e.com/life

ife easier

instructors ... get resources to help you teach.

Visit www.mhhe.com/life

Online Learning Center

A multitude of tools awaits you on *Life*'s Online Learning Center. You'll want to take advantage of our electronic illustrations and photographs from the text; classroom activities; lecture outlines; a message board; and access to PageOut™: Course Website Development Center—All available anytime you want them. Simply visit www.mhhe.com/life.

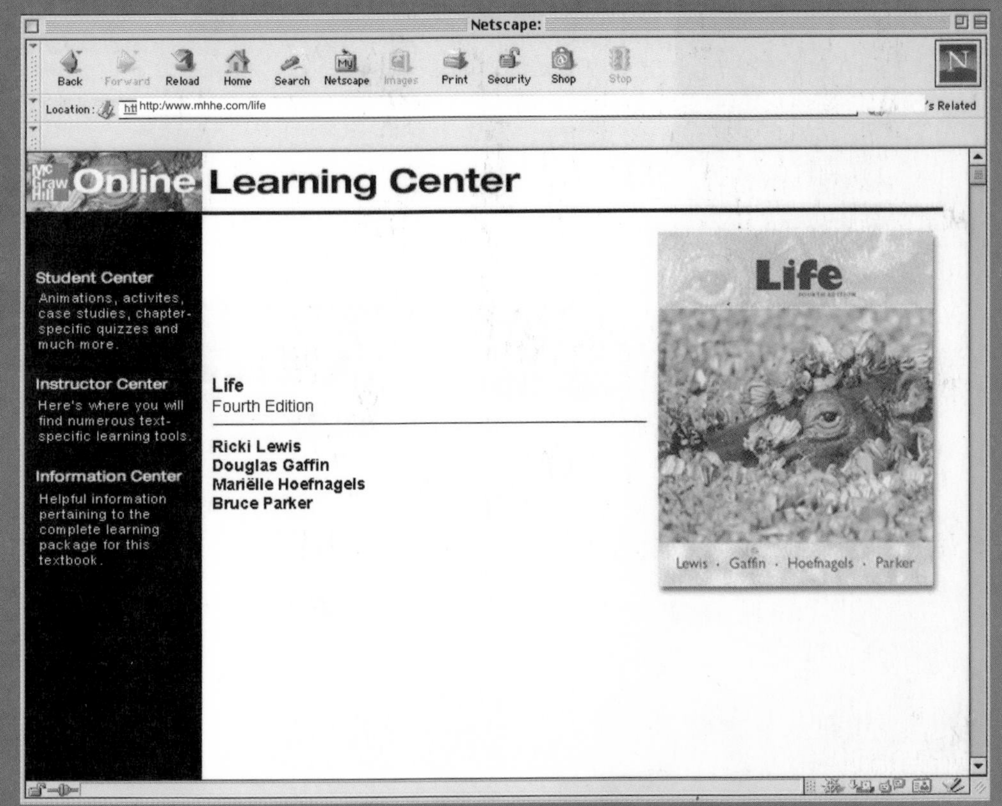

Supplements

- Instructor's Manual with Test Item File
- PageOut™
  - Website authoring software
- 750 Transparencies
- Visual Resource Library CD-ROM
  - Includes all photographs and illustrations
- Life Science Animations CD-ROM
  - 200 animations for classroom presentations
- Online Learning Center at www.mhhe.com/life
  - Now featuring the *Essential Study Partner* 2.0
- BioCourse.com
  - An online biology resource center for study and lecture preparation
- Interactive e-Text CD-ROM
  - An electronic, interactive version of *Life*, Fourth Edition
- Lecture PowerPoint

FOURTH EDITION

Ricki Lewis
The University at Albany
Douglas Gaffin
The University of Oklahoma
Mariëlle Hoefnagels
The University of Oklahoma
Bruce Parker
Utah Valley State College

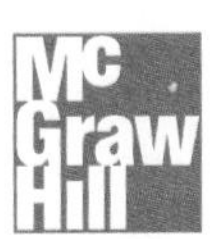

Boston Burr Ridge, IL Dubuque, IA Madison, WI New York San Francisco St. Louis
Bangkok Bogotá Caracas Kuala Lumpur Lisbon London Madrid Mexico City
Milan Montreal New Delhi Santiago Seoul Singapore Sydney Taipei Toronto

McGraw-Hill Higher Education

A Division of The ***McGraw-Hill*** *Companies*

LIFE, FOURTH EDITION

Published by McGraw-Hill, a business unit of The McGraw-Hill Companies, Inc., 1221 Avenue of the Americas, New York, NY 10020.

Some ancillaries, including electronic and print components, may not be available to customers outside the United States.

 This book is printed on recycled, acid-free paper containing 10% postconsumer waste.

1 2 3 4 5 6 7 8 9 0 VNH/VNH 0 9 8 7 6 5 4 3 2 1

ISBN 0–07–027134–8

Publisher: *Michael D. Lange*
Sponsoring editor: *Patrick E. Reidy*
Developmental editors: *Margaret B. Horn/Suzanne Guinn*
Senior marketing manager: *Lisa L. Gottschalk*
Project manager: *Joyce M. Berendes*
Production supervisor: *Kara Kudronowicz*
Coordinator of freelance design: *Michelle D. Whitaker*
Freelance cover/interior designer: *Kathy Cunningham*
Cover photo: *Kevin Schafer/Tony Stone Images*
Photo research coordinator: *John C. Leland*
Photo research: *Mary Reeg*
Supplement producer: *Jodi K. Banowetz*
Compositor: *Precision Graphics*
Typeface: *10/12 Minion*
Printer: *Von Hoffmann Press, Inc.*

The credits section for this book begins on page 936 and is considered an extension of the copyright page.

Library of Congress Cataloging-in-Publication Data

Life / Ricki Lewis . . . [et al.]. — 4th ed.
p. cm.
Rev. ed. of: Life / Ricki Lewis. 3rd. © 1998.
Includes bibliographical references and index.
ISBN 0–07–027134–8
1. Biology. 2. Human biology. I. Lewis, Ricki. II. Lewis, Ricki. Life.

QH308.2 .L485 2002
570—dc21 00–053382
CIP

www.mhhe.com

Dedicated to Our Students

BRIEF CONTENTS

DETAILED CONTENTS

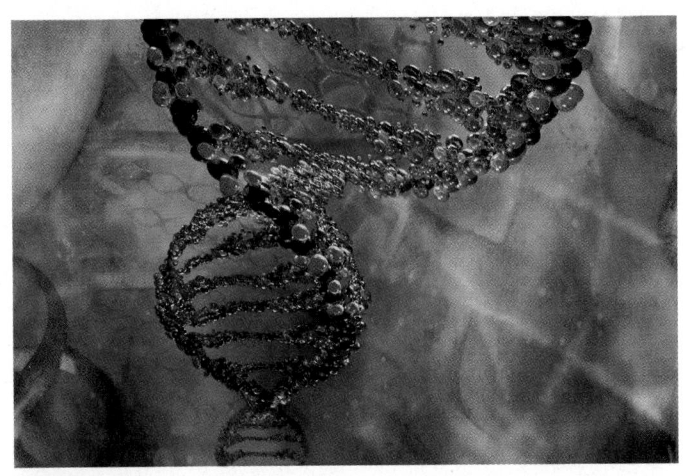

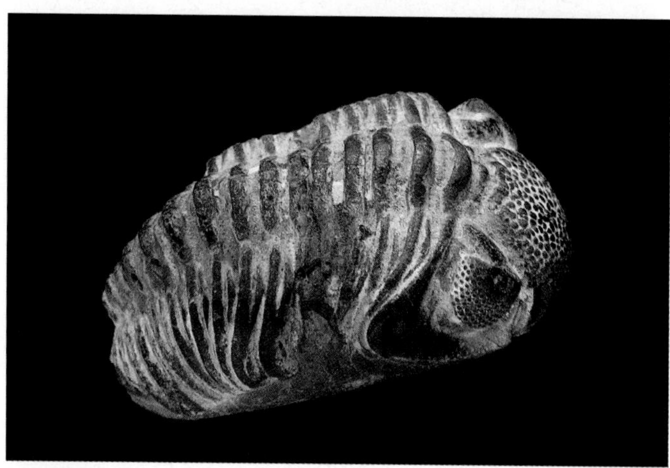

CHAPTER 22
Plantae 412

CHAPTER 23
Fungi 432

CHAPTER 24
Animalia I—Sponges Through Echinoderms 450

CHAPTER 25
Animalia II—The Chordates 476

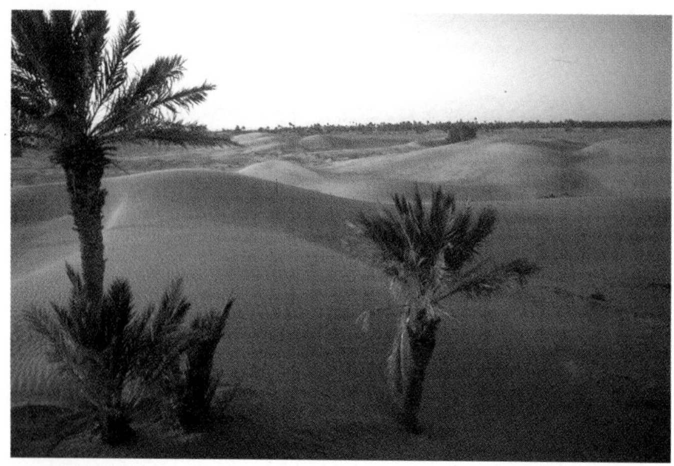

UNIT 5 PLANT LIFE 501

UNIT 7 BEHAVIOR AND ECOLOGY 803

PREFACE

The Changing Face of Biology—And *Life*

To say that the field of biology is changing rapidly is certainly an understatement. Only 50 years ago, James Watson and Francis Crick deciphered the three-dimensional structure of DNA. Now, with just a small DNA sample, biologists can decipher entire genomes—from the simplest bacterium whose genes hold clues to what it is to be alive, to species that seemingly straddle evolutionary leaps, to the most complex plant or animal. On a more practical level, DNA technology and the new life science of genomics have confirmed certain historical references, unraveled the tangled ancestries of wine grapes, and even helped prove the innocence of death row inmates.

Biologists continue to use the molecules of life to reveal new glimpses of the evolutionary relationships that bind all organisms, even species that once thrived in a long-ago, vastly different world. As a result, the way in which biologists classify life is fundamentally different from what it was just a generation ago. Everywhere we look, it's easy to find evidence that these are exciting times for biologists. To reflect these profound shifts in the field, *Life* also has changed.

New Author Team

The fourth edition of *Life* brings together four outstanding biologists. Our team begins with Ricki Lewis. She is well known for her ability to weave together solid biology content with interesting stories, real-life case studies, and applications to student life. With expertise in genetics and science communication, she has published countless articles in magazines, journals, newspapers, and encyclopedias. Her role as contributing editor to *The Scientist* gives her a heads up on much ongoing research, which finds its way into the pages of *Life.* She is also the author of a human genetics textbook and a collection of essays on discovery in the life sciences, and coauthor of human anatomy and physiology textbooks. Ricki has taught a variety of courses at the University at Albany, Empire State College, and Miami University and is a genetic counselor.

Joining forces with Ricki Lewis for the fourth edition of *Life* are three new coauthors, and we are proud to introduce ourselves: Douglas Gaffin and Mariëlle Hoefnagels of the University of Oklahoma, and Bruce Parker of Utah Valley State College. We are all active instructors who use multimedia approaches to teach undergraduate biology to hundreds of majors and nonmajors each semester.

Devotion to, and passion about, teaching unite our team. We thoroughly enjoy telling those interesting stories that are so easy to find at all levels of biology, from molecules to ecology—the stories that, when told correctly, mesmerize even the most reluctant students, causing them to perk up and think "Wow, I never knew that! So that's why . . . !" We all love to watch students get excited about learning a subject they once viewed as too hard or too intimidating. Our enthusiasm for teaching and respect for students have earned us all recognition on our campuses as outstanding teachers.

Our areas of scientific expertise—animal physiology, plant-microorganism interactions, and molecular biology and biochemistry—provide an excellent complement to Ricki's extensive knowledge of genetics. As a result, the fourth edition of *Life* has comprehensive, up-to-date content in all subject areas. But we were careful not to sacrifice Ricki's wonderful way with words.

Life has always had a unique style, reflecting a mix of scientific expertise and journalistic experience. The writing style is neither an authoritative voice talking down to the reader, nor an attempt to water down complicated science—nor a hodgepodge of the two. *Life*'s voice is uniquely clear and exciting. The result is a textbook with substantial content that is accessible to students by mixing in interesting stories and practical applications that make biology relevant to student life.

Our complementary areas of interest and dedication to sharing the wonder of biology with students led the four of us to a united vision for *Life.* We wanted to produce the excellent textbook we were waiting for ourselves—one that is readable, accurate, up-to-date, interesting, and presented in an attractive format that appeals to students. We believe the fourth edition of *Life* meets our goals. Consider the other changes to this edition.

New Content

To keep up with the shifts reverberating throughout biology, and to bring the book more in line with the order in which many instructors teach biology, we overhauled *Life*'s Table of Contents. The first five units have been reorganized and rolled into three (Cell Biology, Genetics and Biotechnology, and Evolution). These new units emphasize the concepts that are common to all life. In the next three units, the book introduces the diversity of life and explores the structures and processes unique to plant and animal life. The final unit considers ecology, concluding with a chapter that addresses environmental challenges today's students may well have to solve.

Life's revamped Table of Contents also introduces several new and reorganized chapters. New chapter 16 is devoted to speciation and extinction, with many fascinating examples of evolution in action. Another new chapter, on viruses and other infectious agents (chapter 19) explores the significance of viruses as emerging pathogens, as useful tools in biotechnology, and also as windows on evolutionary change. To expand coverage of plant life, we added a chapter on transport systems in plants (chapter 27). We moved coverage of animal reproduction and development (chapter 40) to close the unit on animal organ systems, in response to many requests. In addition, animal diversity is now covered in two chapters (24 and 25). Finally, numbered sections throughout reveal at a glance the major themes of each chapter.

Perhaps the greatest change and challenge in *Life* is also the greatest change in the science of life—how to categorize organisms. We completely rewrote Unit 4, The Diversity of Life, to reflect new classification schemes that combine traditional and molecular approaches to taxonomy. Yet we were careful to explain along the way that the molecular data currently providing such a wealth of new information have also thrown the classification of life into upheaval—and acknowledge that the classification schemes we present in this book are provisional. It is important for students to realize that biological facts and concepts are not written in stone. Although we know that the next edition may see even more changes in this unit, we would rather present the current state of taxonomic thought than perpetuate out-of-date classification schemes.

In revising Unit 4, we evaluated and reevaluated the traditional order in which textbooks typically present life's diversity—prokaryotic organisms (bacteria and archaea), followed by protista, fungi, plants, and animals. However, much new scientific evidence suggests that fungi are actually more closely related to animals than to plants. In response to the current state of the science, we placed the chapter on fungi (chapter 23) after plants and before animals, better reflecting how evolution probably unfolded. It is a seemingly minor change, but an important one because it reflects a philosophical shift in how biologists classify life.

New Art and Photo Program

As we examined every word of text in the book, we also scrutinized every piece of art with a critical look at how it works with the text. We added many figures to support the new textual material, and modified many others—then professional biological illustrators rendered each piece anew. The new art is not only visually spectacular, but also pedagogically sound, and it gives *Life* a consistent look from cover to cover. Repeating themes provide continuity, from biochemical reactions to life cycles to feedback loops in animal physiology to evolutionary tree diagrams. Use of color, arrows, and symbols is standardized throughout the text, easing learning. So, for example, DNA, membranes, and other cell structures have a consistent look and color throughout. We have also selected unusual and interesting photos to show students glimpses of the natural world that they may never have seen before. The new art and photos are combined in page layouts that are attractive and interesting—and above all, help students learn.

Highlights on Health, Biotechnology, and Scientific Inquiry

We believe that understanding science and scientific thought is one of the most important things that students should gain from their college experience. *Life* has always emphasized the practical side of biology, and the fourth edition continues that tradition. Each chapter begins with a compelling essay describing a real-life scientific issue, ranging from the worldwide decline of amphibian populations to the evolutionary impact of the varied shapes of male genitalia (in beetles). The content in each chapter supports and expands upon the ideas presented in the opening essays.

Each chapter features one or more boxes highlighting the relevance of the content to health, biotechnology, or scientific inquiry. "Health" boxes provide a human touch. Health 19.1, for example, explores how birds brought influenza and West Nile virus infection to human populations. "Biotechnology" boxes showcase how science segues into practical applications, with looks at such diverse tools as PCR, gene therapy, in vitro evolution, artificial photosynthesis, and molecular taxonomy. A new technique explored in Biotechnology 27.1, for example, is rhizosecretion, a method to coax plants growing in hydroponic culture to secrete useful proteins—some encoded by genes from other species—through their roots. "Investigating Life" features help remove some of the mystique of science, leading the reader through the ways that scientists think when carrying out real investigations and experiments. Investigating Life 14.1, for example, presents compelling evidence of evolution among animals inhabiting a polluted river, taking the reader through the critical experiments and the logic that inspired them step by step. Along different lines, Investigating Life 28.1 invites students to predict the structures of mutant flowers, given a few simple rules governing the interaction between three flower development genes.

New Innovative and Integrative Media Support

The fourth edition of *Life* includes an innovative, comprehensive support package. As we wrote *Life,* we talked a lot about which supplements we would use as instructors. At the top of our list were computer files of textbook art, presented in a format that we could really USE in our multimedia lectures. Most textbooks offer bitmapped files of text art, but small text size and image contrast that is not optimized for large lecture halls often limit the utility of these computer files in the classroom. As instructors, we wanted more flexibility—files we could manipulate ourselves so we could tell our own stories in our own way. As a result, *Life* now offers PowerPoint-compatible, vectorized art files that the instructor can manipulate as he or she sees fit. *Life* is among the first textbooks to offer this feature.

The vectorized art is just one component of an innovative and integrated new program of media support for faculty and students. Instructor presentations will come alive with CD-ROMs that include not only the vectorized art, but also *Life*'s photos and animations. The online Essential Study Partner, which links to *Life's* Online Learning Center, enhances learning, and the new BioCourse.com site rounds out *Life*'s integrated ancillaries.

A Word of Thanks

No single person, no matter how educated, "knows" all of biology. Even an author team whose collective expertise covers most of the field must rely on an almost unimaginable amount of feedback. We greatly appreciate the help of the many reviewers, consultants, and focus group members—committed teachers who went the extra mile to help make this book what it is. We could not have done it without them. We are indebted to Randy Moore and Fred Spiegel for their contributions to the plant life unit. And we are grateful to the students in Dr. Gaffin's Spring 2000 Zoology Capstone Course for their valuable insights as they critiqued portions of the manuscript.

We thank the team at McGraw-Hill who guided us in this new view of *Life*—Michael Lange, Publisher; Patrick Reidy, Sponsoring Editor; Margaret Horn and Suzanne Guinn, Developmental Editors; and Joyce Berendes, Project Manager. We also thank the talented artists and media wizards at Precision Graphics who so beautifully translated our vision. Finally, we hope that both faculty and students will enjoy using our text as much as we loved creating it.We encourage readers to contact us with questions, comments, and suggestions. For at the pace at which biology is progressing, the next edition is just around the corner!

We offer special thanks to the reviewers who spent hours poring over chapter drafts in meticulous detail, spotting errors and inconsistencies, confirming what works and gently critiquing what doesn't, and pointing out sections that we could clarify.

I. Edward Alcamo, *State University of New York—Farmingdale*
Sylvester Allred, *Northern Arizona University*
Harvey J. Babich, *Stern College of Women*
Joseph Bastian, *University of Oklahoma*
Bob Bennett, *Arkansas State University*
Ari Berkowitz, *University of Oklahoma*
Linda Brandt, *Henry Ford Community College*
Jennifer Carr Burtwistle, *Northeast Community College*
Iain M. Campbell, *University of Pittsburgh—Pittsburgh Campus*
Tom Campbell, *Los Angeles Pierce College*
Jeffrey Carmichael, *University of North Dakota*
Margaret Carroll, *Framingham State College*
Ajoy Chakrabarti, *South Carolina State University*
William Chrouser, *Warner Southern College*
Robert Cordero, *Holy Family College*
Frank E. Cox, *Santa Barbara City College*
Penni Croot, *State University of New York—Morrisville*
Charles Denny, *University of South Carolina—Sumter*
Nicholas Despo, *Thiel College*
Randy DiDomenico, *University of Colorado*
Thomas Emmel, *University of Florida*
Lynn Fancher, *College of DuPage*
Deborah Fahey, *Wheaton College*
James Fitch, *Jones County Junior College*
Donald P. French, *Oklahoma State University*
Michael Gleason, *Central Washington University*
Elmer Godeny, *Louisiana State University*
Glenn Gorelick, *Citrus College*
Frederick Gottlieb, *University of Pittsburgh*
Lawrence Gray, *Utah Valley State College*
Patricia Grove, *College of Mount St. Vincent*
Richard Harrington, *Rivier College*
Janet Haynes, *Long Island University*
Julia Hilliard, *Georgia State University*
Jane Johnson, *University of Wisconsin Stevens Point*
Florence Juillerat, *Indiana University—Purdue University*
Judith Kandel, *California State University—Fullerton*
Rebecca Kapley, *Cuyahoga Community College*
Suzanne Kempke, *Armstrong Atlantic State University*
Kerry Kilburn, *Old Dominion University*
Rosemary Knapp, *University of Oklahoma*
Timothy F. Kowalik, *University of Massachusetts Medical School*
Charles Kugler, *Radford University*
Roger Lane, *Kent State University*
Sandra Latourelle, *SUNY—Plattsburgh*
Jacqueline Lee, *Nassau Community College*
Daniel Lim, *University of South Florida*
Yue Lin, *St. John's University*
John Logue, *University of South Carolina*
Cyrus McQueen, *Johnson State College*
Michael Meighan, *University of California—Berkeley*
Thomas Milton, *Richard Bland College*
Jonathan Morris, *Middlesex Community Technical College*
Edward Nelson, *Southeastern Louisiana University*
Janine Nelson, *Tulsa Community College—Northeast Campus*
Mary Theresa Ortiz, *Kingsborough Community College*
Phillip Ortiz, *Skidmore College*
Rhoda Perozzi, *Virginia Commonwealth University*
Maria Menna Perper, *St. Francis College*
Lansing Prescott, *Augustana College (emeritus)*
Darrell Ray, *University of Tennessee at Martin*
Timoth Ruhnke, *West Virginia State College*
Stacia Schneider, *Marquette University*
Elisabeth Schussler, *Louisiana State University*
O. Tacheeni Scott, *California State University—Northridge*
Erik Scully, *Towson University*
Prem P. Sehgal, *East Carolina University*
Michael W. Shaw, *Centers for Disease Control and Prevention*
John Shiber, *Prestonburg Community College*
Jerry Skinner, *Keystone College*
Fred Spiegel, *University of Arkansas*
Carol St. Angelo, *Hofstra College*
Gilbert D. Starks, *Central Michigan University*
Anthea Stavrovlakis, *Kingsborough Community College*
Gregory Stewart, *State University of West Georgia*
Martha Sugermeyer, *Tidewater Community College*
Janice Swab, *Meredith College*
Kevin Swier, *Chicago State University*
Jamey Thompson, *Maysville Community College*
Bruce Voyles, *Grinnell College*
Charlene Waggoner, *Bowling Green State University*
Susan Weinstein, *Marshall University*
Brian Wells, *Northeast Texas Community College*
Mary White, *Southeastern Louisiana University*
Peter Wilkin, *Purdue University North Central*
Courtenay N. Willis, *Youngstown State University*
Anthony Udeogalanya, *Medgar Evers College*
Calvin Young, *Fullerton College*

Art and Design Reviewers

Sylvester Allred, *Northern Arizona University*
Michael Meighan, *University of California—Berkeley*
Bruce Parker, *Utah Valley State College*
Prem Schgal, *East Carolina University*
Calvin Young, *Fullerton College*

In November 1998, at the NABT convention in Reno, NV, a talented group of instructors helped us map out a plan for the revision.

Sylvester Allred, *Northern Arizona University*
Randy DiDomenico, *University of Colorado*
Donald P. French, *Oklahoma State University*
Suzanne Kempke, *Armstrong Atlantic State University*
Manuel Molles, *University of New Mexico*
Barry Palevitz, *University of Georgia*
Bruce Parker, *Utah Valley State College*
Connie Russell, *Oklahoma State University*
D. Courtney Smith, *Ohio State University*
Fred Spiegel, *University of Arkansas*
Linda Tichenor, *University of Arkansas*
Charlene Waggoner, *Bowling Green State University*
Courtenay N. Willis, *Youngstown State University*
Calvin Young, *Fullerton College*

At a focus group in April 1999 in Chicago, we had the opportunity to develop a plan for an extensive new supplements package due to some tremendous advice from a talented and wise group of experienced educators.

Lynn Fancher, *College of DuPage*
Merrill Gassman, *University of Illinois at Chicago*
Sandra Latourelle, *SUNY—Plattsburgh*
Darrel L. Murray, *University of Illinois at Chicago*
Bruce Parker, *Utah Valley State College*
Linda Tichenor, *University of Arkansas*

The following individuals contributed to both the quality of our new supplements and the wide range of outstanding new options for students and faculty.

Jennifer Carr Burtwistle, *Northeast Community College*
Art Animations
Edward Cawley, *Loras College*
Course Integration Guide
Lynn Fancher, *College of DuPage*
Essential Study Partner
Donald P. French, *Oklahoma State University*
Art Animations
Douglas Gaffin, *University of Oklahoma*
Vectorized Art
Sandra Latourelle, *SUNY—Plattsburgh*
Student Study Guide
John Merrill, *Michigan State University*
Online Learning Center
Bruce Parker, *Utah Valley State University*
Instructor's Manual and Test Item File
Nancy Pencoe, *State University of West Georgia*
Instructor PowerPoint Displays
Calvin Young, *Fullerton College*
Art Animations

ABOUT THE AUTHORS

Ricki Lewis has built a multifaceted career around communicating the excitement of life science, especially genetics and biotechnology. She earned her Ph.D. in genetics in 1980 from Indiana University, working with homeotic mutations in *Drosophila melanogaster.* She is an adjunct professor at Miami University and the University at Albany, and has also taught at Empire State College and several community colleges. Ricki has published more than 3,000 articles in publications such as *The Scientist, Genetic Engineering News, BioScience,* and *Discover.* She is a frequent invited speaker, and is a member of the National Association of Biology Teachers, the National Society of Genetic Counselors, and the National Association of Science Writers. Ricki is also a genetic counselor for a large private medical practice, where she helps people make decisions concerning new technologies stemming from genetic research. She lives in upstate New York with chemist husband Larry, three daughters, four cats, two guinea pigs, and a rat, tortoise, and hedgehog. rickilewis@nasw.org

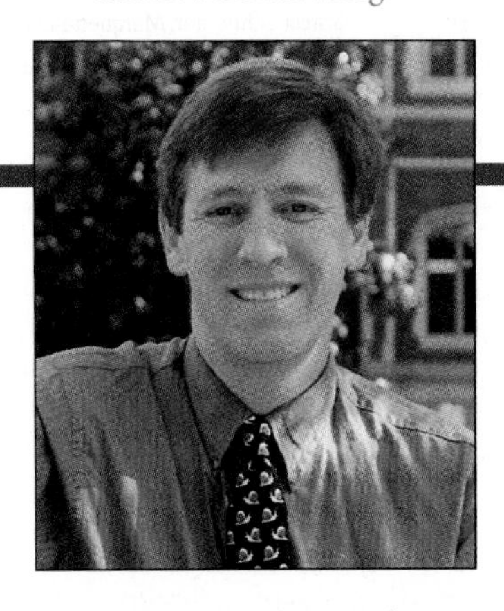

Douglas Gaffin holds a bachelor of science degree from the University of California at Berkeley, and he earned his Ph.D. in zoology from Oregon State University in Corvallis in 1994. His research interests are in sensory neurobiology, where his special focus is on the behavior and sensory physiology of sand scorpions. He has extensive biology teaching experience and has taught students in courses ranging from junior high school to graduate school levels. Doug is currently assistant professor and director of undergraduate studies for the Department of Zoology at the University of Oklahoma, and he has the privilege of teaching introductory zoology to thousands of undergraduates each year. His innovative teaching style and ability to inspire students have been recognized with awards both regionally and nationally. Among other organizations, he is a member of the Society for Neuroscience, the International Society for Neuroethology, the American Arachnological Society, and the National Association of Biology Teachers. In his spare time he enjoys traveling, riding his bike, playing volleyball, and picking the banjo. One of his favorite activities is going to the desert each summer to observe and conduct field research on sand scorpions in their native habitat. ddgaffin@ou.edu

Mariëlle Hoefnagels was raised near San Francisco, and received her B.S. in environmental science (1987) from the University of California at Riverside. After working in a soil analysis lab in Oregon for two years, she earned her master's degree in soil science from North Carolina State University (1991). Her research, on interactions between beneficial fungi and salt marsh plants, led her to return to Oregon to complete her Ph.D. in plant pathology (Oregon State University, 1997). Mariëlle's dissertation work focused on the use of bacterial biological control agents to reduce the spread of fungal pathogens on seeds. She is now a lecturer at the University of Oklahoma, where she teaches nonmajors courses in biology and microbiology, and a course on fungi for advanced botany and microbiology majors. Her current research is on the interactions between plants and beneficial microorganisms in prairie soils, and she particularly enjoys involving undergraduates in her research during the summer. She is a member of the National Association of Biology Teachers, and the American Phytopathological Society. Her hobbies include reading, traveling, photography, and playing volleyball. hoefnagels@ou.edu

Bruce Parker received his Ph.D. in molecular biology/biochemistry from Utah State University in 1988. His areas of expertise include virology, molecular cell biology and biochemistry, and he spent two years in London working on research into viruses that cause cancer, followed by another two years on the same project at St. Jude Children's Research Hospital in Memphis. He has taught general biology for nonmajors and majors at Utah Valley State College since 1992 and has been nominated for Faculty of the Year for six of those years. Bruce currently serves as department chairperson at Utah Valley State College and is included in *Who's Who Among America's Teachers* for 1998. His hobbies include computer programming and amateur radio, when he is not fishing somewhere. parkerbr@uvsc.edu

ILLUSTRATION TEAM

Precision Graphics of Champaign, Illinois, is a specialized composition house. Their own staff of illustrators developed the art program for the fourth edition of *Life*. Each person on the team brought her own skills and strengths to the subject matter. **Connie Balek,** the lead developer at Precision Graphics, not only has a degree in biology, but she also holds a master of fine arts degree in medical and biological illustration from the University of Michigan. **Joanne Bales** has been a medical illustrator at Precision Graphics since 1992 and utilizes her extensive health-care and nursing background in developing illustrations. And **Jan Troutt,** the natural science art director, brings many years of unparalleled experience in rendering art for many of today's leading biology titles. This team collaborated on each piece of art to build an accurate and solid program that will help students learn about life.

THE LEARNING SYSTEM

These pages are a brief guide to tools that *Life* uses to facilitate students' study of biology.

Chapter Opening Vignettes

Each chapter begins with a compelling vignette describing a real-life scientific issue related to the chapter topic.

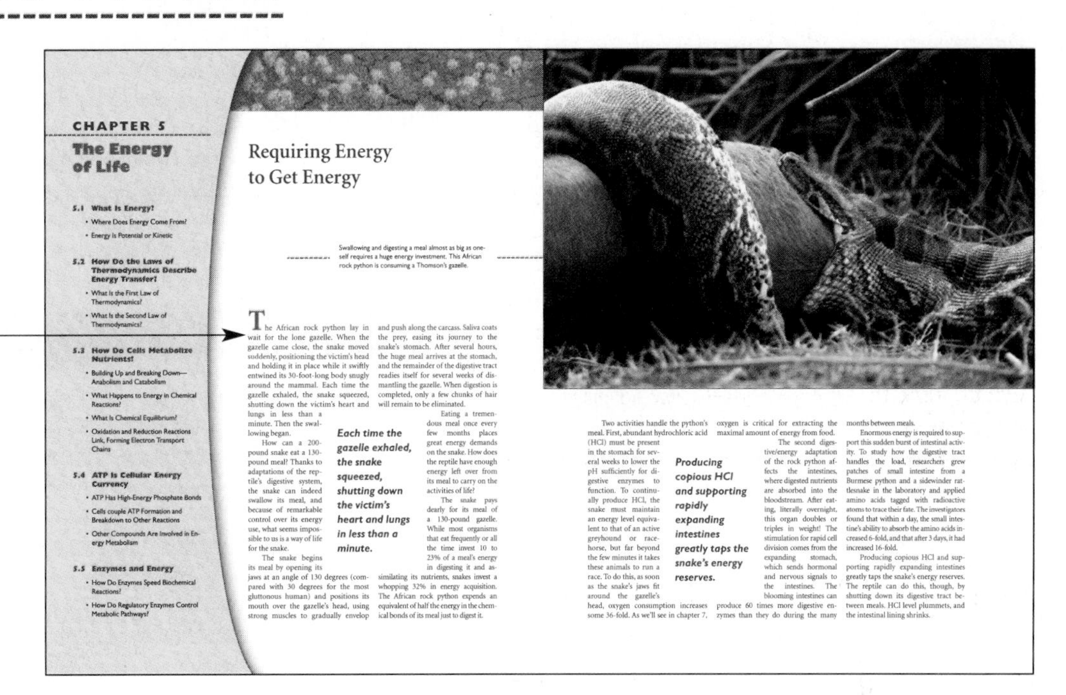

CHAPTER 5

The Energy of Life

Requiring Energy to Get Energy

Swallowing and digesting a meal almost as big as oneself requires a huge energy investment. This African rock python is consuming a Thomson's gazelle.

The African rock python lay in wait for the lone gazelle. When the gazelle came close, the snake moved suddenly, positioning the victim's head and holding it in place while it swiftly entwined its 30-foot-long body snugly around the mammal. Each time the gazelle exhaled, the snake squeezed, shutting down the victim's heart and lungs in less than a minute. Then the swallowing began.

How can a 200-pound snake eat a 130-pound meal? Thanks to adaptations of the reptile's digestive system, the snake can indeed swallow its meal, and because of remarkable control over its energy use, what seems impossible to us is a way of life for the snake.

Each time the gazelle exhaled, the snake squeezed, shutting down the victim's heart and lungs in less than a minute.

The snake begins its meal by opening its jaws at an angle of 130 degrees (compared with 30 degrees for the most gluttonous human) and positions its mouth over the gazelle's head, using strong muscles to gradually envelop and push along the carcass. Saliva coats the prey, easing its journey to the snake's stomach. After several hours, the huge meal arrives at the stomach, and the remainder of the digestive tract readies itself for several weeks of dismantling the gazelle. When digestion is completed, only a few chunks of hair will remain to be eliminated.

Eating a tremendous meal once every few months places great energy demands on the snake. How does the reptile have enough energy left over from its meal to carry on the activities of life?

The snake pays dearly for its meal of a 130-pound gazelle. While most organisms that eat frequently or all the time invest 10 to 23% of a meal's energy in digesting it and assimilating its nutrients, snakes invest a whopping 32% in energy acquisition. The African rock python expends an equivalent of half the energy in the chemical bonds of its meal just to digest it.

Two activities handle the python's meal. First, abundant hydrochloric acid (HCl) must be present in the stomach for several weeks to lower the pH sufficiently for digestive enzymes to function. To continually produce HCl, the snake must maintain an energy level equivalent to that of an active greyhound or racehorse, but far beyond the few minutes it takes these animals to run a race. To do this, as soon as the snake's jaws fit around the gazelle's head, oxygen consumption increases some 36-fold. As we'll see in chapter 7, oxygen is critical for extracting the maximal amount of energy from food.

Producing copious HCl and supporting rapidly expanding intestines greatly taps the snake's energy reserves.

The second digestive/energy adaptation of the rock python affects the intestines, where digested nutrients are absorbed into the bloodstream. After eating, literally overnight, this organ doubles or triples in weight! The stimulation for rapid cell division comes from the expanding stomach, which sends hormonal and nervous signals to the intestines. The blooming intestines can produce 60 times more digestive enzymes than they do during the many months between meals.

Enormous energy is required to support this sudden burst of intestinal activity. To study how the digestive tract handles the load, researchers grew patches of small intestine from a Burmese python and a sidewinder rattlesnake in the laboratory and applied amino acids tagged with radioactive atoms to trace their fate. The investigators found that within a day, the small intestine's ability to absorb the amino acids increased 6-fold, and that after 3 days, it had increased 16-fold.

Producing copious HCl and supporting rapidly expanding intestines greatly taps the snake's energy reserves. The reptile can do this, though, by shutting down its digestive tract between meals. HCl level plummets, and the intestinal lining shrinks.

New Art Program

The accurate and artistically compelling illustrations greatly enhance the student's understanding of difficult processes and concepts.

FIGURE 4.1

Cellular Architecture.

A white blood cell's inner skeleton and surface features enable it to move in the body and to recognize "foreign" cell surfaces—such as those of transplanted tissue. This T lymphocyte rejects foreign tissue.

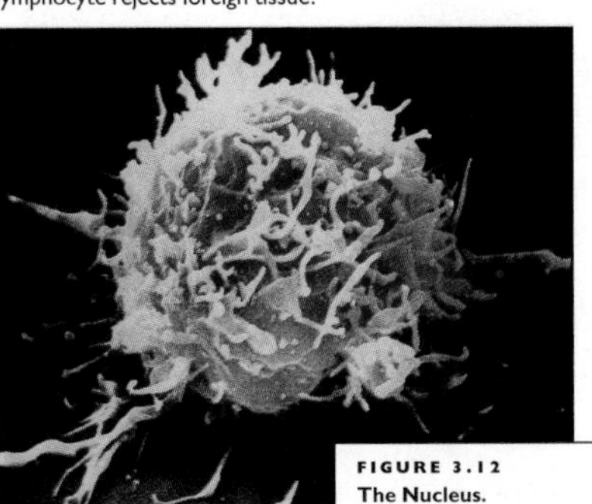

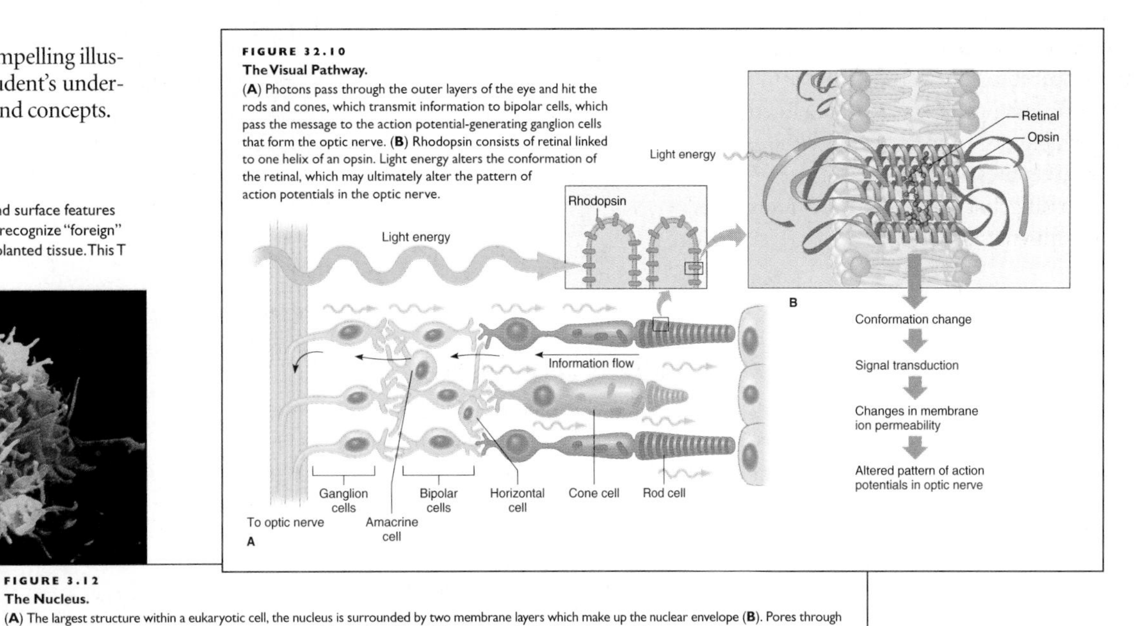

FIGURE 32.10

The Visual Pathway.

(**A**) Photons pass through the outer layers of the eye and hit the rods and cones, which transmit information to bipolar cells, which pass the message to the action potential-generating ganglion cells that form the optic nerve. (**B**) Rhodopsin consists of retinal linked to one helix of an opsin. Light energy alters the conformation of the retinal, which may ultimately alter the pattern of action potentials in the optic nerve.

FIGURE 3.12

The Nucleus.

(**A**) The largest structure within a eukaryotic cell, the nucleus is surrounded by two membrane layers which make up the nuclear envelope (**B**). Pores through the envelope allow specific molecules to move in and out of the nucleus. The darkly staining nucleolus (**C**) is the site of ribosome manufacture and assembly.

A

Cytoplasm
Nuclear pore
Inside nucleus
Nuclear envelope
B

Nuclear pore
Nucleolus
Nuclear membrane
C Nucleus
2 μm

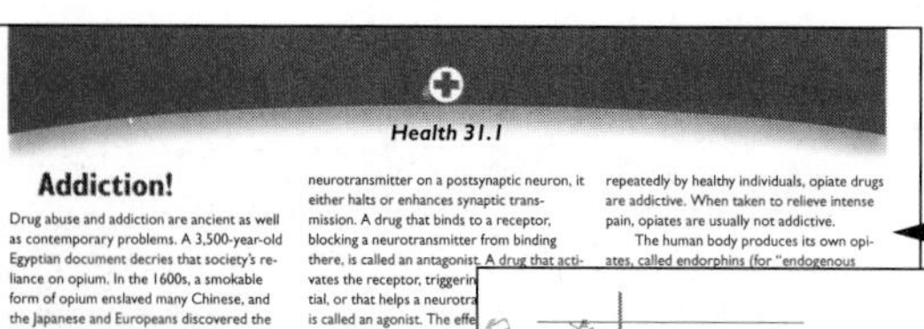

Boxed Readings

These readings highlight the relevance of chapter contents to health, biotechnology, and scientific inquiry.

Health readings discuss health issues of interest to the student.

Biotechnology boxes reveal at a glance how science segues into practical applications.

Investigating Life features help remove some of the mystique of science, leading the reader through ways that scientists think when carrying out real experiments and investigations.

Health 31.1

Addiction!

Biotechnology 39.1

Immunotherapy

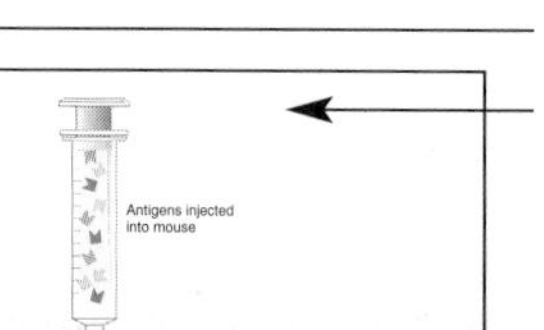

Investigating Life 8.1

Experiments Reveal the Telomere Clock

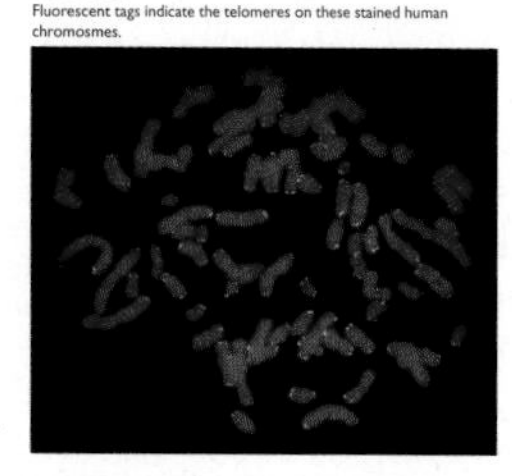

FIGURE 8A
Fluorescent tags indicate the telomeres on these stained human chromosomes.

Chemistry Explains Biology

These brief readings, unique to the chemistry chapter, facilitate understanding of the chapter's material.

Chemistry Explains Biology

Excess Cholesterol

Lipoproteins carry cholesterol in the bloodstream. As their name suggests, lipoproteins consist of lipid and protein. Low-density lipoprotein (LDL) particles carry cholesterol to the arteries. Excess LDL cholesterol that does not enter cells accumulates on the inner linings of blood vessels, eventually impeding blood flow. High-density lipoproteins (HDL), in contrast, carry cholesterol to the liver, where it is removed from the bloodstream. High levels of LDL cholesterol increase risk of heart disease, whereas high levels of HDL cholesterol promote heart health.

Chemistry Explains Biology

Spider Silk

Spider silk is the strongest natural fiber known. Spiders use silk to build webs to capture prey, to store prey, and to protect their eggs.

Silks are proteins, manufactured in sets of glands. A spider usually has several sets of glands, each of which produces a different type of silk. In addition to producing webs, a spider continually secretes a strand of silk called a dragline. Should danger arise, the spider can temporarily escape the web on its dragline.

Dragline silk is best studied in the golden orb weaver spider, *Nephila clavipes*. This especially strong and elastic silk consists of two types of proteins that are dry and practically indestructible once outside the animal's body. Dragline silk proteins include a few types of amino acids that repeat in short sequences. This imparts a conformation of coiled sheets, like the steps of a spiral staircase. Spider silk elegantly illustrates how a protein's shape determines its functions.

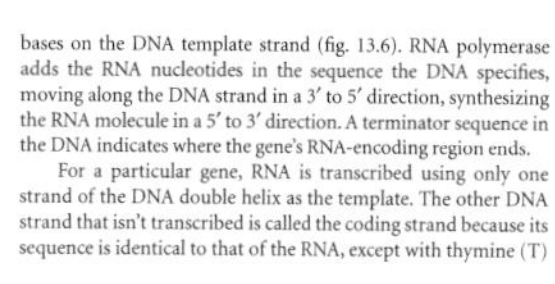

bases on the DNA template strand (fig. 13.6). RNA polymerase adds the RNA nucleotides in the sequence the DNA specifies, moving along the DNA strand in a 3′ to 5′ direction, synthesizing the RNA molecule in a 5′ to 3′ direction. A terminator sequence in the DNA indicates where the gene's RNA-encoding region ends.

For a particular gene, RNA is transcribed using only one strand of the DNA double helix as the template. The other DNA strand that isn't transcribed is called the coding strand because its sequence is identical to that of the RNA, except with thymine (T) in place of uracil (U). Several RNAs may be transcribed from the same DNA template strand simultaneously (fig. 13.7). Since RNA is relatively short-lived, a cell must constantly transcribe certain genes to maintain supplies of essential protein.

To determine the sequence of RNA bases transcribed from a gene, write the RNA bases that are complementary to the template DNA strand, using uracil opposite adenine. For example, if a DNA template strand has the sequence

C C T A G C T A C

then it is transcribed into RNA with the sequence

G G A U C G A U G

The coding DNA sequence is:

G G A T C G A T G

FIGURE 13.7
Transcription of RNA from DNA.
(**A**) Transcription occurs in three states: initiation, elongation, and termination. Initiation is the control point that determines which genes are transcribed and when RNA nucleotides are added during elongation and a terminator sequence in the gene signals the end of transcription. (**B**) Many identical copies of RNA are simultaneously transcribed, with one RNA polymerase starting after another.

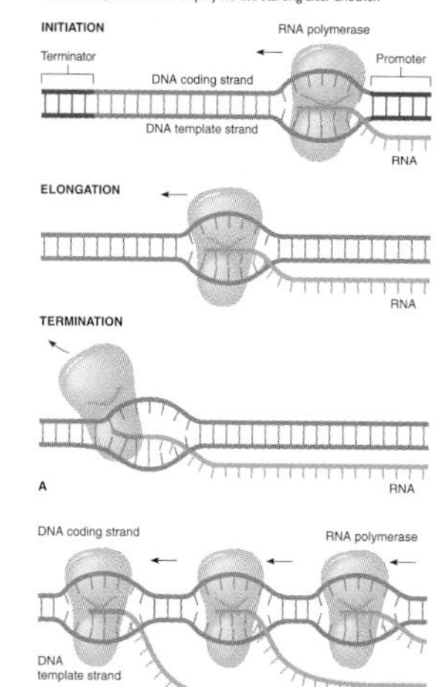

13.1 MASTERING CONCEPTS

1. How do DNA replication, RNA transcription, and translation maintain and use genetic information?
2. How does lactose metabolism in *E. coli* illustrate control of gene expression?
3. How do transcription factors control gene expression in eukaryotes?
4. What are the steps of transcription?

OLC

13.2 RNA Orchestrates Protein Synthesis

RNA carries a gene's information into the cytoplasm, and enables it to be translated into a protein's amino acid sequence. Messenger RNA carries a gene's sequence information; ribosomal RNA is part of ribosomes, which support and bring together amino acids as proteins form; transfer RNA matches specific amino acids to specific mRNA triplets, enabling ribosomes to assemble proteins.

As RNA is synthesized along DNA, it curls into three-dimensional shapes, or conformations, determined by complementary base pairing within the same RNA molecule. These conformations determine how RNA functions. Several types of RNA interact to synthesize proteins (table 13.2).

Types of RNA

Messenger RNA (mRNA) carries the information that specifies a particular protein. Each three mRNA bases in a row forms a genetic code word, or **codon,** that corresponds to a particular amino acid. Because genes vary in length, so do mRNA molecules. Most mRNAs are 500 to 3,000 bases long. Biotechnology 13.1 describes antisense technology, which silences particular genes at the mRNA level.

Ribosomal RNA (rRNA) molecules range from 100 to nearly 3,000 nucleotides long. This type of RNA associates with certain proteins to form a ribosome. Recall from chapter 3 that a ribosome is a structural support for protein synthesis. A ribosome has two subunits that are separate in the cytoplasm but

Mastering Concepts

A short list of questions follows each major text section, to help the student review and understand what was just covered.

Summary Statements

Each major section of the chapter begins with a brief synopsis of the section's material.

Online Learning Center

These icons direct the reader to the *Life* Online Learning Center, which provides self-quizzing, interactive activities, and many other learning tools for students, including the *Essential Study Partner 2.0.* Instructors will find numerous teaching tools, including the Visual Resource Library, Instructor's Manual, and access to PageOut™.

Chapter Summary

The list format of the end of chapter summary makes it easy for students to identify and review key concepts.

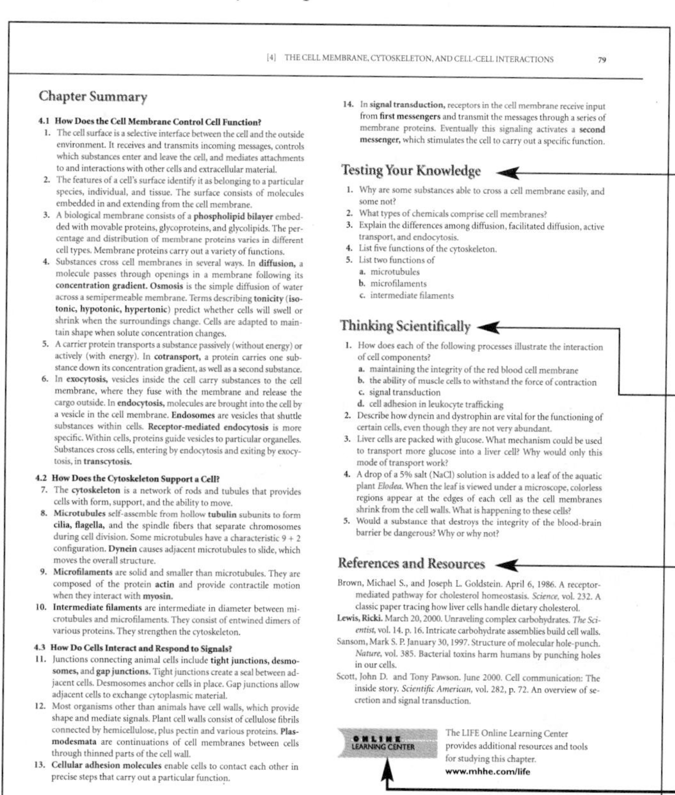

Chapter Summary

4.1 How Does the Cell Membrane Control Cell Function?

1. The cell surface is a selective interface between the cell and the outside environment. It receives and transmits incoming messages, controls which substances enter and leave the cell, and mediates attachments to and interactions with other cells and extracellular material.
2. The features of a cell's surface identify it as belonging to a particular species, individual, and tissue. The surface consists of molecules embedded in and extending from the cell membrane.
3. A biological membrane consists of a **phospholipid bilayer** embedded with movable proteins, glycoproteins, and glycolipids. The percentage and distribution of membrane proteins varies in different cell types. Membrane proteins carry out a variety of functions.
4. Substances cross cell membranes in several ways. In **diffusion,** a molecule passes through openings in a membrane following its **concentration gradient. Osmosis** is the simple diffusion of water across a semipermeable membrane. Terms describing **tonicity** (**isotonic, hypotonic, hypertonic**) predict whether cells will swell or shrink when the surroundings change. Cells are adapted to maintain shape when solute concentration changes.
5. A carrier protein transports a substance passively (without energy) or actively (with energy). In **cotransport,** a protein carries one substance down its concentration gradient, as well as a second substance.
6. In **exocytosis,** vesicles inside the cell carry substances to the cell membrane, where they fuse with the membrane and release the cargo outside. In **endocytosis,** molecules are brought into the cell by a vesicle in the cell membrane. **Endosomes** are vesicles that shuttle substances within cells. **Receptor-mediated endocytosis** is more specific. Within cells, proteins guide vesicles to particular organelles. Substances cross cells, entering by endocytosis and exiting by exocytosis, in **transcytosis.**

4.2 How Does the Cytoskeleton Support a Cell?

7. The **cytoskeleton** is a network of rods and tubules that provides cells with form, support, and the ability to move.
8. **Microtubules** self-assemble from hollow **tubulin** subunits to form **cilia, flagella,** and the spindle fibers that separate chromosomes during cell division. Some microtubules have a characteristic 9 + 2 configuration. **Dynein** causes adjacent microtubules to slide, which moves the overall structure.
9. **Microfilaments** are solid and smaller than microtubules. They are composed of the protein **actin** and provide contractile motion when they interact with **myosin.**
10. **Intermediate filaments** are intermediate in diameter between microtubules and microfilaments. They consist of entwined dimers of various proteins. They strengthen the cytoskeleton.

4.3 How Do Cells Interact and Respond to Signals?

11. Junctions connecting animal cells include **tight junctions, desmosomes,** and **gap junctions.** Tight junctions create a seal between adjacent cells. Desmosomes anchor cells in place. Gap junctions allow adjacent cells to exchange cytoplasmic material.
12. Most organisms other than animals have cell walls, which provide shape and mediate signals. Plant cell walls consist of cellulose fibrils connected by hemicellulose, plus pectin and various proteins. **Plasmodesmata** are continuations of cell membranes between cells through thinned parts of the cell wall.
13. **Cellular adhesion molecules** enable cells to contact each other in precise steps that carry out a particular function.
14. In **signal transduction,** receptors in the cell membrane receive input from **first messengers** and transmit the messages through a series of membrane proteins. Eventually this signaling activates a **second messenger,** which stimulates the cell to carry out a specific function.

Testing Your Knowledge

1. Why are some substances able to cross a cell membrane easily, and some not?
2. What types of chemicals comprise cell membranes?
3. Explain the differences among diffusion, facilitated diffusion, active transport, and endocytosis.
4. List five functions of the cytoskeleton.
5. List two functions of
   a. microtubules
   b. microfilaments
   c. intermediate filaments

Thinking Scientifically

1. How does each of the following processes illustrate the interaction of cell components?
   a. maintaining the integrity of the red blood cell membrane
   b. the ability of muscle cells to withstand the force of contraction
   c. signal transduction
   d. cell adhesion in leukocyte trafficking
2. Describe how dynein and dystrophin are vital for the functioning of certain cells, even though they are not very abundant.
3. Liver cells are packed with glucose. What mechanism could be used to transport more glucose into a liver cell? Why would only this mode of transport work?
4. A drop of a 5% salt (NaCl) solution is added to a leaf of the aquatic plant *Elodea.* When the leaf is viewed under a microscope, colorless regions appear at the edges of each cell as the cell membranes shrink from the cell walls. What is happening to these cells?
5. Would a substance that destroys the integrity of the blood-brain barrier be dangerous? Why or why not?

References and Resources

Brown, Michael S., and Joseph L. Goldstein. April 6, 1986. A receptor-mediated pathway for cholesterol homeostasis. *Science,* vol. 232. A classic paper tracing how liver cells handle dietary cholesterol.

Lewis, Ricki. March 20, 2000. Unraveling complex carbohydrates. *The Scientist,* vol. 14. p. 16. Intricate carbohydrate assemblies build cell walls.

Sansom, Mark S. P. January 30, 1997. Structure of molecular hole-punch. *Nature,* vol. 385. Bacterial toxins harm humans by punching holes in our cells.

Scott, John D. and Tony Pawson. June 2000. Cell communication: The inside story. *Scientific American,* vol. 282, p. 72. An overview of secretion and signal transduction.

ONLINE LEARNING CENTER

The LIFE Online Learning Center provides additional resources and tools for studying this chapter.
www.mhhe.com/life

Testing Your Knowledge

This feature at the end of the chapter tests the student's recall of chapter material. The answers are found on the Online Learning Center.

Thinking Scientifically

These critical thinking questions challenge the student to use concepts of the chapter to solve problems.

References and Resources

These suggested resources can be used for further study of topics covered in the chapter.

TECHNOLOGY SUPPLEMENTS

Life Online Learning Center

This text-specific website provides extensive resources and learning tools for students, including self-quizzing opportunities, interactive activities, bioethics case studies, critical thinking activities, and web links. In addition, the *Essential Study Partner 2.0* is now hosted on this site. Instructors can access the Instructor's Manual, PowerPoint lecture presentations, the Visual Resource Library, PageOut™, a Course Integration Guide, and many other resources. Log on at www.mhhe.com/life

e-Text CD-ROM

The complete *Life* textbook (including art and photos) and study guide PDF files are interlinked and combined with other features for this e-text on CD-ROM.
ISBN 0–07–241257–7

Computerized Test Bank (Microtest)

This computerized test generator contains the complete test item file on a CD in a hybrid format that is compatible with either windows or Macintosh.
ISBN 0–07–027222–0

BioCourse.com

McGraw-Hill's new online life sciences resource for students and instructors includes course success materials, current news and readings, lab information, simulations, animations, and journal search options. Log on at www.biocourse.com

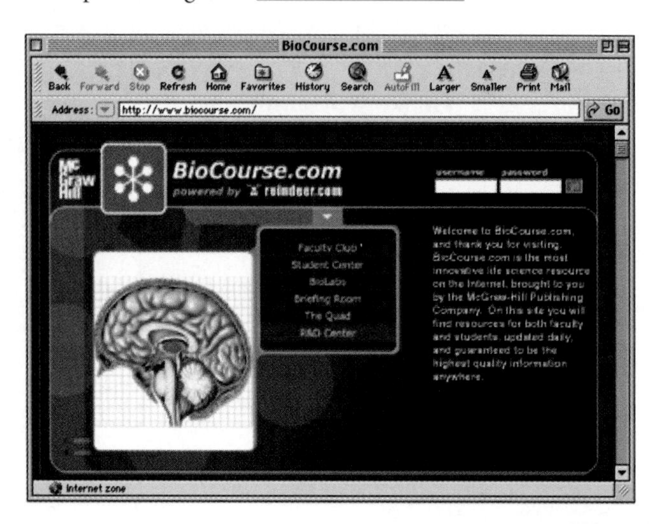

PageOut™

Put together your own customized website using PageOut™, a program designed specifically for instructors wanting to put course information on the web.

Visual Resource Library

Available both on CD-ROM and on the Instructor Center of the Online Learning Center, this resource contains all of the illustrations and photos from the *Life* fourth edition textbook. ISBN 0–07–027221–4

Vectorized (Manipulatable) Art

Selected *Life* text illustrations are available as PowerPoint compatible, vectorized art files that the instructor can manipulate.
ISBN 0–07–246448–8

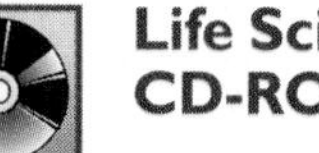

Life Science Animations CD-ROM 2.0

This CD-ROM contains more than 125 animations of important biological concepts and processes.
ISBN 0–07–234296–X

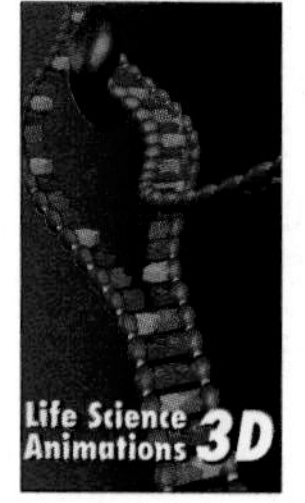

Life Science Animations 3D Videotape

Forty-two key biological processes are narrated and animated with dynamic three-dimensional graphics in full color. ISBN 0–07–290652–9

Life Science Animations Videotape Series

Key physiological processes that are difficult to understand on the static page come to life in this series of five videotapes.

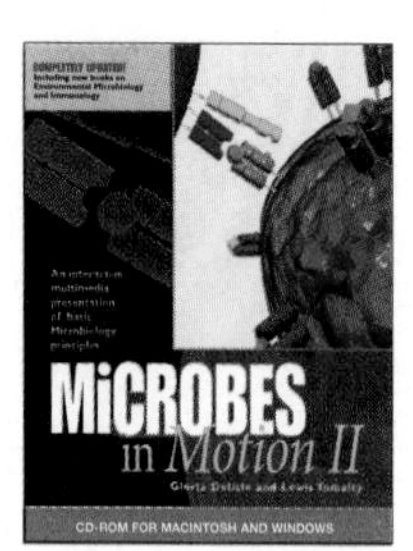

Microbes in Motion CD-ROM, Version 2.0

This interactive CD-ROM allows students to actively explore microbial structure and function. ISBN 0–07–038423–1

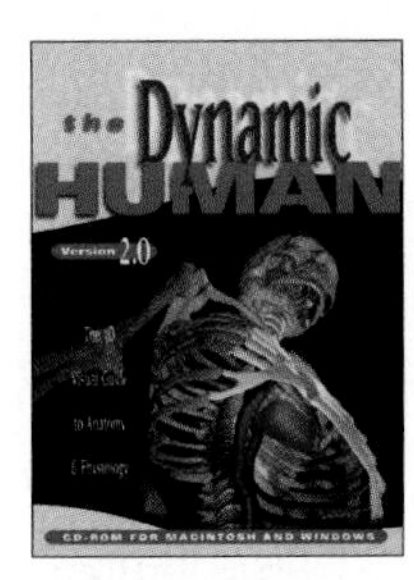

The Dynamic Human CD-ROM, Version 2.0

This guide to anatomy and physiology interactively illustrates the complex relationships between anatomical structures and their functions in the human body. Realistic, three-dimensional visuals are the premier feature of this exciting learning tool.
ISBN 0–07–235476–3

HealthQuest CD-ROM

This interactive CD-ROM allows users to assess their current health and wellness status, determine their health risks and relative life expectancy, explore options, and make decisions to improve the behaviors that impact their health.
ISBN 0–697–29723–3 (Windows)
ISBN 0–07–039335–4 (Macintosh)

Life Science Living Lexicon CD-ROM

Rules of word construction and derivation are carefully explained, in addition to complete definitions of all important terms. ISBN 0–697–37993–0

course*solutions*

McGraw-Hill Course Solutions

Designed specifically to help you with your individual course needs, *Course Solutions* will assist you in integrating your syllabus with *Life, fourth edition,* and state-of-the-art new media tools.

At the heart of the *Course Solutions* you'll find integrated multimedia, a full-scale Online Learning Center, and an enhanced Integration Guide. These unparalleled services are also available as a part of *Course Solutions:* e-Books; Web CT linkage; online animations; McGraw-Hill Course Consultation Service; Visual Resource Library Image Licensing; McGraw-Hill Student Tutorial Service, McGraw-Hill Instructor Syllabus Service; PageOut Lite; PageOut: The Course Website Development Center; and other delivery options.

PRINT SUPPLEMENTS

Instructor's Manual and Test Item File

Bruce Parker has written a new Instructor's Manual and has completely updated and revised the Test Item File.
ISBN 0–07–027224–7

Transparencies

All of the illustrations from the text are included in this set of 750 transparencies. Approximately 50 illustrations from which the labels have been removed are also included. ISBN 0–07–027220–4

Student Study Guide

The study guide, written by Sandra Latourelle, contains activities and questions to help reinforce chapter concepts. ISBN 0–07–027219–0

How to Study Science, Third Edition, by Frederick W. Drewes and Kristen L. D. Milligan

This workbook offers tips on how to take notes and how to overcome science anxiety, in addition to helping students develop critical thinking skills. ISBN 0–697–36051–2

Basic Chemistry for Biology, Second Edition

Carolyn Chapman has written this self-paced book that leads students through basic concepts of inorganic and organic chemistry.
ISBN 0–697–36087–3

The AIDS Booklet, by Frank D. Cox

This booklet describes how AIDS and related diseases are spread so that readers can protect themselves and their friends against this debilitating and deadly disease. ISBN 0–697–29428–5

The Internet Primer: Getting Started on the Internet, by Fritz Erickson and John Vonk

This booklet provides the knowledge you need to access and use some of the Internet's more common features.
ISBN 0–07–303203–4

Critical Thinking Case Study Workbook, by Robert Allen

The 34 case studies in this workbook are designed to immerse students in the process of science and challenge them to solve problems in the same way biologists do. ISBN 0–697–14556–5

Schaum's Outlines: Biology

ISBN 0–07–022405–6

Understanding Evolution, by E. Peter Volpe and Peter Rosenbaum

ISBN 0–697–05137–4

The Online Learning Center with Essential Study Partner 2.0

www.mhhe.com/life

A multitude of teaching and learning tools awaits you on *Life's* Online Learning Center.

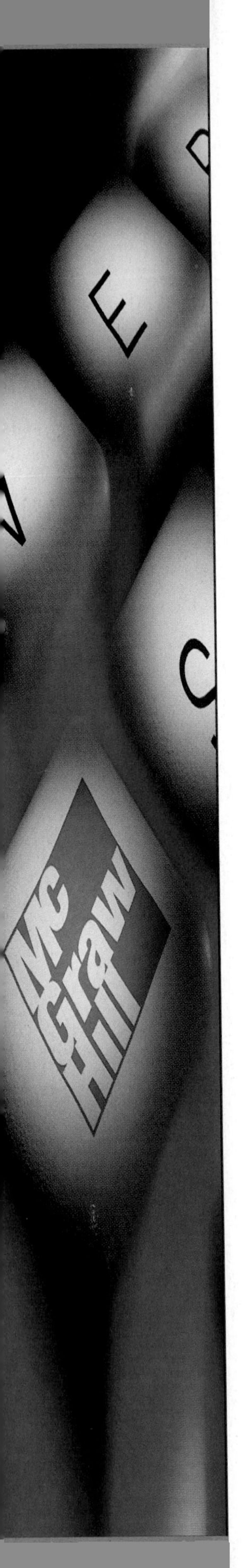

Students, you'll appreciate extensive self-quizzing opportunities; interactive activities; case studies; and related web links in addition to the new Essential Study Partner 2.0—a web-based review of major introductory biology topics—hosted on this site.

Instructors, you'll want to take advantage of our electronic illustrations from the text; classroom activities; lecture outlines; a message board; and access to the PageOut: Course Website Development Center—All available anytime you want them.

Online Learning Center

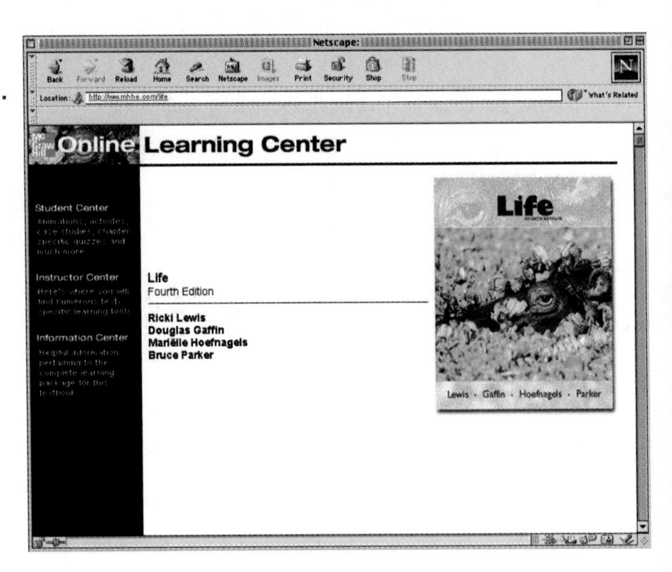

Life's Essential Study Partner 2.0

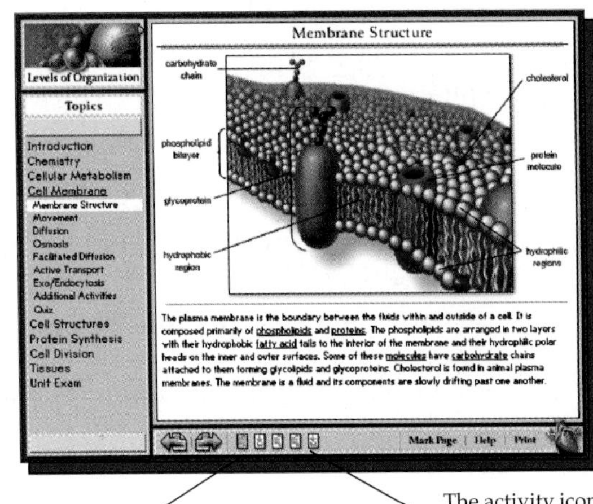

The page icon represents a page of information text.

The activity icon represents an interactive learning activity.

PageOut™

Proven. Reliable. Class-tested.

More than 16,000 professors have chosen **PageOut™** to create course websites. And for good reason: **PageOut™** offers powerful features, yet is incredibly easy to use.

Now you can be the first to use an even better version of **PageOut™**. Through class-testing and customer feedback, we have made key improvements to the grade book, as well as the quizzing and discussion areas. Best of all, **PageOut™** is still free with every McGraw-Hill textbook. And students needn't bother with any special tokens or fees to access your **PageOut™** website.

Customize the site to coincide with your lectures.

Complete the **PageOut™** templates with your course information and you will have an interactive syllabus online. This feature lets you post content to coincide with your lectures. When students visit your **PageOut™** website, your syllabus will direct them to components of McGraw-Hill web content germane to your text, or specific material of your own.

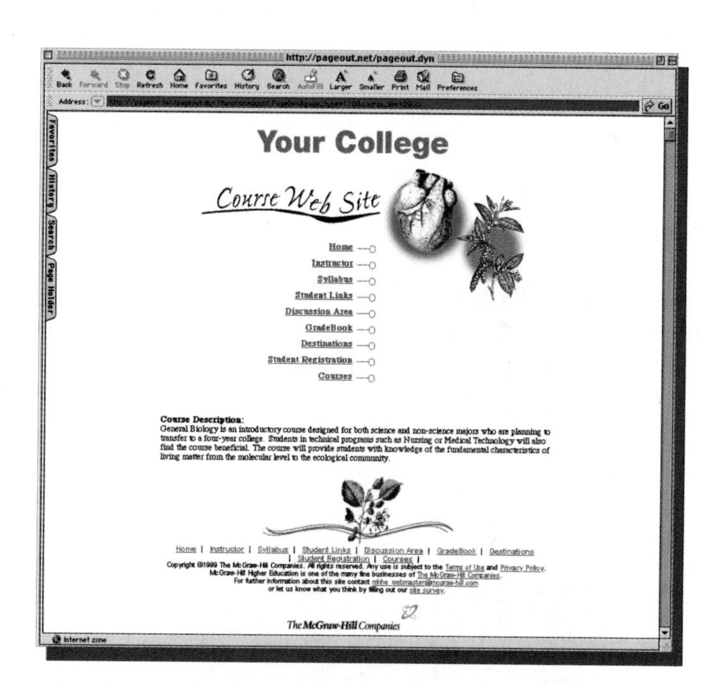

New Features based on customer feedback:

- Specific question selection for quizzes
- Ability to copy your course and share it with colleagues or use as a foundation for a new semester
- Enhanced grade book with reporting features
- Ability to use the **PageOut™** discussion area, or add your own third-party discussion tool
- Password protected courses

Short on time? Let us do the work.

Send your course materials to our McGraw-Hill service team. They will call you for a 30-minute consultation. A team member will then create your **PageOut™** website and provide training to get you up and running. Contact your McGraw-Hill Representative for details.

Contact your McGraw-Hill sales representative for more information or visit *www.mhhe.com*.

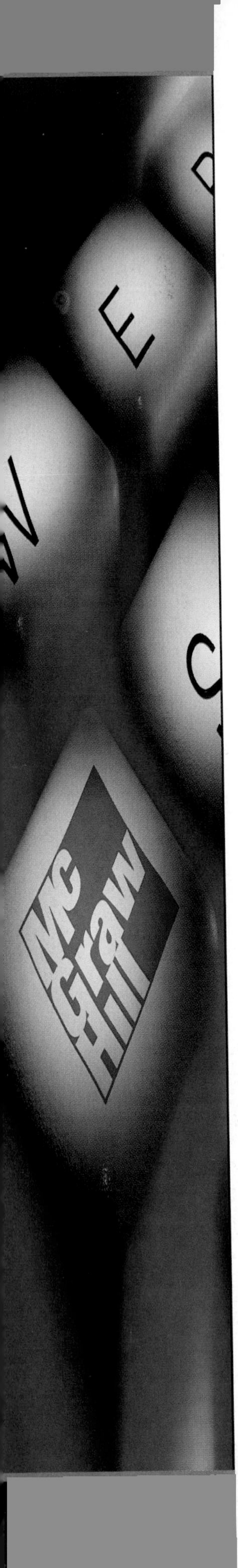

Visual Resource Library CD-ROMs

These CD-ROMs are electronic libraries of educational presentation resources that instructors can use to enhance their lectures. View, sort, search, and print catalog images, play chapter-specific slideshows using PowerPoint, or create customized presentations when you:

- Find and sort thumbnail image records by name, type, location, and user-defined keywords
- Search using keywords or terms
- View images at the same time with the Small Gallery View
- Select and view images at full size
- Display all the important file information for easy file identification
- Drag and place or copy and paste into virtually any graphics, desktop publishing, presentation, or multimedia application

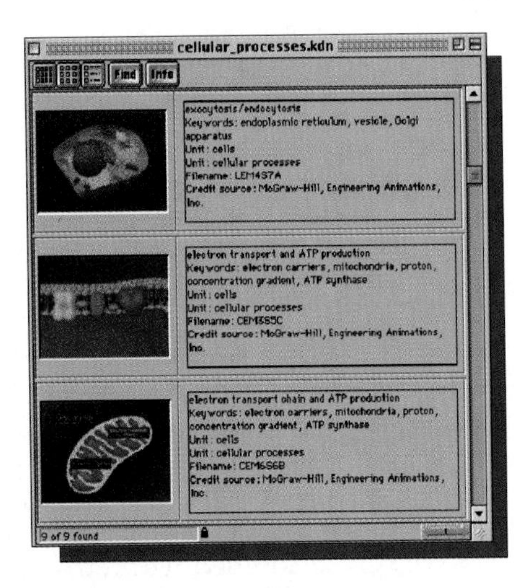

Life Science Animations Visual Resource Library CD-ROM

This instructor's tool, containing more than 125 animations of important biological concepts and processes—found in the *Essential Study Partner* and *Dynamic Human CD-ROMs*—is perfect to support your lecture. The animations contained in this library are not limited to subjects covered in the text, but include an expansion of general life science topics.

Visual Resource Library CD-ROM

This helpful CD-ROM contains ALL 1,500 photographs and illustrations from *Life*. You'll be able to create interesting multimedia presentations with the use of these images, and students will have the ability to easily access the same images in their texts to later review the content covered in class.

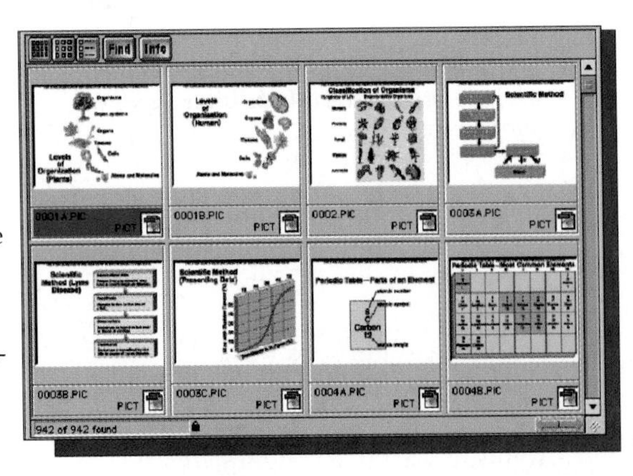

Contact your McGraw-Hill sales representative for more information or visit *www.mhhe.com*.

BioCourse.com

The number one source for your biology course.

BioCourse.com is an electronic meeting place for students and instructors. It provides a comprehensive set of resources in one place that is up-to-date and easy to navigate. You can access BioCourse.com from any of the Online Learning Centers.

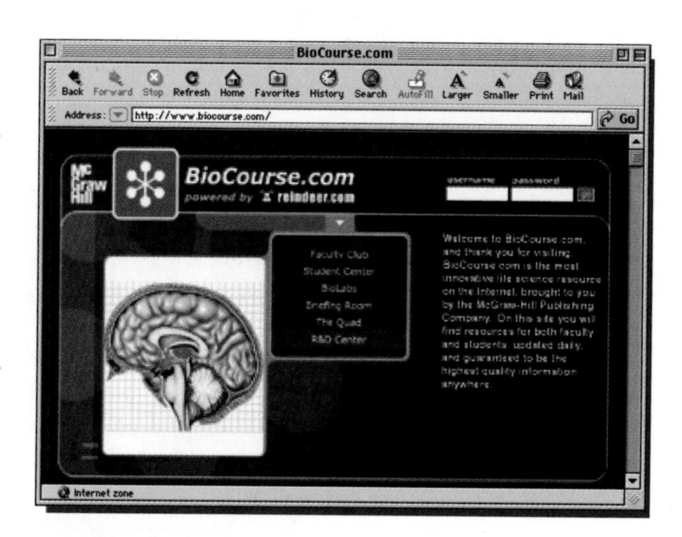

Here is what you will find at BioCourse.com

Faculty Club is an array of information and links to related sites for instructors. Some resources you will find:

- Teaching tips and basic information on pedagogy, assessment, etc.
- Suggestions for classroom and lecture activities.
- Reference searches and literature for faculty.
- Presentation tools.

Student Center contains a wide range of materials to help biology students improve their study skills and achieve success in college and afterward. Examples of materials that will be available:

- Study aids.
- Résumé writing and information on jobs and internships.
- Graduate school options.
- Information for MCAT and other tests.
- Links to content websites by topic.
- A multilingual lexicon.

Briefing Room offers instructors and students up-to-date news articles, a selection of background readings and links to journal search tools and biology magazines. Users can email articles to others, link to search engines, read primary sources online, and access case studies on a wide range of topics.

BioLabs feature materials for lab students and instructors. Some tools you will find include:

For students:

- Dissection techniques.
- Equipment tutorials.
- Safety and setup procedures.

For instructors:

- Lab preparations.
- Lab support.
- Simulations.

Lifelong Learning Content Warehouse is a powerful indexing tool and hierarchical outline of content resources for searching by students and faculty. Users can search by topic through a "content warehouse" featuring text material, activities, visuals, and animations to learn more about a selected topic.

R & D Center features our newest simulations, animations, and other teaching and learning tools. This portion of our site will allow faculty members and students to try out our materials as they are being developed.

Contact your McGraw-Hill sales representative for more information or visit *www.mhhe.com*.

FOURTH EDITION

UNIT 1 • FROM ATOMS TO CELLS

Life is evident on nearly all parts of the planet's surface. The organisms that carpet the earth in many places have great complexity and diversity, yet share similar requirements for being and staying alive, and consist of the same types of chemicals, often organized in the same ways. The animals grazing in the field, and the plants they eat, use atoms and molecules as the foundation of everything they do. Plants rely on sunlight to provide the energy to build stems, roots, and leaves. Ultimately, molecules in the plants are used to make molecules in the moose. This unit explores the chemical foundations of life, cells, and how energy is captured and directed to build and sustain life.

CHAPTER 1

What Is Life?

Life in the Extremes

Redefining the Impossible

"I still have nightmares about being boiled alive," recalls Francis Barany, a biochemist at Cornell University Medical College in New York City. Barany fell into a pool of boiling mud in Yellowstone National Park in 1991, while searching for bacteria. His story vividly illustrates the excitement of biological research, and the possible danger. Barany explored biodiversity and clues to the possible origin of life, while looking for microorganisms that produce interesting chemicals.

A few months later, he was still hobbling about, taking pain medication, and telling everyone about his accident.

The steaming water and mud of hot springs today are found only near volcanoes and certain lakes, where they are topped with colored mats of algae that are home to a few types of thermophiles ("heat loving") microorganisms. Four billion years ago, such hot environments were probably much more common. Could observing how organisms today live in pockets of the environment that resemble conditions of the early earth reveal clues to how life began? Many biologists think so, and so they come to this beautiful park.

Francis Barany wasn't seeking clues to the origin of life when he fell. He was looking for bacteria that make chemicals that do not fall apart under high-temperature conditions, as many chemicals do. He had developed a new method to mass-produce specific pieces of genetic material (deoxyribonucleic acid, or DNA), which would form the basis of tests to diagnose cancer, genetic and infectious diseases. Because the technique required heating the DNA, briefly and repeatedly, so that strands could form against old strands, he needed a DNA-copying enzyme that could tolerate heat. (An enzyme is a protein that speeds the rate of a specific chemical reaction.) Since hot springs microbes reproduce readily, their enzymes must be able to withstand high temperatures.

Indiana University microbiologist Thomas Brock had discovered a mi-

Life in Extremes.
The water in this hot spring in Yellowstone Park is boiling—but certain microorganisms thrive in it.

crobe at Yellowstone in 1965 by dipping glass slides into hot springs, and collecting the heat-loving bacteria that accumulated. It was one of the first observations that life could exist at high temperatures.

A quarter century later, Barany ventured into the hot springs. Here is his story:

> *While collecting my thirteenth sample, the ground simply gave way, collapsed under me, and my left leg sank into the hot mud. It was 70°C—that's 158°F! It blistered the skin right off. I jumped out and hopped onto land, screaming. Someone poured cold water on my leg, and ripped my jeans off. I remember screaming that I had a permit to collect, and to save that thirteenth sample!*

Barany indeed found some novel enzymes in that thirteenth sample. A few months later, he was still hobbling about, taking pain medication, and telling everyone about his accident. But he felt healed enough to marry the woman who pulled him out of the hot springs, and today they have two little girls. Still, Barany often relives in his mind his brush with death, and tells all explorers to exercise caution in their work.

In 1991, it was still somewhat unusual for a biologist to fall into a hot springs while hunting bacteria. Today, the search for "extremophiles," as organisms that live in what to us are hostile conditions are called, is big business for pharmaceutical, chemical, and biotechnology companies. The reason—the microbes' hardy "extremozymes," a new term for enzymes that function under extremes of temperature, pressure, acidity or salinity. Researchers look for extremophiles everywhere. One biotechnology company offers enzymes from life in Arctic pools, Indonesian volcanoes, Costa Rican jungles, the rock beneath the Savannah River in South Carolina, and, of course, Yellowstone National Park.

Extremozymes have varied uses. A detergent that removes lint and fuzz from clothing comes from an organism living in a salty mountain lagoon. Another extremozyme breaks up and bleaches wood pulp, and is used to make paper. Microorganisms can thrive in a sulfuric acid hot spring in Yellowstone, in the heat of a smoldering coal mine, and in places where superheated water bubbles up from cracks in the seafloor. Similar bacteria also live in a cow's stomach, enabling the animal to digest cheaper feed.

Products from microbes living in Antarctic sea ice are used in food processing done at low temperatures to retard spoilage, and in cold-wash laundry detergents. We use extremozymes to stonewash jeans, convict criminals using DNA fingerprinting techniques, and to diagnose infections and produce the antibiotic drugs that treat them. Extremozymes are also often safer than the industrial chemicals that they replace, which may require toxic substances to produce.

Studying extremophiles has also sent biologists back to the drawing board in classifying life into major groups, for life is far more diverse than they had realized. In a more philosophical sense, the finding of life in forbidding places points out starkly how biology is an ever-changing discipline, as is all science. Discoveries compel us to ask more and different types of questions, to alter our thinking, and to continually wonder at just what constitutes this thing we call life.

1.1 What Are the Characteristics of Life?

A distinctive set of characteristics distinguishes the living from the nonliving. Together, they enable an organism to acquire energy to support its growth and functioning, and to reproduce, perpetuating life.

We all have an intuitive sense of what life is: If we see a rabbit on a rock, we "know" that the rabbit is alive and the rock is not. But it's difficult to state just what it is that makes the rabbit alive and the rock not.

Scientists define life in terms of qualities that the living uniquely share. But in the instant before a baby takes its first breath—or after an individual dies—we may ask, as nineteenth-century philosophers called vitalists did, "Does some indefinable 'essence' render a particular collection of structures or processes living?"

Throughout history, the question, "What is life?" has stymied thinkers in many fields. Eighteenth-century French physician Marie Francois Xavier Bichat poetically but imprecisely defined life as "the ensemble of functions that resist death." Others less eloquent and still no more precise have hypothesized that life is a kind of "black box" that endows a group of associated biochemicals with the qualities of life. Still others have tried to break life down into the smallest parts that still exhibit the characteristics of life, and then identify what makes those units different from their nonliving components.

Perhaps we can define life by considering the characteristics that vanish with death. For a human in a hospital, death is defined as cessation of brain activity. But what constitutes death of the more than 30 million other types of organisms?

Five qualities in combination constitute life (table 1.1). An **organism** is a collection of structures that function coordinately and exhibit these qualities. These characteristics alone may also occur in inanimate objects. A computer responds to stimuli, as does a self-flushing toilet, but neither is alive. A fork placed in a pot of boiling water absorbs heat energy and passes it to a hand grabbing it, just as a living organism receives and transfers energy, but this does not make the fork alive. A supermarket is highly organized, but it is obviously not alive! Living organisms have all five characteristics.

Organization

Living matter consists of structures organized in a particular three-dimensional relationship, often following a pattern of structures within structures within structures. Bichat was the first to notice this pattern in the human body. During the bloody French Revolution, as he performed autopsies, Bichat noticed that the body's largest structures, or **organs,** were sometimes linked to form collections of organs, or **organ systems,** but were also composed of simpler structures, which he named **tissues** (from the French word for "very thin"). Had he used a microscope, Bichat would have seen that tissues themselves are comprised of even smaller units called **cells.** All life consists of cells. Within complex cells such as those that make up plants and animals are structures called **organelles** that carry out specific functions. Organelles, and ultimately all living structures, are composed of chemicals (fig. 1.1).

TABLE 1.1 Characteristics of Life

1. Organization
2. Energy use and metabolism
3. Maintenance of internal constancy
4. Reproduction, growth, and development
5. Irritability and adaptation (response to environment)

The chemicals that make up organisms are called biochemicals, although some chemicals abundant in organisms, such as water, are also widespread in the nonliving world. An important biochemical is deoxyribonucleic acid, or DNA. The sequences of DNA's four types of building blocks are like a language of life, encoding information that tells cells how to construct particular proteins. Cells' use of genetic instructions—as encoded in DNA—enables them to specialize and to function in tissues, organs, and organ systems.

At all levels, and in all organisms, structural organization is closely tied to function. Disrupt a structure, and function ceases. Shaking a fertilized hen's egg, for example, stops the embryo within from developing. Conversely, if function is disrupted, a structure will eventually break down. Unused muscles, for example, begin to atrophy (waste away). Biological function and form are interdependent.

Biological organization is apparent in all life. Humans, eels, and evergreens, although outwardly very different, are all nonetheless organized into specialized cells, tissues, organs, and organ systems. Bacteria, although less complex than animal or plant cells, are also composed of highly organized structures.

An organism, however, is more than a collection of smaller parts. The organization of different levels of life imparts distinct characteristics. For example, individual chemicals cannot harness energy from sunlight. But groups of chemicals embedded in the membranes of an organelle within a plant cell use solar energy to synthesize nutrients. Functions that arise as complexity grows are called **emergent properties** (fig. 1.2). These characteristics are not magical; they arise from physical and chemical interactions among a system's component parts.

Energy Use and Metabolism

The organization of life seems contrary to the natural tendency of matter to be random or disordered—that is, energy is required to maintain organization or order. An organism must be able to acquire and use energy to build new structures, repair or break down old ones, and reproduce. These functions occur at the whole-body or organismal level, as well as at the organ, tissue, and cellular levels. The term **metabolism** refers to the chemical reactions within cells that maintain life. Metabolic reactions both build up (synthesize) and break down (degrade) biochemicals.

FIGURE 1.1

Levels of Biological Organization Reveal Common Features of All Life.

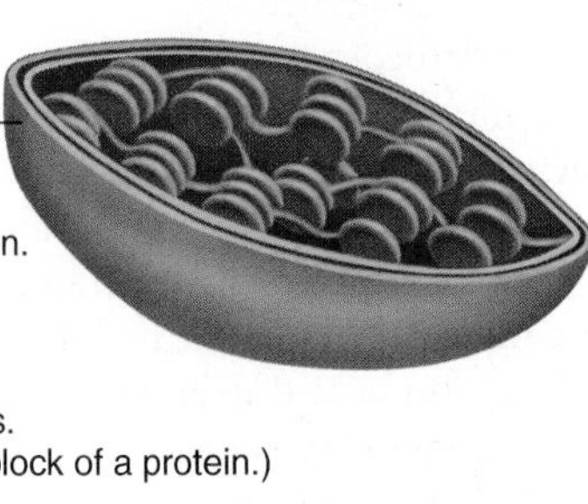

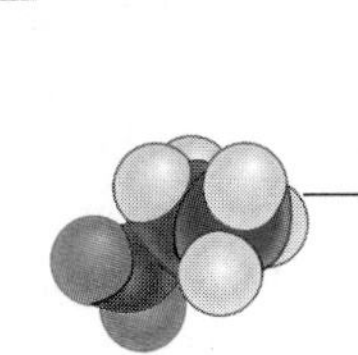

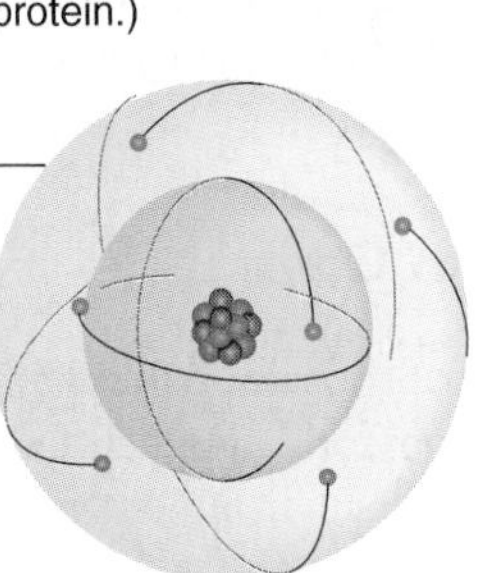

FIGURE 1.2
An Emergent Property—From Tiles to Tubes.
Endothelial cells look like tiles. They adhere to one another to form a sheet. This sheet folds to form a tiny tubule called a capillary, which is the smallest type of blood vessel. The function of these cells does not "emerge" until they aggregate in a specific way.

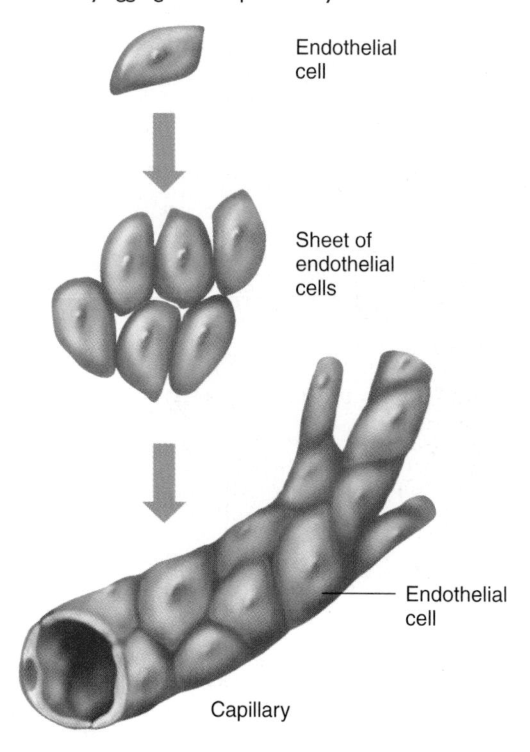

FIGURE 1.3
Life Is Connected.
Organisms extract energy from each other (consumers), and, ultimately, the sun or inorganic chemicals (producers). Decomposers ultimately recycle nutrients to the nonliving environment.

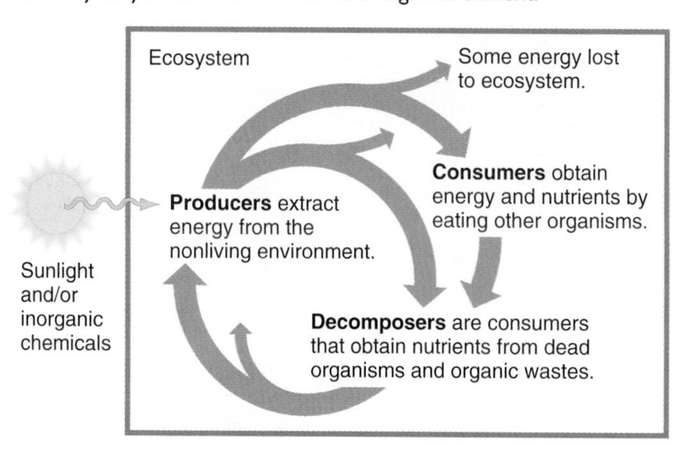

Organisms obtain energy to carry out life processes from the environment (fig. 1.3). **Producers** (also called autotrophs) are organisms that extract energy from the nonliving environment, such as plants capturing light energy from the sun or bacteria deriving chemical energy from rocks. **Consumers** (also called heterotrophs), in contrast, obtain energy by eating nutrients that make up other organisms. **Decomposers** are consumers that obtain nutrients from dead organisms. Fungi, such as mushrooms, are decomposers.

Energy requirements link life into chains and webs of "who eats whom," beginning with producers that capture energy from the nonliving environment and continuing through several levels of consumers and decomposers. This interdependency of life suggests other ways that we can view organization in the living world.

Life's organization and energy use can be described beyond the level of the individual. A **population** includes two or more members of the same type of organism, or **species,** living in the same place at the same time. A **community** includes the populations of different species in a particular region, and an **ecosystem** includes both the living and nonliving components of an area. Finally, the **biosphere** refers to the parts of the planet that can support life, and all of the organisms that live there (see fig. 1.1).

Maintenance of Internal Constancy

A living organism is composed of the same basic chemical elements as nonliving matter, yet it is obviously quite different. The levels of chemicals that make up the environment inside cells must remain within a constant range, even in the face of drastic changes in the outside environment. This ability to keep conditions constant is called **homeostasis.** For example, a cell must maintain a certain temperature and water balance. It must take in nutrients, excrete wastes, and regulate its many chemical reactions to prevent deficiencies or excesses of specific substances. An important characteristic of life, then, is the ability to sense and react to environmental change, thus keeping conditions within cells (and bodies) constantly compatible with being alive.

Reproduction, Growth, and Development

Organisms reproduce—that is, they make other individuals like themselves. These new organisms then grow, which means increasing in size, and develop, which entails adding anatomical detail and taking on specialized functions. Reproduction and development include a bacterium dividing in two, the germination and growth of a tree seedling, and the conception, gestation, and birth of a mammal (fig. 1.4). On the organismal level, reproduction transmits genetic instructions (DNA), which determine the defining characteristics of the organism from one generation to the next. Reproduction is vital for a population of organisms to survive for more than one generation, and on a biosphere level, it is essential for life itself to continue.

Reproduction occurs in two basic ways. In **asexual reproduction** in single-celled or **unicellular** organisms, cellular contents double; then the cell divides to form two new individuals that are genetically identical to the original parent cell. Some many-celled or **multicellular** organisms reproduce asexually. For example, a potato growing on an underground stem can sprout leaves and roots that form a new plant. Asexual reproduction also occurs in some simple animals, such as sponges and sea anemones, when a fragment of the parent animal detaches and develops into a new individual.

In **sexual reproduction,** genetic material from two individuals joins to form a third individual, which has a new com-

FIGURE 1.4
Reproduction Is One of the Most Obvious of Life's Characteristics.
(**A**) These *Escherichia coli* bacterial cells reproduce asexually every 20 minutes under ideal conditions. (**B**) A mighty oak tree begins with a small seedling and each tree produces hundreds of acorns. (**C**) By contrast, mammals such as these deer usually produce only a few offspring per year.

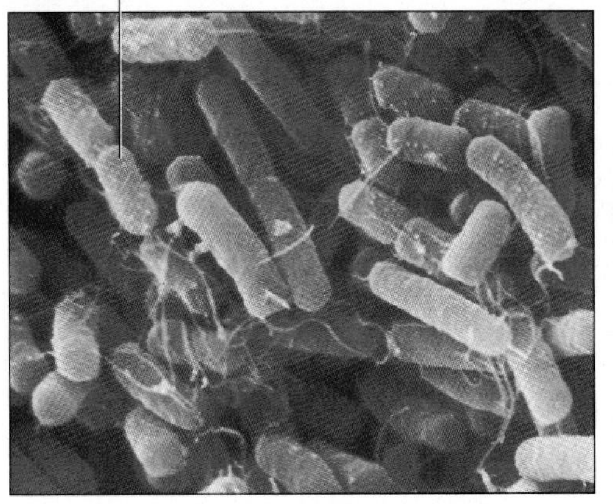

FIGURE 1.5
An Adaptation to Acquire Food.
The superb camouflage of the adder snake, *Bitis peringueyi,* makes it virtually undetectable buried in the sand in the Namib Desert, Namibia (**A**). It is little wonder that the sand lizard, *Aporosaura anchietae,* soon became the meal of the snake (**B**).

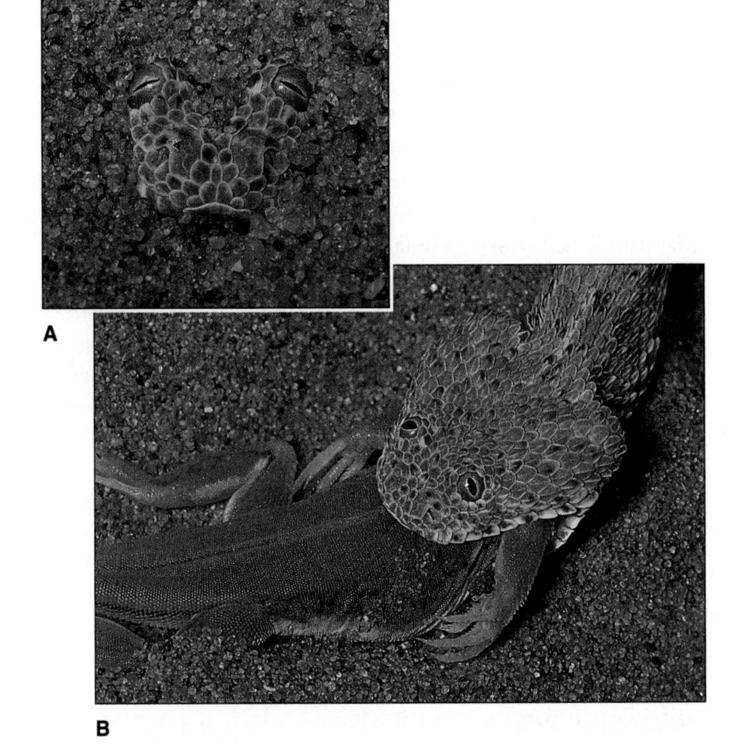

bination of inherited traits. By mixing genetic traits at each generation, sexual reproduction results in tremendous diversity in a population.

Irritability, Adaptation, and Natural Selection

Living organisms sense and respond to certain environmental stimuli while ignoring others. **Irritability** is the tendency to respond immediately to stimuli, and it can be essential for survival. Examples of irritability include a person touching a thorn and quickly pulling that hand away; a dog lifting its ears in response to a whistle; and a plant growing toward the sun.

In contrast to the immediacy of irritability, an **adaptation** is a response that develops over time. It is an inherited characteristic or behavior that enables an organism to successfully reproduce in a given environment. For example, the "extremozymes" of the microorganisms described in the opening essay are adaptations that permit survival at very high temperatures.

Adaptation is usually more complex and specific than irritability. Adaptations are likely to differ from one species to another and within the same species in different environments. Both a human and a dog will pull an extremity away from fire (irritability). But on a hot summer day, the human sweats, while the dog pants (adaptations).

Adaptations can be diverse and very striking. The color patterns of many organisms that enable them to literally fade into the background, such as the snake in figure 1.5, are adaptations. Such a camouflaged organism can hide from predators or await prey unnoticed.

Plants have some interesting adaptations. Most trees, for example, can survive strong winds because of their wide and sturdy

FIGURE 1.6
Leaves of Many Trees Are Adapted to Minimize Wind Damage.
Researcher Steven Vogel subjected various leaves to a wind tunnel and snapped these photos, which reveal a previously unrecognized adaptation to survive a storm. Leaves (**A**) characteristically fold into a more compact form—under moderate (**B**) and high wind (**C**) conditions.

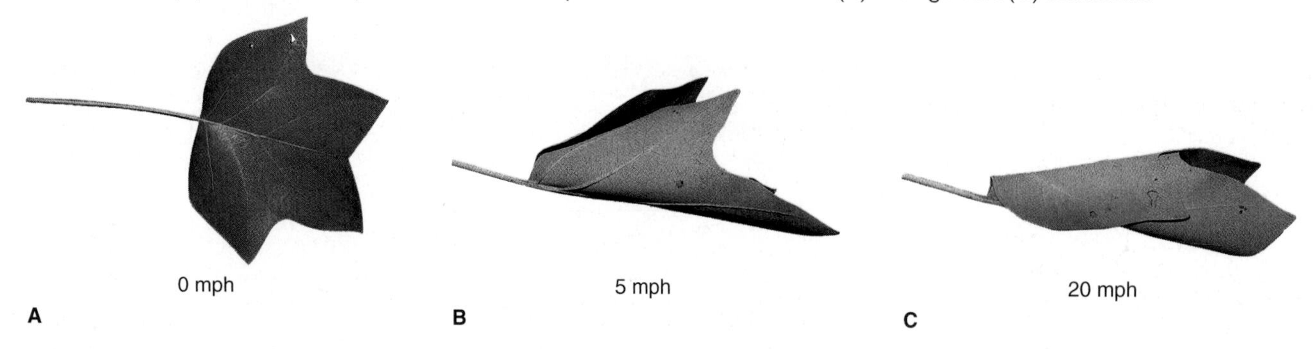

trunks, flexible branches that sway without snapping, and strong, well-spread roots. Figure 1.6 shows another adaptation to wind not readily apparent during the violence of a storm. A researcher who noticed that the fronds of certain algae close into cones and cylinder shapes in turbulent streams studied several types of trees during simulated "wind tunnel storms" to see if similar reactions might occur. Indeed, he found that the leaves of some types of trees do close into tighter shapes, which minimizes wind damage.

Adaptations accumulate in a population of organisms when individuals with certain inherited traits are more likely than others to survive in a particular environment. Trees with leaves that curl to resist wind are more likely to survive for the 20 or more years it takes for most trees to reproduce—even in the face of severe windstorms that strike, on average, every 5 years. Because trees with curling leaves survive more often than trees without this adaptation, more of them survive to reproduce and pass the advantageous trait to future generations. Trees with the adaptation eventually predominate in the population—unless environmental conditions change and protection against the wind becomes less important.

Over time, adaptation can mold the characteristics of a population by enabling individuals with certain combinations of genes to preferentially survive and reproduce. **Natural selection** is the enhanced survival and reproductive success of certain individuals from a population based on inherited characteristics. As individuals that have inherited particularly adaptive traits contribute more offspring, they come to make up more of the population. As the population changes, evolution is occurring. Evolution may be as subtle as the increasing prevalence of a particular trait or as profound as the extinction of a species. Largely because of natural selection, fewer than 1% of the species that have ever existed on earth are alive today. The diversity of life is constantly changing.

1.1 MASTERING CONCEPTS

1. Why is it difficult to define life?
2. What are some of the ways people have tried to define life?
3. What combination of characteristics distinguishes the living from the nonliving?
4. How does life change over time?

OLC

1.2 How Is Life Diverse?

Organisms carry out the same basic functions, but in different ways. Evolution explains the fundamental similarity of all life; biodiversity reveals the many variations on that theme.

The mechanisms of evolution explain the source of the many types of organisms—**species**—that have populated the earth. Life is united in its basic defining characteristics, yet it is also diverse, especially in the adaptive strategies organisms use to maintain those characteristics. **Biodiversity** refers to the many different types of organisms on Earth.

Domains and Kingdoms Describe Life

We humans love to classify things. We organize large stores by departments, sort laundry, and assign students to grades and classes, based on their ages, abilities, and interests.

Similarly, the biological science of **taxonomy** classifies life according to what we know about the evolutionary relationships of organisms—that is, how recently one type of organism shared a common ancestor with another type of organism. The more recently divergence from a shared ancestor occurred, the more closely related we presume the two types of organisms to be. Researchers determine how closely related types of organisms are by anatomical, behavioral, cellular, and biochemical similarities.

Biologists use several levels with increasingly restrictive criteria to describe organisms. Just as a student is assigned to a particular school district, school, grade, and class, an organism is assigned to a specific **domain, kingdom, division** or **phylum, class, order, family, genus,** and species. Taxonomists relatively recently added the domain, as the broadest category, following the discoveries of many types of unicellular organisms called Archaea.

Life's three domains are the Bacteria, Archaea, and Eukarya. Basic differences in cellular constituents and organization distinguish the domains, which are discussed in greater detail in chapters 3, 4, and 18. Figure 1.7 illustrates the relationship between domains and the more familiar and specific kingdoms, with a representative member of each described. Organisms within a kingdom demonstrate the same general strategies for staying alive, but they differ from each other in the details of how they do

FIGURE 1.7

Organizing Life's Diversity.

Before microscopes revealed the microbial world, it was easy to describe life as either plant or animal; the fungi were considered plants. With increasing discoveries and descriptions of microbes, it became clear that two, three, or even four kingdoms (animals, fungi, plants, and bacteria) would not suffice. A five-kingdom system, splitting microorganisms into bacteria and protista, led for a long time, until some years after the 1977 discovery of the Archaea. Identified first in extreme habitats like those of early Earth and therefore considered to be ancient (hence the name), the Archaea share some characteristics with the Bacteria, but also with organisms that have cells with nuclei. Many of their genes have never been described before. The descriptions of Bacteria and Archaea are identical here because most of their differences are at the molecular level. Today, biologists have reorganized the classification of life into three domains, and are still determining the numbers of kingdoms in each domain and their relationship to each other. Overall, biological classification strives to depict evolutionary relationships among organisms, both past and present. Figure 3.4 lists more specific distinctions among the three domains of life.

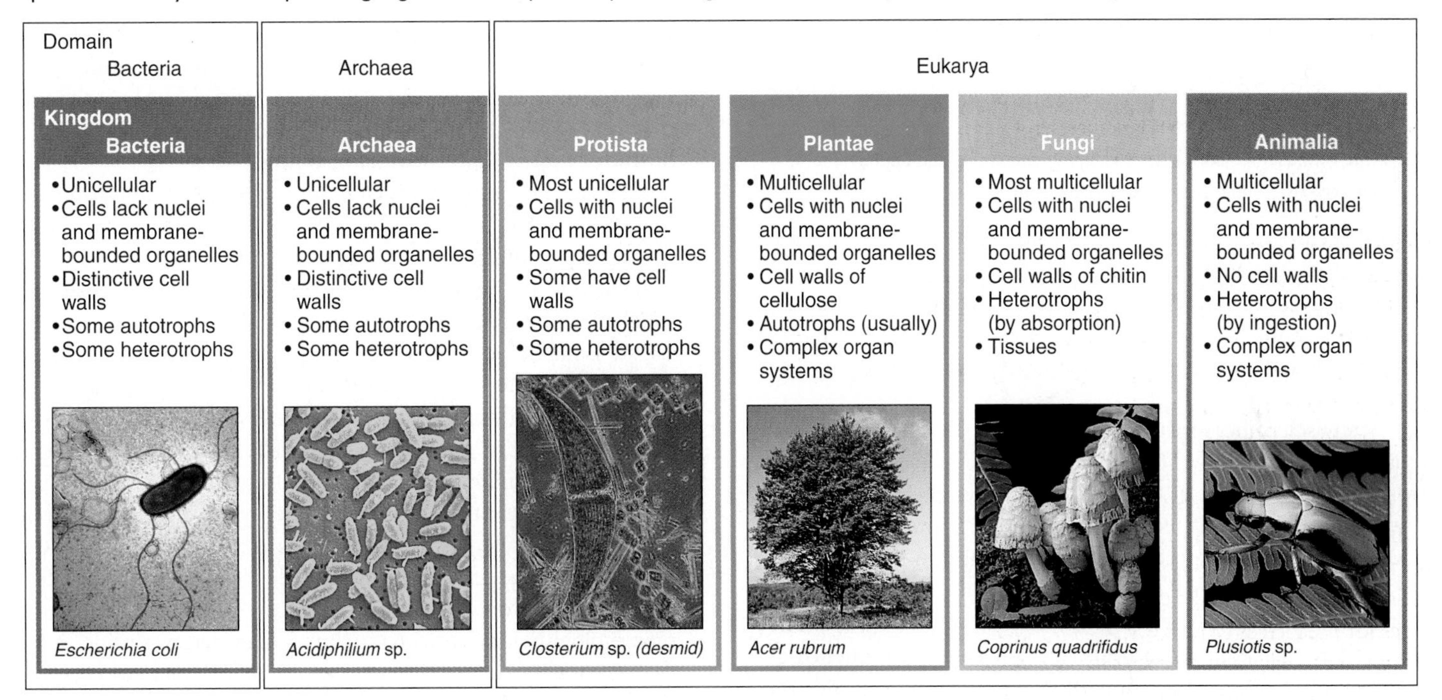

so. Cell complexity, mode of energy acquisition and use, and reproductive mechanisms distinguish the kingdoms. Members of Bacteria and Archaea are all single-celled without nuclei. All eukaryotes have nuclei and many are multicellular, like us. Bacteria and Archaea differ in their chemistry.

The members of a domain or kingdom share many characteristics, yet they are quite diverse—just as students in a school district are alike in that they live in the same area but are nonetheless individuals. A human, squid, and fly, for example, are all members of the animal kingdom, but they are clearly very different from each other and do not share the other more specific taxonomic levels (phylum to species). A human, rat, and pig are more closely related—all belong to the same kingdom, phylum, and class (Mammalia). A human, orangutan, and chimpanzee are even more closely related, sharing the same kingdom, phylum, class, and order (Primates).

Taxonomists give each type of organism an abbreviated name, called its binomial name, that consists of two descriptive words. A human is *Homo sapiens.* Organisms may be classified and combined into smaller and smaller groups based on common features. For example, our full classification is Eukarya Animalia Chordata Mammalia Primates Hominidae Homo *Homo sapiens.* Another familiar organism is Eukarya Plantae Angiospermophyta Monocotyledoneae Liliales Liliaceae Allium *Allium sativum*—otherwise known as garlic. The most restrictive taxon, species, designates organisms that can breed only among themselves and produce fertile offspring or distinctive types of asexually reproducing organisms.

Evolution Is the Backbone of Biology

At the biochemical and cellular levels, different types of organisms show incredible similarities, using the same materials to carry out much the same functions. The most logical explanation for these similarities is that the many species that have existed all descend from a common ancestral type of organism. But how, and why, did life diversify? The answer lies in the informational molecule of life, DNA.

DNA not only encodes the information that cells require to function, but also has a capacity to change. Such a change in the sequence of DNA building blocks is called a **mutation.** DNA's ability to mutate has provided, and continues to provide, the variation upon which natural selection acts. Ever since the first collections of DNA-like molecules became set off from the environment in fatty bubbles and formed the first cells, life has been changing, and diversifying, as the environment altered in ways that allowed only some organisms to survive. Evolution is the consequence of DNA-based changes in species over time. It has been operating since life began, has molded the species that have populated the planet, and continues to act today. Life is always changing.

1.2 MASTERING CONCEPTS

1. What is biodiversity?
2. How are domains related to kingdoms?
3. What is the underlying basis of taxonomic groupings of organisms?
4. How do natural selection and mutation guide evolution?

1.3 How Do Scientists Study the Natural World?

The scientific method is a thinking process that guides our evaluation of what we observe. Curiosity, questioning, and systematic testing enable scientists, and others, to learn how nature works.

Anybody can be a scientist—it basically requires observing and asking questions about nature. Science seeks objective evidence that explains how nature works. Table 1.2 lists some areas that are sometimes confused with science, but are not science because they rely on beliefs and feelings, rather than evaluation of objective data such as measurements.

Many scientific investigations are based on discovery. A microbiologist finds a novel hot springs bacterium; an ecologist catalogs the species that survived severe weather; a botanist discovers a new type of plant in a rain forest. Sometimes a scientist devises a test, or **experiment,** to investigate how or why things are as they are in nature. An experiment tests the validity of a prediction that is based on previous observations or knowledge. The prediction is called a **hypothesis.** Deciphering the sequence of DNA building blocks in a human cell—a decade-long project completed in 2000—is discovery. But devising ways to demonstrate what individual genes do requires experiments.

Often new scientific information comes from connecting observations, which can lead to experimentation, as Investigating Life 1.1 describes. Many links between environmental insults and development of specific cancers, for example, began with noting the types of people at higher-than-normal risk: heavy smokers and lung cancer; people exposed to atomic test blasts in the southwestern United States in the 1950s and blood cancers; young women who painted a radioactive chemical onto watchfaces in the 1920s and bone cancer. Experiments on nonhuman animals and tissues growing in culture then fill in the details of how the environmentally induced cancer develops. The study of disease-related data from real life is called **epidemiology,** and it is often a starting point for experimental approaches.

TABLE 1.2 On Distinguishing Science from Nonscience

WHAT SCIENCE IS	WHAT SCIENCE ISN'T
Understanding how nature works, based on objective evidence that includes reproducible experimental data and measurements and observations	Art Astrology Creationism Extrasensory perception Fortune telling Healing crystals Philosophy Psychic phenomena Reincarnation Religion Telekinesis Telepathy Therapeutic touch

How Do Theories Explain Nature?

When asked to comment on the idea of biological evolution, former U.S. president Ronald Reagan dismissed its validity, saying, "It's only a theory, anyway." In an informal sense, a theory may very well be little more than an opinion. But in many fields of study, "theory" has a distinct meaning.

Unlike a hypothesis, which is applied to initial observations, a **theory** is a systematically organized body of knowledge that applies to a variety of situations. Theories attempt to logically explain natural phenomena, but we cannot know something scientific with absolute certainty. Theories change, as we learn more. The history of science is full of long-established ideas changing as we learned more about nature, often thanks to new technology. People thought that Earth was flat and at the center of the universe before new inventions and data analysis revealed otherwise. Similarly, biologists thought all life was plant or animal until microscopes unveiled a world of organisms invisible to the naked eye. Sometimes evidence for a theory is so consistent, over many years, that it comes to be considered a law. This is the case for the principles that underlie heredity, called Mendel's laws. Although Ronald Reagan scoffed at evolution as "just a theory," most biologists, familiar with the abundant evidence, consider it law.

Theories build on interpretations of observations, evidence, and the results of experiments. Those interpretations arise from application of the **scientific method,** which is a general way of thinking and of organizing an investigation. The scientific method is a framework in which to consider ideas and evidence in a way that can be repeated with the same results. Scientists can take several approaches to answering a particular question. An ecologist, a chemist, a geneticist, and a physicist conduct very different types of investigations. Scientific inquiry consists of everyday activities—observing, questioning, reasoning, predicting, testing, interpreting, and concluding (fig. 1.8).

FIGURE 1.8
The Scientific Method Is a Means of Careful Discovery.

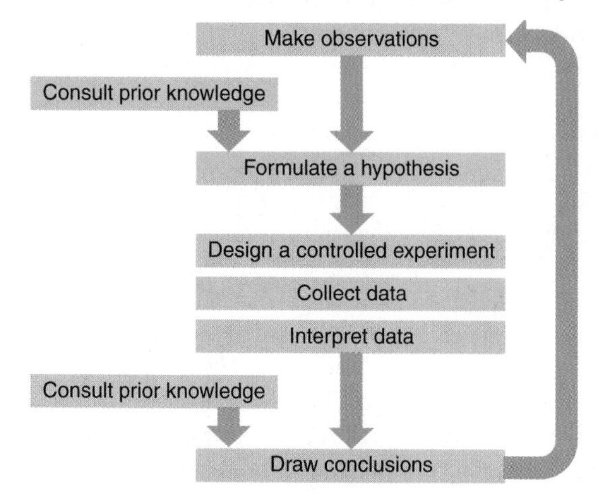

B

C

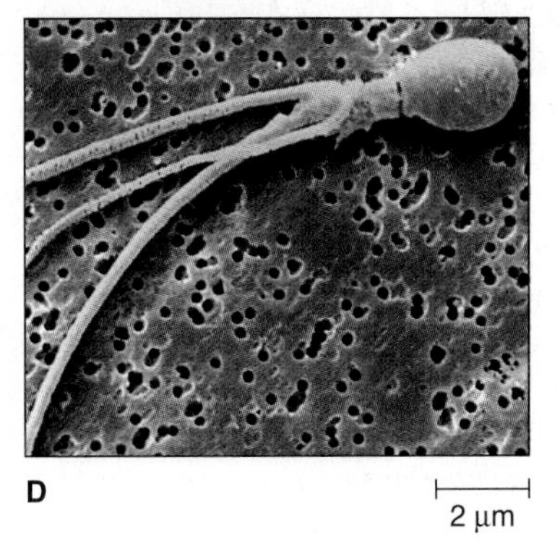

FIGURE 1.9
Theories Are Built on Evidence.
The estrogen mimic theory proposes that estrogen-like chemicals in pesticides cause reproductive abnormalities. (**A**) DDT was sprayed on crops and on some people. (**B**) Bird populations exposed to DDE (a breakdown product of DDT) and other environmental estrogens experience many problems, such as this eagle with a malformed beak. (**C**) Nests containing broken eggs are evidence of the effect of DDT on the proper formation of eggshells, which are too fragile to be properly incubated. (**D**) Abnormal sperm, such as this one with two extra tails, may also be the result of exposure to estrogen-based pesticides. More observations and experiments are needed to confirm the estrogen mimic theory.

It includes thinking, detective work, and seeing connections between seemingly unrelated events. To explore the method of scientific inquiry and the development of a scientific theory, we will consider several aspects of one environmental problem—the effects of chemicals that resemble the hormone estrogen in animals. An animal hormone is a biological messenger that is produced in a gland and is transported in the bloodstream to where it exerts a specific effect on a particular organ. The idea that estrogen-like chemicals in the environment can harm health is called the estrogen mimic theory (fig. 1.9).

Observations The scientific method begins with observations. These observations may be historical accidents:

- From 1949 to 1971, 2 million pregnant women in the United States took an estrogen-based drug called DES, supposedly to prevent miscarriage. Years later, some of their daughters developed a rare vaginal cancer, or less serious reproductive abnormalities.
- In 1980, a large amount of a now-banned pesticide, DDT, was dumped in Lake Apopka in Florida. Exposed male alligators had stunted penises. DDT is broken down to DDE, which functions as an estrogen. Alligators of both sexes had excess estrogen.

Observations can also be based on experimental results, or may arise from making mental connections:

- Estrogen applied to human breast or uterine cells growing in culture stimulates them to divide.
- Women are more likely to develop breast cancer if they begin menstruating early and cease menstruating late—factors that expose them to more estrogen over their lifetimes than women who begin menstruating later and end earlier.
- Since 1938, human sperm counts have dropped, and the incidence of birth defects of the male reproductive system and abnormal sperm have increased. During the same time, use of pesticides containing estrogen-like chemicals has increased.

Background Information Considering existing knowledge is important in scientific inquiry. To understand how estrogen might cause the observed effects, it is important to understand its normal role.

Estrogen molecules enter certain cells and bind to proteins called receptors, which fit them in much the same way that a catcher's mitt fits an incoming baseball. The binding occurs in the nucleus, the part of the cell that houses the genetic material. A cell that is sensitive to estrogen's stimulation may have many thousands of such receptor proteins. Estrogen molecules bound to receptors then activate certain genes, producing the hormone-associated effects, such as cell division.

Some estrogens promote cell division, and some do not. The proportion of cell-division-promoting to non-cell-division-

FIGURE 1.10

Experiments Follow Rules.

(**A**) In nature, the sex of alligators is determined by the incubation temperature of the eggs. Normally, alligator eggs incubated at high temperature hatch males. (**B**) Painting eggs with estrogen or DDE drastically lowers the percentage of male hatchlings. An experiment using alligator eggs to test the estrogen mimic theory must account for possible alternative explanations. The experiment was set up so that researchers did not know which eggs had received which treatment until after results had been tallied. This experimental design prevented unintentional bias, such as looking more carefully for abnormalities in animals that a researcher knows has received estrogen or DDE.

Normal alligator eggs

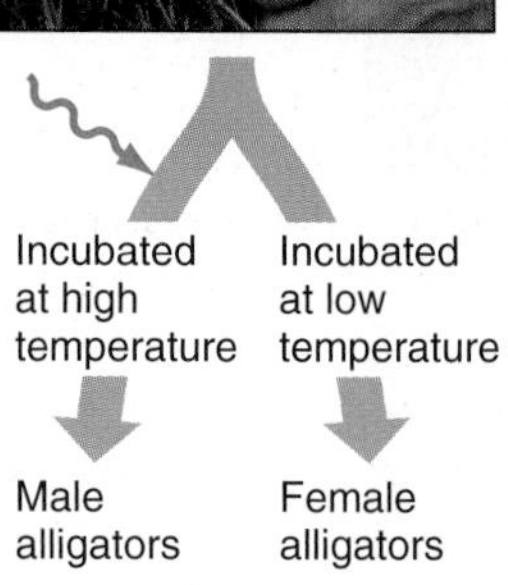

A

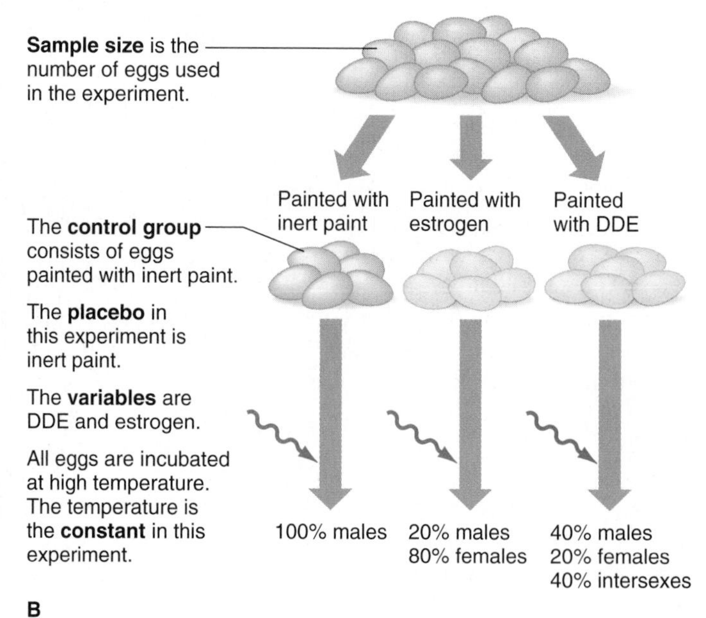

B

promoting estrogens, as well as the ratio of estrogens to other sex hormones, determines whether the reproductive system develops normally. Chemicals acting as estrogen mimics may disrupt this crucial hormonal balance in several ways.

Formulating a Hypothesis Observations in the scientific method lead first to a general question, such as:

> *Does excess estrogen exposure cause reproductive problems in animals, including humans?*

The next step is to formulate a more specific statement to explain an observation, based on previous knowledge. This prediction, or hypothesis, is testable and should examine only one changeable factor, or variable. A hypothesis is more than a question or a hunch; it is based on known facts. Often a hypothesis is posed as an "if . . . then" proposition, which sets up a testable prediction:

> *If exposure to large amounts of estrogen-like chemicals causes reproductive problems in animals, then animal eggs intentionally exposed to estrogen or similar substances should not hatch, or they will hatch to yield abnormal offspring.*

Devising an Experiment Once a hypothesis is posed, a researcher collects information to test it. An experiment can disprove the hypothesis but can never prove it, because there is always the possibility of discovering additional information.

To determine whether DDE in the environment caused the reproductive abnormalities in the alligators, experiments recreated conditions that exposed the eggs to estrogen-like chemicals (fig. 1.10). Investigators collected alligator eggs from a clean lake and incubated them at high temperatures, which in alligators ensures that the hatchlings are male. The researchers painted some eggs with estrogen, some with DDE, and some with an inert substance. This experimental design would test whether just painting an egg can alter development.

Among the estrogen-treated eggs, only 20% produced male hatchlings—a drastic change from the expected 100%. Only 40% of the eggs painted with DDE hatched males, which had excess hormone disturbances in the same proportions as did male alligators from Lake Apopka. Another 40% of the DDE-treated eggs hatched intersexes, which had both male and female reproductive structures. Intersexes are not seen among alligators in clean lakes. The remaining 20% of the DDE-treated eggs produced female hatchlings. All of the eggs

Elephants Calling— The Powers of Observation

If you think only humans use language, consider an elephant's extended family. An elephant clan, led by an elder female, or matriarch, and consisting of females and young males, has quite a vocabulary. Elephant "language" includes calls in infrasound, which are sounds too low for human ears to detect.

In 1984, Cornell University researcher Katy Payne was standing near caged elephants at a zoo when she felt a "throbbing" in the air. The sensation brought back memories of singing in the church choir as a child where the church organ pipes made a similar throbbing. "In the zoo I felt the same thing . . . , without any sound. I guessed the elephants might be making powerful sounds like an organ's notes, but even lower in pitch," she writes in her children's book, *Elephants Calling*.

To test her hypothesis, Payne and two friends used equipment that could detect infrasound to record elephants in a circus and a zoo. Finding infrasound, she moved her study to the Amboseli Plain, a salty, dusty stretch of land at the foot of Mount Kilimanjaro in Tanzania. Her lab was a truck sitting among the elephants, who grew so used to its presence that they regarded it as part of the scenery.

Living among the elephants further sharpened Payne's already highly developed powers of observation. She and her fellow elephant watchers soon became attuned to the subtle communications between mother and calf; the urges of a male ready to mate; and messages to move to find food or water. Writes Payne, "It is amazing how much you can learn about animals if you watch for a long time without disturbing them. They do odd things, which at first you don't understand. Then gradually your mind opens to what it would be like to have different eyes, different ears, and different taste; different needs, different fears, and different knowledge from ours."

FIGURE 1.A
Raoul rarely ventures far from his mother, Renata.

One day, two bulls were fighting for dominance when the youngest family member, baby Raoul, slipped away (fig. 1.A). Finding a hole, Raoul stuck his trunk inside—then leapt back, bellowing, as a very surprised warthog bounded out of his invaded burrow. Raoul's mother, Renata, responded with a roar. Payne described the scene:

Elephants in all directions answer Renata, and they answer each other with roars, screams, bellows, trumpets, and rumbles. Male and female elephants of all sizes and ages charge past each other and us with eyes wide, foreheads high, trunks, tails, and ears swinging wildly. The air throbs with infrasound made not only by elephants' voices but also by their thundering feet. Running legs and swaying bodies loom toward and above us and veer away at the last second.

The family reunited, all the animals clearly shaken. Unable to resist comparing the pachyderms to people, Payne writes, "Renata does not seem angry at Raoul. Perhaps elephants don't ask for explanations."

The kind of observation that leads to new knowledge or understanding, Payne says, happens only rarely. "You have to be alone and undistracted. You have to be concentrating on what's there, as if it were the only thing in the world and you were a tiny child again. The observation comes the way a dream—or a poem—comes. Being ready is what brings it to you."

Here is how the steps of scientific inquiry can be applied to Payne's research.

Observation: The air near the elephant's cage at the zoo seems to vibrate.
Background knowledge: Organ pipes make similar vibrations when playing very low notes.
Hypothesis: Elephants communicate by infrasound, making sounds that are too low for the human ear to hear.
Experiment: Record the elephants with equipment that can detect infrasound.
Results: Elephants make infrasound.
Further observations: Elephants emit infrasound only in certain situations involving communication.
Conclusion: Elephants communicate with infrasound.
Further question: Do different patterns of sounds communicate different messages?

painted with the inert substance hatched normal male alligators.

These experimental results support the hypothesis and lead to the conclusion that DDE alters reproductive structures in alligators. But for a conclusion to become widely accepted, the results must be repeatable—other scientists must be able to perform the same experiments and observe the same results.

Scientific Inquiry Continues Scientific inquiry does not end with a conclusion, because each discovery leads to further questions. Once experiments in painting alligator eggs provided evidence that DDE and estrogen cause reproductive abnormalities, a new question arose:

How do estrogen mimics exert their effects?

This new question suggests a new hypothesis:

If estrogen mimics produce the same or similar effects as estrogen, then the mimics should bind to estrogen receptors.

Chemical experiments support this hypothesis—estrogen receptors bind estrogens and DDE. Various insecticides, herbicides, pesticides, plastics, and fuels also bind to estrogen receptors, suggesting further roads of inquiry.

The cycle of scientific inquiry continues, as puzzle pieces are further connected to reveal a fuller picture of the estrogen mimic story. Because estrogen exposure correlates to increased risk of reproductive abnormalities and cancer, and pesticides introduce estrogen-like compounds into the environment, we can link these facts to pose yet another hypothesis:

Are pesticide residues responsible for an observed increase in breast cancer and reproductive problems in humans?

The experiments and observations discussed so far do not directly address this new hypothesis. The simultaneous increases in incidence of certain reproductive problems and pesticide use do not demonstrate cause-and-effect but rather a correlation—events occurring at the same time that may or may not be causally related. It will take experiments to investigate whether pesticide exposure causes breast cancer and other reproductive problems. If many tests fail to disprove a particular hypothesis, then that hypothesis attains the level of a theory. A scientific theory, then, is an idea, based on hypotheses that have survived strong testing, that has gained acceptance among scientists.

We now take a closer look at experimental design.

How Do Scientists Design Experiments?

Scientists use several approaches when they design experiments in order to make data as valid as possible. The number of individuals or structures that are experimented on is the **sample size.** A useful experiment examines dozens of alligator eggs, or hundreds of cells, because a large sample helps ensure meaningful results. For example, sampling, by chance, only intersexes among the DDE-treated alligator eggs might support the conclusion that DDE exposure causes alligators to always develop as intersexes. Larger samples would reveal that DDE causes intersexes only 40% of the time.

Obtaining a large sample is not always practical or possible. When scientists study very rare disorders, only a few patients may be available. Although scientists prefer to work with a large sample, small-scale (called pilot) experiments are nevertheless valuable because they may indicate whether continuing research is likely to yield valid, meaningful results.

Experimental Controls Distinguishing the unusual requires comparison to the usual. This is why well-designed experiments compare a group of "normal" individuals or components to a group undergoing treatment. Ideally, the only difference between the normal group and the experimental group is the one factor being tested. The normal group is called an **experimental control** and provides a basis of comparison. Designing experiments to include controls helps ensure that a single factor, or **variable,** causes an observed effect.

Experimental controls may take several forms. A **placebo** presents the control group with the same experience as the experimental group. In the alligator egg experiment depicted in figure 1.10, the eggs painted with an inert substance received a placebo. In medical research, a placebo is often a stand-in for a drug being tested—a sugar pill or a treatment already known to be effective.

Another safeguard used in medical experiments is a **double-blind** design, in which neither the researchers nor the participants know who received the substance being evaluated and who received the placebo. The researchers break the "code" of who received which treatment only after the data are tabulated or if one group does so well that it would be unethical to withhold treatment from the placebo group. In the alligator egg study, researchers didn't know which eggs had received which of the three treatments until they compiled the results. This avoided bias, such as looking more carefully for reproductive problems in alligators known to have been exposed to estrogen or DDE. However, it wasn't double-blind—unless the occupants of the eggs could know which treatment they received!

What Are the Limitations of Scientific Inquiry?

The generalized scientific method is neither foolproof nor always easy to implement.

Interpreting Results Experimental evidence may lead to multiple interpretations or unexpected conclusions, and even the most carefully designed experiment can fail to provide a definitive answer. Consider the observation that animals fed large doses of vitamin E live significantly longer than similar animals who do not ingest the vitamin. Does vitamin E slow aging? Possibly, but excess vitamin E causes weight loss, and other experiments associate weight loss with longevity. Does vitamin E extend life, or does the weight loss? The experiment of feeding animals large doses of vitamin E does not distinguish between these possibilities. Can you think of further experiments to clarify whether vitamin E or weight loss extends life?

Another limitation of implementing the scientific method is that researchers may misinterpret observations or experimental results. For example, scientists once concluded that life can arise by heating broth in a bottle and then corking it shut, and observing bacteria in the brew a few days later. The correct explanation was that the cork did not keep bacteria out.

We cannot directly study natural phenomena that occurred only long ago. Many experiments have attempted to re-create the sequence of chemical reactions that might, on early Earth, have formed the chemicals that led to life. Although the experiments produce interesting results and reveal ways that these early events may have occurred, we cannot really know if they accurately re-created conditions at the beginning of life.

Expecting the Unexpected An investigator should try to keep an open mind towards observations, not allowing biases or expectations to cloud interpretation of the results. To do so, scientists must expect the unexpected. But it is human nature to be cautious in accepting an observation that does not fit existing knowledge. The careful demonstration that life does not arise from broth surprised many people who believed that mice sprung from mud, flies from rotted beef, and beetles from cow dung.

Scientific discoveries very much depend upon our willingness to accept unusual ideas. When Edward Jenner noticed in 1796 that young girls who worked with cows did not develop the devastating illness smallpox, he used the observation to invent a smallpox vaccine. He intentionally infected young human volunteers with the cowpox virus and hoped that this would alert their immune systems to fight the related smallpox virus. Jenner had to test his vaccine in private, because people tried to lynch him for the very idea of intentionally infecting people (fig. 1.11). A more recent example of accepting an unusual explanation at the expense of a well-established one is the common belief that stress causes ulcers. Today we know that most ulcers are caused by bacterial infection (see Investigating Life 37.1).

FIGURE 1.11

Unusual Ideas Are Important in Science.

Edward Jenner was at first ridiculed for his idea of using infection with one virus (cowpox) to prevent infection by another (smallpox). Many people at the time were afraid of receiving a smallpox vaccine because they believed it would cause them to grow cow parts, as this cartoon suggested.

Scientific research seeks to understand nature. Because humans are part of nature, we sometimes tend to view scientific research, and particularly biological research, as aimed at improving the human condition. But knowledge without any immediate application or payoff is valuable in and of itself—because we can never know when information will prove useful.

1.3 MASTERING CONCEPTS

1. What is a scientific theory?
2. What are the steps of scientific inquiry?
3. How is scientific inquiry a continuous process?
4. What factors determine whether an experiment will provide meaningful information?
5. What are some limitations of scientific inquiry and experimentation?

OLC

Chapter Summary

1.1 What Are the Characteristics of Life?

1. A living **organism** is distinguished from an inanimate object by the presence of a combination of characteristics.
2. An organism is organized as structures of increasing size and complexity, from biochemicals, to **cells,** to **tissues, organs,** and **organ systems,** to individuals, **populations, communities, ecosystems,** and the **biosphere.**
3. **Emergent properties** arise as the level of organization of life increases and are the consequence of physical and chemical laws.
4. Life requires energy to maintain its organization and functions. **Metabolism** directs the acquisition and use of energy.
5. Organisms must maintain an internal constancy in the face of changing environmental conditions, a state called **homeostasis.**
6. Organisms develop, grow, and reproduce.
7. Organisms respond to the environment, through **irritability** in the short term and by **adaptation** over generations. **Natural selection** eliminates inherited traits that decrease chance of survival and reproduction in a certain environment.

1.2 How is Life Diverse?

8. Biologists classify organisms with a series of names that reflect probable evolutionary relationships.
9. The three domains of life are distinguished by cell structure and organization. Cell complexity, mode of nutrition, and other factors distinguish members of kingdoms.
10. Evolution through natural selection explains why organisms are alike yet diverse and how common ancestry unites all species.

1.3 How Do Scientists Study the Natural World?

11. Scientific theories are ideas based on **hypotheses** that have survived rigorous testing. They attempt to explain natural phenomena.
12. Scientific inquiry, which uses the **scientific method,** is a way of thinking that involves observing, questioning, reasoning, predicting, testing, interpreting, concluding, and posing further questions.
13. Scientific inquiry begins when a scientist makes an observation, raises questions about it, and reasons to construct an explanation, or hypothesis.
14. Experiments test the validity of the hypothesis, and conclusions are based on data analysis. **Experimental controls** ensure that the data is the result of the **variables** in the experiment, and not other factors.
15. **Placebo**-controlled, **double-blind** experiments minimize bias.
16. The scientific method does not always yield a complete answer or explanation or may produce ambiguous results. Discoveries may be unusual, unexpected, or serendipitous.

Testing Your Knowledge

1. List the characteristics of life.
2. Cite two ways that asexual and sexual reproduction differ.
3. How does natural selection act on adaptations in a way that causes evolution to occur?
4. Why do biologists assign taxonomic names to organisms, and what do those names reflect?
5. Describe the relationship between domains and kingdoms.
6. How is a theory more than an opinion, and a hypothesis more than a guess?
7. What is an experiment?

Thinking Scientifically

1. In an episode of the television series *Star Trek,* three members of the crew of the starship USS *Enterprise* meet an enemy who dehydrates each one into a small box. He then rearranges the boxes and crushes one of them. He later attempts to restore the people to life but only two of them reappear. What biological principle does this action illustrate?
2. For the following real-life examples, state whether or not the scientific method was followed. If it was not, indicate which of the following specific faults occurred:
   a. Experimental evidence does not support conclusions
   b. Inadequate controls
   c. Biased sampling
   d. Inappropriate extrapolation from the experimental group to the general population
   e. Sample size too small

   (1) "I ran 4 miles every morning when I was pregnant with my first child," the woman told her physician, "and Jamie weighed only 3 pounds at birth. This time, I didn't exercise at all, and Jamie's sister weighed 8 pounds. Therefore, running during pregnancy must cause low birth weight."

   (2) Eating foods high in cholesterol was found to be dangerous for a large sample of individuals with hypercholesterolemia, a disorder of the heart and blood vessels. It was concluded from this study that all persons should limit dietary cholesterol intake.

   (3) Osteogenesis imperfecta (OI) is an inherited condition that causes easily fractured bones. In a clinical study, 30 children with OI were given a new drug for 3 years. The children all showed increased (improved) bone density, lowered incidence of fractures compared to before treatment began, and less fatigue. The conclusion: The drug is effective in treating OI.

(**4**) Researchers studied HIV in blood and semen from 11 HIV-infected men. In 8, the virus was resistant to several medications, and in 2 of these men, virus from the blood was resistant to the class of drugs called protease inhibitors, but virus from semen was not resistant. The researchers concluded that protease inhibitors do not reach the male reproductive organs.

References and Resources

Blakeslee, Sandra. October 13, 1998. Placebos prove so powerful even experts are surprised. *The New York Times.* Ruling out a placebo effect is crucial in the analysis of new treatments.

Lewis, Ricki. 2001. *Discovery: Windows on the life sciences.* Malden, Massachusetts: Blackwell Science. The "stories" of discovery behind several hot topics in biology today.

Miller, P. J. O. June 22, 2000. Whale songs lengthen in response to sonar. *Nature,* vol. 405, p. 903. Whale communication may be similar to elephant communication.

Rensberger, Boyce. July 7, 2000. The nature of evidence. *Science,* vol. 289, p. 61. The public often confuses scientific evidence with belief.

Sagan, Carl. 1996. *The demon-haunted world.* New York: Ballantine Books. A guide to distinguishing science from nonscience.

Zivin, Justin A. April 2000. Understanding clinical trials. *Scientific American,* vol. 282, p. 69. Clinical trials to evaluate new medical treatments apply scientific inquiry and analysis of data.

The *LIFE* Online Learning Center provides additional resources and tools for studying this chapter.
www.mhhe.com/life

CHAPTER 2

The Chemistry of Life

Nitric Oxide

A Tiny But Versatile Chemical of Life

The chemicals that comprise living organisms are for the most part quite large. Nitric oxide, in contrast, is miniscule. If a typical biological molecule were to be equated to a long freight train, then nitric oxide—abbreviated NO—might be about the size of a foot-long plank of wood in one car of that train, or even shorter.

Before 1980, NO was known mostly for its presence in smog, cigarette smoke, and acid rain. And so it wasn't surprising that when researchers began to suspect that animal cells produce NO in fleeting bursts, and that the tiny chemical could have profound and diverse effects on the functions of an organism, their work was for a time doubted and questioned. Such is the way of scientific investigation. But by the late 1980s, several lines of research had converged that identified NO's role as a messenger in such diverse functions as blood clotting, memory, immune defense, maintenance of blood pressure, and penile erection. NO also acts as a messenger in plants, bacteria, and fungi.

> ***For many scientists, NO came to their attention in 1992, when Science magazine named it the "molecule of the year."***

Nitric oxide is a gas. It consists of just two atoms, one of the element nitrogen, the other of oxygen, that come together in a way that leaves one electron unpaired. This makes NO highly reactive, especially with the larger molecules of life that contain either sulfur or metals. NO also seems not to require special structures to either leave the cells where it is produced or enter others—it appears able to simply move (diffuse) out of one cell and into another. It then flits about for 6 to 10 seconds before colliding with water or oxygen to form chemicals called nitrates and nitrites. In contrast, most molecules must bind to special receptors found on receiving cells, or enter the cell through specific channels.

A type of enzyme called nitric oxide synthase—NOS for short—is required to synthesize NO, which it does by removing the nitrogen and oxygen from a

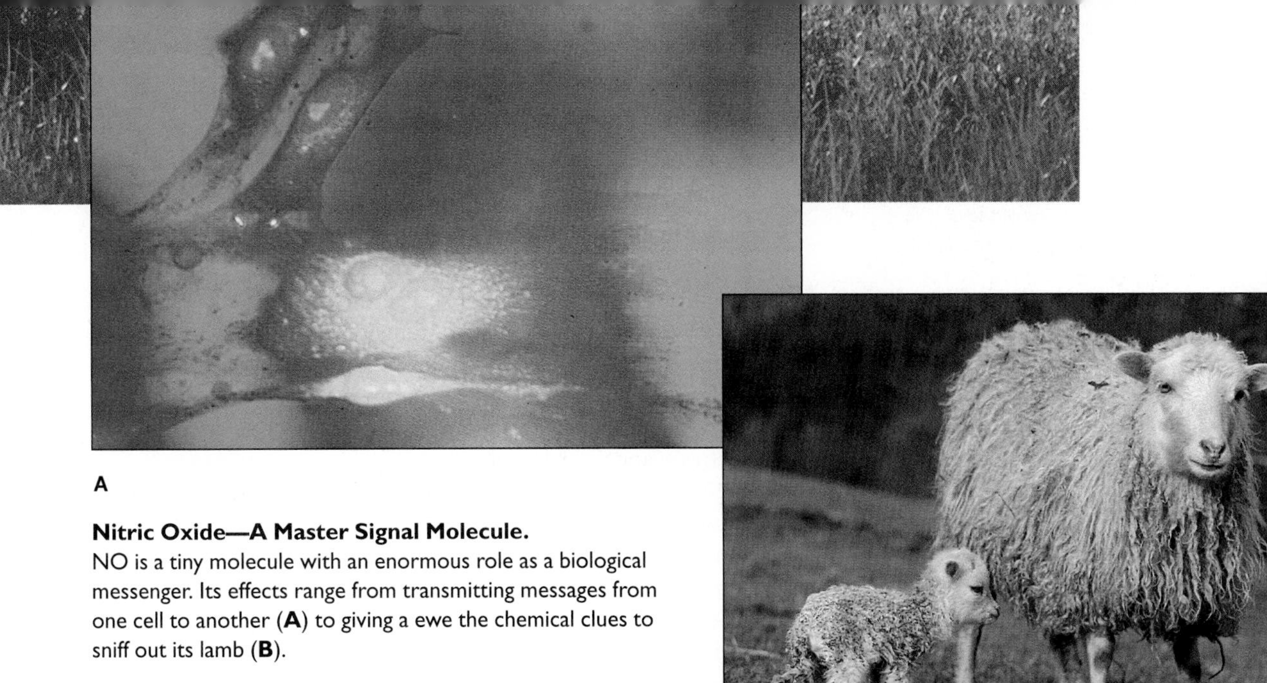

A

Nitric Oxide—A Master Signal Molecule.
NO is a tiny molecule with an enormous role as a biological messenger. Its effects range from transmitting messages from one cell to another (**A**) to giving a ewe the chemical clues to sniff out its lamb (**B**).

B

particular amino acid, leaving a different amino acid as it gives off the reactive gas. In some parts of the body, NOS is present all of the time, quietly pumping out small amounts of NO to carry out everyday sorts of functions, such as maintaining the diameters of blood vessels or secreting neurotransmitters (nerve cell messenger molecules). In other tissues, a different form of NOS is activated only when infectious bacteria invade, and the resulting surge of NO kickstarts the immune response.

For many scientists, NO came to their attention in 1992, when *Science* magazine named it the "molecule of the year." For many nonscientists, NO came to their attention when it hit the headlines as the basis of the functioning of the drug Viagra, used to treat erectile dysfunction (impotence). For those who missed the excitement over Viagra, NO hit the headlines again when three researchers won the 1998 Nobel prize for Physiology or Medicine for unraveling its effects in the human body. Here is a look at a few of the roles of NO in this application, and others.

Erectile Dysfunction A mammal's penis consists of three chambers of spongy tissue that surround blood vessels. When the chambers fill with blood, as they do following sexual stimulation, the organ engorges and stiffens. The stimulation causes nerves as well as the cells that line the interiors of the blood vessels to release NO. The NO then enters muscle cells that form the middle layers of the blood vessels, relaxing them by activating a series of other chemicals. As a result, the vessels widen and blood fills the spongy tissue. The penis becomes erect. Viagra works by mimicking one of the chemicals in the pathway that NO stimulates, called cyclic guanosine monophosphate (cGMP for short). When an enzyme appears on the scene to break down cGMP, it instead latches onto Viagra, leaving the chemical cascade toward erection intact longer.

Recognizing One's Offspring Female sheep learn to recognize their young by smell within the first 2 hours of giving birth. This stimulus is called "learned lamb odor." Because of this "olfactory memory," a mother sheep can easily pick out her lamb from a group of young. But when a drug blocks NOS in new mother sheep, the animals cannot select their offspring from a group. However, the mothers do nurture other lambs, indicating that NO controls offspring recognition, but not parenting behavior. Experiments in mice suggest that NO helps maintain memories, rather than forming them. Blocking NOS in mice renders them incapable of correctly running mazes that they previously could conquer.

Blood Pressure NO also causes blood vessels to widen in parts of the body other than the penis. On a whole-body scale, this activity lowers blood pressure by giving the blood more room to move. Researchers are exploring possible clinical applications of manipulating NO's effect on blood pressure. Giving NO can save the lives of newborns in severe respiratory distress, due to very high blood pressure in the lungs, by lowering the pressure. In an opposite situation, blocking NO production in individuals in septic shock from severe infection can save their lives by raising their dangerously plummeting blood pressure.

As biologists continue to unravel the varied roles of nitric oxide in organisms, drug companies are busy exploring ways to control production of this tiny but important chemical to treat a variety of disorders. We will certainly be hearing more about NO.

2.1 What Is Matter?

Matter consists of pure substances and combinations of those substances. Organisms are built of a subset of the chemicals found in nature.

Organisms consist of **matter** (material that takes up space) and energy (the ability to do work). Chapter 5 discusses the energy of life in detail; this chapter concentrates on the composition of living matter.

Elements

All matter can be broken down into pure substances called **elements.** There are 92 known, naturally occurring elements and at least 17 synthetic ones. Each element has unique properties. The elements may be arranged into a chart, called the **periodic table,** in columns according to their properties.

It is interesting to look at the periodic table positions of the elements that make up organisms. These elements of life appear across the table, indicating that they represent a diverse sampling of all chemicals (fig 2.1). Appendix C contains a complete periodic table.

Twenty-five elements are essential to life. Elements required in large amounts—such as carbon, hydrogen, oxygen, nitrogen, sulfur, and phosphorus—are termed **bulk elements;** those required in small amounts are called trace elements. Many trace elements are important in ensuring that vital chemical reactions occur fast enough to sustain life. Some elements that are toxic in large amounts, such as arsenic, may be vital in very small amounts.

FIGURE 2.1
The Periodic Table of Elements.
Elements 58–71 and 90–103 are omitted for clarity.

1 H																	2 He
3 Li	4 Be											5 B	6 C	7 N	8 O	9 F	10 Ne
11 Na	12 Mg											13 Al	14 Si	15 P	16 S	17 Cl	18 Ar
19 K	20 Ca	21 Sc	22 Ti	23 V	24 Cr	25 Mn	26 Fe	27 Co	28 Ni	29 Cu	30 Zn	31 Ga	32 Ge	33 As	34 Se	35 Br	36 Kr
37 Rb	38 Sr	39 Y	40 Zr	41 Nb	42 Mo	43 Tc	44 Ru	45 Rh	46 Pd	47 Ag	48 Cd	49 In	50 Sn	51 Sb	52 Te	53 I	54 Xe
55 Cs	56 Ba	57 La	72 Hf	73 Ta	74 W	75 Re	76 Os	77 Ir	78 Pt	79 Au	80 Hg	81 Tl	82 Pb	83 Bi	84 Po	85 At	86 Rn
87 Fr	88 Ra	89 Ac	104 Rf	105 Db	106 Sg	107 Bh	108 Hs	109 Mt	110 ***	111 ***	112 ***						

*** These elements have not yet been named.

Bulk biological elements

Trace elements

Possibly essential trace elements

Atoms

An **atom** is the smallest possible "piece" of an element that retains the characteristics of the element. An atom is composed of three major types of subatomic particles: **protons** and **neutrons,** which form a centralized core called the **nucleus,** and **electrons,** which surround the nucleus. Atoms of different elements have characteristic numbers of protons and may vary greatly in size. Hydrogen, the simplest type of atom, has only 1 proton and 1 electron; in contrast, an atom of uranium has 92 protons, 146 neutrons, and 92 electrons.

Protons carry a positive charge, electrons carry a negative charge, and neutrons are electrically neutral. Charge is the attraction between opposite types of particles. Within most atoms, the number of protons equals the number of electrons and the atom is electrically neutral—that is, without a net charge. An electron is vanishingly small compared to a proton or a neutron, yet it exists far away from the nucleus of the atom. If the nucleus of a hydrogen atom were the size of a meatball, the electron belonging to that atom would be about 1 kilometer (0.62 miles) away from it! The term **orbital** refers to the most likely location for an electron relative to its nucleus. Figure 2.2 summarizes the characteristics of the three types of subatomic particles.

The periodic table depicts in shorthand the structures of the atoms of each element. Each element has a symbol, which can come from the English word for that element (*He* for *helium,* for example) or from the word in another language (*Na* for *sodium,* which is *natrium* in Latin). The **atomic number** above the element symbol and name shows the number of protons in the atom, which establishes the identity of the atom. The elements are arranged sequentially in the periodic table by atomic number.

The **atomic mass,** beneath the element symbol, reflects the total number of protons and neutrons in the nucleus of the atom. For biological purposes, we can approximate the mass of a proton (and that of a neutron) to be one. By comparison, the contribution of an electron to an atom's mass is negligible. Therefore, we add the number of protons and neutrons to approximate the atomic mass. Subtracting the atomic number (the number of protons) from the atomic mass (the number of protons and neutrons) yields the number of neutrons.

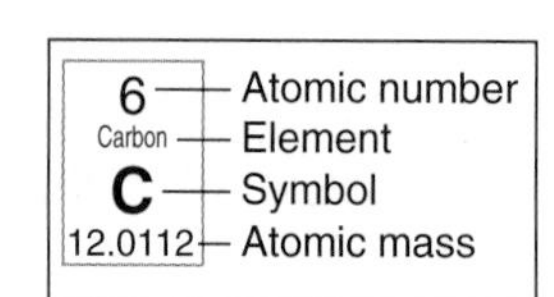

The atoms of an element can exist in different forms, called **isotopes,** that have different numbers of neutrons. Isotopes of an element all have the same charge and characteristics, but different masses. (Weight and mass are related; weight is mass taking gravity into account.) Often one isotope of an element is very abundant, and others are rare. For example, 99% of carbon isotopes have six neutrons, and only 1% are isotopes with seven or eight neutrons. Tin (Sn) has the highest number of naturally occurring isotopes—10. An element's atomic mass in the periodic table is the average mass of its isotopes.

Of the 1,500 known isotopes, 1,236 are unstable, which means they tend to break down into more stable forms, often of other elements. (Atoms with equal numbers of protons and neutrons are more stable.) When these unstable isotopes break down, they emit radioactive energy. Every type of radioactive isotope has a characteristic half-life, which is the time it takes for half of the atoms in a sample to emit radiation, or "decay" to a different, more stable form. The energy is relatively easy to detect, even from very small quantities of isotopes. Such radioactive isotopes have a variety of uses in biomedical and life science research, and we will encounter them in context in later chapters.

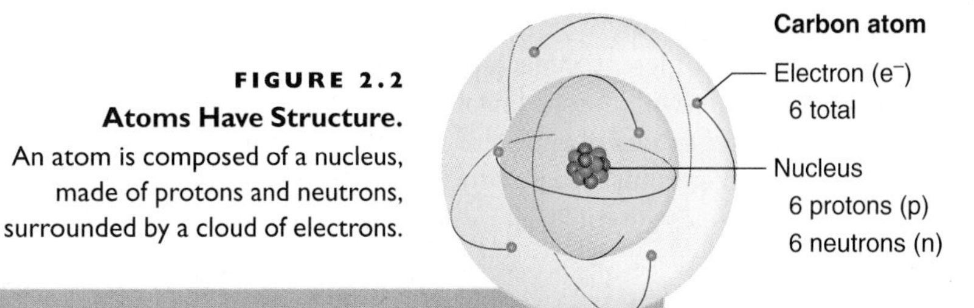

FIGURE 2.2
Atoms Have Structure.
An atom is composed of a nucleus, made of protons and neutrons, surrounded by a cloud of electrons.

Subatomic particles

Particle	Charge	Mass	Function	Symbol	Location
Electron	–	0	Bonding	e^-	Orbitals
Neutron	0	1	Nuclear stability	n	Nucleus
Proton	+	1	Identity	p	Nucleus

Molecules

Atoms of two or more elements joined together form a **compound,** which is a chemical substance with properties distinct from those of its constituent elements. A **molecule** is the smallest unit of a compound. A molecule usually consists of atoms of different elements, but a few exceptions are "diatomic," consisting of two atoms of the same element, such as hydrogen, oxygen, or nitrogen. Table 2.1 reviews these designations of matter.

A compound's characteristics can differ strikingly from those of its constituent elements. Consider table salt, which is the compound sodium chloride. A molecule of salt contains one atom of sodium (Na) and one atom of chlorine (Cl). Sodium is a silvery, highly reactive solid metal, while chlorine is a yellow, corrosive gas. But when equal numbers of these two types of atoms combine, the resulting compound is a white crystalline solid—table salt. The same is true for the swamp gas methane. Its components are carbon, a black sooty solid, and hydrogen, a light, combustible gas.

TABLE 2.1 On the Matter of Matter

DESIGNATION	DEFINITION
Element	A fundamental type of substance
Atom	The smallest piece of an element that retains the characteristics of that element
Compound	A pure substance formed when atoms of different elements bond
Molecule	The smallest piece of a compound that retains the characteristics of that compound

Scientists describe molecules by writing the symbols of their constituent elements and indicating the numbers of atoms of each element in one molecule as subscripts. For example, methane is written CH_4, which denotes 1 carbon atom bonded (joined) to 4 hydrogen atoms. A molecule of the sugar glucose, $C_6H_{12}O_6$, has 6 atoms of carbon (C), 12 of hydrogen (H), and 6 of oxygen (O). A coefficient indicates the number of molecules—6 molecules of glucose is written $6C_6H_{12}O_6$.

Chemistry Explains Biology

Copper Woes

The human body requires certain amounts of trace elements. Inherited disorders called "inborn errors of metabolism" can cause a trace element to be overabundant or lacking—often with severe effects on health. Consider copper.

In Wilson disease, the digestive tract absorbs too much copper from food, causing stomachaches and headaches, changes in one's voice and coordination, and an inflamed liver. Greenish rings around the irises are the key clue to a diagnosis of the disease. A drug, penicillamine, rids the body of excess copper, which turns the urine a coppery color. Symptoms of Wilson disease usually appear during adolescence. The treatment cannot reverse symptoms but prevents the disease from worsening.

In the opposite biochemical situation, Menkes disease, the intestines cannot release copper from food into the bloodstream. Lack of copper causes symptoms that begin shortly after birth. They include yellowish skin, low body temperature, extreme lethargy, seizures, mental retardation, impaired growth, and white, kinky, stubbly hair that gives the disease its other name, "kinky hair syndrome." The white, stubbly hair of a child with Menkes disease provided a clue to the cause. It reminded an astute Australian researcher of the peculiar wool of sheep that graze in his native land on copper-deficient soil. This observation led to the discovery that an inherited copper deficiency causes Menkes disease. Menkes disease cannot be treated, and death occurs in childhood.

In **chemical reactions,** two or more molecules interact with each other to yield different molecules. Chemical reactions are depicted as equations with the starting materials, or **reactants,** on the left and the end products on the right. The total number of atoms of each element must always be the same on either side of the equation. Consider the equation for the formation of glucose in a plant cell:

$$6CO_2 + 6H_2O \longrightarrow C_6H_{12}O_6 + 6O_2$$

In words, this means "six molecules of carbon dioxide and six molecules of water react to produce one molecule of glucose plus six molecules of diatomic oxygen." Note that each side of the equation indicates 6 total atoms of carbon, 18 of oxygen, and 12 of hydrogen.

Organisms are composed mostly of water and carbon-containing molecules, especially **organic molecules,** which contain carbon and hydrogen. Many organic molecules also include oxygen, nitrogen, phosphorus, and/or sulfur. Some organic molecules of life are so large that they are termed **macromolecules.** The plant pigment chlorophyll, for example, contains 55 carbon atoms, 68 hydrogens, 5 oxygens, 4 nitrogens, and 1 magnesium. **Molecular mass** is an indication of a molecule's size. It is calculated by adding the atomic masses of the constituent atoms.

Carbon monoxide (CO) and carbon dioxide (CO_2) are not considered organic because of their simple structure and lack of hydrogen. But like NO, discussed in the opening essay, CO can function as a biological messenger molecule.

FIGURE 2.3
Structures of the Atoms Prevalent in Life.
Shown here are models of the six most common atoms that make up organisms.

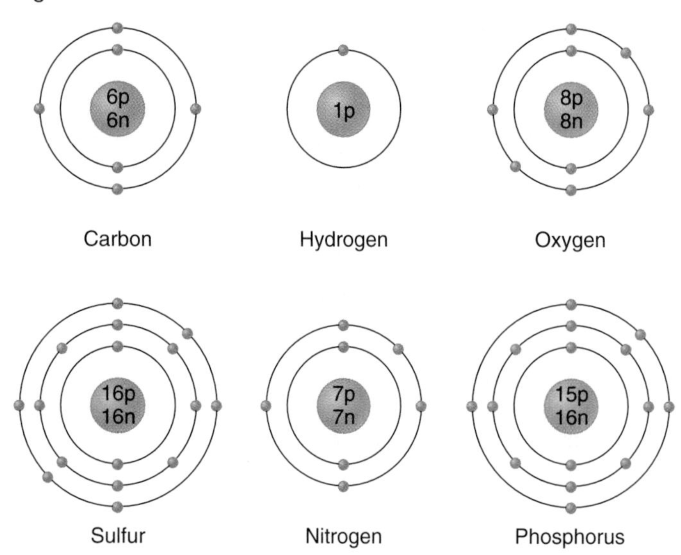

Chemical Bonds

Individual atoms may combine to form molecules through attractive forces called chemical bonds. The type and strength of chemical bonds, and whether they form at all, is determined by the numbers and location of electrons in the interacting atoms. As the numbers of protons in an atom increase, so do the numbers of electrons, which occupy distinct regions around the nucleus. However, since the electrons are constantly in motion, it is impossible to precisely determine the exact location of any single electron at any one point in time. Energy level helps determine an electron's location.

Electrons Determine Reactivity As more electrons are added in increasingly complex atoms, more distance separates the electrons and the nucleus. The farther an electron is from the nucleus, the more energy it has. Levels of energy are defined by a particular distance from the nucleus, called an **energy shell.** Electrons occupy the lowest energy shell available to them and almost always fill one shell before beginning another. If an electron absorbs energy, it shifts to a higher energy level. Upon releasing that energy, the electron returns to the original level. Bioluminescent substances that "glow in the dark" do so because their electrons absorb light energy and are boosted to higher energy levels. To return, the electrons release the energy as a burst of light.

Often electrons are illustrated as dots moving in concentric circles around a nucleus, much like planets moving around a sun, with the orbits symbolizing the energy shells (fig. 2.3). These depictions, called Bohr models, are useful for visualizing the interactions between atoms to form bonds. However, they do not accurately portray the three-dimensional structure of atoms. In a Bohr model, the innermost shell, the one closest to the nucleus, can hold only two electrons. This shell contains one orbital, which is shaped like a sphere, and is called the "s" orbital. The second energy shell contains one larger spherical orbital and three dumbbell-shaped orbitals, arranged along imaginary *x, y,* and *z* axes (fig. 2.4). Each of these four orbitals in

FIGURE 2.4
Electron Orbitals Determine Molecular Shape.
(**A**) The first energy level (designated 1) consists of one spherical (s) orbital containing up to two electrons. (**B**) The second energy level (2) has four orbitals, each containing up to two electrons. One of the orbitals of the second energy level is spherical (s); the other three are dumbbell-shaped and are perpendicular to one another. They are designated p. Therefore, the first energy shell is represented as 1s, and the second energy level as 2s and 2p. The nucleus of the atom is at the center, where the axes intersect. The location of electrons in an atom determines the shape of the molecule formed when atoms bond.

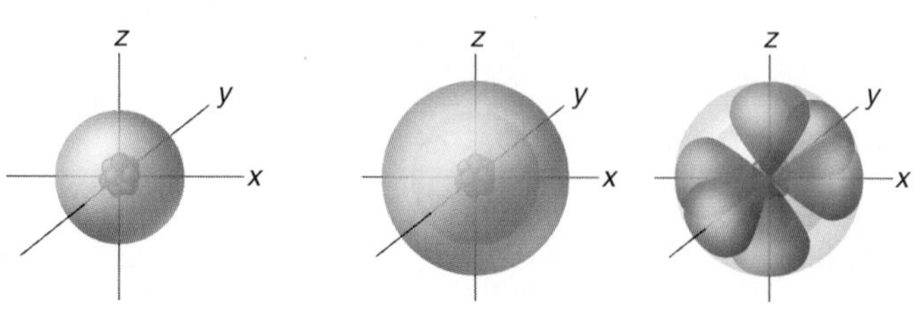

The first shell contains up to 2 electrons in one spherical orbit (1s).

A

The second shell contains up to 8 electrons, 2 electrons in one spherical orbit (2s) and 6 electrons in three perpendicular dumbbell-shaped orbits (2p).

B

the second energy shell can hold two electrons, for a total of eight. Larger atoms have more energy shells, each with its own orbitals. Each shell can accommodate a certain number of orbitals and a certain number of electrons. The outermost shell in an atom is called the **valence shell.** This shell is the one that is most actively involved in chemical reactions, because it is most likely to be only partially filled with electrons. The orientation of the orbitals allows one atom to interact with another.

Atoms are the most stable—that is, least likely to combine with other atoms—when their valence shells are filled. Since most of the valence shells contain spaces for eight electrons, atoms will become most stable with those eight vacancies filled. This tendency to require eight electrons in the valence shell is referred to as the octet rule. To fill that valence shell, atoms will share, borrow, or steal electrons from other atoms to arrive at exactly eight electrons. Chemical bonds are formed through the sharing of electrons or the attraction of two atoms that have, respectively, given up or taken in electrons from each other.

Covalent Bonds—Electron Sharing **Covalent bonds** hold together most of the molecules of life. These strongest of bonds form when two atoms share pairs of valence electrons, and they tend to form among atoms that have three, four, or five valence electrons. Sharing electrons enables both atoms to fill their outermost shells with eight electrons. One electron from each atom spends time around each nucleus, forming a very strong connection between the atoms.

Carbon has four electrons in its outermost shell, and therefore it requires four more electrons to satisfy the octet rule. A carbon atom can attain the stable eight-electron configuration in its outer shell by sharing electrons with four hydrogen atoms, each of which has one electron in its only shell (fig. 2.5). The resulting molecule is the swamp gas methane (CH_4). Figure 2.6 shows several ways to represent its chemical structure.

Covalent bonds are most simply depicted with lines between the interacting atoms. Each bond contains two electrons, one

FIGURE 2.5

Covalent Bonds Form Molecules.

(**A**) Methane (CH_4) is a covalently bonded molecule. One carbon and four hydrogen atoms complete their outermost shells by sharing electrons. Note that the first electron shell is complete with two electrons. (**B**) Methane, also known as "swamp gas," was once a major constituent of Earth's atmosphere.

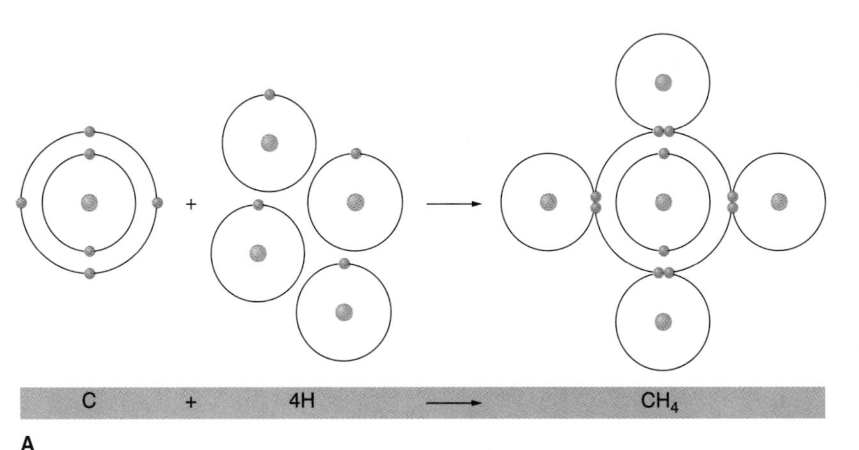

FIGURE 2.6

Different Types of Diagrams Are Used to Represent Molecules.

(**A**) The molecular formula CH_4 indicates that methane consists of one carbon atom bonded to four hydrogen atoms. (**B**) The structural formula shows those bonds as single. (**C**) The electron dot diagram shows the number and arrangement of shared electrons. (**D**) A ball-and-stick model reveals the angles of the bonds between the hydrogens and the carbon. (**E**) A space-filling model shows bond relationships as well as the overall shape of a molecule.

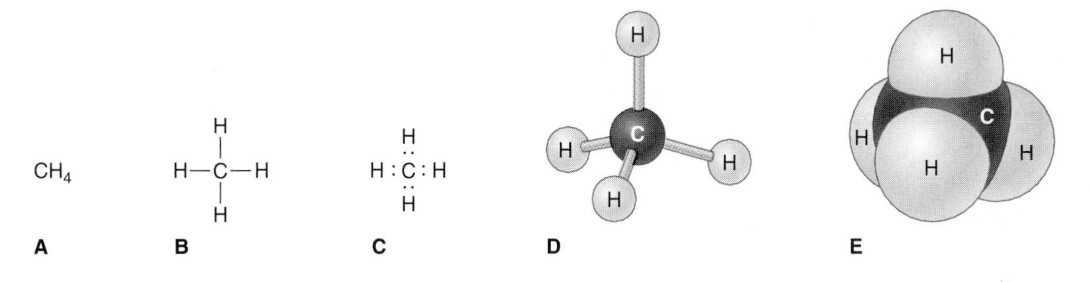

FIGURE 2.7

Carbon Atoms Form Four Covalent Bonds.

Two carbon atoms can bond, forming single, double, or triple bonds. Note that as the number of bonds between carbon atoms increases, the number of bonded hydrogens decreases. (**A**) Ethane, a component of natural gas, is a hydrocarbon built around two singly bonded carbon atoms. Breaking the atoms apart releases enough energy to heat a home or light a fire. (**B**) Ethylene consists of two carbon atoms linked by a double bond. Ethylene is a plant hormone, triggering flowers to drop and fruit to ripen. (**C**) Acetylene, consisting of two carbons held by a triple bond, is a flammable gas used in torches because tremendous heat energy is released when a triple bond is broken. (**D**) Octane is a key ingredient in gasoline and (**E**) benzene is a common solvent. Benzene also shows how a ring structure is possible. The inset shows an abbreviation of the ring that assumes carbon atoms are at the apexes, each with an implied hydrogen atom attached.

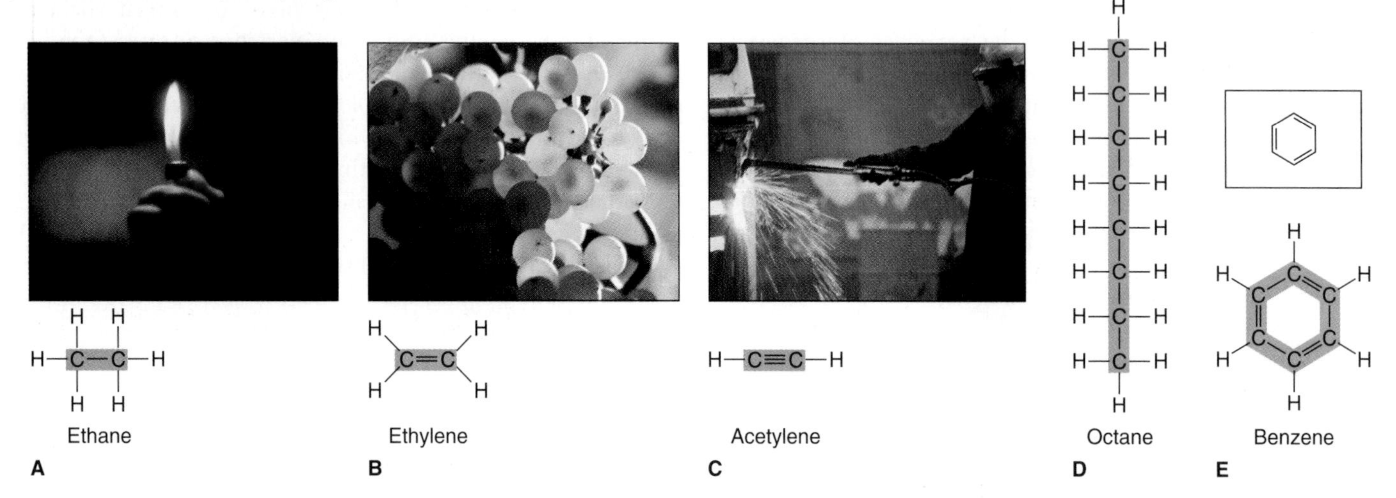

from each atom. Each atom shares the electrons in the bond to help fill its outer octet of electrons. By sharing four electrons with four hydrogen atoms, the carbon in methane fulfills the octet rule and the shell of the hydrogen atom is also filled.

Methane forms from the sharing of single electron pairs in covalent bonds. Carbon atoms can also bond with each other and share two or three electron pairs, forming double and triple covalent bonds, respectively (fig. 2.7). The fact that a carbon atom may form four covalent bonds, which satisfies the octet rule, allows this element to assemble into long chains, intricate branches, and rings. Carbon chains and rings bonded to hydrogens are called **hydrocarbons.** Carbon bonded to other chemical groups yields a great variety of biological molecules.

Methane is held together by **nonpolar covalent bonds,** in which the atoms share all of the electrons in the covalent bond equally. In contrast, in a **polar covalent bond,** electrons draw more towards one atom's nucleus than the other. The term **polar** means there is a difference between opposite ends of a molecule, usually formed by opposite charges. Water (H_2O), which consists of two hydrogen atoms bonded to an oxygen atom, illustrates a molecule with polar covalent bonds (fig. 2.8). The nucleus of each oxygen atom attracts the electrons on the hydrogen atoms more than the hydrogen nuclei do. As a result, the area near the oxygen carries a partial negative charge (from attracting the negatively charged electrons), and the area near the hydrogens is slightly positively charged (as the electrons draw away).

The tendency of an atom to attract electrons is termed **electronegativity.** Oxygen is highly electronegative. Its ability

FIGURE 2.8

Polar Covalent Bonds.

Polar covalent bonds hold together the two hydrogen atoms and one oxygen atom of water (H_2O). Because the oxygen attracts the negatively charged hydrogen electrons more strongly than the hydrogen nuclei do, the oxygen atom bears a partial negative charge and the hydrogens carry a partial positive charge. The resulting partial charges attract one molecule to another in hydrogen bonds.

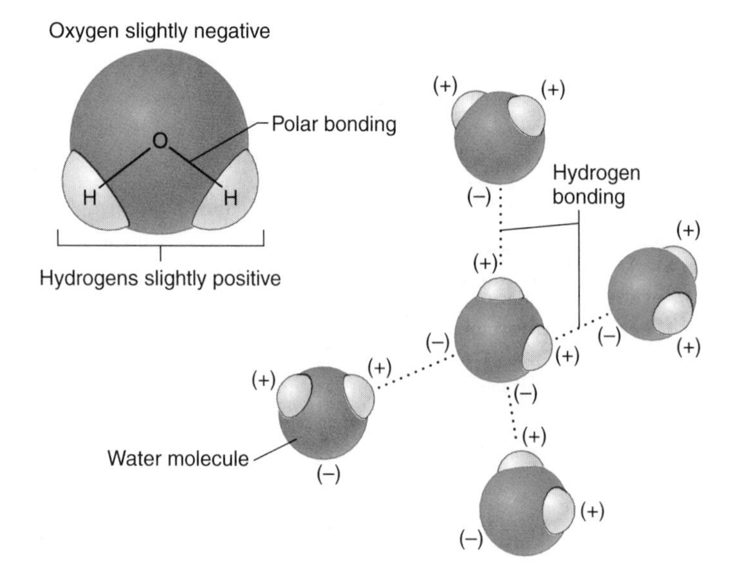

FIGURE 2.9

Table Salt, an Ionically Bonded Molecule.

(**A**) A sodium atom (Na) can donate the one electron in its valence shell to a chlorine atom (Cl), which has seven electrons in its outermost shell. This satisfies the octet rule. The resulting ions (Na^+ and Cl^-) bond to form the compound sodium chloride (NaCl), better known as table salt. (**B**) The ions that constitute NaCl occur in a repeating pattern that produce crystals.

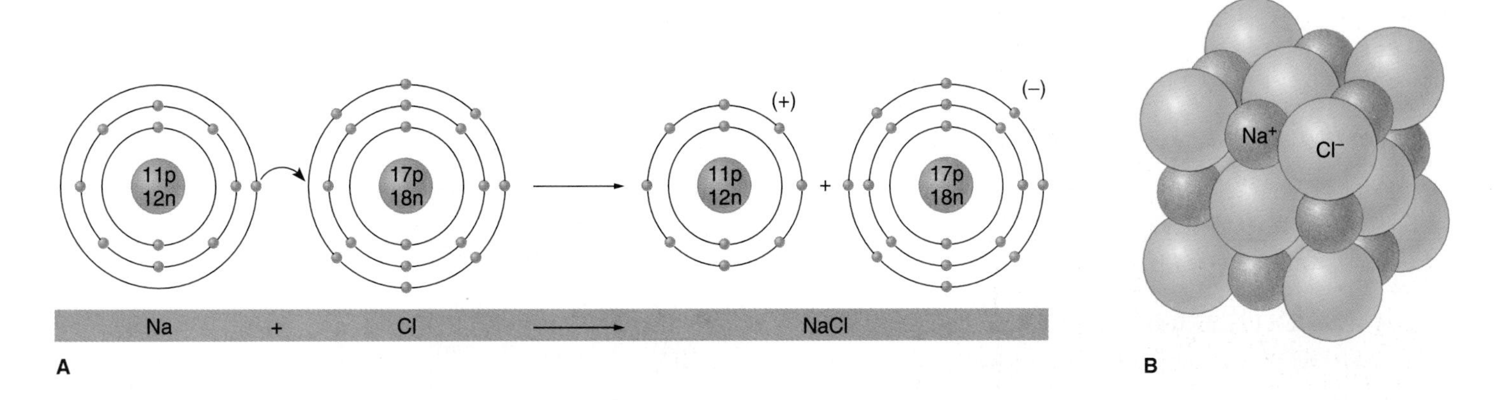

to accept electrons is crucial in helping organisms extract energy from nutrients. Polar covalent bonds form when less electronegative atoms such as carbon and hydrogen bond with electronegative atoms such as oxygen or nitrogen.

Ionic Bonds In a covalent bond, atoms share electrons. Sometimes, the electronegativity of one atom is so great that the atom takes an electron from the other atom. The two atoms then possess opposite charges and attract. This attraction is an **ionic bond.** An atom with one, two, or three electrons in the outermost shell loses them to an atom with, correspondingly, seven, six, or five valence electrons. A sodium (Na) atom, for example, has one valence electron. When it donates this electron to an atom of chlorine (Cl), which has seven electrons in its outer shell, the two atoms bond ionically to form NaCl (fig. 2.9).

Once an atom loses or gains electrons, it has an electric charge and is called an **ion.** Atoms that lose electrons lose negative charges and thus carry a positive charge. Atoms that gain electrons become negatively charged. In NaCl, the oppositely charged ions Na^+ and Cl^- attract in such an ordered manner that a crystal results (fig. 2.9B). A compound composed of oppositely charged ions, like NaCl, is called a **salt.** A salt can consist of any number of ions and is described as a ratio of one type of ion to another. Ionic bonds are strong in a chemical sense, but they often weaken in water.

Most biochemicals contain many negative charges. Many small, positively charged ions are abundant in organisms, such as sodium (Na^+) and potassium (K^+). These ions balance the larger, negatively charged molecules. In the human body, fluids within cells tend to have high concentrations of K^+, whereas fluids outside cells tend to have high concentrations of Na^+. Disruptions of this distribution can cause severe symptoms. Sodium deficiency, for example, causes seizures and coma.

Ions are important in many biological functions. Nerve transmission depends upon passage of Na^+ and K^+ in and out of nerve cells. Muscle contraction depends upon movement of calcium (Ca^{2+}) ions. In many types of organisms, egg cells release calcium ions, which clear a path for the sperm's entry.

Weak Chemical Bonds Weak chemical attractions may also be thought of as a type of bond and are very important in the molecules of life. When two molecules contain atoms that form polar covalent bonds, parts of each atom have small partial charges. The opposite charges on two molecules may attract each other and form **hydrogen bonds** (see fig. 2.8). In these bonds, the partially charged hydrogen on one molecule attracts the partial negative charges on other molecules. These bonds always form between molecules, not individual atoms. Hydrogen bonds are relatively weak, but in large molecules, such as DNA, many of them contribute significantly to holding the molecule together. Water also contains many hydrogen bonds, which provide some of its interesting characteristics.

Different parts of the large and intricately shaped chemicals of life may be temporarily charged because their electrons are always in motion. Dynamic attractions between molecules or within molecules that occur when oppositely charged regions approach one another are called **van der Waals attractions.** They help shape the molecules of life.

Interaction with water is another type of chemical attractive force that helps determine the spatial arrangement of atoms comprising molecules. **Hydrophilic** (water-loving) parts of molecules are attracted to water; **hydrophobic** (water-fearing) parts are repelled by water. Large molecules such as proteins contort so that their hydrophilic regions touch water and their hydrophobic regions shun it. Interactions with water help shape the surfaces of cells. Table 2.2 summarizes the chemical bonds important in life.

TABLE 2.2 Chemical Bonds and Attractive Forces

TYPE	CHEMICAL BASIS	STRENGTH	EXAMPLE
Covalent bonds	Atoms share electron pairs	Strong	Hydrocarbons
Hydrogen bonds	Atoms with partial negative charges attract hydrogen atoms with partial positive charges (due to their participation in a polar covalent bond)	Weak	Water, DNA
Ionic bonds	Atoms donate one or more electrons to other, oppositely charged atoms	Moderate	Sodium chloride
van der Waals attractions	Oppositely charged regions between or within molecules attract	Weak	Protein structure
Water interactions	Hydrophobic parts of molecules repel water; hydrophilic parts are attracted to water	Strong	Protein structure Cell membrane

2.1 MASTERING CONCEPTS

1. What types of chemicals comprise organisms?
2. How are elements and atoms related?
3. Which chemical elements do organisms require in large amounts?
4. What types of particles make up an atom?
5. What do an element's atomic number and atomic mass indicate?
6. What is an isotope?
7. How are compounds and molecules related?
8. How do the number and positions of electrons determine an atom's ability to form bonds?
9. Distinguish among nonpolar and polar covalent bonds, ionic bonds, and hydrogen bonds.
10. How do van der Waals attractions and interactions with water help molecules assume their three-dimensional forms?

OLC

Chemistry Explains Biology

Seeking Salt

Imagine drinking 600 times your weight in just a few hours and excreting most of it as you did so! This is exactly what male moths of species *Gluphisia septentrionis* do to obtain sufficient sodium ions in a salt-poor environment. The moths do this by a behavior called "puddling"—they sip mud puddles. The animals' mouths have sifterlike structures to admit salty water while keeping muck out.

Researchers Scott Smedley and Thomas Eisner discovered the salt-acquiring talent of the male moths by measuring the sodium ion content of puddle water and comparing it to that of the copious urine that spurts from the insects' rectums every few seconds, for hours. They found that the moths collect and concentrate the vital Na^+ in their reproductive organs and then transfer the ions to females during a 5-hour mating act. The female uses some of the ions herself and then passes some to her eggs. Says Smedley, "It's like dad giving the kids a one-a-day vitamin plus minerals."

2.2 How Is Water Important to Life?

Water has distinctive properties that make it essential for life. Water affects life at all levels, from the biochemical to the ecological.

Water has been called the "mater and matrix" of life. "Mater" means mother, and indeed life as we know it could not have begun were it not for this unusual substance. Water is also the matrix, or medium, of life, because it is vital in most biochemical reactions.

Solutions

A molecule of water consists of two atoms of hydrogen and one atom of oxygen. In one second, the hydrogen bonds between a single water molecule and its nearest neighbors form and re-form some 500 billion times! This action of hydrogen bonding makes water fluid. Fluidity is the ability of a substance to flow. The attractions between water molecules, which account for their constant rebonding, is called **cohesion** (fig. 2.10). Water readily forms hydrogen bonds to many other compounds, a property called **adhesion,** which is important in many biological processes. Adhesion enables water to form body fluids that carry vital, dissolved electrically charged (ionic) chemicals called **electrolytes.**

Cohesion and adhesion are important in plant physiology. Movement of water from a plant's roots to its highest leaves depends upon cohesion of water within the plant's water-conducting tubes. Water entering roots is drawn up as water evaporates from leaf cells. Water movement up trees also depends upon the adhesion of water to the walls of the conducting tubes. Water's adhesiveness accounts for **imbibition,** the tendency of substances to absorb water and swell. Rapidly imbibed water swells a seed so that it bursts through the seed coat, stimulating further growth of the embryo within.

Biologists use several terms to describe water's ability to carry other chemicals. A **solvent** is a chemical in which other chemicals, called **solutes,** dissolve. A **solution** consists of one or more chemicals dissolved in a solvent. In an **aqueous solution,**

FIGURE 2.10
Solutions Are Mixtures of Molecules.
Interactions between the solutes and solvents allow solutes to dissolve by forming bonds with the solvent.

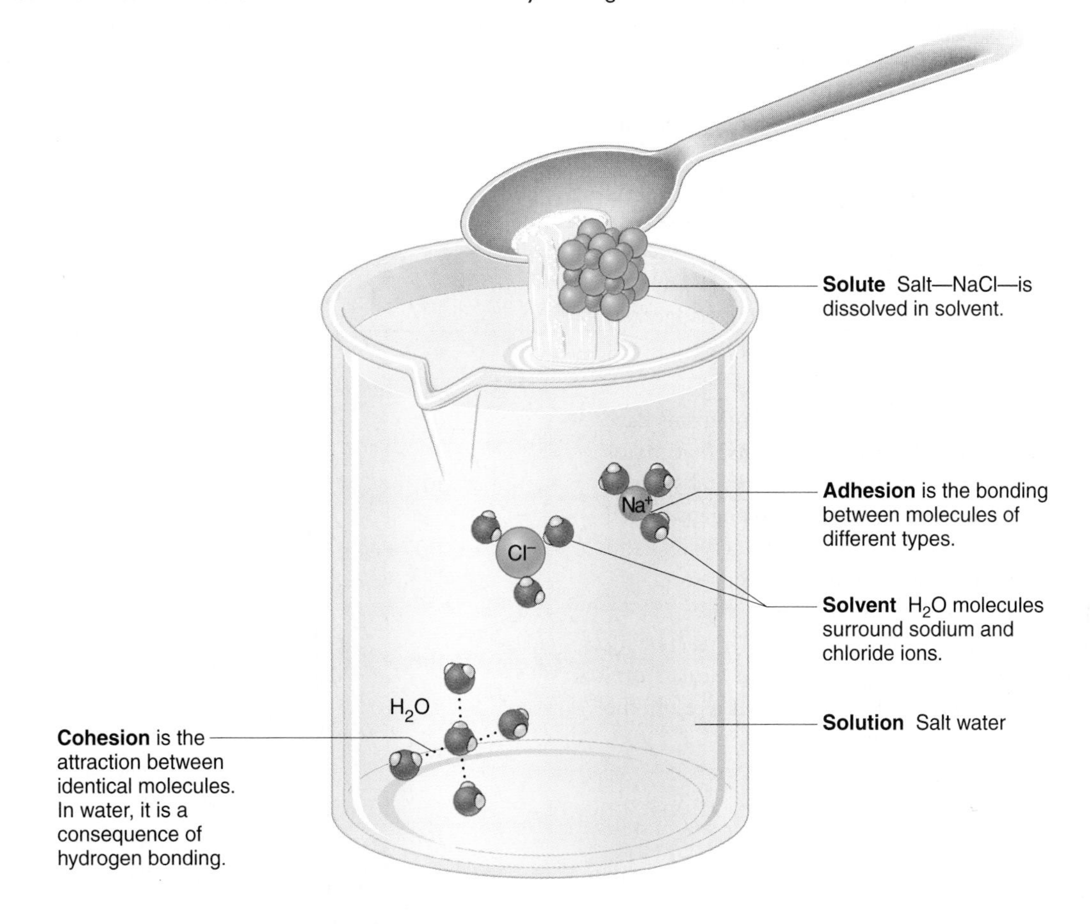

water is the solvent. Water molecules are polar; they have a partial positive charge at one end and a partial negative charge at the other. Ionic bonds can form between these charges on water and oppositely charged ions, such as those in salt. Water molecules surround each of the ions individually, which separates the ions from each other, or dissolves them. Any molecule that has a charge can be dissolved in water. The ability to separate charged atoms or molecules makes water a strong solvent.

Nonpolar molecules, such as fats and oils, do not carry a charge and thus do not usually dissolve in water. This is why oil on the surface of a cup of coffee forms a visible drop and a drop of water on an oily surface beads up.

Acids and Bases

In an aqueous solution, hydrogens from one water molecule are attracted to another water molecule and interact ionically, transforming two molecules of water ($2H_2O$) into a hydronium ion (H_3O^+) and a hydroxide ion (OH^-). (The H_3O^+, H_2O plus H^+, is usually considered the chemical equivalent of H^+.) In pure water, the numbers of free H^+s and OH^-s are equal. However, another chemical dissolved in water can alter the balance of positive and negative charges. Substances that add more H^+ to a solution are **acids.** Common acids include hydrochloric acid (HCl), sulfuric acid (H_2SO_4), phosphoric acid (H_3PO_4), and nitric acid (HNO_3). Substances that decrease the number of H^+s are **bases** and are alkaline (nonacid). In water, bases typically dissociate (come apart) to yield OH^-, which binds with some of the H^+ present in the dissociated water to re-form H_2O. A common base is sodium hydroxide (NaOH), or lye. Shampoo advertised as nonalkaline is actually slightly acidic.

The pH Scale Scientists use a system of measurement called the **pH scale** to gauge how acidic or basic a solution is in terms of its H^+ (hydrogen ion) concentration. The pH scale ranges from 0 to 14, with 0 representing strong acidity (high H^+ concentration) and 14 strong basicity or alkalinity (low H^+ concentration). Each unit on the pH scale represents a 10-fold change in H^+ concentration. The lower the pH, the higher the H^+ concentration becomes. A neutral solution, such as pure water, has a pH of 7 (fig. 2.11). One can think of pH as a relative concentration of OH^- to H^+. At pH 7, there is one OH^- to every H^+. At pH 3 there is one OH^- for every 10,000 H^+.

Many fluids in the human body function within a narrow pH range. Illness results when pH changes beyond this range. The normal pH of blood, for example, is 7.35 to 7.45. Blood pH of 7.5 to 7.8, a condition called alkalosis, makes one feel agitated and dizzy. Alkalosis can be caused by breathing rapidly at high altitudes, taking too many antacids, enduring high fever, or feeling extreme anxiety. Acidosis, in which blood pH falls to 7.0 to 7.3, makes one feel disoriented and fatigued and may impair breathing. This condition may result from severe vomiting, diabetes, brain damage, or lung or kidney disease. If blood pH changes beyond 7.0 or 7.8, acidosis or alkalosis can be fatal.

Buffer Systems Regulate pH in Organisms Organisms have pairs of weak acids and bases, called **buffer systems,** that maintain the pH of body fluids in a comfortable range. Buffer systems keep pH relatively constant by gaining or losing hydrogens, which neutralizes strong acids and bases. Carbonic acid (H_2CO_3) and sodium bicarbonate ($NaHCO_3$) form one such buffer system. It is found inside cells and in some of the fluids outside cells.

A strong acid, such as hydrochloric acid (HCl), reacts with sodium bicarbonate ($NaHCO_3$) and produces the weaker carbonic acid (H_2CO_3) and sodium chloride (NaCl). As a result, the acidity of the fluid decreases:

$$\underset{\text{strong acid}}{HCl + NaHCO_3} \longrightarrow \underset{\text{weak acid}}{H_2CO_3} + NaCl$$

A strong base, such as sodium hydroxide (NaOH), reacts with carbonic acid (H_2CO_3) and produces the weaker base sodium bicarbonate ($NaHCO_3$) and water:

$$\underset{\text{strong base}}{NaOH + H_2CO_3} \longrightarrow \underset{\text{weak base}}{NaHCO_3} + H_2O$$

Other buffer systems maintain the pH of urine and blood within certain ranges.

Outside of organisms, acid solutions affect life on a global scale. Sulfur- and nitrogen-based pollutants react with water in the atmosphere and return to earth as acid precipitation (see chapter 45). Just slightly lowering the pH of a body of water can drastically alter the types of organisms that can survive there. Some organisms, however, are adapted to low-pH environments, such as the bacteria that cause peptic ulcers. They live in the lining of the stomach in humans.

Water and Temperature

Several temperature-related characteristics of water affect life. Water controls temperature in organisms because a great deal of heat is required to raise its temperature—a characteristic called high **heat capacity.** Because of water's high heat capacity, an organism may be exposed to considerable heat before its aqueous body fluids become dangerously warm, and to considerable cold before cooling appreciably. Water's high **heat of vaporization** means that a lot of heat is required to evaporate water. This is why evaporating sweat draws heat out from the body—another important factor in regulating body temperature.

Water's unusual tendency to expand upon freezing also affects life. The bonds of ice are farther apart than those of liquid water. Therefore, ice is less dense and floats on the surface of liquid water. This characteristic benefits water-inhabiting (aquatic) organisms.

FIGURE 2.11
The pH Scale.
The pH scale is commonly used to indicate the strength of acids and bases as a function of the number of hydrogen ions present. The lower the pH, the higher the concentration of free hydrogen ions becomes, and the more acidic the solution. Conversely, the higher the pH, the more free hydroxyl (OH^-) ions there are and the more basic the solution.

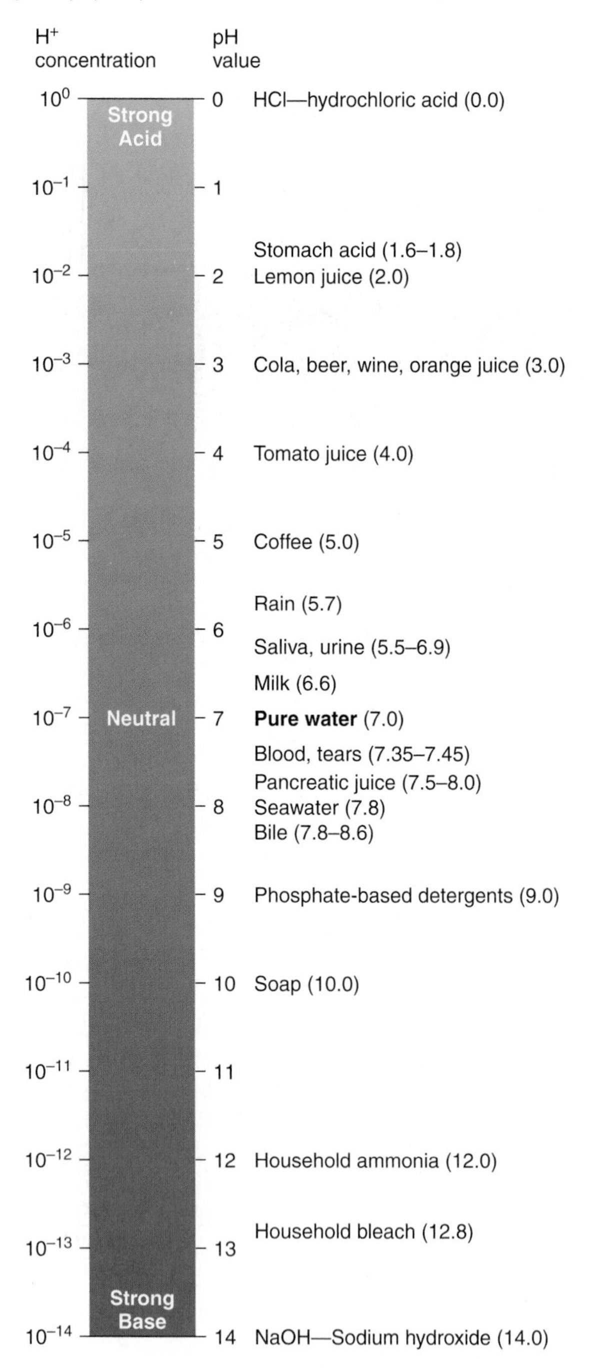

FIGURE 2.12

Survival Without Water.

Tardigrades, also called water bears, are small (less than a millimeter long) and live on aquatic vegetation. Under environmental stress, including freezing, a tardigrade gradually dries up, its body composition changing from 85% water to about 3% water. This creates a state of suspended animation called cryptobiosis that can extend a tardigrade's life span from 1 to 60 years.

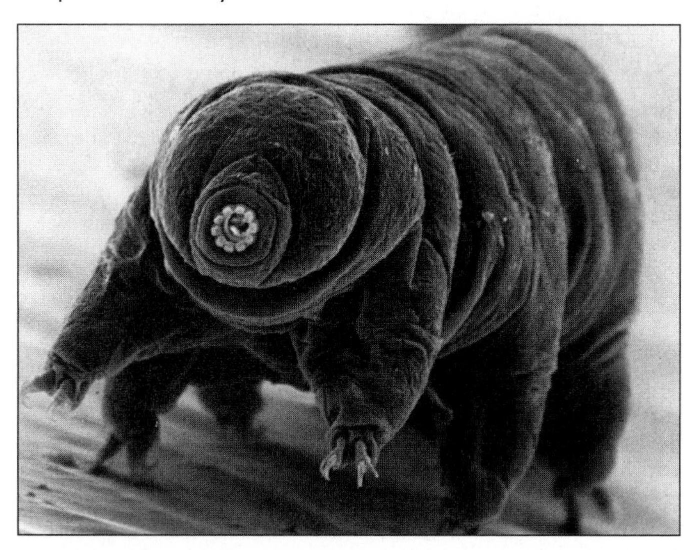

When the air temperature drops sufficiently, a small amount of water freezes at the surface of a body of water, forming a solid cap of ice that retains heat in the water below. Many organisms can then survive in the depths of such shielded lakes and ponds. If water were to become denser upon freezing, ice would sink to the bottom, and the body of water would gradually turn to ice from the bottom up, entrapping the organisms that live there.

On the other hand, organisms do freeze and die because water inside cells expands when it freezes and breaks apart the cells. Many types of organisms have adaptations that enable them to survive freezing, such as producing biochemicals that act as antifreeze, thus keeping body fluids liquid at temperatures low enough to freeze pure water. Another strategy is to temporarily dehydrate, such as does the tardigrade in figure 2.12. The larvae of flies called midges exhibit yet another adaptation to freezing—ice forms between their cells, not inside them.

Table 2.3 lists some of the important properties of water.

The discovery of evidence of water on Mars suggests that life may once have been present there. The presence of water on Jupiter's moon, Europa, has scientists looking for life on that distant world.

2.2 MASTERING CONCEPTS

1. List ways that water is essential to life.
2. What are acids and bases?
3. How does pH measure acidity and basicity?
4. What is the function of buffer systems?
5. Which properties of water enable it to regulate body temperature?
6. How does the fact that water expands upon freezing affect life?

OLC

TABLE 2.3 Properties of Water

PROPERTY	EXAMPLE OF IMPORTANCE TO LIFE
Cohesion	Water flows, forming habitats and a matrix for biological molecules
High heat capacity	Water regulates constant internal body temperature
High heat of vaporization	Evaporation (sweating) cools
Adhesion	Water acts as a solvent
Solid less dense than liquid	Keeps ponds and lakes liquid

2.3 Which Organic Molecules Are Important to Life?

The organic molecules that comprise organisms provide support, energy, protection, and an information system to carry out life functions on a biochemical level.

Organisms are composed of, and must take in large amounts of, four types of organic compounds: carbohydrates, lipids, proteins, and nucleic acids. Many of these compounds are small, single-unit molecules called **monomers.** Linked monomers form **polymers.** For each type of organic compound, the function of the monomer is usually very different from that of the polymer. Vitamins are another group of biologically important organic molecules, but they are required in smaller amounts. Table 2.5 at the end of the chapter summarizes the major organic compounds of life.

Carbohydrates

Carbohydrates consist of carbon, hydrogen, and oxygen, often in the proportion 1:2:1. That is, a carbohydrate has twice as many hydrogen atoms as either carbon or oxygen atoms. Familiar carbohydrates are sugars and starches. Carbohydrates store energy, which is released when their bonds are broken. Some carbohydrates physically support cells and tissues, and others help identify cells.

A monosaccharide is the smallest type of carbohydrate, and it contains three to seven carbons. Monosaccharides can differ from each other by how their atoms are bonded. For example, three six-carbon monosaccharides with the same molecular formula ($C_6H_{12}O_6$) but different chemical structures are glucose (blood sugar), galactose, and fructose (fruit sugar). Three-carbon monosaccharides, called trioses, form as cells break down glucose to release its energy, the subject of chapter 6.

Polysaccharides are huge molecules of hundreds of monomers. Carbohydrates of intermediate length, which consist of 2 to 100 monomers, are called oligosaccharides. (Polysaccharides and large oligosaccharides are also called complex carbohydrates.) The smallest oligosaccharide is a disaccharide, which forms when two monosaccharides link by releasing a molecule of water (H_2O). Figure 2.13 shows how the disaccharide sucrose (table sugar) forms as a molecule of glucose and a molecule of fructose lose a water and join. This type of chemical reaction is called **dehydration synthesis** ("made by losing water"). Dehydration synthesis joins monomers of different kinds to form all of the biological polymers. In the opposite reaction, **hydrolysis** ("breaking with water"), a disaccharide and water react to form two monosaccharides.

Many plants contain abundant sucrose, including sugarcane and sugar beets. Maltose, a disaccharide formed from two glucose molecules, provides energy in sprouting seeds and is used to make beer. Lactose, or milk sugar, is a disaccharide formed from glucose and galactose. Sugars, which are monosaccharides and disaccharides, are sometimes called simple carbohydrates.

FIGURE 2.13

Carbohydrates Can Be Simple or Complex.

Monosaccharides are composed of single molecules, such as glucose or fructose. Monosaccharides generally have the equivalent of one H_2O for every carbon atom. Disaccharides are formed by dehydration synthesis, which binds two monosaccharides and removes water. For instance, glucose and fructose bond to form sucrose. Polysaccharides are long chains formed in a similar way from monosaccharides, such as glucose. Different orientations of these bonds produce different characteristics in the molecules. Recall from figure 2.7 that rings represent carbon atoms at the apexes.

Monosaccharides—simple sugars composed of carbon, hydrogen, and oxygen in the proportions 1:2:1.

Glyceraldehyde $C_3H_6O_3$

Ribose $C_5H_{10}O_5$

Glucose $C_6H_{12}O_6$

Fructose $C_6H_{12}O_6$

Disaccharides—molecules composed of two monosaccharides joined by dehydration synthesis. Hydrolysis converts disaccharides to their component monosaccharides. (The structures of the molecules are simplified to emphasize the joining process.)

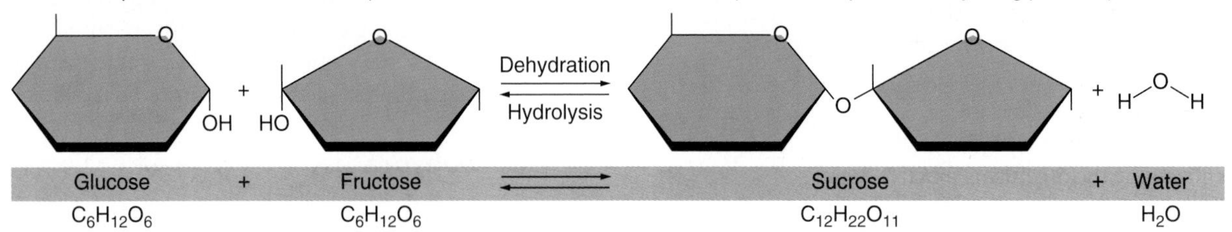

Polysaccharides—also known as "complex carbohydrates," composed of long chains of simple sugars, usually glucose. Their chemical characteristics are determined by the orientation and location of the bond between the monomers.

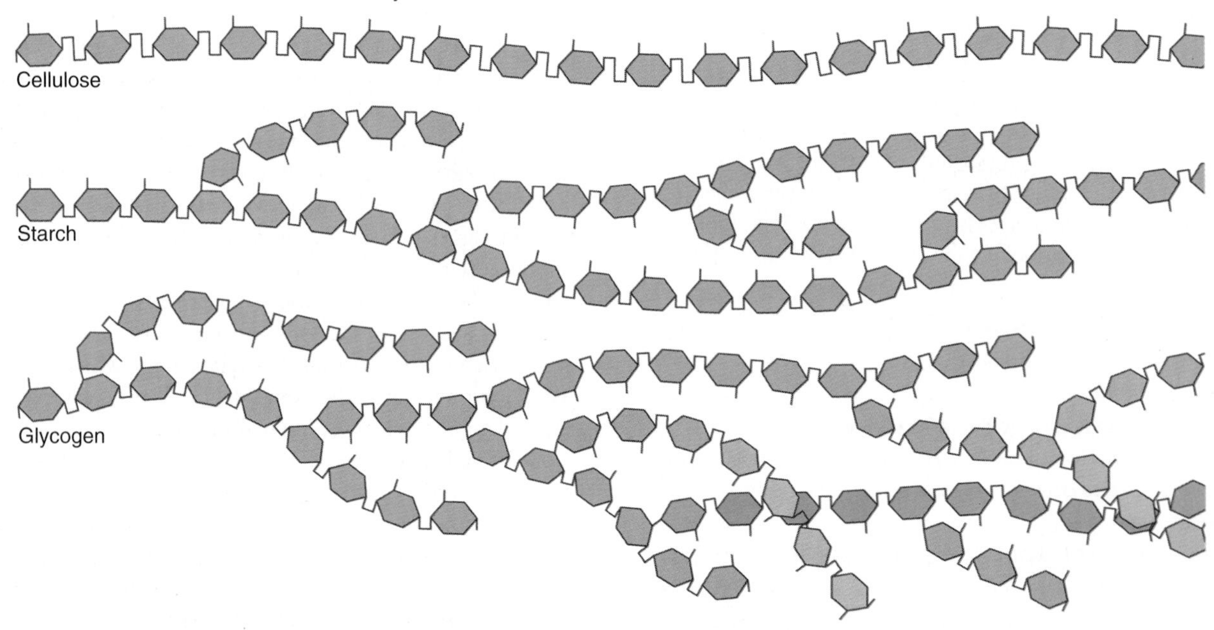

FIGURE 2.14

Complex Carbohydrates Have Multiple Functions.

Carbohydrates can serve structural roles and as sources of energy. (**A**) Glucose monomers join in long chains that form cellulose fibers, a major structural component of plants. (**B**) Chitin forms the hard, outer skeleton (exoskeleton) of insects. (**C**) Animal cells are coated in complex carbohydrate molecules that identify that cell to the rest of the body.

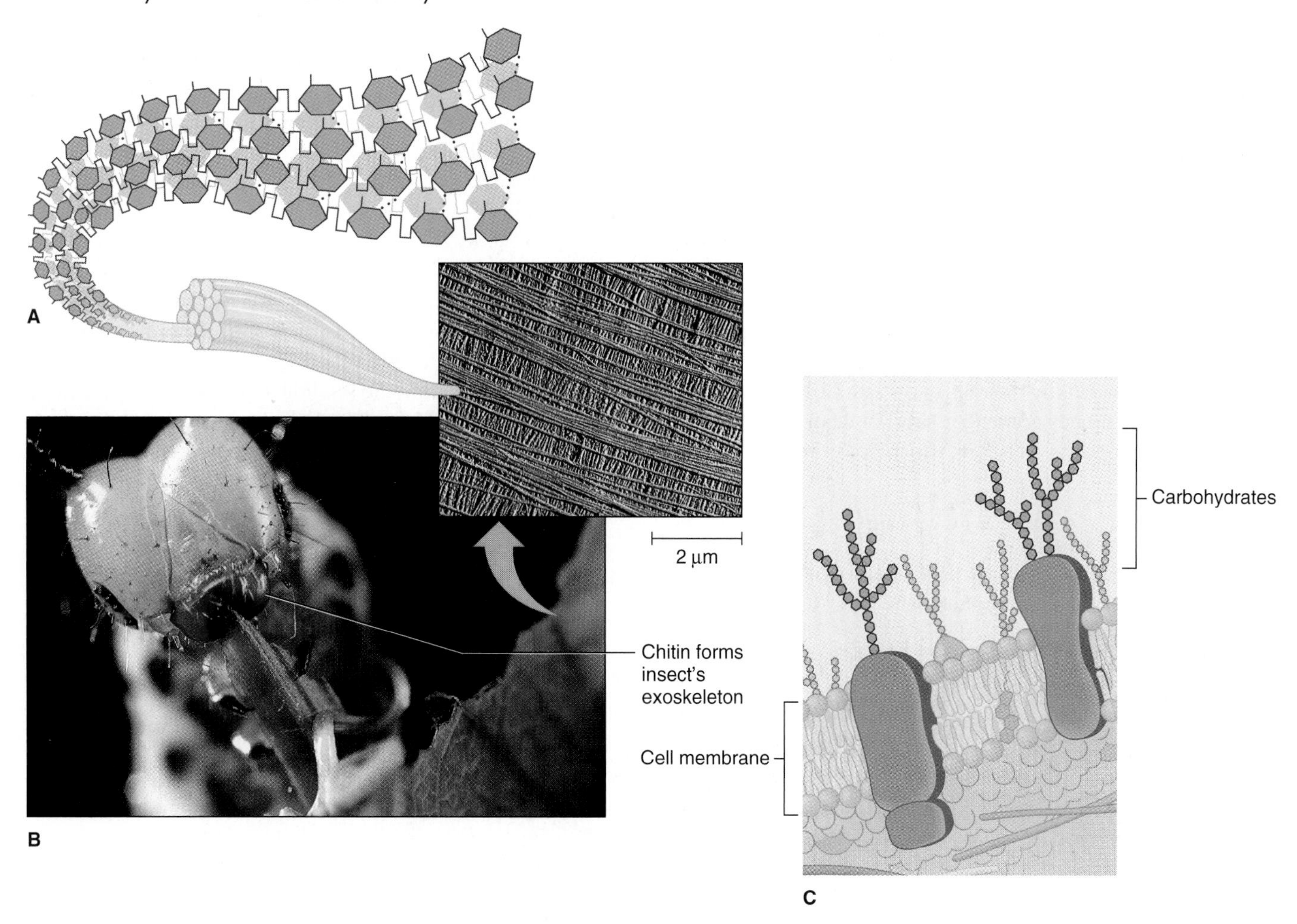

Oligosaccharides attach to proteins that are part of cell membranes, emanating from the cell's surface and creating a topography, like the surface of Earth (fig. 2.14). These glycoproteins on cell surfaces are important in immunity, which is based on the distinctiveness of cell membranes among individuals. Oligosaccharides are also important in enabling proteins called antibodies to assume their characteristic three-dimensional shapes, which is essential to their function in protecting an animal's body from infection.

Polysaccharides ("many sugars") are long polymers of monosaccharides linked by dehydration synthesis (see fig. 2.13). The most common polysaccharides are cellulose, chitin, starch, and glycogen. They are all long chains of glucose, but they differ from each other by their branching patterns. Cellulose forms wood and parts of plant cell walls, and is the most common organic compound in nature. The second most common polysaccharide in nature is chitin. It resembles a glucose polymer, but one OH group in each subunit is replaced with a group that contains nitrogen. Chitin forms the flexible exoskeletons of insects and spiders (fig. 2.14) and forms the cell wall in many types of fungi.

Lipids

Lipids contain the same elements as carbohydrates but with less oxygen. Different types of lipids have different amounts of oxygen. The lipids are diverse molecules that dissolve in organic solvents but not in water. The most familiar lipids are fats and oils.

Lipids are vital to life in many ways. They are necessary for growth and for the utilization of some vitamins. Fats slow digestion, thereby delaying hunger. Some lipids coat leaves, fur, and feathers, making them water repellent. Others cushion organs and help to retain body heat as insulation. Fat is also an excellent energy source, providing more than twice as much energy as equal weights of carbohydrate or protein.

Fat cells aggregate as adipose tissue in animals. White adipose tissue forms most of the fat in human adults. Another type, brown adipose tissue, releases energy as heat and keeps organisms warm, particularly mammals that hibernate. Brown adipose tissue is rare in adult humans, but in newborns it is layered around the neck and shoulders and along the spine.

Lipids enclose all cells as a major component of cell membranes. In our own bodies, lipid-rich cells ensheath nerve cells, which speeds nerve transmission. Human milk is rich in lipids, partly to suit the rapid growth of the brain in the first 2 years of life.

This chapter discusses the triglycerides, fatty acids, sterols, and waxes. Two other types of lipids, phospholipids and glycolipids, are considered in the next two chapters in the context of their functions in cell membranes. A phospholipid is a lipid bonded to a phosphate group (a phosphorus atom bonded to four oxygen atoms). Glycolipids are carbohydrates bonded to lipids.

Triglycerides and Fatty Acids A **triglyceride** consists of a three-carbon alcohol (an organic molecule with an OH) called **glycerol,** from which three hydrocarbon chains called **fatty acids** extend (fig. 2.15). Covalent bonds link the carbons of fatty acids and form tail-like structures of up to 36 carbon atoms. A triglyceride is what is commonly known as "fat."

We describe fatty acids by their degree of saturation, which is a measure of their hydrogen content. A **saturated** fatty acid contains all the hydrogens it possibly can, which occurs when single bonds connect all the carbons. A fatty acid is **unsaturated** if it has at least one double bond and **polyunsaturated** if it has more than one double bond. A **monounsaturated** fatty acid has just one double bond. Olive oil is a monounsaturated fat. Lipids in plants are less saturated than those in animals.

The sites of unsaturation (double bonds) in the fatty acids cause them to form kinks and spread their "tails." This produces an oily consistency at room temperature, and allows membrane phospholipids (fig. 2.16) to be more fluid. The more saturated animal fats tend to be more solid. A food-processing technique called hydrogenation, used to produce margarine, adds hydrogen to an oil to solidify it—in essence, saturating a formerly unsaturated fat.

Sterols **Sterols** are lipids that have four interconnected carbon rings. Vitamin D and cortisone are sterols. A very familiar sterol, cholesterol, is a key part of animal cell membranes (fig. 2.16B). Animal cells use cholesterol as a starting material to synthesize other lipids, including the sex hormones testosterone and estrogen. Liver cells manufacture cholesterol when they break down saturated fats. Eating cholesterol adds to this, and the excess cho-

FIGURE 2.15

Lipids Form Complex Molecules.

A triglyceride is formed by bonding fatty acids to glycerol. This triglyceride is tripalmitin. The double bonds bend the fatty acid tails, making the lipid more fluid. The bend is shown schematically here.

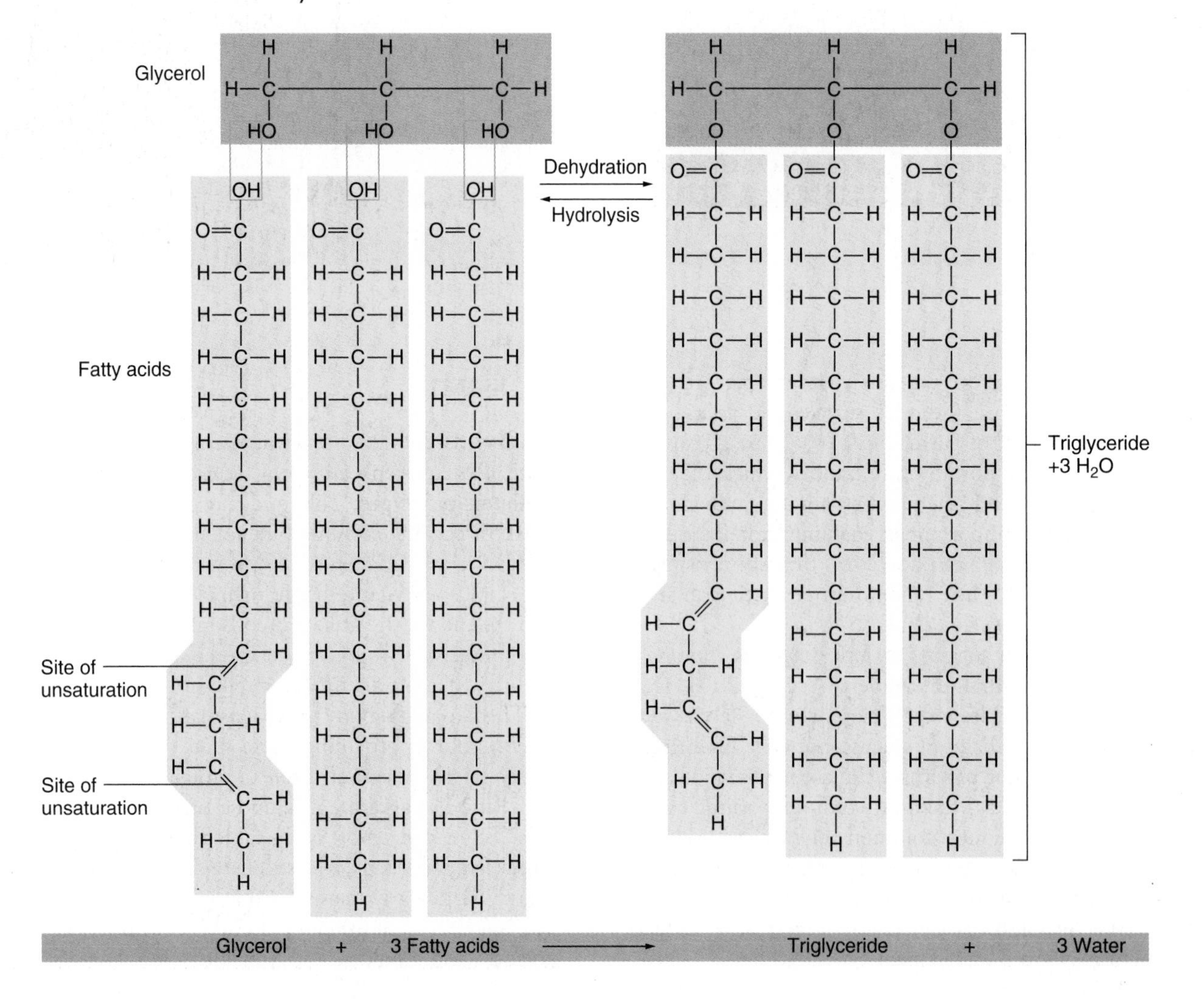

FIGURE 2.16
Lipids Found in Cell Membranes.
Phospholipids (**A**) are the fundamental unit of all biological membranes. Double bonds create "kinks" that make membranes more fluid. In animal cells, cholesterol (**B**) also adds to membrane fluidity in addition to providing raw material for hormone production.

Phospholipid

A

Cholesterol

B

lesterol collects on the inner linings of blood vessels and impedes blood flow. Because the liver essentially converts saturated fat into cholesterol, it is important to limit dietary intake of saturated fats as well as cholesterol.

Waxes Waxes are fatty acids combined with either alcohols or other hydrocarbons that usually form a hard, water-repellent covering. These lipids help waterproof fur, feathers, leaves, fruits, and some stems (fig. 2.17). Jojoba oil, used in cosmetics and shampoos, is unusual in that it is a liquid wax.

Proteins

Proteins consist of monomers of **amino acids** linked to form **polypeptide chains.** A protein consists of one or more polypeptide chains. Amino acids in organisms are of 20 types, even though chemically many others exist. These 20 different types make possible a nearly infinite number of different amino acid sequences. In contrast, a polysaccharide is a polymer of only one or a few types of monomers.

Proteins have diverse biological functions. They enable blood to clot, muscles to contract, oxygen to reach tissues, and nutrients to be broken down to release energy. Enzymes allow biochemical reactions to proceed fast enough to sustain life. Other proteins mark our cell surfaces as distinctly ours. They enable certain cells to migrate and others to attach to one another and build tissues. Table 2.4 lists some of the thousands of kinds of proteins in the human body. Proteins control all the activities of life.

FIGURE 2.17
Lipids Come in Many Forms.
Waxes waterproof the coat of the otter and the cuticles of the grasses growing in the background.

TABLE 2.4 Protein Diversity in the Human Body

PROTEINS	FUNCTION
Actin, myosin, dystrophin	Muscle contraction
Antibodies, antigens, cytokines	Immunity
Carbohydrases, lipases, proteases, nucleases	Digestive enzymes
Casein	Milk protein
Collagen, elastin	Connective tissue
Colony stimulating factors	Blood cell formation
DNA and RNA polymerase	DNA replication, gene expression
Ferritin	Iron transport in blood
Fibrin, thrombin	Blood clotting
Growth factors	Cell division
Hemoglobin, myoglobin	Oxygen transport
Insulin, glucagon	Control of blood glucose level
Keratin	Hair structure
Tubulin, actin	Cell movements
Tumor suppressors	Prevent cancer

Chemistry Explains Biology

Excess Cholesterol

Lipoproteins carry cholesterol in the bloodstream. As their name suggests, lipoproteins consist of lipid and protein. Low-density lipoprotein (LDL) particles carry cholesterol to the arteries. Excess LDL cholesterol that does not enter cells accumulates on the inner linings of blood vessels, eventually impeding blood flow. High-density lipoproteins (HDL), in contrast, carry cholesterol to the liver, where it is removed from the bloodstream. High levels of LDL cholesterol increase risk of heart disease, whereas high levels of HDL cholesterol promote heart health.

Amino Acid and Protein Structure An amino acid (fig. 2.18A) contains a central carbon atom bonded to these four other atoms or groups of atoms:

1. a hydrogen atom;
2. a **carboxyl group** (acid), which is a carbon atom double-bonded to one oxygen and single-bonded to another oxygen carrying a hydrogen (COOH);
3. an **amino group,** which is a nitrogen atom single-bonded to two hydrogen atoms (NH_2);
4. a side chain, or **R group,** which can be any of several chemical groups.

The nature of the R group distinguishes the 20 types of biological amino acids. An R group may be as simple as the lone hydrogen atom in glycine, or as complex as the two organic rings of tryptophan. Figure 2.18B shows three amino acids. The R groups of two amino acids, cysteine and methionine, contain sulfur. Note that due to the composition of amino acids, proteins contain

FIGURE 2.18
Amino Acids Join to Form Peptides.
Amino acids are the monomer subunits of proteins. (**A**) An amino acid is composed of an amino group, an acid (carboxyl) group, and one of 20 R groups attached to a central carbon atom. (**B**) The composition of the R groups contributes different functions to the final protein. (**C**) A dipeptide forms when an OH from a carboxyl group of one amino acid combines with a hydrogen from the amino group of another amino acid, creating a water molecule and linking the carboxyl carbon of the first amino acid to the nitrogen of the other. (**D**) Long chains of amino acids are peptides, which form polypeptides and proteins.

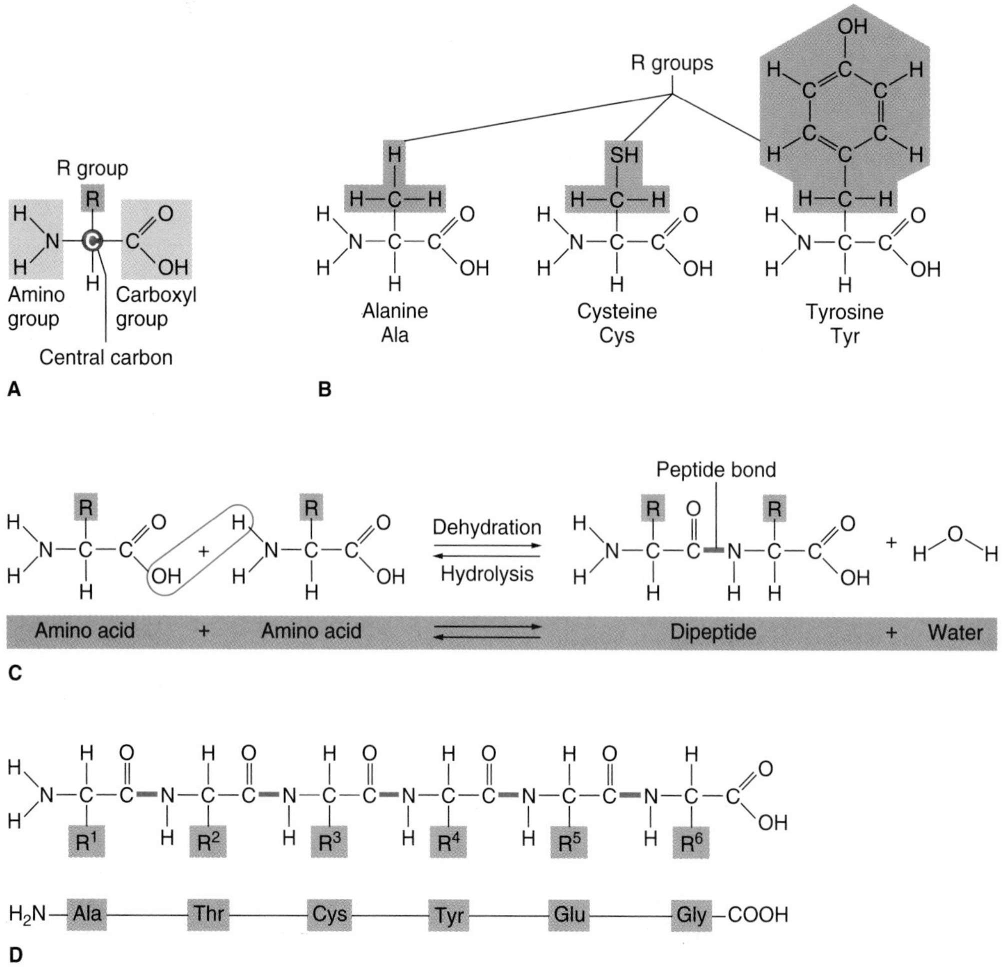

large amounts of nitrogen. This is another way that they differ from carbohydrates and lipids.

Two amino acids join by dehydration synthesis, just as two monosaccharides shed a water molecule to yield a disaccharide. When two amino acids bond, the acid-group carbon of one amino acid bonds to the nitrogen of the other and forms a **peptide bond** (fig. 2.18C). Two linked amino acids form a dipeptide, three a tripeptide; larger chains (fig. 2.18D) with fewer than 100 amino acids are peptides, and finally, those with 100 or more amino acids are polypeptides. Hydrolysis (adding water) breaks a protein into its constituent amino acids.

Protein Folding As a protein is synthesized in a cell, it folds into a three-dimensional structure, known as its **conformation.** The conformation of a protein is determined by the order and kinds of amino acids of which it is composed. The unique final shape of a protein is established by interactions with other proteins and water molecules and from bonds that form within the protein itself. Thousands of water molecules surround a protein, contorting it as its hydrophilic R groups move toward water and its hydrophobic R groups move away from the water toward the protein's interior. Ionic and hydrogen bonds form between the protein "backbone" and certain R groups to contribute to this folding process. The entire structure is further stabilized by the formation of covalent bonds between sulfur atoms in some R groups. Called disulfide bonds, these are abundant in structural proteins such as keratin that forms hair, scales, beaks, wool, and hooves (fig. 2.19). A "permanent wave" curls hair by breaking disulfide bonds in hair keratin and re-forming those bonds in hair that has been wrapped around curlers.

The conformation of a protein may be described at four levels (fig. 2.20). The amino acid sequence of a polypeptide chain is the primary (1°) structure. Hydrogen bonds that occur between parts of the peptide "backbone" create the secondary (2°) structure. These bonds are affected by the R groups, but do not directly involve them.

FIGURE 2.19

A Gallery of Keratin-Based Structures.

Alpha-keratin forms (**A**) the beak of a bird, (**B**) the scales of a snake, and (**C**) the horn of a ram.

A

B

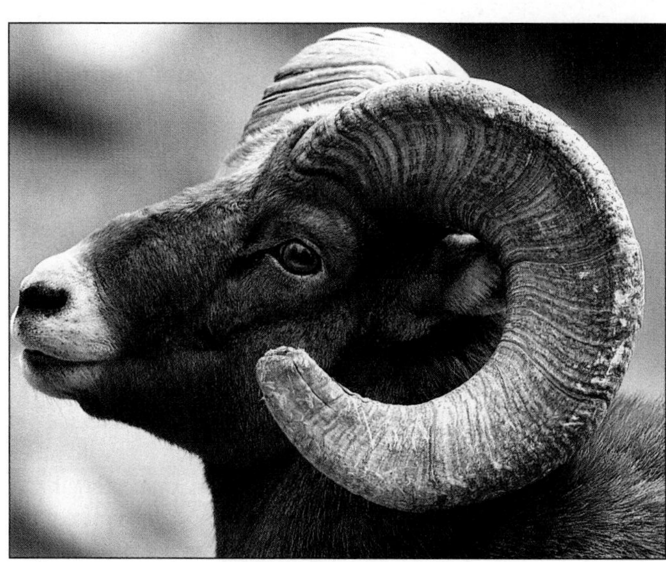

C

Chemistry Explains Biology

Spider Silk

Spider silk is the strongest natural fiber known. Spiders use silk to build webs to capture prey, to store prey, and to protect their eggs.

Silks are proteins, manufactured in sets of glands. A spider usually has several sets of glands, each of which produces a different type of silk. In addition to producing webs, a spider continually secretes a strand of silk called a dragline. Should danger arise, the spider can temporarily escape the web on its dragline.

Dragline silk is best studied in the golden orb weaver spider, *Nephila clavipes*. This especially strong and elastic silk consists of two types of proteins that are dry and practically indestructible once outside the animal's body. Dragline silk proteins include a few types of amino acids that repeat in short sequences. This imparts a conformation of coiled sheets, like the steps of a spiral staircase. Spider silk elegantly illustrates how a protein's shape determines its functions.

FIGURE 2.20

Four Levels of Protein Structure.

(**A**) The amino acid sequence of a polypeptide forms the primary structure, while (**B**) hydrogen bonds between non-R groups create secondary structures such as helixes and sheets. The tertiary structure (**C**) is formed when R groups interact, folding the polypeptide in three dimensions and forming a unique shape. (**D**) If different polypeptide units must interact to be functional, this forms the quaternary structure of a protein.

A Primary structure—the sequence of amino acids

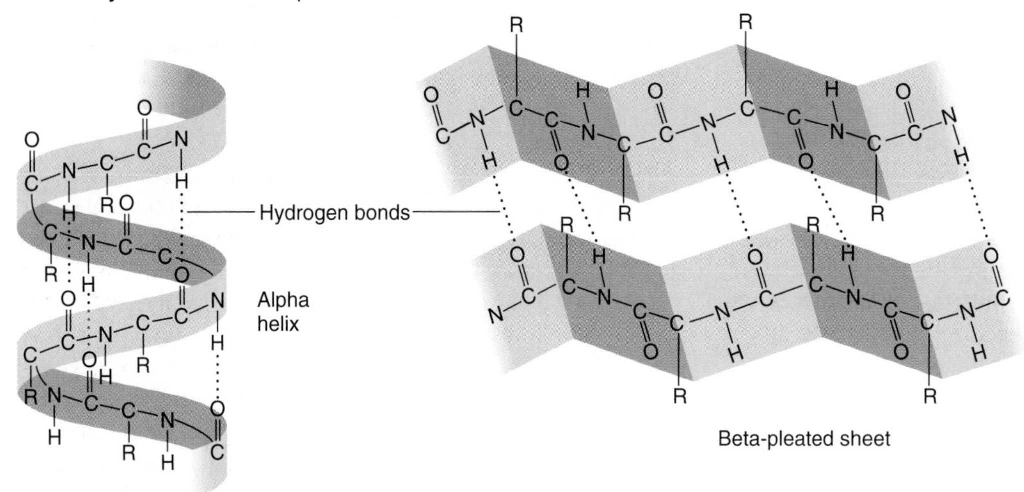

B Secondary structure—hydrogen bonds between nonadjacent carboxyl and amino groups

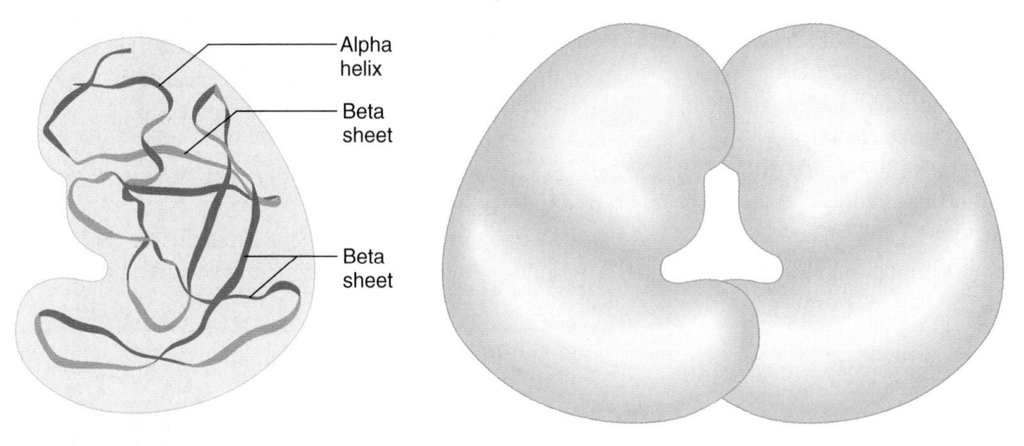

C Tertiary structure—disulfide and ionic bonds between R groups, interactions between R groups and water

D Quaternary structure—hydrogen and ionic bonds between separate polypeptides

The interactions that form the bonds cause the polypeptide to fold into coils, sheets, loops, and a variety of combinations of these shapes. A single polypeptide will usually contain several of these shapes, which are called motifs. Figure 2.20B shows two of the more common motifs; alpha helices and beta-pleated sheets.

Proteins fold into their final tertiary (3°) structures through interactions between R groups and with water. Many proteins are functional at this level. Some polypeptides must interact with other polypeptides to form a functional protein. This level of organization is referred to as quaternary (4°) structure. For example, the blood protein hemoglobin is composed of four polypeptide chains.

A protein's conformation makes possible, and in large part determines, its function. A digestive enzyme holds a large nutrient molecule in just the right way to cause it to break apart. An antibody binds to a very specific part of a molecule on the surface of a bacterium and holds it for a white blood cell to kill. Muscle proteins form long, aligned fibers that slide past one another, shortening their length to create muscle contractions. All functions of a protein are ultimately determined by its primary structure.

Enzymes Among the most important of all biological molecules are enzymes. Enzymes are proteins that speed the rates of specific chemical reactions without being consumed in the process, a phenomenon called **catalysis.** Catalysts may be inorganic (lacking carbon) or organic (carbon-containing). An enzyme is an organic catalyst. Enzymes bind reactants (starting materials) and bring them in contact with each other, so that less energy is required for the reaction to proceed. Without enzymes, many biochemical reactions would proceed far too slowly to support life; some enzymes increase reaction rates a billion times. Most enzymes function under very specific pH and temperature conditions. The "extremozymes" described at the start of chapter 1 can function under very high or low temperature or pH. If the conditions are not optimal, most enzymes will simply stop working.

Enzymes are also very specific in which chemical reaction they catalyze. The key to an enzyme's specificity lies in its **active site,** which is a region to which the reactants, also called **substrates,** bind. A substrate fits into the active site of an enzyme as the enzyme contorts slightly around it, forming a short-lived **enzyme-substrate complex** (fig. 2.21). An enzyme can hold two substrate molecules that react to form one product molecule, or it can hold a single substrate that splits to yield two products. Then the complex breaks down to release the products (or product) of the reaction. The enzyme is unchanged, and its active site is empty and ready to pick up more substrate.

FIGURE 2.21
Enzyme Action.
In this highly schematic depiction, substrate molecules A and B fit into the active site of an enzyme. An enzyme-substrate complex forms as the active site moves slightly to accommodate its occupants. A new compound, AB, is released, and the enzyme is reused. Enzyme-catalyzed reactions can break down as well as build up substrate molecules.

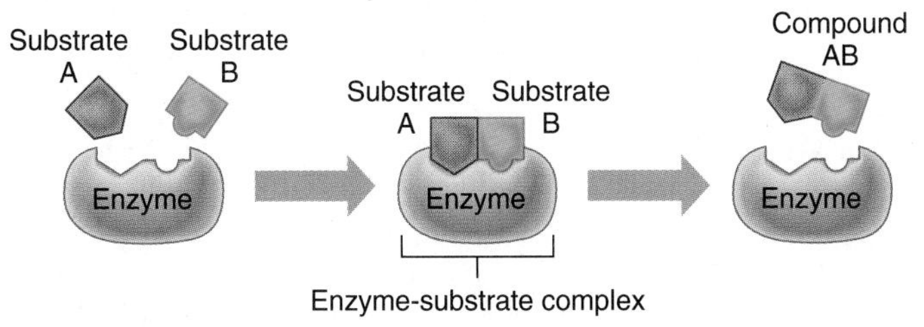

Nucleic Acids

Synthesizing a protein is a more complex task than synthesizing a carbohydrate or fat because of the great variability of amino acid sequences. How does an organism "know" which amino acids to string together to form a particular protein? A protein's amino acid sequence is encoded in a sequence of chemical units of another type of biochemical called a **nucleic acid.**

The two types of nucleic acids are **deoxyribonucleic acid** (DNA) and **ribonucleic acid** (RNA). They are polymers of monomers called **nucleotides** (fig. 2.22A). A nucleotide consists of a five-carbon sugar (deoxyribose in DNA and ribose in RNA), one or more phosphate groups (PO_4), and one of five types of nitrogen-containing compounds called **nitrogenous bases.** The nitrogenous bases are adenine (A), guanine (G), thymine (T), cytosine (C), and uracil (U). DNA contains A, C, G, and T; RNA contains A, C, G, and U. The DNA polymer is a double helix and resembles a spiral staircase in which alternating sugars and phosphates form the rails and nitrogenous bases form the rungs (fig. 2.22B). Chapter 12 discusses nucleic acids in detail.

Long sequences of DNA nucleotides contain information that is copied to RNA molecules for a cell to use to guide assembly of amino acids into polypeptide chains. Each group of three DNA bases in a row specifies a particular amino acid, in a correspondence called the genetic code. A sequence of genetic code specifying a polypeptide is a **gene.** DNA, therefore, is the genetic material.

DNA exists as two strands that are complementary or "opposites" of each other. Each strand is composed of a series of nucleotides. The two strands are held together by specific hydrogen bonds that form between the bases; A with T, C with G. If one strand at a particular location contains an A, the base on the complementary strand must be a T. Because of this unique feature, one strand of DNA contains the information for the other, providing a mechanism for the molecule to replicate.

Even though DNA is the genetic material, RNA is, in some ways, even more important. RNA has several functions, because its single-stranded structure enables it to assume different shapes. In its various guises, RNA enables the information in DNA to be expressed and utilized. RNA can also function as an enzyme. One particular RNA nucleotide serves a vital role in carrying energy that is used in nearly all biological functions. Because of its eclectic roles, RNA, or a molecule similar to it, may have been a bridge between complex groups of chemicals and the first organisms.

Table 2.5 reviews the characteristics of the major types of organic molecules that make up organisms.

2.3 MASTERING CONCEPTS

1. What are the chemical compositions and different types of carbohydrates, lipids, proteins, and nucleic acids?
2. What are the functions of carbohydrates, lipids, proteins, and nucleic acids?
3. Why are proteins extremely varied in organisms, but carbohydrates and lipids are not?
4. What is the significance of a protein's conformation?
5. How are DNA and RNA different in structure and function?

OLC

TABLE 2.5 The Macromolecules of Life

TYPE OF MOLECULE	STRUCTURE	FUNCTIONS	EXAMPLES
Lipids	Diverse, mostly carbon and hydrogen, some oxygen	Membranes Energy storage	Fats, oils, waxes
Carbohydrates	Sugars and polymers of sugars	Energy storage Support	Sugars, starch, glycogen Chitin, cellulose
Proteins	Polymers of amino acids	Transport Muscle action Blood clotting Support Immunity Enzymes	Hemoglobin Myosin, actin Thrombin, fibrin Collagen, elastin Antibodies Chemical reactions in a cell
Nucleic acids	Polymers of nucleotides	Transfer of genetic information	DNA, RNA

Nucleotides—consist of a sugar (ribose or deoxyribose), a phosphate, and one of five nitrogenous bases.

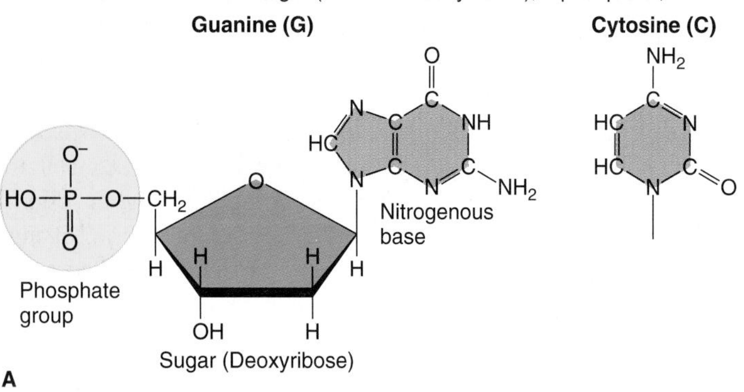

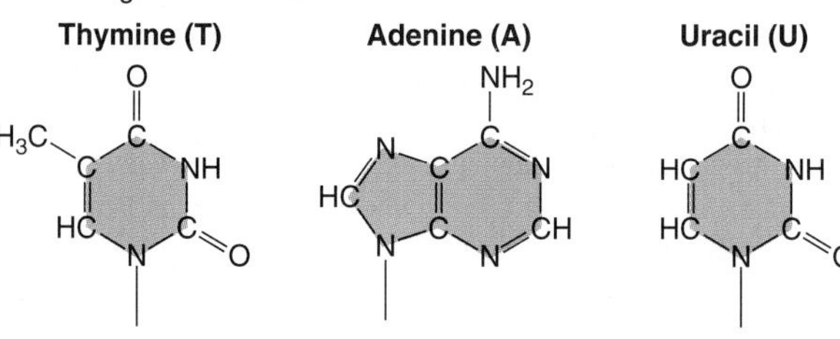

Nucleic acids—nucleotides joined together in long chains to form DNA or RNA. DNA is composed of the nucleotides A, C, T, G. RNA contains the sugar ribose and the nucleotide U instead of T.

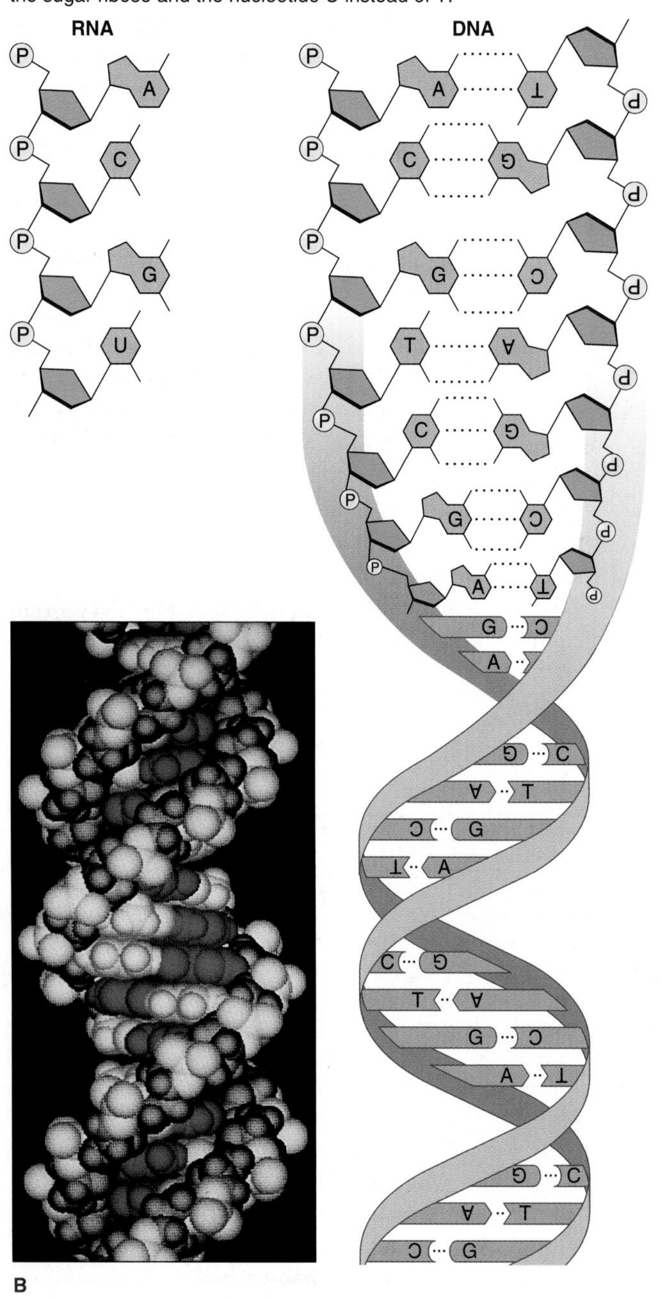

FIGURE 2.22
DNA Structure.
(**A**) The monomer unit of DNA is a nucleotide, which consists of a nitrogenous base (A, T, C, or G), a sugar (deoxyribose), and a phosphate. (**B**) Nucleotide pairs form as complementary bases attract, A with T and G with C. A DNA molecule is double-stranded, with the strands running in opposite orientations—this is why the labels on one strand are upside down. The two strands entwine to form a double helix shape. RNA is usually single-stranded and contains a different sugar—ribose—and a different base—uracil—which replaces thymine.

Chapter Summary

2.1 What Is Matter?

1. Any substance that occupies space is called **matter.** Matter is often equated to chemicals, which form the basis of all life.
2. Matter can be broken down into pure substances called **elements.** Elements are organized in the **periodic table** according to their number of subatomic particles. **Bulk elements,** those essential to life in large quantities, include carbon, hydrogen, oxygen, nitrogen, sulfur, and phosphorus.
3. An **atom** is the smallest unit of an element. Subatomic particles include the positively charged **protons** and neutral **neutrons** that form the **nucleus** and the negatively charged, much smaller **electrons** that circle the nucleus.
4. **Atomic number** is an element's characteristic number of protons, and the **atomic mass** is the mass of its protons and neutrons. **Isotopes** of an element differ by the number of neutrons.
5. **Compounds** are built of bonded atoms of different elements, in a consistent ratio. A **molecule** is the smallest unit of a compound that retains the characteristics of that compound. A compound's characteristics differ from those of its constituent elements.
6. Chemical shorthand indicates the numbers of atoms and molecules in a compound. In a **chemical reaction,** different compounds are broken down and form, but the total number of atoms of each element remains the same.
7. Electrons move constantly; they are most likely to be found in volumes of space called **orbitals,** which contain levels of energy called **shells.** Electrons can absorb energy and move to higher shells.
8. An atom's tendency to fill its outermost or **valence shell** with electrons drives atoms to bond and form molecules.
9. **Covalent bonds** form between atoms that can fill their valence shells by sharing one or more pairs of electrons. These are the strongest of chemical bonds. Carbon atoms form up to four covalent bonds. Atoms in a **nonpolar covalent** bond share all electrons equally. **Electronegative** atoms involved in covalent bonds tend to attract electrons, forming **polar covalent bonds,** resulting in opposite partial charges on different parts of the molecule.

10. Two atoms may donate or receive electrons from each other to fill their valence shell. The resulting atoms possess a charge and are called **ions.** An **ionic bond** forms as two ions are attracted to each other. These are moderately strong bonds.
11. **Hydrogen bonds** form when a hydrogen in one molecule is drawn to part of a neighboring molecule because of unequal electrical charge distribution.
12. **van der Waals attractions** occur between parts of molecules that are temporarily oppositely charged.

2.2 How Is Water Important to Life?

13. Most biochemical reactions occur in an aqueous environment. Water is **cohesive** and **adhesive,** enabling many substances to dissolve in it.
14. **pH** is a measure of H^+ concentration, or how acidic or basic a solution is. Pure water has a pH of 7, which means that the numbers of H^+ and OH^- in water are equal. An **acid** adds H^+ to a solution, lowering the pH below 7. A **base** adds OH^-, raising pH to between 7 and 14. **Buffer systems** consisting of weak acid-and-base pairs maintain the pH ranges of body fluids.
15. Water helps regulate temperature in organisms because of its high **heat capacity** and high **heat of vaporization.**

2.3 Which Organic Molecules Are Important to Life?

16. Most of the large biological molecules are composed of small subunit molecules called **monomers,** which possess characteristics distinct from the resulting **polymers.**
17. Monomers form into polymers by **dehydration synthesis** or are released from polymers by **hydrolysis.**
18. **Carbohydrates** provide energy and support. They consist of carbon, hydrogen and oxygen in the proportions 1:2:1. Monosaccharides are single-molecule sugars such as glucose. Two bonded monosaccharides form a disaccharide. Oligosaccharides are composed of 2 to 100 monomers, whereas polysaccharides are enormous molecules of hundreds of monomers.
19. **Lipids** are diverse organic compounds that provide energy, slow digestion, waterproof the outsides of organisms, cushion organs, and preserve body heat. Lipids include fats and oils, do not dissolve in water, and contain carbon, hydrogen, and oxygen but have less oxygen than carbohydrates. **Triglycerides** consist of **glycerol** and three **fatty acids,** which may be **saturated** (no double bonds), **unsaturated** (at least one double bond), or **polyunsaturated** (more than one double bond). Double bonds make a lipid oily at room temperature, whereas saturated fats are more solid. **Sterols** are lipids containing four carbon rings.
20. **Proteins** have many functions and a great diversity of structures. They consist of 20 types of **amino acids,** each of which consists of a central carbon atom bonded to a hydrogen, an **amino group,** a **carboxyl group,** and an **R group.** Amino acids join by forming **peptide bonds** through dehydration synthesis. A protein's **conformation,** or three-dimensional shape, is vital to its function and is determined by the amino acid sequence (primary structure) and interactions between the non-R group atoms (secondary structure) and ionic, covalent, and hydrophobic interactions between R groups (tertiary structure) in the sequence. A protein with more than one polypeptide has a quaternary structure.
21. **Enzymes** are proteins that accelerate specific chemical reactions under specific conditions and are involved in every aspect of life.
22. **Nucleic acid** sequences determine amino acid sequences. **DNA** and **RNA** are polymers consisting of a sugar-phosphate backbone and sequences of **nitrogenous bases.** DNA includes deoxyribose and the bases adenine, cytosine, guanine, and thymine. RNA contains ribose and has uracil instead of thymine. A **nucleotide,** a nucleic acid monomer, consists of a phosphate, a base, and a sugar. DNA carries genetic information. RNA copies the information to enable the cell to synthesize proteins.

Testing Your Knowledge

1. Define the following terms:
   a. atom
   b. element
   c. molecule
   d. compound
   e. isotope
   f. ion
2. The vitamin biotin contains 10 atoms of carbon, 16 of hydrogen, 3 of oxygen, 2 of nitrogen, and 1 of sulfur. What is its molecular formula?
3. Distinguish among the types of chemical bonds.
4. Distinguish among the terms solute, solvent, and solution.
5. What is the physical basis of pH?
6. Why are buffer systems important in organisms?

Thinking Scientifically

1. If a shampoo is labeled "nonalkaline," would it more likely have a pH of 3, 6, or 12?
2. A topping for ice cream contains fructose, hydrogenated soybean oil, salt, and cellulose. What types of chemicals are in it?
3. Amyotrophic lateral sclerosis, also known as ALS or Lou Gehrig's disease, paralyzes muscles. An inherited form of the illness is caused by a gene (sequence of DNA) that encodes an abnormal enzyme that contains zinc and copper. The abnormal enzyme fails to rid the body of a toxic form of oxygen. Which of the molecules mentioned in this description is a
   a. protein?
   b. nucleic acid?
   c. bulk element?
   d. trace element?
4. A man on a very low-fat diet proclaims to his friend, "I'm going to get my cholesterol down to zero!" Why is this an impossible (and undesirable) goal?

References and Resources

Lewis, Ricki. March 20, 2000. Unraveling complex carbohydrates. *The Scientist,* vol. 14, p. 16. Seemingly simple compared to proteins and nucleic acids, carbohydrates have a unique, three-dimensional complexity.

Rose, George D. January/February 1996. No assembly required. *The Sciences.* A protein must fold properly to function.

Snyder, Solomon H. September 1996. No NO prevents parkinsonism. *Nature Medicine,* vol. 2. Excess NO is linked to Parkinson disease.

Williams, Robert J. P. 1991. The chemical elements of life. *Journal of Chemical Society, Dalton Transactions.* A detailed examination of why and how 24 elements are part of life.

The *LIFE* Online Learning Center provides additional resources and tools for studying this chapter.
www.mhhe.com/life

CHAPTER 3

Cells

Cancer Cells

A Personal Reflection

I never thought I would care very much about the cells composing my thyroid gland. All that changed on August 4, 1993. On that day, a physician-friend, looking at me from across a room, said, "What's that lump in your neck?" And so began my medical journey.

I wasn't very worried, even when a specialist stuck seven thin needles into my neck to sample thyroid cells for testing. Each doctor I'd seen assured me that thyroid tumors are benign (noncancerous) 99% of the time. Still, I consulted my anatomy books to ponder my thyroid. I learned that cancer affecting the outermost cells of the butterfly-shaped gland was nearly always treatable, but cancer affecting cells in the gland's interior, near the blood supply, was much more dangerous.

On that day, a physician-friend, looking at me across a room said, "What's that lump in your neck?"

The doctor's phone call came early on a Monday morning. Knowing that doctors don't like to deliver bad news on a Friday, I was panic struck. I knew I had become a statistic.

I was the 1 in 100 whose thyroid lump defied the odds. Through my tears, my brain registered the words "papillary carcinoma," and I at least knew I had the "good" type of thyroid cancer. Surgery and radiation soon followed, and I was, and am, fine. But I will never forget the terror of discovering that I had cancer.

Cancer Cells Stand Out. Here a cancer cell exhibits the rounded and ruffled appearance common to cancer cells. The smaller cells are normal white blood cells.

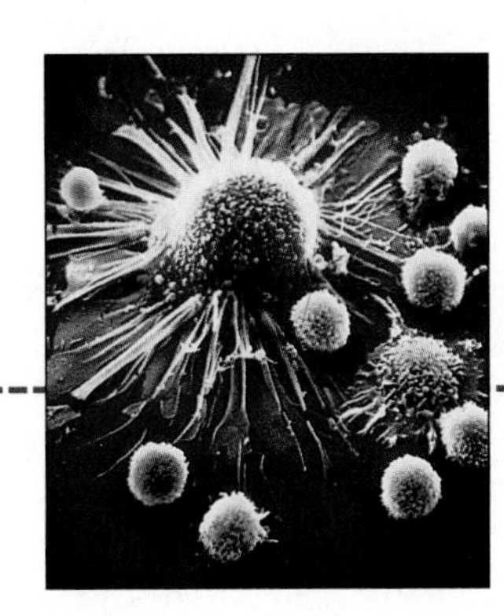

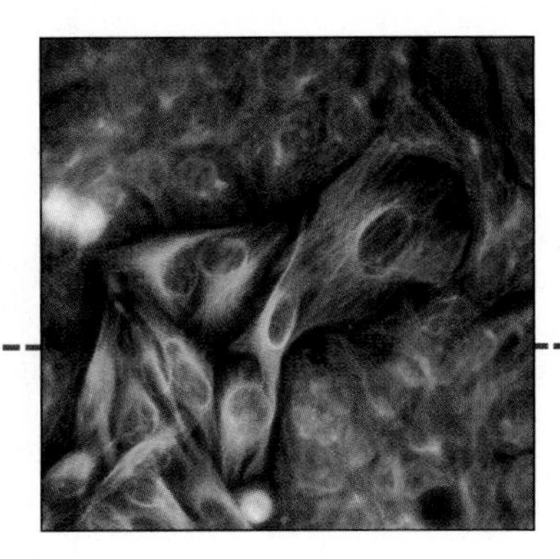

Cancer cells, when stained for the presence of gene variations characteristic of cancer cells only, look very different from surrounding healthy tissue. The orange cells are a melanoma (skin cancer) that is invading normal skin.

Preventing and conquering cancer are compelling reasons to study cell biology. By describing and understanding what is normal, we can begin to battle what is not. And cancer cells are far from normal.

A healthy cell has a characteristic shape, with a boundary that allows entry to some substances, yet blocks others. Not so the misshapen cancer cell, with its fluid surface and less discriminating boundaries (see chapter opening figures). The cancer cell breaches the controls that hold other cells in place, squeezing into spaces where other cells do not, secreting biochemicals that blast pathways through healthy tissue. The cancer even creates its own personal blood supply. The renegade cell's genetic controls differ from those of healthy cells and it transmits these differences when it divides. Cancer cells disregard the "rules" of normal cell division that enable the body to develop and maintain distinct organs. To defy so many biological traditions, the cancer cell uses up tremendous amounts of energy, causing further disruptions.

Nearly one in every three of the people reading this book will develop cancer.

Understanding the biology of cancer cells is our most powerful weapon against them. New combinations of drugs are targeting cancer cells at vulnerable points—altering their surfaces to attract the immune system; dismantling the fibers that enable them to divide often; bolstering the immune system's cancer-fighting biochemical arsenal. Detecting cancer cells' heightened energy use or altered surfaces permits earlier diagnoses, when treatments are more effective. Clinical trials are testing new techniques to monitor recurrence of thyroid cancer, based on several biotechnologies.

We hear so often of those who have died of cancer. Nearly one in every three of the people reading this book will develop cancer. Yet many people living perfectly normal, productive lives either have cancer or have had it. Many times we can defeat this most frightening group of illnesses. Our ability to do so rests solidly on our knowledge of the biology of cells, the units of life.

Ricki Lewis

3.1 Cells Are the Units of Life

The molecules of life aggregate and interact to form larger structures, which also interact to provide the characteristics of life. These chemical assemblies enclosed within a membrane form a cell, the basic unit of life.

A human, a hyacinth, a mushroom, and a bacterium appear to have little in common other than being alive. However, on a microscopic level, these organisms are similar. All organisms consist of microscopic structures called **cells.** Within cells, highly coordinated biochemical activities carry on the basic functions of life, as well as, in some cases, specialized functions (fig. 3.1). The next two chapters introduce the cell, and the chapters that follow delve into specific cellular events.

Cells, as the basic units of life, exhibit all of the characteristics of life. A cell requires energy, genetic information to direct biochemical activities, and structures to carry out these activities. Movement occurs within living cells, and some cells, such as the sperm cell, can move about in the environment. All cells have some structures in common that allow them to reproduce, grow, respond to stimuli, and obtain energy and convert it to a usable form. In all cells, a cell membrane sets the living matter off from the environment and limits size. Complex cells house specialized structures, called **organelles,** in which particular activities take place. The remaining interior of such a cell, as well as the interior of simpler cells, is called **cytoplasm.** The functions of cells are similar to the functions of whole organisms.

But cells can also specialize, containing different numbers of particular types of organelles. For example, an active muscle cell contains many more organelles that enable it to use energy than does an adipose cell, which is little more than a blob of fat. A mesophyll cell in a leaf of a flowering plant is packed with chloroplasts, which are organelles that capture the sun's energy in a process called photosynthesis. The essay that starts chapter 31 explores stem cell technology and tissue engineering, which exploits the process of cell specialization to grow human parts in the laboratory.

Our knowledge of the structures inside cells depends upon technology, because most cells are too small for the unaided human eye to see. Today, cell biologists use sophisticated microscopes to greatly magnify cell contents and many types of stains to color otherwise transparent cellular structures.

Discovering the Cellular Basis of Life

The ability to make objects appear larger probably dates back to ancient times, when people noticed that pieces of glass or very smooth, clear pebbles could magnify small or distant objects. By the thirteenth century, the ability of such "lenses" to aid people with poor vision was widely recognized in the Western world.

Three centuries later, people began using paired lenses. Many sources trace the origin of a double-lens compound microscope to Dutch spectacle makers Johann and Zacharius Janssen. Reports claim that their children were unwittingly responsible for this important discovery. One day in 1590, a Janssen youngster was playing with two lenses, stacking them and looking through them at distant objects. Suddenly he screamed—the church spire

FIGURE 3.1
Specialized Cells.
(**A**) Macrophages (blue) are special cells that "eat" and destroy bacteria (yellow). Long extensions of the macrophages capture the bacteria and draw them inside the cell, where enzymes destroy the bacteria. Macrophages will consume bacteria until they wear out and become "pus" in an infected wound. (**B**) Specialized plant cells are loaded with green chloroplasts, which contain the machinery for photosynthesis.

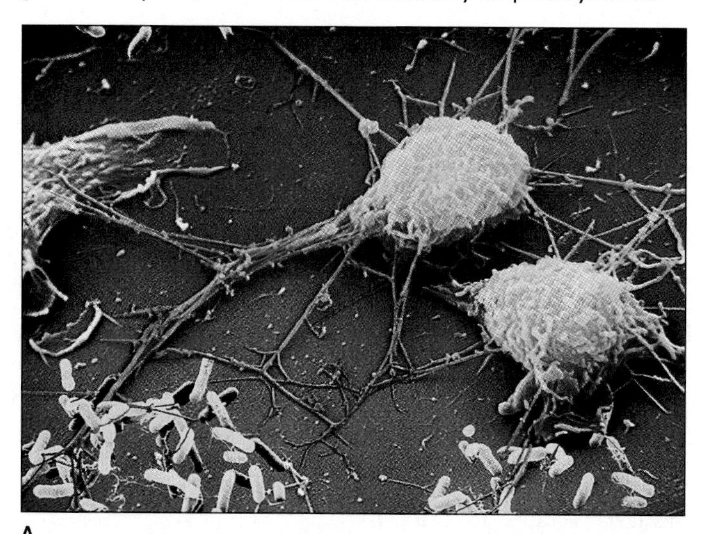

A

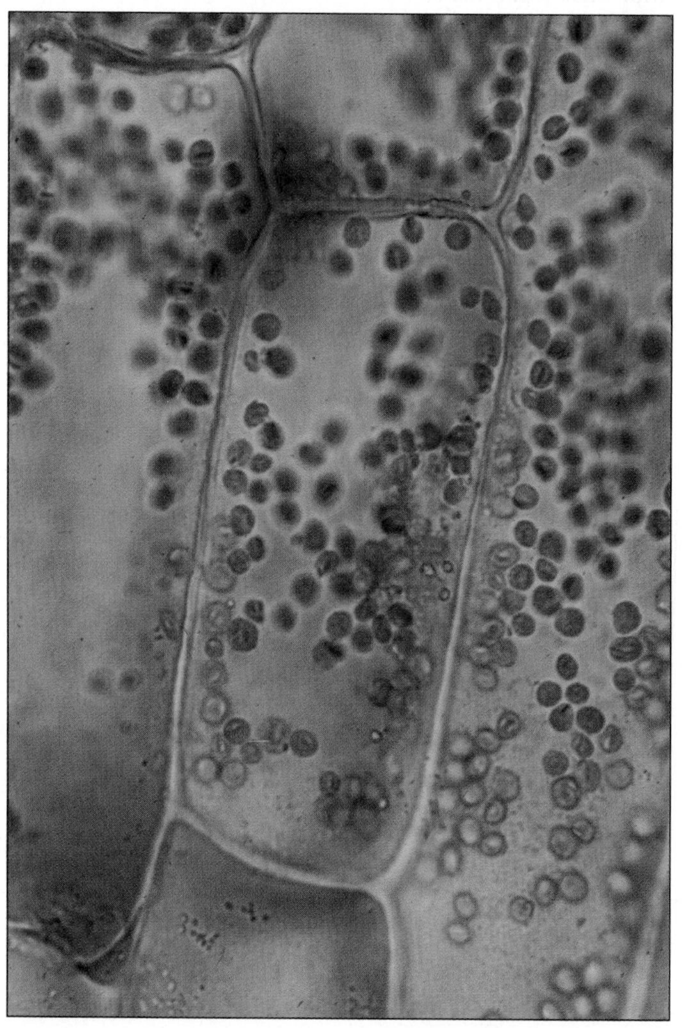

B

looked as if it was coming toward him! The elder Janssens quickly looked through both pieces of glass, and the faraway spire indeed looked as if it was approaching. One lens had magnified the spire, and the other lens had further enlarged the magnified image. This observation led the Janssens to invent the first compound optical device, a telescope. Soon, similar double-lens systems were constructed to focus on objects too small for the unaided human eye to see. The compound microscope was born.

The study of cells—cell biology—began in 1660, when English physicist Robert Hooke melted strands of spun glass to create lenses that he focused on bee stingers, fish scales, fly legs, feathers, and any type of insect he could hold still. When he looked at cork, which is bark from a type of oak tree, it appeared to be divided into little boxes, which were left by cells that were once alive. Hooke called these units "cells," because they looked like the cubicles (cellae) where monks studied and prayed. Although Hooke did not realize the significance of his observation, he was the first person to see the outlines of cells.

In 1673, lenses were improved again, at the hands of Antonie van Leeuwenhoek of Holland. Leeuwenhoek used only a single lens, but it was more effective at magnifying and produced a clearer image than most two-lens microscopes then available. He built more than 500 microscopes! One of his first objects of study was tartar scraped from his own teeth, and his words best describe what he saw there:

> *To my great surprise, I found that it contained many very small animalcules, the motions of which were very pleasing to behold. The motion of these little creatures, one among another, may be likened to that of a great number of gnats or flies disporting in the air.*

Leeuwenhoek opened up a vast new world to the human eye and mind (fig. 3.2). He viewed bacteria and protists that people hadn't known existed. However, he failed to see the single-celled "animalcules" reproduce, and therefore he perpetuated the popular idea at the time that life arises from the nonliving or from nothing. Nevertheless, Leeuwenhoek did describe with remarkable accuracy microorganisms and microscopic parts of larger organisms, including human red blood cells and sperm.

The Cell Theory Emerges

Despite the accumulation of microscopists' drawings of cells made during the seventeenth and eighteenth centuries, the **cell theory**—the idea that the cell is the fundamental unit of all life—did not emerge until the nineteenth century. Historians attribute this delay in the development of the cell theory to poor technology, including crude microscopes and lack of procedures to preserve and study living cells without damaging them. Neither the evidence itself nor early interpretations of it suggested that all organisms were composed of cells. Hooke had not observed actual cells but rather their absence. Leeuwenhoek made important observations, but he did not systematically describe or categorize the structures that cells had in common.

In the nineteenth century more powerful microscopes, with better magnification and illumination, revealed details of life at

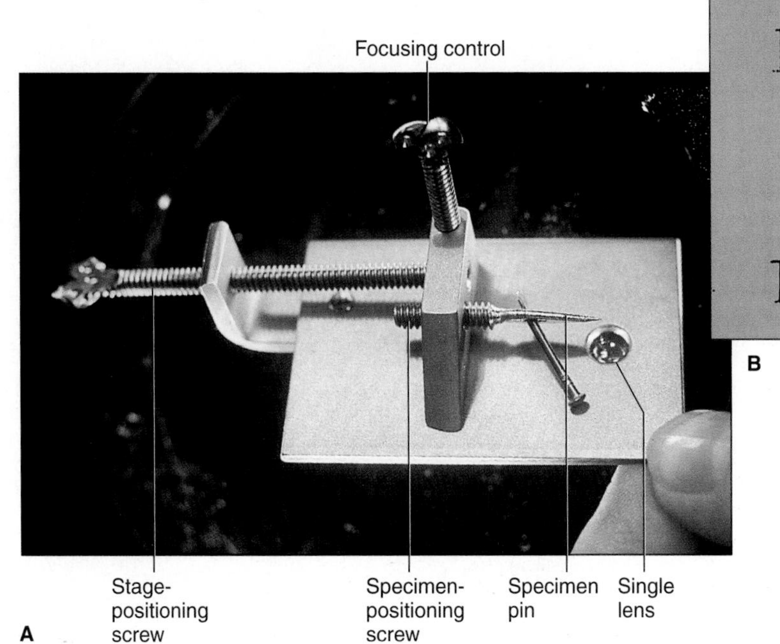

FIGURE 3.2
A First Microscope.
(**A**) Leeuwenhoek made many simple microscopes such as this example, which opened the microscopic world to view. (**B**) He drew what he saw and, in doing so, made the first record of microorganisms.

the subcellular level. In the early 1830s, Scottish surgeon Robert Brown noted a roughly circular structure in cells from orchid plants. He saw the structure in every cell, then identified it in cells of a variety of organisms. He named it the "**nucleus,**" a term that stuck. Today we know that a cell's nucleus houses DNA. Soon microscopists distinguished the translucent, moving material that made up the rest of the cell, calling it the cytoplasm.

In 1839, German biologists Matthias J. Schleiden and Theodore Schwann proposed the cell theory based on many observations made with microscopes. Schleiden first noted that cells were the basic units of plants, and then Schwann compared animal cells to plant cells. After observing many different plant and animal cells, they concluded that cells were "elementary particles of organisms, the unit of structure and function." Schleiden and Schwann described the components of the cell as a cell body and nucleus contained within a surrounding membrane. Schleiden called a cell a "peculiar little organism" and realized that a cell can be a living entity on its own; but the new theory also recognized that in plants and animals, cells are part of a larger living organism.

Many cell biologists extended Schleiden and Schwann's observations and ideas. German physiologist Rudolph Virchow added the important corollary in 1855 that all cells come from preexisting cells, contradicting the still-popular idea that life can arise from the nonliving or from nothingness. Virchow's statement also challenged the popular concept that cells develop on their own from the inside out, the nucleus forming a cell body around itself, and then the cell body growing a cell membrane. Virchow's observation set the stage for descriptions of cell division in the 1870s and 1880s (chapter 8).

Virchow's thinking was ahead of his time, because he hypothesized that abnormal cells cause diseases that affect the whole body. Today, many new treatments for diverse disorders are based on knowledge of the disease process at the cellular level.

The existence of cells is an undisputed fact, yet the cell theory is still evolving. For example, until recently, scientists viewed a complex cell as a structure containing a nucleus and jellylike cytoplasm, with organelles suspended in no particular organization. Researchers are now learning that organelles have precise locations in cells, near other structures with which they interact. So although the cell theory is still considered to be a product of the nineteenth century, we are constantly learning about the organization within cells and about how cells interact to build larger-scale biological organization. Investigating Life 3.1 describes some of the types of microscopes used in cell biology, and figure 3.3 provides a sense of the size of objects that these microscopes can image.

3.1 MASTERING CONCEPTS

1. What is a cell?
2. What types of structures are within complex cells?
3. How did Robert Hooke and Antonie van Leeuwenhoek contribute to the beginnings of cell biology?
4. What is the cell theory?

OLC

FIGURE 3.3

Ranges of the Light, Electron, and Scanning Probe Microscopes.

Biologists use the metric system to measure size. The basic unit of length is the meter (m), which equals 39.37 inches (slightly more than a yard). Smaller metric units measure many chemical and biological structures. A centimeter (cm) is 0.01 meter (about 2/5 of an inch); a millimeter (mm) is 0.001 of a meter; a micrometer (μm) is 0.000001 meter; a nanometer (nm) is 0.000000001 meter; an Angstrom unit (Å) is 1/10 of a nanometer. In the scale below, each segment represents only 1/10 of the length of the segment beneath it. The sizes of some chemical and biological structures are indicated next to the scale.

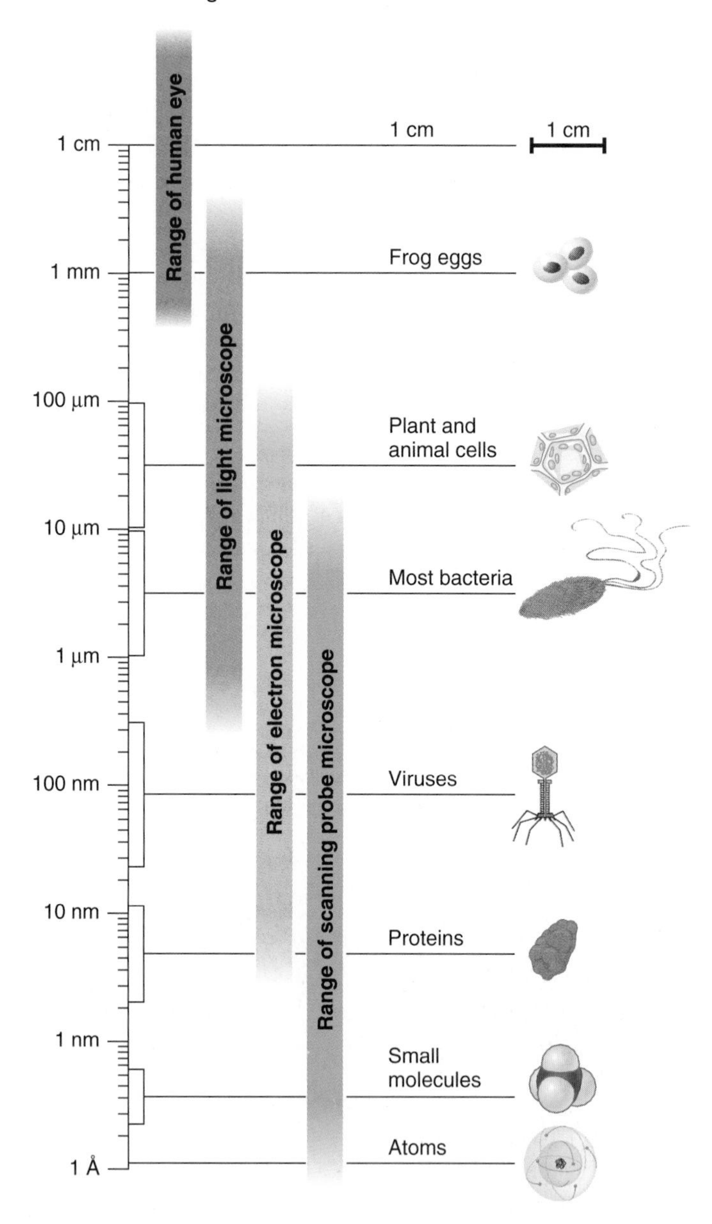

Domain Bacteria	Domain Archaea	Domain Eukarya
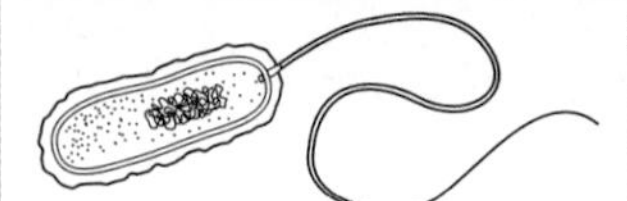		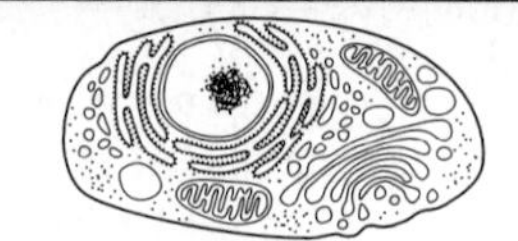
• 1–10 micrometers	• 1–10 micrometers	• 10–100 micrometers
• Cell wall of peptidoglycan	• Cell wall of various molecules; pseudopeptidoglycan, protein	• Cell wall of cellulose or chitin
• No introns present	• Some introns present	• Introns present
• Membrane based on fatty acids	• Membrane based on non-fatty acid lipids (isoprenes)	• Membrane based on fatty acids
• No membrane-bounded organelles	• No membrane-bounded organelles	• Membrane-bounded organelles
• 4-subunit RNA polymerase	• Many-subunit RNA polymerase	• Many-subunit RNA polymerase

FIGURE 3.4
Cells of the Three Domains of Life.
For many years, biologists considered cells to be two types—prokaryotic or eukaryotic—distinguished by absence or presence of a nucleus, respectively. Investigation at the molecular level, however, has revealed that not all prokaryotes are alike. Many biologists today recognize three types of cells: a bacterium, the superficially similar archaeon, and the eukaryotic cell.

3.2 Variations on the Cellular Theme

Cells are of three basic types, two of which lack nuclei and other membrane-bounded organelles. These simpler (prokaryotic) cells are in the domains Archaea and Bacteria. The archaea have unique characteristics, but also share features with bacteria and the third domain, Eukarya. Eukaryotic cells are larger and more complex than prokaryotic cells.

Until recently, biologists recognized two types of organisms, prokaryotes (whose cells lack nuclei) and eukaryotes (whose cells have nuclei). In 1977, University of Illinois physicist-turned-microbiologist Carl Woese detected differences in key molecules in some of the prokaryotes that were great enough to suggest that they were a completely different form of life. He first named them Archaebacteria, which was changed to Archaea when it became apparent that their resemblance to bacteria was only superficial.

The archaea lack nuclei, but have characteristics that resemble either bacteria or eukarya. So far, we do not have a complete enough picture of the archaea to even depict a "typical" cell, as we can for the other two domains (fig. 3.4).

One way that we can distinguish Eukarya from the other domains is by cell size. Most cells of Eukarya dwarf cells of Bacteria and Archaea by 100 to 1,000-fold.

All cells require relatively large surface areas through which they interact with the environment. Nutrients, water, oxygen, carbon dioxide, and waste products must enter or leave a cell through its surfaces. As a cell grows, its volume increases at a faster rate than its surface area, a phenomenon demonstrable with simple calculations (fig. 3.5). Put another way, much of the interior of a large cell is far away from the cell's surface.

The ultimate inability of a cell's surface area to keep pace with its volume can limit a cell's size, and this may be why cells of Bacteria and Archaea are usually small. Another adaptation to maximize surface area is for the cell membrane to fold, just as inlets and capes extend the perimeter of a shoreline. The cells of Eukarya can grow larger than cells of the other domains because their membranous organelles provide needed surface area.

FIGURE 3.5
The Important Relationship Between Surface Area and Volume.
When an object enlarges, its volume grows faster than its surface area. Cells have limited sizes because if they grow too large, the surface areas would be too small to support the large volumes.

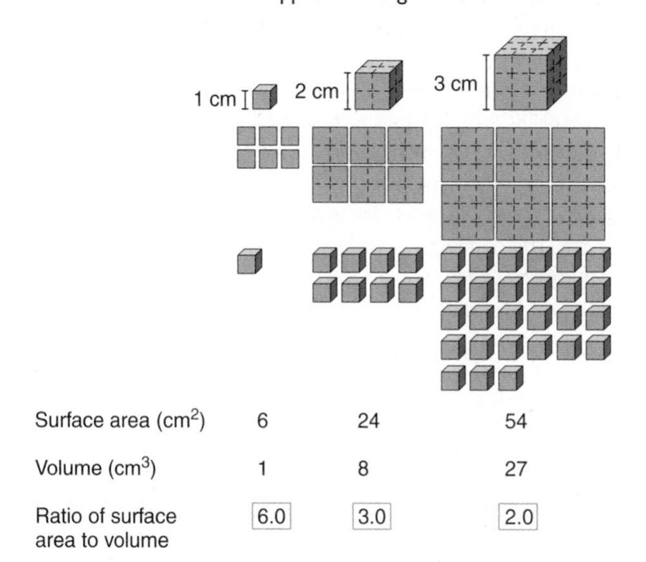

Microscopes Reveal Cell Structure

Studying life at the cellular and molecular levels requires microscopes to magnify structures. Specimens of tissue or single cells usually must be prepared before they can be observed under a microscope. First, specimens are fixed. This means that certain organic chemicals are applied that stop enzyme action, solidify structures, and basically hold the cell and its constituents in place. If the specimen is too thick to be imaged, it must be sectioned into thin slices. Dyes are typically added, alone or in combination, to provide contrast to cell parts that are naturally translucent or transparent.

All microscopes provide two types of power—*magnification* and *resolution* (also called resolving power). A microscope produces an enlarged, or magnified, image of an object. Magnification is the ratio between the size of the image and the object. Resolution refers to the smallest degree of separation at which two objects appear distinct. A compound light microscope can resolve objects that are 0.1 to 0.2 micrometers (4 to 8 millionths of an inch) apart. The resolving power of an electron microscope is 10,000 time greater. Figure 3.A compares these two types of microscopes.

Following is a survey of several types of microscopes.

The Light Microscope The *compound light microscope* focuses visible light through a specimen. Different regions of the object scatter the light differently, producing an image. Three sets of lenses help generate the image. The condenser lens focuses light through the specimen. The objective lens receives light that has passed through the specimen, generating an enlarged image. The ocular lens, or eyepiece, magnifies the image further. Multiplying the magnification of the objective lens by that of the ocular lens gives the total magnification. A limitation of light microscopy is that it focuses on only one two-dimensional plane at a time, so that there is little sense of depth.

FIGURE 3.A
Different Microscopes Reveal Different Details.
An electron microscope offers better magnification and resolution than a light microscope while the scanning tunneling microscope can reveal detail of individual molecules.

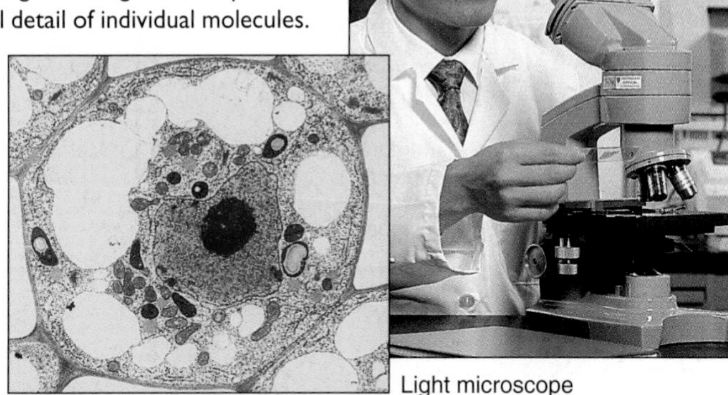

Light microscope

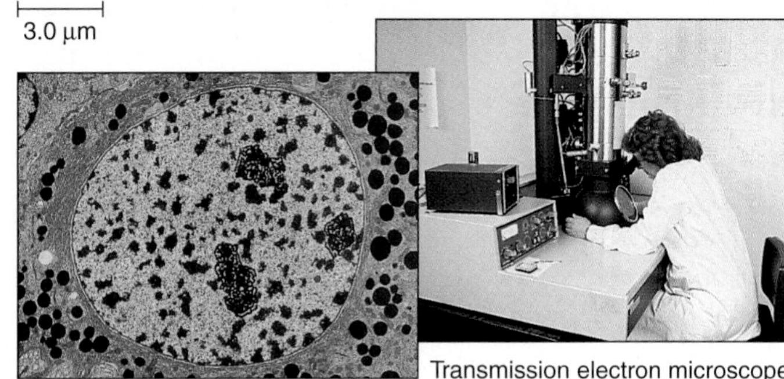

Transmission electron microscope

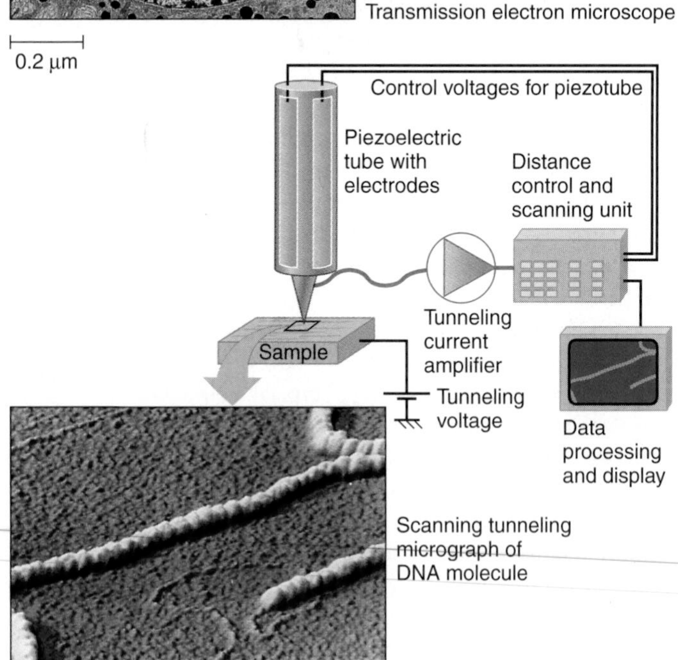

Scanning tunneling micrograph of DNA molecule

A *confocal microscope* is a type of light microscope that enhances resolution by passing white or laser light through a pinhole and a lens to the object. This eliminates the problem of light reflecting from regions of the specimen near the object of interest, which can blur the image. The result is a scan of highly focused light on one tiny part of the specimen. "Confocal" refers to the fact that both the objective and condenser lenses focus on the same small area. Computers can integrate many confocal images of specimens exposed to fluorescent dyes to produce spectacular peeks at living structures. (Fluorescence is an optical phenomenon discussed in chapter 6.) A dividing cell, stained in the right way, reveals vibrant blue chromosomes being pulled to opposite sides of the cell by an apparatus of green fibers.

The Electron Microscope Electron microscopes provide great magnification, better resolution, and better depth than light microscopes. Instead of focusing light, the *transmission electron microscope* (TEM) sends a beam of electrons through a specimen, using a magnetic field rather than a glass lens to focus the beam. Different parts of the specimen absorb electrons differently. When electrons from the specimen hit a fluorescent screen coated with a chemical, light rays are given off, translating the contrasts in electron density into a visible image.

In TEM, the specimen must be killed, chemically fixed, cut into very thin sections, and placed in a vacuum, a treatment that can distort natural structures. The *scanning electron microscope* (SEM) eliminates some of these drawbacks. It bounces electrons off a metal-coated three-dimensional specimen, generating a 3-D image on a screen that highlights crevices and textures.

Scanning Probe Microscopes These microscopes work on a different principle than light or electron microscopes. They move a probe over a surface, and translate the distances into an image—a little like moving your hands over someone's face to get an idea of his or her appearance. There are several types of scanning probe microscopes.

A *scanning tunneling microscope* reveals detail at the atomic level. A very sharp metal needle, its tip as small as an atom, scans a molecule's surface. Electrons "tunnel" across the space between the sample and the needle, creating an electrical current. The closer the needle, the greater the current. An image forms as the scanner continually adjusts the space between the needle and specimen, keeping the current constant over the topography of the molecular surface. The needle's movements over the microscopic hills and valleys are expressed as contour lines, which a computer converts and enhances to produce a colored image of the surface.

Electrons do not pass readily through many biological samples, and so *scanning ion-conductance microscopy* offers an alternative approach. It uses ions instead of electrons, which is useful in imaging muscle and nerve cells, which are specialized to use ions in communication. The probe is made of hollow glass filled with a conductive salt solution, which is also applied to the sample. When voltage passes through the sample and the probe, ions flow to the probe. The rate of ion flow is kept constant, and a portrait is painted as the probe moves. In yet another variation, the *atomic force microscope* uses a diamond-tipped probe that presses a molecule's surface with a very gentle force (fig. 3.B). As the force is kept constant, the probe moves, generating an image. This variation on the theme is useful for recording molecular movements, such as blood clotting and cells dividing.

Figure 3.C contrasts images of human red blood cells taken with the light, electron, and scanning probe microscopes.

FIGURE 3.B
An Atomic Force Microscope "Feels" a Biological Surface.
This device uses a microscopic force sensor called a cantilever, which is attached to a tip that assesses the space to the sample. As the cantilever moves over the surface, the detected distances are converted into an image.

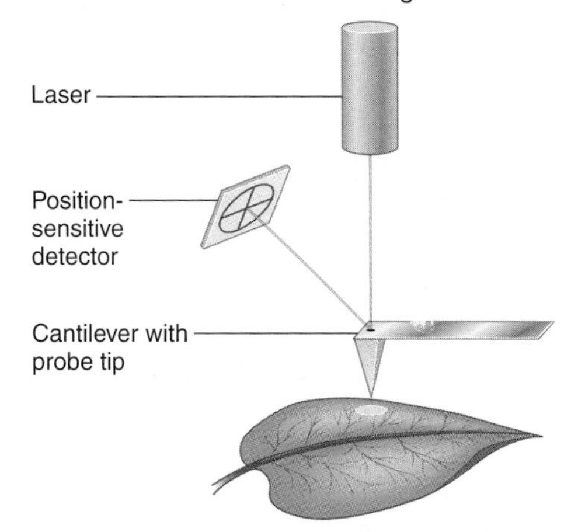

FIGURE 3.C
Three Views of Red Blood Cells.
(**A**) A "smear" of blood visualized under a light microscope appears flat. (**B**) The disc-shaped red blood cells appear with much greater depth with the scanning electron microscope, but material must be killed, fixed in place, and subjected to vacuum before viewing, thus limiting applications. (**C**) The scanning probe microscope captures images of cells and molecules in their natural state, as parts of living organisms.

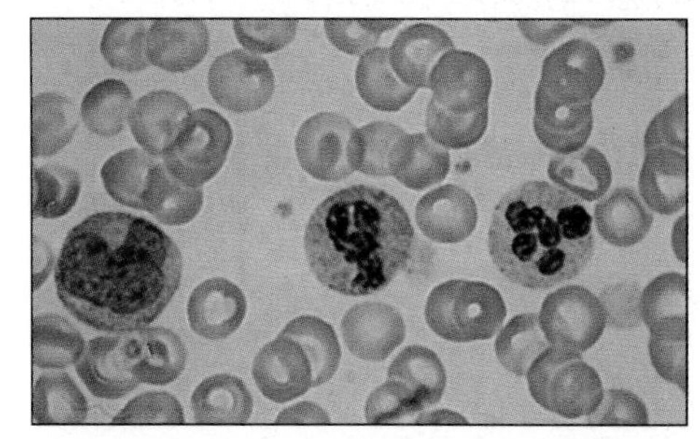
A

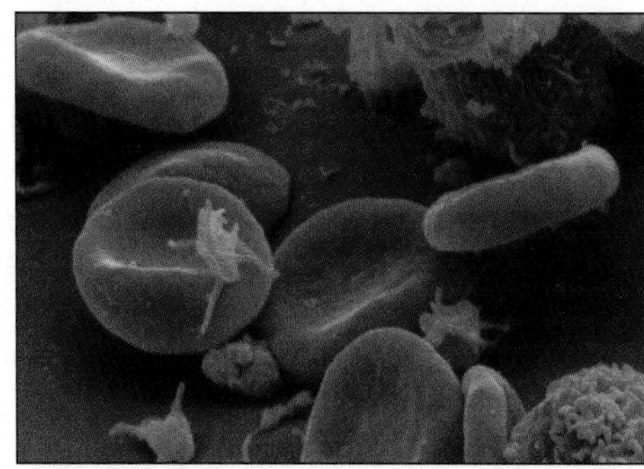
B

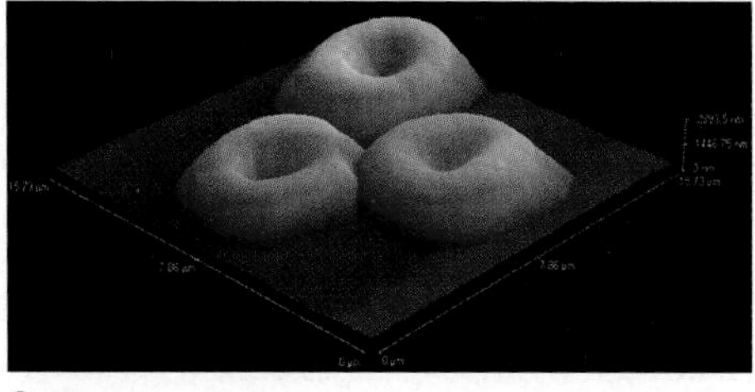
C

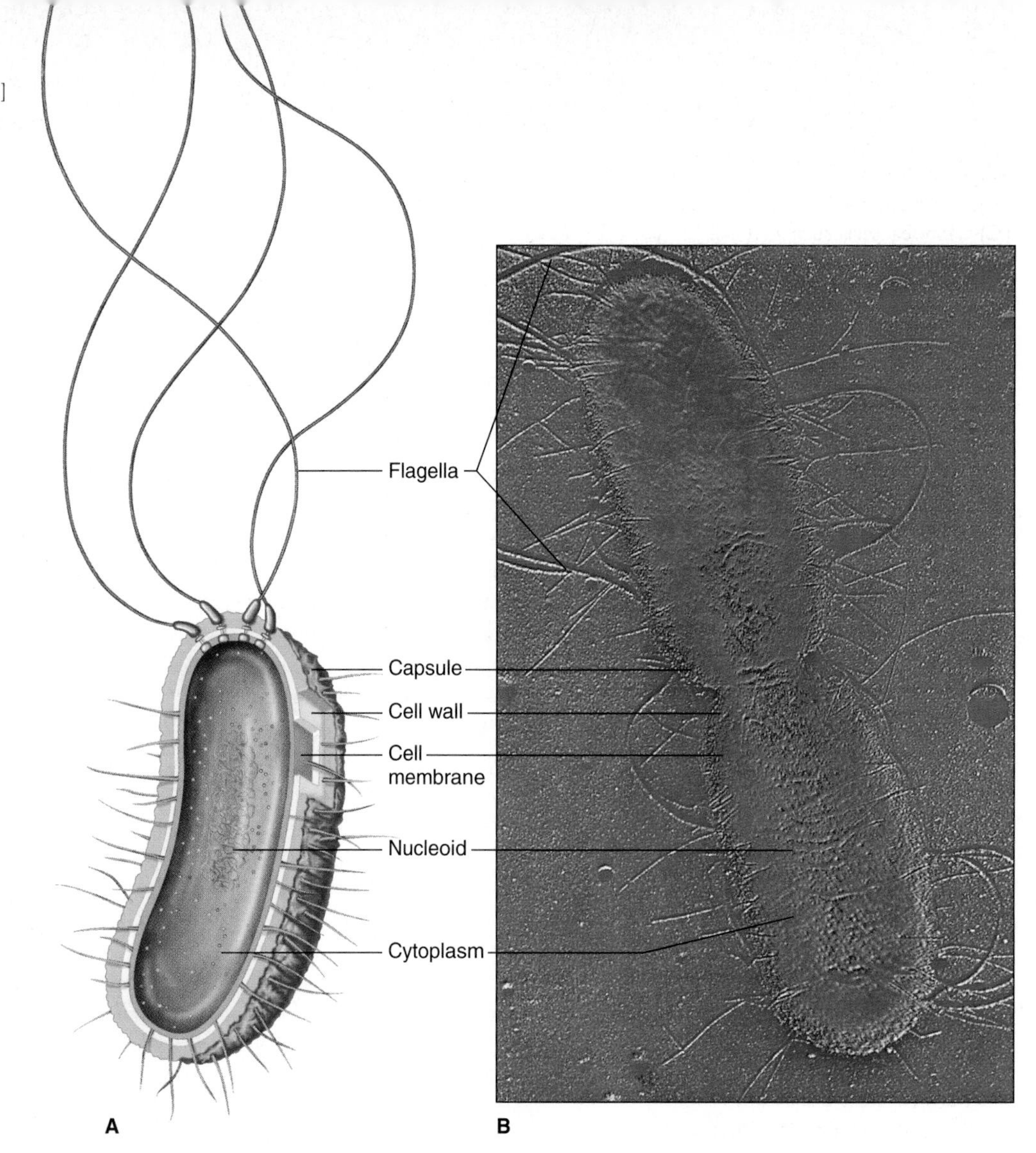

FIGURE 3.6
Anatomy of a Bacterium.
(**A**) Bacteria lack organelles such as a nucleus. Their DNA is suspended in the cytoplasm among the enzymes, nutrients, and fluids that make up the inside of a cell. Bacteria use a rigid cell wall to maintain their shape and some use flagella, long whiplike tails, to provide movement. (**B**) This common bacterium, *E. coli,* is dividing.

Bacterial Cells

All bacteria lack true membrane-enclosed nuclei. These organisms include the familiar bacteria, as well as the cyanobacteria, organisms once known as blue-green algae because of their characteristic pigments and their similarity to the true (eukaryotic) algae. Figure 3.6 depicts the major features of bacterial cells, and figure 3.7 shows a few types of these microorganisms. Bacteria cause many illnesses, but they are also very valuable in food and beverage processing and pharmaceutical production. Bacteria play a critical role in ecosystems as decomposers.

Members of this domain are extremely abundant and diverse—the species that we are familiar with are only a small subset of the entire group. Most bacterial cells are surrounded by rigid **cell walls** that consist of peptidoglycans, which are molecules consisting of amino acid chains and carbohydrates. In some bacteria, polysaccharides on the cell wall form a capsule that protects the cell or enables it to attach to specific types of surfaces. Many antibiotic drugs halt bacterial infection by interfering with the microorganism's ability to build its cell wall.

Microbiologists classify bacteria by cell wall structure and biochemical characteristics such as metabolic pathways and energy sources. Bacterial cells may be round (cocci), rod-shaped (bacilli), spiral (spirilla), comma-shaped (vibrios), or spindle-shaped (fusiform). Bacteria are distinguished as "gram-positive" or "gram-negative" based on how they react to a specific staining procedure described in figure 20.4. Gram-positive bacteria have cell walls that are very thick layers of peptidoglycan. Gram-negative bacteria have thinner peptidoglycan layers plus an outer membrane of protein and lipopolysaccharide.

Unlike nonphotosynthetic bacteria, cyanobacteria contain internal membranes that are continuous with the cell membrane, but are not extensive enough to subdivide the cell into compartments. The cyanobacterium's membranes are studded with pigment molecules that absorb and extract energy from sunlight. In some bacteria, tail-like appendages called flagella, which enable the cell to move, are anchored in the cell wall and underlying cell membrane. (Chapter 4 covers membrane structure and function in detail.)

The main genetic material of a bacterium is a single circle of DNA. Its DNA associates with proteins that are different than those in more complex cells. The part of a bacterial cell where the DNA is located is called the **nucleoid,** and it sometimes appears fibrous under a microscope. Nearby are RNA molecules and **ribosomes,** which are spherical structures consisting of RNA and

FIGURE 3.7
A Trio of Bacteria.
(**A**) *Escherichia coli* (×35,000) inhabits the intestines of some animal species, including humans. (**B**) *Streptococcus pyogenes* infects humans (×900). (**C**) Cyanobacteria such as this *Chroococcus furgidus* (×600) have pigments that enable them to photosynthesize.

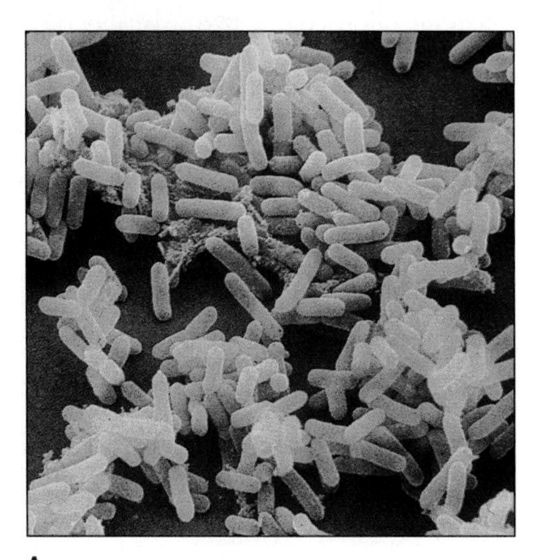
A

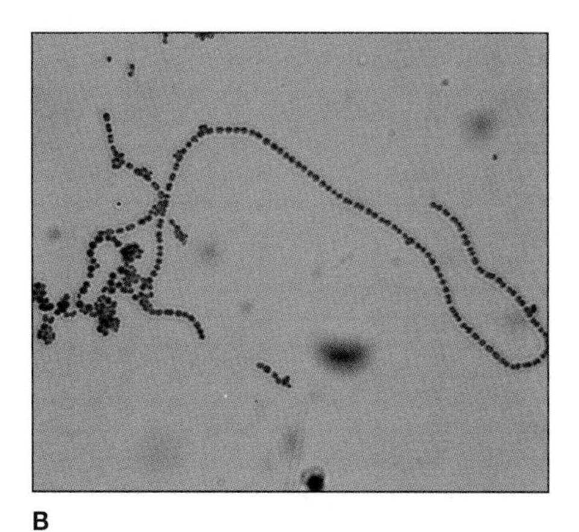
B

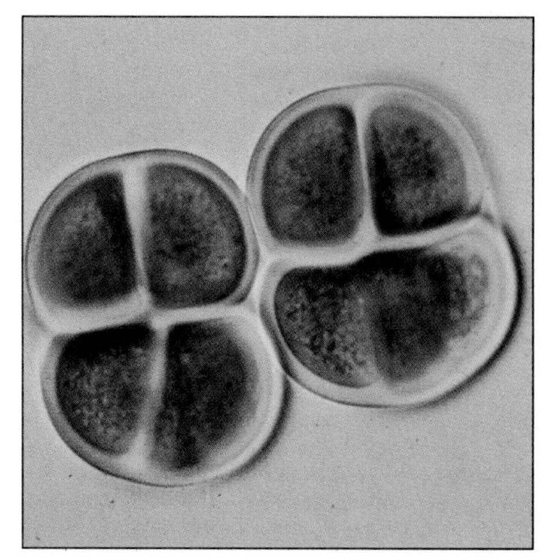
C

protein. The proteins of ribosomes are structural supports for protein synthesis, and the RNA portions of them may help catalyze protein synthesis. In part because the DNA, RNA, and ribosomes in prokaryotic cells are in close contact, protein synthesis is rapid compared to the process in more complex cells, whose cellular components are separated.

Archaean Cells

The first members of Archaea to be described were microorganisms that use carbon dioxide and hydrogen from the environment to produce methane—hence they are called methanogens. Originally archaea were lumped in with bacteria, but Carl Woese delineated their differences from the better-known bacteria and eukarya:

- Their cell walls lack peptidoglycan, found in bacteria.
- They have unique coenzymes to produce methane.
- The base sequence of two types of RNA, called ribosomal RNA and transfer RNA, are distinctly different from these molecules in members of the other domains.

The methanogens have a curious mix of characteristics. These archaea can transport ions within their cells like bacteria do, and share some surface molecules with bacteria. Yet they have proteins, called histones, associated with their genetic material and manufacture proteins more like eukaryotes do. When researchers deciphered all of the genes of a methanogenic archaeon in 1996, they found, as Woese expected they would, that more than half of the genes had no counterpart among bacteria or eukarya. But the fact that slightly less that half of the genes do correspond indicates that the three forms of life recognized today branched off from a shared ancestor long ago (see fig. 1.7).

Researchers have since discovered other types of archaea, and found them in a variety of habitats including swamps, rice paddies, and throughout the oceans (fig. 3.8). Yet we still know hardly

FIGURE 3.8
An Archaeon.
Thermoplasma has characteristics of bacterial and eukaryotic cells. The DNA wraps around proteins much as it does in eukaryotic cells, but there are no nuclei. *Thermoplasma* thrives in the high heat and acidic conditions of smoldering coal deposits. When researcher Gary Darland first discovered these cells, he was so startled by the habitat and mix of features that he named it the "wonder organism."

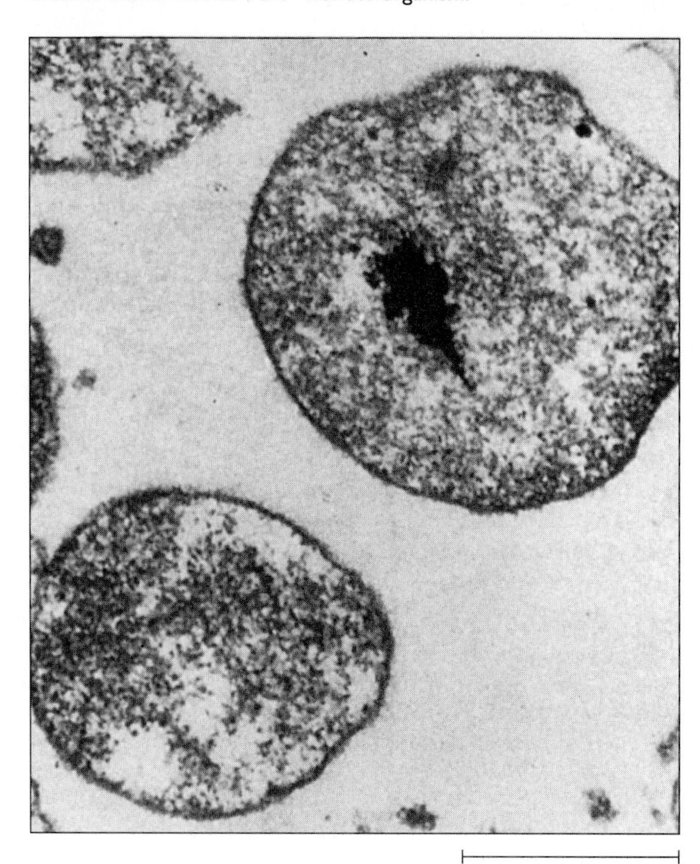

anything about them compared to what we know of the other two domains. Chapter 20 discusses what we do know. But already, the textbooks are literally being rewritten to make room for these little-understood microorganisms. In the years to come, researchers will fill in the details of this domain of life.

Eukaryotic Cells

Plants, animals, fungi, and protista are composed of eukaryotic cells. In these cells, organelles create compartments where specific biochemical reactions can occur. Saclike organelles con-

FIGURE 3.9
A generalized animal cell showing organelles that provide specialized functions for the cell.

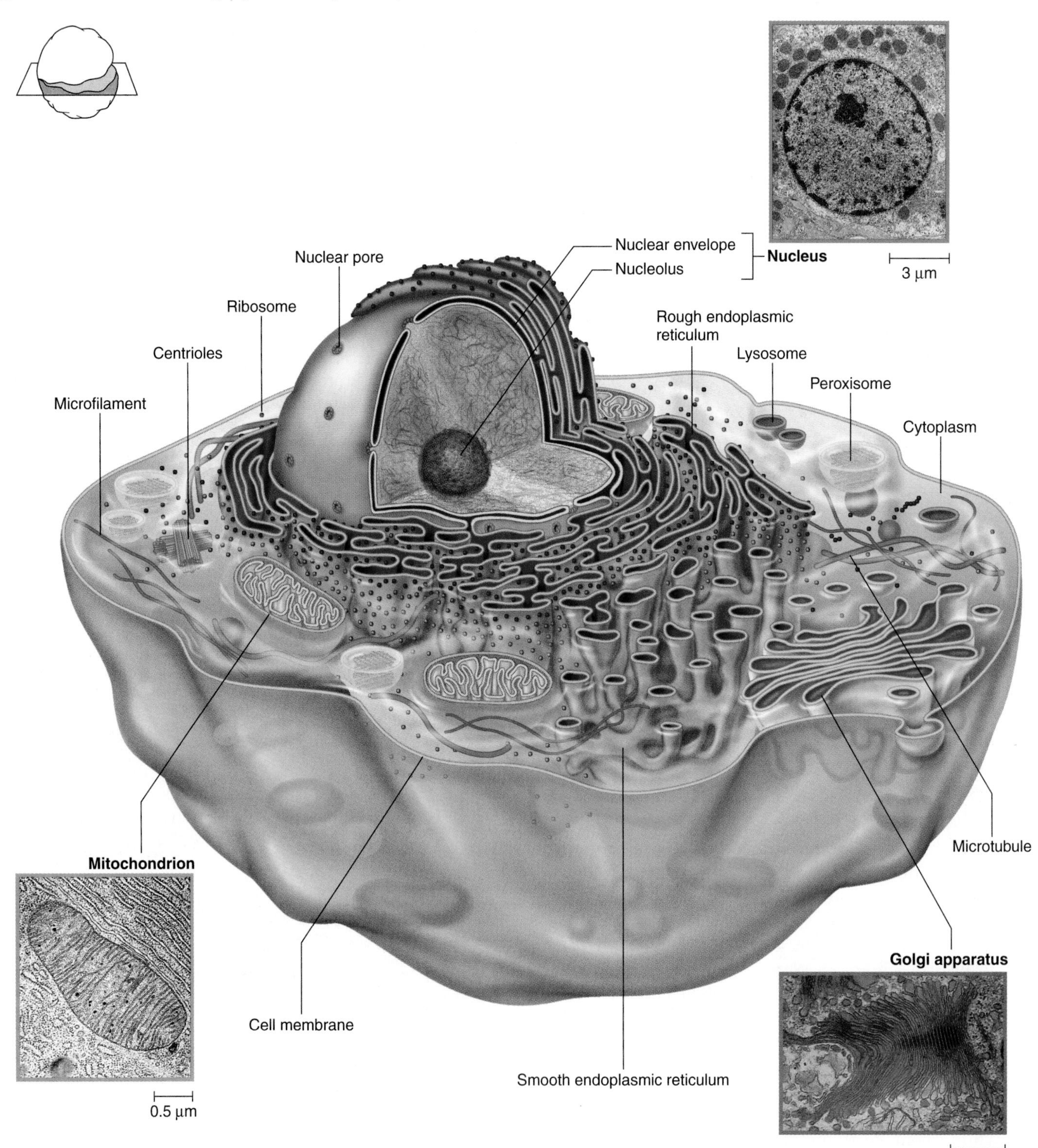

tain biochemicals that might harm other cellular contents. Some organelles consist of membranes studded with enzymes that catalyze chemical reactions on their surfaces. On certain membranes, different enzymes are physically laid out in the order in which they participate in biochemical reactions. In general, organelles keep related biochemicals and structures sufficiently close together to make them function more efficiently. The compartmentalization organelles provide also makes it unnecessary for the entire cell to maintain a high concentration of any particular biochemical. Figures 3.9 and 3.10 depict generalized animal and plant cells.

The most prominent organelle in most cells is the nucleus, which contains the genetic material (DNA) organized with protein into threadlike chromosomes that condense and appear rodlike when the cell divides. The remainder of the cell consists of other organelles and cytoplasm. About half of the volume of an animal cell is organelles; in contrast, some plant cells contain up to 90% water, much of it within a large organelle called a **vacuole.** Plant cell vacuoles have different functions depending upon the specific types of enzymes they contain. All cells, however, are very watery. On the science fiction program *Star Trek,* beings from another planet quite correctly called humans "ugly bags of mostly water."

FIGURE 3.10
A generalized plant cell, which (unlike animal cells) has a cell wall, chloroplasts, and a large vacuole.

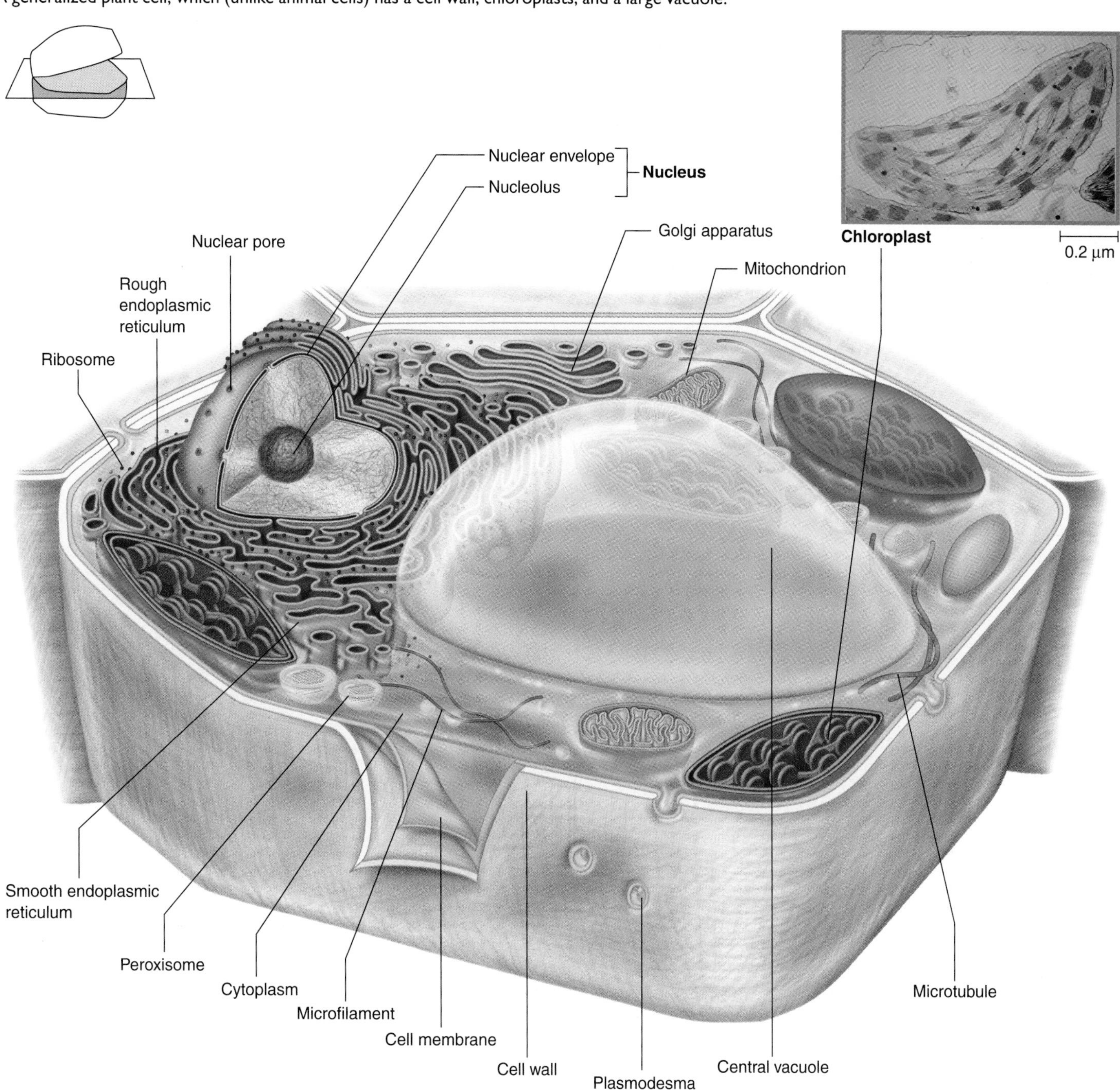

Some eukaryotic cells also contain stored nutrients, minerals, and pigment molecules. Arrays of protein rods and tubules within plant and animal cells form the **cytoskeleton,** which helps to give the cell its shape. Protein rods and tubules also form appendages that enable certain cells to move, and they form structures that are important in cell division. (Chapter 4 examines the cytoskeleton in depth.)

3.2 MASTERING CONCEPTS

1. Why were we unaware of archaea until recently?
2. How do bacteria and archaea differ from eukaryotic cells?
3. What are some major structures of bacterial cells?
4. What are the general functions of organelles?

OLC

3.3 How Do Organelles Divide Labor in a Eukaryotic Cell?

Organelles effectively compartmentalize a cell's activities, improving efficiency and protecting cell contents from harsh chemicals. Organelles enable cells to secrete substances, derive energy from nutrients, degrade debris, and reproduce.

Organelles divide the activities of life at the cellular level, like departments in a large store group related items that are used together. Organelles also interact, providing basic life functions as well as specialized characteristics. Proteins, including enzymes, are key to determining the functions of a cell and how it interacts with other cells in the body. In eukaryotic cells, organelles help to synthesize and process proteins. Instructions for building each protein are stored in the nucleus as genes, which protects them from degradation in the cytoplasm. A series of highly folded membranes comprise the **endomembrane system.** Several compartments within this system house unique enzymes that process proteins made in the first compartment, the **rough endoplasmic reticulum (ER).** Each compartment is connected to the next through **vesicles,** which are small packages of proteins and other molecules surrounded by membrane. Another area, the **smooth endoplasmic reticulum,** synthesizes and modifies lipids. The last compartment in protein processing, the **Golgi apparatus,** finishes the processing and sorts proteins into separate vesicles containing digestive enzymes called **lysosomes.** Other compartments within the cell provide energy, contain special enzymes for oxidation reactions, or simply store useful molecules. Examining a coordinated function, milk secretion, illustrates how organelles interact to produce, package, and release from the cell a complex mixture of biochemicals (fig. 3.11).

How Do Organelles Interact to Secrete Substances?

Special cells in the mammary glands of female mammals produce milk, which is a complex mixture of immune system cells and proteins, fats, carbohydrates, and water in a proportion

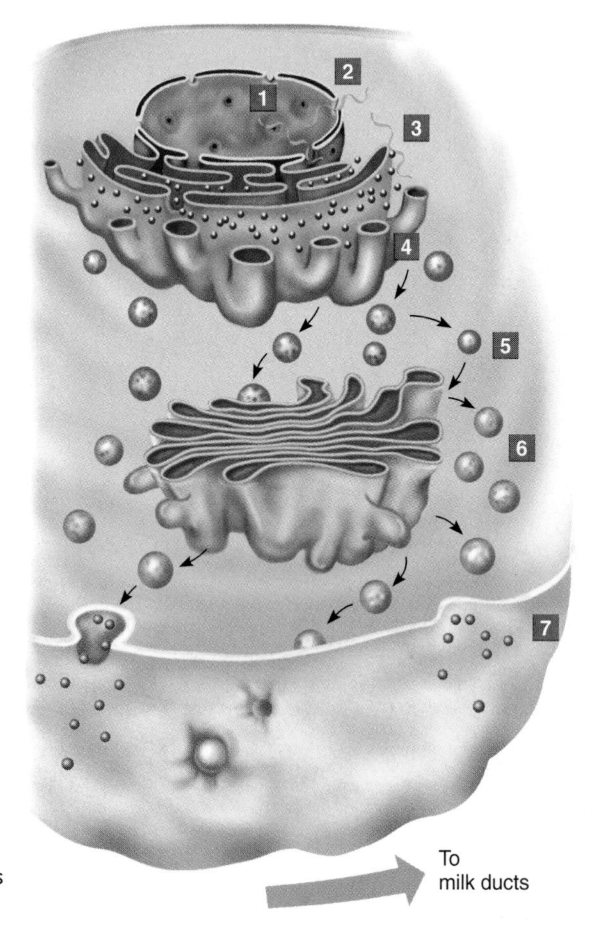

1 Milk protein genes transcribed into mRNA.

2 mRNA exits through nuclear pores.

3 mRNA forms complex with ribosomes and moves to surface of rough ER where protein is made.

4 Enzymes in smooth ER manufacture lipids.

5 Milk proteins and lipids are packaged into vesicles from both rough and smooth ER for transport to Golgi.

6 Final processing of proteins in Golgi and packaging for export out of cell.

7 Proteins and lipids released from cell by fusion of vesicles with cell membrane.

FIGURE 3.11
Secretion.
Milk production and secretion illustrate organelle functions and interactions in a cell from a mammary gland; (*1*) through (7) indicate the order in which organelles participate in this process.

ideal for development of a particular species. Dormant most of the time, these cells increase their metabolic activity during pregnancy and then undergo a burst of productivity shortly after the female gives birth. Organelles form a secretory network that enables individual cells to manufacture milk. Human milk is rich in lipids, which the rapidly growing newborn's brain requires. In contrast, cow milk contains a higher proportion of protein, better suited to a calf's rapid muscle growth. We follow here the production and secretion of human milk.

The Nucleus Secretion begins in the nucleus, where certain genes are copied into another nucleic acid, RNA. These genes encode milk protein and enzymes required to synthesize the carbohydrates and lipids in milk. The RNA molecules move from the interior of the nucleus towards the cytoplasm and then exit the nucleus through holes, called **nuclear pores** (fig. 3.12)**,** in the two-layered **nuclear envelope** that separates the nucleus from the cytoplasm. Nuclear pores are not merely perforations but channels composed of more than 100 different proteins. Traffic through the nuclear pores is busy, with millions of proteins and RNA molecules passing in or out each minute. Proteins tend to enter, and RNA molecules leave. Proteins colorfully called "importins" and "exportins" help regulate this passage of molecules between the nucleus and the cytoplasm.

The Cytoplasm In the cytoplasm, the RNA encounters, but does not actually enter, a maze of enzyme-studded, interconnected membranous tubules and sacs that winds from the nuclear envelope to the cell membrane. This labyrinth is the endoplasmic reticulum (ER) (fig. 3.13). (*Endoplasmic* means "within the plasm" and *reticulum* means "network.") The portion of this membranous system nearest the nucleus is flattened and studded with ribosomes.

Ribosomes consist of proteins and RNA (a different type of RNA than the type that carries information for making a protein) called ribosomal RNA, or rRNA. The parts of ribosomes are assembled in a region within the nucleus called the **nucleolus.** The ribosome-studded part of the ER is called rough ER because of its fuzzy appearance under the electron microscope (fig. 3.13). The RNA that carries the instructions for assembling the protein associates with ribosomes, which begin building the protein. The ribosomes then attach to the surface of the rough ER and complete synthesis of the protein. These proteins may either exit the cell, as does the milk protein casein, or be incorporated into the cell membrane. Proteins synthesized on ribosomes not associated with ER are released into the cytoplasm, where they serve specific functions.

The ER acts as a quality control center for the cell. Its chemical environment enables proteins to fold into the conformations and undergo modifications necessary for their functions. Misfolded proteins are pulled out of the production line and degraded in the ER.

As the rough ER winds from near the nucleus outward toward the cell membrane, the ribosomes become fewer and the diameters of the tubules widen, forming a section called smooth ER (fig. 3.13). Here, fatty acids and other membrane components are synthesized. The smooth ER also houses enzymes that detoxify certain chemicals. The lipids and proteins move along the smooth ER as its tubules narrow and end, then exit the organelle in vesicles that pinch off of the tubular endings of the ER membrane.

A loaded vesicle takes its contents to the next stop in the secretory production line, the Golgi apparatus (fig. 3.14). This organelle is a stack of flat, membrane-enclosed sacs that functions as a processing center. Here, proteins from the ER pass through the series of Golgi sacs, where they complete their intricate folding and become functional. The proteins must visit the parts of the Golgi apparatus in a specific sequence to form properly. The Golgi distinguishes between proteins destined for secretion, packaging them into vesicles, and proteins that will continue into a lysosome. The Golgi apparatus, therefore, compartmentalizes the sequence of steps necessary to produce functional proteins. Also in the Golgi apparatus, simple carbohydrates in the cytoplasm are synthesized, linked to form starches, or attached to proteins to form glycoproteins or to lipids to form glycolipids.

FIGURE 3.12
The Nucleus.
(**A**) The largest structure within a typical eukaryotic cell, the nucleus is surrounded by two membrane layers which make up the nuclear envelope (**B**). Pores through the envelope allow specific molecules to move in and out of the nucleus. The darkly staining nucleolus (**C**) is the site of ribosome manufacture and assembly.

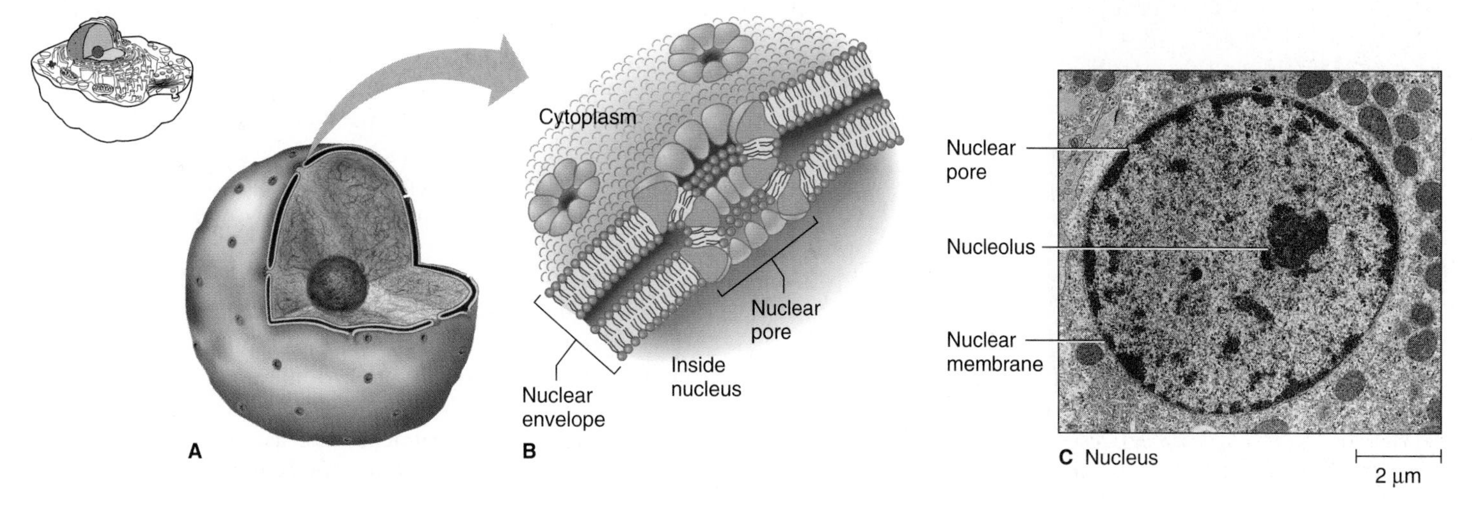

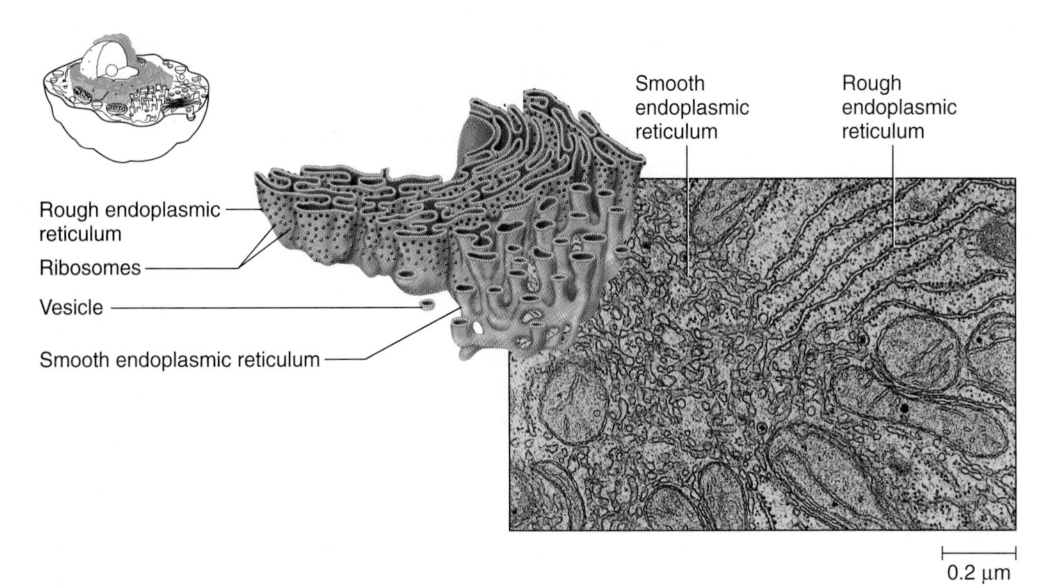

FIGURE 3.13
The Endoplasmic Reticulum.
The rough ER is an extension of the outer membrane of the nuclear envelope and is the site for manufacturing secreted proteins. Ribosomes dot the surface of the ER membrane, giving it the "rough" appearance. The smooth ER is a series of interconnecting tubules and is the site for lipid production and other metabolic processes.

FIGURE 3.14
The Golgi Apparatus.
Seen here in a scanning electron micrograph, the Golgi apparatus is composed of a series of membrane vesicles and flattened sacs. Proteins are sorted and processed as they move through the Golgi apparatus on their way to the cell surface or lysosomes.

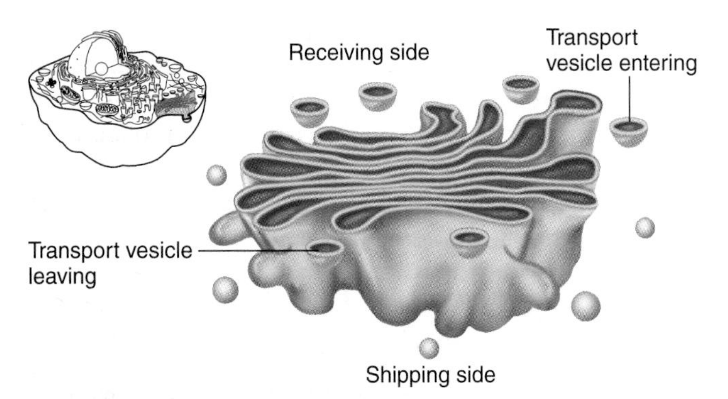

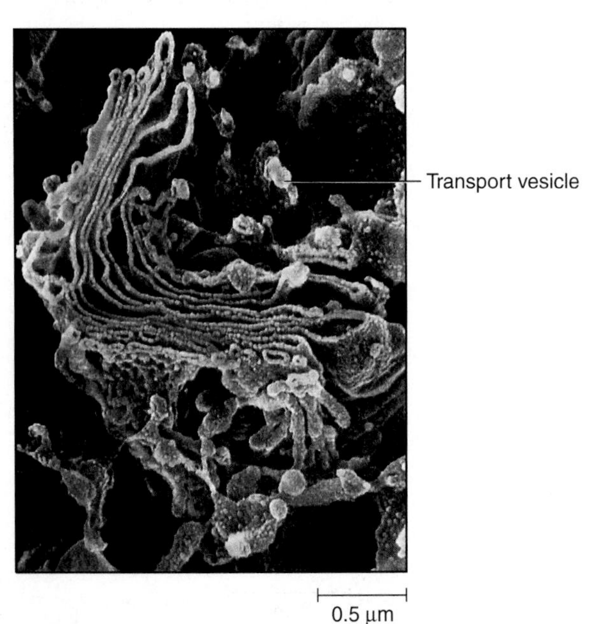

Vesicles budding off of the Golgi apparatus in this example contain milk proteins. The vesicles move toward the cell membrane, where they fuse to become part of the cell membrane and open out facing the exterior of the cell and release the proteins. Fat droplets retain a layer of surrounding membrane when they leave the cell. It isn't surprising that cells that secrete copiously have large ER and numerous Golgi apparatuses.

When a baby suckles, hormones (chemical messengers) released in the mother's system stimulate muscle cells surrounding balls of glandular cells to contract and squeeze milk from them. The milk is released into ducts that lead to the nipple.

Mitochondria Extract Energy from Nutrients

The activities of secretion, as well as the many chemical reactions taking place in the cytoplasm, require a steady supply of energy. In cells of eukaryotes, organelles called **mitochondria** extract energy from nutrient molecules (fig. 3.15). The number of mitochondria in a cell can vary from a few to tens of thousands. A typical liver cell has about 1,700 mitochondria; cells with high energy requirements, such as muscle cells, may have many thousands.

A mitochondrion has an outer membrane similar to those of the ER and Golgi apparatus and an intricately folded inner membrane. The folds of the inner membrane, called **cristae,** contain enzymes that catalyze the biochemical reactions that acquire energy (see chapter 7). This organelle is especially interesting because it contains genetic material, a point we will return to at the chapter's end.

Another unique characteristic of mitochondria is that they are inherited from the female parent only. This is because mitochondria are found in the middle regions of sperm cells but not in the head region, which is the portion that enters the egg to fertilize it. In humans, a class of inherited diseases whose symptoms result from abnormal mitochondria are always passed from mother to offspring. These mitochondrial illnesses usually pro-

FIGURE 3.15
Mitochondria Are the Sites of Energy Reactions.
A transmission electron micrograph of a mitochondrion. Cristae, infoldings of the inner membrane, increase the available surface area containing enzymes for energy reactions.

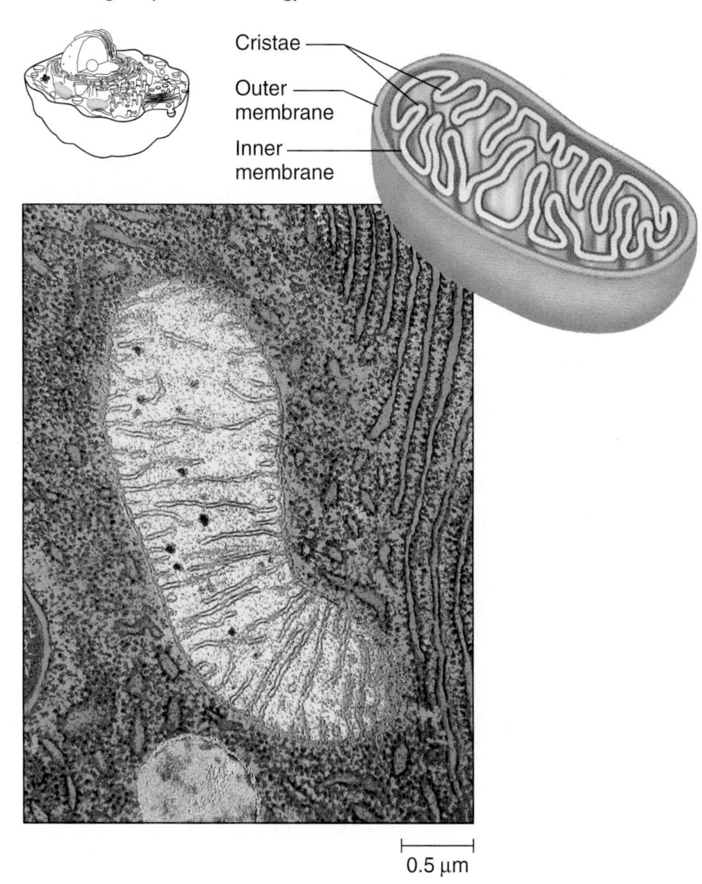

duce extreme muscle weakness, because muscle is a highly active tissue dependent upon the functioning of many mitochondria.

Chloroplasts Provide Plant Cells with Nutrients

In plants, and many protists, an additional organelle carries out photosynthesis to provide the cell with nutrients, mostly glucose. The **chloroplast** (fig. 3.16) is the organelle of photosynthesis within eukaryotes. Like the mitochondria, this organelle contains additional membrane layers. Two outer membrane layers enclose a space known as the **stroma.** Within the stroma is a third membrane system folded into flattened sacs called **thylakoids.** The thylakoids are stacked and interconnected in structures called **grana.** The process of photosynthesis, the subject of chapter 6, occurs in these thylakoids.

Another unique feature that chloroplasts share with mitochondria is the presence of its own genetic information. Chloroplast genes encode proteins and other molecules unique to photosynthesis and the structure of the chloroplasts.

Lysosomes and Peroxisomes Are Cellular Digestion Centers—And More

The cells of eukaryotes break down molecules and structures as well as produce them.

Lysosomes These organelles contain enzymes that dismantle captured bacteria, worn-out organelles, and debris. Lysosomal enzymes also break down some nutrients (fats, proteins, and carbohydrates) into forms that the cell can use, releasing them through the organelle's membrane into the cytoplasm.

FIGURE 3.16
Chloroplasts Are the Sites of Photosynthesis.
A transmission electron micrograph of a chloroplast reveals the stacks of thylakoids that form the grana within the inner compartment, the stroma. Enzymes and light-harvesting proteins are embedded in the membranes of the thylakoids to convert sunlight to chemical energy.

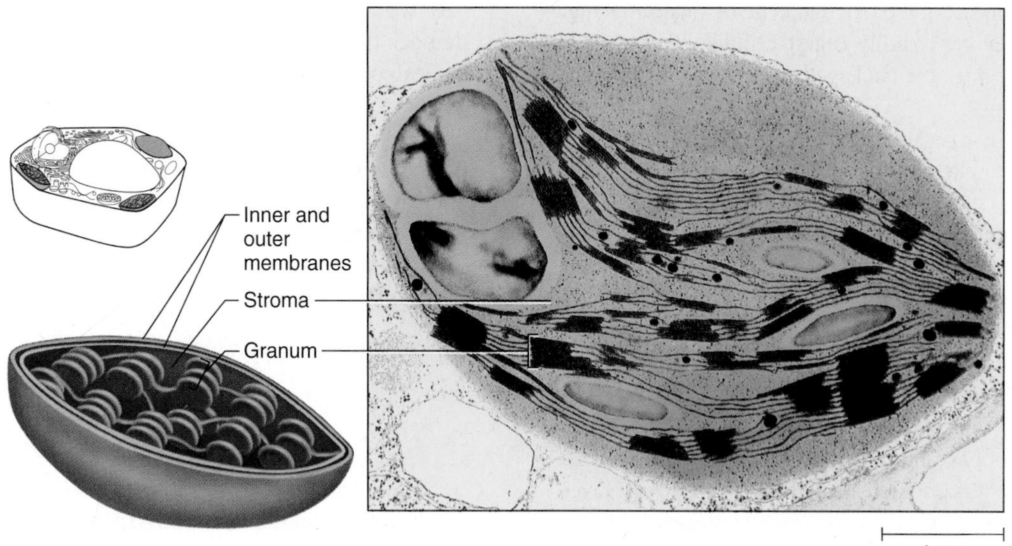

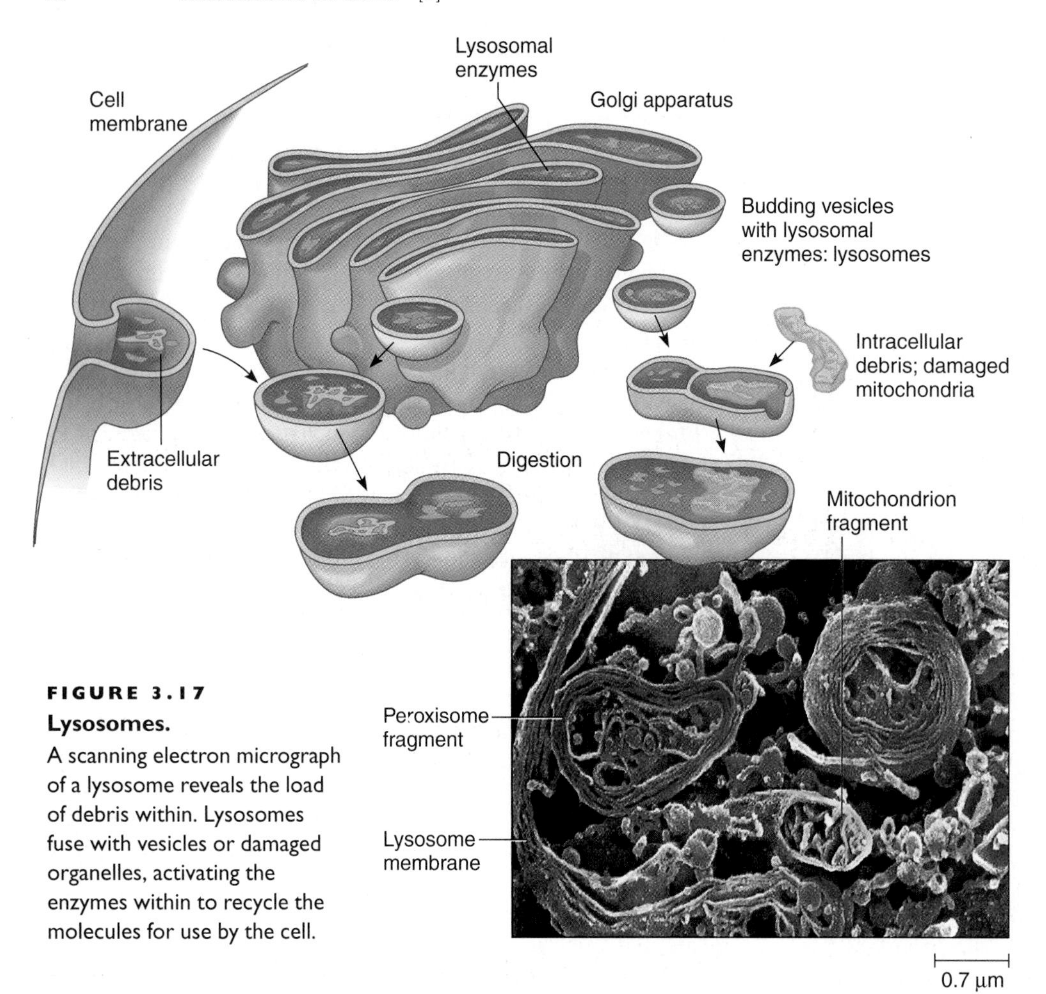

FIGURE 3.17
Lysosomes.
A scanning electron micrograph of a lysosome reveals the load of debris within. Lysosomes fuse with vesicles or damaged organelles, activating the enzymes within to recycle the molecules for use by the cell.

Lysosomes fuse with vesicles carrying debris from outside or from within the cell, and the lysosomal enzymes then degrade the contents (fig. 3.17). Sometimes a lysosome that contains only undigested material fuses with the cell membrane, releasing its contents to the outside.

Lysosomal enzymes are manufactured within the rough ER. The Golgi apparatus then detects and separates enzymes destined for lysosomes by recognizing a particular type of sugar attached to them. Because they can readily digest cellular components, lysosome-specific enzymes are packaged into vesicles that eventually become lysosomes.

Lysosomal enzymes can function only in a very acidic environment and the organelle maintains this environment without harming other cellular constituents. In fact, Belgian cell biologist Christian de Duve discovered lysosomes in 1949 by grinding up cells and noticing the accumulation of a digestive enzyme that did not dismantle structures in intact cells. He deduced that his treatment of the cells had liberated the enzyme from a structure that normally contains it. Years later, the electron microscope revealed lysosomes, which were so named because their enzymes lyse, or cut apart, their substrates.

White blood cells have many lysosomes because these cells engulf debris and bacteria, which must be broken down. Liver cells require many lysosomes to break down cholesterol. In human cells, a lysosome contains more than 40 types of digestive enzymes. The correct balance of these enzymes is important to health. Absence or malfunction of just one type of enzyme can cause a lysosomal storage disease, in which the molecule that is normally degraded accumulates. The lysosome swells, crowding organelles and interfering with the cell's functions. In Tay-Sachs disease, for example, lack of an enzyme that normally breaks down lipid in cells surrounding nerve cells buries the nervous system in lipid. An affected infant begins to lose skills at about age 6 months, then gradually loses sight, hearing, and the ability to move, typically dying within 3 years. Even before birth, the lysosomes of affected cells become hugely swollen.

In plant cells, the large central vacuole serves a function similar to, but not as extensive as, the lysosomes. These enzyme-containing vacuoles soften cells in some fruits, making them more palatable to animals that eat them and spread the seeds.

Peroxisomes **Peroxisomes** are single-membrane-bounded sacs, present in all eukaryotic cells, that contain several types of enzymes. The proteins found in peroxisosomes are first made in the cytoplasm and later transported into vesicles, which become peroxisomes. In some species, exposure to environmental toxins triggers an explosive production of peroxisomes, which helps cells to survive the insult.

Peroxisomes help the cell use oxygen. Specifically, the reactions that some peroxisomal enzymes catalyze produce hydrogen peroxide (H_2O_2). This compound introduces oxygen-free radicals, which are oxygen atoms with unpaired electrons that can damage the cell. To counteract the free-radical buildup, peroxisomes contain abundant catalase, an enzyme that removes an oxygen from hydrogen peroxide. Catalase then combines the oxygen from H_2O_2 with hydrogen atoms removed from various organic molecules, producing harmless water molecules. Liver and kidney cells contain many peroxisomes to help dismantle toxins from the blood (fig. 3.18A).

The peroxisomal enzymes catalyze a variety of biochemical reactions, including:

- synthesizing bile acids, which are used in fat digestion;
- breaking down lipids called very-long-chain fatty acids;
- degrading rare biochemicals;
- metabolizing potentially toxic compounds that form as a result of oxygen exposure.

Abnormal peroxisomal enzymes can harm health. A defect in a receptor on the peroxisome's membrane may affect several of

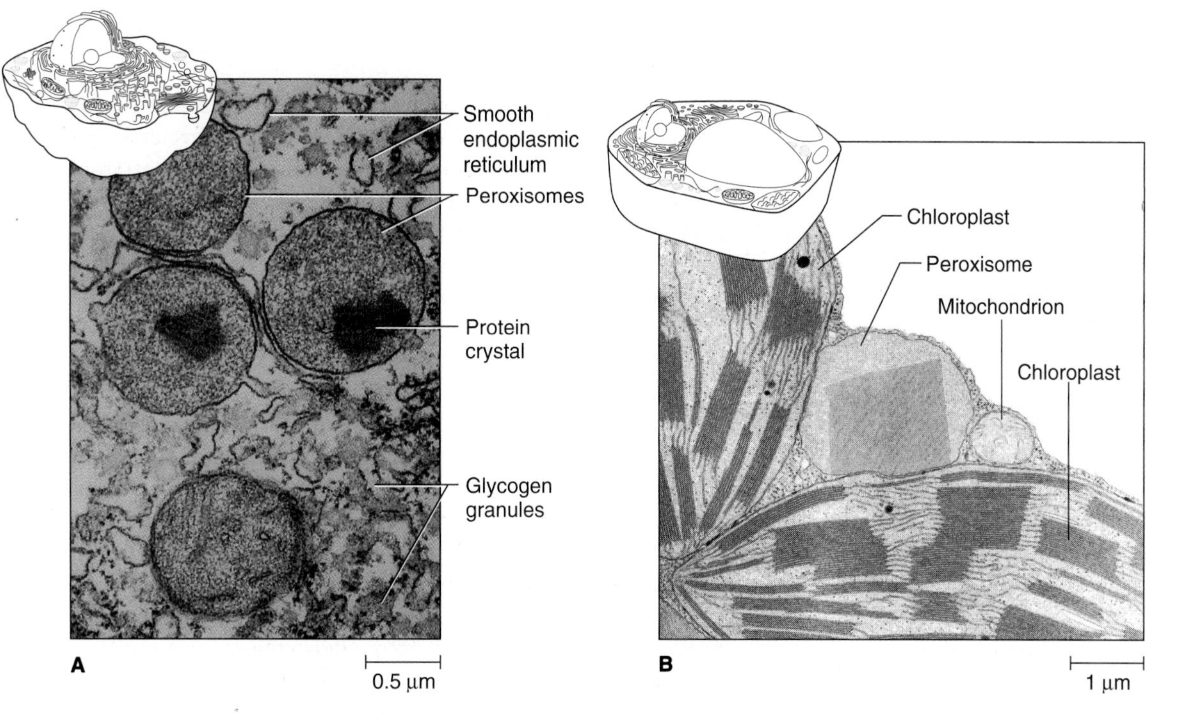

FIGURE 3.18
Peroxisomes.
(**A**) The high concentration of enzymes within results in crystallization of the proteins, giving peroxisomes a characteristic appearance. (**B**) In plants, peroxisomes are involved in many oxidation-reduction reactions that assist in photosynthesis and defense.

the enzymes and cause a variety of symptoms, or a single enzyme type may be abnormal. In adrenoleukodystrophy, for example, one of two major proteins in the organelle's outer membrane is absent. Normally, this protein transports an enzyme into the peroxisome, where it catalyzes a reaction that helps break down a type of very-long-chain fatty acid. Without the enzyme transporter protein, the fatty acid builds up in the cells of the brain and spinal cord, eventually stripping these cells of fatty coverings necessary for nerve conduction. Symptoms include weakness, dizziness, low blood sugar, darkening skin, behavioral problems, and loss of muscular control.

In plants, leaf cells have many peroxisomes. In these cells, a dense area where catalase concentrates often appears as a crystal in electron micrographs (fig. 3.18B). Peroxisomes in plants help to break down organic molecules synthesized in photosynthesis, the process by which plants harness solar energy to synthesize simple carbohydrates.

Table 3.1 summarizes the organelles discussed in this chapter.

Cells also have other, specialized organelles and structures. Some play a role in cell division, others are important in photosynthesis. These organelles are explored in later chapters. The next chapter considers the cell membrane, the cytoskeleton, and how cells interact.

3.3 MASTERING CONCEPTS

1. Which organelles interact to produce and secrete a complex substance?
2. What is the function of the nucleus and its contents?
3. Which organelle houses the reactions that extract chemical energy from nutrient molecules?
4. Which organelles are membrane-bounded sacs containing a variety of enzymes?

OLC

TABLE 3.1 Structures and Functions of Organelles

ORGANELLE	STRUCTURE	FUNCTION
Chloroplast	Three membranes, inner stacks of flattened membrane sacs contain pigments	Photosynthesis
Endoplasmic reticulum	Membrane network; rough ER has ribosomes, smooth ER does not	Site of some protein synthesis and folding; lipid synthesis
Golgi apparatus	Stacks of membrane-enclosed sacs	Sugars linked to form starches or joined to lipids or proteins; proteins finish folding; secretions stored
Lysosome	Sac containing digestive enzymes	Debris degraded; cell contents recycled
Mitochondrion	Two membranes; inner one enzyme-studded	Releases energy from nutrients
Nucleus	Perforated sac containing DNA	Separates DNA from rest of cell
Peroxisome	Sac containing enzymes	Oxygen use, other reactions that protect a cell
Ribosome	Two associated globular subunits of RNA and protein	Scaffold for protein synthesis; RNA may help catalyze protein synthesis
Vesicle	Membrane-bounded sac	Temporarily stores or transports substances

3.4 How Might Complex Cells Have Originated?

The organelles seen in complex cells today likely originated long ago as free-living organisms.

How did eukaryotic cells arise? The **endosymbiont theory** proposes that these complex cells formed as large, nonnucleated cells engulfed smaller and simpler cells. (An endosymbiont is an organism that can live only inside another organism, a relationship that benefits both partners.) The compelling evidence supporting the endosymbiont theory is the striking resemblance between mitochondria and chloroplasts, which are present only in eukaryotic cells, and certain types of bacteria and archaea. These similarities include size, shape, membrane structure, presence of pigments, reproduction by splitting in two, and the close association of DNA, RNA, and ribosomes. The endosymbiont theory was built on evidence from microscopy, but is today being bolstered by genetic evidence, illustrating vividly how our knowledge of life parallels development of new technology.

Investigators first noted in the 1920s that some eukaryotic cell components look like free-living bacteria, and suggested that the former long ago engulfed the latter. But they were ridiculed for having too little evidence. In the 1960s, studies with the electron microscope confirmed the similarities identified four decades earlier between eukaryotic organelles and photosynthetic bacteria. Then genetic studies revealed DNA in mitochondria and chloroplasts, supporting the idea that they were once independent organisms. These converging ideas and observations led biologist Lynn Margulis at the University of Massachusetts at Amherst to propose the endosymbiont theory in the late 1960s.

More recently, DNA sequence evidence suggests that both bacteria and archaea contributed to the origin of eukaryotic cells. One current version of the theory is that archaeal cells with the protein elements of a cytoskeleton and internal membranes enveloped bacterial cells that eventually became mitochondria and chloroplasts (fig. 3.19). Other studies at the genome level are revealing bacteria whose genes are like those of mitochondria, as well as mitochondrial genomes that resemble those of certain bacteria. (A genome is the complete set of genes in an organism.) For example, the genome of the bacterium *Rickettsia prowazekii,* which causes typhus in humans, contains many of the same genes as mitochondria. Looking at the situation from the other direction, the single-celled eukaryote *Reclinomonas americana* has mitochondria whose genes are very much like those of bacteria.

Once we sequence more genomes, more clues to long-ago endosymbiosis may emerge. For now, we can imagine how this merger of organisms might have taken place.

Picture a mat of bacteria and cyanobacteria, thriving in a pond some 2.5 billion years ago. Over many millions of years, the flourishing cyanobacteria pumped oxygen into the atmosphere as a by-product of photosynthesis. Eventually, those organisms that could tolerate free oxygen flourished.

Free oxygen reacts readily with molecules in organisms, producing oxides that can no longer carry out biological functions. One way for a large cell, possibly an archaeon, to survive in an oxygen-rich environment would be to engulf an aerobic (oxygen-utilizing) bacterium in an inward-budding vesicle of its cell membrane. The captured cell would contribute biochemical reactions to detoxify free oxygen. Eventually, the membrane of the enveloping vesicle became the outer membrane of the mitochondrion. The outer membrane of the engulfed aerobic bacterium became the inner membrane system of the mitochondrion, complete with respiratory enzymes. The smaller bacterium found a new home in the larger cell; in return, the host cell could survive in the newly oxygenated atmosphere.

Similarly, archaean cells that picked up cyanobacteria or other small cells capable of photosynthesis obtained the forerunners of chloroplasts and thus became the early ancestors of red algae or of green plants. Once such ancient cells had acquired their endosymbiont organelles, the theory holds, genetic changes impaired the ability of the captured microorganisms to live on their own outside the host cell. Over time the larger cells and the captured ones came to depend on one another for survival. The result of this biological interdependency, according to the endosymbiont theory, is the compartmentalized cells of modern eukaryotes.

Although none of us was present billions of years ago to witness the formation of complex cells, evidence of each of the steps of the endosymbiont theory is seen in present-day organisms. In the absence of a time machine, the elegant endosymbiont theory may be the best window we have to look into the critical juncture in cell history when cells became more complex. As we learn more about the archaea, it will be interesting to see how the endosymbiont theory evolves.

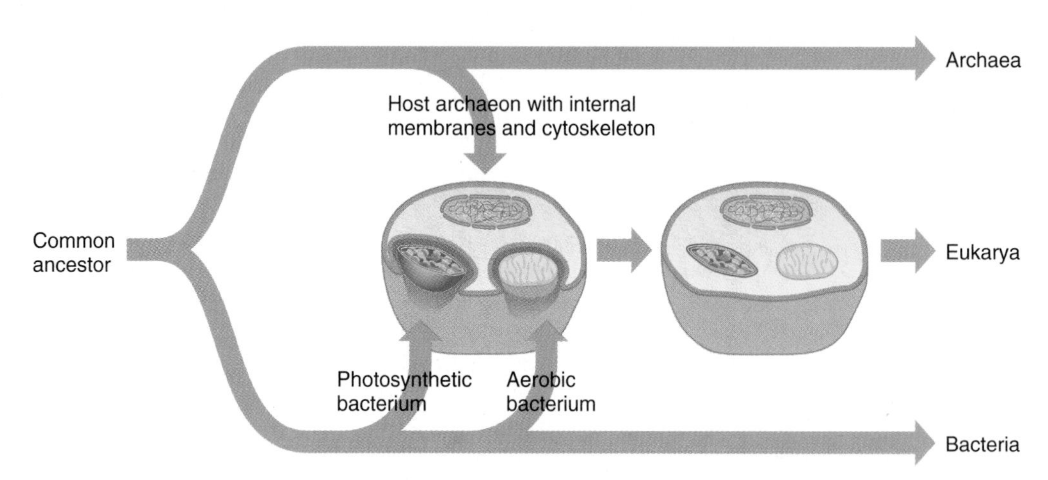

FIGURE 3.19
The Endosymbiont Theory.
Eukaryotic cells may have originated from a long-ago joining of bacterial cells with archaean cells. The captured bacteria eventually became mitochondria and chloroplasts. The archaean host contributed membranes and cytoskeletal elements, which enabled the organism to move and engulf smaller cells. When some of the genes of the captured cells moved to the nucleus-to-be—the archaeon's DNA—the smaller cells became dependent on their host. A complex cell—we think—was so born.

3.4 MASTERING CONCEPTS

1. What is the evidence that certain organelles descend from simpler cells engulfed long ago?
2. How are the hypothesized evolution of mitochondria and chloroplasts related?

OLC

Chapter Summary

3.1 Cells Are the Units of Life

1. **Cells,** the units of life, are the microscopic components of all organisms. Cells exhibit the characteristics of life, but can also specialize.
2. Even the simplest cells are highly organized; complex cells may carry out very specialized functions.
3. Cells were observed first in the late seventeenth century, when Robert Hooke viewed cork with a crude lens. Antonie van Leeuwenhoek viewed many cells under the light microscope he invented. Stains revealed subcellular details.
4. The **cell theory** states that all life is composed of cells, that cells are the functional units of life, and that all cells come from preexisting cells.

3.2 Variations on the Cellular Theme

5. Bacteria are unicellular and lack nuclei and have DNA, **ribosomes** (structures that help manufacture proteins), enzymes for obtaining energy, and various other biochemicals. A cell membrane and usually a rigid **cell wall** enclose the cell contents.
6. Archaea are unicellular and lack nuclei. They share some characteristics with bacteria and eukaryotes but also have unique structures and biochemistry.

3.3 How Do Organelles Divide Labor in a Eukaryotic Cell?

7. Eukaryotic cells sequester certain biochemical activities in organelles. A eukaryotic cell houses DNA in a membrane-bounded **nucleus;** synthesizes, stores, transports, and releases molecules along a network of organelles (**endoplasmic reticulum, Golgi apparatus, vesicles**); degrades wastes and digests nutrients in **lysosomes;** processes toxins and oxygen in **peroxisomes;** extracts energy from digested nutrients in **mitochondria;** and in plants and some protists, extracts solar energy in **chloroplasts.** A cell membrane surrounds eukaryotic cells. Cell walls protect and support cells of many organisms.

3.4 How Might Complex Cells Have Originated?

8. The **endosymbiont theory** proposes that chloroplasts and mitochondria evolved from once free-living bacteria engulfed by larger archaea. Evidence for the endosymbiont theory is that mitochondria and chloroplasts resemble small aerobic bacteria in size, shape, and membrane structure and in the ways their DNA, RNA, and ribosomes interact to manufacture proteins. Gene and genome sequence data also support the endosymbiont theory.

Testing Your Knowledge

1. Identify the cell parts indicated in the diagram in the next column, and describe the functions of each.
2. Until very recently, life scientists thought that there were only two types of cells. How has that view changed?
3. As a cube increases in size from 3 centimeters to 5 centimeters to 7 centimeters on a side, how does the surface area to volume ratio change?
4. Name three structures or activities found in eukaryotic cells but not in bacteria or archaea.
5. Cite two types of evidence in support of the endosymbiont theory.

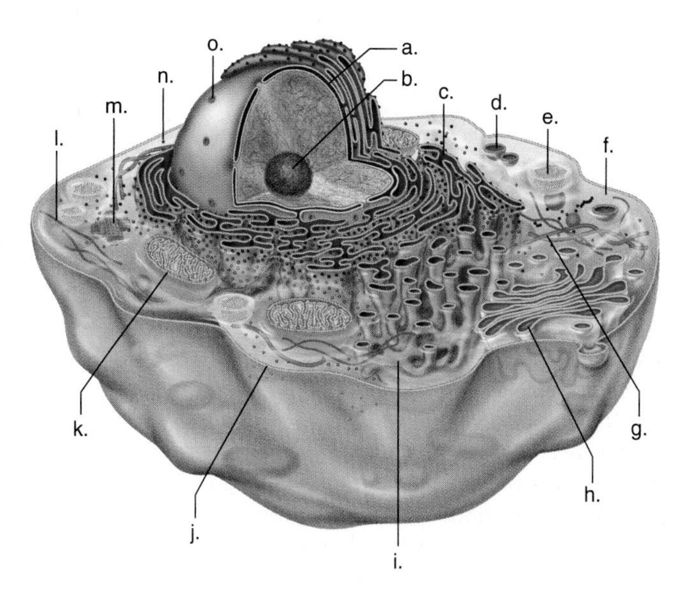

Thinking Scientifically

1. A liver cell has a volume of 5,000 μm^3. Its total membrane area, including the inner membranes lining organelles as well as the cell membrane, fills an area of 110,000 μm^2. A cell in the pancreas that manufactures digestive enzymes has a volume of 1,000 μm^3 and a total membrane area of 13,000 μm^2. Which cell is probably more efficient in carrying out activities that require extensive membrane surfaces, and why?
2. What advantages does compartmentalization confer on a large cell?
3. Why does a muscle cell contain many mitochondria and a white blood cell contain many lysosomes?
4. The amoeba *Pelomyxa palustris* is a single-celled eukaryote with no mitochondria, but it contains symbiotic bacteria that can live in the presence of oxygen. How does this observation support or argue against the endosymbiont theory of the origin of eukaryotic cells?

References and Resources

Bignami, Giovanni F. June 29, 2000. The microscope's coat of arms. *Nature,* vol. 405, p. 999. Thirty years before Leeuwenhoek, another investigator turned magnifying lenses to insects.

Brittle, Elizabeth E., and M. Gerard Waters. July 21, 2000. ER-to-Golgi traffic—this bud's for you. *Science,* vol. 289, p. 403. Proteins guide a vesicle's journey.

Doolittle, W. Ford. February 2000. Uprooting the tree of life. *Scientific American,* vol. 282, p. 90. Many types of cells may have existed at the dawn of life.

Mazzarello, Paolo, and Marina Bentivoglio. April 9, 1998. The centarian Golgi apparatus. *Nature,* col. 392, p. 543. The Golgi apparatus was discovered just over a century ago.

Tromans, Amanda. March 23, 2000. Rafting vesicles. *Nature,* vol. 434. Vesicles link organelle actions.

The *LIFE* Online Learning Center provides additional resources and tools for studying this chapter.
www.mhhe.com/life

CHAPTER 4

The Cell Membrane, Cytoskeleton, and Cell-Cell Interactions

Nicotine Addiction and Cell Biology

Receptors and Reception

In the United States, about 23% of adults and 30% of adolescents smoke cigarettes. In many nations, these figures are even higher. Yet the effects of smoking on health—such as greatly increased risks of developing heart disease, cancer, stroke, and lung disease—are well known. Why, then, do so many people smoke?

The answer is that they are addicted to the nicotine in cigarettes. Cell biology can explain how an enjoyable habit becomes a physical dependency.

Tobacco used for smoking comes from the plant *Nicotiniana tabacum* and has been linked to cancer since the mid-1700s. Recognition of smoking tobacco as an addictive behavior is more recent. A 1964 Surgeon General's report distinguished between "addicting" and "habituating" drugs and defined "addiction" as causing intoxication. On this basis, cigarette smoking was considered merely habituating, a definition that tobacco companies promoted for years. The 1964 report also linked smoking to cancer and lung disease, which frightened some of the 40% of the population that then smoked into quitting. On a worldwide basis, even today, cigarette smoking directly or indirectly causes about 20% of all deaths.

More recent Surgeon General's reports reflect accumulating scientific evidence on the dangers of smoking. The 1988 report stated clearly that nicotine causes addiction, and the 1994 report calls smoking a "totally preventable public health tragedy." It was also in 1994 that the Food and Drug Administration began investigating the possibility of regulating tobacco as a drug, because of its ability to addict. Today, the director of the National Institute on Drug Abuse calls all drug addiction, including nicotine addiction, "a chronic, relapsing brain disorder characterized by compulsive drug seeking." That's a far cry from the billboards depicting happy smokers that once boasted, "Alive with pleasure!"

According to the Diagnostic and Statistical Manual of Mental Disorders, a person addicted to tobacco:

1. must smoke more to attain the same effects (tolerance) over time;
2. experiences withdrawal symptoms when smoking stops, including weight gain, difficulty concentrating, insomnia, restlessness, anxiety, depression, slowed metabolism, and lowered heart rate;
3. smokes more often and for longer than intended;

To increase sales, cigarette companies used ads to suggest smoking was actually good for you.

The cycle of nicotine addiction

Cigarette smoke inhaled.

Nicotine reaches brain cells (neurons).

Nicotine binds to receptors on neurons' cell membranes.

Positive ions can travel through receptors into the neurons.

Ion influx leads to dopamine release.

Dopamine causes feelings of pleasure...

...and suppresses withdrawal symptoms.

Drug-seeking behavior reinforced.

Nicotine is addictive because of effects at the surface of cells. Those effects produce a cycle of nicotine addiction.

4. spends considerable time obtaining cigarettes, and feels compelled to do so;
5. devotes less time to other activities;
6. continues to smoke despite knowing the dangers;
7. wants to stop, but cannot easily do so.

Cigarette smoke contains about 4,000 different chemicals, including carbon monoxide and cyanide. It is the nicotine that causes the addiction, and the addiction supplies enough of the other chemicals to gradually destroy health. Nicotine reaches the brain within seconds of the first inhalation, as it also speeds heart rate and constricts blood vessels, raising blood pressure.

Tracing nicotine's effects on brain cells (neurons) introduces several concepts discussed in this chapter. An activated form of nicotine binds to proteins that form structures called nicotinic receptors that are parts of the cell membranes of certain neurons. These neurons are in a brain region called the nucleus accumbens. The receptors normally bind the neurotransmitter acetylcholine. When sufficient nicotine binds, a channel within the receptor opens and admits positively charged ions into the neuron. When a certain number of ions enter, the neuron is stimulated to release the neurotransmitter dopamine from its other end. Dopamine provides the pleasurable feelings associated with smoking. Addiction stems from two sources—seeking the good feelings of releasing all that dopamine, and avoiding painful withdrawal symptoms.

Cigarette smoke contains about 4,000 different chemicals, including carbon monoxide and cyanide.

Binding nicotinic receptors isn't the only effect of nicotine on the brain. When a smoker increases the number of cigarettes smoked, the number of nicotinic receptors on the brain cells increases. This happens because the way that the nicotine binds impairs the recycling of receptor proteins, so that new receptors are produced faster than they are taken apart. However, after a period of steady nicotine exposure, many of the receptors malfunction and no longer admit the positive ions that trigger nerve transmission. This may be why as time goes on it takes more nicotine to produce the same physical effects.

Experiments with mice demonstrate that it is nicotine's binding to these receptors that reinforces drug-seeking behavior, reinforcing addiction. Mice that lack these receptors (but are otherwise normal) do not push a lever that administers nicotine intravenously to them, as will mice that have the receptors.

Many questions remain concerning the biological effects of tobacco smoking. Why don't lab animals experience withdrawal? Why do people who have successfully stopped smoking often start again 6 months later, even though withdrawal symptoms ease within 3 weeks of quitting? Why do some people become addicted easily, yet others smoke only a few cigarettes a day and can stop anytime?

While scientists try to answer these questions, society must deal with questions of rights and responsibilities that cigarette smoking causes.

4.1 How Does the Cell Membrane Control Cell Function?

Cells must regulate what enters and leaves them; maintain their specific shapes; and interact with other cells. The cell membrane and underlying cytoskeleton make these functions possible.

The 2-year-old's health appeared to be returning mere hours after the liver transplant. Yet, cells of the immune system had already detected the new organ and, interpreting it as "foreign," began to produce molecules to attack it. Even though the donor's liver was carefully "matched" to the little girl—the pattern and types of molecules on the surfaces of its cells was very similar to that on the cells of her own liver cells—the match was not perfect. A rejection reaction would soon destroy the liver and the little girl would need another transplant to survive.

Rejection of a transplanted organ illustrates the importance of cell surfaces in the coordinated functioning of a multicellular organism. The cell surface is one component of a cellular architecture of sorts. It includes structures that give a cell its particular three-dimensional shape and topography, help determine the locations and movements of organelles and biochemicals within the cell, and participate in the cell's interactions with other cells and the extracellular environment.

The cellular architecture consists of surface molecules embedded in the cell membrane, which is the outer covering of a cell. Just beneath the cell membrane protein fibers form part of the cell's interior scaffolding, or **cytoskeleton.** Together, the cell surface and cytoskeleton form a dynamic structural framework that helps to distinguish one cell from another and provide the means for a cell to perform its unique functions (fig. 4.1). The various components of the cellular architecture must communicate and interact for the cell to carry out the processes of life.

FIGURE 4.1
Cellular Architecture.
A white blood cell's inner skeleton and surface features enable it to move in the body and to recognize "foreign" cell surfaces—such as those of transplanted tissue. This T lymphocyte rejects foreign tissue.

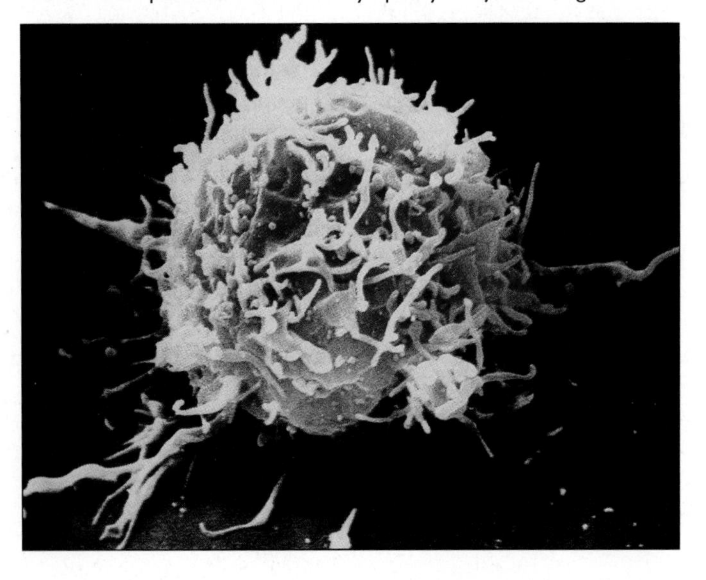

At a conference where most participants do not know one another, name tags are often used to identify people. All cells also have name tags in the form of carbohydrates, lipids, and proteins that protrude from their surfaces. Some surface molecules distinguish cells of different species, like company affiliations on name tags. Other surface structures distinguish individuals within a species.

Surface structures also distinctively mark cells of different tissues in an individual, so that a bone cell's surface is different from that of a nerve cell or a muscle cell. Surface differences between cell types are particularly important during the development of an embryo, when different cells sort and grow into specific tissues and organs.

The special characteristics of different cell types are shaped in part by the substances that enter and leave them. The cell membrane monitors the movements of molecules in and out of the cell. The chemical characteristics and the pattern of molecules that are part of a cell membrane determine which substances can cross it. Archaean cells have interior membranes, and in eukaryotic cells, membranes form organelles.

What Is the Structure of a Cell Membrane?

The structure of a biological membrane is possible because of a chemical property of the phospholipid molecules that compose it. **Phospholipids** are lipid molecules bonded to phosphate groups (PO_4, a phosphorus atom bonded to four oxygen atoms). The phosphate end of a phospholipid molecule is attracted to water (hydrophilic, or "water-loving"); the other end, consisting of two fatty acid chains, is repelled by water (hydrophobic, or "water-fearing"). Because of these water preferences, phospholipid molecules in water spontaneously arrange into the most energy-efficient organization, a **phospholipid bilayer.** In this two-layered, sandwichlike structure, the hydrophilic surfaces are on the outsides of the "sandwich," exposed to the watery medium outside and inside the cell. The hydrophobic surfaces face each other on the inside of the "sandwich," away from water (fig. 4.2).

Cell membranes consist of phospholipid bilayers and the proteins and other molecules embedded in them and extending from them (fig. 4.3). The hydrophobic interior keeps out most substances dissolved in water. However, some proteins embedded in the phospholipid bilayer create passageways through which water-soluble molecules and ions pass. Other proteins are carriers, transporting substances across the membrane.

Proteins within the oily phospholipid bilayer can move laterally within the layer, sometimes at remarkable speed. Because of this movement, the protein-phospholipid bilayer is often called a fluid mosaic.

One way to classify membrane proteins is by their location in the phospholipid bilayer. Membrane proteins may lie completely within the phospholipid bilayer or traverse the membrane to extend out of one or both sides. In animal cells, some membrane proteins, called glycoproteins, contain branchlike carbohydrate molecules which protrude from the membrane's outer surface.

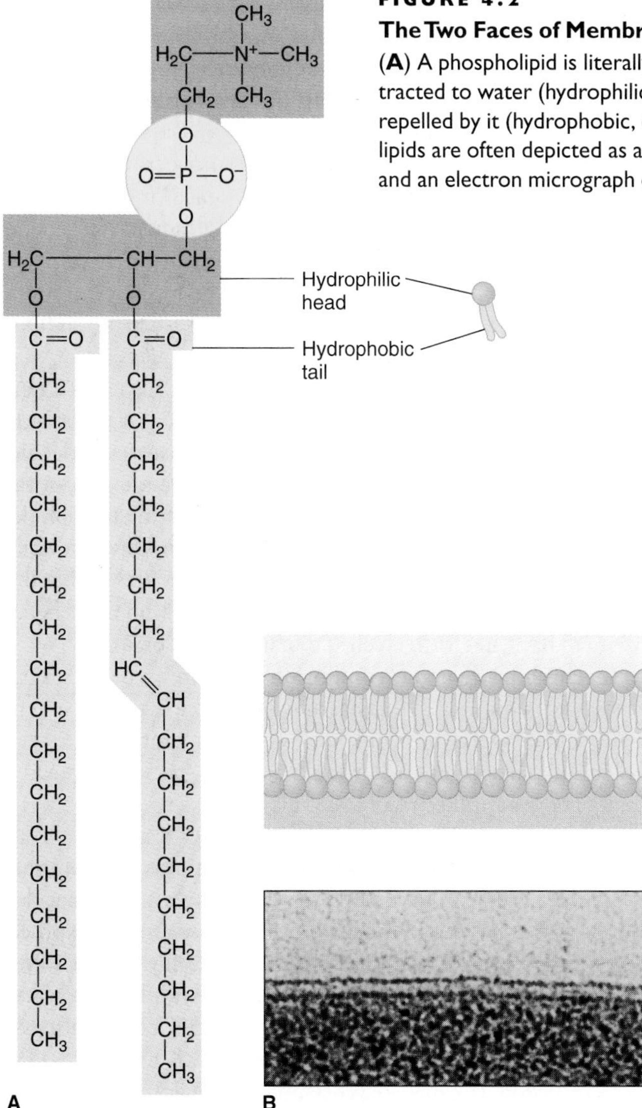

FIGURE 4.2

The Two Faces of Membrane Phospholipids.

(**A**) A phospholipid is literally a two-faced molecule, with one end attracted to water (hydrophilic, or "water-loving") and the other repelled by it (hydrophobic, or "water-fearing"). Membrane phospholipids are often depicted as a circle with two tails. (**B**) A depiction and an electron micrograph of a phospholipid bilayer.

The proteins and glycoproteins that jut from the cell membrane contribute to the surface characteristics that are so important to a cell's interactions with other cells.

The functions of membrane proteins are related to their locations within the phospholipid bilayer. We saw in the case of transplant rejection that one function of cell membrane proteins is to establish a cell's surface as "self." Some membrane proteins exposed on the outer face of the membrane are receptors, binding outside molecules and triggering cascades of chemical reactions in the cell that lead to a specific response. Other membrane proteins enable specific cell types to stick to each other, making cell-to-cell interactions possible. Table 4.1 lists functions of membrane proteins.

TABLE 4.1 Types of Membrane Proteins

PROTEIN TYPE	FUNCTION
Transport proteins	Move substances across membranes
Cell surface proteins	Establish self
Cellular adhesion molecules	Enable cells to stick to each other
Receptor proteins	Receive and transmit messages into a cell

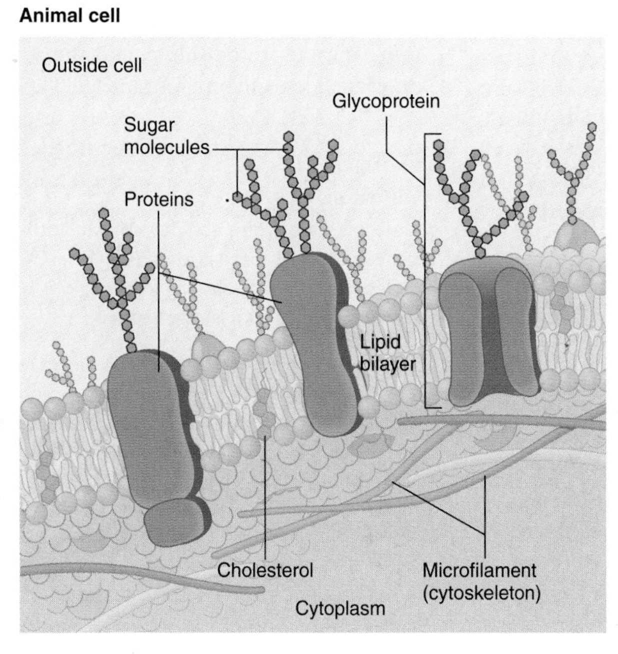

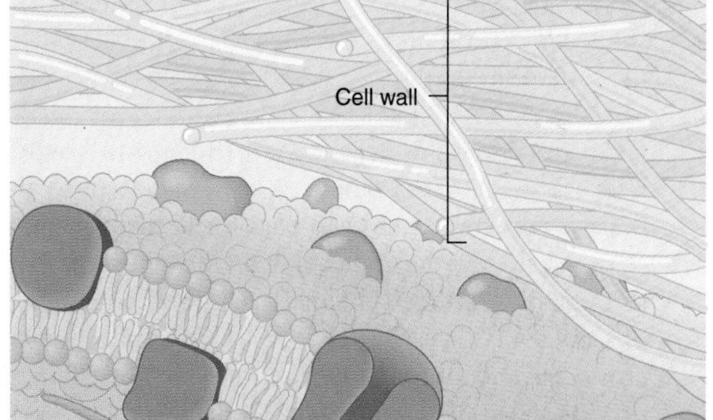

FIGURE 4.3

Anatomy of a Cell Membrane.

In a cell membrane, mobile proteins are embedded throughout a phospholipid bilayer, producing a somewhat fluid structure. An underlying mesh of protein fibers supports the cell membrane. Jutting from the animal cell membrane's outer face are carbohydrate molecules linked to proteins (glycoproteins) and lipids (glycolipids). The typical plant cell is surrounded by a cell wall, a network of cellulose fibers.

Cell membranes have specific protein/phospholipid ratios, and disruption of the ratio can affect health. For example, the cell membranes of certain cells that support nerve cells are about three-quarters myelin, a lipid. The cells enfold nerve cells in tight layers, wrapping their fatty cell membranes into a sheath that provides the insulation that speeds nerve impulse transmission. In multiple sclerosis, the cells coating the nerves that lead to certain muscles lack myelin. The resulting blocked neural messages to muscles impair vision and movement and cause numbness and tremor. In Tay-Sachs disease, the reverse happens. Cell membranes accumulate excess lipid, and nerve cells cannot transmit messages to each other and to muscle cells. An affected child gradually loses the ability to see, hear, and move.

4.1 MASTERING CONCEPTS

1. What are some functions of the cell surface?
2. What types of molecules make up a cell membrane?
3. How are the components of a cell membrane organized?
4. How do the locations of membrane proteins determine their functions?
5. How do membranes differ in different cell types?

How Do Substances Cross Membranes?

Specialized cells function as they do only if certain molecules and ions are maintained at certain levels inside and outside of them. The cell membrane oversees these vital concentration differences. Before considering how cells control which substances enter and leave, recall that an **aqueous solution** is a homogeneous mixture of a substance dissolved in water. The dissolved material is the **solute,** and the liquid in which it is dissolved is the **solvent.** Concentration refers to the relative number of one kind of molecule compared to the total number of molecules present, and it is usually given in terms of the solute. When solute concentration is high, the proportion of solvent (water) present is low, and the solution is concentrated. When solute concentration is low, solvent is proportionately high, and the solution is dilute.

Diffusion Moves Substances from High to Low Concentration The cell membrane is choosy, or selectively permeable—that is, some molecules can pass freely through the membrane (either between the molecules of the phospholipid bilayer or through protein-lined channels), while others cannot. For example, oxygen (O_2), carbon dioxide (CO_2), and water (H_2O) freely cross cell and other biological membranes. They do so by a process called **diffusion,** which is the movement of a substance from a region where it is more concentrated to a region where it is less concentrated, without using energy (fig 4.4). Diffusion occurs because molecules are in constant motion, and they move so that two regions of differing concentration tend to become equal. Heat increases diffusion by increasing the rate of collisions between molecules. An easy way to observe diffusion is to place a tea bag in a cup of hot water. Compounds in the tea leaves dissolve gradually and diffuse throughout the cup. The tea is at first concentrated near the bag, but the brownish color eventually spreads to create a uniform brew.

The natural tendency of a substance to move from where it is highly concentrated to where it is less so is called "moving down" or "following" its **concentration gradient.** A gradient is a general term that refers to a difference in some quality between two neighboring regions. Gradients may be created by concentration, electrical, pH, and pressure differences. Ions such as sodium (Na^+) and potassium (K^+) establish electrical gradients.

Molecules crossing a membrane because of this natural tendency to travel from higher to lower concentration is called simple diffusion because it does not require energy or a carrier molecule. Simple diffusion eventually reaches a point where the concentration of the substance is the same on both sides of the membrane. After this, molecules of the substance continue to flow randomly back and forth across the membrane at the same rate, so that the concentration remains equal on both sides. This point of equal movement back and forth is called **dynamic equilibrium.**

To envision dynamic equilibrium, picture a party taking place in two rooms. Everyone has arrived, and no one has yet left. People walk between the rooms in a way that maintains the same number of partiers in each room, but the specific occupants change. The party is in dynamic equilibrium.

The Movement of Water Is Osmosis The fluids that continually bathe cells of multicellular organisms consist of molecules and ions dissolved in water. Because cells are constantly exposed to water, it is important to understand how a cell regulates water entry. If too much water enters a cell, it swells; if too much water leaves, it shrinks. Either response may affect a cell's ability to function. Movement of water across biological membranes by simple diffusion is called **osmosis.** The concentration of dissolved substances inside and outside the cell determines the direction and intensity of movement.

In osmosis, water is driven to move because the membrane is impermeable to the solute, and the solute concentrations differ on each side of the membrane (fig. 4.5). Water moves across the

FIGURE 4.4
Diffusion.
Molecules and atoms collide with each other, spreading out to have the same volume of space around each one. Diffusion always results in molecules or atoms moving from regions of high concentration toward low until equilibrium is reached.

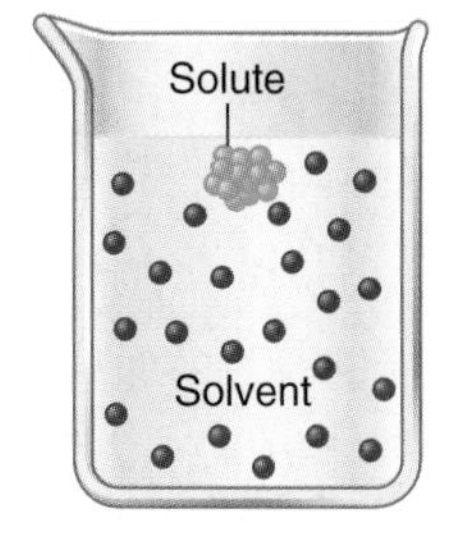

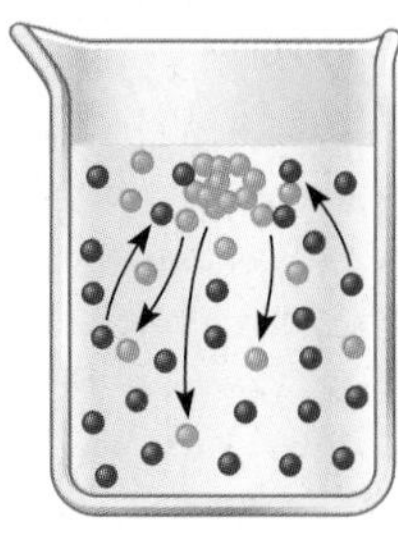
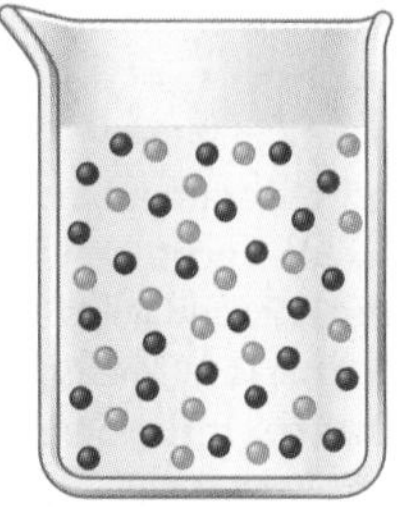

FIGURE 4.5

Osmosis.

An artificial membrane dividing a beaker demonstrates osmosis by permitting water to pass from one chamber to another, but preventing large salt molecules from doing the same. Water will flow from an area of low salt (solute) concentration towards an area of high salt concentration. Eventually, the volume on each side of the membrane will be different, but the final concentrations (amount of solute per unit of volume) will be the same. Dynamic equilibrium is reached when there is no tendency for water to flow in either direction.

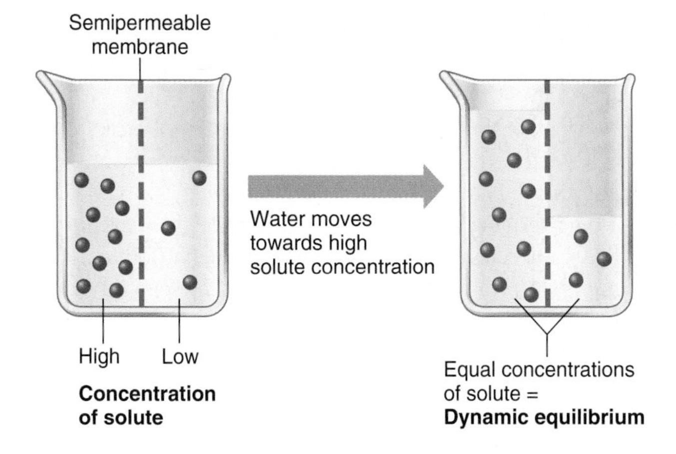

FIGURE 4.6

Diffusion Affects Cell Shape.

A red blood cell changes shape in response to changing plasma solute concentrations. (**A**) A human red blood cell is normally isotonic to the surrounding plasma. When water enters and leaves the cell at the same rate, the cell maintains its shape. (**B**) When the salt concentration of the plasma increases, water leaves the cells to dilute the outside solute faster than water enters the cell. The cell shrinks. (**C**) When the salt concentration of the plasma decreases relative to the salt concentration inside the cell, water flows into the cell faster than it leaves. The cell swells, and may even burst.

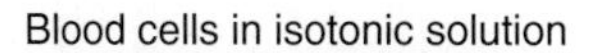

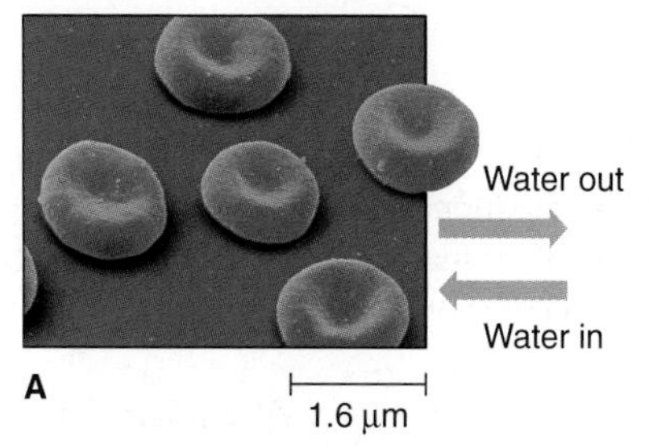

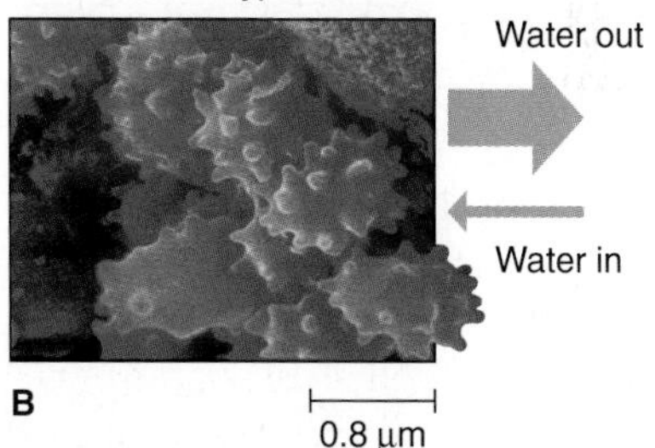

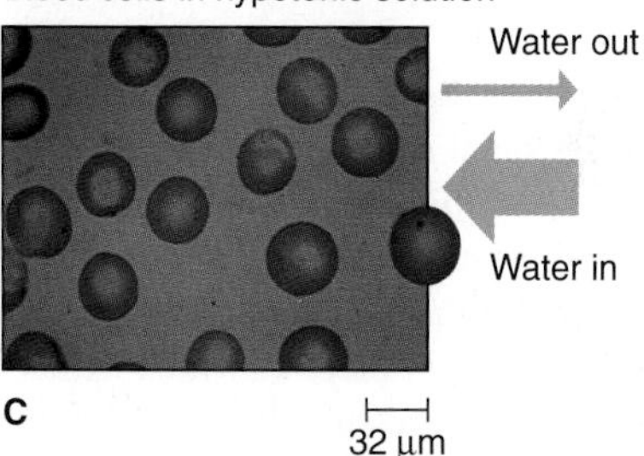

membrane in the direction that dilutes the solute on the side where it is more concentrated.

Variants of the word "tonicity" are used to describe osmosis in relative terms. Tonicity refers to the differences in solute concentration in two compartments separated by a semipermeable membrane. A cell interior is **isotonic** to the surrounding fluid when solute concentrations are the same within and outside the cell. In this situation, there is no net flow of water, a cell's shape does not change, and salt concentration is ideal for enzyme activity.

Disrupting a cell's isotonic state changes its internal environment and shape as water rushes in or leaks out. If a cell is placed in a solution in which the concentration of solute is lower than it is inside the cell, water enters the cell to dilute the higher solute concentration there. In this situation, the solution outside the cell is **hypotonic** to the inside of the cell. The cell swells. In the opposite situation, if a cell is placed in a solution in which the solute concentration is higher than it is inside the cell, water leaves the cell to dilute the higher solute concentration outside. In this case, the outside is **hypertonic** to the inside. This cell shrinks.

Hypotonic and hypertonic are relative terms and can refer to the surrounding solution or to the solution inside the cell. It may help to remember that *hyper* means "over," *hypo* means "under," and *iso* means "the same." A solution in one region may be hypotonic or hypertonic to a solution in another region.

The effects of immersing a cell in a hypertonic or hypotonic solution can be demonstrated with a human red blood cell, which is normally suspended in an isotonic solution called plasma (fig. 4.6). In this state, the cell is doughnut-shaped, with a central indentation. Placing a red blood cell in a hypertonic solution draws water out of the cell, and it shrinks. Placing the cell in a hypotonic solution has the opposite effect. Because there are more solutes inside the cell, water flows into the cell, causing it to swell. Size changes caused by osmosis in plant cells are less dramatic because of the cell wall.

Because shrinking and swelling cells may not function normally, unicellular organisms can regulate osmosis to maintain their shapes. Many cells alter membrane transport activities, changing the concentrations of different solutes on either side of the cell membrane, in a way that drives osmosis in a direction that maintains the cell's shape. This enables some single-celled organisms that live in the ocean to remain isotonic to their salty environment, keeping their shapes. In contrast, the paramecium, a single-celled organism that lives in ponds, must work to maintain its oblong form. A paramecium contains more concentrated solutes than the pond, so water tends to flow into the organism. A

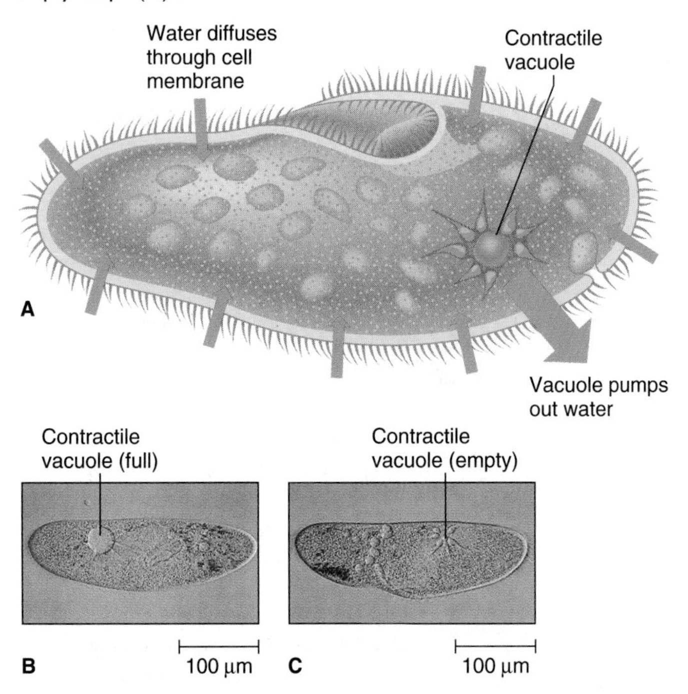

FIGURE 4.7

Aquatic Organisms Must Pump Water.

Paramecia keep their shapes with contractile vacuoles that fill and then pump excess water out of the cells across the cell membrane (**A**). The contractile vacuole moves near the cell membrane as it fills (**B**), and then releases the water to the outside. The organelle then resumes its empty shape (**C**) and moves back to the interior of the cell.

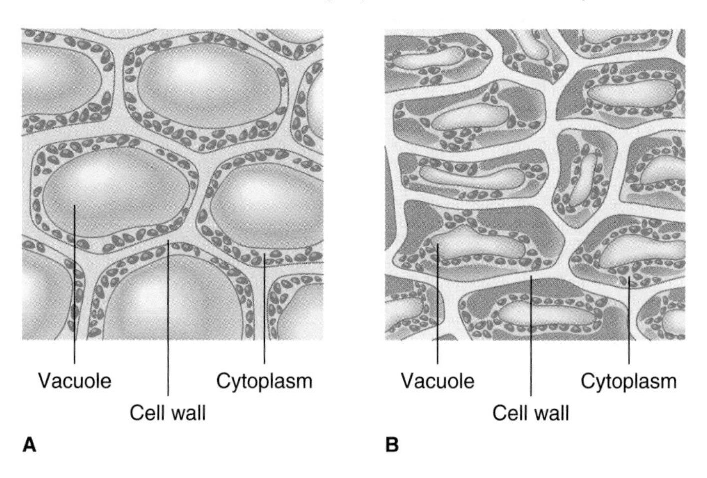

FIGURE 4.8

Plant Cells Keep Their Shapes by Regulating Diffusion.

Like paramecia, plant cells usually contain more concentrated solutes than their surroundings, drawing more water into the cell. (**A**) In a hypotonic solution, water enters the cell and collects in vacuoles. The cell swells against its rigid, restraining cell wall, generating turgor pressure. (**B**) When a plant cell is placed in a hypertonic environment (so that solutes are more concentrated outside the cells), water flows out of the vacuoles, and the cell shrinks. Turgor pressure is low, and the plant wilts.

special organelle called a contractile vacuole pumps the extra water out (fig. 4.7).

Plant cells also face the challenge of maintaining their shapes even with a concentrated interior. Instead of expelling the extra water that rushes in, as the paramecium does, plant cells expand until their cell walls restrain their cell membranes. The resulting rigidity, caused by the force of water against the cell wall, is called **turgor pressure** (fig. 4.8). A piece of wilted lettuce demonstrates the effect of losing turgor pressure. When placed in water, the leaf becomes crisp, as the individual cells expand like inflated balloons.

In the human body, osmosis influences the concentration of urine. Brain cells called osmoreceptors shrink when body fluids are too concentrated, which signals the pituitary gland to release antidiuretic hormone (ADH). The bloodstream transports ADH to cells lining the kidney tubules, where it alters their permeabilities so that water exits the tubules and enters capillaries (microscopic blood vessels) that entwine about the tubules. This conserves water. Without ADH's action, this water would remain in the kidney tubules and leave the body in dilute urine. Instead, it returns to the bloodstream, precisely where it's needed (see chapter 38).

Transport Proteins Move Molecules and Ions The phospholipid bilayer keeps many ions and polar molecules from diffusing across cell membranes. However, such substances can still cross cell membranes with the help of transport proteins. These proteins are abundant and diverse. They typically span the membrane, and come in three varieties—channels, carriers, and pumps (fig. 4.9).

A channel transport protein, as its name implies, forms a pore through which a solute passes. The size of the pore and the charges that line its interior surface determine which molecules or ions can pass through. The transported substance may or may not bind to the protein. Some of these proteins have parts that form gatelike structures that control flow through them under certain conditions. Transport through a channel protein is fast—some 100 million ions or molecules a second may enter or exit the cell.

A carrier protein binds a specific ion or molecule, which contorts the protein in a way that moves the cargo to the other face of the membrane, where it exits. Carrier proteins provide **passive transport** if energy is not expended, or **active transport** if energy drives the movement. Passive transport using a carrier protein is also called facilitated diffusion, because the substance being moved travels down its concentration gradient. The basis of facilitated diffusion is that on the side of the membrane where the substance is more highly concentrated, more molecules contact the carrier protein. Eventually, dynamic equilibrium results, unless some other activity interferes, such as the cell's producing or consuming the substance being transported. Facilitated diffusion can move 100 to 1,000 ions or molecules a second.

Active transport enables a cell to admit a substance that is more concentrated inside than outside. This requires an input of energy. Returning to the party analogy, active transport is like a guest elbowing her way into a room that is already crowded with the majority of the partygoers. Energy for active transport often comes from adenosine triphosphate (ATP), which is discussed further in the next chapter. When a phosphate group is split from ATP, it releases energy that is harnessed to help drive a cellular

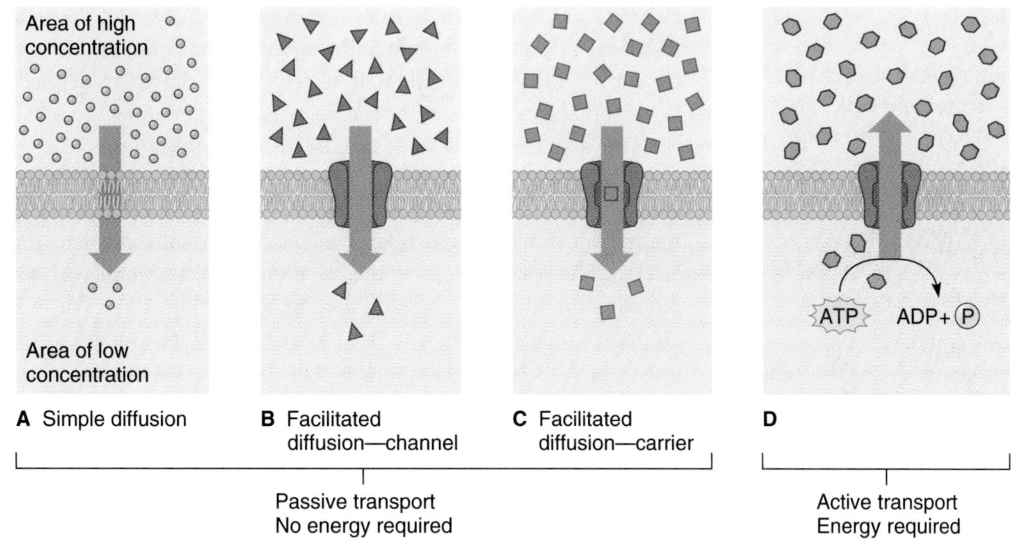

FIGURE 4.9
Transport Moves Substances.
Simple diffusion, facilitated diffusion, and active transport move ions and molecules across cell membranes. In passive transport, molecules move down their concentration gradients by themselves by squeezing through the membrane components (**A**), through a channel protein (**B**), or aboard a carrier protein (**C**), without direct energy input. In active transport (**D**), a molecule or ion crosses a membrane against its concentration gradient, using energy and carrier proteins that function as pumps.

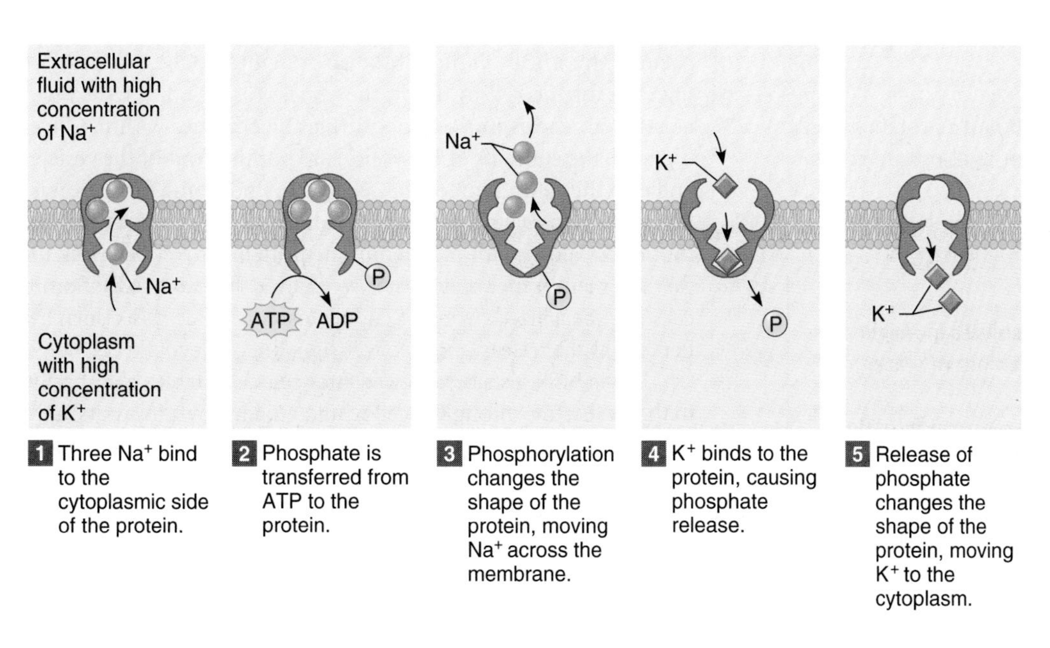

FIGURE 4.10
The Sodium-Potassium Pump.
This "pump," actually a carrier protein embedded in the cell membrane, uses energy (ATP) to move potassium ions (K^+) into the cell and sodium ions (Na^+) out of the cell. The pump first binds Na^+ on the inside face of the membrane (*1*). ATP is split to ADP and a phosphate group is transferred to the carrier or pump. (*2*) This binding alters the conformation of the pump and causes it to release the Na^+ to the outside. The altered pump can now take up K^+ from outside the cell (*3*). Next, the pump releases the bound phosphate (*4*), which again alters the conformation of the carrier protein. This change in shape releases K^+ to the cell's interior (*5*). The pump is also back in the proper shape to bind intracellular Na^+.

function, such as moving a molecule through a membrane against its concentration gradient.

The first active transport system discovered was a carrier protein called the **sodium-potassium pump** found in the cell membranes of most animal cells. Cells must contain high concentrations of potassium ions (K^+) and low concentrations of sodium ions (Na^+) to perform such basic functions as maintaining their volume and synthesizing protein, as well as to conduct more specific activities, such as transmitting nerve impulses and enabling the lungs and kidneys to function.

The sodium-potassium pump contains binding sites for both Na^+ and K^+. A pumping cycle begins with the binding of three Na^+ on the inside of the cell (fig. 4.10). The terminal phosphate of ATP is then transferred to the protein, causing it to change shape and expose the sodium ions to the outside of the cell. Since the sodium ion concentration is lower outside the cell, the ions diffuse away from the pump. This exposes two binding sites for K^+ ions, which are immediately filled. When the pump has bound two K^+ ions, the phosphate is released, causing the pump to return to its original shape. This exposes the K^+ ions to the cytoplasm, where they diffuse away from the pump. The cycle is ready to begin again. The entire process takes only a fraction of a second.

Yet another way to move molecules across membranes is **cotransport,** in which a protein carries one substance as it ferries a different substance down its concentration gradient. Often active transport indirectly powers cotransport. A good example is the process of "sucrose loading" that sends the sugar into certain tissues in plants. An ATP-driven pump in the cell membrane sends hydrogen ions (protons) out of the cell. The protons accumulate in the cell walls that surround the cell membranes, building up energy like a wound spring. Then the protons move down their

FIGURE 4.11
Cotransport.
Energy from ATP is used to pump hydrogen ions out of the plant cell, creating a concentration gradient. The flow of hydrogen ions back into the cell is coupled to the transport of sucrose into a cell by means of a symporter protein channel. The energy of the gradient fuels the active transport.

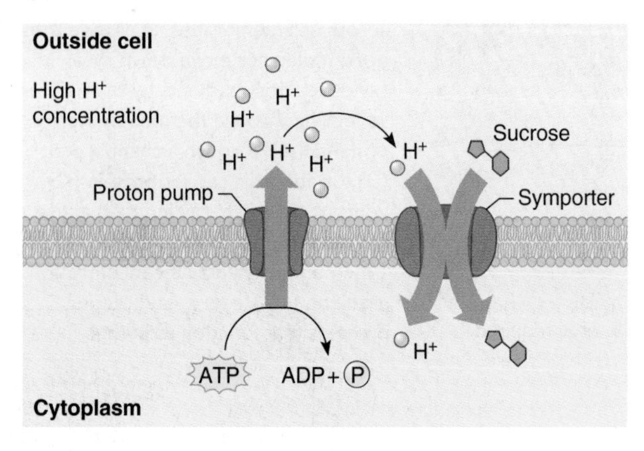

concentration gradient back into the cell aboard a protein called a symporter, using the energy of the gradient to also transport sucrose (fig. 4.11).

4.1 MASTERING CONCEPTS

1. What is diffusion?
2. How do differing concentrations of solutes in neighboring aqueous solutions drive the movement of water molecules?
3. How do facilitated diffusion and active transport differ from each other and from simple diffusion?
4. Describe the mechanisms of passive transport and active transport.

OLC

Exocytosis, Endocytosis, and Transcytosis Most molecules dissolved in water are small, and they can cross cell membranes by simple diffusion, facilitated diffusion, or active transport. Large molecules (and even bacteria) can also enter and leave cells, with the help of vesicles that form from cell membranes.

Exocytosis transports large particles out of cells, such as some components of milk (see fig. 3.11). Inside a cell, a vesicle made of a phospholipid bilayer is filled with substances to be ejected. The vesicle moves to the cell membrane and joins with it, releasing the substance outside the membrane (fig. 4.12). For example, exocytosis in the front tip of a sperm cell releases enzymes that enable the tiny cell to penetrate the much larger egg cell.

Endocytosis allows a cell to capture large molecules on its external surface in a nonspecific way and bring them into the cell. Protein molecules just inside the cell membrane join, forming a vesicle that traps whatever is outside the membrane. The vesicle pinches off and moves into the cell. At times, all that is captured are solutes and water in a process known as pinocytosis (fig. 4.13A).

Phagocytosis is another form of endocytosis that some cells use to capture and destroy debris or smaller organisms such as bacteria (fig. 4.13B). If the substance brought into the cell must be digested for the cell to use it, the vesicle fuses with a lysosome, creating an **endosome** and activating the enzymes within. Once digested, the contents of the vesicle are pumped out of the vesicle and into the cytoplasm for the cell to use. Endosome formation is one way that a cell can capture raw materials. The components of the vesicle, along with the membrane proteins, are cycled back to the cell membrane, and waste is removed by exocytosis. Some cells use specialized forms of endocytosis and exocytosis, such as nerve cells that release and recover signaling molecules.

When biologists first viewed endocytosis in white blood cells in the 1930s, they thought a cell would gulp in anything at its surface. We now recognize a more specific form of the process called **receptor-mediated endocytosis.** A receptor protein on a cell's surface binds a particular biochemical, called a **ligand;** the cell membrane then indents, embracing the ligand and drawing it into the cell (fig. 4.13C).

FIGURE 4.12
Endocytosis and Exocytosis.
Endocytosis brings large particles and even bacteria into a cell. A small portion of the cell membrane buds inward (*1*), entrapping the particles (*2*), and a vesicle forms, which brings the substances into the cell. Biochemicals and particles exit cells by exocytosis. Vesicles surround or take up the structures to be exported (*3*), then move to the cell membrane (*4*) and merge with it, releasing the particles to the outside (*5*).

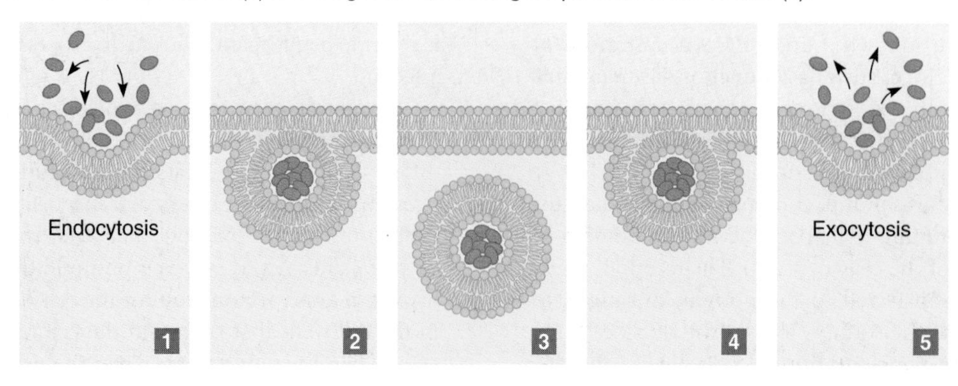

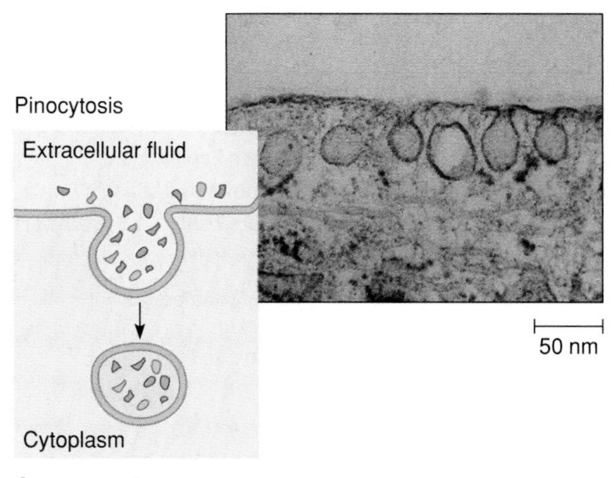

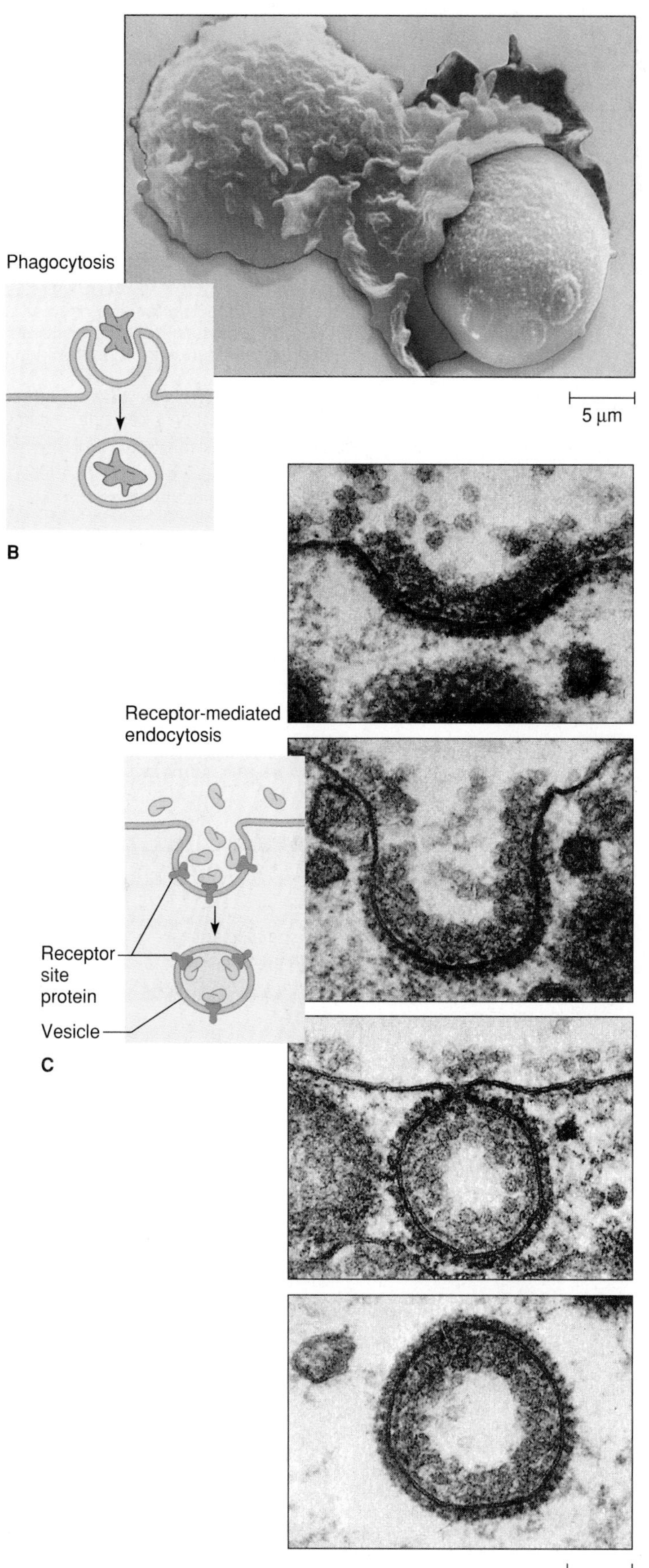

FIGURE 4.13
Three Types of Endocytosis.
(**A**) Pinocytosis captures water containing dissolved substances for the cell. (**B**) Phagocytosis brings large clumps of nutrients into the cell. The white blood cell (in blue) is engulfing a yeast cell. (**C**) Receptor-mediated endocytosis is triggered by the binding of a specific molecule to a receptor protein.

Receptor-mediated endocytosis enables liver cells to take in dietary cholesterol from the bloodstream, where it is carried in lipoprotein particles. One type of cholesterol carrier, a low-density lipoprotein (LDL), binds to receptors clustered in protein-lined pits in the surfaces of liver cells. The cell membranes envelop the LDL particles, forming loaded vesicles and bringing them into the cell, where they move towards lysosomes. Within lysosomes, enzymes liberate the cholesterol from the LDL carriers. The receptors are recycled to the cell surface, where they can bind more cholesterol-laden LDL particles.

Large-scale membrane transport, and movement of substances within eukaryotic cells, requires recognition at the molecular level. How does a cholesterol-containing vesicle "know" how to find a lysosome, or any other organelle? Vesicles move about the cell following specific routes in a process called vesicle trafficking. As is often the case in organisms, proteins add specificity to a generalized process. A vesicle "docks" at its target membrane guided by a series of proteins and then anchors to receptors. Some guiding proteins are within the phospholipid bilayers of the organelle and the approaching vesicle, while others are initially free in the cytoplasm. Certain combinations of proteins join and form a complex that helps draw the vesicle to its destination.

Endocytosis brings a substance into a cell, and exocytosis transports a substance out of a cell. Another process, **transcytosis,** combines endocytosis and exocytosis. Transcytosis is the selective and rapid transport of a substance or particle from one end of a cell to the other. It enables substances to cross barriers formed by tightly connected cells. The most obvious example is the transport of small molecules across the lining of the digestive system into the bloodstream. The digestive system breaks down macromolecules

TABLE 4.2 Movement Across Membranes

MECHANISM	CHARACTERISTICS	EXAMPLE
Diffusion	Follows concentration gradient	Oxygen diffuses from lung into capillaries
Osmosis	Diffusion of water	Water reabsorbed from kidney tubules
Facilitated diffusion	Follows concentration gradient, assisted by carrier protein	Glucose diffuses into red blood cells
Active transport	Moves against concentration gradient, assisted by carrier protein and energy, usually ATP	Salts reabsorbed from kidney tubules
Cotransport	Movement of one substance is coupled to movement of a second substance down its concentration gradient, often countering a proton pump	Sucrose transport into phloem cells
Exocytosis	Membrane-bounded vesicle fuses with cell membrane, releasing its contents outside of cell	Nerve cells release neurotransmitters
Endocytosis	Membrane engulfs substance and draws it into cell in membrane-bounded vesicle	White blood cells ingest bacteria
Transcytosis	Combines endocytosis and exocytosis	Products of the digestive system entering the bloodstream

to monomers. These monomers are transported across the cell membrane of cells (endocytosis), across the cytoplasm to the other side of the cells (trancytosis), and released into the bloodstream (exocytosis).

Table 4.2 summarizes the transport mechanisms that move substances across membranes.

4.1 MASTERING CONCEPTS

1. How do exocytosis and endocytosis transport large particles across cell membranes?
2. How can endocytosis become specialized?
3. Which structures guide vesicles inside a cell?
4. How does transcytosis combine exocytosis and endocytosis?

OLC

4.2 How Does the Cytoskeleton Support a Cell?

Within cells, a vast network of tubules and filaments guides organelle movement, provides overall shape, and establishes vital links to specific molecules that are part of the cell membrane.

The cytoskeleton is a meshwork of tiny protein rods and tubules that molds the distinctive structures of eukaryotic cells, positioning organelles and providing characteristic overall three-dimensional shapes. The protein girders of the cytoskeleton are dynamic structures that are broken down and built up as a cell performs specific activities. Some cytoskeletal elements function as rails, forming conduits for cellular contents on the move; other components of the cytoskeleton, called motor molecules, power the movement of organelles along these rails by converting chemical energy to mechanical energy.

The cytoskeleton includes three major types of elements—**microtubules, microfilaments,** and **intermediate filaments** (fig. 4.14). They are distinguished by protein type, diameter, and

FIGURE 4.14
The Cytoskeleton Is Made of Protein Rods and Tubules.
The three major components of the cytoskeleton are microtubules, intermediate filaments, and microfilaments. Through special staining, the cytoskeleton in this cell glows yellow under the microscope.

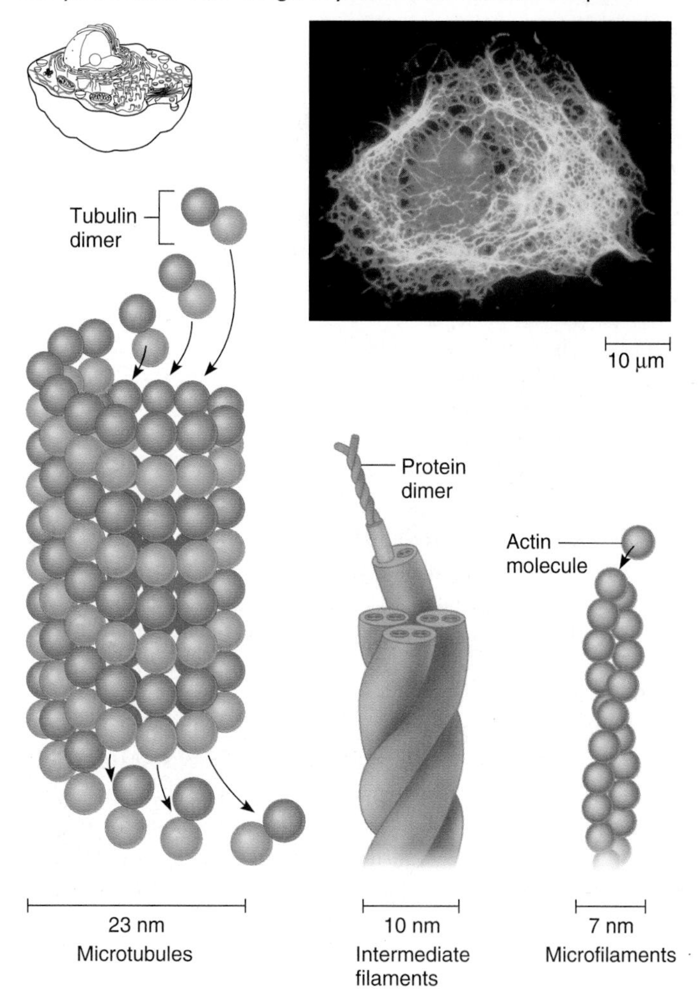

TABLE 4.3 Functions of the Cytoskeleton

1. Moving chromosomes apart during cell division
2. Controlling vesicle trafficking
3. Building organelles by helping to transport their components
4. Enabling cellular appendages and cells themselves to move
5. Connecting cells to each other
6. Secreting and taking up neurotransmitters in nerve cells

how they aggregate into larger structures. Other proteins connect these components to each other, creating the meshwork that provides the cell's strength and ability to resist forces, which maintains shape. Table 4.3 lists some functions of the cytoskeleton, and the following sections discuss the three major components.

Microtubules Are Built of Tubulin

All eukaryotic cells contain long, hollow microtubules that provide many cellular movements. A microtubule is composed of pairs (dimers) of a protein, called tubulin, assembled into a hollow tube 25 nanometers in diameter (fig. 4.14). The cell can change the length of the tubule by adding or removing tubulin molecules.

Cells contain both formed microtubules and individual tubulin molecules. When the cell requires microtubules to carry out a specific function—dividing, for example—the free tubulin dimers self-assemble into more tubules. After the cell divides, some of the microtubules dissociate into individual tubulin dimers. This replenishes the cell's supply of building blocks. Cells are in a perpetual state of flux, building up and breaking down microtubules. Some drugs used to treat cancer affect the microtubules that pull a cell's duplicated chromosomes apart, either by preventing tubulin from assembling into microtubules, or by preventing microtubules from breaking down into free tubulin dimers. In each case, cell division stops.

Microtubules also form locomotor organelles, which move or enable cells to move. The two types of locomotor organelles are **cilia** and **flagella** (fig. 4.15). Cilia are hairlike structures that move in a coordinated fashion, producing a wavelike motion extending from the surface of the cell (fig. 4.16). An individual cilium is constructed of nine microtubule pairs that surround a central, separated pair and form a pattern described as "9 + 2." A type of motor protein called **dynein** connects the outer microtubule pairs and also links them to the central pair. Dynein molecules use energy from ATP to shift in a way that slides adjacent microtubules against each other. This movement bends the cilium (or flagellum). Coordinated movement of these cellular extensions sets up a wave that moves the cell or propels substances along its surface.

Cilia have many vital functions in animal cells. They beat particles up and out of respiratory tubules, and move egg cells in the female reproductive tract. Some single-celled organisms may have thousands of individual cilia, enabling them to "swim" through water.

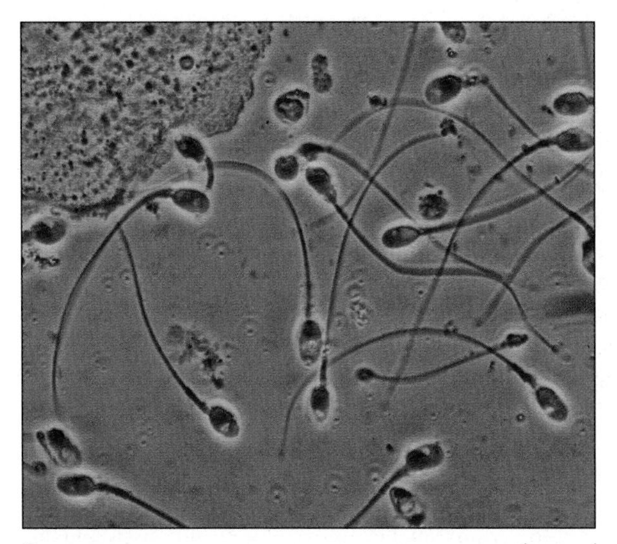

FIGURE 4.15
Cilia and Flagella.
The cilia in (**A**) line the human respiratory tract, where their coordinated movements propel dust particles upward so the person can expel them. The flagella on human sperm cells (**B**) enable them to swim.

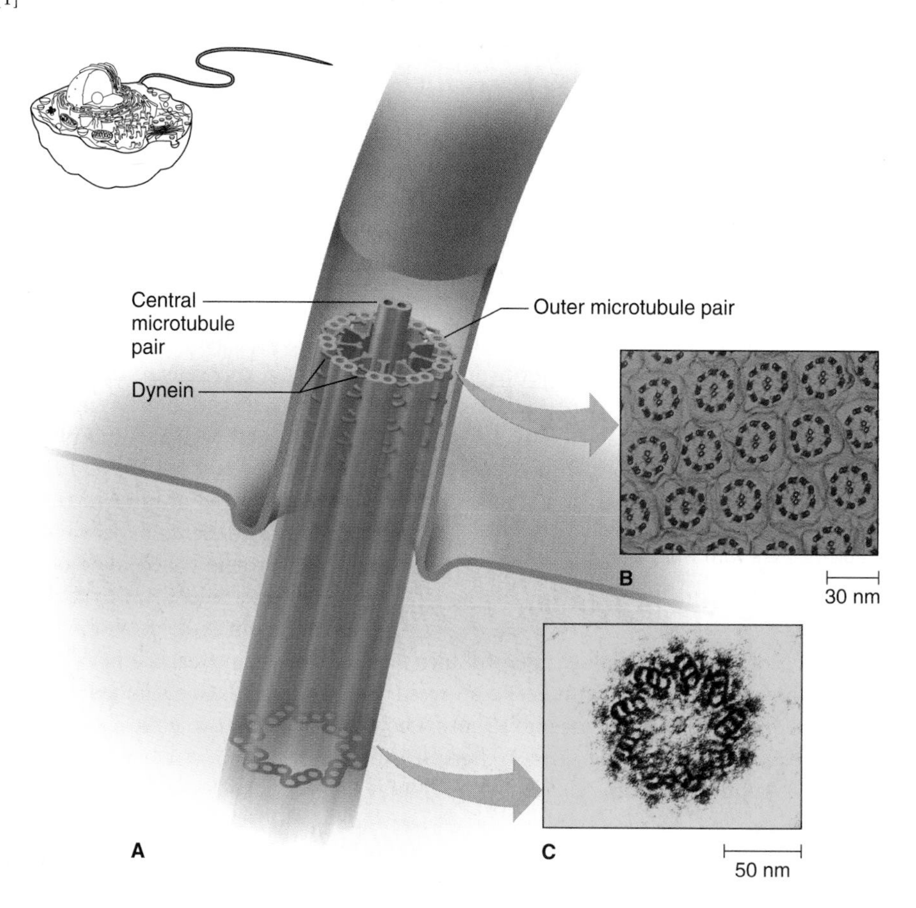

FIGURE 4.16
Microtubules Move Cells. The microtubules that form cilia and the cytoskeletons of flagella have a characteristic "9 + 2" organization. (**A**) Dynein joins the outer microtubule doublets to each other and to the central pair of microtubules. (**B**) A transmission electron micrograph showing a cross section of a group of flagella. (**C**) A structure called a basal body anchors the flagellum to the cell.

Cilia tend to be short and numerous, like a fringe. A flagellum, also built of a 9 + 2 microtubule array, is much longer. Flagella are more like tails, and their whiplike movement enables cells to move. Sperm cells in many species have prominent flagella (see fig. 4.15B). A human sperm cell has only one flagellum, but a sperm cell of a cycad (a type of tree) has thousands of flagella.

Microfilaments Are Built of Actin

Another component of the cytoskeleton is the microfilament (fig. 4.17), which is a long, thin rod composed of the protein actin. In contrast to the microtubules, microfilaments are not hollow and are only about 7 nm in diameter. Microfilaments are vital in providing strength for cells to survive the stretching and compression that often occurs in multicellular organisms. They also help to anchor one cell to another and provide many other functions within the cell through proteins that interact with actin.

One of the best-understood actin-binding proteins is myosin, which uses energy from ATP to move actin filaments. Bundles of myosin slide microfilaments toward each other in muscle cells, generating the motion of muscle contraction (figs. 4.17 and 34.15). Other varieties of myosin-actin interactions move components within the cytoplasm, help cells to move, and distribute the cytoplasm during cell division.

FIGURE 4.17
Muscle Acts by Protein Interactions. Actin microfilaments are interspersed with thicker myosin filaments in muscle tissue. The binding of myosin to actin, followed by a change in the shape of myosin, provides the force behind muscle movement.

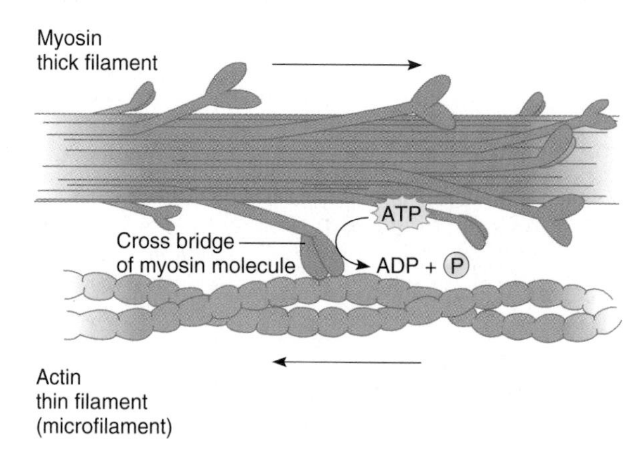

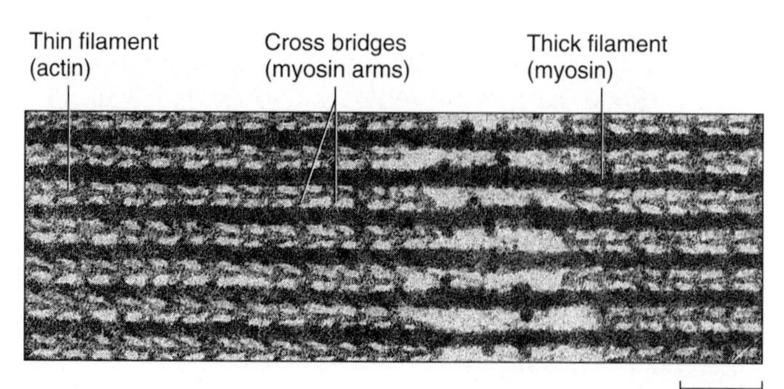

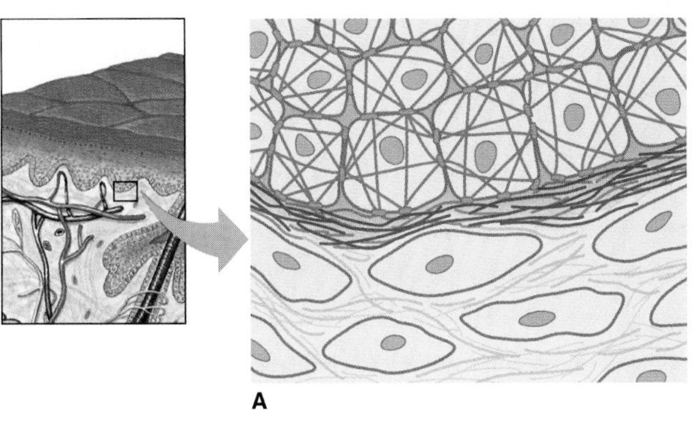

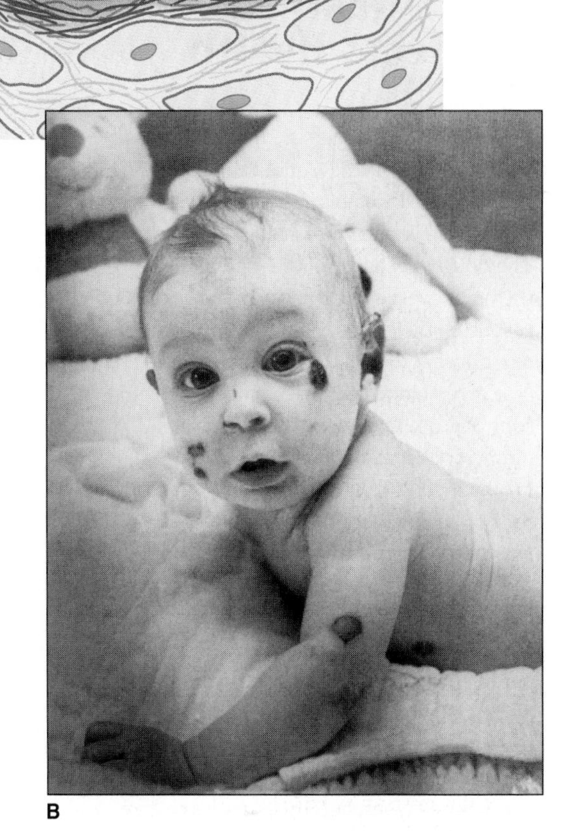

FIGURE 4.18
Intermediate Filaments in Skin.
Keratin intermediate filaments internally support cells in the basal (bottom) layer of the epidermis (**A**). Abnormal intermediate filaments in the skin cause epidermolysis bullosa, a disease characterized by the ease with which skin blisters (**B**).

Intermediate Filaments Provide Scaffolding

Intermediate filaments are so named because their 10 nanometer diameters are intermediate between those of microtubules (25 nm) and microfilaments (7 nm). Unlike microtubules and microfilaments, which consist of a single protein, intermediate filaments are made of different proteins in different specialized cell types. However, all intermediate filaments share a common overall organization of dimers entwined into nested coiled rods (see fig. 4.14). In humans, intermediate filaments comprise only a small part of the cytoskeleton in many cell types, but they are very abundant in skin cells and nerve cells. They form an internal scaffold in the cytoplasm and resist mechanical stress, both functions that maintain a cell's shape.

Intermediate filaments in actively dividing skin cells in the bottommost layer of the epidermis form a strong inner framework (fig. 4.18). It maintains the cells' shapes and firmly attaches them to each other and to the underlying tissue, which is important as this tissue forms a barrier. In an inherited condition called epidermolysis bullosa, keratin intermediate filaments are abnormal, causing the skin to blister easily as tissue layers separate. Intermediate filaments in nerve cells, called neurofilaments, consist of different types of proteins. They surround microtubules and bridge them to each other, stabilizing the extensions of these cells in specific positions, which is important for nerve conduction. The neurofilaments establish and maintain particular diameters of these extensions.

The different components of the cytoskeleton intimately interact through dozens of proteins that interconnect all parts of the cytoskeleton. In multicellular organisms, the cytoskeleton is also important in fostering interactions with noncellular material outside a cell and to other cells. Health 4.1 discusses how the cytoskeleton contributes to the specialized functions of red blood cells, which traverse an endless network of conduits, and muscle cells, which are part of a densely packed, highly active tissue.

4.2 MASTERING CONCEPTS

1. What are some functions of the cytoskeleton?
2. What are the major components of the cytoskeleton?
3. How do the components of the cytoskeleton interact?

OLC

4.3 How Do Cells Interact and Respond to Signals?

Cells must permanently attach to one another to build most tissues; transiently attach to carry out certain functions; and send and receive biochemical messages and respond to them.

All organisms can detect and respond to changes in their environment. A unicellular organism may move toward or away from a particular stimulus. When life became multicellular, about a billion years ago, cell-to-cell communication became crucial too. Such cells "talk" to each other and make contact in thousands of ways. We look now at the structures that join the

Health 4.1

A Disrupted Cytoskeleton Affects Health

Red Blood Cells

Much of what we know about the cohesion between the cell membrane and the cytoskeleton comes from studies of human red blood cells. The doughnut shapes of these cells enable them to squeeze through the narrowest blood vessels on their 300-mile (483-kilometer), 120-day journey through the circulatory system.

A red blood cell's strength derives from rods of proteins called spectrin that form a meshwork beneath the cell membrane (fig. 4.A). Proteins called ankyrins attach the spectrin rods to the membrane. Spectrin molecules are like the steel girders of the red blood cell architecture, and the ankyrins are like the nuts and bolts. If either is absent, the cell's structural support collapses. Abnormal ankyrin in red blood cells causes anemia (too few red blood cells) because the cells balloon out and block circulation and then die. Researchers expect to find defective ankyrin behind problems in other cell types, because it seems to be a key protein in establishing contact between the components of the cell's architecture.

Muscle Cells

Red blood cells and muscle cells are similar in that each must maintain its integrity and shape in the face of great physical force. The red blood cell can dissipate some of the force

FIGURE 4.A

The Red Blood Cell Membrane.

Red blood cells must withstand great turbulent force in the circulation. The cytoskeleton beneath the cell membrane enables these cells to retain their shapes, which are adapted to movement. A protein called ankyrin binds spectrin from the cytoskeleton to the interior face of the cell membrane. On its other end, ankyrin binds a large glycoprotein that helps transport ions across the cell membrane. Abnormal ankyrin causes the cell membrane to collapse—a problem for a cell whose function depends upon its shape.

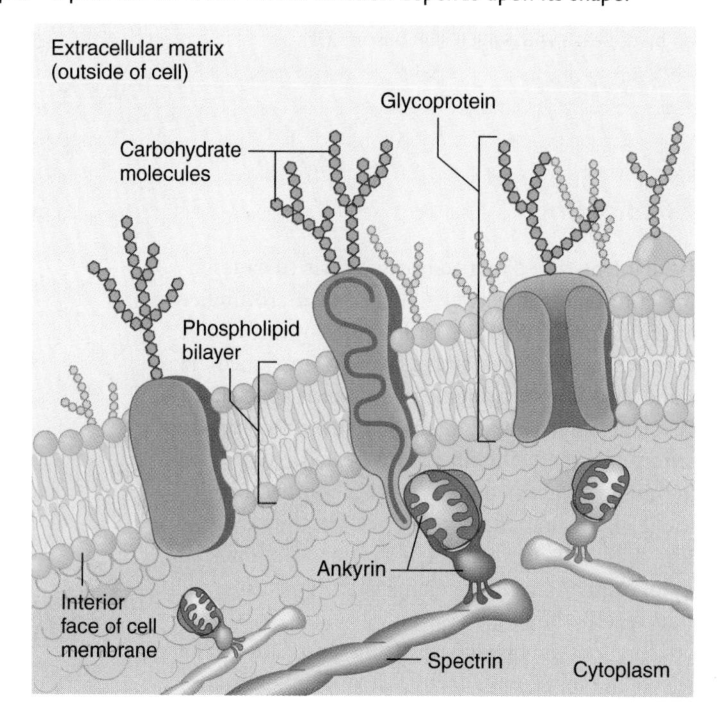

---

cells of multicellular organisms, and then consider two broad types of interactions between cells—cell adhesion and signal transduction.

Animal Cell Junctions Are of Several Types

Animal cells have several types of intercellular junctions (fig. 4.19). In a **tight junction,** the cell membranes of adjacent cells fuse at a localized point. The area of fusion surrounds the two cells like a belt, closing the space between them and ensuring that substances must pass through cells to reach the tissues beyond. Tight junctions join cells into sheets, such as those that line the inside of the human digestive tract.

Extensive tight junctions firmly attach the endothelial cells that form microscopic blood vessels (capillaries) in the human brain, creating some 400 miles of a "blood-brain barrier." This barrier prevents chemical fluctuations, which can damage delicate brain tissue.

The blood-brain barrier readily admits lipid-soluble drugs, because the cell membranes of the endothelium are lipid-rich. The drugs heroin, valium, nicotine, cocaine, and alcohol are lipid-soluble and cross the barrier, which is why they can act so rapidly. Oxygen also enters the brain by directly crossing endothelial cell membranes. Water-soluble molecules must take other routes across the barrier. Insulin moves through the cell by transcytosis. Glucose, amino acids, and iron cross the barrier with the aid of carrier molecules, providing the brain with a constant supply of these nutrients.

because it moves within a fluid. Muscle cells, in contrast, are held within a larger structure—a muscle—where they must withstand powerful forces of contraction, as well as rapid shape changes.

A muscle cell is filled with tiny filaments of actin and myosin, the proteins that slide past one another, providing contraction. A far less abundant protein in muscle cells, dystrophin, is also very important. Dystrophin is critical to muscle cell structure. It physically links actin in the cytoskeleton to glycoproteins that are part of the cell membrane (fig. 4.B). Some of the glycoproteins, in turn, bind to laminin, a cross-shaped protein that is anchored in noncellular material surrounding the muscle cell, called extracellular matrix. By holding together the cytoskeleton, cell membrane, and extracellular matrix, dystrophin and the associated glycoproteins greatly strengthen the muscle cell, which enables it to maintain its structure and function during repeated rounds of contraction. The muscle weakness disorders called muscular dystrophies result from missing or abnormal dystrophin or the glycoproteins it binds to.

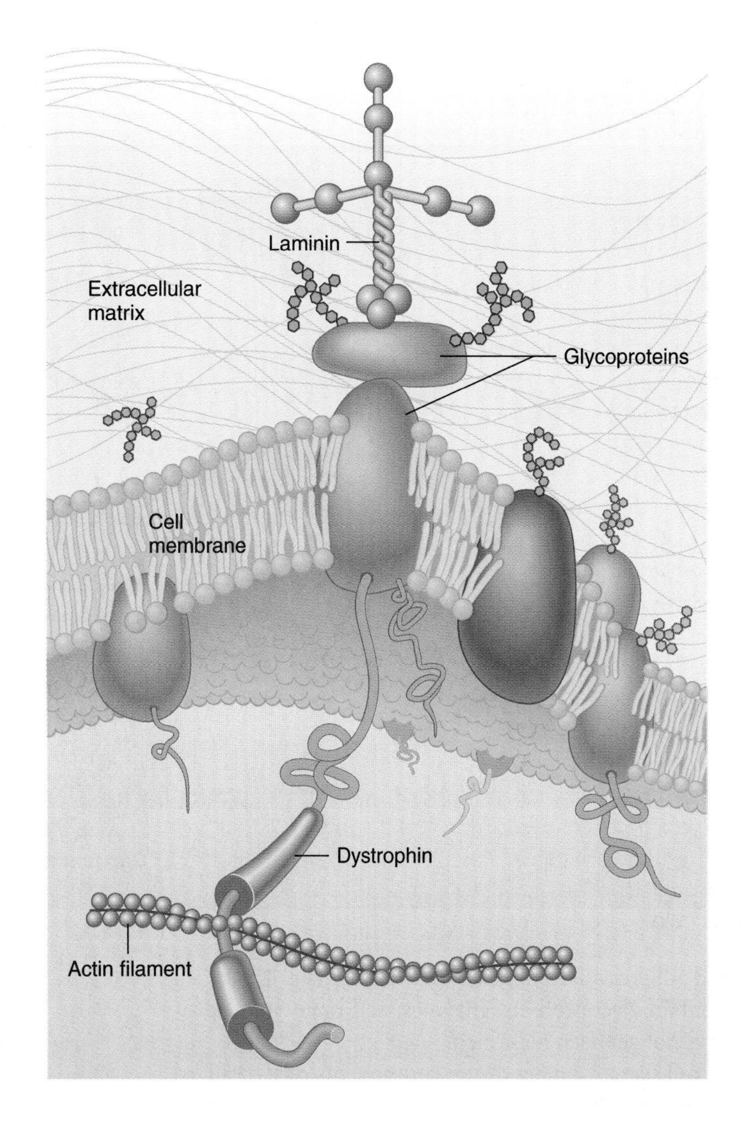

FIGURE 4.B
Cells Need Anchors.
Dystrophin is a membrane protein that stabilizes muscle cell structure by linking the cytoskeleton to the cell membrane. (Actin is present in the cytoskeleton of all cells and is also one of the two major contractile proteins of muscle.)

Another type of intercellular junction in animal cells, called a **desmosome,** links intermediate filaments in adjacent cells in a single spot on both cells (fig. 4.19). Desmosomes hold skin cells in place. A type of connection called a **gap junction** links the cytoplasm of adjacent cells, allowing exchange of ions, nutrients, and other small molecules. Gap junctions join heart muscle cells as well as muscle cells that line the digestive tract.

Cell Walls Are Dynamic Structures

Cell walls surround the cell membranes of nearly all bacteria, archaea, fungi, algae, and plants. Their name is misleading—they are not just barriers that serve only to outline the cells within. Cell walls do impart shape and regulate cell volume, but they also interact with other molecules, helping to determine how a cell in a complex organism specializes.

Cell walls are built of different components in different types of organisms, and may also vary in composition in different parts of the same cell, or at different times in development. Much of the plant cell wall consists of cellulose molecules aligned to form microfibrils that are 10 to 25 nanometers in diameter. Recall that cellulose is a glucose polymer (see fig. 2.15). The microfibrils, in turn, aggregate and twist to form macrofibrils, with diameters of 0.5 micrometers. This fibrous organization imparts great strength, which is increased by molecules of hemicellulose, another polysaccharide, that forms hydrogen bonds to the cellulose microfibrils. Another type of polysaccharide, called pectin, absorbs and holds water, which adds flexibility to the overall

FIGURE 4.19

Animal Cell Connections.

Tight junctions fuse neighboring cell membranes, desmosomes form "spot welds," and gap junctions allow small molecules to move between the cytoplasm of adjacent cells.

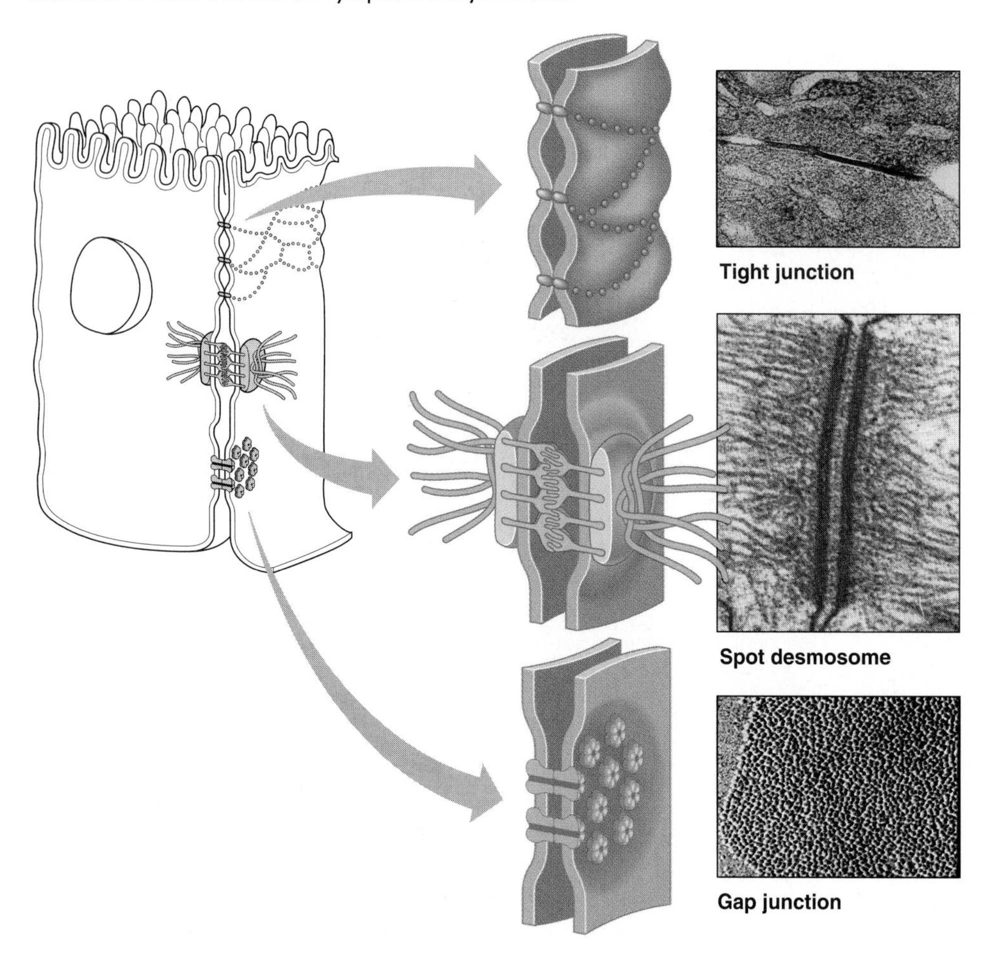

structure and adheres adjacent cells to each other (fig. 4.20). Cell walls also contain glycoproteins, enzymes, and many other proteins that have not yet been identified.

A plant cell secretes many of the components of its wall, so the older layer of a cell wall is on the exterior of the cell, whereas newer layers hug the cell membrane. Some cells have secondary cell walls beneath the initial ones. The region where the walls of adjacent cells meet is called the **middle lamella.** To facilitate cell-to-cell communication and coordination of function, connections of the cell membrane, called **plasmodesmata,** link plant cells. Plasmodesmata form "tunnels" through which the cytoplasm and some of the organelles of one plant cell can interact with those of another (fig. 4.20). They are particularly plentiful in parts of plants that conduct water or nutrients and in cells that secrete oils and nectars.

Experiments reveal the important role of the cell wall in determining cell specialization. Consider a simple organism, the brown alga *Fucus.* It has two cell types that form two tissues: a rootlike rhizoid, and fronds. If the wall is stripped from a *Fucus* cell, it can specialize as either rhizoid or frond, as if its developmental fate has been reset. A cell with its wall intact can become only one type. In plants, whether a cell specializes as root, shoot, or leaf depends upon which cell walls it touches, revealing that a wall affects cells other than the one that it surrounds. Cell walls may therefore help to control the cell-cell interactions that determine how some multicellular organisms develop.

Table 4.4 summarizes intercellular junctions.

Cell Adhesion Directs Cell Movements

Cells stick to each other through a process called adhesion, which is a precise sequence of interactions between proteins that bring cells into contact. One well-studied example of cell adhesion is inflammation—the painful, red swelling at a site of injury or infection. In inflammation, white blood cells move from the circulation to an endangered body part, where they squeeze between the cells of the blood vessel walls to reach the site of injury or infection.

Cellular adhesion molecules (CAMs) help guide white blood cells. Different types of CAMs act in sequence (fig. 4.21). First, CAMs called **selectins** slow the white blood cell they cling to by also binding to carbohydrates on the capillary wall. This has the effect of slowing the cell from moving at 2,500 micrometers per second to a more leisurely 50 micrometers per second. Next, clotting blood, bacteria, or decaying tissue release chemicals that signal the white blood cell to stay and also activate a second type of CAM, called an **integrin.** An integrin links the white blood cell to a third type of CAM, called an **adhesion receptor protein,** that extends from the capillary wall at the injury site. Both types of CAMs then pull the white blood cell between the blood vessel lining cells to the other side, where the damage is.

Cell adhesion is critical to many functions. CAMs guide cells surrounding an embryo to grow toward maternal cells and form the placenta, the supportive organ linking a pregnant woman to the fetus. Sequences of CAMs help establish the connections between nerve cells that underlie learning and memory.

Defects in cell adhesion affect health. Consider the plight of a young woman named Brooke Blanton. She first experienced the effects of faulty cell adhesion as an infant, when her teething sores did not heal. These and other small wounds never accumulated pus, which consists of bacteria, cellular debris, and white blood cells and is a sign of the body fighting infection. Doctors eventually diagnosed Brooke with a newly recognized disorder called leukocyte-adhesion deficiency. Her body lacks the CAMs that

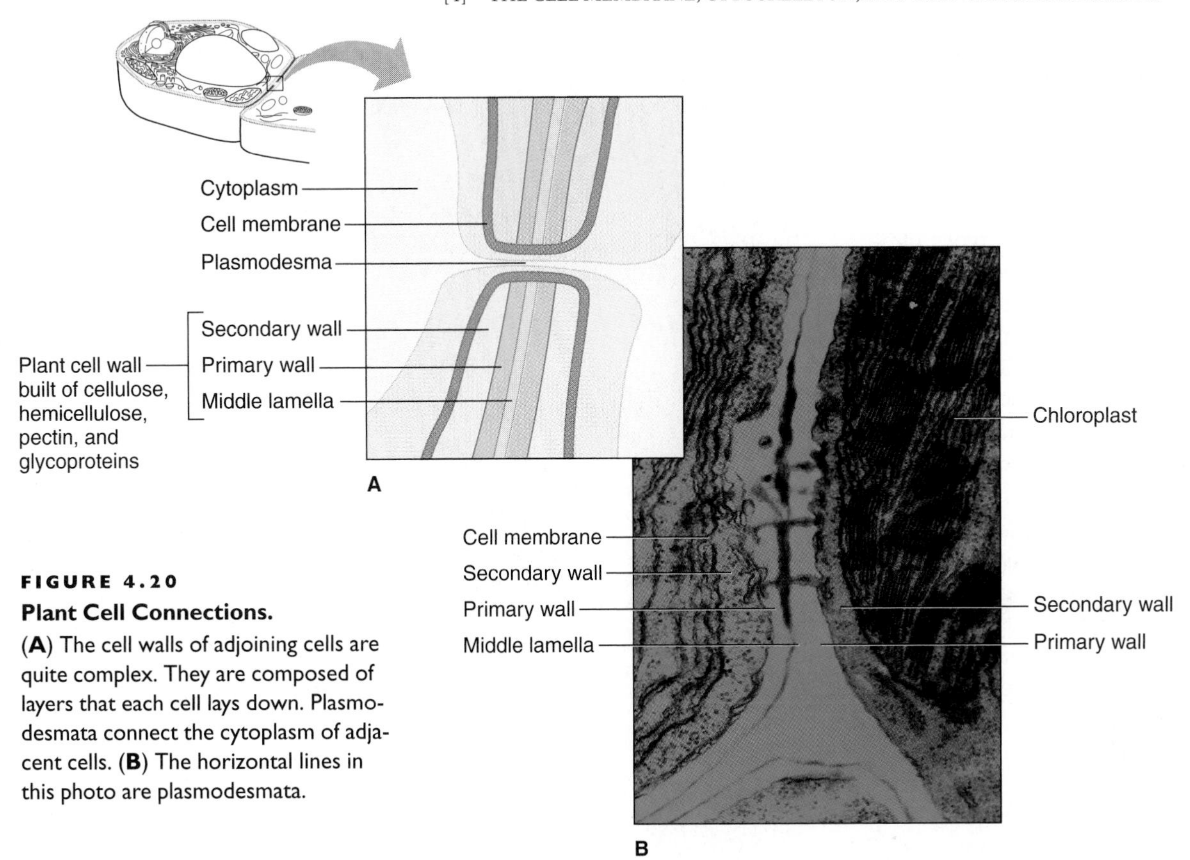

FIGURE 4.20
Plant Cell Connections.
(**A**) The cell walls of adjoining cells are quite complex. They are composed of layers that each cell lays down. Plasmodesmata connect the cytoplasm of adjacent cells. (**B**) The horizontal lines in this photo are plasmodesmata.

enable white blood cells to stick to blood vessel walls. As a result, her blood cells move right past wounds. Brooke must avoid injury and infection, and receive antiinfective treatments for even the slightest wound. More common disorders may reflect abnormal cell adhesion. Lack of cell adhesion eases the journey of cancer cells from one part of the body to another. Arthritis may occur when white blood cells are reined in by the wrong adhesion molecules and inflame a joint where there isn't an injury.

Signal Transduction Mediates Messages

The process by which cells receive information from the outside, amplify it, and then respond, is generally called **signal transduction.** It is an ancient process, because all organisms do it, and many use the same molecules. Plants, fungi, and animals, for example, all use nitric oxide (NO) as one of many signaling molecules (see chapter 2 opening essay).

Organisms of all complexities receive and respond to signals. Bacteria move toward or away from changes in light intensity or the concentration of a particular chemical. A bacterium can detect the presence of a nutrient, and then produce the enzymes required to tap its energy. Environmental extremes can stimulate a bacterium to encase itself in a protective spore, until better conditions arise. Signal transduction in the slime mold *Dictyostelium discoidium,* a protist, provides a glimpse into one way that life may have evolved from the unicellular to the multicellular. When its bacterial food is abundant, the organism exists as single cells. When food becomes scarce, the single cells begin producing cyclic adenosine monophosphate (cAMP), which causes the cells to stream toward each other, and aggregate to form a slug, which can move, possibly finding food (see chapter 21).

The cell membrane is the key site of the chemical interactions that underlie signal transduction. Proteins embedded in the membrane that extend from one or both faces are crucial to the process. The first type of protein in the signal transduction cascade is a receptor, which directly binds to a specific ligand molecule, which is known as the first messenger. The responding receptor contacts a nearby protein, called a regulator, which is the second protein in the pathway. Next, the regulator protein activates a nearby enzyme. The product of the reaction that the

TABLE 4.4 Intercellular Junctions

TYPE	FUNCTION	LOCATION
Tight junctions	Close spaces between animal cells by fusing cell membranes	Inside lining of small intestine
Desmosomes	Spot weld adjacent animal cell membranes	Outer skin layer
Gap junctions	Form channels between animal cells, allowing exchange of substances	Muscle cells in heart and digestive tract
Plasmodesmata	Allow substances to move between plant cells	Weakened areas of cell walls

enzyme catalyzes is called the second messenger, and it lies at the crux of the entire process. The second messenger triggers the cell's response, typically by activating certain genes or enzymes. cAMP is a very common second messenger that mediates many types of messages in cells. Because a single stimulus can trigger production of many second messenger molecules, signal transduction amplifies the incoming information.

Viagra, a drug used to treat erectile dysfunction (impotence), works by altering signal transduction. The drug binds to cyclic guanosine monophosphate (cGMP), a second messenger similar to cAMP. Viagra blocks binding of an enzyme that normally breaks down cGMP. With cGMP around longer, its effect of relaxing the muscle layer of blood vessels in the penis is sustained. The organ remains filled with blood longer.

Cell signaling coordinates the activation of dozens of related processes within a cell by producing groups of molecules that turn on and amplify the effects of other molecules. Signal transduction is extremely complex, with many steps and variants. Figure 4.22 presents a generalized view of the process.

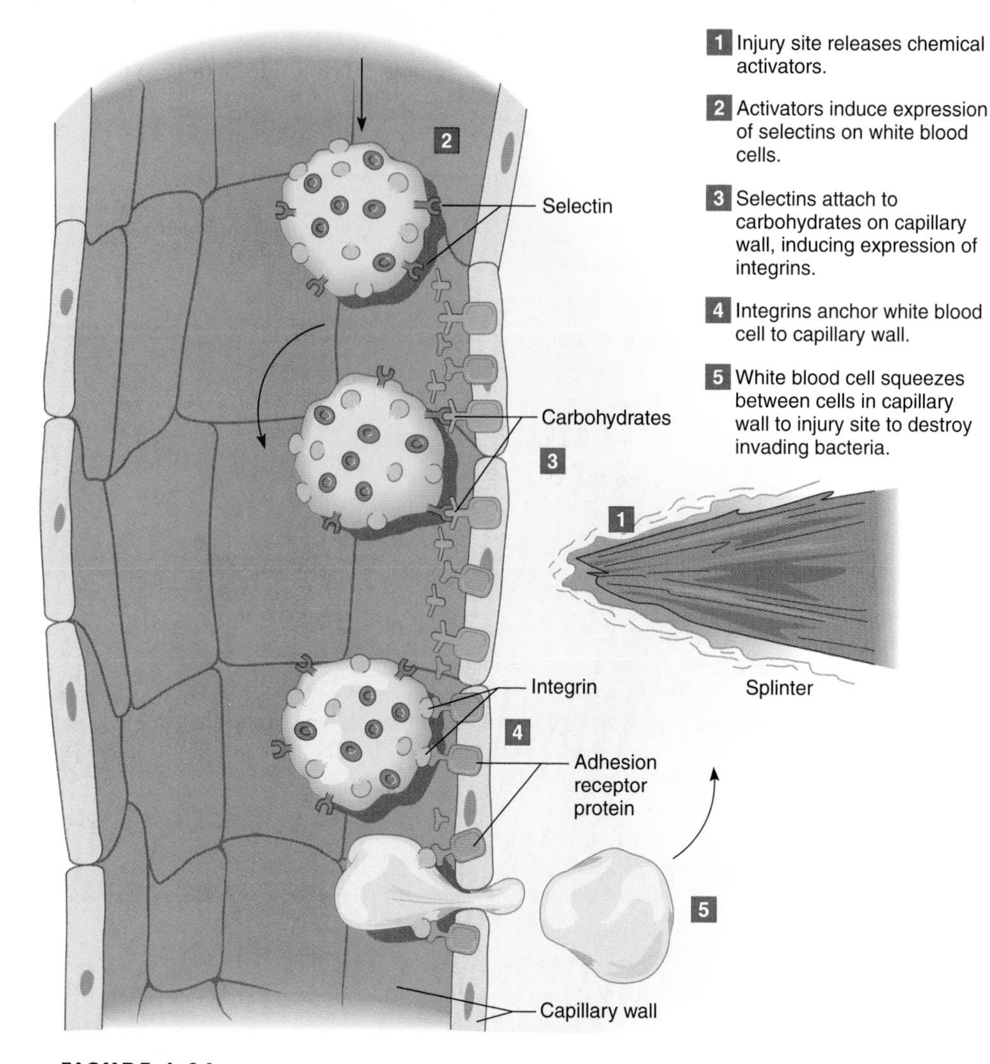

FIGURE 4.21

Adhesion Aids in Defenses.

Cellular adhesion molecules (CAMs), including selectin, integrin, and receptors, direct white blood cells to injury sites.

4.3 MASTERING CONCEPTS

1. What types of structures link cells in animals?
2. What is the composition of a plant cell wall?
3. What functions do cell walls provide?
4. How do cellular adhesion molecules direct a white blood cell to an injury site?
5. How do cells transduce signals?

OLC

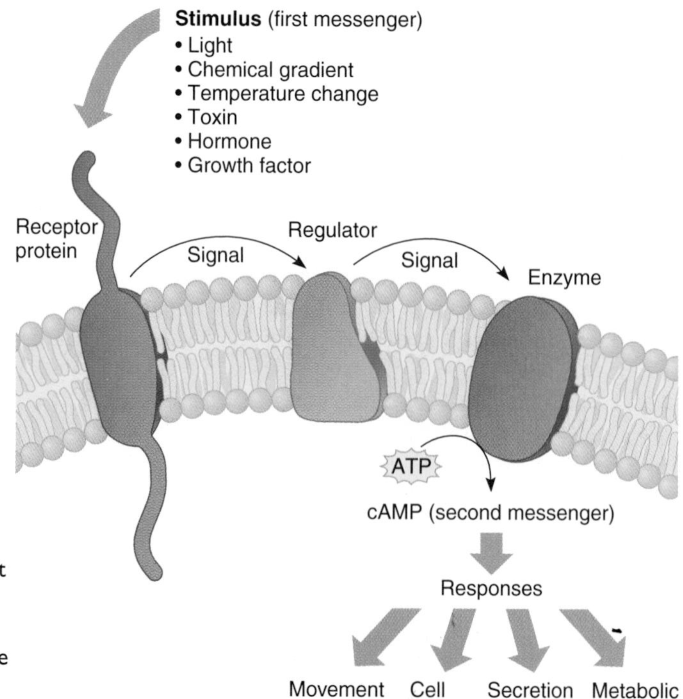

FIGURE 4.22

Signal Transduction.

A first messenger binds a receptor, triggering a cascade of biochemical activity at the cell's surface. An enzyme, adenylate cyclase, catalyzes a reaction inside the cell that circularizes ATP to cyclic AMP, the second messenger. cAMP stimulates various responses, such as cell division, secretion, metabolic changes, and muscle contraction. The splitting of ATP also provides energy.

Chapter Summary

4.1 How Does the Cell Membrane Control Cell Function?

1. The cell surface is a selective interface between the cell and the outside environment. It receives and transmits incoming messages, controls which substances enter and leave the cell, and mediates attachments to and interactions with other cells and extracellular material.
2. The features of a cell's surface identify it as belonging to a particular species, individual, and tissue. The surface consists of molecules embedded in and extending from the cell membrane.
3. A biological membrane consists of a **phospholipid bilayer** embedded with movable proteins, glycoproteins, and glycolipids. The percentage and distribution of membrane proteins varies in different cell types. Membrane proteins carry out a variety of functions.
4. Substances cross cell membranes in several ways. In **diffusion,** a molecule passes through openings in a membrane following its **concentration gradient. Osmosis** is the simple diffusion of water across a semipermeable membrane. Terms describing **tonicity (isotonic, hypotonic, hypertonic)** predict whether cells will swell or shrink when the surroundings change. Cells are adapted to maintain shape when solute concentration changes.
5. A carrier protein transports a substance passively (without energy) or actively (with energy). In **cotransport,** a protein carries one substance down its concentration gradient, as well as a second substance.
6. In **exocytosis,** vesicles inside the cell carry substances to the cell membrane, where they fuse with the membrane and release the cargo outside. In **endocytosis,** molecules are brought into the cell by a vesicle in the cell membrane. **Endosomes** are vesicles that shuttle substances within cells. **Receptor-mediated endocytosis** is more specific. Within cells, proteins guide vesicles to particular organelles. Substances cross cells, entering by endocytosis and exiting by exocytosis, in **transcytosis.**

4.2 How Does the Cytoskeleton Support a Cell?

7. The **cytoskeleton** is a network of rods and tubules that provides cells with form, support, and the ability to move.
8. **Microtubules** self-assemble from hollow **tubulin** subunits to form **cilia, flagella,** and the spindle fibers that separate chromosomes during cell division. Some microtubules have a characteristic 9 + 2 configuration. **Dynein** causes adjacent microtubules to slide, which moves the overall structure.
9. **Microfilaments** are solid and smaller than microtubules. They are composed of the protein **actin** and provide contractile motion when they interact with **myosin.**
10. **Intermediate filaments** are intermediate in diameter between microtubules and microfilaments. They consist of entwined dimers of various proteins. They strengthen the cytoskeleton.

4.3 How Do Cells Interact and Respond to Signals?

11. Junctions connecting animal cells include **tight junctions, desmosomes,** and **gap junctions.** Tight junctions create a seal between adjacent cells. Desmosomes anchor cells in place. Gap junctions allow adjacent cells to exchange cytoplasmic material.
12. Most organisms other than animals have cell walls, which provide shape and mediate signals. Plant cell walls consist of cellulose fibrils connected by hemicellulose, plus pectin and various proteins. **Plasmodesmata** are continuations of cell membranes between cells through thinned parts of the cell wall.
13. **Cellular adhesion molecules** enable cells to contact each other in precise steps that carry out a particular function.
14. In **signal transduction,** receptors in the cell membrane receive input from **first messengers** and transmit the messages through a series of membrane proteins. Eventually this signaling activates a **second messenger,** which stimulates the cell to carry out a specific function.

Testing Your Knowledge

1. Why are some substances able to cross a cell membrane easily, and some not?
2. What types of chemicals comprise cell membranes?
3. Explain the differences among diffusion, facilitated diffusion, active transport, and endocytosis.
4. List five functions of the cytoskeleton.
5. List two functions of
   a. microtubules
   b. microfilaments
   c. intermediate filaments
6. Describe how cells use junctions in different ways.
7. Why are CAMs important?
8. What is signal transduction?

Thinking Scientifically

1. How does each of the following processes illustrate the interaction of cell components?
   a. maintaining the integrity of the red blood cell membrane
   b. the ability of muscle cells to withstand the force of contraction
   c. signal transduction
   d. cell adhesion in leukocyte trafficking
2. Describe how dynein and dystrophin are vital for the functioning of certain cells, even though they are not very abundant.
3. Liver cells are packed with glucose. What mechanism could be used to transport more glucose into a liver cell? Why would only this mode of transport work?
4. A drop of a 5% salt (NaCl) solution is added to a leaf of the aquatic plant *Elodea.* When the leaf is viewed under a microscope, colorless regions appear at the edges of each cell as the cell membranes shrink from the cell walls. What is happening to these cells?
5. Would a substance that destroys the integrity of the blood-brain barrier be dangerous? Why or why not?

References and Resources

Brown, Michael S., and Joseph L. Goldstein. April 6, 1986. A receptor-mediated pathway for cholesterol homeostasis. *Science,* vol. 232. A classic paper tracing how liver cells handle dietary cholesterol.

Lewis, Ricki. March 20, 2000. Unraveling complex carbohydrates. *The Scientist,* vol. 14. p. 16. Intricate carbohydrate assemblies build cell walls.

Sansom, Mark S. P. January 30, 1997. Structure of molecular hole-punch. *Nature,* vol. 385. Bacterial toxins harm humans by punching holes in our cells.

Scott, John D., and Tony Pawson. June 2000. Cell communication: The inside story. *Scientific American,* vol. 282, p. 72. An overview of secretion and signal transduction.

The *LIFE* Online Learning Center provides additional resources and tools for studying this chapter.
www.mhhe.com/life

CHAPTER 5

The Energy of Life

Requiring Energy to Get Energy

Swallowing and digesting a meal almost as big as oneself requires a huge energy investment. This African rock python is consuming a Thomson's gazelle.

The African rock python lay in wait for the lone gazelle. When the gazelle came close, the snake moved suddenly, positioning the victim's head and holding it in place while it swiftly entwined its 30-foot-long body snugly around the mammal. Each time the gazelle exhaled, the snake squeezed, shutting down the victim's heart and lungs in less than a minute. Then the swallowing began.

How can a 200-pound snake eat a 130-pound meal? Thanks to adaptations of the reptile's digestive system, the snake can indeed swallow its meal, and because of remarkable control over its energy use, what seems impossible to us is a way of life for the snake.

The snake begins its meal by opening its jaws at an angle of 130 degrees (compared with 30 degrees for the most gluttonous human) and positions its mouth over the gazelle's head, using strong muscles to gradually envelop and push along the carcass. Saliva coats the prey, easing its journey to the snake's stomach. After several hours, the huge meal arrives at the stomach, and the remainder of the digestive tract readies itself for several weeks of dismantling the gazelle. When digestion is completed, only a few chunks of hair will remain to be eliminated.

Each time the gazelle exhaled, the snake squeezed, shutting down the victim's heart and lungs in less than a minute.

Eating a tremendous meal once every few months places great energy demands on the snake. How does the reptile have enough energy left over from its meal to carry on the activities of life?

The snake pays dearly for its meal of a 130-pound gazelle. While most organisms that eat frequently or all the time invest 10 to 23% of a meal's energy in digesting it and assimilating its nutrients, snakes invest a whopping 32% in energy acquisition. The African rock python expends an equivalent of half the energy in the chemical bonds of its meal just to digest it.

Two activities handle the python's meal. First, abundant hydrochloric acid (HCl) must be present in the stomach for several weeks to lower the pH sufficiently for digestive enzymes to function. To continually produce HCl, the snake must maintain an energy level equivalent to that of an active greyhound or racehorse, but far beyond the few minutes it takes these animals to run a race. To do this, as soon as the snake's jaws fit around the gazelle's head, oxygen consumption increases some 36-fold. As we'll see in chapter 7, oxygen is critical for extracting the maximal amount of energy from food.

Producing copious HCl and supporting rapidly expanding intestines greatly taps the snake's energy reserves.

The second digestive/energy adaptation of the rock python affects the intestines, where digested nutrients are absorbed into the bloodstream. After eating, literally overnight, this organ doubles or triples in weight! The stimulation for rapid cell division comes from the expanding stomach, which sends hormonal and nervous signals to the intestines. The blooming intestines can produce 60 times more digestive enzymes than they do during the many months between meals.

Enormous energy is required to support this sudden burst of intestinal activity. To study how the digestive tract handles the load, researchers grew patches of small intestine from a Burmese python and a sidewinder rattlesnake in the laboratory and applied amino acids tagged with radioactive atoms to trace their fate. The investigators found that within a day, the small intestine's ability to absorb the amino acids increased 6-fold, and that after 3 days, it had increased 16-fold.

Producing copious HCl and supporting rapidly expanding intestines greatly taps the snake's energy reserves. The reptile can do this, though, by shutting down its digestive tract between meals. HCl level plummets, and the intestinal lining shrinks.

5.1 What Is Energy?

All life processes require energy, from a cell dividing or secreting, to cells migrating in an early embryo to build tissues, to a flower blooming, to a bird in flight. Organisms have strategies for harnessing and utilizing energy.

A tortoiseshell butterfly expends a great deal of energy to stay alive. Not only does it fly from flower to flower, but it migrates seasonally over long distances. A less obvious way the insect uses energy is to power the many biochemical reactions of its metabolism. The butterfly obtains this energy from nectar that a flower produces (fig. 5.1); the flowering plant gets its energy from the sun. All animals ultimately extract life-powering energy from plants and from certain microorganisms that capture the energy in sunlight (or certain inorganic chemicals) and convert it into the chemical energy of organic compounds. Because energy is so basic to life, we need to understand just what it is and how organisms use it.

The term *energy* was coined about two centuries ago, when the Industrial Revolution redefined familiar ideas of energy from the power behind horse-drawn carriages and falling water to the power of the internal combustion engine. As people began to think more about harnessing energy, biologists realized that understanding energy could not only improve the quality of life but reveal how life itself is possible. **Bioenergetics** is the study of how living organisms use energy to perform the activities of life.

Energy is the ability to do work—that is, to change or move matter against an opposing force, such as gravity or friction. Because energy is an ability, it is not as tangible as matter, which has mass and takes up space. Energy comes in different forms that can be converted from one to another. Electrons can carry energy. In living systems, **oxidation-reduction reactions** transfer electrons from one molecule to another. Molecules that lose electrons are **oxidized,** whereas those that receive electrons are **reduced.**

FIGURE 5.1
It Takes Energy to Get Energy.
A butterfly expends great amounts of energy flying from flower to flower to obtain food and migrating sometimes thousands of miles. This tortoiseshell butterfly is feeding on nectar from a flower.

Calories are units used to measure energy. A **calorie** (cal) is the amount of energy required to raise the temperature of 1 gram of water from 14.5°C to 15.5°C. The most common unit for measuring the energy content of food and the heat output of organisms is the kilocalorie (kcal), which is the energy required to raise the temperature of a kilogram of water 1°C. The kilocalorie equals 1,000 calories. (Dietary calories are kilocalories.)

The energy transformations that sustain life are similar in all organisms. The most important of these pathways are cellular respiration and photosynthesis, and they are intimately related.

The *energy-requiring* stage of biological energy acquisition and utilization is usually photosynthesis, the subject of chapter 6. During this process, chlorophyll molecules (or similar molecules in certain microorganisms) absorb light energy. They use this energy to reduce carbon dioxide (a low-energy compound) to carbohydrate (a high-energy compound). Oxygen is released as a by-product. Carbohydrate, in turn, fuels the activities of the plant and ultimately all other organisms. During **cellular respiration,** the *energy-releasing* stage of the biological energy process, energy-rich carbohydrate molecules are oxidized to carbon dioxide and water. Cellular respiration liberates the energy necessary to power life. Chapter 7 explores cellular respiration.

Where Does Energy Come From?

Most organisms obtain energy from the sun (fig. 5.2), either directly through photosynthesis or indirectly by consuming other organisms. Even the energy in fossil fuels and in organisms originated as solar energy.

Deep within the sun, temperatures of 10,000,000°C fuse hydrogen atoms, forming helium and releasing electrons and **photons,** which are particles of light energy that travel in waves (see chapter 6). Life depends upon the ability to transform this solar energy into chemical energy before it is converted to heat—the eventual fate of all energy.

The sun's total yearly energy output is about 3.8 sextillion megawatts of electricity, of which Earth intercepts only about two-billionths. Although this is the equivalent of burning about 200 trillion tons of coal a year, most of this energy doesn't reach life. Nearly a third of the sun's incoming energy is reflected back to space, and another half is absorbed by the planet, converted to heat, and returned to space. Another 19% of incoming solar radiation powers wind and other weather phenomena and drives photosynthesis. Of this 19%, only 0.05 to 1.5% is incorporated into plant material—and only about a tenth of that makes its way into the bodies of animals that eat plants. Far less of the original solar energy reaches animals that consume plant-eating animals. Obviously, life on Earth functions using just a tiny portion of solar energy.

Most organisms depend, directly or indirectly, on organisms such as plants that photosynthesize to acquire usable energy.

FIGURE 5.2

Energy Can Take Many Forms.

Energy flows from the sun and is captured by plants to make energy-rich chemicals. Those chemicals, in turn, provide energy for the plants themselves and the organisms that eat plants. In this way energy is continually flowing through one organism to another.

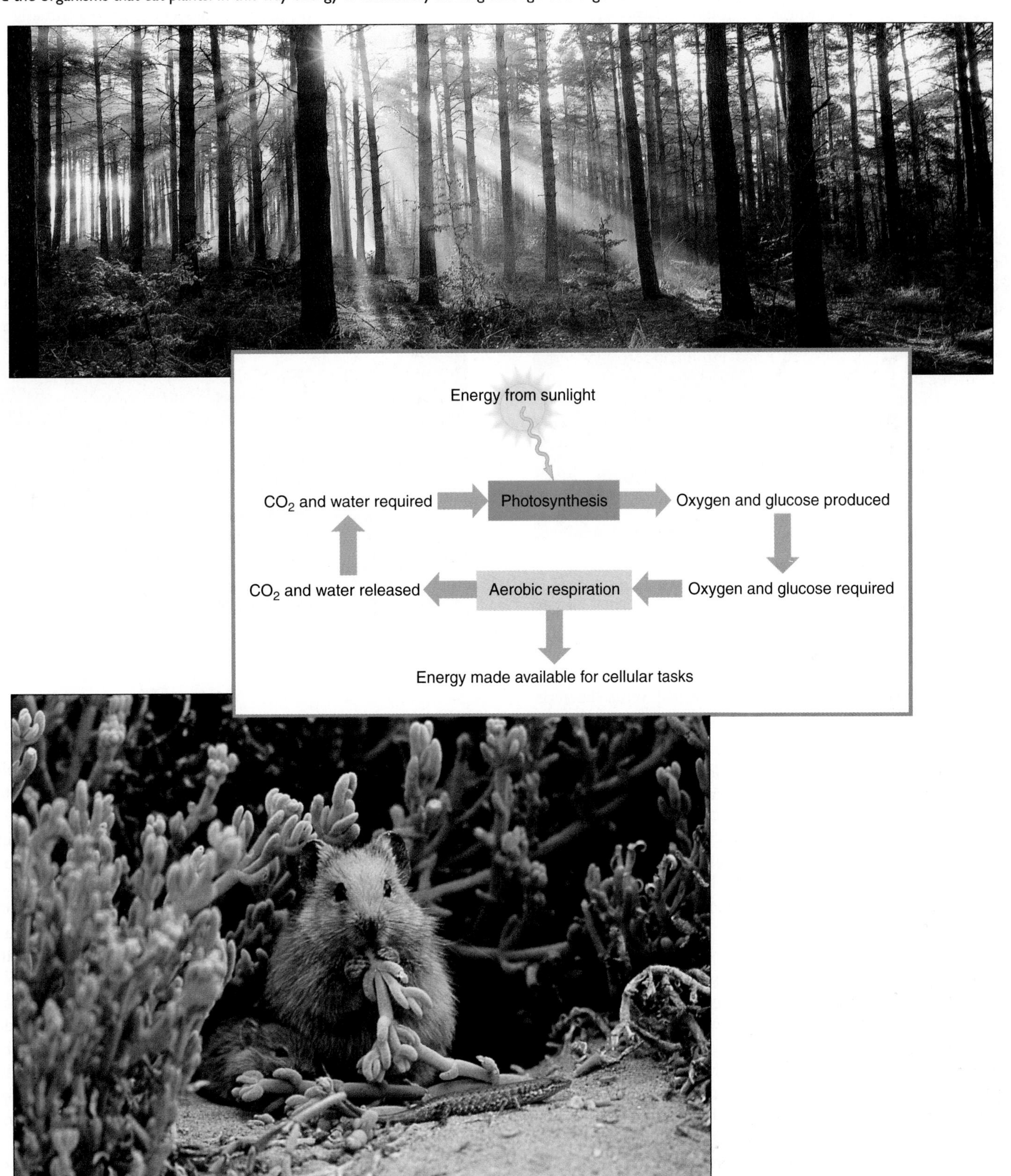

FIGURE 5.3
Geothermal Energy Supports Life.
Chemoautotrophic bacteria support life in deep-sea hydrothermal vents. The bacteria derive energy from inorganic chemicals, such as hydrogen sulfide, that come from minerals. Bacteria, in turn, support other organisms, such as crabs, clams, and the tube worms shown here.

However, a few species can obtain and use geothermal energy and energy from certain inorganic chemicals. Bacteria thrive around dark cracks in the ocean floor where hot molten rock seeps through, creating an intensely hot environment (fig. 5.3). These areas are called deep-sea hydrothermal vents. Bacteria there extract energy from inorganic chemicals surrounding them, specifically from the chemical bonds of hydrogen sulfide (H_2S), a gas present at the vents. The bacteria then add the freed hydrogens to carbon dioxide to synthesize organic compounds that are nutrients for them and the organisms that consume them. Organisms that use inorganic chemicals as the energy source to manufacture nutrient molecules are called **chemoautotrophs.** This term comes from the more general term **autotroph,** which refers to organisms that synthesize their own nutrients.

Energy Is Potential or Kinetic

To stay alive, an organism must continually take in energy through eating or from the sun or the earth and convert it into a usable form. Every aspect of life centers on converting energy from one form to another. In life, two basic types of energy constantly interconvert: potential energy and kinetic energy.

Potential energy is stored energy available to do work. When matter assumes certain positions or arrangements, it contains potential energy. A teaspoon of sugar, a snake about to strike at its prey, a child at the top of a slide, and a baseball player about to throw a ball illustrate potential energy. In organisms, potential energy is stored in the chemical bonds of nutrient molecules, such as carbohydrates, lipids, and proteins.

FIGURE 5.4
Potential and Kinetic Energy.
(**A**) The ball this pitcher is about to throw has potential energy. (**B**) The living world abounds with illustrations of kinetic energy, the energy of action. Here a chameleon's tongue whips out to ensnare a butterfly.

A

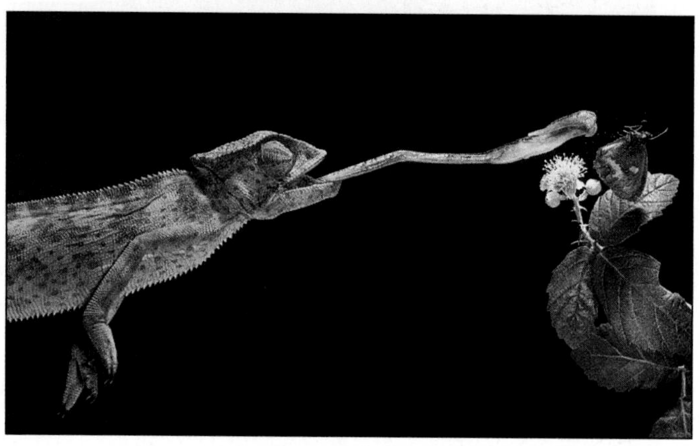

B

Kinetic energy is energy being used to do work. Burning sugar, the snake striking, the child riding down the slide, and the soaring baseball all demonstrate kinetic energy. A chameleon shooting out its sticky tongue to capture a butterfly uses kinetic energy, as does an elephant trumpeting and a Venus's flytrap closing its leaves around an insect. The adder snake in figure 1.4 demonstrates potential energy as it awaits the approach of a sand lizard. Somewhere between figures 1.4A and 1.4B, the snake strikes using kinetic energy to grab its lizard meal. Figures 5.4 and 5.5 illustrate potential and kinetic energy.

Kinetic energy transfers motion to matter. The movement in the pitcher's arm transfers energy to the ball, which then takes flight. Similarly, flowing water turns a turbine, and a growing root can push aside concrete to break through a sidewalk. Put another way, kinetic energy moves objects. Heat and sound are types of kinetic energy because they result from the movement of molecules.

Potential energy can be translated into the kinetic energy of motion, and that burst of energy can be used to do work. Under most conditions, potential and kinetic energy are readily interconvertible, although not at 100% efficiency.

FIGURE 5.5

Potential Energy and Kinetic Energy.

Potential energy in food is converted to kinetic energy as muscles push the cyclist to the top of the hill. The potential energy of gravity provides a free ride by conversion to kinetic energy on the other side.

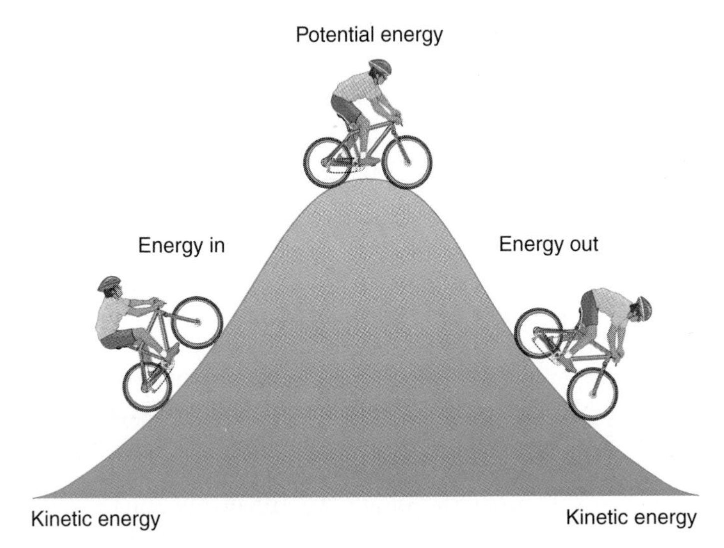

5.1 MASTERING CONCEPTS

1. What is energy?
2. What are some sources of energy?
3. How does solar energy support life?
4. Distinguish between potential and kinetic energy.

5.2 How Do the Laws of Thermodynamics Describe Energy Transfer?

Energy does not arise from nothing: It cannot be created or destroyed, but can change form. The physical laws that explain energy transformation underlie energy use in life, too.

The **laws of thermodynamics** regulate the energy conversions vital for life, as well as those that occur in the nonliving world. These laws operate on a system and its surroundings; the system is the collection of matter under consideration, and the surroundings are the rest of the universe. In figure 5.6, the elephant is the system and its watery background the surroundings. An open system exchanges energy with its surroundings, whereas a closed system is isolated from its surroundings—it does not exchange energy with anything outside the system. Thus, the open system that is the elephant loses heat to its surroundings as it splashes about.

A thermos bottle and its contents illustrate closed and open systems. The uncapped thermos, an open system, loses steam from the heated food to the surrounding air. A closed thermos prevents such an exchange of energy.

The laws of thermodynamics apply to all energy transformations—gasoline combustion in a car's engine, a burning chunk of wood, or a cell breaking down glucose.

What Is the First Law of Thermodynamics?

The laws of thermodynamics follow common sense. The first law of thermodynamics is the law of energy conservation. It states that energy cannot be created or destroyed but only converted to other forms. This means that the total amount of energy in a system and its surroundings remains constant; thus, on a grander scale, the amount of energy in the universe is constant. However, the energy in a living system is constantly changing.

In a practical sense, the first law of thermodynamics explains why we can't get something for nothing. The energy released when a baseball hurtles towards the outfield doesn't appear out of nowhere—it comes from a batter's muscles. Likewise, green plants do not manufacture energy from nothingness; they trap the energy in sunlight and store it in chemical bonds. The energy in sunlight, in turn, comes from nuclear reactions in the sun's matter.

According to the first law of thermodynamics, the amount of energy an organism uses cannot exceed the amount of energy it takes in through the chemical bonds contained in the nutrient molecules of food. Even when starving, an organism cannot use more energy than its tissues already contain. Similarly, the amount of chemical energy that a plant's leaves produce during photosynthesis cannot exceed the amount of energy in the light it has absorbed. No system can use or release more energy than it takes in.

FIGURE 5.6

A System and Its Surroundings.

The elephant obtains energy from the chemical bonds of the nutrient molecules in its food. When the animal moves in the water, it dissipates some energy as heat, sound, and motion into the surroundings.

What Is the Second Law of Thermodynamics?

The second law of thermodynamics concerns the concept of **entropy,** which is a tendency towards randomness or disorder (fig. 5.7). This law states that all energy transformations are inefficient because every reaction results in increased entropy and loses some usable energy to the surroundings as heat. Unlike other forms of energy, heat energy results from random molecule movements. Any other form of energy can be converted completely to heat, but heat cannot be completely converted to any other form of energy. Because all energy eventually becomes heat, and heat is disordered, all energy transformations head towards increasing disorder (entropy). In general, the more disordered a system is, the higher its entropy.

Because of the second law of thermodynamics, events tend to be irreversible (proceed in one direction) unless energy is added to reverse them. Processes that occur without an energy input are termed spontaneous. In natural processes, irreversibility results from the loss of usable energy as heat during energy transformation. It is impossible to reorder the molecules that have dispersed as a result of heat.

The second law of thermodynamics governs cell energetics. Cells derive energy from nutrient molecules and use it to perform such activities as growth, repair, and reproduction. The chemical reactions that transfer this energy are (as the law predicts) inefficient and release much heat. The cells of most organisms are able to extract and use only about half of the energy in nutrients. Although organisms can transform energy—storing it in tissues or using it to repair a wound, for example—ultimately, much of the energy is dissipated as heat because a small amount of energy is lost with each transfer.

FIGURE 5.7
Entropy Represents Disorder.
The destruction of a violent storm symbolizes entropy—extreme disorder. It takes great energy to rebuild, because specific objects must be placed in specific places.

Because organisms are highly organized, they may seem to defy the second law of thermodynamics—but only when they are considered alone, as closed systems. Organisms remain organized because they are *not* closed systems. They use incoming energy and matter, from sources such as sunlight and food, in a constant effort to maintain their organization and stay alive. Although the entropy of one system, such as an organism or a cell, may decrease as the system becomes more organized, the organization is temporary; eventually the system (organism) will die. In addition, life connects or **couples** energy reactions, so that one reaction occurs at the expense of another. In a more general sense, this means that organisms can increase in complexity as long as something else decreases in complexity by a greater amount. Life remains ordered and complex because the sun is constantly decreasing in complexity and releasing energy. The entropy of the universe as a whole is always increasing.

We're familiar with energy transformations that release large amounts of energy at once—explosions, lightning, or a plane taking off. Cellular energy transformations, by contrast, release energy in tiny increments. Cells extract energy from glucose in small amounts at a time via the biochemical pathways of cellular respiration, the subject of chapter 7. If all the energy in the glucose chemical bonds was released at once, it would be converted mostly to heat and produce deadly high temperatures. Instead, cells extract energy from glucose by slowly reducing the overall organization of the molecule in several controlled steps. Each reaction transfers energy from molecule to molecule. Some of this energy is lost as heat, but much of it is stored in the chemical bonds of molecules within the cell.

5.2 MASTERING CONCEPTS

1. On what do the laws of thermodynamics act?
2. What is the first law of thermodynamics?
3. What is the second law of thermodynamics?

5.3 How Do Cells Metabolize Nutrients?

The life of a cell is a complex and continual web of interacting biochemical reactions that build new molecules and dismantle existing ones. Synthesizing the new requires energy; breaking down the old releases energy.

Metabolism, Greek for "change," consists of the chemical reactions that change or transform energy in cells. The reactions of metabolism are organized into step-by-step sequences called **metabolic pathways,** in which the product of one reaction becomes the starting point, or substrate, of another (fig. 5.8). Pathways may branch or form cycles. Enzymes enable metabolic reactions to proceed fast enough to sustain life, a point we return to later in the chapter. Metabolism in its entirety is an enormously complex network of interrelated biochemical reactions organized into chains and cycles.

Substrate Product

A

Substrate Intermediates Product

B

Substrate Intermediate Product Product

C

Substrate

Substrate and product

By-product

D

FIGURE 5.8
Metabolic Pathways.
The chemical reactions of life form straight chains, branched chains, and cycles of metabolic pathways. Depicted are (**A**) an enzyme-catalyzed chemical reaction, (**B**) a straight chain pathway, (**C**) a branched chain pathway, and (**D**) a biochemical cycle. Note that in a cycle, the product of the last reaction is also the starting material of the first. Cycles release by-products that are important in other biochemical pathways.

Building Up and Breaking Down—Anabolism and Catabolism

Metabolism on a familiar, whole-body level, refers to how rapidly an organism "burns" food. On a cellular level, metabolism includes building and breaking down molecules within cells.

Anabolism (synthesis) refers to metabolic pathways that construct large molecules from small ones. Just as erecting a building requires energy (fig. 5.9A), anabolic reactions require energy, because molecules are being built. In anabolism, a small number of chemical subunits join in different ways to produce many different types of large molecules. The pathways of anabolism diverge (branch out) because a few types of precursor molecules combine to yield many different types of products. For example, the 20 different types of amino acids link together in varied sequences to form thousands of different proteins.

FIGURE 5.9
Anabolism Versus Catabolism.
Anabolic reactions (**A**) build complex molecules, like building a barn from bricks and boards, while catabolic reactions (**B**) reverse that process, resulting in smaller components.

Anabolic pathway

Energy in

Energy in

A

Catabolic pathway

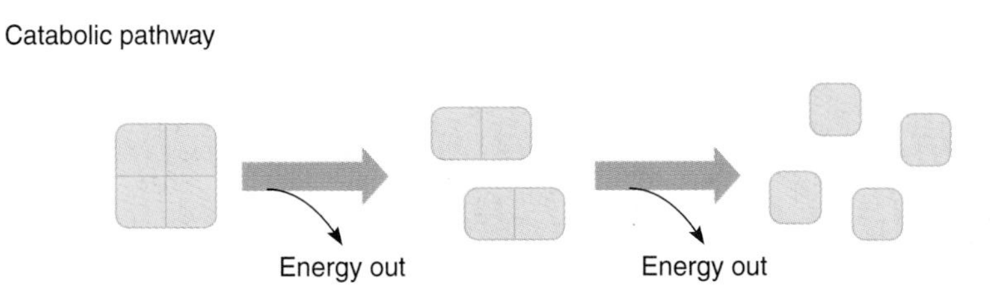

B

Catabolism (degradation) refers to metabolic pathways that break down large molecules into smaller ones. These pathways release energy, just as a collapsing building releases energy in the form of sound, heat, and motion (fig. 5.9B). Catabolic pathways converge (come together) because many different types of large molecules are degraded to yield fewer types of small molecules. Thousands of distinct proteins, for example, are broken down to yield the same 20 types of amino acids. Of the varied foods that organisms consume, all of the nutrients are ultimately broken down or converted to just a few types of building block compounds—glucose, glycerol and fatty acids, and amino acids.

What Happens to Energy in Chemical Reactions?

Each reaction in a metabolic pathway rearranges atoms into new compounds, and each reaction either absorbs or releases energy. According to the first law of thermodynamics, it takes the same amount of energy to break a particular kind of bond as it does to form that same bond—energy is not created or destroyed. The amount of energy stored in a chemical bond (potential energy) is called its **bond energy.** Some chemical bonds are stronger than others—that is, they have greater bond energies. When a stronger bond breaks, it releases more energy. Conversely, forming a stronger bond requires a greater input of energy.

The potential energy of a compound, then, is contained in its chemical bonds. When these bonds break, some of the energy released can be used to do work—for example, to form other bonds. The amount of energy potentially available to form new bonds is called the **free energy** of the reaction that breaks a molecule apart. Chemical reactions change the amount of energy stored in a molecule and potentially available to do work. In these energetic terms, two types of chemical reactions occur—endergonic and exergonic reactions.

In an **endergonic reaction** (energy inward), the products contain more energy than the reactants. Endergonic reactions are not spontaneous—that is, they do not proceed unless there is an input of energy. Entropy decreases in endergonic reactions because the resulting molecules are more complex (fig. 5.10).

An example of an endergonic reaction is combining the monosaccharides glucose and fructose (each $C_6H_{12}O_6$) to form water and the disaccharide sucrose (see fig. 2.13). This synthesis requires energy, and the products of the reaction store more energy in their chemical bonds than do the reactants. Because energy cannot be created or destroyed, this reaction requires an absorption of energy from the surroundings in order for the products to have gained energy.

In an **exergonic reaction** (energy outward), the products contain less energy than the reactants. Energy is released, and the reaction is spontaneous. Entropy increases in exergonic reactions (fig. 5.10).

The breakdown of glucose ($C_6H_{12}O_6$) to carbon dioxide (CO_2) and water (H_2O) is spontaneous and exergonic. The carbon dioxide and water products contain less energy than glucose; some of the energy released can be captured to do work, and the rest is lost as heat.

The reactions of anabolism (synthesis) are endergonic (require energy), and the reactions of catabolism (degradation) are exergonic (release energy).

What Is Chemical Equilibrium?

Most chemical reactions can proceed in both directions—that is, when enough product forms, some of it is converted back to reactants. Arrows between reactants and products going in both directions indicate reversible reactions (fig. 5.11). Such reactions that proceed in both directions reach a point called **chemical equilibrium,** where the reaction goes in both directions at the same rate. However, there may be different amounts of products and reac-

FIGURE 5.10

Endergonic and Exergonic Reactions.

Endergonic reactions store energy, and exergonic reactions release energy.

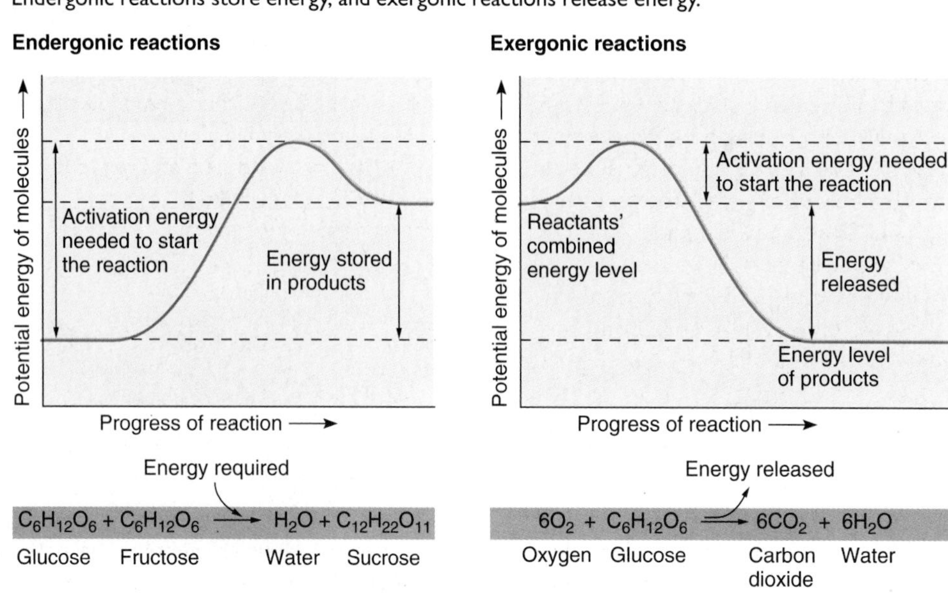

FIGURE 5.11
Chemical Equilibrium.
Reversible chemical reactions proceed toward a "middle point" known as equilibrium. Determined by the concentration of products and reactants, at equilibrium a reaction is equally likely to proceed in either direction. By manipulating these concentrations, living cells can "drive" reactions in a particular direction.

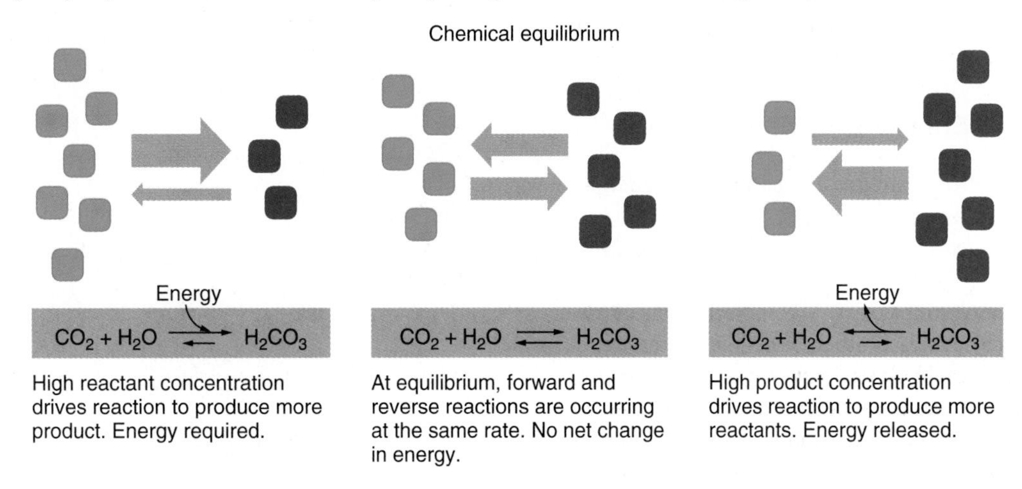

tants at equilibrium—it is their rate of formation that equalizes. At chemical equilibrium, energy is not being gained or lost. When a reaction departs from equilibrium (that is, when either reactants or products accumulate), energy is lost or gained.

Because all activities of life require energy, cells must remain far from chemical equilibrium for their metabolic processes to occur. They do this by continually preventing the accumulation of reactants in metabolic pathways. For example, the large difference in free energy between glucose and its breakdown products, carbon dioxide and water, propels cellular metabolism strongly in one direction—as soon as glucose enters the cell, it is quickly broken down to release energy. This energy allows the cell to stave off equilibrium by continuously forming new products required for life.

Oxidation and Reduction Reactions Link, Forming Electron Transport Chains

Most energy transformations in organisms occur in chemical reactions called oxidations and reductions. In **oxidation,** a molecule loses electrons. The name comes from the observation that many reactions in which molecules lose electrons involve oxygen. Oxidation is the equivalent of adding oxygen because oxygen strongly attracts electrons away from the original atom. Oxidation reactions, such as the breakdown of glucose to carbon dioxide and water, are catabolic. That is, they degrade molecules into simpler products as they release energy.

In **reduction,** a molecule gains electrons. Reduction changes the chemical properties of a molecule. Reduction reactions, such as the formation of lipids, are usually anabolic. They require a net input of energy.

Oxidations and reductions tend to be linked, occurring simultaneously, because electrons removed from one molecule during oxidation join another molecule and reduce it. That is, if one molecule is reduced (gains electrons), then another must be oxidized (loses electrons). Such linked oxidations and reductions form **electron transport chains,** which we will see in the next two chapters.

Many energy transformations in living systems involve carbon oxidations and reductions. Reduced carbon contains more energy than oxidized carbon. This is why reduced molecules such as methane (CH_4) are explosive, whereas oxidized molecules such as carbon dioxide (CO_2) are not. The same principle applies to other compounds: The more reduced they are, the more energy they contain. Anyone who has ever dieted is at least intuitively familiar with this concept of energy content. Saturated fats are highly reduced, and they contain more than twice as many kilocalories by weight as proteins or carbohydrates. Thus, a fatty meal of a bacon double cheeseburger with fries and a shake may contain the same number of kcal (and thus the same amount of energy) as a bathtub full of celery sticks.

5.3 MASTERING CONCEPTS

1. What does metabolism mean in a cellular sense?
2. How are anabolism and catabolism opposites?
3. Distinguish between endergonic and exergonic reactions.
4. What distinguishes a reaction that has reached chemical equilibrium?
5. What are oxidation and reduction reactions, and why are they linked?

OLC

5.4 ATP Is Cellular Energy Currency

ATP, through high-energy phosphate bonds, temporarily stores energy that a cell uses for a wide variety of activities.

Organisms store potential energy in the chemical bonds of nutrient molecules. It takes energy to make these bonds, and energy is released when bonds of these molecules are broken. Much of the released energy of life is stored temporarily in the covalent bonds of **adenosine triphosphate,** a compound more commonly known as **ATP.** Chapter 4 introduced two examples of how a cell uses ATP energy—to power active transport and to move cilia and flagella. ATP also powers metabolism and many other cellular activities. All cells depend on it (see Table 7.1).

ATP Has High-Energy Phosphate Bonds

ATP is composed of the nitrogen-containing base adenine, the five-carbon sugar ribose, and three phosphate groups (each group includes a phosphorus atom bonded to four oxygen atoms and is indicated as PO_4) (fig. 5.12).

The three phosphate groups of an ATP molecule place three negative charges very close to each other. Since like charges repel, this atomic arrangement destabilizes the molecule, and it releases energy when the covalent bonds attaching the phosphates are split off. These phosphate bonds release so much energy that we illustrate them with squiggly lines. However, ATP's "high energy" is not contained completely in its phosphate bonds but in the complex interaction of atoms that make up the entire molecule.

When the endmost (terminal) phosphate group of an ATP molecule is removed, energy is released and the molecule becomes adenosine diphosphate (ADP, which has two phosphate groups rather than three). Still another phosphate bond can break to yield adenosine monophosphate (AMP) and another release of energy. ATP thus provides an energy currency for the cell. When a cell requires energy for an activity, it "spends" ATP by converting it to ADP, inorganic phosphate (symbolized P_i), and energy (fig. 5.12). This reaction is represented as:

$$ATP + H_2O \longrightarrow ADP + P_i + energy$$

In the reverse situation, energy can be temporarily stored by adding a phosphate to ADP, forming ATP. The energy comes from molecules broken down in other reactions. This reaction is represented as:

$$ADP + P_i + energy \longrightarrow ATP + H_2O$$

Investigating Biology 5.1 addresses the role of ATP in bioluminescence, the reaction that gives fireflies a "glow."

ATP is an effective biological energy currency for several reasons. First, converting ATP to ADP + P_i releases about twice the amount of energy required to drive most reactions in cells. The extra energy dissipates as heat. Second, ATP is readily available. The large amounts of energy in the bonds of fats and starches are not as easy to access—they must first be converted to ATP before the cell can use them. Finally, ATP's terminal phosphate bond, unlike the covalent bonds between carbon and hydrogen in organic molecules, is unstable—so it may be broken to release energy.

Just as you can use currency to purchase a great variety of different products, all cells use ATP in many chemical reactions to do different kinds of work. If you ran out of ATP, you would die instantly. Organisms require huge amounts of ATP. A typical adult uses the equivalent of 2 billion ATP molecules a minute just to stay alive. However, ATP is recycled so rapidly that only a few grams are available at any given instant. Organisms recycle ATP at a furious pace, adding phosphate groups to ADP to reconstitute ATP, using the ATP to drive reactions, and turning over the entire supply every minute or so.

FIGURE 5.12
ATP Breakdown and Formation Link Biochemical Reactions. The phosphates bound to ATP (**A**) represent a significant source of stored energy. When a phosphate is removed (**B**), the energy level of ATP (now ADP) is lowered and the released energy can be used to do work. Energy may also be stored by replacing the phosphate to re-form ATP. By adding and removing phosphates (**C**), cells can use an exergonic reaction to fuel an endergonic reaction.

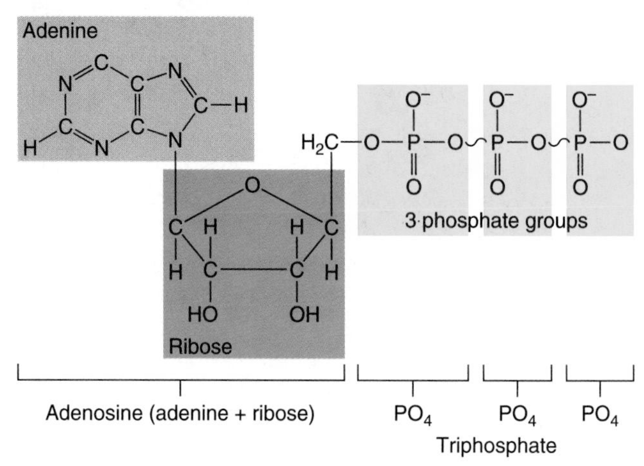

A

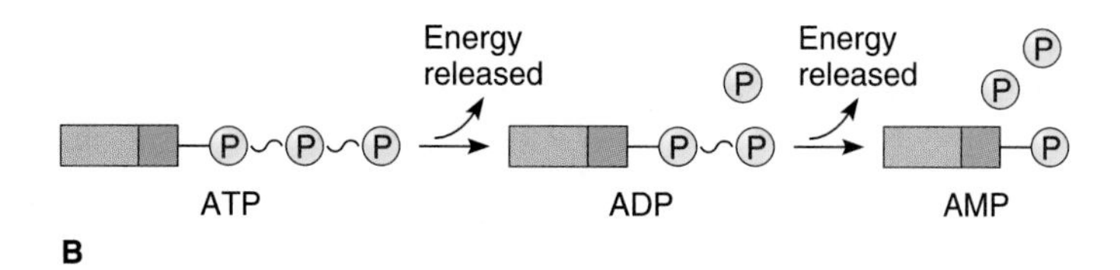

B

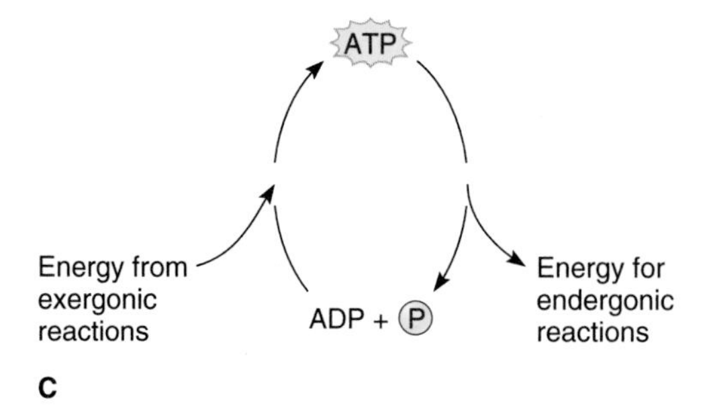

C

Cells Couple ATP Formation and Breakdown to Other Reactions

Cells couple the breakdown of nutrients to ATP production, and they couple the breakdown of ATP to other reactions that occur at the same time and place in the cell. Coupled reactions, as their name implies, are reactions that occur in pairs. One reaction drives the other, which does work or synthesizes new molecules. Consider, once again, the formation of sucrose and water from glucose and fructose. This reaction is not spontaneous; it requires

Investigating Life 5.1

Firefly Bioluminescence

Photosynthesis transforms light energy into chemical energy. In **bioluminescence,** the reverse can happen—chemical energy is converted into light energy. Bioluminescence is common in the oceans, where hordes of glowing microscopic organisms called dinoflagellates lend an eerie bluish cast to fishes, dolphins, or even ships that interrupt their movement.

More familiar, perhaps, is the glow of a firefly's abdomen in late summer. More than 1,900 species of firefly are known, and each uses a distinctive repertoire of light signals to attract a mate. Typically, flying males emit a pattern of flashes. Wingless females, called glowworms, usually are on leaves, where they emit light in response to the male. In one species, *Photuris versicolor,* the female emits the mating signal of another species and then eats the tricked male who approaches her. Some frogs consume so many fireflies that they glow!

In the 1960s, Johns Hopkins University researchers William McElroy and Marlene DeLuca asked Baltimore schoolchildren to bring them jars of fireflies. They then used the insects to decipher the firefly bioluminescence reaction. McElroy and DeLuca found that light is emitted when a molecule called *luciferin* reacts with ATP, yielding the intermediate compound *luciferyl adenylate* (fig. 5.A). The enzyme *luciferase* then catalyzes reaction of this intermediate with molecular oxygen (O_2) to yield *oxyluciferin*—and a flash of light. Oxyluciferin is then reduced to luciferin, and the cycle starts over.

Chemical companies sell luciferin and luciferase, which researchers use to detect ATP. When ATP appears in a sample of any substance, it indicates contamination by an organism. For example, the manufacturers of Coca-Cola use firefly luciferin and luciferase to detect bacteria in syrups used to produce the beverages. Contaminated syrups glow in the presence of luciferin and luciferase because the ATP in the bacteria sets the bioluminescence reaction into motion. Firefly luciferin and luciferase were also aboard the *Viking* spacecraft sent to Mars. Scientists sent the compounds to detect possible life—a method that would only succeed if Martians use ATP.

Although we understand the biochemistry of the firefly's glow, the ways animals use their bioluminescence are still very much a mystery. This is particularly true for the bioluminescent synchrony seen in fireflies in the same trees. When night falls, first one firefly, then another, then more, begin flashing from the tree. Soon the tree twinkles like a Christmas tree. But then, order slowly descends. In small parts of the tree, the lights begin to blink on and off together. The synchrony spreads. A half hour later, the entire tree seems to blink on and off every second. Biologists studying animal behavior have joined mathematicians studying order to try to figure out just what the fireflies are doing—or saying—when they synchronize their glow.

FIGURE 5.A
Fireflies Exhibit a Unique Version of Energy Use.
ATP is used to create flashes of light as energy is transferred to an electron in a specialized molecule called luciferin. The energy is released as light as the electron drops back to its original position.

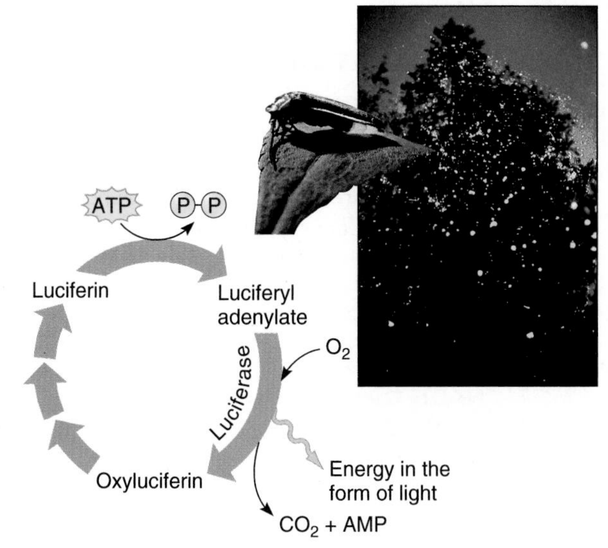

a net input of energy. When ATP provides additional energy, the reactants now have more energy available than the products, and the reaction proceeds. ATP breakdown is therefore coupled to sucrose synthesis.

A cell uses ATP as an energy source in two major ways: to energize a molecule or to change the shape of a molecule. Both activities transfer the terminal phosphate to another molecule, a process called **phosphorylation.** Then the phosphate is released. About 7 kcal of energy is released in splitting 10^{23} molecules of ATP to ADP and phosphate. The presence of a phosphate on a target molecule can make that molecule more likely to bond with other molecules. In this way, ATP is often used to fuel anabolic reactions such as assembling amino acids to make a protein. The other main mechanism of action using ATP changes the shape of a molecule. Adding phosphate to a protein can force that protein into a different shape and removing phosphate allows that molecule to return to its original shape. The cell uses the change to a new shape to move molecules and structures throughout the cell. Muscle contraction is the large-scale effect of millions of small molecules changing shape in a coordinated way. ATP provides the energy.

Other Compounds Are Involved in Energy Metabolism

Several other compounds besides ATP participate in the cell's energy transformations (table 5.1). Nonprotein helpers called **cofactors** assist other chemicals in enabling certain reactions to proceed. Cofactors are often ions. Mg^{2+}, for example, helps to stabilize many important enzymes.

Organic cofactors, which are called **coenzymes,** usually carry protons or electrons. Coenzymes are often nucleotides, as is ATP. But a coenzyme's energy content, unlike ATP's, depends on its ability to donate electrons or protons, not on the presence or absence of a particular phosphate bond. Vitamins are a major source of coenzymes in cells and help to drive metabolic reactions. Vitamins, therefore, do not directly supply energy, but make possible the reactions that extract energy from nutrient molecules.

NAD^+, $NADP^+$, and FAD are other important molecules that function as coenzymes. We introduce them here and return to them in the next two chapters.

Nicotinamide adenine dinucleotide (NAD^+), like ATP, consists of adenine, ribose, and phosphate groups (fig. 5.13). However, NAD^+ also has a nitrogen-containing ring, called nicotinamide, which is derived from niacin (vitamin B_3). The nicotinamide is the active part of the molecule. NAD^+ is reduced when it accepts two electrons and two protons (hydrogens) from a substrate. Both electrons and one proton actually join the NAD^+, leaving a proton (H^+). This reaction is written as:

$$NAD^+ + 2H^+ + 2e^- \longrightarrow NADH + H^+$$

$NADH + H^+$ is reduced and is therefore packed with potential energy. The cell uses that energy to synthesize ATP and to reduce other compounds.

The structure of nicotinamide adenine dinucleotide phosphate ($NADP^+$) is similar to that of NAD^+ but with an added phosphate group. NADPH supplies the hydrogen that reduces carbon dioxide to carbohydrate during photosynthesis. This process "fixes" atmospheric carbon into organic molecules that then serve as nutrients.

Flavin adenine dinucleotide (FAD) is derived from riboflavin (vitamin B_2). FAD, like NAD^+, carries two electrons. However, it also accepts two protons to become $FADH_2$.

The cytochromes are iron-containing molecules that transfer electrons in metabolic pathways. When oxidized, the iron in cytochromes is in the Fe^{3+} form. When the iron accepts an electron, it is reduced to Fe^{2+}. There are several types of cytochromes, all of which carry electrons in cells. In metabolic pathways, cytochromes align to form electron transport chains, with each molecule accepting an electron from the molecule before it and passing an electron to the next. Figure 5.14 shows such a chain; we will see more of them in the next two chapters. Small amounts of energy are released at each step of an electron transport chain, and the cell uses this energy in other reactions. Cytochromes take part in many energy transformations in life, suggesting that this cellular strategy for energy transformation is quite ancient.

5.4 MASTERING CONCEPTS

1. How does ATP supply energy for cellular functions?
2. What are the functions of cofactors and coenzymes in cellular metabolism?
3. What are some other compounds that participate in cellular energy reactions?

OLC

TABLE 5.1 Some Molecules Involved in Cellular Energy Transformations

MOLECULE	MECHANISM
ATP	Compound that phosphorylates molecules, energizing them to participate in other reactions; conversions between ATP and ADP link many biochemical reactions
Cofactors	Substances, usually ions, that stabilize biomolecules
Coenzymes	Vitamin-derived organic cofactors that transfer electrons or protons
NADH	Coenzyme that transfers electrons; used to synthesize ATP
NADPH	Coenzyme that supplies hydrogen and energy to reduce CO_2 in photosynthesis
$FADH_2$	Coenzyme that transfers electrons; used to synthesize ATP in cellular respiration
Cytochromes	Iron-containing molecules that transfer electrons in metabolic pathways

FIGURE 5.13
NADH Carries Electrons.
NAD^+ is a form of nucleotide that is used in many cellular reactions to store and transfer energy-rich electrons. As the electron is added to NAD^+, it becomes reduced as NADH.

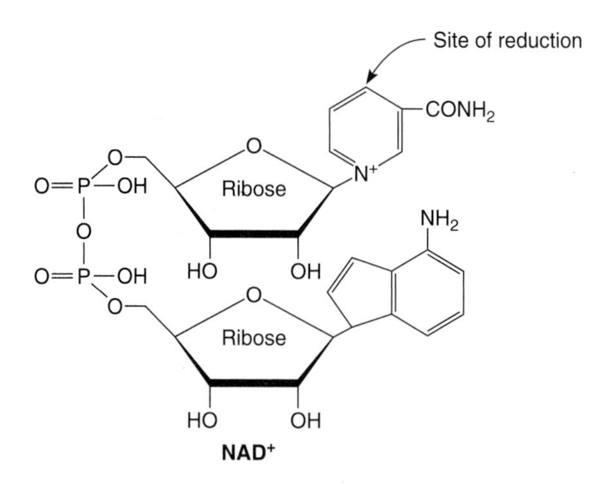

FIGURE 5.14
Electron Transport Chains.
Different cytochrome proteins align within membranes, forming electron transport chains. Electrons pass down the chain in a series of oxidation-reduction reactions, releasing energy in small, usable increments. Electrons are passed between iron molecules attached to the cytochromes. Iron in the ferric form (Fe^{3+}) gains an electron, becoming reduced to the ferrous (Fe^{2+}) form. Enzymes use the energy transfers to pump hydrogens across the membrane and transfer energy ultimately to ATP or other molecules.

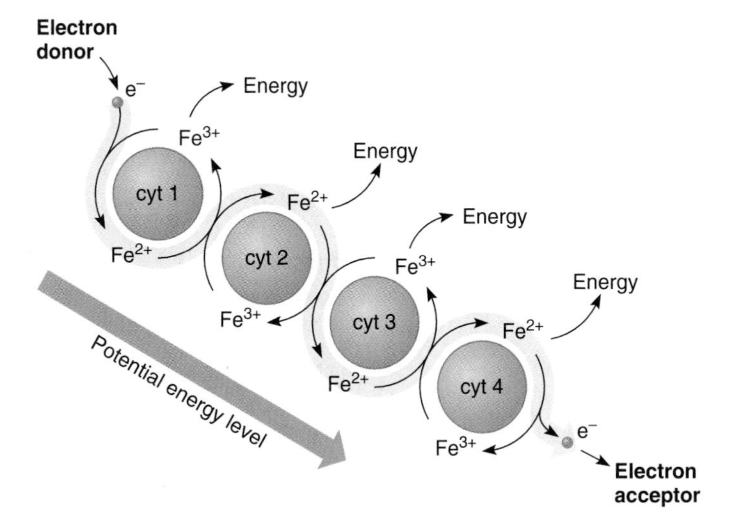

5.5 Enzymes and Energy

Without enzymes, biochemical reactions could not proceed fast enough for life to be possible. Enzymes lower the energy required to start these reactions.

Many chemical reactions require an initial boost of energy, called the **energy of activation.** Sometimes this energy comes from heat in the environment. For example, heat from a lighted match provides the energy of activation to ignite a piece of wood. The amount of heat required to activate most metabolic reactions in cells would swiftly be lethal were it not for enzymes, which lower the required energy of activation (fig. 5.15).

FIGURE 5.15
Enzymes Lower the Activation Energies.
Enzymes speed chemical reactions by increasing the chance that reactants will come together. This decreases the energy needed to start the reaction: the activation energy.

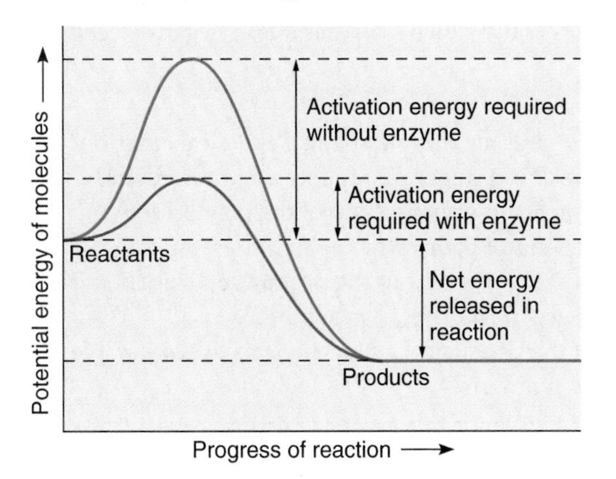

How Do Enzymes Speed Biochemical Reactions?

Recall from chapter 2 that an enzyme is a protein that catalyzes (speeds) specific chemical reactions without being consumed by them. An enzyme catalyzes a chemical reaction by decreasing the energy of activation. The enzyme binds the substrate in a way that enables the reaction to proceed. The conformation of the enzyme is such that the substrate fits into its active site, but not as precisely as a key fits a lock. Rather, the active site contorts slightly, as if it is hugging the substrate (fig. 5.16). Without enzymes, many biochemical reactions would not occur fast enough to support life. The python described in the chapter opener could certainly never process its gazelle meal without the help of digestive enzymes.

Because enzymes are proteins, they are very sensitive to conditions that would disrupt protein structure. Increases in temperature, changes in pH, high or low salt concentrations, or the presence of toxins are all detrimental to the shape of an enzyme. For instance, most enzymes have a very narrow range of temperatures within which they will function correctly. Too high or too low and they simply stop. High enough and they will be forever destroyed. This is one reason it is so useful for organisms to maintain a uniform body temperature.

How Do Regulatory Enzymes Control Metabolic Pathways?

Any living cell is in a constant state of flux, as it rapidly dismantles some molecules and synthesizes others. Disturbances in the balance between these two actions threaten life. Several

FIGURE 5.16
Enzymes are Specific.
(**A**) The shape of the enzyme creates an active site that binds to one or more specific substrates. (**B**) When binding to its substrate, an enzyme changes shape very slightly, as is illustrated here with the binding of glucose to hexokinase.

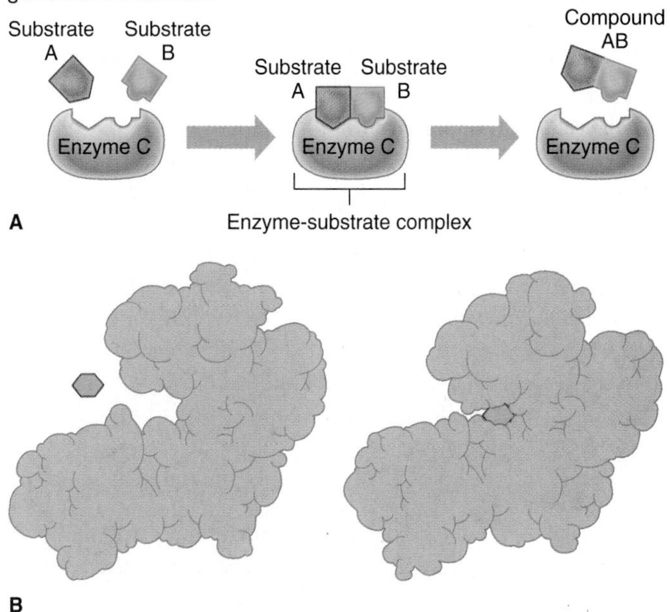

biological mechanisms regulate cellular metabolism and preserve this delicate balance.

Certain enzymes control metabolism by functioning as pacesetters at important junctures in biochemical pathways. The enzyme with the slowest reaction sets the pace for the pathway's productivity, just as the slowest runner on a relay team limits the overall pace for the whole team. This is because each subsequent reaction in the metabolic pathway (like each subsequent relay runner) requires the product of the preceding reaction to continue. The reaction that this enzyme catalyzes is called the rate-limiting step; the enzyme is a **regulatory enzyme** because it regulates the pathway's pace and productivity.

Often a regulatory enzyme is highly sensitive to a specific chemical cue. When this cue is the end product of the pathway the enzyme may "turn off" for a time. This type of regulation is called **negative feedback,** or feedback inhibition. Negative feedback prevents too much of one substance from accumulating—an excess of a particular biochemical effectively shuts down its own synthesis until its levels fall. At that point, the pathway resumes its activity. Negative feedback is somewhat like a thermostat—when the temperature in a building reaches a certain level, the thermostat shuts the heat off for a while. Falling temperature cues the thermostat to ignite the furnace again (see fig. 30.6).

A product molecule can inhibit its own synthesis in two general ways. It may bind to the regulatory enzyme's active site, preventing it from binding substrate and temporarily shutting down the pathway. This is sometimes called competitive inhibition because a substance other than the substrate competes to occupy the active site. Alternatively, product molecules may bind to the regulatory enzyme at a site other than the active site, but in a way that alters the shape of the enzyme so that it can no longer bind substrate. This indirect approach is sometimes called noncompetitive inhibition because the inhibitor does not directly compete to occupy the active site. Both competitive and noncompetitive inhibition are forms of negative feedback. Other molecules may also function as either type of inhibitor.

Figure 5.17 illustrates negative feedback. It shows the sequential pathway that certain bacteria use to synthesize the amino acid histidine. When excess histidine accumulates, it binds (noncompetitively) to the junction between two subunits of the enzyme that regulates the pathway. This temporarily destabilizes the enzyme and impairs catalysis. For a time, histidine synthesis ceases. When levels of histidine in the bacterium fall, the block on the regulatory enzyme is lifted, and the cell can once again synthesize the amino acid.

Substances from outside the body, such as drugs and poisons, can also inhibit enzyme function. A foreign chemical similar in conformation to the substrate may block the enzyme's active site and prevent the substrate from binding. The drug sulfanilamide, for example, competitively inhibits certain enzymes in bacteria and is therefore useful to fight certain infections.

Poisons that bind noncompetitively to a part of the enzyme other than the active site also inhibit enzyme function. Some nerve gases, for example, inhibit the activity of the enzyme acetylcholinesterase, which normally removes the neurotransmitter acetylcholine after it sends its message. When the nerve gas binds to acetylcholinesterase, the enzyme cannot bind its substrate, and nerve transmission ceases by being overloaded by too much acetylcholine. Penicillin, a drug, acts in a similar way, binding irreversibly to an enzyme that bacteria require to build cell walls. Cell walls are not formed correctly and, when the bacterium tries to divide, it falls apart.

Much rarer than negative feedback in organisms is **positive feedback,** in which a molecule activates the pathway leading to its production. Blood clotting, for example, begins when a biochemical pathway synthesizes fibrin, a threadlike protein. The products of the later reactions in the clotting pathway stimulate the enzymes that activate earlier reactions. As a result of positive feedback, fibrin accumulates faster and faster, until there is enough to stem the blood flow. When the clot forms and the blood flow stops, the clotting pathway shuts down—an example of negative feedback.

FIGURE 5.17

Negative Feedback (Feedback Inhibition).

Like other organisms, *E. coli* requires the amino acid histidine to synthesize proteins. This diagram isolates the enzyme-catalyzed, sequential reactions leading to histidine production. When sufficient histidine accumulates, the histidine inhibits the activity of the first enzyme in the pathway, temporarily halting further production of itself. There are two negative feedback mechanisms. In competitive inhibition, the product binds and blocks the active site. In noncompetitive inhibition, a product binds to a different site on the enzyme, but this action alters the active site.

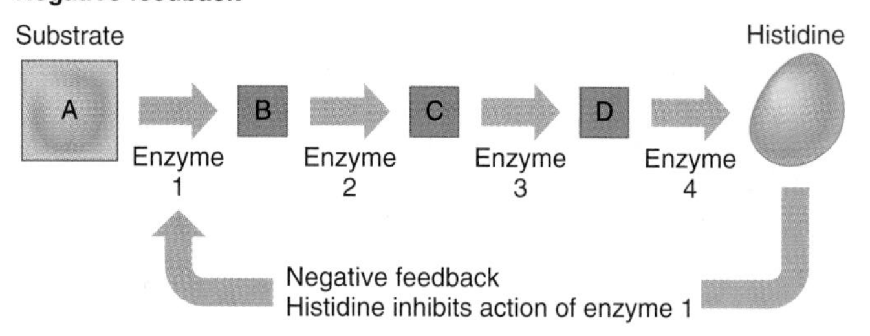

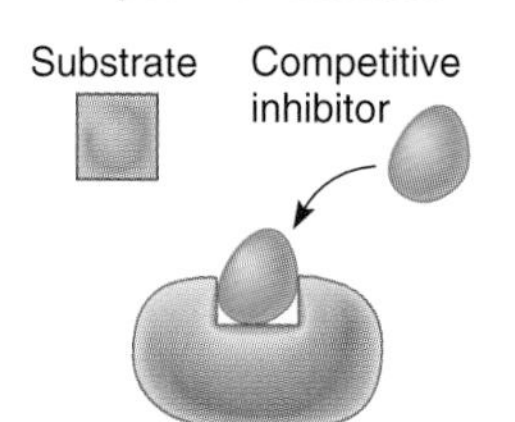

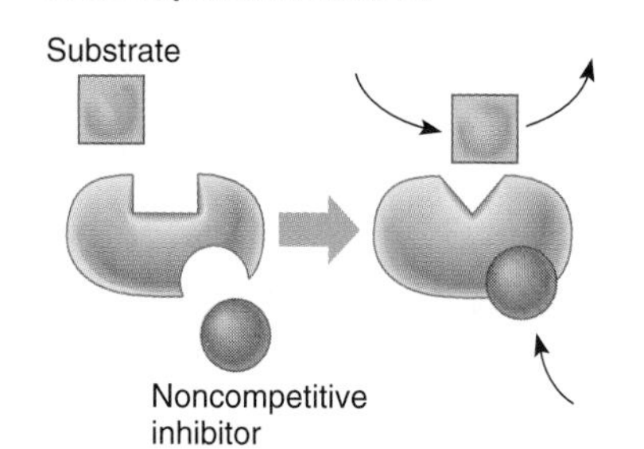

5.5 MASTERING CONCEPTS

1. How does an enzyme speed a chemical reaction?
2. How does temperature affect enzyme activity?
3. How can an enzyme control the productivity of a biochemical pathway?
4. In what two ways can the product of a biochemical pathway turn off its own synthesis?
5. Can the product of a biochemical pathway increase its own synthesis?

OLC

Chapter Summary

5.1 What Is Energy?

1. **Energy** is the ability to do work. In some organisms, photosynthesis stores solar energy in chemical bonds. **Cellular respiration** releases energy.
2. Energy comes from the sun, the wind, moving water, the earth, and the tides and is held in the bonds of molecules of organisms.
3. **Potential energy** is stored energy available to do work. **Kinetic energy** is action.

5.2 How Do the Laws of Thermodynamics Describe Energy Transfer?

4. The **laws of thermodynamics** govern the energy transformations of life.
5. The first law states that energy cannot be created or destroyed but only converted to other forms.
6. The second law states that all energy transformations are inefficient, because every reaction results in increased **entropy** (disorder) and the loss of usable energy as heat.

5.3 How Do Cells Metabolize Nutrients?

7. **Metabolism** is the sum of the energy and matter conversions in a cell. It consists of enzyme-catalyzed reactions organized into often interconnected pathways and cycles.
8. **Anabolism** includes reactions that synthesize molecules. These reactions diverge from a few reactants to many products and require energy.
9. **Catabolism** includes reactions that break down molecules, converge from many reactants to a few products, and release energy.
10. Each reaction of a **metabolic pathway** rearranges molecules into new compounds, which changes the amount of **free energy** available in chemical reactions to do work.
11. In **endergonic reactions,** products have more energy than reactants, and entropy decreases. These reactions are not spontaneous—they require energy input.
12. In **exergonic reactions,** products have less energy than reactants, entropy increases, and energy is released.
13. At **chemical equilibrium,** a reaction proceeds in both directions at the same rate.
14. Many energy transformations in organisms occur via **oxidation** and **reduction** reactions. Oxidation is the loss of electrons from a molecule; reduction is the gain of electrons. Oxidation and reduction reactions occur simultaneously, in pairs, and form **electron transport chains.**

5.4 ATP Is Cellular Energy Currency

15. ATP stores energy in its high-energy phosphate bonds. Many energy transformations involve **coupled reactions,** in which the cell uses the energy released by ATP to drive another reaction. Other compounds that take part in cellular metabolism include **cofactors, coenzymes,** NADH, NADPH, $FADH_2$, and cytochromes.

5.5 Enzymes and Energy

16. Enzymes are proteins that speed spontaneous biochemical reactions to a biologically useful rate by lowering the **energy of activation.**
17. In **negative feedback,** a pathway's product interacts with a **regulatory enzyme** for the pathway, temporarily shutting down its own synthesis when its levels rise. In **positive feedback,** a product stimulates its own production.

Testing Your Knowledge

1. Which units are used to measure energy?
2. What are some sources of energy?
3. Give one example of potential energy and one of kinetic energy.
4. Cite everyday illustrations of the first and second laws of thermodynamics. How do these principles underlie every organism's ability to function?
5. Give an example of entropy.
6. Why isn't heat usable energy?
7. State three differences between anabolic and catabolic reactions.
8. State the differences between endergonic and exergonic reactions.
9. What is chemical equilibrium?
10. Why are oxidation and reduction reactions linked?
11. Cite three reasons why ATP is an excellent source of biological energy.
12. What do ATP, NADH, NADPH, and $FADH_2$ have in common chemically?
13. How does an enzyme speed a chemical reaction?
14. Cite three ways that the product of a biochemical pathway can control its own rate of synthesis.

Thinking Scientifically

1. Some people claim that life's high degree of organization defies the laws of thermodynamics. How is the organization of life actually consistent with the principles of thermodynamics?
2. How might you explain the fact that even species as diverse as humans and yeast use the same biochemical pathways to extract energy from nutrient molecules?
3. Cytochrome *c* is an electron carrier that is nearly identical in all species. No known disorders affecting cytochrome *c* exist. What does this suggest about the importance of this molecule?
4. When a person eats a fatty diet and excess cholesterol accumulates in the bloodstream, cells temporarily turn off their synthesis of cholesterol. What phenomenon described in the chapter does this control of cholesterol illustrate?
5. In 1991, a large volcanic eruption in the Philippines threw great dust clouds into the atmosphere and lowered temperatures around the world for many months. How might this event have affected energy transformations in organisms?

References and Resources

Cossins, Andrew R., and Neil Roberts. January 4, 1996. The gut in feast and famine. *Nature,* vol. 379. Energy adaptations enable a snake to digest a meal that is 65% of its body dry weight.

Diamond, Jared. April 1994. Dining with the snakes. *Discover.* A blow-by-blow description of how a snake eats.

The *LIFE* Online Learning Center provides additional resources and tools for studying this chapter.

www.mhhe.com/life

CHAPTER 6

Photosynthesis

6.1 Life Depends on Photosynthesis

- Photosynthesis Converts Carbon Dioxide to Carbohydrate
- How Might Photosynthesis Have Evolved?

6.2 How Do Cellular Structures Capture and Use Light Energy?

- What Is Light?
- Pigment Molecules Capture Light Energy
- Chloroplasts Are the Sites of Photosynthesis

6.3 The Light Reactions Begin Photosynthesis

- Photosystem II Produces ATP
- Photosystem I Produces NADPH
- Two Photosystems Extract More Energy Than One

6.4 The Carbon Reactions "Fix" Carbon

- How Does the Calvin Cycle "Fix" Carbon?
- How Did Experiments Reveal the Steps of the Calvin Cycle?

6.5 How Efficient Is Photosynthesis?

- How Does Photorespiration Alter the Calvin Cycle?
- C_4 Photosynthesis Decreases Photorespiration
- CAM Photosynthesis Stores Carbon Dioxide at Night

A Jellyfish's "Green Fluorescent Protein" Sheds Light on Photosynthesis—And Much More

In the first century A.D., Roman scholar Pliny the Elder noted an eerie glow in the Bay of Naples, and identified its source—jellyfish. Today, a glowing protein from the jellyfish *Aequorea victoria* is used in cell biology to literally light up any protein. "Green fluorescent protein," or GFP, has enabled researchers to trace such diverse cellular processes as protein production and secretion, contraction of microtubules and sliding of microfilaments, and calcium transport. At the tissue level, GFP reveals the migration of cells during development, and can monitor the course of infection.

When poked, the animal emits a greenish glow, just as Pliny the Elder saw.

The jellyfish uses its GFP to transfer energy. When poked, the animal emits a greenish glow, just as Pliny the Elder saw. The glow is an optic phenomenon called fluorescence, which means that reactive electrons in a molecule excited with one color of light emit light of a different color (the fluoresced light has a longer wavelength). Shine a blue or ultraviolet light onto GFP, and it emits green light for about 10 minutes.

Douglas Prasher and Martin Chalfie at the Woods Hole Marine Biological Laboratory in Massachusetts discovered and described GFP and its gene in 1992. Realizing the potential to hook the GFP gene onto any other gene, from any organism, they applied for a government grant to pursue the idea—but were turned down. Continuing anyway, they attached the GFP gene to genes from a bacterium and a type of worm—and these organisms glowed! GFP as a marker of biological activity is an improvement over chemical dyes for several reasons: it is nontoxic, can be used on live cells, works fast, and is quantitative, indicating the amount of a marked molecule by the intensity of the light emission.

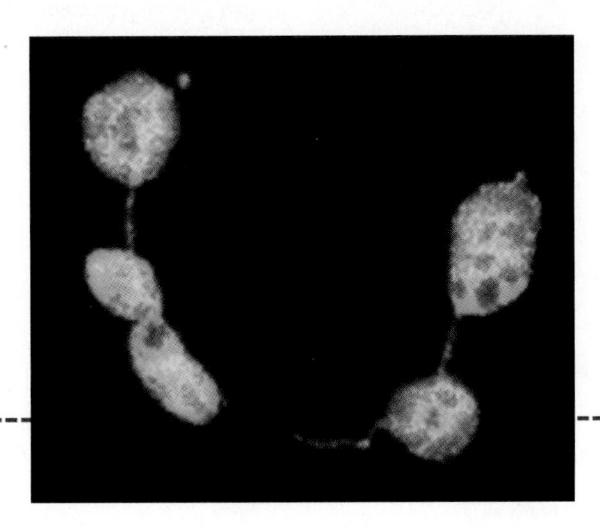

A jellyfish's green fluorescent protein revealed that chloroplasts can connect to each other.

Altering the GFP gene has yielded "new and improved" cell markers. One version functions best at the higher temperatures in a mammal's body compared with that of a jellyfish. A brighter green variety is available, as is a palette of other colors, enabling researchers to trace several molecules at a time. The technology can also confirm and clarify older observations. This is the case for studies on chloroplasts, the plant cell organelles that are the sites of photosynthesis.

Realizing the potential to hook the GFP gene onto any other gene, from any organism, they applied for a government grant to pursue the idea—but were turned down.

An actively photosynthesizing cell in a leaf may contain from 40 to 200 chloroplasts. New chloroplasts arise when one organelle splits in two. Botanists have long thought that chloroplasts are very independent organelles, not interacting with one another. In 1962, Sam Wildman, at the University of California, Los Angeles, filmed what looked like projections forming on one chloroplast and leading toward another in cells from tobacco and spinach. But no one had been able to repeat the finding, and so it was forgotten. Thirty-five years later, GFP revealed that Wildman had seen something real.

In 1997, Maureen Hanson and her colleagues at Cornell University linked GFP to a transit peptide that would take it into the watery region of chloroplasts in tobacco and petunia cells. Sure enough, the chloroplasts glowed greenly, but there was more—strands of fluorescent green extended from one chloroplast to another. Plus, the strands moved! In different live cells they would wave, change shape, extend, and contract. Not every chloroplast nor even every cell had the mysterious protuberances, but the researchers saw enough of them, in various stages of growth, to confirm Wildman's work.

Further experiments revealed that the chloroplast connections were functional as well as structural. When the researchers applied a powerful laser to a particular chloroplast, it turned white, the blast of energy vanquishing its greenness. But when the researchers applied the laser to one of the extensions, both of the chloroplasts that it connected turned white, showing that GFP had migrated from one chloroplast to another through the connections. The next step will be to find out exactly what these extensions that connect chloroplasts do.

FIGURE 6.1
Life Depends on Photosynthesis.
The first signs of life on a barren landscape, the slopes of Mt. St. Helens, are plants. Their roots will retain moisture and nutrients from the environment and establish a foothold for life.

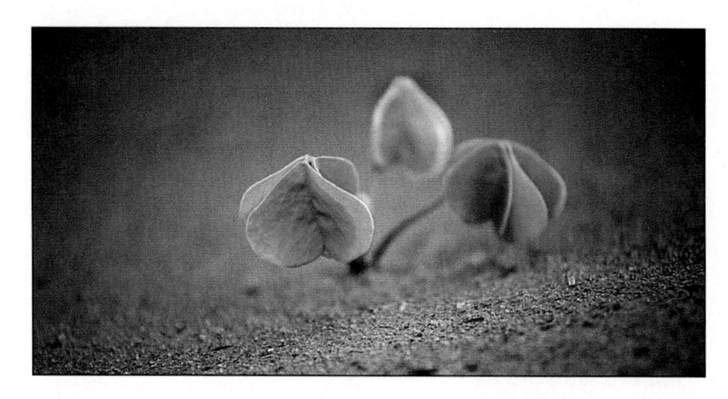

6.1 Life Depends on Photosynthesis

Light ultimately provides the energy that powers nearly all life. Photosynthesis enables plants, algae, and certain microorganisms to harness light energy and convert it to chemical energy.

If asked to designate the most important biochemical pathway, most biologists would not hesitate to cite **photosynthesis,** the process by which plants, algae, and some microorganisms harness solar energy and transduce it into chemical energy. All life on this planet ultimately depends upon photosynthesis.

A world without photosynthesis would not long be a living world. Scientists aptly term the aftermath of an event that would blacken the sky "nuclear winter." Such disasters might include a nuclear holocaust, eruption of a huge volcano, or a meteor impact. After the event, light reaching Earth's surface would be about a tenth of its normal intensity. Plants would die as they broke down scarce carbohydrates for energy faster than they could manufacture more carbohydrates using the energy in sunlight. Animals that ate these plants would go hungry, as would the animals that ate them. A year or even two might pass before enough life-giving light could penetrate that hazy atmosphere, but by then it would be too late. The lethal chain reaction that began with the cessation of photosynthesis would already be well into motion, destroying food webs at their bases. The first organisms to appear after a small-scale disaster are the plants (fig. 6.1).

The products of photosynthesis, glucose and other biomolecules, are often referred to as **photosynthate.** These molecules provide energy for nearly all life—to plants directly and to other organisms indirectly. Plants use about half of all photosynthate as fuel for cellular respiration and a process called photorespiration. They also use photosynthate to manufacture a variety of compounds, including amino acids, starch, sucrose, cellulose, glucose, rubber, quinine, and spices.

Photosynthesis Converts Carbon Dioxide to Carbohydrate

About 10% of the species we know of today photosynthesize. Photosynthesis is a biochemical process that uses energy from the sun to create biologically important molecules. Energy from the sun is captured by specialized pigment molecules in plant cells and used by enzymes in those cells to convert carbon dioxide (CO_2) to glucose ($C_6H_{12}O_6$). Water is used in the process and oxygen (O_2) is released as a by-product. Other organisms use the oxygen to release the energy from glucose. The reactions of photosynthesis can be summarized as follows:

$$6CO_2 + 12H_2O \xrightarrow{\text{light energy}} C_6H_{12}O_6 + 6O_2 + 6H_2O$$

The oxygen liberated in photosynthesis comes from the H_2O and not from the CO_2. This can be demonstrated by exposing a plant to water containing a heavy oxygen isotope designated ^{18}O. The "labeled" isotope appears in the oxygen gas released in photosynthesis, demonstrating that the oxygen came from the water. All organisms readily use the energy in glucose to build energy storage molecules and other biochemicals.

Certain bacteria photosynthesize using hydrogen sulfide (H_2S) in place of H_2O. These microorganisms release sulfur (S_2) rather than oxygen (O_2), along with an energy-rich organic product.

How Might Photosynthesis Have Evolved?

Cells are open systems that can assimilate but not create energy. Before photosynthesis evolved, all organisms were **heterotrophs,** meaning that they assimilated energy-rich organic compounds from the environment, and manufactured other organic compounds from them. As these early heterotrophs oxidized the carbon compounds from their surroundings, they released CO_2 into the environment. Eventually, environmental energy stores began to run out. Because the earliest organisms couldn't use the carbon in CO_2, they faced extinction as soon as they depleted the organic compounds in their habitats.

The evolution of photosynthesis about 3.0 billion years ago gave life a new energy source (fig. 6.2). Photosynthetic organisms, rather than relying on the dwindling supply of energy-rich compounds in the environment, began to use the energy in sunlight. These were the first photosynthetic **autotrophs,** which are organisms that make organic compounds from inorganic compounds such as water and CO_2, rather than obtaining organic compounds directly from the environment. The ability of autotrophs to transduce light energy into chemical energy soon supported most other forms of life.

The rise of photosynthetic organisms radically altered Earth. It decreased the concentration of CO_2 in the atmosphere; lowered global temperature, adding to the polar ice caps and lowering sea level; and filled the atmosphere with a waste product that many other organisms eventually would find essential for life: oxygen. The proportion of oxygen in the atmosphere gradually rose from a tiny fraction of a percent to 20% today (fig. 6.2). Oxygen allowed the evolution of aerobic respiration and more diverse species. All the oxygen in the air we breathe has cycled through photosynthetic organisms.

Our evolutionary look at photosynthesis continues at the end of the next chapter. Investigating Life 6.1 describes some early ideas and experiments that revealed some of the workings of photosynthesis.

FIGURE 6.2

Oxygen Changed the Living World.

Earth's atmosphere was not always rich in oxygen. The evolution of photosynthesis pumped oxygen into the atmosphere, profoundly altering life's diversity. Source: Data from *Teaching About Evolution and the Nature of Science*, 1998, National Academy of Sciences.

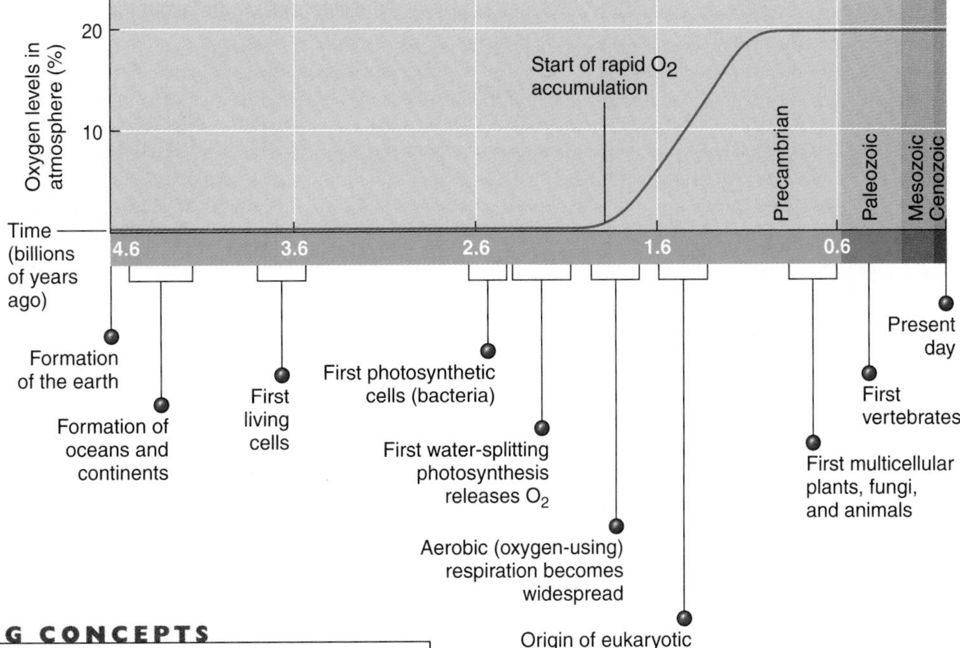

6.1 MASTERING CONCEPTS

1. Why is photosynthesis essential to life on Earth?
2. What is photosynthesis? Describe the process in words and in chemical symbols.
3. How did the origin of photosynthesis alter life on Earth?

6.2 How Do Cellular Structures Capture and Use Light Energy?

Light is a form of energy. Special pigment molecules in plant organelles called chloroplasts capture light energy and channel it to biochemical pathways.

Special pigment molecules in oblong plant organelles called **chloroplasts** capture light energy and channel it to biochemical pathways. Photosynthesis begins with capturing the energy in light.

Each minute, the sun converts more than 120 million tons of matter to radiant energy, releasing much of it outward as waves of radiation. After an 8-minute journey, about two-billionths of this energy reaches Earth's upper atmosphere. Only about 1% of this light is used for photosynthesis—yet this tiny fraction of the sun's power ultimately produces some 3 quadrillion pounds of carbohydrates a year!

What Is Light?

Our understanding of light began about 300 years ago, when Sir Isaac Newton showed that white light passing through a prism, water droplet, or soap bubble separates into a band of colors. Newton also demonstrated that separated light, passed through another prism, can recombine to form white light. Based on these discoveries, Newton proposed that white light is actually a combined spectrum of colors ranging from violet to red. Two centuries later, in the 1860s, Scottish mathematician James Maxwell showed that visible light is, in turn, a small sliver of a much larger spectrum of radiation: the **electromagnetic spectrum** (fig. 6.3).

In 1905, Albert Einstein extended the ideas of Newton and Maxwell and proposed that light consists of packets of energy called **photons.** The intensity (brightness) of light depends on the number of photons (the amount of energy) absorbed per unit of time. Each photon carries a fixed amount of energy determined by how the photon vibrates. The slower the vibration, the less energy the photon carries. Photons travel in waves, and the distance a photon moves during a complete vibration is its **wavelength** (fig. 6.3). The wavelength of visible light is measured in nanometers (nm), or billionths of a meter, and ranges from 390 to 760 nanometers. A photon's energy is inversely proportional to the wavelength of the light. That is, the longer the wavelength, the less energy per photon. Humans perceive light of different wavelengths as distinct colors.

Sunlight consists of about 4% ultraviolet (UV) radiation, 52% infrared (IR) radiation, and 44% visible light. Each of these kinds of light has different energy characteristics and different effects on organisms.

UV radiation contains high-energy photons that drive electrons from molecules, forming ions, which is why UV is called ionizing radiation. UV breaks weak chemical bonds and causes sunburn and skin cancer. Because glass absorbs UV, a window blocks UV radiation and prevents it from burning your skin. The ozone (O_3) in the upper atmosphere also absorbs UV radiation.

IR radiation doesn't contain enough energy per photon to be useful to organisms. Most of its energy is converted immediately to heat. Unlike UV, IR penetrates glass, heating a closed room on a sunny day.

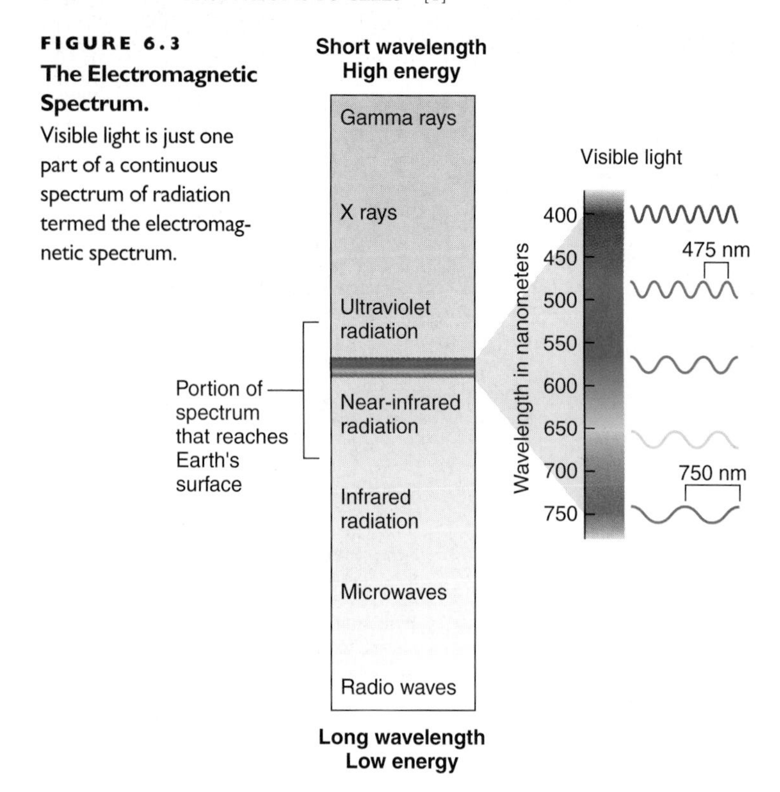

FIGURE 6.3
The Electromagnetic Spectrum.
Visible light is just one part of a continuous spectrum of radiation termed the electromagnetic spectrum.

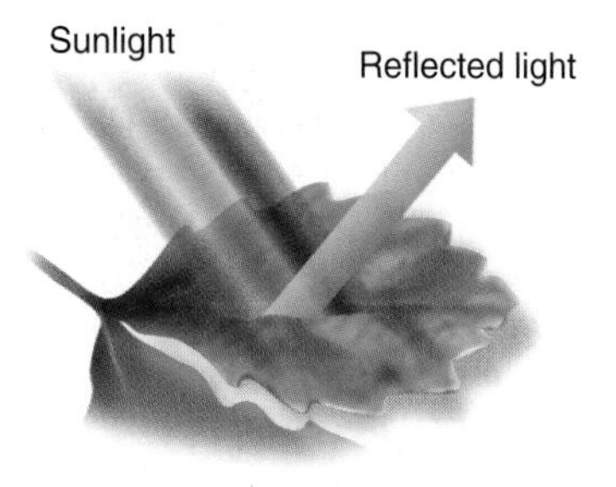

FIGURE 6.4
Leaf Color.
Leaves appear green because chlorophyll molecules in leaf cells reflect green and yellow wavelengths of light and absorb the other wavelengths.

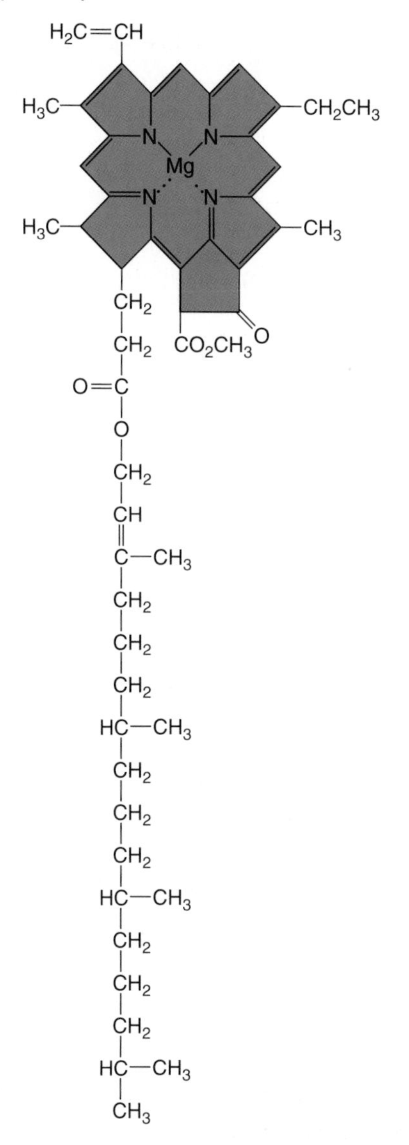

FIGURE 6.5
Chlorophyll.
The green pigment chlorophyll *a* is the dominant pigment in the photosynthetic cells of plants. A long hydrophobic tail anchors the molecule into the chloroplast's thylakoid membranes.

Visible light provides the right amount of energy to power biochemical reactions. Red and blue light are the most effective for photosynthesis. Shorter wavelength radiation, such as X rays and UV, contains enough energy to break chemical bonds and release energy. Visible light, in contrast, contains just enough energy to "excite" or energize molecules. When a photon strikes a pigment molecule in a plant cell, the pigment molecule absorbs it, causing certain electrons to jump to higher energy levels or even out of the atom completely. Such photon-boosted electrons are packed with potential energy, which they release when they fall back to their original positions nearer the nucleus or carry with them when they move to an electron-accepting molecule.

Pigment Molecules Capture Light Energy

Light striking an object is either reflected, transmitted (passed through), or absorbed (fig. 6.4). Only absorbed light can cause an effect. Organisms absorb light energy through pigment molecules. Pigments are colored because they absorb some wavelengths of light and transmit or reflect others. Black pigment absorbs all wavelengths, whereas white absorbs none.

Photosynthetic organisms use a variety of pigment molecules to capture photon energy. **Chlorophyll *a*** is the primary photosynthetic pigment in all photosynthetic organisms except certain bacteria that use a similar molecule called bacteriochlorophyll. Chlorophyll *a* is also present in prokaryotes called cyanobacteria, which may have given rise to today's plants and algae. Chlorophyll *a* absorbs wavelengths corresponding to red and orange (600–700 nm) and blue and violet (400–500 nm) and reflects and transmits green wavelengths. Chlorophyll, therefore, is responsible for the green color of plants that we perceive (fig 6.4).

Chlorophyll *a* is a huge molecule—its chemical formula is $C_{55}H_{22}O_5N_4Mg$ (fig. 6.5). The molecule includes a flat ring-like end that consists of a central magnesium (Mg) atom surrounded

Early Thoughts on Photosynthesis

Today researchers are identifying and describing in molecular detail the various pigments and proteins that interact to capture solar energy and transduce it to chemical energy. However, people have investigated photosynthesis for centuries, with cruder tools, but with no less compelling questions.

An early discussion of photosynthesis appears in Jonathan Swift's eighteenth-century novel *Gulliver's Travels.* In chapter 5 of part 3, "A Voyage to Laputa," Swift writes, "He had been eight years upon a project for extracting sunbeams out of cucumbers, which were to be put into vials hermetically sealed, and let out to warm the air in raw inclement summers." Because the idea of extracting the sun's warmth from a vegetable seemed preposterous, the expression "getting sunbeams from cucumbers" became synonymous with seeking to do the impossible. Historians disagree as to whether Swift was indulging his famed sarcasm or demonstrating a rather deep understanding of photosynthesis, considering how little was known about the process when the novel was published in 1726.

One of the first investigators to explore plant metabolism was Flemish physician and alchemist Jan van Helmont. In the early 1600s, he grew willow trees in weighed amounts of soil, applied known amounts of water, and noted that in 5 years the trees gained more than 100 pounds (45 kilograms), while the soil lost only a few ounces (fig. 6.A). Because he had applied large amounts of water, van Helmont concluded (incorrectly) that plants grew solely by absorbing water. He also noted that burning wood exudes carbon dioxide, but he did not connect this and plant growth. In the 1720s, English botanist and chemist Stephen Hales hypothesized that some substance in the air, rather than water, nourishes plants. Swift may have taken his "sunbeams from cucumbers" imagery from Hales's work.

FIGURE 6.A
Plant Mass Comes From Air.
Jan van Helmont was one of the first scientists to study plant growth. He grew willow trees, carefully recording the amounts of soil and water he used and the weight of the trees after 5 years. He erroneously concluded that the increase in plant matter was due directly to the plentiful water. We now know that the increased mass of the tree came from the carbon in the air.

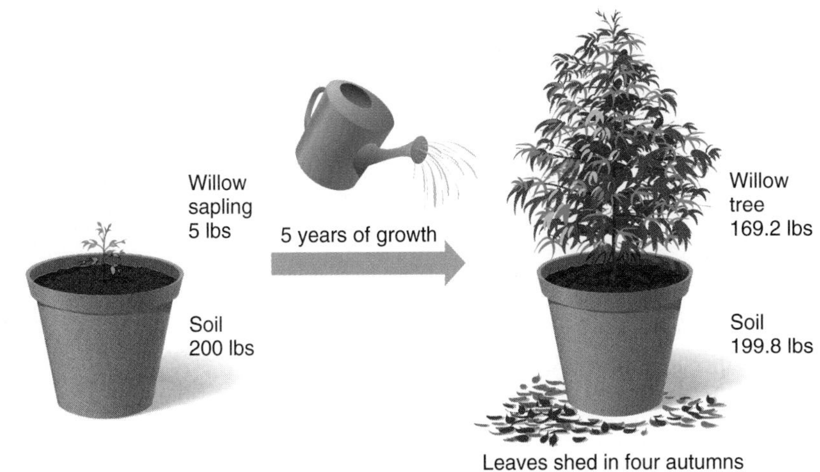

Unitarian minister and chemist Joseph Priestley studied photosynthesis more systematically in the summer of 1771. Priestley knew that if he placed a mouse in an enclosed container with a lit candle, the mouse would die. Hypothesizing that burning somehow "injured" the air so that the mouse couldn't breathe, Priestley looked for a substance that could "purify" the air. A mint plant worked. If a closed chamber held both a burning candle and a mint leaf, a second candle could still burn several days later. Priestley wrote, "I have been so happy as by accident to hit upon a method of restoring air which has been injured by the burning of candles, and to have discovered at least one of the restoratives which nature employs for this purpose. It is vegetation." He also discovered that a mouse could live in a container as long as a plant was present. Priestley concluded that plants purify air, allowing animals to breathe. However, he couldn't always repeat his experimental results—a key part of the scientific method. Part of the problem may have been that Priestley moved his experiments inside to a dark area, where the plant could not photosynthesize and produce oxygen.

Eight summers after Priestley conducted his experiments, Dutch physician Jan Ingenhousz extended knowledge about photosynthesis in his book, *Experiments upon Vegetables, Discovering Their Great Power of Purifying the Common Air in the Sunshine, and of Injuring it in the Shade or at Night.* Ingenhousz conducted more than 500 experiments to identify the components of photosynthesis.

Ingenhousz hypothesized that the sun's warmth powered photosynthesis. He found that when leaves were placed in water in an inverted jar near a fire but shielded from light, "bad" air formed. But leaves from the same plant in a jar exposed to air and sunlight produced "good" air. Ingenhousz called this *dephlogisticated air* (from *phlogiston,* which refers to burning matter). He wrote, "No dephlogisticated air is obtained in a warm room, if the sun does not shine upon the jar containing the leaves. . . . The sun by itself has no power to mend air without the concurrence of plants." He concluded that leaves are the site of "dephlogistication" and that they operate differently in the light than in the dark—two very important contributions to our modern knowledge of photosynthesis. But Ingenhousz also recommended that people place houseplants outdoors at night—otherwise, he asserted, the plants might poison the air!

by four nitrogen atoms and several organic rings. Transfer of energy occurs in this part of the molecule. A long hydrocarbon tail is hydrophobic, and this is why the molecule is anchored in lipids within the chloroplasts of plant cells, rather than dissolved in the watery part of the cell surrounding them.

Organisms that photosynthesize usually have several types of pigments. Having different pigments extends the range of light wavelengths that a cell can harness, because each pigment absorbs a particular range of wavelengths of light, then passes the energy to chlorophylls. These energy-capturing molecules other than the chlorophylls are called **accessory pigments.** When the days become shorter, many plants break down chlorophyll faster than they resynthesize it. The green pigment no longer completely masks the accessory pigments, and the spectacular colors of autumn emerge.

A chart called an absorption spectrum shows the wavelengths that a particular pigment absorbs. The absorption spectrum in figure 6.6 shows at a glance how various accessory pigments absorb wavelengths between 470 and 630 nm that the two main types of chlorophyll cannot. Chlorophylls *a* and *b* have two peaks each, in the red and in the blue parts of the spectrum.

FIGURE 6.6

Other Photosynthetic Pigments.

Accessory pigments absorb wavelengths of light that chlorophylls cannot. These pigments include carotenoids, phycoerythrins, and phycocyanins.

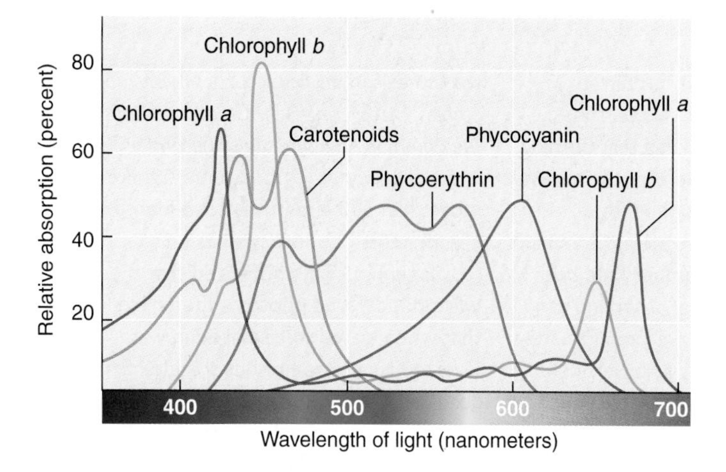

Many of the additional colors found in many plants and algae come from molecules that often extend the range of colors used for photosynthesis. Accessory pigments called phycobilins (phycocyanins and phycoerythrins) impart brilliant colors to algae (table 6.1). Carotenoids, found in many species, include carotenes and their oxidized derivatives, xanthophylls. Carotenoid pigments absorb wavelengths between 460 and 550 nanometers, producing yellow, orange, and red colors. These pigments provide the distinctive colors of carrots, tomatoes, bananas, and squashes. Animals do not manufacture carotenoids, but they eat them in plants and then use the pigments to color various structures. Colorful frogs, fishes, and corals, a squid's ink, flamingo feathers, and egg yolks all owe their brilliant colors to carotenoids, as does the red pigment released when a lobster boils.

Carotenoids also protect plants from damage from oxygen. Recall from the opening essay for chapter 2 on nitric oxide that oxygen tends to form high-energy free radicals that can destroy cellular structures. Many herbicides kill plants by blocking carotenoid synthesis, which removes protection from oxygen free radical damage.

Chloroplasts Are the Sites of Photosynthesis

A chloroplast is a type of **plastid,** which is an organelle that synthesizes or stores nutrients. Other types of plastids include amyloplasts, which store starch, and chromoplasts, which store carotenoids and impart color to flowers and fruits. Chloroplasts are the sites of photosynthesis in plants and algae (fig. 6.7). Most photosynthetic cells contain 40 to 200 chloroplasts, adding up to half a million of the organelles per square millimeter of leaf.

A chloroplast consists of two outside membranes that enclose a gelatinous matrix called the **stroma.** The stroma contains ribosomes, DNA, and enzymes used to synthesize carbohydrates. Suspended in the stroma are folded sacs of **thylakoid mem-**

TABLE 6.1 Accessory Pigments

PIGMENT	ORGANISMS	COLOR
Bacteriochlorophyll	Green bacteria, purple bacteria	Green
Bacteriorhodopsin	Halophilic (salt-loving) archaea	Purple
Carotenoids	Plants, algae, bacteria, archaea	Red, orange, yellow
Chlorophylls	Plants, green algae, cyanobacteria	Bluish green
Fucoxanthin	Brown algae, diatoms, dinoflagellates	Brown
Phycocyanin	Red algae, cyanobacteria	Blue
Phycoerythrin	Red algae	Red
Xanthophylls (oxidized carotenoids)	Plants, algae, bacteria	Red, yellow

branes, which enclose an area called the **thylakoid space.** These membranes are unique to chloroplasts and it is in these membranes that chlorophyll molecules are embedded. Ten to 20 thylakoids may stack into structures called **grana** (singular: granum). Cells that contain chlorophyll (and thus are green) tend to be located on the parts of plants that face the sun. Chlorophylls and accessory pigments in photosynthetic bacteria are also embedded in membranes.

When a chlorophyll molecule absorbs photon energy, its electrons become distributed differently. The molecule becomes excited and unstable, ready to release the just-captured energy. However, a single chlorophyll *a* molecule can absorb only a small amount of photon energy. Several chlorophylls located near each other capture much more energy because they can pass the energy on to other molecules, freeing themselves to absorb other photons as they strike. If 200 chlorophyll molecules aggregate and pass the energy from a photon along, the chance that a second photon will quickly follow is 40,000 times greater than if the first photon hits a single chlorophyll. Chlorophyll molecules aggregate with other pigment molecules to form a **photosystem,** which greatly enhances the efficiency of photosynthesis. Photosynthesis in algae and higher plants occurs along two linked photosystems; one photosystem passes electrons to the other.

FIGURE 6.7

Chloroplasts.

Leaf mesophyll tissue consists of cells with many chloroplasts. The chloroplast is the site of photosynthesis. This double-membraned organelle contains flat, interconnected sacs called thylakoids, which are organized into stacks called grana. Light is absorbed and converted to chemical energy in the grana. The chemical energy is then used in the stroma to manufacture carbohydrates.

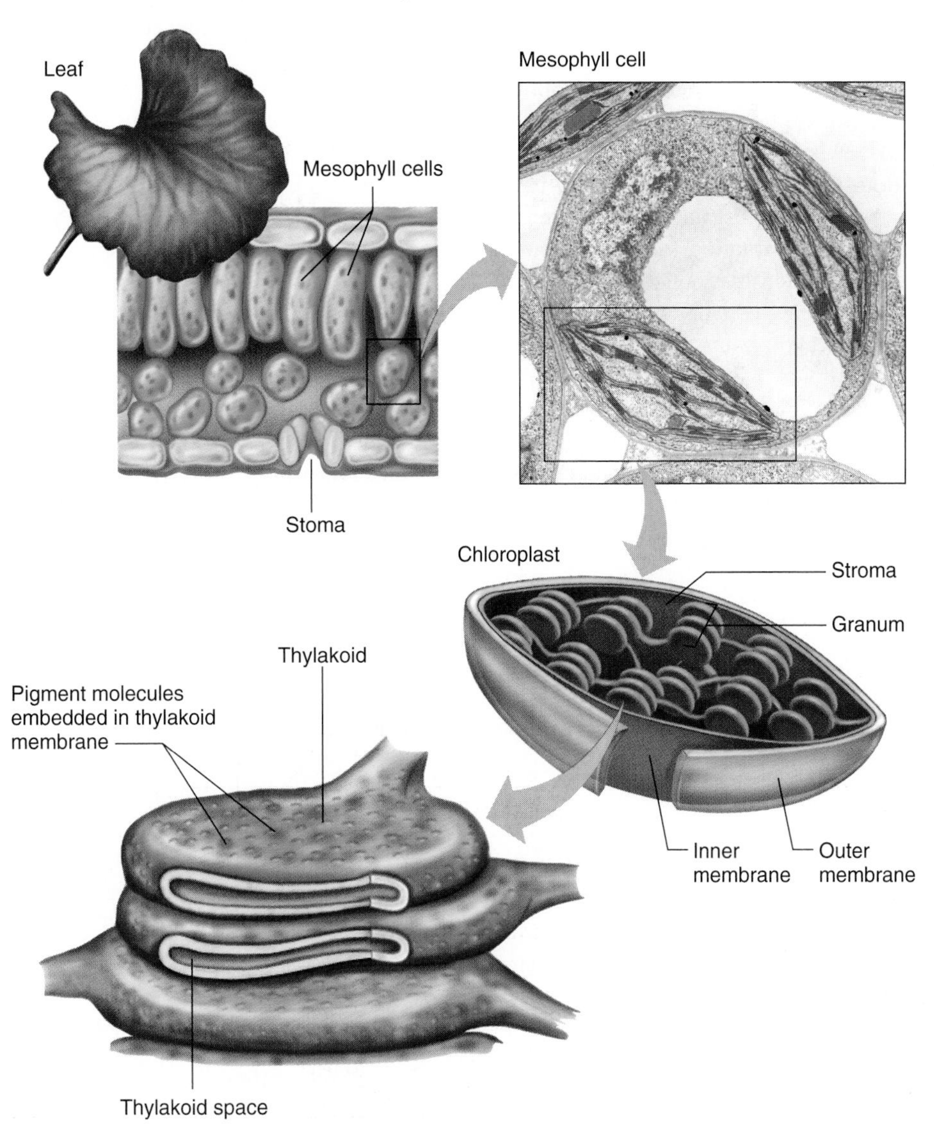

Photosynthesis begins within the pigment complex of a photosystem. The pigment complex has two major parts: an antenna complex and a reaction center. An **antenna complex** is a collection of molecules, including about 300 chlorophyll molecules, 50 accessory pigments, and proteins that anchor the other molecules in the thylakoid membranes (fig. 6.8). Several chlorophylls within the antenna complex act as antennae, which means that they capture photon energy and pass it to other chlorophyll molecules. Within microseconds after the antenna complex absorbs photon energy from sunlight, the energy passes to the **reaction center,** which is a special pair of energy-gathering molecules of chlorophyll *a* coupled with associated proteins.

Each antenna complex contains one reaction center. The electrons of the reaction center's two chlorophyll molecules are less excited than electrons in other chlorophylls. When the photon strikes a reactive (reaction center) chlorophyll, one of its electrons becomes so energized that it is boosted to a higher energy orbital. These boosted chlorophyll electrons can meet any of several fates. Their energy can escape as heat or as an afterglow of light called **fluorescence.** Fluorescent light has a longer wavelength, and therefore less energy, than the light that originally excited the pigment. The green fluorescent protein described in the chapter opening essay illustrates this phenomenon. Chlorophyll fluoresces deep red. Alternatively, photon energy can pass to neighboring molecules. This is what happens in photosynthesis.

Different types of photosynthetic organisms tend to have similar reaction centers, but vary in the organization of antenna complexes. This flexibility reflects adaptation to environmental conditions that transmit different intensities of light.

Eukaryotes and cyanobacteria have two kinds of reaction center chlorophylls because they use two photosystems. The reactive chlorophyll of photosystem I absorbs light energy mostly at 700 nanometers and is therefore called P700 (P stands for pigment). The reactive chlorophyll of photosystem II is called P680 and absorbs energy of 680 nanometers. Although reaction centers comprise less than 1% of the chlorophyll in plants, they are vitally important in photosynthesis because they first capture the light energy.

FIGURE 6.8

The light-harvesting system in chloroplasts collects photons and passes their energy via electrons to the reaction center chlorophylls.

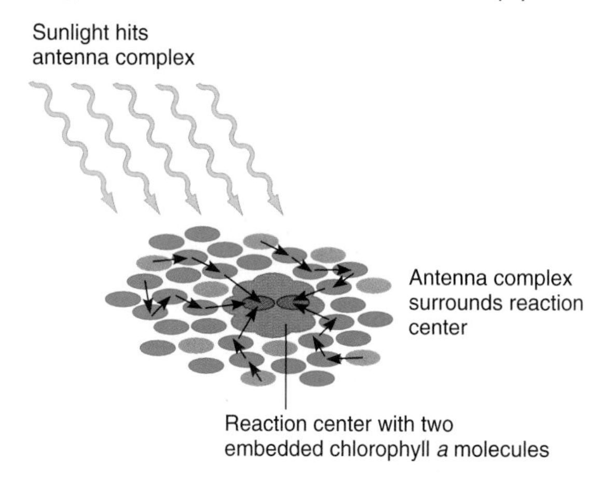

Photosynthetic bacteria other than cyanobacteria use simpler (although still quite complex) collections of pigment molecules to capture photon energy. Consider the purple bacterium *Rhodopseudomonas acidophila,* which lives under algae in murky ponds, gathering stray light rays. In the cell membrane are embedded groups of pigment molecules and proteins that form light-harvesting complexes. Because the cell membrane wraps around the cell several times, it aligns the complexes into layers, which greatly boosts photosynthetic efficiency. A light-harvesting complex can capture and transfer solar energy in mere trillionths of a second. The electrons are passed to reactive bacteriochlorophyll molecules at the center of the complexes, and then to the reactions of carbon fixation.

6.2 MASTERING CONCEPTS

1. What is light?
2. How do pigments capture photon energy?
3. What is the structure of a chloroplast?
4. How are pigment and protein molecules organized in chloroplasts?
5. Why are reaction centers vital to photosynthesis?

OLC

6.3 The Light Reactions Begin Photosynthesis

In plants, two linked photosystems capture light energy and store it in the chemical bonds of ATP and NADPH. These reactions cannot proceed without light.

Photosynthesis occurs in two stages: the **light reactions** and the **carbon reactions.** The light reactions harvest energy from photons and use that energy to synthesize molecules of ATP and NADPH. Electrons from chlorophyll molecules carry the energy and are replaced with electrons removed from water. The hydrogens from water reduce $NADP^+$ and set up a gradient of protons within the chloroplasts. Oxygen gas derived from water is released as a by-product. The carbon reactions then use the ATP and NADPH produced in the light reactions to reduce CO_2 to carbohydrate. The light reactions require light, but the carbon reactions can occur in light or dark. Overall, light is necessary to power photosynthesis and to manufacture many of the enzymes that are essential to the process.

The light reactions take place in four types of clusters of proteins that are embedded in the thylakoid membranes. Two cluster types are photosystems, and these are interspersed with two other types of clusters, which are the electron transport chains. Figure 6.9 shows the light reactions schematically and functionally, and figure 6.10 depicts the same process as it occurs in a thylakoid, providing a more structural view.

Photosystem II Produces ATP

Photosynthesis begins in the cluster of pigment molecules of photosystem II. This may seem illogical, but the two photosystems were named as they were discovered; photosystem II was discovered after photosystem I, but it functions first in the overall process.

Pigment molecules in photosystem II absorb the energy from incoming photons in sunlight. The energy is transferred from one pigment molecule to another until it reaches a chlorophyll *a* reaction center, where it excites an electron. The excited electron is ejected from this chlorophyll *a* molecule and is grabbed by the first electron carrier molecule of the **electron transport chain** that links the two photosystems (fig. 6.10). Photosystems II and I are often called a "Z-scheme" because the electron transport chain connecting them, when diagrammed, resembles the letter Z. The reactive chlorophyll *a* molecules replace their lost electrons when light energy is used to split water (H_2O) into oxygen gas (O_2) and protons (H^+), releasing electrons.

As electrons are passed from carrier to carrier, the energy they lose is used to drive the active transport of protons from the stroma into the thylakoid space. The protons cannot easily

FIGURE 6.9

Overview of Photosynthesis.

Molecules in the thylakoid membranes capture sunlight energy and transfer that energy to molecules of ATP and NADPH. The enzymes of the carbon reactions use this energy to capture carbon dioxide and convert it to glucose.

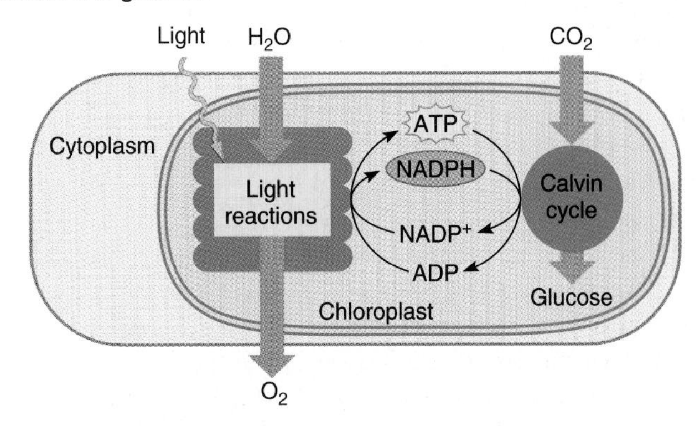

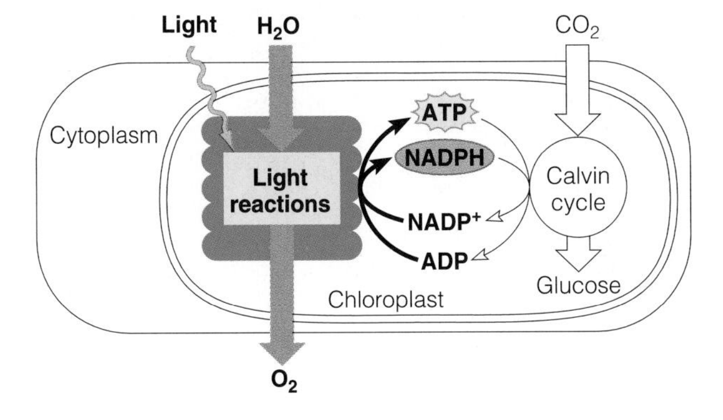

FIGURE 6.10

The Light Reactions of Photosynthesis.

The sun's energy is captured by chlorophyll molecules in photosystem II and transferred to electrons ripped from water molecules. Oxygen is released as a by-product. The energy-rich electrons are passed through a series of carriers that make up an electron transport chain to photosystem I. At each transfer, a small amount of energy is removed and used to pump protons (hydrogen ions taken from water) into the thylakoids. As the thylakoids release the hydrogen gradient, phosphate is added to ADP, forming ATP. In photosystem I, more solar energy is added to the electrons, which are passed to $NADP^+$, creating the energy-rich NADPH. The energy carriers ATP and NADPH are used to fuel the carbon reactions of the Calvin cycle.

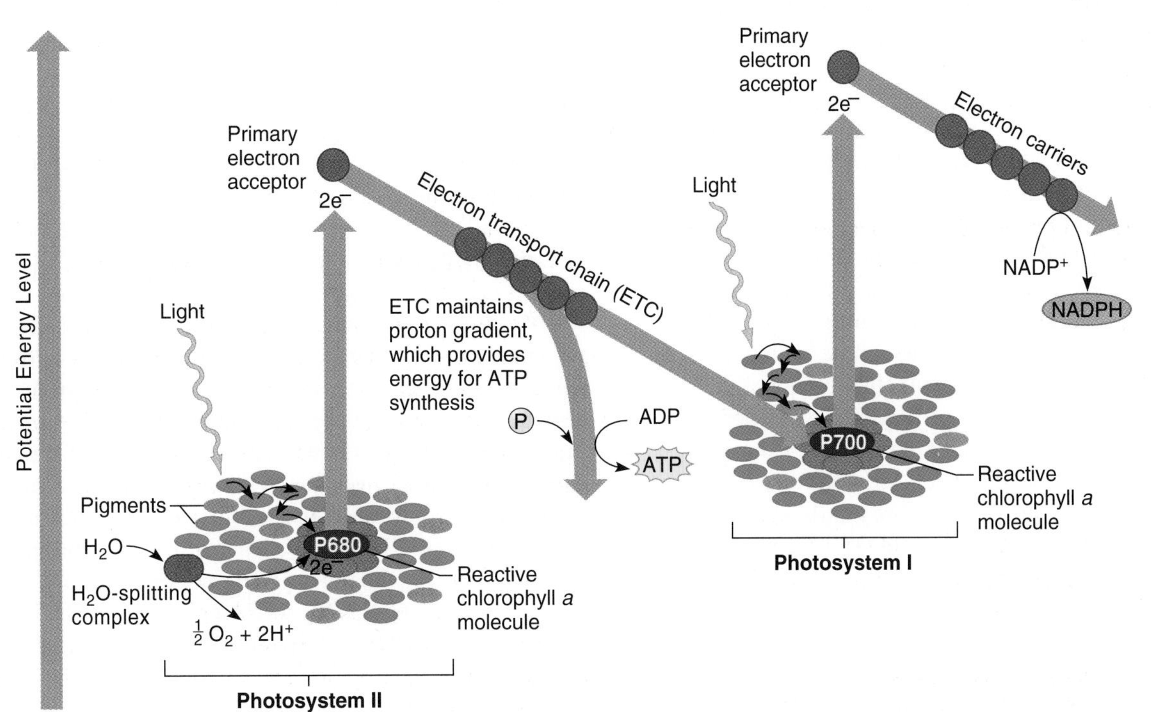

FIGURE 6.11
Chemiosmotic Phosphorylation.
The energy of a proton (H^+) gradient is used to phosphorylate ADP to ATP. As electrons from photosystem II release their energy, enzyme systems pump protons into the thylakoid, forming a tremendous gradient. A membrane channel protein, ATP synthase, releases the protons and uses the force of the gradient to phosphorylate ADP to ATP. The electrons from photosystem II are transferred to photosystem I, where they capture more sunlight energy which is transferred to $NADP^+$ with the electrons to form NADPH. Proton pumping, ATP, and NADPH synthesis occur simultaneously and continuously in the light.

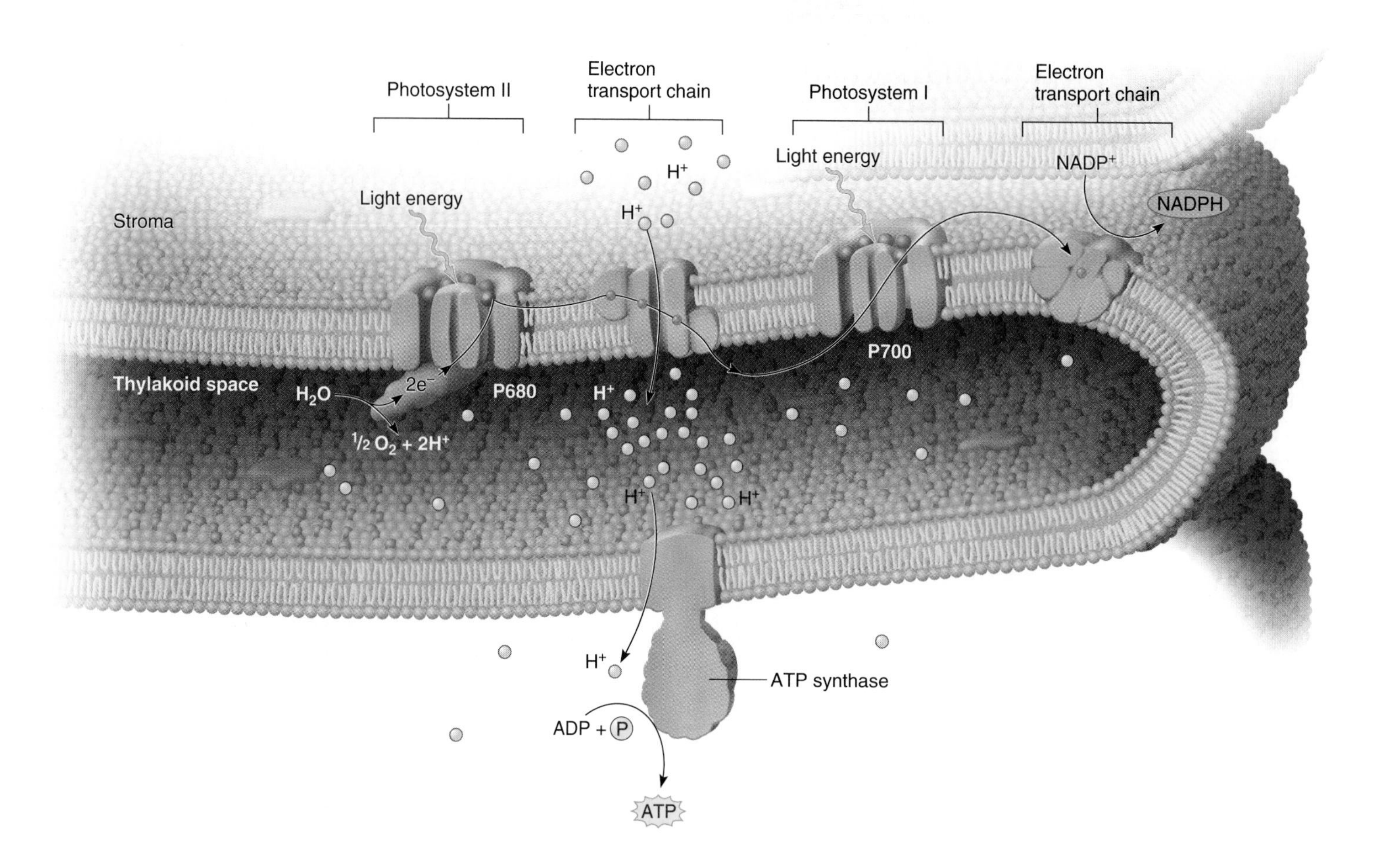

leak back out because the membrane is relatively impermeable to them. The proton concentration increases in the thylakoid, establishing a difference in pH, charge, and chemical concentration between the inside of the thylakoid and the stroma of the chloroplast. This gradient holds such a tremendous amount of potential energy that it is called a "proton motive force." The energy in the gradient is captured by a membrane-bounded enzyme complex known as **ATP synthase** that functions as a channel for the protons (fig. 6.11). As the protons move through the channel, they alter the shape of ATP synthase in a way that adds phosphate to ADP, producing ATP. The ATP molecules formed in this way remain in the stroma, and are used in the carbon reactions of photosynthesis. The coupling of ATP formation to the release of energy from a proton gradient is called **chemiosmotic phosphorylation,** because it is the addition of a phosphate using energy from the movement of chemicals (protons) across a membrane (chemiosmosis). This process is also very important in cellular respiration. Chapter 7 discusses its discovery in mitochondria.

Photosystem I Produces NADPH

Photosystem I functions much as photosystem II does. Photon energy strikes energy-absorbing molecules of chlorophyll *a*, which pass the energy to chlorophyll *a* molecules in the reaction center. The reactive chlorophyll molecules eject electrons to the first electron carrier molecule in a second electron transport chain. The boosted electrons in photosystem I are then replaced with electrons passing down the first electron transport chain from photosystem II. This sequence of electron movement continues until finally the transported electrons of photosystem I reduce a molecule of $NADP^+$ to NADPH. This NADPH, plus the ATP that photosystem II generates, become the energy sources that power the carbon reactions that follow.

Biotechnology 6.1

Mimicking Photosynthesis

No chemist has yet to come up with a series of reactions that harnesses solar energy and converts it to chemical energy as elegantly as nature does in photosynthesis. Although we understand how many parts of the process occur, photosynthesis remains somewhat of a mystery. How can such a straightforward set of reactions require so many large and complex molecules? Can researchers use the chemicals of photosynthesis, or perhaps other molecules, to reconstruct the events in the laboratory? Attempts to do so indicate just how complicated photosynthesis is.

Devens Gust and his coworkers at Arizona State University in Tempe have created a single-photosystem mimic of the light reactions of photosynthesis. It is based on a liposome, which is a bubblelike phospholipid bilayer similar to a cell membrane. First, they assembled an artificial reaction center. It consists of three molecules, covalently bonded to each other (a carotene, a porphyrin, and a fullerene) and placed within the liposome bilayer (fig. 6.B). In a plant cell, the reaction center consists of reactive chlorophyll *a* molecules anchored within the thylakoid membranes.

Artificial photosynthesis starts as light impinges upon the porphyrin, which is a chromophore, a molecule, or part of a molecule that captures photon energy. (Chlorophyll contains a porphyrin, as does hemoglobin, the molecule that transports oxygen in blood.) An electron from the porphyrin becomes excited and jumps over to the fullerene, which is a huge ball of 60 carbon atoms. Meanwhile, the carotene donates an electron to replace the one zapped out of the porphyrin. From the fullerene, the electrons are passed through electron carriers, and they power the buildup of protons inside the liposome cavity, just as happens in the thylakoid space in a chloroplast.

Included in the liposome are embedded molecules of ATP synthase, taken from spinach cells. The buildup of protons within the liposomes forms a gradient that causes them to exit the liposome through the channels in the ATP synthase. The exodus of protons triggers ATP formation.

Being able to mimic parts of photosynthesis in the laboratory will enable researchers to more closely study the process. It may also have practical applications. ATP can be used industrially to power many types of chemical reactions used in the manufacture of diverse products. If the entire process of photosynthesis can be duplicated and scaled up, it could be useful in providing a new, clean energy source or in countering the buildup of carbon dioxide in the atmosphere that may be contributing to global warming.

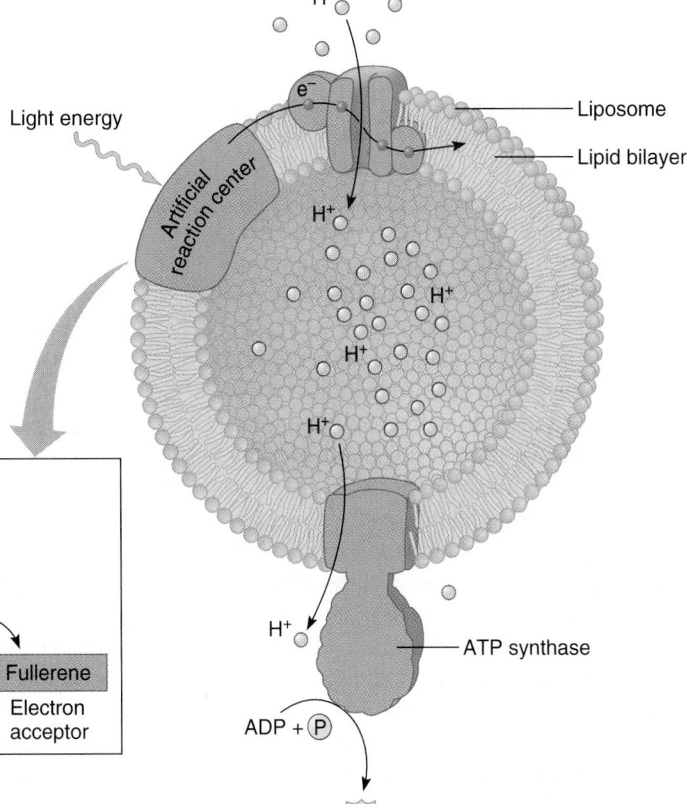

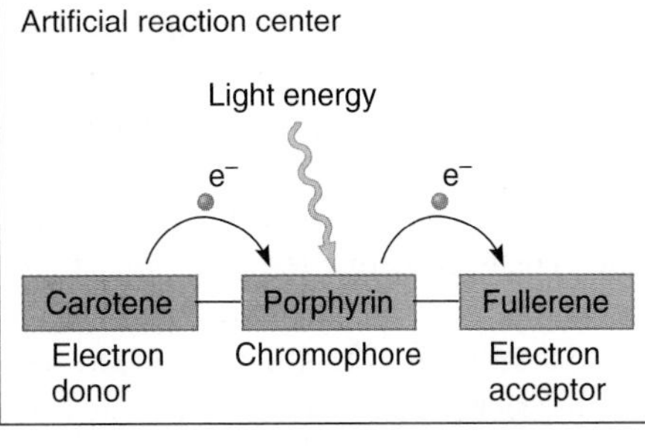

FIGURE 6.B
Studying Photosynthesis.
An artificial thylakoid system developed by chemists at Arizona State University mimics the light reactions. It uses a reaction center consisting of carotene, porphyrin, and fullerene to capture light energy and generate a proton buildup inside a liposome. The protons then diffuse out of channels in ATP synthase molecules embedded in the liposome membrane, triggering ATP formation.

TABLE 6.2 Requirements for the Light Reactions of Photosynthesis

1. Light-harvesting pigment molecules and electron transfer systems must be in close physical proximity.
2. Absorbed light must be in the visible part of the electromagnetic spectrum.
3. Energy must move rapidly from pigments to the reaction center to the electron transport chain. Proton gradient formation and ATP synthesis must also be fast.
4. Electron transport must be unidirectional, establishing a proton gradient.
5. Pigment molecules and electron transport molecules must be used over and over.

Table 6.2 summarizes conditions that must be met for the light reactions to proceed. Biotechnology 6.1 looks at an attempt to re-create photosynthesis in the laboratory. Sometimes people try to do the opposite and stop photosynthesis, which is how many herbicides function. For example, the herbicide DCMU blocks electron flow in photosystem II. The herbicide paraquat, noted for its use in destroying marijuana plants, takes electrons from photosystem I and reacts them with oxygen to produce free radicals, which destroy lipids in the chloroplasts, shutting down photosynthesis.

Two Photosystems Extract More Energy Than One

Why does photosynthesis in eukaryotes require two photosystems? The answer is that two photosystems are more efficient than one—they increase the amount of energy extracted from sunlight.

Photosynthesis in prokaryotes uses only one photosystem and recycles electrons back to the same reaction center from which they originate. This short form of photosynthesis generates chemical energy, but not enough to allow cells to synthesize many molecules. A single photosystem may have been the sole strategy for harnessing solar energy for the first billion years of life. It would have yielded no NADPH or oxygen. Splitting water to replace the light-boosted electrons probably required more energy than these early microbial systems could muster.

Grafting a second, more powerful photosystem onto a version of a simpler microbial system would have enabled plants to extract more energy from sunlight. This second photosystem, photosystem II, acts first. It passes electrons in a straight line rather than cycling them back to the original reaction center until they gain enough energy, as photosystem I does. The two photosystems, connected in a series, produce more energy—much as two batteries increase the voltage in a flashlight. In addition, the newer photosystem II has a unique arrangement of pigments that allows it to harvest shorter wavelengths of light. Electrons move from water to photosystem II to photosystem I, but sunlight hits and reacts with both photosystems simultaneously.

The evolution of photosystem II forever changed life on Earth. No longer did organisms have to rely directly on compounds in the environment as energy sources; they could now harness solar energy. Moreover, the oxygen pumped into the atmosphere by photosynthesis allowed the evolution of aerobic respiration.

6.3 MASTERING CONCEPTS

1. What are the products of the light reactions of photosynthesis?
2. What conditions must exist for the light reactions to proceed?
3. Which events capture and transfer photon energy in photosystem II?
4. How are electrons passed from photosystem II to photosystem I?
5. How are the boosted electrons from photosystem II replaced?
6. How does photosystem II produce ATP?
7. What happens in photosystem I?
8. How do two photosystems increase the efficiency of photosynthesis?

OLC

6.4 The Carbon Reactions "Fix" Carbon

The reactions of the Calvin cycle use the energy stored in the light reactions to reduce carbon dioxide to carbohydrates. These reactions can occur in the light or the dark.

The carbon reactions incorporate, or fix, carbon from CO_2 into organic compounds that cells can use. Photosynthetic organisms get CO_2 from the environment. Algae, bacteria, and some aquatic plants absorb dissolved CO_2 from the water that surrounds them. In plants, atmospheric CO_2 enters through tiny leaf and stem openings called **stomata** (singular: stoma).

How Does the Calvin Cycle "Fix" Carbon?

The carbon reactions can occur in both darkness and light as long as ATP and NADPH are available. These reactions form a metabolic cycle known as the Calvin cycle, named for American biochemist Melvin Calvin. It is also known as the C_3 cycle because a three-carbon compound, phosphoglyceric acid (PGA), is the first stable compound in the pathway. Plants that use only the Calvin cycle to fix carbon from CO_2 are called **C_3 plants.** They include cereals, peanuts, tobacco, spinach, sugar beets, soybeans, most trees, and lawn grasses.

ATP and NADPH, the products of the light reactions, provide the energy necessary to power the carbon reactions. If ATP and NADPH are plentiful, the Calvin cycle continually fixes the carbon from CO_2 into small organic molecules such as PGA. The cell can then use these molecules to synthesize more complex nutrient molecules.

FIGURE 6.12
Carbon Fixation.
The products of the light reactions—ATP and NADPH—drive the carbon reactions (Calvin cycle), shown here in a simplified cycle. The cycle "fixes" carbon from CO_2 into organic molecules. Rubisco catalyzes the first step, the reaction between RuBP and CO_2. The overall cycle of the carbon reactions generates a three-carbon molecule, PGAL, which is converted to glucose and other organic molecules and also regenerates RuBP. This pathway operates in most photosynthesizing plants, but some species have significantly modified cycles.

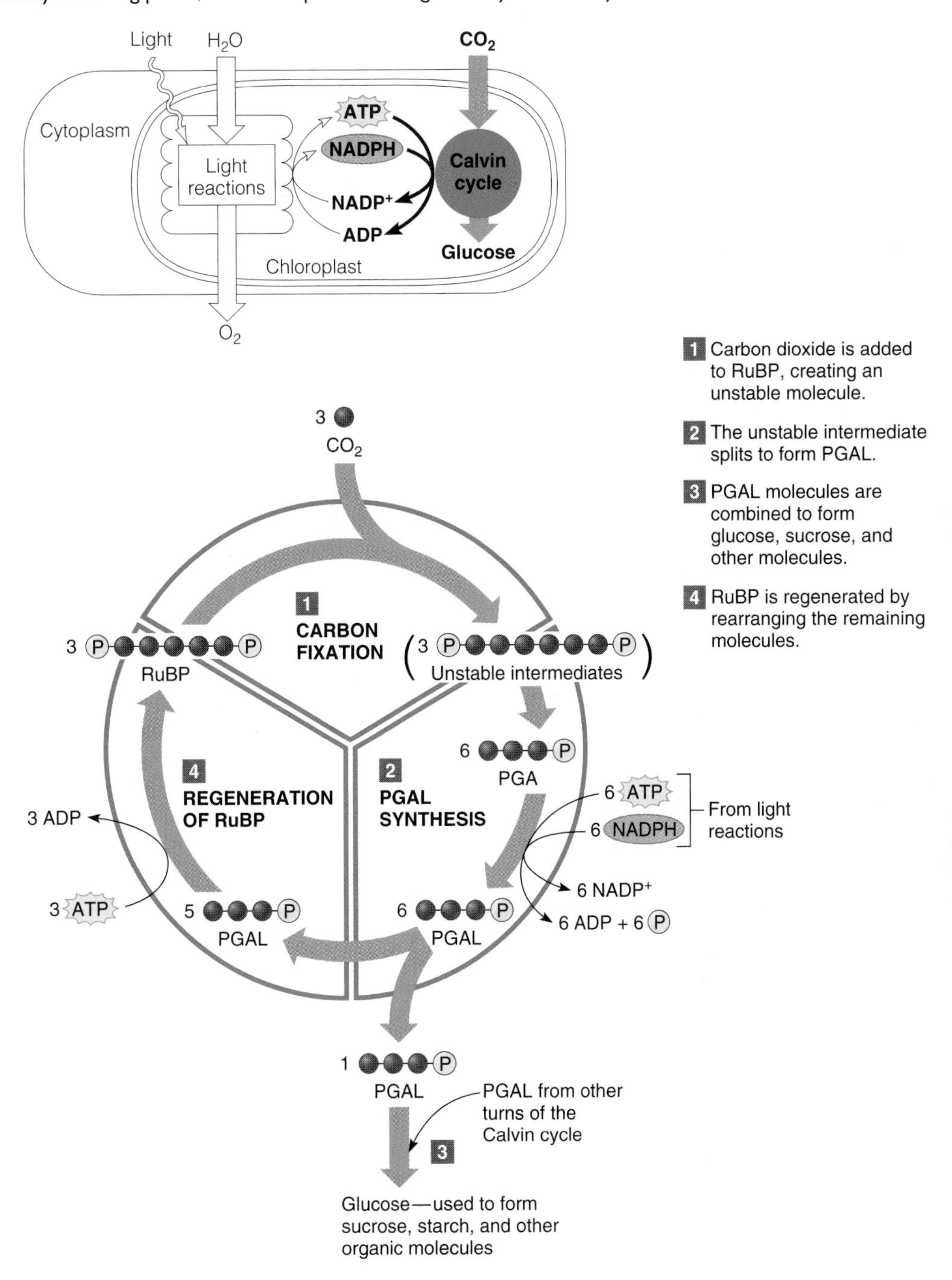

The Calvin cycle begins when CO_2 reacts with **ribulose bisphosphate** (RuBP), a five-carbon sugar with two phosphate groups, to yield an unstable six-carbon molecule (fig. 6.12). This molecule immediately breaks down to two molecules of PGA. The enzyme that catalyzes the reaction between RuBP and CO_2 is RuBP carboxylase/oxygenase, also known as **rubisco.** It uses NADPH and ATP as its energy source. This is one of the most important and abundant proteins in the living world. Further steps in the cycle convert phosphoglyceric acid (PGA) to phosphoglyceraldehyde (PGAL), which is the direct carbohydrate product of the Calvin cycle. PGAL is used in further reactions to produce glucose and other nutrient molecules. Some of the PGAL is rearranged to form RuBP. This regeneration of the five-carbon starting material is important because it perpetuates the cycle.

How Did Experiments Reveal the Steps of the Calvin Cycle?

A series of clever experiments identified the intermediates of the Calvin cycle using ^{14}C, or radioactive carbon. ^{14}C is identical chemically to the common isotope ^{12}C, but it is radioactive because it contains more neutrons.

In the early 1950s, Calvin and his colleagues exposed a dense culture of the green alga *Chlorella* to CO_2 whose carbon was the ^{14}C radioactive isotope. Then Calvin identified the chemicals that progressively incorporated the radioactive carbon "label" and traced the path of the ^{14}C as it moved from CO_2 to carbohydrate. The ^{14}C allowed Calvin to detect very small quantities of product. When Calvin exposed the alga to radioactively labeled CO_2 for 60 seconds, he found that ^{14}C labeled many compounds in the alga. But when he repeated the experiment with a 7-second exposure, most of the ^{14}C appeared in PGA. This suggested that the molecule that initially joins CO_2 is a two-carbon compound, because PGA has three carbons. Calvin searched for this two-carbon acceptor, but he couldn't find it. He then began looking for another molecule that, combined with CO_2, would yield a three-carbon product. He found the elusive acceptor in RuBP. Calvin used a similar approach to decipher the other biochemical reactions of photosynthesis.

6.4 MASTERING CONCEPTS

1. What is the function of the Calvin cycle?
2. What is the role of ribulose bisphosphate in the Calvin cycle?
3. Why is rubisco an important molecule?
4. How does the energy from ATP and NADPH power the Calvin cycle?
5. How did radioactive labels help identify intermediates of the Calvin cycle?

OLC

6.5 How Efficient Is Photosynthesis?

Like all biological processes, photosynthesis is not highly efficient, largely due to a process called photorespiration. Some plants, however, have evolved mechanisms that minimize photorespiration.

The atmosphere is only 0.035% CO_2, yet each year, plants use 200 billion tons of carbon from CO_2 to manufacture organic compounds. That sounds like a large output, but the proportion of photon energy that plants actually harness in the carbohydrate products of photosynthesis is less impressive. The process of energy acquisition is quite inefficient. If each photosystem absorbs the maximum possible number of photons, theoretical calculations yield an efficiency rate of 30%. In reality, this rate is much lower.

On cloudy days, field measurements of the photosynthetic efficiency of individual plants average about 0.1%, while measurements for intensively cultivated plants average about 3%. Some meticulously cared for laboratory-grown plants can reach 25% efficiency. The plant with the greatest natural efficiency—8%—is *Oenothera claviformis,* the annual winter evening primrose grown in Death Valley, California. Sugarcane follows at 7%.

Why do plants waste so much solar energy? One contributing factor is a process that counters photosynthesis called **photorespiration.**

How Does Photorespiration Alter the Calvin Cycle?

Some enzymes, under certain circumstances, use a different substrate and therefore catalyze a different reaction. Often this reaction is less than useful to a cell. One of the best-studied examples of this phenomenon is the enzyme rubisco. Under low carbon dioxide concentrations or high oxygen concentrations, rubisco uses oxygen as a substrate instead of carbon dioxide and starts a process that removes carbon from the carbon reactions. This phenomenon is known as photorespiration. Since this reaction also occurs at higher temperatures, plants living in hot climates face a serious problem: They would fix carbon and release carbon dioxide at nearly equal rates, in a futile cycle. In addition, plants must open their stomata to allow carbon dioxide into their tissues. If they open their stomata in dry, hot climates to fix carbon, they risk losing water at a dangerously high rate.

As plants release oxygen during photosynthesis, the buildup increases the likelihood that photorespiration will occur. Science has yet to discover any benefit for plants from this process. Some plants may lose as much as 30 percent of their fixed carbon through photorespiration, so any mechanism that reduces this process would be useful.

C_4 Photosynthesis Decreases Photorespiration

Photorespiration peaks in hot, dry conditions, when plants begin to close their stomata to conserve water. In this situation, the plants lose carbon. Any plant that can avoid photorespiration has a significant advantage. An adaptation called **C_4 photosynthesis** enables certain plants to reduce photorespiration. This route is called the C_4 pathway, and plants that use it are called C_4 plants, because they use a four-carbon compound to concentrate carbon within special cells.

When biologists repeated Calvin's study labeling carbon from CO_2 in different types of plants, their results usually confirmed Calvin's. But in 1965, a study of sugarcane produced strikingly different results. When sugarcane was exposed to radioactive CO_2 for 1 second, 80% of the radioactivity eventually appeared not in a three-carbon compound, as in the Calvin cycle, but in four-carbon malic acid. Most of this malic acid surfaced in thin-walled **mesophyll cells** (fig. 6.13), which lack most of the enzymes of the Calvin cycle. After about 10 seconds, the radioac-

FIGURE 6.13

C_4 Photosynthesis.

C_4 plants such as corn have a characteristic cellular organization (**A, B**) that physically separates the light reactions and the carbon reactions. A photosynthetic mesophyll cell captures CO_2, combining it with phosphoenolpyruvate (PEP), a three-carbon compound, to yield the four-carbon compound oxaloacetic acid (OAA). OAA is then converted to malic acid and aspartic acid (**C**) which are carried into adjacent bundle-sheath cells, where they are split to release CO_2 and pyruvate. The carbon reactions in the bundle-sheath cells make glucose from the released CO_2, and pyruvate returns to the mesophyll cell, where it is converted to PEP and picks up another CO_2, restarting the C_4 cycle. By concentrating CO_2 in the bundle-sheath cells, C_4 plants reduce the effects of photorespiration.

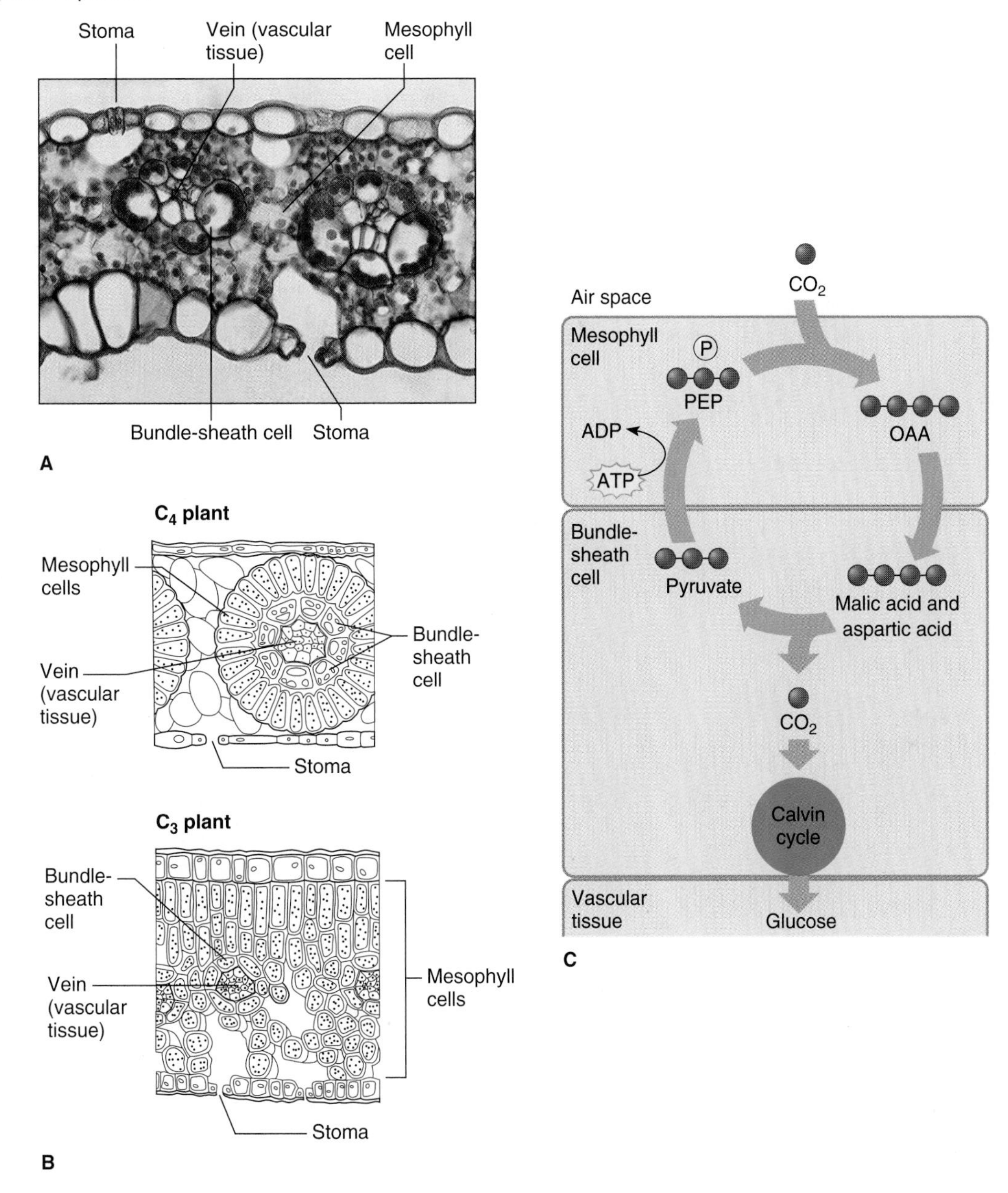

tive label appeared in PGA in the thick-walled **bundle-sheath cells** that surround veins. If CO_2 was being fixed into malic acid, was the Calvin cycle occurring?

Further studies of sugarcane leaves revealed that the plant does fix carbon from CO_2 via the Calvin cycle, but only after first fixing it with another set of reactions. A more complex anatomy in C_4 plants accommodates this extra biochemical pathway. These plants, therefore, have both C_3 and C_4 photosynthesis, but the pathways take place in different cells. C_4 photosynthesis occurs in mesophyll cells, but the Calvin cycle fixes carbon in the bundle-sheath cells of these plants.

In C_4 photosynthesis, CO_2 diffuses into the leaf through stomata and fixes carbon in mesophyll cells. CO_2 combines with a three-carbon compound, phosphoenolpyruvate (PEP), to form the four-carbon oxaloacetic acid (OAA). OAA then is converted to malic acid or aspartic acid, both of which are transported to adjacent bundle-sheath cells. There, the acids split into CO_2 and three-carbon compounds.

FIGURE 6.14

Different Environments Call for Different Approaches to Photosynthesis.

(**A**) At lower temperatures, photorespiration is not significant enough to overload C_3 plants. (**B**) At higher temperatures, C_4 plants use a compartment with high CO_2 concentration to overcome photorespiration, while CAM plants (**C**) store CO_2 at night to reduce photorespiration.

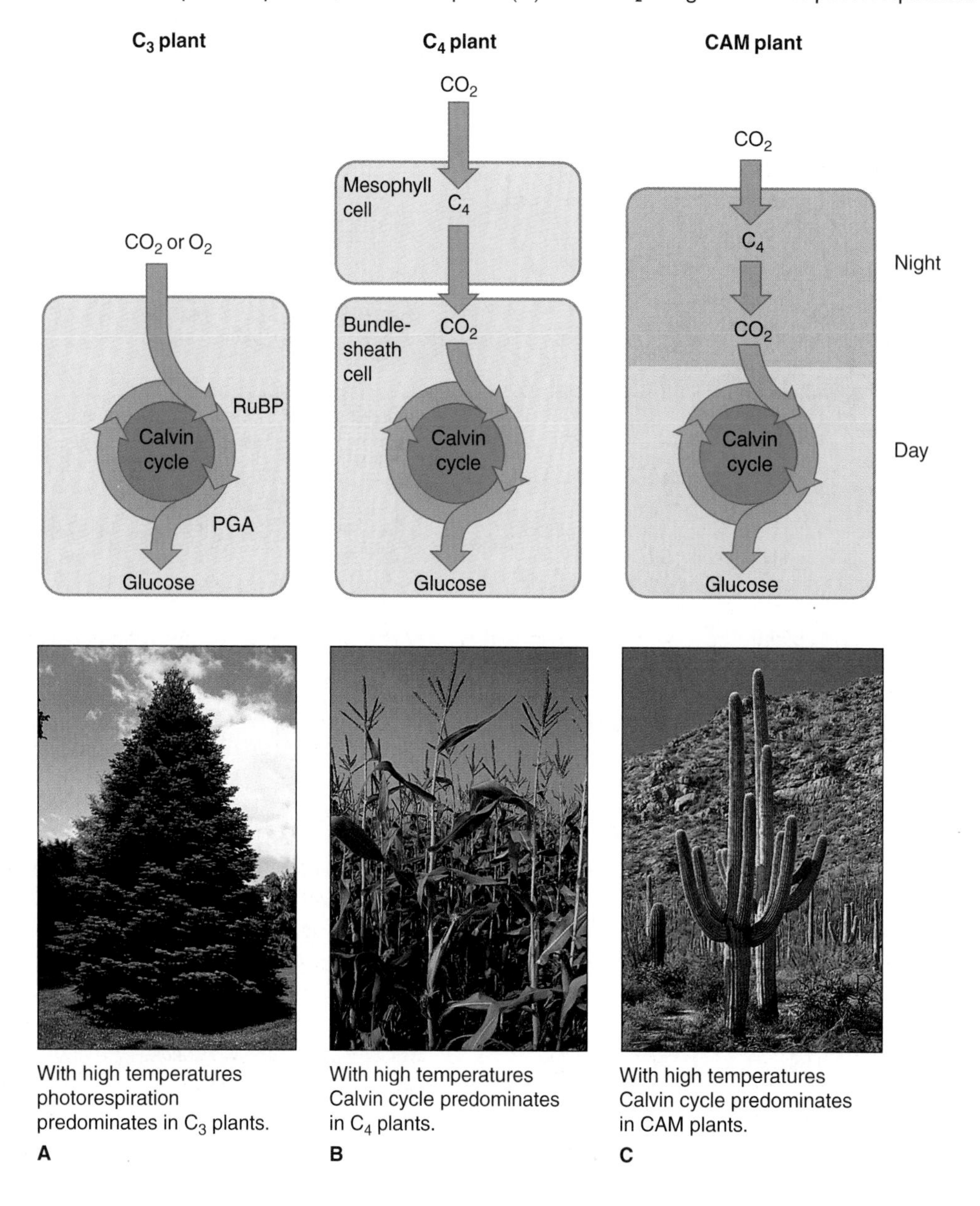

With high temperatures photorespiration predominates in C_3 plants.
A

With high temperatures Calvin cycle predominates in C_4 plants.
B

With high temperatures Calvin cycle predominates in CAM plants.
C

Pumping CO_2 into bundle-sheath cells keeps the internal concentration of CO_2 20 to 120 times greater than normal. This decreases photorespiration and makes C_4 plants more efficient photosynthesizers than C_3 plants in hot, dry, sunny weather. The Calvin cycle fixes the CO_2 released in the bundle-sheath cells. Meanwhile, the three-carbon compound returns to the mesophyll cell, where it is converted to PEP, the initial CO_2 acceptor of C_4 photosynthesis. C_4 can fix CO_2 even when the stomata begin to close, and thus preserve water that would be lost through open stomata. As a result, C_4 plants require only about half as much water as C_3 plants to maintain life.

C_4 photosynthesis evolved in the hot, dry tropics. It probably has persisted because atmospheric CO_2 concentrations have decreased during the last 50 to 60 million years. Today, C_4 plants are all flowering plants growing in hot, open environments. They represent diverse species, suggesting that C_4 photosynthesis may have evolved independently several times. C_4 plants include economically important crops such as corn, millet, and sorghum.

Even though C_4 plants dominate in hot and dry ecosystems because they avoid the photorespiration that affects C_3 plants, in other habitats they are not as abundant. C_4 plants are at an energetic disadvantage in these other environments because they must invest two ATPs for every carbon that goes from the mesophyll cell to the bundle-sheath cell. Still other plants combine aspects of C_3 and C_4 photosynthesis.

TABLE 6.3 Photosynthetic Characteristics of C_3, C_4, and CAM Plants

CHARACTERISTIC	C_3 PLANTS	C_4 PLANTS	CAM PLANTS
Leaf anatomy	Bundle-sheath cells have sparse chloroplasts	Bundle-sheath cells have dense chloroplasts	Large vacuoles
Water (grams) to produce 1 gram of dry photosynthate	450–950	250–350	50–55
Photorespiration	Yes	In bundle-sheath cells	No
Photorespiration rate increases above	15–25°C	30–40°C	About 35°C
Tons of dry photosynthate per hectare per year	20–25	35–40	Low and variable

Source: Randy Moore et al., *Botany.* Copyright © 1995 The McGraw-Hill Companies, Inc.

TABLE 6.4 Factors That Control Photosynthesis

FACTOR	EFFECT
CO_2 concentration	Increasing CO_2 concentration increases the rate of photosynthesis in C_3 plants but not in C_4 plants
Light intensity	The maximum rate of photosynthesis occurs when light is brightest; there is no photosynthesis in the dark
Stage of plant development	Plants in active stages, such as seedlings, require more energy and therefore have higher rates of photosynthesis
Storage or use of photosynthate	The photosynthetic rate increases when injured plants use carbohydrates for energy; it decreases when carbohydrate use diminishes
Temperature	Photosynthetic rate increases at higher temperatures
Water availability	Lack of water closes stomata, inhibiting photosynthesis

CAM Photosynthesis Stores Carbon Dioxide at Night

Just as farmers store a bountiful harvest for use over the long winter, some plants that live in very dry environments take in CO_2 at night, but they do not fix it in the Calvin cycle until daylight. This strategy was discovered in plants of genus *Crassulacea,* and so it is called crassulacean acid metabolism (CAM). Plants with **CAM photosynthesis** include many cacti, pineapple, Spanish moss, orchids, some ferns, and the wax plant. This form of photosynthesis is an adaptation to the day and night temperature and humidity differences in the desert. At night, when temperature drops and humidity rises, stomata open and CO_2 enters the plant. During the hot and dry day, stomata close to conserve water—but the plant already has CO_2 from its nighttime activity. The high CO_2 concentrations produced in this way also reduce photorespiration.

Botanists first observed CAM photosynthesis, indirectly, in the nineteenth century, when they noted that certain plants are acidic at night but become more alkaline during the day. It wasn't until 1958 that botanists discovered the reason for the fluctuating pH—at night, plants open their stomata, take in CO_2, and combine it with the three-carbon compound phosphoenolpyruvate (PEP) to form malic acid. Unlike C_4 metabolism, which occurs in two different cell types, in CAM photosynthesis, malic acid forms and is stored in large vacuoles in the same cells that have chloroplasts. In the daytime, malic acid enters the chloroplast, where it releases CO_2 and sends a three-carbon compound back to the vacuole. The CO_2 is fixed in the Calvin cycle in the chloroplast of the same cell.

The different variations on the photosynthesis theme enable plants to capture and use solar energy in a variety of environments. About 85% of plant species are C_3. Only 0.4% are C_4 plants, but we are very aware of this group because it includes many crop plants. About 10% of plant species use CAM photosynthesis (fig. 6.14). The remaining plant species combine certain reactions of C_3, C_4, or CAM photosynthesis. Some C_3 plants, for example, switch to CAM photosynthesis in times of drought, and some plants use different pathways at different stages in development. Table 6.3 compares C_3, C_4, and CAM plants.

Table 6.4 lists factors that affect photosynthesis.

6.5 MASTERING CONCEPTS

1. How does photorespiration counter photosynthesis?
2. How does C_4 photosynthesis prevent photorespiration?
3. How is CAM photosynthesis an adaptation to the desert environment?
4. How is CAM photosynthesis like C_4 metabolism, and how is it different?

OLC

Chapter Summary

6.1 Life Depends on Photosynthesis

1. **Photosynthesis** is the light-driven reaction of water and carbon dioxide, which produces oxygen, water, and carbohydrate, in the presence of light. The two stages are the **light reactions** and the **carbon reactions.**
2. Plants, algae, and some microorganisms photosynthesize. They are **autotrophs.**
3. All life ultimately depends on photosynthesis.

6.2 How Do Cellular Structures Capture and Use Light Energy?

4. **Photons** move in waves. The longer the **wavelength,** the less energy per photon. Visible light occurs in a spectrum of colors representing different wavelengths. Red and blue light are most effective for photosynthesis.
5. Only absorbed light can affect life, and pigment molecules absorb light. **Chlorophyll *a*** is the primary photosynthetic pigment in plants. **Accessory pigments** absorb wavelengths of light that chlorophyll *a* cannot absorb, extending the range of wavelengths useful for photosynthesis.
6. A chloroplast consists of a gelatinous matrix called the **stroma** that contains stacks of **thylakoid membranes** called **grana.** Pigments aggregate in the grana and absorb light. **Antenna complexes** funnel the energy to reactive chlorophylls that aggregate with proteins to form **reaction centers.**

6.3 The Light Reactions Begin Photosynthesis

7. A **photosystem** consists of an antenna complex and a reaction center. In plants, photosystem II captures light energy and sends electrons from reactive chlorophyll *a* through an **electron transport chain.** The energy drives the active transport of protons into the thylakoid space, where they accumulate and then diffuse out through channels in **ATP synthase.** This movement powers phosphorylation of ADP to ATP. The coupling of the proton gradient and ATP formation is called **chemiosmotic phosphorylation.**
8. Photosystem II passes excited electrons through an electron transport chain to photosystem I, replacing them with electrons from water. The energy from electrons in photosystem I is used to reduce $NADP^+$ to NADPH.
9. In plants, overall the light reactions oxidize water, produce ATP, and reduce $NADP^+$ to NADPH. The two linked photosystems maximize energy extraction.

6.4 The Carbon Reactions "Fix" Carbon

10. The **carbon reactions** use ATP and NADPH to fix carbon into organic compounds.
11. In the Calvin cycle, **rubisco** catalyzes the reaction of CO_2 with **ribulose bisphosphate** (RuBP) to yield two molecules of PGA. ATP and NADPH power the conversion of PGA to PGAL, the immediate carbohydrate product (**photosynthate**) of photosynthesis. This is C_3 photosynthesis.

6.5 How Efficient Is Photosynthesis?

12. Photon energy produces very little carbohydrate. **Photorespiration** contributes to this inefficiency by fixing oxygen instead of carbon from CO_2. Photorespiration wastes CO_2 and energy.
13. **C_4 photosynthesis** resists photorespiration by separating the light and carbon reactions into separate cells. An additional pathway recycles CO_2 to the Calvin cycle and enables these plants to continue photosynthesis even when little atmospheric CO_2 enters their cells.
14. In **CAM photosynthesis,** certain plants that live in hot, dry habitats open their stomata and take in CO_2 at night and store it as malic acid in vacuoles. During the day, they split off CO_2 and fix it in chloroplasts in the same cells storing the malic acid.

Testing Your Knowledge

1. What are the functions of the following molecules in the light reactions?
   a. ATP
   b. NADPH
   c. CO_2
   d. H_2O
   e. chlorophyll *a*
2. Define these terms and arrange them from smallest to largest:
   a. thylakoid membrane
   b. chloroplast
   c. reaction center
   d. photosystem
   e. electron transport chain
3. List the reactants and products of the light reactions of photosynthesis.
4. Why are two photosystems better than one?
5. How does photorespiration counter photosynthesis?
6. How is CAM photosynthesis adaptive to survival in a particular environment?
7. How is C_4 photosynthesis an adaptation based on a spatial arrangement of structures, whereas CAM photosynthesis is temporally based?

Thinking Scientifically

1. Photosynthesis takes place in plants, algae, and some microbes. How does it affect a meat-eating animal?
2. What color would plants be if they absorbed all wavelengths of visible light? Why?
3. How did Priestley and Ingenhousz's experiments follow the scientific method?
4. A mutant strain of corn lacks carotenoids. Seedlings must be grown in the absence of oxygen, and as photosynthesis begins, the oxygen it generates must be removed for the plants to live. Why must the oxygen be removed?

5. When vegetables and flowers are grown in greenhouses in the winter, their growth rate greatly increases if the CO_2 level is raised to two or three times the level in the natural environment. What is the biological basis for the increased rate of growth?

References and Resources

Demmig-Adams, Barbara, and William W. Adams III. January 27, 2000. Harvesting sunlight safely. *Nature,* vol. 403, p. 371. A closer look at an energy-dissipating protein in the light-harvesting complex.

Des Marais, David J. September 8, 2000. When did photosynthesis emerge on Earth? *Science,* vol. 289, p. 1703. The origins of photosynthesis remain murky, but its roots go back quite far.

Gest, Howard. Winter 1991. Sunbeams, cucumbers, and purple bacteria: The discovery of photosynthesis revisited. *Perspectives in Biology and Medicine,* vol. 34, no. 2. The details of photosynthesis are not obvious; some early thinkers developed interesting views of the process.

Guerinot, Mary Lou. January 14, 2000. The green revolution strikes gold. *Science,* vol. 287, p. 241. A strain of rice that manufacturers beta carotene may help combat vitamin A deficiency.

Sharkey, Thomas D. January 21, 2000. Some like it hot. *Science,* vol. 287, p. 435. Temperature affects photosynthesis.

The *LIFE* Online Learning Center provides additional resources and tools for studying this chapter.
www.mhhe.com/life

CHAPTER 7

How Cells Release Energy

"The Greatest Case of Mass Poisoning the World Has Ever Witnessed"

When the World Bank and UNICEF began tapping into aquifers in India and Bangladesh in the late 1960s, they meant well. The cycles of floods and droughts in this area that is criss-crossed by many rivers had brought sewage and industrial waste to the surface water, causing millions of deaths from cholera and other diarrheal diseases. Wells would provide groundwater, supposedly free of the pollution—or so the planners thought.

What they hadn't planned on was a layer of sediment rich in the element arsenic. The chemical has been leaching into the water of at least half of Bangladesh's 4 million wells ever since. "Safe" levels of arsenic in drinking water in many nations is 50 parts per billion (ppb); the World Health Organization advises a level of 10 ppb, based on studies in Taiwan that associated chronic arsenic exposure with cancers. In the contaminated wells, arsenic levels exceed 500 ppb. Effects on health have taken several years to show up.

Arsenic does its damage in several ways. It attacks certain proteins that have sulfur-sulfur bonds. Arsenic's effects on the protein keratin, for example, cause characteristic skin lesions and hair loss. More insidious, however, is arsenic's interference with enzymes that function at two key points in the pathways that extract energy from nutrient molecules.

Arsenic inhibits an enzyme necessary to break the six-carbon sugar glucose into two, three-carbon compounds, and it blocks a second enzyme that acts on another three-carbon compound in mitochondria. The poison also displaces phosphorus in a series of reactions that generate ATP in cellular respiration. In short, the various isotopes of this element shut down the energy extraction pathways. The person weakens and tires and basically

> ***The person weakens and tires and basically falls apart, the body left running on empty.***

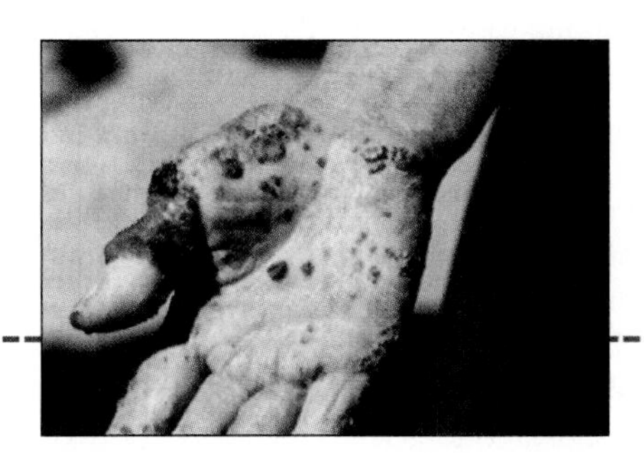

Skin lesions develop after several years from chronic arsenic poisoning.

falls apart, the body left running on empty.

People in India and Bangladesh first noticed signs of chronic arsenic poisoning—tingly, painful darkened sores on the skin—in the middle 1980s. Dipankar Chakraborty, now an environmental scientist at Jadavpur University in Calcutta, initially learned of several cases of the illness, called arsenicosis, on a visit to his parents in 1988 in West Bengal, India. Chakraborty, who had worked with arsenic before, connected the symptoms to exposure in drinking water, and alerted authorities in India and nearby Bangladesh. But the governments ignored him until 1990. Much of the world was not even aware of this enormous environmental problem until 1999.

The statistics are staggering, and predict a disaster that has only just begun. Bangladesh is a country the size of Wisconsin, but with 120 million people crammed within its borders—the population of the United States, by comparison, is 260 million. Average annual income is $260, and 2,308 people occupy each square mile. About 116 million people use the wells, and health agencies estimate that 77 million people have been exposed to dangerous levels of arsenic, for years. Today, hundreds of thousands of people have symptoms of arsenicosis. Over the next several years, more cases will appear. Arsenic-associated cancers of the skin, lungs, and bladder, which take years to develop, are also expected to become more prevalent. Health organizations call the situation in Bangladesh "The greatest case of mass poisoning the world has ever witnessed."

The victims are, understandably, confused. The wells had come to be a status symbol of sorts, indicating one's own water supply, and they had prevented many deaths from diarrheal diseases. Now strangers are telling the people that the wells are slowly killing them. Some of them attribute the poisoning to the will of Allah; others fear it is retribution for having killed snakes to dig the wells. Meanwhile, geologists, chemists, and biologists are just as puzzled. Some remedies have already been tried and failed—chlorine tablets proved toxic, rainwater was difficult to collect, and boiling water was too costly. Scientists have devised all sorts of filters, tried aerating the groundwater, passed water through sand containing iron filings, and dipped alum (an aluminum compound) into the water to precipitate out the arsenic. None of these approaches have succeeded, yet many people continue to use the wells as their water source. There is, as yet, no alternative source of fresh, clean water.

SLOW DEATH FROM ARSENIC POISONING

Skin sores
Hair loss
Fever
Cough
Weakness
Numb arms and legs
Loss of coordination
Abdominal pain
Liver damage
Kidney damage
Cancers

7.1 How Does Cellular Respiration Differ from Breathing?

A respiratory system makes possible gas exchange at the cellular level. Oxygen accepts electrons passed along carrier molecules as energy is liberated from nutrient molecules.

The word "respiration" to most people means "breathing"—as in inhaling and exhaling. Respiration at the cellular level, though, has a different meaning. **Cellular respiration** refers to the biochemical pathways that extract energy from the bonds of nutrient molecules, in the presence of oxygen. Yet cellular respiration and the familiar, whole-body respiration are very much related. Both processes take in oxygen and release carbon dioxide (CO_2), which is called gas exchange (fig. 7.1).

The function of a respiratory system is to filter and convey oxygen to cells, where gas exchange can occur. Specifically, oxygen gas in inhaled air enters red blood cells in the bloodstream, and carbon dioxide leaves through the bloodstream and is exhaled. Cellular respiration also occurs in organisms that do not have respiratory systems, such as plants and fungi.

Recall from chapter 6 that the energy in glucose comes from the process of photosynthesis, which uses carbon dioxide and releases oxygen. In cellular respiration this process is reversed; glucose is ultimately broken down to carbon dioxide and oxygen is consumed. Energy is transferred from nutrient molecules, mostly glucose, to more usable molecules of ATP. The bonds holding together the glucose molecules are rearranged, releasing energy and forming a series of intermediate compounds. The released energy is transferred through electron carrier molecules to other processes in the cell. Ultimately, the energy is used to phosphorylate ADP to form ATP.

Oxygen is the final acceptor of the moving electrons. Without oxygen, most of the energy would remain locked in nutrient molecules. When nutrients are broken down, carbon to carbon bonds are cleaved and each carbon combines with oxygen, forming CO_2, which is a metabolic waste. Cellular respiration, then, explains why our respiratory systems obtain oxygen and get rid of CO_2. Cellular respiration is one of several approaches that organisms use to extract energy from nutrients.

In animals, cellular respiration begins where digestion leaves off. Consider a chipmunk eating a nut. The nut passes through the rodent's digestive system and is broken into clumps of cells. As the nut's cells break apart, they release proteins, carbohydrates, and lipids. After the chipmunk digests these macromolecules into their component amino acids, monosaccharides, and fatty acids and glycerol, they are small enough to enter the blood and lymphatic systems and be transported to the body's tissues. When these smaller nutrient molecules enter the animal's cells, they are broken down further, and some may be converted to glucose. When glucose and other nutrients are broken down, energy is transferred to the high-energy phosphate bonds of ATP.

This chapter describes how cells extract energy from nutrients. The journey from food to energy entails several biochemical pathways, with many chemical names, and may appear overwhelming. But if we consider energy release in major stages, and also one step at a time, the logic becomes clear and, like much of life science, it makes sense.

FIGURE 7.1
"Breathing" Respiration and Cellular Respiration Are Linked. (**A**) The athlete breathes in oxygen (O_2) and exhales carbon dioxide (CO_2), a metabolic waste. Oxygen enters the bloodstream in the lungs and is distributed to all cells. There, in mitochondria, oxygen enables the reactions of cellular respiration to occur, extracting energy from glucose and other nutrient molecules. This energy powers the contractions of the athlete's muscles (**B**).

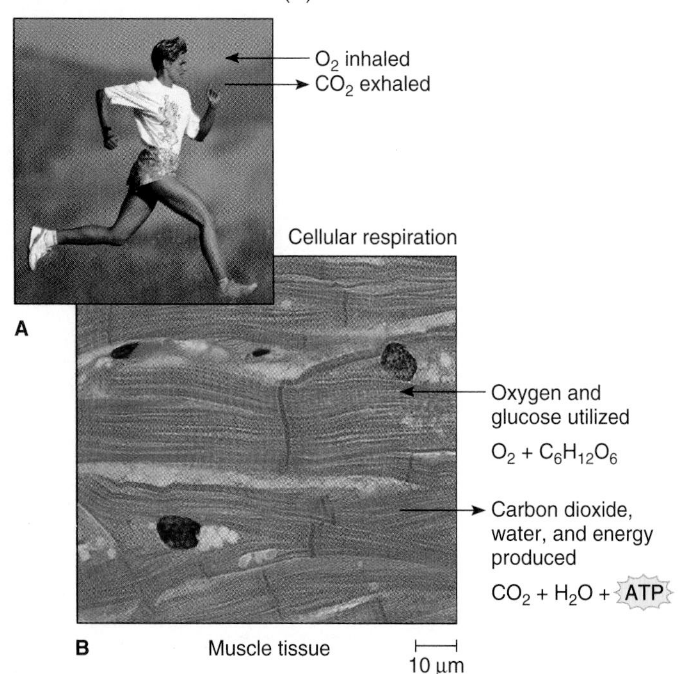

7.1 MASTERING CONCEPTS

1. How does cellular respiration differ from whole-body respiration?
2. What is gas exchange?
3. What is the role of oxygen in cellular respiration?
4. What is the relationship between digestion of foods and cellular respiration?

OLC

7.2 How Do Cells Release Energy?

To extract energy from nutrient molecules, cells use the reactions of glycolysis to split glucose molecules. Other reactions and pathways then capture the energy in the chemical bonds of the breakdown product, pyruvic acid. ATP stores the released energy.

All organisms obtain energy from glucose using a process called **glycolysis** (fig. 7.2). Sources of this glucose include starch, cellulose, other nutrients, or simply molecules of glucose. Even plants must first make glucose via photosynthesis, then extract the energy from glucose. However, glycolysis does not yield as much energy as other processes that use oxygen to completely oxidize glucose to carbon dioxide.

FIGURE 7.2

All Cells Undergo Glycolysis.

The first step in energy release is glycolysis. Subsequent pathways continue the release of energy from nutrient molecules. These are three common pathways.

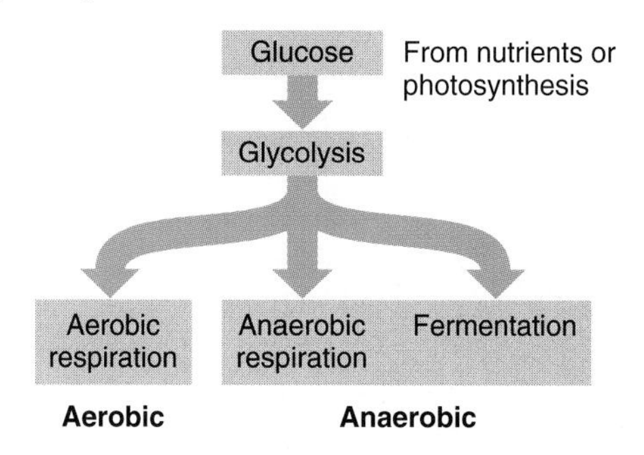

Aerobic respiration refers to the entire conversion of glucose to carbon dioxide in the presence of oxygen. The general equation for aerobic respiration is:

$$\text{glucose} + \text{oxygen} \rightarrow \text{carbon dioxide} + \text{water} + \text{energy}$$

$$C_6H_{12}O_6 + 6O_2 \rightarrow 6CO_2 + 6H_2O + 30\text{ATP}$$

In English, this means: The energy in glucose is transferred to ATP in the presence of oxygen while being broken down to carbon dioxide, with water a by-product. Many organisms, called aerobes, must extract energy in this way to stay alive even though they may be able to function for a short time in the absence of oxygen. Most of the known organisms on Earth are aerobes. The reactions that finish what glycolysis starts are called the **Krebs cycle** and **electron transport.** The enzymes for these reactions are found in the membranes and cytoplasm of bacteria and in the membranes and matrix of mitochondria in more complex species. Organisms called anaerobes obtain energy in the absence of oxygen by using different pathways that also include glycolysis and are referred to as fermentation. Some microorganisms use nitric oxide (NO_3) instead of O_2 as the terminal electron acceptor. This is called anaerobic respiration.

Aerobic respiration occurs in cells containing mitochondria, which includes plants, fungi, and animals. Bacteria and archaea carry out aerobic respiration in much the same way, but without an organelle to extract energy.

An Overview of Glucose Utilization

If all of the energy in glucose were released in one uncontrolled step, enough heat would be liberated (4 kcal per gram) to destroy the cell. To gain the energy in a usable form, cells extract the energy in glucose in small amounts and transfer that energy to ATP. Due to the second law of thermodynamics, this process releases heat. Many organisms use some of this heat to maintain their internal temperatures, and the remainder is dissipated into the surroundings. No one part of a cell becomes hot enough to burst into flames, because the energy is never released all at once. The heat immediately spreads throughout the cell and its surroundings.

Cellular respiration transfers energy from one part of a glucose molecule to another, and then to other molecules as glucose is gradually broken down to carbon dioxide. The first part of respiration, glycolysis (literally "breaking glucose"), occurs in the cytoplasm. It is a series of steps, catalyzed by a specific series of enzymes, that moves the energy in a glucose molecule to concentrate some of that energy into a few electrons. Those electrons are transferred to the energy carriers NADH and ATP. In the process, glucose is split from one six-carbon molecule to two three-carbon molecules of **pyruvic acid.** (The ionized form is called pyruvate.) Next pyruvic acid is transported into the mitochondria, where additional arrangements and energy transfers take place. The second part of respiration transfers two of the carbons in pyruvic acid to a carrier molecule, creating a molecule called acetyl CoA. In the third part of respiration, the two carbons are transferred to another molecule that is then rearranged in the Krebs cycle, which transfers the remaining energy to more molecules of NADH, $FADH_2$, and ATP. The carbons are released as carbon dioxide. The final part of respiration, **electron transport,** transfers energy from NADH and $FADH_2$ to ATP. The energy-rich electrons, plus hydrogen ions from NADH and $FADH_2$, are transferred through a series of membrane proteins and their energy is used to create a gradient of hydrogen ions across the mitochondrial membrane. This gradient holds tremendous potential energy and is used to phosphorylate ADP to ATP. Once the electrons have given up their energy, they are transferred to oxygen, which then combines with hydrogen to yield water. For each gram of glucose, this process captures approximately 35% of the energy in molecules of ATP. Figure 7.3 shows where all of these reactions occur relative to one another in a eukaryotic cell.

FIGURE 7.3

An Overview of Cellular Respiration.

Glucose is broken down to carbon dioxide through a series of enzyme reactions. The energy released phosphorylates ADP to ATP. As the chapter progresses, each of the biochemical pathways will become more detailed. Look to the insets that repeat this diagram with different sections highlighted to follow the part of the overall pathway under discussion.

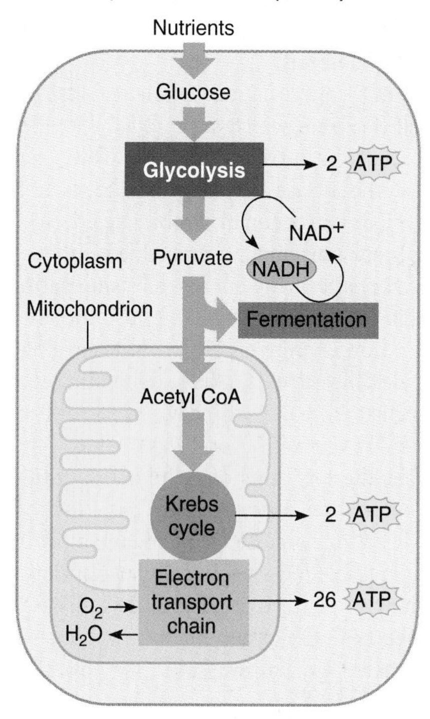

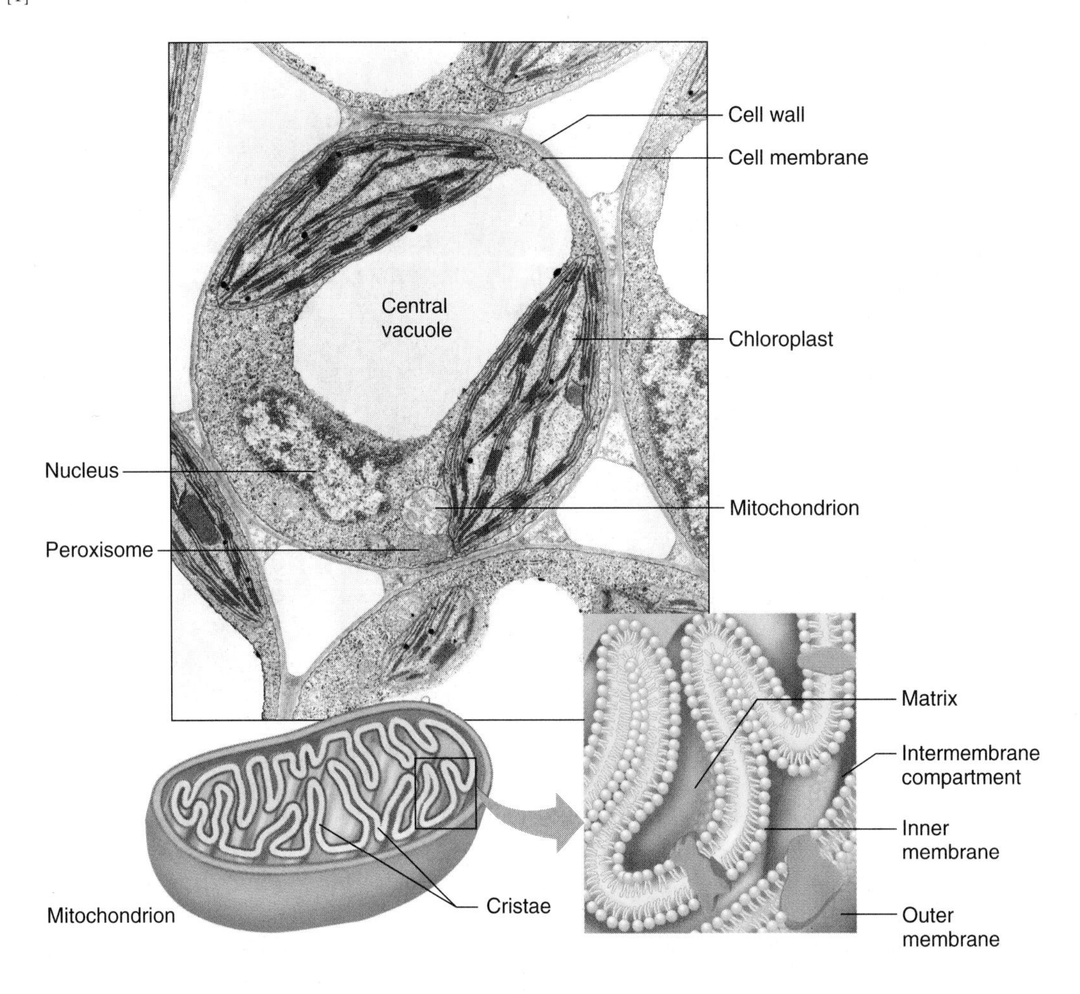

FIGURE 7.4
Cellular Respiration Occurs in Mitochondria.
In eukaryotes, mitochondria provide most of the energy for cellular functions. Enzymes in the matrix and membrane oxidize pyruvic acid to carbon dioxide and transfer the released energy to ATP. The inner membrane of the mitochondria is used to form the proton gradient of chemiosmotic phosphorylation, just as the thylakoid membrane does so in photosynthesis (fig 6.11).

Recall from chapter 3 that a mitochondrion consists of an outer membrane and a highly folded inner membrane, an organization that creates two compartments (fig. 7.4). The innermost compartment, called the **matrix,** is where the Krebs cycle occurs. A second area between the two membranes is called the **intermembrane compartment.**

The inner mitochondrial membrane is folded into numerous projections called **cristae,** which are studded with the enzyme ATP synthase and electron carrier molecules. This extensive folding greatly increases the surface area on which the reactions of the electron transport chain can occur. In aerobic bacteria, which lack mitochondria, respiratory enzymes are embedded in the cell membrane.

The overall result of the energy-releasing pathways is to use energy stored in organic molecules to phosphorylate ADP to ATP. Cellular respiration is quite efficient, with one glucose molecule theoretically yielding 30 to 32 ATPs, depending upon the tissue and organism. Just as we use ordinary currency, the cell can use its energy currency—ATP—in several ways (table 7.1).

In animals, the reactions of cellular respiration begin with glucose. But before this, other carbohydrates in food, such as sucrose (table sugar) or fructose (fruit sugar), must be digested or converted to glucose. Plants begin cellular respiration with compounds that may come from other sources, too. Later in the chapter, we will see how the other major nutrients—proteins and lipids—enter the energy pathways. But for now, we will focus on glucose.

Making and Using ATP

ATP synthesis occurs in two ways. One mechanism uses an enzyme and an energy-rich substrate to add phosphate to a second substrate. The other mechanism uses the energy in a gradient, created by oxidation-reduction reactions, to add phosphate to ADP. Both mechanisms are possible because the resulting molecules contain less energy than the reactants. In the first process, called **substrate-level phosphorylation,** an enzyme oxidizes a molecule, liberating enough energy to directly attach inorganic phosphate to a second substrate. A second reaction transfers this phosphate to ADP. In glycolysis, enough energy is liberated in this reaction to reduce NAD^+ to NADH as well.

TABLE 7.1 Some Functions Fueled by ATP
Cilia and flagella movement
Muscle contraction
Biosynthesis of macromolecules
Enzyme function
Transport of neurotransmitters between nerve cells
Unwinding of DNA double helix during replication
Movement of chromosomes in cell division
Ion transport into mitochondria, chloroplasts, nuclei
Cytoplasmic streaming

The second method of making ATP, **chemiosmotic phosphorylation** (also called oxidative phosphorylation), transfers energy from molecule to molecule and then to a concentration gradient, usually composed of protons. A membrane-bound enzyme system creates and maintains the gradient and sets up a great source of potential energy. The difference in proton concentration on one side of the membrane versus the other is a "force" pushing against that membrane. (A nearly identical mechanism operates in photosynthesis.) The release of that pressure is coupled to the phosphorylation of an ADP, and therefore the gradient is used to make ATP. The process of energy transfer used here is similar to creating electricity using water pressure. Water from rain flows downhill and is trapped by a dam. Tremendous pressure builds on the face of the dam. That pressure is released by forcing the water to turn blades that spin an electric generator. Electricity is made from rain! A mitochondrion uses a similar mechanism to transfer energy from NADH to ATP.

Substrate-level phosphorylation is the simpler and more direct mechanism for producing ATP, but it accounts for only a small percentage of the ATP produced—specifically, that from glycolysis and the Krebs cycle. In the electron transport chain that follows these reactions, ATP is formed through chemiosmotic phosphorylation.

7.2 MASTERING CONCEPTS

1. What is the overall equation that describes cellular respiration?
2. What are the stages of energy acquisition and where do they occur?
3. What are the parts of a mitochondrion?
4. What is substrate-level phosphorylation, and when does it generate ATP?
5. What is chemiosmotic phosphorylation, and when does it generate ATP?

OLC

7.3 How Does Glycolysis Break Down Glucose to Pyruvic Acid?

The reactions of glycolysis start the energy-releasing process by splitting one molecule of glucose into two molecules of pyruvic acid.

Glucose contains considerable bond energy, but cells recover only a small portion of it during glycolysis. In this pathway, glucose splits into two three-carbon compounds. The entire process requires 10 steps, all of which occur in the cytoplasm (fig. 7.5). The first half of the pathway activates glucose so that energy can be extracted from its bonds. The second half of the pathway then extracts this energy.

The First Half of Glycolysis Activates Glucose

The first step of glycolysis uses one molecule of ATP to phosphorylate one molecule of glucose. Phosphorylation activates glucose, enabling the appropriate enzyme to catalyze the next step. Because phosphate is negatively charged, and since charged compounds cannot easily cross the cell membrane, phosphorylation also traps glucose in the cell.

In step 2, the atoms of phosphorylated glucose are rearranged; in step 3 the new molecule (fructose-6-phosphate) is phosphorylated again by a second ATP, forming fructose-1,6-bisphosphate. This compound splits into two three-carbon compounds in steps 4 and 5, and each of the three-carbon products has one phosphate. One of the products, phosphoglyceraldehyde (PGAL), is further metabolized in glycolysis. The other product, dihydroxyacetone phosphate, is converted to PGAL and then is broken down along with the other PGAL. Formation of the two PGALs derived from each glucose molecule marks the halfway point of glycolysis (steps 1 through 5 in fig. 7.5). So far, energy in the form of ATP has been invested, but no ATP has been produced.

The Second Half of Glycolysis Extracts Some Energy

The first energy-obtaining step of the glycolysis pathway occurs in the second half when NAD^+ is reduced to NADH through the coupled oxidation of PGAL (step 6). This oxidation also releases enough energy to add a second phosphate group to PGAL, making it 1,3-bisphosphoglyceric acid. This is one place where arsenic blocks the pathway, as discussed in the chapter opener.

Substrate-level phosphorylation occurs when one of the phosphates of 1,3-bisphosphoglyceric acid is transferred to ADP (step 7). The three-carbon molecule that remains, 3-phosphoglycerate, is then rearranged to form 2-phosphoglycerate (step 8). This compound then loses water and becomes phosphoenolpyruvate (PEP) (step 9). PEP then becomes **pyruvic acid** when it donates its phosphate to a second ADP (step 10). Each PGAL from the first half of glycolysis progresses through this pathway to make two ATPs and one pyruvic acid in the second half. ("Pyruvate" refers to the ionized form of pyruvic acid.)

Because one molecule of glucose yields two molecules of PGAL, and each molecule of PGAL yields two molecules of ATP and one of pyruvic acid, each glucose produces four ATPs and two pyruvic acids. However, because the first half of glycolysis requires two ATPs, the net gain is two ATPs per molecule of glucose.

At the end of glycolysis, a small amount of the chemical energy that started out in glucose ends up in ATP and NADH. However, most of the energy of glucose remains in the bonds of pyruvic acid. This energy is tapped in the mitochondrion to synthesize more ATP. Note that, to this point, the pathway does not require oxygen.

7.3 MASTERING CONCEPTS

1. What are the starting materials and products of glycolysis?
2. Describe the reactions that activate glucose.
3. Describe the reactions that extract energy from the two PGAL molecules derived from glucose.
4. What is the net gain of ATP for each glucose molecule undergoing glycolysis?

OLC

FIGURE 7.5

Glycolysis.

In the glycolysis reactions, glucose is rearranged and split into two three-carbon intermediates, each of which is rearranged further to eventually yield two molecules of pyruvic acid (pyruvate). Along the way, four ATPs and two NADHs are produced. Two ATPs are consumed in activating glucose, so the net ATP yield is two ATP molecules per molecule of glucose.

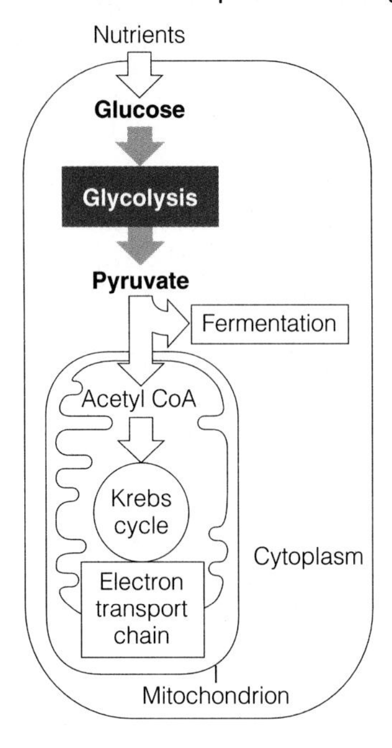

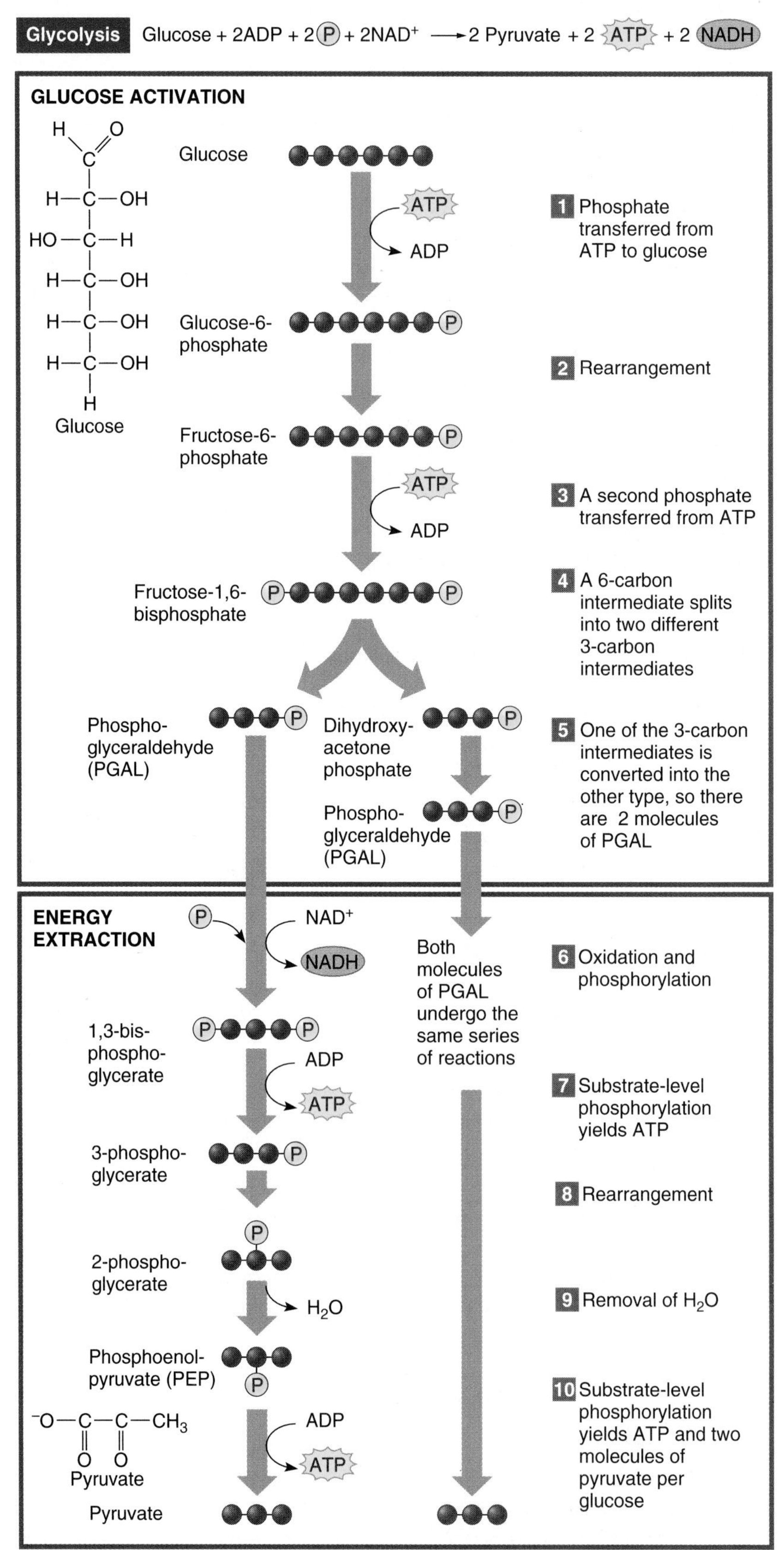

FIGURE 7.6

Transition to the Mitochondria.

Acetyl CoA formation bridges glycolysis and the Krebs cycle. After pyruvic acid enters the mitochondrion, crossing both membranes, it loses CO_2 in a reaction that reduces NAD^+ to NADH. The remaining two carbons combine with coenzyme A to yield acetyl CoA. For every glucose molecule that entered glycolysis, two acetyl CoA molecules now enter the Krebs cycle.

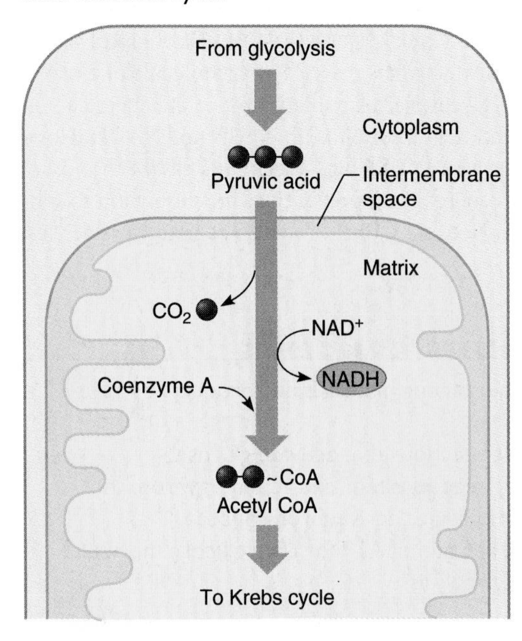

7.4 After Glycolysis—In the Presence of Oxygen

The product of glycolysis—pyruvic acid—may enter mitochondria and be used to form acetyl CoA, and then enter the Krebs cycle and an electron transport chain. These reactions ultimately generate ATP.

The pyruvic acid that is transported into the mitochondrial matrix is not directly used in the Krebs cycle. First a molecule of carbon dioxide is removed as NAD^+ is reduced to NADH. The remaining molecule, called an acetyl group, is transferred to a coenzyme to form **acetyl coenzyme A,** abbreviated **acetyl CoA** (fig. 7.6). This is a second point where arsenic intervenes.

The conversion of pyruvic acid to acetyl CoA links glycolysis and the Krebs cycle. Pyruvic acid is the final product of glycolysis, and acetyl CoA is the compound that enters the Krebs cycle.

The Krebs Cycle Produces ATP and NADH

The Krebs cycle is a cycle because the last step regenerates the reactants of the first step (fig. 7.7). Seven of the eight steps occur in the matrix of the mitochondria. In addition to continuing the breakdown of glucose, the Krebs cycle forms intermediate compounds. The cell then uses the carbon skeletons of these intermediates to manufacture other organic molecules, such as amino acids.

Organisms release enormous quantities of CO_2 generated in the Krebs cycle to the environment. In forests, for example (fig. 7.8) a third of the high levels of CO_2 released to the atmosphere comes from metabolism of nutrients by microorganisms, insects, worms, fungi, plants, and other residents of the top layer of leaf litter. Chapter 43 further explores this cycling of carbon in the environment.

In the first step of the Krebs cycle, the two-carbon acetyl group is transferred to a four-carbon molecule called oxaloacetic

FIGURE 7.7

Products of the Krebs Cycle.

One glucose molecule yields two molecules of acetyl CoA. Therefore, one glucose molecule is associated with two turns of the Krebs cycle. Each turn of the Krebs cycle generates one molecule of ATP, three molecules of NADH, one molecule of $FADH_2$, and two molecules of CO_2, as the carbons acetyl CoA carries are rearranged and oxidized.

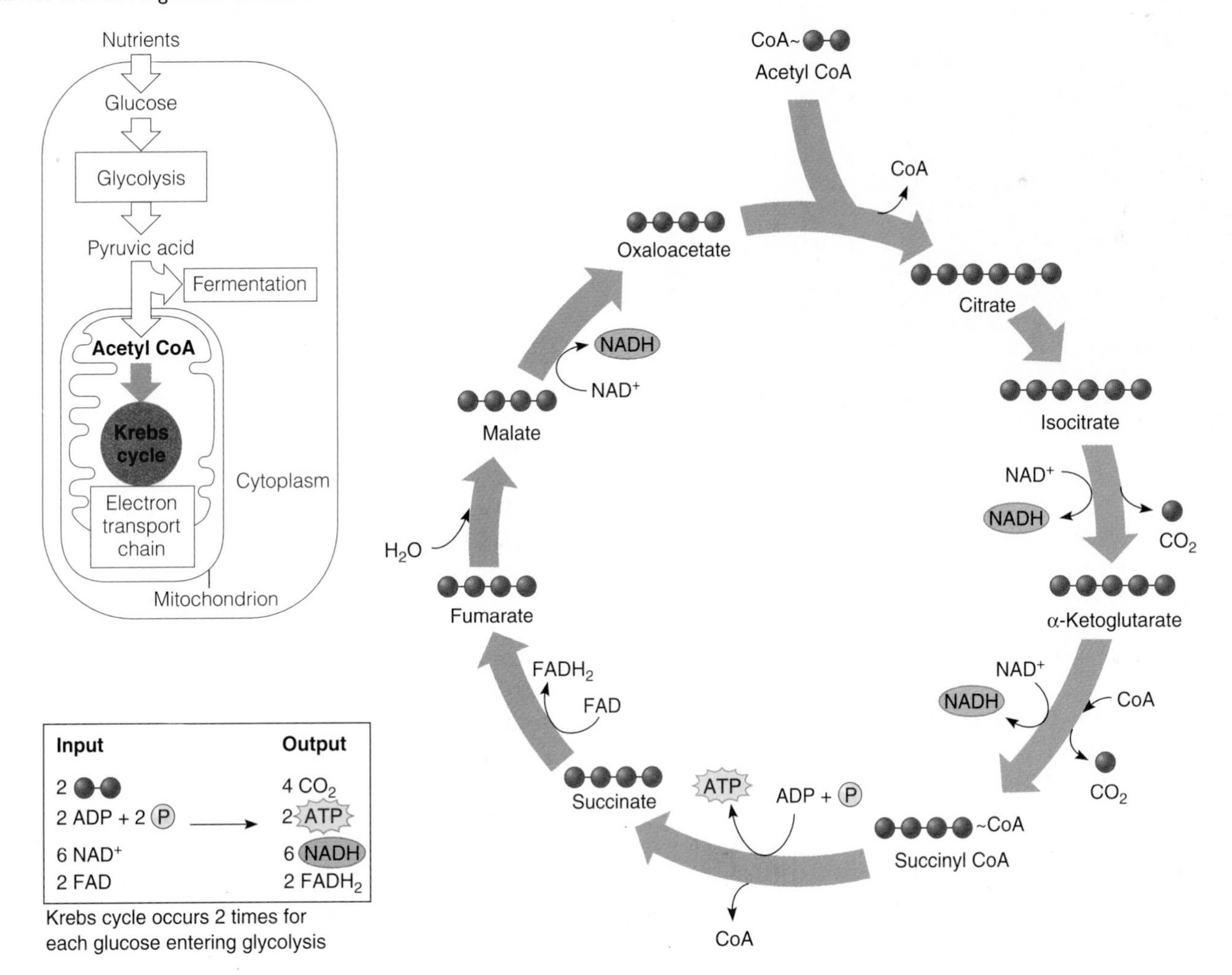

FIGURE 7.8

All Life Releases Carbon Dioxide.

The microbes, insects, worms, fungi, plant roots, and other occupants of soil continually release huge amounts of carbon dioxide (CO_2) from cellular respiration. In forests, a third of the CO_2 released to the atmosphere from soil comes from the top layer of leaf litter. Organisms here metabolize nutrient molecules, releasing CO_2. These nutrients, in turn, derive ultimately from photosynthesis.

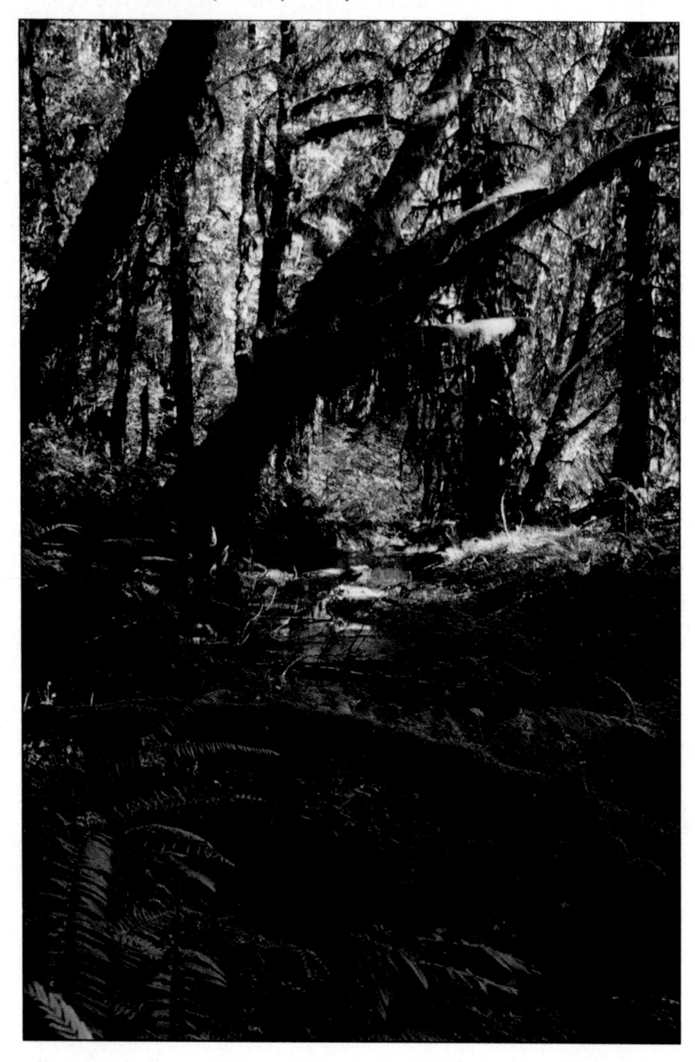

acid. The new six-carbon molecule is called citric acid. For this reason, the Krebs cycle is also called the citric acid cycle. The remaining steps in the Krebs cycle rearrange and oxidize citric acid through several intermediates, with transfer of energetic electrons to the carriers NADH and $FADH_2$, and direct production of ATP. The ATP formed in the Krebs cycle is another example of substrate-level phosphorylation. As the molecules rearrange, the two carbons that entered the cycle are released as carbon dioxide. The molecules in the Krebs cycle are eventually altered to recreate the original acceptor molecule, oxaloacetic acid. The cycle is now ready to repeat.

Since the glycolysis of one molecule of glucose produces two molecules of pyruvic acid, there are two turns of the Krebs cycle per glucose molecule. The potential energy in glucose is transferred through cellular respiration for a net total of four ATP molecules, ten NADH molecules, and two $FADH_2$ molecules, with release of six molecules of CO_2. This yield is possible because the energy in these final molecules is less than that in the original glucose molecule, adhering to the laws of thermodynamics. Not all of the energy is captured in this process—the remainder is lost as heat. However, the cell needs energy in the form of ATP to drive most of it reactions, not NADH. How does the cell transfer the energy from NADH to ATP? An electron transport chain accomplishes this feat.

How Does the Electron Transport Chain Drive ATP Formation?

The electron transport chain is composed of a series of membrane proteins and electron carrier molecules embedded within the inner mitochondrial membrane. These molecules accept electrons, extract some of their energy, use that energy to perform work, and then transfer the electrons to the next set of molecules (fig. 7.9). The work done is to pump hydrogen ions (essentially just protons) out of the matrix of the mitochondrion. By removing the energy from the electrons in a step-wise fashion, cells can extract most of the energy found in NADH and $FADH_2$. This system is another part of the cell that is sensitive to arsenic poisoning.

The final acceptor of the electrons is oxygen, which combines with hydrogen ions to form water. Breathing provides the oxygen for this final electron acceptor. Without oxygen, the cell has no place to put the electrons once they have released their energy, and the entire system is in danger of shutting down. Cyanide is poisonous because it blocks the enzyme that transfers the electrons to oxygen, halting the rest of the system.

The electron transport chain produces a huge amount of potential energy in the proton gradient it creates. The mitochondrion uses an enzyme complex, ATP synthase, to return the protons to the matrix, and uses the energy to phosphorylate ADP to ATP. But figuring out the sequence of events was quite a challenge (table 7.2). Many scientists hypothesized that the electron transport chain forms high-energy compounds directly that fuel ATP synthesis. They searched for these compounds but never found them.

Taking another approach, British researcher Peter Mitchell focused on *where* ATP synthesis occurs rather than on *what* powers it. He looked at the inner mitochondrial membrane, where he knew the carrier molecules of the electron transport chain are located. Did the membrane itself, or perhaps a molecule embedded in it, play a role in ATP synthesis?

To test this idea, Mitchell re-created events that occur in mitochondria. He isolated the inner membranes of mitochondria and made vesicles out of them. When he added oxygen and NADH to the medium containing the vesicles, the electron transport chain functioned as it does in intact cells! But Mitchell noticed something else. The pH of the medium surrounding the vesicles fell, indicating an increase in protons (H^+). (Recall that pH measures hydrogen ion concentration.) Mitchell formulated another hypothesis: Protons pumped out of the vesicles during

FIGURE 7.9

The Electron Transport Chain of the Mitochondria.

Energy-rich electrons removed from NADH and $FADH_2$ slowly release their energy as they are transferred along the inner membrane of the mitochondrion. Membrane-bound enzymes use the energy to pump protons (H^+) out of the mitochondrial matrix and establish a gradient of charge, pH, and atoms across the membrane. Several electron carrier molecules send protons (H^+) from the matrix side of the inner mitochondrial membrane to the intermembrane compartment establishing a pH and electrical charge proton gradient. As the protons pass through a channel in ATP synthase, ADP bound to another part of the enzyme is phosphorylated to form ATP.

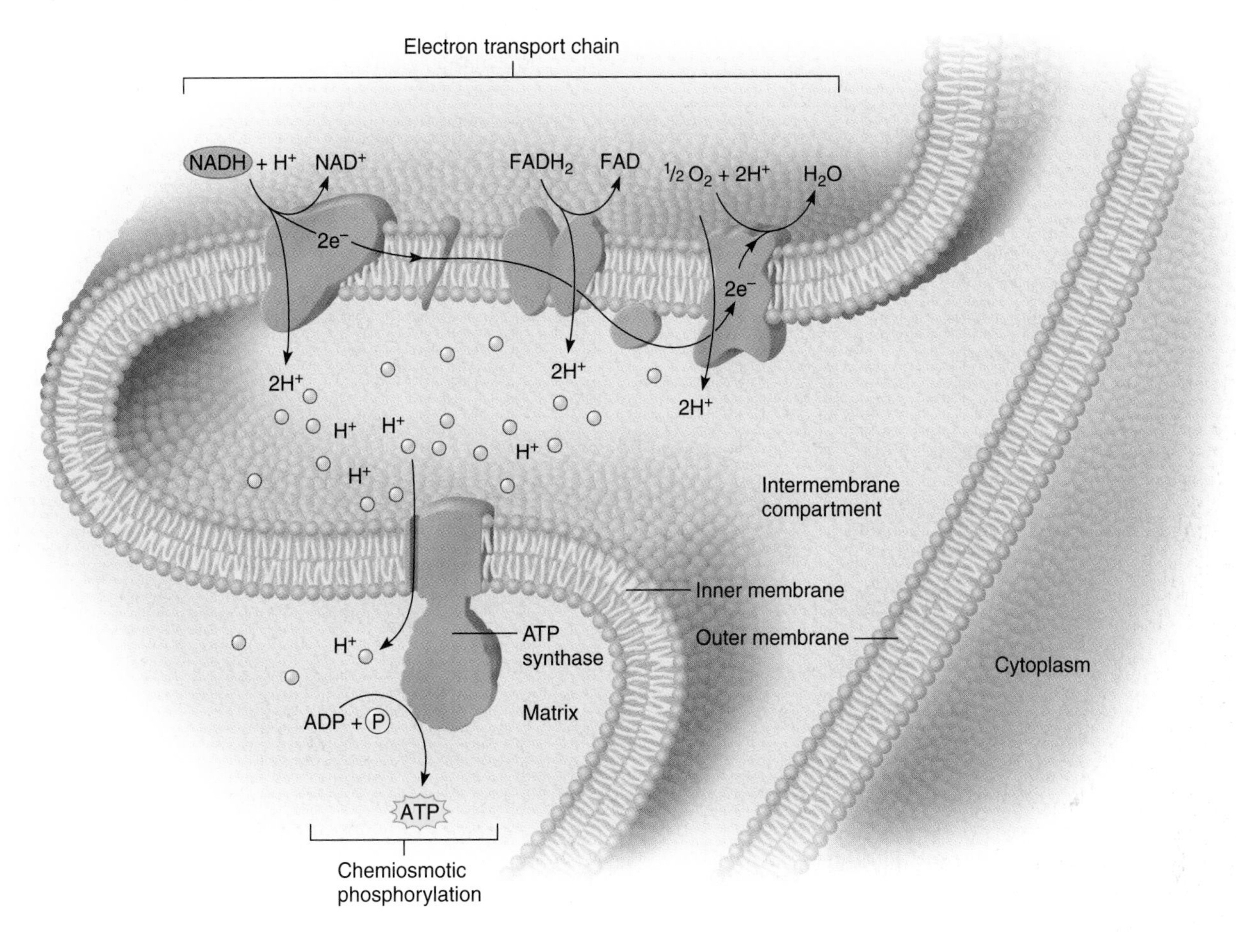

TABLE 7.2 Demonstrating Chemiosmosis—And the Scientific Method

Observation	Mitochondria are the sites of ATP formation.
Hypothesis	If the inner mitochondrial membrane is the site of ATP synthesis, then pieces of membrane cultured in a laboratory and given O_2 and NADH should demonstrate ATP formation.
Experiment	Create vesicles from inner membranes of mitochondria, and add O_2 and NADH to simulate conditions in a cell. See what happens as cellular respiration proceeds.
Results	ATP is produced; pH drops in medium containing vesicles.
Conclusion	A change in concentration of H^+ occurs on the two sides of the inner mitochondrial membrane as ATP forms from ADP.
New hypothesis	If formation of a proton (H^+) gradient causes ATP formation, then a structure or mechanism must physically link the two activities.
Further experiments	Electron microscopy of experimental vesicles reveals clusters of ATP synthase. Further study of the enzyme shows that it both provides a channel in the membrane for H^+ and is the site of phosphorylation of ADP to ATP.
Chemiosmotic theory	A proton gradient across the inner mitochondrial membrane drives ATP synthesis.

FIGURE 7.10
Two Views of ATP Synthase.
(**A**) ATP synthase molecules protrude from an isolated mitochondrial inner membrane on the matrix side, resembling lollipops which are the small dots on the surface of the round membrane fragments in this micrograph. This enzyme, however, is much more complex in structure than the "lollipop" description suggests. It consists of several protein subunits (**B**), some of which researchers have not yet visualized. The ring of protein subunits within the membrane forms the proton channel, which functions like a rotor. The outer portion of the molecule contains the catalytic parts where ADP is phosphorylated to ATP. Cell biologists have called this multi-molecule that is so basic to life "the world's smallest rotary engine."

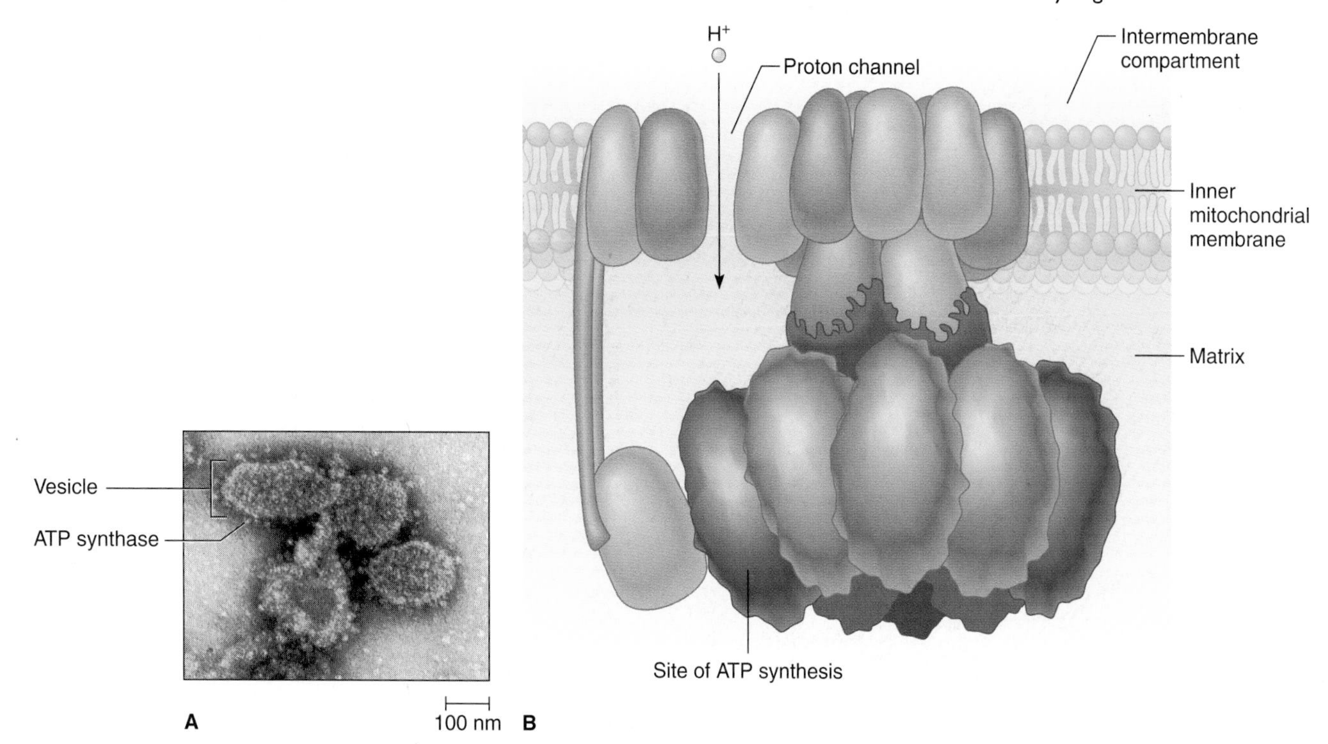

electron transport form a gradient, which provides the energy to drive ATP synthesis.

Further studies using the electron microscope revealed knobby structures protruding from one face of each of Mitchell's experimental vesicles (fig. 7.10). When researchers isolated these structures, they discovered complexes of proteins comprising ATP synthase. This multisubunit enzyme spans the inner mitochondrial membrane, just as it does the thylakoid membrane in a chloroplast, and it catalyzes the phosphorylation of ADP to ATP. The enzyme also forms a channel in the membrane through which protons can flow. Recall from chapter 6 that the portion of the enzyme outside the membrane has sites where ADP and inorganic phosphate (P_i) combine to form ATP. ATP is released only when protons move through the ATP synthase channel. ATP synthase is also found in the respiratory membranes in bacteria and archaea.

Mitchell's theory of ATP synthesis is called chemiosmosis or chemiosmotic phosphorylation. The name derives from the fact that the process uses a chemical gradient as an energy source to add phosphate to ADP, forming ATP. The theory states that the cell uses the energy released from the flow of electrons through the electron transport chain to actively pump protons into the intermembrane compartment. The membrane is not very permeable to protons, so they do not readily leak back across it. Continued electron transport pumps more and more protons into the intermembrane compartment and sets up a proton gradient. As protons move through ATP synthase channels from the intermembrane space to the matrix, ADP is phosphorylated to ATP. Thus, the interaction between electron transport and ATP synthesis is indirect—the electron transport chain produces a proton gradient, and that gradient drives ATP synthesis. Some poisons, such as the insecticide 2,4-dinitrophenol, kill by making the inner mitochondrial membrane permeable to protons. This blocks formation of the proton gradient necessary to drive ATP synthesis. The doomed bugs run out of energy.

How Many ATPs Can One Glucose Molecule Yield?

In following the energy pathways, it is easy to lose track of the overall function of the process—converting the energy in a molecule of glucose into a form the cell can easily use. How productive is cellular respiration? That is, how many ATPs can one molecule of glucose generate? To estimate the yield of ATPs, we can add the presumed maximum net number of ATPs generated from glycolysis, the Krebs cycle, and chemiosmotic phosphorylation. Table 7.3 summarizes this theoretical calculation, as does figure 7.14 near the end of the chapter.

Substrate-level phosphorylation yields two ATPs from glycolysis and two ATPs from the Krebs cycle (one ATP each from

TABLE 7.3 One Glucose Can Yield 30 ATPs

PATHWAYS	COENZYMES REDUCED	ATP YIELD
Glycolysis		
Substrate-level phosphorylation:		2 ATP
Reduction of NAD^+:	2 NADH	
Pyruvic Acid → Acetyl CoA (×2)		
Reduction of NAD^+:	2 NADH	
Krebs Cycle (×2)		
Substrate-level phosphorylation:		2 ATP
Reduction of NAD^+:	6 NADH	
Reduction of FAD:	2 $FADH_2$	
Electron Transport Chain		
Oxidation of 10 NADH × 2.5 ATP/NADH		25 ATP
Oxidation of 2 $FADH_2$ × 1.5 ATP/$FADH_2$		3 ATP
		32 ATP
Energy expended to actively transport NADH from glycolysis into mitochondrion		–2 ATP
	Total	30 ATP

two turns of the cycle). These are the only steps that produce ATP directly. Most of the ATP generated from cellular respiration comes from chemiosmotic phosphorylation. But when we try to determine exactly how much ATP this step produces, some uncertainty enters the picture.

Recent studies on mitochondria have more accurately determined the ATP yield from electron transport. Each NADH would produce 2.5 ATPs, but each $FADH_2$ would produce only 1.5 ATPs, because this molecule enters the electron transport chain a step later. One molecule of glucose yields two NADH molecules from glycolysis, two NADHs from converting two molecules of pyruvic acid to acetyl CoA, and six NADHs and two $FADH_2$s from two turns of the Krebs cycle. This totals 10 NADHs, which would yield 25 ATPs, and two $FADH_2$s, which would yield three more ATPs. Add the four ATPs from substrate-level phosphorylation and the total is 32 ATPs. However, NADH from glycolysis must be actively transported into the mitochondrion, usually at a cost of one ATP for each NADH. This reduces the net production of ATPs from a molecule of glucose to 30.

The number of kilocalories stored in 30 ATPs is about 32% of the total kilocalories stored in the glucose bonds—and this is a theoretical maximum. To put this energy yield into perspective, an automobile uses about 20 to 25% of the energy contained in gasoline's chemical bonds to actually move the car.

The estimate that aerobic respiration yields 30 net ATPs is just that—an estimate. This is because the NADHs produced in cellular respiration are not always used to produce ATP. Respiration efficiency may vary in different cell types; a highly active muscle cell might generate more ATPs than a relatively inactive adipose cell. In addition, the theoretical maximum number of ATPs assumes that all of the protons produced in the electron transport chain are used by ATP synthase. This is not true—the proton gradient plays a part in other processes too, so not all protons are routed through ATP synthase. In addition, many of the intermediates of both glycolysis and the Krebs cycle may be used to form other molecules, such as amino acids.

What Factors Control Cellular Respiration?

A cell is hardly the site of neat pathways and cycles, as diagrams may suggest. It is more like a complex soup, with hundreds of different metabolic reactions occurring at the same time, each with thousands of copies of substrate molecules all mixed together in the cytoplasm. Cellular metabolism consists of many interconnecting pathways occurring simultaneously, subject to feedback controls. The pathways are coordinated so that substances neither build up nor become depleted under ideal conditions.

The cell regulates glycolysis and aerobic respiration so that glycolysis does not proceed so rapidly that it overwhelms the Krebs cycle and the electron transport chain. This control also ensures that ATP is always available. Cells use some of the mechanisms discussed in chapter 5 to regulate the energy pathways.

One important regulatory enzyme is phosphofructokinase, which catalyzes the third step in glycolysis (fig. 7.11). This reaction yields fructose-1,6-bisphosphate, which is the six-carbon intermediate compound that splits into two three-carbon compounds. Negative feedback inhibits phosphofructokinase when NADH or citric acid (an intermediate product of the Krebs cycle) accumulates. (See fig. 5.17 to review negative feedback loops.) NADH and citric acid alter the shape of phosphofructokinase so that it can temporarily no longer catalyze its reaction in glycolysis. ATP buildup also inhibits this enzyme, and excess ADP

FIGURE 7.11

Negative Feedback Regulates Aerobic Respiration.

NADH and citric acid from the Krebs cycle, in addition to ATP at high concentrations, bind to the regulatory enzyme phosphofructokinase (PFK) in a way that temporarily halts its ability to catalyze a reaction in glycolysis (*1*, *2*). (*3*) High levels of ADP restore the enzyme's activity.

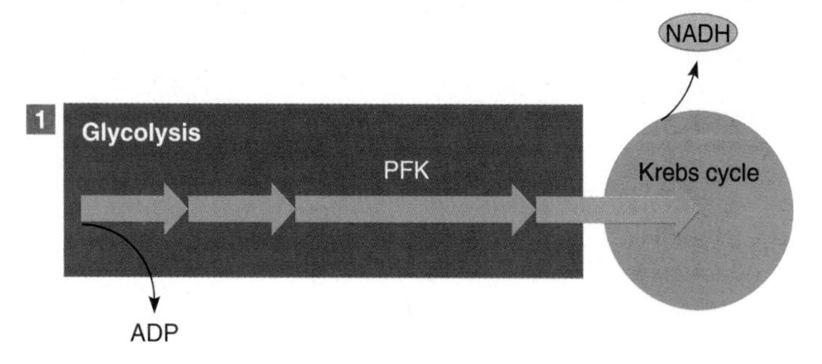

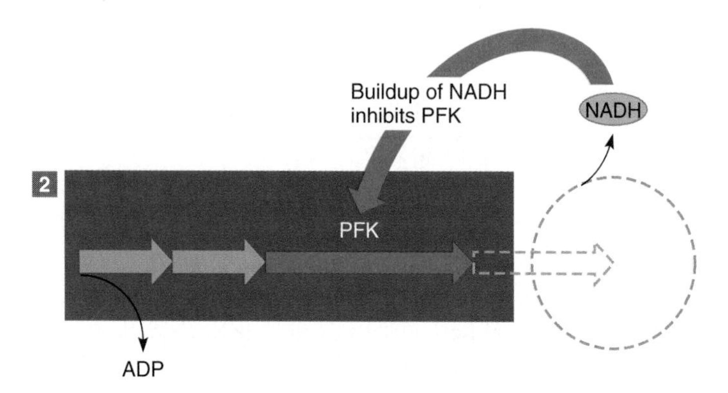

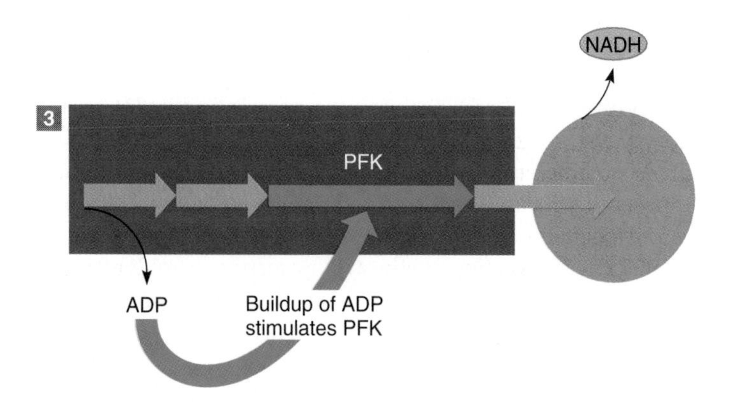

activates it. All of these interactions keep the pace of glycolysis in step with that of subsequent pathways.

Health 7.1 considers energy use on the whole-body level.

Where Do Proteins and Lipids Enter the Energy-Extracting Pathways?

Figure 7.12 shows how the other major nutrients—proteins and lipids—fit into the energy pathways. Amino acids liberated from dietary proteins are usually used to manufacture more proteins. However, when an organism exhausts immediate carbohydrate supplies, cells may use protein as an energy source. When this happens, amino acids are rearranged and broken down. They enter the energy pathways as either pyruvic acid, acetyl CoA, or an intermediate of the Krebs cycle, depending upon the type of amino acid broken down. Ammonia is stripped from the amino groups of the amino acids and eventually excreted.

The fat in food—such as that in the cheese and meat in figure 7.12—is digested into glycerol and fatty acids, which ultimately enter the bloodstream. Glycerol is converted to pyruvic acid and continues from there through acetyl CoA formation, the Krebs cycle, and the electron transport chain. Fatty acids enter cells and are transported into mitochondria, where they are broken down to yield acetyl CoA. From here, the pathways continue as they would for glucose catabolism. The reason that long fatty acid molecules supply a great deal of energy is that they can yield many acetyl (two-carbon) groups for the Krebs cycle.

Plants use lipids to fuel activities such as seed germination. In oily seeds, triglycerides are broken down into glycerol and fatty acids. The fatty acids are then cut into two-carbon pieces that are released as acetyl CoA. This reaction is repeated for every pair of carbons until all of the fatty acids have been processed to form molecules of acetyl CoA.

7.4 MASTERING CONCEPTS

1. What happens to the pyruvic acid generated in glycolysis before it enters the Krebs cycle?
2. How does the Krebs cycle generate CO_2, ATP, NADH, and $FADH_2$?
3. What is the role of oxygen in the electron transport chain of aerobic respiration?
4. How do the electrons captured in NADH and $FADH_2$ power ATP formation?
5. What force fuels ATP formation in aerobic respiration?
6. Why are the energy-releasing reactions and pathways regulated?
7. Where can digested proteins and fats enter the energy pathways?

OLC

7.5 After Glycolysis—In the Absence of Oxygen

Several pathways enable organisms that live in environments without oxygen, or cells far from oxygen, to extract energy from nutrient molecules.

Aerobic cellular respiration via the Krebs cycle and the electron transport chain requires oxygen as the terminal electron acceptor. However, anaerobes do not use oxygen (whether or not it is in the environment) and some cells in aerobic organisms are occasionally without oxygen. Cells can use pathways that do not require oxygen to extract energy from nutrients.

Some anaerobes use fermentation pathways that occur in the cytoplasm. There are many types of fermentation pathways, but all oxidize NADH to NAD^+, which is recycled to keep glycolysis running. However, fermentation is far less efficient than cellular

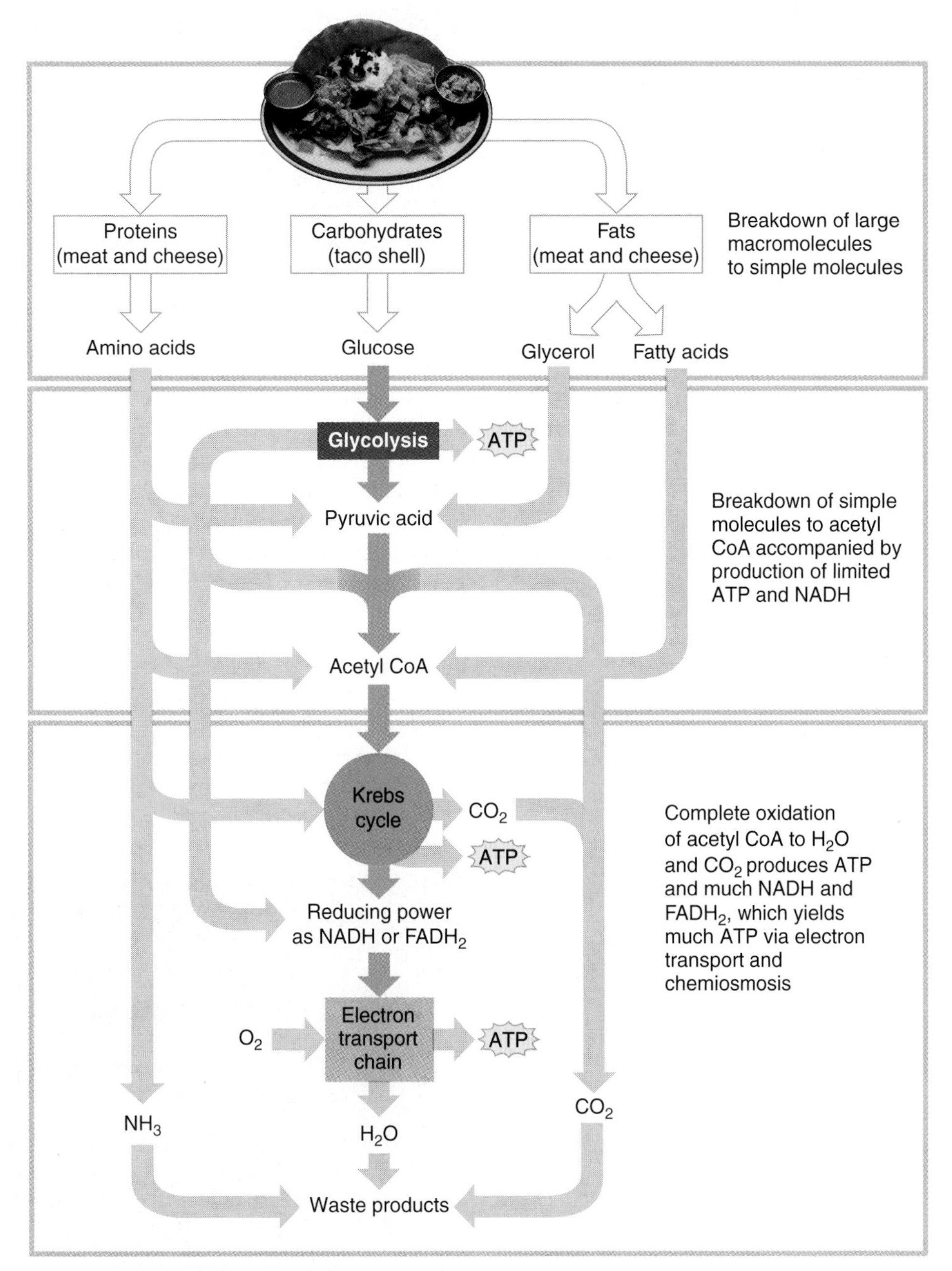

FIGURE 7.12

How Nutrients Enter the Energy Pathways.

Most cells use carbohydrates as a primary source of energy. But amino acids can enter the energy pathways after conversion to pyruvic acid, acetyl CoA, or intermediates of the Krebs cycle. Fats are digested to glycerol and fatty acids, which are catabolized to acetyl CoA.

(aerobic) respiration for energy procurement because it dismantles fewer glucose bonds and does not yield additional ATP besides the ATP generated in glycolysis.

A fermentation pathway includes glycolysis plus one or two reactions in which NADH is oxidized to NAD^+ by reducing pyruvic acid to other biochemicals. This is most commonly accomplished by producing lactic acid or the alcohol ethanol and carbon dioxide (table 7.4). Fermentation also occurs in parts of an organism that are in an anaerobic environment, such as in a plant partially submerged in a pond or in cells deep within a multicellular organism that lack direct access to oxygen.

In **alcoholic fermentation,** yeast cells break down pyruvic acid to ethanol and carbon dioxide, oxidizing NADH to NAD^+

TABLE 7.4 Types of Organisms That Use Fermentation Pathways

ALCOHOLIC FERMENTATION	LACTIC ACID FERMENTATION
Algae	Algae
Bacteria	Animal cells (muscle)
Fungi (yeast)	Bacteria
Plants	Fungi (yeast)
Protists	Protists

Whole-Body Metabolism—Energy on an Organismal Level

Metabolic reactions supply cells with the energy to survive, as well as to accomplish specialized functions in multicellular organisms. However, we are most familiar with whole-body metabolism in animals—such as ourselves.

Energy enters an animal's body as food molecules and exits as heat or activity, (either in movement or in the many internal energy-requiring functions that are part of life) or as undigested food (feces). Usually, energy intake and output are balanced. When they are not, the result can be weight loss or gain.

The energy a vertebrate (an animal with a backbone) requires to stay alive is described as its basal metabolic rate (BMR). In a person, BMR measures the energy (in milliliters of oxygen consumed per kilogram of body weight per minute) required for heartbeat; breathing; nerve, kidney, and gland function; and the maintenance of body temperature when the subject is awake, physically and mentally relaxed, and has not eaten for 12 hours. BMR does not include the energy needed for digestion or physical activity. For an adult human male, the average energy use is 1,750 kilocalories in 24 hours; for a female, 1,450 kilocalories. But because most people do not usually spend 24 hours a day completely at rest and fasting, the number of kilocalories needed to get through a day generally exceeds the basal requirement.

Several factors influence BMR, including age, sex, weight, body proportion, and regular activity level. A sedentary student may only use 1,700 kilocalories in 24 hours and have a BMR of 3 to 3.5. His roommate, who is on the swim team and runs 5 miles each morning, may burn 5,000 kilocalories per day and have a BMR of 4 to 5. When the students sleep, their metabolic rates are likely to fall below basal level; but the swimmer's rate will still be higher because regular exercise elevates metabolism, even during rest.

BMR rises from birth to about age five and then declines until adolescence, when it peaks again. As we age, BMR drops as our energy needs decline. It also falls during starvation as the body attempts to conserve energy. BMR increases during some illnesses, when body temperature rises, and during pregnancy and breastfeeding.

BMR is also related to body composition. For two individuals of the same weight, age, and sex, the one with a greater proportion of lean tissue (muscle, nerve, liver, and kidney) will have a higher BMR because lean tissue consumes more energy than relatively inactive fat tissue.

The thyroid gland in the neck affects BMR. It manufactures the hormone thyroxine, which increases the body's energy expenditure. Too much thyroxine can double the BMR, resulting in excessive consumption of kilocalories and weight loss. Too little thyroxine slows the BMR. The body burns fewer kilocalories, and weight increases.

In general, in warm-blooded animals, the smaller the organism, the higher its BMR. A rat has a higher metabolic rate than a cat, which has a higher rate than a dog, which has a higher rate than a human. This is partly because smaller organisms have higher surface-to-volume ratios and therefore lose more heat to the environment. Consequently, they must burn relatively more fuel to maintain a nearly constant body temperature. Even within a species, smaller individuals have higher metabolic rates.

Whole-body metabolic rate is related to life span in some insects. In houseflies, a measurement called the metabolic potential equals the amount of oxygen the fly consumes in a lifetime. This value is relatively constant from fly to fly, but individuals differ in the rate at which aerobic respiration uses the oxygen. Experiments that compared groups of houseflies with different activity levels showed that the less active the fly, the longer it lives, although all flies consume roughly the same amount of oxygen over their life spans (fig. 7.A).

Does this fact about housefly metabolism suggest that people should forsake the gym and take up residence on the sofa in front of the television? No, because there are differences between human and fly muscles. Researchers who study aging hypothesize that flies succumb to eventual death from excess exercise because toxic by-products of oxygen metabolism destroy their tissues. In humans, exercise strengthens our muscles and cardiovascular systems. Because of differences in whole-body metabolism, what's good for the fly is not necessarily good for the human.

FIGURE 7.A

This fly, caught by the camera while dining on sugar, will live longer if it flies less and faces cool temperatures. An active insect living in a very warm climate takes in the same amount of oxygen as a more sedentary fly but lives a shorter time.

(fig. 7.13). Fermentation by brewer's yeast is used industrially to help manufacture baked goods and alcoholic beverages (fig. 7.13A,C). Depending upon the substance being fermented, the variety of yeast used, and whether carbon dioxide is allowed to escape during the process, yeast fermentation may be used to produce wine or champagne from grapes, the syrupy drink called mead from honey, and cider from apples. Beer is brewed by fermenting grain—barley, rice, or corn.

Using alcoholic fermentation in an industrial process inadvertently led to discovery of lactic acid fermentation. In 1856, a French manufacturer of industrial chemicals who had produced ethanol from fermenting sugar beets noticed that his brew had turned sour-tasting and didn't contain much alcohol. He asked for help from Louis Pasteur, who determined that the type of microorganisms fermenting the beets had changed, and they produced lactic acid instead of alcohol.

Some anaerobic single-celled organisms and some animal cells temporarily deprived of oxygen break down pyruvic acid to the three-carbon compound lactic acid in a single step, oxidizing NADH to NAD^+. **Lactic acid fermentation** occurs in human muscle cells that are working so strenuously that their production of pyruvic acid exceeds the oxygen supply. In this "oxygen-debt" condition, the muscle cells revert to fermentation to extract energy. If enough lactic acid accumulates, the muscle fatigues and cramps. When oxygen is once again present, lactic acid is converted back to pyruvic acid in the liver. Pyruvic acid is then processed as usual in the mitochondria.

Athletic coaches sometimes measure lactic acid levels in the blood to assess the physical condition of swimmers and sprinters. One study reported that lactic acid accumulates to triple the normal levels in the bloodstreams of children who cry vigorously as they are prepared for surgery but not in calm children in the same situation. This suggests that stress may trigger lactic acid fermentation.

Figure 7.14 shows the position of the fermentation reactions in the energy pathways, and reviews all the material in the chapter. Certain bacteria use fermentation pathways other than alcoholic or lactic acid fermentation. These organisms catabolize a variety of organic compounds and release a variety of gases, including carbon dioxide, hydrogen (H_2), hydrogen sulfide (H_2S), and ammonia (NH_3). Some types of *Clostridium* bacteria ferment nearly anything organic—except (unfortunately, from an ecological point of view) plastics!

One less common fermentation pathway proved very useful during World War II. The Allies needed an organic compound called acetone to produce a type of gunpowder. Chemical manufacturers at the time extracted acetone from wood, and supplies were rapidly running out. English chemist Chaim Weizmann

FIGURE 7.13
Fermentation.
The beer and wine industry (**A**) relies on fermentative organisms to produce alcohol. (**B**) Cells use fermentation to produce energy in the absence of oxygen by reducing pyruvic acid and oxidizing NADH to NAD^+, which allows glycolysis to continue. (**C**) In yeast or bacteria, pyruvic acid is broken down to ethanol and carbon dioxide. In animal cells and some microorganisms, lactic acid is the result of oxidizing NADH to NAD^+ in the absence of oxygen.

A

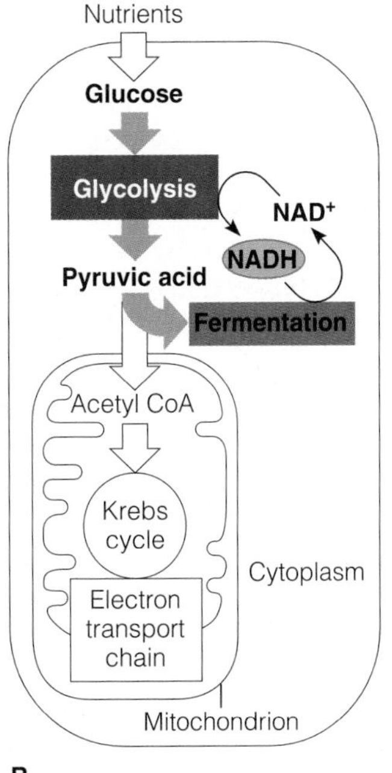

B

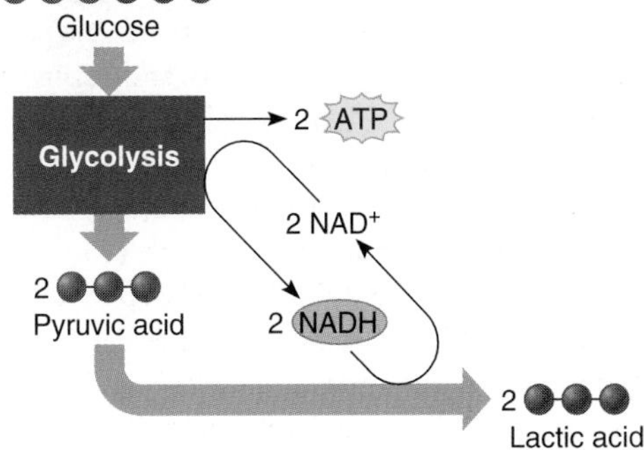

C

FIGURE 7.14
Energy Yield of Respiration.
Breaking down glucose to carbon dioxide can yield as many as 30 ATPs. Fermentation is an alternate pathway to cellular respiration that yields far fewer ATPs per glucose molecule.

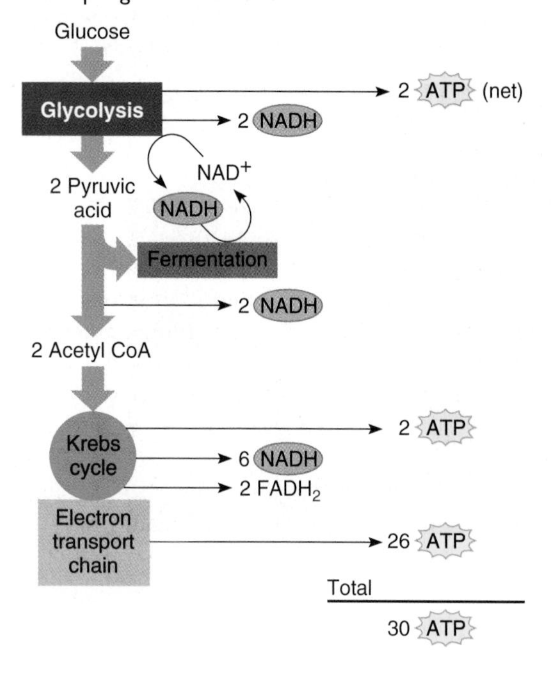

solved the problem by using a fermentation pathway of the anaerobic bacterium *Clostridium acetobutylicum.* The bacteria broke down compounds in grain or molasses into the needed acetone, producing a second valuable chemical, butanol, used to manufacture synthetic rubber.

7.5 MASTERING CONCEPTS

1. What do different fermentation pathways have in common?
2. What are the products of alcoholic fermentation?
3. What are the products of lactic acid fermentation?

OLC

7.6 Possible Origins of the Energy Pathways

Interactions and similarities among the reactions of photosynthesis and the energy-releasing pathways and cycles suggest a sequence in which they might have originated and evolved.

Photosynthesis, glycolysis, and cellular respiration are intimately related (fig. 7.15). The organic products of photosynthesis supply the starting materials for glycolysis. The oxygen produced by

FIGURE 7.15
The Energy Pathways and Cycles Connect Life.
An overview of energy metabolism illustrates how biological energy reactions are interrelated.

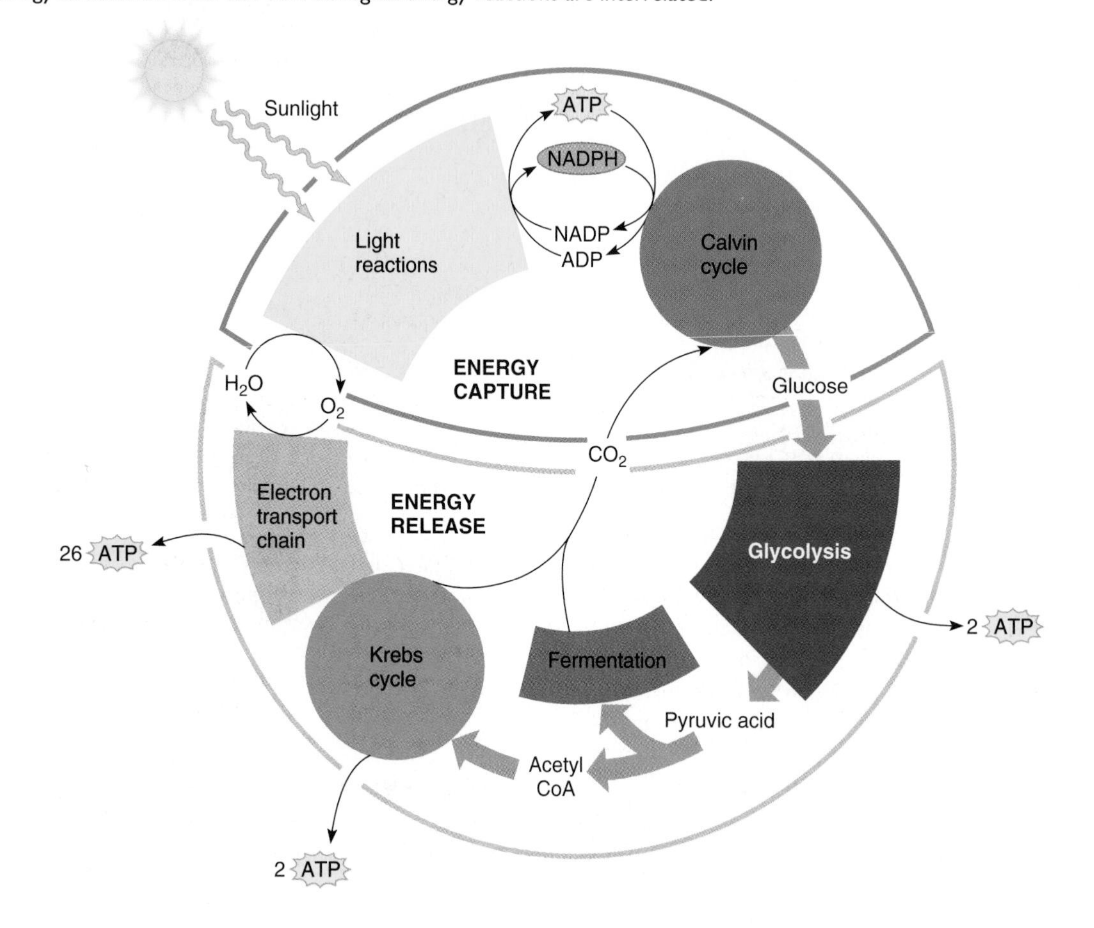

photosynthesis becomes the final electron acceptor in aerobic respiration. CO_2 generated in the Krebs cycle enters the Calvin cycle. Finally, water produced by aerobic respiration is split in photosynthesis, releasing electrons that replace the electrons boosted out of reactive chlorophyll *a* molecules when they absorb photon energy. Together, these biological energy reactions sustain life. How might they have arisen?

Glycolysis is probably the most ancient of the energy pathways because it is common to nearly all cells. The other pathways are more specialized—photosynthesis occurs only in green plants, algae, and some bacteria and archaea; fermentation and anaerobic respiration take place only under certain environmental conditions.

Glycolysis evolved when the atmosphere lacked or had very little oxygen. These reactions enabled the earliest organisms to extract energy from simple organic compounds in the nonliving environment. Photosynthesis may have evolved from glycolysis because some of the reactions of the Calvin cycle are the reverse of some of those of glycolysis. Broadly speaking, glycolysis breaks down a six-carbon compound into two three-carbon compounds; some of the reactions of the Calvin cycle do the opposite.

The reactions of photosynthesis probably arose through a still-unknown chemical rearrangement, based on fossil evidence of cyanobacteria dating to 3.5 billion years ago. Photosynthesis altered life on Earth forever. No longer did organisms have to depend on organic compounds in their surroundings for energy; they now had a way to produce nutrients constantly, from sunlight. Photosynthesis, over time, released oxygen into the primitive atmosphere, paving the way for an explosion of new species capable of using this new atmospheric component. In addition, electrical storms split water in the atmosphere, releasing single oxygen atoms that joined with diatomic oxygen to produce ozone (O_3). As ozone accumulated high in the atmosphere, it blocked harmful ultraviolet radiation from reaching the planet's surface, which prevented some genetic damage and allowed new varieties of life to arise.

Photosynthesis could not have debuted in a plant cell, because such complex organisms were not present on the early earth. The first photosynthetic organisms may have been anaerobic bacteria or archaea or an ancestor of both that used hydrogen sulfide (H_2S) in photosynthesis instead of the water plants use. These first photosynthetic microorganisms would have released sulfur, rather than oxygen, into the environment. Eventually, evolutionary changes in pigment molecules enabled some of these organisms to use water instead of hydrogen sulfide. If a large cell engulfed an ancient photosynthesizing microorganism, it may have become a eukaryotic-like cell that may have been the ancestor of modern plants. (Recall the endosymbiont theory discussed in chapter 3.) Mitochondria might have evolved in a similar way, when larger cells engulfed bacteria capable of using oxygen.

After endosymbiosis, different types of complex cells probably diverged, leading to evolution of a great variety of eukaryotic organisms. Today, the interrelationships of the biological reactions of photosynthesis, glycolysis, and aerobic respiration, and the great similarities of these reactions in diverse species, demonstrate a unifying theme of biology: All types of organisms are related at the biochemical level.

7.6 MASTERING CONCEPTS

1. Which energy pathway is probably the most ancient?
2. Why were the first energy pathways anaerobic?
3. What is the evidence that photosynthesis may have evolved from glycolysis?
4. How did photosynthesis forever alter life on Earth?

OLC

Chapter Summary

7.1 How Does Cellular Respiration Differ from Breathing?

1. The respiratory system provides the oxygen that **aerobic** cellular respiration requires as a final electron acceptor.
2. **Cellular (aerobic) respiration** is a common biochemical pathway that extracts energy from the bonds of nutrient molecules in the presence of oxygen.

7.2 How Do Cells Release Energy?

3. The overall reaction for cellular respiration is:

$$C_6H_{12}O_6 + 6O_2 \rightarrow 6CO_2 + 6H_2O + 30ATP$$

4. All cells begin energy release from nutrients with **glycolysis** and then may follow any of several pathways.
5. Cellular (aerobic) respiration harvests energy gradually. Acetyl CoA formation, the **Krebs cycle,** and an **electron transport chain** follow glycolysis. Glycolysis takes place in the cytoplasm; acetyl CoA formation, the Krebs cycle, and the electron transport chain occur in mitochondria.
6. The two membranes of a mitochondrion enclose the innermost **matrix,** and the **intermembrane compartment** is between the membranes.
7. ATP synthesis occurs by **substrate-level phosphorylation** (phosphate transfer between organic compounds) or by **chemiosmotic phosphorylation** (passage of electrons along carrier molecules through oxidation-reduction reactions, setting up a proton gradient that powers phosphorylation of ADP to ATP).

7.3 How Does Glycolysis Break Down Glucose to Pyruvic Acid?

8. In the first half of glycolysis, glucose is broken down into two molecules of the three-carbon compound PGAL.
9. In the second half of glycolysis, the PGALs are oxidized as NAD^+s are reduced to NADHs, contribute phosphate groups to form two ATPs, and react and are rearranged to form two molecules of **pyruvic acid.**

7.4 After Glycolysis—In the Presence of Oxygen

10. In the mitochondria, pyruvic acid is broken down into **acetyl CoA** in a coupled reaction that also reduces NAD^+ to NADH.

11. Acetyl CoA enters the Krebs cycle, a series of oxidation-reduction reactions that produces ATP, NADH, $FADH_2$, and CO_2. Substrate-level phosphorylation produces ATP in the Krebs cycle.
12. Energy-rich electrons from NADH and $FADH_2$ fuel an electron transport chain. Electrons move through a series of carriers that release energy at each step. The terminal electron acceptor, oxygen, is reduced to form water.
13. Electron transport energy establishes a proton gradient that pumps protons from the mitochondrial matrix into the intermembrane compartment. As protons diffuse back into the matrix through channels in ATP synthase, their energy drives phosphorylation of ADP to ATP.
14. Negative feedback coordinates the rates of glycolysis, acetyl CoA formation, and the Krebs cycle.
15. Amino acids enter the energy pathways as pyruvic acid, acetyl CoA, or an intermediate of the Krebs cycle. Fatty acids and glycerol enter as acetyl CoA.

7.5 After Glycolysis—In the Absence of Oxygen

16. In the absence of oxygen, alcoholic, lactic acid, or other fermentation pathways may run. Fermentation does not produce ATP but oxidizes NADH to NAD^+, which is recycled to glycolysis. **Alcoholic fermentation** reduces pyruvic acid to ethanol and loses carbon dioxide. **Lactic acid fermentation** reduces pyruvic acid to lactic acid.

7.6 Possible Origins of the Energy Pathways

17. The energy pathways are interrelated, with common intermediates and some reactions that mirror the reactions of the others.
18. Glycolysis may be the oldest energy pathway because it is more prevalent—the other pathways are more specialized.
19. Cellular respiration and photosynthesis may have arisen when larger cells engulfed prokaryotes that were forerunners to mitochondria and chloroplasts.

Testing Your Knowledge

1. How are breathing and cellular respiration similar? How are they different?
2. Why is aerobic respiration a more efficient energy-extracting pathway than glycolysis alone?
3. How does substrate-level phosphorylation differ from chemiosmotic phosphorylation?
4. Cite a reaction or pathway that occurs in each of the following locations:
   a. cytoplasm
   b. mitochondrial matrix
   c. inner mitochondrial membrane
   d. intermembrane compartment
5. What is the immediate source of the six carbons in citric acid in the Krebs cycle?
6. At what point does oxygen (O_2) enter the energy pathways of aerobic respiration? What is its role?
7. How are photosynthesis, glycolysis, and cellular respiration interrelated?

Thinking Scientifically

1. The amino acid sequence of the respiratory chain carrier molecule cytochrome *c* is almost identical in all organisms. Disorders caused by abnormal cytochrome *c* are unknown. What do these two observations suggest about the importance of this molecule?
2. Health-food stores sell a product called "pyruvate plus," which supposedly boosts energy. Why is this product unnecessary?
3. A student regularly runs 3 miles (1.9 kilometers) each afternoon at a slow, leisurely pace. One day, she runs a mile as fast as she can. Afterwards she is winded and feels pain in her chest and leg muscles. She thought she was in great shape! What, in terms of energy metabolism, has she experienced?
4. Two men weigh 230 pounds each. One is 6 feet tall and a bodybuilder, with tight, bulging muscles. The other is 5 feet, 4 inches tall and obese. Which man probably has the higher basal metabolic rate? Why?

References and Resources

Bentley, Ronald. Spring 1994. A history of the reaction between oxaloacetate and acetate for citrate biosynthesis: An unsung contribution to the tricarboxylic acid cycle. *Perspectives in Biology and Medicine,* vol. 37. Don't let the title scare you—this is a great description of experiments used to decipher a single step of the Krebs cycle.

Blake, Colin. January 16, 1997. Phosphotransfer hinges in PGK. *Nature,* vol. 385. A look at an enzyme that is crucial to both cellular respiration and photosynthesis.

Levi, Primo. October, 1984. Travels with C. *The Sciences.* An eloquent travelogue of how a single atom of carbon circulates through the living world.

The *LIFE* Online Learning Center provides additional resources and tools for studying this chapter.
www.mhhe.com/life

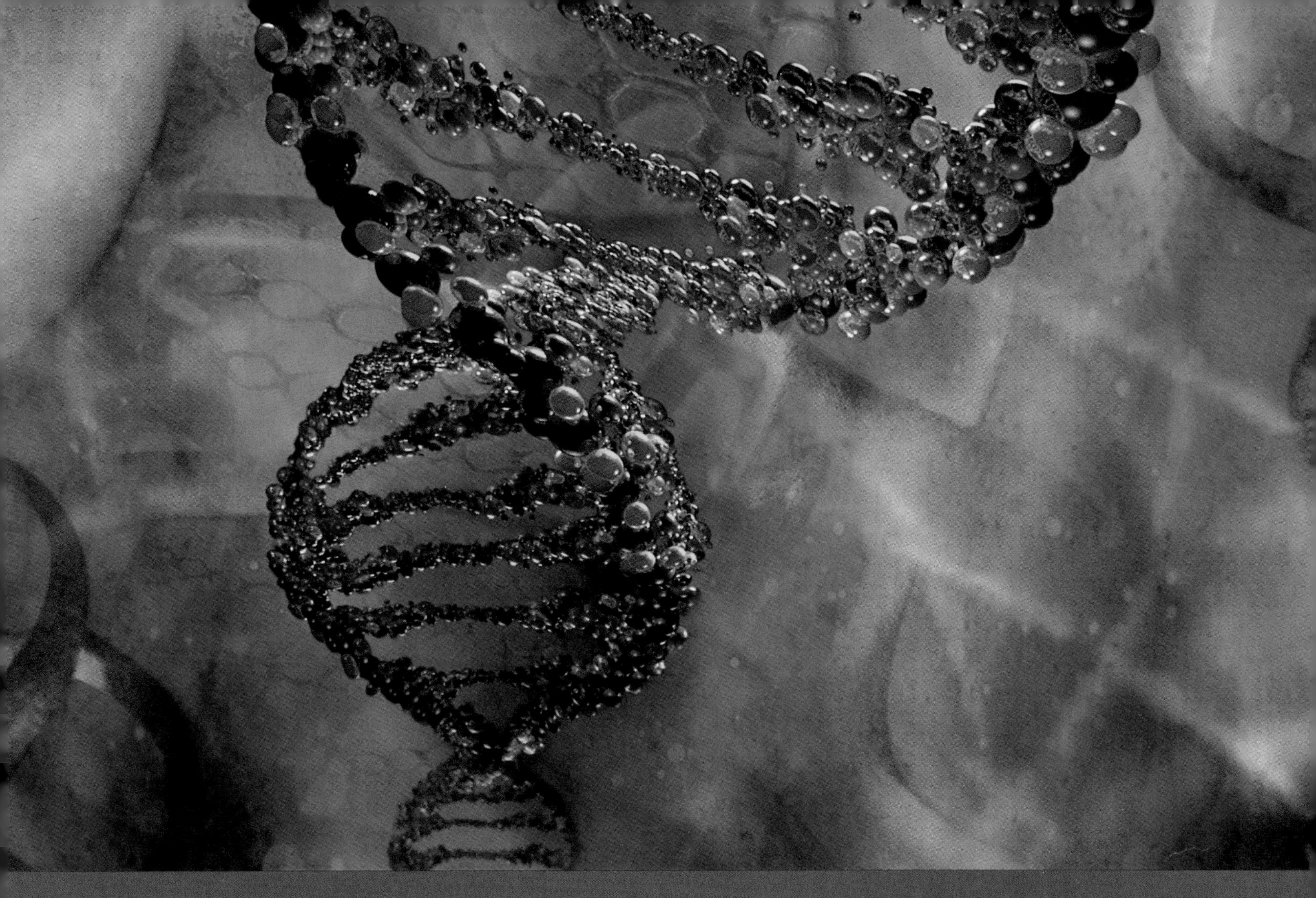

UNIT 2 • GENETICS AND BIOTECHNOLOGY

The field of genetics is only a century old, yet it impacts our lives in many ways. Genetics explains how organisms function at the molecular and cellular levels, how family resemblances arise, and the characteristics of populations. As the basis of biotechnology, genetics affects fields from agriculture to health care to forensics. In a broader sense, genetics provides a living language that reveals and explains the evolution of life on Earth. This unit explores the DNA molecule and how the information encoded in its building block sequences oversees the functioning of organisms.

CHAPTER 8

The Cell Cycle

Starving a Tumor

Probing Angiogenesis

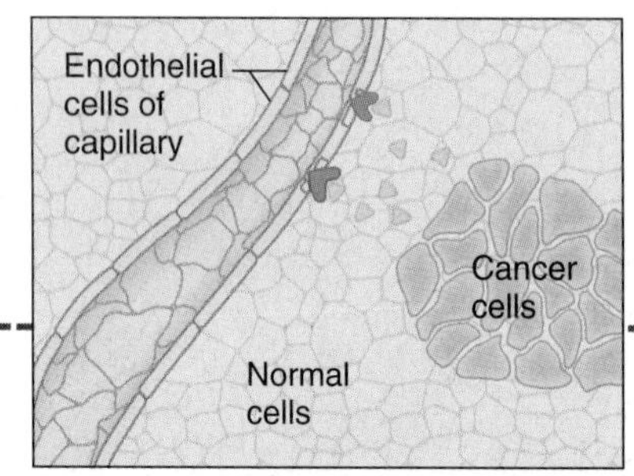

1 Tumor releases factors that bind to endothelial cells of nearby capillaries and stimulate them.

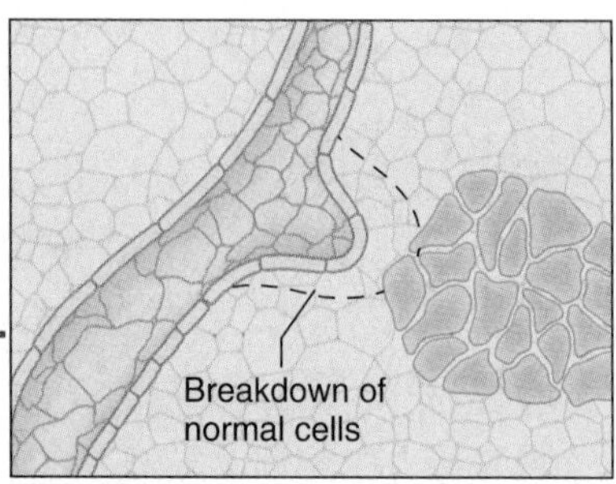

2 Stimulated endothelial cells invade surrounding area (extracellular matrix) and proliferate (divide) toward tumor.

Mitosis is a form of cell division that enables an organism to grow by adding cells. In a process called angiogenesis, capillaries (microscopic blood vessels) form or extend when mitosis occurs in their endothelial cells. In normal development, angiogenesis produces blood vessels that deliver nutrients, hormones, and growth factors to organs, and remove wastes. It is also crucial to healing. After a heart attack, for example, new vessels form in the remaining healthy cardiac muscle to supply blood.

Cancer occurs when cells begin to divide more frequently than the cells from which they derive, and continue to do so. As a cancer spreads, it commandeers otherwise normal angiogenesis, building its own blood supply. Understanding cancer-induced angiogenesis is fueling the development of new drugs that block this support system for the cancer.

The story of angiogenesis inhibitors began in 1971, when Harvard Medical School researcher Judah Folkman reported that when tumors grow larger than a pinhead, they develop their own blood supply. He also noticed, in mice and in humans, that after an initial tumor is removed, secondary tumors, called metastases, often appear and grow throughout the body. Folkman hypothesized that the primary tumor releases chemical factors that inhibit metastases from enlarging. Removing the tumor lifts the inhibition. He further suggested that instead of aiming drugs and radiation at cancer cells, a new approach could strangle tumor-associated angiogenesis. It was an unpopular idea, at first.

A tumor has a great ability to surround itself with blood vessels.

A tumor has a great ability to surround itself with blood vessels. Once a tumor reaches pinhead size, it secretes factors that stimulate nearby capillaries to sprout new branches that extend toward it. Because a tumor harbors some endothelial cells, capillaries may also grow from the tumor itself as these cells divide, assemble into sheets, and roll into tubules. Eventually, capillaries snake out of the tumor. Cancer cells wrap around them, spreading out on this scaffolding into

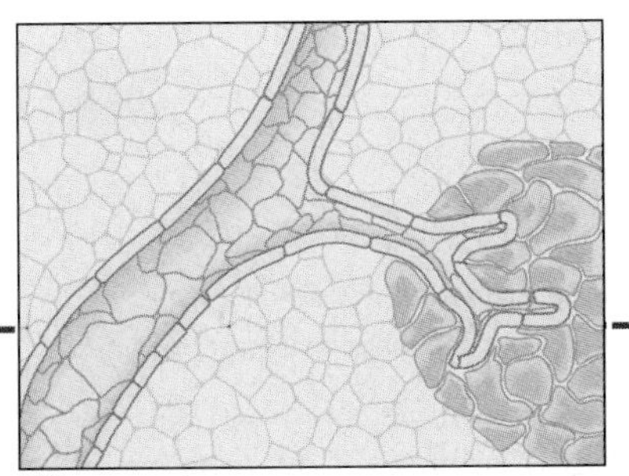

3 New blood vessel extension nourishes tumor.

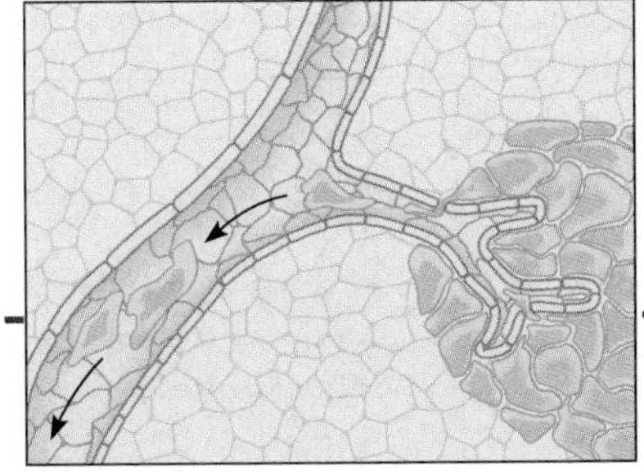

4 Cancer cells leave primary tumor and enter bloodstream, travel, and establish secondary tumors (metastases).

the nearby tissue in the characteristic crablike shape from which the word "cancer" derives. Some cancer cells enter blood vessels and travel to other parts of the body. For a time, maybe even years, the rate of cell division in metastatic tumors equals the rate of cell death, and the growths stay small, adhering to the outsides of the blood vessels that delivered them. But when the primary tumor is removed, the angiogenesis-inhibiting factors vanish, and instead angiogenesis-promoting signals wash over the metastases. As the secondary tumors begin to knit their own blood supplies, the balance of cell division to cell death within them shifts. Mitosis prevails, and the tumors grow.

So renegade were Folkman's ideas that he had difficulty persuading graduate students to do the necessary experiments. Finally in 1991, Michael O'Reilly decided to seek Folkman's hypothesized angiogenesis inhibitors. They took mice with single, large tumors and no metastases and removed the tumors. Within 5 days, the animals had many metastases, and within 15 days, they were dead, their lungs filled with tumors. A biochemical found in the urine in trace amounts before the tumors were removed, but not after, became the first candidate for an angiogenesis inhibitor. Named angiostatin, it was part of a protein called plasminogen that helps blood to clot. The researchers soon discovered angiostatin in human blood.

So renegade were Folkman's ideas that he had difficulty persuading graduate students to do the necessary experiments.

The next experiments were crucial. O'Reilly gave 10 mice bearing single, large tumors injections of angiostatin, and 10 others a placebo, salt water. Sure enough, after removing the primary tumors, the lungs of the mice that received angiostatin remained clear, but the lungs of mice who got salt water rapidly filled with large tumors.

The Harvard researchers then discovered a second angiogenesis inhibitor, which is a fragment of the connective tissue protein collagen. This compound, called endostatin, also blocks endothelial cells from dividing. Experiments in mice at first showed that angiostatin or endostatin shrinks tumors, but only until treatment stops. However, further experiments showed that the effects can be permanent. In 1997, Folkman and O'Reilly gave endostatin to mice with lung cancer. They withdrew the treatment as soon as the tumors shrank, and started it again when the tumors started to regrow, repeatedly. After the sixth cycle, the tumors had shrunk from a huge growth bulging from the animals' sides, to barely visible lumps just under the skin—and remained that way.

Unlike other cancer treatments, angiostatin and endostatin can be used over and over without the targeted tissue becoming resistant, as is the case with conventional drugs. Folkman hypothesizes that this is because endothelial cells do not mutate rapidly, which is how cancer cells become resistant to drugs. Plus, angiostatin and endostatin together make tumors disappear entirely—at least in mice. Several different types of drugs to treat cancer by inhibiting angiogenesis are now being developed.

8.1 The Balance Between Cell Division and Cell Death

The development and continued existence of an animal, and perhaps other organisms, reflects a balance between cell division (mitosis) and cell death (apoptosis).

"In a sense we contain ourselves, wrapped up within ourselves, trillions of times repeated." So wrote noted geneticist Herman J. Muller in 1947, referring to the astonishing fidelity with which a human's trillions of cells each retains the precise genetic information that was present in the fertilized egg. Mitotic cell division, or **mitosis,** is a process that forms two identical cells from one. Cells specialize by expressing subsets of the whole gene set that each contains.

As a multicellular organism develops, cell death occurs too, which helps to carve distinctive structures. A foot might start out as a webbed triangle of tissue, for example, from which digits are carved as certain cells die. This type of cell death that is a normal part of development is called **apoptosis.** It is a precise, genetically programmed sequence of events, as is mitosis. Apoptosis is best studied in animals.

Throughout an animal's life, mitosis and apoptosis are in balance, so that tissue neither overgrows or shrinks. In this way, a child's liver remains much the same shape as she grows into adulthood (fig. 8.1). During early development, mitosis and apoptosis orchestrate the ebb and flow of cell number as new structures form. Later, these processes protect. Mitosis fills in new skin to heal a scraped knee; apoptosis peels away sunburnt skin cells that might otherwise become cancerous.

FIGURE 8.1
Cell Division and Apoptosis Regulate Development.
Mitosis mass-produces cells during development while apoptosis eliminates certain cells, producing the correct shape and size for parts of the body. The evidence of apoptosis can be seen early in development in the structures of a fetus, whose shapes remain essentially the same for life.

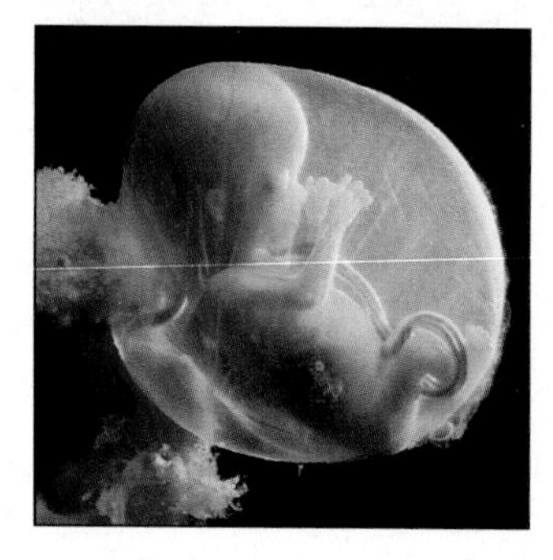

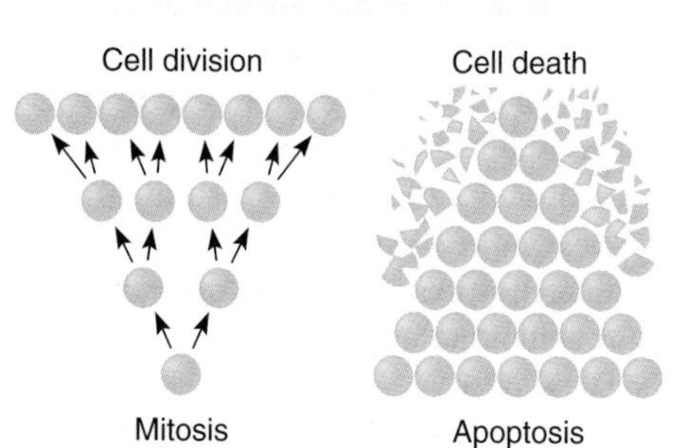

Cancer is a profound derangement of the balance between cell division and death, as we saw in the opening essay about the role of angiogenesis in a cancer's spread. Such abnormal growth occurs when mitosis is too frequent or occurs too many times, or when apoptosis is too infrequent. This chapter explores the opposing but coordinated forces of cell division and death, and considers the consequences of either gone awry.

8.1 MASTERING CONCEPTS

1. Why must mitosis be precise?
2. What is apoptosis?
3. How do mitosis and apoptosis interact?

8.2 The Cell Cycle

A cell has choices: It may divide, specialize but not divide, repair damage, or die. The cell cycle describes the functional stages of a cell's existence.

A cell can be in the process of dividing; not dividing but alive, well, and specialized; or in the throes of death. A series of events called the **cell cycle** describes whether a cell is actively dividing or not (fig. 8.2). The cycle consists of a sequence of activities when the cell synthesizes or duplicates certain molecules, assembles them into larger structures such as membranes and chromosomes, and apportions its contents into two cells. When one cell divides into two, the resulting "daughter" cells must receive complete genetic instructions, as well as molecules and organelles to sustain life and possibly also to specialize.

So vital is the cell cycle to the form and function of a multicellular organism that several **"checkpoints"** control it. Checkpoints are actually interacting proteins that ensure that the proper sequence of events unfolds. They also ensure that the cell cycle can pause briefly so that errors in the sequences of newly formed DNA molecules can be repaired before they are perpetuated.

The cell cycle includes three major stages: mitosis, or **karyokinesis,** when the nucleus is actively dividing; **cytokinesis,** when other cell contents are distributed into the daughter cells; and **interphase,** when the cell is not dividing. Division of the genetic material and other cell contents can overlap in time, depending upon the species. The next chapter considers another form of cell division that results in formation of sex cells.

Interphase—A Time of Great Activity

Biologists in the 1950s and 1960s mistakenly described interphase as a time when the cell is at rest, perhaps because the traditional stains used in microscopy do not bind to the highly unwound interphase DNA. Without visible chromosomes, cells appeared inactive. However, interphase is actually a very active time in the life of a cell. Not only do the basic biochemical life functions continue, but the cell also replicates its genetic material, a process discussed in detail in chapter 14. Interphase is divided into phases of gap (designated "G" and sometimes called "growth") and synthesis ("S").

G_1 Phase During the first gap phase, or **G_1 phase,** the cell carries out the basic functions of life and performs specialized activities. At

FIGURE 8.2

The Cell Cycle.

The cell cycle is divided into interphase, when cellular components replicate, and cell division (mitosis and cytokinesis), when the cell splits in two, distributing its contents into two daughter cells. Interphase is divided into two gap phases (G_1 and G_2), when specific molecules and structures duplicate, and a synthesis phase (S), when the genetic material replicates. Mitosis can be described as consisting of stages: prophase, prometaphase, metaphase, anaphase, and telophase. Several checkpoints control the cell cycle. Of particular importance is the checkpoint called the restriction point, at the end of G_1. It determines a cell's fate—whether it will continue in the cell cycle and divide, enter a stage called G_0 as a quiescent and possibly specialized cell, or die.

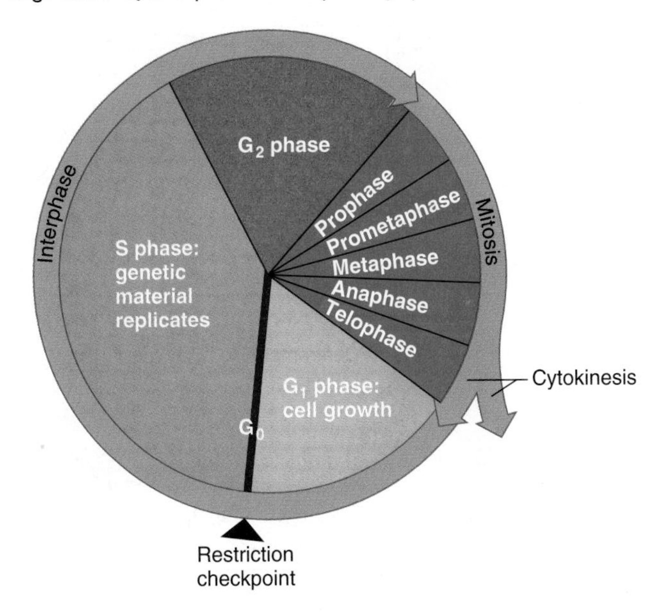

FIGURE 8.3

Cloning Mammals Requires Cell Cycle Synchronization.

At first, researchers thought that using a nucleus that was in G_0 was necessary to clone a mammal, but today cloning is accomplished using donor nuclei from a variety of stages. However, it is important that the cell that donates the nucleus and the recipient cell be in the same stage of the cell cycle.

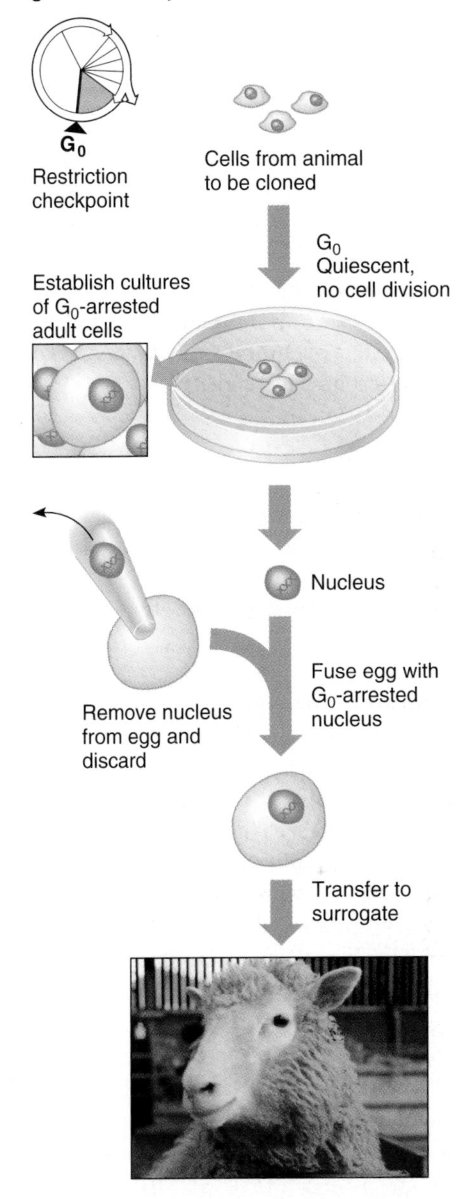

the same time, G_1 is a critical checkpoint that determines a cell's fate. This checkpoint, called the restriction point, "decides" whether a cell goes on to divide, stops to repair damaged DNA, enters a quiescent stage, or dies. A cell in G_1 is exquisitely sensitive to extracellular signals.

Because G_1 is a decision point in the life of a cell, it is the cell cycle period whose duration varies the most among different types of specialized cells. Slow-growing cells, such as human liver cells, may exit G_1 and remain in the quiescent stage for years. Yet human bone marrow cells speed through G_1 in 16 to 24 hours, and early embryonic cells may skip G_1 entirely. Interestingly, the total time spent in the other cell cycle stages is nearly the same in all cells.

G_1 is also a time of preparation for possible impending division. The cell synthesizes proteins, lipids, and carbohydrates. If it divides, the cell will require these molecular supplies to surround the two cells that form from the original cell.

G_0 Phase A cell can exit the cell cycle at G_1 to enter $\mathbf{G_0}$, a quiescent phase. A cell in G_0 can maintain its specialized characteristics, but does not replicate its DNA or divide. Many types of human cells are in G_0, but continue to function.

Understanding how G_0 fits into the cell cycle was crucial in the first cloning of a mammal from an adult cell, which Scottish researchers accomplished in 1997. Cloning produces an individual that is genetically identical to another individual. In the procedure, the nucleus of a specialized cell is transferred to an egg cell whose nucleus has been removed (fig. 8.3). For years such attempts didn't work because the donor nuclei were in S or G_2. When these nuclei enter the egg cells, with their already replicated DNA, they replicate it again, creating a genetic overload that halts development. The DNA of cells in G_0, however, has not yet replicated. The researchers coaxed donor cells from a sheep's mammary gland into G_0 by culturing them and then removing nutrients. It worked! The G_0 donor nucleus, synchronized with the cytoplasm of the recipient egg, supported development. The result was the sheep Dolly. Since then, researchers have been able to alter the cloning protocol to enable nuclei from other cell cycle stages to support development of a clone too.

S Phase **S phase** is a time of great synthetic activity as the cell initiates the immense job of replicating the genetic material. In most human cells, assembling billions of DNA nucleotides during S phase takes from 8 to 10 hours. Many proteins are manufactured during the S phase, including those that are part of the chromosomes and proteins that coordinate the many events taking place in the nucleus and cytoplasm.

G_2 Phase In the **G_2 phase,** the cell synthesizes more proteins, especially abundant tubulin, which will form microtubules (see fig. 4.14). Membrane material is assembled and stored as small empty vesicles beneath the cell membrane. The extra membrane will be used to provide enough material to enclose two cells rather than one. The DNA winds more tightly around its associated proteins, and this start of chromosome condensation signals impending mitosis. Interphase has ended.

Mitosis Distributes Chromosomes

During mitosis, genetic material that replicated in the previous S phase separates. The important structures in the process are the replicated chromosomes and a structure built of microtubules, called the **mitotic spindle,** that separates them into two sets.

Participants in Mitosis A replicated chromosome consists of two identical copies of chromosomal material, called **chromatids.** That is, the DNA sequences of a chromatid pair are identical. Chromatids are joined at a small section of the DNA sequence called the **centromere** (fig. 8.4). The centromere in a replicated chromosome is duplicated too, but it may appear as a constriction. At a certain time during mitosis, the chromatids separate. The number of chromosomes is the same in a cell before mitosis and in each of its daughter cells after mitosis, but the chromosomes are in the replicated form as they enter mitosis.

In order for replicated chromosomes to be evenly distributed into two cells, they must be grouped in a way that they can split equally into two sets. The mitotic spindle accomplishes this task (fig 8.5). It forms rapidly from tubulin subunits that assemble into microtubules.

The spindle grows from two structures called **centrosomes.** These consist of tubulin and other proteins, but biologists still do not know very much about the exact composition. In many animal cells, each centrosome also includes a pair of centrioles (sometimes called basal bodies), which are the structures that give rise to cilia and flagella (see fig. 4.16). However, plant cells and many cells of animal embryos lack centrioles but nonetheless divide, indicating that centrioles are not crucial for mitosis. It is the cloud of proteins that surrounds the centriole, called the pericentriolar material, that triggers microtubules to organize and assemble into a spindle.

FIGURE 8.4
Replicated and Unreplicated Chromosomes.
Chromosomes are replicated during S phase, before mitosis begins. The two genetically identical chromatids of a replicated chromosome attach at the centromere (**A**). (**B**) In the photograph, a human chromosome is in the midst of forming two chromatids. A longitudinal furrow extends from the chromosome tips inward.

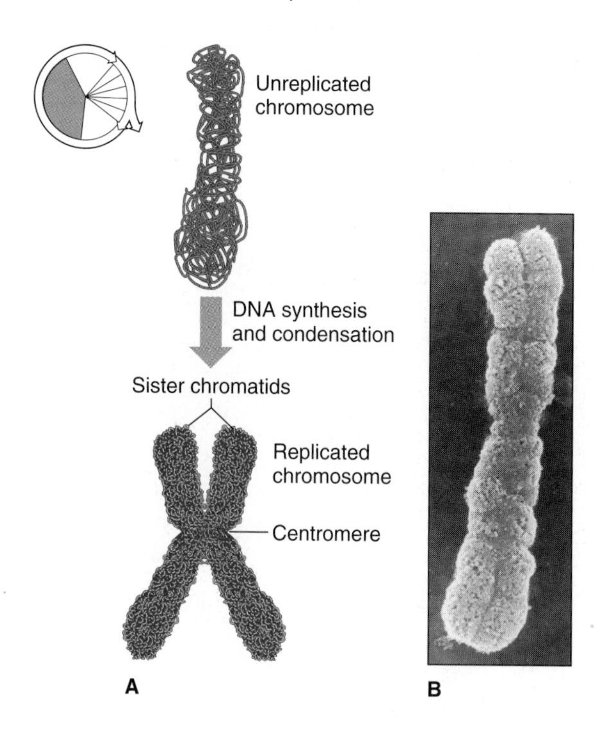

As the spindle begins to take shape, the two points where the microtubules start to extend resemble stars, and are called **asters.** They are initially close to each other, but draw apart as mitosis ensues. The microtubules that extend between the asters are called polar microtubules. Complexes of proteins called **kinetochores** begin to grow on each chromosome's centromere, and these attach the chromosomes to the spindle.

Table 8.1 reviews some terms introduced so far.

The Stages of Mitosis Mitosis (**M phase**) is a continuous process but is considered in stages for ease of understanding. The longer the duration of a particular stage, the more cells are in that stage at any given time, as figure 8.6 shows.

During **prophase,** the first stage of mitosis, DNA coils very tightly around chromosomal proteins, shortening and thickening the chromosomes (fig. 8.7). They are now visible when stained and viewed under a microscope. Chromosomes can separate into two groups more easily when they are condensed.

As prophase begins, microtubules throughout the cell disassemble. At the same time, in animal cells, the centrosome replicates by forming a second centriole at a 90-degree angle to the first. The two centrosomes then migrate toward opposite ends, or poles, of the cell. Microtubules extend from the centrosomes, then back away, and then extend again, like fingers

TABLE 8.1 Miniglossary of Cell Division Terms

TERM	DEFINITION
Asters	Short microtubules extending from centrosomes as spindle forms.
Centriole	A structure that gives rise to cilia and flagella and is part of the centrosome in some cells.
Centromere	A part of a chromosome that joins two chromatids.
Centrosome	A structure consisting of tubulins and other proteins, and sometimes centrioles, that organizes the mitotic spindle.
Chromatid	A single strand of a chromosome. A replicated chromosome consists of two chromatids.
Cytokinesis	Distribution of molecules, cytoplasm, and organelles to daughter cells following division of the genetic material (mitosis).
Karyokinesis	Distribution of replicated genetic material to two daughter cells.
Kinetochore	An aggregation of microtubules on a chromosome where it attaches to the mitotic spindle.

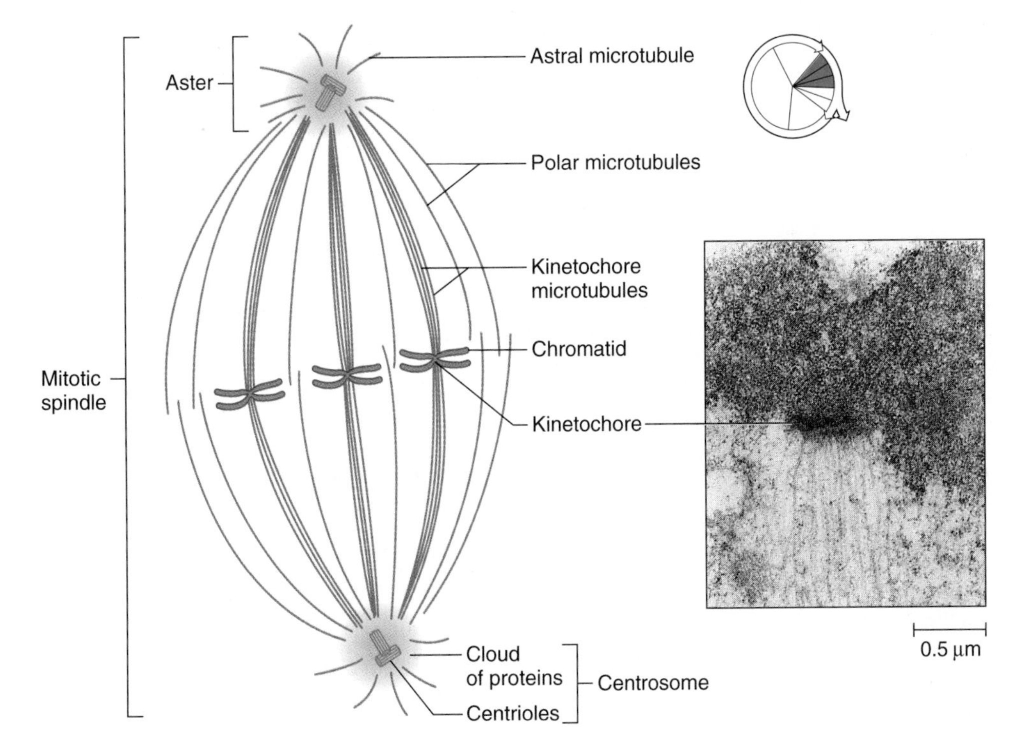

FIGURE 8.5
The Spindle Aligns Chromosomes.
In animal cells, the mitotic spindle consists of highly organized microtubules that form the fibers that grow outward from two centrosomes. The spindle fibers push and pull, aligning chromosomes during the first steps of mitosis.

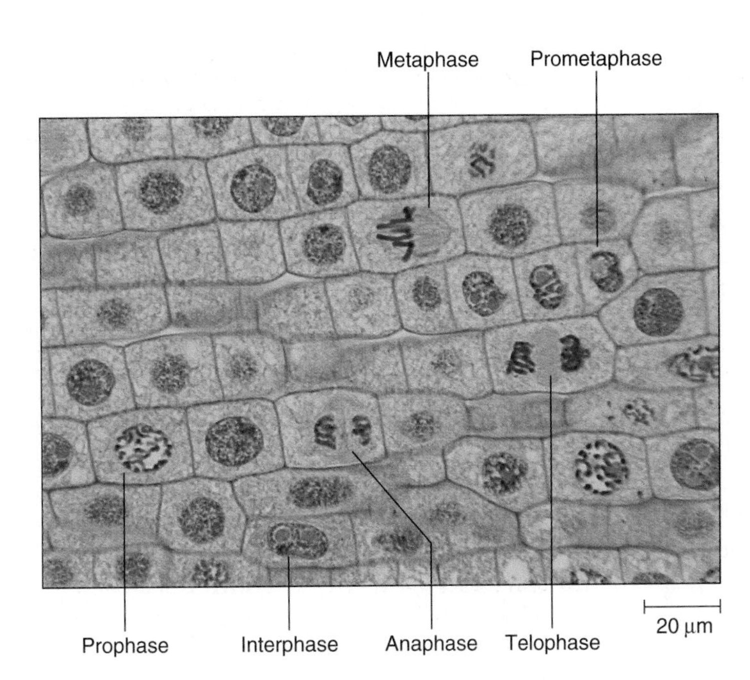

FIGURE 8.6
The Cells of an Onion Root Tip Reveal Stages of the Cell Cycle.
The number of cells in a particular stage reflects the duration of that stage. Interphase takes the longest, and therefore most cells are in interphase.

FIGURE 8.7
Steps of Mitosis.
The illustrations represent an animal cell; the photographs are of a plant cell.

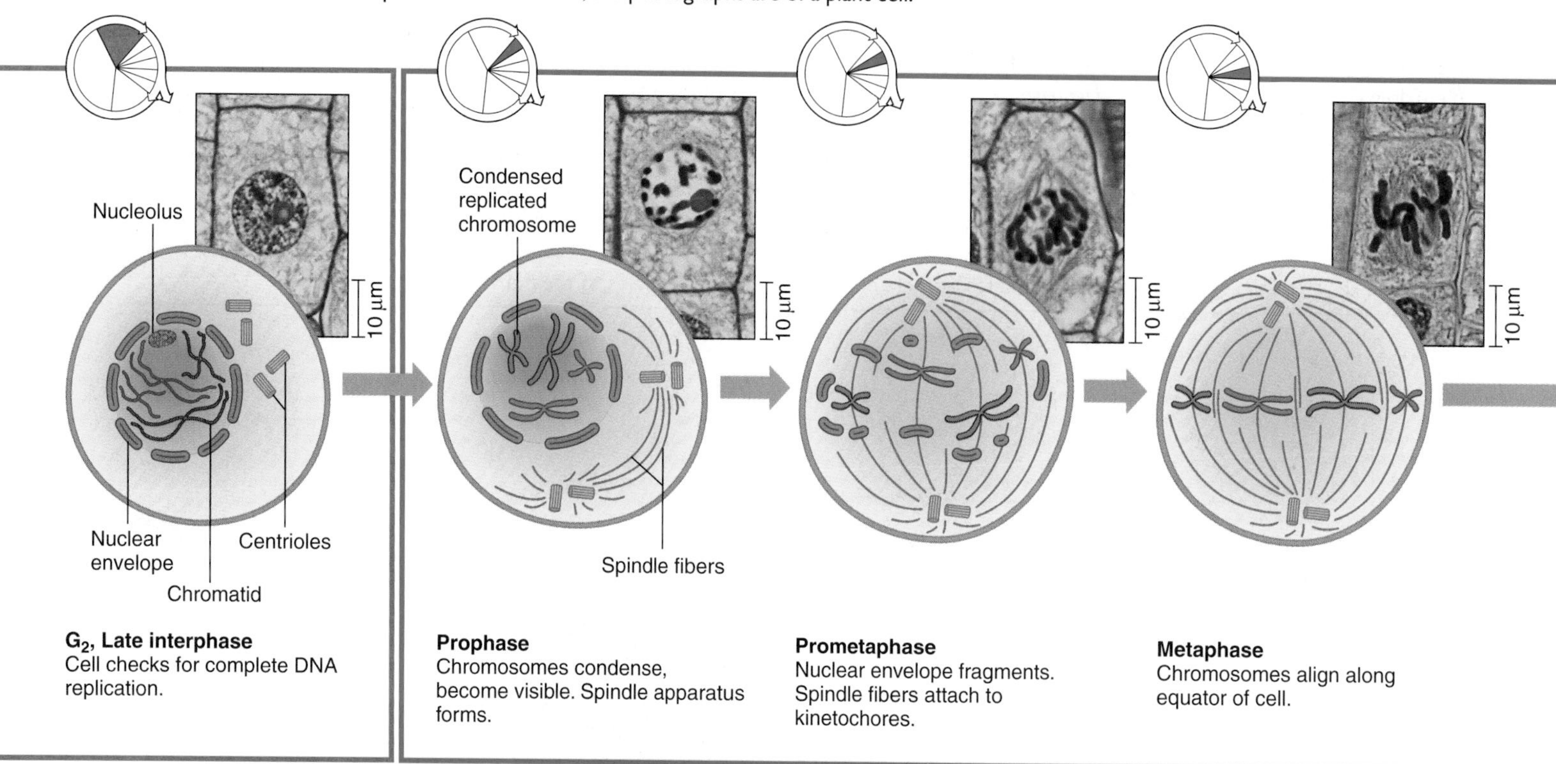

groping for something to hold. When a polar microtubule touches a centromere, it stops elongating. This is the mitotic spindle forming.

During middle prophase, the nucleolus, which is a darkened area in the nucleus where ribosomes are assembled, is no longer visible. It can be seen again during telophase.

Prometaphase occurs immediately after the formation of the spindle. The nuclear membrane breaks into small pieces that move away from the chromosomes and stay just under the cell membrane. The spindle then connects to the centromere of each chromosome.

As the next stage, **metaphase,** begins, the mitotic spindle aligns the chromosomes down the center, or equator, of the cell. Because of this alignment, each resulting cell will contain one chromatid from each chromatid pair.

In the next stage, **anaphase,** the centromeres, one per chromatid, move apart, sending one chromatid from each pair to opposite poles. As the chromatids separate, some microtubules in the spindle shorten and some lengthen in a way that actually moves the poles farther apart, stretching the dividing cell.

In **telophase,** the final stage of mitosis, an animal cell looks like a dumbbell with a set of chromosomes at each end. The mitotic spindle disassembles, and nucleoli and nuclear membranes re-form at each end of the stretched-out cell. Division of the genetic material is now complete. During or just after this karyokinesis, cytokinesis completes as organelles and macromolecules distribute into the two forming daughter cells. Finally, the daughter cells physically separate. They are usually about the same size.

Table 8.2 summarizes the stages of mitosis.

TABLE 8.2 The Stages of Mitosis

STAGE	EVENTS
Prophase	Chromosomes condense
	Microtubules disassemble, re-form, and approach centromeres
	Mitotic spindle forms
	Nucleolus disappears
Prometaphase	Nuclear membrane breaks down
	Spindle fibers attach to centromeres
Metaphase	Chromosomes attached to mitotic spindle align down equator
Anaphase	Centromeres part, sending one set of chromosomes to each pole
	Cytokinesis begins in some cells
Telophase	Mitotic spindle disassembles
	Nuclear membranes form around two nuclei
	Nucleoli reappear
Cytokinesis	Molecules and organelles are distributed to daughter cells
	Daughter cells separate

Cytokinesis Distributes Other Cell Contents

Cytokinesis begins during anaphase or telophase, depending upon the species or cell type (fig. 8.8A). In an animal cell, the first sign of cytokinesis is a slight indentation around the cell at

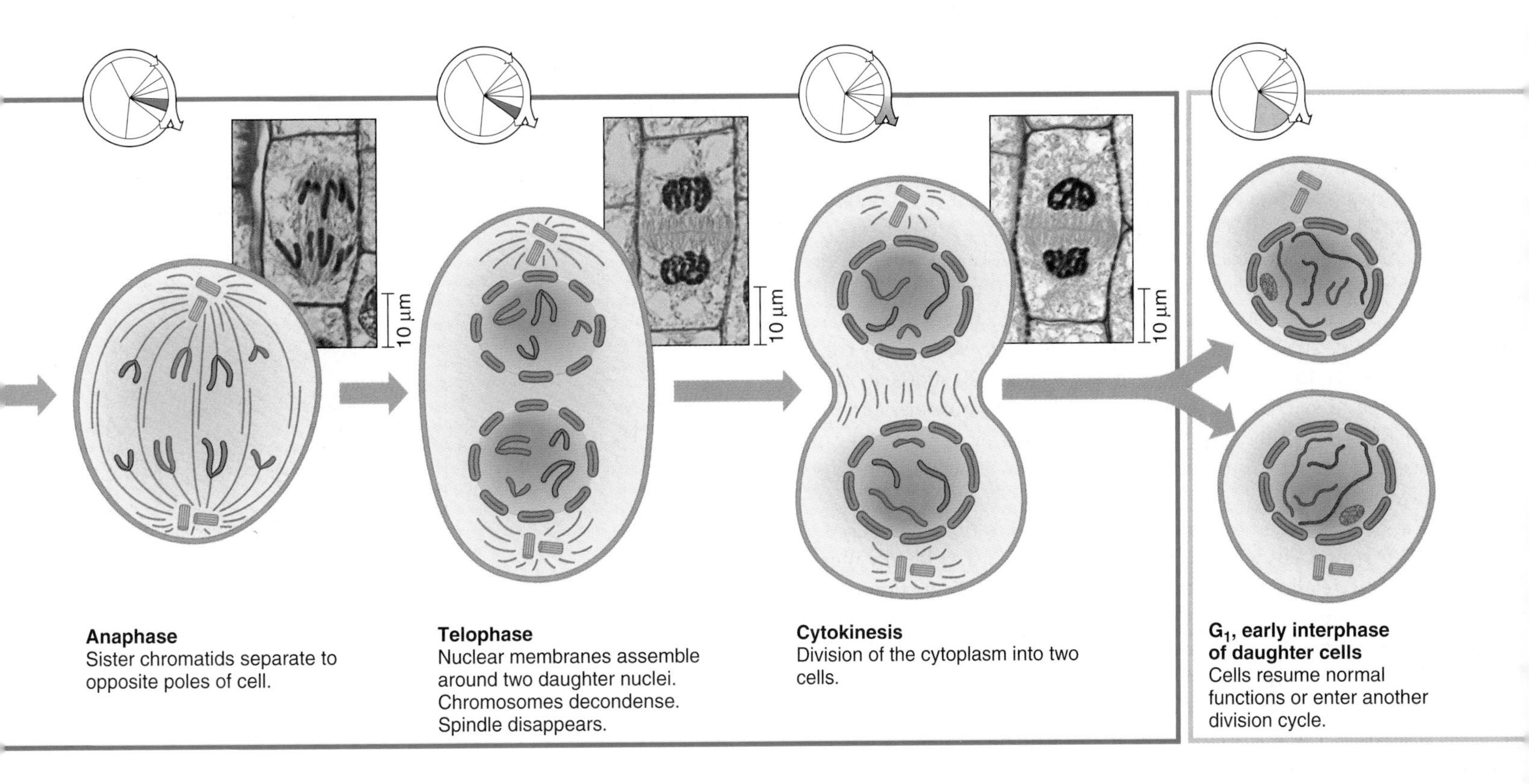

FIGURE 8.8

Cytokinesis Begins as the Chromosomes Separate.

(**A**) A contractile ring consisting of actin and myosin pulls the cell in two. In this dividing kangaroo rat cell, the tubulin of the spindle fibers stain green, and the chromosomes stain purplish-blue (**B**). The contractile ring lies at the center of the spindle halves.

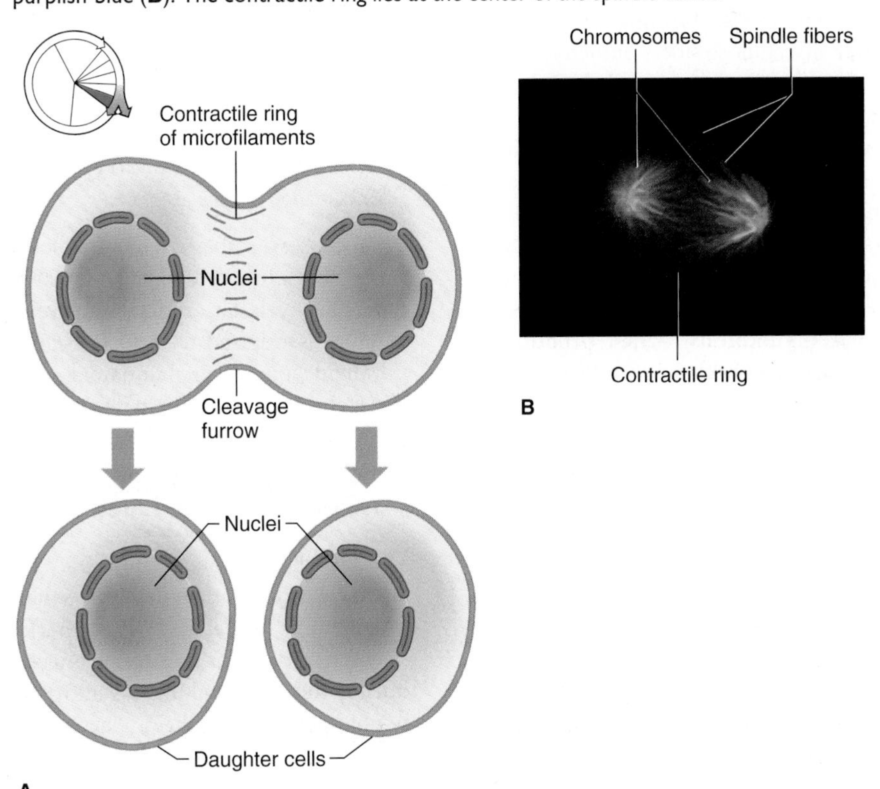

the equator. This indentation, called a **cleavage furrow,** results from a contractile ring of actin and myosin that forms beneath the cell membrane. It contracts like a drawstring, separating the daughter cells.

Actin and myosin molecules normally reside beneath the cell membrane, where they impart a tension to the entire cell surface. Beginning in anaphase, the furrow region becomes more contractile than the rest of the cell surface, which pinches the cell. Two hypotheses once suggested how the pinching might occur—either the poles of the cell relax, or the center contracts. A clever experiment revealed contractility at the center. Researchers used tiny rubber sheets to capture images of cell movement. When dividing cells were placed on the sheets, they left marks where they generated force. The "wrinkles" on the rubber clearly showed the contractile region of the cleavage furrow. Figure 8.8B shows another method to view cytokinesis—staining the chromosomes and the microtubules of the spindle. The contractile ring appears as a gap between the microtubules, although it is actually actin and myosin.

The location of asters determines the location of the cleavage furrow. Cells with abnormal numbers of asters vividly illustrate the relationship between asters and cleavage furrows. The normal number of asters is two. Cells lacking asters do not develop a furrow and yield one large cell with two nuclei. Cells with three or four asters develop extra cleavage furrows and yield three or four daughter cells, respectively. Extra asters occur when more than one sperm fertilizes an egg, a highly unusual and abnormal situation.

Plant cells must construct a new cell wall that separates the two daughter cells. A structure consisting of microtubules, called a phragmoplast, forms between the daughter cells and entraps vesicles containing structural materials (cellulose fibers, other polysaccharides, and proteins) for the cell wall. The first sign of cell wall construction is a line separating the forming cells, called a cell plate (fig. 8.9). The layer of cellulose fibers embedded in surrounding material makes a strong and rigid wall that gives plant cells their rectangular shapes.

Usually, cytokinesis completes shortly after karyokinesis. In some situations, cytokinesis does not follow karyokinesis, resulting in a mass of multinucleated cells called a syncytium. In some plants, endosperm tissue, which nourishes the developing embryo in a seed, consists of cells with many nuclei sharing a common cytoplasm. Some algae, slime molds, and fungi undergo mitosis without immediate cytokinesis, which produces multinucleated cells extending a meter or more in length.

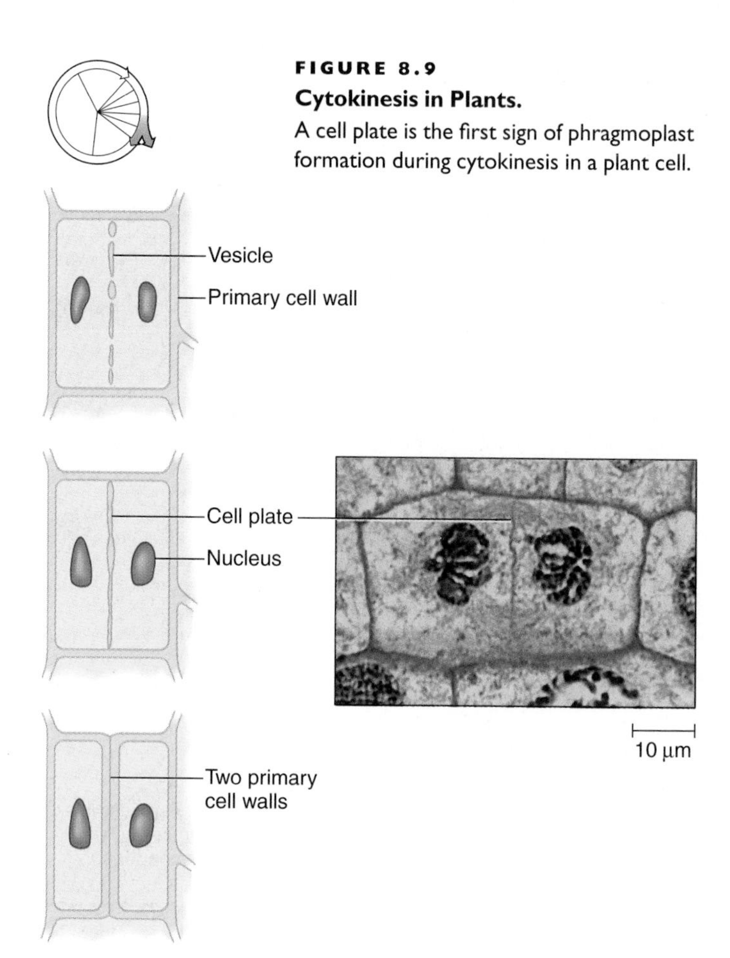

FIGURE 8.9
Cytokinesis in Plants.
A cell plate is the first sign of phragmoplast formation during cytokinesis in a plant cell.

8.2 MASTERING CONCEPTS

1. What events occur during each of the stages of the cell cycle?
2. What happens during interphase?
3. How does the mitotic spindle form?
4. What happens during each stage of mitosis?
5. Distinguish between karyokinesis and cytokinesis.

OLC

8.3 How Is the Cell Cycle Controlled?

Biochemical checkpoints—a clock provided by shortening chromosome tips, and signals from outside and inside the cell—regulate the cell cycle. Stem cells maintain the growth and specialization of a tissue.

What causes a somatic cell to divide or to cease dividing? Mechanisms that control the cell cycle and cell death operate at the molecular, cellular, and tissue levels. Understanding how and when cells respond to signals to divide, remain specialized but not divide, pause to repair damage, or die may reveal how abnormal states such as cancer arise.

Checkpoints Keep the Cell Cycle on Track

The checkpoints of mitosis are difficult to describe, for they are more complex than the arrows that are used to represent them on diagrams of the cell cycle. A checkpoint is actually a group of interacting proteins that controls a cell's fate at a particular point during the cycle, taking inventory on certain key events that might span several stages. Checkpoints have been meticulously

FIGURE 8.10
Cell Cycle Control.
Cell cycle checkpoints ensure that events occur in the correct sequence. Many types of cancer result from deranged checkpoints—caused by missing or faulty proteins.

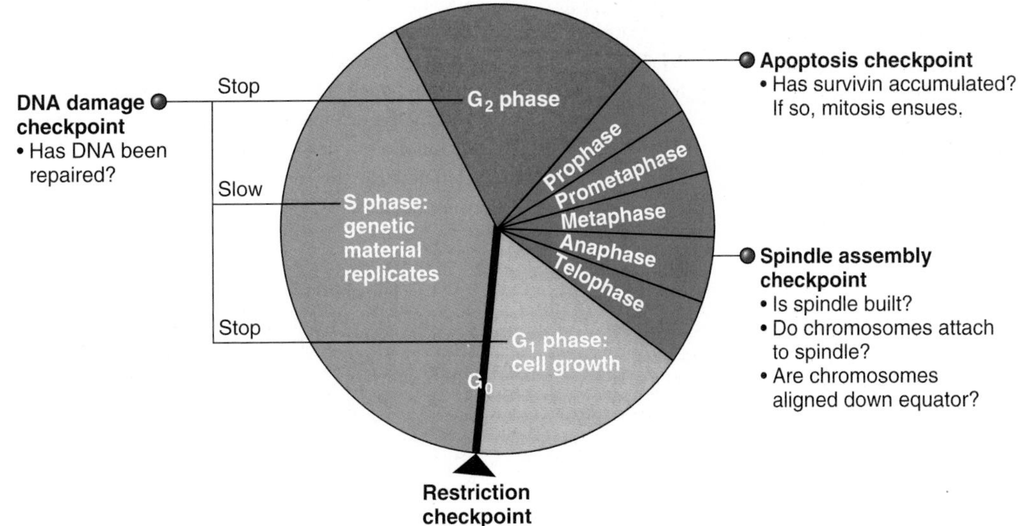

deciphered by looking at many different types of mutations in a variety of organisms, most notably yeasts, which form characteristically small colonies when their cell cycles are impaired.

Figure 8.10 illustrates three of the possibly dozens of different checkpoints that together ensure that chromosomes are faithfully replicated and apportioned into daughter cells. The "DNA damage checkpoint" has several effects—stopping cells in G_1 and G_2 and slowing DNA replication during S phase. As this checkpoint temporarily pauses the cell cycle, it also turns on genes that manufacture proteins that repair damaged DNA. The cell gains the time to recover from an injury. An "apoptosis checkpoint" turns on as mitosis begins. This checkpoint consists of a protein called survivin that appears to be necessary to override a signal for the cell to die, keeping it in M phase. Later in mitosis, the "spindle assembly checkpoint" oversees construction of the spindle and the binding of chromosomes to it. We have probably not yet discovered all of the cell cycle checkpoints.

Telomeres Provide a Cellular Clock

In the 1960s, biologist Leonard Hayflick discovered that mammalian cells grown in the laboratory obey an internal "clock" that allows them to divide a maximum number of times. A fibroblast (connective tissue cell) from a human fetus, for example, divides from 35 to 63 times, but a fibroblast from an adult divides only 14 to 29 times. Hayflick noted that the younger the individual providing the cell, the greater the number of times it could divide. It is as if cells "know" how many times they have divided and how many more divisions they can undergo.

In the early 1990s, cell biologists discovered that the long-sought cellular clock resides in chromosome tips, called **telomeres.** The chromosomes of humans and nearly all vertebrates studied so far end in hundreds to thousands of repeats of a specific six-nucleotide DNA sequence. At each cell division, the telomeres in a cell collectively lose 50 to 200 DNA nucleotides from their ends. This means that chromosomes gradually shorten each time a cell divides. After about 50 divisions, a key amount of lost telomere DNA signals cell division to cease. The cell may remain functional but not divide further, or it may die.

More interesting than the many cells whose telomeres tick down are the few cell types whose telomeres do *not* shrink. In general, cells in highly proliferative tissues maintain their telomeres, which enables them to divide beyond the 50-or-so division limit. In humans, such cells include those lining the small intestine, bone marrow cells, and certain blood cells. The cells that give rise to human sperm cells also "ignore" the cellular clock, perhaps because they must continually produce huge numbers of sperm and develop into the cells that give rise to the next generation of sperm. Cancer cells maintain long telomeres. This may be part of how they continue to divide far beyond the capacity of the cells they derive from.

The mechanism of telomere shortening is somewhat indirect—cells with shrinking telomeres do not manufacture an enzyme called **telomerase.** When cells do make telomerase, their telomeres stay long. Telomerase can continually add DNA to chromosome tips because it includes a small piece of RNA that binds to chromosome ends, serving as a model, or template, for additional DNA nucleotides to add on (fig. 8.11). In humans, telomerase levels are high in sperm-forming cells, blood cells, and digestive system lining cells. In cancer cells, the level of telomerase rises as the disease progresses. Much research focuses on finding ways to inactivate telomerase in cancer cells. Investigating Life 8.1 discusses key experiments that revealed the composition of telomeres.

In plants telomeres are also short DNA repeats, but they do not shrink as part of a mitotic clock, as they do in the somatic (nonsex) cells of vertebrates. Plant cells produce telomerase, and many divide beyond 50 divisions. Lack of a telomere clock in plants makes sense

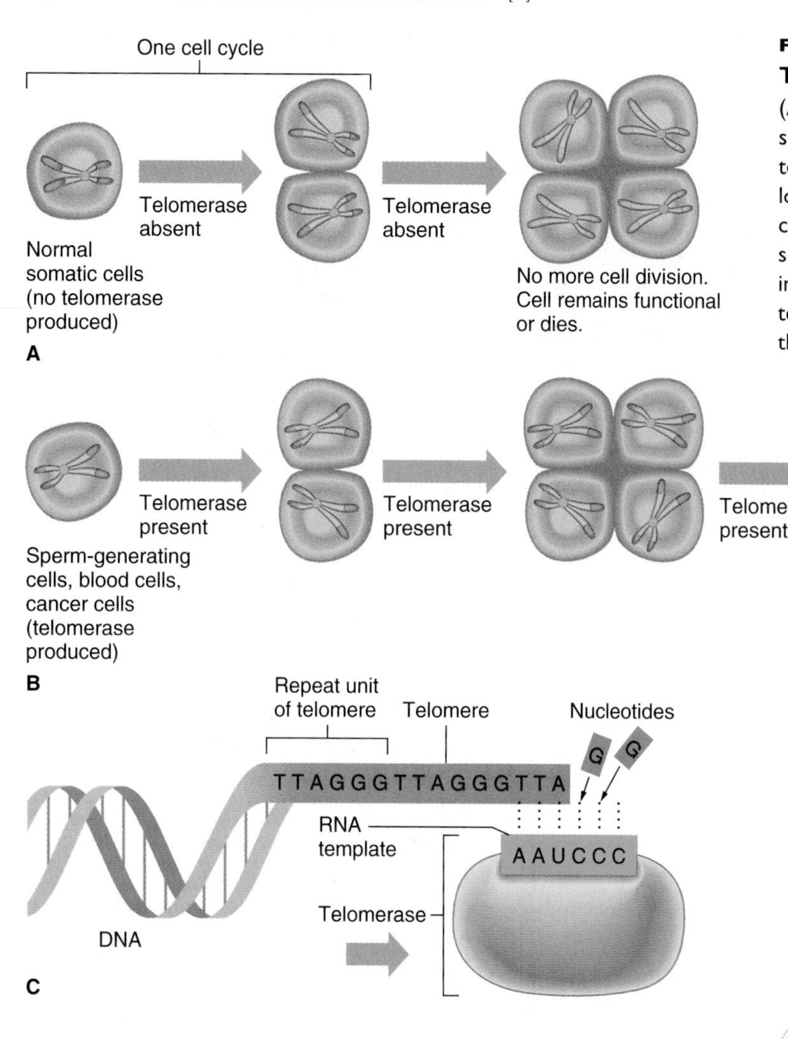

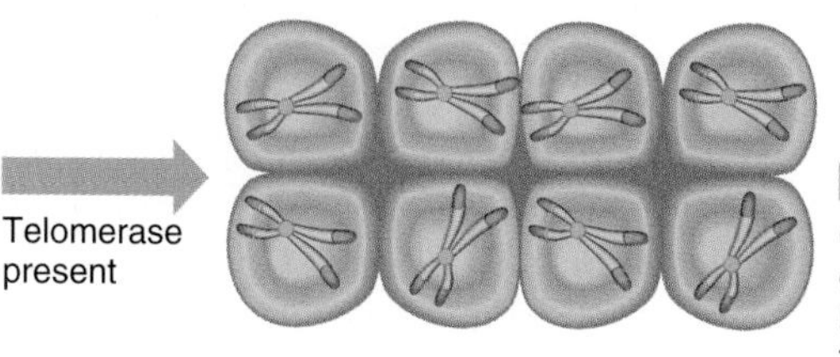

FIGURE 8.11
Telomeres.
(**A**) In normal somatic (nonsex) cells of vertebrates, telomeres shorten with each cell division because the cells do not produce telomerase. When the telomeres shrink to a certain point, the cell no longer divides. (**B**) Sperm-generating cells, blood cells, and cancer cells produce telomerase and continually extend their telomeres, resetting the cell division clock. (**C**) Telomerase contains RNA, which includes a portion (the nucleotide sequence AAUCCC) that acts as a template for the repeated DNA sequence (TTAGGG), which forms the telomere.

for the way that they develop. Unlike animals that set aside special cells destined to start a new individual, many plant parts have abundant meristem tissue, which has an unlimited ability to divide. If plants could not keep their telomeres long, the chromosomes of meristem cells would wear down and the cells cease to divide—and the plant cease to grow. Telomerase level is higher in a plant's most proliferative parts, but this is because researchers measure the enzyme from a piece of tissue, sampling many cells at once. Highly renewable tissue (such as root tips and seedlings) have a higher proportion of meristem cells than do less proliferative structures such as leaves, and so their overall telomerase levels are higher.

What Signals a Cell to Divide?

Signals from outside the cell and from within affect the cell cycle. Crowding can slow or halt cell division. Normal cells growing in culture stop dividing when they form a one-cell-thick layer (a monolayer) lining their container. If the layer tears, remaining cells bordering the tear grow and divide to fill in the gap but stop dividing once it is filled (fig. 8.12). The term **contact inhibition** refers to the inhibiting effect of crowding on cell division. Cancer cells display a characteristic lack of contact inhibition.

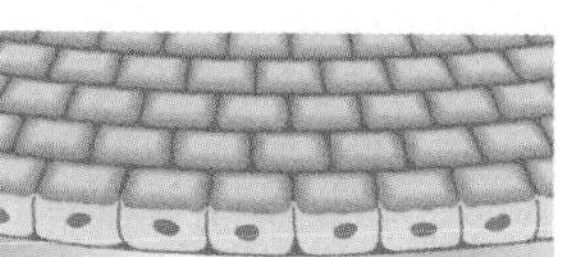

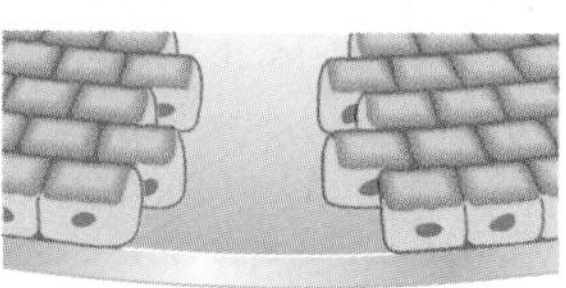

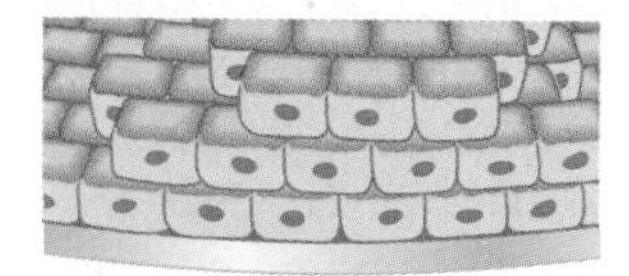

FIGURE 8.12
Contact Inhibition.
(**A**) Normal animal cells in culture divide until they line their container in a one-cell-thick sheet (a monolayer). If the monolayer is damaged (**B**), the cells bordering the removal site grow and divide, (**C**) filling the gap. However, the cells do not pile up on each other because of contact inhibition. In contrast, (**D**) cancer cells pile up on each other.

Experiments Reveal the Telomere Clock

Chromosome tips, or telomeres, have long fascinated biologists. Since the 1930s, geneticists have noticed that chromosomes missing their tips behave strangely—they stick together, forming clumps and bridges, and may vanish altogether as a cell divides. Without something that would function much like the plastic end of a shoelace, a chromosome would lose material from its ends whenever the DNA replicated. That something is a telomere, a short piece of DNA, repeated many times, that caps chromosomes. An enzyme, telomerase, includes an RNA template that tacks the telomere sequence onto the chromosomes of highly proliferative cells. When telomerase isn't manufactured, the chromosome tips are whittled away with each cell division to a point that somehow tells the cell to cease dividing.

Telomeres began yielding their secrets thanks to a pond-dwelling ciliate (a type of protistan) called *Tetrahymena thermophila.* In the 1970s, Elizabeth Blackburn and Joseph Gall, at Yale University, took advantage of a peculiarity of *Tetrahymena* to sample large amounts of telomeric DNA. This organism has two nuclei, one small and one large. When the larger one divides, its chromosomes shatter into up to 10,000 pieces—each of which ends in two telomeres. One such cell, then, would have 20,000 telomeres! Compare that to the 92 telomeres in a human cell (fig. 8.A). Blackburn and Gall were able to collect enough material to determine that each telomere consists of the same DNA building block sequence, repeated 50 to 70 times.

Over the next several years, similar telomeres were found in yeast, the simplest eukaryote, indicating that these chromosome caps are ancient. (Bacteria and archaea lack them because their DNA is circular.) In 1989, human telomeres were found to consist of nearly the same DNA sequence found in *Tetrahymena,* and a year later, researchers linked telomere shrinkage in human somatic (nonsex) cells to increasing numbers of cell divisions. In the mid-1980s Elizabeth Blackburn, then at the University of California at Berkeley, and her graduate student Carol Greider, identified telomerase. In the years since, many investigators have described the components of telomerase—the 6-base RNA, a "reverse transcriptase" enzyme that makes DNA using the RNA as the template, and an associated protein. Medical information came too. Apparently telomerase is turned off in most somatic cells, but is expressed in cancer cells. And conversely, the chromosomes in cells from people suffering from accelerated-aging disorders whittle down too quickly.

What was missing in all this work was evidence that ticking-down telomeres really function as a mitotic clock. Those key experiments came in 1998, when researchers examined what happens to cells robbed of telomerase, and to cells given extra telomerase.

Carol Greider, then at Cold Spring Harbor Laboratory on Long Island, bred mice that lack the RNA template of telomerase, and looked at the animals' proliferative tissues, which normally divide many times. As the researchers predicted, the mice were infertile and had shrunken reproductive organs and spleens and degenerating bone marrow.

At about the same time, researchers at a biotechnology company and the Texas Southwestern Medical Center added telomerase to normal human somatic cells growing in culture. The cells were relevant to health—pigmented cells from the retina that break down in macular degeneration (a common cause of aging-related visual loss); fibroblasts that, when aged, lose collagen and contribute to skin wrinkles; and cells that line blood vessels (vascular endothelium, discussed in the opening essay and important in heart disease). The telomerase-boosted cells received a new lease on life, dividing beyond their normal limits.

Interestingly, mouse cells lacking telomerase could still become cancerous, and human cells with extra telomerase did not become cancerous—precisely the opposite of what would be predicted. These findings indicate that cancer is not a simple matter of keeping one molecule—telomerase—turned on. It's clear that we still have a lot to learn about telomere biology.

FIGURE 8.A
Visualizing Telomeres.
Fluorescent tags indicate the telomeres on these stained human chromosmes.

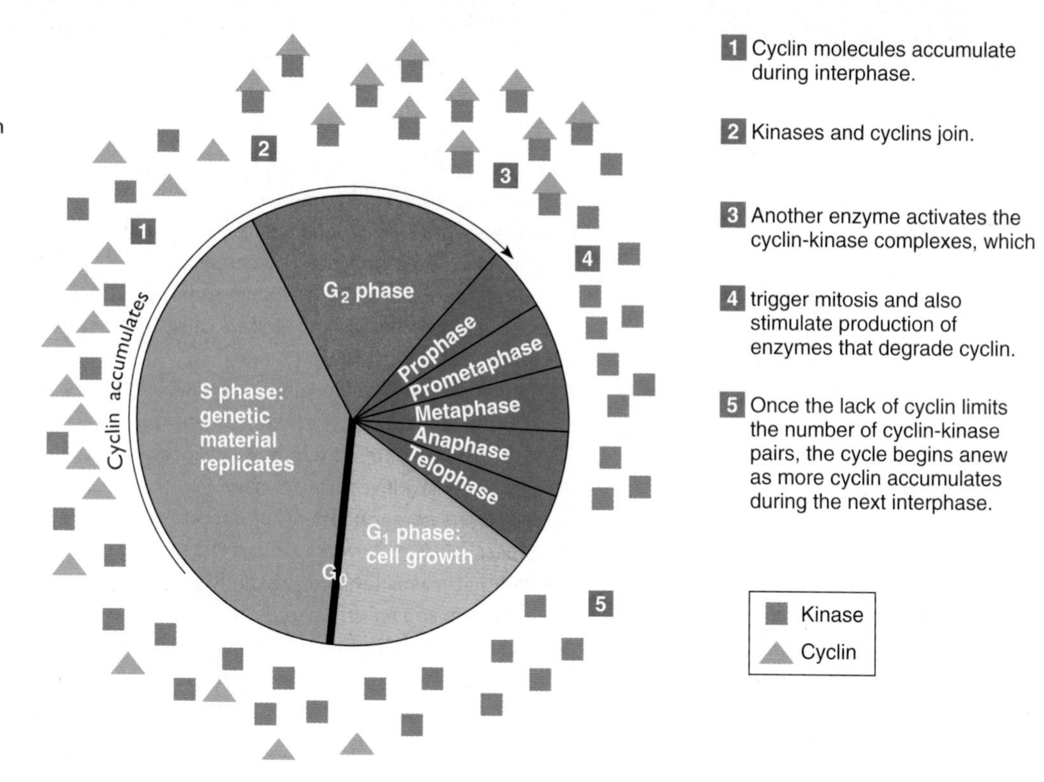

FIGURE 8.13
Cyclins and Kinases Pair to Regulate the Cell Cycle.
Kinase molecules are always present in the cell, and they join cyclin molecules that accumulate during interphase.

Certain hormones and growth factors are biochemicals that signal a cell to divide. In an animal, a hormone is manufactured in a gland and travels in the bloodstream to another part of the body, where it exerts a specific effect (see chapter 33). Cell division is one such consequence of a hormonal signal, as we saw in chapter 1 for estrogen. At a certain time in the monthly hormonal cycle in the human female, estrogen levels peak, stimulating the cells lining the uterus to divide, and build tissue in which a fertilized egg can implant. If an egg is not fertilized, another hormonal shift triggers cell death that breaks the lining down, resulting in menstruation. In plants, hormones coordinate growth and development, often by stimulating cell division in roots, shoots, seeds, fruits, and young leaves (see chapter 29).

Growth factors are proteins that stimulate local cell division. At a wound site, growth factors increase the rate of cell division, which replaces damaged tissue. For example, epidermal growth factor (EGF) stimulates epithelium (covering tissue) to divide, which fills in new skin underneath a scab. Salivary glands produce EGF, which aids healing when an animal licks its wounds. Biotechnologists have put EGF to a fascinating new use. Sheep engineered to produce extra EGF shed their wool about every six weeks. The animals are fitted with baggy coats, and when their wool falls out, the farmer needs only collect it from the coats—shearing is unnecessary.

Biochemicals inside cells also stimulate cell division. Pairs of proteins called **kinases** and **cyclins** activate genes whose protein products carry out cell division. Cyclins form the checkpoint that operates in G_1 to determine whether or not a cell divides.

A kinase is a type of enzyme that activates other proteins by adding phosphates to them. The kinase controlling the cell cycle is present in all cells of eukaryotes at all times. In contrast, levels of cyclin fluctuate, as its name implies. At one point in the cell cycle, cyclin levels plummet as if something is rapidly degrading it.

The relationship between kinase and cyclin controls the cell cycle (fig. 8.13). First, kinase molecules bind to cyclin molecules, which have accumulated during the previous interphase. An enzyme activates this kinase-cyclin complex, making mitosis inevitable and stimulating production of enzymes that break down cyclin. As cyclin levels fall, levels of cyclin-degrading enzymes also drop, and cyclin accumulates again. When enough accumulates that cyclin again combines with the always-present kinase, cell division begins anew. Hence, a negative feedback loop controls the ebb and flow of cyclin.

Control at the Tissue Level—Stem Cells and Cell Populations

Tissues must maintain their specialized natures, as well as retain the capacity to generate new cells as the organism grows or repairs tissues damaged by injury or illness. Many tissues contain a few cells, called **stem cells,** that can divide and thereby replenish the tissue. When a stem cell divides to yield two daughter cells, one remains a stem cell, able to divide again, while the other no longer divides and specializes to perform certain functions. Figure 8.14 shows stem cells in the basal (deepest) layer of the epidermal skin layer in a human. These basal cells readily divide, but the cells above them, which divided from the basal stem cells, do not. The cells in the upper skin layers die via apoptosis—they flake off after a brisk toweling. This organization of stem cells pushing more specialized cells upward also occurs in the folds of the small intestinal lining.

Recently researchers have discovered stem cells in the human brain. This was a surprise, because most neurons are in G_0 and

FIGURE 8.14
Stem Cells.
In some tissues, only cells in certain positions divide. The outer layer of human skin, the epidermis, has actively dividing stem cells in its basal (deepest) layer that push most of their daughter cells upward, yet they maintain a certain number of stem cells.

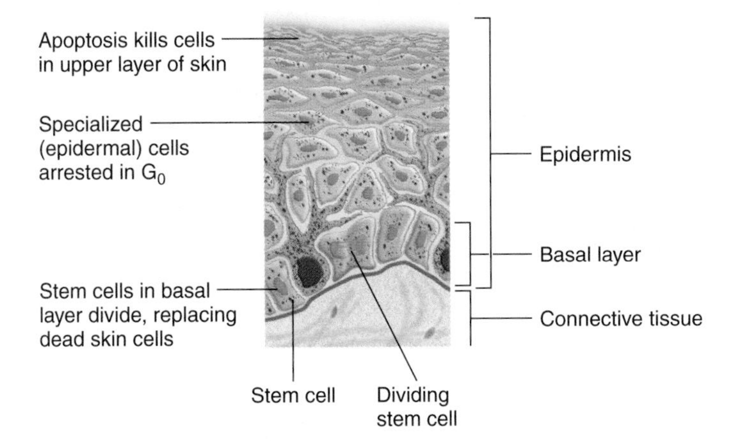

therefore do not divide. Neural stem cells can replace neurons and neuroglia (cells that support, nourish, and interact with neurons) as they die, or restore function following injury. Experiments show that neural stem cells cultured in the laboratory with appropriate biochemical signals can divide to yield a neuron or neuroglial cell, plus another stem cell. Too few neural stem cells might cause neurodegenerative disorders such as Alzheimer's disease; too many may lead to cancer. Biotechnologists are exploring ways to develop these cells into implants to treat various brain disorders.

Tissues can be described by the percentage of cells or cell populations in particular stages of the cell cycle. In a renewal cell population, cells are actively dividing. Renewal cell populations in animals maintain linings that are continually shed and rebuilt from dividing cells, such as the inside surface of the digestive tract. They replace many millions of cells each day.

In an expanding cell population, up to 3% of the cells are dividing. The remaining cells are not actively dividing, but they can divide to repair an injury. The fast-growing tissues of young organisms, as well as adult kidney, liver, pancreas, and bone marrow tissues, consist of expanding cell populations.

Cells that are highly specialized and no longer divide make up static cell populations. Nerve and muscle cells form static cell populations and remain in G_0. These cells enlarge rather than divide. A single nerve cell may grow to a meter in length, but within an organism it normally does not divide. Cells in static cell populations are not dividing, nor are they dying. They simply remain as they are, specialized.

8.3 MASTERING CONCEPTS

1. What are cell cycle checkpoints?
2. How does telomere length serve as a cell division clock in some organisms?
3. How does crowding affect cell division?
4. Which biochemicals outside and inside cells affect cell division rate?
5. What is the function of stem cells in a tissue or organ?

OLC

8.4 How Is Cell Death a Part of Life?

Cells that perish as a normal part of the life of an organism exhibit a characteristic choreography, almost a dance of death.

The word "apoptosis" comes from the Greek for a tree shedding leaves. Similarly, apoptosis in an animal's body refers to cell death that is part of normal development. Biologists had observed what they thought to be a programmed form of cell death for nearly a century—the webbing that vanishes as a chicken's foot forms but not in a duck's foot, or the resorption of a tadpole's tail (fig. 8.15).

FIGURE 8.15
Apoptosis Carves Toes.
A developing chicken foot undergoes extensive apoptosis (**A**), but a developing duck's foot doesn't (**B**), retaining the webbing.

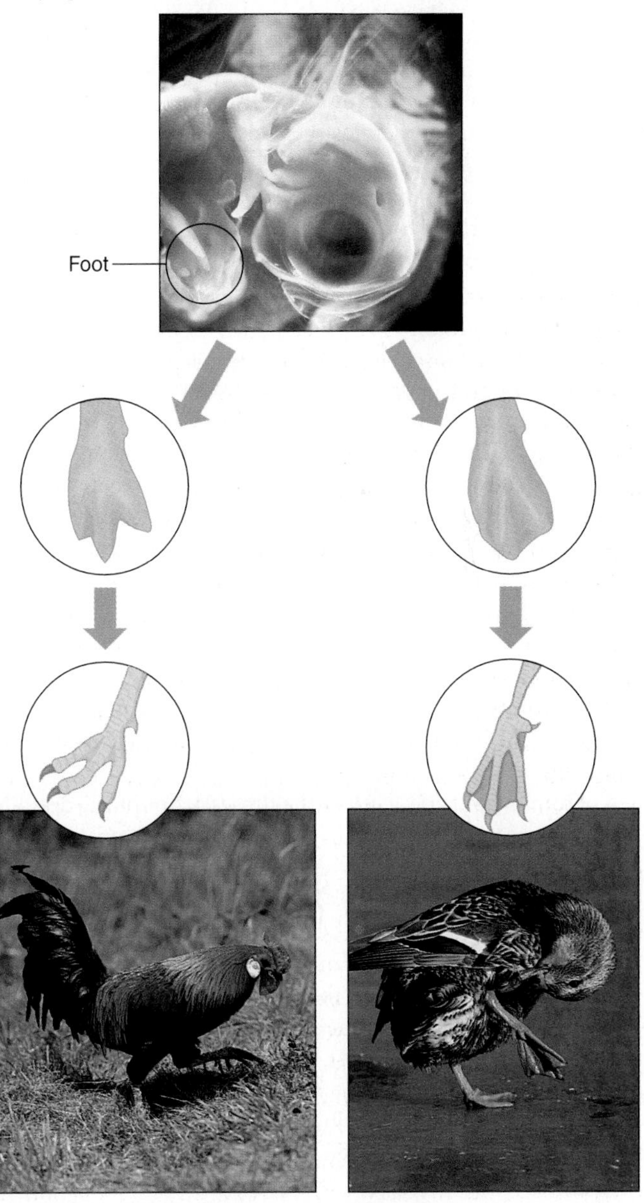

In the 1990s researchers began to unravel the genetic controls of the process, starting with the discovery that in the tiny roundworm *Caenorhabditis elegans,* exactly 131 of its 1,090 larval cells die, as a normal part of development, before the adult forms. Soon researchers discovered versions of these "death genes" in many animal species, including our own.

What Happens During Apoptosis?

Like mitosis, the steps of apoptosis are highly regulated, specific, and sequential. Overall, apoptosis rapidly and neatly dismantles a cell into membrane-bounded pieces that a phagocyte (a cell that engulfs and destroys another) can mop up. In contrast is necrosis, which is cell death that occurs in response to injury. Necrosis is messy—the cell swells, bursts, and causes great inflammation.

Apoptosis begins when a "death receptor" on the doomed cell's membrane receives a signal to die. (It can also occur in response to an insult, such as ultraviolet radiation, X rays, or toxic chemicals.) Within seconds of death receptor binding, a chemical cascade begins as enzymes called **caspases** become activated. The caspases go to work activating each other and snipping apart various cell components. An early act of destruction is to shut down other proteins that normally prevent apoptosis, and in so doing release fragments of those proteins that keep the caspase cascade functioning—an example of positive feedback.

Then the caspases let loose as molecular killing machines. They destroy intermediate filaments that support the nuclear membrane, which causes the chromatin within to condense. The caspases also demolish the enzymes that replicate and repair DNA, then activate enzymes that chew DNA up. The killer enzymes tear apart the cytoskeleton. Then they ready the cell for the phagocytic mop-up to come. They disable cellular adhesion molecules so that the cell can no longer cling to others, and send a particular phospholipid from the cell membrane's inner face to its outer surface, where it attracts phagocytes. Meanwhile, mitochondria get in on the action. The electron transport chains come apart, and some of the electron carriers that are released into the cytoplasm activate more caspases!

From the outside, a cell in the throes of apoptotic death has a characteristic appearance (fig. 8.16). It rounds up as contacts with other cells are cut off, and the cell membrane undulates, forming bulges called blebs. The nucleus bursts, releasing the chromatin, now broken into same-sized pieces. Then the cell shatters, but pieces of freed membrane quickly encapsulate the fragments, which prevents them from causing inflammation. Within an hour of when the death receptor first received the bad news, the event is all over. The phagocytes descend, and the cell is gone.

So far apoptosis as described above appears to be part of animal life. However, many of the genes are also seen in such diverse organisms as yeasts, other fungi, and slime molds. Plant cells die too, but apparently not in precisely the way that animal cells meet their programmed fate. Instead, plant cells are digested by enzymes in their own vacuoles, and often the remains are useful, such as vacated cell walls providing support. Plants also use a form of cell death to limit infection, the affected area meeting its end so it cannot spread the infection.

FIGURE 8.16
Death of a Cell.
A cell undergoing apoptosis loses its characteristic shape, forms blebs, and finally falls apart. Caspases destroy the cell's insides. Phagocytes digest the remains.

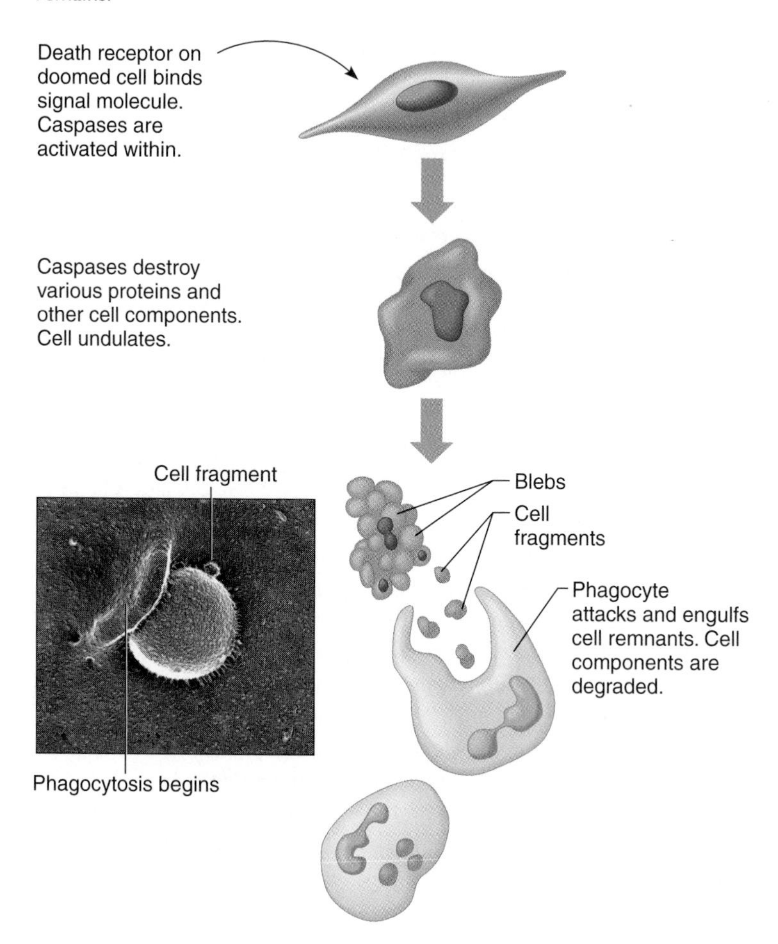

Why Do Cells Die?

On an organismal level, obviously if some cells didn't die yet mitosis continued unchecked, we would all be massive lumps. On an organ level, apoptosis eliminates excess cells to essentially carve out a working structure.

In the human brain, connections between nerve cells are essential for proper functioning. To establish these links from a large mass of neurons, certain cells release nerve growth factors

(NGFs), which attract other neurons. Once a "receiving" neuron sends an extension toward a neuron beckoning it, the growth signal ceases, and no other neurons approach it. Apoptosis eliminates neurons not responding to NGF. This intentional death of nearly half of all brain cells sculpts a highly organized organ from what would otherwise be a meaningless tangle of neurons.

Apoptosis also fine-tunes the human immune system so that it can distinguish "self" from "nonself." In the fetus, the thymus gland produces many immature T cells that recognize an enormous diversity of cell surfaces, only a few of which are "self." Apoptosis kills the T cells that do not recognize self. If this were not the case, the T cells would attack the body—resulting in an autoimmune disorder. As the immune system matures, the bone marrow continually produces many more blood cells than enter the circulation. A type of white blood cell called a neutrophil, for example, is churned out in huge numbers in the bone marrow, but relatively few make it into the circulation to arrive at the scene of infection. Apparently the routine killing of neutrophils assures a constant, fresh supply.

A second function of apoptosis is protective, to detect and weed out cells that could grow uncontrollably. In fact, apoptosis appears to be a "default" option in the cell cycle, which must be overcome at a checkpoint in order for mitosis to proceed (see fig. 8.10). A good example of how apoptosis protects is the skin peeling that follows a sunburn. A protein called p53 stands guard during G_1 as a checkpoint. If a cell has already undergone many cell divisions and ultraviolet radiation in sunlight damages its DNA, it is repaired, by an enzymatic process discussed in chapter 12. But if the cell is "young" in terms of the number of mitoses it has undergone, or the damage very severe, p53 sends it along a pathway toward apoptosis. The alternative is too many mitoses—skin cancer.

Like any biological process, apoptosis, when abnormal, harms health. The examples above show that too little apoptosis can cause autoimmune disorders and cancer. Deficient apoptosis may also lie behind cardiovascular disease by failing to remove cells that pile up on blood vessel inner linings. On the other hand, excess apoptosis upsets cell number balance too. A stroke in the brain, for example, triggers apoptosis in the cells surrounding the directly affected area that is robbed of oxygen, extending the damage as cells die. Too much apoptosis may also be responsible for neurodegenerative conditions such as Alzheimer's disease. Liver failure can reflect excess apoptosis. Iron overload, alcoholism, cancer, and viral infection (hepatitis) are all associated with too much apoptosis in the liver.

8.4 MASTERING CONCEPTS

1. What happens to a cell undergoing apoptosis?
2. What are the functions of apoptosis?
3. How can abnormal apoptosis affect health?

OLC

8.5 Cancer—When the Cell Cycle Goes Awry

Cancer is a derangement in cell cycle control. A cancer cell lacks a mitotic "brake," and has other characteristic features. Mutant genes cause cancer.

One out of three of us will develop cancer. In this group of disorders, certain body cells lose normal control over the rate and number of divisions they undergo, or they may ignore apoptosis signals. Instead, the cell continues to divide, more often or more times than the surrounding cells, and grows into a mass called a malignant tumor, or moves in the blood. Cancer begins with a single cell that breaks through its death and division controls.

Each of us probably has cancer cells from time to time, because cell division occurs so frequently that an occasional cell is bound to escape the control. In animals, certain cells, sometimes organized into immune systems, destroy cancer cells. Plants can develop abnormal growths called crown galls that are similar to cancer.

How Do Cancer Cells Differ from Normal Cells?

Cancer cells differ from normal cells in many ways. Overall, these distinctions give a cancer cell physical characteristics and behaviors like those of embryonic cells.

Given sufficient nutrients and space, cancer cells can divide uncontrollably and eternally. The cervical cancer cells of a woman named Henrietta Lacks vividly illustrate these characteristics. Shortly before she died in 1951, she donated some of her cancer cells to a laboratory at Johns Hopkins University. Lacks's cells grew so well, dividing so often, that they quickly became a favorite of cell biologists seeking cells to culture that would divide beyond the normal 50-division limit. Still used today, "HeLa" (for Henrietta Lacks) cells grow so vigorously that if just a few of them contaminate a culture of different cells, within days they completely take over!

Cancer cells divide more times than the cells from which they derive. They are not necessarily the fastest-dividing cells. Even the speediest cancer cells, which complete a cell cycle every 18 to 24 hours, do not divide as often as some normal human embryo cells. A cancer cell divides faster than the normal cell type it arises from, or it divides continuously at the normal rate.

A cancer's growth rate depends upon the type of cell affected. The smallest detectable fast-growing tumor is half a centimeter in diameter and can contain hundreds of millions of cells, dividing at a rate that produces a million or so new cells an hour. If 99.9% of such a tumor's cells are destroyed, a million would still be left to proliferate. Other cancers are very slow to develop. Lung cancer, for example, may take three to four decades to develop (fig. 8.17). However, any tumor's growth rate

FIGURE 8.17

Cancers Take Many Years to Spread.

Lung cancer due to smoking begins with irritation of respiratory tubes. Ciliated cells die (but can be restored if smoking ceases), basal cells divide, and then, if the irritation continues, cancerous changes may appear.

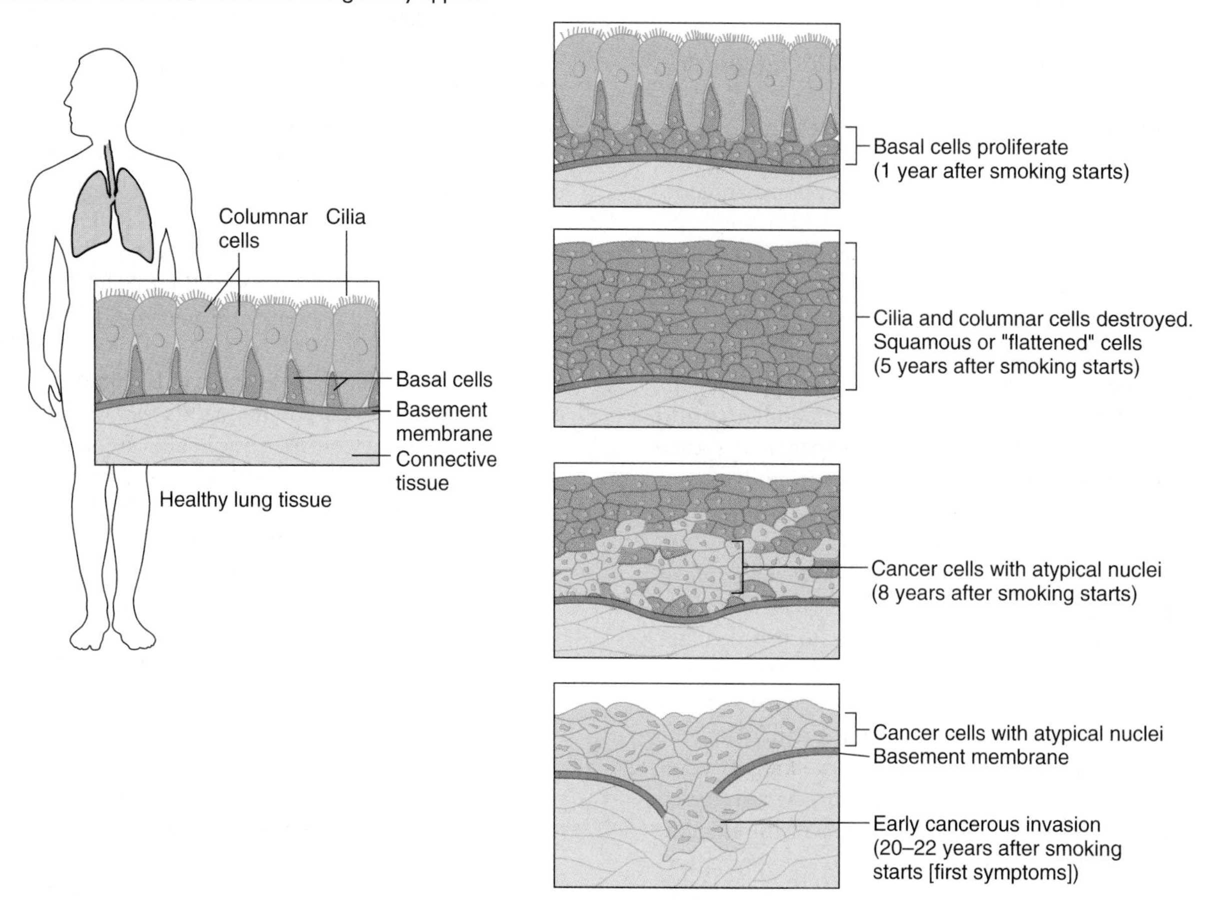

is slower at first because fewer cells are dividing. Once a tumor forms, it may grow faster than surrounding tissue simply because a larger proportion of its cells are actively dividing.

When a cell becomes cancerous, it passes its loss of cell cycle control to its descendants. This is called heritability. It does not mean that the cancer is inherited, but that the characteristics of cancer are passed to daughter cells. A cancer cell is also transplantable. This means that a cancer cell injected into a healthy animal can spread the disease as the errant cell divides and forms more cancerous cells.

A cancer cell looks different from a normal cell. It is rounder because it is less adhesive and the cell membrane more fluid. Tumors form because of the lack of contact inhibition. Cancer cells often undergo genetic changes and "dedifferentiate," losing some of the specializations of their parent cells.

Cancerous cells have surface structures that enable them to squeeze into, or invade, any available space. They anchor themselves to tissue boundaries and secrete chemicals that cut paths through healthy tissue. As the opening essay describes, cancer cells send signals that lure blood vessels to grow toward them, and even grow their own vessels. Once metastasized, cancer cells stimulate angiogenesis in new places. Meanwhile, as a cancer spreads, it mutates, and a treatment that shrank the original tumor may have no effect on this secondary growth. Table 8.3 reviews the characteristics of cancer cells. The figure in the opening essay for chapter 3 shows a cancer cell.

What Causes Cancer?

Cancer arises from mistimed or misplaced cell division or absence of normal apoptosis. If cells in an adult liver divide at the rate or to the extent that embryo liver cells divide, the resulting overgrowth may lead to liver cancer. Because proteins such as growth factors, kinases, survivin, and cyclins control the cell cycle, and because genes encode proteins, genes play a key role in causing cancer.

Two major classes of genes contribute to causing cancer. **Oncogenes** are versions of genes that normally trigger cell divi-

TABLE 8.3 Characteristics of Cancer Cells

Loss of cell cycle control
Heritability
Transplantability
Genetic mutability
Dedifferentiation
Loss of contact inhibition
Angiogenesis
Invasiveness
Ability to spread (metastasis)

sion, but they are overexpressed and accelerate the cell cycle. For example, one human oncogene normally active at the site of a wound stimulates production of growth factors, which prompt cell division that heals the damage. When the oncogene is active at a site other than a wound, it still stimulates growth factor production and cell division, but, because there is no damaged tissue to replace, the new cells form a tumor.

In contrast to oncogenes are **tumor suppressor genes,** whose protein products normally prevent a cell from dividing or promote normal cell death. Inactivation or removal of a tumor suppressor gene can cause cancer. For example, a childhood kidney cancer is caused by the absence of a gene that normally halts mitosis in the rapidly developing kidney tubules in the fetus. Cell division does not cease on schedule, and the child's kidney retains pockets of cells that divide as frequently as if they were still in the fetus. In the child, these cells form tumors. The p53 gene that can send a cell on the road to apoptosis is abnormal in a very rare syndrome in which family members are at extremely high risk of developing a variety of cancers. Chapter 13 explores mutations that predispose to cancer in more detail.

We know of hundreds of different oncogenes and tumor suppressor genes. Often a cancer results from a series of cancer-promoting genetic changes. Some forms of colon cancer, for example, occur because of a sequence of genetic abnormalities that include oncogene activation and tumor suppressor inactivation as well as possible environmental influences, such as diet and exercise habits.

Oncogenes may be activated or tumor suppressor genes inactivated in response to environmental triggers, such as sun exposure or cigarette smoking. Environmental influences that raise cancer risk include exposure to certain chemicals, radiation, and viruses as well as nutrient deficiencies.

Chemical carcinogens (cancer-causing chemicals) were recognized as long ago as 1761, when nasal cancer was linked to use of snuff, an inhaled form of tobacco. In 1775, British physician Sir Percival Potts suggested that the high rate of skin cancer in the scrotum among chimney sweeps was due to their exposure to a chemical in soot. Other associations between cancers and environmental triggers followed. In 1795, physicians noted that lip cancer was more common among pipe smokers, and in 1879, lung cancer was linked to working in silver, cobalt, and uranium mines. It wasn't until 1950 that researchers established the link between lung cancer and cigarette smoking.

Mitosis is the division of somatic, or nonsex, cells, and it makes growth possible. The next chapter considers another type of cell division, meiosis, that makes sexual reproduction possible.

8.5 MASTERING CONCEPTS

1. How are the cell cycles of cancer cells abnormal?
2. How do cancer cells differ from normal cells?
3. What are the roles of genes and environmental factors in causing cancer?

OLC

Chapter Summary

8.1 The Balance Between Cell Division and Cell Death

1. Cell division (**mitosis**) and cell death (**apoptosis**) shape organs, select certain cell types to persist in development, and protect tissues by ridding the body of cells that could become cancerous.

8.2 The Cell Cycle

2. The **cell cycle** is a sequence of events that describes whether a cell is dividing its genetic material (mitosis or **karyokinesis**); dividing its cytoplasm, organelles, and macromolecules (**cytokinesis**); or preparing to divide (**interphase**).
3. Interphase includes two gap periods, $\mathbf{G_1}$ and $\mathbf{G_2}$, when the cell makes proteins, carbohydrates, and lipids, and a synthesis period (**S**), when it replicates genetic material. G_1 includes a **checkpoint** that sets a cell's fate—to divide, stay specialized but not divide, or die. $\mathbf{G_0}$ is a "time out" from the cell cycle.
4. Each replicated chromosome consists of two complete sets of genetic information, called **chromatids,** attached at a section of DNA called a **centromere** that also replicates.
5. Microtubules assemble to build the **mitotic spindle.** It arises from paired **centrosomes.** Chromosomes attach to the mitotic spindle with microtubule assemblies called **kinetochores.**
6. Mitosis consists of five stages. In **prophase,** the chromosomes condense and become visible when stained, the nuclear membrane disassembles, and the mitotic spindle forms. In **prometaphase,** the nuclear membrane fragments and moves out of the way. In **metaphase,** spindle fibers align replicated chromosomes down the cell's equator. In **anaphase,** the chromatids of each replicated chromosome separate, sending a complete set of genetic instructions to each end of the cell. In **telophase,** the spindle breaks down and nuclear membranes form.

7. In cytokinesis in an animal cell, a **cleavage furrow** forms and a contractile band draws the two cells apart. In a plant cell, a phragmoplast provides space for a new cell wall to be laid down, separating the cells. Cytokinesis usually begins during anaphase or telophase. Lack of cytokinesis results in a huge cell with many nuclei.

8.3 How Is the Cell Cycle Controlled?

8. Checkpoints are interacting proteins that maintain the sequence of cell cycle events and control cell fate.
9. Shrinking **telomeres** track the number of divisions a cell has undergone, and when telomeres reach a certain length, division ceases. Cancer cells and certain rapidly dividing cells retain long telomeres and divide continually.
10. **Contact inhibition** prevents normal cells from dividing. Extracellular signals (hormones and growth factors) and intracellular signals (**cyclins** and **kinases**) control cell division rate and number.
11. **Stem cells** actively divide, replenishing tissues. Different cell populations include specific proportions of cells in different stages of the cell cycle.

8.4 How Is Cell Death a Part of Life?

12. Apoptosis is a form of programmed cell death. **Caspases** carry out the destruction. Apoptosis is neat compared to necrosis. An apoptotic cell rounds up, the cell membrane forms blebs, the nuclear membrane breaks down, chromatin condenses, and DNA is cut into many equal-sized pieces.
13. Apoptosis shapes structures, and protects by getting rid of cells that could become cancerous.

8.5 Cancer—When the Cell Cycle Goes Awry

14. Cancer can result from excess cell division or deficient apoptosis. A cancer cell divides more often or more times than surrounding cells, has an altered surface, loses specialization, and divides to yield other cancer cells. A malignant tumor infiltrates nearby tissues and metastasizes if it reaches the bloodstream.
15. Cancer can result from an overexpressed **oncogene** or an inactivated **tumor suppressor gene** and may be sensitive to environmental triggers.

Testing Your Knowledge

1. Identify the stage of mitosis the following illustrations depict:

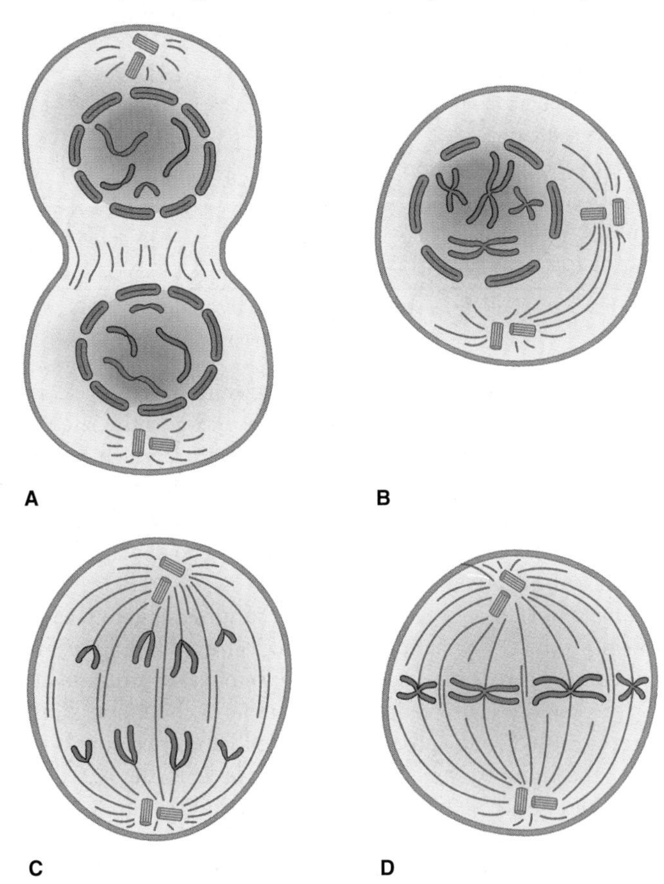

2. What might be the consequence of blocked mitosis? Blocked apoptosis?
3. Describe the events that take place during mitosis.
4. Why is G_1 a crucial time in the life of a cell?
5. Why isn't interphase a time of cellular rest, as biologists once thought?
6. List four ways that cancer cells differ from normal cells.
7. How do biochemicals from inside and outside the cell control the cell cycle?

Thinking Scientifically

1. A cell from a newborn human divides 19 times in culture and is then frozen for 10 years. Upon thawing, how many times is the cell likely to divide?
2. Cytochalasin B is a drug that blocks cytokinesis by disrupting the microfilaments in the contractile ring. What effect would this drug have on cell division?
3. How might the observation that more advanced cancer cells have higher telomerase activity be developed into a test that could help physicians treat cancer patients?
4. A researcher removes a tumor from a mouse and breaks it into cells. He injects each cell into a different mouse. Although all of the mice in the experiment are genetically identical and were raised in the same environment, the animals develop cancers that spread at different rates. Some mice die quickly, some linger, and others recover. What do these results indicate about the cells that made up the original tumor?
5. Why can combining a traditional cancer treatment with an angiogenesis inhibitor be more effective than either treatment alone?

References and Resources

Carr, Antony M. March 10, 2000. Piecing together the p53 puzzle. *Science,* vol. 287, p. 1765. p53 protein controls an important cell cycle checkpoint.

Hall, Stephen S. January 30, 2000. The recycled generation. *The New York Times Magazine.* Can telomerase applied to stem cells cloned from embryo copies of ourselves make us immortal?

Lewis, Ricki. December 1998. Telomere tales. *BioScience,* vol. 48, p. 981. Long of interest to cell biologists, telomeres are now impacting on medicine.

The August 28, 1998 issue of *Science* magazine has several articles on apoptosis.

The *LIFE* Online Learning Center provides additional resources and tools for studying this chapter.
www.mhhe.com/life

CHAPTER 9

Meiosis

Of Aphids and a Mosaic Child

Parthenogenesis

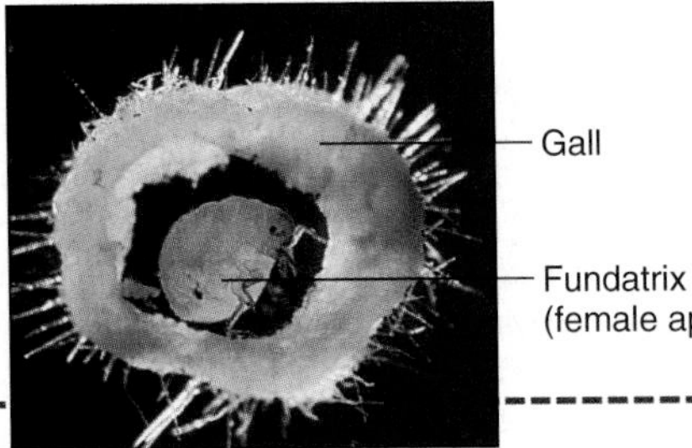

This cross section of a protective gall on a sumac tree shows a fundatrix, a female aphid who will reproduce clones of herself by parthenogenesis.

Parthenogenesis is Greek for "virgin birth" and refers to the unusual situation of an oocyte (immature egg cell) becoming activated, doubling its genetic material, and then dividing mitotically to yield a viable offspring. It is reproduction without a mate, and occurs in salamanders, lizards, snakes, turkeys, roundworms, flatworms, and various pond dwellers. The most familiar parthenogenotes are the males (drones) in certain bee, wasp, and ant societies, which develop from unfertilized eggs. Stimuli provoke parthenogenesis, allowing calcium to enter an oocyte. For example, wasp eggs start developing when they are mechanically stimulated while exiting the female's abdomen. Certain female salamanders mate with males from a related species, and the sperm activate their eggs, but the male cells degenerate before they can contribute paternal nuclei. Researchers can induce parthenogenesis in some species by pricking an oocyte, which admits calcium.

Parthenogenesis is advantageous when an organism is well adapted to its environment, because it can lead to a sudden population explosion. Parthenogenesis explains how a pond forming overnight from a puddle becomes rapidly overrun with little swimming animals. In a changing environment, though, parthenogenesis becomes a liability. The organisms are genetically alike, and all perish if conditions become harsh. Because environments change, parthenogenesis is rare.

The Complicated Sex Life of an Aphid

The first report of parthenogenesis is attributed to Swiss naturalist Charles Bonnet, who noticed that aphid eggs develop without benefit of sperm. Aphids are insects that live on trees, mosses, ornamental flowers, and garden vegetables. Their unusual reproductive cycle begins in June, when small females called fundatrices hatch on tree bark and move to the leaves. There they burrow into the leaves, nestling into sacs called galls. As each fundatrix grows and matures within a gall, embryos begin to develop in her abdomen, without the aid of a male aphid (parthenogenetically).

By August, the gall that at first housed a single fundatrix has swollen to accommodate a thriving matriarchy of the original female and her daughters, all pale and wingless, and even some grand-

The fundatrix's daughters look like her—pale and wingless. Her granddaughters, destined to migrate, are dark and winged. Curiously, this colony is a clone—all of the individuals are genetically identical. However, the third generation expresses a different set of genes than the previous two generations, causing these individuals to look different.

First polar body
Second polar body
X
X
DNA replication
Y
Fertilization
XX
XY

A Most Unusual Child.
"F.D." arose from two initial cells, one of which was a polar body (a product of meiosis that is not an oocyte) that replicated its DNA, rather than being fertilized by a sperm. A second polar body was fertilized. The two initial cells remained together, producing a child with two cell populations. He is a partial parthenogenote.

daughters. One female may even contain her daughters and granddaughters. By autumn, the granddaughters emerge from the gall, mature, winged, and dark. They take flight, landing on mosses to live out their last few days. Sacs of embryos develop in the granddaughters, again without male input. As death nears, the granddaughters deposit groups of young on the moss.

Individuals of this fourth generation reproduce by parthenogenesis, grow, and live with their young protected beneath the waxy coat of the moss. Come springtime of the next year, or sometimes the year after that, the young aphids fly up from the mosses. But the aphid's curious multigenerational reproductive cycle isn't over yet.

The winged aphids alight on sumac trees, where they release offspring—this time of two types, male and female. The males and females mate. This rare sexual generation introduces genetic variability; individuals from different colonies, and therefore descended from different fundatrices, mate and pass their genes to the next generation in new combinations. The sexual females deposit fertilized eggs on sumac bark. Sometime later, the eggs hatch as a new generation of fundatrices, and the complex sequence of reproductive events begins anew.

The Case of F.D.

Mammal parthenogenotes develop only partway. A mouse embryo derived from an activated egg shrivels up and ceases developing. Humans fare even worse. If an oocyte somehow doubles its genetic material and starts to develop, it doesn't form an embryo, but a tumorlike mass called a teratoma. If a sperm doubles its DNA and divides, it forms a different type of abnormal growth. But biologists thought it might be possible for a mammal to be a partial parthenogenote—that is, part of the body is derived from an unfertilized maternal cell. The birth of a most unusual child in 1991, whom the medical journals call "F.D.," showed that this can indeed happen.

Shortly after F.D.'s birth, his parents noticed that his head was lopsided, with the left side skewed and drooping. Photographs of the boy from one side look completely normal, yet from the other side, he has a jutting chin, sunken eye, and oddly clenched jaw. Not obvious are a cleft palate and other abnormal throat structures. F.D. is relatively healthy, although he is slightly developmentally delayed and has aggressive outbursts.

Because chromosome abnormalities are often associated with unusual facial structures, doctors took a blood sample to examine the child's chromosomes. To their surprise, they found two X chromosomes in the blood cells, indicating a female. But F.D. looked like a boy!

The doctors continued to look at the sex chromosomes in different types of cells in F.D., and found some to be XX, and some to be XY. What had happened?

F.D.'s body consists of two cell populations, one XX and one XY. One possible explanation for the child's beginnings is a detour in meiosis in the mother. As an oocyte formed by meiotic cell division, cells called polar bodies were also produced—this is normal. What wasn't normal was that one polar body evidently replicated its DNA to become an XX cell, and another was fertilized by a sperm. The two cells remained attached, and developed into F.D., such that half the body is chromosomally male, and the other half female. Fortunately, F.D. looks and feels male, and can one day become a father because his reproductive system descended from the XY cell.

9.1 Why Reproduce?

Reproduction is not essential to the life of an individual, but it is for the perpetuation of a species. Different reproductive strategies yield offspring identical to a parent, or offspring that combine the traits of two parents. Either approach is adaptive under certain conditions.

Organisms must reproduce—generate other individuals like themselves—for a species to survive. The amoeba in figure 9.1A demonstrates a straightforward way to reproduce. It replicates its genetic material and then splits in two, doubling and then apportioning the cellular contents of one individual into two. This ancient form of reproduction, called **binary fission,** is still common among single-celled organisms.

In an unchanging environment, the mass production of identical individuals, as in binary fission, makes sense. However, environmental conditions are rarely constant in the real world. If all organisms in a species were well suited to a hot, dry climate, the entire species might perish in a frost. A population of individuals of the same species with a diversity of characteristics is better able to survive in a changeable environment. Binary fission cannot create or maintain this diversity; **sexual reproduction** can. Sexual reproduction is a process that shuffles and recombines inherited traits from one generation to the next. The persistence of sexual reproduction over time and in many diverse species attests to its success in a changing world.

The kittens in figure 9.1B differ from each other because they were conceived sexually. In contrast are the identical amoebae in figure 9.1A, which are the products of **asexual reproduction,** or reproduction without sex. The opening essay describes another form of reproduction, **parthenogenesis,** in which an offspring is derived solely from a female parent. Parthenogenesis is rare.

Sexual reproduction has two essential qualities: It introduces new combinations of genes from different individuals (the parents), and it increases the number of individuals by producing offspring. The cells from the parents that combine to form the first cell of the offspring are called **gametes.** The nuclei of gametes contain only one set of chromosomes and are said to be **haploid,** abbreviated **1*n*.** When two haploid gametes merge, they reconstitute the double, or **diploid (2*n*),** number of chromosome sets in the cells of the offspring. In humans, sperm and egg cells (oocytes) are the gametes. Some species, including ciliates and certain fungi, produce gametic (haploid) nuclei, but do not enclose them in separate cells.

9.1 MASTERING CONCEPTS

1. Under what circumstances are asexual and sexual reproduction adaptive?
2. How do asexual and sexual reproduction differ?
3. What are gametes?
4. How do haploid and diploid nuclei differ?

OLC

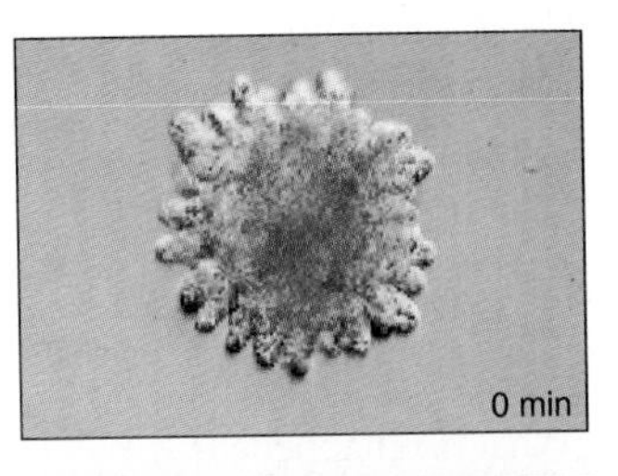

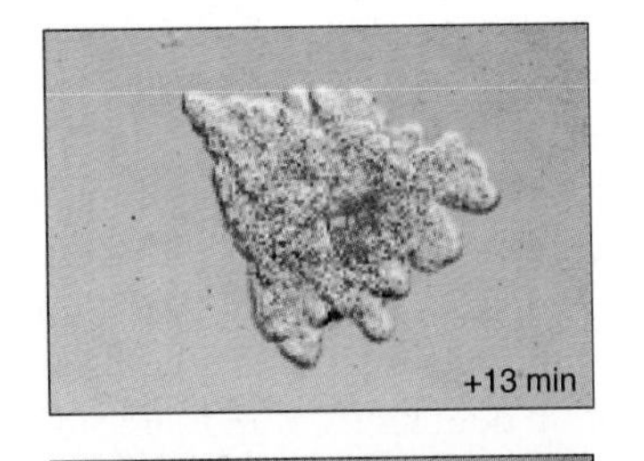

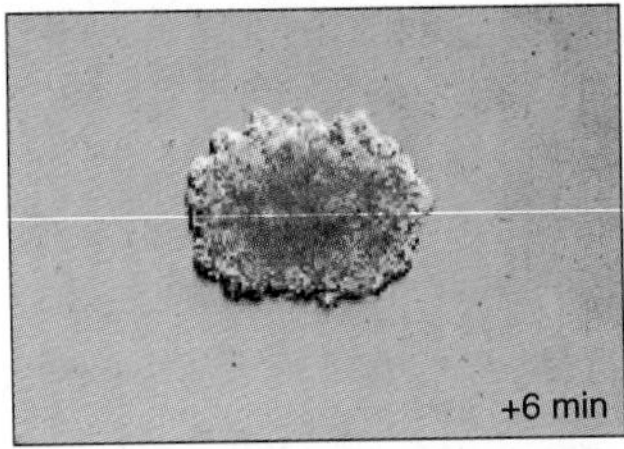

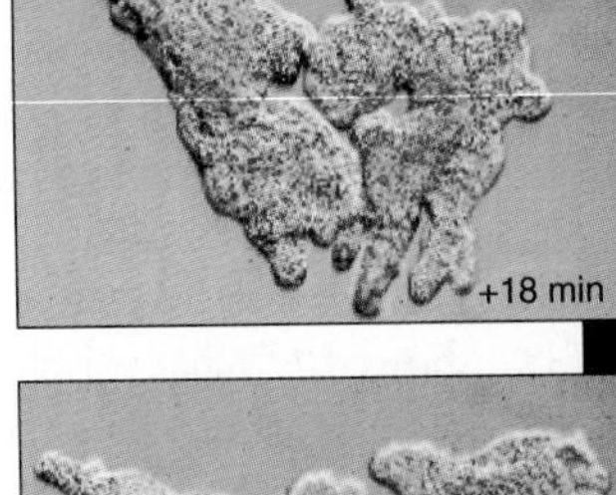

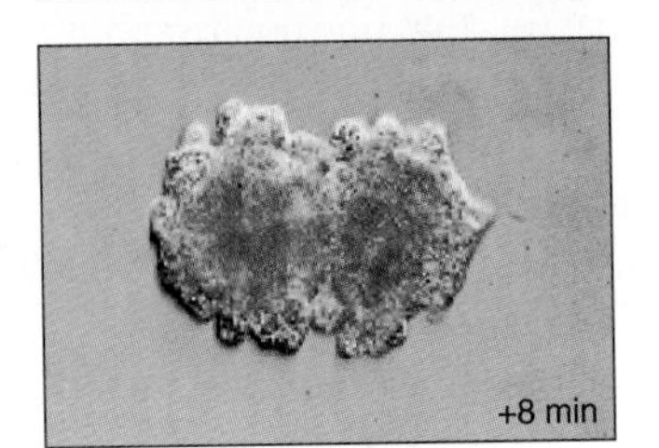

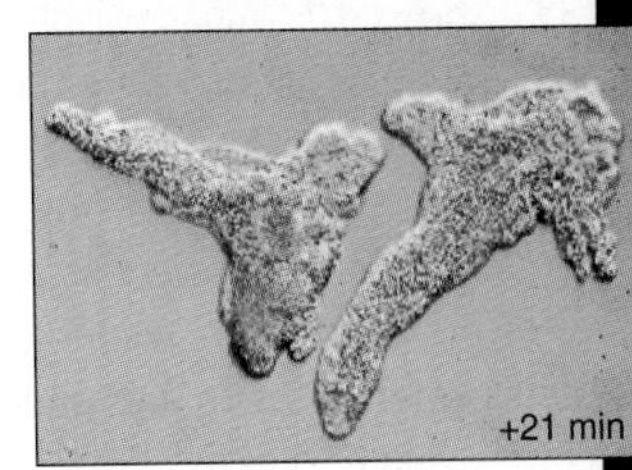

A

B

FIGURE 9.1
Asexual and Sexual Reproduction.
(**A**) The single-celled *Amoeba proteus* follows an asexual reproductive strategy by splitting in two (binary fission). (**B**) These kittens are obviously not exactly alike. The reason—each receives different combinations of the parents' genes, thanks to sexual reproduction.

9.2 Variations on the Sexual Reproduction Theme

Organisms may reproduce asexually, sexually, or have alternating asexual and sexual phases. Some microorganisms exchange genetic material without forming a new individual.

Organisms reproduce in many ways. Some species exhibit distinctly different phases. Consider the yeast *Saccharomyces cerevisiae,* which is a single-celled eukaryote that reproduces both asexually and sexually. A single diploid cell of the yeast can replicate its genetic material and "bud," yielding genetically identical diploid daughter cells. This is asexual reproduction, because each offspring has one parent. Alternatively, yeast can form a structure called an ascus, which gives rise to specialized haploid cells. Two of these haploid cells can fuse, restoring the diploid chromosome number. This is sexual reproduction, because two individuals produce a third.

Sexual life cycles describe the proportion and timing of an individual's existence that is haploid and/or diploid (table 9.1). The two main events of a sexual life cycle are meiosis and fertilization. **Meiosis** is a form of cell division that halves the genetic material. **Fertilization** is the joining of haploid nuclei that result from meiosis, which reconstitutes the diploid condition. In humans, meiosis leads to formation of the sperm or egg cells, and fertilization occurs following sexual intercourse. Mitosis is also a major part of sexual life cycles, producing growth of the body, or somatic growth.

Sexually reproducing species vary in the timing and extent of mitosis. In green algae, cellular slime molds, and some fungi, mitotic growth occurs between meiosis and fertilization, resulting in a body that is haploid. Yet in ciliates, brown algae, water molds, and animals, the period of mitotic activity occurs between fertilization and meiosis, producing a diploid body.

Land plants and some other organisms have a more complex life cycle that includes multicellular diploid and haploid stages. These stages are called generations, and because they alternate within a life cycle, sexually reproducing plants are said to undergo **alternation of generations** (fig. 9.2).

The diploid generation, or sporophyte, produces haploid spores through meiosis. Haploid spores divide mitotically to

TABLE 9.1 Sexual Life Cycles

PERIOD OF SOMATIC GROWTH	RESULT	TYPES OF ORGANISMS
Between meiosis and fertilization	Somatic body haploid	Some green algae Cellular slime molds Fungi (ascomycetes, chytrids)
Between fertilization and meiosis	Somatic body diploid	Ciliates Brown algae Water molds Animals
Between both meiosis and fertilization, and fertilization and meiosis	Alternation of haploid and diploid somatic generations	Some algae Land plants Yeast

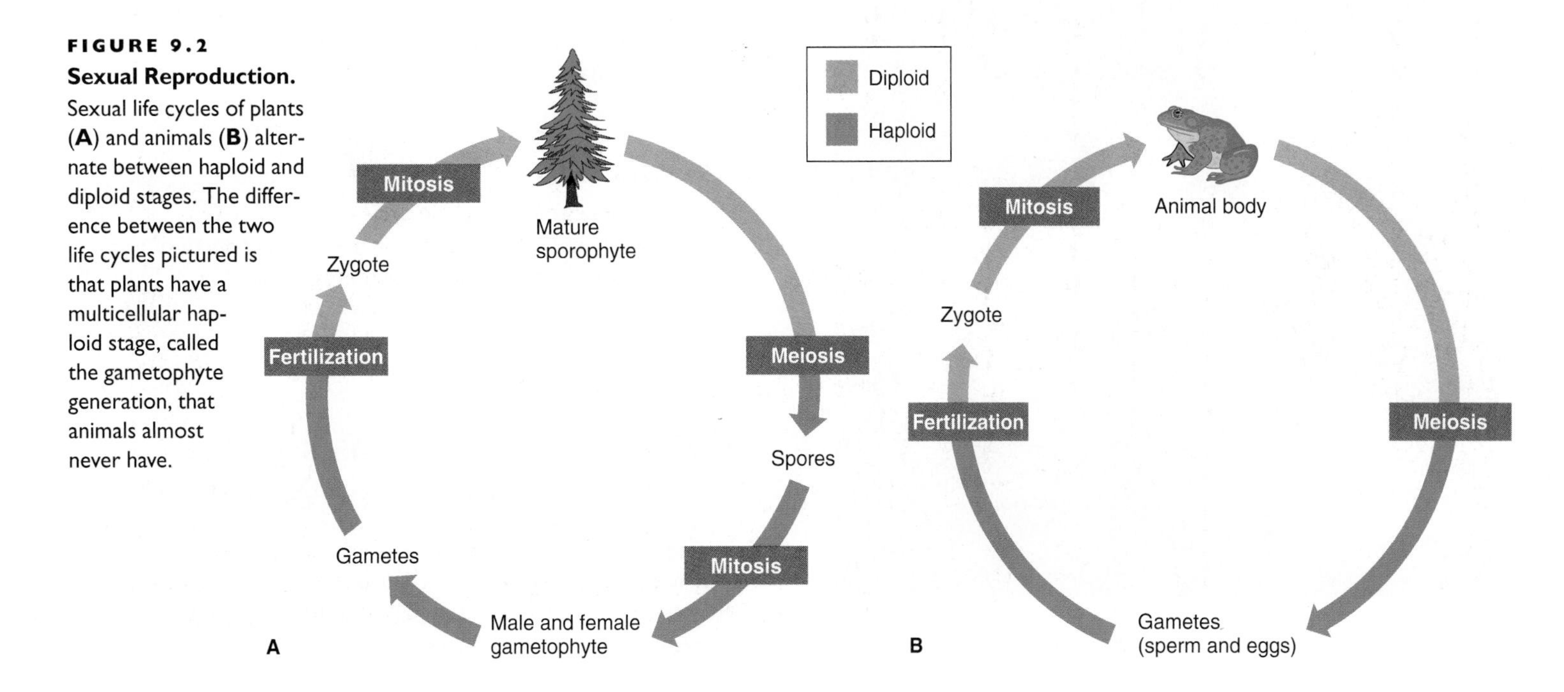

FIGURE 9.2
Sexual Reproduction. Sexual life cycles of plants (**A**) and animals (**B**) alternate between haploid and diploid stages. The difference between the two life cycles pictured is that plants have a multicellular haploid stage, called the gametophyte generation, that animals almost never have.

produce a multicellular haploid individual called the gametophyte. Eventually, the gametophyte produces haploid gametes—eggs and sperm—which fuse to form a fertilized ovum, or **zygote.** The zygote grows into a sporophyte, and the cycle begins anew. Some plants can also reproduce asexually when a "cutting" from a larger plant grows into a new organism. Chapter 28 discusses flowering plant reproduction.

Learning how diverse organisms reproduce and exchange genetic material today can provide clues to how sexual reproduction may have evolved. The earliest process that combines genes from two individuals appeared about 3.5 billion years ago, according to fossil evidence, in a form of bacterial gene transfer called **conjugation** that is still prevalent today (see chapter 20). In conjugation, one bacterial cell uses an outgrowth called a sex pilus to transfer genetic material to another bacterial cell (fig. 9.3). Bacterial conjugation is sex without reproduction because it alters the genetic makeup of an already existing organism rather than producing a new individual.

Paramecium is a unicellular eukaryote that clearly separates reproduction and sex. Paramecia reproduce asexually by binary fission, yet two organisms can conjugate by aligning their oral cavities and forming a bridge of cytoplasm between them. Nuclei are transferred across this bridge, resulting in new gene combinations but no new offspring.

FIGURE 9.3
Conjugation.
The bacterium *Escherichia coli* usually reproduces asexually by binary fission. But *E. coli* can also transfer genetic material to another bacterium using a projection of the cell membrane, called a sex pilus, to insert DNA into another cell. DNA transfer between bacterial cells, called conjugation, is similar to sexual reproduction in that it produces a new combination of genes. It is not reproduction, however, because a new individual does not form.

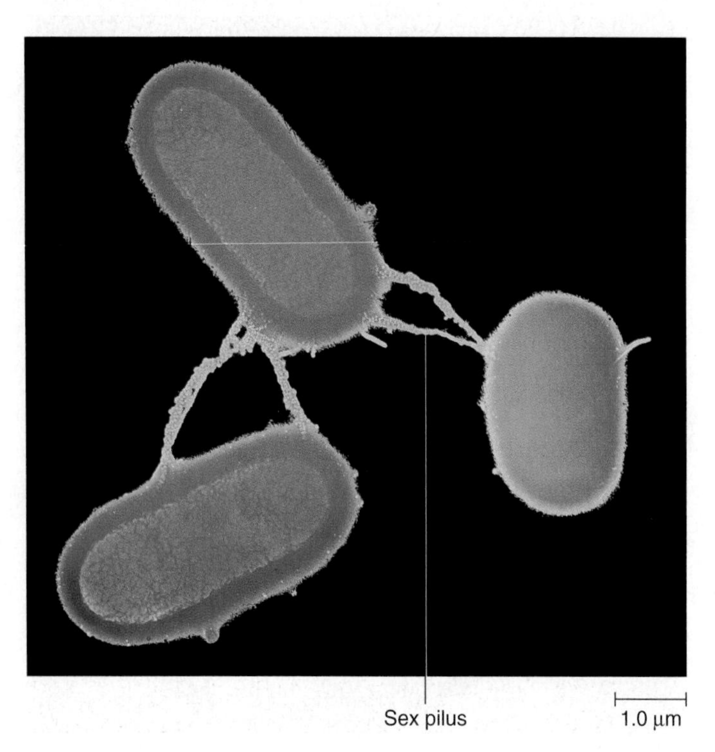

Conjugation is sexual in the sense that it exchanges genetic material, but it is not reproduction, because the number of individuals does not increase. Unicellular green algae of the genus *Chlamydomonas* exhibit a simple form of true sexual reproduction that may be similar to the earliest sexual reproduction, which may have begun about 1.5 billion years ago.

In the more common asexual phase, *Chlamydomonas* cells are haploid (fig. 9.4A). The cells are of two mating types, designated plus and minus. Under certain environmental conditions, a plus and minus cell join and form a single diploid cell. This cell then undergoes meiosis, yielding four haploid cells—two plus and two minus. The next generation begins when plus and minus cells join, thus meeting the two requirements of sexual reproduction—the number of individuals increases, and the genetic contributions of two parents are mixed and reapportioned in the offspring.

When reproduction depends upon the meeting and fusion of two (or more) types of cells, it is critical that the different types recognize each other. The two types of *Chlamydomonas* cells have different patterns of molecules on their surfaces, and they probably differ in other ways obvious to them. Sexual reproduction in *Chlamydomonas* begins when molecules on the tail-like flagella (fig. 9.4B) of each type attract and draw the plus and minus cells into physical contact. Certain regions of the two cell membranes touch, like soap bubbles coalescing. The plus cell then extends a fingerlike appendage toward the minus cell and establishes a bridge of cytoplasm that further weds the two cells.

In more complex organisms, the two types of mating cells may differ more in appearance. Still, despite the differences—such as the enormity of eggs compared to sperm—each cell plays the same role as an emissary carrying genetic material to the next generation. Meiosis halves the chromosome number. An additional series of steps in many species, called maturation, strips much of the cytoplasm and organelles from developing male gametes and, conversely, concentrates these supplies in female gametes. The cells that undergo meiosis and give rise to gametes are called **germ cells.** Meiosis and maturation together constitute **gametogenesis.** The remainder of the chapter considers gametogenesis in humans.

9.2 MASTERING CONCEPTS

1. What are some sexual life cycles?
2. How can sex occur with and without reproduction?
3. What are the two stages of gametogenesis?

9.3 How Does Meiosis Halve the Chromosome Number?

Sexual reproduction is possible because meiosis, a form of cell division, replicates the genetic material once, but distributes it into two cells twice. The result is four haploid nuclei, which may be packaged into cells called gametes.

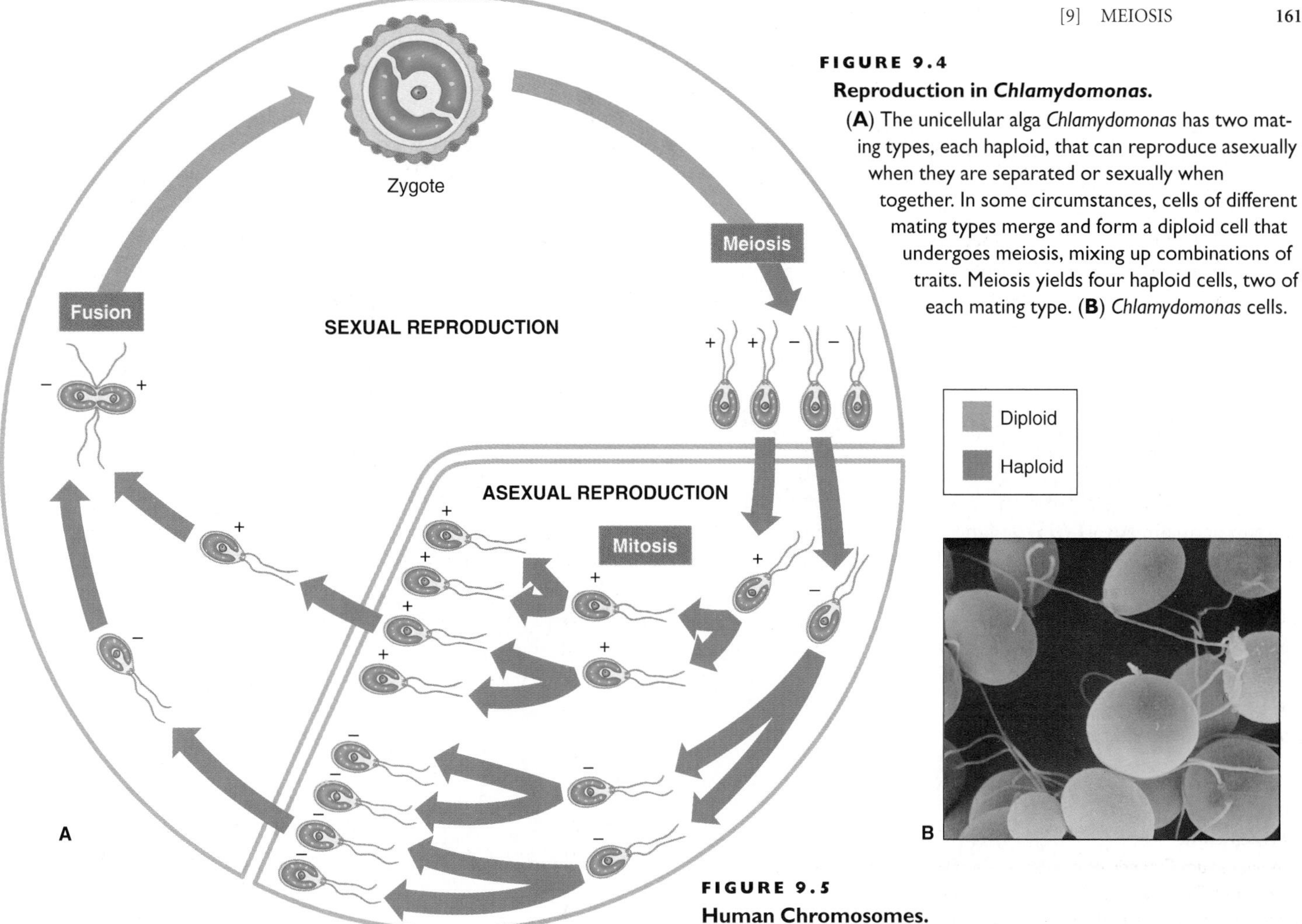

FIGURE 9.4
Reproduction in *Chlamydomonas*.
(**A**) The unicellular alga *Chlamydomonas* has two mating types, each haploid, that can reproduce asexually when they are separated or sexually when together. In some circumstances, cells of different mating types merge and form a diploid cell that undergoes meiosis, mixing up combinations of traits. Meiosis yields four haploid cells, two of each mating type. (**B**) *Chlamydomonas* cells.

FIGURE 9.5
Human Chromosomes.
A chart called a karyotype displays in size order the 23 pairs of chromosomes in a somatic cell from a human male. Note the X and Y sex chromosomes. They pair during meiosis, perhaps because they have several regions in common. A somatic cell from a female has two X chromosomes.

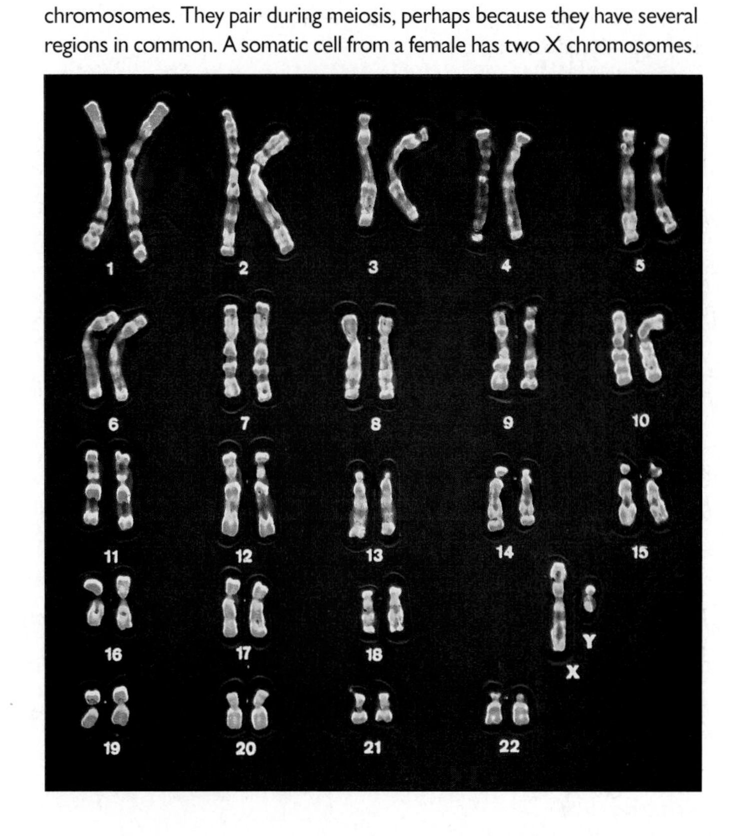

In meiosis, the genetic material replicates only once, but divides twice. The first division is called **reduction division** (or meiosis I) because it reduces the number of chromosomes—in humans, from 46 to 23. The second division, called the **equational division** (or meiosis II), produces four cells from the two formed in the first division.

Before meiosis begins, DNA replicates during S phase of interphase. The cell in which meiosis begins has **homologous pairs** of chromosomes, or homologs for short. Homologs look alike and carry the genes for the same traits in the same sequence. A diploid human cell has 23 pairs of homologs. Of these, 22 pairs do not determine sex and are called **autosomes;** the other chromosomes (the X and the Y) carry genes that determine sex and are known as **sex chromosomes** (fig. 9.5).

One homolog in a pair comes from the person's mother, and the other comes from the father. Different colors distinguish the parental origins of members of homologous pairs in the illustrations, and different pairs are distinguished by size (fig. 9.6). When meiosis begins, the replicated DNA of each homolog forms two chromatids joined by a centromere. The chromosomes are not yet condensed enough to be visible when stained.

FIGURE 9.6
A Schematic Overview of Meiosis.
Meiosis is a form of cell division in which certain cells are set aside to give rise to haploid gametes. In humans, the first meiotic division reduces the number of chromosomes to 23, all in the replicated form. In the second meiotic division the cells essentially undergo mitosis. The result of the two divisions of meiosis is four haploid cells. Homologous pairs of chromosomes are indicated by size, and parental origin of chromosomes by color.

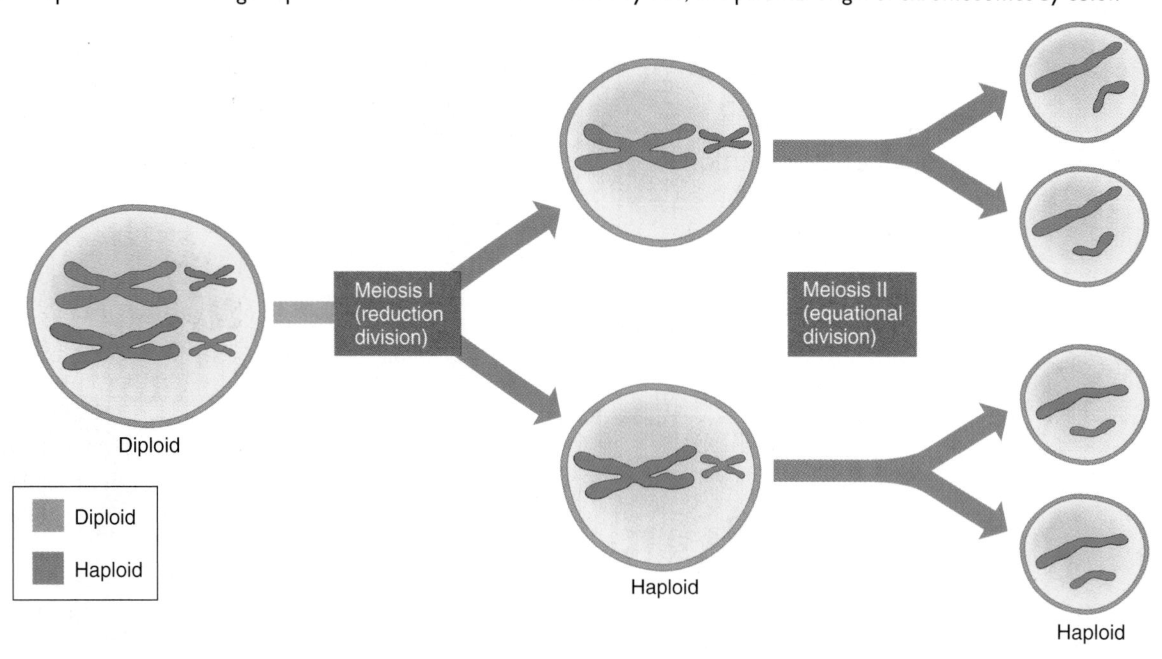

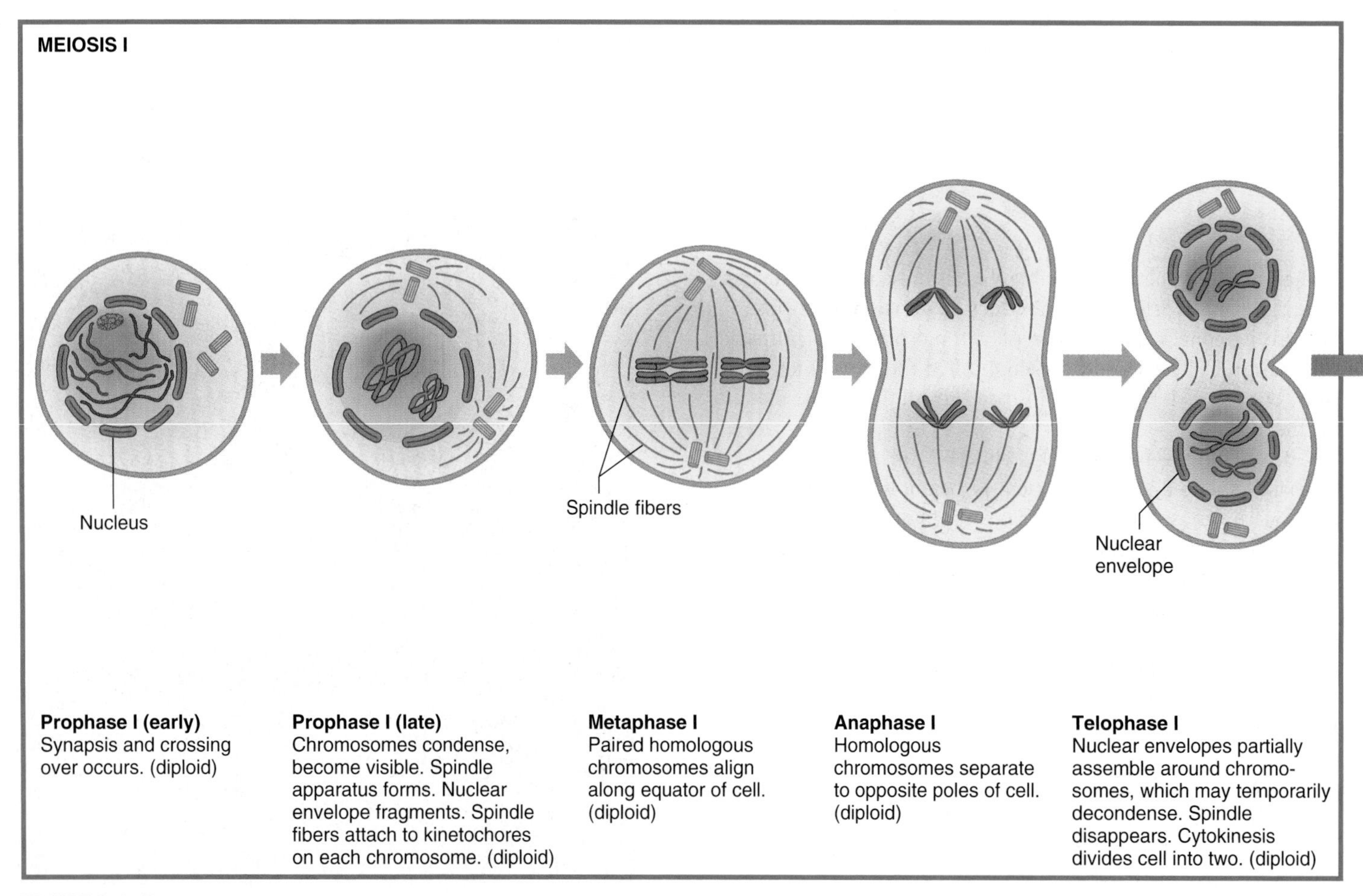

FIGURE 9.7
An Overview of Meiosis.

Prophase I (so called because it is the prophase of meiosis I) follows interphase. Early in prophase I, replicated chromosomes condense and become visible when stained (fig. 9.7). Toward the middle of prophase I, the homologs line up next to one another, gene by gene, a phenomenon called synapsis. A complex of RNA and protein connects the paired chromosomes.

Toward the end of prophase I, the synapsed chromosomes separate but remain attached at a few points along their lengths, called chiasmata (fig. 9.8). Here, the homologs exchange chromosomal material in a process called **crossing over.** Because each homolog comes from a different parent, crossing over results in chromosomes that have some genes from the mother and some from the father. New gene combinations arise when the parents carry different forms of the same gene, which are called **alleles.** The mixing up of trait combinations resulting from crossing over is one reason why siblings (except for identical twins) are never exactly alike genetically.

Consider a simplified example of how crossing over mixes trait combinations. Suppose that one homolog carries genes for hair color, eye color, and finger length. One of the chromosomes in the homolog pair—perhaps the one that came from the father—carries alleles for blond hair, blue eyes, and short fingers. The homolog from the mother carries alleles for black hair, brown eyes, and long fingers. After crossing over, one of the chromosomes might bear alleles for blond hair, brown eyes, and long fingers, and the other bear alleles for black hair, blue eyes, and short fingers.

Meiosis continues in metaphase I, when the homologs align down the center of the cell (see fig. 9.7). A spindle forms, and each chromosome attaches to a spindle fiber stretching to the opposite pole. The chromosomes' alignment during metaphase I is important in generating genetic diversity. Within each homolog pair, the maternally-derived and paternally-derived members attach to each pole at random. The greater the number of chromosomes, the greater the genetic diversity, because different combinations of maternal and paternal homologs move to each pole. It is a little like a double line of schoolchildren. Imagine the many different ways that 23 pairs of students could form a double line, while maintaining specific pairs.

For two pairs of homologs, four (2^2) different metaphase configurations are possible. For three pairs of homologs, eight (2^3) configurations can occur. Our 23 chromosome pairs can thus line up in 8,388,608 (2^{23}) different ways! (And a class of 23 pairs of students could line up in this many ways.) The random distribution of homologs in a cell in metaphase I is called **independent assortment** (fig. 9.9). It accounts for a basic law of inheritance, which is discussed in the next chapter.

Homologs separate in anaphase I (see fig. 9.7), and they complete their movement to opposite poles in telophase I. In most species, the cell divides in two after telophase I to produce two

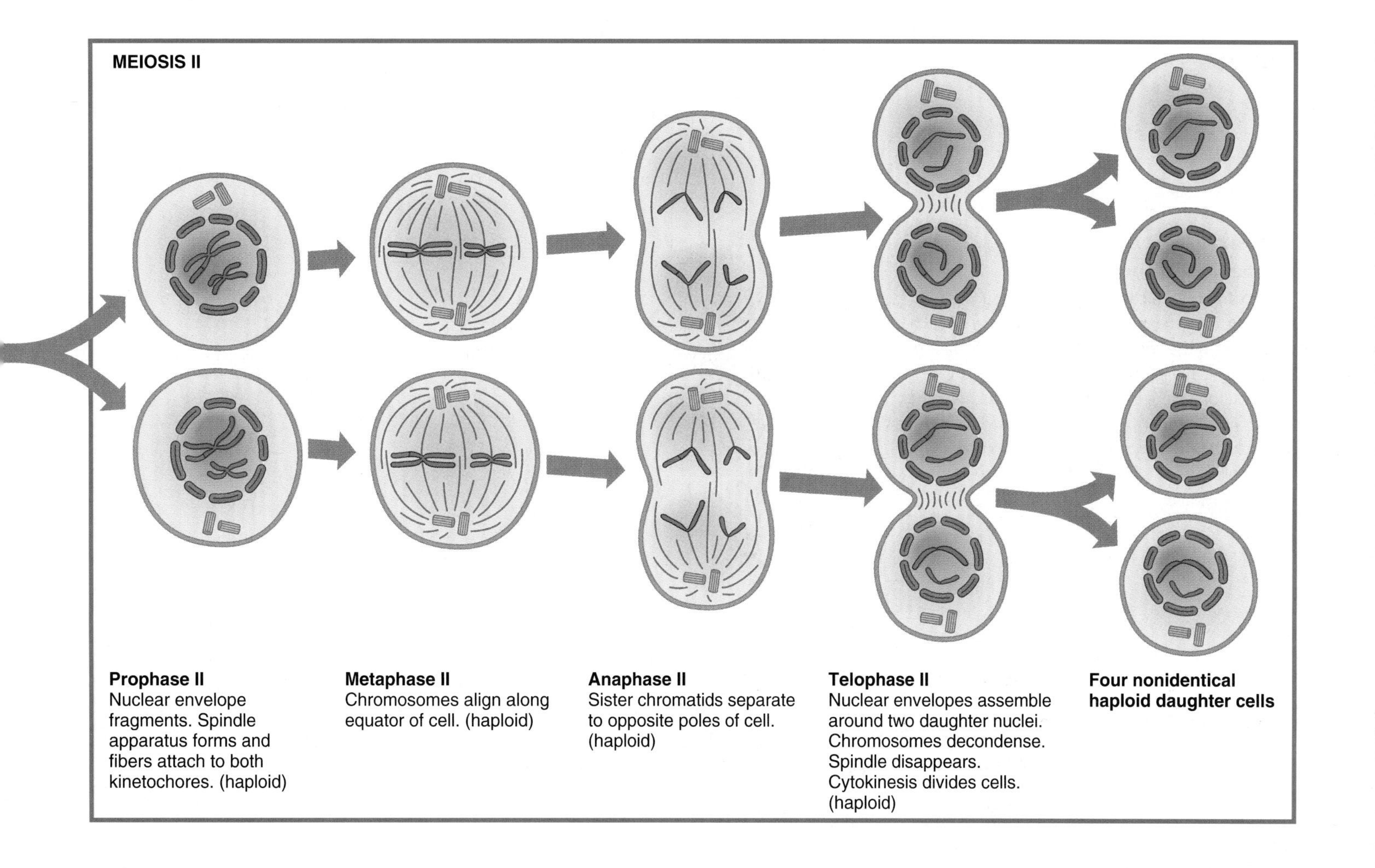

FIGURE 9.8

Crossing Over Recombines Genes.

Crossing over helps to generate genetic diversity by mixing up parental traits. (**A**)The capital and lowercase forms of the same letter represent different forms (alleles) of the same gene.
(**B**) Chiasmata are visible points of crossing over between homologous chromosomes.

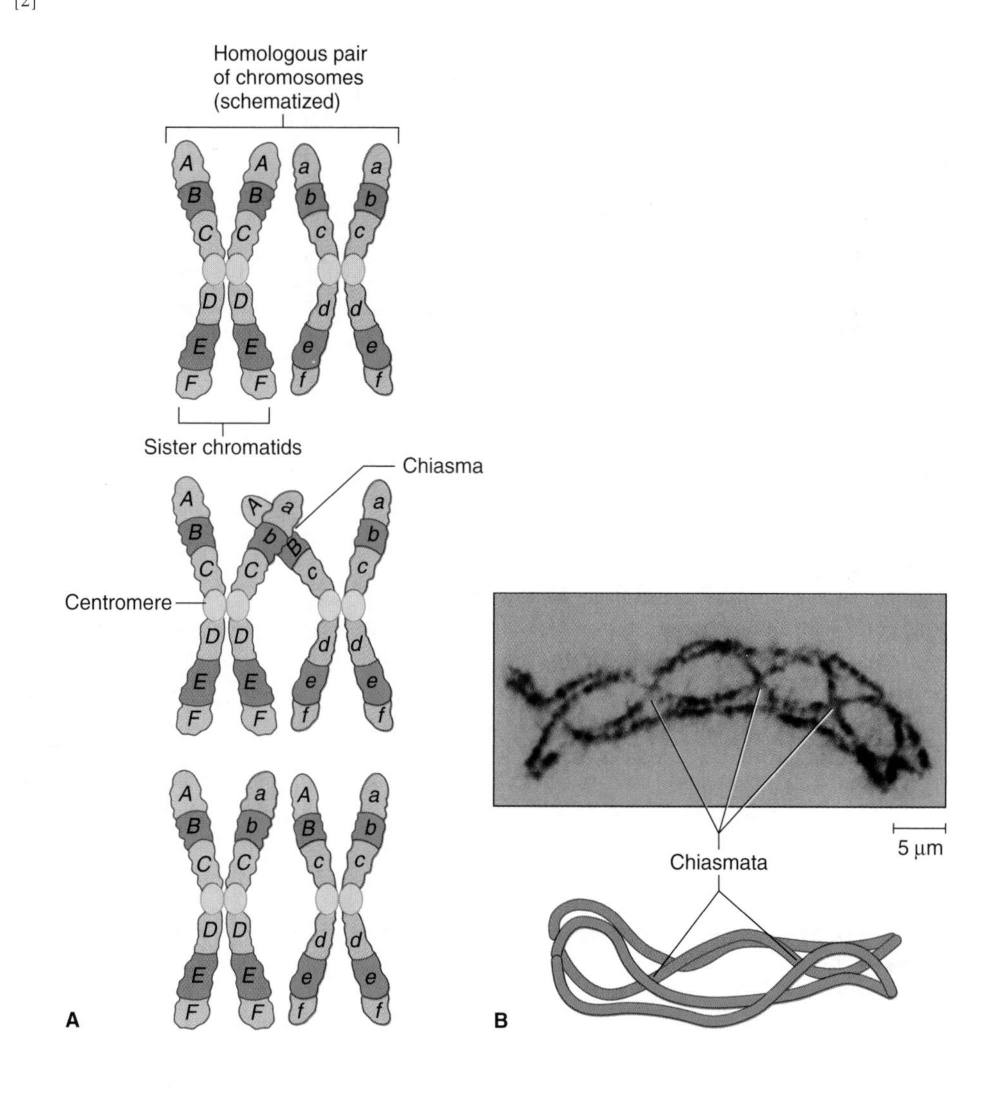

FIGURE 9.9

Independent Assortment.

The pattern in which homologs align during metaphase I determines the combination of chromosomes in the daughter cells. Two pairs of chromosomes can align in two different ways to produce four different possibilities in the daughter cells. The potential variability generated by meiosis skyrockets when one considers all 23 chromosome pairs and the effects of crossing over.

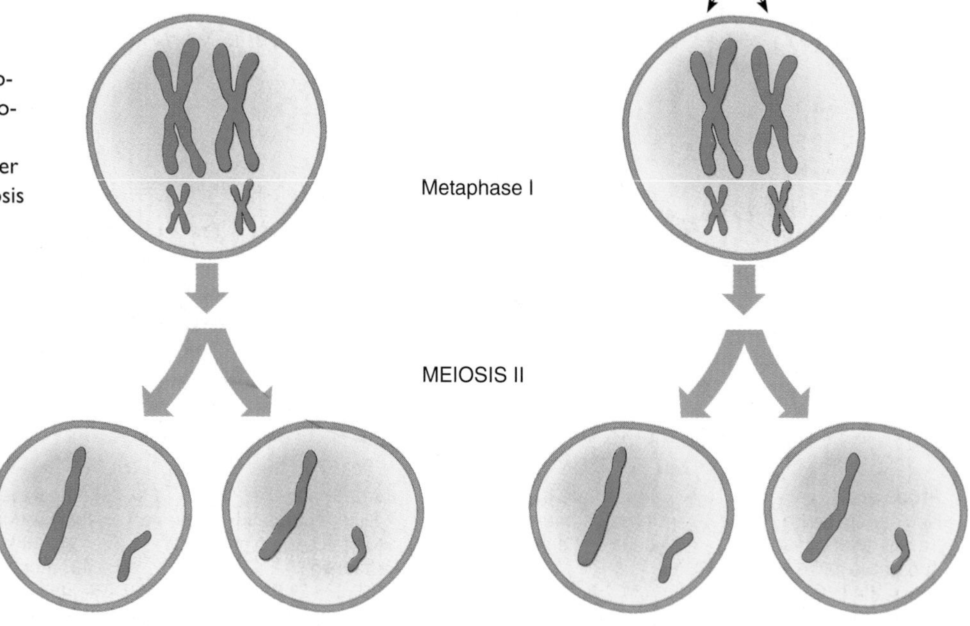

haploid cells. During a second interphase, the chromosomes unfold into very thin threads. (Some species skip a second interphase.) Proteins are manufactured, but the genetic material does not replicate a second time. The double division of meiosis is what halves the chromosome number to produce four haploid cells.

Prophase II marks the start of the second meiotic division (see fig. 9.7). The chromosomes again condense and become visible. In metaphase II, the replicated chromosomes align down the center of the cell. In anaphase II, the centromeres part and the chromatids move to opposite poles. In telophase II, nuclear envelopes form around the four nuclei containing the separated sets of chromosomes.

Cytokinesis separates the two groups of two nuclei into individual cells. In most species, cytokinesis occurs between the first and second meiotic divisions, forming first two cells; a subsequent cytokinesis after telophase II yields a total of four cells. In some species, cytokinesis occurs only following telophase II; the four nuclei enclosed in one large cell are separated into four smaller haploid cells.

Table 9.2 summarizes the stages of meiosis, and table 9.3 and figure 9.10 compare mitosis to meiosis.

Meiosis generates astounding genetic variety. Any one of a person's 8,388,608 (2^{23}) possible chromosome combinations can combine with any one of the 8,388,608 combinations of a partner, raising potential variability to more than 70 trillion ($8,388,608^2$) genetically different individuals! Crossing over contributes even more genetic variability to gametes.

9.3 MASTERING CONCEPTS

1. How do the numbers of DNA replications and cell divisions halve the chromosome number in meiosis?
2. Which events of meiosis generate genetic variability?
3. What happens during the stages of meiosis I and meiosis II?

OLC

9.4 The Sculpting of Gametes

Both sperm and oocyte arise from meiosis, but look about as different as two cells can look. A sperm is lean and motile, and an oocyte is huge and nutrient-packed.

TABLE 9.2 The Stages of Meiosis

STAGE	EVENTS
INTERPHASE	DNA replicates
MEIOSIS I (reduction division)	(Halves chromosome number)
Prophase I	Replicated chromosomes condense
	Homologs synapse and cross over
Metaphase I	Homologs align down equator
	Spindle forms and attaches each chromosome to one pole, with members of homologous pairs attached to opposite poles
Anaphase I	Homologs move apart
Telophase I	Homologs arrive at opposite poles
INTERPHASE	
MEIOSIS II (equational division)	(Doubles number of daughter cells)
Prophase II	Chromosomes condense
Metaphase II	Chromosomes align down equator
Anaphase II	Centromeres part, separated chromatids move to opposite poles
Telophase II	Nuclear envelopes re-form as each of the two cells from meiosis I forms two cells

TABLE 9.3 Comparison of Mitosis and Meiosis

MITOSIS	MEIOSIS
One division	Two divisions
Two daughter cells per cycle	Four daughter cells per cycle
Daughter cells genetically identical	Daughter cells genetically different
Chromosome number in daughter cells same as in parent cell	Chromosome number in daughter cells half that in parent cell
Occurs in somatic cells	Occurs in germ cells
Occurs throughout life cycle	In humans, completed only after sexual maturity
Used for growth, repair, and asexual reproduction	Used for sexual reproduction, producing new gene combinations in new individuals

FIGURE 9.10
Differences Between Mitosis and Meiosis.

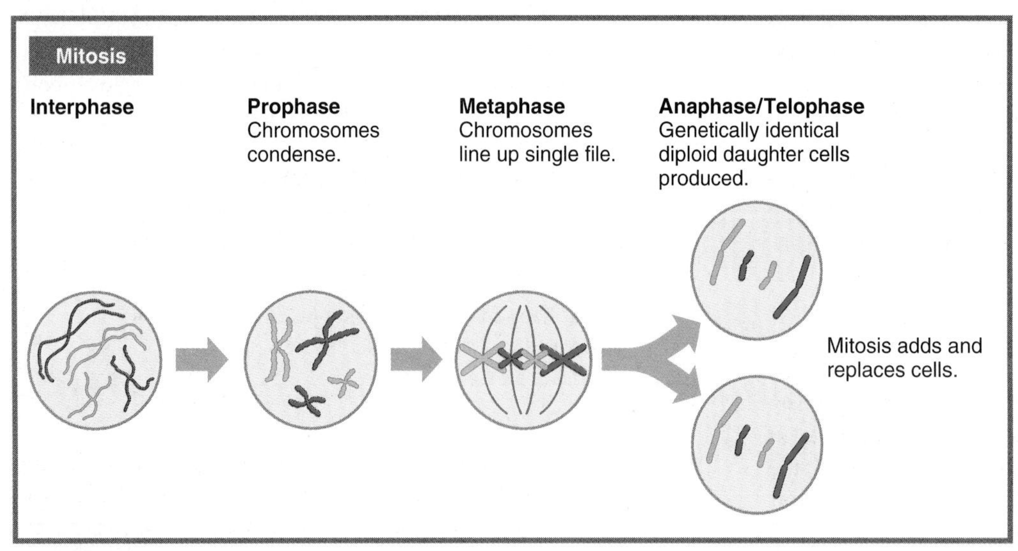

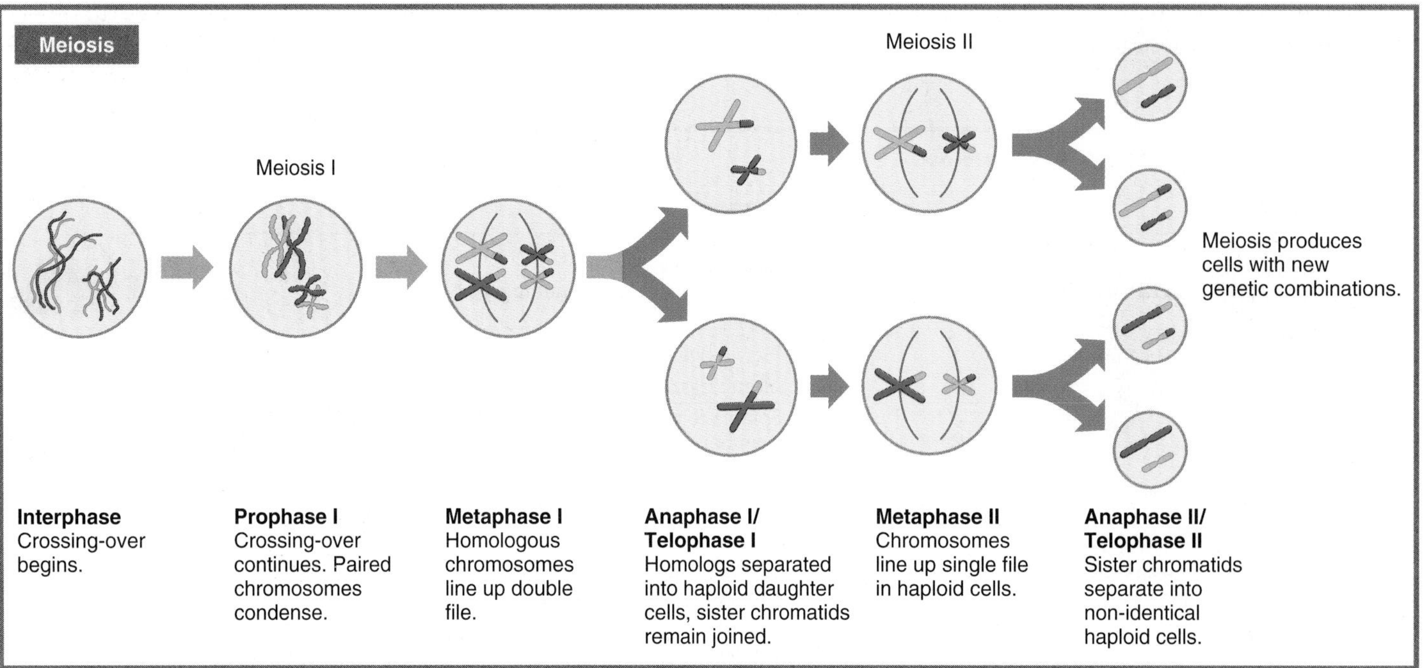

Sperm and egg each have a haploid set of chromosomes, but in many species they look quite distinct because each receives different amounts of other cellular components. A sperm is motile and lightweight; an egg is huge by comparison and packed with nutrients and organelles (fig. 9.11). So far we have considered meiosis in a general sense. We now look at the embellishments that transform the products of meiosis into sperm or egg.

Sperm Development

Sperm cells have long intrigued biologists, perhaps in part because the cells are so interesting to watch under a microscope (fig. 9.12A,B). Sperm cells look a little like microorganisms; indeed, when Antonie van Leeuwenhoek first viewed sperm in 1678, he thought they were parasites in the semen. By 1685, he modified his view and suggested that each sperm is a seed containing a preformed being (fig. 9.12C) and requiring a period of nurturing in the female to develop into a new life.

In the 1770s, Italian biologist Lazzaro Spallanzani took an experimental approach to studying sperm by placing toad semen onto filter paper. He observed that the material that passed through the tiny holes in the filter paper—seminal fluid minus the sperm—could not fertilize eggs. Even though he had shown indirectly that sperm are required for fertilization, Spallanzani,

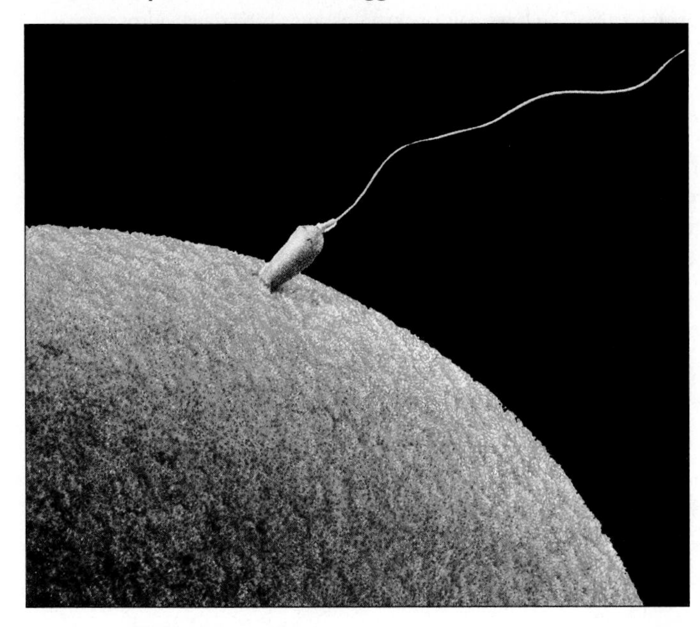

FIGURE 9.11
A Human Sperm Contacts an Egg.

like Leeuwenhoek before him, concluded that the sperm cells were contaminants. Several researchers in the nineteenth century finally showed that sperm were not microbial invaders but human cells that play a key role in reproduction.

Formation and specialization of sperm cells is called **spermatogenesis** (fig. 9.13). The process begins with a diploid cell that divides mitotically, yielding diploid daughter cells destined to become sperm cells. This initial cell is a **spermatogonium.** Several spermatogonia may remain attached to each other by bridges of cytoplasm, and they undergo meiosis simultaneously. First, the spermatogonia accumulate cytoplasm and replicate their DNA. They are now called **primary spermatocytes.** Next, during reduction division (meiosis I), each primary spermatocyte divides, forming two equal-sized haploid cells called **secondary spermatocytes.** These cells remain attached by bridges of cytoplasm.

In meiosis II, each secondary spermatocyte divides, yielding two equal-sized **spermatids** that are still attached. Each spermatid then develops specialized structures, including the characteristic sperm tail, or flagellum. The base of the tail has many mitochondria and ATP molecules, forming an energy system that enables the sperm to swim inside the female reproductive tract.

FIGURE 9.12
Human Sperm.
(**A**) A sperm contains distinct regions that assist in the job of delivering DNA to an egg. (**B**) Scanning electron micrograph of human sperm cells. (**C**) When human sperm were first seen in the microscope, they were thought to be infectious microbes. This 1694 illustration by Dutch histologist Niklass Hartsoeker presents another once-popular hypothesis about the role of sperm—some thought they were carriers of a preformed human called a homunculus.

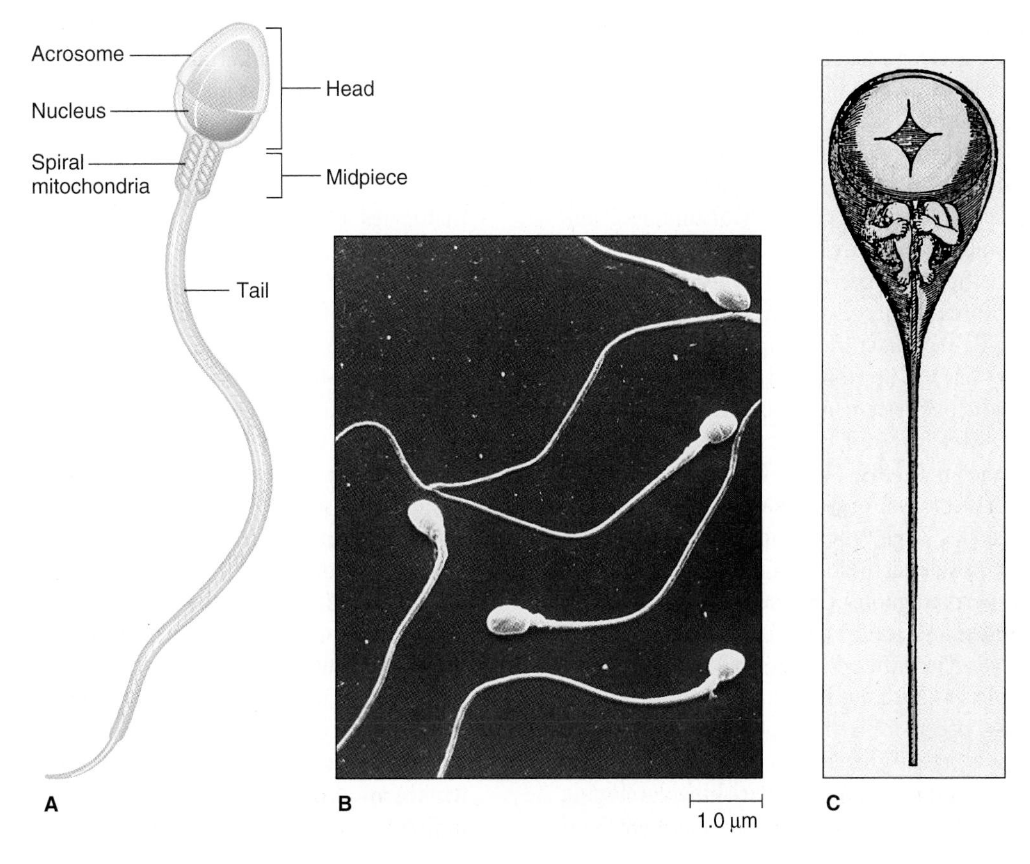

FIGURE 9.13

Sperm Formation (Spermatogenesis).

In humans, primary spermatocytes have the normal diploid number of 23 chromosome pairs. The large pair of chromosomes represents autosomes (nonsex chromosomes). The X and Y chromosomes are sex chromosomes.

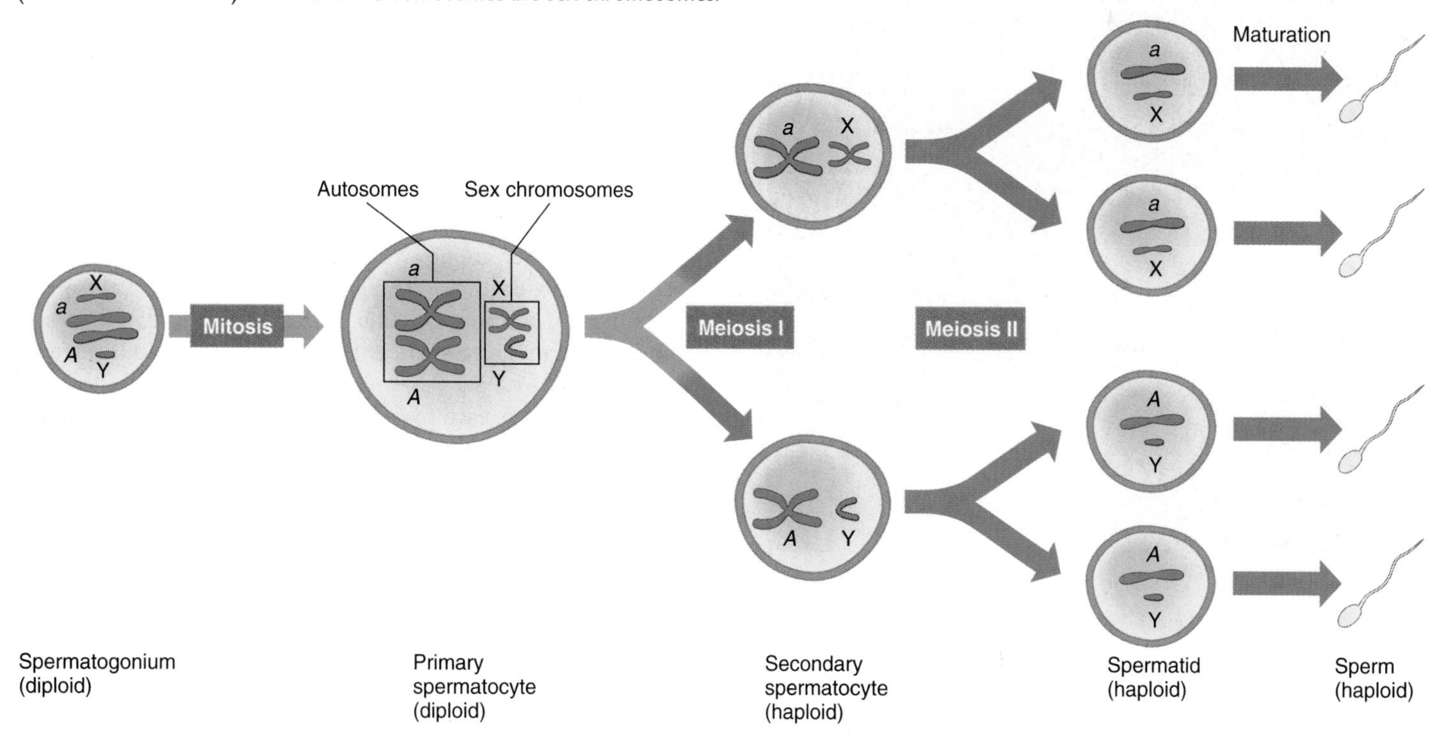

(A sperm, which is a mere 0.006 millimeters long, must travel about 178 millimeters within the female reproductive system to reach an egg!) After the spermatids form, some of the cytoplasm connecting the cells falls away, leaving mature, tadpole-shaped **spermatozoa,** or sperm.

A mature sperm has three functional parts (see fig. 9.12A): a haploid nucleus, a locomotion system, and enzymes to penetrate an egg. Each sperm cell consists of a tail, body or midpiece, and head region. A small organelle on the front end, the **acrosome,** contains enzymes that help the sperm cell penetrate the egg's outer membrane. Within the large sperm head, DNA wraps around proteins. A male manufactures trillions of sperm in his lifetime, yet only a very few will approach an egg.

In the male reproductive system, meiosis begins in the **seminiferous tubules.** If stretched out, the seminiferous tubules would be about 12.5 meters (41 feet) long. They are packed into the paired **testes,** which are oval organs that, with surrounding tissues, form the testicles (fig. 9.14). Within a seminiferous tubule, spermatogonia reside farthest from the lumen (the central cavity). When a spermatogonium divides, one daughter cell moves towards the lumen and accumulates cytoplasm, becoming a primary spermatocyte. The other daughter cell remains in the tubule wall. There, it acts as a stem cell, remaining unspecialized but continually giving rise to cells that specialize into sperm. Large sustentacular cells extend the entire width of a seminiferous tubule, surrounding, supporting, and nourishing developing sperm cells. Interstitial cells fill in spaces between the tubules and secrete male sex hormones.

The developing sperm cells move toward the lumen of the seminiferous tubule, and by the time they reach it, they are spermatids. This sequential and synchronized sperm development can be compared to students at a high school. Just as the student body consists of freshmen, sophomores, juniors, and seniors, all present at the same time but usually in different classes, a seminiferous tubule houses spermatogonia (freshmen), primary spermatocytes (sophomores), secondary spermatocytes (juniors), and spermatids (seniors).

The **epididymis** is a tightly coiled tube leading from each testis that stores spermatids. Here the spermatids complete their differentiation into spermatozoa. The entire process, from spermatogonium to spermatozoan, takes 74 days in the human.

The epididymis contracts during ejaculation, sending sperm cells into the **vasa deferentia,** which are tubes that join the urethra. The sperm, along with secretions from several accessory glands, form semen, which exits the body through the penis.

Meiosis in the male has some built-in protections against birth defects. Spermatogonia exposed to toxins may be so damaged that they never differentiate into mature sperm. Toxins can render sperm cells unable to swim. Toxic drugs carried in the seminal fluid but not actually damaging sperm can endanger an embryo or fetus if they enter a pregnant woman's reproductive tract. They can harm the uterus or enter a woman's circulation and reach the placenta, the organ connecting the woman to the fetus. Cocaine can affect a fetus by another route—it attaches to thousands of binding sites on the sperm, without harming the cells or impeding their movements. Therefore, sperm can ferry cocaine to a fertilized ovum.

FIGURE 9.14
Meiosis Produces Sperm Cells.
Diploid cells divide through mitosis in the linings of the seminiferous tubules. Some of the daughter cells then undergo meiosis, producing haploid spermatocytes which differentiate into mature sperm cells.

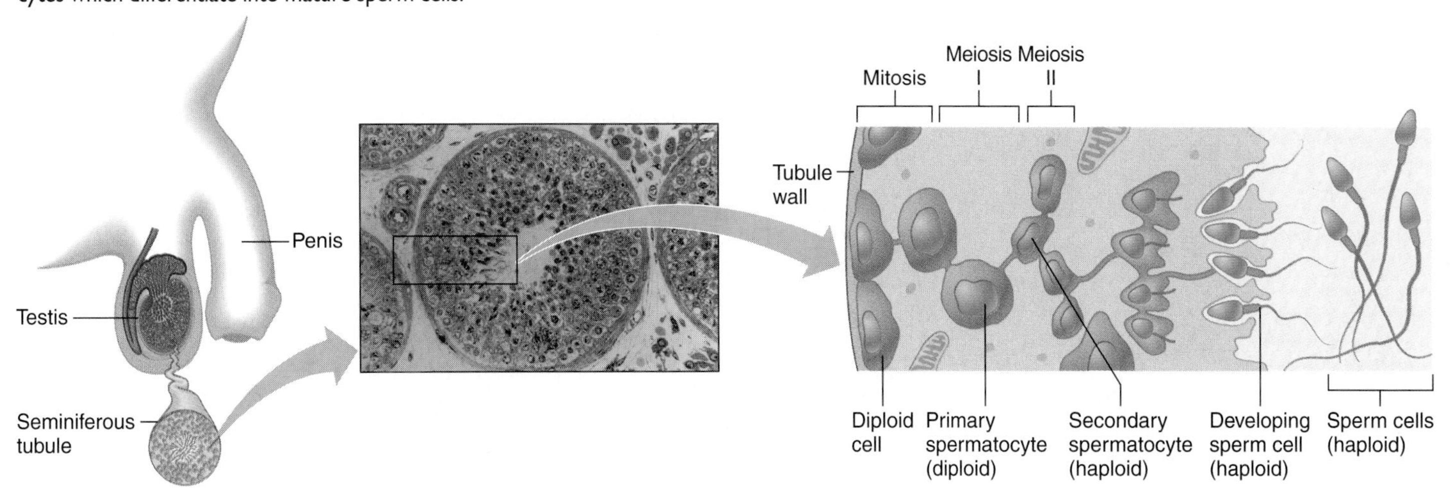

Ovum Development

Meiosis in the female is called **oogenesis** (egg making). It begins, like spermatogenesis, with a diploid cell, an **oogonium.** Unlike spermatogonia, oogonia are not attached to each other, but each is surrounded by a layer of follicle cells. Each oogonium grows, accumulates cytoplasm, and replicates its DNA, becoming a **primary oocyte.** The ensuing meiotic division in oogenesis, unlike that in spermatogenesis, produces cells of different sizes.

In meiosis I, the primary oocyte divides into a small cell with very little cytoplasm, called a **polar body,** and a much larger cell called a **secondary oocyte** (fig. 9.15). Each cell is haploid but with replicated chromosomes. In meiosis II, the tiny polar body may divide to yield two polar bodies of equal size or may decompose. The secondary oocyte, however, divides unequally in meiosis II to

FIGURE 9.15
Ovum Formation (Oogenesis).
In humans, primary oocytes have the normal diploid number of 23 chromosome pairs. Meiosis in females is uneven, concentrating most of the cytoplasm into one large cell, an oocyte (or egg). The other products of meiosis, called polar bodies, contain the other three sets of chromosomes and are discarded by the body.

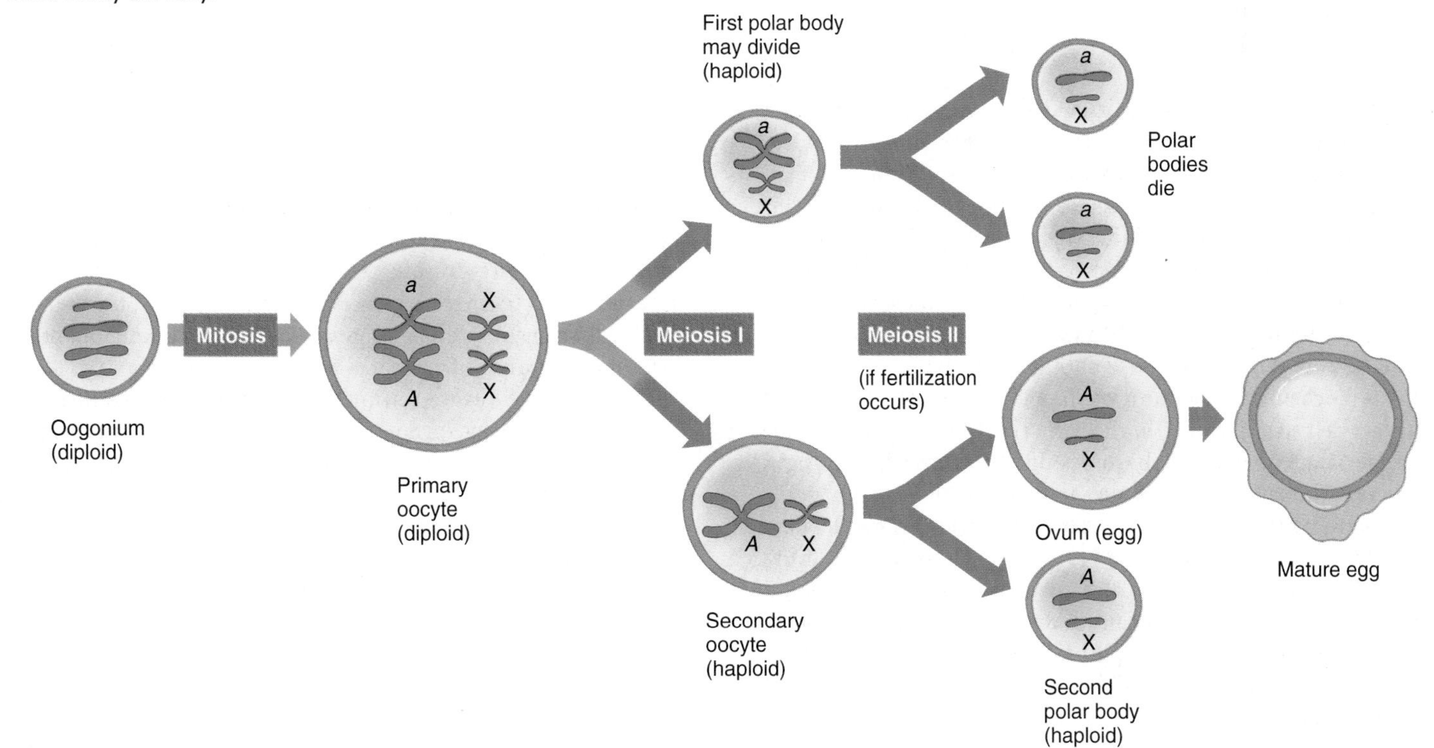

FIGURE 9.16

The Making of Oocytes.

Oocytes develop within the ovary in protective follicles. An ovary contains many oocytes in various stages of maturation. After puberty, each month the most mature oocyte in one ovary bursts out, an event called ovulation.

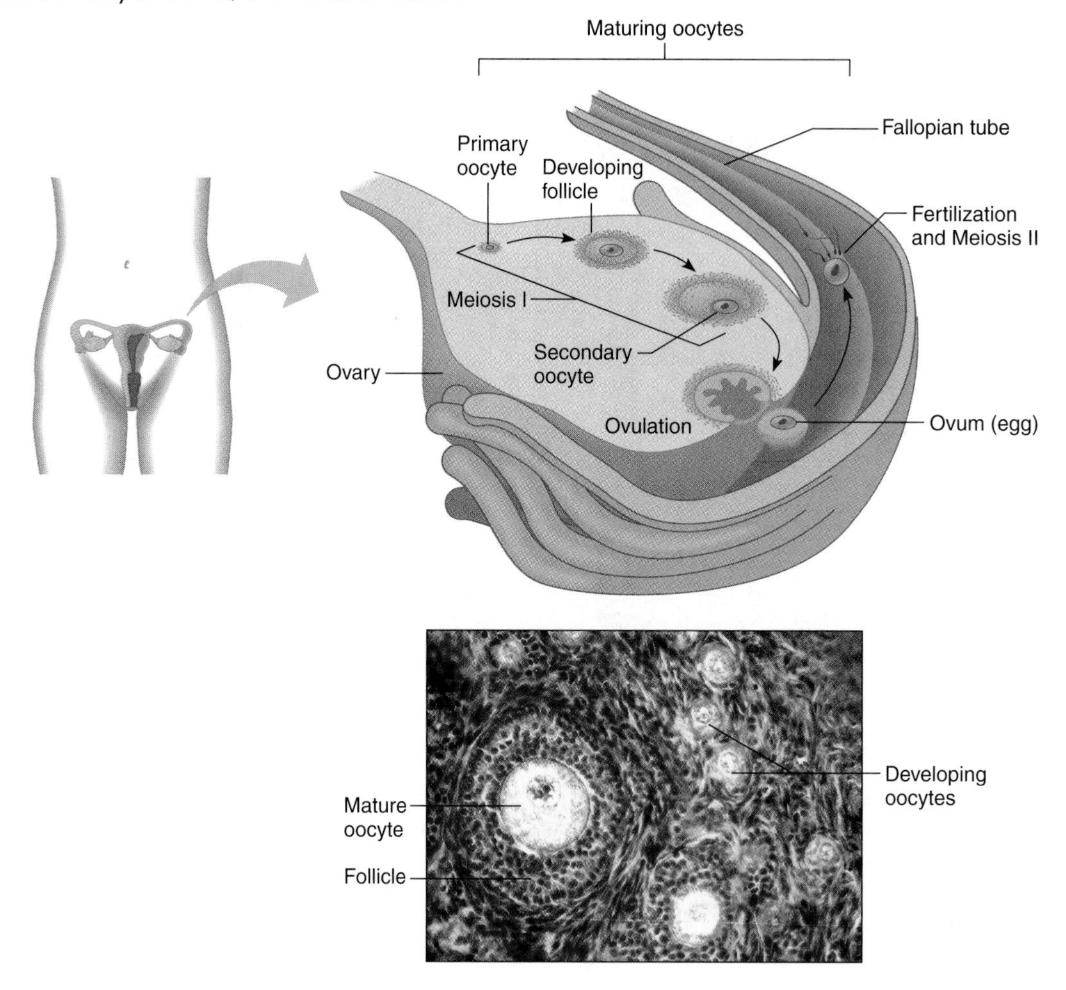

produce a small second polar body and the mature egg cell or ovum, which contains a large amount of cytoplasm. Therefore, a cell undergoing meiosis in a female typically yields four cells, only one of which can become an ovum. Oocytes develop in the paired **ovaries,** within protective structures called follicles (fig. 9.16). Fallopian tubes lead to the uterus, the sac where an embryo can develop.

The secondary oocyte, in receiving most of the cytoplasm of four cells, is packed with biochemicals and organelles that the zygote (fertilized ovum) will use until its genes begin to function. These biochemicals include proteins and RNA molecules that influence cell specialization in the early embryo. Some amphibian oocytes amass a trillion ribosomes during meiosis, providing a zygote with a large supply of tools for protein synthesis.

A woman's body absorbs polar bodies, which normally play no further role in development. However, rarely, sperm can fertilize polar bodies, and a mass of tissue that does not resemble an embryo grows, until the woman's body rejects it. A fertilized polar body, sometimes called a "blighted ovum," accounts for about 1 in 100 spontaneous abortions.

In the female's body, polar bodies provide a way to package maximal cytoplasm and organelles with one haploid nucleus, forming an ovum. Polar bodies are the basis of a new medical technology called polar body biopsy. A test to analyze specific genes in polar bodies can determine whether or not an oocyte contains a disease-causing gene. This test is used to identify "healthy" oocytes that have been removed from a woman's body and will be fertilized in laboratory glassware. If a woman knows that she is a carrier of a particular inherited condition, and a polar body has the disease-causing form of the gene, then it can be inferred that the oocyte does not have the gene, because the homologs that carry the genes separate as the polar body and oocyte form (fig. 9.17).

The timetable for oogenesis differs greatly from that of spermatogenesis. Three months after conception, the ovaries of a female fetus contain 2 million or more primary oocytes. From then on, the number of primary oocytes declines. A million are present by birth, and about 400,000 remain by the time of puberty. At birth, the oocytes arrest in prophase I. After puberty, one

FIGURE 9.17

Polar Body Biopsy.

The fact that an oocyte shares a woman's divided genetic material with a much smaller companion, the polar body, is the basis for a new technology to select oocytes that are free of certain disease-causing genes.

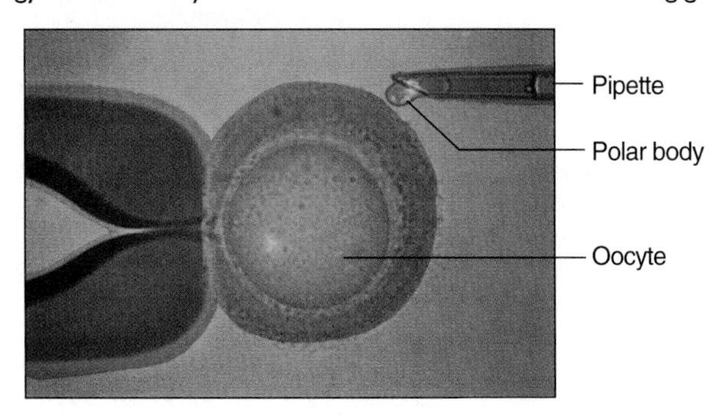

or a few oocytes complete meiosis I each month. These oocytes stop meiosis again, this time at metaphase II. Specific hormonal cues each month prompt an ovary to ovulate, or release a secondary oocyte. If a sperm penetrates the oocyte membrane, meiosis in the oocyte completes, and the two nuclei approach slowly and combine to form the fertilized ovum. Therefore, female meiosis does not even finish unless fertilization occurs. If the secondary oocyte is not fertilized, it degenerates and leaves the body in the menstrual flow.

The stage of meiosis that occurs when a sperm fertilizes an oocyte varies with species. In many worms, fertilization occurs just as the primary oocyte forms, whereas in foxes and dogs, it occurs in an older primary oocyte. In shellfish, octopuses, and squids, fertilization takes place during metaphase I, and in the sea urchin, it follows the completion of meiosis.

A human female will only ovulate about 400 oocytes between puberty and menopause (after menstrual periods cease), and only a very few of these will likely contact a sperm cell. Only one in three fertilized ova will continue to grow, divide, and specialize to eventually form a new human life.

Figure 9.18 compares spermatogenesis and oogenesis. Biotechnology 9.1 investigates routes to fertilization other than sexual intercourse. Chapter 40 explores human reproduction and development further.

9.4 MASTERING CONCEPTS

1. What are the stages of sperm development and maturation?
2. What are the parts of a mature sperm cell?
3. Where do sperm develop?
4. What are the stages of development and maturation of an oocyte?
5. How does the timetable for meiosis differ in males and females?

OLC

FIGURE 9.18

Spermatogenesis and Oogenesis in the Human Compared.

The steps of sperm and oocyte formation are similar, and the products of each are haploid. But the male and female gametes look very different.

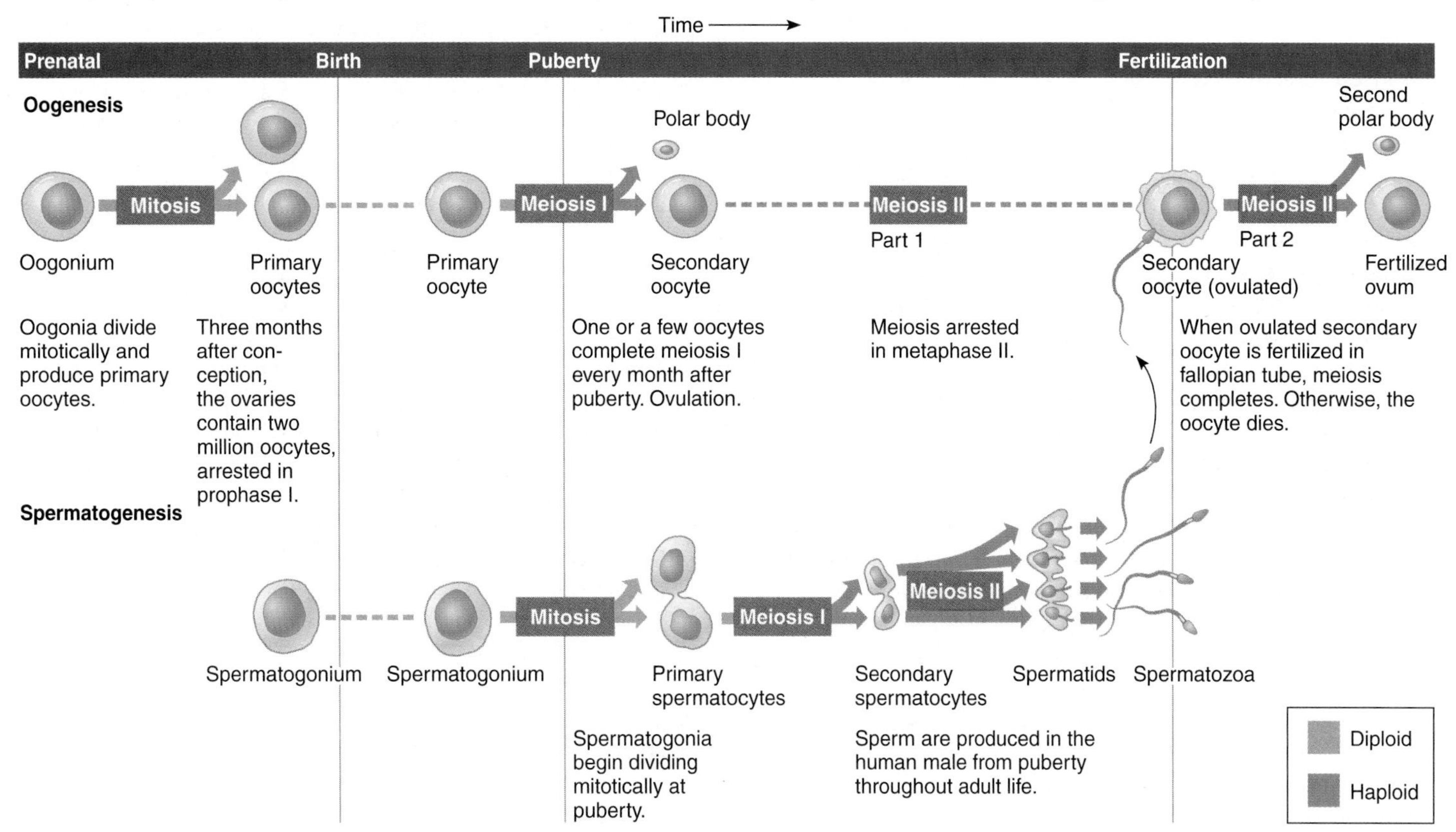

Biotechnology 9.1

Assisted Reproductive Technologies

Conception requires the meeting and merging of sperm and oocyte, which naturally occurs in the woman's reproductive tract. Abnormal gametes or blockages that impede this meeting of cells can result in infertility (inability to conceive). Assisted reproductive technologies can help couples conceive. The procedures usually involve a laboratory technique and sometimes participation of a third individual. These techniques are often costly and may take several attempts, and some have very low success rates. Most assisted reproductive technologies were developed in nonhuman animals. For example, the first artificial inseminations were performed in dogs in 1782, and the first successful in vitro fertilization was accomplished in 1959, in a rabbit. Here is a look at some procedures.

Donated Sperm—Artificial Insemination

In artificial insemination, a doctor places donated sperm in a woman's reproductive tract. A woman might seek artificial insemination if her partner is infertile or carries a gene for an inherited illness or if she desires to be a single parent. More than 250,000 babies have been born worldwide as a result of this procedure. The first human artificial inseminations by donor were done in the 1890s. For many years, physicians donated sperm, and this became a way for male medical students to earn a few extra dollars. By 1953, sperm could be frozen and stored for later use. Today, sperm banks freeze and store donated sperm and then provide it to physicians who perform artificial insemination.

A woman or couple choosing artificial insemination can select sperm from a catalog that lists the personal characteristics of the donors, including blood type; hair, skin, and eye color; build; and even educational level and interests. Of course, not all of these traits are inherited. Although artificial insemination has helped many people to become parents, it, and other assisted reproductive technologies, have led to occasional dilemmas (table 9.4).

In the future, an alternative to artificial insemination may be a transplant of sperm-producing tissue in a male. Researchers have restored fertility to male mice with transplants of spermatogonia from fertile mice.

A Donated Uterus—Surrogate Motherhood

A man's role in reproductive technologies is simpler than a woman's. A man can be a genetic parent, providing half of his genetic self in sperm, but a woman can be both a genetic parent (providing the oocyte) and a gestational parent (providing the uterus).

If a man produces healthy sperm but his partner's uterus is absent or cannot maintain a pregnancy, a surrogate mother may be able to help. She is artificially inseminated with the man's sperm and carries the pregnancy, but the couple raises the child. The surrogate is both the genetic and the gestational mother.

Another type of surrogate mother lends only her uterus; she receives a fertilized egg from a woman who has healthy ovaries but lacks a functional uterus. For example, Arlette Schweitzer's daughter, Christa, had been born without a uterus but had healthy ovaries. Christa's oocytes were fertilized in laboratory glassware with her husband's sperm, and the resulting embryos were implanted into Arlette's uterus. The result—twins Chad and Chelsea.

TABLE 9.4 Assisted Reproductive Dilemmas

1. A physician in California used his own sperm to artificially inseminate 15 patients and told them that he had used sperm from anonymous donors.
2. A plane crash killed the wealthy parents of two early embryos stored at −320°F (−195°C) in a hospital in Melbourne, Australia. Adult children of the couple were asked to share their estate with two eight-celled balls.
3. Several couples in Chicago planning to marry discovered that they were half-siblings. Their mothers had been artificially inseminated with sperm from the same donor.
4. Two Rhode Island couples sued a fertility clinic for misplacing embryos.
5. Several couples in California sued a fertility clinic for implanting their oocytes or embryos in other women without consent from the donors. One woman is requesting partial custody of the resulting children if her oocytes were taken and full custody if her embryos were used, even though the children are of school age and she has never met them.
6. A man sued his ex-wife for possession of their frozen embryos as part of the divorce settlement.

In Vitro Fertilization

In in vitro fertilization (IVF), which means "fertilization in glass," sperm meets oocyte outside the woman's body. The fertilized ovum divides two or three times and is then introduced into the oocyte donor's (or another woman's) uterus. If all goes well, a pregnancy begins.

A woman might undergo IVF if her ovaries and uterus work but her fallopian tubes, which conduct fertilized ova to the uterus, are blocked. To begin, she takes a hormone that hastens maturity of several oocytes. Using a laparoscope to view the ovaries and fallopian tubes, a physician removes a few of the largest oocytes and transfers them to a laboratory dish. He or she then adds chemicals similar to those in the female reproductive tract, and sperm.

If a sperm cannot penetrate the oocyte in vitro, it may be sucked up into a tiny syringe and injected using a tiny needle into the female cell (fig. 9.A). This variant of IVF, called intracytoplasmic sperm injection (ICSI), is very successful, resulting in a 68% fertilization rate. It can help men with very low sperm counts, high numbers of abnormal sperm, or injuries or illnesses that prevent them from ejaculating. Minor surgery removes testicular tissue, from which viable sperm are isolated and injected into oocytes. A day or so later, a physician transfers some of the resulting balls of 8 or 16 cells to the woman's uterus. The birthrate following IVF is about 17%, compared with 31% for natural conceptions.

Gamete Intrafallopian Transfer

One reason that IVF rarely works is the artificial fertilization environment. A procedure called GIFT, which stands for gamete intrafallopian transfer, circumvents this problem by moving fertilization to the woman's body.

In GIFT, a woman takes a superovulation drug for a week and then has several of her largest oocytes removed. A man donates a sperm sample, and a physician separates the most active cells. The collected oocytes and sperm are deposited together in the woman's fallopian tube, at a site past any obstruction so that implantation can occur. GIFT is 26% successful.

In zygote intrafallopian transfer (ZIFT), a physician places an in vitro fertilized ovum in a woman's fallopian tube. This is unlike IVF because the site of introduction is the fallopian tube and unlike GIFT because fertilization occurs in the laboratory. Allowing the fertilized ovum to make its own way to the uterus seems to increase the chance that it will implant. ZIFT is 23% successful.

Oocyte Banking and Donation

Can oocytes be frozen and stored, as sperm are? If so, a woman wishing to have a baby later in life, when fertility declines, could set aside oocytes while she is young. Stored oocytes would also benefit women undergoing a medical treatment that could damage oocytes, who work with gamete-damaging toxins, who are carriers for certain genetic diseases, or who have entered menopause prematurely.

Unfortunately, freezing oocytes isn't as easy as freezing sperm. Oocytes are frozen in liquid nitrogen at −86°F to −104°F (−30°C to −40°C) when they are at metaphase of meiosis II. At this time, the chromosomes are aligned along the spindle, which is sensitive to temperature extremes. If the spindle comes apart as the cell freezes, the oocyte may lose a chromosome, which can devastate development. Frozen oocytes may retain a polar body and form a diploid gamete.

Alternative approaches attempt to overcome the limitations of freezing oocytes. For example, researchers can culture oocyte-packed, 1-millimeter-square sections of ovaries in the laboratory, which requires a high level of oxygen and a complex combination of biochemicals.

Healthy young women may also donate healthy oocytes to other women. The donor must receive several daily injections of a superovulation drug, have her blood checked each morning for 3 weeks, and then undergo laparoscopic surgery to collect the oocytes. The prospective father's sperm and the donor's oocytes are then placed in the recipient's uterus.

FIGURE 9.A
Fertilizing an Egg.
Intracytoplasmic sperm injection (ICSI) enables some infertile men and men with spinal cord injuries and other illnesses to become fathers. A single sperm cell is injected into the cytoplasm of an oocyte.

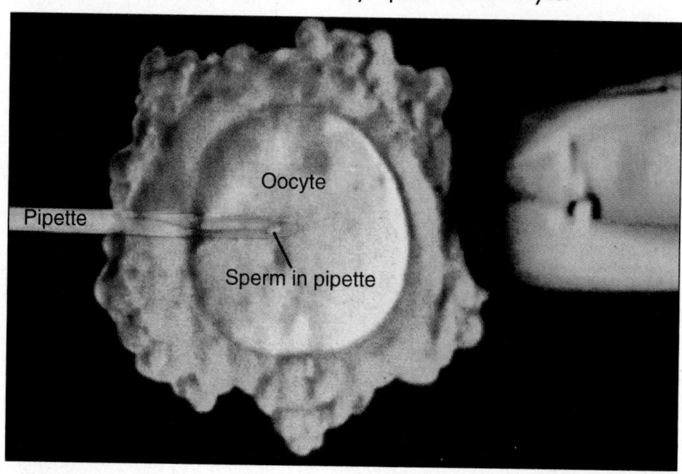

Chapter Summary

9.1 Why Reproduce?

1. Reproduction is essential to species' survival.
2. **Asexual reproduction,** such as **binary fission,** can be successful in an unchanging environment.
3. **Sexual reproduction** mixes traits and therefore provides species' protection in a changing environment. Sexual reproduction also increases the number of organisms. It occurs when **haploid gametes** fuse, restoring the **diploid** state.

9.2 Variations on the Sexual Reproduction Theme

4. **Meiosis** halves the genetic material, and **fertilization** is the joining of haploid nuclei. Meiosis and fertilization are two key events in sexual life cycles.
5. Somatic growth occurs between meiosis and fertilization, or between fertilization and meiosis.
6. Sexual reproduction in plants involves an **alternation of generations** and generations with multicellular haploid and diploid phases.
7. **Conjugation,** a form of gene transfer in some microorganisms, is sexual because one individual transfers genetic material to another, but it is not reproduction because no additional individual forms. *Chlamydomonas* undergoes cell fusion, which may have been a forerunner of sexual reproduction.
8. In animals and some other organisms, cells undergoing meiosis are **germ cells.** After meiosis, haploid gametes specialize in a process called maturation.

9.3 How Does Meiosis Halve the Chromosome Number?

9. Meiosis halves the number of chromosomes in somatic cells, producing haploid gametes. A species' chromosome number stays constant because in gamete-producing cells, the DNA replicates once, but the cells divide twice.
10. In meiosis I, **reduction division,** the chromosome number is halved. In meiosis II, **equational division,** the two products of meiosis I undergo essentially a mitotic division. Each division proceeds through stages of prophase, metaphase, anaphase, and telophase.
11. Meiosis provides genetic variability by partitioning different combinations of genes into gametes through **independent assortment. Crossing over,** which occurs in prophase I as **homologous pairs** synapse, increases the variability.

9.4 The Sculpting of Gametes

12. **Spermatogenesis** begins with **spermatogonia,** which accumulate cytoplasm and replicate their DNA to become **primary spermatocytes.** After meiosis I, the cells are haploid **secondary spermatocytes.** In meiosis II the secondary spermatocytes divide to each yield two **spermatids,** which differentiate along the male reproductive tract to yield sperm.
13. Sperm develop within the **testes** in the walls of the **seminiferous tubules** and mature in the **epididymis.** Sperm join secretions and leave the body through the **vasa deferentia.**
14. In **oogenesis, oogonia** replicate their DNA, becoming **primary oocytes.** In meiosis I, the primary oocyte divides, apportioning cytoplasm to one large **secondary oocyte** and a much smaller **polar body.** In meiosis II, the secondary oocyte divides, yielding the large ovum and another small polar body. Oogenesis occurs in the **ovaries.**
15. The development of a sperm cell takes 74 days. Meiosis in the female begins before birth and completes at fertilization.

Testing Your Knowledge

1. What is the evidence that sexual reproduction has been successful on an evolutionary scale?
2. Distinguish among asexual reproduction, parthenogenesis, and sexual reproduction.
3. Describe how a plant life cycle may include a multicellular haploid and a diploid phase.
4. What are the two main events of a sexual life cycle?
5. Why is conjugation sexual but not reproductive?
6. Define the following terms:
   a. crossing over
   b. gamete
   c. haploid
   d. homolog
   e. synapsis
   f. diploid
7. How many sets of chromosomes are present in each of the following cell types?
   a. an oogonium
   b. a primary spermatocyte
   c. a spermatid
   d. a secondary oocyte
   e. a polar body derived from a primary oocyte
8. How are the timetables different for oogenesis and spermatogenesis in humans?

Thinking Scientifically

1. A dog has 39 pairs of chromosomes. Considering only independent assortment (the random lining up of maternally and paternally derived chromosomes), how many genetically different puppies are possible from the mating of two dogs? Is this number an underestimate or an overestimate? Why?
2. Why is it extremely unlikely that a human child will be genetically identical to a parent?
3. Many male veterans of the Vietnam War claim that their children born years later have birth defects caused by a contaminant in the herbicide Agent Orange used as a defoliant in the conflict. What types of cells would the chemical have to have affected in these men to cause birth defects years later? Explain your answer.
4. Why is a fertilized polar body unable to support the development of an embryo, while an oocyte, which is genetically identical to it, can?
5. How do the structures of the male and female human gametes aid them in performing their functions?

References and Resources

Bell, Graham. February 1992. Dividing they stand. *Natural History.* Paramecia reproduce without sex and have sex without reproduction.

Gould, Stephen Jay. 1980. Dr. Down's syndrome. In *The Panda's Thumb.* New York: W. W. Norton. An entertaining essay on the importance of meiosis.

Moran, Nancy A. April 1992. Quantum leapers. *Natural History.* Sumac gall aphids shift from a parthenogenetic to a sexual lifestyle.

Roach, Mary. January 1997. Sperm futures. *Discover.* Sperm in the news.

The *LIFE* Online Learning Center provides additional resources and tools for studying this chapter.
www.mhhe.com/life

CHAPTER 10

How Inherited Traits Are Transmitted

From Mendel to Medical Genetics

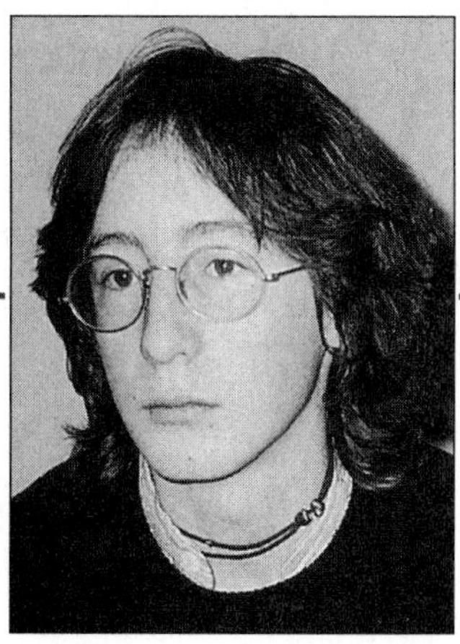

Heredity is apparent in appearances and talents, as illustrated by John (left) and Julian (right) Lennon.

Inherited similarities can be startling, such as the physical likeness between John Lennon and son Julian and in their distinctive singing voices. Interest in heredity is probably as old as humankind itself, as people throughout time have wondered at their similarities—and fought over their differences.

Awareness of heredity appears as early as an ancient Jewish law that excuses a boy from circumcision if relatives bled to death from the ritual. Nineteenth-century biologists thought that body parts controlled trait transmission and gave the units of inheritance, today called genes, such colorful names as pangens, idioblasts, bioblasts, gemmules, or just characters. An investigator who used the term *elementen* made the most lasting impression on what would become the science of genetics. His name was Gregor Mendel.

Mendel's laws are truly scientific laws—they apply as much to the traits he studied in peas as they do to human families with inherited disease.

Mendel spent his early childhood in a small village in what is now the Czech Republic, near the Polish border. His father was a farmer, and his mother was the daughter of a gardener, so Mendel learned early how to tend fruit trees. He did so well in school that at age 10 he left home to attend a special school for bright students, supporting himself by tutoring. After a few years at a preparatory school, Mendel became a priest at the Augustinian monastery of St. Thomas in Brno. But this was not a typical monastery—the priests were also teachers, and they did research in natural science. It was perfect for Mendel.

The young man eagerly learned how to artificially pollinate crop plants to control their breeding. He desperately wanted to teach natural history, but had difficulty passing the necessary exams, a victim of test anxiety. At age 29, he did such a good job as a substitute teacher

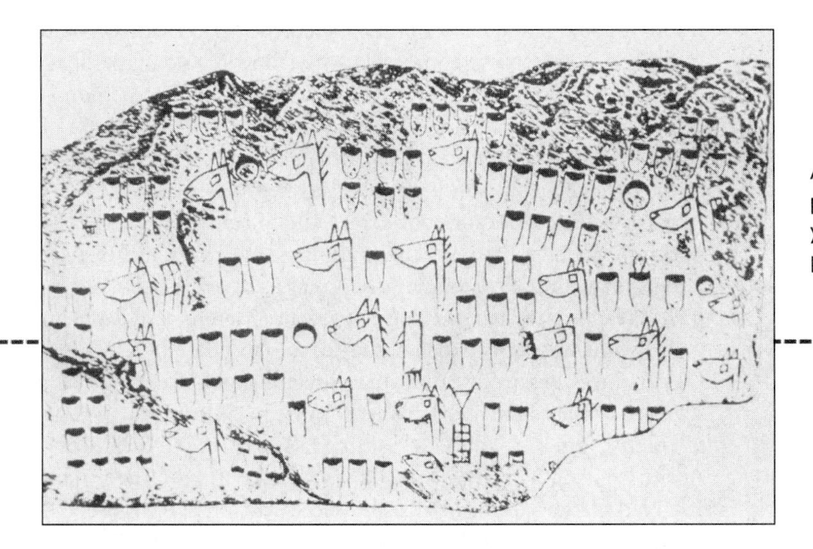

Ancient cultures recognized inheritance patterns. A horse breeder in Asia 4,000 years ago etched a record of his animals' physical characteristics in stone.

that he was sent to earn a college degree. At the University of Vienna, courses in the sciences and statistics fueled his long-held interest in plant breeding. Then Mendel began to think about experiments to address a compelling question that had confounded other plant breeders—why did certain traits disappear, only to reappear a generation later? Mendel confronted this puzzle by breeding hybrids and applying the statistics he had learned in college.

From 1857 to 1863, Mendel crossed and cataloged some 24,034 plants, through several generations. He deduced that consistent ratios of traits in the offspring indicated that the plants transmitted distinct units, or "elementen." He derived two laws that explain how inherited traits are transmitted, the subject of the first half of this chapter. Mendel described his work to the Brno Medical Society in 1865 and published it in the organization's journal the next year. The remarkably easy to read paper discusses plant hybridization, the reappearance of traits in the third generation, and the joys of working with peas, plus Mendel's data.

Mendel's laws are truly scientific laws—they apply as much to the traits he studied in peas as they do to human families with inherited disease. Mendel's inferences can best be appreciated by realizing what he did not know. He deduced how genes are passed on chromosomes without knowing what these structures are!

Finally, his treatise was published in English in 1901. Then three botanists (Hugo DeVries, Karl Franz Joseph Erich Correns, and Seysenegg Tschermak) rediscovered the laws of inheritance independently and eventually found Mendel's paper. All credited Mendel, who came to be regarded as the "father of genetics." As is the way of science, many others repeated and confirmed Mendel's work. However, researchers in this new field of genetics also identified situations that seem to disrupt Mendelian ratios, the subject of the second half of this chapter.

In the twentieth century, researchers discovered the molecular basis of Mendel's traits. The "short" and "tall" plants reflect expression of a gene that enables a plant to produce the hormone gibberellin, which elongates the stem. One tiny change to the DNA, and a short plant results. Likewise, "round" and "wrinkled" peas arise from the *R* gene, whose encoded protein connects sugars into branching polysaccharides. Seeds with a mutant gene cannot attach the sugars, water exits the cells, and the peas wrinkle.

Today genetics and DNA are familiar, and the entire set of genetic instructions to build a person—the human genome—has been deciphered. The average person probably knows more about DNA than about the rules of heredity discovered by an Augustinian monk more than a century ago. But the study of heredity on an organismal level is still intriguing. New parents don't usually ask about their newborn's DNA sequences, but note such obvious traits as sex, hair color, and facial features. When we select mates we don't usually consider DNA sequences, but more tangible manifestations of our genes. And so our look at genetics begins the traditional way, with Gregor Mendel, but we can appreciate his genius in light of what we know about DNA.

10.1 Tracing the Inheritance of One Gene

Mendel's law of segregation states that gene variants (alleles) separate during meiosis as chromosomes are packaged into gametes. Patterns of single gene transmission and expression depend upon dominance relationships and whether a gene is on an autosome or a sex chromosome.

The beauty of studying heredity is that different types of organisms can be used to demonstrate underlying principles common to all species. However, some organisms are easier to work with than others, as we'll see in subsequent chapters of this unit. Peas (*Pisum sativum*) proved a good choice for probing trait transmission for several reasons—peas are easy to grow, develop quickly, and have many traits that take two easily distinguishable forms. It was this last quality of "differentiating characters" that particularly interested Gregor Mendel, the Augustinian monk whose experiments from 1857 through 1863 revealed the two basic laws of heredity. Figure 10.1 illustrates the seven traits that Mendel followed.

Experiments with Peas

Mendel's first experiments dealt with single traits that have two expressions, such as "short" and "tall." He set up all combinations of possible artificial pollinations, manipulating fertilizations as figure 10.2 shows: tall with tall; short with short; and tall with short, which produces hybrids. Mendel noted that short plants crossed to other short plants were "true-breeding," always producing short plants.

The crosses of tall plants to each other were more confusing. Tall plants were only sometimes true-breeding. Some tall plants, when crossed with short plants, produced only tall plants, but certain other tall plants crossed with each other yielded surprising results—about one-quarter of the plants in the next generation were short. It appeared as if in some tall plants, tallness could mask shortness (fig. 10.3). The trait that masks the other is said to be **dominant;** the masked trait is **recessive.** Wrote Mendel, "In the case of each of the seven crosses the hybrid character resembles that of one of the parental forms so closely that the other either escapes observation completely or cannot be detected with certainty. . . . The expression 'recessive' has been chosen because the characters thereby designated withdraw or entirely disappear in the hybrids, but nevertheless reappear unchanged in their progeny." Mendel conducted up to 70 hybrid crosses with each of the seven traits he studied.

To investigate these non-true-breeding tall plants further, Mendel set aside the short plants that had arisen from crossing tall plants, and crossed the remaining tall offspring to each other. Again, he saw the ratio of one-quarter short to three-quarters tall, and yet further experiments showed two-thirds of those tall plants to be non-true-breeding. He wrote ". . . of those forms which possess the dominant character in the first generation, two-thirds have the hybrid character, while one-third remains consistent with the dominant character."

FIGURE 10.1
Traits Mendel Studied.
Gregor Mendel studied the transmission of seven traits in the pea plant. Each trait has two easily distinguished expressions, or phenotypes.

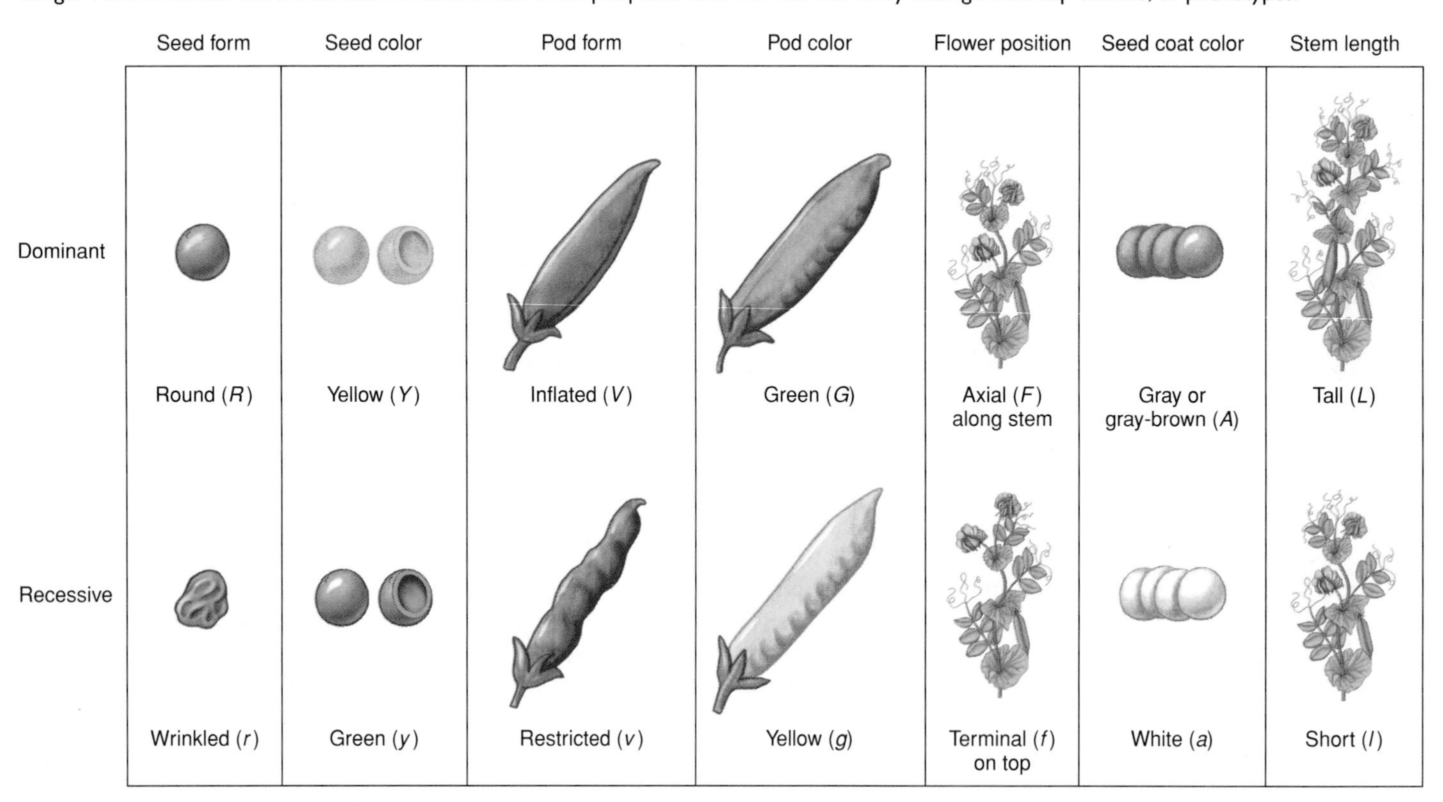

FIGURE 10.2

Mendel's Experimental Approach for Breeding Peas.

Gregor Mendel set up specific crosses of pea plants, so he could observe the appeareance of traits in the next generation.

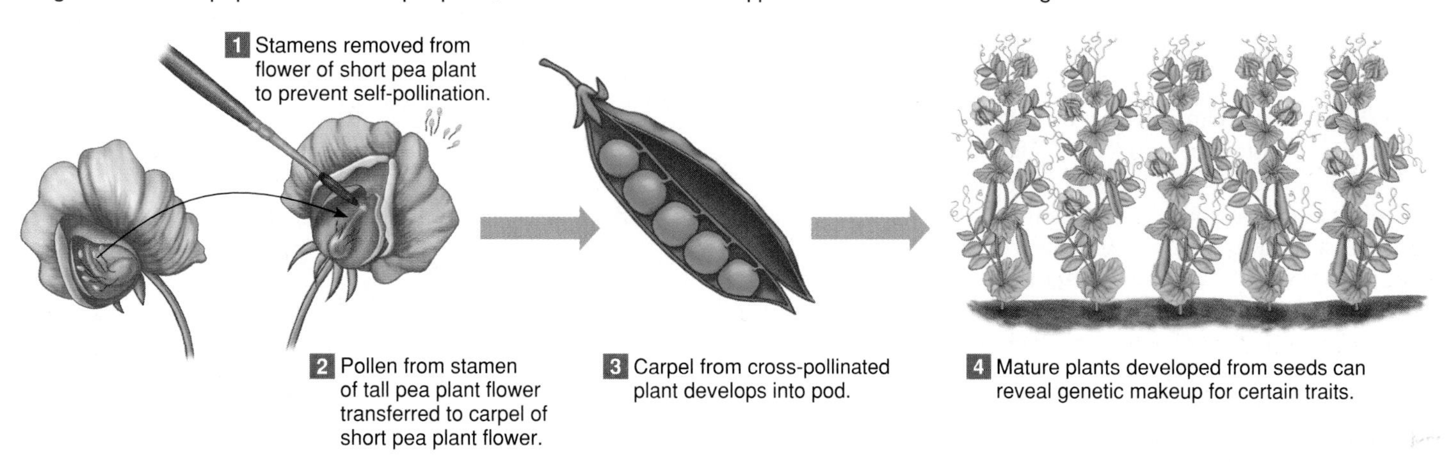

In these initial experiments with one trait, Mendel confirmed that hybrids hide one expression of a trait, which reappears when hybrids are crossed. But Mendel went farther, trying to explain how this happened. He suggested that gametes distribute "elementen,"—what we now call genes—because these cells physically link generations. Sets of elementen would separate from each other as gametes form. When gametes join at fertilization, the elementen would group into new combinations. Mendel reasoned that each element was packaged in a separate gamete, and if opposite-sex gametes combine at random, then he could mathematically explain the different ratios of traits produced from his pea plant crossings. Mendel's idea that elementen separate in the gametes would later be called the **law of segregation.** Mendel eventually turned his energies to monastery administration when other scientists did not recognize the importance of his work.

FIGURE 10.3

Mendel Crossed Short and Tall Pea Plants.

(**A**) When Mendel crossed short pea plants with short pea plants, all of the progeny were short. (**B**) Some tall plants crossed to tall plants yielded only tall plants. (**C**) Certain tall plants crossed with short plants produced all tall plants. (**D**) Other tall plants crossed with short plants produced some tall plants and some short plants.

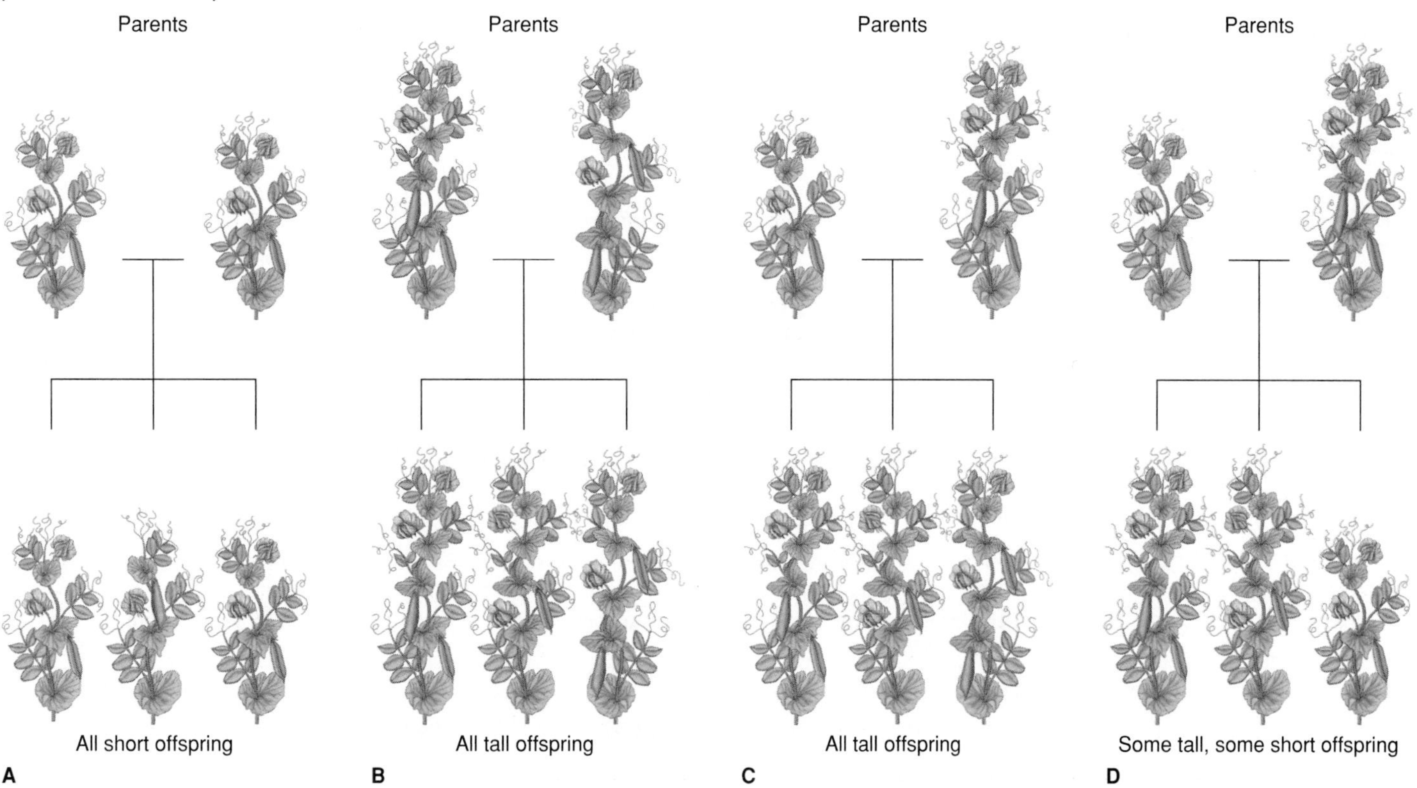

When Mendel's ratios were demonstrated again and again in several species in the early 1900s, at the same time that chromosomes were being described for the first time, it quickly became apparent that elementen and chromosomes had much in common. Both paired elementen and paired chromosomes separate at each generation and pass—one from each parent—to offspring. Both elementen and chromosomes are inherited in random combinations. Chromosomes provided a physical mechanism for Mendel's hypotheses. In 1909, Mendel's elementen were renamed genes (Greek for "give birth to"), but for the next several years, the units of inheritance remained a mystery. By the 1940s, scientists began investigating the gene's chemical basis. We pick up the historical trail at this point in chapter 12.

Studying genetics brings its own terminology. Table 10.1 summarizes the important terms, grouped by similarities.

How Do Biologists Describe Single Genes?

Mendel observed two different expressions of a trait—for example, short and tall. Traits are expressed in different ways because a gene can exist in alternate forms, or **alleles** (fig. 10.4).

FIGURE 10.4
Homologous Chromosomes.
A diploid cell in an organism has two alleles for every gene located on a homologous pair of chromosomes. One allele came from each of the organism's parents. This organism is homozygous for the dominant allele (*AA*) for trait A and heterozygous (*Bb*) for trait B, and homozygous for the recessive allele (*dd*) for trait D.

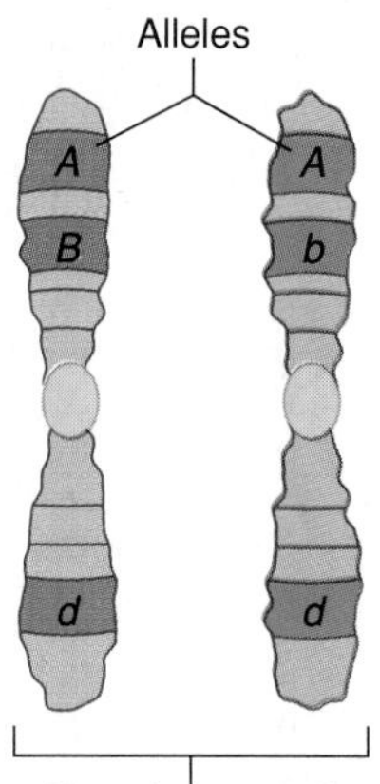

TABLE 10.1 A Glossary of Genetic Terms

TERM	DEFINITION
Gene	A sequence of DNA that encodes a protein
Chromosome	A dark-staining body in a cell's nucleus that consists of a continuous double helix of DNA plus associated proteins
Allele	An alternate form of a gene
Dihybrid	An individual who is heterozygous for two particular genes
Monohybrid	An individual heterozygous for a particular gene
Genotype	The allele combination in an individual
Phenotype	The observable expression of an allele combination
F_1	The first filial generation; offspring
F_2	The second filial generation; offspring of offspring
P_1	The parental generation
Dominant	An allele that masks the expression of another allele
Recessive	An allele whose expression is masked by another allele
Heterozygous	Possessing different alleles of a gene
Homozygous	Possessing identical alleles of a gene
Segregation	Mendel's first law; alleles of a gene separate into equal numbers of gametes
Independent assortment	Mendel's second law; a gene on one chromosome does not influence the inheritance of a gene on a different (nonhomologous) chromosome because meiosis packages chromosomes randomly into gametes
Autosomal	A gene located on an autosome (a chromosome that does not determine sex) or a trait resulting from such a gene's expression.
X-linked	A gene located on the X chromosome or a trait that results from the activity of a gene on the X chromosome
Wild type	The most common phenotype or allele for a gene in a population
Mutant	A phenotype or allele resulting from a change (mutation) in a gene
Mutation	A change in a gene

Because a gene is a long sequence of DNA, it can vary in many ways. It therefore isn't surprising that a gene can have different expressions. We return to this idea in chapter 13. An individual with two identical alleles for a gene is **homozygous** for that gene. An individual with two different alleles is **heterozygous.** Mendel's "non-true-breeding" plants were heterozygous, also called "hybrid."

When a gene has two alleles, it is common to symbolize the dominant allele with a capital letter and the recessive with the corresponding small letter. If both alleles are recessive, the individual is homozygous recessive. Two small letters, such as *tt* for short plants, symbolize this. An individual with two dominant alleles is homozygous dominant. Two capital letters, such as *TT* for tall pea plants, represent homozygous dominance. Another possible allele combination is one dominant and one recessive allele—*Tt* for non-true-breeding tall pea plants, or heterozygotes.

An organism's appearance does not always reveal its alleles. Both a *TT* and a *Tt* pea plant are tall, but the first is a homozygote and the second a heterozygote. The **genotype** describes the organism's alleles, and the **phenotype** describes the outward expression of an allele combination. A pea plant with a tall phenotype may have genotype *TT* or *Tt*. A **wild type** phenotype is the most common expression of a particular gene in a population. A **mutant** phenotype is a variant of a gene's expression that arises when the gene undergoes a change, or **mutation.**

When analyzing genetic crosses, the first generation is the parental generation, or P_1; the second generation is the first filial generation, or F_1; the next generation is the second filial generation, or F_2, and so on. If you considered your grandparents the P_1 generation, your parents would be the F_1 generation, and you and your siblings are the F_2 generation.

How Does Mendel's First Law Reflect Meiosis?

Mendel's observations on the inheritance of single genes reflect the events of meiosis. When a gamete is produced, the two copies of a particular gene separate, as the homologs that carry them do. In a plant of genotype *Tt*, for example, gametes carrying either *T* or *t* form in equal numbers during anaphase I. When gametes meet to start the next generation, they combine at random. That is, a *t*-bearing oocyte is neither more nor less attractive to a *t*-bearing sperm than to a *T*-bearing sperm. These two factors—equal allele distribution into gametes and random combinations of gametes—underlie Mendel's law of segregation (fig. 10.5).

Mendel's crosses of short and tall plants make more sense in terms of meiosis. He crossed short plants (*tt*) with true-breeding tall plants (*TT*). The resulting seeds grew into F_1 plants of the same phenotype: tall (genotype *Tt*). Next, he crossed the F_1 plants with each other in a monohybrid cross, where the inheritance of one trait is tracked by crossing two hybrids. The three possible genotypic outcomes of such a cross are *TT, tt,* and *Tt*. A *TT* individual results when a *T* sperm fertilizes a *T* oocyte; a *tt* plant results when a *t* oocyte meets a *t* sperm; and a *Tt* individual results when either a *t* sperm fertilizes a *T* oocyte, or a *T* sperm fertilizes a *t* oocyte.

FIGURE 10.5

Mendel's First Law—Gene Segregation.

During meiosis, homologous pairs of chromosomes (and the genes that compose them) separate from one another and are packaged into separate gametes. At fertilization, gametes combine at random to form the individuals of a new generation. Green and blue denote different parental origins of the chromosomes.

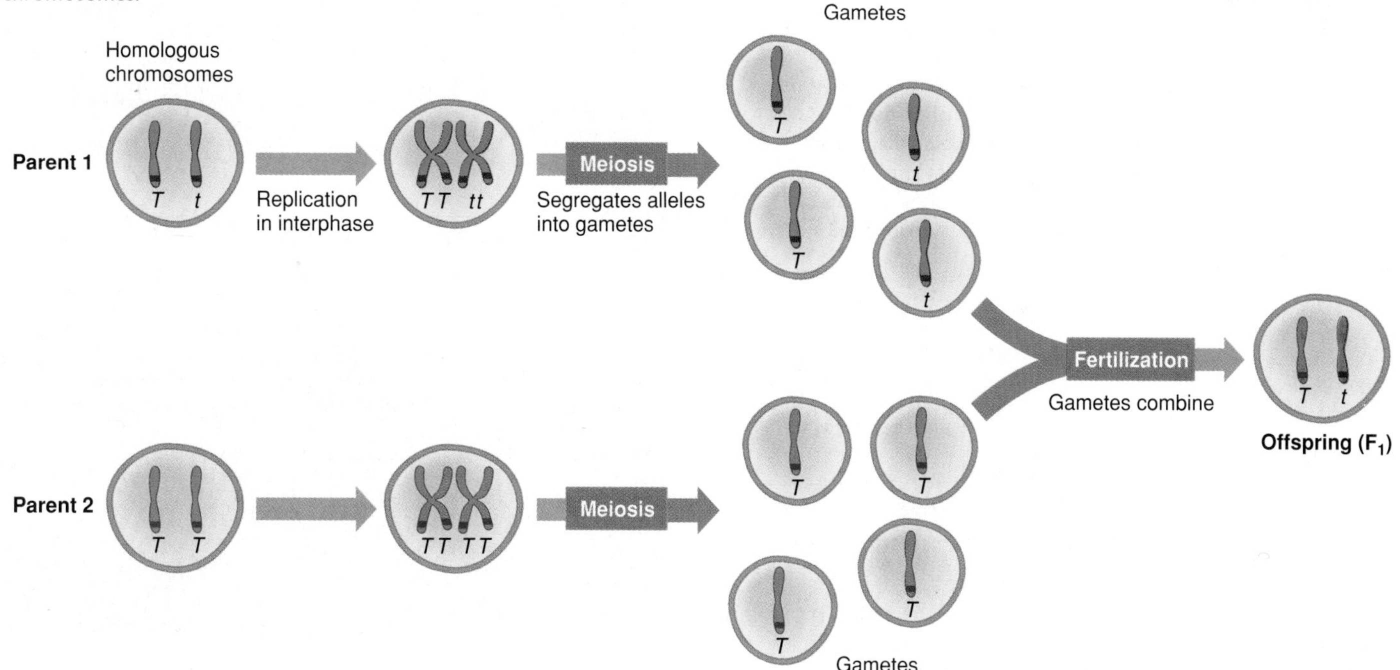

TABLE 10.2 Mendel's Law of Segregation: Crossing Heterozygotes Produces a 3:1 Dominant to Recessive Phenotypic Ratio

EXPERIMENT	TOTAL	DOMINANT	RECESSIVE	RATIO
1. Seed form	7,324	5,474	1,850	2.96:1
2. Seed color	8,023	6,022	2,001	3.01:1
3. Seed coat color	929	705	224	3.15:1
4. Pod form	1,181	882	299	2.95:1
5. Pod color	580	428	152	2.82:1
6. Flower position	858	651	207	3.14:1
7. Stem length	1,064	787	277	2.84:1
				Average = 2.98:1

Because two of the four possible gamete combinations produce a heterozygote, and each of the others produces a homozygote, the genotypic ratio expected of a monohybrid cross is 1 *TT*: 2 *Tt*: 1 *tt*. The corresponding phenotypic ratio is three tall plants to one short plant, a 3:1 ratio. Mendel saw these results for all seven traits that he studied, although as Table 10.2 shows, the ratios were not exact. Today we use a diagram called a **Punnett square** to derive these ratios (fig. 10.6). Experiments yield numbers of offspring that approximate these ratios.

Mendel distinguished the two genotypes resulting in tall progeny—*TT* from *Tt*—with additional crosses. He bred tall plants of unknown genotype with short (*tt*) plants. If a tall plant crossed with a *tt* plant produced both tall and short progeny, Mendel knew it was genotype *Tt;* if it produced only tall plants, he knew it must be *TT.* Crossing an individual of unknown genotype with a homozygous recessive individual is called a test cross. The homozygous recessive is the only genotype that can be identified by the phenotype—that is, a short plant is only *tt.* The homozygous recessive is therefore a "known" that can reveal the unknown genotype of another individual when the two are crossed. Figures 10.7 and 10.8 show the results of monohybrid crosses in two other familiar species.

FIGURE 10.6
A Punnett Square.
A diagram of gametes and how they can combine in a cross between two particular individuals is helpful in following the transmission of traits. The different types of female gametes are listed along the top of the square; male gametes are listed on the left-hand side. Each compartment within the square contains the genotype that results when gametes that correspond to that compartment join. The Punnett square here describes Mendel's monohybrid cross of two tall pea plants. Among the progeny, tall plants outnumber short plants 3:1. Can you determine the genotypic ratio?

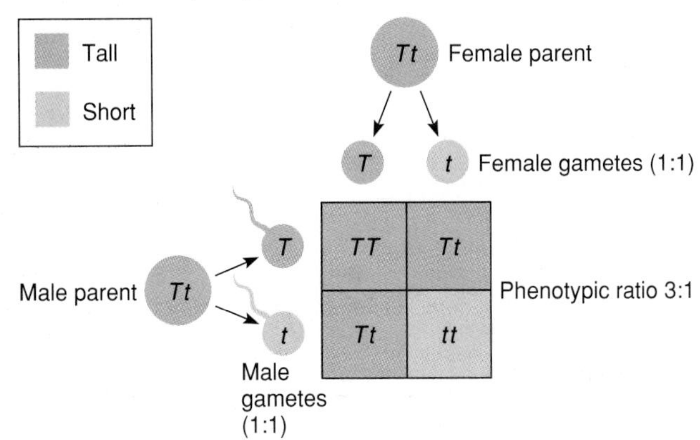

FIGURE 10.7
Monohybrid Crosses.
(**A**) Each kernel on an ear of corn represents the progeny of a single cross. When a plant with a dominant allele for purple kernels and a recessive allele for yellow kernels is self-crossed, the resulting ear has approximately three purple kernels for every yellow kernel.
(**B**) Albinism can result from a monohybrid cross in a variety of organisms. A heterozygote, or "carrier," for albinism has one allele that directs the synthesis of an enzyme needed to manufacture the skin pigment melanin and one allele that fails to make the enzyme. Each child of two carriers has a one in four chance of inheriting the two deficient alleles and being unable to manufacture melanin.

A

B

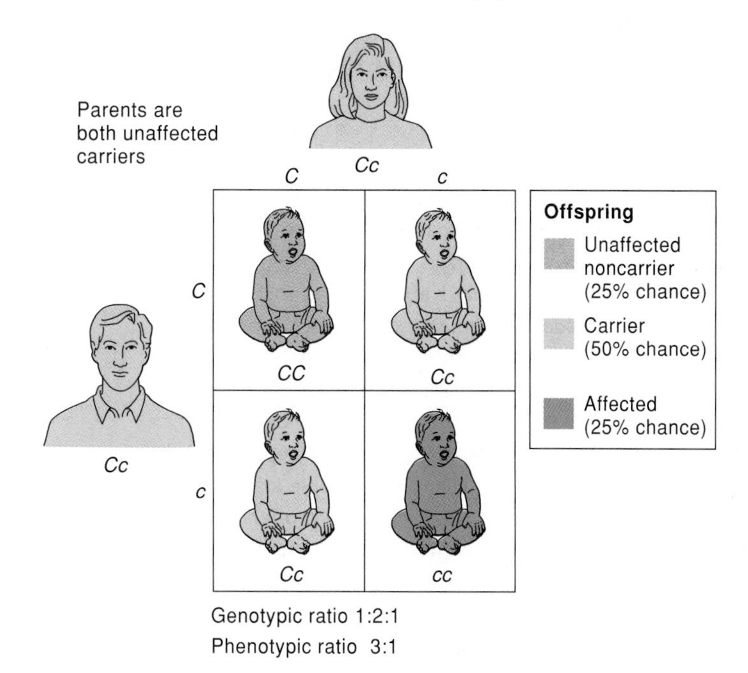

FIGURE 10.8
Carriers and Genotypic Ratios.
A 1:2:1 genotypic ratio results from a monohybrid cross, whether in peas or people. When parents are carriers for the same autosomal recessive trait or disorder, such as cystic fibrosis (CF), each child faces a 25% risk of inheriting the condition (a sperm carrying a CF allele fertilizing an oocyte carrying a CF allele); a 50% chance of being a carrier like the parents (a CF-carrying sperm with a wild type oocyte, and vice versa); and a 25% chance of inheriting two wild type alleles. In CF, abnormal ion channels in cells lining the respiratory tract and pancreas cause accumulation of very sticky mucus. Symptoms include great difficulty in breathing and digesting, and frequent lung infections.

Mendel's laws apply to all diploid species, and characteristics determined by single genes are termed Mendelian traits. Traits carried on the X or Y chromosome are generally called sex-linked, or more specifically X-linked or Y-linked. The next chapter will discuss these traits. All other genes, and the traits they specify, are termed **autosomal.**

Mendelian disorders in humans are rare, each affecting only one person in many thousands. About 2,500 Mendelian disorders are known, and some 2,500 other conditions are suspected to be Mendelian, based on their recurrence patterns in large families. Table 10.3 lists some other interesting Mendelian traits in humans.

We describe Mendelian traits by their transmission pattern in families, called the mode of inheritance. An **autosomal recessive** trait affects both sexes, because its gene is carried on an autosome. Such a trait can skip generations, because heterozygotes (carriers) show no symptoms. Individuals whose children have an autosomal recessive condition must both be carriers. In contrast, an **autosomal dominant** trait can affect both sexes because it is autosomal, but an affected individual must have an affected parent, unless the trait arose by mutation. If a generation arises in which no individuals inherit a disease-causing autosomal dominant gene, transmission of that autosomal trait stops.

Pedigrees Show Modes of Inheritance

Charts called **pedigrees** depict family relationships and phenotypes and can reveal the mode of inheritance of a particular trait. A pedigree consists of lines connecting shapes (fig. 10.9).

TABLE 10.3 Unusual Mendelian Traits
A compendium called *Mendelian Inheritance in Man* lists a wide variety of traits thought to be inherited, many by single genes. Here are a few odder ones.
Uncombable hair
Misshapen toes or teeth
Pigmented tongue tip
Inability to smell freesia flowers, musk, or skunk
Lack of teeth, brows, nasal bones, thumbnails
White forelock
Tone deafness
Double toenails
Magenta urine after eating beets
Hairy nosetips, knuckles, palms, soles
Extra, missing, protruding, fused, shovel-shaped, or snowcapped teeth
Grooved tongue
Duckbill-shaped lips
Flared ears
Egg-shaped pupils
Three rows of eyelashes
Spotted nails
Whorl in the eyebrow
Hairs that are triangular in cross section or have multiple hues
Sneezing fits in bright sunlight

Health 10.1

Autosomal Recessive and Autosomal Dominant Inheritance

Xeroderma Pigmentosum

When she was three years old, Katie Mahar said "the sun is a monster." For her, because of an autosomal recessive disorder, sunlight is extremely dangerous.

Caren and Dan Mahar had no reason to suspect that their youngest child would have a serious inherited medical condition; their other three children were healthy. But one day when Katie was about a month old, her parents put her under a tree in the backyard to enjoy the warm spring weather. Almost immediately, Katie broke out everywhere in spots. As the baby shrieked in pain, the spots turned to blisters, and later to scabs.

At first the Mahars thought the skin reaction was an isolated incident, perhaps an allergic reaction to something in their yard. But it happened every time Katie encountered sunlight. Even a shaft of light entering through a window and falling on the little girl could cause painful blisters to form instantaneously.

Eventually, Katie was diagnosed with *xeroderma pigmentosum,* a disruption of her DNA's ability to repair damage caused by ultraviolet radiation in sunlight. Because this ability to repair normally helps prevent skin cancer, Katie is highly susceptible. She is one of less than 1000 people in the world with the condition. Xeroderma pigmentosum is inherited as an autosomal recessive trait—that is, each of Katie's parents is a carrier. The parents were unaware of any other affected relatives because never before had two mutant alleles of the gene been passed to the same individual.

Today, Katie lives a shaded, indoor existence, which her parents hope will enable her to live long and without pain. She is liberally smeared with sunblock up to eight times a day to protect against accidental sun exposures. To go to a doctor, she bundles up at night and travels in a car with blocked windows. These measures may prevent her from developing skin cancer, which affects more than half of all children with xeroderma pigmentosum by the time they enter their teens. As figure 10.A shows, she is a beautiful child.

FIGURE 10.A
Although Katie has the recessive trait leading to xeroderma pigmentosum, her skin is free of lesions thanks to changes in her lifestyle that avoid the sun.

Autosomal recessive
inheritance
I
II

Vertical lines represent generations; horizontal lines connect parents; siblings are connected to their parents by vertical lines and joined by a horizontal line. Squares indicate males, circles females, and diamonds individuals of unknown sex. Colored shapes indicate individuals who express a particular trait, and half-filled shapes are known carriers.

The different modes of inheritance have characteristic pedigree patterns as discussed in Health 10.1. An autosomal recessive condition can skip generations; an autosomal dominant condition cannot skip generations. Both modes of inheritance affect both sexes. Figure 10.9 depicts transmission of an autosomal recessive trait, albinism, to two of three children of two carriers. Figure 10.10 depicts transmission of an autosomal dominant condition. A pedigree can be inconclusive—that is, either autosomal dominant or recessive—if the phenotype only mildly affects health, so that autosomal recessive individuals can have children, and therefore generations aren't skipped.

Pedigrees can be difficult to construct and interpret for several reasons. People sometimes hesitate to supply information because symptoms embarrass them. Adoption, children born out of

FIGURE 10.9
A Pedigree for an Autosomal Recessive Trait.
Autosomal recessive traits can skip generations.

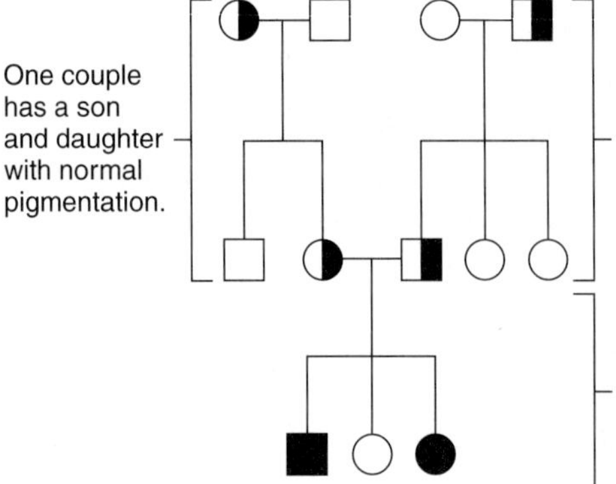

FIGURE 10.B

Rudimentary nails, shown here on the second and fifth toes, are characteristic of distal symphalangism. Another symptom of distal symphalangism is brachydactyly, or short fingers.

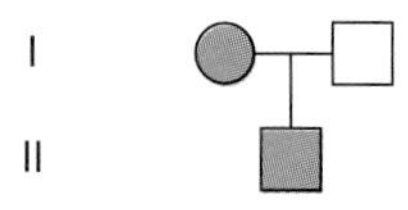

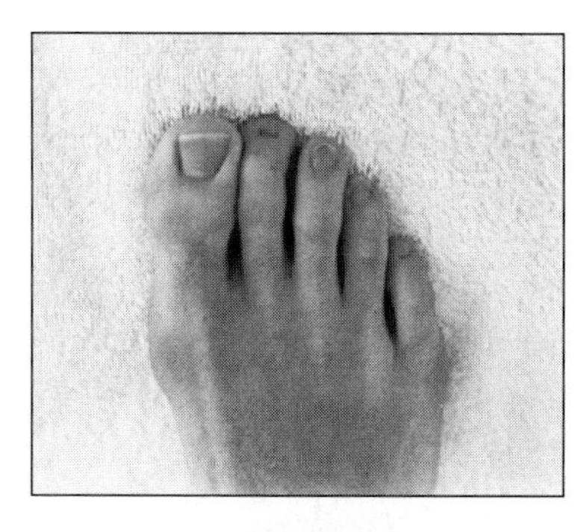

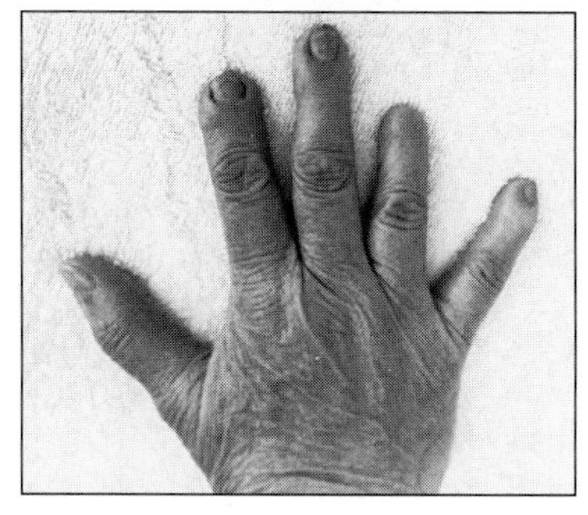

Like many families with a chronically ill member, the Mahars have adapted their way of life to make Katie's easier. The children play in the yard with Katie at night, and the jungle gym is in the garage to allow occasional daytime play. Caren and Dan run a special camp where children with the disorder can turn night into day so that they, like other children, can enjoy the outdoors (Camp Sundown).

Autosomal Recessive Inheritance

- Parents are carriers or are affected.
- Can skip generations.
- Affects males and females.
- In humans, usually early onset.

Distal symphalangism

As a high school student, James Poush saw Mendel's laws in action when he explored his own family for a science fair project. His work was the first full report on an autosomal dominant condition called distal symphalangism (fig. 10.B).

James had never thought that his stiff fingers and toes with their tiny nails were odd. Others in his family had them, too. But when he studied genetics, he realized that his quirk might be a Mendelian trait. After much detective work, James identified 27 affected individuals among 156 relatives, and he went on to analyze a second family with the disorder.

James concluded that the trait is autosomal dominant. Of 63 relatives with an affected parent, 27 (43%) expressed the phenotype, close to the 50% expected of autosomal dominant inheritance. The figure is lower, James reasons, because of underreporting and a few cases where a person inherited the dominant gene but didn't express the phenotype (a phenomenon discussed later in the chapter).

The dominant allele responsible for distal symphalangism affects people to different degrees. Some relatives did not realize they had inherited the variation until James pointed it out to them!

Autosomal Dominant Inheritance

- At least one parent is affected.
- Does not skip generations.
- Affects males and females.
- In humans, often adult onset.
- No carriers, but affected individuals before symptom onset may appear to be carriers.

FIGURE 10.10

A Pedigree for an Autosomal Dominant Trait.

Autosomal dominant traits do not skip generations.

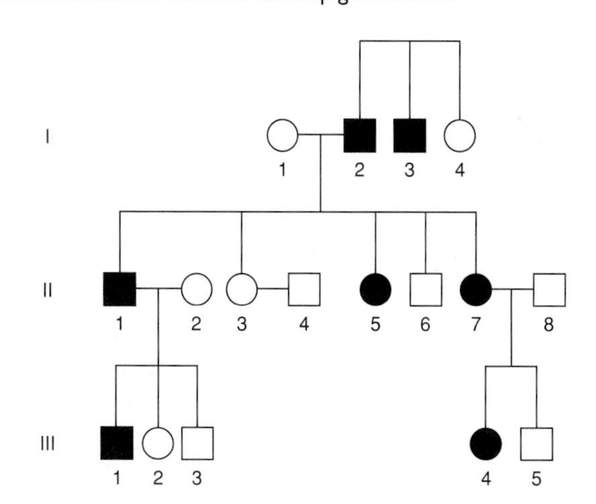

wedlock, serial marriages and the resulting blended families, and assisted reproductive technologies complicate tracing family ties. Many people cannot trace their families back far enough to reveal modes of inheritance.

Table 10.4 describes some autosomal recessive and autosomal dominant disorders.

10.1 MASTERING CONCEPTS

1. How did Mendel's experiments reveal how single genes are transmitted?
2. How do Mendel's observations reflect events of meiosis?
3. What is the law of segregation?
4. Distinguish between a heterozygote and a homozygote; phenotype and genotype; dominant and recessive; wild type and mutant.
5. What are the genotypic and phenotypic ratios expected of a monohybrid cross?
6. How are Punnett squares and pedigrees helpful in following transmission of single genes?

OLC

TABLE 10.4 Some Mendelian Disorders in Humans

DISORDER	SYMPTOMS
Autosomal Recessive	
Ataxia telangiectasis	Facial rash, poor muscular coordination, involuntary eye movements, high risk for cancer, sinus and lung infections
Cystic fibrosis	Lung infections and congestion, poor fat digestion, male infertility, poor weight gain, salty sweat
Familial hypertrophic cardiomyopathy	Overgrowth of heart muscle causes sudden death in young adults
Gaucher disease	Swollen liver and spleen, anemia, internal bleeding, poor balance
Hemochromatosis	Body retains iron; high risk of infection, liver damage, excess skin pigmentation, heart and pancreas damage
Maple syrup urine disease	Lethargy, vomiting, irritability, mental retardation, coma, and death in infancy
Phenylketonuria	Mental retardation, fair skin
Sickle cell disease	Joint pain, spleen damage, high risk of infection
Tay-Sachs disease	Nervous system degeneration
Autosomal Dominant	
Achondroplasia	Dwarfism with short limbs, normal size head and trunk
Familial hypercholesterolemia	Very high serum cholesterol, heart disease
Huntington disease	Progressive uncontrollable movements and personality changes, beginning in middle age
Lactose intolerance	Inability to digest lactose causes cramps after eating this sugar
Marfan syndrome	Long limbs, sunken chest, lens dislocation, spindly fingers, weakened aorta
Myotonic dystrophy	Progressive muscle wasting
Neurofibromatosis (I)	Brown skin marks, benign tumors beneath skin
Polycystic kidney disease	Cysts in kidneys, bloody urine, high blood pressure, abdominal pain
Polydactyly	Extra fingers and/or toes
Porphyria variegata	Red urine, fever, abdominal pain, headache, coma, death

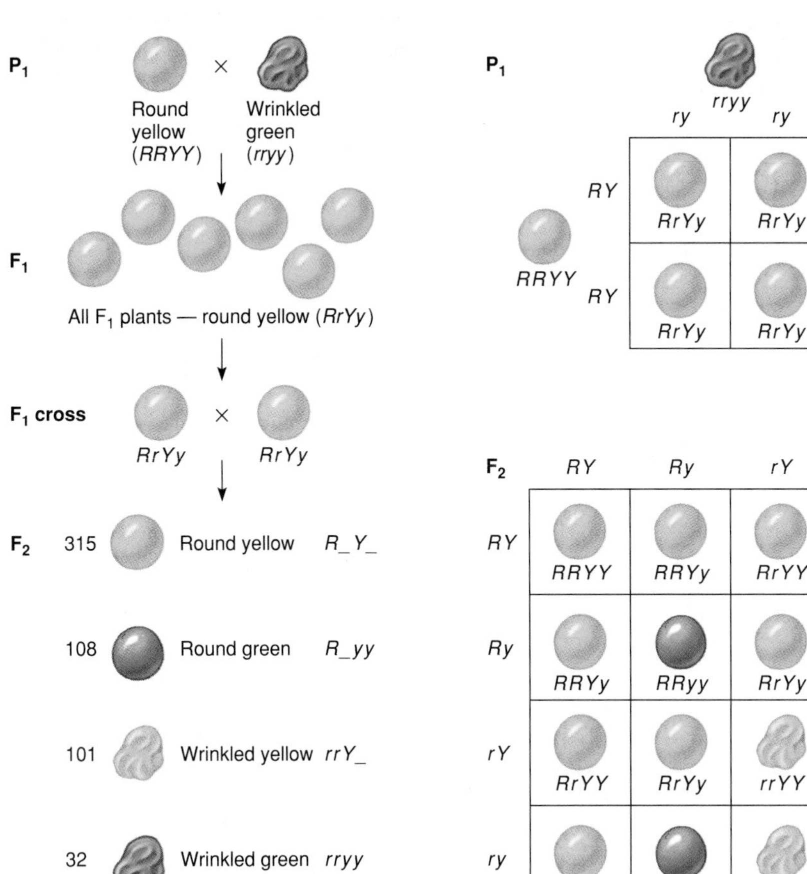

FIGURE 10.11
Plotting a Dihybrid Cross.
A Punnett square can be used to represent the random combinations of gametes produced by dihybrid individuals. An underline in a genotype (in the F_2 generation) indicates that either a dominant or recessive allele is possible. The numbers in the F_2 generation are Mendel's experimental data.

10.2 Tracing the Inheritance of Two Genes

Mendel's law of independent assortment states that a gene transmitted on one chromosome does not influence transmission of a gene on a different chromosome. This law is used to predict the proportions of progeny classes when more than one trait is considered.

The law of segregation follows the inheritance pattern of two alleles of a single gene. In a second set of experiments with pea plants, Mendel examined the inheritance of two different traits, each of which has two different alleles: he looked at seed shape, which may be either round or wrinkled (determined by the *R* gene), and seed color, which may be either yellow or green (determined by the *Y* gene). When Mendel crossed plants with round, yellow seeds with plants with wrinkled, green seeds, all the progeny had round, yellow seeds (fig. 10.11). Therefore, he concluded that round is completely dominant to wrinkled, and yellow is completely dominant to green.

Next, Mendel took F_1 plants (genotype *RrYy*) and crossed them with each other. This is a dihybrid cross, because the individuals are heterozygous for two genes. Mendel found four types of seeds in the F_2 generation: round, yellow (315 plants); round, green (108 plants); wrinkled, yellow (101 plants); and wrinkled, green (32 plants). This is an approximate phenotypic ratio of 9:3:3:1. The mathematically-minded Mendel deduced that the traits would have to be transmitted independently for this ratio to occur.

To determine the genotypes of plants that produce the 9:3:3:1 phenotypic ratio, Mendel took each plant from the F_2 generation and crossed it with wrinkled, green (*rryy*) plants. These test crosses established whether each F_2 plant was true-breeding for both genes (*RRYY* or *rryy*), true-breeding for one but heterozygous for the other (*RRYy, RrYY, rrYy,* or *Rryy*), or heterozygous for both genes (*RrYy*). Based upon the results of the dihybrid cross, Mendel proposed what is now known as the **law of independent assortment.** It states that a gene for one trait does not influence the transmission of a gene for another trait. That is, genes carried on different chromosomes are randomly packaged into gametes with respect to each other. This second law is true only for genes on different chromosomes, such as pea shape and color.

Mendel had again inferred a principle of inheritance based on meiosis. Independent assortment occurs because chromosomes from each parent combine in a random fashion (fig. 10.12). In Mendel's dihybrid cross, each parent produces equal numbers of gametes of four different types: *RY, Ry, rY,* and *ry.* (Note that each of these combinations has one gene for each trait.) A Punnett square for this cross predicts that the four types of seeds—round, yellow (*RRYY, RrYY, RRYy,* and *RrYy*); round, green (*RRyy, Rryy*);

FIGURE 10.12
Meiosis Produces Genetic Variation.
The independent assortment of genes carried on different chromosomes results from the random alignment of chromosome pairs during metaphase of meiosis I. An individual of genotype *RrYy*, for example, manufactures four types of gametes, containing the dominant alleles of both genes (*RY*), the recessive alleles of both genes (*ry*), and a dominant allele of one with a recessive allele of the other (*Ry* or *rY*). The allele combination depends upon which chromosomes are packaged together into the same gamete—and this happens at random.

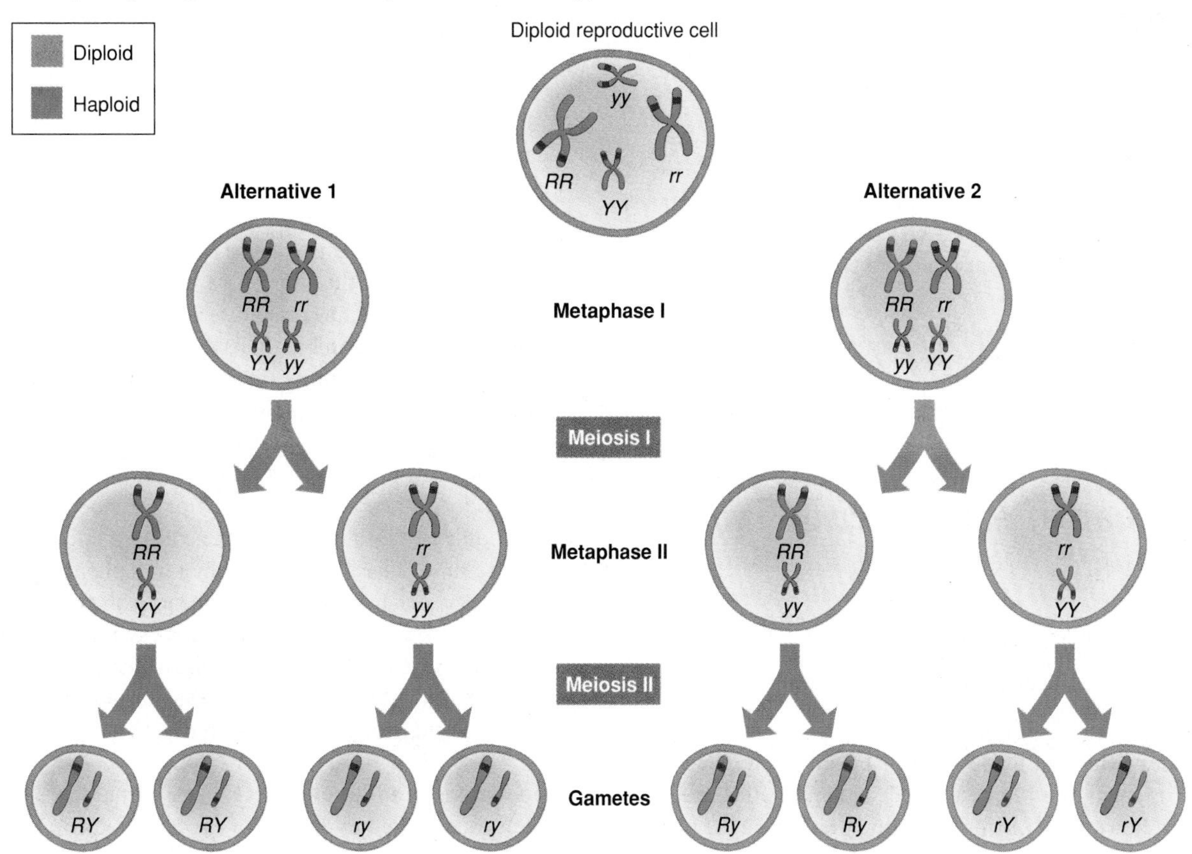

FIGURE 10.13

The Product Rule.

Each parent is tall with round, yellow peas, and is a trihybrid (*RrYyTt*). Multiplying probabilities derived from Punnett squares reveals the chance that the offspring plant is also a trihybrid. (Shaded boxes indicate a dominant phenotype.)

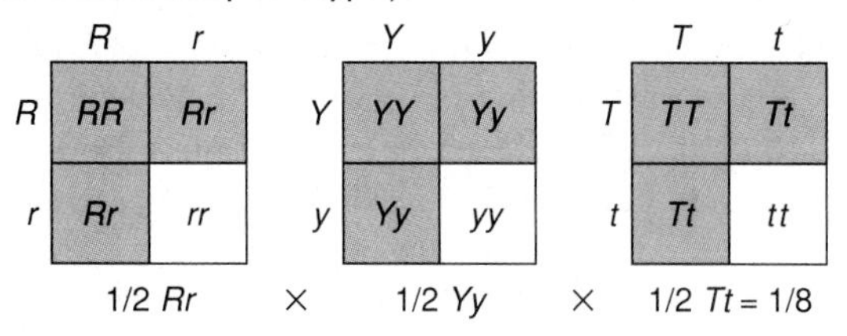

wrinkled, yellow (*rrYY, rrYy*); and wrinkled, green (*rryy*)—will occur in the ratio 9:3:3:1, just as Mendel found.

Punnett squares become cumbersome when analyzing more than two genes—a Punnett square for three genes has 64 boxes; for four genes, 256 boxes. An easier way to predict genotypes and phenotypes is to use the mathematical laws of probability on which Punnett squares are based. Probability predicts the likelihood of an event. The **product rule** states that the chance that two independent events will both occur—such as two alleles both being inherited—equals the product of the individual chances that each event will occur.

The product rule can predict the chance of obtaining a wrinkled, green (*rryy*) plant from dihybrid (*RrYy*) parents. Consider the dihybrid individual one gene at a time. A Punnett square for *Rr* crossed with *Rr* shows that the probability that two *Rr* plants will produce *rr* progeny is 25%, or 1/4. Similarly, the chance of two *Yy* plants producing a *yy* individual is 1/4. According to the product rule, the chance of dihybrid parents (*RrYy*) producing homozygous recessive (*rryy*) offspring is 1/4 multiplied by 1/4, or 1/16. Now consult the 16-box Punnett square for Mendel's dihybrid cross. Only 1 of the 16 boxes is *rryy*. Figure 10.13 applies the product rule to three traits.

10.2 MASTERING CONCEPTS

1. What is the law of independent assortment?
2. How is Mendel's second law based on meiosis?
3. What are the expected genotypic and phenotypic ratios of a dihybrid cross?
4. How can the product rule be used to predict the results of crosses involving more than one gene?

10.3 How Can Gene Expression Appear to Alter Mendelian Ratios?

Sometimes Mendel's laws do not seem to be operating, because expected ratios of progeny classes do not occur. Allele interactions and effects of other genes and the environment can alter phenotypes, but the laws still operate.

Mendel's crosses yielded easily distinguishable offspring. A pea is either yellow or green, round or wrinkled; a plant is either tall or short. At times, however, offspring traits do not occur in the proportions Punnett squares or probabilities predict. It may appear that Mendel's laws do not apply—but they do. The underlying genotypic ratios are there, but the nature of the phenotype or other genes or the environment alter how traits appear.

Allele Interactions

Mendelian ratios can be disrupted if progeny classes are absent, a phenotype has several effects, or the phenotype varies or is altered by another gene.

Lethal Allele Combinations Genes begin to function soon after fertilization. Some allele combinations halt development very early, so that an entire phenotypic class of offspring never appears. An allele that causes such early death is called a lethal allele.

Lethal alleles are seen—or not seen—in plants of certain genotypes that die at fertilization, during seed development, or as seedlings. In humans, lethal alleles can cause spontaneous abortion (miscarriage). When a man and woman each carry a recessive lethal allele for the same gene, each pregnancy has a 25% chance of spontaneously aborting—a figure that represents the homozygous recessive class.

Sometimes a double dose of a dominant allele is lethal. This is the case for Mexican hairless dogs. Inheriting one dominant allele confers the coveted hairlessness trait (genotype *Hh*), but inheriting two dominant alleles (genotype *HH*) is lethal to the embryo. Breeders cross hairless to hairy ("powder-puff" genotype *hh*) dogs, rather than hairless to hairless, to avoid losing the lethal homozygous dominant class—a quarter of the pups, as figure 10.14 indicates.

Multiple Alleles A diploid cell has a "place," or locus, for two alleles of a given gene, one on each of a pair of homologous chromosomes. The alleles can be identical, producing a homozygous genotype, or different, producing a heterozygote. Even though an individual diploid cell has only two alleles for a particular gene, a gene can exist in more than two allelic forms. This is because a change, or mutation, in any of the hundreds or thousands of DNA bases of a gene constitutes a different allele. The change is detectable if the protein product the new allele encodes changes in a way that affects the phenotype.

Multiple alleles can complicate interpretation of Mendel's laws because in different combinations they cause different variations of the phenotype. For example, in rabbits, four alleles determine coat color (table 10.5). Allele *c* is always recessive; c^h is dominant to *c*; c^{ch} is dominant to c^h and *c*; and *C* is dominant to all alleles. In humans, the cystic fibrosis gene can mutate in hundreds of different ways and produce phenotypes ranging from frequent bronchitis and pneumonia to life-threatening, near-constant infections.

The more alleles a gene has, the more phenotypes and genotypes are associated with it. For a gene with two alleles, such as round (*R*) versus wrinkled (*r*) peas, three genotypes are possible—*RR, Rr,* and *rr*. A gene with three alleles has six genotypes. As rabbit coat color illustrates, a gene with four alleles has ten associated genotypes and five or more phenotypes.

Penetrance and Expressivity The same allele combination can produce different degrees of the phenotype in different indi-

FIGURE 10.14
Lethal Alleles.
(**A**) This Mexican hairless dog has inherited a dominant allele that makes the animal hairless. Inheriting two such dominant alleles is lethal during embryonic development. (**B**) Breeders cross Mexican hairless dogs to hairy ("powder-puff") dogs to avoid dead embryos and stillbirths that represent the *HH* genotypic class.

A

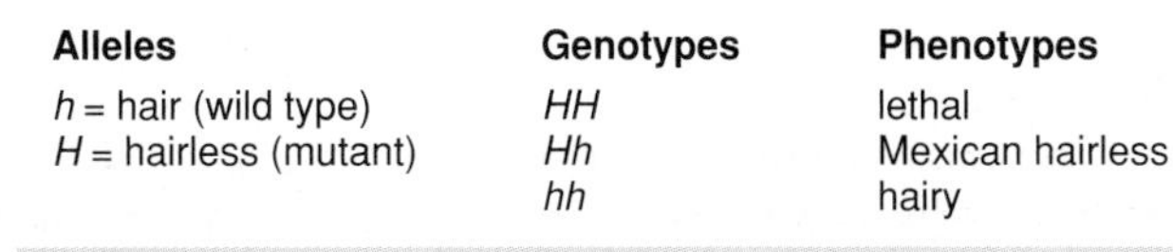

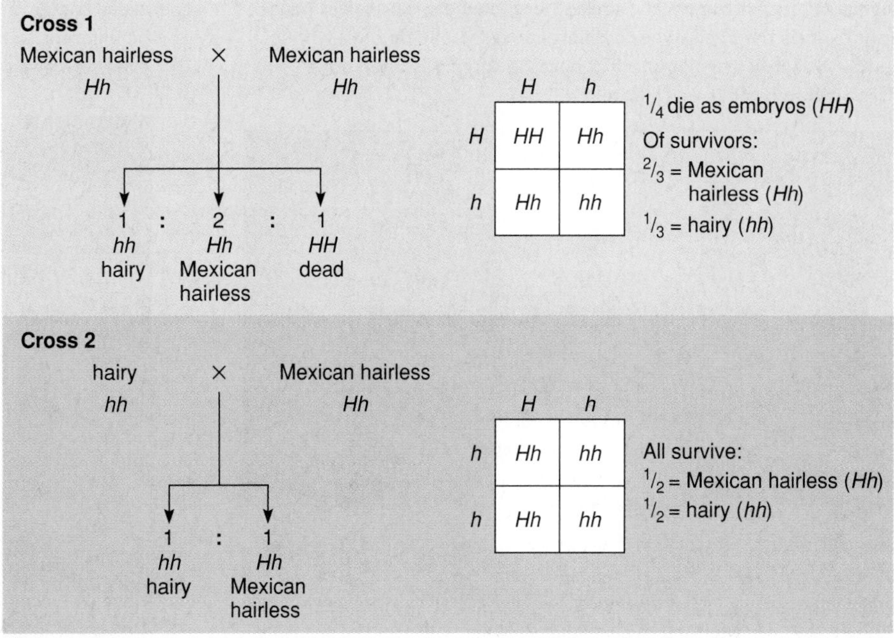

B

viduals, even siblings, because of outside influences such as nutrition, toxic exposures, illnesses, and other genes. Most disease-causing allele combinations are completely penetrant, which means that all individuals who inherit that genotype have symptoms. A genotype is **incompletely penetrant** if some individuals who inherit the disease-causing genotype do not express the phenotype. Polydactyly, having extra fingers or toes, is an incompletely penetrant trait seen in many types of mammals. Cats that inherit the dominant polydactyly allele have more than four toes on at least one paw, yet others who are known to have the allele (because they have an affected parent and offspring) have the normal number of toes. The penetrance of a gene is described numerically. If 80 of 100 cats that inherit the dominant polydactyly allele have extra digits, the allele is 80% penetrant. Figure 10.15 shows a person with polydactyly.

A phenotype is **variably expressive** if the symptoms vary in intensity in different individuals. One cat with polydactyly might have an extra digit on two paws; another might have two extra digits on all four paws; a third cat might have just one extra toe. Penetrance refers to the all-or-none expression of a genotype; expressivity refers to the severity of a phenotype.

TABLE 10.5 Phenotypes and Genotypes for Rabbit Coat Color

PHENOTYPE	POSSIBLE GENOTYPES	PHENOTYPE	POSSIBLE GENOTYPES
Gray	CC, Cc^{ch}, Cc^{h}, Cc	Himalayan	$c^{h}c^{h}$, $c^{h}c$
Chinchilla	$c^{ch}c^{ch}$	Albino	cc
Light gray	$c^{ch}c^{h}$, $c^{ch}c$		

FIGURE 10.15

Incomplete Penetrance and Variable Expressivity.

The dominant allele that causes polydactyly is incompletely penetrant and variably expressive. Some people who inherit the allele have the normal number of fingers and toes. Those in whom the allele is penetrant express the phenotype to different degrees. Some of us who inherit polydactyly may not know it, because surgery is often done soon after birth to remove extra fingers or toes.

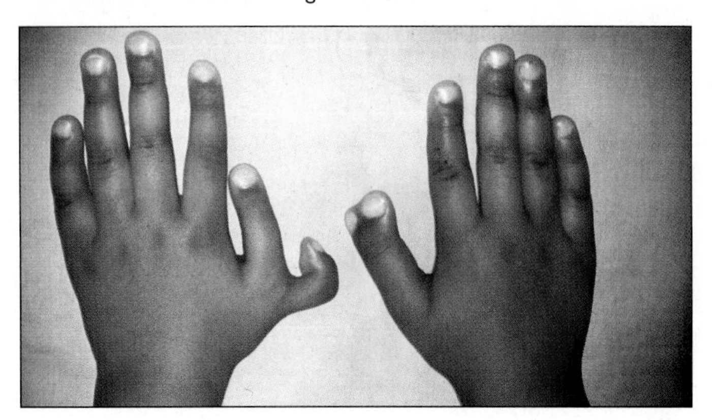

FIGURE 10.16

Pleiotropy.

(**A**) King George III suffered from the autosomal dominant disorder porphyria variegata—and so did several other family members. (**B**) Because of pleiotropy, the family's varied illnesses and quirks appeared to be different, unrelated disorders. In King George, symptoms appeared every few years in this order.

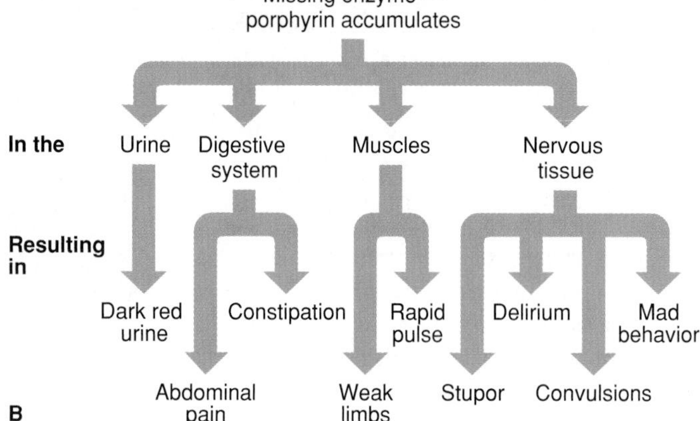

Pleiotropy A Mendelian disorder or trait with many symptoms or expressions is **pleiotropic,** and arises when a single protein is important in different biochemical pathways, or affects more than one body part or process. For example, the long limbs, spindly fingers, caved-in chest, weakened aorta, and lens dislocation of Marfan syndrome seemed unrelated, until researchers discovered the same connective tissue protein abnormality in all of these structures.

Pleiotropic conditions can be difficult to trace through families, because individuals with different subsets of symptoms may appear to have different disorders. In fact, some historians blame the American Revolution on a pleiotropic blood disorder, porphyria variegata, that afflicted England's King George III.

Starting at age 50, King George III experienced recurrent bouts of a long list of symptoms (fig. 10.16) and then he would recover. Because the king ripped off his wig and clothing and ran about at the peak of a fever, he was eventually deemed unfit to rule and was dethroned.

Doctors were permitted to do very little to the royal body and based their diagnoses on the king's raving comments. In the twentieth century, researchers realized that George's red urine indicated porphyria variegata. Lack of a liver enzyme results in buildup of compounds called porphyrins that are parts of certain enzymes. It still isn't understood how this inborn error causes the neurological and other symptoms. Careful study of George's relatives revealed that some of them had different subsets of the symptoms, explaining why the family's medical problems looked like different disorders. Medical charts showed that several of George's relatives had porphyria symptoms, but because they had different combinations of the many symptoms, in the past no one suspected a single genetic cause.

Epistasis One gene masking or otherwise affecting the expression of another is called **epistasis,** and it may appear to disrupt operation of Mendel's laws. Epistasis is different from a dominant allele masking a recessive allele of the same gene. Instead, the masking happens between two different genes. Epistasis is responsible for certain blood type inconsistencies between parents and offspring.

A person's ABO blood type is determined by the *I* gene, which has three alleles—I^A, I^B, and *i*. Figure 10.17 shows the various genotypes that are the basis of the four ABO blood types of A, B, AB, and O. The *I* gene encodes enzymes that insert two types of molecules onto the surfaces of red blood cells. Such cell surface molecules are termed antigens. A person with antigen A attached has type A blood; antigen B has type B blood; both A and B has type AB blood; and neither A nor B has type O blood. However, the A and B antigens can attach only if a glycoprotein links them to the cell surface. If a gene called *H* is mutant, the carbohydrate portion of that glycoprotein is missing, and antigens A and B cannot attach. As a result, a person who is genotype *hh* has blood that tests as type O, even though he or she may have an ABO genotype indicating type A, B, or AB. Confusion over paternity might arise if a type O offspring is not possible given the parents' ABO genotypes. The epistatic relationship between the *H* and *I* gene is called the Bombay phenotype. (The soap

FIGURE 10.17
Epistasis.
If a person is genotype *hh,* the ABO blood type will appear to be O, because the *I* gene's product depends upon the presence of the *H* gene's product. The _ in a genotype indicates that the allele can be *H* or *h*.

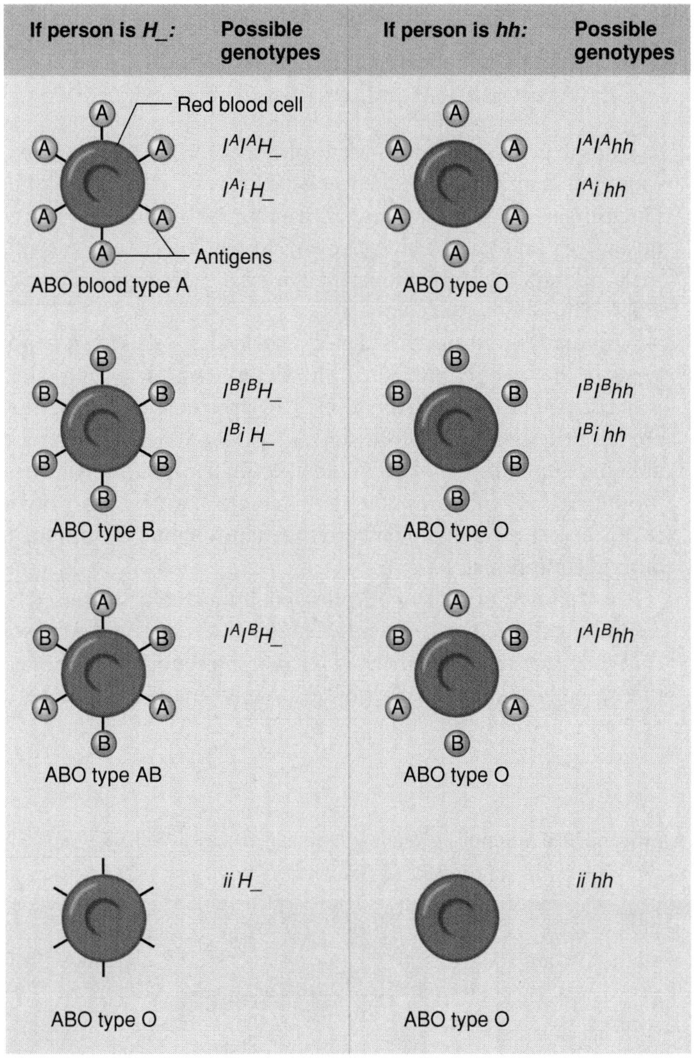

If person is *H_*:	Possible genotypes	If person is *hh*:	Possible genotypes
ABO blood type A (Red blood cell; Antigens)	$I^AI^AH\_$ $I^Ai\,H\_$	ABO type O	I^AI^Ahh $I^Ai\,hh$
ABO type B	$I^BI^BH\_$ $I^Bi\,H\_$	ABO type O	I^BI^Bhh $I^Bi\,hh$
ABO type AB	$I^AI^BH\_$	ABO type O	I^AI^Bhh
ABO type O	$ii\,H\_$	ABO type O	$ii\,hh$

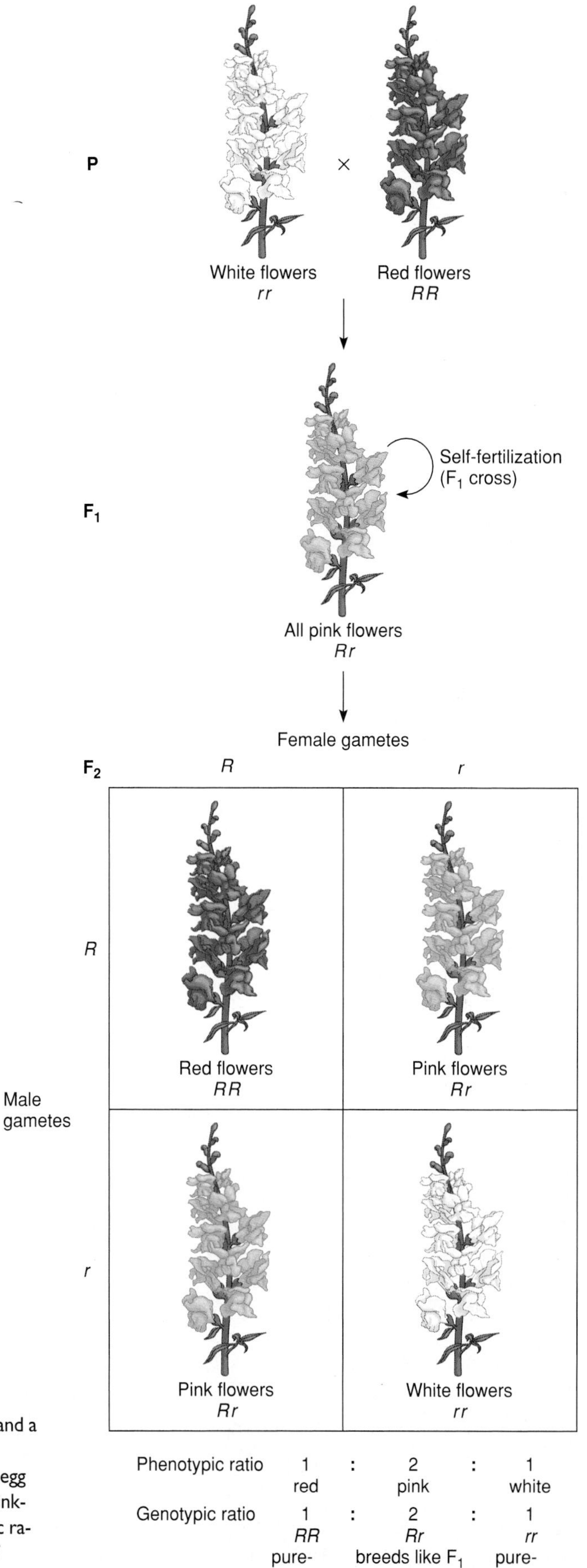

FIGURE 10.18
Incomplete Dominance in Snapdragon Flowers.
A cross between a homozygous dominant plant with red flowers (*RR*) and a homozygous recessive plant with white flowers (*rr*) produces a heterozygous plant with pink flowers (*Rr*). When *Rr* pollen fertilizes *Rr* egg cells, one-quarter of the progeny are red-flowered (*RR*), one-half are pink-flowered (*Rr*), and one-quarter are white-flowered (*rr*). The phenotypic ratio of this monohybrid cross is 1:2:1 (instead of the 3:1 seen in cases of complete dominance) because the heterozygous class has a phenotype different from that of the homozygous dominant class.

FIGURE 10.19
Codominance.
Even though the I^A and I^B alleles of the I gene are codominant, they still follow Mendel's law of segregation. These Punnett squares follow the genotypes that could result by crossing a person with type A blood with a person with type B blood. Is it possible for parents with type A and type B blood to have a child who is type O? All possible crosses of type A and type B individuals are shown at the right.

Genotypes	Phenotypes	
	Antigens on surface	ABO blood type
I^AI^A	A	Type A
I^Ai	A	
I^BI^B	B	Type B
I^Bi	B	
I^AI^B	AB	Type AB
ii	None	Type O

Type A × Type B:

Type B \ Type A	I^A	I^A
I^B	I^AI^B AB	I^AI^B AB
I^B	I^AI^B AB	I^AI^B AB

Type B \ Type A	I^A	i
I^B	I^AI^B AB	I^Bi B
I^B	I^AI^B AB	I^Bi B

Type B \ Type A	I^A	I^A
I^B	I^AI^B AB	I^AI^B AB
i	I^Ai A	I^Ai A

Type B \ Type A	I^A	i
I^B	I^AI^B AB	I^Bi B
i	I^Ai A	ii O

opera *General Hospital* once used it as a plot device!) People with an *hh* genotype are extremely rare.

Different Dominance Relationships Some genes show **incomplete dominance,** with the heterozygous phenotype intermediate between those of the two homozygotes. For example, the piebald trait in domestic cats confers varying numbers of white spots and shows incomplete dominance. Cats of genotype *SS* have many spots, *ss* cats have no spots, and *Ss* cats have an intermediate number of spots. Another example of incomplete dominance is the snapdragon plant (fig. 10.18). A red-flowered plant of genotype *RR* crossed with a white-flowered *rr* plant gives rise to a pink-flowered *Rr*. An intermediate amount of pigment in the heterozygote confers the pink color.

Different alleles that are both expressed in a heterozygote are termed **codominant.** The two dominant alleles, named A and B, of the *I* gene for blood type, are actually codominant. People of blood type AB have both antigens A and B on the surfaces of their red blood cells. Even though the I^A and I^B alleles are codominant, they segregate between generations (fig. 10.19).

Environmental Influences

Some gene expressions are exquisitely sensitive to the environment. Temperature influences gene expression in some familiar animals—Siamese cats and Himalayan rabbits have dark ears, noses, feet, and tails because these parts are colder than the animals' abdomens. Heat affects the abundance of pigment molecules that provide coat color.

Temperature-sensitive alleles also lead to striking phenotypes in the fruit fly (fig. 10.20). In flies mutant for a gene called proboscipedia, mouthparts develop as antennae when the flies are raised at low temperatures and as legs when the flies are raised at high temperatures. When raised at room temperature, the flies' mouthparts are mixtures of leg and antennal tissue. A phenotype that is expressed only under certain environmental conditions is termed conditional.

A trait may appear to be inherited, but actually be caused by something in the environment. Such an environmentally caused trait that appears to be inherited is called a **phenocopy.** It may either resemble a Mendelian disorder's symptoms or mimic inheri-

FIGURE 10.20
A Conditional Mutant.
A temperature-sensitive mutant gene in the fruit fly *Drosophila melanogaster* transforms normal mouthparts (center) into leg structures at high temperatures (left) and into antennal structures at low temperatures (right).

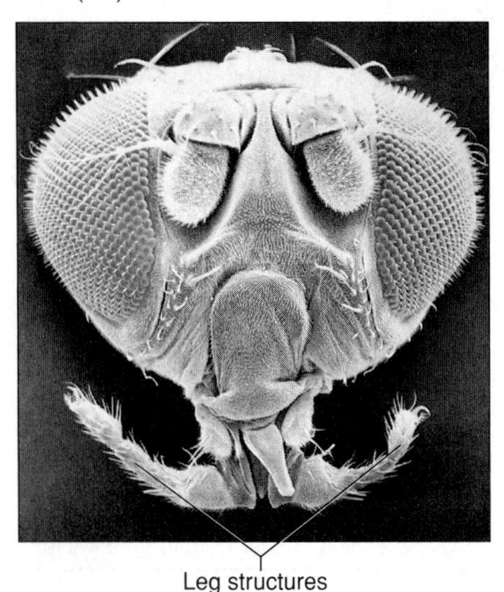

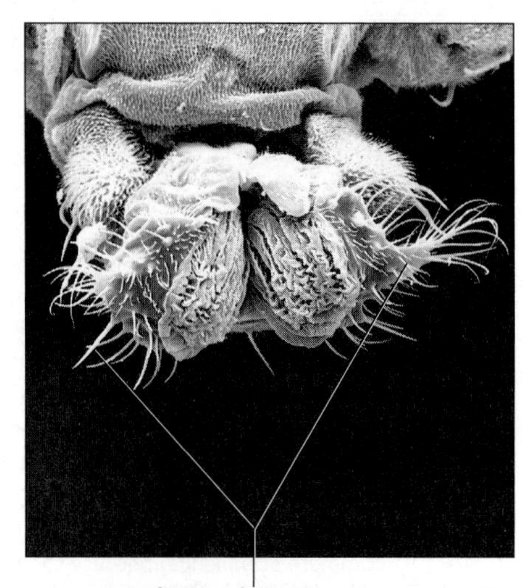

tance by occurring in certain relatives. For example, the drug thalidomide causes grossly shortened limbs in children whose mothers took it while pregnant. This birth defect is a phenocopy of an inherited illness called phocomelia. An infection can also appear to be a Mendelian disorder. Children who have AIDS may contract the infection from HIV-positive parents. But these children acquire AIDS by viral infection, not by inheriting a gene. Unlike an inherited disorder, an infection does not recur with a predictable frequency. Its recurrence depends upon exposure to the infective agent.

Genetic Heterogeneity—More Than One Way to Inherit a Trait

Another complication of Mendel's laws is that different genes can produce similar or identical phenotypes, a phenomenon called **genetic heterogeneity.** Because of genetic heterogeneity, Mendel's laws can appear to not be operating. For example, 132 forms of deafness are transmitted as autosomal recessive traits in humans. If a man who is heterozygous for a deafness gene on one chromosome has a child with a woman who is heterozygous for a deafness gene on a different chromosome, then that child faces only the general population risk of inheriting either form of deafness—not the 25% risk that Mendel's law predicts for a monohybrid cross. This is because the parents are heterozygous for *different* genes.

Genetic heterogeneity can occur when genes encode different enzymes that participate in the same biochemical pathway. For example, eleven biochemical reactions lead to blood clot formation. Clotting disorders may result from abnormalities in genes specifying any of these enzymes, causing several types of bleeding disorders.

Table 10.6 summarizes phenomena that appear to alter Mendelian inheritance.

10.3 MASTERING CONCEPTS

1. How do lethal alleles and epistasis decrease the number of phenotypic classes, and multiple alleles, incomplete dominance, and codominance increase the number of phenotypic classes?
2. How do penetrance and expressivity differ?
3. How does pleiotropy differ from incomplete penetrance and variable expressivity?
4. How can the environment affect a phenotype?
5. How does a phenocopy differ from an inherited trait?
6. What is genetic heterogeneity?

10.4 Complex Traits

Traits that "run in families," but do not adhere to Mendel's laws, are termed "complex." They are determined by more than one gene, and also sometimes by the environment. Many behavioral characteristics are complex.

A woman who is a prolific writer has a daughter who becomes a successful novelist. An overweight man and woman have obese children. A man whose father was an alcoholic is himself an alcoholic. Are these characteristics—writing talent, obesity, and alcoholism—inherited or imitated? It can be difficult to tell. For single-gene disorders, we can predict the probability that a certain family member will inherit the condition. Some traits and diseases, though, "run in families," appearing in a few relatives with no apparent pattern, because they are caused by more than one gene, the environment, or both.

Characteristics that do not follow Mendel's laws but have an inherited component are termed **complex traits.** These may be

TABLE 10.6 Factors That Alter Mendelian Phenotypic Ratios

PHENOMENON	EFFECT ON PHENOTYPE	EXAMPLES
Lethal alleles	A phenotypic class dies very early in development	Spontaneous abortion
Multiple alleles	Produces many variants of a phenotype	Rabbit coat color, cystic fibrosis
Penetrance	Some individuals inheriting a particular genotype do not have the associated phenotype	Polydactyly
Expressivity	A genotype is associated with a phenotype of varying intensity	Polydactyly
Pleiotropy	The phenotype includes many symptoms, with different subsets in different individuals	Porphyria variegata, Marfan syndrome
Epistasis	One gene masks another's phenotype	Bombay phenotype
Incomplete dominance	A heterozygote's phenotype is intermediate between those of two homozygotes	Snapdragon flower color, piebald trait
Codominance	A heterozygote's phenotype is distinct from and not intermediate between those of the two homozygotes	ABO blood types
Conditional mutations	An environmental condition affects a gene's expression	Temperature-sensitive coat colors
Phenocopy	An environmentally caused condition whose symptoms and recurrence in a family make it appear to be inherited	Infection, environmentally caused birth defect
Genetic heterogeneity	Genes on different chromosomes produce same phenotype	Deafness, clotting disorders

FIGURE 10.21
Skin Color Follows a Polygenic Pattern of Inheritance.
Multiple copies of genes, or different versions of genes, combine to increase the pigment in skin cells. Different numbers of the contributing genes produce the intermediate phenotypes.

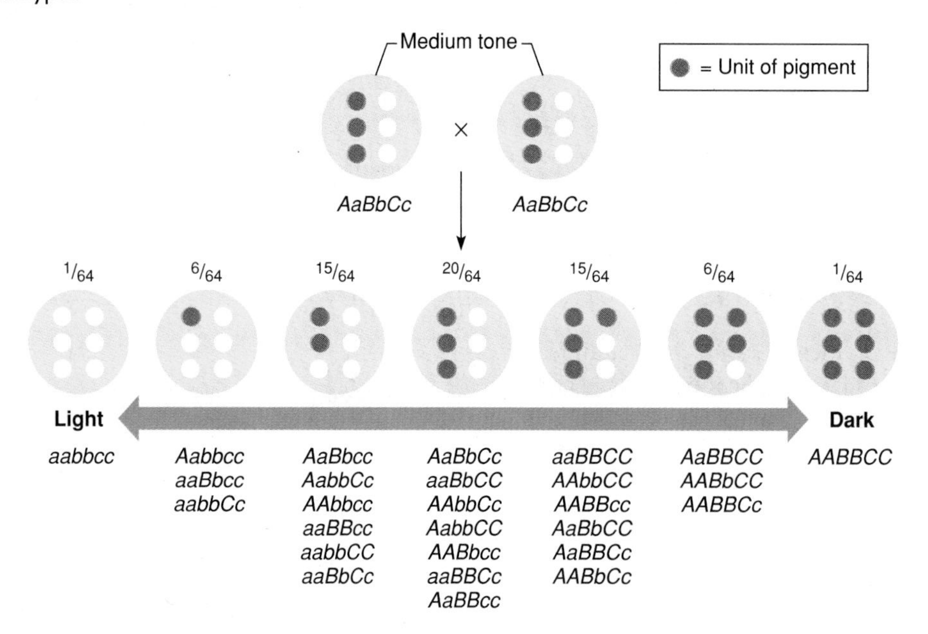

polygenic (determined by more than one gene) or **multifactorial** (determined by one or more genes plus the environment). The term "environment" encompasses many influences, including position in the uterus, experiences, and exposure to infectious agents.

For polygenic (also called quantitative) traits, several genes each contribute to the overall phenotype in equal, small degrees. Their combined actions produce a continuum, or continuously varying expression, of the trait. Figure 10.21 shows how a polygenic model can account for nuances of skin color. The genes that correspond to a polygenic trait follow Mendel's laws individually (unless they are on the same chromosome), but they don't produce typical ratios because they each contribute to the phenotype, and they are neither dominant nor recessive with respect to each other. Instead, their effect is additive.

Height and skin color are familiar examples of polygenic traits in humans (fig. 10.22). Eye color is also polygenic and continuously varying, determined by enzymes that control production and distribution of the pigment melanin. Darker eyes have more clumps of melanin, which absorb incoming light, than do lighter eye colors. In plants, continuously varying traits include flower color, petal length, blade length, and stomata density on the leaves. When the frequencies of all the phenotypes associated with a polygenic trait are plotted on a graph, they form a characteristic bell-shaped curve (fig. 10.22).

Measuring the Input of Genes and the Environment

Multifactorial traits are determined by one or more genes and the environment. For example, cardiovascular health (or disease) reflects the actions of genes controlling blood pressure, clotting ability of the blood, and lipid levels in the blood, as well as environmental factors such as diet, exercise habits, and stress. Body weight and intelligence are other multifactorial traits.

Information from population studies and from family relationships helps describe the inherited component of multifactorial traits. **Empiric risk** is a prediction of recurrence based on a multifactorial trait's incidence in a specific population. In general, empiric risk increases with the severity of the disorder, the number of affected family members, and how closely related the person is to affected individuals.

Empiric risk helps, for example, to predict the likelihood that a neural tube defect (NTD) will recur. A NTD is an opening or lesion in the brain or spinal cord in an embryo that occurs on about the 28th day of prenatal development in humans. In the United States, the overall population risk of carrying a fetus with a NTD is about 1 in 1,000. However, if a person has a NTD, the risk to a sibling is 3%, and if two siblings are affected, the risk to a third child is even greater. The statistics come from direct observations of the prevalence of NTDs. These birth defects have a variety of causes, including mutation and folic acid deficiency. The environmental risk of NTDs may be greatly reduced by supplementing a pregnant woman's diet with folic acid.

Heritability is another description of a complex trait. It estimates the proportion of phenotypic variation in a group that can be attributed to genes. Because the environment can change, so too can heritability. For example, the heritability of human skin color in a Canadian population is greater in the winter, when sun exposure is less likely to darken skin.

One way to measure heritability is to consider pairs of individuals who are related in a certain way, and compare the frequency with which Mendel's laws predict they should share

FIGURE 10.22
Height Is Polygenic.
Many genetics classes demonstrate polygenic (continuously varying) traits by having students line up by height. Such an exercise reveals the input of the environment, because recent lineups have more tall students than lineups done early in the twentieth century. However, all such lineups display the characteristic bell-shaped curve.

a particular trait to the actual frequency. Such a calculation depends on a measurement called the **correlation coefficient,** which is the proportion of genes that two people related in a particular way share (table 10.7). For example, parents and children, as well as siblings to each other, theoretically share 50 percent of their genes, because of the mechanism of meiosis. If the heritability of a trait is very high, then out of a group of 100 pairs of siblings, nearly 50 would be predicted to have it. But for height, which has a greater environmental component, only 40 of a group of 100 sibling pairs are the same height. Heritability for height among this group is .40/.50 (observed/expected), or 80 percent. That is, about 20 percent of their height is attributed to environmental factors, such as diet.

Separating Genetic from Environmental Influences

Multifactorial inheritance analysis does not easily lend itself to the scientific method, which ideally examines one variable. Two types of people, however, have helped geneticists tease apart the genetic and environmental components of complex traits—adopted individuals and twins. (In the future, information about DNA sequences from the human genome project will aid researchers in identifying inherited components of complex traits.)

TABLE 10.7 Relatives Share Genes

RELATIONSHIP	PERCENT SHARED GENES
Sibling to sibling (non-identical)	50% (1/2)
Parent to child	50% (1/2)
Uncle/aunt to niece/nephew	25% (1/4)
First cousin to first cousin	12.5% (1/8)

Adopted Individuals A person adopted by nonrelatives shares environmental influences, but not genes, with his or her adoptive family. Conversely, adopted individuals share genes, but not the exact environment, with their biological parents. Therefore, biologists assume that similarities between adopted people and their adoptive parents reflect environmental influences, whereas similarities between adopted people and their biological parents mostly reflect genetic influences. Information on both sets of parents can reveal to what degree heredity and the environment contribute to a trait.

Many adoption studies use the Danish Adoption Register, a list of children adopted in Denmark from 1924 to 1947. One study examined causes of death among biological and adoptive parents and offspring. If a biological parent died of infection before age 50, the adopted child was five times more likely to die of infection at a young age than a similar person in the general population. This may be because inherited variants in immune system genes increase susceptibility to certain infections. In support of this hypothesis, the risk that adopted individuals would die young from infection did not correlate with adoptive parents' death from infection before age 50. The study also revealed that environment affects longevity. If adoptive parents died before age 50 of cardiovascular disease, their adopted children were three times as likely to die of heart and blood vessel disease as a person in the general population. Can you think of a factor that might explain this correlation?

Twins **Monozygotic** (MZ), or identical, twins are always of the same sex and have identical genes, because they develop from one fertilized ovum. **Dizygotic** (DZ), or fraternal, twins are no more similar genetically than any other two siblings, although they share the same prenatal environment because they develop at the same time from two fertilized ova.

A trait that occurs more frequently in both members of identical twin pairs than in both members of fraternal twin pairs is at least partly controlled by genes. The **concordance** of a trait is the degree to which it is inherited, and it is calculated as the percentage of twin pairs in which both members express the trait (fig. 10.23). Twins can be used to calculate heritability, which equals approximately double the difference between MZ and DZ concordance values for a trait.

Diseases caused by single genes, whether dominant or recessive, are always 100% concordant in MZ twins—that is, if one twin has it, so does the other. However, among DZ twins, concordance is 50% for a dominant trait and 25% for a recessive trait, the same Mendelian values that apply to any two siblings. For a trait determined by several genes, concordance values for MZ twins are significantly greater than for DZ twins. Finally, a trait molded mostly by the environment exhibits similar concordance values for both types of twins.

Using twins to study genetic influence on complex traits dates to 1924, when German dermatologist Hermann Siemens compared school transcripts of identical versus fraternal twins. Noticing that grades and teachers' comments were much more alike for identical twins than for fraternal twins, he concluded that genes contribute to intelligence. Siemens also suggested that a better test would be to study identical twins who were separated at birth and then raised in very different environments.

Twins separated at birth provide natural experiments for distinguishing nature from nurture. Much of what they have in common can be attributed to genetics, especially if their environments have been very different, which the famous cartoon in figure 10.24 shows. The twins' differences reflect differences in their upbringing.

Since 1979, more than 100 sets of twins and triplets separated at birth have visited the laboratories of Thomas Bouchard at the University of Minnesota. Each pair of twins undergoes a 6-day battery of tests that measure physical and behavioral traits, including 24 blood types, handedness, direction of hair growth, fingerprint pattern, height, weight, functioning of all organ systems, intelligence, allergies, and dental patterns. Researchers videotape the twins' facial expressions and body movements in different circumstances and probe their fears, vocational interests, and superstitions.

Identical twins separated at birth and reunited later can be remarkably similar, even when they grow up in very different adoptive families. Idiosyncrasies are particularly striking. A pair of identical male twins raised in different countries practicing different religions were astounded, when they were reunited as adults, to find that they both laugh when someone sneezes and flush the toilet before using it. Twins who met for the first time in their thirties responded identically to questions; each paused for 30 seconds, rotated a gold necklace she was wearing three times, and then answered the question. Coincidence, or genetics?

The "twins reared apart" approach is not a perfectly controlled way to separate nature from nurture. Identical twins share an environment in the uterus and possibly in early infancy that may affect later development. Siblings, whether adoptive or biological, do not always share identical home environments.

FIGURE 10.23
Concordance.
A trait that is more often present in both members of monozygotic twin pairs than it is in both members of dizygotic twin pairs has a significant inherited component.

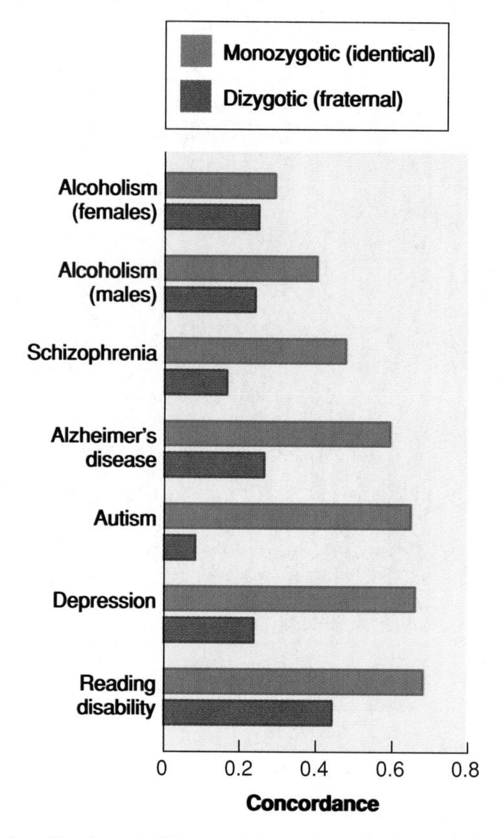

Source: Robert Plomin, et al., "The genetic basis of complex human behaviors," *Science,* 17 June 1994, vol. 264, pp. 1733–1739. Copyright 1994 American Association for the Advancement of Science.

FIGURE 10.24
The New Yorker Collection 1981. Charles Addams from cartoonbank.com.

Separated at birth, the Mallifert twins meet accidentally.

Differences in age, sex, general health, school and peer experiences, temperament, and personality affect each individual's perception of such environmental influences as parental affection and discipline.

Adoption studies, likewise, are not perfect experiments. Adoption agencies often search for adoptive families who have socioeconomic or religious backgrounds similar to those of the biological parents. Thus, even when different families adopt separated twins, their environments may not be as different as they might be for two unrelated adopted people. However, at the present time, twins reared apart and adopted individuals provide intriguing insights into the body movements, psychological quirks, interests, behaviors, and other personality traits that seem to be rooted in our genes.

The next three chapters explore genetics at the subcellular and molecular levels.

10.4 MASTERING CONCEPTS

1. Define complex, polygenic, continuously varying, and multifactorial traits.
2. What do empiric risk, heritability, and coefficient of relationship indicate?
3. How do adopted individuals and twins help geneticists distinguish between genetic and environmental influences on traits?

OLC

Chapter Summary

10.1 Tracing the Inheritance of One Gene

1. Using pea plant crosses, Mendel followed transmission of one or two traits at a time. The genes for the traits he studied were carried on different chromosomes, and each trait had two forms, or **alleles.**
2. Mendel's **law of segregation** states that inherited "elementen" (genes) separate in meiosis. Each individual receives one copy of each gene from each parent. Single genes cause Mendelian traits.
3. An allele whose expression masks another is **dominant;** an allele whose expression is masked by a dominant allele is **recessive.**
4. A **heterozygote** has two different alleles of a gene. A **homozygous** recessive individual has two recessive alleles. A homozygous dominant individual has two dominant alleles.
5. The combination of alleles is the **genotype,** and the expression of a genotype is its **phenotype.** A **wild type** allele is the most common in a population. A change in a gene is a **mutation** and may result in a **mutant** phenotype.
6. The parental generation is designated P_1, the next generation is the first filial generation, or F_1, and the next is the second filial generation, or F_2.
7. A monohybrid cross yields a genotypic ratio of 1:2:1 and a phenotypic ratio of 3:1. A test cross breeds an individual of unknown genotype to a homozygous recessive individual.
8. **Punnett squares,** based on the principles of probability, can be used to predict the outcomes of genetic crosses.
9. An **autosomal recessive** trait can appear in either sex, is passed from parents who are carriers or are affected, and can skip generations. An **autosomal dominant** trait affects both sexes and is inherited from one affected parent. **Pedigrees** can trace traits in families and reveal mode of inheritance.

10.2 Tracing the Inheritance of Two Genes

10. Mendel's **law of independent assortment** was derived by observing transmission of two or more characters whose genes are on different chromosomes. Because maternally and paternally derived chromosomes (and the genes they carry) assort randomly in meiosis, different gametes receive different combinations of genes.
11. A dihybrid cross yields a 9:3:3:1 phenotypic ratio.
12. The **product rule** can be used as an alternative to Punnett squares for following inheritance of more than one trait at a time.

10.3 How Can Gene Expression Appear to Alter Mendelian Ratios?

13. In some crosses, the ratio of progeny phenotypes does not seem to follow Mendel's laws. Lethal allele combinations cease development, eliminating a progeny class. A gene can have multiple alleles because its DNA sequence may be altered in many ways.
14. In an **incompletely penetrant** genotype, phenotype is not expressed in all individuals that inherit it. Phenotypes that vary in

intensity are **variably expressive.** The environment can influence expression of conditional mutations.

15. A **pleiotropic** gene has several expressions.
16. In **epistasis,** one gene masks the effect of another.
17. Different dominance relationships influence phenotypic ratios. Heterozygotes of **incompletely dominant** alleles have phenotypes intermediate between those of the two homozygotes. **Codominant** alleles are both expressed.
18. A **phenocopy** appears to be inherited because it occurs repeatedly within a family, but is environmentally caused.
19. **Genetic heterogeneity** occurs when different genes are associated with the same phenotype.

10.4 Complex Traits

20. **Complex traits** do not follow Mendel's laws but have an inherited component. A **polygenic trait** is determined by more than one gene and varies continuously in its expression. The environment and genes cause **multifactorial traits.**
21. **Empiric risk** measures the likelihood of a multifactorial trait recurring based on its prevalence. The risk rises with genetic closeness to an affected individual, severity of the phenotype, and number of affected relatives.
22. **Heritability** estimates the proportion of variation in a multifactorial trait due to genetics. It is calculated by comparing observed to expected frequencies of a trait among pairs of individuals who are related in a certain way.
23. Characteristics adopted people share with their biological parents are mostly inherited, whereas similarities between adopted people and their adopted parents mostly reflect environmental influences.
24. **Concordance** measures the expression of a trait in **monozygotic** or **dizygotic** twins. It equals the percentage of twin pairs in which both members express the trait. A high concordance value indicates that genes predominantly cause a trait.

Testing Your Knowledge

1. State how Mendel's two laws derive from meiotic events.
2. Distinguish between genotype and phenotype; homozygote and heterozygote; mutant and wild type; recessive and dominant.
3. A man and a woman each have dark eyes, dark hair, and freckles. The genes for these traits assort independently. The woman is heterozygous for each of these traits, but the man is homozygous. The dominance relationships of the alleles are as follows:

   B = dark eyes; b = blue eyes
   H = dark hair; h = blond hair
   F = freckles; f = no freckles

   a. What is the probability that their child will share the parents' phenotype?
   b. What is the probability that the child will share the same genotype as the mother? as the father?

   Use probability (the product rule) or a Punnett square to obtain your answers. Which method do you think is easier?
4. Explain how each of the following appear to disrupt Mendelian ratios:
   a. pleiotropy
   b. variable expressivity
   c. incomplete penetrance
   d. lethal alleles
5. A man with type AB blood has children with a woman who has type O blood. What are the chances that a child they conceive will have type A blood? type B? AB? O?
6. A male cat has short hair, a stubby tail, and extra toes. A female cat has long hair, a long tail, and extra toes. The genes and alleles are as follows:

   L = short hair; l = long hair
   M = stubby tail (Manx); m = long tail
   Pd = extra toes; pd = normal number of toes

   The two cats have kittens. One has long hair, a long tail, and no extra toes. Another has short hair, a stubby tail, and extra toes. The third kitten has short hair, a long tail, and no extra toes. What is the genotype of the father?

Thinking Scientifically

1. Why would Mendel's observations have differed if the genes he studied were part of the same chromosome?
2. A white woman with fair skin, blond hair, and blue eyes and a black man with dark brown skin, hair, and eyes have fraternal twins. One twin has blond hair, brown eyes, and light skin, and the other has dark hair, brown eyes, and dark skin. What Mendelian principle does this real-life case illustrate?
3. Many plants have more than two sets of chromosomes. How would having four (rather than two) copies of a chromosome more effectively mask expression of a recessive allele?
4. In an attempt to breed winter barley that is resistant to barley mild mosaic virus, agricultural researchers cross a susceptible domesticated strain with a resistant wild strain. The F_1 plants are all susceptible, but when the F_1 plants are crossed with each other, some of the F_2 individuals are resistant. Is the viral resistance gene recessive or dominant? How do you know?

References and Resources

Bhattacharyya, Madan K., et al. January 12, 1990. The wrinkled-seed character of pea described by Mendel is caused by a transposon-like insertion in a gene encoding starch-branching enzyme. *Cell,* vol. 60, pp. 115–127. Using modern techniques, scientists are discovering the molecular basis of Mendel's traits.

Lester, Diane R., et al. August 1997. Mendel's stem length gene (Le) encodes a gibberellin 3-β-hydroxylase. *The Plant Cell,* vol. 9, pp. 1435–1443. The molecular and cellular basis of Mendel's short and tall traits involves regulation of a growth hormone.

Lewis, Ricki. 2000. *Human Genetics: Concepts and Applications,* 4th ed. St. Louis: McGraw-Hill. A textbook with many examples of human Mendelian traits.

Lewis, Ricki. December 1994. The evolution of a classic genetic tool. *BioScience.* Computers have greatly increased the information content of pedigrees.

The *LIFE* Online Learning Center provides additional resources and tools for studying this chapter.
www.mhhe.com/life

CHAPTER 11

Chromosomes

The Art and Science of Displaying Chromosomes

Viewing the Molecules of Heredity

Today, many couples expecting a child, about midway through pregnancy, can see a photograph of the fetus's chromosomes, displayed in a size-ordered chart called a karyotype. This is quite remarkable, considering the long history of developing the technology to view chromosomes.

Since the late nineteenth century, microscopists have tried to illustrate human chromosomes. However, obtaining an image of a cell where the chromosomes were sufficiently spread apart so that they did not touch was so difficult that estimates of the number of human chromosomes ranged from 30 to 80! In 1923, Theophilus Painter published sketches of human chromosomes, concluding that the number in a somatic cell was 48. But his subjects were patients at a mental hospital, and they may very well have had abnormal chromosome numbers.

Chromosomes are easiest to see if the cell is dividing, when they are the most condensed and likely to soak up a dye. From early on, geneticists used colchicine, from the chrysanthemum plant, to arrest cells in mitosis. Then in 1951, an accident revealed how to untangle the spaghetti-like mass of chromosomes. A technician mistakenly washed white blood cells in a salt solution that was less concentrated than the cells' interiors. Water rushed into the cells, swelling them and separating the chromosomes. Other improvements followed, including ways to drop cells onto microscope slides prepared with stains. By 1956, everyone agreed that the human somatic cell chromosome number was 46, and that the number in gametes was 23.

But viewing 46 separate chromosomes did not provide very useful information, especially since many of them were similar in size. The first stains used, in the late nineteenth century, weren't too helpful in distinguishing chromosomes either. These were alkaline compounds that bound uniformly to the acidic DNA, darkening all the chromosomes about equally. As a result, for many years, chromosomes were grouped into broad, size-based classes. Researchers and physicians could distinguish only profound abnormalities, such as entire missing or extra chromosomes.

In the 1970s, Swedish researchers developed more specific stains that home in on DNA sequences that are rich in the base pairs A and T or C and G. These stains, especially when used together, created banding patterns unique to each chromosome type. Then in the late 1970s researchers discovered how to synchronize the cell cycles of white blood cells in culture, stopping them in early mitosis. This treatment revealed many more bands, making it easier to detect smaller abnormalities.

Chromosome bands revealed by stains and synchronization were, however, not specific—all chromosomes bound the same materials, but in different patterns. A much more specific technique is FISH, which stands for "fluorescence in situ hybridization." FISH uses DNA probes, which are pieces of DNA attached to molecules of a dye that glows when hit with light. When the probe binds (hybridizes) its DNA sequence complement among

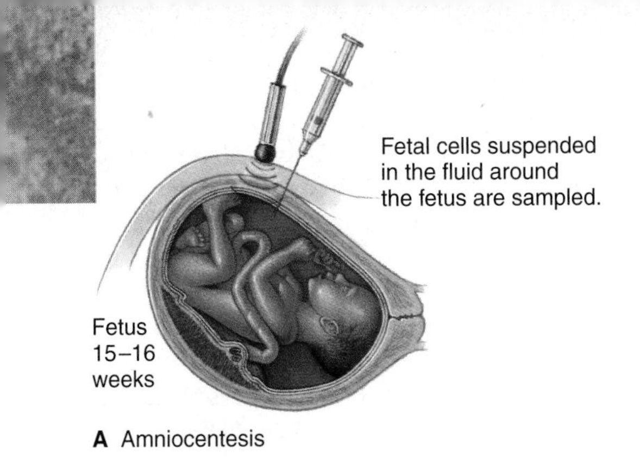

A Amniocentesis

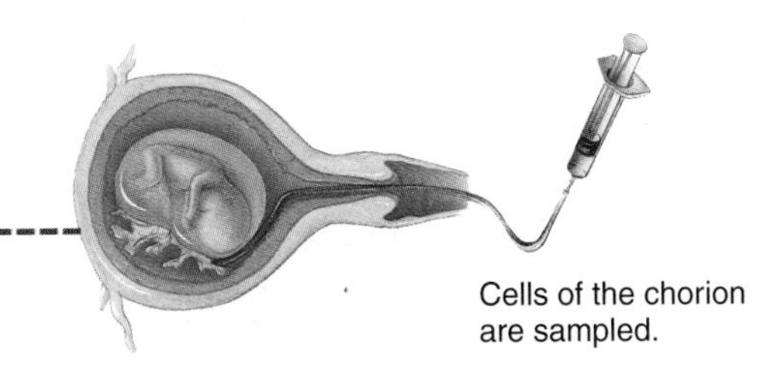

B Chorionic villi sampling

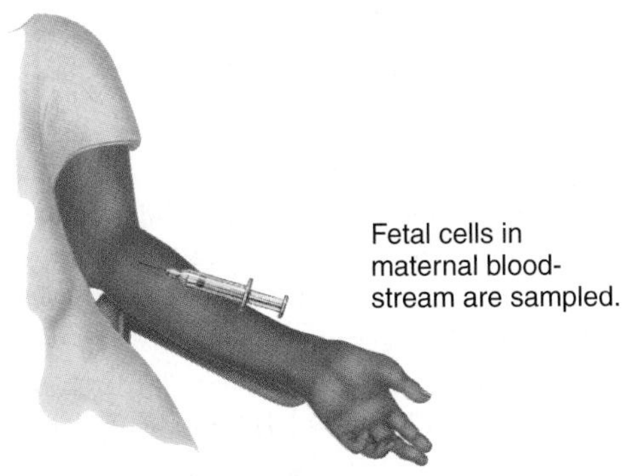

C Fetal cell sorting

1 2 3 4 5
6 7 8 9 10 11 12
13 14 15 16 17 18
19 20 21 22
Sex chromosomes

D Fetal karyotype (normal female)

Three Ways of Checking a Fetus's Chromosomes.
(**A**) Amniocentesis harvests fetal cells shed during development into the amniotic fluid by drawing out some of that fluid. (**B**) Chorionic villi sampling removes some of the cells that would otherwise develop into the placenta. Since these cells came from the fertilized ovum, they have the same chromosomal constitution as the fetus. (**C**) Improved techniques at identifying and extracting specific cells allow researchers to detect fetal cells in a sample of blood from the woman. (**D**) For all three techniques, the harvested cells are allowed to reach metaphase, where chromosomes are most visible, and then broken open on a slide. The chromosomes are stained or their DNA probed, then arranged into a karyotype.

chromosomes spread out on a microscope slide and light is applied, a flash of color results, revealing the location of a particular gene. By using many different probes and dyes, researchers can "paint" each chromosome a unique color.

Karyotypes have also changed over the years. Until a decade ago, a researcher or technician would examine cells using a microscope and photograph one with well-spread chromosomes, then construct the size-order chart using scissors and tape. Today, computerized karyotype devices scan ruptured cells in a drop of stain or treated with DNA probes, and select one in which the chromosomes are the most visible and well spread. Image-analysis software then recognizes the band pattern of each chromosome pair, size-orders them into a chart, and prints it. If the software recognizes an abnormal band pattern, it supplies similar karyotypes from a database, and provides clinical information on associated symptoms.

Today, fetal chromosomes are checked in three ways. The most common and oldest method, done routinely, is *amniocentesis.* A needle inserted into the uterus after the 15th week of pregnancy removes fetal fibroblasts from fluid surrounding the fetus. These cells are cultured and karyotypes prepared.

Chorionic villi sampling (CVS) checks the chromosomes in cells of the chorionic villi, which are fingerlike projections surrounding a fetus that develop into the placenta, the organ that connects to the woman. Because the chorionic villi and the fetus descend from the same fertilized ovum, a chromosomal abnormality in a villus cell should also be present in the fetus. A catheter passed through the vagina is used in CVS. This test is done after the 10th week of pregnancy.

Fetal cell sorting is a newer technique that is safer than amniocentesis or CVS because it detects rare fetal cells in blood taken from a woman's arm, rather than invading her reproductive tract. The test traces its roots to 1957, when a pregnant woman died when cells from a very early embryo lodged in a major blood vessel in her lung. Doctors found that the blockage contained cells with a Y chromosome, which could not possibly have come from the woman, because women do not have Y chromosomes. Male fetal cells were subsequently found to often enter the maternal circulation, and presumably, female fetal cells do too, although they usually cannot be distinguished from maternal cells. In fetal cell sorting, a device called a fluorescence-activated cell sorter recognizes surface characteristics unique to fetal cells and collects them. Then, these cells are karyotyped. Fetal cell sorting may not be routinely available for a few more years.

11.1 What Is a Chromosome?

A chromosome is a long, continuous strand of DNA, plus several types of associated proteins, and RNA. A species has a characteristic chromosome number, and chromosomes are distinguished by size, centromere position, and banding and DNA probe patterns. Chromosome stability and integrity are essential to trait transmission.

A **chromosome** is a very long, continuous molecule of DNA—and more. Several biochemicals associate with the DNA, including RNA and various proteins that help replicate the DNA and transcribe it into RNA. Other proteins serve as scaffolds around which DNA tightly entwines and coils particularly tightly during mitosis (see fig. 8.4). Chromosomes also include proteins specific to certain cell types. We distinguish chromosomes by size, shape, and the pattern in which they stain to form colored bands or bind fluorescent dyes.

Distinguishing Chromosomes

Each species has a characteristic number of chromosomes. The number does not reflect the complexity of the organism nor necessarily how closely related two species are. A mosquito has 6 chromosomes; a grasshopper, rice plant, and pine tree 24; a dog has 78; a carp 104. The 46 chromosomes in a human somatic cell are size-ordered into **karyotype** charts for ease of study, as the opening essay discusses.

Stains applied to chromosomes create patterns of bands, which differ among the chromosome types. Most stains do this by binding preferentially to tightly wound DNA with many repetitive sequences, called **heterochromatin,** which stains darkly. In contrast is lighter-staining **euchromatin,** which harbors more unique sequences and is looser (fig. 11.1). Heterochromatin comprises the telomeres (chromosome tips) and centromeres and may help maintain chromosomes' structural integrity. Euchromatin encodes proteins and may be more loosely wound so that its information is accessible.

Centromere position is also used to distinguish chromosomes. Recall from chapter 8 that a centromere is a characteristically located constriction, consisting of repetitive DNA sequences, to which spindle fibers attach in mitosis. A chromosome is **telocentric** if the centromere is very close to a chromosome tip; **acrocentric** if the centromere pinches off only a small amount of material; **submetacentric** if the centromere establishes one long arm and one short arm; and **metacentric** if the centromere divides it into two arms of approximately equal length (fig 11.2). The long arm of a submetacentric or acrocentric chromosome is termed *q*, and the short arm *p*. Some chromosomes also have bloblike ends called satellites that extend from a stalklike bridge from the rest of the chromosome.

Artificial Chromosomes

We can sequence genes and analyze genomes, but the basis of a chromosome's integrity remains somewhat mysterious. What are the minimal building blocks necessary to form a chromosome that persists throughout the repeated rounds of DNA replication and reassortment that constitute the life of a cell?

To remain stable, a chromosome requires three basic parts:

- telomeres;
- origins of replication, which are sites where DNA replication begins;
- centromeres.

What would it take to construct a chromosome? There are two ways to tackle this question—pare down an existing chromosome to see how small it can get and still hold together, or build up a new chromosome from DNA pieces.

To cut an existing chromosome down to size, researchers swap in a piece of DNA that includes telomere sequences. New telomeres form at the insertion site, a little like prematurely ending a sentence by adding a period. This technique forms small chromosomes, but they can't be easily isolated from cells for further study. The alternative approach, building a chromosome, was challenging because researchers did not know the sequences of origin of replication sites.

Huntington Willard and colleagues at Case Western Reserve University solved the problem. To create "human artificial chro-

FIGURE 11.1

Anatomy of a Chromosome.

Dark-staining chromosomal material (heterochromatin) was once thought to be nonfunctional. Even though heterochromatin does not encode as many proteins as the lighter-staining euchromatin, it is important; it stabilizes the chromosome.

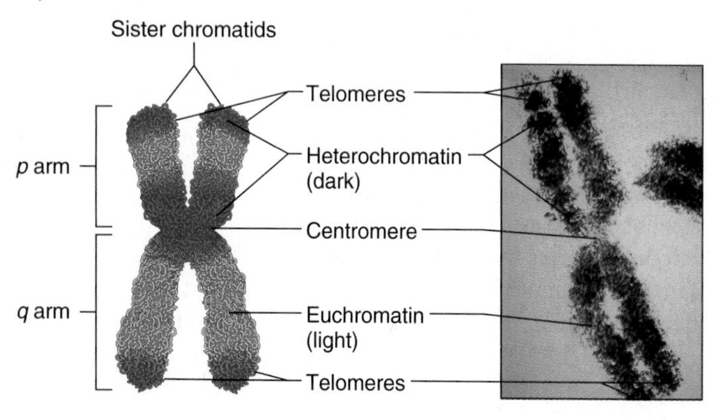

FIGURE 11.2

Centromere Position Is Used to Distinguish Chromosomes.

(**A**) A telocentric chromosome has the centromere very close to one end. (**B**) An acrocentric chromosome has the centromere near an end. (**C**) A submetacentric chromosome's centromere creates a long arm (*q*) and a short arm (*p*). (**D**) A metacentric chromosome's centromere creates equal-sized arms.

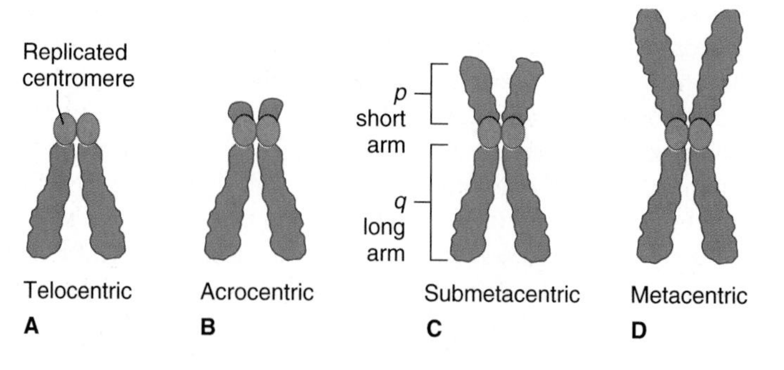

mosomes," they sent separately into cultured cells telomere DNA, repetitive DNA known to form a crucial part of centromeres, and, because they didn't know the origin of replication sites, random pieces of DNA from the human genome. In the cells, these pieces associated in a correct orientation and formed functional and stable structures about 5 to 10 times smaller than the smallest natural human chromosome. The hardy human artificial chromosomes, or "HACs", withstand repeated rounds of cell division.

11.1 MASTERING CONCEPTS

1. What are the chemical components of a chromosome?
2. Cite several ways that chromosomes are distinguished from each other.
3. What are structural and functional differences between heterochromatin and euchromatin?
4. What are the minimal requirements for building a chromosome?

OLC

11.2 "Linking" Mendel's Laws to Chromosomes

Genes carried on the same chromosome do not follow Mendel's second law, because they do not independently assort in meiosis. Maps are constructed based on how often crossovers occur between two linked genes—the more crossovers, the farther apart the genes.

At the turn of this century, genetic researchers are building human artificial chromosomes and finally learning the minimal requirements for these bearers of genetic information. But at the turn of the last century, when chromosomes were first visualized against the backdrop of rediscovering Mendel's laws, they were quite mysterious.

Experiments with Peas and Flies Reveal That Genes Are Linked on Chromosomes

In 1902, German biologist Theodor Boveri and U.S. graduate student Walter Sutton independently realized that chromosomes transmit inherited traits. The association of particular chromosomes with particular traits, including abnormalities, constitutes the field of **cytogenetics.** It began with the study of corn chromosomes, but today is a medical specialty.

As biologists in the early decades of the twentieth century cataloged traits and the chromosomes that transmit them in several species, it soon became clear that the number of traits far exceeded the number of chromosomes. Fruit flies, for example, have four pairs of chromosomes, but dozens of different bristle patterns, body colors, eye colors, wing shapes, and other characteristics. How might a few chromosomes control so many traits? The answer: Chromosomes contain many genes which are **linked,** or inherited together, and do not assort independently, just as the cars of a train arrive at the same destination at the same time, whereas automobiles headed for the same place do not arrive exactly together. The seven traits that Mendel followed in his pea plants were transmitted on different chromosomes. Had the same chromosome carried these genes near each other, Mendel would have generated markedly different results in his dihybrid crosses.

The different inheritance pattern of linked genes was first noticed in the early 1900s, when William Bateson and R. C. Punnett observed offspring ratios in pea plants which were different from the ratios Mendel's laws predicted. They looked at different traits, crossing true-breeding plants with purple flowers and long pollen grains (genotype *PPLL*) with true-breeding plants with red flowers and round pollen grains (genotype *ppll*). Then they crossed the F_1 plants, of genotype *PpLl,* with each other. Surprisingly, the F_2 generation did not show the expected 9:3:3:1 phenotypic ratio for an independently assorting dihybrid cross (fig. 11.3).

Two types of F_2 peas—those with the same phenotypes and genotypes as the parents, *P_L_* and *ppll*—were more abundant than predicted, while the other two progeny classes—*ppL_* and *P_ll*—were less common. Bateson and Punnett hypothesized that the prevalent parental allele combinations reflected genes

FIGURE 11.3

Expected Results of a Dihybrid Cross.
(**A**) When genes are not linked, they assort independently. The gametes then represent all possible allele combinations. The expected phenotypic ratio of a dihybrid cross would be 9:3:3:1. (**B**) If genes are linked by being on the same chromosome, there will only be two allele combinations expected in the gametes. The expected phenotypic ratio would be 3:1, the same as a monohybrid cross.

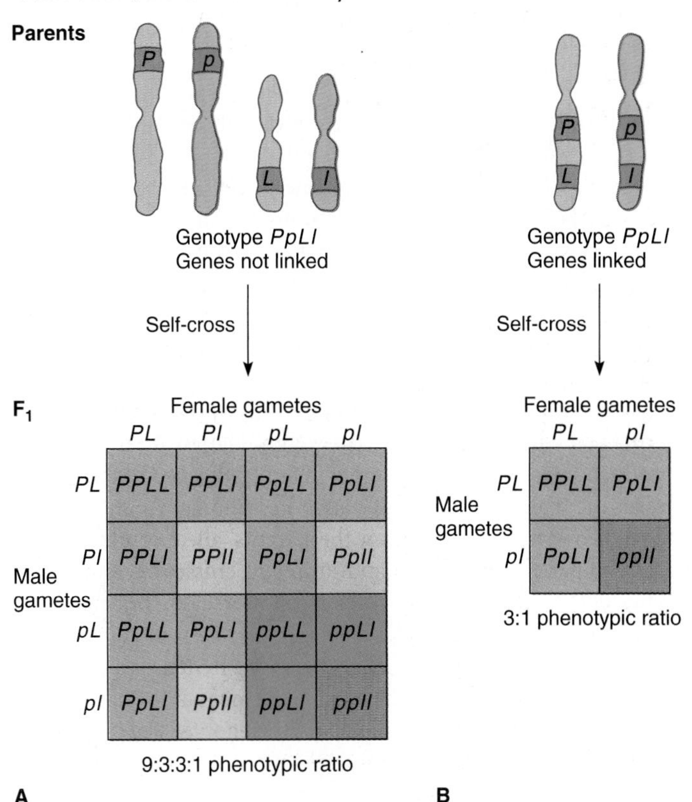

FIGURE 11.4
Crossing Over.
Genes linked closely on the same chromosome are usually inherited together. Linkage between two genes can be interrupted if the chromosome they are located on crosses over with its homolog at a point between the two genes. This packages recombinant arrangements of the genes into gametes.

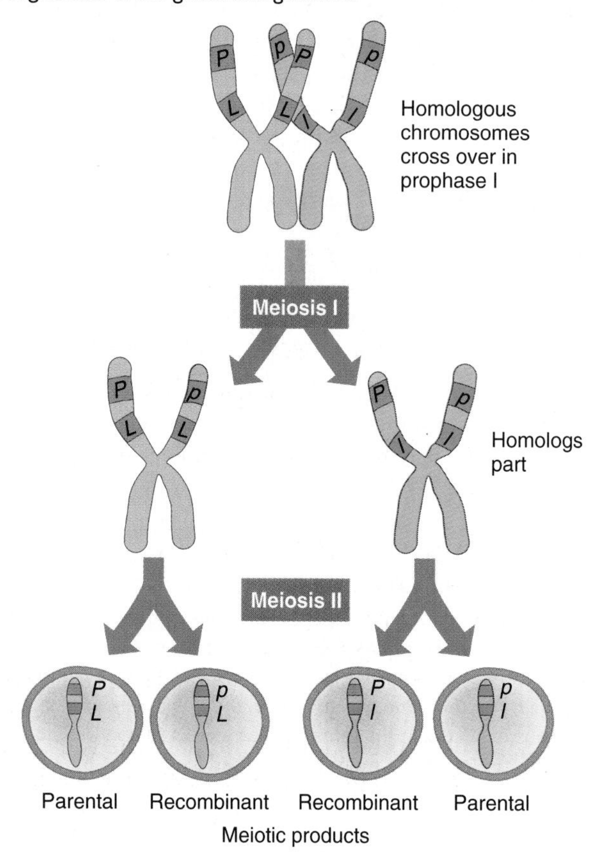

transmitted on the same chromosome, and the genes therefore did not separate during meiosis.

If the two genes were on the same chromosome, why did the researchers see non-parental trait combinations at all? These offspring classes arise because of another meiotic event, crossing over. Recall that crossing over is an exchange between homologs that mixes up maternal and paternal gene combinations in the gametes (see fig. 9.8). Figure 11.4 follows the fate of alleles during crossing over.

Recombinant chromosomes result from the mixing of maternal and paternal alleles into new combinations in the meiotic products. **Parental** chromosomes retain the gene combinations from the parents. "Parental" and "recombinant" are relative terms, however, depending on the parents' allele combinations. Had the parents in Bateson and Punnett's crosses been of genotypes *ppL_* and *P_ll* (phenotypes red flowers, long pollen grains and purple flowers, round pollen grains), then *P_L_* and *ppll* would be the recombinant rather than the parental classes.

Two other terms describe the arrangement of linked genes in heterozygotes. Consider a pea plant with genotype *PpLl.* These alleles can be on the same chromosome in two different positions. If the two dominant alleles travel on one chromosome and the two recessive alleles on the other, the genes are said to be in coupling. In

FIGURE 11.5
Allele Configuration Is Important.
Parental chromosomes can be distinguished from recombinant chromosomes only if the allele configuration of the two genes is known—they are either in coupling (**A**) or in repulsion (**B**).

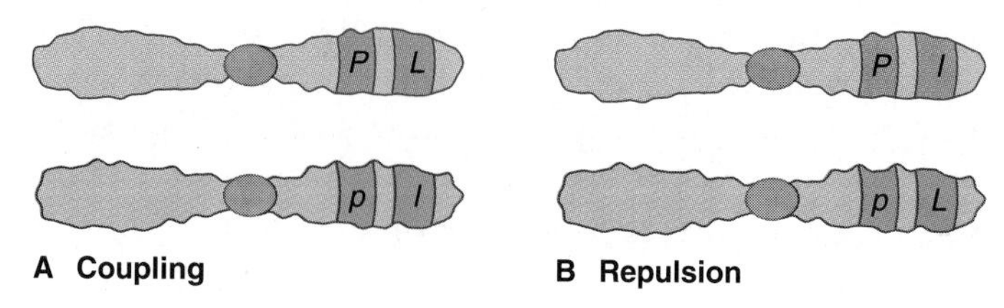

the opposite configuration, when one dominant allele is near one recessive allele on a chromosome, the two genes are in repulsion (fig. 11.5).

Knowing the allele configuration is important in predicting trait transmission. Alleles in coupling (the same type on the same homolog) tend to be transmitted together. However, the same type of alleles on different homologs separate at meiosis and may not be passed on together. This is valuable information when tracing traits, because alleles in coupling tend to stick together, whereas alleles in repulsion separate with each generation. Therefore, alleles in coupling transmit dominant expressions of the two traits, or recessive expressions of both. We return to medical implications of this distinction shortly.

As Bateson and Punnett were studying linkage in peas, Thomas Hunt Morgan at Columbia University was breeding the fruit fly *Drosophila melanogaster* to determine whether pairs of traits were linked. As data accumulated, the trait pairs fell into four groups. Within each group, crossed dihybrids did not produce the proportions of offspring Mendel's second law predicts. This indicated four sets of linked genes—precisely the number of chromosome pairs in the fly. The traits fell into four groups because the genes specifying traits that are transmitted together are on the same chromosome.

Morgan wondered why the size of the recombinant classes varied among the crosses. Might the differences reflect different physical relationships between the genes on the chromosome? Alfred Sturtevant, Morgan's undergraduate assistant, explored this idea. In 1911, he developed a theory and technique that would profoundly affect the fledgling field of genetics in his day and the medical genetics of today. Sturtevant proposed that the farther apart two genes are on a chromosome, the more likely a crossover is to occur between them—

FIGURE 11.6
Breaking Linkage.
Crossing over is more likely to occur between the widely spaced linked genes A and B or between A and C than between the more closely spaced linked genes B and C, because there is more room for an exchange to occur.

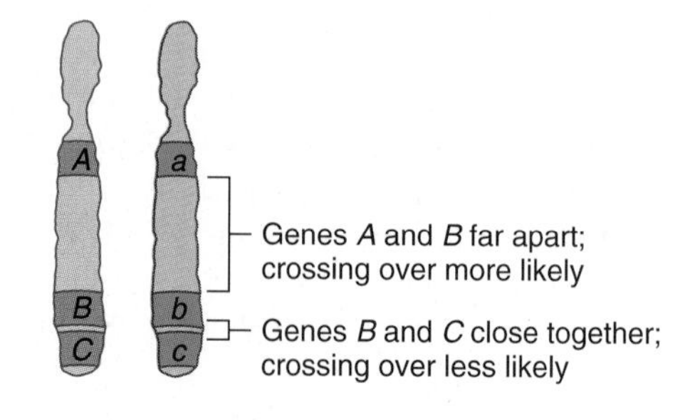

simply because there is more space between the two genes (fig. 11.6). This idea became the basis for mapping genes on chromosomes.

Linkage Maps

Geneticists use the correlation between crossover frequency and the distance between genes to construct **linkage maps,** which are diagrams of gene order and spacing on chromosomes. By determining the proportion of recombinant offspring, investigators can infer the frequency with which two genes cross over and how far apart the genes are on the chromosome. Genes that occupy opposite ends of the same chromosome cross over often, generating a large recombinant class. In contrast, a crossover would rarely separate genes lying very close on the chromosome, and they would not generate a large recombinant class (fig. 11.6). It is a little like the ease of walking between two people standing at opposite ends of a room compared to the ease of walking between two people who are slow dancing.

In 1913, Sturtevant published the first genetic linkage map, depicting the order of five genes on the X chromosome of the fruit fly. Researchers rapidly mapped genes on all four fly chromosomes. In humans, researchers mapped the X chromosome first. It is simpler to map than the autosomes because human males have only a single X chromosome. This means that X chromosome alleles in the human male are never masked and are therefore easier to study.

By 1950, genetic researchers began to contemplate the daunting task of mapping genes on the 22 human autosomes. The gene mapper needs a clue to begin matching a certain gene to its chromosome. To locate genes on autosomes, helpful clues come from individuals who have particular traits and particular chromosome abnormalities. That is, if people with a certain set of symptoms have an abnormality affecting the same part of a certain chromosome, then the gene causing the disorder is probably located on that chromosome.

Finding disease-chromosome links is rare. More often, researchers used the sorts of experiments Sturtevant conducted on his flies—calculating the percentage of recombination (crossovers) between two genes. Because individual humans do not produce hundreds of offspring, as fruit flies do, cytogeneticists had to devise other ways to detect linkage and recombination in humans. One way is to observe the same traits in many families and pool the data to obtain enough information to establish gene linkages. That is, if two specific traits nearly always occur together in many families, then the genes are probably linked and are usually inherited together, unless crossing over separates them.

Linkage of specific pairs of genes is the basis of genetic marker technology, which can predict the likelihood that a person will develop a particular inherited illness. A genetic marker is a detectable DNA sequence that all family members who have the illness share but that healthy relatives do not have. The DNA sequence "marks" the presence of the disease-causing gene because the two are closely linked—similar to the concept of guilt by association.

The linkage maps of the mid-twentieth century provided the rough drafts upon which DNA sequence information was added later in the century. Today, linkage maps and genetic marker technology are obsolete, because we know much of the human genome sequence—and several others, as the opening essay to chapter 13 describes.

11.2 MASTERING CONCEPTS

1. What is the relationship between genes and chromosomes?
2. How do patterns of inheritance differ when pairs of genes are linked, versus when they are not linked?
3. How can linkage be used to map genes on chromosomes?

OLC

11.3 How Do Chromosomes Determine Sex?

Presence or absence of a sex chromosome determines maleness or femaleness in some species, often reflecting the actions of individual genes. In other organisms, certain combinations of sex chromosomes, autosomes, or genes determine sex, and in still others, the environment may influence whether one develops as a male or a female.

Sex determination refers to the mechanism by which an individual develops as a male or a female. In many species, males and females differ in the types of sex chromosomes that they have. These chromosomes may determine sex because they include genes that direct development of reproductive structures of one sex, while suppressing development of structures of the other sex.

The sex with two different sex chromosomes is the **heterogametic** sex, and the other, with two of the same sex chromosomes, is the **homogametic** sex. In all mammals, the male is the heterogametic sex. In birds, moths, and butterflies, the female is the heterogametic sex.

A Diversity of Sex Determination Mechanisms

Sex determination mechanisms are diverse in the living world. Grasshoppers, for example, have only one type of sex chromosome, designated X. If two sex chromosomes are present (XX), the insect is a female; if only one is present (XO), it is a male.

In the fruit fly *Drosophila melanogaster,* the X-to-autosome ratio is important. Individuals with a pair of X chromosomes and a pair of each autosome are normal females; those with one X chromosome and a pair of each autosome are normal males. Although fertile male fruit flies are XY, it is not the Y chromosome that determines maleness but the ratio of the one X chromosome to the paired autosomes. An XO fruit fly is a sterile male (in humans, an XO is female), and an XXY fruit fly is female (in humans, an XXY is male).

Plants exhibit a variety of sex determination mechanisms. These include X-to-autosome ratios, active Y chromosomes, and autosomal genes determining sex.

Sex determination is more complex in some other species. Certain beetles have 12 X chromosomes, 6 Y chromosomes, and 18 autosomes. In other species, the number of total sets of chromosomes determines sex. In bees, a fertilized ovum becomes a diploid female; an unfertilized ovum develops into a haploid

FIGURE 11.7
Environment Affects Sex Determination.
For the slipper limpet, position in the stack, not chromosomes, determines sex.

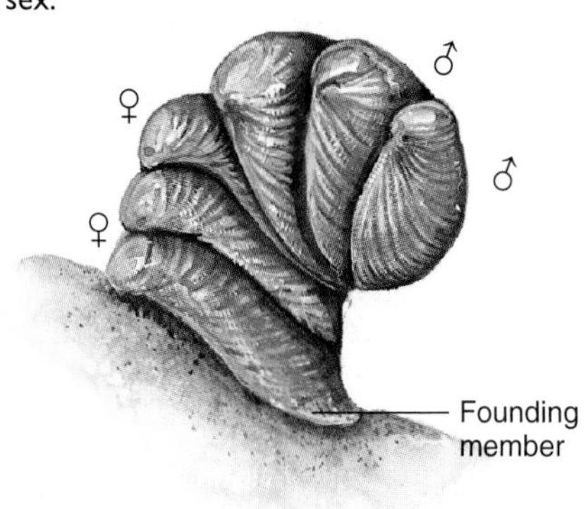

male. The queen bee actually determines sex: The oocytes that she allows to be fertilized become females, and those she doesn't become males.

In some species, the environment determines sex. For certain turtles, the temperature of the land the eggs are laid on determines sex; eggs laid in the sun hatch females, and those laid in the shade produce males. Chapter 1 discussed the efffect of pollution on sex determination in alligators.

Some marine bivalves and fish can even change sex in adulthood (fig. 11.7). For example, in the slipper limpet *Crepidula fornicata,* position determines sex. These organisms aggregate in groups anchored to rocks and shells on muddy sea bottoms. The founding member is a small male. As he grows, he becomes a she, and a new small male enters the top of the colony. When this second member grows large enough to become female, a third recruit enters. Thus, one's size and position in the pile determines sex.

Sex Determination in Humans

In mammals, the sexes have equal numbers of autosomes, but males have one X and one Y chromosome, and females have two X chromosomes. Figure 11.8 depicts the sex chromosomes in humans.

The human X chromosome carries more than 1,000 genes. The Y chromosome is much smaller than the X and carries far fewer genes. A quarter of the genes on the Y chromosome correspond to genes on the X chromosome, and are located at the ends of the Y chromosome. These areas are called pseudoautosomal regions, and crossing over can occur between these corresponding genes on the X and Y chromosomes. The other genes on the Y chromosomes are considered functionally to fall into two other groups—one consisting of genes similar in sequence to genes on the X chromosome, the other including genes that are unique to the Y chromosome.

Researchers identified the part of the Y chromosome that determines maleness by studying certain interesting people—men who have two X chromosomes and women who have one X and one Y chromosome, the reverse of normal. A close look at these individuals' sex chromosomes revealed that each of the XX males actually had a small piece of a Y chromosome, and each of the XY

FIGURE 11.8
The Human X and Y Chromosomes.

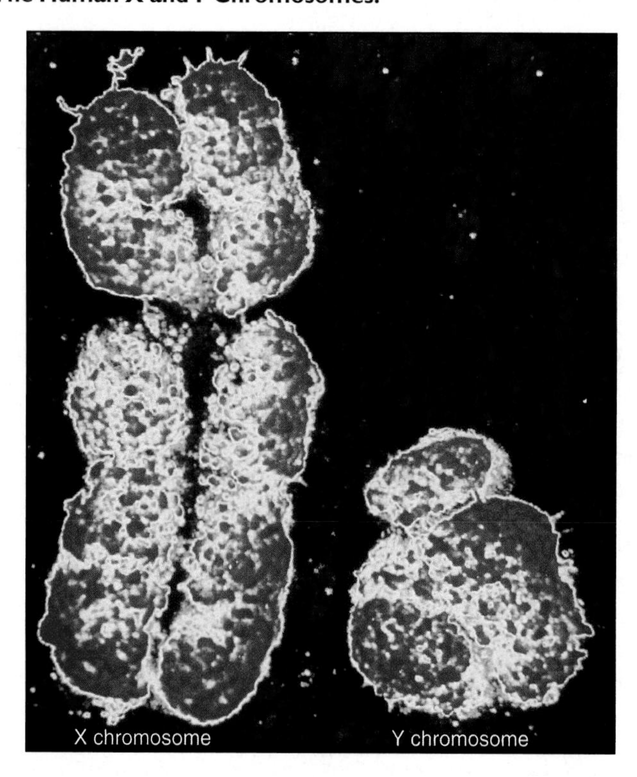

females lacked a small part of the Y chromosome. The part of the Y chromosome present in the XX males was the same part that was missing in the XY females. It was only a tiny part of the Y chromosome, about half a percent of its total structure, but it proved important because it includes a gene, called *SRY* (for sex-determining region of the Y), that specifies maleness.

The *SRY* gene produces a protein that switches on other genes that direct the embryo to develop male structures (see fig. 40.15). Rudimentary testes begin to secrete the male sex hormone testosterone, and cascades of other gene activities follow, shrinking breast tissue and promoting swellings that develop into male sex organs. The *SRY* protein also activates a gene that encodes a protein that destroys rudimentary female structures.

11.3 MASTERING CONCEPTS

1. What is sex determination?
2. How do chromosomes determine sex?
3. What are some nonchromosomal mechanisms of sex determination?

OLC

11.4 Inheritance of Genes on Sex Chromosomes

In humans, X-linked recessive traits pass from carrier mothers to affected sons and are more severe in males, because they have only one X chromosome to a female's two. X-linked dominant traits are rare. In female mammals, one X chromosome is inactivated in each somatic cell early in development.

Health 11.1

X-Linked Recessive and X-Linked Dominant Inheritance

Ichthyosis

The middle-aged man had rough, brown, scaly skin since early childhood. Because no one else in the family had the condition, no one suspected that it might be inherited. Then the man's daughter had a son. By the time the boy was a year old, his skin had begun to appear dried out. As the condition worsened, the family suspected that he had the same problem that his grandfather had.

A medical geneticist diagnosed ichthyosis, a rare X-linked recessive condition. Lack of an enzyme interferes with the removal of cholesterol from skin cells. The outermost skin layer cannot peel off as it normally would as the cells fill with keratin and die, and instead accumulate, producing the rough, scaly appearance. The mother, who is a carrier because her father and son are affected, produces enough of the enzyme so that her own skin is normal.

Congenital Generalized Hypertrichosis

In X-linked dominant inheritance, a female who inherits a disease-causing allele generally has symptoms, but a male has a much more severe phenotype, because he has no normal gene counterpart on a second X chromosome.

In congenital generalized hypertrichosis (CGH) a person has many more hair follicles than normal. Affected males have dense hair on the face and upper body. The hair growth is milder and spottier in females because of hormonal differences and the effects of a normal gene on the second X chromosome.

Researchers studied a large Mexican family that had 19 relatives with CGH. The pattern of inheritance was distinctive for X-linked dominant inheritance, which is quite rare. In one portion of the pedigree, an affected man passed the trait to all four of his daughters, but to none of his nine sons. Because sons inherit the X chromosome from their mother, and only the Y from their father, they could not have inherited CGH from their affected father.

The mutant gene that causes CGH is atavistic, which means that it controls a trait also present in ancestral species. A version of the gene is probably present in chimpanzees and other hairy primates. Sometime in our distant past, the wild type form of the gene must have mutated in a way that enables humans to grow dense hair only on their heads and in areas dictated by sex hormones.

Unfortunately, men with this condition have been displayed in circus side shows, advertised as "ape men" or "werewolves."

Genes transmitted on the X chromosome are **X-linked,** and those on the Y chromosome, **Y-linked.** Y-linked traits in humans are extremely rare because the Y chromosome contains very few genes. Most of the well-studied X-linked traits are recessive. Health 11.1 discusses an X-linked recessive and an X-linked dominant medical condition.

X-Linked Recessive Inheritance

Predicting the inheritance pattern of an X-linked trait requires understanding the differences in gene expression between the sexes. Any gene on the X chromosome of a male mammal is expressed in his phenotype, because he lacks a second allele for that gene, which could mask its expression. An allele on an X chromosome in a female may or may not be expressed, depending upon whether it is dominant or recessive and on the nature of the allele on the second X chromosome. The male is **hemizygous** for X-linked traits because he has half the number of genes the female has. He either has the trait or does not—he cannot be a carrier.

A male always inherits his Y chromosome from his father and his X chromosome from his mother (fig. 11.9). A female inherits one X chromosome from each parent. If a mother is heterozygous for an X-linked gene, her son has a 50% chance of inheriting either allele from her. X-linked genes are therefore passed from mother to son. Because a male does not receive an X chromosome from his father, a man cannot pass an X-linked trait to his son.

FIGURE 11.9

Sex Determination in Humans.

An oocyte contains a single X chromosome. A sperm cell contains either an X chromosome or a Y chromosome. If a Y-bearing sperm cell with an *SRY* gene fertilizes an oocyte, the zygote is a male (XY). If an X-bearing sperm cell fertilizes an oocyte, then the zygote is a female (XX).

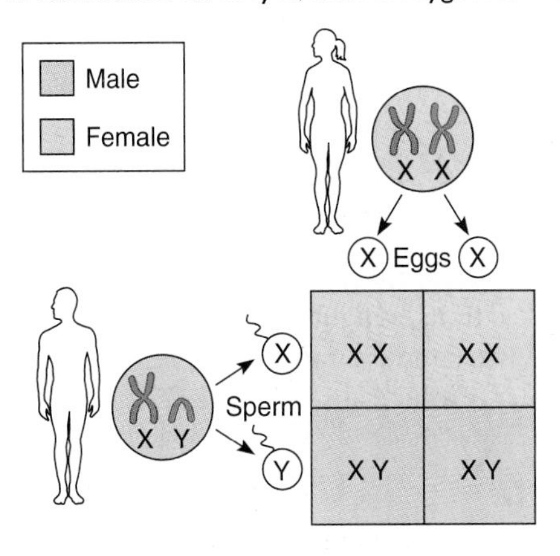

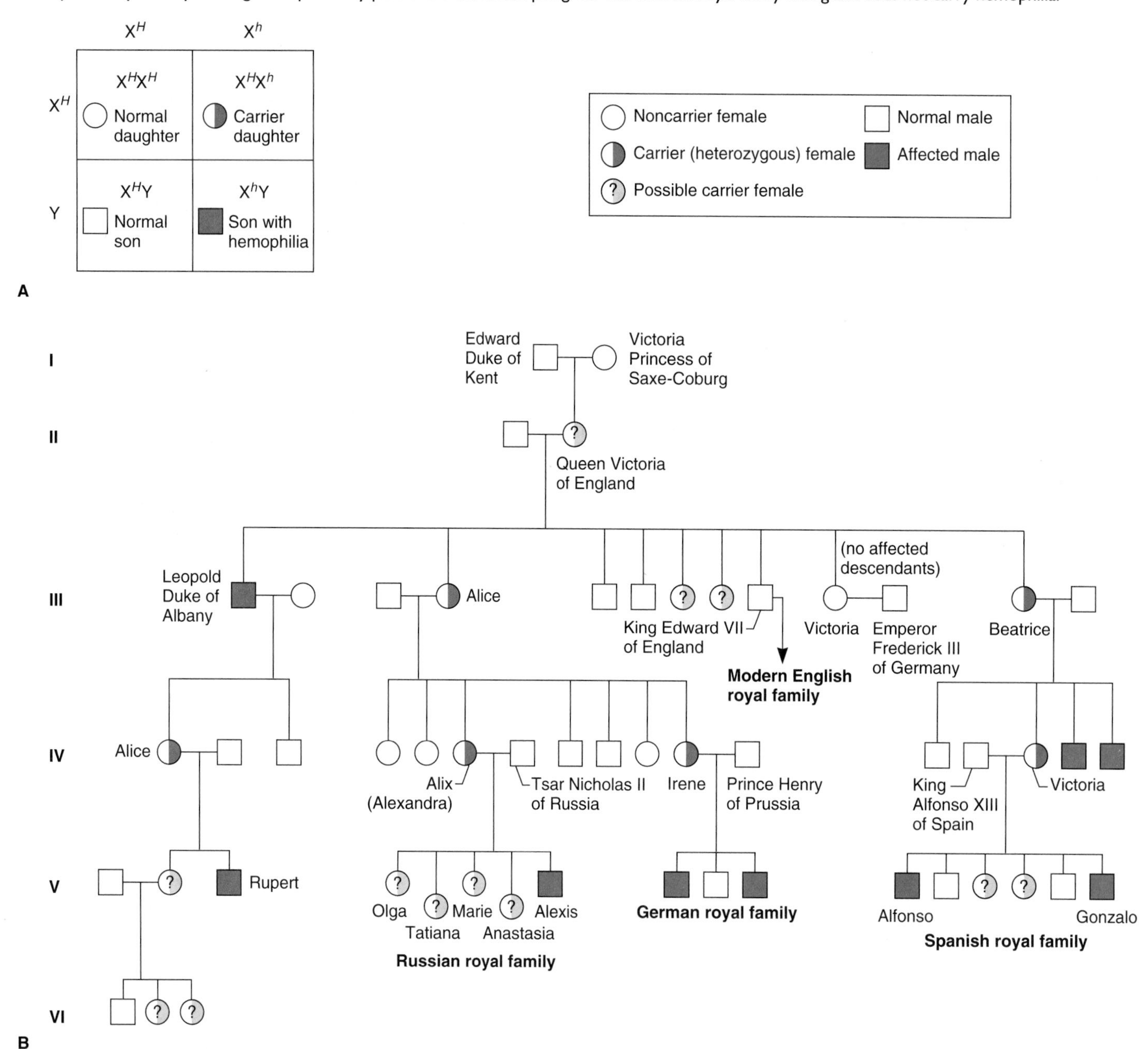

FIGURE 11.10
Hemophilia A.
(**A**) This X-linked recessive disease usually passes from a heterozygous woman (designated $X^H X^h$, where X^h is the hemophilia-causing allele) to heterozygous daughters or hemizygous sons. (**B**) The disorder has appeared in the royal families of England, Germany, Spain, and the former Soviet Union. The mutant allele apparently arose in Queen Victoria, who was either a carrier or produced oocytes in which the gene mutated. She passed the allele to Alice and Beatrice, who were carriers, and to Leopold, who had a case so mild that he fathered children. In the fourth generation, Alexandra was a carrier and married Nicholas II, Tsar of Russia. Alexandra's sister Irene married Prince Henry of Prussia, passing the allele to the German royal family, and Beatrice's descendants passed it to the Spanish royal family. This figure depicts only part of the enormous pedigree. The modern royal family in England does not carry hemophilia.

Punnett squares and pedigrees are used to depict transmission of X-linked traits, as figure 11.10 shows for the inheritance of hemophilia A in a famous family. Hemophilia A is an X-linked recessive disorder in which absence or deficiency of a protein clotting factor greatly slows blood clotting. Cuts are slow to stop bleeding, bruises happen easily, and internal bleeding may occur without the person's awareness. Receiving the missing clotting factor can control the illness.

The risk that a carrier mother will pass hemophilia A to her son is 50%, because he can inherit either her normal allele or the mutant one. The risk that a daughter will inherit the hemophilia allele and be a carrier like her mother is also 50%.

Consider a woman whose brother has hemophilia A but whose parents are healthy. The woman's chance of having inherited the hemophilia allele on the X chromosome she received from her mother and being a carrier is 50%, or one-half. The

Investigating Life 11.1

Of Preserved Eyeballs and Duplicated Genes—Color Blindness

John Dalton, a famous English chemist, saw things differently than most people. In a 1794 lecture, he described his visual world. Sealing wax that appeared red to other people was as green as a leaf to Dalton and his brother. Pink wildflowers were blue, and Dalton perceived the cranesbill plant as "sky blue" in daylight, but "very near yellow, but with a tincture of red," in candlelight. The Dalton brothers, like 7% of males and 0.4% of females today, had the X-linked recessive trait of color blindness.

Dalton was very curious about the cause of his color blindness, so he made arrangements with his personal physician, Joseph Ransome, to dissect his eyes after he died. Ransome snipped off the back of one eye, removing the retina, where the cone cells that provide color vision nestle among the more abundant rod cells that impart black-and-white vision. Because Ransome could see red and green normally when he peered through the back of his friend's eyeball, he concluded that it was not an abnormal filter in front of the eye that altered color vision.

Fortunately, Ransome stored the eyes in dry air, where they remained relatively undamaged. In 1994, Dalton's eyes underwent DNA analysis at London's Institute of Ophthalmology. The research showed that Dalton's remaining retina lacked one of three types of pigments, called photopigments, that enable cone cells to capture certain incoming wavelengths of light.

Each photopigment consists of a vitamin A-derived portion called retinal and a protein portion called an opsin. The presence of opsins—because they are controlled by genes—explains why color blindness is inherited.

John Dalton wasn't the only person interested in color blindness to have work done on his own tissue. Johns Hopkins University researcher Jeremy Nathans does so today—but not posthumously. Nor is he color blind.

Nathans identified three opsin genes—one on chromosome 7, the other two on the X chromosome. But his opsin genes were not entirely normal, and this gave him a big clue as to how the trait arises and why it is so common. On his X chromosome, Nathans has one red opsin gene and two green genes, instead of the normal one of each (the colors refer to the wavelengths of light a protein captures). Because the red and green genes have similar DNA sequences, Nathans reasoned, they can misalign during meiosis in the female (fig. 11.A). The resulting oocytes would then have either two or none of one opsin gene type. An oocyte lacking either a red or a green opsin gene would, when fertilized by a Y-bearing sperm, give rise to a color-blind male.

FIGURE 11.A

(**A**) The sequence similarities between the opsin genes responsible for color vision may cause chromosome misalignment during meiosis in the female. Offspring may inherit too many, or too few, opsin genes. A son inheriting an X chromosome missing an opsin gene would be color blind. A daughter, unless her father is color blind, would be a carrier. (**B**) A missing gene causes X-linked color blindness.

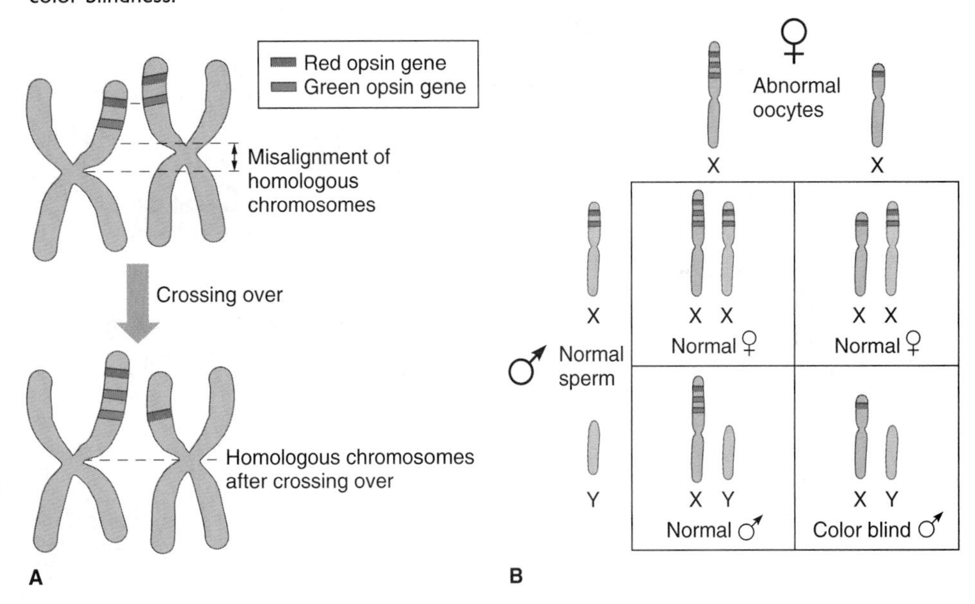

chance that the woman will conceive a son with hemophilia is one-half multiplied by one-half. This is because the chance that she is a carrier is one-half and, if she is a carrier, the chance that she will transmit the X chromosome bearing the hemophilia allele to a son is also one-half.

Females usually do not exhibit X-linked traits because the wild type allele on their second X chromosome masks the expression of the mutant allele. A female has the trait only if she inherits a recessive mutant allele from each parent. Unless she has a specific biochemical or genetic test, a woman usually doesn't know she carries an X-linked recessive trait unless she has an affected son. Color blindness is an X-linked trait seen in females more often than some other X-linked traits because it does not harm health sufficiently to prevent a man from fathering children. It is discussed in Investigating Life 11.1.

Even though the Y chromosome has very few protein-encoding genes, researchers can still use its DNA sequences to trace father-to-son inheritance. Comparing Y chromosome sequences has enabled researchers to identify Thomas Jefferson's descendants through his slave Sally Hemings, and to show that a priestly group of Jewish men called cohanim share genetic as well as Biblical ancestry, possibly extending back to Moses' brother Aaron.

X Inactivation—Equalizing the Sexes

Female mammals have two alleles for every gene on the X chromosome to a male's one. A mechanism called **X inactivation** helps balance this inequality by inactivating one X chromosome in each female cell during early prenatal development.

Whether the turned-off X chromosome comes from the mother or father is random. As a result, a female expresses the paternal X chromosome genes in some cells and the maternal genes in others.

A specific part of the X chromosome, the X-inactivation center, shuts off the chromosome one gene at a time, leaving only a few genes active. Once an X chromosome is inactivated in one cell, all the cells that form when that cell divides have the same inactivated X chromosome (fig. 11.11). Because the inactivation occurs early in development, females have patches of tissue that differ in their ex-

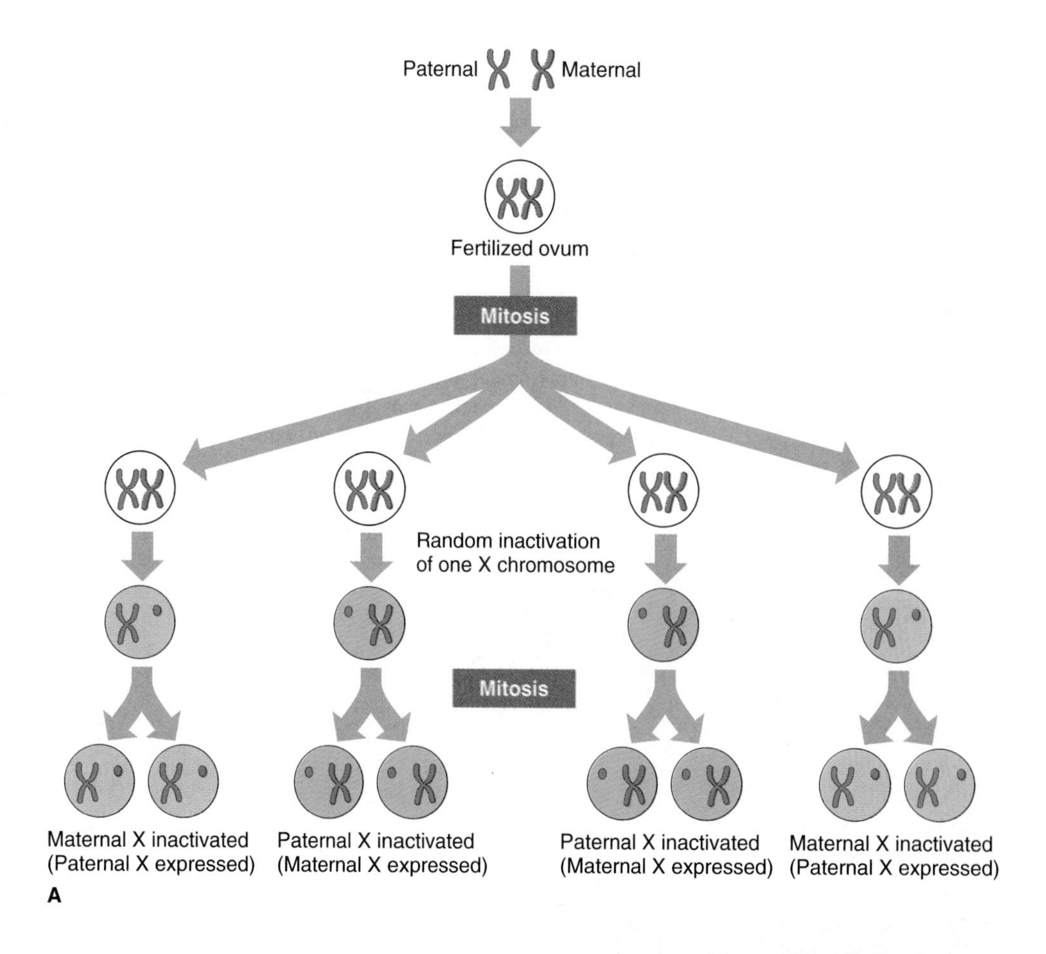

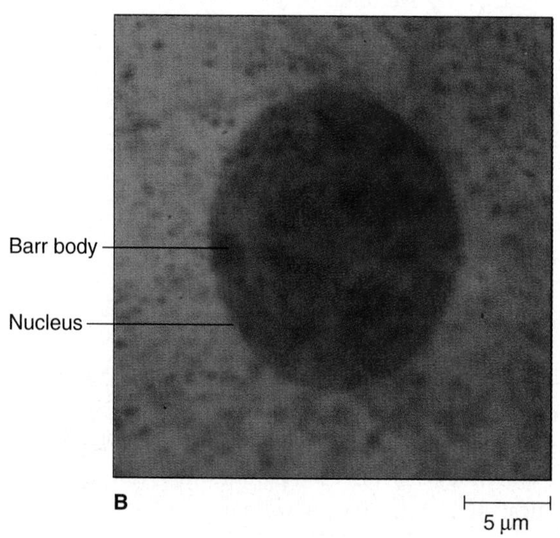

FIGURE 11.11
X Inactivation.
(**A**) At about the third week of embryonic development in the human female, one X chromosome at random in each diploid cell is inactivated, and all daughter cells of these cells have the same X chromosome turned off. The inactivated X may come from the mother or father, resulting in a female who is mosaic at the cellular level for X chromosome gene expression. (**B**) A cell of a normal female human has one Barr body (inactivated X chromosome).

pression of X-linked genes. A female is thus a genetic mosaic for any heterozygous genes on the X chromosome because some cells express one allele, and other cells express the other allele.

For most traits, the cells with active wild type alleles produce enough of the specified protein for the female to be healthy. Rarely, a female who is heterozygous for an X-linked gene can have symptoms if, by chance, the wild type allele is inactivated in many of her cells that the condition affects. A carrier of hemophilia might experience slow blood clotting, and a carrier of muscular dystrophy, muscle weakness. Figure 11.12 explains how X inactivation of either of two alleles of a coat color gene causes the distinctive appearance of calico cats, which are always female. The earlier X inactivation occurs, the larger the patches.

FIGURE 11.12
Calico and Tortoiseshell Cats Reveal Patterns of X Inactivation.
A striking exhibition of X inactivation occurs in the calico cat. Each orange patch is made up of cells descended from a cell in which the X chromosome carrying the coat color allele for black was inactivated; each black patch is made of cells descended from a cell in which the X chromosome carrying the orange allele was turned off. A different gene accounts for the white background. (**A**) X inactivation happened early in development, resulting in large patches. (**B**) X inactivation occurred later in development, producing smaller patches and a "tortoiseshell" coloration pattern.

A

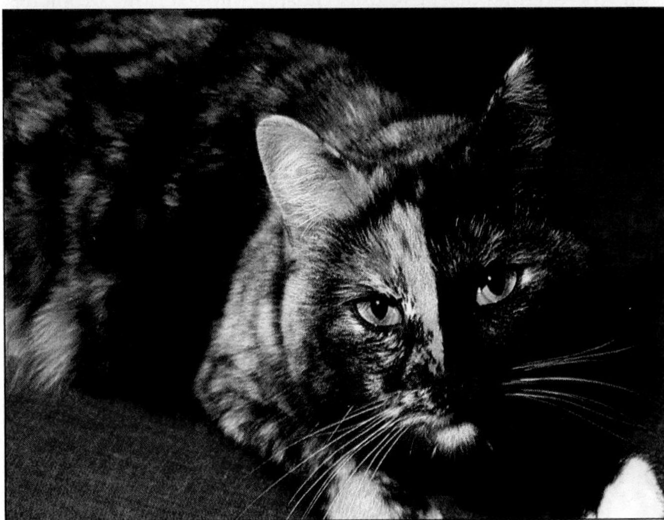
B

X inactivation is directly observable in cells, because the turned-off X chromosome absorbs a stain much more readily than an active X chromosome does (fig. 11.11b). In the nucleus of a female cell in interphase, the dark-staining, inactivated X chromosome is called a Barr body. Murray Barr, a Canadian researcher, first noticed the bodies in 1949 in the nerve cells of female cats. A normal male cell has no Barr bodies because the one X chromosome remains active. Cells of a female with extra X chromosomes have extra Barr bodies.

11.4 MASTERING CONCEPTS

1. How are genes on the X chromosome inherited?
2. Why are X-linked genes expressed differently in males and females?
3. How does X inactivation in mammals equalize the contributions of X-linked genes between the sexes?

OLC

11.5 When Chromosomes Are Abnormal

Chromosomal health is a matter of balance. Extra or missing chromosomes, or their parts, or inverted or rearranged segments, can drastically alter the phenotype.

An altered chromosome number drastically affects the phenotype because a chromosome consists of so many genes. Even small abnormalities in chromosome structure can have devastating effects on health. Table 11.1, at the end of the chapter, reviews several types of chromosome abnormalities.

Extra Chromosome Sets—Polyploidy

An error in meiosis can produce a gamete with an extra complete set of chromosomes, a condition called **polyploidy** ("many sets"). For example, if a sperm with the normal one copy of each chromosome fertilizes an oocyte with two copies of each chromosome, the resulting zygote (fertilized ovum) will have three copies of each chromosome, a type of polyploidy called triploidy.

Most human polyploids cease developing as embryos or fetuses, but occasionally one survives for a few days after birth, with defects in nearly all organs. However, about 30% of flowering plant species tolerate polyploidy well, and many crops are polyploids. The wheat in pasta is tetraploid—it has four sets of seven chromosomes—and the wheat species in bread is a hexaploid—it has six sets of seven chromosomes.

Extra and Missing Chromosomes—Aneuploidy

A **euploid** ("true set") cell has the normal chromosome number for the species. An **aneuploid** ("not true set") cell, by contrast, has a missing or extra chromosome. Symptoms or

FIGURE 11.13
Extra and Missing Chromosomes—Aneuploidy.
Unequal division of chromosome pairs can occur at the first or second meiotic division. (**A**) A single pair of chromosomes is unevenly partitioned into the two cells arising from the first division of meiosis in a male. The result: two sperm cells that have two copies of the chromosome, and two sperm cells that have no copies of that chromosome. When a sperm cell with two copies of the chromosome fertilizes a normal oocyte, the zygote is trisomic for that chromosome; when a sperm cell lacking the chromosome fertilizes a normal oocyte, the zygote is monosomic for that chromosome. Symptoms depend upon which chromosome is involved. (**B**) This nondisjunction occurs at the second meiotic division. Because the two products of the first division are unaffected, two of the mature sperm are normal and two are aneuploid. Oocytes can undergo nondisjunction as well, leading to zygotes with extra or missing chromosomes when normal sperm cells fertilize them.

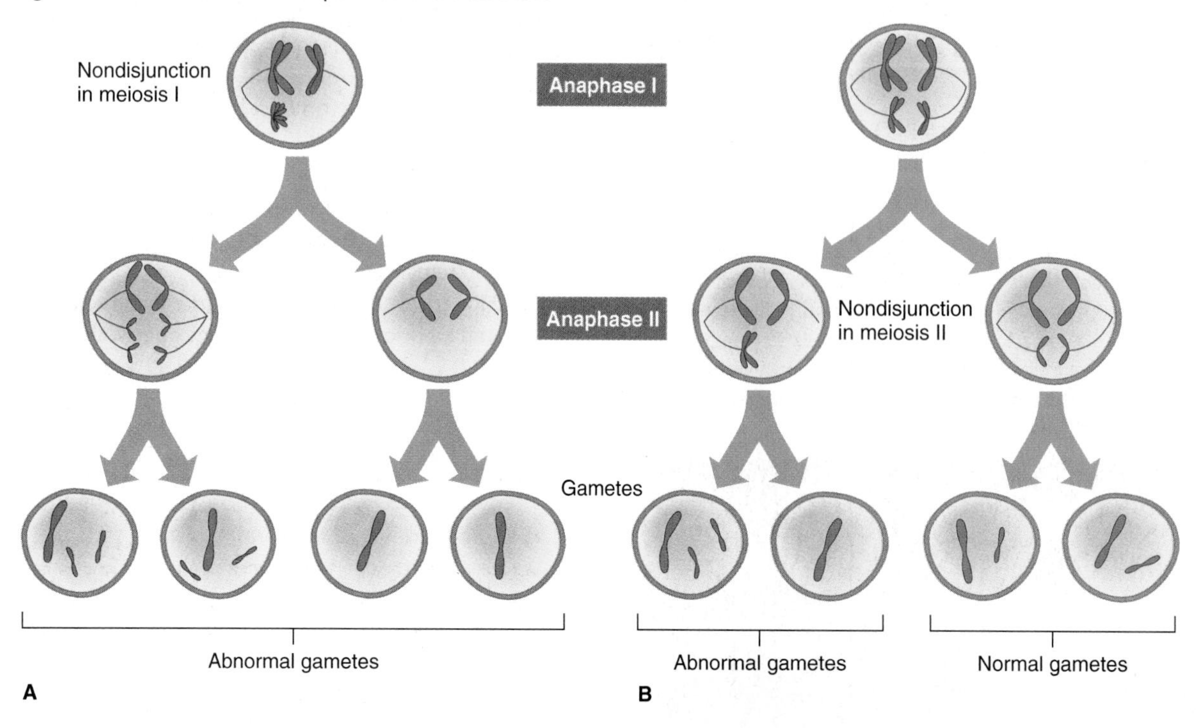

altered characteristics resulting from aneuploidy depend upon which chromosome is affected. In humans, autosomal aneuploidy often results in mental retardation, because so many genes contribute to brain function. Sex chromosome aneuploidy in humans is less severe.

Extra genetic material is less dangerous than missing material. This is why most children born with the wrong number of chromosomes have an extra one (called a **trisomy**) rather than a missing one (a **monosomy**). Trisomies and monosomies are named according to the affected chromosome, and the syndromes they cause were originally named for the investigator who first described them. An extra chromosome 21, for example, is called trisomy 21, and it causes Down syndrome.

Aneuploidy occurs due to a meiotic error called **nondisjunction** (fig. 11.13). Recall that in normal meiosis, pairs of homologous chromosomes separate, and each of the resulting cells (sperm or egg) contains only one member of each pair. In nondisjunction, a chromosome pair fails to separate, either at the first or second meiotic division. The result is a sperm or oocyte with either two copies of a particular chromosome or none at all, rather than the normal one copy. In humans, when such a gamete fuses with its opposite at fertilization, the resulting zygote has either 45 or 47 chromosomes instead of the normal 46. Most aneuploids cease developing before birth and account for about 50% of all spontaneous abortions. In plants, aneuploidy may produce a seed (an embryo and its food supply) that does not germinate.

Aneuploidy and polyploidy can also arise during mitosis and produce groups of somatic cells with a chromosomal aberration. Having only a few cells with an abnormal chromosome number may not impair health. A normally functioning human liver, for example, may have patches of polyploid cells. If mitotic aneuploidy occurs in a very early embryo, however, many cells descend from the original defective one, causing more serious problems. Some people who have mild Down syndrome, for example, are really chromosomal mosaics—that is, some of their cells carry an extra chromosome 21, and some are normal.

Following is a look at some syndromes in humans resulting from aneuploidy.

Autosomal Aneuploids An individual with trisomy 21, the most common autosomal aneuploid, has distinctive facial features, including a flat face, slanted eyes, straight, sparse hair, a protruding tongue, and thick lips. He or she also has an abnormal pattern of hand creases, loose joints, and poor reflex and muscle tone. Children reach developmental milestones (such as sitting, standing, and walking) slowly, and toilet training may take several years. Intelligence varies greatly; some have profound mental impairment, whereas others attend college (fig. 11.14).

FIGURE 11.14

Trisomy 21 Down Syndrome.

(**A**) Wendy Weisz enjoys studying art at Cuyahoga Community College. (**B**) A karyotype for trisomy 21 Down syndrome shows the extra chromosome 21.

A

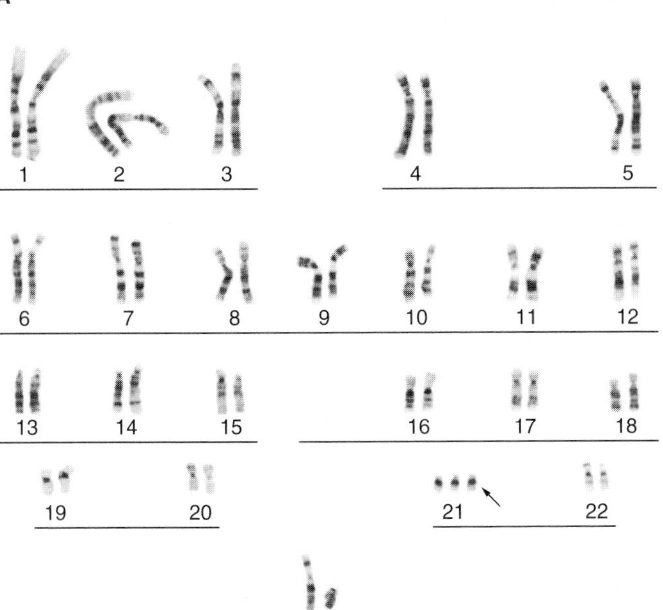

B

Prenatal tests can detect trisomy 21, but they cannot predict the severity of the syndrome. Nearly 50% of affected children die before their first birthdays, often of heart or kidney defects or infection. Many people with trisomy 21 who live past age 40 develop Alzheimer disease.

The likelihood of giving birth to a child with trisomy 21 increases dramatically as a woman ages, based on empirical evidence. For women under 30, the chances of conceiving a child with the syndrome are 1 in 3,000. For a woman of 48, the incidence jumps to 1 in 9. The increased likelihood of nondisjunction in older oocytes may account for this age association. However, about 40% of trisomy 21 cases result from aneuploid sperm.

Trisomies 13 and 18 are the next most common autosomal aneuploids. Affected fetuses usually cease developing before birth. An infant with trisomy 13 has an underdeveloped face, extra and fused fingers and toes, heart defects, small adrenal glands, and a cleft lip or palate. An infant with trisomy 18 suffers many of the problems seen in trisomy 13, plus a few distinctive features. These include a peculiar positioning of the fingers and flaps of extra abdominal skin called a "prune belly." Trisomies undoubtedly occur with other chromosomes but are naturally aborted very early.

Sex Chromosome Aneuploids People with the same chromosomal abnormality often share similar physical characteristics, as in trisomy 21. At a 1938 medical conference, a physician, Henry Turner, reported on seven young women, aged 15 to 23, who were short and sexually undeveloped, and had folds of skin on the backs of their necks. Other physicians soon began recognizing patients with this disorder, named Turner syndrome and today called XO syndrome. At first doctors thought that the odd collection of traits reflected a hormonal insufficiency; in 1954, researchers discovered that the cells of Turner syndrome patients lack Barr bodies and therefore have only one X chromosome. This deficit causes the symptoms, particularly the failure to mature sexually.

More than 99% of XO fetuses do not survive to be born, and the other 1% account for the 1 in 2,000 newborn girls with XO syndrome. Many affected young women do not know they have a chromosomal abnormality until they lag in sexual development. They are usually of normal intelligence and, if treated with hormone supplements, lead fairly normal lives, but they are infertile.

About 1 in every 1,000 to 2,000 females has an extra X chromosome in each cell, a condition called triplo-X. Symptoms are tallness, menstrual irregularities, and a normal-range IQ but that is slightly lower than that of other family members. However, a woman with triplo-X may produce some oocytes bearing two X chromosomes, which increases her risk of giving birth to triplo-X daughters or XXY sons.

Males with an extra X chromosome (XXY) have Klinefelter or XXY syndrome. They are sexually underdeveloped, with rudimentary testes and prostate glands and no pubic or facial hair. They also have very long arms and legs and large hands and feet, and they may develop breast tissue. Individuals with XXY syndrome may be slow to learn, but they are usually not mentally retarded unless they have more than two X chromosomes, which is rare. The syndrome occurs in 1 out of every 500 to 2,000 male births. Like XO syndrome, XXY syndrome varies greatly, and some affected individuals do not realize anything is amiss until well into adulthood.

One male in 1,000 has an extra Y chromosome, a condition called Jacobs or XYY syndrome. It once made headlines, and was the subject of an episode of the television program *Law and Order.* In 1965, researcher Patricia Jacobs published results of a survey among inmates at a high-security mental facility in Scotland. Of 197 men, 12 had chromosome abnormalities, 7 of them an extra Y! Jacobs hypothesized that the extra Y chromosome might cause these men to be violent and aggressive. When other facilities reported similar incidences of XYY men, *Newsweek* magazine ran a cover story on "congenital criminals." In the early 1970s, hospital nurseries in several nations began screening for XYY boys. Social workers and psychologists offered "anticipatory guidance" to parents in raising their toddling criminals. However, a correlation between two events does not necessarily signify cause and effect. By 1974, geneticists and others halted the program.

Today, we know that 96% of XYY males are apparently normal, the only consistent symptoms being great height, acne, and speech and reading problems. An explanation for the high prevalence of XYY among prison populations may be more psychological than biological. Teachers, employers, parents, and others may expect more of these physically large boys and men than of their peers, and a small percentage of them may cope with this stress with aggression.

Medical researchers have never reported a sex chromosome constitution of one Y and no X. When a zygote lacks any X chromosome, so much genetic material is missing that it probably cannot sustain more than a few cell divisions.

Smaller-Scale Chromosome Abnormalities

Because chromosomes consist of so many genes, even small changes—material missing, extra, inverted, or moved—can affect the phenotype. Figure 11.15 illustrates smaller chromosomal changes.

Sometimes only part of a chromosome is missing (a deletion) or extra (a duplication). Large deletions and duplications appear as missing or extra bands on stained chromosomes. Cri du chat syndrome (French for "cat's cry"), for example, is associated with deletion of the short arm of chromosome 5. The child has an odd cry similar to the mewing of a cat and pinched facial features, mental retardation and developmental delay. The chromosome region responsible for the catlike cry is distinct from the region causing the mental and developmental symptoms, indicating that a deletion causes a syndrome by removing more than one gene.

Duplications, like deletions, are more likely to cause symptoms if they affect large amounts of genetic material. For example, duplications within chromosome 15 are common, but they do not produce a distinctive phenotype (seizures and mental retardation) unless they repeat several genes.

Chromosome Rearrangements Deletions and duplications arise from chromosome rearrangements, which force chromosomes to pair in meiosis in ways that delete or duplicate genes. Rearrangements include inversions and translocations.

FIGURE 11.15
Chromosomal Mistakes.
Chromosome abnormalities include deletions, duplications, and inversions. The letters represent genes.

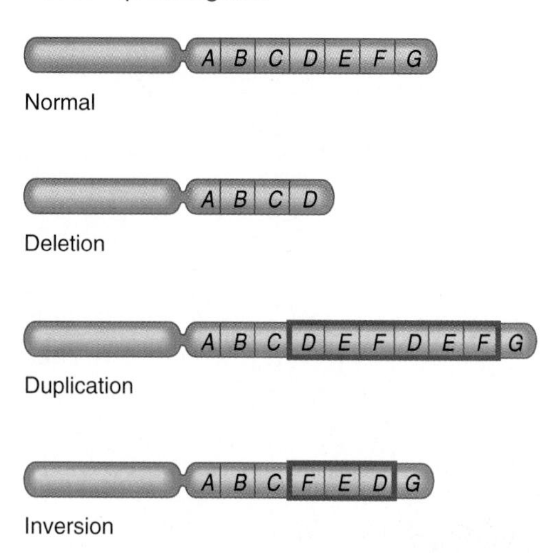

In an inversion, part of the chromosome flips and reinserts, rearranging the gene sequence. In humans, 5 to 10% of inversions harm health, probably because they disrupt vital genes.

An adult who is heterozygous for an inversion in one chromosome can be healthy, because all the genes are present, but he or she has reproductive problems. This is because in meiosis the inverted chromosome and its noninverted homolog twist around each other in a way that generates chromosomes with missing or extra genes, if crossing over occurs. Spontaneous abortions or birth defects may result.

In a translocation, different (nonhomologous) chromosomes exchange parts or combine. Certain viruses, drugs, and radiation can cause translocations, but we often do not know how they arise.

There are two types of translocations. In a **Robertsonian translocation,** the short arms of two nonhomologs break, leaving sticky ends that then join the two long arms into a new, large chromosome. The person with the large chromosome may not have symptoms if a crucial gene has not been deleted or damaged, but he or she may produce sperm or oocytes with too many or too few genes. This can lead to spontaneous abortion or birth defects. One in twenty cases of Down syndrome arises because a person has a Robertsonian translocation involving chromosome 21.

In a **reciprocal translocation,** two nonhomologs exchange parts (fig. 11.16). If this action does not break any genes, then a person who has both translocated chromosomes is healthy and is called a translocation carrier. He or she has the normal amount of genetic material, but it is rearranged. A translocation carrier produces some gametes that have duplications or deletions of genes in the translocated chromosomes. This occurs if the sperm or oocyte receives one reciprocally translocated chromosome but not the other, causing a genetic imbalance. The

FIGURE 11.16

A Reciprocal Translocation.

In a reciprocal translocation, two nonhomologous chromsomes exchange parts. In (**A**), genes *C, D,* and *E* on the blue chromosome exchange positions with genes *M* and *N* on the red chromosome. (**B**) shows a reciprocal translocation that is highlighted using FISH, which is described in the opening essay. The pink chromosome with the dab of blue, and the blue chromosome with a small section of pink, are the translocated chromosomes.

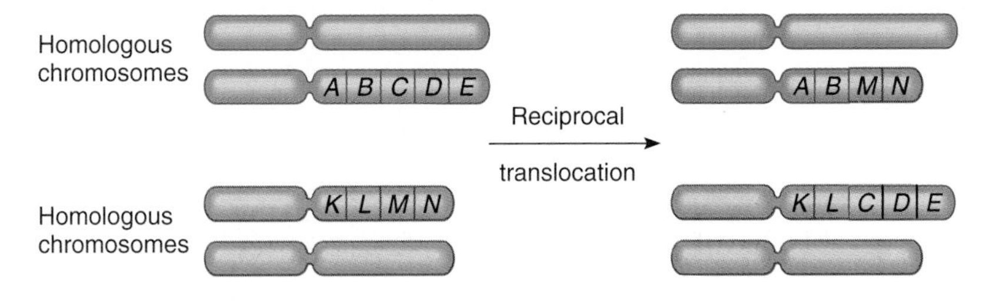

A

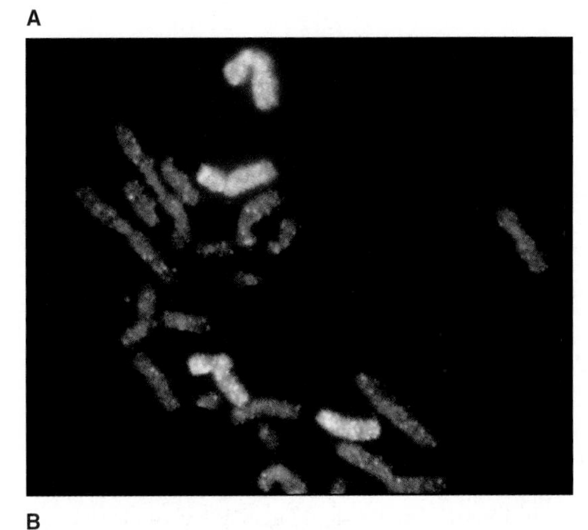

B

FIGURE 11.17

Fragile X Syndrome.

A fragile site on the tip of the long arm of the X chromosome (**A**) is associated with mental retardation and a characteristic long face that becomes pronounced with age (**B**).

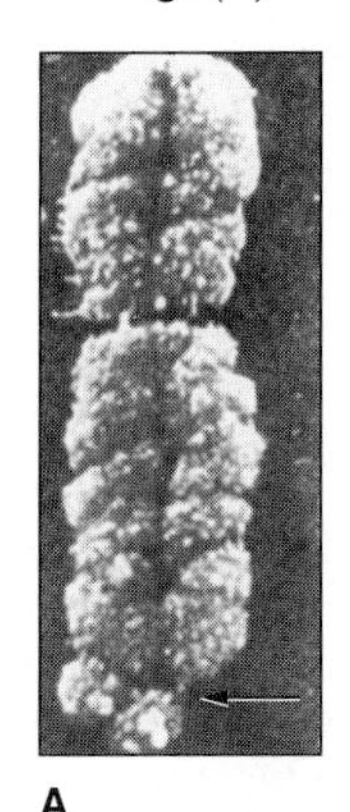

A

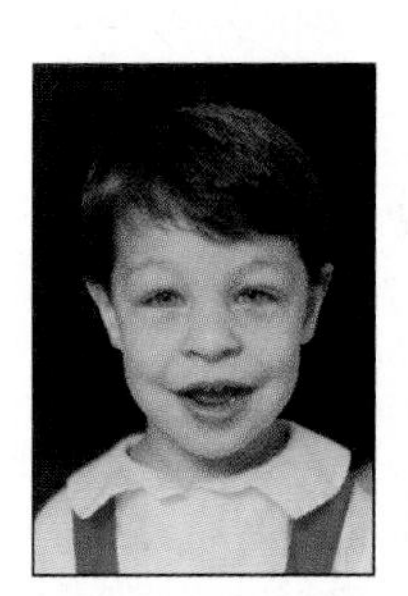

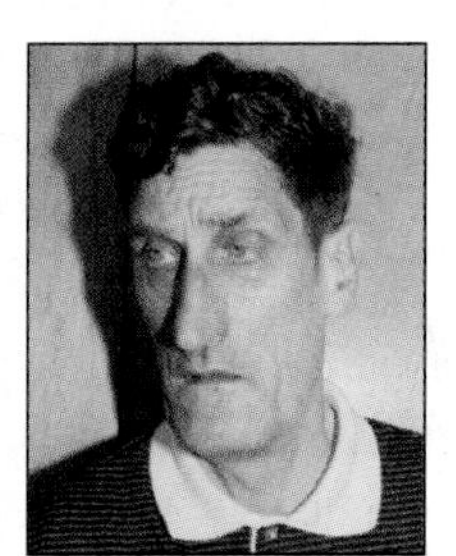

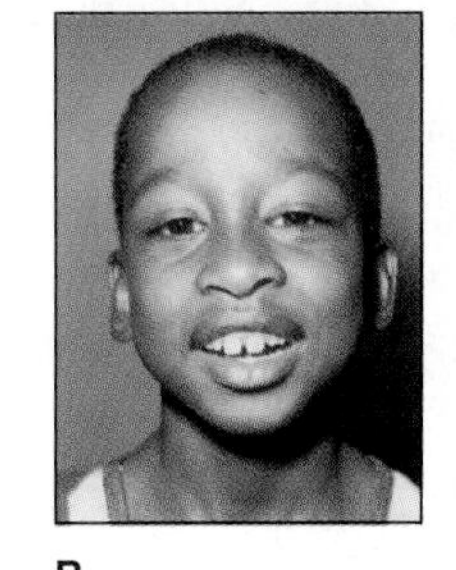

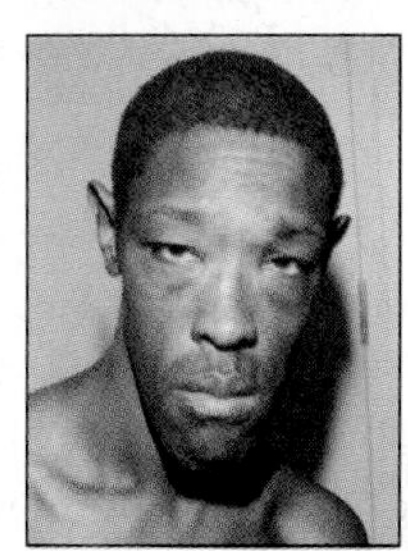

B

phenotype depends upon the particular genes that the rearrangement disrupts.

A chromosomal aberration in a class by itself causes fragile X syndrome, a form of mental retardation that affects 1 in 1,500 males and 1 in 2,500 females. When chromosomes from affected people are cultured in media low in certain nutrients, an X chromosome tip dangles, making it prone to breakage—hence the name (fig. 11.17). A person with fragile X syndrome has protruding ears and a long jaw. Fragile X syndrome is caused by a gene that expands with each generation, and is discussed further in chapter 13.

11.5 MASTERING CONCEPTS

1. How do polyploidy and aneuploidy differ?
2. What is the meiotic basis of polyploidy and aneuploidy?
3. How do monosomies and trisomies differ?
4. How can deletions, duplications, inversions, and translocations cause symptoms?
5. How do inversions and translocations cause reproductive problems?

OLC

TABLE 11.1 Chromosome Abnormalities

TYPE OF ABNORMALITY	DEFINITION
Polyploidy	Extra full sets of chromosomes
Aneuploidy	An extra or missing chromosome
Deletion	Part of a chromosome missing
Duplication	Part of a chromosome present twice
Inversion	Piece of chromosome reversed
Translocation	Piece of one chromosome type on another

Chapter Summary

11.1 What Is a Chromosome?

1. A **chromosome** is a continuous double-stranded molecule of DNA, with RNA and associated proteins that provide scaffolding or help carry out replication or transcription.
2. A **karyotype** is a size-ordered chart of chromosome pairs, which are distinguished by size, centromere position, and banding patterns of dark-staining **heterochromatin** and light-staining **euchromatin.**
3. The minimal requirements for chromosomal integrity are a centromere, telomeres, and origins of replication.

11.2 "Linking" Mendel's Laws to Chromosomes

4. Each species has a characteristic number of chromosomes. **Cytogenetics** associates abnormal chromosomes to phenotypes.
5. Genes on the same chromosome are **linked;** rather than demonstrating independent assortment, they produce a large number of **parental** genotypes and a small number of **recombinant** genotypes.
6. Genotype predictions and **linkage maps** are derived from knowing allele configurations (coupling or repulsion) and crossover frequencies, which are directly proportional to distances between genes.

11.3 How Do Chromosomes Determine Sex?

7. Sex determination mechanisms set development on a course toward malenss or femaleness.
8. Sex determination mechanisms are diverse. In chromosomal strategies, the sex with two different sex chromosomes is **heterogametic,** and the sex with two of the same type of sex chromosome is **homogametic.**
9. In humans, the male is heterogametic, and the female is homogametic. The *SRY* gene on the Y chromosome controls other genes that stimulate development of male structures and suppress development of female structures.

11.4 Inheritance of Genes on Sex Chromosomes

10. An **X-linked** trait passes from mother to **hemizygous** son because the male inherits his X chromosome from his mother and his Y chromosome from his father. Y-linked traits are very rare.
11. **X inactivation** shuts off one X chromosome in the cells of female mammals, equalizing the number of active X-linked genes in each sex. Early in development, each female cell inactivates one X chromosome. A female is mosaic for X-linked heterozygous genes on the X chromosome because the X chromosomes are inactivated at random with respect to their parental origin.

11.5 When Chromosomes Are Abnormal

12. **Polyploid** cells have extra full chromosome sets, and **aneuploids** have extra or missing individual chromosomes. A **trisomy** (one extra chromosome) is less harmful than a **monosomy** (one absent chromosome). Sex chromosome aneuploidy is less severe than autosomal aneuploidy. **Nondisjunction,** an uneven division of chromosomes in meiosis, causes aneuploidy.
13. Chromosomal rearrangements disrupt meiotic pairing, which can delete or duplicate genes. An inversion flips gene order, affecting the phenotype if it disrupts a vital gene. A **Robertsonian translocation** fuses the long arms of two nonhomologs. In a **reciprocal translocation** two nonhomologs exchange parts.

Testing Your Knowledge

1. Why is FISH a more precise way of identifying chromosomes than stains?
2. Human chromosome 3 is metacentric, and chromosome 22 is acrocentric. What are two ways that these chromosomes differ?
3. What are the minimal requirements to construct a chromosome?
4. How are linked genes inherited differently than genes that are located on different chromosomes?
5. How are X-linked genes inherited differently in male and female humans?
6. How is sex determination different in fruit flies than it is in humans?
7. What does X inactivation accomplish?

Thinking Scientifically

1. Why doesn't the inheritance pattern of linked genes disprove Mendel's laws?
2. A normal-sighted woman with a normal-sighted mother and a color-blind father marries a color-blind man. What are the chances that their son will be color blind? their daughter?
3. A fetus dies in the uterus. Several of its cells are examined for chromosomal content. Approximately 75% of the cells are diploid, and 25% are tetraploid. What has happened, and when in development did it probably occur?
4. Why are there no male calico cats?
5. A fetus has an inverted chromosome. What information might reveal whether the expected child will have health problems stemming from the inversion?
6. Patricia Jacobs concluded in her 1965 *Nature* article on XYY syndrome, "It is not yet clear whether the increased frequency of XYY males found in this institution is related to aggressive behavior or to their mental deficiency or to a combination of these factors." Why is this wording unscientific? What harm could it (or did it) do?

References and Resources

Copel, J. A., et al. August 12, 1999. Prenatal screening for Down syndrome—a search for the family's values. *The New England Journal of Medicine,* vol. 341, p. 521. Blood tests are often done to detect increased risk of a fetus with trisomy 21 (Down syndrome), instead of amniocentesis or CVS in women under age 35, but are less accurate.

Hill, Emmeline, et al. March 23, 2000. Y-chromosome variation and Irish origins. *Nature,* vol. 404, p. 351. Y chromosome DNA sequences reflect Irish ancestry.

Lewis, Ricki. June 12, 2000. Chromosome 21 reveals sparse gene content. *The Scientist,* vol 14, p. 1. Compared to other chromosomes, 21 has little genetic information.

Roberts, Lori M., et al. October 1999. New solutions to an ancient riddle: Defining the differences between Adam and Eve. *The American Journal of Human Genetics,* vol. 65, pp. 933–942. The sexes differ chromosomally.

Wade, Nicholas. May 9, 1999. Group in Africa has Jewish roots, DNA indicates. *The New York Times,* p. F1. Y chromosome sequences and anthropological evidence reveal that the Lemba of Africa are descended from Jewish people.

The *LIFE* Online Learning Center provides additional resources and tools for studying this chapter.
www.mhhe.com/life

CHAPTER 12

DNA Structure and Replication

DNA Analysis Solves a Royal Mystery

One night in July 1918, Tsar Nicholas II of Russia and his family met gruesome deaths at the hands of Bolsheviks in a mountain town in Siberia called Ekaterinburg. Captors led the tsar, Tsarina Alexandra, their four daughters and one son, the family physician, and three servants to a cellar and shot and bayoneted them. The executioners then stripped the bodies and loaded them onto a truck. They planned to hurl the victims down a mine shaft, but the truck broke down. Instead, the killers placed the bodies in a shallow grave and poured sulfuric acid over their faces so they could not be identified.

Instead, the killers placed the bodies in a shallow grave and poured sulfuric acid over their faces so they could not be identified.

In July 1991, two amateur historians found the grave and alerted the government that they might have unearthed the long-sought bodies of the Romanov family. Forensic researchers soon determined that the 1,000 pieces of bone at the scene came from nine individuals. The sizes of the skeletons indicated that three were children. The porcelain, platinum, and gold in some of the teeth suggested that some of the people were royalty. But the bullets, bayonets, and acid had so destroyed the remains that some conventional forensic tests were not possible. However, one very valuable type of evidence survived—DNA.

First, detecting Y-chromosome-specific DNA sequences distinguished male from female skeletons. Then mitochondrial DNA sequences identified the mother and children, because mitochondria are maternally inherited. But the researchers still had to connect the skeletons with living royals. That was indeed possible: The tsarina's DNA had some key sequences that matched those of Prince Philip, a modern member of the royal family.

DNA analysis identified the remains of the murdered members of the Romanov family—and an interesting genetic phenomenon.

The next step was to prove that a particular male skeleton with metallic teeth was the tsar. This proved more challenging, because of a surprise in his DNA.

The problem involved nucleotide position 16169 of a mitochondrial gene that is highly variable in sequence among individuals. About 70% of bone cells examined from the remains had cytosine (C) at this position, but the remaining 30% of the cells had thymine (T) at this site. Skeptics at first suspected contamination or a laboratory error, but when the odd result was repeated, researchers realized that this historical case had revealed a genetic phenomenon not seen before in human DNA. The bone cells apparently harbored two populations of mitochondria, one type with C at this position, the other with T. Possibly it could be explained by looking at the DNA of other family members.

The DNA of a living blood relative of the tsar, Countess Xenia Cheremeteff-Sfiri, had only T at nucleotide site 16169. Xenia is the great-granddaughter of Tsar Nicholas II's sister. However, DNA of Xenia and the murdered man with the fancy teeth matched at every other site. DNA of another living relative, the Duke of Fife, who is the great-grandson of Nicholas's maternal aunt, matched Xenia at the famed 16169 site. A closer relative, Nicholas's nephew Tikhon Kulikovsky, refused to lend his DNA, citing anger at the British for not assisting the tsar's family during the Bolshevik revolution.

But the story wasn't over. It would take an event in yet another July, in 1994, to clarify matters.

Attention turned to Tsar Nicholas's brother, Grand Duke of Russia Georgij Romanov. Georgij had died at age 28 in 1899 of tuberculosis. His body was exhumed in July 1994, and researchers sequenced the troublesome mitochondrial gene in bone cells from his leg. They found a match! Georgij's mitochondrial DNA had the same double-base site as the man murdered in Siberia. That man was, therefore, Tsar Nicholas II. The researchers calculated the probability that the remains are truly those of the tsar, rather than resembling Georgij by chance, as 130 million to 1. The murdered Russian royal family can finally rest in peace, thanks to DNA analysis.

Solving the mystery of the Romanov family by comparing DNA sequences is but one example of a biotechnology. These tools today are important in such diverse fields as forensics, agriculture, veterinary medicine, development of new pharmaceuticals, the food and beverage industry, and diagnosis of genetic and infectious diseases. Biotechnologies will increasingly become part of health care now that most of the sequence of the human genome is known.

Many textbooks introduce biotechnology—sometimes called genetic engineering—in its own chapter, after discussing the principles of genetics. Because technology derives from scientific principles, this book presents biotechnologies in boxed readings that accompany discussion of the science on which they are based. This approach more accurately reflects how ideas and discoveries become practical tools.

12.1 How Did Experiments Identify and Describe the Genetic Material?

A series of elegant experiments revealed DNA, and not protein, to be the genetic material, and deciphered its structure.

Today, DNA is one of the most familiar molecules on the planet, the subject matter of movies and headlines (fig. 12.1). Fictional dinosaurs are reconstructed from DNA preserved in an ancient mosquito's gut. Real-life trials hinge on DNA evidence, cloned animals raise questions about the role of DNA in determining who we are, and DNA-based discoveries are yielding many new genetic tests and treatments.

Far more important than DNA's role in society today is its role in life itself. Of all the characteristics that distinguish the living from the nonliving, the one most important to the continuance of life is the ability to reproduce. At the cellular level, reproduction duplicates a cell—be it a simple single-celled organism on Earth billions of years ago, or a cell lining a person's intestine or part of a leaf today. At the molecular level, reproduction depends upon a biochemical that has dual abilities: to direct the specific activities of that cell and to manufacture an exact replica of itself so that the instructions are perpetuated.

DNA is the multifunctional molecule that replicates as well as orchestrates cellular activities by controlling protein synthesis. But the recognition of DNA's vital role in life was a long time in coming.

FIGURE 12.1

DNA Is Highly Packaged.

DNA bursts forth from this treated bacterial cell, illustrating just how much DNA is tightly wound into a single cell.

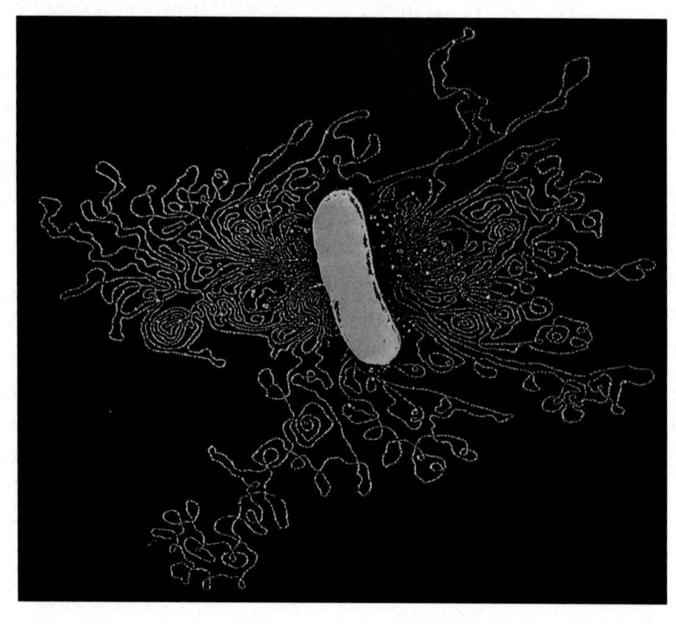

DNA Transmits Traits

Swiss physician and biochemist Friedrich Miescher was the first investigator to chemically analyze the contents of a cell's nucleus. In 1869, he isolated the nuclei of white blood cells obtained from pus in soiled bandages. In the nuclei, he discovered an unusual acidic substance containing nitrogen and phosphorus. Miescher and others went on to find it in cells from a variety of sources. Because the material resided in cell nuclei, Miescher called it nuclein in his 1871 paper; subsequently it was called a nucleic acid.

Miescher's discovery, like those of his contemporary Gregor Mendel (see the opening essay in chapter 10), was not appreciated for years. Instead, most investigators researching inheritance focused on the association between inherited disease and proteins.

In 1909, English physician Archibald Garrod was the first to associate inheritance and protein. Garrod noted that people with inherited "inborn errors of metabolism" lacked certain enzymes. Other researchers added supporting evidence: they linked abnormal or missing enzymes to unusual eye color in fruit flies and nutritional deficiencies in bread mold variants. But how do enzyme deficiencies produce traits? Experiments in bacteria would answer the question and return, eventually, to Miescher's nuclein.

In 1928, English microbiologist Frederick Griffith inadvertently contributed the first step in identifying DNA as the genetic material. Griffith studied pneumonia in mice caused by a bacterium, *Diplococcus pneumoniae.* He identified two types of bacteria. Type S bacteria form smooth colonies, because they are encased in a polysaccharide capsule. When injected into mice, type S bacteria cause pneumonia. Type R bacteria form rough-shaped colonies, and when injected into mice do not cause pneumonia. Therefore, the smooth polysaccharide coat seemed to be necessary for infection.

When Griffith heated type S bacteria ("heat-killing" them) and injected them into mice, they no longer caused pneumonia. However, when he injected mice with a mixture of type R bacteria plus heat-killed type S bacteria—neither able to cause pneumonia alone—the mice died of pneumonia (fig. 12.2). Their bodies contained live type S bacteria encased in polysaccharide. What had happened?

In the 1940s, U.S. physicians Oswald Avery, Colin MacLeod, and Maclyn McCarty offered an explanation. They hypothesized that something in the heat-killed type S bacteria entered and "transformed" the normally harmless type S strain into a killer. Was this "transforming principle" a protein? Treating the solution from the type S strain with a protein-destroying enzyme (a protease) failed to inhibit the type R strain from being transformed into a killer. Therefore, a protein was not responsible for transmitting the killing trait. Treating the solution from the heat-killed S bacteria with a DNA-destroying enzyme (DNase) first, however, prevented the killing ability. Could DNA transmit the killing trait (fig. 12.3)?

Avery, MacLeod, and McCarty confirmed that DNA transformed the bacteria by isolating DNA from heat-killed type S bacteria and injecting it along with type R bacteria into mice. The mice died, and their bodies contained active type S bacteria. The conclusion: Type S DNA altered the type R bacteria, enabling them to manufacture the smooth coat necessary to cause infection.

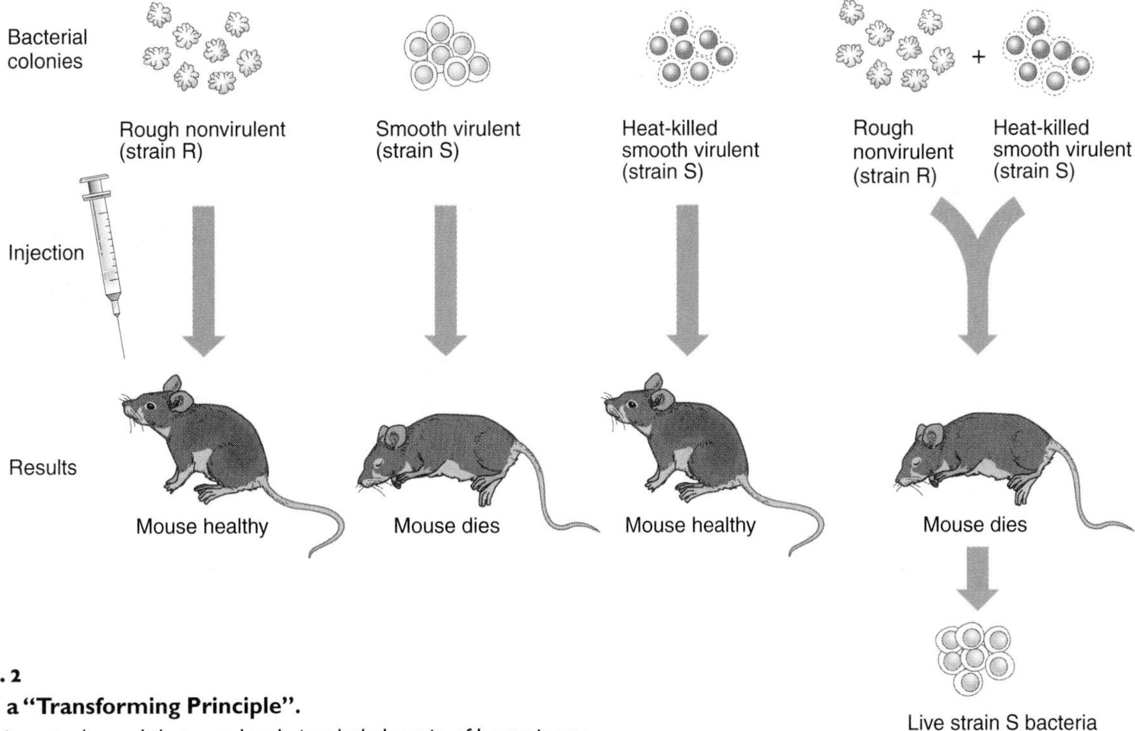

FIGURE 12.2
Discovery of a "Transforming Principle".
Griffith's experiments showed that a molecule in a lethal strain of bacteria can transform nonkilling bacteria into killers.

FIGURE 12.3
DNA is the "Transforming Principle".
Avery, MacLeod, and McCarty identified Griffith's transforming principle as DNA. By adding enzymes that either destroy proteins (protease) or DNA (DNase) to the types of solutions that Griffith used in his experiments, they demonstrated that DNA transforms bacteria—and that protein does not.

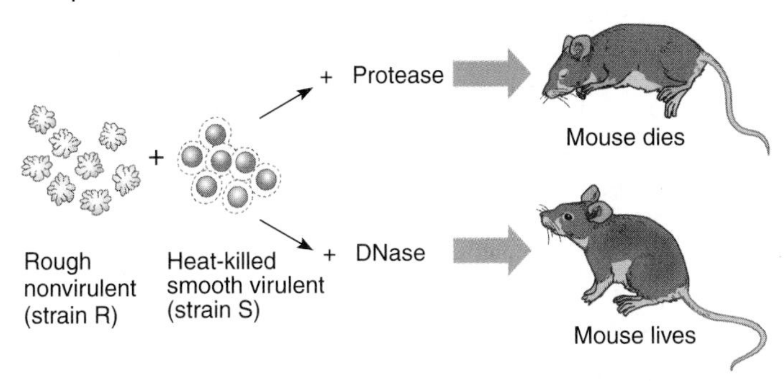

The Genetic Material—DNA, Not Protein

Biologists at first were rather hesitant in accepting DNA as the biochemical of heredity. More was known about proteins than about nucleic acids, and it was thought that protein, with its 20 building blocks, was more versatile and therefore more likely to be able to encode many traits than DNA, with its mere 4 types of building blocks. In 1950, U.S. microbiologists Alfred Hershey and Martha Chase showed that DNA—not protein—is the genetic material.

Hershey and Chase used a very simple system—*Escherichia coli,* a bacterium, and T4, a virus that infects bacteria (called a bacteriophage). Most bacterial viruses consist of only a protein coat and a nucleic acid (DNA in this case) core. We now know that when the virus infects the bacterial cell, it injects its DNA, and the protein coat is left attached loosely to the bacterium (fig. 12.4). The viral DNA then uses the bacterial cell's protein synthetic machinery to manufacture more of itself. New virus particles burst from the cell. Much of this information was not available in 1950.

Hershey and Chase wanted to know which part of the virus controls its replication—the DNA, or the protein coat. They knew that virus grown in the presence of radioactive sulfur becomes radioactive, and that the radioactivity is emitted from the protein coats, but not from the DNA. This is because protein contains sulfur, but DNA does not. But if virus is grown in the presence of radioactive phosphorus, the radioactivity comes from the viral DNA but not from the protein. This is because DNA contains phosphorus, but protein does not. (Recall that Miescher had identified phosphorus in nuclein nearly a century earlier.)

In further experiments, Hershey and Chase "labeled" two batches of virus, one with radioactive sulfur (which marked protein) and the other with radioactive phosphorus (which marked DNA) (fig. 12.5). They used each type of labeled virus to infect a separate batch of bacteria and allowed several minutes for the virus particles to bind to the bacteria and inject their DNA into them. Then they agitated each mixture in a blender, which knocked the remaining viruses and empty protein coats from the surfaces of the bacteria. They poured the mixtures into test tubes, and centrifuged them (spun them at high speed). This settled the

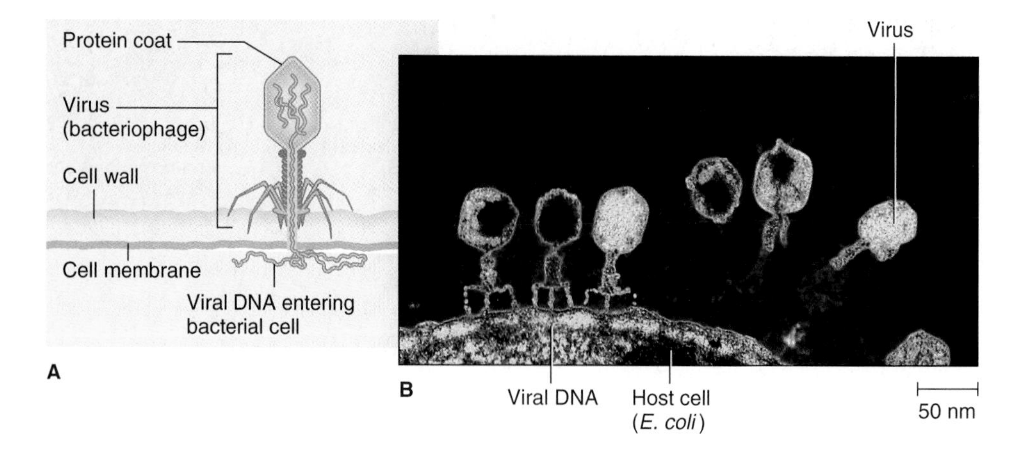

FIGURE 12.4
A Virus (Bacteriophage) Infects a Bacterium.
(**A**) A bacteriophage is a virus that consists only of a nucleic acid in a protein coat. The virus uses the protein coat to attach and inject its DNA into a bacterial cell. (**B**) This photo shows several bacteriophages infecting a bacterium.

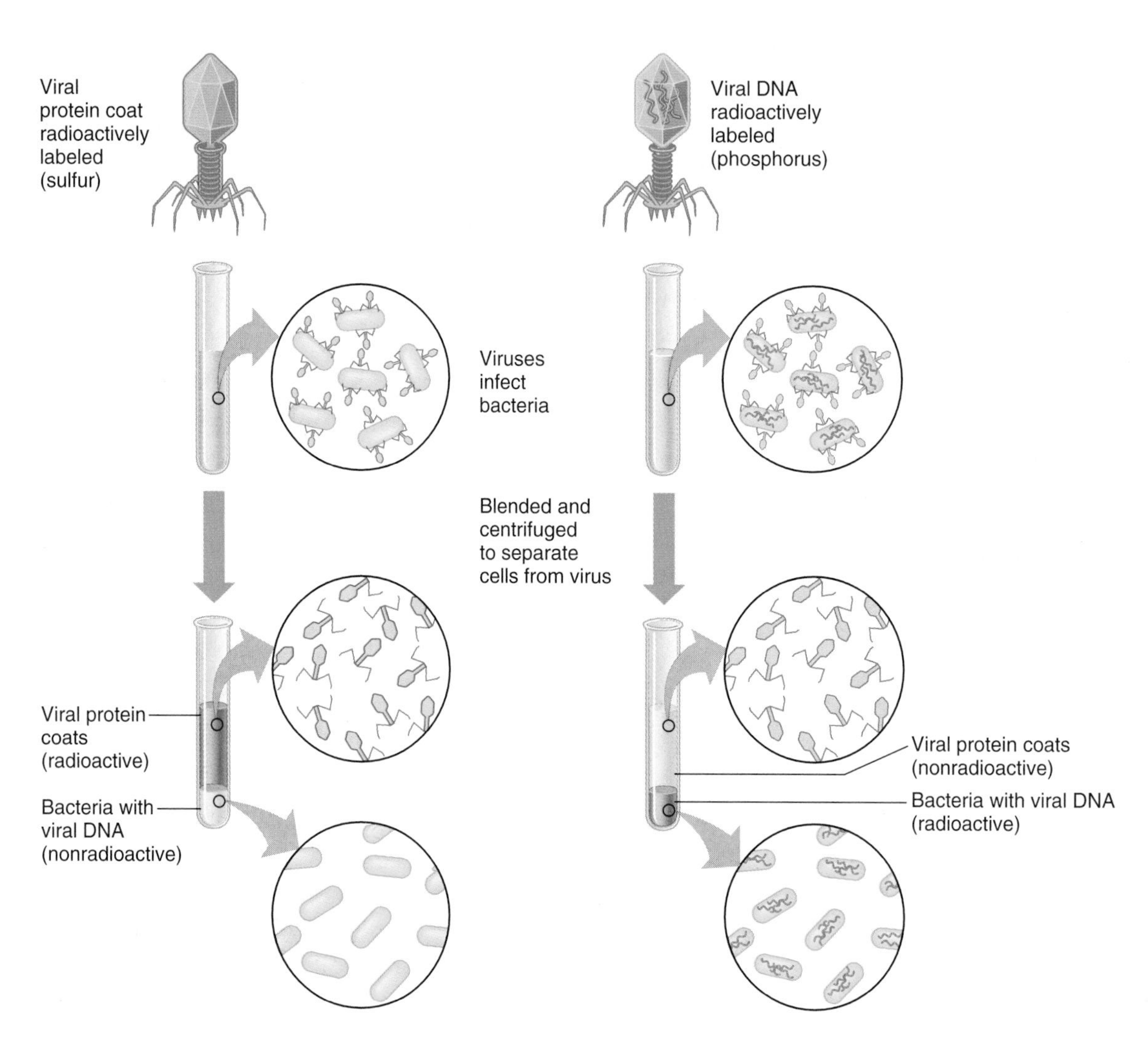

FIGURE 12.5
Hershey-Chase Experiment.
Hershey and Chase used different radioactive isotopes to distinguish the viral protein coat from the genetic material (DNA). These "blender experiments" showed that DNA is what the virus transfers to the bacterium.The experiment showed DNA was sufficient for reproduction of viruses.

infected bacteria at the bottom of each test tube, because they were heavier than the liberated viral protein coats.

Hershey and Chase examined the contents of the bacteria that had settled to the bottom of each tube. In the test tube containing sulfur-labeled virus, the virus-infected bacteria were not radioactive, but the fluid portion of the material in the tube was. In the other tube, where the virus contained radioactive phosphorus, the infected bacteria were radioactive, but the fluid was not. The "blender experiment" therefore showed that the part of the virus that could enter the bacteria and direct them to mass-produce more virus was the part with the phosphorus label—the DNA. The genetic material, therefore, was DNA and not protein.

Deciphering the Structure of DNA

Early in the twentieth century, Russian-American biochemist Phoebus Levene continued Miescher's chemical analysis of nucleic acids. In 1909, Levene identified the five-carbon sugar **ribose** in

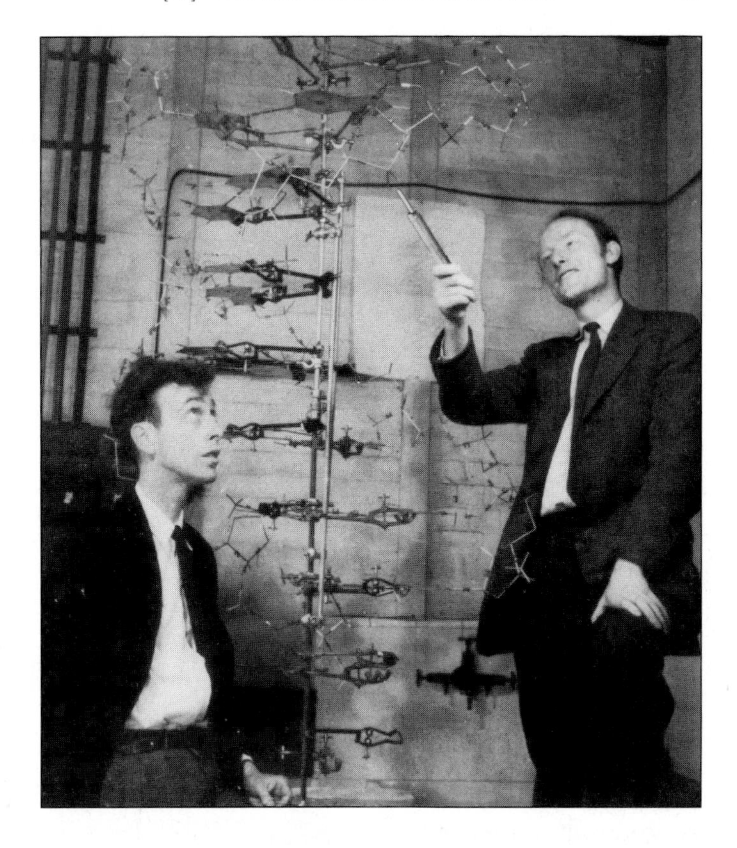

FIGURE 12.6
The "Secret of Life."
Using data provided by Maurice Wilkins and Rosalind Franklin, James Watson and Francis Crick were the first to make an accurate model of the molecular structure of DNA. Watson and Crick, shown here, announced in a pub that they had discovered the "secret of life." For their now-famous discovery, Wilson, Watson and Crick shared the Nobel Prize in 1962. Unfortunately, Rosalind Franklin had died in 1958, and by the rules of the award, could not be included.

some nucleic acids; in 1929, he discovered a similar sugar, **deoxyribose.** Levene's work revealed a major chemical distinction between the two types of nucleic acid, RNA (ribonucleic acid) and DNA (deoxyribonucleic acid).

Levene then determined that the three parts of a nucleic acid—sugars, nitrogen-containing groups, and phosphorus-containing components—occur in equal proportions. He deduced that a nucleotide building block must include one of each component. Levene also found that while nucleotides always contain the same sugars and phosphates, they may contain any one of four different nitrogen-containing bases. For several years thereafter, scientists erroneously thought that the nitrogenous bases occur in equal amounts. If this were the case, DNA bases could not encode much information, just as a sentence could not say much if it was restricted to equal numbers of certain letters.

In the early 1950s, two lines of evidence converged that would reveal DNA's structure. Austrian-American biochemist Erwin Chargaff showed that DNA contains equal amounts of the bases adenine (A) and thymine (T) and equal amounts of the bases guanine (G) and cytosine (C). Next, two English researchers, physicist Maurice Wilkins and chemist Rosalind Franklin, bombarded DNA with X rays, using a technique called X-ray diffraction to determine the three-dimensional shape of the molecule. The X-ray diffraction pattern revealed a regularly repeating structure of nucleotides. Wilkins and Franklin provided key clues to DNA's three-dimensional structure.

In 1953, U.S. biochemist James Watson and English physicist Francis Crick combined these clues to guide them in building a replica of the DNA molecule using ball-and-stick models (fig. 12.6). Their model included equal amounts of G and C and of A and T, and it had the sleek symmetry of the X-ray diffraction pattern. Their model was the now familiar double helix (table 12.1). Watson, Crick, and Wilkins were awarded the Nobel Prize in 1962 for their ground-breaking discovery. Unfortunately, Rosalind Franklin had died in 1958 and could not be included in the award.

12.1 MASTERING CONCEPTS

1. When was DNA first described?
2. What evidence linked enzyme deficiencies to inherited traits?
3. How did researchers demonstrate that DNA is the genetic material and protein is not?
4. What are the components of DNA?
5. What evidence enabled Watson and Crick to decipher the structure of DNA?

OLC

TABLE 12.1 The Road to the Double Helix

INVESTIGATOR	CONTRIBUTION	DATES
Friedrich Miescher	Isolated nuclein in white blood cell nuclei	1869
Frederick Griffith	Killing ability in bacteria can be transferred between strains	1928
Oswald Avery, Colin MacLeod, and Maclyn McCarty	DNA transmits killing ability in bacteria	1944
Alfred Hershey and Martha Chase	The part of a virus that infects and replicates is its nucleic acid and not its protein	1950
Phoebus Levene, Erwin Chargaff, Maurice Wilkins, and Rosalind Franklin	DNA components, proportions, and positions	1909–early 1950s
James Watson and Francis Crick	DNA's three-dimensional structure	1953

12.2 What Is the Three-Dimensional Structure of DNA?

The DNA molecule is a double helix with sugar-phosphate rails and pyrimidine-purine pairs as rungs. The base sequence specifies the amino acid sequence of a particular protein. DNA must be very highly coiled to fit into the nucleus.

Genes and proteins can carry information because they consist of sequences—genes of four types of DNA bases and proteins of 20 types of amino acids. Genes consist of thousands of base pairs, and proteins of hundreds of amino acids.

How Does DNA Structure Encode Information?

If the DNA double helix is pictured as a twisted ladder, the rungs are A–T and G–C base pairs joined by hydrogen bonds, and the rails that make up the DNA's sugar-phosphate "backbone" are alternating units of deoxyribose and phosphate (PO_4) joined with covalent bonds. A nucleotide consists of one deoxyribose, one phosphate, and one base, as figure 2.22 shows. Figure 12.7 offers other views of DNA.

DNA base sequences encode a gene's information. Biotechnology 12.1 discusses DNA fingerprinting, which is based on detecting DNA sequence differences among individuals. Adenine and guanine are **purines,** which have a double organic ring structure. Cytosine and thymine are **pyrimidines,** which have a single organic ring structure (fig. 12.8). The pairing of A with T and G

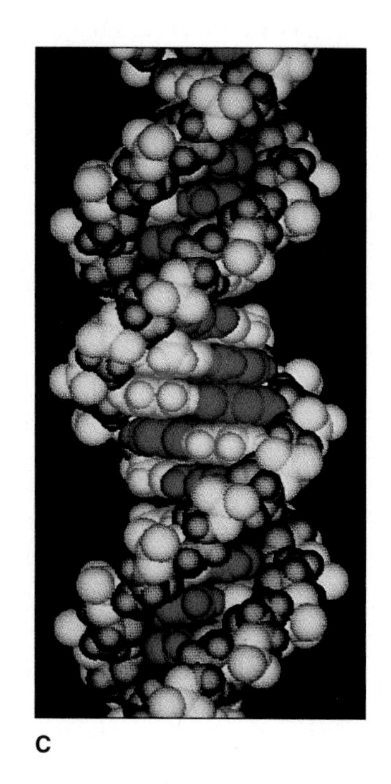

FIGURE 12.7
Different Ways to Represent the DNA Double Helix.
(**A**) The helix is unwound to show the base pairs in blue and the sugar-phosphate backbone in green and yellow. (**B**) This representation shows the helical structure of DNA. The sugar-phosphate rails are identical in all DNA molecules, and run in opposite directions. (**C**) This photo is a space-filling model of a DNA molecule that shows the three-dimensional relationships of the component atoms.

FIGURE 12.8
DNA Bases.
Guanine and adenine are purines, each composed of a six-membered organic ring plus a five-membered ring. Cytosine and thymine are pyrimidines, each built of a single six-membered ring. In a nucleotide, one nitrogenous base is joined to the sugar, deoxyribose, and a phosphate.

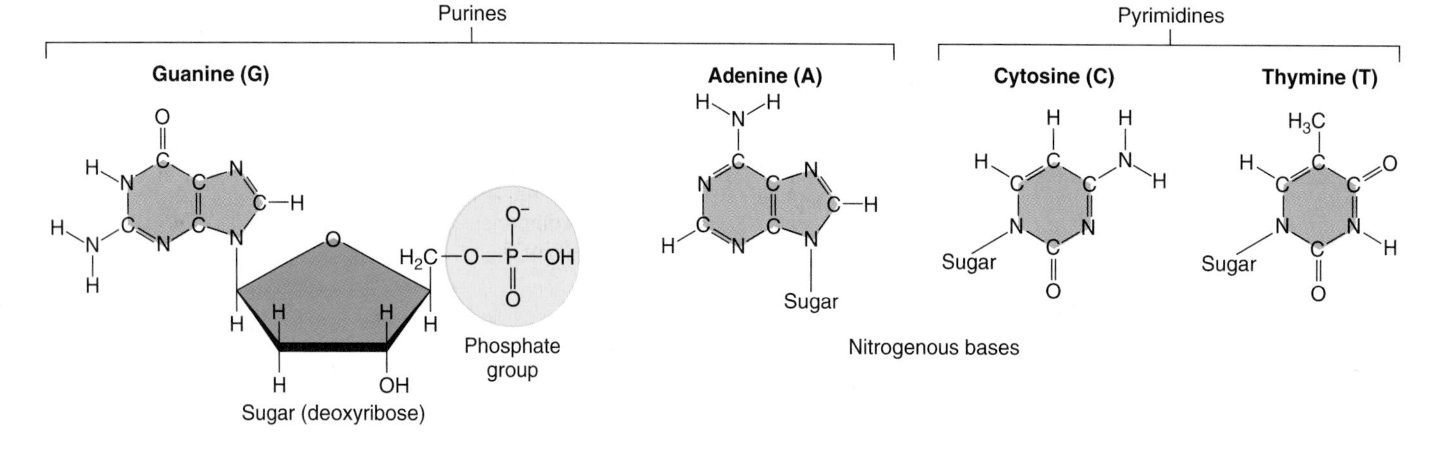

with C through hydrogen bonds, maintains the helix's width because each pair includes a single-ringed structure and a double-ringed structure. These specific purine-pyrimidine couples are termed **complementary base pairs** (fig. 12.9). Complementary base pairing is the basis of gene function, as chapter 13 discusses.

The two chains of the DNA double helix run opposite to each other, somewhat like artist M. C. Escher's depiction of drawing hands (fig. 12.10). This head-to-tail arrangement, called **antiparallelism,** is apparent when the deoxyribose carbons are numbered consecutively from right to left, according to chemical convention. Where one chain ends in the 3′ carbon, the opposite chain ends in the 5′ carbon. (The 3′ and 5′ designate the opposite ends of each strand of the DNA molecule.)

DNA Is Highly Coiled

If the DNA bases of all 46 human chromosomes were typed as A, C, T, and G, the 3 billion letters would fill 4,000 books of 500 pages each! How can a cell only one-millionth of an inch across contain so much material? The explanation is that DNA is wrapped around proteins, much as a very long length of thread is wound around a wooden spool.

To form a chromosome in a eukaryotic cell, a stretch of 146 base pairs of DNA wraps twice around a structure of eight proteins, called histones, to form a **nucleosome,** which is 10 nanometers (nm, billionths of a meter) in diameter. A continuous thread of DNA connects nucleosomes like beads on a string, which in turn fold into structures that are 30 nanometers in diameter. Like thread, DNA must unwind to function (fig. 12.11). As sections of DNA unwind, the DNA in the more widely

FIGURE 12.9
DNA Base Pairs.
The key to the constant width of the DNA double helix is the pairing of purines with pyrimidines. Specifically, adenine pairs with thymine through two hydrogen bonds, and cytosine pairs with guanine through three hydrogen bonds.

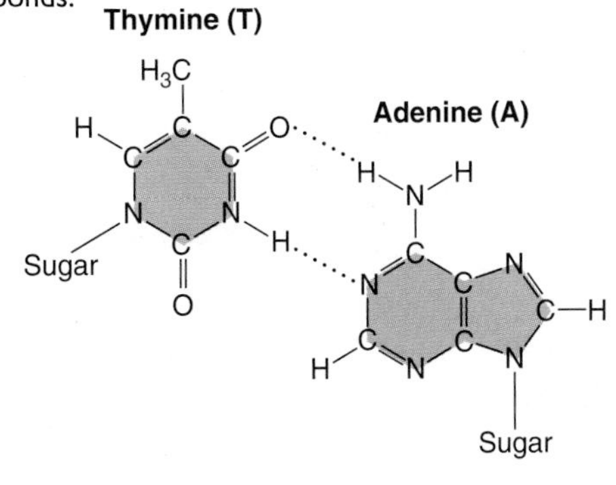

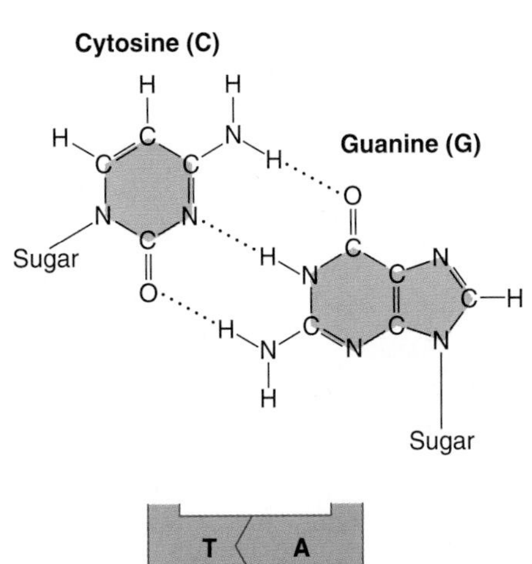

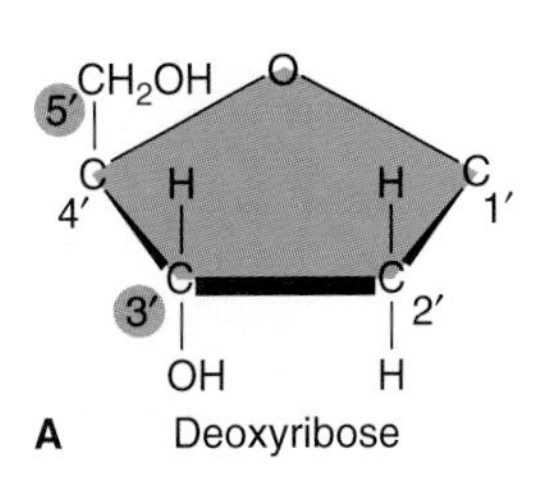

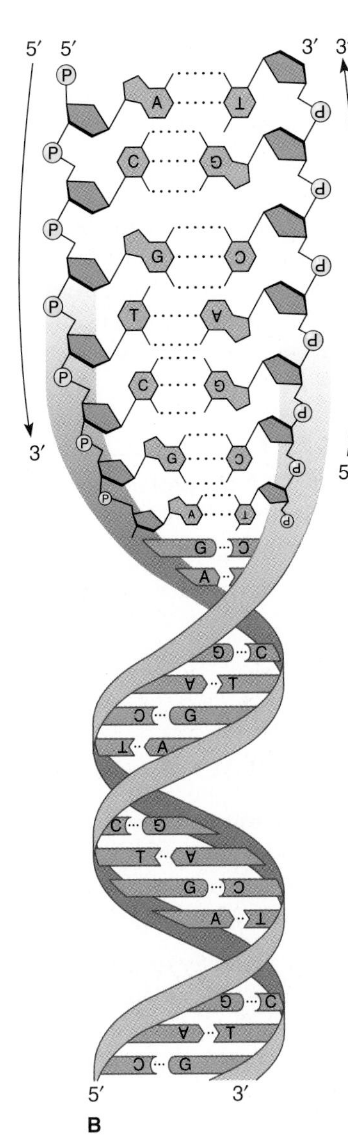

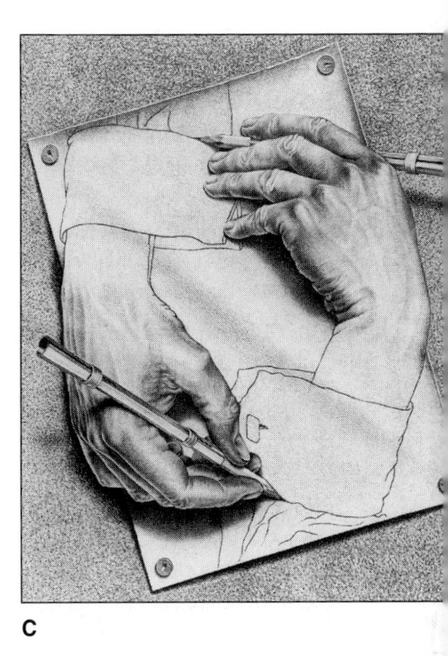

FIGURE 12.10
DNA Strands Are Antiparallel.
(**A**) Chemists assign numbers to differently positioned carbons in organic molecules. (**B**) The two strands of the DNA double helix run opposite one another in orientation. This arrangement is called antiparallelism. (**C**) The spatial relationship of these two hands resembles that of the two DNA chains that make up the DNA double helix.

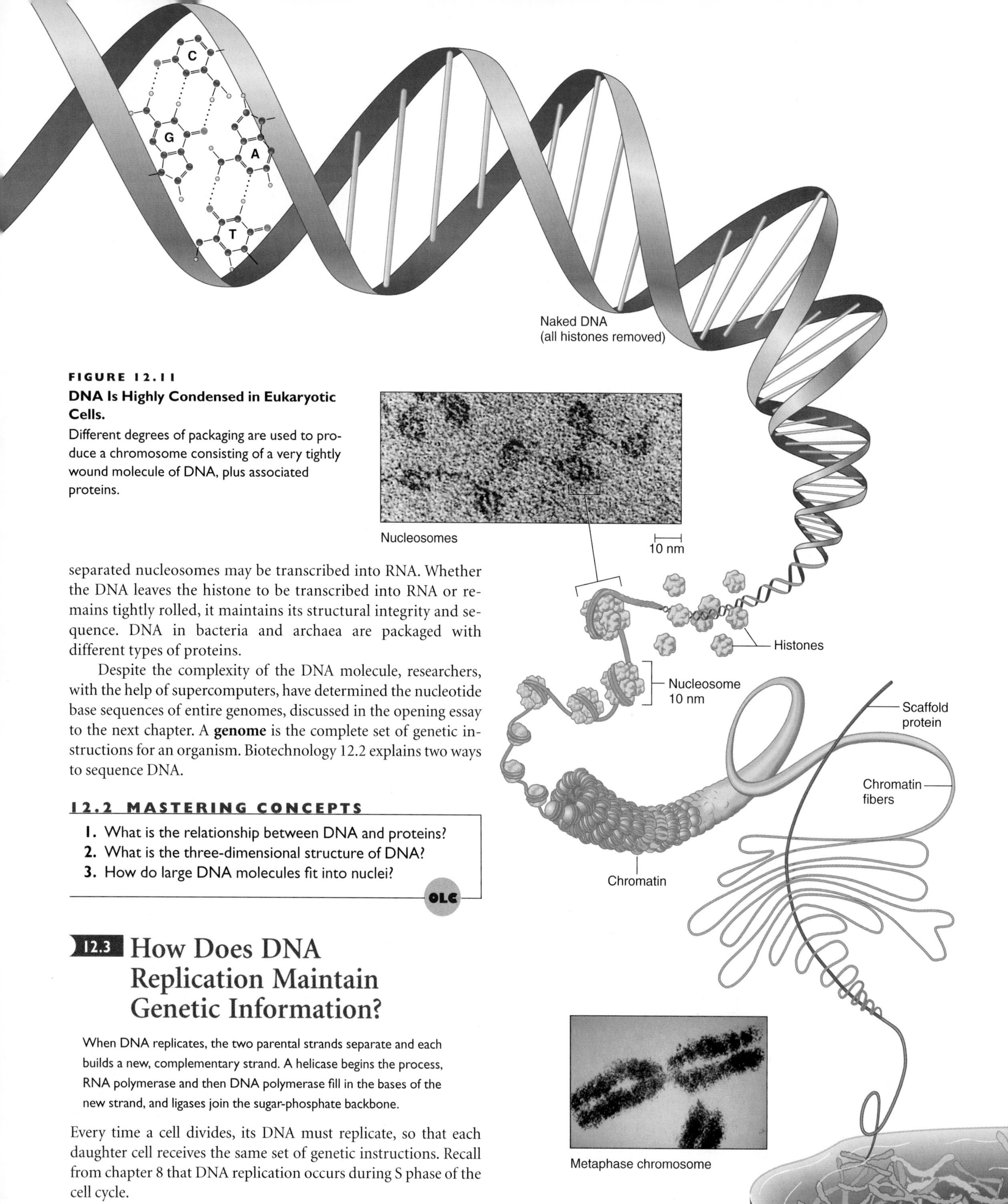

FIGURE 12.11
DNA Is Highly Condensed in Eukaryotic Cells.
Different degrees of packaging are used to produce a chromosome consisting of a very tightly wound molecule of DNA, plus associated proteins.

separated nucleosomes may be transcribed into RNA. Whether the DNA leaves the histone to be transcribed into RNA or remains tightly rolled, it maintains its structural integrity and sequence. DNA in bacteria and archaea are packaged with different types of proteins.

Despite the complexity of the DNA molecule, researchers, with the help of supercomputers, have determined the nucleotide base sequences of entire genomes, discussed in the opening essay to the next chapter. A **genome** is the complete set of genetic instructions for an organism. Biotechnology 12.2 explains two ways to sequence DNA.

12.2 MASTERING CONCEPTS

1. What is the relationship between DNA and proteins?
2. What is the three-dimensional structure of DNA?
3. How do large DNA molecules fit into nuclei?

OLC

12.3 How Does DNA Replication Maintain Genetic Information?

When DNA replicates, the two parental strands separate and each builds a new, complementary strand. A helicase begins the process, RNA polymerase and then DNA polymerase fill in the bases of the new strand, and ligases join the sugar-phosphate backbone.

Every time a cell divides, its DNA must replicate, so that each daughter cell receives the same set of genetic instructions. Recall from chapter 8 that DNA replication occurs during S phase of the cell cycle.

Biotechnology 12.1

DNA Fingerprinting: An Application of Understanding DNA Structure

DNA FINGERPRINTING

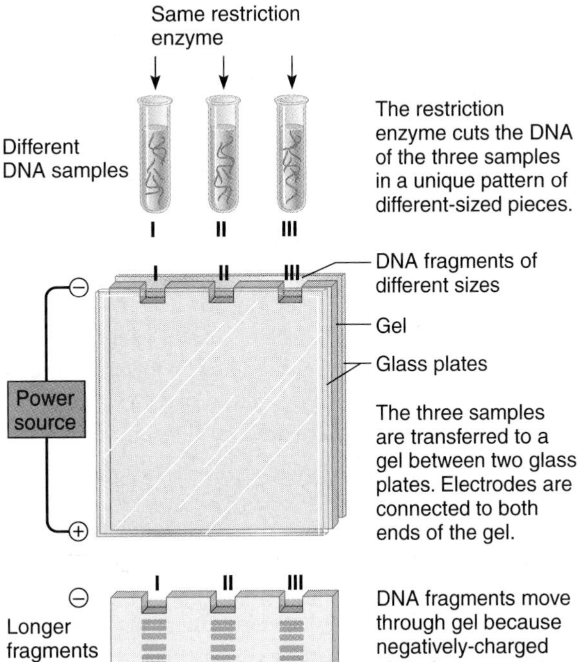

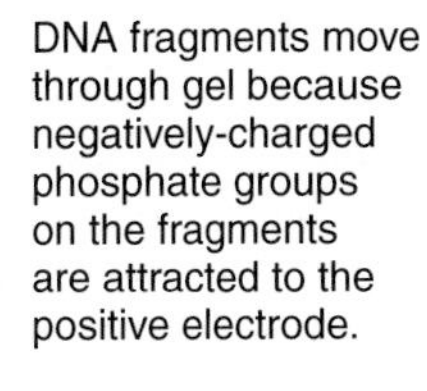

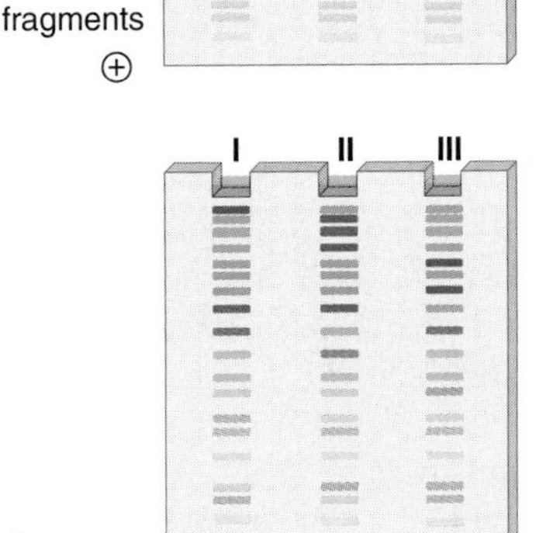

FIGURE 12.A

DNA Fingerprints from a Murder Case.

(**A**) Different DNA samples produce different patterns of fragments when cut with the same restriction enzyme. (**B**) DNA from bloodstains on the defendant's clothes matches the DNA fingerprint of the victim but differs from the DNA fingerprint of the defendant. This is evidence that the blood on the defendant's clothes came from the victim, not the defendant.

DNA fingerprinting is a biotechnology that compares DNA sequences between individuals, which can reveal or disprove genetic relationships. A faster way to compare DNA sequences than actually determining the base order uses restriction enzymes, which are bacterial enzymes that cut DNA at specific sequences. If the DNA of two individuals differs at a cutting site for a restriction enzyme, then that enzyme cuts their DNA into different-sized pieces. The pieces are then separated by size order, which makes single base differences stand out (fig. 12.A).

The power of DNA fingerprinting in humans stems from the fact that there are many more ways for the 3 billion bases of the genome to vary than there are people. The results are often given in terms of probability—that is, how likely it is that two DNA fingerprints match because the sources are the same (such as semen in a raped woman's body and white blood cells from a suspect) rather than because two individuals resemble each other by chance. To make this distinction, the investigator consults population databases on allele frequencies to calculate the likelihood of a particular combination occurring in a certain population. If the allele combination being examined is fairly common, then a DNA match could happen by coincidence, just like two people with blond hair and blue eyes are not necessarily related. DNA fingerprinting typically examines a few DNA sequences that vary greatly among individuals to avoid coincidences.

To generate the statistics that make DNA fingerprinting results meaningful, allele frequencies are multiplied, an application of the product rule. For example, if rare DNA sequence variants at five sites in the genome are exactly the same for sperm cells collected from a rape victim and blood from a suspect, then chances are very high that the same man provided both samples.

In one variant of DNA fingerprinting, individuals are distinguished by the numbers of repeats of particular short DNA sequences. This approach was recently used to show that President Thomas Jefferson fathered a child, Eston, with his slave Sally Hemings.

In the Courtroom

In 1986, DNA fingerprinting was unheard of outside of scientific circles. So rapist Tommie Lee Andrews thought he was being very meticulous in planning his crimes. He picked his victims months before he attacked, and watched

continued

Biotechnology 12.1 *concluded*

them so that he knew exactly when they would be home alone. On a balmy Sunday night in May 1986, Andrews lay in wait for Nancy Hodge, a young computer operator at Disney World, at her home. Surprising her when she was in the bathroom, he covered her face, then raped and brutalized her.

Andrews was very careful not to leave fingerprints, threads, or hairs, but he had not counted on DNA fingerprinting. Thanks to a clear-thinking crime victim and scientifically informed lawyers, Andrews was soon at the center of a trial not only of himself, but also of the technology that would eventually help convict him. When Andrews was arrested for another assault, DNA fingerprinting was done on his white blood cells, and the pattern matched the DNA from the sperm sample taken from Hodge. The match left little doubt that Tommie Lee Andrews was guilty.

DNA fingerprinting has rapidly become a standard forensic tool. It saves time, money, and reputations. In the United States and England, a third of all rape suspects are released because DNA analysis vindicates them. Dozens of wrongly convicted people have been released from prison thanks to DNA evidence that revealed that they could not possibly have committed the crimes for which they stood trial. The U.S. military has a "genetic dog tag" program that will make it possible to identify remains if teeth or conventional fingerprints are not available.

In the Garden

Vitis vinifera—better known as the grape—was domesticated more than 6,000 years ago. Traditionally, grape tasters distinguish 14,000 to 24,000 cultivated strains, or cultivars, based on growth characteristics; vine branching pattern; leaf, fruit, and flower shape; and pollen grain diameter. Researchers currently cataloging grape cultivars by DNA fingerprint pattern have discovered that several popular varieties descend from the same parental strains, one of which is not itself a good wine grape. DNA fingerprinting is enabling researchers to follow new combinations of characteristics, and to objectively identify cultivars to settle patent disputes.

In yet another application, DNA fingerprinting has revealed a close evolutionary relationship among sugar beets (*Beta vulgaris*), three cultivated strains (fodder beets, leaf beets, and garden beets), and a wild subspecies (*B. vulgaris maritima*). Not surprisingly, the native sugar beets and the cultivars are very similar genetically, whereas the wild strain is much more variable. Researchers are using this information to identify valuable characteristics in the wild strain, such as cold tolerance and disease resistance, that could be bred or genetically engineered into sugar beets.

Showing That DNA Replication Is Semiconservative

Watson and Crick had a flair for the dramatic, ending their report on the structure of DNA with the tantalizing statement, "It has not escaped our notice that the specific pairing we have postulated immediately suggests a possible copying mechanism for the genetic material." They envisioned the immense molecule unwinding, exposing unpaired bases that would attract their complements, and neatly knitting two double helices from one. This route to replication is called semiconservative, because each new DNA molecule conserves half of the original genetic material.

Not everyone thought DNA replication was so straightforward. One critic was Max Delbrück, another of the founders of molecular biology. When Watson shared his and Crick's results with Delbrück before publication of their paper, Delbrück wrote back that " . . . for a DNA molecule of molecular weight 3,000,000 there would be about 500 turns around each other. These would have to be untwiddled to separate the strands." Although "untwiddled" is hardly a scientific term, the problem at hand was clear—separating the DNA strands was like having to keep two pieces of thread the length of a football field from tangling.

Because separating the DNA strands seemed so difficult, some researchers suggested that the molecule replicated some other way. Replication might be **conservative,** with one double helix specifying creation of a second double helix. Or replication might be **dispersive,** with a double helix shattering into pieces that would then join with newly synthesized DNA pieces to form two molecules. Figure 12.12 distinguishes among the hypothesized mechanisms of DNA replication.

Delbrück suggested experiments that might reveal how DNA replicates. If newly synthesized DNA from a virus could incorporate radioactive phosphorus, then the radiation could be detected by its ability to expose photographic film, and the new generation of DNA thus identified. By 1956, though, results were not clear enough to show that replication was either semiconservative, conservative, or dispersive. But just over a year later, a variation on the approach definitively answered the question of how DNA replicates. The elegant series of experiments not only supported one hypothesis, but disproved the other two—about as good as the scientific method gets.

A young researcher named Matthew Meselson decided to label newly synthesized DNA with a common isotope of nitrogen (^{14}N), which would then be distinguishable from older DNA synthesized using a less common heavier isotope, ^{15}N. (Recall from chapter 2 that an isotope is an atom with a different number of neutrons than is usual for that element.) Meselson and coworker Franklin Stahl examined DNA replication in bacteria. The idea was that DNA that incorporated the heavy nitrogen could be separated from DNA that incorporated the normal lighter nitrogen by its

FIGURE 12.12

DNA Replication Is Semiconservative.

Density shift experiments distinguished the three hypothesized mechanisms of DNA replication. DNA molecules containing light nitrogen are designated "LL" and those with heavy nitrogen, "HH". Molecules containing both isotopes are designated "LH".

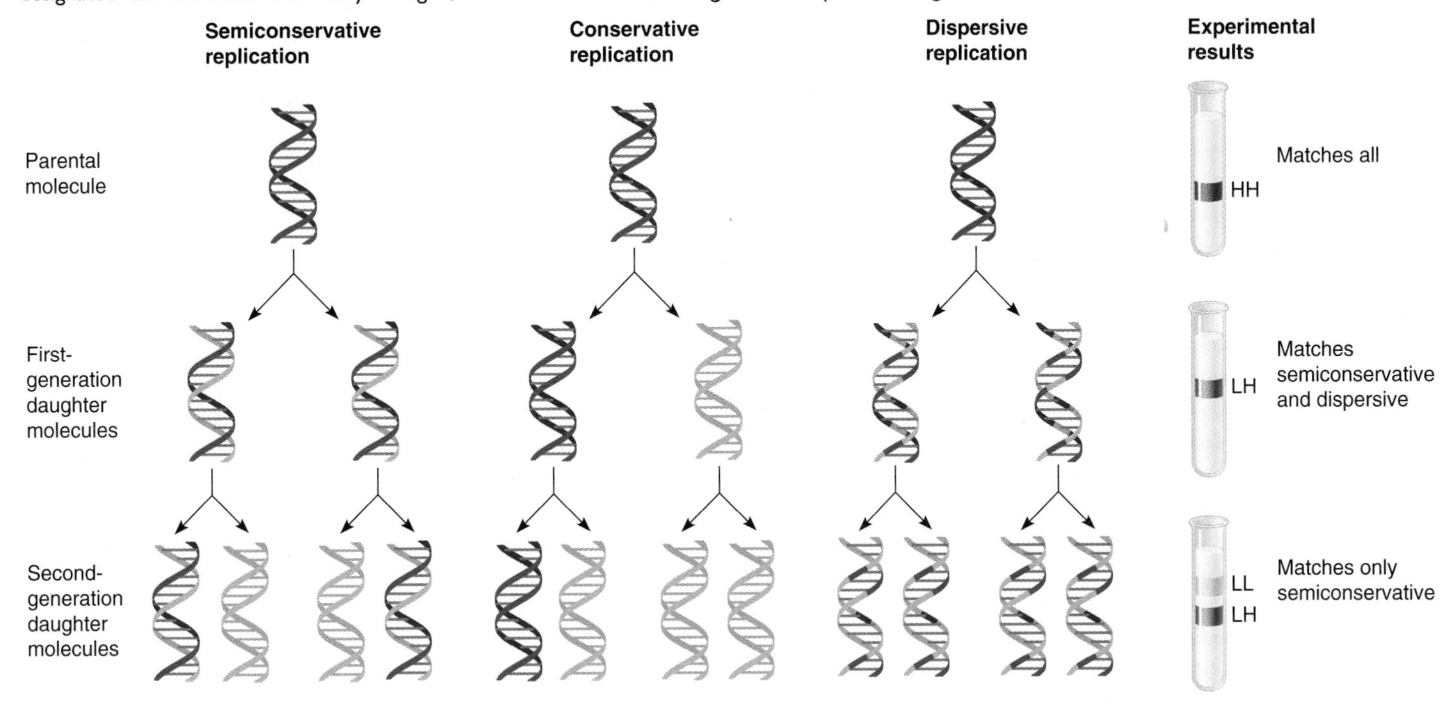

greater density. DNA in which one strand of the double helix was light and one heavy would be of intermediate density. In what came to be known as density shift experiments, Meselson and Stahl grew cells on media that would enable them to trace replication. They then broke open the cells, extracted DNA, and spun the DNA in a centrifuge. The heavier DNA settled near the bottom of the centrifuge tube, the light DNA rose to a higher level and the "heavy-light" double helices settled in the middle area of the tube. The "shift" referred to the fact that bacterial cultures were shifted to media with either of the two types of nitrogen.

The first step was to grow *E. coli* on medium containing ^{15}N for several generations. The bacteria had completely heavy DNA (fig. 12.12). Meselson and Stahl knew this because only "heavy-heavy" molecules appeared in the density gradient. They then shifted the bacteria to medium containing ^{14}N, allowing enough time for the bacteria to divide only once (about 30 minutes). The proportions of heavy and light nitrogen over the next two replications would reveal the type of mechanism.

When the researchers collected the DNA this time and centrifuged it, the double helices were all of intermediate density, indicating that they contained half ^{14}N and half ^{15}N. This pattern was consistent with semiconservative DNA replication—but it was also consistent with a dispersive mechanism. In contrast, the result of conservative replication would have been one band of material completely labeled with ^{15}N corresponding to one double helix, and one totally "light" band containing ^{14}N only, corresponding to the other double helix.

Meselson and Stahl definitively distinguished among the three possible routes to replication, supporting the semiconservative mode and disproving the others, by extending the experiment one more generation. For the semiconservative mechanism to hold up, each hybrid (half ^{14}N and half ^{15}N) double helix present after the first generation following the shift to ^{14}N medium would part and assemble a new half from bases labeled only with ^{14}N. This would produce two double helices with one ^{15}N (heavy) and one ^{14}N (light) chain, plus two double helices containing only ^{14}N. The density gradient would appear as one heavy-light band and one light-light band. This is indeed what Meselson and Stahl saw.

The conservative mechanism would have yielded two bands in the third generation, indicating three completely light double helices for every completely heavy one. The third generation for the dispersive model would have been a single large band, somewhat higher than the second generation band because additional ^{14}N would have been randomly incorporated.

Semiconservative DNA replication was demonstrated in other species in a similar manner to Meselson and Stahl's density shift experiments. Introducing a radioactive label into growth medium and then removing it and monitoring cell division allowed researchers to see replicated chromosomes in which half of each chromatid displays the radioactivity (by exposing photographic film) and the other does not. These experiments extended Meselson and Stahl's results by demonstrating semiconservative replication in the cells of more complex organisms and at the whole-chromosome level.

Biotechnology 12.2

Two Routes to DNA Sequencing

The Sanger Method

Modern DNA sequencing instruments utilize a basic technique Frederick Sanger developed in 1977. The overall goal is to generate a series of DNA fragments of identical sequence that are complementary to the sequence of interest. These fragments differ in length from each other by one end base:

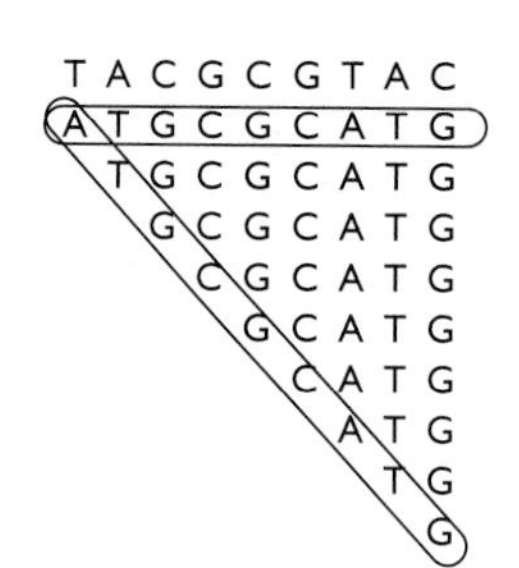

Note that the entire complementary sequence appears in the sequence of end bases of each fragment. If the complement of the gene of interest can be cut into a collection of such pieces, and the end bases distinguished with a radioactive or fluorescent label, then a technique called polyacrylamide gel electrophoresis can be used to separate the fragments by size. Then once the areas of overlap are aligned, reading the labeled end bases in size order reveals the sequence of the complement. Replacing A with T, G with C, T with A, and C with G establishes the sequence of the gene in question.

Sanger invented a way to generate the DNA pieces. In a test tube, he included the unknown sequence and all of the biochemicals needed to replicate it, including supplies of the four nucleotide bases. Some of each of the four types of bases were chemically modified at a specific location on the base sugar to contain no oxygen atoms instead of one—in the language of chemistry, they were dideoxyribonucleotides rather than deoxyribonucleotides. A radioactive nucleotide, usually "C," is also included in each reaction. DNA synthesis halts when DNA polymerase encounters a "dideoxy" base, leaving only a piece of the newly replicated strand.

Sanger repeated the experiment four times, each time using a dideoxy version of A,

FIGURE 12.B

Determining the Sequence of DNA.

In the Sanger method of DNA sequencing, complementary copies of an unknown DNA sequence are terminated early because of the incorporation of dideoxynucleotides terminators. A researcher or computer deduces the sequence by placing the fragments in size order. Radioactive labels are used to visualize the sparse quantities of each fragment.

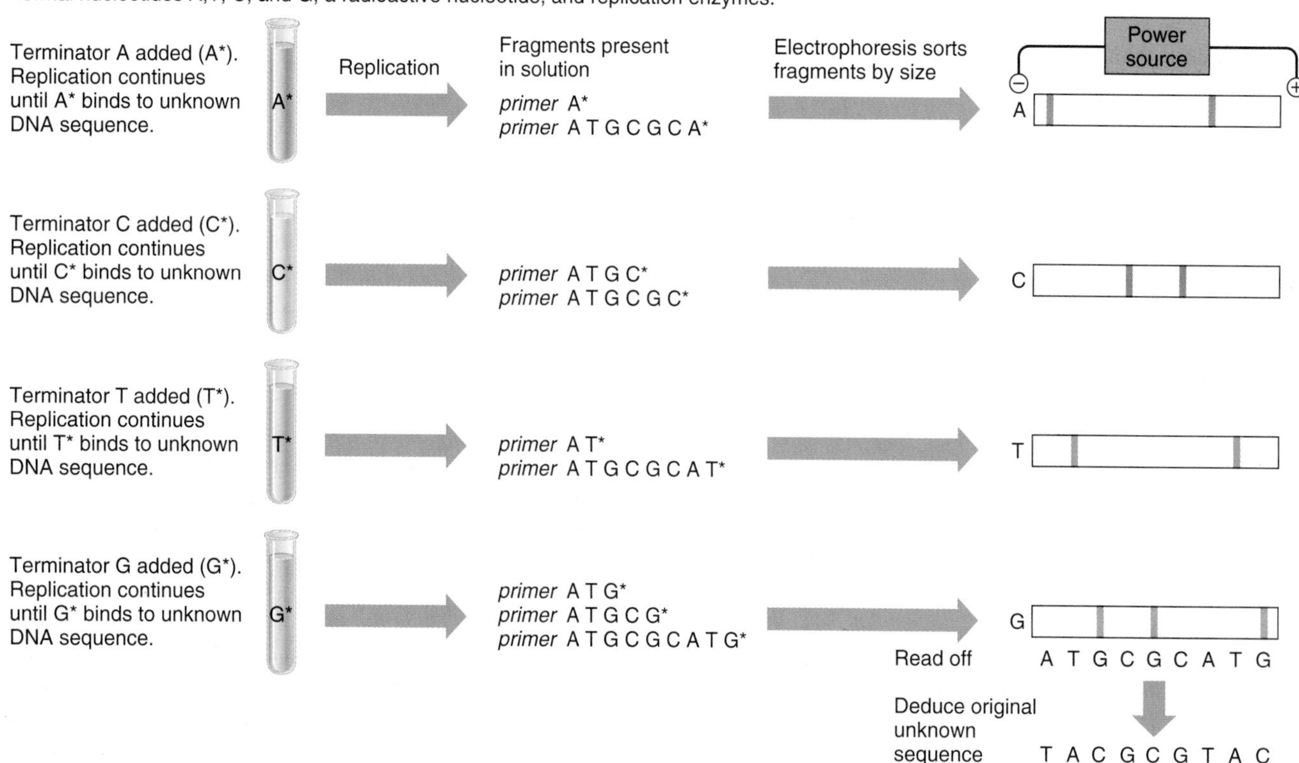

continued

T, C, then G. The four experiments were run in four lanes of a gel (fig. 12.B). Today fluorescent labels are used, one for each of the four base types, allowing a single experiment to reveal the sequence. The data appear as a sequential readout of the wavelengths of the fluorescence from the labels.

Sequencing on a DNA Chip

The best automated gene sequencers claim output of up to 7,200 bases per hour. A new technique, sequencing-by-hybridization, promises to sequence 32,000 bases an hour. The method utilizes a small glass square on which short DNA fragments of known sequence are immobilized. The DNA-studded glass square is called a DNA microchip or a DNA microarray. In one version, the 4,096 possible 6-base combinations (hexamers) of DNA are placed onto a 1-centimeter-by-1-centimeter microchip. Copies of an unknown DNA segment incorporating a fluorescent label are then also placed on the microchip. The copies stick (hybridize) to immobilized hexamers whose sequences are complementary to the DNA segment's sequences. Under laser light, the bound hexamers fluoresce. Because the researcher (or computer) knows which hexamers occupy which positions on the microchip, a scan of the chip reveals which 6-base sequences comprise the unknown sequence. Then, software aligns the identified hexamers by their overlaps. This reconstructs the complement of the entire unknown sequence. Figure 12.C depicts a simplified version of sequencing-by-hybridization.

FIGURE 12.C
Sequencing by Hybridization.
A labeled DNA segment of unknown sequence base pairs to short, known DNA sequences immobilized on a small glass microchip. Identifying the small, bound sequences, overlapping them, and then reading off their sequences, reveals the unknown DNA sequence.

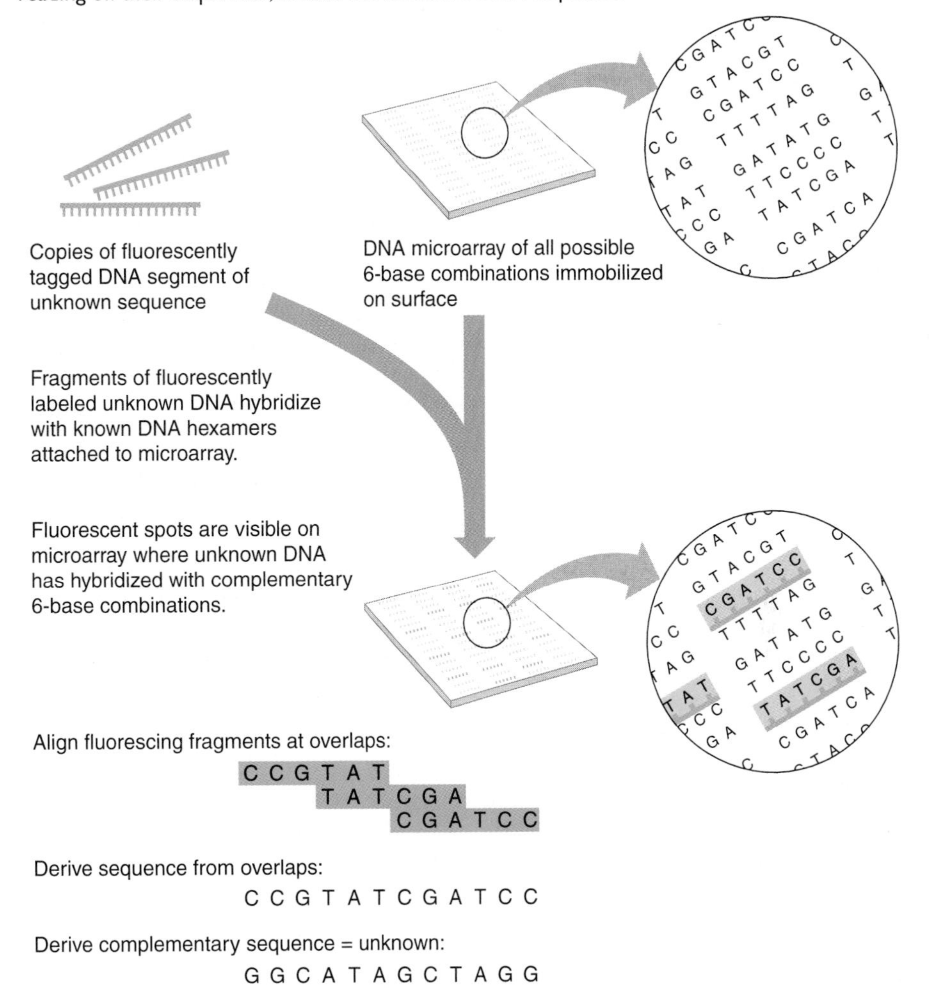

Steps and Participants in DNA Replication

A contingent of enzymes carries out DNA replication, as figure 12.13 whimsically shows. Enzymes called helicases unwind and hold apart replicating DNA so that other enzymes can guide the assembly of new DNA strands.

DNA replicates at hundreds of points along a chromosome, and then the newly made pieces merge to make one complete DNA molecule from each parent strand. While replicating, DNA resembles a fork in a road. Fittingly, the open portions of a replicating double helix are called **replication forks.** A human chromosome has many replication forks, which is necessary to replicate so much DNA in a relatively short time.

DNA replication begins when a helicase breaks the hydrogen bonds that connect a base pair. This first step occurs at an **origin of replication site.** Another type of enzyme, called primase, builds a short complementary piece of RNA, called an **RNA primer,** at the start of each DNA segment to be replicated. This is necessary because the major replication enzyme, **DNA polymerase,** can only add bases to an existing strand. Next, the RNA primer attracts DNA polymerase, an enzyme that draws in DNA nucleotides to complement the exposed bases on the parental strand and adds new bases one at a time, starting at the RNA primer. The new DNA strand grows and forms hydrogen bonds between the complementary bases. Special binding proteins keep the two strands apart.

DNA polymerase also "proofreads" as it goes, excising mismatched bases and inserting correct ones. At the same time,

FIGURE 12.13
An Army of Enzymes Replicates DNA.
This advertisement for a company that sells DNA-cutting enzymes depicts the number of participants (enzymes) that help to replicate and repair DNA. What is wrong with this cartoon?

another enzyme removes the RNA primer and replaces it with the correct DNA bases. Enzymes called **ligases** knit together the sugar-phosphate backbone. Figure 12.14 depicts the replication process. Biotechnology 12.3 discusses a widely used technology that uses DNA polymerase, called the polymerase chain reaction, or PCR, to mass produce a target DNA sequence.

DNA polymerase works directionally, adding new nucleotides to the exposed 3′ end of the deoxyribose in the growing strand. Replication proceeds in a 5′ to 3′ direction, because this is the only chemical configuration in which DNA polymerase can add bases. In order to follow this 5′ to 3′ "rule," DNA replication proceeds continuously (in one piece) on one strand but discontinuously (in short 5′ to 3′ pieces) on the other strand (fig. 12.15). The short pieces on the discontinuous strand are called Okazaki fragments.

12.3 MASTERING CONCEPTS

1. Why must DNA replicate?
2. What is semiconservative replication?
3. How did Meselson and Stahl demonstrate that DNA replication is semiconservative?
4. What are the steps of DNA replication?

OLC

12.4 How Does DNA Repair Itself?

Three mechanisms repair errors in base sequence that arise during DNA replication. Abnormal or absent DNA repair causes chromosome breakage and greatly increases the risk of cancer.

Any manufacturing facility tests a product in several ways to see whether it has been assembled correctly. Production mistakes are rectified before the item goes on the market—most of the time. The same is true of a cell's DNA production.

DNA replication is incredibly accurate—only about 1 in 100,000 bases is incorporated incorrectly. In addition to the "proofreading" capabilities of the DNA polymerase, repair enzymes further assure the accuracy of DNA replication.

Ultraviolet radiation and errors in replication can cause damage that must be repaired. Ultraviolet radiation damages DNA by causing an extra covalent bond to form between adjacent pyrimidines on the same strand, particularly thymines. The linked thymines are called thymine dimers. This extra

FIGURE 12.14
Overview of DNA Replication.

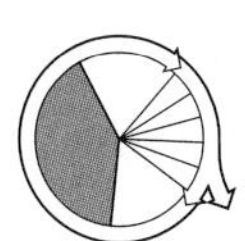

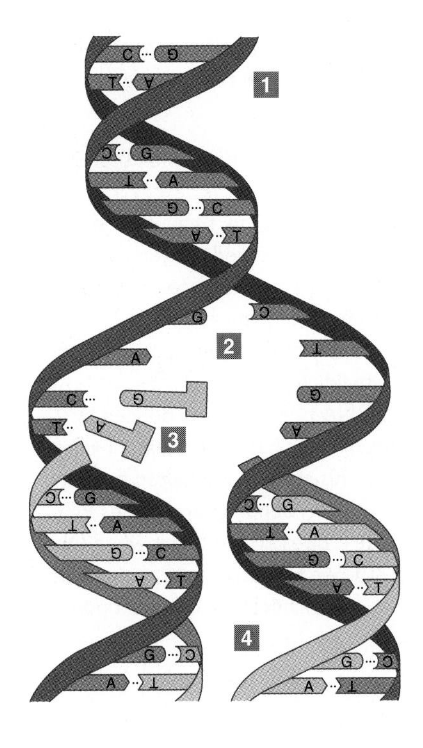

bond kinks the double helix, and is enough of an upset to disrupt replication and allow insertion of a noncomplementary base. In a type of repair called **photoreactivation,** enzymes called photolyases absorb energy from the blue part of visible light and use it to break the extra bond of a pyrimidine dimer (fig. 12.16).

Another type of DNA self-mending, **excision repair,** cuts the bond between the deoxyribose and the base and removes the pyrimidine dimer and surrounding bases. Then, a DNA polymerase fills in the correct nucleotides, using the exposed template as a guide.

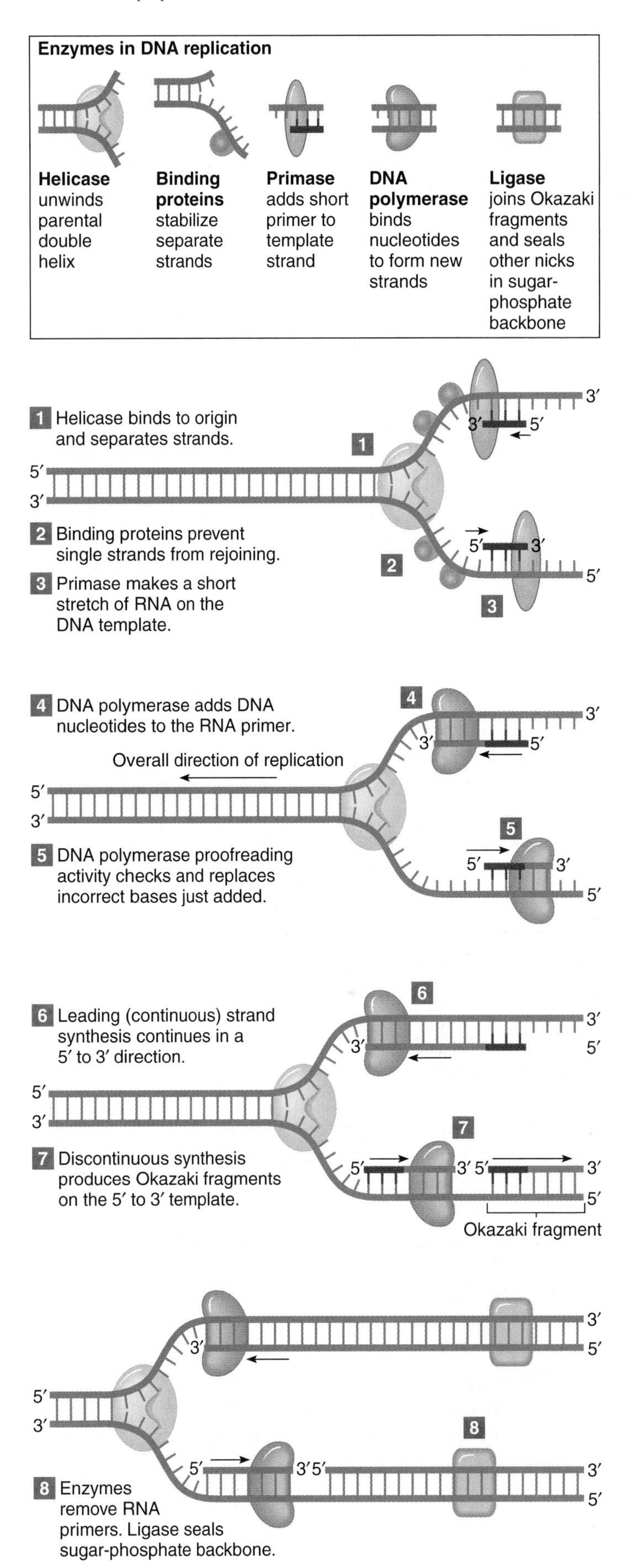

FIGURE 12.15
DNA Replication Takes Many Steps.

Biotechnology 12.3

PCR: An Application of Understanding DNA Replication

Every time a cell divides, it replicates all of its DNA. A technology called the polymerase chain reaction (PCR) uses the cell's DNA copying machinery to rapidly produce millions of copies of a specific DNA sequence of interest. Mass-producing specific DNA sequences is valuable in research, and also in developing highly sensitive medical tests, as table 12.2 describes.

PCR is useful in any situation where a small amount of DNA or RNA would provide information if it was mass-produced. Applications span a variety of fields. In forensics PCR is used routinely to establish blood relationships, identify remains, and to help convict criminals or exonerate the falsely accused. When used to amplify the nucleic acids of microorganisms and viruses and other parasites, PCR is important in agriculture, veterinary medicine, environmental science, and human health care. In genetics, PCR is both a crucial laboratory tool to map and identify genes as well as the basis of many tests to diagnose inherited disease, sometimes even years before symptoms arise.

Inspiration on a Starry Night

PCR was born in Kary Mullis's mind on a moonlit night in northern California in 1983. As he drove up and down the hills, Mullis, a molecular biologist, was thinking about the incredible precision and power of DNA replication. Suddenly, a way to tap into that power popped into his mind. He excitedly explained his idea to his girlfriend and then went home to think it through further. "It was difficult for me to sleep with deoxyribonuclear bombs exploding in my brain," he wrote much later.

The idea behind PCR was so stunningly straightforward that Mullis had trouble convincing his superiors at Cetus Corporation that he was really onto something. He spent the next year using the technique to amplify a well-studied gene so he could prove that his brainstorm was not just a flight of fancy. One by one, other researchers glimpsed Mullis's vision. After he convinced his colleagues at Cetus, Mullis published his landmark 1985 paper and filed patent applications, launching the era of gene amplification. However, the only compensation he received from Cetus was a $10,000 bonus—the company later sold the technology for $300 million. Mullis did have the consolation of winning a Nobel prize for his invention.

Surprisingly Simple

PCR rapidly replicates a selected sequence of DNA in a test tube. The requirements include:

1. Knowing parts of a target DNA sequence to be amplified.
2. Two types of lab-made, single-stranded, short pieces of DNA called primers. These are complementary in sequence to opposite ends of the target sequence.
3. A hefty supply of the four types of DNA nucleotide building blocks.
4. TaqI, a DNA polymerase produced by *Thermus aquaticus,* a bacterium that inhabits hot springs. This enzyme is adapted to its natural host's hot surroundings and makes PCR easy because it does not fall apart when DNA is heated. (Other heat-tolerant polymerases can be used, too, as described in the opening essay for chapter 1.)

In the first step of PCR, heat is used to separate the two strands of the target DNA. Next, the temperature is lowered and the two short DNA primers and TaqI DNA polymerase are added. The primers bind by complementary base pairing to the separated

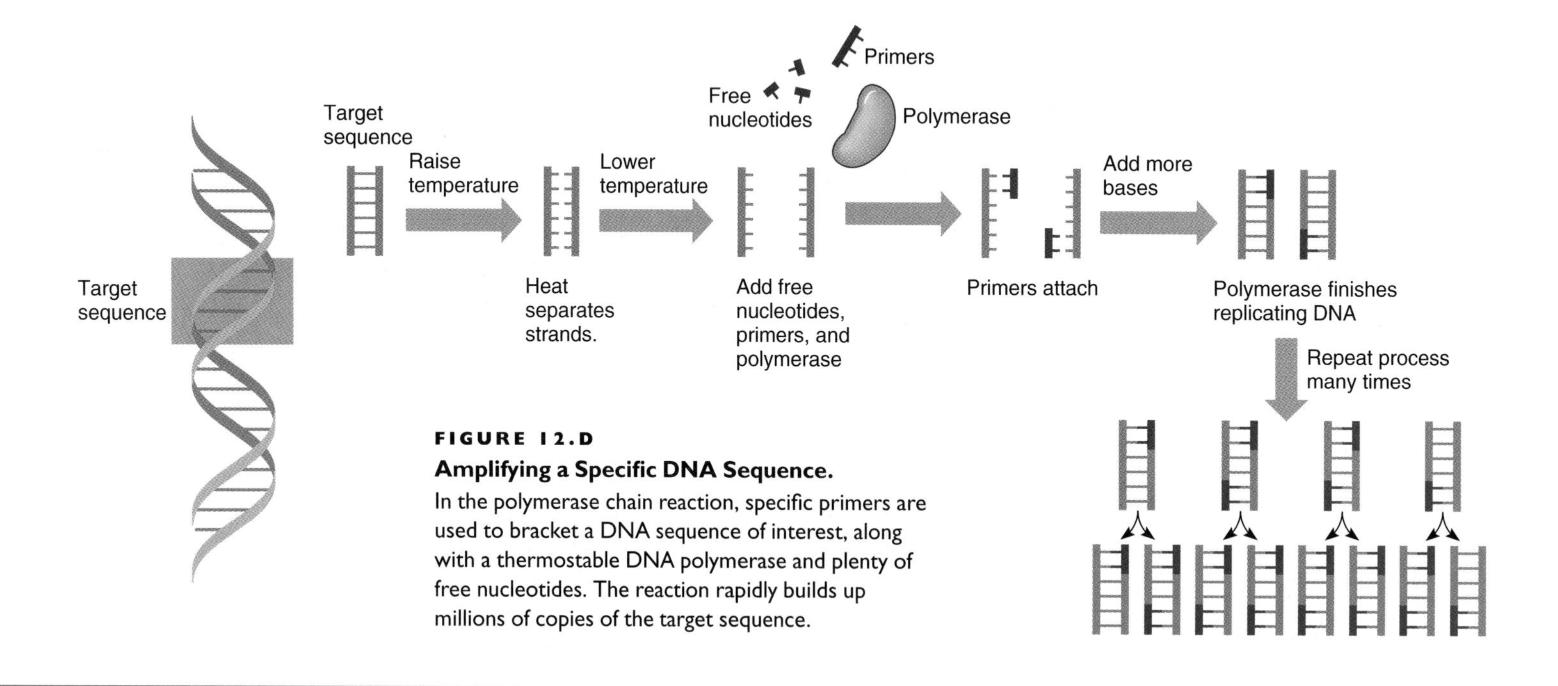

FIGURE 12.D
Amplifying a Specific DNA Sequence.
In the polymerase chain reaction, specific primers are used to bracket a DNA sequence of interest, along with a thermostable DNA polymerase and plenty of free nucleotides. The reaction rapidly builds up millions of copies of the target sequence.

target strands. In the third step, more DNA nucleotide bases are added. The DNA polymerase adds bases to the primers and builds a sequence complementary to the target sequence. The newly synthesized strands then act as templates in the next round of replication, which is initiated immediately by raising the temperature to separate the strands. All of this is done in an automated device called a thermal cycler that controls the key temperature changes.

The pieces of DNA accumulate geometrically. The number of amplified pieces of DNA equals 2^n, where n equals the number of temperature cycles. After just 20 cycles, a one million-fold increase in the number of copies of the original sequence accumulates in the test tube (fig. 12.D).

PCR's greatest strength is that it works on crude samples of rare and minute sequences, such as a bit of brain tissue on the bumper of a car, which in one criminal case led to identification of a missing person. PCR's greatest weakness, ironically, is its exquisite sensitivity. A blood sample submitted for diagnosis of an infection contaminated by leftover DNA from a previous run, or a stray eyelash dropped from the person running the reaction, can yield a false result.

TABLE 12.2 PCR Applications

PCR HAS BEEN USED TO AMPLIFY:

- Genetic material from HIV in a human blood sample when infection was so recent that antibodies were not yet detectable.
- A bit of DNA in a preserved quagga (a relative of the zebra) and a marsupial wolf, which are recently extinct animals.
- Genes from microorganisms that cannot be grown or maintained in culture for study.
- Mitochondrial DNA (mtDNA) from various modern human populations. Comparisons of mtDNA sequences indicate that *Homo sapiens* originated in Africa, supporting fossil evidence.
- Similar genes from several species. Comparing the extent of similarity reveals evolutionary relationships among species.
- DNA from the brain of a 7,000-year-old human mummy, which indicates that native Americans were not the only people to dwell in North America long ago.
- Genetic material from saliva, hair, skin, and excrement of organisms that we cannot catch to study. The prevalence of a rare DNA sequence among all of the bird droppings from a certain species in an area can be extrapolated to estimate the population size.
- DNA in the digestive tracts of carnivores, to reveal food web interactions.
- DNA in deteriorated road kills and carcasses washed ashore, to identify locally threatened species.
- DNA in products illegally made from endangered species, such as powdered rhinoceros horn, sold as an aphrodisiac.
- DNA sequences in animals indicating that they carry the bacteria that cause Lyme disease, providing clues to how the disease is transmitted.
- DNA from genetically altered bacteria that are released in field tests, to follow their dispersion.
- DNA from one cell of an 8-celled human embryo to diagnose cystic fibrosis.
- Y-chromosome-specific DNA to determine sex of oocytes fertilized in the laboratory.
- A human papillomavirus DNA sequence present in, and possibly causing, an eye cancer.
- DNA from poached moose meat in hamburger.
- DNA from human remains in Jesse James's grave, to make a positive identification.
- DNA from maggots in a rotting human corpse, to extrapolate the time of murder by determining which particular species of insect deposited larvae.
- DNA in artificial knee joints, indicating infection.
- DNA from semen on a blue dress of a White House intern, which helped identify the person with whom she had a sexual encounter.
- DNA from cat hairs on a murder victim, matched to DNA from a suspect's cat Fluffy.

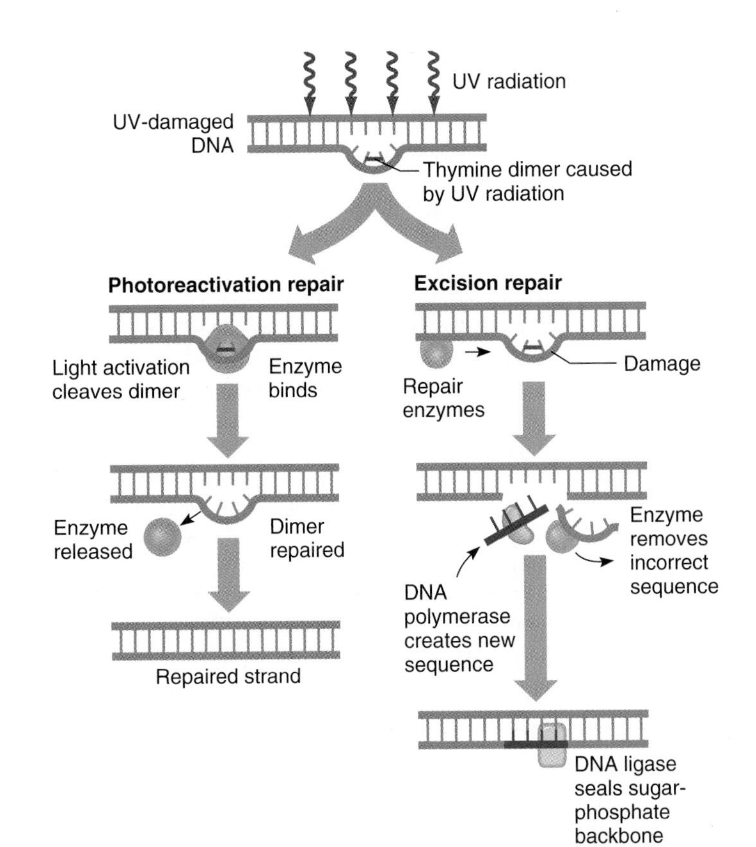

In a third type of mechanism called **mismatch repair,** enzymes proofread newly replicated DNA for small loops that form where the two strands are not precisely aligned. Such mismatching tends to occur in regions of very short DNA sequence repeats, called microsatellites, that are scattered throughout the genome. Their lengths can vary from person to person, but within an individual they are all the same length. Mismatch repair maintains microsatellite length in an individual.

FIGURE 12.16
Two Types of DNA Repair.
DNA damaged by UV light is repaired by photoreactivation, in which a pyrimidine dimer is split, or by excision repair, in which the pyrimidine dimer and a few surrounding bases are removed and replaced.

Several inherited disorders reflect faulty DNA repair. Such conditions usually produce chromosome breaks and a high susceptibility to cancer following exposure to ionizing radiation or chemicals that affect cell division. Xeroderma pigmentosum, described in Health 10.1, is caused by a defect in excision repair.

Disorders of DNA repair reveal how essential precise DNA replication is to cell survival and proliferation. Once these processes are completed, the cell begins to use the informational content of DNA. This is the subject of the next chapter.

12.4 MASTERING CONCEPTS

1. Why is DNA repair necessary?
2. How does ultraviolet radiation damage DNA?
3. What are different types of DNA repair?
4. What are the symptoms of disorders that result from defective DNA repair?

OLC

Chapter Summary

12.1 How Did Experiments Identify and Describe the Genetic Material?

1. DNA encodes the information necessary for a cell's survival and specialization and must be able to replicate for a cell to divide.
2. Many experiments described DNA and showed it to be the genetic material. Miescher identified DNA in white blood cell nuclei. Garrod connected heredity to symptoms resulting from enzyme abnormalities.
3. Griffith determined that a substance transmits a disease-causing trait to bacteria; Avery, MacLeod, and McCarty showed that the transforming principle is DNA; Hershey and Chase confirmed that the genetic material is DNA and not protein.
4. Levene described the proportions of nucleotide components. Chargaff discovered that A and T, and G and C, occur in equal proportions. Wilkins and Franklin provided X-ray diffraction data. Watson and Crick combined these clues to propose the double helix conformation of DNA.

12.2 What Is the Three-Dimensional Structure of DNA?

5. The rungs of the DNA double helix consist of hydrogen-bonded complementary base pairs (A with T, and C with G). The rails are chains of alternating **deoxyribose** and phosphate, which run **antiparallel** to each other.
6. DNA is highly coiled around proteins, forming **nucleosomes.**

12.3 How Does DNA Replication Maintain Genetic Information?

7. Density shift experiments showed that DNA replication is **semiconservative,** and not **conservative** or **dispersive.**
8. To replicate, DNA unwinds locally at several **origins of replication. Replication forks** form as hydrogen bonds break. Primase builds a short **RNA primer,** which is eventually replaced with DNA. Next, **DNA polymerase** fills in DNA bases, and **ligase** seals the sugar-phosphate backbone.
9. Replication proceeds in a 5′ to 3′ direction, necessitating that the process be discontinuous in short stretches on one strand.

12.4 How Does DNA Repair Itself?

10. **Photoreactivation** splits pyrimidine dimers.
11. **Excision repair** cuts out the damaged area and replaces it with correct bases.
12. **Mismatch repair** scans newly replicated DNA for mispairing and corrects the error.
13. Repair disorders break chromosomes and raise cancer risk.

Testing Your Knowledge

1. If a cell contains all the genetic material it must have to synthesize protein, why must the DNA also replicate?
2. State the functions of the following enzymes that participate in DNA replication or repair:
   a. primase
   b. DNA polymerase
   c. ligase
   d. helicase
   e. photolyase
3. What part of the DNA molecule encodes information?
4. Write the complementary DNA sequence of each of the following base sequences:
   a. T C G A G A A T C T C G A T T
   b. C C G T A T A G C C G G T A C
   c. A T C G G A T C G C T A C T G
5. List the steps of DNA replication.
6. Choose an experiment mentioned in the chapter and analyze how it follows the scientific method.

Thinking Scientifically

1. To diagnose encephalitis (brain inflammation) caused by West Nile virus infection, a researcher needs a million copies of a viral gene. She decides to use the polymerase chain reaction on a sample of cerebrospinal fluid, which bathes the person's infected brain. If one cycle of PCR takes 2 minutes, how long will it take the researcher to obtain her million-fold amplification if she starts with a single copy?
2. Give an example from the chapter of different types of experiments used to address the same hypothesis. Why might this be necessary?
3. The experiments that revealed DNA structure and function used a variety of organisms. How can such diverse organisms demonstrate the same genetic principles?
4. A person with deficient or abnormal ligase or excision repair may have an increased cancer risk and chromosomes that cannot heal breaks. The person is, nevertheless, alive. How long would an individual lacking DNA polymerase be likely to survive?
5. HIV infection was once diagnosed by detecting antibodies in a person's blood or documenting a decline in the number of the type of white blood cell that HIV initially infects. Why is detection using PCR more accurate?

References and Resources

Foster, E. A., et al. November 5, 1998. Jefferson fathered slave's last child. *Nature,* vol. 396. DNA analysis showed that President Jefferson's slave Sally Hemings had his child.

Mullis, Kary B. April 1990. The unusual origin of the polymerase chain reaction. *Scientific American.* How PCR arose from a brainstorm.

Piper, Anne. April 1998. Light on a dark lady. *Trends in Biological Sciences,* vol. 23. A tribute to Rosalind Franklin, by her best friend.

Watson, James D. 1968. *The Double Helix.* New York: New American Library. An exciting, personal account of the discovery of DNA structure.

The *LIFE* Online Learning Center provides additional resources and tools for studying this chapter.
www.mhhe.com/life/

CHAPTER 13

Gene Function

On Genomes

The Human Genome Project, which has revealed the DNA sequences of all the genes in a human cell, promises to revolutionize health care. From a biologist's viewpoint, however, even more exciting is the sequencing of genomes from many species. To obtain this information, researchers cut several copies of the same genome into many pieces, and then automated devices and computers sequence and overlap the pieces, deriving the continuous sequences of chromosomes. One way that researchers discover gene functions is by identifying similarities in databases of sequenced genes from other organisms.

Several dozen species have already had their entire genetic selves laid bare, and the greatest lesson we've learned so far is how much we do not know. DNA sequences that have remained relatively unchanged over the ages among diverse species indicate much shared, if distant, ancestry. By comparing the genomes of species that lie at the boundaries of great evolutionary leaps, researchers can investigate the compelling questions posed below:

What Is the Minimum Number of Genes Required for Life? The smallest cell known to be able to reproduce is the bacterium-like *Mycoplasma genitalium.* It infects cabbage, citrus fruit, corn, broccoli, honeybees, and spiders, and causes respiratory illness in chickens, pigs, cows, and humans. Researchers call its tiny genome the "near-minimal set of genes for independent life." Comparing *Mycoplasma* genes to those of other organisms offers an approximation of how this most micro of microorganisms uses what it's got:

% of genome	Function
30	Maintaining its cell membrane
24	Expressing genes as proteins
22	Unknown: doesn't match any known gene
9	DNA replication
8	Obtaining energy
4.5	Evading immune attack by host cell
4	Making and recycling nucleic acids

(They add up to more than 100 percent because there is some overlap in function.) According to this genome, only 300 or so genes may be necessary for life.

What Are the Fundamental Distinctions Among the Three Domains of Life? Home for the archaean *Methanococcus jannaschii* is the bottom of a 2,600-meter-tall "white smoker" chimney deep in the Pacific Ocean, at high temperature and pressure and without oxygen. Less than half of

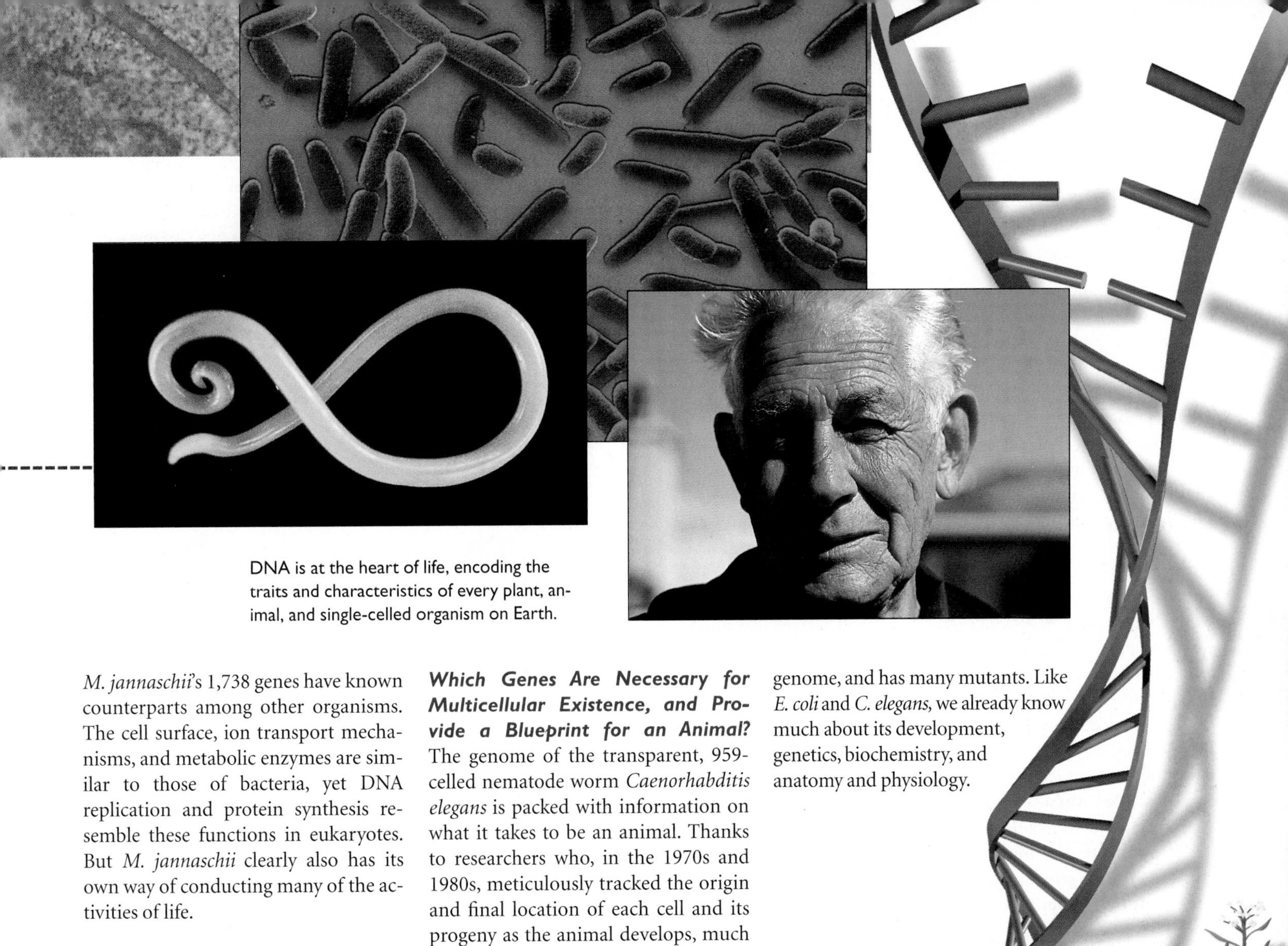

DNA is at the heart of life, encoding the traits and characteristics of every plant, animal, and single-celled organism on Earth.

M. jannaschii's 1,738 genes have known counterparts among other organisms. The cell surface, ion transport mechanisms, and metabolic enzymes are similar to those of bacteria, yet DNA replication and protein synthesis resemble these functions in eukaryotes. But *M. jannaschii* clearly also has its own way of conducting many of the activities of life.

How Does a Bacterium Work? In 1966, Francis Crick suggested that by identifying all the genes in an *Escherichia coli* cell, we would understand how it works. Crick's idea became reality in 1997, with the sequencing of the *E. coli* genome, but having the sequence was only a first step. In 1997, researchers knew the functions of only 2,000 of the 4,288 protein-encoding genes—more than half were a mystery.

What Is the Genetic Organization of the Simplest Eukaryote? The title of journal article unveiling the genome of *Saccharomyces cerevisiae,* "Life with 6,000 Genes," was deceptive for this not-so-simple unicellular yeast. A third of its genes have counterparts among mammals, including those for more than 70 disease-causing genes in humans. Many of the yeast's genes are duplicated or clustered, similar to the organization in multicellular organisms.

Which Genes Are Necessary for Multicellular Existence, and Provide a Blueprint for an Animal? The genome of the transparent, 959-celled nematode worm *Caenorhabditis elegans* is packed with information on what it takes to be an animal. Thanks to researchers who, in the 1970s and 1980s, meticulously tracked the origin and final location of each cell and its progeny as the animal develops, much of the biology of the worm was already known before its 97 million DNA bases were revealed late in 1998. Its signaling pathways, cytoskeleton, immune system, DNA replication, and certain nerve-cell proteins are similar to ours! Comparisons to the yeast genome reveal requirements for multicellularity. Many of the worm's genes encode cell surface receptors for hormones, allowing cell-to-cell communication not seen in the much simpler yeast.

Which Genes Are Distinct to Plants? *Arabidopsis thaliana* is the ideal model organism for plant biologists. Until the early 1980s, most species used for basic research were of agricultural value. But *Arabidopsis,* a member of the mustard family, offers such compelling advantages for research that it is the first flowering plant to have its genome sequenced. It has a short life cycle, produces abundant seeds, has a small genome, and has many mutants. Like *E. coli* and *C. elegans,* we already know much about its development, genetics, biochemistry, and anatomy and physiology.

13.1 How Does a Cell Access Genetic Information?

A cell accesses specific genes by transcribing their DNA sequences into RNA sequences. A cell's characteristics reflect control of transcription.

DNA replication preserves genetic information by ensuring that each new cell receives a complete set of genetic instructions. A cell uses the information encoded in DNA sequences to manufacture proteins. First, the process of **transcription** copies the DNA sequence comprising a gene into an RNA molecule that is complementary to one strand of the DNA double helix, called the template strand. Then the process of **translation** uses the information copied into RNA to manufacture a specific protein by aligning and joining certain amino acids. Watson and Crick, shortly after publishing their structure of DNA in 1953, expressed this relationship between nucleic acids and proteins as a directional flow of information they called the "central dogma" (fig. 13.1). Francis Crick explained the central dogma in a talk he gave in 1957: "The specificity of a piece of nucleic acid is expressed solely by the sequence of its bases, and this sequence is a code for the amino acid sequence of a particular protein."

RNA is of crucial importance in the flow of genetic information. This eclectic nucleic acid differs from DNA in several ways: It contains the sugar ribose instead of deoxyribose, it has the nitrogenous base uracil in place of thymine; and it can be single-stranded. RNA can also be a catalyst, a role not known for DNA (fig. 13.2 and table 13.1).

Transcription is highly regulated. In a unicellular organism, some genes must be transcribed all the time because they specify proteins that are essential to life, such as the enzymes

FIGURE 13.1
DNA to RNA to Protein.
The central dogma of biology states that information stored in DNA is copied to RNA (transcription), which is used to assemble proteins (translation). DNA replication perpetuates genetic information. This figure repeats through the chapter, with the part under discussion highlighted.

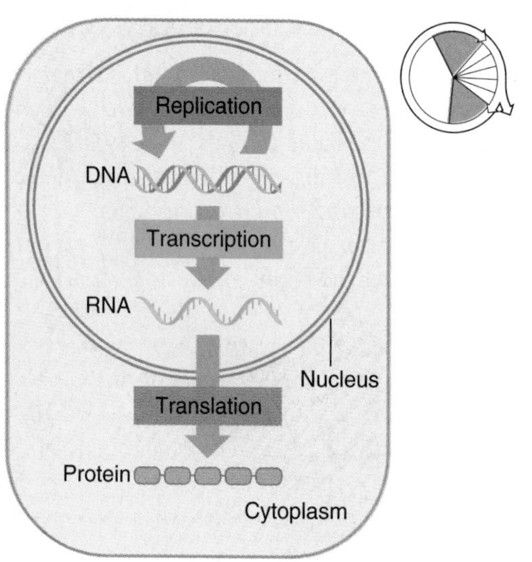

FIGURE 13.2
DNA and RNA Differ Functionally and Structurally.
DNA and RNA have different functions (**A**). DNA is double-stranded; RNA is usually single-stranded (**B**). DNA nucleotides include deoxyribose, whereas RNA nucleotides have ribose (**C**). Finally, DNA nucleotides include the pyrimidine thymine, whereas RNA has uracil (**D**).

DNA
Stores RNA- and protein-encoding information, and transfers information to daughter cells

RNA
Carries protein-encoding information, helps to make proteins

A

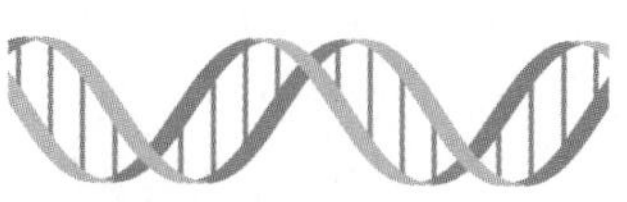
Double-stranded

Generally single-stranded

B

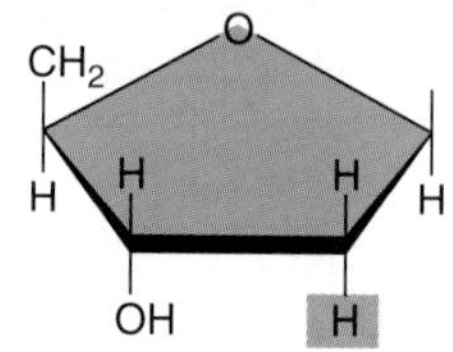

Deoxyribose as the sugar

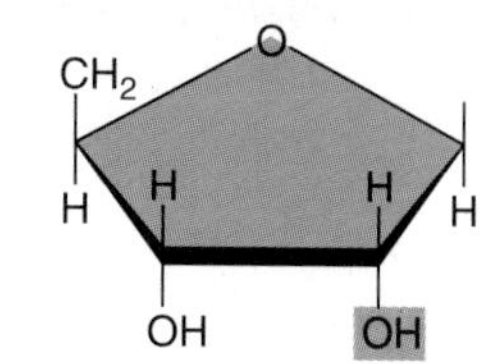

Ribose as the sugar

C

Bases used:

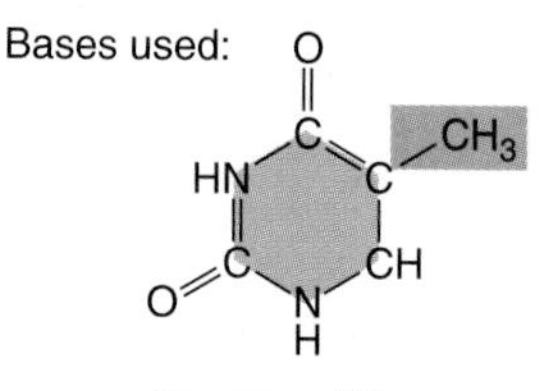

Thymine (T)
Cytosine (C)
Adenine (A)
Guanine (G)

Bases used:

Uracil (U)
Cytosine (C)
Adenine (A)
Guanine (G)

D

TABLE 13.1 How RNA and DNA Differ

RNA	DNA
Usually single-stranded	Usually double-stranded
Has uracil as a base	Has thymine as a base
Ribose as the sugar	Deoxyribose as the sugar
Carries protein-encoding information	Maintains protein-encoding information
Can be catalytic	Not catalytic

that catalyze the reactions of glycolysis. Other genes that must be transcribed continuously include those coding for essential proteins that are very short-lived. Other genes are transcribed only under certain conditions to enable the unicellular organism to survive a change in the environment. A multicellular organism presents even greater transcriptional complexity. How does a muscle cell "know" to repeatedly transcribe the genes encoding actin and myosin? How does a cell in a leaf "know" to express the genes for the enzymes that catalyze the reactions of photosynthesis?

Cells have built-in controls of transcription. We look first at these controls, and then follow step-by-step as a gene's information is first transcribed into an accessible form and then translated into protein. Finally we look at changes in a gene's structure and sometimes function—mutation.

Bacteria Use Operons to Turn Genes On or Off

It wasn't long after the structure of DNA was published that researchers began to unravel the controls of gene expression. In 1961, French biologists François Jacob and Jacques Monod described the remarkable ability of *E. coli* to produce the exact enzymes they require to metabolize the sugar lactose only when lactose is present in the cell's surroundings. What "tells" a simple bacterial cell to transcribe the genes whose products metabolize lactose at precisely the right time?

The lactose itself is the trigger. A modified form of the sugar attaches to a protein, called a **repressor,** that in the absence of lactose binds to and suppresses the DNA sequence that signals the cell to transcribe the three enzymes that break down the sugar (fig. 13.3). When lactose removes the repressor, the **promoter,** which is a DNA sequence that controls transcription of other genes, turns on the genes encoding the enzymes. Lactose, in a sense, causes its own dismantling.

Jacob and Monod named the genes (and their controls) that produce the enzymes required for lactose metabolism an **operon.** Soon, geneticists discovered operons for the metabolism of other nutrients, and in other bacteria. Some operons, like the lactose operon, negatively control transcription by removing a block. Others act positively, producing factors that turn on transcription. As Jacob and Monod stated in 1961, "The genome contains not only a series of blueprints, but a coordinated program of protein synthesis and means of controlling its execution." Operons were originally described in bacteria, but the genome sequence of the roundworm *C. elegans* revealed that nearly a quarter of its genes are organized into operon-like groups, too.

FIGURE 13.3

The Lactose Operon.

(**A**) In the lactose operon, a group of genes that encode proteins needed for the utilization of the sugar lactose are under the control of a single promoter. (**B**) In the absence of lactose, a repressor protein binds to the operator region of the DNA, preventing transcription of the genes. (**C**) In the presence of lactose, the repressor binds to lactose exclusively, allowing transcription to occur. This allows the cell to only make these enzymes when they are needed.

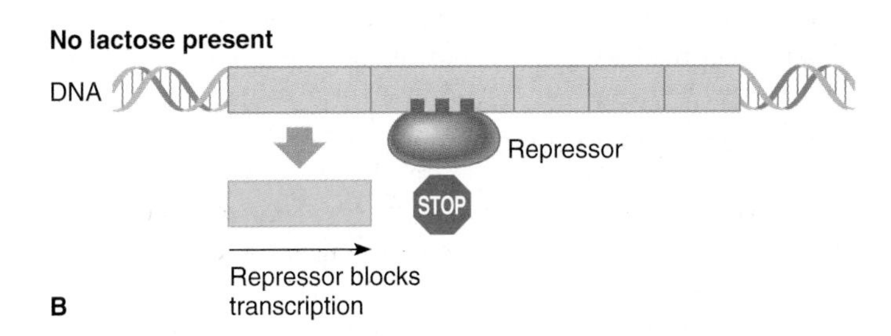

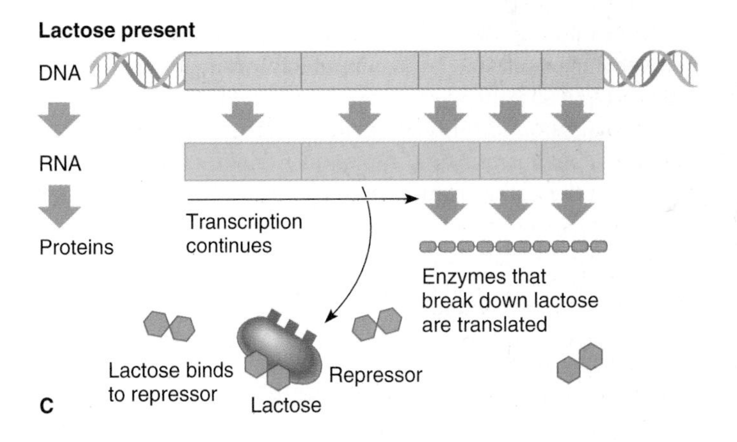

Multicellular Organisms Use Transcription Factors to Turn Genes On or Off

In bacteria, operon controls function like switches, turning gene transcription on or off. In multicellular eukaryotes like ourselves, genetic control is more complex because different cell types express different subsets of genes. To manage such complexity, groups of proteins called **transcription factors** come together, forming an apparatus that binds DNA and initiates transcription at specific sites on the chromosome. The transcription factors, activated by signals from outside the cell, set the stage for transcription to begin by forming a pocket for **RNA polymerase**—the enzyme that actually builds an RNA chain.

Several types of transcription factors are required to transcribe a eukaryotic gene. Because transcription factors are proteins, they too are gene-encoded. The DNA sequences that transcription factors bind may be located near the genes they control, or as far as 40,000 bases away. DNA may form loops so that the genes encoding proteins that act together come near each other. Proteins in the nucleus may help bring certain genes and their associated transcription factors in close proximity, much as books on a specialized topic might be grouped together in a library for easier access.

Hundreds of transcription factors are known, and in humans, defects in them lie behind some diseases, including cancers. Many transcription factors have regions in common called motifs that fold into similar three-dimensional shapes, or conformations. These motifs generally enable the transcription factor to bind DNA. They have very colorful names, such as "helix-turn-helix,"

FIGURE 13.4

Binding to DNA.

Transcription factors are proteins that bind to DNA and regulate gene expression by turning on or blocking transcription of particular genes. Motifs are regions named for the characteristic shapes of the DNA binding regions found in these proteins. A common example is shown here.

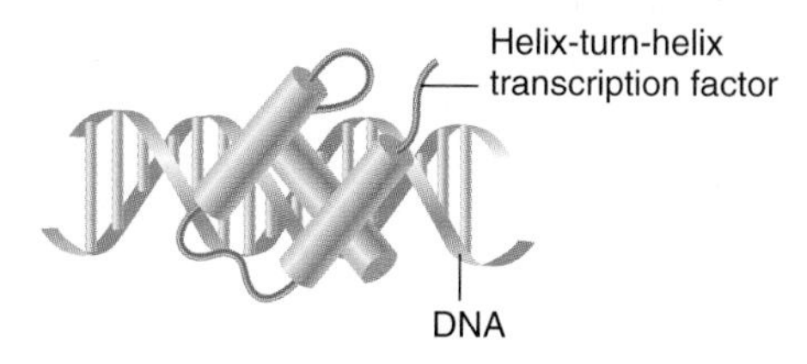

"zinc fingers," and "leucine zippers," that reflect their distinctive shapes. Figure 13.4 shows how one transcription factor associates intimately with DNA, thereby controlling gene expression.

Steps of Transcription

How do transcription factors and RNA polymerase "know" where to bind to DNA to begin transcribing a specific gene? Transcription factors and RNA polymerase are attracted to a **promoter,** which is a special sequence that signals the start of the gene. Figure 13.5 shows one order in which transcription factors bind, to set up a site to receive RNA polymerase. The first transcription factor to bind, called a TATA binding protein, is attracted to a DNA sequence called a TATA box, which consists of the base sequence TATA surrounded by long stretches of G and C. Once the first transcription factor binds, it attracts others in groups and finally RNA polymerase joins the complex, binding just in front of the start of the gene sequence.

Complementary base pairing underlies transcription, just as it does DNA replication. First, enzymes unwind the DNA double helix, and RNA nucleotides bond with exposed complementary

FIGURE 13.5

Setting the Stage for Transcription to Begin.

(**A**) The promoter region of a gene has specific sequences recognized by proteins that initiate transcription. (**B**) A binding protein recognizes the TATA region and binds to the DNA. This allows other transcription factors to bind. (**C**) The presence of the necessary transcription factors allows RNA polymerase to bind and begin making RNA.

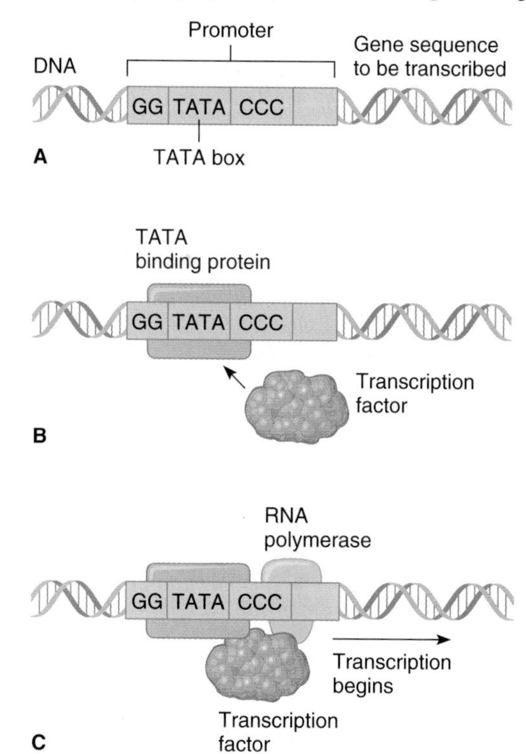

FIGURE 13.6

The Relationship Among RNA, the DNA Template Strand, and the DNA Coding Strand.

The RNA sequence is complementary to that of the DNA template strand and therefore is the same sequence as the DNA coding strand, with uracil (U) in place of thymine (T).

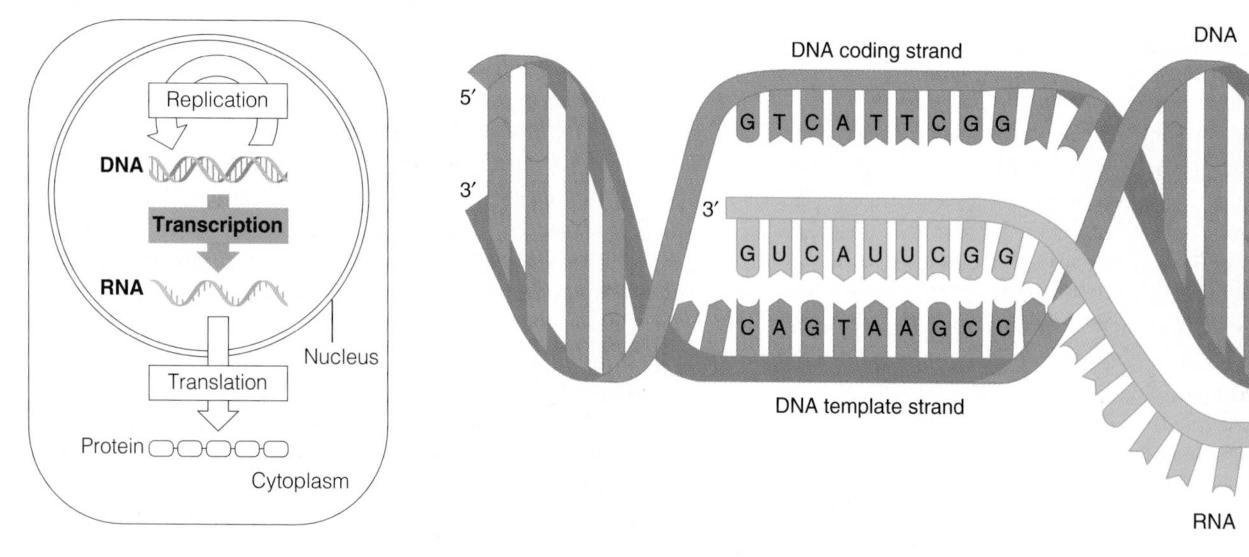

bases on the DNA template strand (fig. 13.6). RNA polymerase adds the RNA nucleotides in the sequence the DNA specifies, moving along the DNA strand in a 3′ to 5′ direction, synthesizing the RNA molecule in a 5′ to 3′ direction. A terminator sequence in the DNA indicates where the gene's RNA-encoding region ends.

For a particular gene, RNA is transcribed using only one strand of the DNA double helix as the template. The other DNA strand that isn't transcribed is called the coding strand because its sequence is identical to that of the RNA, except with thymine (T) in place of uracil (U). Several RNAs may be transcribed from the same DNA template strand simultaneously (fig. 13.7). Since RNA is relatively short-lived, a cell must constantly transcribe certain genes to maintain supplies of essential protein.

FIGURE 13.7

Transcription of RNA from DNA.

(**A**) Transcription occurs in three stages: initiation, elongation, and termination. Initiation is the control point that determines which genes are transcribed and when. RNA nucleotides are added during elongation, and a terminator sequence in the gene signals the end of transcription. (**B**) Many identical copies of RNA are simultaneously transcribed, with one RNA polymerase starting after another.

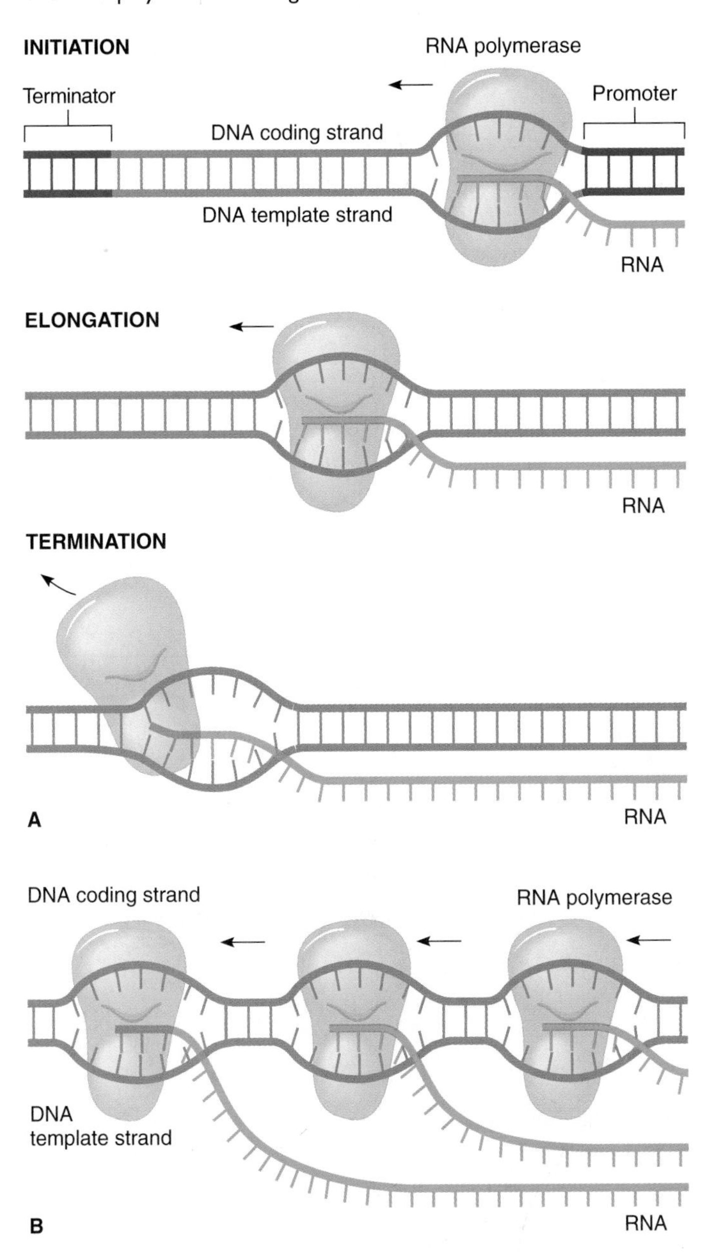

To determine the sequence of RNA bases transcribed from a gene, write the RNA bases that are complementary to the template DNA strand, using uracil opposite adenine. For example, if a DNA template strand has the sequence

C C T A G C T A C

then it is transcribed into RNA with the sequence

G G A U C G A U G

The coding DNA sequence is:

G G A T C G A T G

13.1 MASTERING CONCEPTS

1. How do DNA replication, RNA transcription, and translation maintain and use genetic information?
2. How does lactose metabolism in *E. coli* illustrate control of gene expression?
3. How do transcription factors control gene expression in eukaryotes?
4. What are the steps of transcription?

OLC

13.2 RNA Orchestrates Protein Synthesis

RNA carries a gene's information into the cytoplasm, and enables it to be translated into a protein's amino acid sequence. Messenger RNA carries a gene's sequence information; ribosomal RNA is part of ribosomes, which support and bring together amino acids as proteins form; transfer RNA matches specific amino acids to specific mRNA triplets, enabling ribosomes to assemble proteins.

As RNA is synthesized along DNA, it curls into three-dimensional shapes, or conformations, determined by complementary base pairing within the same RNA molecule. These conformations determine how RNA functions. Several types of RNA interact to synthesize proteins (table 13.2).

Types of RNA

Messenger RNA (mRNA) carries the information that specifies a particular protein. Each group of three mRNA bases in a row forms a genetic code word, or **codon,** that corresponds to a particular amino acid. Because genes vary in length, so do mRNA molecules. Most mRNAs are 500 to 3,000 bases long. Biotechnology 13.1 describes antisense technology, which silences particular genes at the mRNA level.

Ribosomal RNA (rRNA) molecules range from 100 to nearly 3,000 nucleotides long. This type of RNA associates with certain proteins to form a ribosome. Recall from chapter 3 that a ribosome is a structural support for protein synthesis. A ribosome has two subunits that are separate in the cytoplasm but

Biotechnology 13.1

Antisense Technology Silences Gene Expression

Is it wasteful to transcribe only one side of the DNA double helix for any given gene? It may seem so, but the unexpressed side may provide a tool for controlling gene expression. An RNA sequence complementary to a gene's messenger RNA (mRNA) could block that gene's expression by physically binding to the mRNA, making protein synthesis impossible. Use of complementary nucleic acid sequences to block expression of other sequences is called antisense technology. The name derives from the fact that mRNA is sometimes called "sense" RNA. Its complement is therefore called "antisense" RNA (fig. 13.A).

Antisense technology as first conceived in the mid-1980s used synthetic short pieces of nucleic acids, usually about 20 bases long, called oligonucleotides. But the "oligos" proved unstable, quickly dismantled by an enzyme called RNase. A more successful approach is to engineer organisms from the beginning of development so that each cell includes a DNA sequence that encodes the antisense RNA sequence. This works well for plants, but is banned in humans. (Such an organism is called transgenic, and is discussed in Biotechnology 13.3.) Researchers at a California biotechnology company fashioned the "FlavrSavr" tomato with this transgenic approach. An antisense sequence silenced a ripening (softening) enzyme, and the tomatoes remained firm and red in the supermarket for days longer than their unaltered counterparts. However, FlavrSavr failed in the marketplace because of other, undesirable traits, and the public's fear of genetically altered crops.

An alternative to oligos is to synthesize related chemicals that retain the base sequence that provides the technique's specificity, but to alter the sugar-phosphate backbone in a way that makes the molecules more stable. One drug being developed treats an eye infection caused by cytomegalovirus, which often occurs with AIDS. The antisense compound targets the mRNA for a protein that the virus must manufacture in infected human cells in the eye to survive. Patients take it as weekly eyedrops. Another experimental antisense treatment counters the inflammation of Crohn's disease, a form of inflammatory bowel disease in which a cellular adhesion molecule (see chapter 4) is overproduced. A synthetic antisense compound blocks the mRNA encoding the adhesion protein, quelling the inflammation in the intestines.

Antisense technology has tremendous potential—theoretically it can be used to squelch activity of any gene, if the RNA can be coaxed to stick around long enough, and if it can be delivered to the appropriate tissue. With genome projects revealing gene sequences on a daily basis, and the human genome sequence known, there will be no shortage of targets for antisense technology. All that remains is perfecting the approach.

FIGURE 13.A
Silencing Gene Expression.
Antisense technology uses a complementary nucleic acid sequence, or a similar molecule, to bond with mRNA, thereby preventing protein synthesis. RNase dismantles the blocked mRNA.

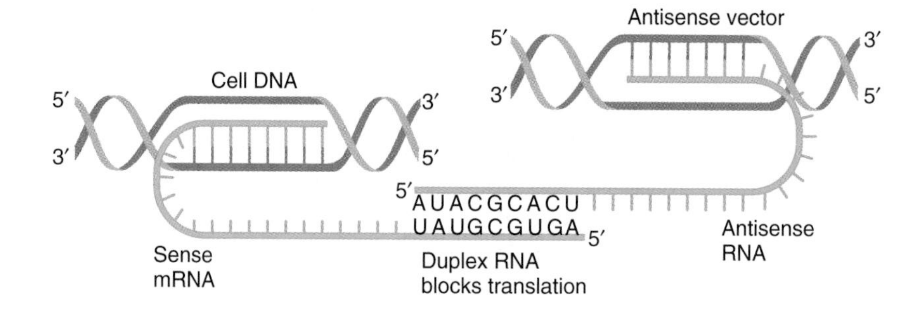

TABLE 13.2 Major Types of RNA

MOLECULE	TYPICAL SIZE (NUMBER OF NUCLEOTIDES)	FUNCTION
mRNA	500–3,000	Codons encode amino acid sequence
rRNA	100–3,000	Associates with proteins to form ribosomes, which structurally support and catalyze protein synthesis
tRNA	75–80	Binds mRNA codon on one end, amino acid on the other, linking a gene's message to the amino acid sequence it encodes

FIGURE 13.8
The Ribosome.
A ribosome from a eukaryotic cell, shown here, has two subunits, containing 82 proteins and four rRNA molecules altogether.

5,080 RNA bases (in 2 or 3 molecules) ~49 proteins

1,900 RNA bases (in a single molecule) ~33 proteins

join at the initiation of protein synthesis (fig. 13.8). Certain rRNAs catalyze formation of the peptide bonds between amino acids, and others help to correctly align the ribosome and mRNA.

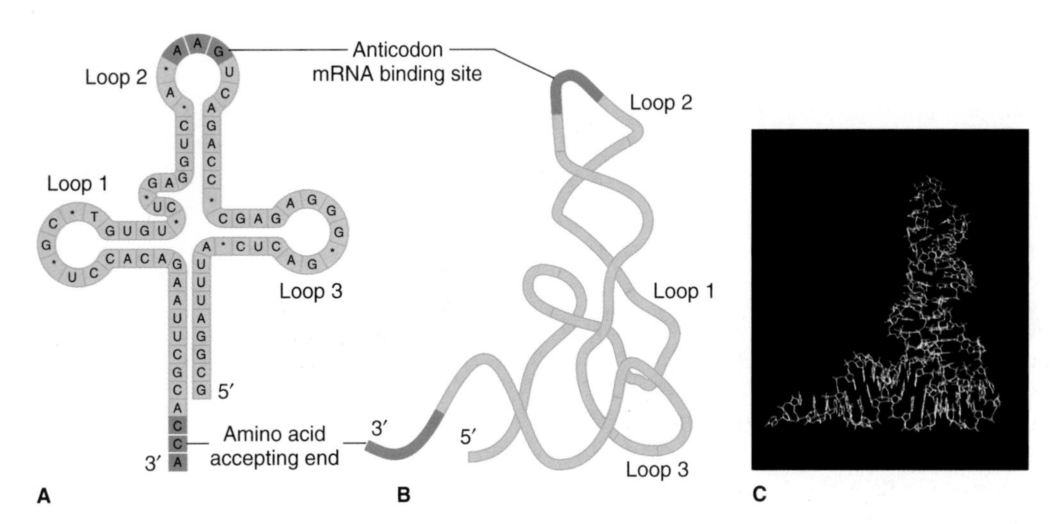

FIGURE 13.9
Transfer RNA.
(**A**) Certain nucleotide bases within a tRNA molecule hydrogen bond with each other to give the molecule a "cloverleaf" conformation that can be represented in two dimensions. The darker bases at the top form the anticodon, the sequence that binds a complementary mRNA codon. Each tRNA terminates with the sequence CCA, where a particular amino acid covalently bonds. Three-dimensional representations of a tRNA (**B**) and (**C**) depict the loops that interact with the ribosome to give tRNA its functions in translation.

Transfer RNA (tRNA) molecules are "connectors" that bind mRNA codons at one end and specific amino acids at the other. A tRNA molecule is only 75 to 80 nucleotides long. Some of its bases form weak (hydrogen) bonds with each other, which fold the tRNA into a characteristic shape (fig. 13.9). One loop of the tRNA has three bases in a row that form the **anticodon,** which is complementary to an mRNA codon. The end of the tRNA opposite the anticodon forms a strong (covalent) bond to a specific amino acid. A tRNA with a particular anticodon always carries the same amino acid. For example, a tRNA with the anticodon sequence AAG always picks up the amino acid phenylalanine. Special enzymes attach amino acids to the tRNAs that bear the appropriate anticodons.

RNA Is Processed

In bacteria and archaea, RNA is translated into protein as soon as it is transcribed from DNA because a nucleus does not physically separate the two processes. In cells of eukaryotes, mRNA must first exit the nucleus to enter the cytoplasm, where protein synthesis occurs. RNA is altered before it participates in protein synthesis in these more complex cells.

In eukaryotic cells, after mRNA is transcribed, a short sequence of modified nucleotides, called a cap, is added to the 5′ end of the molecule. At the 3′ end, 100 to 200 adenines are added, forming a "poly A tail." The cap and poly A tail "signal" the cell as to which mRNAs should exit the nucleus.

In addition to these modifications, not all of an mRNA is translated into an amino acid sequence in eukaryotic cells. Parts of mRNAs called **introns** are transcribed but are later removed. The ends of the remaining molecule are spliced together before the mRNA is translated. The parts of mRNA that are translated are called **exons** (fig. 13.10). Small catalytic RNAs associate in groups with proteins and cut introns out and knit exons together to form the mature mRNA that exits the nucleus.

Introns range in size from 65 to 100,000 bases. While the average exon is 100 to 300 bases long, the average intron is about 1,000 bases long. Many genes are riddled with introns—the human collagen gene, for example, contains 50 of them. The number, size, and organization of introns vary from gene to gene.

Once regarded as an oddity because their existence was not expected, introns are now known to make up large parts of the genomes of complex organisms. They are not common among bacteria and archaea. Introns are thought to function in export of mRNA from the nucleus, or in enabling DNA sequences to cross over. Facilitating crossing over may be important in evolution by quickly creating new genes.

Just as newly replicated DNA is proofread for errors, so too may enzymes check newly transcribed RNA molecules for errors. Messenger RNAs that are too short may be stopped from exiting the nucleus. A proofreading mechanism also monitors tRNAs, ensuring that the correct conformation forms and that the CCA amino acid–binding end is there.

FIGURE 13.10
Messenger RNA Processing—the Maturing of the Message.
Several steps carve the mature mRNA. First, a large region of DNA containing the gene is transcribed. Then a modified nucleotide cap and poly A tail are added, and introns are spliced out. Finally, the mature mRNA is transported out of the nucleus.

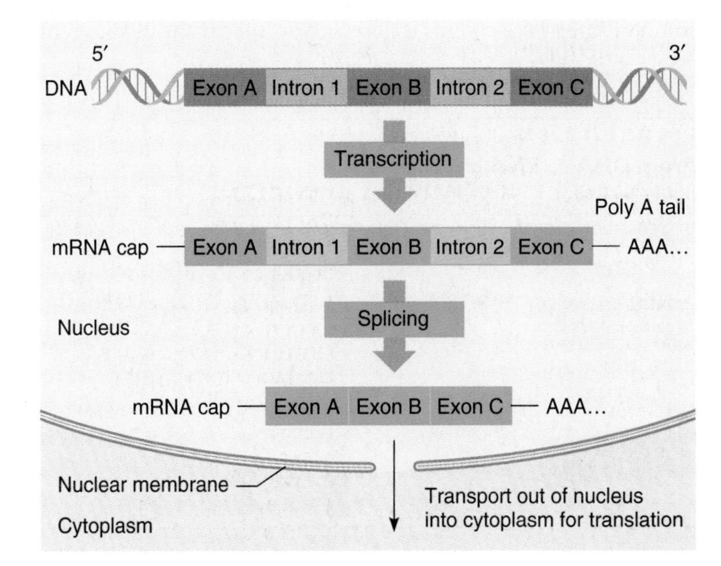

13.2 MASTERING CONCEPTS

1. Why is RNA conformation important?
2. How are amino acids brought to mRNA codons?
3. What are the components and function of ribosomes?
4. What are the steps of transcription and mRNA maturation?

OLC

13.3 How Does a Cell Build a Protein Using Genetic Information?

To synthesize a protein, tRNA molecules carrying specific amino acids base pair with mRNA molecules. Ribosomes align the amino acids. A protein must fold into a certain conformation to function.

Transcription copies the information encoded in a DNA base sequence into the complementary language of mRNA. The next step is translating this "message" into the specified sequence of amino acids. Particular mRNA codons (three bases in a row) correspond to particular amino acids (fig. 13.11). This correspondence between the chemical languages of mRNA and protein is called the **genetic code.** In the 1960s, many researchers collaborated to crack the code and determine which mRNA codons correspond to which amino acids. They used a combination of logic and experiments and began by posing certain questions.

The Genetic Code Connects Gene to Protein

Question 1—How many RNA bases specify one amino acid? Because the number of different protein building blocks (20) exceeds the number of different mRNA building blocks (4), each codon must contain more than one mRNA base. A genetic code in which codons consist of only one mRNA base could specify only four different amino acids, one corresponding to each of the four bases: A, C, G, and U. A code consisting of two-base codons could specify only 16 different amino acids. A code with a minimum of three-base codons is necessary to specify the 20 different amino acids that make up biological proteins. (Actually, more than one codon type can specify the same amino acid, a point reexamined in section 13.4)

Francis Crick and his coworkers conducted experiments that confirmed the triplet nature of the genetic code. They added one, two, or three bases within a gene with a known sequence and protein product. Altering the sequence by one or two bases greatly disrupted the coded order of specified amino acids, known as the reading frame. This produced a different amino acid sequence. However, adding or deleting three bases in a row added or deleted only one amino acid in the protein product (fig. 13.12). They deduced that the code is triplet.

Question 2—Does the genetic code overlap? Consider the hypothetical mRNA sequence AUCAGUCUA. If the genetic code is triplet and does not overlap (that is, each three bases in a row forms a codon, but any one base is part of only one codon), then this sequence contains three codons: AUC, AGU, and CUA. If the code overlaps, the sequence contains seven codons: AUC, UCA, CAG, AGU, GUC, UCU, and CUA (fig. 13.13).

An overlapping code would pack maximal information into a limited number of bases, but would constrain protein structure because certain amino acids would always follow certain others. For example, the amino acid the first codon specifies, AUC, would always be followed by an amino acid whose codon begins with UC. Experiments that determine the sequences of proteins show that no specific amino acid always follows another. The code rarely overlaps. It does, however, in some viruses.

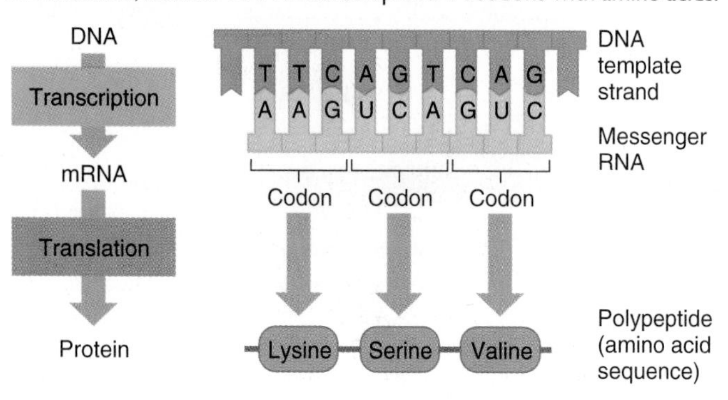

FIGURE 13.11
From DNA to RNA to Protein.
Messenger RNA is transcribed from a locally unwound portion of DNA. In translation, transfer RNA matches up mRNA codons with amino acids.

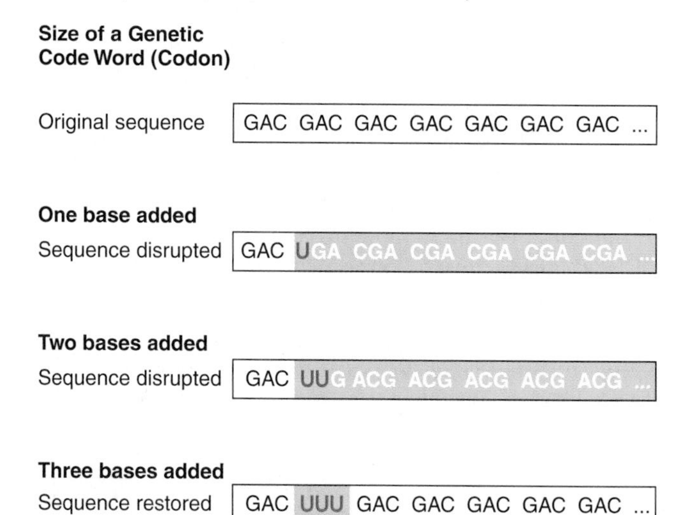

FIGURE 13.12
Three at a Time.
Adding or deleting one or two nucleotides to a DNA sequence disrupts the encoded amino acid sequence. However, adding or deleting three bases does not disrupt the reading frame. Therefore, the code is triplet. This is a simplified representation of the Crick experiment.

FIGURE 13.13
The Genetic Code Does Not Overlap.
An overlapping genetic code may seem economical, but it is restrictive, dictating that certain amino acids must follow others in a protein's sequence. This does not happen; therefore, the genetic code is nonoverlapping.

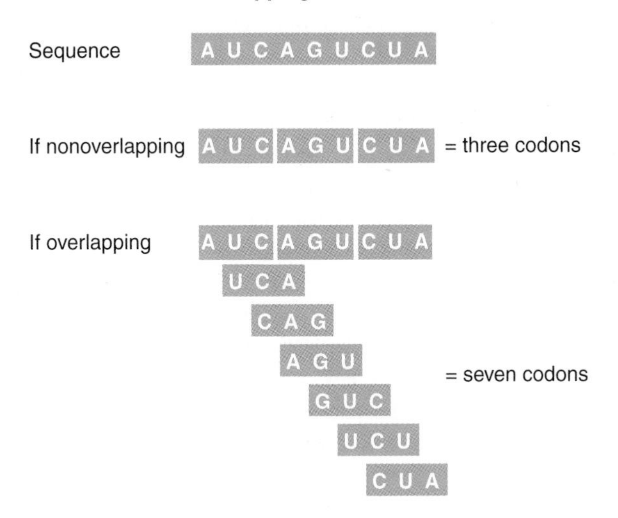

Question 3—Can mRNA codons signal anything other than amino acids? Chemical analysis eventually showed that the genetic code contains directions for starting and stopping translation. The codon AUG signals "start," and the codons UGA, UAA, and UAG each signify "stop." Another form of "punctuation" is a short sequence of bases at the start of each mRNA, called the leader sequence, which enables the mRNA to hydrogen bond with rRNA in a ribosome.

Question 4—Do all species use the same genetic code? The fact that all species use the same mRNA codons to specify the same amino acids is part of the abundant evidence that all life on Earth evolved from a common ancestor. The genetic code is universal, with the exception of DNA in the mitochondria and chloroplasts of certain single-celled organisms. The ability of a cell from one species to translate mRNA from another species makes possible recombinant DNA technology, discussed in Biotechnology 13.2.

Question 5—Which codons specify which amino acids? In 1961, researchers at the National Institute of Health began deciphering which codons specify which amino acids using an elegant series of experiments. First they synthesized mRNA molecules in the laboratory. Then they added them to test tubes containing all the chemicals and structures needed for translation, which they had extracted from *E. coli* cells. Which amino acids would each synthetic RNA specify?

The first synthetic mRNA tested had the sequence UUU UUU. . . . In the test tube, this was translated into a peptide consisting entirely of one amino acid type: phenylalanine. Thus was revealed the first entry in the genetic code dictionary: the codon UUU specifies the amino acid phenylalanine. The number of phenylalanines always equaled one-third of the number of mRNA bases, confirming that the genetic code is triplet and does not overlap. The next three experiments revealed that AAA codes for the amino acid lysine, GGG for glycine, and CCC for proline.

The next step was to synthesize chains of alternating bases. Synthetic mRNA of the sequence AUAUAU . . . introduced codons AUA and UAU. When translated, the mRNA yielded an amino acid sequence of alternating isoleucines and tyrosines. But was AUA isoleucine and UAU tyrosine, or vice versa? Another experiment answered the question.

An mRNA of sequence UUUAUAUUUAUA . . . encoded alternating phenylalanine and isoleucine. Because the first experiment showed that UUU codes for phenylalanine, the researchers deduced AUA must code for isoleucine. If AUA codes for isoleucine, they reasoned, looking back at the previous experiment, then UAU must code for tyrosine. Table 13.3 summarizes some of these experiments.

By the end of the 1960s, researchers had deciphered the entire genetic code (table 13.4). Many research groups contributed to this monumental task. Some of the more exuberant personalities organized an "amino acid tie club" and inducted a new member whenever someone added a new piece to the puzzle of the genetic code, anointing him (there were no prominent hers) with a tie emblazoned with the structure of the specified amino acid.

Steps of Protein Synthesis

Protein synthesis requires mRNA, tRNAs carrying amino acids, ribosomes, energy-storing molecules such as adenosine triphosphate (ATP) and guanosine triphosphate (GTP), and various protein factors. These pieces come together at the beginning

TABLE 13.3 Deciphering RNA Codons and the Amino Acids They Specify

SYNTHETIC RNA	ENCODED AMINO ACID CHAIN	PUZZLE PIECE
UUUUUUUUUUUUUUUUUU	Phe-Phe-Phe-Phe-Phe-Phe	UUU = Phe
AAAAAAAAAAAAAAAAAA	Lys-Lys-Lys-Lys-Lys-Lys	AAA = Lys
GGGGGGGGGGGGGGGGGG	Gly-Gly-Gly-Gly-Gly-Gly	GGG = Gly
CCCCCCCCCCCCCCCCCC	Pro-Pro-Pro-Pro-Pro-Pro	CCC = Pro
AUAUAUAUAUAUAUAUAU	Ile-Tyr-Ile-Tyr-Ile-Tyr	AUA = Ile or Tyr UAU = Ile or Tyr
UUUAUAUUUAUAUUUAUA	Phe-Ile-Phe-Ile-Phe-Ile	AUA = Ile UAU = Tyr

Biotechnology 13.2

Recombinant DNA

The fact that all species use the same genetic code means that one type of organism can express a gene from another. Recombinant DNA technology takes advantage of this fact by using bacteria or other cells growing in culture to mass-produce a gene of interest and its protein product. Drugs produced in this way are free of viral contamination that other sources, such as donated tissue or cadavers, might contain. Another advantage is that the proteins are the human versions, and therefore are less likely to provoke allergic reactions than are drugs derived from nonhuman animals. Several dozen drugs are produced using this technology, including insulin, blood clotting factors, immune system biochemicals, and fertility hormones. But applications aren't all medical. The indigo dye used to make blue jeans blue, for example, comes from *E. coli* given another bacterial gene, rather than using the endangered plant that once supplied the dye.

Constructing recombinant DNA molecules requires several general components:

- enzymes that cut the donor and recipient DNA (restriction enzymes);
- DNA circles to carry the donor DNA (a cloning vector);
- recipient cells (bacteria or other cultured cells);
- a way to separate cells harboring the gene of interest, then coax them to express the foreign gene, and collect the desired protein.

Restriction enzymes are used to cut a gene from its normal location and insert it into a vector. The natural function of restriction enzymes is to protect bacteria by cutting up DNA from infecting viruses. Each of the hundreds of types of restriction enzymes cuts double-stranded DNA at a specific base sequence, generating single-stranded ends that "stick" to each other by complementary base pairing (fig. 13.B).

A common type of cloning vector is a plasmid, which is a small circle of double-stranded DNA found in some bacteria, yeasts, plants, and other organisms (fig. 13.C). The term "cloning" refers to making many identical copies of a particular DNA sequence. Bacteriophages (viruses that infect bacteria) and retroviruses (viruses that can integrate DNA into the host's genome) are also used as vectors. They are manipulated to transport DNA but not cause disease.

To create a recombinant DNA molecule, a restriction enzyme cuts DNA from a donor cell at sequences known to bracket the gene of interest, which leaves sticky ends of a certain base sequence. Next, a plasmid is cut with the same restriction enzyme, leaving the same sticky ends. When the cut plasmid and donor DNA are mixed, the single-stranded sticky ends of some plasmids base pair with those of the donor DNA (fig. 13.D).

(continued)

FIGURE 13.B
Recombining DNA.
A restriction enzyme makes "sticky ends" in DNA by cutting it at specific sequences. (**A**) The enzyme EcoRI cuts the sequence GAATTC between G and the A. (**B**) This staggered cutting pattern produces "sticky ends" of sequence AATT. The ends attract through complementary base pairing. (**C**) DNA from two sources is cut with the same restriction enzyme. Pieces join, forming recombinant DNA molecules.

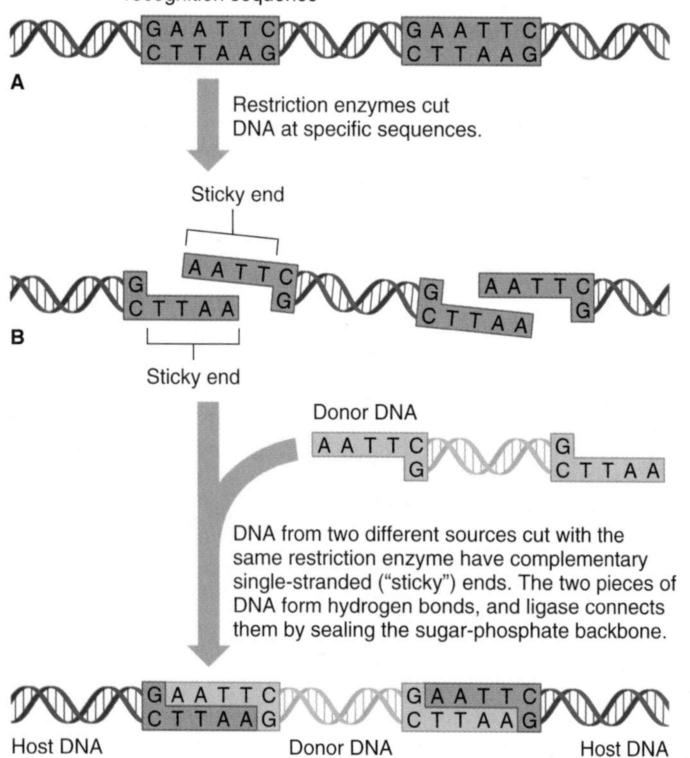

FIGURE 13.C
Plasmids.
Plasmids are small circles of DNA found naturally in the cells of some organisms. A plasmid, along with any other DNA inserted into it, can replicate independently of the chromosomal DNA. For this reason, plasmids make excellent cloning vectors—structures that carry DNA from cells of one species into the cells of another.

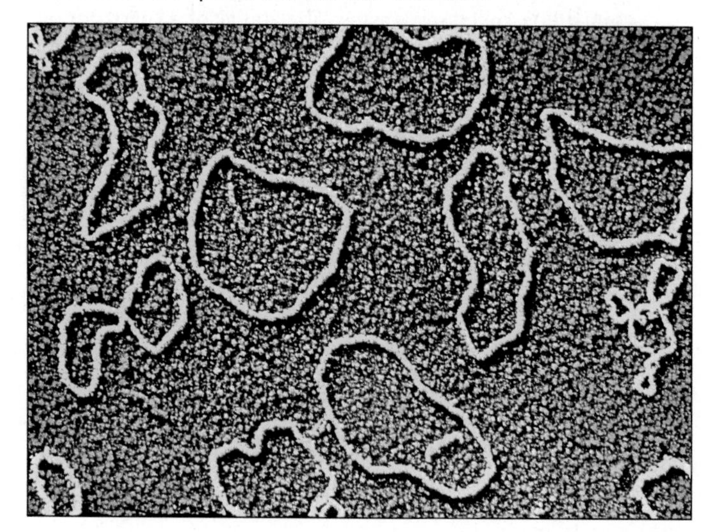

A recombinant DNA experiment is planned so that recombinant DNA molecules can be separated from molecules consisting of just donor DNA or just plasmid DNA. One way to do this is to use a plasmid that contains two genes that each enable a cell to grow in the presence of a different antibiotic drug. When the gene of interest inserts into the plasmid, creating the recombinant DNA molecule, it inactivates one of the antibiotic-resistance genes. In this way, a cell that dies in the presence of one antibiotic but not the other is known to harbor the foreign gene. (The researcher keeps some of the cells aside so as not to kill off the desired ones.) Cells that grow in the presence of both antibiotics carry plasmids that have not incorporated the foreign gene; cells killed by both antibiotics lack plasmids.

When cells containing the recombinant plasmid divide, so does the plasmid. Within hours, the original cell gives rise to many harboring the recombinant plasmid. The enzymes, ribosomes, energy molecules, and factors necessary for protein synthesis transcribe and translate the plasmid DNA and its stowaway foreign gene, producing the desired protein.

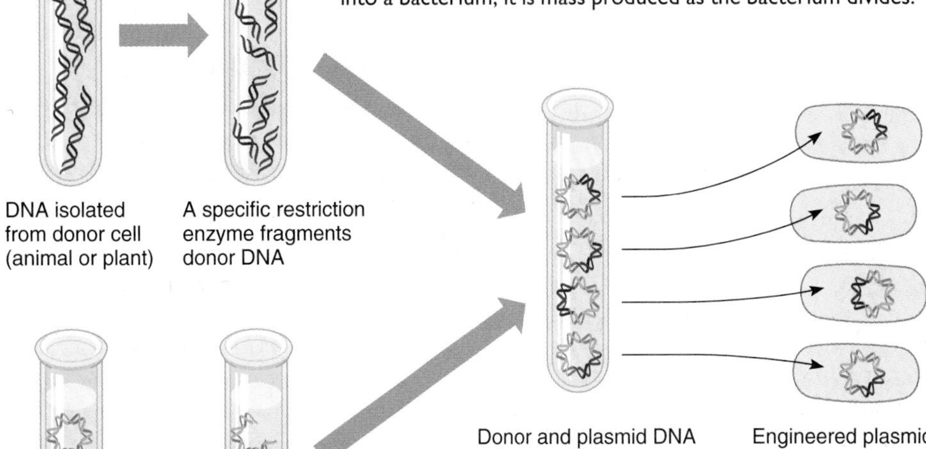

FIGURE 13.D
Recombinant DNA.
To construct a recombinant DNA molecule, DNA isolated from a donor cell and a plasmid are cut with the same restriction enzyme and mixed. Some of the sticky ends from the donor DNA hydrogen bond with the sticky ends of the plasmid DNA, forming recombinant DNA molecules. When such an engineered plasmid is introduced into a bacterium, it is mass produced as the bacterium divides.

TABLE 13.4 The Genetic Code

FIRST LETTER	SECOND LETTER: U		C		A		G		THIRD LETTER
U	UUU UUC	Phenylalanine (Phe)	UCU UCC UCA UCG	Serine (Ser)	UAU UAC	Tyrosine (Tyr)	UGU UGC	Cysteine (Cys)	U C
	UUA UUG	Leucine (Leu)			UAA	"stop"	UGA	"stop"	A
					UAG	"stop"	UGG	Tryptophan (Trp)	G
C	CUU CUC CUA CUG	Leucine (Leu)	CCU CCC CCA CCG	Proline (Pro)	CAU CAC	Histidine (His)	CGU CGC CGA CGG	Arginine (Arg)	U C
					CAA CAG	Glutamine (Gln)			A G
A	AUU AUC AUA	Isoleucine (Ile)	ACU ACC ACA ACG	Threonine (Thr)	AAU AAC	Asparagine (Asn)	AGU AGC	Serine (Ser)	U C
	AUG	Methionine (Met) and "start"			AAA AAG	Lysine (Lys)	AGA AGG	Arginine (Arg)	A G
G	GUU GUC GUA GUG	Valine (Val)	GCU GCC GCA GCG	Alanine (Ala)	GAU GAC	Aspartic acid (Asp)	GGU GGC GGA GGG	Glycine (Gly)	U C
					GAA GAG	Glutamic acid (Glu)			A G

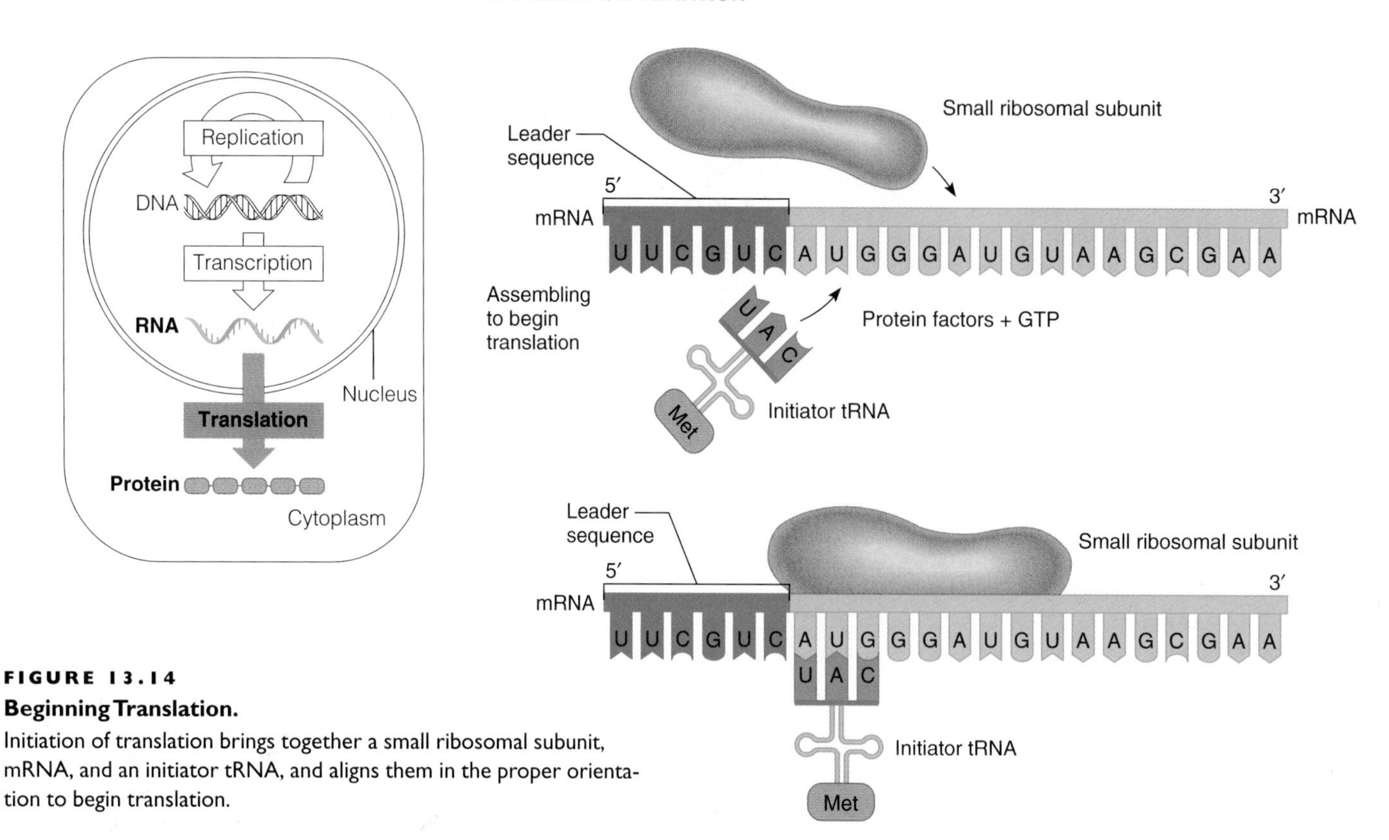

FIGURE 13.14
Beginning Translation.
Initiation of translation brings together a small ribosomal subunit, mRNA, and an initiator tRNA, and aligns them in the proper orientation to begin translation.

of translation in a stage called initiation (fig. 13.14). First, the mRNA leader sequence hydrogen bonds with a short sequence of rRNA in a small ribosomal subunit. The first mRNA codon to specify an amino acid is usually AUG, which attracts an initiator tRNA that carries the amino acid methionine. This methionine signifies the start of a polypeptide. The small ribosomal subunit, the mRNA bonded to it, and the initiator tRNA with its attached methionine form the initiation complex.

To start the next stage of translation, called elongation, a large ribosomal subunit attaches to the initiation complex. The codon adjacent to the initiation codon, which is GGA in figure 13.15A, then bonds to its complementary anticodon, which is part of a free tRNA that carries the amino acid glycine. The two amino acids (methionine and glycine in the example), which are still attached to their tRNAs, align. A peptide bond forms between them, and the first tRNA is released. It will pick up another amino acid and be used again. The ribosome and its attached mRNA are now bound to a single tRNA, with two amino acids extending from it. This begins the polypeptide.

Next, the ribosome moves down the mRNA by one codon. A third tRNA enters, carrying its amino acid (cysteine in fig. 13.15B). This third amino acid aligns with the other two and forms a peptide bond to the second amino acid in the growing chain. The tRNA attached to the second amino acid is released and recycled. The polypeptide continues to build, one amino acid at a time, as new tRNAs deliver their cargo (fig. 13.15C). The tRNAs carry anticodons that correspond to the mRNA codons.

Elongation halts at an mRNA "stop" codon (UGA, UAG, or UAA) by the binding of specific proteins, called release factors. No tRNA molecules correspond to these codons (see fig. 13.15D). The last tRNA is released from the ribosome, the ribosomal subunits separate from each other and are recycled, and the new polypeptide is released (fig 13.15E).

Protein synthesis is economical. A cell can produce large amounts of a particular protein from just one or two copies of a gene. A plasma cell in the human immune system, for example, manufactures 2,000 identical antibody molecules per second. To mass-produce on this scale, RNA, ribosomes, enzymes, and other proteins must be continually recycled. Transcription always produces multiple copies of a particular mRNA, and each mRNA may be bound to dozens of ribosomes, as figure 13.16 shows. As soon as one ribosome has moved far enough along the mRNA, another ribosome will attach. In this way, many copies of the encoded protein will be made from the same mRNA.

Protein Folding

For many years, biochemists thought that protein folding was straightforward; the amino acid sequence dictated specific attractions and repulsions between parts of a protein, contorting it into its final form as it emerged from the ribosome. But these

TRANSLATION ELONGATION

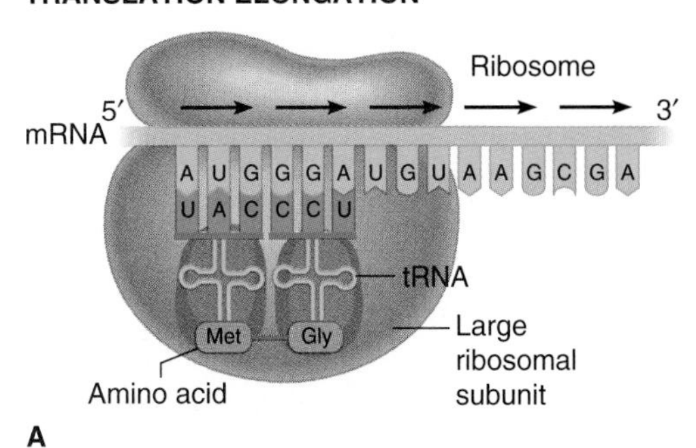

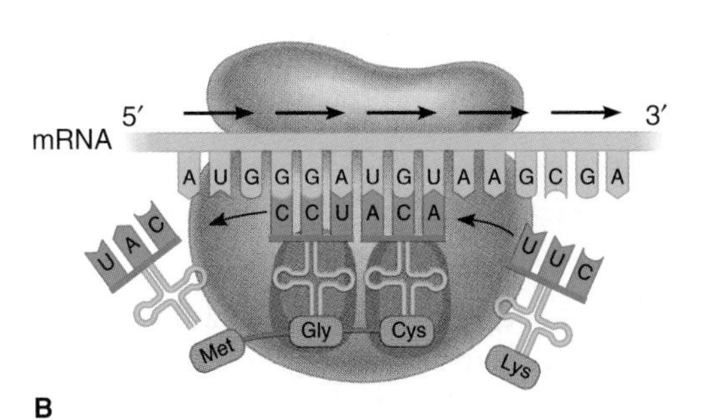

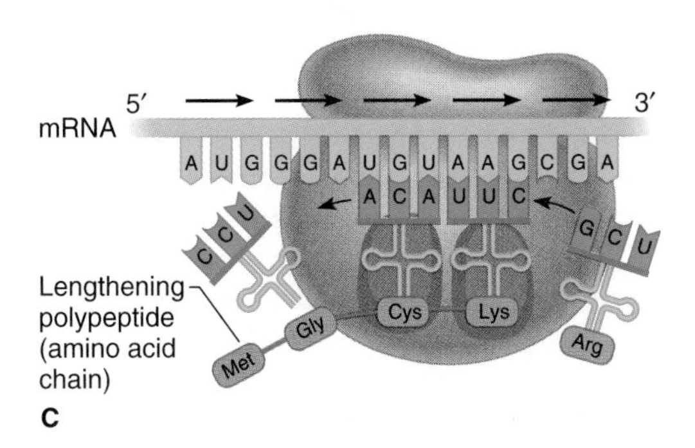

TRANSLATION TERMINATION

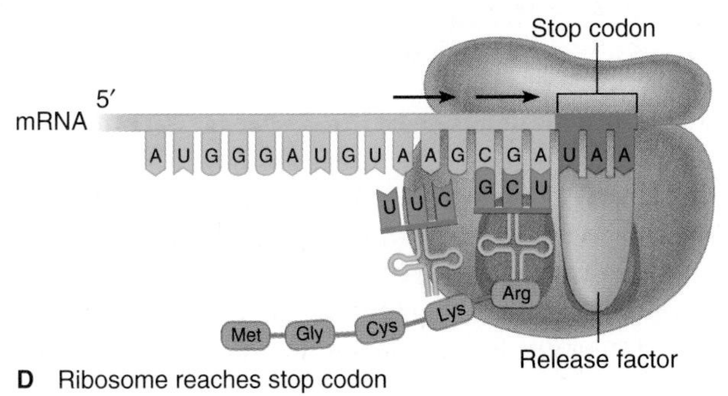

D Ribosome reaches stop codon

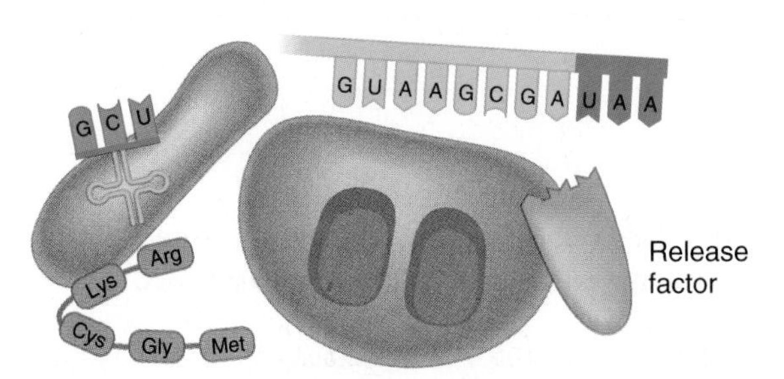

E Once stop codon is reached, elements disassemble

FIGURE 13.15

Building a Polypeptide.

A large ribosomal subunit binds to the initiation complex, and a tRNA bearing a second amino acid (glycine, in this example) forms hydrogen bonds between its anticodon and the mRNA's second codon (**A**). The methionine brought in by the first tRNA forms a peptide bond with the amino acid brought in by the second tRNA, and a third tRNA arrives, in this example carrying the amino acid cysteine (**B**). A fourth amino acid is linked to the growing polypeptide chain (**C**), and the process continues until a termination codon is reached. (**D**) A protein release factor binds to the stop codon, releasing the completed protein from the tRNA and (**E**) freeing all of the components of the translation machine.

attractions and repulsions are not sufficient to fold the polypeptide into the highly specific form essential to its function. A protein apparently needs help to fold correctly.

An amino acid chain may start to fold as it emerges from the ribosome. Localized regions of shape form, and possibly break apart and form again, as translation proceeds. Experiments that isolate proteins as they are synthesized show that other proteins oversee the process of proper folding. These accessory proteins include enzymes that foster chemical bonds and chaperone proteins, which stabilize partially folded regions that are important to the molecule's final form.

Just as repair enzymes check newly replicated DNA for errors and RNAs are proofread, proteins scrutinize a folding protein to detect and dismantle incorrectly folded regions. Errors in protein folding can cause illness. Some mutations that cause cystic fibrosis, for example, prevent the encoded protein from assuming its final form and anchoring in the cell membrane, where it normally controls the flow of chloride ions. Alzheimer disease is associated with a protein called amyloid that forms an abnormal, gummy mass instead of remaining as distinct molecules, because of improper folding.

The spongiform encephalopathies, such as "mad cow disease" and similar conditions in sheep and humans, are caused by abnormal aggregation of protein particles called prions. The prion disorders are discussed further in chapter 19.

In addition to folding, certain proteins must be altered further before they become functional. Sometimes enzymes must shorten a polypeptide chain for it to become active. Insulin, which is 51 amino acids long, for example, is initially translated as the polypeptide proinsulin, which is 80 amino acids long. Some polypeptides must join others to form larger protein molecules (see fig. 2.20). The blood protein hemoglobin, for example, consists of four polypeptide chains.

The linguistic nature of the flow of genetic information makes it ideal for computer analysis. The view of DNA sequences as a language emerged in the 1960s, as experiments revealed the linear relationship between nucleic acid sequences in genes and amino acid sequences in proteins. Yet the "rules" by which DNA sequences specify protein shapes are not well understood, even as we routinely decipher the sequences of entire genomes.

FIGURE 13.16
Making Multiple Copies of a Protein.
Several ribosomes can translate the same protein from a single mRNA at the same time. (**A**) The ribosomes have different-sized polypeptides dangling from them—the closer a ribosome is to the end of a gene, the longer its polypeptide. Chaperone proteins help fold the polypeptide into its characteristic conformation. (**B**) In the micrograph, the ribosomes on the left have just begun translation and the polypeptides are just barely visible. Further along in translation, the polypeptides are longer. The chaperones are not visible.

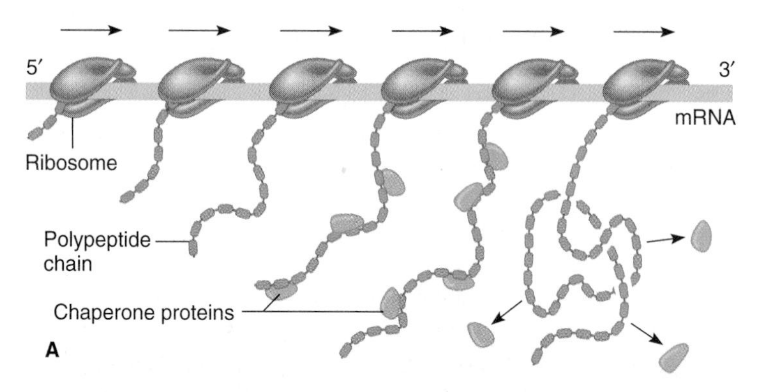

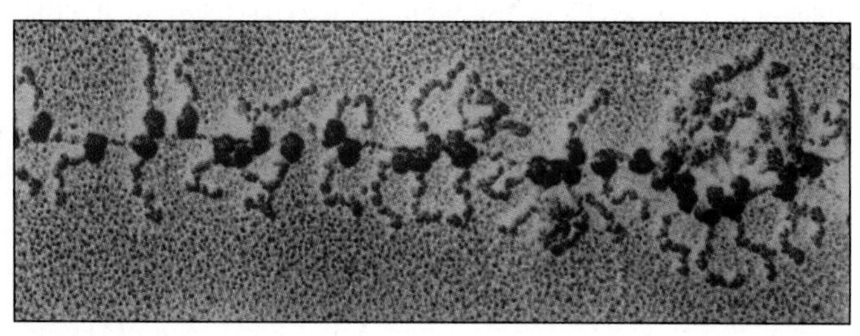

Researchers today send newly discovered gene and protein sequences to a clearinghouse, such as GenBank in the United States. A researcher can compare a newly discovered sequence to those on file to identify regions in common that might suggest the molecule's function. For example, proteins embedded in cell membranes are often anchored by seven distinctive loops formed by certain amino acids hydrogen bonding with each other. A new protein with these seven loops is probably a membrane protein. Searching such databases is a crucial part of genomics (see chapter opening essay).

13.3 MASTERING CONCEPTS

1. What is the genetic code?
2. How did logic and experiments help determine that the genetic code is triplet and nonoverlapping?
3. How did researchers determine which codons specify which amino acids?
4. What are the steps of protein synthesis?
5. Why is protein folding important?

13.4 Mutation—Genetic Misinformation

A change in a gene's nucleotide sequence can alter the gene's function and affect the phenotype. Mutations may occur spontaneously as an error in DNA replication, or be induced.

A gene can change, or mutate, in many ways. This is hardly surprising, given DNA's informational content. Think of how many ways the words on this page can be altered!

A **mutation** is a physical change in the genetic material. The change can be a single DNA base substituted for another, the addition or deletion of one or more DNA nucleotides, or even a transfer of nucleotides between chromosomes. A mutation can occur in the part of a gene that encodes amino acids, in a sequence that controls transcription, in an intron, or at a site critical to intron removal and exon splicing. Whether or not a mutation causes a mutant (different) phenotype depends upon how the alteration affects the gene's product. Table 13.5 lists some general ways that mutations cause inherited illness in humans. Many mutations are "silent"—DNA changes that do not affect the phenotype. Others affect traits that make individuals different without causing disease. Mutation provides variety, which makes evolution possible—the subject of the next unit.

Discovering Mutation in Sickle Cell Disease

The first genetic illness traced to a specific mutation was sickle cell disease. In the 1940s, scientists showed that sickle-shaped red blood cells accompanied this inherited form of anemia (weakness and fatigue caused by a deficient number of red blood cells). In 1949, a team led by Linus Pauling discovered that hemoglobin (the oxygen-carrying molecule in red blood cells) functioned differently depending on whether it came from

TABLE 13.5 Effects of Mutations

EFFECT	EXAMPLE
Prevents a protein from forming	Lack of dystrophin causes muscle cells to collapse in Duchenne muscular dystrophy
Lowers amount of a protein	Blood clots very slowly due to too little clotting factor in hemophilia A
Alters a protein	Skin blisters because amino acid substitution alters protein filaments that hold skin layers together, in epidermolysis bullosa
Adds a function to a protein	Addition of bases to the Huntington disease gene adds a stretch of amino acids to the protein product that gives it a new function that somehow leads to brain degeneration

healthy people or people with sickle cell disease. When Pauling placed the hemoglobin in a solution in an electrically charged field (a technique called electrophoresis), each type of hemoglobin moved to a different position. Hemoglobin from sickle cell carriers (heterozygotes) showed both movement patterns.

The researchers suspected that a physical difference accounted for the difference in electrophoretic mobility between normal and sickle hemoglobin. But how could they identify the part of hemoglobin that sickle cell disease alters? Hemoglobin is very large, consisting of four globular-shaped polypeptide chains, each surrounding an iron atom that is at the center of an organic group called a heme.

Protein chemist V. M. Ingram tackled the problem by cutting normal and sickle hemoglobin with a protein-digesting enzyme and then separating the pieces, staining them, and displaying them on paper. The patterns of fragments differed, which meant, Ingram deduced, that the two molecules must differ in amino acid sequence. That is, one piece of the cut-up molecule occupied a different position for the two types of hemoglobin. Ingram concentrated on this peptide (just 8 amino acids long, rather than the 146 of a full beta globin sequence) to find the site of the mutation. It was in the sixth amino acid in the primary sequence—valine appeared in the sickle cell hemoglobin where glutamic acid appeared in the normal chain. At the DNA level, the mutation is a change in asingle base—a CAC replaced the normal CTC (fig. 13.17).

FIGURE 13.17

Sickle Cell Disease Results from a Single Base Change.

Hemoglobin carries oxygen throughout the body. When normal (**A**), the protein forms an orderly block of adjacent proteins that enables the cell containing it to assume a rounded shape. In sickle-cell disease (**B**), a single DNA base change replaces one amino acid in the protein with a different one. The result is a protein that assembles with others in long, curved rods which change the shape of the red blood cell. Sickled cells obstruct blood vessels, causing symptoms that arise from blocked circulation.

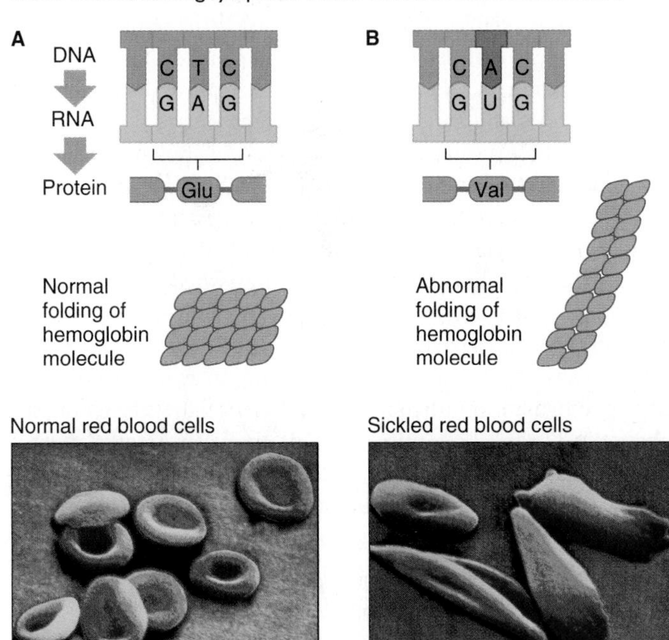

What Causes Mutation?

A mutation can form spontaneously or be induced by chemical or radiation exposure. An agent that causes mutation is called a **mutagen.**

Two healthy people of normal height have a child with an autosomal dominant form of dwarfism called achondroplasia. How can this be, if neither parent is affected? The boy's achondroplasia arose from a new, or *de novo,* mutation that occurred by chance in his mother's or father's gamete. Such a spontaneous mutation usually originates as a DNA replication error.

Sometimes spontaneous mutation stems from chemical changes in DNA's nitrogenous bases. Free nitrogenous bases exist in two slightly different forms called tautomers. For a short time, each base takes on an unstable tautomeric form. If such an unstable base inserts by chance into newly forming DNA, an error will be perpetuated. Figure 13.18 shows how a spontaneous mutation affects DNA replication.

The spontaneous mutation rate varies in different organisms and for different genes. Mutations occur more frequently in bacteria and viruses than in complex organisms because their DNA replicates more often. The larger a gene, the more likely it is to

FIGURE 13.18

Spontaneous Mutation.

DNA bases are very slightly chemically unstable and, for brief moments, they exist in altered forms. If a replication fork encounters a base in its unstable form, a mismatched base pair can result. After another round of replication, one of the daughter cells has a different base pair than the one in the corresponding position in the original DNA.

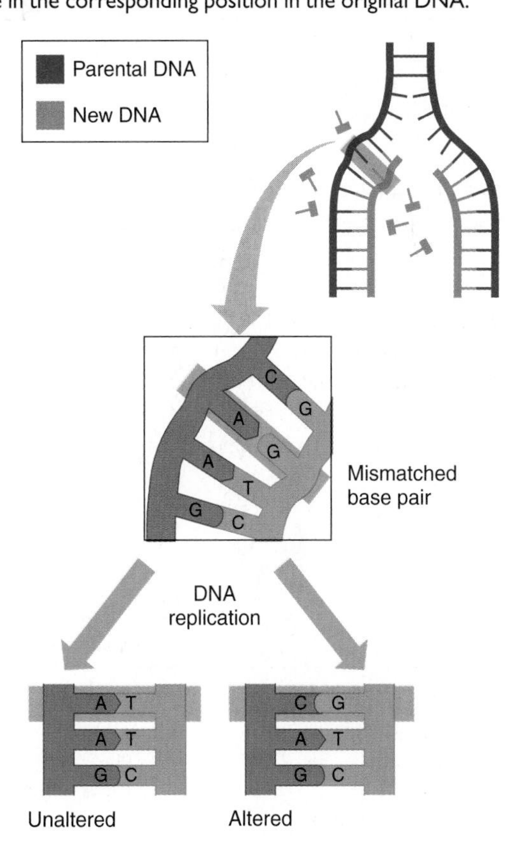

spontaneously mutate. A gene's sequence also influences mutation rate. Genes with repeated base sequences, such as GCGCGC . . . , are more likely to mutate because bases from a single strand may pair with each other as the double helix unwinds to replicate. It is as if the molecules that guide and carry out replication become "confused" by short, repeated sequences, as an editor scanning a manuscript might miss the spelling errors in the words "happpiness" and "bananana."

Genes containing repeated sequences may mutate by misaligning during meiosis. For example, the alpha globin genes are especially prone to mutation because they are repeated in their entirety next to each other on each human chromosome 16. If mispairing occurs in meiosis, chromosomes may result that have more or fewer than the normal two copies of the gene (fig. 13.19). This type of mutation is also responsible for color blindness, as Investigating Life 11.1 describes.

The spontaneous mutation rate for most genes is quite low—each human gene, for example, has about a 1 in 100,000 chance of mutating spontaneously. For researchers who wish to study mutations to learn how genes normally function, the spontaneous mutation rate is not high enough to efficiently provide variants, so they use chemicals or radiation to cause mutations.

FIGURE 13.19
Gene Duplication and Deletion.
The repeated nature of the alpha globin genes makes them prone to mutation by mispairing during meiosis. A person missing one alpha globin gene can develop anemia.

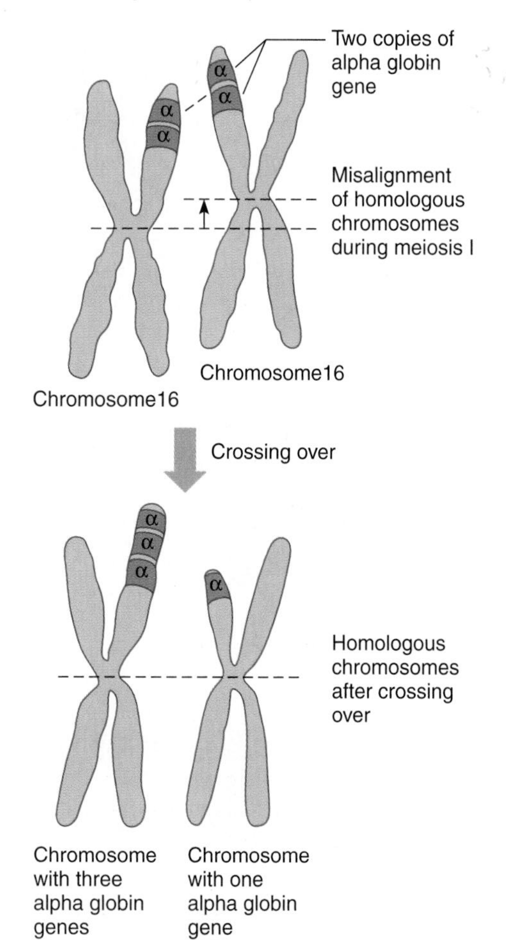

The resulting induced mutation rate is much higher. For example, chemicals called alkylating agents alter bases, enabling them to base pair in different ways that allow the wrong base to be incorporated on the other strand. The mismatch is a mutation. The mutation rate can go as high as 1 in 100! Other chemical mutagens add or remove several nucleotides in a row, and radiation breaks DNA and causes thymine dimers to form.

Researchers test the ability of substances to induce mutation by exposing cells growing in culture. The ability of sodium nitrite in smoked meats and certain pesticides and food additives to cause mutation was detected this way. A test called the Ames test routinely examines chemicals for the ability to induce mutation in bacteria. Because about 90% of mutagens cause cancer, this test is one of many used to screen new drug candidates for their ability to cause cancer.

Induced mutation also occurs from environmental exposures. Natural exposures that cause mutation include cosmic rays from space, ultraviolet light in sunlight, and radiation from rocks. Unnatural mutagenic agents and exposures include chemical weapons such as mustard gas, fallout from atomic bomb tests, and nuclear accidents, such as the one at the Chernobyl Nuclear Power Station in Ukraine in April 1986. Increased rates of certain cancers among children and workers exposed to released radiation may indicate an increased mutation rate. Other evidence for an elevated mutation rate among exposed humans is in the form of repetitive DNA sequences that are of different sizes in children and their parents. Normally they are the same size and only mutation can account for the change from one generation to the next.

Types of Mutations

Mutations can affect somatic cells or sex cells. Mutations also can alter DNA to different extents. Table 13.6 summarizes the different types of mutations using an analogy to a sentence.

A **germinal mutation** occurs during the DNA replication period preceding meiosis. The change appears in the resulting gamete and all the cells that descend from it following fertilization. A **somatic mutation** occurs during the DNA replication period just prior to mitosis, and the genetic change is perpetuated in the daughter cells. Because a somatic mutation occurs in somatic cells rather than gametes, it affects a subset of a multicellular organism's cells.

There are several types of mutations. A **point mutation** is a change in a single DNA base. Two types of point mutations, called missense and nonsense, are distinguished by their extent and consequences. A **missense mutation** changes a codon that normally specifies a particular amino acid into one that encodes a different amino acid. If the substituted amino acid alters the protein's conformation sufficiently at a site critical to its function, the phenotype changes. A missense mutation can greatly affect a gene's product if it alters a site controlling intron removal. The encoded protein has additional amino acids, even though the causative mutation only affects a single DNA base.

A **nonsense mutation** is a single base change that changes a codon specifying an amino acid into a "stop" codon. This shortens the protein product, which can profoundly influence the phenotype. If a missense mutation changes a normal stop codon into a codon that specifies an amino acid, the resulting protein would contain extra amino acids and may not function properly.

In genes, the number three is very important, because triplets of DNA bases specify amino acids. Adding or deleting bases by any number other than a multiple of three devastates a gene's function. It disrupts the reading frame, and therefore it also disrupts the sequence of amino acids. Such a change is called a **frameshift mutation.** Even adding or deleting multiples of three can alter a phenotype if an amino acid is added that alters the protein's function, or a crucial amino acid is removed.

Missing genetic material—even a single base—can greatly alter gene function. A DNA deletion can be so large that detectable sections of chromosomes are missing, or it can be so small that only a few genes or parts of a gene are lacking. The mutation that causes severe cystic fibrosis deletes only a single codon, and the resulting protein lacks just one amino acid, but it cannot function.

A recently discovered type of mutation is an expanding repeat, in which the number of copies of a three-nucleotide sequence increases over several generations. Consider myotonic dystrophy, an autosomal dominant form of muscular dystrophy. A grandfather might experience only mild weakness in his forearms, his daughter might have moderate arm and leg weakness, but her children, if they inherit the gene, have severe muscle impairment. The disease actually worsens with each generation because the gene expands! The gene, on chromosome 19, has an area rich in the trinucleotide CTG that repeats. A person who does not have myotonic dystrophy has from 2 to 28 repeats, whereas a person with the disorder has from 28 to 50 copies of the sequence.

Expanding genes underlie several well-known inherited disorders, including Huntington disease and fragile X syndrome. Most of these disorders affect brain function. In Huntington disease, expanded repeats of GTC cause extra glutamines (an amino acid) to be incorporated into the gene's protein product. The abnormal protein forms fibrous clumps in the nuclei of certain brain cells, which causes the symptoms of uncontrollable movements and personality changes. Biotechnology 13.3 describes how genetically engineered mice helped researchers trace the beginnings of this neurological disorder.

TABLE 13.6 Types of Mutations

A sentence of three-letter words can serve as an analogy to demonstrate the effects of mutations on gene sequence:

Wild type	THE ONE BIG FLY HAD ONE RED EYE
Missense	TH**Q** ONE BIG FLY HAD ONE RED EYE
Nonsense	THE ONE BIG
Frameshift	THE ONE **Q**BI GFL YHA DON ERE DEY
Deletion of three letters	THE ONE BIG HAD ONE RED EYE
Duplication	THE ONE BIG FLY **FLY** HAD ONE RED EYE
Insertion	THE ONE BIG **WET** FLY HAD ONE RED EYE
Expanding mutation	P_1 THE ONE BIG FLY HAD ONE RED EYE
	F_1 THE ONE BIG FLY **FLY FLY** HAD ONE RED EYE
	F_2 THE ONE BIG FLY FLY **FLY FLY FLY FLY** HAD ONE RED EYE

Many types of organisms have some genes that can move from one chromosomal location to another. About 10% of the human genome consists of these "jumping genes," or **transposable elements.** These pieces of DNA can block gene expression by inserting into a gene and preventing its transcription.

Some viruses are, in a sense, transposable elements because they insert their genetic material into the host cell's chromosomes. Transposable elements in bacteria carry genes conferring resistance to certain antibiotics.

Natural Protection Against Mutation

DNA proofreading and repair, and the perusal of newly made RNA molecules for accurate conformation (such as the cloverleaf shape of a tRNA) are some of the built-in protections against mutation. The nature of the genetic code also minimizes the effects of mutation.

The genetic code has excess information—61 types of codons specify 20 types of amino acids (the other three codon types signal "stop"). Different codons that encode the same amino acid are termed **degenerate codons.** This redundancy protects against the effects of mutation because many alterations of the third codon position do not alter the specified amino acid and are therefore "silent." For example, both CAA and CAG mRNA codons specify glutamine. A change from one to the other does not alter the designated amino acid, so it would not alter a protein containing that amino acid and would not affect the phenotype.

Mutations in the second codon position often cause one amino acid to replace another with a similar conformation, which may not disrupt the protein's form drastically. For example, if a GCC codon mutates to GGC, glycine replaces alanine; both are very small amino acids.

A mutation's effect on the phenotype depends upon how it alters the conformation of the gene product. A mutation that replaces an amino acid with a very dissimilar one may not affect the phenotype if the change occurs in a part of the protein that is not critical to its function. Certain mutations in the beta globin gene, for example, do not cause anemia or otherwise affect how a person feels, but they may slightly alter how hemoglobin migrates in an electric field.

Some proteins are more vulnerable to disruption by mutation than others. For example, collagen, a major constituent of connective tissue, is unusually symmetrical. The slightest change in its DNA sequence can greatly disrupt its overall structure and lead to a disorder of connective tissue.

Although mutations can cause disease, they also provide much of the variation that makes life interesting. The next unit explores how genetic variation provides the raw material for evolution.

13.4 MASTERING CONCEPTS

1. What is a mutation?
2. What causes a spontaneous mutation?
3. What is the difference between a germinal and a somatic mutation?
4. How do different types of mutations alter DNA?
5. What are some natural protections against mutation?

OLC

Biotechnology 13.3

Transgenes, Gene Therapy, and DNA Microarrays

Transgenic Technology

Recombinant DNA technology adds a gene to single cells. The equivalent in a multicellular organism is transgenic technology. Genes of interest are introduced into a gamete or fertilized ovum, and the organism that develops carries the foreign gene, or "transgene," in every cell, and expresses it in appropriate tissues. Figure 13.E shows some ways that researchers introduce transgenes—zapping the recipient cell with electricity to open temporary holes that admit "naked" DNA; shooting or injecting DNA in; sending it in inside a fatty bubble called a liposome; or hitching DNA to an infecting virus. In many flowering plants, transgenes are delivered in a Ti plasmid, which is a ring of DNA that normally resides in certain bacteria. A variation on the transgenic theme, called gene targeting, uses the natural tendency of a gene to bind its complement to deliver a transgene to its normal location on a chromosome.

Transgenic organisms have diverse applications:

Model Organisms A mouse harboring a gene that affects human health can reveal how a disease begins. This was the case for Huntington disease (HD), a neurological disorder beginning in middle age that causes uncontrollable movements and personality changes. In 1979, researchers discovered clumps of an unfamiliar protein, detectable only after autopsy, in the nuclei of certain brain cells. In 1997, mice transgenic for the HD gene showed the telltale protein clumps in cytoplasm of brain cells before symptoms arose, and once the clumps appeared in the nuclei, symptoms began in the mice. Now researchers can use these mice to try to halt the disease at its earliest stages.

"Pharming" Transgenic farm animals can secrete human proteins in their milk or semen, yielding abundant, pure supplies of otherwise rare substances that act as drugs. For example, transgenic goats produce human tissue plasminogen activator (tPA), a "clot buster" drug used to limit damage in heart attacks and stroke; sheep manufacture enzymes used to treat hereditary emphysema and secrete human clotting factors. Transgenic pharming can produce interesting combinations (fig. 13.F). For example, transgenic rabbits secrete the salmon version of calcitonin, a hormone used to treat various bone disorders in humans. The transgene includes a human gene encoding a protein (alpha-lactoglobulin) that responds to signals to secrete milk, and a sheep gene segment (promoter) that turns on transcription of the calcitonin and alpha-lactoglobulin genes.

Agriculture Transgenic crops resist pests, survive harsh environmental conditions, or contain nutrients that they otherwise wouldn't. Tobacco plants can be altered to express the glow of a firefly's luciferase gene, or produce human hemoglobin. Transgenic fish given growth hormone reach market size very quickly, and those given antifreeze genes can live in colder climes. "Golden rice" contains genes from petunia and bacteria that enable it to produce beta-carotene (a vitamin A precursor) and extra iron, which may make it very valuable in preventing malnutrition.

Gene Therapy

Gene therapy replaces a nonfunctioning gene in somatic cells. The last decade of the twentieth century began with a gene therapy success, but ended with a tragic failure. Both events occurred, coincidentally, in mid-September.

In 1990, 4-year-old Ashanti DaSilva received her own white blood cells bolstered with a gene whose absence caused her inherited immune deficiency. She and another child treated soon after did well, but had to receive booster treatments as their replaced white blood cells naturally died off. Others treated as babies received genetically altered umbilical cord stem cells, which have provided a longer-lasting cure, with healthy cells gradually replacing their deficient ones.

Like antisense technology (see Biotechnology 13.1), gene therapy makes sense theoretically, but has had limited clinical success. The therapeutic gene must be isolated and delivered to the cell type that needs correction, then be expressed long enough to ameliorate symptoms—without alerting the immune system. Unfortunately, this is what happened to 18-year-old Jesse Gelsinger, who died in a gene therapy experiment.

In September 1999, Gelsinger received a massive infusion of viruses carrying a gene to correct his inborn error of metabolism, called ornithine transcarbamylase deficiency. He died in days from an overwhelming immune system reaction. The great tragedy of Jesse's case was that he could control his symptoms with diet and drugs, and had volunteered for the gene therapy to help researchers develop a cure for others who died of the condition as newborns. Jesse's death led to a temporary halt to several gene therapy protocols, and stricter rules for conducting experiments.

DNA Microarrays Monitor Gene Expression and Interaction

A DNA microarray, also known as a "DNA chip," is a collection of genes or gene pieces

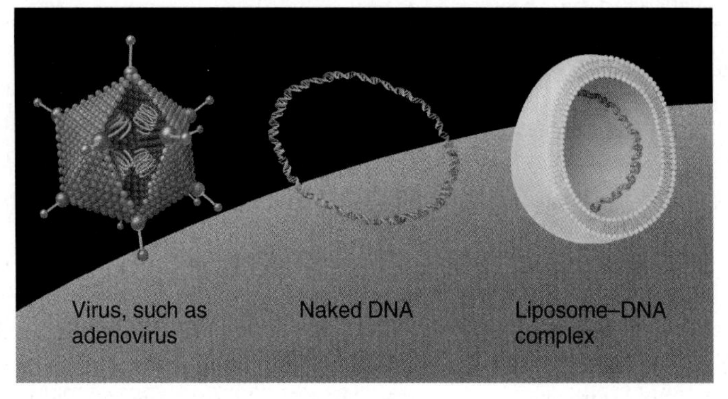

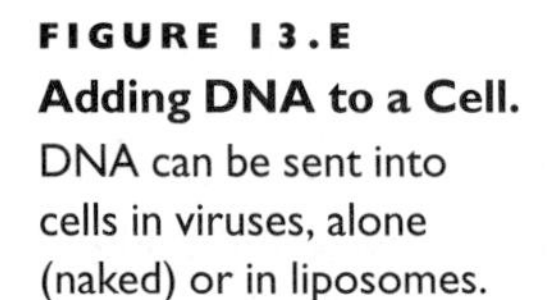
FIGURE 13.E
Adding DNA to a Cell. DNA can be sent into cells in viruses, alone (naked) or in liposomes.

FIGURE 13.F
Combining Genes.
Transgenic technology can combine gene segments from different organisms to produce a gene that does exactly what researchers want it to do. DNA regions from a sheep, a salmon, and a human are combined to make a transgenic rabbit that produces calcitonin for treating bone disorders.

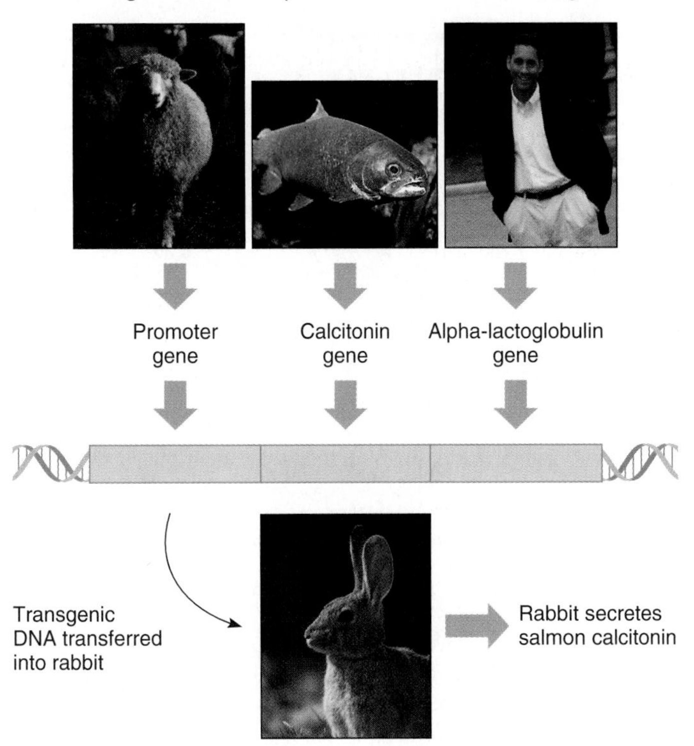

placed in defined positions on a small piece of glass, nylon, silicon (an element), or silica (sand). When fluorescently tagged DNA from a tissue sample, such as blood, is applied to the chip, dots of light fluoresce where the sample DNA matches the chip's DNA. Computer software, given the functions or identities of the genes probed, then can analyze the significance of the pattern of colored dots on the square.

Overall, a DNA microarray provides a peek at gene expression. By creatively choosing which genes to target and which to test, researchers can ask and answer a seemingly endless list of questions, from basic biology to clinical applications.

Scrutinizing several genes at once, or multiple activities of a particular gene in different tissues, is called genomics, and it is a great departure from the one-gene one-function focus of genetics research in the last century. Already genomics has spawned specialties. In functional genomics, researchers infer a gene's function by identifying where in an organism and when in development it is expressed, using tissue-specific gene expression microarrays. Functional genomics can be used to study normal physiology, or the progression of a disease. Another new, related field is toxicogenomics, which identifies an individual's inherited susceptibilities to environmental contaminants or pollutants.

Yet another application of microarray technology is pharmacogenomics, which tailors drug treatments to individuals based on their genotypes. Pharmacogenomics may profoundly alter the ways that physicians choose drugs to treat cancer, infection, cardiovascular disease, and many other conditions. Consider the types of information that pharmacogenomics can provide to select treatment for the blood cancer acute lymphoblastic leukemia (ALL). An "ALL" DNA chip scans genes in a patient's blood sample that reveal the cancer subtype; whether the person's cells will admit a particular drug and, if they do so, whether it will be metabolized in a safe and effective way; and how the person's immune system is likely to respond to both the cancer and the drug. Similar "infection DNA chips" identify which antibiotics are most likely to work in a particular individual with a particular infection.

The science of genetics began with the study of the transmission of natural variants—mutations—in Mendel's peas. Then mutant fruit flies, bread mold, corn, bacteria, and other organisms led researchers to unravel the nature and structure of chromosomes and the genes that they carry. The twentieth century began with the rediscovery of Mendel's laws and the quest to identify and describe the gene, and closed with the almost routine sequencing not only of individual genes, but of entire genomes. Along the way, modern biotechnology was born and increasingly affects our lives. Figure 13.20 summarizes the major biotechnologies.

Biologists, however, are now reconsidering their expectations that deciphering the chemical nature of the gene and genomes would reveal much about the workings of life, for genes do not act in a vacuum. They interact, with each other and with the environment. Knowing that a person has inherited a specific mutation may not be useful information, unless the natures of other genes and environmental exposures are also known. The complexity of interacting genes is a prime example of how science proceeds—the more we learn, the more we realize there is to know.

FIGURE 13.20
Biotechnology.
Understanding gene function has spawned several biotechnologies. Antisense technology uses RNA complementary to mRNA to selectively block gene action. Transgenic technology introduces a genetic change into a gamete or fertilized ovum, so that the change is perpetuated in all cells of the resulting individual.

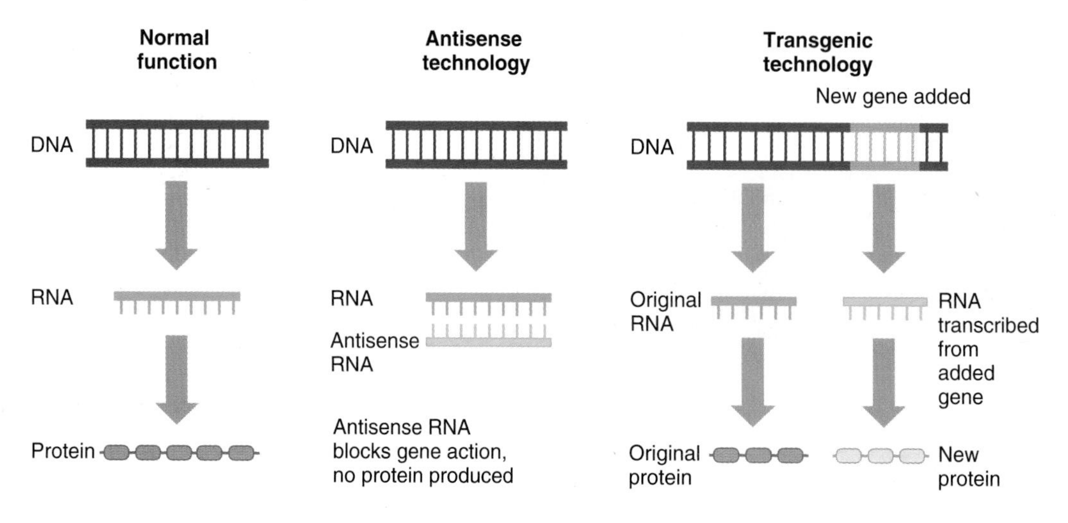

Chapter Summary

13.1 How Does a Cell Access Genetic Information?

1. A gene's information must be **transcribed** into RNA before it is **translated** into a sequence of amino acids. RNA is a single-stranded nucleic acid similar to DNA, but it contains uracil and ribose rather than thymine and deoxyribose.
2. In bacteria, **operons** coordinate expression of grouped genes whose encoded proteins participate in the same metabolic pathway. In multicellular organisms, **transcription factors** regulate which genes are transcribed and when in a particular cell type. These factors have certain common regions called motifs.

13.2 RNA Orchestrates Protein Synthesis

3. Transcription begins when transcription factors help **RNA polymerase** bind to a promoter on the DNA template strand, and then builds an RNA molecule. After transcription, **introns** are cut out of RNA and the remaining **exons** spliced together.
4. Several types of RNA participate in translation. **Messenger RNA** (mRNA) carries a protein-encoding gene's information. **Ribosomal RNA** (rRNA) associates with certain proteins to form ribosomes, which support and help catalyze protein synthesis. **Transfer RNA** (tRNA) has an **anticodon** sequence complementary to a particular mRNA **codon** on one end and a particular amino acid at the other end.

13.3 How Does a Cell Build a Protein Using Genetic Information?

5. Each group of three consecutive mRNA bases is a codon that specifies a particular amino acid. The correspondence between codons and amino acids constitutes the **genetic code.** Of the 64 different codons, 61 specify amino acids and three signal the end of translation. **Degenerate** codons encode the same amino acid. The genetic code is nonoverlapping, triplet, and identical in all species. Scientists deciphered the code by exposing synthetic RNA molecules to the contents of *E. coli* cells and noting which amino acids formed peptide chains. Other experiments confirmed the triplet nature of the code.
6. Translation requires tRNA, ribosomes, energy storage molecules, enzymes, and protein factors. An initiation complex forms when mRNA, a small ribosomal subunit, and a tRNA usually carrying methionine join. A large ribosomal subunit joins the small one. Next, a second tRNA binds by its anticodon to the next mRNA codon, and its amino acid bonds with the methionine the first tRNA brought in. tRNAs continue to add amino acids, elongating a polypeptide. The ribosome moves down the mRNA as the chain grows.
7. When the ribosome reaches a "stop" codon, it separates into its subunits and is released, and the new polypeptide breaks free. Chaperone proteins help fold the polypeptide, and it may be shortened or combined with others.

13.4 Mutation—Genetic Misinformation

8. A **mutation** adds, deletes, alters, or moves nucleotides. A phenotype that a mutation alters is mutant. A gene can mutate spontaneously, particularly if it contains regions of repetitive DNA sequences. **Mutagens** are chemicals or radiation that increase the mutation rate.
9. A **germinal mutation** originates in meiosis and affects all cells of the progeny. A **somatic mutation** originates in mitosis and affects a subset of cells. A **point mutation** alters a single DNA base, and may be **missense** (substituting one amino acid for another) or **nonsense** (substituting a "stop" codon for an amino acid–coding codon). Altering the number of bases in a gene (a **frameshift mutation**) may disrupt the reading frame, altering the amino acid sequence of the gene's product. Expanding triplet repeat mutations cause some in-

herited illnesses. **Transposable elements** move, possibly disrupting a gene's function.

10. Some mutations are silent. A mutation in the third position of a **degenerate codon** can substitute the same amino acid. A mutation in the second codon position can replace an amino acid with a similarly shaped one. A mutation in a nonvital part of a protein may not affect function.

Testing Your Knowledge

1. Define and distinguish between transcription and translation.
2. List the differences between RNA and DNA.
3. Where do DNA replication, transcription, and translation occur in a eukaryotic cell?
4. List the three major types of RNA and their functions.
5. List the sequences of the mRNA molecules transcribed from the following template DNA sequences:
   a. TTACACTTGCTTGAGAGTT
   b. ACTTGGGCTATGCTCATTA
   c. GGAATACGTCTAGCTAGCA
6. Given the following partial mRNA sequences, reconstruct the corresponding DNA template sequences:
   a. GCUAUCUGUCAUAAAAGAGGA
   b. GUGGCGUAUUCUUUUCCGGGUAGG
   c. AGGAAAACCCCUCUUAUUAUAGAU
7. Refer to the figure below to answer the following questions:
   a. Label the mRNA and the tRNA molecules, and draw in the ribosomes.
   b. What are the next three amino acids to be added to peptide *b*?
   c. Fill in the correct codons in the mRNA complementary to the template DNA strand.
   d. What is the sequence of the DNA coding strand (as much as can be determined from the figure)?
   e. Is the end of the peptide encoded by this gene indicated in the figure? How can you tell?
   f. What might happen to peptide *b* after it is terminated and released from the ribosome?
8. To answer the following questions, refer to this template DNA sequence: GCAAAACCGCGATTATCATGCTTC.
   a. What is the sequence of the coding DNA strand?
   b. What is the mRNA sequence?
   c. What amino acid sequence does the template DNA sequence specify?
   d. What would the amino acid sequence be if the genetic code were completely overlapping?
   e. If the 15th DNA base mutates to thymine, what happens to the amino acid chain?
9. A protein-encoding region of a gene has the following DNA sequence:

   G T A G C G T C A C A A A C A A A T C A G C T C

   Determine how each of the following mutations alters the amino acid sequence:
   a. substitution of a T for the C in the 10th DNA base
   b. substitution of a G for the C in the 19th DNA base
   c. insertion of a T between the 4th and 5th DNA bases
   d. insertion of a GTA between the 12th and 13th DNA bases
   e. deletion of the first DNA base

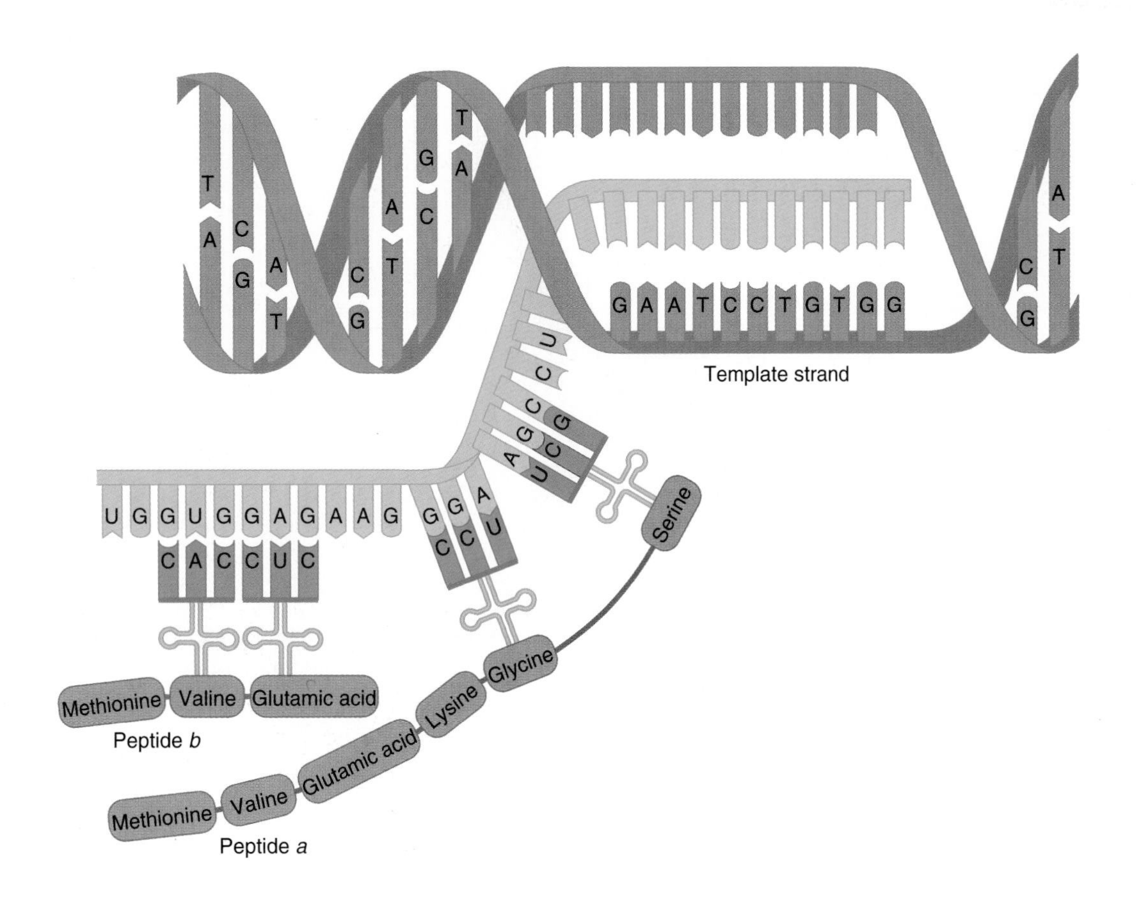

Thinking Scientifically

1. When the sequencing of the human genome was announced in June 2000, most media reports proclaimed that the "human genetic code" had been "cracked." How is this statement incorrect—in two ways?
2. How can a mutation alter the sequence of DNA bases in a gene but not produce a noticeable change in the gene's polypeptide product? How can a mutation alter the amino acid sequence of a polypeptide yet not alter the phenotype?
3. Which biotechnology might be able to accomplish the following goals? More than one answer may be possible.
   - **a.** Develop a mouse that has a gene causing breast cancer in humans in each of its cells to serve as a model for that disease.
   - **b.** Shut off HIV genes integrated into the chromosomes of people with HIV infection (which leads to AIDS).
   - **c.** Create bacteria that produce human growth hormone, used to treat extremely short stature.
4. In the past, certain drugs were used to treat certain types of cancers. Many patients would waste precious time taking drugs that were ineffective, or too toxic. How might DNA microarray technology refine the treatment of cancer?

References and Resources

Gilbert, Walter. February 9, 1978. Why genes in pieces? *Nature,* vol. 271. A classic and insightful look at the enigma of introns.

Ingram, V. M. 1957. Gene mutations in human hemoglobin: The chemical difference between normal and sickle cell hemoglobin. *Nature,* vol. 180, p.1326. This classic paper explains the molecular basis for sickle cell disease.

Lewis, Ricki. July 24, 2000. Keeping up: genetics to genomics in four editions, *The Scientist,* vol 14. p. 46. Writing four editions of a human genetics textbook chronicles the evolution of genetics to genomics.

Lewis, Ricki. February 1996. On cracked codes, cell walls, and human fungi. *The American Biology Teacher.* Newspapers, magazines, and films often misuse the term "genetic code."

Varki, Ajit. September 2000. A chimpanzee genome project is a biomedical imperative. *Genome Research,* vol. 10, p. 1065. AIDS, wisdom tooth impaction, and difficult childbirth are all human conditions not seen in chimps. Knowing their genome could answer questions about ourselves.

The *LIFE* Online Learning Center provides additional resources and tools for studying this chapter.
www.mhhe.com/life

UNIT 3 · EVOLUTION

Over many millions of years,
this planet has been home to a startling
diversity of microorganisms, fungi, plants, and animals.
Some exist now only as fossil remains that paint partial portraits
of long-ago living communities. The preserved cast of a trilobite reveals
an organism that lived in an ocean among thousands of other animals
more than 300 million years ago. This unit explores the processes
that produced so much diversity and how those
mechanisms, at the same time,
unite all life.

CHAPTER 14

The Evolution of Evolutionary Thought

On the Evolution of Male Genitalia

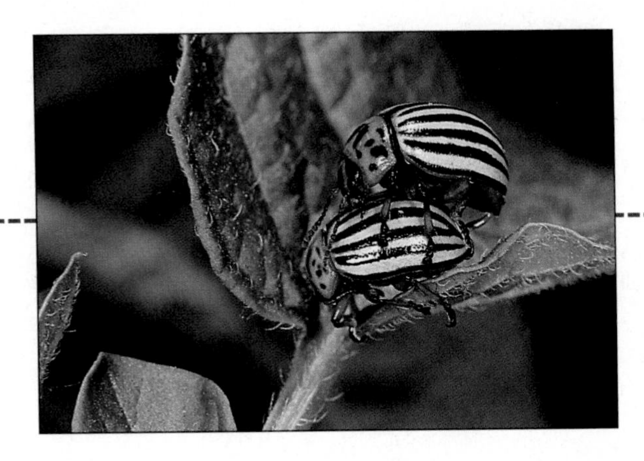

Reproduction is essential for species survival. These are mating Colorado beetles (*Leptinotarsa decemlineate*).

The beauty of the theory of evolution is that it explains many observations of the living world. So it is for the great variety of shapes of male genitalia. A cat's penis has spines; a monkey's winglike flaps; a human's a foreskin; and a dog's a sheath. But in a contest for penile diversity, insects would surely win, with the extensions that emanate from their organs bearing such vivid names as epiprocts, cerci, gonotremes, and titillators. The genitalia of the male stonefly *Brachyptera occidentalis* are among the most complex, resembling a Swiss army knife with a series of distinctly shaped parts that unfold in a specific sequence. For years entomologists have meticulously drawn pictures of intricate insect genitalia to help distinguish among species.

A stag's antlers or rhinoceros's horn, used to vanquish other males in contests over females, are results of sexual selection.

Why are the sex organs of certain male insects so specialized? Biologists have proposed at least two hypotheses, based upon the function of these body parts. The first hypothesis is that species-specific male genitalia help ensure that he fits a she like a lock fits a key, assuring sperm delivery and preventing matings between members of different species.

A second explanation for the different shapes of penises among insects, and especially within species, is that sex organ adornments are just that—structures that attract the opposite sex. As such, penises are agents of what Charles Darwin called sexual selection, the persistence of traits that contribute to attraction, courtship, mating, or otherwise gaining access to a mate. A stag's antlers or rhinoceros's

Sexual selection has molded the intricate shapes of male genitalia in some insect species, such as these three stonefly *(Brachyptera)* species.

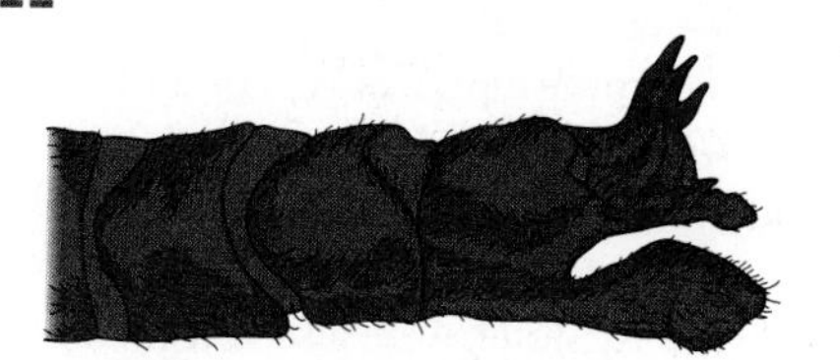

horn, used to vanquish other males in contests over females, are results of sexual selection. Darwin found abundant evidence for sexual selection among the insects—the tone of a song, pattern of veins on a wing, or shape of horns on a beetle's head might make the difference in securing a successful pairing. He tended toward the anthropomorphic, writing in *The Descent of Man* in 1871 that "sexual selection . . . implies the possession of considerable perceptive powers and of strong passions. . . ."

We cannot know about the innate passion of beetles and butterflies, but we do know that when it comes to evolution, sex is of utmost importance. Sex determines which genes make it to the next generation, and genetic change in populations over time is the essence of evolution. Darwin described sexual selection at the whole-body or behavioral level. Today we know that it extends to the cellular level as well, at the time of fertilization. In perhaps the ultimate example of sexual selection, among insect species where females mate with several males, a second male's sperm can destroy sperm already deposited from an earlier sexual encounter.

Goran Arnqvist, a researcher at the University of Umea in Sweden, investigated whether the distinct shapes of male genitalia among insects promote species recognition, or reflect sexual selection—or both. He hypothesized that the mating systems of species might determine how the male genitalia evolve. As data, Arnqvist used thousands of published illustrations of the male sexual organs of insects, available from decades of entomological research. In an approach called "geometric morphometrics," he identified 19 points of physical comparison among the organs and, as a control, also compared nonsexual body parts such as wings and legs.

Arnqvist considered many pairs of closely related insects that differed predominantly in whether the females had one mate (monandrous) or several (polyandrous). The selected pairs spanned four broad types of insects—mayflies, butterflies and moths, flies and mosquitoes, and beetles. He predicted that in monandrous species, where the female must "get it right" because she only mates with one male, one penis would look pretty much like another. But in polyandrous species, where a male competes with others to mate, male sex organ shape should vary in response to sexual selection by females preferring more ornamental males.

Results were striking—penises of polyandrous species were indeed more than twice as variable compared with those of monandrous species. The control experiment was very important—wings and legs did not show a difference in variability between monandrous and polyandrous species. Therefore, penis shape is sexually selected, at least in some insect species. Arnqvist's inspired experiment exemplifies the power of evolutionary theory. It is an intellectual framework that enables us to explain why nature is as it is.

14.1 Evolution Is a Framework for Understanding How Life Works

Evolution is genetic change over time in a population. Various thinkers attempted to explain the appearances and disappearances of different types of organisms over vast time periods. In the nineteenth century, people began to notice how changes on Earth may have influenced life.

Evolution is genetic change in a population over time. A population is a group of interbreeding organisms (members of a species) that live in the same area. Evolution occurs when the frequencies (percentages) of alleles change from one generation to the next. As the chapters in this unit will repeatedly demonstrate, evolution is a process that is ongoing and everywhere, and obvious in many ways. It serves as such a compelling conceptual framework for many observations about life that noted geneticist Theodosius Dobzhansky entitled a much-quoted article "Nothing in biology makes sense except in the light of evolution."

Evolution is a continuing process that explains the history of life on Earth, as well as the diversity of organisms today both in terms of the number of species and of variation within species. Evolution also explains the great unity of life, why organisms diverse in many ways nonetheless use the same genetic code, the same reactions to extract energy from nutrients, even the same or very similar enzymes and other proteins. Shared ancestry—that is, descent from a common ancestor—explains the similarities among species. Natural selection—the differential survival and/or reproductive success of individuals with particular genotypes in response to environmental challenges—accounts for much of the diversity of life. Natural selection operates on individuals, but evolution occurs in populations. The result is a planet packed with spectacular living diversity, of millions of variations of the same underlying biochemical theme.

Biological evolution includes large-scale events, such as new species appearing and existing ones dying out. Such large changes are called **macroevolution.** However, evolution also includes changes in individual allele frequencies within a population, termed **microevolution.** Often macroevolutionary change is the consequence of accumulating microevolutionary changes. Macroevolutionary events tend to span very long time periods, whereas microevolutionary events happen so rapidly that we can sometimes observe them over periods of just a few years. Investigating Life 14.1 discusses how adaptations spread within a population of animals, demonstrating short-term microevolution. The discussion of bacterial antibiotic resistance at the chapter's end provides another compelling example of evolution happening right now.

How Have People Explained Life's Diversity?

The term **evolution** has been around for a very long time (fig. 14.1), although it has had many meanings. Up to the mid 1700s, most people used the term to describe the process of development from embryo to adult. Greek philosopher Aristotle (384 B.C. to 322 B.C.) applied the term to mean the progression from imperfection to perfection based upon innate potential. Although Aristotle recognized that all organisms are related in a hierarchy of simple to complex forms, he believed that all organisms were created with an infinite potential to become a perfect version of that particular type or "essence." By "essence" Aristotle meant that all members of a species were identical in form and capacity. This idea was to influence scientific thinking for nearly 2,000 years.

Several ideas, not original with Aristotle, were also considered to be fundamental principles of science well into the 1800s. Among them was the concept of a "special creation," the sudden appearance of organisms on Earth. This creative event was assumed to have been planned and purposeful, and people believed that species were fixed and unchangeable and that Earth was relatively young (some still believe this). The idea of a special creation also implied that there could be no extinctions. Scientists studying nature worked under the belief that they were studying the work of God as seen in the laws and diversity of life.

As scientists observed the many different varieties of organisms, they struggled to reconcile the accepted fixity of species consistent with a special creation with evidence that species could in fact change. Fossils, discovered at least as early as 500 B.C., were at first believed to be accidents of nature, oddly shaped crystals or faulty attempts at life to form through spontaneous generation in rocks. By the mid-1700s, the increasingly obvious connection between organisms and fossils argued against the idea that they were mere accidents or bore only a coincidental resemblance to

FIGURE 14.1
Science Evolves.
Significant contributions were made by many scientists, over many years, to develop the foundation Darwin used to describe natural selection as the mechanism for evolution.

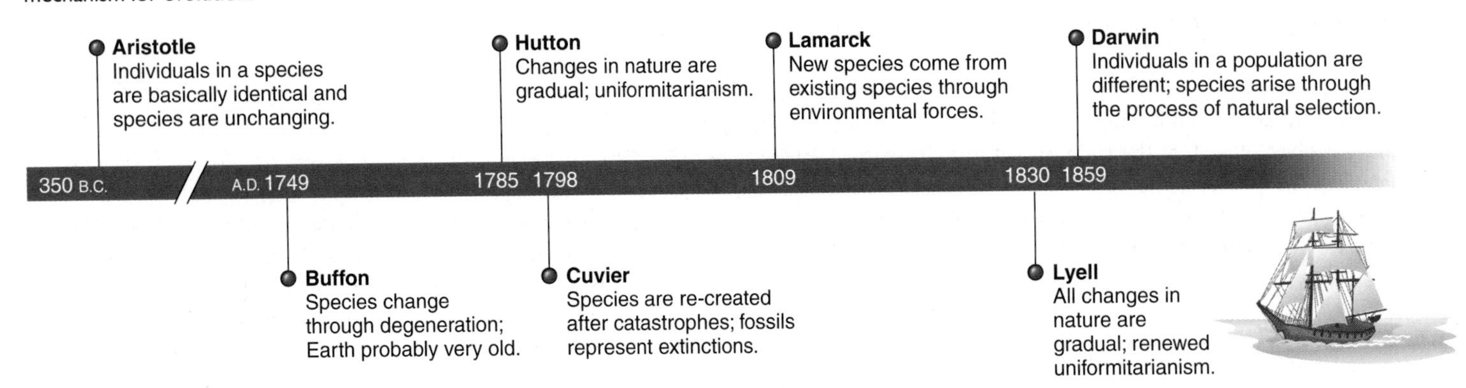

Investigating Life 14.1

Pollution Drives Evolution by Selecting Preexisting Variants

The rapid adaptation of certain small aquatic invertebrates (animals without backbones) to environmental pollution illustrates microevolution in action. Individuals whose genetic makeups render them vulnerable to pollution die out or do not produce fertile offspring, eventually disappearing from the population. The gene(s) for pollution resistance, already present before exposure to the pollution, become(s) more common in the population. This happened in Foundry Cove, part of the Hudson River about 50 miles (80 kilometers) north of New York City. Foundry Cove has a toxic history. During the Revolutionary War, a forge at the site manufactured metal chains that were placed in the river to stop British ships. During the Civil War, bullets were produced near the cove. Then, in the 1950s, manufacturing facilities at Foundry Cove made batteries. With such a record of heavy metal manufacture, it's not surprising that the sediments at the bottom of the river contained up to 25% cadmium.

Curiously, the polluted river area swarmed with invertebrate life, as did neighboring coves. But when animals from the nearby regions with cleaner sediments were moved to the polluted areas, they died. Had the Foundry Cove animals inherited the ability to survive heavy metal poisoning? That is, did the population once include animals with differing abilities to survive and reproduce in the presence of certain chemicals? Did years of exposure to pollution select the resistant animals, who came to constitute most of the population? To find out, biologists devised an experiment. They took animals from the areas with polluted sediments and bred them in the laboratory for several generations. They then returned the descendants to the polluted cove, where they survived—indicating that the toxin resistance is genetic.

Next, researchers hypothesized that the population had evolved in a way that enabled the animals to withstand the pollution. This may have taken as little as 30 years. To test their hypothesis, researchers tried to re-create the cove's history in the laboratory (fig. 14.A). They sampled invertebrates from a clean cove and exposed them to cadmium-rich mud in the laboratory. Then they bred the survivors with each other, building a population where most individuals could tolerate cadmium. When placed in polluted waters, the animals survived. Evolution had occurred, and was occurring, due to the more likely survival and reproduction of animals that had inherited genes enabling them to live, and produce fertile offspring, in the presence of heavy metal.

FIGURE 14.A
Observing Evolution.
Evolution can occur rapidly enough for us to observe it, as the animal life in Foundry Cove illustrates.

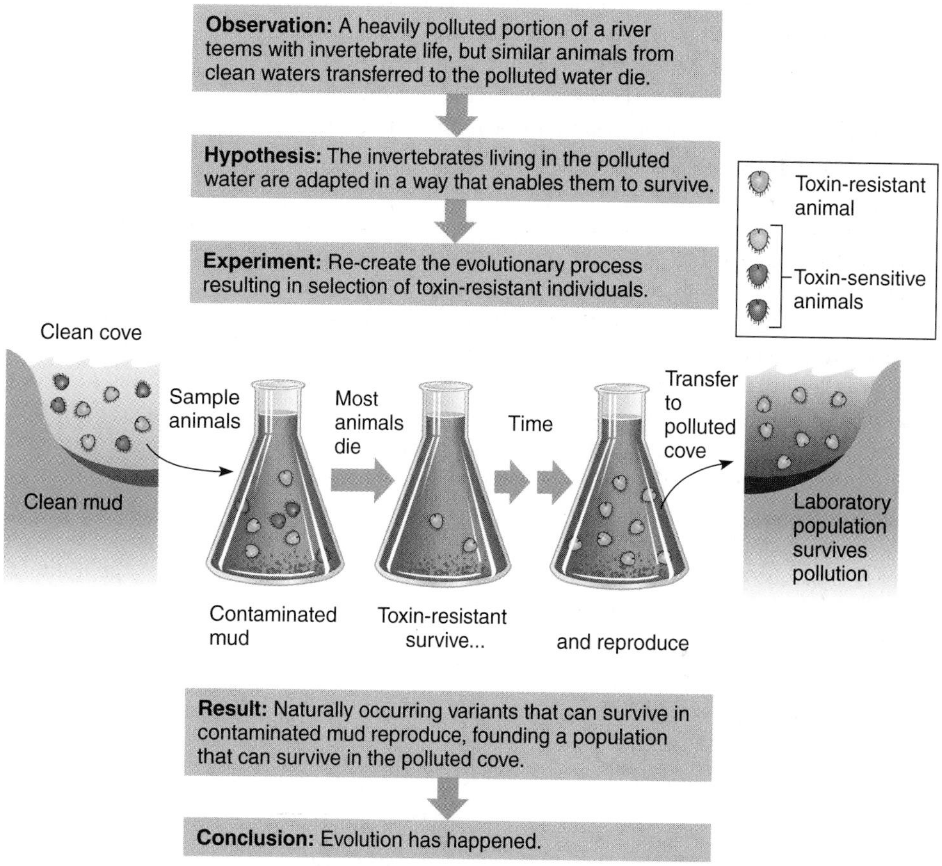

Observation: A heavily polluted portion of a river teems with invertebrate life, but similar animals from clean waters transferred to the polluted water die.

Hypothesis: The invertebrates living in the polluted water are adapted in a way that enables them to survive.

Experiment: Re-create the evolutionary process resulting in selection of toxin-resistant individuals.

Toxin-resistant animal
Toxin-sensitive animals
Clean cove
Clean mud
Sample animals
Contaminated mud
Most animals die
Toxin-resistant survive...
Time
and reproduce
Transfer to polluted cove
Laboratory population survives pollution

Result: Naturally occurring variants that can survive in contaminated mud reproduce, founding a population that can survive in the polluted cove.

Conclusion: Evolution has happened.

life. To reconcile this idea, early scientists suggested that fossils represented organisms destroyed during the biblical flood. However, some of the fossils depicted organisms no one had ever seen before. Since the people of the day believed firmly that species created by God could not become extinct, these fossils presented quite a paradox. The conflict between ideology and observation widened as geologists discovered that certain fossils were always found in the same layers of rock. A different layer of rock would reveal an entirely different group of fossils, all now extinct.

In 1749, French naturalist Georges Buffon (1707–1788) became one of the first to openly suggest that individuals within a species were changing. According to Buffon, new species were actually degenerate forms of existing species. Since everything else in nature seemed to degenerate, why not organisms as well?

He was quite wrong, but he did suggest that species *were* changing, which was a radical idea at the time. By moving the discussion into the public, he made possible new consideration of evolution and its causes. Buffon also suggested that Earth would be much older than a few thousand years based upon how long it would take to cool after moving away from the sun.

Clues in the Earth

In the early 1700s and 1800s, geology was a more accepted science than biology, so it was geology that attracted most of the attention of the natural scientists of the day. Most scientists thought that rapid, catastrophic events, such as floods and earthquakes, were responsible for most geological formations. In 1785, physician James Hutton (1726–1797) proposed that the forces that formed the earth would have acted in a gradual, yet uniform, way. His ideas became known as **uniformitarianism,** but were not widely accepted at the time.

The father of paleontology, Georges Cuvier (1769–1832), was an outspoken opponent of uniformitarianism. He frowned on the practice of making predictions from a limited set of data, which he believed Hutton had done. Cuvier's observations of the action of water on coastlines convinced him that a series of catastrophes had caused the geographic changes, an idea called **catastrophism.** Cuvier also described the anatomical similarities among organisms. Because of his knowledge of anatomy, he was able to identify many fossils from just a few bones. He was also the first to recognize that older, simpler fossils appeared at the lower layers of rock (fig. 14.2)—a concept known as the **principle of superposition.** Although he had to accept that certain species must have become extinct, he refused to believe that they were not originally formed through creation. He argued that catastrophes would destroy most of the organisms in an area, but then new life would be created or arrive from surrounding areas. To Cuvier, each rock layer represented a separate destructive event, followed by influx of new life.

In 1809, French taxonomist Jean-Baptiste de Lamarck (1744–1829) proposed a radical new theory of the formation of new species. The recognition of fossils as evidence of extinct species caused many people to acknowledge that species were changing, but no one had suggested a mechanism for such changes. Lamarck proposed that the environment could exert a powerful influence on populations, guiding changes. He applied an accepted theory of the day, called the inheritance of acquired characteristics, to provide a way for traits to be passed to the next generation. Whereas Cuvier had worked mostly with fossils of more complex animals, such as elephants, Lamarck was the expert of the time on invertebrates, such as clams and mussels. By comparing fossils with living examples of these animals, Lamarck made a breakthrough in thinking—he concluded that one species becomes extinct by becoming a new, different species.

Lamarck reasoned that organisms that used one part of their body repeatedly would increase their abilities, very much like today's weight lifters developing strong arms. He proposed that the resulting changes in individuals would give them the ability to get more food in a changing environment. Other individuals would die if they did not similarly change. But to explain how those traits were passed to the next generation, Lamarck applied the theory of the inheritance of acquired characteristics. To illustrate his point, he suggested that this was why all of the sons of a blacksmith were born strong. The mechanism is absurd in light of what we know today about genetics, but Lamarck had the distinction of being the first to propose a mechanism for evolution and to suggest that animals would change or become extinct in response to their interactions with their environment.

Geologist Charles Lyell (1797–1875) renewed the argument for uniformitarianism in 1830 in a lengthy work on principles of geology, suggesting that natural processes are slow and steady. One obvious conclusion from his contribution is that gradual changes in some organisms are represented in the fossil layers. Although many scientists at the time held to the views of Cuvier, Lyell was so persuasive that some people began to support the idea of gradual geologic change. With these new theories and ideas, people were beginning to accept the concept of evolution, but could not understand how it could result in the formation of new species. Charles Darwin was schooled in geology, and became convinced that Lyell's explanations of geologic change were correct. Ultimately, Darwin recognized their application to the changing diversity of life on Earth.

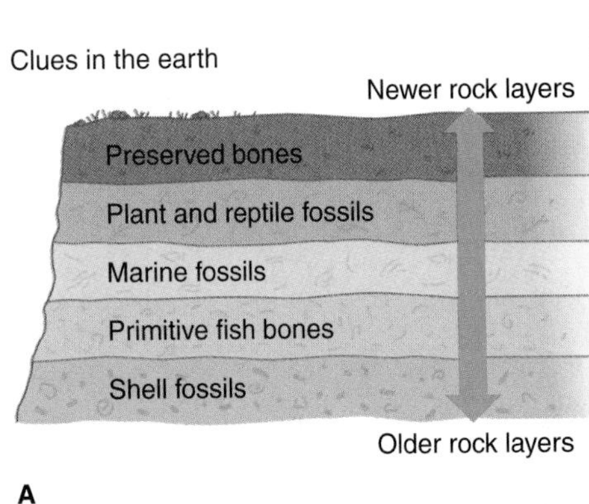

FIGURE 14.2
Rock Layers Reveal Earth History and Sometimes Life History.
(**A**) Layers, or strata, of sedimentary rock formed from sand, mud, and gravel deposited in ancient seas. The rock layers on the bottom are older than those on top. Rock strata sometimes contain fossil evidence of organisms that lived when the layer was formed and provide clues about when the organism lived. (**B**) Sediment layers are visible along the Grand Canyon. Hiking there is like taking a journey through time. Although the rim now rises over 2,000 meters (6,500 feet) above sea level, it has repeatedly been submerged and uplifted.

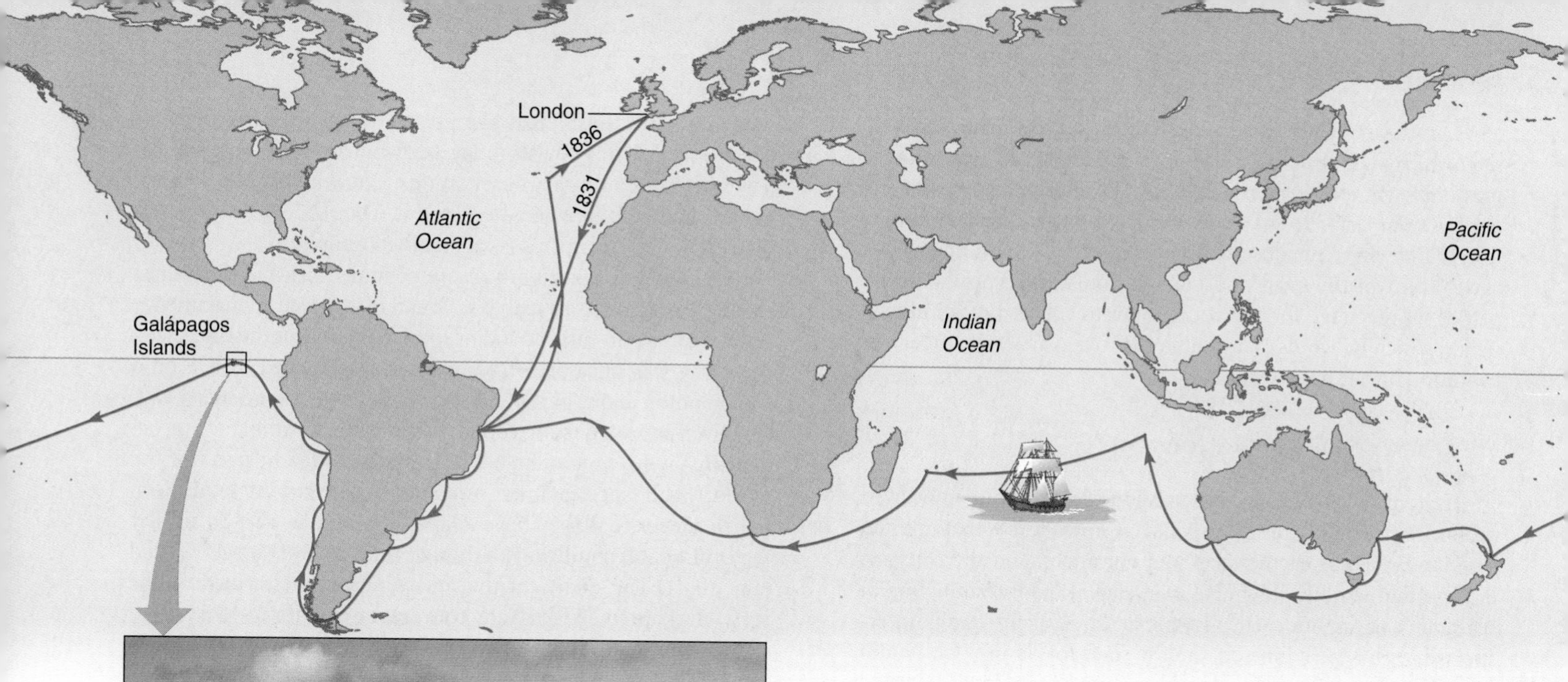

FIGURE 14.3

The Voyage of the *Beagle*.

Darwin formulated many of his theories about organisms and evolution through observing life and geology in many places in the world during the journey of the HMS *Beagle*. Many of the ideas of natural selection had their origins in the observations Darwin made of the different species inhabiting the Galápagos Islands.

14.1 MASTERING CONCEPTS

1. What is evolution?
2. Distinguish between macroevolution and microevolution.
3. What were some of the ways that people thought species arose and diversified before Charles Darwin published his theory of evolution by natural selection?
4. How does the principle of superposition provide a framework for evaluating evidence of past life?
5. Distinguish among the theories of Lamarck, Cuvier, and Buffon.

OLC

14.2 Charles Darwin's Voyage on the HMS *Beagle*, and Afterward

Darwin explored the Galápagos Islands, noting in detail the distribution and diversity of organisms. He formulated his theory of the origin and diversification of species by natural selection based on these observations and the ideas of others. Objection to Darwin's theory centered on his consideration of humans as animals that are related by descent to other types of animals, although this was only a small part of his treatise.

Charles Darwin (1809–1882) was the son of a physician and grandson of noted physician and poet Erasmus Darwin. Young Charles was a poor student. He preferred to wander the countryside examining rock outcroppings, collecting shells, and observing birds and insects. Under family pressure, he began to study medicine but abandoned it because he could not stand to watch children undergoing surgery without anesthesia. He grew bored with his studies, leading his father to comment, "You're good for nothing but shooting guns, and rat catching, and you'll be a disgrace to yourself and all of your family." At the urging of his family, Darwin finally attended Cambridge and completed studies to enter the clergy.

While Darwin reluctantly explored the vocations his family preferred, he also followed his own interests. He joined several geological field trips and met several eminent geology professors. Eventually Darwin was offered a position as captain's companion and naturalist aboard the HMS *Beagle*. Before the ship set sail for its 5-year voyage 2 days after Christmas in 1831 (fig. 14.3), the botany professor who had arranged Darwin's position gave the young man the first volume of Lyell's *Principles of Geology*. Darwin picked up the second two volumes in South America.

Darwin read Lyell's volumes while battling continual seasickness. By the time he finished reading, Darwin was an avid proponent of uniformitarianism. Lyell's ideas meshed with his observations prior to boarding the HMS *Beagle.* The combination of Darwin's keen observational skills, Lyell's ideas, and a voyage to some of the most unusual and undisturbed places on the planet set the stage for the young man to collect bountiful evidence, which he would, years later, use to construct the theory of evolution by natural selection.

Darwin's Observations

Darwin recorded his observations as the ship journeyed around the coast of South America. He noted forces that uplifted new land, such as earthquakes and volcanoes, and the constant erosion that wore it down. He marveled at the intermingling of sea and land in the earth's layers, at forest plant fossils interspersed with sea sediments, and at shell fossils in a mountain cave. The layers of earth revealed gradual changes in life. Darwin suggested that fossils, like sediments, appeared in a chronological sequence, with those most like present forms in the uppermost layers. Like paleontologists today, Darwin tried to reconstruct the past from contemporary observations, and wondered how an organism had arrived where he found it. The study of the physical distribution of organisms is called **biogeography.**

Darwin was particularly aware of similarities and differences among organisms, on a global scale and in localized environments. If there was a one-time period of special creation, as the Bible held, then why was one sort of animal or plant created to live on a mountaintop in one part of the world, yet another type elsewhere?

Even more puzzling to Darwin was the resemblance of organisms living in similar habitats but in different parts of the world. Such species have undergone **convergent evolution,** which means that they have adapted in similar ways to similar environmental conditions although they are not closely related (fig. 14.4). The giant carnivorous dinosaurs described at the start of chapter 17 illustrate convergent evolution—they lived on different continents and were therefore probably not closely related, but their fossilized bones are remarkably alike. Large, flightless birds that inhabit Africa, Australia, and South America

FIGURE 14.4
Convergent Evolution.
(**A**) Plants in different parts of the world have evolved similar mechanisms for protecting themselves, as this ocotillo from California and allauidi from Madagascar illustrate. (**B**) The North American wolf and Tasmanian wolf look very similar and live in similar environments, yet their methods of nurturing their unborn young are vastly different. The North American wolf is a placental mammal. The Tasmanian wolf, now extinct, was a marsupial that nurtured young in the mother's pouch. (**C**) Each of these three animals has a back fin, flippers, a tail with two lobes, and a streamlined body, all adaptations to aquatic habitats. Yet the animal on the left is an ancient reptile (an ichthyosaur), the middle animal is a modern fish, and the animal on the right is a dolphin, a mammal.

A

B

C

are contemporary evidence of convergent evolution. The ostrich lives on the plains and deserts of Africa and stands 8 feet tall (2.4 meters) and weighs 345 pounds (156 kilograms). The emu is Australian and stands 5 1/2 feet (1.7 meters) tall and weighs 100 pounds (45 kilograms), and the rhea lives on the grasslands of South America, reaching 5 feet (1.5 meters) and weighing 50 pounds (23 kilograms). The similarities in their habitats provided the selective force that led to flightlessness in different places on Earth.

Darwin kept detailed notes on organisms he saw and on geological formations. He wrote after the voyage that wherever he found a barrier—a desert, river, or mountain—different types of organisms often populated the areas on each side. He also noted that islands lacked the types of organisms that could not travel to them. Why would a Creator have put only a few types of organisms on islands, Darwin wondered?

In the fourth year of the voyage, the HMS *Beagle* spent a month in the Galápagos Islands. Although Darwin spent half of this time shipbound due to illness and visited only 1 of the 11 islands, the notes and samples he brought back would form the seed of his theory of evolution.

Toward the end of the voyage, Darwin began to assimilate all he had seen and recorded. Pondering the great variety of residents of South America, their relationships to fossils and geology, he began to think that these were clues to how species originate.

After the Voyage

Darwin returned to England in 1836, and by 1837 began assembling his notes in earnest. He published the first account of the voyage, *A Naturalist's Voyage on the Beagle,* in 1839. During that time he had spoken with other naturalists and geologists whose observations and interpretations helped mold his thoughts on the evolutionary process as it was revealed in the diverse life of South America.

In March 1837, Darwin consulted an ornithologist (an expert on birds) who was very excited by the finches the *Beagle* brought back from the Galápagos Islands. The ornithologist could tell from beak structures that some of the birds ate small seeds, some large seeds, others leaves or fruits, and some insects. In all, the ornithologist described 14 distinct types of finch, each different from the finches on the mainland. A special creation of slightly different finches on each island did not make sense; Darwin thought it more likely that the different varieties of finch on the Galápagos had descended from a single ancestral type of bird that had flown to the islands and, finding a relatively unoccupied new habitat, flourished. Gradually, the finch population branched in several directions, with different groups eating insects, fruits and leaves, and seeds of different sizes, depending on what natural resources each island had to offer. Figure 14.5 describes more recent experiments that used Galápagos finches. Darwin noted similar changes in the length of the Galápagos tortoise's neck. He called this gradual change from an ancestral type "descent with modification."

In September 1838, Darwin read a work that enabled him to understand the diversity of finches on the Galápagos Islands. Economist and theologian Thomas Malthus's *Essay on the Principle of Population,* written 40 years earlier, stated that food availability, disease, and war limit the size of a human population. Wouldn't populations of other organisms face similar limitations? If so, then individuals who could not obtain essential resources would die. The insight Malthus provided was that individual members of a population were not all the same, as Aristotle had taught. Individuals better able to obtain resources were more likely to survive and to reproduce. This would explain the observation that more individuals are produced in a generation than survive—they do not all obtain enough vital resources. Over time, environmental challenges would "select" out the more poorly equipped variants, and gradually the population would change. Darwin called this "weeding out" of less adaptive variants **natural selection.** The process is sometimes also viewed as selecting for certain traits that are adaptive in a particular environment. He got the idea of natural selection from thinking about **artificial selection,** the breeding strategies used to produce varieties of organisms with specific combinations of inherited traits, using naturally occurring extreme versions of inherited traits. Artificial selection is responsible for many of the breeds of domestic dogs and cats (see Investigating Life 15.1). Darwin himself raised pigeons, and artificially selected several new breeds.

Natural selection explained the diversity of finches on the Galápagos. Originally, some finches flew from the mainland to one island. When that first island population grew too large for all individuals to obtain enough small seeds, those who could eat nothing else starved. But birds who could eat other things, perhaps because of an inherited quirk in beak structure, survived. Since the new food was plentiful, these once-unusual birds gradually came to make up more of the population. Because each of the islands had slightly different habitats, different varieties of finches predominated on each one. The divergence of several new types from a single ancestral type is called **adaptive radiation.** The opening essay for chapter 15 describes adaptive radiation occurring in a laboratory experiment.

FIGURE 14.5

Finch Beak Shape Reflects Natural Selection.

Since the early 1970s, Princeton University researchers Peter and Rosemary Grant have continued Darwin's observations of changes in the finch populations of the Galápagos. The average beak size in a population—an inherited trait—can change appreciably, altering dietary possibilities, in as short a time as a year. By meticulously capturing, banding, and recapturing birds among populations of the medium ground finch *Geospiza fortis* and monitoring their beak sizes, the Grants discovered that following a very dry season, when seeds were sparse and the small, easy-to-crack ones were eaten fast, birds with large beaks survived because only they were strong enough to open the large, tough-to-crack seeds so they could consume the contents.

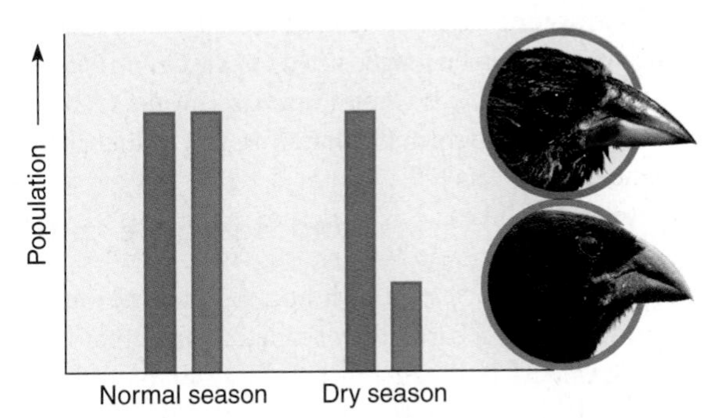

Darwin described his theory of evolution by natural selection in a 35-page sketch in 1842 and 2 years after that as a 230-page analysis, but did not publish either account. He continued to work feverishly on his ideas, and in 1858, with his treatise expanded but still not yet published, he was disturbed to receive a 4,000-word manuscript from British naturalist Alfred Russell Wallace. Its title was rather long-winded: *On the Tendency of Varieties to Depart Indefinitely from the Original Type.* It was as if Wallace had peered into Darwin's mind. Actually, Wallace, like Darwin, had observed the principles of evolution demonstrated among the diverse species of South America.

Darwin submitted his paper, along with Wallace's, to the Linnaean Society meeting later that year. In 1859, Darwin finally published 490 pages of the even longer-titled *On the Origin of Species by Means of Natural Selection, or Preservation of Favoured Races in the Struggle for Life.* It would form the underpinning of life science.

The *Origin of Species* and Response to It

Darwin's *Origin of Species,* as it became known, was not remotely a travelog, but a carefully reasoned treatise that he called "one long argument." In the first five of the 14 chapters, he set out the observations underlying the theory of natural selection: how artificial selection molds breeds of domesticated animals, and how nature similarly selects preexisting variants. Chapter 3 of Darwin's book describes the Malthus-inspired struggle for existence, and chapter 4 details how the environment influences natural selection. The fifth chapter guesses at the sources of this natural variation, but Darwin did not know about the work of his contemporary, Gregor Mendel. The sixth chapter counters expected arguments against the theory, and the remaining chapters detail the observations that natural selection explains. Darwin revised his book six times by 1872.

Although some members of the scientific community happily embraced Mr. Darwin's efforts, other people were less appreciative. Some of his ideas were perceived to clash with religious beliefs. At the time, many people believed that all life arose from separate special creations that occurred in the year 4004 B.C., and that once created, organisms did not change, although ever-more-perfect forms could be created. This was a sharp contrast to the geologic ages that Darwin hypothesized as the backdrop for fossil formation and distribution.

The nonscientific view of the diversity of life also held that nature is harmonious and purposeful. Darwin's world instead was purposeless, with evolutionary change occurring because individuals in a population had an advantage in a particular environment, due to inherited differences. Nor was the struggle to obtain resources harmonious. Perhaps most disturbing to many people was that Darwin removed humans from the pedestal on which many thinkers, such as Aristotle, had placed them. Humans were just one more species competing for resources that would enable them to survive and reproduce. More shocking was that Darwin's theory held that humans are related by common descent to all other species. The idea that humans were not special was heretical, a view that came to a head in a courtroom in the small town of Dayton, Tennessee, in the sweltering summer of 1925.

In what became know as the "Scopes monkey trial," a 24-year-old substitute science teacher, John Thomas Scopes, was tried for violating a state statute that banned the teaching of evolution (fig. 14.6). Scopes had actually answered an ad placed by the American Civil Liberties Union, volunteering to test the law, and he played a rather low-key role in the proceedings. Still, the trial quickly captured national attention, as the prosecution ignored the many examples in Darwin's teachings, focusing on the idea that humans evolved from a more primitive type of animal. Dayton became festooned in monkey memorabilia, and on the first day of the trial, two apes and a three-foot human calling himself "the missing link" arrived on the scene, to be used as evidence against Darwin's theory. In retrospect, especially in light of the enormous volume of evidence for evolution from many fields, the prosecution's arguments may seem ridiculous. But at the time, they made sense to many people who were uncomfortable with the idea of a human being a mere animal. Scopes was convicted, fined $100, and quickly returned to a life of relative ob-

FIGURE 14.6

The Scopes Trial Pitted Religion Against Evolution.
This sign commemorates the famous trial. It is outside the Rhea County Courthouse in Dayton, Tennessee, where the trial took place.

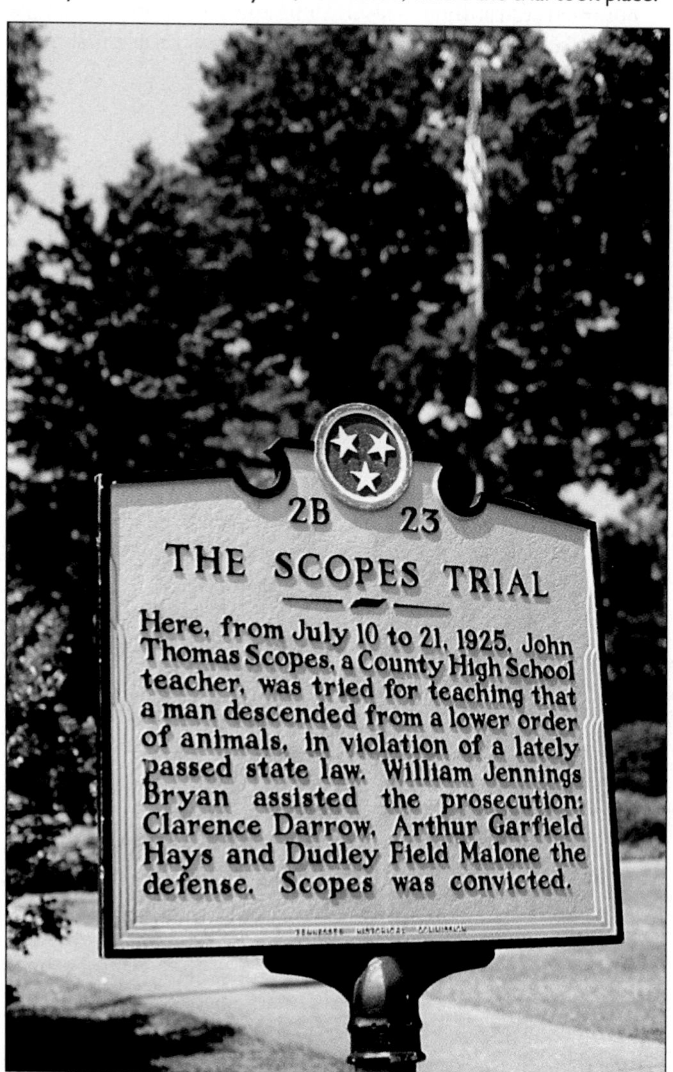

scurity. Unfortunately, objection to the theory of evolution by natural selection—a theory supported by so much evidence that some scientists consider it a law or principle—continues today.

Natural Selection—The Mechanism Behind Darwinian Evolution

Darwin observed that organisms within species vary, and that certain variants are more prevalent than others in different environments. He envisioned this natural variation as the raw material of evolution. Depending upon environmental conditions and competition for resources, certain variants would be less likely to have offspring healthy enough to produce offspring of their own. However, the traits that contribute to reproductive success need not directly affect reproduction. Any trait that ensures an individual's survival can make reproduction more likely.

Sexual selection is a type of natural selection that directly affects traits that increase an individual's chance of reproducing. The chapter opening essay describes the striking effect of sexual selection on male genitalia in some insect species. Sexually selected traits include elaborate feathers in male birds, horns in species ranging from beetles to giant elk, and the courtship songs many insects and birds sing to attract mates (fig. 14.7). So important is successful reproduction that an adaptation that gives a male a mating advantage will be perpetuated, even though that trait may result in his death!

In sexual selection, traits are associated with reproductive success, but the individual does not select a mate with evolution as its goal—evolution is not purposeful. For example, female barn swallows select mates whose tail feathers are symmetric in length. Experiments show that male barn swallows with even tails resist parasitic infections better than birds with more ragged rears, and are therefore more likely to survive to produce fertile offspring.

The direction of natural selection can change. Natural selection does not lead to perfection, but reflects adaptation to a prevailing environmental condition. Selective forces act upon preexisting characteristics; they do not create new, adapted variants. The furry

FIGURE 14.7
Sexual Selection.
The individual who is most successful in attracting a mate and reproducing is the most fit, according to Darwin, even if a male dies in the process. (**A**) The male bowerbird builds intricate towers of sticks and grass to attract a mate. (**B**) Diverse organisms, such as these Hercules beetles, use horns or antlers to battle rivals for access to the hornless sex. (**C**) The male bird of paradise displays bright plumes and capes in his quest for sexual success. (**D**) A female long-horned, wood-boring beetle chooses her mate based on the size of his territory, a succulent saguaro fruit. A better territory means more food for offspring.

A

B

C

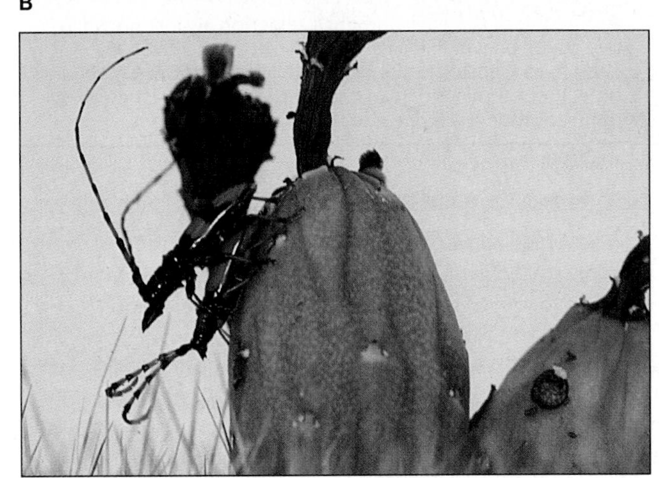
D

fox that survives a frigid winter to bear pups in the spring is not "perfect" and would certainly be less adapted to an unusually mild winter or hot summer. Similarly, bighorn sheep whose dwindling numbers grace the crags of the northern Rockies were far more plentiful at the height of the most recent ice age, when they flourished in the cooler climate.

A phenotype that seems to be "perfect" in one set of circumstances may be a liability in another. Natural selection can take the same population in different directions under different environmental conditions. Consider the finches on the tiny Galápagos Island of Daphne Major (see fig. 14.5). In a very dry season in the early 1980s, birds with large beaks were more likely to survive because they could eat the large, tough seeds that were left when the small amounts of small seeds were depleted. In 1983, 8 months of extremely heavy rainfall allowed many small seeds to accumulate. Over the next 2 years, finches with small beaks, who could easily eat the tiny seeds, came to predominate in the population.

A population may undergo sexual selection under one circumstance and natural selection for survival under another. This happens in guppy populations. When predatory fish are present for an extended time, the guppies are mostly drab, an adaptation that enables them to blend in with the murky stream bottom and escape predation. Natural selection (being eaten) weeds out the colorful guppies. When the stream lacks predators for a time, however, vibrantly colored fish gradually come to make up more of the population. Here, the color reflects sexual selection, because female guppies "prefer" colorful mates.

Table 14.1 summarizes Darwin's main ideas.

TABLE 14.1 Darwin's Main Ideas

OBSERVATIONS OF NATURE

1. Organisms are varied, and some variations are inherited. Within a species, no two individuals (except identical siblings) are exactly alike.
2. More individuals are born than survive to reproduce.
3. Individuals compete with one another for the resources that enable them to survive.

INFERENCES FROM OBSERVATIONS

4. Within populations, the inherited characteristics of some individuals make them more able to survive and produce fertile offspring in the face of certain environmental conditions.
5. As a result of the environment's selection against nonadaptive traits, only individuals with adaptive traits live long enough to transmit traits to the next generation. Over time, natural selection can change the characteristics of populations, even giving rise to new species.

14.2 MASTERING CONCEPTS

1. What did Darwin observe about the distribution of organisms and geology that led him to develop his theory of the origin of species driven by natural selection?
2. What is the raw material of evolution?
3. How did Darwin's ideas challenge prevailing beliefs about life's diversity and the status of humans?
4. How can natural selection favor different traits within the same population at different times?

OLC

14.3 Evolution Today—Epidemiology

The principles of Darwinian evolution operate today. Emerging and reemerging infectious diseases, epidemics, and antibiotic resistance all reflect selective pressures.

So far this chapter's view of evolution has been historical in flavor. Biological evolution, however, is continual and ongoing and is evidenced today in several familiar ways. We look now at two areas of evolution in action—the changeable nature of infectious diseases and the growing resistance of bacteria to antibiotic drugs.

Emerging Infectious Diseases

Epidemiology, the study of infectious disease origin and transmission in populations, includes microevolution. Anyone who has ever had more than one cold (upper respiratory infection) is a victim of evolutionary change. Surface features of the envelopes of cold-causing viruses change faster than our immune systems can fight them. Similarly, rapid genetic changes that alter the pattern of proteins on the surfaces of influenza viruses render one season's flu vaccine ineffective the next year.

"Flu" viruses usually pass from fowl such as chickens and ducks, through pigs, to people. In 1997 to 1998, a flu variant arose in Hong Kong that passed directly from chickens to a few dozen people, raising fears that a new deadly strain would become a worldwide killer (see chapter 19). Fortunately, a large epidemic never materialized, but millions of chickens were slaughtered in an attempt to prevent spread of the virus. Such a changing pattern of an infectious disease illustrates evolution in action.

Infectious diseases change in severity and type over easily observed spans of time because the causative microorganisms or viruses have very short generation times. A human generation takes 20 to 30 years; a bacterial generation takes 20 to 40 minutes. Given the success of antibiotic drugs in shortening the course of infections and vaccines in preventing them, we grew complacent about our power to defeat these diseases. We banished one scourge, smallpox, completely, although some nations harbor frozen samples of the virus. The resurgence of some diseases, including measles, dengue fever, cholera, diphtheria, and tuberculosis, and the appearance of apparently new ones, especially AIDS, has made us all too aware that the microorganisms and viruses that cause infectious diseases in humans are ever-evolving (Health 14.1).

Over the past quarter-century, toxic shock syndrome, Legionnaires' disease, Lyme disease, and AIDS have arisen (fig. 14.8). The bacteria or viruses causing these "new" illnesses may not be new at all; rather, genetic change may have led to emergence of the associated illness. A mutation may enable an infection to spread by a different route, affect a new species, or produce different symptoms. For example, hantavirus caused an outbreak of a fatal respiratory syndrome in the southwest United States in 1994, yet in Korea in the past, and today in the former Yugoslavia, a variant of the virus causes a hemorrhagic fever that results in bleeding and kidney failure. The infection is changing as the virus diverges into two types.

Crowd Diseases

When Europeans first explored the New World, they inadvertently brought lethal weapons—bacteria and viruses to which their immune systems had adapted. The immune systems of Native Americans, however, had never encountered these pathogens before and were thus unprepared.

Smallpox, a viral infection (fig. 14.B), decimated the Aztec population in Mexico from 20 million in 1519, when conquistador Hernán Cortés arrived from Spain, to 10 million by 1521, when Cortés returned. By 1618, the Aztec nation had fallen to 1.6 million. The Incas in Peru and northern populations were also dying of smallpox. When explorers visited what is now the southeast United States, they found abandoned towns where natives had died from smallpox, measles, pertussis, typhus, and influenza.

The diseases that so easily killed Native Americans are known as "crowd" diseases, because they arise with the spread of agriculture and urbanization and affect many people. Crowd diseases swept Europe and Asia as expanding trade routes spread bacteria and viruses along with silk and spices. More recently, air travel has spread crowd diseases. Returning soldiers introduced penicillin-resistant gonorrhea from southeast Asia to the United States during the Vietnam war. More recently, travelers transported cholera by jet from Peru to Los Angeles.

Crowd diseases tend to pass from conquerors who live in large, intercommunicating societies to smaller, more isolated and more susceptible populations, and not vice versa. When Columbus arrived in the New World, the large populations of Europe and Asia had existed far longer than American settlements. In Europe and Asia, infectious diseases had time to become established and for human populations to adapt to them. In contrast, an unfamiliar infectious disease can quickly wipe out an isolated tribe, leaving no one behind to give the illness to invaders.

Crowd diseases are often zoonoses, which are diseases that pass to humans from other types of animals. Typically, a zoonosis begins when a person picks up a microbe from constant close contact with an animal, such as a bird, cow, monkey, or sheep. The microbe randomly mutates in a way that enables it to infect humans, possibly causing symptoms. For example, two apparently new viral diseases of humans that emerged in the 1990s are carried in fruit bats, but passed by certain mammals to people. In Australia, people contracted hendra virus from horses; in Singapore and Malaysia, the related nipah virus was passed from coughing pigs! Both viruses cause encephalitis, a swelling of the membranes that enclose the brain.

Fortunately, most crowd diseases vanish quickly, in any of several ways. Medical researchers may develop vaccines or treatments to stop the transmission. People may alter their behaviors to avoid contracting the infection, or the disease may be so devastating that infected individuals die before they can pass the disease on. Sometimes, we don't know why a disease vanishes or becomes less severe.

We may be able to treat and control newly evolving infectious diseases one at a time, with new drugs and vaccines. But the process that continually spawns new genetic variants in microbe populations—resulting in evolution—means that the battle against infectious disease will continue.

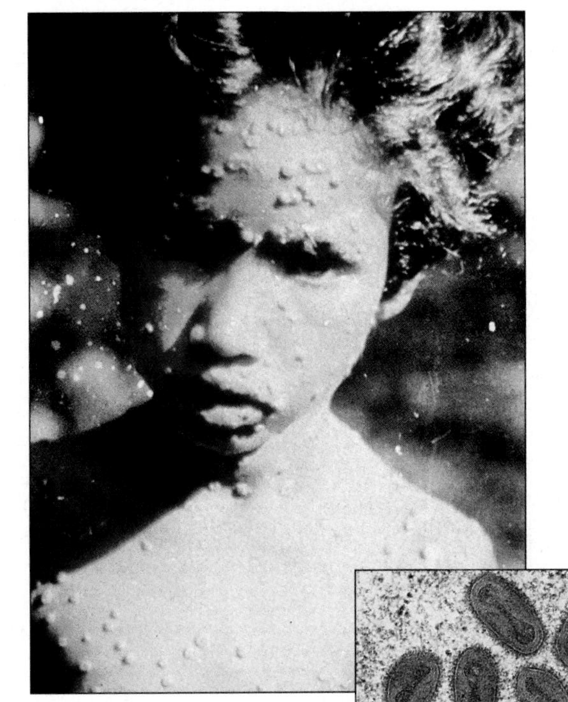

FIGURE 14.B
Death of a Disease.
This boy is one of the last victims of smallpox, which has not infected a human since 1977. Because many doctors are unfamiliar with smallpox, and people are no longer vaccinated against it, an outbreak would be a major health disaster. For these reasons, the former Soviet Union developed smallpox as a weapon, even though it had signed the Biological Weapons Convention in 1972 pledging "not to develop, produce, stockpile or otherwise acquire or retain" such agents of war. The inset shows smallpox viruses.

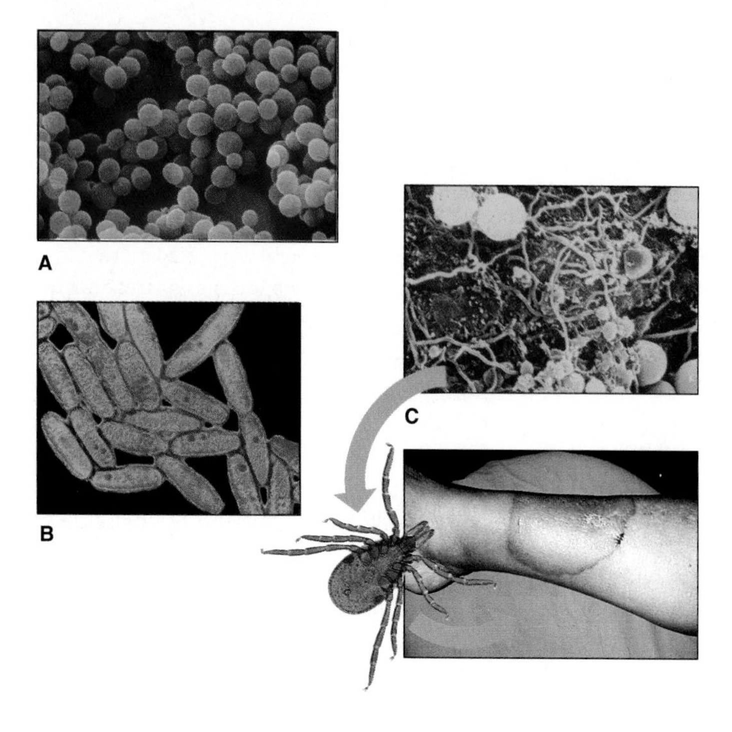

FIGURE 14.8
A Trio of Recently Recognized Bacterial Diseases.
(**A**) Toxic shock syndrome first appeared in the early 1980s, mostly in women who used a certain brand of tampon. A material in the tampons entrapped the common bacterium *Staphylococcus aureus,* provoking it to produce a toxin that causes vomiting, diarrhea, plummeting blood pressure, skin peeling, and death. (**B**) Legionnaires' disease is a form of pneumonia caused by the bacterium *Legionella pneumophila,* which was implicated in the 1970s after it struck American Legion conventioneers in Philadelphia by traveling through air-conditioning ducts. (**C**) Lyme disease produces a rash, aches, and sometimes arthritis and nervous system complications. A spiral-shaped bacterium, *Borrelia burgdorferi,* causes the illness and is transmitted to humans by the bite of a deer tick. The tick is about the size of a period.

HIV infection is changing too. On a population level, more people are living longer with the infection. On an individual level, natural selection acts on the "swarm" of viral particles that infect a body. As the virus replicates, its genetic material mutates. Viral genetic diversity is low as infection begins, as rapidly replicating viruses predominate. Then during the 2- to 15-year latency period, the person's immune system selects out—destroys—certain viral variants. Gradually, hardier viral variants accumulate and symptoms appear. In HIV infection, the virus usually eventually "wins" the evolutionary battle.

Another factor that can favor once-rare disease-causing bacterial or viral variants is a change in the environment, such as deforestation or drought. This was the case for Lyme disease, a bacterial infection that causes a flulike illness and arthritis (joint inflammation). When the first recognized cases of the disease appeared in Lyme, Connecticut, in the early 1970s, physicians thought it was a new form of infectious juvenile arthritis. But the disease was actually commonplace in colonial times, when communities were more heavily forested. The bacterium that causes the illness, *Borrelia burgdorferi,* lives in ticks that feed on deer that inhabit forests. As colonists cleared land to build settlements, the habitat of the deer, ticks, and bacteria shrank, and the arthritis became rare. Then in the northeast people began abandoning farms. Forests regrew, deer returned, and ticks once again transferred *Borrelia burgdorferi* to people enjoying the outdoors. Lyme disease returned.

A mutation that enables a microorganism to produce a new toxin, or a change in the environment that favors a microorganism that produces a toxin, can result in an infectious disease. Each year in the United States, a few people die after eating hamburger containing a strain of *E. coli,* a normally harmless bacterium, that produces a toxin that causes bloody diarrhea and kidney failure. Researchers at Cornell University developed a very simple way to avoid the problem, based on understanding how evolution works. Feeding cows barley changes the environment in their digestive tracts in a way that selects against the toxin-producing strain. Hamburger made from such cattle lacks the toxic strain of *E. coli.*

In a general sense, the types of symptoms that define infectious diseases are actually adaptations that confer a survival advantage to the infectious agents. Sneezing, bleeding, and diarrhea spread viruses and bacteria. Ebola virus vividly illustrates how an infectious agent "uses" a human body to reproduce.

Evolving Infectious Illness Caused by Ebola Virus

Ebola virus was discovered as the cause of a hemorrhagic (bleeding) fever in humans in 1976 in what was then Zaire, with a 90% fatality rate. Outbreaks occurred in Africa every few years, and then the illness would mysteriously vanish. The infection spreads by contact with body fluids of an infected individual. Symptoms progress rapidly over a 2-week or shorter period from headache to rash, vomiting blood, and a massive breakdown of internal tissues until the victim "bleeds out," expelling tissues and secretions from all orifices.

Given this frightening picture, when in 1989 a few monkeys imported from the Philippines and housed in an animal facility in Reston, Virginia, began dying of an airborne infection identified as Ebola, public health authorities feared the worst—a deadly virus transmitted from monkeys to humans *in the air.* But this illness seemed to be different. The monkeys died of extreme lethargy and failure to eat or drink, not the dramatic "bleeding out" of a human Ebola victim. The few humans infected with what came to be known as "Ebola-Reston" did not get very sick and fully recovered. The virus had changed from its African form, or perhaps was a different variant from the Philippines. But the deadly African form was still very much in existence.

In the spring of 1995, Ebola hemorrhagic fever swept the city of Kikwit, Zaire (now Congo). The virus infected 315 people before it could be contained, with a 79% fatality rate—lower than earlier outbreaks probably because of swift medical attention. As in 1976, the infection spread by body fluid contact.

In 1996, Ebola-Reston emerged again, this time in laboratory monkeys imported from the Philippines being kept at a facility in Texas. Like the Reston incident, the virus was airborne and produced different symptoms. Genetic analysis showed that the virus in the Texas monkeys matched that from the Virginia monkeys in 98.9% of its DNA sequence.

The two types of Ebola virus—from Africa causing hemorrhagic fever in humans and from the Philippines causing lethargy in monkeys—illustrate evolution in action in its most stripped-down form, a changing virus. Two strains diverged from a common viral ancestor.

The Rise of Antibiotic Resistance

The use of antibiotic drugs revolutionized medical care in the twentieth century and enabled people to survive many once-deadly bacterial infections (fig. 14.9). These drugs are naturally occurring compounds, or chemicals based on them. However, some infectious diseases are becoming resistant to the antibiotic drugs that once vanquished them. The reason for the rise in antibiotic resistance is evolution.

Any natural population of organisms includes variants. If an antibiotic dismantles a component of a bacterial cell wall, for example, a resistant mutant might have a slightly different molecule in its cell wall that the antibiotic cannot alter. When a person takes an antibiotic, the drug kills all the susceptible bacteria and leaves behind those that can resist it. The resistant bacteria multiply, increasing their numbers a million-fold in a day. The antibiotic does not create the bacterial resistance, but creates a situation where an already existing variant can flourish. If this sounds a lot like natural selection, it should. It is.

Antibiotic-resistant bacteria appeared just four years after the drugs entered medical practice in the late 1940s, and researchers responded by discovering new drugs. But the microbes kept pace, as evolution tends to do. Today, 40% of hospital *Staphylococcus* infections resist all antibiotics but one, and already some laboratory strains are resistant to all antibiotics. We may be able to slow antibiotic resistance by using these powerful drugs only when it is appropriate.

The evolution of bacterial antibiotic resistance is similar to the evolution of pollution resistance among invertebrates in Foundry Cove, as figure 14.A depicts. Other examples of the evolution of resistance include weeds that can survive in the presence of herbicides and cancer cells that continue to divide in the presence of chemotherapeutic drugs.

New and returning infectious diseases, and the growing resistance of bacteria to antibiotic drugs, offer compelling evidence that evolution, driven by natural selection and accumulating genetic change, is an ongoing process.

14.3 MASTERING CONCEPTS

1. How can a genetic change alter the characteristics of an infectious disease?
2. How can an environmental change alter an infectious disease?
3. What factors have contributed to the rise in resistance of bacteria to antibiotic drugs?

OLC

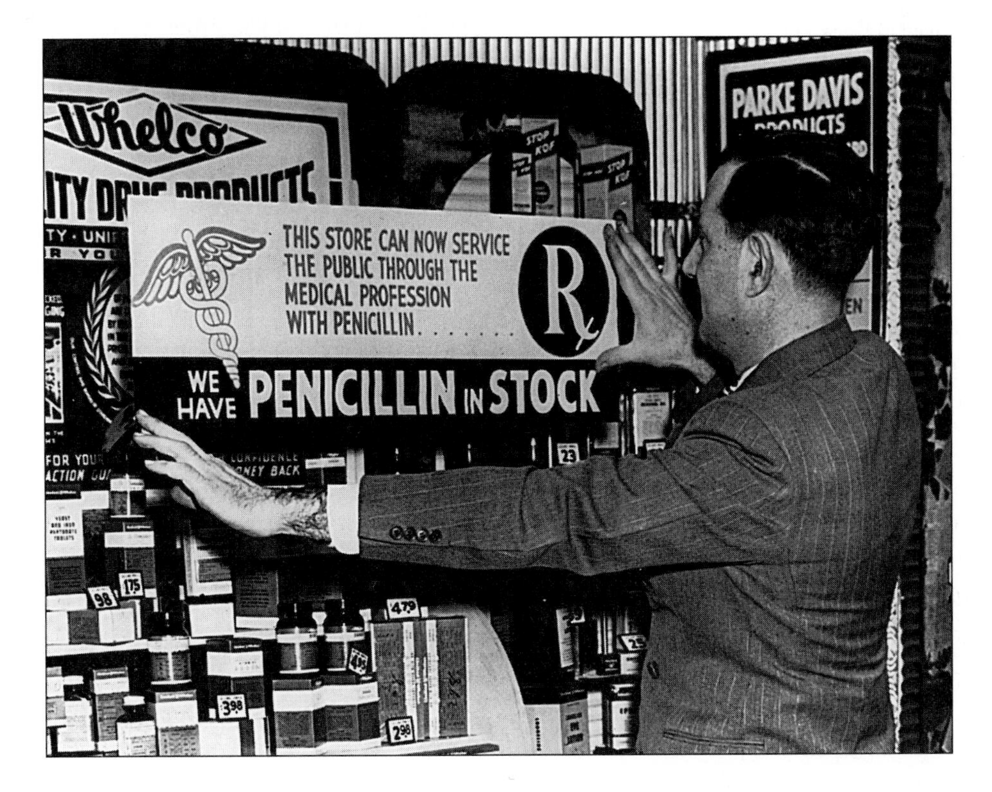

FIGURE 14.9
Penicillin, a fungus-produced antibiotic compound, revolutionized health care during World War II by enabling people to survive otherwise fatal bacterial infections. Penicillin destroys the cell walls of susceptible bacteria. Today, many antibiotics are losing their effectiveness because bacteria have become drug resistant. The reason for antibiotic resistance lies in the basic concepts of evolution.

Chapter Summary

14.1 Evolution Is a Framework for Understanding How Life Works

1. Biological **evolution** is change in allele frequencies in populations. Evolution has occurred in the past and is constant and ongoing.
2. Evolution consists of large-scale, species-level changes (**macroevolution**) as well as gene-by-gene changes (**microevolution**).
3. Before Darwin, attempts to explain life's diversity were human-centric and subjective. Lamarck was the first to propose a mechanism of evolution, but it was based on acquired rather than inherited traits.
4. Geology laid the groundwork for evolutionary thought. Some people explained the distribution of rock strata with the idea of **catastrophism** (a series of floods). The more gradual **uniformitarianism** (continual remolding of Earth's surface) became widely accepted. The **principle of superposition** states that lower rock strata are older, suggesting a time frame for fossils within them.

14.2 Charles Darwin's Voyage on the HMS Beagle, and Afterward

5. During the voyage of the HMS *Beagle,* Darwin observed the distribution of organisms in diverse habitats (**biogeography**) and their relationships to geological formations. He noted that similar adaptations can lead to **convergent evolution.** After much thought, and considering input from other scientists, he synthesized his theory of the origin of species by means of natural selection.
6. The bountiful evidence in Darwin's treatise is organized as "one long argument." However, people who believed Earth to be young, humans to be unique, and nature to move toward perfection, had difficulty with his ideas.
7. Darwin's theory was based on the observations that populations include individuals that vary for inherited traits; that many more offspring are born than survive; and that life is a struggle for use of limited resources. According to the theory of **natural selection,** individuals least adapted to their environments are less likely to leave fertile offspring, and therefore their genes will diminish in the population over time. The genes of those better adapted to the particular environment will persist. Natural selection caused the diversification, or **adaptive radiation,** of finches on the Galápagos Islands.
8. **Sexual selection** is a form of natural selection in which certain inherited traits make an individual more likely to mate and produce viable offspring.

14.3 Evolution Today—Epidemiology

9. **Epidemiology** explores connections between genetics, evolution, and infectious diseases in populations.
10. Emerging and returning infectious illnesses and increasing resistance of bacteria to antibiotic drugs illustrate evolution in action.
11. Changes in the severity of illness or types of symptoms of viral infections can reflect evolution.

Testing Your Knowledge

1. How can evolution explain the seeming contradiction that life is very diverse, but that all species share many features, especially at the biochemical and cellular levels?
2. What is the difference between microevolutionary and macroevolutionary change?
3. How did James Hutton, Georges Cuvier, Georges Buffon, Jean-Baptiste de Lamarck, Charles Lyell, and Thomas Malthus influence Charles Darwin's thinking?
4. Why must natural selection work on inherited, rather than on acquired, traits in order to drive evolution?
5. What sorts of traits lead to "fitness" in the Darwinian sense?
6. How do the appearances of new infectious diseases or resurgences of old ones illustrate continuing evolution?

Thinking Scientifically

1. The giant anteater lives in South America, the scaly anteater lives in tropical parts of Asia and Africa, and the spiny anteater lives in Australia. The three animals are not at all closely related, but resemble each other closely. They have long, sticky tongues that they use to eat ants, no teeth, large salivary glands, and long, bald snouts. Which principle that Darwin observed do these animals illustrate?
2. Which theory best explains Darwinian evolution—catastrophism or uniformitarianism? Give a reason for your answer.
3. An early and major objection to Darwin's concept of natural selection as the mechanism behind evolution was that it refutes the pleasant idea that we humans are more advanced and special than other species. How does Darwinian evolution do this?
4. You have a pet cocker spaniel and believe that if you snip off the end of his tail, and then breed him, his puppies will be born with snipped-off tails. Is this idea consistent with Darwinian evolution? Why or why not?
5. Many articles about the rise of antibiotic-resistant bacteria claim that overuse of antibiotics creates resistant strains. How is this statement incorrect?

References and Resources

Alibek, Ken. 1999. *Biohazard.* New York: Delta. The fact that researchers in the former Soviet Union could select especially virulent bacteria and viruses to create bioweapons is evidence of evolution.

Darwin, Charles. 1958 (1859). *On the origin of species by means of natural selection.* New York: Mentor Books. Darwin's writing is very detailed but offers fascinating examples and interpretations of the natural world.

Dobzhansky, Theodosius. 1973. Nothing in biology makes sense except in the light of evolution. *The American Biology Teacher,* vol. 35. A concise explanation of why evolution unites all fields of biology.

Jones, Steve. 2000. *Darwin's Ghost: The Origin of Species Updated.* New York: Random House. Cell and molecular biology fit beautifully into the scheme of Darwinian evolution.

Lewis, Ricki. August 21, 2000. An eclectic look at emerging infectious disease. *The Scientist,* vol. 14, p. 1. Bats and migrating pigs bring new viral diseases to people.

Mayr, Ernst. 1982. *The growth of biological thought.* Cambridge, Massachusetts: The Belknap/Harvard Press. A comprehensive treatment of the thoughts and events that led to current biological concepts.

Scott, Eugenie. May 5, 2000. Not (just) in Kansas anymore. *Science,* vol. 288, p. 813. When school boards allow religion to clash with science, students are unprepared to understand many biological principles.

The February 1999 issue of *Natural History* has several articles on microevolution occurring today, in a variety of species.

The *LIFE* Online Learning Center provides additional resources and tools for studying this chapter.
www.mhhe.com/life

CHAPTER 15

The Forces of Evolutionary Change—Microevolution

Observing Microevolution in a Microbial Microcosm

A group of children let loose on an empty playground tends to spread out in a way that allows everyone to do something—usually a few children will climb on each piece of equipment, rather than all of them piling onto one swing set or slide. So it is with populations of organisms facing an environment that offers diverse places to live. In the Galápagos Islands, Charles Darwin inferred that finches had arrived and diversified as they adapted to different habitats. The Hawaiian Islands present a similar showcase for such adaptive radiation, with a spectacular diversity of fruit fly species.

The microcosms that formed within stationary dishes of broth are analogous to the different islands of the Galápagos or Hawaii.

Adaptive radiation of animals such as birds and insects can take many years. But the same process can be easily observed in laboratory experiments, using microorganisms. Paul Rainey and Michael Travisano at the University of Oxford did just that, with bacteria called *Pseudomonas fluorescens.* Bacteria offer many advantages in tracking population dynamics and even evolution, including a fast generation time, huge populations, and the ability to be stored in suspended animation so that researchers can compare different generations at the same time. The inherited trait that Rainey and Travisano looked at was the shape that a colony of bacteria takes on a plate of nutritive medium. A colony forms as an original bacterial cell reproduces asexually, but the shape can change if mutation occurs.

The researchers set up a simple experiment beginning with an "ancestral" phenotype of smooth colonies, adding

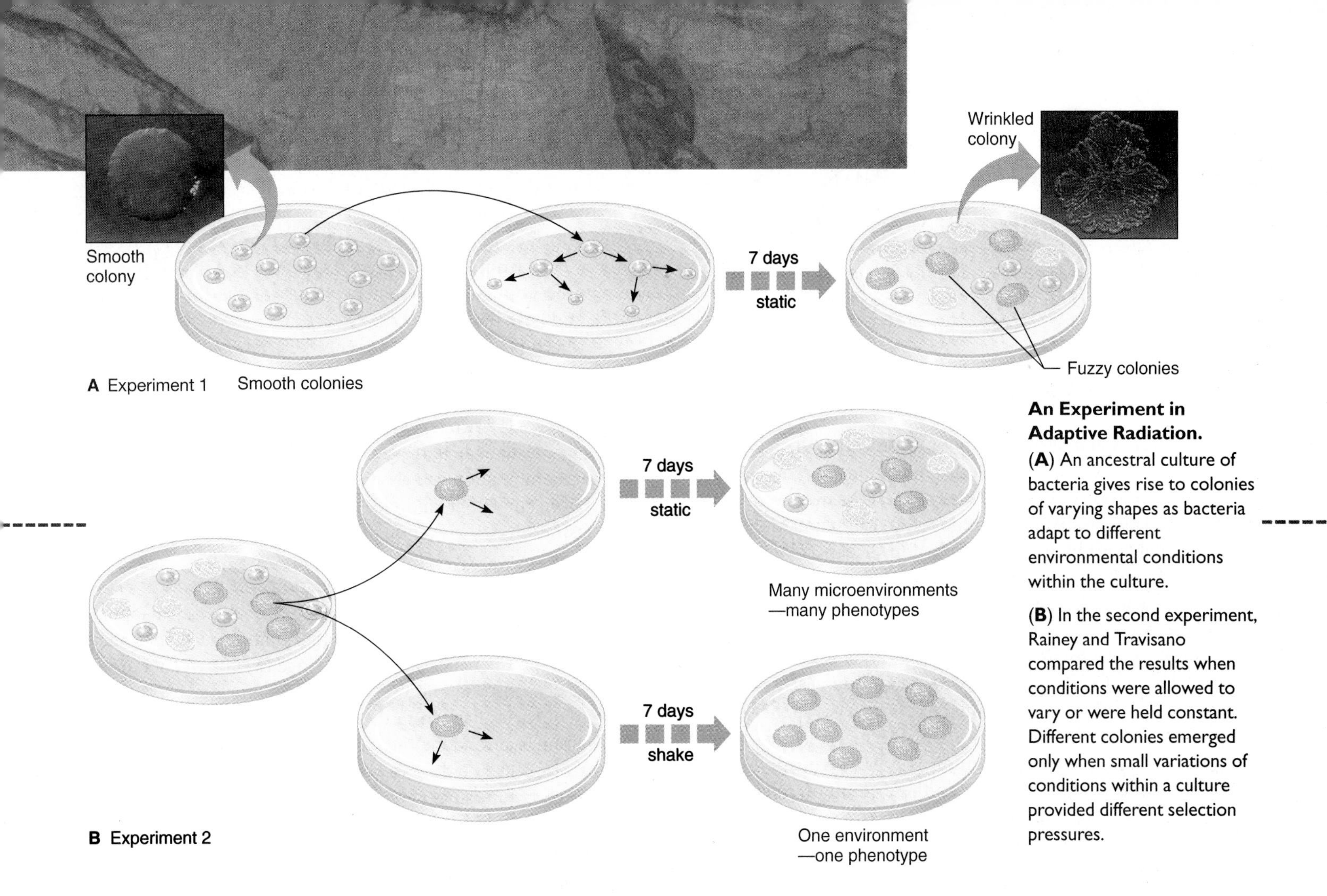

An Experiment in Adaptive Radiation.

(**A**) An ancestral culture of bacteria gives rise to colonies of varying shapes as bacteria adapt to different environmental conditions within the culture.

(**B**) In the second experiment, Rainey and Travisano compared the results when conditions were allowed to vary or were held constant. Different colonies emerged only when small variations of conditions within a culture provided different selection pressures.

bacteria to a small dish containing a broth of nutrients. The key to the experiment was to keep the dish absolutely still. Usually, bacterial cultures are shaken to evenly distribute the nutrients and oxygen. Keeping them still created "microcosms," areas within the broth that differ from each other in nutrient and oxygen content. (A similar phenomenon occurs in a large swimming pool with the filter turned off. Pockets of warm or cold water form as the circulation stops.) In the cultures, areas near the top of the containers would contain more dissolved oxygen than areas near the bottom, because they are closer to the air. The microcosms that formed within stationary dishes of broth are analogous to the different islands of the Galápagos or Hawaii.

The bacterial cultures sat still for 7 days. When Rainey and Travisano looked at the dishes, they could see the diversity of colony shapes immediately. Not all of the colonies had smooth edges—some were wrinkled, others fuzzy. From this initial observation that diverse habitats cause diverse populations to emerge, the researchers hypothesized that the formation of microcosms selected different genetic variants that had formed randomly by spontaneous mutation. To test this hypothesis, further experiments were necessary. If the habitat diversification selected the variants, then destroying that diversification should lead to a uniform population.

In the second experiment, Rainey and Travisano plucked an individual colony from the dishes harboring the three variants, and transferred part of it to a new dish that was shaken, and the other part to a second new dish that was held still. After two 7-day periods, the static dish held all three types of bacterial colonies—smooth, wrinkled, and fuzzy. But the dish that had been shaken had only the type of colony originally placed on it. The researchers repeated this protocol 10 times, with the same results. The conclusion: An environment containing different niches (habitats) supports genetic diversity, whereas an environment that is the same throughout selects only one type of bacterium. Diversifying bacterial cultures modeled adaptive radiation.

Evolution also occurs when individuals of different phenotypes compete for resources. This too happened among the bacteria in Rainey and Travisano's lab. By letting the cultures grow for many days, the researchers observed that the cells forming the wrinkled colonies tended to occupy the broth/air interface, where they stuck to each other to form a tightly knit mat. For a time, the living mat consumed most of the oxygen, to the detriment of the bacteria that formed smooth and fuzzy colonies. But eventually the bacteria that grew in wrinkled colonies reproduced too much for their own good—the mat got so heavy that it sank! Deprived of oxygen at the dish bottom, the wrinkled bacterial colonies began to die out. The populations of smooth and fuzzy colony-forming bacteria resurged to take their place. Microevolution was occurring, because allele frequencies were changing within the total population, over time, in response to environmental change.

15.1 Why Is Evolution Much More Likely Than Not?

Evolution occurs at the population level as gene (allele) frequencies change. Algebra can be used to represent Hardy–Weinberg equilibrium, the unlikely situation in which evolution does not occur. Hardy–Weinberg equilibrium provides a background against which microevolution can be detected.

The "raw material" of evolution is inherited variation. Genes affect individuals, but influence evolution at the level of the population.

All the genes and alleles in a population constitute its **gene pool,** and allele movement between populations is called **gene flow.** The proportion of different alleles for each gene determines the characteristics of that population. A Swedish population, for example, might include a large proportion of hair color alleles conferring blondness; a population of Asians would have very few, if any, such alleles but would have many alleles conferring darker hair.

Shifting allele frequencies in populations are the small steps of change that collectively drive evolution. (Frequencies are percentages expressed as decimal fractions.) Given the large number of genes in any organism and the many factors that can alter allele frequencies, evolution is not only possible but unavoidable. This chapter focuses on these common phenomena that make evolution very much a part of life—past, present, and future.

Conditions for Evolution *Not* to Happen: Hardy–Weinberg Equilibrium

Mathematics is used to describe the highly unlikely situation when allele frequencies do not change from one generation to the next, called **Hardy–Weinberg equilibrium.** It serves as a basis of comparison to reveal when microevolution is occurring. Hardy–Weinberg equilibrium is only possible if a population is large, its members mate at random and produce fertile offspring, and there is no migration, mutation, or natural selection—conditions that in actuality do not occur together.

In 1908 mathematician G. H. Hardy and physician W. Weinberg independently proposed that the expression $p + q = 1$ could be used to represent the frequency of all of the alleles $(p + q)$ for a particular gene in a population of diploid organisms, if only two alleles exist for that gene (fig. 15.1). This expression means that p and q represent all of the alleles in the population for a particular trait. If 70% of the alleles in a ferret gene pool confer dark fur, and the gene has only two alleles, then the allele frequency for the dark fur allele p is 0.7, and that for the alternate allele q, which confers tan fur, is 0.3.

Once we know allele frequencies, we can calculate genotype frequencies. This is because the proportion of alleles equals the proportion of gametes, as a consequence of Mendel's law of segregation. If p represents the frequency of the dominant allele (*D* for dark fur), then the proportion of the population that is homozygous *D* (genotype *DD*) equals p multiplied by p, or p^2, which equals 0.7 times 0.7, or 0.49, in our example. Likewise, the proportion of population members that are homozygous recessive (*dd*) equals 0.3 times 0.3 (q^2), or 0.09. To calculate the proportion of the population that is heterozygous, subtract the total proportion of homozygotes ($0.49 + 0.09 = 0.58$) from 1, which gives 0.42. That is, 4.2% of the ferrets in the population are heterozygous for the fur color gene (*Dd*), and have dark fur.

In algebraic terms, $2pq$ represents the heterozygous class. Note that $2pq$ equals 2 times 0.7 times 0.3, or 0.42—the same result as inferring the proportion of heterozygotes by subtracting the frequencies of the two types of homozygotes (*DD* and *dd*) from 1. "$2pq$" means that a heterozygote has one p allele and one q allele, and this combination can arise in two ways—p from the mother and q from the father, and vice versa. This is a population-level demonstration of segregation—recall that a Punnett square for a monohybrid cross yields heterozygotes in two ways.

Hardy and Weinberg used an equation—$p^2 + 2pq + q^2 = 1$—to represent the proportions of genotypes that make up a population. If conditions of Hardy–Weinberg equilibrium are met, allele and genotype frequencies will not change in future generations. We can demonstrate Hardy–Weinberg equilibrium by deriving allele frequencies back from the genotype frequencies, assuming that mating is at random, there is no migration or mutation, and that natural selection doesn't weed out any phenotypes.

Ferrets comprising the population of 49% *DD* (dark), 9% *dd* (tan), and 42% *Dd* (dark) animals manufacture gametes. The proportion of *D* alleles equals that of the homozygous dominant (*DD*) class (0.49) plus one-half of the gametes from the heterozygotes (*Dd*), which equals one-half of 0.42, or 0.21. Therefore, the proportion of *D* alleles is $p = 0.49 + 0.21 = 0.70$. If p equals 0.7, then q equals 1 minus p, or 0.3. We are back at the beginning. Hardy–Weinberg equilibrium persists, and evolution is not occurring. In the next generation, the same proportion of ferrets will have dark fur. These calculations also reveal how heterozygotes maintain recessive alleles in a population, even if homozygous recessive individuals are at a disadvantage.

Figure 15.1 organizes these calculations, depicting the random mating that underlies Hardy–Weinberg equilibrium. The "male" and "female" columns show every possible combination of gametes, and the other columns show how genotype frequencies of the next generation are derived from the random matings.

What Conditions Cause Evolutionary Change?

Microevolution occurs when the frequency of an allele in a population changes. This may happen when

- mutation introduces new alleles (*mutation*);
- individuals migrate between populations (*migration*);
- individuals remain in closed groups, mating among themselves within a larger population (*nonrandom mating*);
- some phenotypes are better adapted to a particular environment than others (*natural selection*);
- allele frequencies change due to chance (*genetic drift*).

FIGURE 15.1

Hardy–Weinberg Equilibrium—An Idealized State When Evolution Does Not Occur.

At Hardy–Weinberg equilibrium, allele frequencies remain constant from one generation to the next.

p = frequency of D (dominant allele) = dark fur = 0.7
q = frequency of d (recessive allele) = tan fur = 0.3

Algebraic Expression	What It Means
$p + q = 1$	Frequency of all dominant alleles plus frequency of all recessive alleles for this gene.
$p^2 + 2pq + q^2 = 1$ ($DD + 2Dd + dd = 1$)	For a particular gene, the frequencies of all the homozygous dominant individuals (p^2) plus heterozygotes ($2pq$) plus all homozygous recessives (q^2) add up to all of the individuals in the population.

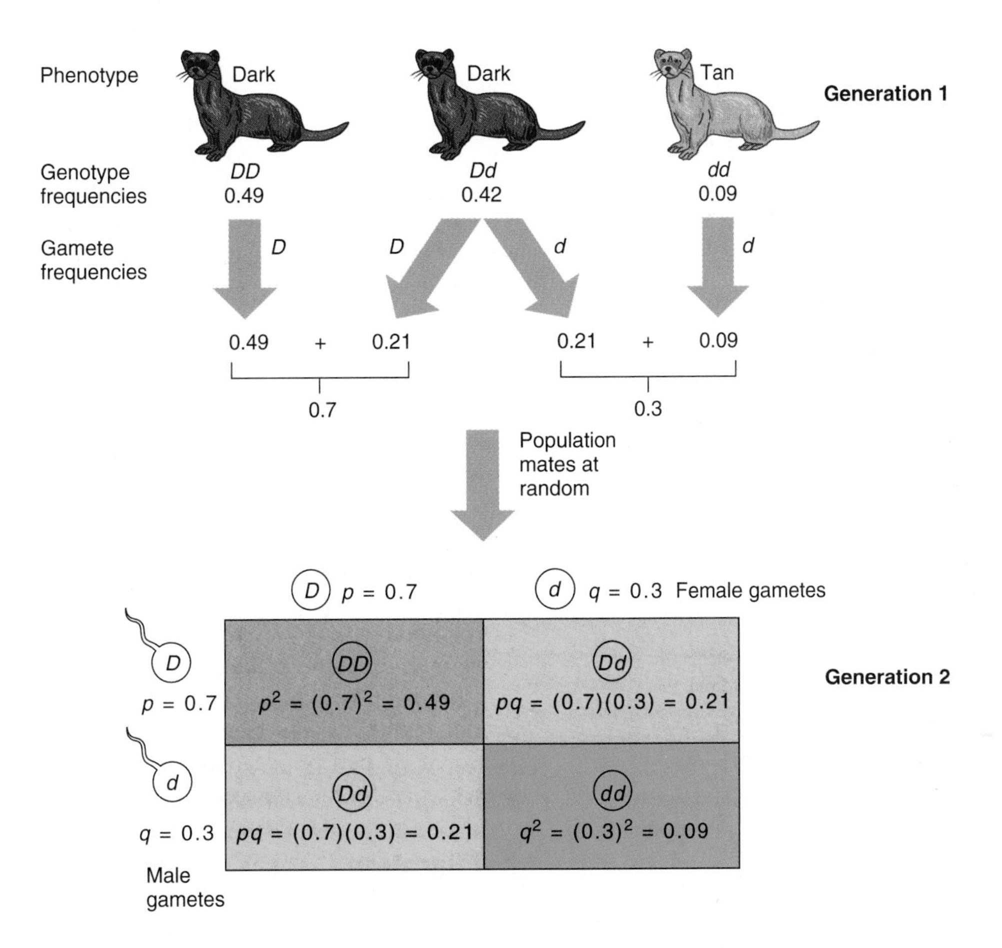

All possible crosses			F_1 genotype produced			Proportion of crosses
Male		Female	*DD* +	*Dd* +	*dd* =	
0.49 *DD*	×	0.49 *DD*	0.2401			0.2401
0.49 *DD*	×	0.42 *Dd*	0.1029	0.1029		0.2058
0.49 *DD*	×	0.09 *dd*		0.0441		0.0441
0.42 *Dd*	×	0.49 *DD*	0.1029	0.1029		0.2058
0.42 *Dd*	×	0.42 *Dd*	0.0441	0.0882	0.0441	0.1764
0.42 *Dd*	×	0.09 *dd*		0.0189	0.0189	0.0378
0.09 *dd*	×	0.49 *DD*		0.0441		0.0441
0.09 *dd*	×	0.42 *Dd*		0.0189	0.0189	0.0378
0.09 *dd*	×	0.09 *dd*			0.0081	0.0081
			0.49	0.42	0.09	1.0000
			DD +	*Dd* +	*dd* = 1	

Because mutation, migration, nonrandom mating, natural selection and genetic drift are common, the conditions necessary for Hardy–Weinberg equilibrium (unchanging gene frequencies) are rarely, if ever, met in natural populations. We give great consideration to selecting mates—it is hardly a random process. Migration disrupts allele frequencies as we move and mix, yet at the same time we create population pockets of ethnicity by living in the same areas and having children with people like ourselves. We counter natural selection by medically correcting phenotypes that would otherwise prevent some of us from having children. A cougar with poor vision cannot hunt and thus starves; a human with poor vision simply wears glasses, and life goes on.

Let's look now at how frequencies in real populations change. We will focus especially on human populations, for we know the most about ourselves.

15.1 MASTERING CONCEPTS

1. Why doesn't evolution occur if Hardy–Weinberg equilibrium exists?
2. What are the conditions for Hardy–Weinberg equilibrium?
3. Why doesn't Hardy–Weinberg equilibrium occur in real populations?

OLC

15.2 How Does Mate Choice Influence Evolution?

Nonrandom mating is one way for evolution to occur. Human behavior and migration make mating nonrandom. As a result, some alleles become more or less common in future generations.

Many factors influence mate choice, including geographical restrictions, access to the opposite sex, sexual selection, and behavior. Among humans, cultural factors are particularly important in choosing partners.

Nonrandom Mating

The random mating that is a requirement for Hardy–Weinberg equilibrium refers to genes considered one at a time—something we hardly think about when choosing a mate. Humans don't generally have sex at random; we choose partners based on appearance, ethnic background, intelligence, and many other factors. Economics, culture, and even political events affect mating patterns. Agriculture offers an extreme example of nonrandom mating, where certain individuals contribute disproportionately to the next generation. An animal or plant with valuable characteristics may be extensively bred. Semen from one prize bull, for example, is used to artificially inseminate thousands of cows. Breeds of popular domesticated animals also illustrate nonrandom mating (Investigating Life 15.1).

Occasionally, a man in a human population fathers an unusually large number of children and a mutant gene reveals his activity. In the Cape population of South Africa, for example, a Chinese immigrant known as Arnold had a very rare autosomal dominant disease that causes the teeth to fall out before age 20. Arnold had seven wives. Of his 356 living descendants, 70 have the dental disorder. The frequency of this allele in the Cape population is exceptionally high, compared to elsewhere, thanks to Arnold's considerable contribution to the gene pool.

The high frequency of people with albinism among Arizona's Hopi Indians also reflects nonrandom mating. This autosomal recessive condition, marked by lack of pigmentation in skin and hair, affects about 1 in 12,500 people in the general U.S. population, but about 1 in 200 Hopi Indians. The reason is cultural. Hopi men with albinism often stay back and help the women, rather than risk severe sunburn working in the fields with the other men. They contribute disproportionately to the next generation because they have greater contact with the women.

Historical events often influence mating patterns. When one group of people is subservient to another, genes tend to flow from ruling class to underclass because ruling class males have sex with underclass females. Historical records and studies of DNA sequences on the Y chromosome, which enable researchers to track male gene transmission, support this observation.

Despite our personal mating choices and the effects of cultural and historical trends, many traits do mix randomly in each human generation. This may be because we are unaware of these characteristics, and so cannot choose or reject them, or it may be because we do not consider them important in mate selection. The genes that encode such traits may exhibit Hardy–Weinberg equilibrium. We hardly choose life partners, for example, on the basis of blood type! However, sometimes the opposite occurs, and people with the same alleles have children together more often than they would by chance. This happens when families with a particular inherited illness meet each other through activities sponsored by patient organizations, and marriages occur between carriers for the same condition.

Migration

Large cities, with their pockets of ethnicity, defy Hardy–Weinberg equilibrium by their very existence. Waves of immigration built the population of New York City, for example. The original Dutch settlers of the 1600s lacked many of the gene variants present in today's metropolis; English, Irish, Slavic, African, Hispanic, Italian, Asian, and many other types of immigrants introduced them.

We can trace migration-driven microevolution by correlating allele frequencies in present-day populations to events in history. For example, the frequency of ABO blood types in certain parts of the world reflects Arab rule. The ABO distribution is very similar in northern Africa, the Near East, and southern Spain, precisely the regions the Arabs ruled until 1492. Even over half a millennium later, the proportions of the ABO alleles have not changed significantly.

Dogs and Cats—Products of Artificial Selection

The pampered poodle and graceful greyhound may win in the show ring, but they are poor specimens in terms of genetics and evolution. Human notions of attractiveness can lead to bizarre breeds that may never have evolved naturally. Behind carefully bred traits lurk small gene pools and extensive inbreeding—all of which may harm the health of highly prized and highly priced show animals. Purebred dogs suffer from more than 300 types of inherited disorders.

The sad eyes of the basset hound make this dog a favorite in advertisements, but these runny eyes can be quite painful (fig. 15.A, top right). Short legs make the dog prone to arthritis, the long abdomen encourages back injuries, and the characteristic floppy ears often hide ear infections. The eyeballs of the Pekingese protrude so much that a mild bump can pop them out of their sockets. The tiny jaws and massive teeth of pugs and bulldogs cause dental and breathing problems, as well as sinusitis, bad colds, and their notorious "dog breath" (fig. 15.A, top left). Folds of skin on their abdomens easily become infected. Larger breeds, such as the Saint Bernard, have bone problems and short life spans. A Newfoundland or a Great Dane may suddenly die at a young age, its heart overworked from years of supporting a large body (table 15.1).

We artificially select natural oddities in cats too. One of every 10 New England cats has six or seven toes on each paw, thanks to a multitoed ancestor in colonial Boston (fig. 15.A, bottom left). Elsewhere, these cats are quite rare. The sizes of the blotched tabby populations in New England, Canada, Australia, and New Zealand correlate with the time that has passed since cat-loving Britons colonized each region. The Vikings brought the orange tabby to the islands off the coast of Scotland, rural Iceland, and the Isle of Man, where these feline favorites flourish today.

A more modern breed appealing to cat fanciers is the American Curl cat, whose origin is traced to a stray female who wandered into the home of a cat-loving family in Lakewood, California, in 1981. This cat passed her unusual, curled-up ears to kittens in several litters (fig. 15.A, bottom right). A dominant gene that makes extra cartilage grow along the outer ear causes the trait. Cat breeders attempting to fashion this natural peculiarity into an official show animal are hoping that the gene does not have other, less lovable effects. Cats with floppy ears, for example, are known to have large feet, stubbed tails, and lazy natures.

All these examples make one genetic truth clear: You may be able to breed desired characteristics into a dog or cat, but you can't always breed other traits out.

FIGURE 15.A

Pets Exhibit a Variety of Unusual Traits Due to the Genetic Variations Selected Through Breeding.

The bulldog (top left) was selected for its flattened face and fierce demeanor. Some traits are simply attractive or unusual, such as curled ears (bottom right). Multi-toed cats (bottom left) make better mousers and the mournful expression of the basset hound (top right) accompanies its heightened sense of smell. All of these traits originally occurred naturally due to genetic variation.

Bulldog

Basset hound

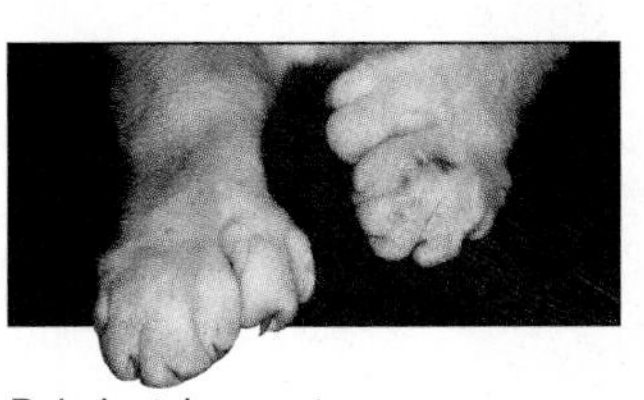
Polydactylous cat

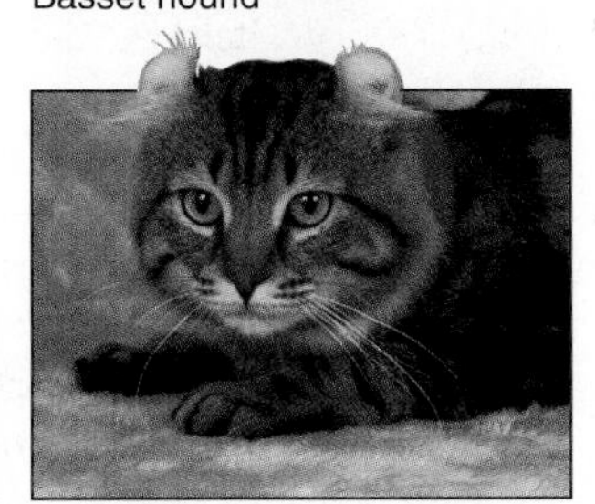
American Curl cat

TABLE 15.1 Purebred Plights

BREED	HEALTH PROBLEMS
Cocker spaniel	Nervousness
	Ear infections
	Hernias
	Kidney problems
Collie	Blindness
	Bald spots
	Seizures
German shepherd	Hip dysplasia
Golden retriever	Lymphatic cancer
	Muscular dystrophy
	Skin allergies
	Hip dysplasia
	Absence of one testicle
Great Dane	Heart failure
	Bone cancer
Labrador retriever	Dwarfism
	Blindness
Shar-pei	Skin disorders

Directional gene flow also occurred when nomadic peoples with a hunter-gatherer lifestyle encountered a more stable group of people. For example, in the eighteenth century, nomadic European Caucasians called trekboers migrated to the Cape area of South Africa. The men had children with the native women of the Nama tribe and settled in the area. The resulting mixed society remained fairly isolated and led to the present-day "Richtersveld coloureds" of the Cape region.

Because geographical barriers greatly influence migration patterns, allele frequencies sometimes differ between relatively close geographical regions, such as on either side of a mountain range. In Europe, geographical barriers are important factors in creating **clines,** which are allele frequencies that progressively change across a region. For example, the allele frequencies for many genetic diseases vary among different European populations as a result of geographical barriers.

Genetic Drift

When a small group of individuals is separated from a larger population, or mates only among themselves within a larger population, allele frequencies may change as a result of chance sampling from the whole. This change in allele frequency that occurs when a small group is considered separately from the larger whole is termed **genetic drift.** It can be compared to reaching into a bag of jellybeans and, by chance, grabbing only green and yellow ones. The allele frequency changes that occur with genetic drift are random and therefore unpredictable, just as reaching into a jellybean bag a second time might yield mostly black and orange candies.

A common cause of genetic drift in human populations is the **founder effect,** which occurs when small groups of people leave their homes to found new settlements. The new colony may have different allele frequencies than the original population, amplifying some traits while diminishing or even losing others (fig. 15.2).

The Amish people are an extreme example of this type of genetic drift. They have a higher incidence of certain traits than do other populations because people with these traits or who carry the genes for them marry within the group.

Genetic drift is also striking in the Dunker community of Germantown, Pennsylvania. The Dunkers left Germany between 1719 and 1729 to settle in the New World. Today, the frequencies of some genotypes are different among the Dunkers than among either their non-Dunker neighbors and/or the people living in their ancestral German village. For example, the frequency of type A blood in the United States is 40% and in Germany 45%. Yet among the Dunkers, the frequency of type A blood is 60%. The original settlers included a disproportionate number of individuals with type A or AB blood.

Genetic drift also may result from a **population bottleneck.** This occurs when many members of a population die, and a few remaining individuals mate, eventually restoring their numbers. The new population has lost much of the genetic diversity that was present in the larger ancestral population.

Population bottlenecks sometimes occur when people (or other organisms) colonize islands and then face disaster. An extreme example of this type of genetic drift occurred among the Pingelapese people of the Eastern Caroline islands in Micronesia. Five percent of the current population of 3,000 today have "Pingelapese blindness," an autosomal recessive combination of color blindness, nearsightedness, and cataracts (clouding of the lens). Elsewhere only 1 in 50,000 to 100,000 people are affected. The prevalence of this condition among the Pingelapese traces back to a typhoon that decimated the population in 1775. Only 20 people survived, and four generations later, due to unavoidable inbreeding, the blindness began to show up.

The world's cheetahs are currently undergoing a population bottleneck (fig. 15.3). Until 10,000 years ago, these cats were prevalent in many areas. Today, just two isolated populations live in South and East Africa, numbering only a few thousand animals. The amino acid sequences of several proteins among the cheetahs are uniform, indicating descent from a common ancestor and therefore a recent population bottleneck. The South African cheetahs are so genetically alike that even unrelated animals can accept skin grafts from each

FIGURE 15.2

The Founder Effect.

(A) When some members of a population leave to found a new group, allele frequencies can shift. **(B)** This Amish child from Lancaster County, Pennsylvania, has inherited Ellis-van Creveld syndrome. He has short-limbed dwarfism, extra fingers, heart disease, and fused wrist bones, and he had teeth at birth. Ellis-van Creveld is autosomal recessive and occurs in 7% of the people of this Amish community—a high figure, because they marry among themselves.

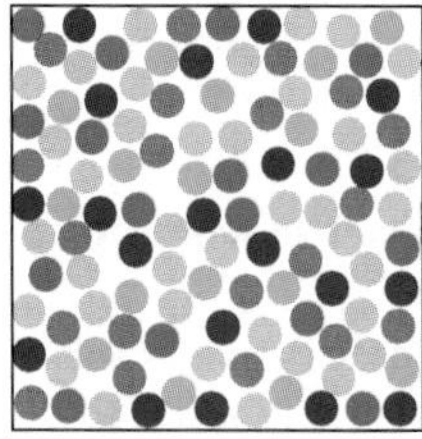

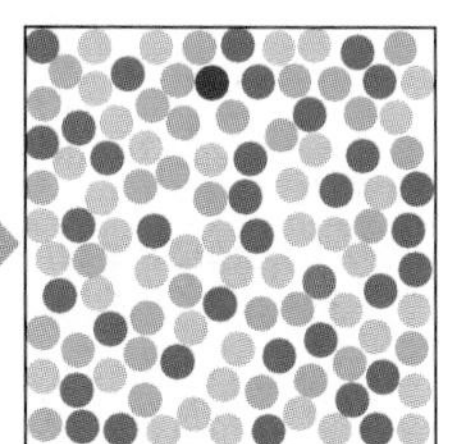

A

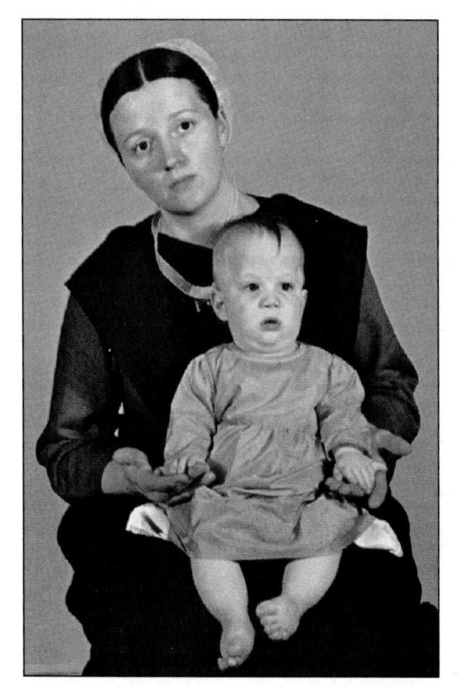

B

FIGURE 15.3

The Bottleneck Effect.

A population bottleneck occurs when the size of a genetically diverse population drastically falls and a few individuals mate, restoring it. The rebuilt population loses some genetic diversity because alleles are lost in the bottleneck event. The two dwindling cheetah populations in South and East Africa vividly illustrate bottlenecks. Cheetahs are difficult to breed in zoos because sperm quality is poor and many newborns die—both due to lack of genetic diversity.

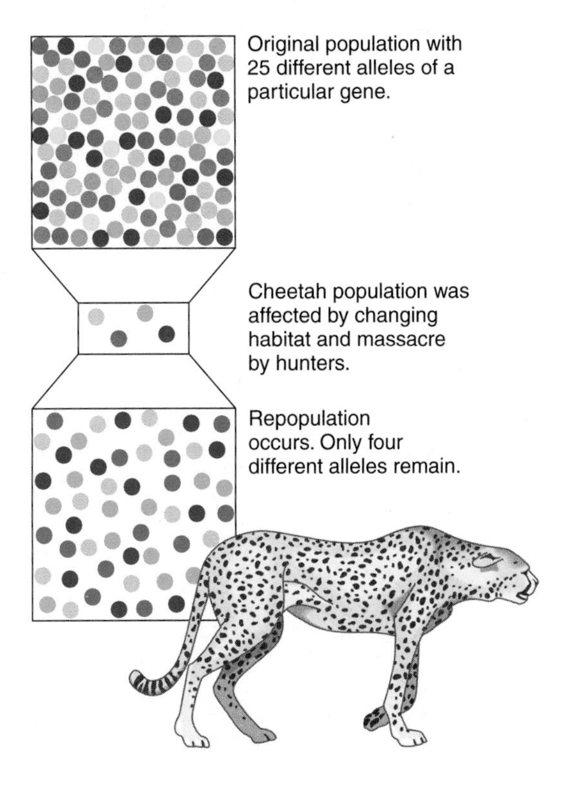

other. Researchers attribute the genetic uniformity of cheetahs to two bottlenecks—one that occurred at the end of the most recent ice age, when habitats changed drastically, and another when humans slaughtered many cheetahs during the nineteenth century.

Evidence of a population bottleneck is apparent among the Finns, who descend from settlers who came to Finland 4,000 years ago. Isolated by both geography and the harsh climate, alleles peculiar to the settlers stayed in the population. Today, the descendants not only have a distinctive appearance and language, but a unique spectrum of inherited disease. Cystic fibrosis, common in many populations, is practically nonexistent among the Finns, yet they have high incidence of 30 or so genetic disorders that are extremely rare elsewhere. Analysis of mitochondrial DNA, which reflects maternal lineage, and from the Y chromosome, which reflects paternal heritage, provides molecular evidence of a population bottleneck. That is, the Finns show much less genetic variation within their group than do other Europeans.

15.2 MASTERING CONCEPTS

1. How do nonrandom mating, migration, and genetic drift disrupt Hardy–Weinberg equilibrium?
2. What is a founder effect?
3. What is a population bottleneck?

OLC

15.3 How Does Mutation Fuel Evolution?

Deleterious alleles in a population constitute the genetic load, and arise from mutation or are perpetuated in heterozygotes. Mutations influence evolution by introducing new genetic variants.

Recall from chapter 13 that a mutation is a change in the DNA. Mutations are the raw material for evolution, because natural selection acts on phenotypes. Sexual reproduction ensures that a population includes individuals with different phenotypes. Should an environmental condition harm individuals of one particular phenotype, others may survive. Mutations help to provide this variability. The genetic makeup of populations, and ultimately species, changes as natural selection permits differential survival of variants that are adapted to a particular environment.

Most mutations are neither beneficial nor useful, with no effect on phenotype; some are harmful (deleterious), resulting in defects in protein production that can lead to disease. Most of these harmful genetic traits are recessive and, therefore, the alleles persist in a population through heterozygotes. In evolutionary terms, these alleles constitute a **genetic load** that may be the target of some future natural selection on a given population. The potential protection that genetic diversity offers is why inbreeding is so detrimental. As related individuals mate, heterozygosity is diminished.

15.3 MASTERING CONCEPTS

1. Are all mutations deleterious?
2. What is the role of mutation in evolution?
3. What is the genetic load?
4. How are deleterious recessive alleles maintained in populations?

OLC

15.4 How Does Natural Selection Mold Evolution?

Natural selection may favor one phenotype, two extreme phenotypes, or an intermediate phenotype. Balanced polymorphism maintains a deleterious allele when a heterozygote is unusually resistant to a specific, usually infectious, illness.

Allele frequencies may change in response to environmental change. This is natural selection in action. Different types of natural selection are distinguished by their effects on phenotypes (fig. 15.4).

Types of Natural Selection

In **directional selection,** a changing environment selects against one genotype, allowing another to gradually become more prevalent. For example, populations of approximately 100 insect species have undergone color changes enabling them to blend into polluted backgrounds. This adaptive response of

FIGURE 15.4

Types of Natural Selection.

Directional selection (**A**) results from selection against one extreme phenotype. In disruptive selection (**B**) two extreme phenotypes each have a selective advantage, and both persist. Stabilizing selection (**C**) maintains an intermediate expression of a trait by selection against extreme variants.

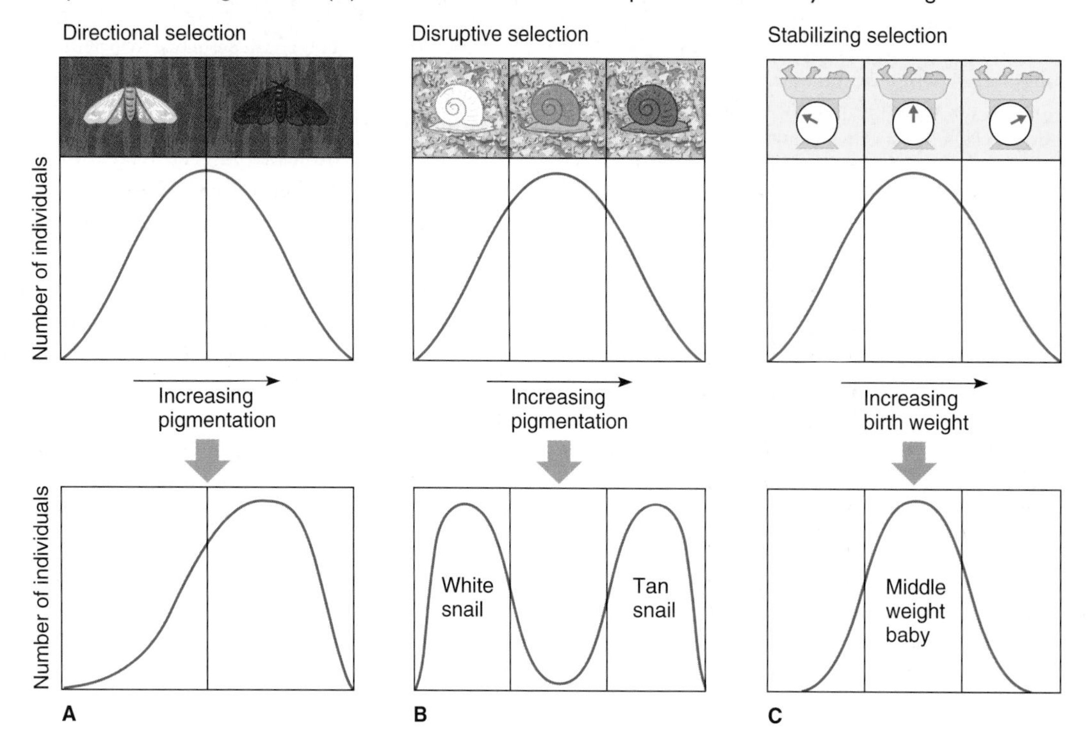

darkening is called **industrial melanism.** The rise of antibiotic resistance among infection-causing bacteria also reflects directional selection, as does the increase in pesticide-resistant plants. Biotechnology 15.1 discusses an intentional form of directional selection conducted in the laboratory.

In **disruptive selection** (sometimes called diversifying selection), two extreme expressions of a trait are the most fit, and come to predominate. For example, in a population of marine snails that live among tan rocks encrusted with white barnacles, the animals are either white and camouflaged while near the barnacles, or tan and hidden while on the bare rock. The snails that are not white or tan, or that lie against the opposite-colored background, are more often seen and eaten by predatory shorebirds.

In a third form of natural selection, called **stabilizing selection,** extreme phenotypes are less adaptive, and an intermediate phenotype has greater survival and reproductive success. Human birth weight illustrates this tendency to stabilize. Newborns who are under 5 pounds (2.27 kilograms) or over 10 pounds (4.54 kilograms) are less likely to survive than babies weighing between 5 and 10 pounds.

Balanced Polymorphism

Stabilizing selection can result in **balanced polymorphism,** which maintains a genetic disease in a population even though the illness clearly diminishes the fitness of affected individuals. (Polymorphism means "multiple forms," and refers specifically to genetic variants.) The inherited disease persists because carriers (heterozygotes) have some health advantage over individuals who have two copies of the wild type allele. Balanced polymorphism explains some fascinating links between certain inherited and infectious diseases. Figure 15.5 depicts the example discussed below, and table 15.2 lists some others.

Sickle Cell Disease and Malaria Sickle cell disease is an autosomal recessive disorder that causes anemia, joint pain, a swollen spleen, and frequent, severe infections. Homozygous individuals usually do not feel well enough, or live long enough, to reproduce. Sickle cell disease carriers, who do not have symptoms, are resistant to malaria, which is an infection by any of four species of the protistan genus *Plasmodium.* Mosquitoes transmit the parasite to humans, causing the agonizing cycle of severe chills and fever of malaria (see chapter 21).

Discovering the sickle cell–malaria link took clever medical sleuthing. In 1949, British geneticist Anthony Allison found that the frequency of sickle cell carriers in tropical Africa was unusually high in regions where malaria raged. Blood tests of children hospitalized with malaria revealed that nearly all of them were homozygous for the wild type sickle cell allele, giving them normal blood cells. The few sickle cell carriers among them had the mildest cases of malaria. Was malaria somehow selecting against the wild type sickle cell allele? In the United States, where malaria is very rare, sickle cell disease is also less common, suggesting a relationship between the two disorders.

Biotechnology 15.1

In Vitro Evolution—Not Quite Natural Selection

Most people associate evolution with the highly visible changes of macroevolution, such as the formation or extinction of species. Evolution works on phenotypes, and therefore ultimately affects the informational molecules of life—the proteins and nucleic acids that underlie those phenotypes. Researchers can witness evolution in action by monitoring sequence changes among very large numbers of these molecules, in a process called in vitro selection or evolution. Three groups of researchers invented the technique in 1989.

Overall, in vitro evolution amplifies those molecules in a group that have certain desired characteristics. Over several rounds of selecting these molecules, they come to comprise a greater proportion of the population. By subtly altering the selection criteria and/or changing the selected population slightly at each iteration, eventually the process yields molecules with specified properties.

In vitro evolution experiments usually focus on RNA, for three reasons:

- RNA is informational; it has a sequence.
- RNA is catalytic (it functions as an enzyme in some reactions).
- RNA is single-stranded and can fold into many intricate three-dimensional conformations.

These characteristics make RNA the prime candidate for the molecule that began life, which is discussed in chapter 18.

An in vitro evolution experiment usually begins with a pool of randomly varied RNA sequences (fig. 15.B). A typical effort might use 10 quadrillion molecules of 27 or fewer RNA nucleotides, generated synthetically. (Computer simulation can also be used.) Next, certain molecules are selected by passing the pool over a column to which particular other types of molecules are bound, or by testing to see which ones can catalyze a specific chemical reaction. All RNAs that do not bind or catalyze are discarded, and the now-enriched population is amplified using the polymerase chain reaction (PCR, see Biotechnology 12.3). Changes can be introduced into the PCR primers, which results in minor variations of the amplified sequences. This enriched pool is then passed through the binding/catalyzing selection again, and again, up to 10 times. The ultimate result is the desired molecule.

In vitro evolution is like biological evolution by means of natural selection in that a subgroup is selected from a large group of variants. But it differs from natural selection in important ways. First, in in vitro selection, someone sets the criteria. In nature, it is environmental change that is detrimental to some phenotypes but not others. Secondly, natural selection acts on phenotypes, not directly on nucleic acids. That is, a sequence of DNA that is part of a chromosome but does not affect the phenotype would not be subject to natural selection.

In addition to modeling biological evolution to a limited extent, two powerful applications of in vitro evolution target basic research and practical use. For studies on the origin of life, the technique is revealing enzymatic activities of RNA that have not been observed in organisms. We are learning what this eclectic nucleic acid can do, and might once have done to jump-start life, before proteins took over the catalytic capabilities. In the practical area of drug discovery, in vitro selection allows researchers to guide the functions of nucleic acids and proteins. The technique is so sensitive that, for example, one experiment yielded a molecule that could distinguish caffeine from theophylline, two chemicals that differ from each other by just a methyl (CH_3) group. By fashioning molecules that bind tenaciously and preferentially to a particular cell surface receptor, the products of these lab experiments might become powerful new drugs.

FIGURE 15.B
In vitro evolution selects molecules that have desired characteristics.

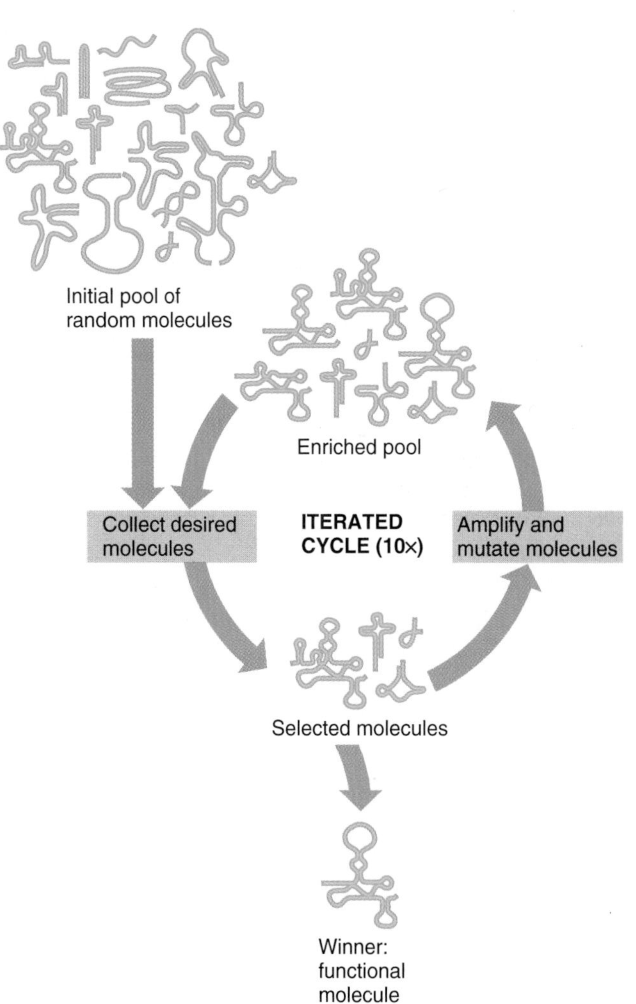

FIGURE 15.5

Balanced Polymorphism.

Being a carrier for a specific inherited illness can protect against another type of condition. Maps that compare frequency distributions of both disorders are evidence of this phenomenon. (**A**) The classic example of balanced polymorphism is sickle cell disease heterozygosity protecting against malaria. (**B**) Evidence is starting to accumulate for a similar relationship between people who have one or two alleles conferring resistance to HIV infection and protection against bubonic plague, an infectious disease that ravaged Europe in the Middle Ages.

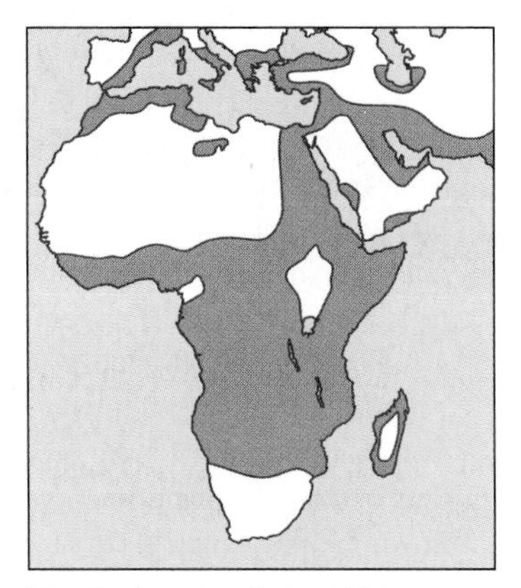

Distribution of malaria, 1920s

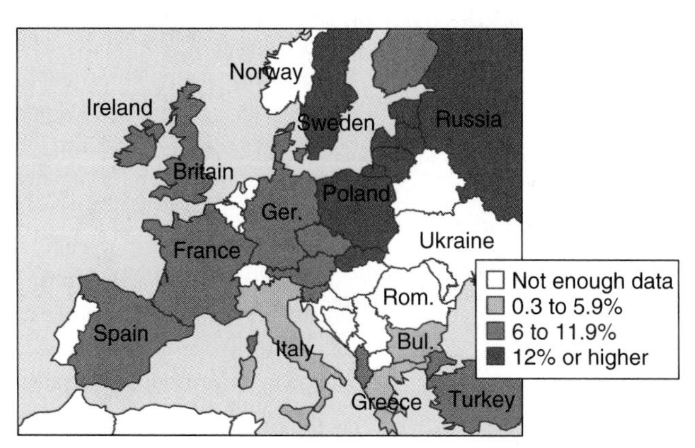

Distribution of AIDS-resistant gene

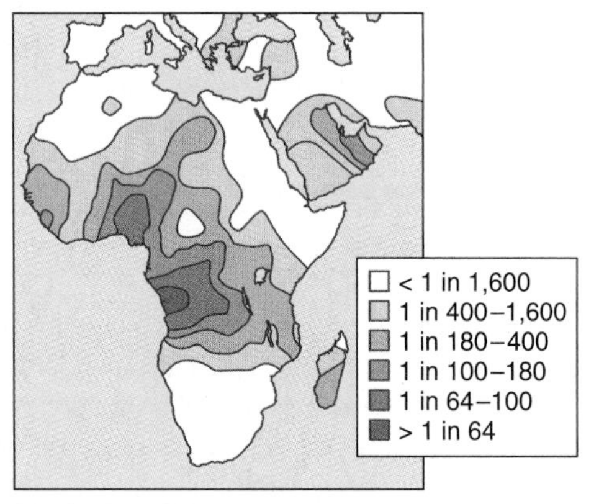

Distribution of sickle cell disease carriers

A

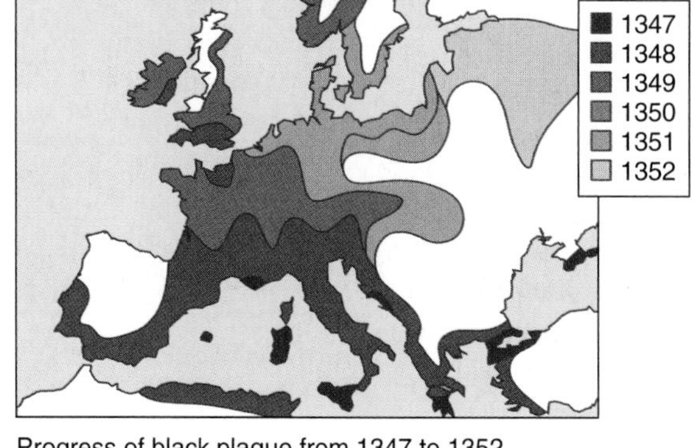

Progress of black plague from 1347 to 1352

B

TABLE 15.2 Balanced Polymorphism

PERSON WHO HAS OR CARRIES	IS PROTECTED FROM	POSSIBLY BECAUSE
Cystic fibrosis	Diarrheal disease	Carriers have too few functional chloride channels in intestinal cells, blocking toxin
G6PD deficiency	Malaria	Red blood cells inhospitable to malaria parasite
Phenylketonuria (PKU)	Spontaneous abortion	Excess amino acid (phenylalanine) in carriers inactivates ochratoxin A, a fungal toxin that causes miscarriage
Sickle cell disease	Malaria	Red blood cells inhospitable to malaria parasite
Tay-Sachs disease	Tuberculosis	Unknown
Non-insulin-dependent diabetes mellitus	Starvation	Associated tendency to gain weight once protected against starvation during famine.

Further historical evidence supports the hypothesis that being a sickle cell carrier in a malaria-ridden environment confers a selective advantage. The rise of sickle cell disease parallels cultivation of crops that provide breeding grounds for the malaria-carrying *Anopheles gambiae* mosquito. About 1000 B.C., sailors from southeast Asia traveled in canoes to East Africa, bringing new crops of bananas, yams, taros, and coconuts. When the jungle was cleared to grow these crops, mosquitoes came, offering a habitat for the malaria parasite in the early part of its life cycle.

When an infected mosquito feeds on a human who does not have or carry sickle cell disease, the malaria parasite enters the red blood cells, eventually bursting them and releasing the parasite throughout the body. In all the red blood cells of a person with sickle cell disease, and in about half of the cells of a carrier, the abnormal beta globin chains adhere to one another to form aggregates that bend the cell into the characteristic sickle shape, which also thickens the blood (see fig. 13.7). For a reason still unknown, the sickled cells are inhospitable to the parasite.

When malaria first invaded East Africa, sickle cell carriers, who remained healthier, had more children and passed the protective allele to approximately half of them. Gradually, over 35 generations, the frequency of the sickle cell allele in East Africa rose from 0.1% to a spectacular 45%. However, whenever two carriers produced a child who suffered from sickle cell disease—a homozygote—the child paid the price for this genetic protection.

A cycle set in. Settlements with large numbers of sickle cell carriers escaped debilitating malaria. Their residents were therefore strong enough to clear even more land to grow food—and support the disease-bearing mosquitoes. Even today, sickle cell disease is more prevalent in agricultural societies than among people who hunt and gather their food.

Cystic Fibrosis and Diarrheal Infections The cellular defect that underlies cystic fibrosis (CF) protects against certain diarrheal diseases caused by bacterial infection, such as cholera and typhoid fever.

Some strains of *Vibrio cholerae* bacteria produce a toxin that causes severe diarrhea. The toxin opens chloride channels in cells lining the small intestine. As salt (NaCl) leaves the cells, water follows to dilute the salt. Water rushing out of intestinal cells leaves the body as diarrhea, because the large intestine cannot reabsorb the fluid fast enough. A victim can lose his or her body weight in water in a week. The relentless dehydration leads to shock, kidney failure, and heart failure.

The 1989 identification of the CF gene and its protein product CFTR, a chloride channel regulator in certain secretory cells, suggested why the genetic disease is so prevalent. The cholera toxin opens chloride channels, allowing chloride and water to leave cells. The abnormal CF protein does just the opposite; it closes chloride channels, trapping salt and water in cells and drying out mucus and other secretions. A person with CF is very unlikely to contract cholera, because the toxin cannot open the chloride channels in the small intestine cells.

Carriers of CF therefore enjoy the mixed blessing of balanced polymorphism. They do not have enough abnormal chloride channels to suffer the labored breathing and clogged pancreas of CF, but they do have a sufficient defect to prevent cholera toxin from opening chloride channels.

Typhoid fever is a different diarrheal disease that CF protects against. The infecting bacterium itself, *Salmonella typhi,* rather than a toxin, enters cells lining the small intestine—but only if functional CFTR channels are present. Cells of people with two copies of the most common CF-causing allele, whose CFTR proteins never reach the cell surface, admit none of the bacteria; cells of carriers admit some.

During the devastating cholera and typhoid fever epidemics that have peppered history, individuals who carried mutant CF alleles had a selective advantage, and they disproportionately transmitted those alleles to future generations. CF endures.

Given the large number of genes in any organism, the changes in allele frequencies that constitute microevolution must occur nearly all the time. Figure 15.6 summarizes these changes. The next chapter considers the easier-to-see macroevolutionary changes of speciation and extinction.

15.4 MASTERING CONCEPTS

1. Distinguish among directional, disruptive, and stabilizing selection.
2. How can being a carrier for an inherited disease be beneficial?

OLC

FIGURE 15.6

Factors That Alter Allele Frequencies and Thereby Contribute to Evolution.

(**A**) In Hardy-Weinberg equilibrium, allele frequencies stay constant. (**B**) Nonrandom mating increases some allele frequencies and decreases others because individuals with certain phenotypes are more attractive to the opposite sex. (**C**) Migration removes alleles from or adds alleles to populations. (**D**) Genetic drift samples a portion of a population, altering allele frequencies. (**E**) Mutation creates new alleles. (**F**) Natural selection operates when environmental conditions prevent individuals of certain genotypes from reproducing successfully.

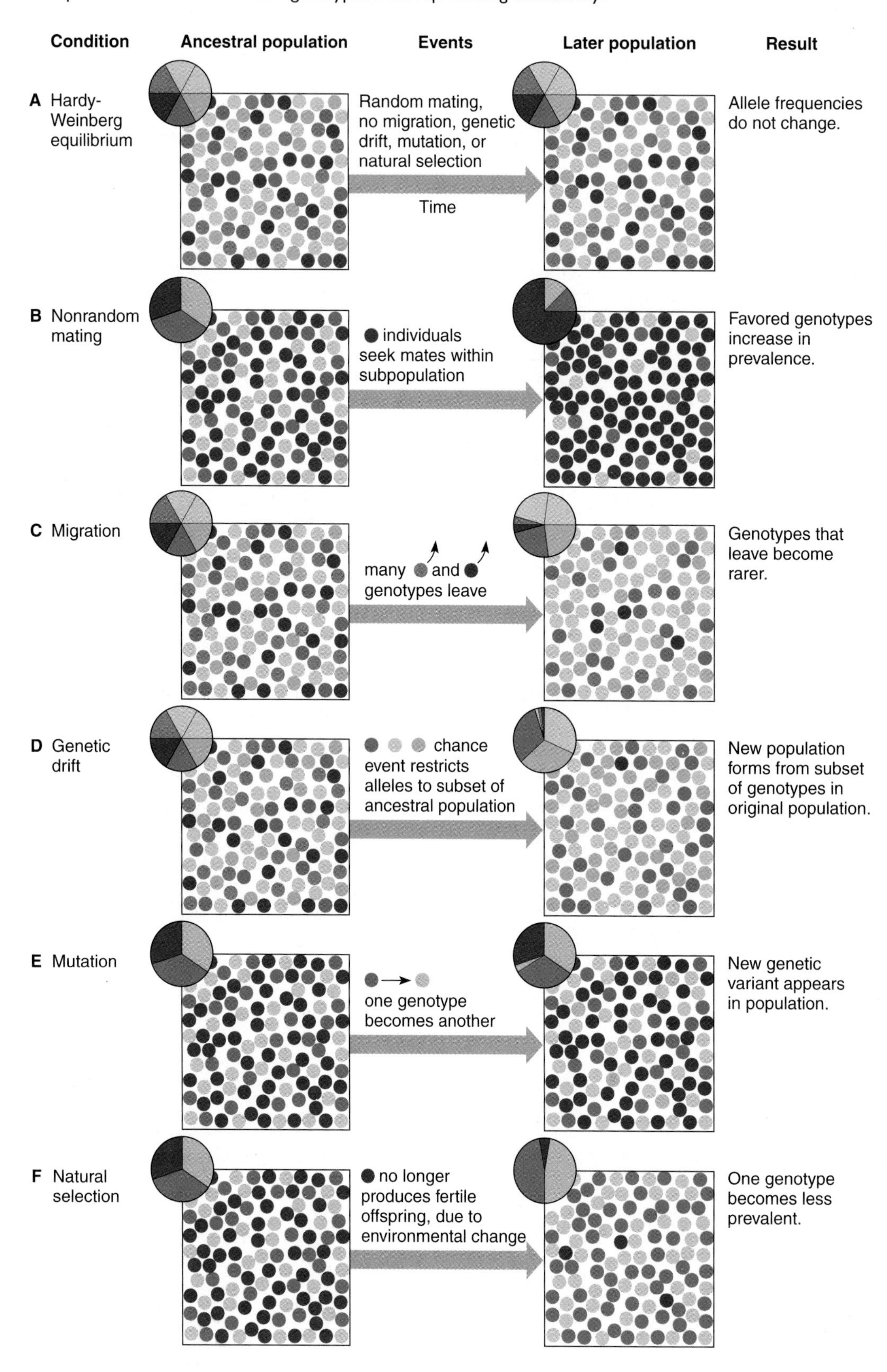

Chapter Summary

15.1 Why Is Evolution Much More Likely Than Not?

1. A **gene pool** includes all the genes in a population. Allele movement between populations is **gene flow.** Inherited characteristics of the individuals in a population reflect allele frequencies.
2. In **Hardy–Weinberg equilibrium,** evolution is not occurring because allele frequencies do not change from generation to generation. In this idealized state, we can calculate the proportion of genotypes and phenotypes in a population by inserting known allele frequencies into an algebraic equation: $p^2 + 2pq + q^2$. The equation also can reveal allele frequency changes when we know the proportion of genotypes in a population.

15.2 How Does Mate Choice Influence Evolution?

3. When allele frequencies change, evolution occurs.
4. Nonrandom mating causes certain alleles to predominate because a particular phenotype is more attractive to the opposite sex.
5. In **genetic drift,** small populations separate from larger ancestral populations and establish a new gene pool, with different allele frequencies. The **founder effect** and **population bottlenecks** are forms of genetic drift.

15.3 How Does Mutation Fuel Evolution?

6. The deleterious alleles in a population constitute its **genetic load.**
7. Mutation alters allele frequencies by changing one allele into another and providing new phenotypes for evolution to act on.
8. Harmful recessive alleles are selected against in homozygotes, but heterozygotes maintain them and mutation introduces them into the gene pool.

15.4 How Does Natural Selection Mold Evolution?

9. In **directional selection,** an extreme phenotype becomes more prevalent in a population. **Industrial melanism** is an example.
10. In **disruptive selection,** extreme expressions survive at the expense of intermediate forms.
11. In **stabilizing selection,** an intermediate phenotype has an advantage.
12. **Balanced polymorphism** is a form of stabilizing selection that maintains deleterious recessive alleles because heterozygotes are protected against another medical condition.

Testing Your Knowledge

1. Define the following:
   a. population
   b. gene pool
   c. gene flow
   d. genetic load
   e. genetic drift
2. Explain how the algebraic expression $p^2 + 2pq + q^2$ is used to represent the proportions of different genotypes in a population.
3. The fraggles are a population of mythical, mouselike creatures that live in underground tunnels and chambers beneath a large vegetable garden that supplies their food. Of the 100 fraggles in this population, 84 have green fur and 16 have gray fur. A dominant allele *F* confers green fur and a recessive allele *f* confers gray fur. Assuming Hardy–Weinberg equilibrium is operating, answer the following questions:
   a. What is the frequency of the gray allele *f*?
   b. What is the frequency of the green allele *F*?
   c. How many fraggles are heterozygotes (*Ff*)?
   d. How many fraggles are homozygous recessive (*ff*)?
   e. How many fraggles are homozygous dominant (*FF*)?
4. One spring, a dust storm blankets the usually green garden of the fraggles in gray. Under these conditions, the green fraggles become very visible to the Gorgs, who tend the gardens and try to kill the fraggles to protect their crops. The gray fraggles, however, blend into the dusty background and find that they can easily steal radishes from the garden. How might this event affect microevolution in this population of fraggles?
5. What are two reasons that deleterious recessive alleles persist in populations, even though they prevent individuals from reproducing when present in two copies?
6. Give examples of directional, disruptive, and stabilizing selection.

Thinking Scientifically

1. How do the following situations or practices disrupt Hardy–Weinberg equilibrium?
   a. Couples who find out that they are heterozygotes for the same illness decide to not have children together.
   b. Several dozen young adults in a large midwestern city discover they are half-siblings. Each was conceived by artificial insemination, with sperm from the same donor.
   c. Members of a very close-knit Amish community are forbidden to marry outside the community.
   d. A new viral illness kills only people who have a certain blood type.
2. A thin-shelled crab can more readily move to escape a predator than can a thick-shelled crab, but it is more vulnerable to predators that drill through the shell. As a result of these opposing forces, shell thickness for many types of crabs has remained within a narrow range, over a long time. What type of natural selection does crab shell thickness illustrate?
3. Which factors contributing to evolution discussed in this chapter do the following science fiction film plots illustrate?
   a. In *When Worlds Collide,* Earth is about to be destroyed. One hundred people, chosen for their intelligence and fertility, leave to colonize a new planet.
   b. In *The Time Machine,* set in the distant future on Earth, one group of people is forced to live on the planet's surface and another group is forced to live in caves. After many years, they look and behave differently. The Morlocks, who live belowground, have dark skin, dark hair, and are very aggressive, whereas the Eloi, who live aboveground, are blond, fair-skinned, and meek.
   c. In *Children of the Damned,* genetically identical beings from another planet impregnate all the women in a small town.
   d. In *The War of the Worlds,* Martians cannot survive on Earth because they are vulnerable to infection by terrestrial microbes.
4. How is in vitro evolution similar to the Foundry Cove experiments discussed in Investigating Life 14.1?

References and Resources

Kolata, Gina. May 26, 1998. Scientists see a mysterious similarity in a pair of deadly plagues. *The New York Times.* The link between HIV resistance and plague could be due to balanced polymorphism.

Landweber, Laura F., Peter J. Simon, and Thor A. Wagner. February 1998. Ribozyme engineering and early evolution. *BioScience,* vol. 48. Selection is easy to observe at the molecular level.

Majerus, Michael E. N. 1998. *Melanism: Evolution in Action.* Oxford University Press. Famous experiments that demonstrated industrial melanism in England may have been seriously flawed, but directional selection does occur.

McKusick, Victor A. March 2000. Ellis-van Creveld syndrome and the Amish. *Nature Genetics,* vol. 24, p. 203. McKusick did some of the first genetic studies on the Amish. The work continues today.

Pier, Gerald B., et al. May 7, 1998. *Salmonella typhi* uses CFTR to enter intestinal epithelial cells. *Nature,* vol. 393. The defective cell surface chloride channel that underlies cystic fibrosis may protect against certain diarrheal diseases.

Sachs, Oliver. 1998. *The Island of the Colorblind.* New York: Random House Vintage Books. The story of the Pingelapese blindness.

The *LIFE* Online Learning Center provides additional resources and tools for studying this chapter.
www.mhhe.com/life

CHAPTER 16

Speciation and Extinction

Islands Provide Windows on Evolution

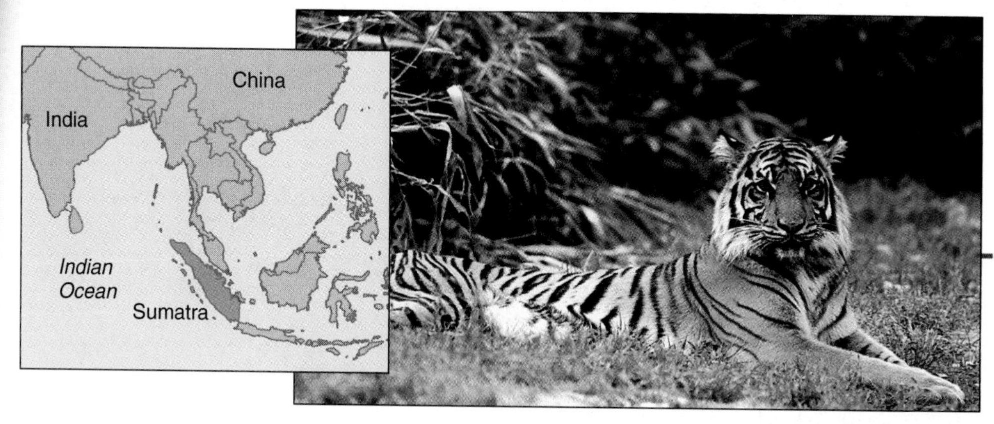

Sumatran tiger

Groups of islands isolated from a mainland provide peeks at evolution. Islands that broke apart from large landmasses sequestered parts of gene pools by genetic drift. Islands that form from undersea volcanoes become populated from the air or from organisms arriving aboard natural rafts. The particular environments of an island present specific selective pressures. In this way, microevolution is gradually shaping the tiger population on the island of Sumatra, which is part of Indonesia. Sumatra became separated from other islands when the sea rose 6,000 to 12,000 years ago.

Tigers belong to one species—*Panthera tigris*—but their scattered populations are genetically so divergent that they might be on their way to becoming distinct species, and so they are said to exist as five subspecies. The Sumatran tiger is the smallest, the male weighing 220 to 310 pounds (99.9 to 140.7 kilograms) and growing to 7 to 8 feet (2.1 to 2.4 meters) in length. The largest subspecies is the Siberian tiger, with males weighing 470 to 675 pounds (213 to 307 kilograms) and reaching 9 to 11 feet (2.7 to 3.4 meters) in length. These distinctions in body size may reflect adaptation to the other species in the very different environments of a tropical island and Siberia. The other subspecies come from Bengal, South China, and Indochina. DNA sequence comparisons indicate that the small Sumatran tiger is genetically distinct from the others, and therefore is on its way to becoming a separate species.

DNA sequence comparisons indicate that the small Sumatran tiger is genetically distinct from the others, and therefore is on its way to becoming a separate species.

Sumatran tigers are significantly smaller than their counterparts in Siberia, Bengal, South China, and Indochina. Their isolation has permitted natural selection to move their gene pools away from those of the other tiger subspecies.

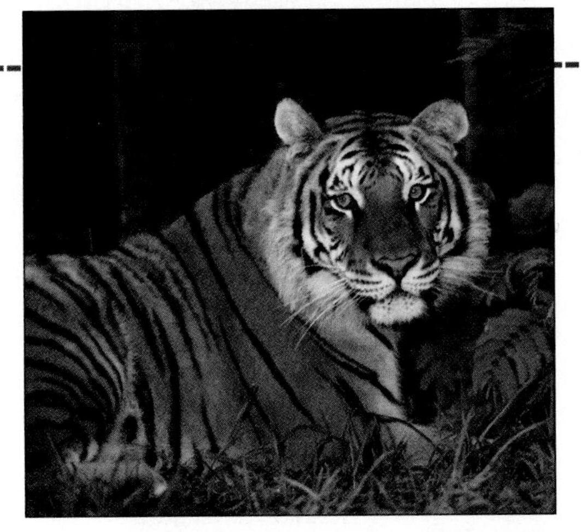

Siberian tiger

Africa
Mauritius
Madagascar
Indian Ocean

Humans caused extinction of the dodo. Most of what we know about these birds comes from remains of 400 individuals and this 1848 description from fossils: "These birds were of large size and grotesque proportions, the wings too short and feeble for flight, the plumage loose and decomposed, and the general aspect suggestive of gigantic immaturity. . . ." So rapid and complete was their extinction that the vague descriptions given of them by early navigators were long regarded as fabulous or exaggerated, and these birds, almost contemporary of our great-grandfathers, became associated in the minds of many persons with the Griffin and the Phoenix of mythological antiquity.

Other islands have interesting evolutionary tales to tell. If we know when an island formed, which species were the first to colonize it, which species live there today, and when humans arrived and brought organisms with them, we can try to reconstruct events. Consider the island of Mauritius, which rose in volcanic fury from the depths of the Indian Ocean about 8 million years ago. By the sixteenth century, the island teemed with dense, tall forests, colorful birds, scurrying insects, and basking reptiles, all well adapted to the island and to one another.

One inhabitant of Mauritius was the flightless, 3 foot (1 meter) tall, 30 pound (14 kilogram) dodo bird, described by one scientist as "a magnificently overweight pigeon". The dodo ate the hard fruits of the abundant *Calvaria* tree.

Life flourished on Mauritius until the 1500s, when European sailors arrived. The men ate dodo meat, and their pet monkeys and pigs ate dodo eggs. Nearly all the *Calvaria* trees died. Rats and mice swam ashore from ships and attacked native insects and reptiles. The sailors' Indian myna birds inhabited nests of the native echo parakeet, while Brazilian purple guava and oriental privet plants crowded the seedlings of native trees that had been undisturbed for centuries. Soon, the forest that had been home mostly to birds, insects, and reptiles became a mammalian haven. By the mid-1600s, only 11 of the original 33 species of native birds remained. The dodo was exterminated by 1681, although a few birds were taken to zoos. The animal became the first recorded extinction due to direct human intervention.

Today, although 685 of the 765 original species of land plants survive on Mauritius, humans have introduced 730 others. Half the island is planted with sugar-cane. Introducing organisms to islands not only drives out native species, but causes what some ecologists call "McDonaldization." That is, different islands, once valued for their unique assemblages of life, are coming to resemble one another. Many of the same species of plants and birds introduced to Mauritius, for example, were also brought to the Seychelles Islands, 930 miles (1,500 kilometers) to the northwest. Because of human intervention, a sameness is settling over many islands that were, not too long ago, stunning showcases of evolution in action.

16.1 What Is a Species?

A mushroom, bird, and tree are obviously different species; a leopard, lion, and panther are perhaps not as different in appearance, but are clearly separate species because they cannot produce fertile offspring with each other. Species designations identify types of organisms and represent their evolutionary relationships.

If a visitor from another planet with a body about the size of ours landed on a populated area of Earth and viewed the inhabitants, the variety of terrestrial residents might be obvious. A mushroom is clearly distinct from a bird or a tree, all of which are not at all like microscopic life-forms. Some types of organisms would clearly appear to be more closely related to certain others, as figure 16.1 depicts. If the visitor returned to the same place every 10 million years or so, it might notice that the types of organisms change—new ones have appeared, and some existing ones vanished, although many remain the same. But the visitor would get a different view if it arrived soon after one of a dozen or so points in Earth's history marked by mass extinctions followed by explosive speciation. The panorama of

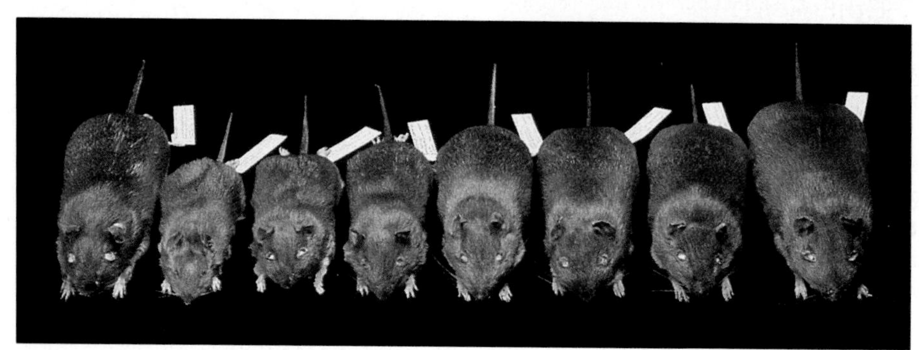

FIGURE 16.1

Species Are Distinctive Types of Organisms. The more recently different species diverged from a common ancestor, the more characteristics they share. The bacteria, tree, and bird in (**A**) are about as unlike as three types of organisms can be—their last common ancestor must have lived before cells diverged into the three basic types (bacterial, archaean, and eukaryotic). The three cats in (**B**) shared their last common ancestor much more recently. They have many similar characteristics. The spiny rats in (**C**) are so alike that researchers have difficulty telling them apart. Often they do so by comparing the rats' penises, which vary in shape from species to species.

life on Earth continuously changes, but not at a constant or even predictable rate.

The Evolving Species Concept

The first investigator to use the term "species" and to combine it with the broader classification "genus" to denote a biological name was Swedish botanist Carolus Linnaeus (1707–1778), who is discussed in greater detail in chapter 18. Linnaeus relied on observable (to us) characteristics, defining species as "all examples of creatures that were alike in minute detail of body structure." From childhood, he made meticulous lists, comparing, categorizing, and describing every organism he could find, eventually publishing them in 10 editions of huge volumes called "Systema Naturae."

Linnaeus's species designations organized the great diversity of life and helped scientists communicate with each other, although he did not consider the role of evolution. Linnaeus thought that each species was created separately, and that species could not change. Therefore, species could not appear or disappear, nor were they related to each other. It was Charles Darwin who connected species diversity to evolution, writing that "our classifications will come to be, as far as they can be so made, genealogies."

In the 1940s, Harvard biologist Ernst Mayr amended the work of Linnaeus and Darwin by considering reproduction and genetics. Mayr defined a **biological species** as a population, or group of populations, whose members can interbreed and produce fertile offspring. Microevolution would become macroevolution as a population became divided and natural selection led to sufficient genetic divergence between the groups so that if they once again came into contact, they could no longer produce fertile offspring.

Today most biologists accept Mayr's more restrictive species definition, which is much less subjective than Linnaeus's observations. It might be obvious to us that a tree and a rodent are too different to produce offspring together, but could we determine that two groups of fruit flies that appear to be identical to us belong to two different species? Under the system of Linnaeus this would be impossible, but using Mayr's definition, if the two groups can produce fertile offspring together, they belong to one species. Mayr's definition does not rely on physical appearance to assign species, which removes much of the potential confusion. The reasons that two groups cannot successfully mate are irrelevant to the definition. Today most species are assigned based upon ability to produce fertile offspring. New species arise when members of a population can no longer successfully interbreed. The formation of new species is called **speciation;** new species always come from existing species.

Mayr's definition of a species, however, raises three difficulties. How are organisms classified that have the *potential* to interbreed, but do not do so in nature? How are organisms that do not reproduce sexually assigned species designations? How is the species of a fossil recognized? DNA sequence analysis has allowed scientists to fill in some of these gaps and make more detailed comparisons of the relatedness of one organism to another, and is very useful in assigning organisms to species when no other information is available. In this way, DNA comparisons are redefining the concept of a species, yet at the same time raising more questions. How much difference in DNA sequence designates two organisms as different species? Scientists are now using many different sources and types of information to determine species and species relationships.

Tree Diagrams Describe the Comings and Goings of Species

Linnaeus thought that all species appeared when they were created and did not change from their original state. Unlike Linnaeus, Lamarck envisioned species as changeable, but he proposed that they did so in a straight line, with each new species an improvement on the previous one. Darwin, on the other hand, proposed that evolution would occur in a branched fashion, with one major species giving rise to others as they occupy and adapt to new habitats. In fact, the only figure he included in the *Origin of Species* was his diagram showing this branching pattern of one species giving rise to many others. Figure 16.2 shows an example of the current view of evolution as a series of branches that form a treelike diagram.

A branching pattern of lines represents populations that diverge genetically to split off a new species or subspecies (a

FIGURE 16.2

Different Views of Species Relationships.

(**A**) Linnaeus thought that each species was created once, and has not changed since. (**B**) According to Lamarck, and many others, evolution has been a transformation of one type of organism into another, sometimes demonstrating increasing complexity or a perceived progression toward perfection. (**C**) Evolutionary trees are diagrams that represent how species are related to each other by common descent, based on many kinds of evidence. The short lines represent extinct species. The tiger tree below is hypothetical—we need to compare more gene sequences to untangle the evolutionary relationships among the five extant subspecies.

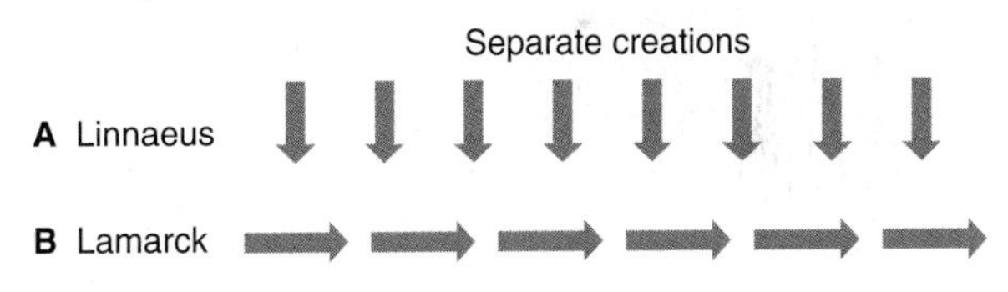

C Evolutionary tree diagram

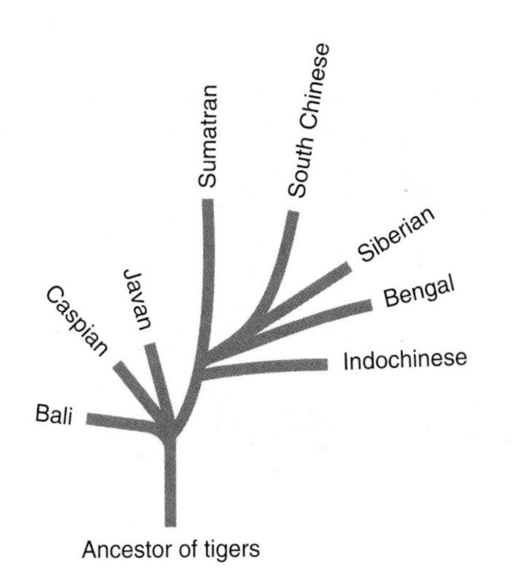

species-in-the-making). Discontinuation of a line indicates extinction. For example, the five subspecies of tigers described in the chapter opening essay descend from an original ancestral form, much as the branches on a tree arise from the same trunk. Yet this family tree also has three branches that stop, indicating the extinct Bali, Caspian, and Javan tigers.

For decades, biologists have used evolutionary trees, called **phylogenies,** to depict species' relationships based on descent from shared ancestors. More recently, molecular data have enabled researchers to refine approximate timescales on evolutionary tree diagrams originally made from fossil data. Much of the molecular data consists of comparing sequences of DNA found in mitochondria. This "mtDNA" represents only the mother, because sperm do not usually pass these organelles to oocytes during fertilization. Also, mtDNA mutates more rapidly than DNA from a cell's nucleus. Comparing the number of mtDNA nucleotide base differences between pairs of species provides a way to estimate elapsed time because we can estimate the rate at which this DNA mutates. That is, the more divergent the mtDNA, the longer ago the two species in question shared an ancestor. Chapter 17 discusses evolutionary trees and mtDNA dating in greater detail.

Morphology Versus Molecules—What Is a Platypus?

Describing physical characteristics of organisms, as Linneaus did, enables us to sort out species relationships to an extent. Molecular data can confirm phylogenies based on comparing obvious traits, or sometimes indicate the need to reevaluate the biological classification of an organism, as happened for *Ornithorhynchus anatinus,* the duck-billed platypus.

The platypus's characteristics reflect adaptations for swimming in Australian streams—a flattened, bald snout that it uses to scoop up small invertebrates from stream bottoms, a broad tail, and webbed feet (fig. 16.3A). The animal can extend claws beneath the webbing to walk on dry land and even dig. The young drink their mother's milk for about 4 months. The platypus is the only mammal besides the two species of echidna that lays eggs—the three species have traditionally been considered to comprise a group of primitive mammals called monotremes. A female platypus typically lays one to three eggs that hatch 10 days later.

Mammalogist W. K. Gregory spent his life studying platypuses, and in 1947 developed the "marsupionta hypothesis" that considered them mammals despite the egg-laying. Platypuses share with other mammals brain structure, glands, and milk secretion. The distinctive snout, tail, and feet, Gregory maintained, reflected specialized adaptations to an aquatic lifestyle, rather than a primitive state compared to other mammals.

In recent years researchers have sequenced the entire mitochondrial genomes of many mammalian species, and this has allowed a revision of the classification of the monotremes (fig. 16.3B). Instead of a tiny side branch with the echidnas, the DNA sequencing places both egg-layers along the lineage of marsupials, the pouched mammals. The data revealed that the monotremes are more like marsupials at the genetic level than had been suspected based on their unusual appearance.

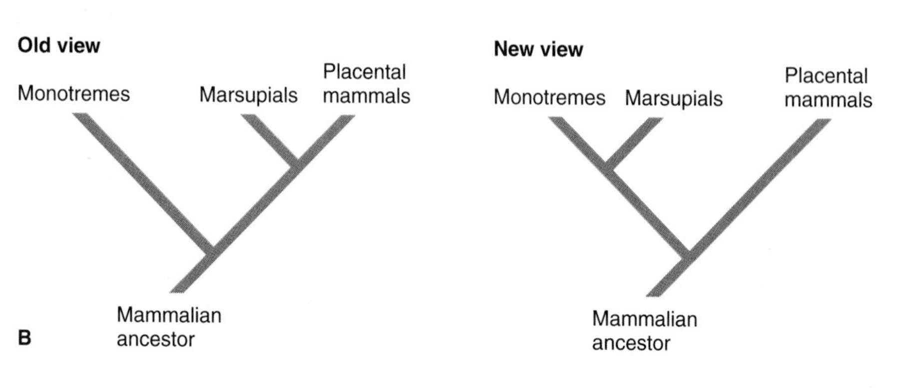

FIGURE 16.3
Placing the Platypus.
(**A**) The platypus shares features with many other animals, making it difficult to classify. (**B**) Molecular data can be used to refine species designations. Sequencing and comparing mitochondrial genomes from 19 mammalian species revealed that the monotremes are a side branch of the marsupials, rather than a separate lineage. The three groups are distinguished by the way that offspring are born. Monotremes lay eggs, marsupials have pouches, and placental mammals nurture fetuses within the female, where an organ called a placenta delivers nutrients and removes wastes. In tree diagrams, the groups that reach the top are modern species, and groups that are more closely related are closer together.

One lesson learned from recategorizing the platypus is that our definitions of species can change as our ability to detect differences among types of organisms changes. To define a duck-billed platypus requires consideration of its morphology (form), ecology (adaptations to the environment), and molecules (DNA sequences).

16.1 MASTERING CONCEPTS

1. How does today's definition of species differ from Linnaeus's definition?
2. What are some of the challenges in defining and describing species?
3. How are tree diagrams used to represent speciation and extinction?
4. What types of information are used to hypothesize how species are related to one another by descent from shared ancestors?

OLC

16.2 How Do Reproductive Differences That Distinguish Species Arise?

Many characteristics must match for two individuals to successfully mate and produce fertile offspring. Reproduction barriers occur at all points along the continuum of development.

In keeping with Mayr's biological species concept, new species form when sexually reproducing individuals of a population can no longer produce viable, fertile offspring with each other. The precise point of the incompatibility varies along the continuum of reproduction and development, from the courtship behaviors that bring the sexes together, through the many steps of fertilization and formation of an embryonic offspring, to the ability of the next generation to itself reproduce. For ease of study, biologists define reproductive isolating mechanisms—the source of the separation that underlies speciation—as prezygotic or postzygotic. (A zygote is the fertilized egg.) Figure 16.4 summarizes these mechanisms.

Early Obstacles

Prezygotic reproductive isolation acts before the nuclei of male and female gametes join. The earliest prezygotic isolating mechanisms prevent mating, and usually reflect the circumstances of place or time. In ecological isolation, members of two populations mate in different habitats. In temporal isolation, they have different mating seasons. In behavioral isolation, the organisms exhibit different mating rituals.

Two species of toad descended from a single ancestral type illustrate prezygotic reproductive isolation. *Bufo americanus* breeds in the early spring in small, shallow puddles or nearby dry creeks, whereas *Bufo fowleri* breeds in the late spring in large pools and streams. Each type of toad also has a unique mating call. The two species are thus ecologically, temporally, and behaviorally isolated. A more extreme example of behavioral isolation is a male moose attempting to mate with a dairy cow. This happens in New England occasionally when female moose are scarce. The males apparently respond to the next best thing—a dairy cow. The act doesn't proceed beyond fruitless attempts at mounting.

Sometimes individuals of two species may be in the same place at the same time, exchange the right behavioral cues, and even have compatible enough reproductive structures to permit mating or pollination, but obstacles to successful fertilization occur at the point of contact between the gametes. For example, in abalone, a type of shelled marine invertebrate, sperm penetrate oocytes using a protein called lysin that cuts specifically shaped holes in the envelope surrounding the female cell. The oocyte surface bears a receptor that binds only the lysin from the same species. If two abalone mate, but the sperm's lysin does not physically fit the oocyte's lysin receptors, then fertilization cannot occur. The two species remain reproductively isolated, and therefore separate.

Once a sperm enters an oocyte, prezygotic reproductive isolation can still operate if the chromosome numbers and types are not compatible. As an extreme example consider a dog and a cat. Even if a dog and a cat could mate (unlikely because of behavioral isolation), and sperm could enter oocytes (unlikely because enzymes that a dog's sperm release cannot penetrate a cat's oocyte), the canine's 39 different chromosomes could not pair with the feline's 19!

Chromosomal incompatibility can occur within populations, too, and drive speciation. This happened when a population of the plant *Clarkia rubicunda,* common along the coast of central California, underwent a bottleneck event. A severe drought in the Golden Gate Bridge region in San Francisco nearly decimated the local population of *C. rubicunda.* The only survivors had several chromosomal abnormalities. These plants cross-fertilized among themselves and established a new population in which a certain chromosomal aberration became the norm. When the drought ended, *C. rubicunda* plants encroached from the surrounding regions, but they could not reproduce with the Golden Gate group. The genetic material was organized too differently in the two groups, although both types of plants descended from the same ancestors. A new species, *C. franciscana,* had arisen.

Later Obstacles

Postzygotic reproductive isolation may act any time after fertilization, arresting development or producing a malformed, weak, or infertile offspring. Some cross-species hybrids, such as a mule (the offspring of a female horse and a male donkey), are always infertile. A mule is infertile because one parent's gamete has one more chromosome than the other. Meiosis is impossible. Different species may produce hybrids under unusual circumstances, such as captivity. For example, sometimes in zoos a male lion will mate with a female tiger, and a liger results. Likewise, when zebras and horses mate, they produce zebroids that look like a mixture of the two. Ligers and zebroids are infertile.

Rarely, hybrids are fertile but not very healthy. In these cases, postzygotic reproductive isolation may take a few generations to become evident. When groups within a population

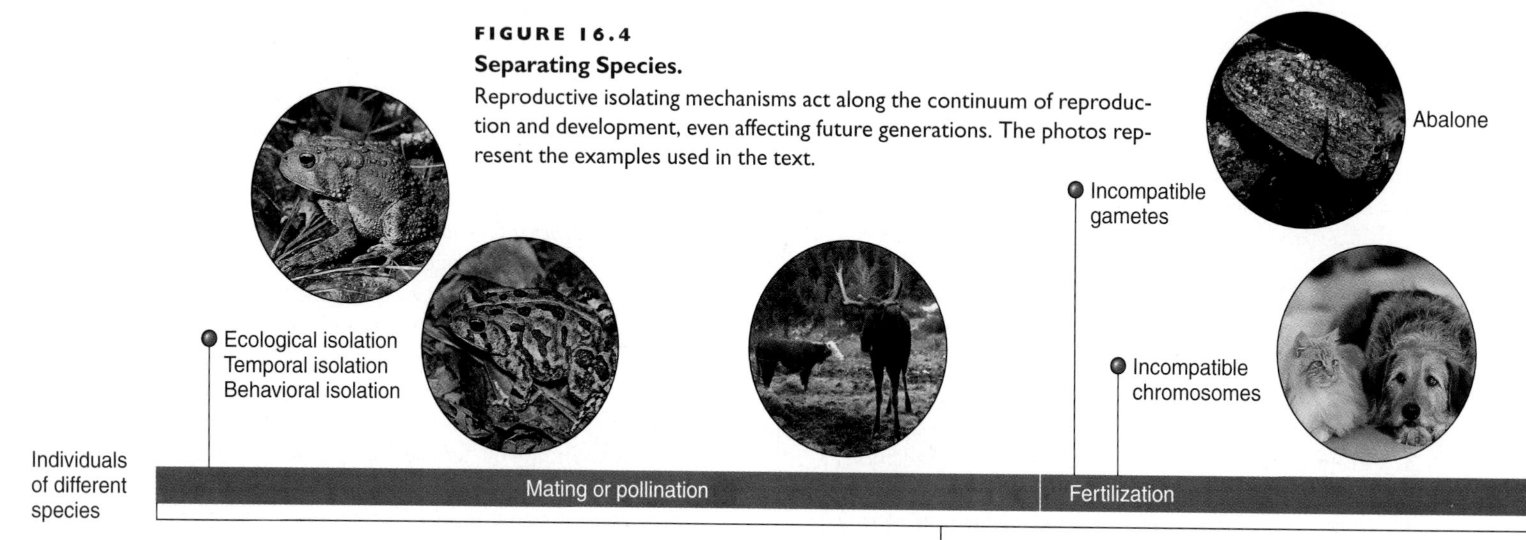

FIGURE 16.4

Separating Species.

Reproductive isolating mechanisms act along the continuum of reproduction and development, even affecting future generations. The photos represent the examples used in the text.

begin to diverge genetically, certain combinations of linked genes become associated with each subgroup, and perhaps offer some selective advantage. Mating between members of different groups then disrupts these advantageous linkages because of crossing over, and resulting offspring are less fit. With each generation, fewer hybrids survive, and ultimately the population drifts into two genetically distinct species.

16.2 MASTERING CONCEPTS

1. How can reproductive isolation occur in different ways?
2. Name three modes of prezygotic reproductive isolation that prevent mating.
3. Describe two cellular or molecular level mechanisms of prezygotic reproductive isolation.
4. What are two ways that postzygotic reproductive isolation may occur?

OLC

16.3 Speciation and Space

Species may arise when a geographical barrier separates a single population into two that then diverge genetically; when groups are neighbors and share a border but mate mostly among themselves; within the same area if diverse resources are available or if genetic differences are great.

Speciation is intimately tied to geography, because the spatial distribution of organisms determines their access to each other for mating purposes. Biologists consider three geographical circumstances under which speciation can occur—populations that are physically separated, that inhabit neighboring areas, or that share a habitat.

Allopatric Speciation Reflects a Geographic Barrier

In **allopatric speciation,** a geologic event or structure physically separates a population into two groups. Agents of allopatric speciation include volcanoes, earthquakes, storms, tidal waves, glaciers, floods, and formation or destruction of mountains or bodies of water. After the event, natural selection, mutation, nonrandom mating, genetic drift, and migration act independently on the geographically separated groups. The action of these forces may result in one or more means of reproductive separation. If after the geographic barrier is lifted descendants of the members of the original two populations can no longer interbreed, one species has branched into two.

The bluish-gray pupfish, which inhabits a warm spring at the base of a mountain near Death Valley, Nevada, illustrates one way that allopatric speciation occurs. The spring was isolated from other bodies of water about 50,000 years ago, preventing genetic exchange between the fish trapped within and those in the original population. In that time, the gene pool has shifted sufficiently so that a Death Valley pupfish cannot mate with a pupfish from another spring. It has become a distinct species.

The geographic routes to allopatric speciation may be obvious, or not so obvious. In the Amazon jungle, the source of much of the area's striking biodiversity seems clear—the huge rivers that are constantly changing the terrain as they seasonally dry out and flood, or meander and create ephemeral pools and islands. When Alfred Wallace, who discovered the principles of evolution in South America just about when Darwin was doing so in the Galápagos, arrived in the area in 1848 at age 25, he immediately identified the Amazon, Negro, and Madeira rivers as providing answers to "the problem of the origin of species."

A study of mitochondrial DNA sequences in tamarin monkeys supports Wallace's hypothesis that rivers divide populations, driving speciation. Where the Amazon is very wide, monkeys on one side of the river are brown, but on the other side they are

white. Yet populations on either side of the river where the banks are close together have both types of coat colors, presumably because the animals have easy access to each other (fig. 16.5).

The Amazon's rivers aren't the only force driving allopatric speciation. Consider another mitochondrial DNA study, of 11 species of similar-appearing rats. Their DNA sequences are quite divergent, indicating that the rats last shared ancestors several million years ago. Researchers studied two populations of each rat species, hypothesizing that, like the tamarins, populations on each side of a wide river would differ genetically. The population pairs indeed differed, but not in the expected way—for each of the 11 species, animals that lived upriver differed from those that lived downriver. It was as if some invisible barrier was cutting across the river, blocking gene flow between groups of rats.

A geological theory called **plate tectonics** explained the puzzling speciation pattern of the Amazon rats. According to the theory, Earth's surface consists of several rigid layers, called tectonic plates, that can move, like layers of ice on a lake. A plate boundary west of the Amazon River was pushing the continent westward, against the Andes Mountains. This generated a ripple effect outward, with some parts of the area rising, others falling. The shifting plates affected the formation and laying down of soil, which in turn affected the types of plants that came to comprise the forests, creating unique habitats and boundaries. When a geologist drew a map of the hidden rocky ridges underlying the Amazon tributaries, the boundaries directly coincided with the geographic boundaries delineating genetic differences between rat populations and species.

Parapatric Speciation Occurs in Neighboring Regions

In **parapatric speciation,** neighboring populations share a border zone. Most individuals mate within their own populations, but a few may venture outside the group and into the border zone for sexual partners. The genetic divergence that results

FIGURE 16.5
Allopatric Speciation.
Tamarin monkey populations on different sides of the Amazon River where the banks are widely separated are diverging toward speciation. The populations can still mix where the river is narrow.

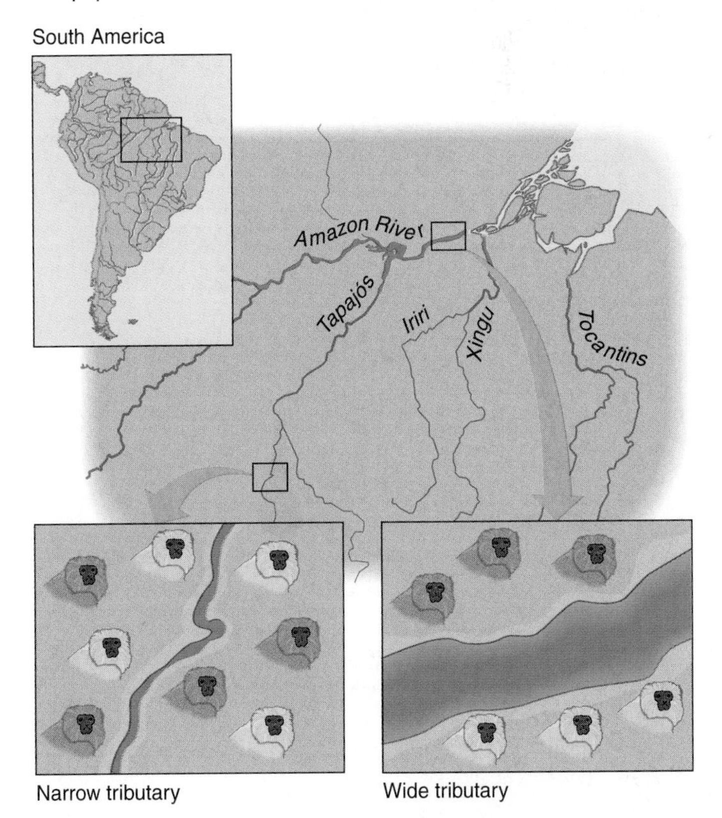

between the group sharing the border and the two original populations can be an initial step toward speciation.

Consider the little greenbul (*Andropadus virens*), a small green bird that lives in the tropical rain forest of Cameroon, West Africa (fig. 16.6). The birds also inhabit neighboring patches of grassland. Where forest and grassland meet, a type of border area, called an **ecotone,** is formed. From 1990 until 1996, researchers captured birds from six tropical rain forest sites and six grassland sites, and cataloged traits known to be subject to natural selection in birds. These included weight, beak depth, and the lengths of leg bones, upper jaws, and wings. The birds in the grassland patches were quite different from their rain forest counterparts. Their greater weight, deeper bills, and longer legs and wings had adapted them to living where fewer trees offer protection.

The little greenbuls from the ecotones can still mate with those from the rain forest, and so speciation has not yet occurred. But what is happening is that the forces of natural selection are greater than the gene flow between the two populations, taking the groups, gradually, farther apart. They will likely one day become two species.

FIGURE 16.6
Parapatric Speciation.
The little greenbul of Cameroon (**A**) lives in the rain forest, the grasslands, and the border zone between them (**B**). Their different-sized body parts (**C**) indicate that the two populations are diverging genetically, and one day may become separate species.

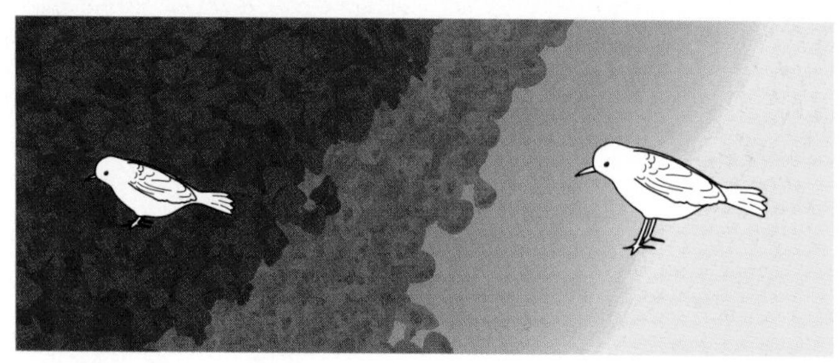

Sympatric Speciation Occurs in a Shared Habitat

In **sympatric speciation,** populations diverge genetically while in the same physical area. Often sympatric speciation reflects the fact that a habitat that appears to us to be uniform—such as a body of water—actually consists of many minienvironments. Fishes called cichlids, for example, have diversified into many species within the same large lakes in Cameroon. Their speciation is probably the consequence of ecological isolation. Some populations feed exclusively on the lake bottom, whereas others prefer the regions near roots of aquatic plants, or closer to the surface. Because the members of each population do not come into contact, no genetic exchange occurs. As changes in the genes of each population accumulate, they become more reproductively isolated. Over many years, the populations have become distinct species.

Sympatric speciation of crayfish seems to be ongoing in North America (fig. 16.7). In one Wisconsin lake, fishermen introduced the rusty crayfish *Orconectes rusticus* in 1960 as bait. A few of the rusty crayfish mated with the smaller resident species, the blue crayfish *O. propinquus.* Analysis of mitochondrial DNA sequences showed that 89% of the matings were between rusty crayfish females and the smaller blue crayfish males, suggesting that something about those males exerted sexual selection pressure on the females. The result is apparently a third population that may be on its way to becoming a distinct species. Laboratory experiments that place all three types of crayfish in the same tanks reveal that the hybrids are better at getting food and finding shelter under rocks than either species from which it derived—perhaps in nature it will eventually take over. Investigating Life 16.1 describes an experiment that monitored formation of a third species of sunflower from two others.

Often a large-scale genetic change precipitates sympatric speciation. This was the case for the chromosomal abberations that suddenly became predominant among *Clarkia rubicunda* plants in San Francisco, discussed in the preceding section. An even more drastic chromosomal change is **polyploidy,** when the number of sets of chromosomes increases. Polyploidy can occur when meiosis fails, producing, for example, diploid sex cells in a diploid individual. If diploid sex cells in a plant self-fertilize, a tetraploid individual (with four sets of chromosomes) results.

When extra chromosome sets derive from the same species, the organism is called an **autopolyploid.** In the rose,

Sunflower Speciation in the Laboratory

Plant biologists can watch speciation occur among sunflowers in a greenhouse experiment. The diploid sunflower *Helianthus anomalus,* which lives among sand dunes in Utah and Arizona, arose about 100,000 years ago—a fairly recent arrival. Examination of its chromosomes suggests that *H. anomalus* is a hybrid of two other, much older, species, *H. annuus* and *H. petiolaris* (fig. 16.A). Researchers knew that hybrids form from the two older species in the wild, but the new plants are usually infertile because of chromosome rearrangements that make fertilization and embryo development impossible. A new species could arise, theoretically and naturally, if several hybrid individuals had the same chromosome rearrangements and would therefore be compatible with each other.

To entice hybrid formation and speciation to occur, researchers bred the two ancestral plants—*H. annuus* and *H. petiolaris*—in a greenhouse. They followed three particular hybrid lines, breeding them in various ways with each other and with the parental species. After five generations, the investigators examined the genetic makeup of the hybrids to determine which genes the new hybrids retained. The result: Certain genes, obviously essential, remained from the ancestral species in each hybrid.

The researchers extrapolated from their greenhouse experiment what might have happened in the deserts of the U.S. Southwest: a group of individual plants arose with the same chromosome rearrangements—so they were now the norm and not aberrant—and genes necessary for survival and reproduction remained. This set the stage for speciation.

FIGURE 16.A
Producing New Species.
Two species of sunflower (top) were crossed to create a third (bottom)—in a greenhouse.

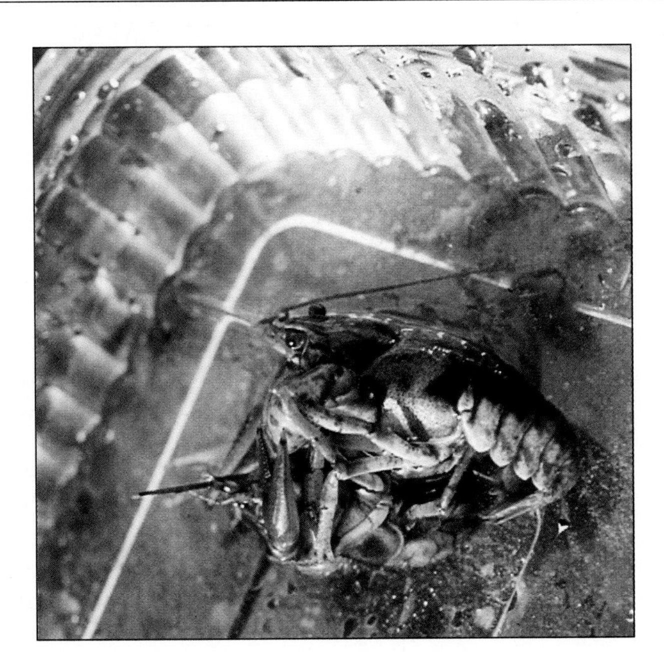

FIGURE 16.7
Sympatric Speciation.
In midwestern lakes, different species of crayfish occupy the same area. Here, a female rusty crayfish from a species introduced as bait in 1960 mates with a blue crayfish, a species that has lived in the area for a long time. The hybrid offspring are so vigorous that they may be evolving into a separate species.

autopolyploids have 14, 21, 28, 35, 42, or 56 chromosomes—presumably all descended from an ancestral species having 7 chromosomes. In contrast, **allopolyploids** form when gametes from two different species fuse. For example, an "Old World" species of cotton has 26 large chromosomes, whereas a species found in Central and South America has 26 small chromosomes. The type of cotton commonly cultivated for cloth is an allopolyploid of the Old World and American types. It has 52 chromosomes—26 large and 26 small.

Polyploidy may occur in the wild in response to unusually low temperatures. Agriculturalists induce polyploidy intentionally because such plants have larger leaves, flowers, and fruits. The drug colchicine, an extract of the autumn crocus plant, induces polyploidy by dismantling the mitotic spindle, which normally aligns and then separates the chromosomes in dividing cells. Many new crop varieties are induced polyploids, including alfalfa, apples, bananas, barley, potatoes, and peanuts.

The genetic differences between a polyploid and the plant it arises from are so great that geographic isolation is not even necessary for speciation. Nearly half of all flowering plant species are natural polyploids, which indicates the importance of this form of reproductive isolation in plant evolution.

Polyploidy is rare among animals. An example is the gray tree frog *Hyla versicolor,* a tetraploid probably derived from the identical-looking diploid *Hyla chrysoscelis.* However, recent molecular evidence suggests that something similar to polyploidy can occur when genes duplicate, rather than entire chromosomes. If the result is reproductive isolation, a new species forms. Genome doubling in this manner might have led to a burst of speciation 300 million years ago among the ray-finned fishes, producing today's 25,000 species. Researchers using the zebrafish (*Danio rerio*) as a model organism discovered this phenomenon. At first they noticed that several genes were duplicated, and eventually they realized that the entire genome is doubled in some species. Gene duplication is a mechanism of microevolutionary change because it can enable an organism to "try out" a new function (that is, manufacture a new protein), while retaining the old one.

FIGURE 16.8

Flycatchers Illustrate Allopatry and Sympatry.

In Wales, where the pied flycatcher and collared flycatcher live in separate areas, the males of the two species look very similar. But where they live sympatrically elsewhere in Europe, the males are less alike, perhaps so females can tell them apart when they seek mates. Source: Data from Roger K. Butlin and Tom Tregenza, "Is Speciation No Accident?" in *Nature*, Vol. 38, June 5, 1997.

Difficulties in Examining Speciation and Space

Classifying speciation as allopatric, parapatric, or sympatric again introduces the limitations of one species attempting to discern what is happening with others. Geographic barriers fuel allopatric speciation—but how can we be certain that we can detect barriers that are important to other species? Another problem is that of scale. Is a patch of forest an environment that would foster sympatric speciation, because it appears to us to be the same throughout? To an insect on the forest floor, the area around it is far different from a habitat in the treetops.

Allopatric speciation has been considered the most common mechanism because the evidence for it is the most abundant and compelling. For example, of the 18,818 known species of fishes, 36% live in freshwater habitats, although these places account for only 1% of Earth's surface. The countless numbers of lakes, ponds, streams, and rivers provide very diverse habitats, and certainly more boundaries to genetic exchange, compared with the vast but few oceans. In the oceans, biodiversity is highest where water meets land, or where waters of differing temperatures or velocities meet.

Another complication in identifying and distinguishing species is that the degree to which two species appear different from each other may depend upon whether they live divided by a geographic barrier or in the same area. Consider the European pied flycatcher *Ficedula hypoleuca* and the European collared flycatcher *F. albicollis.* In Wales, the ranges of the two types of birds are physically separate (allopatry), and the males look much like one another—they are black and white (fig. 16.8). But in central and eastern Europe, where the ranges coincide, male pied flycatchers are brown but collared flycatcher males have white patches. The hypothesis: Where the ranges overlap, selection favors distinct coloration that enables females to more easily tell the males apart.

16.3 MASTERING CONCEPTS

1. How does the physical relationship of the ranges of two populations affect speciation?
2. Distinguish among allopatric, parapatric, and sympatric speciation.
3. What are some difficulties that are encountered in considering the effects of geography on speciation?

OLC

16.4 Speciation and Time

Speciation is ongoing, as populations diverge genetically. It may be gradual or fast.

Speciation can happen quickly or very gradually—or at any rate or temporal pattern in between. At the same time, one species can be changing very rapidly, and another hardly at all. On a global level, Earth has seen times of relatively little change in the living landscape, but also periods of bursts of speciation, as well as times of mass extinctions.

Evolution May Be Gradual or Proceed by "Leaps and Starts"

Darwin envisioned evolution as one type of organism gradually transforming into another through a series of intermediate stages. The pace as Darwin saw it was slow, although not necessarily constant, and came to be known as **gradualism.** Many steps in species formation, however, did not leave fossil evidence, and so we do not know of many intermediate or transitional forms. Reasons for the incompleteness of the fossil record are many, including poor preservation of biological material, natural forces that destroyed fossils, and simply the fact that we haven't discovered all there is to discover. Chapter 17 explores these reasons in greater detail.

Another explanation for the absence of some predicted transitional forms in the fossil record is that such "missing links" never existed! Perhaps some biological changes occurred too quickly, perhaps due to a single genetic step that had a profound effect, to have left much evidence. Just as a criminal who robs a bank fast is less likely to be caught on a video camera than one who takes his time, some evolutionary change may have happened too fast to be caught in the act.

In 1944, paleontologist George Gaylord Simpson suggested that about 10% of the gaps in the fossil record represent the sudden evolution of a species from an ancestral form followed by periods of little change. (A paleontologist studies evidence of past life.) But at the time, strict Darwinian gradualists did not accept Simpson's theory of evolution by "leaps and starts."

In 1972, two young paleontologists, Stephen Jay Gould and Niles Eldredge, again raised the idea of swift speciation interspersed with long periods of stability (stasis). They concluded that evolution occurs by leaps and starts from their investigations of the fossil record—Gould for land snails in Bermuda, and Eldredge for trilobites (ancestors of insects) in New York. They coined the term **punctuated equilibrium** to describe a pattern of fast evolutionary change interrupting long periods of stasis. The fossil record, they argued, lacks some transitional forms because they never existed in a particular location or there were simply too few organisms to leave fossils.

The fossil record reveals that both punctuated equilibrium and gradualism occur—and many variations in between. In evolutionary tree diagrams, the angle of branch points distinguishes the two tempos (fig. 16.9). A small angle indicates gradualism, and a horizontal offshoot depicts rapid change.

The fossil records of bryozoa, which are sea-dwelling animals that resemble lacy plants, and of cyanobacteria (once called blue-green algae) illustrate the difference between punctuated equilibrium and gradualism. Fossil bryozoa appear unchanged for millions of years and then "suddenly" (over a period of 100,000 to 200,000 years) split to yield new species that coexist with the

FIGURE 16.9
Plotting Evolution.
An evolutionary tree diagram can depict rates and times of speciation and extinction events.

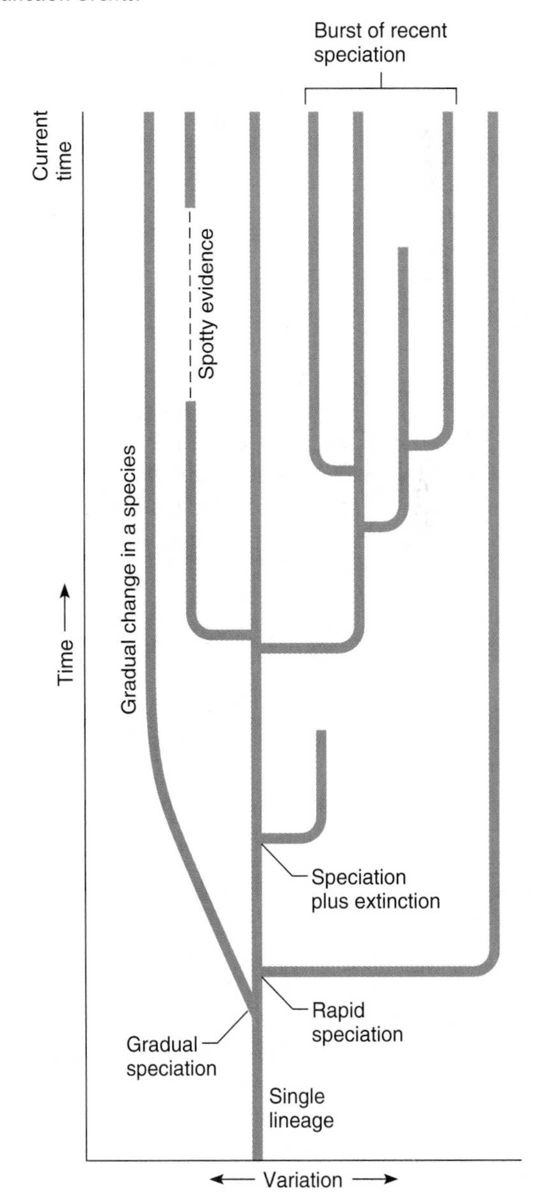

FIGURE 16.10

The Pace of Evolution—Speciation and Stasis.

(**A**) According to the fossil record, bryozoa changed little over millions of years and then suddenly split, forming several new species. As a result, modern species (top) often differ from fossil species (bottom). (**B**) In contrast, many species of modern cyanobacteria are remarkably like their ancestors; compare modern *Lyngbya* (top) to the 95-million-year-old *Palaeolyngbya* (bottom).

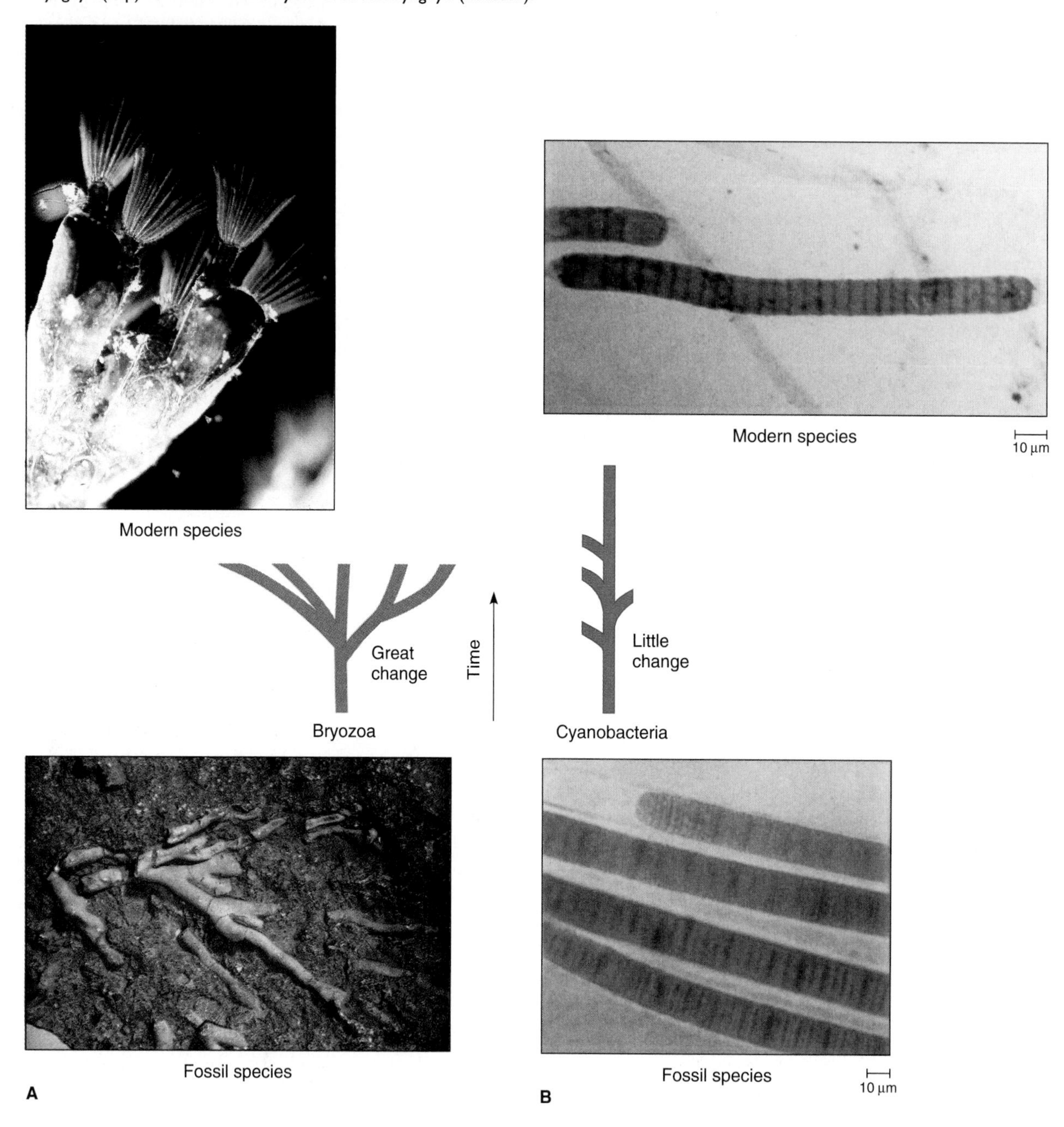

original species. This is punctuated equilibrium (fig. 16.10). In contrast, cyanobacteria appear much the same today as they appeared a billion years ago. This is extreme gradualism or stasis. The cyanobacteria may have changed little because they are extraordinarily well adapted to environmental changes. Their superb DNA repair systems, for example, may have enabled them to survive environmental disasters that released DNA-damaging radiation.

The examples of evolution observable over a human life span described in previous chapters illustrate the fast part of punctuated equilibrium. Factors that speed evolution include small populations, genes with high mutation rates, chromosomal aberrations and polyploidy, a short generation time, such as in bacteria, and rapid environmental change. A mutation in a single gene can also have evolutionary impact. A mutation that changes the timing of early developmental events may cause obvious changes in the adult. For example, an inherited delay in pigmentation of an embryo that greatly alters the adult's external color pattern can be important if survival depends upon protective coloration. Similarly, a single genetic "switch" that alters the timing of mitosis could have produced the prolonged brain growth characteristic of our own species.

An unusual experiment demonstrated how a single genetic change can alter the phenotype in a way that may be of evolutionary significance. By blocking expression of one gene, researchers forced a chicken's foot to develop to resemble a duck's foot (fig. 16.11). A chicken's foot is scale-covered and has four distinct digits. In contrast, a duck's foot is webbed and smooth. These differences enable a duck to swim and fly long distances over water, which a chicken cannot do.

Bursts of Speciation—Adaptive Radiation

Speciation can occur in bursts, called **adaptive radiations,** when a population faces an environment with abundant and diverse resources. Adaptive radiation can also occur if some members of a population inherit a structure or ability that gives them an advantage, such as a large beak that enables a bird to eat seeds too large for others of its kind.

Adaptive radiation occurs at all levels, from the microscopic to the global. Chapter 15's opening essay describes a laboratory experiment that demonstrated rapid microevolution of one strain of *Pseudomonas* into three when an ancestral strain was placed in medium that was allowed to develop microenvironments. Figure 16.9 shows four species that branch within a short time period from a shared ancestor. Such a group of species related by recent common descent is called a **clade.**

The finches and tortoises that Darwin observed in the Galápagos illustrate the adaptive radiation that is common in island groups (archipelagos). A stunning island example of adaptive radiation occurred in the Greater Antilles, which consists of Cuba, Jamaica, Puerto Rico, and Hispaniola. On each of the islands, small lizards called anoles have diverged in precisely the same ways to adapt to different parts of the trees on which they live (fig. 16.12). Adaptive radiation occurred separately on each island. In a familiar

FIGURE 16.11

From Chicken Legs to Duck Legs with a Single Genetic Switch.
The leg on the left is a normal, untreated chicken leg. The leg on the right, from the same animal, has blocked cell surface receptors that prevent certain cells from receiving the genetic message to die, which would normally carve digits out of webbing. The blocked leg also developed feathers instead of scales. The experiment shows that a single genetic change can exert a profound effect on phenotype—one great enough to have influenced evolution.

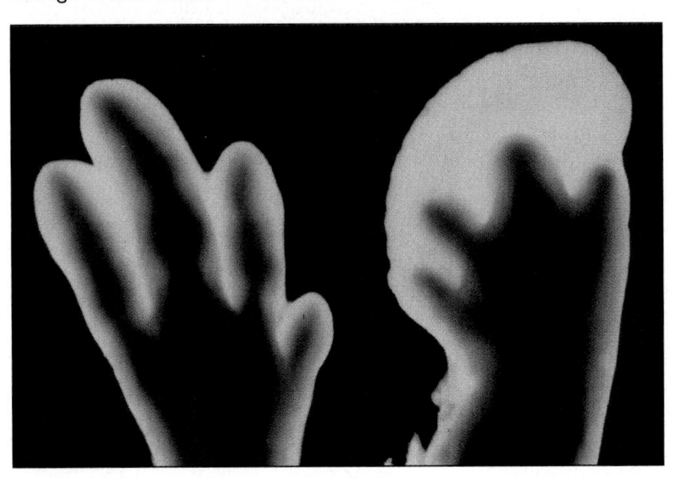

FIGURE 16.12

Adaptive Radiation on Islands.
Adaptive radiation of anoles has occurred on the islands of the Greater Antilles in a convergent fashion. On each island, different species of the lizards have adapted to living in different parts of trees, in strikingly similar ways.

Species example	Tree location
	**Tree crown** Large body, large toe pads Cuba—*Anolis equestris* Hispaniola—*A. ricordii* Jamaica—*A. garmani* Puerto Rico—*A. cuvieri*
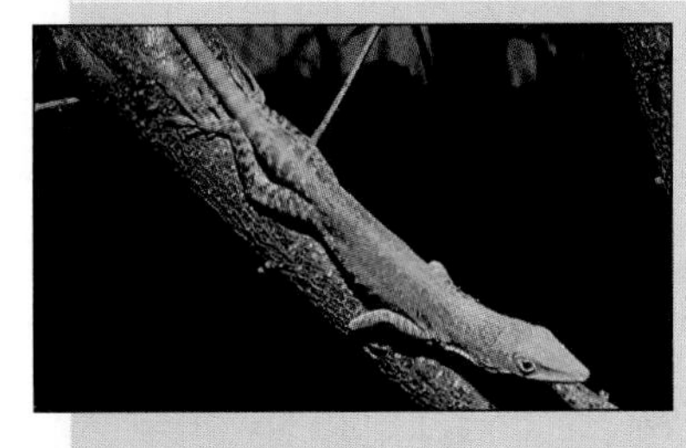	**Upper trunk/canopy** Large toe pads, can change color Cuba—*Anolis porcatus* Hispaniola—*A. chlorocyanus* Jamaica—*A. grahami* Puerto Rico—*A. evermanni*
	**Twig** Short body, slender legs and tail Cuba—*Anolis angusticeps* Hispaniola—*A. insolitus* Jamaica—*A. valencienni* Puerto Rico—*A. occultus*
	**Midtrunk** Long forelimbs, vertically flattened body Cuba—*Anolis loysiana* Hispaniola—*A. distichus* Jamaica—none found Puerto Rico—none found
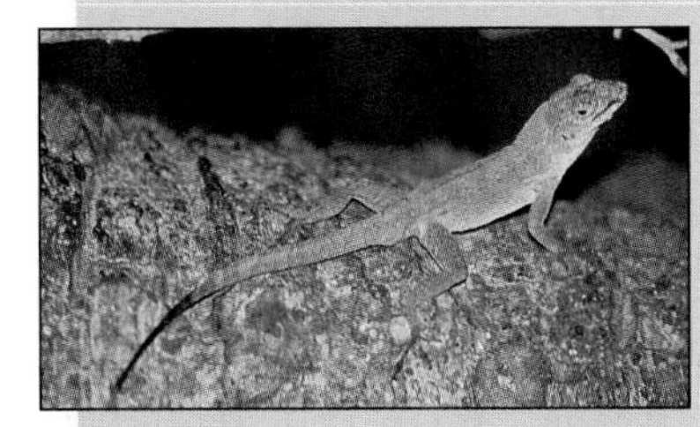	**Lower trunk/ground** Stocky body, long hind limbs Cuba—*Anolis sagrei* Hispaniola—*A. cybotes* Jamaica—*A. lineatopus* Puerto Rico—*A. gundlachi*
	**Grass/bush** Slender body, very long tail Cuba—*Anolis alutaceus* Hispaniola—*A. olssoni* Jamaica—none found Puerto Rico—*A. pulchellus*

FIGURE 16.13

Adaptive Radiation on a Global Scale.

With many ecological niches vacated when dinosaurs became extinct, mammals could flourish. The Mesozoic and Cenozoic are eras of time, discussed further in chapter 17.

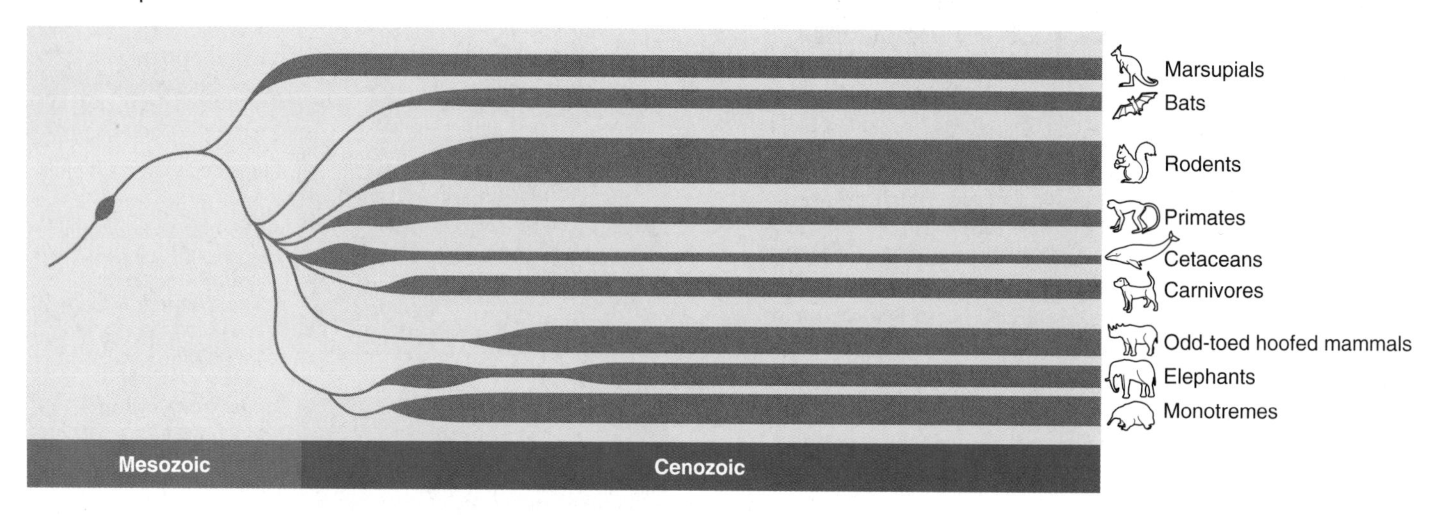

global scale example, mammals underwent an enormous adaptive radiation when the extinctions of the dinosaurs opened up many new habitats (fig. 16.13).

Sometimes the cause of adaptive radiation is not as obvious as it might seem. Consider glaciers. The burst of speciation of North American songbirds has been attributed to glaciers that 2.5 million years ago split many ranges. But mitochondrial DNA sequence comparisons indicate that songbird species diverged just *before* the glaciers formed. Similar evidence for other types of organisms that experienced adaptive radiation at about the same time as the songbirds—such as antelopes and rodents and our own ancestors in Africa—indicates that they, too, began to diverge before the ice age. From this information, researchers concluded that global cooling that ushered in the ice age triggered rapid speciation.

16.4 MASTERING CONCEPTS

1. How can gradualism and punctuated equilibrium both occur?
2. How can a single genetic change cause rapid evolution?
3. What factors can lead to adaptive radiation?

OLC

16.5 How Do Species Become Extinct?

Extinctions occur continuously as populations lose genetic diversity or cannot adapt to changing environmental conditions. Mass extinctions reflect a global disturbance.

Like speciation, extinction happens continually. It occurs on local as well as global scales.

Extinctions on a Small Scale

Many factors can cause extinction. Populations may dwindle, and a species eventually vanish, because individuals are not adapted well enough to a changing environment to leave enough fertile offspring to sustain their numbers. Extinction may also be a matter of bad luck—sometimes not a single population of a species survives a volcanic eruption or asteroid impact.

A decrease in genetic diversity lies behind some extinctions. The danger of genetic similarity in a population is dual. A new pathogen to which most or all the members of a population are susceptible may leave so few survivors that the host population can no longer reproduce. Secondly, genetic uniformity—often the result of inbreeding—tends to bring together deleterious lethal alleles for the same gene in the same individuals, which can drastically weaken the organisms. A field experiment on the Glanville fritillary butterfly (*Melitaea cinxia*) demonstrated the role of genetic uniformity in promoting extinction (fig. 16.14).

In Finland, fritillary butterflies live in metapopulations, which are many small populations that live apart in the same large, dry meadows. Because of a high reproductive rate and inbreeding, some of these smaller groups consist exclusively of close relatives. In a typical meadow, about 200 populations vanish and 114 new ones arrive each year. In the experiment, researchers captured female butterflies and identified the alleles for eight highly variable genes. Those populations with the least variability for these genes were more likely to become extinct. Before these doomed populations vanished, the researchers charted the signs of impending disaster: fewer eggs hatching, fewer larvae surviving, increased time in cocoons, and a shortened female life span. The extinction of a population can serve as a model for the mechanism of extinction of an entire species.

The common extinctions that happen gradually as populations shrink are termed **background extinctions.** From consulting the fossil record, paleontologists calculate that most species live from 1 to 10 million years before extinction, and that the rate of background extinctions is very roughly 0.1 to 1.0 species per year per million species. Earth has also witnessed at least a dozen periods of **mass extinctions,** when biodiversity and the sheer numbers of organisms plummeted (table 16.1). Although most

A

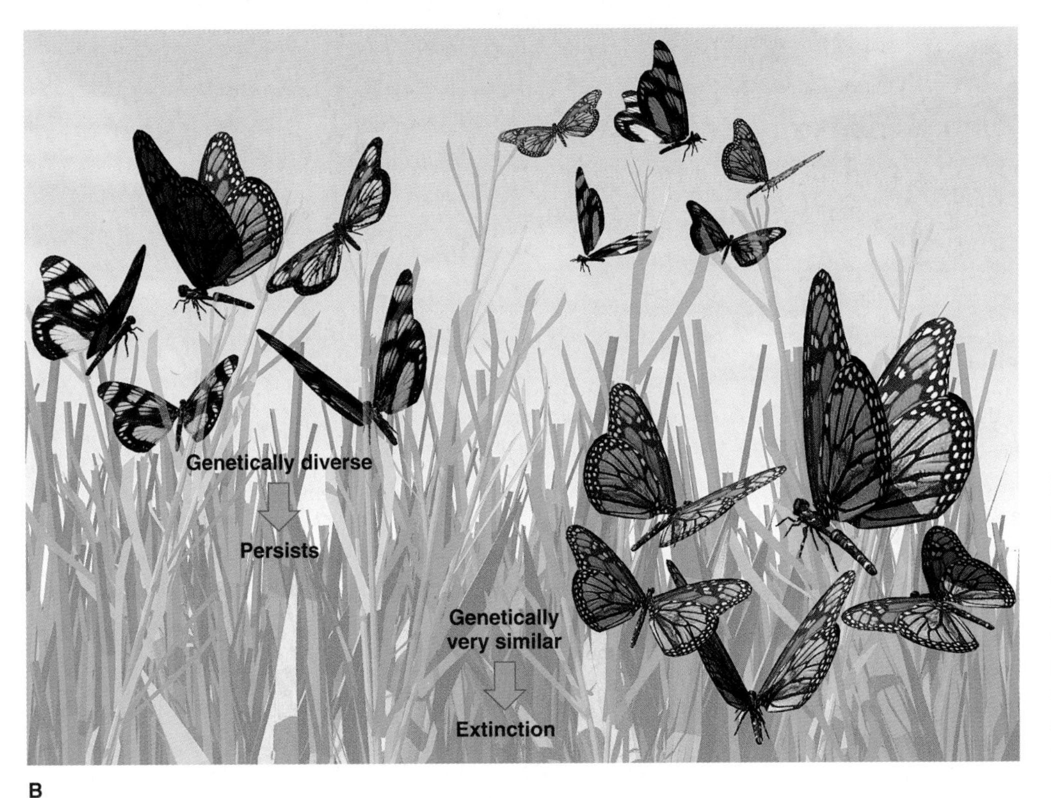

B

FIGURE 16.14
Inbreeding Contributes to Extinction at the Population Level.
(**A**) Genetically similar individuals may be vulnerable to the same pathogens.
(**B**) A genetically diverse population is more likely to survive environmental change than a population of genetically similar individuals. Such genetic uniformity arises from inbreeding.

extinctions overall are background, mass extinctions have had a great impact on Earth history because they have periodically opened vast new areas for adaptive radiation to occur.

What Causes Mass Extinctions?

During mass extinctions, many species disappeared over relatively short expanses of time (fig. 16.15). Paleontologists study clues in Earth's sediments to understand the catastrophic events that contributed to mass extinctions. Two general hypotheses have emerged in recent years to explain these events, although several processes have probably contributed to mass extinctions.

The **impact theory** suggests that a meteorite or comet crashed to Earth, sending dust, soot, and other debris into the sky, blocking the sun and setting into motion a deadly chain reaction. Without sunlight, plants, unable to photosynthesize, died. The animals that ate plants, and the animals that ate those animals, then perished. An extraterrestrial object that landed in the ocean would also have been devastating, because it would have mixed water layers. Oxygen-poor deeper waters would rise in the turbulence, and upper-dwelling organisms adapted to oxygen in their usual surroundings would die of oxygen starvation.

Evidence for the impact theory of the extinction at the end of the Cretaceous period includes centimeter-thin layers of earth that are rich in iridium, an element rare on Earth but common in meteorites. Quartz crystals in iridium deposits are cracked at angles that suggest an explosion. The impact theory of mass extinction may explain why layers of rock unusually devoid of fossils lie near an iridium layer. The absence of fossils may reflect an absence of life following a catastrophic impact event. Another type of chemical known to form in meteorites lies right above iridium deposits—large molecules of carbon called fullerenes.

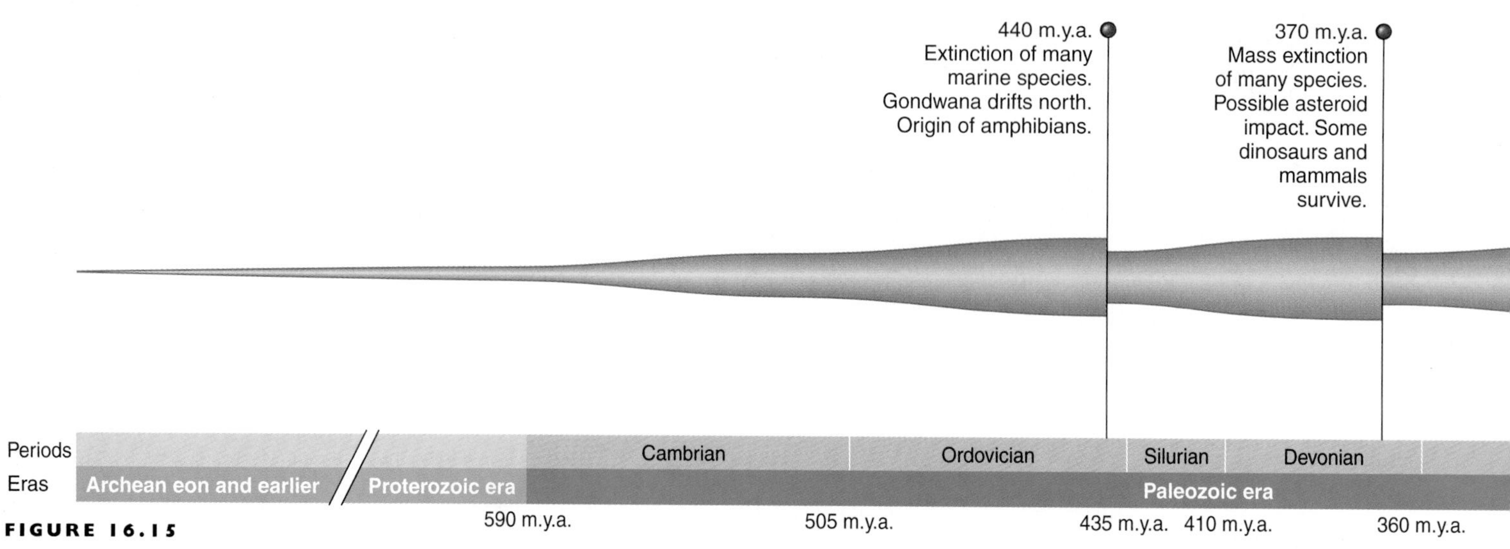

FIGURE 16.15
Catastrophes Affect Evolution.
Earth has seen several mass extinctions followed by adaptive radiation. Each sudden narrowing of the line representing life in this illustration represents an extinction and the ensuing adaptive radiation. Figure 17.6 describes the time frames of the five eras and the periods that comprise them.

Alternatively, the restlessness of the planet's rocks may explain some mass extinctions. Tectonic plates continually drift away from oceanic ridges, where new molten rock bubbles forth. Older regions of tectonic plates sink into Earth's interior at huge trenches.

According to the **plate tectonics** theory, the movement of tectonic plates caused continents to drift apart, then come back together. Oceans were mixed and separated. The result was dramatic environmental change that profoundly affected life. Organisms that had thrived in certain habitats were competing with unfamiliar species for limited resources. Weather conditions changed; ice ages and droughts killed many. Shifting continents altered shorelines, diminishing shallow sea areas packed with life. The shrinking habitats of large, meat-eating dinosaurs that accompanied continent formation, described in chapter 17's opening essay, vividly illustrate the effects of changing landmasses on the distribution of species, called **biogeography.**

Clues from fossils and biogeography help paleontologists reconstruct scenarios of past mass extinctions. These events probably date to the dawn of life, as the earliest cell-like structures lost the competition for resources to more efficient successors. An early mass extinction reflected the increase in atmospheric oxygen as photosynthesis arose. The oxygen released was toxic to some anaerobic species, and many died out as the more efficient energy users flourished.

About 440 million years ago, a huge continent, Gondwana, formed and covered the South Pole, causing an ice age and severely disrupting life in the seas. Glaciers formed, drawing water from the oceans and destroying many habitats. Then, 370 million years ago, geological upheaval occurred again. Sulfur-containing

TABLE 16.1 Mass Extinctions

TIME	SPECIES AFFECTED	SUGGESTED CAUSE
3 b.y.a.* (Precambrian)	Many anaerobic bacteria	Oxygen in atmosphere
505 m.y.a.† (end of Cambrian)	Trilobites, other marine invertebrates	Meteorite impact
440 m.y.a. (Ordovician)	Marine invertebrates	Supercontinent Gondwana formed
370 m.y.a. (end of Devonian)	Most fish and invertebrates	Meteorite impact, Gondwana moved, asteroid shower
250 m.y.a. (Permian)	90% of marine species	Pangaea formed
200 m.y.a. (end of Triassic)	75% of marine invertebrates	Meteorite impact
140 m.y.a. (end of Jurassic)	Many marine species	Not known
90 m.y.a. (mid-Cretaceous)	Many dinosaurs	Rise of flowering plants
65 m.y.a. (Cretaceous/Tertiary boundary)	Dinosaurs, marine species	One or more meteorite impacts
11,000 years ago	Large mammals	Drought, infectious disease, hunting, climate warming

*Billion years ago.
†Million years ago.

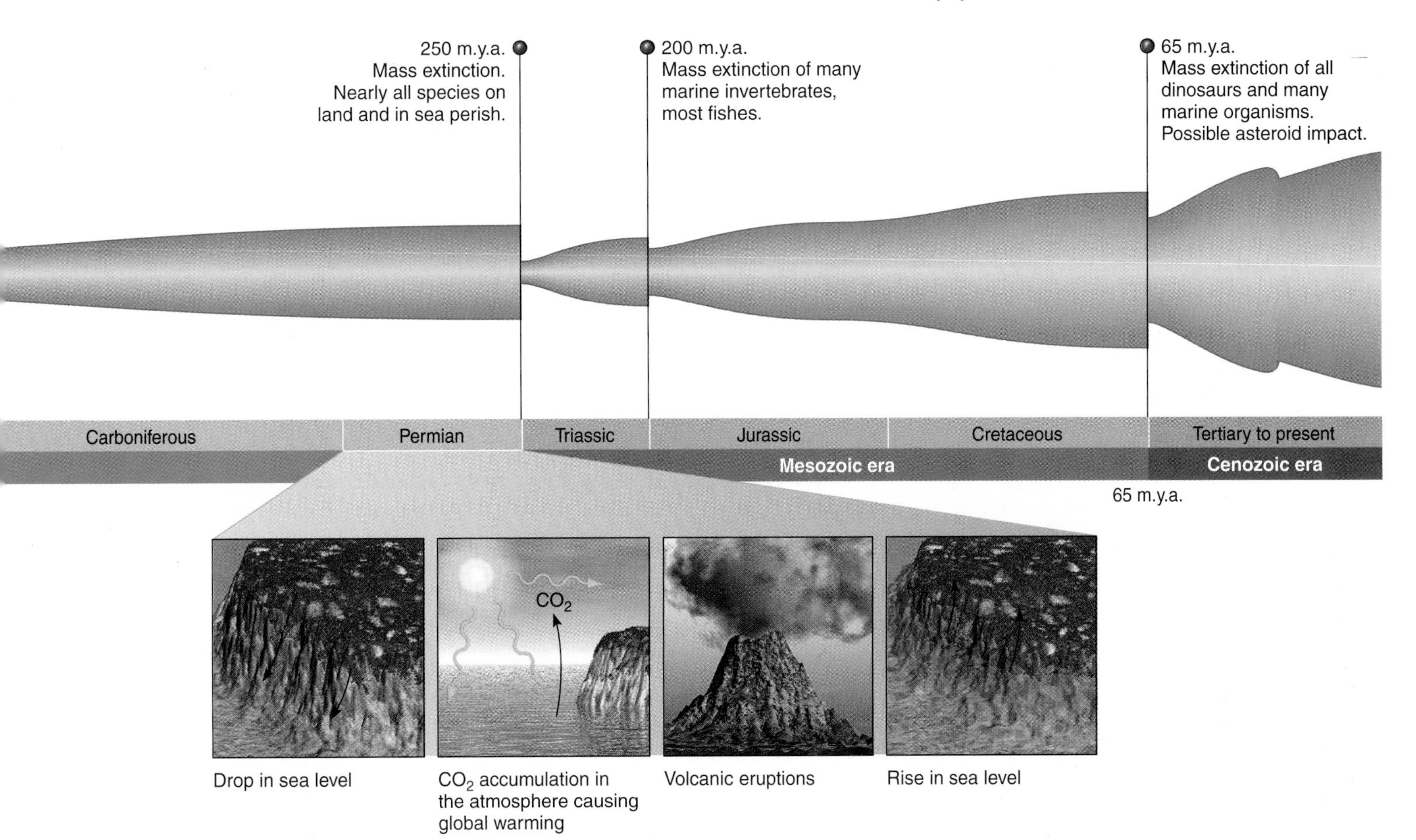

minerals in disturbed rock layers in the Canadian Rockies indicate that ancient ocean layers mixed at about this time. Iridium in these layers suggests a possible meteorite impact, and rock formations suggest glaciation. Many fish species and nearly 75% of all marine invertebrates perished.

The end of the Permian period, about 250 million years ago, saw what paleontologists call "the mother of mass extinctions." Geography was much different then, with today's landmasses forming the supercontinent Pangaea, and 70% of the planet's surface a single ocean, called Panthalassa (fig. 16.16). Over about 8 million years, 90% of marine species vanished in several waves of death. Most vulnerable were animals that clung to the seafloor, such as corals, sea lilies, brachiopods, and bryozoa. On the land, many types of insects, amphibians, and reptiles disappeared. Abundant fossils of fungal spores suggest that habitats for these organisms had opened up.

Paleontologists hypothesize that the Permian extinctions were the result of a sequence of events (see fig. 16.15). The first might have been a drop in sea level, which dried out coastline communities of the one landmass of the time, Pangaea. Next carbon dioxide accumulated in the atmosphere from oxidation of organic molecules, which raised global temperature and depleted oxygen dissolved in water. Rocks from that time show that the sea bottom was depleted of oxygen. These events could have devastated sea life. A long series of volcanic eruptions beginning 255 million years ago, lasting a few million years, further altered global climate. Finally, sea level rose, drowning coastline communities.

Forty million years after the Permian extinctions, the environment changed drastically again, and 75% of marine invertebrate species vanished. A crater in Quebec, Canada, roughly half the size of Connecticut, may be evidence of a meteorite impact that had devastating repercussions at this time. Alternatively, shifting tectonic plates could have caused these extinctions.

FIGURE 16.16

The World at the Time of the Permian Extinctions.

About 255 million years ago, Earth consisted of a large landmass, Pangaea, and a huge ocean, Panthalassa.

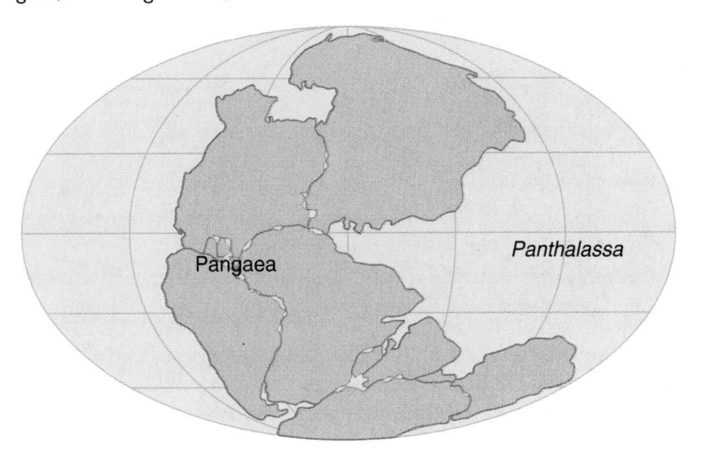

About 65 million years ago, an asteroid (a large meteorite) 6.2 miles (10 kilometers) in diameter crashed into Chicxulub, Mexico, part of the Yucatan peninsula. This was a time in Earth history called the Cretaceous/Tertiary, or K/T boundary. The impact left an offshore crater that filled with debris and clay, and a huge semicircle of sinkholes onshore. The area of damage was 112 miles (180 kilometers) in diameter, and so disturbed the crust that few geologic clues to life remained. However, deep-sea drilling rigs off the coast of southern New Jersey have produced cores that show an abrupt absence of microfossils (remains and evidence of microscopic life) in sediment layers that correspond to the time of the impact.

The loss of life from the Chicxulub impact was staggering. Nearly 75% of all species, including the dinosaurs and many birds, perished. Biologists estimate that photosynthesis was almost nonexistent for 3 years, as debris thrown into the sky blocked the sunlight. In the seas, plankton, which provides microscopic food for many larger marine dwellers, died as well. The ammonites, once-abundant marine organisms, vanished, as did many types of sea urchins and about 10% of the kinds of bony fishes. However, these and other mass extinctions may not have occurred as one great cataclysmic event. Although the impact would have obviously killed many organisms outright, evidence of changes in ocean chemistry that span 2 million years around the time of the impact suggest that the background extinction rate may have already been elevated before the event.

Mass extinctions in more recent times may reflect human influences. About 50,000 years ago in Australia occurred massive die-offs of "megafauna"—large animals. About 85% of the genera (plural for genus) of animals larger than 20 pounds (44 kilograms), including mostly marsupials but also some large birds and reptiles, became extinct at this time. Humans may have lent a hand in the extinctions because this was a time of only moderate climatic change, and it coincides with when our ancestors arrived on the scene. The connection might be only coincidence, but extinctions in other times and places also correlate to human occupation. For example, between 11,000 and 12,500 years ago, many animal species vanished from the North American plains, including saber-toothed tigers, mastodonts, huge sloths, and giant birds. Nearly all of the known 135 species driven to extinction then were megafauna.

How might humans have caused extinctions in the past? It is unlikely that our ancestors could have hunted enough to have decimated entire groups of organisms, but they might have caused or hastened extinctions by transmitting infectious diseases as they traveled, or by destroying habitats.

The role of *Homo sapiens* in causing extinctions is evident today. In California, several species of shrimplike crustaceans have recently become extinct as the transient puddles called vernal pools that they live in have been filled in with concrete. In Hawaii, 100 of 135 native bird species have vanished since humans arrived in A.D. 400, and 24 others are endangered. The Mississippi River and its tributaries once housed nearly 300 species of freshwater mussels; today 35 are extinct, 56 endangered. A study of archeological records from the past 6,000 years correlates the decrease in mussel biodiversity with the arrival of European settlers, and their sewage, paper mills, tanneries, plows, and dams. People began noticing fewer invertebrates in the streams and rivers a century ago. The list goes on and on as the problem worsens.

Today ecologists are documenting an alarming increase in background extinction rates for many types of organisms to 20 to 200 per million species per year. But only time will tell if we are on the brink of another mass extinction, or are just at a peak in the many ebbs and flows of biodiversity that have characterized life on Earth.

16.5 MASTERING CONCEPTS

1. What factors can cause or hasten extinction?
2. Distinguish between background extinction and mass extinctions.
3. Describe the major mass extinctions.
4. What is the evidence that humans have influenced extinctions?

OLC

Chapter Summary

16.1 What Is a Species?

1. Linnaeus's species designations helped scientists communicate. Darwin added evolutionary meaning and Mayr added the requirement for reproductive isolation to define **biological species.**
2. Diagrams of branching lines represent evolutionary relationships among species. Such evolutionary trees are called **phylogenies. Speciation** is the formation of a new species.
3. Comparing sequences of mtDNA is increasingly being used to clarify and supplement species relationships based on visible characteristics.

16.2 How Do Reproductive Differences That Distinguish Species Arise?

4. **Prezygotic reproductive isolation** occurs before or during fertilization. It includes obstacles to mating in space, time, and behavior; molecular mismatches between gametes; and incompatible chromosomes.
5. **Postzygotic reproductive isolation** results in weak or infertile offspring, or may act more gradually by disrupting beneficial gene linkages.

16.3 Speciation and Space

6. **Allopatric speciation** occurs when a geographic barrier separates populations, which then diverge genetically to the point that members of the two populations can no longer produce fertile offspring together.
7. **Parapatric speciation** occurs when two populations live in neighboring areas but share a border zone. Genetic divergence between the two groups exceeds gene flow, driving speciation.
8. **Sympatric speciation** enables populations that occupy the same area to diverge, via drastic genetic change such as **polyploidy** or the use of different resources.

16.4 Speciation and Time

9. Evolutionary change occurs at many rates, from slow and steady **gradualism,** to the "leaps and starts" of **punctuated equilibrium.**
10. In **adaptive radiation,** an ancestral species rapidly branches into several new species, reflecting sudden availability of new habitats or resources or a particularly beneficial adaptation. A group of species related by recent common descent is a **clade.**

16.5 How Do Species Become Extinct?

11. Extinction is the disappearance of species.
12. Decreased genetic diversity may lead to extinction of populations and possibly species.
13. **Background extinctions** are ongoing. **Mass extinctions** result from global changes.

Testing Your Knowledge

1. What did Linnaeus, Darwin, and Mayr contribute to the meaning of the term "species"?
2. Describe the relationships among the species that the letters represent in the following evolutionary tree diagrams:

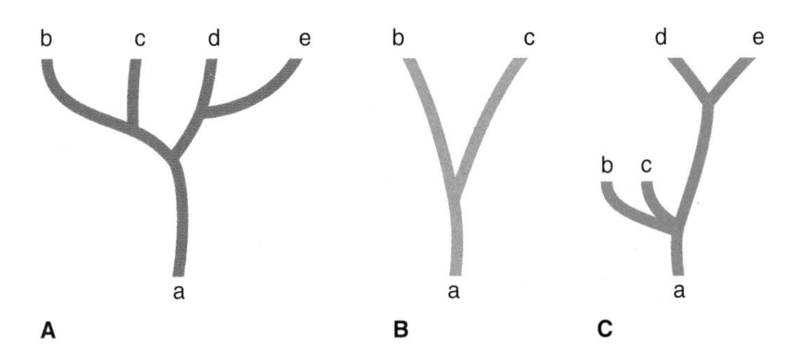

3. Explain how researchers compared mitochondrial DNA sequences to:
   a. detect speciation among tigers
   b. clarify the relationship between the platypus and other mammals
   c. detect divergence of populations of tamarin monkeys and rats in South America
   d. identify the cause of adaptive radiation of songbirds in North America
4. Whenever two particular types of animals mate, pregnancy occurs, but the embryos stop developing when they are just balls of cells. Is this prezygotic or postzygotic reproductive isolation?
5. Distinguish among allopatric, parapatric, and sympatric speciation and give an example of each.

Thinking Scientifically

1. A researcher sets up an experiment in which two genetically very different populations of fruit fly are placed in two enclosed areas that are connected by a narrow tube, through which the flies can fit. Suggest two scenarios that can occur in the entire setup, over time.
2. Each island of the Greater Antilles is home to a similar-appearing assortment of anoles (fig. 16.12). However, mtDNA analysis showed that the anoles from different islands that resemble each other are not alike genetically. How is this possible?
3. Give three examples from the chapter of events that may not have occurred as we initially thought they did.
4. Investigation of fossilized eggshells from a huge bird (*Genyormis newtoni*) shows that it became extinct about 50,000 years ago in Australia, a time that coincided with human colonization. What types of evidence would support or refute the hypothesis that humans hunted the bird into extinction?

References and Resources

Alvarez, L., et al. June 6, 1980. Extraterrestrial cause for the Cretaceous-Tertiary extinction. *Science,* vol. 208. The original paper describing how an impact event may have killed the dinosaurs.

Penny, David, and Masami Hasegawa. June 5, 1997. The platypus put in its place. *Nature,* vol. 387. DNA sequence analysis alters the placement of the platypus on the mammalian family tree.

Stebbins, G. Ledyard, and Francisco J. Ayala. August 28, 1981. Is a new evolutionary synthesis necessary? *Science,* vol. 213. A classic paper placing rapid evolution in the framework of Darwinian gradualism.

Wilson, E. O. Spring 2000. Vanishing before our eyes. *Time,* Special Issue. Mass extinctions are happening right now.

The *LIFE* Online Learning Center provides additional resources and tools for studying this chapter.
www.mhhe.com/life

CHAPTER 17

Evidence for Evolution

Are Birds Dinosaurs?

Imagine being asked to build a story from the following characters:

1. a pumpkin that turns into a coach;
2. a prince who hosts a ball;
3. a poor but beautiful young woman with a pleasant voice and dainty feet who has two mean stepsisters who are ugly and have large feet.

Chances are that unless you're familiar with the fairy tale "Cinderella," you wouldn't come up with that exact story. In fact, 10 people given the same pieces of information might construct 10 very different tales. But if more information was available, the stories might become more alike. So it is with the evidence that we have of the evolution of life on Earth—pieces of a puzzle in time, some out of sequence, many missing. As we fit the puzzle pieces together, pictures and stories emerge.

Over the years, the similarities that researchers have identified between dinosaurs and birds have added up.

Investigating how birds are related to dinosaurs is one such story of life. Discoveries of dinosaur fossils with clear evidence of feathers, plus similarities between bone structure among birds and certain types of dinosaurs, support the theory that birds are not only descended from dinosaurs, but, as some biologists insist, *are* dinosaurs. Perhaps the dictionary definition of "dinosaur" as an extinct reptile will have to change!

The hypothesis that birds are intimately related to small, bipedal (standing on two legs) meat-eating dinosaurs, called theropods, began with the discovery of a fossil of a theropod with feathers in Bavaria, dating to about 150 million years ago. It was 1860, the year after Darwin published *On the Origin of Species.* The animal, named *Archaeopteryx lithographica,* was the size of a blue jay, with a mix of features seen in birds and reptiles today—wings and feathers, but also a toothed jaw and a long, bony tail. In 1870, Thomas Henry Huxley reported to the Geological Society of London that he had identified 35 features of theropod skeletons and ostriches that only those two types of organisms shared, which he interpreted to be evidence of a close relationship. Others dismissed Huxley's ideas, largely because dinosaurs were not known to have flown. How could they, then, have given rise to birds? By 1916, the hypoth-

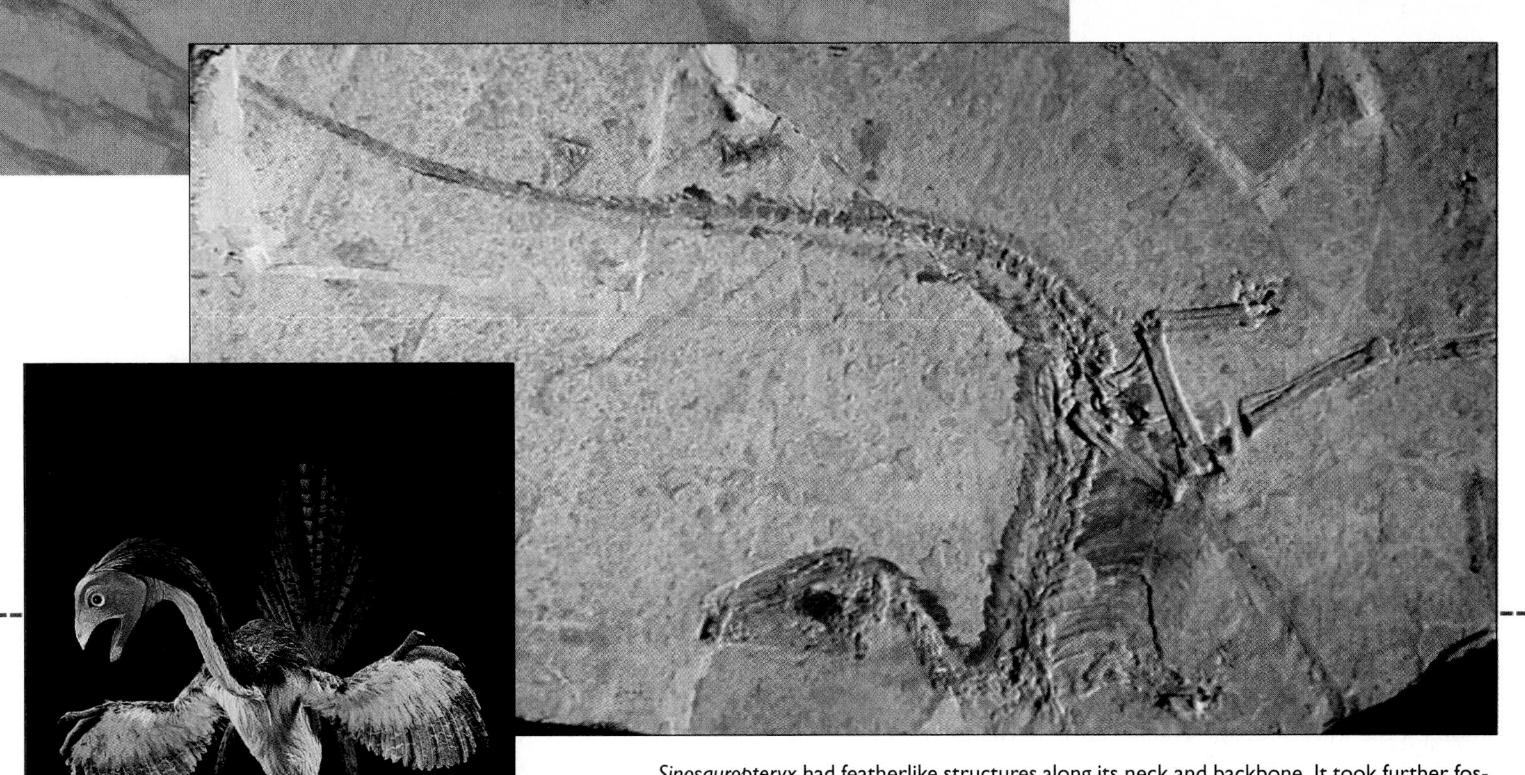

Sinosauropteryx had featherlike structures along its neck and backbone. It took further fossil finds to convince some researchers that dinosaurs really had feathers.

The theropod dinosaur *Protoarchaeopteryx* lived about 145 million years ago in China. It had feathers that resembled down on its body and tail, but also larger barbed feathers at the end of the tail and perhaps elsewhere. Feathers are only one characteristic linking birds to dinosaurs. This is a model of what clues suggest the animal might have looked like.

esis that birds descended from dinosaurs seemed even less likely, when Gerhard Heilmann, a Danish physician and fossil collector, noted that theropods lack clavicles, which form a bird's wishbone. Fossils of theropods with clavicles had simply not yet been discovered.

In the 1960s, Yale University paleontologist John Ostrom bolstered the bird-dinosaur link with an exhaustive comparison of bones, including the clavicles of theropods, discovered in the 1920s. His work focused on *Deinonychus,* a theropod that lived in Montana about 115 million years ago. In the 1970s and 1980s, application of a new technique called cladistics, in which the characteristics used to determine species relationships are based on shared ancestors, supported Ostrom's work.

Over the years, the similarities that researchers have identified between dinosaurs and birds have added up. We have tended to think of birds in terms of their most obvious distinctions from other types of animals—feathers. But many of the anatomical characteristics that would make flight possible were present in dinosaurs that preceded birds by many millions of years, and likely had functions other than a prelude to flight. For example, certain theropods had hollow bones, a bipedal stance in which they stood on their toes, a horizontal back, long arms, and a short tail. Rare finds of dinosaur nests reveal behavioral similarities to birds, particularly the preserved skeletons of theropods over their eggs, as if protecting them when disaster struck. The dinosaurs also laid only one or two eggs, which is more like a modern bird than most other reptiles, which deposit large clutches of eggs.

For several years, about all that was missing to definitively connect birds to dinosaurs was evidence of feathered dinosaurs. That came in the mid-1990s.

In 1996, researchers discovered fossils of *Sinosauropteryx* in Liaoning, a region of China rich in dinosaur fossils. About the size of a turkey, this animal had a fringe of downlike structures along its neck and back and along its flanks, although they were apparently not as organized as today's plumages. In 1998, a bonanza fossil find introduced *Protarchaeopteryx* and the somewhat later *Caudipteryx,* also from China. The 1998 discoveries were even more feathery than the fossil of *Sinosauropteryx.*

Although we now have clear evidence that some dinosaurs had feathers, how flight came about is still very much a mystery. The traditional view of animals gliding from tree to tree until they were truly airborne doesn't seem likely, because *Archaeopteryx* fossils are not found in treed areas and these reptile-birds lacked other adaptations to life in the trees. Instead, flight may have begun in small, light dinosaurs running at high speeds on open land. Eventually, with the aid of feathers, they took off.

17.1 Reconstructing the Stories of Life

Evolutionary biologists assemble clues from fossils as well as from structures and informational molecules in modern species to paint portraits of past life. Cladograms are diagrams that group species based on specialized characteristics that only they have inherited from a recent shared ancestor.

The millions of species alive today did not just pop into existence, but are the culmination of continuing genetic change to organisms that lived in the past. Many types of clues enable us to hypothesize about how modern species evolved from those no longer living, and also to understand how the organisms that live today are related to each other.

Clues in the Earth and in Molecules

Traditionally, evidence for evolution came mostly from **paleontology,** the study of fossil remains or other clues to past life. Increasingly that evidence is being supplemented with measurements on corresponding body parts between pairs of modern as well as extinct species, and with clues to evolutionary relationships in protein and nucleic acid sequence similarities.

Fossil evidence is diverse, revealing ancient structures, such as flowers from a plant that lived about 138 million years ago, or an eggshell from just 14,000 years ago (fig. 17.1A and B). Figure 17.1C is a fossil packed with functional information. It is a lump of excrement left by an extinct ground sloth about 20,000 years ago in a cave in Las Vegas. The location and types of bones found near the lump helped researchers identify the type of animal that deposited it, and DNA analysis helped to flesh out the scene. The few intestinal cells in the dung yielded DNA similar to that of modern sloths, and plant cells in the dung revealed that the ancient animal dined on mint leaves, grapes, yucca plants, and grasses.

Evolutionary biology today is a mix of old and new approaches. For example, researchers recently compared the sizes, shapes, and positions of 680 bones in the ray-finned fish *Amia calva* (also known as a dogfish or cypress trout), the only living fish of its genus, to bones preserved in hundreds of fossils, to decipher evolutionary relationships in this fish group. Harder to classify, though, are the millions of types of nematodes (roundworms), for which researchers have identified 15,000 species.

FIGURE 17.1

A Gallery of Spectacular Fossils.

(**A**) *Archaefructus* is one of the earliest fossils of flowers and fruits, from the Liaoning Province of northeast China about 138 million years ago. (**B**) This emu (*Genyornis newtoni*) eggshell from Lake Eyre, central Australia, is unchanged chemically from its original state, from about 14,000 years ago. (**C**) A ground sloth, *Nothrotheriops shastensis*, dropped this coprolite—fossilized excrement—about 20,000 years ago in a cave in Las Vegas.

Nematodes live everywhere, including in the bodies of many types of organisms. Researchers often use DNA sequence comparisons to tell the worm species apart, plus morphological features and which types of organisms the worms parasitize. The different criteria align fairly well—that is, worms that have the most similar DNA sequences also tend to parasitize the same types of hosts.

Systematics Assembles the Stories of Life

The varied types of evidence for evolution are part of a larger field called **systematics,** which attempts to explain how species are related to each other, and how they arose. Before 1980 or so, systematists constructed evolutionary tree diagrams by comparing as many characteristics as possible among species. Those organisms with the most characteristics in common would be neighbors on the tree's branches. But simply comparing what we can see doesn't necessarily reflect shared ancestry. For example, traditional evolutionary tree diagrams may not account for convergent evolution, which is the evolution of a similar structure or function separately in two types of organisms in response to a similar environment (see figs. 14.4 and 16.12).

More meaningful than an evolutionary tree diagram based on apparent similarities is a **cladogram,** which is another type of treelike diagram built using specific features common to only one group of organisms. The group is called a **clade,** and the distinguishing feature is a **derived character.** A clade indicates monophyletic evolution, or a single pathway. For example, feathers is a trait suitable as the basis of a clade because they are found only among birds. Not useful is a trait such as the flippers of a penguin and a porpoise, which arise from different structures (a wing and a leg, respectively) and are adaptations to swimming.

Figure 17.2 depicts the subtle distinction between a traditional evolutionary tree diagram and a cladogram, highlighting vertebrates (animals with backbones). The traditional evolutionary tree clearly separates mammals and birds from reptiles. In contrast, the cladogram places them on a continuum that includes reptiles.

To construct a cladogram, a researcher begins by selecting traits that are of evolutionary import—that is, traits that probably reflect descent from a shared ancestor. As an analogy, consider a

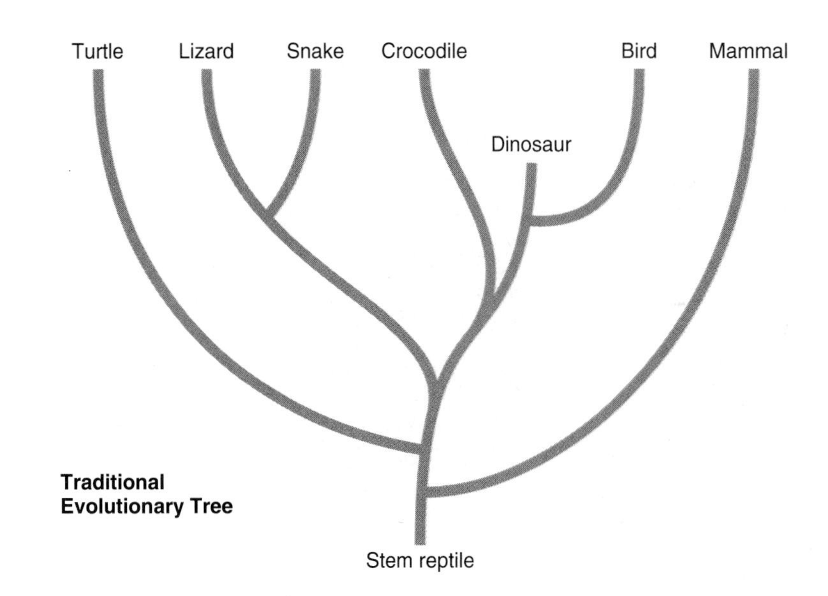

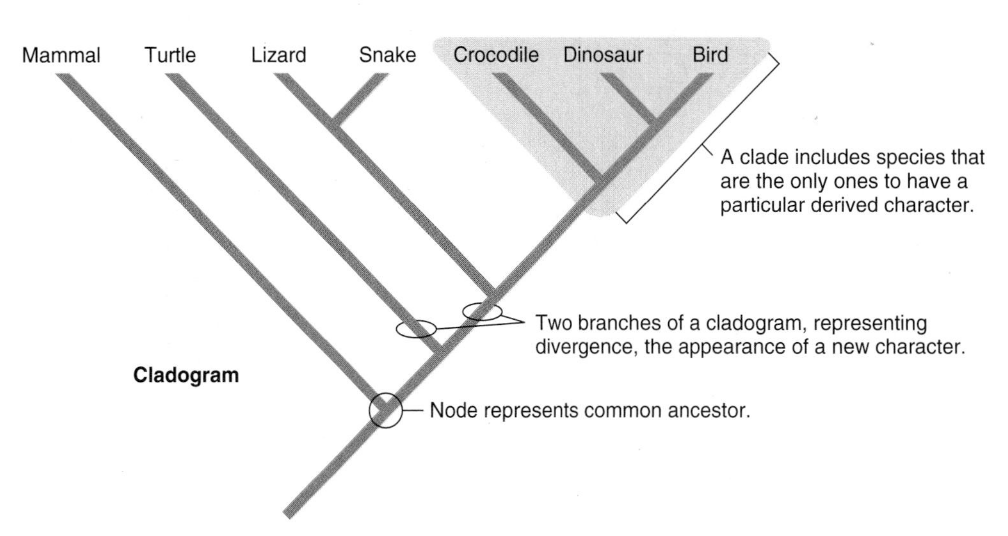

FIGURE 17.2
Two Types of Evolutionary Tree Diagrams.
A traditional evolutionary tree places mammals apart from modern reptiles; a cladogram has them on a continuum. The traditional tree is based on morphological (structural) differences. The cladogram is based on evolutionary relatedness, as presumed from derived characters that only the members of a clade share.

class of 24 students on a Monday morning. Six of them show up wearing identical T-shirts from a concert the night before. It is more likely that those students got their shirts from a recent shared experience—the concert—than that they all just happened to pick out the same shirts from vast collections to wear on the same day. Similarly, shared characters reflect a shared origin.

The next step is to make a chart listing which organisms under consideration have which traits. Then a tree is built, with those species sharing the most derived characters occupying the farthest and closest branches. The nodes indicate where a new feature arises from a common ancestor. Figure 17.3 demonstrates how to construct a cladogram, using the example from the chap-

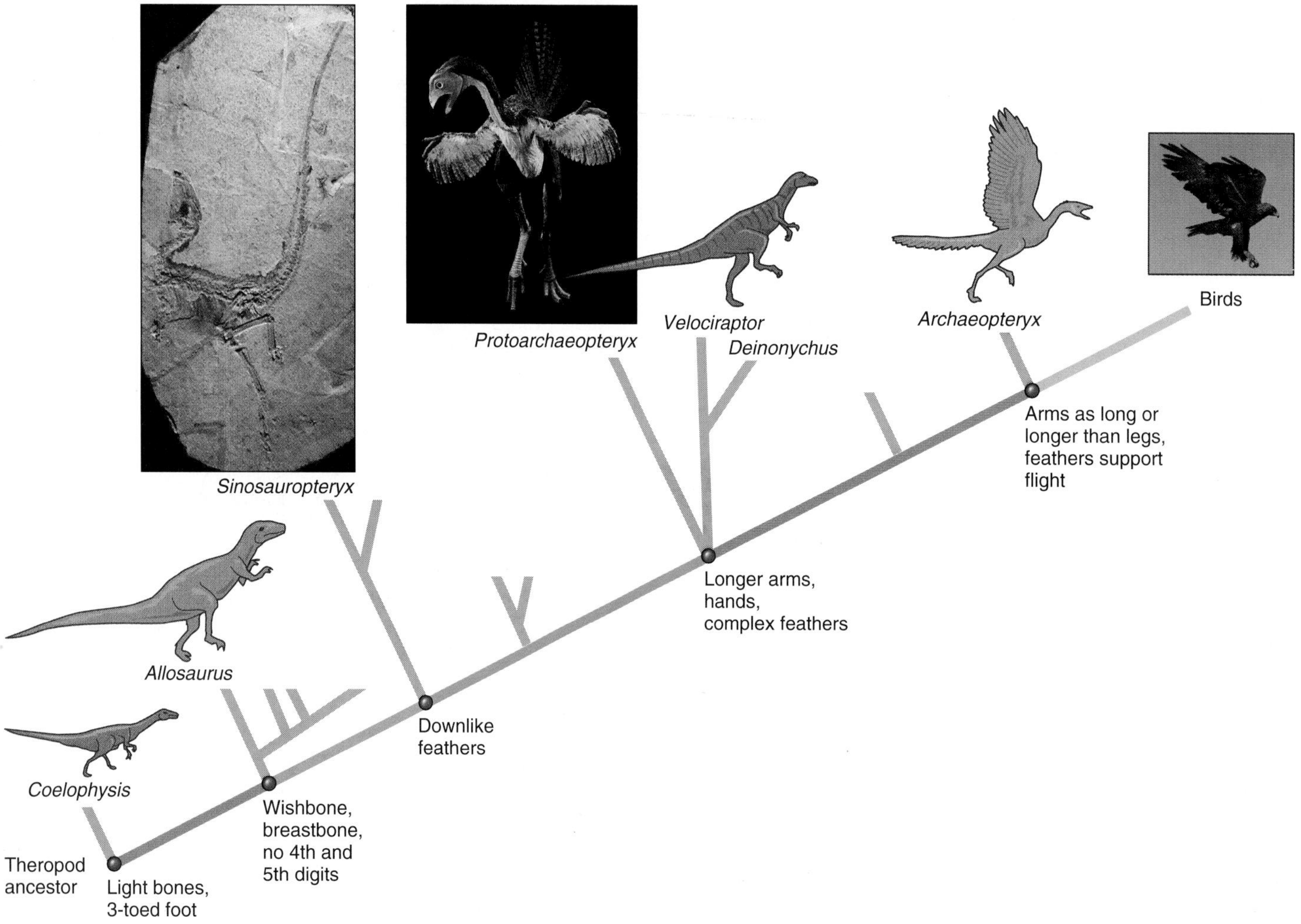

FIGURE 17.3
Constructing a Cladogram.
To build a cladogram, tally up the derived characters that species share. Pairs that share the greatest number of characters are depicted as the closest on the cladogram.
Source: Cladogram data based on "When Is a Bird Not a Bird?" in *Nature,* June 25, 1998, p. 729.

	Coelophysis	*Allosaurus*	*Sinosauropteryx*	*Protoarchaeopteryx*	*Archaeopteryx*	Birds
• Light bones • 3-toed foot	✓	✓	✓	✓	✓	✓
• Wishbone • Breastbone • Loss of 4th and 5th digits		✓	✓	✓	✓	✓
• Downlike feathers			✓	✓	✓	✓
• Longer arms • Hands • Complex feathers				✓	✓	✓
• Arms as long or longer than legs • Feathers support flight					✓	✓

Sinosauropteryx
Protoarchaeopteryx
Velociraptor
Deinonychus
Archaeopteryx
Birds
Arms as long or longer than legs, feathers support flight
Longer arms, hands, complex feathers
Allosaurus
Downlike feathers
Coelophysis
Wishbone, breastbone, no 4th and 5th digits
Theropod ancestor
Light bones, 3-toed foot

ter opening essay. It depicts key dinosaurs on the path to "birddom," using derived characters that are important in the acquisition of flight. We will return to this cladogram later in the chapter.

A problem with cladistics is that the diagrams can become enormously complicated when many species and derived characters are used. Mathematically, several trees can accommodate any one data set. Typically computers are used to derive trees, and the one that requires the fewest steps—called the most parsimonious—is selected to be the closest to reality. Cladograms, because they are based on limited and sometimes ambiguous information, are not peeks into the past, but are tools that researchers can use to construct hypotheses about the origins and relationships of different types of organisms. Investigating Life 17.1 describes an example of how molecular evidence and the resulting cladogram led to a reassessment of the classification of hippos.

17.1 MASTERING CONCEPTS

1. What is systematics?
2. What types of information supplement fossils in investigating evolutionary relationships?
3. How is a cladogram constructed?

17.2 What Can Fossils Reveal?

Fossils can reveal the immediate environment of an ancient organism, clues to a scene, or even global change. Fossils form in a variety of ways. The time that a fossilized organism lived can be estimated relatively from its position in rock strata, or more definitively by following the extent of breakdown of one radioactive isotope into another, which reflects passage of time.

Fossils have meaning only in a context. Where they are found, what they are found with, the conditions they are in, and their composition all provide clues to life.

The Role of Geology in Interpreting Fossil Evidence

To provide maximal information, fossil evidence is considered in the larger context of its environment—time and place. This information may be highly specific to how a particular organism lived and died, or reveal an event in a specific location from ages ago. Fossil evidence can also provide insights into time and place in a more sweeping sense.

As an example of the environmental information in an individual fossil, return to the preserved sloth dung of figure 17.1c. Intact cells found in the sloth's droppings allowed researchers to identify the plants that comprised the sloth's last meal, indicating, compared to similar evidence from longer ago, that the climate was changing, an event that might have contributed to the species' eventual extinction.

A spectacular fossil find in Patagonia, Argentina, reveals a long-ago disaster in an extensive valley. Here in 1998, researchers discovered a floodplain literally littered with preserved dinosaur eggs, naming it Auca Maheuvo. Thousands of eggs contain about-to-hatch babies, frozen in time and place as reddish-brown mud and silt covered them when shallow streams flooded about 89 million years ago (fig. 17.4). The shape of the preserved teeth place them in a group of plant-eaters called sauropods. The newborns would have been only 15 inches long, and would attain an adult length of 45 feet. The Auca Maheuvo fossils include the first skeletons of dinosaur embryos found in the southern hemisphere, as well as preserved skin.

Fossils can provide information on a global scale, too. Consider three types of dinosaurs that lived about 90 million years ago. All were large, upright meat-eaters (theropods) with similar skeletons (fig. 17.5). *Carcharodontosaurus* lived in Africa, *Acrocanthosaurus* in North America, and *Giganotosaurus* in Argentina.

Is it coincidence or convergent evolution that explains the similarities among dinosaurs that lived on three continents at about the same time? A more likely explanation is that they didn't always live on separate landmasses. That is, it is possible that all three giants

FIGURE 17.4

A Field of Dinosaur Embryos Trapped in Mud and Time. Thousands of sauropod eggs were buried as streams overflowed this Argentina plain (**A**), which was a valley about 89 million years ago. Preserved embryonic teeth (**B**) were so tiny that a mouthful of 32 of them would fit into an "o" in this sentence. The embryonic skin preserved in (**C**) from the Argentinian dinosaurs resembles that from a modern-day 13-week-old Nile crocodile—it does not yet have the hard parts, called osteoderms, that formed the characteristic armor of the adults.

A

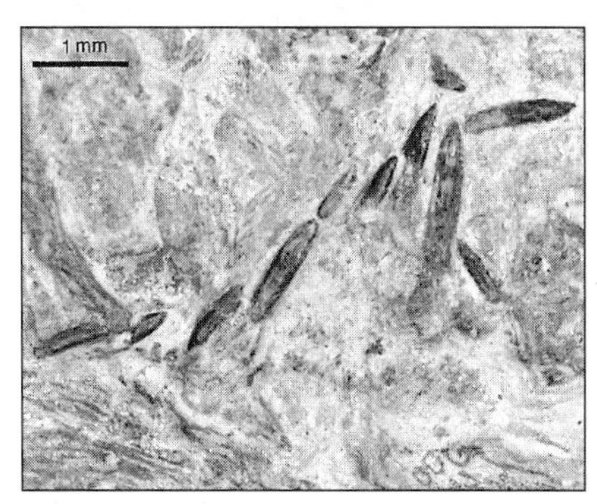

B

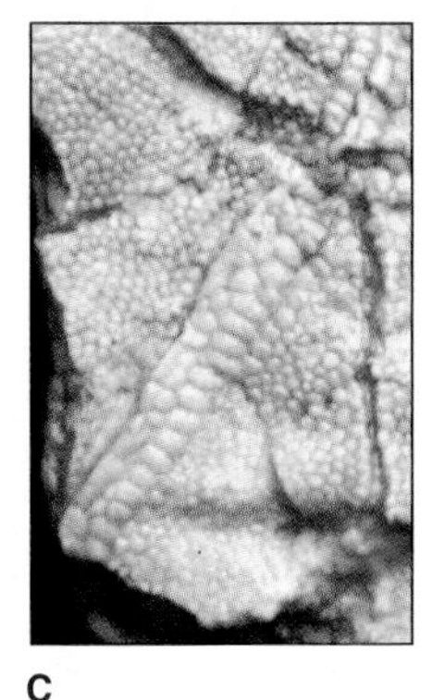

C

Investigating Life 17.1

Hippos and Whales—Closer Relatives Than We Thought?

Classifying cetaceans—a group that includes whales, porpoises, and dolphins—has always been a problem. The cetaceans have many adaptations to life in the water that complicate comparisons to terrestrial vertebrates, and according to fossil evidence, they went from land to water very rapidly, in just a few million years. About a century ago, biologists placed the cetaceans closest to the ruminants (which are hoofed, grazing mammals such as deer, cows, sheep, and giraffes), based on a variety of skeletal similarities. The ruminants are part of a larger group, the artiodactyls or even-toed ungulates, that also includes hippos, pigs, peccaries, camels, and llamas. All of these mammals first appeared in modern form about 48 to 50 million years ago.

The most defining feature of the artiodactyls is a very mobile heel joint, something that cannot be observed in the legless cetaceans. The axis of symmetry that splits their feet is also nonexistent in the footless whales, porpoises, and dolphins. Artiodactyls are classified by a triple row of cusps (raised areas) on the back molars, a trait that may have disappeared in the cetaceans because of dietary differences from their land-dwelling relatives. The traditional family tree, based largely on superficial resemblances, placed the cetaceans closest to the ruminants, and then most closely related to the hippos, which in turn were closest to pigs and peccaries (figure 17.A). In commonsense terms, a hippo looks more like a pig than it does a porpoise!

But molecules told a different story. Starting in 1994, evidence began to accumulate that showed striking biochemical similarities between hippos and cetaceans. DNA sequences from the nucleus and the mitochondria support a tree that places the cetaceans, ruminants, and hippos as one clade, or monophyletic group. This means that they are set off together by at least one derived character. Very definitive molecular evidence came in 1997 in the form of highly distinct DNA sequences called "short interspersed elements," or SINES for short. These sequences are so unusual that it is very unlikely for them to appear in different species by chance alone—inheritance from a shared ancestor is much more probable. The SINES were integrated into the genomes of cetaceans, ruminants, and hippos—but notably absent in the genomes of pigs, peccaries, camels, and llamas.

As the molecular evidence continues to mount that hippos are more closely related to whales than to the similar-appearing pigs, paleontologists and evolutionary biologists are looking for clues that could explain the apparent contradiction between the two types of trees—fossils of the animals that gave rise to the cetaceans.

FIGURE 17.A

Classification Based on Molecules.

Molecular evidence led to reconsideration of the relationships among hippos, pigs, and whales. Several types of molecular measures indicate that cetaceans (whales, dolphins, and porpoises), ruminants (deer, cows, and sheep), and hippos form a monophyletic group, or clade.

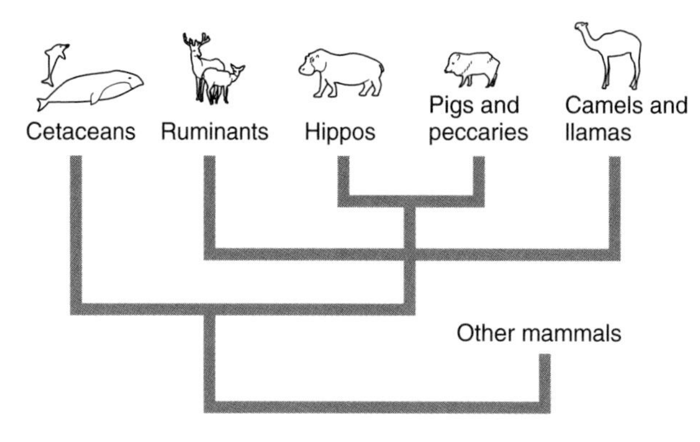

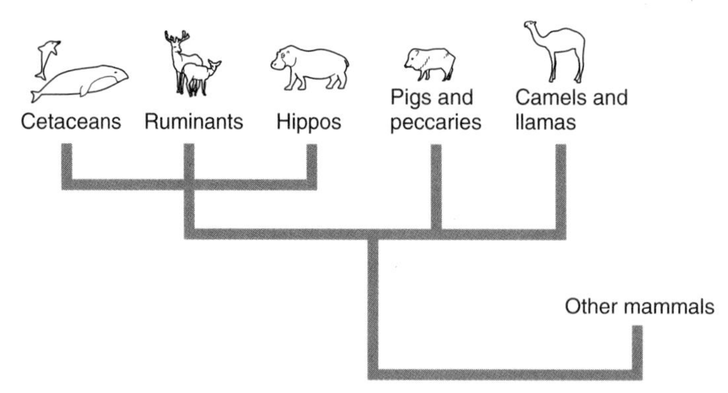

FIGURE 17.5

Fossil Distribution Yields Clues to Changing Geography.

The formation of continents explains the geographical distribution of fossils of certain theropods. About 140 million years ago, *Allosaurus* lived in many areas. As the world's large continent broke apart, populations of dinosaurs became separated and diverged genetically from each other, eventually becoming several species, three of which are shown here. The dots on the modern globe represent location of fossils of the dinosaurs that lived 90 million years ago. The inset is the portion of figure 17.3 that includes the dinosaurs highlighted.

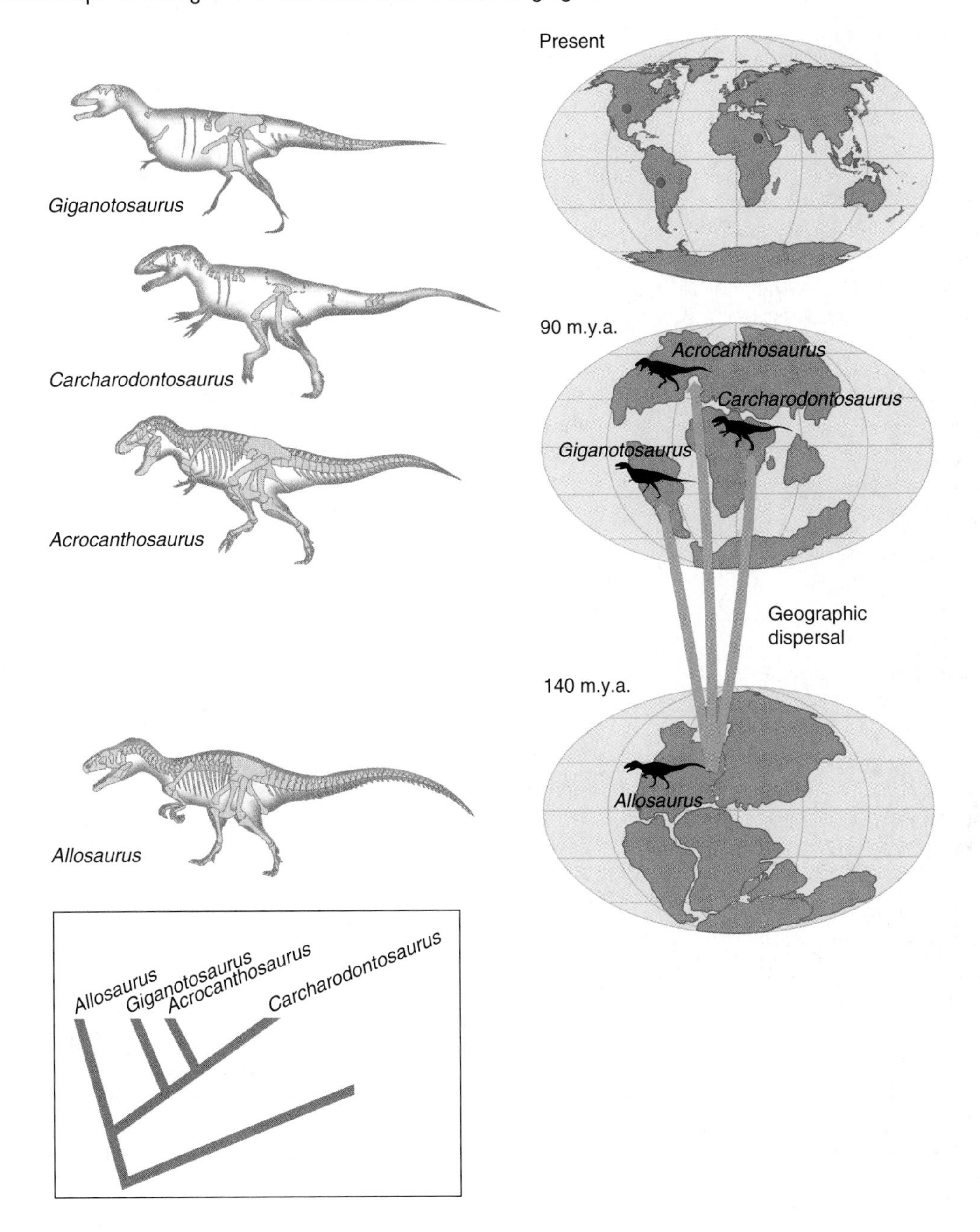

descended from a common ancestor in whom their derived characters arose, and that ancestor lived at a time when the landmasses of Earth were united. The common ancestor may have been *Allosaurus.*

Placing the fossil evidence for these giant carnivores into a bigger picture requires consulting the **geological timescale,** which uses mass extinctions to define a series of eras. Within the five eras—Archean, Proterozoic, Paleozoic, Mesozoic, and Cenozoic—time is further divided into periods and epochs. The time representation is very skewed, reflecting the incompleteness of our knowledge, so that the vastness of the older eras is not accurately depicted. That is, the Cenozoic era is but a flicker of time compared with all of Earth's existence, but it is the time about which we know the most, since it is the most recent, and ongoing. By contrast, the Archean and Proterozoic eras, sometimes collectively called the Precambrian time, are the longest, yet we know the least about them. (A distinction between Precambrian and Cambrian time was made because evidence of Precambrian life was once extremely rare.)

Figure 17.6 shows landmasses at different spans along the geological timescale. *Allosaurus* dwelled while Earth had one

FIGURE 17.6
A Changing World.
The distribution of landmasses on Earth has changed with time, due to shifting tectonic plates.

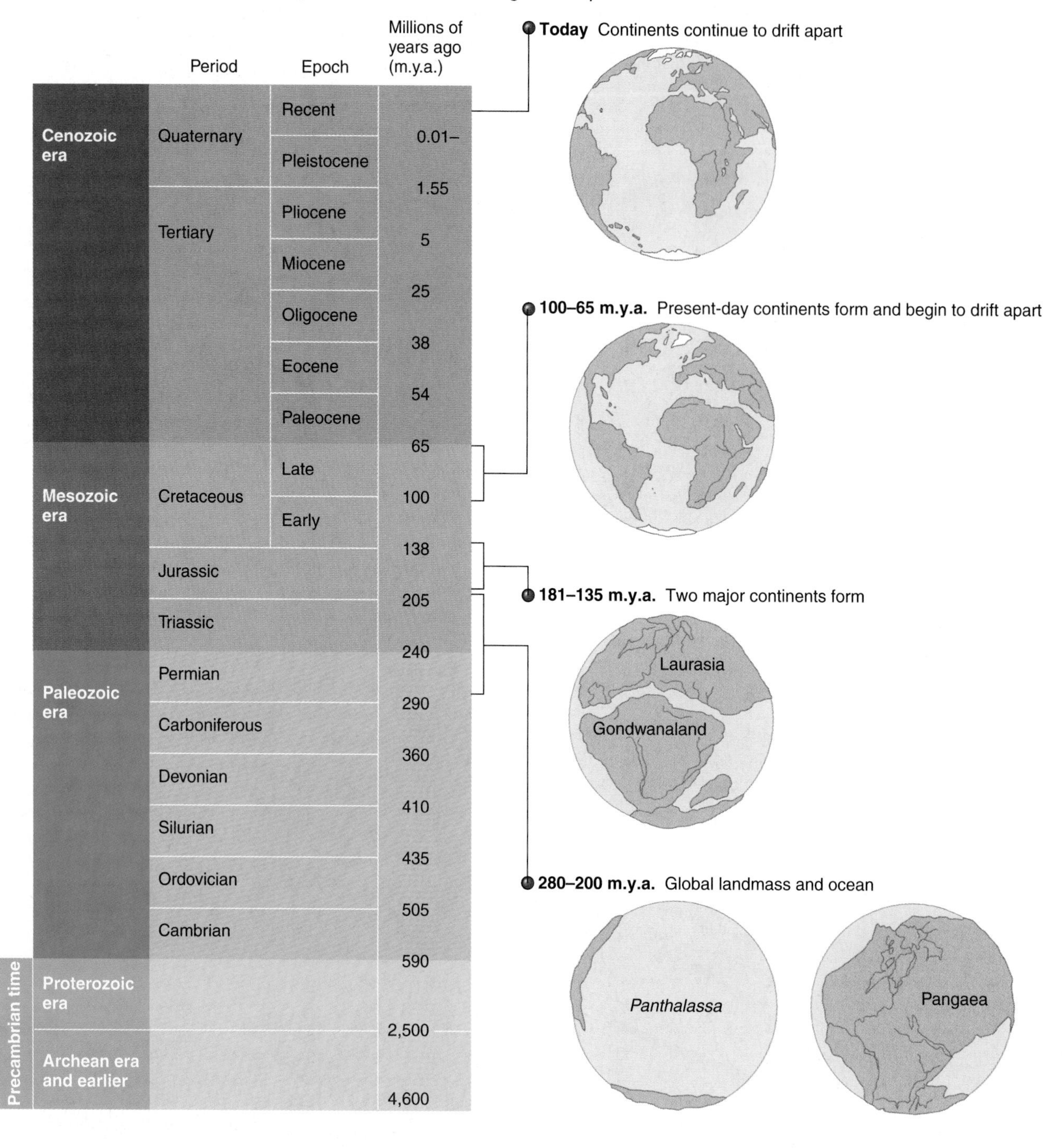

large landmass, Pangaea. Fossils of this giant indicate that it lived in many places, which were accessible because all the land was attached. But as Pangaea broke apart from 135 to 65 million years ago, these theropods became separated into three populations, or groups of populations. By this time, the last land bridges connecting northern and southern lands had sunk, and oceans separated the southern continents. The forces of allopatric speciation came into play as the gene pools of the giants diverged, but certain key features—the semiupright stance, distinctive shapes of facial bones, and behaviors associated with a carnivorous lifestyle—persisted. Eventually arose the three species that we recognize today from their preserved bones.

How Did Fossils Form?

A fossil is any evidence of an organism from more than 10,000 years ago. It may be part of the original organism, such as a thin film of plant cuticle that remained after a leaf was buried under sediments and the water squeezed out (fig. 17.7A). This is an example of a **compression fossil.** Coal, oil, and natural gas are also actual remains of plants and are called fossil fuels. An **impression fossil** is an outline that formed when hard parts of a dead organism pressed against soft sediment, which then hardened after the organism decayed (fig. 17.7B). Footprints and worm borings are impression fossils.

Many fossils formed when minerals slowly replaced tissue. Picture an inch-long horn coral that died on an ancient sea bottom in what is now Indiana and was covered by sand and mud. Gradually, the sediment hardened into rock. Meanwhile, the hard part of the horn coral dissolved, leaving an impression of its shell. Millions of years later, a person walking along land that was once that sea bottom sees the impression. Perhaps the mold was filled in

FIGURE 17.7
How Fossils Form.
(**A**) A compression fossil of a leaf preserves part of the plant. (**B**) An impression fossil reveals anatomical details as an imprint. Many shelled animals and ferns left imprint fossils. (**C**) This horn coral is a cast. Once-living material dissolved and was replaced by mud that hardened into rock. (**D**) Most fossils of our immediate ancestors are mineralized bones and teeth, usually found in fragments. (**E**) Calcium phosphate raining down in the ocean preserved this ball of cells—an early animal embryo, species as yet unknown. Researchers at first thought these balls were algae, but the telltale constancy of the ball size, irrespective of the number of cells, was a giveaway that these were actually animal embryos. They come from a time just before animals greatly diversified.

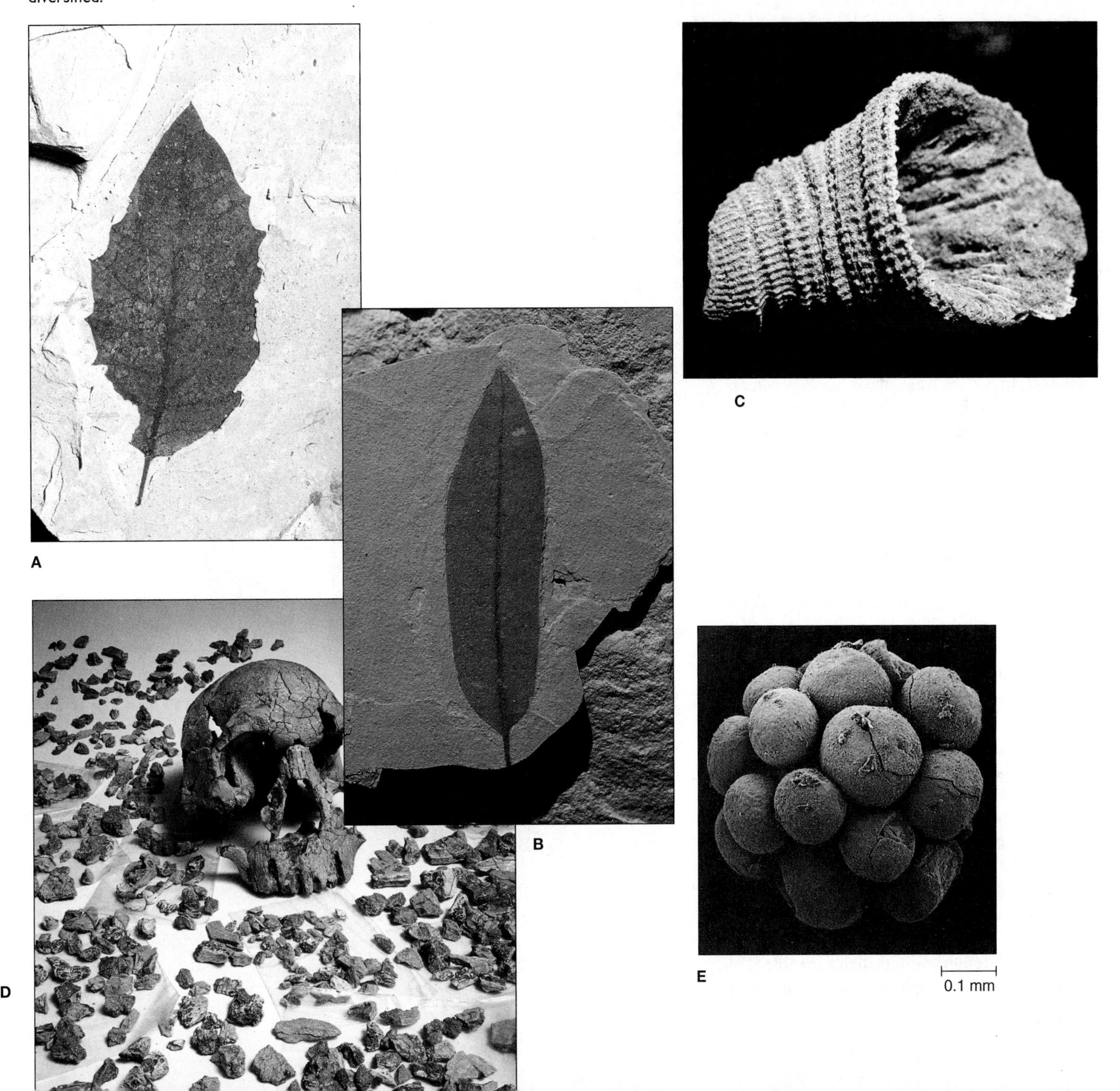

with more mud, which also hardened into rock. The explorer may then find a cast of the horn coral, a rocky replica of the ancient animal (fig. 17.7C). Evidence of our recent relatives often consists mostly of teeth, which are even harder than bone and are the likeliest anatomical parts to be preserved (fig. 17.7D). Bones became fossilized when minerals replaced cells and intercellular materials.

In a process called **petrifaction,** minerals percolate into the tissues of trees, producing petrified wood that reveals cellular structures when sliced thin. In a type of petrifaction called phosphatization, calcium phosphate replaces soft parts soon after death, preserving subcellular structures such as nuclei in exquisite detail. Phosphatization has preserved small objects, less than 2 millimeters in diameter. German researchers discovered the process in the 1970s as the mechanism of preservation of certain microscopic crustacean (shellfish) larvae. Then in the mid-1990s, several research groups discovered many phosphatized remains of algae and the embryos of sponges and other simple animals in the Doushantuo formation in south-central China (fig. 17.7E). This rich community flourished from 570 to 580 million years ago, a time of great paleontological import because it predates two periods that left an enormous assemblage of fossils—the Ediacaran (about 565 million years ago), when bizarre-looking animals unlike any modern types lived, and the Cambrian period (from about 543 to 530 million years ago), when evidence of life is so abundant that it is often called an "explosion."

Unusual circumstances may have preserved biological material. For example, gold miners found a perfectly preserved baby mammoth in the ice of the Arctic circle in Siberia. Elsewhere, sticky tree resin entrapped insects and frogs and hardened them in translucent amber tombs. Cloning extinct organisms from DNA preserved in amber is a popular science fiction plot. Biotechnology 17.1 discusses what cloning a wooly mammoth might entail.

The most striking fossils formed when sudden catastrophes, such as mudslides and floods, rapidly buried organisms in an oxygen-poor environment. Without oxygen, tissue damage was minimal, and scavengers could not reach the dead, enabling even soft-bodied organisms to leave detailed anatomical portraits. Such a sudden burial apparently happened in the Patagonian floodplain that preserved thousands of dinosaur eggs and embryos (see fig. 17.4).

Estimating the Age of a Fossil

Scientists can estimate when an organism that left fossils lived in two ways. **Relative dating** places a fossil in the context of other fossils, and is usually based on the principle of superposition, with lower rock strata presumed to be older. The farther down a fossil is, the longer ago the organism it represents lived—a little like a memo at the bottom of a pile of papers on a desk being older than one nearer the top. Relative dating provides an age relative to other fossils in strata above and below.

Researchers obtain more precise dates, called **absolute dating,** by using the constant and measurable rate of decay of certain radioactive minerals (fig, 17.8). **Radiometric dating** is a type of absolute dating that uses natural radioactivity as a "clock." Isotopes of certain elements are naturally unstable, which causes them to emit radiation. As the isotopes release radiation (or "radioactively decay"), they change into different isotopes of the same element, or they become a different element. (Recall from chapter 2 that an isotope is a form of an element distinguished by a different number of neutrons.) Each radioactive isotope decays to its other form at a characteristic and unalterable rate called its **half-life.** The half-life is the time it takes for half of the isotopes in a sample of the original element to decay into the second form. If we know the half-life of a radioactive isotope and the amounts of its "before" and "after" forms in a rock (a fossil) or tissue sample, we can deduce when the sample formed.

Two radioactive isotopes often used to assign dates to fossils are potassium 40 and carbon 14. Potassium 40 radioactively decays to

FIGURE 17.8

Carbon 14 Dating.

Living organisms accumulate radioactive carbon 14 as they assimilate carbon through photosynthesis or ingestion of other organisms. Although the carbon 14 is decaying and being released as nitrogen 14, there is a steady supply of carbon 14 through metabolism. After death, the replenishment ceases, and the proportion of carbon 14 to carbon 12 in the organism begins to shift. Measuring the proportion allows scientists to determine how long ago the organism that was fossilized died.

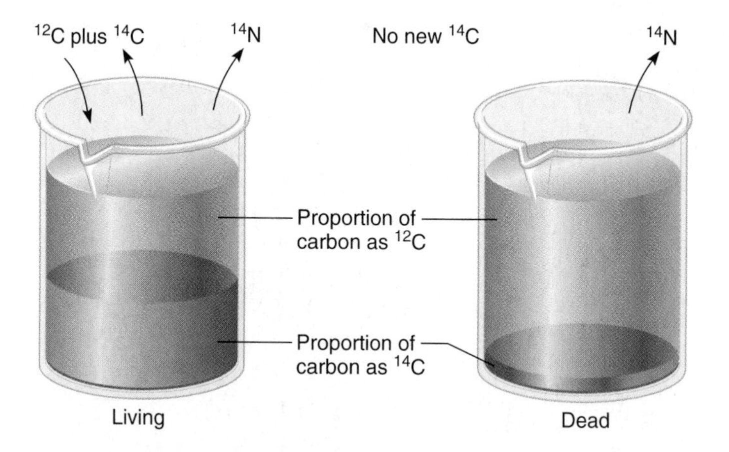

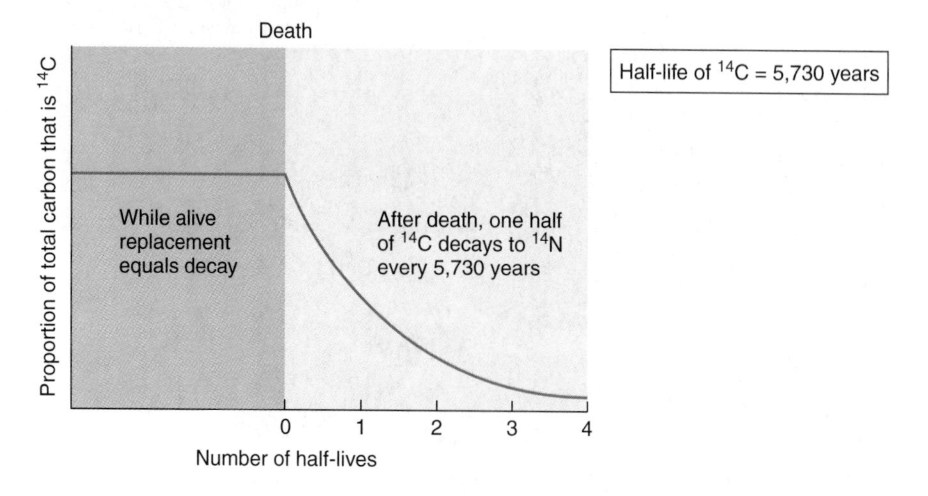

Biotechnology 17.1

Can We Bring Back the Woolly Mammoth?

High in the cliffs of Siberia, rare bones and tufts of hair poking from the bare rock may be the remains of woolly mammoths. These elephant ancestors roamed the grasslands here from 1.8 million years ago until about 11,000 years ago. Starvation following the last ice age drove their extinction, although a few isolated populations survived until about 3,800 years ago (fig. 17.B).

Today biologists from Japan are searching for mammoths that were flash-frozen, so that some of their cell nuclei might have been preserved, and can possibly yield DNA. Once the researchers have mammoth DNA in hand, they hope to take two approaches to recreating the mammoth. If sperm are found, the researchers plan to use those bearing X chromosomes to artificially inseminate a modern elephant. After a 600-day pregnancy, the elephant should give birth to a female that is half elephant, half mammoth. That offspring, some 13 or so years later, can then be inseminated by more mammoth sperm, producing a baby that is three-quarters mammoth, and so on. All of this depends upon whether the hybrid animal is fertile. It was in this time-consuming manner that conservation biologists "brought back" Przewalski's horse, repeatedly breeding members of the dwindling population to other horses to build a breed that is genetically mostly the now-extinct Przewalski's variety.

A faster way to recreate a woolly mammoth is to clone one. This would require a nucleus from a somatic cell, which could then be delivered into an elephant's oocyte whose nucleus has been removed. The offspring would be all mammoth, and likely fertile, in contrast to the half-breed that would result from artificial insemination. Males and females would need to be cloned to begin a population, and from 10 to 20 different genetic variants would be necessary to sustain a population.

The idea to clone a mammoth is an outgrowth of the effort to collect DNA from extinct organisms preserved in amber, which is hardened resin from certain pine trees. The mix of chemicals in amber entomb whatever happens to fall into it when it is the consistency of maple syrup. Alcohols and sugars in the resin dry out the specimen, and other organic molecules act as fixatives, keeping cellular contents in place. The resin itself seals out oxygen and bacteria, which would otherwise decompose tissue before it could be preserved. Finally, organic molecules called terpenes link together, hardening the resin over a period of 4 to 5 million years.

The novel and film *Jurassic Park* described the cloning of dinosaurs from blood in mosquitoes trapped in amber—not a very likely scenario. But in reality, the first successful extraction of bits of DNA from amber occurred in 1990, from a 17-million-year-old magnolia leaf entombed in amber. Those researchers started "the million-plus club" to encourage others to obtain ancient DNA (fig. 17.C). This challenge has sent researchers into the back rooms of museums, literally dusting off specimens of pressed leaves, insects stuck with pins into Styrofoam, and old bones and pelts, in search of nucleic acid clues to life in the past.

FIGURE 17.B
Frozen in Time.
"Dima" the baby mammoth was about 9 months old when she perished about 40,000 years ago in Siberia near the Arctic Circle. She stood about 3 feet (1 meter) tall and weighed about 140 pounds (64 kilograms).

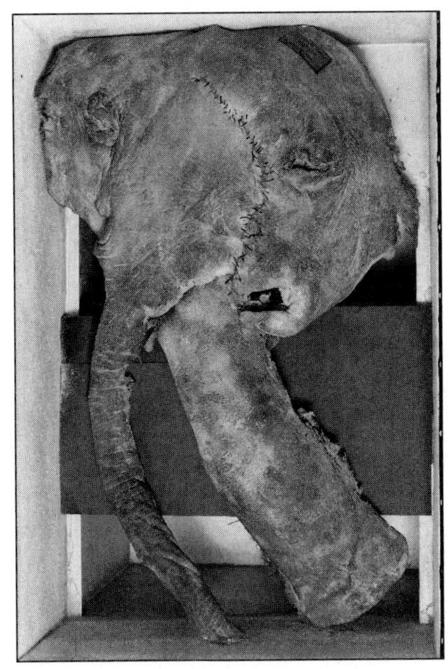

FIGURE 17.C
Ancient DNA.
Researchers have extracted DNA from these organisms.
Source: Data from Roger Lewin, "Patterns in Evolution" in *Scientific American*, 1997.

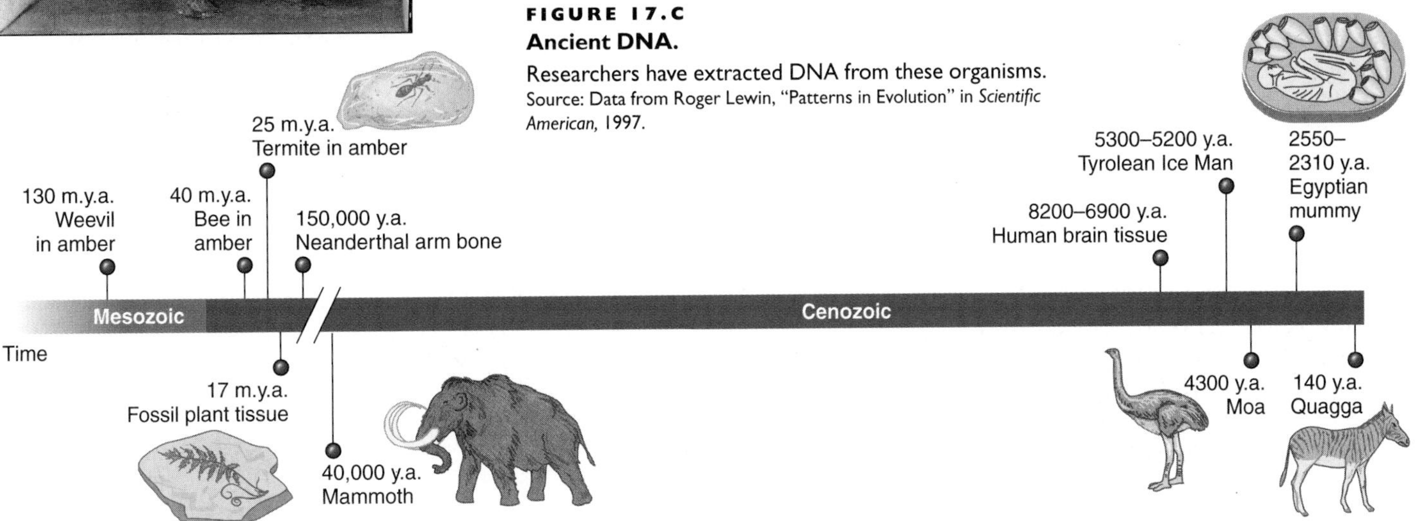

argon 40 with a half-life of 1.3 billion years, so it is valuable in dating very old rocks containing traces of both isotopes. As volcanic rocks form, they trap a certain amount of potassium when they cool. Since a known proportion of the trapped potassium is potassium 40, scientists use the proportion of potassium 40 to argon 40 as a measure of the age of the rock. They then extrapolate and estimate the age of any fossils found in sedimentary layers above or below the volcanic rock. Chemical analyses can detect argon 40 in amounts small enough to correspond to fossils that are about 300,000 years old or older.

Fossils up to 40,000 years old are radiometrically dated by measuring the proportion of carbon 12 to the rarer carbon 14 (see fig. 17.8). Carbon 14 is a radioactive isotope that forms naturally in the atmosphere when nitrogen gas is bombarded with cosmic rays from space. Less than 1% of atmospheric carbon (in CO_2) is carbon 14. Organisms accumulate carbon 14 during photosynthesis or by eating organic matter. When an organism dies, however, its intake of carbon, including carbon 14, stops. Carbon 14 decays to the more stable nitrogen 14 with a half-life of 5,730 years. As carbon 14 decays, the ratio of carbon 12 to carbon 14 changes. The ratio is then used to determine how long ago death occurred. For example, radioactive carbon dating was used to determine the age of fossils of vultures that lived in the Grand Canyon. The birds' remains have about one-fourth the carbon 14 to carbon 12 ratio of a living organism. Therefore, about two half-lives, or about 11,460 years, passed since the animals died—it took 5,730 years for half of the carbon 14 to decay, and another 5,730 years for half of what was left to decay to nitrogen 14.

One limitation of potassium-argon and carbon 14 dating is that they leave a gap, resulting from the different half-lives of the radioactive isotopes. Carbon 14 dates extend up to 40,000 years ago, but potassium-argon dates do not begin reliably until 300,000 years ago. Several new techniques cover the missing years.

17.2 MASTERING CONCEPTS

1. What can fossils reveal about life?
2. What are some of the ways that fossils form?
3. Distinguish between relative and absolute dating of fossils.

OLC

17.3 Comparing Structures

Comparing bones in various dinosaurs and birds vividly illustrates evolutionary change. Homologous structures are retained from a shared ancestor, whereas analogous structures reflect adaptation to a similar environment. Vestigial organs and similarities among embryos are also evidence of evolution.

Some clues to the biological past come from the biological present. Such techniques compare anatomy, physiology, and biochemistry among modern species. In the best of situations, comparative information is available from modern species as well as from fossils.

Comparing Anatomical Parts Can Reveal Evolutionary Relationships

A compelling explanation for anatomical similarities among different modern species is that the organisms descended from a common ancestor. For example, all vertebrate skeletons consist of the same parts, are made of the same materials, and support the body (fig. 17.9). An ancestor to modern vertebrates must have originated this skeletal organization, which then gradually became modified as more recent organisms (fishes, amphibians, reptiles, birds, and mammals) adapted to their environments.

FIGURE 17.9
Vertebrate Skeletal Organization Reflects Common Ancestry.
Although a human walks erect, a lion walks on all fours, a bird flies, and a seal swims, all have skeletons that are similarly organized and composed of the same type of tissue.

A return to the dinosaur-bird saga illustrates the power of comparing anatomical structures—bones in this case—in establishing a relationship between two types of organisms, as well as in vividly revealing how species change over time. Figure 17.10 depicts changes in three parts of the skeletons of five key dinosaurs represented in the cladogram of figure 17.3. Specifically, the pubic bone in the pelvis shifted gradually from pointing forward in the theropod *Coelophysis* to pointing backward in modern birds. The hand narrowed as the number of digits decreased, and the wristbone changed into a form that would ultimately support the flapping of wings. Finally, the clavicles fused, thinned, and narrowed into wishbones as dinosaurs became birds. All of these changes contributed to the evolution of flight.

Structures that are similar in different types of organisms because they are inherited from a recent common ancestor, such as the vertebrate skeleton, are termed **homologous.** The structures may or may not have similar functions, but they share a common origin. The middle ear bones of mammals, for example, originated as bones that supported the jaws of primitive fishes, and they still exist as such in simpler vertebrates. These bones are homologous and reveal our shared ancestry with fishes.

FIGURE 17.10
Comparative Anatomy.
Comparing pelvic, wrist and hand bones, and clavicles reveals the adaptations that underlie the evolution of flight.

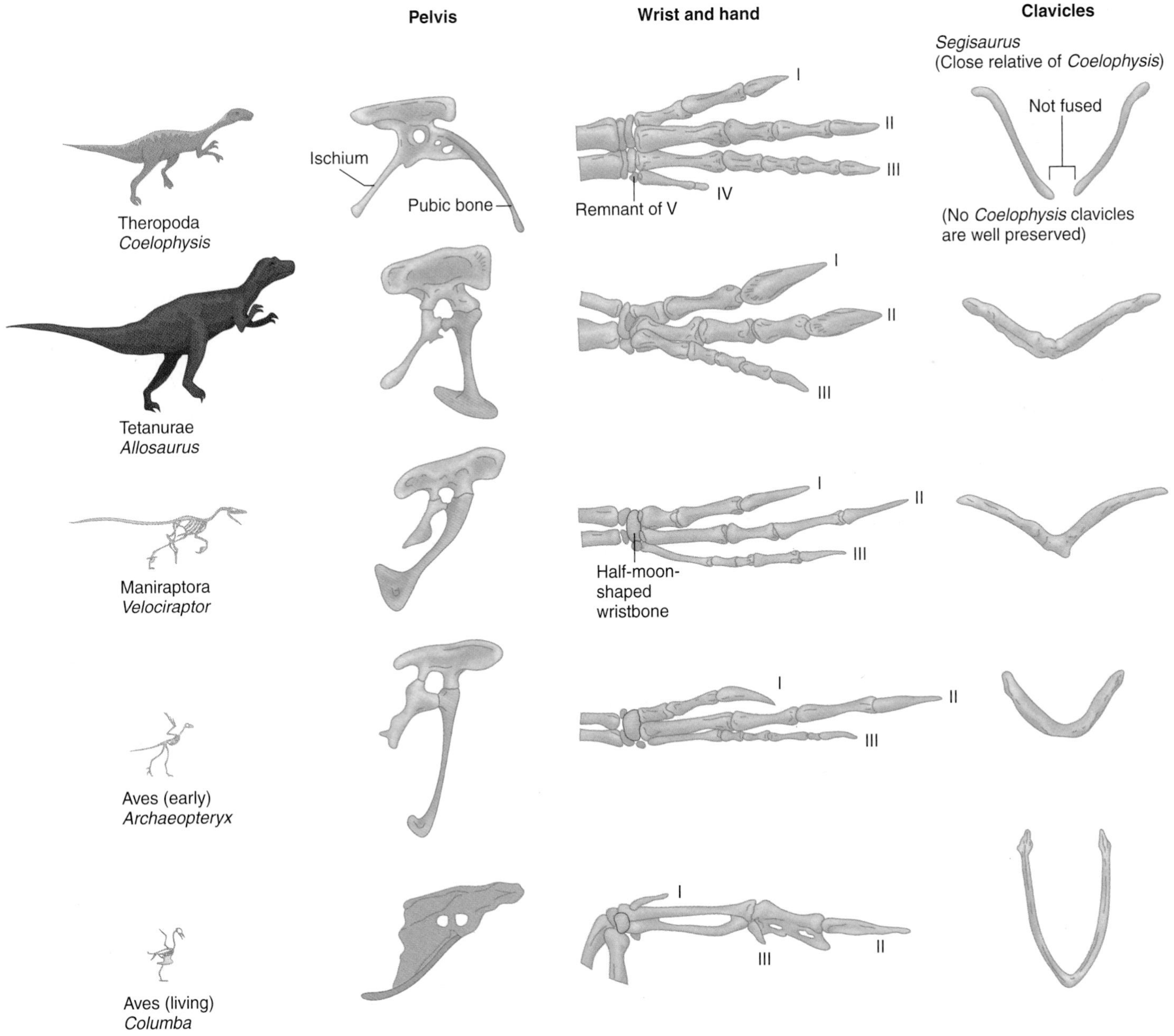

Structures that are similar among different species but that evolved independently, perhaps in response to similar environmental challenges, are termed **analogous.** Bats, birds, and insects, for example, have wings, but the bird's wing is a modification of vertebrate limb bones, whereas the insect's wing is an outgrowth of the cuticle that covers its body (fig. 17.11).

Determining whether particular body parts are homologous or analogous can be difficult. In general, though, analogous structures tend to resemble one another only superficially, whereas the similarities in body parts between two species related by recent common descent—homologies—tend to be complex and numerous.

Considering the original definitions of the terms, coined by British anatomist Sir Richard Owen, can help distinguish between them. According to Owen, homology is "the same organ . . . under every variety of form and function." Analogy, however, refers to "a part or organ . . . which has the same function. . . ."

Vestigial Structures Reflect the Past

Evolution is not a perfect process. As environmental changes select against certain structures, others are retained, sometimes persisting even if they are not used. A structure that seems to have no function in one species, yet is homologous to a functional organ in another species, is termed **vestigial.** Darwin compared vestigial structures to silent letters in a word—they are not pronounced, but they offer clues to the word's origin.

Humans have several vestigial organs. In the digestive system, the appendix serves no known function, but may once have helped process different types of foods. The tiny muscles that are attached to our hairs made fur stand on end in our hairier ancestors, helping them to conserve heat. In us they serve only as the basis of gooseflesh! Human embryos have tails and gill pouches, structures that persist in the adults of some other types of animals. A trio of muscles above our ears that most of us can't use

FIGURE 17.11
Homology and Analogy.
A bat wing, mouse forelimb, and human arm are homologous structures, all built of bone, and presumably inherited from a recent common ancestor. The wings of bats, birds, and butterflies are analogous structures because they are not made of the same materials, nor are they organized in the same way, indicating that they are not inherited from a recent common ancestor.

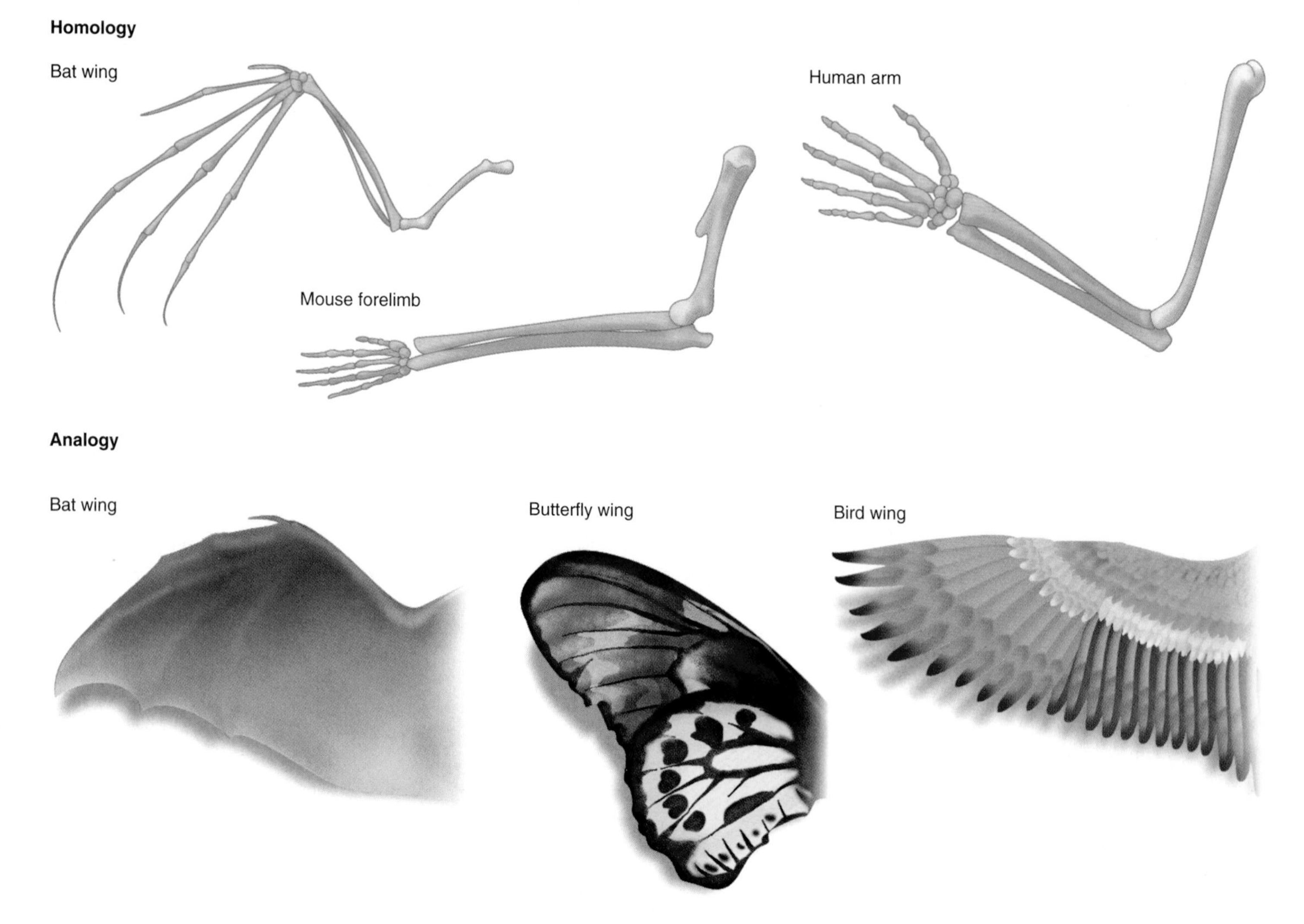

help other mammals move their ears in a way that improves hearing. For humans to possess these structures, we must have the genes for making them. Each structure, then, links us to other animals that still use these features.

In whales and some snakes, tiny leg bones are vestigial, retained from vertebrate ancestors who used their legs (fig. 17.12A). The vestigial leg bones in whales is one factor that may have led to their misclassification (see Investigating Life 17.1).

The reason for snakes' lack of limbs stems from the unusual expression of certain genes, called HOX genes, in early development. In many species, these genes direct the development of the correct body parts in embryos, including limb formation, by specifying protein gradients that tell cells which genes to express, molding their specialization. The HOX genes that stimulate limbs to form in mouse and chicken embryos do so by being turned on in cells located at certain key points in the middle layer of the embryo. In snake embryos, in contrast, these genes are turned on all along the length of the middle layer, and so there is no special signal telling the embryo where to start sprouting limbs. Without this developmental signal, the snake's legs remain mere buds, vestiges of what would otherwise be (fig. 17.12B). A fossil snake, *Pachyrhachis problematicus,* may have been a transitional form in snake evolution—it has hindlimbs but lacks forelimbs. Perhaps it underwent an initial mutation in a HOX gene, which was later compounded by a second mutation that blocked formation of hindlimbs too.

FIGURE 17.12
Vestigial Organs.
(**A**) Snakes have tiny femurs (leg bones) that are vestigial, but detectable only in the skeleton (**B**).

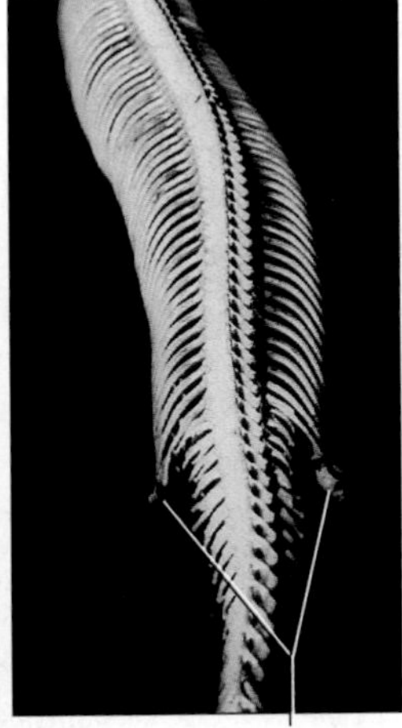

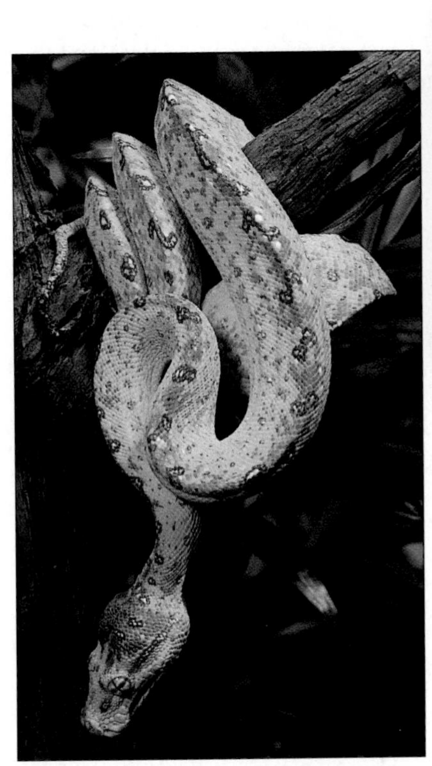

Embryos Provide Evolutionary Clues

Discovering occupied dinosaur eggs and being able to manipulate genes in modern embryos brings prenatal organisms into the realm of the study of evolutionary biology. Since related organisms share many physical traits, they must also share developmental processes that produce those traits. By comparing embryos at different stages, it should be possible to reconstruct some of the steps that led to differences among species. In times past, however, comparing embryos led to some confusion because embryos of many types of vertebrates look alike, particularly early in development (see fig. 40.2). Darwin suggested that the striking similarities of vertebrate embryos reflect shared adaptations to their similar environments—floating in a watery bubble, either in a uterus or in an egg. Natural selection also plays a role in the resemblance of vertebrate embryos, because those that have a devastating mutation never develop far enough to pass on the gene. Over time, similar functions carried out by similar structures came to persist in the embryos.

In 1874, German naturalist Ernst Haeckel published drawings of embryos from different vertebrate species, including fishes, reptiles, birds, and mammals, that appeared nearly identical. Shortly thereafter, Haeckel admitted to a bit of artistic license, saying that he blackened the chick's eye, for example, and curled the bird embryo's tail to make it look more like a human's. He tended to blur or leave out distinctions among the embryos, and did not include species designations. The biggest problem, though, was that Haeckel did not represent scale. That is, although embryos might start out with basically the same parts, different subsequent growth rates distinguish species as gestation proceeds, such as the very fast growth of the head in the primate fetus. Differential gene action also accounts for retention of juvenile features in some species, such as salamanders, and the presence or absence of hair.

Despite Haeckel's admission and many doubters, biology textbooks continued to print his illustrations for generations. A few years ago, a group of developmental biologists photographed embryos and fetuses of a variety of vertebrate species throughout development, finally laying to rest the issue of Haeckel's fudging (fig. 17.13). The data show that there really are similarities in embryonic structures, supporting the concept of common ancestry.

17.3 MASTERING CONCEPTS

1. What can comparing homologous structures reveal about evolution?
2. How can distinguishing between homologous and analogous structures be confusing?
3. What is a vestigial organ?
4. Why do vertebrate embryos appear similar, but then become very different adult animals?

OLC

FIGURE 17.13

Embryo Resemblances.

Vertebrate embryos appear alike early in development, reflecting the similarities of basic processes as cells divide and specialize, as the figure shows for five species. As development continues, and parts grow at different rates in different species, the embryos do not look as similar. The drawings show the similarities that are sometimes observed in photos by yolk sacs or other membranes.

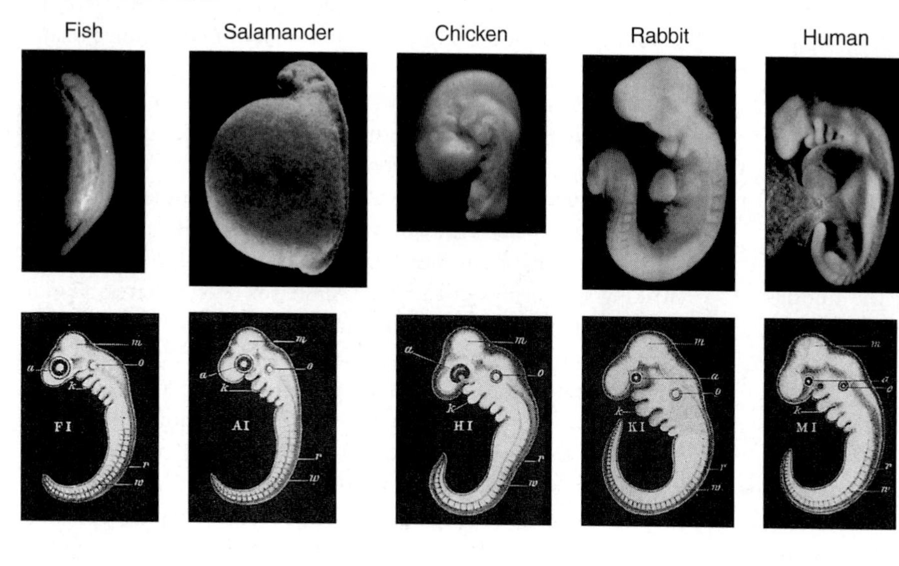

17.4 What Can Molecular Evidence Reveal?

Molecular evidence for evolution includes similarities at the gene, protein, chromosomal, and genome levels. Gene sequence differences among species can be placed in a time frame derived from mutation rates.

If a major problem with the fossil record is its incompleteness, then the major problem with molecular evidence for evolution is just the opposite—there is so much information that researchers can't analyze it fast enough! With entire genomes being sequenced on a monthly basis, it is becoming routine to locate a gene, then scan massive databases to study its function in other species. The opening essay for chapter 13 describes these genome projects and what they reveal.

The techniques used to study molecular evolution are based on information—they compare gene and protein building block sequences, or chromosome banding patterns, among different species. The logic is that it is highly unlikely that two unrelated species would evolve precisely the same gene (nucleotide) or protein (amino acid) sequences by chance. It is more likely that the similarities were inherited from a common ancestor and that differences arose by mutation after the species diverged from the ancestral type. Interpreting molecular sequence data is also less subjective than deciding whether two similar-appearing structures are homologous or analogous.

An underlying assumption of molecular evolution studies is that the greater the molecular similarities between two modern species, the closer their evolutionary relationship. Chromosome banding patterns and gene or protein sequences that are identical or very similar among different species are said to be **highly conserved.**

Comparing Chromosomes

Karyotypes (chromosome charts) can provide clues to how the organization of genomes has changed over time. Researchers compare the number of chromosomes, chromosome band patterns, and gene orders on chromosomes to measure species relatedness. For example, human chromosome banding patterns most closely match those of chimpanzees, then gorillas, and then orangutans. If both copies of human chromosome 2 were broken in half, we would have 48 chromosomes instead of 46, and the 48 would very closely resemble those of the three species of apes. Of our 23 chromosome pairs, 18 appear virtually identical to their counterparts in the apes. The remaining differences are largely due to inversions or translocations of chromosome segments.

Chromosome patterns can also be compared between species that are not as closely related as are humans and other primates (table 17.1). For example, the banding pattern of chromosome 1 in humans, chimps, gorillas, and orangutans matches the pattern in two small chromosomes in the African green monkey. Because fossil evidence indicates that monkeys are more ancient than apes, the two small chromosomes may have originated in the ancestor of green monkeys and then fused as primates continued to evolve. Another example of chromosome conservation is that all mammals have identically banded X chromosomes. Figure 17.14 shows similarities between the karyotypes of a human and a cat.

Chromosome band similarities can be striking, but they do not measure species relatedness as precisely as do a gene's nucleotide or a protein's amino acid sequence. Chromosome band patterns resulting from traditional staining may superficially resemble each other in different species, without corresponding on a gene-by-gene basis. Incorporating DNA probes into chromosome

TABLE 17.1 Percent of Common Chromosome Bands Between Humans and Other Species

Chimpanzees	99+%
Gorillas	99+%
Orangutans	99+%
African green monkeys	95%
Domestic cats	35%
Mice	7%

FIGURE 17.14
Conserved Regions of Human and Cat Chromosomes.
In each pair, the chromosome on the left is from a cat, and the chromosome on the right is from a human. The colored boxes indicate apparently corresponding areas. The chromosomes in (**A**) have similar banding patterns generated from traditional stains, which do not target specific genes but rather chromosome regions with generally similar DNA base content. The chromosomes in (**B**) do not look alike, but DNA probes to specific genes indicate that the parts do indeed share many specific DNA sequences. (**C**) One of the authors with her cat, Nirvana.

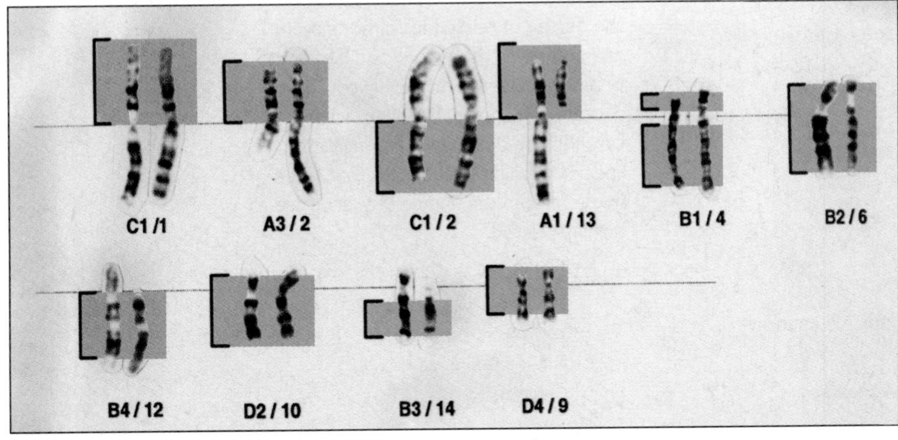

A

C

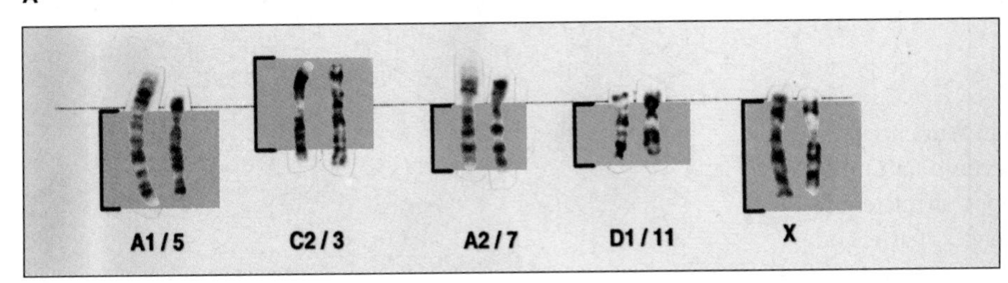

B

banding enables researchers to compare individual genes (see the opening essay in chapter 11). The goal is to identify regions of **synteny,** or identical sequences of genes along parts of chromosomes. For example, 11 genes are closely linked on human chromosome 21, on mouse chromosome 16, and on chromosome U10 in cows. However, several genes on human chromosome 3 have counterparts in mice and cows on chromosomes corresponding to human chromosome 21. Perhaps a mammal ancestral to all three species carried all of these genes on one chromosome. In humans, the genes dispersed to chromosomes 3 and 21. Chromosome similarities are thus like puzzle pieces of the past.

Comparing Protein Sequences

The fact that all species use the same genetic code to synthesize proteins argues for a common ancestry to all life on Earth. Additional evidence for evolution is that many different types of organisms use the same proteins, with only slight variations in amino acid sequence. The keratin genes that encode a sheep's wool protein, for example, have counterparts on human chromosome 11.

Comparisons of amino acid sequences among different species often support fossil and anatomical evidence of evolutionary relationships, but sometimes they reveal startling information. In the early 1960s, for example, comparison of a blood protein called serum albumin led, after considerable argument, to a rearrangement of the accepted human family tree (fig. 17.15). In those days before DNA sequencing, Morris Goodman, at Wayne State University, examined how rabbit antibodies (immune system proteins) distinguished among albumin from humans, chimpanzees, gorillas, orangutans, and gibbons. The test, based on highly specific molecular recognition, suggested that we are rather closely related to other primates: The rabbit antibodies reacted in the same way to albumin from humans, chimps, and gorillas, but not as strongly to albumin from orangutans and even less strongly to that of gibbons.

Another study done in 2000 found 7 of 20 selected proteins to be identical in amino acid sequence between humans and chimps. One of these proteins is cytochrome *c,* which is part of the electron transport chain in cellular respiration. In addition, 20 of its 104 amino acids occupy identical positions in the cytochrome *c* of all eukaryotes. Figure 17.16 shows that the more closely related two species are, the more alike is their cytochrome *c* amino acid sequence.

Comparing DNA Sequences

Amino acid sequence similarities of course reflect DNA sequence similarities, because genes encode proteins. Because different DNA triplets can specify the same amino acid, however, DNA sequences can provide more information than can amino

FIGURE 17.15

Humans Are Very Closely Related to Gorillas and Chimpanzees.

Molecular evidence indicates a recent shared ancestry among gorillas, chimpanzees, and humans.

Traditional classification

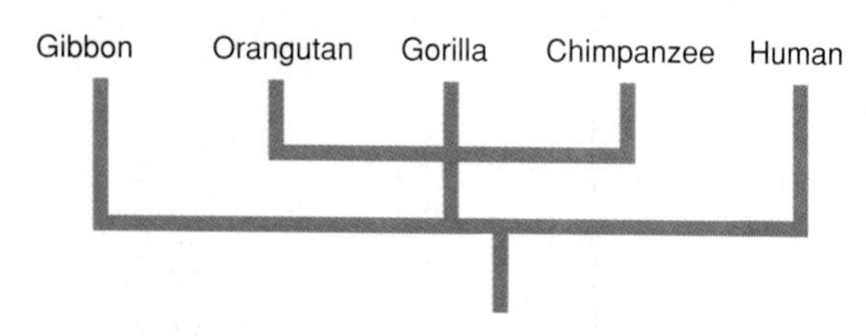

Modern classification

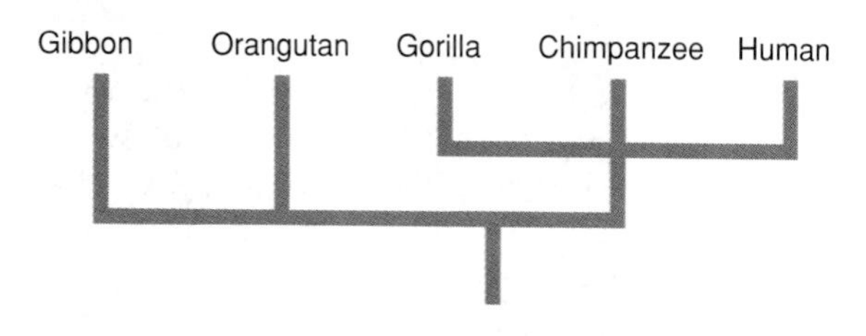

acid sequences. However, comparing amino acid sequences may be a better measure of evolutionary change, because natural selection acts only on phenotypes, and phenotypes ultimately reflect protein function. That is, two DNA sequences that encode the same amino acid sequence because of the degeneracy of the genetic code would be subject to the same natural selection pressures. Still, DNA differences can be used to trace origins.

Similarities in two species' DNA sequences can be assessed for a short stretch of nucleotides, for a single gene, for families of genes with related structures and/or functions, and for entire genomes. Sometimes underlying gene sequence similarities are expressed as striking phenotypes. People with Waardenburg syndrome, for example, have a characteristic white forelock of hair, wide-spaced, light-colored eyes, and hearing impairment. The causative gene is very similar in sequence to one in cats that have white coats and blue eyes and are deaf (fig. 17.17). Horses, mice, and minks also have this combination of traits. The phenotype may result from abnormal movements of pigment cells in the embryo's outermost layer.

When researchers began to compare DNA sequences among different types of organisms, in the 1970s, they used a rather crude technique called DNA hybridization. It uses complementary base pairing to estimate how similar are the genomes of two species. DNA double helices from two species are unwound and mixed. The rate at which hybrid DNA double helices—that is, DNA molecules containing one helix from each species—form is a direct measure of how similar they are in sequence. The faster the DNA from two species forms hybrids, the more of the sequence they share, and the more closely related they are presumed to be (fig. 17.18). A frequently cited DNA hybridization study from 1984 showed, for example, that total DNA differs in 1.5% of its base pairs from chimpanzee DNA, in 2.3% from gorilla DNA, and in 3.7% from orangutan DNA.

A 98.5% genetic similarity between humans and chimps may seem disturbingly high, but the true value is probably even greater!

FIGURE 17.16

Amino Acid Sequence Similarities Are a Measure of Evolutionary Relatedness.

Similarities in amino acid sequence for the respiratory protein cytochrome *c* in humans and other species parallel the degree of relatedness among them. (**A**) compares the sequence differences of nine species for this highly conserved protein. (**B**) shows the differences in cytochrome *c* sequence among four species visually. Amino acids that differ from those in the human sequence are highlighted red.

Cytochrome *c* Evolution

Organism	Number of amino acid differences from humans
Chimpanzee	0
Rhesus monkey	1
Rabbit	9
Cow	10
Pigeon	12
Bullfrog	20
Fruit fly	24
Wheat germ	37
Yeast	42

A

B

Human — Chimpanzee — Honeybee — Rice

FIGURE 17.17
Similar Genes Specify Similar Phenotype
A gene in mice, cats, humans, and other species causes light eye color, hearing or other neurological impairment, and a fair forelock of hair. The genes that cause these similar phenotypes may be homologous.

FIGURE 17.18
The Rate of DNA Hybridization Reflects the Degree of Evolutionary Relatedness.
This highly schematic diagram shows why DNA from a human hybridizes more rapidly with chimpanzee DNA than it does with chicken DNA. The * refers to sites where chimp or chicken DNA is different from human DNA.

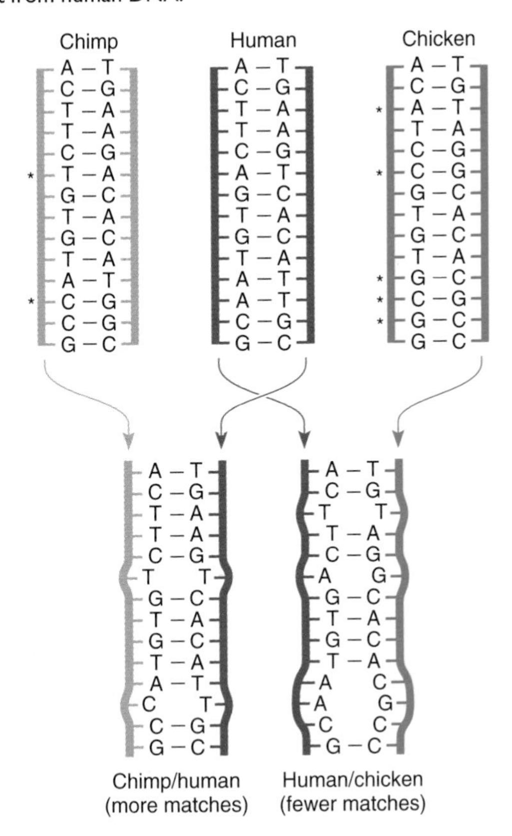

This is because part of that key 1.5% difference arises from variations in nucleotide base sequences (called polymorphisms) *within* each species. Part of the difference is also in noncoding regions of the DNA, and in degenerate codons. That is, only some of the differences are meaningful in terms of phenotype. One team of researchers combined data from DNA hybridization experiments and many individual gene sequences, and concluded that "uniquely human features" derive from less than 1% of our genomes; the rest we presumably share with chimps! This is as closely related as the tigers described in chapter 16's opening essay are to each other. Yet common sense tells us that humans and chimps are hardly interchangeable.

Why can we recite poetry and build ships and form governments, and chimps can't? Why don't chimps get AIDS or develop cancer at the rates that humans do? Some researchers think that single genes with profound effects will be found to underlie the key differences, such as genes that control the rate of brain growth in the fetus and child, or the number and proximity of hair follicles, accounting for a chimp's fur. A single change to the larynx may be all that was needed to give humans speech. With these ideas in mind, several biotechnology companies are scrutinizing human and chimp genes that encode slightly different amino acid sequences, because these may be proteins on which natural selection has acted. It is here that the answers to the distinctions between humans and chimps will emerge, offering important clues to our origins long ago, as well as to our health problems today.

Molecular Clocks

A clock measures the passage of time by moving its hands through a certain portion of a circle in a specific and constant interval of time—a second or a minute. Similarly, a polymeric molecule (one with many building blocks) can be used as a **molecular clock** if its components are replaced at a known and constant rate. Molecular clock studies, however, are not quite as straightforward as glancing at a wristwatch.

If biologists know the mutation rate for the bases in a particular gene, plus the number of differences in the DNA sequences for that gene in two species, they can estimate the time when the organisms diverged from a common ancestor. For example, many human and chimpanzee genes differ in about 4 to 6% of their bases, and substitutions occur at an estimated rate of 1% per 1 million years. Therefore, about 4 to 6 million years have passed since the two species diverged, based on analysis of these genes.

A complication of molecular clock studies is that DNA base substitutions (microevolution) occur in different genes at different rates. This is because some DNA sequences, such as highly repeated ones, are more prone to replication errors than others. Once mutations have occurred, natural selection acts on them to different extents, depending upon the phenotype and the environment. Also, a particular gene may exert a drastic effect on evolution, yet others may exert little effect. Therefore, to construct molecular tree diagrams incorporating molecular clocks, researchers often consider DNA sequences that are not subject to natural selection, such as genes encoding ribosomal RNAs. Timescales based on fossil evidence and molecular clocks can be superimposed on molecular tree diagrams constructed from sequence data.

A special type of molecular clock measures mutations in mitochondrial DNA. Recall from chapters 3 and 7 that mitochondria are organelles that house the biochemical reactions of cellular energy acquisition. Mitochondria contain about 17,000 DNA bases, which encode RNA molecules and a few enzymes. Mitochondrial DNA (mtDNA) is valuable in tracking recent evolutionary time because its molecular clock "ticks" 5 to 10 times faster than the nuclear DNA clock—that is, mtDNA mutates 5 to 10 times faster than DNA in the nucleus. This is because mitochondria lack some of the DNA repair enzymes found in the nucleus.

Just as with other measures of evolutionary relatedness, the more similar the mtDNA sequence between two individuals, the more recently they are presumed to share a common ancestor. Mitochondrial DNA in humans, however, presents intriguing possibilities because it is inherited from mothers only, in oocytes. The part of the sperm cell that enters the oocyte at fertilization usually does not contribute mitochondria, thus preventing males from contributing mitochondria to offspring.

Mitochondrial DNA comparisons are used to trace human origins. For example, many studies show that people from Africa have the most numerous and diverse mitochondrial mutations. This suggests that Africans have existed longer than other modern peoples, because it takes time for mutations to accumulate. Additional evidence in support of the African origin of all modern humans comes from cladograms in which the groups of people whose DNA sequences form the bases of the diagrams are from Africa. This idea that early humans originated in Africa and then migrated, eventually replacing hominids elsewhere, is called the single origin hypothesis. An alternative explanation, the multiregional hypothesis, suggests that separate interbreeding populations of early humans lived throughout the world and all evolved into modern humans.

Possibly, some combination of the single origin and multiregional models is closest to the truth. Perhaps African peoples migrated to Asia and Europe, where they met and mixed with other early humans, forming the gene pools from which we ultimately arose.

How Can We Reconcile Fossil, Biogeographical, and Molecular Evidence?

Just as fossil evidence has limitations, so does molecular evidence of evolution. Molecular data paint gene-by-gene comparisons, but do not account for gene interactions. Nor can molecular information tell us directly about extinct life, unless we can obtain preserved DNA and use it to produce a phenotype in a living organism. It also cannot explain the evolution of many obvious physical differences between organisms whose genes are very similar.

What molecular evidence *can* tell us is how species are related to each other. Fossil and biogeographical evidence introduce the element of time—that is, when species most likely diverged from common ancestors, against the backdrop of other events happening on Earth.

It is exciting when molecular evidence supports other kinds of evidence for evolution. This is so for the hundreds of species of cichlid fishes that have spread during the past several hundreds of thousands of years in the African great lakes. As new lakes formed, they isolated populations of the fishes, and the animals eventually accumulated sufficient genetic changes to constitute new species. The approximate times when DNA sequences diverged, according to mitochondrial DNA sequence comparisons, coincide with geographic evidence of when earthquakes occurred. The earthquakes may have formed the lakes that separated the ancient gene pools of cichlid fishes.

Sometimes different types of evolutionary evidence do not agree. For example, most fossils of simple animals date to 500 to 600 million years ago, with a very few dating back a billion years. DNA and protein sequence data on modern descendants of these early animals, however, indicate that they diverged into several species at least a billion years ago. A possible explanation for the disagreement between molecular and fossil evidence is that simple animals may have been abundant a billion years ago, as the sequence data suggest, but left few fossils from that time. The recent discoveries of never-before-seen microscopic algae and animal embryos from south-central China (see fig. 17.7E) predating the Cambrian explosion suggest that the molecular data were indeed on the right track.

As the stories of life told in the rocks and in the sequences of informational biomolecules accumulate, converge, and overlap, we will learn more about where we have come from. Perhaps this knowledge will give us a better understanding of where we are headed—subjects of the next chapter.

17.4 MASTERING CONCEPTS

1. What types of structures and molecules can be compared to yield information on evolution?
2. What is the conceptual basis for molecular evolution studies?
3. What are the advantages and limitations of molecular evolution studies?
4. What is the basis of applying a timescale to molecular sequence information?

OLC

Chapter Summary

17.1 Reconstructing the Stories of Life

1. Evidence for evolutionary relationships comes from **paleontology** (the study of past life) and comparing anatomical and biochemical characteristics of species.
2. **Systematics** examines species relationships. **Cladograms,** based on **derived characters,** have largely replaced evolutionary tree diagrams based on less objective physical similarities.

17.2 What Can Fossils Reveal?

3. Fossils can provide information on individual organisms, on a particular locale, or on global changes.
4. A fossil may form when mineral replaces tissue gradually or after a sudden catastrophe, or may be indirect evidence such as footprints.
5. In **petrifaction,** minerals replace living tissue. Phosphatization is a form of petrifaction that replaces very small structures with calcium phosphate.
6. Fossil age is estimated in relative and absolute terms. The rock layer a fossil is in provides a **relative date.** The ratio of a stable radioactive isotope to its breakdown product gives an **absolute date,** which is a more precise range of time when an organism lived. This is a type of **radiometric dating.**

17.3 Comparing Structures

7. **Homologous structures** are inherited from a shared ancestor, are made of the same materials, but may differ in function. **Analogous structures** are similar in function due to convergent evolution.
8. **Vestigial structures** and similar embryonic structures in different species reflect actions of genes retained from ancestors.

17.4 What Can Molecular Evidence Reveal?

9. Molecular evolution considers similarities and differences among sequences of chromosome bands, a protein's amino acids, a gene's DNA bases, or genomes. Presumably, these sequences contain so many bits of information that it is unlikely that similarities happened by chance. More likely is descent from a shared ancestor.
10. Similar chromosome band patterns may not reflect similarity at the gene level. DNA probes can reveal **synteny,** or corresponding sections. Many genes and proteins are **highly conserved.**
11. A **molecular clock** estimates the time when two species diverged from a common ancestor by comparing DNA or protein sequences. Molecular clocks based on mitochondrial DNA are used to date recent events because this DNA mutates faster than nuclear DNA.

Testing Your Knowledge

1. What evidence was important in placing the following species in the same clade?
   a. birds and certain dinosaurs
   b. whales and hippos
   c. humans, chimps, and gorillas
2. This cladogram shows the evolutionary relationships among a tree species with leaves preserved in amber, *Hymenaea protera;* three living relatives; and six other types of plants:
   a. Which organism is *H. protera*'s closest relative?
   b. Is *H. protera* more closely related to tobacco or to palm?
   c. Which organism depicted is ancestral to all the others?

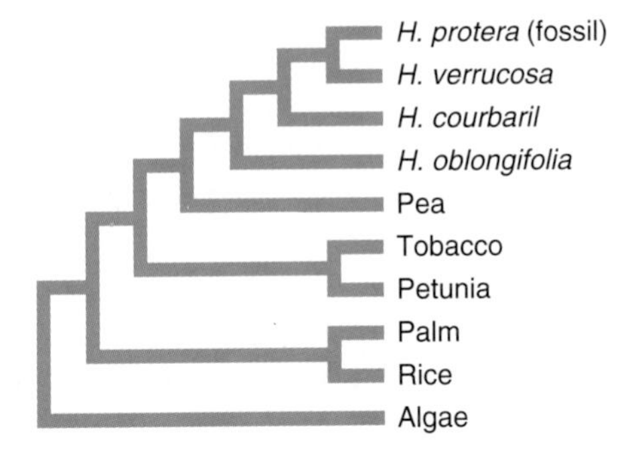

3. Why can't potassium/argon dating be used on preserved woolly mammoths?
4. Suggest a single type of genetic change that could have a drastic effect on evolution.
5. How does the type of data that molecular sequence data provide differ from the type of information that relative and absolute dating provide?

Thinking Scientifically

1. Give an example of how molecular and anatomical evidence can complement or support each other.
2. If DNA could be extracted from dinosaurs, how could the sequence be used to learn more about the origins of *Carcharodontosaurus, Giganotosaurus,* and *Acrocanthosaurus?*
3. Some genes are more alike between human and chimp than other genes are from person to person. Does this mean that chimps are humans or that humans with different alleles are different species? What other explanation fits the facts?
4. Inheritance can be traced through females by tracking mitochondrial DNA sequences. Suggest a way that male inheritance can be followed.
5. Why is the DNA sequence of one gene a less accurate indicator of the evolutionary relationship between two species than a DNA hybridization experiment comparing large portions of the two genomes?
6. What present environmental conditions might preserve organisms to form tomorrow's fossils?

References and Resources

Palevitz, Barry, and Ricki Lewis. February 1, 1999. Short shrift to evolution? *The Scientist,* vol. 13, p. 8. A look at *Science* or *Nature* in any given week provides mountains of evidence for evolution—as this essay demonstrates.

Rensberger, John M., and Mahito Watabe. August 10, 2000. Fine structure of bone in dinosaurs, birds and mammals. *Nature,* vol. 406, p. 619. Bony evidence supports the theory that birds and dinosaurs are closely related.

Sun, Ge, et al. November 27, 1998. In search of the first flower: A Jurassic angiosperm, *Archaefructus,* from northeast China. *Science,* vol. 282, p. 1692. This fossil exhibits the defining characteristics of flowering plants.

Tchernov, Eitan, et al. March 17, 2000. A fossil snake with limbs. *Science,* vol. 287. p. 2010. A 95-million-year-old fossil snake shows clear evidence of leg bones.

Varki, Ajit. September 2000. A chimpanzee genome project is a biomedical imperative. *Genome Research,* vol. 10, p. 1065. A "box" in this article clearly explains how the degree of similarity between the human and chimp genomes is calculated.

The *LIFE* Online Learning Center provides additional resources and tools for studying this chapter.
www.mhhe.com/life

CHAPTER 18

The Origin and History of Life

Alternate Views of Life's Origins

It is an inarguable fact that none of us was present on Earth to witness the origin, or origins, of life. However, experiments that attempt to re-create what might have happened when life began overwhelmingly support the theory that the first cells arose as collections of chemicals gradually formed that could carry information, extract energy from the environment, reproduce, and change. The first part of this chapter addresses this evolution of the chemical to the biological, which nearly all biologists accept. But people have suggested alternative explanations for how life arose. One idea, spontaneous generation, was long ago laid to rest. Another, that life arrived from space, is resurging in light of interpretations of recent evidence.

To understand how life began, we must begin with chemistry.

Spontaneous Generation

The concept of spontaneous generation held that the living came rather suddenly from the nonliving. For centuries people thought that life sprang from nowhere—maggots in meat, frogs in a puddle-turned-pond, cockroaches in a kitchen. It took many experiments that monitored the comings and goings of life in closed environments to show that life did not appear unless preexisting life was present.

In 1668, Italian physician Francesco Redi put meat into four containers open to the air and in four sealed containers. When maggots appeared only in the open flasks, Redi, thinking that flies had laid eggs on the meat, repeated the experiment covering the open flasks with gauze. As he had predicted, keeping out the flies kept out the maggots. But the question of spontaneous generation arose anew with the invention of microscopes. Could bacteria appear from nowhere? English naturalist John Needham thought that they could—when he boiled mutton broth for a few minutes in sealed glass vessels, bacteria appeared. But 20 years later, Italian biologist Lazzaro Spallanzani repeated the experiments, boiling the cultures longer and then sealing the flasks. Under these conditions, bacteria did not appear.

Still people held on to the idea of spontaneous generation, claiming that boiling killed a "vital principle" required for life. Finally, a century later, Louis Pasteur laid the question to rest by boiling meat broth in flasks that had S-shaped curved open necks that allowed air (and the vital principle, whatever it was) in, but kept out dust particles that carry bacterial spores. No bacteria grew. People finally believed what they were seeing—life does not arise spontaneously on a daily basis. Conditions on Earth now are probably too hostile for the chemical reactions that would eventually produce a living cell. And without cells, there can be no life.

From Panspermia to Cosmic Ancestry—Life from Space

In a 1908 book entitled *Worlds in the Making,* Swedish physical chemist Svante Arrhenius suggested that life came to

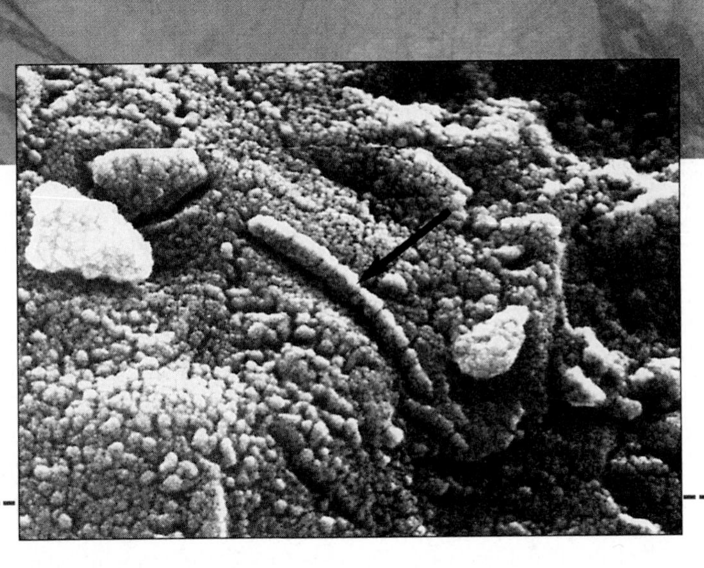

Possible evidence of life on Mars isn't little green men or monsters, but squiggles that resemble microfossils on Earth. A carbonate globule in Mars meteorite ALH84001 has structures (arrow) that look like fossilized bacteria from the Columbia River basin in Washington State.

Earth from spores that are abundant in the cosmos. He later broadened the idea, calling it "panspermia," and proposing that the life-carrying spores arrived on interstellar dust, comets, asteroids, and meteorites. The spores survived the frigid vacuum of space, and were propelled on their intergalactic journey by the energy of radiation. A version of panspermia from the late 1970s proposed that instead of spores, which are protected cells, only biochemical instructions for life arrived from space.

The reasoning behind panspermia is that some people believe Earth has not been in existence long enough for the great diversity of life to have evolved. Seeding the planet, either with cells or biochemical blueprints, would have saved time. Panspermia appeared in yet another guise in the 1990s, called "cosmic ancestry," but this time it derives from several types of possible evidence.

Martian Meteorites Alan Hills 84001 (ALH84001) is a softball-sized rock that was discovered in Antarctica in 1984. It is a meteorite that formed on Mars about 4.5 billion years ago, and landed on Earth some 13,000 years ago. The rock bears bits of organic compounds called polycyclic aromatic hydrocarbons (PAHs), which geologists hypothesize formed between 3.6 and 4 billion years ago, when water trickled through cracks. In 1996, researchers dissolved these PAHs in acetic acid, and saw grains emerge of the mineral magnetite, ordered into hexagonal structures that looked exactly like the remains of certain bacteria. The meteorite markings did not at all resemble magnetite precipitated from nonliving materials. Another type of possible evidence for life on Mars came two years later, when the Mars *Global Surveyor* detected a 300-mile stretch of hematite on the red planet, a mineral that forms in the presence of warm water. But the debate continues today as to whether the aligned magnetite grains, other patterns in the rock that resemble microfossils, and evidence for warm water on a long-ago Mars reflect ancient life there that could have come here.

Clues in Chemical Handedness While scientists were debating whether ALH84001 contains martian life, another type of possible evidence of life from space was found in the well-studied Murchison meteorite, a 100-kilogram rock that fell in Australia in 1969. Researchers at Arizona State University discovered mostly left-handed forms of an amino acid in powder from the meteorite, which is consistent with the fact that the amino acids in organisms are always of the left-handed form. In contrast, the same amino acids synthesized in a laboratory occur in equal amounts of left- and right-handed forms (Investigating Life 18.1 describes this phenomenon). The single-handedness of amino acids in the meteorite could mean that it carried proteins of biological origin. Alternatively, magnetized dust could have selectively destroyed right-handed amino acids as the meteor headed for Earth. Either way, amino acids from space may have provided some of the "building blocks" necessary for life to begin. Finding biomolecules in meteorites also indicates that their formation is common in the universe.

Life in Extreme Environments Life on Earth must have begun under harsh conditions—conditions not unlike those on the outer planets today. Therefore, might life have come from such an environment beyond Earth? Candidates for such hardy species are ancestors of the various extremophiles, described in Chapter 1's opening essay. A large, reddish bacterium called *Deinococcus radiodurans*, for example, provides evidence that life can survive the intense radiation in space. This organism's spectacular DNA repair system enables it to tolerate 1,000 times the radiation level that a person can—it even lives in nuclear reactors! *Deinococcus radiodurans* was discovered in a can of spoiled ground meat at the Oregon Agricultural Experiment Station in Corvallis in 1956, having survived the radiation used to sterilize the food. Its genome sequence, deciphered in 1998, may contain clues to how it survives. Other extremophiles live in rocks; in hot springs; under the tremendous heat and pressure of chimneylike "black smokers" that rise from the Pacific seafloor; and in pockets of water within icy Antarctic lakes—surroundings not unlike the icy insides of a comet. Bacteria may have traveled through space protected in similar pockets of water and landing here long ago on a meteorite, comet, or dust.

Spontaneous generation is no longer a valid scientific theory, and the evidence for cosmic ancestry is sparse. Yet even if the evidence mounts, it would not reveal how life *began*—just how it might have gotten here. To understand how life began, we must begin with chemistry.

18.1 How Might Life Have Originated?

Clues from geology, paleontology, and biochemistry provide evidence for the gradual formation of the first living cells from interacting collections of chemicals on long-ago Earth. RNA or a similar molecule that could encode information, replicate, and change may have bridged the chemical and the biological worlds.

Reconstructing life's start is like reading all the chapters of a novel except the first. A reader can get an idea of the events and setting of the opening chapter from clues throughout the novel. Similarly, scattered clues from life through the ages reflect events that may have led to the origin of life. This evidence includes experiments that simulate chemical reactions that may have occurred on early Earth; fossilized microorganisms that might have been among Earth's first residents; exploring how clays or minerals might have molded biochemical building blocks into polymers (chains) important in life; and identifying ancient gene and protein sequences retained in the cells of modern organisms.

When and Where Might Life Have Arisen?

The study of the chemistry that led to life actually begins with astronomy and geology (fig. 18.1). Earth and the other planets of the solar system formed about 4.55 billion years ago, as solid matter condensed out of a vast expanse of dust and gas. The red-hot ball had cooled enough to form a crust by about 4.2 to 4.1 billion years ago, when the temperature at the surface ranged from 500°C to 1000°C (932°F–1832°F) and atmospheric pressure was ten times what it is now.

During the planet's first 500 to 600 million years, comets, meteors, and possibly asteroids bombarded the surface, repeatedly boiling off the seas and vaporizing rocks to carve the features of the fledgling world. The oldest rocks that remain today, and house the oldest known fossils, are from Greenland, in an area called the Isua geological formation. But most of the initial crust was torn down and built up again into sediments, heated and compressed, or dragged into Earth's interior at deep-sea trenches and possibly recycled to the surface. The oldest hints of life are or-

FIGURE 18.1
A Time Line of Early Earth.
Originally a violent place, the surface eventually supported life. (Billions of years ago = b.y.a.)

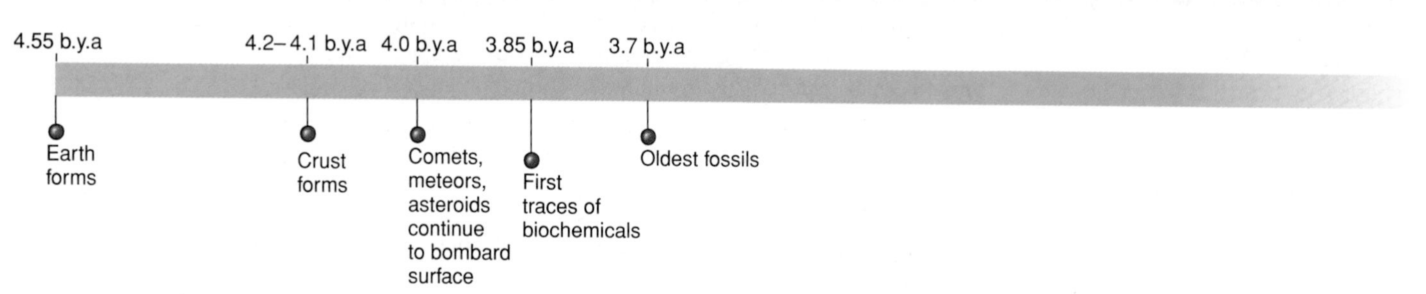

ganic deposits in quartz crystals that are rich in the carbon isotopes found in organisms, dating to about 3.85 billion years ago. Some of the oldest actual fossils are from 3.7-billion-year-old rock in Warrawoona, Australia, and Swaziland, South Africa. They are large formations of cyanobacteria-like cells called stromatolites.

The clues from geology and paleontology suggest that from 4.2 to 3.85 billion years ago, life—or more likely something that is more complex than an organized collection of organic molecules yet less complex than an actual cell—arose. This event may have happened in many places, in many forms, and at many times—we can't know. But somehow, an entity that could reproduce and adapt gained a foothold on the young planet, and set out on the road of diversification that became life.

Charles Darwin envisioned life's beginnings in a "warm little pond," but the geological evidence paints a clearly chaotic picture of volcanic eruptions, seismic upheaval, cosmic bombardment, and ultraviolet radiation. Still, there were probably protected pockets of the environment where chemicals could aggregate and perhaps interact. One such early environment might have been areas near deep-sea hydrothermal vents. Here, in a zone where hot water meets cold water, a life-sustaining warmth prevailed where prebiotic chemical collections could have been continually exposed to a rich brew of minerals spewed from Earth's interior. These minerals may have provided energy, functioned as catalysts, and formed physical molds, or templates, on which the building blocks of life could have linked to build larger molecules. When these larger molecules attained the ability to make more of themselves, and to change, the road to cells would have been well under way.

So harsh and unsettled was the early environment of Earth that cell-like groups of chemicals may have formed many times, only to be torn apart by heat, crushing debris from space, or radiation. Life may have begun quickly—in 10 million years or less—because this is the time that it takes the entire water supply to circulate through thermal vent systems, where the intense heat of Earth's interior would have dismantled the delicate molecules of life. Figure 18.2 summarizes the major steps in chemical evolution, discussed below.

FIGURE 18.2
Chemical Evolution.
An overview of the stages of chemical evolution leading to the origin of life. Figure 18.6 elaborates on this flowchart.

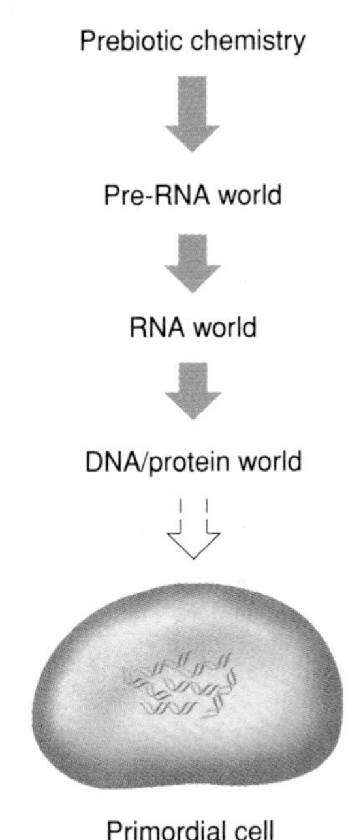

Prebiotic Simulation Experiments Show How Life Might Have Arisen

Early Earth was not only very different geologically from what it is today, but it was different chemically too. In the first half of the twentieth century, researchers thought about what the atmosphere might have been like about 4 billion years ago.

The era of **prebiotic simulations**—experiments that attempt to re-create chemical conditions on Earth before life arose—began in 1924, when Soviet chemist Alex I. Oparin suggested that life began as simple organic molecules linked into chains, which interacted to eventually form cells. Five years later, English chemist John B. S. Haldane added that early Earth had a different atmosphere than it does today. Oparin refined these ideas further in a 1938 book, *The Origin of Life*, where he hypothesized that a hydrogen-rich, or reducing, atmosphere was necessary for organic molecules to form. The hydrogen atoms could have become linked to carbon skeletons to form increasingly complex organic molecules. Oparin thought that this long-ago atmosphere included methane (CH_4), ammonia (NH_3), water (H_2O), and hydrogen (H_2), similar to the atmospheres of the outer planets today. Due to the extreme conditions on Earth immediately after cooling, little free oxygen (O_2) would have been available. Without oxygen, certain chemical reactions would have been possible that form amino acids and nucleotides. More recent geological evidence suggests that the early atmosphere may have contained abundant carbon dioxide (CO_2). The atmosphere today is rich in carbon dioxide, nitrogen, water, and oxygen.

In the fall of 1951, a new graduate student in chemistry at the University of Chicago, Stanley Miller, sat at a seminar listening to his mentor, Harold Urey, talk about Oparin's work. Both Urey and Miller questioned why no one had tried to test experimentally whether or not Oparin's atmosphere could indeed give rise to amino acids and/or nucleotides. Soon after, over Urey's protests that he might never see any results, Miller built a glass enclosure to contain Oparin's four gases, through which he passed electric discharges to simulate lightning (fig. 18.3). Next, he condensed the gases in a narrow tube and passed them over an electric heater, a laboratory version of a volcano. After a few failures because he had initially added the spark after the condenser and had the heat too high, Miller saw the condensed liquid turn yellowish. Chemical analysis showed that he had made glycine, the simplest type of amino acid in organisms. When he

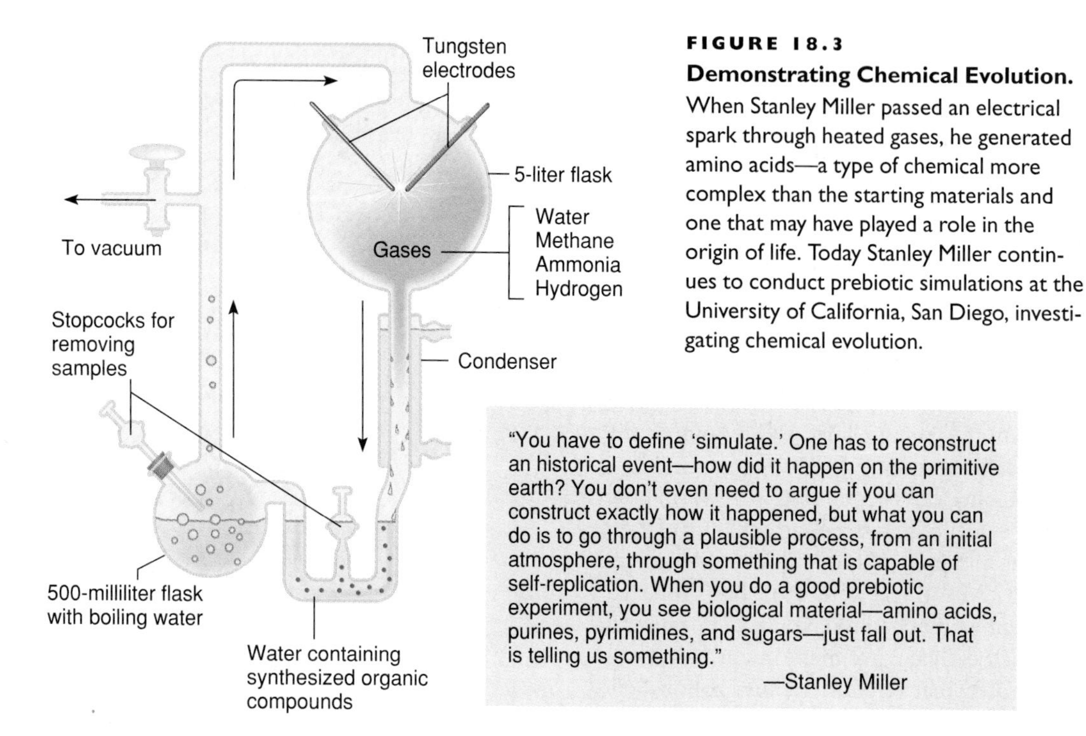

FIGURE 18.3
Demonstrating Chemical Evolution.
When Stanley Miller passed an electrical spark through heated gases, he generated amino acids—a type of chemical more complex than the starting materials and one that may have played a role in the origin of life. Today Stanley Miller continues to conduct prebiotic simulations at the University of California, San Diego, investigating chemical evolution.

let the brew cook a full week, the solution turned varying shades of red, pink, and yellow-brown, which turned out to be several more amino acids, some found in life. A prestigious journal published the work—which Urey gallantly refused to put his name on—and the 25-year-old-graduate student made headlines in newspapers and magazines reporting that he'd created life in a test tube.

A few amino acids do not an organism make, but "the Miller experiment" would go down in history as the first prebiotic simulation. Miller and many others confirmed and extended the results by altering conditions or using different starting materials. For example, methane and ammonia could form clouds of hydrogen cyanide (HCN) which, in the presence of ultraviolet light and water, produced amino acids. Prebiotic "soups" that included phosphates yielded nucleotides, including the biological energy molecule ATP. Other experiments produced molecules important in energy metabolism, carbohydrates, and phospholipids similar to those in biological membranes.

More recent prebiotic simulations mimic deep-sea hydrothermal vents (fig. 18.4). One laboratory version of a type of vent called a black smoker places mineral-rich lava with seawater containing dissolved carbon dioxide under high temperature and pressure, and simple organic compounds form. In another thermal vent model, nitrogen compounds and water are mixed with

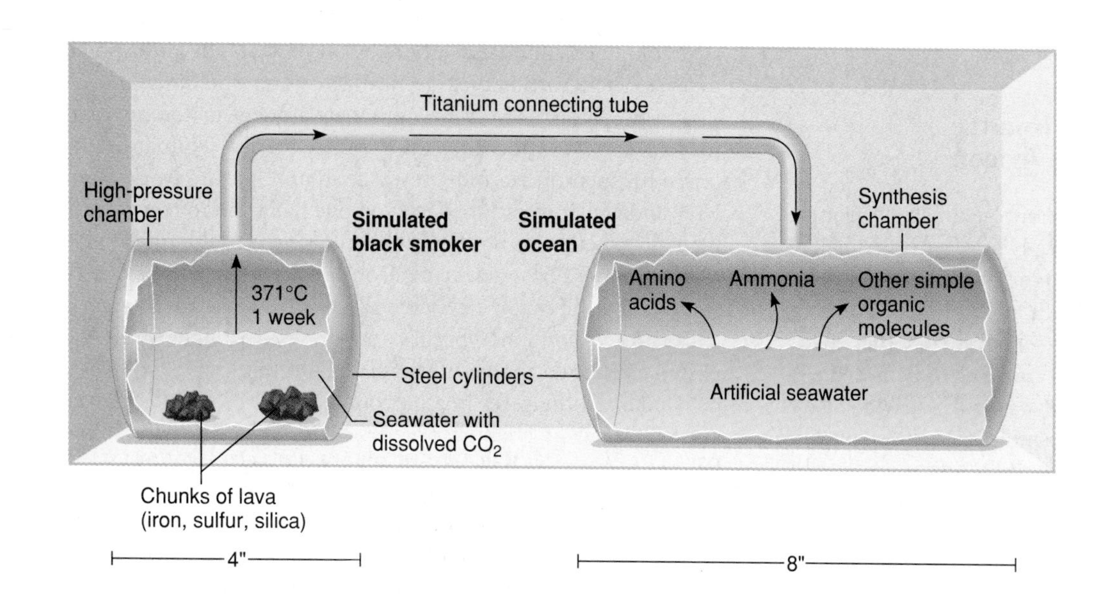

FIGURE 18.4
Life in the Deep Ocean.
A prebiotic simulation of a black smoker, which is a type of deep-sea hydrothermal vent, mixes pieces of mineral-rich lava with seawater. When heated under very high pressure, instead of vaporizing, the water remains in a "superheated" liquid state. The superheated water leaches minerals from the lava, and with carbon dioxide from the seawater moves through a tube to a second chamber. There, the chemicals react to produce simple organic molecules.

an iron-containing mineral under high temperature and pressure. The iron catalyzes reactions that produce ammonia (NH_3)—one of the components of the original Miller experiment.

Once the building blocks of macromolecules were present, whether in warm little ponds or deep-sea hydrothermal vents or in some other place, they had to have polymerized (linked into chains). This may have happened on clays or other minerals that provided ample, dry surfaces. Continuing the culinary theme of prebiotic "soup," this stage in the origin of life is sometimes described as "crepes," which are French pancakes that are made by pouring liquid dough onto a hot surface, where it dehydrates to form the solid pancake. Similarly, organic contents of primordial soups may have dried out and polymerized on hot mineral or clay surfaces that served as molds for creating certain molecular shapes.

The reason for the "crepe" following the "soup" is chemical. In an aqueous (watery) environment, hydrolysis is more likely to occur than its opposite reaction, dehydration synthesis (see fig. 2.13). Hydrolysis breaks down chains of amino acids and nucleotides faster than they can form, countering the buildup of long chains (polymers). Clays reverse the chemistry, favoring polymerization. In addition, some clays include minerals, such as iron pyrite ("fool's gold"), that could provide energy for forming chemical bonds by releasing electrons. Carbon monoxide contacting such minerals can react to yield pyruvate, the product of glycolysis.

Prebiotic simulations conducted on a common type of clay called montmorillonite demonstrate that these surfaces could have been the site of formation of the first RNA molecules. Not only do the positive charges on the clay's surface attract and hold negatively charged RNA nucleotides, but they foster formation of the phosphate bonds that link the nucleotides into chains, and even draw in other nucleotides to form a complementary strand (fig. 18.5). For example, a strand of 10 adenines would attract 10 uracils, which would then link. So far in experiments, 55 nucleotides can link and attract a complementary strand. Four billion years ago or thereabouts, clays might have been fringed with an ever-increasing variety of growing polymers. Some of these might have become the macromolecules that would eventually build cells.

FIGURE 18.5

From "Soup" to "Crepes."

Clays may have provided templates on which nucleotides could polymerize. In one hypothesized scenario, the dry environment enabled the rate of dehydration synthesis (condensation) to exceed that of hydrolysis, and gradually chains of nucleotides grew. Energy came from the sun, and catalytic capability from iron pyrite ("fool's gold").

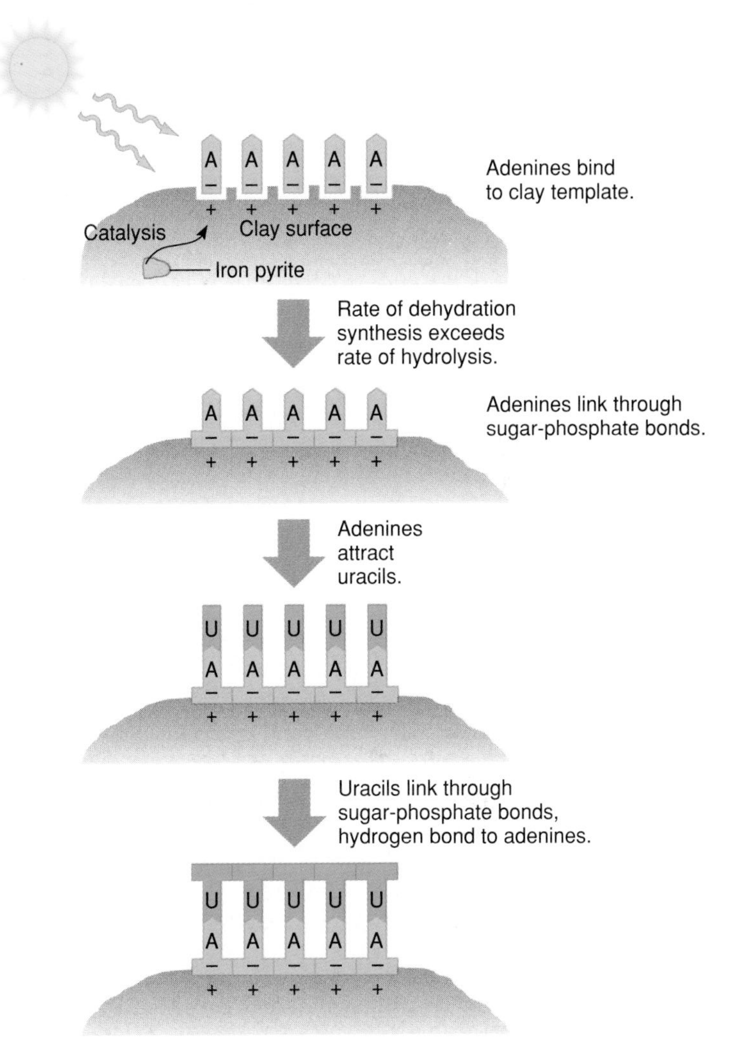

An RNA World May Have Preceded Cellular Life

Life requires an informational molecule. That molecule may have been RNA, or something like it, because RNA is the most versatile molecule that we know of—it carries information, uses it to manufacture proteins and control gene expression, and can function as a catalyst (table 18.1). In 1986, Harvard molecular biologist Walter Gilbert wrote in a one-page paper, "One can contemplate an RNA world, containing only RNA molecules that serve to catalyze the synthesis of themselves." The term **"RNA world"** stuck, and has come to describe how RNA, or more likely a similar molecule that was more chemically stable, jump-started life by providing the key ability to replicate (fig. 18.6).

RNA, however, may have been too unstable to have lasted long enough to support formation of cells. A hardier, similar molecule, or even several different types of nucleic acid–type molecules, may have played this pivotal role, forming a pre-RNA world. What might this pre-RNA have been? Some researchers think that it might have been similar to a synthetic molecule called a peptide nucleic acid (PNA) that consists of four types of nitrogenous bases emanating from a backbone built of the simple amino acid glycine. (Recall that glycine was an ingredient of Miller's "soup.") A single strand of PNA strongly binds to a single strand of RNA, DNA, or another PNA, and it is stable if one of its ends is blocked. Of course we can't know if PNA itself was on early Earth, but its characteristics suggest that a molecule other than the familiar DNA and RNA, perhaps one that is no longer around, might have been important at the dawn of life.

TABLE 18.1 Roles of RNA (Today)

1. RNA contains a gene's information in mRNA.
2. RNA (tRNA) brings amino acids to mRNA and forms part of ribosomes (rRNA).
3. RNA catalyzes certain chemical reactions, including its own synthesis, cutting other RNAs, and fostering peptide bonds.
4. RNA nucleotides are cofactors that enable enzymes in the energy reactions to function.
5. A small piece of RNA starts DNA replication. That is, DNA requires RNA for its synthesis, but not vice versa.
6. Because it is single-stranded, RNA can fold into many shapes, making many functions possible.

As time passed, pieces of RNA or a chemical cousin would have continued to form and accumulate, growing longer, becoming more complex in sequence, and changing as replication errors led to mutations that went uncorrected because there were as yet no repair systems. Nucleic acid pieces might even have flitted about, joining one another to create different combinations, as occurs today in the "jumping genes" or transposons that can move from one chromosome to another. Then natural selection, on a chemical level, exerted an effect, in the sense that some members of this accumulating community of molecules were more stable in a particular environment than others. Sums up Stanley Miller, "The origin of life is the origin of evolution, which requires replication, mutation, and selection. Replication is the hard part. Once a genetic material could replicate, life would have just taken off."

RNA might have begun encoding proteins, just short chains of amino acids at first. Eventually, an RNA molecule grew long enough to encode the enzyme **reverse transcriptase,** which copies RNA into DNA. With the formation of DNA from RNA came the ability to retain information, and the chemical blueprints of life could be laid down. More protein enzymes began taking over some of the functions of catalytic RNAs.

Meanwhile, lipids would have been entering the picture, as they formed and aggregated around the sensitive nucleic acids and proteins, setting them off from the harsh environment much as cell membranes do today. Under certain temperature and pH conditions, and with the necessary precursors, phospholipids could have formed membrane-like structures, some of which left evidence in ancient sediments. Experiments using lipids show that pieces of membrane can indeed grow on structural supports,

FIGURE 18.6
A Theory of Life's Beginnings.
Molecules assemble, leading to formation of the first cell.

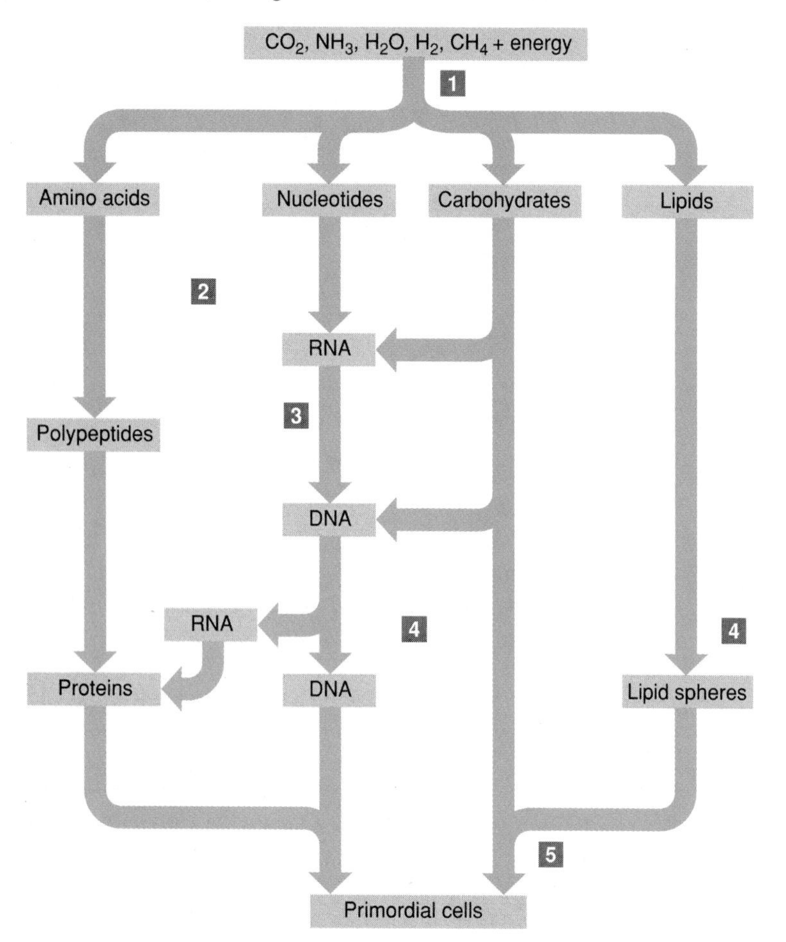

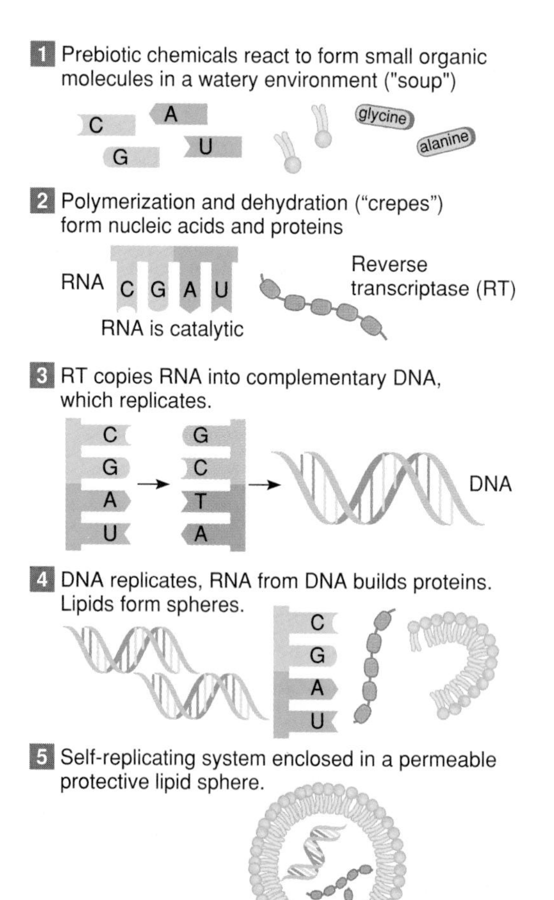

and break free when they reach a certain size and form a bubble, called a liposome. Might an ancient membrane bubble have enclosed a neighboring collection of nucleic acids and proteins to form a cell-like assemblage?

These ancient aggregates of RNA, DNA, proteins, and lipids—precursors of cells but not yet nearly as complex—were named **progenotes** by Carl Woese, who described the domain Archaea. They are also called protocells or protobionts. The next step in the origin of life was the evolution of metabolism.

How Did Metabolism Begin?

Metabolism is the ability of organisms to acquire and use energy to maintain the organization necessary for life. In a simple sense, this means converting one kind of molecule to another. The capacity of nucleic acids to mutate may have enabled progenotes to become increasingly self-sufficient, giving rise eventually to the reaction pathways of metabolism. Here is how it might have happened (fig. 18.7).

FIGURE 18.7
Evolution of Metabolic Pathways.
As early cells developed the ability to enzymatically convert more substances into nutrients, metabolic pathways may have emerged. Natural selection would have favored progenotes that could use more diverse food sources.

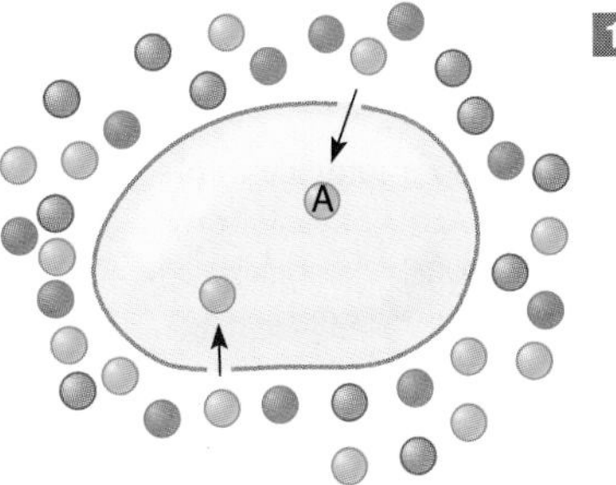

1 Progenote can absorb nutrient A for energy.

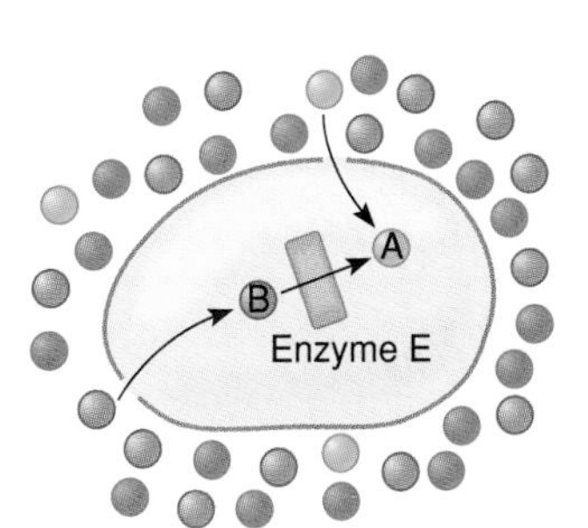

2 When concentration of A decreases, only progenotes with enzyme E, which can make A from B, will survive. If A is available, this progenote has two energy sources.

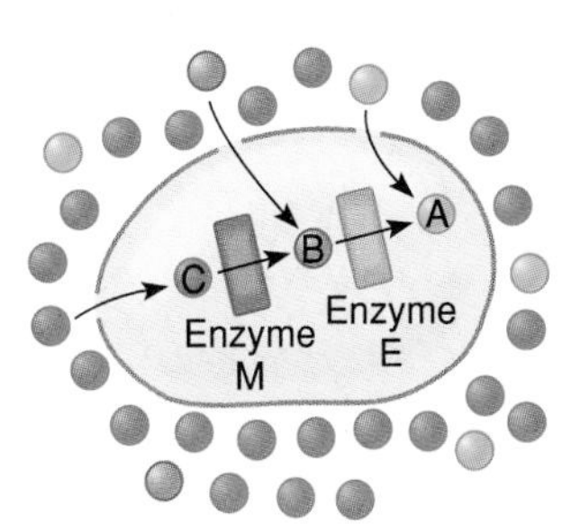

3 As the concentrations of A and B decreases, natural selection favors progenotes with enzyme M, which can make B from C, and enzyme E, which can make A from B. When A or B is available, they can be utilized by the progenote also.

Imagine a progenote that fed on a molecule, nutrient A, that was abundant in its environment. As the progenote reproduced, it began to use up nutrient A. But the ancient seas probably held more than one genetic variety of progenote. One type might have had an enzyme that could convert another nutrient, B, into the original nutrient A. The second progenote type would have a nutritional advantage, because it could extract energy from food source B as well as from A.

Soon the first type of progenote, totally dependent on nutrient A, would die out as its food vanished. For a while, the variant progenote that could convert nutrient B to A would flourish. But, in time, nutrient B would also become scarce. Meanwhile, perhaps a new type of progenote would have arisen with an additional enzyme that converted a nutrient C to B (and then, using the first enzyme, B to A). In time, this progenote would flourish; then another, and yet another. Over time, an enzyme-catalyzed sequence of connected chemical reactions—a biochemical pathway linking D to C to B to A—would arise. More pathways would form. As intermediates of one pathway spawned others, metabolism evolved.

Many millions of different associations of proteins, nucleic acids, and lipids must have formed spontaneously and randomly over many millions of years. Somehow a certain combination, or combinations, persisted and diversified. Investigating Life 18.1 discusses how a chance event might have accounted for the handedness of nucleic acids and proteins in organisms. With so many chance experiments of nature, it seems almost inevitable that, eventually, a cell-like structure, with the ability to replicate and change, would emerge. Biological evolution would begin.

18.1 MASTERING CONCEPTS

1. What happened to Earth before life began?
2. What types of information can prebiotic simulations provide?
3. Why is RNA the molecule most likely to have been pivotal in life's beginnings?
4. How might metabolic pathways have originated and evolved?

OLC

18.2 Highlights in the History of Life

Cells formed and changed and gradually gave rise to the last common ancestor of the three domains of life. Eukaryotes appeared, and multicellular life formed and diverged. We know about only those chapters in the history of life that have been represented in the fossil record and left clues in modern organisms and their molecules.

If a person was able to observe the first 3 billion (3,000 million) years or so of life on Earth, the scene wouldn't be terribly exciting. For a long time, most of the planet's surface was water. When land emerged, some of it became covered, gradually, with mats of cyanobacteria that harnessed solar energy, beginning the oxygenation of the atmosphere. But the abundant bare rock may have been misleading. In the ancient seas, in widespread spots at first, single-celled organisms were reproducing, diversifying, and

Handedness in Biomolecules

Before there was life, there were collections of chemicals. But a living entity is more than a subset of chemicals. One way that the chemistry of cells differs from that of nonliving matter is in the symmetry of certain organic molecules.

Nineteen of the 20 amino acids of life, and the nucleic acid sugars ribose and deoxyribose, are asymmetric. They can exist in two forms that are mirror images of each other but are not superimposable—like a right and left hand (fig. 18.A, top). This situation, called chirality, occurs when a central carbon bonds to four different chemical groups (fig. 18.A , bottom). When chiral molecules are synthesized in the laboratory, half of any given batch is one configuration, half the other. But in organisms, the 19 amino acid types are all left-handed (the exception, glycine, does not have four different groups around the central carbon); ribose and deoxyribose are always right-handed. How, and why, is this so?

The handedness of the proteins and nucleic acids of life has been the subject of murder mysteries and science fiction. Biologists and chemists have puzzled over it for 150 years.

French physicist Jean Baptiste Biot discovered chirality in 1815. He passed beams of light through certain organic chemicals in solution and found that some solutions of a particular compound moved the light in a counterclockwise direction, some in a clockwise direction, and some not at all. He suggested that a quality of the molecules affected the way they bend polarized light. French chemist Louis Pasteur extended Biot's observation. He synthesized crystals of various organic compounds and then examined them under a microscope. He discovered two types of crystals that differed subtly in their handedness. When he separated a 50:50 mixture of the two types of crystals by handedness, he found that the right-handed crystals rotated light in a clockwise direction, and the left-handed crystals rotated light counterclockwise. Solutions with equal numbers of types cancel each other out, and the path of light doesn't rotate.

Ten years after the crystal-picking experiment, Pasteur discovered that chiral organic molecules in organisms are only one form. Today, life scientists still argue about what this asymmetry means, and how and when it happened. Two hypotheses are:

1. Meteorites or comets carrying just one form of chiral molecules seeded Earth with the precursors of life. The observation that the Murchison meteorite has predominantly the left-handed form of the amino acid alanine is evidence for this explanation.
2. Chirality was part of the origin of life. As various combinations of primordial self-replicating molecules formed, one type, including either a left-handed or right-handed version of a key constituent, had an advantage—such as a better ability to catalyze or replicate, perhaps—and therefore outcompeted other forms and persisted.

Perhaps the answer to why the biomolecules of life are chiral will come from space explorations. If amino acids are left-handed and nucleic acid sugars right-handed because of a chance choice long ago, then life elsewhere might be the opposite. But if the handedness on Earth is a prerequisite for life, then it should be this way beyond Earth—if life there is as we know it.

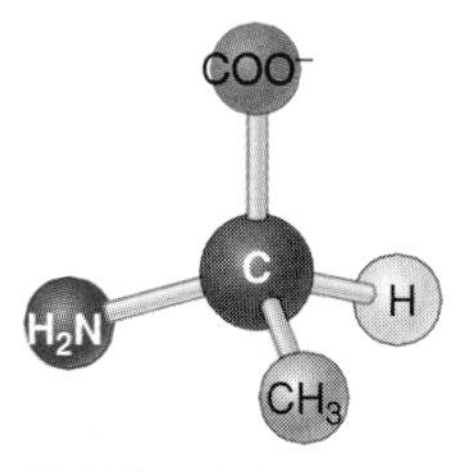

FIGURE 18.A
Living Systems Recognize "Handedness."
Molecules that are composed of identical atoms can be mirror images of each other, like human hands. Although chemically identical, only one orientation of such a chiral molecule functions in cells. Of the 20 amino acids of life, 19 are left-handed, and the 20th not chiral. The sugars ribose and deoxyribose are right-handed.

interacting. A time would come when life would seem to explode everywhere.

It would take a textbook many thousands of pages long to capture all of the events that passed from the rise of the first cells to life today—if we even knew them. Instead, here, we highlight key events in the history of life, and refer to others mentioned in preceding chapters. Table 18.2 and figures 18.8 and 18.9 summarize the major events of life history, which are discussed as the chapter continues.

Cells Emerge and Evolve in the Archean and Proterozoic Eras

Evidence from prebiotic simulations, geology, and paleontology indicate that living cells appeared in a slice of time of about 100 million years between 3.8 and 3.7 billion (3,800 and 3,700 million) years ago (fig. 18.9). Those first organisms were simpler than any cell known today. Several types of early cells probably prevailed for millions of years, competing for resources and possibly sharing

TABLE 18.2 The Geological Timescale

ERA	PERIOD	EPOCH	MILLIONS OF YEARS AGO	LIFE
CENOZOIC	Quaternary	Recent		?
			0.01	
		Pleistocene		*Homo*
			1.8	
		Pliocene		*Australopithecus*
			5	
	Tertiary	Miocene		
			24	Hominoids
		Oligocene		*Aegyptopithecus*
			38	
		Eocene		Mammals and flowering plants diversify
			54	
		Paleocene		First primates, hoofed mammals, insects, birds diversify
MESOZOIC	Cretaceous		65	First mammals; widespread dinosaurs and flowering plants
			138	Diverse and abundant dinosaurs and flowering plants
	Jurassic			First birds
			205	First dinosaurs
	Triassic			Therapsids and thecodonts; forests of cycads, ginkgos, and conifers
PALEOZOIC	Permian		240	Fewer amphibians, more reptiles
				Cotylosaurs and pelycosaurs
			290	Reptiles arise
	Carboniferous			Ferns and conifers; amphibians arise and diversify
			360	Bony fishes, corals, crinoids, land plants diversify, land invertebrates
	Devonian			
			410	First land plants and animals
	Silurian			
			435	Algae, invertebrates, graptolites, jawless fishes
	Ordovician			
			505	"Explosion" of sponges, worms, jellyfish, "small shelly fossils"; ancestors of all modern animal groups appear
	Cambrian			
PRECAMBRIAN	PROTEROZOIC		590	Ediacaran organisms
			670	
			700	
			1500	First fossils of eukaryotes
	ARCHEAN		2500	
				First fossils (cyanobacteria)
			3500	Progenotes
				RNA world
				Pre-RNA world
			4550	Earth forms

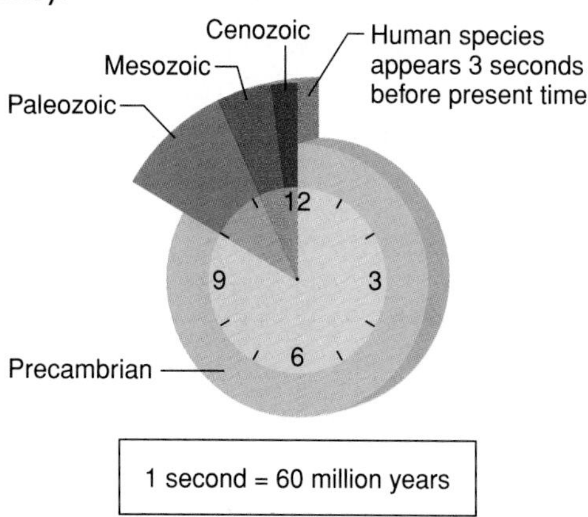

FIGURE 18.8
Precambrian Time Accounts for Five-Sixths of Earth's History.
The time when we think life on Earth has been abundant—the Paleozoic, Mesozoic, and Cenozoic eras—accounts for only one-sixth of the planet's history.

genetic material. Carl Woese describes this time as "a diverse community of cells that survived and evolved as a biological unit." Eventually, a type of cell arose that was the last shared ancestor of Archaea, Bacteria, and Eukarya.

The earliest fossil evidence indicates that cells with pigments that could harness solar energy were present by 3.5 billion (3,500 million) years ago, although this process was not as complex as photosynthesis today. We know very little about the earliest life until about 3 billion (3,000 million) years ago, when the geological tumult had calmed somewhat and unicellular organisms inhabited nearly every part of the planet. This was the Archean era, which ended 2.5 billion (2,500 million) years ago. Fossils from Rhodesia from 2.8 billion (2,800 million) years ago contain breakdown products of chlorophyll, a testament to the existence of photosynthesis by that time. The continuing photosynthesis gradually added oxygen to the atmosphere, which forever altered the pattern of life on Earth. Natural selection would begin to favor organisms that could use the accumulating oxygen to maximize energy harnessing from nutrients, while anaerobic species would persist in pockets of the environment away from oxygen.

Eukaryotic cells emerged in the Proterozoic era, which lasted from the end of the Archean era until 590 million years ago. Recall from chapter 3 that a modern version of the endosymbiont theory proposes that eukaryotic cells arose from ancestral cells, perhaps archaea, that acquired mitochondria by engulfing aerobic bacteria and chloroplasts by taking in photosynthesizing bacteria. The host archaean cell contributed membrane material, cytoskeletal elements, and some genes and other cellular constituents. Biologists are continually refining the endosymbiont theory as sequenced microbial genomes (see chapter 13's opening essay) reveal species relationships at the DNA level. Figure 18.10 recaps figure 3.19, and shows some organisms that might hold clues to how endosymbiosis occurred. Like the origin of life, endosymbiosis may have happened more than once, and mitochondria and chloroplasts, which have genomes, continue to evolve.

The earliest fossils of eukaryotic cells date from 1.9 to 1.4 billion (1,900 to 1,400 million) years ago, and come from China (fig. 18.11). These organisms may have been unicellular algae. Australian fossils consisting of an organic residue 1.69 billion (1,690 million)

FIGURE 18.9
A Glimpse of the Sequence of Appearance of Different Life-Forms on Earth.
We know little about the earliest types of organisms, but they were single-celled. Life has increased in complexity, yet simpler organisms, such as bacteria and archaea, still flourish. Note how recently humans appear.

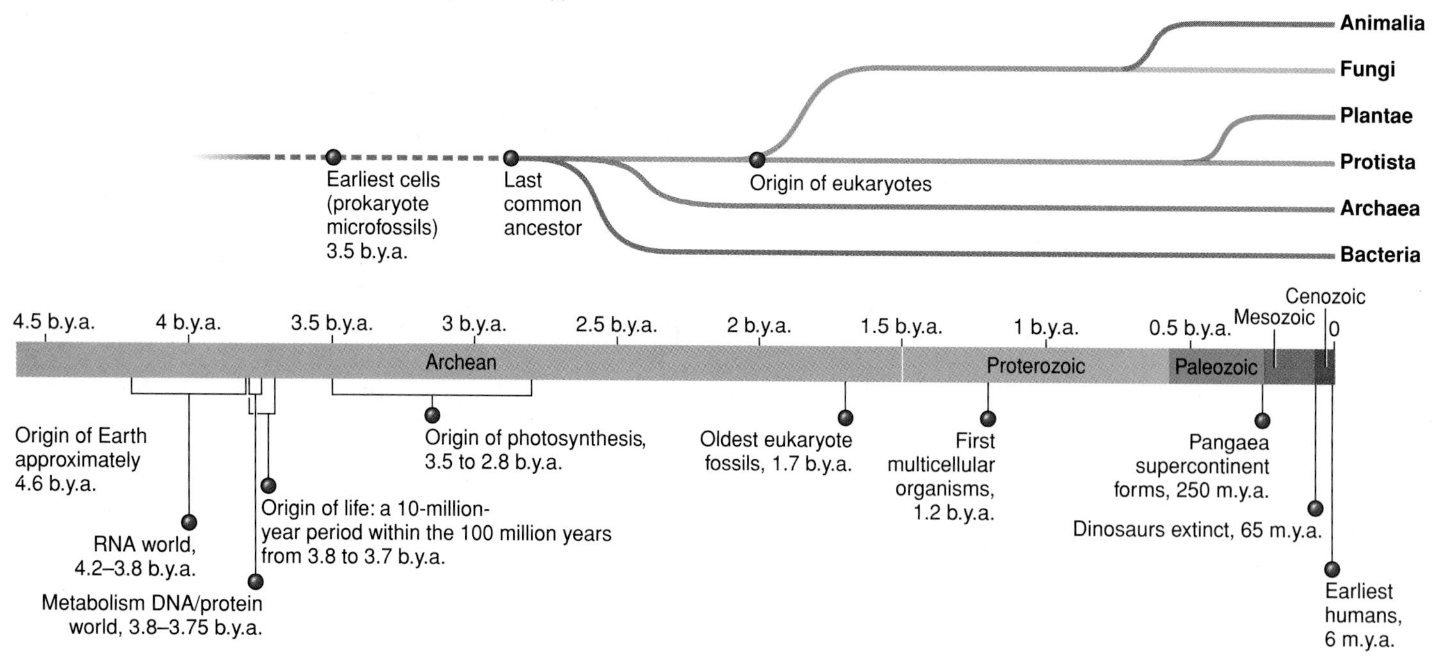

FIGURE 18.10

The Endosymbiont Theory Revisited.

(**A**) Resemblance between aerobic bacteria and mitochondria, cyanobacteria and chloroplasts, and other observations formed the basis of the endosymbiont theory. (**B**) Modern examples of microorganisms living within cells of another support the theory such as the algae living inside this protistan. (**C**) Compelling evidence for endosymbiosis also comes from genome sequences. The simple bacterium *Rickettsia prowazekii* carries out energy reactions very much like a mitochondrion does. *Reclinomonas americana* (**D**) is a unicellular eukaryote whose mitochondrial genome is packed with bacteria-like operons, supporting the bacterial origin of mitochondria.

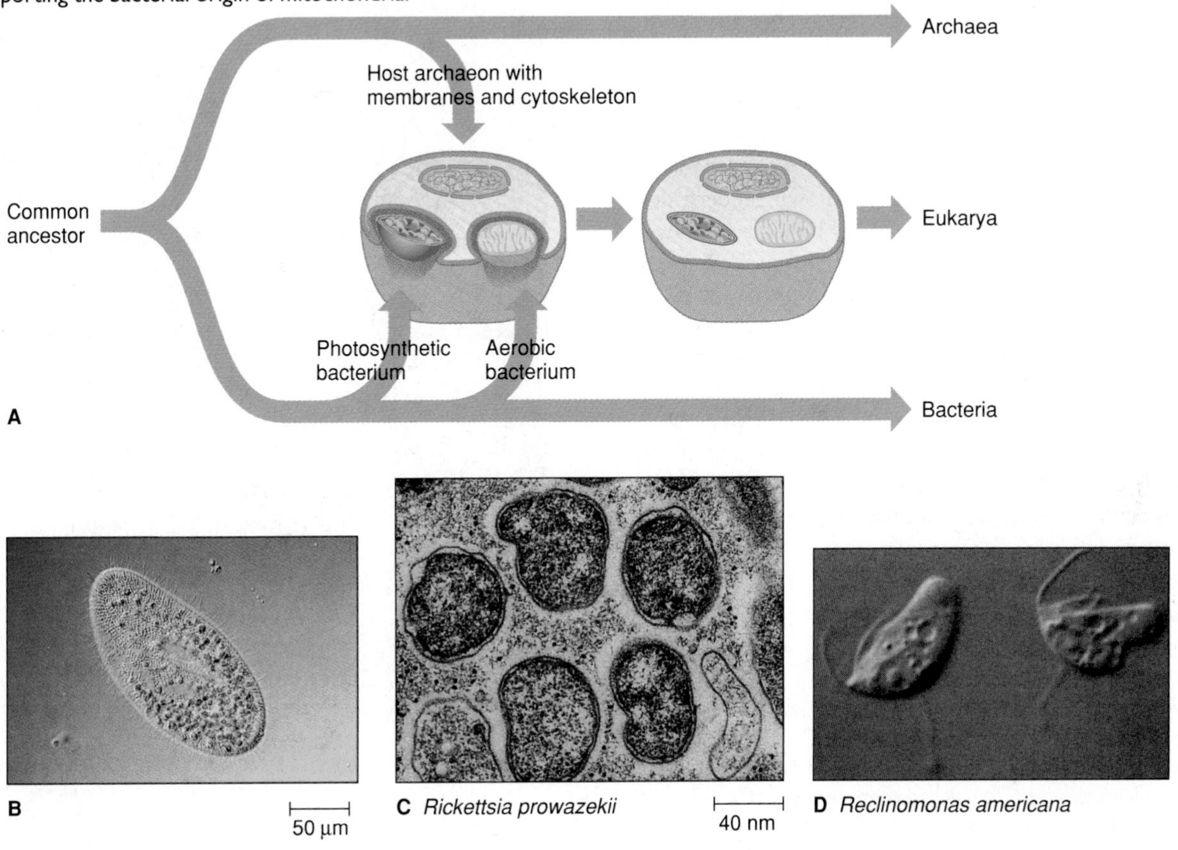

FIGURE 18.11

Fossil Evidence of Cells.

A very ancient eukaryote, *Grypania spiralis,* left this spiral of organic matter in a 1.4-billion-year-old rock in China.

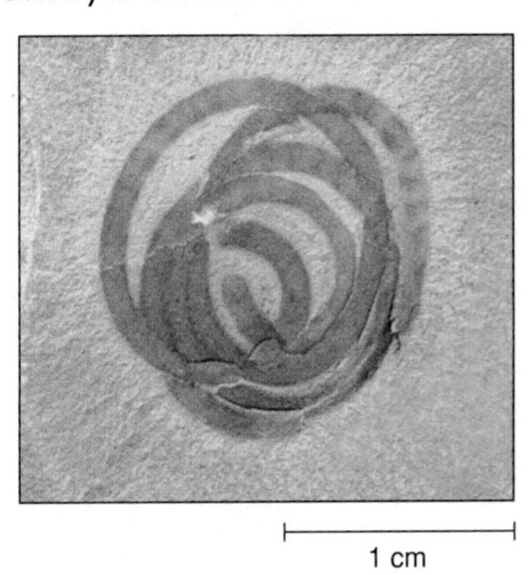

years old are chemically similar to eukaryotic membrane components and may come from a very early unicellular eukaryote.

About 1.2 billion (1,200 million) years ago, multicellular life appeared. Exactly how life proceeded from the single-celled to the many-celled is a mystery. Perhaps many unicellular organisms came together and joined, then took on different tasks to form a multicellular organism. Alternatively, a large single-celled organism may have divided into many subunits and diverged in gene expression and function. The earliest fossils of multicellular life are from a red alga that lived about 1.25 billion (1,250 million) to 950 million years ago in Canada. Abundant fossil evidence of multicellular algae comes from Siberia, dating from a billion years ago.

The Strange Ediacarans Preceded the Cambrian Explosion

There is a saying that important events tend to "divide time," and this has been the case with paleontologists' interpretations of the fossil record. They often divide natural history into the

Precambrian and everything that came during and after the Cambrian (590 until 505 million years ago), which was the first period of the Paleozoic era. The reason for this distinction is that fossils of all the major phyla of animals appear rather suddenly in sediments from the Cambrian. But keeping in mind the special set of circumstances that must be present for fossilization to occur, it seems highly unlikely that the late Precambrian was lifeless, only that it didn't leave much of a record. Charles Darwin wrote that the Cambrian explosion was misleading, and that the Precambrian seas must have "swarmed with living creatures."

As the Proterozoic era became the Paleozoic, from about 600 to 544 million years ago, many parts of Earth were home to organisms called Ediacarans, which have no known modern descendants. The Ediacarans have long puzzled biologists because of their appearances, preserved in sandstones and shales as casts and molds, that are notably different from life as we know it today. Their bodies were apparently soft and very flat—figure 18.12 shows one representative organism, *Dickinsonia,* that was a meter in diameter but less than 3 millimeters in body thickness. This flattened form may have been an adaptation for extracting maximal oxygen from the ancient seas. Ediacaran bodies were built of sections of tubing, branches, ribs, discs, and fronds, a little like a toy in which a child pieces together similar-shaped objects to create a creature. Ediacarans had no complex internal organ systems, no obvious body openings, and they probably could not move very far. Biologists have interpreted fossils to be everything from worms to ferns to fungi, as figure 18.13 demonstrates.

FIGURE 18.12
Complexity Increases.
A "typical" Ediacaran organism, *Dickinsonia,* grew to a meter in diameter, but only 3 millimeters thick. It had segments and two different ends, and internal detail that paleontologists have interpreted as remnants of a simple circulatory or digestive system. But just what it was remains unclear. One group of paleontologists couldn't agree on whether it most resembled a jellyfish, a flatworm, or a fungus! Like other Ediacarans, *Dickinsonia* might not have been anything like the species we are familiar with today.

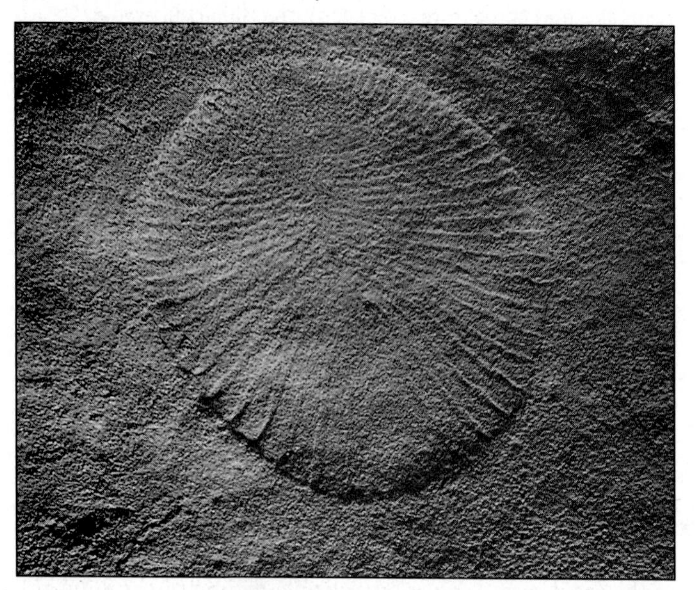

FIGURE 18.13
Plant or Animal?
It isn't clear from this fossil of the Ediacaran *Spriggina* just what type of organism it was!

The Ediacarans vanished from the fossil record about 544 million years ago, by which time the Cambrian explosion was in full swing. Their disappearance has been attributed to sudden mass extinction, but another explanation is that the very specific conditions under which they were preserved changed. Recall from fig. 17.7E that the height of the Ediacaran reign—from 570 to 580 million years ago—was also the time that phosphatization preserved large numbers of algae, sponges, and unidentified animal embryos in the seas. So instead of Ediacaran life suddenly giving way to familiar animal phyla, these organisms may have more gradually vanished, opening up habitats for the organisms that left the phosphatized remains, and others. As the Ediacarans disappeared, the Cambrian seas filled with a spectacular diversity of life.

Life "Explodes" in the Paleozoic Era

The Cambrian seas held remnants of the Ediacaran world, plus abundant algae, sponges, jellyfish, and worms. Most notable were the earliest known organisms with hard parts, such as shellfish, insectlike trilobites, nautiloids, scorpion-like eurypterids, and brachiopods, which resembled clams. The early Cambrian seas were home to diverse wormlike, armored animals, some of which would die out, others of which would continue to evolve, giving rise to modern mollusks, various worm phyla, and arthropods.

Fossils from the Cambrian are, from a geological time frame, so suddenly different from what came before that paleontologists revert to plain English in recognition, calling the period's clues to the past simply "small, shelly fossils." Bodies with hard parts eased fossilization. And once again, an unusual geologic circumstance discovered in 1909 preserved a slice of life from this time. In an area of British Columbia called the Burgess Shale, enormous numbers of organisms were buried in a mudslide in the middle of the Cambrian, including soft-bodied invertebrates not seen elsewhere, and animals with skeletons. The Burgess Shale animals were abundant, very diverse, and preserved in exquisite detail (fig. 18.14).

The Ordovician period followed the Cambrian period, lasting from 505 to 435 million years ago. The seas continued to support huge communities of algae and invertebrates. Organisms called graptolites were common, named for their fossils, which resemble pencil markings. The first vertebrates to leave fossil evidence, the jawless fishes, appeared at this time.

At the end of the Ordovician period and during the Silurian period, which lasted from 435 to 410 million years ago, fossilized spores indicate that life ventured onto land, in the form of primitive plants that may have resembled modern liverworts (fig. 18.15). The first vascular plants (with specialized tissue for water and nutrient transport) had bare stems from which spores scattered, and underground branches, but they lacked leaves and roots. The tallest was only about 10 centimeters high. Figure 23.3 shows a time line of plant evolution. The first animals to leave fossils on land also date from the Silurian. They resembled scorpions.

The Devonian period of 410 to 360 million years ago was the "Age of Fishes." The seas continued to support most life, including the now prevalent invertebrates, plus fishes with skeletons of cartilage or bone. Corals and animals called crinoids that resembled

FIGURE 18.14

A Cambrian Scene.

This mid-Cambrian sea bottom scene, depicting what is now British Columbia, was reconstructed from fossils in the Burgess Shale. Life in the seas was incredibly diverse, including many species with no known modern descendants.

FIGURE 18.15

Plants Settle the Land.

Plants likely evolved from freshwater algae, and left the first traces of life on land in the late Ordovician. This spore from that time is from *Tetrahedraletes medinensis,* a plant related to the liverworts.

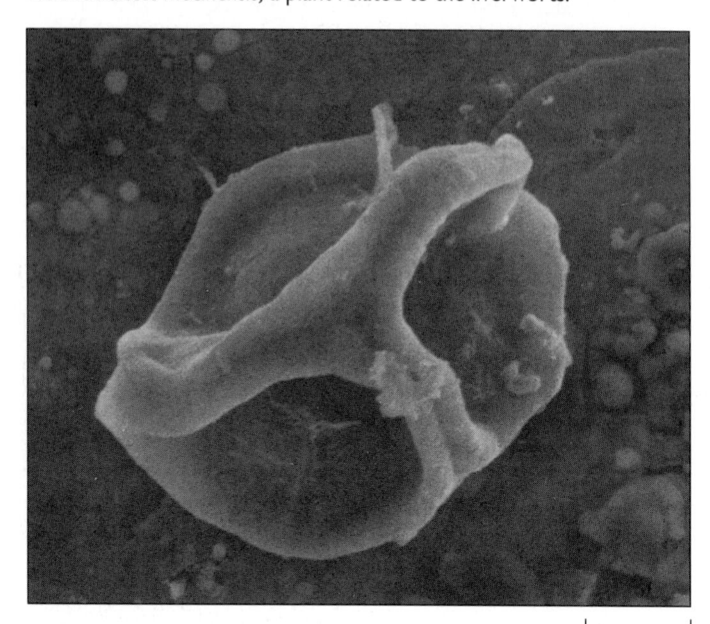

flowers were abundant. The many types of spores indicate that by this time, plants were conquering the land and diversifying. Scorpions, millipedes, and other invertebrates lived on the land.

The fresh waters of the Devonian were home to the lobe-finned fishes (fig. 18.16). These animals had fleshy, powerful fins and could obtain oxygen through gills as well as through primitive lunglike structures. The lobe-finned fishes first appeared about 410 million years ago and persist today as lungfish and coelacanths.

By 360 million years ago, the first four-limbed animals, or tetrapods, appeared. The earliest tetrapods, with their adaptations to living in water and on land, provide a study in transition. The 3-foot-long (0.9-meter-long) *Acanthostega* had a fin on its tail like a fish and used its powerful tail to move underwater, but it also had hips, legs, and toes, useful for navigating on land. Its rib cage wasn't strong enough to support internal organs without the buoyancy of water, but preserved footprints indicate that the animal could venture briefly onto land. Its feet were especially interesting—flat to move easily through the water but with eight toes! A contemporary of *Acanthostega,* called *Ichthyostega,* was more like an amphibian than a fish. Its legs were more powerful and its rib cage stronger, yet it had a skull shape and finned tail reminiscent of fish ancestors.

As this "Age of Amphibians" progressed, the descendants of the lobe-finned fishes were spending more time on land, but their legs were too weak to support full-time walking. They had to return to the water for frequent dips to wet their skins and to lay eggs, so that the embryos within could get sufficient moisture and nutrients.

Between 300 and 350 million years ago, as amphibians flourished, some types arose that coated their eggs with a hard shell.

FIGURE 18.16

From an Aquatic to a Terrestrial Existence.

(**A**) Lobe-finned fishes lived during the Devonian period and had an adaptation that would permit their descendants to venture onto the land—lungs. (**B**) *Acanthostega* lived as the Devonian period ended. It stayed mostly in the water but had adaptations that permitted it to spend short periods of time on land—including legs with eight toes! (**C**) A contemporary, *Ichthyostega,* could spend longer periods of time on land because its rib cage was stronger. It retained the skull shape and finned tail of its ancestors, however.

A *Coelacanth*

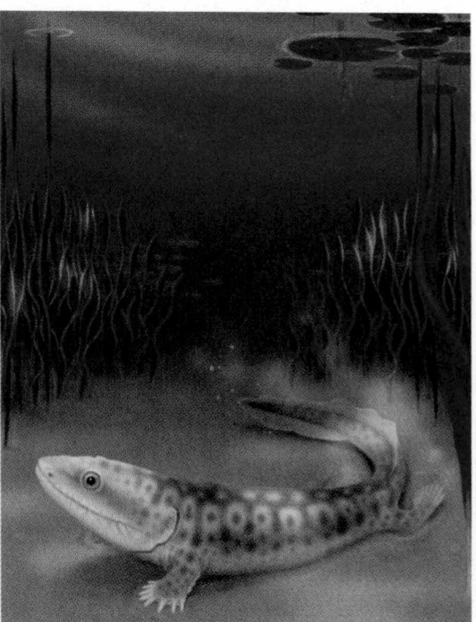

B *Acanthostega*

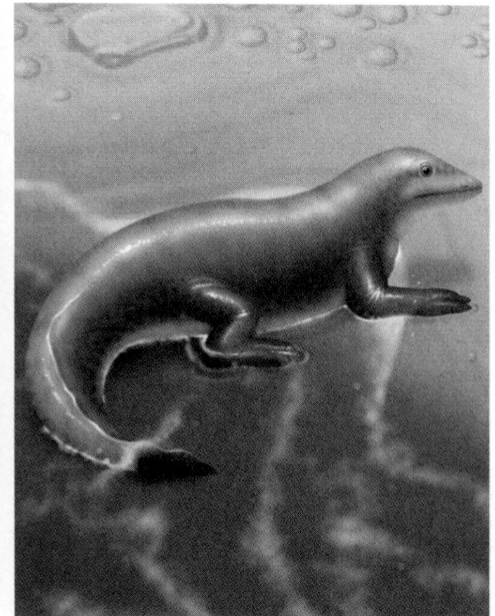

C *Ichthyostega*

These animals branched from the other amphibians to eventually give rise to reptiles, birds, and mammals. The first animals capable of living totally on land, primitive reptiles, appeared about 300 million years ago. Earth then was one large, green landmass within a vast ocean, the waters and swamps brimming with life. The air, warm year-round, was alive with the sounds of dragonflies, grasshoppers, and crickets, some of them giant versions of their familiar modern descendants. Invertebrates flourished in the sediments.

The swamps of this time, the Carboniferous, had fernlike plants and majestic conifers towering to 130 feet (40 meters). By the end of the period, many of the plants had died, buried beneath the swamps to form, over the coming millennia, coal beds (fig. 18.17).

FIGURE 18.17

The Carboniferous, or Coal Age.

(**A**) About 300 million years ago, forests were so lush that they were eventually preserved in massive coal beds. The fernlike plants in the foreground are very ancient seed-bearing gymnosperms. The thin tree on the right is an extinct ancestor of the modern horsetail, and the thicker trunks on the left, also extinct, gave rise to modern club mosses and ground pines. (**B**) Fossils from the Carboniferous period are found in coal and shale, demonstrating the beauty and diversity of the plant life during that period.

A

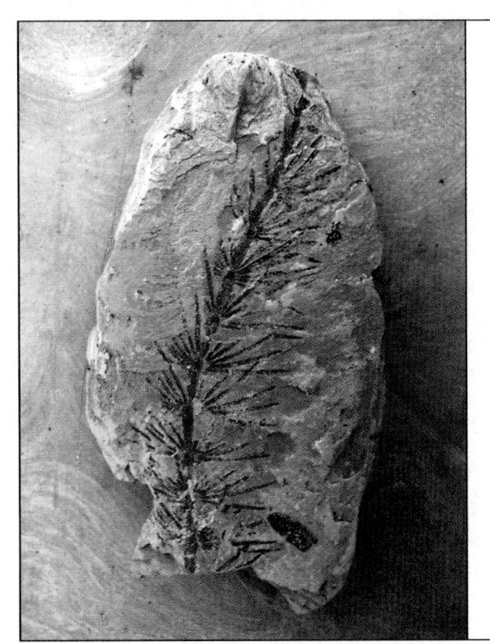

B

Because of these bountiful reminders of the ancient first forests, this time is also called the Coal Age. Today, a split piece of coal will sometimes reveal an impression left by an ancient fern.

The Paleozoic era ended dramatically with the Permian period (290 to 240 million years ago), when half the families of vertebrates and more than 90% of species in the shallow seas became extinct (see fig. 16.15). Reptiles were becoming more prevalent, as amphibian populations shrank.

The reptile introduced a new biological structure, the **amniote egg,** in which an embryo could develop completely on dry land. An amniote egg contains nutrient-rich yolk and protective extraembryonic membranes. The amnion encloses the embryo in a sac of fluid that is chemically very similar to seawater, which nurtured so many ancient species. Amniote eggs persist today in reptiles, birds, and a few mammals.

The Permian period foreshadowed the dawn of the dinosaur age. Cotylosaurs were early Permian reptiles that gave rise to the dinosaurs, as well as to modern reptiles, birds, and mammals. They coexisted with their immediate descendants, the pelycosaurs, or sailed lizards.

Reptiles and Flowering Plants Thrive During the Mesozoic Era

During the Triassic period from 240 to 205 million years ago, small animals called thecodonts flourished, occupying a place in time between the Permian cotylosaurs and the great dinosaurs to come. Thecodonts shared the forest of cycads, ginkgos, and conifers with other animals called therapsids. These were reptiles, but their posture and teeth were mammal-like (fig. 18.18A). By the end of the Mesozoic era, the therapsids would evolve into small, hair-covered animals, most of whom lived on the forest floor—the first mammals.

At the close of the Triassic period, thecodonts and therapsids were becoming rarer, as much larger animals began to infiltrate a wide range of habitats. These new, well-adapted animals were the dinosaurs, and they would dominate for the next 120 million years. The dinosaurs had characteristics of the reptiles that preceded them and of the birds that would follow.

By the Jurassic period of 205 to 138 million years ago, the dinosaurs were everywhere. Ichthyosaurs swam in the seas. *Protoarchaeopteryx* and then *Archaeopteryx* glided through the air and apatosaurs, stegosaurs, and allosaurs roamed the land in the second half of the period (see chapter 17's opening essay and figure 17.3). The first flowering plants (angiosperms) appeared, as figure 17.1A of *Archaefructus* vividly shows, but forests still consisted largely of tall ferns and conifers, club mosses, and horsetails.

The Cretaceous period (138 to 65 million years ago) was a time of great biological change, as the continents drifted apart. Figure 17.5 shows how the allosaurs diverged at this time. Flowering plants spread in spectacular diversity. Although this was the beginning of the end for the dinosaurs, many species soared to quite healthy numbers, as the field of dinosaur eggs and embryos deposited in Patagonia 89 million years ago shows (see figure 17.4). Duck-billed maiasaurs, for example, traveled in herds of thousands in what is now Montana (fig. 18.18B). Huge herds of apatosaurs migrated on the plains of Alberta to the Arctic, northern Europe, and Asia, which were then joined. By the end of the period, triceratops were so widespread that some paleontologists call them the "cockroaches of the Cretaceous."

The reign of the dinosaurs ended about 65 million years ago, as section 16.5 discussed. We do not know for certain what caused their demise, but it is clear that their disappearance opened up habitats for many other species, including the mammals that eventually gave rise to our own.

18.2 MASTERING CONCEPTS

1. In what time interval did cells likely originate?
2. When and how did eukaryotic cells arise?
3. What were the Ediacarans like?
4. What types of organisms flourished in the Cambrian?
5. How did life diversify after the Cambrian?

18.3 The Age of Mammals

Mammals first appeared in the Mesozoic, but radiated extensively in the Cenozoic, as the climate changed. The hominoids were ancestral to apes and humans; extinct hominids were animals that were ancestral to humans only.

Considering the parade of life that has graced Earth for nearly 4 billion years, our own species seems rather insignificant—and certainly a late arrival on the scene! Still, our look at the history of life concludes with our own beginnings.

Mammals Have Radiated During the Cenozoic Era

The dawn of the Cenozoic era 65 million years ago was a time of great adaptive radiation for mammals, according to the fossil record (see fig. 16.13). Molecular evidence, however, suggests that mammals lived at the end of the Mesozoic, perhaps as long ago as 100 million years in the past. Researchers are not certain whether this discrepancy reflects the incompleteness of the fossil record, or some unknown event that led to the rapid evolution of certain genes. But whenever it was that mammals first appeared, by 65 million years ago, they were clearly geographically widespread.

At the start of the Cenozoic era, diverse hoofed mammals grazed the grassy Americas. Many may have been marsupials (pouched mammals) or the more primitive egg-laying monotremes, ancestors of the platypus (see fig. 16.3). Their young were helpless, born tiny, hairless, and blind. They crawled along the mother's fur to sweat glands modified to form tiny nipples, which secreted milk.

Then a new type of mammal appeared whose young were better protected and therefore more likely to survive, reproduce, and persist. These newcomers were the placental mammals. The young remain within the female's body longer, where the placenta nurtures them. Mammary glands are well developed. Fossil evidence indicates that the placental mammals burst rather suddenly onto the scene. Within just 1.6 million years during the

Tertiary period (65 to 1.8 million years ago), 15 of the 18 modern orders of these mammals appeared.

Placental mammals eventually replaced most of the marsupials in North America, then invaded South America, thanks to changing geography. Until 2 to 3 million years ago, South America was an island. During the Tertiary period, several types of large marsupials thrived there because the placental mammals, which had originated in North America, could not reach the enormous southern island.

Then the Bering land bridge rose and connected Asia to North America. Many mammals, including our ancestors, probably journeyed from Asia to Alaska and southward through North America. The isthmus of Panama formed and provided a route for northern placental mammals to invade the southern marsupial communities. The first animals to arrive, rodents and ground sloths, crossed when the land bridge was only a string of islands. Arrival of the placental mammals and their use of natural resources drove nearly all the South American marsupials to

A

FIGURE 18.18
Life in the Mesozoic.
(**A**) The dawn of the Mesozoic era saw many small animals, called thecodonts, that had characteristics of reptiles and mammals. Mammals did not suddenly appear with the demise of the dinosaurs; they just became more prevalent. (**B**) Duck-billed maiasaurs left fossil evidence on a hill in Choteau, Montana, named Egg Mountain. A sudden volcanic eruption 80 million years ago preserved in time their community, including nine inhabited nests. "Dinosaur upchuck" near the nests suggested that parents fed offspring by chewing food for them, much as birds do today.

B

FIGURE 18.19

Placental Mammals Migrated.

About 3 million years ago, the appearance of the Panama land bridge between the separate continents of North and South America provided a route for the migration of animals between the two huge landmasses. Many species wandered in both directions. The southward invasion of placental mammals drove to extinction many marsupial species living in the former island continent.

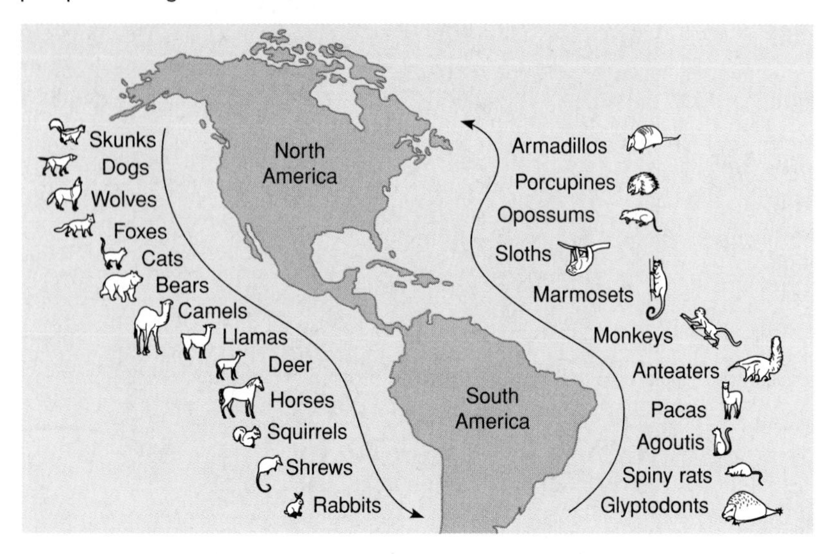

extinction (fig. 18.19). Many of the species found there today, including peccaries, llamas, alpacas, deer, tapirs, and jaguars, are descendants of immigrants from the north. A few species, including the opossum, armadillo, and porcupine, traveled in the opposite direction.

Geology and the resulting climate changes molded the comings and goings of species throughout the Cenozoic. The era began with formation of new mountains and coastlines as tectonic plates shifted. The wet warmth of the first epoch, the Paleocene (from 65 to 54 million years ago), which opened up many habitats for mammals, continued into the Eocene (from 54 to 38 million years ago), providing widespread forests and woodlands. Grasslands began to replace the forests by the end of the Eocene when the temperature and humidity dropped. Extinctions of some mammals paralleled the changing plant populations as the forests diminished, but grazing mammals thrived throughout the Oligocene, Miocene, and Pliocene.

The Road to Humanity

Our species, *Homo sapiens* ("the wise human"), probably first appeared during the Pleistocene epoch, about 200,000 years ago. The earliest of our order, Primates, were rodentlike insect eaters that lived about 60 million years ago. The ability of primates to grasp and to perceive depth provided the flexibility and coordination necessary to dominate the treetops. Primates diversified into many new species.

About 30 to 40 million years ago, a monkeylike animal about the size of a cat, *Aegyptopithecus,* lived in the lush tropical forests of Africa. Although the animal probably spent most of its time in the trees, fossils of limb bones indicate that it could run on the ground too. Fossils of different individuals found together indicate that these were social animals. *Aegyptopithecus* had fangs that it might have used for defense. This animal and a few monkeylike contemporaries are possible ancestors of gibbons, apes, and humans.

From 22 to 32 million years ago lived the first **hominoids,** primates ancestral to apes and humans only. Most hominoid evidence comes from Africa, but an animal living in Europe at that time, named *Dryopithecus,* was also a hominoid (fig. 18.20). The way the bones fit together suggests that this ape lived in the trees but could walk farther than *Aegyptopithecus.* It was also more dextrous than its predecessors.

Several species of apes lived 11 to 16 million years ago in Europe, Asia, and the Middle East. These apes were about the size of a human 7-year-old and had small brains and pointy snouts. Possibly in response to competition in the treetops and forests shrinking due to climate shifts, these apes ventured onto the grasslands. One species of Miocene ape survived to give rise to humans and African apes.

Hominoid and **hominid** (ancestral to humans only) fossils from 4 to 19 million years ago are scarce. ("Hominid" is being replaced with "hominine," a term based more on cladistics than on just fossil evidence.) This was the time that the stooped, large-brained ape became the upright, smaller-brained ape-human. A rare 10-million-year-old fossilized face from northern Greece has small canine teeth and thick tooth enamel, which are more humanlike characteristics. This animal could have been an immediate forerunner of hominids.

Four million years ago, **bipedalism**—the ability to walk upright—opened vast new habitats on the plains to hominids. Several species of *Australopithecus* lived at this time. These ani-

FIGURE 18.20

Early Hominids.

The "oak ape" *Dryopithecus* lived from 32 to 22 million years ago in Europe. The fossils were found with oak leaves.

mals had flat skull bases, a feature seen in all modern primates except humans. They stood from 4 to 5 feet (1.2 to 1.5 meters) tall and had humanlike teeth and brains about the size of a gorilla's. The angle of preserved pelvic bones, plus the finding of fossils near those of grazing animals, suggests that this ape-human had left the forest.

Fossils have fleshed out our knowledge of the australopithecines. These include a child's skull discovered in a limestone outcropping in Taung, South Africa, in 1925; the skeleton (dubbed Lucy) of a 20-year-old female found in 1974; and astonishingly clear footprints discovered in Laetoli, Tanzania, in 1976 (fig. 18.21). In 1999, paleontologists discovered three skeletons of a previously unknown hominid, which they named *Australopithecus garhi.* It lived about 2.5 million years ago, in eastern Ethiopia, and was possibly a contemporary of members of our own genus, *Homo.* The fossils revealed that *A. garhi* hunted and butchered large animals for food.

Two million years ago, australopithecines coexisted with *Homo habilis*—a more humanlike primate who lived communally in caves and cared intensely for young. "Habilis" means handy, and this primate is the first to evidence extensive tool use. *H. habilis* coexisted with and eventually was replaced by *Homo erectus.* Fossilized teeth and jaws suggest that *H. erectus* ate meat. They were the first primates to have an angled skull base that would have permitted a greater range of sounds, making speech possible. *H. erectus* fossils have been found in China, tropical Africa, and southeast Asia, indicating that they may have been the first hominids to leave Africa. The distribution of fossils suggests that *H. erectus* lived in families of male-female pairs (most primates have harems). They left fossil evidence of cooperation, social organization, and tool use, including the use of fire (fig. 18.22).

FIGURE 18.21
An Early Walk.
Much of what we know about the australopithecines comes from two parallel paths of footprints, preserved in volcanic ash in the Laetoli area of Tanzania. Archaeologist Mary Leakey and her team discovered them by accident in 1976. The 27-meter- (89-feet-) long trail of footprints, from 3.6 million years ago, were probably made by a large and small individual walking close together, and a third following in the steps of the larger animal in front. The shape of the prints indicates that their feet and gait were remarkably like ours.

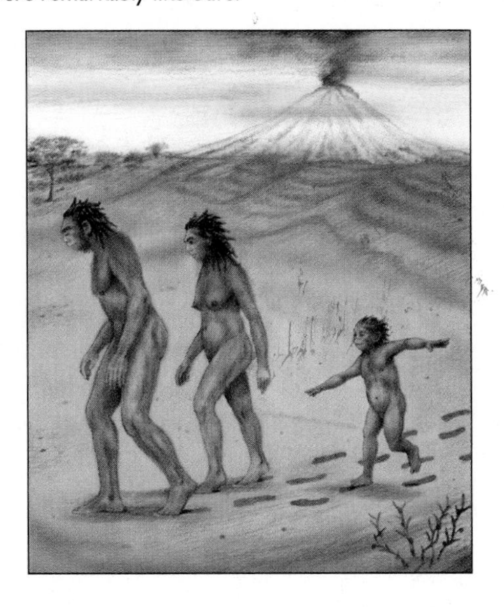

FIGURE 18.22
Handy Hominid.
Homo erectus made tools out of bone and stone, used fire, and dwelled communally in caves from 1.6 million years ago to possibly as recently as 35,000 years ago.

H. erectus lived from the end of the australopithecine reign some 1.6 million years ago to possibly as recently as 35,000 years ago, and probably coexisted with the first *Homo sapiens.* Several pockets of ancient peoples may have been dispersed throughout the world at that time.

The Neanderthals, contemporaries of *H. erectus* and of *Homo sapiens,* appeared in Europe about 150,000 years ago. By 70,000 years ago, they had spread to western Asia. The Neanderthals had slightly larger brains than us, gaps between certain teeth, very muscular jaws, and large, barrel-shaped chests. They take their name from Neander Valley, Germany, where their fossils were discovered in 1856. Other skeletons buried with flowers suggest that the Neanderthals recognized important events.

Molecular evidence has added to what we know of the Neanderthals from their fossils. In 1997, a graduate student ground up a bit of arm bone from the original French Neanderthal skeleton, and extracted and amplified a 100-base sequence from a mitochondrial gene. Then he compared it with the corresponding region among 986 modern *Homo sapiens.* Neanderthal DNA had three times the number of differences seen between pairs of the most unrelated modern humans. The conclusion: it is highly unlikely that Neanderthals and modern humans ever interbred, because they are too genetically dissimilar.

From 30,000 to 40,000 years ago, the Neanderthals coexisted with the lighter-weight, finer-boned, and less hairy Cro-Magnons, who had high foreheads and well-developed frontal brain regions. The first Cro-Magnon fossils found were the skeletons of five adults and a baby arranged in what appeared to be a communal grave in a French cave. Nearby were seashells pierced in a way that suggested they may have been used as jewelry. Intricate art decorated the cave walls.

Hominid Health

We tend to look at the transition from a hunter-gatherer lifestyle to agriculture and then urbanization as steps forward, but clues in preserved skeletons indicate that as far as health goes, this "progress" may actually be a step backward. Hunting and gathering required activity and mobility to obtain a diet that was low in fat and high in fiber and vitamins. The isolated bands of people that lived this way, before 4,000 years ago, had little contact with others and so the chances of contracting infectious diseases were low. Plus, the wild plants that they ate often provided phytochemicals that treated parasitic infections.

In contrast, urbanization led to "sedentism" (inactivity) and dietary changes (fig. 18.B). The resulting malnutrition and spread of infectious diseases are seen in the preserved bones of children. When a child starves or suffers from severe infection, the ends of the long bones stop growing. When health returns, growth resumes, but leaves behind telltale areas of dense bone.

Several types of studies vividly point out the health differences between then and now. Tooth decay, for example, is an excellent indicator of health. Among teeth from hunter-gatherers, 3% indicate decay, whereas among farming people the percentage jumps to 8.7, and among city dwellers it is 17%. Before agriculture, life expectancy was about 25 years; after, it fell to 19 years. Coprolites (preserved excrement) from native peoples who lived in Arizona 9,000 years ago did not contain parasites, but once the people began to farm, their excrement contained pinworms, tapeworms, and hookworms. The people picked up these infections by living near their excrement, using it to fertilize crops, storing grain, and no longer eating plants that contained antiparasite chemicals.

Paleopathologists, who study the diseases of ancient peoples, have discovered some fascinating glimpses of the past. The australopithecine Lucy, the original Neanderthal skeleton, and Ötzi, the 5,300-year-old "ice man," all had signs of arthritis. An australopithecine child from 3 million years ago had periodontal disease. In a Florida swamp, the skeleton of an adolescent male from 8,000 years ago revealed that he had spina bifida and a shriveled right leg. The fact that he was carefully buried indicated that his community took care of their ill. A 5,000-year-old skeleton from Jordan had pits in the spine characteristic of tuberculosis. And a child's jawbone from about A.D. 1500 found in the southwest United States had the spongy appearance that is a sign of scurvy, a disease that results from deficiency of vitamin C. Despite increased exposure to parasites and other infections, the constant and controllable food supplies allowed human populations to grow explosively.

FIGURE 18.B
Lifestyle Determines Health.
As modern humans went from a hunter-gatherer lifestyle to agriculture and then urbanization, malnutrition, infectious disease, and other conditions became more prevalent. The hunter-gatherer way of life offered regular exercise, a low-fat and high-fiber diet rich in vitamins, and isolation from groups that could spread infection.

Fossils indicate that Neanderthals and Cro-Magnons, two types of humans, coexisted, but did not interbreed, in the region of the Mideast that links Africa and Eurasia. What kept them from mating with each other? Perhaps it was incompatible gametes, or a geographical separation. Maybe they migrated, occupying the same areas at different times. Another theory is behavioral—perhaps the great facial differences between the large-boned Neanderthals and more delicate Cro-Magnons made it impossible for them to exchange mating cues. Whatever happened, by 30,000 years ago, the Neanderthals were no longer, and the Cro-Magnons presumably continued on the path to humanity.

Evolution continued, taking on a cultural aspect as social and communication skills improved. Cave art from about 14,000 years ago indicates that our ancestors had developed fine hand coordination and could use symbols. By 10,000 years ago, people had migrated from the Middle East across Europe. In many places agriculture had begun to replace a hunter-gatherer lifestyle, with medical consequences (see Health 18.1).

We know a little of what humans were like 5,300 years ago from Ötzi, a man whose remains hikers discovered in the Ötztaler Alps of northern Italy in 1991 (fig. 18.23). Ötzi was apparently tending sheep high on a mountaintop when he fell into a ditch and was covered by a sudden snowstorm, quickly freezing to death. He was dressed for the harsh weather, and berries found with him place the season as late summer or early fall. Ötzi not only had an illness, but he carried a remedy with him! An autopsy revealed the whipworm *Trichuris trichiura* in his large intestine, which caused anemia and cramps. About his waist, Ötzi wore two small sacs that hung from a leather thong. Each sac contained the birch fungus *Piptoporus betulinus,* which produces an oil that kills whipworms, and also causes diarrhea that can flush the infection out of the body. Limited DNA analysis of Ötzi's tissues indicate that he belongs to the gene pool of people living in the area today. His genetic similarity to modern humans isn't surprising, considering that the history of life on Earth began more than 3 billion years ago. The 5,300 years separating Ötzi's existence from ours is but a flicker of evolutionary time.

18.3 MASTERING CONCEPTS

1. How did changing conditions in the Cenozoic lead to the adaptive radiation of mammals?
2. Distinguish between hominoids and hominids.
3. What were our immediate ancestors like?

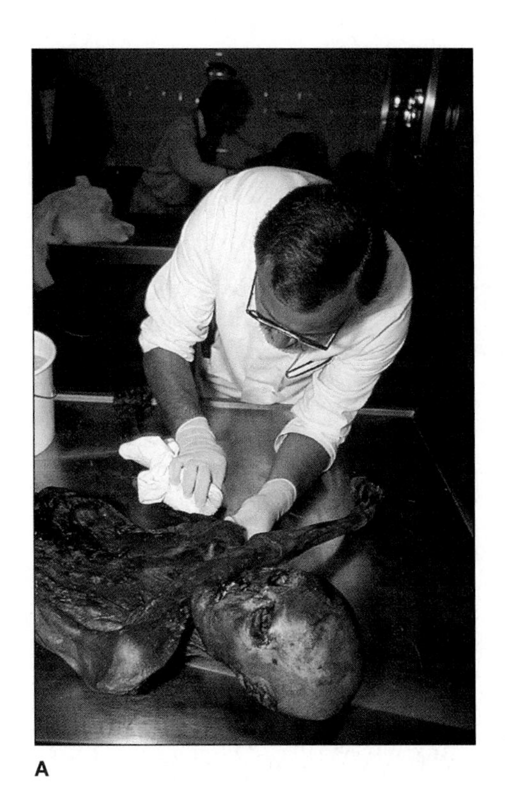

A

B

FIGURE 18.23
The "Ice Man" Ötzi.
(**A**) The oldest known intact human, at 5,300 years of age, lay frozen in the Alps between Austria and Italy, where hikers found him in 1991.
(**B**) He wore well-made clothing, had intricate arrows indicating familiarity with ballistics and engineering, and carried mushrooms with antibiotic properties. He had tattoos, indentations in his ears that suggest he wore earrings, evidence of a haircut, and a hat. This depiction was derived from the evidence on and near the preserved body. Ötzi has been periodically defrosted so that researchers could sample his bone, tooth enamel, and DNA for genome sequencing.

Chapter Summary

18.1 How Might Life Have Originated?

1. The solar system formed about 4.55 billion years ago, and life left evidence on Earth by about 700 million years after that. Conditions when life originated differed from today.
2. When life began, Earth was geologically unstable, and the atmosphere was high in hydrogen and low in oxygen.
3. **Prebiotic simulations** combine simple chemicals that include the elements in organisms in the presence of energy. Organic building blocks form, some of which are also in organisms.
4. Amino acids may have polymerized into peptides and nucleotides into nucleic acids on hot clay or mineral surfaces that served as templates, structural supports, and supplied energy and catalysis.
5. The **RNA world** theory proposes that RNA or a similar but more stable molecule preceded formation of the first cells because it could replicate, encode information, change, and catalyze reactions. Proteins became enzymes. **Reverse transcriptase** could have copied RNA's information into more permanent DNA.
6. Phospholipid sheets that formed bubbles around proteins and nucleic acids may have formed cell precursors, or **progenotes.**
7. Metabolic pathways may have originated when **progenotes** mutated in ways that enabled them to use alternate or additional nutrients.

18.2 Highlights in the History of Life

8. Life began in the seas. The earliest fossils are of cyanobacteria from 3.7 billion years ago. The oldest eukaryotic fossils, of algae, date from 1.9 billion years ago. Evidence of multicellularity dates to 1.2 billion years ago.
9. The Ediacarans were soft, flat organisms unlike modern species that lived in the late Precambrian and early Cambrian periods.
10. The Cambrian explosion introduced evidence of many species, notably those with hard parts. Amphibian-like animals ventured onto land about 360 million years ago. By 300 million years ago, reptiles had appeared and then diverged, eventually evolving also into birds and mammals. Invertebrates, ferns, and forests flourished. The Paleozoic era ended with mass extinctions.

18.3 The Age of Mammals

11. Dinosaurs prevailed throughout the Mesozoic era, when forests were largely cycads, ginkgos, and conifers. In the middle of the era, flowering plants became prevalent. When the dinosaurs died out 65 million years ago, resources opened up for mammals.
12. Molecular evidence dates the origin of mammals to 100 million years ago, and fossil evidence indicates their adaptive radiation beginning 65 million years ago. Placental mammals eventually replaced many marsupial species.
13. Mammalian species changed throughout the Cenozoic as the climate cooled and dried.
14. *Aegyptopithecus* and other primates preceded the **hominoids,** which were ancestral to apes and humans. **Hominids** were ancestral to humans only. Four million years ago several species of *Australopithecus* existed and were gradually replaced with *Homo habilis* and then *Homo erectus.* Cro-Magnons were *Homo sapiens,* as are modern humans. The Neanderthals lived at the same time as early *Homo sapiens,* but were probably a separate species.

Testing Your Knowledge

1. About how soon after Earth formed did life first appear, according to fossil evidence?
2. Cite a chemical and a geological reason why life may have originated in as short a time as 10 million years.
3. What chemical elements had to have been in primordial "soup" to generate nucleic acids and proteins?
4. Cite four reasons why RNA or a similar molecule is a likely candidate for the most important biochemical in the origin of life.
5. List the major life events of the five geological eras and trace human evolution through time.

Thinking Scientifically

1. What are the limitations, if any, of the recent evidence used to support cosmic ancestry?
2. How was the scientific method used to show that life does not arise spontaneously?
3. Louis Pasteur is often credited with disproving the theory of spontaneous generation. Stanley Miller often receives credit for "brewing" primordial soup. What previous discoveries and hypotheses influenced Pasteur and Miller?
4. What is the chemical evidence that life may have originated in different ways?
5. Why does molecular evidence usually yield an earlier date of a speciation event than fossil evidence?
6. What might have happened to evolving life between the first types of progenotes, and the last common ancestor that the three domains of life shared?

References and Resources

Bear, Greg. 1999. *Darwin's Radio.* New York: The Ballantine Publishing Group. A novel about modern humans evolving into a new species.

Cody, George D., et al. August 25, 2000. Primordial carbonylated iron-sulfur compounds and the synthesis of pyruvate. *Science,* vol. 289, p. 1337. A prebiotic simulation yields the product of glycolysis.

Hoffman, Hillel J. February 2000. The rise of life on earth. *National Geographic,* vol. 197, p. 34. A view of Earth at the time of life's debut.

Lewis, Ricki. 2000. *Discovery: Windows on the Life Sciences.* Malden, Mass.: Blackwell Science. Chapter 2 offers an extensive analysis of evidence about events at the dawn of life.

Tattersall, Ian. January 2000. Once upon a time, *H. sapiens* may not have been the only hominid. *Scientific American,* vol. 282, p. 56. Modern humans may have emerged the sole survivor among several types of ancestral hominids.

Wong, Kate. April 2000. Who were the Neanderthals? *Scientific American,* vol. 282, p. 98. How similar to us were the Neanderthals?

The *LIFE* Online Learning Center provides additional resources and tools for studying this chapter.
www.mhhe.com/life

UNIT 4 • THE DIVERSITY OF LIFE

Walk across Earth and the moon
a few billion years ago and your boots get
covered with dust. Walk across the moon today and
your boots still get covered with dust. But between 4 and 5 billion
years ago something different happened on Earth. Walk across Earth today
and your boots sink into a carpet of grass, moss, or pine needles. Depending on
where you are, you may run into a redwood tree, an alligator, or a passing
elephant. You swat at a mosquito, which harbors a universe of
microorganisms in its gut. Even these microbes may have
troubles of their own, in the form of viruses.
Welcome to Earth—planet of life.

CHAPTER 19

Viruses and Simple Infectious Agents

Ocean Viruses

Viruses seem to be everywhere. These assemblies of protein and nucleic acid, far simpler than cells, but nonetheless able to cripple an organism the size of an elephant or a redwood, are responsible for some of the worst diseases in humans, and cause a variety of conditions in plants. They are also abundant in the oceans. A teaspoon of seawater might contain 10 million to 100 million viruses.

Viruses were known to infect dolphins, seals, clams, salmon, crabs, oysters, and mussels, but their effects on microscopic ocean life weren't discovered until a decade ago. In the oceans, viruses outnumber bacteria 10 to 1. They are especially abundant in the top millimeter of the oceans, where bacteria and plankton (single-celled algae and other protistans) support vast food webs. The number of viral particles in this top layer of all the planet's ocean waters is 3 followed by 30 zeroes.

Mathematical models predict that each day, viruses rip apart 20% of the ocean's bacteria, releasing yet more viruses to the water—plus abundant cellular contents.

More important than viral numbers is their rapid turnover—population sizes fluctuate wildly, and any given viral particle exists for only a few days. This means that ocean viruses infect cells and reproduce within them with great speed. Mathematical models predict that each day, viruses rip apart 20% of the ocean's bacteria, releasing yet more viruses to the water—plus abundant cellular contents.

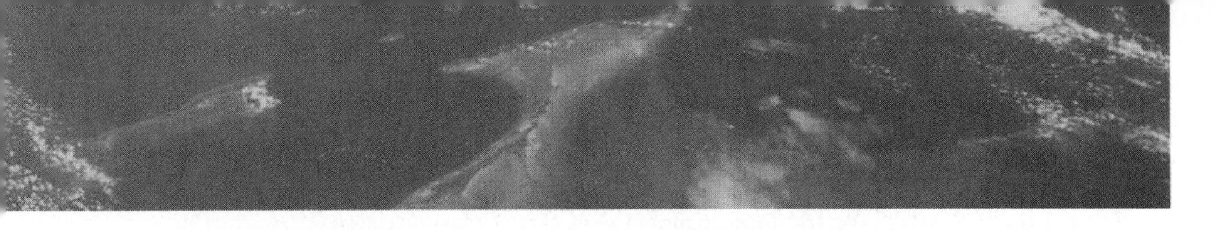

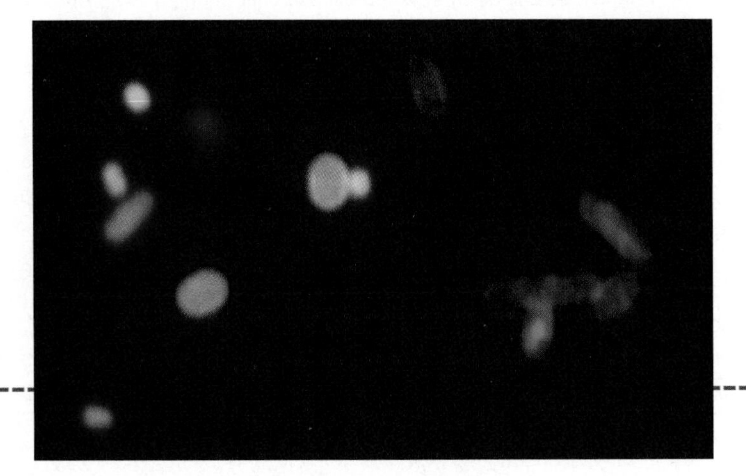

Ocean Bacteria and Viruses. The blue halos around the red dots are hundreds of viruses surrounding bacteria.

The viruses also maintain biodiversity of various species by quickly diminishing any bacterial or plankton population that grows too swiftly.

The unrelenting viral attack on bacteria and plankton is important for the ocean ecosystem, because it releases a steady diet of nutrients that other organisms use. The viruses also maintain biodiversity of various species by quickly diminishing any bacterial or plankton population that grows too swiftly. For example, viruses may be responsible for the unexplained disappearances of phytoplankton "blooms," which are population explosions that can tinge the water with red pigments.

Curtis Suttle and Amy Chan, researchers at the University of British Columbia, demonstrated experimentally how important viruses are to ocean ecosystems. They built a microcosm: a tank of ocean that includes a realistic spectrum of microscopic species—and then removed the viruses. The result: Diverse types of plankton stopped growing. The populations apparently needed the viruses to obtain nutrients, and to keep their sizes and interactions in balance.

Viruses aren't all bad!

19.1 What Is a Virus?

A virus is not alive and is much simpler than a cell, consisting of a nucleic acid wrapped in protein and perhaps a fatty envelope. Inside cells of specific species, a virus can take over protein synthesis and mass-produce itself, with powerful effects.

A **virus** is a small, infectious agent that can pass through filters that are designed to trap most bacteria. It is not a cell, but simply a genetic informational molecule—DNA or RNA—enclosed in a protein coat. As such, a virus straddles the boundary between the chemical and the biological, a structure that is more than an assembly of macromolecules, yet less than a cell.

Biologists do not consider a virus to be alive, because it does not metabolize, respond to stimuli, or even reproduce on its own. A virus requires a living host cell to manufacture more of itself, which is really all it can do. In the process, the virus usually damages or destroys the host cell, which is how disease begins. The symptoms of a viral infection reflect the type of cell that is invaded. All kinds of organisms are susceptible to viral infection.

How Were Viruses Discovered?

People recognized viral infections many centuries before they knew what caused the illnesses. Once the early microscopists had discovered bacteria, these microorganisms were thought to cause all infectious diseases. In 1892, Russian botanist Dmitri Ivanovsky investigated a disease that stunted and mottled the leaves of tobacco plants (fig. 19.1A). To isolate the bacteria that he thought caused the disease, Ivanovsky passed extracts of damaged leaves through a porcelain filter designed to trap bacteria. But when he applied the filtered material to healthy plants, they showed the same signs of infection. Although he could not grow bacteria from any of his samples, Ivanovsky concluded that a bacterium must be the cause, and that it had somehow slipped through the filter.

In 1895, Dutch botanist Martinus Beijerinck repeated Ivanovsky's experiments. Taking them one step further, Beijerinck showed that the filtered fluid could retain its potency after being passed from one plant to another. This result suggested that the infective substance was in some way reproducing, ruling out the possibility that it was a toxic chemical. In 1898, Beijerinck concluded that the cause of the tobacco disease was a living entity and called it a "filterable agent," or virus, which means "poison" in Latin. That same year, researchers isolated the first filterable virus in animals, the cause of hoof-and-mouth disease. In 1901, scientists identified a virus as the cause of yellow fever in humans. In 1939, researchers used an electron microscope to capture the first images of what Beijerinck investigated: tobacco mosaic virus (TMV) (fig. 19.1B).

How Are Viruses Studied?

In the early 1900s, microbiologists learned how to culture viruses in layers of cells growing in petri dishes. Today, researchers still study viruses by observing their effects on cells that are grown in culture. By diluting virus preparations, researchers can isolate single strains of viruses from small, localized infections seen in cell cultures called **plaques** (fig. 19.2). The viruses that comprise a plaque descend from one original viral particle, and therefore represent a single

FIGURE 19.1
Tobacco Mosaic Virus.
(**A**) Tobacco mosaic virus (TMV) infection causes a characteristic mottling (spotting) of the leaf. (**B**) An electron micrograph of the tobacco mosaic virus (×400,000). (**C**) Illustration of TMV structure. The virus is intricate and highly symmetrical, but built of only a few protein units in a repeated organization.

A

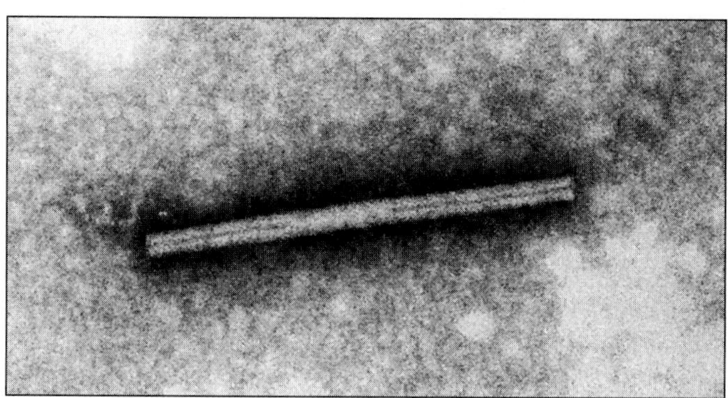

B

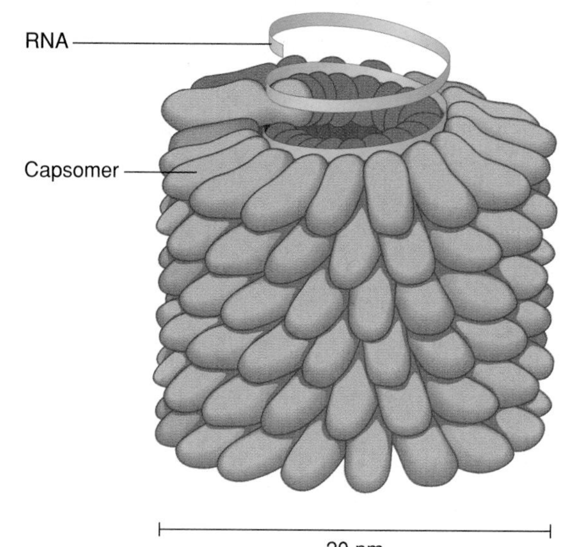

C

FIGURE 19.2
Viral Plaques.
Each spot in the culture of cells represents a zone of killed cells that started from one virus infecting one cell. This dish is about 15 centimeters in diameter.

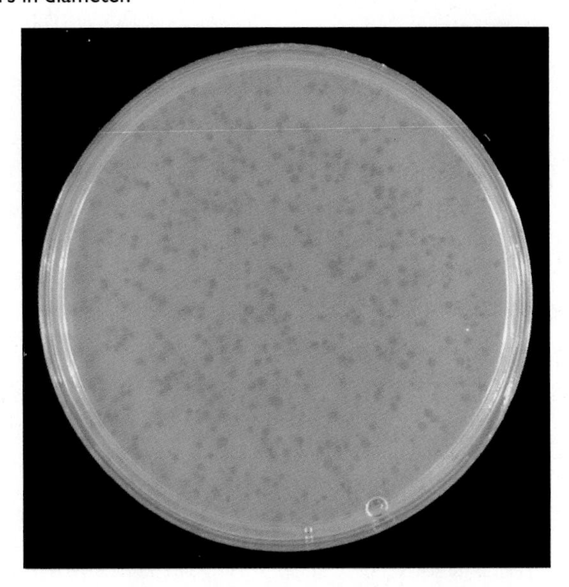

genetic strain. Large-scale cell cultures are used to mass-produce viruses for use in research and to manufacture vaccines. Researchers have sequenced the complete genomes of many viruses, making it possible to study the structure and function of each viral protein.

Viruses are important in studying basic biological processes, and they also have practical uses. Although viruses are not alive, their ability to borrow a cell's protein synthetic machinery to reproduce has provided valuable glimpses into genetic control and cellular organization. Some viruses have also revealed much about the control of cell growth and the development of cancer. Because they are so adept at entering target cells, viruses are often used as vectors in gene therapy, for which they are stripped of their infection-causing genes and loaded with the genetic instructions to repair a specific defect in a host cell.

Despite our use of virally derived molecular tools, viruses remain, in several ways, an enigma. Too small and simple to even qualify as being alive, they can nevertheless kill a complex, trillion-celled human in just days. Their genomes encode a mere handful of proteins compared with other genomes, yet by usurping biochemicals from their host cells, viruses rapidly reproduce, at the ultimate expense of the infected cell. At the same time, the combination of rapid reproduction and lack of genetic repair systems renders some viruses highly changeable, enabling them to keep ahead of host defenses in an evolutionary arms race of sorts. We still have much to learn about viruses.

How Are the Components of a Virus Organized?

A virus is much simpler than a cell, lacking a nucleus, organelles, and even cytoplasm. Only a few types of viruses even contain enzymes. Because of its stripped-down structure, a virus is often referred to as a particle. A virus is also much smaller than a cell (fig. 19.3A). An average virus is about 80 nanometers in diameter, compared with about 5,000 nanometers for a bacterium and 50,000 nanometers for an animal cell. It would take millions of aggregated viral particles to form a dot the size of a printed period.

All viruses contain genetic material that carries instructions to synthesize their constituent molecules. The protein coat, called the **capsid,** is composed of protein complexes called **capsomers,** which assemble in an ordered fashion around the genetic material. The capsomers form the capsid much like pentagons and hexagons join to form the pattern on a soccer ball. The complete infectious viral particle that can exist outside of a cell is called a **virion** (fig. 19.3B). Many virions are spherical or icosahedral in shape—an icosahedron is a polyhedron built of triangular sections. Other virions are rod-shaped, oval, or filamentous. The way that the capsid proteins associate with the genetic material and with each other determines a virion's overall form, which is one characteristic used to distinguish viruses.

The major criterion for classifying viruses is whether the genetic material is DNA or RNA. Either type of viral genetic material may be single-stranded or double-stranded. The type of genetic material determines the steps that are necessary for the virus to reproduce. A viral genome may contain as few as three to more than 200 genes—not much compared with other genomes. Nearly all viral proteins assist in recognizing and entering host cells, replicating genetic material, or helping the virus remain hidden within a cell. Some viruses that infect animal cells have genes that encode proteins that help in evading the host's immune defenses. All viruses obtain at least some of the enzymes they require to reproduce from the host cell.

Not only are viral genomes small, but they are highly efficient. Many copies of a particular protein may form the capsid. Like a greenhouse made of identical glass panes, a virus assembles identical capsomer proteins as modular units, getting maximal structure out of a minimal set of genes. The tobacco mosaic virus in figure 19.1B and the adenovirus and herpesvirus in fig. 19.3 illustrate the repeated nature of the capsid.

In addition to the protein coat, some viruses have a layer of membrane, derived from the host cell membrane, called an **envelope.** A virus's overall structure can be quite intricate and complex. For example, some viruses that infect bacteria, termed bacteriophages, have parts that resemble tails, legs, and spikes, and they look like the spacecrafts once used to land on the moon. Some of these surface molecules enable viruses to attach to target cells and inject their genetic material.

19.1 MASTERING CONCEPTS

1. What is a virus?
2. How were viruses discovered?
3. What are the relative sizes of viruses, bacterial cells, and eukaryotic cells?
4. How are viruses masters of genetic efficiency?
5. How do the components of a virus assemble?

OLC

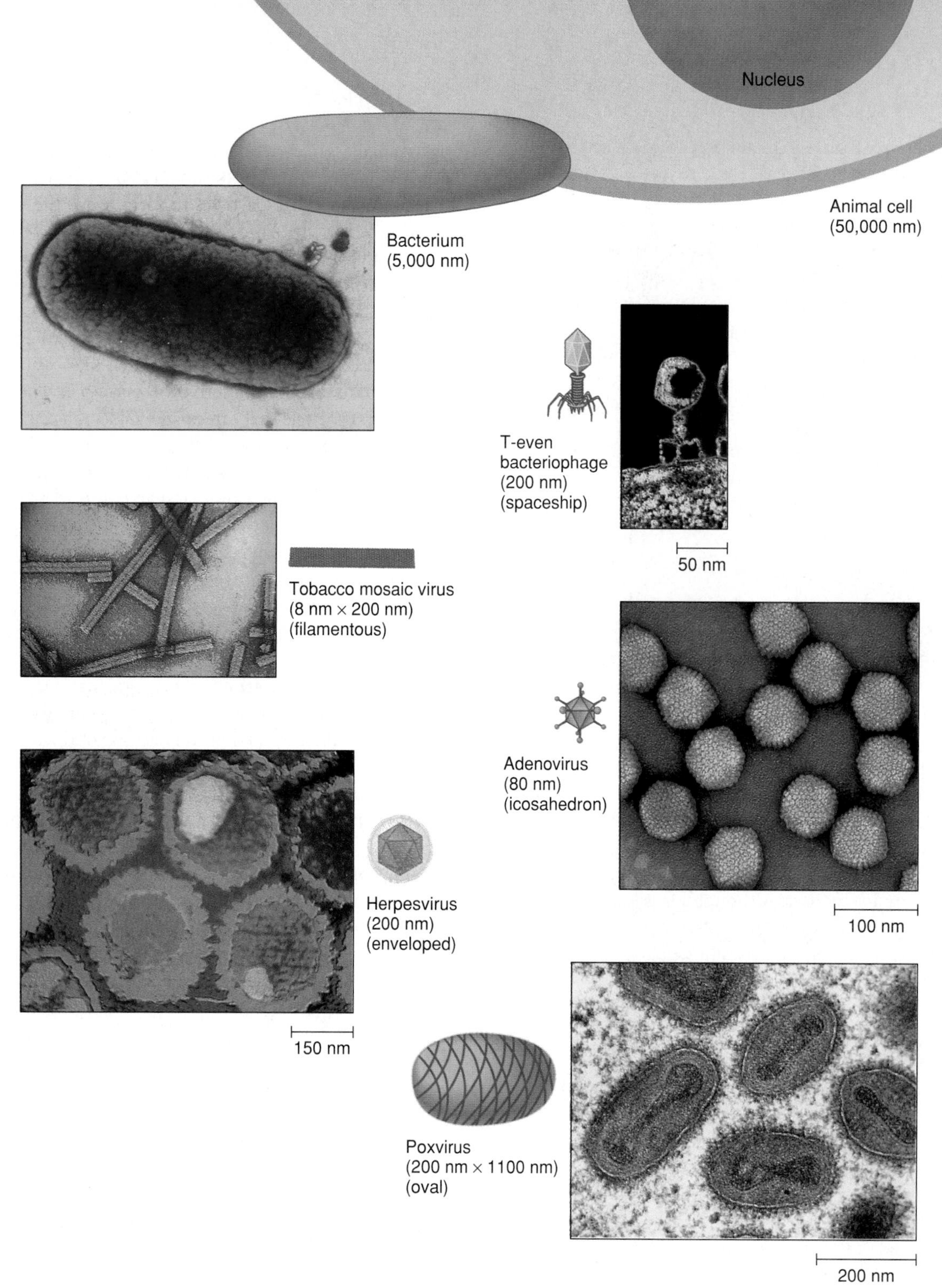

FIGURE 19.3

Viruses Are Small Compared to Cells.

Each type of virus has a very characteristic structure, which is tiny compared to a bacterial or eukaryotic cell. The organization of proteins determines the shape of the virus.

19.2 How Do Viruses Infect Cells and Reproduce?

A viral infection begins as a virus binds to a specific molecule on a cell surface, and then injects its genetic material or is engulfed. The virus may set into motion its own reproduction, eventually bursting the cell to release viral progeny, or nestle into the host's DNA and persist for some time. The steps of viral replication depend upon whether the genetic material is RNA or DNA, single-stranded or double-stranded.

A virion must bind to a specific target cell, enter it, and deliver its genetic material in order to manufacture more of itself. Alternatively, some viruses can integrate into the host's chromosome or remain as an extra piece of DNA within the cell and be perpetuated, sometimes for many years and even from generation to generation.

Viruses Bind Target Cells and Introduce Their Genetic Material

A virus attaches to a host cell by adhering to a surface molecule that the cell uses for some other function. For instance, a bacteriophage called lambda has a surface molecule that is shaped like the sugar maltose, which is an energy source for bacteria. The viral molecule binds to the receptor for maltose on the bacterial cell surface. The human immunodeficiency virus (HIV), which causes AIDS, binds to two molecules on helper T cells, which are part of the immune system (fig. 19.4). The required binding between a viral surface molecule and a molecule on the host cell's surface determines the specificity of the infection—that is, generally the virus infects only a cell within which it can reproduce.

The second step in viral infection, after binding, is entry of the viral genetic material into the cell (fig. 19.5). Many bacteriophages do this by acting like small hypodermic syringes, injecting their genetic material through the cell wall and cell membrane. Animal cells, which lack cell walls, often engulf virus particles and bring them into the cytoplasm through endocytosis (see fig. 4.12). Once inside an animal cell, virus particles are unwrapped, releasing the genetic material. Often lysosomal enzymes release the viral genetic material to the cytoplasm. Even viruses with envelopes require specific molecule-to-molecule attachments to infect, and the envelope may assist the virus in entering the cell.

Plants can fall victim to viral infections too. Viruses can cause a "mosaic disease" in which leaves or flowers become mottled with light green or yellow areas. Other viral infections of plants cause areas of cell death. To infect a plant cell, a virus must penetrate waxy coats and thick cell walls (fig. 19.6). Some viruses do this by taking advantage of mechanical damage to a leaf, such as from farm equipment, the wind, or a nibbling animal. Seeds and pollen may carry viruses. Worms in soil pass viruses to roots, as do fungal parasites to leaves. Most viral infections of plants begin with the bite of an insect that has sucking mouthparts, such as a leafhopper or aphid. When a leafhopper bites an infected plant, it sucks out fluid that contains virions. As the insect moves from plant to plant, it infects each new leaf that it punctures.

Within a plant, virions easily spread from cell to cell through the plasmodesmata, which are bridges of cytoplasm between plant cells (see fig. 4.20). The virus usually enters a cell in a leaf's upper epidermal layer, reproduces there, then crosses plasmodesmata to the internal leaf cells, called mesophyll, where photosynthesis occurs. Viruses can use special cellular proteins to cross through the plasmodesmata. The infection spreads as viruses invade the tissues that distribute sap.

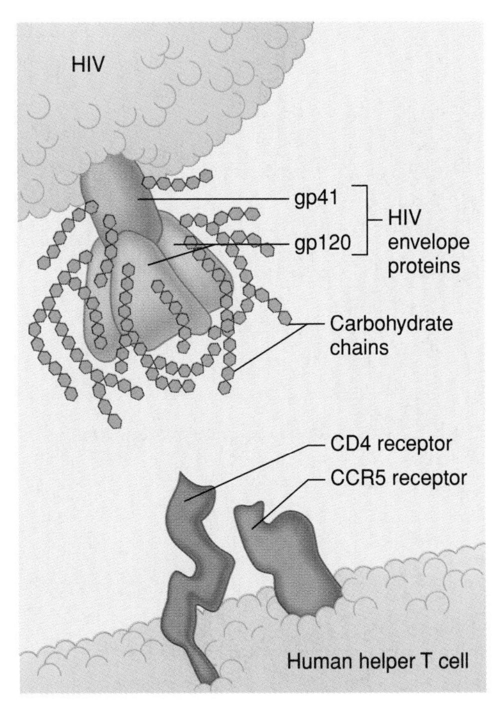

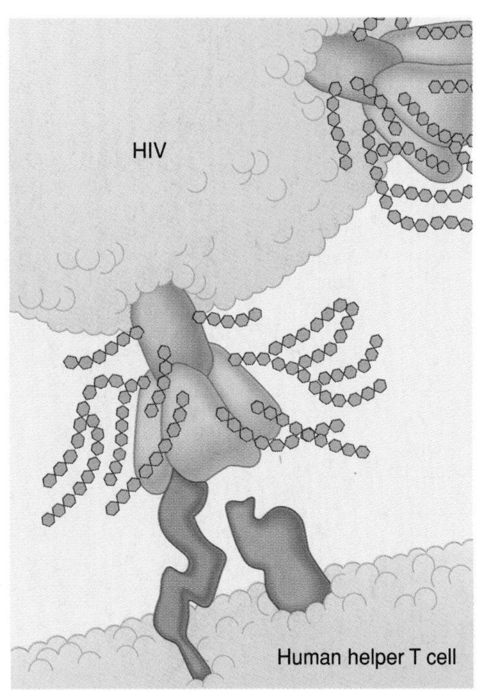

FIGURE 19.4
HIV Binds to a Helper T Cell.
(**A**) The part of HIV that binds to helper T cells is called gp120 (gp stands for glycoprotein). (**B**) When the carbohydrate chains that shield the protein portion of gp120 move aside as they approach the cell surface, the viral molecule can bind to a CD4 receptor. A receptor called CCR5 is also necessary for HIV to dock at a helper T cell. Once bound to the cell surface, the viral envelope fuses with the cell membrane, enabling the virus to enter. A few lucky individuals lack CCR5, and cannot be infected by HIV. (The size of HIV here is greatly exaggerated. Figure 19.9 shows the accurate proportions.)

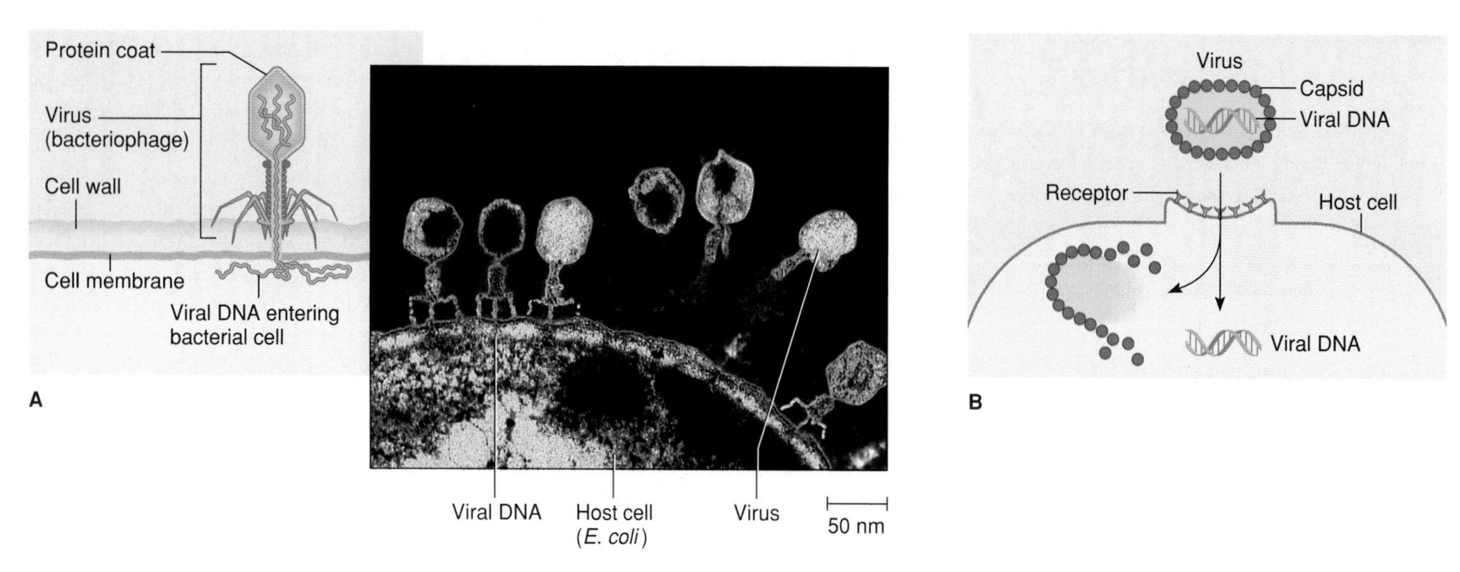

FIGURE 19.5
Viral Entry into Cells.
(**A**) Bacteriophages inject DNA directly into their host cell. (**B**) Animal viruses often enter the cell intact. The protein coat then dissolves to release the genetic material into the cytoplasm.

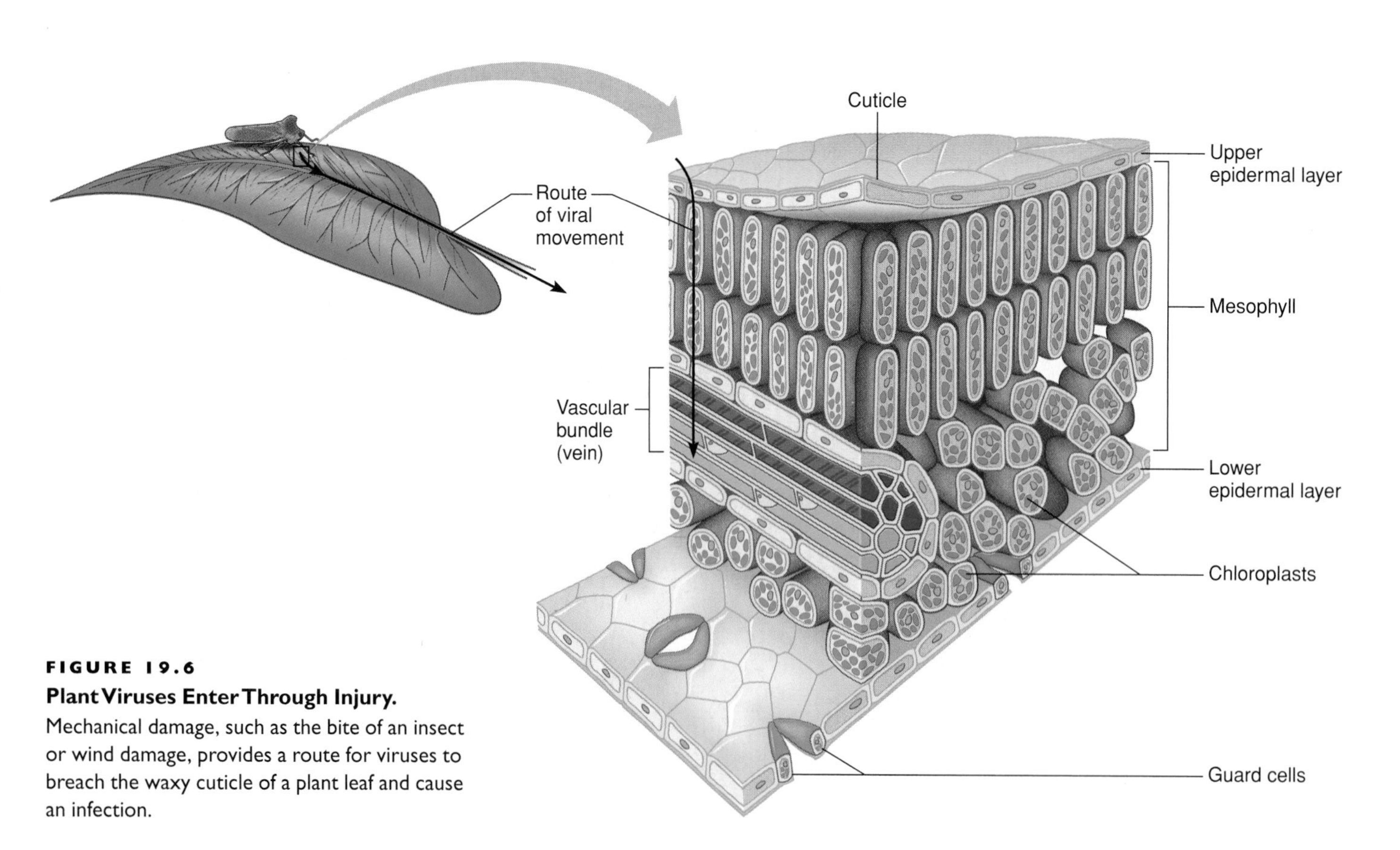

FIGURE 19.6
Plant Viruses Enter Through Injury.
Mechanical damage, such as the bite of an insect or wind damage, provides a route for viruses to breach the waxy cuticle of a plant leaf and cause an infection.

FIGURE 19.7
A Generalized Viral Replication Cycle.

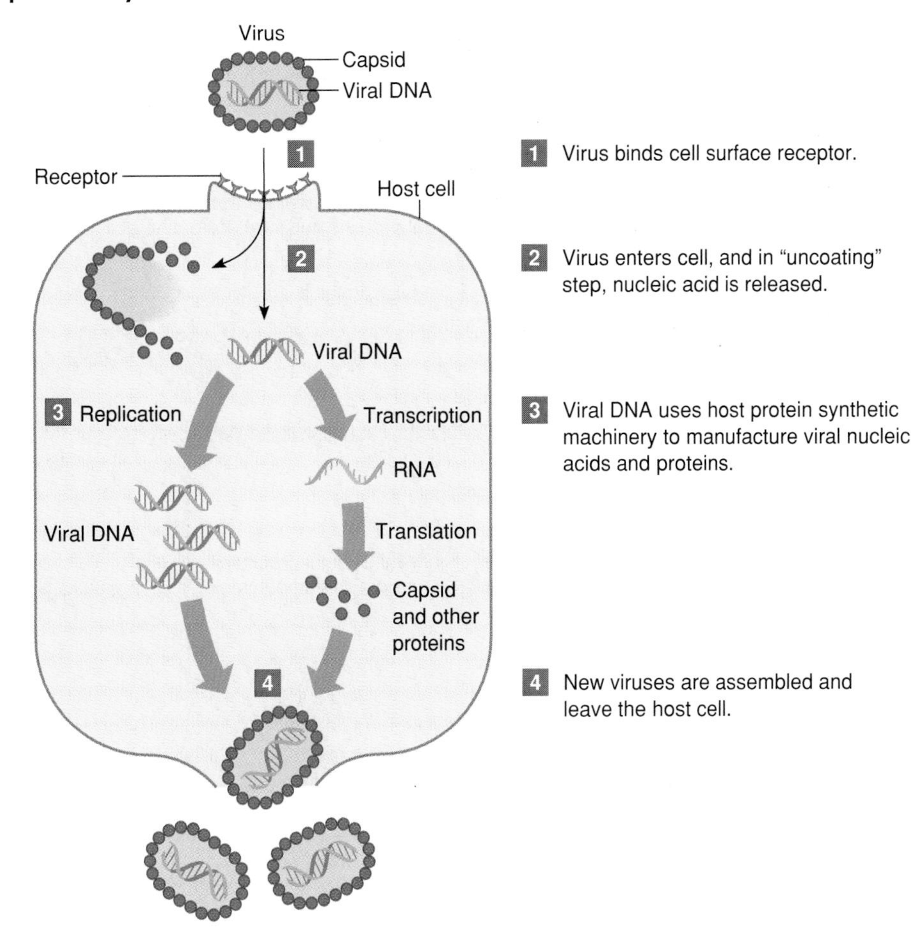

1 Virus binds cell surface receptor.

2 Virus enters cell, and in "uncoating" step, nucleic acid is released.

3 Viral DNA uses host protein synthetic machinery to manufacture viral nucleic acids and proteins.

4 New viruses are assembled and leave the host cell.

Whatever the host species or cell type, the result of viral infection is the same—viral genetic material is replicated and commandeers the cell's protein synthetic machinery. Some of these proteins help the virus to reproduce, whereas others are the actual components of the capsids. As the virus reproduces, it damages or destroys the host cell. A viral infection in plants can produce unexpected results. The beautiful color patterns in the leaves and flowers of some plants are the result of viral infections.

Two Viral Lifestyles: Lysis and Lysogeny

Although viruses use a host cell for reproduction, the way new progeny virions are made is very different from the method cells use to divide. Instead of doubling all of its components and then dividing, as a cell does, a virus reproduces by manufacturing many copies of the virion proteins and nucleic acids at one time. These components are assembled all at once, producing many virions. Because one virion produces hundreds of progeny, this process is often referred to as viral replication (fig. 19.7).

Viruses follow two major strategies to replicate. All viruses replicate, at some point, through what is known as a **lytic infection** that often lyses, or bursts, the cell. In contrast is a **lysogenic infection,** in which viruses "hide" in host cells as molecules of genetic material until conditions are right for them to replicate. Bacteriophages offer the simplest examples of these types of viral infections, but most animal, fungal, and plant viruses use similar strategies.

A lytic infection (fig. 19.8) begins when the viral genetic material enters the host cytoplasm. Many bacteriophages have double-stranded DNA. The bacterial proteins that control transcription bind to the viral genes and transcribe them, and soon protein synthesis begins. The promoters of many of the viral genes are so active that the host cell transcribes only those genes, whereas some of the newly made viral proteins actually block transcription of the host's genes and destroy its DNA. The host cell continues to replicate viral DNA and produce capsomer proteins, until, eventually, enough components accumulate so that hollow virions begin to be assembled. Then proteins insert viral DNA into the cores, forming complete **progeny virions,** the new viral particles. The host cell swells as hundreds of viruses assemble simultaneously. As a final insult, a viral protein breaks open the bacterial cell, releasing a flood of progeny virions and killing the cell. Within hours, all of the bacteria in the culture are dead. Considering the fact that just one virion infecting one cell starts this overwhelming infection process, it isn't surprising that a viral infection can kill cells, or cause symptoms, so quickly.

In eukaryotic cells, viruses follow a similar route to replication. However, they are able to use the organelles of these more complex

FIGURE 19.8

Lysis Versus Lysogeny.

(**A**) After invading a cell, viral genetic information is used by the cell's enzymes to produce many copies of the components of the virion. Lysis of the cell occurs when the virus particles assemble within the cell, causing it to burst. (**B**) Some viruses insert their DNA into the host chromosome. It may remain integrated there, even through many generations, in what is known in bacteria as lysogeny. An environmental change may trigger a lysogenic virus to become lytic.

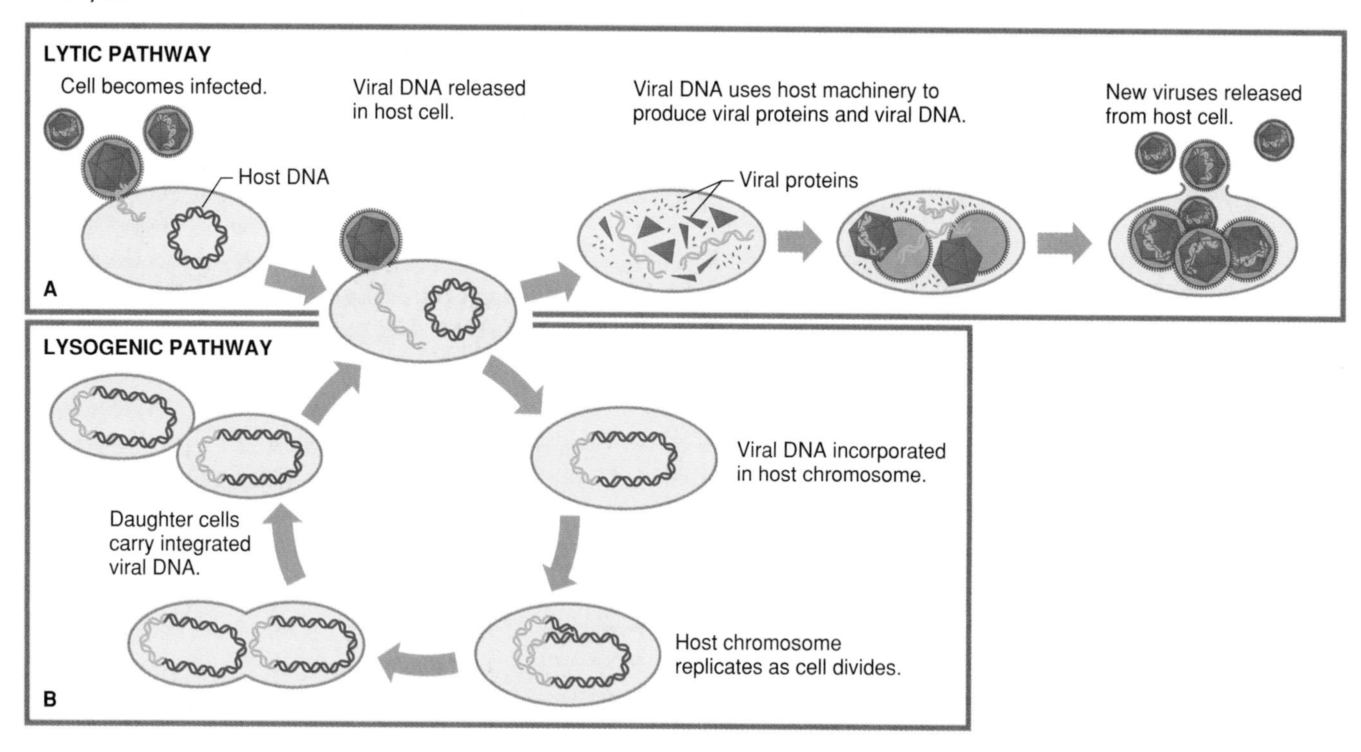

host cells to manufacture larger and more intricate proteins, glycoproteins, and lipids. For example, some viruses surround themselves with a membrane envelope by using the secretory network formed by the endoplasmic reticulum, the Golgi apparatus, and associated vesicles (see fig. 3.11) to manufacture certain viral proteins that lodge in the cell membrane. As progeny virions exit the cell, they bind to these embedded proteins, wrapping bits of the cell membrane around themselves, thereby creating envelopes. As more and more virions carry off segments of the cell membrane, the cell eventually loses its integrity and lyses. Viruses that lack envelopes use a variety of other pathways to leave the cell, including bursting out in the manner of bacteriophages.

An alternative strategy to lysis that some bacteriophages exhibit is **lysogeny,** in which the viral DNA integrates into the DNA of the host and remains "hidden" for some time. A lysogenic virus cuts the host cell DNA with an enzyme, then uses other enzymes to join its DNA to the host DNA molecule. When the cell divides, it replicates the viral genes too. Lysogeny does not damage the host cell, and at the same time ensures survival of the virus, because the cell cannot identify and remove the stowaway foreign DNA. Only a few viral proteins are transcribed and translated, and most of these proteins function as a switch to determine whether and when the virus should become lytic and exit the cell. At some signal, such as stress from DNA damage or lack of nutrients for the cell, these viral proteins cut the viral DNA out of the host genome and proceed to a lytic infection cycle. The resulting progeny virions then infect other cells, where they may enter a lytic or lysogenic state, depending upon the condition of the new host cells.

Some animal viruses can remain dormant as DNA molecules within a cell until conditions make it possible, or necessary, to replicate. Viruses in this state are **latent.** They may remain in cells for a long time, without producing progeny virions. An example of a latent virus is the herpesvirus that causes "cold sores." After the infection, the herpesvirus DNA remains in the infected cells indefinitely. When the cell becomes stressed or damaged, the DNA is transcribed and viral proteins are made. New virions are assembled and leave the cell to infect other cells. Cold sores often recur at the site of the original infection. Epstein-Barr virus (EBV)—the virus that causes mononucleosis—is another latent virus. Because it is so efficient at remaining undetected, more than 80% of the human population carries EBV. Because latent viruses persist by signaling their host cells to divide continuously, they may cause cancer.

HIV has a unique variation of lysis (fig. 19.9). The HIV genome is incorporated into the host DNA as a part of its replication cycle. Shortly after infection, many HIV virions are produced and released. Eventually, a persistent infection is established where the infected cells produce small quantities of virus particles. Infected individuals have almost no symptoms, yet HIV is

FIGURE 19.9
Replication of HIV.
HIV infects a cell by integrating first into the host chromosome, then producing more virions.

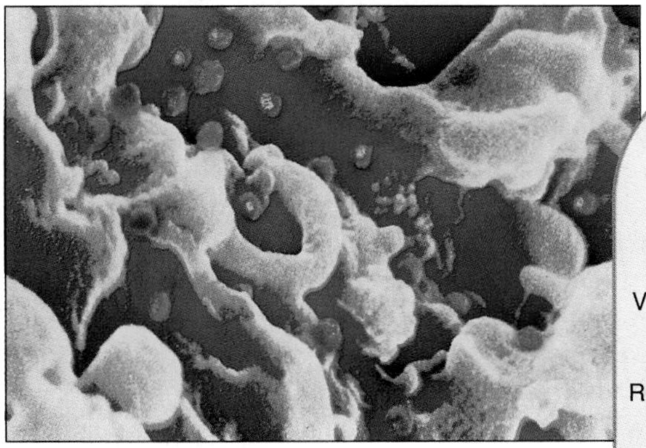

Scanning electron micrograph of HIV on a host cell.

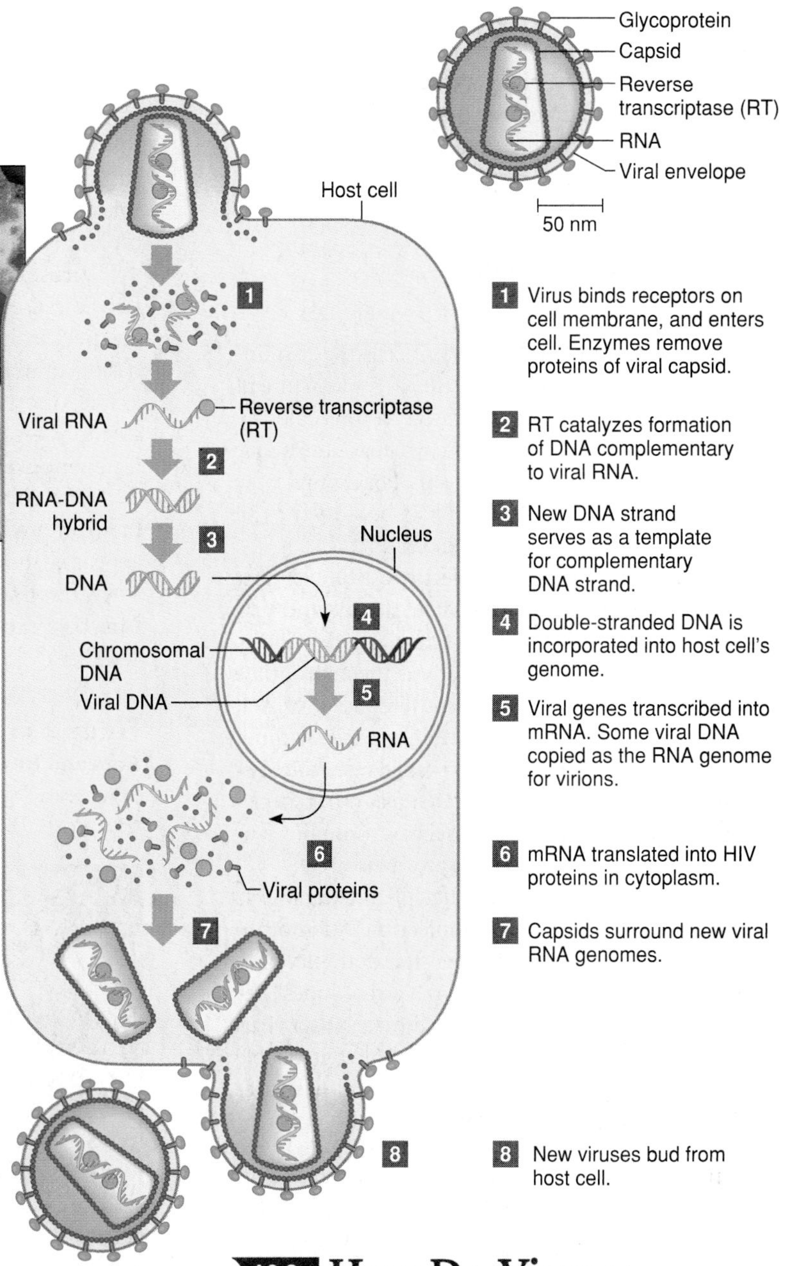

1. Virus binds receptors on cell membrane, and enters cell. Enzymes remove proteins of viral capsid.
2. RT catalyzes formation of DNA complementary to viral RNA.
3. New DNA strand serves as a template for complementary DNA strand.
4. Double-stranded DNA is incorporated into host cell's genome.
5. Viral genes transcribed into mRNA. Some viral DNA copied as the RNA genome for virions.
6. mRNA translated into HIV proteins in cytoplasm.
7. Capsids surround new viral RNA genomes.
8. New viruses bud from host cell.

present in their bloodstreams. This persistent phase can last for many years. Finally, in response to some as yet unidentified stimulus, the virus enters a productive stage once more. So many viruses are produced that nearly all of the target cells in the body are killed. Vital immune system cells are destroyed, leaving the body unable to defend itself from infections. AIDS is the result.

19.2 MASTERING CONCEPTS

1. What factors determine whether a virus infects a cell?
2. How do viruses infect bacterial, animal, and plant cells?
3. Distinguish among lysis, lysogeny, and latency.
4. What is unique about the infection of cells by HIV?

OLC

19.3 How Do Viruses Cause Disease?

The symptoms of a viral illness arise from direct effects of the virus on certain cells, and from the immune system's response. A virus can infect only specific cells of particular species. Some viruses "jump" from one species to another. Sometimes viruses persist, without causing harm, in reservoir species. Organisms have evolved defenses against viral infection.

The specific symptoms of a viral infection reflect the types of host cells that are destroyed, and the host's responses. For example, rhinoviruses grow most efficiently in the mucous-producing cells in a person's nose, throat, and lungs, causing the symptoms of the common cold. Papillomaviruses infecting cells lining the reproductive tract cause growths called genital warts. The virus that causes hoof-and-mouth disease infects cells in the limbs and mouths of cattle.

Symptoms of viral infections felt at the whole-body level are generally caused by molecules that the immune system releases. The immune response renders the body inhospitable to the infectious agent, raising body temperature, causing the aches and pains of inflammation, and increasing mucous secretions, among other actions. Unfortunately, these responses can also make the host rather miserable!

Viruses Infect Specific Cells of Certain Species

Viruses are very specific about the types of cells they can invade. They infect only cells that have a specific target attachment molecule, or receptor, on their surfaces. Some target molecules are on a very small subset of cells in an organism, whereas others may be found in an entire group of related organisms. For example, the rabies virus can infect humans, skunks, and bats because they all share some common target molecules on their cells.

The kinds of organisms a virus can infect constitute its **host range.** The types of cells a virus can infect contain the enzymes and other molecules that the virus requires to effectively reproduce. However, a virus will enter any cell that contains the appropriate attachment molecule, whether it can efficiently replicate or not. Sometimes a virus can infect and replicate within a species without causing symptoms. Such carriers of a virus are called **reservoirs** because they act as a source of that virus that can infect other species and produce disease in them. Often insects such as mosquitoes or leafhoppers are reservoirs (see fig. 19.6) and spread disease.

Identifying the reservoir for a virus can help epidemiologists develop strategies for preventing spread of infection. Attempts to limit outbreaks of Ebola hemorrhagic fever, for example, have been stymied because the reservoir—where the virus "hides" between surfacing in human settlements—is unknown. Ebola virus is passed in body fluids and causes a rapidly fatal uncontrollable bleeding in humans. In contrast, smallpox virus vanished after vaccines successfully kept it out of human bodies. With only one host—us—the virus had no other cells to infect. It exists today only in certain laboratories.

Another feature of viruses is their ability to change as they infect new species. This is rare, but can produce frightening results. For instance, HIV is believed to be the result of a similar chimpanzee virus becoming able to infect humans.

A virus can "jump" species, widening its host range to include another species, only if the cells of that species have receptors to which the virus can bind, and enter the cell. Then, the new host must have the enzymes necessary to replicate the viral genetic material. If a species' cells are not compatible with the virus in these ways, then the virus cannot infect it and replicate efficiently. For this reason, a pet cat cannot transmit the feline form of the AIDS virus to its owner, nor can a tomato plant pass bushy-stunt disease to a gardener. However, a person can get the flu from ducks, which serve as reservoirs. Influenza is an example of a **zoonosis,** which is an infection that jumps from another animal species to humans.

Influenza is perhaps the most familiar viral zoonosis, able to infect the cells that line the respiratory tracts of various species of birds and mammals. A well-documented infectious route is from human to pig to duck and back to humans. Along the way, because these viruses lack genetic repair systems, mutations accumulate, making the virus appear different to the human immune system from one year to the next. Epidemiologists must track the base sequences of influenza virus genetic material—which is single-stranded RNA—among people, pigs, and various fowl species the world over, continually, to predict the coming year's strains and develop vaccines. Health 19.1 takes a closer look at the role of birds in flu epidemics and in a more recent outbreak of encephalitis caused by the West Nile virus.

Viruses that acquire new types of hosts can cause new infectious diseases. Many zoonoses probably never affect more than a few people, because they often are transmitted only from originating animal to human, but not from human to human. For example, most outbreaks of monkeypox—a disease similar to smallpox—soon vanish because people must catch it from infected monkeys, not from other people.

Often birds or bats are the agents of a new zoonosis because they can travel long distances, spreading an illness that might disappear if left in just one host population. In Australia, for example, a new respiratory illness in humans came from horses, courtesy of fruit bats carrying what is now called Hendra virus (fig. 19.10). The disease killed several thoroughbred horses and their trainer in

FIGURE 19.10

Bats and Birds Can Transport Viruses to New Hosts.
In Australia, when people moved closer to bats, a new respiratory illness arose. The virus jumped from pregnant bats to horses to people.

From the Birds: Influenza and West Nile Virus

Influenza has caused much human suffering and death over the centuries. Flu also affects horses, ferrets, seals, and pigs, but the disease's origins trace to waterfowl, the natural viral reservoir.

Flu probably began about 4,500 years ago in China, where the virus moved from wild to domesticated ducks. In the seventeenth century, the Chinese brought domesticated ducks to live among rice paddies, where the birds ate insects and crabs, but not rice. In this way, flu-infested ducks came to live close to people, as well as near pigs and chickens. Occasionally pigs would snort about in infected duck droppings, triggering a curious and crucial molecular mixing. The cells that line a pig's throat bear receptors for both the avian (bird) and human versions of flu viruses. In pig throats long ago, epidemiologists hypothesize, an avian flu virus mutated in a way that enabled it to infect humans. Today we know that avian and human flu viruses commonly intermingle in pigs and exchange genetic segments, generating new viruses that can take our immune systems by surprise, causing epidemics. But genetic exchange isn't the only contributing factor to worldwide epidemics, known as pandemics. Consider the special conditions of wartime, when the sickest people, taken to hospitals, spread the illness to many others.

The flu pandemic of 1918 killed more people in the United States than both world wars, the Korean war, and the Vietnam war combined. Unlike flu outbreaks in more recent times that are deadly mostly to only the very young and very old, the 1918 flu killed mostly people ages 20 to 40 (fig. 19.A). The first reported case came from Camp Funston, Kansas, on March 4, 1918. It spread to several cities, and then U.S. troops brought it overseas. Over the summer, the virus became more virulent, and starting in August, it hammered Europe, where the death rate peaked in the fall. The 1918 flu spread and killed fast. A young man might awaken in the morning with a stuffed nose and die by midnight, his lungs filled with fluid. Nurses would estimate how close a man was to death by the color of his feet, which would turn black as ravaged lungs failed to deliver oxygen. Researchers today are trying to determine why the 1918 flu was so deadly by studying viruses in lung tissue from several preserved victims.

Flu pandemics also occurred in 1957, 1968, and 1977. Two decades later, epidemiologists feared a possible repeat of pandemics past when a new flu variant arose with a frightening twist—it appeared to jump from birds directly to humans.

On May 10, 1997, a three-year-old boy in Hong Kong died of a fierce flu caused by a virus never before seen in humans, but known in birds. In 1983, the killer strain had passed from wild ducks to chickens in warehouses in Pennsylvania, where it caused mild digestive symptoms in the birds. Then, a single mutation enabled the virus to kill the chickens. When two other Hong Kong children fell ill with the chicken virus, apparently infected by chickens kept as pets in their preschools, panic set in. Not only did Hong Kong have a huge chicken population, but it seemed possible that the virus could also pass from person to person. Plus, the usual route to vaccine development, which uses chicken eggs, was obviously not possible. To avert an epidemic, the government killed every chicken in Hong Kong. Fortunately and inexplicably, the outbreak never went beyond 18 people, with six deaths.

Influenza isn't the only virus to stop over in birds. When 67 people suddenly developed encephalitis in the late summer of 1999 in New York City, seven of them dying, researchers at first suspected the mosquito-borne St. Louis encephalitis virus. But when local crows and then a bald eagle at the Bronx Zoo died, an astute veterinarian sent samples to be tested for viruses and the true culprit was identified—West Nile virus. This virus, discovered in 1937 in Uganda and then traveling in birds to cause outbreaks in Romania, Russia, and the Middle East, arrived for the first time in North America with an Israeli goose. The virus replicates in many common bird species and lives in a mosquito reservoir.

West Nile virus survived its first winter in North America, and birds have since carried it to several other states. A few more people have become ill, mostly older individuals or those with suppressed immune systems. Others may develop only short-lived flulike symptoms and not realize that they have been infected. Prevention methods include avoiding standing water, where mosquitoes breed, and using insecticides. Despite such precautions, epidemiologists expect the disease to spread throughout the United States.

FIGURE 19.A
Guarding Against the Flu.
The 1918 flu epidemic spread so quickly that cities took drastic steps. Policemen in London wore surgical masks to avoid infection.

a suburb of Melbourne in the fall of 1994. After similar cases were reported, epidemiologists identified what looked like a large version of a known virus, equine morbillivirus. Because the cases were in different geographic regions, virus-trackers suspected a bird or bat source. They focused on bats for two reasons—they are more closely related to horses and humans than are birds, and the expanding suburbs of Australian cities have been encroaching on bat habitats.

Bats became implicated from people who cared for injured bats that they found in their backyards. Blood from the bat-rescuers contained antibodies to the newly described virus, indicating that they had been infected, but had recovered. Then, a woman took a bat that had become caught in a barbed wire fence to a public health department, where the virus was found in the animal's uterine fluid. A picture of disease transmission began to emerge. Bats roost in barns, and when they give birth, uterine fluids, rich in the virus, drop onto horses, or near them. Some of the horses become infected by being dripped on or by eating bat placentas. When the horses fall ill and stop eating, their trainers force-feed them, becoming scratched—and infected—in the process.

A close relative of Hendra virus, called Nipah virus, also transmits a disease—encephalitis (brain inflammation)—to humans via fruit bats. The unaffected but infected bats pass the virus to pigs, which develop a violent cough, by which they infect people who slaughter the animals for food. An initial outbreak in Malaysia in 1998 and 1999 infected 269 people, killing 108. Then, as Malaysian pigs were brought southward into Singapore, so was Nipah virus encephalitis. More than a million pigs were destroyed in an attempt to contain the outbreak.

Some viruses with broad host ranges can cause severe disease in one species, while having little or no effect on another, related species. Consider the Asian elephant and African elephant. For years, Asian elephants were dying in zoos from an unknown, fast-acting illness. Recently researchers performed an autopsy on Kumari, a baby Asian elephant that died in the National Zoo in Washington, D.C., following sudden, severe internal bleeding. The linings of her blood vessels contained dark round particles, evidence of a herpesvirus infection. Researchers used the polymerase chain reaction (PCR) to identify telltale herpes genes, and discovered a previously unknown type of herpesvirus. Kumari had contracted it from an African elephant kept in the compound. The same virus that kills an Asian elephant causes only mild skin or genital warts in African elephants. A different herpesvirus has precisely the opposite effect, killing African elephants but causing only mild illness in their Asian cousins. Microbiologists do not understand how a virus's host range develops, but clearly zookeepers need to separate their elephants!

Defenses Against Viral Infections

Halting a viral infection presents quite a challenge to an organism, because viruses use the host's enzymes to replicate. Therefore, interfering with the virus would also affect the host cells. In bacteria, enzymes called restriction endonucleases (or restriction enzymes) degrade viral DNA by binding to and cutting very specific base sequences. The cell's own DNA is protected from its restriction endonucleases with methyl (CH_3) groups.

Animal cells do not make restriction endonucleases, but have immune systems that attack viruses in several ways. Antibodies are proteins that can coat viral particles, hindering their attachment to target cells. Infected cells release chemicals that activate neighboring cells to attack viruses when they enter. Virally infected cells may burst before progeny viruses can be released, thanks to the immune system. We also use antiviral drugs that interfere with enzymes or other proteins that are unique to viruses and vital to their replication. The anti-AIDS drugs AZT and ddC, for example, block critical viral enzymes used in their replication, and protease inhibitors are drugs that block the ability of the virus to complete processing its proteins. But antiviral drugs can do little more than slow the replication of the virus. Ultimately, our own immune systems must rid us of the virus. Chapter 39 discusses the immune system in greater detail.

19.3 MASTERING CONCEPTS

1. How do viruses cause symptoms, directly and indirectly?
2. What is a virus's host range?
3. What is a viral reservoir?
4. What is required for a virus to "jump" to infecting another species?
5. How do organisms evade viral infection?

OLC

19.4 Other Infectious Agents

Viroids are infectious RNA molecules that harm plants. Prion diseases result when abnormal prion proteins bend normal prion proteins into an alternate shape.

The idea of something as comparatively simple as a piece of DNA or RNA wrapped in protein—a virus—causing devastating illness may seem unlikely. But some infectious agents are even simpler than viruses.

A Viroid Is an Infectious RNA Molecule

A **viroid** is a highly wound circle of RNA that causes a certain disease in a particular plant species. It differs from a virus in that it lacks a protein coat—a viroid is, essentially, naked RNA. The RNA coils so tightly, bonding with itself to form double-stranded RNA, that host cell enzymes that would normally degrade RNA cannot do so.

Viroids are passed from plant to plant in a mechanical fashion, as parts of one plant touch another, or as insects feed from plant to plant. Farm machinery that harvests plants is a major source of viroid spread. Alternatively, viroids can be transmitted from generation to generation. The diseases that they cause have vividly descriptive names, such as "cucumber pale fruit,"

FIGURE 19.11
Plants Can Fall Victim to an Infectious Particle Called a Viroid.
Shown here are a tomato with "tomato bunchy top," caused by a viroid (*left*), and a healthy tomato plant (*right*).

"grapevine yellow speckle," "coconut cadang-cadang," and "dapple peach." Figure 19.11 shows another viroid disease, "tomato bunchy top."

Viroid infections can ruin crops. Avocado sunblotch, for example, spreads readily and causes yellow blotches to form on the fruits, which cannot be sold. Chrysanthemum stunt disease nearly decimated the market for that flower in the United States in the early 1950s. Today, researchers use biotechnology to control viroid diseases. Seed banks use a version of the polymerase chain reaction that detects RNA to identify viroids. Researchers in India created transgenic potato plants that carry a gene from yeast that enables the plants to selectively destroy double-stranded RNA. The genetically engineered potatoes resist infection by the potato spindle tuber viroid, which in the past has destroyed crops by severely shriveling the plants.

A Prion Is One Protein That Takes Variant Forms

Another type of infectious agent simpler than a cell is a **prion,** which stands for "proteinaceous infectious particle." A prion is a normal cellular protein whose function is not understood. In a certain conformation, however, prions can cause disease. Prions are known in more than 80 types of mammals, but are well studied in only a few. A prion disease called scrapie has been familiar to shepherds for centuries. It causes such intense itching that afflicted animals rub themselves raw—hence the name.

Prion protein—PrP for short—can cause disease because it can exist in more than one conformation (fig. 19.12). (Recall that a protein's conformation is its three-dimensional shape.) Some shapes are "normal," yet others, collectively called the "scrapie" forms, cause conditions called **transmissible spongiform encephalopathies.** These diseases are highly species-specific. They riddle the brain with holes and may also cause a buildup of dense collections, called fibrils, of the abnormal PrP protein. A prion disease results from a chain reaction in which a scrapie PrP triggers normal PrP molecules to change (fig. 19.13).

FIGURE 19.12
Prions Change Shape.
A prion disease may begin when a single scrapie PrP contacts a normal PrP and causes it to switch into the abnormal conformation. As the change spreads, the disease occurs in susceptible species.

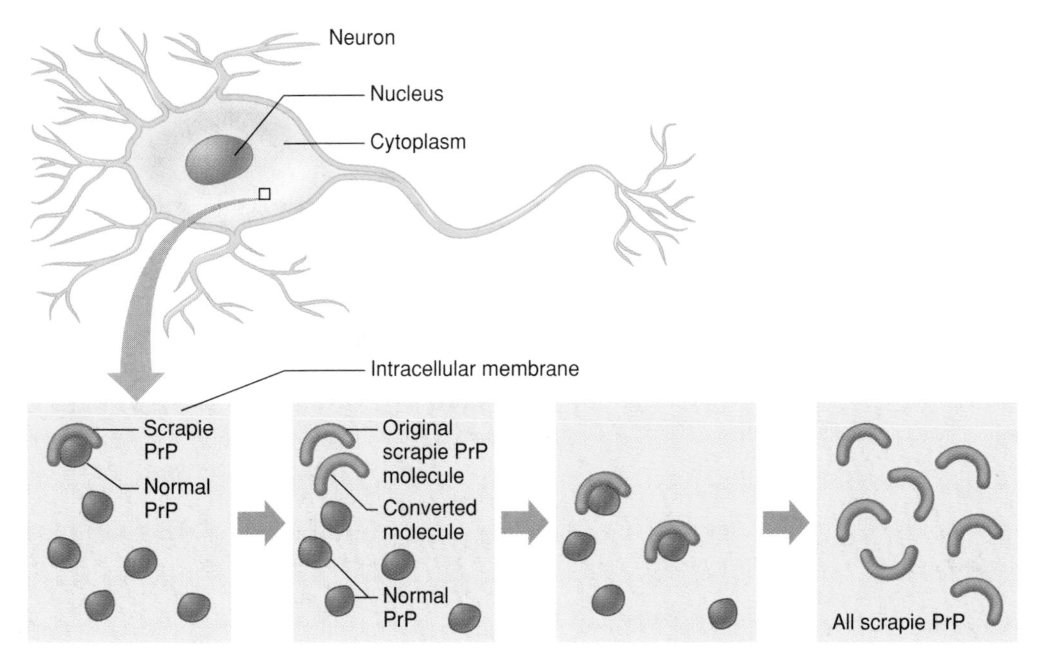

FIGURE 19.13

The Effects of BSE, a Prion Disease.

In this image, the holes and clumps of protein fibrils are evident in a brain from a cow that died of bovine spongiform encephalopathy (BSE), also known as "mad cow disease."

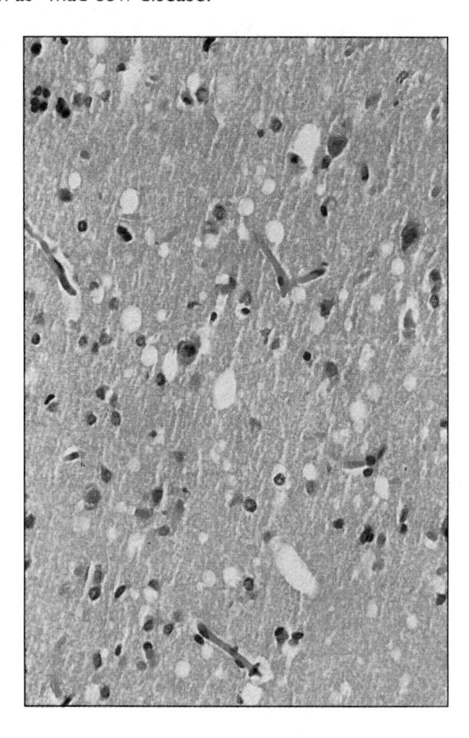

Like a virus, a prion is tiny, has a specific host range, and can be carried in reservoir species that do not become ill. But other evidence indicates that a prion is most definitely not a virus. Heat, radiation, and chemicals that would degrade a virus do not harm a prion. More telling is that the host's immune system does not attack a prion—inflammation does not occur nor are antibodies produced. Whatever a prion is, the host's body does not recognize it as "foreign," as it would a flu virus. This suggests that a prion is somehow familiar to the body.

The transmissible spongiform encephalopathies cause loss of coordination, wasting, and holes in brain tissue. These disorders result from ingesting infected tissue, which can occur in a variety of ways:

- animals eating tissue from affected animals;
- transplants or hormones from infected tissue;
- cannibalism;
- humans eating other animals that have a prion disease.

The first transmissible spongiform encephalopathy recognized was scrapie. Prions entered the realm of human health care with a mysterious illness called kuru that affected the Foré people of New Guinea, until they abandoned a centuries-old cannibalism ritual. A young U.S. government physician, D. Carleton Gajdusek, began unraveling the story of kuru in 1957. He connected scrapie, kuru, and the very rare but similar Creutzfeldt-Jakob disease (CJD) in humans, which in 1972 caught the attention of a young physician, Stanley Prusiner. Prusiner has been investigating prions ever since, and proposed the explanation for their mechanism of action shown in figure 19.12. Both men won the Nobel prize for their work on prion diseases, which are still not completely understood.

CJD was known to affect one in a million people, usually over age 60, either sporadically (no known cause) or from contamination by surgical instruments or transplanted tissue. In 1996, evidence began to mount that a "new variant" of CJD could be transmitted to humans who ate nerve tissue from cows suffering from the prion disease bovine spongiform encephalopathy (BSE), which quickly became known as "mad cow disease." New-variant CJD, which became known as variant CJD (vCJD) as its newness faded, seems to have been limited to a few dozen cases in the United Kingdom, as Health 19.2 recounts.

Prions may be more common than has been realized. Similar infectious proteins are known in yeast, where they have metabolic functions different from those associated with the prions of mammals. Prions may also illustrate a general disease mechanism of a cascade of proteins being changed into abnormal forms that cause them to clump.

19.4 MASTERING CONCEPTS

1. What is a viroid?
2. What is a prion?
3. How do scientists think prions cause disease?

OLC

Creutzfeldt-Jakob Disease (CJD)—The Old and the New

CJD has always been an extremely rare disease that was either inherited, or acquired from contaminated transplants, growth hormone, or medical devices that touched the brain. In the 1970s, the seeds were sown for another route of transmission of this prion disease. In the United Kingdom, food processors changed the way that animal tissue was converted into animal feed, heating the material differently and adding different types of animal fats. Between 1981 and 1982, calves ate this food, and became infected with what would become known as bovine spongiform encephalopathy, or BSE. Farmers noted the first "mad cows" in 1985—animals that grew fearful, then aggressive, then had difficulty standing and collapsed. The animals rapidly lost weight and died. The disease obviously took several years to develop and produce symptoms. It was first noted in dairy cattle, because beef cattle were slaughtered before they lived long enough to become ill.

Between 1994 and 1996, a few young people in the United Kingdom and one in France became suddenly ill with frightening symptoms. One woman noticed that her son's loss of coordination was eerily like that of the "mad cows" that she saw staggering in news reports. Two young women who worked in a meat pie factory and a butcher shop developed the illness too. Altogether, 34 people became ill with what looked like CJD. But on examination of their brain tissue, the pattern of holes was much more reminiscent of kuru and BSE. The British government, while denying for a time that there could be a link between BSE and this variant CJD, nevertheless was concerned enough to kill 4 million cattle. Figure 19.B chronicles the BSE epidemic.

In retrospect, epidemiologists hypothesize that the people who developed variant CJD contracted the prion infection from eating brain, spinal cord, and digestive tracts of cows, which before an explicit ban in 1989 had been added to sausage, hamburger, and processed meat. Studies in mice since then have established a fairly clear link between BSE and variant CJD:

- The PrP proteins that cause both illnesses have the same types and patterns of sugars attached to them.
- The patterns of holes in affected brains from people with variant CJD are most like those in cows affected with BSE, and not like brains from people who died of CJD traced to the older sources of contamination.
- The abnormal prions from cows with BSE, from people with variant CJD, and from cats and zoo animals that have similar disorders, cause the same illness in mice, with long incubation periods and similar distributions of brain holes.

Although new cases are still being reported, the outbreak of variant CJD in the United Kingdom subsided because of bans begun in the late 1980s on using animal parts that could carry abnormal prions in animal feed and in human food. Research also indicates that prion diseases are not spread in blood or blood products. Still, because BSE or scrapie, or an as yet unknown type of abnormal prion disease, can arise spontaneously, some people remain wary of eating beef.

continued

(concluded)

FIGURE 19.B

Chronology of an Epidemic Zoonosis.

Cases of humans infected with variant Creutzfeldt-Jakob disease (vCJD) began to appear as the bovine spongiform encephalopathy (BSE) epidemic was winding down. Several years had elapsed between the removal of infected cattle and the first cases in humans, illustrating the slow onset of this disease. It is now clear that BSE was the source of the new form of CJD. Epidemiologists are not sure what the future holds for the incidence of vCJD.

From Richard Johnson and Clarence Gibbs, Jr., *New England Journal of Medicine,* vol. 339, no. 27, February 2000. Copyright © 2000 Massachusetts Medical Society. Reprinted by permission.

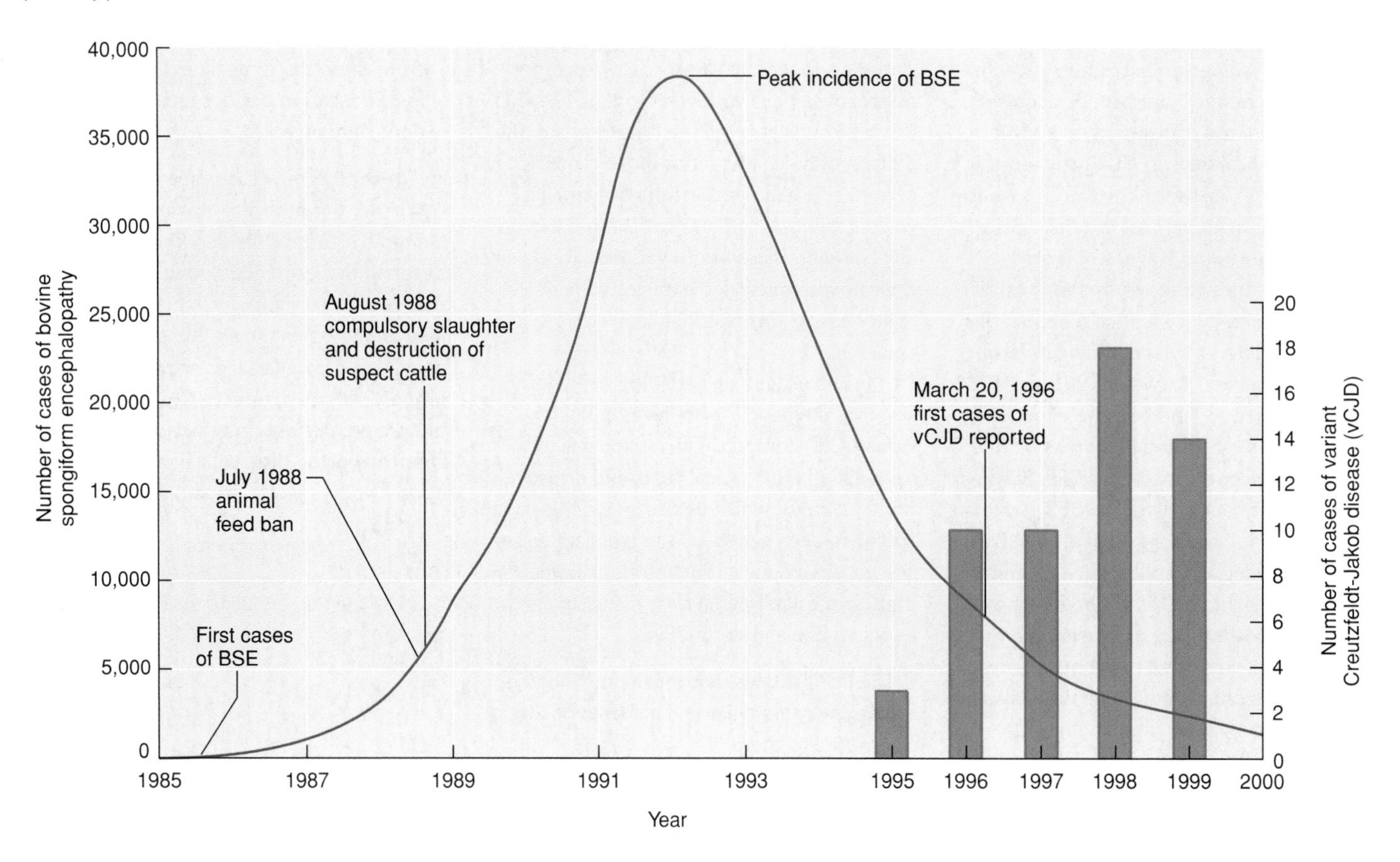

Chapter Summary

19.1 What Is a Virus?

1. A **virus** is a nucleic acid in a protein coat and perhaps an **envelope.** It must be in a cell to reproduce.
2. Viruses were discovered when scientists passed fluid from infected organisms through filters that held back bacteria, and then demonstrated that the strained material was infectious.
3. A virus's protein coat is a **capsid;** the viral particle outside a cell is a **virion.** Capsids have a variety of shapes.
4. A viral **envelope** includes virally encoded proteins and lipids, some of which may have come from host cell membranes.

19.2 How Do Viruses Infect Cells and Reproduce?

5. A virus binds to a surface molecule on a host cell, then injects its nucleic acid or is engulfed. Insects assist viruses in infecting plant cells. The viruses then spread via plasmodesmata that link plant cells.
6. In a **lytic infection, progeny virions** are manufactured, assembled, and released. In a **lysogenic infection,** the bacteriophage nucleic acid integrates into the host DNA and remains hidden. Some animal viruses can exist hidden within a host cell in a **latent** infection.
7. The steps of viral replication are determined by whether the genetic material is RNA or DNA, and whether it is single-stranded or double-stranded.

19.3 How Do Viruses Cause Disease?

8. Viruses cause disease by harming certain cell types, and indirectly by stimulating immune system responses.
9. The types of species that a virus infects are its **host range.** A virus may be carried in a **reservoir** species, in which it doesn't cause symptoms.
10. A **zoonosis** is an infection that humans acquire from another type of animal. Birds and bats spread some zoonoses.
11. Protection against viral infection includes restriction endonucleases in bacteria and immune systems in animals.

19.4 Other Infectious Agents

12. **Viroids** are infectious RNAs that affect plants.
13. A **prion** is a protein that can assume different conformations, some of which can convert other prions to the scrapie form and thereby cause **transmissible spongiform encephalopathies.**

Testing Your Knowledge

1. How does a virus differ from:
   - **a.** a cell?
   - **b.** a toxin?
   - **c.** a prion?
   - **d.** a viroid?
2. What are the four types and configurations of nucleic acids in viruses?
3. Describe how a virus's envelope can include lipid molecules that are also part of host cell membranes.
4. Distinguish among lysis, latency, and lysogeny.
5. Explain how different viruses can affect different parts of a multicellular organism's body.
6. Why is a virus that causes an illness in a pet turtle unlikely to cause illness in the turtle's owner?
7. What event would make a zoonosis extremely dangerous?
8. What are the causes of
   - **a.** kuru
   - **b.** BSE
   - **c.** monkeypox
   - **d.** AIDS
9. What are two ways in which prions are similar to viruses?

Thinking Scientifically

1. How can a virus have an enormous capsid, yet very few genes?
2. Suggest two ways that a virus can evade the host's immune system.
3. Hantavirus pulmonary syndrome (HPS) is a flulike illness that can kill a person within days. In 1993, a small outbreak occurred in the southwestern United States. By studying the spread of other diseases, epidemiologists concluded that HPS was spread in rat excrement, and not from person to person. In larger outbreaks of hantavirus infection that occurred in Korea among U.S. military personnel during the war there in the 1950s, and again among soldiers in that nation in 1988, the condition was called Korean hemorrhagic fever. It had different symptoms—bleeding and kidney failure. Explain how the same virus at different times can spread in different ways and cause different symptoms.
4. Smallpox once killed millions of people, but thanks to widespread vaccination, it vanished by 1977. The virus, called variola, is kept as deep-frozen samples in two laboratories, although it is possible—some say likely—that many nations have kept samples for use as potential bioweapons. By 1993, researchers knew the genome sequence of variola, yet ironically, most doctors have never seen a case of smallpox. Using this information, provide arguments for and against destroying all remaining samples of this virus.
5. Since the 1970s, people have no longer been vaccinated against smallpox, because the disease has vanished. The vaccine also protected against a related viral illness, monkeypox. In the Democratic Republic of Congo, epidemiologists recorded 37 monkeypox cases between 1981 and 1986. But in 1996 and 1997, more than 92 cases were recorded. Explain why monkeypox might be on the rise.

References and Resources

Hill, Andrew F., et al. August 29, 2000. Species-barrier independent prion replication in apparently resistant species. *Proceedings of the National Academy of Sciences,* vol. 97, p. 10248. Prions "hide" in reservoir species.

Lewis, Ricki. August 21, 2000. An eclectic look at infectious diseases. *The Scientist,* vol. 14. p. 1 A short review of Nipah and Hendra virus outbreaks.

Lewis, Ricki. April 17, 2000. West Nile Virus—Part 2? *The Scientist,* vol. 14, p. 1. Mosquitoes carrying this virus that causes encephalitis survived the New York City winter, and the disease has since spread.

Lewis, Ricki. July 6, 1998. Classic technique reveals HIV in action. *The Scientist,* vol. 12. p. 1. As HIV approaches a T cell, its docking site is revealed.

Rhodes, Richard. 1997. *Deadly feasts.* New York: Simon & Schuster. A fast-paced book that recounts the connected tales of kuru, BSE, and new-variant CJD.

True, Heather L. and Susan L. Lindquiest. September 28, 2000. A yeast prion provides a mechanism for genetic variation and phenotypic diversity. *Nature,* vol. 407, p. 477. Yeast prions have different functions than mammalian prions—they control gene expression.

The *LIFE* Online Learning Center provides additional resources and tools for studying this chapter.
www.mhhe.com/life

CHAPTER 20

Bacteria and Archaea

Behaving Bacteria

Like familiar animals and plants, microorganisms encounter challenges that they must overcome if they are to survive. A soil bacterium, for example, will likely die if it cannot quickly adjust to the sudden flood of a heavy rain following a long dry spell. Microorganisms must also be able to respond to fluctuations in supplies of oxygen or nutrients, by either moving on or forming tough, resistant survival structures. Although these responses are not as easy to observe as the behaviors that many animals display, they serve many of the same functions. Here are some ways that bacteria "behave."

When a particular nutrient is available everywhere, a bacterium locomotes in a "tumbling" movement that results from clockwise rotation of its flagellum.

Chemotaxis Bacteria can detect and approach beneficial chemicals, or avoid harmful ones. Such directed movement toward or away from a chemical, called chemotaxis, is based on signal transduction (see fig. 4.22). The signal is a change in concentration of the chemical. When a particular nutrient is available everywhere, a bacterium locomotes in a "tumbling" movement that results from clockwise rotation of its flagellum. The tumbling continually changes the direction of movement, keeping it random. Alternatively, a bacterium can "run," moving in one direction for a short time. Having more runs than tumbles sends a bacterium in a specific direction. Nutrient molecules binding to receptors on the cell surface can activate a signaling pathway, and the organism runs more than it tumbles—taking a jagged path to the meal.

Bioluminescence *Vibrio fischeri* bacteria live in the open ocean, and also in a light-emitting organ in the squid

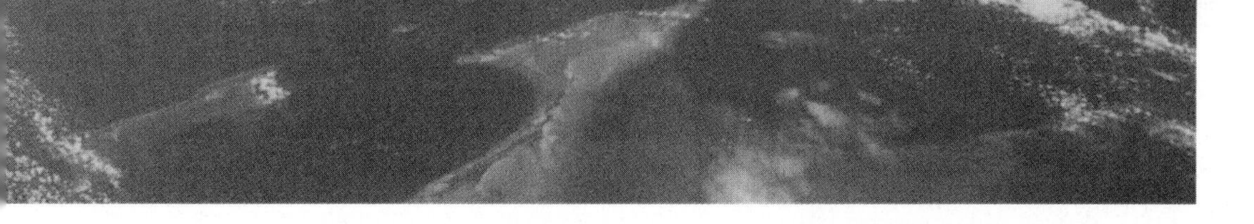

Chemotaxis.
When nutrients are absent or evenly distributed, bacteria alternate "tumbles" and "runs," moving about randomly. When nutrients are present in a gradient, bacteria replace tumbles with longer runs that bring them closer to the food source.

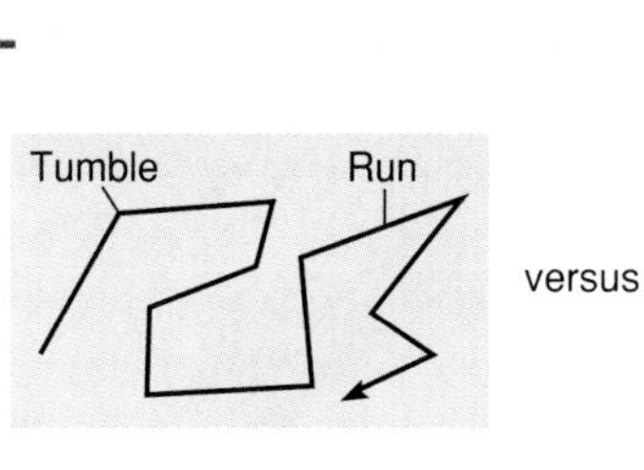

versus

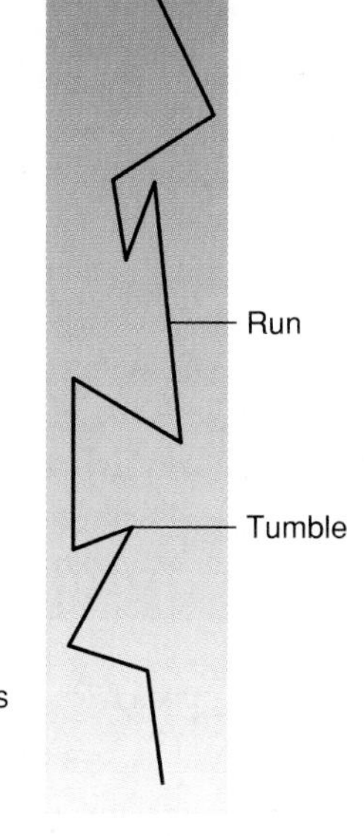

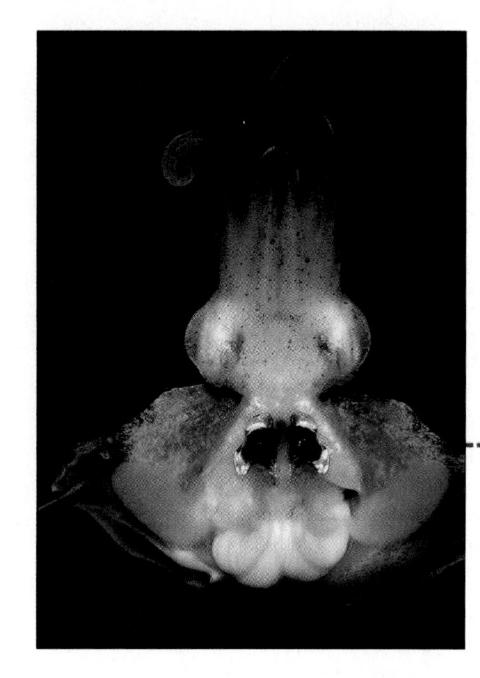

Bioluminescence.
The squid *Euprymna scolopes* escapes predation thanks to glowing bacteria (*Vibrio fischeri*) in its light-emitting organ.

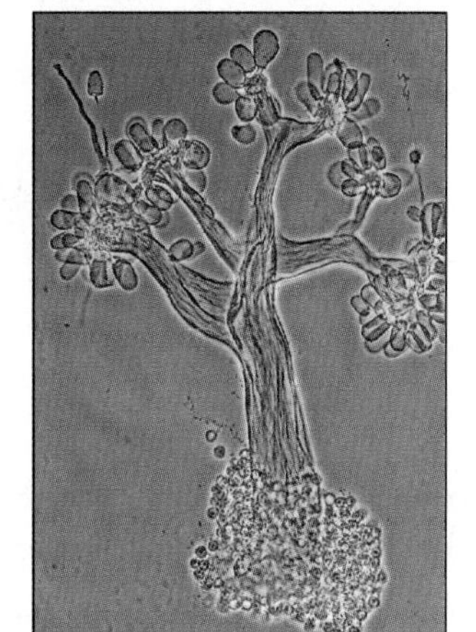

Spore Formation.
When nutrients and water are scarce, cells of myxobacteria aggregate and specialize to produce a fruiting body, such as this stalk of *Chondromyces crocatus*. The structures clustered at the branch tips each contain hundreds or thousands of desiccation-resistant spores that might be transported to a more hospitable environment.

Euprymna scolopes. Inside the squid, the bacteria produce light, either in response to a biochemical signal from the squid, or because the bacteria are packed tightly together. In the open ocean, where the bacteria are spread out, they do not glow. The bioluminescence is adaptive to both bacteria and squid—an example of symbiosis. The bacteria are sheltered and fed. The nocturnal squid evades predators or hides from prey as the glow from its bacterial passengers mimics moonlight coming from above.

Spore Formation Soil-dwelling myxobacteria glide along decaying organic matter or dung, absorbing nutrients and secreting a slime trail that other myxobacteria follow. When they run out of water and nutrients, they emit signals that draw thousands of them together, forming a structure called a fruiting body. Because of the fruiting body's size, the human eye can see it as a colored speck on a leaf or bark. The fruiting body contains spores, which are thick-walled structures adapted for dispersal or protection from a harsh environment. The aggregated spores can travel on wind or in water, or stick to the coat of a passing mammal, finding a new location where food and water might be available.

20.1 Prokaryotes Are a Biological Success Story

Prokaryotes (bacteria and archaea) are abundant, diverse, and everywhere—they number about 5×10^{30}. We actually know little about these organisms that are essential to life on Earth.

If all of the animals and plants on Earth were to become extinct, microbial life would nevertheless continue. But if all the prokaryotes on the planet were to vanish, all the rest of life would cease too. The prokaryotes—single-celled organisms lacking nuclei and membrane-bounded organelles—were the first life-forms and remain essential to life in many ways.

Prokaryotes in a Global Context

Prokaryotes have had a tremendous impact on Earth's natural history. The first cells were probably more like existing prokaryotes than any other type of organism. Along the road of evolution, prokaryotes were probably the precursors of the chloroplasts and mitochondria of eukaryotic cells (see fig. 18.10). Ancient photosynthetic prokaryotes also contributed oxygen gas (O_2) to Earth's atmosphere, creating a protective ozone layer and paving the way for aerobic respiration.

Even now, the seemingly simple prokaryotes lie at the crux of the continual cycling of chemical elements between organisms and the nonliving environment (see chapter 43). Besides their role in decomposing dead organic matter, only bacteria can regularly convert atmospheric nitrogen gas (N_2) into biologically useful forms. **Nitrogen fixation** is a process in which bacteria reduce N_2 to ammonia (NH_3), which other organisms can incorporate into organic molecules. Some of these bacteria, such as those in the genus *Rhizobium,* induce the formation of nodules in the roots of certain host plants (fig. 20.1). Inside the nodules, *Rhizobium* cells share the nitrogen that they fix with their hosts, offering the host plant a growth advantage in nutrient-poor soils. In exchange, the bacteria receive nutrients and protection from their hosts. Other nitrogen-fixing bacteria live independent of plant hosts.

Biologist Mark Wheelis sums up the importance of the **Bacteria** and **Archaea,** the two prokaryotic domains of life: "The earth is a microbial planet, on which macroorganisms are recent additions—highly interesting and extremely complex in ways that most microbes aren't, but in the final analysis relatively unimportant in a global context." Bacteria and archaea were once thought to comprise half of all living matter. The percentage is probably much higher. We have likely underestimated the *diversity* of the prokaryotic domains because these organisms are not easy to observe. But when we try to survey the *abundance* of prokaryotes, their numbers are astounding. Three researchers from the University of Georgia—William Whitman, David Coleman, and William Wiebe—did just that.

Where Do Prokaryotes Live? Nearly Everywhere!

By dividing Earth into types of habitats and consulting many studies that evaluated the number of prokaryotic residents in each place, the researchers estimate that more than 5 million trillion trillion prokaryotes live on or in the planet—that's

5,000,000,000,000,000,000,000,000,000,000

prokaryotes. Although many people tend to think of prokaryotes in terms of those that cause disease in humans, the number of prokaryotes in all animals combined probably accounts for less than 1% of the total prokaryotes on Earth. The proportion of all prokaryotes that inhabit plants is also low.

Prokaryotes live in an incredible diversity of places—within rocks and ice, in thermal vents (see fig. 5.3), nuclear reactors (see the opening essay to chapter 18) and hot springs (see the opening essay to chapter 1), and in extremely low pH conditions (see chapter 2). They live as far up as 40 miles (69.4 kilometers) into the atmosphere, and as far down as 7 miles (11.3 kilometers) beneath the ocean's surface. Prokaryotes live in fish guts, polar bear hairs, and plant roots. Nearly 94% of all prokaryotes live in the terrestrial and marine subsurface, which extends to 26.2 feet (8 meters) beneath land and

FIGURE 20.1
Nitrogen-Fixing Bacteria.
(**A**) *Rhizobium* infects these sweet clover roots, producing root nodules where nitrogen fixation occurs. (**B**) Cross section of a root nodule, showing bacteria inside the plant's cells.

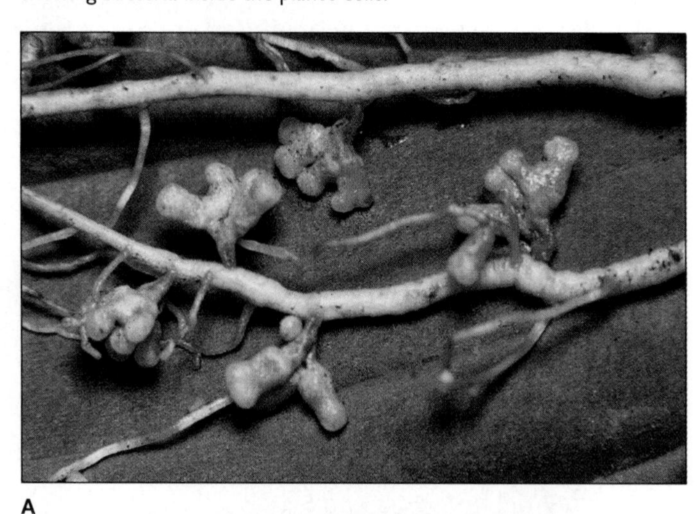
A

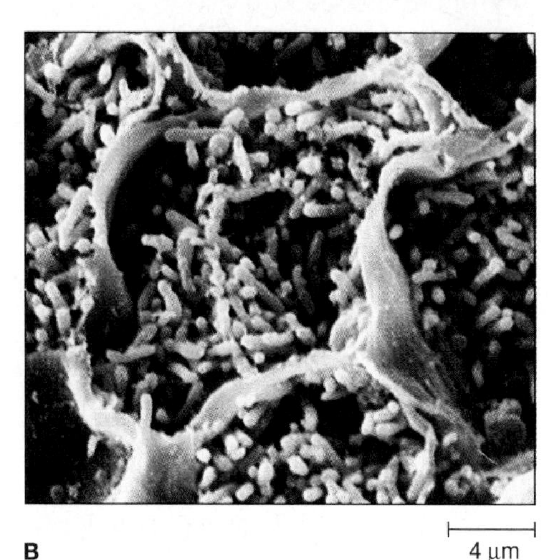

B

3.9 inches (10 centimeters) beneath marine sediments. The next most populated habitats are the soil and the open ocean. Table 20.1 lists various prokaryote homes.

Most of what we know about prokaryotes comes from a few species that microbiologists are able to culture in the laboratory. Prokaryotes are grown in petri dishes on a layer of agar (a gelatin-like seaweed extract) that includes nutrient combinations that support particular types of microorganisms. However, culturable microorganisms hardly represent the natural diversity of the prokaryotes. Carl Woese, who named and defined the domain Archaea, describes such lab-variety microorganisms as "a zoo of laboratory freaks that perform the physiological feats required of them." Laboratory culture may not accurately mimic the natural nutrient supply, and does not replicate the diversity of microbial species in a natural habitat.

20.1 MASTERING CONCEPTS

1. List several ways that prokaryotes are vital to life on Earth.
2. What are some of the places where prokaryotes live?
3. Why have microbiologists underestimated the diversity of prokaryotic life?

OLC

20.2 What Are the Parts of a Prokaryotic Cell?

Basic structural features of prokaryotes include genetic material, extensions that help them adhere to surfaces and move, and protective layers.

Like the cells of other organisms, all prokaryotic cells are bounded by a cell membrane and contain cytoplasm, DNA, and ribosomes. At about 1 to 10 micrometers long, however, a typical prokaryotic cell is 10 to 100 times smaller than most eukaryotic cells. Prokaryotes also lack the membrane-bounded organelles found in eukaryotic cells. Figures 3.6 and 20.2 depict bacterial cell structures; a given cell may have some or all of the structures pictured.

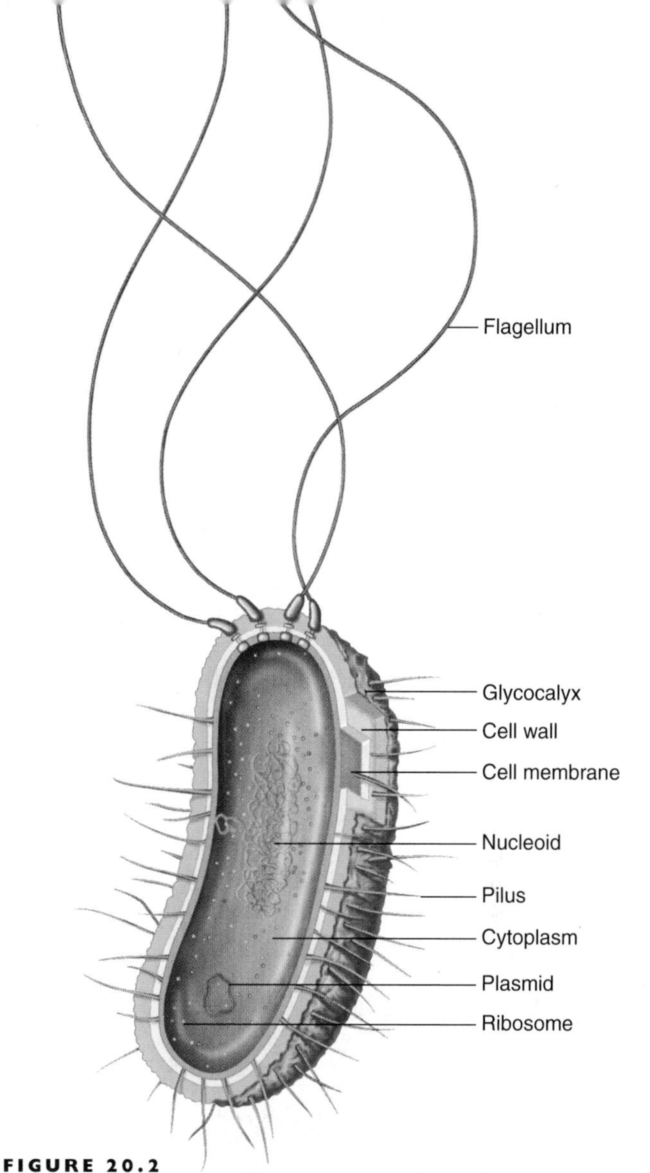

FIGURE 20.2
Bacterial Cell Structure.
A bacterial cell has DNA associated with some proteins and RNA molecules in an area called the nucleoid, but it lacks a true membrane-bounded nucleus. Ribosomes are free in the cytoplasm. Some cells have flagella (which provide movement), pili (which serve as attachment points to surfaces or other cells), and a slimy glycocalyx around the cell wall. A plasmid is a circle of DNA apart from the cell's chromosome.

TABLE 20.1 A Selection of Prokaryote Habitats

HABITAT	ESTIMATED DENSITY OF PROKARYOTIC CELLS	ESTIMATED NUMBER OF PROKARYOTES WORLDWIDE
Ocean and soil subsurface	$0.34 - 220 \times 10^6$ cells/cm^3	4.9×10^{30}
Soil (all ecosystems combined)	$1.0 \times 10^6 - 1.0 \times 10^9$ cells/gram	2.6×10^{29}
Open ocean (water)	$0.5 - 5 \times 10^5$ cells/ml	1.0×10^{29}
Lakes, fresh and saline	1.0×10^6 cells/ml	2.3×10^{26}
Cattle rumen	2.1×10^{10} cells/ml	29.0×10^{23}
Termite hindgut	2.7×10^6 cells/termite	6.5×10^{23}
Human colon	3.2×10^{11} cells/gram	3.9×10^{23}
Bird cecum (chicken, duck, turkey)	9.5×10^{10} cells/gram	2.4×10^{21}
Human skin	3.0×10^8 cells/adult	1.8×10^{18}

Internal Structures

A prokaryotic cell's DNA is typically a single large, circular chromosome that, along with some RNA and a few proteins, is located in a region of the cell called the **nucleoid.** Unlike the nucleus of a eukaryotic cell, the nucleoid is not surrounded by a membranous envelope.

In addition to chromosomal DNA, many prokaryotic cells contain circles of double-stranded DNA called **plasmids** (fig. 20.3A). Plasmids may include genes that make possible their transfer and replication, and may also include genes that provide a survival advantage in a particular environment, such as the ability to resist a drug or toxin. Other plasmid genes may enable a cell to cause disease, or alter its metabolism. Plasmids may integrate into chromosomal DNA. Free or integrated, plasmid DNA is replicated when the cell divides. Plasmids remain in the host cell as long as they confer an advantage. Sometimes a plasmid is lost if it is not replicated, or it may remain in a cell's progeny for many generations. Recombinant DNA technology uses plasmids, which can easily pass between prokaryotic cells, to ferry genes from one kind of prokaryote to another.

FIGURE 20.3

Internal Structures of Prokaryotic Cells.

(**A**) Plasmids are rings of DNA that can transfer genes between prokaryotes. (**B**) Bacteria that cause the disease anthrax, *Bacillus anthracis*, survive environmental extremes by forming thick-walled endospores.

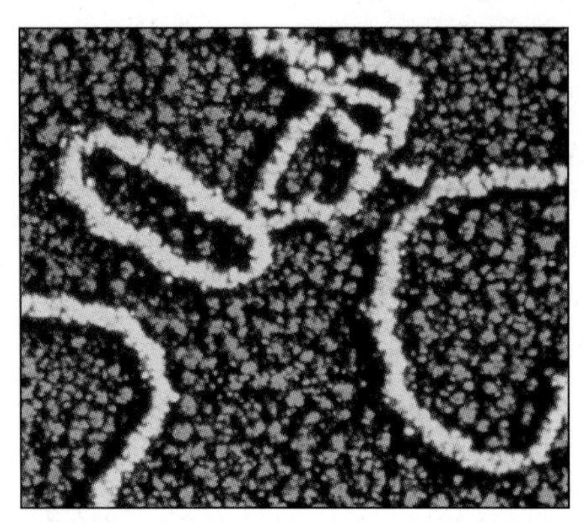

A Plasmids

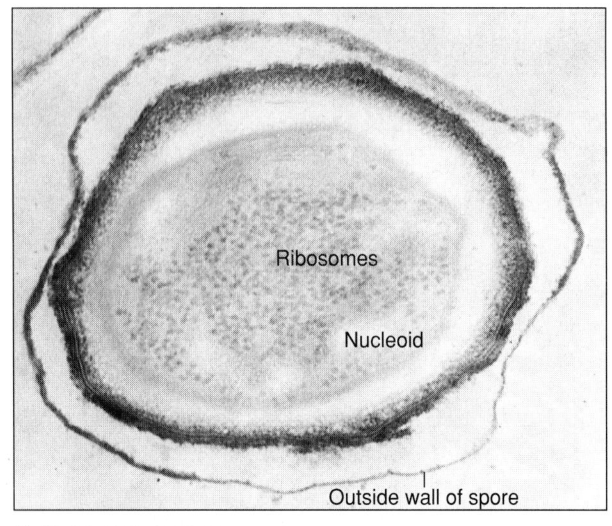

B Endospore structure

Prokaryotic ribosomes, where proteins are assembled, are structurally different from eukaryotic ribosomes. Certain antibiotics exploit this structural difference. Streptomycin, for example, binds specifically to bacterial ribosomes, inhibiting protein synthesis without harming eukaryotic host cells.

Some prokaryotes can form structures called **endospores** that enable them to survive harsh conditions (fig. 20.3B). An endospore is a walled structure that forms around the DNA and a small amount of cytoplasm. The normal cellular form returns when environmental conditions improve. Endospores, which are much tougher than the myxobacterial spores described in the chapter opening essay, can withstand boiling and drying. Because many pathogenic (disease-causing) bacteria form endospores, food manufacturing processes typically include a 10- to 15-minute superheated steam treatment to destroy them.

External Structures

Most prokaryotic cells are surrounded by a **cell wall.** The chemical composition of the cell wall differs between bacteria and archaea. Bacterial cell walls contain a complex, cross-linked polysaccharide called **peptidoglycan,** which is absent from cell walls in archaea. The antibiotic penicillin inhibits bacteria but not archaea because it interferes with the final steps in peptidoglycan synthesis.

The **Gram stain** reaction (fig. 20.4) highlights differences in cell wall architecture among bacteria, and provides a way to distinguish between two major groups. To determine whether a type of bacterium is gram-negative or gram-positive, cells are first stained with a purplish-blue dye called crystal violet. They are then exposed to iodine, which helps retain the stain. Next, cells are washed with alcohol, which removes the crystal violet from some cells. Finally, cells are exposed to a second dye, safranin, which imparts a reddish-pink color. Gram-positive cells, which have walls made primarily of a thick layer of peptidoglycan, retain the crystal violet and appear purplish blue. In contrast, gram-negative cells, because they have a much thinner layer of peptidoglycan, lose the violet hue, become colorless, and then take up safranin and appear pink. Gram staining is commonly used in identifying unknown bacteria.

Besides the thin inner layer of peptidoglycan, the cell walls of gram-negative bacteria have an outer covering of lipid, protein, and polysaccharide (fig. 20.4). Certain components of this outer membrane are responsible for the toxic properties of many medically important gram-negative bacteria, including *Salmonella.*

Some cells have short, hairlike projections called **pili** (singular: pilus). Attachment pili enable the cells to attach to objects (fig. 20.5A). The bacterium that produces a toxin that causes cholera, for example, attaches to a human's intestinal wall using pili. Sex pili, which occur only in certain bacterial groups, can aid in the transfer of DNA from cell to cell.

Another external structure found on some cells is a **flagellum,** which is an extension that rotates, moving the cell (fig. 20.5B). As described in the chapter opening essay, prokaryotes use their flagella to move toward or away from a stimulus, a response called **chemotaxis.** A flagellum is thin, only about 10 to 30 nanometers

FIGURE 20.4

Gram Stain.

The Gram stain is used to distinguish bacteria on the basis of cell wall structure. Gram-positive cells have a thick layer of peptidoglycan that retains the purple dye. Gram-negative cells have a thinner peptidoglycan layer, coated with an outer membrane. The inset shows a stained smear that contains both gram-positive and gram-negative bacteria.

Gram-negative bacteria—stain pink

Cytoplasm

Outer membrane (removed during gram staining procedure)

Peptidoglycan layer

Cell membrane

Gram-positive bacteria—stain purple

Peptidoglycan layer

Cell wall

Cell membrane

4.0 μm

FIGURE 20.5

External Structures of Prokaryotic Cells.

(**A**) Pili are extensions that enable cells to attach to objects, surfaces, and other cells. This is *Escherichia coli*. (**B**) The flagella on this *Legionella pneumophila* enable it to move. (**C**) A glycocalyx is a sticky layer surrounding the cell wall that enables cells such as this *Bacteroides* to adhere to surfaces.

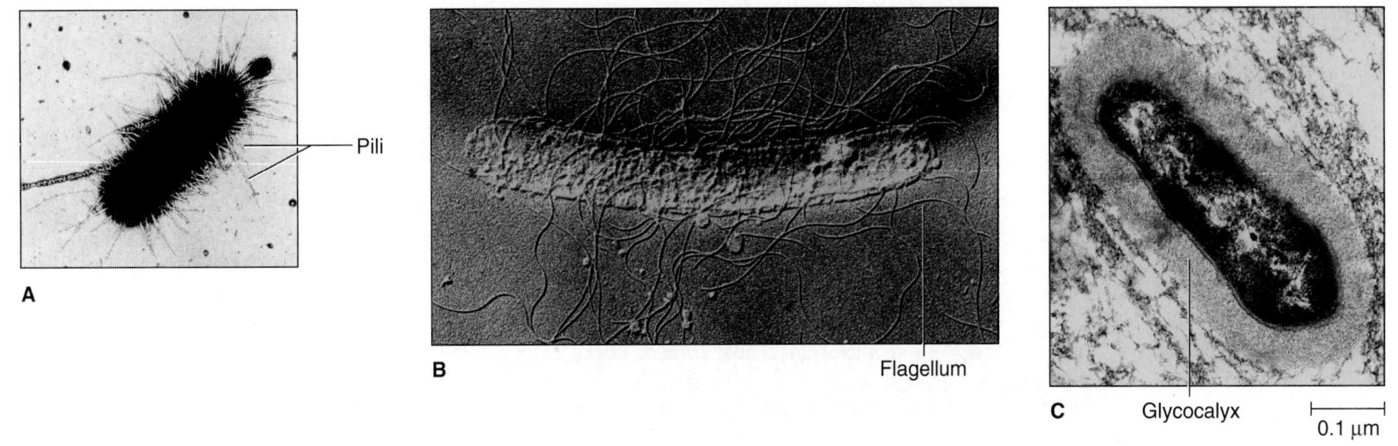

in diameter. It is very different chemically from a flagellum on an animal cell, such as a human sperm cell (see figs. 4.15 and 4.16).

The prokaryotic cell wall may be surrounded by a thick or thin sticky layer called a **glycocalyx,** which is composed of proteins and/or polysaccharides (fig. 20.5C). A loose glycocalyx is called a **slime layer,** and a firm glycocalyx is a **capsule.** The glycocalyx has many functions, including attachment to various surfaces, resistance to drying, and, in animal pathogens, protection from host immune system cells.

20.2 MASTERING CONCEPTS

1. Where is prokaryotic DNA located?
2. What is the advantage of endospores to species that make them?
3. What is the basis of Gram staining?
4. What are the functions of pili, flagella, and a glycocalyx?

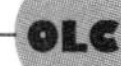

20.3 How Do Microbiologists Distinguish and Describe Prokaryotes?

Classification of prokaryotes is undergoing a quiet revolution as molecular characteristics are considered along with more traditional ways of grouping organisms.

People have been examining prokaryotes since Antonie van Leeuwenhoek first scraped them from his teeth and viewed them under his microscope in 1673. GenBank, the U.S. government database that stores DNA sequences of organisms, lists about 14,000 prokaryotic species (including 1,000 or so archaea). However, microbiologists estimate that there are 100,000 to 10,000,000 species of prokaryotes.

We still know very little about prokaryotes compared with what we know about large, multicellular organisms, partly due to our own limitations. Prokaryotes are tiny, making them difficult to visualize. They may live in inaccessible habitats. Also, many microorganisms are difficult, if not impossible, to culture in numbers sufficient to study them. In spite of these hurdles, microbiologists have devised taxonomic groupings based on characteristics that they could observe (table 20.2). These traditional groupings do not necessarily reflect evolutionary relationships, and they are slowly being revised as more molecular sequences are discovered.

Traditional Methods of Classifying Prokaryotes

Cell structure, or morphology, is one important criterion for classifying prokaryotes. Individual cells can take a variety of forms. Three of the most common are **cocci** (spherical), **bacilli** (rod-shaped), and **spirilla** (spiral) (fig. 20.6). The arrangement of the cells in pairs, tetrads, grapelike clusters (*staphylo-*), or chains (*strepto-*) also is sometimes important. *Staphylococcus,* for example, causes infections in humans; its spherical cells form clusters. Other morphological characteristics include the presence or absence of certain structures such as flagella, pili, endospores, or a glycocalyx.

Besides direct microscopic observation of cells, staining is used to divide prokaryotes into groups based on differences in cell wall structure. For example, the Gram stain technique, described earlier, is one of the methods by which bacteria are placed into major subgroups.

The lack of structural diversity among prokaryotes is amply compensated by their astonishing metabolic diversity. The methods by which prokaryotes acquire carbon and energy form another basis for their classification. The most fundamental distinction reflects carbon source. Recall from chapter 6 that **autotrophs** acquire carbon from inorganic sources such as carbon dioxide (CO_2), and **heterotrophs** use more reduced and complex organic molecules,

TABLE 20.2 A Partial Selection of Prokaryote Groups, Using Traditional Taxonomic Criteria

GROUP	FEATURES	EXAMPLE
Purple and green sulfur bacteria	Bacterial photosynthesis using H_2S (not H_2O) as electron donor	*Chromatium vinosum, Chlorobium limicola*
Cyanobacteria	Photosynthesis releases O_2; some fix nitrogen; free-living or symbiotic with plants, fungi, or protists	*Nostoc, Anabaena*
Spirochetes	Spiral-shaped; some pathogens of animals	*Borrelia burgdorferi* (causes Lyme disease) *Treponema pallidum* (causes syphilis)
Enteric bacteria	Rod-shaped, facultative anaerobes in animal intestinal tracts	*Escherichia coli, Salmonella* spp.
Vibrios	Comma-shaped, facultative anaerobes common in aquatic environments	*Vibrio cholerae* (causes cholera)
Gram-positive endospore-forming bacteria	Aerobic or anaerobic; rods or cocci	*Bacillus anthracis* (causes anthrax) *Clostridium tetani* (causes tetanus)

FIGURE 20.6
Prokaryotes Can Be Classified by the Shapes of Their Cells.
Micrococcus (**A**) are spherical (cocci). *Bacillus megaterium* (**B**) are rods (bacilli). *Rhodospirillum rubrum* (**C**) are spiral-shaped (spirilla).

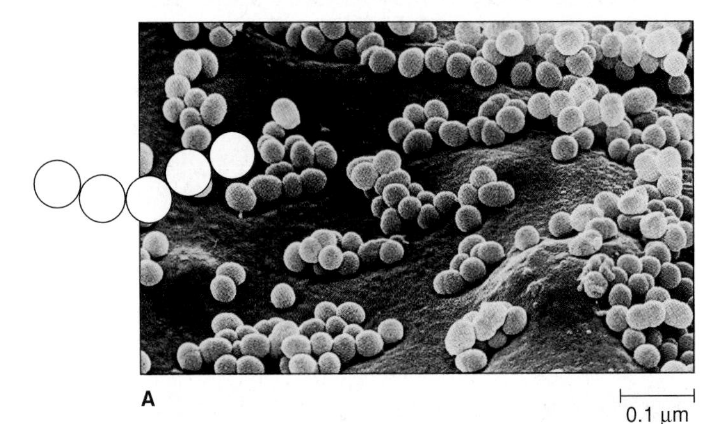

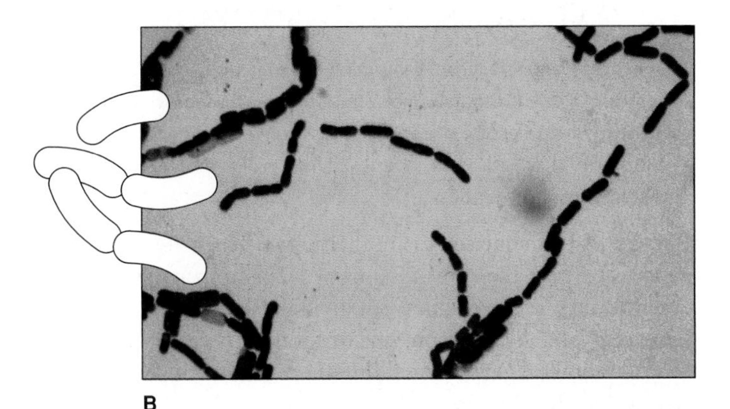

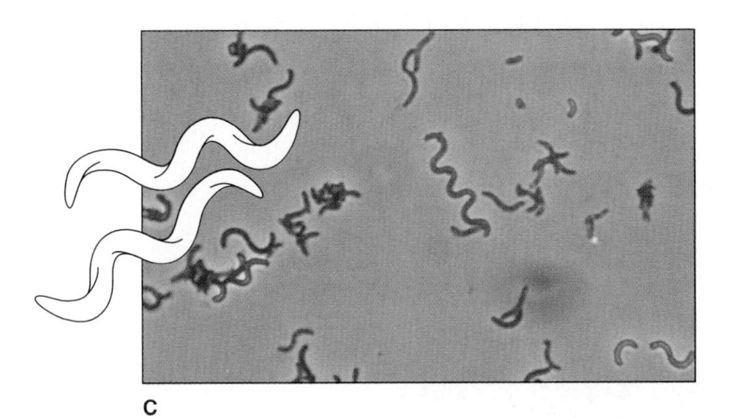

typically from other organisms. An autotroph or heterotroph may also be a **phototroph** or **chemotroph,** which refers to the energy source. Phototrophs derive energy from the sun, and chemotrophs get energy by oxidizing inorganic or organic chemicals.

Combining these terms describes how specific microorganisms function and fit into the environment. Cyanobacteria, for example, are photoautotrophs—which means that they use sunlight (photo) for energy and CO_2 (auto) for carbon. Many pathogenic (disease-causing) bacteria are chemoheterotrophs, because they use organic molecules from their hosts to acquire both carbon and energy. Chemoautotrophic bacteria acquire carbon from CO_2 and energy by oxidizing inorganic nitrogen- or sulfur-containing molecules. The metabolic diversity of prokaryotes allows them to live in many environments that lack eukaryotes.

Prokaryotes are also classified by physiological tests such as temperature and pH optima, and oxygen requirements. For example, **obligate aerobes** require oxygen for generating ATP. For **obligate anaerobes** such as *Clostridium tetani* (the bacterium that causes tetanus), oxygen is toxic, and they live in habitats that lack it. **Facultative anaerobes,** which include *E. coli* and *Salmonella,* can either use oxygen or not.

Finally, prokaryotes may also be classified based on where they live. Table 20.3 lists some of these designations. But classification based on habitat can be confusing, because archaea and bacteria may share habitats. For example, both the bacterium *Thermus aquaticus* and the archaeon *Thermoplasma volcanium* are thermophiles, living in hot conditions. Another complication of this method is that one organism might fall into more than one category.

Molecular Data Reveal Evolutionary Relationships

Traditional classification systems are still in widespread use, because they are based on characteristics that are easily observed. One problem with describing microorganisms based on their appearance and methods of energy acquisition, however, is that members of the groups may be only distantly related to one another. Molecular data are leading scientists closer to a classification system that reflects evolutionary relationships among all organisms, including prokaryotes.

Starting in the late 1970s, microbiologists began to compare organisms by ribosomal RNA sequences, which tend to change little over evolutionary time and therefore might reveal common ancestries. Such studies compare **signature sequences,** which are

TABLE 20.3 Identifying Prokaryotes by Habitat

DESIGNATION	ENVIRONMENT	SAMPLE HABITATS
Acidophile	Low pH (1.0–5.4)	Hot springs
Halophile	Extreme salt (3.5–30%)	Ocean, Dead Sea, evaporation ponds
Methanogen (methane is a metabolic by-product)	Anaerobic	Aquatic sediments, human gastrointestinal tract
Thermophile	Extreme heat (50°C–110°C)	Compost heaps, boiling springs

short stretches of nucleotides that are unique to certain types of organisms. If two varieties of bacteria share sequences that no others have, then one explanation is that they descended from a recent shared ancestor. Such information is used to build phylogenetic trees. Recall from chapter 3 that RNA sequencing of prokaryotes was how Woese distinguished archaea from bacteria.

RNA analysis can also be valuable in studying microorganisms that are difficult to culture. This was the case for *Epulopiscium fishelsoni,* discovered in 1993 in the intestine of a large fish in the Red Sea. Microbiologists thought it was a eukaryote, because it was thousands of times the size of *E. coli* and even larger than the unicellular eukaryote *Paramecium* (fig 20.7). *E. fishelsoni* survives for only 20 minutes outside of the fish, so laboratory culture did not look promising. However, because of its large size, researchers were able to isolate the organism, retrieve its DNA, and use the polymerase chain reaction (see Biotechnology 12.3) to amplify the sequences that encode rRNA. The signature sequences were unmistakably those of bacteria, not eukaryotes. *E. fishelsoni* is a giant bacterium.

By knowing the types of proteins that a prokaryote manufactures, researchers are obtaining unprecedented glimpses into how these organisms live. Consider *Treponema pallidum,* which causes syphilis. Microbiologists have long wondered why this bacterium is remarkably comfortable in the human body, taking up residence for decades, yet it dies within days in laboratory culture. The genome sequence, published in 1998, revealed that *T. pallidum* cannot manufacture many nucleotides, fatty acids, and enzyme cofactors, but it does have a large supply of transport proteins to obtain these essentials from its host. Laboratory culture simply can't provide what a warm human body can. Today, the sequencing of entire genomes is prompting researchers to reevaluate prokaryote taxonomy (see Investigating Life 20.1).

20.3 MASTERING CONCEPTS

1. What are some of the challenges encountered in studying prokaryotes?
2. What are the basic shapes of prokaryotic cells?
3. What are the different ways that bacteria and archaea acquire energy and carbon?
4. Why do groupings of prokaryotes based on rRNA signature sequences sometimes not match groupings based on other characteristics?

OLC

20.4 Prokaryotes Transfer Genes Vertically and Horizontally

Genes are transferred from generation to generation, but also from cell to cell horizontally, in several ways. Horizontal gene transfer may complicate our ability to discover what the earliest cells were like, as well as to understand how prokaryotes are related to each other today.

Genes pass from one cell to another in two general ways. **Vertical gene transfer** is the transmission of DNA from a parent cell to daughter cells as the original cell divides. **Horizontal gene transfer** occurs laterally, from one cell to another. That is, vertical gene transfer entails a generational difference (inheritance), but horizontal gene transfer does not.

FIGURE 20.7
Giant Bacterium.
Epulopiscium fishelsoni is a million times larger than many other prokaryotes, and it is even larger than the single-celled eukaryote *Paramecium* (×200).

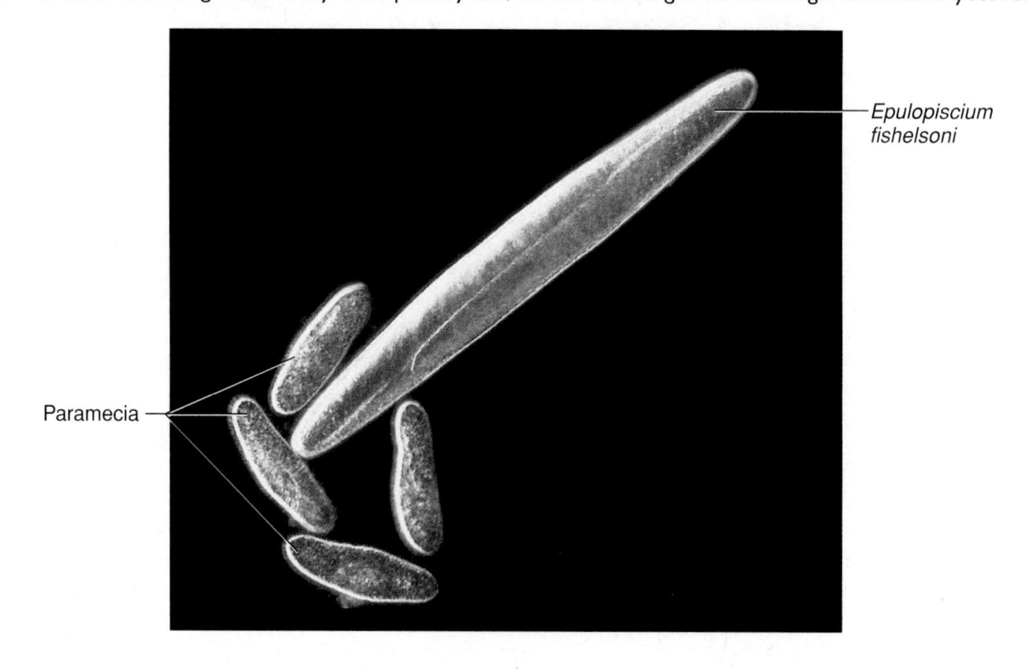

Investigating Life 20.1

Reconciling Molecular Versus Traditional Characteristics

When microbiologists began incorporating molecular sequence data into classification schemes, they hoped that the groups defined by appearance or metabolism would match those indicated by RNA sequences. However, this isn't always so, because the characteristics on which many classifications have been based do not necessarily reflect shared ancestry, as sequence similarities are presumed to do.

Two Situations

1. *Both types of data agree.* This is the case for the spirochetes. These bacteria are characterized by their corkscrew shape as well as twisted flagella, called axial filaments, that lie beneath a sheath and enable the bacteria to move on surfaces (see fig. 14.8C). All spirochetes defined by shape and axial filaments also fall into the same group defined by rRNA sequences.

2. *The types of data disagree.* Mycoplasmas are the smallest known cells, so tiny that their genome of about 650 genes is termed the "minimal functional gene set." One biotechnology company is trying to synthesize a living cell based on subtracting genes systematically from a *Mycoplasma* genome (see the opening essay to chapter 13). These organisms were classified as gram-negative because they lack cell walls, but they in fact have more in common, molecularly speaking, with gram-positive bacteria. The mycoplasmas may have descended from bacteria that had cell walls, and for some unknown reason, lost them.

Molecular data also differ from traditional taxonomic groupings for *Bacteroides* and *Flavobacterium.* Both genera contain gram-negative bacteria, but *Bacteroides* lives in the absence of oxygen in the human digestive tract, accounting for 30% of the bacteria in feces. Flavobacteria are aerobes and they were grouped separately from *Bacteroides.* Yet they share key rRNA sequences, and some members of each group produce certain lipids that are not seen in other prokaryotes.

Bacteria and Archaea Revisited

Nowhere is the discrepancy between what we can see and what molecules tell us as great as in distinguishing the domains Bacteria and Archaea. The tendency to lump together all organisms that lack nuclei and membrane-bounded organelles hid much of the diversity in the microbial world for many years. Although microbiologists haven't yet scrutinized many archaean species intensely, and many distinctions probably have not yet been discovered, these cells do differ from bacterial cells in some key ways (see fig. 3.4):

- Archaea exhibit different rRNA and tRNA sequences, ribosomal proteins, and antibiotic sensitivity patterns.
- Archaea have certain cell wall components and membrane lipids not seen in bacteria.
- An archaean genome contains more introns and repeated sequences than does a bacterial genome.
- Some archaea are thin rectangles (fig. 20.A), a form not seen among bacteria, indicating that members of the two domains differ in their cellular organization and possibly composition.

Although we have a long way to go in understanding the invisible microbial world, it is clear that archaea are not bacteria.

FIGURE 20.A
Rectangular Cells.
Archaea can take forms not seen in bacteria, indicating a different type of cell structure.

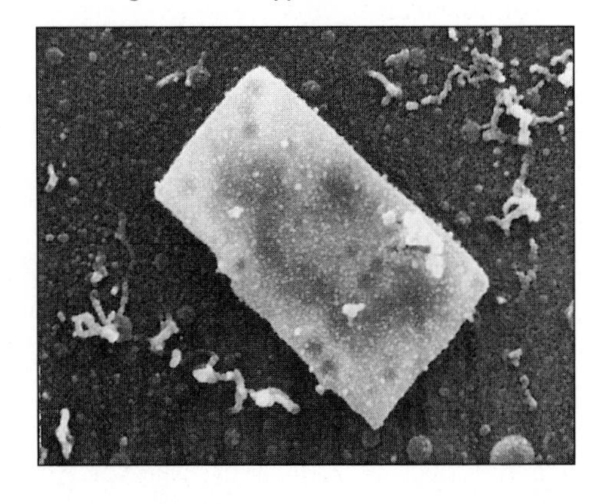

Binary Fission Provides Vertical Gene Transfer

Prokaryotes reproduce by **binary fission,** a process that replicates DNA and distributes it and other cellular constituents into two daughter cells. Binary fission therefore provides vertical gene transfer, from one generation to the next.

In prokaryotic cells, DNA is attached to a point on the inner face of the cell membrane (fig. 20.8). In binary fission, the DNA begins to replicate, and the cell membrane grows between the two DNA molecules and separates them. Then the cell membrane dips inward, pinching off two daughter cells from the original one. Formation of cell walls completes the process.

Binary fission superficially resembles mitosis, but it is different because it lacks spindle fibers and many of the types of proteins that are associated with chromosomes in more complex organisms. Also, prokaryotic cells lack the linear chromosomes whose positions define the stages of mitosis.

Horizontal Gene Transfer Occurs by Three Routes

In horizontal gene transfer, DNA from one cell enters another, either by itself, or as part of a plasmid or bacteriophage. This type of gene transfer has profound implications in fields as diverse as origin of life research, medicine, biodiversity, evolution, taxonomy, and biotechnology.

Horizontal gene transfer among prokaryotes occurs in three ways:

Transformation In transformation, a prokaryote takes up naked DNA without cell-to-cell contact. Frederick Griffith discovered and

FIGURE 20.8
Prokaryotic Cell Division.
(**A**) In binary fission, the cell membrane grows and then indents as the DNA replicates, separating one cell into two. (**B**) This false-color image shows two new *E. coli* cells on the verge of separating from each other. The cytoplasm (green) has already divided, and the cell walls are nearly complete.

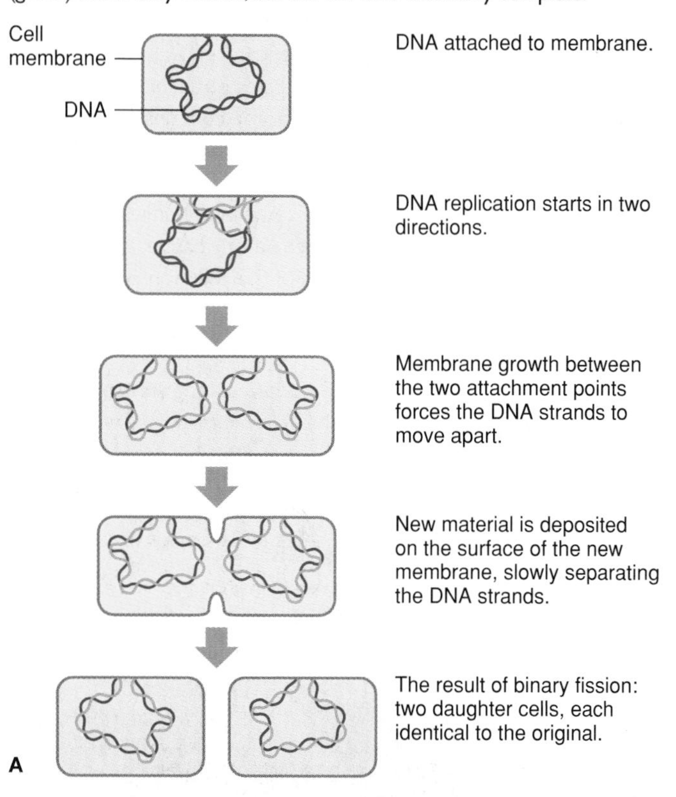

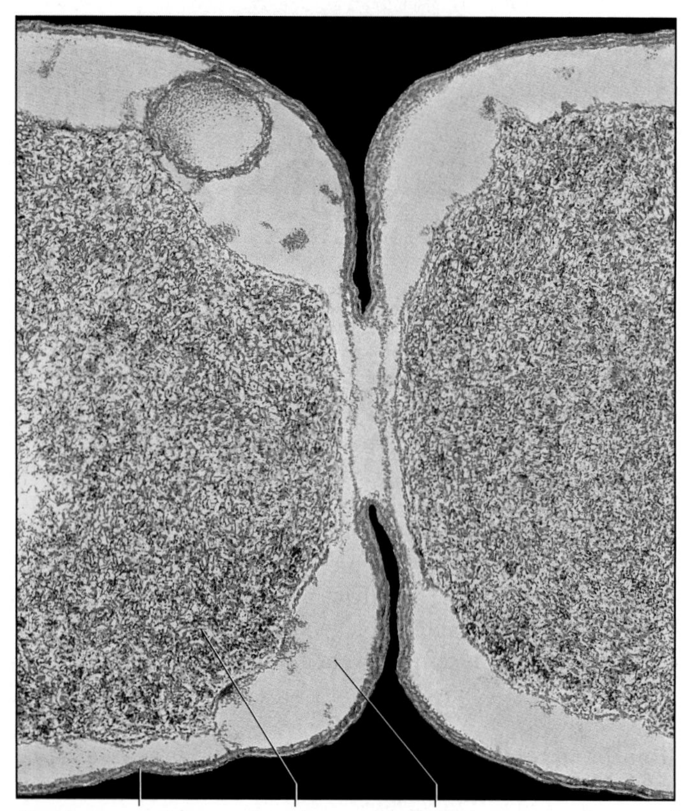

used transformation to demonstrate that a "transforming principle" transfers virulence of *Pneumococcus* bacteria in mice (fig. 20.9 and see fig. 12.2). The transforming principle was, of course, DNA. In Griffith's experiment, the dead virulent cells released their DNA, which the nonvirulent cells picked up. In nature, transformation occurs when prokaryotes die and pieces of their DNA or plasmid DNA enter other cells. Natural transformation is probably rare. In natural ecosystems, when prokaryotes take up DNA from their dead neighbors, it is more often dismantled and its building blocks recycled. Plus, not all cells are naturally "competent" to take up naked DNA.

FIGURE 20.9
Transformation.
In transformation, a cell picks up DNA fragments (**A**) or plasmids (**B**) released from another cell. If the receiving cell does not break down the new DNA, it may acquire useful new genes.

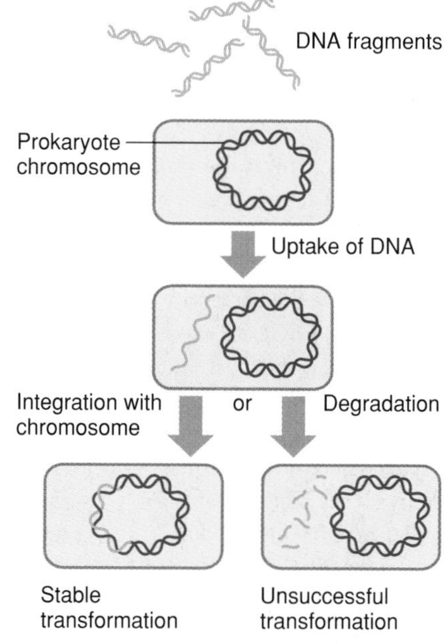

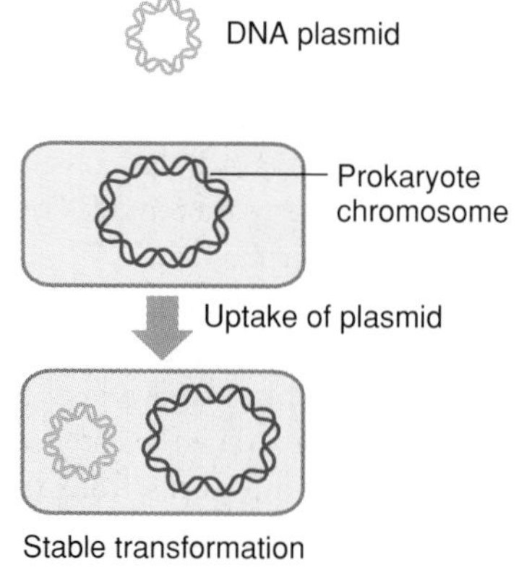

Transduction In transduction, a bacteriophage mistakenly packages host cell DNA along with bacteriophage DNA, then includes that bacterial DNA when it injects its own DNA as it attempts to infect another bacterial cell (fig. 20.10). Considering that there are billions of bacteriophages per milliliter of water in lakes, ponds, rivers, and oceans, transduction is probably very common. Because viruses can infect only specific hosts, however, transduction probably does not transfer DNA between very different species.

Conjugation In conjugation, a prokaryote that carries a set of genes called a sex factor on chromosomal or plasmid DNA passes these genes to a cell lacking them, along a bridge of cytoplasm called a **sex pilus** (fig. 20.11). Sometimes, if the sex factor is chromosomal, the cell copies its DNA and sends it to the recipient cell along with the

FIGURE 20.10

Transduction.

In transduction, a bacteriophage (virus) picks up DNA from one cell and transfers it to another.

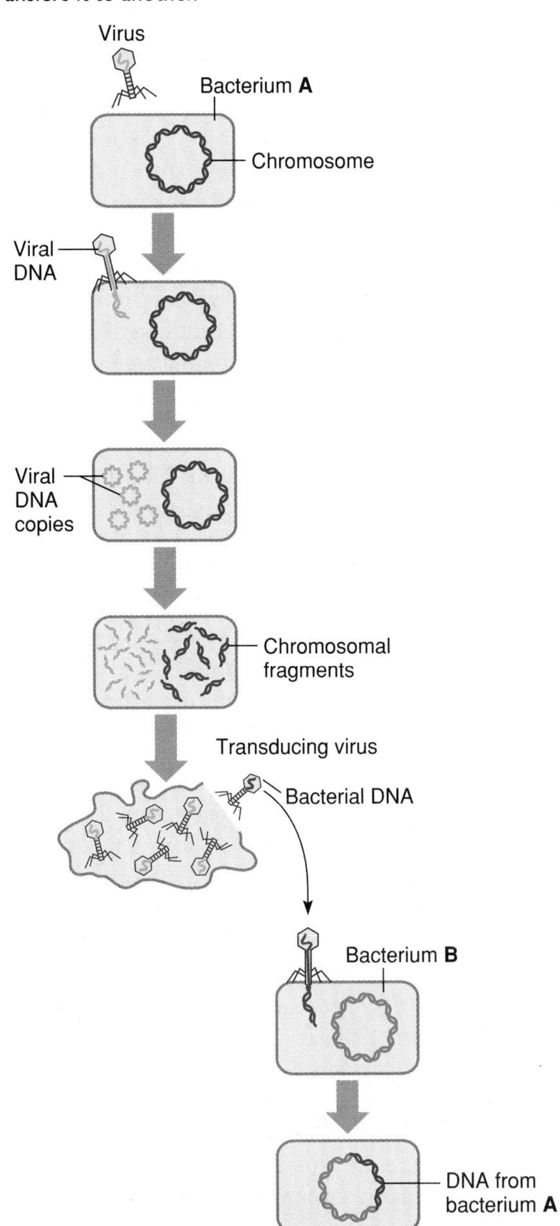

FIGURE 20.11

Conjugation.

In conjugation, a copy of some of one cell's DNA moves into another via a cytoplasm bridge (sex pilus). The new DNA may replace corresponding sequences, creating new gene combinations.

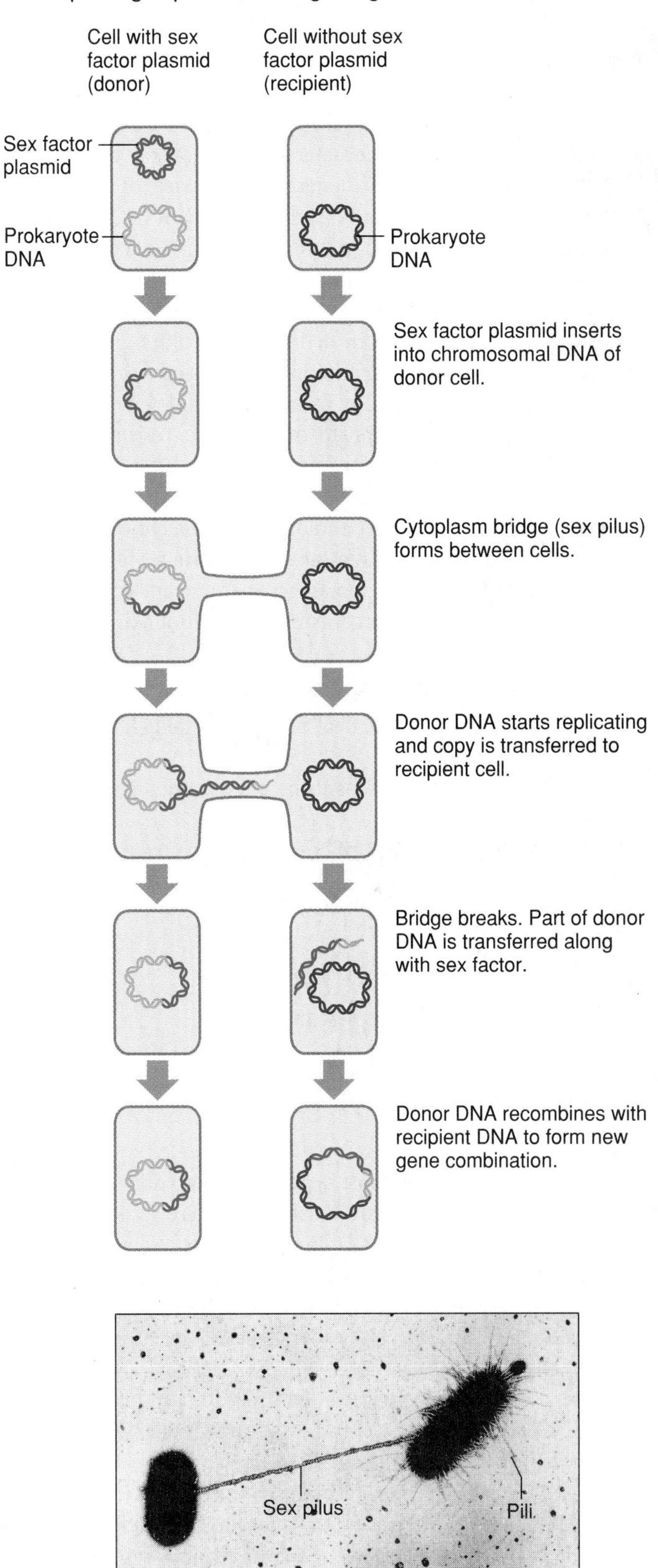

genes of the sex factor. The cytoplasm bridge occasionally collapses before all of the sending cell's DNA is transferred. The new genes arriving in the recipient cell trade places with their counterparts there, which enzymes then destroy. As a result, the receiving cell retains some of its own genes and some from the other cell, creating a new gene combination. Conjugation is sometimes called "bacterial sex" because sex refers to a process that generates new gene combinations.

Now that we are aware of how many trillions of prokaryotes inhabit Earth, it seems likely that horizontal gene transfer is very prevalent. In fact, all three mechanisms occur in soil, both freshwater and marine aquatic ecosystems, and in a variety of organisms. Horizontal gene transfer is perhaps most significant when it occurs among unrelated organisms. The long-ago swapping of gene pieces among the earliest cells may make it difficult for us to unravel life's beginnings. Today, promiscuous sharing of genes among species may hamper the ability to even define species, although relationships can still be inferred from infrequently transferred genes.

On the medical technology front, horizontal gene transfer is partly responsible for the spread of antibiotic-resistance genes among bacteria. Yet at the same time, biotechnologists take advantage of the process when they send genes of interest into bacteria in recombinant DNA technology (see Biotechnology 13.2). Horizontal gene transfer is a great concern in agricultural biotechnology, where it may spread genes from genetically modified bacteria to other microorganisms, plants, fungi, or animals.

20.4 MASTERING CONCEPTS

1. Distinguish between vertical and horizontal gene transfer.
2. What are the three mechanisms of horizontal gene transfer?
3. What is the significance of horizontal gene transfer?

OLC

20.5 How Prokaryotes Affect Humans

Few species of prokaryotes cause disease. Bacteria and their toxins can be used as weapons. Many industries use prokaryotic products or processes.

Humans never encounter most types of prokaryotes, and those that we do contact usually are harmless. We are most aware of those few species of bacteria that are pathogenic, as well as those prokaryotes that we harness for their products and processes. So far, no archaea are known to cause disease in humans.

Traditionally, microbiologists have used a set of rules called **Koch's postulates** to determine whether an organism causes a specific illness. The approach is named after Robert Koch (1843–1910), a German physician and microbiologist. A suspected pathogen is isolated from an infected, ill organism, grown in pure culture, and transferred to a healthy individual of the same species. If it, too, develops symptoms, and the microorganism can be recovered from its body, then an infection connection is indicated. Today, Koch's postulates are increasingly being supplemented with gene-detection methods to verify the causes of infectious diseases.

How Do Bacteria Infect Humans?

Bacterial infections may be spread in air, carried in arthropod (insect and spider) vectors, transmitted by direct contact, or passed in water or food.

Airborne Bacteria that can withstand prolonged drying can be transmitted in air. For example, *Legionella pneumophila* causes legionellosis, a form of pneumonia. This infection first came to public attention, as Legionnaires' disease, when a group of American Legion conventioneers staying at a Philadelphia hotel in 1976 suddenly developed cough, fever, headache, and pneumonia. Several of them died, as epidemiologists frantically traced the outbreak to an aerobic, gram-negative, rod-shaped bacterium that spread through the hotel's air conditioning system. Identification of this bacterium explained several earlier mysterious outbreaks of respiratory illness. Some bacteria need not be dried out to be carried in the air, such as the bacterium that causes tuberculosis.

Arthropods The arthropods include insects (mosquitoes, lice, flies, fleas, etc.) and ticks that may transmit bacteria. Arthropods transmit diseases when they bite one host, ingest the disease-causing microorganism, then pass on the disease when they bite a second host. A tick-borne bacterial infection is Lyme disease, caused by a spirochete, *Borrelia burgdorferi,* that lives in deer ticks. Lyme disease can begin with a bull's-eye-shaped rash, then produces malaise, flulike symptoms, and sometimes arthritis. The arthritis and weakness can persist for years and be quite disabling. Another example of an arthropod-borne bacterial disease is bubonic plague, caused by *Yersinia pestis.* It usually passes from rats to fleas to humans, although it can also be contracted through fleas from dogs and cats. The plague-causing bacteria proliferate in the host's immune system cells that normally destroy bacteria, enabling them to survive in the host and be passed to others.

Direct Contact Some bacteria enter the host directly, through intact or broken skin. In the sexually transmitted disease gonorrhea, bacteria attach by pili to cells lining the human urethra. White blood cells engulf the bacteria, which persist inside the cells, where they are shielded from immune system biochemicals (fig. 20.12A). Meanwhile, the area in the urethra where the bacteria attach becomes inflamed, and fibrous tissue forms, blocking urine flow. Painful urination is often one of the first signs of gonorrhea. Syphilis and chlamydia are two other examples of bacterial diseases that are spread by direct sexual contact.

Food- and Waterborne Many familiar bacterial infections fall into the food- and waterborne category. Several types of bacteria cause food poisoning, usually producing symptoms of abdominal cramps and diarrhea. Each year in the United States, more than 4 million people develop food poisoning caused by bacteria, and 1,200 die. Millions of additional cases are of undetermined origin.

Several *Salmonella* species cause a common type of food poisoning called salmonellosis (fig. 20.12B). The bacteria often proliferate in eggs, raw chicken, or undercooked beef. Children have contracted salmonellosis from pet reptiles. Cholera and *E. coli*–associated diarrheal illnesses are food- or waterborne diseases whose symptoms are caused by toxins.

Table 20.4 summarizes common routes of bacterial infection.

FIGURE 20.12

Pathogenic Bacteria.

(**A**) *Neisseria gonorrhoeae* causes the sexually transmitted disease gonorrhea by inhabiting the host's white blood cells (×500). (**B**) *Salmonella enteritidis* causes food poisoning (gastroenteritis).

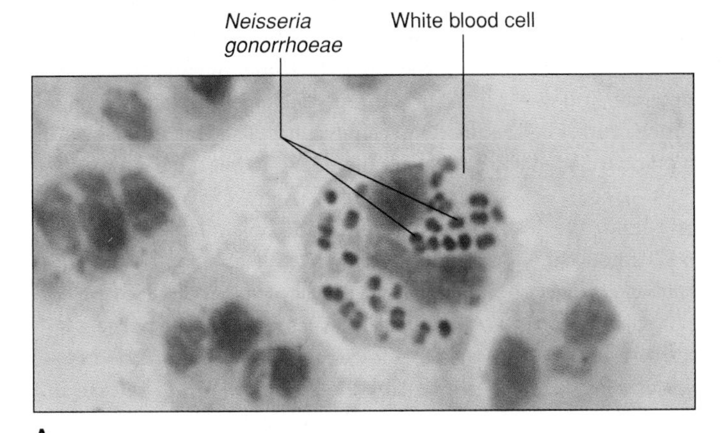

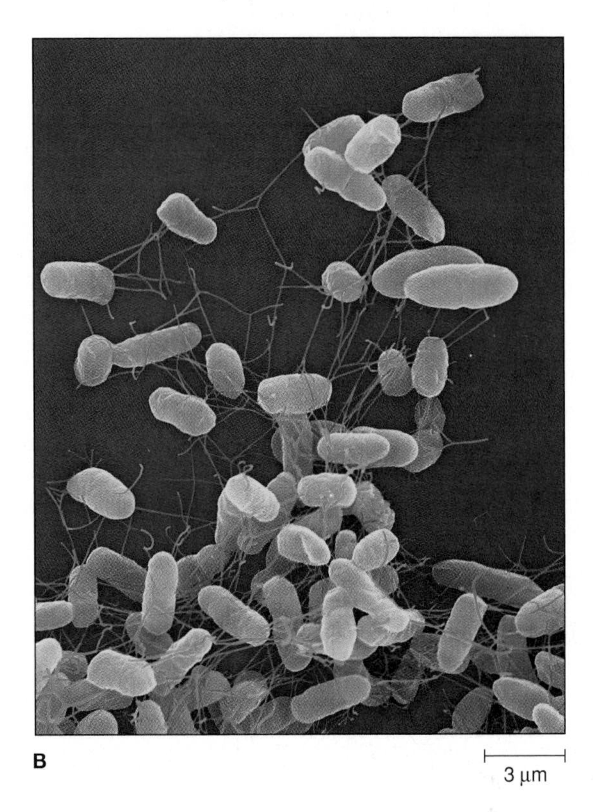

Bacteria as Bioweapons

It may be hard to believe that people have ever even considered using pathogens to intentionally cause disease, but this is the essence of bioweaponry. Today what was once called "germ warfare" is even more threatening because of the ability to genetically alter pathogens, making them more toxic or resistant to vaccines or drug treatments. Although many nations have taken measures to prevent bioterrorism, history suggests that it can happen, as Biotechnology 20.1 relates.

A biological weapon delivers highly virulent and incapacitating viruses, bacteria, fungi, or their toxins. A millionth of a gram of *Bacillus anthracis,* the bacterium that causes anthrax, can kill a person; a gram of aerosolized botulinum toxin, produced by *Clostridium botulinum,* can kill 1.5 million people. If delivered through the air, a microbial bioweapon would be tasteless, odorless, and invisible. A population might not even know it was under attack until infection had already occurred. Although treatments might be available, response might be slow because most physicians are unfamiliar with the "exotic" disorders that serve as the basis for bioweapons.

TABLE 20.4 Routes of Transmission of Bacterial Infection

ROUTE	DISEASE	BACTERIUM
Airborne	Legionellosis	*Legionella pneumophila*
	Pertussis	*Bordetella pertussis*
	Diphtheria	*Corynebacterium diphtheriae*
	Tuberculosis	*Mycobacterium tuberculosis*
Arthropod	Lyme disease	*Borrelia burgdorferi*
	Bubonic plague	*Yersinia pestis*
Direct contact	Gonorrhea	*Neisseria gonorrhoeae*
	Anthrax	*Bacillus anthracis*
Food- or waterborne	Salmonellosis (food poisoning)	*Salmonella enteritidis*
	Cholera	*Vibrio cholerae*
	Diarrhea and kidney failure	*Escherichia coli*

Industrial Microbiology Borrows Prokaryotes' Products and Processes

For centuries, humans have used prokaryotes to help make food. Bacteria are instrumental in the production of vinegar, sauerkraut, pickles, olives, yogurt, and cheese (fig. 20.13). They also help produce enormous quantities of vitamins, enzymes, and other useful chemicals such as ethanol and the solvent acetone.

FIGURE 20.13

Bacteria in Food Production.

Bacteria of genus *Lactococcus* are used to manufacture cheddar cheese from fermenting milk. The name of the cheese comes from Cheddar, a village in England where the cheese originated.

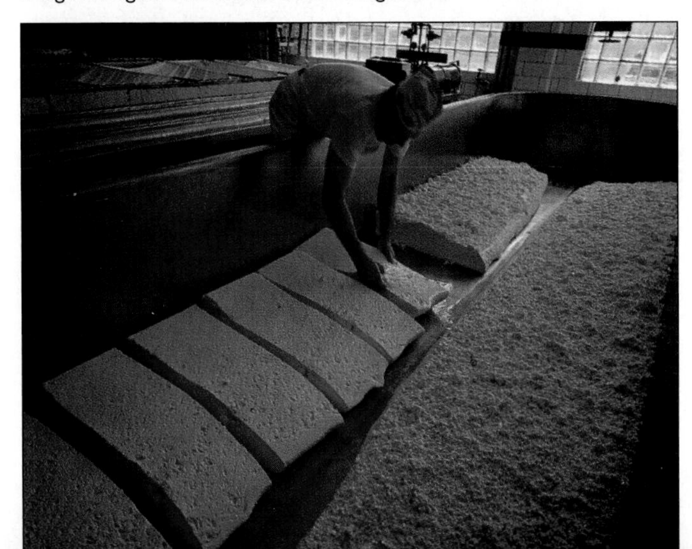

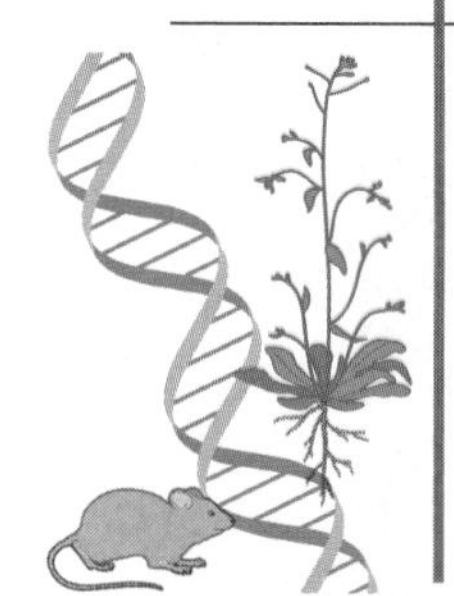

Biotechnology 20.1

Bioweaponry: A Brief History

Biological weapons have been around since medieval warriors catapulted plague-ridden corpses over city walls to kill the inhabitants. During the French and Indian War, the British gave Indians blankets intentionally contaminated with secretions from smallpox victims. Although germ warfare was banned by international law in 1925, from 1932 until 1942 Japan field-tested bacterial bioweapons, killing thousands.

In 1973, the Soviet Union established an organization called Biopreparat. Thousands of workers in 50 facilities prepared anthrax bombs and other bioweapons under the guise of manufacturing legitimate drugs, vaccines, and veterinary products. In 1979, an accident occurred in a city then called Sverdlovsk. At Military Compound Number 19, a miscommunication between shift workers in charge of changing safety air filters resulted in the release of a cloud of dried anthrax spores over the city. Within weeks, 66 people died of anthrax. The government officially announced that the deaths were due to eating infected meat. In each person, about 10,000 spores stuck to their lungs, entering their bloodstreams. Soon respiratory symptoms began, but health returned for a few days. Then, the spores germinated and the bacteria began dividing and releasing toxins, causing death. In September 1992, Boris Yeltsin officially halted bioweapon research.

In the United States, a small-scale bioweapons effort began in 1942, although it was overshadowed by the Manhattan Project to develop the atomic bomb. A facility at Fort Detrick in Frederick, Maryland, stored 5,000 bombs loaded with anthrax spores; a production facility for the bombs was located in Terre Haute, Indiana; and Mississippi and Utah had test sites. President Richard Nixon halted the program in 1969, because he thought that conventional and nuclear weapons were enough of a deterrent to war. On April 10, 1972, political leaders in London, Moscow, and Washington signed the Biological Weapons Convention, an effort to prevent bioterrorism whose protocols are being strengthened today. By 1973, the U.S. government destroyed all bioweapons and records of their production. Today, a sealed, four-story building at Fort Detrick contains fermentation tanks once used to grow anthrax bacilli and other pathogens.

Bioterrorism incidents so far have been rare. The only known United States case was the deliberate contamination of restaurants with *Salmonella typhimurium* by the Rajneeshee cult in The Dalles, Oregon, in 1984. Cult members added the bacteria to salad bars, and more than 750 people became ill with food poisoning—no one died. The goal was to keep people from voting in a local election. In March 1995, the Aum Shinrikyo sect released the nerve gas sarin in a Tokyo subway, following failed attempts by the group in 1993 to release anthrax spores from a building and botulinum toxin from a vehicle, and a 1992 trip to Zaire to obtain Ebola virus for weaponization.

Bacteria and viruses subverted for use as weapons have been called "the poor man's bomb," because they are inexpensive and fairly easy to produce, with potentially devastating consequences. Leaders of the biotechnology community are very concerned with this misuse of their field, and are taking measures to ensure that the tools of microbiology and molecular biology are never used for purposes of war or terrorism.

These industrial applications of microbiology take advantage of natural metabolic diversity among prokaryotes.

Today, bacteria are also mass-producing human proteins such as insulin and blood-clotting factors, using genes inserted into their genomes by recombinant DNA technology (see Biotechnology 13.2). The heat-, acid-, and salt-tolerant enzymes in archaean and bacterial cells from extreme environments have many potential applications in industry. For example, a bacterial enzyme helped improve the efficiency of PCR, the polymerase chain reaction (see Biotechnology 12.3). Because PCR requires frequent temperature shifts, researchers previously had to frequently add new DNA polymerase as heat dismantled it. Thus, PCR was inefficient until its discoverer incorporated DNA polymerase from the hot springs bacterium *Thermus aquaticus* (see the opening essay to chapter 1). This heat-tolerant bacterial enzyme works continuously.

Prokaryotes are also used in water and waste treatment. Sewage treatment plants in most communities, for example, harbor countless aerobic and anaerobic microbes that degrade organic wastes. **Bioremediation** uses microorganisms to naturally metabolize pollutants. For example, when the Exxon *Valdez* dumped oil into Prince William Sound in Alaska, biologists fed native shoreline bacteria inorganic nitrogen and phosphorus. The microbes consumed much of the oil, degrading it into CO_2, water, and biomass. Microorganisms can also detoxify chlorinated compounds such as polychlorinated biphenyls (PCBs) in soil, and remove heavy metals such as cadmium, lead, copper, zinc, silver, and mercury from water.

20.5 MASTERING CONCEPTS

1. What are Koch's postulates?
2. How do infectious bacteria enter human hosts?
3. What are bioweapons?
4. What are some industrial uses of prokaryotes?

Chapter Summary

20.1 Prokaryotes Are a Biological Success Story

1. Prokaryotes are important in many ways. They contribute gases to the atmosphere, form the bases of food webs, and **fix nitrogen.** More than a billion years ago, they gave rise to certain organelles.
2. Prokaryotes are very abundant and diverse and occupy a great variety of habitats.
3. Many prokaryotes are studied in the laboratory on plates of agar supplemented with appropriate nutrients, but others remain poorly studied because they cannot be cultured.

20.2 What Are the Parts of a Prokaryotic Cell?

4. Like other organisms, prokaryotic cells contain DNA and ribosomes and are bounded by a cell membrane.
5. DNA along with some RNA and proteins associate in an area called the **nucleoid. Plasmids** are circles of DNA in addition to the chromosome.
6. Some bacteria wait out harsh conditions protected in **endospores.**
7. A **cell wall** surrounds most prokaryotic cells. **Gram staining** reveals differences in cell wall architecture to group bacteria.
8. **Pili** are projections that allow cells to adhere to surfaces or other cells. **Flagella** provide movement.
9. A **glycocalyx** outside the cell wall provides adherence or protection from host immune system cells.

20.3 How Do Microbiologists Distinguish and Describe Prokaryotes?

10. Traditionally, microbiologists have classified prokaryotes based on cell morphology, metabolic capabilities, physiological tests, or habitat.
11. Bacteria are spherical **cocci,** rod-shaped **bacilli,** spiral-shaped **spirilla,** or variations of these. Archaea may have other shapes.
12. Prokaryotes can be classified based on how they acquire carbon and energy. **Autotrophs** acquire carbon from inorganic sources, and **heterotrophs** obtain carbon from other organisms. A **phototroph** derives energy from the sun, and a **chemotroph** by oxidizing organic or inorganic chemicals.
13. **Obligate aerobes** require oxygen, **facultative anaerobes** can live whether or not oxygen is present, and **obligate anaerobes** cannot function in the presence of oxygen.
14. Microbiologists are currently using molecular data, including **signature sequences** that are unique to certain groups of organisms, to reconsider traditional taxonomic classification of prokaryotes.

20.4 Prokaryotes Transfer Genes Vertically and Horizontally

15. **Binary fission** is division of a prokaryotic cell to yield two daughter cells. It is **vertical gene transfer** because DNA goes from one generation to the next.
16. In **transformation,** cells take up pieces of naked DNA from the environment. In **transduction,** a virus transfers DNA from one cell to another. In **conjugation,** a sex pilus transfers DNA from one cell to another. These are routes of **horizontal gene transfer.**

20.5 How Prokaryotes Affect Humans

17. **Koch's postulates** are used to demonstrate whether a type of microorganism causes a particular set of symptoms.
18. Pathogenic bacteria are transmitted in air, by direct contact, via arthropod vectors, or in food and water.
19. Bioweapons are highly virulent bacteria, viruses, or fungi. A bioweapon attack would be difficult to detect, trace, and treat.
20. Prokaryotes are used in the manufacture of many foods, drugs, and other chemicals. They are also used for treating sewage and cleaning the environment.

Testing Your Knowledge

1. Give five examples that illustrate how bacteria and archaea are important to other types of organisms.
2. Distinguish between:
   - **a.** a phototroph and a chemotroph
   - **b.** bacteria and archaea
   - **c.** a halophile and a methanogen
   - **d.** a gram-negative and a gram-positive bacterium
   - **e.** transformation and transduction
   - **f.** an autotroph and a heterotroph
   - **g.** vertical and horizontal gene transfer
   - **h.** an obligate anaerobe and a facultative anaerobe
3. What structures do prokaryotic cells have that are not found in eukaryotic cells?
4. What are the criteria that microbiologists use in classifying prokaryotes?
5. How can the polymerase chain reaction be useful in studying prokaryotes that do not survive in laboratory culture?
6. List Koch's postulates and explain why they are important.

Thinking Scientifically

1. What kind of change in the environment would cause a bacterium to change its movement from the pattern on the left to the pattern on the right?

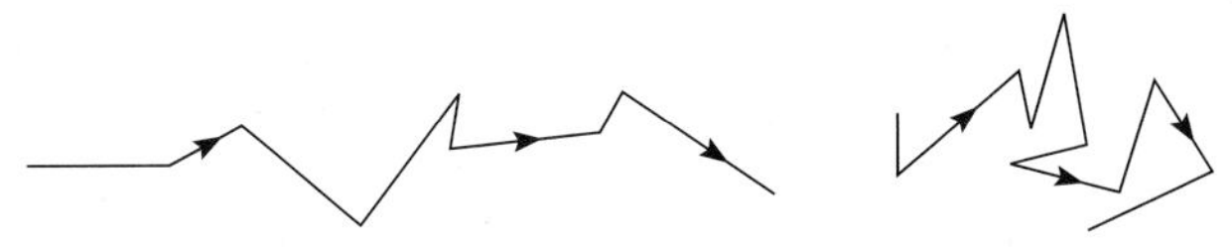

2. Coral living off the coast of Florida have developed an infection that researchers are calling a "plague." What steps should they take to identify a causative microorganism?
3. Why would a type of bacterium that forms spores make a more effective bioweapon than one that does not form spores?
4. The genetic material of different strains (varieties) of *Mycobacterium tuberculosis,* which causes tuberculosis, has different stretches of nucleotides called insertion sequences. Are these more likely to have been acquired by vertical or horizontal gene transfer? Cite a reason for your answer.
5. Would a gorilla or a human of the same approximate size have more bacteria on the skin? Cite a reason for your answer.
6. A young child develops a very high fever and an extremely painful sore throat. Knowing that the child could have an infection with a strain of *Streptococcus* that could be deadly, the physician seeks a very specific diagnosis. What three approaches might the doctor (or a laboratory) use to tell whether this infection is viral or bacterial and, if the latter, identify the bacterium?

References and Resources

Losick, Richard, and Dale Kaiser. February 1997. Why and how bacteria communicate. *Scientific American,* p. 70. Bacteria send and respond to signals.

Mayr, Ernst. August 1998. Two empires or three? *Proceedings of the National Academy of Sciences,* vol. 95, p. 9720. A prominent biologist argues why all prokaryotes are pretty much the same.

Stover, C. K., et al. August 31, 2000. Complete genome sequence of *Pseudomonas aeruginosa* PA01, an opportunistic pathogen. *Nature,* vol. 406, p. 959. This bacterium's large genome and complex metabolism enables it to evade antibiotics and disinfectants.

Whitman, William B., et al. June 1998. Prokaryotes: The unseen majority. *Proceedings of the National Academy of Sciences,* vol. 95, p. 6578. Bacteria and archaea are much more abundant than previously imagined.

Woese, Carl. September 1998. Default taxonomy: Ernst Mayr's view of the microbial world. *Proceedings of the National Academy of Sciences,* vol. 95, p. 11043. Woese answers Mayr.

The *LIFE* Online Learning Center provides additional resources and tools for studying this chapter.
www.mhhe.com/life

CHAPTER 21

Protista

Trout Whirling Disease

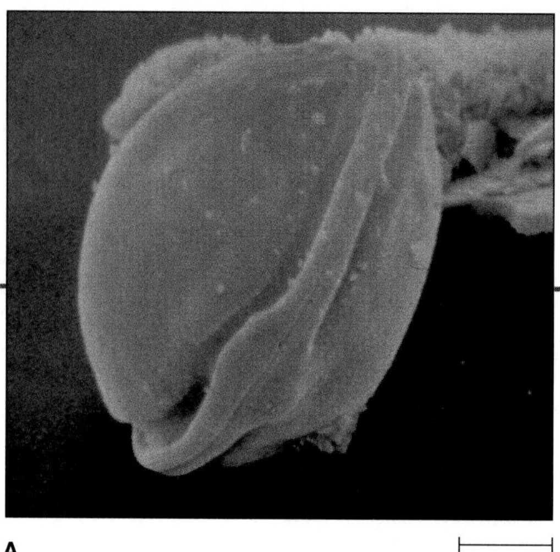

A

4 μm

Kingdom Protista includes a very diverse group of mostly unicellular eukaryotes. Some of them are parasitic. In humans, protista cause malaria and paralytic shellfish poisoning, and account for several of the serious infections that plague people with AIDS. Trout and other salmonid fishes suffer from a protist-caused infection called whirling disease.

Whirling disease begins when a protist called *Myxobolus cerebralis* enters the skin of a rainbow trout that is less than 2 months old. The parasite finds its way to cartilage in the head and spine, and begins a rapid series of nuclear divisions. As the parasites grow, holes form in the fish's tissues. Within months, the cavities fill with the feeding stage of the parasite, called trophozoites, as well as spores that can persist for years. When the parasites invade the auditory-equilibrium organ behind the eye, the fish's coordination vanishes, and the animal chases its tail feverishly, in no particular direction. The fish whirls about frantically, and may finally rest on the river bottom, whereupon it frequently becomes the meal of a larger fish. Birds such as herons and kingfishers pluck exhausted whirling fish from near the water's surface. Fish that survive develop deformities as their infected cartilage hardens into bone.

When the parasites invade the auditory-equilibrium organ behind the eye, the fish's coordination vanishes, and the animal chases its tail feverishly, in no particular direction.

Whirling disease is passed in several ways. Birds that have eaten infected trout pass the spores to other aquatic habitats when they defecate after

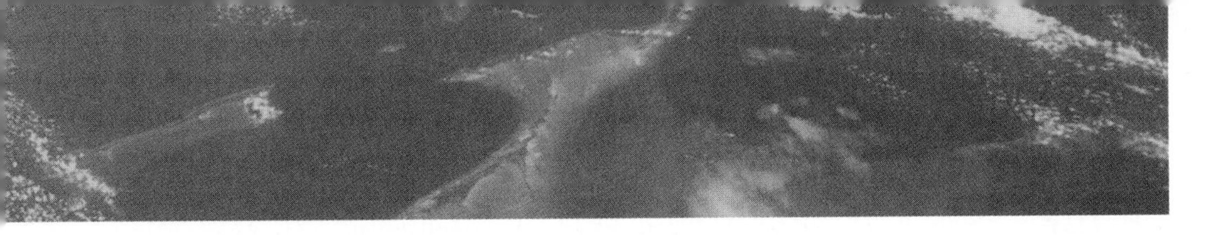

Fish Disease.

Whirling disease is caused by a protist called *Myxobolus cerebralis* (**A**). Rainbow trout that develop whirling disease develop extreme deformities. Below, a healthy fish (**B**) is compared to a deformed one (**C**).

B

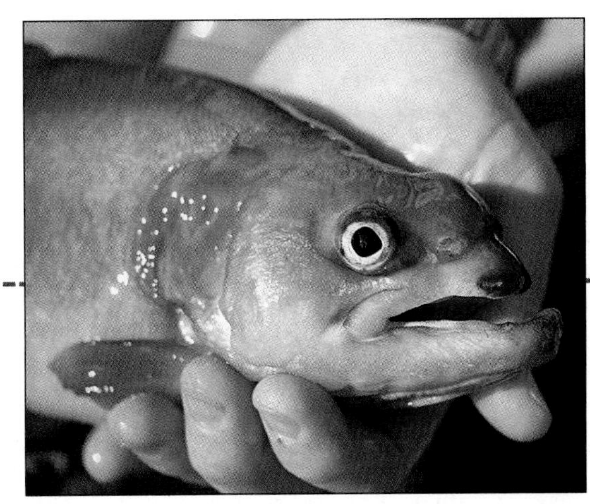

C

they fly away. A whirling trout torn apart by another fish releases thousands of spores into the water. Essential to the life cycle of *Myxobolus cerebralis* are aquatic worms of the species *Tubifex tubifex.* They eat the spores, which develop into a form that, when released, can infect more fish.

People first became aware of the disease in Europe in 1900 when rainbow trout were introduced. The parasite may have been passed from brown trout, which it infects without causing symptoms. It came to the United States in the late 1950s from Europe and today, whirling disease is affecting rainbow trout populations in 22 states. In some areas, rainbow trout populations have plummeted by 75%. Along one river in Montana, rainbow trout, which once numbered several thousand per mile, are now down to fewer than 50 per mile.

Along one river in Montana, rainbow trout, which once numbered several thousand per mile, are now down to fewer than 50 per mile.

Understanding the biology of *Myxobolus cerebralis* may help wildlife biologists develop ways to control outbreaks of whirling disease. Current research efforts include:

- Determining why brown trout and coho salmon become infected but do not develop symptoms, whereas rainbow trout, chinook salmon, and cutthroat trout develop whirling disease. Resistance genes could then be bred into the susceptible fish.
- Devising ways to raise young fish in areas free of *Myxobolus cerebralis.*
- Mating tubifex worms that spread the parasite with strains that do not. Preliminary experiments have shown that this strategy can reduce the number of *Myxobolus* organisms released by up to 80%.
- Fish hatcheries are testing treatment of young rainbow trout with antibiotics and ultraviolet radiation to kill the parasites.
- Ecologists are identifying the environmental conditions that lead to "hot spots" for whirling disease. For example, Little Prickly Pear Creek is a tributary of the Missouri River that is a spawning ground. Because the rainbow trout population here includes so many young, the infection has been particularly devastating.

21.1 Kingdom Protista Lies at the Crossroads Between the Simpler and the Complex

The Protista are a very diverse group of organisms whose classification is currently in flux, as molecular data sometimes support and sometimes challenge past hypotheses. These are the simplest of the eukaryotes, the organisms that gave rise to plants, fungi, and animals. The most ancient Protista lack mitochondria or have mitochondria that resemble bacteria.

Between the unicellular organisms of the Bacteria and Archaea domains and the multicellular eukaryotes lies a kingdom of very diverse organisms, the **Protista,** that represent a bridge of sorts from simple to complex life-forms. Most protista are single eukaryotic cells. The simplest and most ancient lack mitochondria, and provide glimpses into what the earliest eukaryotes might have looked like.

Protista are of particular interest to evolutionary biologists because current forms may retain clues to important milestones in eukaryote history. For example, a group of protista called jakobids have mitochondria that resemble bacteria more than those of any other type of eukaryote. Therefore, jakobids may look like organisms that lived shortly after cells acquired aerobic bacteria as endosymbionts (see figs. 3.19 and 18.10). At the other end of the evolutionary spectrum are the choanoflagellates, protista that may be similar to the forerunners of animals. These organisms bear an uncanny resemblance to the "collar cells" in sponges, which are the simplest animals (fig. 21.1). Since some choanoflagellates form colonies in which the individual organisms seem to interact to move and obtain food, they might represent a step on the road to the earliest multicellular organisms.

Textbooks and taxonomists have traditionally considered the protista in terms of the more familiar organisms that they resemble—the plantlike algae, animal-like protozoa, and funguslike slime molds. These designations, however, probably do not reflect evolutionary relationships. As with other major groupings of organisms, biologists are in the process of reconsidering the classifications within the kingdom to reflect nucleic acid sequence, which is a more objective measure of relatedness than comparing possibly superficial appearances. In fact, some new classification systems propose splitting the Protista into more than one kingdom. The groupings presented in this chapter should therefore be considered provisional, because they will undoubtedly be modified as researchers gather more information.

The **algae** (an informal term for photosynthetic protista) illustrate the ongoing taxonomic revision of the kingdom. These organisms are often described together for convenience, but they do not form a natural grouping. In fact, photosynthetic organisms are found among at least five different groups of protista: the euglenida (basal eukaryotes), dinoflagellates (alveolates), golden and brown algae (stramenopiles), red algae, and green algae. These groups are defined by their photosynthetic pigments, carbohydrate storage molecules, and other fundamental characteristics.

Because the diversity of the protista is tremendous, this chapter provides just a small sampling of some interesting members. The chapter begins with the **basal (founding) eukaryotes,** which do not encompass a single taxonomic group. Rather, they include several distantly related groups that may reveal stages in the acquisition and evolution of endosymbiont bacteria into mitochondria. That is, some of these protista lack mitochondria or anything resembling them, and others have fully developed energy-extracting organelles. The remainder of the chapter is devoted to four groups of the so-called **crown eukaryotes,** which diverged later: alveolates, stramenopiles, red algae, and green algae.

Basal Eukaryotes That Lack Mitochondria

One group of basal eukaryotes that lack mitochondria is the **Parabasalia.** These flagellated organisms, found only in association with animals, include *Trichomonas vaginalis,* which causes a

FIGURE 21.1

Choanoflagellates Resemble Choanocytes in Sponges.
Choanoflagellates (**A**) are protista that greatly resemble choanocytes (collar cells) in sponges (**B**). The photo depicts a member of genus *Salpingoeca.* The organism's flagellum and tentacles from the collar are partly visible near the left side. Movement of the flagellum produces water currents that carry food particles into the theca, which envelops the cell.

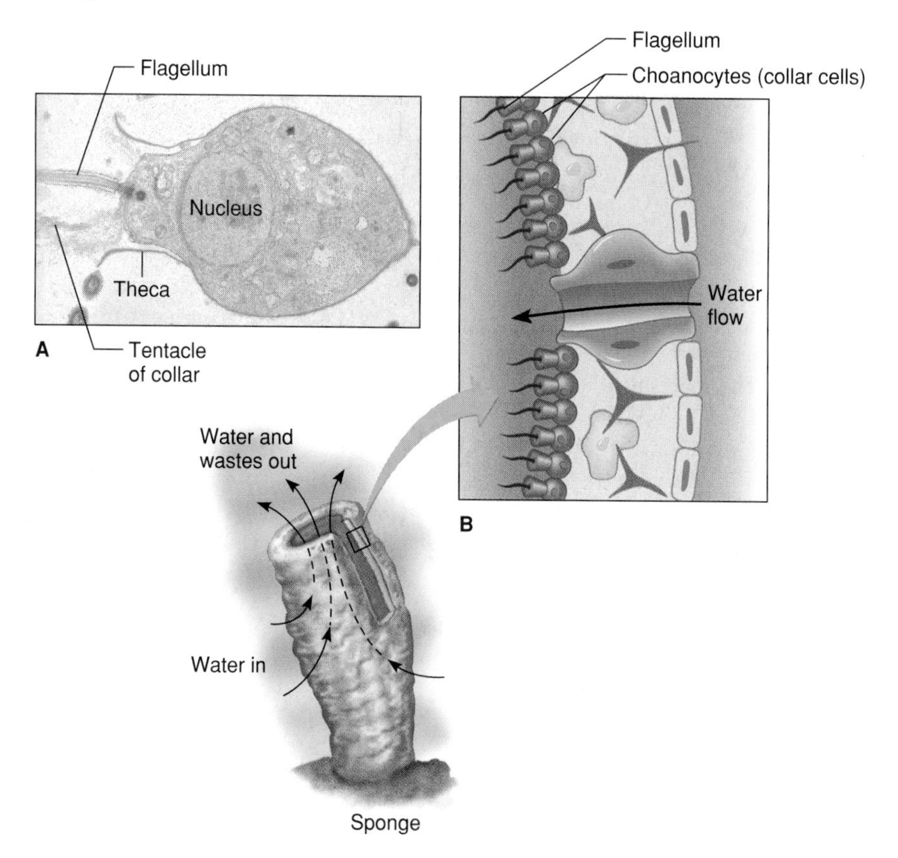

sexually transmitted disease in humans (fig. 21.2). Other parabasalia live in the human mouth and small intestine with no apparent ill effects. Some species of parabasalia live in the intestines of termites and roaches, where they in turn harbor bacteria that digest cellulose. It is this bacterium-within-protista living organization that enables termites to "digest" wood (fig. 21.3).

FIGURE 21.2
Trichomonas vaginalis.
This protistan is a flagellated basal eukaryote that causes the sexually transmitted disease trichomoniasis.

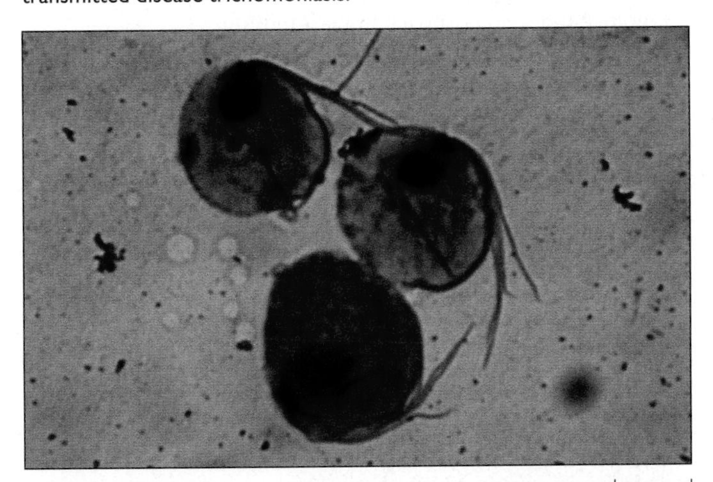

FIGURE 21.3
Life Inside a Termite.
A termite cannot digest the cellulose in its wood meal without symbiotic protista, such as this *Trichonympha,* which in turn harbor symbiotic bacteria. Exposing termites to high oxygen or high temperature kills their protist symbionts, and the insects soon die, with guts full of undigested wood.

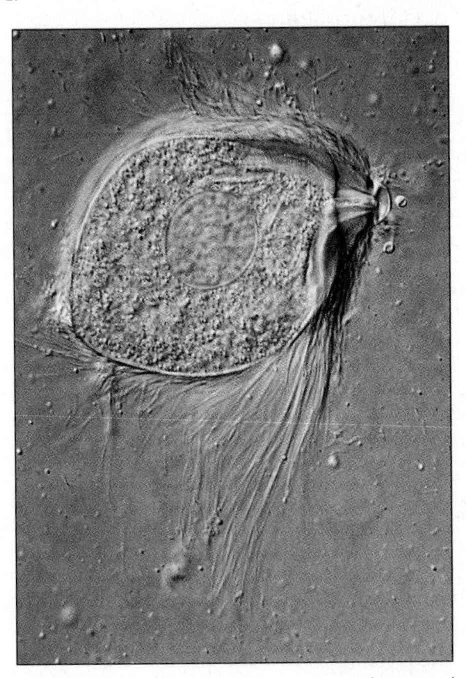

Another group of modern basal eukaryotes that might resemble the very early eukaryotes is the **Diplomonadida.** These organisms, which live in stagnant freshwater environments or in intestines, also have flagella and lack mitochondria. One species, *Giardia lamblia,* causes "hiker's diarrhea," also known as giardiasis (fig. 21.4). People may drink the nonmotile, thick-walled **cysts**

FIGURE 21.4
Giardia lamblia.
This basal eukaryote can cause giardiasis, also called hiker's diarrhea. The organism attaches to the lining of a person's small intestine (**A**) and leaves an impression when it detaches (**B**).

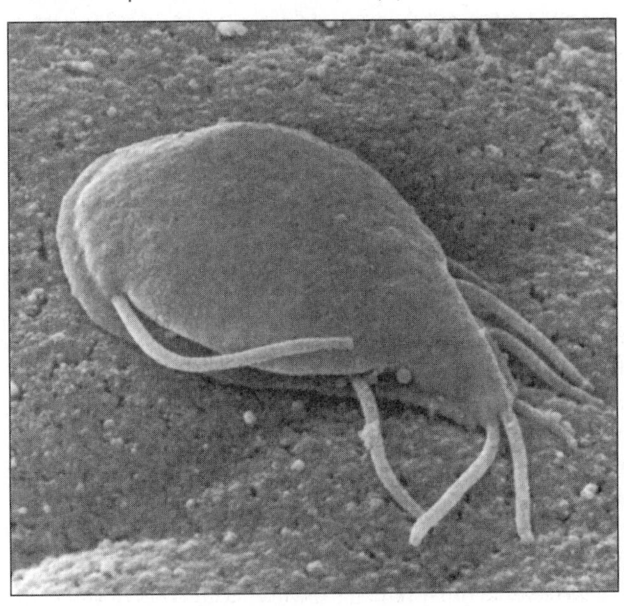

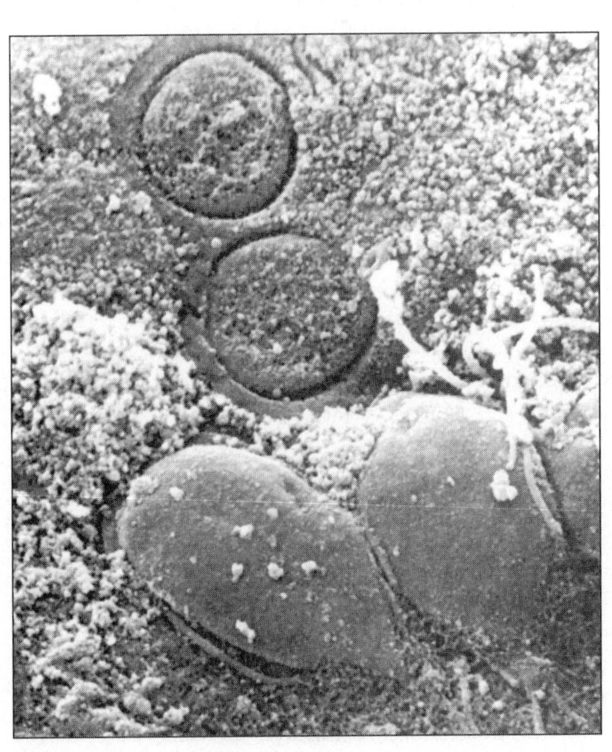

of the organism in unboiled stream water, which then become the feeding, active form in the small intestine. As *Giardia* cells divide, they impair the ability of the intestine to absorb nutrients, and diarrhea and cramping result. To make matters worse, the parasite produces a toxin that causes stomach pain, intestinal gas, weight loss, and lack of appetite.

The human digestive tract may also house *Entamoeba histolytica,* another basal eukaryote that lacks mitochondria. This organism, which causes amoebic dysentery, exists as cysts that form in fecal matter as it dries out, and as active feeding cells that emerge in the small intestine and latch onto the lining. Acute illness results in 15 to 20 bloody, runny bowel movements per day, and death may result from heart failure, a perforated intestine, or sheer exhaustion. Many people are infected painlessly, however, and pass hundreds of cysts in each bowel movement.

Basal Eukaryotes with Mitochondria

Human intestines house several types of **amoebae,** most of which do not make their presence known. Amoebae move in a characteristic way, allowing their cytoplasm to flow into projections called pseudopods ("false feet") (fig. 21.5). Certain stages in the life cycle of other protista are called amoeboid if they locomote in this manner. Most such organisms are included among the basal eukaryotes whose cells contain mitochondria. One example is the familiar *Amoeba proteus,* an important laboratory protist. This aquatic organism eats by engulfing other protista and bacteria within its pseudopods.

The slime molds are often grouped together as funguslike protista, but they actually form two separate groups whose relationship to each other remains unclear. Both types of slime molds can be found on forest floors and in deposits of cow manure. In addition, each type of organism exists as single, amoeboid cells and also as large masses that respond as if they are a single multicellular organism. The major difference between the two types is reflected in their names: the acellular and the cellular slime molds.

Acellular slime molds exist as enormous single cells containing thousands of nuclei. This multinucleated mass is called a **plasmodium** (not to be confused with genus *Plasmodium,* which includes the protista that cause malaria). When the habitat is moist, the plasmodia migrate in soil or along the forest floor, over leaves, debris, and rotting logs, consuming bacteria as they move (fig. 21.6). In times of drought or famine, they halt and form stalks topped with reproductive structures that ultimately release single amoeboid or flagellated cells. Two of these cells may fuse, forming a zygote that develops into a new plasmodium.

In contrast, individual cells of a **cellular slime mold** retain their separate identities throughout the life cycle. The cells exist as amoebae, feeding on bacteria. When food grows scarce, these amoebae secrete chemical attractants, which stimulate the neighboring amoebae to aggregate into a sluglike structure called a pseudoplasmodium (fig. 21.7). Cellular differentiation distinguishes cells at the front and back ends of the slug. The structure

FIGURE 21.5
"False Feet."
Pseudopods are temporary projections from the cells of amoeboid organisms. A pseudopod enables the organism to move or take in food. This organism, *Amoeba proteus,* is consuming *Euglena,* another protist (small green cell at right).

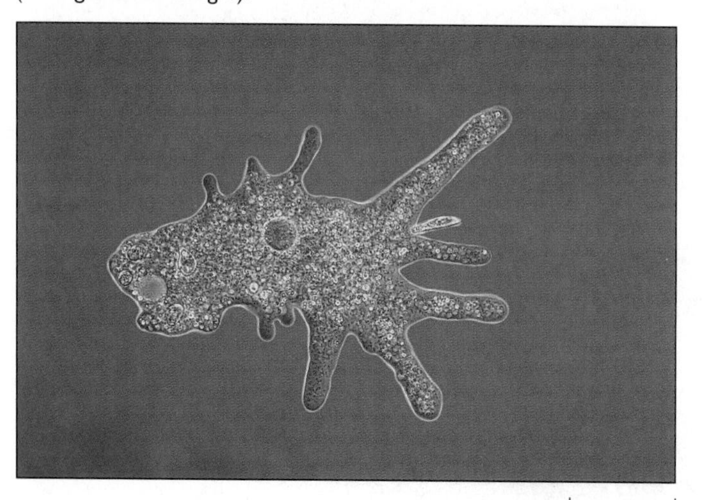

100 μm

FIGURE 21.6
Acellular Slime Mold.
Streaming masses of the acellular slime mold *Physarum* can move.

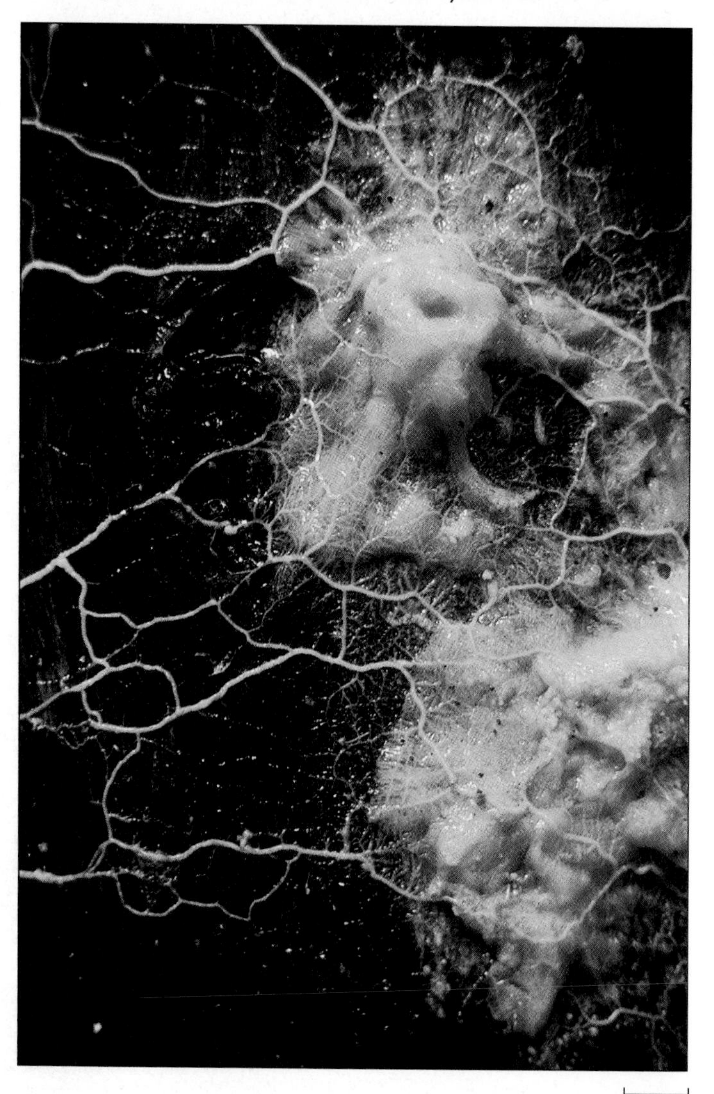

50 μm

moves toward light, stops, and forms a stalk topped by a fruiting body that contains spores. The cells of the stalk perish, but the spores survive to be spread to new habitats by water, soil animals, or birds. If environmental conditions are favorable, the spores form haploid amoebae. The cycle renews.

Another group of basal eukaryotes is the **Euglenida.** The euglenida include about 1,000 species, some of which have chloroplasts. Their elongated cells are covered with protein strips that form a rigid or elastic structure called a pellicle, a feature that is unique to this group. The pellicle may provide shape changes necessary for these organisms to creep along pond bottoms. The cells have one or two flagella, which the organisms use for swimming. *Euglena* (fig. 21.8) nicely illustrates one of the reasons that protist taxonomy has long been confusing—it has characteristics

FIGURE 21.7
Life Cycle of the Cellular Slime Mold *Dictyostelium discoideum*.
The amoeboid cells of this species are usually haploid. Only the asexual portion of the life cycle is shown, although a sexual phase may also occur.

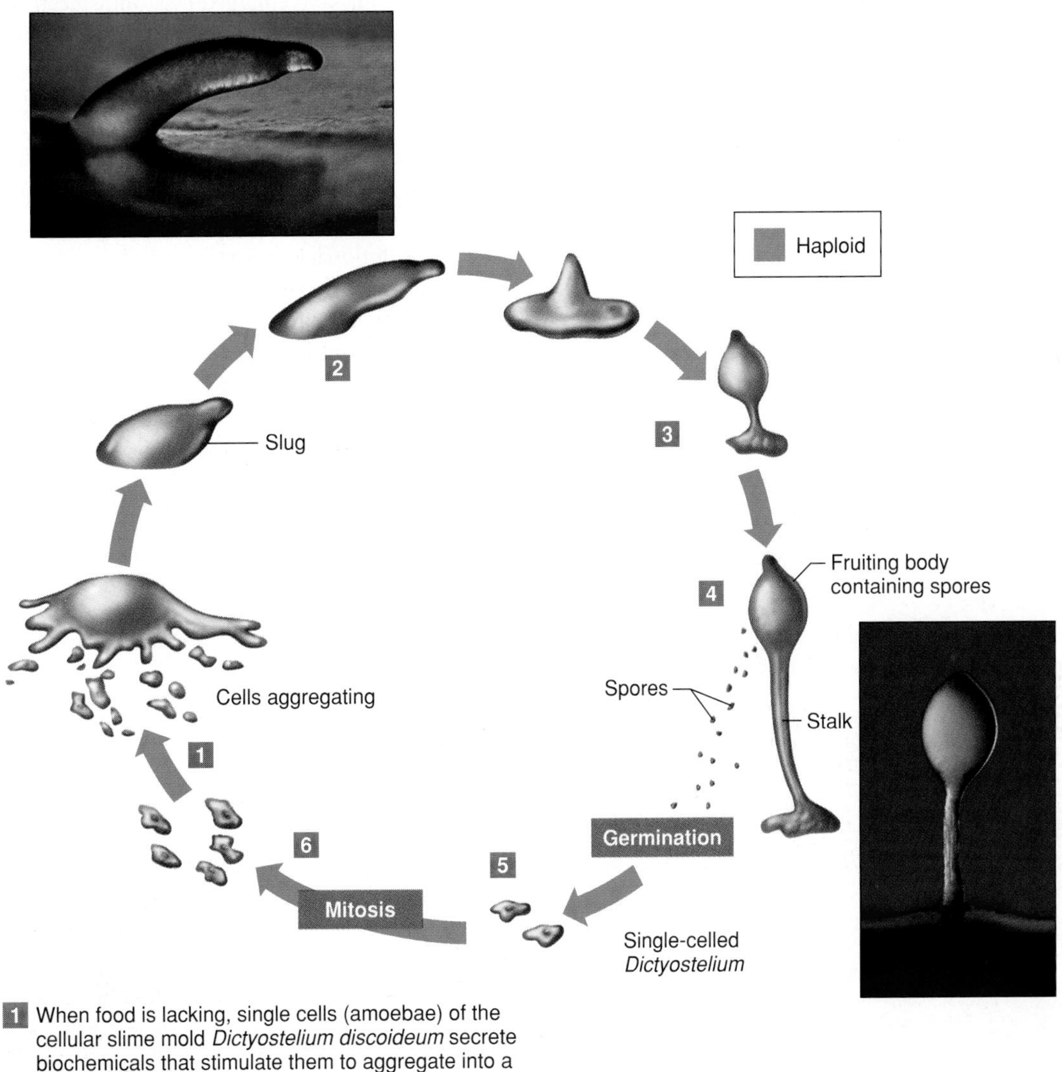

1 When food is lacking, single cells (amoebae) of the cellular slime mold *Dictyostelium discoideum* secrete biochemicals that stimulate them to aggregate into a multicellular slug.

2 The slug travels toward light.

3 Then the slug halts, forming a stalk with a fruiting body on top containing spores.

4 The stalk cells perish,
5 leaving spores that may be carried to new, more favorable habitats, where they germinate.

6 If food grows scarce, the switch to multicellularity occurs again.

FIGURE 21.8

Euglena.

The pond-dwelling *Euglena* has a flagellum and chloroplasts, but can also ingest food particles.

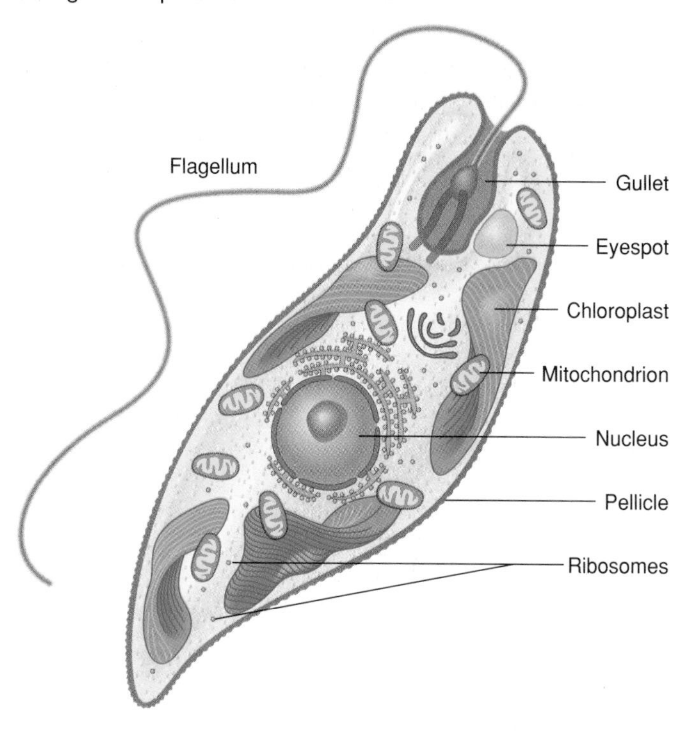

of both plants and animals. Like plants, these organisms photosynthesize. Like animals, they are motile and can ingest food.

Closely related to the euglenida are the **Kinetoplastida.** These flagellated basal eukaryotes are named for a structure that is unique to them, called a **kinetoplast.** The kinetoplast is a mass of mitochondrial DNA, usually near the flagellum; it is part of a single long mitochondrion that spans the length of the cell. Included among the kinetoplastida are the **trypanosomes,** which cause several diseases transmitted to humans by blood-seeking biting flies. For example, tsetse flies transmit African sleeping sickness, caused by *Trypanosoma brucei.* Another parasite, *Leishmania,* causes an infection called leishmaniasis. A fatal form of this illness causes fever, anemia, and enlarged liver and spleen. Two other forms affect the skin and the mucous membranes of the nose, mouth, and throat, resulting in extreme deformities (fig. 21.9).

21.1 MASTERING CONCEPTS

1. How are the protista simple yet complex?
2. Why are the jakobids and choanoflagellates of evolutionary import?
3. How might the basal eukaryotes provide information on the origin of mitochondria?
4. How are the acellular and cellular slime molds both different and similar?
5. What are some types of basal eukaryotes, both with and without mitochondria?

OLC

FIGURE 21.9

Trypanosomes.

These protista share a unique organelle, the kinetoplast (**A**). Trypanosomes (**B**) cause Chagas disease, African sleeping sickness, and leishmaniasis. The mucocutaneous form of leishmaniasis has deformed this man's face (**C**).

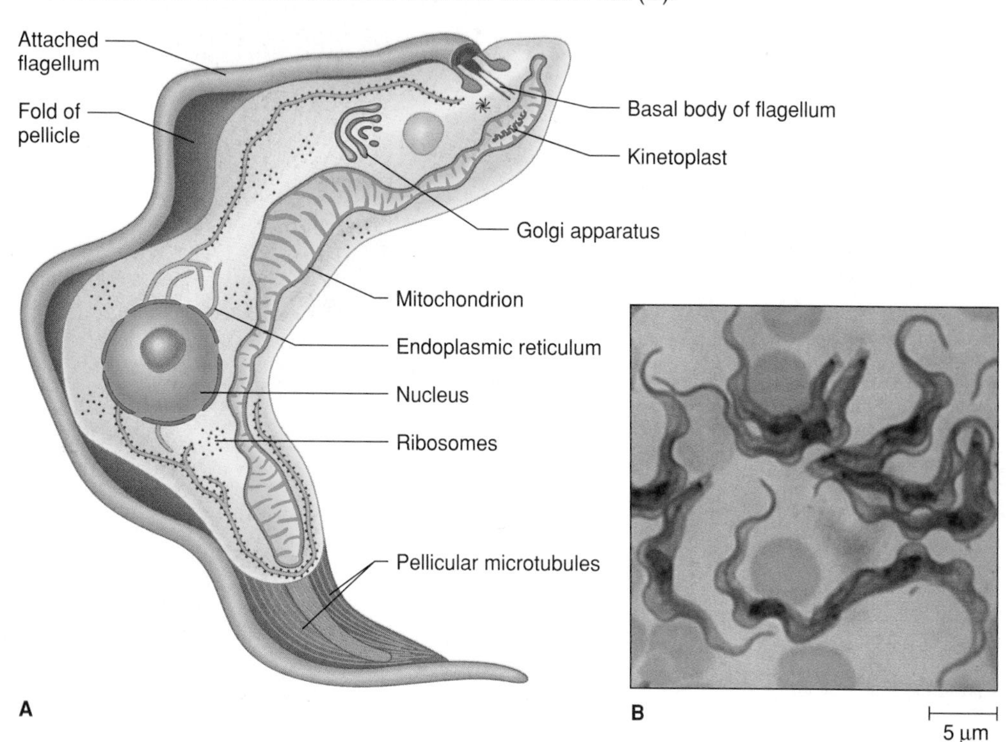

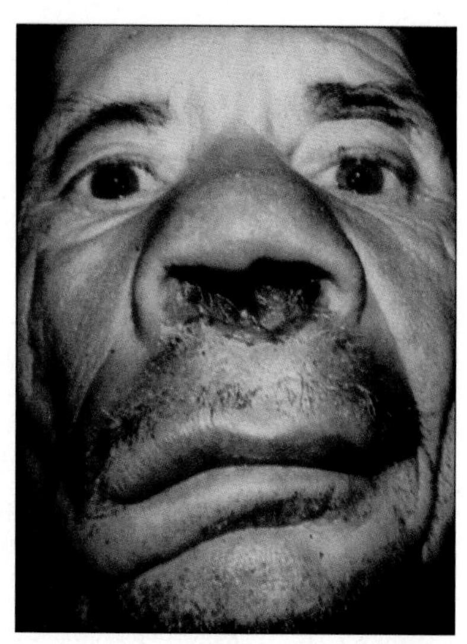

21.2 The Alveolates Include Dinoflagellates, Apicomplexa, and Ciliates

Membranous sacs beneath the cell membrane, tubular mitochondrial cristae, and DNA sequence similarities define the alveolates. The group includes the dinoflagellates, apicomplexa, and ciliates.

The **alveolates** are crown eukaryotes that have in common a series of flattened sacs, or alveoli, just beneath the cell membrane. Many species have three membrane layers—the cell membrane and two membranes formed by the alveoli. These organisms also possess mitochondrial cristae that are tubular. (The cristae in other protista may be disc-shaped or flat.) Molecular sequence evidence from rRNA and from mitochondrial DNA also places these protista together. The alveolates include three major groups (dinoflagellates, ciliates, and apicomplexa), and several groups with fewer members. Among these smaller groups is the foraminifera, whose durable shells have left microscopic fossils over millions of years. These tiny remnants can provide important clues to the age and environment in which a rock formed.

Dinoflagellates

The **dinoflagellates** take their name from the Greek *dinein,* which means to whirl. They have two flagella, of different lengths, one of which is longitudinal and the other transverse in orientation (fig. 21.10). The transverse flagellum beats in a way that moves the cell ahead with a whirling motion. Another characteristic of many dinoflagellates is that their cell walls are divided into 2 to 100 overlapping cellulose plates. The shape and pattern of these plates is the main basis for classification of the dinoflagellates.

Dinoflagellates are important ecologically because they help form plankton, the microscopic food that supports the ocean's vast food webs. About half of the 2,000 or so species are photosynthetic; the remainder obtain nourishment from other organisms, either as predators or parasites. Some species are bioluminescent; others are symbionts of giant clams, jellyfish, sea anemones, mollusks, or corals.

Many dinoflagellates can enter a dormant cyst stage, with a cell wall and lowered metabolic demands. Cysts enable the dinoflagellate to survive a temporarily harsh environment, or provide a means of transport from one host to another. When environmental conditions improve, food becomes available, or an appropriate host arrives, the dinoflagellate bursts from its protective cyst. Sometimes, populations of the dinoflagellates *Gymnodinium* and *Gonyaulax* "bloom," with many cysts bursting at once, producing a "red tide" effect (fig. 21.10C). The dinoflagellates also may release a neurotoxin, which becomes concentrated as fish eat it and pass it up food chains (see fig. 43.18). In humans, the neurotoxin inhibits sodium ion transport in the nerve cells and causes the numb mouth, lips, face, and limbs that result from paralytic shellfish poisoning.

FIGURE 21.10
Dinoflagellate Features.
Gonyaulax tamarensis exhibits classic dinoflagellate structure—note the characteristic transverse and longitudinal flagella (**A** and **B**). Blooming populations of dinoflagellates are responsible for red tides (**C**).

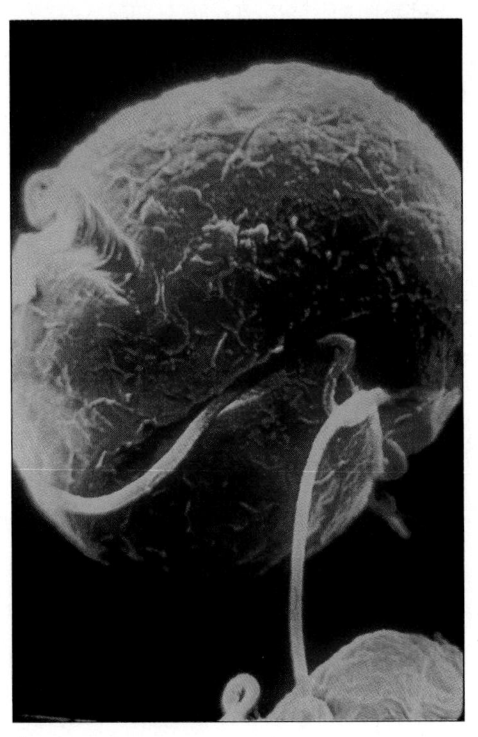

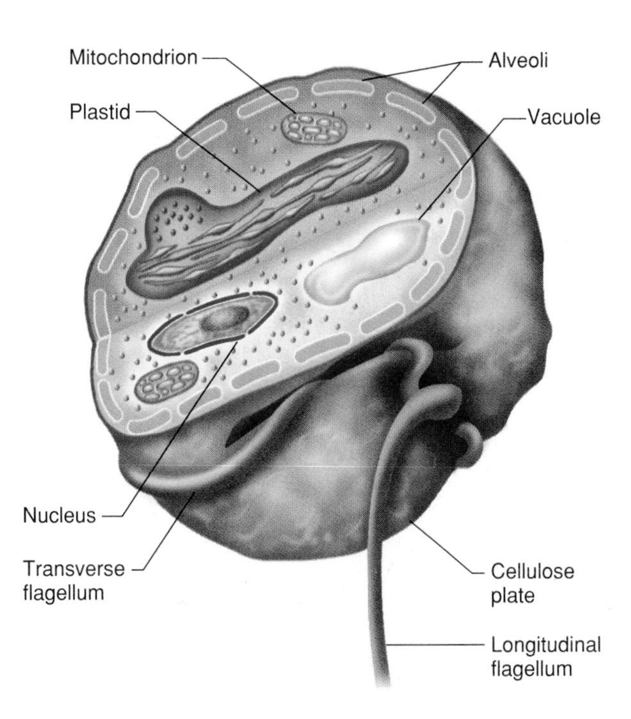

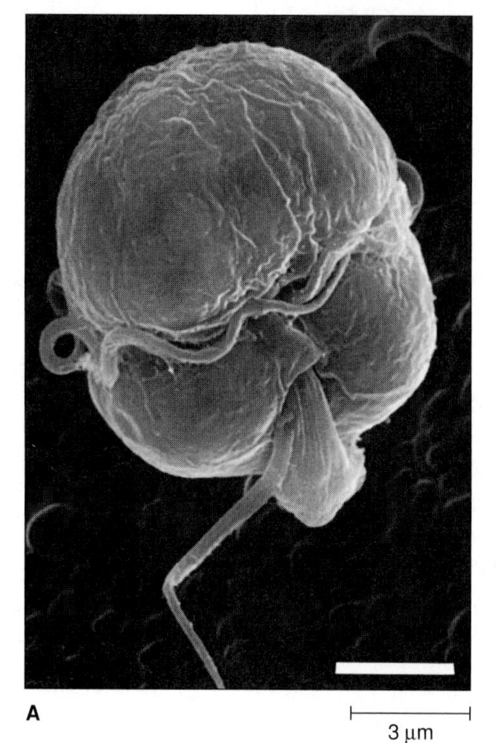

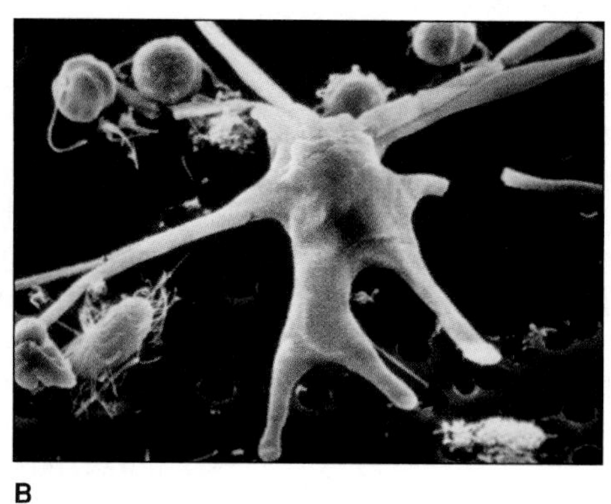

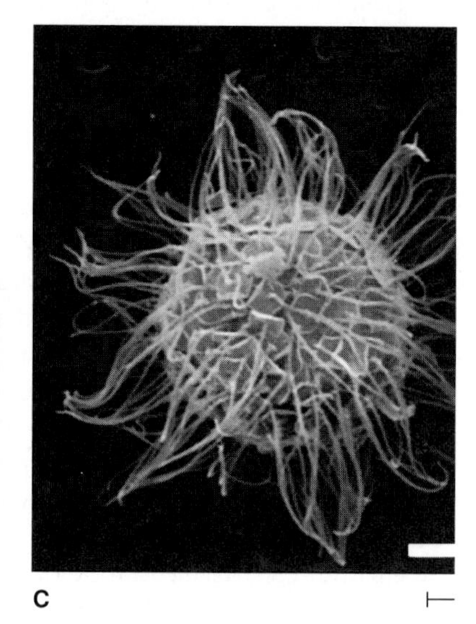

FIGURE 21.11
Forms of *Pfiesteria*.
Pfiesteria has at least 24 stages to its life cycle, which, when drawn out, looks like a complicated road map for a large city! This figure shows (**A**) the free-swimming form that feeds on fish, an amoeboid cell (**B**), and a cyst (**C**).

Another dinoflagellate that produces a toxin and can make people ill is *Pfiesteria piscicida,* discovered in 1988 in the waters off the coast of North Carolina. *Pfiesteria* has 24 stages to its life cycle, but assumes three basic forms—flagellated cells, amoeboid cells, and cysts (fig. 21.11). Researchers suspect that these organisms normally exist in nontoxic forms in estuaries, but that excrement and secretions from schooling fish such as Atlantic menhaden trigger a switch to the toxin-producing forms. Just 300 *Pfiesteria* cells per milliliter of water (about 20 drops) is enough to cause bloody sores on fish or kill them. In some areas of extensive "fish kills," the density of dinoflagellates reaches 250,000 cells per milliliter! The toxin release takes only a few hours, and then the poison breaks down rapidly, but the fish die over the next several days. The ephemeral nature of the toxin has made it very difficult to study, as Health 21.1 describes.

Apicomplexa

The 4,000 species of **apicomplexa** are obligate parasites of animals. Among other features, these organisms all possess a group of organelles collectively called an **apical complex** (fig. 21.12). The apical complex consists of characteristic organelles which apparently help the organism attach to and invade the cells of its host.

The apicomplexa exhibit a great range of life cycles, many of which are quite complicated and involve multiple hosts. Reproduction can be asexual and sexual, but only some of the stages may occur in a particular species. The form of the organism that enters (infects) a host is called a sporozoite. Within the host, it divides by mitosis repeatedly, giving rise to daughter cells called merozoites. Some of the merozoites differentiate into sexual cells that divide, producing gametes that join to form zygotes, which undergo yet another series of divisions to re-form sporozoites. Depending on the species, the parasites may spread when

FIGURE 21.12
Apical Complex.
A longitudinal section through the invasive stage of an apicomplexan (*Toxoplasma gondii*) illustrates various organelles comprising the characteristic apical complex (AC). N = Nucleus. Bar is 0.5 μm (Supplied by DJP Ferguson, Oxford University, with permission from *The Biologist*).

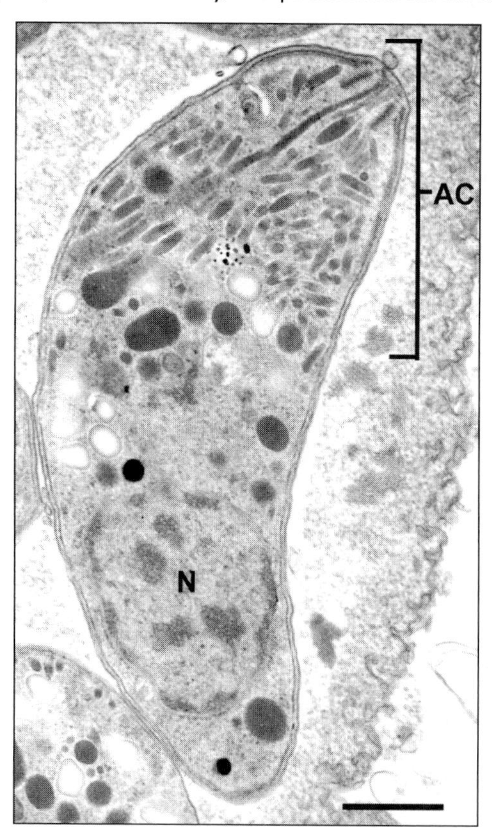

A "New Clinical Syndrome" Linked to *Pfiesteria*

In the fall of 1996, people who fished along the rivers and estuaries of the eastern shore of the Chesapeake Bay began to notice sores on many fish, rotting ulcers that looked like they had been punched out of the animals. The fish swam erratically, then died. The problem worsened in the following spring and summer, and by August, several huge fish kills had occurred (fig. 21.A). Light and electron microscopy of water samples revealed *Pfiesteria piscicida* and other related but unknown species, collectively called "toxic *Pfiesteria* complex." The fish kills, and the presence of the dinoflagellates, were eerily similar to outbreaks in North Carolina in 1992 and 1993 (fig. 21.B).

Also in late 1996 and through 1997, the Chesapeake fishers developed symptoms, which intensified in parallel to the worsening fish kills. They experienced fatigue, headache, breathing difficulty, diarrhea, weight loss, and rashes. Most disturbing were the neurological effects. People with high exposure to the water became confused and unable to concentrate. Memory impairment caused some people to forget their destinations while driving, that they had mailed packages or done other chores, and to place certain critical supplies on their boats. Some people even forgot their names and phone numbers.

A research team from the University of Maryland and Johns Hopkins University investigated and described this apparent new illness by comparing 24 people exposed to the waters during the time of the fish illness, with eight watermen who worked on the ocean side of the bay, where *Pfiesteria* isn't found. The participants answered a battery of questions, and underwent many medical tests. The large number of tests were necessary to demonstrate an effect, because *Pfiesteria* was in the news at the time, and people tended to exaggerate their symptoms.

The link of the "new clinical syndrome" to *Pfiesteria* is still considered a hypothesis, until researchers can isolate and describe the toxin. But evidence for the link is compelling:

1. Symptoms in humans began when the fish began to develop sores.
2. Symptoms in humans vanish in 3 to 6 months after stopping contact with infested water.
3. Toxic *Pfiesteria* complex was identified in waters from the estuaries where fish were dying, at that time.
4. Lesions in fish could be induced in the lab by exposure to infested water.
5. Laboratory rats experienced learning difficulties after exposure to *Pfiesteria.*
6. A dose-response effect exists. That is, people with more intense or prolonged exposures have worse symptoms.

Meanwhile, researchers at North Carolina State University, led by Joann Burkholder, who discovered *Pfiesteria,* are strengthening the link between illness and the dinoflagellate by using a DNA probe to rule out infection by other organisms. (The illnesses are actually not infections at all, but poisonings.) They are also carefully monitoring the affected environments to determine the precise conditions that precipitate outbreaks. Burkholder and some of her coworkers have been sickened by exposure to the toxin.

FIGURE 21.A
***Pfiesteria* Fish Lesions.**
Pfiesteria causes gaping sores on fish.

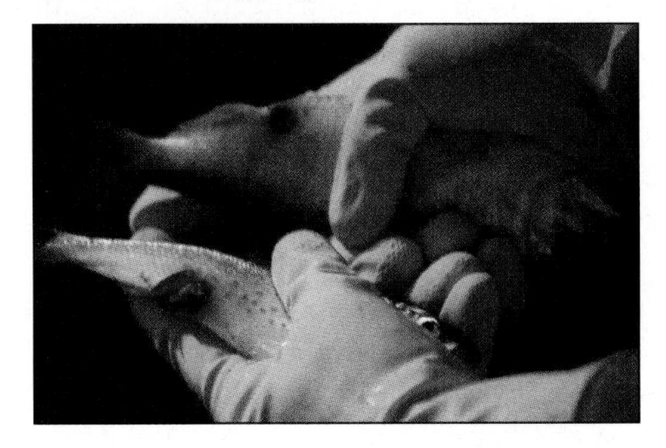

FIGURE 21.B
Infested Waters.
Fish kills caused by *Pfiesteria* and related dinoflagellates have struck several mid-Atlantic coastal states.

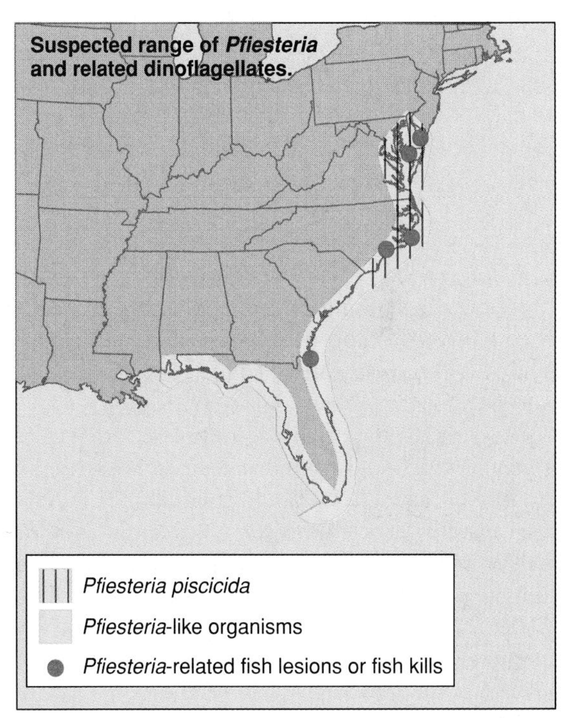

Source: Data from Environmental Protection Agency.

FIGURE 21.13
***Plasmodium* Causes Malaria.**
A mosquito ingests *Plasmodium* gametocytes (sexual forms) that join and form sporozoites, which the insect transmits while feeding. The sporozoites migrate to the liver, where they divide to form merozoites, which may continue to cycle in red blood cells for some time. At some point, gametocytes form from merozoites, and mosquitoes ingest them, starting the cycle anew.

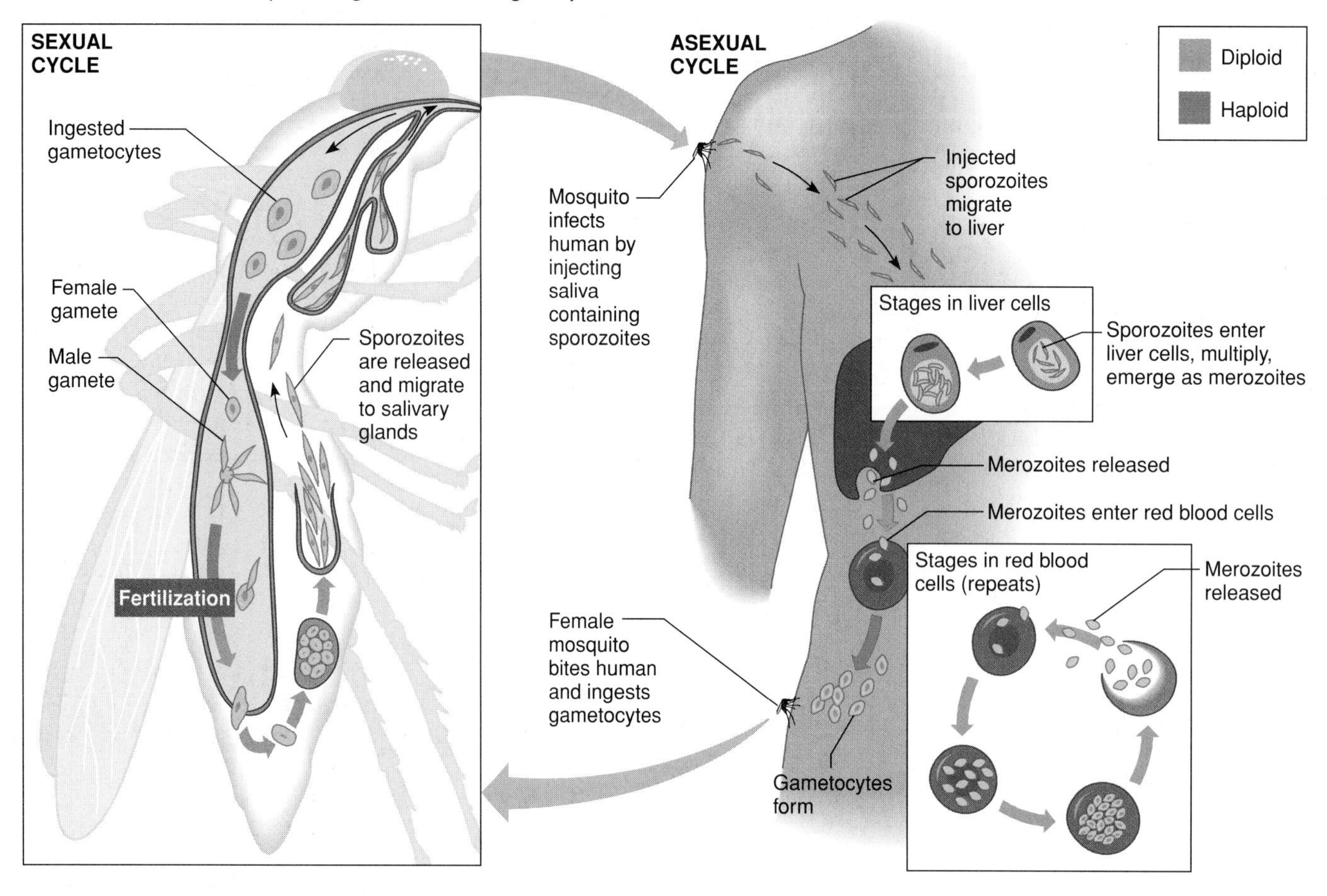

a new host ingests infective cells liberated by sick animals, or when a blood-sucking arthropod transmits them to new hosts. Figure 21.13 shows a simplified life cycle for *Plasmodium,* which causes malaria, the subject of Health 21.2.

Two other important apicomplexa produce mild illness in people with healthy immune systems, but severe symptoms in people who have AIDS. Cryptosporidiosis, a waterborne infection caused by *Cryptosporidium parvum,* causes mild diarrhea in healthy individuals, but severe diarrhea, weight loss, fever, abdominal pain and sometimes respiratory distress in people with AIDS. *Toxoplasma gondii* causes toxoplasmosis, which usually produces symptoms that are so mild (fatigue) that a person can be infected and not know it. In people with AIDS, however, the parasite quickly spreads to the brain, lungs, heart, liver, and eyes. Healthy pregnant women can become unknowingly infected (via cat feces, for example) and pass on the disease to the fetus, where it can result in low birth weight and other serious problems.

Ciliates

As their name implies, the **ciliates** are characterized by their abundant cilia, which are hairlike structures that often number in the thousands (fig. 21.14). (Recall from chapter 4 that cilia are built of microtubules; see figs. 4.15 and 4.16.) Some ciliates are completely covered with cilia, whereas others have them in bands or tufts. When the cilia beat, they propel the organism, and can also sweep food (bacteria, algae, and other ciliates) into the ciliate's gullet. Structures called food vacuoles then surround and transport the captured meal inside the cell. A permanent anal pore releases the wastes. Another feature of the ciliates is a contractile vacuole, which pumps excess water out of the cell (see fig. 4.7). Many ciliates also have two types of nuclei, a small haploid micronucleus and a larger, diploid macronucleus.

The 8,000 or so species of ciliates live in diverse habitats. Some ciliates, for example, live in the stomachs of ruminant mammals, where they house bacteria that break down cellulose in the grasses that the animals eat. Others attach to the bodies of aquatic invertebrates, or are free-living in moist soil, marine environments, or fresh water. A drop of pond water, for example, is likely to reveal many ciliates. With luck, an observer might find a scintillating scene—the voracious *Didinium nasutum* attacking and eating the larger *Paramecium* (fig. 21.15). *Didinium* is in perpetual motion, but it is only when it smacks into a *Paramecium* that it releases threadlike structures called trichocysts that capture it and reel it in. *Didinium* typically attaches to its prey in the center, then maneuvers the larger protist into its gullet.

Malaria

Malaria is a serious, even deadly, infection that has left its mark on human history. Today, in many tropical parts of the world, it continues to threaten human health. Worldwide, more than 300 million people a year suffer from malaria. In tropical Africa, more than a million children die each year of the infection.

Symptoms and Course of Infection

Malaria starts with chills and violent trembling and then progresses to high fever and delirium. The person sweats profusely, is completely exhausted, and has a dangerously enlarged spleen. The disease strikes in a relentless cycle, with symptoms returning every 2 to 4 days. The waxing and waning symptoms reflect the parasite's life cycle. The frequent crises cause anemia as the person's red blood cell supplies plummet and strain the liver and spleen, both of which process broken-down red blood cells.

A cycle of malaria begins when an infected female mosquito of any of 60 *Anopheles* species feeds on a human (see fig. 21.13). The insect's saliva contains an anticlotting agent, as well as sporozoites, which are small, haploid cells of *Plasmodium falciparum, P. vivax, P. malariae,* or *P. ovale*. The sporozoites enter the human host's liver cells, where they multiply rapidly, eventually emerging as merozoites. Some merozoites reinfect liver cells, and others infect blood cells. For the next 2 to 3 days, the merozoites enlarge and divide, finally bursting from the red blood cells. The subsequent infection of other red blood cells, and the release of merozoites into the bloodstream 48 to 72 hours later, is often synchronized throughout the victim's body. This causes the recurrent chills and fever of malaria.

Meanwhile, a few *Plasmodium* cells specialize into gametocytes (sexual forms). When mosquitoes ingest the gametocytes from an infected person's blood, the gametocytes unite in the insect's stomach. After several additional steps, sporozoites form. These move to the mosquito's salivary glands, ready to enter a new host when the insect seeks its next blood meal.

Of the *Plasmodium* species that can cause malaria, *P. falciparum* is the most widespread and virulent, and causes most fatal cases. Recovery is more likely when other species cause the disease. *Plasmodium vivax* and *P. ovale,* however, can remain dormant in the liver for months or years. A person infected with these species may feel well for months, then become ill again.

The type of vegetation in an area determines whether the mosquito and its parasite will thrive. Agriculture often creates conditions that encourage malaria by replacing dense forests with damp rice fields, a haven for moisture-loving mosquitoes. One reason that malaria is relatively uncommon in the United States is that in many places warm weather doesn't last long enough for the mosquitoes to survive and perpetuate the infective cycle. The disease was, however, once widespread in the southern United States.

History

For centuries, people have recognized the link between swamps and recurring malaria. The name of the disease comes from seventeenth-century Italy, where people living near smelly swamps outside of Rome developed mal'aria—"bad air." As long ago as the fourth century B.C., however, Greeks noted that people living near swamps had bouts of fever and enlarged spleens. It took centuries to identify the apicomplexan that causes malaria.

In 1880, French army surgeon Charles Louis Alphonse Laveran saw the parasites in a sick person's blood (fig. 21.C). By the end of the century, researchers had discovered the role the *Anopheles* mosquito plays in transmitting the disease. In 1902, British army surgeon Sir Ronald Ross, using birds as models, deciphered the entire complex cycle of malaria.

FIGURE 21.C
Emerging Parasites.
Plasmodium cells burst from these red blood cells in a laboratory culture dish.

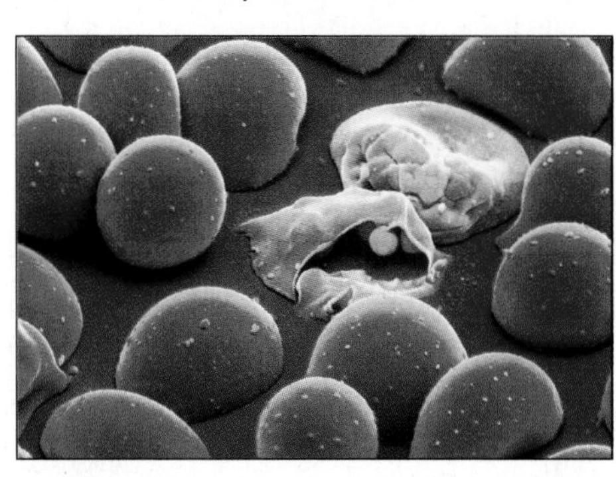

Eradication

People were able to treat malaria long before they understood the infective cycle. In the sixteenth century, Peruvian natives gave Jesuit missionaries on their way to Europe their secret malaria remedy—the powdered bark of the cinchona tree. It was not until 1834, though, that French chemist Pierre Joseph Pelletier extracted the active ingredient from cinchona bark: quinine, which is still used in some forms, in addition to many other drugs developed to keep pace with *Plasmodium*'s evolving resistance to various drugs. In 1955, the World Health Organization announced an eradication campaign against malaria; by 1976, it admitted failure. Malaria had actually spread through developing nations as people cleared land for farming.

Although we have moderately effective ways to prevent and treat malaria, stemming the illness is very challenging, for biological as well as sociological reasons. Not only does *Plasmodium* continue to develop drug resistance, but nations troubled by poverty and civil unrest struggle to distribute drugs to prevent or treat malaria. Still, certain measures can lower the risk of contracting malaria:

- a drug called mefloquine can prevent malaria when administered weekly for 2 years and can kill parasites within 48 hours;
- bednets soaked in insect repellent keep mosquitoes away;
- screens on windows and doors help keep mosquitoes out;
- wearing long pants and long sleeves, especially during the evening, helps prevent mosquito bites.

FIGURE 21.14

Anatomy of *Paramecium*.

Structures in *Paramecium* include a micronucleus and a macronucleus; trichocysts associated with cilia, which may help the organism to feed or defend itself; an oral groove with a "mouth" into which cilia wave bacteria for food; food vacuoles for storage; contractile vacuoles to maintain solute concentrations; and an anal pore, which releases wastes.

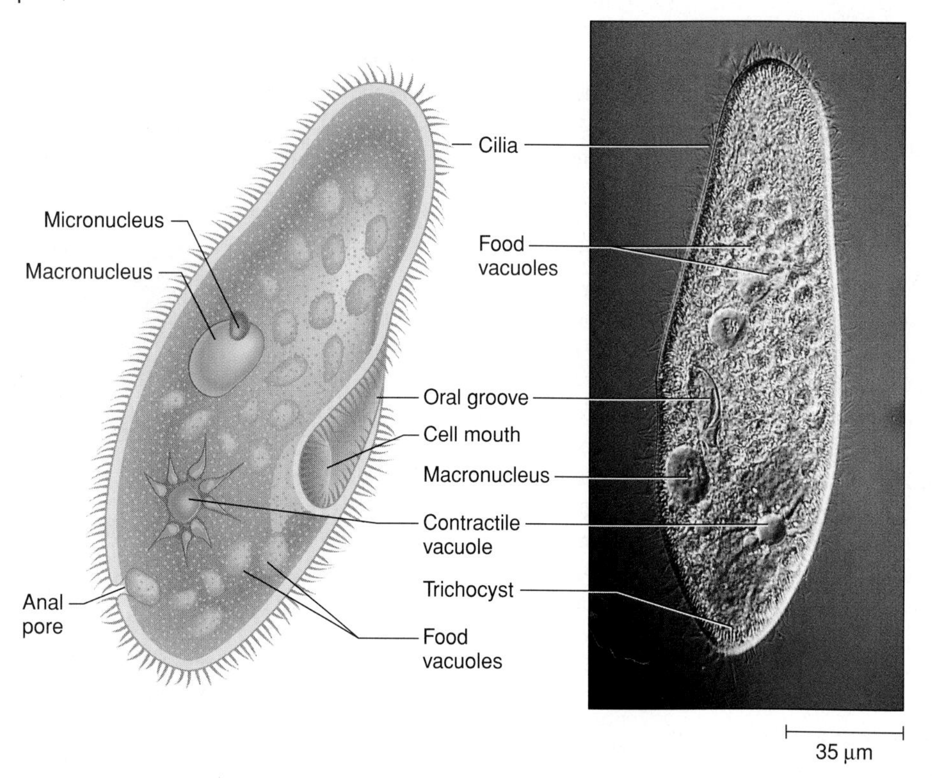

FIGURE 21.15

Protistan Predation.

Didinium consumes a *Paramecium*.

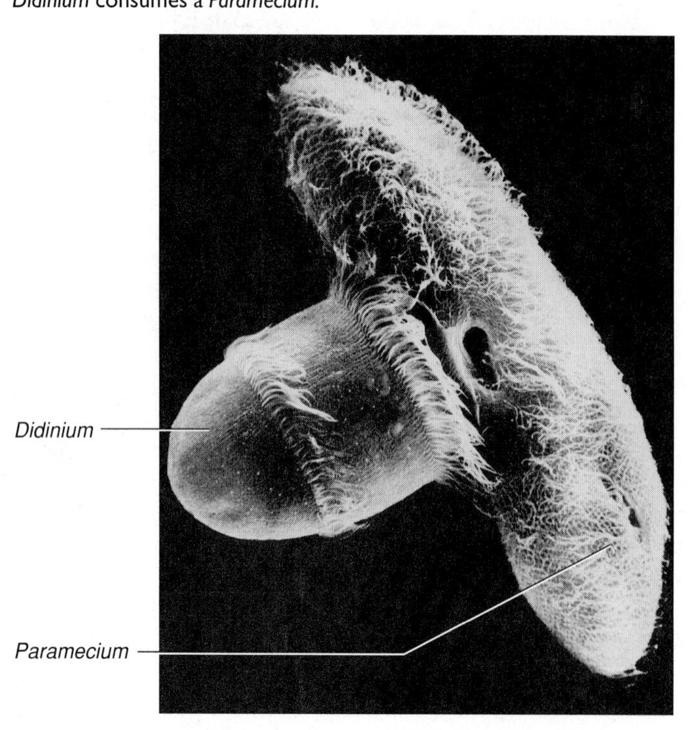

21.2 MASTERING CONCEPTS

1. What are the general characteristics of alveolates?
2. Describe a dinoflagellate.
3. What are the distinguishing characteristics of apicomplexa?
4. How do ciliates move and eat?

OLC

21.3 The Stramenopiles Include Water Molds, Diatoms, and Some Algae

Three-part tubular hairs, pigments not found in plants, a storage carbohydrate called laminarin, and DNA sequence similarities characterize the stramenopiles. The group includes water molds, diatoms, brown algae, and golden algae.

Like the alveolates, the **stramenopiles** are another grouping of protista that unites organisms once thought to be very dissimilar—but molecular evidence says otherwise. Included are the filamentous, heterotrophic water molds, which were once classified with the fungi; diatoms; and brown and golden algae, which were once thought to be plants. In size, its members range from the microscopic diatoms to giant kelp forests that hug the shorelines in temperate areas. Stramenopiles are also called Chromista, in recognition of the pigments in many member species.

The diversity among the stramenopiles, however, belies some important shared features. At some point in their life cycles, most have cells with two flagella, one of which is covered with tubular hairs. The photosynthetic stramenopiles have chlorophyll *c* and other pigments not found in plants. Finally, instead of the starch found in plants, stramenopiles use a storage carbohydrate called laminarin.

Water Molds

Water molds, or oomycetes, absorb nutrients from their surroundings. This mode of nutrition, along with their filamentous growth form, is why the water molds were once classed with the fungi. In addition, like a group of fungi called chytrids, water molds produce flagellated zoospores which aid their dispersal in moist environments. They are unlike fungi, however, in many respects. Water molds are diploid, for example, whereas fungi are often haploid. Fungi have cell walls containing chitin, but water mold cell walls are composed of cellulose. The storage carbohydrates of water molds are also unlike those of fungi.

The water molds live in fresh water and moist soil, and on living plants and animals. The best-known water molds are those that have thwarted agriculture, causing such diseases as white rust, which affects radishes, and downy mildew, which affects grapes and lettuce. In the 1870s, downy mildew of grapes, caused by *Plasmopara viticola,* nearly destroyed the French wine industry. The water mold *Phytophthora infestans,* which means "plant destroyer," causes late blight of potato (fig. 21.16) and also affects tomatoes. This disease was responsible for the 1845–47 potato famine in Ireland, when more than a million people starved. The Irish potato famine followed several rainy seasons, which fostered the rapid and devastating spread of the mold.

FIGURE 21.16
Plant Killer.
(**A**) *Phytophthora infestans* is a water mold that has biflagellated zoospores. (**B**) It causes "late blight" of potatoes, and was responsible for the Irish potato famine in the late 1800s.

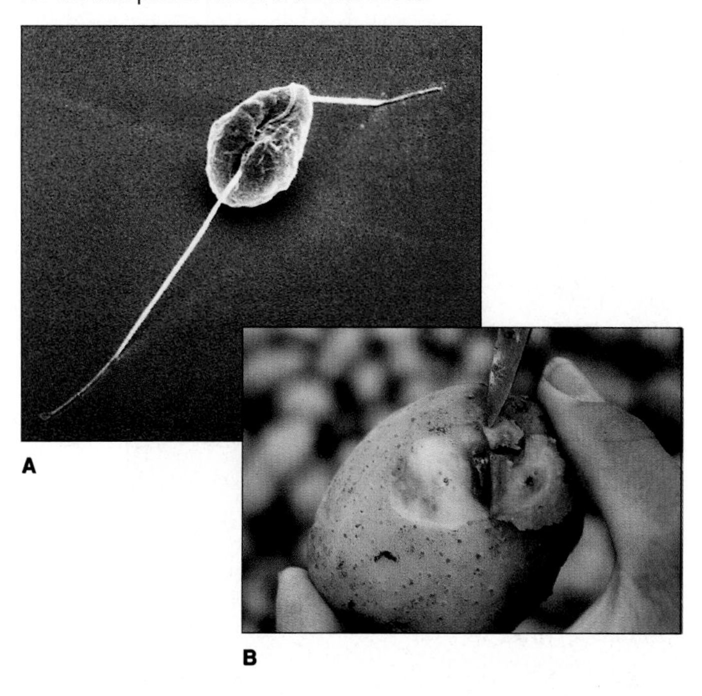

Diatoms

Diatoms have two-part silica walls called frustules that give them a variety of intricate shapes (fig. 21.17). The abrasiveness of frustules makes diatoms useful ingredients in swimming pool filters, polishes, toothpaste, and many other products. Diatoms also impart the distinct quality of paint used in reflective roadway signs.

Diatoms are an abundant form of phytoplankton (microscopic, free-floating photosynthetic organisms): A liter of seawater may contain a million diatoms. These organisms undergo a form of asexual reproduction that decreases the average cell size at each generation. When cells reach a certain minimum average size, a sexual form of reproduction occurs and restores the original cell size.

FIGURE 21.17
Diatoms.
The silica walls of these photosynthetic protista exhibit a dazzling variety of forms.

Brown and Golden Algae

The **brown algae** (Phaeophyta) owe their distinctive color to a photosynthetic accessory pigment called fucoxanthin. Brown algae are multicellular seaweeds that live along shorelines in cool climates all over the world. They are also found in freshwater habitats. These are the most complex algae and the largest of the stramenopiles, some reaching 30 meters in length. Some brown algae

FIGURE 21.18
A Giant Kelp.
A holdfast organ (not shown) anchors this brown alga to surfaces when tides and currents are strong. The bladder is a gas-filled float that enables the upper parts of the organism to rise above the holdfast.

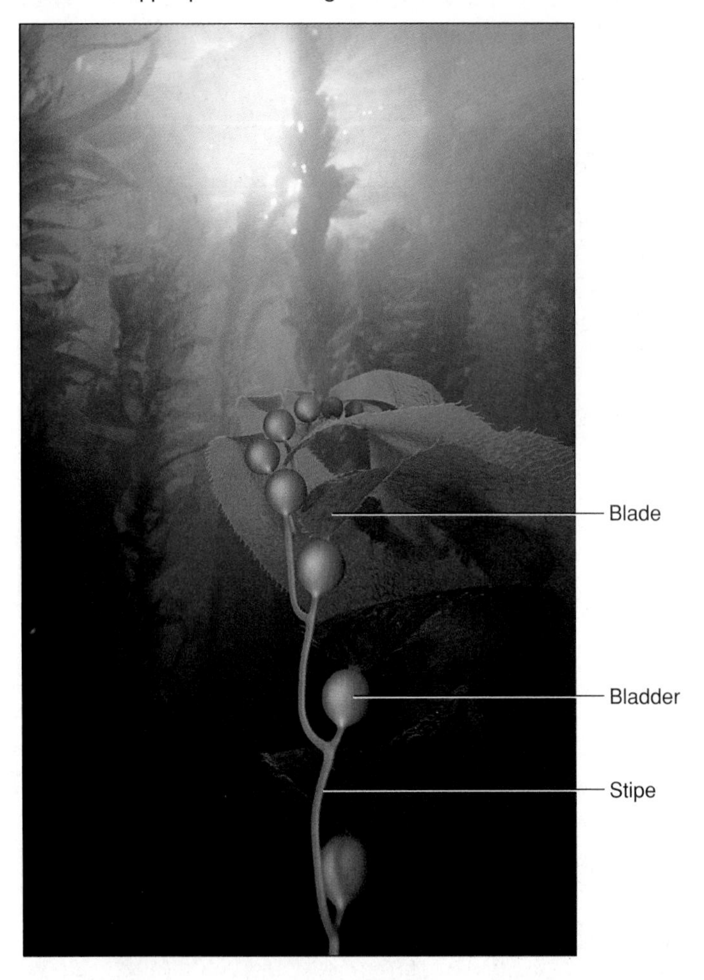

have a distinctive body form that consists of a holdfast organ, a stemlike region called a stipe, a balloonlike area called a bladder, and a blade emanating from each bladder (fig. 21.18).

The vast majority of the **golden algae** (Chrysophyta) are photosynthetic, and they are a significant source of food for zooplankton in freshwater and marine ecosystems. When light or nutrient supplies dwindle, however, many golden algae can consume bacteria or diatoms—they are facultative heterotrophs. Although most are unicellular, filamentous and colonial forms do exist. *Dinobryon* is an example of a freshwater genus of golden algae, in which individual vase-shaped cells stack end-to-end to produce branched or unbranched chains. The evolutionary relationships among the golden algae, and between the golden and brown algae, remain unclear.

21.3 MASTERING CONCEPTS

1. What are the generalized characteristics of the stramenopiles?
2. Why were water molds once classified as fungi?
3. How are diatoms different from brown and golden algae?

21.4 The Red Algae Contain Unique Photosynthetic Pigments

The red algae are genetically distinct from other types of algae. They have a modified form of starch, cellulose cell walls, and chlorophyll *a*. They also produce unique pigments that enable them to photosynthesize at great depths.

The **red algae** (Rhodophyta) form another group within the crown eukaryotes, and are genetically distinct from other types of algae. About 3,900 species of red algae, most of which are marine organisms, belong to this group. The red algae are typically multicellular with branches up to a meter long (fig. 21.19). They are somewhat similar to green plants in that they store carbohydrates as a modified form of starch, have cell walls containing cellulose, and produce chlorophyll *a*. Unlike plants, however, red algae also have photosynthetic pigments called phycobilins, which include a red form (phycoerythrin) and a blue form (phycocyanin). These pigments allow a seaweed far below the ocean's surface to absorb wavelengths of light that chlorophyll in plants cannot capture. These accessory pigments pass solar energy to chlorophyll *a* molecules in the alga's chloroplasts.

The slippery feel of red algae comes from a mucilaginous matrix that forms the outer layer of the cell walls. One polysaccharide in the matrix of some species is agar, a gelatinous substance used as a culture medium for microorganisms, as an inert ingredient in medications, to gel canned meats, and to thicken ice cream and yogurt. Another red algal product is the polysaccharide carageenan, which is used to emulsify fats in chocolate bars and to stabilize paints, cosmetics, and creamy foods. The Japanese have cultivated and eaten nori, a red alga, for more than 300 years.

21.4 MASTERING CONCEPTS

1. How are red algae similar to and different from green plants?
2. What useful products are made from red algae?

FIGURE 21.19
A Red Alga, *Bossiella*.
The pigment phycoerythrin provides the characteristic color of a red alga.

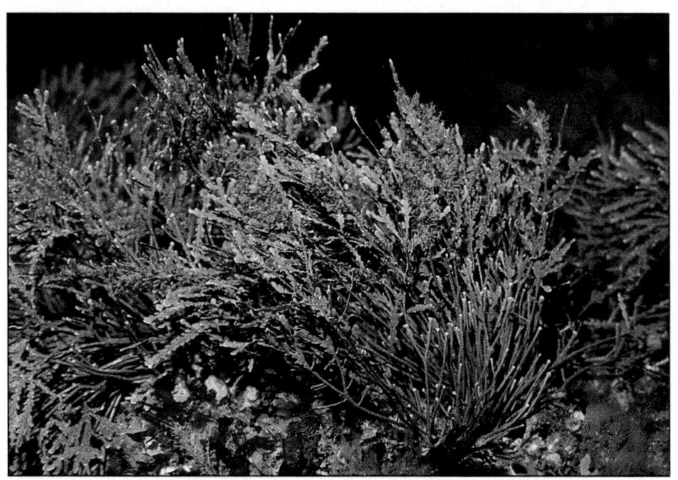

21.5 The Green Algae

The green algae have pigments and alternation of generations like plants, yet differ in their aquatic habitats. Algae in general span several taxonomic groups, but the green algae are the most closely related to plants.

"Algae" is a general descriptive term that refers to relatively simple organisms that live in water and manufacture their own food (aquatic autotrophs). Algae appear in many classifications that are based on nucleic acid sequences. Even the cyanobacteria, which are photosynthetic bacteria, were once called "blue-green algae."

Algae are not plants. Although plants are also autotrophs, the vast majority of them are terrestrial. However, the **green algae** (Chlorophyta) are the most plantlike of the algae because they use chlorophyll *a* and *b*, use starch as a storage carbohydrate, and have cell walls containing cellulose. Like plants, many green algae also have **alternation of generations,** with alternating haploid (gametophyte) and diploid (sporophyte) phases in their life cycles (fig. 21.20). Unlike plants, however, green algae may be unicellular; plants are multicellular. Also, although algae may have rootlike and stemlike structures, the cells of multicellular algae are far less specialized than those of higher plants. Another feature that sets green algae apart from higher plants are pyrenoids, which are protein-rich regions in the chloroplasts that are centers for starch formation.

Green algae are distinguished from each other by body form, the number and arrangement of their flagella, details of cell division, and habitat. The 7,500 or so species live in marine as well as freshwater environments, even in snow. They also live on and in soil and forest plants, coat rocks near waterfalls, and dwell within polar bear hairs!

The body forms of green algae are very diverse (fig. 21.21). One well-studied green alga is *Chlamydomonas,* which tinges

FIGURE 21.20

Reproduction of a Green Alga, the Sea Lettuce *Ulva*.

(*1*) A diploid sporophyte undergoes meiosis (*2*) to yield haploid reproductive male and female spores. The spores divide mitotically to form male and female gametophytes (*3*), which look like each other (and also like the sporophyte). The gametophytes produce male and female gametes (*4*), which fuse (*5*) to yield a diploid zygote (*6*) that grows by mitosis and matures to form a sporophyte. The cycle begins anew.

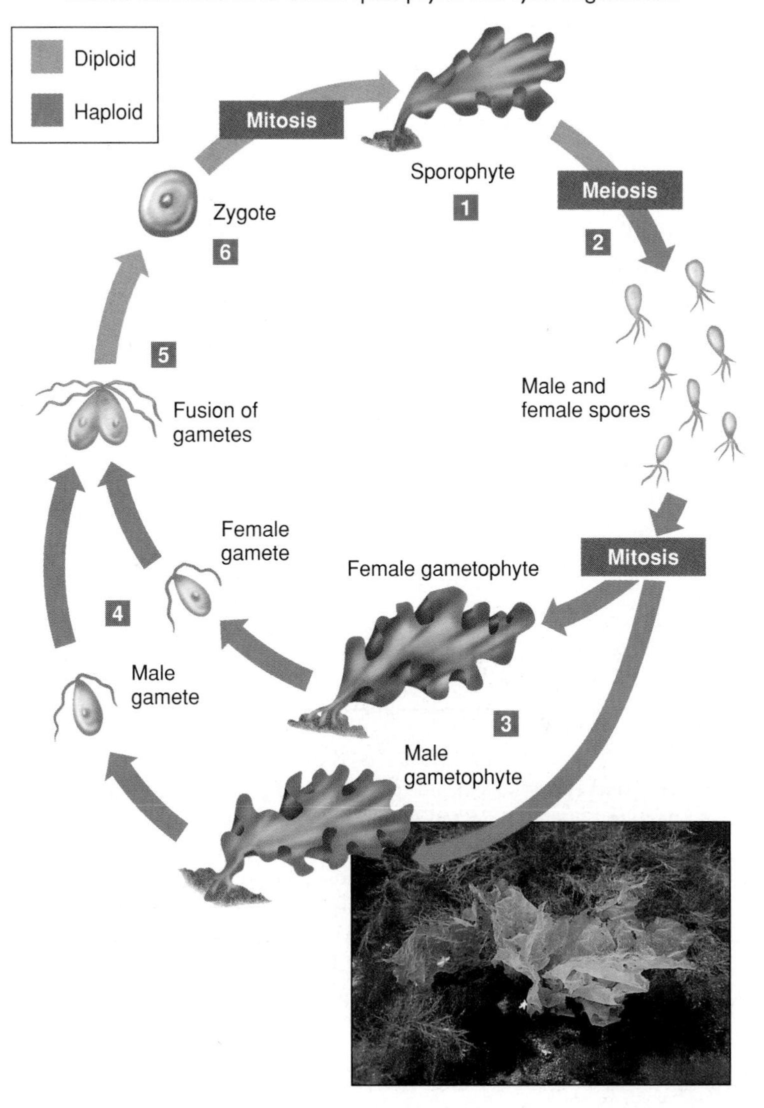

FIGURE 21.21

Diversity of Green Algae.

Green algae have a variety of body forms, from solitary microscopic cells to complex multicellular forms.

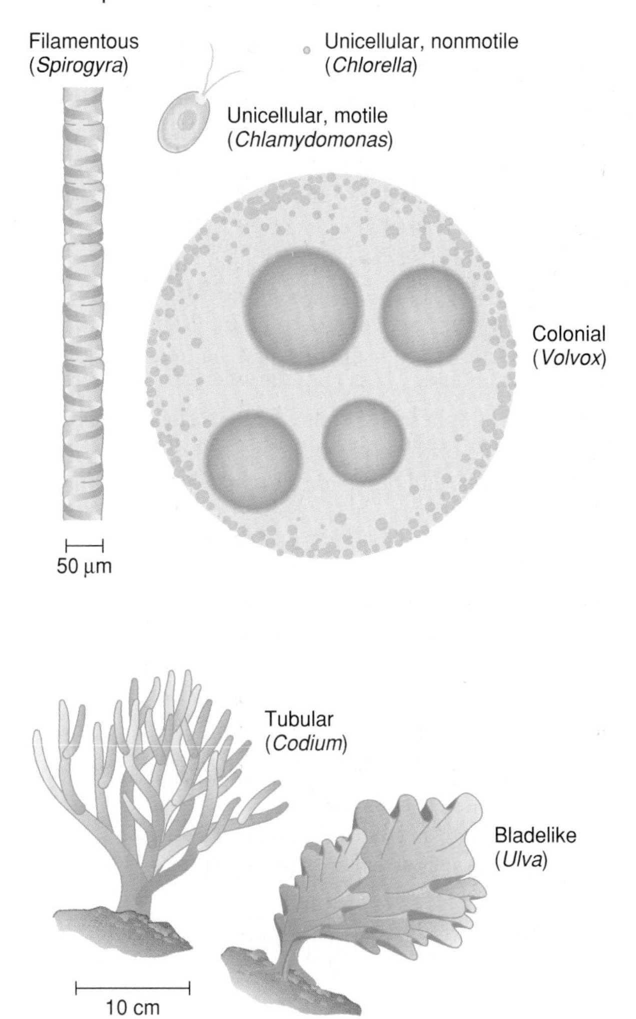

TABLE 21.1 Summary of Protist Groups

PROTIST GROUP	SOME DISTINGUISHING FEATURES	EXAMPLES
BASAL EUKARYOTES		
Nonmitochondrial	Cells lack mitochondria	*Trichomonas, Giardia, Entamoeba*
Mitochondrial		
Slime molds	In response to starvation, cells aggregate into movable mass	Acellular and cellular slime molds
Euglenida	Pellicle, some photosynthetic, one or two flagella	*Euglena*
Kinetoplastida	Kinetoplasts	Trypanosomes
CROWN EUKARYOTES		
Alveolates	Alveoli beneath cell membrane	
Dinoflagellates	Longitudinal and transverse flagella; cellulose plates	Red tide organisms, *Pfiesteria*
Apicomplexa	Obligate animal parasites; apical complex; complex life cycles	*Plasmodium, Cryptosporidium, Toxoplasma*
Ciliates	Multiple cilia in bands, tufts, or covering entire cell	*Paramecium, Didinium*
Stramenopiles	Biflagellate cells; some flagella with tubular hairs; chlorophyll *c* in photosynthetic members; laminarin	
Water molds (oomycetes)	Filamentous, heterotrophic	*Phytophthora*
Diatoms	Silica shells	
Brown algae	Fucoxanthin (pigment)	Giant kelp
Golden algae	Facultative heterotrophs	*Dinobryon*
Red Algae	Multicellular, phycobilins (pigments)	*Bossiella*
Green Algae	Alternation of generations; chlorophyll *a* and b; pyrenoids	*Ulva, Chlamydomonas, Volvox*

puddles green after a heavy rain. The 600 or so recognized species are distinguished by the shapes of their cells and whether they are solitary or form groups. *Chlamydomonas* cells may exist in a thin-walled vegetative state, or form thick-walled cysts that may appear reddish due to carotenoids. To survive drought, *Chlamydomonas* cells pull in their flagella and become cysts. When water returns, the cells regrow flagella and reproduce. Figure 9.4A shows how *Chlamydomonas* reproduces sexually and asexually.

Volvox is a green alga that is colonial. From 500 to 600,000 *Volvox* cells form hollow balls that may be barely visible to the human eye. The cells wave their flagella in a coordinated fashion, which moves the sphere. Reproductive cells cluster at one end of the sphere. Some of these reproductive cells divide asexually to yield new colonies, and others produce gametes that, after fertilization, also divide to produce a daughter colony. New colonies remain within the parental ball of cells until they burst free.

Table 21.1 reviews and summarizes the protista.

A cell in itself is extremely complex and highly organized. The second half of this unit explores a giant leap in complexity among life-forms—the acquisition and elaboration of multicellularity. With multicellularity came another layer of complexity, the interactions of cells to build tissues and organs.

21.5 MASTERING CONCEPTS

1. What characteristics of green algae distinguish them from plants?
2. How does *Chlamydomonas* differ from *Volvox*?

OLC

Chapter Summary

21.1 Kingdom Protista Lies at the Crossroads Between the Simpler and the Complex

1. Most **protista** are unicellular eukaryotes; a few are multicellular.
2. Some protista may resemble early eukaryotes or more complex plants, animals, and fungi.
3. Classification of protista is changing as RNA and DNA sequence data are considered along with traditional traits.
4. The **basal eukaryotes** include protista that may reveal stages in the evolution of endosymbiont bacteria into mitochondria. They may be free-living or parasitic.
5. Basal eukaryotes lacking mitochondria include *Giardia lamblia,* which exists as a nonmotile, thick-walled **cyst** or a flagellated feeding form, and **Parabasalia,** some of which cause amoebic dysentery.
6. The **acellular** and **cellular slime molds,** along with some **amoebae, Euglenida,** and **Kinetoplastida,** represent the diversity of basal eukaryotes that have mitochondria.

21.2 The Alveolates Include Dinoflagellates, Apicomplexa, and Ciliates

7. The **alveolates** share flattened sacs beneath the cell membrane and tubular mitochondrial cristae.
8. **Dinoflagellates** have two different-sized flagella at right angles that generate a whirling movement. Many have outer plates made of cellulose.
9. The **apicomplexa** are obligate parasites of animals. They share an **apical complex** of organelles that helps them attach to or penetrate host cells. Apicomplexa cause malaria and toxoplasmosis.
10. The **ciliates** are complex cells with cilia, food and contractile vacuoles, trichocysts, and two types of nuclei.

21.3 The Stramenopiles Include Water Molds, Diatoms, and Some Algae

11. The **stramenopiles** are diverse organisms that have unique types of flagella, laminarin, and photosynthetic pigments not found in plants.
12. The **water molds** are filamentous heterotrophs that live in moist or wet environments. Some are pathogens of plants.
13. **Diatoms** are microscopic phytoplankton with silica walls called frustules.
14. Fucoxanthin colors **brown algae,** which are large, multicellular seaweeds. The **golden algae,** which may be related to the brown algae, are photosynthetic but can eat other microorganisms when light or nutrient supplies decline.

21.4 The Red Algae Contain Unique Photosynthetic Pigments

15. Like plants, **red algae** have cellulose cell walls and chlorophyll *a*. Red algae produce phycobilin pigments that expand their photosynthetic range.
16. Red algae provide products such as agar and carageenan.

21.5 The Green Algae

17. **Green algae** are aquatic, store carbohydrates as starch, are photosynthetic and use the same pigments as plants. Many have **alternation of generations.** Unlike plants, green algae lack true roots, stems, and leaves, and they have less specialized cells.
18. The green algae have diverse body forms, ranging from microscopic, one-celled organisms to large, multicellular seaweeds.

Testing Your Knowledge

1. Give three examples of how nucleic acid sequencing data have challenged traditional taxonomic classifications of protista.
2. How are the jakobids, choanoflagellates, and green algae of evolutionary importance?
3. Name an organism that uses each form of locomotion:
   a. beating cilia
   b. amoeboid motion
   c. flagella movement
4. Name the storage carbohydrates found in plants, diatoms, brown algae, and green algae.
5. For each of the following structures, cite at least one type of organism that has it:
   a. trichocysts
   b. plasmodium
   c. frustules
   d. apical complex
   e. kinetoplast
   f. pellicle
   g. pyrenoid
   h. cyst
   i. laminarin
   j. pseudopod
6. Name three types of pigments found in protista, and list an organism that has each.
7. Which protist causes each of the following diseases?
   a. malaria
   b. grape downy mildew
   c. amoebic dysentery
   d. late blight of potato
   e. African sleeping sickness
   f. toxoplasmosis
   g. whirling disease
   h. leishmaniasis
   i. hiker's diarrhea

Thinking Scientifically

1. Give one example where molecular evidence unites organisms once thought to be dissimilar, and one example where such evidence indicates that organisms once thought to be alike due to their appearance are actually not closely related.
2. What strategies enable cellular slime molds and dinoflagellates to survive a temporarily harsh environment?
3. How do stramenopiles and green algae differ from plants?
4. A strange event occurred on a South African beach. Half a million spiny rock lobsters, which are a delicacy, became stranded on the sand, and police had to set up roadblocks to prevent people from plucking free dinners from the sea. The lobsters had fled the ocean due to a "red tide" which tinged the water red. What type of protistan might have caused the lobster stranding? Are they safe to eat?
5. How is it adaptive for a red alga to have pigments other than chlorophyll?
6. How might *Volvox's* colonial lifestyle be adaptive?

References and Resources

Burkholder, JoAnn M. August 1999. The lurking perils of *Pfiesteria. Scientific American,* p. 42.

Poser, Charles M. and George W. Bruyn. 1999. *An Illustrated History of Malaria.* New York: Parthenon. A look at malaria outbreaks and epidemics over the past 500 years.

Webster, Donovan. September 2000. Malaria kills one child every 30 seconds. *Smithsonian,* vol. 31, p. 32. Malaria, caused by a protista, is the world's #1 killer.

The *LIFE* Online Learning Center provides additional resources and tools for studying this chapter.
www.mhhe.com/life

CHAPTER 22

Plantae

Nature's Botanical Medicine Cabinet

In Europe they are called phytomedicines, in the United States and Canada, herbal remedies or botanicals. Plant-based medicines have been a staple of many cultures for millennia. Before chemists began synthesizing drugs in the 1930s, pharmaceuticals came entirely from nature—mostly plants. Today, 25% of prescription drugs in the United States are derived from plants, and the proportion is higher in Europe (see chapter opening table). Many other plant products are sold as food supplements. In the United States, it is not required that the plant products undergo rigorous testing for safety and efficacy, and therefore manufacturers are not permitted to claim that the products treat specific illnesses. Still, many people take *Ginkgo biloba* to improve memory, St. John's wort to fight depression, and *Echinacea* to bolster immunity. Clinical trials are ongoing to test whether anecdotal reports are consistent with scientific evidence of efficacy for these and other plant extracts.

Medicinal use of botanical substances dates from ancient times. Remains of prehistoric settlements include parts of poppy plants that once yielded opiates used to dull pain. Native peoples in East Africa chewed twigs from the Indian neem tree to prevent tooth decay, an effect of a chemical that in the plants repels desert locusts. Writings from ancient Egypt and China list several plant-based medicines. Other animals may use plant biochemicals too. A Navajo Indian story told of a bear teaching people to use the flowering plant *Ligusticum* to heal damaged skin. In an experiment, Kodiak bears in captivity chewed the roots of the plant and rubbed the mash on sore paws. Chimpanzees and certain birds eat plants that contain potent compounds that kill bacteria or parasitic worms.

The tale of two drugs used to treat cancer illustrates how a phytochemical becomes a pharmaceutical.

Many societies have used extracts of the Madagascar periwinkle *Catharanthus roseus* to treat fever, sore throat, eye inflammation, bleeding, and wasp stings. In 1957, researchers ground up the pink flowers and dissolved them in alcohol, then gave the preparation to rats. The animals' white blood cell counts fell, suggesting that the active ingredient might hamper other rapidly dividing cells—such as cancer cells. Researchers isolated a chemical called vinblastine from the flowers, and later another called vincristine, both of which became very successful drugs

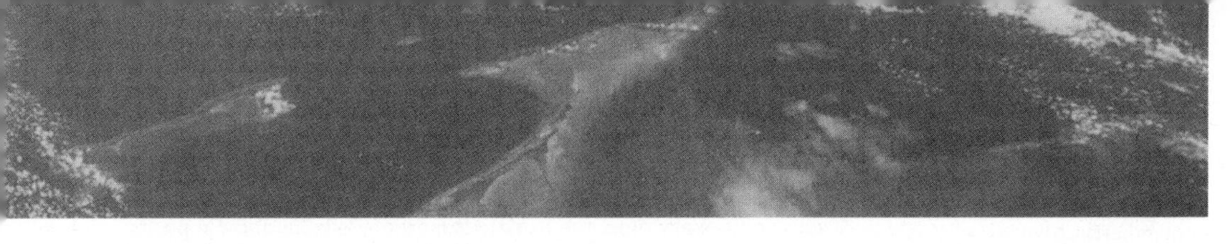

Medicinal Plants.
The rosy periwinkle from Madagascar (*Catharanthus roseus* **left**) and Pacific yew (*Taxus brevifolia* **right**) are just two of many plants that provide medicines.

used to treat leukemias and lymphomas. The plant manufactures 70 different types of organic compounds called alkaloids, six of which have anticancer effects. Vincristine and vinblastine work by preventing microtubules from assembling into mitotic spindles, crippling a cell's ability to divide. Side effects stem from the drugs' actions on normal cells that also happen to divide often.

In the early 1980s, natural products chemists searching among thousands of phytochemicals identified one from the bark of the Pacific yew tree, *Taxus brevifolia*. Unlike vincristine and vinblastine, this chemical, called paclitaxel (brand name Taxol) paralyzes microtubules once they form, and mitotic spindles cannot break down. Daughter cells can't separate, and division ceases. Paclitaxel can extend the lives of women whose ovarian or breast cancer has not responded to other treatments, and it was approved in 1992 and 1994 for these diseases. But an environmental concern arose—the Pacific yew was endangered, and centuries-old forests were being destroyed to manufacture the valuable drug.

At first, the prospect of creating the compound synthetically was daunting—it took 25 separate chemical reactions. Fortunately, a combination of biological and chemical know-how found a way around the paclitaxel shortage. Researchers discovered that several common yew species can provide chemicals that, with a few added steps in the laboratory, yield anticancer compounds. And so instead of Pacific forests, English hedgerows and French graveyard foliage became the source of the drug. Paclitaxel is also derived from plant cells growing in culture and from a fungus.

DRUGS from PLANTS

Common Name	Scientific Name	Active Compound	Use
Autumn crocus	*Colchicum autumnale*	Colchicine	Gout
Belladonna	*Atropa belladonna*	Atropine	Muscle relaxant
Burn plant	*Aloe vera*	Aloin	Soothes skin
Coca	*Erythoxyolon coca*	Cocaine	Local anesthetic; addictive, psychoactive drug
Feverbark tree	*Cinchona*	Quinine	Malaria
Ipecac	*Cephaelis ipecacuanha*	Cephaeline emetine	Induces vomiting
Marijuana	*Cannabis sativa*	Tetrahydrocannabinol	Glaucoma, antinausea for chemotherapy, increases appetite in people with AIDS
Mayapple	*Podophyllum peltatum*	Podophyllin	Genital warts
Purple foxglove	*Digitalis purpurea*	Digitoxin	Congestive heart failure
Red hot chili peppers	*Capsicum ohinense*	Capsaicin	Local anesthetic
St. John's wort	*Hypericum perforatum*	Hypericum	Depression
White willow	*Salix alba*	Aspirin	Pain, fever, inflammation
Yam	*Dioscorea*	Diosgenin	Oral contraceptives

22.1 Introducing the Plants

Plants are diverse multicellular organisms that photosynthesize, have cellulose cell walls, and exhibit alternation of generations. They evolved from green algae about 480 million years ago, and many special features of modern plants reflect their ancient conquest of the land. Plants are classified by the presence or absence of transport tissues, seeds, flowers and fruits, and by molecular evidence.

The planet would not be what it is today were it not for plants. When plants settled and conquered the land hundreds of millions of years ago, they set into motion a complex series of changes that would profoundly affect both the living and nonliving worlds. The explosion of photosynthetic activity altered the atmosphere, lowering carbon dioxide levels and raising oxygen content. Because plants are autotrophs, they came to form the bases of the intricate food webs that sustain life. Tall trees provided diverse and nutrient-packed habitats for many types of animals, while flowers fed insects and birds. Leaf litter accumulating on forest floors created a rich soil for countless microorganisms, insects, and worms, and when washed into streams and rivers, fueled a spectacular diversification of fishes and other aquatic animals.

In our lives plants provide food, fuel, clothing, shelter, and medicines, as the chapter opening essay describes. Yet the plants that we cultivate are only a tiny percentage of the vastly diverse kingdom Plantae. This chapter introduces the diversity of plants, and the chapters in unit 5 expand upon this information.

What Are the Distinguishing Features of Plants?

Plants are multicellular eukaryotes that have cellulose-rich cell walls and use starch as a nutrient reserve. The vast majority of plants are photoautotrophic, using photosynthesis to convert solar energy into chemical energy in glucose. The cells of photosynthetic plants have chloroplasts that contain chlorophyll *a* and *b*, and other pigments. Unlike the green algae from which they arose, most plants live in terrestrial habitats.

Plants and certain algae have a characteristic type of life cycle termed **alternation of generations.** Figure 22.1 repeats figure 9.2, which depicts how a plant alternates between a diploid stage called a **sporophyte** ("spore-making body") and a haploid stage called a **gametophyte** ("gamete-making body"). The sporophyte develops from a zygote that forms when gametes come together at fertilization. In turn, certain sporophyte cells undergo meiosis and produce haploid **spores,** which divide mitotically to form the gametophyte. It is the gametophyte that produces gametes that fuse at fertilization, starting the cycle anew.

Complex plants, distinguished by specialized tissues that transport water and nutrients, have a sporophyte that is larger and more visible than the gametophyte. In the flowering plants, for example, the gametophyte is microscopic, and among the conifers, it is visible but small. A blade of grass, a maple tree, and a rose bush are all sporophytes. In contrast, simpler plants such as mosses have a more prominent gametophyte stage, and a reduced sporophyte. However, there are exceptions. Botanists thought that the Killarney fern (*Trichomanes speciosum*) was extinct because changing weather conditions proved challenging for the sporophyte, which became rare. The gametophyte, however, persisted in many parts of Europe because it was better adapted to the drier environment that has prevailed since the 1700s.

Classification of plants is fairly straightforward compared with some other groups of organisms, although scientists disagree on some aspects (fig. 22.2). Plants can be classified by

FIGURE 22.1
Sexual Life Cycles.
Over evolutionary time, the haploid gametophyte generation in plants (**A**) has become less conspicuous, and the diploid sporophyte generation has become dominant. In contrast, gametes are the only haploid cells in the life cycles of most animals (**B**).

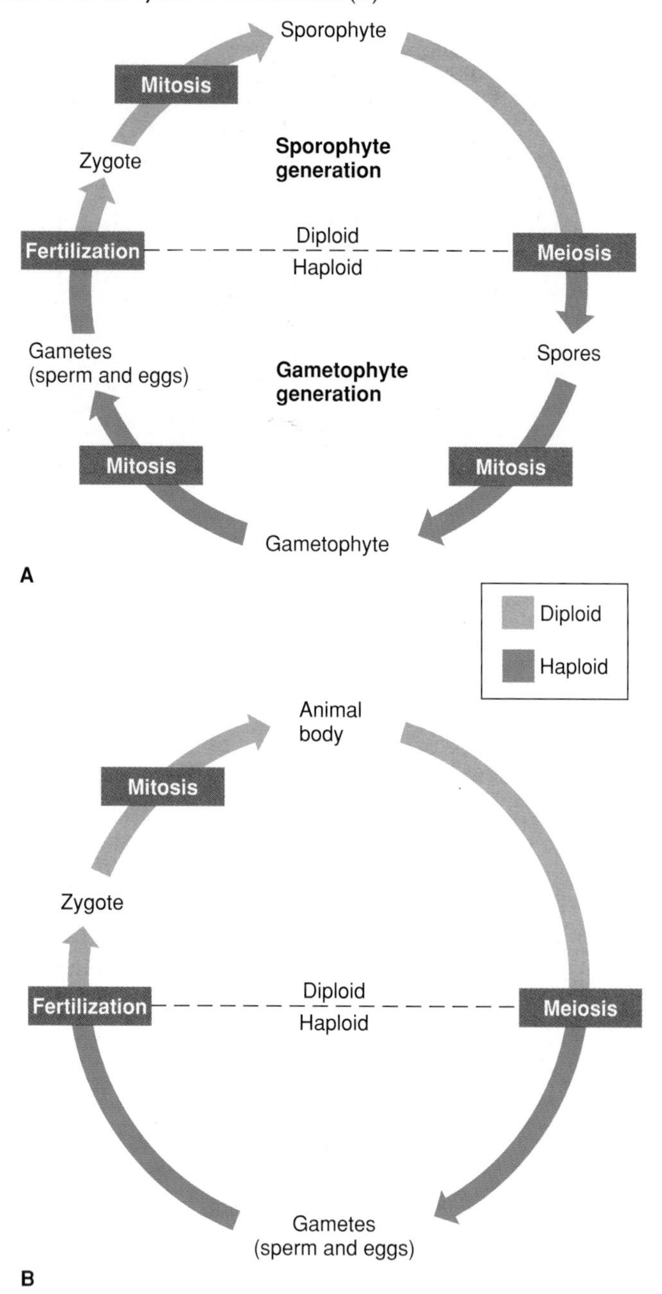

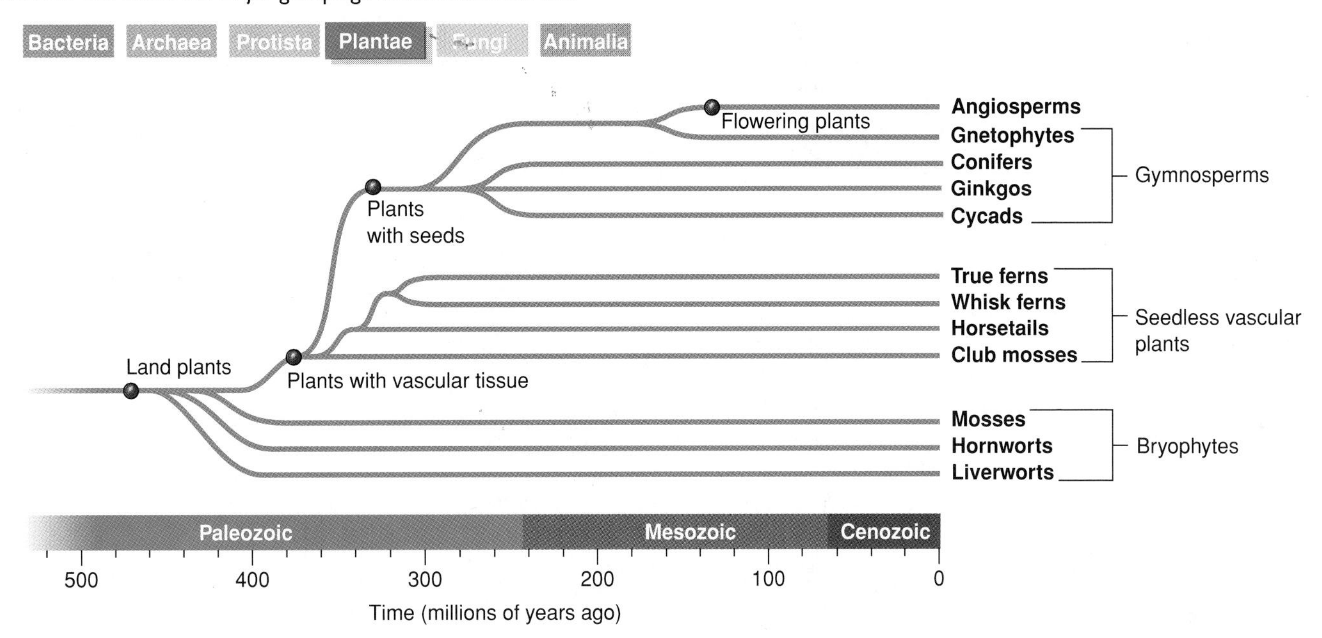

FIGURE 22.2
Plant Diversity.
Although botanists have traditionally classified plants according to the presence or absence of vascular tissue, seeds, and flowers, evolutionary relationships between and within the major groupings remain controversial.

whether or not they have vascular tissue to transport water and nutrients. The simplest plants, which lack vascular tissue and other complex parts such as roots, stems, and leaves, are the **bryophytes,** and they include liverworts, hornworts, and mosses. These are the nonvascular plants. In contrast, the vascular plants have vascular tissue, and may be seedless or seed-producing. Seedless vascular plants include club mosses, horsetails, whisk ferns, and true ferns. The seed-producing plants may have uncovered or "naked" seeds (the **gymnosperms**) or enclose the seeds in fruits (the **angiosperms**). The gymnosperms include cycads, ginkgos, conifers, and gnetophytes. The angiosperms are the flowering plants.

Origin and Evolution of Land Plants

As was the case for the other kingdoms, researchers hypothesize how plants diversified by examining fossil evidence and clues in nucleic acid sequences of modern forms. Molecular evidence, new fossil finds, and increased attention to the simpler plants trace the origin of plants to freshwater green algae called **charophytes.** These algae have many of the complex biochemicals seen among the plants, but their reproduction occurs in water—they did not require adaptations to permit gametes to form and meet on dry land. Sometime before about 480 to 470 million years ago, in the middle of the Paleozoic era, these algae gave rise to an organism probably similar to a modern liverwort. These first plants were aquatic, like the algae from which they descended.

By the end of the Devonian period, about 360 million years ago, the plant way of life was firmly established beyond water's edge (fig. 22.3). Life on dry land required several adaptations that enabled plants to survive and reproduce without a constant water supply. Plants were able to conquer dry land because of:

- a dominant sporophyte that protects the gametophyte and zygote. Some land plants, such as mosses, have a dominant gametophyte. But as the invasion of land continued, the sporophyte became dominant, and the gametophyte smaller.
- vascular tissues that transport water and nutrients rapidly and efficiently throughout the larger plant body. As described in chapter 27, **xylem** tissue carries dissolved ions and water, and **phloem** tissue carries sugars produced in photosynthesis.
- development and elaboration of root systems that anchored plants and tapped water and nutrients from the soil, and of leaves that captured solar energy.
- the complex polymer **lignin,** which strengthens and supports cell walls, enabling plants to grow tall and upright, and to form branches, an important adaptation in the intense competition for sunlight in thick forests.
- a waterproof cuticle and stomata that minimize water loss and control gas exchange.

Fossil evidence for plants tends to come in bursts, reflecting periods of rapid diversification as new environments were conquered, as well as fortuitous conditions for preserving signs of life. The earliest plant fossils, resembling liverworts, date from 476 to 432 million years ago.

A spectacular diversity of fossil plants appears from about 402 to 256 million years ago. Vascular plants were beginning to take over. Fossils from about 360 million years ago represent extinct plants (*Cooksonia* and *Rhynia*) that appeared to be transitional forms between bryophytes and the first vascular plants. Like bryophytes, these plants lacked well-defined leaves, roots,

FIGURE 22.3

Diversification of Plants Occurred Rapidly After They Conquered Dry Land.

Plants have increased in complexity since their earliest days on land. Many of the plants shown are known only from fossils. Note that flowering plants are relative newcomers.

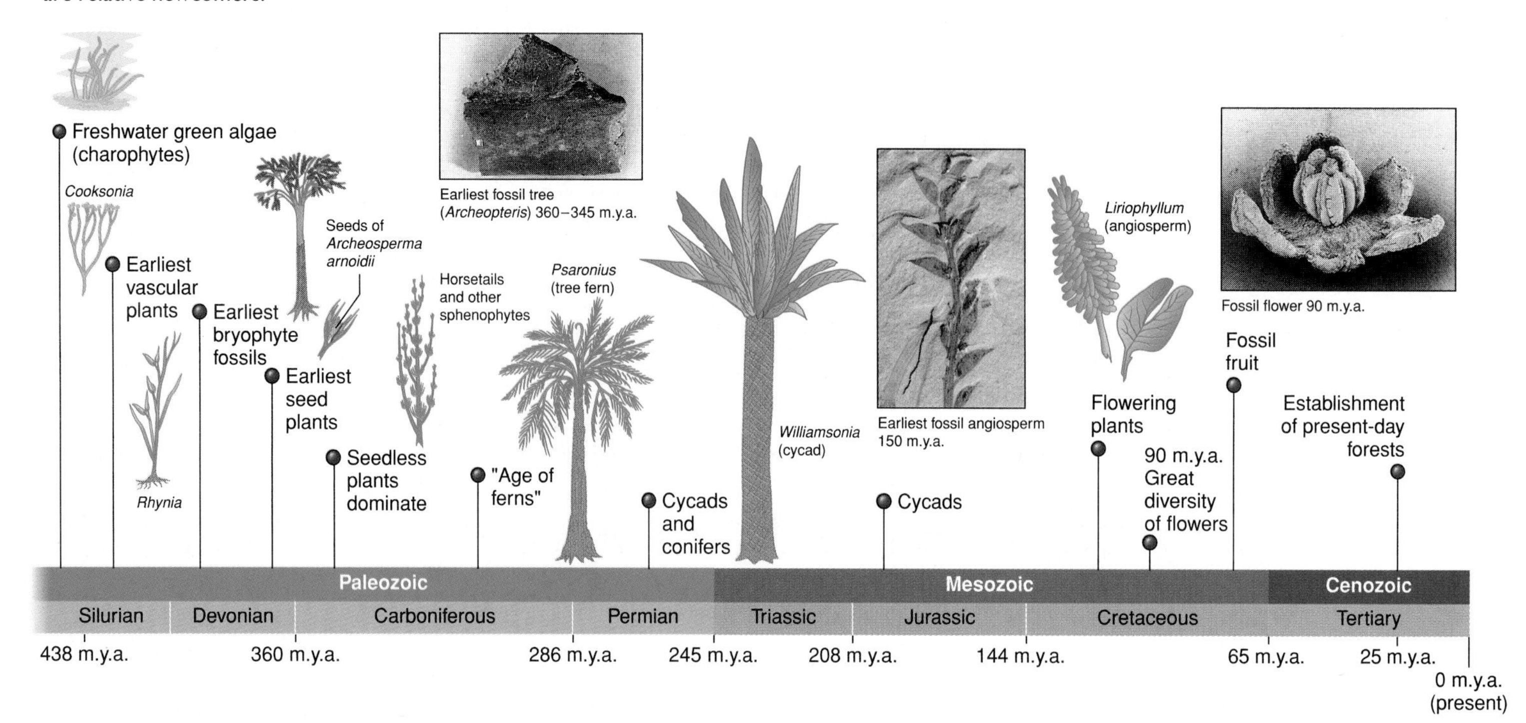

and vascular tissue, but like vascular plants, they show the first signs of a dominant sporophyte stage. Seeds also appeared during that time, and the Mesozoic saw great forests of seed-bearing cycads, conifers, and other gymnosperms.

The angiosperms dominate today's world. Botanists place angiosperm origins at about 200 million years ago, but the earliest fossils with unambiguous angiosperm traits date from about 150 million years ago (see fig. 17.1A). Their primitive flowers lacked petals and were probably wind-pollinated. Fossils indicate that by 125 million years ago, flower parts had developed elaborate spiral arrangements. By 90 million years ago, an evolutionary race of sorts was taking place, as flowers and the insects that pollinate them took turns adapting to each other's shapes.

By 55 million years ago, fruits had evolved, which protect developing seeds and aid in their dispersal. Animals ate the fruits and spread seeds farther than they had ever traveled before. The plants faced new selective pressures, and continued to diversify. We will return to the remarkable success of the angiosperms at the end of the chapter.

22.1 MASTERING CONCEPTS

1. How are plants vital to life on Earth?
2. What is alternation of generations?
3. What are the major types of plants?
4. Which adaptations enabled plants to thrive on dry land?
5. What are some of the main events in the evolution of plants?

OLC

22.2 Bryophytes Are the Simplest Plants

The liverworts, hornworts, and mosses are simple plants that lack transport vessels. They also lack roots, so the entire plant absorbs nutrients and water directly from its surroundings.

The earliest plants were likely simple organisms with no organized vascular systems. They may have resembled modern bryophytes. Figure 22.4 reviews the bryophytes' basal location among the plants. Although the term "bryophyte" is commonly used, the figure shows

FIGURE 22.4

Bryophytes.

The earliest land plants probably resembled existing bryophytes, a collective term for mosses, hornworts, and liverworts.

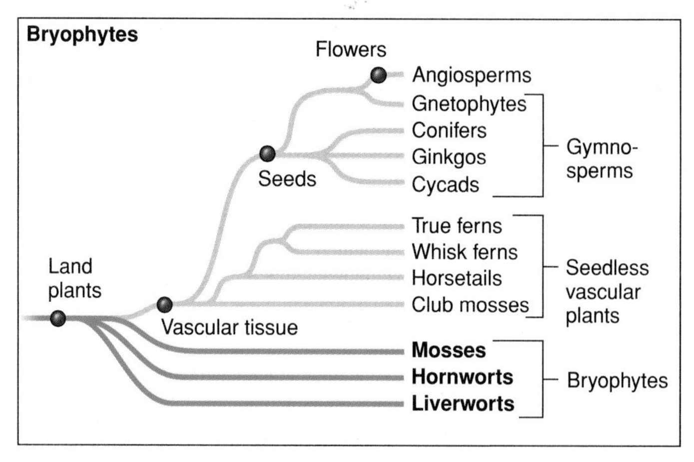

that these plants do not form a natural grouping. They are discussed together, however, because they share some important features.

Bryophyte Characteristics and Life Cycle

Bryophytes are small, compact, green plants that do not have vascular tissue. They also lack supporting tissue, but the polymer lignin hardens them. Bryophytes grow close to the ground, where they absorb water and nutrients. They also obtain nutrients from dust, rainwater, and chemicals dissolved in water at the soil's surface. Hairlike extensions called **rhizoids** along a bryophyte's lower surface anchor the plant and absorb water and minerals. Although bryophytes do not have true leaves and stems, many have structures that are functionally equivalent to these organs.

Most reproduction in bryophytes occurs asexually when a plant breaks apart and each piece grows into a new plant. The Japanese use tiny fragments of *Polytrichum* moss to create their lavish moss gardens. In nature, moss fragmentation is so extensive that a cubic meter of arctic snow can contain more than 500 bryophyte fragments, each of which can form a separate plant.

Bryophytes also reproduce sexually. The bryophytes are unlike other land plants in that the most conspicuous generation is the haploid gametophyte. Gametes form by mitosis in multicellular structures called **antheridia** (male) and **archegonia** (female). Each flask-shaped archegonium produces one egg, and each saclike antheridium produces many sperm (fig. 22.5). A protective sheath of cells surrounds each gamete-producing structure. Because sperm swim through water to eggs, bryophytes require free water for sexual reproduction. Sperm released from an antheridium do not swim randomly but are attracted to a biochemical that a nearby archegonium produces. The sperm fertilizes an egg

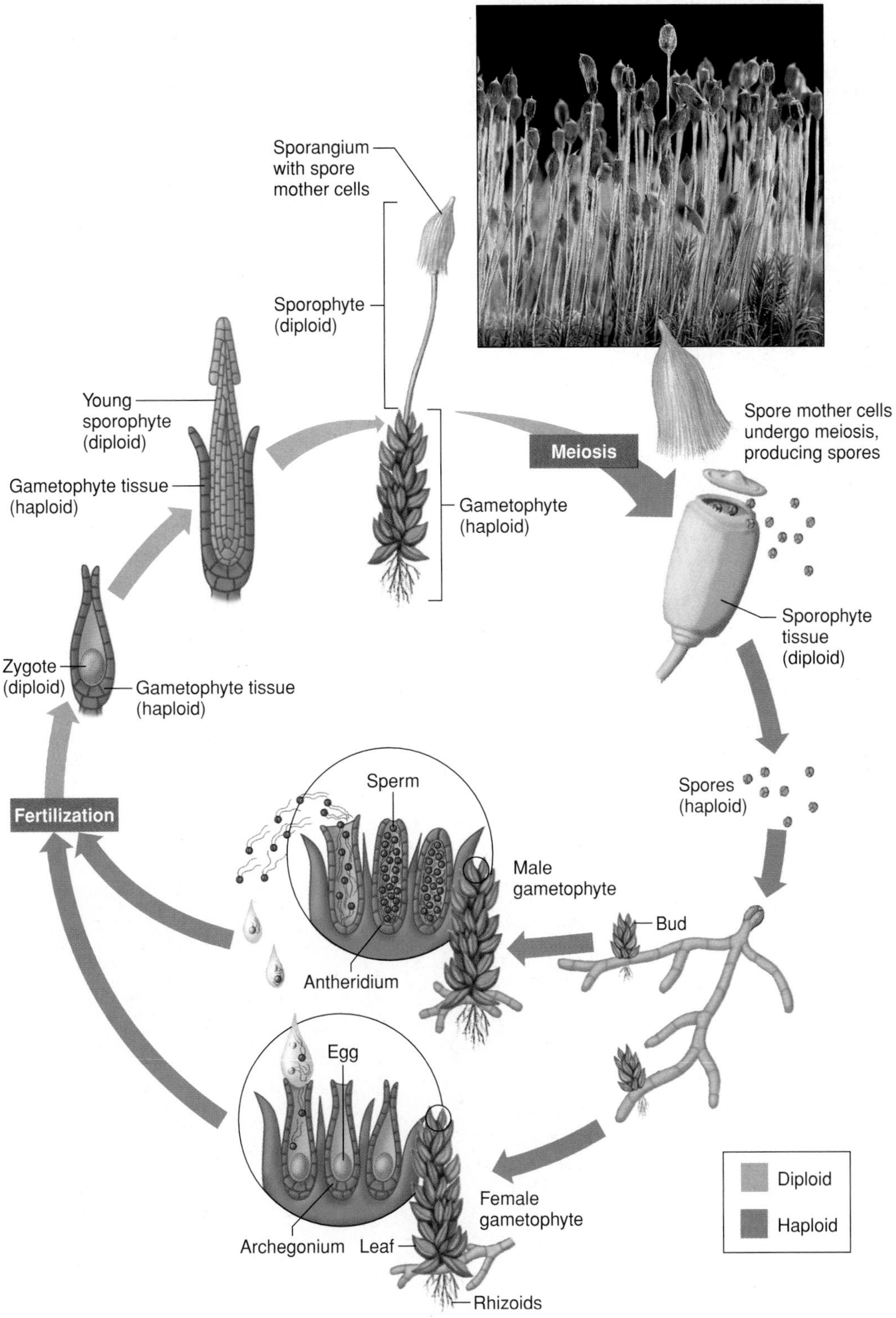

FIGURE 22.5
Alternation of Generations in a Bryophyte.
In mosses, the gametophyte is dominant, as is typical for bryophytes. The sporophyte generation, although distinct, is attached to and dependent on the gametophyte. In the sporophyte, spore mother cells in sporangia undergo meiosis, which yields haploid spores that generate the male and female gametophytes. Antheridia house sperm, and archegonia contain eggs, at the gametophyte tips. Gametes join to form a zygote, which develops into a new sporophyte.

FIGURE 22.6
Liverworts.
(**A**) The gametophyte of *Calypogeia muelleriana* resembles small leaves. (**B**) Note the ribbonlike shape of *Fossombronia cristula.* (**C**) Gemmae cups on a liverwort, *Lunularia,* are involved in asexual reproduction. Raindrops splash gemmae from the cups. In their new habitat, each gemma can form a new plant.

and forms a diploid zygote, thus beginning the diploid (sporophyte) generation of the life cycle. During this phase of the life cycle, which is less conspicuous than the gametophyte, spore mother cells in a caplike structure called a **sporangium** undergo meiosis, producing the haploid spores that give rise to the gametophyte.

Bryophyte Diversity

The 16,000 or so species of bryophytes are classified into three divisions (the equivalent of phyla): liverworts, hornworts, and mosses. Although all bryophytes are seedless, nonvascular plants, molecular data suggest they are not monophyletic (that is, they only share primitive characters and not distinctive, derived characters; see chapter 17). The liverworts are apparently the bryophytes most closely related to ancestral forms that evolved from green algae. The mosses are the bryophytes most closely related to the vascular plants.

Liverworts (division Hepaticophyta) take their name from the resemblance of some species to a human liver, which has lobes. "Wort" means herb. In the fifteenth and sixteenth centuries, people thought that a plant that resembled a part of the human body could treat a disorder there, and so liverworts were thought to heal liver ailments.

The 6,000 species of liverworts vary in size and shape. A leaflike liverwort may be less than half a millimeter in diameter, or a broader liverwort may extend more than 8 inches (20 centimeters) in diameter. Liverwort gametophytes are of two general shapes—leafy, and thallose, which are flat and resemble ribbons (fig. 22.6A and B).

Liverworts reproduce asexually with pieces of tissue called **gemmae.** These form in small "splash cups" on the upper surfaces of the gametophyte. Falling raindrops detach the gemmae, which can then grow into new plants (fig. 22.6C). Liverworts also reproduce sexually, when swimming sperm fertilize eggs produced by a female gametophyte.

Hornworts (division Anthocerotophyta) are the smallest group of bryophytes, with only about 100 species recognized. A hornwort's sporophyte is shaped like a tapered horn, hence the name (fig. 22.7). Unlike the archegonia of other bryophytes, hornwort archegonia are not discrete organs but are embedded in the plant. Also, like many green algae (but unlike most land plants), hornwort cells have only one chloroplast each, a clue to their aquatic heritage.

FIGURE 22.7
The Hornwort *Anthoceros.*
The tapered hornlike structures are sporophytes, below which the flat gametophytes are visible.

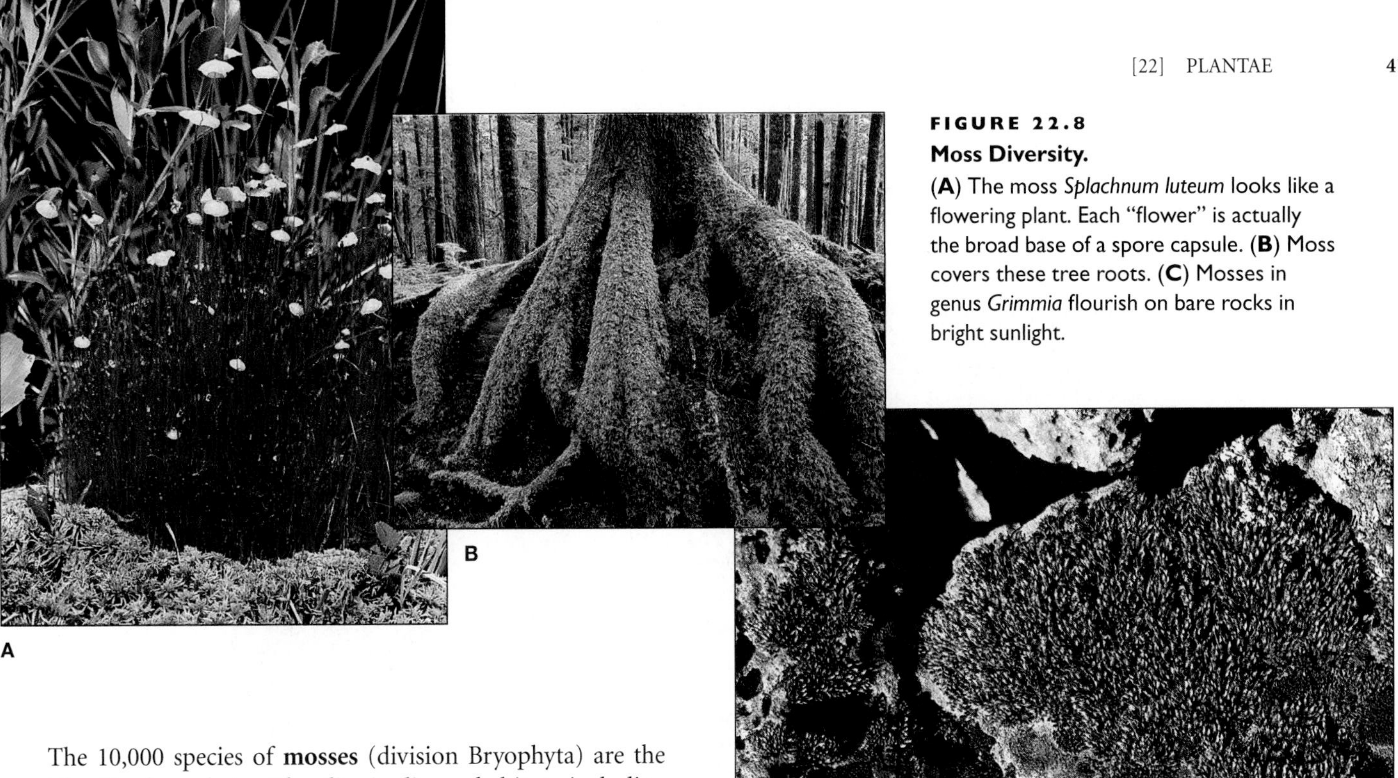

FIGURE 22.8
Moss Diversity.
(**A**) The moss *Splachnum luteum* looks like a flowering plant. Each "flower" is actually the broad base of a spore capsule. (**B**) Moss covers these tree roots. (**C**) Mosses in genus *Grimmia* flourish on bare rocks in bright sunlight.

The 10,000 species of **mosses** (division Bryophyta) are the most familiar bryophytes. They live in diverse habitats, including waterfalls, lava beds, at the mouths of caves, and on rocks (fig. 22.8). Mosses often grow alongside flowering plants.

A moss gametophyte may resemble leaves, but unlike the complex leaves of vascular plants, they are usually only one cell thick. The tiny brown or green moss sporophyte does not resemble the gametophyte at all.

We use mosses in diverse ways. They are used to stuff furniture, condition soil, and absorb oil spills. Florists use peat moss as a damp cushion around plants designated for shipment, and aboriginal people use *Sphagnum* moss for diapers and as a disinfectant. *Sphagnum* also was used to pack wounds during the Civil War. In some areas, peat moss may also be an important source of renewable energy.

22.2 MASTERING CONCEPTS

1. In what ways are bryophytes simple and similar to the earliest plants?
2. How do bryophytes reproduce?
3. What are some habitats of bryophytes?
4. Describe the three types of bryophytes.

22.3 Seedless Vascular Plants

The club mosses, horsetails, whisk ferns, and true ferns retain features that initially enabled plants to live on dry land—a protective cuticle and stomata. They do not have seeds, but they do have true leaves, stems, and roots.

The earliest vascular plants are extinct, but their successful strategies for adapting to land persist in their descendants. The more than 250,000 species of vascular plants are divided into two groups: those that do not produce seeds and those that do. Figure 22.9 shows the evolutionary position of these plants.

Seedless Vascular Plant Characteristics and Life Cycle

Seedless vascular plants share several features with bryophytes, including the same types of pigments, the same basic life cycle of alternation of generations, and stored starch as a primary food reserve. But the seedless vascular plants differ from bryophytes in that they contain vascular tissue. The evolution of vascular tissue enabled plants to live in drier land habitats more effectively than bryophytes could. Unlike bryophytes, seedless vascular plants have a well-developed cuticle that minimizes water loss, as well as stomata, which are tiny pores that regulate gas exchange.

FIGURE 22.9
Seedless Vascular Plants.
The seedless vascular plants include club mosses, horsetails, whisk ferns, and true ferns.

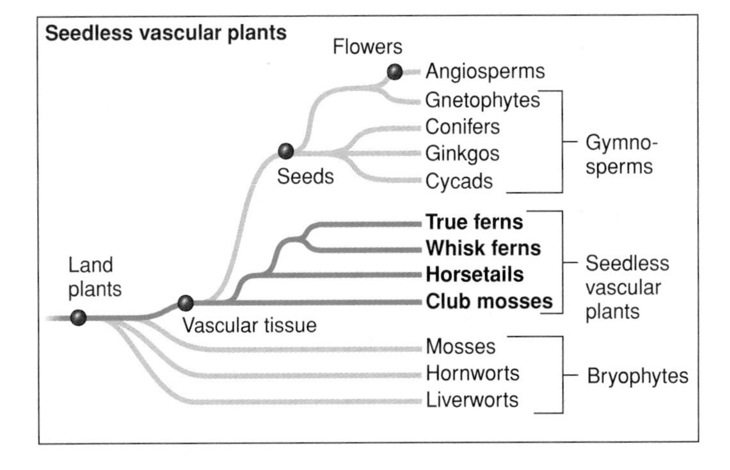

FIGURE 22.10

Alternation of Generations in a Fern.

The diploid sporophyte generation is most visible, although the heart-shaped haploid gametophyte still forms a tiny separate stage. Sporangia on sporophyte fronds house spore mother cells that produce spores through meiosis. The haploid spores develop into gametophytes. Archegonia and antheridia on each gametophyte produce eggs and swimming sperm. These gametes join to form a zygote, which develops into the sporophyte. The large sporophyte quickly dwarfs the tiny gametophyte.

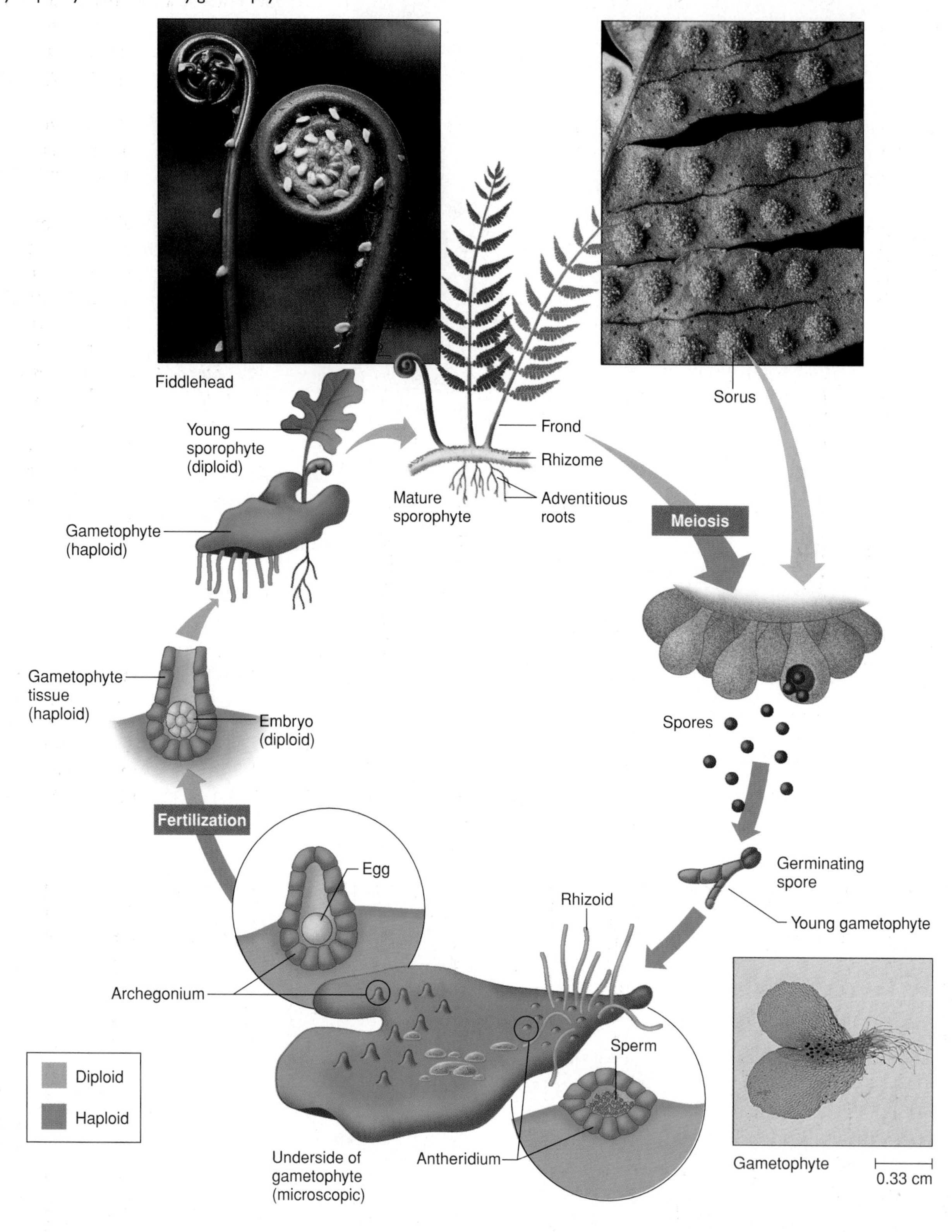

Seedless vascular plants such as ferns have a life cycle similar to that of bryophytes (fig. 22.10). The sporophyte (diploid phase) produces spores by meiosis. Each spore germinates and grows into a gametophyte that produces gametes by mitosis. Archegonia produce eggs and antheridia produce swimming sperm. These gametes fuse to form diploid zygotes. In the vascular plants, the gametophyte generation is much less conspicuous than the sporophyte.

Seedless Vascular Plant Diversity

The seedless vascular plants include the club mosses, horsetails, whisk ferns, and true ferns (fig. 22.11). Most varieties of these plants live in tropical areas.

The more than 1,000 species of **club mosses** (division Lycophyta), also called lycopods, live in various habitats worldwide. Most of the species are included in two genera, the club mosses (*Lycopodium*) and the spike mosses (*Selaginella*), which take their names from their club- or spike-shaped reproductive structures. Lycopods have leaves, stems, and roots. In some species, the roots grow from **rhizomes** (horizontal, underground stems) extending from a central point, and they may be quite extensive. One such ring structure of a *Lycopodium* species began growing in 1839 and is still expanding!

An interesting *Selaginella* species is the resurrection plant (fig. 22.12). During drought, this plant rolls up in a tight, dried-up ball. When rain comes, its branches expand and the plant photosynthesizes.

Horsetails (division Sphenophyta) include only one living genus, *Equisetum*, with 15 species. Some of these plants have branched stems that look somewhat like horses' tails. These plants grow worldwide, usually along streams or at the borders of forests. Their rhizomes are highly branched. Horsetails have aerial stems that are green and photosynthetic and can poison livestock that eat them. The stems bear the reproductive (spore-producing) structures at their tips.

Horsetails are also called scouring rushes because their epidermal (surface)

A

B

C

D

FIGURE 22.11

Diversity of Seedless Vascular Plants.

The seedless vascular plants include (**A**) the club moss *Lycopodium selage,* (**B**) horsetails such as this *Equisetum fluvatile,* (**C**) the whisk fern *Psilotum nudum,* and (**D**) the narrow beech fern (*Phegopteris connectilis*), a true fern.

FIGURE 22.12
Resurrection Plant.
The scale-leaf spike moss *Selaginella lepidophylla* forms a dried-up ball (**A**) in times of drought, losing up to 97% of its moisture. Yet it can expand and turn green once rain comes (**B**).

A

B

tissue contains abrasive silica particles. Native Americans used horsetails to polish bows and arrows, and early colonists and pioneers used them to scrub pots and pans.

Whisk ferns (division Psilophyta) are simple vascular plants, lacking roots. Most species have no obvious leaves. Instead of roots, whisk ferns have rhizomes. Since no fossils of whisk ferns have been discovered, their relationship with other land plants remains debatable.

The name "whisk fern" comes from the highly branched, broomlike stems of *Psilotum,* which resemble whisk brooms (see fig. 22.11C). *Psilotum,* the largest of the two genera in this group, is widespread in subtropical regions of the southern United States and Asia. These plants are easily cultivated and grown in greenhouses worldwide. The other genus of whisk ferns, *Tmesipteris,* is rarely cultivated and grows only on islands in the South Pacific.

True ferns (division Pterophyta) are by far the largest group of seedless vascular plants, with 12,000 living species. Ferns are primarily tropical plants, but some grow in temperate regions, and others grow in deserts. North or south of the tropics, the number of fern species decreases because of decreasing moisture. In Guam, for example, about 12% of the species of vascular plants are ferns, but in California, only about 2% are ferns. Ferns were especially widespread and abundant during the Carboniferous period, when their huge fronds dominated warm, moist forests (see fig. 18.17A). They form most coal deposits.

The leaves, or **fronds,** of ferns are their most obvious feature (see fig. 22.11D). Some genera of ferns have leaves that are the largest and most complex in the plant kingdom. One species of tree fern has leaves up to 30 feet (9 meters) long and 16 feet (5 meters) wide, which is nearly the size of a two-car garage! At the other extreme, the aquatic fern *Azolla* has tiny leaves, about the diameter of a pencil eraser.

Each year, new fronds of true ferns grow from rhizomes. The new leaves are at first lopsided, because their upper surfaces grow more slowly than their lower surfaces. This growth pattern coils the new leaves into structures called fiddleheads. As each growing season begins, new fiddleheads arise near the growing tip of the rhizome.

Most fern fronds seasonally produce dark spots on their undersides called sori (singular: sorus). Sori are collections of sporangia that are often enclosed in a protective covering. Sori can produce as many as 50 million spores per plant each season. The drying sporangium catapults the spores away from the plant.

22.3 MASTERING CONCEPTS

1. How do the two major groups of vascular plants differ from each other?
2. What do seedless vascular plants have in common with bryophytes?
3. What are some types of seedless vascular plants?

22.4 Seed-Producing Vascular Plants

Evolution of seeds continued the trend of adapting to life on land. The gymnosperms have "naked" seeds. Angiosperms enclose their seeds in fruits. Both groups are very diverse.

The ancestors of modern seedless vascular plants dominated Earth's vegetation for more than 250 million years, eventually giving way to the seed plants (see figure 22.3). Two major groups of vascular plants produce seeds: gymnosperms (naked seed plants) and angiosperms (flowering plants). One of the most significant events in the evolution of vascular plants was the origin of the seed, because it eased plants' survival on dry land.

Like many other plants, a seed-bearing plant has two spore types. One type develops into a female gametophyte, giving rise to eggs. The other type becomes a **pollen grain,** which develops into a male gametophyte. After fertilization, the zygote develops into an embryo, which is packaged with nutritive materials into a tough outer coat to form a seed.

Fossil evidence of the first seeds dates to 360 million years ago, and the first pollen dates to at least 280 million years ago. Seed plants arose as the cool, dry Permian period followed the wet, swampy Devonian and Carboniferous periods. Many seeded plant species arose and then became extinct as the environment continued to change. A group called the seed ferns dominated for about 70 million years and then disappeared as towering gymnosperm trees flourished, particularly cycads, conifers, and cycadeoids, which are extinct, cycadlike plants. So dominant were gymnosperm trees by the Jurassic that botanists call this period the Age of Cycads, in contrast to the more familiar Age of Dinosaurs. Today, fewer than 750 types of cycads and conifers remain.

FIGURE 22.13

Gymnosperms.

The gymnosperms, or "naked seed" plants, include cycads, ginkgos, conifers, and gnetophytes.

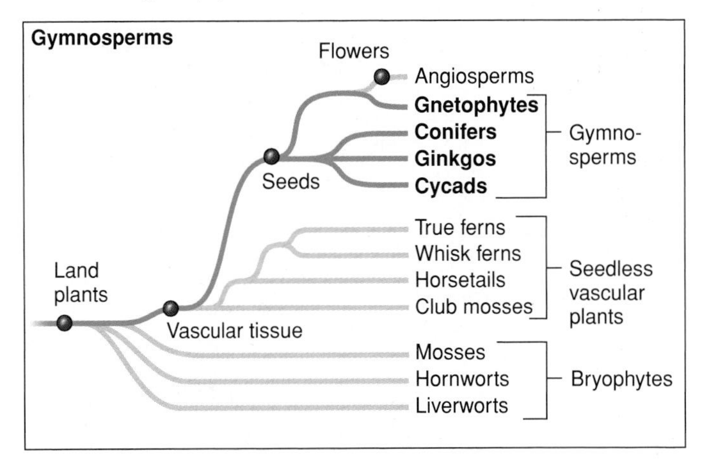

Gymnosperms Are "Naked Seed" Plants

The term "gymnosperm" derives from the Greek words *gymnos,* meaning "naked," and *sperma,* meaning "seed." The seeds of these plants are termed "naked" because they are not enclosed in fruits. Most gymnosperms are woody trees or shrubs, although a few species are more vinelike. Figure 22.13 shows the evolutionary position of gymnosperms. Note that like the bryophytes, the gymnosperms do not form a natural grouping. However, the term "gymnosperm" remains in common use, and these plants are discussed together because they share some important characteristics.

Gymnosperms and angiosperms differ from the seedless plants in that they do not require water for sperm to swim in to reach the egg. Only the cycads and the ginkgo, both gymnosperms, produce swimming sperm. A watery medium, however, is not necessary for fertilization—animals, the wind, or moving water can carry pollen to the female gametophyte. Many gymnosperms produce huge amounts of pollen, which is an adaptation to increase the likelihood of fertilization when pollen dispersal is inefficient. A male pine cone annually releases about 2 million pollen grains!

Living gymnosperms are remarkably diverse in reproductive structures and leaf types. Leaves range from simple, flat blades to needles, large fernlike leaves, and tiny, reduced leaves.

Gymnosperm Diversity The 721 species of gymnosperms are grouped into four divisions: cycads, the maidenhair tree (*Ginkgo*), conifers, and the gnetophytes.

Cycads (division Cycadophyta) include 185 species that live primarily in tropical and subtropical regions. These trees have palmlike leaves unlike the leaves of other living gymnosperms, but they do produce large cones (fig. 22.14A). Although cycads are planted as ornamentals, only two species are native to the United States. Both are in the genus *Zamia,* and in the United States neither lives in the wild outside of Florida. Cycads dominated Mesozoic era landscapes, but today many species are near extinction. This is because of their slow growth, low reproductive rate, and shrinking habitats.

Cycad plants are male or female, and they reproduce every 2 to 3 years. These plants are well adapted to entice insects to pollinate the female. A male plant has 15 to 20 cones, only one or two of which are ready to release pollen at any given time. Those cones emit an odor and warmth that attract insects, which arrive, breed, and feed. At this time, the female cones produce a toxin that discourages visits from hungry insects. Sometime later, the female cones warm up and release a pleasant aroma too, attracting newly hatched insects that are conveniently covered in pollen. So fine-tuned are these adaptations that insect larvae may enter a dormant state until the cycad manufactures its perfume once again. Adaptations that correspond between species are said to have co-evolved.

The maidenhair tree, *Ginkgo biloba,* is the only living representative of division Ginkgophyta and is the oldest living tree species. It is a popular cultivated tree that no longer grows wild in nature. The **ginkgo's** distinctive, fan-shaped leaves have remained virtually unchanged for 80 million years (fig. 22.14B). All living ginkgo trees descend from plants grown in the temple gardens of China and Japan.

A ginkgo tree is either male or female. The seeds are very fleshy, resembling berries or small plums, but they produce a foul odor and may irritate a person's skin. In some parts of Asia, however, pickled ginkgo seeds are a delicacy.

Extracts of *Ginkgo biloba* appear to have medicinal value. The Chinese have used ginkgo tea to treat asthma and bronchitis for more than 5,000 years. *Ginkgo biloba* extracts increase blood flow, particularly to the brain, and it appears to improve mental function in elderly persons with mild dementia. Many healthy people take it to help memory and concentration, but these effects have not been demonstrated in clinical trials.

Conifers (division Coniferophyta) have needlelike or scalelike leaves and produce cones. At any given time, a few needles are shed, so that a conifer is not truly "evergreen" but turns over its needle supply every few years. Bristlecone pine needles are exceptionally long-lived—a needle may last 25 to 30 years!

A

B

C

FIGURE 22.14

Gymnosperm Diversity.

(**A**) *Zamia furfuracea* is a cycad that produces large cones. (**B**) *Ginkgo biloba* is also called the maidenhair tree. Plants are either male or female. Male trees are preferred for cultivation, because seeds on female trees emit an odor of rot. (**C**) The bristlecone pine (*Pinus longaeva*) is one of the most ancient plants known. It is a conifer. (**D**) *Ephedra* is a gnetophyte that is the source of the stimulant and decongestant ephedrine.

D

The pines (*Pinus*) are the largest group of conifers and are the most abundant trees in the northern hemisphere. Included in this genus is the bristlecone pine (*Pinus longaeva*, fig. 22.14C) of the western United States, some of which are among the oldest known plants. One tree, named Methuselah, is about 4,725 years old. Other conifers include fir, larch, spruce, juniper, and the coastal redwood. The Pacific yew is a source of paclitaxel, the drug used to treat certain cancers, as described in the chapter opening essay.

Gnetophytes (division Gnetophyta) include some of the most distinctive (if not bizarre) of all seed plants: *Ephedra, Gnetum,* and *Welwitschia.* Molecular evidence suggests that these plants are most closely related to the flowering plants.

Ephedra, known as Mormon tea, or Ma Huang, contains a powerful decongestant, ephedrine, that also stimulates the central nervous system, and is sometimes abused as the drug "herbal ecstasy" (figure 22.14D). Another compound from this plant, pseudoephedrine, has the decongestant effect without affecting the nervous system, and is used in many preparations that treat the common cold.

Welwitschia is a slow-growing plant that lives in African deserts, where it gets most of its water from fog. Mature plants have a single pair of large, strap-shaped leaves that persist throughout the life of the plant.

A Closer Look at the Pine Life Cycle Like other plants, pines have an alternation of generations. Cones are the reproductive structures of pines (fig. 22.15). Large female cones bear

FIGURE 22.15

Pine Life Cycle.

Alternation of generations occurs in all plants, but in gymnosperms (and angiosperms) the gametophyte generation consists of just a few cells (lower right part of cycle). As in all plants, male and female gametophytes produce sperm and egg cells. Each female cone scale has two ovules (only one is visible in the figure), each of which produces a gametophyte with two or three archegonia, but only one egg per ovule is fertilized. Pollen, the male gametophyte, delivers a sperm cell to an egg via a pollen tube. More than a year after pollination, each fertilized egg (zygote) completes its development into a seed, which is the young sporophyte. Once the sporophyte is a mature tree, it produces male and female cones to complete the life cycle.

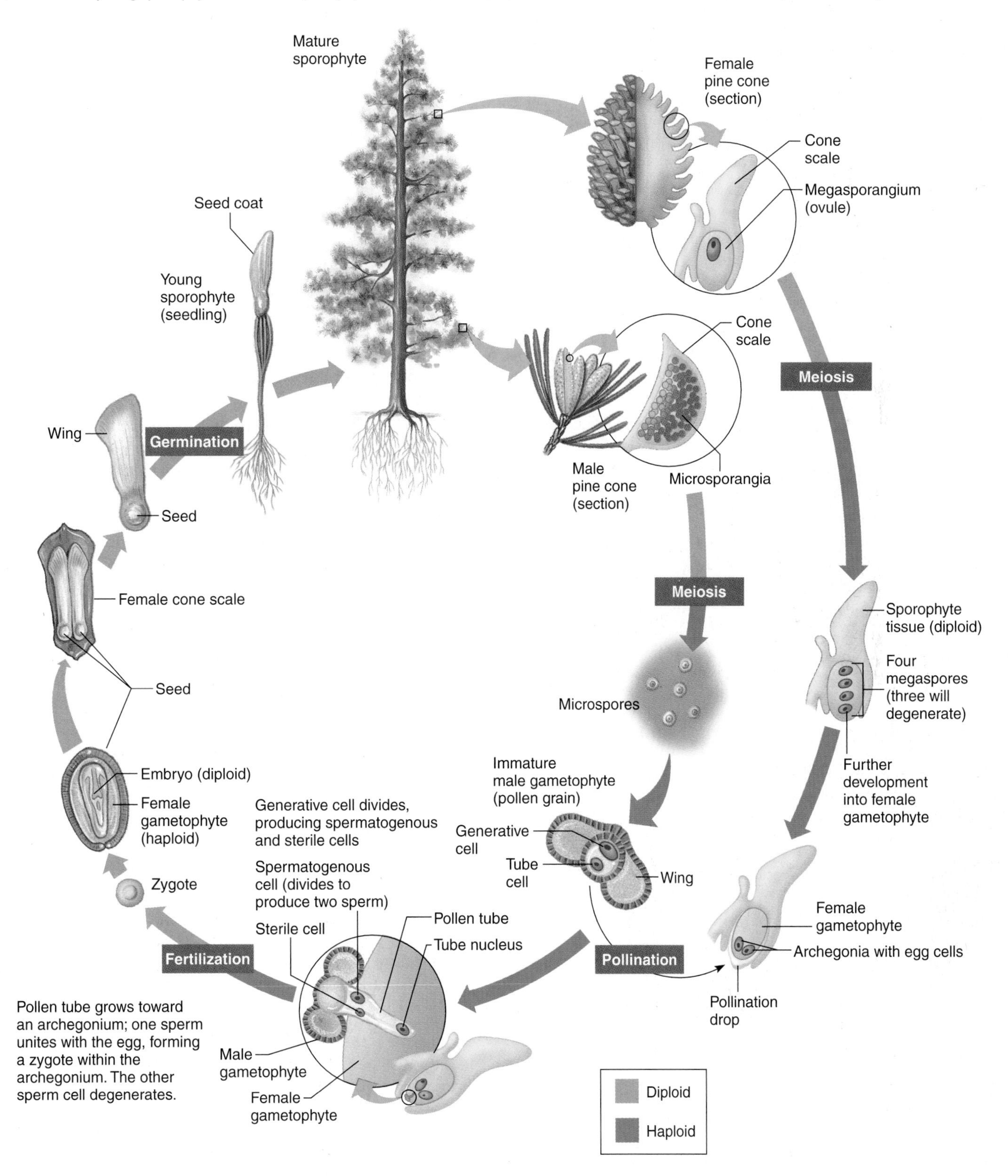

two ovules (megasporangia) on the upper surface of each scale. Through meiosis, each ovule produces four haploid structures called megaspores, three that degenerate and one that continues to develop into a female gametophyte. Over many months, the female gametophyte undergoes mitosis. Finally, two to six archegonia form, each of which houses an egg ready to be fertilized.

Small male cones bear pairs of microsporangia on thin, delicate scales. Through meiosis, these microsporangia produce microspores which eventually become pollen grains (immature microgametophytes). The microsporangia burst, releasing millions of winged pollen grains. Tapping a male cone in the spring releases a cloud of pollen. Most of these pollen grains, however, never reach a female cone.

Pollination occurs when airborne pollen grains drift between the scales of female cones and adhere to drops of a sticky secretion. Female cones are above male cones on most species, which makes it unlikely that pollen will drift upwards to land on female cones on the same tree. This is an adaptation that encourages outcrossing (mating between different individuals), which fosters new combinations of genes in the next generation.

After pollination occurs, the cone scales grow together, and a structure called a **pollen tube** begins growing through the ovule toward the egg. But the pollen is not yet mature—before the pollen tube reaches the egg, the pollen grain must divide twice more to become a mature microgametophyte. Two of the cells become active sperm cells, and one of them fertilizes the egg cell. The whole process is so slow that fertilization occurs about 15 months after pollination.

Within the ovule, the haploid tissue of the megagametophyte nourishes the developing embryo. Following a period of metabolic activity, the embryo becomes dormant and the ovule develops a tough, protective seed coat. It may remain in this state for another year. Eventually, the seed is shed, to be dispersed by wind or animals. If conditions are favorable, the seed germinates, giving rise to a new tree.

Angiosperms Are Flowering Plants

Angiosperms (division Anthophyta or Magnoliophyta) are plants that produce flowers (fig. 22.16). The angiosperm life cycle is complex, with alternation of generations and double fertilization. Chapter 28 considers angiosperm reproduction in detail, and figures 28.4 and 28.5 depict the life cycle. Investigating Life 22.1 describes rare types of flowering plants that "eat" other types of organisms.

Angiosperm Diversity More than 95% of modern plants are angiosperms, (fig. 22.17) and about half of the more than 230,000 species live in tropical rain forests. Except for conifer forests and moss-lichen tundras, angiosperms dominate major terrestrial zones of vegetation. Only about 5% of flowering plants are aquatic.

FIGURE 22.16
Angiosperms.
The angiosperms are distinguished by flowers, and seeds that are enclosed in fruits.

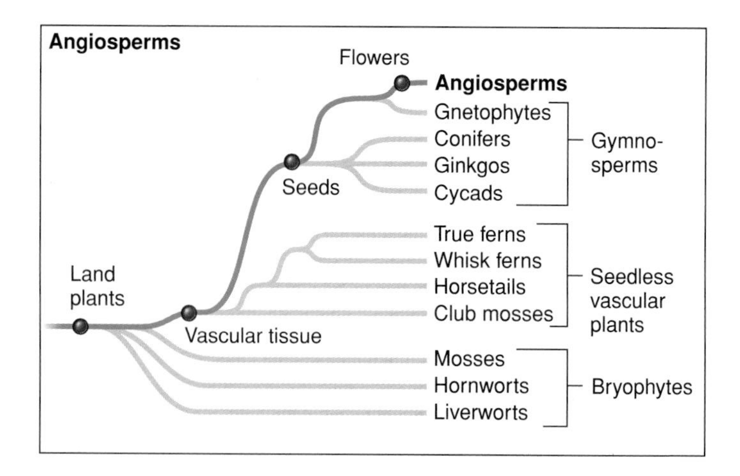

The two major classes of angiosperms are **monocotyledons** (monocots) and **dicotyledons** (dicots), which are distinguished by whether there are one or two embryonic seed leaves (fig. 22.18). The groups also differ by the other characteristics described in Table 22.1 (many of these structural differences are further described in chapter 26). However, these differences are far from absolute, and molecular evidence indicates that several types of plants thought to be dicots based on their seed leaves are actually more closely related to monocots. The monocots apparently are monophyletic (forming a clade), but the dicots are a more complex group.

Why Have Angiosperms Been So Successful? Flowering plants dominate the contemporary landscape, yet they are relatively recent arrivals. What accounts for the success of the an-

TABLE 22.1 Monocots and Dicots

CHARACTERISTIC	MONOCOTS	DICOTS
Number of seed leaves (cotyledons)	1	2
Number of flower parts is a multiple of	3	4 or 5
Presence of wood (secondary growth)	No	Yes
Configuration of leaf veins	Parallel	Reticulate (meshlike)
Pattern of vascular bundles in stem	Throughout	Ring
Root system	Fibrous	Taproot

FIGURE 22.17

Gallery of Angiosperms.

The angiosperms exhibit an astonishing variety of flowers and fruit. (**A**) In bananas (genus *Musa*), flowers occur in clusters. Yellow flowers and green developing fruits are visible in this photograph. (**B**) Red maple (*Acer rubrum*) produces small, bright red flowers that develop into winged fruits after fertilization. (**C**) The red passion vine (*Passiflora coccinea*) is a tropical plant with very showy flowers. (**D**) Cattails (*Typha latifolia*) are familiar wetland plants. The brown cylindrical "tails" are actually spikes of tiny brown flowers.

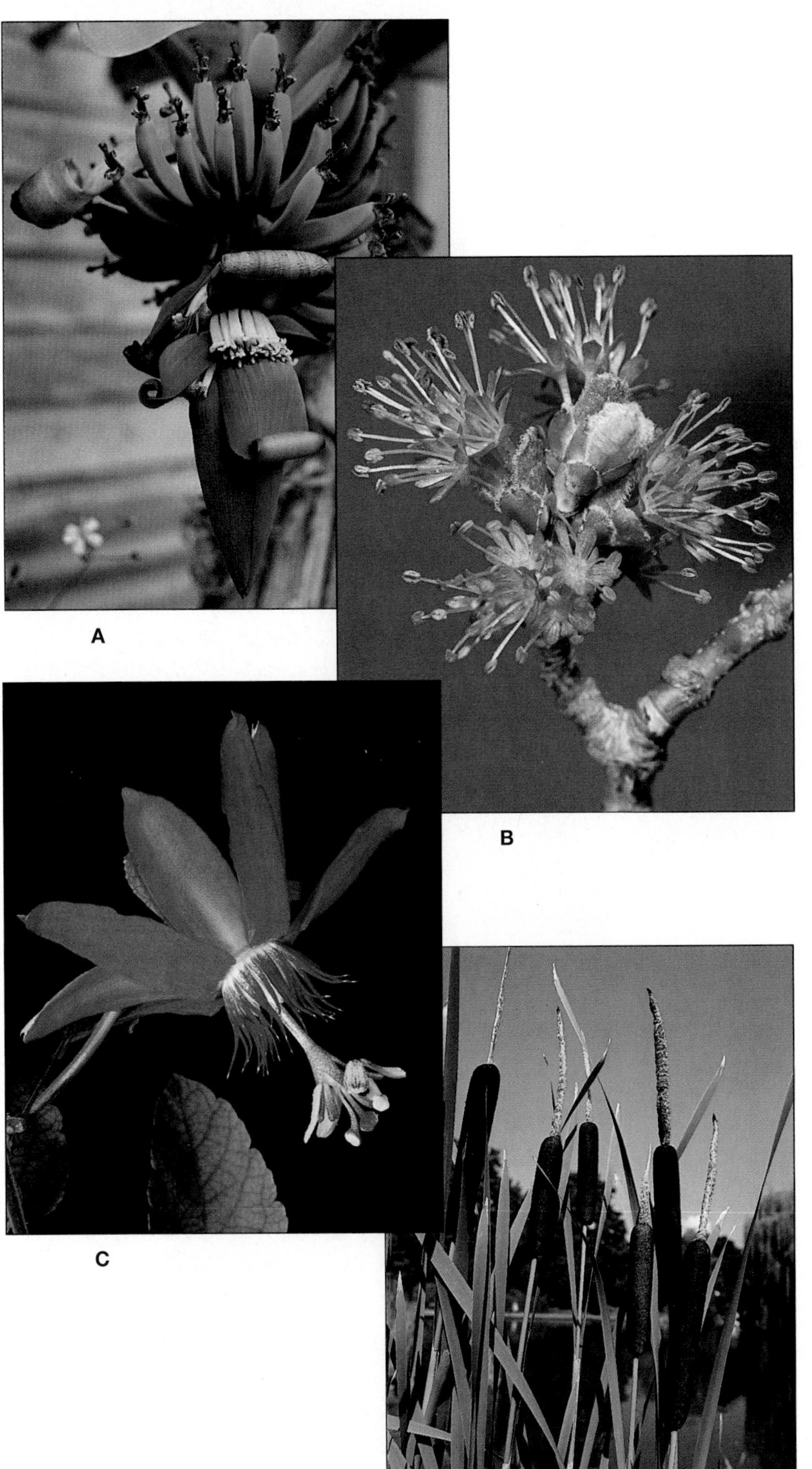

FIGURE 22.18

Monocots and Dicots.

Monocots have a single seed leaf (not visible in this figure), whereas dicots have two. Other characteristics distinguish these groups. Molecular evidence indicates that the monocots share a common ancestor, whereas the dicots form a more complex group.

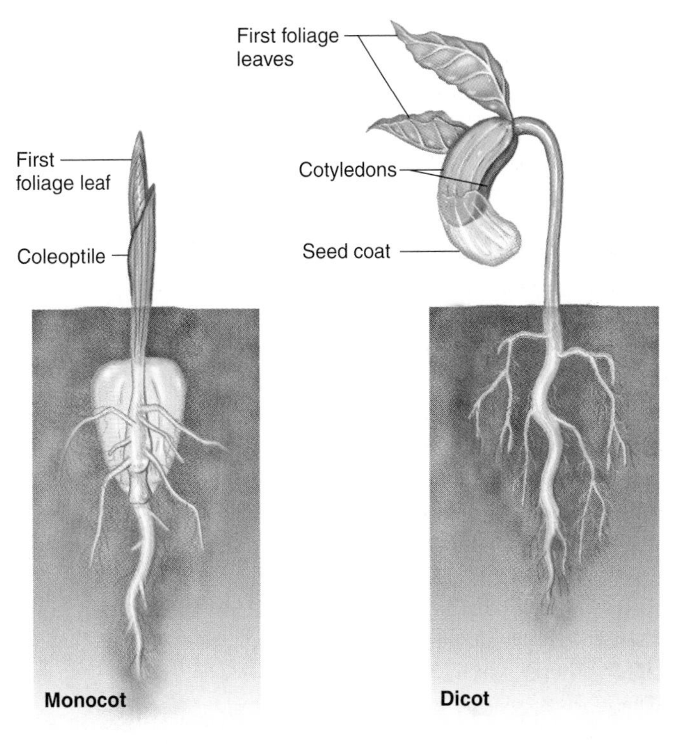

giosperms? Coevolution with animal pollinators may have played a part. Insects, reptiles, birds, and mammals transport pollen more efficiently than wind, which most gymnosperms rely on. Enclosing seeds within fruits also encourages widespread dispersal by animals. Another hypothesis to explain the success of the flowering plants is that the dietary habits of dinosaurs benefited angiosperms while harming gymnosperms.

The enormous sauropods, such as *Apatosaurus* and *Brachiosaurus,* roamed in great herds, browsing on the tops of conifers and other gymnosperms. This nibbling would not have killed large trees or harmed the smaller seedlings, which the giants could not reach. But by the late Cretaceous, smaller dinosaurs had

Investigating Life 22.1

Carnivorous Plants

To a human observer, the sundew plant is a magnificent member of a swamp community. To an insect, however, the plant's beckoning club-shaped leaves, bearing tiny nectar-covered tentacles, are treacherous. Once the insect alights on the plant and begins to enjoy its sweet, sticky meal, the surrounding hairlike tentacles begin to move. Gradually they fold inward, entrapping the helpless visitor and forcing it down toward the leaf's center. Here, powerful digestive enzymes dismantle the insect's body and release its component nutrients. After 18 hours, the leaves open. All that remains of the previous day's six-legged guest is a few bits of indigestible matter.

The Venus's flytrap (*Dionaea muscipula*) is another insect-eating plant, found on the coasts of North and South Carolina. Its two-sided leaves have highly sensitive trigger hairs that capture insects, which are then digested. (These plants, which are endangered in the wild because of habitat loss, are commercially produced by cloning and sold to the general public.) The pitcher plant (*Sarracenia*) entices insects to explore a hornlike structure that collects rainwater. When the hapless insect falls in, digestive enzymes go to work. By summer's end, pitcher plants contain many leftover insect parts. Figures 22.A–C show the sundew, Venus's flytrap, and pitcher plant.

Only about 450 plants display a carnivorous lifestyle, which is an adaptation that permits survival in nutrient-poor soils. These plants function actively, as the Venus's flytrap does in snapping shut, or passively, such as the pitcher plant's trap. All carnivorous plants were thought to attract and attack insects, until the recent discovery of a small flowering plant from South America and tropical Africa whose victims are ciliates and other protista.

Genlisea aurea is an unusual plant among unusual plants (figs. 22.D–G). It lives in white sands and on outcroppings of moistened rock, with a small rosette of leaves close to the ground and yellow or purple flowers extending up a foot or so from the leaves. Underground are extensive outgrowths that look like roots, but are actually highly

A

B

C

FIGURE 22.A–C
Carnivorous Plants.
(**A**) The sundew plant grows in the swamps of upstate New York and the Pine Barrens of New Jersey. Its sticky leaves are insect traps. (**B**) The leaves of the Venus's flytrap snap shut to catch prey. This plant lives only in North and South Carolina. (**C**) Leaves of the pitcher plant *Sarracenia* trap and kill insects, which the plant digests.

FIGURE 22.D–G

A Ciliate-Eating Plant.

(D) The rootlike structures in this *Genlisea* plant are actually underground leaves, highly modified to trap ciliated protists. **(E)** Ciliates swarm around a *Genlisea* leaf. **(F)** A close-up view of the leaf reveals the tiny ciliate traps and the long hairs **(G)** that prevent escape.

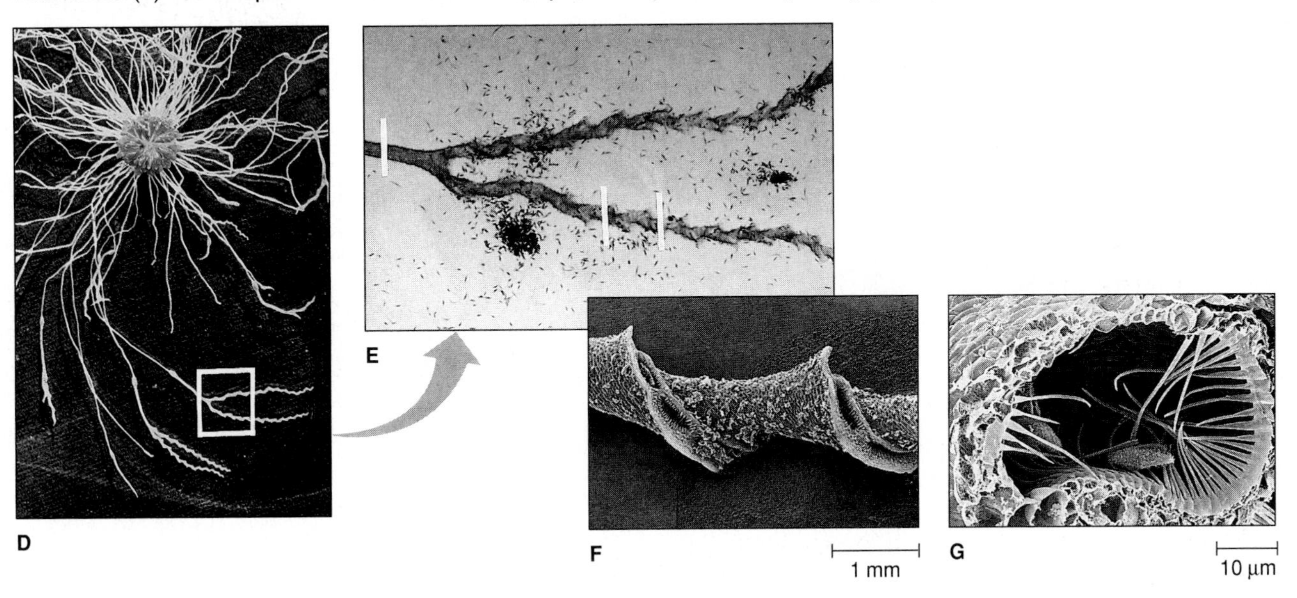

modified leaves. Rather than photosynthesizing, as leaves typically do, these structures have long hairs that ensure that ciliates swimming toward the plant cannot retreat.

Botanists had known for some time that *Genlisea aurea* was a carnivorous plant, but no one had observed its prey, assumed to be an insect. But the size of the opening between the traps, and the absence of insect evidence suggested another type of meal. So, researchers grew the plant in a greenhouse, and fed it something that seemed the right size—ciliates (a type of protist). The results were striking: The ciliates headed right for the underground leaves, which are booby-trapped with glands that secrete digestive enzymes. The ciliates did not approach roots of other plants, sending researchers looking for a chemical attractant, which they indeed found. Further experiments showed that when ciliates were fed radioactive food, the radioactivity appeared 2 days later in the rosette of leaves. Finally, sampling from the field identified several types of ciliates in the vicinity of *Genlisea*'s unusual underground traps.

---

begun to replace the large herbivores (fig. 22.19). The low browsers probably devastated the gymnosperms by eating small seedlings before the plants could reach maturity and produce seeds. Because the first angiosperms were smaller and herbaceous (nonwoody), they probably grew and reproduced more rapidly than the woody gymnosperms and therefore were more likely to produce seeds before they were eaten. Furthermore, small dinosaurs eating gymnosperms opened up new habitats for angiosperm invasion and evolution. When the dinosaurs became extinct in the Tertiary period, angiosperms exploded in abundance and diversity.

Perhaps dinosaurs' dining habits and coevolution with other types of animals influenced the rapid evolution of angiosperms. Whatever the precise triggers, the spurt of angiosperm diversity in the Cretaceous period set the stage for their spectacular global diversification.

22.4 MASTERING CONCEPTS

1. What are the characteristics of gymnosperms?
2. What are the four divisions of gymnosperms?
3. What happens during and after pollination in gymnosperms?
4. Why is it adaptive for male cones to be below female cones on a tree?
5. How do angiosperms differ from gymnosperms?
6. What are the types of angiosperms?
7. How might angiosperms have come to dominate terrestrial habitats?

OLC

FIGURE 22.19
The Rise of Angiosperms.
Tall gymnosperms flourished when dinosaurs were giants, because the reptiles didn't eat the lower portions of the plants. As smaller dinosaurs appeared and began eating younger gymnosperms, the plants began to die out, opening habitats for angiosperms.

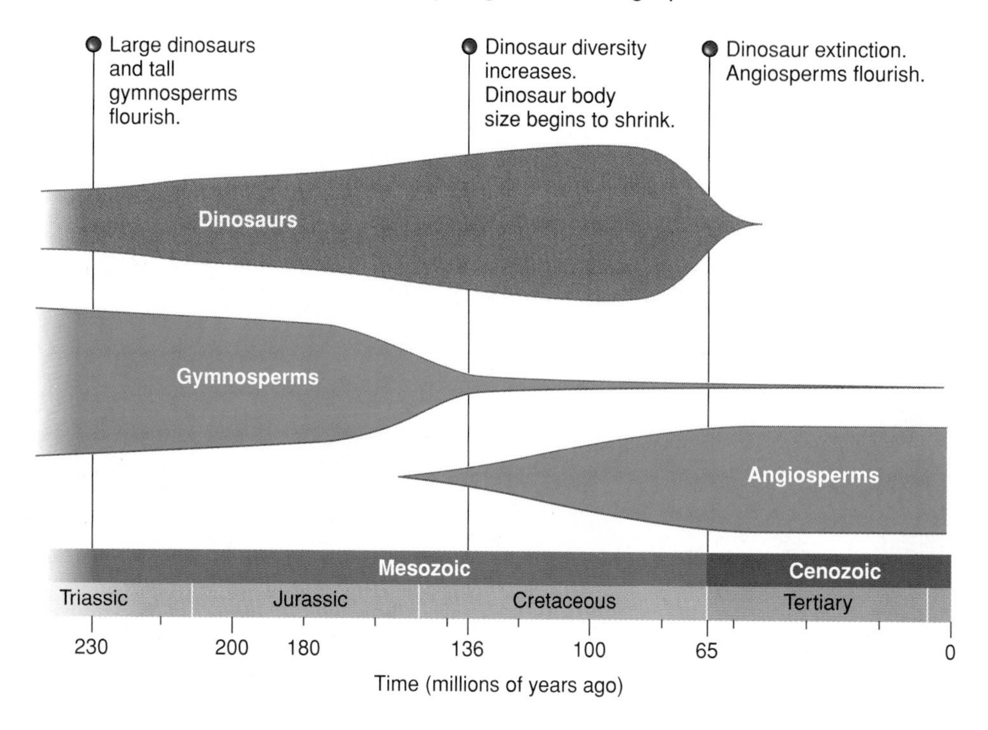

Chapter Summary

22.1 Introducing the Plants

1. Plants are multicellular eukaryotes that have cellulose cell walls and use starch as a carbohydrate reserve. Most photosynthesize.
2. Plants have alternation of a **sporophyte** (diploid) phase and a **gametophyte** (haploid) phase.
3. Plants are classified by presence or absence of **vascular tissue, seeds,** and flowers and fruits. **Bryophytes** lack vascular tissue. Vascular plants have vascular tissue and may be seedless or produce seeds. **Gymnosperms** have naked seeds, and **angiosperms** have seeds enclosed in fruits.
4. Plants originated about 480 million years ago from a type of green alga, **charophytes.** By 360 million years ago, plants had conquered dry land. Adaptations that made this possible include a dominant sporophyte, vascular tissue (**xylem** and **phloem**), roots and leaves, **lignin,** a waterproof cuticle, and stomata. By 200 million years ago, the angiosperms appeared and by 90 million years ago had diversified greatly.

22.2 Bryophytes Are the Simplest Plants

5. **Bryophytes** are small green plants lacking vascular tissue, supportive tissue, and true leaves and stems. **Lignin** hardens bryophytes and **rhizoids** anchor them to the ground, where they absorb water and nutrients.
6. In bryophytes, the gametophyte is dominant, and sexual reproduction requires water for sperm to travel through. Sperm from an **antheridium** travel to eggs in an **archegonium.**
7. The three divisions of bryophytes are **liverworts, hornworts,** and **mosses.**

22.3 Seedless Vascular Plants

8. Seedless vascular plants have the same pigments, reproductive cycles, and starch storage mechanisms as bryophytes, but they also have vascular tissue and a dominant sporophyte generation.
9. Seedless vascular plants include **club mosses, horsetails, whisk ferns** and **true ferns.**

22.4 Seed-Producing Vascular Plants

10. The seed-producing vascular plants are the gymnosperms and angiosperms. Neither requires water for sperm to meet eggs.
11. Gymnosperm seeds are not enclosed in fruits. They have diverse leaves and reproductive structures.
12. Gymnosperms include **cycads, ginkgos, conifers,** and **gnetophytes.**
13. In gymnosperms, male cones produce **pollen** and females cones produce eggs inside ovules. Pollen germination yields a **pollen tube,** through which a sperm cell fertilizes the egg. The resulting embryo remains dormant in a seed until germination.
14. Angiosperms are mostly terrestrial. The two major groups are **monocotyledons** and **dicotyledons.** The success of the angiosperms may reflect coevolution with animals, and dinosaur foraging habits.

Testing Your Knowledge

1. Describe a plant that was or is intermediate between each of the plants in the pairs described below, in terms of evolution.
   a. algae and the first plants
   b. bryophytes and vascular plants
   c. true ferns and angiosperms
2. What do all plant life cycles have in common?
3. What are two characteristics used to distinguish the major groups of plants?
4. Describe asexual and sexual reproduction in bryophytes.
5. List four adaptations that enabled plants to live on land.
6. How do angiosperms differ from gymnosperms?
7. Why are angiosperms considered to be successful? What adaptations and factors may have contributed to angiosperm success?

Thinking Scientifically

1. List the botanical extracts in five household products or drugs.
2. Why are bryophytes much smaller than most vascular plants?
3. Which evolutionary developments enabled seedless vascular plants to invade new habitats?
4. What advantage did the development of seeds offer in early plant evolution?
5. Some scientists hypothesize that a geological catastrophe, such as an asteroid impact, killed the dinosaurs. If this is true, how might angiosperms have survived such a catastrophe?
6. Michael Crichton's novel and its film adaptation, *Jurassic Park,* featured large herbivorous dinosaurs and huge ferns. What organisms might have been featured if the film had been called Cretaceous Park? What if it had been Tertiary Park?
7. Why might carnivorous plants live primarily in damp areas?

References and Resources

Gandolfo, M. A., et al. August, 1998. Oldest known fossils of monocotyledons. *Nature,* vol. 394, p. 532. About 90 million years ago, many types of flowers lived.

The May 1999 issue of *Natural History* magazine is devoted to the flower. It has many interesting articles on plant diversity.

Swerdlow, Joel L. April 2000. Medicines in nature. *National Geographic,* vol. 197, p. 98. Many drugs are based on botanical products.

Wandersee, James H., and Elisabeth E. Schussler. February 1999. Preventing plant blindness. *The American Biology Teacher,* vol. 61, p. 84. Many *Homo sapiens* are guilty of being unaware of plants.

The *LIFE* Online Learning Center provides additional resources and tools for studying this chapter.
www.mhhe.com/life

CHAPTER 23

Fungi

Amphibian Killers

When populations of adult amphibians in various places around the world began to plummet a few years ago, researchers at first sought an environmental explanation. Might the thinning ozone layer have allowed ultraviolet radiation to reach the surface and weaken and kill the animals, perhaps by damaging their skins? Or could pesticides be the culprit? The pattern of spread of the dying frogs and toads, however, suggested an infection. The pathogen turned out to be an unexpected organism—a type of chytrid, a simple fungus.

The research that led to discovering a chytrid with an unusual taste for amphibians began in 1993, when veterinarian pathologist Lee Berger, at the University of Queensland in Australia, noticed what looked like spore casings on the skins of dead sharp-snouted day frogs that had died in the Melbourne Zoo. The frogs had come from high up in the tropical rain forest. Berger and others found similar casings on the skins of amphibians that had died in Panama, and by 1997, they had appeared among the dead in zoos in Philadelphia, Chicago, Washington, D.C., and New York City.

At first, investigators suspected that a protist was the cause, because of the shape of the telltale objects on the amphibian skin. Tests for protist DNA came up negative. Researchers could not detect any viruses, nor did any type of bacterium

In the outer skin layer, cells die or erode, and some thicken with extra keratin, while cells in the deeper layer divide more often than normal.

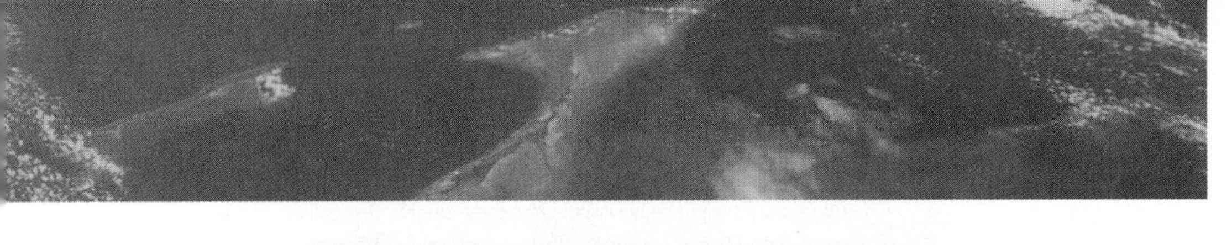

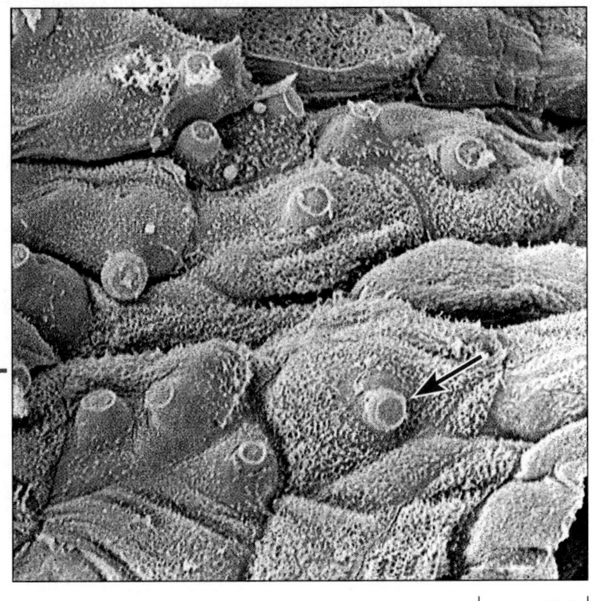

Amphibian Infection.
A chytrid is partly responsible for the decline in amphibian populations. Fungal extensions (small arrow) can be seen poking through these frog skin cells.

affect all of the dying species. Then closer examination with the electron microscope revealed fungal hyphae poking through the skin cells. In addition, DNA sequences were closest to those of known chytrids. The presence of unique, membrane-bounded lipid globules clinched the diagnosis of a chyrtid.

What was so unusual about this chytrid, called *Batrachochytrium dendrobatidis,* is that it attacks vertebrates—its relatives commonly parasitize plants or insects. The lethal illness that it causes in amphibians is called "cutaneous chytridiomycosis," in recognition of the infection site—the skin. In the outer skin layer, cells die or erode, and some thicken with extra keratin, while cells in the deeper layer divide more often than normal. The chytrid secretes enzymes that break down the keratin, then sucks up the liberated nutrients, as fungi do. These pathogens coat the legs and undersides of amphibians, impairing their ability to breathe through their skin. They also clog the pelvic patch, an area where amphibians take in water, which disrupts water balance. The nonkeratinized parts of the animals—the eyes, mouth, tongue, nose, and intestine—remain chytrid-free. Tadpoles only become infected in their mouthparts, which is the sole body part that contains keratin.

These pathogens coat the legs and undersides of amphibians, impairing their ability to breathe through their skin.

Experiments revealed that the chytrids are responsible for the amphibian die-offs, and are not just innocent bystanders. Berger took skin scrapings from infected frogs and added them to water. When he added healthy frogs, they sickened and died, showing the characteristic skin lesions. Although many biologists suspect that there may be additional causes of amphibian decline, it is clear that this newly discovered fungus plays an important role.

Wildlife biologists still do not know how or why the chytrid invasion of amphibians began. However it happened, it is a recent phenomenon, because examination of amphibian samples stored between 1987 and 1995 showed no evidence of the chytrids. One possible explanation is that an environmental problem, such as increased ultraviolet radiation or pollution, weakened amphibians so that a formerly nonpathogenic chytrid became a threat. An alternate hypothesis is that biologists unwittingly spread the chytrids from Australia to Panama or vice versa, and to zoo populations.

23.1 What Are Fungi?

Fungi are eukaryotes whose absorptive mode of nutrient acquisition profoundly influences other types of organisms and ecosystems. Their bodies grow toward their food. Molecular evidence generally supports traditional classification based on sexual structures, and also indicates that kingdom Fungi is the sister kingdom to the animals.

Fungi live nearly everywhere—in soil, in and on plants and insects, in water, in dung, and in the intestines of diverse animals. Cryptoendolithic ("hidden within rock") fungi live inside rocks in Antarctica. Microscopic fungi infect diatoms, while a single fungus extends for 1,500 acres beneath Washington State, coating tree roots (fig. 23.1).

Fungi are vitally important in the biosphere because they secrete digestive enzymes that break down living or once-living material. A fungus sends out extensions that probe the surroundings for food, shifting its direction of growth to maximize nutrient procurement. As decomposers, fungi release carbon, oxygen, nitrogen, and phosphorus from dead plants and animals in soil, recycling them to plants. Much of the 85 billion tons of carbon recycled to the environment each year passes through fungi. These organisms are, in a sense, the garbage processors of the planet. So vital are they to plants that fungal diversity in the soil largely determines the spectrum of species comprising plant communities, which in turn affects the types of animals present.

Despite the widespread distribution and varying habitats of fungi, we know little of their diversity. Mycologists—biologists who study fungi—estimate that they have identified only about 5% of the million or more species that exist. Fungi familiar to us include mushrooms and the black or colorful growths on food abandoned in the refrigerator. Fungi cause Dutch elm disease, chestnut blight, athlete's foot, diaper rash, ringworm, and yeast infections. They are used to manufacture cheese, bread, alcoholic beverages, dyes, plastics, soaps, toothpaste, drugs, soy sauce, flavorings, fuel, and food coloring. Health 23.1 at the end of the chapter describes a few fungi that cause illness.

Which Characteristics Do the Fungi Uniquely Share?

The fungi provide a powerful example of the value of molecular sequence evidence in sorting out major groups that descend from a common ancestor (fig. 23.2). Traditionally, mycologists classified fungi by sexual structures. Fungi in phylum Zygomycota (**zygomycetes**) produce thick-walled, black structures; Ascomycota (**ascomycetes**) house spores in characteristic sacs; and Basidiomycota (**basidiomycetes**) release spores from club-shaped structures. Fungi whose sexual status was unknown or unclear were formerly placed in a group called deuteromycetes. As molecular data reveal more about evolutionary relationships among fungi, biologists are gradually reclassifying these asexual fungi, usually as ascomycetes.

Molecular sequence analysis confirms that zygomycetes, ascomycetes, and basidiomycetes comprise distinct groups, and is aiding in placing the homeless deuteromycetes within them.

FIGURE 23.1
Fungi Range Greatly in Size.
(**A**) Some fungi are so small that they inhabit the cells of other organisms. The dark structures in these cotton cells are the fungus *Gigaspora margarita*. (**B**) Some fungi are enormous. One individual of *Armillaria gallica* in Michigan extends underground for 37 acres. Here, *A. gallica* parasitizes a white birch tree (*Betula papyrifera*).

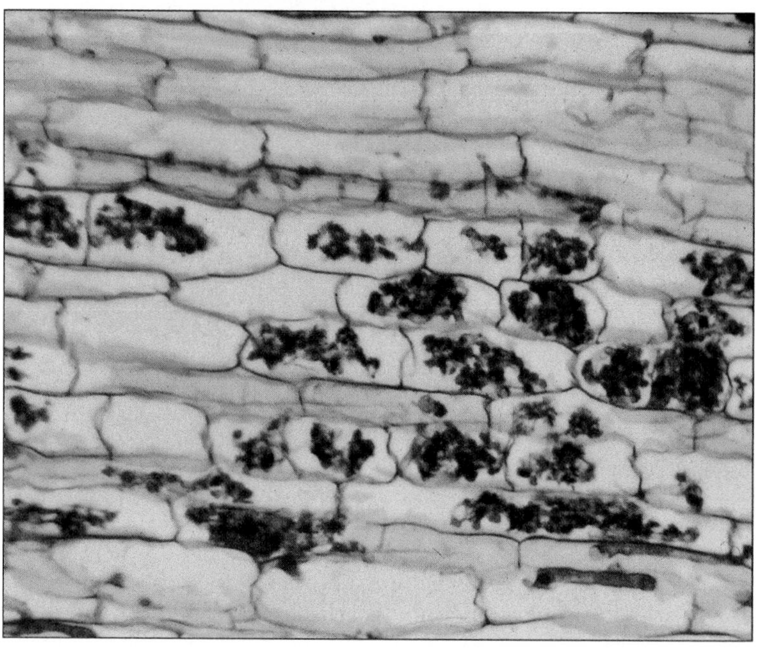

A

B

Molecular data have also established that the **chytridiomycetes** (chytrids) are fungi and not protista, where some biologists formerly placed them (see the chapter opening essay). Molecular evidence has reclassified several zygomycetes with the chytrids. On a broader level, molecular evidence clearly places the fungi closer to animals than to plants (see fig. 18.9), a finding that might surprise those who erroneously think that mushrooms are plants!

FIGURE 23.2

Fungal Phyla.

The fungal kingdom contains four phyla, which are distinguished mainly on the basis of spore type. Broken lines indicate uncertain relationships among phyla. The time at which fungi originated remains uncertain, but they evidently colonized land around the same time as plants.

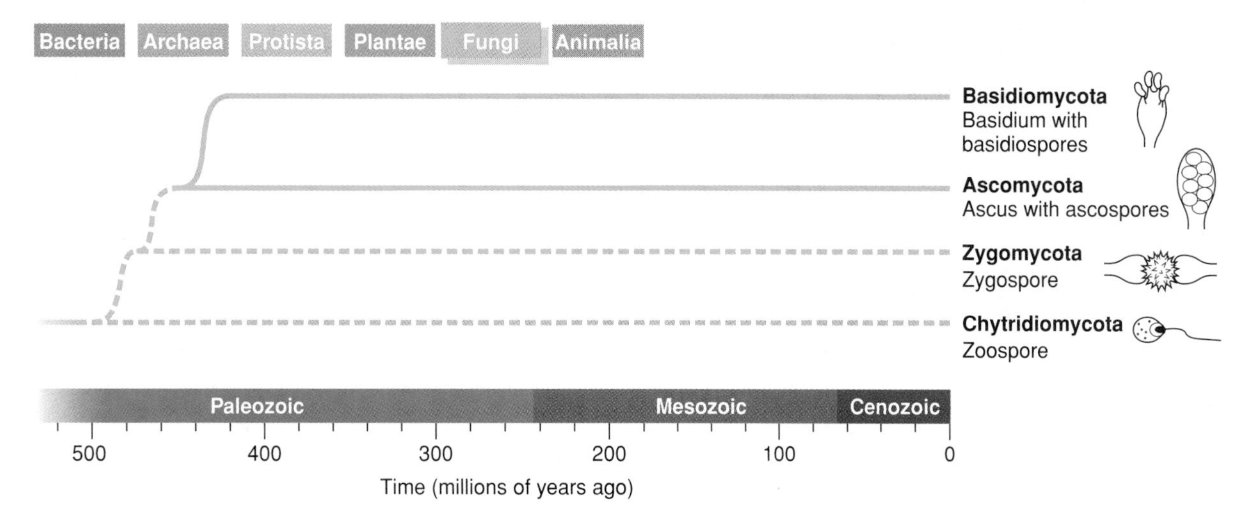

Fungi share several characteristics:

1. Cell walls are composed primarily of the modified carbohydrate chitin (see fig. 2.14). Plant cell walls, in contrast, consist mainly of cellulose.
2. The storage carbohydrate of fungi is glycogen, the same as for animals.
3. Fungi are heterotrophs, absorbing nutrients from other organisms. They do this in several ways. As **saprotrophs,** fungi secrete enzymes that break down dead organic matter to release sugars and amino acids. Fungi may also be parasites, breaking down nutrients within a living host. Some fungi are symbionts; about 80% of land plants have symbiotic fungi.
4. Like members of other eukaryotic kingdoms, many fungi have both asexual and sexual stages to their life cycles. Unlike other eukaryotes, however, in most fungi the zygote is the only diploid cell. Meiosis occurs in the zygote to yield haploid nuclei, which subsequently divide mitotically as the organism grows. Some fungi remain haploid throughout most of the remainder of their life cycles. Alternatively, two individuals may fuse to form a **dikaryotic** stage in which each cell retains two separate nuclei (see sections 23.4 and 23.5). Dikaryotic cells are unique to the fungi.
5. Fungi have some unique cell processes. For example, fungal cells use an unusual pathway for the manufacture of the amino acid lysine, and their mitochondrial RNA uses the codon UGA differently from other organisms. Also, fungal mitosis is slightly different from that in other organisms.

Fungi are also distinguished by what they cannot do. Unlike plants, they do not photosynthesize. Fungi lack the organized nervous tissue of animals, but they sense their environment and can react to it in very visible ways, as the carnivorous fungus *Arthrobotrys anchonia* in figure 23.3 illustrates. Some fungi communicate via biochemicals called pheromones, which are well studied among the insects.

FIGURE 23.3

A Carnivorous Fungus.

A sightless nematode worm burrowing through decomposing wood detects a sweet odor, and the worm approaches it. But the promise of food is a lure, set by the fungus *Arthrobotrys anchonia* (× 750). When the worm touches the sweet-smelling threadlike loops emanating from the fungus, it sticks and instantly becomes trapped. The loops constrict, swiftly crushing the worm. The fungus threads itself into its prey, releasing digestive enzymes and eating it from within.

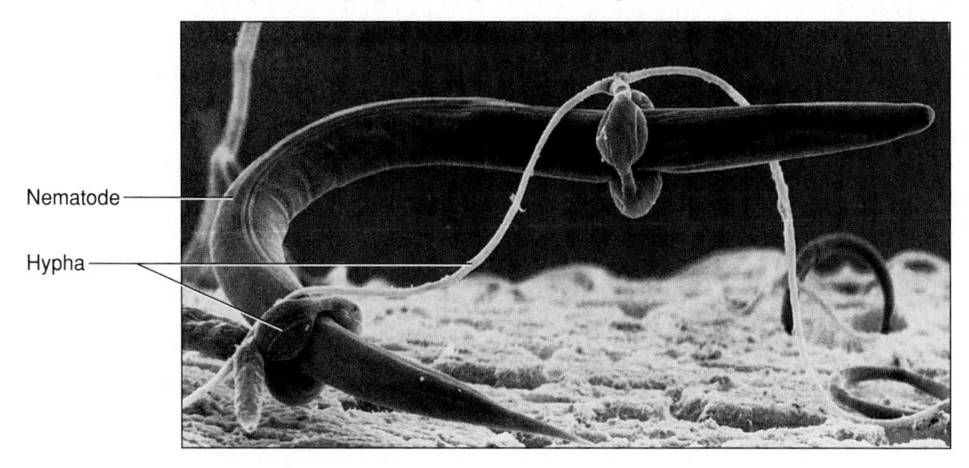

FIGURE 23.4
Yeast Cells.
A yeast is a unicellular fungus. This cell is reproducing by budding.

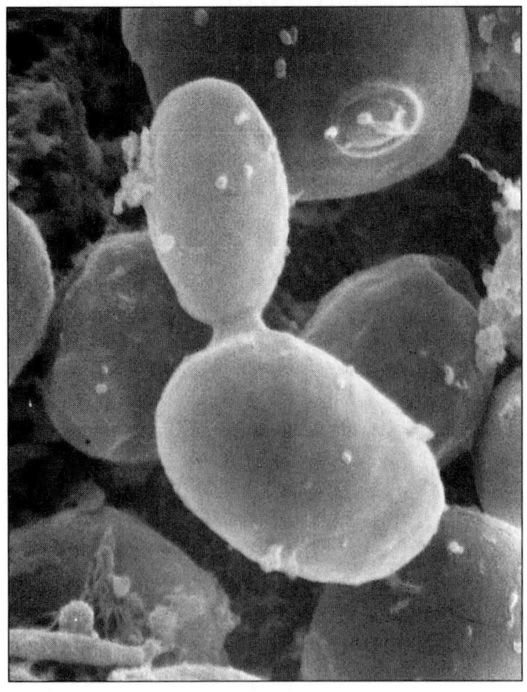

FIGURE 23.5
The Fungal Body.
(**A**) The fruiting body, a reproductive structure, is typically aboveground. It arises from the mycelium, which is a collective term for the extensive network of hyphae that penetrates the fungus' food source. Like the mycelium, the fruiting body is made of hyphae, but they are tightly aligned to form a solid structure. (**B**) Hyphae may be nonseptate with a continuous, multinucleate cytoplasm, or septate, with crosswalls delimiting individual cells.

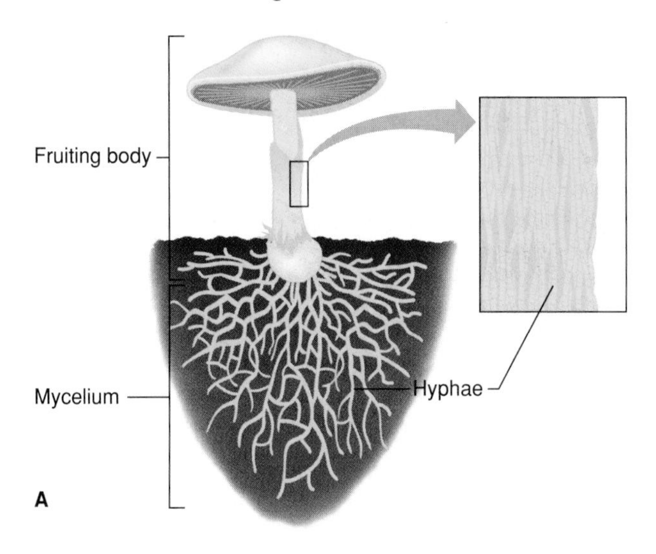

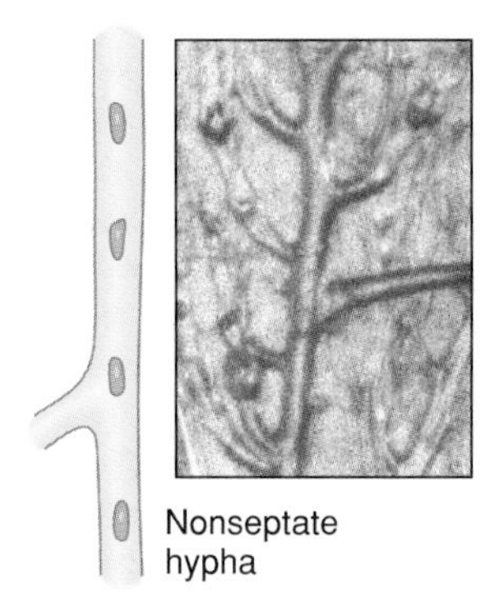

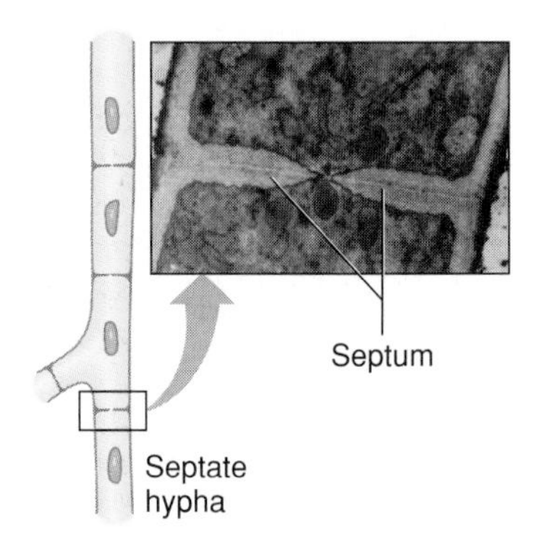

What Are the Parts of a Fungus?

The nonreproductive part of a fungus is called a **thallus,** which is a general term that refers to a relatively undifferentiated body form that lacks roots, stems, and leaves. Most fungi are multicellular, but many have a unicellular form called a **yeast.** A few types of fungi are strictly unicellular, although the single cells may aggregate into long strands that resemble the threads of multicellular fungi. The bread yeast *Saccharomyces cerevisiae* is a unicellular fungus (fig. 23.4).

The bulk of a multicellular fungus is the **mycelium,** which is composed of numerous threads called **hyphae** (singular: hypha)(fig. 23.5A). Often the mycelium is beneath the soil or within its food. The visible, familiar part of a fungus—a mushroom, puffball, or truffle, for example—is actually made of hyphae aggregated into a sexual structure called a **fruiting body.** A fungus may have an enormous number of hyphae that grow rapidly toward a food source. Hyphal growth may add up to a kilometer of new fungus daily! A hypha grows from its tip, and its diameter ranges from 0.5 to 100 micrometers. Thousands of microscopic hyphae may align, branch, or fuse, forming visible strands.

Hyphae are of two types (fig. 23.5B). In a **nonseptate hypha,** cytoplasm streams continuously along the length and encircles many nuclei, which are not separated into distinct cells. In a **septate hypha,** dividers called septa partition off cells, each of which may have one or more nuclei. The septa have pores that enable cytoplasm to flow from cell to cell.

Fungi also take the form of **spores.** A fungal spore is either a reproductive structure or a survival structure encased within a tough protective wall. Spores serve two functions. They may be carried to new locations, where food may be more plentiful, and they enable the organism to survive a harsh environment. Spores germinate when favorable conditions return.

Fungi produce and release so many spores that in some areas they outnumber pollen grains in the air; the record for fungal spore density is 5.5 million of them per cubic foot! Spores are also disseminated in flowing water, in rain, from the digestive tracts of animals, and on insects; some are forcibly released from fungal structures. Consider *Pilobolus,* a fungus that lives in cow dung (fig. 23.6). Its spores must pass through a cow's digestive system to germinate. To complete its life cycle, therefore, the fungus must send its spores onto grass, where cows will eat them and pass them, undigested, in their dung. The fungus grows a stalk topped with a black mass of spores. The stalk extends toward the sun, and

FIGURE 23.6
***Pilobolus*, the "Shotgun Fungus."**
(**A**) *Pilobolus* has a light-sensing organ that causes pressure to build, flinging spores onto the grass, where cows eat them. (**B**) The cows pass the spores in their dung, where they germinate to complete the cycle.

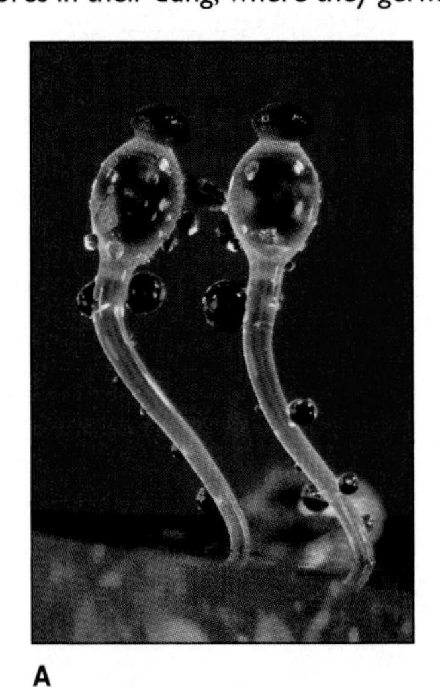

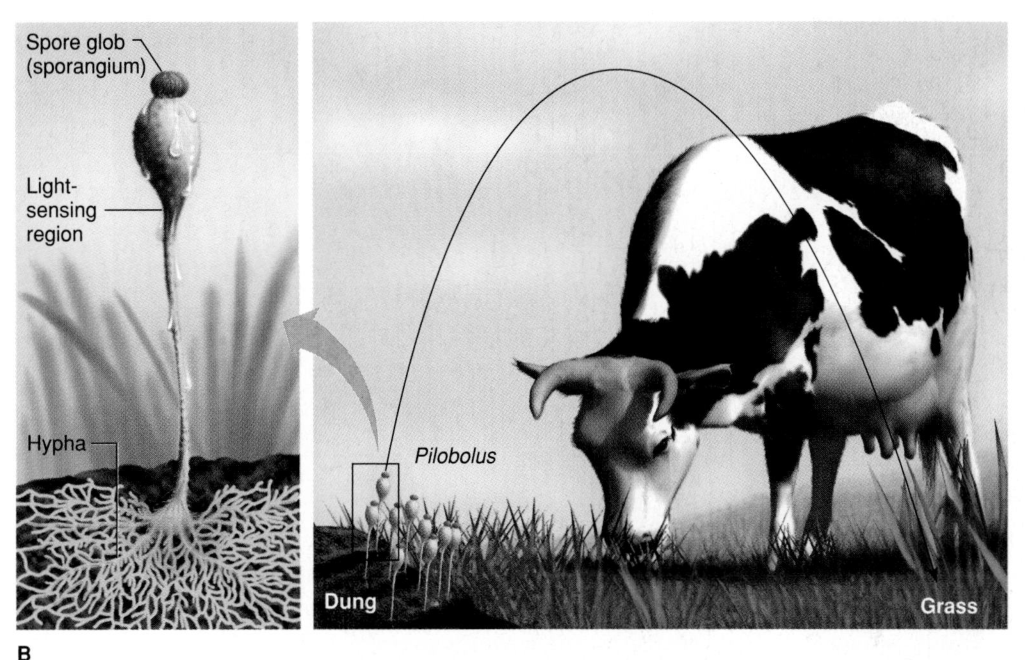

exposure to light causes water pressure to build until the stalk literally blows its top. The spores, accompanied by a sticky substance, shoot to a height of 6 feet (1.83 meters) and may travel as far as 8 feet (2.44 meters). This is far enough to get them out of the dung and onto the grass to continue the life cycle.

The Place of Fungi in Life History

The fungal fossil record is sparse, largely because of the lack of hard parts in these organisms. Fossils are often difficult to classify because the sexual structures (the basis for grouping fungi) are rarely preserved. Fossils usually represent spores and hyphae, and most are microscopic. The oldest fossils are from about 600 million years ago, from northern Russia, but the fossil trail could go as far back as 900 million years ago. Signs of fungi appear with insect fossils from the Silurian (435 to 410 million years ago) and with land plants from the Devonian (410 to 360 million years ago), entwined around and in vines, stems, and roots. By the end of the Devonian, all four fungal phyla were established.

Recall from unit 3 that molecular sequence evidence is often used to supplement the fossil record by inferring relationships from modern species. Such evidence indicates that fungi and animals diverged from a shared ancestor about a billion years ago. This ancestor was likely similar to the choanoflagellates, discussed in chapter 21. Molecular evidence also shows that the ascomycetes and basidiomycetes are sister groups, meaning that they share derived characters that no other groups have. They diverged most recently from a shared ancestor, as figure 23.2 indicates. Very little is known about the evolutionary relationships of the chytrids and zygomycetes within the kingdom.

23.1 MASTERING CONCEPTS

1. Why are fungi important in the living world?
2. Which characteristics define the fungi?
3. What are the four fungal phyla?
4. Describe the major parts of a fungus.
5. To what other kingdom are the fungi most closely related?

OLC

23.2 Chytridiomycetes Are Flagellated Fungi

Chytrids are microscopic and widespread, but little is known about them. They are unique among fungi in having a flagellated stage. Chytrids degrade the major carbohydrate coverings of organisms.

The chytrids are sometimes referred to as primitive fungi, and may provide a glimpse of what the earliest members of this kingdom were like. These microscopic fungi are found everywhere—in puddles, mud, soil, bogs, in marine and fresh water, on decaying insects and inside a wide variety of plants, winged insects, worms, algae, and other fungi. Their body forms vary from the unicellular to slender nonseptate hyphae (fig. 23.7A). Biologists have yet to discover the details of most chytrid life cycles.

Chytrids differ from other fungi in that they produce **zoospores** (motile spores), each with a single, posterior flagellum, which propels them toward light or food (fig. 23.7B). These fungi

FIGURE 23.7
Chytrids.
(**A**) Hyphae and sporangia of the chytrid *Allomyces.* Flagellated zoospores will emerge from the rounded sporangia at the tips of the hyphae.
(**B**) Zoospores from the chytrid *Blastocladiella.*

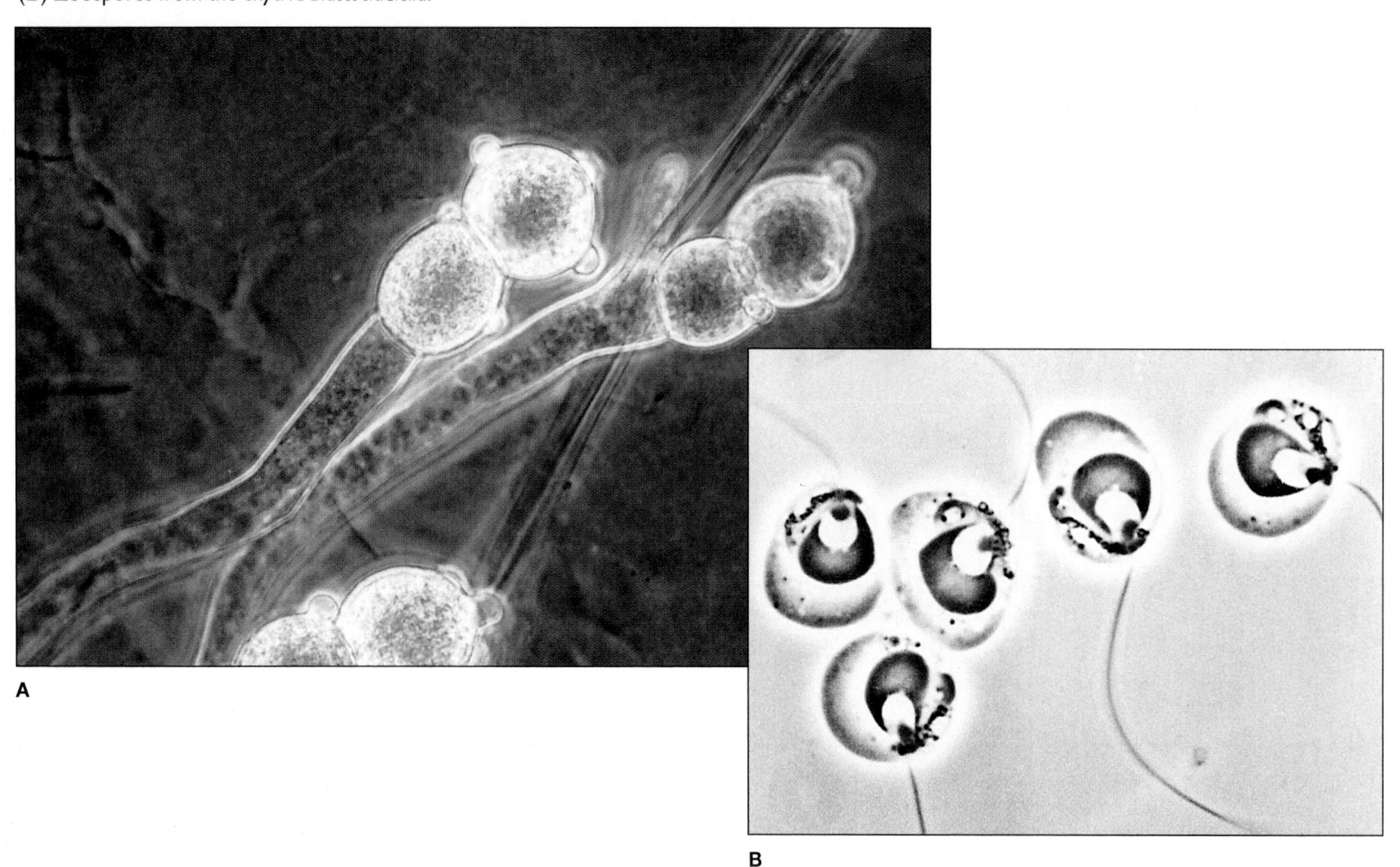

are powerful decomposers, secreting enzymes that degrade the three most common outsides of organisms—cellulose, chitin, and keratin, as the chapter opening essay discusses. One ecosystem where resident chytrids are particularly valuable is a ruminant's digestive tract. Anaerobic chytrids here start digesting cellulose in a cow's grassy meal, paving the way for bacteria to continue the process.

Some chytrids cause disease in familiar organisms. *Synchytrium endobioticum,* for example, causes "black wart" of potatoes. This pest is difficult to banish because the spores live as long as 40 years within thick chitin walls, surviving pesticides and crop rotation. Another chytrid, *Coelomomyces,* is used to control populations of the mosquitoes that cause malaria and yellow fever in humans.

23.2 MASTERING CONCEPTS

1. Where are chytrids found?
2. What are some characteristics of chytrids?

OLC

23.3 Zygomycetes Are Prolific, Nonseptate Fungi

The bread molds reproduce asexually from lightweight spores and reproduce sexually when haploid nuclei from cells of different mating types merge.

The 900 or so species of zygomycetes account for only about 1% of identified fungi, but include the familiar black mold, *Rhizopus stolonifer,* that grows on breads, fruits, and vegetables. In addition to forming black fuzz on refrigerated leftovers, the zygomycetes can be found on decaying plant and animal matter in soil, in plant roots, and as parasites of insects. These fungi are known for their spectacular growth rates. The zygomycetes have nonseptate hyphae.

Reproduction among the zygomycetes is both asexual and sexual (fig. 23.8), and both strategies involve spores. The asexual spores are haploid and are adapted for easy dispersal, as the explosive spores of *Pilobolus* show. Each spore gives rise to a hypha which grows into a mycelium. Within days, the new mycelium sprouts its own spore sacs, and the cycle begins anew. These

FIGURE 23.8

Zygomycete Reproduction.

In asexual reproduction, spore sacs release haploid spores that give rise to hyphae, which in turn produce more spores. In sexual reproduction, haploid nuclei from cells of different mating types merge to yield a diploid zygote. Meiosis occurs within the resulting zygospore, which germinates to yield a spore sac filled with haploid spores. The inset shows the zygospore between two vacated gametangia, a characteristic of this group of fungi.

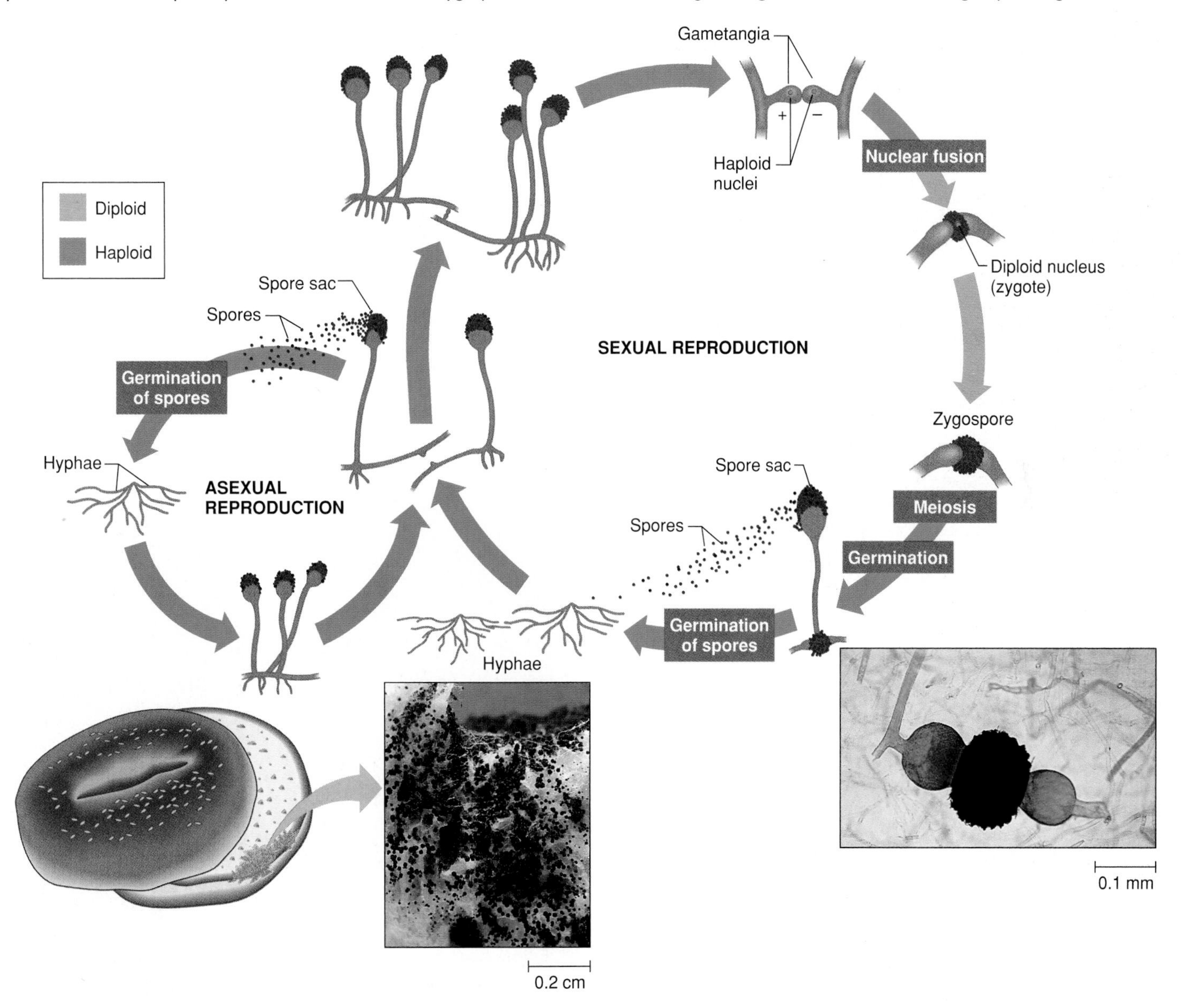

asexual growths can be seen on moldy bread, fruit, or in horse dung. Asexual spores of the zygomycete *Entomophthora muscae* appear as a white substance on and around dead flies, such as those that accumulate on window sills. Flies become infected when they explore the corpses (fig. 23.9).

The zygomycetes take their name from their mode of sexual reproduction—"zygos" is Greek for "yoke," which means "to join." This is what happens when hyphae of different mating types align, approach, and part repeatedly, as if in a dance. (In the complicated world of fungal sex, "mating types" are somewhat analogous to gender, but a given fungal species may have one to hundreds of different mating types instead of only two.) Then their haploid nuclei merge into a new structure called a **zygospore.** After the merger, the cells from which they arose, called gametangia, appear as vacated areas that hug the zygospore (fig. 23.8 inset). These structures are characteristic of the zygomycetes. The diploid zygospore nucleus undergoes meiosis, and a haploid hypha with a spore sac at its tip emerges. It eventually releases haploid spores. Each spore then gives rise to a hypha that grows into a haploid mycelium.

23.3 MASTERING CONCEPTS

1. Where are zygomycetes found?
2. How do zygomycetes reproduce asexually and sexually?
3. Give an example of a zygomycete.

OLC

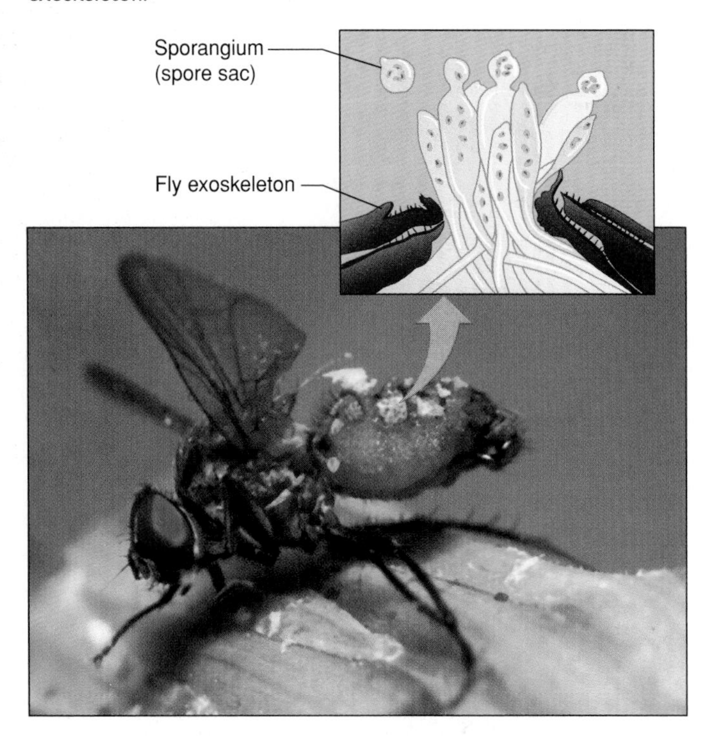

FIGURE 23.9
Fly Killer.
Entomophthora muscae is a zygomycete that grows on insects. The inset shows sacs containing asexual spores bursting from the fly's exoskeleton.

TABLE 23.1 Familiar Ascomycetes

SPECIES	USE OR DISEASE CAUSED
Aspergillus flavus	Produces aflatoxin, a carcinogen
Candida albicans	Thrush, diaper rash, yeast infections
Claviceps purpurea	Produces toxin that causes hallucinations
Cryphonectria parasitica	Chestnut blight
Morchella esculentum	Edible morel
Neurospora crassa	Bread mold, model organism for genetics research
Ophiostoma ulmi	Dutch elm disease
Penicillium chrysogenum	Antibiotic penicillin
Penicillium roqueforti and *camembertii*	Cheeses
Pneumocystis carinii	Pneumonia in people with immune deficiencies
Saccharomyces cerevisiae	Brewing and baking, model organism for research
Tuber magnatum	Edible truffle

23.4 The Ascomycetes Are Sac Fungi

The sac fungi include the delectable morels and truffles, but also species that cause disease. Their cells can have two nuclei, but most are haploid for most of the life cycle. Ascomycetes reproduce asexually by budding, and sexually with ascospores.

Ascomycetes are known for their eclectic tastes—under the right conditions, they will absorb nearly anything containing carbon, even paint and jet fuel! Some ascomycetes recycle plant material in soil, and others live with insects.

Ascomycetes cause various diseases of plants and animals and are used in the baking, brewing, and pharmaceutical industries (table 23.1). The fruiting bodies of two types of ascomycetes are prized for their unique and delicious taste—truffles and morels (fig. 23.10). Truffles grow on underground tree roots and are difficult to detect. For thousands of years people have used dogs and pigs to root these delicacies out of the ground. Squirrels and chipmunks disperse the spores of uprooted truffles. The French have attempted to farm truffles. Natural or farmed, they sell for $500 a pound! Morels grow aboveground and are easy to spot, although care must be taken to eat only nontoxic species. Morels grow in shaded areas that are moist and full of nutrients, such as orchards, swamps, flowerbeds, and forests, especially after a fire.

FIGURE 23.10
Edible Ascomycetes.
Familiar ascomycetes include (**A**) truffles (*Tuber melanosporum*) and (**B**) morels (*Morchella esculenta*).

Reproduction among ascomycetes is both asexual and sexual. For example, the well-studied yeast *Saccharomyces cerevisiae* reproduces asexually by budding (see fig. 23.4), but also reproduces sexually. In multicellular ascomycetes, asexual spores pinch off from the ends of exposed hyphae. In sexual reproduction, compatible mating types attract using pheromones. Their hyphae

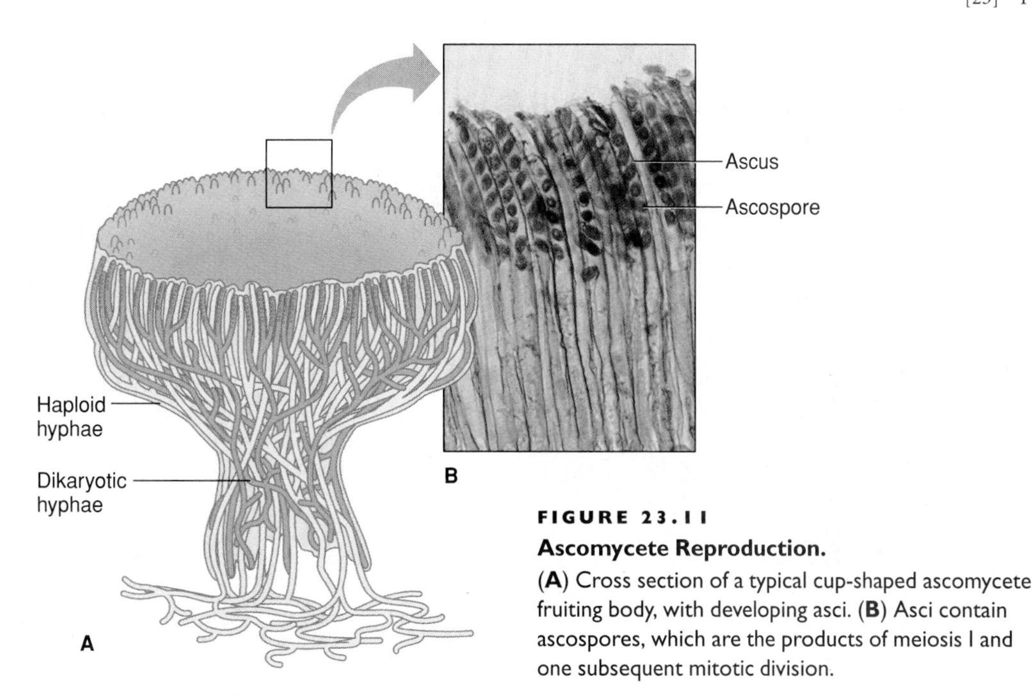

FIGURE 23.11
Ascomycete Reproduction.
(**A**) Cross section of a typical cup-shaped ascomycete fruiting body, with developing asci. (**B**) Asci contain ascospores, which are the products of meiosis I and one subsequent mitotic division.

fuse, but unlike in zygomycetes, the individual nuclei from the two parents do not immediately merge. The result is a dikaryotic cell, which occurs only in ascomycetes and basidiomycetes. When dikaryotic cells divide, the two nuclei undergo mitosis separately, each retaining its genetic identity.

In ascomycetes, the sexual fruiting body consists of haploid tissue (often in a cuplike or bottlelike shape) that houses a fertile layer of dikaryotic cells. Eventually the two nuclei in each dikaryotic cell fuse to form a diploid zygote, which immediately undergoes meiosis to form four haploid nuclei. Usually each haploid nucleus divides once mitotically to yield a total of eight haploid **ascospores,** so named because they form in saclike structures called **asci** (singular: ascus) as shown in figure 23.11. After dispersal, ascospores either persist in the environment or germinate to yield new haploid fungi.

The 30,000 ascomycete species include 500 that are single-celled yeasts, many that are multicellular, and some that are both. Most ascomycetes are haploid for much of their life cycles. Molecular data have placed *Penicillium* and *Candida* within the ascomycetes. They were formerly considered deuteromycetes because sexual structures were not observed.

Ascomycetes and basidiomycetes are sister groups. In addition to sharing DNA sequences, these fungi also have septate hyphae, and before sexual reproduction enter a dikaryotic stage.

23.4 MASTERING CONCEPTS

1. What are some places where ascomycetes live?
2. What are some examples of ascomycetes?
3. Why are ascomycetes and basidiomycetes considered to be sister groups?
4. How do ascomycetes reproduce asexually and sexually?

OLC

23.5 The Basidiomycetes Are Club Fungi

The club fungi include many familiar species, and may grow quite large. They have two genetically different nuclei per cell for most of the life cycle. Spore-bearing structures on mushrooms carry out sexual reproduction.

The 25,000 or so species of basidiomycetes account for about a third of all recognized fungi. They are named club fungi because during sexual reproduction, their hyphae tips form club-shaped, spore-bearing swellings called **basidia** (singular: basidium). Basidia often grow in pores or along strips of tissue called gills that are on the underside of the mushroom. Basidiomycete species differ from each other by the shape of the basidia, how many spores a basidium produces, and the organization of the gills or pores.

Familiar basidiomycetes include mushrooms, toadstools, puffballs, stinkhorns, shelf fungi, rusts, bird's nest fungi, and smuts (fig. 23.12). In extreme cases, these fungi may grow quite large—a puffball from Canada is 9 feet (2.75 meters) in circumference, a mushroom in England 100 pounds (45.4 kilograms), and the underground mycelium of *Armillaria gallica* growing in northern Michigan covers 37 acres (15 hectares). Some mushrooms are edible, some are deadly, and others are hallucinogenic. Smuts and rusts are pathogens of cereal crops. The extensive mycelia of some basidiomycetes grow outward from their food source in a circle beneath the ground, depleting the soil of nutrients. Mushrooms poke up above these spreading mycelia, creating the "fairy rings" of folklore.

Reproduction among the basidiomycetes is asexual and sexual. Asexual reproduction occurs by budding, fragmentation, and

FIGURE 23.12
A Gallery of Club Fungi (Basidiomycetes).
(**A**) Puffballs (genus *Lycoperdon*), (**B**) stinkhorns (*Phallus impudicus*), (**C**) turkey tail bracket fungi (*Trametes versicolor*), and (**D**) bird's nest fungi (order Nidulariales).

A

B

C

D

production of asexual spores. Figure 23.13 shows how a mushroom is a key player in the sexual reproductive cycle. Lining the mushroom's gills (or pores) are dikaryotic basidia, each housing two nuclei of compatible mating types. The haploid nuclei in each basidium fuse, giving rise to a diploid zygote, which immediately undergoes meiosis to yield four haploid nuclei. Each nucleus migrates to a **basidiospore** which, after being shed, germinates to produce a haploid hypha. Two hyphae of compatible mating types fuse to create a secondary mycelium whose cells are once again dikaryotic. This mycelium grows, producing a mushroom, and the cycle begins anew.

23.5 MASTERING CONCEPTS

1. Which features distinguish the basidiomycetes from other fungi?
2. What are some familiar basidiomycetes?
3. How do basidiomycetes reproduce?

OLC

23.6 Fungi Interact with Other Organisms

Many fungi wrap their extensions around plant roots, or penetrate root cells, bringing the plant minerals and receiving photosynthate. A fungus in intimate association with a green alga or a cyanobacterium forms a lichen. Ants cultivate certain fungi, with the help of bacteria that keep harmful fungi out.

The natural histories of fungi and plants are intimately intertwined. Fungi accompanied plants when they moved onto the land, and today remain an integral part of the underground root systems that sustain plant communities. Fungi can also associate with certain types of bacteria and algae to form a kind of compound organism called a **lichen.** Yet another fascinating fungal interaction is their cultivation by certain species of ants.

Mycorrhizal Fungi Live on or in Roots

Fungal hyphae in association with roots constitute **mycorrhizae** (literally, fungus roots). Nearly all land plants have mycorrhizae, and some, such as orchids, cannot live without their mycorrhizal associates. The symbiotic relationship benefits both partners—the plant obtains water and minerals that the hyphae absorb from the surroundings, and the fungus gains carbohydrates from the plant's photosynthetic activity. The relationships among root fungi and plants can become complex, with a single plant attracting several types of fungi, and the extensive hyphae

FIGURE 23.13

Reproduction in Basidiomycetes.

A mushroom is a short-lived reproductive structure of a basidiomycete (*1*). Basidia form on thin sheets of tissue called gills on the underside of a mushroom cap (*2*). Each dikaryotic basidium contains two nuclei, representing two compatible mating types. (*3*) The nuclei fuse, creating a diploid zygote, which immediately undergoes meiosis to form four haploid nuclei. The nuclei migrate into separate basidiospores, which the mushroom sheds (*4*). The spores germinate and new haploid hyphae grow (*5*). Hyphae of compatible mating types soon unite and form a secondary mycelium that is dikaryotic (*6*)—that is, with two nuclei per cell. This dikaryotic fungus grows (*7*) and forms a mushroom (*8*). The cycle begins anew.

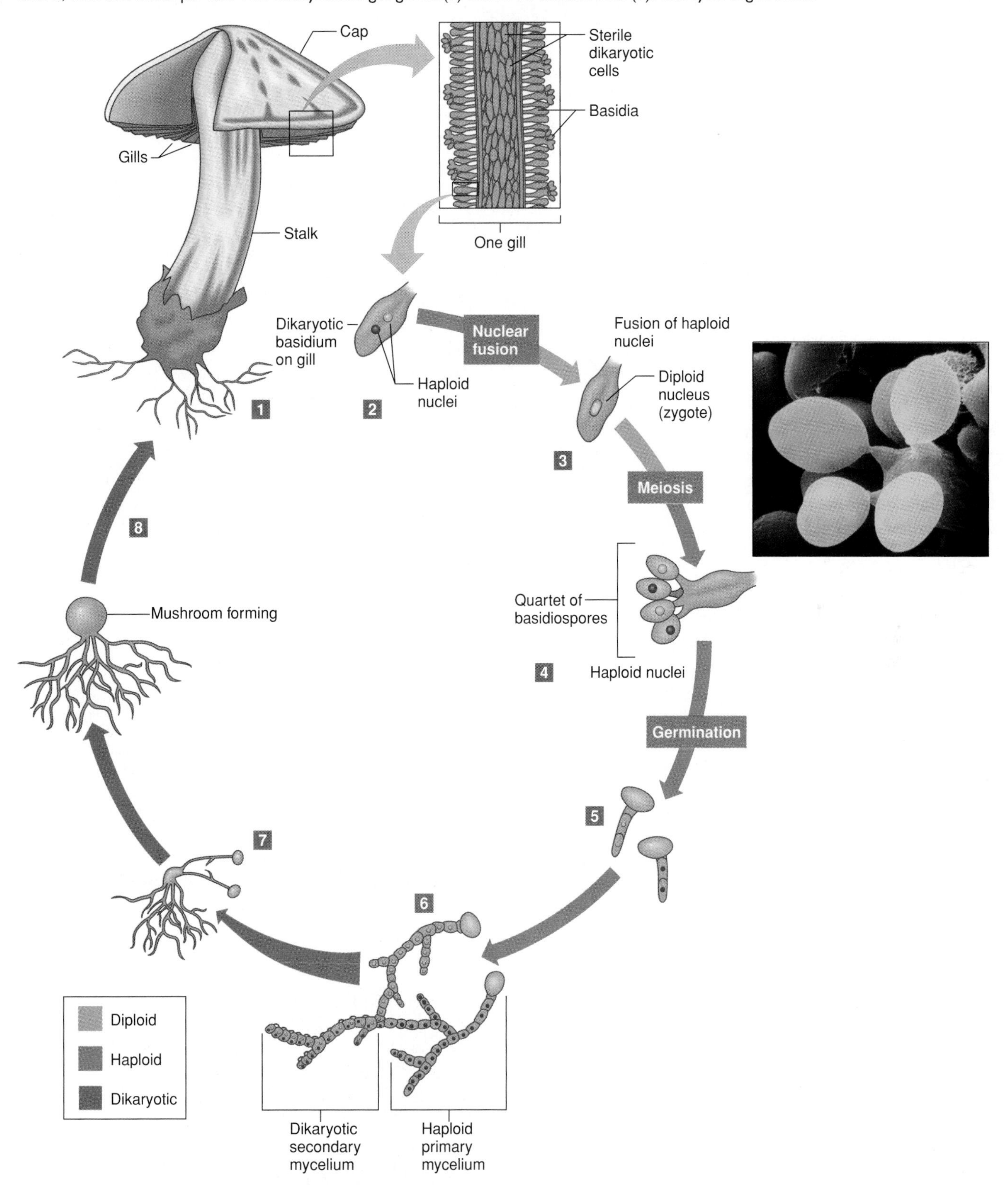

from different fungi interconnecting the roots of many plants.

Mycorrhizae are of two types. Ectomycorrhizal fungi (fig. 23.14) wrap around roots and may extend between root cells, but do not greatly penetrate the plant body. These fungi are found mostly in temperate trees and shrubs, such as beech, oak, birch, and conifers. Ascomycetes and basidiomycetes form ectomycorrhizae. In the other, more common type, endomycorrhizal fungi pierce the cells of roots in grasses, tomatoes, oranges, apples, beans, corn, wheat, and other crop plants, where they may occupy much of the host cell (see fig. 23.1A). Most endomycorrhizal fungi are zygomycetes.

FIGURE 23.14
Ectomycorrhizae.
(**A**) The roots of this lodgepole pine (*Pinus contorta*) seedling are colonized by an unidentified basidiomycete, which has produced a mushroom. The mycelium-covered mycorrhizal root tips are recognizable by their pale creamy to light brown color and Y-shaped branching pattern. (**B**) Hyphae of an ectomycorrhizal fungus wrap around a host's root tip.

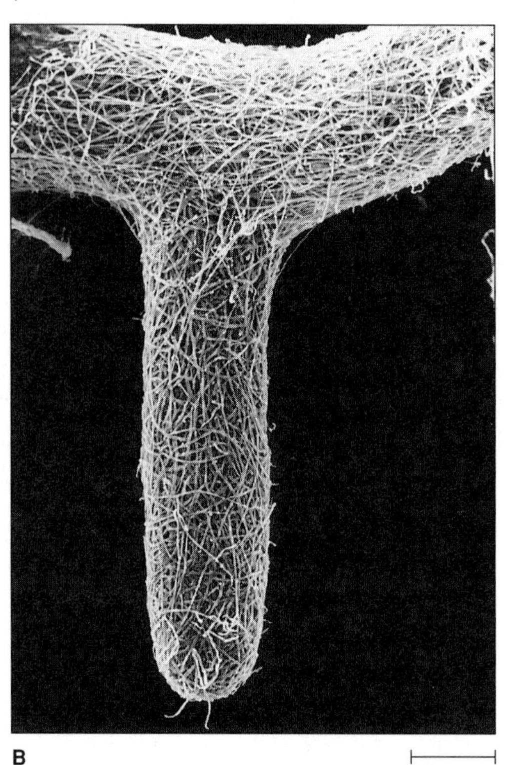

Lichens Are Distinctive Dual Organisms

A lichen is a dual organism that forms when certain fungi house green algae or cyanobacteria (fig. 23.15). Mycologists recognize about 13,500 types. Lichens are typically flattened growths that are green, orange, yellow, or black. Many lichens look quite different from either constituent, suggesting that the composite organism synthesizes compounds that neither partner does alone (fig. 23.16). Some green algae, for example, produce certain types of alcohols only when they are part of a lichen.

The algal or cyanobacterial part of a lichen contributes the ability to photosynthesize. The fungus contributes other synthetic capabilities. Experiments suggest that a lichen benefits the fungus at the expense of the alga. Grown alone, the alga flourishes compared with its growth rate as part of the lichen. In contrast, the fungal component cultured alone forms slow-growing, small colonies compared to its enhanced growth when it is part of a lichen.

Lichens grow on trees and rocks, in tiny crevices between mineral grains in rocks, on the backs of beetles, in the driest deserts, and in the wettest tropical rain forests. They dry out to survive extreme environmental conditions, springing back when moisture returns. They break rock down into soil, which can then support plant growth.

One type of habitat is hostile to lichens—polluted areas. Lichens absorb toxins, but cannot excrete them. Toxin buildup hampers photosynthesis, and the lichen dies. Disappearance of native lichens is a sign that encroaching pollution is disturbing the environment, and in fact lichens are used to monitor pollution.

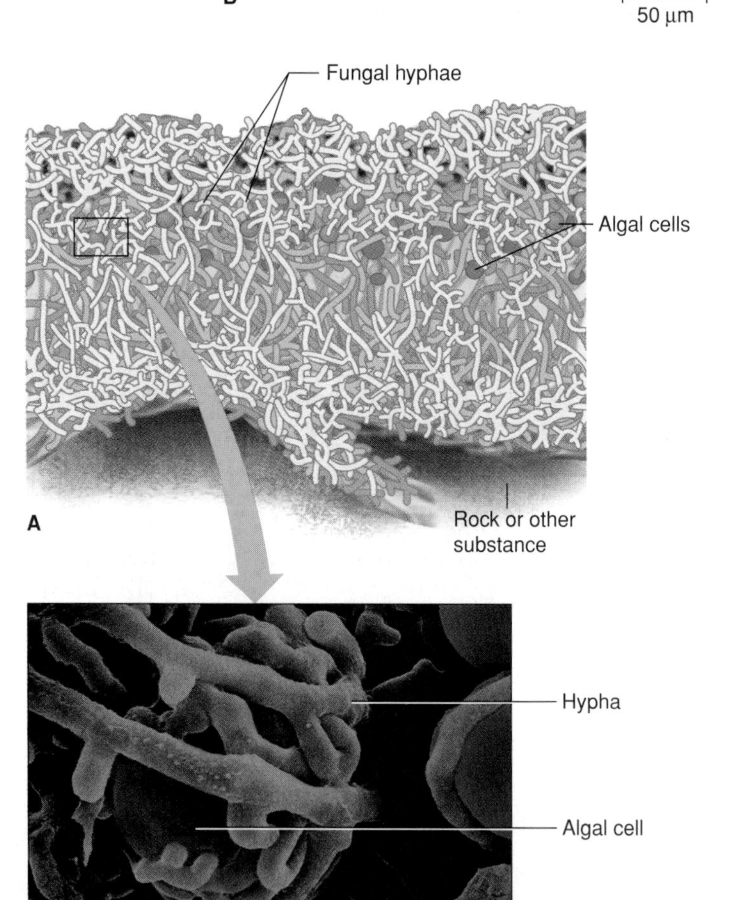

FIGURE 23.15
Anatomy of a Lichen.
Fungi associate intimately with certain green algae or cyanobacteria to form a composite organism called a lichen. This figure shows a cross section of a lichen encrusting a rock (**A**). Note that fungal hyphae wrap tightly around algal cells (**B**).

Lichens have left their mark on human history. The "manna lichen" *Lecanora esculenta* blows in the wind in the Near East, sometimes accumulating in drifts so large that people use it to make bread. Might this lichen have been the "manna from heaven" that sustained the starving Israelites in the desert during biblical times? More recently, Native Americans have used lichens to tan leather, poison arrowheads, treat infections and stomachaches, fashion fabric, and as a food source during famine.

Ants, Bacteria, and Fungi

The leaf-cutter ants of Central and South America and the southern United States have long been known for their cultivation of the basidiomycete *Lepiota* in special chambers in their underground colonies. A single colony may contain 2,000 such chambers. Into each, the ants transport a paste made from discs that they cut from live leaves, mixed with their saliva. The fungi grow on the paste, and both adult and larval ants eat the hyphae so quickly that mushrooms never get a chance to form—although they will when grown in a laboratory. The ants truck out old paste and bring in fresh supplies, and transport fungi to new chambers.

The ants and their fungal gardens constitute a mutualistic symbiosis—the ants eat, and the fungi have a home. Both benefit. But how do the ants keep out competitors or pathogens? Mycologists thought that the ants simply ate all interlopers, but recently a third partner in this symbiosis has been discovered. *Streptomyces* bacteria coat parts of the ants' cuticles and secrete a potent antibiotic that kills an ascomycete, *Escovopsis,* a virulent pathogen of the cultivated fungus.

This particular symbiosis-turned-ecosystem has four interacting species that we know of—ants that eat basidiomycetes, ascomycetes that feed on the basidiomycetes, and bacteria that enable the ant and its food to survive. Considering how few fungal species we know of, nature is certain to have many other examples of these intriguing organisms living with and within diverse organisms. Table 23.2 reviews the phyla of kingdom Fungi.

FIGURE 23.16
The British Soldier Lichen *Cladonia cristatella.*
A lichen looks like a single organism when viewed at a large scale.

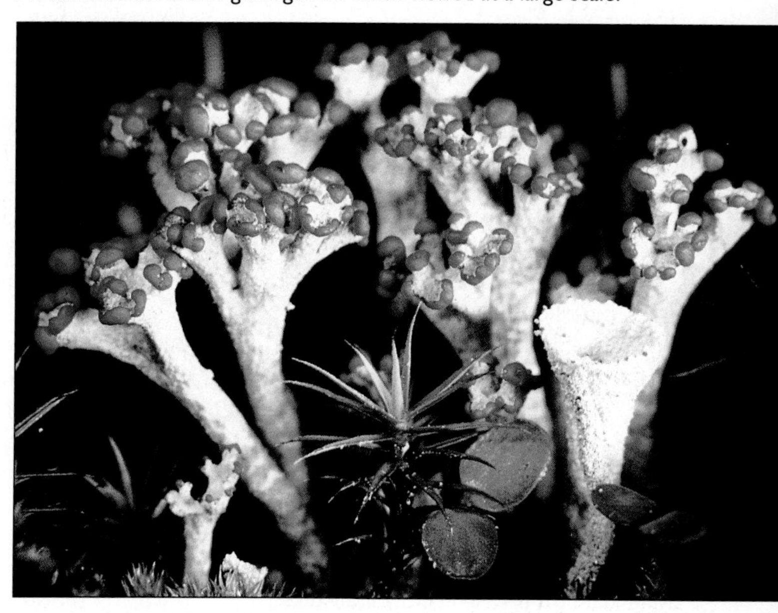

23.6 MASTERING CONCEPTS

1. What are three examples of interactions between fungi and other organisms?
2. What does the plant and the fungus gain from a symbiotic relationship beneath the soil?
3. What is a lichen?
4. How do ants, fungi, and bacteria interact in leaf-cutting ant colonies?

OLC

TABLE 23.2 Kingdom Fungi

PHYLUM	DISTINGUISHING FEATURES
Chytridiomycota (chytrids)	Flagellated zoospores (motile spores)
Zygomycota (bread molds)	Nonseptate hyphae Hyphae are haploid throughout life cycle; dikaryotic phase absent Three-part sexual structure of zygospore and vacated gametangia
Ascomycota (sac fungi)	Septate hyphae Haploid for much of life cycle Dikaryotic cells restricted to fruiting body Sexual reproduction yields haploid ascospores in asci
Basidiomycota (club fungi)	Septate hyphae Dikaryotic for much of life cycle Sexual reproduction yields basidia that bear haploid basidiospores

Health 23.1

Three Fungi That Affect Human Health

Pathogenic fungi cause about 50 recognized conditions in humans, ranging in severity from superficial skin outbreaks such as black piedra of the hair, to deeper skin infections such as ringworm, to systemic (bodywide) infections that harm major organs. Many opportunistic infections, which gain a foothold in people with suppressed immune systems such as those with AIDS or undergoing cancer chemotherapy, are fungal. A fungus infection that would be mild in an otherwise healthy individual is more severe and widespread in an immunosuppressed person. Here we look at three fungi that affect human health.

A Rain of Fungal Spores—Valley Fever

Valley fever is a systemic fungal infection caused by *Coccidioides immitis,* a fungus that lives in parts of Arizona, Nevada, California, New Mexico, Texas, and Utah. The fungus' spores, so small that they penetrate protective masks, are normally quiescent, lying undisturbed in dust and rarely entering human lungs. The spores burst into the air when rain follows a drought. People can then inhale the spores and become ill. The infection doesn't pass from person to person but from the disturbed environment to human lungs.

While usually extremely rare, Valley fever began increasing in prevalence in 1991, particularly in California's Kern County, in the San Joaquin Valley. Two simultaneous events triggered the outbreak—the end of a six-year drought and increased construction. Beating rain and excavation exposed and stirred up dust and released many millions of fungal spores (fig. 23.A). The 428 annual cases of Valley fever reported in California surged to more than 4,000 in the early 1990s.

At 4:31 A.M. Monday morning, January 17, 1994, a major earthquake shook central and southern California. As mountains rose, freeways fractured, gas lines and oil mains broke, and buildings collapsed. Dust was forcefully expelled into the air, releasing trillions of spores. The more than 1,000 aftershocks following the earthquake created repeated showers of fungal spores.

Hundreds of thousands of people became ill with Valley fever, technically known as coccidiomycosis. Symptoms could be as mild as those of an upper respiratory infection that lasts a few weeks and then clears up, or as serious as a prolonged high fever, bad headache, bumpy rash, swollen feet, fatigue, joint pain, pneumonia, or meningitis (inflammation of the membranes around the brain). Fortunately, antifungal drugs treated most cases of the illness.

FIGURE 23.A
***Coccidioides Immitis* Causes Valley Fever.** These microscopic, thick-walled structures, called spherules, each contain hundreds of fungal spores.

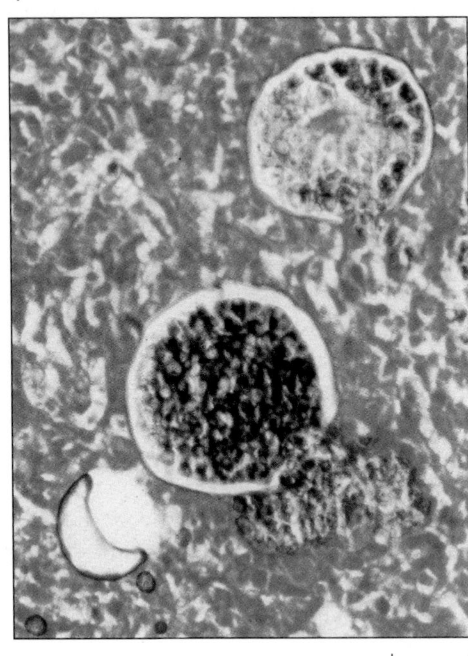

Histoplasmosis

A common systemic fungal infection is histoplasmosis. *Histoplasma capsulatum,* which causes the infection, grows within the macrophages (immune system cells) that normally fight infection. These macrophages then harm rather than help by spreading their fungal stowaways throughout the body.

Bird droppings transmit histoplasmosis, which is widespread in Mississippi, Kentucky, Tennessee, and Ohio. It produces symptoms in humans and bats, but not in the birds that

Chapter Summary

23.1 What are Fungi?

1. Fungi are widespread and profoundly affect ecosystems because they are decomposers and parasites.
2. The four fungal phyla are Chytridiomycota (**chytrids**), Zygomycota (**zygomycetes**), Ascomycota (**ascomycetes**), and Basidiomycota (**basidiomycetes**).
3. Fungal characteristics include chitin cell walls, glycogen, and a **saprotrophic** or parasitic mode of acquiring nutrients. Fungi have both asexual and sexual reproduction, and some have unique **dikaryotic** cells with two nuclei. Fungi do not photosynthesize.
4. A fungal body includes a **mycelium** built of threads called **hyphae,** which may form a **fruiting body.** Some fungi occur both as hyphae and as unicellular yeasts. Hyphae may be continuous (**nonseptate**)

transmit it, because their high body temperatures prevent symptoms from occurring. Most people experience fever, cough, and joint pain that clears up on its own. Exploring caves or working with bird droppings increases the risk of contracting this mild infection.

LSD and Ergotism

Lysergic acid diethylamide (LSD) is a drug that became widely known in the 1960s, when it was used recreationally because of its hallucinogenic effects. LSD is a derivative of lysergic acid, found in the ascomycete *Claviceps purpurea*. LSD was first synthesized in 1938, and its mind-altering effects first described in 1943, but the reputation of this fungus to affect human perceptions goes much farther back in time.

Claviceps purpurea parasitizes rye and other grasses, causing a disease of plants called ergot (fig. 23.B). People who eat bread made from infected grain can develop the convulsions, gangrene, and psychotic delusions of ergotism, also known as St. Anthony's fire because of a burning sensation in the hands. This illness killed tens of thousands of people in Europe during the Middle Ages. Some of the women in Salem, Massachusetts, who were burned at the stake as witches in the late 1690s may have been suffering from ergotism after eating infected grain; people thought their uncontrollable movements meant they were in the throes of demonic possession. More recently, in 1951, an outbreak of ergotism in a French town made 30 infected people race about insisting that snakes and devils were chasing them—5 people died.

FIGURE 23.B
***Claviceps Purpurea* Infects This Grain.**
Ergotism may result if the dark fungal structures become mixed in with grain used to mill flour.

C. purpurea and its extracts affect the human body in several ways—dilating the pupils and producing sleeplessness, tremors, dry mouth, elevated temperature, profuse sweating, and lack of appetite. Even short-term users of LSD may experience flashbacks, in which the sensations and perceptions of a "trip" recur without actually ingesting more of the chemical. Long-term use of LSD can lead to psychoses, including depression and schizophrenia. It is an extremely dangerous drug.

---

or divided into cells (**septate**). Fungi reproduce using **spores.** The nonreproductive part is a **thallus.**

5. Molecular evidence indicates that ascomycetes and basidiomycetes are sister groups, and that fungi diverged from animals, their closest relatives, about 1 billion years ago.

23.2 Chytridiomycetes Are Flagellated Fungi

6. Chytrids are microscopic fungi that produce flagellated, motile **zoospores.** They may resemble primitive fungi.
7. Chytrids decompose major biological carbohydrate coverings.

23.3 Zygomycetes Are Prolific, Nonseptate Fungi

8. Zygomycetes include bread molds. They grow and reproduce rapidly.
9. Asexual reproduction occurs via haploid, thin-walled spores.
10. Sexual reproduction occurs when gametangia of different mating types fuse their nuclei into a **zygospore.** The zygospore undergoes meiosis, then generates a spore sac. Spores germinate to yield hyphae, continuing the cycle.

23.4 The Ascomycetes Are Sac Fungi

11. Ascomycetes may be multicellular, unicellular, or both. They include important pathogens of plants and animals and have many uses in industry.
12. Hyphae are septate and haploid for most of the life cycle, although there is a brief dikaryotic stage.
13. Asexual reproduction occurs as asexual spores which are released from hyphal tips, or by budding.
14. In sexual reproduction, ascomycetes produce haploid **ascospores** in saclike **asci.**

23.5 The Basidiomycetes Are Club Fungi

15. Basidiomycetes include mushrooms and other familiar fungi. Many important plant pathogens are basidiomycetes.
16. Basidiomycetes are dikaryotic for most of the life cycle.
17. Asexual reproduction occurs by budding, fragmenting of a thallus, or with asexual spores.
18. Sexual reproduction occurs when haploid nuclei in **basidia** along the gills of mushrooms fuse, then undergo meiosis. The resulting haploid **basidiospores** germinate, and hyphae of compatible mating types fuse to regenerate the dikaryotic state.

23.6 Fungi Interact with Other Organisms

19. Symbiotic associations between fungi and roots are called **mycorrhizae.**
20. A **lichen** is a compound organism, with unique characteristics, that consists of a fungus in intimate association with a cyanobacterium or a green alga.
21. Leaf-cutter ants cultivate basidiomycetes that they eat, and bacteria on the ants kill a parasitic ascomycete.

Testing Your Knowledge

1. Identify each structure drawn below from this list:
   - **a.** mycelium
   - **b.** ascus
   - **c.** zoospore
   - **d.** nonseptate hypha
   - **e.** basidium with basidiospores
   - **f.** septate hypha
   - **g.** zygospore

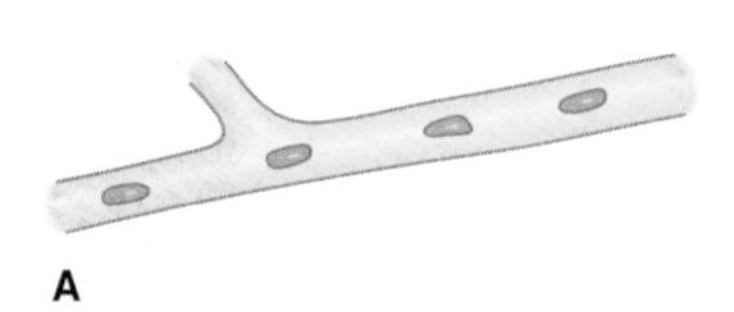

A

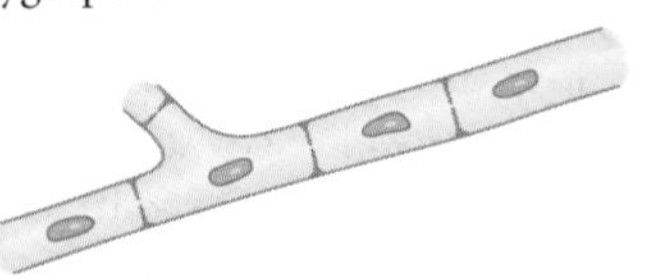

E

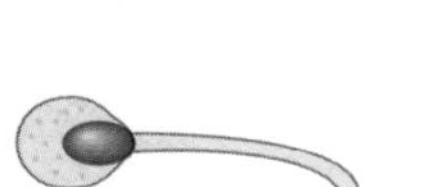

B

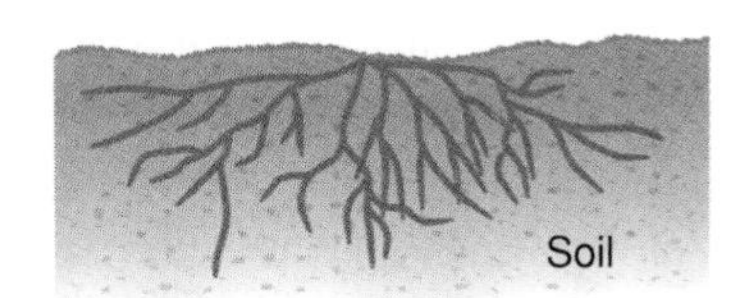

F

C

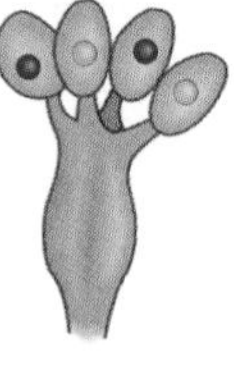

G

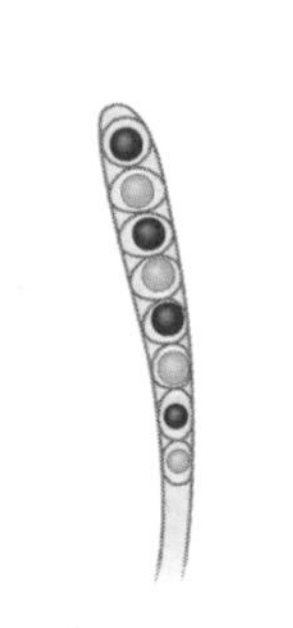

D

2. Which characteristics do all types of fungi share?
3. List the characteristics that distinguish the four phyla of fungi.
4. Distinguish between:
   - **a.** nonseptate and septate hyphae
   - **b.** a yeast and a mycelium
   - **c.** an ascomycete and a basidiomycete
   - **d.** ectomycorrhizae and endomycorrhizae
   - **e.** a saprotroph and a parasite
   - **f.** a zoospore and a zygospore
   - **g.** diploid and dikaryotic
5. Give an example of reclassification of a fungus based on molecular evidence.

Thinking Scientifically

1. Describe the relationships that connect:
   a. roots and mycorrhizal fungi
   b. leaf-cutter ants, their fungal food, bacterial symbionts, and pathogenic fungi
2. Penicillin, an antibiotic derived from a fungus, kills certain bacteria by destroying their cell walls. Why doesn't penicillin harm fungal cell walls?
3. Describe how experiments showed that
   a. chytrids are killing amphibians.
   b. fungi benefit more from a lichen relationship than do algae.
   c. bacteria help leaf-cutting ants cultivate one fungus while killing another.
4. What are two functions of spores?

References and Resources

Alexopoulos, Constantine, J., C. W. Mims, and M. Blackwell. 1996. *Introductory Mycology,* 4th ed. New York: John Wiley & Sons. An excellent textbook.

Heist, Annette. Joyous mushrooms, September 1999. *Natural History,* vol. 108, p. 48. A tale of fungus and feminism.

Kendrick, Bryce. 1992. *The Fifth Kingdom,* 2nd ed. Newburyport, Mass.: Focus Information Group, Inc. A very readable overview of the fungi.

Lewis, Ricki. June 1994. A new place for fungi. *BioScience.* Fungi are more closely related to animals than they are to plants.

Murawski, Darlyne A. August 2000. Fungi. *National Geographic,* vol. 198, p. 58. A tour through kingdom Fungi.

The *LIFE* Online Learning Center provides additional resources and tools for studying this chapter.
www.mhhe.com/life

CHAPTER 24

Animalia I—Sponges Through Echinoderms

Life on a Lobster's Lips

Discovering and describing never before seen organisms is nothing new to biologists. Finding *Symbion pandora* was unusual not only because of its habitat on the mouthparts of the Norwegian lobster *Nephrops norvegious,* but because its characteristics do not fit into any known animal phylum. Danish researchers discovered the animal in late 1995.

Until biologists can learn more about *Symbion pandora*'s distinguishing characteristics, it has been assigned to a new phylum, Cycliophora. The name in Greek means "small wheel" and refers to the circular, ciliated ring that is the animal's mouth. "Pandora" refers to the Greek myth of Pandora's box, in recognition of a stage in which a larva harbors a different feeding stage within its body.

S. pandora is less than a millimeter long. The life cycle is complex, with short swimming stages as well as lengthier immobile (sessile) stages. The animal is easiest for a human observer to see when it feeds, and has a saclike body that adheres by suction to a lobster's mouthparts. The gut continues from the ciliated mouth ring, opening as an anus near the mouth. In addition to large female and smaller male feeding forms, a larval stage moves but doesn't feed. Larvae are liberated from a degenerating female attached to a lobster. The end of molting signals sexual reproduction, and the male, for reasons unknown, has two penises.

"Pandora" refers to the Greek myth of Pandora's box, in recognition of a stage in which a larva harbors a different feeding stage within its body.

What exactly is *S. pandora*? The unusual internal budding that gives rise to larvae is reminiscent of two types of animals, the entoprocta (an obscure group) and the ectoprocta (which includes lacy-looking "moss animals" called bryozoa). Because of the limited information that researchers have, gene and protein sequence comparisons will be important in figuring out whether these curious organisms reside in one of the existing animal phyla, or will remain in their own group.

The researcher who codiscovered *S. pandora,* Reinhardt Kristensen, says that "this is only the beginning," and that many types of animals await our detection and description. Kristensen discovered another animal phylum, the Loricifera, in 1983. These animals live between grains of sand. It's intriguing to think of how much we do not know about this most familiar of biological kingdoms.

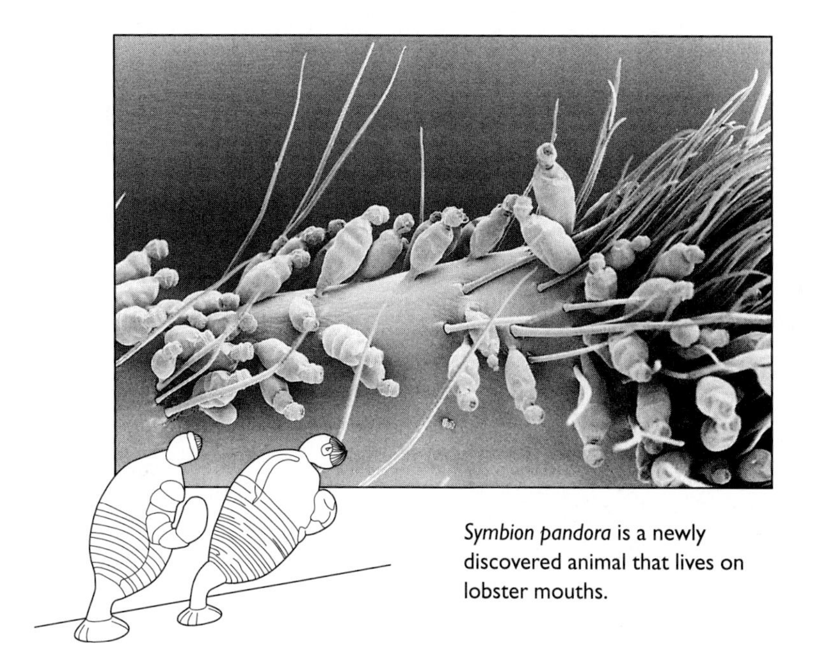

Symbion pandora is a newly discovered animal that lives on lobster mouths.

24.1 What Is an Animal?

Many people equate animals with the furry and familiar. The animal kingdom actually includes more than a million very diverse species. Animals probably evolved from choanoflagellates, and their first fossils date from 580 to 570 million years ago.

The animal kingdom includes an extremely diverse group of organisms, of which we are a member—despite our tendency to use such phrases as "humans and animals," or to equate animals solely with mammals. Biologists have catalogued more than a million animal species, yet possibly tens of millions more exist. About 45,000 known animal species are **vertebrates** (animals with backbones). Because most vertebrates are relatively easy to see, it is unusual to discover a new species. **Invertebrates** (animals without backbones) comprise most of the kingdom. Phylum Arthropoda, for example, includes more than 750,000 identified species of insects, crustaceans, spiders, and other organisms.

An animal is a multicellular organism with eukaryotic cells that lack cell walls. Most animals have diploid cells, sexual reproduction and perhaps asexual reproduction, and have bodies organized into cells, tissues, organs, and possibly organ systems (see fig. 1.1). Animals pass through immature developmental stages—embryos and/or larvae—and are motile at some phase of the life cycle.

The immediate ancestors of animals may have been the choanoflagellates. Recall from figure 21.1 that these protista bear an uncanny resemblance to certain sponge cells, and sponges are the simplest and most ancient of animals (fig. 24.1). Choanoflagellates can form colonies, and their ribosomal RNA sequences place them closest to the sponges, and far from plants and fungi.

FIGURE 24.1
Animal Beginnings?
An immediate animal ancestor may have resembled a choanoflagellate.

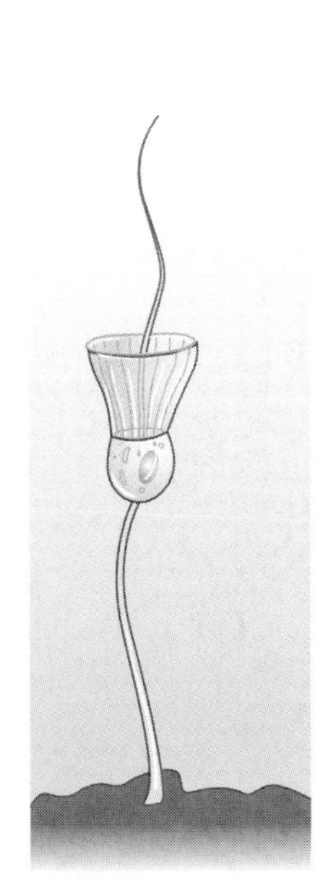

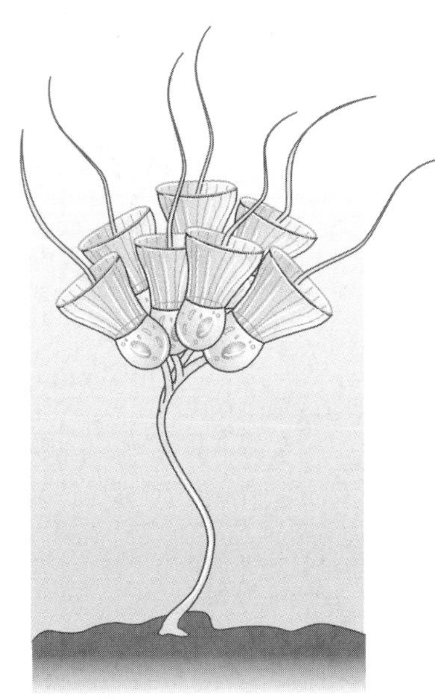

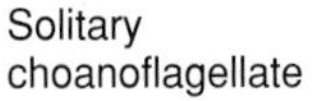

By 580 to 570 million years ago, the first animals lived in the seas. Recall from chapter 17 that unusual fossilization conditions led to the preservation of many embryos from this time, which most likely represent sponges, corals, jellyfishes, and sea anemones, or organisms similar to these (see figs. 17.7C and 18.12). These first animals lived alongside the mysterious Ediacarans, organisms that bear little resemblance to familiar forms today. By the time of the Cambrian explosion, from about 544 to 505 million years ago, the Ediacarans had vanished, and the seas brimmed with animals that left "small, shelly fossils," worms, and a great diversity of other animals (see fig. 18.14).

24.1 MASTERING CONCEPTS

1. What are some major types of animals?
2. From what type of organism did animals likely originate?
3. What were the earliest animals that we know of?

24.2 What Characteristics Define and Distinguish Animals?

Animals eat other organisms, move, and have cells that lack walls and secrete extracellular matrix. Biologists classify animals by body symmetry, presence or absence of a body cavity, mechanisms of reproduction and development, and molecular sequences.

Traditionally, biologists have classified animals on the basis of such characteristics as energy (food) sources, body complexity and form, and mechanisms of reproduction and development. More recently, cell and molecular level information has been incorporated to confirm or alter older classifications, as is true for the other kingdoms we have considered. For example, animal cells are unique in secreting an extracellular matrix that consists of collagen, proteoglycans, glycoproteins, and integrins. Depending upon its exact composition, this matrix enables some animal cells to move, others to assemble into sheets, and yet others to immerse themselves in supportive surroundings, such as bone or shell. Animals are not the only multicellular organisms, but they are the only ones to secrete an extracellular matrix.

Animals Obtain Nutrients from Others

Animals are heterotrophs, which means that they obtain carbon and energy from nutrients that are part of or released by other organisms, in contrast to manufacturing nutrients through photosynthesis. In an ecological context, animals are consumers because they eat others. An animal may be a parasite; a detritivore that eats dead organisms or nutrients in soil; or a predator. A carnivore is a predator that eats other animals, and an herbivore eats plants. We will return to these designations in chapter 43.

Body Complexity and Form

Recall from chapter 1 that multicellular organisms consist of cells, which associate with others of the same or similar type to form tissues, which in turn build organs, which may be linked into organ systems. These levels of organization are reflected in the levels of complexity of animal phyla.

One way to divide the animal kingdom is on the basis of cellular organization. Animals in subkingdom Parazoa have individual cells that do not work together in a coordinated fashion (that is, their cells do not form tissues; see chapter 30). Subkingdom Parazoa contains only one extant phylum (Porifera, or sponges). In sponges, cells aggregate to form a body, and different cells have distinct structures and functions, but they do not interact as they would in a true tissue. In contrast, all other animals (subkingdom Eumetazoa) have true tissues. For example, some cnidaria (jellyfishes, sea anemones, and corals) have cells linked into "nerve nets" that coordinate movement. In more complex eumetazoa, tissues come to form organs that carry out specific functions. The simplest animals to show this level of complexity are the flatworms (phylum Platyhelminthes), which demonstrate organ systems in their reproductive structures.

Biologists also describe animal bodies by symmetry, which is the organization of body parts around an axis (fig. 24.2). Many sponges are bloblike and therefore lack symmetry—they are asymmetrical. In a **radially symmetrical** animal, any plane passing from the oral end (the mouth) to the aboral end (opposite the mouth) divides the body into mirror images. Hydras have radial symmetry. In a **bilaterally symmetrical** animal, such as a crayfish, only one plane divides the animal into mirror images. Bilaterally symmetrical animals have anterior (head) and posterior (tail) ends. These animals move through their environment in a headfirst manner. This behavior is linked evolutionarily with **cephalization,** the tendency to concentrate neural elements such as sensory organs and a brain at the anterior end of the organism. Bilaterally symmetrical body forms also have back sides (dorsal) and belly or undersides (ventral).

Animal Reproduction and Development

Animals reproduce sexually, although there are variations on this theme. Some species can reproduce asexually, such as the budding of a hydra. Earthworms are **hermaphroditic,** producing sperm and eggs in the same individual. Among the social insects, such as bees and wasps, certain members are haploid. But for the most part, an animal's somatic cells are diploid, and the gametes (sperm and oocytes) are haploid. A sperm cell fertilizes an oocyte, restoring the diploid state in the zygote that results. The zygote—the first cell of the new organism—undergoes a series of rapid mitoses called cleavage divisions, forming the developing organism, or embryo. Chapter 40 discusses prenatal animal development in detail.

From the zygote, the embryo develops as a solid ball of cells that hollows out to form a **blastula.** The space inside is called the blastocoel. Soon complex foldings generate a structure called a **gastrula,** which has three primary germ layers, from which specialized tissues develop. These layers, called the **ectoderm, mesoderm,** and **endoderm,** are discussed further in chapter 40. Animal species have either ectoderm and endoderm and are termed **diploblastic,** or all three layers, and are **triploblastic** (fig. 24.3). The immature form of a jellyfish, for example, is diploblastic and consists of two layers—ectoderm and endoderm—that sandwich a layer of noncellular material. A human embryo, in contrast, is triploblastic and has three layers—ectoderm and endoderm, sandwiching the cellular mesoderm.

FIGURE 24.2
Animal Body Forms.
(A) Some sponges have an asymmetrical body form. **(B)** Hydra has a radially symmetrical body form. Any plane passing through the oral-aboral axis divides the animal into mirror images. **(C)** A crayfish has a bilaterally symmetrical body form. Only one plane divides the animal into mirror images.

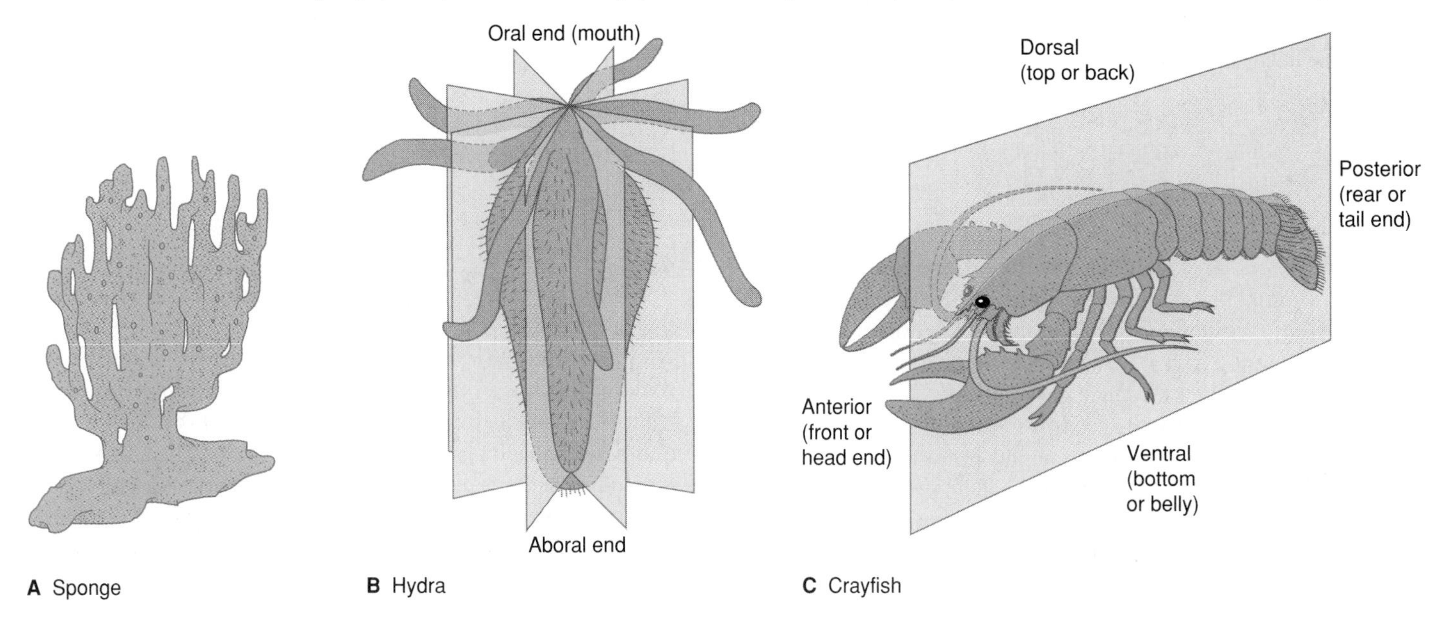

FIGURE 24.3
Gastrulation Produces Embryonic Tissue Layers.
Early during development of many animal embryos, a hollow ball of cells called a blastula undergoes gastrulation, which generates primary germ layers. In diploblastic animals these are the ectoderm (outer—blue) and endoderm (inner—green) layers. In triploblastic animals a third layer, the mesoderm (red), forms between the other two.

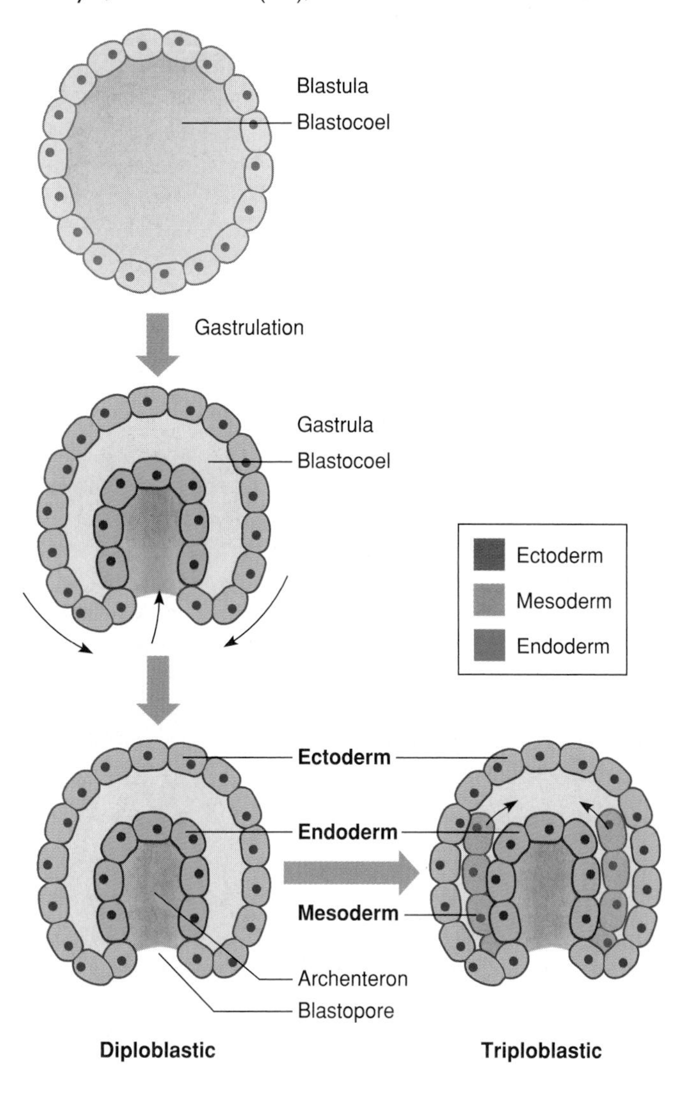

FIGURE 24.4
Body Cavities.
Presence or absence of a body cavity distinguishes animals. Certain worms are acoelomates (**A**), which lack body cavities. Some other worms are pseudocoelomates (**B**), with body cavities that are not completely lined with mesoderm. Vertebrates and many other types of animals are coelomates whose body cavities are completely lined with mesoderm (**C**).

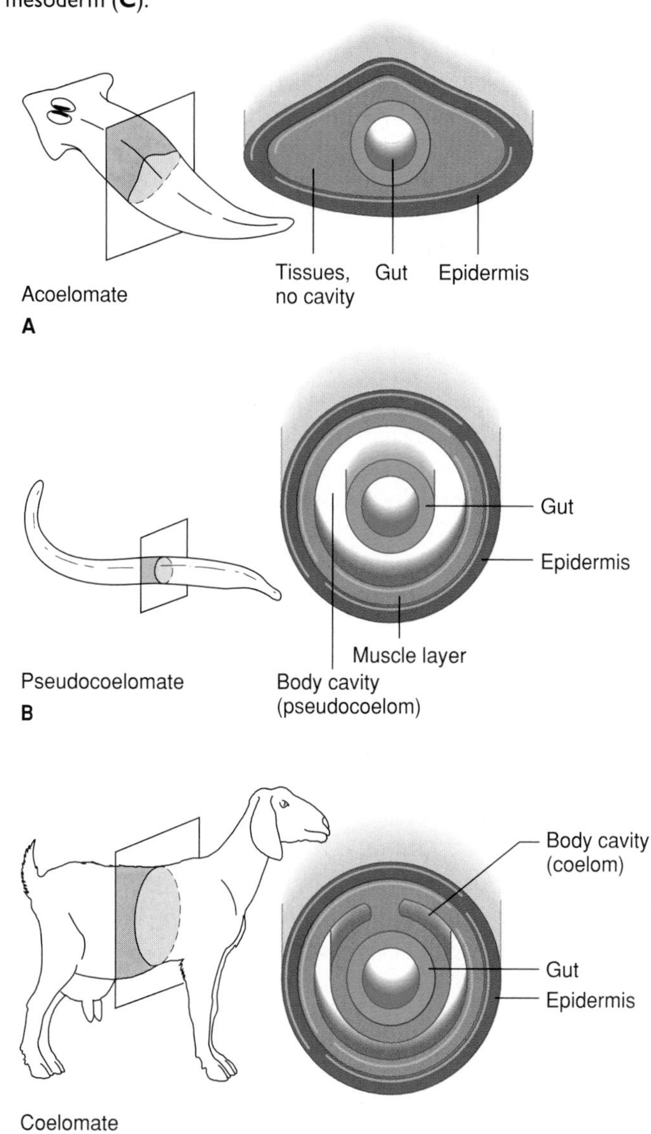

Development from fertilized egg to adult happens in either of two ways. Animals that undergo **direct development** hatch from eggs or are born resembling adults of the species. A newborn elephant, for example, looks like an adult elephant. Animals that directly develop have no larval stage. In contrast, an animal that develops indirectly may spend part of its life as a **larva,** which is an immature stage that does not resemble the adult of the species. Larvae undergo a developmental process called **metamorphosis,** in which they change greatly. Tadpoles and caterpillars are familiar larvae. Many aquatic animals metamorphose, some developing through several larval stages.

Major lineages in the animal kingdom reflect distinctions in the development of a body cavity. Animals with triploblastic development can be classified based on the presence or absence of a **coelom,** which is a fluid-filled body cavity that forms completely within the mesoderm (fig. 24.4). Flatworms (phylum Platyhelminthes) are triploblastic but lack a coelom, and hence are called **acoelomates.** Some animals form a body cavity called a pseudocoelom ("false coelom") that is lined partly with mesoderm and partly with endoderm. Roundworms (phylum Nematoda) display

FIGURE 24.5
Protostomes Versus Deuterostomes.
Early developmental differences characterize protostomes and deuterostomes.

	Fate of blastomere	Cleavage pattern Early cleavage (8-celled stage)	Fate of blastopore region (Gastrula stage)
Protostomes Examples: Mollusks Annelids Arthropods	Remove a blastomere; Development arrested (Determinate cleavage)	Spiral cleavage	Blastopore forms into mouth
Deuterostomes Examples: Echinoderms Chordates	Remove a blastomere; Normal larva develops; (Indeterminate cleavage); Normal larva develops	Radial cleavage	Blastopore forms into anus

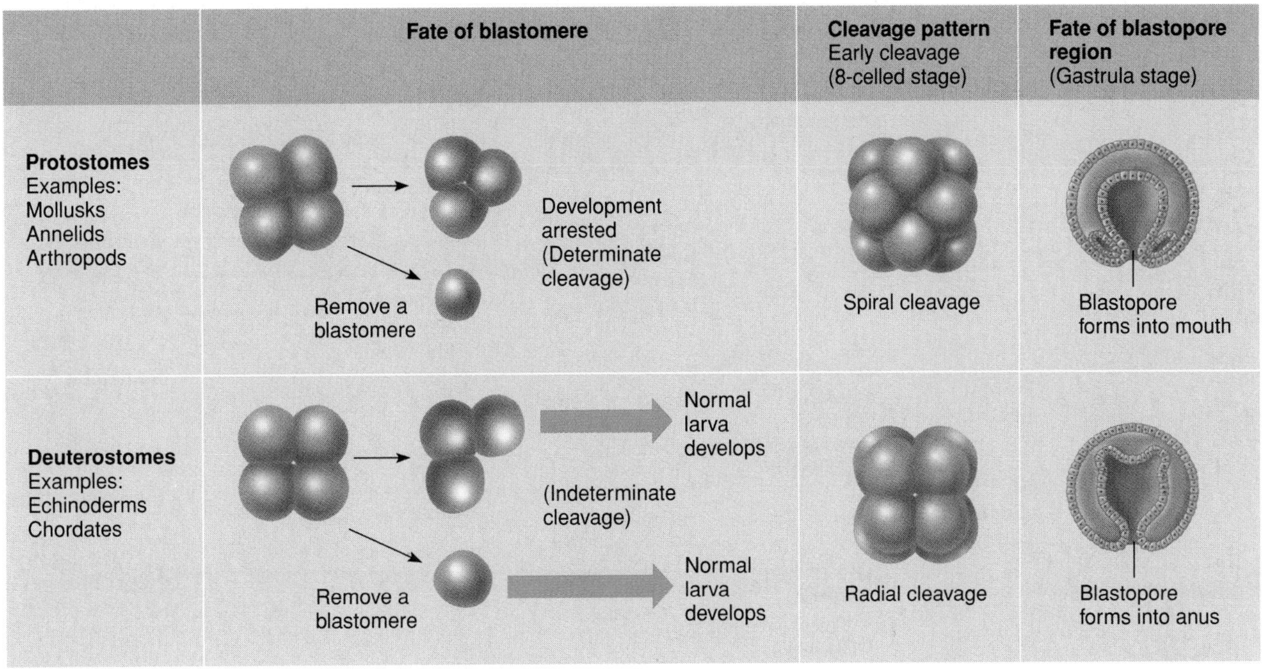

this type of cavity development and are termed **pseudocoelomates.** Animals that have a true coelom are called **coelomates** and include several groups of complex animals, including snails, insects, starfishes, and humans.

Among the coelomates are two major lineages, called **protostomes** and **deuterostomes.** Protostomes include mollusks (snails, clams, slugs, octopuses, and others), annelids (earthworms, leeches, and others), and arthropods. Deuterostomes include echinoderms (spiny-skinned marine organisms) and chordates, the group to which vertebrates belong.

Protostomes and deuterostomes show three major distinctions very early in development (fig. 24.5). The first major difference between protostomes and deuterostomes is the developmental potential of the first cells (blastomeres) resulting from early cleavage divisions. In protostomes, if an early cell is removed, development ceases, leaving an incomplete mass. This phenomenon, called **determinate cleavage,** occurs because the fate of each cell of the embryo is predetermined. In deuterostomes, if one cell is removed from the two- or four-celled stage, both the remaining three cells and the separated fourth cell form complete, genetically identical embryos. This is called **indeterminate cleavage,** because the fate of each cell is not predetermined at this early stage—that is, a cell separated this early retains the potential to yield a complete embryo.

The second difference is in the pattern of cells at the eight-celled stage. In protostomes, the top layer of four cells is offset from the bottom layer, giving the embryo a spiral appearance. The pattern of cell division that gives rise to this form is called **spiral cleavage.** In contrast, the two four-celled layers in the eight-celled deuterostome are directly aligned, demonstrating **radial cleavage.**

The third distinction between protostomes and deuterostomes emerges later in development, once the embryo has begun to fold into a gastrula. In protostomes, an indentation called the **blastopore** develops into a mouth; the word protostome means "first mouth" in Greek. In deuterostomes, the blastopore is a different shape and develops into an anus. Deuterostome means "second mouth."

The major types of animals are distinguished by these characteristics of body form and development (fig. 24.6). To fully describe all of the animal phyla would require a book—and a long one. Here we consider animal phyla that include more than 5,000 species, from simplest to most complex.

24.2 MASTERING CONCEPTS

1. What are the distinguishing characteristics of animals?
2. What types of body symmetry are seen among animals?
3. What is the range of body complexity among animals?
4. In what three ways do protostomes and deuterostomes differ?

OLC

FIGURE 24.6
Animal Classification.
Classification of animal phyla reflects complexity of body form and developmental characteristics.

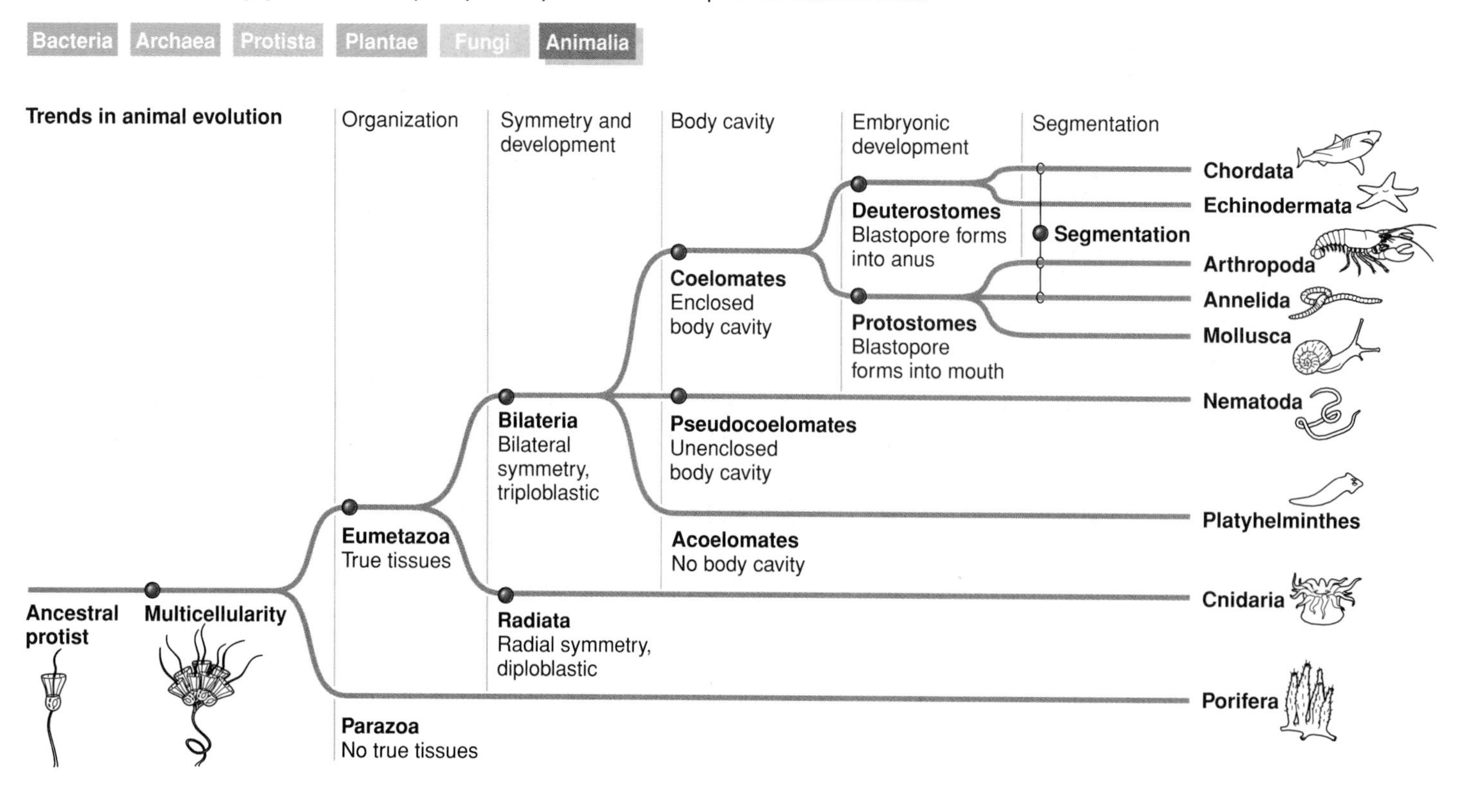

24.3 The Sponges (Porifera) Are the Simplest Animals

Sponges are more than the immobile lumps that they seem. They move, albeit slowly; recognize different sponge species; have cell types but not true tissues; strain food particles from water; and have skeletons.

Sponges belong to phylum Porifera, which means "pore-bearers"—an apt description of these simple animals. They are considered parazoa because they have embryonic and larval stages, but they lack true tissues. Figure 24.7 shows their location within the kingdom. Sponges are either asymmetrical or radially symmetrical. They are all aquatic and mostly marine, and are suspension feeders, capturing food particles suspended in the water. Sponges were thought for a long time to be sessile (unable to move), but an observant biologist showed otherwise, as Investigating Life 24.1 discusses.

The sponge body form is quite different from that of other animals. The body wall consists of cells, called pinacocytes, embedded in a gelatinous matrix. The mesohyl region of a sponge contains amoebocytes, which are multifunctional, mobile cells. Amoebocytes can digest, store and transport nutrients, divide, or

FIGURE 24.7
Phylum Porifera.
(**A**) Location of phylum Porifera (sponges) within the animal kingdom. Note the lack of symmetry in this sponge (**B**).

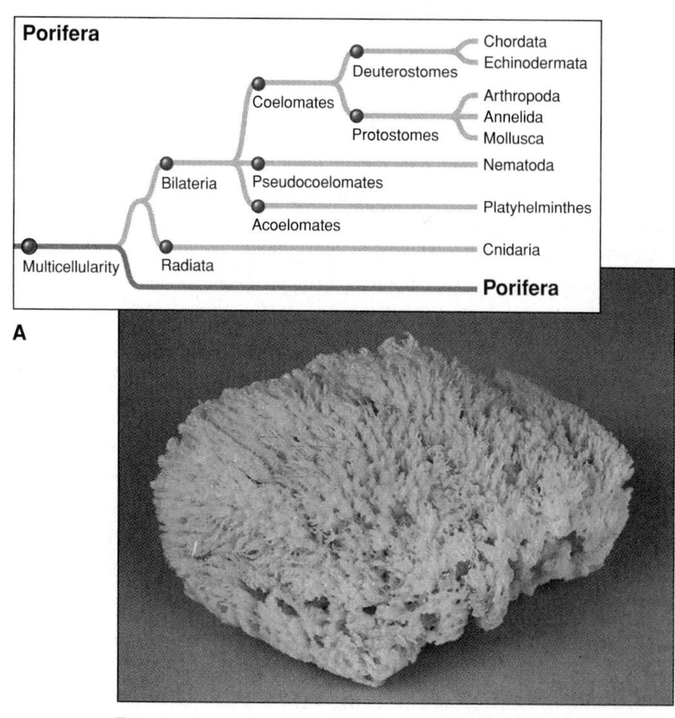

Investigating Life 24.1

Sponges on the Move

Sponges have been maligned in the pages of many textbooks, described as sessile when, in actuality, they can and do move—perhaps too slowly for a human observer to easily note, but movement nonetheless.

The idea that sponges can locomote was first proposed in 1933 by Maurice Burton, curator of the invertebrate collection at the British Museum. When he reported at a meeting of the Zoological Society of London that sponges could move, a British tabloid, in jest, ran a story about a pet sponge that moved about a house seeking water. After World War II, Burton left the museum for the ocean, where he meticulously watched and recorded the movements of sponges, publishing his work in 1949. Still, sponges couldn't shake their reputation as stationary lumps. Others who saw them move attributed the activity to water currents or a response to adding new cells.

In the 1990s, a graduate student at the University of North Carolina revisited the sponge movement story with a high-tech tool—a laundry marker. Interested in studying sponges as a simple model of animal development, Calhoun Bond collected freshwater species from ponds and streams near the Chapel Hill campus, and filmed them in their laboratory containers, marking their positions on tanks with the marker. The animals moved, if ever so slowly, about 2 millimeters a day. In Bermuda and the Caribbean, Bond collected large marine species. Once in tanks for a few weeks, the sponges would form a flattish lump with a protrusion at one end. When the protrusion hit a wall, it would drag the rest of the animal up the surface.

He soon found a speedster among sponges, *Haliclona loosanoffi,* which moves at twice the rate of others. He placed the animals under a microscope and on a deformable rubber coating that would record the "steps." He found that the sponge locomotes from one end like an amoeba. Some cells at the leading edge actually detach, then rejoin the animal. The interior of the sponge moves too, as amoebocytes carry spicules to the leading edge and build new skeleton as the animal moves. The canals are constantly being built, torn down, and changed.

Bond hypothesizes that the sponge's motility is adaptive in helping to rapidly close wounds and search for nutrients—even at 2 millimeters a day.

---

secrete skeletal components called spicules. Sponges also have cells called archeocytes, which can differentiate into the other two somatic cell types.

The simplest sponges are hollow and feed by drawing water through many body surface pores, called ostia, into the central cavity, or spongocoel. To accomplish this, flagellated cells called choanocytes, or "collar cells," produce a current that moves or propels food and water into the spongocoel. (These are the cells that choanoflagellates resemble.) Particles of organic matter and small unicellular organisms move to the cell body and are ingested in a food vacuole, where they are partially broken down. The food particles then pass to amoebocytes, which complete digestion. Water and wastes exit the sponge through a large hole at the top, the osculum (fig. 24.8).

In more complex sponges, body wall folding increases the surface area-to-volume ratio. This increased ratio allows the sponge to move water, cap-

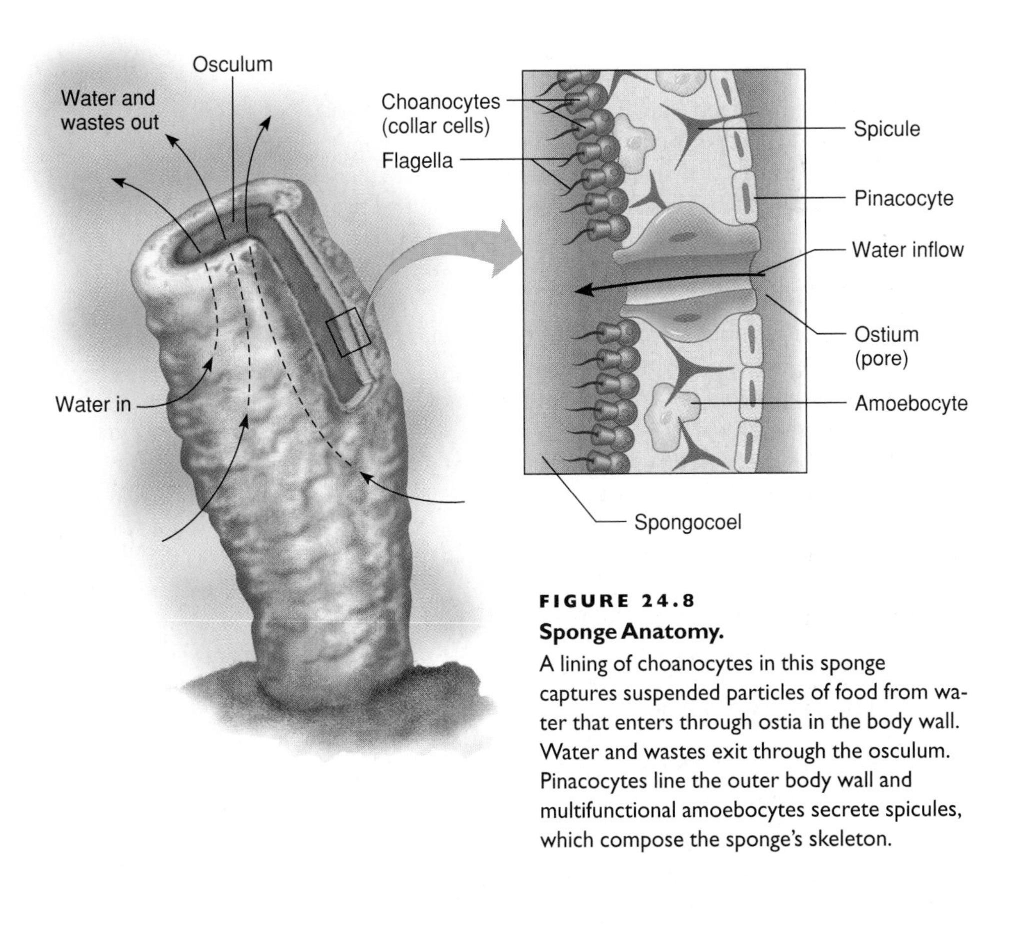

FIGURE 24.8
Sponge Anatomy.
A lining of choanocytes in this sponge captures suspended particles of food from water that enters through ostia in the body wall. Water and wastes exit through the osculum. Pinacocytes line the outer body wall and multifunctional amoebocytes secrete spicules, which compose the sponge's skeleton.

ture food, and excrete wastes more efficiently, which preserves energy and enables the sponge to grow.

Sponge skeletons consist of individual slivers of glassy or limy **spicules** within a mesh of a protein called spongin. Amoebocytes secrete spicules, which come in many shapes, resembling toothpicks, umbrellas, dumbbells, and sunbursts. Biologists classify sponges by skeleton type. Natural bath sponges have no spicules and only spongin, and therefore are soft enough to use to bathe.

A sponge moves by producing and transporting spicules to the leading edge of the body. Its body is in a constant state of flux, with new skeleton forming as remnants of old skeletal material are left behind.

Sponges can reproduce either sexually with gametes or asexually by budding or regenerating fragments. Freshwater and some marine sponges asexually produce "resting bodies" called gemmules when their habitat dries up or freezes. Gemmules contain amoebocytes and lie within spicules. When the environment becomes habitable, amoebocytes leave gemmules through a small opening and grow into new sponges.

Although sponges can grow anew from single cells, different species clearly recognize each other. If cells from different sponge species are isolated and mixed, they sort themselves out and regenerate two sponges, one from each species.

Sponges are abundant in diverse marine environments. They fend off predators with their spicules and with toxic chemicals. Sponges frequently house symbionts that avoid predators in their spiky host. Fishes, for example, hide in spongocoels, and worms and shrimps occupy sponge canals. The decorator crab attaches sponges to its shell to camouflage itself from predators!

Sponges may be predators. Nearly 3.1 miles (5,000 meters) beneath the sea, near the lip of a mud volcano in the Barbados Trench, live hundreds of sponges of the family Cladorhizidae, each about the size of a large dog. Unlike other sponges, these cannot suck up nutrients. Instead, they stab passing crustaceans with their spicules and consume them. The sponges also feed on at least two types of bacteria that live between and within their cells.

24.3 MASTERING CONCEPTS

1. How are sponges distinct from other types of animals?
2. What are the major cell types of a sponge?
3. How is a sponge's form an adaptation to acquiring nutrient-containing water?
4. How do gemmules protect sponges in harsh environments?

OLC

24.4 Cnidaria Are Radially Symmetrical, Aquatic Invertebrates

The diverse cnidaria include jellyfishes, hydras, and corals.
Cnidaria have sessile as well as motile stages, and many can sting.

Cnidaria are aquatic animals that have radial symmetry (fig. 24.9). The phylum includes hydras, jellyfishes, sea anemones, and corals.

FIGURE 24.9
Phylum Cnidaria.
(**A**) Cnidaria are radially symmetrical, diploblastic animals. The phylum includes sea anemones (**B**) and the Portuguese man-of-war (**C**), which is actually a colony of individual cnidaria. Other members include hydras, jellyfishes, and corals.

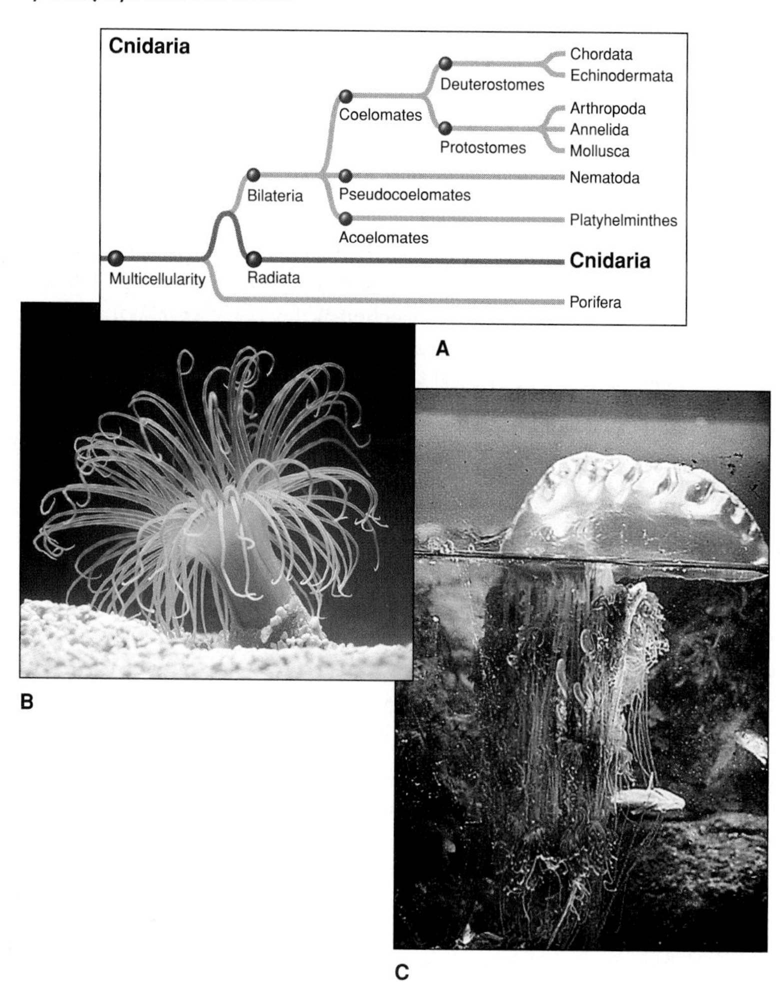

Characteristics of Cnidaria

Many members of phylum Cnidaria, including the sea nettle, sea wasp, fire coral, and Portuguese man-of-war, can sting animals that venture too close. They sting by discharging microscopic capsules called **nematocysts** from cells called **cnidocytes** that are located on the tentacles that surround the mouth. These animals use different kinds of nematocysts for defense, locomotion, attachment, and capturing prey. Figure 24.10 shows the location and function of nematocysts in the body wall of a hydra.

Cnidaria are mostly marine. Many, such as corals, are sessile, but others are free-swimming. All cnidaria are predators, and many are prey for other predators. Small cnidaria consume zooplankton, and large ones eat animals as large as fishes. Cnidaria usually obtain food passively, by digesting whatever happens to swim near the stinging tentacles.

The radial symmetry of a cnidarian is built around an oral-aboral axis that extends from the mouth (the only opening) to a point opposite the mouth. Cnidarian embryos are diploblastic. Their ectoderm is called an epidermis, and their endoderm a gastrodermis. These layers sandwich a jellylike, noncellular substance called mesoglea. In hydra, the mesoglea is little more than a thin gluelike layer, but in jellyfishes, the thick layer is their "jelly." Digestion and gas exchange occur outside of cells in the **gastrovascular cavity,** which comprises much of the organism. The animal ejects wastes through its mouth.

The two types of cnidarian body forms are a generally sessile **polyp** and a free-swimming **medusa.** Hydra and sea anemones are polyps. Adult jellyfishes are medusae. Cnidarian species that undergo alternation of generations have both polyp and medusa stages. The medusa is the mature, sexually reproducing stage, with male and female individuals. Taxonomists classify cnidaria partly on the basis of the dominant stage in the life cycle.

The life cycle of a cnidarian begins as male and female medusae release eggs and sperm into the water (fig. 24.11). Fertilization occurs, and a cilia-fringed larva called a planula develops. The planula attaches to a surface and metamorphoses into the polyp form. The polyp then buds asexually, generating a colony called a strobila. In response to a trigger—perhaps an environmental change or reaching a maximal colony size—medusae are produced asexually. The cycle begins anew.

Cnidaria may aggregate into colonies. The Portuguese man-of-war (*Physalia*) is a free-swimming colony of individual medusae and polyps. The floating portion, or pneumatophore, is a gas-filled structure that keeps the colony at the ocean's surface. Corals are colonies of polyps that reproduce asexually and sexually.

Some cnidaria have a hydrostatic skeleton, which is a fluid-filled compartment that provides support. Other cnidaria secrete a limy exoskeleton or have an endoskeleton of protein or the carbohydrate chitin. As successive generations of colonial coral polyps die and new individuals grow, their limy exoskeletons form magnificent coral reefs.

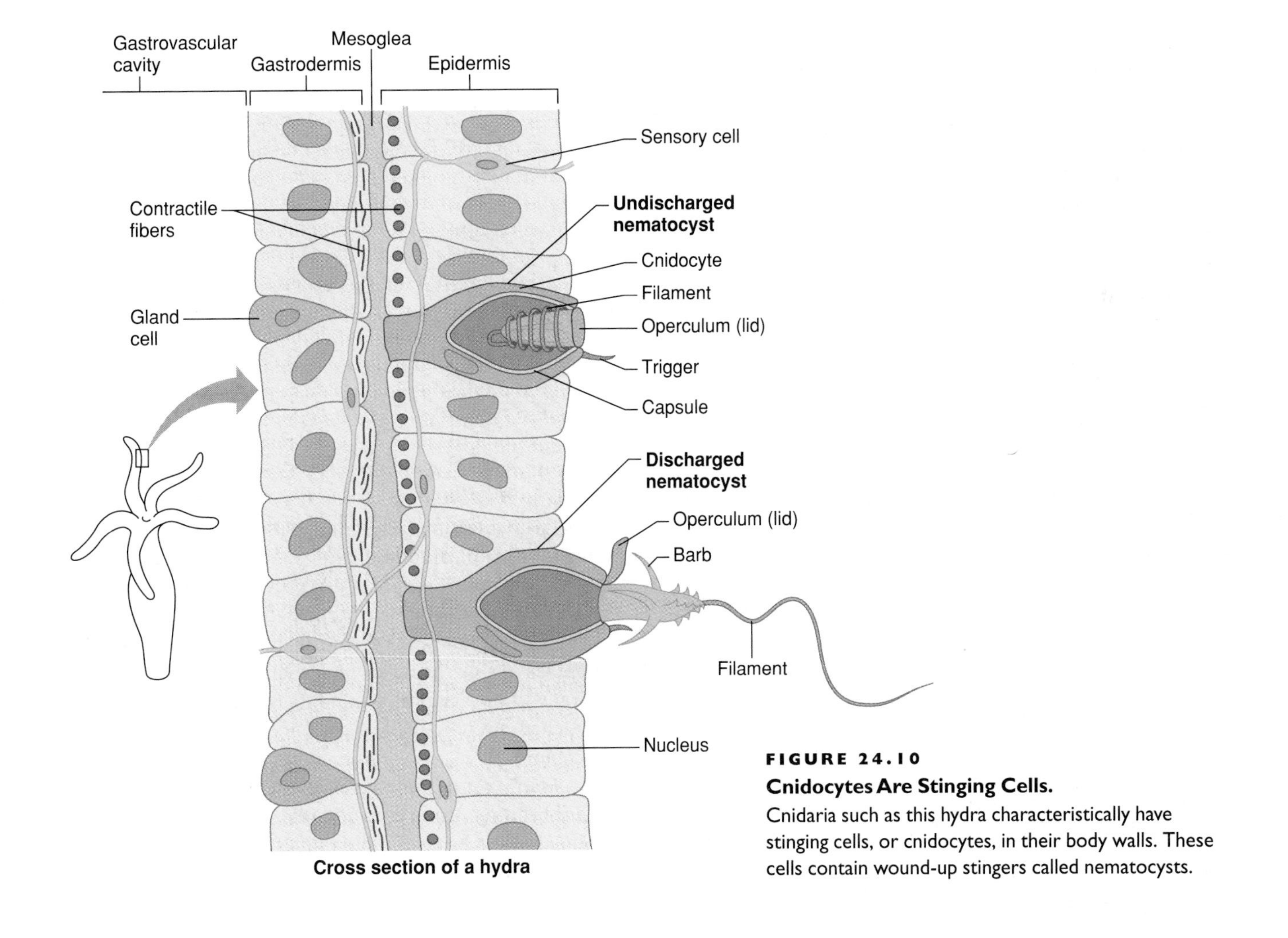

FIGURE 24.10
Cnidocytes Are Stinging Cells.
Cnidaria such as this hydra characteristically have stinging cells, or cnidocytes, in their body walls. These cells contain wound-up stingers called nematocysts.

FIGURE 24.11
The Cnidarian Life Cycle.
(**A**) Male and female medusae release sperm and eggs, and after fertilization, a planula larva develops. It attaches to a surface and becomes a polyp, which reproduces asexually to form a colony (strobila). New medusae form asexually from the polyp colony. (**B**) and (**C**) show the anatomy of the polyp and medusa stages.

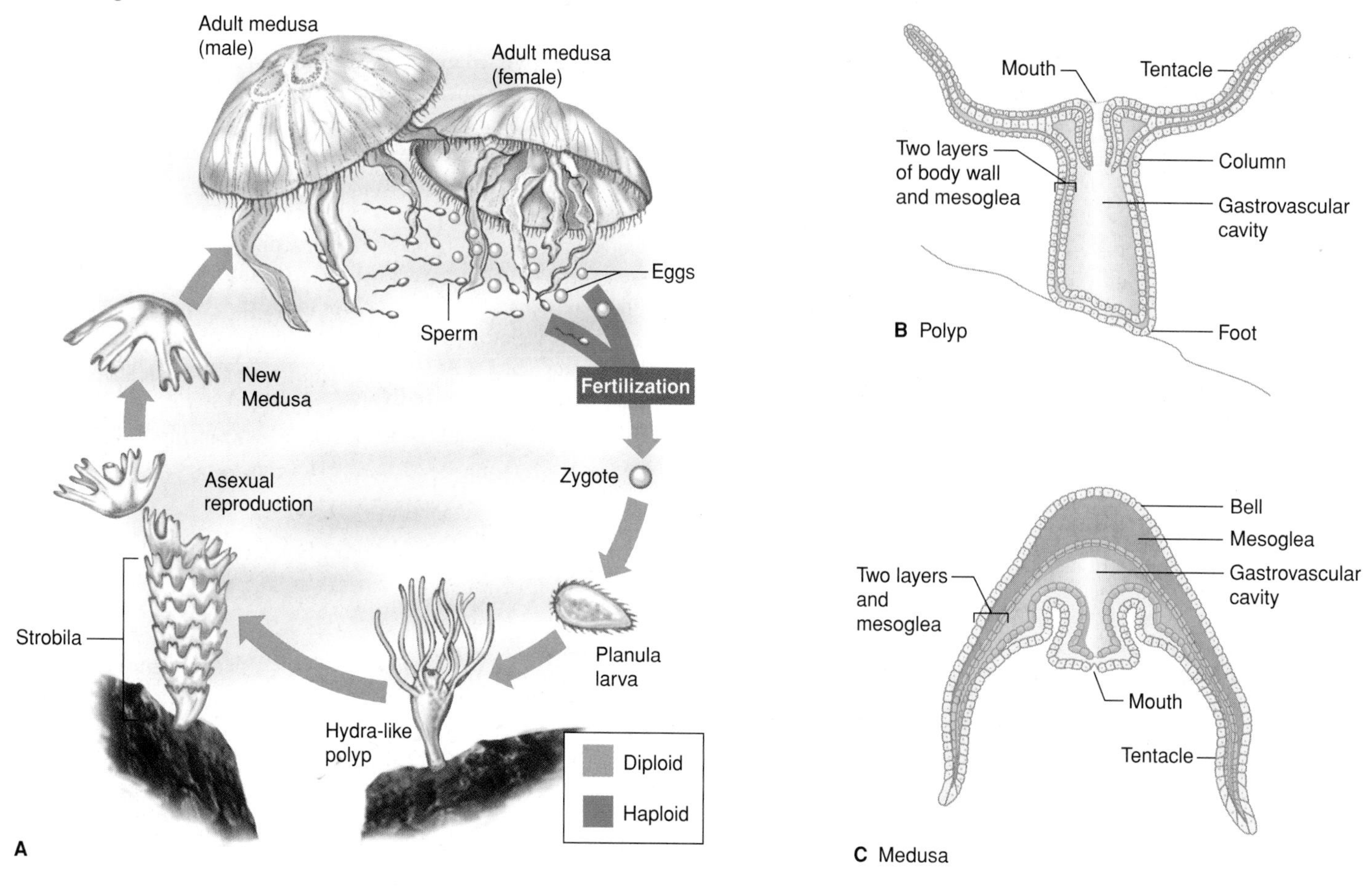

A Closer Look at Corals

Corals have the distinction of being the only type of organism to have a type of ecosystem named after it! The extensive calcium carbonate secretions of corals house 35,000 to 60,000 species, including other cnidaria (fig. 24.12). The skeletons form enormous growths. The Great Barrier Reef in Australia is 1,250 miles (2,012 kilometers) long. Overall, corals cover 232,000 square miles (600,880 square kilometers) in the oceans. The reefs are like huge office buildings, with different visitors present at different times of day, plus many permanent residents. Turtles, sea horses, thousands of different fishes, octopuses, eels, lobsters, and sea slugs all dart about coral reefs.

The coral animal itself is a small, soft-bodied polyp. Many polyps form a living layer atop the coral, where they grab passing food with their tentacles and soak up sunlight. Within polyps live a variety of dinoflagellates and algae that impart spectacular colors, and more importantly provide oxygen from photosynthesis. Sponges help filter particles from the water, allowing sunlight to penetrate, and parrotfish and sea urchins keep other algal populations in check, keeping them from blocking the light. A coral without its algal symbionts is a dead coral—it is white or bleached. A cubic inch of coral may be home to half a million symbionts.

Many corals are endangered. Increasing tourism in the Caribbean has contributed to the destruction of these delicate animals. Visitors take coral samples home and construction work sends soil into the ocean, blocking sunlight. Nitrogen and

FIGURE 24.12
Coral.
(**A**) An individual coral is a soft-bodied organism. More familiar are the reefs that form from their calcium carbonate secretions (**B**).

A

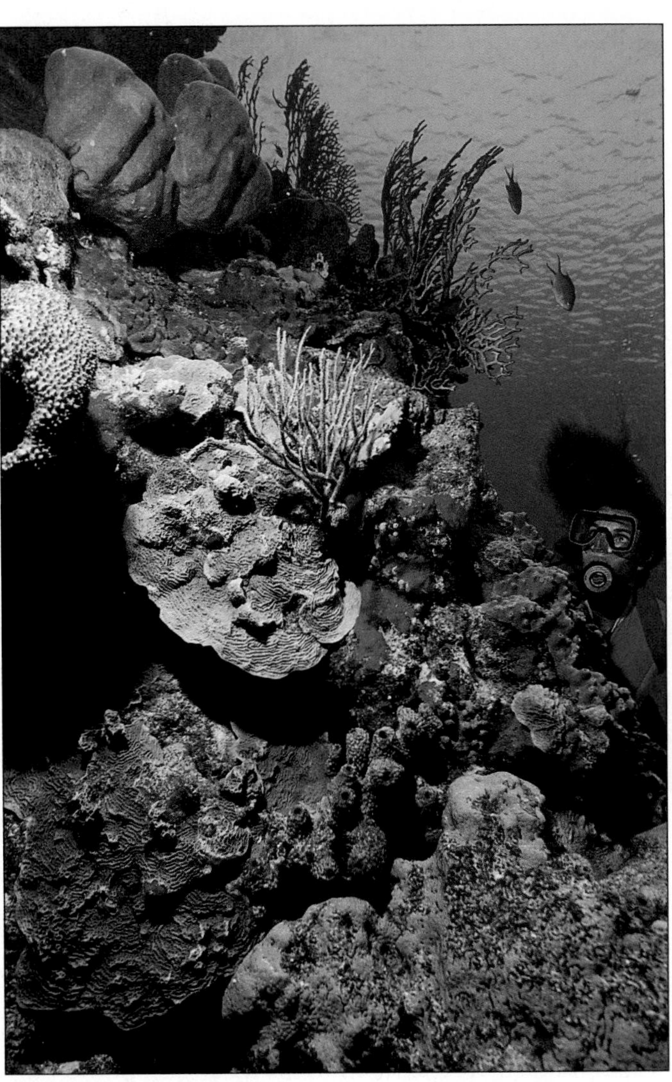
B

phosphorus from fertilizer cause algae to overgrow on the coral, blocking the light even more. Water movements from El Niño have also eroded some reefs. In addition, bacterial infections are killing corals.

24.4 MASTERING CONCEPTS

1. What are the characteristics of cnidaria?
2. What are some examples of cnidaria?
3. How do cnidaria eat?
4. How do cnidaria reproduce?
5. What is the role of symbionts in coral?

24.5 Flatworms (Platyhelminthes)

Planaria, flukes, and tapeworms belong to phylum Platyhelminthes. Their bodies are adapted to a parasitic or free-living lifestyle. Flatworms lack circulatory and respiratory systems, but they have concentrations of nerve cells and excretory structures.

Phylum Platyhelminthes, the flatworms, includes planaria, flukes, and tapeworms. Flatworms are triploblastic, unsegmented acoelomates with bilateral symmetry (fig. 24.13). Although flatworms lack a coelom, some biologists suggest that these animals secondarily lost their body cavities, and therefore group the flatworms with the protostomes. We have adopted the more traditional approach in our placement of these animals.

Variations within phylum Platyhelminthes reflect adaptations to either a free-living or parasitic existence. Parasitic flatworms have greater reproductive capacity, as evidenced by their greater number or size of reproductive organs compared to those of free-living species. Many flatworms parasitize humans and domestic animals and are usually contracted by eating undercooked fish, beef, or pork.

Flukes are flatworms that parasitize vertebrates. They feed on their host's blood and other tissues through an anterior mouth surrounded by an adhesive oral sucker that attaches to the host. A muscular pharynx pulls food into two large, unbranched, saclike intestinal ceca for digestion and absorption.

Tapeworms parasitize vertebrate intestines. They lack mouths and absorb food through the tegument, which is a protective body covering also found in flukes. This covering has microscopic, finger-shaped projections that increase the surface area for absorption. What looks like the head of a tapeworm is actually a holdfast organ, the scolex, that helps the worm attach to the host's intestine. Because tapeworms absorb digested food directly from their hosts, they lack digestive systems.

Free-living flatworms usually are predators or scavengers in aquatic ecosystems. A planarian, for example, has a structure called an eversible pharynx that flicks out of the mouth to catch prey or eat dead animals. This worm has a highly branched gut for extracellular digestion and ejects wastes through its mouth (fig. 24.14). The digestive system is incomplete because there is no anus.

FIGURE 24.13

Tapeworms.

(**A**) Tapeworms, members of phylum Platyhelminthes (flatworms), are bilateral acoelomates. (**B**) The dog tapeworm *Taenia pisiformis* has many adaptations for life as an internal parasite, including hooks and suckers on its scolex (holdfast organ) that enable it to latch onto its host's intestine.

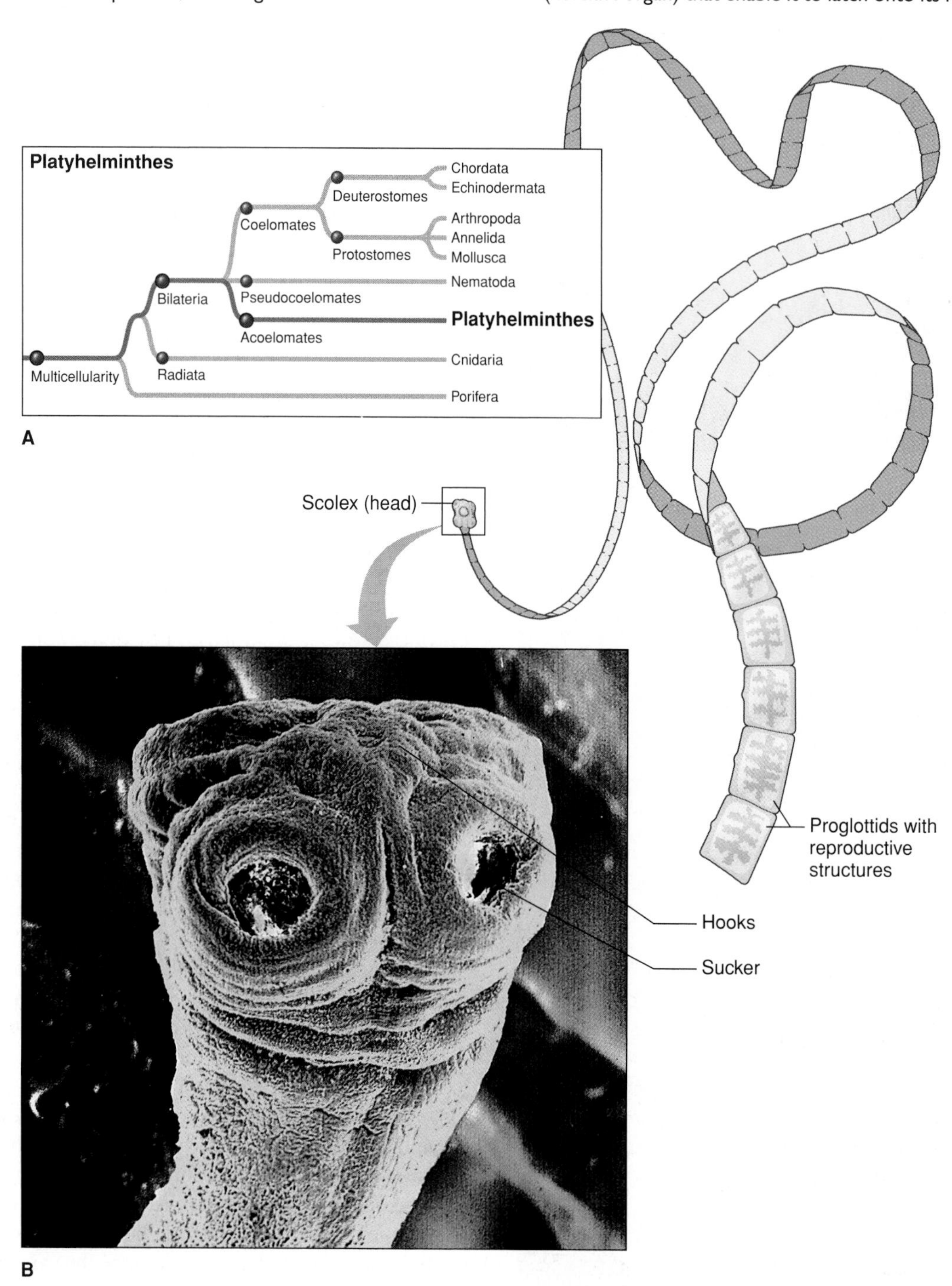

FIGURE 24.14

A Planarian Flatworm.

(**A**) The pharynx is a tube attached to the mouth that passes food to intestinal ceca. (**B**) Flame cells form an excretory system. (**C**) A planarian has a brain, ladderlike nerve cords, and eyespots. (**D**) Free-living flatworms are hermaphroditic.

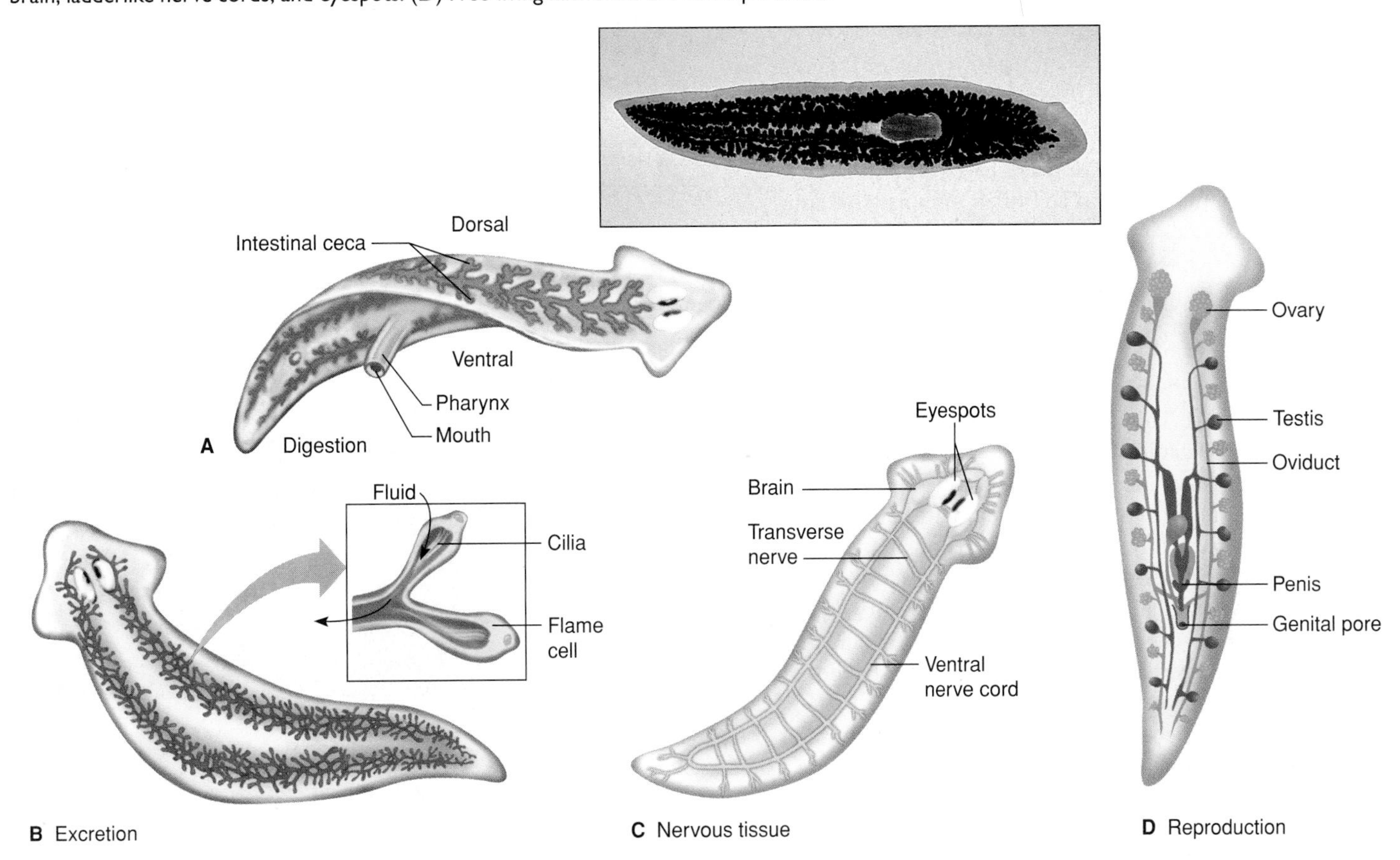

Some flatworms have concentrations of nerve cell bodies in their anterior ends, which form a rudimentary brain, and nerve cords. Others have a simple nerve net. Chemosensory (taste and smell) cells in some freshwater planaria cluster on earlike structures called auricles. Eyespots detect light but do not form images.

Protonephridia are structures in flatworms that maintain internal water balance and form a simple excretory system. Because of their shape and beating cilia, protonephridial cells are often called flame cells, resembling a flickering candle. Most excretion occurs by simple diffusion through the body wall. Flatworms have no circulatory or respiratory systems—diffusion accomplishes these functions.

Many flatworms reproduce both sexually and asexually. Free-living species may simply pinch in half and regenerate the missing parts asexually. Free-living flatworms are hermaphroditic, but they cross-fertilize by mating with each other. Young that resemble adults hatch from eggs, except in some marine species that have ciliated larvae.

Most flukes are hermaphroditic and cross-fertilize inside the host. Eggs pass out of the host with urine or feces and must land in water to hatch. The intermediate host—the animal that houses the larval form of the parasite—is usually a snail. Another organism may serve as a second intermediate to the fluke before it moves to its final host to develop into an adult.

Tapeworms are reproductive machines. They are constructed of many structures, called proglottids, each of which houses a complete set of male and female reproductive structures. The animals can self-fertilize or cross-fertilize. After eggs mature and are fertilized, individual proglottids break off and leave the host along with the feces. When another appropriate host swallows the proglottids, eggs hatch and start a new generation. These worms can grow to be quite large. The tapeworm *Diphyllobothrium pacificum,* for example, passes as eggs from sea lion feces into the water, where they pass up food webs to eventually reach fish, which people eat. Raw fish can transmit live worms, which can grow up to 16 feet (almost 5 meters) long!

24.5 MASTERING CONCEPTS

1. What are some adaptations of flatworms for a parasitic existence?
2. What types of specialized cells and tissues do flatworms have?
3. How do flatworms reproduce?

OLC

24.6 Roundworms (Nematoda)

A chunk of soil houses hundreds, maybe thousands, of roundworms. Nematodes are unsegmented, lack respiratory and circulatory organs, but are distinguished from the simpler flatworms by their complete digestive systems.

Roundworms of the phylum Nematoda are cylindrical worms without segments. They are tapered at both ends and have a surrounding cuticle (fig. 24.15). Most are barely visible to the unaided eye and are free-living in soil or in sediments of aquatic ecosystems. Nematodes are extremely abundant. One sample of sediment from Loch Ness, the largest lake in Great Britain, yielded 27 species of nematodes, some never seen before! Some roundworms parasitize plants or animals and are acquired in a variety of ways (table 24.1). Many developmental biologists use the nematode *Caenorhabditis elegans* as a model organism to trace cell lineages (see Thinking Scientifically 40.1). It is perhaps the best-studied of all animals.

Roundworms have only longitudinal (lengthwise) muscles, which limits their range of movement. They have complete digestive systems (one-way flow from mouth to anus), and are thus more complex than flatworms, but they still have poorly developed anterior ends and lack circulatory or respiratory organs. A roundworm's hydrostatic skeleton consists of a fluid-filled pseudocoelom. An external layer of tissue, called the hypodermis, secretes the cuticle.

Most species of nematodes have separate sexes. Females of parasitic species produce large numbers of eggs that are extremely resistant to harsh environmental conditions such as dryness and exposure to harsh chemicals. Most nematodes undergo direct development, with most parasitic forms living in only one host. Mosquitoes can transmit certain roundworm infections.

FIGURE 24.15
Roundworms.
(**A**) Roundworms are pseudocoelomates and are among the most abundant animals. (**B**) Millions of humans harbor the giant intestinal roundworm *Ascaris lumbricoides*. The parasite is acquired through contact with human feces containing eggs, which hatch in the intestine. This is a mature male giant intestinal roundworm. Note the hooked posterior end.

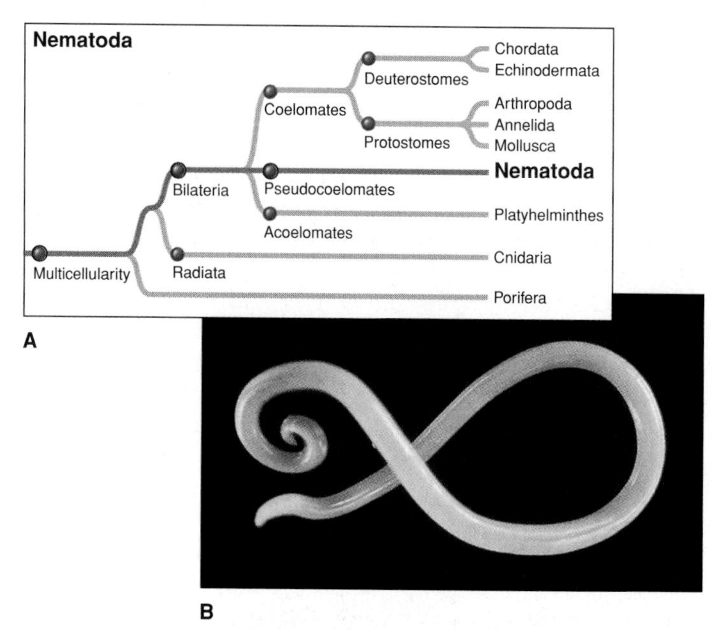

24.6 MASTERING CONCEPTS

1. Describe a roundworm.
2. How is a roundworm more complex than a flatworm?
3. How do roundworms reproduce?

OLC

TABLE 24.1 Nematode Parasites of Humans and Domestic Animals

WORM (COMMON NAME)	HOST	MODE OF INFECTION	DISTRIBUTION
Enterobius vermicularis (pinworm)	Humans only	Inhaling eggs; fingers to anus	Worldwide;* most common worm parasite in United States
Necator americanus (New World hookworm)	Humans only	Larvae burrow into feet from contaminated soil; can cause anemia	New World;† southeastern United States
Ascaris lumbricoides (giant human intestinal roundworm)	Humans only	Eating infective stage eggs in contaminated food	Worldwide;* Appalachia and southeastern United States
Trichinella spiralis (trichina worm)	Humans, pigs, dogs, cats, rats	Eating undercooked meat with encysted larvae; can be fatal	Worldwide;* throughout United States
Wuchereria bancrofti (filarial worm)	Humans	Mosquito; can cause elephantiasis (severe swelling)	Tropics
Dirofilaria immitis (heartworm)	Dogs	Mosquito; can be fatal	Worldwide

*Distribution is scattered throughout the world; appears to varying degrees in different places.
†Distribution is scattered throughout the New World; appears to varying degrees in different places.

24.7 Mollusks Include Clams, Snails, Octopuses, and Squids

Mollusks have all organ systems. Although they appear quite different from each other, all mollusks share certain basic body parts.

Phylum Mollusca is the second largest phylum in the animal kingdom, with many species living in the oceans, in fresh water, and on land. The giant squid is the largest invertebrate known, ranging up to 65 feet (20 meters) in length, including its massive arms.

Mollusks are among the most complex invertebrates and possess all the organ systems of vertebrates. The mollusk circulatory system is usually open—that is, blood circulates throughout the body cavity instead of within vessels. Aquatic species have gills for gas exchange, and terrestrial species have one lung derived from a space called the mantle cavity. Mollusks are protostomes, and most are bilaterally symmetrical (fig. 24.16).

Mollusks have several body structures in common (fig. 24.17), although these animals are quite diverse in appearance—compare a giant clam, a snail, and an octopus! One of these common body parts is the **mantle,** which is a dorsal fold of tissue that secretes a shell. Many mollusks have two-part shells and are therefore called **bivalves.** Mollusks locomote by using a ventrally located organ called a **muscular foot.** An area called the **visceral mass** contains the digestive and reproductive systems. Many mollusks have a chitinous, tonguelike structure called a radula that they use to scrape algae or plant matter into their mouths.

Classification of mollusks often is based on the shapes of the shell and foot. Most bivalves, for example, have a hatchet-shaped foot for digging into the sediment. **Gastropods** are very diverse mollusks that live in marine, freshwater, and terrestrial environments. They include snails with spiral shells and slugs with small internal shells, and have a broad, flat foot for crawling. Snails and slugs secrete a mucous trail they glide on, which prevents the epidermis from peeling away as the animals move. A snail's body twists as it develops and becomes asymmetrical. **Cephalopods** are mollusks that have reduced or absent shells, such as octopuses and squids.

The molluscan nervous system varies from simple and ladderlike to very complex and cephalized. An octopus's nervous system is so sophisticated that neurobiologists studying how learning occurs use the animal in laboratory experiments. The octopus also has a very highly developed visual system.

Most mollusks have separate sexes. Bivalves shed eggs or sperm into the water, where external fertilization occurs. Gastropods such as snails, slugs, limpets, whelks, and conches have copulatory organs and fertilize eggs internally. A male cephalopod reaches into his mantle cavity with an armlike appendage to draw out a package of sperm and then delivers the sperm package to the female's mantle cavity, where fertilization occurs.

Many marine mollusks have a ciliated larval stage called a trochophore, which may develop into a larva with a tiny mantle, shell, foot, and swimming organ. The trochophore larva settles to the bottom of the sea, where it develops into an adult.

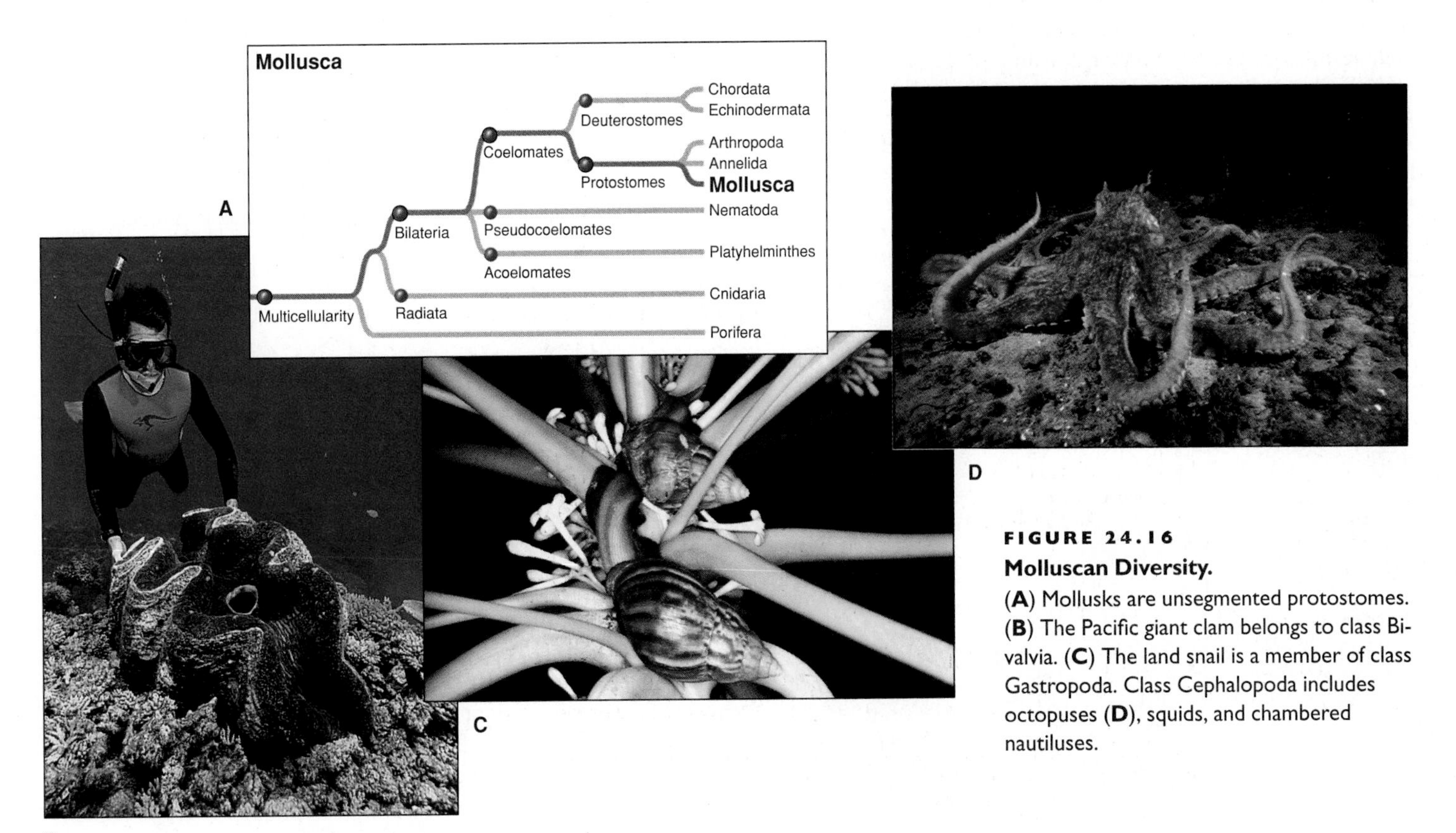

FIGURE 24.16
Molluscan Diversity.
(A) Mollusks are unsegmented protostomes. **(B)** The Pacific giant clam belongs to class Bivalvia. **(C)** The land snail is a member of class Gastropoda. Class Cephalopoda includes octopuses **(D)**, squids, and chambered nautiluses.

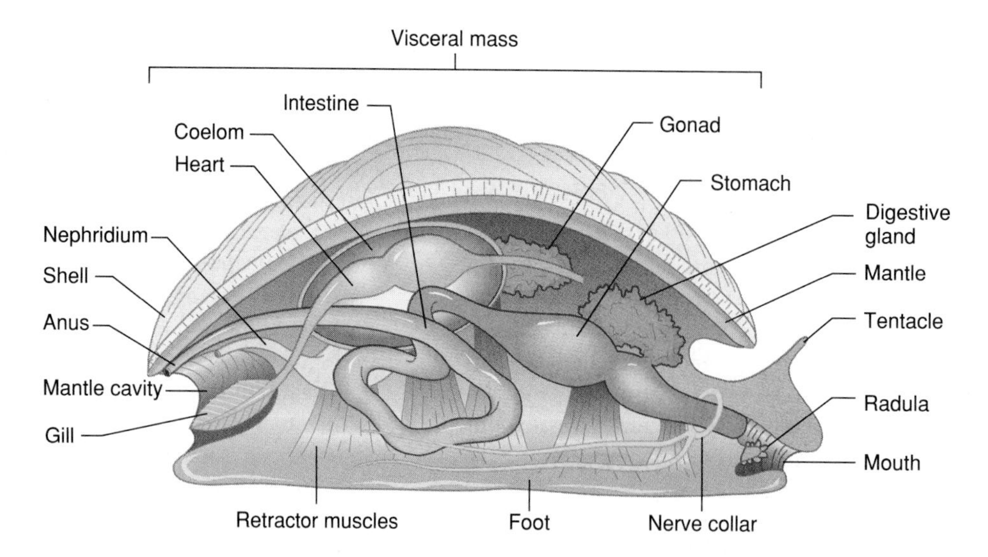

FIGURE 24.17
Generalized Molluscan Body Plan.
The digestive, circulatory, excretory, and reproductive systems are all contained within the visceral mass. The muscular foot provides locomotion and the mantle secretes a protective shell.

Freshwater mussel larvae are parasites, clamping onto gills or fins of fishes. When the larvae are developed, they detach, settle to the bottom, and mature into free-living adults. In cephalopods, terrestrial gastropods, and some other mollusks, all larval development occurs inside the egg. The hatchlings resemble adults.

Mollusks feed in diverse ways. Aquatic bivalves such as oysters, clams, scallops, and mussels are suspension feeders, eating organic particles and small organisms they strain out of the water. Mollusks concentrate pollutants or toxins produced by dinoflagellates and may become poisonous to human seafood eaters. Slugs consume vegetation so voraciously that they can rapidly defoliate certain garden plants. Some marine snails drill holes in bivalve shells to eat them. Cephalopods are active predators. Their keen eyes, closed circulatory systems, and ability to move by "jet propulsion," squirting water out of modified feet, allow them to detect and catch fast-moving prey such as fishes. The octopus and chambered nautilus scavenge dead fishes.

24.7 MASTERING CONCEPTS

1. What structures do mollusks share?
2. Why are mollusks complex invertebrates?
3. How do mollusks locomote, reproduce, and feed?

OLC

FIGURE 24.18
Annelid Diversity.
(**A**) Annelids are segmented protostomes. (**B**) Oligochaetes, such as this earthworm (*Lumbricus terrestris*), show the metamerism (repeated body parts) characteristic of phylum Annelida. (**C**) *Spirobranchus giganteus*, the Chistmas-tree worm, is a sedentary, tube-dwelling polychaete that uses its double crown of radioles to filter organic particles from seawater. (**D**) The third class of annelids, the hirudinea, includes the leeches, which have suckers and lack parapodia and setae.

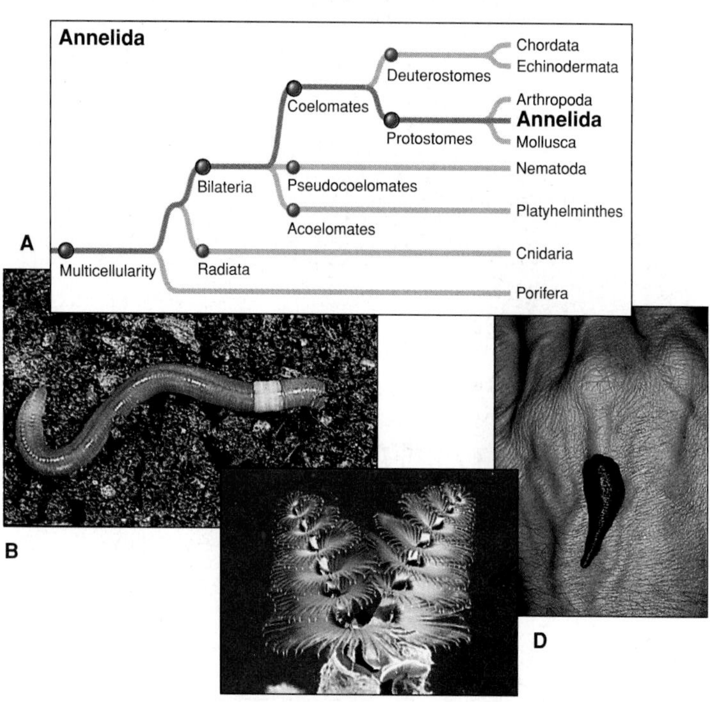

24.8 Annelids Are Segmented Worms

One type of annelid tolerates incredible temperature fluctuations, hugging deep-sea hydrothermal vents; another type lives nestled into ice blocks. These protostomes are segmented worms, such as leeches and the familiar earthworm.

Segmented worms belong to the phylum Annelida (fig. 24.18). Figure 24.16 shows the place of annelids on the protostome branch of the animal family tree. The annelids' main identifying characteristic is their body form of repeated segments, called metamerism. Annelids, like mollusks, have trochophore larvae. This is a distinguishing characteristic of protostomes.

Phylum Annelida includes three easily recognized classes. Oligochaetes, such as earthworms, have a few small bristles called setae on their sides (fig. 24.19). A part of the outer covering called

FIGURE 24.19
Anatomy of an Earthworm.
Earthworms show the body segmentation, both internal and external, that is typical of annelids. Groups of cell bodies (ganglia) above the pharynx fuse to form a primitive brain that connects by nerve rings to the ventral nerve cord. The blood system is closed and the anterior end of the dorsal blood vessel contracts rhythmically, forcing blood through five aortic arches that connect with the ventral blood vessel. Setae are bristles that provide traction for locomotion. Pairs of nephridia in each segment remove waste from the coelomic fluid and blood.

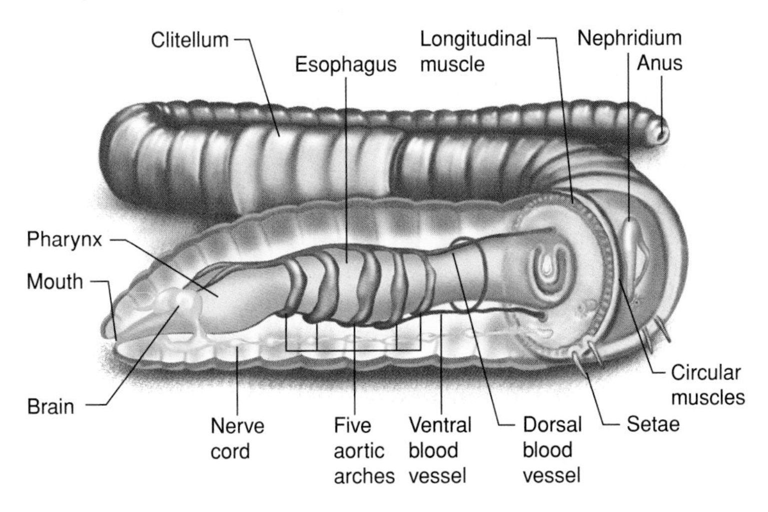

the clitellum secretes mucus when the worms copulate and a cocoon that protects the animal early in development.

Marine segmented worms, the polychaetes, have pairs of fleshy appendages called parapodia that they use for locomotion. Parapodia are located on the sides of segments and have many long setae embedded in them. Setae anchor annelids and keep them from slipping backwards when they move. The third class of annelids, the hirudinea, include the leeches. They lack parapodia and setae and have suckers and demarcations called annuli between their segments.

Annelids feed in several ways. Earthworms and several other annelids are deposit feeders, which take in large quantities of soil or aquatic sediment and strain out organic material for food. Many polychaetes are suspension feeders, using their crowns of ciliated structures called radioles to filter small organisms and organic particles from seawater. Some polychaetes are predators with formidable jaws. Many leeches suck blood from vertebrates, but most eat small organisms such as arthropods or other annelids. A blood-thinning chemical from leeches is used to reattach severed human digits and ears.

Annelids have complex organ systems. Polychaetes have various external structures that carry out gas exchange. Oligochaetes and hirudinea respire more simply, by diffusion through the body wall. An annelid's blood is confined to vessels, called a **closed circulatory system.** The coelom serves as a hydrostatic skeleton; muscles work against it as the worm crawls, burrows, or swims. The nervous system consists of a brain connected around the digestive tract to a ventral nerve cord, with lateral nerves running through each segment.

Leeches and oligochaetes reproduce differently than polychaetes. Leeches and oligochaetes are hermaphroditic and also cross-fertilize. Two individuals copulate, each discharging sperm that its partner temporarily stores. Eggs and sperm then are shed into a cocoon that each worm's clitellum secretes. In earthworms, the clitellum is visible at all times, but in leeches it is only visible during breeding season. Juvenile leeches that resemble adults develop within the cocoon and hatch from the fertilized eggs.

In contrast, polychaetes have separate sexes; eggs and sperm are usually shed into the ocean, where fertilization occurs. The *Palolo* worm has a particularly interesting method of reproduction, using specialized posterior body segments. On one night with a full moon in October or November, these segments detach, swarm to the surface of the ocean, and break open, releasing many gametes. The anterior part of the body survives and can regenerate the posterior portion the following year.

One type of polychaete, the Pompeii worm *Alvinella pompejana,* has the distinction of being the animal that tolerates the largest known environmental temperature gradient. Adhering to the sides of deep-sea hydrothermal vents off the coast of Mexico, the anterior end of the worm might be 81°C (178°F), while its posterior end is 21°C (70°F)! The temperature gradient occurs because at the vent sites, 300°C (572°F) water spews from Earth's interior and hits the seawater, which is 2°C (36°F). The worms, which are about 6 centimeters (2.4 inches) long, spend most of their time in papery tubes that they secrete, poking their heads out, but can travel as far as a meter to eat bacteria (fig. 24.20).

FIGURE 24.20
The Pompeii Worm.
The polychaete Pompeii worm *Alvinella pompejana* lives along deep-sea hydrothermal vents, where it tolerates extreme temperature gradients. Biologists don't know how the worms do it!

Other polychaetes thrive in cold. *Hesiocaeca methanicola* lives on the bottom of the Gulf of Mexico in methane that is so cold that it forms solid blocks. The worms consume bacteria that produce nutrients by metabolizing the methane. If the temperature rises slightly, the ice (methane hydrate) turns into gas. But while it is solid, it swarms with polychaete worms.

24.8 MASTERING CONCEPTS

1. Describe the annelid body form.
2. What are the distinguishing features of the three classes of annelids?
3. How do annelids feed, respire, exchange gases, respond to the environment, and reproduce?
4. Where do the classes of annelids live?

24.9 Arthropods Include Insects, Crustaceans, and Arachnids

The majority of animal species are arthropods—they are everywhere. Arthropods are segmented with jointed appendages and external skeletons of chitin.

If biological success is judged in terms of diversity, perseverance, and sheer numbers, then the phylum Arthropoda certainly qualifies as the most successful group of organisms. Biologists hypothesize that more than 75% of all animal species are arthropods, the phylum that includes insects, crustaceans, and arachnids.

Arthropods are protostomes with bilateral symmetry. They have segmented bodies, skeletons composed of chitin, and jointed appendages (arthropoda means "jointed foot"). These animals molt their **exoskeletons** to grow, and many metamorphose.

Arthropods have **open circulatory systems.** Their blood, called **hemolymph,** circulates freely through the body cavity, or hemocoel. The heart is dorsal and has holes that blood enters. In many terrestrial arthropods, the circulatory system plays only a limited role in gas exchange. The arthropod respiratory system consists of body wall holes called spiracles that open into a series of branching tubes—the tracheae and smaller tracheoles. The smallest tracheoles serve individual cells. The tracheal system very efficiently transports oxygen and carbon dioxide to and from tissues. Excretion is accomplished by unique organs called **Malpighian tubules** (see fig. 38.12).

An arthropod's nervous system is similar to an annelid's, with a dorsal brain and a ventral nerve cord. Many arthropods have large **compound** (faceted) **eyes.**

Different types of arthropods vary in the number of major body regions and appendages; whether mouthparts chew or pierce; and whether certain appendages, such as antennae or legs, are branched. An appendage with two branches is biramous; with one lobe, uniramous. Many arthropods have three major body regions: head, thorax (chest), and abdomen (fig. 24.21).

FIGURE 24.21
The Arthropod Body Form.
The grasshopper displays typical arthropod characteristics.

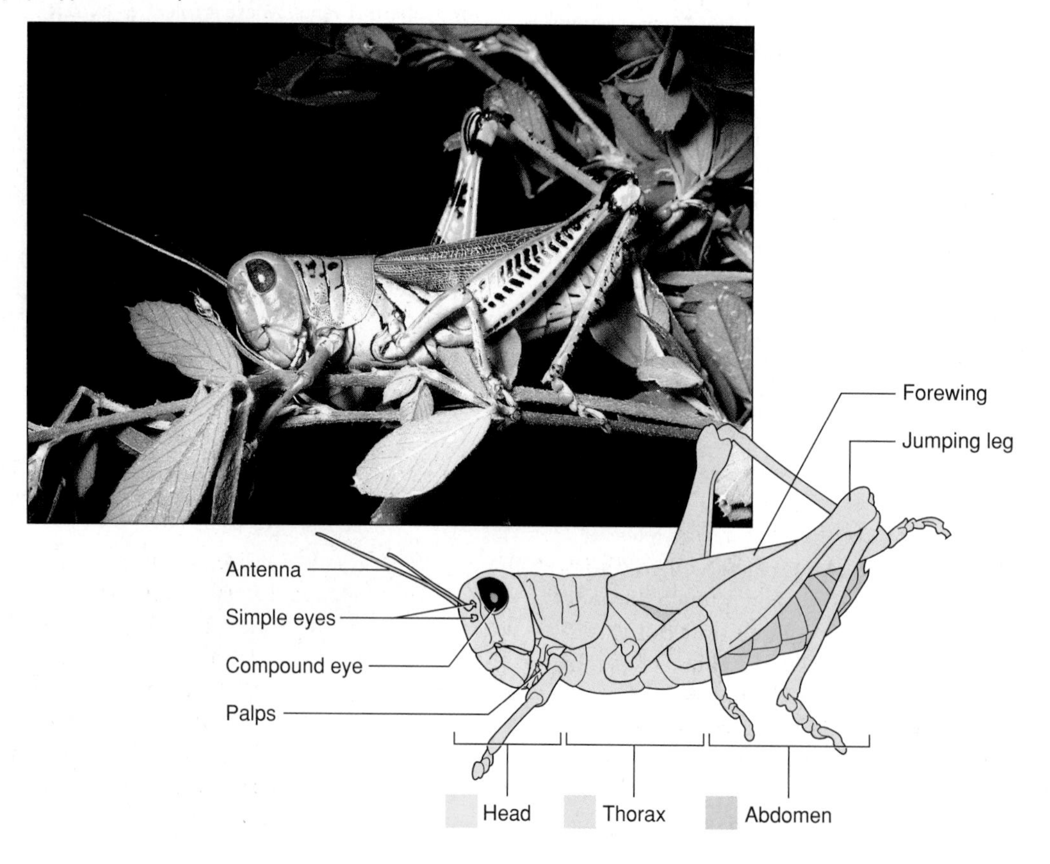

FIGURE 24.22

Arthropod Diversity.

(**A**) Arthropods are segmented protostomes with exoskeletons and jointed appendages. Subphylum Chelicerata includes (**B**) horseshoe crabs and (**C**) arachnids, such as this spider. Subphylum Crustacea includes many familiar animals such as lobsters (**D**). Subphylum Uniramia includes the myriapods, such as this centipede (**E**) and the insects, such as this molting cicada (**F**) and beetle (**G**), which before metamorphosis was a grub (**H**).

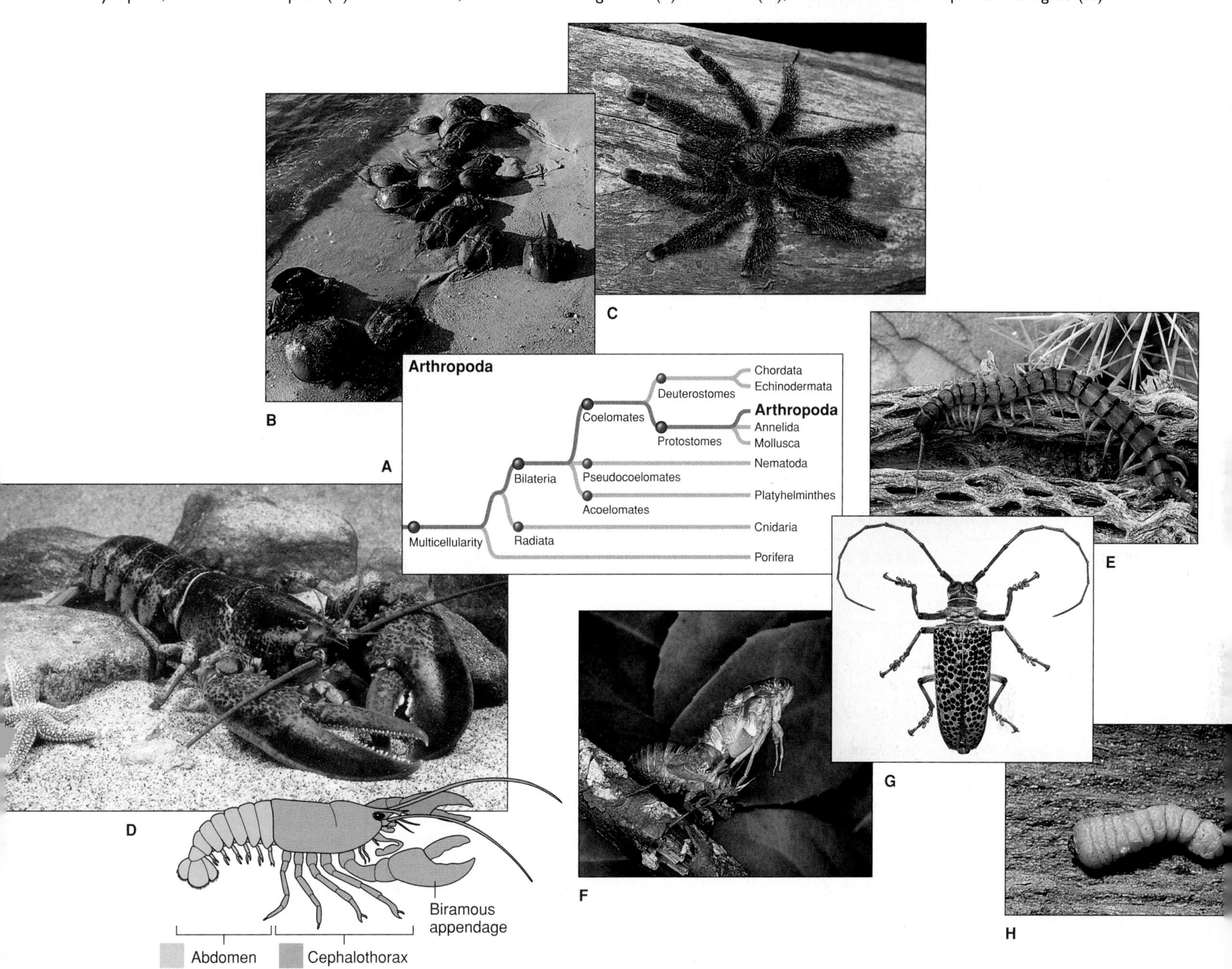

Because of its great diversity, phylum Arthropoda is divided into four subphyla (fig. 24.22). The evolution of subphyla and classes within the phylum has caused controversy among biologists. Older references say that modern arthropods descended from animals called trilobites that dominated the oceans of the early Paleozoic era, but were extinct before the Cenozoic era began. Trilobites had biramous appendages and a three-lobed body. Today, however, biologists hypothesize that trilobites formed a dead-end subphylum.

Subphylum Chelicerata includes marine horseshoe crabs and the mainly terrestrial **arachnids,** such as mites, ticks, scorpions, and spiders. They have piercing mouthparts, four pairs of legs, and they lack antennae. Arachnids and horseshoe crabs have two major body segments: a cephalothorax and an abdomen.

Subphylum Crustacea includes a wide variety of mostly aquatic arthropods, including lobsters, crayfishes, crabs, and shrimp, and many smaller forms such as waterfleas (*Daphnia*) and copepods. Barnacles are unusual crustaceans in that they are sessile. Isopods, commonly known as pillbugs, are terrestrial crustaceans. **Crustaceans** have gills, biramous appendages, mandibles (jaws), two pairs of antennae, and two or three major body segments.

Subphylum Uniramia includes **insects** and myriapods, which are millipedes and centipedes. Myriapods are terrestrial, as are most insects, but some insects live in fresh water. The ocean is about the only place devoid of insects. As their name suggests, uniramians have uniramous appendages, one pair of antennae, and mandibles. Insects have a head, thorax, and abdomen, and myriapods have a head and trunk. Most insects also have six legs and usually two pairs of wings. Millipedes ("thousand leggers") do not really have a thousand legs, but they differ from centipedes because they have two pairs of legs per body segment rather than one pair. Insects range in size from wingless soil species less than 1 millimeter long to fist-sized beetles, foot-long walkingsticks, and others with foot-wide wingspans. Fossil dragonflies were even larger—one had a wingspan of about 30 inches (about 76 centimeters)!

Table 24.2 lists some arthropods that threaten human health.

TABLE 24.2 Arthropods of Medical Importance in the United States

ARTHROPOD	EFFECT ON HUMAN HEALTH
SPIDERS	
Latrodectus species (black widows)	Venomous bite
Loxosceles reclusa (brown recluse, violin spider, or fiddleback)	Venomous bite
MITES	
Trombiculid mites (chiggers)	Dermatitis
Sarcoptes scabiei (itch mite or scabies)	Dermatitis
TICKS	
Ixodes dammini (deer tick)	Transmits Lyme disease
Dermacentor species (dog tick, wood tick)	Transmits Rocky Mountain spotted fever
SCORPIONS	Venomous sting (not dangerous in most U.S. species)
CENTIPEDES	Venomous bite (not dangerous in U.S. species)
INSECTS	
Mosquitoes	Female transmits disease (encephalitis, filarial worms)
Horseflies, deerflies	Female has painful bite
Houseflies and relatives	Many transmit bacteria, viruses, worms to food or water
Fleas	Dermatitis; transmits plague, tapeworms
Bees, wasps, ants	Venomous stings (single sting not dangerous unless person is allergic)

24.9 MASTERING CONCEPTS

1. What are the distinguishing characteristics of arthropods?
2. What types of organisms are arthropods?
3. Describe an arthropod's circulatory, digestive, and nervous systems.

24.10 The Echinoderms: Life Based on Five-Part Symmetry

With their spiny skins and five-part body plans, adult echinoderms look very different from all other animals. But their embryos and larvae reveal closeness to the chordates, the topic of the next chapter.

The phylum Echinodermata includes sea stars (starfishes), brittle stars, sea urchins, sand dollars, sea lilies, and sea cucumbers. Their name means "spiny skin," in recognition of just one of several very distinct traits that characterize this group of organisms. Figure 24.23 shows the relationship of echinoderms to other animals and illustrates some characteristics of this group.

Echinoderm Characteristics

Biologists classify echinoderms by variations in the **endoskeleton** (the inner supportive framework) and general body form. Some echinoderms, such as the sea cucumber, have small, separate calcareous (calcium-containing) structures called ossicles embedded in their leathery skin. Others, such as the sea star, have a more elaborate system of calcareous spines and endoskeletal plates that are strong but lightweight. Sea urchins and sand dollars exhibit yet another endoskeletal variant—their plates are completely fused into a shell called a test.

Echinoderms have a variety of body forms and locomote in different ways. Sea stars and brittle stars have five arms, a form called pentaradial symmetry. Feather stars and sea lilies have branched arms that are adapted for suspension feeding, while sea urchins and sand dollars have no arms and "walk" on stiltlike spines. Echinoderms are among the most colorful of sea inhabitants. Most are bottom dwellers that crawl on coral reefs or other hard surfaces, but echinoderms live at all depths and latitudes. Sand dollars and some sea cucumbers burrow into soft sediments. Some brittle stars can swim using their appendages.

A unique echinoderm feature is pedicellariae, which are tiny pincers around the base of the spines on sea stars and sea urchins. Pedicellariae clean the skin surface of debris and larvae. Another unique characteristic is the echinoderm's **water vascular system,** in which seawater enters a series of enclosed canals that end in suction-cup-like structures called **tube feet** (fig. 24.24). When muscles contract, tube feet extend, forcing water in. When the muscles relax and water exits, the sucker at the bottom of each tube foot clamps to the seafloor. The wavelike pumping of water in and out of the tube feet allows the animal to locomote in slow, gliding movements. Echinoderms are like living hydraulic systems that tap the power of moving water to locomote.

FIGURE 24.23

Echinoderm Characteristics.

(**A**) Echinoderms' place in the animal kingdom is closest to the chordates. (**B**) A sea urchin's spines illustrate the source of the phylum's name. (**C**) Look closely at this underside of an Indonesian sea urchin—note the pentaradial (five-part) symmetry. (**D**) The aboral surface of a sea star shows spines with pedicellariae (pincers) at their tips.

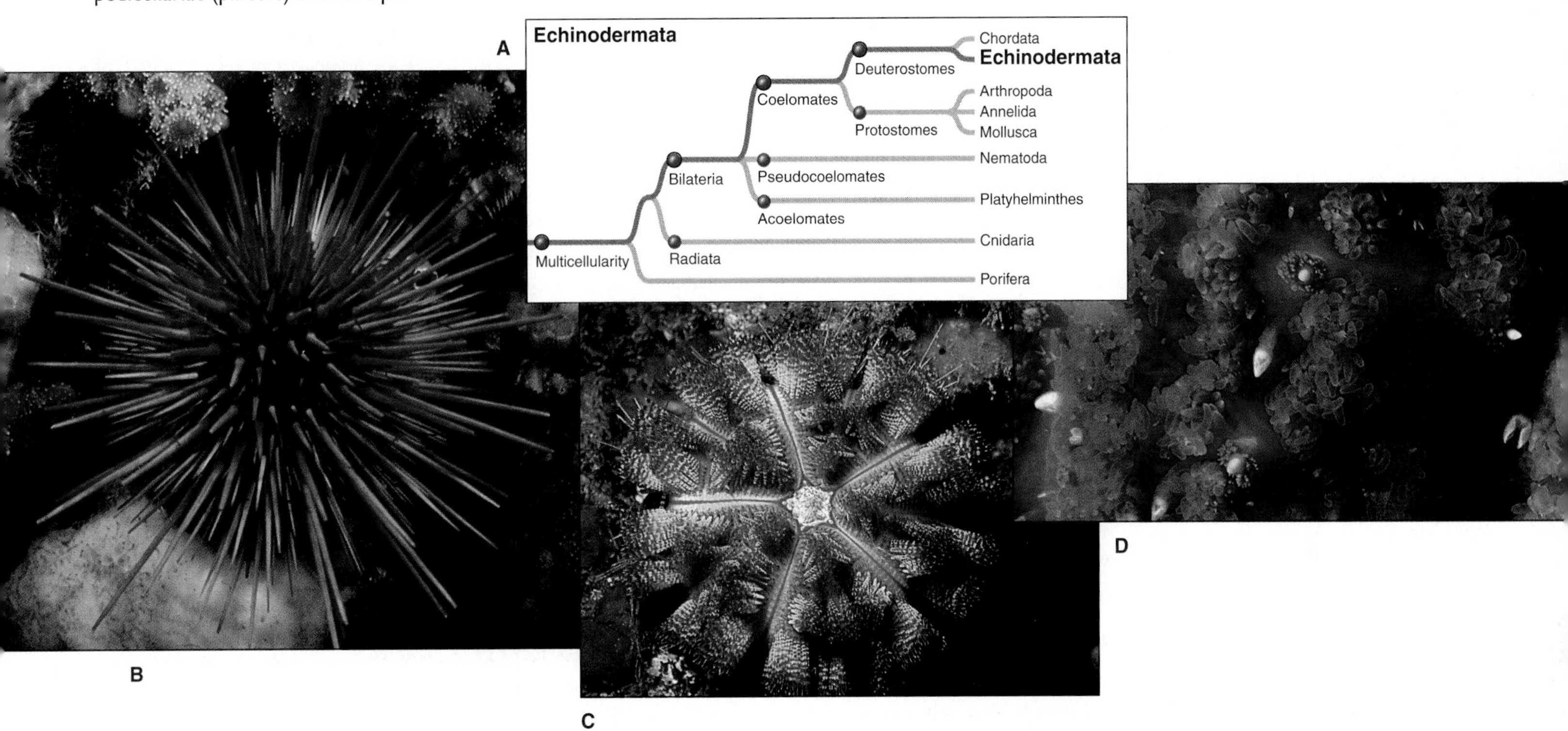

Tube feet have other functions. A sea star uses its tube feet to open bivalve shelves so it can eat the soft flesh within. The sea star attaches its tube feet to prey and steadily pulls until the muscles of the bivalve tire and the shell opens. The sea star then everts its stomach through its mouth into the bivalve, secretes digestive enzymes, and absorbs the liquefied food. Tube feet also function as gills, sucking up oxygen from the water, and serve as sense organs because they house sensory cells.

A sea cucumber's tentacles are modified tube feet used to catch food suspended in the water. The animal then inserts the food-covered tentacles into the mouth. A brittle star's tube feet lack suckers and are used mainly for feeding, whereas the sea urchin's tube feet are mainly respiratory.

Due to their relatively sedentary lifestyle, echinoderms lack heads, but nerves extend down their arms and tentacles. They have reduced circulatory systems. Sea stars, for example, respire

FIGURE 24.24

Tube Feet.

(**A**) This sea cucumber uses tube feet to locomote on the ocean floor. (**B**) The bright yellow tube feet shown in this ventral view also help it to breathe and feel.

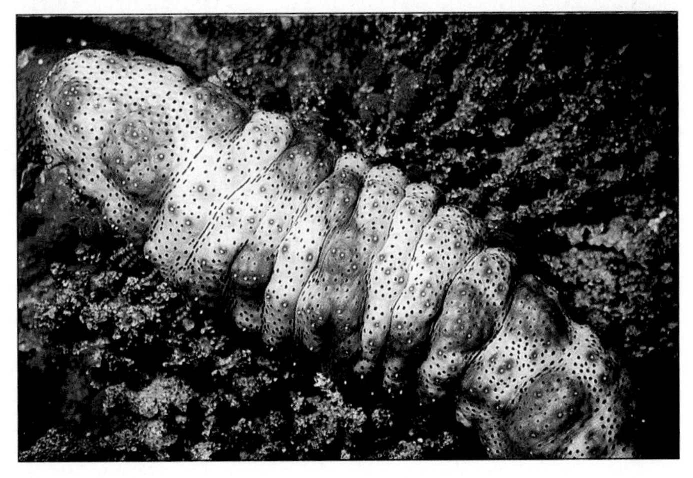

by means of tiny skin gills. In sea cucumbers, a network of passageways called a respiratory tree projects internally from the cloaca (a common opening used for both waste disposal and reproduction). This tree exchanges gases and eliminates nitrogenous wastes. Some echinoderms lack excretory systems, yet others have complete ones.

Usually echinoderms reproduce sexually, with male and female gametes combining in the sea, and coming from separate individuals. Some starfishes and other echinoderms can regenerate lost parts such as arms. In fact, some animals reproduce asexually by splitting in two and regenerating the missing arms.

Echinoderm Evolution and Development

Because of their trademark spiny skin, the echinoderms have left a rich fossil record. The first fossils date from about 535 million years ago. These ancient animals had tube feet and the hydraulic systems that powered them, but they did not yet have pentaradial symmetry. Of the 25 main groups of echinoderms that lived before the dinosaurs reigned, only 5 remain today.

Echinoderms were initially classified with jellyfishes and corals, based on superficial resemblance. Early in the twentieth century, however, embryologists noted striking similarities to the chordates (the topic of the next chapter). Echinoderms, chordates, and a few other phyla are the only deuterostomes, sharing the pattern of early cell divisions and embryonic stages, and bilaterally symmetrical larvae. And what strange larvae they are!

Echinoderm larvae look nothing like the adult forms (fig. 24.25). Plus, adults of different species that look very much alike may have totally different-looking larvae. Metamorphosis is quite dramatic. Within as short a time as 30 minutes, a collection of cells on one side of a larva assembles into a five-sided, small disc. All of a sudden, the disc everts and consumes the remainder of the larva. The animal is now a tiny juvenile, a replica of the adult form. Within a year, it will be mature enough to reproduce.

Comparing DNA sequences has revealed the close evolutionary tie between the echinoderms and the chordates. Study of the genes that regulate development are even more telling. Several genes that control development of the head and limbs in vertebrates, for example, have counterparts in the echinoderms, where they control functioning of the tube feet. It is as if the echinoderms used the same set of fundamental genetic instructions as chordates, but in their own way, resulting in the unusual five-sided forms that set this phylum apart from all others.

Biologists recognize 35 animal phyla—or 36 if the lobster lip dwellers described in the chapter opening essay are counted (table 24.3). Animals are alike in that they eat, move, lack cell walls, and secrete extracellular matrix, but their body forms are distinctive enough to form the basis of designating phyla. This makes sense—a hydra seems nothing like a nematode or a spider or a sea star. But at the genetic level, these distinctions break down. Animals may turn out to be more like each other than they are different, with the same sets of regulatory genes becoming activated at different times and in different tissues to sculpt the myriad of body forms in this kingdom.

24.10 MASTERING CONCEPTS

1. What are some characteristics unique to echinoderms?
2. What are some functions of tube feet?
3. How do echinoderms eat, respire, and reproduce?
4. How are echinoderms similar to chordates?

A

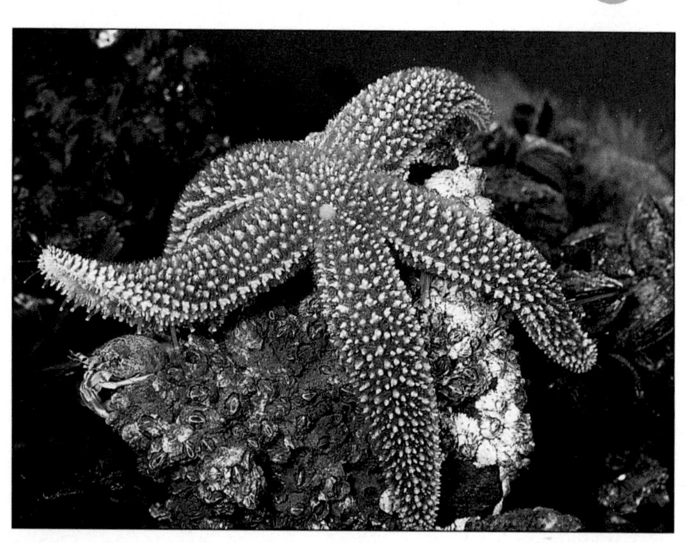

B

FIGURE 24.25

Echinoderm Metamorphosis.

An echinoderm larva (**A**) typically does not resemble the adult organism (**B**), as this starfish illustrates. The larvae help reveal the evolutionary placement of echinoderms with other bilaterally symmetrical animals.

TABLE 24.3 Some Animal Phyla

PHYLUM	EXAMPLES	NO. OF KNOWN SPECIES	BODY SYMMETRY	BODY FORM	DEVELOPMENT	SPECIFIC STRUCTURES/ CHARACTERISTICS (NOT NECESSARILY IN ALL SPECIES)
Porifera	Sponges	9,000+	Asymmetrical, radial			Archeocytes, pinacocytes, amoebocytes Mesohyl Skeleton of spicules and spongin
Cnidaria	Jellyfishes Hydras Corals Sea anemones	9,000+	Radial		Diploblastic	Polyp and medusa stages Mesoglea Gastrovascular cavity Nematocysts
Platyhelminthes	Planaria Flukes Tapeworms	20,000	Bilateral	Acoelomate No segments	Triploblastic	Intestinal ceca Tegument Protonephridia (flame cells)
Nematoda	*C. elegans*	12,000+	Bilateral	Pseudocoelomate No segments	Triploblastic	Pseudocoel Complete digestive system
Mollusca	Clams Squids Snails Octopuses	100,000	Bilateral	Coelomate No segments	Triploblastic Protostome	Mantle Muscular foot Visceral mass Radula
Annelida	Earthworms Leeches Polychaetes	15,000	Bilateral	Coelomate Segments	Triploblastic Protostome	Setae Clitellum Parapodia Closed circulatory system
Arthropoda	Insects Arachnids Crustaceans Myriapods	1,000,000	Bilateral	Coelomate Segments	Triploblastic Protostome	Hemolymph in hemocoel Spiracles Jointed appendages Exoskeleton
Echinodermata	Sea stars Brittle stars Sea urchins Sand dollars Sea lilies Sea cucumbers	6,500	Bilateral larvae Pentaradial adults	Coelomate No segments	Triploblastic Deuterostome	Ossicles Pedicellariae Tube feet Water vascular system

Chapter Summary

24.1 What Is an Animal?

1. Animals are multicellular eukaryotes whose cells do not have cell walls. **Vertebrates** have backbones; the more abundant **invertebrates** do not.
2. Animals have immature stages, and are motile at some point in the life cycle.
3. An immediate animal ancestor was likely a choanoflagellate.
4. The earliest fossil evidence of animals is from 580 to 570 million years ago.

24.2 What Characteristics Define and Distinguish Animals?

5. Animal cells secrete an extracellular matrix.
6. Animals are heterotrophs and may be carnivores, herbivores, or detritivores.
7. Animal bodies exhibit degrees of organization into tissues, organs, and organ systems.
8. Body symmetry may be **radial** with an oral end and aboral end, or **bilateral,** with a **cephalized** (head) end.
9. An animal zygote undergoes cleavage divisions to form a **blastula** and then a **gastrula,** which has **ectoderm, endoderm,** and in some

species, **mesoderm.** Those with two layers are **diploblastic;** with three layers, **triploblastic.**

10. Animals may undergo **direct development** or **metamorphose** from a **larva** to an adult. They reproduce sexually and sometimes asexually.
11. A **coelom** is a body cavity lined with mesoderm. Animals may be **coelomate, acoelomate,** or **pseudocoelomate.**
12. Major coelomate lineages differ in developmental pattern. **Protostomes** have **spiral** and **determinate cleavage,** and a **blastopore** developing into a mouth. **Deuterostomes** have **radial** and **indeterminate cleavage** and a blastopore developing into an anus.

24.3 The Sponges (Porifera) Are the Simplest Animals

13. Sponges are asymmetrical or radially symmetrical suspension feeders that move slowly.
14. Pinacocytes, amoebocytes, archeocytes, and choanocytes are sponge cells. They do not form true tissues.
15. A sponge's skeleton consists of **spicules** within spongin.
16. Sponges reproduce sexually and may bud asexually using gemmules.

24.4 Cnidaria Are Radially Symmetrical, Aquatic Invertebrates

17. Cnidaria include jellyfishes, corals, and Portuguese men-of-war.
18. These animals capture prey with stinging **nematocysts** released from **cnidocytes.**
19. Cnidaria have radial symmetry, are diploblastic, and partially digest food in a **gastrovascular cavity.**
20. A cnidarian body form is a **polyp** or a **medusa.** Fertilization in water produces a planula, which attaches and becomes a polyp that reproduces asexually to generate a strobila, which yields medusae.

24.5 Flatworms (Platyhelminthes)

21. Flatworms include planaria, flukes, and tapeworms.
22. Flatworms are triploblastic, lack coeloms, and have bilateral symmetry. **Protonephridia** maintain internal water balance.
23. Most flatworms are parasitic and have great reproductive capacity.
24. Flatworms reproduce asexually and sexually, and some are hermaphrodites.

24.6 Roundworms (Nematoda)

25. Roundworms include parasitic and free-living species in soil or aquatic sediments.
26. Nematodes are bilaterally symmetrical, cylindrical, and unsegmented.
27. Nematodes have pseudocoeloms, complete digestive systems, and separate sexes.

24.7 Mollusks Include Clams, Snails, Octopuses, and Squids

28. The diverse mollusks have **mantles, muscular feet, visceral masses,** and complex organ systems.
29. Mollusks are protostomes with bilateral symmetry and have trochophore larvae.
30. Sexes are separate in the mollusks.

24.8 Annelids Are Segmented Worms

31. Annelids include oligochaetes, characterized by setae (bristles) and a clitellum; polychaetes, with parapodia as appendages; and the hirudinea (leeches) with annuli between their segments.
32. Annelids feed in diverse ways and have organ systems.
33. Leeches and oligochaetes are hermaphroditic; polychaetes have separate sexes.

24.9 Arthropods Include Insects, Crustaceans, and Arachnids

34. Arthropods are protostomes with bilateral symmetry, and they are segmented with jointed appendages and a chitinous exoskeleton.
35. Arthropods have **open circulatory systems,** spiracles to respire, and a nervous system.
36. Four subphyla include trilobites; horseshoe crabs and **arachnids; crustaceans;** and **insects** and myriapods.

24.10 The Echinoderms: Life Based on Five-Part Symmetry

37. Echinoderms are spiny-skinned marine animals with pentaradial symmetry that move using **water vascular systems** to power **tube feet.**
38. Similarities in embryonic development and molecular evidence place echinoderms closest to chordates.
39. Echinoderm larvae look very different from adults and display bilateral symmetry.

Testing Your Knowledge

1. Cite the taxonomic criteria used to distinguish each of the following:
   a. animals from other organisms
   b. protostomes from deuterostomes
   c. a tapeworm from a nematode from an earthworm
2. Distinguish between or among the following:
   a. radial, bilateral, and pentaradial symmetry
   b. direct development and metamorphosis
   c. spiral and radial cleavage
   d. determinate and indeterminate cleavage
   e. biramous and uniramous appendages
   f. diploblastic and triploblastic embryos
   g. coelomate, acoelomate, and pseudocoelomate body forms
3. Identify an animal that has the following cell types:
   a. cnidocytes
   b. flame cells
   c. choanocytes
4. Identify an animal that has the following tissues:
   a. epidermis and gastrodermis
   b. pseudocoel and hypodermis
   c. hemolymph
5. On an episode of the television series *The X-Files,* two FBI agents chase a bizarre creature that has the form of a human but with a scolex atop its head. From what animal phylum did the writer get inspiration for including a scolex?
6. Compare how a sponge and a sea urchin feed.

Thinking Scientifically

1. Compare the major animal phyla in the order in which the chapter presents them, listing the new complexities seen in each group.
2. Give an example from the animal kingdom where molecular sequence information confirmed a classification based on physical resemblance.
3. In a new medical technology called preimplantation genetic diagnosis, a single cell removed from an eight-celled human embryo is tested for the presence of a disease-causing gene. If the disease isn't present, the remaining seven-celled embryo is placed in the woman's uterus, where it continues development. What property of early human development makes this procedure possible?
4. The Chincharro people lived on the northern coast of Chile from 5500 to 500 B.C. Many of them died at very young ages due to a number of parasitic diseases. Their preserved excrement (coprolites) and mummies contain eggs and other remains of tapeworms, roundworms, and flukes. How can a researcher tell these worms apart?

References and Resources

Bischof, Barbie. December 1997/January 1998. Cities beneath the sea. *Natural History,* vol. 106, p. 40. Corals are living communities.

Bischof, Barbie. December 1997/January 1998. Reefs in Crisis. *Natural History,* vol. 106, p. 46. Infections and encroaching construction and tourists threaten coral reefs.

Bond, Calhoun. December 1997/January 1998. Keeping up with the sponges. *Natural History,* vol 106, p. 22. Sometimes the most powerful tool of the biologist is simple observation.

Morris, Simon Conway. December 14, 1995. A new phylum from the lobster's lips. *Nature,* vol. 378, p. 661. Researchers identified an organism living on lobster mouthparts that has such an unusual combination of traits that they placed it in a new phylum.

The *LIFE* Online Learning Center provides additional resources and tools for studying this chapter.
www.mhhe.com/life

CHAPTER 25

Animalia II—The Chordates

Vietnam's Diverse Vertebrates

There aren't many places on Earth where biologists can discover new species of large animals, but Vietnam is certainly one of them. Since the country's jungles were opened for exploration in 1982, researchers have described dozens of species of insects, amphibians, reptiles, and mammals that live there. Sometimes an animal is new to science, but familiar to natives. This was the case for *Sus bucculentus*, a long-snouted, warty-faced, short-legged pig that biologists thought had been extinct for more than a century. But in 1998, a researcher found one in the Annamite Range along the Laos/Vietnam border. The pig was a popular dinner dish among the natives! DNA sequence comparisons between someone's dinner and a museum specimen revealed that the old and new pigs were of the same species.

> ***Sometimes an animal is new to science, but familiar to natives. This was the case for Sus bucculentus, a . . . pig that biologists thought had been extinct for more than a century.***

In the Tam Dao reserve near Hanoi, biologists have cataloged 108 new species of snakes, accounting for 4% of the world's total of 2,700 species. Other reptiles are abundant too, including blind snakes, legless lizards, snakes that eat lizards, green tree snakes, giant salamanders, and frogs that resemble moss. Their camouflage is astounding—frogs blend in with bark, and a lizard's face has the uncanny mosaic green pattern of a leaf. The reason for the incredible biodiversity at Tam Dao is that here, three drastically

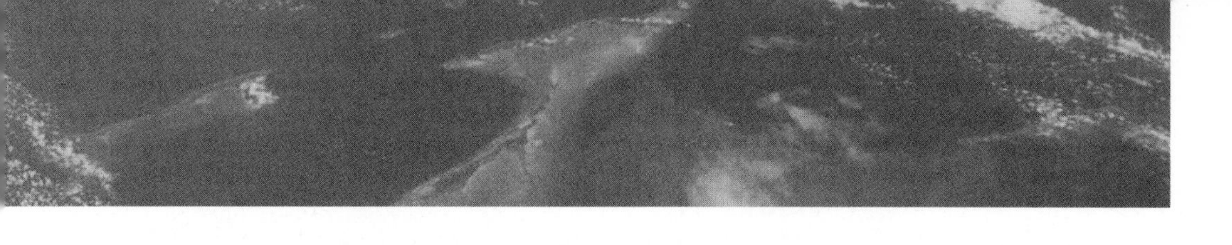

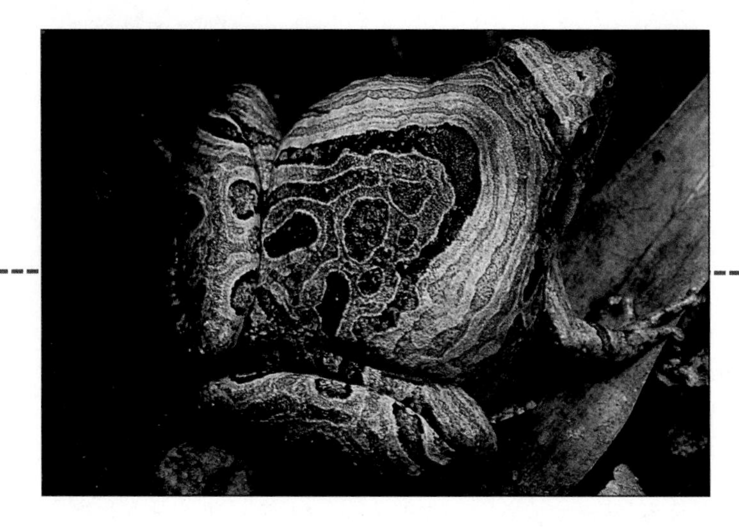

Many new species of amphibians and reptiles are being discovered in the forests and jungles of Vietnam. The Asian painted frog *Microhyla pulchra* blends in with this branch.

different types of ecosystems intersect—the eastern Himalayan alpine forest, the south China temperate forest, and the southeast Asian tropics. The many peaks, valleys, furrows, and ridges create microecosystems, small isolated areas where the genetic divergence that can become speciation is commonplace.

Trees tend to grow taller here than elsewhere in the world, and they are farther apart, not linked by vines as they are in other forests. Very different types of animals have adapted to this forest topography in strikingly similar ways—they have evolved into gliders. Lizards, snakes, frogs, and squirrels have extensive webbing between their toes and skin flaps on their abdomens that effectively turn them into living parachutes, able to fly short distances from tree to tree.

The pig was a popular dinner dish among the natives!

The teams of biologists combing Vietnam's forests and jungles are cataloging species, taking population counts, and studying diets and reproductive strategies. The hope is that the attention of biologists will attract tourists interested in natural history, which will bring money that will be used in environmental conservation of these rare habitats and their residents.

The green skin pattern of the water dragon *Physignathus cocincinus* looks remarkably like a leaf.

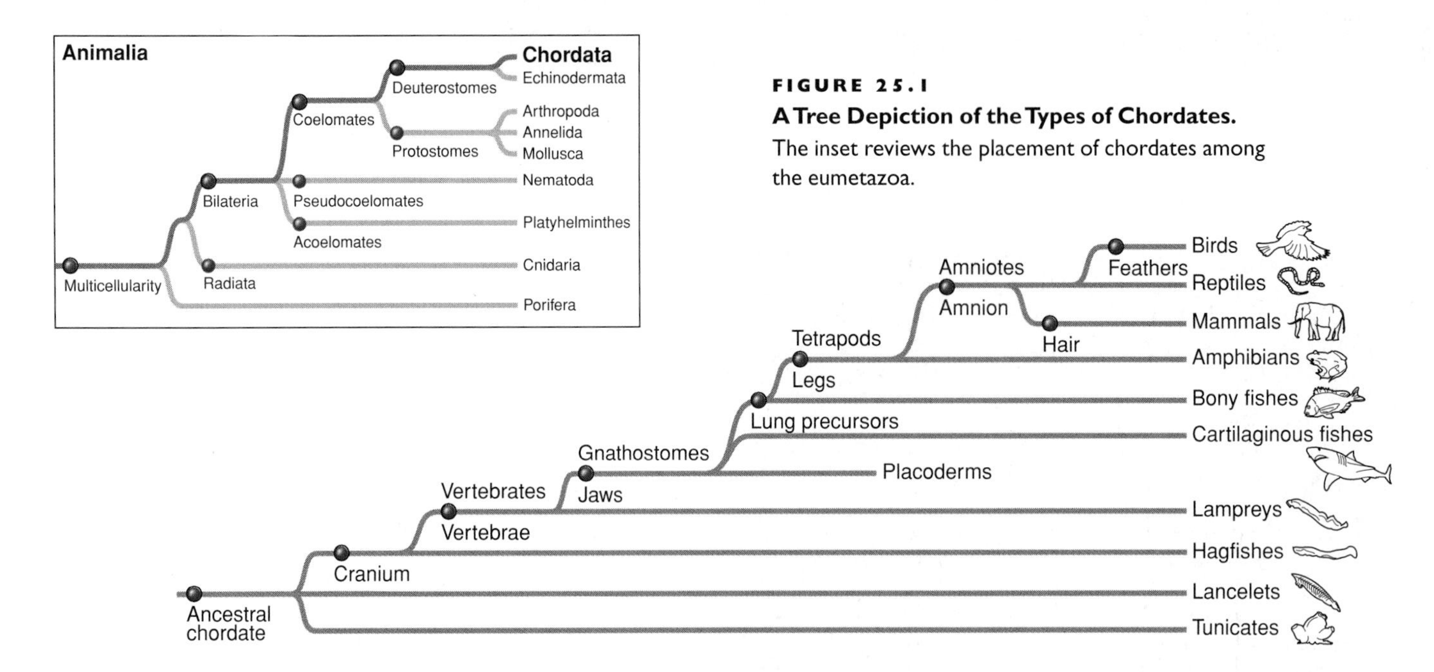

FIGURE 25.1
A Tree Depiction of the Types of Chordates.
The inset reviews the placement of chordates among the eumetazoa.

25.1 What Are Chordates?

Many familiar animals—fishes, amphibians, reptiles, birds, and mammals—belong to phylum Chordata. They share four key characteristics.

The chordates are a diverse group of about 55,000 species that include many familiar animals (fig. 25.1). They are divided into three subphyla. The tunicates (subphylum Urochordata) and the lancelets (subphylum Cephalochordata) are invertebrates. Members of the third subphylum, Vertebrata, have either cartilage or bony inner skeletons (endoskeletons) and are further subdivided based on acquisition of specialized features, such as jaws, lungs, limbs, protective membranes around fetuses, hair and feathers. The phylum includes two small groups of jawless fishes (the hagfishes and lampreys), cartilaginous and bony fishes, and the **tetrapods** (four-limbed animals), which are amphibians, reptiles, birds, and mammals.

Chordates have a three-layered embryo, bilateral symmetry, and a coelom. They also share four major characteristics at some point during their life histories (fig. 25.2):

1. The term Chordata derives from the distinguishing characteristic of the **notochord.** This flexible rod within a fibrous sheath, which is derived from embryonic mesoderm, extends dorsally (along the back) down the length of the animal's body, and it provides support and also allows side-to-side movement. Some of the invertebrate chordates retain the notochord throughout life, and it aids in support and locomotion as well as providing points for muscle attachments. The notochord in vertebrates is gen-

FIGURE 25.2
A Chordate Has Four Defining Characteristics.
(*1*) A notochord is a flexible rod that extends down the body dorsally. It is present at some point in every chordate's life. (*2*) The nerve cord in a chordate is characteristically dorsal to the digestive tract. (*3*) Chordates have pharyngeal gill pouches or slits. (*4*) Chordates also have a postanal tail. The illustration depicts these characteristics among four representative chordates.

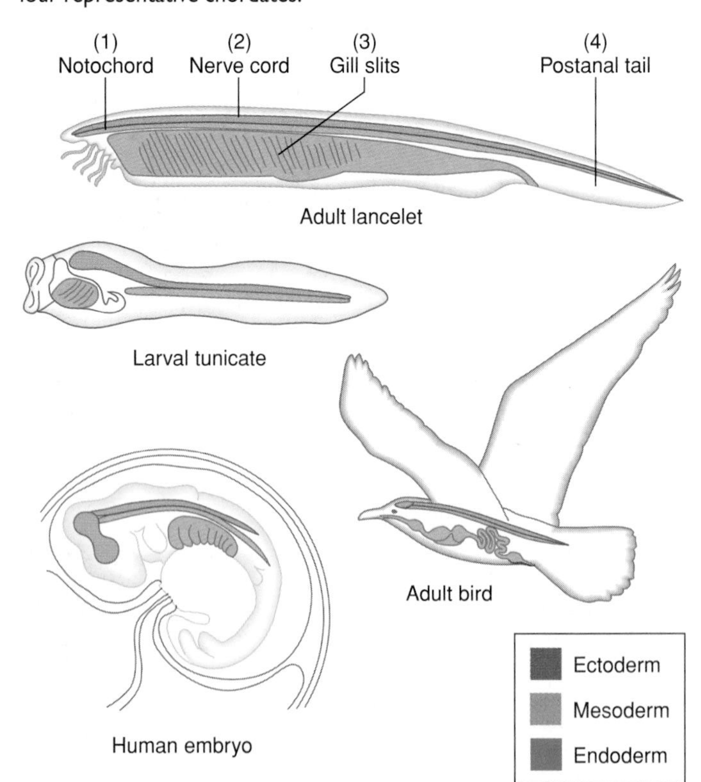

erally prominent early in their development, but it stops growing, leaving only vestiges between the vertebrae of the adult animal.

2. Chordates have a **dorsal, hollow nerve cord,** which is derived from a plate of embryonic ectoderm. In many species the nerve cord enlarges at the anterior end to form a brain. Nerve cords in chordates are dorsal to the notochord, and they are hollow. In contrast, nerve cords in other animals are ventrally located and solid. In most vertebrates, bony structures protect the nerve cord—the vertebrae and cranium (skull).
3. Chordate embryos develop **pharyngeal pouches** where the endoderm and ectoderm grow toward each other. In invertebrate chordates, the pouches are used to filter food particles from water. In aquatic chordates, such as fishes, the two pockets meet and open, forming gill slits within a muscular tube called the **pharynx.** These slits, which may develop further into **gills,** are respiratory organs. In tetrapods the embryonic pharyngeal pouches develop into various structures, including the middle ear cavity, the Eustachian tube, tonsils, and the parathyroid glands.
4. A **postanal tail** consists of the posterior tip of the notochord plus muscles and other tissues that form an appendage that enables juveniles of certain chordates to move freely in water. In many species, adults retain and use the tail. The human vestige of this structure is the coccyx, or tailbone.

The chordates originated at least 570 million years ago. By 540 million years ago, chordate ancestors swam in the ancient oceans. The earliest chordate fossils date from about this time, when soft bodies began to develop hard parts, such as bones and teeth, that could leave evidence of their existence. Animals in a small phylum called Hemichordata (the acorn worms) may be the descendants of the immediate ancestors of the chordates, forming a link of sorts to the echinoderms. The hemichordates share some characteristics with the chordates (gill slits and a dorsal tubular nerve cord), yet their larvae resemble both those of tunicates and echinoderms. As such, they may represent a "transitional form" in the evolution of animals.

25.1 MASTERING CONCEPTS

1. What are the three subphyla of chordates?
2. What are the four defining characteristics of chordates?
3. About when did chordates originate?

OLC

25.2 The Tunicates and Lancelets Are "Protochordates"

Tunicates and lancelets are small, cylindrical, aquatic invertebrates that extract oxygen and trap nutrients from water using siphons and a ciliated pharynx. They are the modern organisms that most closely resemble ancestral chordates.

Members of subphylum Urochordata (the tunicates) and subphylum Cephalochordata (the lancelets) are the modern organisms that most resemble the ancestral chordates. Because of their primitive characteristics, they are termed protochordates.

The Tunicates

Tunicates are named for their nonliving protective covering, or tunic, which consists of cellulose and protein. The presence of cellulose is unusual because this carbohydrate is typically found in plants, not animals. Tunicates may live individually or in groups that share a tunic. Urochordata means "tail-chordate," in recognition of the fact that the tunicate larva retains the notochord only in its tail. Figure 25.3 locates the tunicates among the chordates, and shows a representative organism. Many of the approximately 3,000 species of tunicates are brightly colored. Most tunicates are hermaphroditic and reproduce sexually by external cross-fertilization, but some tunicates reproduce asexually by budding.

A generalized adult tunicate body resembles a bag with two projections, or siphons. The animal extracts oxygen and food from the water that flows through it. Water, pulled in by cilia on the pharynx, enters through the incurrent siphon, then moves across mucus-covered gill slits in the extensive pharynx. Oxygen dissolved in the water diffuses into blood vessels near the pharynx, and carbon dioxide diffuses out. Suspended food particles become trapped in the mucus, and cilia on the pharynx carry the food to the digestive organs. The water continues into a cavity called the atrium, then exits through the excurrent siphon. The best-studied tunicates, the ascidians, are also called sea squirts because they can forcibly eject water from their excurrent siphons if disturbed.

Tunicate larvae have all four chordate characteristics, but after a period of metamorphosis most adults retain only gill slits. The notochord disappears and the nerve cord shrinks to nearly nothing. The free-swimming larvae, which resemble tadpoles, move vigorously, dispersing sediments and thereby affecting nutrient distribution in aquatic ecosystems. Many adults are sessile and attach to boat bottoms, dock pilings, mollusk shells, algae, or corals. Tunicates range in size from a few millimeters to 30 centimeters in length.

Three classes of tunicates are recognized. Ascidiacea (the sea squirts) are one. Members of the second class, Thaliacea, are transparent and live in the ocean, singly or as colonies. Their mostly hollow bodies are barrel-shaped, with an incurrent siphon at one end and an excurrent siphon at the other, which differs from the closer location of the siphons in sea squirts. The thaliacean body has bands of muscles that enable the animal to move, respire, and strain food particles from the water. Some thaliacea can bioluminesce in response to phytoplankton "blooms." Members of the third class of tunicates, class Larvacea, take their name from their resemblance to tadpoles. They have an unusual way of feeding. A larvacean builds a "house" around its mouth, a hollow sphere made of mucus and fibers. A filterlike structure within the house strains out food particles. About six times a day, the filter clogs, and the animal just moves away from it and secretes another.

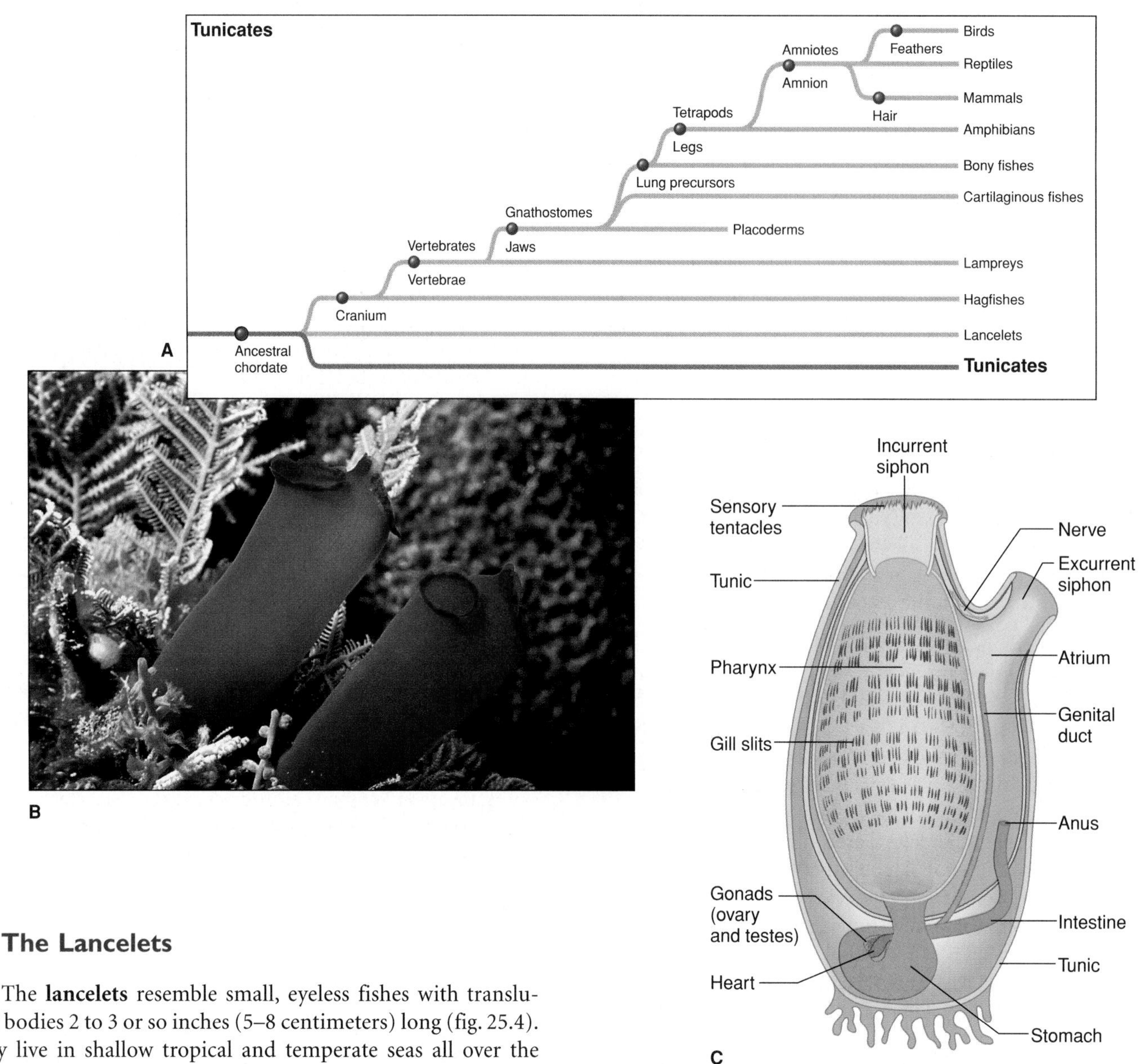

FIGURE 25.3
Tunicates.
(**A**) Tunicates, members of subphylum Urochordata, are one type of invertebrate chordate. The adult (**B**) is sessile (still) and shows none of the chordate-defining characteristics that the free-swimming larva does. The tunic and incurrent and excurrent siphons are major features of the tunicate body (**C**).

The Lancelets

The **lancelets** resemble small, eyeless fishes with translucent bodies 2 to 3 or so inches (5–8 centimeters) long (fig. 25.4). They live in shallow tropical and temperate seas all over the world, assuming a characteristic positioning with their tails buried in the sand or sediments and their anterior ends extending into the water. Some parts of the oceans contain more than 5,000 lancelets per square meter. The fossil record is mostly silent on the lancelets much as it is with the tunicates, because of their lack of hard parts. Today biologists recognize 25 species of lancelets.

Like tunicates, the lancelets filter the water for nutrients, capturing suspended food particles by using ciliated structures extending from their mouths. Cilia on the gills coupled with the mucus secreted by the pharynx effectively trap and move food particles into the digestive tract for processing and absorption. The filtered water passes out through the gill slits, leaving the body through the atriopore, which is the counterpart to the sea squirt's excurrent siphon.

Lancelets have a circulatory system that lacks a heart, but is closed—that is, the vessels are all connected. Pulsing of the vessels moves the blood. A lancelet's blood distributes nutrients, but is not complex enough to handle gas exchange. The nervous system is a nerve cord with a slight swelling at the anterior end, plus sensory receptors on the body. Sexes are separate, with gametes released through the atriopore. Fertilization occurs in the water,

FIGURE 25.4
Lancelets.
(**A**) Lancelets, members of subphylum Cephalochordata, are invertebrate chordates whose adults retain chordate-defining characteristics. (**B**) This lancelet is in its suspension-feeding posture, with its head sticking up out of the substrate and its tail buried. The internal anatomy of the lancelet reveals the chordate characteristics (**C**).

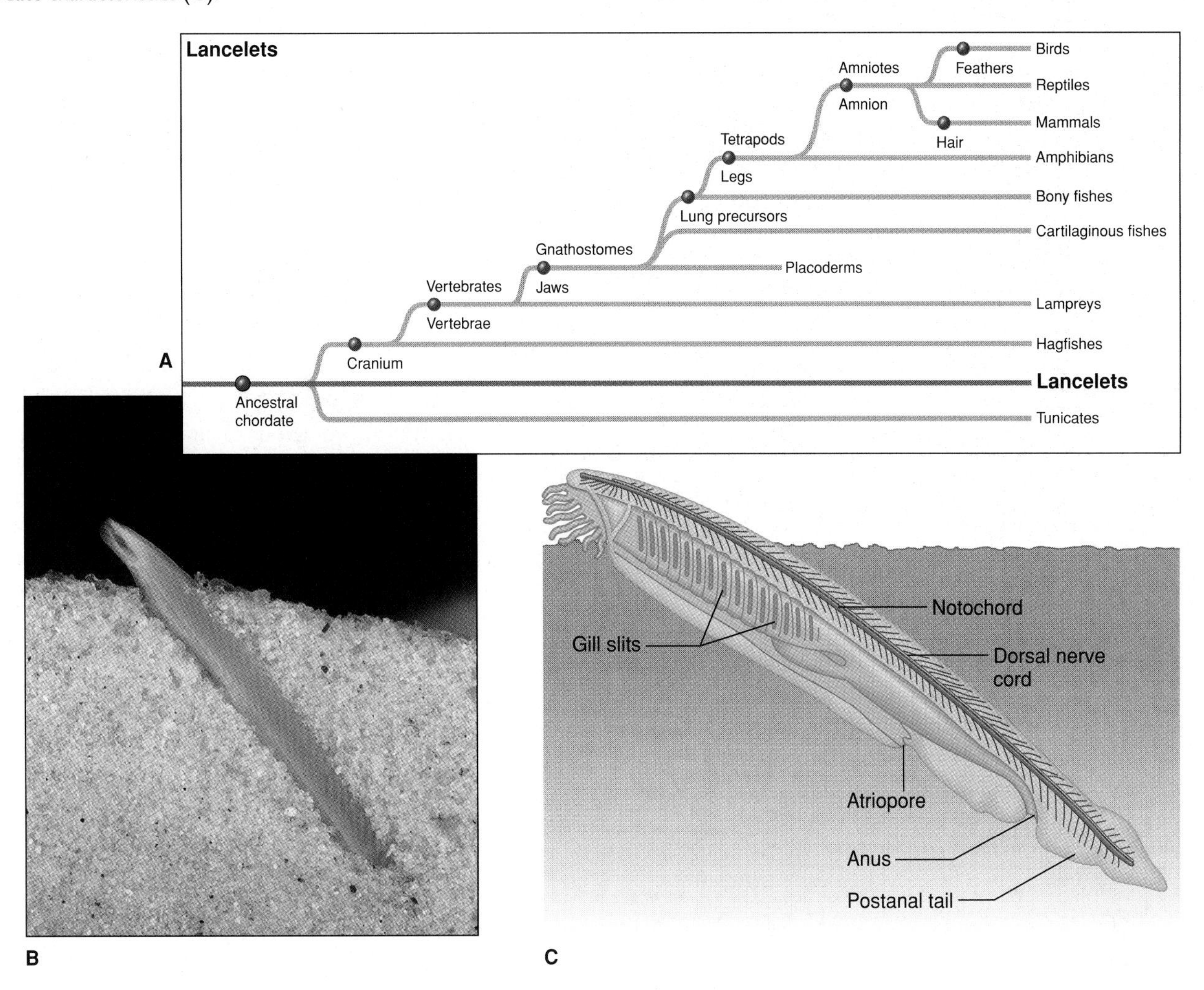

resulting in embryos that develop into ciliated free-swimming larvae.

Lancelets were long considered the model chordates, because they very clearly display all four major chordate characteristics, as well as inklings of the organ systems that appear in the vertebrates. Lancelets were once hailed as the direct ancestor of vertebrates, but because they lack the sophisticated sensory organs, brain, and mobility of vertebrates, it is more likely they branched from an early shared ancestor. Today, lancelets are the modern organism that probably most closely resembles the ancestor to the vertebrates.

A clue to the primitive status of the lancelets is that they share with the insects many genes that determine body segmentation. Biologists had thought that the segmentation of insects and the multisubunit structure of the backbone of vertebrates represented convergent evolution, two separate solutions to the challenge of building a bilateral body. The identification of segmentation genes in the lancelets, however, argues instead for a homologous relationship—that is, that arthropods and chordates acquired the ability to form segmented bodies from a shared ancestor.

25.2 MASTERING CONCEPTS

1. In a tunicate, how does the pharynx provide nutrition as well as gas exchange?
2. What are the types of tunicates?
3. Why are tunicates and lancelets considered to be chordates?
4. What is the likely evolutionary relationship between lancelets and vertebrates?

OLC

25.3 The Vertebrates

In addition to their namesake, the vertebrae of the backbone, most vertebrates share paired appendages, an endoskeleton that can include jaws, and organ systems. The evolution of lungs and limbs within this group made life on land possible for chordates.

The **vertebrates** are distinguished by the presence of a vertebral column (backbone), and a combination of other characteristics. For most, these include paired limbs, an endoskeleton (on the inside) that includes a cranium surrounding a brain, and more complex organ systems. Many of these features are explored in detail in unit 6 so they are just introduced here.

If the vertebrates are compared with the invertebrate chordates, three evolutionary trends emerge:

1. With increasing complexity, the notochord became less pronounced, giving way to a backbone built of linked hard parts, the **vertebrae.** The backbone provides a point of attachment for muscles and gives the animal a greater range of movement.
2. Jaws developed from gill supports. This greatly expanded the ways that the animals feed, opening up new behaviors that led to the use of new habitats.
3. The evolution of lungs and limbs enabled some vertebrates —the tetrapods—to live on land.

Because of their endoskeletons, vertebrates have left a rich fossil record. A recent analysis of nearly 700 highly conserved genes used to estimate the time of origin of the different types of vertebrates is, happily, very consistent with this fossil evidence. Figure 25.5 depicts an interpretation of this molecular dating information.

The 50,000+ modern (extant) species of vertebrates vary in several ways. Vertebrates occupy a diversity of areas, ranging from high mountains to great ocean depths, from dry hot deserts to wet tropical rain forests and icy polar regions. They locomote in diverse ways—many swim, others walk, jump, run, crawl, or fly. Some vertebrates lay eggs **(oviparous),** some retain their eggs and give birth to live offspring **(ovoviviparous),** and others produce live young directly **(viviparous).**

There are eight classes of vertebrates: four are fishes, plus amphibians, reptiles, birds, and mammals. The four fish classes include an extinct group, jawless fishes, cartilaginous fishes, and bony fishes. Reptiles, birds, and mammals are called **amniotes** because they have membranes (the amnion, chorion, and allantois) that protect a developing fetus.

25.3 MASTERING CONCEPTS

1. What is the function of a vertebral column?
2. Which characteristics define vertebrates?
3. List the evolutionary trends seen among the vertebrates.
4. What are the major groups of vertebrates?

OLC

25.4 Fishes—Vertebrate Life in the Waters

Fishes are the most abundant and diverse of the vertebrates, and account for half of the classes that comprise the phylum. Fishes have well-developed organ systems and are exquisitely adapted to an aquatic lifestyle. Their skeletons are composed of cartilage or bone and they are considered as two groups: the jawless and the jawed fishes.

Fishes are the most diverse and abundant of the vertebrates, with nearly 25,000 known living species. They vary greatly in size, shape, and color. Fishes exploit and occupy fresh, estuarine, and marine waters worldwide. Most of these animals are **ectotherms,** which means that the environment controls the body temperature. While fishes cannot tolerate hot springs, they can live in frigid waters by synthesizing antifreeze compounds that maintain their circulation. In these diverse habitats, fishes eat in a variety of ways. Most sharks are voracious carnivores, while others like the tilapia are herbivores that feed on aquatic plants. Parrotfish nibble on coral reefs, some lampreys are parasitic, and hagfishes are scavengers. Very large fishes, such as the whale shark, can consume the smallest of prey, microscopic plankton. The organs and organ systems of fishes are well developed. Fishes use structures called gills to extract oxygen from the water (see fig. 36.4). Oxygen diffuses into the blood through the membranes that form the gills. A two-chambered heart pumps blood through an elaborate labyrinth of arteries and veins. Fish senses are well developed. These animals can see, hear, and sense vibrations and electrical fields that help them to detect predators, prey, and potential mates.

The Jawless Fishes

Modern fishes have traditionally been distinguished by easy to observe features, such as whether they are jawless or jawed, and whether the skeleton is made of cartilage or bone. As with other types of organisms, biological classification is now embracing molecular sequence data. Consider the jawless fishes, which include about 60 species of hagfishes and 15 species of lampreys (fig. 25.6). The jawless fishes were grouped together because of what they lack—jaws, fins, bones, and scales. Their bodies resemble worms and are less than a meter long. But DNA sequence information reveals that the two types of fishes are actually distinct phylogenetic groups.

Hagfishes are the more primitive of the two types of jawless fishes. They are the simplest animals to have a cranium (a braincase) but because they lack supportive cartilage around the nerve cord, they are not technically vertebrates. The hagfishes diverged from the vertebrate line about 530 million years ago, and they haven't changed very much in about 330 million years, earning the nickname "living fossil." These fishes are also called "slime hags," in recognition of the 200 or so slime glands that release a sticky white substance when the animal is disturbed. Dining is a varied experience. Tentacles near the mouth with sensory cells and a tongue help locate food. Hagfishes can

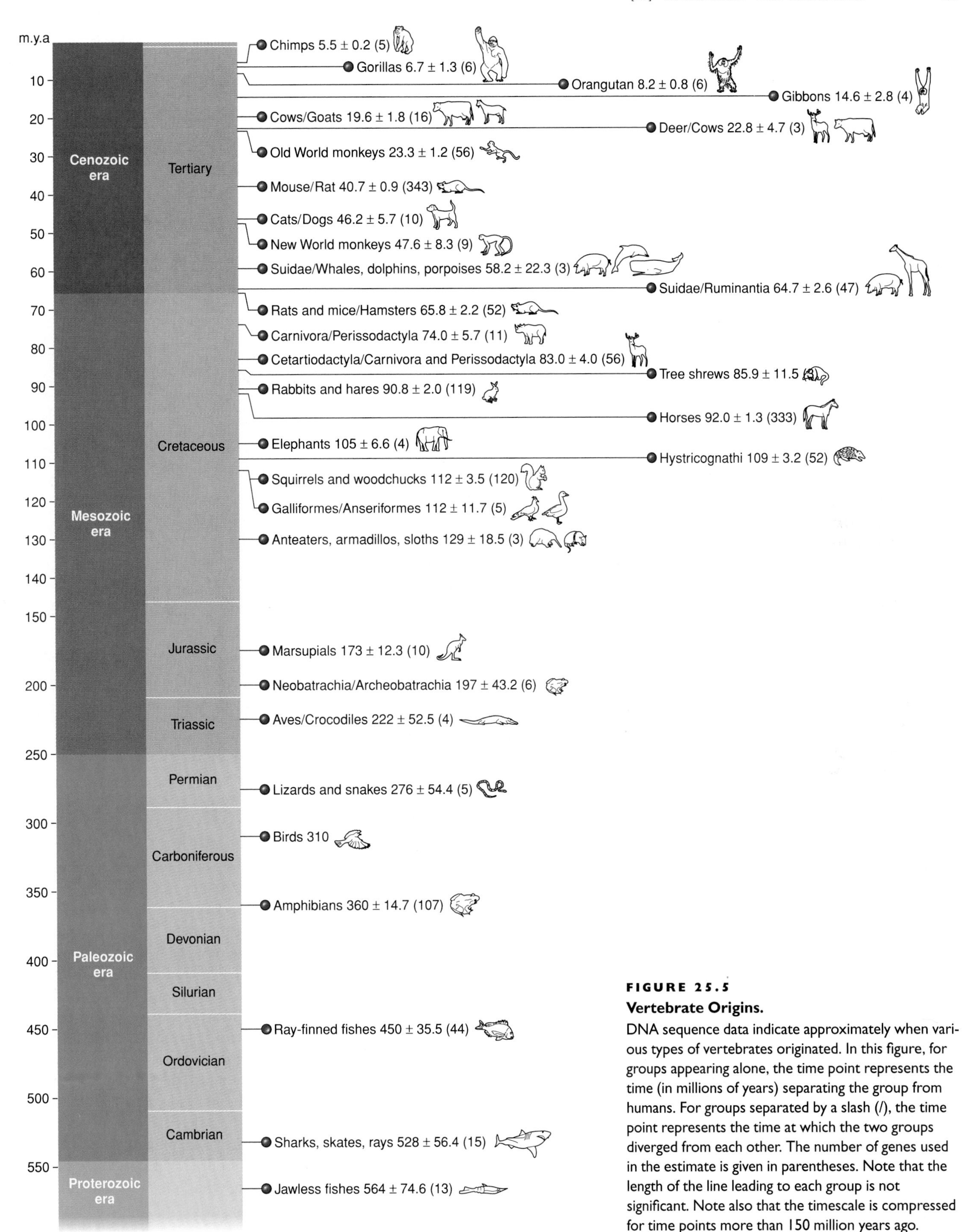

FIGURE 25.5

Vertebrate Origins.

DNA sequence data indicate approximately when various types of vertebrates originated. In this figure, for groups appearing alone, the time point represents the time (in millions of years) separating the group from humans. For groups separated by a slash (/), the time point represents the time at which the two groups diverged from each other. The number of genes used in the estimate is given in parentheses. Note that the length of the line leading to each group is not significant. Note also that the timescale is compressed for time points more than 150 million years ago.

FIGURE 25.6

The Jawless Fishes.

(**A**) Hagfishes and lampreys have been grouped together because they lack jaws, fins, eyes, bones, and scales. But DNA sequence evidence indicates that they are two distinct groups, with hagfishes the more ancient. Hagfishes have craniums and scattered cartilage, but lack cartilage on the nerve cord, which the lampreys have. (**B**) The half-meter-long hagfishes are shown feeding on the carcass of a decomposing whale. (**C**) Some lampreys have distinctive sucker mouths.

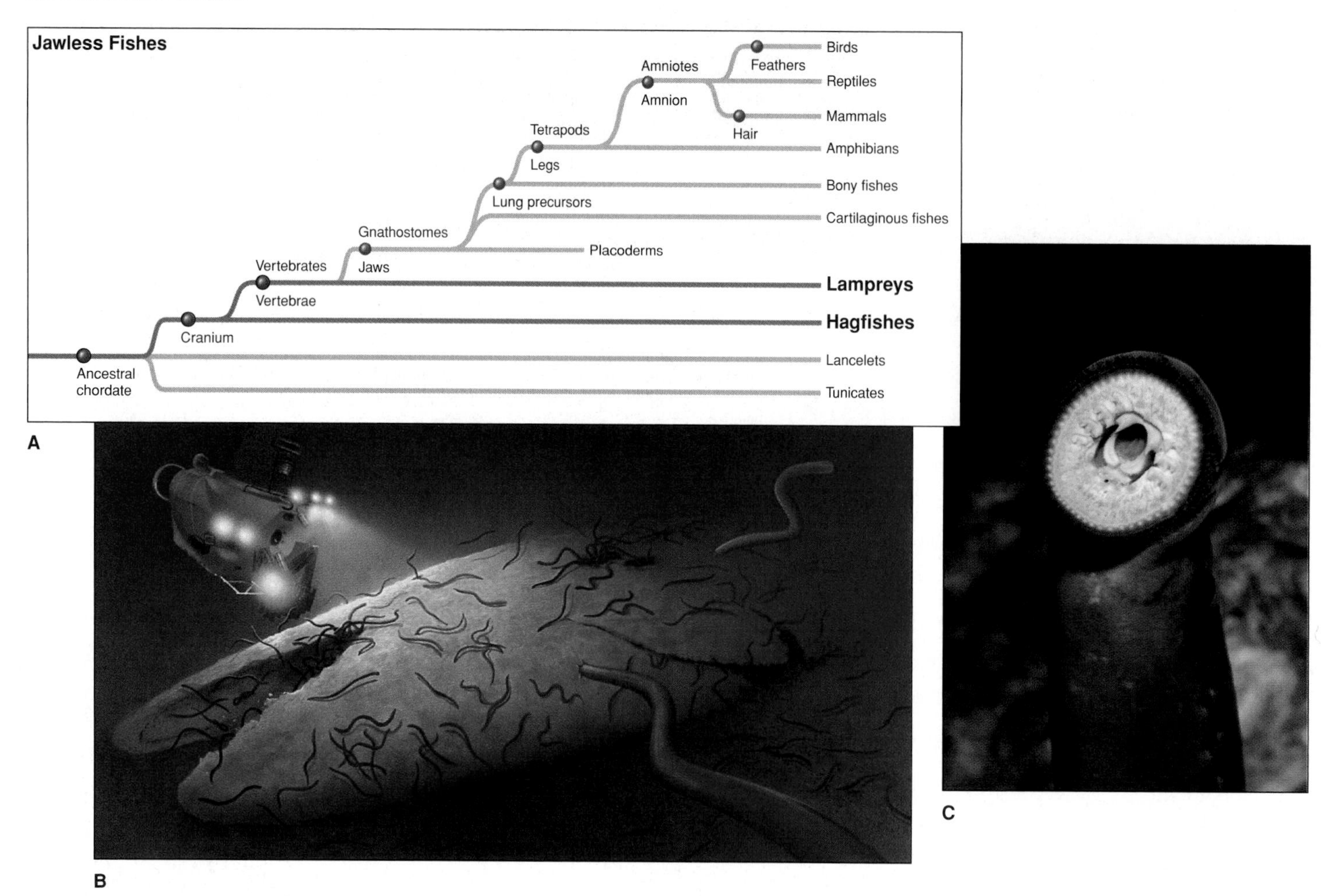

spear a polychaete worm on a spike that lies above their two dental plates, which they use to grab food that is already dead. Often a large carcass fills with hungry hagfishes, which look like eels. (Eels are bony fishes.)

Unlike hagfishes, which are marine, **lampreys** are found in the sea and in fresh water. Lampreys are the first organisms to have cartilage around the nerve cord, and so are the simplest true vertebrates. Though most lampreys are filter feeders, a few species have a characteristic oral disc equipped with projections that enables them to suck on fish. With the opening of the Welland Canal in 1829, lampreys could venture beyond their natural Lake Ontario range. In the other Great Lakes they have been largely responsible for the decline in populations of lake trout and whitefish.

The Jawed Fishes

Scant fossil evidence reveals extinct fishes that lived around the time that jawed fishes evolved. The **ostracoderms** lived in the Cambrian, vanishing when fishes with jaws, which were better able to feed, radiated during the Silurian. Ostracoderms were jawless filter feeders that lived on ocean bottoms. They had notochords and brains, and their bodies were encased in plates consisting of bone and dentin, the material in our teeth. Another ancient fish was the **placoderm,** which had jaws and paired fins, and a notochord that included some bony tissue. This animal also dwelled on the ocean bottom but, instead of waiting passively for a meal to float by, it scavenged or pursued and ate other animals. Placoderms enjoyed more expansive diets than the jawless ostracoderms because the structures that support gills had enlarged into jaws. They may have been the first jawed fishes. Some placoderms were giants, about 13 feet (4 meters) long (fig. 25.7).

The jawed fishes include nearly all familiar fishes. The **cartilaginous jawed fishes** (class Chondrichthyes), with fewer than 900 recognized species, have large fins, 5 to 7 gill slits on either side of the pharynx, and toothlike scales. The group includes chimaeras, skates, rays, and sharks (fig. 25.8). Chimaeras (also called ratfishes) are a small group with characteristics intermediate between those of sharks and the bony fishes. Skates and rays are flat bottom-dwellers that consume shellfish and various marine worms with their flattened teeth. Some can sting, and others are electric.

Sharks are well known for their sharp senses, such as responding rapidly to blood in the water. Their heads are dotted with pores that bear receptors that sense electrical fields around animals—one shark ate a researcher's camera because it gave off an electrical field! A shark can detect vibrations from a fish in distress up to a mile away, thanks to an adaptation called a **lateral line system.** This is a network of canals extending along the fish and over the head that houses receptor organs that detect vibrations and pass the information through the nervous system, which sends the shark in the direction of the unfortunate source. Sharks also have sophisticated immune systems, with spleens, thymus glands, and the same types of protective white blood cells and antibody proteins found in our own immune systems. Sharks have a rich fossil record because of their teeth, a few of which fall out daily.

The **bony fishes** (class Osteichthyes) include 96% of existing species of fishes. They have skeletons reinforced with mineral deposits of calcium phosphate to form bone. An organ called a **swim bladder** permits bony fishes to adjust their buoyancy to maintain position in the water, if conditions warrant stillness. Like sharks, these fishes also have a lateral line system that they use in detection of prey, predators, objects, and in some species, social partners.

The diversity and abundance of the bony fishes are testament to their superb adaptations to a watery world. Several key features enable fishes to swim, maximize oxygen acquisition, and conserve heat, as the bluefin tuna of figure 25.9 illustrates. The tuna is a powerful swimmer, thanks to a sleek body shape with fins and finlets that decrease drag, an effect that human swimmers try to imitate by wearing tight-fitting suits, bathing caps, and shaving their legs. The tuna can retract certain fins to increase speed at crucial times, and its eyeballs lie flat against the head, further decreasing drag. The thin, lightweight, yet strong and flexible tail is crescent-shaped, and the animal moves rapidly in a side-to-side motion to propel itself forward, swimming up to 20 miles (32.2 kilometers) per hour.

FIGURE 25.7
Development of Jaws.
(**A**) Jaws developed from bones that support gills. (**B**) Placoderms had jaws and paired fins. Although extinct, they probably shared an ancestor with the modern jawed fishes.

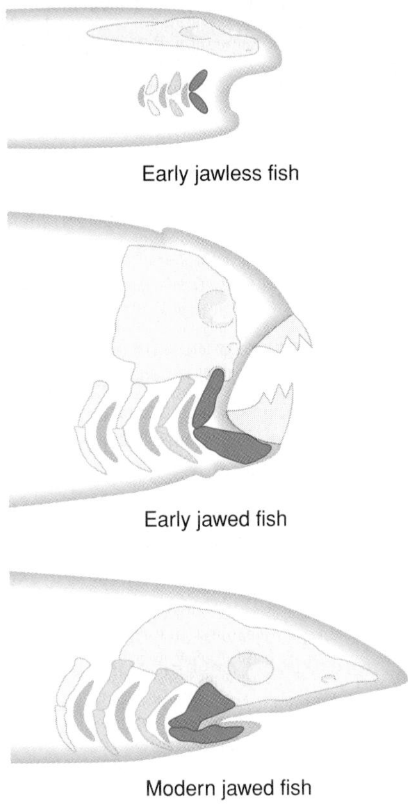

FIGURE 25.8
The Cartilaginous Fishes.
(**A**) The placement of cartilaginous fishes on the chordate tree. This group includes skates and rays (**B**), and sharks (**C**).

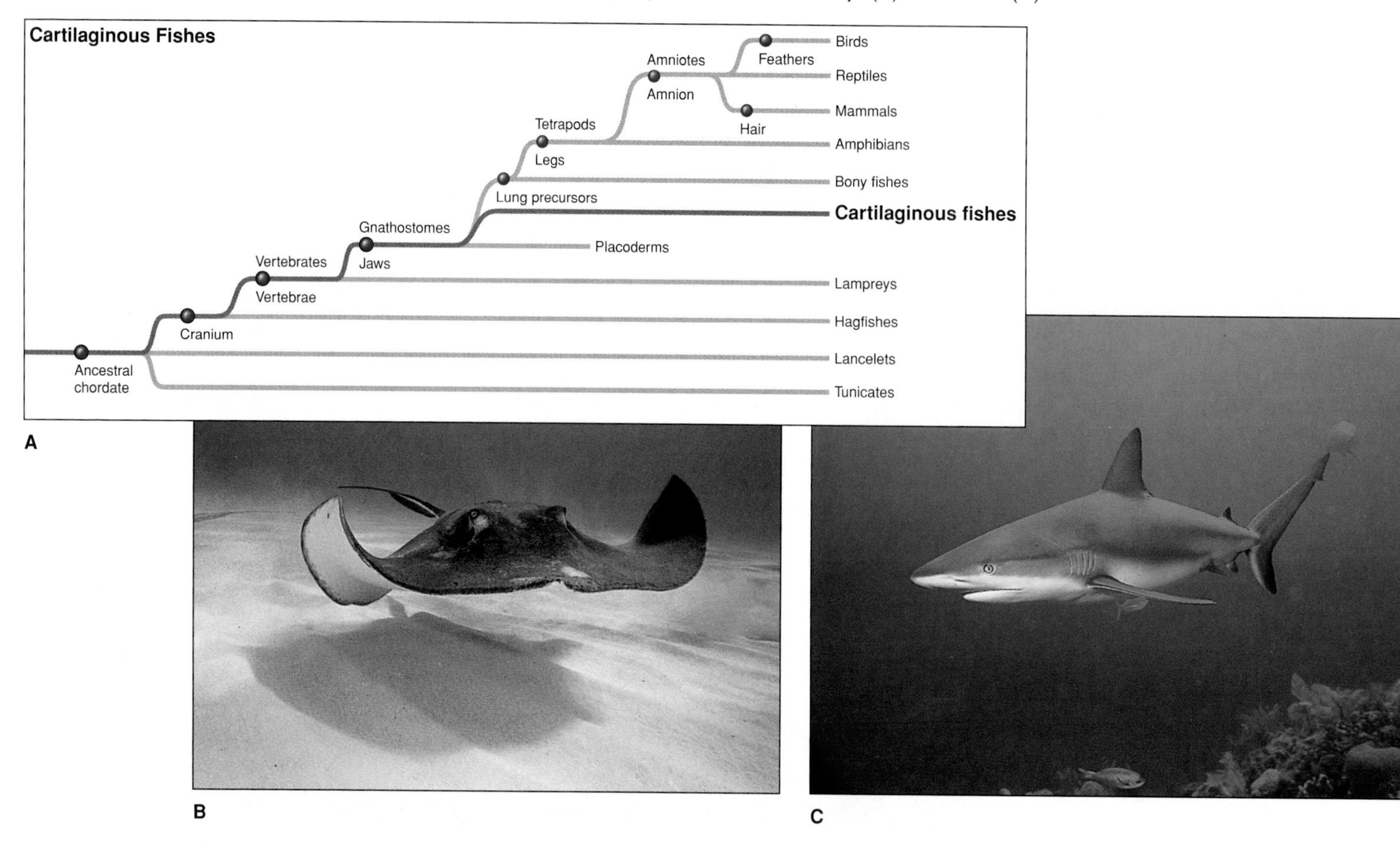

The tuna must swim forward rapidly to force water over its gills, because its hard head prevents it from sucking in water, as some other fishes can do. The tuna's gills are huge and lushly supplied with tiny blood vessels, enabling it to extract maximal oxygen from the water. Its muscles are packed with myoglobin, a protein that readily binds oxygen, and the circulatory system is adapted to meet requirements for oxygen and heat. Heart rate increases with exertion, an ability that mammals have but many other fishes do not. Though not endothermic, the tuna gains a measure of temperature control because in many places in the body, particularly the head, blood vessels from the surface bearing colder blood come near vessels from the interior carrying warmer blood. The heat is transferred to the cooler blood.

Bony fishes are divided into three groups (fig. 25.10). The **ray-finned fishes** include nearly all familiar fishes—eels, minnows, catfish, trout, tuna, salmon, and many others. The **lungfishes** are represented by only three living species, one per continent in the waters of Africa, Australia, and South America. They are so named because during droughts, they can survive by burrowing into the mud beneath stagnant water, gulping air, and temporarily slowing their metabolism. DNA sequence analysis reveals that the lungfishes are the bony fishes most closely related to the tetrapods. The **lobe-finned fishes** have only one surviving member, the coelacanth. These animals originated during the Devonian, about 360 million years ago, and were thought to have become extinct 60 million years ago, the date of the most recent fossils. Then a fisherman caught one off the coast of South Africa in 1938. Researchers subsequently discovered a small population of coelacanths in caves near undersea volcanoes in the Comoros archipelago of the western Indian Ocean. At least one other population exists. On September 18, 1997, a biologist's wife saw one in a fish market in Indonesia, and just a year later, her husband had the thrill of spending several hours with a live, but dying, coelacanth that had been caught in a shark net! Tissue samples carefully taken before the animal expired are certain to provide clues that will help in clarifying its evolutionary status. The coelacanth's nasal organs and eyes are separated from the ear and brain, a primitive characteristic not known in any other modern animal. Yet it moves its fins more like they are limbs, and has ears that are similar to those of tetrapods.

25.4 MASTERING CONCEPTS

1. What characteristics are used to classify different types of fishes?
2. How are fishes adapted to a life in water?
3. What are the major types of ancient and modern fishes?

OLC

FIGURE 25.9

Built to Swim.

A bluefin tuna, which can exceed 10 feet (3 meters) and 1,400 pounds (636 kilograms), gets maximal oxygen from water and conserves heat.

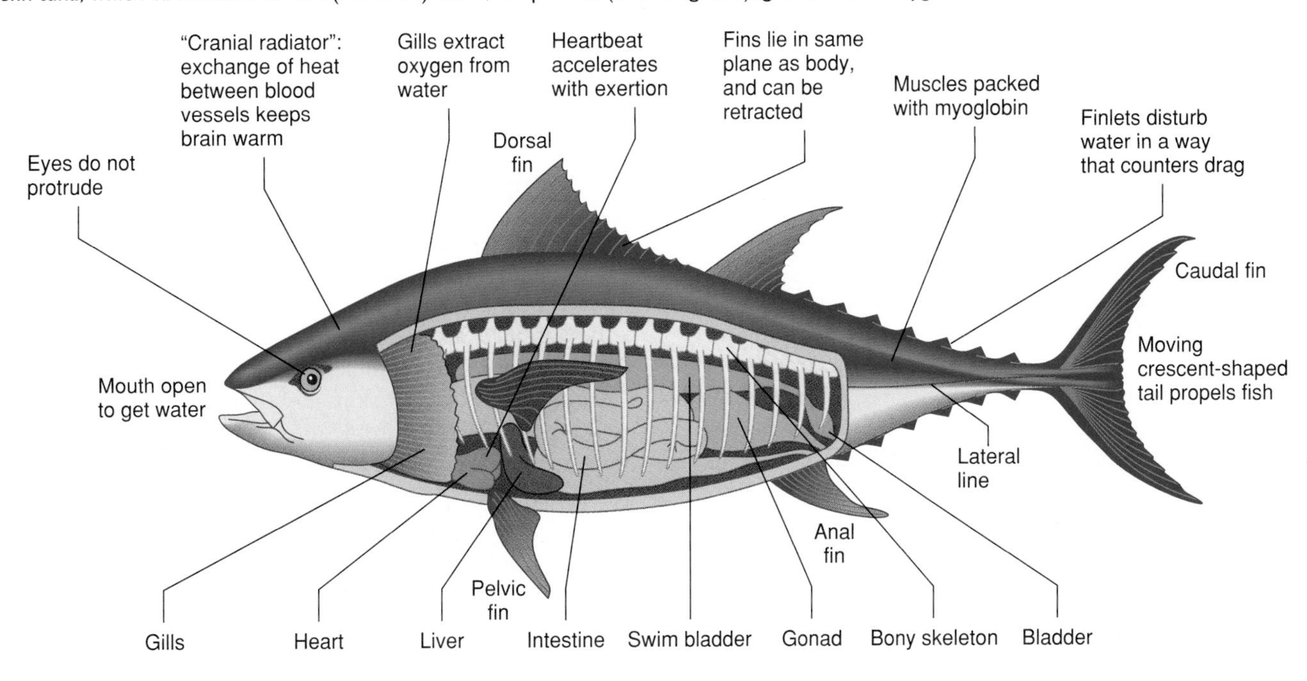

FIGURE 25.10

The Bony Fishes.

(**A**) The bony fishes have lung precursors and skeletons reinforced with mineral deposits of calcium phosphate. This group includes the lungfishes (**B**), lobe-finned fishes (**C**), and ray-finned fishes (**D**).

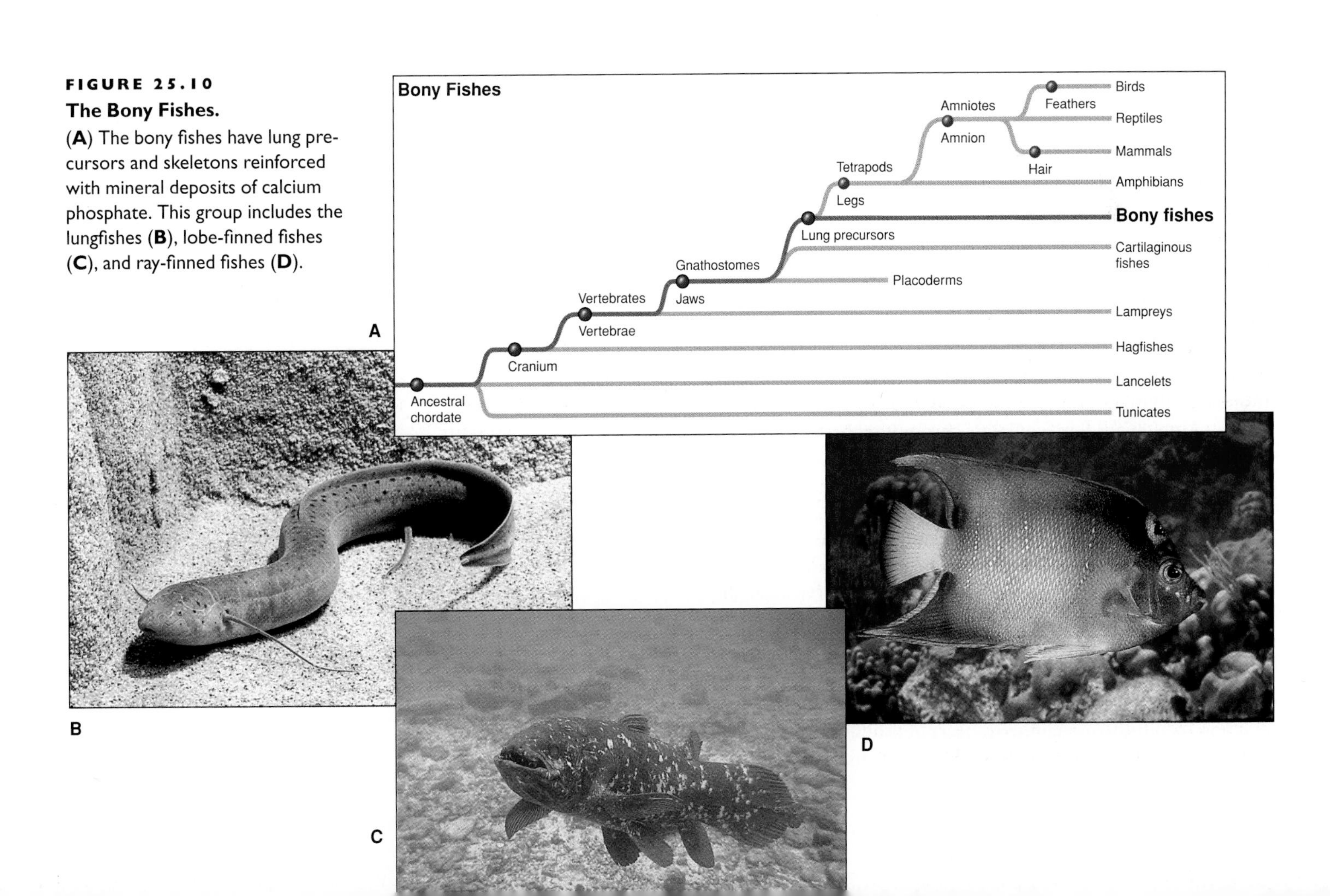

25.5 The Amphibians Lead a Dual Lifestyle

Salamanders, newts, frogs, toads, and the wormlike caecilians are doubly adapted to life in water and on land. In the amphibians, gills gave way to lungs, circulatory systems grew more efficient, the skeleton became denser, and senses attuned to terrestrial existence sharpened. Amphibians reproduce in several ways, and some care for young.

The word **"amphibian"** is Greek for "double-life," referring to the defining ability of these vertebrates to live in water and on land. The amphibians are a life-form seemingly in transition, retaining certain characteristics of the fishes, yet able to live on land like other tetrapods. It's easy to envision how gulping lungfishes might have foreshadowed the lungs of amphibians, or how the fleshy fins of lobe-finned fishes might have become legs (fig. 25.11).

Amphibian ancestors faced many challenges in conquering land (see fig. 18.16). The terrestrial habitat provided new opportunities—space, protection, and food—but there were trade-offs. The animals faced wider swings in temperature. Air had more oxygen than water, but delicate gills collapsed without the buoyancy of water. With improved lungs, circulatory systems grew more complex and powerful. The skeleton became denser and better able to withstand the force of gravity. Different senses were valuable on land—the fishes' lateral line system of detecting vibrations in deep water was no longer of much use, and remained, in rudimentary form, in only a few aquatic amphibian species. Natural selection favored acute hearing and sight, with tear glands and eyelids eventually forming to keep eyes moist. Yet even with these adaptations, amphibians then and now must return to water to breed, because their eggs lack shells and protective membranes to nourish them and keep them from drying out. Amphibians today face different challenges. The opening essay to chapter 23 describes how chytrids are causing an infectious disease that is decimating amphibian populations worldwide.

Amphibian Characteristics

The amphibians have several defining characteristics. The skin is smooth, moist, and thin, and is permeated with blood vessels. Although the thinness promotes rapid water loss, it also allows respiration. Amphibian skin contains **chromatophores,** which are highly branched cells that have pigment concentrated in their centers. Under certain conditions, the pigment spreads into the cell branches, which has the overall effect of coloring the skin.

Amphibians have a bony skeleton, and most have four limbs. The mouth is typically large, with two nostrils and small teeth in one or both jaws. They respire through lungs, some through gills, and many supplement oxygen intake by gas exchange through the skin. The excretory system includes paired kidneys and the waste product is urea, discussed further in chapter 38. All amphibians have three-chambered hearts. This promotes more effective circulation than the two-chambered heart of fishes, but oxygen-rich and oxygen-poor blood still mix in the one ventricle. Amphibians have separate sexes. Frogs and toads have external fertilization; salamanders and **caecilians** (wormlike amphibians) have internal fertilization. Their name of "double-life" might also apply to their characteristic metamorphosis, with larval forms of some species looking quite unlike their adult counterparts.

Amphibian Diversity

Modern amphibians are classified into three orders (fig. 25.12). Most of the 3,900 extant species are frogs and toads (order Anura). The class also includes salamanders and newts (order Urodela), and the wormlike caecilians (order Gymnophiona).

About 3,450 amphibian species are frogs or toads, the latter distinguished by drier skins and more time spent on land. Frogs and toads live in the tropics or in temperate regions, where they are active during warm seasons, and hibernate beneath forest litter or aquatic sediments during the winter. The adults have large mouths, heads fused to their trunks, and lack tails and scales. Female frogs lay their eggs directly in the water, as the male frogs clasp the backs of the females and release sperm, which fertilize the eggs externally. The eggs hatch into tadpoles (larvae) that respire using gills. Tadpoles are strictly herbivores, feeding on algae. Eventually the tadpoles metamorphose, developing legs, lungs, and carnivorous tastes.

Frogs and toads are remarkably variable in size—the largest is the 12 inch (30 centimeter) long West African *Conraua goliath*, so

FIGURE 25.11
Fins Become Legs.
Ichthyostega was an animal that ventured onto land using modified fins.

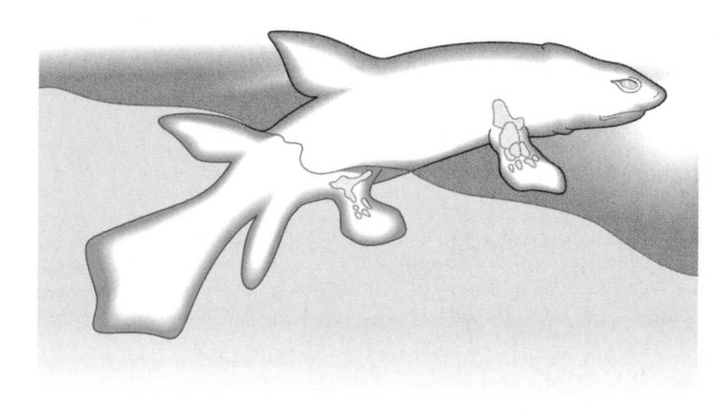

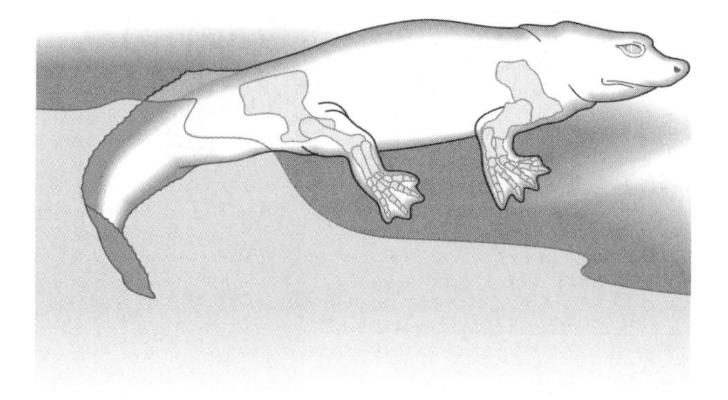

FIGURE 25.12
Amphibian Diversity.
(**A**) Amphibians were the first chordates to develop legs and venture onto land. The moss frog (**B**) was recently discovered in the Tam Dao reserve in Vietnam, where more than a dozen new species of amphibians and reptiles have been identified since 1982. Its skin resembles moss. (**C**) Salamanders are small, juvenile-appearing amphibians. (**D**) Caecilians are legless amphibians that resemble earthworms or snakes.

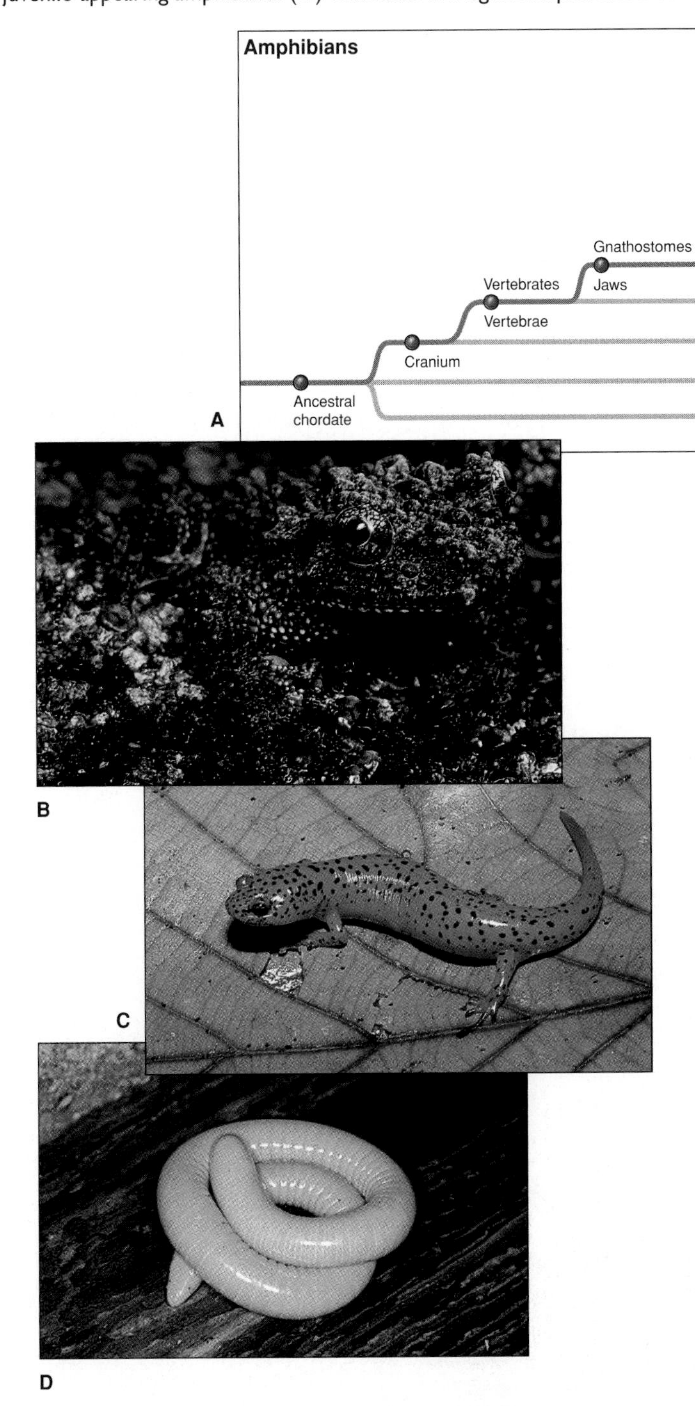

large that it can swallow a duck! At the other end of the size spectrum is *Psyllophryne didactyla*, which lives in the Brazilian rain forest and is less than 0.4 inches (1 centimeter) long. Frogs and toads have remarkable adaptations to avoid predation. The "moss frog" recently discovered in Vietnam (see chapter opening essay and fig. 25.12B) is the epitome of camouflage, looking just like wet moss on a rock. Other frogs display vibrant colors that warn of toxins secreted from glands in the skin. We use some of those toxins as antibiotic drugs. Frogs and toads also evade the fate of becoming a meal by being master jumpers, expert at playing dead, and inflating their mouths so that a snake can't swallow them.

Order Urodela includes about 370 species of salamanders and newts. They live in temperate and tropical habitats, where they eat small invertebrates. Salamanders and newts have tails and most are small—but the largest, the Japanese giant salamander, is about 5 feet (1.5 meters) long. They hold their four legs at a characteristic 90-degree angle from the body. Aquatic salamanders lay their eggs in water, which hatch out free-swimming larvae that have external gills and a finlike tail. Terrestrial salamanders lay their eggs in depressions in the soil or under logs. They undergo direct development, with offspring hatching from eggs and resembling adults. In many terrestrial species, the adults guard the young.

In salamanders that undergo metamorphosis, the larval and adult forms more closely resemble each other than is true for frogs. Salamanders exhibit **paedomorphosis,** which means that the adults have features that were present in larval stages of their ancestors. The familiar mudpuppy, for example, swims on pond bottoms and does not metamorphose, retaining gills and appearing fishlike. In some amphibian species, metamorphosis occurs only in response to certain environmental signals. The Mexican axolotl, for example, respires primarily through gills when it lives in ponds. Drought triggers metamorphosis. The adults, equipped with lungs, pull themselves from drying ponds and walk on land in search of other ponds, where they reproduce.

The 160 species of caecilians are the third type of amphibian. Lacking limbs, these animals look much like giant earthworms. Most species burrow under the soil in the tropical forests of South America, Africa, and southeast Asia, but a few inhabit shallow freshwater ponds. Caecilians are carnivores, eating insects and worms.

25.5 MASTERING CONCEPTS

1. Which characteristics of amphibians are retained from ancestors that lived in the water, and which are adaptations to life on land?
2. What are some amphibian adaptations to avoid predation?
3. How do amphibians reproduce?
4. What are the major types of amphibians?

25.6 The Reptiles Fully Conquer Land

Armed with leathery eggs harboring built-in food supplies and protective membranes to seal in moisture; more efficient respiratory, circulatory, and excretory systems; and better brain power, reptiles conquered land. Life no longer required an aquatic environment.

Reptiles evolved from amphibians during the Carboniferous period. They dominated animal life during the Mesozoic era, until their decline beginning 65 million years ago, as chapter 18 explains. Although we tend to think of reptiles as large—such as crocodiles, pythons, and tortoises—members of most of the 7,000 extant species are small. If one measures success by diversity, then lizards and snakes win by a landslide, accounting for almost 95% of modern reptile species. Other reptiles include the crocodilians (crocodiles and alligators), turtles and tortoises, an obscure group called "worm-lizards" that burrow and are more closely related to snakes and lizards than to other reptiles, and the tuatara, a lizardlike animal that lives on only a few islands near New Zealand. Figure 25.13 is a phylogenetic tree that depicts the evolutionary relationships among reptiles, birds, and mammals, and figure 25.14 shows some representative reptilian species.

Certain reptiles have lost particular specialized characteristics through evolutionary time, which can complicate classifying them. Figure 17.12, for example, shows that the ancestors of snakes had legs. Another example is the turtle. These reptiles had been considered primitive because they lack holes on the sides of

FIGURE 25.13

Phylogenetic Relationships of the Tetrapods.
This expanded tree shows how the amniotes are split into the synapsids (which led to the three groups of mammals), the anapsids (the turtles), and the diapsids (which gave rise to crocodiles, dinosaurs, and birds). The question mark after "Anapsids" reflects the state of flux of the classification, in light of recent mtDNA evidence that places turtles with diapsids. Another possible modification is that some vertebrate biologists group birds with reptiles.

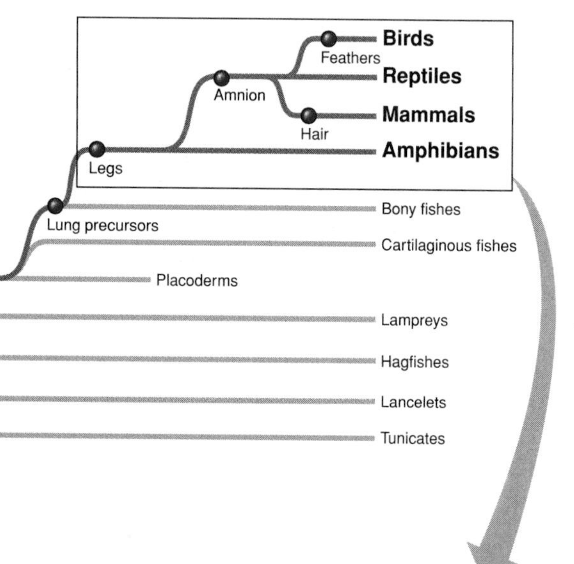

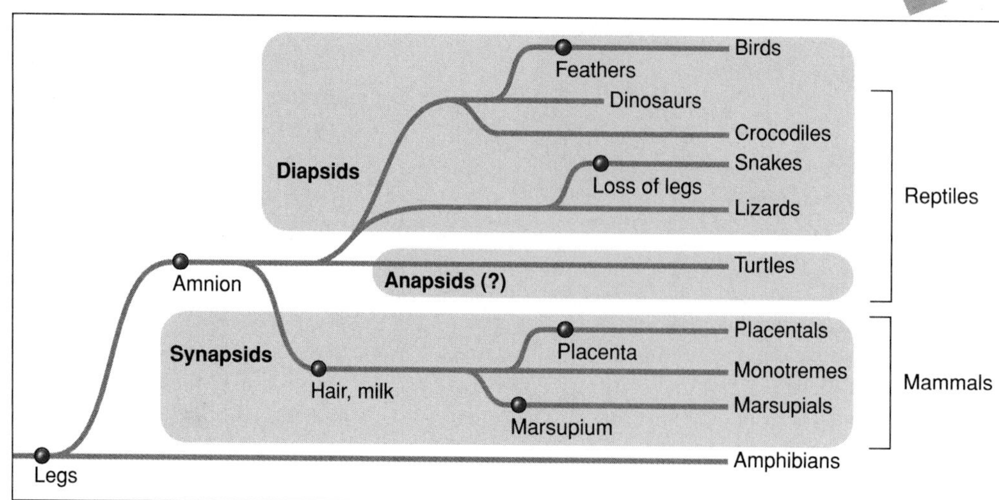

FIGURE 25.14

Reptilian Diversity.

(**A**) The reptiles are the first group to develop an amniote egg, which helped make life solely on land possible. They include turtles and tortoises (**B**), lizards and snakes (**C**), crocodiles and alligators (**D**), and a few other groups including the extinct dinosaurs.

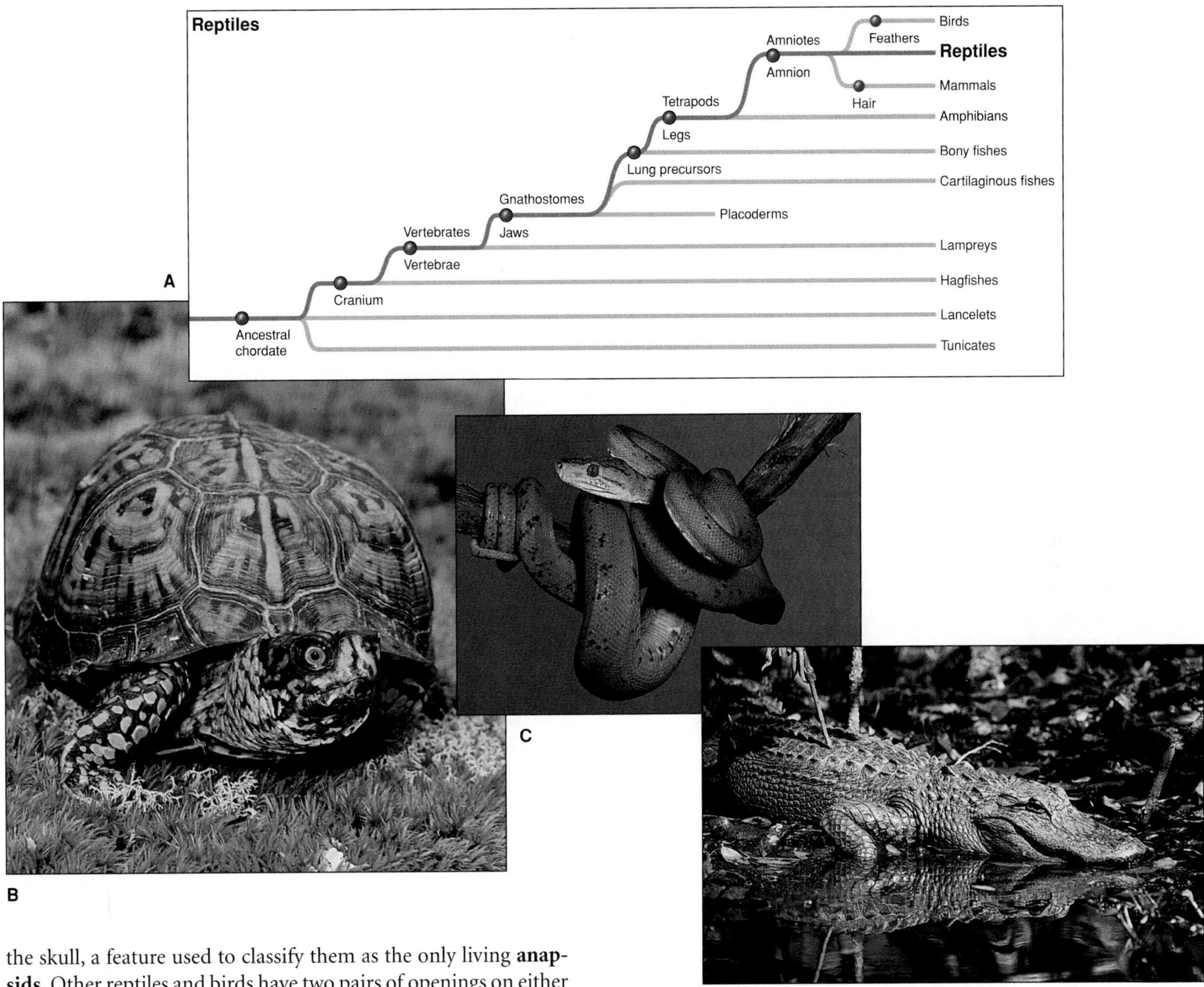

the skull, a feature used to classify them as the only living **anapsids.** Other reptiles and birds have two pairs of openings on either side of the skull, behind the eye orbits, and are termed **diapsids.** However, analysis of mitochondrial DNA reveals that turtles belong with the other reptiles as diapsids—meaning that they "lost" the skull holes over evolutionary time. Mammals and their immediate ancestors have a single pair of openings on either side of the skull behind the eye orbits, and are called **synapsids.**

Reptiles are ectothermic and are thus confined to deserts, tropical, and temperate regions of the Earth. Most live on dry land. Several adaptations that persist today enabled the first reptiles to become independent of water. Their excretory systems, including paired kidneys, enable them to conserve water, as do their tough, scaly coverings composed of keratin. The eggs of reptiles are also adapted to dryness. They have leathery and hard outer layers that surround an **amniote egg** (fig. 25.15). This type of egg was a milestone in evolution. It includes a nutritive yolk as well as internal membranes that retain moisture and add support, enabling an embryo to survive without being in water. Internal fertilization, in which a male uses a copulatory organ to deliver sperm inside a female's body, freed the reptiles from having to deposit sperm in water, not a very efficient means of fertilization. Though some snakes and lizards are viviparous, most female reptiles lay their eggs on land and, depending on the species, either leave the eggs to their fate or guard them until they hatch.

An important adaptation to a terrestrial existence is increased respiratory capacity, thanks to more highly developed lungs associated with enhanced circulation from a three-chambered heart. Crocodiles and alligators have four-chambered hearts. The increasing number of heart chambers from the two in fishes, to three in amphibians and most reptiles, and four in birds and mammals

FIGURE 25.15

The Amniote Egg.

The amniote egg freed reptiles from having to return to water to breed, as amphibians must do. The embryo is encased in a hard, protective shell, and is supported internally by three membranes—the amnion, allantois, and chorion. These are discussed further in chapter 40. Yolk supplies nutrients.

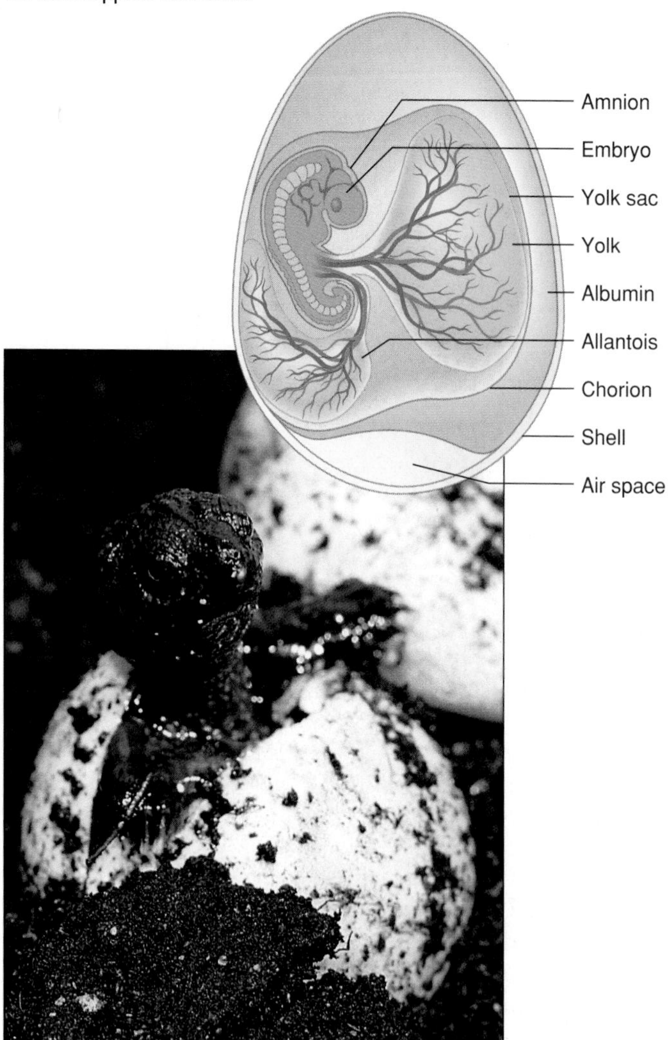

reflects more complete separation of oxygen-rich from oxygen-poor blood, and is discussed further in chapter 35.

The reptiles have developed sophisticated strategies to obtain food as well as avoiding becoming food. A lizard might escape a confrontation with a predator minus its tail, but that is better than being eaten. Camouflage and the ability to hold perfectly still for long times enable snakes to surprise their prey, then subdue it by injecting venom or wrapping coils around the victim and squeezing the life out of it. Snakes then demonstrate a skill unique to them—they can unhinge their jaws, which allows them to swallow animals much wider than they are. The opening photo to chapter 5 shows an African rock python swallowing a gazelle! Certain large, ground-dwelling snakes grow strong enough to ambush big prey by being chronically constipated. The Gaboon viper *Bitis gabonica*, for example, defecates only once every few months. The feces accumulate at the back end of the snake, giving the snake a secure anchor for pulling in its large prey. In contrast, thin and much more active tree-dwelling snakes keep their weight down and activity level up by defecating frequently.

25.6 MASTERING CONCEPTS

1. What are some types of reptiles?
2. What are the characteristics of reptiles?
3. How are reptiles adapted to life on land?

25.7 Birds—Vertebrate Life in the Sky

Birds are quite distinctive as the only organisms that have feathers, although others can fly. Just as the fish is a living machine adapted to water, a bird's body is a collection of adaptations to a life spent partially in flight.

Feathers and the ability to fly, plus hard-shelled eggs have helped the vertebrate class Aves succeed and flourish (fig. 25.16). **Birds** come in many sizes, from the hovering tiny nectar-feeding hummingbirds to the enormous flightless ostriches and emus. All 9,000+ species share feathers, four-chambered hearts, and **endothermy,** which is heat generated from their metabolism. Most birds fly. The powerful heart and lungs supply the oxygen that supports the high metabolic demands of flight. The distinctive shapes of birds' beaks and feet reveal their varied habitats and diets (fig. 25.17).

Scales and egg laying are two characteristics that birds retain from their reptilian ancestors. Some biologists consider birds to be reptiles—modern-day dinosaurs, specifically. The avian egg, with its hard calcified shell, protects the developing embryo inside, and the adults care for, feed, and otherwise protect their hatchlings. Babies that hatch with feathers and eyes open, such as ducks and chickens, are called **precocial.** Young that hatch helpless and totally dependent on the parents, such as robins, are called **altricial.**

Birds are anatomically adapted for flight (fig. 25.18). Tapered bodies offer a streamlined profile, and hollow bones reduce weight. Feather-covered wings help to provide lift in the air while highly developed muscles power flight. The sternum (breastbone) of flying birds resembles the shape of the keel, or axis, of a sailboat, and serves as the major structural support to which two powerful flight muscles attach. When the bird contracts one muscle (the pectoralis), the wing moves downward; when that muscle relaxes and the bird contracts the other (the supracoracoideus), the wing is raised. Coordinating these actions makes flight possible. Birds with this keel type of sternum are called carinates, and include birds that fly. However, penguins too are carinates—they "fly" underwater in search of prey. Flightless birds, known as ratites, have sternums that are not keeled. Emus, ostriches, and rheas are ratites.

Birds are the only modern animals that have feathers. These structures enable birds to fly, provide insulation, and are impor-

FIGURE 25.16

The Relationship of Birds to Other Chordates.

Birds diverged fairly recently from reptiles, and some biologists consider them to actually be reptiles—modern-day dinosaurs.

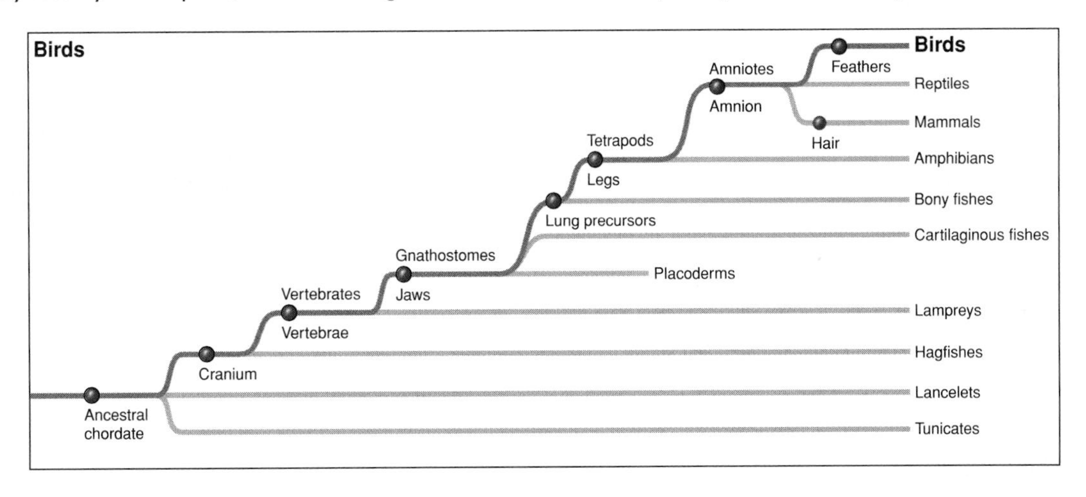

FIGURE 25.17

Birds' Beaks Are Related to Their Functions.

The shapes of birds' beaks are adapted to their diets, as Charles Darwin noted on his journey through the Galápagos Islands.

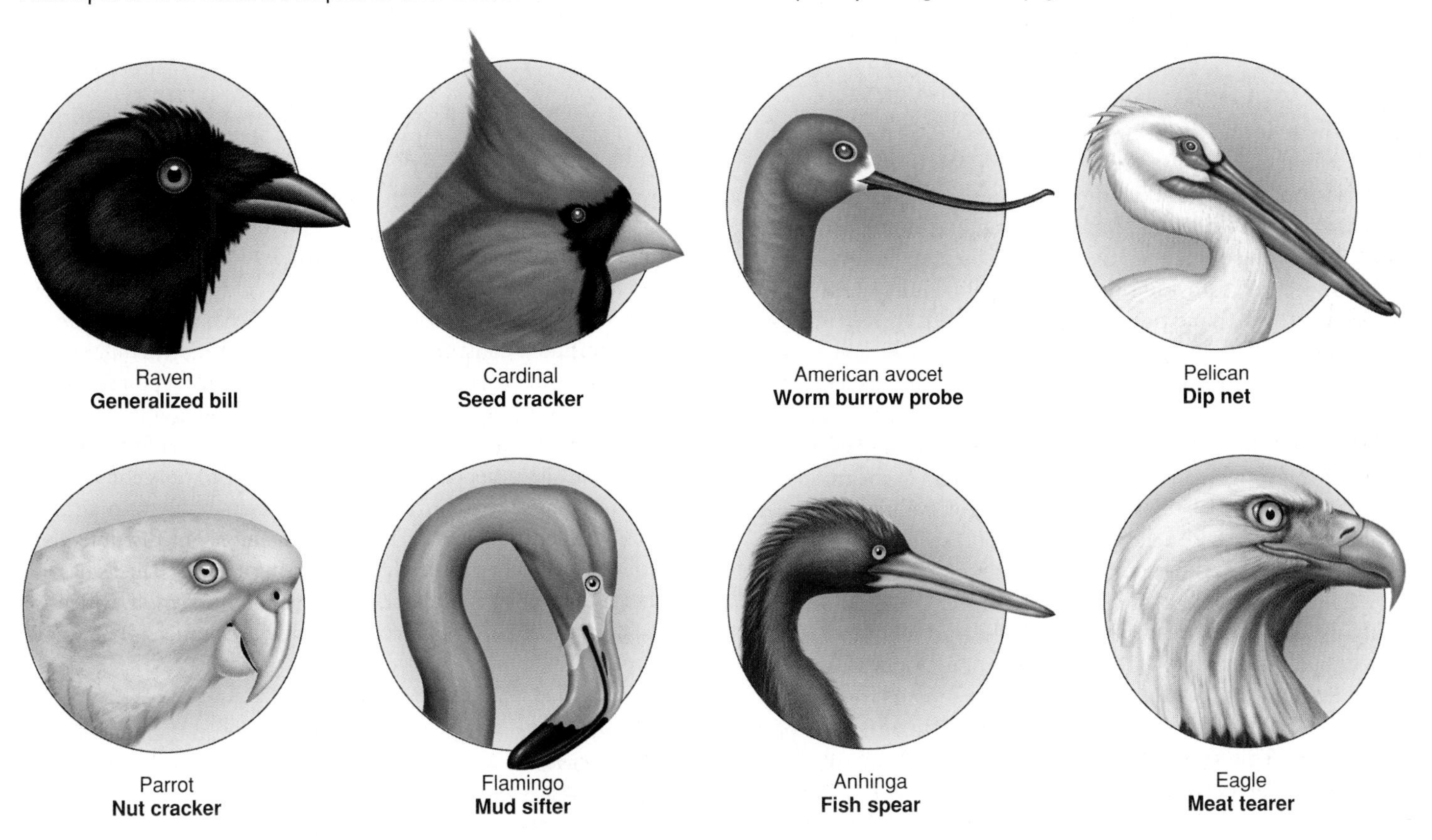

FIGURE 25.18

Adaptations for Flight.

(**A**) Feathers provide a streamlined contour and lift. Hollow bones (**B**) keep a bird's body lightweight. Alternate contraction of flight muscles anchored to the keel of the breastbone powers flight (**C**).

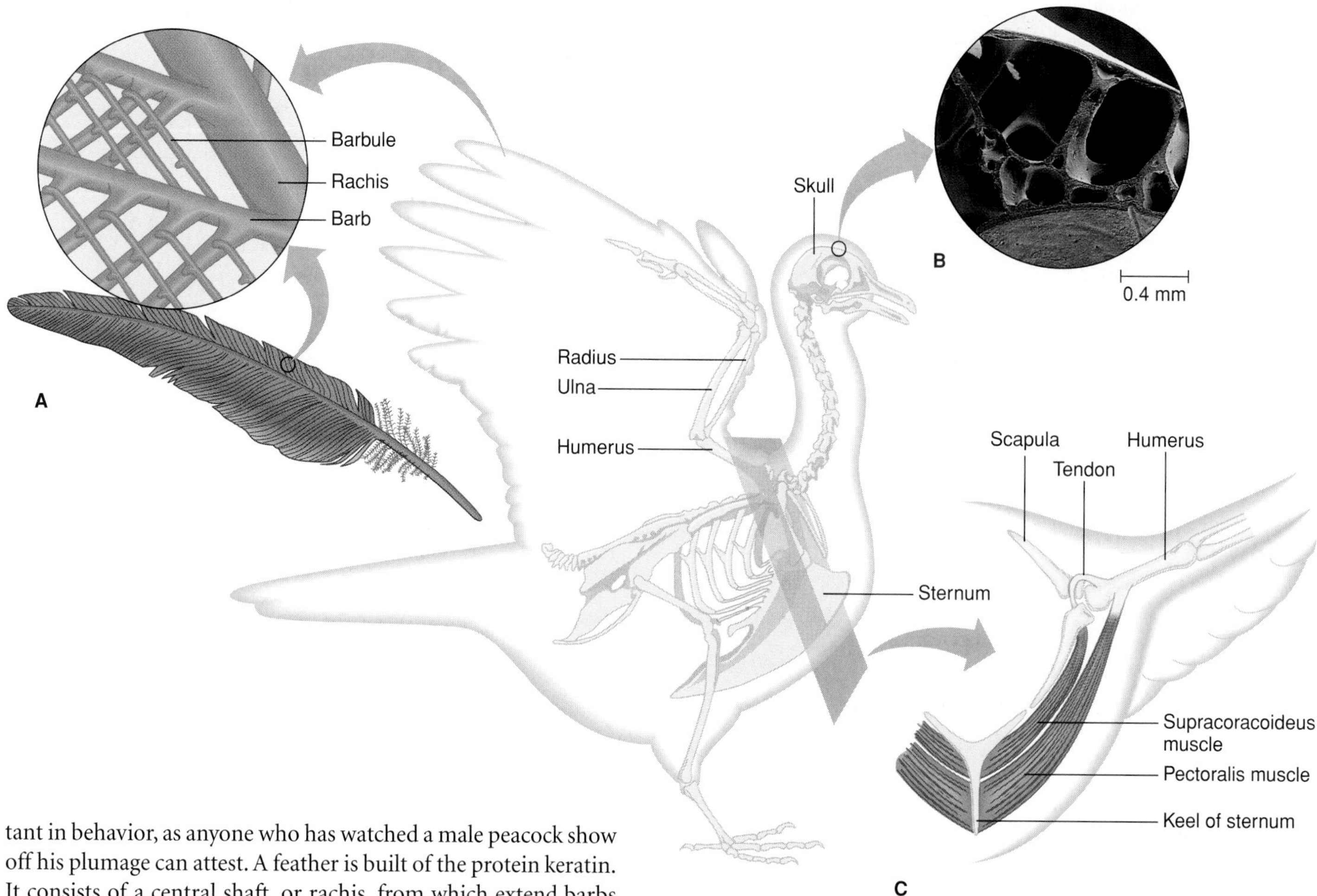

tant in behavior, as anyone who has watched a male peacock show off his plumage can attest. A feather is built of the protein keratin. It consists of a central shaft, or rachis, from which extend barbs and from them, barbules. Different-sized and different-shaped feathers serve distinct functions. Flight feathers are long and thin, contour feathers provide overall shape, filoplumes are used in displays, and down provides warmth (fig. 25.19).

Feather colors arise in two ways. Some colors come from biochromes, which are the same pigment molecules seen in other types of organisms. Melanin provides dark speckles. Carotenoids, derived from the diet, impart the pink in pink flamingoes, as well as yellows, oranges, and reds. Porphyrins, similar chemically to the hemoglobin that makes blood red but containing a copper atom instead of an iron atom, provide the magenta wings of the African turaco and other deep and intense colors.

In contrast to biochromes are structural colors, which are patterns of a feather's nanostructure that scatter light in ways that produce a hue, usually blue. Consider the iridescence of a hummingbird's throat. On each barbule are several layers of hollow, platelike structures that are partitioned into air spaces of a precise size. This structure reflects incoming light rays in a way that crosses them, which produces the iridescence. At the same time, surface melanin scatters the outgoing blue light, while deeper melanin absorbs red and yellow wavelengths. The overall effect is the shininess of the bird's throat. The precision of structural colors reflects natural selection. By studying feather nanostructure with the electron microscope, ornithologists (biologists who study birds) have shown that a small change in structure can have enormous consequences to a bird. For example, a difference in the spacing of air pockets in the barbule of a male plum-throated cotinga (*Cotinga maynana*) of less than five-millionths of an inch can discolor his throat plumage. The result: He can't find a mate.

Because feathers are so vital, their care and upkeep is a major part of being a bird. Birds preen their feathers by running them through their beaks, which reconnects and aligns disrupted barbules on the rachis, somewhat like closing a zipper. Ducks have oil glands that keep their feathers waterproof. Birds molt, which is a highly regulated loss of old feathers and development of new ones.

25.7 MASTERING CONCEPTS

1. What characteristics do birds retain from their reptilian ancestors?
2. How are birds adapted for flight?
3. What are the functions of feathers?
4. How do feather colors arise?

OLC

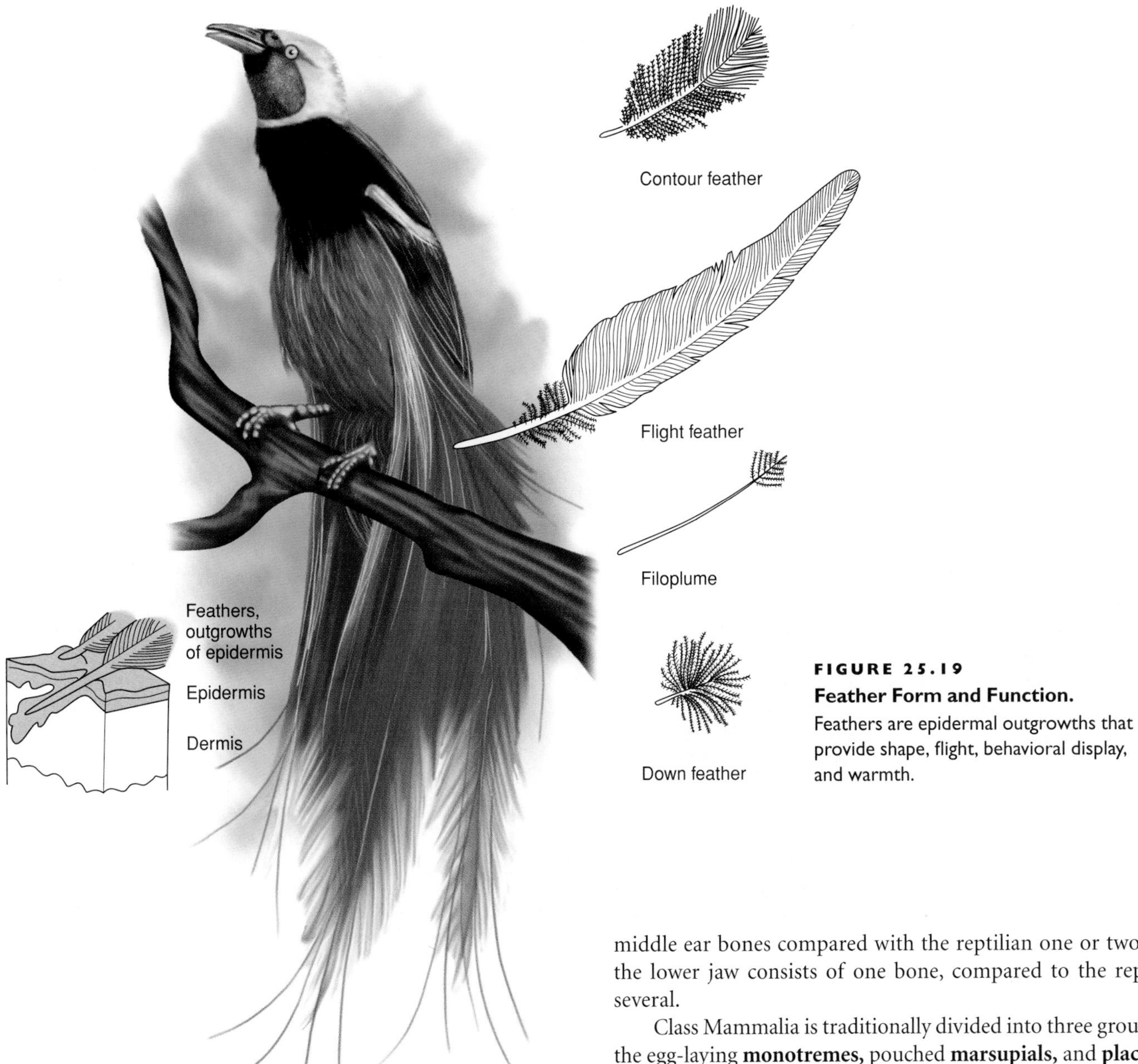

FIGURE 25.19

Feather Form and Function.

Feathers are epidermal outgrowths that provide shape, flight, behavioral display, and warmth.

25.8 Mammals Are Furry Milk-Drinkers—And More

Blueprints for a mammal: hair, mammary glands, red blood cells that lack nuclei, sweat and oil glands, two sets of teeth, and complex brains. We share the class with a few egg-layers and pouched mammals, and many other types of placental mammals.

Mammals are furry animals that as infants are nourished by their mothers' milk. Only 4,500 species are recognized, but we are perhaps most aware of them because we *are* them. (How often do people equate the word "animal" with "mammal"?) Because fur and breasts do not leave fossils, mammalian remains are recognized in other ways. Specifically, a mammal has three middle ear bones compared with the reptilian one or two, and the lower jaw consists of one bone, compared to the reptile's several.

Class Mammalia is traditionally divided into three groups—the egg-laying **monotremes,** pouched **marsupials,** and **placental mammals** (see fig. 25.13)—but as the cladogram in figure 16.3 indicates, DNA sequence evidence indicates that the monotremes may have branched from the marsupial lineage. The monotremes include the duck-billed platypus and spiny anteaters of Australia. Marsupials such as kangaroos and opossums have a pouch or **marsupium** in which the immature young nurse and develop. Female placental mammals carry their young inside the uterus, where a **placenta** connects the maternal and fetal circulatory systems to nourish and remove wastes from the developing offspring. Most mammals are placental.

Mammals probably arose from reptiles (fig. 25.20). Figure 18.8a depicts a thecodont, which had characteristics of both mammals and reptiles. Figure 16.13 shows the adaptive radiation of mammals, and section 18.3 discusses their evolution.

The word "mammal" is derived from the Latin *mammae* for "breast," and refers to the hallmark milk-secreting **mammary glands** of the female. These glands are rudimentary in males. Mammals are also distinguished from other vertebrates by hair,

FIGURE 25.20
Mammalian Diversity.
(**A**) Mammals differ from other vertebrates in that they have hair and mammary glands. Platypuses (**B**) retain some characteristics thought to have been present in the earliest mammals. Mammals have successfully invaded a variety of habitats, including the air (**C**—bats) and the sea (**D**—dolphins). This mammal, a human baby (**E**), is deriving nourishment from his mother's mammary gland.

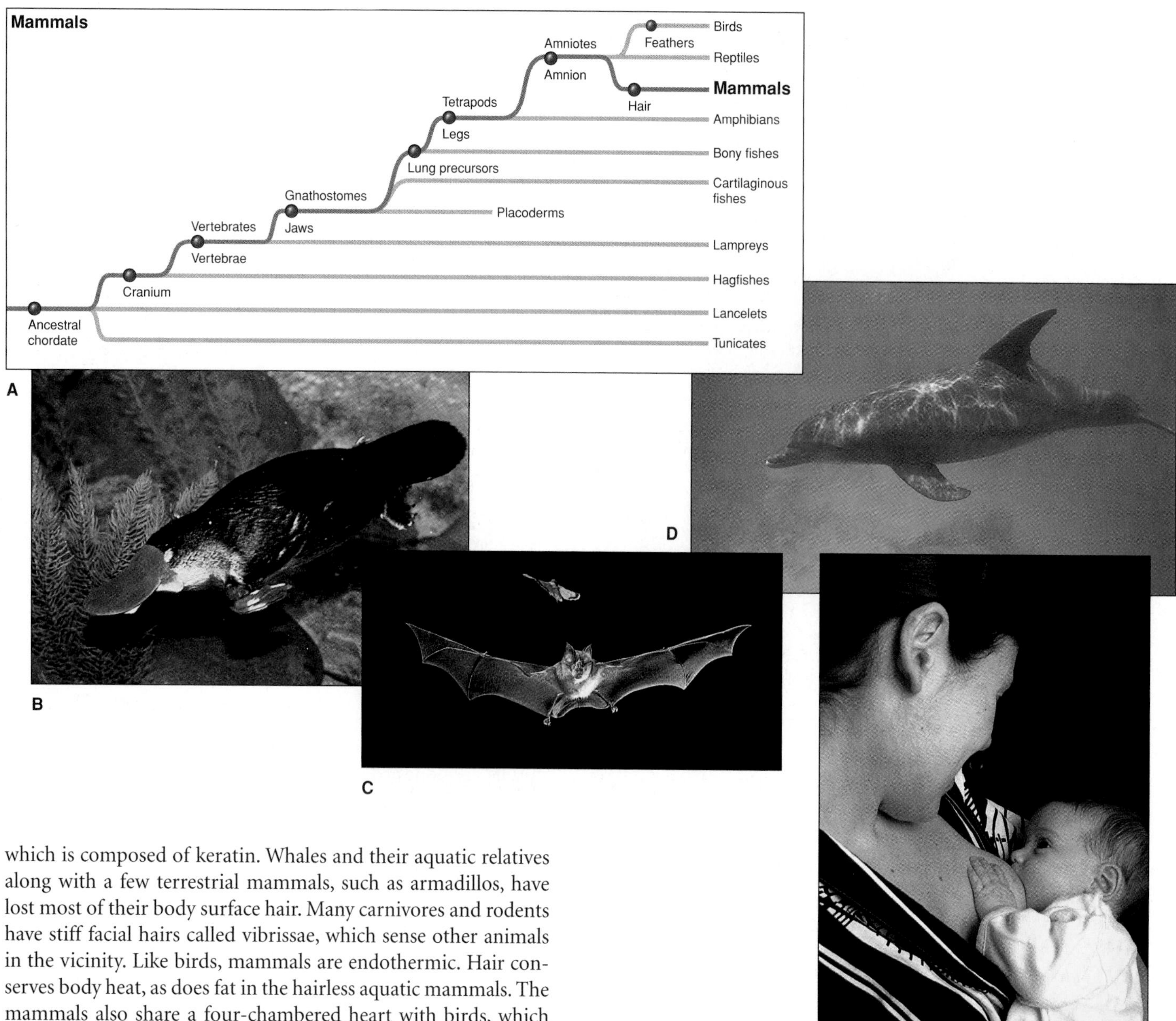

which is composed of keratin. Whales and their aquatic relatives along with a few terrestrial mammals, such as armadillos, have lost most of their body surface hair. Many carnivores and rodents have stiff facial hairs called vibrissae, which sense other animals in the vicinity. Like birds, mammals are endothermic. Hair conserves body heat, as does fat in the hairless aquatic mammals. The mammals also share a four-chambered heart with birds, which evolved independently in the two groups.

Mammals are the only animals that have mature red blood cells that lack nuclei. Teeth are distinctive too. Mammals typically have two sets of teeth, rather than continually replacing them as many other vertebrates do. Much as beak shape reveals eating habits of birds, tooth shape indicates diet in mammals. Carnivores have sharp canine teeth for tearing tough meat, whereas herbivores have well-developed incisors and molars for biting and mashing fibrous plant tissues. Sebaceous (oil) and sweat glands are another unique characteristic of mammals. Sebaceous glands lubricate hair and retain moisture in skin, and sweat glands cool bodies by evaporating secretions. Sebaceous and sweat glands vary in location on the bodies of different types of mammals. The cerebral cortex in mammals is very well developed, enabling these animals to learn, remember, and even think. As a result, many mammals are unlike other organisms in that they can purposefully respond to environmental stimuli.

Of the 26 orders of mammals, five are extinct (fig. 25.21). The order that includes humans is primates. These mammals are adapted to life in the trees (an arboreal existence). Primates' acute vision, excellent coordination, and ability to grasp branches are essential for obtaining food and, in some species, using tools.

FIGURE 25.21

The Extant Orders of Mammalia.

Shown here are 20 extant orders of mammals; five orders (not shown) are extinct. The relationship between monotremes and marsupials is under debate, and some biologists break the marsupials into two orders, which would bring the total number of extant orders to 21.

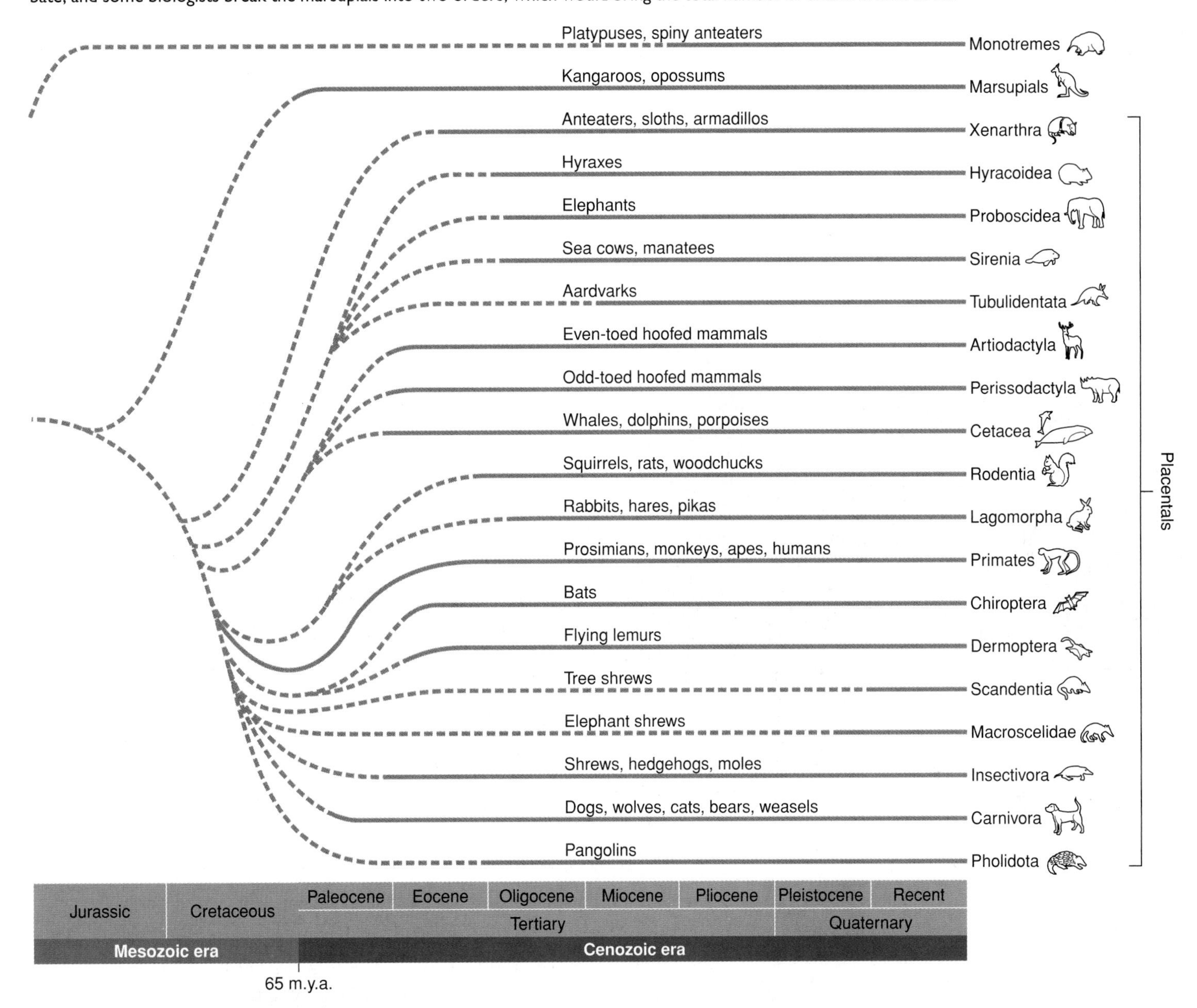

Characteristics that taken together distinguish primates from other mammals include:

- a large brain;
- color vision;
- binocular vision;
- five digits on each limb;
- nails instead of claws;
- an "opposable" thumb, meaning that it folds against the other fingers, enabling the animal to grasp.

Figure 25.22 shows some members of order Primates. Suborder Prosimii includes the more primitive primates, such as lemurs, bush babies, tarsiers, and lorises. These are small animals that are arboreal and look like squirrels. Suborder Anthropoidea ("resembling man") includes three subgroups that are monophyletic (each is a clade). The New World monkeys have a tail described as prehensile, which means that it can wrap around objects, almost like a third arm. They also have flattened noses, nonopposable thumbs, and they lack cheek pouches and calloused rear ends (ischial callosities). The New World monkeys include the organ grinder, spider, and howler monkeys. Old World monkeys, in contrast, have close-set nostrils, cheek pouches, ischial callosities, and opposable thumbs, but lack prehensile tails. The group includes the mandrill and the rhesus and proboscis monkeys. The third group of Anthropoidea includes gibbons, orangutans, chimpanzees, gorillas, and humans. These primates lack cheek pouches and have tails reduced to a coccyx.

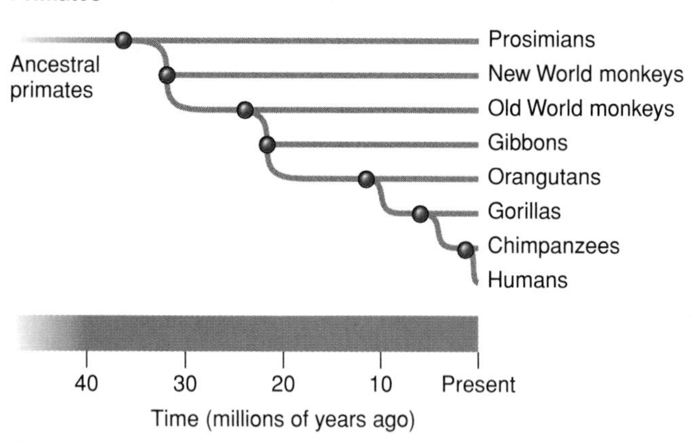

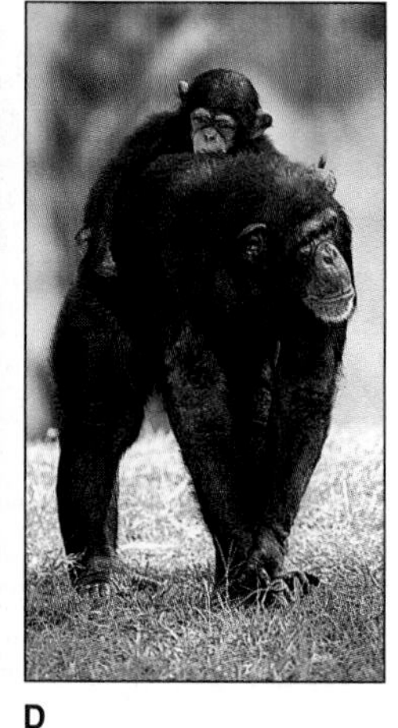

D

We have a tendency to judge the diversity of the living world according to what we can see. The chapters in this unit should indicate that the slices of life with which we are familiar are very small, indeed. We humans are, biologically speaking, organisms like any other. Perhaps we can alter the environment to an extent greater than, for example, a slime mold or a polychaete worm, and in some instances we can counter natural selection. But we are a species like any other, descended from ancestors with which we share many characteristics, yet with our own specific set of features. It is intriguing to think about where the human species is headed, which species will vanish, and how life will continue to diversify.

25.8 MASTERING CONCEPTS

1. Which characteristics define a mammal?
2. How do monotremes, marsupials, and placental mammals differ in how they reproduce?
3. What types of animals are primates?
4. How are primates adapted to forest life?

FIGURE 25.22
Primate Diversity.
(**A**) This phylogenetic tree shows proposed relationships among the primates and their approximate times of divergence. Representative primates include (**B**) white-faced capuchins (a New World monkey), (**C**) gibbons, and (**D**) chimpanzees.

Chapter Summary

25.1 What Are Chordates?

1. Phylum Chordata includes the tunicates, lancelets, and the vertebrates. Vertebrates include fishes, amphibians, reptiles, birds, and mammals.
2. **Chordates** share four characteristics: a **notochord,** a **dorsal hollow nerve cord, pharyngeal pouches** or slits, and a **postanal tail.**
3. The hemichordates may bridge echinoderms and chordates.

25.2 The Tunicates and Lancelets Are "Protochordates"

4. The **tunicates** and **lancelets** are the only invertebrate chordates.
5. Tunicates have a covering that contains cellulose, and retain the notochord only in the larval tail. They obtain food and oxygen with a siphon system.
6. Lancelets resemble eyeless fishes. They have rudimentary circulatory and nervous systems.

25.3 The Vertebrates

7. **Vertebrates** are distinguished from other animals by a backbone, paired limbs, an endoskeleton, and organ systems.
8. Compared with invertebrate chordates, vertebrates have a diminished notochord, a central nervous system, and in the tetrapods, lungs and limbs.
9. Vertebrates are distinguished from each other by habitat, mode of locomotion, and how they have young. Reptiles, mammals, and birds are **amniotes,** possessing membranes that protect the fetus.
10. Vertebrates include four fish classes, amphibians, reptiles, birds, and mammals.

25.4 Fishes—Vertebrate Life in the Waters

11. Fishes are abundant, diverse, and include four classes.
12. Aquatic adaptations include a streamlined body, lightweight bones, fins and tails, and ways to maximize extraction of oxygen from water. They are **ectotherms,** with the environment controlling body temperature.
13. **Hagfishes** and **lampreys** are two types of jawless fishes. The hagfishes are more primitive, and secrete slime. Lampreys have a sucking mouth.
14. **Ostracoderms** were jawless fishes, now extinct. The **placoderms** are an extinct class that may have originated jaws.
15. The **cartilaginous jawed fishes** include chimaeras, skates, rays, and sharks. Sharks detect vibrations from prey with a **lateral line system.**

16. The **bony jawed fishes** account for 96% of extant fish species. They include **ray-finned fishes, lungfishes,** and **lobe-finned fishes**. Bony fishes have lateral line systems and also **swim bladders,** which enable them to control their buoyancy.

25.5 The Amphibians Lead a Dual Lifestyle

17. **Amphibians** retain some characteristics of fishes, yet are adapted to life on land. They have denser bones, lungs, four limbs, different senses emphasized, and more powerful circulatory systems. **Chromatophores** enable some of them to change color.
18. Amphibians include frogs and toads (order Anura), salamanders and newts (order Urodela), and caecilians (order Gymnophiona).
19. Amphibians have many adaptations to avoid predators, reproduce, and respire in a variety of ways.

25.6 The Reptiles Fully Conquer Land

20. **Reptilian** adaptations to life on land include efficient excretory systems, amniote eggs, better respiration and circulation, internal fertilization, and a cerebral cortex.
21. Reptiles include dinosaurs, lizards and snakes, crocodilians, turtles and tortoises, and worm-lizards.

25.7 Birds—Vertebrate Life in the Sky

22. **Birds** retain scales and egg-laying from reptilian ancestors, but are highly specialized for flight.
23. Hollow bones, streamlined bodies, and feathered wings are adaptations for flight.
24. Feather colors derive from biochromes or structural colors.

25.8 Mammals Are Furry Milk-Drinkers—And More

25. The **mammals** include **monotremes, marsupials,** and **placental mammals.**
26. Defining characteristics include fur, milk secretion, non-nucleated red blood cells, two sets of teeth, oil and sweat glands, and a highly developed cerebral cortex.
27. Primates have excellent vision and coordination, a large brain, five digits per limb, and opposable thumbs.

Testing Your Knowledge

1. What are the four distinguishing characteristics of chordates?
2. How do tunicates and lancelets differ from tetrapods?
3. List evolutionary trends seen in vertebrates compared with invertebrate chordates.
4. List five adaptations each that enabled:
   a. fishes to live in water
   b. amphibians to live on land
   c. birds to fly
   d. primates to live in trees
5. How is an amphibian's skin both an advantage and a liability?
6. In which group of vertebrates does parental care of young first appear?
7. How are opposable thumbs, prehensile tails, and large cerebral cortexes adaptive to a primate's lifestyle?

Thinking Scientifically

1. Tunicates and lancelets did not leave fossil evidence because their parts were not hard. The distinguishing characteristics of mammals—hair and breasts—also are not hard, yet mammals have a rich fossil record. Explain why this is so.
2. How are a fish's and a bird's skeletons similar in structure and function?
3. Fishes are adapted to life in water, and tetrapods to life on land. Cite two criteria for assessing which group has been more successful.
4. Give two examples in which DNA sequence evidence contradicts traditional classification of vertebrates based on easy-to-observe characteristics.
5. Why do some biologists consider hagfishes not to be vertebrates? Birds to be reptiles?
6. Lungfishes, thecodonts, and hemichordates are each thought to be transitional forms. Which two types of animals does each resemble?
7. Why does a group of organisms that has lost a specialized characteristic complicate biological classification? Cite two examples of this phenomenon. Suggest a way that biologists can overcome this confusion.
8. If you found a fossil and weren't sure whether it was from a reptile or a mammal, how might you tell the difference?

References and Resources

Hofrichter, Robert. 2000. *Amphibians: The World of Frogs, Toads, Salamanders and Newts.* New York: Firefly Books. Fantastic photos of amphibians.

McRae, Michael. June 1999. Tam Dao: Vietnam's sanctuary under siege. *National Geographic,* vol. 195, p. 80. Biologists are identifying dozens of new species of amphibians, reptiles, and mammals in the forests and jungles of Vietnam.

Mermet, Gilles. June 2000. Supermodels with 6 legs. *Smithsonian,* vol. 31, p. 60. A photo gallery of insects.

Serena, Melody. April 2000. Duck-billed platypus: Australia's urban oddity. *National Geographic,* vol. 197, p. 118. The platypus lived with dinosaurs, and persists today.

Zimmer, Carl. March 3, 2000. In search of vertebrate origins: Beyond brain and bone. *Science,* vol. 287, p. 1576. Lancelets may hold clues to transitional forms between invertebrates and vertebrates.

The *LIFE* Online Learning Center provides additional resources and tools for studying this chapter.
www.mhhe.com/life

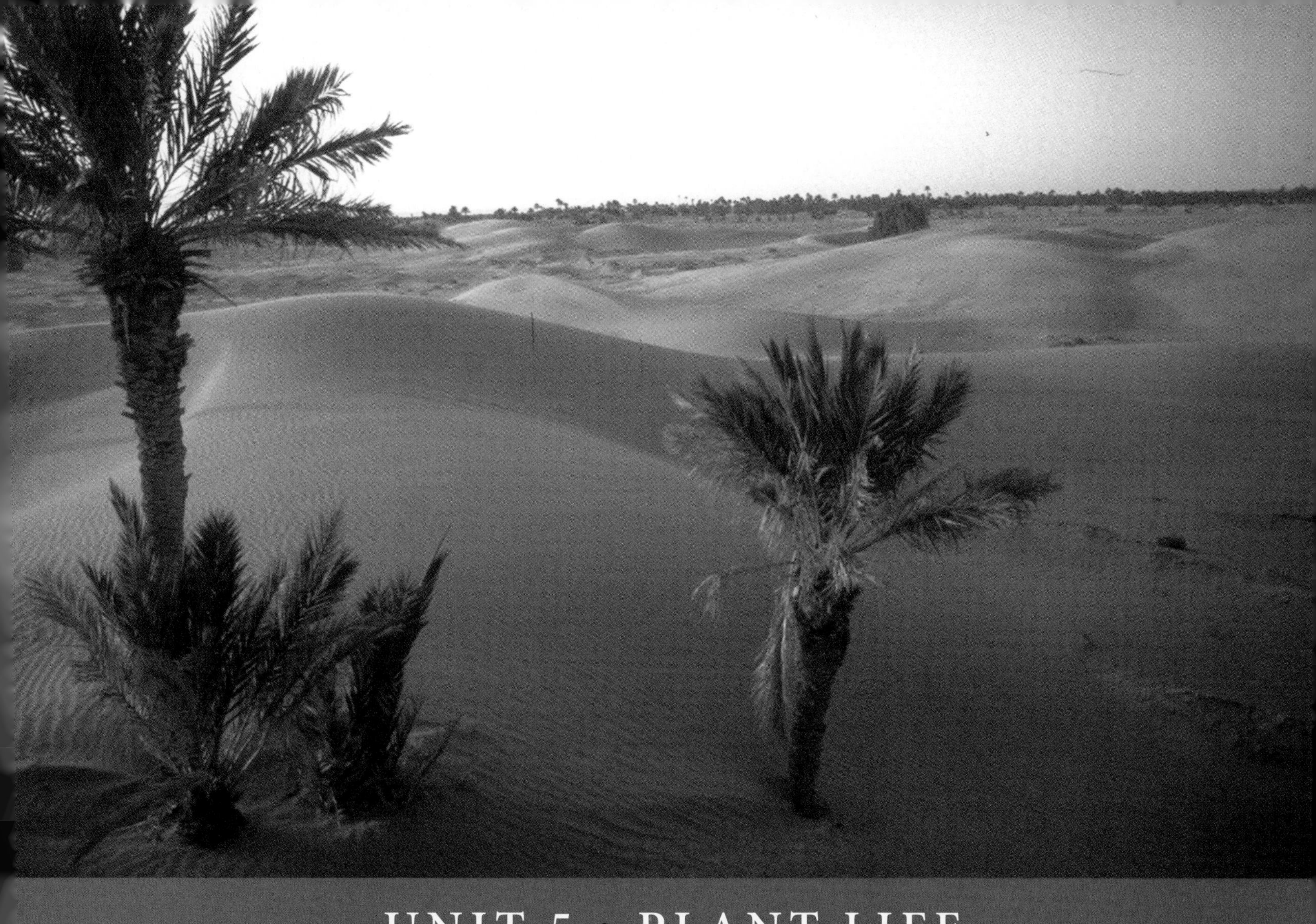

UNIT 5 • PLANT LIFE

In some of the world's driest deserts, huge expanses of sand offer little relief against extreme heat and drought. An oasis is a cool, wet break in the desert, a place where groundwater rises close enough to the surface to support plant and animal life. For people living in the oasis, the trunks and leaves of date palms provide raw materials for buildings and handicrafts. The date itself is a sweet and nutritious fruit that feeds humans and livestock alike. Farmers desire improved varieties of these long-lived and resilient trees, but conventional propagation is slow and unpredictable. Modern biotechnologies such as tissue culture and transgenesis may soon give nature a boost in this unforgiving environment.

CHAPTER 26

Plant Form and Function

Versatile Paper

Our uses of paper seem endless. We use it to package food, as currency, to communicate, to decorate walls, and even to mop up various bodily secretions. This versatile product consists of cellulose fibers from plant cell walls, mixed with water and pressed through a fine mesh. The fibers entangle, adjacent cellulose molecules bond, the water evaporates, and paper forms.

The word "paper" comes from "papyrus," which refers to the sedge plant **Cyperus papyrus,** ***used by the ancient Egyptians to make paper.***

Manufacturing paper today demands a constant supply of wood pulp from millions of trees. Over the centuries, however, paper has been made from cotton, hemp, jute, bamboo, sugarcane, wheat, rice straw, aspen, beech, birch, fir, gum, oak, pine, hemlock, and spruce.

The word "paper" comes from "papyrus," which refers to the sedge plant *Cyperus papyrus,* used by the ancient Egyptians to make paper. They took strips from the interiors of stems, flattened them, and arranged them crosswise. They added water, then pressed the material so that the fibers would adhere, and allowed it to dry. The paper was smoothed and rolled up. About 500 B.C., the Greeks and Romans discovered papyrus as a paper source too. Then a Chinese man named Ts'ai Lun, who served in the emperor's court in A.D. 105, invented modern paper from the inner bark of the mulberry tree, *Broussonetia papyrifera.* He ground up the bark and passed it through a cloth mesh held between sticks of bamboo. People used different plants to manufacture different qualities of paper.

Arabs who took Chinese prisoners learned from them the art of papermaking, and soon Baghdad had a thriving paper industry. Paper came to Japan in A.D. 610 and the Near East in 800. The Crusades and the Moorish

Making Paper.
To make paper, plant material is broken apart, soaked in water, spread into sheets, and dried. The Chinese invented papermaking about 2000 years ago. The sheet of paper on the right was handmade with paper mulberry (*Broussonetia papyrifera*). The long fibers are easily visible in the finished sheets.

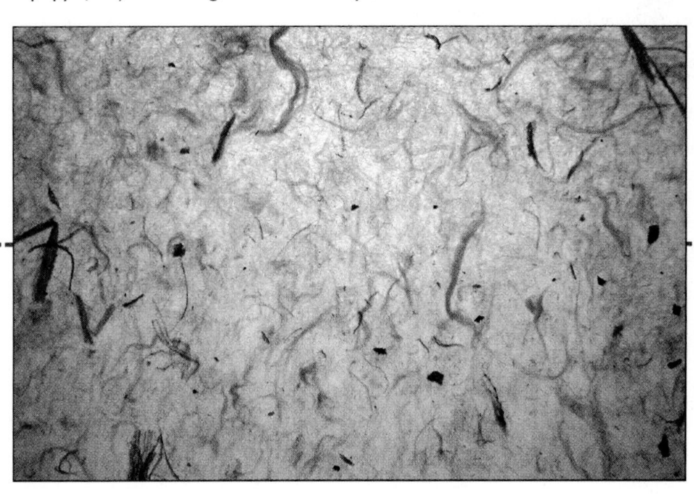

conquest of north Africa and Spain brought papermaking to Europe, first in Spain in 1150. Meanwhile, the Aztecs and Mayans used fibers from native American plants to make paper.

The first paper mill in the United States was built near Philadelphia in 1690. Paper for newspapers came from linen or cotton rags—some fine documents today are still made from cloth. When demand for paper rose sharply as more people began to read regularly, supplies of rags dwindled, and new sources were needed. For a short while paper was made from the wrappings around Egyptian mummies, but soon these supplies diminished too.

Today each issue of the Sunday Times uses 150 acres (0.61 square kilometers) of trees.

Around 1800, machines took over the difficult task of making paper. In 1840, attention turned to wood pulp—the Germans were the first to use it to make paper, and in 1854 the British began to use wood. To make paper, the plant material is gently broken apart in water, forming a slurry of fibers. The material is compressed into a sheet, then dried. Several chemical steps were added to the papermaking process to improve the transformation of wood pulp into paper. On August 23, 1873, *The New York Times* published its first all-wood issue. Today each issue of the Sunday *Times* uses 150 acres (0.61 square kilometers) of trees.

Currently, people in the United States use almost 200,000 tons of paper each day—enough to cover 1,350 square miles (an area the size of Long Island). In time, the increasing manufacture of these products will outstrip the wood pulp supply.

How can we conserve wood resources? Many publishers now print magazines and books on recycled paper, and napkins, towels, and paperboard come from recycled paper. To recycle paper, old newspapers, magazines, and junk mail are placed into huge tanks called pulpers, to which solvents and detergents are added to remove inks. The fibers are then reassembled into new paper.

Another approach to preserving wood is to find alternative sources of paper. One jeans manufacturer converts leftover fabric to paper, which the company uses for stationery and labels. Hemp fibers make an excellent, long-lasting paper that doesn't require bleaching, and the plant is easy to cultivate. Early drafts of government documents, including the Declaration of Independence and the Constitution, were made of hemp. However, laws in the United States ban growing *Cannabis sativa*, the source of hemp fibers, because a high-narcotic variety of the same species is the source of marijuana.

One large American publishing company took a clever approach to saving paper—it trimmed 2.5 centimeters from the width of toilet paper rolls in its building. The employees still used the same number of rolls each month. Trimming all rolls of toilet paper in the United States would save a million trees a year. Other inventive ideas could undoubtedly save millions more.

26.1 Plant Tissues

Plant bodies consist of four basic tissue types. Cells in meristems divide, increasing the length and girth of the plant. Ground tissue is specialized for storage, photosynthesis, and physical support. Dermal tissue protects the plant, while allowing gas exchange and water and mineral uptake. Water and dissolved materials move within a plant in vascular tissue.

A cactus, an elm tree, and a dandelion look very different from one another, yet each consists of the same basic parts. Most people are familiar with the vegetative, or nonreproductive, plant organs—the roots, stems, and leaves that support each other. Through photosynthesis, the aboveground part of a plant, or **shoot,** produces carbohydrates. A portion of this sugar supply nourishes the **roots,** which are usually belowground. In turn, roots absorb water and minerals that are transported to the shoots.

The plant organs are composed of specialized tissues that make and store food, acquire and transport water and dissolved nutrients, grow, provide support, and protect the plant from predators. Specialized cells give these tissues their unique properties. This chapter begins by exploring the tissue and cell types that perform the functions of plant life, then describes how these cell and tissue types work together to produce the diversity of stems, leaves, and roots found in herbaceous (nonwoody) and woody plants. The emphasis is on flowering plants—angiosperms—the most diverse and abundant plants.

Plants consist of four basic tissue types (fig. 26.1 and table 26.1). Meristematic tissue adds new cells that enable a plant to grow

FIGURE 26.1
Parts of a Flowering Plant.
(**A**) A plant consists of a root system and a shoot system. Roots, stems and leaves are vegetative organs; flowers and fruits are reproductive structures. (**B**) Four tissue types build plant organs.

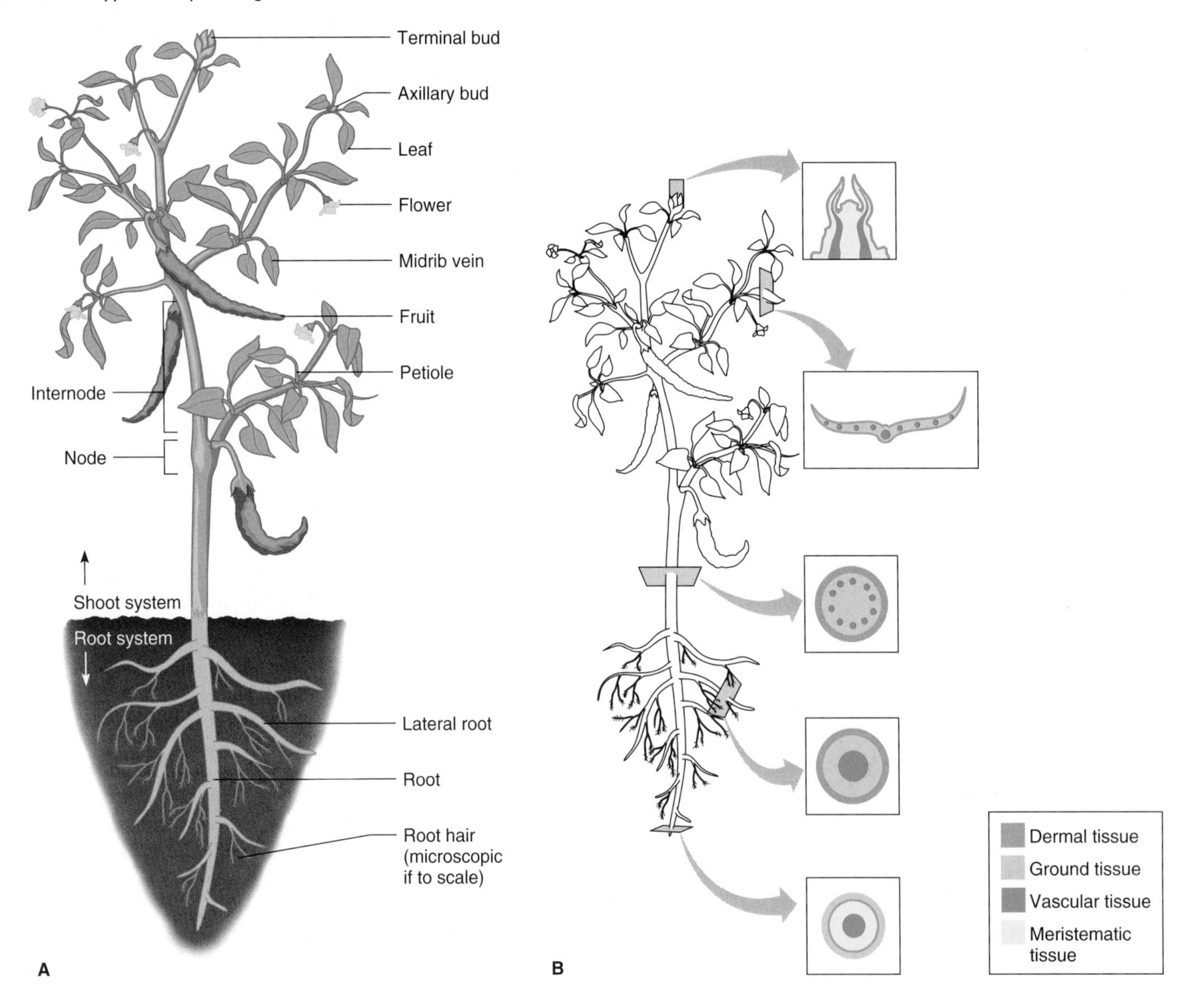

TABLE 26.1 Tissue Types in Angiosperms

TISSUE	FUNCTION
Meristem	Cell division and growth
Ground	Bulk of interiors of roots, stems, and leaves
Dermal	Protects plant; controls gas exchange
Vascular	Conducts water, dissolved minerals, and photosynthate

and specialize. Ground tissue makes up the bulk of the living plant tissue and stores nutrients or photosynthesizes. Dermal tissue covers and protects the plant, controlling gas exchange with the environment. Vascular tissues, including xylem and phloem, conduct water, minerals, and photosynthate (sugars) throughout the plant.

Meristems

Meristems are localized regions in a plant that undergo cell division and are the ultimate source of all the cells in a plant. These regions of active cell division account for the elongation of root and stem tips, the growth of buds, and the thickening of some stems and roots. They function throughout a plant's life; because of them, plants need never stop growing.

Apical meristems are near the tips of roots and shoots in all plants. Figure 26.2 depicts the location of the apical meristem in a shoot, and figure 26.15 in a root. Cells in the apical meristems are small and unspecialized. When the meristematic cells divide, the root or shoot tip lengthens in what is called **primary growth.** Apical meristems give rise to three other types of meristems—the ground meristem, protoderm, and procambium—which produce ground tissue, dermal tissue, and vascular tissue.

In contrast to apical meristems, which lengthen a plant, **lateral meristems** grow outward to thicken the plant. This process, called **secondary growth,** does not occur in all plants. Wood forms from secondary growth, which is discussed at the end of this chapter.

In some plants, intercalary meristem occurs between areas that are more developed. Grasses, for example, tolerate grazing (and mowing) because their leaves have intercalary meristems that

FIGURE 26.2
Primary Growth of a Dicot's Shoot.
Looking at the tissue layers that comprise the growing tip of a shoot reveals how apical meristems give rise to primary meristems, which form mature, specialized tissues visible in older parts of the shoot.

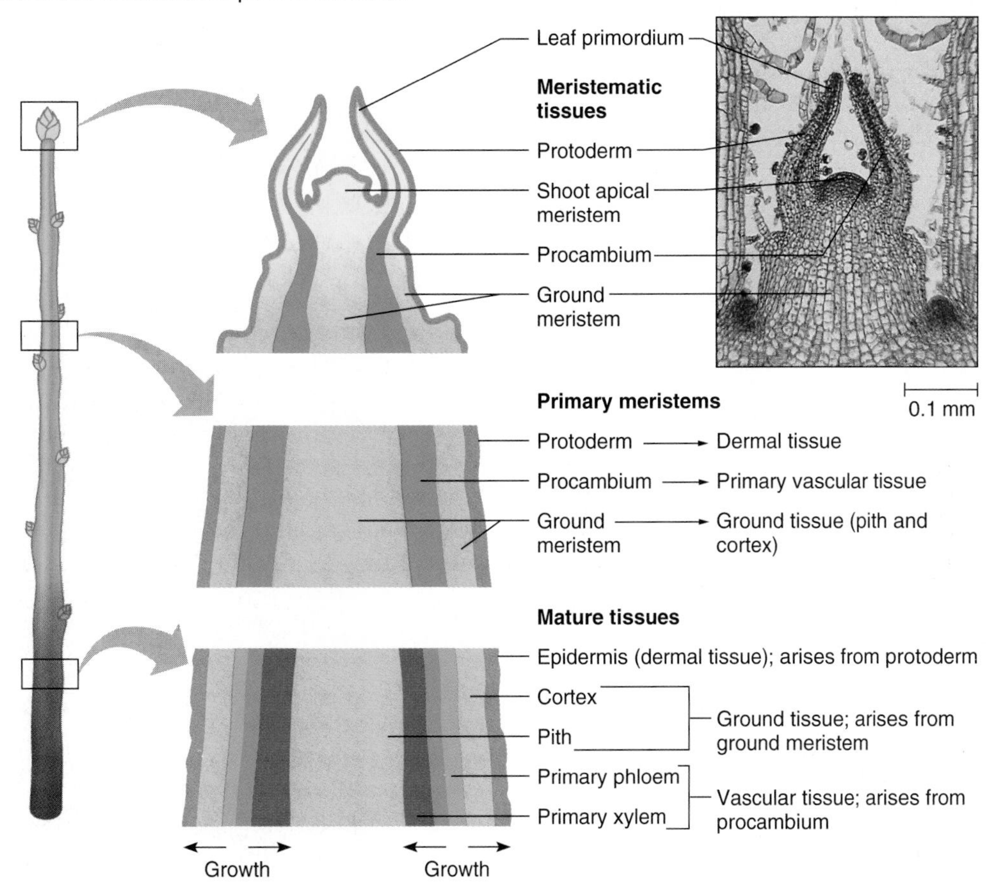

TABLE 26.2 Meristem Types

TYPE	FUNCTION
Apical meristem	Growth at root and shoot tips
Lateral meristem	Growth outward, thickening plant
Intercalary meristem	In grass stems, allows rapid regrowth of mature stems

divide to regrow the plant when the tip of the stem is munched off. Table 26.2 summarizes meristem types.

Ground Tissue

Ground tissue makes up most of the primary body of a flowering plant, filling much of the interior of roots, stems, and leaves. These cells have many functions, including storage, support, and basic metabolism. Ground tissue consists of three cell types: parenchyma, collenchyma, and sclerenchyma.

Parenchyma cells are the most abundant plant cells. They are relatively unspecialized, although they can divide, which enables the tissue to become specialized in response to injury or a changing environment.

Parenchyma cells store the edible biochemicals in plants, such as the starch in a potato or a kernel of corn. These cells may also store fragrant oils, salts, pigments, and organic acids. Parenchyma cells of oranges and lemons, for example, store citric acid, which gives them their tart taste. These cells also conduct vital functions, such as photosynthesis, cellular respiration, and protein synthesis. Chlorenchyma cells are parenchyma cells that photosynthesize. Their chloroplasts impart the green color to leaves (fig. 26.3A).

Collenchyma cells are elongated living cells that differentiate from parenchyma and support the growing regions of shoots. Collenchyma cells have unevenly thickened primary (outer) cell walls that can stretch and elongate with the cells (see chapter 4 to review plant cell wall characteristics). As a result, collenchyma provides support without interfering with the growth of young stems or expanding leaves (fig. 26.3B).

Sclerenchyma cells have thick, nonstretchable secondary cell walls (a trilayered structure inside the outer, or primary, cell wall). These cells, which are usually dead at maturity, support parts of plants that are no longer growing. Two types of sclerenchyma form from parenchyma: sclereids and fibers.

Sclereids have many shapes and occur singly or in small groups. Small groups of sclereids create a pear's gritty texture (fig. 26.3C). Sclereids may form hard layers, such as in the hulls of peanuts. Fibers are elongated cells that usually occur in strands that vary from a few to a few hundred millimeters long (26.3D). Many textiles are sclerenchyma fibers. Humans now cultivate more than 40 families of plants for fibers and have fashioned cords from fibers since 8000 B.C. The fibers of *Agave sisalana,* commonly known as sisal, or the century plant, are used to make brooms, brushes, and twines. Linen comes from the fibers of *Linum usitatissimum,* or flax.

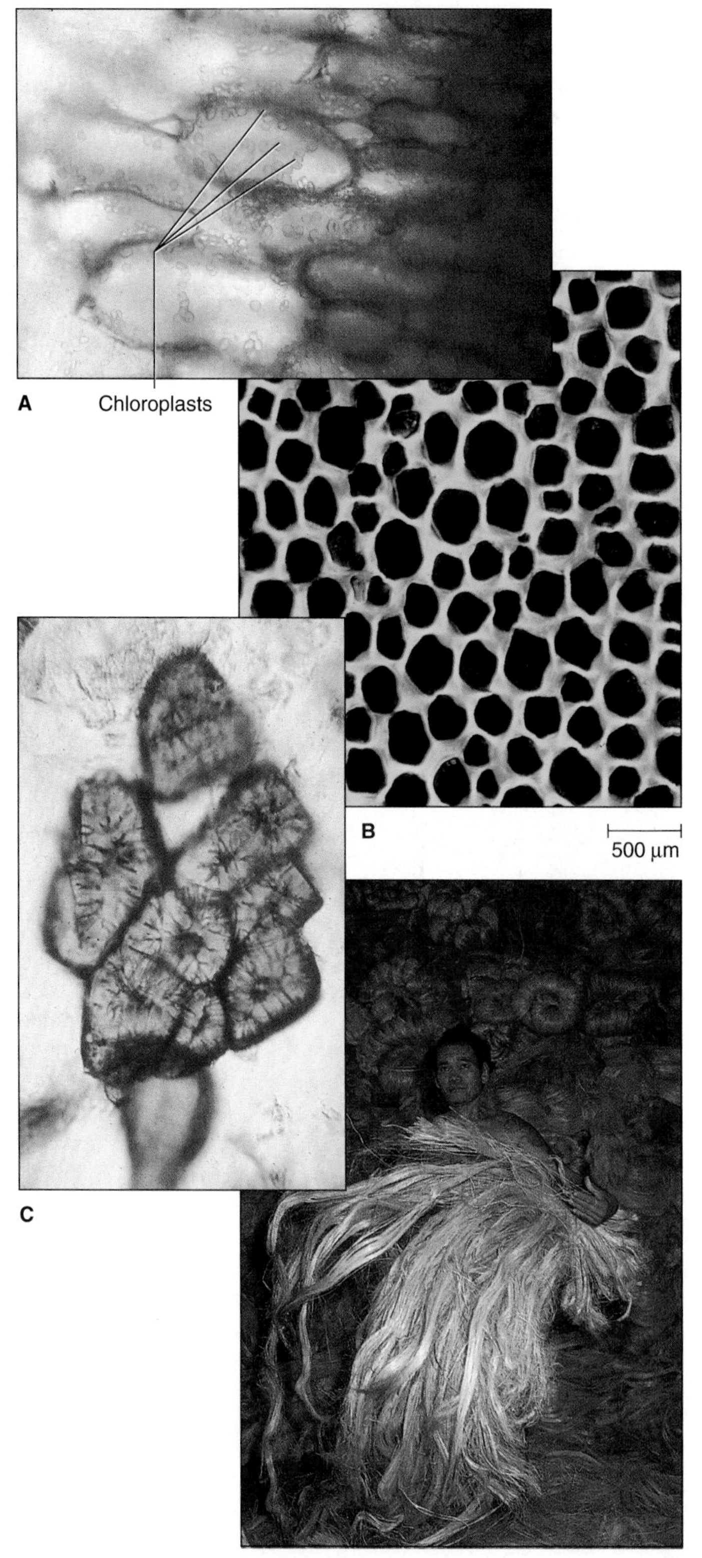

FIGURE 26.3
Plant Cells.
(**A**) Parenchyma cells are usually unspecialized and store carbohydrates and other important biochemicals. Chlorenchyma is a type of parenchyma that photosynthesizes. (**B**) Collenchyma differentiates from parenchyma and supports growing shoots with thick, elastic cell walls. The collenchyma tissue pictured forms the "string" of a celery stalk. (**C**) Sclerenchyma supports plant parts that are no longer growing. This cell type includes these sclereids that provide the gritty texture of pear flesh and long fibers (**D**), such as these from Manila hemp (*Musa textiles*), which is used to make rope.

Dermal Tissue

Dermal tissue covers the plant. The **epidermis,** usually only one cell thick, covers the primary plant body. Epidermal cells are flat, transparent, and tightly packed. Special features of the epidermis provide a variety of functions.

The **cuticle** is an extracellular covering over all the aerial parts of a plant; it protects the plant and conserves water (fig. 26.4A). The cuticle consists primarily of cutin, a waxy material that epidermal cells produce. This covering retains water and prevents desiccation. As a result, plants can maintain a watery internal environment—a prerequisite to survival on dry land. The cuticle and underlying epidermal layer also are a first line of defense against predators and infectious agents. In many plants, a smooth white layer of wax covers the cuticle; when it is thick, it is visible on leaves and fruits. The layer on the undersides of wax palm leaves may be more than 5 millimeters thick. It is harvested and used to manufacture polishes and lipstick.

An impermeable cuticle covers the tightly packed epidermal cells, but plants must exchange water and gases with the atmosphere. They do this through specialized pores, called **stomata** (singular: stoma) (see fig. 6.7 and fig. 26.4B). **Guard cells** surround the pores and control their opening and closing, which regulates gas and water exchange.

When a plant has sufficient water, potassium ions (K^+) move into guard cells. Because the concentration of K^+ becomes greater inside than outside the cells, water enters them by osmosis. This swells the guard cells, which opens the stoma between them. When too much water evaporates from the plant, cells in the leaves lose turgor pressure, which causes release of a hormone, abscisic acid (see chapter 29). This hormone mobilizes calcium ions, which ultimately lowers K^+ concentration inside the guard cells (fig. 26.5). Water exits, the guard cells collapse onto each other, and the stoma closes.

Stomata control the amount of carbon dioxide that diffuses into a leaf for photosynthesis. They also regulate the amount of water that evaporates from leaves, a process called transpiration, which is discussed in the next chapter. Stomata may be very numerous. The underside of a black oak leaf, for example, has 100,000 or so stomata per square centimeter! Because stomata help plants conserve water, they are an essential adaptation for life on land.

Trichomes are outgrowths of the epidermis in almost all plants (see figs. 26.4C–E). These structures have many functions, including deterrence of predators. Hook-shaped trichomes may impale marauding animals. In some plants, predators may inadvertently break off the tips of trichomes, releasing a sticky substance that traps the animal. For example, trichomes of the stinging nettle have spherical tips that break off and penetrate a predator's body, injecting their poisonous contents into the wounds. Trichomes of carnivorous plants such as the Venus's flytrap secrete enzymes to digest trapped animals (see Investigating Life 22.1). These trichomes then absorb the digested prey.

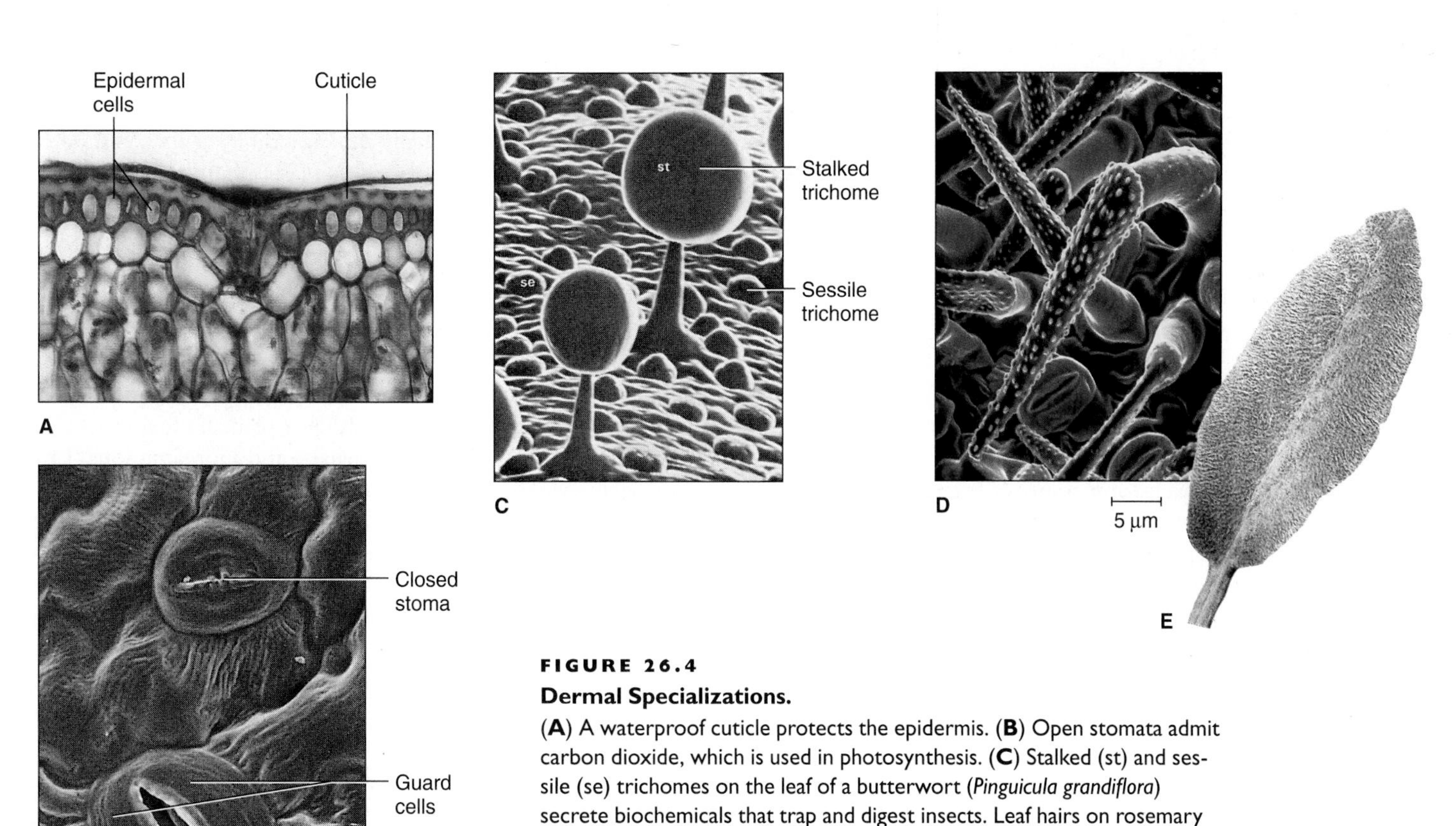

FIGURE 26.4
Dermal Specializations.
(**A**) A waterproof cuticle protects the epidermis. (**B**) Open stomata admit carbon dioxide, which is used in photosynthesis. (**C**) Stalked (st) and sessile (se) trichomes on the leaf of a butterwort (*Pinguicula grandiflora*) secrete biochemicals that trap and digest insects. Leaf hairs on rosemary (*Rosmarinus officinalis*) (**D**) and lamb's ear (*Stachys byzantina*) (**E**) slow air moving over the leaf surface, reducing water loss.

FIGURE 26.5
Opening and Closing of Stomata.
Relative concentrations of potassium ions (K^+) inside and outside the guard cells that surround a stoma determine whether the stoma is opened or closed. When intracellular concentrations are higher, water enters and the guard cells swell, opening the stoma. When intracellular concentrations are lower, water leaves and the stoma closes as the guard cells collapse.

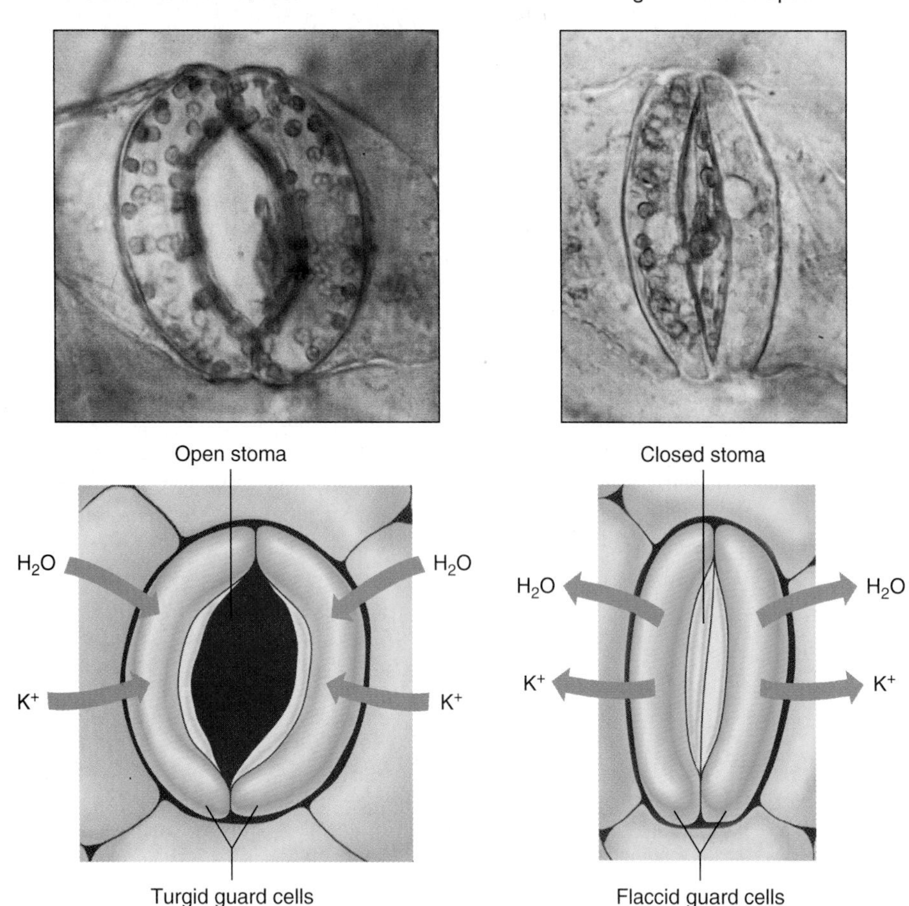

Xylem Xylem, from the Greek for wood, transports water and dissolved minerals from the roots to all parts of the plant. This water replaces water lost in transpiration through stomata.

The two kinds of conducting cells in xylem are called **tracheids** and **vessel elements** (fig. 26.6). Both are elongated, dead at maturity, and have thick walls. The fact that these cells are dead means that organelles do not block water flow. The thick cell walls prevent tracheids and vessel elements from collapsing as water rises through the plant under negative pressure (suction). They also provide support that helps keep the plant upright.

Tracheids, the least specialized conducting cells, are long and narrow, overlapping at their tapered ends. Water moves from tracheid to tracheid through thin areas in cell walls called pits. The water movement is slow because of the small diameter and end walls of the tracheids. Tracheids are probably ancient cells, because they are the only means of water conduction in most nonflowering vascular plants, which are more ancient than flowering plants.

Vessel elements are more specialized than tracheids. Unlike tracheids, vessel elements are short, wide, barrel-shaped cells. Vessel elements stack end to end, and their end walls usually disintegrate, forming hollow tubes, or vessels, that may extend from a centimeter to 3 meters long. Vessel elements are like cellulose pipes, and water in them moves much faster than in the narrower tracheids (fig. 26.6). The greater width of vessel elements and the fact that water can pass directly from one "cell" to another accounts for its more efficient water conduction, a point we return to in the next chapter.

Leaf hairs, a class of trichomes, are epidermal structures that slow the movement of air over the leaf surface (figs. 26.4D–E). This action reduces water loss. **Root hairs** are trichomes that increase the root surface area for absorbing water and minerals.

Many trichomes are economically important. Cotton fibers, for example, are trichomes from the epidermis of cotton seeds. Burrs, on which the fastener Velcro is based, are also trichomes. Menthol comes from peppermint trichomes, and hashish, a powerful narcotic, is purified resin from *Cannabis* trichomes.

Vascular Tissue

Vascular tissues are specialized conducting tissues that transport water, minerals, carbohydrates, and other dissolved compounds throughout the plant. The two types of vascular tissue, **xylem** and **phloem,** form a continuous distribution system (see fig. 27.1). The cell types making up vascular tissue are described here, and their function is further explored in chapter 27.

Phloem Phloem transports dissolved organic compounds, primarily carbohydrates, throughout a plant. Phloem sap also contains hormones, alkaloids, viruses, and inorganic ions. Unlike xylem, which transports water upward under negative pressure, like a drinking straw, phloem transports substances under positive pressure, which is like water flowing through a hose when the spigot is turned on. Thus, water and dissolved sugars can move through phloem in all directions. Also, the conducting cells of phloem, unlike those of xylem, are alive at maturity. Their cell walls have thin areas perforated by many structures called sieve pores, and solutes move through these pores from cell to cell.

Different vascular plants have different types of phloem cells. **Sieve cells** are long, tapering cells with overlapping ends that are usually found in gymnosperms and seedless vascular plants. Phloem sap moves from cell to cell through sieve pores, which permeate all walls of the sieve cells. **Sieve tube members** are mostly in angiosperms, and they are more specialized than sieve cells. In most sieve tube members, the pore areas are aggre-

FIGURE 26.6

Two Types of Xylem Cells.

Xylem transports water and dissolved minerals from roots to shoots. Tracheids (**A**) are long, narrow, and less specialized than the barrel-shaped vessel elements (**B**). Both types of xylem structures consist of dead cells.

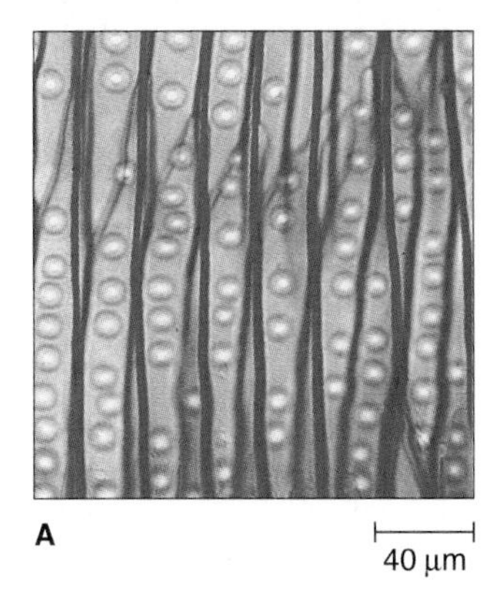

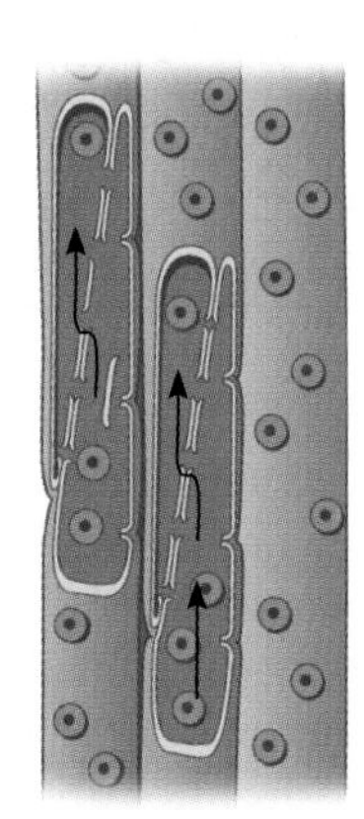

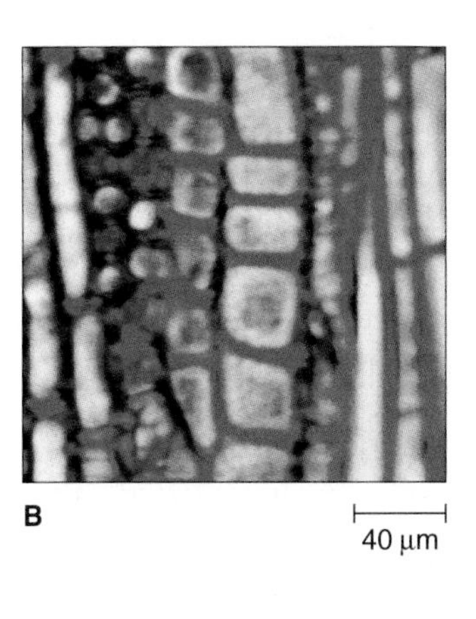

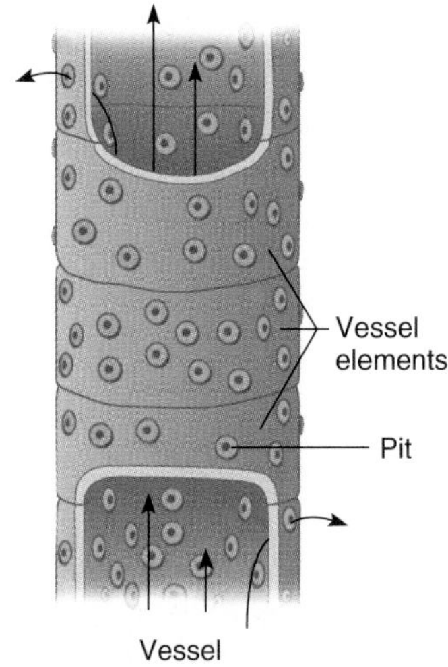

FIGURE 26.7

Phloem Cells.

Phloem consists of living cells, and transports an aqueous solution of carbohydrates, hormones, ions, and various other substances under positive pressure in all directions. In flowering plants, sieve tube members with perforated end walls form sieve tubes that carry phloem sap. A sieve plate is visible in the upper left corner of the photo.

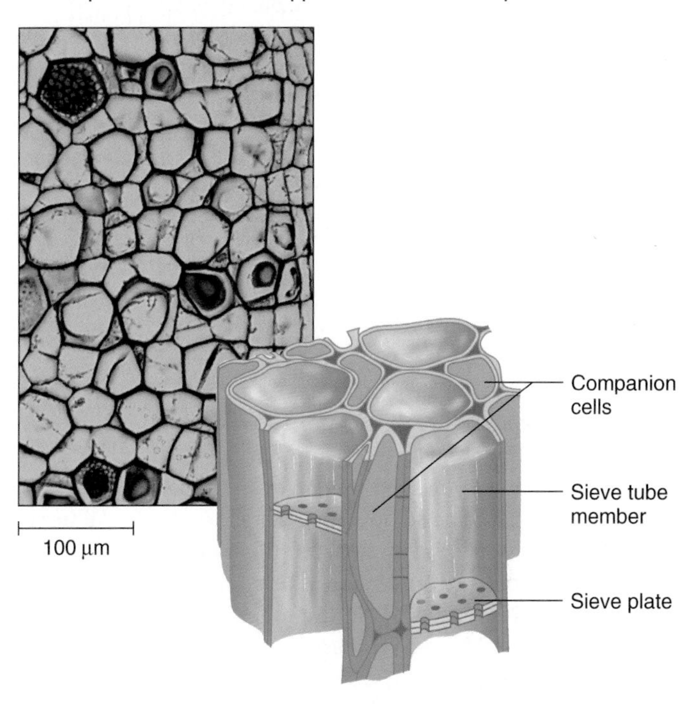

gated into **sieve plates,** usually at the ends of the cells (fig. 26.7). Many sieve tube members lined up end to end comprise a single functional unit, the sieve tube, much as vessels in xylem are made up of vessel elements arranged end to end. Near sieve tube members are **companion cells,** which help transfer carbohydrates into and out of the sieve tube members, as described in chapter 27.

Table 26.3 reviews the major tissue and cell types in plants.

26.1 MASTERING CONCEPTS

1. What is the function of meristematic tissue?
2. What cell types make up ground tissue?
3. How does dermal tissue protect a plant but enable gas exchange to occur?
4. What are the functions of the two types of vascular tissue in plants?
5. What types of cells make up each of the two types of vascular tissue?

OLC

26.2 Anatomy of a Plant

A plant's tissues are organized into stems, leaves, and roots, each of which can be modified and associate to produce a great variety of specialized body forms.

The organs of a plant are composed of meristematic, dermal, ground, and vascular tissues. Natural selection sculpts different organizations of these basic plant tissues as certain forms become better adapted to particular environments. We look now at the basic parts of a flowering plant.

TABLE 26.3 Plant Tissue and Cell Types

TISSUE TYPE	CELL TYPES	FUNCTION
Meristem	Undifferentiated	Produce new cells that add to plant's length or girth
Ground	Parenchyma	Photosynthesis, storage, respiration
	Collenchyma	Elastic support
	Sclerenchyma	Nonelastic support
Dermal	Epidermal	Protection, regulation of gas exchange in stems and leaves, absorption of water and minerals in roots
Vascular (xylem)	Tracheids or vessel elements	Water and mineral conduction, support
Vascular (phloem)	Sieve cells or sieve tube members	Conduction of organic molecules
	Companion cells	Transfer carbohydrates to/from sieve tube members

Stems

The central axis of a shoot system is the **stem.** Stems support leaves, produce and store nutrients, and transport nutrients and water between roots and leaves. Stems may provide food for other species—asparagus, for example, is an edible stem.

Stems consist of **nodes** (areas of leaf attachment) and **internodes** (portions of the stem between the nodes) (see fig. 26.1). The region of the stem just above the point where the leaf attaches is the leaf axil. Axillary buds are undeveloped shoots that form in leaf axils. Although axillary buds can elongate to form a branch or flower, many remain small and dormant.

Stems grow and differentiate at their tips, with new cells originating at the shoot's apical meristem (see fig. 26.2). The shoot elongates as cells divide and become specialized into ground tissue, vascular tissue, or dermal tissue.

In most plants, stems also elongate in the internodal regions, and separate nodes may be easily distinguished. However, some plants have stems called rosettes that do not elongate. Rosettes have short internodes and overlapping leaves. A cabbage head is a rosette made of large, tightly packed leaves.

The epidermis surrounding a stem is a transparent, unicellular layer. It contains stomata, but fewer than are in a leaf's epidermis. The epidermis of a stem also may have protective trichomes.

Vascular tissues in the stems of nonwoody flowering plants are organized into groups called **vascular bundles** that branch into leaves at the nodes. Phloem forms on the outer portions of a bundle, whereas xylem forms to the inside. Often, thick-walled sclerenchyma fibers associate with vascular bundles and strengthen the vascular tissue.

Vascular bundles are organized differently in different types of plants. Consider the two groups of flowering plants: monocotyledons (monocots for short), which have one first, or "seed," leaf; and dicotyledons (dicots), which have two seed leaves (see fig. 22.18). Monocots such as corn have vascular bundles scattered throughout their ground tissue, whereas dicots such as sunflowers have a single ring of vascular bundles (fig. 26.8).

The ground tissue that fills the area between the epidermis and vascular tissue in a stem is called the **cortex.** This area is mostly parenchyma, but may include a few supportive collenchyma strands. Some cortical cells are photosynthetic and store starch. The centrally located ground tissue in dicots is called **pith.**

Many plant stems are modified for special functions such as reproduction, climbing, protection, and storage (fig. 26.9). Some specialized stems are listed below.

- Stolons, or runners, are stems that grow along the soil surface. New plants form from their nodes. Strawberry plants develop stolons after they flower, and several plants can arise from the original one.
- Thorns often are stems modified for protection, such as on hawthorn plants.
- Succulent stems of plants such as cacti are fleshy and store large amounts of water.
- Tendrils support plants by coiling around objects, sometimes attaching by their adhesive tips. Tendrils enable a plant to maximize sun exposure. The stem tendrils of green bean plants readily entwine around anything they can touch. Grapevines also have many tendrils. (Leaves may also be modified into tendrils.)
- Tubers are swollen regions of underground stems that store nutrients. Potatoes are tubers produced on burrowing stolons.
- Rhizomes are underground stems that produce roots and new shoots. Ginger is a spice that is derived from a rhizome.

Leaves

In addition to the stem, a shoot system has **leaves.** Like stems, leaves consist of epidermal, vascular, and ground tissues. Leaves are the primary photosynthetic organs of most plants. They provide an enormous surface area for the plant to capture solar energy. For example, a maple tree with a 1-meter-wide trunk has approximately 100,000 leaves, with a total surface area that would cover the area of six basketball courts (about 2,500 square meters).

Leaves are extremely diverse in form. The leaves of some tropical palms may be 65 feet (20 meters) long, whereas the leaves of *Wolffia*, an aquatic plant, are no larger than a pinhead. A mature American elm may have several million leaves, whereas the

FIGURE 26.8
Stem Anatomy.
Cross sections of (**A**) a monocot stem (corn) and (**B**) a dicot stem (sunflower) (×10). Notice the scattered vascular bundles in the monocot stem and the ring of vascular bundles in the dicot stem.

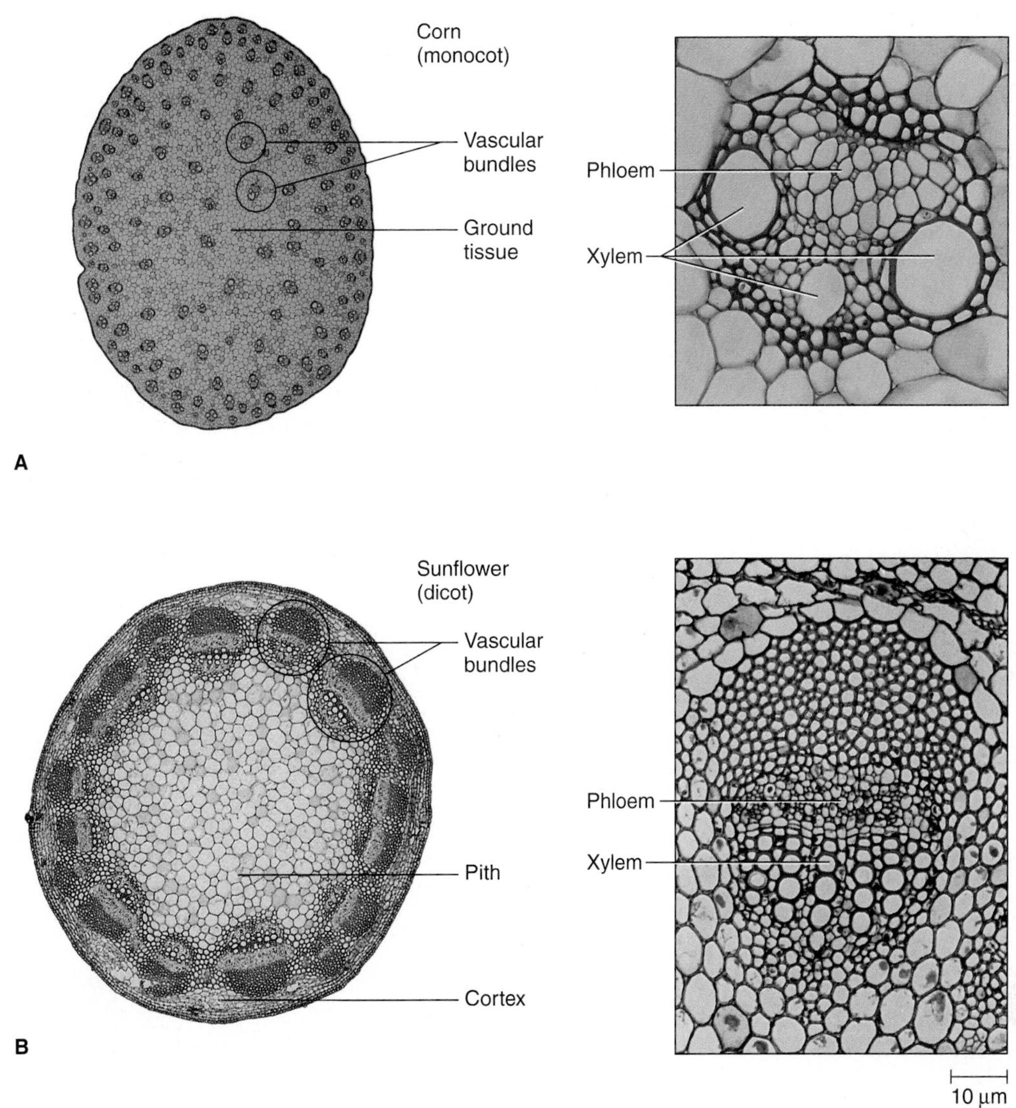

desert gymnosperm *Welwitschia mirabilis* produces only two leaves during its entire lifetime. Leaves may be needlelike, feathery, waxy, or smooth. No two species have identical leaves.

Botanists categorize leaves according to their basic forms (fig. 26.10). Most leaves consist of a flattened **blade** and a supporting, stalklike **petiole.** The large vein down the center is called the midrib. Leaves may be simple or compound. Simple leaves have flat, undivided blades. Elm, maple, and zinnia have simple leaves. Compound leaves are divided into leaflets and are further distinguished by leaflet position. Pinnate compound leaflets are paired along a central line and include ash, rose, and walnut. Palmate compound leaflets all attach to one point at the top of the petiole, like fingers on a hand, and include lupine, four-leaf clover, and shamrock.

Leaf epidermis covers the leaf and consists of tightly packed transparent cells. With the exception of guard cells, the epidermis is usually nonphotosynthetic. It may contain many stomata—more than 11 million in a cabbage leaf, for example. Water loss is minimized in many species because stomata are most abundant on the shaded undersides of leaves that are horizontal. The floating leaves of lily pads are unusual in having stomata on the upper surfaces only. Vertical leaves, such as those of grasses, have equal densities of stomata on both sides.

Xylem and phloem in leaves are connected to the stem's vascular tissue at the stem's nodes. Inside the leaves, the vascular tissue branches into an intricate network of veins. Sclerenchyma fibers and parenchyma cells support leaf veins. Leaf veins may be of two types: netted, with minor veins branching off from larger, prominent midveins, or parallel, with several major parallel veins connected by smaller minor veins (fig. 26.11). Most dicots have netted veins, and many monocots have parallel veins. Vein endings are the blind ends of minor veins, where water and solutes move in and out.

FIGURE 26.9

Modified Stems.

(**A**) Stolons of the beach strawberry (*Fragaria chilensis*) run parallel to the ground. (**B**) The thorns that protect this honey locust are outgrowths of the stem. (**C**) The stem of the fishhook barrel cactus is highly modified to store water; its thorns are actually modified leaves. (**D**) Tendrils may be stems modified to coil around objects, supporting and anchoring plants. (**E**) The potato is a tuber. Sprouts grow from its "eyes" and form new plants. (**F**) The rhizome of an iris is an underground stem.

FIGURE 26.10
Leaf Forms.
(**A**) This simple leaf of *Thottea* attaches to the stem by its petiole. (**B**) This pinnate compound leaf of a *Mimosa* plant has many leaflets. (**C**) The horse chestnut plant has palmate compound leaves, with five leaflets.

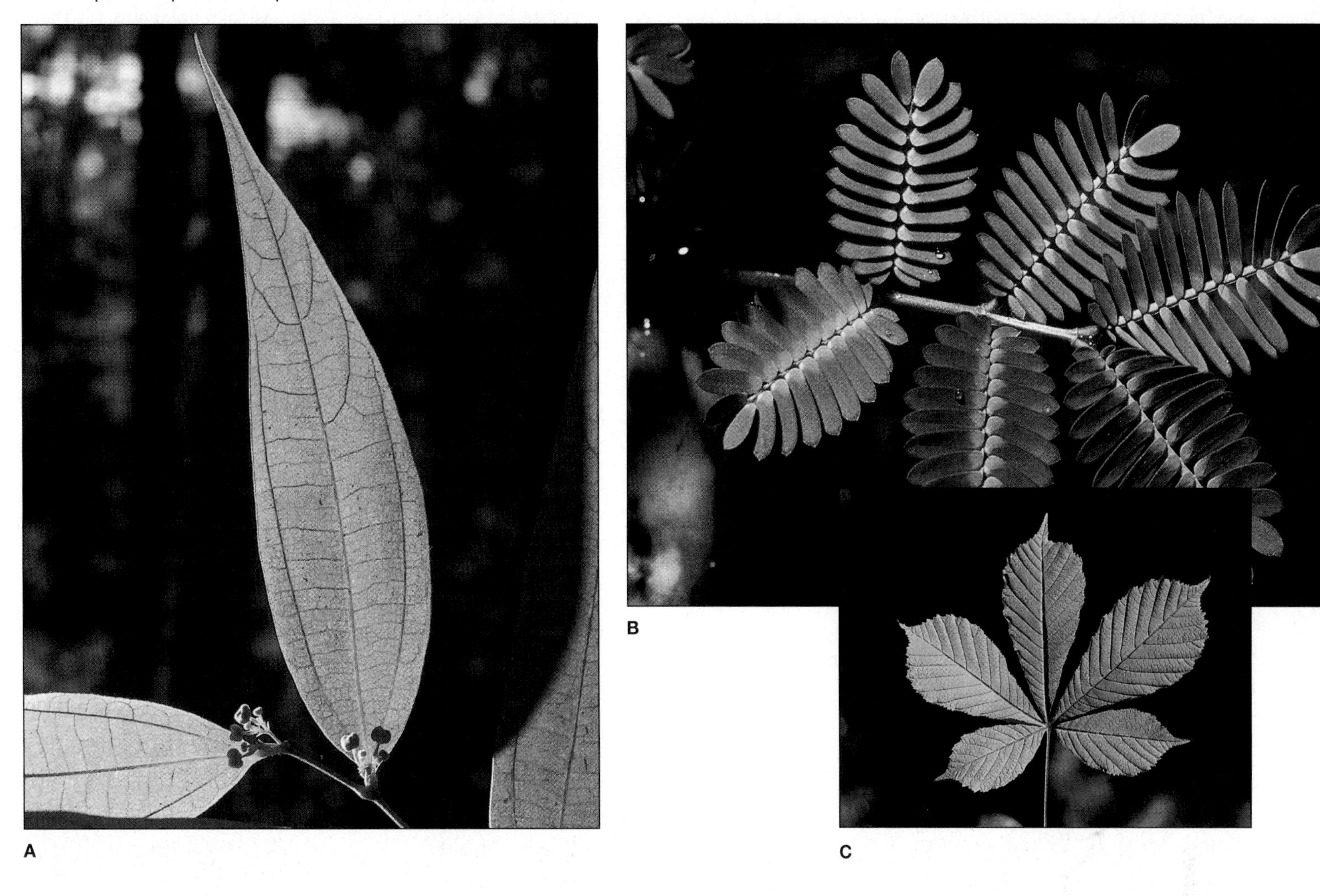

FIGURE 26.11
Leaf Venation in Dicots and Monocots.
(**A**) Leaves of dicots, such as this pumpkin plant, have netted venation. (**B**) Leaves of many monocots, such as this lily, have prominent parallel veins interconnected by many tiny veins.

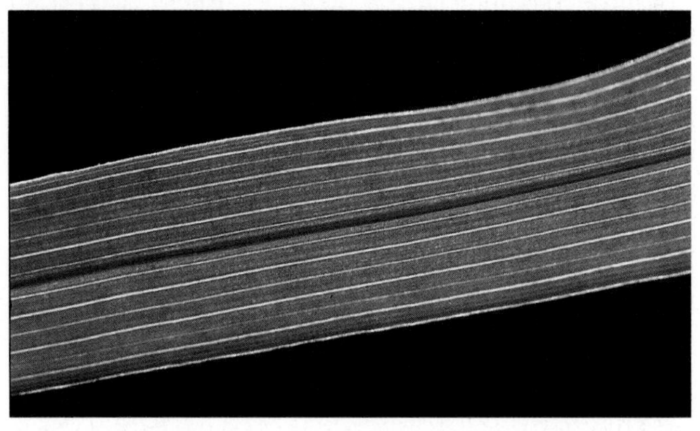

Leaf ground tissue, which is called **mesophyll,** is made up largely of parenchyma cells (see fig. 6.7). Some of these cells are chlorenchyma and therefore photosynthesize and produce sugars. Horizontally oriented leaves have two types of chlorenchyma (fig. 26.12). The long, columnar cells along the upper side of a leaf, called palisade mesophyll cells, are specialized for light absorption. Below the palisade layer are spongy mesophyll cells, which are irregularly shaped chlorenchyma cells separated by large air spaces. These cells are specialized for gas exchange and can also photosynthesize.

In addition to photosynthesizing, leaves provide support, protection, and nutrient procurement and storage, with the following specializations:

- Tendrils are modified leaves that wrap around nearby objects, supporting climbing plants. Pea plants growing in a garden will "grab" a fence with leaf tendrils. (Both leaves and stems can be modified into tendrils.)
- Spines of plants such as cacti are leaves modified to protect the plant from predators.
- Bracts are floral leaves that protect developing flowers. They are colorful in some plants, such as poinsettia. Other floral parts, such as sepals and petals, are also modified leaves.
- Storage leaves are fleshy and store nutrients. Onion bulbs are the bases of such leaves.
- Insect-trapping leaves are found in about 200 types of carnivorous plants and attract, capture, and digest prey, as Investigating Life 22.1 explains.
- Cotyledons are embryonic leaves that may store carbohydrates, which supply energy for germination.

Anyone who has ever raked a lawn knows well that leaves have a limited life span. Leaf abscission is the normal process by which a plant sheds its leaves. Deciduous trees shed leaves at the end of a growing season. Evergreens retain leaves for several years but shed a few leaves each year.

Leaves are shed from an abscission zone, which is a region at the base of the petiole (fig. 26.13). In response to environmental cues such as shortening days or cooler temperatures, a separation layer forms in the abscission zone, isolating the dying leaf from the stem. Eventually, wind, rain or some other disturbance, such as a scurrying squirrel, breaks the dead leaf from the stem. Leaf abscission is important because it removes injured or dying leaves from the plant.

FIGURE 26.12
Anatomy of a Leaf.
Leaf mesophyll consists of an upper palisade layer and spongy mesophyll below. Stomata are concentrated on the lower leaf surface. Leaf veins deliver water and minerals and carry off the products of photosynthesis.

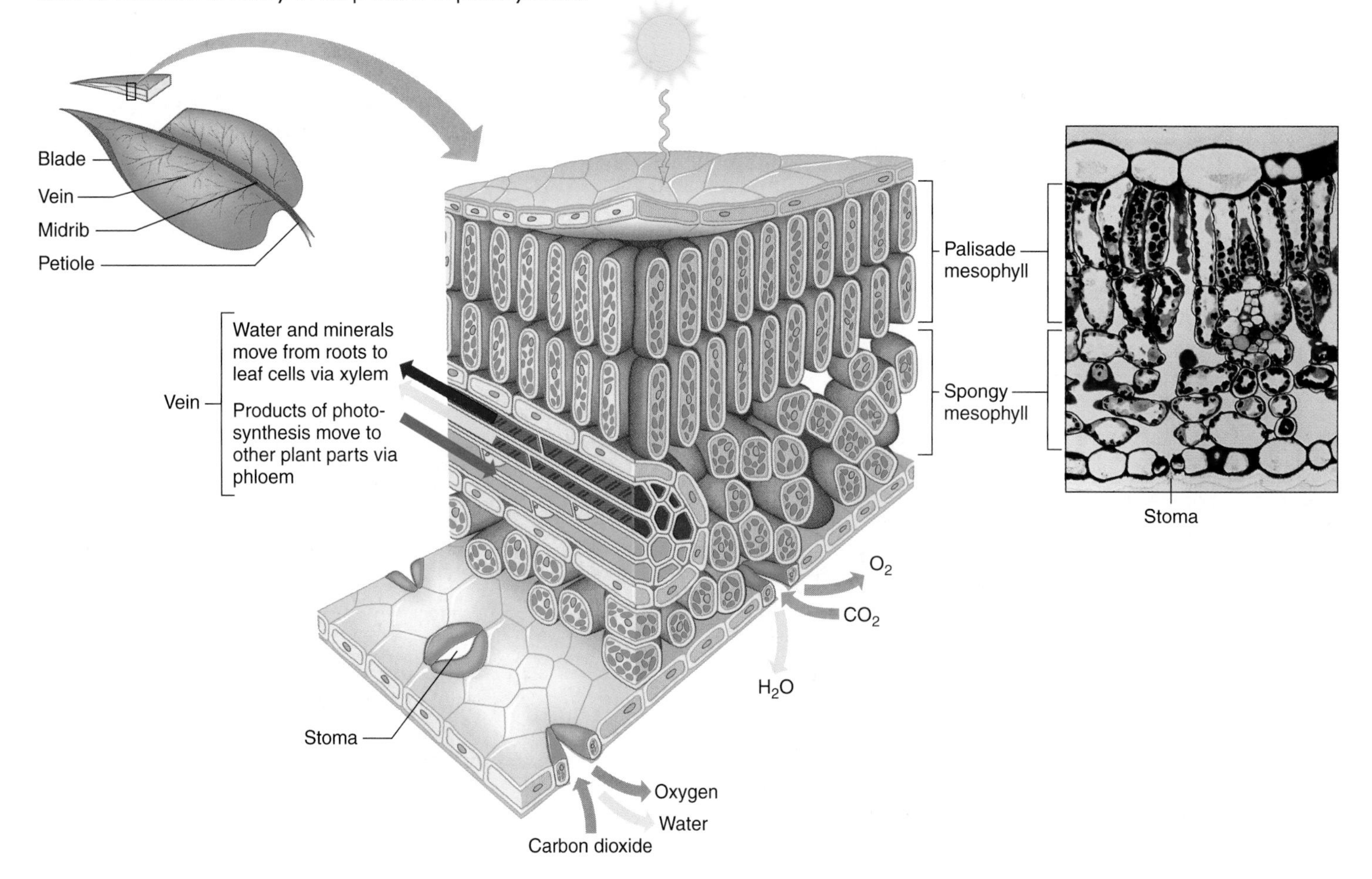

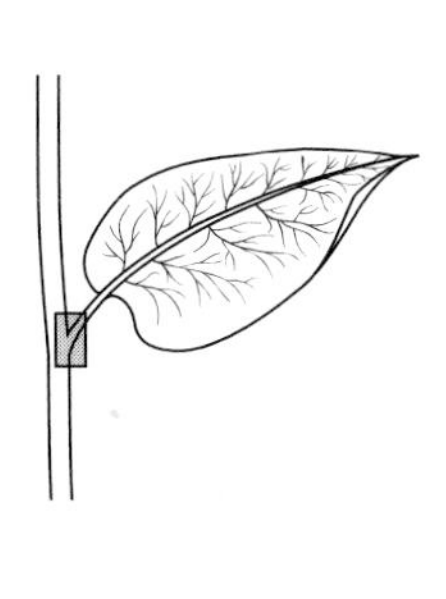

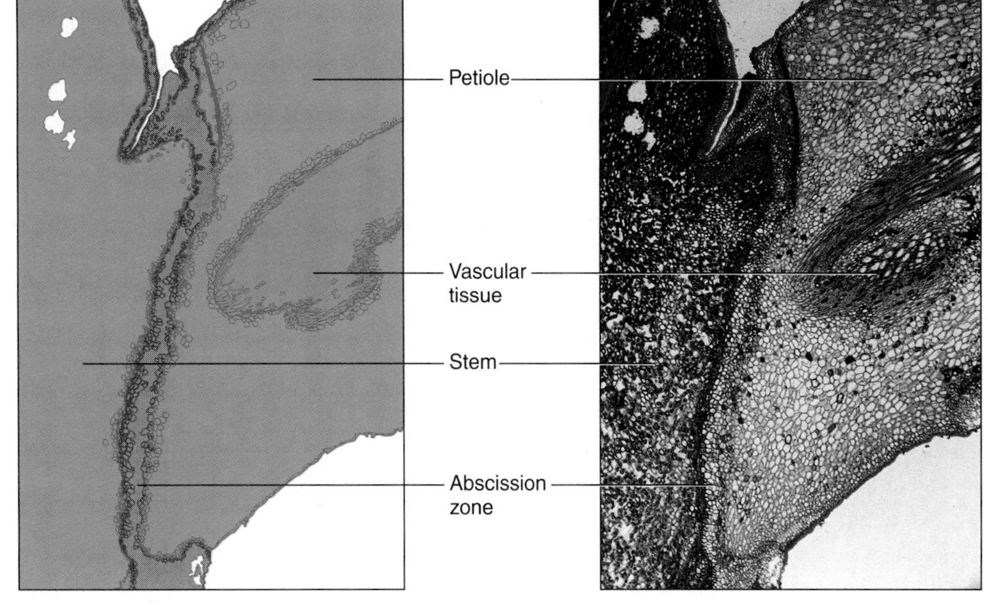

FIGURE 26.13
Leaf Abscission.
The abscission zone is a region of separation that forms near the petiole base of a leaf. This zone minimizes risk of infection and nutrient loss when the leaf is shed.

Roots

Plants are immobile, but they are biologically active, especially underground, where roots grow. Roots are so indispensable to plant growth and photosynthesis that annual root production often consumes more than half of a plant's energy and may account for a substantial portion of its bulk. Roots anchor plants and absorb, transport, and store water and nutrients. They absorb oxygen from between soil particles; roots pushing through firmly packed or water-logged soil may die from lack of oxygen because they cannot respire.

The first root to emerge from a seed is the radicle. The lifespan of the radicle differs in two types of root systems. In a **taproot system,** the radicle enlarges to form a major root that persists throughout the life of the plant (fig. 26.14A). Taproots grow fast and deep, maximizing support and enabling a plant to use minerals and water deep in the soil. Engineers once found a mesquite root growing 174 feet (53 meters) below the earth's surface! Most dicots develop taproot systems.

A **fibrous root system,** conversely, has a short-lived radicle that is replaced with adventitious roots, which form on stems (fig. 26.14B). Because they are relatively shallow, fibrous root systems rapidly absorb minerals and water near the soil surface and prevent soil erosion. Most monocots have fibrous root systems.

Like stems, roots grow at their tips as a result of cell division at the apical meristem, which is located just behind the root tip (fig. 26.15). Toward the tip, cells produced at the apical meristem differentiate into the **root cap,** whose cells slough off as the root grows through the soil. Root cap cells produce a slimy substance, mucigel, which lubricates the root as it grows and protects the tip from desiccation. Toward the base of the root, the apical meristem produces cells that elongate and differentiate into xylem, phloem, and other root tissues. Root hairs, which are extensions of root epidermal cells, often give this zone of cell maturation a fuzzy appearance (fig. 26.16). Figure 26.15D distinguishes the zones of cell division, elongation, and maturation of a root.

FIGURE 26.14
Two Main Patterns of Root Organization.
(**A**) The taproot system of a dandelion, a dicot. (**B**) The fibrous root system of barley, a monocot.

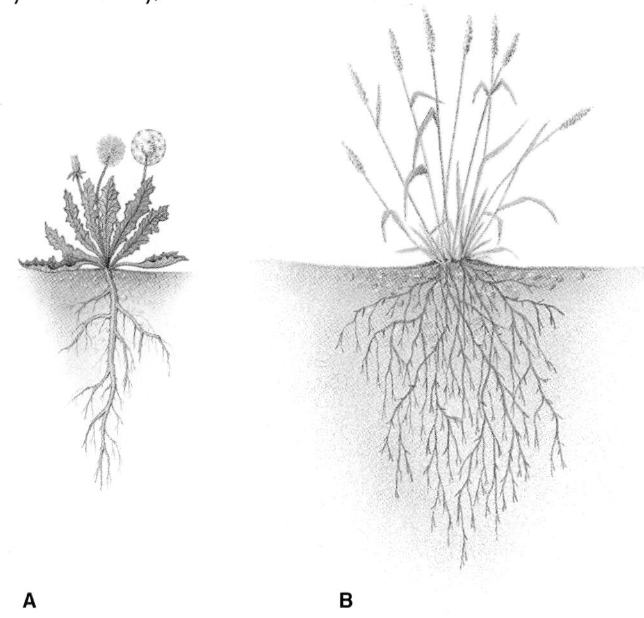

The root epidermis surrounds the entire primary root except the root cap. Root epidermal cells have a very thin cuticle or none at all, and are thus well adapted for absorbing water and minerals (a topic that is covered in more detail in chapter 27).

Interior to the epidermis is the cortex, which makes up the majority of the primary root's bulk. It consists of loosely packed, interconnected parenchyma cells that may store starch or other materials. The air spaces between the cells allow for aeration and water movement. The **endodermis** is the innermost ring of the

FIGURE 26.15

Anatomy of a Primary Root.

Cross sections of (**A**) a monocot root (corn) and (**B**) a dicot root (buttercup). A close-up of the dicot's vascular cylinder appears in (**C**), along with a diagram of the root's internal structures (**D**). The apical meristem (**E**) produces root cap cells toward the root tip and meristematic tissues toward the base. These meristems produce cells that mature into ground tissue, vascular tissue, and the root epidermis.

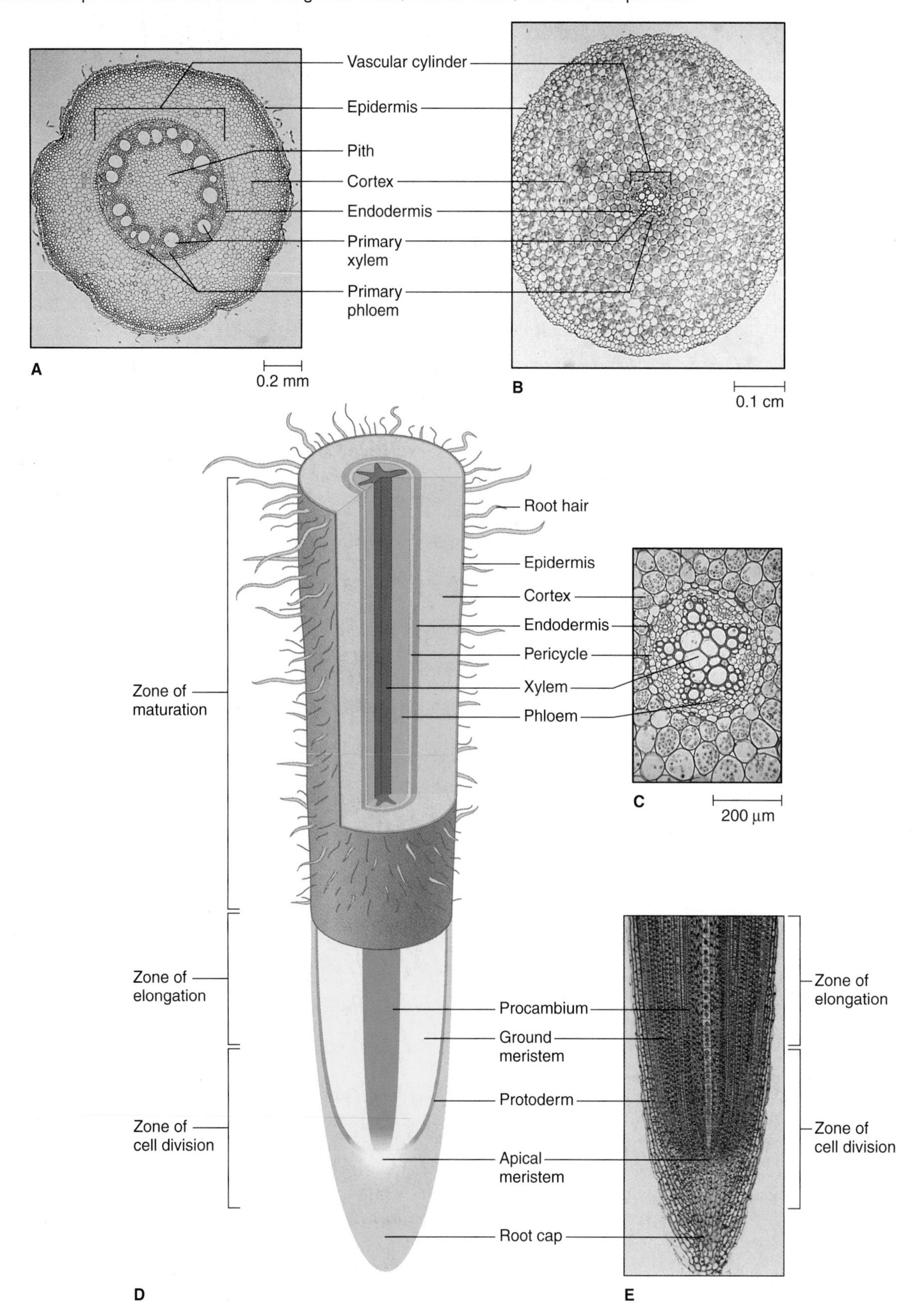

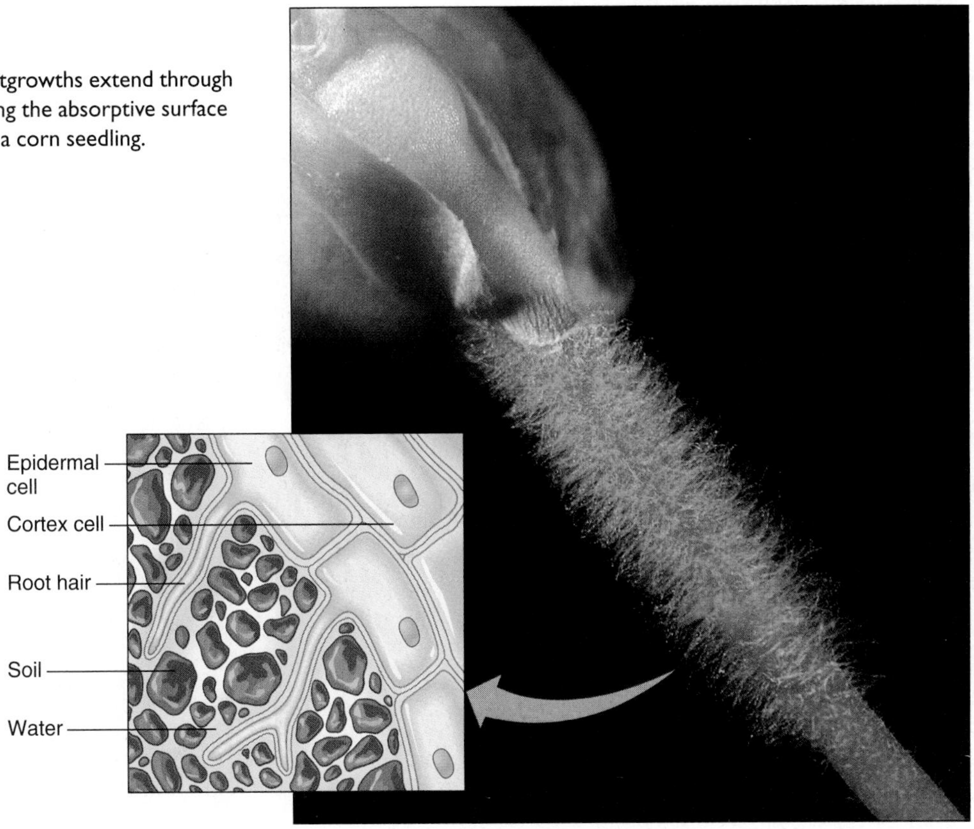

FIGURE 26.16
Root Hairs.
These epidermal cell outgrowths extend through the soil, greatly increasing the absorptive surface area of the root. This is a corn seedling.

cortex. It includes a single layer of tightly packed cells, whose walls contain a waxy, waterproof material called suberin. The waxy deposits form a barrier, called the **Casparian strip,** which ensures that all materials entering the vascular cylinder pass through the endodermal cells first (fig. 26.17).

Inside the endodermis is a layer of cells called the **pericycle,** which produces branch roots that grow through the cortex and epidermis and reach into the soil. The root's vascular cylinder is interior to the pericycle. In many roots, the vascular cylinder consists of a solid core of xylem, with ridges that project to the pericycle. Phloem strands are generally located between the "arms" of the xylem core. In other roots, a ring of vascular bundles surrounds a central core of parenchyma cells.

Like stems and leaves, roots are modified for special functions (fig. 26.18):

- Storage is a familiar root specialization. Beet, carrot, and sweet potato roots store carbohydrates, and desert plant roots may store water.
- Pneumatophores are specialized roots that form on plants growing in oxygen-poor environments, such as swamps. Black mangrove trees have pneumatophores. These roots form underground and grow up into the air, allowing oxygen to diffuse in.
- Aerial roots are adventitious roots that form and grow in the air. Mistletoe and orchids have aerial roots.
- Thick, enormous buttress roots at the base of a tree provide support, as do prop roots that arise from the stem, as seen in corn.

FIGURE 26.17
The Casparian Strip.
The Casparian strip blocks movement of water and dissolved minerals between endodermal cells, ensuring that all materials entering the vascular cylinder first pass through living cells.

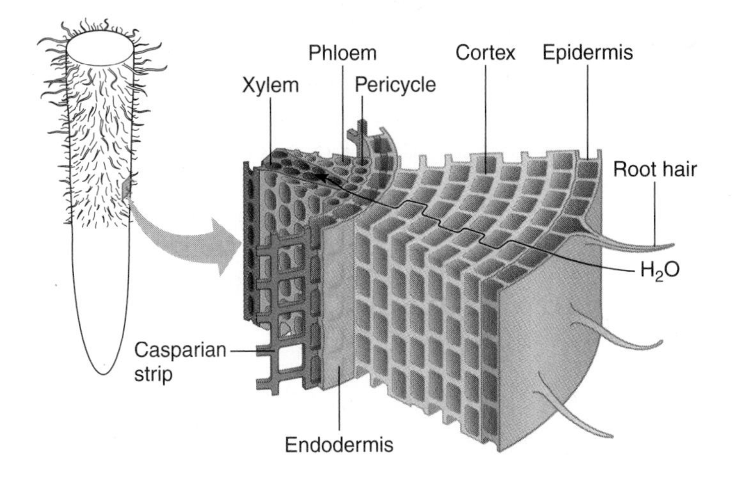

A plant's roots may interact with other organisms. Recall from chapter 23 that many roots form mycorrhizae with beneficial fungi. The fungi absorb water and minerals from soil, while the plants provide carbohydrates to the fungi. Roots of legume plants, such as peas, are often infected with bacteria of genus *Rhizobium.* The roots form nodules in response to the infection. The

FIGURE 26.18
Root Modifications.
(**A**) The banyan tree has aerial roots growing out of its branches. (**B**) This tropical fig tree has buttress roots so enormous they resemble a trunk. (**C**) Prop roots on corn arise from the stem and support the plant.

A

B

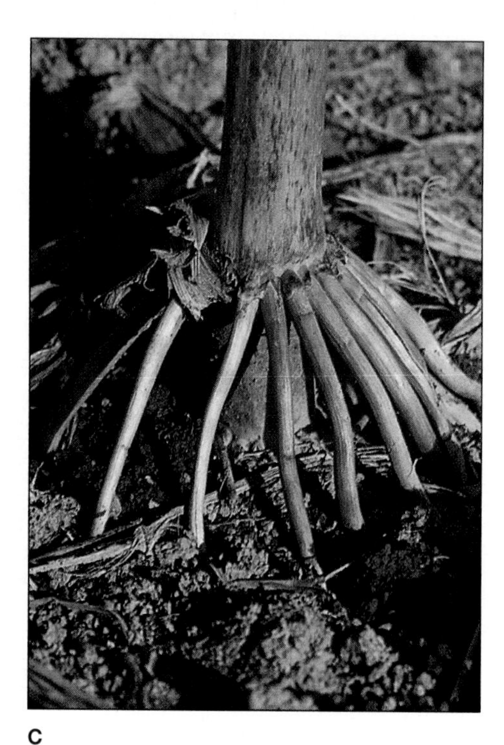
C

bacteria function as built-in fertilizer, providing the plant with nitrogen "fixed" into compounds it can use.

26.2 MASTERING CONCEPTS

1. What are the parts and tissues of a stem?
2. What are the functions of stems?
3. What are the structures and functions of leaves?
4. How do the two types of root systems differ?
5. What are the regions and structures of a root?
6. What are some special modifications of stems, leaves, and roots?

OLC

26.3 How Does a Plant Increase in Girth?

Wood and bark support the stems, branches, and roots of many plants. These supportive tissues arise from two lateral meristems, the vascular and cork cambia, that allow the plant to grow outward.

The tallest plants can intercept the most light. However, continued elongation poses a problem, because primary tissues cannot adequately support tall plants. Lateral meristems, which increase the girth of stems and roots by secondary growth, address this problem. These meristems are called the vascular cambium and cork cambium.

Secondary growth can be impressive. A 2,000-year-old tule tree in Oaxaca, Mexico, is 148 feet (45 meters) in circumference and only 131 feet (40 meters) tall. A 328-foot-tall (100-meter) giant sequoia in northern California is more than 23 feet (7 meters) in diameter. To support all of this extra tissue, a plant's transport systems must also become more complex and powerful.

Vascular Cambium

The **vascular cambium** is a ring of meristematic tissue that produces most of the diameter of a root or stem. Generally, it forms only in plants that exhibit secondary growth—primarily woody dicots and gymnosperms.

In roots and stems that undergo secondary growth, the vascular cambium forms a thin layer between the primary xylem and phloem. In stems with discrete vascular bundles, the vascular cambium extends between the bundles to form a ring (fig. 26.19). The cells of the vascular cambium produce secondary xylem toward the inside of the cambium, and secondary phloem on the outer side. Overall, the vascular cambium produces much more secondary xylem than secondary phloem.

Secondary xylem is more commonly known as wood. The vascular cambium divides to produce wood during the spring and summer. During the moist days of spring, wood is made of large cells and is specialized for conduction of water. During the drier days of summer, the vascular cambium produces summer wood that has small cells and is specialized for support. These seasonal differences in wood cell sizes generate visible demarcations called growth rings (fig. 26.20). Secondary xylem can be

FIGURE 26.19

Secondary Growth Produces Wood.

The secondary growth of a woody stem involves the activities of two lateral meristems—vascular cambium and cork cambium. The microscopic vascular cambium produces secondary xylem to the inside and secondary phloem to the outside.

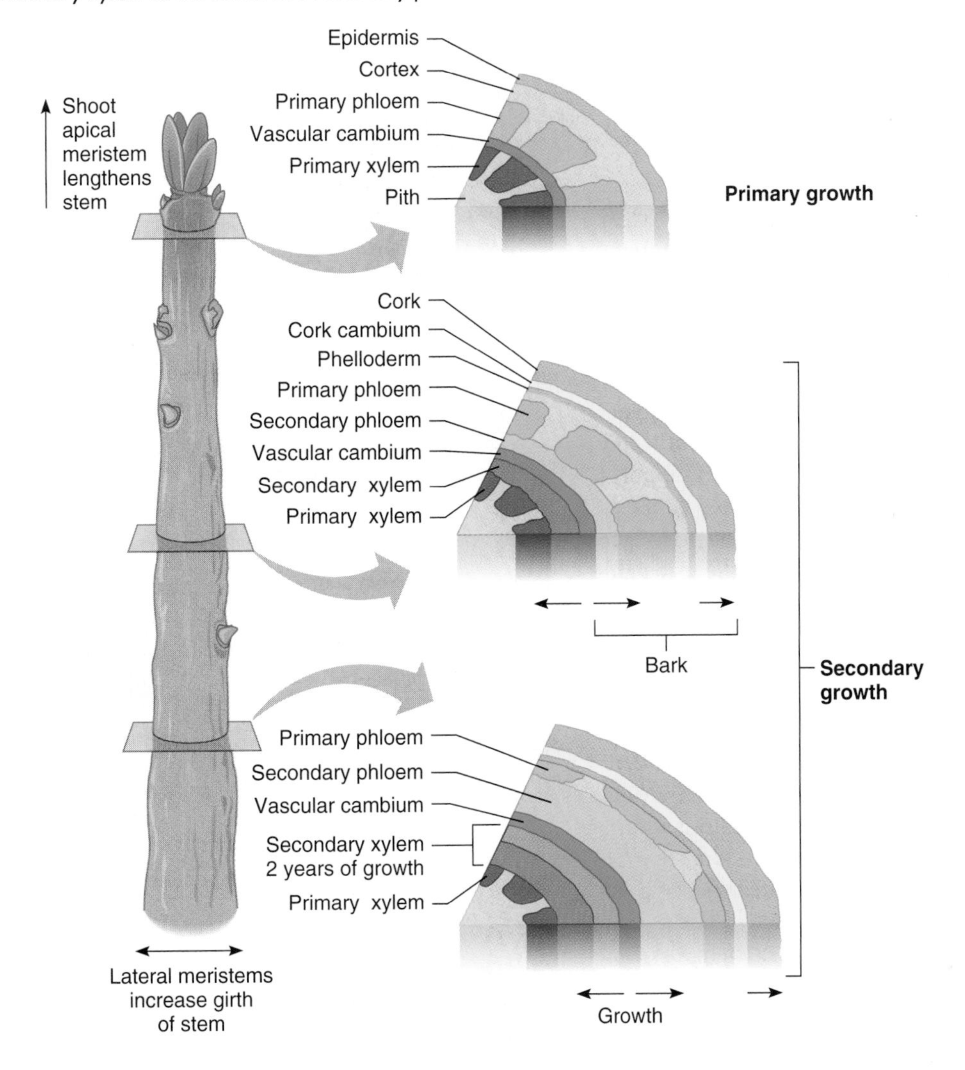

FIGURE 26.20

Anatomy of a Secondary Stem.

(**A**) Wood is secondary xylem, and bark is all the tissue outside the vascular cambium. (**B**) Differences in available soil moisture when wood forms in spring and summer result in different-sized cells (**C**), which are visible as growth rings.

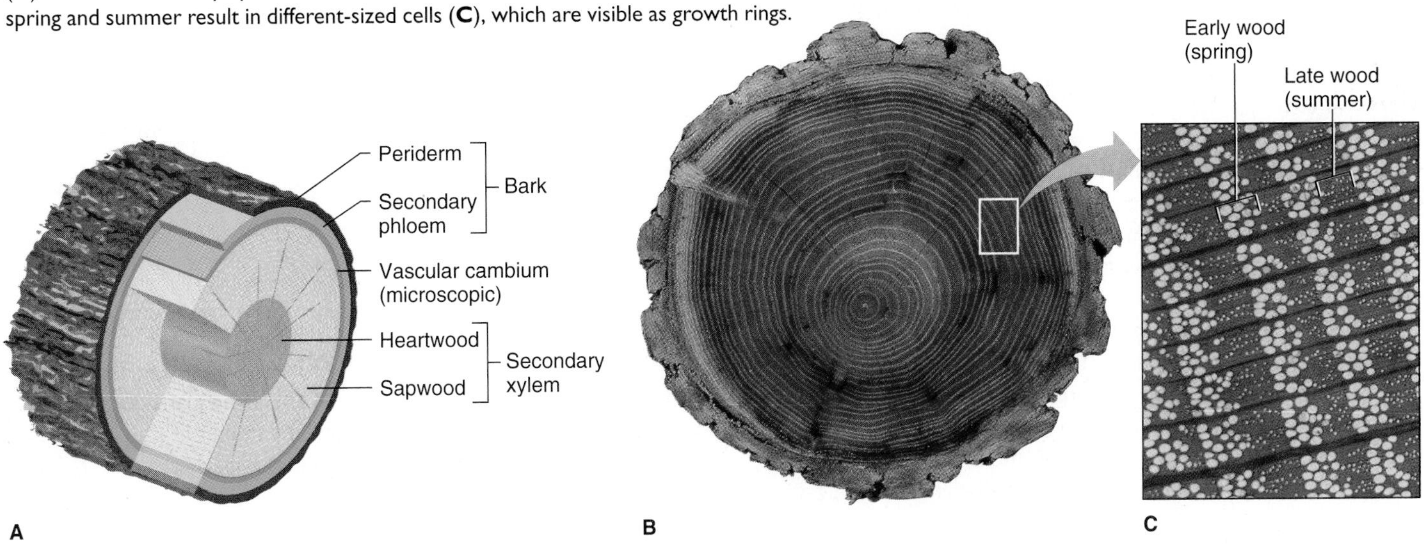

Periderm
Secondary phloem
Bark
Vascular cambium (microscopic)
Heartwood
Sapwood
Secondary xylem
Early wood (spring)
Late wood (summer)

A B C

used to measure the passage of time because the larger spring wood cells appear light colored, and summer wood cells are smaller and darker colored. The contrast between the summer wood of one year and the spring wood of the next creates the characteristic annual tree ring. The most recently formed ring is next to the microscopic vascular cambium. Tropical species, which have secondary growth all year long, do not have regular tree rings. Investigating Life 26.1 describes how tree ring data are aligned to correlate to historical events.

The older, innermost increments of secondary xylem gradually become unable to conduct water; this nonfunctioning wood is called **heartwood.** The **sapwood,** located nearest the vascular cambium, transports water and dissolved minerals.

Woods differ in hardness. Dicots such as oak, maple, and ash are often called hardwood trees, and gymnosperms such as pine, spruce, and fir are called softwood trees. Dicot wood contains tracheids, vessels, and supportive fibers, whereas wood from softwood trees contains 90% tracheids. As a result, dicot wood is usually stronger and denser than wood from gymnosperms (the soft, light wood from the balsa tree, a dicot, is a notable exception). The terms "hardwood" and "softwood" are not scientific terms. They are used mainly in industry to reflect the observation that dicot wood is usually harder than wood from gymnosperms.

The tissues outside the vascular cambium are collectively called **bark.** Secondary phloem forms the live, innermost layer of bark that transports phloem sap within the tree (the primary phloem is crushed as the stem or root continues to grow outward). The outer layer of bark is called the **periderm,** a protective layer of tissue that replaces the epidermis as the girth of a stem or root increases.

Cork Cambium

The **cork cambium** is the second type of lateral meristem, and it gives rise to cork to the outside and phelloderm to the inside (see fig. 26.19). Together, the phelloderm, cork cambium, and cork make up the periderm. Cork cells are waxy, densely packed cells covering the surfaces of mature stems and roots. They are dead at maturity and form waterproof, insulating layers that protect plants. The cork used to stopper wine bottles comes from a cork oak tree that grows in the Mediterranean. Every 10 years, harvesters remove much of the cork cambium and cork, which grows back. In contrast to cork, phelloderm cells are live parenchyma cells.

This chapter has described many structural features of flowering plants, and noted some of the differences between the two main groups of angiosperms, monocots and dicots. Figure 26.21 illustrates some of these differences (including some mentioned earlier, in chapter 22). In the chapters that follow, the functions, growth, and development of these structures will be described in greater detail.

FIGURE 26.21
Monocots and Dicots Compared.
Although many variations occur, most monocots and dicots differ from each other in flower and leaf morphology, stem anatomy, root system architecture, and seed structure.

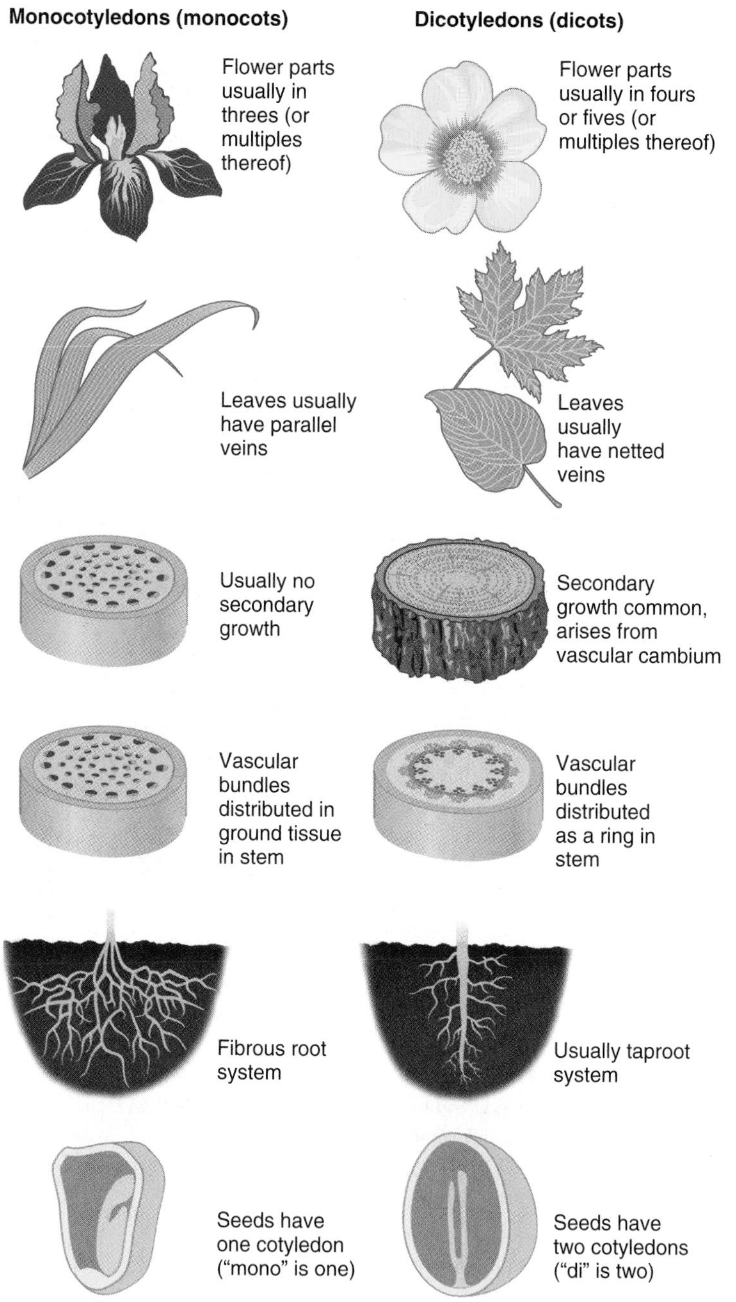

26.3 MASTERING CONCEPTS

1. How does secondary growth affect the plant body?
2. What is the function of vascular cambium?
3. What are the components of bark?

OLC

Investigating Life 26.1

Tree Rings Reveal the Past

Studying tree ring patterns—a technique called dendrochronology—is helping to fill gaps in our knowledge of ancient civilizations by providing information on past climatic events. Rings arise in trees that grow in temperate climates, where wood cells that form in the spring are larger than those that form in summer, creating a distinctive pattern in the wood grain.

Tree rings also provide information on climate. The larger the ring, the more plentiful the rainfall that year. A fire leaves behind a charred "burn scar." Two rings fall very close together when voracious caterpillars or locusts eat the season's early leaves. With fewer leaves, the tree cannot obtain sufficient nutrition from photosynthesis, so the plant does not produce much secondary xylem. The tree rings appear narrower than in a better year. Tree ring structures and patterns also provide clues about light availability, altitude, temperature, and length of the growing season in times past.

Researchers at Cornell University combined dendrochronology with other types of evidence to "anchor" a nearly 1,500-year sequence in the history of a little-understood society—that of ancient Turkey—into an absolute time frame. They used wood and charcoal samples from the Midas burial mound that were once juniper trees. The evidence may actually rewrite some ancient history.

When researchers can align overlaps in the patterns among several trees, tree ring data can span many years. As long as one of the trees has a ring next to the vascular cambium, so that the dendrochronologist can determine the most recent year, such comparisons can yield a very accurate "master chronology." For example, the oldest known living tree is a bristlecone pine (*Pinus longaeva*) growing in the White Mountains of California. It is more than 4,760 years old. Combining its tree ring information with information from older, dead trees in the area has provided rainfall data going back 8,200 years!

The juniper remains from the Midas Mound in ancient Turkey, however, had no link to the present, and so the Cornell researchers had a "floating" dendrochronology, a sequence of events with no anchoring end. But the researchers used two clever approaches to assign absolute dates.

First, in a technique called wiggle-matching, the researchers correlated irregularities in the tree ring data to irregularities in radioactive carbon dating and established an approximate date. Then they aligned an unusual wideness in the tree rings to similar anomalies in patterns from trees in the United States and Europe, which had been linked to a volcanic eruption on the island of Thera, near Crete, in 1628 B.C. With the new tree ring data, anthropologists now have a time frame in which to place objects found with the wooden remains in ancient Turkey. Tree rings may have much to tell us about early history.

FIGURE 26.A

Researchers Infer the Age of Wooden Structures by Studying Tree Ring Patterns.
To establish a chronology, dendrochronologists overlap patterns. In the study depicted here, a tree cut down in 1950 establishes a starting date (**A**). This tree's rings are compared to those of a tree cut down and used as a wooden beam in a newer house (**B**) and then to those of a tree used as a beam in an older house (**C**). The tree ring data pictured begin shortly after 1840 and extend until 1950. When a starting date can be established, scientists can use historical tree ring data to infer the age of centuries-old wood (**D**).

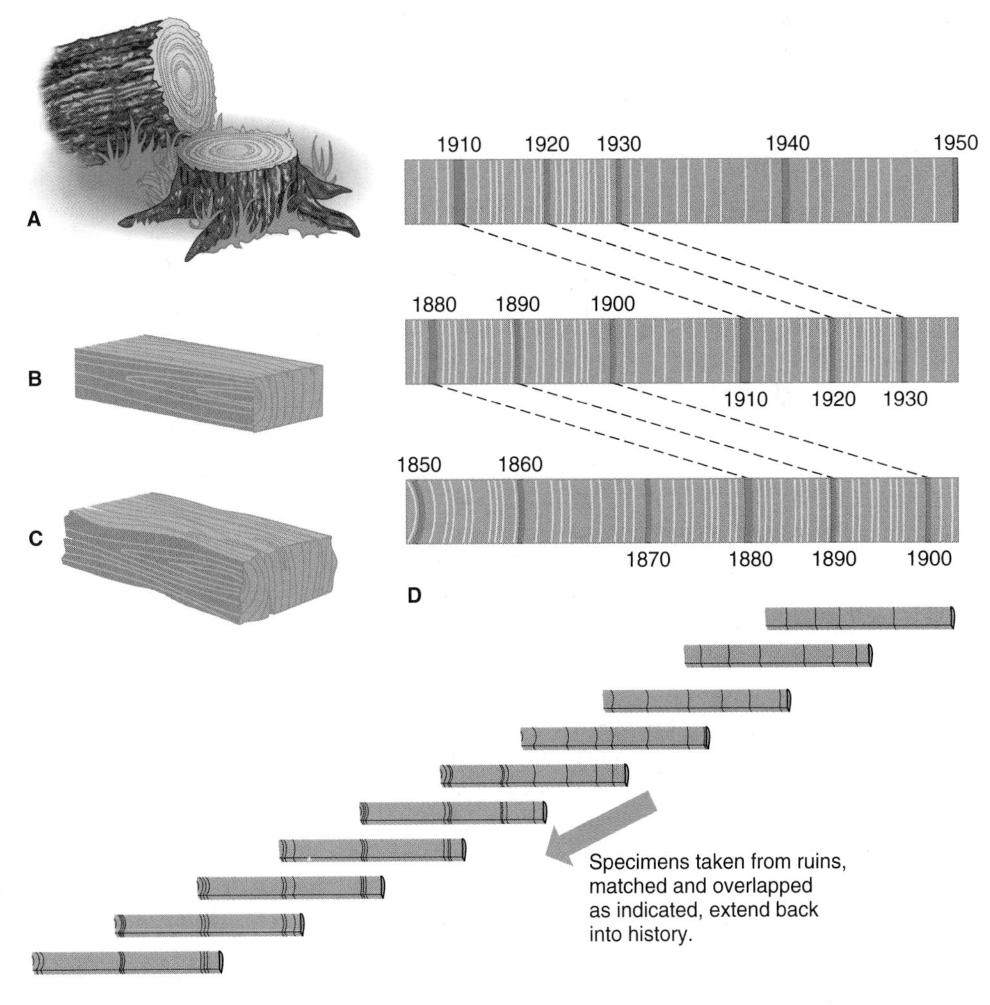

Chapter Summary

26.1 Plant Tissues

1. The tissues of a flowering plant are meristems, ground tissue, dermal tissue, and vascular tissue, which includes phloem and xylem. The plant body consists of a **shoot** and a **root.**
2. **Meristems** are localized collections of cells that divide throughout the life of the plant. **Apical meristems** located at the plant's tips provide **primary growth,** and **lateral meristems** add girth, or **secondary growth.**
3. Most of the plant body is **ground tissue.** It includes **parenchyma** cells, which can divide and store substances. A chlorenchyma cell is a type of parenchyma cell that photosynthesizes. **Collenchyma** supports growing shoots and **sclerenchyma** supports plant parts that are no longer growing. Sclerenchyma includes sclereids, which form hard coverings, and fibers, which form strands.
4. **Dermal tissue** includes the **epidermis,** a single cell layer covering the plant. The epidermis secretes a waxy **cuticle** in aerial plant parts. Gas and water exchange occur through epidermal pores (**stomata**), which are bounded by **guard cells.** Trichomes are epidermal outgrowths, and include **root hairs.**
5. **Vascular tissue** is specialized conducting tissue. **Xylem** transports water and dissolved minerals from roots upwards. Xylem cells are elongated with thick walls and are dead. They include the long, narrow, less-specialized **tracheids** and the more specialized, barrel-shaped **vessel elements.**
6. **Phloem** transports dissolved carbohydrates and other substances throughout a plant and includes **sieve cells** and, in flowering plants, the more specialized **sieve tube members.** Pores of sieve tube member cells cluster at **sieve plates,** allowing more efficient nutrient transport. **Companion cells** help transfer carbohydrates.

26.2 Anatomy of a Plant

7. A **stem** is the central axis of the shoot and consists of **nodes,** where leaves attach, and **internodes** between leaves, where the stem elongates. **Vascular bundles** in the stem contain xylem and phloem, which are scattered in monocots and form a ring in dicots. Between a stem's epidermis and vascular tissue lies the **cortex,** made of ground tissue. **Pith** is ground tissue in the center of a stem. Stem modifications include stolons, thorns, succulent stems, tendrils, and tubers.
8. Simple **leaves** have undivided **blades,** and compound leaves form leaflets, which may be pinnate (with a central axis) or palmate (extend from a common point). Leaves are the sites of photosynthesis.
9. Leaf epidermis is tightly packed, transparent, and nonphotosynthetic. Veins may be in either netted or parallel formation. Leaf ground tissue includes palisade and spongy **mesophyll** cells.
10. Leaf modifications include tendrils, spines, bracts, storage leaves, insect-trapping leaves, and cotyledons.
11. Leaves are shed from an abscission zone in response to environmental cues.
12. Roots absorb water and dissolved minerals. A plant's first root is the radicle. **Taproot systems** have a large, persistent major root, whereas **fibrous root systems** are shallow, branched, and shorter-lived. A **root cap** protects the tip as an apical meristem replaces cells lost during rapid extension. The apical meristem also produces cells that differentiate into the root's epidermis, cortex, and vascular tissues.
13. The root cortex consists of storage parenchyma and **endodermis.** Some roots are specialized for storage or adapted to low-oxygen environments.

26.3 How Does a Plant Increase in Girth?

14. Secondary tissues increase a plant's girth.
15. Two lateral meristems, the **vascular cambium** and **cork cambium,** produce outward growth. The vascular cambium produces secondary xylem and secondary phloem. The cork cambium produces cork and other tissues that, along with secondary phloem, comprise **bark.**

Testing Your Knowledge

1. Which tissue provides a plant's continuous growth?
2. Which plant tissue is the most abundant?
3. What are the functions of the following substances?
   a. cutin
   b. mucigel
   c. suberin
4. How are the two types of xylem cells similar to the two types of phloem cells?
5. List two ways that a plant's anatomy is adapted to existence on land.
6. Describe how leaves and roots maximize surface area.
7. What are some of the ways in which we classify leaves?
8. Cite a stem specialization and a leaf specialization that provide protection.
9. Corn is a monocot and cucumber a dicot. How do these plants differ in the following?
   a. stem structure
   b. leaf venation
   c. root organization
10. Moving from the outermost bark layer to the center of a tree's trunk, which tissues are encountered, and what type of meristem produced each?

Thinking Scientifically

1. If an overanxious tomato picker tears off some tomato plant stems and leaves, the plant regrows these parts in a few weeks. Which tissue type is responsible for this regrowth?
2. What plant structures are adaptations to conserve and transport water?
3. Why can phloem transport materials in all directions, while xylem can transport materials upward only?
4. How does a tree "know" when to shed its leaves?
5. Paper "fibers" actually include fibers, tracheids, and vessel elements. Describe each of these.

References and Resources

Renfrew, Colin. June 27, 1996. Kings, tree rings and the Old World. *Nature,* vol. 381, pp. 733–734. Tree ring dating supplements historical information.

Roach, Mary. June 1996. Bamboo solution. *Discover,* p. 92. Bamboo has many uses, but it is not strong enough to support a building greater than one story tall.

Shanley, Patricia. October 1999. To market. *Natural History,* vol. 108, p. 44. People in Brazil use and sell many parts of many types of plants.

Smith, William K., et. al. December 1997. Leaf form and photosynthesis. *BioScience,* vol. 47, p. 785. Leaf structure and orientation reflect adaptations to maximize production of photosynthate.

The *LIFE* Online Learning Center provides additional resources and tools for studying this chapter.
www.mhhe.com/life

CHAPTER 27

Transport Systems in Plants

Fog Drip

The coastal redwood *Sequoia sempervirens* lives in a narrow band of forest along the coast of southwest Oregon and northern and central California. It is at the center of an ecosystem that was once much larger, warmer, and wetter, until encroaching glaciers invaded its boundaries during the Cretaceous period. Before then, other redwoods shared the land, including the giant sequoia (*Sequoiadendron*) now found only in California, and the dawn redwood (*Metasequoia*) that is now native only to China. The coastal redwood population is now only 4% of what it was then.

The majestic tree is both large and long-lived. The tallest coastal redwood living today reaches to 370 feet (101 meters), with a diameter of 15 feet (4.6 meters). The tree is mature at about 400 to 500 years, and the oldest has survived for 2,200 years. It resists infection by insects and fungi, and survives fires.

Fog consists of suspended particles of liquid water. . . . Ferns could not survive without the redwoods' fog drip.

The coastal redwood requires vast amounts of moisture, some thousand tons for every ton of its weight. Much of this water comes from an unusual source—fog. Thanks to these trees, the ecosystem captures up to 50 inches (127 centimeters) of water annually that it would otherwise not receive.

Fog consists of suspended particles of liquid water. It forms as moisture-rich air from the north falls over the west coast, and some of the water condenses. As the day heats up inland, rising warm air pulls in cooler air from over the ocean. The fog contacts the many needle-covered branches and limbs of the redwoods, dripping to the forest floor, unimpeded because the lower third of the tree lacks branches. The air in the area cools and humidity rises as the tree's

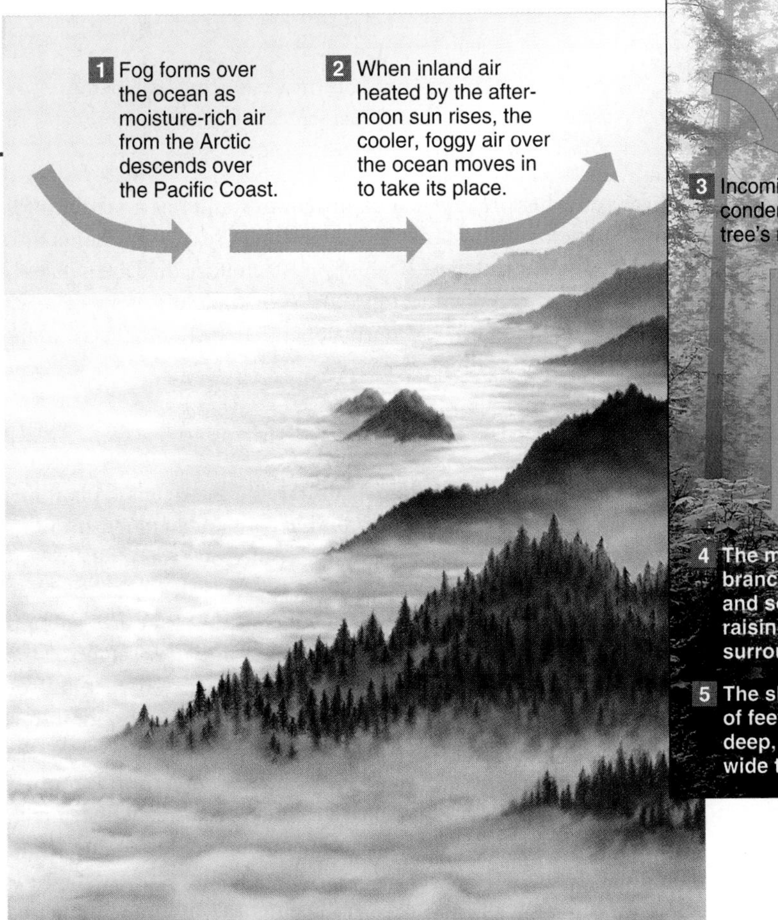

Fog Water Recycling.
Water carried in fog condenses on the surface of a giant redwood tree and drips to the forest floor. The tree's shallow roots absorb some of the water, which moves up the massive trunk and returns to the atmosphere through the tree's needles.

enormous surface area captures the moisture in the fog and passively delivers it—thanks to gravity—to the soil below.

Ferns could not survive without the redwoods' fog drip. At the same time, the trees' massive root systems, with several main branches so close to the surface that some of them actually break through, suck up the water, which is pulled upward through xylem tissue to the leaves and released as water vapor. The root system is not deep but extends well beyond the area covered by each tree. The tree absorbs about 35% of the fog drip. It is the tree's ability to tap so much of the moisture in fog that may explain its great height. By keeping the immediate environment near-saturated, less water leaves the plant, and there is less pressure to pull up more water rapidly from the roots.

Coastal redwoods, as suppliers of water, are critical components of these forest ecosystems. But they may become endangered, because their strong, pliable, insect-free and knot-free wood commands a very high price. It isn't surprising that areas cleared of these trees often experience drought.

27.1 Complex Multicellular Organisms Require Complex Transport Systems

The first forms of life on Earth were unicellular, simple enough that all cells could maintain contact with the environment. As life evolved into more complex forms, adaptations accumulated that maximized exchange with the environment. Among the plants, these adaptations included development of transport systems.

Organisms engage in constant exchange with the environment, acquiring nutrients that provide energy and chemical building blocks, and releasing the wastes that accumulate as part of being alive. The single-celled organisms that were the first forms of life were small enough to handle this back and forth with their surroundings. The first multicellular forms, too, were probably sufficiently simple that each cell had easy access to the environment.

The earliest plants were probably descendants of green algae, perhaps resembling liverworts, as chapter 22 discusses. These organisms were aquatic, likely bathed in chemicals released from dead and decaying organisms, and diffusion was sufficient to transport materials into and out of each cell. When plants conquered land, none at first grew very tall, forming carpetlike growths that could soak up water and dissolved nutrients from soil, dust, or rain. Over time, leaves greatly increased the photosynthetic capabilities of plants, and root systems grew more complex, expanding the range of nutrient and water sources available to plants. Long taproots plunged toward the water table, while other roots extended laterally, soaking up the nutrients released into the soil from decaying organisms.

As organisms in general grew larger and more complex, bodies folded and contorted and developed extensions that could increase the surface area, which otherwise could not keep pace with increasing volume. Today, the members of the three multicellular kingdoms of life exhibit their particular solutions to this challenge of maximizing the surface area necessary to maintain contact with the environment. Fungi grow countless long, slender strands of hyphae that permeate the soil and absorb its bounty of nutrients. Animals and plants have internal networks of tubules and tubes that deliver water and nutrients. In many animals, circulatory systems built of a vast expanse of blood vessels and a pumping heart distribute nutrients, and other tubules comprising excretory systems process and rid the body of wastes.

In vascular plants, xylem transports water and dissolved minerals, and phloem distributes photosynthate, the products of photosynthesis (fig. 27.1). In roots, both types of vascular tissue are bundled together within a layer of parenchyma; together, these tissues are called the stele. In many plants, the xylem is at the center of the stele, and phloem is in strands toward the outside. In stems, the stele diverges to form discrete vascular bundles and associated ground tissue such as pith (see chapter 26).

27.1 MASTERING CONCEPTS

1. Why did the first organisms on Earth not require intercellular transport systems?
2. How did more complex bodies change in ways that maximize contact with the environment?
3. Which tissues comprise plant transport systems?

27.2 Plants Require Water and Nutrients

All plants require at least 16 essential minerals, obtained from the atmosphere or soil. Bacteria convert atmospheric nitrogen to forms that plants can take up and use.

Like any other organism, a plant requires water as a medium for and participant in many of its metabolic reactions. Plants also need essential **nutrients,** which are elements that are vital for metabolism, growth, and reproduction. A plant gets most of its water from the soil. The waterproof cuticle and stomata are adaptations for conserving water (see fig. 26.4).

Essential Nutrients for Plants

In the mid 1800s, botanists discovered the essential plant nutrients by growing plants in water that contained known amounts of particular elements, a technique called hydroponics. In this way, they identified 16 elements vital to all plants (fig. 27.2). Nine of these are macronutrients, which means that they are required in fairly large amounts—at least 0.1% of the dry weight of the plant. The macronutrients are carbon, hydrogen, oxygen, nitrogen, potassium, calcium, magnesium, phosphorus, and sulfur. The micronutrients, required in much smaller amounts, are chlorine, iron, boron, manganese, zinc, copper, and molybdenum. Figure 27.2 lists some of their functions. In general, macronutrients are required for production of structural and storage carbohydrates such as cellulose and starch, enzyme synthesis and control, altering osmotic potential, and as participants in metabolic reactions. Just as is the case for all organisms, nutrient deficiencies can harm plants, as figure 27.3 shows.

Certain plants require elements other than the 16 that all plants use. Horsetails and certain grasses, for example, accumulate silicon in their cells, and soybeans use nickel. Some other plants accumulate lead, cadmium, copper, zinc, or cobalt. Most interesting is the locoweed, *Astragalus*—more than 1% of its dry weight is selenium. Cattle that eat the selenium-laden plants stagger about and seem intoxicated, and farmers usually have to destroy them. Locoweed accumulates selenium because this element reduces the toxic effect of phosphorus, which is often present in high levels in selenium-rich soil.

Accumulating minerals that may be toxic to other organisms could be an adaptation to avoid predation, but humans have also taken advantage of that capability. A biotechnology called phytoremediation uses plants with toxic tastes to naturally remove

FIGURE 27.1

Plant Transport Systems.

In vascular plants, xylem transports water and dissolved minerals absorbed from soil, and phloem distributes the products of photosynthesis to fruits, roots, and other plant parts.

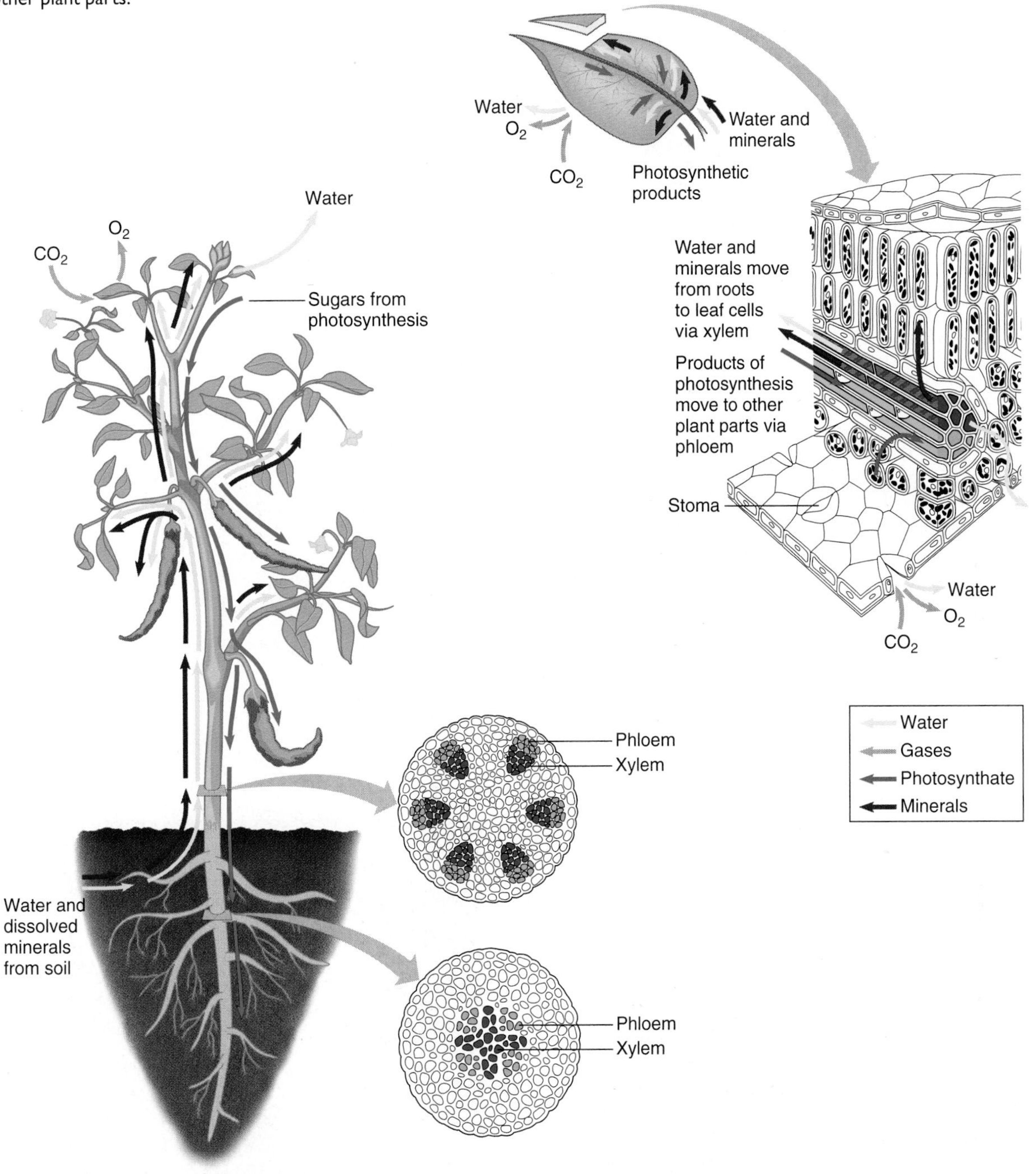

heavy metals and other dangerous chemicals from the environment. Transgenic technology (see Biotechnology 13.3) is being used to boost plants' detoxifying capabilities, while taking advantage of their role in the environment. For example, researchers have engineered tobacco plants with a bacterial gene to produce the only known enzyme that completely detoxifies trinitrotoluene, an organic compound more familiarly known as TNT, or dynamite. In the United States, the Environmental Protection Agency has identified 22 sites where weapons were manufactured, tested, or dumped, leaving TNT in the topsoil. Although the bacterial enzyme itself is powerful, it is easier to cover such areas with transgenic plants than it is to permeate the soil with bacteria. (Tobacco is used in botanical research as an experimental model organism, not to manufacture cigarettes.)

FIGURE 27.2

Essential Nutrients for Plants.

The nine most abundant elements in plants are called macronutrients. The micronutrients occur in much lower concentrations but are also essential for plant survival. The table lists the micronutrients present in all plants; there are others too.

Macronutrients	Percent dry weight	Selected functions
Carbon	45	Part of organic compounds
Oxygen	45	Part of organic compounds
Hydrogen	6	Part of organic compounds
Nitrogen	1.5	Part of nucleic acids, amino acids, coenzymes, chlorophyll, ATP
Potassium	1.0	Controls opening and closing of stomata, activates enzymes
Calcium	0.5	Cell wall component, activates enzymes, second messenger in signal transduction, maintains membranes
Magnesium	0.2	Part of chlorophyll, activates enzymes, participates in protein synthesis
Phosphorus	0.2	Part of nucleic acids, sugar phosphates, ATP, coenzymes, phospholipids
Sulfur	0.1	Part of cysteine and methionine (amino acids)
Micronutrients	**Percent dry weight**	**Selected functions**
Chlorine	0.01	Water balance
Iron	0.01	Chlorophyll synthesis, cofactor for enzymes, part of electron carriers
Boron	0.002	Growth of pollen tubes, sugar transport, regulates certain enzymes
Zinc	0.002	Hormone synthesis, activates enzymes, stabilizes ribosomes
Manganese	0.005	Activates enzymes, electron transfer, photosynthesis
Copper	0.0006	Part of plastid pigments, lignin synthesis, activates enzymes
Molybdenum	0.00001	Nitrogen fixation

FIGURE 27.3

Nutrient Deficiencies Produce Distinctive Symptoms.

(**A**) Phosphorus deficiency causes dark green or purple leaves in seedlings. (**B**) Iron deficiency causes chlorotic (yellowed) leaves, but the veins remain green. A deficiency of manganese produces similar symptoms.

A

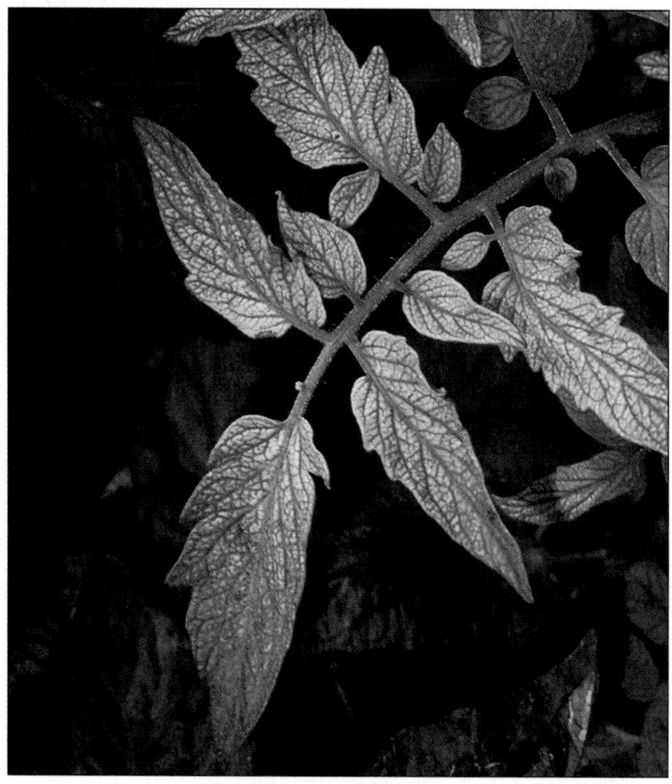

B

Plants also assimilate elements that are present in their particular environments, but that may or may not help or harm the plant. For example, plants growing near gold mines have gold in their tissues (which prospectors use as clues to promising deposits), plants drenched in cadmium-containing fertilizers contain that metal, and plants growing along superhighways take up lead from gasoline exhaust. Near nuclear test sites, plants absorb radioactive strontium—as do people who drink the milk from cows that graze on grasses exposed to nuclear fallout.

Nutrients Come from the Atmosphere and Soil

Plants obtain carbon, oxygen, and hydrogen atoms from carbon dioxide gas (CO_2) and water (H_2O). CO_2 comprises 0.035% of the atmosphere. This gas enters the leaf through open stomata, dissolves in the water film surrounding the cells, and diffuses into the cytoplasm. As CO_2 is used in photosynthesis, its concentration drops in the cell relative to its surroundings, and more diffuses in from air spaces between the chlorenchyma cells. The gas continues to diffuse into the cells as long as its concentration in the air spaces exceeds its concentration in the cytoplasm. For CO_2 uptake to continue, the air in the intercellular spaces must be continuous with the air outside the leaf. That is, the stomata must be open.

The other nutrients come from soil, which consists of small particles of rocks and minerals mixed with decaying organisms and organic molecules. It is full of living organisms as well, which may compete with plant roots for nutrients. The characteristics of soil change with depth (fig. 27.4), and the composition and texture vary from place to place.

FIGURE 27.4
Soil.
Soil consists of layers called horizons, designated A through C. The top layer is litter, which covers the rich topsoil. Water passing through the topsoil deposits clay and minerals in the next layer, and below this is a layer of partially weathered rock.

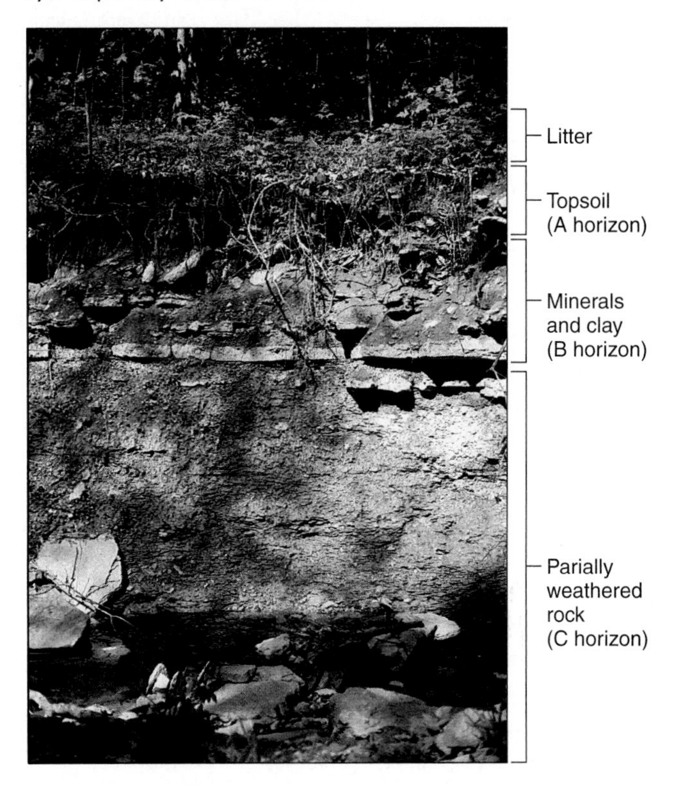

The mineral and rock particles in soil are the products of weathering of rocks on the earth's surface. The texture of a soil depends upon the size of these particles. Sandy soils are rich in coarse particles, whereas silty soils have finer particles. Clays are the soils with the finest particles. Many soils are a mixture of sand, silt, and clay, and soil scientists consider the relative amounts of each to classify the soil. The size of the particles in a soil also determines its water-holding capacity. The finer the soil particles, the more surface area per unit of soil volume is available to hold water.

Some of the minerals that plants extract from the soil come from rocks weathered into soil particles. Others are released during the decomposition of dead organisms.

Most of the decaying material in soil comes from plants and forms a layer of pieces, called litter, on the soil surface. Complex organic chemicals that are hard to digest along with very small fragments of leaves and stems form a material called **humus** in the upper layer of soil, the **topsoil.** This layer is a major source of water and nutrients for most plants growing in the soil. The topsoil layer is also called the A horizon, and it usually extends 4 to 12 inches (10–30 centimeters) below the surface. Below topsoil is the B horizon, which has less organic matter, although roots extend to this depth. The C horizon, at about 36 to 48 inches (90–120 centimeters) below the surface, consists almost entirely of partially weathered pieces of rocks and minerals. It is an interface between the bedrock below and the soil above.

Roots Tap Water and Dissolved Minerals

Unlike animals that can easily move in search of water, plants must take up water and the minerals in it with their roots, which is why these plant organs are usually very extensive. Roots offer a spectacular example of biological maximization of surface area. Not only do the roots themselves often reach well beyond the area covered by the aboveground plant body, but they also form millions of root hairs that increase surface area for water absorption (see fig. 26.16). Fungi associated with roots, called **mycorrhizae,** extend this absorptive surface even more (see figs. 23.1 and 23.14). Mycorrhizae increase the growth rate of onions 30-fold! This is very much a mutualistic relationship—recall that the plant benefits from the increase in water and minerals assimilated by and transferred from the fungus, and the fungus receives photosynthate.

Inside the roots, the cortical cells provide yet more surface area, because water has to pass through the cell membranes to reach the xylem (see fig. 26.17). These cell membranes also provide the selectivity that makes it possible for plants to have different concentrations of elements than are present in the soil. The selectivity comes from membrane proteins that admit only certain ions. For example, a soil might be very rich in aluminum, but

Biotechnology 27.1

Rhizosecretion

Tobacco plants that secrete a jellyfish protein, a protein normally made in a human's placenta, and a bacterial enzyme vividly illustrate a new biotechnology, called rhizosecretion (fig. 27.A). This technique taps into the ability of roots to secrete proteins, and gives them the ability to secrete specific desired products.

Recall from Biotechnology 13.3 that a transgenic organism harbors a gene from another species in each of its cells. Plant transgenesis is already a thriving technology, with seeds, fruits, tubers, and roots expressing foreign genes. By adding a genetic "tag" to transgenes, researchers can guide where the products will be secreted—such as into the cytoplasm, in chloroplasts, or into the apoplast, which consists of the cell walls and the extracellular spaces that are continuous with the outside environment.

But plant transgenesis has a major problem—it is difficult to collect the desired products from cells bounded by tough cell walls. Plus, the product is often impure. Extraction and purification are both time-consuming and costly, factors that limit the commercialization of these processes. Still, a plant cell can add sugars and other organic side groups to protein products that recombinant bacteria cannot.

The secret to perfecting plant transgenesis may simply be to use a part of the plant that will naturally cooperate. That is what researchers from Rutgers University and the Institute of Cell Biology and Genetic Engineering in Kyiv, Ukraine, did. Realizing that roots are adept at secreting the carrier proteins that harness minerals from soil, they created tobacco plants transgenic for the jellyfish green fluorescent protein (see the opening essay for chapter 6), a human placental protein, and an enzyme from a thermophilic bacterium. The transgenes included DNA sequences that directed protein production to occur in the roots. The researchers grew the plants in hydroponic media, and were able to collect impressive amounts of the products. Because roots do not naturally secrete many types of proteins, purification was fairly easy—and the plants pumped out the products continuously.

Not only should rhizosecretion be straightforward to scale up, but the hydroponic setup can be converted into a bioreactor. That is, transgenic roots will be able to produce a foreign enzyme that catalyzes a desired reaction. Adding the correct precursor chemicals (including the enzyme's substrate) should result in synthesis and secretion of the desired product into the medium.

FIGURE 27.A
Rhizosecretion.
Transgenic plant technology has been problematic because of the difficulty of extracting protein products from plant cells. Enlisting roots to secrete—rhizosecretion—may solve the problem. A transgenic plant that expresses the gene for a desired protein in its roots can secrete the products directly into the surrounding medium.

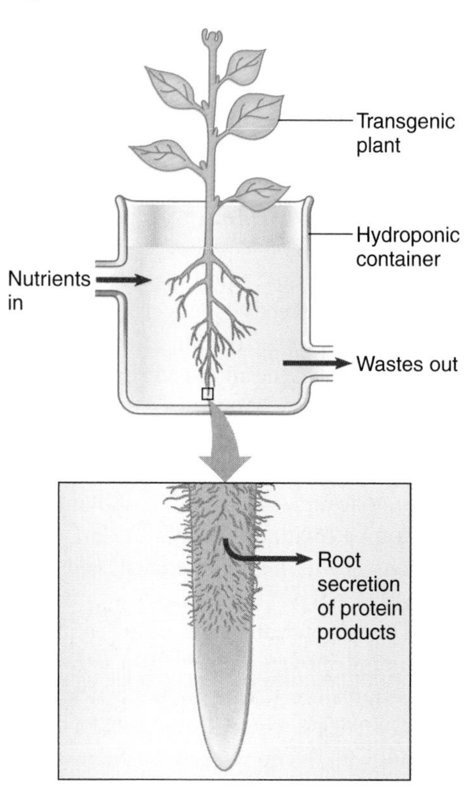

the plants that grow in it have little of the element because the number of carrier proteins limits uptake. On the other hand, plants often have higher concentrations of essential micronutrients than are present in the soil.

Minerals may enter a root cell by facilitated diffusion or active transport, depending on the relative concentrations inside and outside the cell (see fig. 4.9). Specific minerals traverse specific membrane channels, with the assistance of carrier proteins. To enhance mineral uptake, some plants secrete proteins to help them acquire scarce or insoluble elements. For example, monocots such as grains secrete small proteins called siderophores that bind to iron in soil, rendering it better able to enter root cells. The ability of roots to secrete proteins is exploited in a new biotechnology called rhizosecretion (see Biotechnology 27.1).

How Plants Obtain Nitrogen

Plants combine nitrogen with Krebs cycle intermediates to manufacture certain amino acids, which in turn are used to build proteins, nucleic acids, hormones, chlorophyll, secondary metabolites, and other biochemicals. The nitrogen that is abundant in the atmosphere, however, is in the diatomic form—N_2. It is held together by three covalent bonds, which require a tremendous input of energy to break—more than a plant can muster. Instead, plants must acquire nitrogen as nitrates (NO_3^-), nitrites (NO_2^-), or ammonium ion (NH_4^+).

Several types of microorganisms "fix" atmospheric nitrogen into these more accessible forms. Certain types of bacteria, including those in the genus *Rhizobium,* are stimulated by signals from plants of the legume family (peas, peanuts, soybeans, and alfalfa) to

FIGURE 27.5

Bacteria Provide Usable Nitrogen to Plants.

In response to signals from leguminous plants such as beans, *Rhizobium* triggers the development of root nodules (**A**). The bacteria enter through root hairs (**B**), and live symbiotically within the plant's cells (**C**).

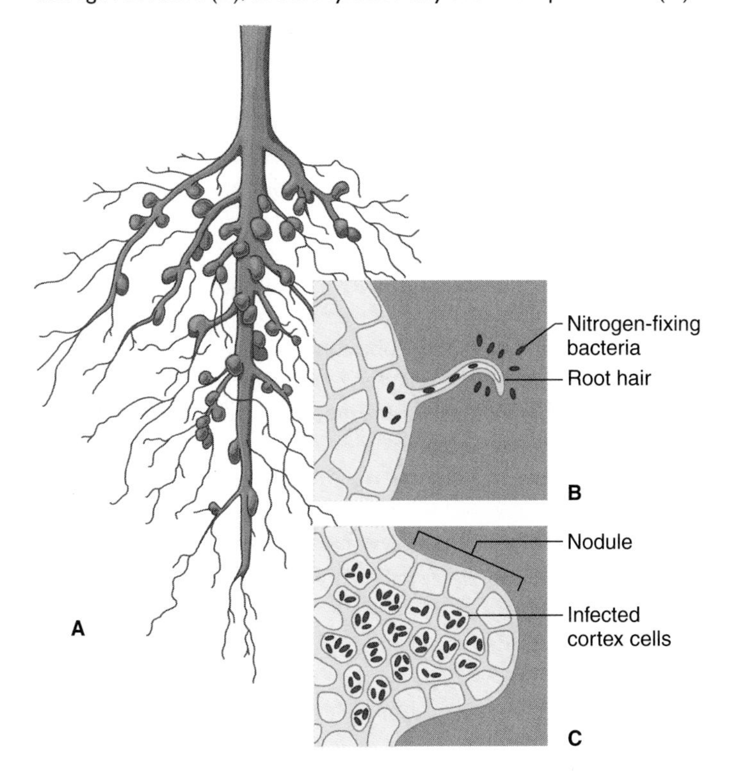

associate in growths called **nodules** on the plants' roots, where they use **nitrogen-fixing** enzymes to convert N_2 to usable forms (figs. 20.1 and 27.5). Cyanobacteria and *Azotobacter* are other types of bacteria that provide nitrogen to plants; some of these microorganisms form symbiotic relationships, while others are free-living.

Rhizobium reduces atmospheric N_2 to ammonium ion (NH_4^+), which also comes from fertilizer and from other types of bacteria that live on decaying organisms. Certain autotrophic bacteria, including *Nitrosomonas* and *Nitrobacter,* oxidize NH_4^+ into nitrites and nitrates. The form of nitrogen that plants actually take up, usually NH_4^+ or NO_3^-, depends on the plant species and soil conditions such as pH. The nitrogen cycle is depicted in figure 43.15.

A few plants acquire nitrogen from more unusual sources. For example, carnivorous plants, which often live in nitrogen-poor soils, obtain nitrogen compounds from the insects that they consume (see Investigating Life 22.1).

27.2 MASTERING CONCEPTS

1. What are the 16 essential macronutrients and micronutrients that plants require, and why are they considered essential?
2. Where do plants obtain nutrients?
3. How are roots adapted to maximize surface area?
4. How do bacteria help plants to acquire nitrogen from the environment?

OLC

27.3 Water and Dissolved Minerals Are Pulled Up to Leaves

Water molecules exiting leaf stomata in transpiration pull up water from farther down in the plant, creating a hydraulic system that continually brings water and dissolved nutrients absorbed in the roots upward—even hundreds of feet.

Because leaves need a constant supply of water, water must be transported from the roots, which can be quite a long distance away. This process has three interrelated steps: transpiration of water from leaves, replacement of that water, and uptake of water in the roots.

Water Vapor Is Lost Through Transpiration

When a plant is actively photosynthesizing, heat builds up, both from the sun and from the plant's metabolism. This heat causes water in the cell walls to evaporate into the intercellular spaces, cooling the leaf and establishing a gradient with a higher concentration of water vapor in the intercellular spaces than in the air surrounding the leaf. Water moves by osmosis down the gradient, and is lost to the outside as long as the stomata are open—this is **transpiration,** discussed in chapters 6 and 26. More than 90% of the water that enters the leaf eventually is lost via transpiration; a small amount escapes through the cuticle. Transpirational loss can be great. A corn plant loses 1.3 gallons (5 liters) of water a day—imagine what a coastal redwood at full maturity, discussed in the chapter opening essay, loses when there isn't much fog!

Leaf cells will quickly shrink, and if water lost in transpiration is not replaced, the leaf will wilt. The leaves on a branch pulled off a tree and left in the sun on a warm day rapidly wilt. In contrast, a branch left on the tree, or cut off and placed in a container of water, retains firm leaves. What is happening is that as water molecules evaporate from the leaves' open stomata, they are immediately replaced by water molecules that are drawn up through the xylem, either from the container of water or via the roots and stem if the plant is intact.

Water Moves Through Xylem

Water is transported from the roots up the stem to the leaves through the xylem, a special water-conducting tissue. Xylem makes up wood and is also present in the petioles and veins of leaves and in strands that extend nearly to the tips of roots and shoots. Labeling water with a radioactive tracer or a dye reveals the path from xylem in the roots through the veins of leaves. Environmental conditions affect the rate of water movement in xylem. The warmer the day, the faster water moves. The relative humidity of the surrounding air and wind speed also affect the rate of water movement .

Recall from chapter 26 that the water conducting cells of the xylem, called vessel elements and tracheids in flowering plants, are only the remaining cell walls of long, tapering cells that died when they reached maturity (see fig. 26.6). This system of microscopic pipes transports a dilute aqueous solution, the **xylem sap,** that consists of water and dissolved minerals. Because the functional cells of the xylem are dead, the metabolic activity of the plant is *not* what drives the movement of water.

How then, does water in the xylem get from roots to the mesophyll in the leaves? As transpiration removes water from the leaf, the solutes in the cytoplasm of chlorenchyma cells and in the water layer surrounding these cells become more concentrated than they were before transpiration began. The water in the xylem in the veins of the leaf has a lower concentration of these solutes. Therefore, water diffuses down a concentration gradient, from the xylem in the veins to the mesophyll tissues in the leaf.

As water leaves the xylem in the vein, the water molecules, because of their polarity and ability to form hydrogen bonds (see chapter 2), attract water molecules further down the water column in the xylem, pulling them toward the mesophyll of the leaf. This attraction between water molecules is the property of cohesion, and it enables water to be drawn up to the top of a tall tree. Water in the roots is thus pulled up through the xylem, as if being sucked through a straw, rather than being pushed from below. The mineral-laden water in xylem forms a continuous hydraulic system that rises in columns through the plant body. The movement of water from the roots to the leaves and out through the stomata is called the transpiration stream, and the **cohesion-tension theory** explains its mechanism based on the properties of water in a tube (fig. 27.6).

Water must overcome powerful gravitational forces to ascend a plant. Any column of water has mass, and will collapse under the force of gravity unless a greater force holds it up. The force required can be formidable. For example, a 300-foot (about 92 meters) tree can lose hundreds of gallons of water from its leaves on a hot day. Yet even the tree's highest leaves get sufficient moisture, because of water's cohesion as well as another property, its adhesion. Recall from chapter 2 that adhesion is a molecule's attraction to another type of substance—in this case, water forms

FIGURE 27.6

Water and Dissolved Minerals Are Pulled to Treetops.

Transpiration of water from leaves creates a force that pulls water up a plant's body from the roots. The cohesiveness and adhesiveness of water make this possible.

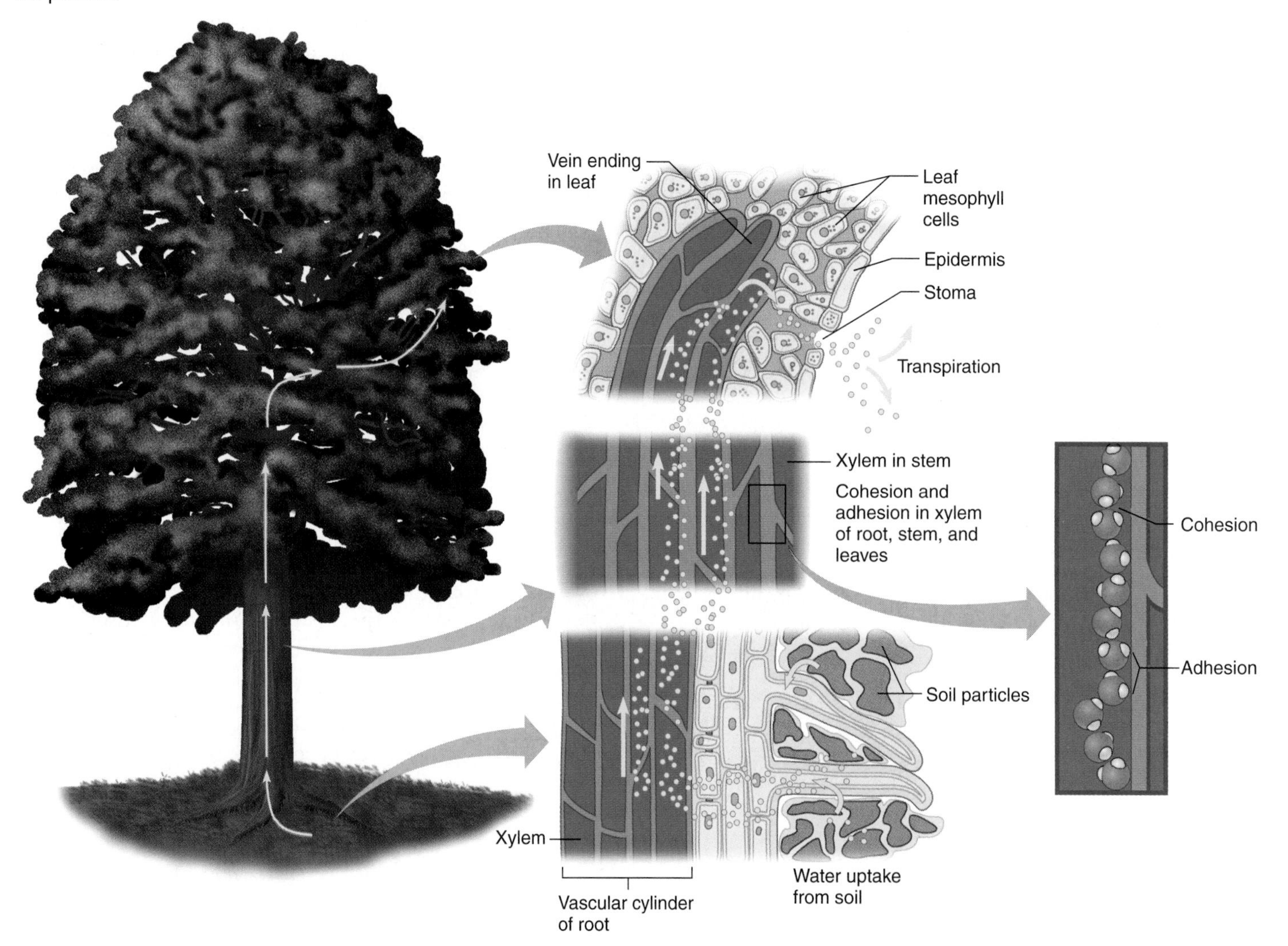

hydrogen bonds with the walls of the xylem tubes. In other words, water molecules stick to each other, and also adhere to the walls of vessel elements and tracheids. These combined properties of water hold a water column up against the force of gravity.

How high a given mass of water can rise is a function of the diameter of the tube in which it is contained (fig. 27.7). The thinner the diameter of the tube, the higher the water column. This is why it is easier to suck up a soda through a narrow straw than through a wide one, and why a finger placed over the upper end of a filled thin straw will retain the straw's contents, but attempting the same with a wide straw will result in a spill. Likewise, the thin diameters of tracheids and vessel elements, which range from a few to several tens of micrometers, support water columns that reach to the tallest treetops.

Xylem tubes are remarkably strong. Problems with water transport are generally due not to damaged xylem, but to an interruption in the water stream, a condition called **cavitation.** A water column being pulled through a xylem tube by the transpiration stream is under stress and is actually being slightly stretched. If too hard a force is applied to pull water through the xylem, the aqueous solution in the column may vaporize and create a bubble of gas. If the bubble fills the xylem tube across its diameter, the continuous water column breaks, which halts further flow in that tube. Cavitation is more likely to occur in vessel elements, with their larger diameters, than in tracheids. Plants are particularly subject to cavitation when the transpiration rate is high and the replacement rate of water in the roots is relatively low. If too many xylem tubes cavitate, then too little water will reach the crown of the plant, and the crown will begin to dry out.

As long as the soil supplies sufficient water to replace water moving through the transpiration stream, the plant will flourish. However, if replacement of water in the xylem slows because the soil dries out or cavitation occurs, the plant must temporarily shut down the transpiration stream or it will wilt and die. It does this by closing the stomata in the leaves, because that is the primary route through which the plant loses water. However, a consequence of closing stomata is also the shutdown of photosynthesis because CO_2 can no longer enter the leaf. Under drought conditions, plants must balance the opposing requirements for CO_2 and water.

Water Enters Plants in Roots

As evaporation from leaf surfaces pulls water up the stem, additional water enters roots from the soil. The concentration of solutes in the soil is generally lower than the concentration of solutes in the cells of a plant's root. Therefore, at the interface between a root and the soil, water moves into root cells by osmosis. The water-absorbing interface between the roots of most plants and the soil is a region just basal to the growing tip of the root, where the epidermis is fringed with many root hairs (see fig. 26.16). Root hairs greatly increase the surface area of the root that contacts particles in the soil. Because the cell walls of root hairs are hydrophilic, the water in them is continuous with both the water in the soil and the water in the cytoplasm.

Water and the dissolved minerals it carries can move to the xylem in two ways (fig. 27.8). In the apoplastic pathway, water moves along the cell walls and the extracellular spaces. This water and its solutes remain between the cells of the epidermis and the cortex of the root, an area called the apoplast.

In the symplastic pathway, water and solutes move from cell to cell through the plasmodesmata that link the cytoplasm of the cells of the epidermis and the cortex (see fig. 4.20). This continuous cytoplasm of many cells constitutes the symplast. Some of the minerals dissolved in soil water enter cells of the root, then also pass from cell to cell through the plasmodesmata.

FIGURE 27.7

A Column of Fluid Rises Higher in a Thin Tube Than in a Wider One.

(**A**) Water rises up a tube because water molecules stick to the tube walls (adhesion) and each other (cohesion). The thinner the tube diameter, the taller the water column it can support. (**B**) Xylem cells vary in diameter but are narrow enough to supply water to the tops of very tall trees.

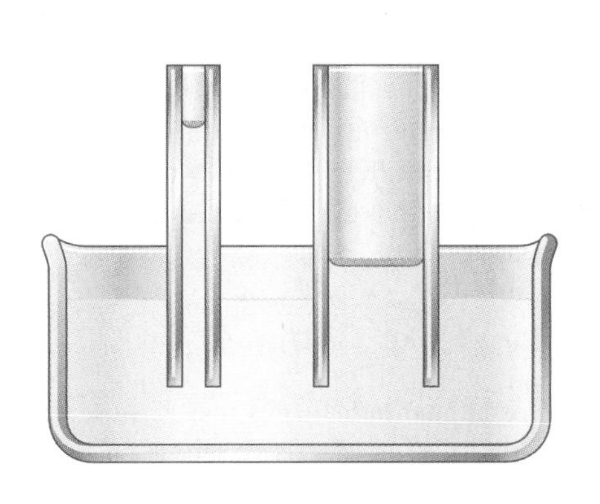

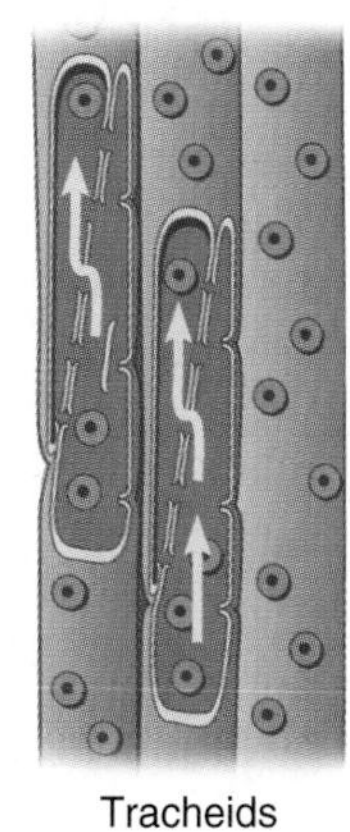

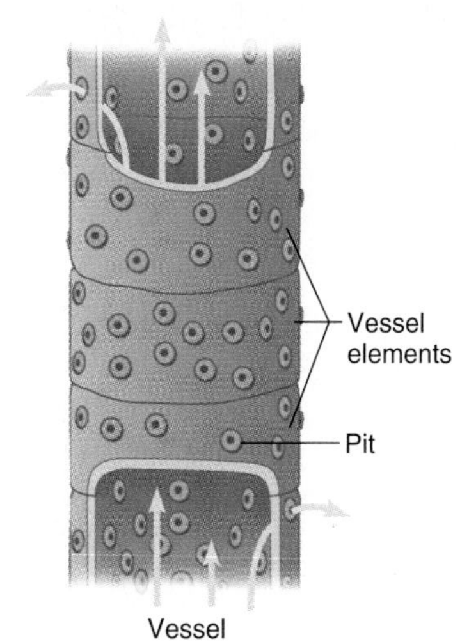

FIGURE 27.8

Two Routes into Roots.

Water and dissolved minerals travel through a root's cortex either apoplastically (between cells) or symplastically (through cells). All of this incoming material must pass through the cells of the endodermis to reach the xylem.

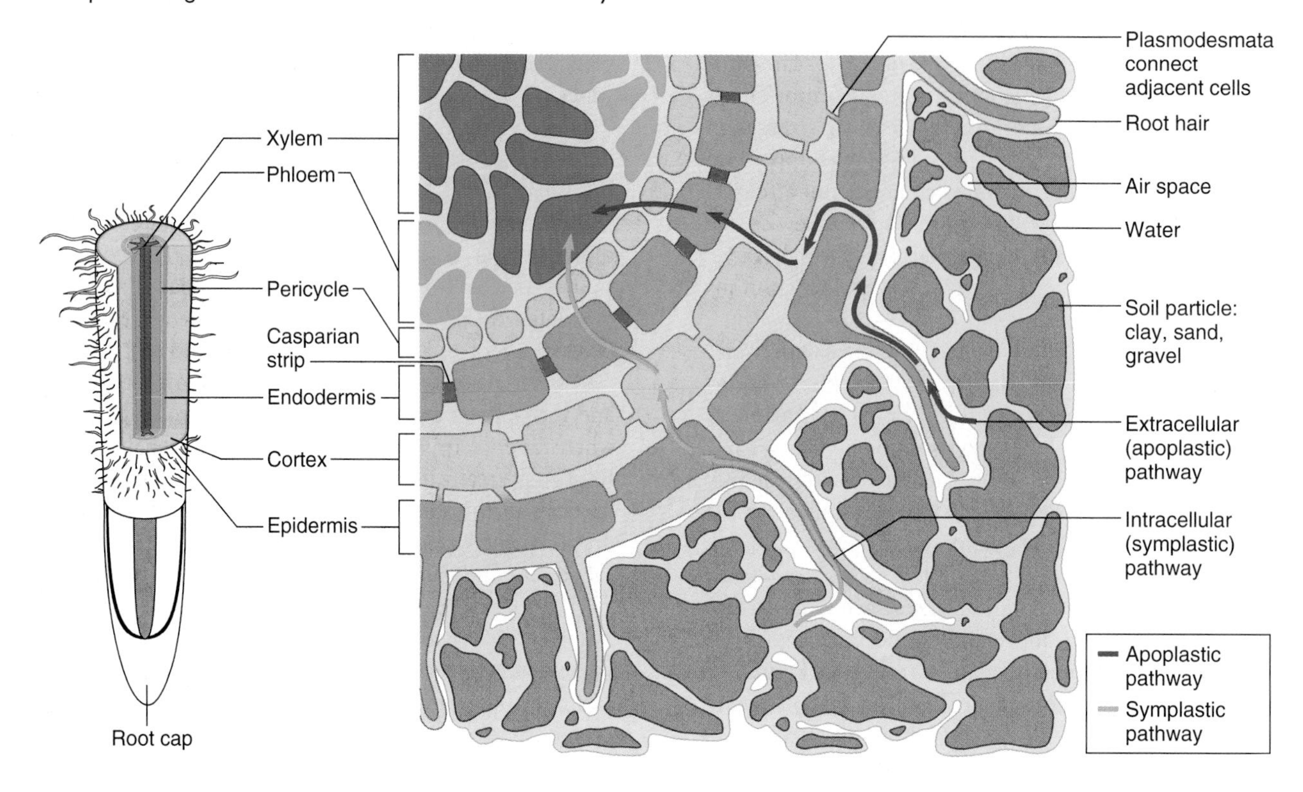

When the incoming water and minerals contact the endodermis—the innermost layer of the cortex—and its impermeable **Casparian strip** (see fig. 26.17), apoplastic movement is blocked, which forces the water and minerals that had gone around cells to now enter them. Because endodermal cell membranes have receptors, carrier proteins, and channels for only certain minerals, the endodermis exercises some selectivity in which substances enter the xylem. Water and dissolved minerals that cross the endodermis continue into the pericycle, which consists of a layer of living cells at the outside of the stele. The water and minerals then enter the xylem. The overall effect of this cellular organization is that all the water inside the endodermis must have passed through a cell membrane to get into the stele.

27.3 MASTERING CONCEPTS

1. Describe how transpiration occurs.
2. How do the properties of water and xylem cells enable xylem to pull up water from roots?
3. How do roots take up water and dissolved minerals?

OLC

27.4 Photosynthate Is Pushed to Heterotrophic Cells

Pressure from incoming fluid from the xylem pushes phloem sap—which contains photosynthate—to heterotrophic cells.

When water is plentiful enough to allow photosynthesis to proceed, and sufficient dissolved minerals arrive in the transpiration stream, a chlorenchyma cell will produce a surplus of photosynthate that can then be transported to the heterotrophic, or nonphotosynthesizing, cells of the plant. Heterotrophic cells include those in the root, the dividing cells of shoot apical meristems, and cells in flowers and fruits. If these cells do not receive enough photosynthate to generate the ATP they require, the plant may die or fail to reproduce. In a multicellular organism, all parts ultimately depend upon others.

Phloem Sap Composition

Photosynthate in leaf cells may have several fates. It may be stored in plastids, usually as starch; used to synthesize fats and

FIGURE 27.9

Aphids Reveal Sieve Tube Pressure.

Botanists learned about the composition of phloem sap with the help of aphids. These insects excrete drops of "honeydew" (**A**), which is derived from phloem sap, when their stylets (mouthparts) penetrate phloem cells (**B**) in the host plant.

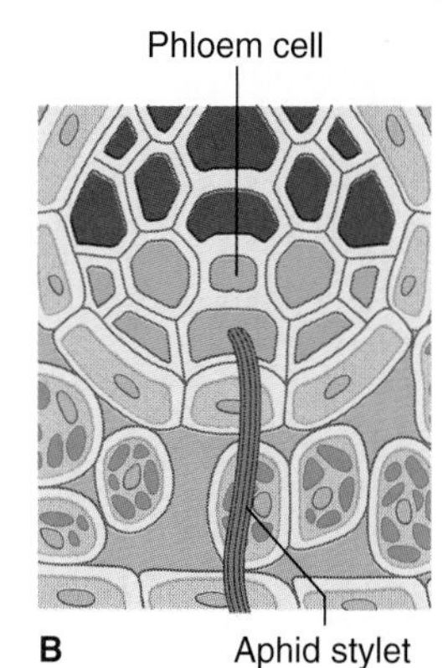

proteins; or transported from the leaf mesophyll through the phloem. The photosynthate in phloem is dissolved in an aqueous solution called the **phloem sap,** which includes water and minerals from the xylem. The carbohydrates in phloem sap are mostly simple sugars, because polysaccharides such as starch are too insoluble to transport. Most of the dissolved photosynthate is the disaccharide sucrose (see fig. 2.13), or a derivative of sucrose. Phloem sap also contains amino acids, hormones, enzymes, and messenger RNAs that are translated into proteins in cells distant from the site of their transcription.

Researchers learned about the composition of phloem sap with the help of aphids, the insects discussed in the opening essay to chapter 9. Aphids feed on phloem sap, excreting a sticky, sweet material euphemistically called "honeydew" (fig. 27.9). Ants eat the honeydew, and even "farm" aphids to keep a steady supply of the treat. As soon as one drop of honeydew falls from an aphid's anus, another begins to form. The honeydew is very rich in sucrose and other carbohydrates—indirect evidence that the aphid is ingesting the phloem sap. Carefully cutting an aphid away from its mouthparts reveals direct evidence that phloem sap is under pressure. If the mouthparts are allowed to remain in a sieve tube, sugary drops continue to form at and fall from them. Because the mouthparts are just a passive tube, drops can only form in them if the phloem sap is under pressure. These insects are living spigots.

Radioactive tracers can be used to demonstrate phloem transport. Expose a leaf to CO_2 labeled with carbon 14, and the radioactive carbon is incorporated first into organic compounds in chlorenchyma cells—this is photosynthate. Some of the radioactive photosynthate then moves from the chlorenchyma cells to the veins of the leaf, and travels through phloem to all parts of the plant, both above and below the leaf. Heterotrophic cells take up the photosynthate from phloem in sites throughout the plant.

The Pressure Flow Theory Explains Photosynthate Distribution

The major transport structures of phloem in flowering plants are the microscopic sieve tubes (see fig. 26.7). Unlike xylem, the sieve tube members that comprise sieve tubes are alive. The end walls, or sieve plates, of adjacent sieve tube members are perforated and look like miniature sieves that allow the carbohydrates in photosynthate to easily pass through. Sieve tube members carry on metabolism but lack nuclei. The final cell division that gives rise to the cell that develops into a sieve tube member also produces a companion cell, which retains its nucleus. The complex of the sieve tube element and its companion cell forms the functional unit of the phloem sieve tube system.

Phloem sap flows through the sieve tubes under pressure, following the laws of physics. When a container holding a fluid under pressure is opened, the fluid moves out of the container through that opening, and will flow until the force on the fluid inside the container equals the force outside it. If more fluid is added to the container as it leaves, then the fluid flows through the container. The part of the container where fluid is released is called a **sink,** and the part where fluid is added is a **source.**

Xylem is critical to the function of phloem, and the two tissues are adjacent to each other in vascular bundles. Water diffusing in from xylem creates the force that drives phloem sap through sieve tubes. Sucrose from photosynthate enters the sieve tube members in leaves by cotransport, powered by ATP from companion cells (see fig. 4.11). Because sucrose becomes so much more concentrated in the sieve tubes than in the adjacent xylem, osmosis drives water from the xylem into the phloem sap in the sieve tubes. But because liquids cannot be compressed, when the phloem sap increases in volume from the fluid input from the xylem, pressure is generated.

When the pressure is relieved at a sink somewhere along the length of the sieve tube, more phloem sap under pressure flows through the sieve tube toward the sink. This movement mechanism is called the **pressure flow theory** (fig. 27.10). Common sinks are fruits, flowers, and roots. Here, heterotrophic cells take up the sucrose (and other compounds in the phloem sap) through facilitated diffusion or active transport. As the sucrose is unloaded from the sieve tubes, the concentration of solutes in the phloem sap becomes more and more dilute, until water diffuses from the sieve tube to the surrounding tissue (often xylem). Movement of water out of the sieve tube relieves the pressure, and the pressurized phloem sap in the sieve tube continues to flow toward the sink. In contrast to water and dissolved minerals that are pulled up xylem, photosynthate is pushed toward sinks.

27.4 MASTERING CONCEPTS

1. What are the components of phloem sap?
2. How do researchers know that phloem sap is under pressure?
3. How is phloem sap pushed through plants?

OLC

FIGURE 27.10

Photosynthate Moves Under Pressure from a Source to a Sink.

Sugar produced in green "source" organs such as leaves moves via phloem to roots, fruits, and other nonphotosynthetic "sinks." According to the pressure flow theory, phloem sap is pushed from sources to sinks.

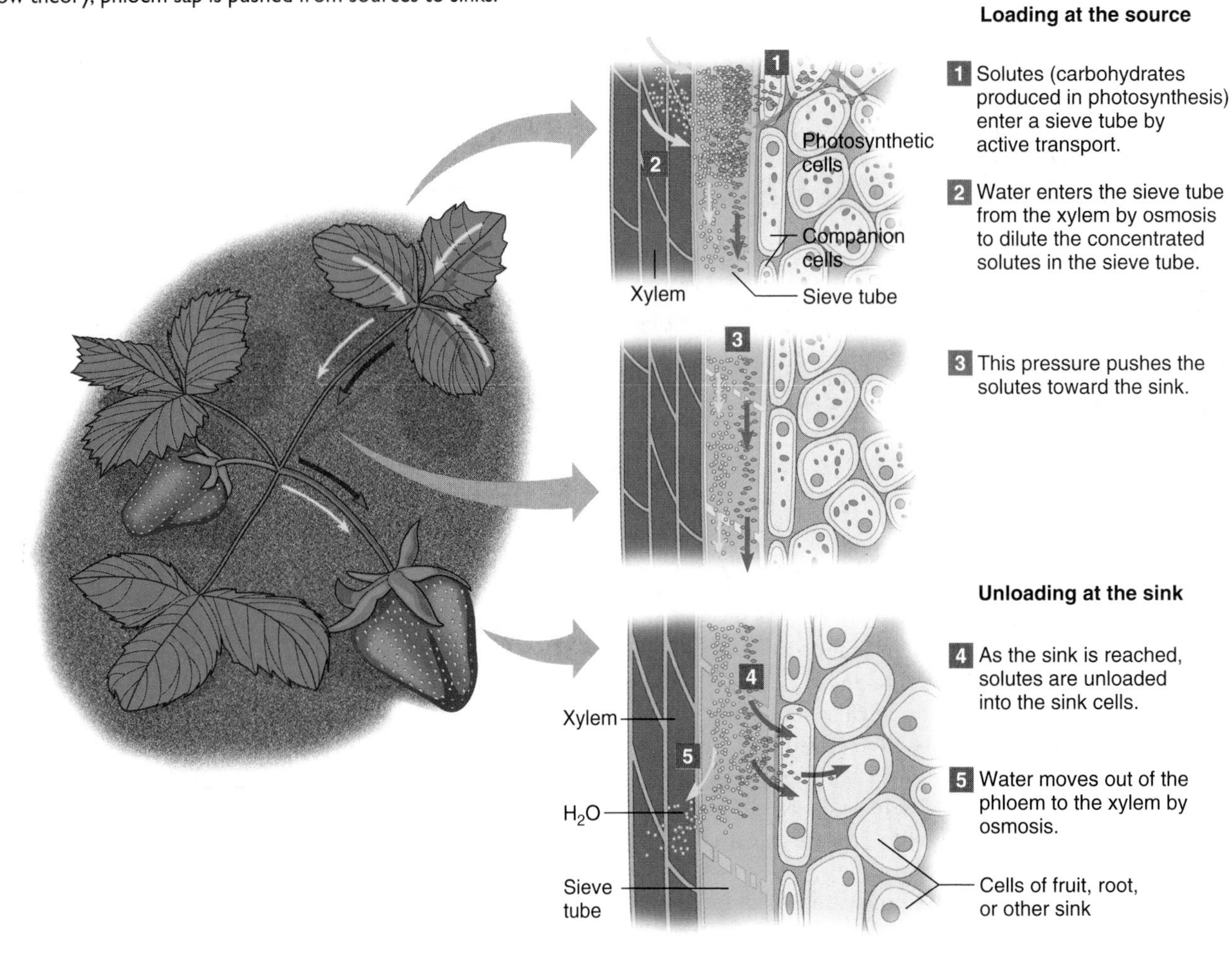

Chapter Summary

27.1 Complex Multicellular Organisms Require Complex Transport Systems

1. The first life-forms were simple enough to directly exchange materials with the environment.
2. More complex multicellular organisms developed folds and transport systems, which maximized surface area and contact with the environment.
3. In plants, xylem transports water and dissolved minerals, and phloem distributes photosynthate. Both tissue types are bundled together in roots to form the stele.

27.2 Plants Require Water and Nutrients

4. In all plants, the macronutrients are carbon, hydrogen, oxygen, nitrogen, potassium, calcium, magnesium, phosphorus, and sulfur. The micronutrients are chlorine, iron, boron, manganese, zinc, copper, and molybdenum. Certain plants assimilate other elements.
5. Plants get carbon and oxygen from CO_2 in the atmosphere, and hydrogen from water. The other elements come from soil.
6. Soil consists of rock and mineral particles mixed with decaying organic molecules and organisms. Soil layers are called horizons. **Topsoil** contains **humus** and is a major source of water and nutrients for plants.
7. Root branches, root hairs, **mycorrhizae,** and root cortical cells provide abundant surface area to absorb water and dissolved minerals. Some plants secrete proteins to enhance mineral uptake.
8. Several types of bacteria **fix nitrogen** into forms that plants can use.

27.3 Water and Dissolved Minerals Are Pulled Up to Leaves

9. Water and dissolved minerals (**xylem sap**) are pulled up through xylem to replace water lost through **transpiration** in leaves. This is called the **cohesion-tension theory.**

10. Xylem consists of microscopic tubes made from dead tracheids and vessel elements.
11. Water's properties of cohesion and adhesion make xylem transport possible.
12. A break in the flow of water through xylem is called **cavitation.**
13. Water enters roots by osmosis because the solute concentration in the soil is less than that of cells of the root.
14. The endodermis with its impermeable **Casparian strip** controls which minerals enter the nearby xylem. Water and dissolved minerals move to xylem by the apoplastic pathway (extracellularly) and by the symplastic pathway (through cells).

27.4 Photosynthate Is Pushed to Heterotrophic Cells

15. Photosynthate may be used or stored in its cell of origin, or moved in phloem to heterotrophic cells in roots, flowers, or fruits.
16. **Phloem sap** includes photosynthate and water and minerals from xylem.
17. Phloem sap flows through sieve tubes from a **source** to a **sink,** with pressure generated by continual influx of water from xylem. This mechanism is called the **pressure flow theory.**

Testing Your Knowledge

1. What is the difference between a macronutrient and a micronutrient?
2. How do plants obtain carbon and nitrogen?
3. How do the cells of xylem and phloem differ?
4. Describe how cavitation harms a plant.
5. Distinguish between cohesion and adhesion.
6. At which points in the transport of water and nutrients in plants do the following processes occur?
   a. osmosis
   b. active transport
   c. cotransport
7. Describe the paths of water and dissolved minerals into xylem in a root.
8. How does xylem flow influence phloem flow?
9. Distinguish between a source and a sink.
10. What are the applications of phytoremediation and rhizosecretion?

Thinking Scientifically

1. Give three examples of plant adaptations that maximize surface area.
2. How can cutting down coastal redwoods for lumber harm ferns?
3. How did vascular systems influence plant evolution?
4. How might transgenic technology (see Biotechnology 13.3) be used to endow plants with the ability to fix nitrogen?
5. On a planet with an atmosphere similar to Earth's, which elements would have to be present in the soil for earthly plants to grow?
6. To collect sap to make maple syrup, a metal tube is placed into the sapwood—at a certain time of day in only some parts of the United States—and sap flows out. What is this sap? What force pushes it out of the tree?
7. Rice plants given a gene encoding a soybean protein called ferritin triple the amount of iron that they can store. How might this engineered rice be of value, and how might it be harmful?
8. Explain how chlorenchyma and root cortical cells interact.

References and Resources

Finer, John J. May 1999. Plant protein secretion on tap. *Nature Biotechnology,* vol. 17, p. 427. Hydroponically growing roots can be used as bioreactors.

Hooker, Brian S., and Rodney S. Skeen. May 1999. Transgenic phytoremediation blasts onto the scene. *Nature Biotechnology,* vol. 17, p. 428. Tobacco plants carrying a bacterial transgene can detoxify TNT.

Lewin, Roger. April 3, 1999. Ruling passions. *New Scientist,* p. 34. A system of tubules to transport vital substances is not unique to plants.

Strauss, Evelyn. January 1, 1999. RNA molecules may carry long-distance signals in plants. *Science,* vol. 283. Phloem transports mRNA as well as photosynthate.

The *LIFE* Online Learning Center provides additional resources and tools for studying this chapter.
www.mhhe.com/life

CHAPTER 28

Reproduction of Flowering Plants

Imperiled Pollinators

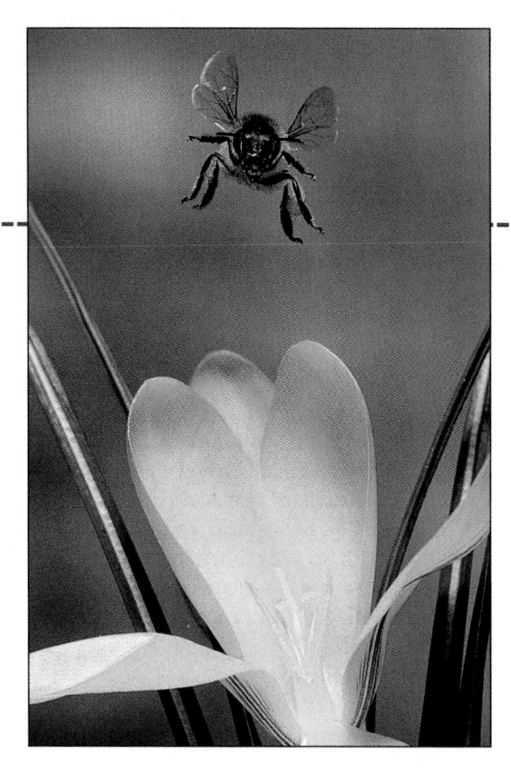

Nearly all life in terrestrial ecosystems depends, ultimately, on plants, because of photosynthesis. Yet animals make much of plant life possible by transferring pollen from one flower or plant to another. Bees are particularly proficient pollinators and may store pollen from dozens of plant species.

European honeybees (*Apis mellifera*) pollinate many flowering plant species in the United States. Europeans introduced these bees in Jamestown, Virginia, in 1621. Because these honeybees can use nectar (a sweet substance that attracts pollinators to flowers) and pollen from a variety of plants in a variety of habitats, they rapidly displaced native bee species, reaching a population pinnacle of nearly 6 million colonies in 1947.

European honeybees (Apis mellifera) pollinate many flowering plant species in the United States.

Several adaptations enabled honeybees to flourish. With their superb senses they spot rare flowers peeking out of otherwise barren landscapes more readily than can other pollinators. Honeybees have easily satisfied tastes. They flock to many angiosperm species, even in disrupted areas, while other bees look for specific plants. But the honeybees' wide-ranging tastes make them less-than-efficient pollinators, because they often deposit pollen from one species on plants of another species. Still, the large numbers of honeybees

Pollinating Bees
Honeybees (*Apis mellifera*) pollinate flowers of crocus, clover, and many other important plants. Many flowering plants depend for reproduction on healthy populations of pollinating animals. Unfortunately, honeybee populations are declining, partly because of mites (*Varroa jacobsoni*) infesting their hives. Reddish-brown mites are visible on the bees shown at right.

maintain many flowering plant populations.

Clearly, honeybees are important for reproduction in both crops and wild populations of flowering plants. But bee populations worldwide are plummeting.

Several factors have contributed to the decline of honeybee populations in North America. Increased use of pesticides in agriculture began killing bees shortly after World War II and had depleted more than a million colonies by 1972. A study of *Cereus* cactus in an area along the Mexico and Arizona border revealed the effects of herbicide use to clear land to build new homes. Where herbicide was used, 27% of the plants monitored were pollinated and 5% developed fruit. Without herbicide, 60 to 100% of the cacti were pollinated and 75 to 100% developed fruit.

Bee experts—called apiculturists—predict that the falling bee populations will soon lead to diminished crop yields.

Pesticide use is only one threat to bee pollinators. Tracheal mites first appeared in Florida bee colonies in 1984 and rapidly spread throughout the United States, killing up to half of the bees in individual hives. Another type of mite, the *Varroa* mite, began invading hives in Wisconsin in 1987 and likewise spread, eventually destroying 60% of honeybee colonies in the United States. Caterpillars with an appetite for beeswax plagued many hives. Finally, Africanized bees invaded at the end of the 1980s, bringing deadly mites to many honeybee hives.

Bee experts—called apiculturists—predict that the falling bee populations will soon lead to diminished crop yields. European apiculturists have more experience than North American bee authorities, and they recommend that we can avert a crisis by preserving, or restoring, areas of native vegetation. An experiment on Staten Island in New York City suggests that the Europeans are correct. Staten Island is home to a huge landfill. Ecologists have begun to plant small sections of it with native trees and shrubs. In a short time, honeybees as well as other types of bees have come back and are pollinating the new plants. The closer the new areas are to undisturbed woodlands, the more bees gather.

28.1 Asexual Reproduction

Cloning may be new and science-fictionish for mammals, but it's been known in plants for centuries.

Flowering plants (angiosperms) dominate many terrestrial landscapes. From grasslands to deciduous forests, garden plots to large-scale agriculture, angiosperms have been extremely successful. They have also been useful to humans, as sources of food, shelter, clothing, furniture, and fuel.

The angiosperms owe their widespread distribution to several adaptations. First, pollen (which also occurs in gymnosperms) enables sperm to fertilize the egg in the absence of free water. Second, angiosperms and gymnosperms produce seeds, which protect the embryo during dormancy and nourish the developing seedling. Third, only angiosperms have flowers, which promote pollination and develop into fruits that help disperse the seeds far from the parent plant.

Although some plants reproduce asexually, the majority are sexual (see chapter 22 for life cycles of other vascular and nonvascular plants). This chapter begins by briefly examining asexual reproduction, then moves on to gamete formation, fertilization, and seed development in flowering plants.

Many plants reproduce asexually and form new individuals by mitotic cell division. **Asexual reproduction** does not involve meiosis, gametes, and fertilization, but it is still a complex biological process that we do not fully understand. Recall from chapter 9 that in asexual reproduction, a parent organism produces progeny that are genetically identical to it and to each other—they are clones.

Asexual reproduction is advantageous when environmental conditions are stable and plants (or other organisms) are well adapted to their surroundings, because the clones will also be well suited to the environment. Perhaps the most stable and predictable environment is a laboratory, where researchers can control conditions. Biotechnology 28.1 describes several techniques, based on asexual reproduction, that may help scientists develop new, useful plant varieties.

Plants often reproduce vegetatively (asexually) by forming new plants from portions of their roots, stems, or leaves (fig. 28.1). For example, buds form on the roots of cherry, pear, apple, and black locust plants, and when they sprout, aerial shoots grow upward. If these shoots, called "suckers," are separated from the parent plant, they become new individuals. New strawberry plants arise asexually from nodes on aboveground stems called stolons (fig. 28.1A). A few strawberry plants will become many within a few years. Some plants use leaves to reproduce asexually. When the leaves of a "maternity plant" (*Kalanchoe daigremontiana*) lie on a moist surface, roots and shoots develop at their edges. These plantlets become new individuals when the parent's leaves shed them.

Sometimes new plants derived from asexual reproduction remain attached to the parent and produce huge individuals. For

FIGURE 28.1
Asexual Reproduction.
(**A**) Strawberry plants (*Fragaria ananassa*) reproduce asexually, with new plants growing from nodes on horizontal stems called stolons. They also reproduce sexually. (**B**) The "maternity plant" (*Kalanchoe daigremontiana*) grows plantlets on its leaves. (**C**) Quaking aspen (*Populus tremuloides*) trees that cover more than 100 acres in Utah are clones.

A

B

C

Biotechnology 28.1

New Routes to Plant Reproduction

Traditional plant breeding introduces new varieties through sexual reproduction: a sperm cell carrying genes for certain traits fertilizes an egg cell carrying a different set of genes. Most plant biotechnologies instead start with somatic cells from nonsexual parts of the plant, including leaves, stems, or embryos. These cells contain a complete set of genetic instructions. Plants regenerated from somatic cells do not have the unpredictable mixture of characteristics of sexually derived plants because the somatic cells are clones.

Protoplast fusion and cell culture are plant biotechnologies that use somatic cells.

Protoplast Fusion

A protoplast is a plant cell stripped of its cell wall. Protoplast fusion combines these cells from different species and regenerates a mature plant hybrid (fig. 28.A). Cells may fuse spontaneously or after treatment with digestive enzymes, calcium, polyethylene glycol (an antifreeze), electricity, or a laser. A single gram (about 1/28 of an ounce) of plant tissue can yield up to 4 million protoplasts. A plant regenerated from protoplasts of different species is called a somatic hybrid.

Not all protoplast fusions yield mature plants, and not all that do are agriculturally useful. A protoplast fusion of radish and cabbage, for example, may yield a plant that grows radish leaves and cabbage roots! Protoplast fusion is most successful when the parent cells come from closely related species. Consider fusion of a protoplast from a potato plant that is normally killed by the herbicide triazine with a protoplast from wild black nightshade, a relative that resists the herbicide. The resulting hybrid grows well in triazine-treated soil.

Cell Culture

When tiny pieces of plant tissue called explants are nurtured in a dish with nutrients and hormones, the cells lose their specialized characteristics after a few days and form a white lump called a callus. The callus grows, its cells dividing, for a few weeks. Then certain callus cells develop into either a tiny plantlet with shoots and roots, or a tiny embryo (fig. 28.B). Such an embryo is called a somatic embryo because it derives from somatic, rather than sexual, tissue. Researchers are not sure how or why the calli of some species give rise to somatic embryos, others form plantlets, and still others never develop beyond a lump of tissue. Callus growth is unique to plants. The human equivalent would be a cultured skin cell multiplying into a blob of unspecialized tissue and then sprouting tiny humans or human embryos!

Embryos or plantlets grown from a single callus are usually clones, but sometimes they differ from each other because mutations can occur in individual cells, which are then grown into separate plants. Researchers can to some extent control whether somatic embryos or plantlets are identical or variant by altering the nutrients and hormones in the culture medium. Either clones or variants may have uses.

Identical cultured somatic embryos are the basis of artificial seed technology. A natural seed is a plant embryo and its food supply packaged in a protective coat. An artificial seed is a somatic embryo suspended in a gel containing nutrients and hormones. A biodegradable polymer coat provides protection and shape. Researchers can package somatic embryos with pesticides, fertilizer, nitrogen-fixing bacteria, and even microscopic

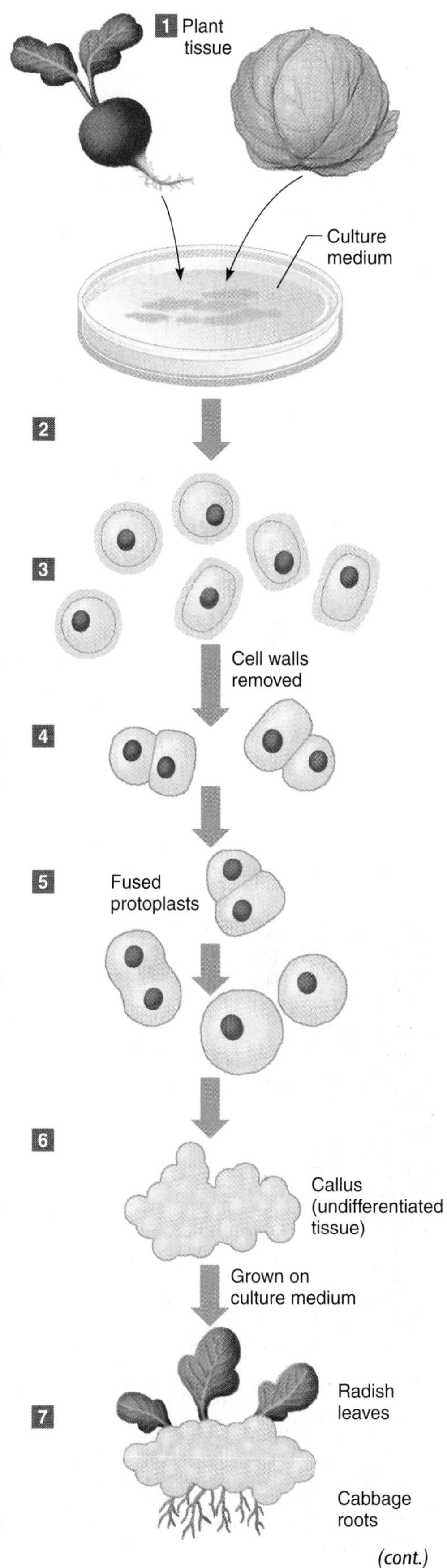

FIGURE 28.A
Protoplast Fusion—Making One Cell from Two.
Protoplast fusion can yield plants with the genetic traits of two species. Tissues from different plants (*1*) are placed on a tissue culture medium (*2*), and the cell walls are dissolved (*3*). The resulting protoplasts are mixed and stimulated to fuse (*4*). Among the fusion products may be hybrid protoplasts (*5*), which are selected and grown in culture into an unspecialized lump of tissue called a callus (*6*). Sometimes, plants regenerated from the callus exhibit new combinations of traits (*7*). The radish leaves and cabbage roots were not exactly what researchers sought!

(cont.)

Biotechnology 28.1 (continued)

parasite-destroying worms! So far, however, artificial seeds are too costly to be practical.

The occasional variant plantlets that grow from a callus are called somaclonal variants because they derive from a single somatic cell and were therefore originally clones. Normally, new plant varieties that arise from spontaneous mutation are one in a million. Somaclonal variants occur much more frequently. In one experiment, researchers chopped up a tomato leaf and cultured it into a callus and regenerated 230 plantlets. Thirteen of the plantlets were new variants—one had light orange fruits; two lacked joints between stems and tomatoes, which makes harvesting easier; and two others had a high solids content, which is important in food processing. Other intriguing somaclonal variants include stringless celery, crunchier carrots, and popcorn with a built-in buttery taste.

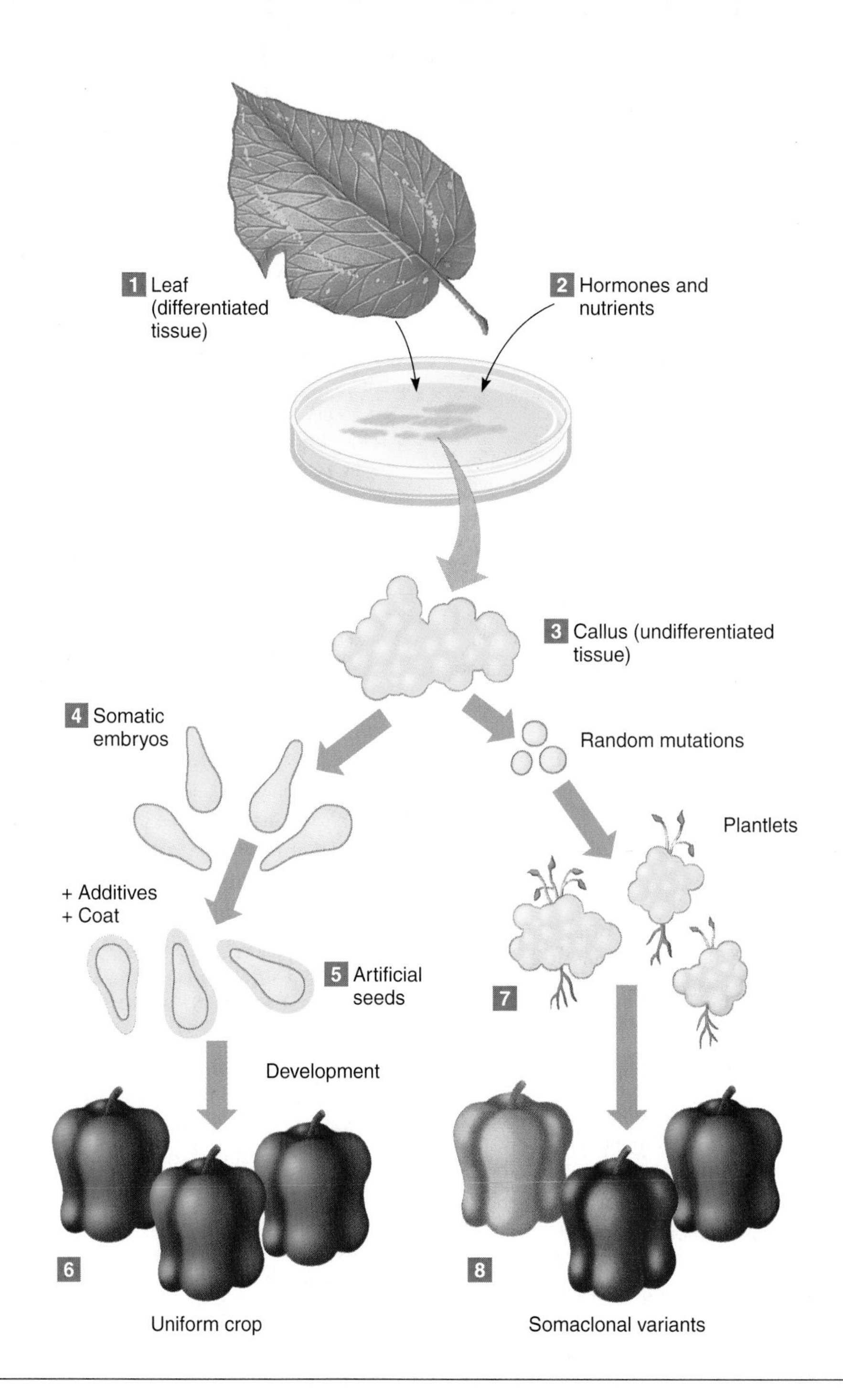

FIGURE 28.B
The Fate of a Callus—Embryos or Plantlets.
New plants grown from cell culture are valuable either for their uniformity or their variation. Leaf tissue (*1*) is cultured with hormones and nutrients in a dish (*2*), giving rise to a callus (*3*). Under certain conditions, the callus produces somatic embryos (*4*), which when encapsulated become artificial seeds (*5*). When planted, artificial seeds grow into genetically identical plants, producing a uniform crop (*6*). Under other conditions, callus cells undergo random mutations, producing somaclonal variant plantlets (*7*), which may yield new crop varieties (*8*).

example, in the Wasatch Mountains of Utah, a quaking aspen tree (*Populus tremuloides*) named *Pando* (Latin for "I spread") covers more than 100 acres of land! That plant, produced by asexual reproduction, consists of 47,000 tree trunks (each with an ordinary tree's usual complement of leaves and branches) and weighs more than 13.2 million pounds (6 million kilograms), making it arguably the world's most massive organism. It may also be the oldest—the clone may have started 1 million years ago!

28.1 MASTERING CONCEPTS

1. When are sexual and asexual reproduction each adaptive?
2. How does a plant reproduce asexually?

28.2 Sexual Reproduction

Flowers are sexual structures that house and nurture gametes and entice pollination, which is followed by fertilization.

The major groups of multicellular organisms—fungi, animals, and plants—all have different sexual life cycles. All involve meiosis, but the proportion of the life cycle spent as haploid or diploid cells varies among the kingdoms. Fungal life cycles are the most variable; some fungi are primarily haploid, a few are diploid, and many are dikaryotic (see chapter 23). In contrast, animal bodies are diploid. Eggs and sperm, the products of meiosis and gamete maturation, are usually the only haploid animal cells.

In plants, eggs and sperm form as a direct result of mitosis, not meiosis. Recall from chapter 22 that in sexually reproducing plants, the diploid generation, or **sporophyte,** produces haploid spores through meiosis (fig. 28.2). The haploid spores divide mitotically to produce a multicellular haploid **gametophyte,** which produces haploid eggs or sperm (gametes) by mitosis. These haploid gametes fuse to form a diploid zygote. The zygote grows and develops as its cells divide mitotically, eventually becoming a mature sporophyte, and the cycle begins anew.

In angiosperms, **flowers** are reproductive organs that are specialized for bringing together eggs and sperm, but they have other functions. A flower may protect its seeds as they develop. Also, certain parts of the flower develop into a **fruit,** which provides a mechanism for seed dispersal.

Flowers Are Reproductive Organs

Complete flowers have four types of structures, all of which are modified leaves. Each type of structure forms a whorl, or circle, around the end of the flower stalk (fig. 28.3). The calyx is the outermost whorl of a flower. It consists of green, leaflike **sepals,** which enclose and protect the inner floral parts. Inside the calyx is the corolla, which is a whorl of **petals.** In many flowers, large, colorful petals attract pollinators. The calyx and the corolla do not play a direct role in sexual reproduction and are therefore considered accessory parts of the flower.

The two innermost whorls of a flower are essential for sexual reproduction. The whorl within the corolla consists of male reproductive structures called **stamens,** which are stalklike filaments that bear pollen-producing oval bodies called **anthers** at their tips. The whorl at the center of a flower consists of the female reproductive structures, collectively called the pistil. The pistil, in turn, is made of

FIGURE 28.2

Alternation of Generations in Flowering Plants.

The flowering plant life cycle includes an alternation of haploid and diploid generations. Megaspores and microspores divide mitotically to produce megagametophytes and microgametophytes, which produce eggs and sperm.

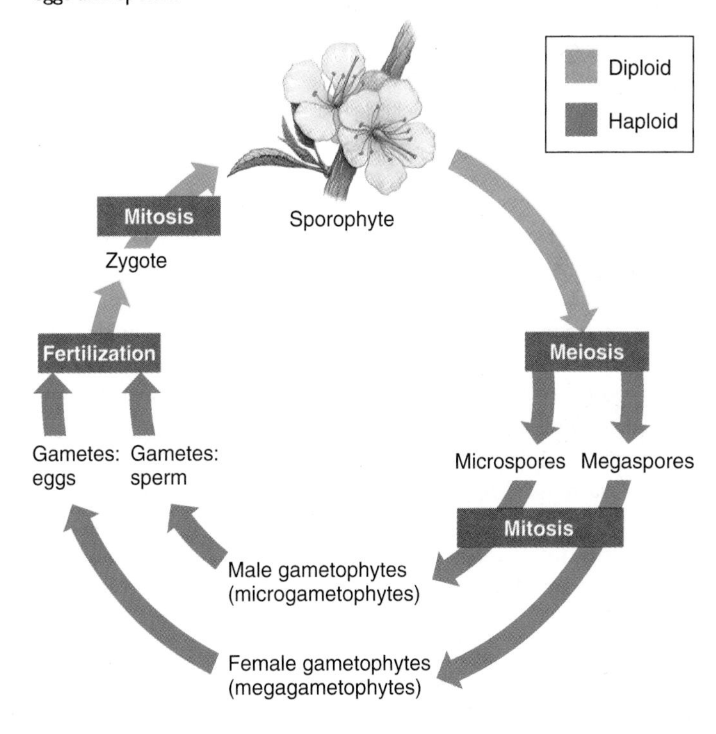

FIGURE 28.3

Parts of a Flower.

A flower is made up of four levels, or whorls. The outermost whorl (whorl 1) is sepals, the next whorl (whorl 2) is petals, the next whorl (whorl 3) is stamens, and the innermost whorl (whorl 4) is the pistil, which consists of one or more carpels. The pistil of the flower shown here has only one carpel.

Whorl 1
Calyx, made up of all sepals

Whorl 2
Corolla, made up of all petals

Whorl 3
Stamens, male reproductive organs

Whorl 4
Carpels, female reproductive organs

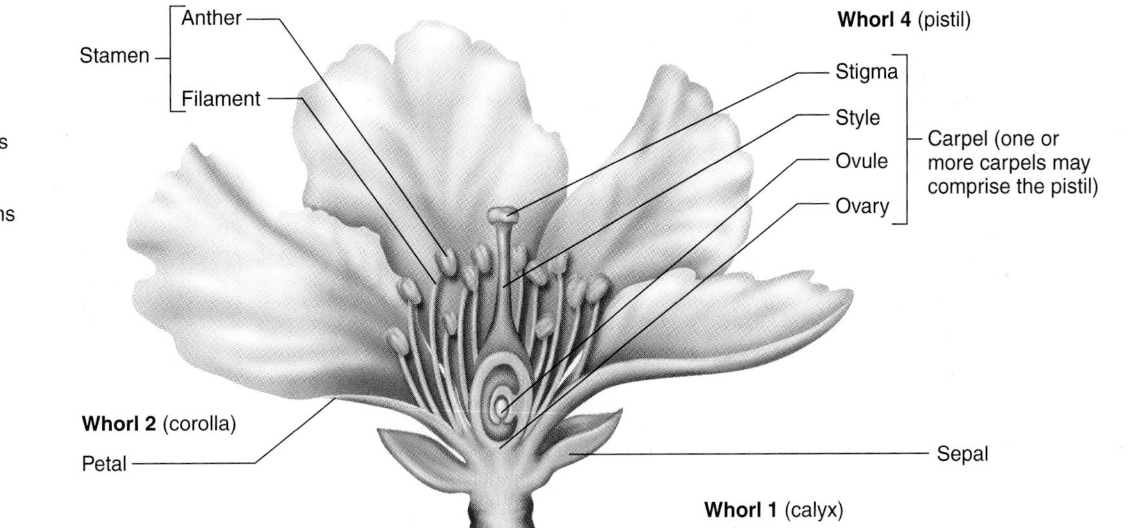

Unraveling the Genetic Controls of Flower Formation

All multicellular organisms have genes that operate early in development, and ensure that the right parts form in the right places. These are called homeotic genes. Figure 10.20 shows how a mutation in such a gene mixes up development in the fruit fly *Drosophila melanogaster,* causing legs to grow in place of antennae. These types of genes also control development of flowers, where having the right parts in the right places is critical.

Developmentally, flower formation is straightforward. Meristematic tissue in a stem tip differentiates into structures of one of a flower's four whorls—sepal, petal, stamen, or carpel. If single genes control "decisions" on such specialization, then mutations in those genes should produce flowers with parts in the wrong places but in predictable patterns. This is indeed the case.

Many plant geneticists work with *Arabidopsis thaliana,* a small plant in the mustard family. In the early 1980s, California Institute of Technology researcher Elliot Meyerowitz alerted colleagues that he was looking for abnormal *Arabidopsis* flowers. He and his coworkers also exposed seeds to mutagens to produce mutant flowers. Soon, they had a collection of unusual flowers that fell into three categories. Instead of the normal sequence, from outside inwards, of sepals-petals-stamens-carpels (fig. 28.C), the mutant flowers were either carpels-stamens-stamens-carpels, sepals-sepals-carpels-carpels, or sepals-petals-petals (fig. 28.D).

Geneticists like to figure out how genes interact to produce certain phenotypes. Meyerowitz hypothesized how a few mutations could produce each of the three mutant phenotypes he observed. Here are his "rules," followed by explanations of how the mutant classes arise:

Rule 1: Three groups of genes control flower formation: *A, B,* and *C.* Each class of genes is expressed in certain parts of the flower:

- In sepals, *A* genes are expressed.
- In petals, *A* and *B* genes are expressed.
- In stamens, *B* and *C* genes are expressed.
- In carpels, *C* genes are expressed.

Put another way, *A* genes are expressed in the outer two whorls (sepals and petals), *B* genes are expressed in the middle two whorls (petals and stamens), and *C* genes are expressed in the inner two whorls (stamens and carpels).

Rule 2: When an *A* gene is inactive, a *C* gene replaces it.
Rule 3: When a *C* gene is inactive, an *A* gene replaces it.

After geneticists devise rules, they induce mutations to see if the model accurately predicts the phenotypes that arise. Meyerowitz inactivated each class of gene to see what effect this had on the phenotype. The results supported his hypothesis.

FIGURE 28.C
Two Views of a Wild-type (Normal) Flower of *Arabidopsis thaliana.*
Sepals, petals, stamens, and the stigma are clearly visible in normal (wild-type) flowers.

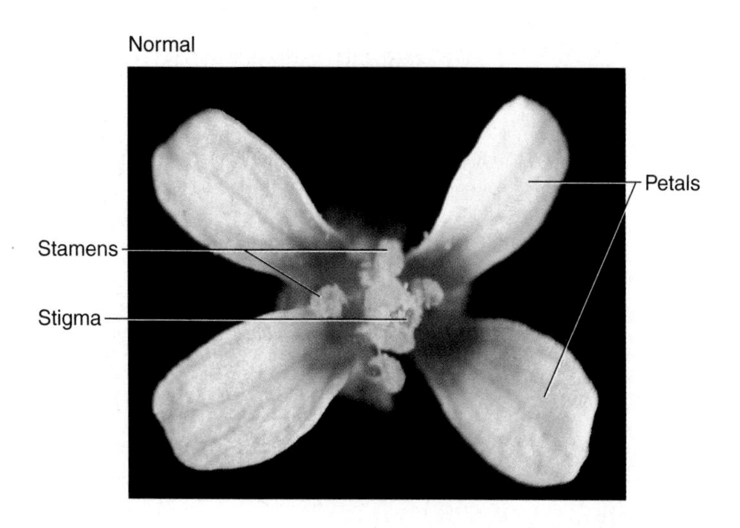

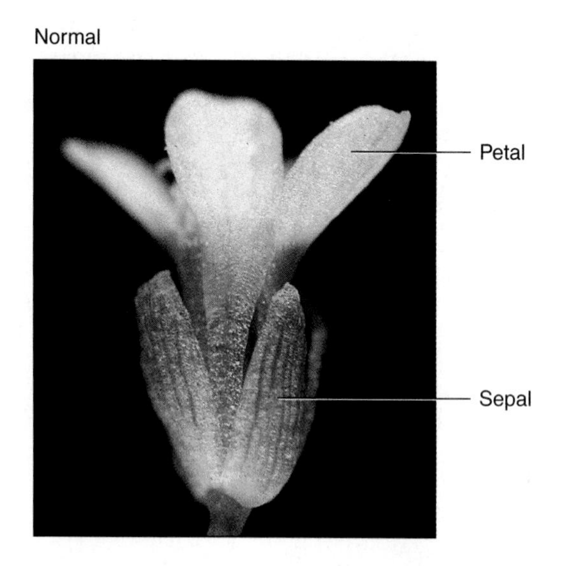

one or more **carpels,** which are leaflike structures enclosing the **ovules.** The bases of carpels and their enclosed ovules make up the **ovary.** The upper part of the carpels are stalklike **styles** that bear structures called **stigmas** at their tips. Stigmas receive pollen.

Not all flowers have both male and female parts. A flower is considered "perfect" if it has stamens and carpels, and "imperfect" if it has only one or the other. An individual plant with male and female flowers is called **monoecious** (literally, "one house"). A plant

Class A Mutations

(Remember that if an *A* gene is inactivated, a *C* gene takes over.)

- The first whorl (normally sepals) sees *C* instead of *A* and develops as carpels.
- The second whorl (normally petals) sees *C* and *B* instead of *A* and *B* and develops as stamens.
- The third whorl (normally stamens) still sees *B* and *C* and develops normally as stamens.
- The fourth whorl (normally carpels) still sees *C* and develops normally as carpels.

The resulting flower whorl pattern is carpels-stamens-stamens-carpels. The flower lacks sepals and petals.

Class B Mutations

- The first whorl (normally sepals) still sees *A* and develops normally as sepals.
- The second whorl (normally petals) only sees *A* and therefore also forms sepals.
- The third whorl (normally stamens) sees only *C* instead of *B* and *C* and therefore develops as carpels.
- The fourth whorl (normally carpels) still sees only *C* and develops normally as carpels.

The resulting flower whorl pattern is sepals-sepals-carpels-carpels. The flower lacks petals and stamens.

Class C Mutations

(Remember that if a *C* gene is inactivated, an *A* gene takes over.)

- The first whorl (normally sepals) still sees *A* and develops normally as sepals.
- The second whorl (normally petals) still sees *A* and *B* and therefore forms petals.
- The third whorl (normally stamens) instead of *B* and *C* sees *B* and *A* and develops petals.
- The fourth whorl (normally carpels), for reasons unknown, does not develop.

Deletion of *C* also apparently lifts control on whorl formation. This mutant flower develops many whorls, in the pattern sepals-petals-petals.

Meyerowitz then made double mutants, inactivating two gene classes at a time. Again, his predictions of flower structure were accurate. Other researchers soon found that the same broken genetic rules underlie mutant flowers in other species. The picture is even more complex than explained here, because yet other genes control the *A, B,* and *C* genes.

Not surprisingly, like homeotic genes of animals, the *A, B* and *C* genes encode transcription factors, which are proteins that activate other genes. Although the DNA sequences in plants and animals are not the same, these two types of organisms do share a fundamental aspect of their development—control by a very few genes.

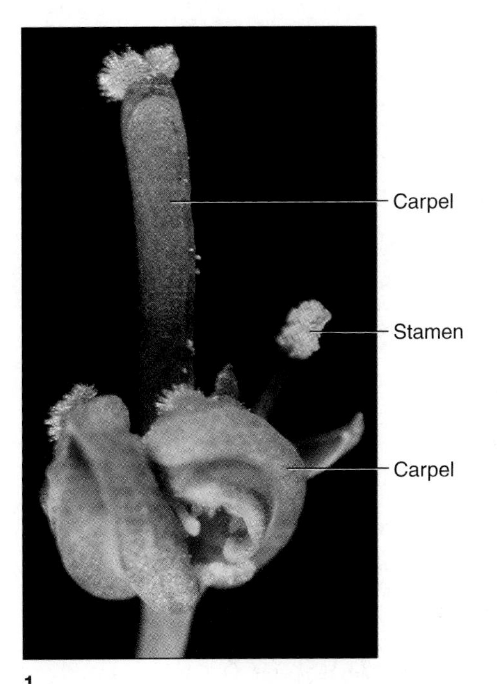

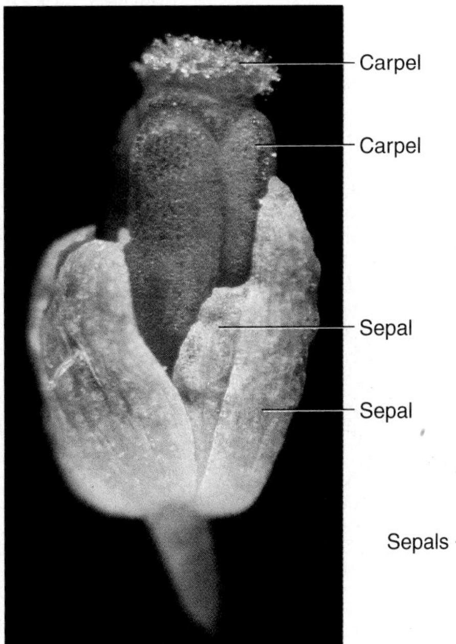

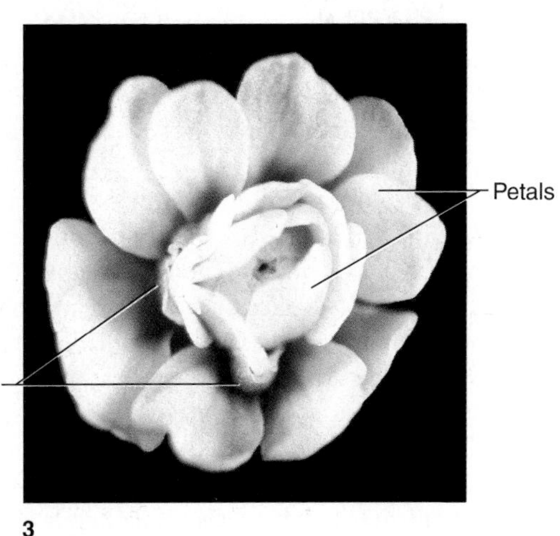

FIGURE 28.D
Mutant Flowers.
(*1*) A mutation in an *A* gene yields a flower that consists of carpels-stamens-stamens-carpels (only one stamen is visible). (*2*) A mutation in a *B* gene yields a flower that consists of sepals-sepals-carpels-carpels. (*3*) A mutation in a *C* gene yields a flower that consists of several repeats of sepals-petals-petals.

species with opposite sex flowers on different plants is called **dioecious** ("two houses"). Corn is monoecious; hemp is dioecious.

Investigating Life 28.1 describes how mutant genes misdirect development so that flowers have their parts in the wrong places.

Gamete Formation

In flowering plants, the gametophyte generation is very small, consisting of only a few cells. Because the female egg-producing

FIGURE 28.4

Major Steps in the Life Cycle of a Flowering Plant.

A sperm cell fertilizes an egg in the ovary of a flower. Then the ovule develops into a seed, which contains the embryo. The ovary matures into a fruit enclosing the seed. The seed is dispersed—perhaps with the help of a hungry or passing animal—and germinates, developing into the mature sporophyte.

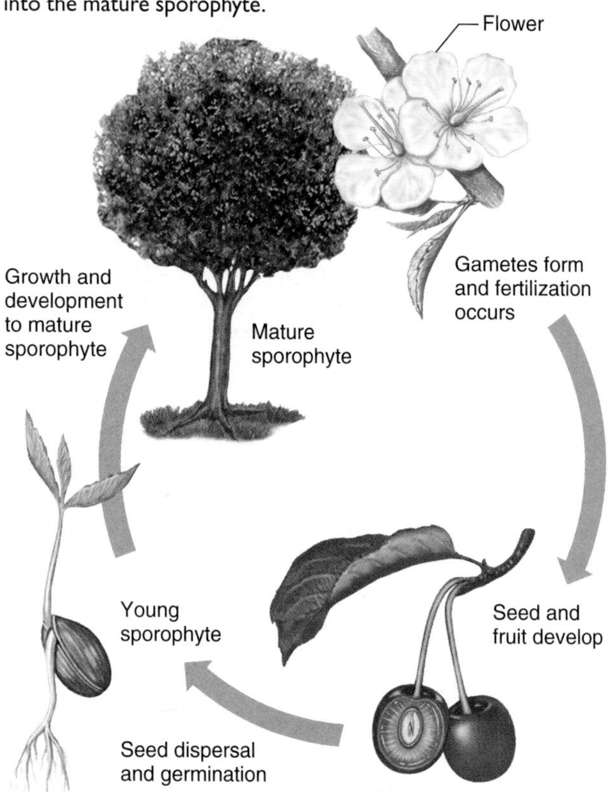

FIGURE 28.5

Sperm and Egg Formation in a Flowering Plant.

Pollen, which gives rise to sperm, is produced in pollen sacs on stamens. Eggs are produced in ovules inside the ovary.

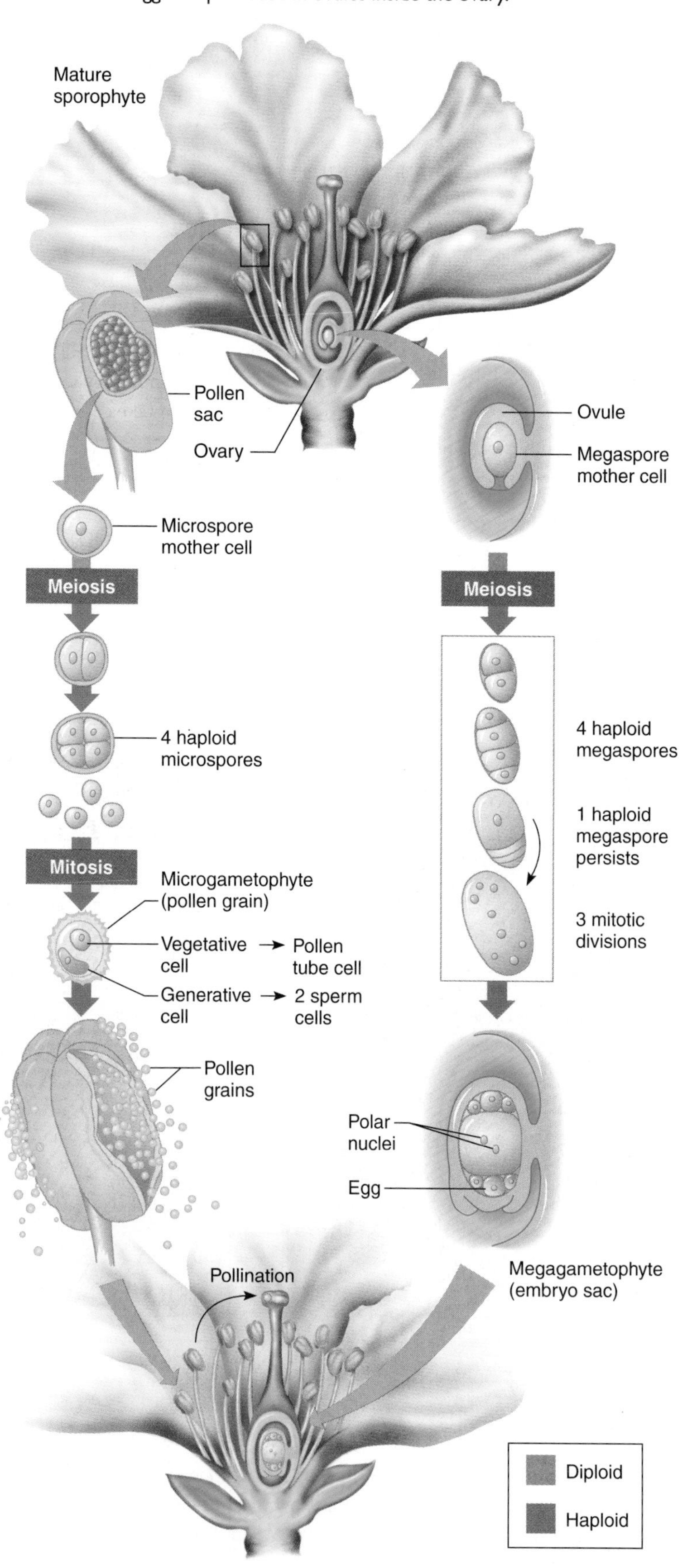

gametophytes are often larger than the male sperm-producing gametophytes, the female structures are called **megagametophytes,** and the male structures are called **microgametophytes.** Female and male gametophytes arise from two types of spores called **megaspores** and **microspores.** Figure 28.4 shows the place of gamete formation and fertilization within the entire angiosperm life cycle.

Microspores form in the anthers at the tips of the stamens (fig. 28.5). Each anther contains four **pollen sacs,** which contain microspore mother cells. These cells divide meiotically to produce four haploid microspores. Each microspore then divides mitotically and produces a two-celled, thick-walled structure called a **pollen grain,** which is the young male gametophyte. A mature pollen sac opens to release millions of pollen grains.

The two haploid cells in the pollen grain are called the generative cell and the vegetative cell. The generative cell divides mitotically to form two sperm cells, either before or after the pollen sac sheds the pollen grains. The vegetative cell gives rise to a pollen tube when the pollen grain reaches a stigma. The pollen tube transports sperm cells to the ovule.

Megaspores form in the ovary of the flower. The ovary may contain one or more ovules, inside of which a megaspore mother cell divides meiotically to produce four haploid megaspores. In many species, three of these cells quickly disintegrate, leaving one large megaspore. The megaspore undergoes three mitotic divisions to form a female gametophyte with eight haploid nuclei but only seven cells—one large, central cell has two nuclei called **polar nuclei.** In a mature megagametophyte, also called an **embryo sac,** one of the cells with a single nucleus is the egg.

Pollination Brings Pollen to the Stigma

Pollination is the transfer of pollen from an anther to a receptive stigma. Some angiosperms are self-pollinating, meaning that pollen grains are transferred from the anther to a stigma on the same plant. Other angiosperms are outcrossing species. In these plants, pollen grains from one flower are carried to a stigma on a different plant's flower.

Outcrossing is beneficial in a changing environment because it produces a genetically variable population that can adapt to changing conditions. But how do flowers promote outcrossing? Some angiosperms are dioecious, with male and female flowers on separate plants, making self-pollination impossible. Other plants, such as the magnolia, ensure cross-pollination by releasing pollen only after its eggs have been fertilized by pollen from other individuals. Still other plants use biochemical signals to prevent their pollen from germinating on their own stigmas, a phenomenon called self-incompatibility.

Animals often transfer pollen from plant to plant. Common pollinators are insects, birds, and bats (fig. 28.6); rarer pollinators

FIGURE 28.6
Animals as Pollinators.
Animal pollinators include insects, birds, bats, and many others that brush past flowering plants. Bumblebees (**A**) and hummingbirds (**B**) are efficient pollinators. Note the shape of a hummingbird's bill—suited to poking into a flower. Certain moths, such as the striped morning sphinx moth (*Hyles lineata*) (**C**), strikingly resemble hummingbirds, a case of convergent evolution reflecting similar adaptations for efficient pollination. The moth has a long proboscis that it uses to gather nectar from flowers. (**D**) The bat *Leptonycteris curasoae* rubs itself on a flower of a saguaro cactus, *Cereus giganteus*.

A

B

C

D

FIGURE 28.7
Ultraviolet Flower Markings.
Bees can detect ultraviolet wavelengths of light, so this black-eyed Susan (**A**), which appears fully yellow to us, has a dark patch in the center to bees (**B**).

A

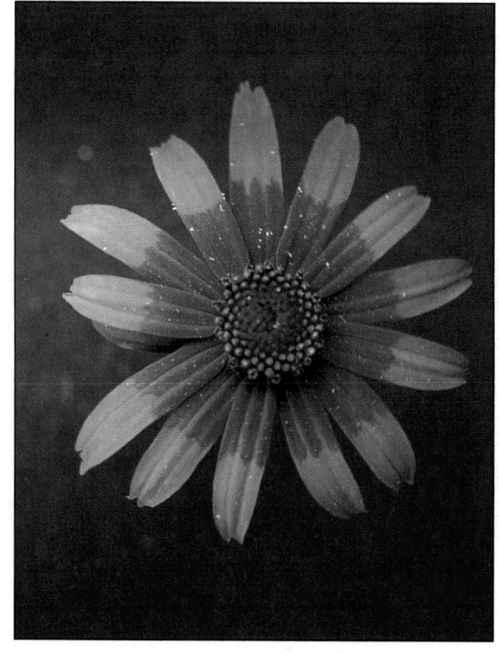
B

include geckos, gliding possums, lemurs, flying foxes, and even an occasional human hiker who brushes against flowers and transports the pollen to other plants of the same species. Many pollinators benefit from their association with plants—they obtain food in the form of pollen or nectar, they may seek shelter among the petals, or they use the flower for a mating ground. Nectar is a sweet-tasting substance that has no known function in plants other than attracting pollinators.

Flower color, shape, and odor attract particular animals to particular flowers. For example, birds are attracted to red flowers, a color that insects cannot distinguish. Different insects have characteristic floral preferences as well. Bees are attracted to blue or yellow sweet-smelling blooms, whereas beetles respond to dull-colored flowers with spicier scents. Because bees can detect ultraviolet light, bee-pollinated flowers often have markings that are visible only under this wavelength of light (fig. 28.7). A most intriguing flower is that of *Lisianthius nigrescens*—it looks like jet-black tubular bells. Like a black hole that absorbs all light in deep space, so does this flower absorb many wavelengths of light, including ultraviolet. It reflects none, appearing black. Researchers have yet to discover the plant's pollinator.

Most moths pollinate white or yellow heavily scented flowers, which are easy to locate at night, when these insects are most active. The flowers may have flat surfaces that their pollinators use as landing strips. In the tropics, where more flowers are open at night, bats are important pollinators.

Some flowers produce heat to attract pollinators. Skunk cabbage (*Symplocarpus foetidus*), a member of the lily family, maintains its flowers from 59°F to 72°F (15°C–22°C), even when the temperature drops to below freezing (fig. 28.8). The heat emanat-

FIGURE 28.8
Hot Plants.
Jean-Baptiste de Lamarck described the heat given off by the arum lily (*Arum italicum*) in 1778. This eastern skunk cabbage (*Symplocarpus foetidus*) also generates heat, which vaporizes scents and may increase the activity of pollinators.

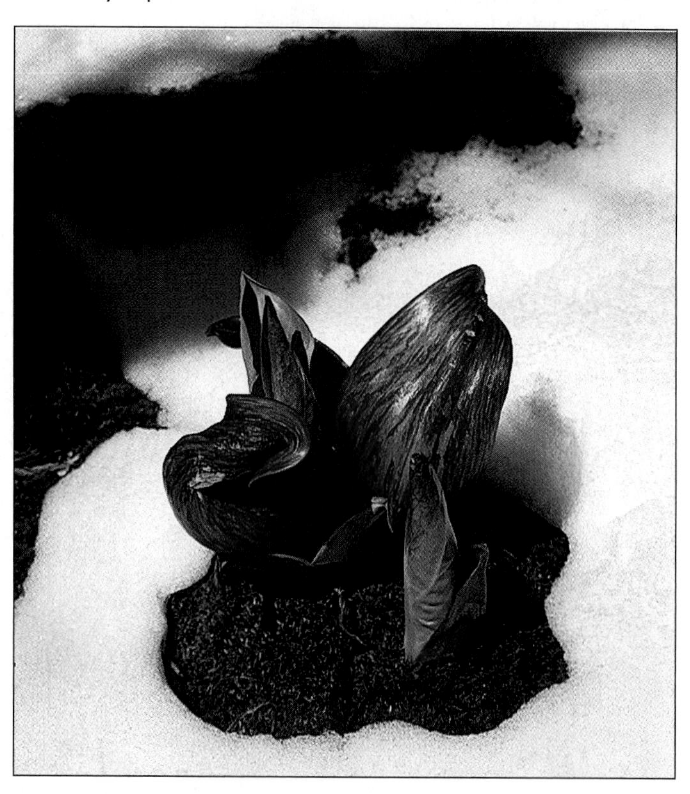

ing from the flowers has two functions. First, it volatilizes aromatic molecules, releasing a skunklike odor that attracts pollinators. Secondly, pollinators use the heat to power their activity. Many insect pollinators become sluggish when the temperature drops and muscle function slows.

As the skunk cabbage illustrates, not all floral scents are pleasing to the human nose. In South Africa, the "carrion flowers" of stapelia plants smell remarkably like rotting flesh, a repulsive scent to humans but a highly attractive scent to flies. As if the stench were not sufficient to beckon the insects, the stapelia's leaves are wrinkled and reddish brown, resembling decaying meat.

Many types of angiosperms (and most gymnosperms) use wind rather than animals to carry pollen. Oaks, cottonwoods, ragweed, and grasses shed large quantities of pollen that breezes disperse; this pollen output provokes allergic reactions in many human sufferers. The wind, however, does not carry pollen very far. This is why wind-pollinated plants must grow closely together and manufacture abundant pollen. Wind pollination is also far less precise than animal pollination. The wind drops pollen where and when it slackens, whereas an animal delivers pollen directly to a particular type of plant. The flowers of wind-pollinated angiosperms are small, greenish, and odorless. Rather than invest energy in developing perfume and showy flowers, it is more efficient for these plants to invest energy in pollen production.

With so much pollen carried by animals and the wind, certainly pollen is deposited on stigmas of other species. Experiments show that matching pollen to the correct species is not a passive process. In *Arabidopsis,* signals or structures from the cell walls of the stigma "grab" pollen from that species, and that species only. Researchers allowed *Arabidopsis* plants to bind pollen, and then attempted to dislodge it by spinning the flowers in a centrifuge. The pollen wouldn't budge. But when *Arabidopsis* was exposed to pollen from petunia plants, the pollen wouldn't even bind. One researcher described the role of the cell wall in this species-recognition process as "a playground where a lot of molecules can interact and talk with each other."

Double Fertilization Yields Zygote and Endosperm

After a pollen grain lands on a stigma of the correct species, the vegetative cell gives rise to a pollen tube (fig. 28.9). The pollen grain's two sperm cells enter the pollen tube as it grows through the tissues of the stigma and the style towards the ovary. When the pollen tube reaches an ovule, it discharges its two sperm cells into the embryo sac. One sperm fuses with the egg nucleus and forms a diploid zygote. After a series of mitotic cell divisions, the zygote becomes the embryo.

The second sperm fuses with the two polar nuclei and forms a triploid cell. Recall that a triploid cell has three complete sets of chromosomes. The triploid nucleus divides to form a tissue called **endosperm,** which is stored food for the developing embryo. Familiar endosperms are coconut milk and the starchy part of a kernel of corn. Notice that the egg and the polar nuclei both are fertilized, a phenomenon termed **double fertilization.**

Double fertilization is characteristic of all angiosperms. To date, it has been found in only one other group of plants—the

FIGURE 28.9
Double Fertilization.
Pollen sticks to a stigma on a flower of the same species. A pollen tube grows from the vegetative cell toward the ovule, and transports two sperm. One sperm fertilizes the egg to form a zygote, and the other fertilizes the polar nuclei to yield the endosperm.

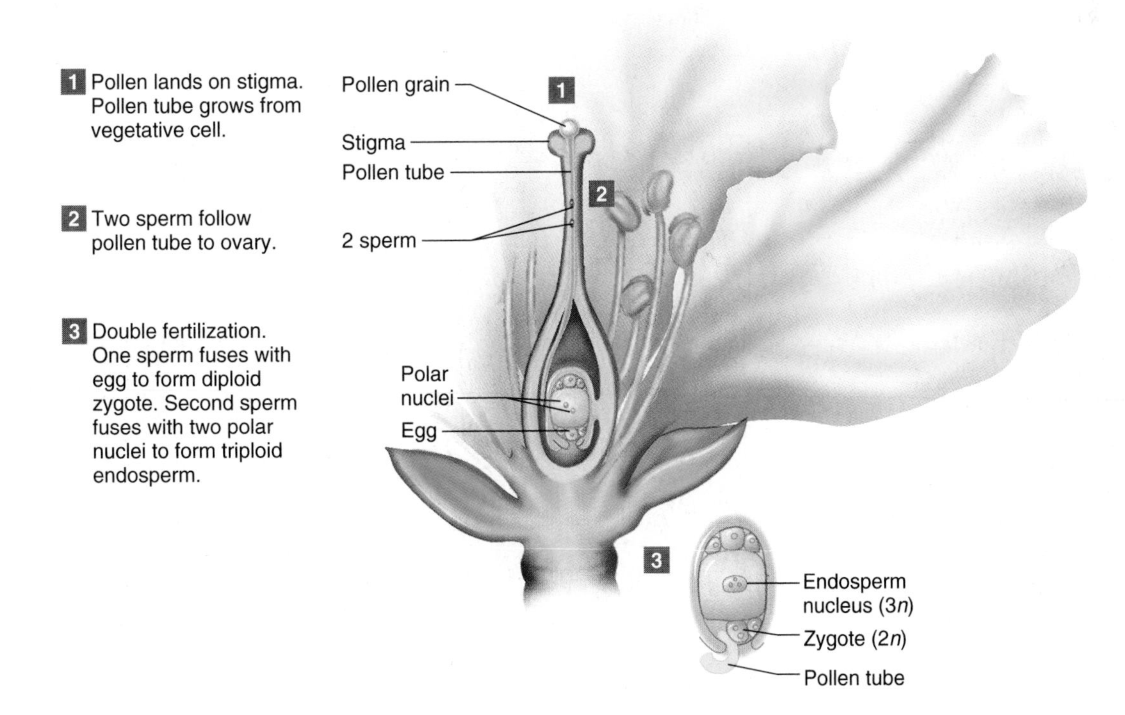

gnetophytes. As discussed in chapter 22, the gnetophytes form a group of gymnosperms that may be most closely related to angiosperms.

28.2 MASTERING CONCEPTS

1. How do female gametophytes differ from male gametophytes?
2. What are the parts of a flower?
3. How do female and male gametes of an angiosperm form?
4. How does pollen move from one flower to another?
5. Describe the events of double fertilization.

OLC

28.3 Seeds and Fruits

Seeds protect the earliest stages of sporophyte development. Fruits protect seeds from drying out and being destroyed, while providing a vehicle for their dispersal.

The seed protects and nourishes the new plant, and the fruit aids in seed dispersal.

Seed Development and Dormancy

Immediately after fertilization, the ovule contains an embryo sac with a diploid zygote and a triploid endosperm, both of which are encased in layers of maternal tissue (fig. 28.10). Initially, the

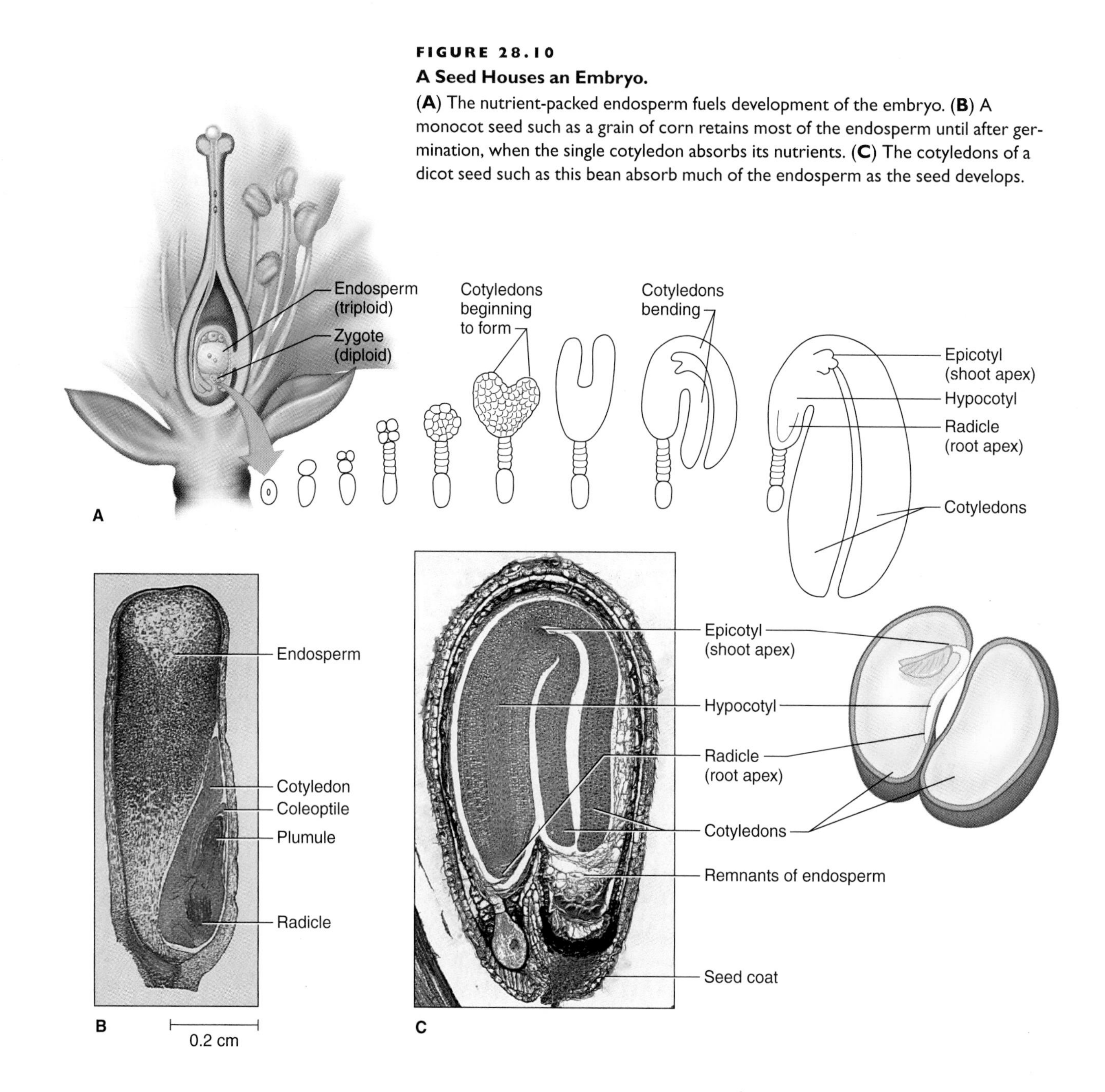

FIGURE 28.10
A Seed Houses an Embryo.
(**A**) The nutrient-packed endosperm fuels development of the embryo. (**B**) A monocot seed such as a grain of corn retains most of the endosperm until after germination, when the single cotyledon absorbs its nutrients. (**C**) The cotyledons of a dicot seed such as this bean absorb much of the endosperm as the seed develops.

endosperm cells divide more rapidly than the zygote and thus form a large multicellular mass. This endosperm serves as the nutrient supply for the developing embryo.

The developing embryo forms **cotyledons,** or seed leaves. Recall that angiosperms that have one cotyledon are monocots, and those with two cotyledons are dicots (figs. 22.18 and 26.21). In monocots, the cotyledon does not absorb the endosperm but transfers nutrients from it to the nonleaf part of the embryo during germination. In many dicots, the cotyledons become thick and fleshy as they absorb the endosperm.

The shoot apical meristem forms at the tip of the epicotyl, which is the stemlike region above the cotyledons. The stemlike region below the cotyledons is called the hypocotyl. The root apical meristem differentiates near the tip of the embryonic root, or radicle. When one or more embryonic leaves form on the epicotyl, the epicotyl plus its young leaves is called a plumule. (You can see the tiny plumule of an embryonic peanut plant if you carefully part the two halves of a peanut.) In monocots, a sheathlike structure called the coleoptile covers the plumule.

At a certain point in embryonic development, in response to hormonal signals, cell division and growth stop, and the embryo becomes dormant. A tough outer layer, the **seed coat,** protects the dormant plant embryo and its food supply. Together, the plant embryo, stored food, and seed coat make up the **seed.**

Seed dormancy is a crucial adaptation that enables seeds to postpone development when the environment is unfavorable, such as during a drought or frost. Some seeds can delay development for centuries! The oldest seed to germinate was a 1,288-year-old lotus seed from China, which germinated in 1996 in just 4 days. Favorable conditions trigger growth to resume when young plants are more likely to survive.

Fruit Formation

A flower begins to change as seeds form. When a pollen tube begins growing, the stigma produces large amounts of ethylene, a simple organic molecule that is a plant hormone, discussed in the next chapter. (A hormone is a biochemical produced in one part of an organism that exerts a physiological effect on some other part of the organism.) Ethylene triggers senescence (aging) of the flower. Floral parts that are no longer vital—usually all parts except the ovary—wither and fall to the ground.

In many angiosperms, the ovary, and in some plants also surrounding tissues, grows rapidly and extensively to form a structure that contains one or more seeds—this structure is a fruit. Some fruit parts, such as the pulp of an apple, derive from the region where the flower attaches to the shoot, called the **receptacle.** Figure 28.11 shows how the parts of a flower give rise to specific parts of fruits. In a green bean, for example, the seeds form from the ovules and the pod is the ovary.

Fruit parts derived from the ovary consist of three layers. The skin is the exocarp; the middle part, which can be very fleshy, is the mesocarp, and the interior layer, which hugs the seed(s), is the endocarp. The layers are distinctive in fleshy fruits, but in dry fruits they are joined into one layer, called the pericarp.

Botanists classify fruits based on fleshiness or dryness, and origins from floral parts (table 28.1). Simple fleshy fruits develop from one or more united carpels. These fruits are of three types. A drupe is a fleshy fruit with a hard pit that surrounds one or more carpels that have a single seed each. A peach is a drupe. A berry (such as a tomato) is fleshy and each of its one or more carpels has many seeds. A pome is a specialized type of simple fleshy fruit in which much of the pulp comes from the receptacle. Pomes

FIGURE 28.11
Fruit Parts Derive from Flower Parts.
(**A**) After a flower is pollinated and the egg is fertilized, the flower begins to develop into a fruit to house the seeds. The fleshy part of an apple develops from the receptacle, which grows to enclose the ovary and seeds within. (**B**) Flowering plants differ markedly in the ways that fruit parts develop from flowers.

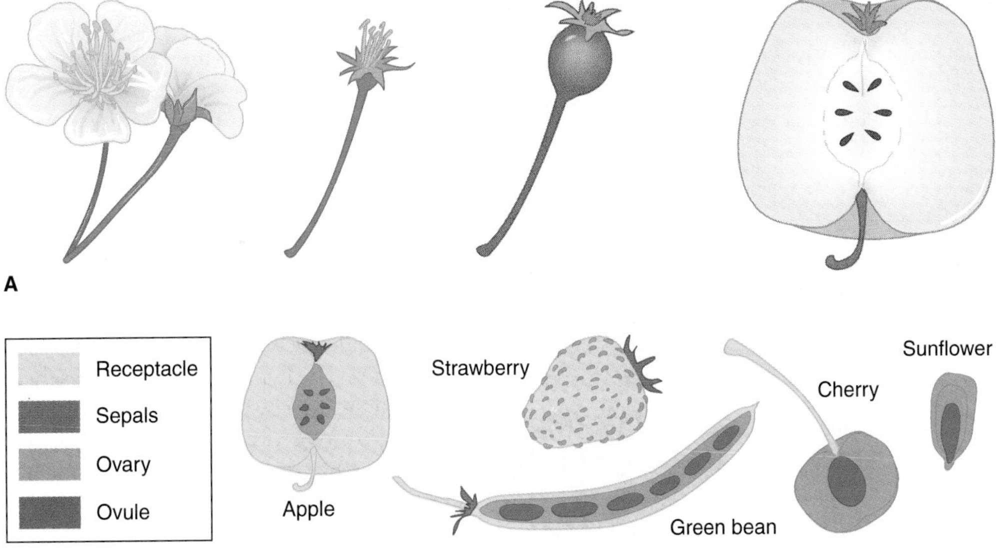

TABLE 28.1 Types of Fruits

Fleshy fruits	Characteristics	Examples
Simple	**One or more united carpels**	
Drupe	Fleshy fruit, hard pit, one or more carpels with single seed each	Olive Cherry Peach Plum Coconut
Berry	One or more carpels, each with many seeds; inner part fleshy	Grape Tomato Pepper Eggplant
Pome	Tough core is ovary; pulp derived from flower base area (receptacle)	Apple Pear
Complex	**Multiple separate carpels**	
Aggregate	Derives from one flower with many separate carpels	Blackberry Strawberry Raspberry Magnolia
Multiple	Develops from tight clusters of flowers	Pineapple
Dry fruits	**Characteristics**	**Examples**
Dehiscent	Mature fruit splits and releases seeds	Bean Pea Peanut Radish Milkweed
Indehiscent	Mature fruit remains around seed(s)	
	Hard, thick pericarp	Hickory Acorn Chestnut
	Thin pericarp	Parsley Carrot Maple Sunflower

include apples and pears. Complex fleshy fruits include aggregate fruits that develop from one flower with many separate carpels, and multiple fruits, which are actually several fruits that develop from a cluster of flowers and remain attached.

Dry fruits are of two types. Dehiscent fruits split down one or more seams to release seeds. They include the legumes, such as peas and peanuts. Indehiscent fruits may have hard, thick pericarps that remain around the seed. Nuts belong to this group. The pericarp may also be thin, as in maple or sunflower fruits.

Seed Dispersal and Germination

Fruits protect seeds from desiccation and destruction, and facilitate seed dispersal. Animals, which often depend on fruits and seeds for food, can carry seeds far from a parent plant, as can wind and water.

Colored berries attract birds and other animals, which carry the ingested seeds to new locations, where they are released in the animals' feces. Some seeds germinate only after passing through the intestines of birds or mammals. Animals such as squirrels that hide seeds to eat later also unknowingly disperse seeds when they forget some of their cache locations.

Birds and mammals spread seeds from place to place when fruits bearing hooked spines attach to their feathers or fur. The fruit of the burdock plant, for example, has barbed hooks that cling to a passing deer or a hiker's jeans. The inspiration for Velcro, a fuzzy fabric that sticks to other fabrics, came from the strong attachment of burdock fruits (fig. 28.12).

Wind and, less commonly, water can also distribute seeds. Wind-dispersed fruits such as those of dandelions and maples have wings or other structures that enable them to ride on air currents and land far from the original plant. Coconuts are water-dispersed fruits that travel long distances before colonizing distant islands. (Store-bought coconuts are actually seeds with a watery endosperm, with the green husk that forms the rest of the fruit already removed.)

Germination is the resumption of growth and development after dormancy is broken. It usually requires water, oxygen, and a source of energy. First, the seed imbibes (absorbs) water. In some seeds, imbibition causes the embryo to release hormones that stimulate the endosperm's breakdown. The starch in the endosperm is broken down to sugars that the embryo can use for energy. Imbibition also swells a seed, eventually rupturing the seed coat and exposing the plant embryo to oxygen. At this point, an embryo may resume growth; however, seeds of some plants normally germinate only after additional stimulation, such as exposure to light of a certain intensity or several days at a particular temperature.

After the growing embryo bursts out of the seed coat, further growth and development depend upon the root and shoot apical meristems. The hypocotyl, with its attached radicle, emerges first from the seed. In response to gravity, the radicle grows downward and anchors the plant in the soil. Root systems develop rapidly, providing the continual water and mineral supply that plants require to grow.

The shoot emerges from the soil in different ways, depending on the species. In most dicots, the elongating hypocotyl forms an arch that breaks through the soil and straightens in response to

FIGURE 28.12
Seed Dispersal.
(**A**) Dandelion fruits have fluff that enables them to float on a breeze. (**B**) Some seed pods have explosive methods of scattering seeds. Seeds pop out of this orchid's pod. (**C**) This cedar waxwing helps disperse the seeds of winterberry, a type of holly. (**D**) Velcro resembles the burdock fruit, which is specialized for hitching a ride on passing animals. (**E**) This tumbleweed plant looks calm and firmly rooted, but if a brisk wind comes along, the entire plant can blow away, shedding seeds.

light, pulling the cotyledons and epicotyl up out of the soil. In most monocots, the coleoptile-protected epicotyl begins growing upward shortly after root growth starts. The single cotyledon remains underground (fig. 28.13).

A plant begins photosynthesizing as its shoot emerges from the soil. By the time the plant has exhausted its embryonic food reserves, it is already producing its own food photosynthetically. If conditions are favorable, the seedling continues to grow until it reaches reproductive maturity, which may take weeks, years, or even decades, depending on the species. At that time it too will develop flowers, continuing the life cycle. During its life span, the plant may encounter fluctuating environmental conditions; the next chapter describes some of the ways the plant may respond.

28.3 MASTERING CONCEPTS

1. How does seed development differ in monocots and dicots?
2. What is a fruit?
3. How are fruits classified?
4. What happens during seed germination?

OLC

FIGURE 28.13

Seeds Yield New Plants.

Seed structure, germination, and development in (**A**) a monocot (corn) and (**B**) a dicot (green bean).

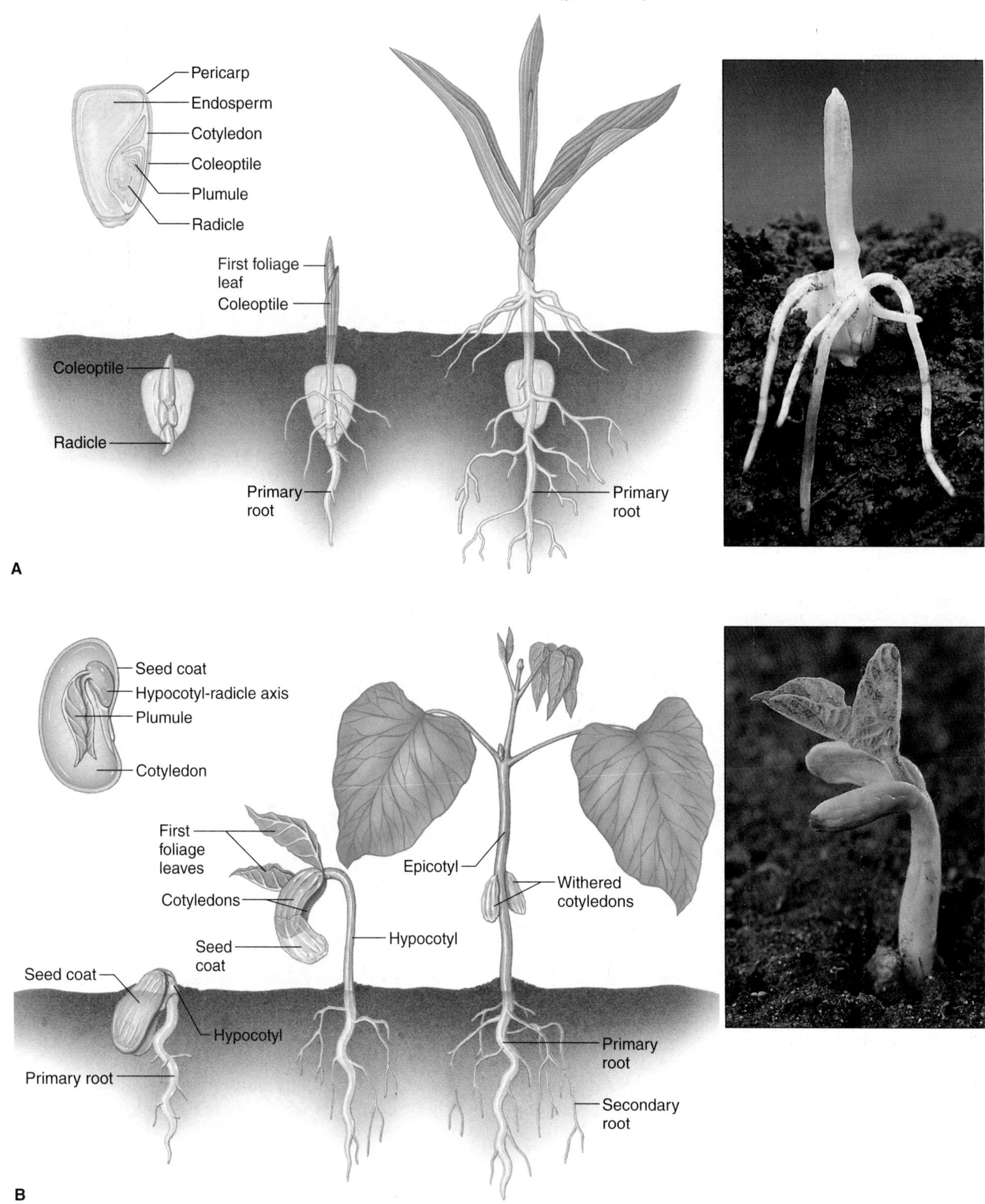

Chapter Summary

28.1 Asexual Reproduction

1. In **asexual reproduction,** a parent plant gives rise to clones, which can develop from roots, stems, or leaves.
2. Asexual reproduction is advantageous in a stable environment where plants are well adapted to their surroundings.

28.2 Sexual Reproduction

3. **Flowers** are reproductive structures built of whorls of parts. The calyx, made of **sepals,** and the corolla, made of **petals,** are accessory parts. Inside the corolla, the male parts consist of **stamens** and their pollen-containing **anthers.** At the center of the flower are the female parts, including the **ovary.** The **stigma** extends from the ovary and captures pollen.
4. A perfect flower has male and female parts; an imperfect flower has either. In a **monoecious** plant, opposite-sex flowers are on the same plant. In a **dioecious** plant, a plant has flowers of only one sex.
5. Female gametophytes are **megagametophytes,** and male structures are **microgametophytes.** They arise from **megaspores** and **microspores,** respectively.
6. Male and female structures produce gametes. In the anther, **pollen sacs** contain microspore mother cells. Each divides meiotically to yield four haploid microspores, which each divide mitotically to yield a haploid generative cell and a haploid vegetative cell; these two cells and their covering are a **pollen grain.** The pollen grain is the immature male gametophyte. Sperm cells arise from the generative cell and a pollen tube grows from the vegetative cell.
7. In the ovary, megaspore mother cells divide meiotically to yield four haploid cells, one of which persists as a haploid megaspore that divides mitotically three times. The resulting megagametophyte, or **embryo sac,** contains seven cells. One is the egg.
8. Animals or wind transfer pollen from the anthers of one plant to its own or another plant's stigma. Flower structures and odors are adapted to encourage animal or wind **pollination.**
9. The stigma retains pollen from the same species.
10. Once on a stigma, a pollen grain grows a pollen tube, and its two sperm move through the tube towards the ovary. In the embryo sac, one sperm fertilizes the egg to form the zygote, and the second sperm fertilizes the **polar nuclei** to form the **endosperm.** This phenomenon is termed **double fertilization.**

28.3 Seeds and Fruits

11. A **seed** is an embryo, endosperm, and **seed coat.** The endosperm nourishes the developing embryo.
12. As the embryo grows, **cotyledons** develop. In dicots, the cotyledons absorb the endosperm and become fleshy.
13. Cells in apical meristems divide to promote the growth of shoot and root in the embryo.
14. The epicotyl is the stemlike region above the cotyledons; along with its leaves it is a plumule. The hypocotyl is the stemlike region beneath the cotyledons. The radicle is the embryonic root.
15. Seeds enter a dormancy period in which the embryo postpones development.
16. After fertilization, nonessential floral parts fall off, and hormones may influence the ovary and sometimes other plant parts to develop into a **fruit.** Fleshy fruits have distinct exocarp, mesocarp, and endocarp layers. In dry fruits, these join to form a pericarp. Drupes, berries, and pomes are types of fruit.
17. Seed **germination** requires oxygen, energy, and water. When the embryo bursts from the seed coat, the plant's primary growth ensues.

Testing Your Knowledge

1. Why doesn't sexual reproduction generate clones?
2. Give an example of asexual reproduction in a plant.
3. How does a flowering plant's life cycle differ from that of other multicellular organisms?
4. What is a "perfect" flower?
5. State whether the following floral parts are male, female, or accessory structures:
   - **a.** sepal
   - **b.** carpel
   - **c.** stigma
   - **d.** calyx
   - **e.** style
   - **f.** petal
   - **g.** stamen
   - **h.** corolla
   - **i.** pistil
   - **j.** anther
6. Which cells or structures develop from the generative cell and the vegetative cell?
7. In what two ways is pollination by animals more efficient than pollination by wind?
8. List three adaptations in plants that promote outcrossing.
9. Name tissues or cells in an angiosperm that are haploid, diploid, and triploid.
10. Distinguish between or among:
    - **a.** microspore, megaspore
    - **b.** drupe, berry, pome
    - **c.** exocarp, mesocarp, endocarp, pericarp

Thinking Scientifically

1. Why is pollination by animals less efficient in the early morning?
2. Plant reproductive biotechnologies can yield many identical plants, or encourage variation. Cite two examples of when each would be desirable.
3. Outline the steps of the scientific method that Elliot Meyerowitz used to test his hypothesis that three genes control flower development.
4. Chefs consider a plant food a fruit or a vegetable according to how it is prepared and eaten and how sweet it tastes. How does this differ from the biological definition of a fruit?
5. A few types of plants, such as dandelions, can produce seeds asexually, in a process called apomixis. How might a plant produce a seed without fertilization, and what would be an advantage of such a system?

References and Resources

Bradley, Desmond, et al. January 3, 1997. Inflorescence commitment and architecture in *Arabidopsis. Science,* vol. 275, p. 80. Mutations alter flower structure.

Conniff, Richard. September 2000. So tiny, so sweet . . . so MEAN. *Smithsonian,* vol. 31, p. 72. Hummingbirds are relentless pollinators.

Kao, Teh-hui, and Andrew G. McCubbin. February 24, 2000. A social stigma. *Nature,* vol. 403, p. 840. Genes prevent plants from pollinating themselves.

Nicholson, Rob. May 1999. The blackest flower in the world. *Natural History,* p. 60. Botanists are puzzled as to why a flower would not reflect any light.

The *LIFE* Online Learning Center provides additional resources and tools for studying this chapter.
www.mhhe.com/life

CHAPTER 29

Plant Responses to Stimuli

Plants Attack Caterpillars

Ravenous insect larvae can devastate a plant by consuming its leaves. But plants may not be the passive organisms that they appear to be. Their defenses go beyond waxy coverings and thorns, to a biochemical arsenal that helps them survive damage from animals. For example, lectins are plant compounds that bind to the lining of an insect's digestive tract, blocking absorption of nutrients and starving the pest. Plants also produce enzymes called protease inhibitors that dismantle herbivores' digestive enzymes.

A lipid-derived weapon called jasmonic acid is a hormone that plants such as tomato and legumes produce only when they are being eaten. The reaction begins when a caterpillar damages leaves. Cells in the leaf are stimulated to synthesize and secrete an 18-amino-acid-long peptide hormone called systemin into the apoplast (the cell wall and intercellular material). Systemin travels in the phloem to target cells in other parts of the plant, where it binds to receptors in the cell membranes. In response to the binding, certain membrane lipids react, producing jasmonic acid. This is the key biochemical in the signaling pathway. Jasmonic acid then activates genes that direct cells to produce the protease inhibitors that destroy the insects' digestive enzymes. Plus, the tomato plant shares its success—jasmonic acid volatilizes, forming a gas that wafts over to nearby plants and

. . . the tomato plant shares its success — jasmonic acid volatilizes, forming a gas that wafts over to nearby plants and announces that hungry caterpillars are in the area.

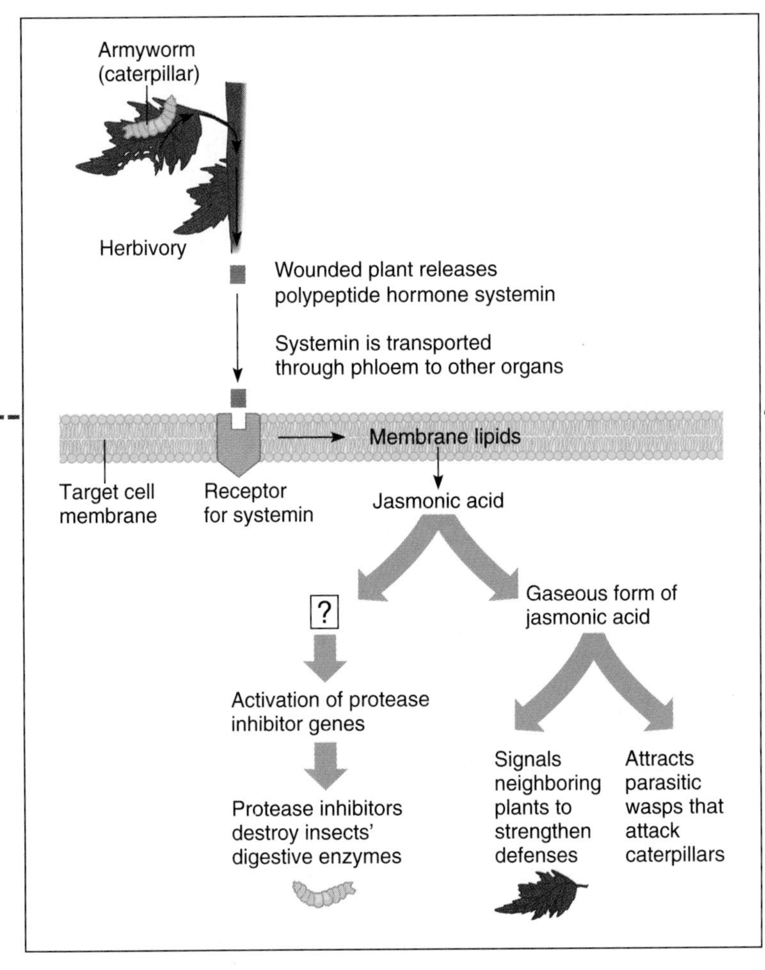

Hormones and Plant Defenses. Attack by voracious caterpillars causes some plants to release a series of hormones that culminate in production of protease inhibitors, which destroy the insects' digestive enzymes. Plants can also signal neighboring plants and indirectly vanquish the caterpillars by attracting parasitic wasps to the scene.

announces that hungry caterpillars are in the area. Laboratory experiments indicate that other plants that receive the signal begin to manufacture their own supplies of protease inhibitors, keeping them available for the arriving hungry hordes.

Jasmonic acid also has an indirect effect that harms leaf-munching insects—it attracts natural parasites of certain caterpillars.

Jasmonic acid also has an indirect effect that harms leaf-munching insects—it attracts natural parasites of certain caterpillars. Jennifer Thaler, a researcher at the University of California at Davis, investigated this subtle effect in the laboratory. She monitored tomato plants; beet armyworms, which are inch-long green caterpillars with a taste for tomato leaf; and parasitic wasps that deposit their eggs in the caterpillars, which are then destroyed from the inside out. Thaler set up 103 field plots, each containing four to six tomato plants. She sprayed half of the plants with jasmonic acid and half with water, and then beet armyworms arrived. Three weeks later, the treated plants had twice as many wasps as did the untreated plants. Was the jasmonic acid somehow attracting the exact species of insect that would kill the armyworms?

To demonstrate that it was the plants that were recruiting the wasps, and not that the armyworms on treated tomato leaves were weaker due to the protease inhibitors, Thaler ran another series of experiments keeping the armyworms in cups placed under tomato leaves, so that they could not suffer any ill effects of the protease inhibitors. These "sentinel caterpillars" were fed an artificial diet and were not in contact with the leaves, but the wasps could reach them. And they did. A day after setting the cups out, the caterpillars beneath the jasmonic acid–treated tomato plants had 37% more visits by parasitic wasps seeking a place to deposit their eggs. Thaler's conclusion: The plant produces a chemical—one that provides a "leafy" odor—that attracts the wasps.

29.1 How Do Plants Grow?

A plant can't quickly go after what it needs or run to avoid danger. Plants respond to environmental change mostly through growth that causes movement—activities orchestrated by hormones.

Plants respond to their environment. Shoots grow toward light, and against gravity; roots do just the opposite. Many plants leaf out in spring, produce flowers and fruits, then return to dormancy in autumn—all in response to seasonal changes. Other responses are more immediate, such as the closing of stomata to reduce transpiration during hot times of day. Plants may even send signals to each other, warning of such dangers as insect infestations, as the chapter opening essay describes.

Plant responses are usually much more subtle (to us) than those of many animals—plants cannot run away from unfavorable conditions. Instead, plants adjust to the environment. Their responses are partly mediated by chemicals called hormones. Classically, a **hormone** is a biochemical synthesized in small quantities in one part of an organism and transported to another, where it stimulates or inhibits a response from target cells. This definition must be modified slightly for plants, however, because some plant hormones also act close to or at their site of synthesis.

In many ways, hormones mediate the interactions between a plant's environment and its genome. Recall that an organism's DNA sequence provides blueprints for organizing specialized cells into tissues and organs, and these genetic instructions can include several options. External factors, such as light intensity or temperature, may trigger the release of a hormone that directly or indirectly alters the expression of certain genes. The result may be a different phenotype, a finely tuned growth response that enables the plant to adjust to its surroundings.

Plants have five major classes of plant hormones: auxins, gibberellins, cytokinins, ethylene, and abscisic acid (table 29.1). These chemicals interact in complex ways to regulate many aspects of plant growth, flower and fruit development, senescence (aging), and responses to environmental extremes. Predicting a particular plant's response to a hormone may not be straightforward—it depends on many factors, including the concentration of the hormone, the sensitivity of the cells in a tissue to the hormone, and the presence of other hormones. Hormones also elicit different effects in plants of different species or developmental stages.

Auxins

Auxins, from the Greek "to increase," were the first plant hormones that scientists described. In the late 1870s, decades before researchers determined the chemical structures of plant hormones, Charles Darwin and his son Francis learned that a plant-produced "influence" caused plants to grow toward light (fig. 29.1). They grew oats and canary grass and noted that seedlings bent toward the light, but not if the coleoptiles (protective structures at the tips of grass shoots) were covered. The Dar-

TABLE 29.1 Major Classes of Plant Hormones

CLASS	SELECTED ACTIONS	SYNTHESIS SITE	ROUTE OF TRANSPORT
Auxins	Elongate cells in seedlings, shoot tips, leaves, embryos Adventitious root growth of cuttings Inhibit leaf and fruit abscission Stimulate synthesis of ethylene Inhibit growth of lateral buds	Developing leaves and seeds, shoot tips	Cell to cell, usually downward
Gibberellins	Elongate and divide cells in seeds, roots, shoots, young leaves Seed germination Stimulate flowering	Young shoot, developing seeds	Xylem and phloem
Cytokinins	Stimulate cell division in seeds, roots, young leaves, fruits Delay leaf senescence Allow lateral buds to grow	Root tips	Xylem (roots to shoots)
Ethylene	Hastens fruit ripening Leaf and flower senescence Leaf and fruit abscission	All parts, especially under stress, aging, or ripening	Diffusion
Abscisic acid	Inhibits shoot growth Closes stomata Induces and maintains seed dormancy Stores proteins in seeds	Mature leaves, plants under stress	Xylem and phloem

FIGURE 29.1
Plants Bend Toward Light.
Charles Darwin and his son Francis demonstrated that a plant-produced "influence" promotes bending toward light. (**A**) Shoots normally bend toward a unidirectional light source. (**B**) When the Darwins blocked the coleoptile at the tip of an oat seedling, the shoot no longer bent toward the light. (**C**) Blocking the shoot beneath the tip did not have this effect. Therefore, something in the growing tip enables the plant to sense and respond to light.

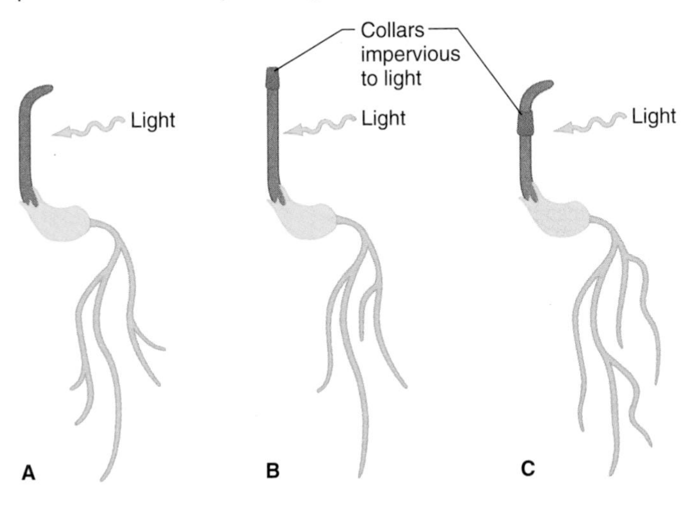

wins wrote in *The Power of Movement in Plants,* a book published in 1881: "When seedlings are freely exposed to a lateral light, some influence is transmitted from the upper to the lower part, causing the latter to bend."

The Darwins were describing auxins, a group of plant hormones that stimulate cell elongation in grass seedlings and herbs. Among other effects, auxins stimulate cells to elongate by altering the plasticity of cell walls, so that the walls stretch by taking in more water. Specifically, the hormone causes proteins in the cell membrane to pump protons (H^+) into the cell wall. Here, the protons activate enzymes that separate the cellulose microfibrils, enabling the wall to expand and elongate against turgor pressure. More microfibrils form, stabilizing the new, longer cell shape. When this happens to several cells close together, the plant bends (fig. 29.2).

Three auxins occur naturally. The most active, indoleacetic acid (IAA), is produced in shoot tips, embryos, young leaves, flowers, fruits, pollen, and coleoptiles. The hormone moves in one direction from its site of synthesis toward the bases of leaves and stems, or to root tips. Carrier molecules in parenchyma cells encircling vascular bundles transport auxins from cell to cell, by passive diffusion. Auxins act rapidly, spurring noticeable growth in a grass seedling in minutes.

For many years auxins have been used to stimulate the growth of adventitious roots in cuttings, which is important in the vegetative propagation of various plants. Synthetic compounds having auxinlike effects, such as 2,4-D (2,4-dichlorophenoxyacetic acid), are important commercially. 2,4-D is used extensively as an herbicide. Such a hormone-mimic is harmful because the plant cannot completely break it down, and it accumulates to lethal levels. For reasons that are not completely understood, these herbicides are selective, killing broad-leaf weeds but not grasses.

Gibberellins

In 1926, Japanese botanists studying "foolish seedling disease" in rice discovered **gibberellins,** another class of plant hormone that causes shoot elongation. Plants suffering from foolish seedling disease are infected by a fungus that causes them to grow rapidly, becoming so spindly that they fall over and die. The botanists discovered that a chemical extract of the fungus also produced the disease symptoms. In 1934, scientists isolated the active compound and named it gibberellin. We now know of at least 84 naturally occurring gibberellins.

Gibberellins are present in all plant parts, in varying amounts. They have several functions. Gibberellins stimulate shoot elongation in trees, shrubs, and a few other plants (fig. 29.3A). They induce both cell division and elongation and also stimulate seed germination by "awakening" seeds from dormancy. Elongation induced by gibberellins occurs after about a 1-hour delay, which is much slower than auxin-induced growth. These hormones move throughout plants in xylem and phloem. They are used in agriculture to stimulate stem elongation in seedless grapes (fig. 29.3B).

Cytokinins

As early as 1913, scientists knew that a biochemical stimulates plant cells to divide. It was not until 1964 that researchers discovered a naturally occurring **cytokinin,** in corn kernels. Since then, researchers have isolated other cytokinins and have synthesized several artificial forms.

Cytokinins earned their name because they stimulate cytokinesis (the division of the cell after the genetic material has replicated and separated). The effects of cytokinins are similar to some of the effects of auxins and gibberellins—they promote cell division and growth and participate in development, differentiation, and senescence. In flowering plants, most cytokinins affect roots and developing organs such as seeds, fruits, and young leaves.

The actions of cytokinins and auxins apparently compete in an effect called **apical dominance** (fig. 29.4). Both types of hormones are present in a gradient along the plant body, with cytokinins more concentrated in the roots, and auxins more concentrated in shoot tips. The intact apical (terminal) bud of a plant normally secretes auxins, which move downward and suppress the growth of lateral buds. In contrast, cytokinins move upward in the xylem and stimulate bud sprouting. If the shoot tip is cut off, the concentration of auxins decreases, and cytokinins increase. Apical dominance is relieved and lateral buds grow; this is why gardeners often promote bushier growth by pinching off plants' tips. Cytokinins also retard leaf senescence, and they can be used commercially to extend the shelf lives of leafy vegetables.

FIGURE 29.2

Auxin Promotes Cell Lengthening.

In diffuse light (**A**), seedlings grow straight upwards but in directional light (**B**), the stem tips bend toward the light. (**C**) Auxin builds up on the shaded side of the seedling, promoting acidification of the cell walls and subsequent lengthening of the shaded cells.

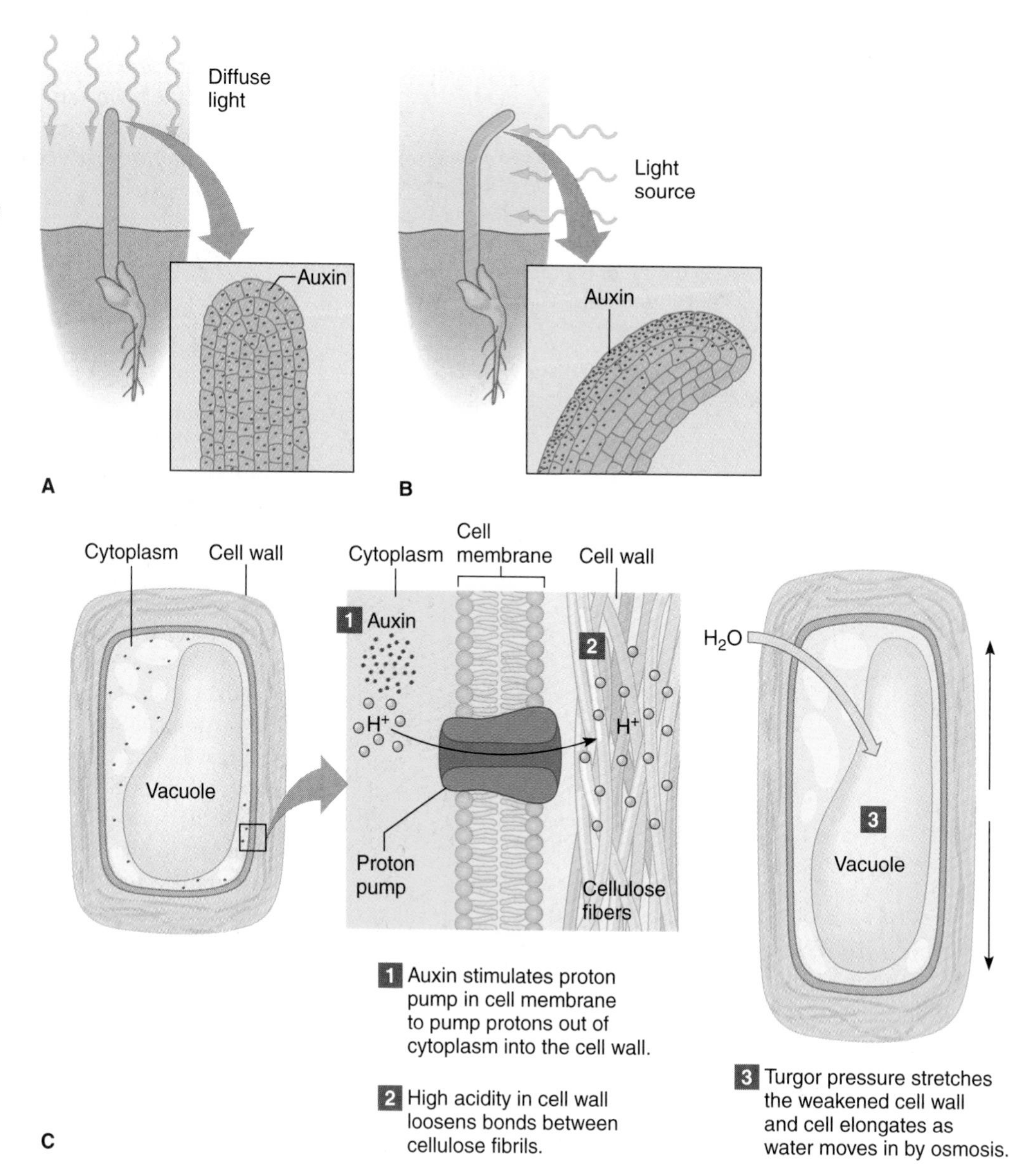

FIGURE 29.3

Gibberellins Induce Shoot Elongation.

(**A**) Gibberellins applied to cyclamen plants (marked GA-1250 in photo) elongate the flower stems, causing them to shoot upward relative to controls, marked C. (**B**) Gibberellins applied to grapes lengthen the stems, giving the grapes more room to grow. Treated grapes are on the right.

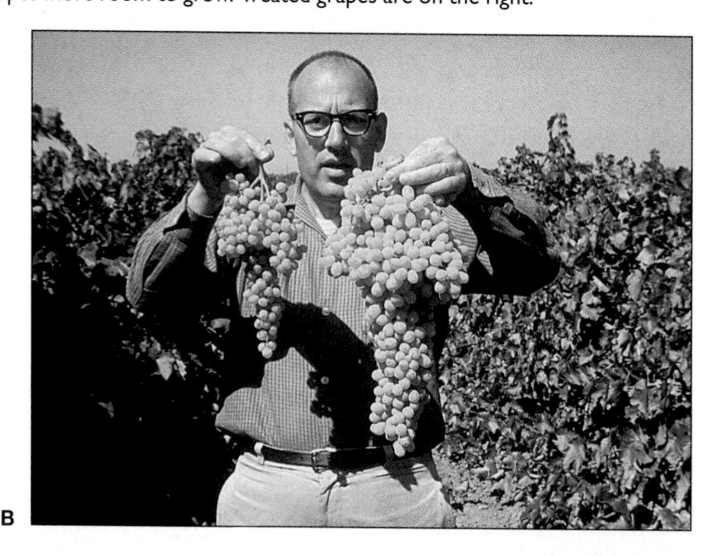

FIGURE 29.4

Apical Dominance.

(**A**) High concentrations of auxin enable the apical meristem to grow at the expense of lateral (axillary) buds. (**B**) Removing the apical meristem changes the auxin-cytokinin balance, which lifts this control and stimulates the lateral buds to grow.

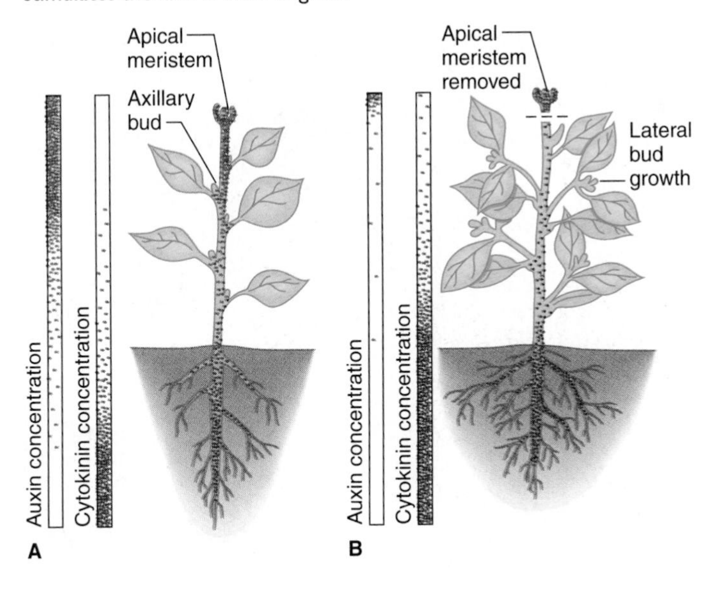

Ethylene

The effects of **ethylene** were noticed long before its discovery. In 1910, scientists in Japan observed that bananas ripened prematurely when stored with oranges. Apparently, the oranges produced something that induced rapid ripening. By 1934, scientists realized ethylene gas was hastening ripening. Later, the source of the ethylene was traced to fungi growing on the oranges.

All parts of flowering plants synthesize ethylene, particularly the shoot apical meristem, nodes, flowers, and ripening fruits. Like many hormones, ethylene has several effects. In most species, ethylene gas causes flowers to fade and wither (fig. 29.5).

Ethylene ripens fruit in many species. Ripening is a complex process that includes pigment synthesis, fruit softening, and breakdown of starches to sugars. A tomato picked from the garden when it is hard and green with streaks of pale orange will turn, aided by ethylene, into a soft, red, succulent tomato. The dark spots on a ripening banana peel are concentrated pockets of melanin, produced under the influence of ethylene. Because ethylene is a gas, its effects can be contagious. The expression "one bad apple spoils the whole batch" refers to the way ethylene released from one apple can hasten the ripening, and eventual spoiling, of others nearby.

Ethylene also helps ensure that a plant will survive injury or infection. When a damaged plant produces the hormone, it hastens aging of the affected part so that it can be shed before the problem spreads to other regions of the plant. This effect was noticed in Germany in 1864, when ethylene in a mixture of gases used to light streetlamps caused nearby trees to lose their leaves.

Abscisic Acid

Can plants manufacture substances that inhibit growth? By the 1940s, botanists thought so. Twenty years later, researchers isolated such a compound.

Abscisic acid, or ABA, inhibits the growth-stimulating effects of many other hormones. It inhibits seed germination, countering the effects of gibberellins. ABA also closes stomata, which helps plants conserve water during drought. This hormone promotes leaf, flower, and fruit abscission (shedding), although it remains unclear whether ABA is directly involved or stimulates the production of ethylene, which also causes abscission. ABA is produced in higher amounts in response to stresses such as drought or frost, and promotes protective responses. In fact, it is used commercially to inhibit the growth of nursery plants so that shipping is less likely to damage them.

FIGURE 29.5

Ethylene's Effects on Flowers.

All four of these petunia flowers were treated with ethylene gas (2 parts per million) for 18 hours. The wild-type flowers on the left withered, but the flowers on the right, genetically engineered to be insensitive to ethylene, remained fresh. The fresh flowers have mutant ethylene receptor genes.

29.1 MASTERING CONCEPTS

1. What are the five major classes of plant hormones?
2. What are the functions of the five types of plant hormones?
3. Give an example of how plant hormones interact.

29.2 How Do Plants Move?

A plant can grow toward light, with or against gravity, or encircle an object in response to touch. Such tropisms are directed movements.

The term **tropism** refers to plant growth toward or away from environmental stimuli, such as light or gravity. Tropisms result from differential growth, in which one side of the responding organ grows faster than the other, curving the organ. When an organ curves towards the stimulus, as a stem grows toward light, it is called a positive tropism. When an organ curves away from a stimulus, it is a negative tropism. Hormones are involved in many tropisms. Tropisms are named for the stimuli that elicit the responses. Phototropism, for example, is a growth response to unidirectional light, and gravitropism is a growth response to gravity. Table 29.2 lists some tropisms.

Phototropism—Response to Unidirectional Light

In **phototropism,** cells on the shaded side of a stem elongate more than cells on the lighted side of the stem (fig. 29.6). Auxins from the shoot apex control the rapid elongation of cells along the shaded side of coleoptiles (see fig. 29.2).

In the 1950s, Winslow Briggs and his colleagues were the first to discover precisely how auxins control phototropism. First, they determined that the amount of auxins produced by coleoptiles grown in the light is the same as the amount produced by coleoptiles grown in the dark—that is, they established that light does not destroy auxins. They then discovered that they could collect more auxins from the shaded side of coleoptiles than from the lighted side, which suggested that light causes auxins to migrate to the shaded side of the stem. More recent experiments support this finding. Auxins labeled with radioactive carbon (^{14}C) and exposed to unidirectional light move to the shaded side of coleoptiles. Cells in the shade then elongate more than cells in the light, curving the coleoptile toward the light.

FIGURE 29.6
Phototropism.
These autumn crocuses (*Colchicum purpura*) show strong phototropism when the sun is off to the side. Phototropism occurs because auxin moves to the shaded side of a shoot, where it promotes cell elongation. This hormonal action bends the plant toward the light.

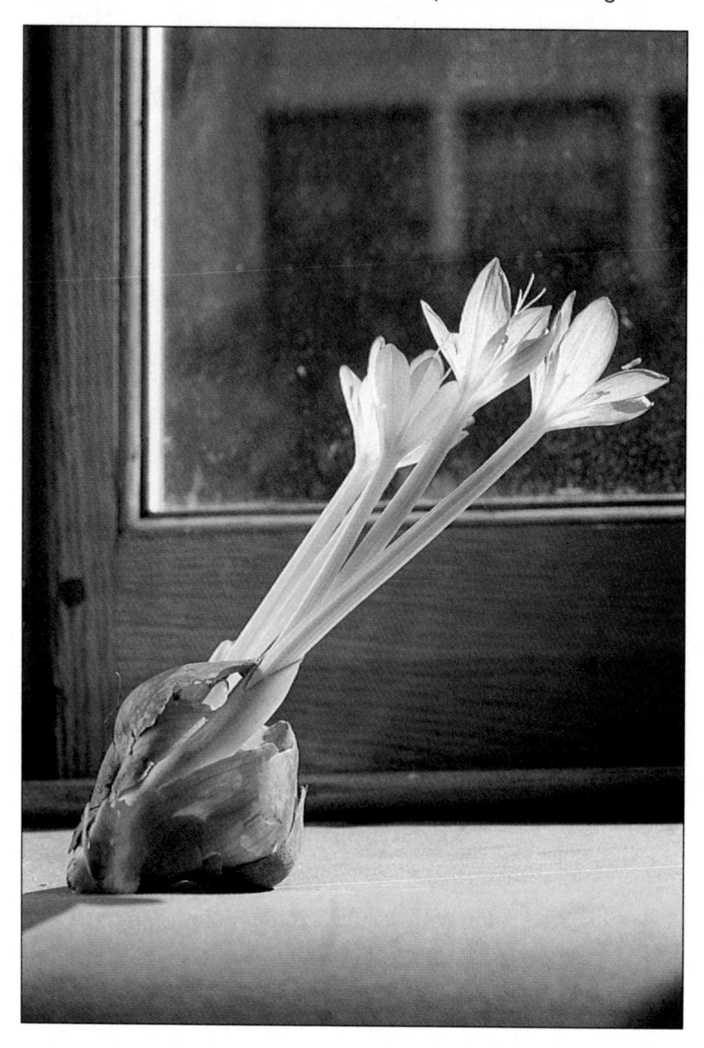

TABLE 29.2 Types of Tropisms

STIMULUS	TROPISM
Light	Phototropism
Gravity	Gravitropism
Touch	Thigmotropism
Temperature	Thermotropism
Chemical	Chemotropism
Water	Hydrotropism

How does a plant detect light coming from one direction? Only blue light with a wavelength less than 500 nanometers effectively induces phototropism. The yellow pigment flavin is probably the photoreceptor molecule for phototropism. Somehow, flavin alters auxin transport to the shaded side of the stem or coleoptile. Similar pigment molecules may also control phototropism in other organisms, including fungi.

Gravitropism—Response to Gravity

Even as a seed germinates, its shoot seeks light by growing upward, while its root heads downward into the soil. The seedling's response to gravity is called **gravitropism,** another tropism that Charles Darwin and his son Francis investigated.

In the late 1800s, it was already known that roots would not respond to gravity if their root caps were cut off, suggesting that something in the root cap is necessary for gravitropism. Despite more than a century of subsequent research, the role of the root cap in sensing gravity is still not completely understood. One hypothesis for how a plant senses gravity centers on cells in the root that have modified starch-containing plastids, called amyloplasts, that function as **statoliths,** or gravity detectors. These granules normally sink to the bottoms of the cells, somehow telling the cells which direction is down. Figure 29.7 shows how turning a root causes the statoliths to move, reorienting the plant to grow downward. Movement of the statoliths is somehow coupled to calcium ion and auxin movement in a way that directs root growth downward. The complete story is still not known, and other organelles and hormones may also be involved.

One approach to studying gravitropism is to attempt to counteract the influence of gravity. A clinostat is a slowly revolving wheel that cancels the effect of gravity for plants mounted on it (fig. 29.8). On a clinostat, roots no longer grow down, nor do stems grow up. Another strategy is to study plant growth in minimal gravity, as when plants grow aboard an orbiting spacecraft (Investigating Life 29.1).

FIGURE 29.7
Gravity Influences Plant Growth.
(**A**) Gravitropism causes shoots and stems to grow up and roots to grow down no matter how their seeds are oriented. (**B**) Starch-filled plastids (amyloplasts) that function as statoliths may be the plant's mechanism for detecting gravity. (**C**) When a root is moved sideways, the statoliths move to occupy the new "down" side. This action somehow signals changes that set the root back on its normal downward course.

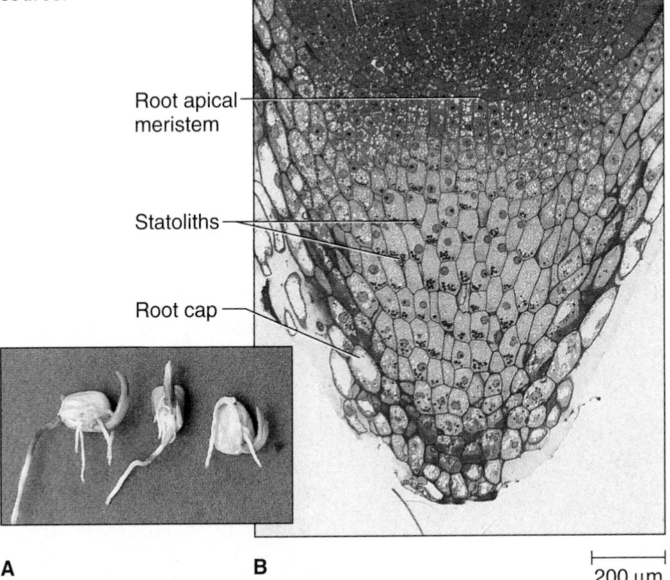

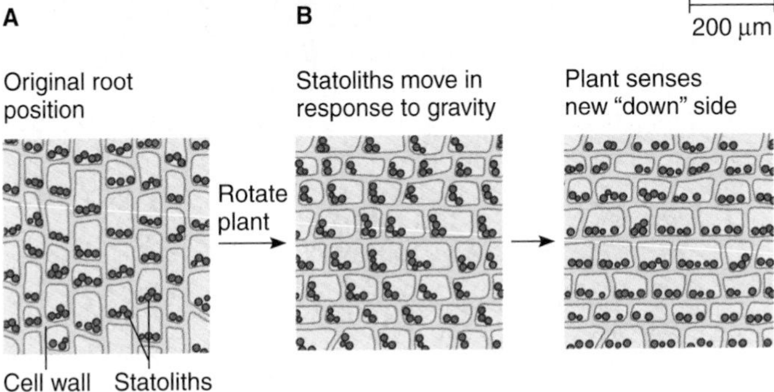

Thigmotropism—Response to Touch

The coiling tendrils of twining plants such as morning glory and bindweed exhibit **thigmotropism,** a response to touch. When the tendrils hang free, they often grow in a spiral fashion, which increases their chances of contacting an object they can cling to (fig. 29.9). A plant detects contact with an object with specialized epidermal cells, which induce differential growth of the tendril. In only 5 to 10 minutes, the tendril completely encircles the object. Thigmotropism is often long-lasting. Stroking the tendril of a garden pea plant for only a few minutes induces a curling response that lasts for several days. Auxin and ethylene control thigmotropism.

Interestingly, tendrils seem to "remember" touch. Tendrils touched in the dark do not respond until they are illuminated. They apparently store sensory information in the dark but do not respond until light is present.

Nastic Movements

Plant movements that are not oriented with respect to a stimulus are called **nastic movements.** A nastic movement resulting

FIGURE 29.8
A Clinostat.
Rotating plants slowly on a wheel makes them unable to sense gravity. As a result, roots no longer grow down and stems up. This classic experiment demonstrated the effects of gravity on plant growth.

Plants in Space

Plants will play a key part in space colonization, providing food and oxygen. But how will organisms that have evolved under constant gravity function in its near absence beyond Earth?

Researchers are studying the effects of microgravity on a variety of species by sending them on space shuttle voyages. These experiments can more realistically assess plant growth and development in space than simulations conducted on Earth, which use a rotating device called a clinostat to diminish gravitational force. So far, it appears that lack of gravity greatly affects plants, in everything from subcellular structural organization to the functioning of the organism as a whole.

Subcellular Responses to Microgravity

Plant cells grown in space have fewer starch grains and more abundant lipid-containing bodies than their Earthly counterparts, indicating a change in energy balance. Organelle organization is also grossly altered; endoplasmic reticula occur in randomly spaced bunches and mitochondria swell. Nuclei enlarge, and chromosomes break. Chloroplasts have enlarged thylakoid membranes and small grana.

Interesting effects occur in amyloplasts, which are starch-containing granules in certain root cells. On Earth, amyloplasts aggregate at the bottoms of these cells, which tells roots which way is up. But in space, amyloplasts occur throughout the cytoplasm. For many years, botanists hypothesized that sinking starch granules in these cells are gravity receptors. An alternative hypothesis holds that cells sense gravity by detecting the difference in pressure on the cell membrane at the top versus the bottom of the cell. This difference results from protoplasm sinking in response to gravity, so there is less of a difference as gravitational force falls. The actual "sensing" may be carried out by a cell membrane protein called an integrin, and the starch granules may amplify the difference in pressure by sinking in the presence of gravity. Whatever the precise mechanism of gravity sensation, a root tip cannot elongate normally in microgravity.

Cell Division

Microgravity halts mitosis, usually at telophase, which produces cells with more than one nucleus. Oat seedlings germinated in space have only one-tenth as many dividing cells as seedlings germinated on Earth. Microgravity also disrupts the spindle apparatus that pulls chromosome sets apart during mitosis.

Cell walls formed in space are considerably thinner than their terrestrial counterparts, with less cellulose and lignin. Microgravity also inhibits regeneration and alters cell distribution (fig. 29.A). A decapped root will regenerate in 2 to 3 days on Earth but not at all in space. Lettuce roots have a shortened elongating zone when grown in space.

Growth and Development

Germination is less likely to occur in space than on Earth because of chromosome damage, but it does happen. Early growth seems to depend upon the species—bean, oat, and pine seedlings grow more slowly than on Earth, and lettuce, garden cress, and cucumbers grow faster. Many species, including wheat and peas, cease growing and die before they flower. In 1982, however, *Arabidopsis* successfully completed a life cycle in space—indicating that human space colonies containing plant companions may indeed be possible.

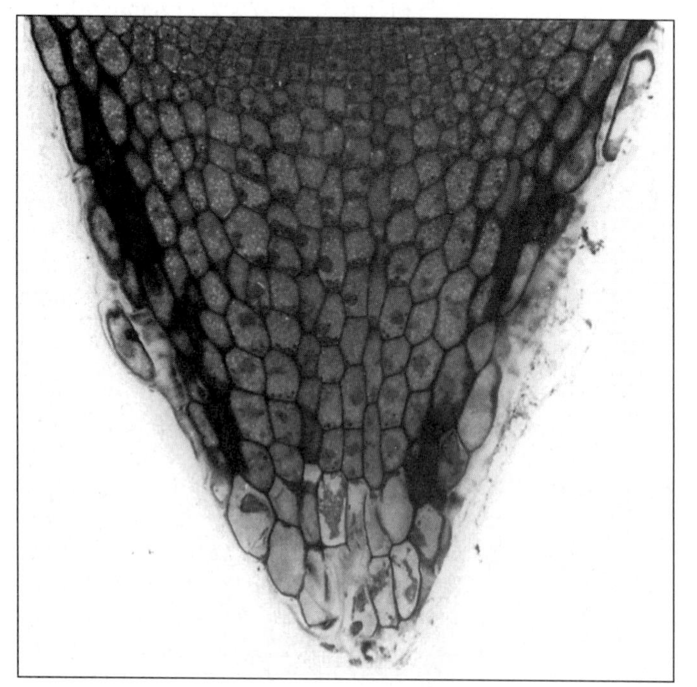

1

FIGURE 29.A

Plant Growth in Space.

(*1*) On Earth, a root whose cap has been removed regenerates an organized, functional root cap. (2) In the microgravity of space, however, regrowth of a decapped root is disorganized. Clearly, gravitational cues help direct normal regeneration.

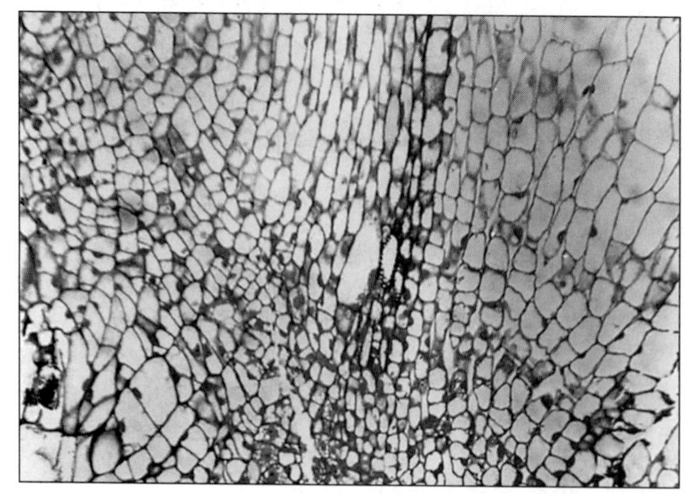

2

FIGURE 29.9
Thigmotropism.
A tendril's epidermis is sensitive to touch. This tendril of a passion vine wraps around a blackberry stem.

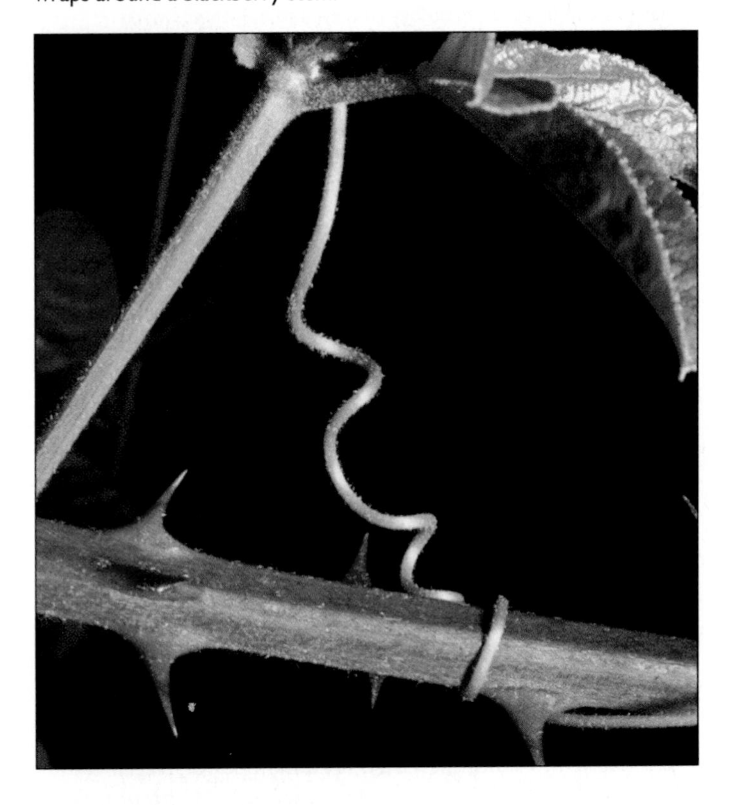

from physical contact or mechanical disturbance is **thigmonasty.** Thigmonastic movements depend upon a plant's ability to transmit a stimulus rapidly from touch-sensitive cells to responding cells elsewhere. Consider leaf movement in the "sensitive plant," *Mimosa.* When the leaves of *Mimosa* are touched, the leaflets immediately fold and the petiole droops (fig. 29.10). Touch causes the membranes of parenchyma cells called motor cells at the base of the leaflet to become more permeable to K^+ and other ions. As ions move out of the motor cells, the osmotic potential in the surrounding area decreases, causing water to move out of the motor cells by osmosis. The water loss shrinks the motor cells and causes thigmonastic movement. This electrical stimulation and response is considerably faster than a hormone-induced action. Reversal of the process unfolds the leaves in approximately 15 to 30 minutes.

Thigmonastic movements are protective. In some plants, closing leaflets expose sharp prickles and stimulate cells at the leaflet bases to secrete noxious substances called tannins, which discourage hungry animals. The Venus's flytrap is famous for its dramatic thigmonastic response (see Investigating Life 22.1). Unlike *Mimosa,* in which thigmonastic movements result from reversible changes in turgor pressure, the Venus's flytrap's movements result from increased cell size, which begins when the cell walls are acidified to pH 4.5 and below. The leafy traps each consist of two lobes, and each lobe has three sensitive "trigger" hairs overlying motor cells. When a meandering animal touches two of these hairs, it signals the plant's motor cells, which then initiate H^+ transport to epidermal cell walls along the trap's outer surface. Acidification promotes softening of the cell wall and osmotic swelling of the outer epidermal cells along the central portion of the leaf. Since epidermal cells along the inner surface of the leaf do not change volume, the flytrap shuts. Empty traps usually open after 8 to 12 hours.

FIGURE 29.10
Thigmonasty.
Mimosa pudica is called the "sensitive plant" because of its thigmonastic movements. Leaves are erect in undisturbed plants (**A**), but touching a leaf (**B**) causes the leaflets to fold and the petiole to droop (**C**).

Photonasty, the nastic response to daily rhythms of light and dark, is also known as "sleep movement." This is an example of a circadian (day-based) rhythm, discussed further in section 29.4. The prayer plant *Maranta* is an ornamental houseplant that exhibits such a response. Prayer plant leaves lie horizontally during the day, which maximizes their interception of sunlight. At night, the leaves fold vertically into a configuration resembling a pair of hands in prayer (fig. 29.11).

The prayer plant's leaves move in response to light and dark as the turgor pressure changes in motor cells at the base of each leaf. In the dark, K^+ moves out of cells along the upper side and into cells along the lower side of a leaf base. This moves water, via osmosis, into cells along the lower side of the leaf base, swelling them, as cells along the upper side lose water and shrink. As the cellular volume changes, the leaf stands vertically. At sunrise, the process reverses and the leaf again lies horizontally. Changes in leaf position can conserve water and heat.

FIGURE 29.11
Photonasty.
The prayer plant, *Maranta,* exhibits photonasty. When the sun goes down, the prayer plant's leaves fold inward.

Sorrel and legumes such as beans have similar sleep movements that occur at the same time each day. Carolus Linnaeus, the Swedish botanist who proposed a widely used taxonomic scheme in the eighteenth century, made clever use of these regular movements. He filled wedge-shaped portions of a circular garden with plants that had sleep movements at different times. By checking to see which plants in his so-called *horologium florae* (flower clock) were "asleep," Linnaeus could tell the time of day.

29.2 MASTERING CONCEPTS

1. What is the physical basis of a tropism?
2. How does auxin cause phototropism?
3. How is gravitropism adaptive?
4. What is a nastic movement?
5. Describe two types of nastic movements.

OLC

29.3 How Do Plants Respond to Seasonal Changes?

Many aspects of plant physiology are keyed to day length, which reflects seasonal changes. The ratio of two forms of a photopigment provides critical cues to day length.

Seasonal changes affect plant responses in many ways. For instance, autumn in temperate regions brings cooler nights and shorter days, which produce beautifully colored leaves, dormant buds, and decreased growth. In the spring, buds resume growth and rapidly transform a barren forest into a dynamic, photosynthetic community. These seasonal changes illustrate the complex interactions among environmental signals, hormones and other biochemicals, and the plant's genes.

Flowering Is a Response to Photoperiod

Flowering reflects seasonal changes. Many plants flower only during certain times of the year. Clover and iris flower during the long days of late spring and summer, whereas poinsettias and asters bloom in the short days of early spring or fall.

Studies of how seasonal changes influence flowering began in the early 1900s. W. W. Garner and H. A. Allard at a U.S. Department of Agriculture research center in Maryland were studying tobacco, which flowers during late summer in Maryland. One group of tobacco mutants did not flower, but continued to grow vegetatively into autumn. These mutants became large, leading Garner and Allard to name them Maryland Mammoth. Since these oversized mutants had the potential for increasing tobacco crop yields, Garner and Allard moved their Mammoth plants into the greenhouse to protect them from winter cold, and continued to observe their growth. To their surprise, the mutants finally flowered in December!

Could the plants somehow measure day length? To test this hypothesis, Garner and Allard set up several experimental plots of the tobacco, each planted approximately a week apart. All of the plants flowered at the same time, despite the fact that the staggered planting times resulted in plants of different ages and sizes. Garner and Allard suggested that the plants were exhibiting **photoperiodism,** the ability to measure seasonal changes by day length.

Photoperiodism attunes plant responses to a changing environment. Plants measure and respond to a critical day length rather than other climatic factors, such as rainfall or temperature, because weather is unpredictable from year to year. Day length is consistent due to the position of Earth as it moves around the sun.

Botanists classify plants into groups, depending upon response to photoperiod (duration of daylight) (fig. 29.12). **Long-day plants** flower when light periods are longer than a critical length, usually 9 to 16 hours. These plants typically bloom in the spring or early summer and include lettuce, spinach, beets, clover, corn, and iris. **Day-neutral plants** do not rely on photoperiod to stimulate flowering. These include roses, snapdragons, cotton, carnations, dandelions, sunflowers, tomatoes, cucumbers, and

FIGURE 29.12

Flowering Responses to Day Length.

Long-day plants flower in response to light periods longer than a critical period. Flowering in day-neutral plants does not rely on response to day length. Short-day plants respond to a light period shorter than a critical length.

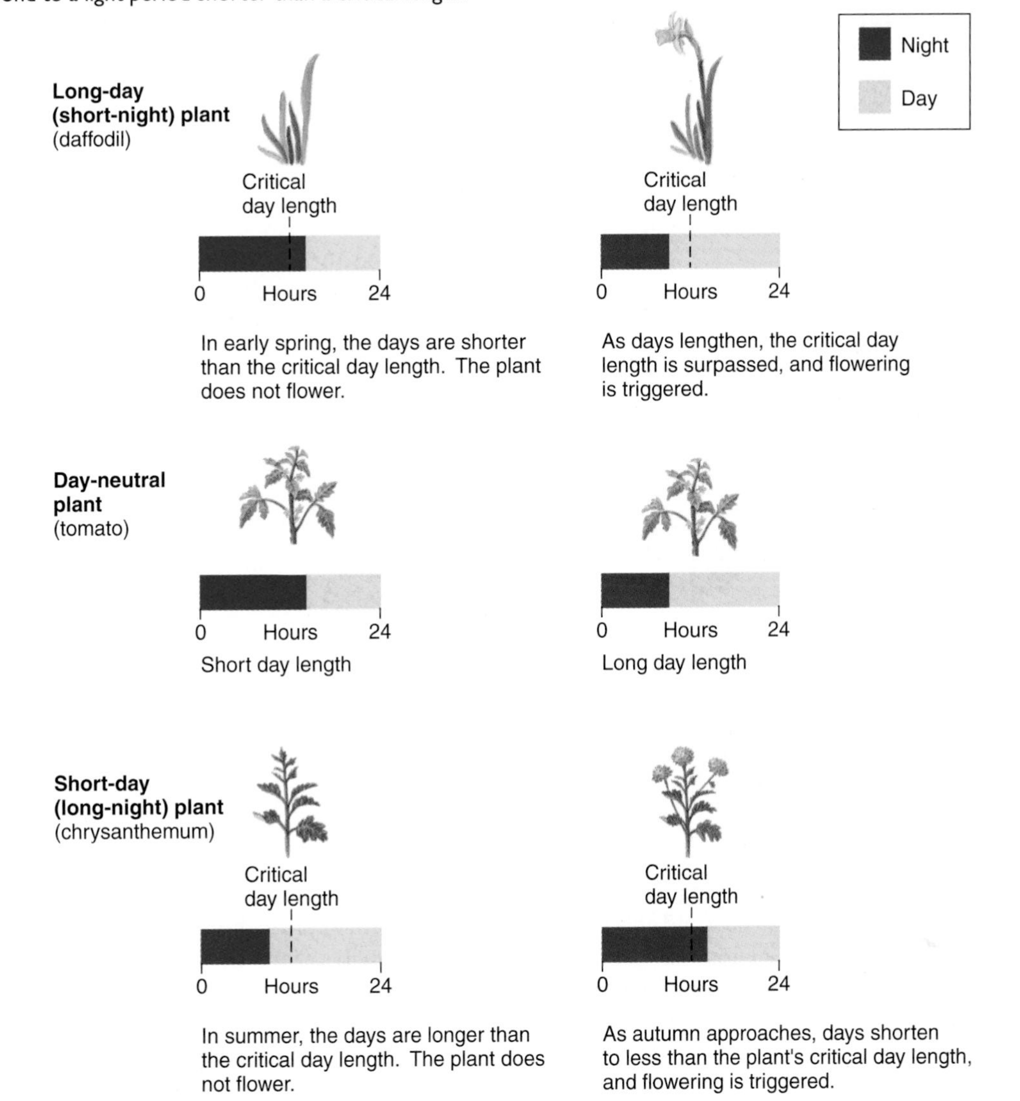

many weeds. **Short-day plants** require light periods shorter than some critical length. These plants usually flower in late summer or fall. For example, ragweed plants flower only when exposed to 14 hours or fewer of light per day. Asters, strawberries, poinsettias, potatoes, soybeans, and goldenrods are short-day plants. Many short-day plants will not grow in the tropics, where days are always too long to induce flowering.

The measuring system in plants is in the leaves: remove a plant's leaves, and it does not respond to changes in photoperiod. Response to photoperiod is also remarkably sensitive. Henbane, a long-day plant, flowers when exposed to light periods of 10.3 hours, but not when the light period is 10.0 hours. These flowering responses may ensure that flowers of different sexes open within a few days of each other.

Do Plants Measure Day or Night?

Plant physiologists Karl Hamner and James Bonner continued Garner's and Allard's work by studying the photoperiodism of the cocklebur, a short-day plant requiring 15 or fewer hours of light to flower. Hamner and Bonner used controlled-environment growth chambers to manipulate photoperiods. They were startled to discover that plants responded to the length of the dark period rather than the light period. The cocklebur plants flowered only when the dark period exceeded 9 hours.

Hamner and Bonner also discovered that flowering did not occur if a 1-minute flash of light interrupted the dark period, even if darkness exceeded the required 9 hours. In the reverse experiment, darkness interrupting the light period had no effect on

FIGURE 29.13

Night-time Light Flashes Inhibit Flowering in Short-Day Plants.

Length of day and night influence flowering in long-day plants such as clover and short-day plants such as cocklebur. Long-day plants require a dark period shorter than a critical length. Short-day plants require an uninterrupted dark period longer than a critical period. Interrupting the dark period of a short-day plant inhibits flowering.

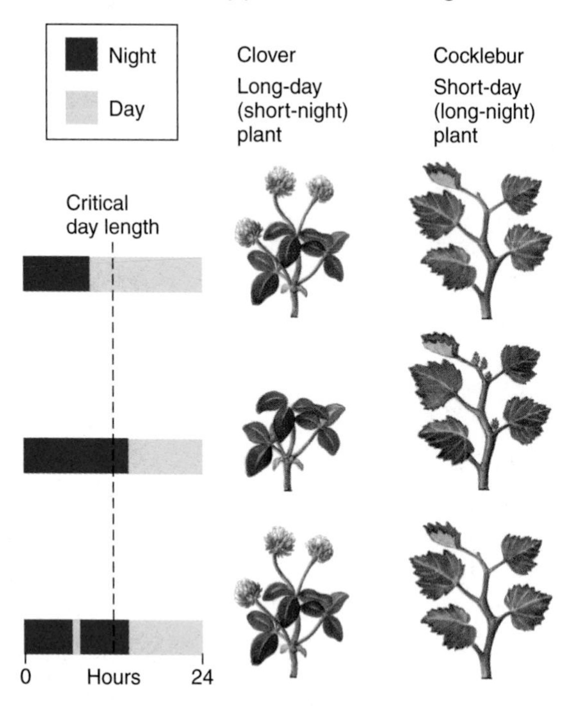

FIGURE 29.14

The Blue Pigment Phytochrome Comes in Two Forms.

(**A**) P_r and P_{fr} absorb slightly different wavelengths of red light; as a result, both appear blue. (**B**) P_r absorbs red light and is rapidly converted into P_{fr}; the reverse occurs rapidly when P_{fr} absorbs far-red light. P_{fr} also converts slowly to P_r in the absence of light. P_{fr} has a variety of biological effects.

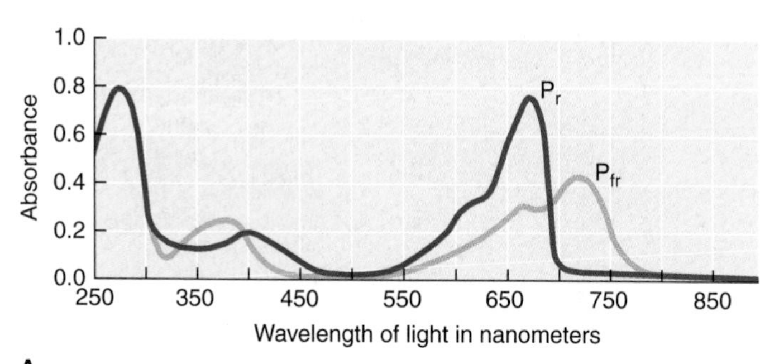

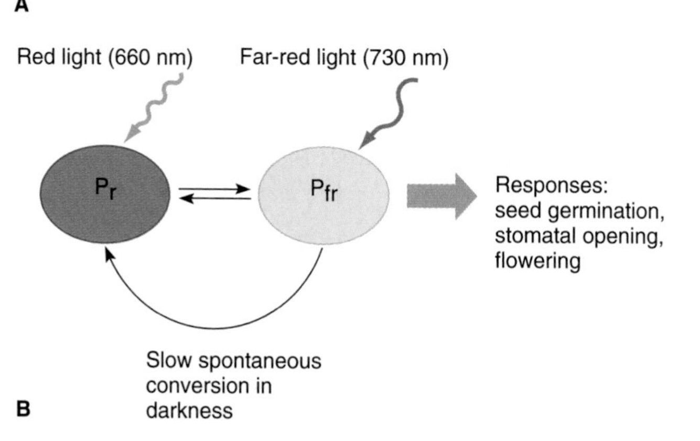

flowering. Furthermore, a long-day plant flowering on a photoperiod of 16 hours light to 8 hours dark also flowered on a photoperiod of 8 hours light to 16 hours dark if a 1-minute light flash interrupted the dark period. Other experiments with long- and short-day plants confirmed that flowering requires a specific period of uninterrupted darkness, rather than uninterrupted light (fig. 29.13). Thus, short-day plants are really long-night plants, because they flower only if their uninterrupted dark period exceeds a critical length. Similarly, long-day plants are really short-night plants.

Phytochrome Is a Pigment That Controls Photoperiodism

How do plants measure the lengths of night and day? The "clock" is a blue pigment molecule called **phytochrome.**

Phytochrome exists in two interconvertible forms that have very similar structures (fig. 29.14). The P_r form absorbs red wavelengths of light (600 nanometers). The P_{fr} form absorbs in the far-red portion of the electromagnetic spectrum (730 nm). P_r is converted to P_{fr} nearly instantaneously in the presence of red light, but the reverse transformation, P_{fr} to P_r, is slow (unless plants are exposed to far-red light). It is the different amounts of red and far-red wavelengths in sunlight throughout a 24-hour period that the plant senses with its phytochrome system. Because sunlight has more red than far-red, in the daytime, P_{fr} production is rapid, which somehow tells the plant that it is day. As night falls, the P_{fr} form of phytochrome is slowly converted back to P_r.

Often in science the existence of a structure or phenomenon is predicted based on indirect evidence, and later confirmed by discovery or experiment. With phytochrome, researchers hypothesized involvement of a pigment in controlling flowering because light (or its absence) was clearly part of the response. Then, a series of experiments showed that a flash of red light during the dark period could inhibit flowering, but that a subsequent flash of far-red light could cancel this effect (fig. 29.15). The presumed pigment was therefore sensitive to red and far-red wavelengths. In 1959, researchers isolated and identified phytochrome.

Photoperiod and Phytochrome Influence Other Responses

Phytochrome also affects seed germination. In seeds of lettuce and many weeds, red light stimulates germination, and far-red light inhibits it. Like the flowering response shown in figure 29.15, seeds alternately exposed to red and far-red light are affected only by the last exposure. Therefore, the phytochrome system can "inform" a seed whether sunlight is available for photosynthesis and thus promote germination under favorable conditions, or inhibit germination until good conditions prevail. If seeds are buried too deeply in the soil, P_{fr} is absent (due to lack of sunlight needed to convert P_r to P_{fr}), and germination does not occur.

FIGURE 29.15

The Last Flash Matters.

(**A**) A long-day plant flowers when daylight exceeds a critical length. (**B**) A short-day plant flowers when daylight is less than a critical length. (**C**) Interrupting night with a flash of red light shortens continuous darkness, and a long-day plant flowers, but a short-day plant does not because the time of uninterrupted darkness is too short. (**D**) A flash of far-red light closely following a flash of red cancels the effect of the red, so the results are the same as (**B**). Three flashes (**E**) — red, far-red, red—has the same effect as one flash of red, and four flashes (**F**) —red, far-red, red, far-red—has the same effect as (**D**). The conclusion: The last flash matters because it determines the prevalent form of phytochrome, which plants use to detect the length of the night.

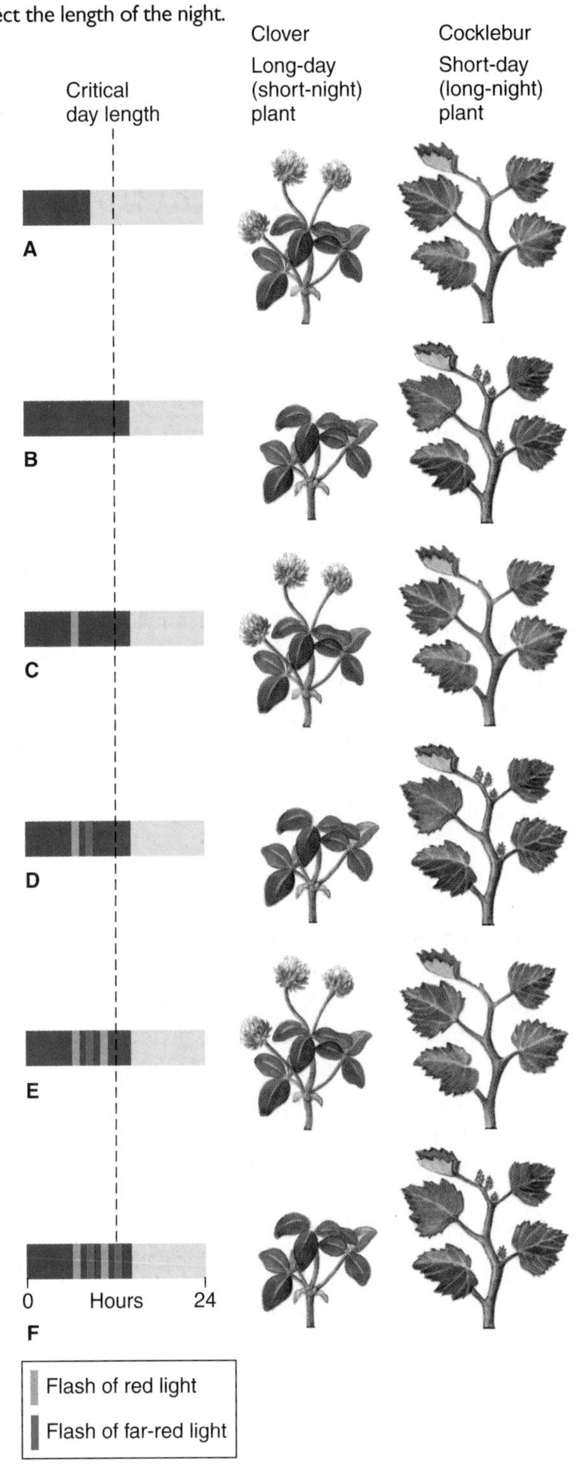

Clover
Long-day (short-night) plant

Cocklebur
Short-day (long-night) plant

Critical day length

A

B

C

D

E

F

0 Hours 24

Flash of red light
Flash of far-red light

Phytochrome also controls early seedling growth. Seedlings grown in the dark are etiolated—they have abnormally elongated stems, small roots and leaves, a pale color, and a spindly appearance (fig. 29.16). Bean sprouts used in Chinese cooking are etiolated. Etiolated plants rapidly elongate before exhausting their food reserves. Normal growth replaces etiolated growth once the plants are exposed to red light, even if for just one minute.

Phytochrome may also help direct shoot phototropism, in addition to the influence of auxin. Light coming from only one direction would presumably create a gradient of P_r and P_{fr} across the stem. P_{fr} would be most abundant on the illuminated side of the stem, because in daylight P_r is rapidly converted to P_{fr}. P_r would be most abundant on the shaded side of the stem. This phytochrome gradient could bend a shoot as P_r promotes stem elongation and P_{fr} inhibits it.

FIGURE 29.16

Growing Up in the Dark.

An overlying log was lifted to reveal these etiolated seedlings, which germinated and grew in the absence of light. Note the pale, elongated stems and tiny leaves.

Senescence and Dormancy

Senescence, or aging, is also a seasonal response of plants. Aging occurs at different rates in different species. The flowers of plants such as wood sorrel and heron's bill shrivel and die only a few hours after blooming. Slower senescence is seen in the colorful changes in leaves in autumn.

Whatever its duration, senescence is not merely a gradual cessation of growth but an energy-requiring process brought about by new metabolic activities. Leaf senescence begins during the shortening days of summer, as nutrients are mobilized. By the time a leaf is shed, most of its nutrients have long since been transported to the roots for storage. Fallen leaves are little more than cell walls, wastes, and remnants of nutrient-depleted protoplasm.

Destruction of chlorophyll in leaves is part of senescence. In autumn, the yellow, orange, and red carotenoid pigments, previously masked by chlorophyll, become visible. Senescing cells also produce pigments called anthocyanins. Loss of chlorophyll, visibility of carotenoids, and production of anthocyanins combine to create the spectacular colors of autumn leaves.

Before the onset of harsh environmental conditions such as cold or drought, plants often become **dormant** and enter a state of decreased metabolism (fig. 29.17). Like leaf senescence, dormancy entails structural and chemical changes. Cells synthesize sugars and amino acids, which function as antifreeze, preventing or minimizing cold damage. Growth inhibitors accumulate in buds, transforming them into winter buds covered by thick, protective scales. These changes in preparation for winter are called acclimation.

Growth resumes in the spring as a response to changes in photoperiod and/or temperature. Lengthening spring days awaken birch and red oak trees from dormancy. Fruit trees such as apple and cherry have a cold requirement to resume growth. This is why a warm period in December, before temperatures have really plummeted, will not stimulate apple and cherry trees to bud, but a similar warm-up in late February will induce growth. The exact mechanism by which photoperiod or cold breaks dormancy is unknown, although hormonal changes are probably involved.

FIGURE 29.17
Dormancy.
Plants enter a seasonal state of dormancy, which enables them to survive harsh weather. Note the protective scales on these buds.

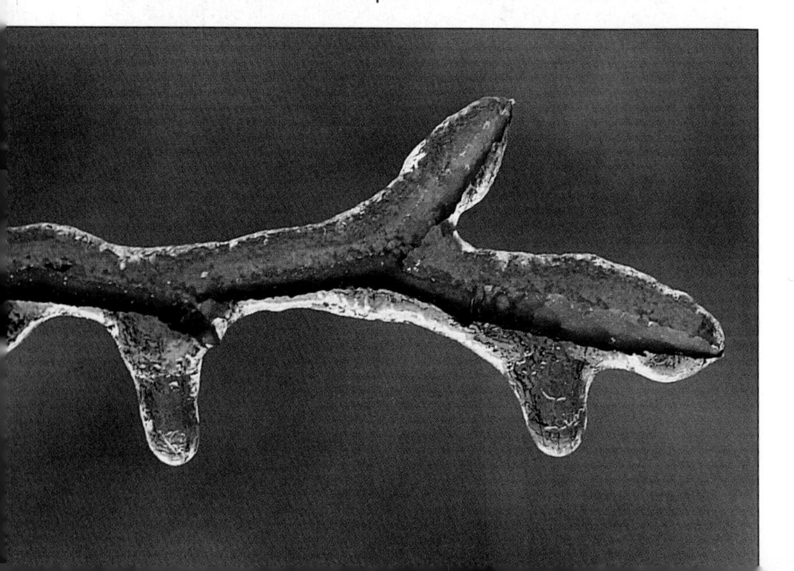

In some plants, factors other than photoperiod or temperature trigger dormancy. In many desert plants, for example, rainfall alone releases the plant from dormancy. In contrast, potatoes require a dry period before renewing growth.

29.3 MASTERING CONCEPTS

1. Define the three types of plants in relation to photoperiod.
2. How do biologists know whether photoperiodism is a response to duration of light or dark?
3. Explain how phytochrome controls photoperiodism and seed germination.
4. What events occur during senescence and dormancy?

29.4 Plant Circadian Rhythms

Biological clocks control activities that cycle according to regular schedules. Circadian rhythms, such as solar tracking, are based on the length of a day. Biological clocks are internally controlled but refined by the environment.

The photoperiodic flowering response occurs in response to seasonal stimuli such as days becoming longer or shorter. Other rhythmic responses are not seasonal but daily, and these are called **circadian rhythms** (from the Latin *circa* = approximately and *dies* = day). Consider the common four-o'clock, which opens its flowers only in late afternoon, or the yellow flowers of evening primrose, which open only at nightfall. Similarly, the photonastic movements of prayer plants occur at the same time every day. Other circadian rhythms in plants include stomatal opening and nectar secretion. These daily rhythms allow plants to synchronize their activities. Flowers of a particular species can open when pollinators are most likely to visit; for example, bat-pollinated flowers of genus *Cereus* must open at night when bats are active. Such daily rhythms also occur in many protista, fungi, and animals.

Control of Circadian Rhythms

The biological clocks that regulate circadian rhythms are controlled both internally (by genes that encode "clock" proteins) and externally, by environmental factors, but the precise mechanism (or mechanisms) remains unknown. Evidence for internal control is that in many species, circadian rhythms do not exactly coincide with a 24-hour day, but may be a few hours longer or shorter. In addition, circadian rhythms often continue under laboratory conditions of constant light or dark. They are ingrained.

Environmental factors, such as a change in photoperiod, can affect or reset a plant's circadian rhythms. This environmentally controlled resynchronization of the biological clock is called entrainment. However, entrainment to a new environment is limited. If the new photoperiod differs too much from a plant's biological clock, the plant reverts to its internal rhythms. Also, a plant maintained in a modified photoperiod over a long period of time reverts to its natural rhythms when placed in constant light.

FIGURE 29.18
Heliotropism.
Sunflowers turn to face the sun in a circadian response called heliotropism.

Solar Tracking Is a Circadian Rhythm

The sight of a field of sunflowers with their glorious heads turned to face the sun is so striking that anyone lucky enough to see it simply stares (fig. 29.18). The response of flowers turning to face the sun, called heliotropism, is circadian because it is the sun that defines day and night. Heliotropism, more commonly known as solar tracking, is seen among flowers that grow in Arctic and alpine environments, where days are short. Snow buttercups (*Ranunculus adoneus*) and mountain avens (*Dryas integrifolia*) also display heliotropism.

By turning toward the sun, heliotropic plants maximize sun exposure, which is used not only for photosynthesis, but to warm pollinators, which become more active. Charles Darwin described solar tracking, noting that it is both fast and reversible. Phototropism, in contrast, is slower and irreversible.

A bit of folklore inspired a series of recent experiments that traced heliotropism to auxin activity. In the nineteenth century, the French noticed that plants shielded by bottles of red wine did not point toward the sun, as others did that were not blocked. Reasoning that the red color blocked blue wavelengths of light, Candace Galen, a biologist at the University of Missouri at Columbia, devised a laboratory version of the wine phenomenon, depicted in figure 29.19.

Galen grew buttercups shielded with a red acrylic filter that blocked blue light; with no filter; and with a filter that allowed blue wavelengths through. The flowers not exposed to blue light did not orient toward the sun. To determine how the flowers sensed blue light, Galen followed a simple approach—she lopped off different parts of the plant, to see which one had to be missing for the plant to be unable to detect blue light. The implicated area turned out to be the upper tip of the main stem (fig. 29.19B). Finally, she used an opaque substance to cover different parts of the main stem, and so discovered that blocking the tip eliminated heliotropism (fig. 29.19C). By examining the plants under a microscope, she saw that the cells on the side of the middle of the stem away from the sun were larger than the cells facing the sun, suggesting that auxin is involved in the response. The conclusion: The top of the stem detects blue light and somehow passes this information to cells in the middle of the stem, where auxin accumulates on the shaded side, causing the plant to bend toward the sun.

This chapter has focused on plants' responses to environmental stimuli. These responses are easy to describe—leaves turn to intercept the sun, or flowers bloom. More difficult is untangling the many complex interactions that produce the responses. Plant hormones, which help control such fundamental functions as cell division and elongation, surely play some part in flowering, which in many plants occurs in response to photoperiod. But the precise interaction between phytochrome and plant hormones remains unknown. Furthermore, the hormones described in this chapter are by no means the only influences on plant growth and development. As described in Investigating Life 29.2, many other mechanisms also play roles in growth, cell division, and plant responses to the environment.

29.4 MASTERING CONCEPTS

1. What is a circadian rhythm?
2. What is the evidence that genes and the environment influence biological clocks?
3. How does heliotropism (solar tracking) occur?

OLC

FIGURE 29.19

Experiments Reveal That Blue Light Controls Heliotropism Through Auxin.

(**A**) Blocking blue light destroys the ability of buttercups to track the sun. (**B**) The tip of the main stem just below the flower is the plant part that detects blue light. Even decapitated plants turn toward the sun, as long as their stems are intact. (**C**) By applying an opaque coating (Liquid Paper) to the tops, middles, and bottoms of stems, researcher Candace Galen showed that the tip detects the blue light. Microscopy revealed that an auxin response occurs in the middle of the stem. The tip must therefore communicate with the cells in the middle section.

A

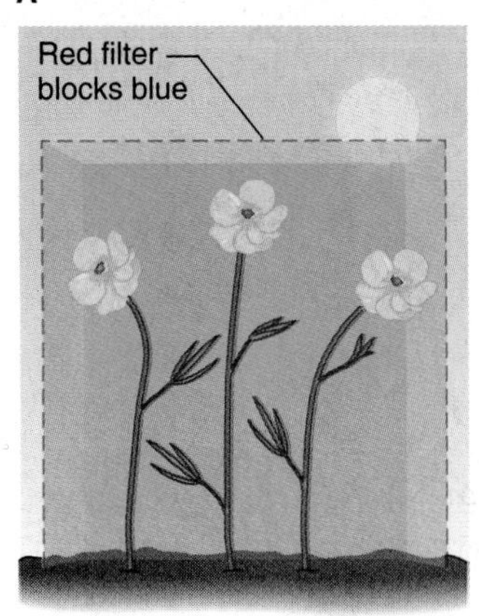

Experimental

Control 1

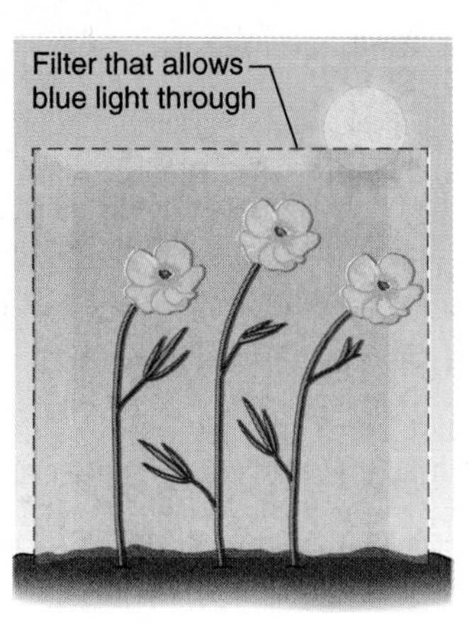

Control 2

Experiment 1
Conclusion: Blue light is required for solar tracking.

B

Decapitated

Control

Experiment 2
Conclusion: Solar tracking ability resides in stem.

C

Block tip of stem

Block middle of stem

Block bottom of stem

Experiment 3
Conclusion: Solar tracking ability resides in upper stem.

Plant Communication Molecules

The tools of molecular biology are revealing previously unrecognized internal signals that are important in plant growth and development (table 29.3). Here are a few.

Steroids

We usually associate steroid hormones with their muscle-building effects in humans. Plants have steroids too. Botanists recognize a plant steroid called brassinolide, but until recently didn't know its function.

Understanding the role of brassinolide began when geneticists working with *Arabidopsis thaliana* identified a gene similar to one that encodes an enzyme in mammals essential for producing male steroid sex hormones. Plants with mutations in this gene cannot use cues from light to regulate growth and development. As a result, plants are stunted and too green (fig. 29.B). When researchers gave mutant plants brassinolide, the plants developed normally. Therefore, brassinolide helps control growth and chlorophyll production.

Peptides

Like animals, plants use short amino acid chains as communication molecules. Researchers isolated the first one in 1991—a single microgram of systemin (discussed in the chapter opening essay) from 30,000 tomato plants! The second plant communication peptide discovered, ENOD40, enables certain root cells to ignore signals from auxin that would normally stop their growth. Under ENOD40's influence, these cells divide, forming root nodules in the presence of nitrogen-fixing bacteria.

Corn plants with crinkled leaves and kernels led researchers to a gene encoding a receptor for a kinase. Kinases are enzymes that add phosphate groups to other molecules, and they are vital in signal transduction, the biochemical pathways that cells use to receive messages. Kinases activate certain other types of molecules. The crinkled kernel results from a failure to communicate—cells forming a protective layer on the embryo cover only half of the embryo, because of a block in the signal to migrate during development. Kinase receptors also control stem thickness in *Arabidopsis* and resistance to bacterial infection in rice, and prevent self-pollination in broccoli and cauliflower. Pollen sends a signal to carpels on the same plant, which blocks pollen grains from sticking to them.

Methylation

Turning off certain genes at certain times fuels development. In humans and mice, a process called methylation places methyl (CH_3) groups on certain genes, which inactivates them. *Arabidopsis* mutants that cannot methylate sufficiently are deranged. Mutant plants grow up to 27 extra leaves, take 47 days to flower compared with the normal 26 days, and develop five times the normal number of flower stalks!

Methylation may function as an internal clock in plants, because its absence in *Arabidopsis* mutants mistimes development. Meristems do not receive messages to cease division, and plant parts continue to form when they shouldn't, resulting in such conditions as extra petals. In addition, in the same region of a mutant plant, some cells differentiate as leaves, and others as flowers.

FIGURE 29.B
Lack of Steroid Hormones Can Stunt Plant Growth.
Compared to the normal *Arabidopsis thaliana* plant at left, the plant on the right is severely stunted because it cannot produce a steroid hormone needed to grow and develop in response to light. The dwarfed plant is dark green because of excess chlorophyll.

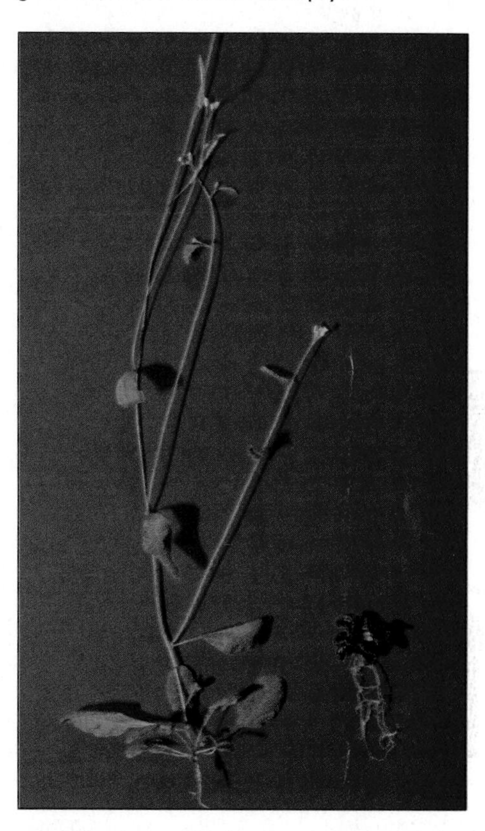

TABLE 29.3 Plant Communication Chemicals

BIOCHEMICAL	CLASS	FUNCTION
Brassinolide	Steroid	Cell division and elongation
		Chlorophyll synthesis and distribution
ENOD40	Peptide	Overrides auxin in roots to form nodules
Jasmonic acid	Carboxylic acid	Triggers synthesis of defense proteins
		Controls seed germination and root growth
Kinase receptor	Protein	Formation of protective layer on corn embryo
		Stem thickness
		Resistance to bacterial infection
		Prevents self-pollination
Methyl groups	CH_3	Regulate development by silencing certain genes

Chapter Summary

29.1 How Do Plants Grow?

1. Plants respond to the environment with changes in growth and movement, mediated by the action of **hormones.**
2. **Auxins** stimulate cell elongation in shoot tips, embryos, young leaves, flowers, fruits, and pollen. Auxins are more concentrated at the main shoot tip, which blocks growth of lateral buds **(apical dominance).**
3. **Gibberellins** stimulate cell division and elongation but act more slowly than auxins.
4. **Cytokinins** stimulate mitosis in actively developing plant parts.
5. **Ethylene** speeds ripening.
6. **Abscisic acid** inhibits the growth-inducing effects of other hormones.

29.2 How Do Plants Move?

7. A **tropism** is a growth response toward or away from an environmental stimulus, usually caused when different parts of an organ or structure grow at different rates.
8. In **phototropism,** light sends auxin to the shaded portion of the plant, stimulating growth towards the light.
9. Shoot growth is a negative **gravitropism,** and root growth is a positive gravitropism. The positions of amyoplasts in cells apparently help plants detect gravity.
10. **Thigmotropism** is a response to touch.
11. **Nastic movements** are not oriented toward a stimulus. **Thigmonasty** is response to contact. Nastic response to light and dark is **photonasty,** caused by osmotic changes that differentially alter cell volume.

29.3 How Do Plants Respond to Seasonal Changes?

12. Plants sense seasonal and other environmental changes. **Photoperiodism** is the ability of a plant to measure length of day and night. Flowering can depend upon photoperiodism.
13. **Short-day plants** flower only when the duration of light is less than a critical length. **Long-day plants** require a light period longer than a critical length. **Day-neutral plants** do not use light or dark cues to flower. The type of plant determines the season when it flowers.
14. Plants may respond to length of darkness rather than length of daylight.
15. A plant pigment, **phytochrome,** controls response to light. One form, P_r, absorbs red light to become P_{fr}. P_{fr} promotes flowering of long-day plants and inhibits flowering of short-day plants. It is reconverted to P_r by absorbing far-red light. The ratio between these two forms provides information about daylight because sunlight has more red than far-red light.
16. Phytochrome controls seed germination and early seedling growth, and helps direct shoot phototropism.
17. **Senescence** is an active and passive cessation of growth. Growth becomes **dormant** during cold or dry times and resumes when environmental conditions are more favorable.

29.4 Plant Circadian Rhythms

18. Internal biological clocks control daily responses, or **circadian rhythms.** Environmental changes can alter, or entrain, these clocks.
19. Heliotropism (solar tracking) is a circadian rhythm in which flowers face the sun when cells in the upper stem respond to blue light.

Testing Your Knowledge

1. What are some examples of ways that plants respond to the environment?
2. How do gibberellins and auxins differ in their actions?
3. Describe how ion movement can cause plant movement.
4. What effects do red light and blue light have on plants?
5. How do auxins and cytokinins interact in apical dominance?
6. How does a tropism differ from a nastic movement?
7. What are three effects of phytochrome?
8. What are three factors that can release a plant from dormancy?
9. What is the difference between photoperiodism and a circadian rhythm such as photonasty?

Thinking Scientifically

1. Describe the steps of the scientific method that Jennifer Thaler used to study plant defense against insects and that Candace Galen used to identify the plant part responsible for heliotropism.
2. Poinsettias are short-day plants. What can be done to encourage these plants to flower in December?
3. Why does mixing leftover fruit salad with fresh fruit hasten rotting of the mix?
4. Tomato plants produce jasmonic acid only when they are being eaten by insects. What are two reasons that it would not be adaptive to produce this chemical all the time?
5. Although spinach and ragweed each require 14 hours of sunlight to flower, spinach flowers in the summer and ragweed (as hay fever sufferers can attest) flowers in the fall. Explain the seasonal difference in flowering.

References and Resources

Galen, Candace. May 1999. Sun stalkers. *Natural History,* p. 49. Plants that display heliotropism detect blue light, which triggers auxin activity.

Lewis, Ricki. December 11, 1995. Chronobiology researchers say their field's time has come. *The Scientist,* vol. 9, p. 1. Diverse species exhibit circadian rhythms.

Thaler, J. S. June 17, 1999. Jasmonate-inducible plant defenses cause increased parasitism of herbivores. *Nature,* vol. 399, p. 686. A field experiment shows that plants attack insect pests both directly and indirectly.

Trewaveas, Anthony. April 1999. How plants learn. *Proceedings of the National Academy of Sciences,* vol. 96, p. 4216. Signal transduction enables plants to respond to a changing environment.

The *LIFE* Online Learning Center provides additional resources and tools for studying this chapter.
www.mhhe.com/life

UNIT 6 • ANIMAL LIFE

A quiet drama is being played out on a moonless summer's night on a sand dune in the middle of the Mojave Desert. A desert sand scorpion sits in patient ambush for a burrowing cockroach to betray itself through its movements. With exquisitely sensitive sensors on its legs, the scorpion detects the cockroach's subtle vibrations in the sand. Methodically, using triangulation to the waves that spread under its feet, the scorpion orients and moves closer to the source of the disturbance. Then, with a quick lurch, a grab, and a neurotoxic sting, dinner is served. Such a drama could only be seen in the animal kingdom, for only animals have the tissues required—muscles and nerves.

CHAPTER 30

Animal Tissues and Organ Systems

Regenerative Medicine

Organs in a human body harbor pockets of cells that are reservoirs for future growth and repair. These stem cells, so called because they can differentiate into particular cell types, have been studied for decades, but their value in replacing or rejuvenating worn body parts is just now being realized. Researchers are investigating stem cells' potentials in the laboratory, using the simplest tissues therapeutically, and creating hybrids of synthetic materials and cells.

Mesenchymal stem cells (MSCs) are tucked into bone marrow and other places, but migrate to the site of an injury. Here, responding to signals, they differentiate into connective tissue, blood, cartilage, bone, fat cells, and the outer layers of blood vessels—whatever can repair the damage. Until recently, it wasn't clear whether the differentiated cells sprang from several progenitors—bone from one type of stem cell, cartilage from another—or if a single cell could, in response to different combinations of signals, follow different fates.

To answer this question, researchers exposed MSCs that descended from single cells to different mixtures of hormones, growth factors, vitamins, and serum, specific for differentiation as fat, cartilage, or bone cell. A few weeks later, future fat cells bulged with fat, the cartilage cells contained characteristic proteins, and bone cells stockpiled the minerals and enzymes of bone tissue. The conclusion: A single mesenchymal stem cell can indeed develop as any of several cell types.

MSCs and other types of stem cells are the basis of "regenerative medicine." Experiments in nonhuman animals demonstrate several promising approaches. Stem cells that give rise to cardiac muscle can mend damaged hearts. Stem cells that give rise to cartilage can repair injuries in this bloodless and therefore hard-to-heal tissue. One form of regenerative medicine already in practice is storing stem cells from umbilical cord blood. These cells can be used to treat blood or immune system disorders in the child, or be donated to help others. The cells are much less likely to provoke an immune system rejection than more mature, specialized cells. People being treated for cancer can store stem cells separated from their own blood or bone marrow before undergoing intensive radiation and chemotherapy.

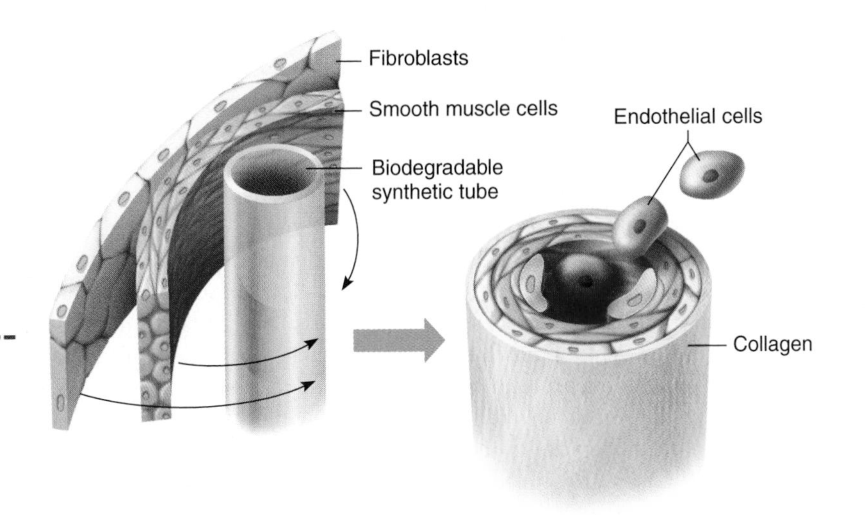

Tissue Engineering.
To construct new blood vessels, sheets of smooth muscle cells and fibroblasts are wrapped around a synthetic tube mold. Later, the tube is removed and the inner surface of the vessel is "seeded" with endothelial cells that proliferate to form a smooth lining. The fibroblasts secrete an outer noncellular layer of collagen that will reinforce the vessel and anchor the vessel to overlying structures.

When their blood counts fall in response to the therapy, the patients can receive infusions of their own stem cells, enabling them to tolerate higher doses of the treatments, which are more likely to be effective against the cancer.

Researchers do not yet even know the limits of regenerative medicine. Certain cells may harbor a developmental plasticity that biologists never imagined possible. For example, bone marrow cells, given certain signals, can become something else, such as brain cells, liver cells, or muscle cells. This means that physicians may one day be able to heal hard-to-reach places in the brain, liver, and muscles through the blood.

Future applications will supply tissues that consist of more than one cell type, and, eventually, organs.

The first applications of regenerative medicine will use single cell types or blood components. Future applications will supply tissues that consist of more than one cell type, and, eventually, organs. To do this, biologists must learn how cells aggregate and interact to form tissues, then how those tissues build organs. This requires not only discovering and replicating the right mix of biochemicals to coax stem cells into becoming nerve or muscle, cartilage or bone, but also identifying the cell densities, mechanical forces, and spatial organizations necessary to form larger-scale parts of animal bodies.

In the meantime, a related field, tissue engineering, combines living cells and synthetic substances to create replacement parts. The challenge is to identify and reproduce the characteristics that enable the part to function in the body. For example, a blood vessel must:

- be strong enough to withstand the force of circulating blood and pressure from surrounding tissue;
- be flexible;
- be smooth enough to prevent formation of blood clots;
- not evoke an immune response.

Several biotechnology companies have developed blood vessels by wrapping various tissues around synthetic tubes and surrounding the structure with collagen, the major protein of connective tissue. However, adding collagen disturbs the fibroblasts that normally secrete it, resulting in a rather ragged tube. An innovative graduate student improved the approach by letting the engineered blood vessel secrete its own outer covering.

Nicolas L'Heureux knew that the vessel required an inner smooth lining of sheetlike cells called endothelium; a middle layer of smooth muscle and elastic connective tissue; and an outer layer of fibroblasts. So he grew smooth muscle cells and fibroblasts into sheets, then wrapped the sheets around a biocompatible polymer tube to form a tubule. He seeded the inner surface of the muscle tube with endothelial cells, which divided and formed a one-cell-thick inner lining, just as in a natural blood vessel. Then the fibroblasts secreted collagen, coating the vessel. The result: sleek blood vessel replacements that may one day help people who need grafts in their legs or new coronary arteries.

30.1 Animal Tissue Types

An animal's body is built of differentiated cells as well as stem cells that retain the potential to specialize. Cells associate to form four major tissue types: epithelial tissue, connective tissue, nervous tissue, and muscle tissue.

Animal bodies are built of tissues. Most **tissues** consist of two or more cell types that interact. The number of cell types reflects the complexity of the animal. A human body has more than 200 cell types; a sponge has only four. In complex animals, two or more tissues interact to form **organs,** and organs may be linked, either physically or functionally, into **organ systems.**

This initial chapter introduces the four basic tissue types of animals, focusing on the human, and concluding with a look at a sample organ system, the integument. The other chapters in the unit consider organ systems one at a time. Each includes an interesting invertebrate's system and the corresponding system in the human.

Cells Specialize and Associate to Form Tissues

Sperm fertilizes egg, and soon the DNA from two individuals mingles in a new nucleus. Hours later, that initial cell divides to become two, then later, it does so again and the animal is four cells, and so on. The cells form a ball, and then the ball hollows out. Up until this point, one cell is pretty much like another, in terms of appearance and position. Then changes ensue. A few of the cells collect on the inner face of the ball, spread out to form sheets, and then these sheets fold, first forming two layers, then a third in the middle (see fig. 24.3). From these three primary germ layers—ectoderm (outer), mesoderm (middle), and endoderm (inner)—develop all of the specialized structures of the animal. Chapter 40 explores these steps of development in detail.

Cell differentiation begins as different genes are turned on or off in the new organism's genome. A cell destined to be part of the heart muscle produces the proteins actin and myosin, and if this is a human, by about the fourth week, those cells will beat in unison. Another cell might express genes enabling it to produce neurotransmitters, eventually becoming a nerve cell.

Not all cells specialize. **Stem cells** are less differentiated than others, retaining the ability to become more specialized should conditions warrant (see the chapter opening essay). These cells enable a tissue to continue to grow and to repair damage. Figure 8.14 depicts stem cells in the basal layer of human skin. Stem cells are a basic feature of animal bodies. Recall from chapter 24 that even sponges, the simplest animals, have archeocytes, which can give rise to either a pinacocyte or an amoebocyte.

The 200+ types of cells in a human body (and those of other vertebrates) fall into four broad categories:

- **Epithelial tissue** covers and lines organs and forms most of the functional tissue.
- **Connective tissue** provides support, adhesion, insulation, and attachment.
- **Nervous tissue** forms rapid communication networks between cells and distant body regions.
- **Muscle tissue** contracts, powering the movements of life.

Epithelial Tissue Forms Layers

An animal's body has many surfaces, from the most obvious, the skin or integument, to those on the outsides and insides of organs. Epithelial tissues line structures with one or more layers of tightly packed cells. They cover organs and line the inside of hollow organs and body cavities. These tissues always have a free or apical surface that is exposed either to the outside or to a space within the body. The other side, or basal surface, is anchored to other tissue by a noncellular layer called a **basement membrane.** Epithelial tissues lack blood vessels; nutrients diffuse in and wastes out. Epithelial tissues participate in a variety of functions, including protection, absorption, and secretion.

The shapes of the cells and how they are layered is used to classify epithelial tissues (fig. 30.1). The three most common cell shapes are squamous (flattened), cuboidal (cube-shaped), and columnar (tall, thin). Cell layering is characterized as either simple (a single layer of cells) or stratified (more than one layer). For example, simple squamous epithelium is a single layer of flat cells. In contrast, stratified columnar epithelium consists of multiple layers of cells with the outermost layer made of tall column-shaped cells. The apical layer is used to describe epithelial tissue. Stratified squamous epithelia, for example, usually have cuboidal cells on their basal surfaces, but the cell layers become flattened—squamous—by the time they reach the apical side. In the upper respiratory tract, the epithelium is a single layer of columnar cells, but the staggered location of nuclei in different parts of the cells gives the impression of layers. Such epithelium is termed "pseudostratified."

The structure of an epithelial tissue is closely tied to its function. Simple squamous epithelium lines the microscopic air sacs in the lungs, where gases diffuse across its narrow width. Simple cuboidal epithelium lines kidney tubules, where it regulates the composition of fluids. Most of the digestive tract's lining is simple columnar epithelium. This type of epithelium often is fringed with extensions called microvilli that increase the surface area available to absorb nutrients. The multiple layers and regenerative abilities of stratified squamous epithelium enable it to resist abrasion, such as occurs at various body openings and the outer layer of the skin (epidermis).

Connective Tissues Include Blood, Bone, and Cartilage

Much of a vertebrate's body consists of connective tissues. These tissues fill spaces, attach epithelium to other tissues, protect and cushion organs, and provide mechanical support. All connective tissues consist of cells embedded in a nonliving substance called an extracellular matrix.

A **fibroblast** is a common type of connective tissue cell. Fibroblasts manufacture and secrete two main types of protein fibers that become part of the matrix—**collagen,** a flexible white

FIGURE 30.1
Types of Epithelium.
Epithelial tissues are composed of tightly packed cells in single or multiple layers that rest atop a basement membrane. Epithelial tissues cover body surfaces and line hollow organs and body cavities.

Types of epithelium	Properties and examples
Simple squamous Basement membrane; Connective tissue	Single layer of flattened cells; important in areas of filtration and diffusion; examples: lining of heart and blood vessels, alveoli of lungs, glomeruli of kidneys.
Simple cuboidal 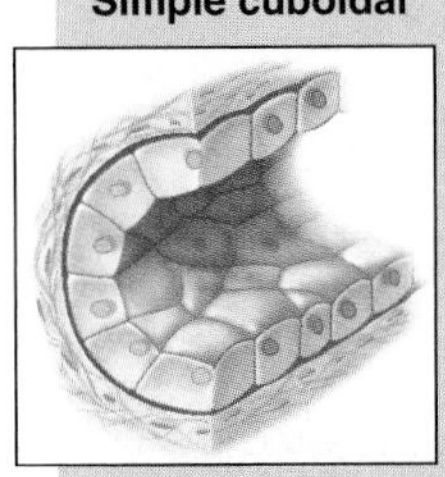	Single layer of cube-shaped cells; important in secretion and absorption; examples: glandular tissue, lining of the kidney tubules.
Simple columnar 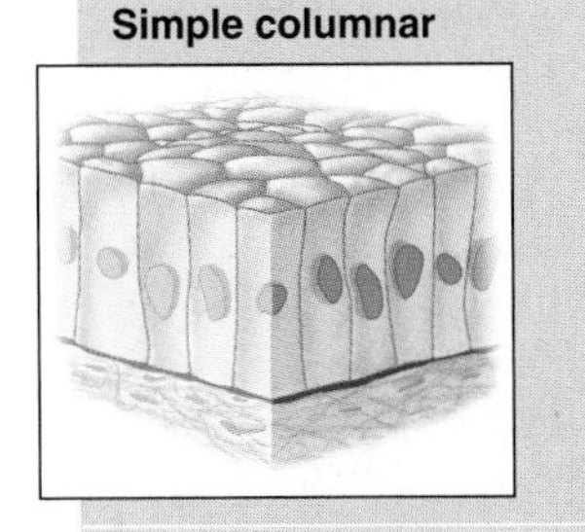	Single layer of column-shaped cells; important in absorption and secretion; examples: lining of the stomach, intestines, and oviducts.
Stratified squamous 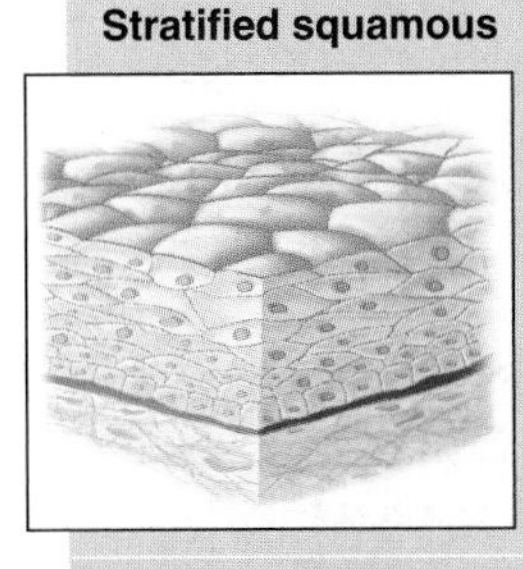	Multiple layers of cells with flattened cells on exposed surface; important in areas subject to abrasion; example: outer layer of skin, lining of body openings.
Pseudostratified ciliated columnar 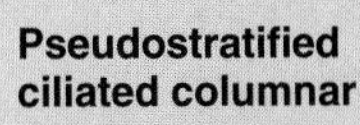 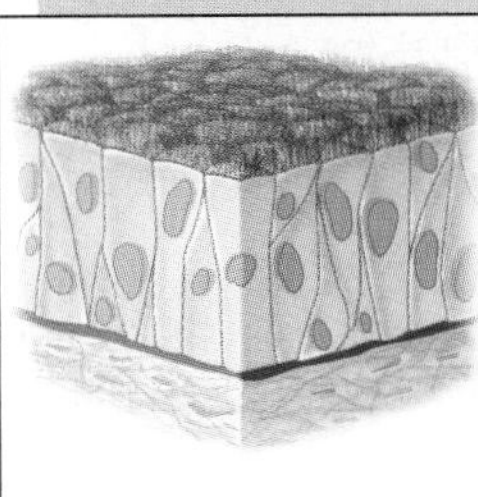	Single layer of cells that appears stratified due to staggered nuclei; cilia beat rhythmically to move dust and microorganisms out of airways; example: lining of the upper respiratory tract.

protein that resists stretching, and **elastin,** a yellowish protein that stretches readily, like a rubber band. The matrix also includes a thin gel of proteoglycans, which are complex carbohydrates linked to proteins.

The major types of connective tissues are loose connective tissue, dense connective tissue, adipose tissue, blood, cartilage, and bone (fig. 30.2). Matrix composition, types of fibers, cell specializations, and the proportion of cells to matrix distinguish each type of connective tissue. For example, blood has a very high proportion of matrix, whereas adipose tissue (fat) has a high proportion of cells.

Loose connective tissue is the "glue" of the body. It consists of widely spaced fibroblasts and a few **adipocytes** (fat cells) within a meshwork of collagen and elastin fibers. In contrast, **dense connective tissue** consists of more tightly packed tracts of collagen. Dense connective tissue forms ligaments, which bind bones to each other, and tendons, which connect muscles to bones, and much of the middle layer of skin.

Adipocytes, which comprise **adipose tissue,** form from fibroblasts that accumulate enormous amounts of lipid (fat), which pushes the nuclei to the cell periphery. Adipose tissue insulates, cushions joints, protects organs, and stores energy.

Blood is a complex mixture of different cell types suspended in a liquid matrix called **plasma** (see chapter 35). Blood cells circulate through the body, whereas other connective tissue cells do not migrate. **Red blood cells** (erythrocytes) transport oxygen and constitute the bulk of the blood cells. **White blood cells** (leukocytes) are less numerous than red blood cells and are of several varieties. White blood cells protect against infection and help clear the body of cells that have worn out or become abnormal. Blood also contains cell fragments called **platelets,** which release chemicals that promote blood clotting.

Cartilage is a connective tissue that cushions organs and forms a structural framework that keeps tubular organs from collapsing, such as in the ear, nose, and respiratory passages. In joints, cartilage can sustain weight while allowing bones to move against one another. Unlike most connective tissues, cartilage consists of a single cell type, the **chondrocyte.** These cells lodge within oblong spaces called **lacunae** that are embedded in a collagen matrix called ground substance. Cartilage grows as chondrocytes secrete collagen. Some cartilage also contains elastin. Cartilage lacks a blood supply, which is why it is hard to heal.

Bone also consists of cells within a matrix but with an important addition—mineral salts. The mineral hydroxyapatite, which contains calcium and phosphate, constitutes most of the mineral part, or phase, of bone. A labyrinth of tunnels forms as the hard mineral is deposited. The organic phase of bone consists almost entirely of collagen. Cells called **osteocytes** occupy lacunae (spaces) in the bone. Long narrow passageways, called **canaliculi,** connect lacunae. Osteocytes extend through canaliculi and touch each other.

Bone, which has an ample blood supply, is an unusual tissue in that it is continually under reconstruction. Several cell types (other than osteocytes) remodel bone. **Osteoblasts** secrete bone matrix. **Osteoclasts** degrade bone matrix. Stem cells called **osteoprogenitor cells** line bone passageways, and they differentiate into osteoblasts if growth or injury requires additional bone to form. Chapter 34 considers cartilage and bone further.

FIGURE 30.2

Types of Connective Tissue.

Connective tissues are highly diverse in form and function. However, they all consist of cells embedded in a nonliving matrix.

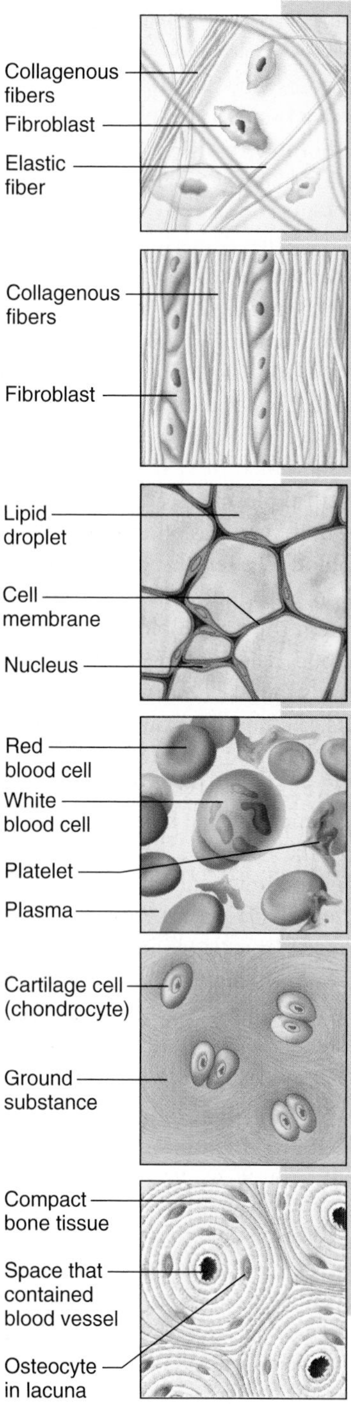

Types of connective tissue	Cells	Matrix composition	Cells to matrix	Site
Loose connective tissue	Fibroblasts, adipose tissue, white blood cells	Loose elastin and collagen networks	Low	Under skin
Dense connective tissue	Fibroblasts	Dense elastin and collagen networks	Low	Ligaments and tendons
Adipose tissue	Adipocytes	Cells with abundant lipid	High	Beneath skin, between muscles, around heart and joints
Blood	Red and white blood cells, platelets	Plasma	Low	In vessels throughout the body
Cartilage	Chondrocytes	Fine fibers of collagen	Low	Ears, joints, bone ends, respiratory passages, embryonic skeleton
Bone	Osteoclasts, osteoblasts, osteocytes, osteoprogenitor cells	Collagen, minerals	Low	Skeleton

Nervous Tissue Connects and Integrates the Body

Nerve cells, called **neurons,** along with associated cells called **neuroglia** make up nervous tissue. This tissue forms a communication network within an animal's body that enables it to receive information from the outside environment and respond to it.

Neurons can be quite long, with intricate branches. A typical neuron consists of an enlarged portion called the **cell body** (which contains the nucleus), a branch called the **axon,** and several branches called **dendrites.** Most dendrites receive information from other neurons in the form of chemical neurotransmitters. Environmental stimuli such as light, heat, or pressure may activate neurons specialized as sensory receptors, located in the skin and in sense organs. Arrival of a neurotransmitter or sensory stimulation alters the permeability of the dendrite's cell membrane and triggers an electrochemical impulse that then passes to another neuron or to a muscle or gland cell (fig. 30.3).

Several types of neuroglia, as well as connective tissue, are found around and between neurons. One abundant type of neuroglia, Schwann cells, have very fatty cell membranes that wrap around axons and form insulating sheaths of the lipid myelin, which speeds nerve impulse conduction. Other neuroglia form scaffoldings that support highly branched neurons, and some supply nutrients and growth factors to neurons or remove ions and neurotransmitters that accumulate in the spaces between neurons. Chapters 31 and 32 examine nervous tissue.

FIGURE 30.3
Nervous Tissue.
Neurons and several types of neuroglia make up nervous tissue. A neuron has several dendrites that receive messages from neurotransmitters, and usually one axon, which relays impulses to other cells. Some neurons are wrapped in a sheet of myelin that is part of Schwann cells, a type of neuroglia. This fatty insulation speeds nerve message transmission.

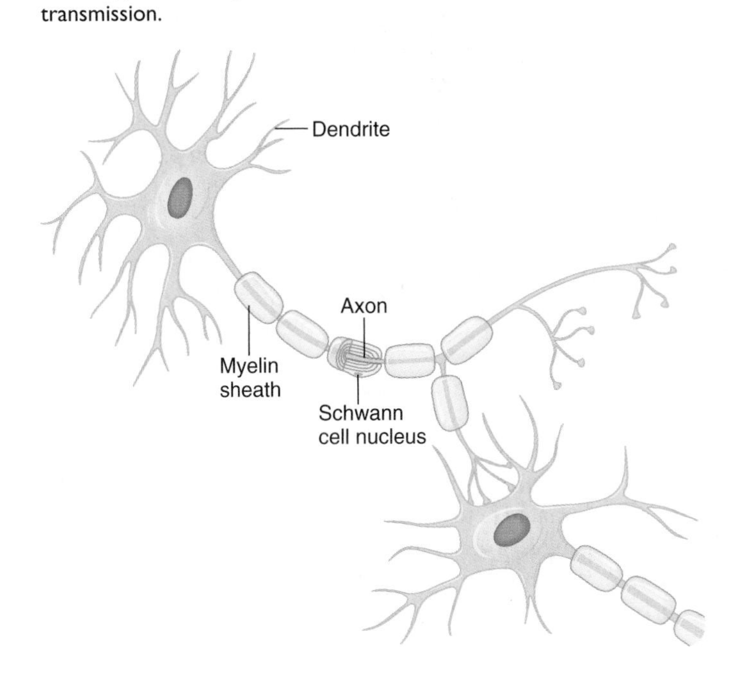

Muscle Tissue Provides Movement

Muscle tissue provides voluntary and involuntary movements. Muscle cells contract when two types of filaments, composed of the proteins **actin** and **myosin,** slide past one another, thus shortening their total length. Other, less abundant proteins are necessary for muscle contraction, too. Muscle cells have many mitochondria, which provide the energy for contraction.

Contractile cells are of three types (see fig. 34.10). A fiber of **skeletal muscle** consists of one long cell with many nuclei. It appears striped, or striated, when viewed under a microscope because the actin and myosin proteins align. Skeletal muscle provides voluntary movements. **Cardiac muscle,** found only in the heart, is striated, but the cells have single nuclei. Disc-like structures join cardiac muscle cells to each other. **Smooth muscle** is not striated, and its involuntary contractions are slower than those of other contractile cells. Smooth muscle cells pulsate along the digestive tract, helping to move food along.

Figure 30.4 summarizes the four tissue types in a human.

30.1 MASTERING CONCEPTS

1. How do cells specialize?
2. What is the function of stem cells?
3. What are the four types of tissues in humans?

30.2 Tissues and Organs Build Animal Bodies

The simplest animal, a sponge, has only four types of cells; a human has more than 200, organized into tissues that in turn are organized into complex organ systems. The organ systems of a human provide communication, structure and support, energy, protection, and continuity. These systems interact in many ways.

Among bacteria, archaea, and most protista, the cell is the body. With the evolution of multicellularity came the opportunity for cells to diverge, to take on specific characteristics that would enable them to divide the labor of being alive by associating into tissues and organs. Fungi and plants have distinctive cell types, but when it comes to cell specialization, the more complex animals have many more types than do members of the other two multicellular kingdoms.

Tissues and Organs Increased in Complexity with Evolution

Chapters 24 and 25 provide an overview of the evolution of body complexity among the animals, as reflected in modern species. Figure 24.1 depicts what the first animals were probably like—a cluster of choanoflagellates, one cell exactly like another. The sponges have a few specialized cell types, including the choanocytes that are perhaps a major clue to animal origins. The more complex cnidaria have more specialized cells that are organized into rudimentary organ systems. Muscle cells interact; cavities, tubules, and passageways form simple digestive and excretory

FIGURE 30.4
Human Tissue Types.
This diagram shows some examples of the locations of the four primary tissue types in the human body.

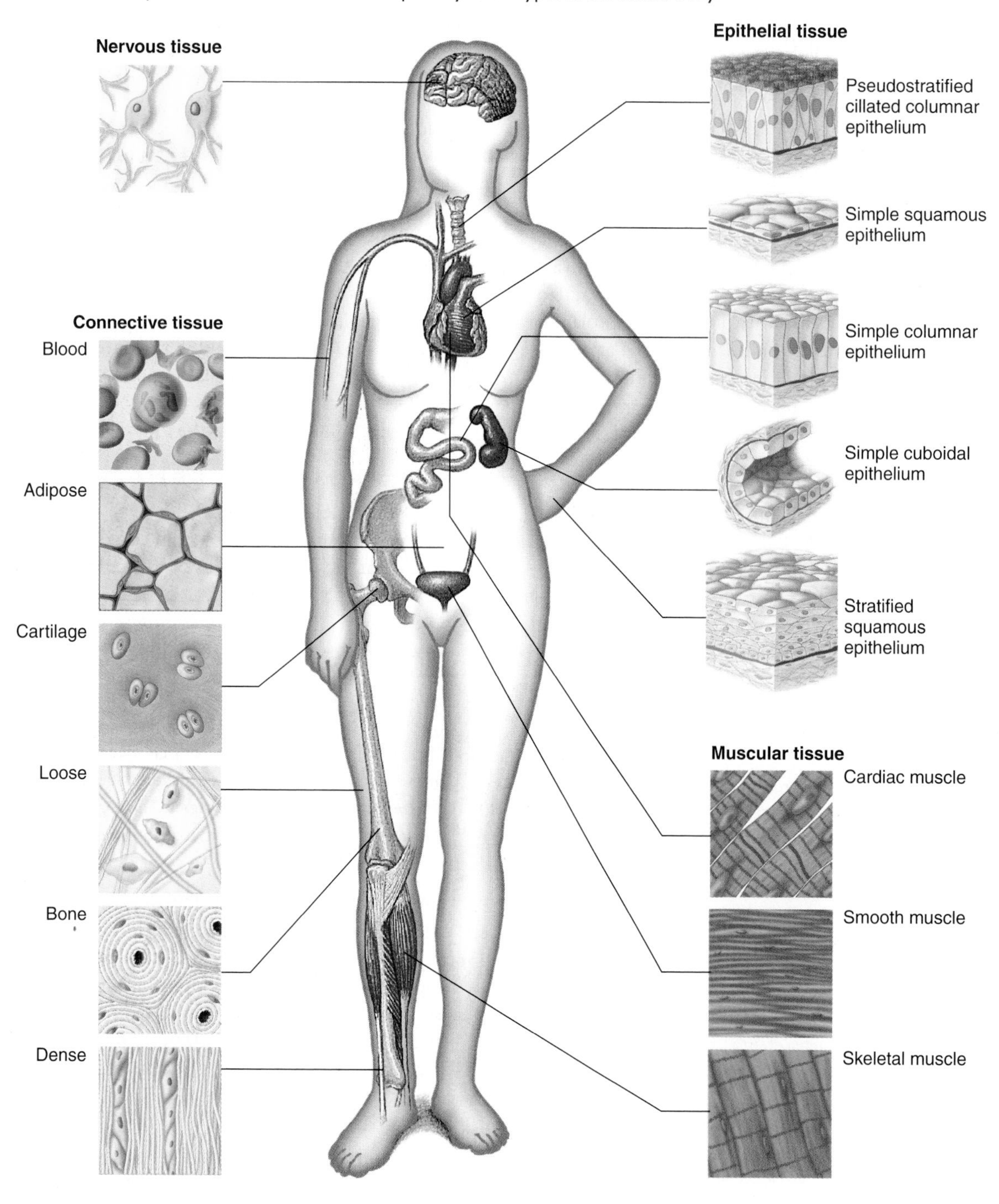

tracts; and sensory cells tune the animal in to its environment. The planarian flatworm introduces a ladderlike nerve cord, a more complex digestive tract, and an excretory system based on highly specialized flame cells.

Organ systems are yet more sophisticated in the mollusks, annelids, arthropods, echinoderms, and chordates. Senses sharpened; numbers of blood vessels and nerves increased; digestive systems became more specialized; and efficiency of circulatory systems improved. With the evolution of these organ systems also came additional connections between them, enabling an animal to better coordinate its functions.

Following is a brief overview of the organ systems of a human. The remaining chapters in this unit elaborate on each one.

The Organ Systems of a Human

Figure 30.5 shows the organ systems of a human grouped by general function. However, the organ systems interact in many ways.

Communication The nervous and endocrine systems integrate and coordinate the activities of all organ systems.

The **nervous system** in a human is a vast interconnected network of trillions of neurons, surrounded, nourished, and supported by several types of neuroglia. Organization within the nervous system ranges from simple sequences of neurons, to the central relay center that is the spinal cord, to the almost incomprehensible complexity of the brain. The sense organs that provide information about the outside world are also part of the nervous system. Some neurons are specialized sensory receptors that detect changes from the environment or inside the body, while others receive the impulses transmitted from sensory receptors and either relay, interpret, or act on them. Yet other neurons carry impulses from the brain or spinal cord to muscles or glands, which contract or secrete products in response. Nervous tissue infiltrates many

Figure 30.5
Human Organ Systems.

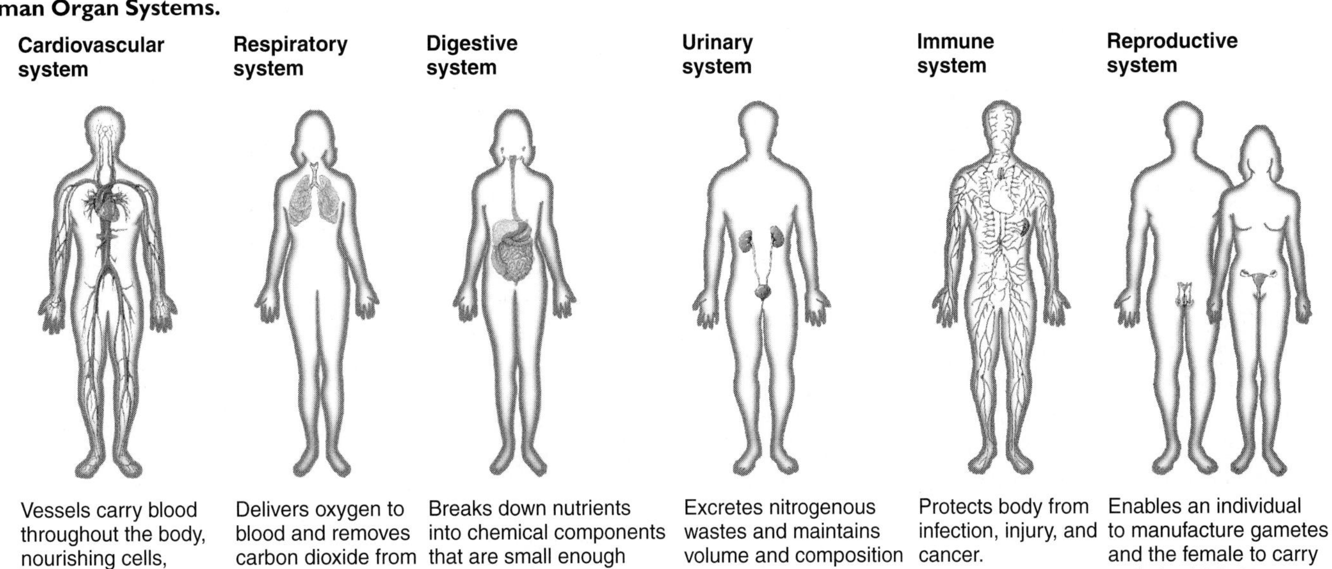

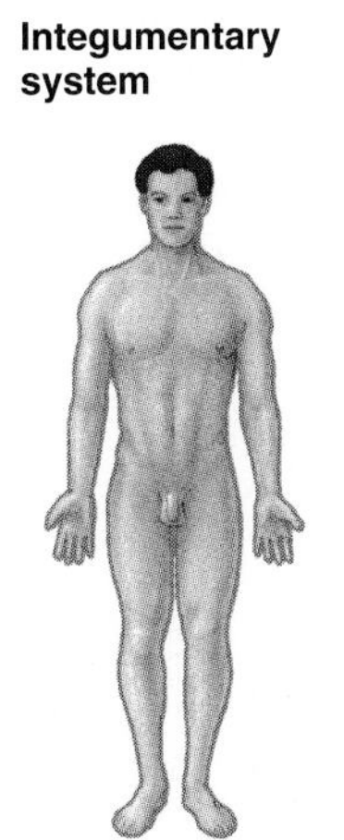

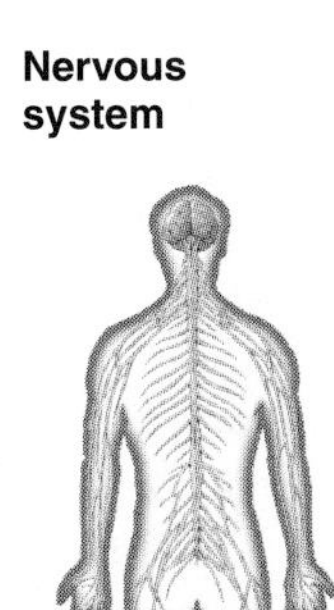

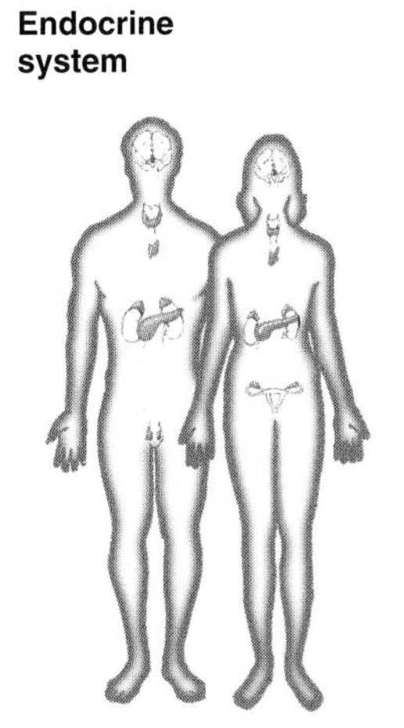

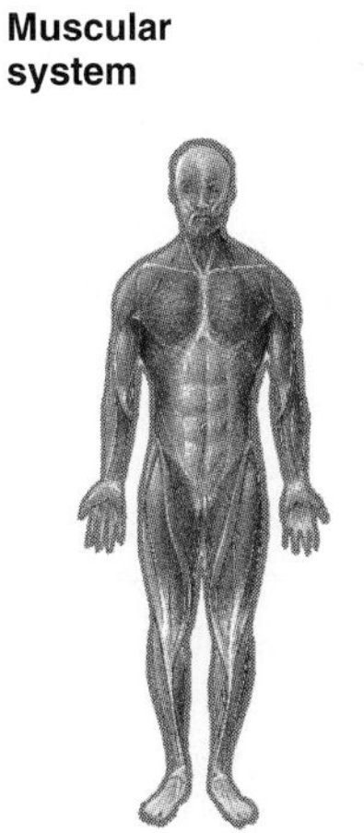

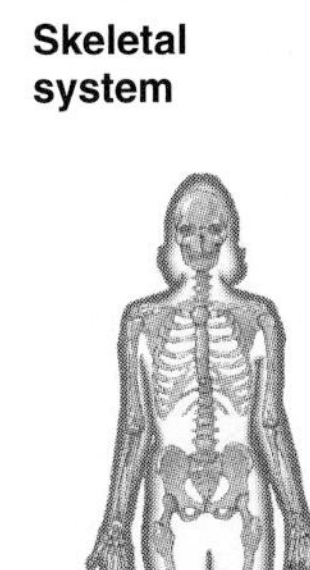

organs. The digestive tract, for example, has its own mini–nervous system, which explains why and how our emotions can affect appetite and digestion.

The **endocrine system** includes glands that secrete hormones (communication biochemicals). Most hormones travel within the circulatory system to target tissues, where they affect metabolism in a way that produces a characteristic response. For example, the pituitary gland in the brain produces prolactin, which signals the breasts of a woman who has just given birth to secrete milk. The hormones of the endocrine system affect development, reproduction, mental health, metabolism, and many other functions.

Structure and Support The skeletal and muscular systems enable us to stand upright and to move, and also power movements within the body.

The **skeletal system** consists of bones, which are built of bone cells, their secretions, and deposited minerals, as well as ligaments (sheets of connective tissue) and cartilage that bind bones to each other. Bones provide frameworks and protective shields for softer tissues, and serve as attachments for muscles. The marrow within certain bones is also a vital tissue, producing the components of blood and storing inorganic salts.

Individual muscles are the organs that comprise the **muscular system.** Muscles contract when the protein filaments that comprise them slide past one another. This contraction provides the forces that cause movements of major body parts. Muscles also help maintain posture, keep food moving through the digestive tract, enable the lungs and heart to work, and are a major source of body heat.

Getting Energy The **respiratory system** acquires oxygen from the atmosphere and releases carbon dioxide (CO_2); the **cardiovascular system** delivers CO_2 to the lungs for exhalation and carries oxygen throughout the body; and the **digestive system** provides nutrients. The oxygen that the respiratory system collects and the cardiovascular system delivers is used to extract maximal energy from the chemical bonds of nutrient molecules.

All three of these systems maximize surface area through a generalized organization of parts within parts within parts. The respiratory system is a series of linked tubes that collect, filter and then deliver air to the paired lungs, where oxygen enters capillaries (tiny blood vessels) in millions of microscopic air sacs. CO_2 leaves the capillaries here and is ultimately exhaled. The freshly oxygenated blood returns to the heart and is distributed through a vast system of blood vessels throughout the tissues, providing them with the oxygen that they require to obtain energy to carry on life processes. Nerve and muscle tissues, which use large quantities of oxygen and nutrients, are packed with capillaries, whereas tissues with low metabolic demands, such as cartilage and the epidermis, lack capillaries.

The digestive system is also a series of tubes and chambers that conduct food into the body, breaking it apart mechanically and chemically along the way. Finally, at the small intestine, the molecules are small enough to be absorbed into capillaries, which then distribute the nutrients throughout the body.

Protection The body protects itself in several ways. The **integumentary system,** for example, discussed in detail in section 30.4, is a barrier between an animal and its environment. In addition, the compositions of body fluids must remain within certain ranges for life to continue in the face of a changeable environment, and the body must have a way to fight infection, injury, and cancer.

The **urinary system** filters the blood, removing toxins and wastes, reabsorbing valuable substances, and maintaining the concentrations of a variety of ions, keeping the electrolytes in body fluids in balance. The system consists of the paired kidneys, which are each packed with tubular units called nephrons, along which filtered blood travels as the contents are adjusted; two tubes, called ureters, into which the kidney tubules drain; a urinary bladder to store urine; and the urethra, the tube leading to the outside. All of the tissue types are represented in the urinary system. Epithelium lines tubules, connective tissue fills in spaces in the kidneys, nerves provide the sense of having to urinate, and muscles propel urine from the body.

The **immune system** is a huge contingent of several types of highly specialized cells, vessels that transport them, plus organs where these cells are produced and collect. This system is remarkably diverse—it can launch an attack against nearly any virus or bacterium, and it "remembers," which is how the body develops immunity to certain infections. The immune system recognizes cells of the body as "self," and launches an attack against anything that is "nonself," such as certain viruses, microorganisms, and cancer cells.

Biological Continuity The **reproductive systems** are not vital to the functioning of a human body in the way that the other organ systems are, but reproduction is essential in a larger sense, to perpetuate an individual's genes. The reproductive systems of male and female consist of organs where gametes form; tubules that transport gametes to parts of the body where the opposite types can meet; and various glands whose secretions enable the gametes to mature and function. The female body also has a built-in system to nurture a developing offspring.

The reproductive system illustrates how the different organ systems are, in a sense, not separate at all. Consider the uterus, the pear-shaped sac that houses the embryo and fetus. It is innervated, which is why a woman feels cramps when it contracts; hormones stimulate these contractions. The majority of the uterus is muscle. The entire system is richly supplied with blood vessels, which also deliver cells and biochemicals of the immune system.

30.2 MASTERING CONCEPTS

1. How have organ systems grown more complex through evolutionary time?
2. What are the general functions of the organ systems of the human body?
3. How are organs and organ systems built of tissues?
4. What are some examples of how organ systems interact?

OLC

30.3 Organ System Interaction Promotes Homeostasis

An animal's body must maintain concentrations of the components of body fluids within a certain range, and regulate other functions. Organ systems interact to provide this internal constancy, or homeostasis.

The environment changes. Light and dark cycle, temperatures rise and fall, food and water may be available, or not. In order to function in a changing environment, an animal's body must maintain a stable internal environment. Concentrations of nutrients and oxygen in body fluids must be within a certain range, and internal temperature and blood pressure must remain within certain parameters too. This state of internal constancy is termed **homeostasis.** Organ systems interact in ways that promote homeostasis. One of the great challenges in transplanting an organ is for the new body part to integrate and function with the rest of the body to maintain homeostasis. Health 30.1 explores organ transplants.

Maintenance of body temperature illustrates the coordination and control that is necessary to maintain homeostasis. A part of the brain functions as a thermostat of sorts, sensing when body temperature deviates from the normal 98.6°F (about 37°C). If body temperature drops, the brain triggers activities that conserve or generate heat. Small groups of muscles contract—the body shivers. Simultaneously, blood vessels beneath the skin constrict, keeping warm blood away from the body's surface to minimize heat loss. In the opposite situation, a rising temperature, homeostatic changes promote loss of body heat. Sweat pours out, cooling the skin as it evaporates. Blood vessels in the skin dilate, releasing heat from deeper tissues to the environment. Heart rate increases, sending more blood to the surface vessels to allow heat to be dissipated.

Maintenance of blood pressure is another bodily function controlled homeostatically. Rising blood pressure triggers receptors in the walls of blood vessels leading away from the heart to signal the brain to signal the heart to slow its contractions, sending less blood into the already stressed vessels. The pressure drops. If blood pressure falls too low, the brain center signals the heart to speed up, sending out more blood.

Most homeostatic responses demonstrate **negative feedback,** which is an action that counters an existing condition, as discussed in chapter 5 in a biochemical sense (see fig. 5.17). When temperature or blood pressure is low, homeostasis raises it, and vice versa. Figure 30.6 summarizes the negative feedback in a familiar situation—maintaining room temperature. Only a few biological functions demonstrate **positive feedback,** in which an

FIGURE 30.6
Negative Feedback Systems Promote Homeostasis.
(A) In negative feedback systems, some variable is controlled within limits. Sensors monitor changes in the controlled variable and activate an effector if the value drifts too high or too low. The effector's response counteracts the original change in the controlled variable (negative feedback). **(B)** Some thermostats contain a bimetal strip that is sensitive to temperature change. As the room warms the strip bends and eventually turns off the heater. As the room cools the strip straightens, turning on the heater.

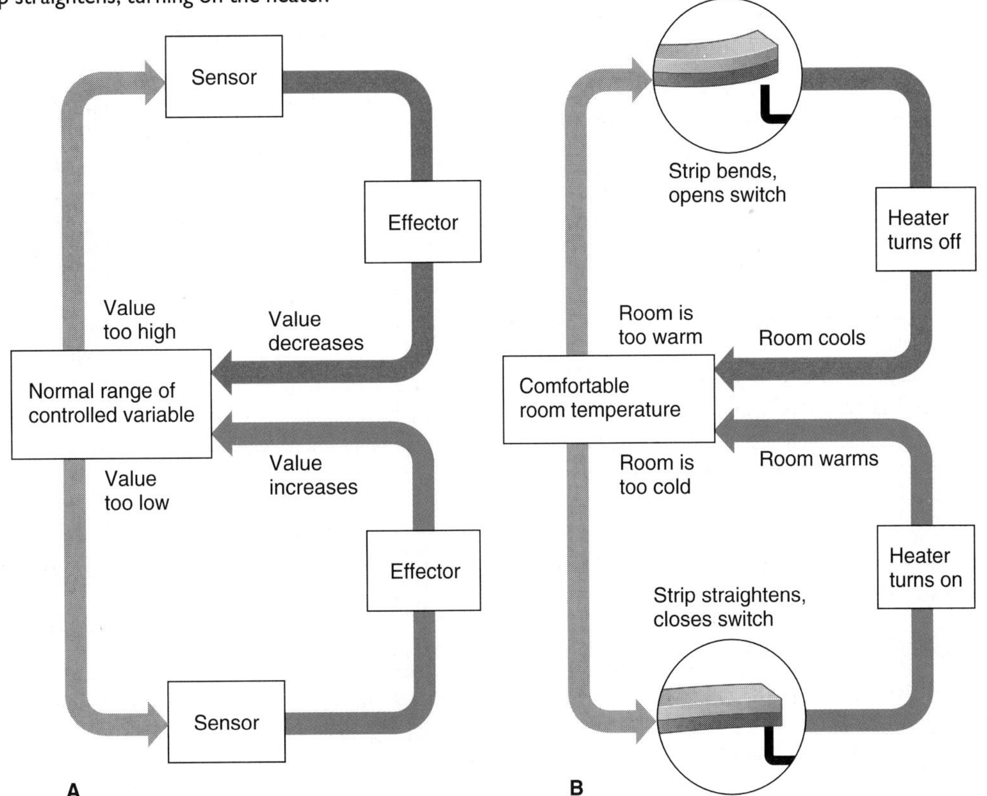

Health 30.1

Organ Transplants

Damaged or diseased tissues and organs can be repaired and/or regrown to differing extents. Medical technology can help this process by replacing damaged parts. Transplanted organs include corneas, pancreases, kidneys, skin, livers, lungs, bone marrow, parts of the digestive tract, and hearts. The greatest challenge in an organ transplant—after obtaining the organ—is to prevent the recipient's immune system from rejecting the foreign material. Physicians minimize rejection by selecting patients based on cell surface compatibility (see chapter 4), giving drugs to suppress the immune system, and by stripping the molecules from donor organs' cell surfaces that might provoke an immune attack.

Recipients of transplanted organs face other risks. One is graft-versus-host disease, in which the transplanted cells release biochemicals that attack the body. Another risk is infection, resulting from the use of immunosuppressive drugs.

Types of Transplants

The relationship of donor to recipient distinguishes transplant types (fig. 30.A). In an autograft, tissue from one part of a person's body is transferred to another part of the body. Technically, this is not truly a transplant, because only one person is involved. A skin graft from the thigh to replace burned skin on the chest is an example of an autograft. (A graft is a general term referring to placing something in the body. A transplant is a type of graft.) Another type of autograft is a coronary bypass, in which a blood vessel from a patient's leg or chest is used as a new artery in the heart, where it restores blood flow.

An allograft, the most common type of transplant, is from a member of the same species. Allografts may come from blood relatives, whose cell surfaces are likely to be similar to the recipient's, or from nonrelatives who, by chance, have similar cell surfaces. Kidney and bone marrow transplants are commonly performed allografts.

A rare type of transplant is an isograft. This is a transplant from one identical twin to another. Because such twins are genetically identical, the recipient does not reject the organ. The only successful ovary transplants have been done between identical twins.

A xenograft is a transplant of tissues or organs from one species to another. ("Xeno" means foreign.) A common type of xenograft is heart valves from pigs, used to replace malfunctioning ones in humans. Table 1 lists some experimental xenotransplants.

A transplant may be an entire organ, such as a kidney or a heart, or part of one. For example, people with diabetes receive transplants of clusters of pancreatic cells that produce insulin, the hormone that their bod-

FIGURE 30.A

Transplant Types.

An autograft is within an individual. An isograft is between identical twins. An allograft is between members of the same species, and a xenograft is between members of different species.

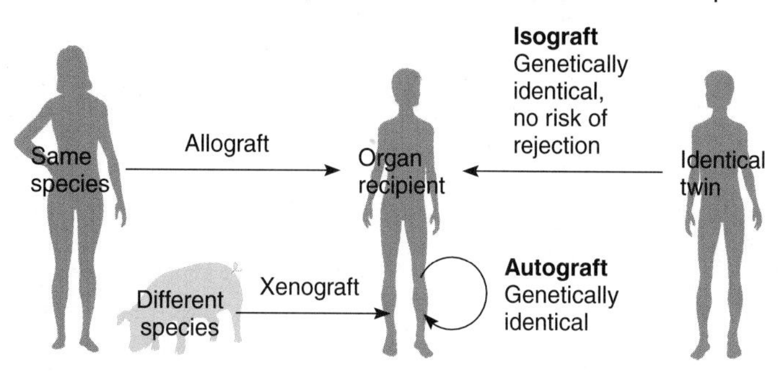

TABLE 1 Xenografts

YEAR	ORGAN TRANSPLANTED
1984	Baboon heart to California newborn, lived 20 days
1993	Baboon liver to man in Pittsburgh, lived 30 days
1996	Baboon bone marrow to California AIDS patient, transplant failed, patient lived
1997	Pig liver "bridges" support four patients awaiting human organs, two survive
1998	Fetal pig cells to brains of people with Parkinson's disease, Huntington's disease, seizure disorders

activity stimulates itself to keep going. Blood clotting and milk secretion are examples of positive feedback.

30.3 MASTERING CONCEPTS

1. What is homeostasis?
2. Why is homeostasis important?
3. Distinguish between negative and positive feedback.

OLC

30.4 A Sample Organ System—The Integument

Skin, horns, hair, feathers, and antlers are all parts of the integument, the organ system that protects an animal's body from the environment.

The chapters of this unit examine the organ systems depicted in figure 30.5. To illustrate how cells and tissues make up organs, and ies cannot make. The cells come from cadavers and are infused into a vein in the recipient's liver, where they secrete insulin directly into the bloodstream, the normal route of hormone delivery. Some transplants are parts of donated organs. One liver, for example, can be transplanted into several recipients, whose bodies regenerate complete organs. Figure 30.B shows two very unusual transplant examples.

Controversial Transplants

Until tissue engineering and stem cell biology can be combined so that researchers can craft organs in the laboratory, an organ shortage will prevail. In the United States, 10 people die every day awaiting organs. This situation has stimulated a search for alternate ways to replace organs, but possible solutions are controversial.

Human fetal tissues and organs are potential sources of transplant material. From a biological viewpoint, healthy fetal tissue is more suitable for transplant than adult material because the cells do not yet have many cell surface antigens that could provoke an immune rejection. Plus, more cells in a fetal organ are still capable of division than in an adult counterpart. Fetal brain tissue, for example, can partially repopulate brain tissue depleted in people with Parkinson's disease, restoring some function.

Interest in using nonhuman animals for spare parts—xenografts—has been around for decades, but resurged in the mid 1990s when researchers were able to create transgenic pigs that bear human antigens on their cell surfaces that dampen the immune response (see Biotechnology 13.3). This makes pig parts much less likely to cause an explosive immune reaction that occurs when tissue from one species is placed in another. The pig is a good candidate for heart transplants because its heart is about the same size as that of a human, and its cardiovascular system is quite similar to ours. Pigs are also potential donors of livers and kidneys. Experimental procedures have attached people to pig livers or kidneys for a few days, circulating their blood through the foreign organs, while they awaited human organs.

A possible danger of xenotransplants is that people may acquire viruses from the organ donors. Recall from chapter 19 that influenza viruses can "jump" species, from ducks to pigs to humans, and the outcome in the new host is unpredictable. AIDS, for example, may have its origin in chimpanzees, who do not become as ill as humans do. So far, it is known that a virus called PERV—for "porcine endogenous retrovirus"—can infect human cells in culture. However, a study of several dozen patients who had received implants of pig tissue, for a variety of reasons, revealed that none showed evidence of PERV years later. That study, though, looked only at blood. We still do not know what effect pig viruses can have on a human body. Because many viral infections take years to cause symptoms, a new infectious disease in the future could be the trade-off for using xenotransplants to solve the organ shortage.

FIGURE 30.B
Transplants.
In 1984, a California newborn, Baby Fae (left), lived 20 days with a baboon heart. She was born with an underdeveloped left half of the heart. (right) Surgeons from Kleinert, Kutz and Associates Hand Care Center and the University of Louisville begin the process of attaching a donor hand to Matthew Scott's arm at Jewish Hospital in 1999. A year later, a follow-up report in *The New England Journal of Medicine* described the successful outcome, although soon afterward, Scott claimed that the hand did not function.

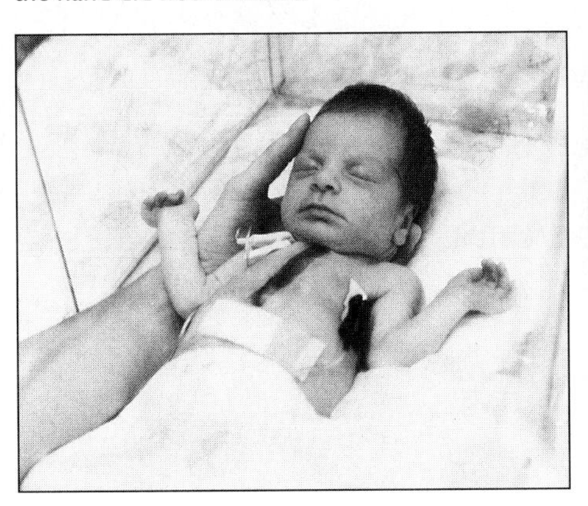

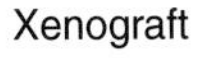
Xenograft

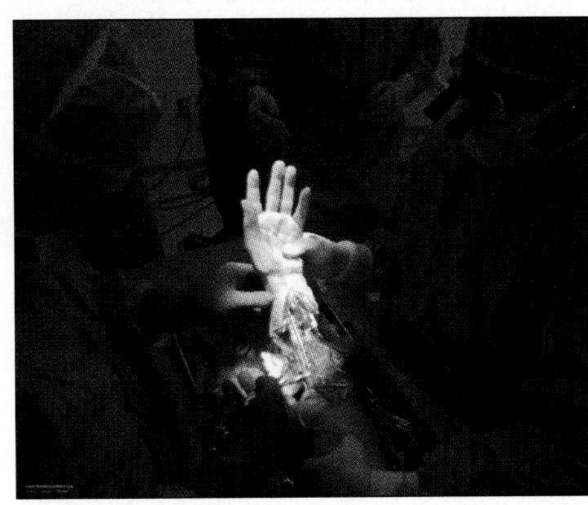

Allograft

to introduce organ systems, we look now at the integumentary system—the coverings of animal bodies. Figure 2.19 shows various integumentary structures based on the protein keratin (fig. 30.7).

The integumentary system has several functions. As the boundary between an animal and its surroundings, the integument obviously protects. In many complex animals, immune system cells in the integument add to the barrier function by recognizing and destroying infectious microorganisms and initiating an inflammatory response to injury. A second function of an integument is sensory. Sensory receptors in the skin enable us to distinguish hot from cold and a pinch from a caress. Skin also prevents water loss and helps to regulate body temperature. Glands that are part of the skin secrete sweat, oils, and scents.

A Closer Look at Human Skin

The structure of human skin illustrates how cells aggregate and form tissues. Human skin has several layers (fig. 30.8). The

FIGURE 30.7
Integumentary Diversity.
Integumentary systems are varied in detail, but similar in their protective function.

outermost layer, the **epidermis,** is mostly stratified squamous epithelium. Scattered cells deep in the epidermis, called **melanocytes,** produce melanin pigment, which imparts color to skin. Melanin absorbs ultraviolet wavelengths of light that can damage the DNA in skin cells, protecting against some types of skin cancer. Although melanocytes are not abundant, they can add a great deal of color to skin because they have long extensions that grow between the other skin cells, spreading the pigment throughout the upper skin layers.

Epidermal cells produce and accumulate the protein keratin, which hardens them into a protective scaliness. Stem cells in the deepest epidermal layer continually push daughter cells upward, so that the outermost skin layer constantly renews itself.

The inner layer of skin, the **dermis,** is mostly dense connective tissue, plus various epithelial derivatives. The dermis includes a specialized type of epithelium, called myoepithelium, which can contract. For example, arrector pili muscles move hairs and propel secretions from glands. The dermis also contains blood vessels and nervous tissue. Neurons here stimulate glands to secrete and convey information from sensory receptors to the brain for interpretation. Blood vessels carry nutrients and oxygen to the dermis. They constrict to conserve heat and dilate to dissipate heat.

FIGURE 30.8
Human skin.
Human skin consists of the epidermis and the dermis, with the subcutaneous layer beneath it. A basement membrane separates the epidermis and the dermis. Associated structures include hair follicles, sweat glands, and sebaceous (oil) glands. Skin color results from the pigment melanin released from melanocytes.

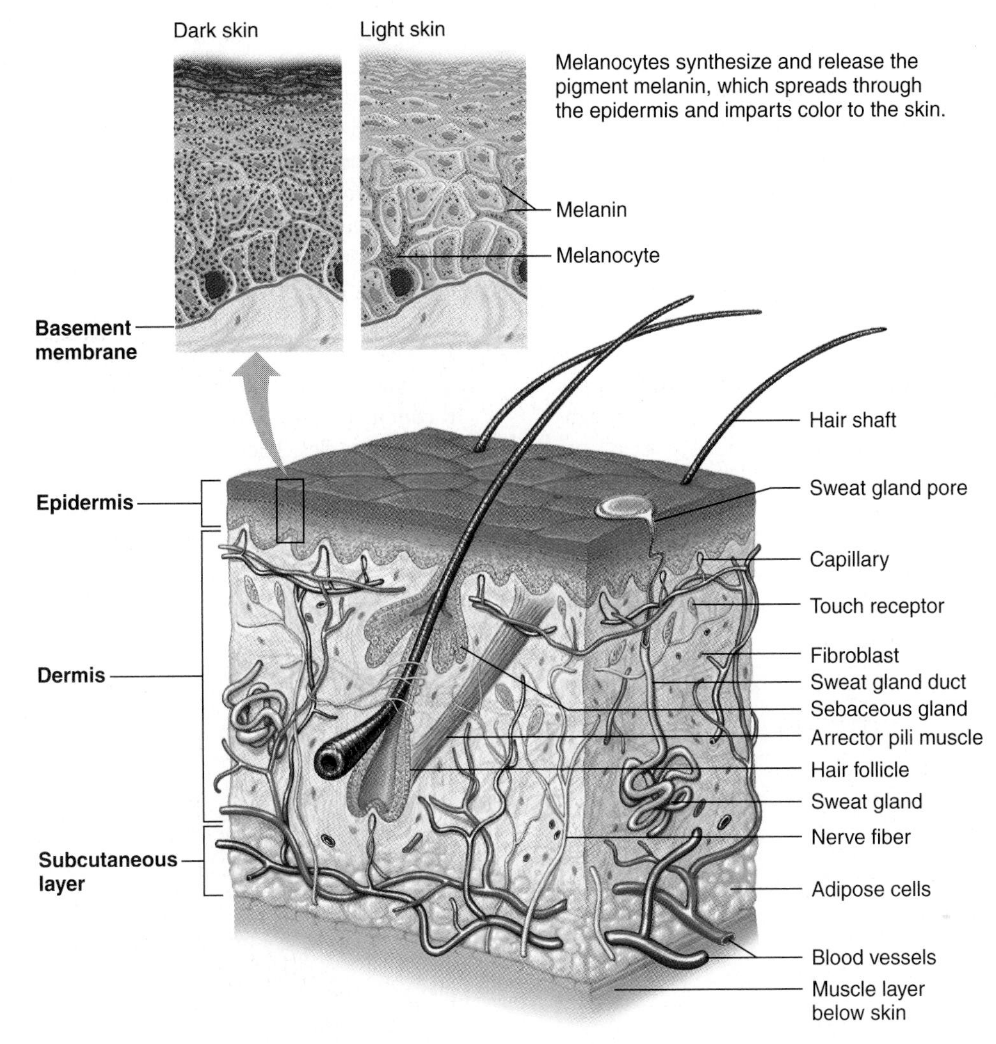

The epidermis is separated from the dermis by the basement membrane. Here, a protein called laminin binds the layers, much as a staple seals two sheets of cardboard together, as discussed in Health 4.1. Beneath the dermis lies a subcutaneous layer that consists of loose connective tissue and adipose tissue. The subcutaneous layer is not technically part of the skin. Altogether, the skin is an extensive organ system, comprising about 15% by weight of an adult's body.

The integrity of the epidermis and dermis is crucial to the skin's ability to protect and retain fluids. This is one reason severe burns threaten life. Figure 4.18 illustrates an inherited disorder, called epidermolysis bullosa, that causes the skin to blister from even slight touching, due to abnormalities in different skin proteins.

Specializations of the Integument

Many animals have integuments consisting of an epidermis and a dermis. Animals also share specialized integumental structures that are similar in composition, but they appear as different as a hair, a feather, and a great variety of claws, hooves, nails, horns, beaks, and scales. These structures can have widely different functions. Hair insulates, horns defend, feathers provide locomotion and insulate, and claws enable an animal to scamper up a tree to escape a predator.

A hair grows from a group of cells in the epidermis called a **hair follicle,** which is anchored in the dermis and nourished by blood vessels there. Epidermal cells at the base of a hair follicle divide, pushing daughter cells up. These cells stiffen with keratin and die (fig. 30.9). Hairs are really just dead epidermal cells.

A feather is also an outgrowth of the epidermis and like a hair, a feather is keratinized. Feathers develop from **feather follicles,** which are extensions of the epidermis dipping into the dermis, much like hair follicles. Feathers have different roles, depending upon their size, structure, and location (see fig. 25.19).

Nails, claws, and hooves are keratinized, stratified squamous epithelium that is part of the epidermis. They differ in their position on the animal's digits (fig. 30.10). Nails are plates of very tightly packed keratinized cells on the fingers and toes of primates. Claws are curved and project beyond the digit in some amphibians and in most birds, reptiles, and mammals. Hooves are large keratinized plates on the ends of digits in deer, elk, and moose.

An animal's dermis may be distinctive for its pattern of collagen fibers, glands, or outgrowths of bone. Dermis is composed largely of sheets of collagen fibers. How tightly these fibers pack against each other, and their orientation, determines the integument's flexibility. In aquatic vertebrates the collagen sheets are very ordered, which assists in locomotion through water. The dermal collagen in terrestrial vertebrates is less ordered, perhaps because movement on land depends more on muscle action than on the integument.

Consider a shark. A shark's skin is extremely tight, which enables the animal to swim rapidly through water without wrinkling, which would cause some turbulence. In a shark's skin, collagen bundles lie at 45-degree angles to each other, much like in woven cloth (fig. 30.11). This pattern is functional in both sharks and garments. When the collagen or material fibers are pulled from above or the side, they do not deform much, if at all. If pulled from an oblique angle, however, the patch of collagen or material readily moves. This provides flexibility to the animal—and to a person moving about in a garment. Collagen bundles in human skin are not nearly as highly ordered, and as a result our skin sags and wrinkles easily, especially as dermal collagen stores fall as we age.

The dermis of birds differs from that of sharks and humans. It contains many blood vessels, sensory nerve endings, and smooth muscles. These tissues heat the dermis in the breast area of a bird, which is adaptive for brooding eggs.

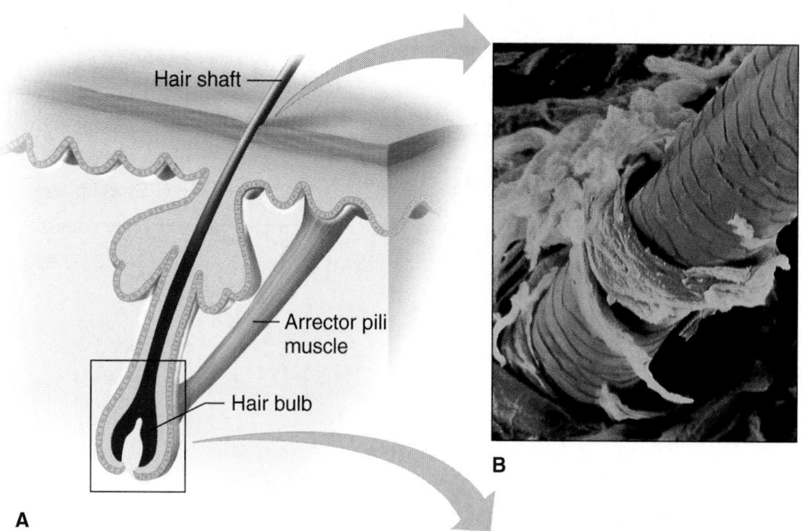

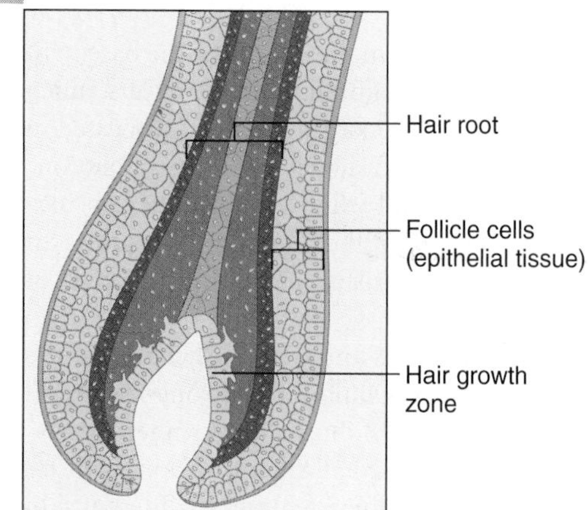

FIGURE 30.9
Structure of Hair and Follicle.
(**A**) The hair follicle is an invagination of the epidermal portion of skin. (**B**) This scanning electron micrograph shows a human hair emerging from the epidermis (×340). Note the scaliness of the cells that make up the hair. (**C**) Longitudinal drawing of the hair bulb region showing the actively dividing epithelial cells that form the base of the hair.

FIGURE 30.10

Integumentary Specializations.

Nails (**A**), hooves (**B**), claws (**C**), feathers (**D**), quills (**E**), and rhinoceros horns (**F**) are all keratinized epidermal outgrowths. Antlers (**G**) and true horns (**H**) are derived from skin and underlying bone. Antlers are shed each year and have a velvety, blood-rich integument, which covers the bone for part of the year. True horns are not shed and a keratinized skin layer covers the bone.

FIGURE 30.11

Shark Skin.

The collagen fibers in a shark's skin are very tightly arrayed, in a pattern similar to that of a woven cloth. This organization results in a skin that is not easily deformed but remains flexible.

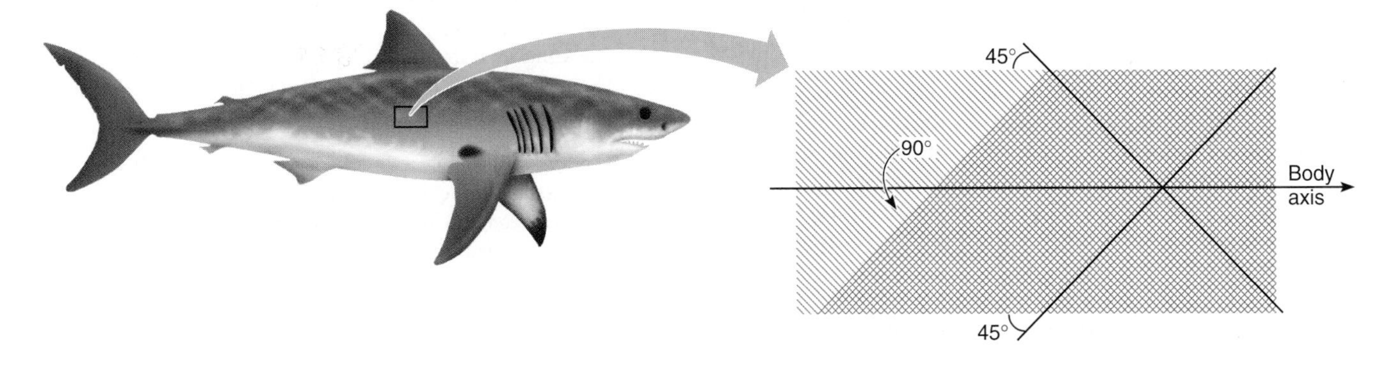

Perhaps the most distinctive dermal derivative is baleen, which hangs in plates from the roof of certain whales' mouths. Baleen screens microscopic organisms (krill) from the water, which the whale consumes. Baleen plates originate in the dermis and extend down into the epidermis, where they are covered with keratin.

Glands are part of the dermis. **Sweat glands** are epidermal invaginations into the dermis. Humans have two types of sweat glands. One type is present before puberty, produces a thin secretion, and is not associated with a hair follicle. This type of sweat gland cools the body when its secretion evaporates. The second type of sweat gland in humans is near a hair follicle, produces a thick, pungent form of sweat, and develops at or after puberty.

Not all mammals have sweat glands, and they appear in different places in different species. Humans and chimpanzees have widespread sweat glands. Elephants lack them. Rabbits have sweat glands on their lips, deer at the base of the tail, and duck-billed platypuses on the snout.

Scent glands are derived from sweat glands, and they too occur in different parts of the body in different species. Rodents, cats, and dogs have scent glands near the anus; camels on the back; rabbits on the chin; alligators on the lower jaw; and cattle on the feet and legs. Scents are important in communication. Many animals, including humans, have **sebaceous glands,** which secrete an oily substance called sebum that softens the skin and hair. Mammary glands, which produce milk in mammals, are also derived from sweat glands.

Horns and antlers form when bone pushes the integument upwards, creating a protrusion (fig. 30.10). In horns, the integument forms a tough, keratinized coating on the bone. Sheep, cows, antelope, goats, wildebeests, and bison have horns. These outgrowths enlarge and persist for the lifetime of the animal. In some larger species, both sexes have horns, but these structures are typically larger in the males, who use them for displays of strength. In smaller species, females may lack horns. A rhinoceros's horns are different from others in that they are outgrowths of the epidermis only, and lack the core of bone seen in other horns.

Antlers differ from horns in that they are covered with an integument that is rich in blood vessels, creating a velvety texture. The skin falls away, leaving bare bone, which is the mature antler. Male deer, elk, and moose grow antlers, which are shed yearly.

30.4 MASTERING CONCEPTS

1. What functions does an integumentary system provide?
2. What are the layers of the human integumentary system?
3. What are some epidermal and dermal specializations?

OLC

Chapter Summary

30.1 Animal Tissue Types

1. Specialized cells express different genes. These cells aggregate and function together to form **tissues.**
2. **Stem cells** retain the ability to specialize, and enable tissues to grow and heal.
3. **Epithelium** is lining tissue. It may be simple (one layer), stratified (more than one layer), or pseudostratified (one layer appearing as more than one). Epithelial cells may be squamous (flat), cuboidal (cube-shaped), or columnar (tall and thin). Keratin hardens squamous epithelium. Epithelial tissue protects, senses, and secretes.
4. Connective tissues consist of cells within a matrix. **Fibroblasts** secrete **collagen** and **elastin. Loose connective tissue** has loose collagen strands, whereas **dense connective tissue** has tightly packed collagen fibrils. **Adipose tissue** consists of **adipocytes** swelled with lipid. The matrix of **blood** is **plasma,** which carries **red blood cells, white blood cells,** and **platelets. Cartilage** consists of **chondrocytes** in **lacunae,** in a collagen matrix. In **bone, osteocytes** communicate through **canaliculi. Osteoclasts** and **osteoblasts** tear down and rebuild bone tissue. **Osteoprogenitor cells** can become osteoblasts. Bone matrix consists of collagen and minerals.
5. **Neurons** and **neuroglia** constitute nervous tissue. A neuron has a **cell body,** an **axon,** and **dendrites** and functions in communication. Neuroglia support neurons.
6. Muscle tissue provides movement when filaments of **actin** and **myosin** slide past one another. Contractile cells include **skeletal, cardiac,** and **smooth muscle** cells.

30.2 Tissues and Organs Build Animal Bodies

7. The **nervous** and **endocrine systems** coordinate all others, via neurotransmitters and hormones.
8. The **skeletal** and **muscular systems** provide support and locomotion.
9. The **respiratory system** obtains oxygen and the **circulatory system** delivers it to tissues, where it is used to extract maximal energy from nutrient molecules. The **digestive system** provides nutrients.
10. The **integumentary system** provides a physical barrier between an animal and its surroundings.
11. The **urinary system** removes wastes from the blood and reabsorbs useful substances.
12. The **immune system** protects against infection, injury, and cancer.
13. The **reproductive systems** are essential for the perpetuation of an individual's genes.

30.3 Organ System Interaction Promotes Homeostasis

14. **Homeostasis** is the maintenance of internal constancy.
15. **Negative feedback** reduces or increases the level of a substance or parameter to within a normal range.

30.4 A Sample Organ System—The Integument

16. The integument in many animals consists of an **epidermis** over a **dermis,** plus specialized structures, such as hairs, feathers, claws,

and glands. A **basement membrane** joins the epidermis to the dermis, and a subcutaneous layer underlies the dermis. **Melanocytes** provide pigment.

Testing Your Knowledge

1. Identify and describe the four basic tissue types.
2. State the type of epithelium illustrated below:

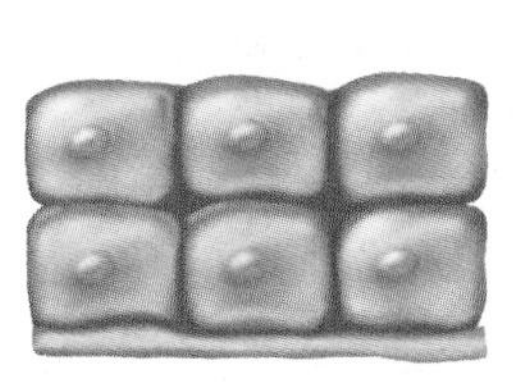

A

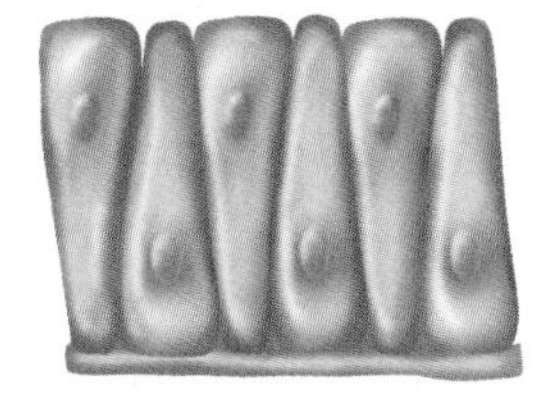

B

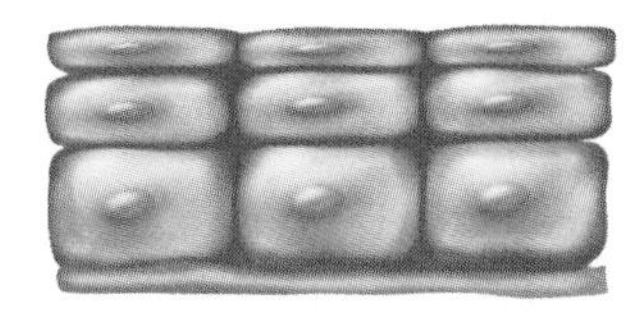

C

3. What is homeostasis and why is it important?
4. Distinguish between
   - **a.** neurons and neuroglia
   - **b.** negative and positive feedback
   - **c.** an axon and a dendrite
5. What are two ways that the integuments of a shark and a deer differ?
6. Match the following cell types and structures to the tissues they are a part of:

Cell Types	Tissue
a. platelets	**1.** cartilage
b. chondrocytes	**2.** blood
c. neurons	**3.** loose connective tissue
d. osteocytes	**4.** muscle
e. fibroblasts	**5.** adipose tissue
f. neuroglia	**6.** bone
g. adipocytes	**7.** nervous tissue
h. actin and myosin filaments	**8.** dense connective tissue
i. melanocytes	**9.** epithelium

7. How are hair follicles and feather follicles similar, and how are they different?

Thinking Scientifically

1. There are three types of the inherited disorder epidermolysis bullosa. One type affects keratin in the epidermis; one affects collagen in the dermis; and the third affects laminin. Which form do you think is most severe, and why?
2. What materials—living or synthetic—might you use to create a tissue engineered version of:
   - **a.** blood
   - **b.** cartilage
   - **c.** bone
3. Which of the organ systems are highly folded to maximize surface area, and why is this important?
4. List three ways that organ systems interact.
5. Collagen injections can temporarily fill out facial skin, diminishing wrinkles. Into which skin layer should collagen be injected? Why isn't the treatment permanent?

References and Resources

Jones, Jon W., et al. August 17, 2000. Successful hand transplantation: one-year follow-up. *The New England Journal of Medicine,* vol. 343, p. 469. A man received a hand transplant from a cadaver.

Lewis, Ricki. October 16, 2000. Porcine possibilities. *The Scientist,* vol. 14, p. 1. Pig tissue transmits retroviruses to human cells, but the effects aren't known.

Lewis, Ricki. March 6, 2000. A paradigm shift in stem cell research? *The Scientist,* vol. 14, p. 1. The developmental potential of many tissues is far greater than biologists had thought.

Lewis, Ricki. April 12, 1999. Human mesenchymal stem cells differentiate in the lab. *The Scientist,* vol. 13, p. 1. Given the proper combination of nutrients, a mesenchymal stem cell may follow any of several fates.

The February 25, 2000 issue of *Science,* vol. 287, has many excellent articles on stem cells.

The April 1999 issue of *Scientific American* has several articles on tissue engineering.

The *LIFE* Online Learning Center provides additional resources and tools for studying this chapter.
www.mhhe.com/life

CHAPTER 31

The Nervous System

Mercury

A Toxic Legacy

Cases of mercury poisoning are fortunately rare, for it is a terrible way to die. The Mad Hatter, a character in Lewis Carroll's *Alice in Wonderland,* made light of this serious type of poisoning that was once an occupational hazard of hatmakers. Mercuric nitrate was sprayed onto the felt to make hats shiny, and the workers would inhale the toxic fumes. After many years of exposure, symptoms arose, including twitching hands and difficulty breathing. It is also a risk that chemists face when they work with mercury-containing compounds. These include elemental mercury and two forms of the element in organic compounds—methylmercury and dimethylmercury.

On August 14, 1996, Dartmouth College chemist Karen Wetterhahn was working with dimethylmercury, when a few drops spilled onto her latex glove. She cleaned up the spill, then took off the glove. Although she recorded the accident in her notebook, she didn't think too much about it—until January, when symptoms of mercury poisoning began.

At first, Dr. Wetterhahn felt tingles in her fingers and toes, and developed poor balance and slurred speech.

At first, Dr. Wetterhahn felt tingles in her fingers and toes, and developed poor balance and slurred speech. Within days, her visual field narrowed and she saw flashes of light, and her hearing diminished. By the end of January, following chemical analysis of her blood, urine, and hair, the diagnosis was in: mercury poisoning. The level of the deadly compound in her blood exceeded that required for toxicity 80-fold. By the end of February, she no longer responded to light, sound, or light touch, although her eyes opened, she responded to painful stimuli, and she had bouts of crying. She never regained consciousness, and died on June 8, 1997.

Dimethylmercury is a "supertoxin" that heads straight for the brain. It rapidly crosses the blood-brain barrier, then binds to sulfur-containing

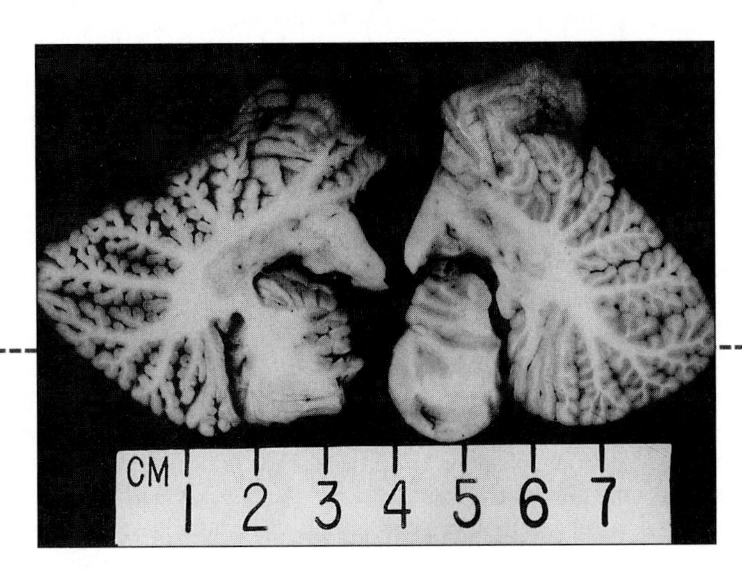

Autopsy of Karen Wetterhahn's Brain.
Dr. Karen Wetterhahn's brain (left) was missing much tissue compared with the brain of a healthy woman of the same age.

amino acids on and in neurons, destroying them. The neurons, which communicate with each other, gradually die as the more supportive cells, called neuroglia, overgrow. On autopsy, Dr. Wetterhahn's brain was a mere vestige of normal. Her cerebral cortex, which controls thinking, sensation, and perception, was reduced to a thin lining, and her cerebellum, which controls balance, appeared eaten away.

The neurons, which communicate with each other, gradually die as the more supportive cells, called neuroglia, overgrow.

Some good, however, came out of Dr. Wetterhahn's tragic death. Between the time she was diagnosed and when she lost consciousness, she did everything possible to prevent others from suffering her fate. Her department chairman immediately notified the American Chemical Society, which published a prominent warning in its most widely read journal. An independent laboratory tested all gloves in Wetterhahn's lab, finding that dimethylmercury penetrates them in 15 seconds! And although symptoms don't appear for months (weeks for the methyl form), the compound begins to destroy the brain almost immediately. Just a month after the accident, all of Wetterhahn's body fluids contained mercury. Attempts to use drugs to trap and remove the mercury are invariably futile.

Karen Wetterhahn wasn't the first to die of mercury poisoning. The world learned of the toxic effects of organic mercury compounds in the 1960s, when many severely deformed children were born to women who had eaten fish that had swum in contaminated waters in Minimata, Japan. Another mass-poisoning occurred in Iraq in 1972, when farmers used methylmercury as a fungicide to treat grain. The grain should have been planted, but instead it was milled and used to make bread. People ate the tainted bread, and 100 died. As a result, the World Health Organization banned use of the compound in agriculture.

Chemists, however, continued to use dimethylmercury to standardize nuclear magnetic resonance devices, a common analytical tool. Karen Wetterhahn was preparing such a standard when the deadly drop fell onto her latex glove. Hers was one of only a handful of known cases of dimethylmercury poisoning. The other victims were laboratory workers in England who originally synthesized the compound in 1865; two secretaries in Canada who inhaled fumes from a leaking vial in a warehouse; and a chemist working without adequate protection in the former Czechoslovakia. Thanks to Dr. Wetterhahn's efforts to alert her colleagues as she approached death, perhaps mercury poisoning will never happen again. Said her doctor, "She was concerned that people immediately increase efforts to use better and safer handling techniques, and that physicians learn to pick up the signs earlier. She was so sick, yet concerned above all about her colleagues."

31.1 The Nervous System Maintains Homeostasis

A nervous system enables an organism to detect, react to, and survive threats in the environment. Networks of neurons and neuroglia build nervous systems.

The lynx is one of the few inhabitants of the snowy midwinter northern Montana plain. The cat is acutely attuned to its frigid environment, able to hear the quiet swish of a snowshoe hare on the icy cover before it sees its well-camouflaged prey (fig. 31.1). The lynx watches. The hare doesn't realize it is being observed until the lynx bursts from behind a tree. The cat's snowshoelike feet enable it to quickly overtake and capture its fleeing prey.

The interaction between the lynx and the hare is orchestrated by each animal's nervous system, a vast collection of interconnected neurons and their associated neuroglial cells. The cat's adaptations as a predator—inch-long claws, knifelike fangs, compact powerful body, and snowshoe feet—require a nervous system that enables it to spot and attack prey. Similarly, the hare's nervous system triggers its swift flight—although too late in this case.

Overall, the function of the nervous system, and of all animal organ systems, is to maintain a range of internal constancy, or homeostasis, by detecting and reacting to changes in the environment. The nervous system maintains homeostasis by regulating virtually all other organ systems. The lynx's nervous system controls its respiratory system, which brings oxygen to the circulatory system, which delivers the oxygen plus nutrients to all cells. The nervous system also controls the excretory system, as cell wastes move through the circulatory system to the kidneys. The nervous system enables muscles to act on bones so that the lynx can move; keeps the digestive system providing nutrients to cells for energy conversion and work; causes the endocrine system to secrete adrenaline, which sharpens the animal's awareness and coordinates the many systems for rapid action; and keeps the immune system on alert against infection.

The lynx's pursuit of the hare demonstrates three major roles of the nervous system—sensory input, sensory integration, and motor response. Sensory information arrives as sound and visual images. The lynx compares this information to past hunting experiences and decides how to act. The motor systems coordinate muscles to move the lynx into position to catch the hare. These events are initiated at the conscious level of the brain. Meanwhile, the heart beats, the lungs inhale and exhale, and the cat maintains its balance through sensory input, decision making, and motor output at the subconscious level.

Neurons make possible many sensations, actions, emotions, and experiences. Networks of interacting nerve cells control mood, appetite, blood pressure, coordination, and perception of pain and pleasure. The unique ability of neurons to communicate rapidly enables animals to be aware of and react to the environment and to screen out unimportant stimuli, to learn, and to remember. Yet despite their diverse functions, all neurons communicate in a similar manner. This chapter first explores how neurons function, and then considers the evolution and functioning of nervous systems.

31.1 MASTERING CONCEPTS

1. What are the components of a nervous system?
2. How does a nervous system help maintain homeostasis?

OLC

FIGURE 31.1

Nervous Systems in Action.

Networks of nerve cells control this lynx's attack—and the hare's attempt to escape. Sensory organs such as eyes and ears receive sensory input and relay it to the brain and spinal cord, which integrates the information and produces appropriate motor responses.

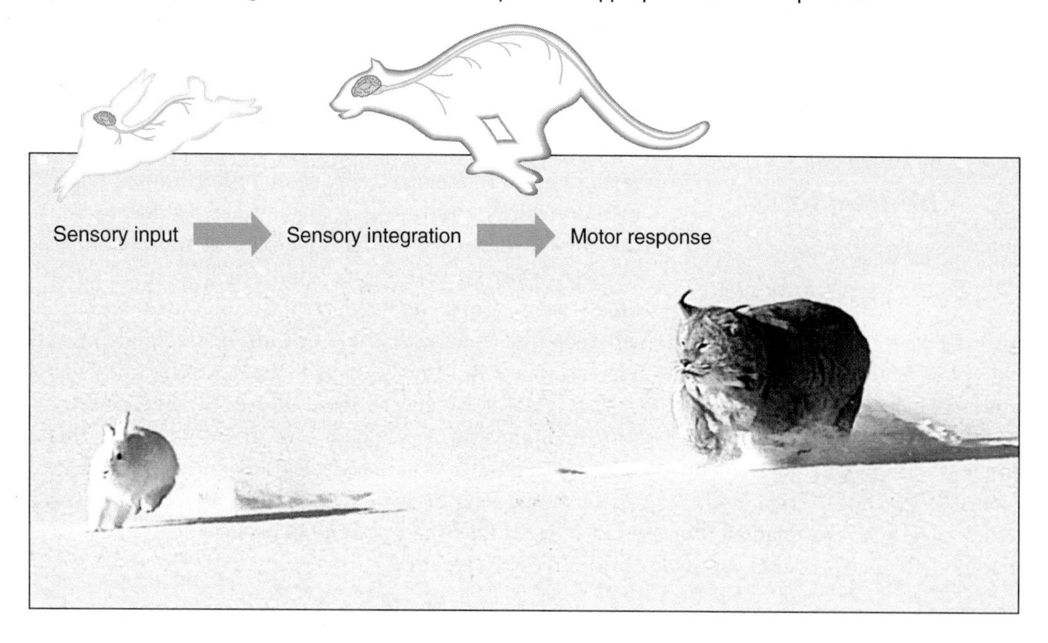

31.2 Neurons Are Functional Units of a Nervous System

Sensory neurons bring information toward the central nervous system (CNS), motor neurons function in the reverse direction, and interneurons connect neurons in the CNS. A resting neuron is negatively charged inside compared with outside due to the distribution of sodium and potassium ions, and proteins. During an action potential, which indicates a neural impulse, a neuron's interior near the cell membrane becomes fleetingly positively charged, a condition that spreads down the axon.

All neurons have the same basic parts, but they vary considerably in shape and size.

Anatomy of a Neuron

A typical **neuron** consists of a rounded central portion with many emanating long, fine extensions, a form well adapted to receiving, integrating, and conducting messages over long distances (fig. 31.2). The central portion of the neuron, the **cell body,** does most of the neuron's metabolic work. It contains the usual organelles: a nucleus, extensive endoplasmic reticulum, mitochondria to supply energy, and ribosomes to manufacture proteins necessary to convey messages.

A neuron's extensions are of two types. The shorter, branched, and more numerous extensions are **dendrites.** They receive information from other neurons and transmit it toward the cell body. The many branching dendrites can receive input from many other neurons. The second type of extension from the cell body is the **axon,** which conducts the message away from the cell body and towards another cell. Because a nerve's message may have to reach to a cell quite far away, an axon is usually longer than a dendrite—sometimes surprisingly so. An axon that permits a person to wiggle a big toe, for example, extends from the base of the spinal cord to the toe. An axon is usually thicker than a dendrite, and a neuron usually has only one axon. Bundles of axons from several cells are sometimes called **nerves.**

To picture the relative sizes of a typical neuron's parts, imagine that the cell body is the size of a tennis ball. The axon might then be 1 mile (1.6 kilometers) long and half an inch (1.27 centimeters) thick. The dendrites would fill an average-size living room.

One way to classify neurons is into three groups according to their general functions (fig. 31.3). A neuron that brings information

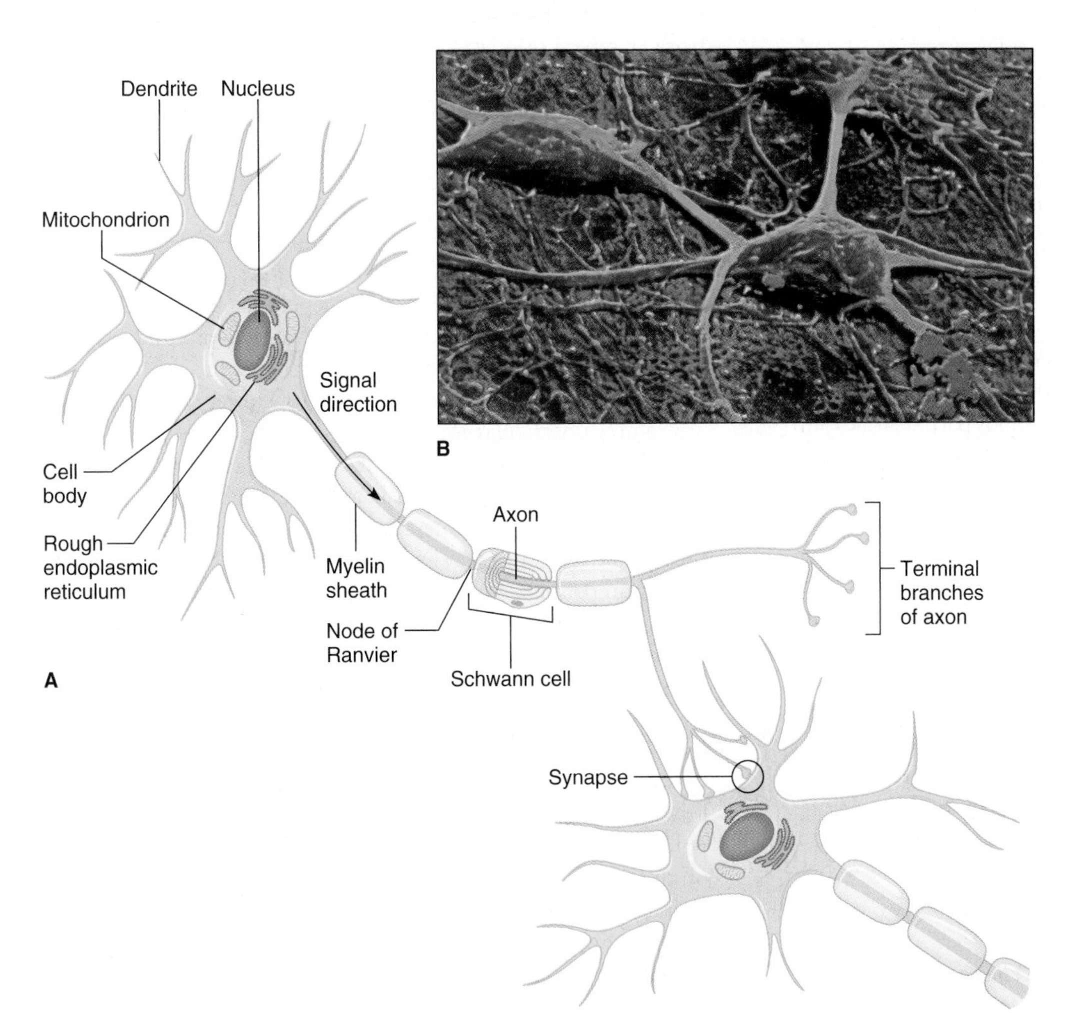

FIGURE 31.2
Parts of a Neuron.
(**A**) A neuron consists of a rounded cell body, "receiving" branches called dendrites, and a "sending" branch called an axon. The space between an axon terminal of one neuron and a dendrite of an adjacent neuron is a synapse. Many axons are encased in fatty myelin sheaths. Unmyelinated regions between adjacent myelin sheath cells are called nodes of Ranvier. (**B**) These unmyelinated neurons from the cortex of a human brain are magnified 500 times. Note their entangled axons and dendrites.

FIGURE 31.3
Categories of Neurons.
Sensory neurons transmit information from sensory receptors in contact with the environment to the central nervous system. Motor neurons send information from the central nervous system to muscles or glands. Interneurons connect other neurons. Note that neuron shape varies considerably. The cell body of the sensory neuron is near one end of the cell and has a single branch that exits the cell body. The motor neuron's cell body is at one end of the cell and has several dendrites and a single axon.

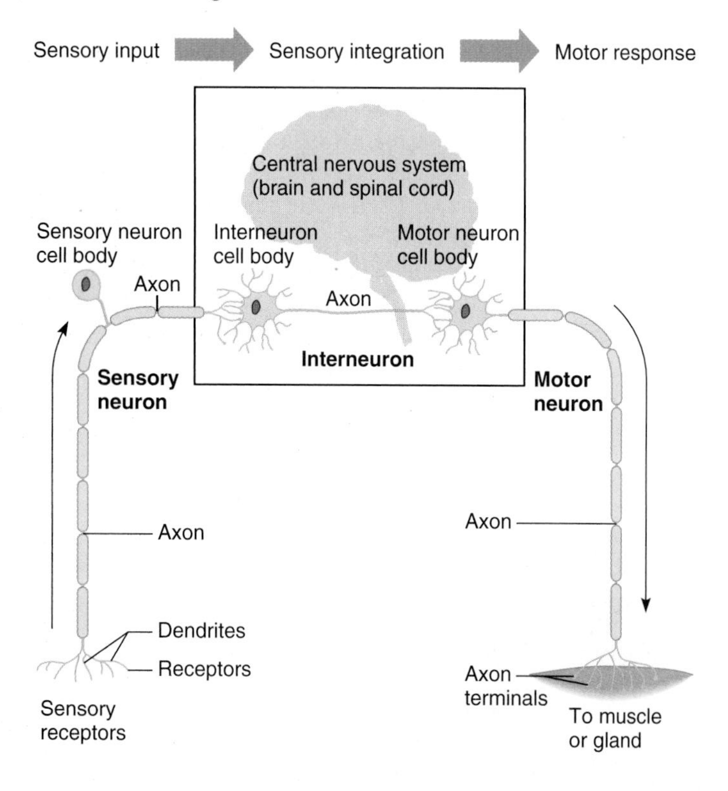

toward the **central nervous system** (CNS) (the brain and spinal cord) is called a **sensory** (or afferent) **neuron.** It has remote, or far away, dendrites that carry its message from a body part, such as from the skin, toward the cell body, located just outside the spinal cord. A sensory neuron delivers its message to another neuron whose dendrites are located nearby within the spinal cord.

A **motor** (or efferent) **neuron** conducts its message outward, from the CNS toward muscle or gland cells. It has a long axon to reach the effector (the muscle or gland) and short dendrites. When a motor neuron stimulates a muscle cell, it contracts, and when a neuron stimulates a gland, it secretes. A third type of neuron, an **interneuron,** connects one neuron to another within the CNS to integrate information from many sources and coordinate responses. Large, complex networks of interneurons receive information from sensory neurons, process and store this information, and generate the messages that the motor neurons carry to effector organs.

A Neuron at Rest Has a Negative Charge

The message that a neuron conducts is called a neural impulse. This is an electrochemical change that occurs when ions move across the cell membrane. A measurement called an **action potential** describes the ionic changes that are the neural impulse. A neural impulse is the spread of electrochemical change (action potentials) along an axon. To understand how and why ions move in an action potential, it helps to be familiar with the **resting potential,** the state a neuron is in when it is not conducting an impulse.

The membrane of a resting neuron is polarized. That is, the inside carries a slightly negative electrical charge relative to the outside. This separation of charge creates an electrical "potential" that measures around –70 millivolts. (A volt measures the difference in electrical charge between two points. The minus sign indicates that the inside of the cell is negative when compared with the outside of the cell.) The charge difference across the membrane results from the unequal distribution of ions (fig. 31.4).

How is this unequal distribution of ions established and maintained? First, the cell membrane is selectively permeable; it admits some ions but not others. Ions move through the membrane through small channels. Some channels are always open, but others open or close like gates, depending on proteins that change shape. A few of these gates are voltage regulated—whether a gate opens or closes depends upon the electrical charge of the membrane. Other gates open and close in response to certain chemicals. Some membrane channels are specific for sodium ions (Na^+), and others are specific for potassium ions (K^+).

Another property of the membrane that establishes and maintains ion distribution is the sodium-potassium pump. Recall from chapter 4 (see fig. 4.10) that this pump is a mechanism that uses cellular energy (ATP) to transport Na^+ out of and K^+ into the cell. The sodium-potassium pump actively transports Na^+ and K^+ against their concentration gradients, resulting in a concentration of K^+ 30 times greater inside the cell than outside, and a concentration of Na^+ 10 times greater outside than inside.

Ions distribute themselves in response to two forces. First, ions follow an electrical gradient. Like charges (negative and negative; positive and positive) tend to repel one another. Opposite charges (negative and positive) attract. Second, ions follow a concentration gradient and diffuse from an area in which they are highly concentrated toward an area of lower concentration. Therefore, a particular ion enters or exits the cell at a rate determined by its permeability and concentration gradient.

Three mechanisms establish and maintain the resting potential. First, the sodium-potassium pump concentrates K^+ inside the cell and Na^+ outside. The pump ejects three Na^+ for every two K^+ it pumps in. Second, large, negatively charged proteins and other negative ions are trapped inside the cell because the cell membrane is not permeable to them. Third, the membrane in the resting state is 40 times more permeable to K^+ than to Na^+.

Because of the concentration gradient and high permeability, K^+ is able to diffuse out of the cell. As K^+ moves through the membrane to the outside of the cell, it carries a positive charge, leaving behind large, negatively charged molecules. A charge or potential is therefore established across the membrane: positive on the outside and negative on the inside. The magnitude of the charge is determined by the balance of opposing forces acting on K^+. The concentration gradient drives K^+ outward, and the negative charge inside the cell holds K^+ in. When these two opposing forces are equal, no net movement of K^+ occurs, and the cell is in equilibrium, at the resting potential.

FIGURE 31.4

The Resting Potential.

(**A**) A voltmeter measures the difference in electrical potential between two electrodes. When one electrode is placed inside an axon at rest and the other is placed outside, the electrical potential inside the cell is approximately –70 millivolts (mV) relative to the outside due to the separation of positive (+) and negative (–) charges along the membrane. (**B**) At rest the concentration of Na^+ ions is greatest inside the cell while the concentration of K^+ ions is greatest inside the cell. Large negatively charged proteins are confined to the cell interior and help contribute to the net negative cell interior. The Na^+/K^+ pump uses ATP to continuously move three Na^+ ions out of and two K^+ ions into the cell. One class of K^+ channel is open at rest and the diffusion of K^+ out of the cell is balanced by the electrical flow of K^+ back into the negative cell interior.

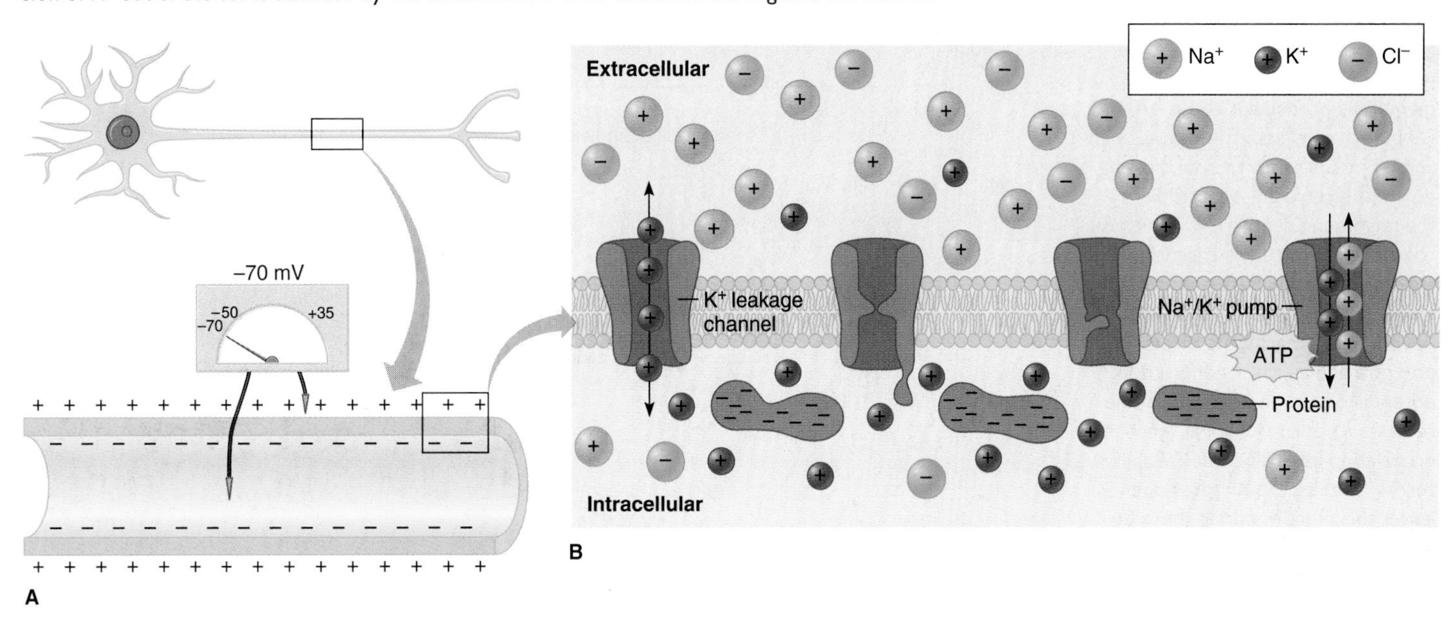

The importance of the sodium-potassium pump in maintaining the resting potential becomes evident when a metabolic poison such as cyanide disables the pump. K^+ slowly diffuses out and Na^+ in, destroying the concentration gradients. Nerve transmission is then impossible because a charge no longer exists across the membrane. Death occurs in minutes.

It is curious that a neuron uses more energy while resting than it does conducting an impulse. Presumably, expending energy to maintain the resting potential allows the neuron to respond more quickly than it could if it had to generate a potential difference across the membrane each time it received a stimulus. This is analogous to holding back the string on a bow to be continuously ready to shoot an arrow.

A Neuron Transmitting an Impulse Undergoes a Wave of Depolarization

During an action potential, Na^+ and K^+ quickly redistribute across a small patch of the cell membrane, creating an electrochemical change that moves like a wave along the nerve fiber. An action potential begins when a stimulus (a change in pH, a touch, or a signal from another neuron) changes the permeability of the membrane so that some Na^+ begins to leak into the cell (fig. 31.5). As Na^+ enters the neuron, the interior becomes less negative, or depolarizes because of the influx of Na^+. When enough Na^+ enters to depolarize the membrane to a certain point, called the threshold potential, the sodium gates sensitive to charge in that area of the membrane open, increasing permeability to Na^+. The threshold potential is about –50 mV. Driven by both the electrical gradient and the concentration gradient, enough Na^+ enters the cell to positively charge the cell interior near the membrane. Na^+ influx continues until the positive charge peaks.

At this peak of the action potential, membrane permeability changes again. Permeability to Na^+ halts as sodium gates close, but permeability to K^+ suddenly increases as delayed potassium gates open. Now Na^+ cannot enter in large numbers; however, exit of K^+ begins, driven by both electrical and concentration gradients. K^+ flows outward because it is more concentrated inside than outside and because the inside of the membrane is now positively charged due to the influx of Na^+.

The loss of positively charged K^+ restores the negative charge to the interior of the cell, repolarizing the cell membrane. The electrical potential fleetingly drops below the resting value, because the K^+ gates stay open slightly longer than the Na^+ gates. Quickly, the ion concentrations return to normal as the sodium-potassium pump "resets" the membrane potential to the resting state. Figure 31.4 further describes the opening and closing of these ion channels.

While the Na^+ and then the K^+ gates are open, a second action potential cannot begin. Still, an action potential takes only 1 to 5 milliseconds, and it spreads rapidly to adjacent areas of membrane, away from the stimulus. This capacity to rapidly transmit neural impulses makes the nervous system an effective communication network.

The characteristic changes in membrane permeability that constitute the action potential travel along the neuron, usually from dendrite to cell body and down an axon. The action potential

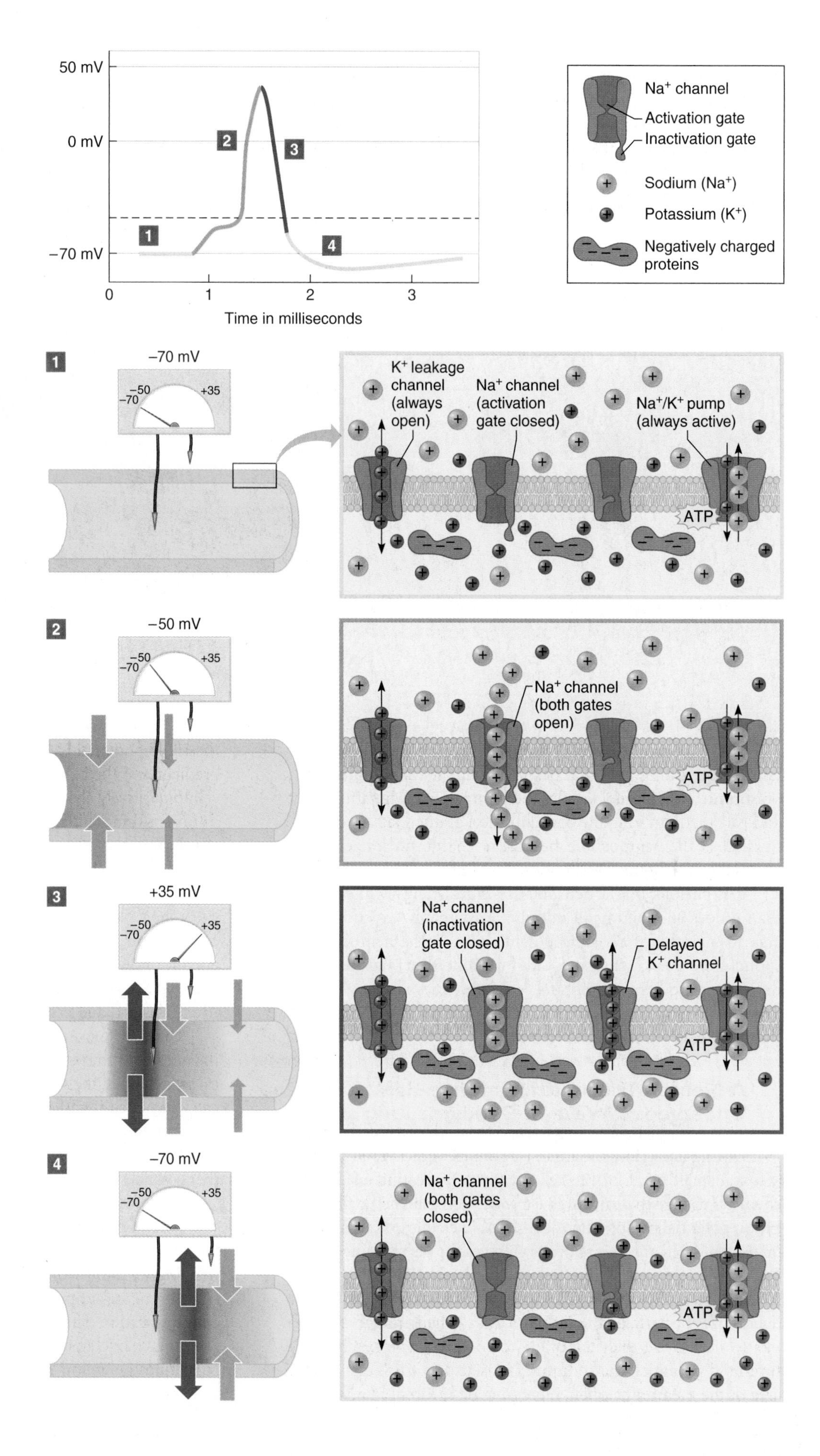

FIGURE 31.5
The Generation of an Action Potential.
(*1*) Shown is a diagram of the neuronal membrane at rest. The K^+ leakage channel is open and K^+ is in equilibrium between its diffusional force outward and its electrical force inward. The Na^+ channel activation gate is closed, restricting the movement of Na^+ into the cell; the membrane potential is about –70 mV. (*2*) Na^+ channel activation gates open and Na^+ rushes into the cell, flowing down its electrochemical gradient; the membrane depolarizes. This local change in potential will trigger opening of additional voltage-sensitive Na^+ channels. If enough Na^+ channels open, the cell reaches its threshold potential and an all-or-none action potential occurs and the membrane reverses its polarity. (*3*) Na^+ channel inactivation gates close after a split second and delayed K^+ gates open. K^+ leaves the axon, restoring the polarized condition of the resting potential. (*4*) Na^+ channel activation gates close and the membrane briefly hyperpolarizes as the resting membrane is reestablished.

FIGURE 31.6

Sheaths of Fatty Myelin Encase Many Axons of Vertebrate Nerve Cells.

This covering forms when a Schwann cell (**A**) winds around an axon in a spiral so that several layers of its lipid-rich cell membrane surround the axon. The Schwann cell's cytoplasm and nucleus are often squeezed into the periphery of the sheath. The unmyelinated sections of axon between Schwann cells are called nodes of Ranvier. (**B**) A myelinated axon shown in cross section.

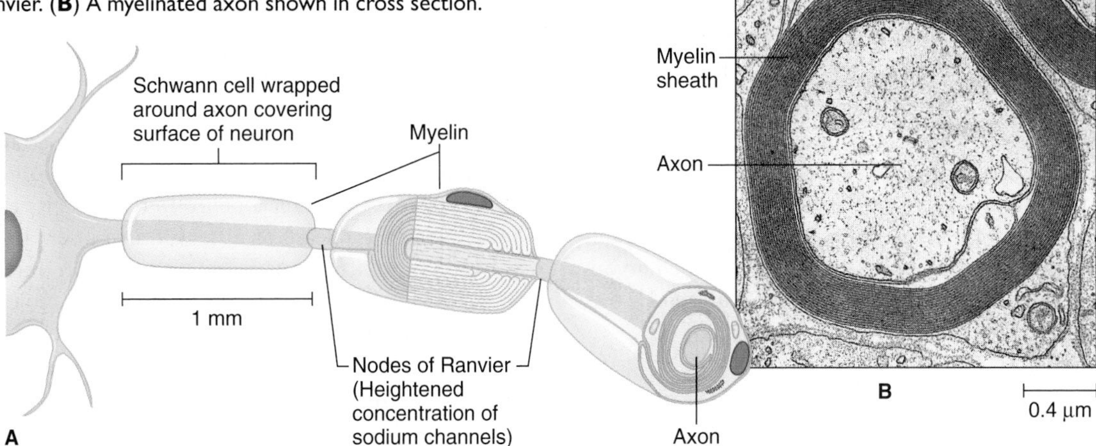

spreads because some of the Na^+ rushing into the cell at a particular point moves to the neighboring part of the neuron and causes it to reach threshold. This triggers an influx of Na^+ there, carrying the action potential forward. An action potential is an all-or-none phenomenon—it either happens or it doesn't.

Interneurons discern the intensity of a stimulus from the frequency of action potentials from sensory neurons. Whereas a light touch to nerve endings in the skin might produce 10 impulses in a given time period, a hard hit might generate 100 impulses, intensifying the sensation. Although all action potentials are the same, neurons also distinguish the type of stimulation. We can tell light from sound because the neurons that light stimulates transmit impulses to a different place in the brain than sound-generated impulses reach.

A Fatty Sheath Speeds Impulse Conduction

Not all nerve fibers conduct action potentials at the same speed. Conduction speed depends on certain characteristics of the fiber. The greater the diameter of the fiber, the faster it conducts an action potential; however, thin vertebrate nerve fibers can conduct action potentials very rapidly when they are coated with a fatty material called a **myelin sheath.**

Outside the brain and spinal cord, **Schwann cells,** which contain enormous amounts of lipid, form myelin sheaths. Each Schwann cell wraps around a small segment of an axon many times, forming a whitish coating (fig. 31.6). Between Schwann cells is a short region of exposed axon called a **node of Ranvier.** Some neurons in the brain and spinal cord are wrapped in myelin that neuroglial cells called **oligodendrocytes** produce.

In a myelinated axon, sodium channels are concentrated at the nodes of Ranvier. When an action potential travels along the axon, it "jumps" from node to node in a type of transmission called **saltatory conduction** (fig. 31.7). The impulse appears to leap from node

FIGURE 31.7

Saltatory Conduction.

In myelinated axons, the inward diffusion of Na^+ can occur only at the nodes of Ranvier. Thus, action potentials appear to "jump" from one node to the next, which speeds up impulse transmission along the axon.

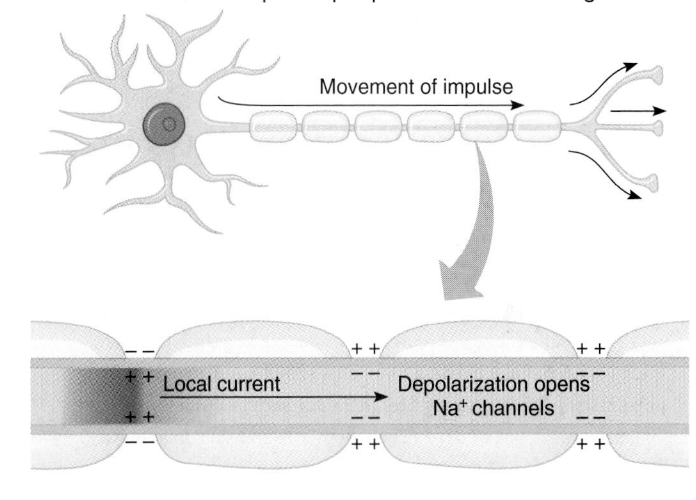

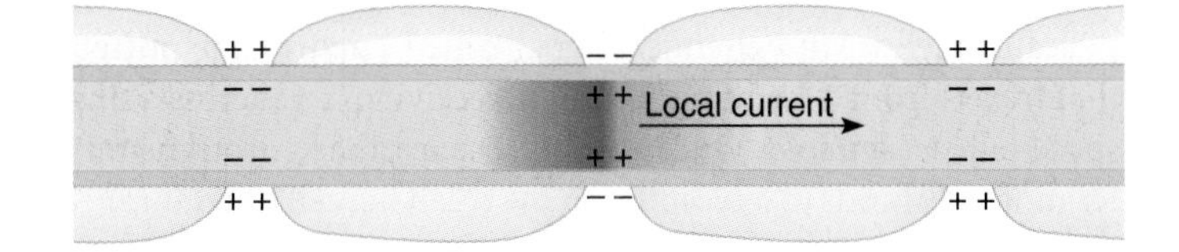

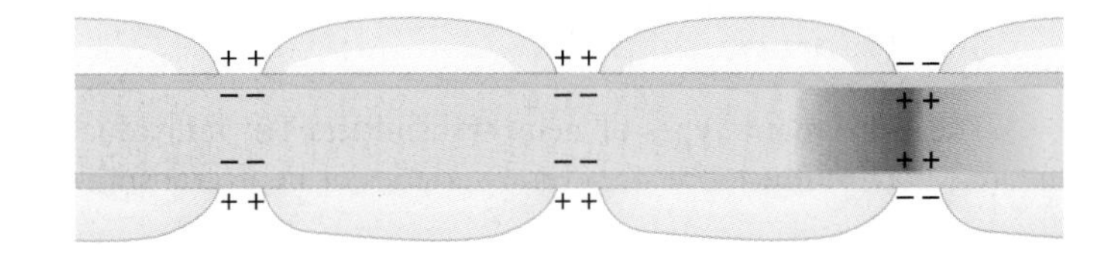

to node because the myelin insulation prevents ion flow across the membrane, but a small electrical current spreads instantly between nodes. Because an action potential moves faster when it jumps from node to node, saltatory conduction speeds nerve impulse transmission. Myelinated axons may conduct action potentials 100 times faster than unmyelinated axons, at speeds of up to 394 feet (120 meters) per second (an astounding 270 miles [435 kilometers] an hour)! This means that a sensory message travels from the toe to the spinal cord in less than 1/100 of a second.

Myelinated fibers are found in pathways that transmit impulses over long distances. They make up the white matter of the nervous system. Cell bodies and interneurons that lack myelin usually specialize in interpreting multiple messages. These unmyelinated fibers, which make up the gray matter of the nervous system, form much of the nerve tissue in the brain and spinal cord.

31.2 MASTERING CONCEPTS

1. What are the parts of a neuron?
2. What are the three types of neurons, and how do they differ from each other?
3. What three mechanisms establish and maintain the resting potential?
4. How do changing cell membrane ion permeabilities generate and transmit a neural impulse?
5. How do action potentials indicate stimulus intensity?
6. How do fatty coverings on neurons speed neural impulse transmission?

OLC

31.3 Neurotransmitters Pass the Message from Cell to Cell

Neurotransmitters are chemicals that are released in vesicles from knobs on axons in response to an action potential, diffuse across the space between neurons, and bind to specific receptors on the receiving cell membrane. Neurotransmitter binding alters the permeability of the membrane in a way that stimulates or blocks depolarization. A neuron integrates incoming excitatory and inhibitory messages to determine whether to transmit an impulse.

To form a communication network, a neuron must convey an action potential to another neuron or to a muscle or gland cell. Most neurons do not touch each other, so the action potential cannot travel directly from cell to cell. Instead, the action potential causes release of a chemical signal, called a **neurotransmitter,** that travels from a "sending" cell to a "receiving" cell across a tiny space called a **synapse.** Once across the space, the neurotransmitter molecule binds to a receptor on the receiving cell's membrane, altering the permeability in a way that either provokes or prevents an action potential from occuring.

Types of Neurotransmitters and Neuromodulators

There are many types of neurotransmitters. An individual neuron may produce one or several varieties of neurotransmitters. The **peripheral nervous system** (PNS; the part outside the brain and spinal cord) uses three neurotransmitters: acetylcholine, noradrenaline, and adrenaline. The central nervous system uses more than 40 neurotransmitters, including serotonin, dopamine, and gamma amino butyric acid (GABA).

Neurotransmitters can be classified chemically. Adrenaline, noradrenaline, dopamine, and serotonin are called monoamines, and they are modified amino acids. Unmodified amino acids can also function as neurotransmitters, and these include glycine, glutamic acid, aspartic acid, and GABA. Certain peptides (short chains of amino acids) function as **neuromodulators,** which means that they alter a neuron's response to a neurotransmitter, or block the release of a neurotransmitter. The enkephalins are five-amino-acid-long peptides that are produced in the CNS and relieve acute pain. Endorphins are longer molecules that provide more extended pain relief. Yet another neuromodulator is substance P. It is found throughout the nervous system, and may be the molecule upon which enkephalins and endorphins act to relieve pain.

Neurotransmitter function is enormously complex, because the nervous system is enormously complex. A neurotransmitter's effect depends upon its concentration and the types and numbers of receptors and ion channels on the receiving cells' membranes. Plus, neurotransmitters affect each other's levels. The same neurotransmitter can even have opposite effects on different types of cells.

Consider acetylcholine. In the PNS, acetylcholine controls muscle contraction, as discussed in chapter 34. In the brain's cerebral cortex, however, acetylcholine's role is not as straightforward. Here, this neurotransmitter coordinates function by enhancing nerve transmission in a horizontal direction, while at the same time inhibiting transmission in a vertical direction. The cortex is organized into columns, and horizontal transmission is necessary to coordinate functioning. Acetylcholine can both inhibit and stimulate because it can bind two types of receptors. One type, found on certain brain cells called basket cells, binds acetylcholine and then decreases levels of GABA production, which otherwise inhibits activity. The result is a stimulation of horizontal nerve transmission because GABA's inhibitory effect is lifted. The other type of acetylcholine receptor, located on brain cells called bipolar cells, increases GABA production. This strengthens the inhibition of vertical transmission.

Neurotransmission is very much a matter of balance, and too much or too little of a neurotransmitter can cause disease. Excess acetylcholine, for example, causes seizures. In contrast, death of acetylcholine-producing cells causes the forgetfulness and difficulty thinking that characterize Alzheimer disease.

Table 31.1 lists drugs that alter specific neurotransmitter levels, and table 31.2 describes illnesses that result from neurotransmitter imbalances.

Neurotransmitter Action at the Synapse

The end of an axon has tiny branches that enlarge at the tips to form **synaptic knobs.** These knobs contain many synap-

TABLE 31.1 Mechanisms of Drug Action

DRUG	NEUROTRANSMITTER AFFECTED	MECHANISM OF ACTION	EFFECT
Tryptophan	Serotonin	Stimulates neurotransmitter synthesis	Sleepiness
Reserpine	Noradrenaline	Packages neurotransmitter into vesicles	Limb tremors
Curare	Acetylcholine	Decreases neurotransmitter in synaptic cleft	Muscle paralysis
Valium	GABA	Enhances receptor binding	Decreases anxiety
Nicotine	Dopamine	Stimulates release of neurotransmitter	Increases alertness
Cocaine	Dopamine	Binds dopamine transporters	Euphoria
Tricyclic antidepressants	Noradrenaline	Blocks reuptake	Mood elevation
Monoamine oxidase inhibitors	Noradrenaline	Blocks enzymatic degradation of neurotransmitter in presynaptic cell	Mood elevation
Ritalin	Dopamine	Binds dopamine transporters	Treats attention deficit disorder

TABLE 31.2 Disorders Associated with Neurotransmitter Imbalances

CONDITION	IMBALANCE OF NEUROTRANSMITTER IN BRAIN	SYMPTOMS
Alzheimer disease	Deficient acetylcholine	Memory loss, depression, disorientation, dementia, hallucinations, death
Depression	Deficient noradrenaline and/or serotonin	Debilitating, inexplicable sadness
Epilepsy	Excess GABA leads to excess noradrenaline and dopamine	Seizures, loss of consciousness
Huntington disease	Deficient GABA	Personality changes, loss of coordination, uncontrollable movement, death
Hypersomnia	Excess serotonin	Excessive sleeping
Insomnia	Deficient serotonin	Inability to sleep
Myasthenia gravis	Deficient acetylcholine at neuromuscular junctions	Progressive muscular weakness
Parkinson disease	Deficient dopamine	Tremors of hands, slowed movements, muscle rigidity
Schizophrenia	Deficient GABA leads to excess dopamine	Inappropriate emotional responses, hallucinations

tic vesicles, which are small sacs that hold neurotransmitter molecules. An action potential passes down the axon of the **presynaptic neuron,** the cell sending the message. When the action potential reaches the membrane near the space, or **synaptic cleft,** the permeability of the membrane changes, allowing calcium ions to enter the cell. Calcium ions cause the loaded vesicles to move toward the synaptic membrane, fuse with it, and dump their neurotransmitter contents into the synaptic cleft by exocytosis (fig. 31.8).

Neurotransmitter molecules diffuse across the cleft and attach to protein receptors on the membrane of the receiving neuron, the **postsynaptic neuron.** A particular neurotransmitter fits only into a specific receptor type, as a key fits only a certain lock. When the neurotransmitter contacts the receptor, the shape of the receptor changes. This binding opens channels in the postsynaptic membrane, admitting specific ions and changing the probability that an action potential will occur.

Disposal of Neurotransmitters

If a neurotransmitter stayed in the synapse, its effect on the receiving cell would be continuous, perhaps causing it to fire unceasingly and bombard the nervous system with stimuli. Two mechanisms control the amount of neurotransmitter in a synapse: It is either destroyed by an enzyme, or taken back into the presynaptic axon soon after its release, an event called reuptake. Poisonous nerve gases and certain insecticides work by blocking reuptake of the neurotransmitter acetylcholine. The resulting excess acetylcholine activity stimulates skeletal muscle to contract continuously, and the person convulses and dies. The twitching legs of a cockroach sprayed with certain insecticides also demonstrate the effects of blocking acetylcholine breakdown.

Several classes of drugs work by blocking reuptake of serotonin. Consider the antidepressants called "selective serotonin reuptake inhibitors" (SSRIs). These include Prozac, Zoloft, Paxil,

FIGURE 31.8

Transmission Across a Chemical Synapse.

(**A**) Arrival of a wave of depolarization—the action potential—at an axon of a presynaptic neuron triggers release of neurotransmitter from vesicles. The neurotransmitter crosses the synapse and binds with receptors in the postsynaptic neuron's cell membrane. This action opens ion channels in a way that increases or decreases the likelihood that an action potential will be generated. (**B**) This micrograph of a synapse shows synaptic vesicles and a synaptic cleft.

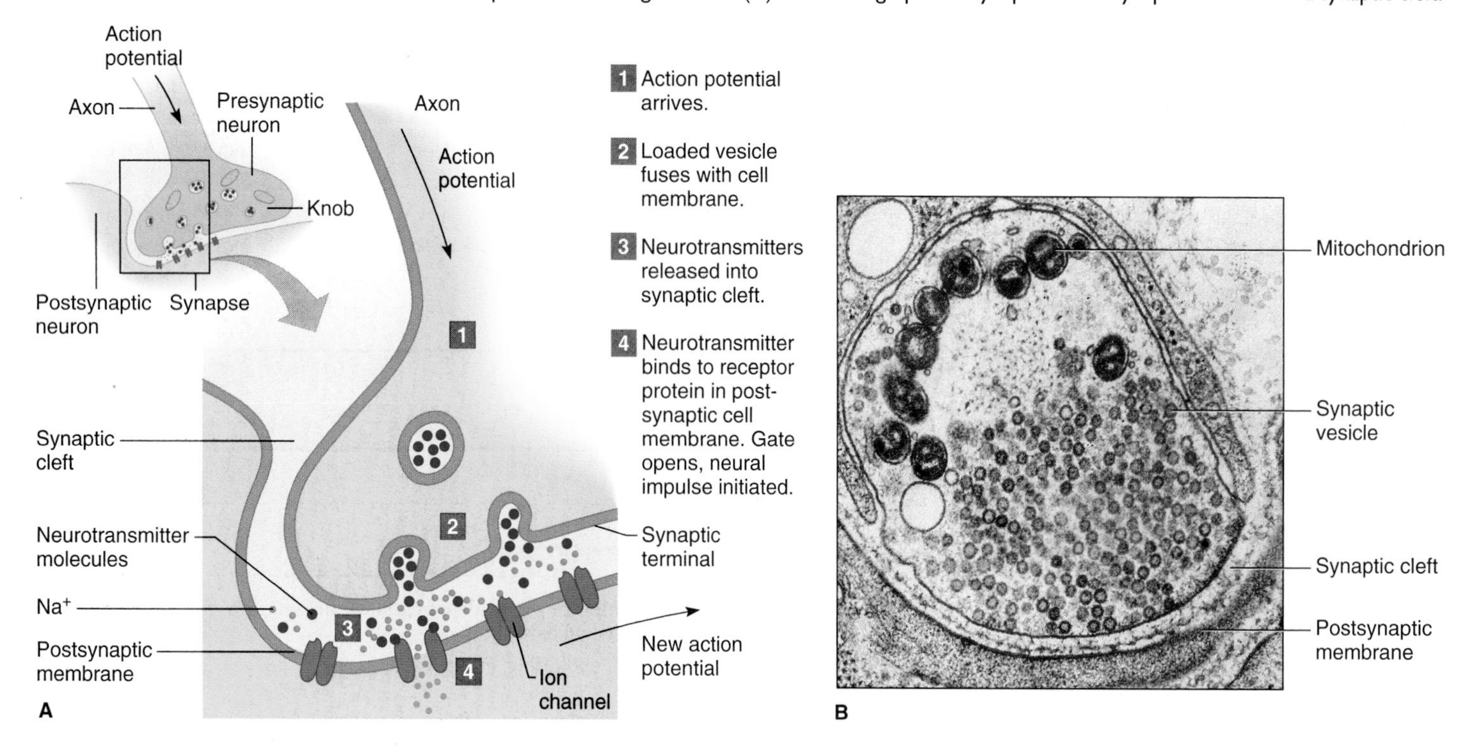

and several newer drugs. By blocking reuptake of serotonin, they enable the neurotransmitter to accumulate in the synapse, offsetting a deficit that presumably causes the symptoms (fig. 31.9). Increasing serotonin availability is also a way to treat migraine headaches and obesity due to overeating. The reverse situation, excess serotonin, causes sleepiness. This is why people become sleepy after eating turkey. The meat contains abundant tryptophan, an amino acid that is required for the body to synthesize serotonin.

Two recent experiments examining serotonin reuptake demonstrated the similarity of nervous systems among vertebrates and invertebrates. Knowing that people who raise clams feed them serotonin to get them to release sperm and eggs at the same time, a biologist fed clams Prozac, the first SSRI to be used as a drug in humans. A tiny dose of the drug triggered sperm and egg release, giving clam farmers a much cheaper way to increase reproduction rates at their facilities. In another experiment, a researcher gave antidepressants to barnacles and mussels. The excess serotonin sped the development of the larvae, causing them to detach from surfaces. The drugs may be used as antifoulant agents to keep barnacles off ships! Yet another study is examining the effects of increasing serotonin on aggression in lobsters.

A Neuron Integrates Incoming Messages

The nervous system has two types of synapses, excitatory and inhibitory. The combination of excitatory and inhibitory synapses provides finer control over a neuron's activities.

Excitatory synapses depolarize the postsynaptic membrane, and inhibitory synapses increase the polarization, or hyperpolarize it. A neurotransmitter that acts at an excitatory synapse increases the probability that an action potential will be generated in the second neuron by slightly depolarizing it. For example, when acetylcholine binds to the receptors at an excitatory synapse, channels open that admit Na^+ into the postsynaptic cell. In a millisecond, half a million sodium ions flow in. If enough Na^+ enters to reach a threshold level of depolarization, it triggers an action potential in the postsynaptic cell.

On the other hand, a neurotransmitter may inhibit an action potential in the postsynaptic cell by making the cell's interior more negative than the usual resting potential. In this case, extra Na^+ must enter before the membrane depolarizes enough to generate an action potential.

A single neuron in the nervous system may receive input from tens of thousands of other neurons, some excitatory and others inhibitory. Nearly half of a neuron's receiving surface adjoins synapses. Whether that neuron transmits an action potential depends on the sum of the excitatory and inhibitory impulses it receives (fig. 31.10). If it receives more excitatory impulses, the postsynaptic cell is stimulated; if inhibitory messages predominate, it is not. A neuron's evaluation of impinging nerve messages, which determines whether an action potential is "fired," is termed neural or **synaptic integration.**

Synapses markedly increase the informational content of the nervous system. The human brain has a trillion neurons, each of

FIGURE 31.9
Anatomy of an Antidepressant.
Selective serotonin reuptake inhibitors (SSRIs) are antidepressant drugs that act as their name states—they block the reuptake of serotonin, making more of the neurotransmitter available in the synaptic cleft. This restores a neurotransmitter deficit that presumably causes the symptoms. Overactive or overabundant reuptake receptors can cause the deficit. The precise mechanism of SSRIs is not well understood.

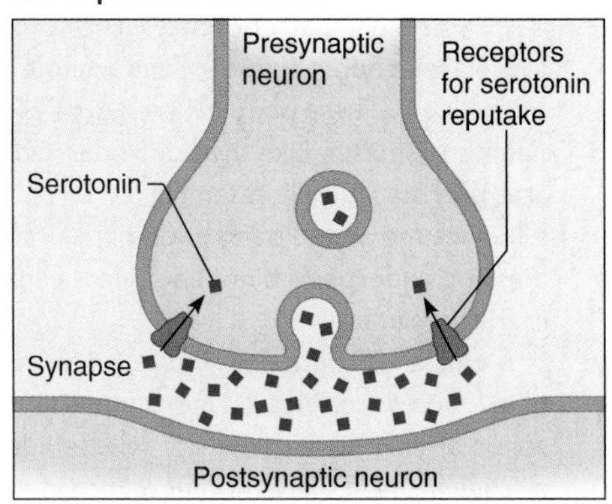

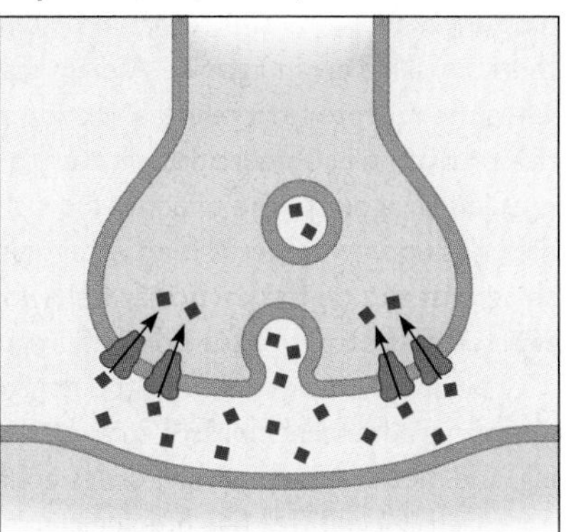

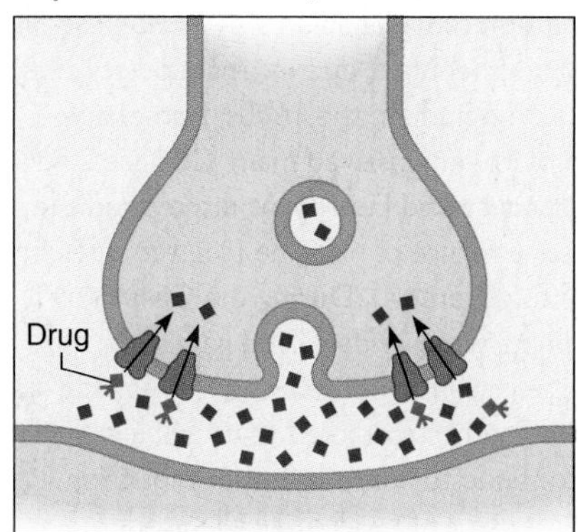

which can be viewed as carrying bits of information. But if a synapse is also considered a unit of information, then the informational capacity of the brain increases a thousandfold, because a typical brain neuron has synaptic connections to a thousand other neurons, each sending or receiving messages hundreds of times per second. Health 31.1 discusses how drug addiction interferes with neurotransmission.

FIGURE 31.10
Integration of Neural Impulses.
The synaptic terminals from many neurons converge on the cell body of a single postsynaptic neuron. Some of these terminals are inhibitory while others are excitatory—the postsynaptic cell integrates the contributions from all active inputs and produces an action potential, or not, based on this calculation.

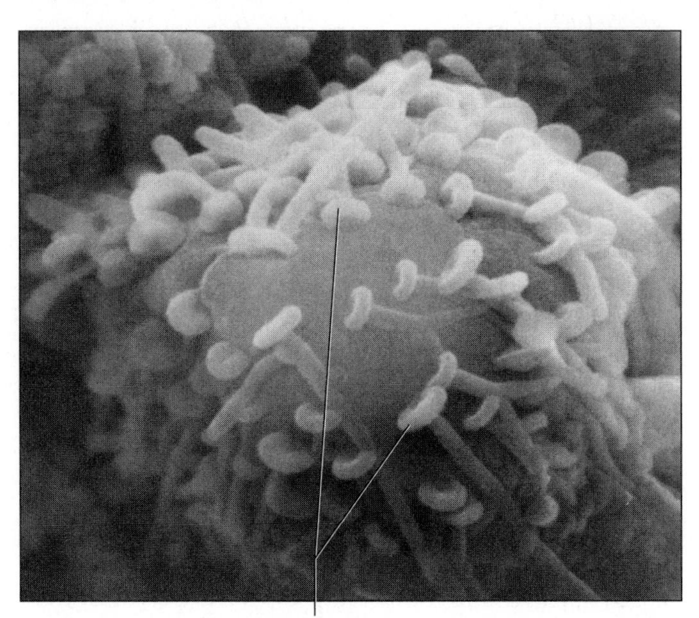

31.3 MASTERING CONCEPTS

1. How do neurotransmitters send action potentials from a neuron to another neuron, a muscle, or a gland cell?
2. What happens to a neurotransmitter after it is released?
3. What is neural integration?

31.4 Evolutionary Trends in Nervous Systems

Nervous systems in different animal phyla reflect adaptations to particular environments. In invertebrates, these systems increase in complexity, from simple net and ladder organizations, to the anterior concentration of neurons to form a rudimentary brain. The vertebrate nervous system consists of a centralized brain and spinal cord and peripheral nerves.

Multicellularity posed a profound new challenge to organisms—how would cells communicate with each other? In a single-celled organism, the cell membrane serves as a selective gateway to the outside environment. In a multicellular organism, cells maintain this gatekeeper function; they react to messages that must pass between cells to coordinate the cells' functions. These messages are electrical and chemical changes that transmit information by stimulating or inhibiting cell surface receptor molecules.

All nervous systems consist of organizations of interacting cells. Comparing nervous systems among diverse modern species reveals that nervous systems probably increased in complexity as new species evolved. A nervous system might be a loose network of relatively few cells, or a structure as complex as the human brain.

Different organizations of nervous systems are adaptive in particular environments. Radially symmetrical animals such as sea anemones respond to stimuli that can come from any direction. These animals usually have simple, diffuse nerve nets that make all parts of their bodies equally receptive to stimuli.

Other animals have more intricate nerve cell organizations in certain areas of the body. Bilaterally symmetrical animals have

Addiction!

Drug abuse and addiction are ancient as well as contemporary problems. A 3,500-year-old Egyptian document decries that society's reliance on opium. In the 1600s, a smokable form of opium enslaved many Chinese, and the Japanese and Europeans discovered the addictive nature of nicotine (see the opening essay to chapter 4). During the Civil War, morphine was a widely used painkiller; cocaine was introduced a short time later to relieve veterans addicted to morphine. Today, we continue to abuse drugs intended for medical use. LSD was originally used in psychotherapy, but was abused in the 1960s as a hallucinogen (see Health 23.1). PCP was an anesthetic before being abused in the 1980s. Why people become addicted to certain drugs lies in the complex interactions of neurons, drugs, and individual behaviors.

The Role of Receptors

Eating hot fudge sundaes is highly enjoyable, but we usually don't feel driven to consume them repeatedly. Why do certain drugs compel a person to repeatedly use them in steadily increasing amounts—the definition of addiction? The biology of neurotransmission helps to explain how we, and other animals, become addicted to certain drugs.

Understanding how neurotransmitters fit receptors can explain the actions of certain drugs. When a drug alters the activity of a neurotransmitter on a postsynaptic neuron, it either halts or enhances synaptic transmission. A drug that binds to a receptor, blocking a neurotransmitter from binding there, is called an antagonist. A drug that activates the receptor, triggering an action potential, or that helps a neurotransmitter to bind, is called an agonist. The effect of a drug depends upon whether it is an antagonist or an agonist and on the particular behaviors the affected neurotransmitter normally regulates.

Neural pathways that use the neurotransmitter noradrenaline control arousal, dreaming, and mood. Amphetamine drugs enhance noradrenaline activity, heightening alertness and mood. Amphetamine's structure is so similar to that of noradrenaline that it binds to noradrenaline receptors and triggers the same changes in the postsynaptic membrane.

Cocaine has a complex mechanism of action, both blocking reuptake of noradrenaline and binding to molecules that transport dopamine to postsynaptic cells. Cocaine's rapid and short-lived "high" reflects its rapid stay in the brain—its uptake takes just 4 to 6 minutes, and within 20 minutes the drug loses half its activity.

Opiates in the Human Body

Opiate drugs, such as morphine, heroin, codeine, and opium, are potent painkillers derived from the poppy plant. These drugs alter pain perception, making pain easier to tolerate, and they also elevate mood. When taken repeatedly by healthy individuals, opiate drugs are addictive. When taken to relieve intense pain, opiates are usually not addictive.

The human body produces its own opiates, called endorphins (for "endogenous morphine"). These peptides are a type of neurotransmitter. Like the poppy-derived opiates they structurally resemble, endorphins influence mood and perception of pain. Opiates and endorphins bind the same receptors in the human brain.

Humans produce several types of endorphins, which are released in response to stress or pain. Endorphins are released during a "runner's high," as a mother and child experience childbirth, and during acupuncture.

Endorphins help explain why some people addicted to an opiate drug such as heroin experience withdrawal pain when they stop taking the drug. Initially, the body interprets the frequent binding of heroin to its endorphin receptors as excess endorphins. To bring the binding down, the body slows its own production of endorphins. Then, when the person stops taking heroin, the body is caught short of opiates (heroin as well as endorphins). The result is pain.

Drug addiction is a powerful force, its causes rooted in both biology and psychology and its effects an economic, sociological, and political problem. Fighting an addiction is far more difficult than "just saying no" to temptation, as one might to a hot fudge sundae.

---

nervous tissue centralized or concentrated in certain areas of the body, where large numbers of nerve cells maintain highly intricate interconnections. This increases the number and complexity of possible responses. Animals became increasingly cephalized, with accumulation of nervous tissue into a brain and the development of sensory structures to form a head. Thus, nervous tissue concentrates in the end of the animal that takes in the most sensory information from the environment. Other nervous tissue is part of the peripheral nervous system, which transmits information to the CNS from receptors and from the CNS to effectors such as muscles and glands.

Invertebrates Have Nerve Nets, Ladders, or Simple Brains

The simplest nervous systems are found in the phylum Cnidaria, which possess diffuse networks of neurons, called **nerve nets.** In body walls of hydra, jellyfishes, and sea anemones, nerve nets synapse with muscle cells near the body surface, enabling each animal to move its tentacles. A stimulus at any point on the body spreads over the entire body surface. Cnidaria can maintain balance and detect touch, certain chemicals, and light.

This diffuse nervous system organization and the ability of each neuron to conduct impulses in both directions allow cnidaria to react to stimuli that approach from any direction. If an animal is floating in the ocean where danger and food may approach from any direction, a nerve net is adaptive. Figure 31.11 illustrates some nervous systems of invertebrates.

Flatworms have rudimentary brains that consist of two clusters of nerve cell bodies, called **ganglia,** and two nerve cords that extend down the body and connect to each other to form a "ladder" type of nervous system. Motor structures and the neurons

FIGURE 31.11
Invertebrate Nervous Systems.
As invertebrate bodies increased in complexity, so did their nervous systems.

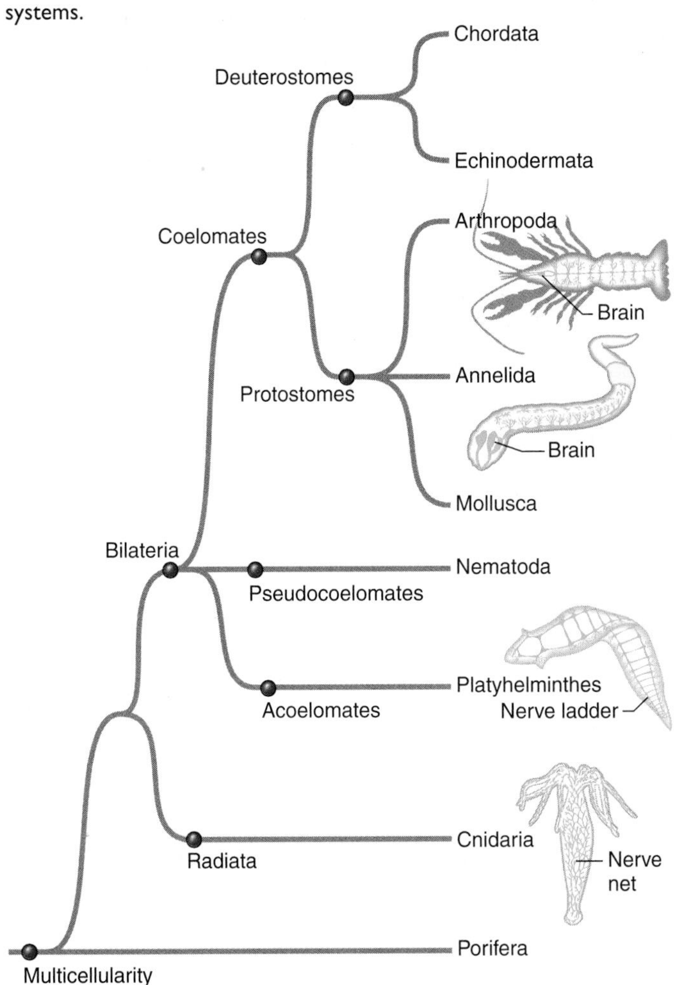

that control them are paired, which allows the worm to move in a coordinated forward motion. Paired symmetrical sense organs at the head end allow the flatworm to detect a stimulus and crawl toward it. Bilaterally paired receptors enable the animal to determine which direction stimulation is coming from by comparing the stimulus intensity at each receptor. The receptor with the strongest response is closest to the stimulus.

Segmented worms have a more elaborate ladder organization and a larger brain. Two nerve cords run down the body on the ventral side beneath the digestive tract. These cords are solid and fused to each other. Peripheral nerves branch from the central fused ladder, and neuron clusters along the nerve cords coordinate functions in particular regions of the body through interneurons.

The nervous systems of arthropods (including insects, spiders, lobsters, and scorpions) also consist of a brain and a ventral, fused nerve cord. The highly cephalized arthropod nervous systems have well-developed sensory organs for touch, smell, hearing, chemical reception, and balance. These animals have complex behaviors and even rudimentary learning.

Vertebrates Have Central and Peripheral Nervous Systems

Less variation is seen among vertebrate nervous systems than among invertebrate systems, which is consistent with the fact that invertebrates are more diverse than vertebrates. The vertebrate nervous system is a modification of the symmetry seen among the simpler worms and insects. A tubular **spinal cord** extends lengthwise and expands at the anterior end to form the **brain.** The cord conducts information to and from the brain and centralizes rapid involuntary responses.

Figure 31.12 shows the major divisions of the vertebrate nervous system. The PNS consists of **cranial nerves** that exit the CNS from the brain, **spinal nerves** that extend from the spinal cord, and collections of cell bodies. The PNS is divided into the sensory (or afferent) pathways that carry information to the CNS and the motor (or efferent) pathways that transmit action potentials from the CNS to muscle or gland cells. The motor pathways in turn consist of the **somatic** (voluntary) **nervous system,** which leads to skeletal muscles, and the **autonomic** (involuntary) **nervous system,** which goes to smooth muscle, cardiac muscle, and glands. Finally, the autonomic nervous system consists of the **sympathetic nervous system,** which mobilizes the body to respond to threatening environmental stimuli, and the **parasympathetic nervous system,** which controls more mundane functions such as digestion, respiration, and heart rate at rest. The autonomic nervous system is well developed in reptiles, birds, and mammals but less so in fishes and amphibians.

The most complex nervous systems are in primates. The remainder of the chapter focuses on the structures and functions of the human nervous system.

31.4 MASTERING CONCEPTS

1. Contrast the complexity of invertebrate and vertebrate nervous systems.
2. How do the organizations of nervous systems in invertebrates reflect their lifestyles?
3. What are the subdivisions of the vertebrate nervous system?

OLC

31.5 The Human Central Nervous System

The spinal cord receives impulses from the rest of the body, conducts reflexes, and communicates with the brain. Structures in the hindbrain control vital functions and connect to other regions; neurons in the midbrain process and integrate certain sensory input; in the forebrain the cerebrum receives sensory input and directs motor responses, and the hypothalamus regulates secretion of certain hormones.

The human CNS is essentially like that of many other vertebrates. Figure 31.13 compares the brains of several vertebrate species.

FIGURE 31.12

Hierarchical Organization of the Vertebrate Nervous System.

The peripheral nervous system consists of all nervous tissue outside the central nervous system. Most generally, it can be divided into motor and sensory pathways. The motor pathways are further subdivided into the somatic nervous system, which innervates skeletal (voluntary) muscle, and the autonomic nervous system, which stimulates (involuntary) effectors. The autonomic nervous system is further subdivided into the sympathetic nervous system, which predominates in threatening situations, and the parasympathetic nervous system, which predominates in restful circumstances.

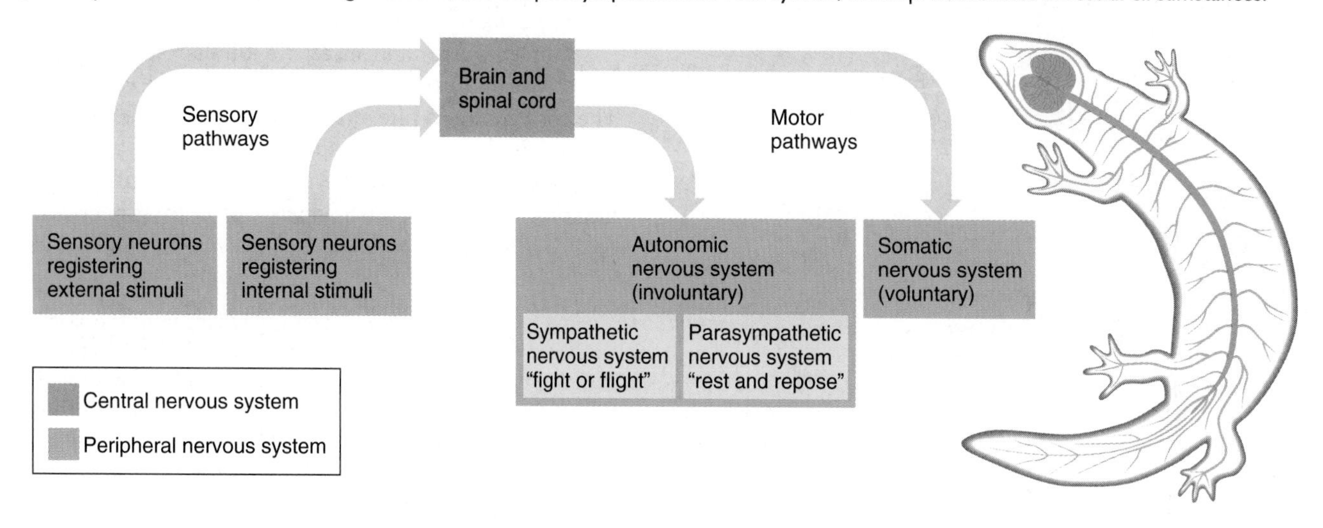

FIGURE 31.13

Vertebrate Brains.

The vertebrate brain forms in the embryo as a sheet of ectoderm that folds into a tube, and then swells at the anterior end. Then the cells specialize into neurons or neuroglia, and associate and interact to form three distinct regions—the forebrain, midbrain, and hindbrain. These divisions persist in the adult, although some areas are more highly developed in certain species compared with others. (A fish's nervous system forms slightly differently, with a solid tube hollowing out rather than a sheet folding to form a hollow cylinder.)

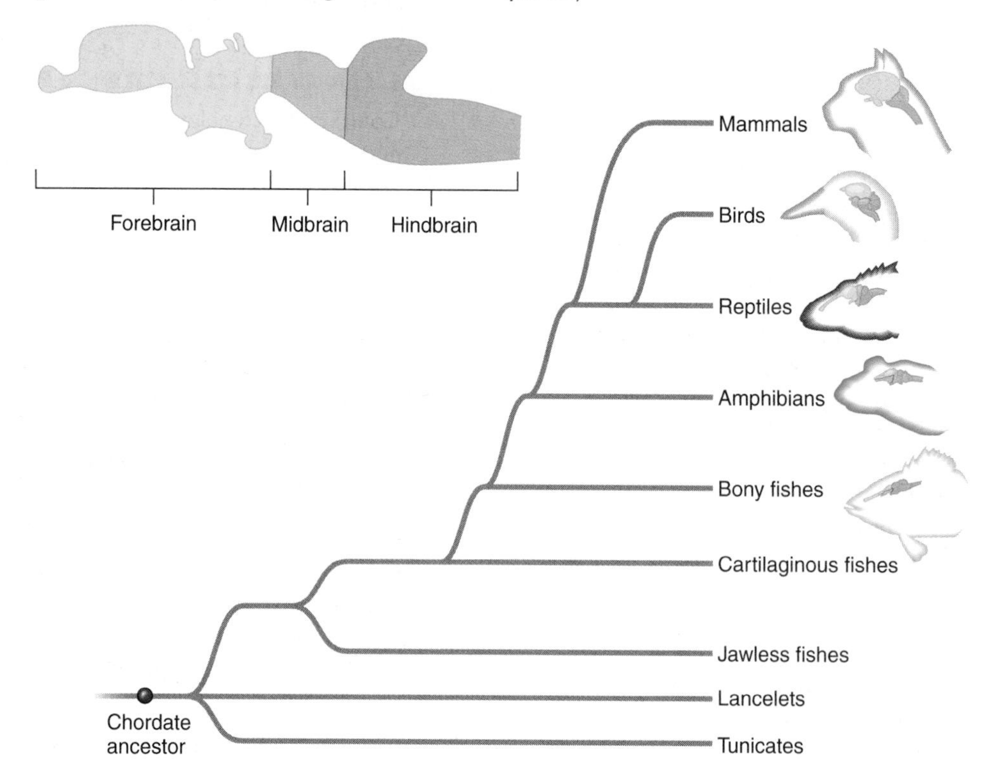

The Spinal Cord

The spinal cord is a tube of neural tissue encased in the bony armor of the vertebral column (backbone). The vertebral column is composed of individual vertebrae. It protects the delicate nervous tissue and provides points of attachment for muscles.

The spinal cord extends about 17 inches (43 centimeters) from the base of the brain to an inch or so below the last rib. It carries impulses to and from the brain and is the site where spinal reflex neurons interact. A reflex is a rapid, involuntary response to a stimulus that may come from within or outside the body.

The spinal cord communicates with and receives information from the rest of the body through 31 pairs of spinal nerves, numbered according to their position along the cord, as fig. 31.14 depicts. Sensory information travels to the rear of the spinal cord, and instructions for activity of skeletal muscles pass outward from the

FIGURE 31.14
Overview of the Human Nervous System.
(**A**) Dorsal view of the brain, spinal cord, and spinal nerves. (**B**) Lateral view of the spinal cord showing the location of the 31 pairs of spinal nerves, which are named and numbered according to the point at which they leave the cord. There are 8 pairs of cervical nerves, 12 pairs of thoracic nerves, 5 pairs of lumbar nerves, 5 pairs of sacral nerves, and 1 pair of coccygeal nerves. (**C**) Expanded view of a part of the spinal cord from the thoracic region. The cord runs through the intervertebral canal, the tunnel formed within the stacked vertebrae of the vertebral column. The spinal nerves exit laterally between the vertebrae. Note that membranes called meninges surround the spinal cord. (**D**) Cross section of spinal cord viewed from the top. The central region of gray matter, which in cross section resembles the letter H, contains short interneurons, unmyelinated fibers, motor neuron cell bodies, and neuroglial cells. The white matter surrounding the H is made up of myelinated nerve tracts.

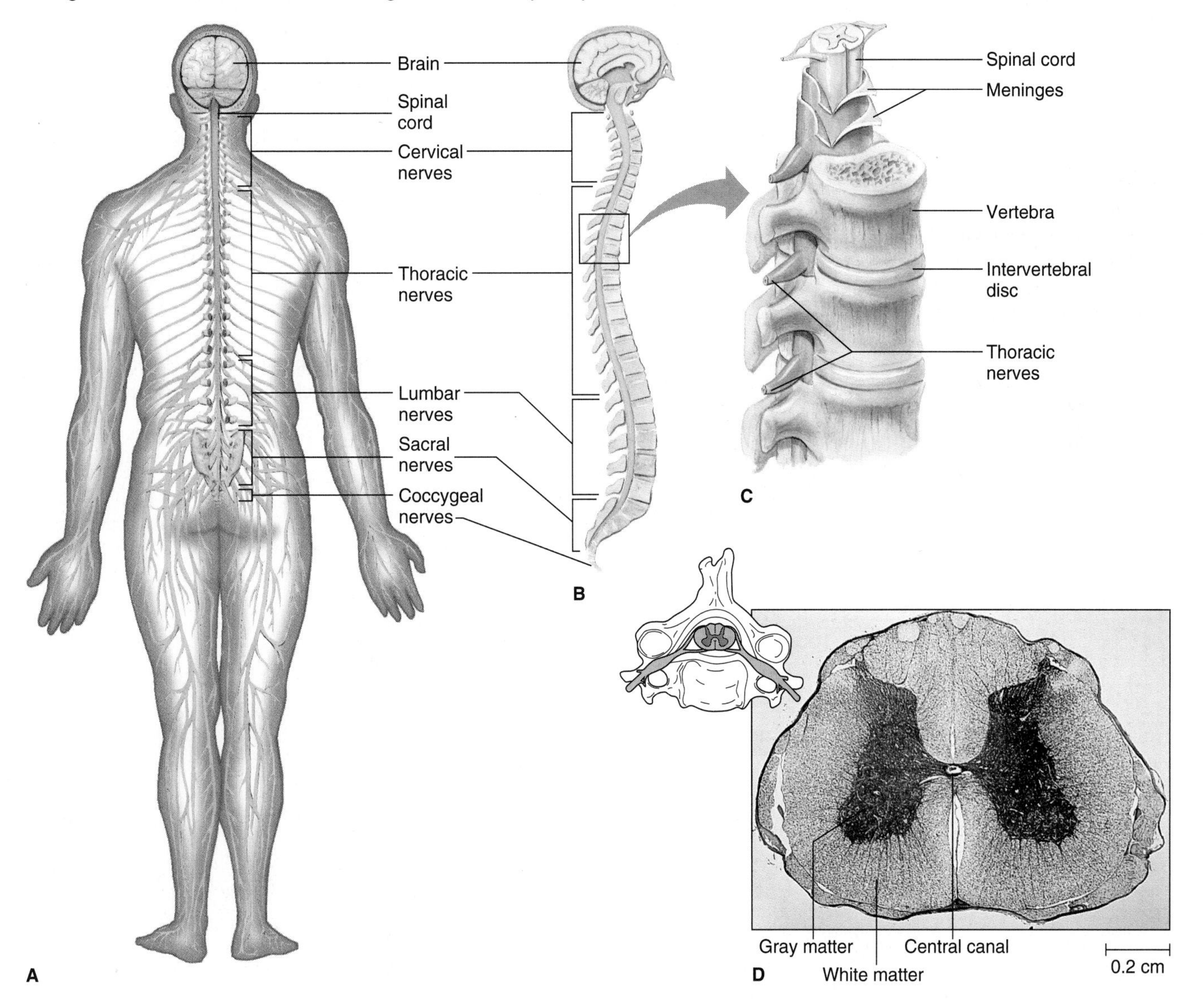

front of the cord. The sensory and motor nerve fibers form a single cable, which passes through the opening between the vertebrae.

The spinal cord conducts information to and from the brain via myelinated fibers that form **white matter** at the periphery of the cord. Specific types of information travel in particular tracts (axons and dendrites) of the white matter. Ascending tracts carry sensory information to the brain, and descending tracts carry motor information from the brain to muscles and glands. The **gray matter** interior of the spinal cord consists of motor neuron cell bodies, unmyelinated fibers, interneurons, and neuroglial cells.

The spinal cord handles reflexes without interacting with the brain (although impulses must be relayed to the brain for awareness to occur). Two or more neurons may carry out reflex reactions. A simple neural connection of sensory neuron to interneuron to motor neuron allows the body to respond more rapidly than it could if the information had to travel to the brain. Reflexes are often protective.

A **reflex arc** is a neural pathway that links a sensory receptor and an effector such as a muscle. The three neurons that cause you to pull your hand away from a painful stimulus such as a thorn make up a reflex arc (fig. 31.15). The arc begins when a sensory receptor, perhaps a sensory neuron whose dendrite extends all the way to your fingertip, detects the sharp thorn. An action potential is generated along its axon and is transmitted to the dorsal (back) side of the spinal cord. Within the spinal cord, the sensory neuron's axon synapses with an interneuron; within the gray matter, the interneuron synapses with a motor neuron. The motor neuron's axon exits the spinal cord on the ventral side, and its action potential stimulates a skeletal muscle cell to contract. When a sufficient number of muscle fibers contract, you pull your hand away from the thorn. The original sensory neuron synapses with other interneurons, too. Some of these send action potentials to the brain, perhaps prompting you to yell in pain. Health 31.2 discusses the spinal cord injury of a well-known individual, Christopher Reeve.

FIGURE 31.15
A Reflex Arc.
A reflex arc links a sensory receptor to an effector. A sensory receptor—perhaps a sensory neuron dendrite that ends in the skin—senses an environmental stimulus and relays an action potential to its cell body in the spinal cord. This neuron sends the information to an interneuron, which relays the action potential to a motor neuron; this neuron, in turn, stimulates a muscle cell to contract. Motor neurons may activate other neurons or gland or muscle cells.

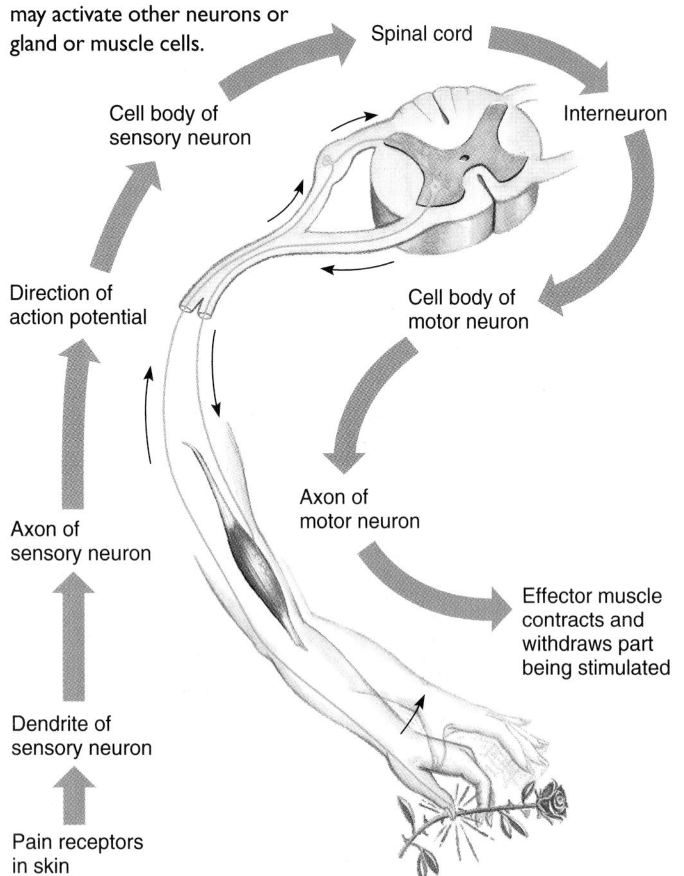

Major Brain Regions

The human brain weighs 3 pounds (1.36 kilograms) and looks and feels like grayish pudding. The brain requires a large and constant energy supply to oversee organ systems and to provide the qualities of "mind"—learning, reasoning, and memory. At any given time, brain activity consumes 20% of the body's oxygen and 15% of its blood glucose. Permanent brain damage occurs after just 5 minutes of oxygen deprivation.

Brain anatomy can be considered in terms of the three subdivisions that first appear in the early embryo—the **hindbrain,** the **midbrain,** and the **forebrain** (fig. 31.16 and see fig. 31.13). Within each region lie major structures and a few smaller (but nonetheless important) structures. The three largest parts are the **brain stem,** the **cerebellum,** and the **cerebrum.** The brain stem and the cerebellum, which lie in the hindbrain, are considered more primitive brain structures because they also appear in species less complex than humans. The cerebrum, located in the forebrain, is most highly developed in the primates.

FIGURE 31.16
The Human Brain.
A midsagittal section through the human skull (dividing the brain into right and left halves) shows the relationship of the brain with the overlying bone structures. The three major areas of the vertebrate brain are the hindbrain, the midbrain, and the forebrain.

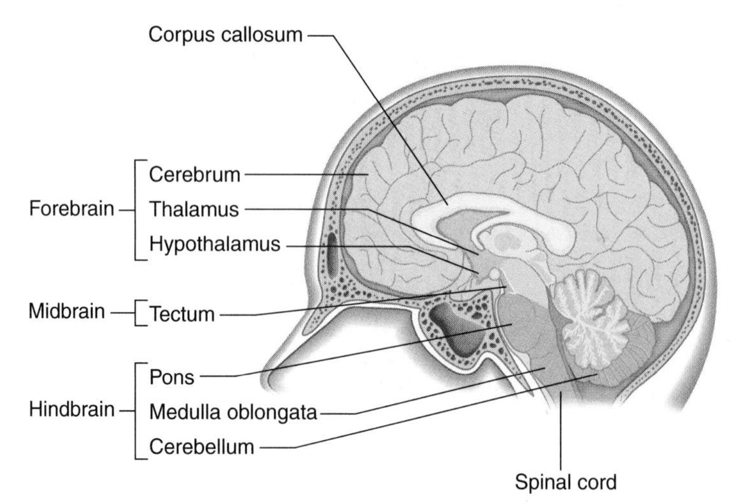

Spinal Cord Injuries

Christopher Reeve's life changed forever in a split second on May 27, 1995. Reeve, best known for his portrayal of Superman in four films, was one of 300 equestrians competing on that bright Saturday in Culpeper County, Virginia. He and his horse, Buck, were poised to clear the third of 15 hurdles in a 2-mile event. The horse's front legs went over the hurdle, and Reeve's back arched as he propelled himself forward, in sync with the horse, as they had done many times in practice. But this time was different.

Buck stopped, his back legs never clearing the fence. Reeve hurled forward, pitching over the horse's head, striking his head on the fence. He landed on the grass—unconscious, not moving or breathing.

Reeve had broken the first and second cervical vertebrae, between the neck and the brain stem. Someone performed mouth-to-mouth resuscitation until paramedics inserted a breathing tube and then stabilized him on a board. At a nearby hospital, Reeve received methylprednisolone, a drug that diminishes the extremely damaging swelling that occurs as the immune system responds to the injury. If given within 8 hours of the accident, this drug can save a fifth of the damaged neurons. Reeve was then flown to a larger medical center, where he was sedated and fluid suctioned from his lungs.

The first few days following a spinal cord injury are devastating. At first, the vertebrae are compressed and may break, which sets off action potentials in neurons, many of which soon die. The massive neuron death releases calcium ions, which attract tissue-degrading enzymes. Then white blood cells arrive and produce inflammation that can destroy healthy as well as damaged neurons. Axons tear, myelin coatings are stripped off, and vital connections between nerves and muscles are cut. The tissue cannot regenerate.

On the whole-body level, a spinal cord injury prevents motor impulses from descending from the brain, causing paralysis (fig. 31.A). Blockage of sensory input upward halts control of bowel and bladder function, blood pressure, muscle action, and sensation of temperature change. In the days following the injury, the person may experience bleeding, inflammation, swelling, and pain. Later, muscle spasms, leakage of cerebrospinal fluid, and infection become problems.

A week after the accident, Reeve could move a few muscles in his chest and back, which indicated that his spine had not been severed. Three weeks after the accident, he could sit and had some sensation in his upper body. On June 5, he underwent surgery to implant U-shaped wires in his neck to limit further damage. A month later he went to a rehabilitation facility to begin therapy.

Progress has been slow. It wasn't until January 1996 that Reeve could breathe unaided for longer than a few minutes, which he compared to climbing a very steep hill. A year after the injury, he could breathe unaided for close to an hour, which indicated that some healing had occurred. By 1997, Reeve was providing voiceovers, directing, and acting in roles requiring a wheelchair-bound character. By 1998, Reeve was able to breathe on his own for 90 minutes. Today he continues to act and direct.

Each year, about 12,000 people in North America sustain spinal cord injuries (table 31.3). Worldwide, a few million people live with these injuries, 55% with partial paralysis due to damage in the lower back, the rest with

TABLE 31.3 Causes of Spinal Cord Injuries

CAUSE	PERCENTAGE OF CASES
Motor vehicle accidents	44
Falls	22
Sports (diving, football, horseback riding)	18
Violence	16

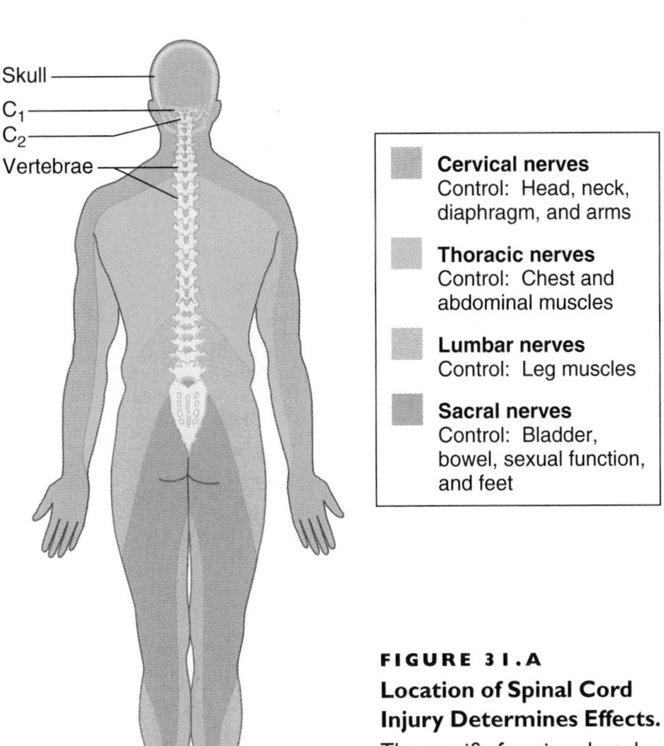

FIGURE 31.A Location of Spinal Cord Injury Determines Effects. The specific functions lost due to a spinal cord injury reflect the site of the damage.

Continued

(concluded)

damage to the upper spine and near-total paralysis.

Several new treatments are on the horizon for spinal cord injuries. They work in three ways:

1. *Limiting damage during the acute phase.* When 17-year-old Chinese gymnast Sang Lan misjudged a practice vault at the Goodwill Games in New York City in the summer of 1998, fracturing two cervical vertebrae, prompt administration of an experimental drug called GM1 ganglioside limited damage. This carbohydrate, normally found on neuronal cell membranes, blocks the actions of amino acids that function as excitatory neurotransmitters, which cuts the deadly calcium ion influx into cells. It also blocks apoptosis (programmed cell death) and stimulates synthesis of nerve growth factor.

2. *Restoring or compensating for function.* A new drug called 4-aminopyridine blocks potassium channels on neurons. This boosts electrical transmission and compensates for the myelin-stripping effects of the injury. Being developed for patients injured at least 18 months previously, this drug can restore some sexual, bowel, and bladder function.

3. *Regeneration.* Neurobiologists have known since 1981 that regeneration of damaged spinal cord cells should be possible. Experiments then showed that spinal axons can grow in the PNS, but not in the CNS. In 1988, researchers discovered a protein in the spinal cord that inhibits regeneration. Blocking this protein may allow some regeneration. In rats with damaged spinal cords, a neural cellular adhesion molecule blocks the inhibitor, restoring some walking ability. Using another approach, investigators have effectively patched severed rat spinal cords with implants of peripheral nervous tissue from the chest area and added growth factors. These animals too regained partial walking ability.

Clinical trials are underway to see if implants of neural stem cells can regenerate spinal cord neurons. One day, neural stem cells taken from an injured person's brain—which one researcher calls "brain marrow"—might be expanded in culture and used to "patch" a severed spinal cord.

The Hindbrain and Midbrain Include the Brain Stem, Cerebellum, and Pons The hindbrain is located toward the back of the skull. It contains the brain stem, which includes the **medulla oblongata** (often just "medulla"), the cerebellum, and the **pons.** The brain stem is a continuation of the spinal cord within the skull. It is aptly named for its appearance: it supports the brain as a stem supports a head of cauliflower. But it does much more. The brain stem controls vital functions.

The section of brain stem closest to the spinal cord is the medulla oblongata, which regulates essential physiological processes including breathing, blood pressure, heart rate, and muscle contractility. The medulla adjusts these activities to suit the body's varying requirements, increasing heart and breathing rates and blood pressure during vigorous activity and slowing them during rest. The medulla also contains reflex centers for vomiting, coughing, sneezing, urinating, defecating, swallowing, and hiccoughing.

The medulla is a thoroughfare, passing most messages that enter or leave the brain. As axons and dendrites carrying sensory or motor messages traverse this region, most of them cross from one side of the body to the other. In this way, the left side of the brain receives impressions from and directs the motor activity of the right side of the body, and vice versa.

The cerebellum ("little brain") is an outgrowth of and is connected to the brain stem. The cerebellum is gray on the outside, white with gray patches on the inside, and divided into two hemispheres. As the second largest structure in the human brain, it accounts for almost one-eighth of the brain's mass.

The neurons of the cerebellum refine motor messages and coordinate muscular movements subconsciously. The cerebellum receives sensory input from the cerebral cortex (the outer portion of the cerebrum) and the PNS. It then compares the action the cerebrum intended with the actual movement and makes corrections so that the two agree. For example, the cerebellum acts when we try to bring the tips of our two index fingers together without looking. The cerebellum quickly recognizes and corrects the initial miss.

Many of our conscious activities have subconscious components that the cerebellum governs. If you had to consciously plan every movement in catching a ball, for instance, you would probably not be quick enough to prevent it from striking your head. Training and practice result in an automatic program of movements, many of which the cerebellum controls at a subconscious level.

The area above the medulla is the pons, which means "bridge." This is a suitable name for this oval mass, because white matter tracts in the pons form a two-way conduction system that connects higher brain centers with the spinal cord and connects the pons to the cerebellum. In addition, gray matter in the pons controls some aspects of breathing.

The medulla and pons have changed very little through vertebrate evolution. The cerebellum, however, is greatly enlarged in more complex vertebrate species.

A narrow region above the pons, the midbrain, is also part of the brain stem. The midbrain receives sensory information from touch, sound, visual, and other receptors and passes it to the forebrain. Neurons in the midbrain interpret sensations as perceptions. Much of this integrating activity occurs in a thickened area of gray matter at the roof of the midbrain called the **tectum.** In amphibians and fishes, the tectum is the largest part of the brain.

The Forebrain The part of the brain that has changed the most through vertebrate evolution is the forebrain. It includes two major regions: the telencephalon, which includes the cerebrum, and the diencephalon, which lies in front of the midbrain and includes the thalamus, hypothalamus, pineal gland, and pituitary gland. (Chapter 33 discusses the latter two structures.) Forebrain structures are important in complex behaviors such as learning, memory, language, motivation, and emotion.

The **hypothalamus** weighs less than 4 grams and occupies less than 1% of the brain volume, but it regulates many vital functions, including body temperature, heartbeat, water balance, and blood pressure. Groups of nerve cell bodies in the hypothalamus control hunger, thirst, sexual arousal, and feelings of pain, pleasure, anger, and fear. The hypothalamus also regulates hormone secretion from the pituitary gland at the base of the brain. Thus, the hypothalamus links the nervous and endocrine systems, the body's two communication systems.

The **thalamus** is a gray, tight package of nerve cell bodies and their associated neuroglial cells located beneath the cerebrum. It acts as a relay station for sensory input, processing incoming information and sending it to the appropriate part of the cerebrum.

The **reticular formation** is not really localized to hindbrain, midbrain, or forebrain but is a diffuse network of cell bodies and nerve tracts that extends through the brain stem and into the thalamus. The name *reticular* ("little net") alludes to the formation's role in screening sensory information so that only certain impulses reach the cerebrum. If the reticular formation did not do so, the senses would be overwhelmed—you would be acutely aware of every cough, sneeze, and rustle of paper from those around you, every scent, every movement, in the classroom; you would even be aware of the touch of your clothing against your skin.

The reticular formation is also called the reticular activating system because it is important in overall activation and arousal. When certain neurons within the reticular formation are active, you are awake. When other neurons inhibit them, you sleep. The thalamus is a gateway between the reticular formation and regions of the cerebrum. On a cellular level, sleep is a state of synchrony, when many neurons in all of these regions are inhibited from firing action potentials at the same time.

The electrical activity of the brain—the brain waves detectable in an electroencephalogram (EEG) tracing—differs during wakefulness and sleep. We experience two types of sleep—REM sleep, named for the rapid eye movements that occur under closed lids, and nonREM sleep. During REM sleep, the nervous system is quite active, although movement is suppressed. The autonomic nervous system is alert, and heart rate and respiration are elevated. Brain temperature rises as its blood flow and oxygen consumption increase. Bursts of action potentials zip along certain brain pathways more frequently than during nonREM sleep or even during wakefulness. REM sleep is a time of vivid dreaming. NonREM sleep proceeds through four continuous stages, from light sleep to deep sleep, and the corresponding EEG wave tracings become less frequent but higher in amplitude (size).

Sleep follows a predictable pattern—70 to 90 minutes of nonREM followed by 5 to 15 minutes of REM, repeated many times a night. If a person is deprived of REM sleep and therefore doesn't dream, the next night's pattern compensates to deliver more REM sleep.

We know little about sleep, but we do know that it is essential. Prions, the infectious proteins discussed in chapter 19, can cause a condition called fatal insomnia. It begins with inability to fall asleep. As the months progress, the person suffers hallucinations and delusions. After about a year, he or she cannot distinguish reality from the dream state. Death soon follows

The Cerebrum Is the "Mind"

The human brain has a large, highly developed cerebrum that controls the qualities of "mind," including intelligence, learning, perception, and emotion. The cerebrum is divided into two **cerebral hemispheres,** which make up 80% of total human brain volume.

The outer layer of the cerebrum, the **cerebral cortex,** consists of gray matter that integrates incoming information. In mammals, the cerebral cortex increases in size and complexity as species become more complex. Through evolutionary time, the cortex expanded so rapidly that it folded back on itself to fit into the skull, forming characteristic convolutions.

The cerebral cortex contains sensory, motor, and association areas (fig. 31.17). Sensory areas receive and interpret messages

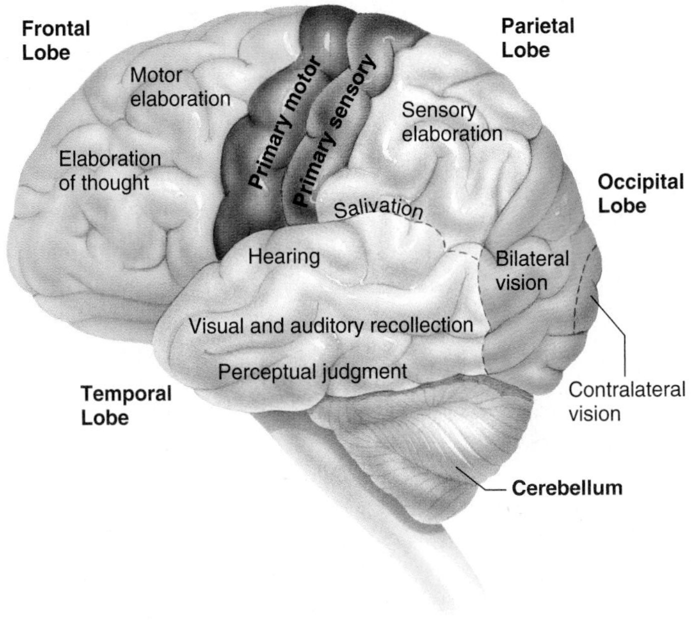

Frontal Lobe
Motor elaboration
Primary motor
Primary sensory
Parietal Lobe
Sensory elaboration
Elaboration of thought
Salivation
Occipital Lobe
Hearing
Bilateral vision
Visual and auditory recollection
Perceptual judgment
Temporal Lobe
Contralateral vision
Cerebellum

FIGURE 31.17

Cerebral Specialization.

The cerebrum contains four main sections visible from the surface—the frontal, temporal, occipital, and parietal lobes. Experiments have shown that different parts of the cerebrum carry out different specific functions. Each hemisphere of the cerebral cortex contains a band of tissue, the primary sensory area, which receives sensory input from the muscles, joints, bones, and skin. Each also has a band of tissue called the primary motor area that controls voluntary muscles. Association areas are located in the front of the frontal lobe and in parts of the occipital, parietal, and temporal lobes.

from sense organs about temperature, body movement, pain, touch, taste, smell, sight, and sound. Motor areas send impulses to skeletal muscles. Association areas do not appear to be either sensory or motor, but they are the seats of learning and creativity.

A band of cerebral cortex extending from ear to ear across the top of the head, called the primary motor cortex, controls voluntary muscles. Just behind it is the primary sensory cortex, which receives sensory input from the skin. Nearly every part of the body is represented in both the sensory cortex and motor cortex. The surface area devoted to a particular body part is proportional to the degree of sensitivity and motor activity in that area. For example, the hands, tongue, and face are extensively represented in both the sensory and motor regions of the cortex (fig. 31.18).

Each cerebral hemisphere has some specific functions, yet the hemispheres work together and are interconnected by a thick band of nerve fibers called the **corpus callosum.** In most people, parts of the left hemisphere are associated with speech, linguistic skills, mathematical ability, and reasoning, while the right hemisphere specializes in spatial, intuitive, musical, and artistic abilities. Under normal conditions, both cerebral hemispheres gather and process information simultaneously.

Memory

Why is it that you cannot remember facts your biology professor mentioned an hour ago, but you can easily recite lyrics to a song you haven't heard in years? Memory is closely associated with learning. Whereas learning is the acquisition of new knowledge, memory is the retention of learning with the ability to access it at a later time. Although we usually think of learning and memory as human abilities, we all know of pets who recognize people. Even an animal as simple as a sea snail can remember.

Researchers have recognized two types of memory, short-term and long-term, for many years, but they now think that the two types differ in characteristics other than duration. Research today focuses on how neurons in different parts of the brain encode memories and how short-term memories are converted to long-term memories.

FIGURE 31.18
Sensory and Motor Cortex.
Different body parts are innervated to different degrees. The amount of surface area devoted to a particular body part depends not upon the size of the part but upon how precisely the nervous system must control it. A distorted human figure superimposed on this diagram of the primary sensory and motor areas indicates the areas that are finely innervated, such as the face and hands.

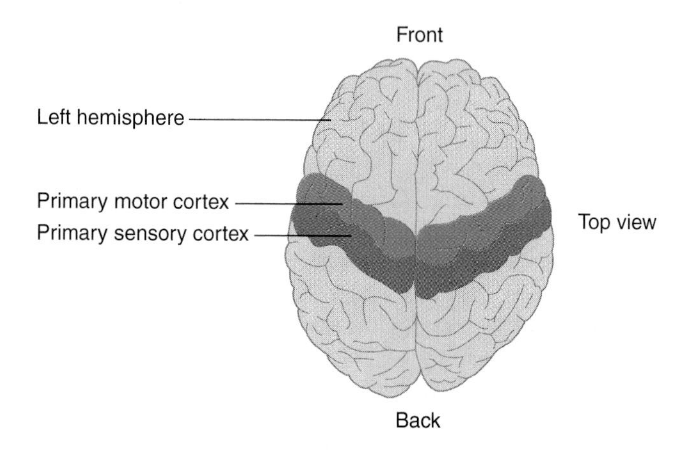

Primary motor cortex
Fingers
Little
Ring
Middle
Index
Thumb
Neck
Brow
Eyeball and eyelid
Face
Lips
Vocalization
Jaw
Swallowing
Tongue
Mastication
Salivation
Hand
Wrist
Elbow
Shoulder
Trunk
Hip
Knee
Ankle
Toes
Left hemisphere

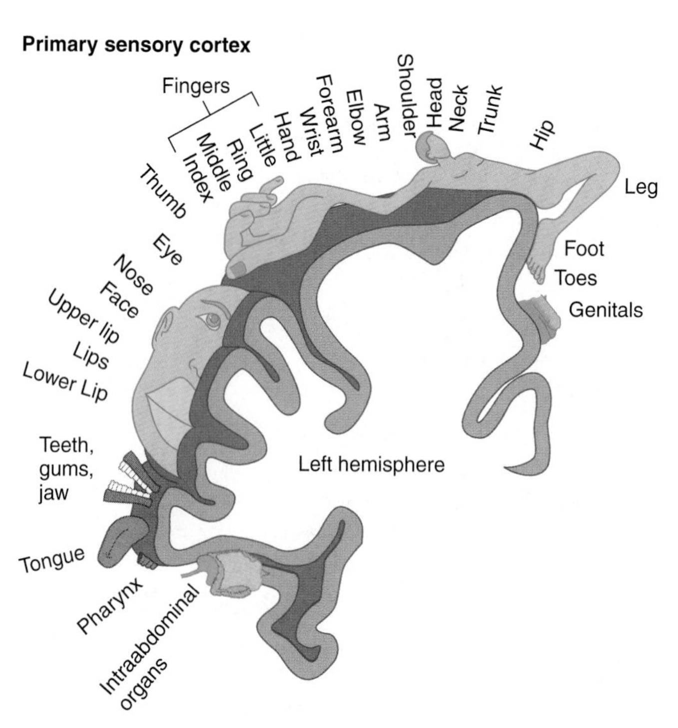

Short-term memories are thought to be electrical in nature. Neurons may connect in a circuit so that the last in the series restimulates the first. As long as the pattern of stimulation continues, you remember the thought. When the reverberation ceases, so does the memory—unless it enters long-term memory.

Long-term memory probably changes neuron structure or function in a way that enhances synaptic transmission. A long-term memory may be a certain synaptic connection pattern. Such synaptic patterns fulfill two requirements of long-term memory. First, there are enough synapses in a human's brain to encode an almost limitless number of memories. Each of the 10 billion neurons in the cortex can make tens of thousands of synaptic connections to dendrites of other neurons, forming 60 trillion synapses. Second, a certain pattern of synapses can persist for years.

According to a theory called long-term synaptic potentiation, frequent stimulation of the neurons in an area of the cerebral cortex called the **hippocampus** strengthens their synaptic connections. This strengthening could occur because the action potentials triggered in postsynaptic cells in response to the repeated stimuli undergo greater depolarization. For example, certain neurons are consistently stimulated when a 2-year-old first learns what an orange is. The child's senses tell the brain that the object is round, orange, and rough in texture and that it smells and tastes sweet. With repetition, the stimulation of the same sets of neurons communicating different aspects of "orange" become, somehow, associated into a single memory. Soon, just the smell of an orange or a drawing of one summons up a mental image—the memory—of the real thing. Glutamate is the neurotransmitter involved in formation of long-term memories.

31.5 MASTERING CONCEPTS

1. Describe the functions of the neurons that form a reflex arc.
2. What are the major structures in the hindbrain, midbrain, and forebrain?
3. How do REM and nonREM sleep differ?
4. Distinguish between the functions of sensory and motor areas of the cerebral cortex.
5. How might neurons encode memories?

31.6 The Human Peripheral Nervous System

The somatic nervous system stimulates voluntary muscles, whereas the autonomic nervous system stimulates smooth muscle, cardiac muscle, and gland secretion, which are involuntary responses. The sympathetic nervous system dominates under stress; the parasympathetic nervous system dominates during rest.

The PNS, which consists of nerve cells outside the CNS, is divided into sensory pathways and motor pathways. The motor pathways, in turn, are divided into the somatic (voluntary) nervous system and the autonomic (involuntary) nervous system.

The Somatic Nervous System

The nerves of the somatic nervous system send impulses to the muscles, sense organs, and glands associated with the body's surface. They also conduct action potentials to the CNS from muscles and glands.

Somatic nerves are classified by their site of origin. Twelve pairs of cranial nerves arise from the brain or brain stem. Thirty-one pairs of spinal nerves exit the spinal cord and emerge between the vertebrae. Eleven of the cranial nerve pairs innervate portions of the head or neck, while the exception, the vagus nerve, leads to internal organs. Each pair of spinal nerves innervates a section of the body near its point of departure from the spinal cord. Loss of feeling in a particular patch of skin can indicate damage to the spinal nerves that lead to that area.

The Autonomic Nervous System

The autonomic nervous system enables internal organs to function properly without our conscious awareness by receiving sensory information and transmitting impulses to smooth muscle, cardiac muscle, and glands. It is subdivided into the sympathetic and parasympathetic nervous systems. When an individual is under stress, the sympathetic nervous system dominates; during more relaxed times, the parasympathetic nervous system is in charge. Although both the sympathetic and parasympathetic nervous systems often innervate the same organ, they usually have opposite effects on it (fig. 31.19). For example, the parasympathetic nervous system stimulates salivation, intestinal activity, and pupil constriction. In contrast, the sympathetic nervous system inhibits salivation and intestinal activity and dilates the pupils.

The sympathetic nervous system prepares the body to face emergencies—it accelerates heart rate and breathing rate; shunts blood to where it is required most, such as the heart, brain, and the skeletal muscles necessary for "fight or flight" (as opposed to "rest and repose"); and dilates airways, easing gas exchange.

The parasympathetic nervous system is activated when the body returns to rest. After an emergency, heart rate and respiration slow and digestion resumes. Different organs recover at different rates, because the organization of the parasympathetic nerves permits independent organ control.

The activities of the autonomic nervous system are involuntary, but we can control them to an extent, if we are aware of them. Biofeedback is a practice that uses a mechanical device to detect and amplify an autonomically controlled body function, such as blood pressure or heart rate. Once made aware of an unhealthy measurement, a person can consciously try to bring the body function within a normal range.

31.6 MASTERING CONCEPTS

1. What are the pathways of the peripheral nervous system?
2. Which division of the autonomic nervous system dominates during stress, and which division dominates when the situation returns to normal?

OLC

FIGURE 31.19

The Autonomic Nervous System.

Neurons from the parasympathetic and sympathetic divisions of the autonomic nervous system innervate the internal organs. Note that in the parasympathetic division, the nerves emerging from the central nervous system are long and those near the organ are short. The reverse is true in the sympathetic nervous system. Each system has an opposite effect on the same organs.

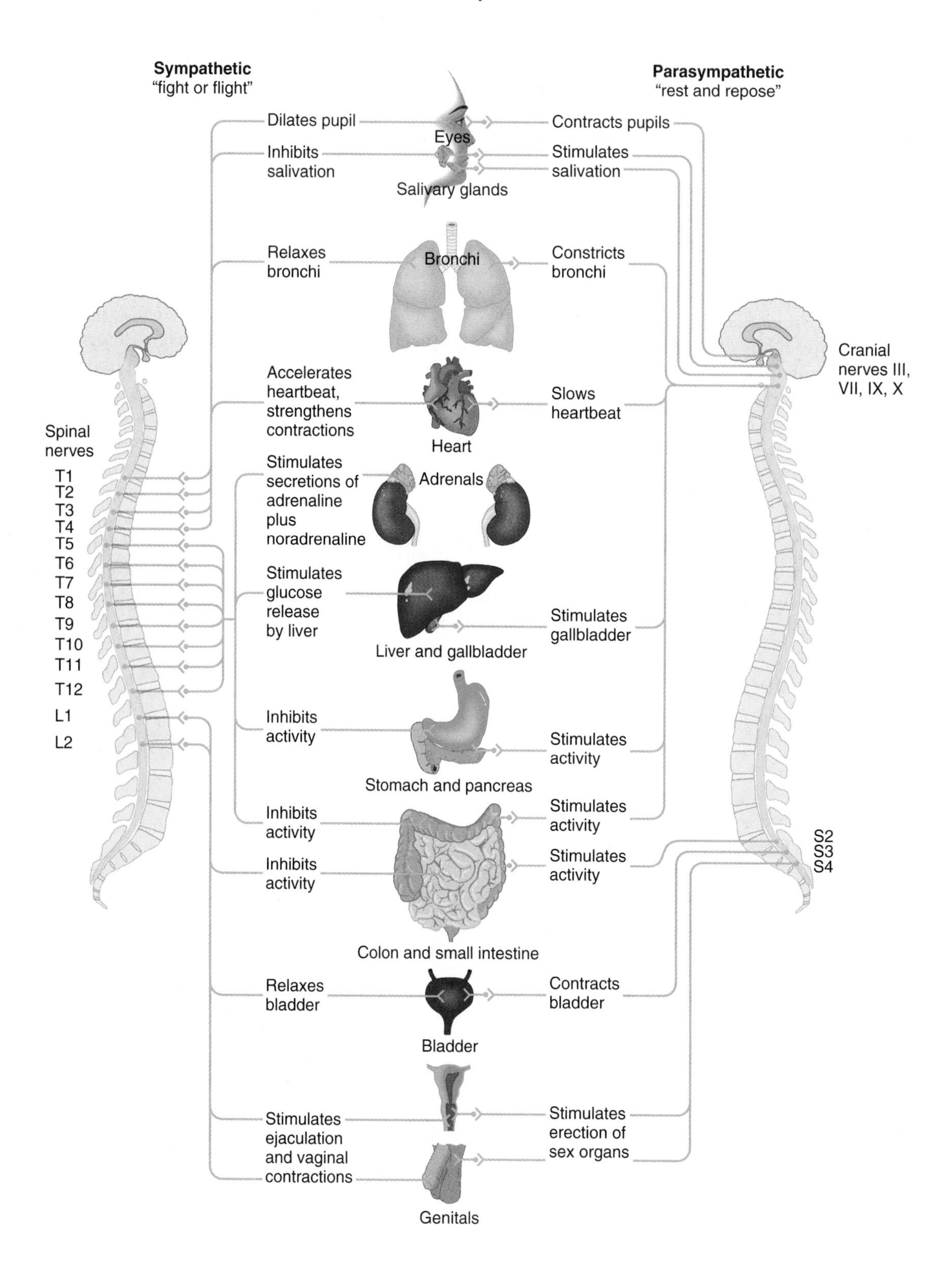

31.7 Protection of the Nervous System

Because it coordinates all organ systems, the nervous system must be protected from disease and injury. Bones, cushioning fluid and membranes, selective entry into the brain, redundancy, and limited regeneration accomplish this.

Because the nervous system interacts so closely with other organ systems, injuries or illnesses can be devastating. Many adaptations protect the nervous system or enable the body to survive damage to it.

Several structures protect the CNS. First, bones of the skull and vertebral column shield the delicate nervous tissue from bumps and blows. Trilayered membranes called **meninges** jacket the CNS for further protection. The **blood-brain barrier,** formed by specialized brain capillaries, helps protect the brain from extreme chemical fluctuations. The tilelike cells that form these capillaries fit so tightly together that only certain chemicals can cross into the **cerebrospinal fluid** that bathes and cushions the brain and spinal cord. This fluid, made by cells that line ventricles (spaces) in the brain, further insulates the CNS from injury.

Despite these protections, disease or injury can harm the nervous system. Many nervous system problems are difficult to treat because mature neurons usually cannot divide to repair damaged tissue. In addition, the blood-brain barrier blocks many drugs used to combat CNS infections. How, then, can the nervous system compensate for damage?

Neurons of the PNS can regenerate. Schwann cells at the tip of a damaged peripheral nerve cell extend and provide a pathway for new growth. A crushed peripheral nerve can grow back, reforming its synaptic connections.

In the CNS, healing is minimal. Damaged unmyelinated neurons can sprout new axons, but these do not extend as far as the old ones. Crushed or damaged neurons in a spinal tract are replaced with a scar consisting of neuroglial cells, connective tissue, and random axons, rather than with a new set of regrown, connected axons. However, pockets of neural stem cells line some areas of the brain's ventricles, which may open up new treatment possibilities for those with neurodegenerative conditions such as Alzheimer disease.

Surviving neurons can help maintain overall functions when some neurons die. Undamaged neighboring neurons extend new terminals to the cells that previously synapsed with the destroyed cell, much as office workers assume the duties of a departed coworker.

The brain, like other organs, loses efficiency over time. In the aging brain, some synaptic connections deteriorate, some postsynaptic receptors become less sensitive to neurotransmitters, and levels of some neurotransmitters fall. By age 60, the human brain decreases in weight by about 3 ounces (85 grams). Although an older person's memory may be less reliable and learning more difficult, he or she can continue to be mentally alert and enjoy the sensations, feelings, and perceptions the nervous system makes possible for many years.

31.7 MASTERING CONCEPTS

1. List three ways that the nervous system is protected from damage.
2. To what extent can the nervous system regenerate?
3. How does the brain change with age?

OLC

Chapter Summary

31.1 The Nervous System Maintains Homeostasis

1. Nervous systems maintain homeostasis by enabling animals to respond to environmental changes.
2. A nervous system receives information, integrates it, and may respond to it, often with movement.

31.2 Neurons Are Functional Units of a Nervous System

3. A neuron has a **cell body; dendrites,** which receive impulses and transmit them toward the cell body; and an **axon,** which conducts impulses away from the cell body.
4. A **sensory neuron** carries information toward the brain and spinal cord (**central nervous system,** or CNS). A **motor neuron** carries information from the CNS and stimulates an effector (a muscle or gland). An **interneuron** conducts information between two neurons and coordinates responses.
5. A neural impulse, measured as an **action potential,** is an electrical change. In a neuron at rest, K^+ concentration is 30 times greater inside the cell than outside, and the Na^+ concentration is 10 times greater outside than inside. The resulting diffusional gradients combined with negatively charged proteins within the cell give the interior a negative charge. When an action potential begins, membrane channels open and allow Na^+ in, creating a positive charge inside. The positively charged Na^+ depolarizes the membrane. At the peak of depolarization, Na^+ channels close. Repolarization occurs as K^+ leaves the cell, restoring the negatively charged **resting potential.** The action potential spreads along the nerve fiber.
6. Myelination increases the speed of neural impulse transmission. A **myelin sheath** forms around some nerve fibers as fatty **Schwann cells** wrap around the fiber. The gaps between these insulating cells are **nodes of Ranvier.** In a myelinated fiber, the neural impulse "jumps" from one node to the next, an action called **saltatory conduction.**

31.3 Neurotransmitters Pass the Message from Cell to Cell

7. The several types of **neurotransmitters,** which are chemicals that transmit a neural impulse, include modified and unmodified amino acids. **Neuromodulators** alter a neuron's response to a neurotransmitter. Neurotransmitters and neuromodulators interact in complex ways.

8. An action potential reaching the end of an axon causes vesicles in the **presynaptic neuron** to approach the cell membrane and release neurotransmitters into the **synaptic cleft.** These chemicals diffuse across the cleft and bind to receptors on the **postsynaptic neuron.**
9. Used neurotransmitter is enzymatically destroyed or reabsorbed into the presynaptic cell.
10. If the neurotransmitter is excitatory, it slightly depolarizes the postsynaptic membrane, making an action potential more probable. An inhibitory neurotransmitter hyperpolarizes the membrane, making an action potential less likely. **Synaptic integration** sums excitatory and inhibitory messages, finely controlling neuron activity.

31.4 Evolutionary Trends In Nervous Systems

11. Nervous systems are groups of interacting cells that help coordinate the activities of animals. The form of a nervous system is adapted to a species' environment.
12. The simplest invertebrate nervous systems are **nerve nets,** which detect stimuli from any direction. Flatworms have simple brains, paired sense organs, and two nerve cord ladders that allow localized motor control. Annelids have a more elaborate brain and ladder organization, with acute senses. Arthropods have even more complex sensory organs and behaviors.
13. The vertebrate nervous system includes the CNS and the **peripheral nervous system** (PNS), which includes all neural tissue outside of the CNS.

31.5 The Human Central Nervous System

14. The human spinal cord is a tube of neural tissue encased in the vertebral canal. The **white matter** on the periphery of the cord conducts impulses to and from the brain. The spinal cord is a reflex center. A reflex is a quick, automatic, protective response that travels through a **reflex arc.** A reflex arc usually consists of a sensory receptor, a sensory neuron, a spinal interneuron, a motor neuron, and an effector, such as a muscle or gland.
15. The brain has three regions. The **hindbrain** includes the **medulla oblongata,** which controls many vital functions; the **cerebellum,** which coordinates unconscious movements; and the **pons,** which bridges the medulla and higher brain regions and connects the cerebellum to the cerebrum. The **midbrain** processes visual and auditory sensory information; the **tectum** integrates this information. The **forebrain** consists of the telencephalon, which includes the **cerebrum** and the diencephalon, which contains the **thalamus,** a relay station between lower and higher brain regions; the **hypothalamus,** which regulates vital physiological processes and regulates levels of some pituitary hormones; and the pineal and pituitary glands. The **reticular formation** filters sensory input and is important for arousal.
16. The cerebrum has an inner layer of white matter and an outer layer of convoluted **gray matter,** which comprises the **cerebral cortex,** where information is processed and integrated. The cerebrum's two **hemispheres** receive sensory input from and direct motor responses to the opposite side of the body. The left hemisphere specializes in language and analytical reasoning, and the right hemisphere regulates spatial, intuitive, and creative abilities.
17. Short-term memory may depend on temporal electrical activity in neuronal circuits. Short-term memories may consolidate into long-term memories, which depend on permanent chemical or structural changes in neurons. Memories form in the **hippocampus.**

31.6 The Human Peripheral Nervous System

18. The PNS consists of the **somatic** (voluntary) **nervous system** and the **autonomic** (involuntary) **nervous system.** The somatic nervous system includes cranial and spinal nerves that transmit sensations from sensory receptors or stimulation to voluntary muscles. The autonomic nervous system receives sensory information and conveys impulses to smooth muscle, cardiac muscle, and glands. Within the autonomic nervous system, the **sympathetic nervous system** controls physical responses to stressful events, while the **parasympathetic nervous system** dominates during rest.

31.7 Protection of the Nervous System

19. Bones of the skull and vertebrae, **cerebrospinal fluid,** the **blood-brain barrier,** and **meninges** protect the CNS. PNS neurons can regenerate. The CNS has stem cells. Undamaged neurons can sometimes take over some functions of damaged neurons.

Testing Your Knowledge

1. In the illustration below, label:
   - **a.** dendrites
   - **b.** axons
   - **c.** cell bodies
   - **d.** a Schwann cell
   - **e.** a myelin sheath
   - **f.** a synapse

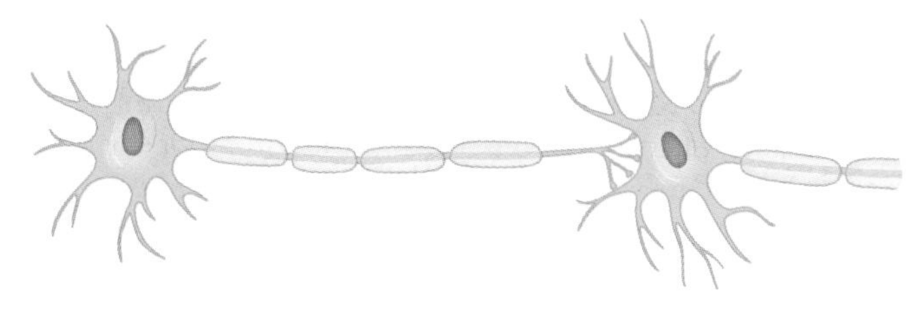

2. How do the functions of sensory, motor, and interneurons differ?
3. Describe the distribution of charges in the membrane of a resting neuron.
4. What ionic events cause the wave of depolarization and repolarization constituting an action potential?
5. How does myelin alter conduction of an action potential along a nerve fiber?
6. In what part of the brain do the qualities of "mind" lie?
7. How does the structure of the cerebral cortex provide different degrees of sensitivity in different body parts?

Thinking Scientifically

1. If a student gets only 3 hours of sleep for several days, how might her sleeping pattern differ from normal on her next full night of sleep?
2. All human brains are about the same size, contain the same major structures, and function in similar ways. How, then, does each of us develop a distinct personality?
3. How might defects in cell adhesion and signal transduction affect neural impulse transmission?
4. In myasthenia gravis, progressive muscle weakness results from destruction of a neurotransmitter at the junctions between neurons and muscle cells. Which neurotransmitter is it?
5. What symptom might result from an overdose of an SSRI drug? (Don't try this!)
6. In 1999, British researchers examined the brain of Albert Einstein, which had been hidden for decades. Although the brain was of normal size, the researchers identified a part of the parietal lobe that was about 15% wider than normal. This area controls mathematical reasoning, imagery, and the ability to visualize objects in space.

A particular groove in the area appeared much reduced, leading the researchers to speculate that this might have allowed more synaptic connections to form than normal. What additional information would help to determine whether Einstein's brain distinctions could have accounted for his genius?

References and Resources

Fox, Cynthia. September 18, 2000. Making neurons. *The Scientist,* vol. 14, p. 1. Neural stem cell research is exploding.

Kulig, Ken. June 4, 1998. A tragic reminder about organic mercury. *The New England Journal of Medicine,* vol. 339, p. 1692. The sad case of Karen Wetterhahn illustrated how mercury destroys the nervous system.

Lemonick, Michael D. June 28, 1999. Was Einstein's brain built for brilliance? *Time.* Einstein's brain has some differences from normal that may have accounted for his exceptional abilities in mathematics.

Sapolsky, Robert M. May 28, 1998. The stress of Gulf War syndrome. *Nature,* vol. 393, p. 308. Neuropsychological symptoms of this condition may reflect altered activity of acetylcholine in brain synapses.

Seger, Idan. May 21, 1998. Sound grounds for computing dendrites. *Nature,* vol. 393, p. 207. Dendrites form treelike patterns, with specific shapes characteristic of certain brain regions.

Tsien, Joe Z. April 2000. Building a brainier mouse. *Scientific American,* vol. 282, p. 62. A genetically boosted strain of mice is revealing the molecular basis of learning and memory.

The *LIFE* Online Learning Center provides additional resources and tools for studying this chapter.
www.mhhe.com/life

CHAPTER 32

The Senses

Diverse Ways of Sensing the World

Most animals sense the environment with specialized cells that receive, transduce, and relay information on to other neurons in the course of producing an appropriate response. Species exhibit many variations on this theme.

Fire Beetles Use Thermoreceptors

The black, half-inch-long beetles of subgenus *Melanophila* mate on conifers smoldering from forest fires and lay their eggs beneath the scorched bark. Larvae live in the tree for a year, eating dead insects. Then they leave and pupate.

The beetles use pit organs, located near their middle legs, to detect the distinctive infrared electromagnetic radiation a forest fire emits. The pits house clusters of receptor cells that are exposed only during flight. The atmosphere doesn't absorb the range of wavelengths from the fire, enabling the beetles to detect them from many miles away.

Each sensory neuron in a fire beetle's pit organ lies within a sphere formed from inner layers of the cuticle, with a dendrite sticking out of the domed top. Only infrared wavelengths trigger action potentials in the pit organ sensory neurons, but they don't do so directly. The radiation causes the bonds of the cuticle to vibrate, deforming the dome enough to bend the dendrite, which fires an action potential in response. The alerted beetle uses this information to orient toward the fire.

Blowflies Have a Gyroscopic Sense

A gyroscope is used in navigation to keep a ship on course. The blowfly *Calliphora vicina* has a gyroscope of sorts in its halteres, small structures behind the wings. The halteres sense the animal's position and rotation in space, providing feedback to the nervous system that enables the fly to keep on track, avoid obstacles, and resist turbulence.

Each haltere consists of a base, attached to the cuticle, that houses five clusters of neurons specialized as strain receptors. From the base an extension leads to a club-shaped expansion at the tip. Turbulence deforms the clubs, which conduct this mechanical energy to the strain receptors. The ventral cord integrates input from the halteres and directs the animal to fly in the right direction. Eleven small muscles at the base of the halteres move them counter to the wings, which beat 150 times per second. Two of the muscles synapse with neurons from the visual system, so that what a fly sees determines where it goes.

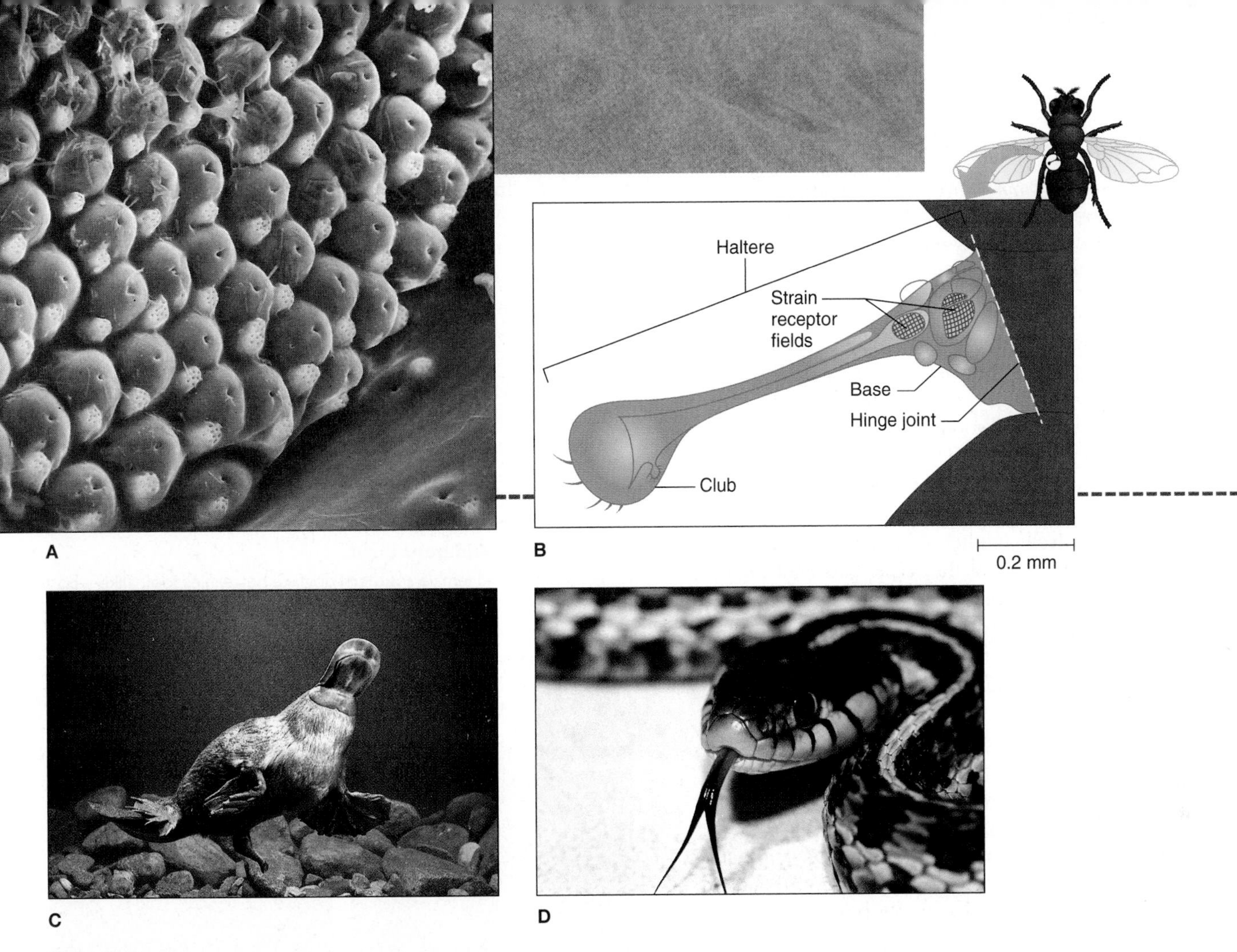

Eclectic Senses.
(**A**) Fire beetles sense infrared radiation that a forest fire gives off using domed structures located in pit organs near their middle legs. The radiation, transduced to heat, deforms the domes in a way that presses on neurons within them, triggering action potentials. The beetles lay their eggs in bark of burned conifers. (**B**) Blowflies use their halteres to sense air movements, and integrate this information to orient themselves in space. (**C**) The platypus uses its ears and eyes when its head is above water, but uses touch and electroreception to find food on the rocky river bottom. (**D**) A snake's forked tongue detects chemical gradients, enabling the animal to follow a scent trail.

Platypus Electroreception

A platypus eats half its body weight daily as it navigates along the river bottoms of Australia, nudging rocks aside with its bill to find crayfish, worms, insect larvae, frogs, and fish. But the bill doesn't sense its targets by touch alone—it can also detect the weak electrical fields that prey emit in water as a result of their muscle activity. The sensors in the platypus bill are specialized neurons called electroreceptors.

These electroreceptors were unknown until Australian researchers placed hungry platypuses in pools with live and dead batteries. The animals explored only the live batteries, suggesting that they are attracted to electricity. Next, the researchers recorded platypuses' brain waves while stimulating their bills electrically and detected a response. A close look at the bills revealed threadlike ends of nerve fibers deep within the pits that form mucous glands. These were the electroreceptors. The tiniest flick of the smallest shrimp's tail can stimulate the electroreceptors to trigger 20 to 50 action potentials per second!

The Snake's Forked Tongue

The forked tongue of the serpent is a metaphor for deceit. However, snakes and some other reptiles actually use their tongues to detect scents. A forked tongue can detect a scent that diminishes or strengthens over a distance by sensing at two points simultaneously. The forks bring odor molecules into the mouth, where they pass through openings in the palate and contact sensory cells that detect chemicals. These cells reside in the snout in structures called vomeronasal organs. They transmit action potentials to a center in the brain that compares the dual sensory input and interprets whether the trail is weakening or strengthening. The snake integrates this information along with the nature of the chemical stimulus to determine whether it is approaching a potential mate or a meal.

32.1 Sensory Systems Are Adaptive

Perhaps no part of an animal's body illustrates natural selection as vividly as do the sense organs. Sensory systems are well suited to a particular species' surroundings, enabling an animal to receive important environmental cues and to respond to them in a way that promotes survival.

Every animal inhabits a world of its own whose characteristics are determined by the animal's sense organs. The sensory abilities of animals are tremendously diverse. Consider a person and a dog walking through the New England woods in the fall. The human's sensations are primarily visual, but the dog detects odors that paint a picture human eyes miss—fox urine left as a calling card near an oak and the scent in a shrub revealing a deer's bed.

If this is an evening stroll, another animal, the little brown bat, experiences yet another view of the scene. Bats emit high-frequency pulses of sound and analyze the resulting echoes to picture their surroundings in a way that we cannot, because our ears cannot detect such high-frequency sounds. Although a bat emits sound pulses that are about as intense as a passing train, to our ears these are silent screams.

Animal species have diverse sensory systems to sense the range of stimuli most critical for survival and reproduction. Cells in a frog's eyes respond selectively to small moving objects, which help the animal to locate food. Male and female butterflies that look alike to us appear quite different to one another (fig. 32.1). The ears of female cricket frogs respond to the frequencies in the croaks of males of the species but cannot hear other frequencies.

Sensory receptor cells are sensory neurons or specialized epithelial cells that detect and pass stimulus information to sensory neurons, which in turn pass the information to the central nervous system (CNS). Generally, a stimulus alters the shape of a protein embedded in a sensory receptor's cell membrane in a way that changes the membrane's permeability. If the stimulus is strong enough, then resulting changes in ion movement across the membrane trigger action potentials in the sensory neurons. Figure 32.2 and table 32.1 depict the information flow from environmental stimulus to the brain that underlies sensation and **perception,** which is the animal's interpretation of the sensation.

Animals live in virtually all habitats on Earth, and are active at all times of the day and night. In all of these habitats and conditions, animals must maintain homeostasis. This requires that they respond to changes in their external and internal environments. Natural selection has molded sensory receptors in every group of animals in ways that reflect their environments, the time of day when they are active, the foods they eat, their reproductive mechanisms, and other characteristics. The link between sensory systems and natural selection is best seen in systems that assist the reproduction of a species. For example, female silk moths (*Bombyx*

FIGURE 32.1

Butterflies in a Different Light.

To us, the male and female southern dogface butterflies look alike—both are yellow and black (**A**). The insects, however, can see in the ultraviolet range of light; to them, only the male has reflective patches on his wings (**B**). This marking is important in courtship behavior. However, this photograph is still a human visual perception of a butterfly under UV illumination. A butterfly probably sees much less precise outlines.

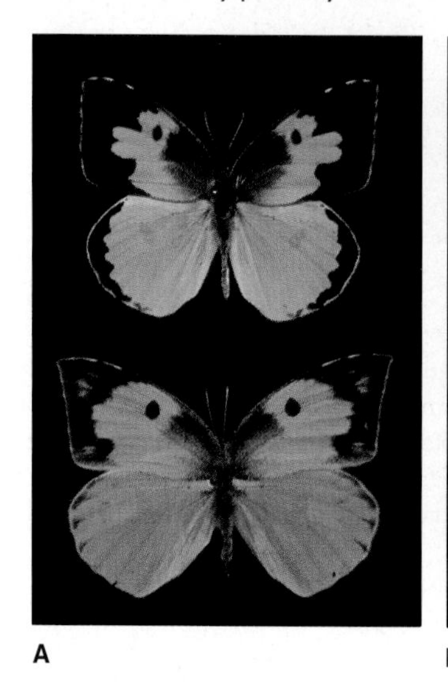

A

B

Sensory receptor cell
Specialized neuron or epithelial cell responds to a specific stimulus

Sensory neuron
One or more sensory neurons relay action potentials to the central nervous system. In some cases, sensory receptor cells also serve this function

Central nervous system (brain and spinal cord)

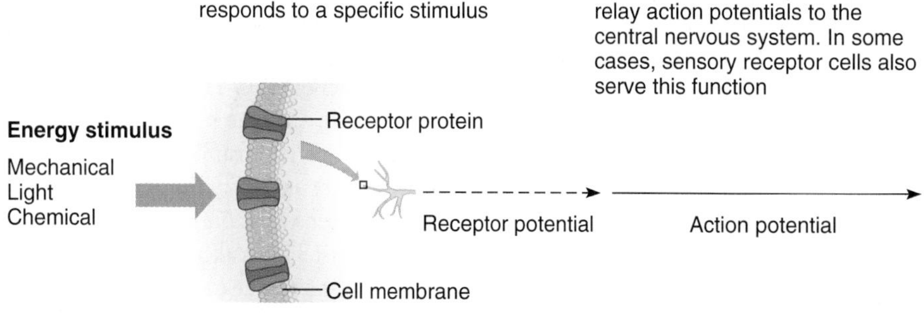

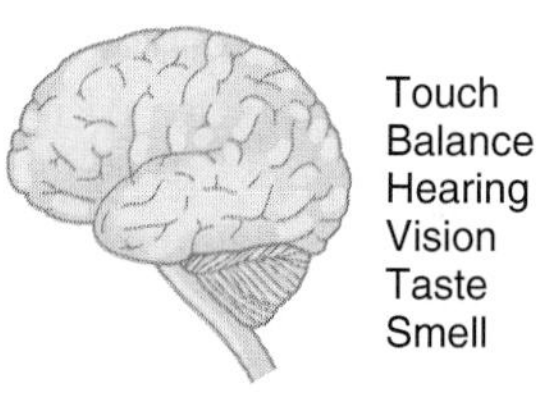

Stimulus contorts receptor protein, which alters membrane permeability. Receptor potential varies with strength of stimulus. An action potential is generated in some receptor cells.

A neurotransmitter released by the receptor cell triggers a receptor potential in the sensory neuron. An action potential is triggered if threshold is reached.

The brain receives sensory input and integrates the information.

FIGURE 32.2

Sensory Information Flow.

This figure depicts a generalized view of information flow in familiar senses.

TABLE 32.1 Experiencing an Apple

SENSORY RECEPTOR ⟶	SENSORY NEURON ⟶	CNS ⟶	SENSATION +	PERCEPTION
Olfactory cells in nose	Olfactory nerve	Cerebral cortex	A sweet smell	Smell of an apple
Taste bud receptor cells	Sensory fibers in taste buds	Cerebral cortex	A sweet taste	Taste of an apple
Rods and cones in retina of eye	Optic nerve	Midbrain and visual cortex	A small, round, red object	Sight of an apple
Hair cells in cochlea of inner ear	Auditory nerve	Midbrain and cortex	A crunching sound	Sound of biting into an apple

mori) release volatile chemical substances called **pheromones** (one is called bombykol) that attract males up to several miles away. Sensory receptors in the male antennae are so sensitive to these odors that a single molecule is enough to trigger an action potential in the cell. Proteins that are part of the receptor detect the pheromone. A single mutation can alter the protein structure and its sensitivity to the odorant molecule, making a male moth unable to sense the pheromone. He would not find the female, not mate, and not pass this particular gene on to the next generation.

32.1 MASTERING CONCEPTS

1. In what ways are sensory systems among animal species the same? How are they different?
2. How has natural selection molded sensory systems?

32.2 Sensory Systems, Although Diverse, Operate by the Same General Principles

Proteins in the cell membranes of sensory receptors bind specific stimuli and contort, changing the membrane permeability and firing an action potential. These action potentials follow specific neural pathways to the brain, where their input is interpreted as a perception. Sensory receptors also amplify incoming stimuli, signal the strength of stimuli, and detect changes in input.

The brain receives sensory input, integrates the information, and interprets it by consulting memories to form a perception. Sensory receptors, then, are portals through which nervous systems experience the world.

Sensory Receptors Detect, Transduce, and Amplify Stimuli

If all of the action potentials that sensory neurons transmit from sensory receptors are alike, how can an animal distinguish sight from sound? The answer is that sensory receptors are selective and they use different neural pathways. They absorb a particular type of energy—light energy, for example—and convert it into the electrochemical energy of an action potential through a process called transduction. The action potentials that a specific type of receptor generates travel to the brain over a specific sensory pathway. Different qualities of sensation, such as light, sound, or pressure, are the brain's interpretation of the input from different pathways. Receptors also amplify input, so that a single photon (packet of light) or a single odor molecule produces an impulse that has inherently more energy than the original stimulus.

In addition to transmitting information about the type of stimulus, sensory receptors send messages about the strength of the stimulus. One way stimulus intensity is signaled to the brain is in neural impulse frequency; the more intense the stimulus, the more action potentials are generated in a given time period. In response to an environmental change, specialized epithelial receptor cells produce **receptor potentials.** A receptor potential is a local change in the electrical potential of a cell membrane that precedes generation of an action potential. Like an action potential, redistribution of ions triggers a receptor potential. Unlike an action potential, however, a receptor potential is not "all-or-none."

The magnitude of a receptor potential varies with the strength of the stimulus. For example, a loud sound causes greater depolarization than a soft noise, and therefore results in a larger receptor potential. The change in the receptor potential, in turn, influences the likelihood that an action potential will be generated. In some receptors, the strength and duration of the receptor potential determine the rate at which action potentials are generated. The brain interprets this as an increase in stimulus intensity. Figure 32.3 shows the generation of receptor potentials and action potentials in a Pacinian corpuscle, which is a touch receptor in the deep layers of mammalian skin.

A second way information about stimulus strength reaches the brain is through variations in the number of sensory neurons carrying the message. As the stimulus strengthens, more receptors are stimulated, and these activate more sensory neurons.

Sensory Adaptation

Many sensory receptors detect changes in input—if a sensation remains constant, it becomes less noticeable. This mechanism, called **sensory adaptation,** keeps the nervous system from becoming overly sensitive. The strong smell of a fish market, for example, may be overpowering to anyone who has just entered, but to the people working there, it is hardly noticeable. Similarly, a tub filled with hot water may seem too hot at first, but it soon becomes tolerable, even pleasant. Without sensory adaptation, we would be distinctly aware of the touch of clothing and every sight and sound. Concentrating on a single stimulus, such as a person speaking, would be difficult. Some sensory receptors, however, are very slow to adapt—and this is also protective. For example, being very aware of continuous pain helps ensure action to remove its source.

Types of Sensory Receptors

Humans have multiple sensory receptors, providing a rich tapestry of sensations. A person's skin contains about 4 million pain receptors, 500,000 pressure receptors, 150,000 receptors specific for cold, and 16,000 specific for heat. Specialized receptors in the joints, tendons, ligaments, and muscles, called proprioceptors, provide information that, when combined with information from sense organs such as the eyes and ears, gives us a feeling of where certain body parts are relative to the rest of the body. Sensory receptors are classified by the type of energy they detect. Thus, receptor cells responsive to chemicals are called **chemoreceptors** and those sensitive to mechanical energy are **mechanoreceptors.** Table 32.2 shows the major types of sensory receptors and the type of energy to which each is most sensitive.

Malfunctioning or absent senses can greatly disrupt our lives. Health 32.1 examines an unusual condition that mixes sensations and perceptions. The remainder of the chapter explores in more detail the senses of chemoreception, photoreception, and mechanoreception.

32.2 MASTERING CONCEPTS

1. What is the source of the specificity of sensory receptors?
2. How do sensory receptors detect differences in the strength of stimuli?
3. What are the major types of sensory receptors?

FIGURE 32.3

Receptor Potentials Trigger Action Potentials.

The skin of mammals contains mechanoreceptors called Pacinian corpuscles that detect touch and pressure. The corpuscle is actually connective tissue wrapped around a sensory neuron. When pressure deforms the connective tissue, the neuron undergoes a local flow of electrical current called a receptor potential. Strong pressure causes stronger receptor potentials, until a threshold is reached that triggers an action potential.

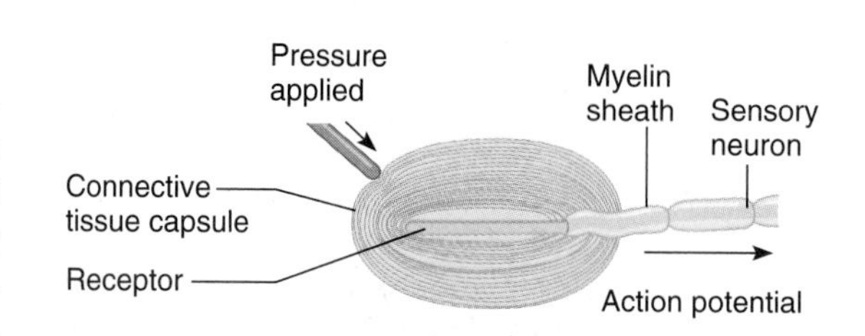

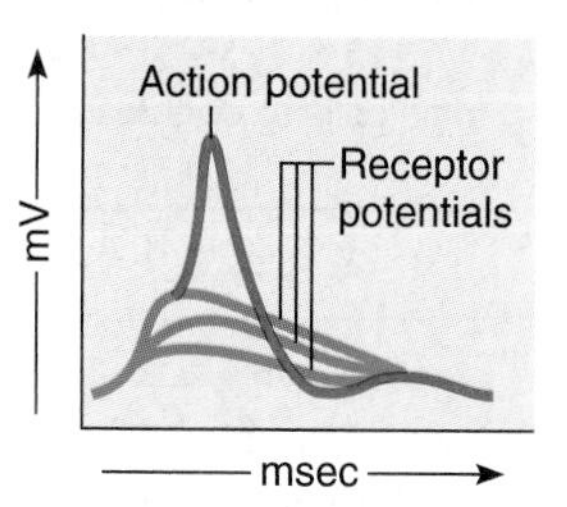

32.3 Chemoreception

The ability to detect and respond to specific chemicals is the most ancient of all of the senses. Invertebrates have chemoreceptors on various parts of their bodies that provide smell and/or taste. Vertebrates have receptor cells in olfactory epithelium that bind odorant molecules and send signals to the brain that are interpreted as smell. Vertebrate taste receptor cells in the tongue bind food molecules.

Chemoreception is the ability to detect specific chemicals in the environment. It is the most ancient sense, even predating the evolution of animals. Prokaryotes and protista use chemoreception when they sense concentrations of external chemicals and use these cues to approach food or move away from danger.

Chemoreception requires that the stimulus molecule be dissolved in an aqueous solution (such as saliva or the moist lining of a nasal passage). In addition, the molecule must interact with a receptor on a cell membrane. Most, if not all, animals can detect and respond to chemicals in their environments, although different animals have receptors tuned to different important chemical stimuli.

TABLE 32.2 Some Types of Sensory Receptors

TYPE OF RECEPTOR	STIMULUS	EXAMPLES
Thermoreceptor	Heat	Fire beetles detect burned conifers in which to lay eggs
Mechanoreceptor	Touch, vibration	Pacinian corpuscles in mammalian skin Lateral line system in bony fishes Hearing in mammals
Chemoreceptor	Airborne molecules	Snake's forked tongue compares odor strength at two points to detect scent trail
	Dissolved molecules	Taste in vertebrates
Photoreceptor	Light	Eyespots in planarian flatworms Light cells in coverings of simple invertebrates Visual systems in many taxa
Pain receptor	Chemicals released from injured cells	Release of endorphins and enkephalins from brain stem of vertebrates
Electroreceptor	Electrical fields	Platypus detects electrical currents generated by prey in water

Health 32.1

Mixed-Up Senses—Synesthesia

"The song was full of glittering orange diamonds."
"The paint smelled blue."
"The sunset was salty."
"The pickle tasted like a rectangle."

To 1 in 500,000 people with a condition called synesthesia, sensation and perception mix, so that the brain perceives a stimulus to one sense as coming from another. Most commonly, letters, numbers, or periods of time evoke specific colors. These associations are involuntary, are very specific, and persist over a lifetime. For example, a person might report that 3 is always mustard yellow, or Thursday brown.

We do not know what causes synesthesia, although it does seem to be inherited and more common in women. One of the authors of this book (R. L.) has it—she has always perceived days of the week and months as specific colors. People have reported the condition to psychologists and physicians for at least 200 years. Synesthesia has been attributed to an immature nervous system that cannot sort out sensory stimuli or altered brain circuitry that routes stimuli to the wrong part of the cerebral cortex.

PET (positron emission tomography) scanning reveals a physical basis to synesthesia. Brain scans of six nonsynesthetes were compared with those of six synesthetes who reported associating words with colors. Cortical blood flow was monitored while a list of words was read aloud to both groups. Interestingly, cortical blood flow was greatly elevated in the synesthetes compared with the nonsynesthetes. Furthermore, while blood flow was increased in word-processing areas for both groups, the scans revealed that areas important in vision and color processing were also lit up in those with synesthesia.

Chemoreception in Invertebrates

Chemoreceptors occur on various parts of invertebrate bodies, depending on where they are likely to encounter biologically important molecules. Chemosensory cells on a flatworm's head detect a potential meal of algae, dead animals, or other live invertebrates. Crayfishes have chemoreceptors on their mouthparts, appendages, and antennae that allow them to taste food. Lobsters and insects taste with receptors on their legs. Scorpions search for food and potential mates by "tasting" the ground surface, using comblike structures (similar to antennae) on the lower surface of their bodies. An octopus samples food with taste receptors on its tentacles.

Arthropods have chemoreceptors that provide the senses of smell and taste. Many insects use chemicals to communicate, by detecting pheromones that others of the same species emit. Pheromones, discussed further in chapter 41, signal such important information as alarm, food availability, readiness to mate, and the locations of sites to lay eggs. An insect's body bears many chemoreceptors that intercept pheromones in the air. Figure 32.4 shows the clustering of these receptors on an insect's antennae. Thousands of hairs on the antennae house sensory neurons, whose dendrites detect pheromones. A male emperor moth can detect one molecule of a female's scent 3 miles away!

A

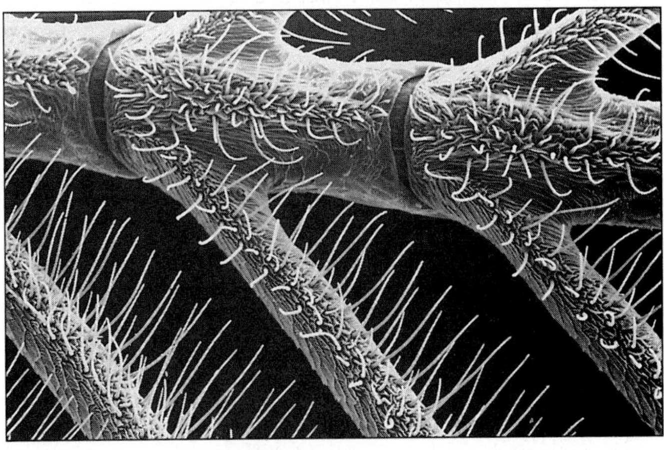

B

FIGURE 32.4
Moth Pheromone Chemoreception.
(**A**) The antennae of the male silkworm moth *Bombyx mori* are large, paired appendages that effectively trap wafting chemicals. (**B**) The 50,000 or so sensory hairs detect pheromones that the female emits. Each sensory hair includes a fluid-filled chamber in which the pheromones dissolve.

Chemoreception in Vertebrates

Vertebrate sensory systems—like those of invertebrates—reflect species-specific adaptations. A shark samples water that continually flows through its paired nostrils on the underside of its snout to smell blood 500 yards away. A catfish tastes with its whiskers. Male red-sided garter snakes, which collect airborne molecules with their tongues, are exquisitely sensitive to pheromones that females secrete through their skins. The emissions from just a few females (or a paper towel spiked with the pheromone) can provoke hundreds of males into a mating frenzy. Bloodhounds are used to trace the scent of a missing person because they can smell the skin cells that specific individuals shed. We now focus on the human senses of taste and smell.

Smell Chemoreceptors in humans, like those of other vertebrates, provide the senses of smell and taste. Taste is intimately related to smell, as anyone whose sense of taste has been dulled by a stuffy nose can attest. For millions of people who cannot smell normally, eating is a tasteless experience. This is because about 75 to 80% of flavor derives from smell.

Smell, or **olfaction,** is the ability of specialized olfactory receptor cells to detect certain molecules, called odorant molecules. These "smell cells" are located in an inch-square patch of tissue high in the nasal cavity called the olfactory epithelium (fig. 32.5). To trigger the sense of smell, an odorant molecule must bind to receptor proteins on cilia in this patch of tissue. Odorant molecules, usually in gaseous form, pass over the olfactory epithelium during breathing and dissolve in the mucous layer covering the cilia and diffuse to the receptors. Many types of molecules participate in olfaction, including about 500 different types of membrane-bound receptor proteins that bind specific odorant molecules, information revealed from human genome project results. In addition, multiple signaling proteins inside the receptor cell translate the chemical signal (binding of the odorant to the receptor protein) into the electrochemical language of the nervous system.

Part of each olfactory receptor cell passes through the skull and synapses with neurons in the brain's olfactory bulb. Different receptor cells likely have different combinations of odorant receptor proteins, which generate specificity in the pattern of activated neurons that stimulate the brain. Olfactory receptor cells with the same receptor proteins are not necessarily near each other, but their axons meet in the olfactory bulb. From here, sensory neurons relay the message to the cerebral cortex, which interprets the message as an odor and identifies it. The brain associates a particular pattern of neurons that activate the same odorant receptor proteins to a particular smell. For example, if there are 10 types of odor receptors, banana smell might stimulate receptors 2, 4, and 7; garlic, receptors 1, 4, and 9.

Humans have about 12 million olfactory receptor cells, each with 10 to 20 cilia that increase the surface area for receiving odorant molecules. We can detect a single molecule of the chemical that gives a green pepper its odor in 3 trillion molecules of air! Other species have even more acute senses of smell. A bloodhound has 4 billion receptor cells, and its olfactory epithelium, if it were

FIGURE 32.5

How Smells Signal the Brain.

(**A**) The powerful sense of smell derives from an inch-square section of sensitive ciliated olfactory receptor cells in the nasal cavity. (**B**) This close-up of the olfactory epithelium shows the cilia that fringe receptor cells. An olfactory receptor cell physically binds an odorant molecule (such as from garlic) at its ciliated end and transmits an action potential to cells in the olfactory bulb. The axons of these neurons form the olfactory tract, which passes the information to the brain for interpretation as a particular smell.

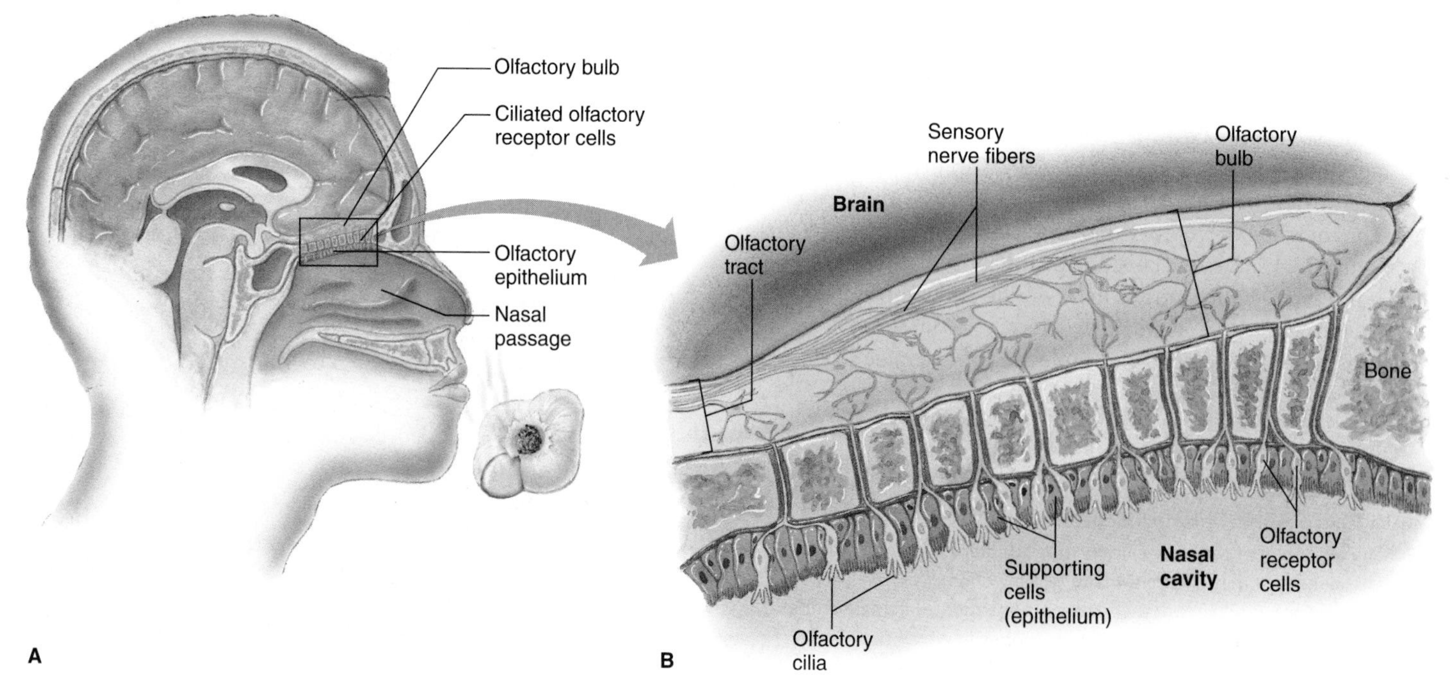

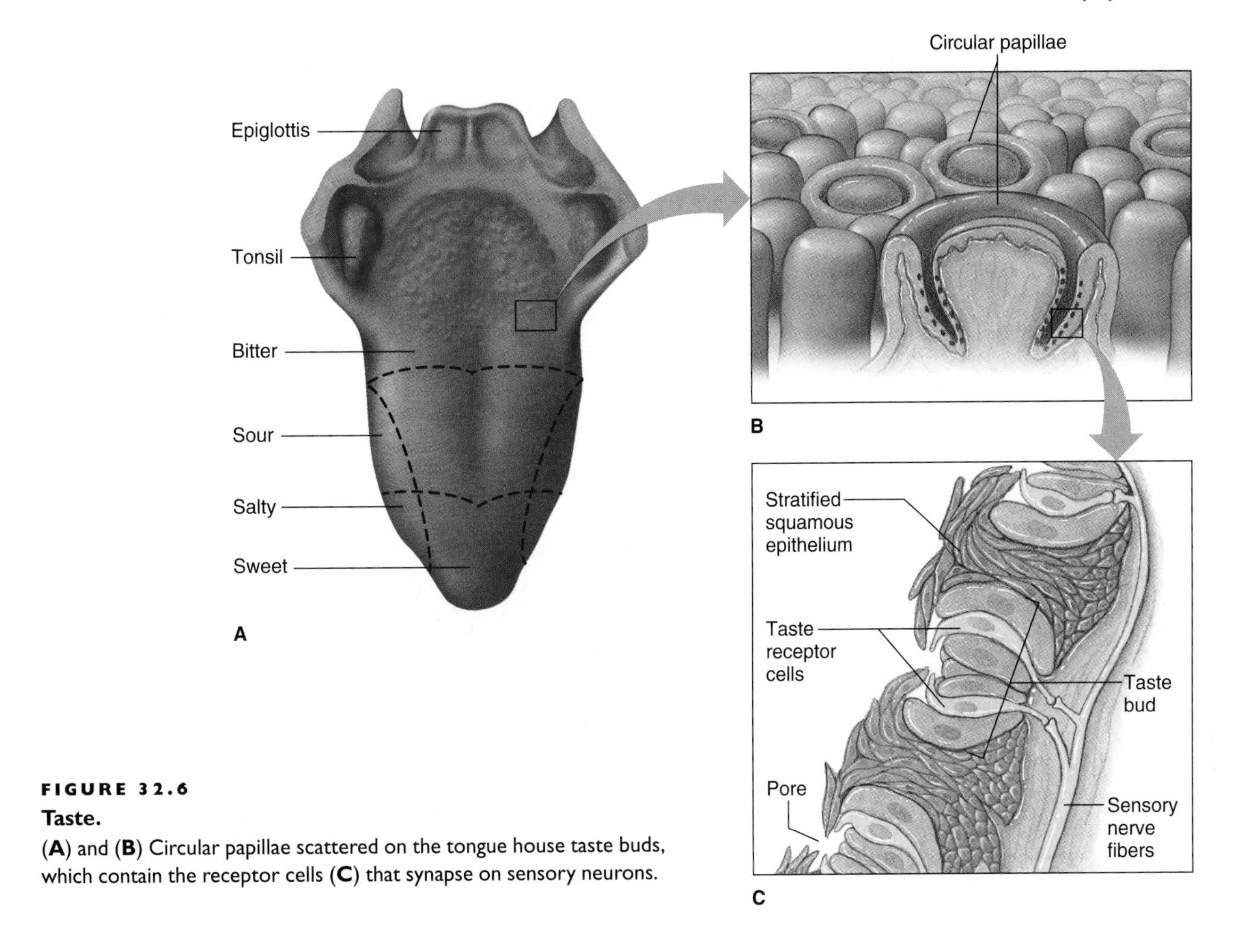

FIGURE 32.6
Taste.
(**A**) and (**B**) Circular papillae scattered on the tongue house taste buds, which contain the receptor cells (**C**) that synapse on sensory neurons.

unfolded, would cover 59 square inches, compared with our 1 to 1 1/2 square inches.

In addition to stimulating the olfactory bulb, sensory information from olfactory receptors also travels to the limbic system, the brain center for memory and emotion. This is why we may become nostalgic over a scent from the past. A whiff of the perfume Grandma used to wear may bring back a flood of memories. Olfactory input to the limbic system explains why odors can alter mood so easily. For example, the scent of freshly cut hay or rain on a summer's morning makes many people feel good.

Taste In humans, taste (**gustation**) is centered primarily in the mouth in clusters of cells called taste buds that resemble the sections of an orange. Humans have about 10,000 taste buds; cattle have up to 25,000. Although taste buds are lightly scattered around all the places that food and drink are likely to touch, including the cheeks, throat, and roof of the mouth, they are concentrated on the tongue in the grooves around certain of the bumps, or papillae, on the tongue's upper surface.

Each human taste bud contains 50 to 150 receptor cells, which generate action potentials when food molecules bind to them. The receptor cells synapse at the base of the taste bud onto sensory neurons that relay information to the medulla of the brain. The chemoreceptors derive from epithelial cells of the tongue and live for only about 3 days before being replaced by new ones.

The human tasting experience is not limited to the four primary taste sensations of sweet, sour, salty, and bitter. Rather, our sense of taste depends upon the pattern of activity across all taste neurons. Most taste buds sense all four of the "primary" tastes, but each responds most strongly to one or two (fig. 32.6). Although individual "tongue maps" vary considerably, we can make some generalizations. Receptors that respond most strongly to sugars concentrate on the tip of the tongue, and receptors at the back are most sensitive to bitter substances. Sour receptors are most common toward the middle edges of the tongue, and a little further forward are those most sensitive to salts.

The sense of taste also reflects what happens to food as animals chew it. Most foods are chemically complex, and so they stimulate different receptors. To track the actual act of tasting, chemists collected samples of air from participants' nostrils as they bit into juicy red tomatoes. An analytical technique called mass spectrometry revealed that chewing activates a sequence of chemical reactions in the tomato as its tissues are torn, releasing first aromatic hydrocarbons, then after a 30-second delay, products of fatty acid breakdown, and, finally, several alcohols. This gradual release of stimulating molecules is why we experience a series of flavors as we savor a food.

Inherited differences account for some of our food preferences. A single gene, for example, determines whether we perceive broccoli as mild or bitter. The same gene may control whether we find diet sodas to taste like the calorie-laden variety, or bitterly poor alternatives.

32.3 MASTERING CONCEPTS

1. What are the functions of chemoreceptors in invertebrates?
2. Which cells and molecules impart our sense of smell?
3. How does a taste bud function?

32.4 Photoreception

Animals detect light with special light-sensitive cells called photoreceptors. Photoreceptors are grouped into several different types of eyes in invertebrates. In the vertebrate eye, light focuses onto the retina, where it stimulates rod photoreceptor cells to provide black and white vision, and cone cells to provide color vision.

Animals detect light with **photoreceptors,** which include a pigment molecule associated with a membrane. When the pigment absorbs light, its structure changes. This alters the charge across the associated membrane, which may generate an action potential.

Vision is interesting from an evolutionary standpoint, because the pairing of photoreceptor cells and pigment molecules appears to be quite ancient, yet the visual abilities of different species are adapted to particular ways of life. Nearly all multicellular species with photoreceptor cells use a pigment molecule called **rhodopsin,** which bends in response to light to provoke a change in nerve activity. Even some unicellular organisms, such as the alga *Chlamydomonas,* use rhodopsin. A photosynthetic prokaryote, *Halobacterium halobium,* uses a rhodopsin variant for photosynthesis.

Photoreception in Invertebrates

Different species use rhodopsin in very different structures that confer varying degrees of visual acuity. Invertebrates have several types of eyes. A turbellarian flatworm's photoreceptor cells are gathered into two cup-shaped eyes. These simple structures enable the flatworm to detect shadows, which is sufficient light sensation for the animal to orient itself in its environment (fig. 32.7). Similarly, a starfish has cells at the tip of each arm that detect differences in light intensity. Some insects have **ocelli,** which are simple, single-lens eyes that can detect the presence of light but are incapable of forming images. Most adult insects have paired **compound eyes** that consist of closely packed photosensitive units called **ommatidia.** Each ommatidium contains a lens that transmits light to its own or nearby photoreceptor cells, generating a tiny view of the world. The animal's nervous system then integrates the many images from each ommatidium. Animals with many ommatidia may see an image as clearly as we do.

Some arthropods have combinations of ocelli and compound eyes, or they have different photoreceptors at different stages of development. A crayfish's compound eyes, located on two movable eyestalks, may contain from 25 to 14,000 ommatidia. Earlier in life, the wormlike crayfish larva has ocelli that enable it to detect light at the water's surface, where its single-celled food lives. Insects have compound eyes of up to 28,000 ommatidia each and may also have up to three ocelli, each consisting of 500 to 1000 receptor cells grouped under a lens. Ocelli may complement visual information from the compound eyes. While compound eyes detect movement, enabling insects to capture moving prey, ocelli detect shadows, which may help insects to escape approaching predators.

The cephalopods (octopuses, squids, and cuttlefishes) have visual systems based on another type of eye, the single-lens eye. An opening in the eye, the **pupil,** admits light, which a lens focuses

FIGURE 32.7
Invertebrate Eyes.
Invertebrates have several types of eyes, three of which are: eye cups (**A**), compound eyes built of many ommatidia (**B**), and single-lens eyes (**C**).

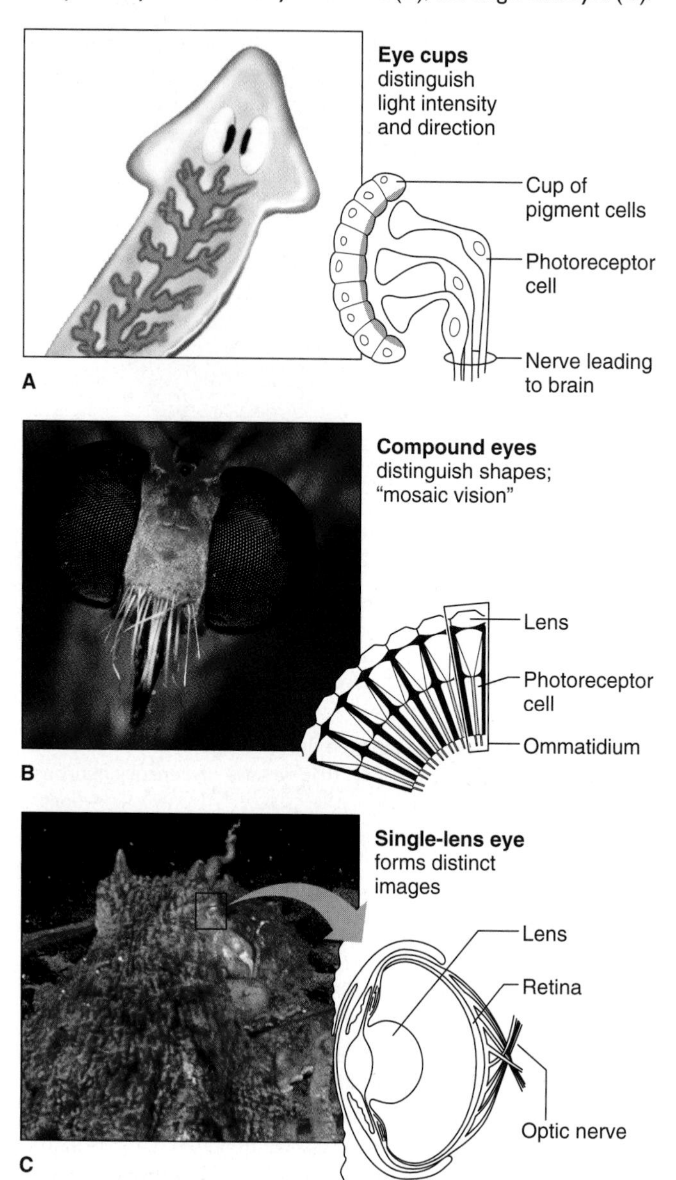

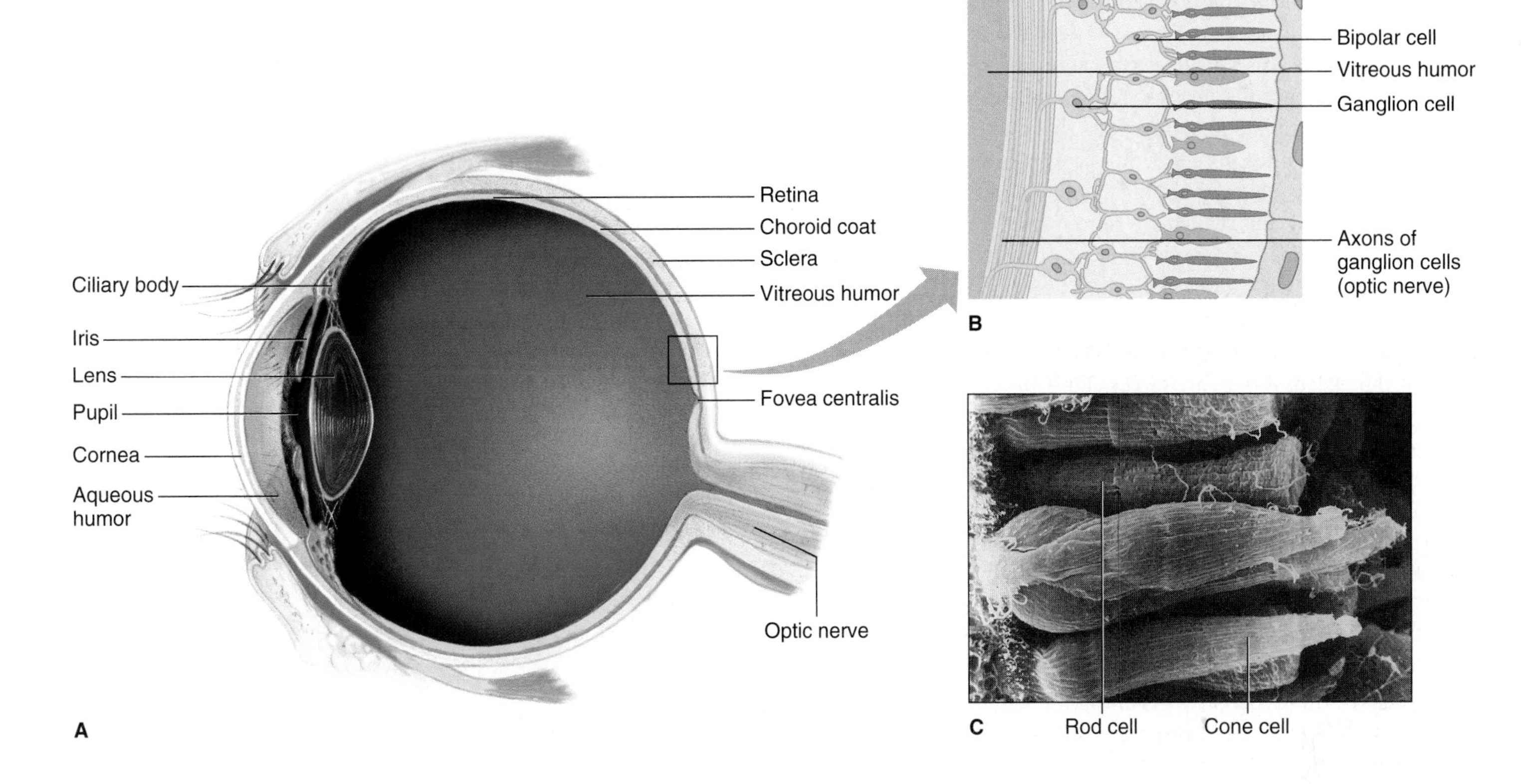

FIGURE 32.8
Anatomy of the Vertebrate Eye.
(**A**) The parts of the eyeball collect, transmit, and focus light onto photoreceptor cells at the back of the eyeball. (**B**) The retina includes layers of photoreceptor cells (rods and cones) that transduce and transmit light energy to bipolar cells, which transmit the information to ganglion cells, whose axons form the optic nerve. (For simplicity, horizontal and amacrine cells are omitted.) (**C**) Rods and cones contain highly folded membranes studded with visual pigments that absorb photon energy and transduce it to an electrochemical message.

onto photosensitive cells at the back of the eye that are clustered into a structure called the **retina.** Action potentials travel along nerves to the brain, where the visual information is interpreted. Cephalopod vision is excellent, and their visual systems are much like our own—both function similar to the way that a camera works.

Photoreception in Vertebrates

The eyes of vertebrates work like a camera. Both a camera and the vertebrate eye have a **lens** system that focuses incoming light into an image on a light-detecting surface. In a camera, this surface is the film; in the eye, it is a sheet of photoreceptors, the retina. The retina is, in a sense, a window into the CNS, but it does more than simply transmit information. Its intricate circuits of interacting neurons accentuate some input, while dampening other input, thereby controlling the information flow to a certain extent.

Anatomy of the Eye The eyeball is a fluid-filled sphere made up of three distinct layers. The white of the eye is the outermost layer, or **sclera,** which protects the inner structures (fig. 32.8). Toward the front of the eye, the sclera is modified into the **cornea,** a transparent curved window that admits light. This curve bends incoming light rays, which helps to focus them on the photoreceptor cells.

The middle layer of the eyeball, the **choroid,** is rich in blood vessels that nourish the eye. The choroid contains a dark pigment that absorbs light and prevents it from reflecting off the retina, much as black paint inside a camera absorbs light. Like the sclera, the choroid is specialized into different structures. At the front of the eye, the choroid thickens into a highly folded structure called the ciliary body, which houses the ciliary muscle. This muscle has filaments that loosely hold the lens in a way that enables it to alter its curvature, bending incoming light so that it focuses on the center of the visual field.

In front of the ciliary body, the choroid becomes the thin, opaque **iris,** which provides eye color. Like the diaphragm in a camera, the iris regulates the amount of light entering the eye by altering the size of the hole in its middle, the pupil. In bright light, the pupil constricts, shielding the retina from excess stimulation. In dim light, the pupil dilates, letting more light strike the retina.

The third and innermost layer of the eyeball, the retina, is itself built of several cell layers (see fig. 32.8). Five major classes of neurons comprise the retina. Three of them are organized into a pathway. Photoreceptors (rods and cones) absorb light energy and transduce it into a graded change in the cell's membrane potential. The photoreceptors then pass the message to a layer of bipolar neurons, so called because their cell bodies are centrally located. The bipolar neurons in turn transmit the message to **ganglion cells** (the first cells in the chain to generate action potentials), whose axons form the **optic nerve** that leads to the brain. The point where the optic nerve exits the retina is called the optic disc. This region is also called the blind spot, because it is devoid of photoreceptors, and light that hits this small region cannot be sensed. The two other types of neurons in the retina, called horizontal and amacrine cells, form lateral connections that modify the information sent along the photoreceptor-bipolar-ganglion pathways.

The human eye contains about 125 million rod cells and 7 million cone cells. **Rod cells,** which are concentrated around the edges of the retina, provide black-and-white vision in dim light and enable us to see at night. **Cone cells,** which are concentrated toward the center of the retina, detect color. An indentation directly opposite the lens, the **fovea centralis,** contains only cones. The number of cones in this area is a measure of the acuity of an animal's vision. Humans have about 150,000 cones per square millimeter, whereas some predatory birds have a million. Birds of prey owe their incredible eyesight to a second fovea. Other animals have different adaptations. For example, a cat's eyes glow in the dark because an extra layer (the tapetum) behind the retina reflects back the light that enters the retina. This increases visual acuity, especially at night.

The distribution of rods and cones helps explain why human vision works as it does. To see detail in bright light, it is best to look directly at the object, because this focuses light on the central retina area that is dense with cones. At night, however, we can see an object more clearly from the corner of the eye. From this perspective, the light bouncing off the object stimulates the rod-rich region of the eye. Only rods are sensitive enough to respond to dim light. In fact, just five photons of light appear to the human eye as a flash, thanks to the rods.

The eyeball is filled with fluid. Behind the lens, making up most of the volume of the eyeball, is the jellylike mass of **vitreous humor.** Between the cornea and the lens is the watery **aqueous humor.** Both humors help bend light rays to focus them on the retina. The aqueous humor cleanses and nourishes the cornea and the lens and presses against the sclera, which maintains the shape of the eyeball. A fresh supply of aqueous humor is secreted and absorbed every 4 hours. Some aqueous humor diffuses to the back of the eye and bathes the rods and cones.

Focusing the Light Vision begins when light rays pass through the cornea and lens and are focused on the retina. Because most objects that we see are considerably larger than our retinas, the light rays must bend to form an image on the fovea centralis. The cornea, the lens, and the humors of the eye bend the light rays.

The lens changes shape to focus onto the retina light that reflects from objects. This adjustment to suit the distance between the eye and the viewed object is called **accommodation.** To focus on a very close object, the ciliary muscle contracts, rounding the lens so that it can bend incoming light rays at sharper angles. Close work often causes eyestrain because the ciliary muscle must remain contracted to view nearby objects. When viewing a distant object, the ciliary muscle relaxes and the lens flattens. That is why gazing into the distance can relieve eyestrain.

The eyeball must be a certain shape for the cornea and lens to focus light rays precisely on the retina. Unfortunately, not all of us have perfectly shaped eyeballs. Fortunately, though, corrective lenses (eyeglasses and contact lenses) can alter the path of light. Figure 32.9 shows how these lenses correct vision.

Transducing Light Energy to Neural Messages Pigment molecules within the rods and cones change conformation upon absorbing photon energy. This response sets into motion a signal transduction pathway (see chapter 4) that converts and amplifies light into receptor potentials.

FIGURE 32.9

Correcting Vision.

(**A**) A normally shaped eyeball focuses light rays on the retina. (**B**) If the eyeball is elongated, light rays converge in front of the retina, impairing ability to see distant objects (nearsightedness). A concave eyeglass lens alters the point of focus to the retina. (**C**) In astigmatism, the lens or cornea is shaped so unusually that incoming light rays do not focus evenly. An irregular-shaped eyeglass lens can compensate for the eyeball's abnormal shape and focus light rays on the retina. (**D**) A short eyeball focuses light beyond the retina, and the person has difficulty seeing close objects (farsightedness). A convex lens corrects this problem.

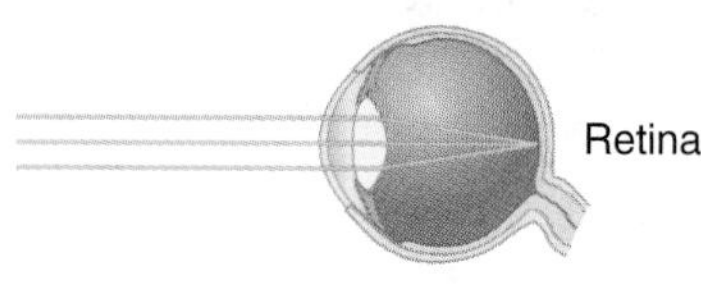

A Normal sight
Rays focus on retina

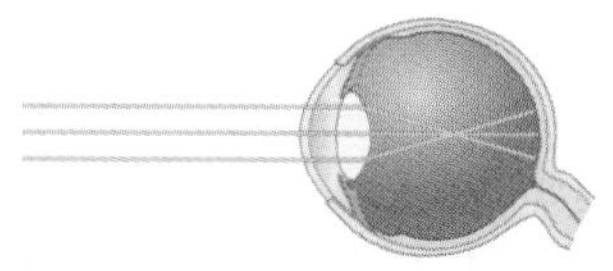

B Nearsightedness
Rays focus in front of retina

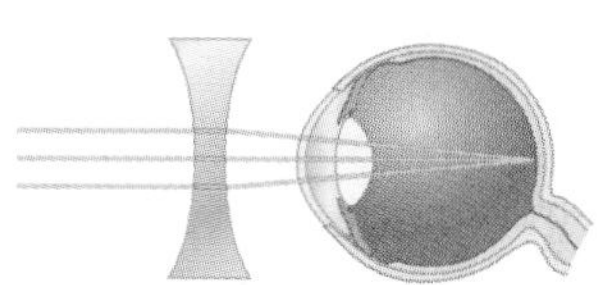

Concave lens corrects nearsightedness

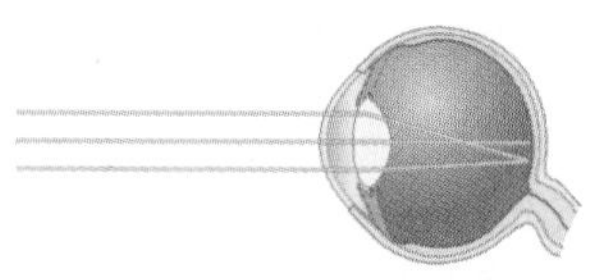

C Astigmatism
Rays do not focus equally

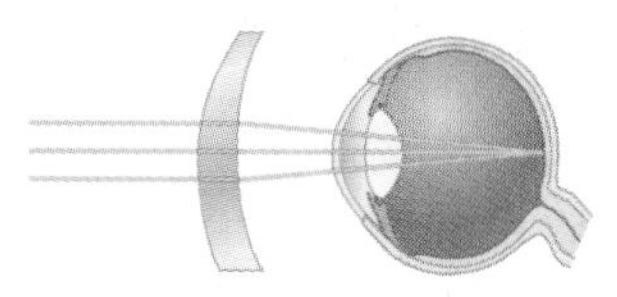

Uneven lens corrects astigmatism

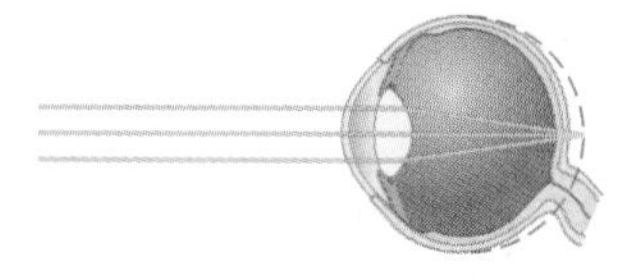

D Farsightedness
Rays focus behind retina

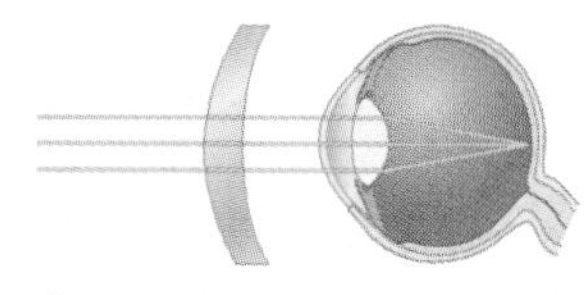

Convex lens corrects farsightedness

FIGURE 32.10
The Visual Pathway.
(**A**) Photons pass through the outer layers of the eye and hit the rods and cones, which transmit information to bipolar cells, which pass the message to the action potential-generating ganglion cells that form the optic nerve. (**B**) Rhodopsin consists of retinal linked to one helix of an opsin. Light energy alters the conformation of the retinal, which may ultimately alter the pattern of action potentials in the optic nerve.

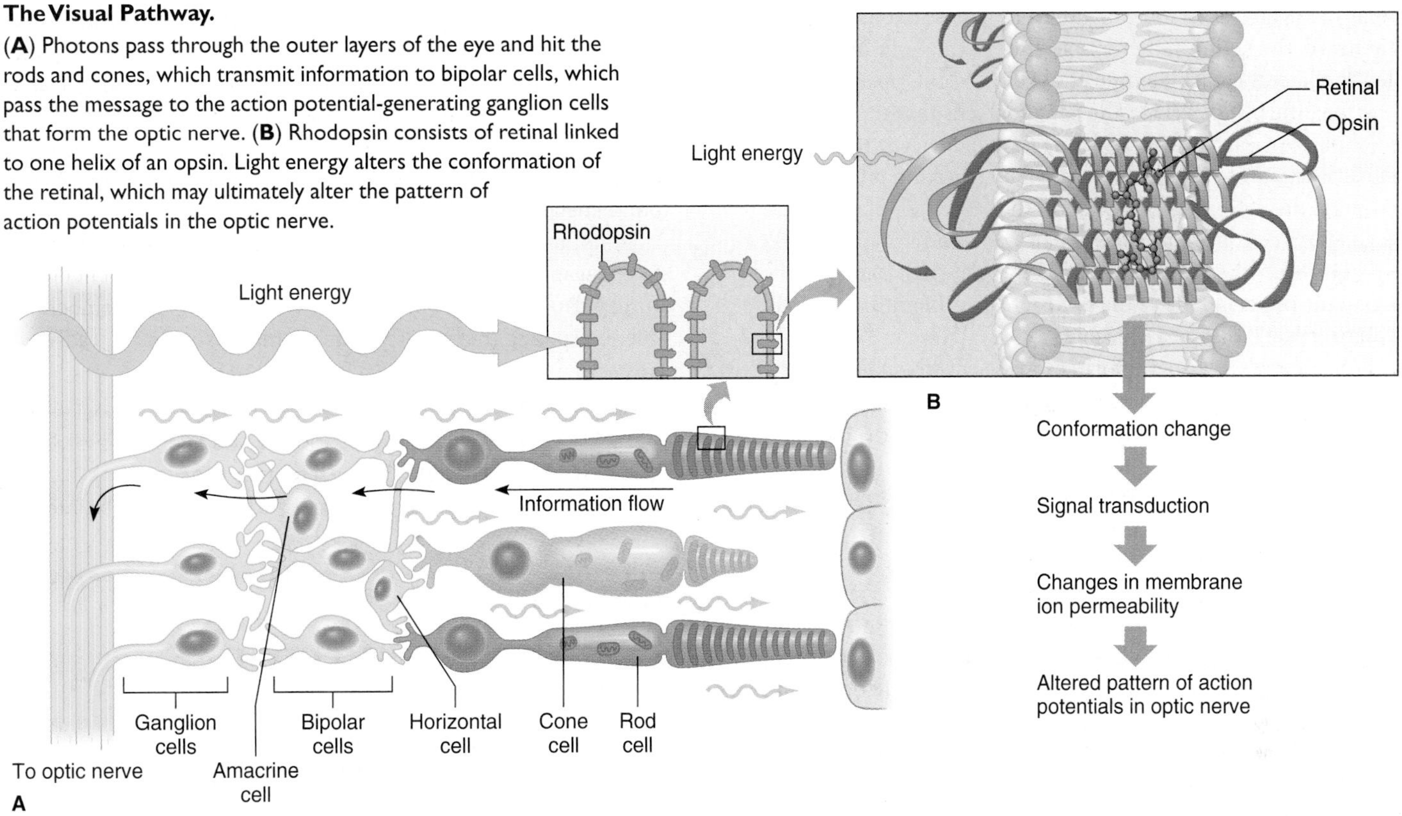

Light energy
Rhodopsin
Information flow
Ganglion cells
Bipolar cells
Horizontal cell
Cone cell
Rod cell
To optic nerve
Amacrine cell
A
Light energy
Retinal
Opsin
B
Conformation change
Signal transduction
Changes in membrane ion permeability
Altered pattern of action potentials in optic nerve

Rod cells provide black-and-white-vision. A rod cell can respond to a single photon of light, and the response is then amplified along the signal transduction pathway. One end of the cell consists of about 2,000 interconnected discs of lipid bilayer, derived from the cell membrane, that are studded with rhodopsin molecules (fig. 32.10). The other end of the cell houses the typical organelles that supply energy and continually produce rhodopsin and other proteins necessary for its function.

A rhodopsin molecule consists of a small organic group called **retinal** that is bonded to one of seven helices that make up the other part, an enzyme called **opsin.** (Many molecules that take part in signal transduction have a seven-membered membrane-spanning component.) It is the retinal, a derivative of vitamin A, that actually absorbs light. (Beta-carotene, the plant pigment that gives carrots and other vegetables their orange color, is converted in our bodies to vitamin A.) Upon absorbing light, retinal elongates, which distorts the opsin, triggering a cascade of signal transduction. This activity changes the ion permeability of the cell membrane, which regulates neurotransmitter release.

Color vision is possible because the cone cells contain different pigments, and each pigment absorbs light of a different range of wavelengths. Although an individual cone cell contains only one type of pigment, the wavelengths that each type of cone absorbs overlap. In humans, "blue" cones absorb shorter wavelengths of light, "green" cones absorb medium wavelengths, and "red" cones absorb long wavelengths. Examination of the retinas of different people reveals that individuals have unique patterns of cone types, all apparently able to provide color vision. Some parts of the retina are even normally devoid of one particular type, yet the brain integrates information from all over to "fill in the gaps," creating a continuous overall image. People who lack a cone type, though, due to a mutation, are color blind. Genes that control production of pigments sensitive to red and green wavelengths are on the X chromosome, but the gene that regulates synthesis of blue-sensitive pigments is on an autosome. The different types of color blindness correspond to which cone types are absent.

The molecules that provide color vision also consist of opsin bound to retinal. When a cone absorbs light, its pigment molecule contorts, generating a receptor potential. The brain interprets the ratio of activities among the different cone types as light of a particular color, integrating thousands of incoming signals in a process of image analysis. The result is not just a trio of colors, but a rainbow of hues.

As primates, we humans enjoy a more multicolored world than many other mammals. This is because the visual systems of nonprimate mammals funnel input from groups of photoreceptor cells into the CNS. That is, several photoreceptors signal the same bipolar neurons, which in turn pool their input to ganglion cells. Primates are the only mammals to have three types of cones (others have two), and it appears that primates excel in color vision because the rods and cones connect individually to neural pathways to the brain.

The Visual Cortex Integrates Information Vision is a complex sense. About a third of the human brain's circuitry is devoted to seeing.

Once the rods and cones have sent their input through the bipolar neuron layer to the ganglion cells, whose axons form the optic nerve, the action potentials go first in the brain to an area called the lateral geniculate nucleus (LGN) in the posterior part of the thalamus (fig. 32.11). The ganglion cells that make up the medial parts of each retina, which receive light from the central part of the field of view (due to the light-inverting effect of the lens), send their axons to the LGN on the same side of the brain. The ganglion cells from the lateral parts of each retina, which receive light from the lateral field of view, send their axons to the LGN on the opposite side of the brain, crossing behind the eyes in a structure called the optic chiasma.

From the LGN, visual information is sent to neurons at the rear of the brain that comprise the primary visual cortex. Here, different neuron groups interpret the incoming information in different ways. To do this, different subgroups of neurons respond to different aspects of an object, such as its color, shape, size, and texture.

Overall, vision is not simply a linear pathway of light to retina to brain, but an enormously intricate network of amplified and interacting signals, which the brain interprets with astonishing speed and accuracy. Because of the curved shape of the lens, the image projected onto the retina is actually upside down and backward. The brain processes this information so that we perceive a right-side-up world. In many mammals, the brain also integrates the overlapping visual fields coming from each eye, creating the three-dimensional images necessary for depth perception. This is possible when an animal's eyes are located on the front of its head. Animals whose eyes are on the side of the head, such as fishes and rabbits, do not experience nearly the same degree of depth perception.

FIGURE 32.11

From the Eyes to the Brain.

The optic nerve passes the light stimulus first to the lateral geniculate nucleus, then to the primary visual cortex, where the information is processed and integrated.

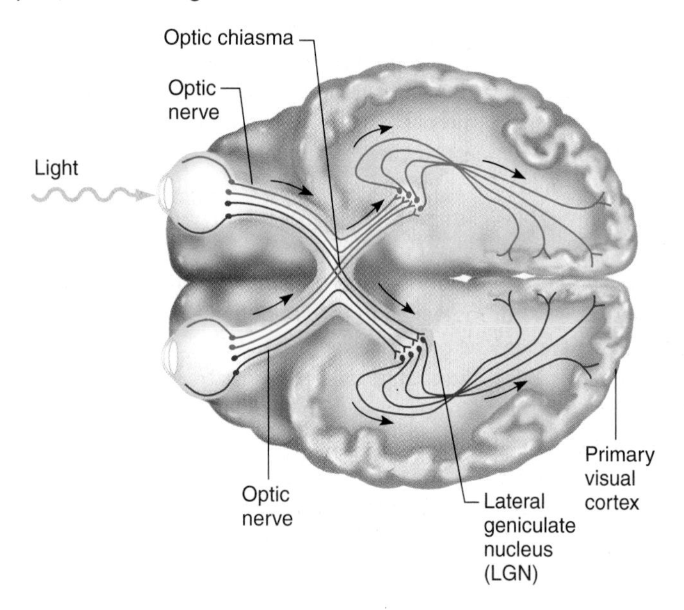

The visual cortex fuses images formed at discrete moments of time into a fluid perception of motion. Inability to do this is like watching a movie in extremely slow motion, so that each scene change is painfully obvious. It's hard to appreciate the importance of this ability, but one woman did when a stroke damaged part of her visual cortex, destroying her ability to perceive motion. She saw movement as a series of separate, static images. Her deficit had profound effects on her life. She could not pour a drink, because she could not tell when the cup would overflow. She could not cross a street because she could not detect cars moving toward her. Many animals can detect prey only if it moves. If she were a member of a different species, the inability to detect motion could be deadly.

32.4 MASTERING CONCEPTS

1. Name two types of light-sensing organs in invertebrates.
2. What are the parts of the vertebrate eye?
3. How does the human eye transduce incoming light into an image?

OLC

32.5 Mechanoreception

Mechanoreceptors enable an organism to respond to physical stimuli. Invertebrates and vertebrates have mechanoreceptors grouped into specialized organs for the senses of hearing, equilibrium and balance, and touch. Structures in the vertebrate ear transmit sound waves that contact hair cells, triggering action potentials on the auditory nerve, providing hearing. Other structures in the ear control balance and equilibrium. Receptors in the skin provide the sense of touch.

Mechanoreceptors enable an organism to respond to physical stimuli (touch) and sound waves (hearing), and to maintain balance and equilibrium.

Mechanoreception in Invertebrates

The sense of hearing operates when pressure waves hit and move receptors. In insects, hairlike structures called **setae** at the bases of the antennae vibrate in response to sound. The movement stimulates sensory cells that transmit the information to the brain.

Some insects also have drumlike **tympanal organs,** which are thin membranes stretched over large air sacs that form resonating chambers for sound. Sensory cells beneath the membrane detect sound pressure waves and relay this information to the brain. Crickets and katydids have tympanal organs on their legs, while grasshoppers and some moths have them on their abdomens. These organs are usually on both sides of an insect's body, which enables it to sense not only the presence but also the direction of the sound.

Setae and tympanal organs are highly sensitive. In fact, insects can detect much higher pitches than the human ear can hear. The ability of certain moths to escape predatory bats vividly displays the sensitivity of insect hearing. Moths can detect the high-pitched cries of bats and dive down when the sound approaches. This confuses the bat's **echolocation,** the ability to locate airborne insects by bouncing ultrasound waves off them.

When moths fly down near the ground, the bat's echolocation bounces off of the moths and the ground at about the same time. Thus, the bat can no longer pick the moths out.

Mechanoreception also includes a sense of balance and equilibrium. In many animals, structures called **statocysts** provide this sense. A statocyst is a fluid-filled cavity containing sensory hairs and bits of mineral called statoliths. When the animal's body tilts, the statoliths touch the sensory hairs, stimulating the nervous system. In response, the animal can orient itself with respect to gravity. Such diverse invertebrates as jellyfishes, snails, squids, crayfishes, and earthworms use statocysts for balance. Recall that statoliths also function in a plant root's ability to detect gravity (see fig. 29.7). Invertebrate statocysts are similar to human structures that carry out the same functions, discussed in a later section.

Many invertebrates are startled by touch, which they sense using specialized mechanoreceptors. For example, an insect's body is laden with contact-sensitive hairs and with sensors that detect strain on its exoskeleton and joints. Some insect antennae function as both "feelers" and chemoreceptors. Spiders have modified strain gauges in their legs that feel the quivering of their webs when an insect becomes entangled. Scorpions have similar structures that detect the small ground-surface vibrations from nearby prey.

Mechanoreception in Vertebrates

Hearing The clatter of a train, the sounds of a symphony, a child's wail—what do they have in common? All sounds, regardless of the source, originate when something vibrates. The vibrating object creates repeating pressure waves in the surrounding medium, such as air, water, or even the ground. The size and energy of these pressure waves determine the intensity or loudness of the sound, while the number of waves (cycles) per second determines the sound's frequency or pitch. The more cycles per second, the higher the pitch.

Vertebrates display interesting variations on the theme of hearing. For example, Panamanian golden frogs lack outer and middle ears, yet they can detect and precisely localize the calls from other frogs. Perhaps they can hear because sound waves that hit the sides of the frog's body move the lungs, which then transmit vibrations to the intact inner ears for processing. Barn owls can pinpoint the height a sound emanates from because one ear is higher than the other and one points up, while the other points down. Bats' ears can swivel and their receptor cells are sensitive to their own high-pitched calls that reflect off of nearby objects, such as their flying insect meals. Dolphins also use echolocation. High-frequency clicks and trills, generated in nasal sacs in their heads, are focused by an "acoustical lens" called the melon, located in their foreheads. In dolphins, the auditory canal typical of land-dwelling mammals is closed. As such, the sound waves that bounce off of objects, such as fishes, pass to the brain by way of the lower jaw and the middle ear. Using this sophisticated sense, dolphins can discriminate nearly identical objects.

In humans, sound transduction and perception begins with the fleshy outer part of the ear (the pinna) that traps sound waves and funnels them down the **auditory canal** (ear canal) to the **tympanic membrane,** or eardrum (fig. 32.12). The sound pressure waves set up vibrations in the eardrum, which moves three small bones, the maleus, incus, and stapes, located in the middle ear. These bones transmit the incoming sound and amplify it 20 times.

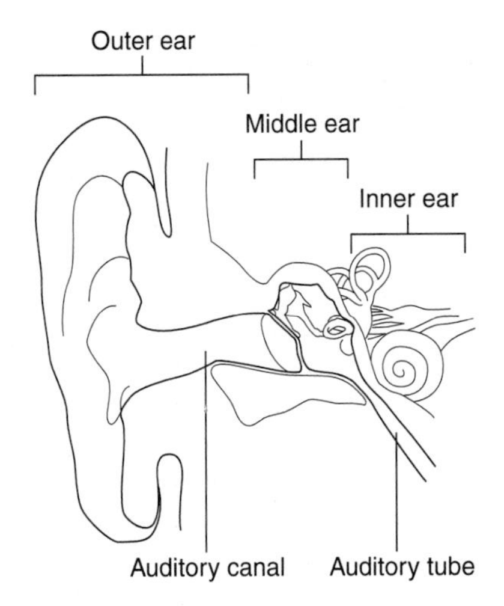

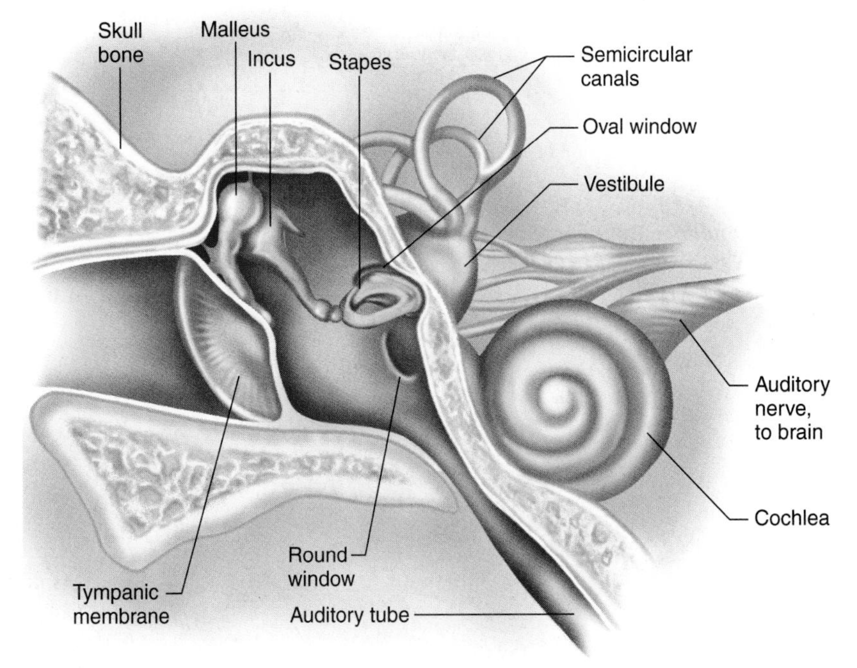

FIGURE 32.12
Anatomy of the Ear.
Sound enters the outer ear and, upon reaching the middle ear, impinges upon the tympanic membrane, which vibrates three bones (malleus, incus, and stapes). The inner ear houses the organs of hearing and balance. The vibrating stapes hits the oval window, and hair cells in the cochlea convert the vibrations into action potentials, which are sent along the auditory nerve to the brain. Hair cells in the semicircular canals and in the vestibule sense balance. The auditory tube connects the middle ear to the rear of the nasal cavity, equalizing air pressure.

From the middle ear, the vibrations of the stapes are transmitted through the **oval window,** a membrane that opens into the inner ear. Running between the middle ear and the air passageways in the back of the nose is a canal called the **eustachian** (or auditory) **tube.** This tube allows us to adjust pressure differences on the inside of the tympanic membrane if the outside pressure should change. "Ear popping" occurs when the internal and external pressures re-equilibrate and the tympanic membrane 'pops' back into shape, such as when we yawn during an airplane takeoff.

The inner ear is a fluid-filled chamber that houses both balance and hearing structures. Two parts of the inner ear, the semicircular canals and the vestibule, control balance and are discussed in the next section. The remaining portion of the inner ear is the snail-shaped **cochlea.** The spirals of the cochlea are three fluid-filled canals called the vestibular, cochlear, and tympanic canals. The vestibular and tympanic canals actually form a continuous U-shaped tube. Between them lies the cochlear canal. When the last bone of the middle ear, the stapes, moves, it pushes on the oval window, transferring the vibration to the fluid in the vestibular canal. The pressure of each vibration is dissipated by the movement of the round window at the other end of the U. The vibration of the fluid also moves the **basilar membrane,** which forms the lower wall of the cochlear space, and initiates the change of mechanical energy to receptor potentials.

Sound is translated into the universal action potential language of the nervous system within the cochlea. Here, specialized **hair cells** (the mechanoreceptors) lie between the basilar membrane below and another sheet of tissue, the **tectorial membrane,** above (fig. 32.13). The basilar membrane is narrow and rigid at the base of the cochlea and widens and becomes more flexible closer to the tip. Because of this variation in width and flexibility, different areas of the basilar membrane vibrate more intensely when exposed to different frequencies. The high-pitched tinkle of a bell stimulates the narrow region of the basilar membrane at the base of the cochlea, while the low-pitched tones of a tugboat whistle stimulate the wide end.

When a region of the basilar membrane vibrates, the hair cells are displaced relative to the tectorial membrane, and this initiates action potentials in fibers of the **auditory nerve.** This nerve carries the impulses to the brain, which interprets the input from different regions as sounds of different pitches (fig. 32.14). The louder the sound, the greater the vibration of the basilar mem-

FIGURE 32.13
Hearing.
Hearing is the transduction of vibrations to nerve impulses. The cochlea, shown in anatomical context in (**A**) and in cross section in (**B**), is a spiral-shaped structure consisting of three fluid-filled canals. When the stapes moves against the oval window, the fluid in the vestibular canal vibrates. The vibrations press on the hair cells that lie between the basilar and tectorial membranes (**C**). The hair cells push against the tectorial membrane and this triggers action potentials in the auditory nerve. Cilia fringe the tops of the hair cells.

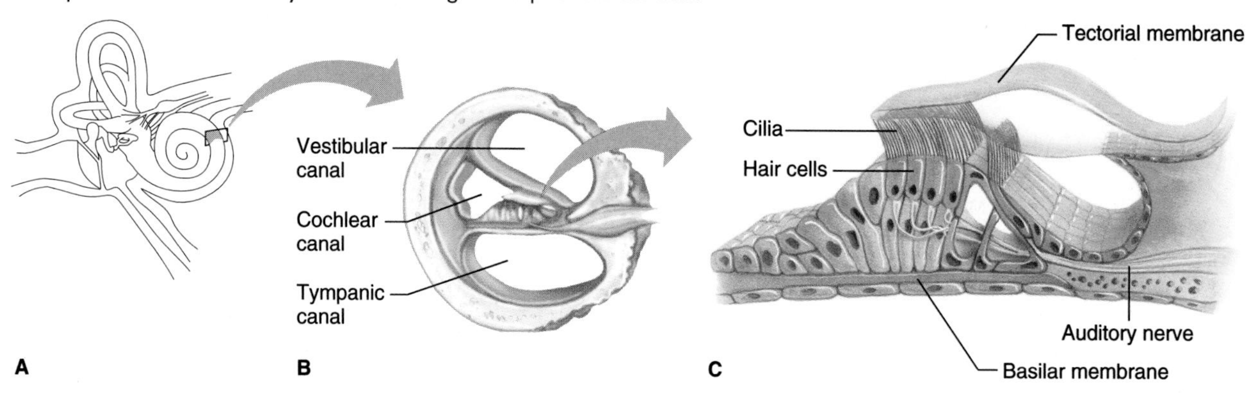

FIGURE 32.14
Correspondence Between Cochlea and Cortex.
Sounds of different frequencies (pitches) excite different sensory neurons in the cochlea. These neurons, in turn, send their input to different regions of the auditory cortex.

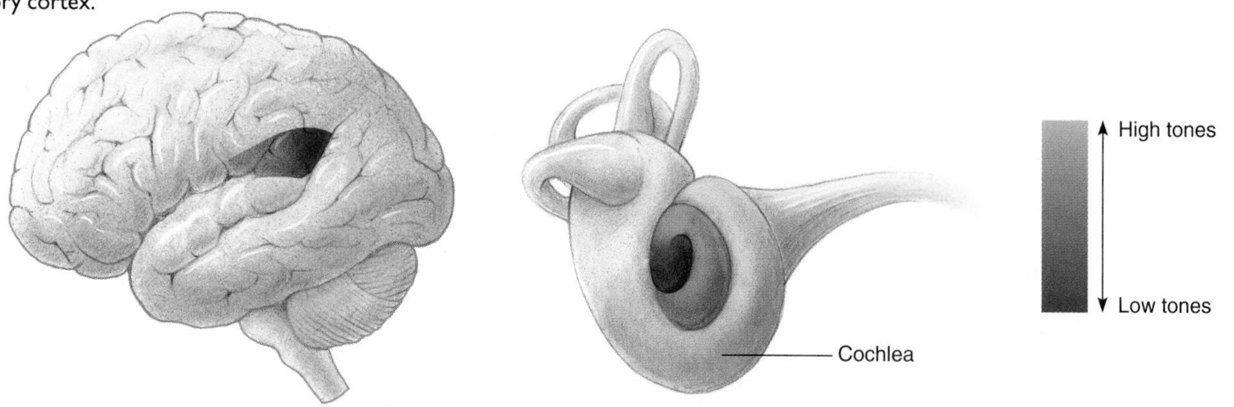

TABLE 32.3 Hearing Impairment

PROBLEM	CAUSE	RESULT
Conductive deafness	Impaired transmission of sound through middle ear due to infection, earwax buildup, damaged eardrum, or fusion of middle ear bones	Inability to clearly hear all pitches of sound
Sensory (neural) deafness	Inability to generate action potentials in the cochlea, blocked communication between cochlea and brain, or brain's inability to interpret sensory messages	Inability to hear some pitches
Tinnitus	Damaged hair cells in inner ear due to infection, brain tumor, earwax buildup, or loud noise	Ringing in the ears

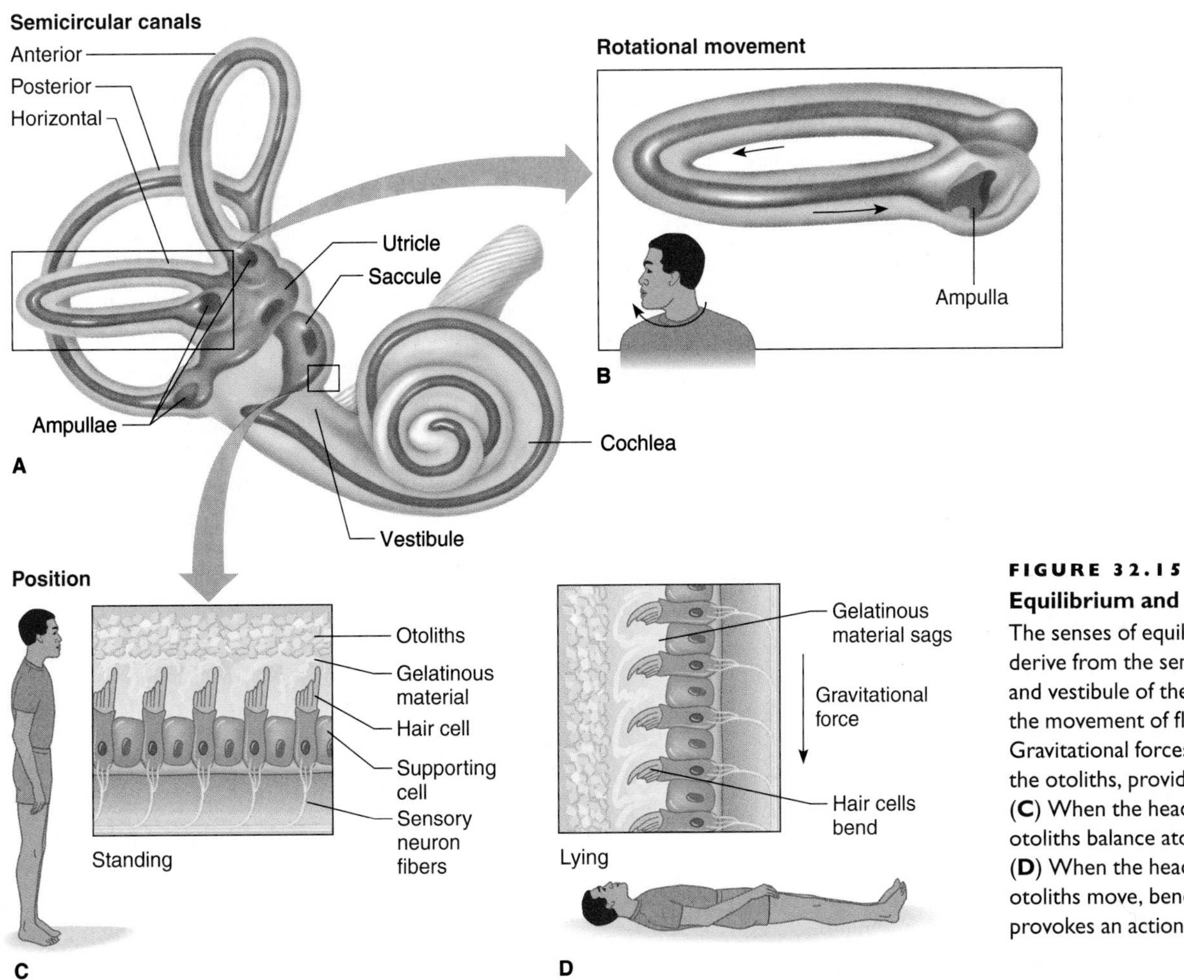

FIGURE 32.15
Equilibrium and Balance.
The senses of equilibrium and balance derive from the semicircular canals and vestibule of the inner ear (**A**) and the movement of fluid in them (**B**). Gravitational forces on the head move the otoliths, providing a sense of balance. (**C**) When the head is upright, the otoliths balance atop the cilia of hair cells. (**D**) When the head is horizontal, the otoliths move, bending the cilia. This provokes an action potential.

brane. As a result, individual hair cells fire more action potentials and more cells are provoked to fire. The brain interprets the resulting increase in the rate and number of neurons firing as an increase in amplitude, or loudness.

Hearing loss may be temporary or permanent. It can result from conductive deafness (blocked transmission of sound waves through the middle ear) or from sensory deafness (damage to the nervous system) (table 32.3). Many cases of hearing loss result from exposure to loud noise. Sound intensity is measured in decibels, with each 10-decibel increase representing a 10-fold increase in sound intensity. Damage to the inner ear's hair cells begins to occur at about 80 decibels, which is as loud as heavy traffic. The degree of damage depends both upon decibel level and duration of exposure to the sound.

Many rock stars, and their fans, of the 1960s have hearing loss today. The damage begins as the hair cells develop blisterlike bulges that eventually pop. The tissue beneath the hair cells swells and softens until the hair cells and sometimes the neurons leaving the cochlea become blanketed with scar tissue and degenerate.

Equilibrium and Balance The **semicircular canals** and the **vestibule** of the inner ear regulate the sense of balance (fig. 32.15).

The semicircular canals tell us when the head is rotating and help us maintain the position of the head in response to sudden movement. The enlarged bases of the semicircular canals, the ampullae, are lined with small, ciliated hair cells. A caplike structure called a cupula covers the hair cells. Because the semicircular canals are perpendicular to each other, the fluid that fills a canal may or may not flow back and forth in response to a movement, depending on its direction. When the fluid in a canal moves, it bends the cilia on the hair cells in that ampulla, which stimulates action potentials in a nearby cranial nerve. The brain interprets these impulses as body movements.

Information from the vestibule senses the position of the head with respect to gravity. In addition, the vestibule senses changes in velocity when traveling in a straight line. When riding in a car, for example, we can sense acceleration and deceleration. The vestibule functions in a similar way to the semicircular canals. It contains two pouches, the **utricle** and the **saccule,** both of which are filled with jellylike fluid and lined with ciliated hair cells. Granules of calcium carbonate, called **otoliths,** float on the fluid. The granules in the utricle move in response to acceleration or to tilting of the head or body, bending the cilia on the hair cells. When the cilia bend in one direction, the rate of sensory impulses to the brain increases. A shift in the opposite direction inhibits the sensory neuron. The brain interprets this information as a change in velocity or a change in body position.

Motion sickness results from contradictory signals. The inner ear signals the brain that the person is not accelerating. At the same time, the eyes detect passing scenery and signal the brain that the person is moving. The result of these mixed signals is nausea.

Touch Different species have different structures that sense touch—a fish uses its pressure-sensitive lateral lines, cats and rats use their whiskers, and nearly all vertebrates have diffuse touch receptors covering their bodies. Human sensitivity to touch comes from several types of receptors in the skin—Pacinian and Meissner's corpuscles and free nerve endings (fig. 32.16). Both types of corpuscles are encapsulated, resembling an onion—they consist of tissue surrounding a single nerve fiber. A touch pushes the flexible sides of the corpuscle inwards, generating an action potential in the nerve fiber. Firm pressure, such as a bear hug, stimulates Pacinian corpuscles. On the other hand, a light touch stimulates Meissner's corpuscles, particularly a moving touch, as in a gentle caress.

Besides touch, other specialized receptors in the skin detect environmental stimuli. For example, **thermoreceptors** respond to temperature changes. Heat receptors are most sensitive to temperatures above 77°F (25°C), and become unresponsive at temperatures above 113°F (45°C). Approaching this upper limit also triggers pain receptors, causing the intense pain of a burn. Cold receptors are most sensitive when the temperature ranges from 50°F (10°C) to 68°F (20°C). Temperatures below this lower limit stimulate pain receptors, causing the sensation of freezing. At the more common intermediate temperatures, the brain integrates input from many cold and heat receptors. Thermoreceptors adapt fast, which is why we quickly become comfortable after jumping into a cold swimming pool, or submerging into a steaming hot tub.

FIGURE 32.16
Touch.
Touch receptors include free nerve endings, which respond to touch, pressure, and pain; Meissner's corpuscles, which respond to light touch; and Pacinian corpuscles, which detect firm pressure.

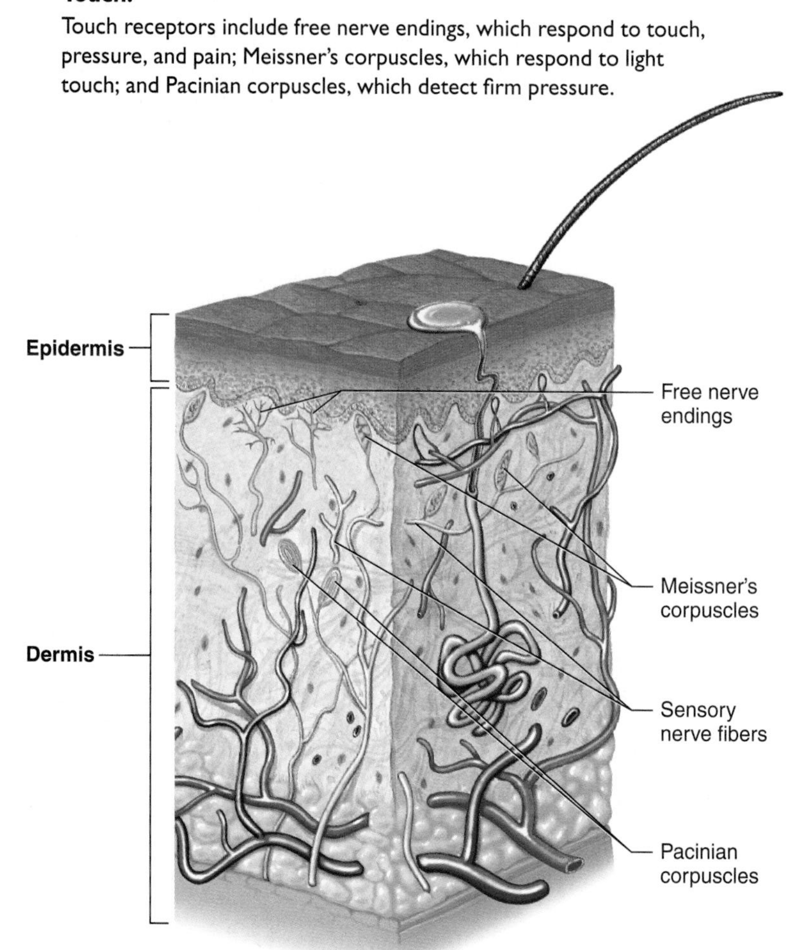

Pain receptors are located nearly everywhere in the body, except the brain. They are specialized to detect mechanical damage, temperature extremes, chemicals, and blood deficiency. A particular pain sensation may be the consequence of activation of different types of receptors. A muscle cramp, for example, reflects the mechanical compression of the tightly contracting muscle, as well as the cutoff of oxygen to part of the tissue. Applying heat to a cramped muscle dilates the blood vessels, bringing in more oxygen and relieving some of the pain.

The fact that pain receptors from different parts of the body can share neural pathways to the brain causes a phenomenon called referred pain, when pain appears to come from a body part other than the one that is injured. For example, a person suffering a heart attack may initially feel pain in the left arm, because pain receptors in the heart and in the skin of the left shoulder and inside of the left arm share neural pathways. Similarly, people who have undergone amputations often report pain in the removed limb, because neural pathways that connected pain receptors in the area remain. This is called a phantom limb.

Senses are not only essential to our survival, but they help make life enjoyable. Fortunately, because our senses both complement and overlap, we can still live full, functional lives if one of our senses is impaired; we can compensate with one sense for another. A person who cannot see a garden can smell the flowers and touch the plants to gain a detailed perception of it. A person who cannot hear can read lips or use sign language to converse. Thanks to the blend of our senses, the world is an exciting, multidimensional place.

32.5 MASTERING CONCEPTS

1. What senses do mechanoreceptors provide?
2. What are the parts of the ear, and how do they transmit sound?
3. How do structures in the ear provide equilibrium and balance?
4. Which structures provide the sense of touch?

Chapter Summary

32.1 Sensory Systems Are Adaptive

1. **Sensory receptors** detect, transduce, and amplify stimuli, allowing an animal to perceive its environment.
2. Natural selection has molded sensory systems.

32.2 Sensory Systems, Although Diverse, Operate by the Same General Principles

3. A sensory receptor selectively responds to a single form of energy and converts it to **receptor potentials,** which change membrane potential in proportion to stimulus strength.
4. Sensory receptors amplify stimuli.
5. If a stimulus remains constant, a sensory receptor ceases to respond (**sensory adaptation**).
6. Sensory receptors are classified by the type of energy they detect: chemicals, light, or mechanical energy.

32.3 Chemoreception

7. Invertebrates have **chemoreceptors** to detect food, escape danger, and communicate.
8. In humans, **olfaction** occurs when odorant molecules bind to receptors in the olfactory epithelium of the nasal passages. The brain perceives a smell by evaluating the pattern of olfactory receptor cells that bind to odorant molecules.
9. Humans perceive taste (**gustation**) when chemicals stimulate receptors within taste buds.

32.4 Photoreception

10. Invertebrate **photoreceptors** contain pigments (usually **rhodopsin**), associated with membranes. Light stimulation alters the pigment and changes the charge across the membrane, which may generate an action potential. Visual systems range from simple eye cups to **compound eyes** composed of **ommatidia,** to lens systems.
11. Humans perceive vision with a complex **lens** system. The human eye contains three layers. The outer layer, the **sclera,** protects. It forms the transparent **cornea** in the front of the eyeball. The next layer, the **choroid coat,** is pigmented, located toward the rear of the eye, and absorbs light. In the front of the eye, the choroid coat forms the ciliary body, which controls the shape of the lens that focuses light on the photoreceptors, and the opaque **iris.** The **pupil** constricts or dilates to adjust the amount of light entering the eye.
12. The innermost eye layer is the multilayered **retina.** Beneath a pigment layer lie photoreceptors: **rods** for black-and-white vision in dim light and **cones** for color vision in brighter light. These cells synapse with bipolar cells that form the middle retinal layer. The bipolar cells, in turn, synapse with **ganglion cells** whose fibers leave the retina as the **optic nerve,** which carries the neural messages to the brain for interpretation. Light activates rhodopsin, altering **retinal,** which ultimately alters ion permeability of the receptor cell membrane. Three types of cones each contain a pigment that maximally absorbs light of a range of particular wavelengths. The brain interprets the ratio of the activities of the three cone types as a color.

32.5 Mechanoreception

13. Invertebrate hearing depends on **setae** and **tympanal organs,** and balance and equilibrium depend on statolith crystals within **statocysts.**
14. **Mechanoreceptors** bend in response to sound, to movement, or touch. In human hearing, sound enters the **auditory canal,** vibrating the **tympanic membrane.** These vibrations are transmitted through the middle ear and amplified by three bones, the malleus, incus, and stapes. The movement of these bones changes the pressure in fluid within the **cochlea,** which in turn vibrates the **basilar membrane.** As the basilar membrane moves, it pushes hair cells against the **tectorial membrane,** which signals the brain to perceive the pitch of the sound through the location of the moving **hair cells.** The brain determines the sound's loudness from the frequency of action potentials and the number of stimulated hair cells.
15. The **semicircular canals** and the **vestibule** in the inner ear sense body position and movement. Fluid movement within these areas stimulates sensory hair cells, and the brain interprets this information, providing a sense of equilibrium.
16. Pacinian corpuscles, Meissner's corpuscles, and free nerve endings are mechanoreceptors that detect touch. Pacinian corpuscles respond to intense pressure, Meissner's corpuscles to gentle pressure, and free nerve endings to touch, pressure, and pain.

Testing Your Knowledge

1. What is a sense organ?
2. What are three general types of senses?
3. How can a pigment associated with a protein transduce an environmental stimulus into a sensation?
4. Distinguish between sensation and perception.
5. Distinguish between ocelli and ommatidia.
6. How does the nervous system detect different smells?
7. In what way are the senses of smell and taste similar?
8. List the structures of the human eye and their functions.
9. In humans, otolith movement provides equilibrium and balance. What is the corresponding structure in an invertebrate?

10. Why doesn't burning your tongue on a hot slice of pizza permanently damage your sense of taste?
11. How does vision in humans differ from that of an insect?
12. How do we see colors other than red, blue, or green?
13. In what ways do the three major structures of the inner ear (cochlea, semicircular canals, and vestibule) function similarly?
14. Sensation results from an anatomical pattern of many sensory cells that transmit impulses to a nerve leading to a specific part of the brain. Describe the organization of neurons in three human senses that fit this pattern of funneling incoming information from many sensory neurons to a nerve fiber leading to the brain.

Thinking Scientifically

1. People can inherit an inability to smell certain substances, such as freesia flowers, jasmine, skunk scent, and hydrogen cyanide. How could the hypothesis that odors are encoded in combinations of stimulated receptors explain these conditions?
2. Cite two examples of how people who lack one sense can compensate by relying more heavily on another.
3. People who are hearing impaired due to cochlea damage do not suffer from motion sickness. Why not?
4. We have relatively few sensory systems. How, then, do we experience such a diversity of sensory perceptions?
5. Humans love sucrose (table sugar), but armadillos, hedgehogs, lions, and seagulls do not respond to it. Opossums love lactose (milk sugar) but rats avoid it, and chickens hate the sugar xylose, while cattle love it. In what way might these diverse tastes in the animal kingdom be adaptive?

References and Resources

Clyne, Peter J., et al. March 10, 2000. Candidate taste receptors in *Drosophila. Science,* vol. 287, p. 1830. The genes that control tasting ability have been identified in the fruit fly.

Firestein, Stuart. April 6, 2000. The good taste of genomics. *Nature,* vol. 404, p. 552. Sequencing genomes revealed a fifth basic taste sensation.

Hughes, Howard C. 1999. *Sensory exotica: A world beyond human experience.* Cambridge: MIT Press. A look at sensory systems among diverse animal species.

Lewis, Ricki. September/October 2000. Total color blindness: A rare gene defect. *Biophotonics International,* vol. 7, p. 38. Total color blindness results from an abnormal ion channel in cone cell photoreceptors.

In 1999 *Scientific American* published a supplement called "Science's Vision: The Mechanics of Sight." It has many useful articles about the research and tools that have revealed the details of how we see.

Seielstad, Mark. October 2000. Whiffs of selection. *Nature Genetics,* vol. 26, p. 131. Human genome data reveals that natural selection has carved our sense of smell to avoid toxins.

The *LIFE* Online Learning Center provides additional resources and tools for studying this chapter.
www.mhhe.com/life

CHAPTER 33

The Endocrine System

History of an Illness

Diabetes Mellitus

Insulin Treatment.
(**A**) A mother holds her son, who was 3 years old and weighed 15 pounds when this photo was taken in December 1922. He suffered from diabetes mellitus. (**B**) The same boy 2 months later; after insulin treatment, his weight increased to 29 pounds.

The sweet-smelling urine that is the hallmark of diabetes mellitus was noted as far back as an Egyptian papyrus from 1500 B.C. In A.D. 96 in Greece, Aretaeus of Cappadocia described the condition as a "melting down of limbs and flesh into urine."

Type I (juvenile-onset) diabetes mellitus results from failure of the pancreas to secrete the hormone insulin. Normally, insulin enables cells to take up glucose from the bloodstream. Without the hormone, excess glucose enters the urine.

Complications from the inability to adequately control blood glucose level include nerve damage, retinal damage, and heart disease. Diabetes causes weakness and weight loss as the body breaks down tissues to obtain energy. Eventually toxins enter the bloodstream, and coma and death follow. For many years "treatment" - was a starvation diet and exercise—which hastened death. Then a series of telling experiments changed everything.

Diabetes causes weakness and weight loss as the body breaks down tissues to obtain energy. Eventually toxins enter the bloodstream, and coma and death follow.

In October 1920, a young surgeon named Frederick Banting was lecturing medical students at the University of Toronto. While describing an experiment performed in 1889 that demonstrated that removing a dog's pancreas caused it to dramatically weaken, lose weight, and die, Banting wondered whether a pancreatic extract might restore such a dog's health—and also wondered if such an approach could help people with the same symptoms. Banting asked his department chair for help and received a tiny lab, an assistant named Charles Best, and 10 dogs.

In May 1921, Banting and Best's experiments began. They removed the

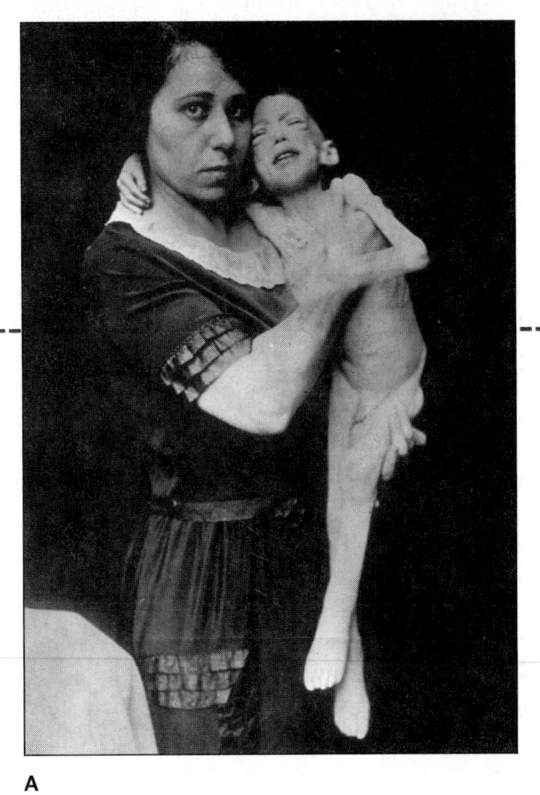

A

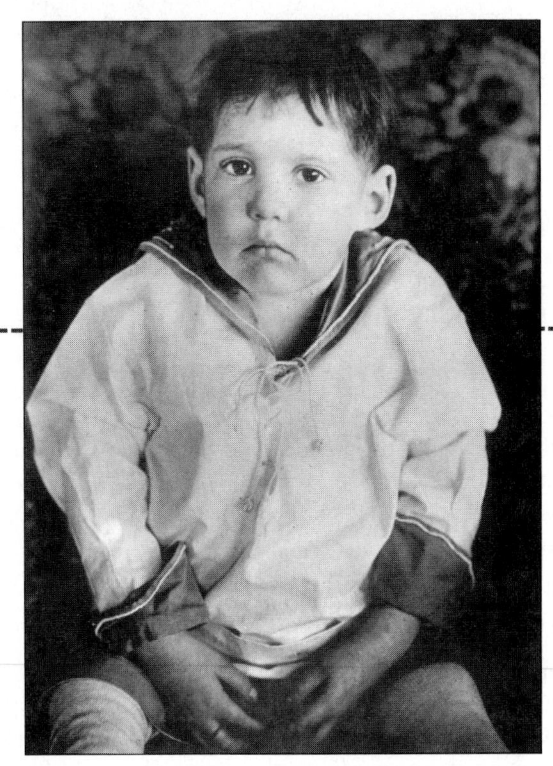

B

pancreas from one dog and tied off the duct from the second dog's pancreas that releases digestive enzymes. Within a week the first dog lay dying; the second couldn't digest very well. Banting and Best removed the dried-up pancreas from the second dog, mashed it up, and extracted fluid from it. They then injected the fluid into the dog dying from diabetes. An hour later, the diabetic dog was walking about wagging its tail!

But the treated dog died the next day. Further experiments showed that dogs require daily insulin to prevent diabetes. Banting soon thought of a better source of insulin—calf fetuses removed from cows at slaughter and discarded. Fetal pancreases produce insulin but not digestive enzymes, which makes the insulin more effective.

For many years "treatment" was a starvation diet and exercise—which hastened death. Then a series of telling experiments changed everything.

The next step was to give insulin to people. The first volunteer was a physician-friend of Dr. Banting who was dying of diabetes—the treatment saved his life. Then a 14-year-old boy was saved. Banting conducted a large-scale trial on residents of an institution for injured and ill soldiers, again achieving astounding success. Soon, people with diabetes were flocking to Toronto. By mid-1922, insulin treatment became widely available.

Biochemists produced purer insulin from cows and pigs, but some people were allergic to the nonhuman hormones. In 1982, geneticists stitched the human insulin gene into *E. coli*, and soon pure and plentiful human insulin became available, thanks to recombinant DNA technology.

Today, people with diabetes have choices—human insulin, implanted insulin pumps, and implants of insulin-producing cell clusters. Despite these advances, we still cannot exactly duplicate the function of the pancreas in secreting its hormone product.

33.1 Hormones Are Chemical Messengers and Regulators

Endocrine systems consist of glands and the hormones they secrete, as well as scattered hormone-secreting cells. A hormone moves in the bloodstream and binds to receptors on or in target cells, and the stimulated cells carry out the associated effect. The endocrine system interacts with the nervous system to control vital functions, reproduction, and maintain homeostasis.

As animals became more complex, communication systems evolved that enabled cells to "talk" to each other, as well as for individuals to interact. The **endocrine system**—glands and the hormones they secrete—helps accomplish this task.

Roles of Endocrine Systems

A **hormone** is a biochemical that certain cells release into the body fluids, which carry it to another part of the body where it alters other cells' activities. The endocrine system helps maintain homeostasis by integrating and coordinating many diverse physiological functions. Preparing the body for reproduction, for example, involves far more than initiating sperm or egg formation. Many developmental changes, some activating long-dormant genes, must ready the animal's body for reproductive maturity. In many animals, growth, maturation, and reproduction must coincide with the time of year when climatic conditions are favorable and food is available. Hormones help coordinate the timing of these events, which is crucial to their success.

The endocrine system also coordinates with the nervous system to perform another vital function—communication within the animal's body. Although the endocrine system communicates much more slowly than the nervous system, its effects are generally more prolonged. These two systems work together to produce a variety of responses, from sexual and reproductive behavior, to control of growth and development, to adjusting the delicate chemical balance of body fluids. Their functions overlap, and they control each other.

Traditionally, hormones are considered **endocrine gland** products that are secreted into the tissue spaces and transported in the circulation to exert their effects on distant cells. In recent years, however, some researchers have extended the term "hormone" to **prostaglandins,** which exert their effects locally. Similar to these communication biochemicals are **pheromones,** which are substances that an individual secretes that stimulate a physiological or behavioral response in another individual of the same species. Pheromones are similar to hormones in that they affect reproduction and behavior, but they differ because they transmit information to other individuals, rather than within an organism. They are discussed further in chapter 41.

Hormone systems likely arose as scattered hormone-secreting cells in the simpler animals. As animal life grew more complex, such cells aggregated to form glands. Today, endocrine systems include networks of glands, but also scattered hormone-secreting cells. As technology reveals more widely dispersed hormone-producing cells, the list of endocrine system components will grow.

The fact that chemical regulation is present in diverse modern organisms indicates that it is an ancient adaptation. For example, the messenger molecule **cyclic AMP** (cyclic adenosine monophosphate), or cAMP, detects a hormone's arrival at a cell in diverse species. In slime molds, cAMP initiates aggregation of single cells into a multicellular "slug" form. In the simple aquatic organism *Hydra,* cAMP conveys signals to regenerate damaged tissue. In insects, it controls metamorphosis from larva to adult. In our own bodies, cAMP relays signals that control cell division rate.

Components of Endocrine Systems

The basis of hormone function is molecular recognition—a recurrent theme in biology. Many types of hormones may circulate to all cells of an animal's body, but only certain cells respond to a particular hormone. This specificity exists because cells have receptor molecules that only bind certain chemicals. A cell that binds a particular hormone is called that hormone's **target cell.** A target cell's receptor physically fits the molecular shape of the hormone, and the hormone-receptor complex initiates the cell's response (fig. 33.1).

An endocrine gland consists of groups of cells organized into cords or plates, connected with a large blood supply to carry away the hormones that the gland produces. Whereas endocrine secretions eventually move into the circulatory system, **exocrine glands,** such as sweat, salivary, and digestive glands, secrete their nonhormonal products into ducts that release the substances onto the body's interior or exterior surface (fig. 33.2). Also important as chemical communicators are **neurosecretory cells,** which release hormones that travel to exert their effects, rather than the more locally acting neurotransmitters.

FIGURE 33.1
Chemical Messenger Systems Are Based on Molecular Recognition.
Endocrine glands and neurosecretory cells broadcast their chemical signals into tissue fluids or the general circulation. However, only target cells receive and respond to the message because they express specific molecules that interact with the signal molecule.

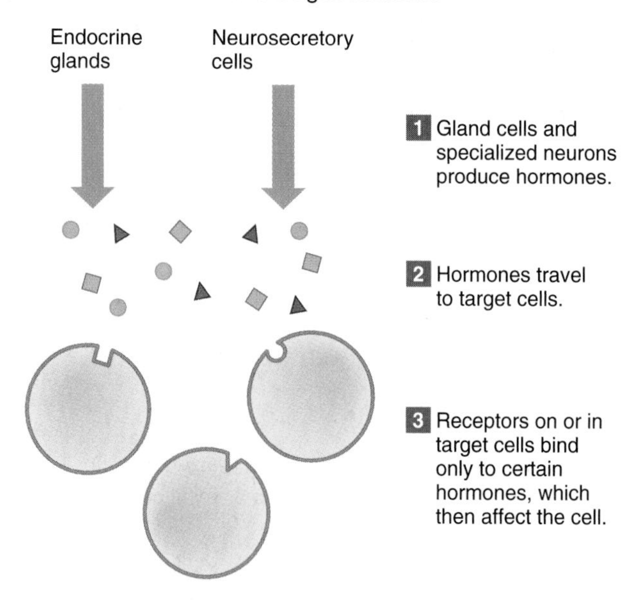

FIGURE 33.2
Endocrine Versus Exocrine Glands.
(**A**) Exocrine glands release their secretions through ducts. (**B**) Endocrine glands release their hormones directly into the circulatory system.

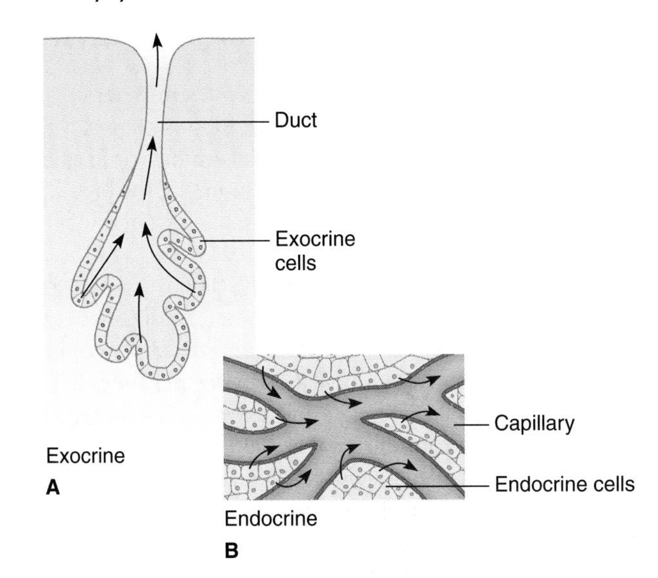

Not all hormones originate in cells within a gland, however, and not all hormones travel very far to evoke a response. Individual cells scattered throughout a tissue or organ may produce hormones that exert localized effects. For example, the stomach and intestinal linings contain cells that secrete "gut hormones" that regulate movement of the digestive tract and secretion rates of digestive biochemicals. Similarly, scattered cells in the mammalian heart secrete atrial natriuretic hormone, which regulates blood pressure. The kidney and placenta also contain hormone-secreting cells, and white blood cells produce and release hormones.

33.1 MASTERING CONCEPTS

1. What is the overall function of an endocrine system?
2. How do neural and hormonal regulation differ?
3. What is the basis of recognition in chemical communication systems?
4. What is the evidence that hormonal systems are ancient?

OLC

33.2 How Hormones Exert Their Effects

Peptide hormones cannot cross the cell membrane's lipid bilayer, and instead bind to receptors that trigger second messenger cascades. Steroid hormones can cross the lipid bilayer, and bind to receptors inside cells, ultimately activating expression of particular genes. Hormone levels remain within specific ranges by negative feedback mechanisms and, less commonly, positive feedback loops.

One way to classify hormones is as either water-soluble or lipid-soluble. Water-soluble hormones include **peptide hormones,** which are short chains of amino acids. Lipid-soluble hormones include **steroid hormones.** These two general types of hormones have different mechanisms of action. Peptide hormones bind to cell surface receptors, whereas steroid hormones bind to receptors that are inside their target cells (fig. 33.3).

FIGURE 33.3
Two Types of Hormones.
(**A**) Peptide hormones are water-soluble and bind to cell surface receptors. (**B**) Steroid hormones are lipid-soluble. They pass through cell membranes and bind to receptors in the cytoplasm or nucleus.

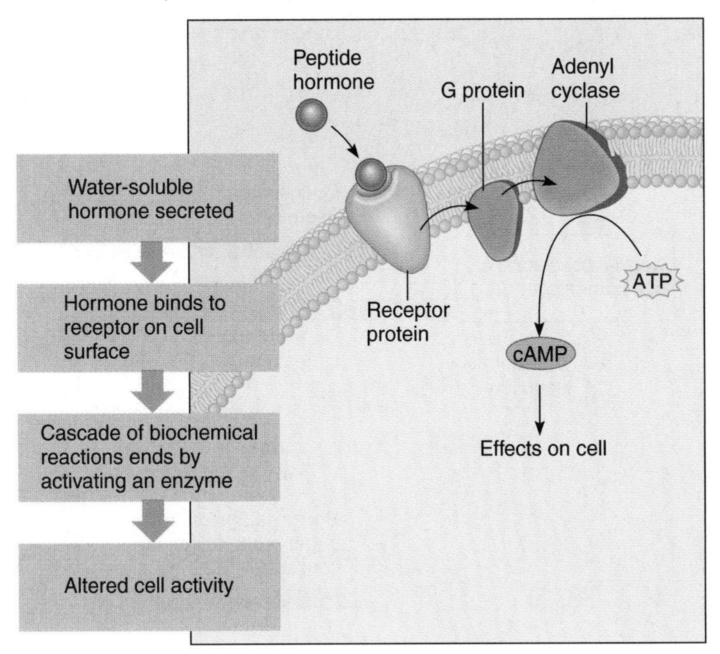

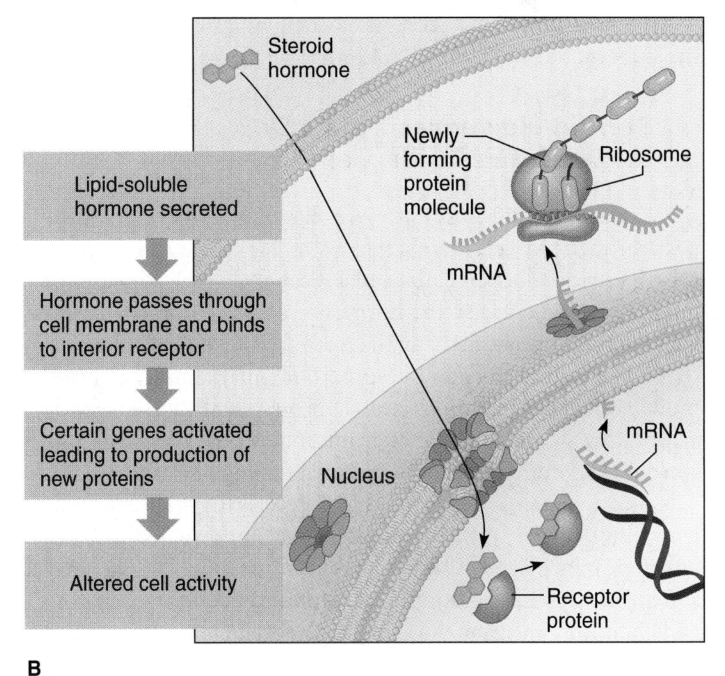

Peptide Hormones Bind to Cell Surface Receptors

Peptide hormones are water-soluble, so the watery plasma of the bloodstream easily carries them to virtually all of an animal's cells. Because peptide hormones are lipid-insoluble, however, they cannot cross cell membranes. Thus, a peptide hormone generally does not enter the target cell itself (one exception may be thyroid hormone), but instead it binds to one of the hundreds or even thousands of receptors on the target cell that it fits as a key fits a lock. When a peptide hormone binds to its receptor, it initiates the target cell's response. For example, specific ion channels in the membrane might open. This would alter the concentration of ions within the cell, which might change the rate of certain cellular activities.

Usually when a hormone and a membrane receptor bind to form a complex, it activates a second messenger that triggers the cellular response. (Chapter 4 introduced second messengers.) This intracellular molecule is called a second messenger because it must first be activated by a hormone before it activates the enzymes that produce the effects associated with the hormone.

Cyclic AMP is a common second messenger (see fig. 4.22). When a hormone binds to a receptor, it activates a coupling protein (G protein), which in turn activates another membrane-associated enzyme called adenyl cyclase on the inner face of the membrane. Adenyl cyclase then catalyzes conversion of ATP into cAMP. (Recall that ATP is the energy currency of the cell.) In turn, cAMP activates specific enzymes within the cell. These enzymes catalyze the reactions that produce the response to the hormone, such as cell division, secretion, contraction, or metabolic changes. Different hormones can stimulate cAMP formation. The specific biochemical events following cAMP formation trigger the effects characteristic of different peptide hormones.

Parathyroid hormone is one type of peptide hormone. Cells of the parathyroid glands in the neck synthesize and secrete the hormone into the bloodstream. The hormone eventually binds to receptors on specific kidney cells, where it converts ATP to cAMP. The cAMP sets into motion a series of biochemical reactions that cause the kidney tubules to reabsorb calcium from the forming urine.

Steroid Hormones Bind to Receptors Inside Cells

Because steroid hormones are fat-soluble, they can easily pass through the cell membranes of target cells. Once inside a cell, the steroid hormone binds to a receptor in the cytoplasm or nucleus. In the nucleus, the hormone-receptor complex stimulates particular genes to direct the manufacture of particular proteins. The steroid hormone testosterone, for example, activates muscle cell genes that direct muscle protein synthesis. The body synthesizes steroid hormones from cholesterol—which is one reason why some cholesterol is essential for good health.

Abuse of steroid drugs by athletes reveals the many effects of a hormone on the body. Steroid users may improve performance and appearance in the short term, but in the long run they may suffer. Steroids hasten adulthood, stunting height and causing early hair loss, and may cause males to develop breast tissue and females to develop a deepened voice, hairiness, and a male physique. Steroids can damage the kidneys, liver, and heart and cause atherosclerosis. In males, the body mistakes synthetic steroids for the natural hormone and lowers its own production of testosterone, causing infertility later.

Feedback Loops Control Hormone Levels

To maintain homeostasis, an animal's body must strictly regulate the levels of specific hormones in the bloodstream at a given time. Feedback interactions between the hormone (or its effects) and the gland often regulate levels of hormones in the blood. In a **negative feedback** loop, the accumulation of a certain biochemical switches off hormone synthesis (fig. 33.4).

A negative feedback loop controls blood glucose levels (fig. 33.5). After a meal rich in carbohydrates, the digestive system breaks down the complex carbohydrates into the simple carbohydrate glucose, which enters the circulation in the small intestine. The resulting rise in blood sugar level stimulates beta cells in the pancreas to secrete the hormone insulin into the bloodstream. Then, insulin stimulates target cells to admit glucose, which the cell then stores or breaks down to release energy. As glucose leaves the bloodstream to enter cells, the pancreas slows insulin secretion. Rising insulin levels thus eventually turn off insulin production as the hormone completes its "job." Negative feedback also controls levels of the hormone glucagon, but with effects opposite those of insulin. Alpha cells in the

FIGURE 33.4

Feedback Loops Regulate Hormone Action.

In this generalized depiction, cells release hormone A to lower the levels of an accumulating biochemical, and hormone B to offset a deficit.

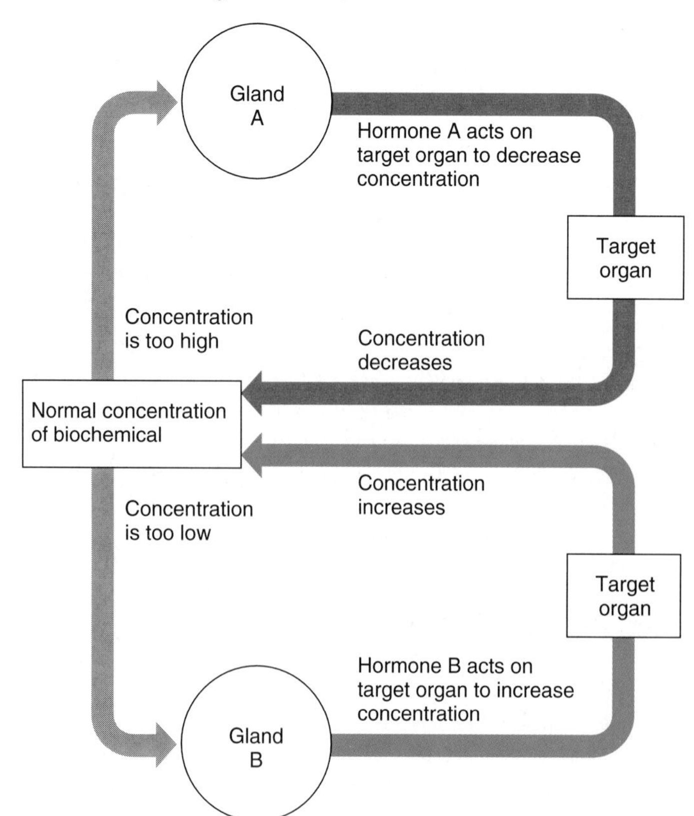

FIGURE 33.5

The Pancreas Regulates Blood Glucose Level.

Beta cells secrete insulin, which admits glucose into cells. Alpha cells release glucagon, which increases blood glucose levels.

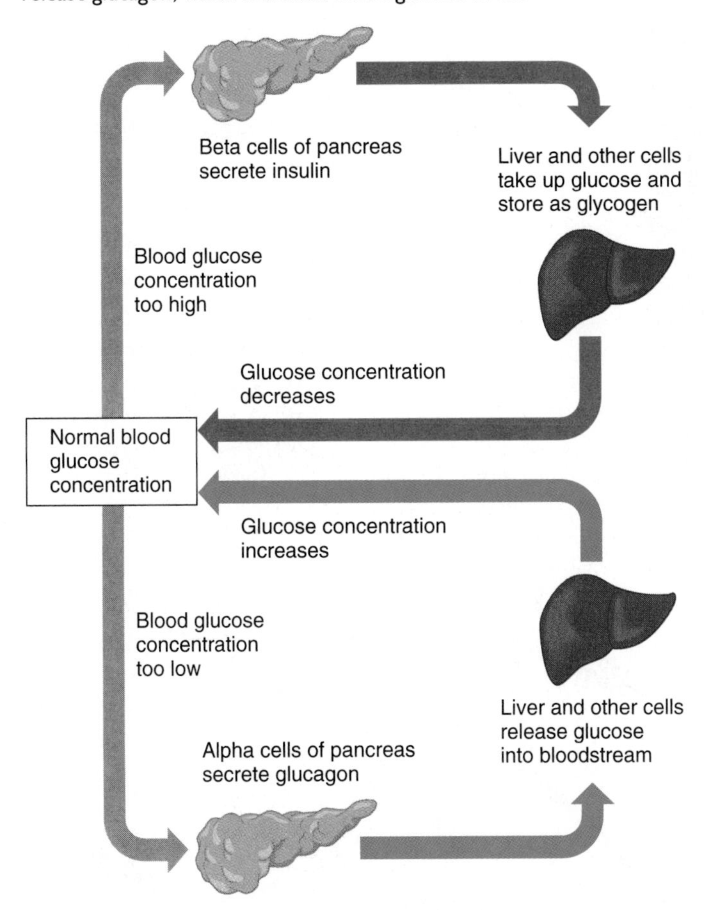

pancreas secrete glucagon, in response to lowering levels of glucose in the bloodstream.

In a **positive feedback** loop, an accumulating biochemical increases its own production. For example, at the onset of labor, the uterus contracts, releasing the hormone oxytocin and prostaglandins. These substances intensify the uterine contractions, and the contractions further stimulate oxytocin and prostaglandin production. When the baby is born, other hormonal changes stop the cycle.

33.2 MASTERING CONCEPTS

1. How do water-soluble hormones function?
2. How do steroid hormones exert their effects?
3. Distinguish between the ways that negative and positive feedback loops control hormone secretion.

OLC

33.3 Diversity of Endocrine Systems

Changing levels of two hormones control insect metamorphosis. Complex hormonal cycles regulate reproduction in vertebrates. The hypothalamus and the pituitary gland control endocrine systems in vertebrates.

Hormones control and coordinate timing and expression of complex biological functions that may entail many biochemical reactions. All animals have some hormonal control. Neurosecretory cells accomplish this in simpler invertebrates; similar cells organized into glands control hormone synthesis in more complex invertebrates and in vertebrates.

Hormones in Invertebrates

Among invertebrates, hormones control a variety of functions, including regeneration in jellyfishes and flatworms; gonad maturation in flatworms, mollusks, and arthropods; and regulation of blood sugar level in segmented worms. Molting of the outer covering (exoskeleton) of crustaceans, such as crabs and lobsters, is controlled by a molt-inhibiting hormone produced in the sinus gland located in the animal's eyestalks. The eyestalks also produce hormones that adjust body coloration and regulate blood sugar levels. A familiar hormone-orchestrated event of invertebrate life is a larval insect's metamorphosis into its adult form (fig. 33.6). The physiological and biochemical changes that sculpt a butterfly from a caterpillar, a fly from a maggot, or a beetle from a grub are among the most profound in the animal kingdom.

An insect larva is a streamlined eating machine. The larva is equipped to eat but lacks such structures as sex organs and wings. It grows at a spectacular rate, increasing its size several thousandfold in a few days. This creates a problem—how can the young insect contain its increasingly bulging body? The solution is to molt—to periodically shed its cuticle (outer covering) to allow for rapid growth.

After several larval molts, the insect secretes a cocoon and becomes a pupa. Larval cells die, as other cells that were set aside in tiny disclike packets in early larval life obtain energy from the degenerating larval cells. The reawakened cells divide and project outwards, like a painting springing into three-dimensional life. Within a few days, the cells form small adult body parts—legs, antennae, wings. Eventually the pupa splits, and an adult insect emerges.

The adult insect is wet and compacted as it emerges from the cocoon, its wings plastered against its abdomen. It takes a few hours to dry out and for fluid to flow into veins in the wings and expand them. Soon the insect is free to embark on an existence very different from its life as a larva. Rather than living to eat, the adult insect lives to mate.

Insect metamorphosis is a complex, highly coordinated, and intricately timed process. Changing levels of two hormones—**juvenile hormone** and **molting hormone** (a cholesterol derivative called ecdysone)—control the process. In the early larva, juvenile hormone predominates. Neurosecretory cells of the brain produce juvenile hormone, which is released from structures called the corpora allata. Periodic increases in molting hormone levels trigger molting between larval stages. Then, just before the pupa forms, the neurosecretory cells that produce juvenile hormone shut off, and molting hormone predominates. Molting hormone then oversees metamorphosis. Molting hormone is produced from the prothoracic gland in the brain, under the influence of brain hormone.

Hormones and Glands in Vertebrates

It is difficult to generalize about hormones and glands across vertebrate species, because the same hormone may have different

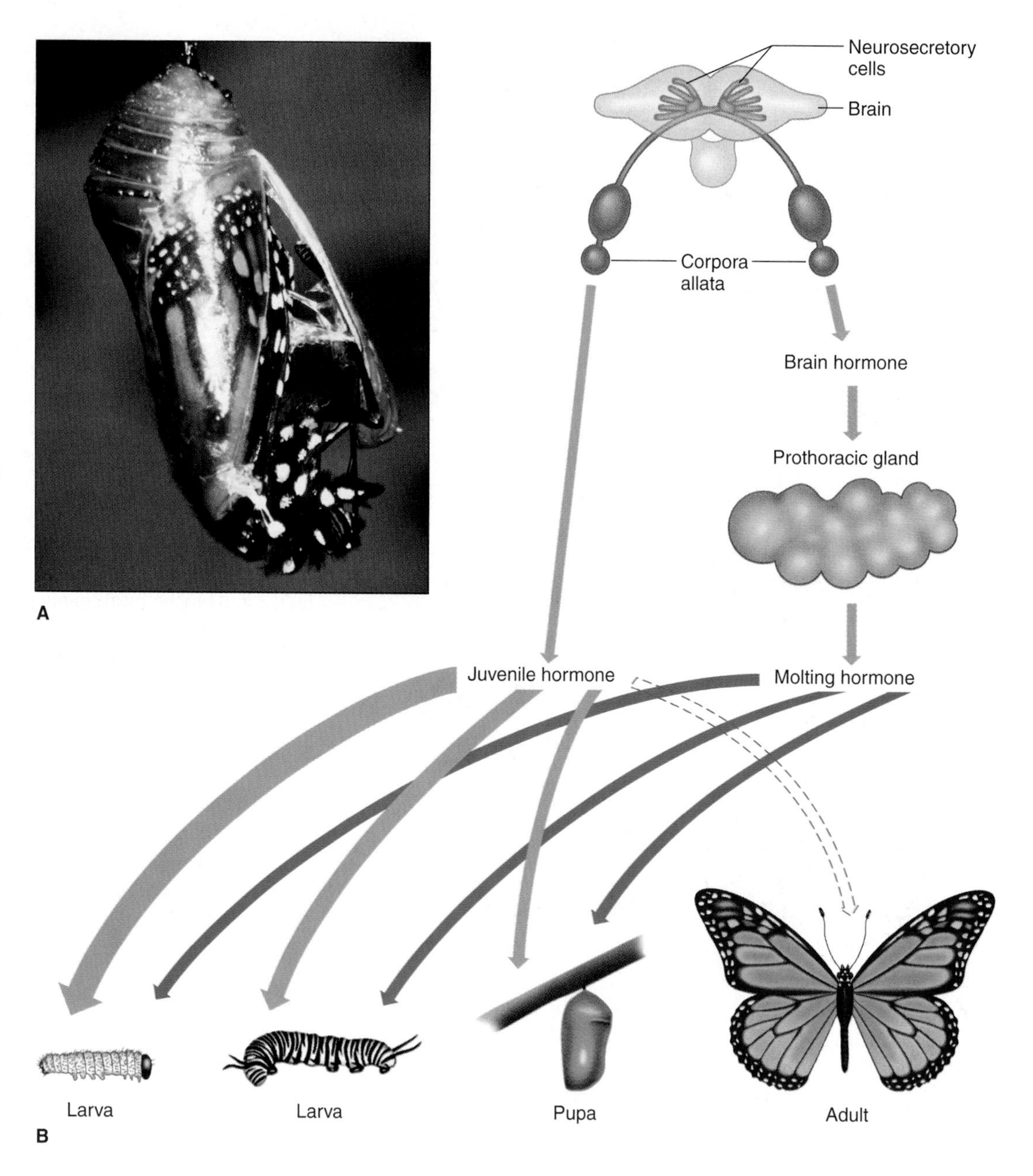

FIGURE 33.6
Metamorphosis.
(**A**) A monarch butterfly emerges from its chrysalis (pupa of a butterfly). (**B**) Two hormones control insect metamorphosis. Juvenile hormone, produced by neurosecretory cells of the brain and released from the corpora allata, predominates in the larva. Molting hormone is released periodically from the prothoracic gland under the control of brain hormone. The episodic increase in molting hormone stimulates molting of the larval stages. The decline of juvenile hormone levels toward the end of larval life results in a molt into a pupa and then metamorphosis into the adult.

functions in different species. This may reflect differences in lifestyles. In humans, for example, the thyroid gland hormone thyroxine sets the pace of metabolism, but in frogs, thyroxine controls metamorphosis. Add thyroxine to water, and a tadpole becomes a frog too soon. Block a tadpole's thyroxine secretion, and it remains a tadpole. In birds, which have metabolic requirements different from those of humans and frogs, thyroxine controls heat production and oxygen consumption. Other diversities in hormone function are harder to explain. For example, antidiuretic hormone (ADH) causes the human kidney to reabsorb water into the blood. In the prairie vole, a rodent, the same hormone is responsible for a male's monogamy (dedication to a single mate) and caring for its young!

Sexual Seasons Reproductive behavior illustrates how hormonally controlled processes can vary among vertebrate species. The human female is unusual among animals in that she is sexually receptive at all times. Because of her monthly cycle, she is referred to as a cyclic ovulator. Her sexual receptivity is not limited to the time when she ovulates.

In contrast, most female mammals are seasonal ovulators. They go through periods of estrus (heat) when they are both sexually receptive and ovulating. The correlation of sexual behavior and peak fertility may be an adaptation to maximize the chance of conceiving. During estrus in the seasonal ovulator, estrogen production surges. Higher estrogen levels induce her to display her interest in potential mates by her scent, appearance, posture, and behavior. When she is not in estrus, she is unreceptive to mating behavior of any sort. In some species, females are actually physically incapable of copulation when they are not in estrus. The African bush baby, for example, grows a covering of skin over her vagina in the sexual off-season. The male adjusts to the female pattern. In some species, males manufacture sperm only when females are in estrus.

Even greater reproductive economy is seen among induced ovulators. In rabbits and domestic cats, copulation induces ovulation. The female mink ovulates when her mate bites her on the back of the neck.

In mice, a male can disrupt a female's hormones even after she becomes pregnant. If a female is less than 4 days pregnant when she meets a "new" male (a male other than the father of the embryos), his presence causes her to abort her embryos and to become sexually receptive to him. To trigger this "pregnancy block," the male emits a pheromone that begins a hormonal cascade in the female's body, culminating with the drastic lowering of blood levels of the hormone prolactin. When prolactin levels are inadequate, embryos cannot remain implanted in the uterus.

The Hypothalamus Oversees Neuroendocrine Control All vertebrate endocrine systems are regulated in large part by the **hypothalamus,** which is the neuroendocrine control center in the brain, and the **pituitary,** which is a nearby pea-sized gland. The hypothalamus directly or indirectly controls most other vertebrate endocrine glands (fig. 33.7).

FIGURE 33.7
Endocrine System Control.
The hypothalamus and the pituitary gland control the functioning of the endocrine system.

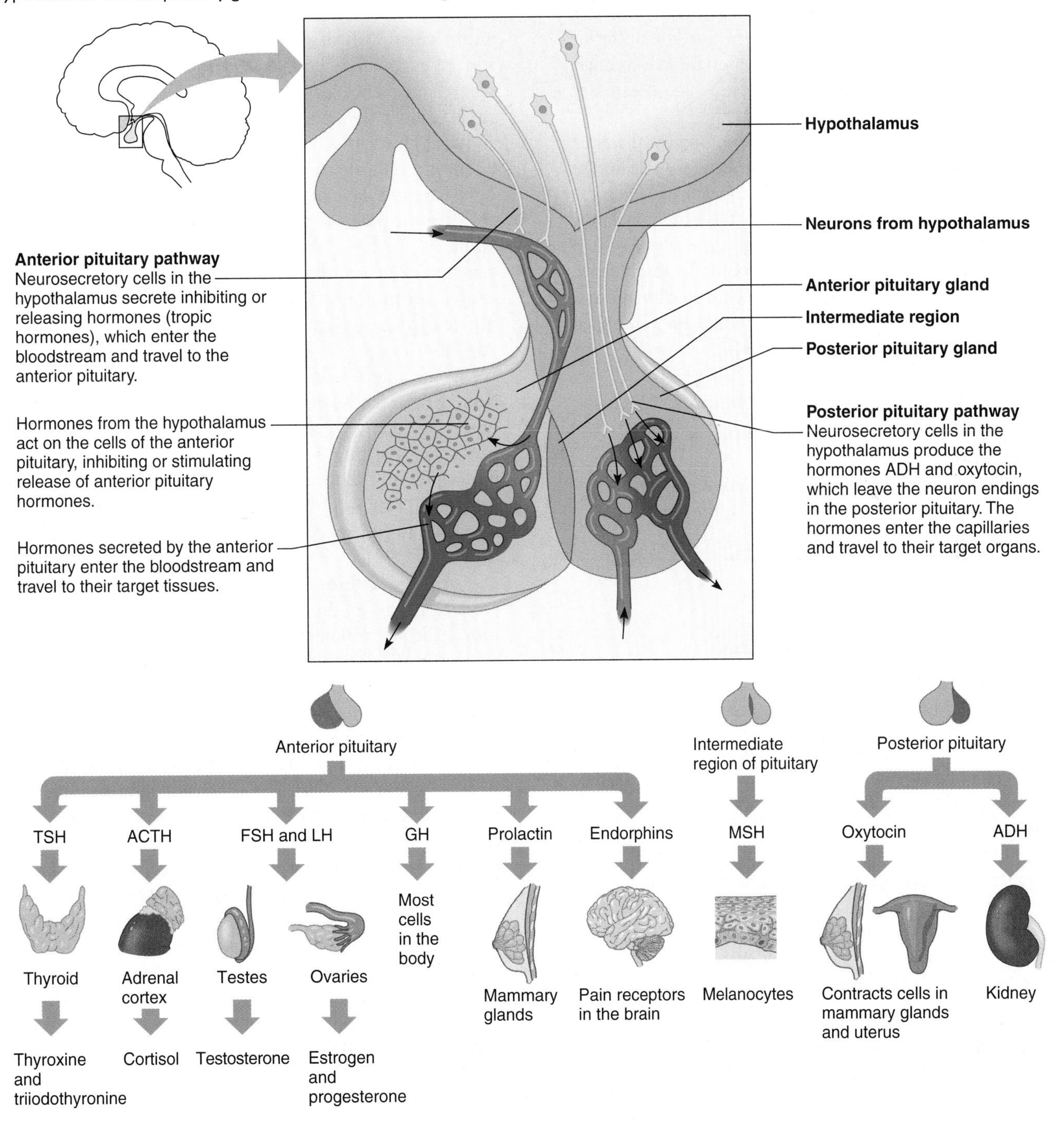

The hypothalamus links the nervous and endocrine systems by controlling pituitary secretions. The pituitary gland is attached to a stalk extending from the hypothalamus and is actually two distinct glands, the anterior (front) and posterior (back) pituitary. Each half of the pituitary releases a different set of hormones and is regulated by the hypothalamus in a different manner. Between the anterior and posterior lobes of the pituitary is an intermediate region known to produce at least one hormone.

The hypothalamus develops from the central nervous system and maintains a close relationship with the brain, with nerve fibers touching many parts of the brain. In the embryo, the posterior pituitary develops from the hypothalamus. Later, neurosecretory cells in the functional hypothalamus synthesize antidiuretic hormone (ADH) and oxytocin, the hormones that the posterior pituitary releases. ADH and oxytocin move through the stalk along the axons of the cells that produced them to the posterior lobe of the pituitary, where they are stored within the cell endings. When neural activity in the brain stimulates the neurosecretory cells, they release their hormones.

The hypothalamus also controls the anterior lobe of the pituitary, but in a different manner. Other neurosecretory cells in the hypothalamus secrete hormones that either stimulate or inhibit the production of anterior pituitary hormones. These hypothalamic hormones reach the anterior pituitary through a specialized system of blood vessels. Hormones that one gland produces to influence another gland's hormone production are called **tropic hormones.** When the hypothalamus produces such hormones, they are called **releasing hormones** or **inhibiting hormones.**

33.3 MASTERING CONCEPTS

1. Explain how changing levels of two hormones control insect metamorphosis.
2. How do ovulation patterns vary in different mammals?
3. How do the hypothalamus and the pituitary gland interact?

OLC

33.4 A Closer Look at the Endocrine System in Humans

Releasing hormones from the hypothalamus stimulate the anterior pituitary to secrete growth hormone, thyroid-stimulating hormone, adrenocorticotropic hormone, prolactin, and sex hormones. The posterior lobe stores and releases antidiuretic hormone and oxytocin, and the intermediate region secretes melanocyte-stimulating hormone. The thyroid, parathyroid and adrenal glands, and the pancreas secrete hormones that control metabolism. Cycling levels of hormones control the female reproductive system. In males constant hormone levels foster sperm development, and testosterone promotes male characteristics and controls secretion from the prostate gland.

Vertebrate endocrine systems vary, but consist of the same basic glands and hormones, summarized in table 33.1. Here we consider the human system as an example (fig. 33.8).

The Pituitary Is Two Glands in One

The pituitary gland is really two glands in one, both structurally and functionally.

Hormones of the Anterior Pituitary In humans, the anterior pituitary gland produces hormones that directly affect body parts and hormones that affect other endocrine glands. **Growth hormone (GH)** promotes growth and development of all tissues by increasing protein synthesis and cell division rates. GH stimulates cells to take up more amino acids, to mobilize fats, and to release glucose from the liver to supply energy. GH also increases the cell division rate in cartilage and bone, making the body grow taller.

Levels of GH peak in the preteen years and, together with rising levels of sex hormones, cause adolescent growth spurts. Abnormal levels of GH have noticeable effects. A severe deficiency of the hormone during childhood leads to pituitary dwarfism, which produces extremely short stature. This condition can be treated with human growth hormone manufactured using recombinant DNA technology. Excess GH causes overgrowth of many tissues. In a child, a pituitary tumor produces a pituitary giant. In an adult, excess GH does not affect overall height because the growth regions in the long bones are no longer active, but it causes acromegaly. This is a thickening of the bones, most noticeably in the hands and face, that can respond to GH (fig. 33.9).

Several anterior pituitary hormones affect reproduction. **Prolactin** stimulates milk production in a woman's breasts after she gives birth. In males and in women who are not breast-feeding, an inhibitory hormone produced in the hypothalamus suppresses prolactin synthesis in the anterior pituitary. The inhibition is overcome in nursing mothers by nerve impulses generated when the infant sucks on the nipples.

The **endorphins** are natural analgesics (painkillers) that are also found in the anterior pituitary. These peptides have morphinelike (opiatelike) effects and are released in response to prolonged painful stimuli. Endorphins bind to opioid receptors in the brain to initiate their analgesic and possibly other effects.

The remaining anterior pituitary gland hormones are tropic hormones that influence hormone secretion by other endocrine glands. The gonadotropic hormones affect the ovaries and testes, the organs where gametes are produced. In the human female, **follicle-stimulating hormone (FSH)** initiates development of ovarian follicles, oocyte maturation, and release of estrogen from the follicles. In the male, FSH promotes the development of the testes and manufacture of sperm cells. In the female, **luteinizing hormone (LH)** stimulates an ovary to release an oocyte each month. In the male, LH stimulates the testes to produce the hormone testosterone, which is necessary for sperm development. LH in the male is also called interstitial cell stimulating hormone (ICSH).

TABLE 33.1 Summary of Major Hormone Functions in the Human Endocrine System

GLAND	HORMONE	TYPE	TARGET	EFFECTS	REGULATED BY
Hypothalamus	Releasing hormones	Peptides	Anterior pituitary	Stimulate or inhibit release of hormones from anterior pituitary	
	Inhibiting hormones	Peptides			
Anterior pituitary	Growth hormone (GH)	Protein	All cells	Increases cell division, protein synthesis	Hypothalamus
	Thyroid-stimulating hormone (TSH)	Glycoprotein	Thyroid gland	Stimulates secretion of thyroxine and triiodothyronine	Hypothalamus, Thyroxine blood level
	Adrenocorticotropic hormone (ACTH)	Protein	Adrenal cortex	Stimulates secretion of glucocorticoids and mineralocorticoids	Hypothalamus , Glucocorticoid blood level
	Prolactin	Protein	Cells in mammary glands	Stimulates milk secretion	Hypothalamus
	Follicle-stimulating hormone (FSH)	Protein	Ovaries	Stimulates follicle development, oocyte maturation, release of estrogen	Hypothalamus
	Luteinizing hormone (LH)	Protein	Ovaries, testes	Promotes ovulation; stimulates late sperm cell development, testosterone synthesis	Hypothalamus
Intermediate pituitary	Melanocyte-stimulating hormone (MSH)	Peptide	Skin	Increases skin pigmentation	Hypothalamus
Posterior pituitary	Antidiuretic hormone (ADH)	Peptide	Kidneys, smooth muscle cells of blood vessels	Helps maintain composition of body fluids	Concentration of solutes in body fluids
	Oxytocin	Peptide	Uterus, cells of mammary glands	Stimulates muscle contraction	Nervous system
Thyroid gland	Thyroxine, triiodothyronine	Amines	All cells	Increase metabolic rate	TSH
	Calcitonin	Peptide	Bone	Increases rate of calcium deposition in bone	Blood calcium level
Parathyroid glands	Parathyroid hormone	Peptide	Bone, digestive organs, kidneys	Releases calcium from bone, increases calcium absorption in digestive tract and kidneys	Blood calcium level
Adrenal cortex	Glucocorticoids	Steroids	All cells	Increase glucose levels in blood and brain	ACTH
	Mineralocorticoids	Steroids	Kidneys	Maintain blood volume and electrolyte balance	Blood potassium level
Adrenal medulla	Catecholamines	Amines	Blood vessels	Raise blood pressure, constrict blood vessels, slow digestion	Nervous system
Pineal gland	Melatonin	Amine	Other endocrine glands	Regulates effects of light-dark cycle on other glands	Light-dark cycle
Pancreas	Insulin	Peptide	All cells	Increases cellular glucose uptake and conversion of glucose to glycogen	Blood glucose level
	Glucagon	Protein	All cells	Stimulates breakdown of glycogen to glucose	Blood glucose level
	Somatostatin	Peptide	Small intestine	Regulates nutrient absorption rate	Blood glucose level
Ovaries	Progesterone	Steroid	Uterine lining	Controls monthly secretion patterns	FSH and LH
	Estrogen	Steroid	Uterine lining	Increases rate of mitosis	FSH and LH
Testes	Testosterone	Steroid	Skin, muscles, sperm-producing cells	Maintains secondary sexual characteristics, promotes sperm development	FSH and LH

FIGURE 33.8

Some Human Endocrine Glands.

The endocrine system includes several glands that contain specialized cells that secrete hormones, as well as cells scattered among the other organ systems that also secrete hormones. (The parathyroids are not actually visible, as depicted here—they are behind the thyroid.)

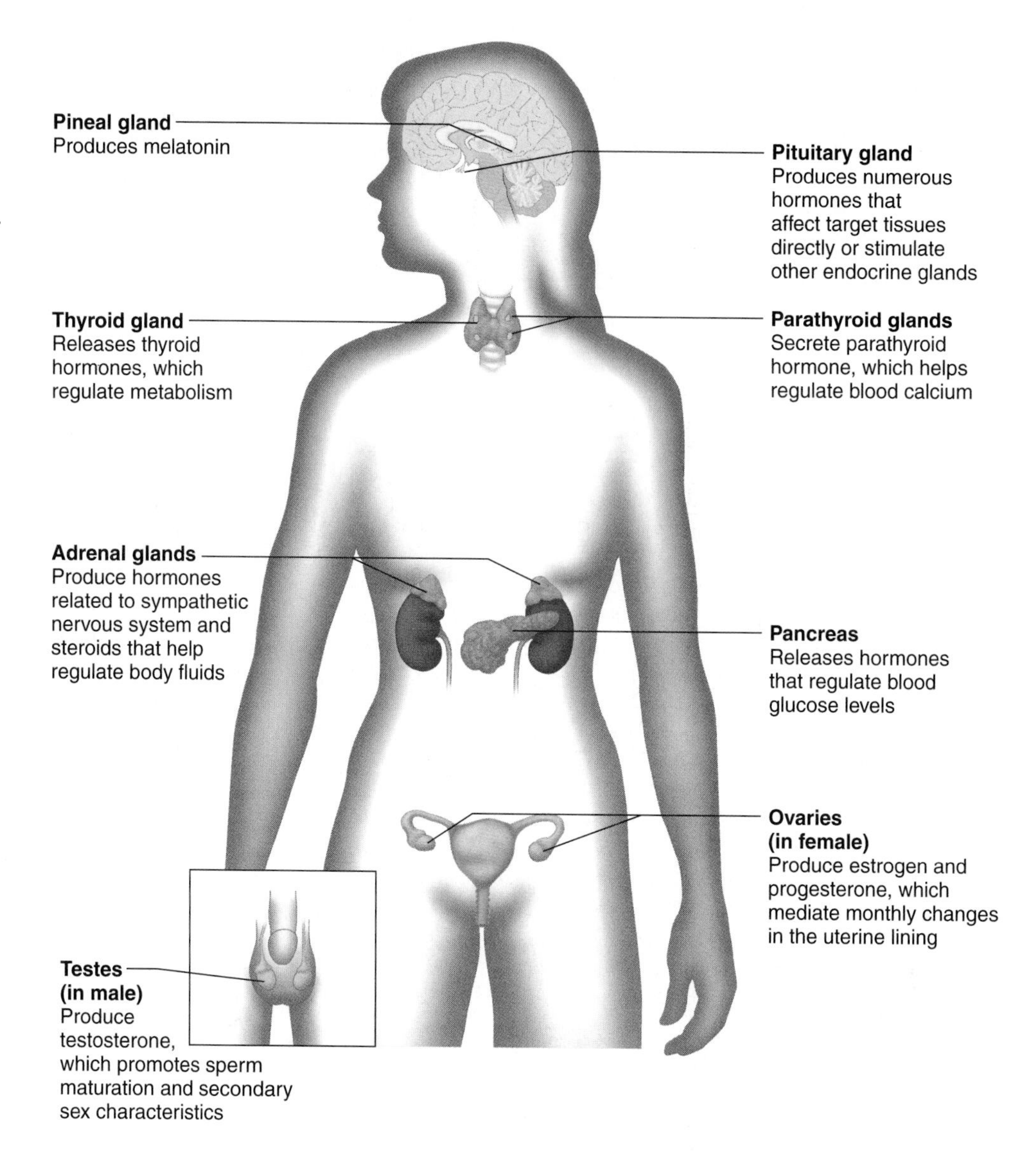

FIGURE 33.9

Growth Hormone Overproduction That Begins in Adulthood Causes Acromegaly.

Adult limb bones no longer respond to growth hormone; however, the bones of the hand and face still respond. When GH is overproduced, these bones enlarge considerably, as the changes in the facial features of this woman show as she ages from (**A**) 16 years to (**B**) 33 years to (**C**) 52 years.

The anterior pituitary also releases **thyroid-stimulating hormone (TSH),** which causes the thyroid gland in the neck to release two hormones that control metabolism. Another anterior pituitary hormone, **adrenocorticotropic hormone (ACTH)**, stimulates release of the **glucocorticoid hormones** from the outer portion (cortex) of the **adrenal glands,** which are located on top of the kidneys. The glucocorticoids increase blood glucose level during stress.

Hormones of the Posterior Pituitary Two hormones are synthesized in the hypothalamus and stored in and released from the posterior pituitary. One, **antidiuretic hormone (ADH),** contracts smooth muscle cells lining blood vessels, which helps to control blood pressure. It is called antidiuretic ("diuretic" means increasing urine flow) hormone because it also makes the kidneys' collecting ducts more permeable to water so that the body can reabsorb more water.

Control of ADH secretion helps maintain the chemical balance of body fluids. Specialized cells in the hypothalamus called osmoreceptors monitor the amount of water in the blood. If the blood has too little water, the osmoreceptor cells stimulate the posterior pituitary to release more ADH. As ADH increases water reabsorption within the kidneys, diluting the blood, the osmoreceptor cells signal the ADH-manufacturing cells in the hypothalamus to decrease their activity.

Oxytocin is the other hypothalamus-produced hormone stored in and released from the posterior pituitary. This hormone contracts cells in the breasts, causing them to release milk when a baby nurses, and contracts cells in the uterus, which pushes a baby out during labor. Synthetic oxytocin can induce labor or accelerate contractions in a woman who is exhausted from a long effort at giving birth.

Melanocyte-stimulating hormone (MSH) is produced in the region of the pituitary between the anterior and posterior lobes. MSH causes the color changes that many vertebrates undergo. When the hormone binds to receptors on melanocytes in the skin, pigment granules disperse, darkening the skin.

In fishes, amphibians, reptiles, and some mammals, MSH is an adaptation that provides camouflage. In humans, the intermediate lobe is poorly developed, and significant levels of MSH are normally found only in children and pregnant women. Some researchers hypothesize that excess sunlight may trigger MSH secretion, which protects the skin against sun damage.

33.4 MASTERING CONCEPTS

1. Which pituitary hormones are tropic hormones?
2. Which hormones does the anterior pituitary produce, and what are their functions?
3. What are the functions of ADH and oxytocin?
4. What is the function of melanocyte-stimulating hormone?

OLC

Glands That Regulate Metabolism

The thyroid gland, parathyroid glands, adrenal glands, and the pancreas secrete hormones that control metabolism.

The Thyroid Gland Sets the Metabolic Pace The **thyroid** is a two-lobed gland located at the front of the larynx and trachea in the neck. It produces two hormones, **thyroxine** and **triiodothyronine,** that increase the rate of cellular metabolism in target cells. Dietary iodine is required to produce these hormones. The thyroid gland actively absorbs and concentrates iodine, which reaches levels 25 times higher than blood levels.

The lobes of the thyroid consist of sections called follicles that are lined with cells that secrete thyroxine and triiodothyronine. The follicles fill with a sticky substance that consists of the two hormones bound to a glycoprotein called thyroglobulin, which keeps the hormones temporarily out of the bloodstream when their levels in the blood rise too high. The endocrine cells secrete thyroxine and triiodothyronine upon stimulation by TSH from the pituitary, which in turn is stimulated by thyrotropin-releasing hormone (TRH) from the hypothalamus, in response to low blood levels of the hormones (fig. 33.10).

FIGURE 33.10
Control of Thyroid Gland Function.
Thyrotropin-releasing hormone (TRH) from the hypothalamus triggers secretion of thyroid-stimulating hormone (TSH) from the anterior pituitary, which causes the thyroid gland to secrete the thyroid hormones thyroxine and triiodothyronine. Blood levels of these two hormones affect the rate of their synthesis in a negative feedback loop.

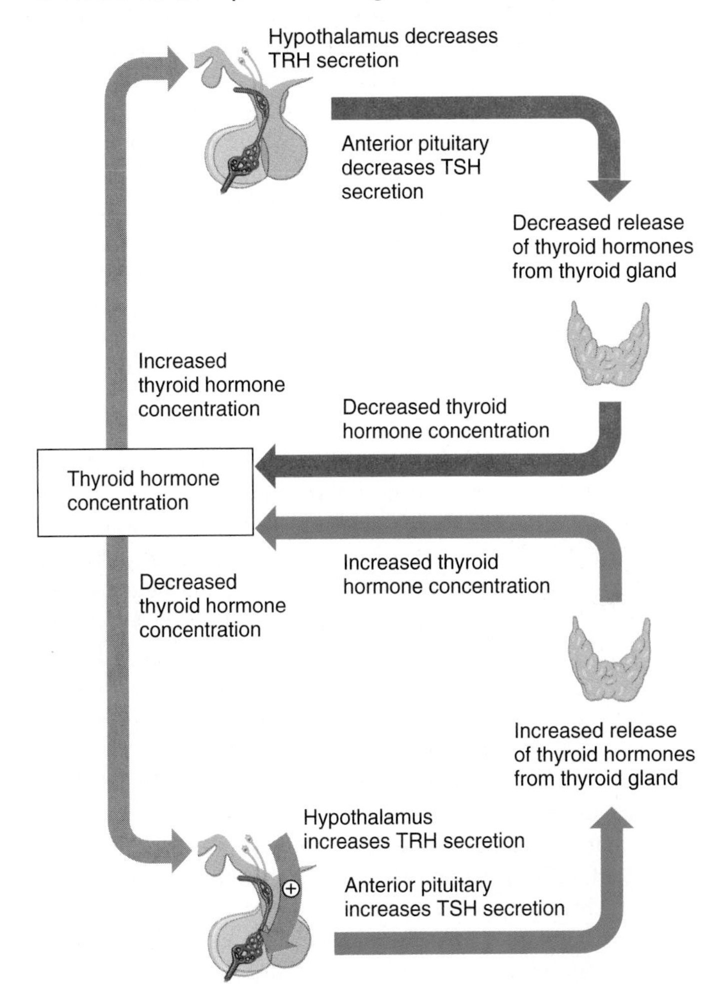

Thyroid hormones bind to many different cell types, where they speed metabolism by increasing supplies of enzymes required for cellular respiration. The specific nature of this activity depends upon the cell type. Under thyroid stimulation, the lungs exchange gases faster, the small intestine absorbs nutrients more readily, and fat levels in cells and in blood plasma diminish.

An underactive thyroid gland (hypothyroidism) slows metabolic rate. Because the body burns fewer calories, weight increases despite a poor appetite. Heartbeat slows, and blood pressure and body temperature fall. Hypothyroidism beginning at birth produces cretinism, leaving the child physically and mentally retarded. If cretinism is detected at birth and treated with thyroxine before 3 months of age, however, the child develops normal intelligence. Hypothyroidism beginning in adulthood is termed myxedema and causes lethargy, a puffy face, and dry, sparse hair. It, too, is treatable with thyroxine.

Hypothyroidism due to lack of iodine in the diet causes the thyroid gland to swell and form a lump in the neck called a **goiter.** Ancient Egyptian doctors treated this type of goiter with seaweed, not realizing that iodine in the seaweed reversed the condition by enabling the thyroid to produce more hormone. Today iodine-deficient goiter is rare in nations where iodine is added to table salt.

An overactive thyroid, called hyperthyroidism, can also cause a goiter. Other symptoms reflect accelerated metabolism, including a very short attention span, irritability and hyperactivity, and elevated heart rate, blood pressure, and temperature. Appetite is great, but the high metabolic rate keeps weight off. The chapter 3 opening essay discusses thyroid cancer.

The thyroid produces a third hormone, **calcitonin,** which decreases blood calcium level under certain conditions. Calcitonin stimulates adenyl cyclase activity in its target tissues, which include the kidneys, bones, and the small and large intestines. The effect of calcitonin is to inhibit bone resorption, which is the process by which bones release calcium to the bloodstream. For example, calcitonin is important during pregnancy by keeping the developing fetus from "taking" too much calcium from the woman's skeletal system. Levels of calcitonin greatly increase during pregnancy and lactation.

The Parathyroid Glands Control Calcium Level Embedded in the back of the thyroid gland are four small groups of cells that form the **parathyroid glands.** These glands secrete **parathyroid hormone (PTH),** which maintains calcium levels in blood and tissue fluid by releasing calcium from bones and by enhancing calcium absorption through the digestive tract and kidneys (fig. 33.11). Parathyroid hormone action opposes that of calcitonin and also inhibits the kidneys from reabsorbing phosphate.

Calcium is vital to muscle contraction, neural impulse conduction, blood clotting, bone formation, and the activities of many enzymes. Underactivity of the parathyroids can swiftly be fatal. Excess parathyroid hormone causes calcium to leave bones faster than it accumulates, resulting in easily broken bones. This condition, called osteoporosis, is most common in women who have reached menopause (cessation of menstrual periods). The estrogen decrease that accompanies menopause makes bone-forming cells more sensitive to parathyroid hormone, which depletes bone mass.

FIGURE 33.11
Regulation of Blood Calcium Level.
Calcitonin lowers blood calcium level with actions on the bones, kidneys, and intestines. Parathyroid hormone (PTH) acts in reverse, stimulating bone to release calcium (Ca^{2+}) and the kidneys to conserve calcium. PTH indirectly stimulates the intestine to absorb calcium. The resulting increase in blood calcium concentration inhibits secretion of PTH.

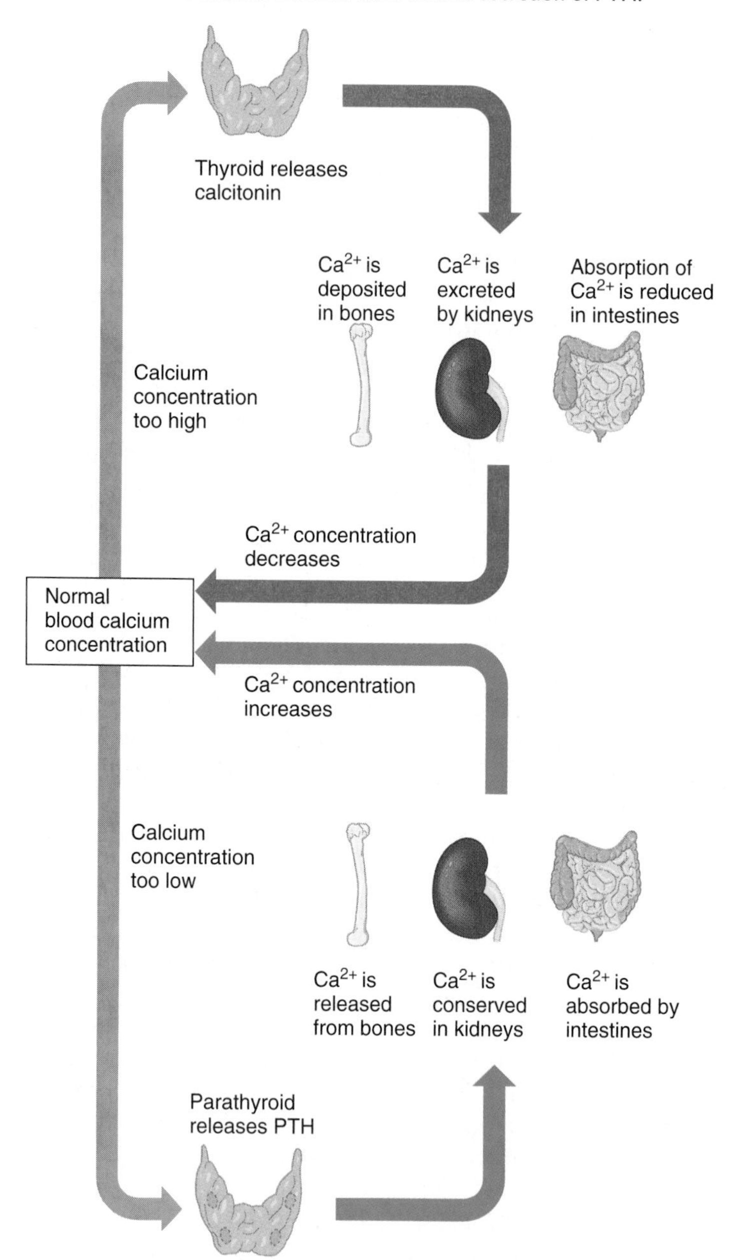

The Adrenal Glands Secrete Catecholamines and Steroid Hormones The paired adrenal glands ("ad" means near, "renal" means kidney) sit on top of the kidneys. The inner portion of each gland is the **adrenal medulla,** and the outer portion is the **adrenal cortex.** These sections differ in structure, function, and embryonic origin and can be considered separate but physically attached glands.

The adrenal medulla hormones, **epinephrine** (also known as adrenaline) and **norepinephrine** (also called noradrenaline),

FIGURE 33.12
The Adrenal Glands.
(**A**) The adrenal medulla secretes the catecholamines epinephrine and norepinephrine, which ready the body for an immediate "fight-or-flight" response during emergencies. (**B**) The adrenal cortex secretes mineralocorticoids and glucocorticoids, which enable the body to survive prolonged stress. The adrenal cortex also secretes small amounts of sex hormones (not shown).

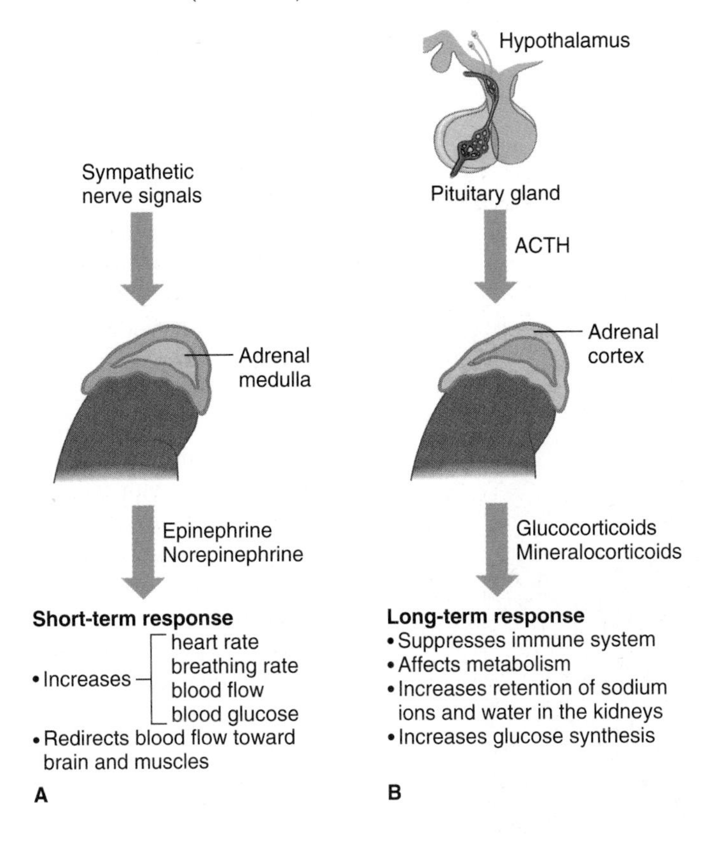

help the body respond to and survive emergencies by increasing heart and breathing rates and blood flow. These hormones, called catecholamines, are under the control of the sympathetic nervous system. Figure 33.12 lists their specific effects.

The adrenal cortex secretes three types of steroid hormones: **mineralocorticoids, glucocorticoids,** and sex hormones. The mineralocorticoids and glucocorticoids coordinate during prolonged stress to mobilize energy reserves while stabilizing blood volume and composition.

The mineralocorticoids maintain blood volume and electrolyte balance by stimulating the kidneys to return sodium ions and water to the blood while excreting potassium ions. The major mineralocorticoid, aldosterone, maintains the level of sodium ions in the blood by increasing the amount the kidneys reabsorb. Since water reabsorption follows sodium ion reabsorption, aldosterone also affects blood volume. This action is particularly important in compensating for fluid loss from severe bleeding.

The glucocorticoids, the most important of which is cortisol, are essential in the body's response to prolonged stress. Whereas the adrenal medulla braces the body against an immediate danger, the glucocorticoids take over for longer periods, for instance during an infection. Glucocorticoids affect carbohydrate, protein, and lipid metabolism. They break down proteins into amino acids and then stimulate the liver to synthesize glucose from parts of these freed amino acids. The newly synthesized glucose supplies energy for healing after the immediate supply of glucose is depleted. Glucocorticoids also indirectly constrict blood vessels, which slows blood loss and prevents tissue inflammation, important responses to injury.

A deficiency in adrenal cortex hormones causes Addison's disease. Lack of mineralocorticoids causes an ion imbalance that decreases water retention so that the person has low blood pressure and is dehydrated. Symptoms include weight loss, mental fatigue, weakness, and impaired resistance to stress. Lack of glucocorticoids in Addison's disease darkens the skin. Replacing the deficient hormones can treat Addison's disease. Former president John F. Kennedy had the condition.

Excess glucocorticoids causes Cushing's syndrome, which redistributes body fat. A person with the syndrome may have thin legs but excess fat in the face, behind the shoulder blades, and in the abdomen. The person bruises easily and heals slowly.

The adrenal cortex secretes small amounts of both male and female sex hormones. The cortex produces more male sex hormone (testosterone) than female sex hormones (estrogen and progesterone). Adrenal hormones are particularly important in postmenopausal women.

The Pancreas Regulates Nutrient Utilization The **pancreas** is a large gland located between the spleen and the small intestine. It produces digestive enzymes and hormones. Dispersed among the cells that produce digestive enzymes are clusters of cells called **islets of Langerhans** (fig. 33.13), which are the endocrine portion of the pancreas. The hormones they secrete are polypeptides that regulate the body's use of nutrients.

The islets produce three hormones, each secreted by a different type of cell: **insulin** (Latin for "island"), **glucagon,** and **somatostatin.** Figure 33.5 illustrated how the actions of insulin and glucagon oppose one another in regulating blood glucose levels. Somatostatin controls the rate at which the blood absorbs nutrients.

Insulin deficiency, or an inability of the body to use insulin, causes **diabetes mellitus,** the subject of this chapter's opening essay. Because the body cannot use or store glucose properly, sugar builds up in the blood and appears in the urine. Untreated diabetes leads to a condition similar to starvation. Fifteen percent of affected individuals have type I, or insulin-dependent, diabetes, which usually begins in childhood. Symptoms include thirst, blurred vision, weakness, fatigue, irritability, nausea, and weight loss. Untreated, a severe insulin deficiency can result in a lethal diabetic coma. Insulin-dependent diabetes may be an autoimmune disease, in which the body attacks insulin-producing beta cells as if they are foreign.

The more common form of the disorder, non-insulin-dependent diabetes (also known as type II), usually begins in adulthood. The body produces insulin but is unable to use it, perhaps due to defective insulin receptors on cell surfaces or an abnormality in a protein that regulates insulin activity. Symptoms of non-insulin-dependent diabetes mellitus include fatigue, itchy skin, blurred vision, slow healing, and poor circulation. Diet and exercise can help about 90% of these people; the remainder use

FIGURE 33.13
The Pancreas.
The pancreas lies above the small intestine and beneath the stomach. It consists of many lobes and produces digestive enzymes and hormones. The expanded drawing shows the hormone-secreting islets of Langerhans next to enzyme-secreting exocrine cells.

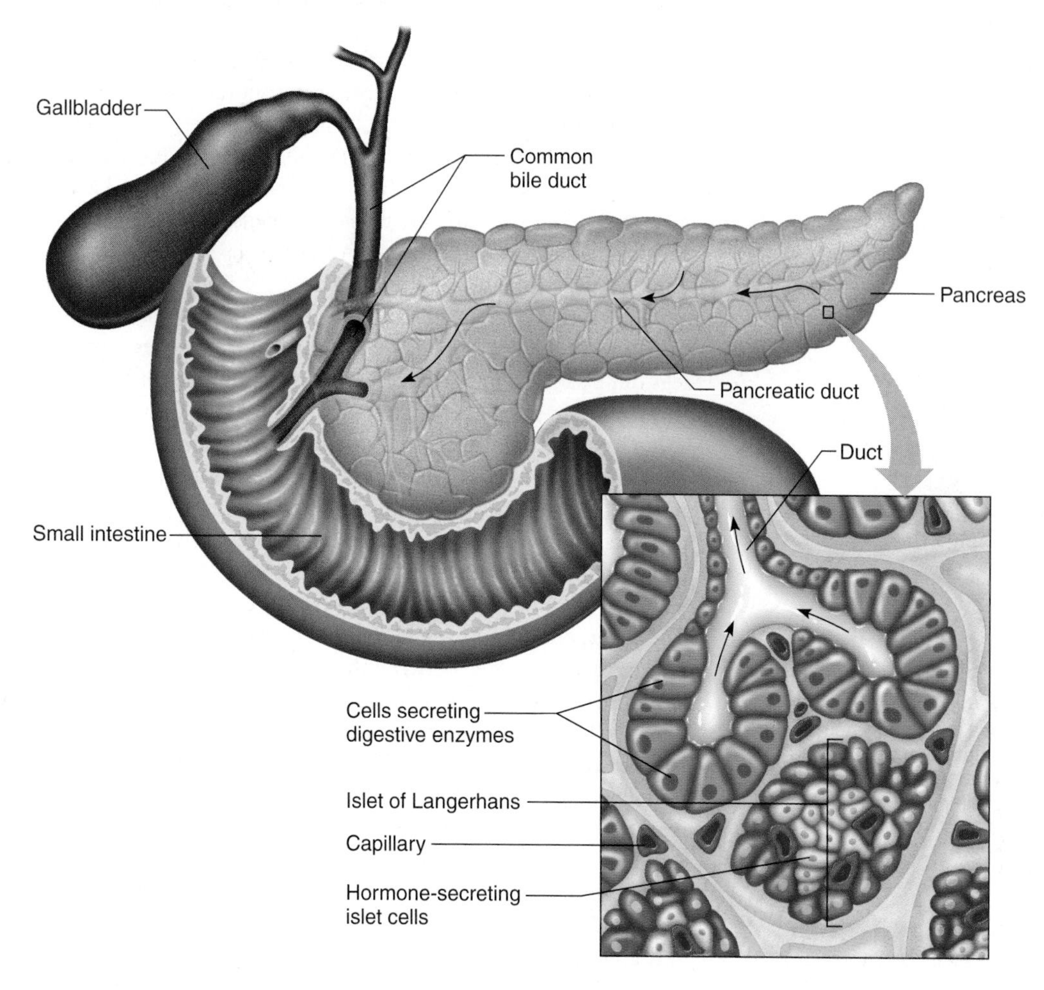

drugs to enable the body to more effectively use insulin. Another form of diabetes mellitus occurs during pregnancy, when placental hormones interfere with insulin action.

Hypoglycemia is a deficient level of glucose in the blood, indicating excess insulin production. A person with this condition feels weak, sweaty, anxious, and shaky. A healthy person might experience transient hypoglycemia following very strenuous exercise. Following a diet of frequent, small meals low in carbohydrates and high in protein can help relieve symptoms of hypoglycemia. This prevents insulin surges that lower blood sugar level. Figure 33.14 outlines steps in glucose use at the cellular level that may be altered by diabetes mellitus and hypoglycemia.

The Pineal Gland Secretes Melatonin The **pineal gland** is a small oval structure in the brain near the hypothalamus. It produces **melatonin,** a hormone that helps regulate reproduction in certain mammals by inhibiting the anterior pituitary hormones that regulate gonadal activity.

Melatonin production depends upon the pattern of light and dark an individual is exposed to. Light inhibits melatonin synthesis; darkness stimulates it. Because melatonin synthesis depends upon the absence of light, changing levels of the hormone may signal an organism to prepare for changing seasons—perhaps by growing a heavier coat or initiating mating behavior.

Melatonin influences other endocrine glands, and in this sense affects metabolism. Melatonin levels cue vertebrates to molt, mate, migrate, and hibernate. In rodents, melatonin injections lower the levels of thyroid hormone, melanocyte-stimulating hormone, and sex hormones.

The functions of the human pineal gland are not well understood, possibly because we alter natural light-dark cycles with artificial lighting. Mood swings may be linked to abnormal melatonin secretion patterns, particularly a form of depression called seasonal affective disorder (SAD). Exposing such individuals to additional hours of daylight can elevate their moods. Even though we know very little about the role of melatonin in the human body, this hormone has been "hyped" as a cure for many conditions. We do know that melatonin levels decrease with age, and from this observation arose the hypothesis that taking extra melatonin might prevent diseases associated with aging. One study did find that melatonin helps older people with insomnia to sleep, but less sleep may be a normal part of aging. A book published in 1995 called the hormone an "amazing natural pill that can reset the body's aging clock and restore youth." The book claimed that melatonin "rejuvenates sexual organs" (although in some rodent studies melatonin shrunk gonads) and "is an extremely useful tool in helping to prevent heart disease" (although clinical trials in human subjects have never been conducted). A long-held theory is that the pineal gland in more primitive vertebrates, such as reptiles, serves as a "third eye," a light-sensitive brain region that perhaps receives photons through the top of the skull. Such an ancestry may explain the role of this gland in monitoring environmental dark-light cues.

Endocrine Control of Reproduction

The **gonads**—the **ovaries** in females, and the **testes** in males—are the glands that manufacture gametes. The hormones

FIGURE 33.14

Diabetes and Hypoglycemia.

Diabetes mellitus and hypoglycemia disrupt the endocrine system's control of blood glucose level.

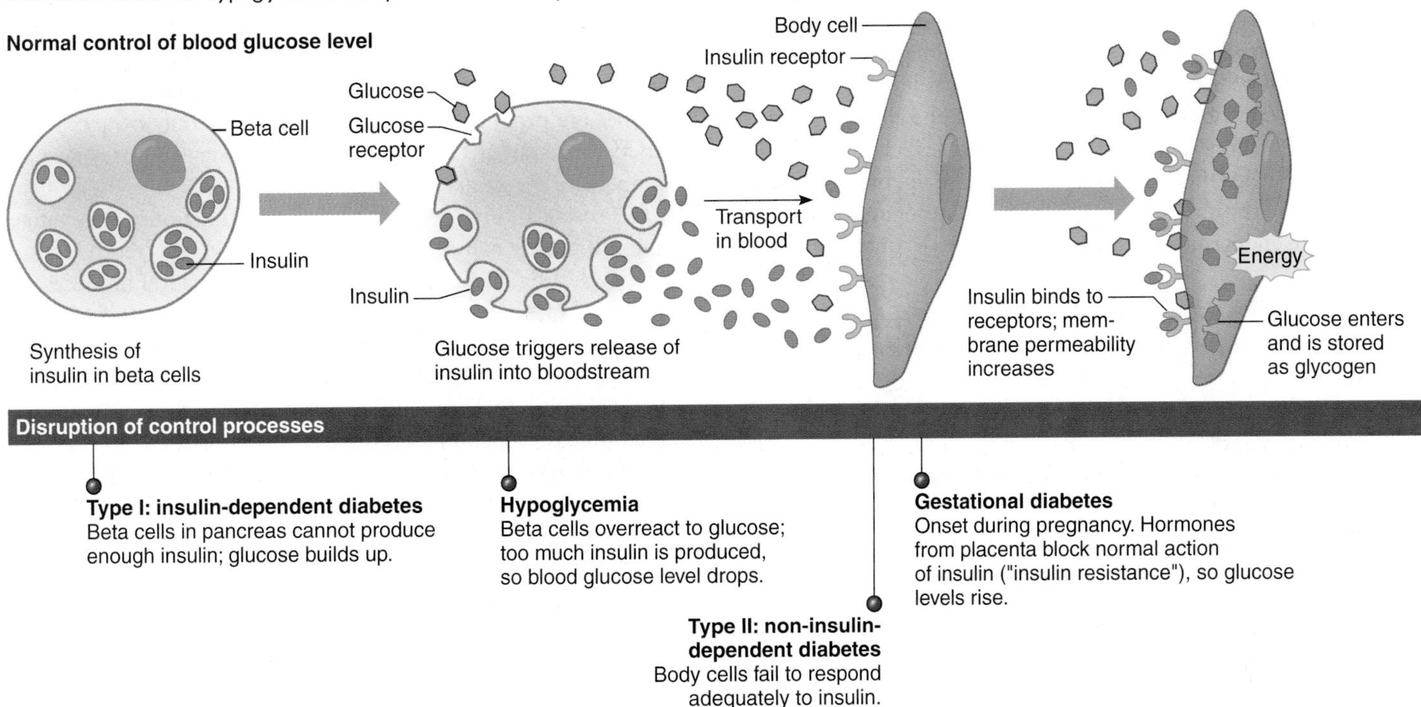

that the gonads secrete enable the gametes to mature, are responsible for secondary sexual characteristics, and are under negative feedback control by the pituitary gland and the hypothalamus (fig. 33.15).

In a woman of reproductive age, the levels of several sex hormones cycle on a monthly basis. Low blood levels of **estrogen** and **progesterone** trigger the hypothalamus to produce and secrete gonadotropin-releasing hormone (GnRH), which then stimulates the release of luteinizing hormone (LH) and follicle-stimulating hormone (FSH) from the anterior pituitary. LH and FSH then stimulate oocyte maturation and division and growth of follicle cells in the ovary. The growing follicle cells release estrogen and progesterone, which, in turn, exert negative feedback control on both the hypothalamus and pituitary. The cycling of female reproductive hormones and their complex effects on oocyte maturation, oocyte release (ovulation), and uterine lining changes are discussed in more detail in chapter 40.

In the human male, reproductive hormone levels do not cycle to the same extent as in females. GnRH stimulates the anterior pituitary to release FSH and LH. FSH and LH travel in the bloodstream to the testes. There, FSH stimulates the early stages of sperm formation, and LH completes sperm production and stimulates the testes to synthesize the male steroid hormone **testosterone.** Testosterone and another hormone, called inhibin, prevent overproduction of sperm in a negative feedback loop. Testosterone stimulates development of male characteristics in the embryo and promotes later

FIGURE 33.15

Control of Reproductive Hormones.

(**A**) In females, a monthly surge of LH and FSH expels an oocyte from an ovary. Different hormonal events follow, depending upon whether a fertilized ovum implants in the uterine lining. If not, falling levels of progesterone (released from the ovary) initiate the breakdown of the uterine lining. Because progesterone and estrogen inhibit both the hypothalamus and pituitary, lowered levels of these hormones allow LH and FSH levels to increase again and the cycle begins anew. (**B**) Male sex hormones do not cycle, but are still subject to negative feedback controls.

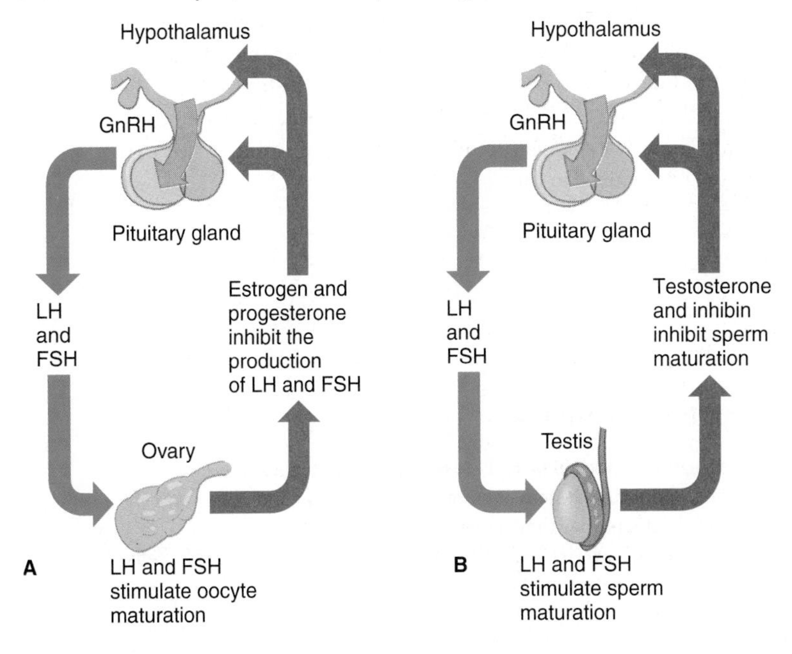

development of male secondary sexual characteristics, including facial hair, deepening of the voice, and increased muscle growth.

Testosterone also controls growth and activity of the prostate gland, a walnut-sized, two-lobed structure beneath the urinary bladder. The prostate produces a thin, milky fluid with an alkaline pH. Contractile cells within the gland eject the fluid into the urethra, where it contributes to the seminal fluid; the alkalinity neutralizes acid in the sperm's metabolic wastes. The prostate secretion also helps the sperm to swim. Because the prostate does not secrete into the bloodstream, it is technically not an endocrine gland, but an endocrine secretion—testosterone—controls it.

33.4 MASTERING CONCEPTS

1. Describe the hormonal fluctuations of the menstrual cycle.
2. What are the functions of testosterone and the prostate gland?
3. What are the three thyroid hormones, and what do they control?
4. What is the function of parathyroid hormone?
5. How do the functions of hormones secreted by the adrenal cortex and the adrenal medulla differ?
6. What are the functions of insulin, glucagon, and somatostatin?
7. How do darkness and light affect melatonin secretion?

OLC

33.5 Prostaglandins Provide Localized Control

Prostaglandins are short-lived molecules that, unlike hormones, function near their sites of synthesis. They are also chemical regulators with a variety of functions, and interact with hormones.

Prostaglandins are biochemicals that exert profound effects on various tissues and organs, often by altering hormone levels. Prostaglandins are lipids that appear locally and transiently when cells are disturbed. When a cell membrane is disrupted by an injury, binding of a hormone, or an immune system attack, the damaged membrane releases certain fatty acids into the cytoplasm. These fatty acids provide the substrate for enzymes that catalyze prostaglandin formation.

Prostaglandins do not fit the classical definition of a hormone. They function at their synthesis site, rather than traveling in the bloodstream to a target tissue. Different types of stimuli initiate prostaglandin secretion and hormone release, compared with the specificity of hormone action. Also, prostaglandins are in every mammalian tissue type thus far examined, whereas hormones are produced in significant amounts only by specialized endocrine system cells. Nonetheless, prostaglandins, like hormones, are chemical regulators.

Prostaglandins have a variety of functions: they affect smooth muscle contraction, secretion, blood flow, reproduction, blood clotting, respiration, transmission of neural impulses, fat metabolism, the immune response, and inflammation. Certain prostaglandins stimulate the smooth muscle in blood vessel walls to adjust blood flow and in the walls of airways within the lungs to alter oxygen delivery. Still other prostaglandins contract uterine muscles. Depending on the circumstance, these uterine contractions may assist sperm in their journey to the oocyte, propel a fetus out of the uterus, or cause menstrual cramps.

Learning how different prostaglandins exert specific effects has explained the mechanisms of certain drugs and led to some interesting new medical applications. For example, some prostaglandins promote inflammation, the immune system's attempt to fight infection at the site of a wound by sending in fluid and white blood cells. Aspirin relieves the pain of inflammation by inactivating an enzyme required to synthesize several prostaglandins. Aspirin also blocks synthesis of prostaglandins that lower the threshold of pain receptors in the nervous system.

Prostaglandins' functions in blood clotting and controlling blood vessel diameter may also play a part in heart attacks. Daily doses of aspirin may reduce heart attack risk by altering prostaglandin activity. Prostaglandins that dilate blood vessels may treat hypertension (high blood pressure), which results from constricted blood vessels.

33.5 MASTERING CONCEPTS

1. How are prostaglandins similar to hormones, yet different from them?
2. What are some functions of prostaglandins?
3. How does prostaglandin function explain aspirin's effects?

OLC

Chapter Summary

33.1 Hormones Are Chemical Messengers and Regulators

1. The **endocrine system** includes several glands and scattered cells and the **hormones** they secrete into the bloodstream.
2. A hormone exerts a specific physiological effect on **target cells,** which have receptors for it.
3. The nervous and endocrine systems interact to maintain homeostasis. The nervous system acts faster and more locally than the endocrine system. **Neurosecretory cells,** which are neurons that secrete hormones, are a physical link between the two systems.
4. Diverse species have hormonelike regulation, and many use the signaling molecule **cAMP.**
5. Local-acting hormones include those that regulate digestion, and atrial natriuretic hormone.

33.2 How Hormones Exert Their Effects

6. **Peptide hormones** are water-soluble and bind to the surface receptors of target cells, stimulating conversion of ATP to cAMP on the inner face of the cell membrane. The cAMP then triggers a specific metabolic effect.
7. The lipid-soluble **steroid hormones** cross target cell membranes, bind to receptors in the cytoplasm or nucleus, and activate genes to direct synthesis of proteins that provide the cell's response.
8. In a **negative feedback loop,** excess of a hormone or the product of a hormone-induced response suppresses further synthesis or release of that hormone until levels return to normal.
9. In a **positive feedback loop,** the hormone causes an event that increases its production.
10. Ions or nutrient levels near the endocrine cells, input from the nervous system, and other hormones control feedback loops.

33.3 Diversity of Endocrine Systems

11. Simpler invertebrates have neurosecretory cells. More complex invertebrates have interacting hormones.
12. Changing levels of **juvenile hormone** and **molting hormone** control insect metamorphosis.
13. The same hormones may function differently in different species.
14. The **hypothalamus** produces **releasing hormones,** which travel in neurosecretory cells to the anterior lobe of the **pituitary gland.**

33.4 A Closer Look at the Endocrine System in Humans

15. Anterior pituitary hormones include **growth hormone,** which stimulates cell division, protein synthesis, and growth in all cells; **thyroid-stimulating hormone,** which prompts the **thyroid gland** to release **thyroxine** and **triiodothyronine,** which regulate metabolism; **adrenocorticotropic hormone,** which stimulates the **adrenal cortex** to release hormones that enable the body to cope with a serious threat; **prolactin,** which stimulates milk production; and the sex hormones, which control sex cell development.
16. The hypothalamus manufactures two hormones that are stored in and released from the posterior lobe of the pituitary gland—**antidiuretic hormone,** which regulates body fluid composition, and **oxytocin,** which contracts the uterus and milk ducts.
17. In many vertebrate species, the region between the anterior and posterior lobes of the pituitary secretes **melanocyte-stimulating hormone,** which affects skin coloring.
18. The **parathyroid glands** secrete **parathyroid hormone,** which increases blood calcium level by releasing calcium from bone and increasing its absorption in the gastrointestinal tract and kidneys. **Calcitonin,** which the thyroid gland secretes, lowers the level of calcium in the blood.
19. The **adrenal cortex** secretes **mineralocorticoids** and **glucocorticoids,** which mobilize energy reserves during stress and maintain blood volume and blood composition. The **adrenal medulla** secretes **epinephrine** and **norepinephrine,** which ready the body to cope with an emergency.
20. The endocrine portion of the **pancreas** secretes **insulin,** which stimulates cells to take up glucose; **glucagon,** which increases blood glucose levels; and **somatostatin,** which controls the rate at which the blood absorbs nutrients.
21. The **pineal gland** may regulate the responses of other glands to light-dark cycles through its hormone, **melatonin.**
22. The **ovaries** secrete **estrogen** and **progesterone,** hormones that stimulate development of female sexual characteristics and with GnRH, **FSH,** and **LH** control the menstrual cycle. The **testes** secrete **testosterone,** which stimulates sperm cell production, the development of secondary sexual characteristics, and controls prostate gland function.
23. **Prostaglandins** are lipids that form enzymatically when disturbed cell membranes release fatty acids. They function at the site where they are released, and are diverse.

Testing Your Knowledge

1. Why doesn't a hormone affect all body cells?
2. How do the mechanisms of peptide and steroid hormone function differ?
3. How do hormones regulate their own levels?
4. Describe how thyroglobulin, thyroxine, triiodothyronine, TSH, and TRH interact.
5. Which hormone fits each description below?
   - **a.** Too little of this hormone causes fatigue.
   - **b.** A woman who is breast-feeding produces this hormone.
   - **c.** This hormone increases blood calcium level.
   - **d.** This hormone decreases blood glucose level.
   - **e.** Synthetic steroid drugs mimic the muscle-building effects of this hormone.
6. How are the following glands really two glands in one?
   - **a.** the thyroid
   - **b.** the pituitary
   - **c.** the adrenal glands

Thinking Scientifically

1. In an episode of *The X-Files,* a monster sucks out people's pituitary glands. What symptoms might they develop?
2. How might mutations in different genes cause deficits or malfunctions of the same hormones?
3. A queen honeybee secretes a substance from a gland in her mouthparts that inhibits the development of ovaries in worker bees. Is this substance most likely a hormone, prostaglandin, or pheromone? Cite a reason for your answer.
4. Imagine you are a researcher who has just found a small glandlike structure in a human. How would you determine whether it is an endocrine gland? If it is, how might you identify the gland's role in the endocrine system?

References and Resources

Berreby, David. June 9, 1998. Studies explore love and the sweaty T-shirt. *The New York Times,* p. F1. An unusual experiment suggests that humans may have sex pheromones.

Lewis, Ricki. January 18, 1999. New tests monitor thyroid cancer. *The Scientist,* p. 1. New follow-up tests are based on knowing how thyroid hormones are regulated.

Utiger, R. D. April 20, 2000. Treatment of acromegaly. *The New England Journal of Medicine,* vol. 342, p. 1210. A genetically engineered drug blocks production of growth hormone in people who have acromegaly.

The *LIFE* Online Learing Center provides additional resouces and tools for studying this chapter.
www.mhhe.com/life

CHAPTER 34

The Musculoskeletal System

On the Origins of Biomineralization

In 1997, French researchers attempted a bold experiment. They collected nacre, a substance better known as mother-of-pearl, from the inside surface of the shell of the mollusk *Pinctada maxima* and ground it into a fine powder. Then they mixed the powder with the blood of each of eight women suffering from loss of bone in their upper jaws, and injected the material at the sites of bone loss. Six months later, the jaw holes had filled, and the women's immune systems had not rejected the molluscan grafts. But the new jaw material wasn't nacre—it was human bone. Somehow, the nacre had stimulated osteoblasts to secrete new bone material around the particles of molluscan material, which were slowly dissolving. Osteoclasts shaped the new bone so that it did not deform the jaw, but the cells did not eat away at the nacre.

Then they mixed the powder with the blood of each of eight women suffering from loss of bone in their upper jaws, and injected the material at the sites of bone loss.

The researchers got the idea for this experiment from Mayan skulls discovered in 1931. The skulls had teeth that were composed of nacre, yet X-ray images revealed that the roots were growing into the jawbone as if they were normal, human teeth. What did long-ago Mayan dentists know that we don't?

To find out, the researchers placed chips of nacre on a layer of human osteoblasts growing in culture. The bone cells divided, attached to the nacre chips, and new bone material collected at the interfaces. Next, the researchers placed nacre chips and bone chips about a millimeter apart on the osteoblast layer. After a few weeks, the chips grew together, yet maintained their distinctive characteristics—the bone looked confluent, but the nacre retained the bricks-and-mortar configuration that it has when lining a mollusk's shell. Bone chips alone or nacre chips alone didn't grow. Clearly, the nacre and

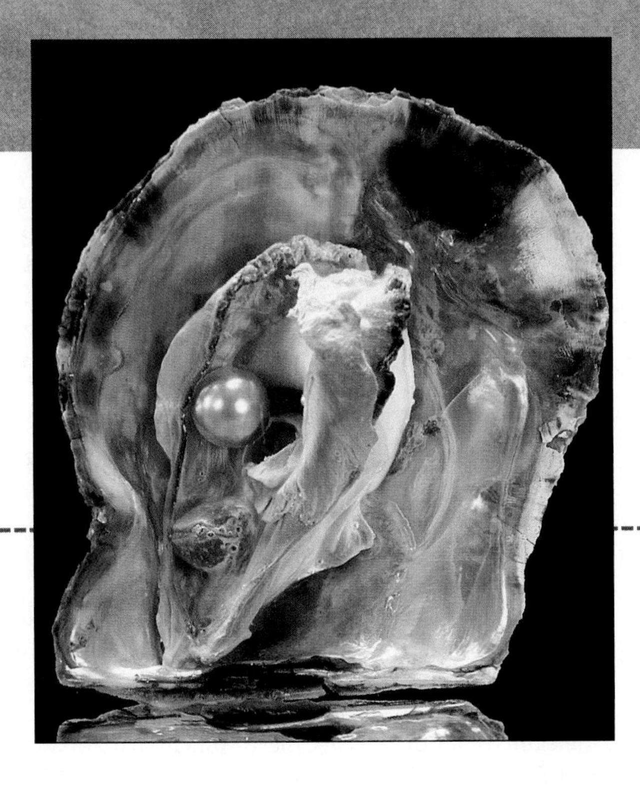

Mother of Pearl
Nacre, better known as mother-of-pearl, is obtained from the inside of the shell of *Pinctada maxima.*

bone were interacting, despite their natural presence in animals that have very different skeletons. And in another set of experiments performed in 2000, hip fractures in sheep were healed with implants of mesenchymal stem cells (see the chapter 30 opening essay) grown in the nooks and crannies of coral.

A possible explanation for why and how a human accepts material from a mollusk, or a sheep accepts a coral-based graft, takes us back to the time just before the Cambrian explosion, when the mysterious soft-bodied Ediacaran animals swam in the seas (see fig. 18.14). Recall that a major difference between the doomed Ediacarans and the ancestors of modern animals that flourished in the Cambrian was the advent of hard parts—skeletons. The undersea rain of phosphates responsible for some of the earliest known animal fossils (see fig. 17.7E) attests to the fact that the environment was rich in minerals that organisms could tap, if they had the molecular tools to do so.

The skulls had teeth that were composed of nacre, yet X-ray images revealed that the roots were growing into the jawbone as if they were normal, human teeth. What did long-ago Mayan dentists know that we don't?

For organisms to develop skeletons—either those worn on the outside or inside—required control over the deposition of minerals present in the aquatic environment in supersaturated solutions. Researchers hypothesize that animals that preceded those with hard parts had enzymes that prevented minerals from forming on them, a natural tendency that would have encrusted them out of existence. This is similar to a protein present in saliva that prevents too much mineral from building up on teeth. Skeletons arose when molecules evolved that could turn off that control, but in a regulated manner, so that minerals could combine with organic materials, a process called biomineralization.

The first animals to incorporate minerals were the "small shelly fossils" that were so abundant during the early Cambrian period. Then, the hypothesis continues, other molecules formed and persisted that could further biomineralization, sculpting the many animal body forms that appeared, rather suddenly, during the Cambrian explosion.

What does this scenario have to do with human jaws that accept mother-of-pearl replacement parts, or sheep with coral in their healed hips? Researchers were at first stymied, because nacre and coral are pure calcium carbonate, whereas bone mineral is predominantly calcium phosphate. The two hard materials are not the same, so they could not have been inherited from a recent shared ancestor. An alternate explanation is that *Pinctada maxima* and humans share the molecular signaling pathways that enable them to incorporate minerals from the environment.

A larger lesson in the story of nacre and human bone is that we never know when different aspects of the science of life will interact. This story brings together signal transduction (see chapter 4), the evolution of animals (see chapter 18), and the skeletal system (this chapter).

34.1 A Musculoskeletal System Helps an Animal to Respond to Its Environment

Muscles may operate alone or with skeletal systems to enable an animal to move, which helps to maintain homeostasis. Natural selection has molded species-specific characteristics of musculoskeletal systems.

The dancers line up nervously, each perfectly still, waiting for the choreographer's signal. At the cue, they suddenly come alive, jumping and turning in unison, their graceful movements precisely timed to the music (fig. 34.1A). Their ability to move depends upon connected bones and muscles, just as the leap of a gray tree frog does (fig. 34.1B).

Muscle attached to bone is only one type of musculoskeletal system. An earthworm's skeleton, for example, isn't bone but fluid-filled compartments. Together with perpendicular muscles in its body wall, this skeleton helps the worm cling tenaciously to its burrow.

The first organisms were probably able to move by a mechanism similar to the one that powers our muscles today—protein filaments that slide past each other, shortening and thereby contracting the cells containing them. In multicellular animals, contractile cells are organized into muscle tissues, which form organs called muscles.

Some muscles operate independently of skeletal structures. A vertebrate's digestive tract muscles, for example, mix and propel food without skeletal support. When many large animals move, however, muscles pull against rigid bones to form lever systems, which increase the strength and efficiency of movement. The skeletal and muscular systems have evolved as closely allied, interacting organ systems.

Musculoskeletal systems are under perpetual biological renovation. A person beginning an intense weight-lifting program, for example, will gradually develop larger muscles and stronger bones in response to increased use. In the opposite situation, astronauts lose bone density if they are in a prolonged weightless environment, because their musculoskeletal systems don't have to work as hard in the absence of gravitational force.

An internal skeleton can grow as the animal grows. Because bones are constantly broken down and rebuilt, a bone retains its shape yet enlarges as an individual develops. In contrast, arthropods such as insects and certain crustaceans periodically shed their external skeletons and grow new ones, which accommodate their enlarging bodies.

At the species level, the skeletal system's capacity to change is striking, especially among vertebrates. The skeletons of the hummingbird and the hippo, the cheetah and the sloth, and the salmon and the whale are distinctive, yet they are composed of the same types of cells and have similar spatial patterns. The skeleton, in turn, influences the distribution of the musculature. The characteristics of each species' musculoskeletal system reflect its common ancestry and the selective forces in its environment.

34.1 MASTERING CONCEPTS

1. How do the skeletal and muscular systems interact?
2. What are some examples of bones and muscles changing in response to changing conditions?

OLC

FIGURE 34.1

Muscles and Bones Assist in Movement.

(**A**) The graceful movements of these dancers are possible because muscles contract against bones in a lever system. (**B**) The leap of this gray tree frog also depends on the musculoskeletal system.

A

B

34.2 Types of Skeletons

A skeleton is a supportive framework. A hydrostatic skeleton consists of constrained fluid. An exoskeleton is an external braced framework, and an endoskeleton is an internal braced framework. Vertebrate endoskeletons are built of cartilage and/or bone, organized into axial and appendicular parts.

A skeleton is a supporting structure or framework. It gives an animal's body shape, protects internal organs, and provides a firm surface for muscles to pull against, facilitating movement. Several types of skeletons perform these functions for different species.

Hydrostatic Skeletons Consist of Constrained Fluid

The simplest type of skeleton is a **hydrostatic skeleton** ("hydro" means water), which consists of liquid within a layer of flexible tissue. In a hydra, for example, the gastrovascular cavity fills with water, providing rigidity. Hydrostatic skeletons are also found in jellyfishes, sea anemones, slugs, and annelid worms. The tension of the constrained fluid supports and helps to determine shape, as it does in a water-filled balloon. Combined with muscle action, a hydrostatic skeleton can provide locomotion.

Movement of an earthworm through soil illustrates how a hydrostatic skeleton makes locomotion possible. The earthworm has two layers of muscle in the body wall (fig. 34.2). One layer encircles the body (circular muscle) and the other runs lengthwise (longitudinal muscle). In segments where the circular muscles contract against the fluid in the body cavity (coelom), the worm lengthens and thins out, just as a water-filled balloon would if squeezed at one end. In alternating groups of segments, the longitudinal muscles contract, which shortens and widens those areas. These wider segments adhere to the surrounding soil with their setae, whereas the thinner segments elongate. The overall effect is that the animal pulls itself through the burrow.

Similar to a hydrostatic skeleton in that it consists of tissue that cannot be compressed is a **muscular hydrostat,** which is a group of muscles that generate a supportive force without a skeleton. An elephant's trunk is a muscular hydrostat. Its many functions are easily seen in a cooperative zoo elephant. The tongues of reptiles and mammals and the tentacles of squids and octopuses are other examples of muscular hydrostats.

Exoskeletons Protect from the Outside

A more complex skeleton is a **braced framework,** which consists of solid structural components strong enough to resist pulling forces. Muscles can attach to the surfaces of the framework, facilitating movement.

An **exoskeleton** ("exo" means outside) is a braced framework that protects an organism from the outside, much like a suit of armor. Many groups of invertebrates have exoskeletons, including arthropods (such as lobsters and insects) and mollusks (such as clams). Figure 34.3 shows the jointed exoskeletons of a flea and a lobster, which consist of the complex carbohydrate chitin.

FIGURE 34.2
Earthworm Locomotion.
(**A**) The earthworm has a hydrostatic skeleton. (**B**) Coordinated contraction of the longitudinal and circular muscles moves the worm.

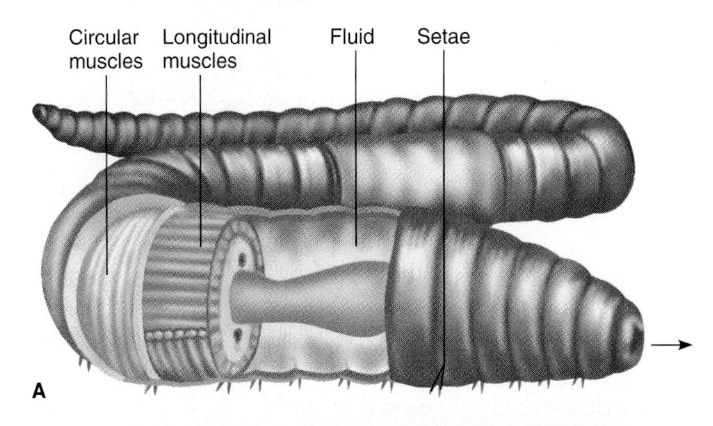

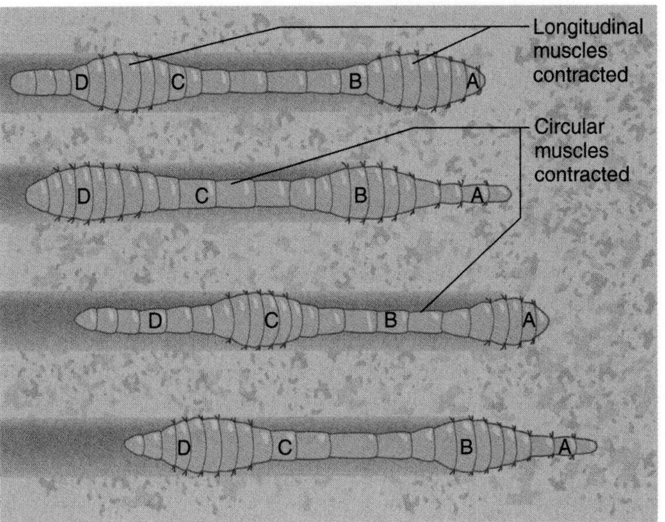

Clams and snails use readily available minerals to produce hard, calcium-containing shells, much as the ancient animals described in the chapter opening essay did. Rigid shells and spike-like structures thicken as the organism grows, remaining with it for life.

Arthropod exoskeletons, however, have a shortcoming in common with armor—the animal may outgrow them. Animals must periodically shed, or molt, the exoskeleton and grow a slightly larger new one. Until the new exoskeleton has formed and hardened, the animal is in greater danger of being eaten. Seafood lovers may be familiar with the results of molting: A soft-shelled crab has just begun to harden its exoskeleton, and a hard-shelled crab has completed the process. During the summer the cast-off

FIGURE 34.3
Exoskeletons.
A flea and a lobster have exoskeletons consisting of chitin.

exoskeletons of certain insects, such as cicadas, are shed intact, leaving behind an eerie cast of the animal.

Endoskeletons Support, Protect, and Provide Movement

An **endoskeleton** ("endo" means inner) is an internal braced framework and is found in sponges, echinoderms, and vertebrates (fig. 34.4). An endoskeleton offers several advantages over an exoskeleton—it does not restrain growth and, because it consumes less of an organism's total body mass, it increases mobility.

Recall from chapter 25 that some fishes, including sharks, have endoskeletons made of cartilage. The endoskeletons of most vertebrates, including humans, are composed primarily of bone.

The skeletons of vertebrates are organized in very similar ways. If you were to line up the skulls of a variety of species, they would all be readily recognizable as skulls, yet with species-specific adaptations. Figure 17.9 compares the skeletons of four vertebrates—a human, a bird, a seal, and a lion.

Functions of Endoskeletons The vertebrate endoskeleton has several functions:

- *Support*—The skeleton is a framework that supports an animal's body against gravity. To a large extent it provides the body's shape.
- *Lever system for movement*—The vertebrate skeleton is a system of muscle-operated levers. Typically, the two ends of a skeletal (voluntary) muscle attach to different bones

FIGURE 34.4
Endoskeleton Diversity.
A sponge's spicules (**A**), a ray's cartilaginous skeleton (**B**), and a meerkat's bony skeleton (**C**) are variations on the endoskeletal theme.

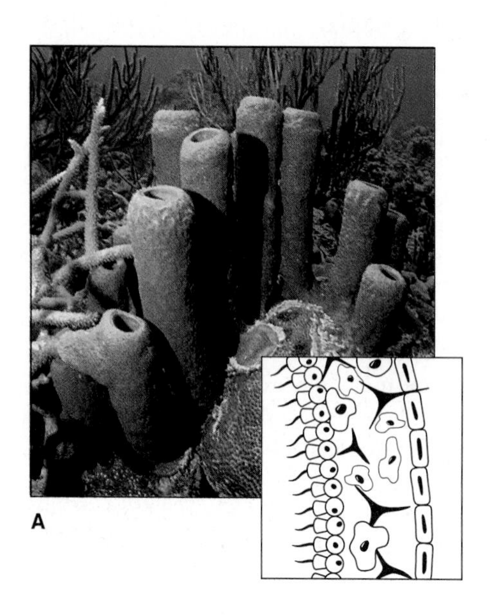

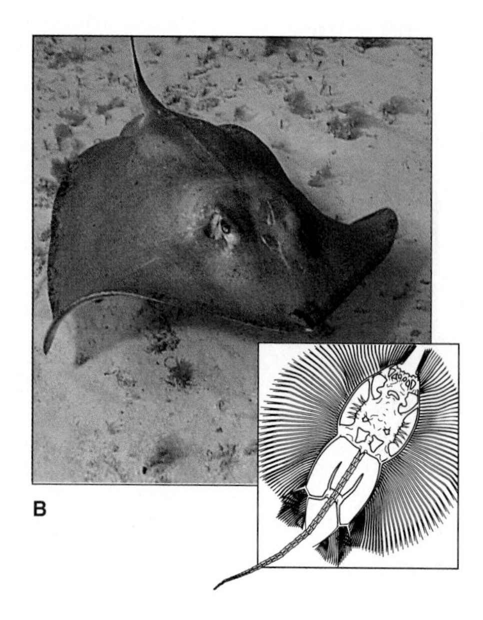

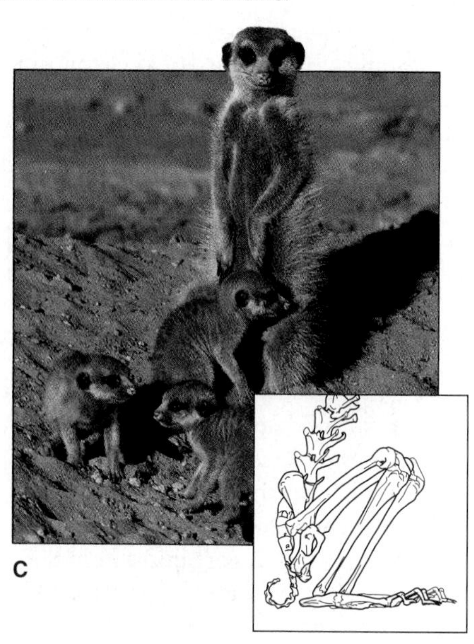

that connect in a structure called a **joint.** When the muscle contracts, one bone is pulled towards the other.

- *Protection of internal structures*—Some bones form cages that protect soft internal organs. The backbone surrounds and shields the spinal cord, the skull protects the brain, and ribs protect the heart and lungs.
- *Production of blood cells*—Many bones, such as the long bones of the human arm and leg, contain and protect red marrow, a tissue where red blood cells, white blood cells, and platelets are produced.
- *Mineral storage*—The skeleton stores calcium and phosphorus, minerals required for a number of important activities. About 99% of the calcium in the human body is stored in the bones. Some of the remaining 1% is carried in blood plasma. Calcium is vital for the activity of some enzymes, muscle contraction, synaptic function, blood clotting, cell adhesion, and cell membrane permeability. To maintain the blood concentrations of calcium necessary for these functions, the body constantly shuttles calcium between blood and bone. Phosphorus is an essential component of nucleic acids, as well as ATP.

Variations on the Vertebrate Skeletal Theme The general properties and organization of the vertebrate musculoskeletal system can vary to accommodate animals that live in vastly different environments—for example, the owl and the seal (fig. 34.5).

Bird skeletons are lightweight, which eases flight (see fig. 25.17). Natural selection has modified the owl's forelimb into a wing. The digits are fused, forming a shape better able to support flight. The owl's forelimb bones—the radius, ulna, and humerus—are thin but sturdy. Its bones, joints, and muscles connect in a way that promotes flapping movements.

FIGURE 34.5

Bones Are Adapted to Different Environments.

The length of the humerus in the barred owl (left) and the northern fur seal (right) is the same, but the two bones differ markedly in width and density. These differences are adaptations to the animals' vastly different environments.

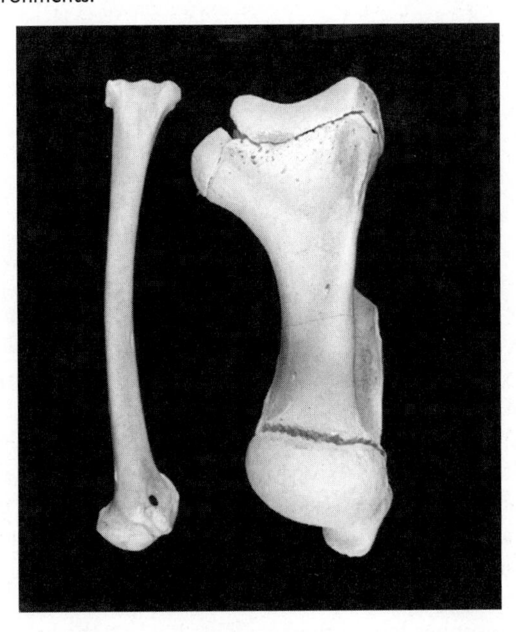

The fur seal's humerus is as long as the owl's, but it is at least twice its width and denser. The seal's bone weighs more than five times that of the owl. The short, heavy forelimb of the fur seal is an adaptation to its habitat, which includes both sea and land. The seal's heavy limb bones enable the animal to forage on the ocean bottom for food without expending enormous muscular energy to fight the buoyancy of water. The bone's density provides the strength required to move against water resistance, a force considerably greater than the air resistance birds encounter. At the same time, the joints of the seal's musculoskeletal system provide enough flexibility to permit the animal to drag itself ashore.

Structure of the Vertebrate Skeleton The bones of the vertebrate skeleton are grouped into two parts. The **axial skeleton,** so named because it is located in the longitudinal central axis of the body, consists of the skull, vertebral column, ribs, and sternum (breastbone). The **appendicular skeleton** consists of the limbs and bones that support them. The clavicles (collarbones) and scapulae (shoulder blades) form the **pectoral girdle,** which supports the forelimbs, and the **pelvic girdle** supports the hindlimbs. Both parts of the skeleton effectively absorb the tremendous shocks generated in locomotion. Figure 34.6 depicts the human axial and appendicular skeletons. The number of bones differs among vertebrate species, but the organization is similar. We consider the human to illustrate the major parts of the skeleton.

The human axial skeleton shields soft body parts. The skull protects the brain and many of the sense organs. It consists of hard, dense bones that fit like puzzle pieces at immovable boundaries called sutures. Sutures also help dissipate the shock of a blow to the head. Inside the skull, nooks and crannies fit snugly around the brain and eyes. All the head bones are attached with immovable joints except for the lower jaw and the middle ear. These bones must move to enable chewing, speech, and hearing. Spaces within the bones of the head, called sinuses, help lighten the skull and create a resonating voice chamber. The foramen magnum, a hole at the base of the skull, allows the nervous tissue of the brain to join the spinal cord.

The **vertebral column** protects the spinal cord down to the lower back (fig. 34.6 inset). The position of the vertebral column and its flexible, multipart construction promote locomotion. A human vertebral column consists of 33 vertebrae. The 7 **cervical vertebrae** in the neck are the smallest and allow the widest range of motion so that we can turn our heads in many directions. In the upper back are 12 heavier **thoracic vertebrae.** Indentations on each side help anchor the ribs in place. The 5 **lumbar vertebrae** in the small of the back are the largest of the spinal bones. The 5 fused pelvic vertebrae make up the **sacrum,** which forms a wedge between the two hipbones. The final 4 vertebrae fuse to form the **coccyx,** or tailbone, which has no known function in humans.

FIGURE 34.6
The Human Skeleton.
The human skeleton is divided into the axial skeleton and the appendicular skeleton. The axial skeleton in humans includes the skull, vertebral column, and rib cage. The bones of the limbs and those that support them make up the appendicular skeleton.

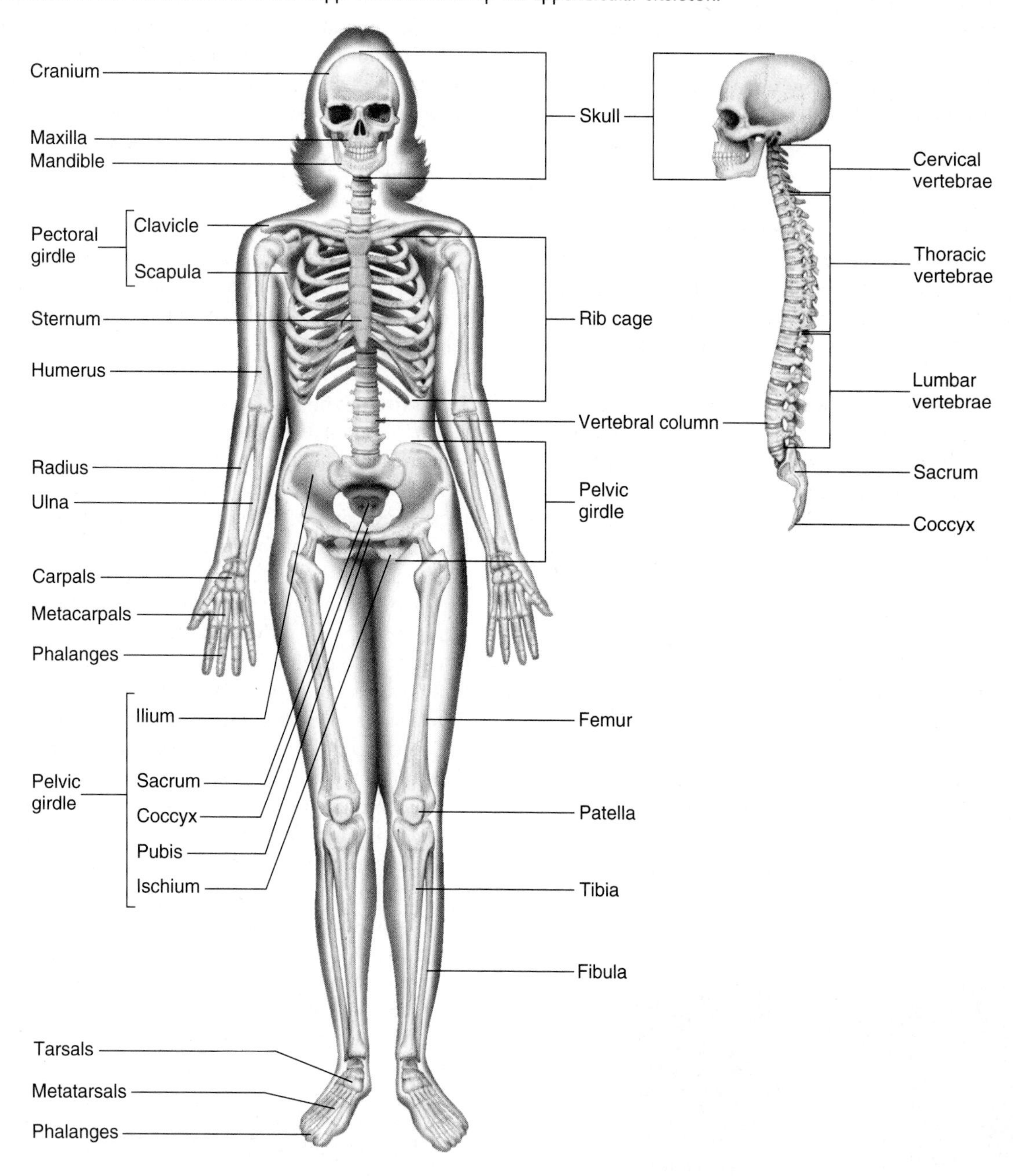

The rib cage protects the heart and lungs. Humans have 10 pairs of ribs attached to the sternum, or breastbone, and two additional, unattached pairs that "float." All the ribs also attach to the vertebral column. The flexibility of the cartilage between the ribs and other bones allows muscles to elevate the ribs, a movement important in breathing.

The long and short bones of the appendicular skeleton function as lever systems, which, when combined with muscles, power movement. The greater the number of bones connected by movable joints, the finer and more variable the movements an animal can make.

The lower limb bones attach to the pelvic girdle. This girdle is made up of the two hipbones, each of which is actually three separate but fused bones. The hipbones join the sacrum in the rear and meet each other in front, creating a bowl-like pelvic cavity. (The term "pelvis" is Latin for "basin.") The bony pelvis protects the lower digestive organs, the bladder, and some of the reproductive structures (especially in the female). The front of the female pelvis is broader than the male's, and it is larger and has a wider bottom opening. These differences in the female pelvis are adaptations for childbirth. In the later months of pregnancy, hormones loosen the ligaments holding the pelvic bones together, allowing them greater flexibility when the newborn passes through.

Cells and Tissues of the Vertebrate Skeleton The vertebrate skeletal system consists primarily of cartilage and bone. These two types of connective tissue are specialized to serve as an internal framework. Recall from chapter 30 that cartilage consists of chondrocytes that secrete collagen. Bone cells include osteoblasts that secrete bone matrix, osteoclasts that break down bone matrix to remodel the tissue, and mature cells, the osteocytes. Figure 34.7 compares cartilage and bone.

Cartilage has many functions in the skeleton. During embryonic development, the skeleton first forms in cartilage, providing a mold for the bone that will later replace it (see Health 34.1). Cartilage is widely dispersed in the adult body. The flexibility of elastic cartilage helps the rib cage expand during inhalation. Another type of cartilage, in the pads between the bones that make up the backbone, can withstand compression. In the appendages, smooth cartilage coverings allow the surfaces of adjoining bones to slide smoothly past one another.

The chondrocytes, which are housed in lacunae within a matrix, secrete proteins that give the tissue its firmness and flexibility. Strong networks of collagen fibers enable the tissue to resist breakage and stretching even when bearing great weight. Elastin provides flexibility. The protein network in cartilage also entraps a great deal of water, which provides support and firmness. The high water content of the matrix makes cartilage an excellent shock absorber, which is its primary function between many bones. The water motion within cartilage that accompanies body movement also cleanses the bloodless tissue and bathes it with dissolved nutrients.

Bone packages maximal strength into a lightweight form. One cubic inch of bone can bear loads up to 19,000 pounds (approximately 8,618 kilograms). The strength of bone rivals that of light steel, but a steel skeleton would weigh four or five times as much.

Bone derives strength from collagen and hardness from minerals. A 20-pound (9-kilogram) weight could dangle from a collagen strand a mere millimeter thick without causing it to snap. Collagen's strength is an asset to bone, but it is flexible and elastic. The hardness and rigidity of bone comes from the minerals, primarily calcium and phosphate, which precipitate out of body fluids coating the collagen fibers. If the minerals were removed from bone, it could be tied in a knot like a garden hose; however, removal of the collagen would make the bone brittle and easily shattered.

Embedded in the bone matrix are living cells responsible for growth and repair. Nutrients and wastes cannot diffuse through the matrix to and from the cells; instead, blood reaches the cells through small canals that penetrate the matrix. The predominant bone cells, osteocytes, lie in spaces (lacunae) in concentric rings. A group of such rings forms an **osteon** (fig. 34.8). Osteons surround central portals, called **osteonic canals,** that contain the blood supply. Communicating canals link adjacent osteonic canals, and narrow passageways, the canaliculi, connect lacunae. Osteocyte extensions reach through the canaliculi, maintaining contact between widely separated osteocytes. Materials pass from the blood, and from osteocyte to osteocyte, through a "bucket brigade" of up to 15 cells. Still other canals connect the entire labyrinth to the outer surface of the bone and to the marrow cavity within. A layer of connective tissue called the periosteum surrounds the entire complex structure of bone.

Bones are lightweight as well as strong because they are porous rather than solid. Most of the irregularly shaped flat bones and almost all of the bulbous tips of long bones are made of **spongy bone.** This type of bone is hard, but has many large spaces between a web of bony struts that align with the lines of stress, increasing the bone's strength. The spaces within spongy bone are filled with red marrow, a nursery for blood cells and platelets that releases more than a million cells a second. Spongy bone is covered

FIGURE 34.7

Cartilage and Bone.

(**A**) Cartilage consists of chondrocytes in lacunae (spaces) within a matrix of collagen and elastin. (**B**) Bone consists of osteons, which are rings of osteocytes within lacunae in a matrix of collagen and mineral. Osteocytes form concentric circles around a central passageway, the osteonic canal, that contains the blood supply.

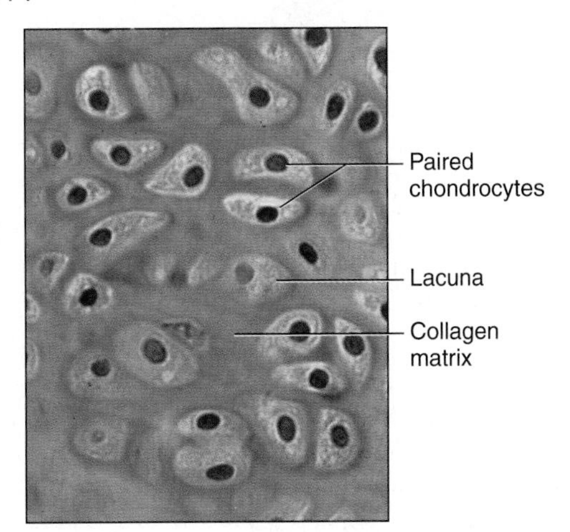

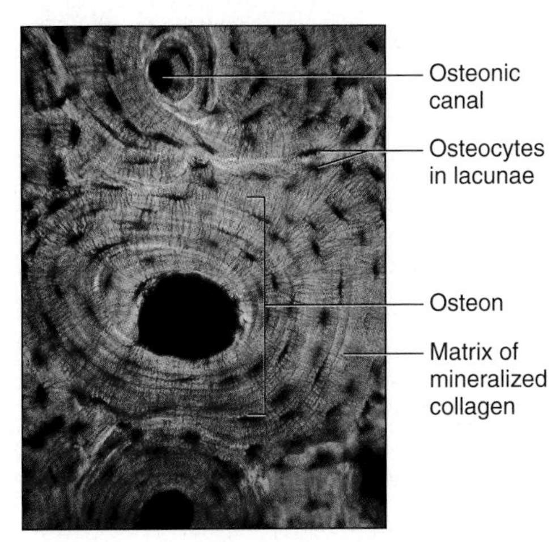

The Human Skeleton over a Lifetime

Before Birth

Cartilage and bone interact throughout development, sculpting the ever-changing skeleton. Most bones originate in the embryo as cartilage models that bone tissue gradually replaces (fig. 34.A). Cartilage models are shaped like the bones they will become. A layer of connective tissue surrounds the models.

In a 6-week embryo, cells just beneath this connective tissue secrete a "collar" of compact bone around the central shaft. Meanwhile, cartilage cells and their lacunae within the shaft enlarge, squeezing the cartilage matrix into thin spicules. The compressed cartilage matrix hardens with calcium salts, which kills the cartilage cells. The cartilage matrix degenerates, and capillaries replace it. Bone cells enter and secrete matrix, establishing the first internal region of new bone.

During Childhood

A baby's skull bones still contain dense connective tissue (rather than bone) at the sutures. The regions between the bones pulsate from underlying blood vessels. The skull bones of the newborn, not yet fused, can compress as the infant squeezes through the birth canal. The soft spots in the head usually fill in with bone tissue by 18 months of age.

Within a few months of birth, bone growth becomes concentrated near the ends of the long bones in thin disks of cartilage called the epiphyseal plates. These cartilage disks enable bones to elongate as the child grows. The cartilage cells in the plate divide, pushing daughter cells toward the shaft, where they calcify. The dividing cartilage cells keep the disk relatively constant in thickness, but the calcified daughter cells become part of the shaft, elongating it. Growth continues until the late teens, when bone tissue begins to replace the cartilage plates. By the early twenties, bone growth is complete. Only a line remains marking the position of a cartilage disk.

Fracture Repair

A series of events occurs when broken bones heal. The immediate reaction to a fracture is bleeding, followed rapidly by blood clot formation at the site of the break. Then dense connective tissue replaces the blood clot. Next, cartilage cells enter the connective tissue and form a fibrous "callus" that fills the gap the injury left. New spongy bone replaces the cartilage, closing the gap. Exercising the injured bone stimulates bone cells in the healing area to secrete collagen, which compacts the newly formed bone.

Osteoporosis

Decreasing bone mass due to loss of calcium salts is a normal part of aging, but it can lead to osteoporosis (holes in the bones), in which bones are likely to break (fig. 34.B). Ten percent of women have severe enough osteoporosis by age 50 to suffer bone fractures. Nearly 25% of all women develop osteoporosis by age 60, and by age 90, one-third of all women and one-sixth of all men have it. Osteoporosis causes shrinking stature, chronic back pain, and frequent fractures.

FIGURE 34.A
Bone Growth.
During prenatal development, bone gradually replaces the cartilage skeleton. A cartilage "model" for most bones appears in the 6-week embryo. Bone forms first around the central shaft as cartilage cells die within a calcifying matrix. Later, bones elongate from the epiphyseal plates, which are cartilage disks located between the shaft and the knob at the end of the bone. The developmental stages listed under each bone are for the femur, other bones may develop at different rates.

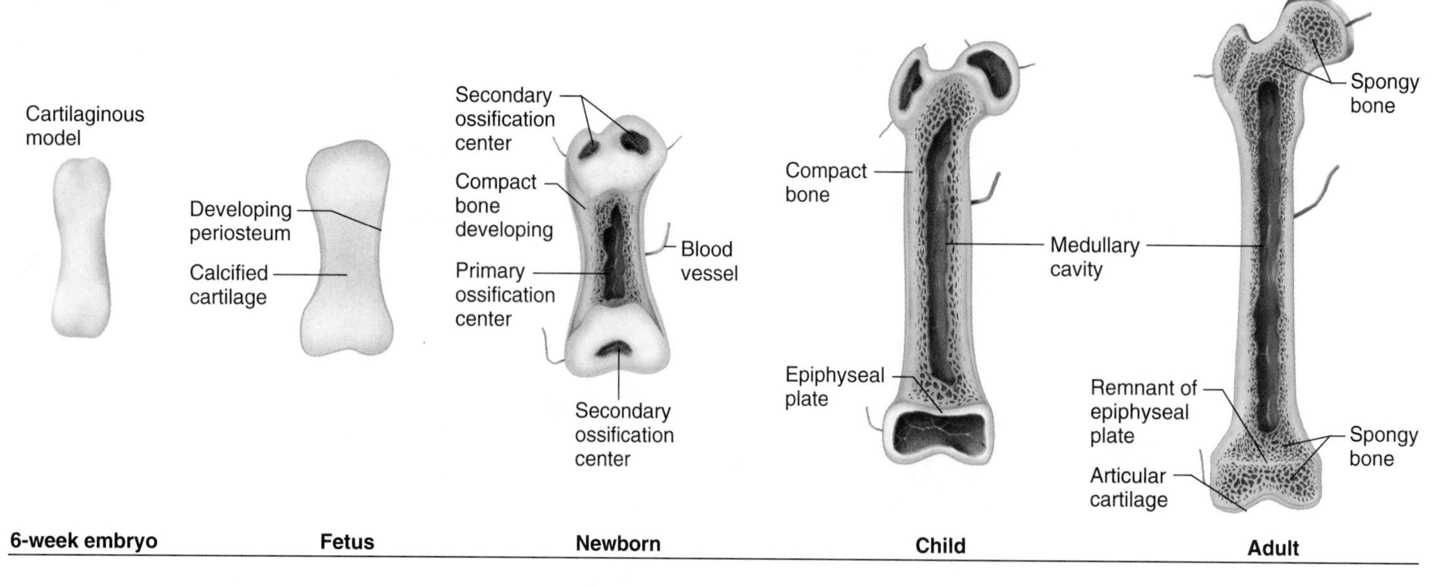

FIGURE 34.B
Osteoporosis.
In osteoporosis, calcium loss weakens bones. The vertebra on the left is normal; the vertebra on the right has been weakened by osteoperosis.

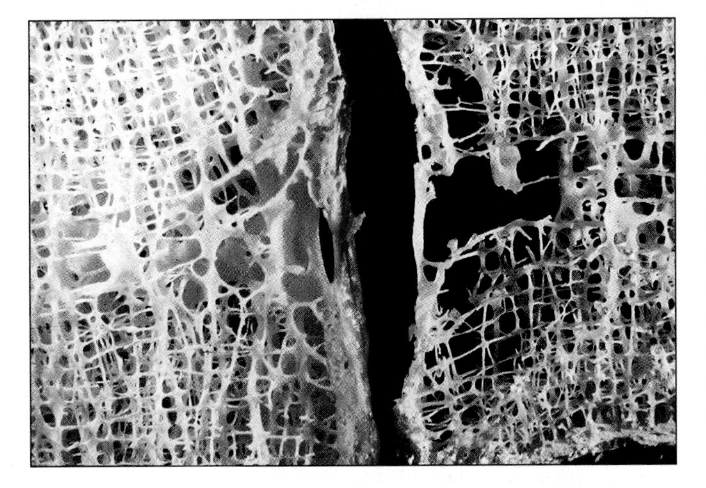

Females are more likely to suffer from osteoporosis than males for several reasons. Women's longer life span contributes to their overrepresentation among those with osteoporosis. Furthermore, the bone mass of the average woman is about 30% less than a man's, so her bones are more easily depleted of calcium stores. A decline in the level of estrogen also makes a woman more prone to osteoporosis. For this reason, doctors advise all women to take 1,000 to 1,500 milligrams of calcium daily and to exercise regularly.

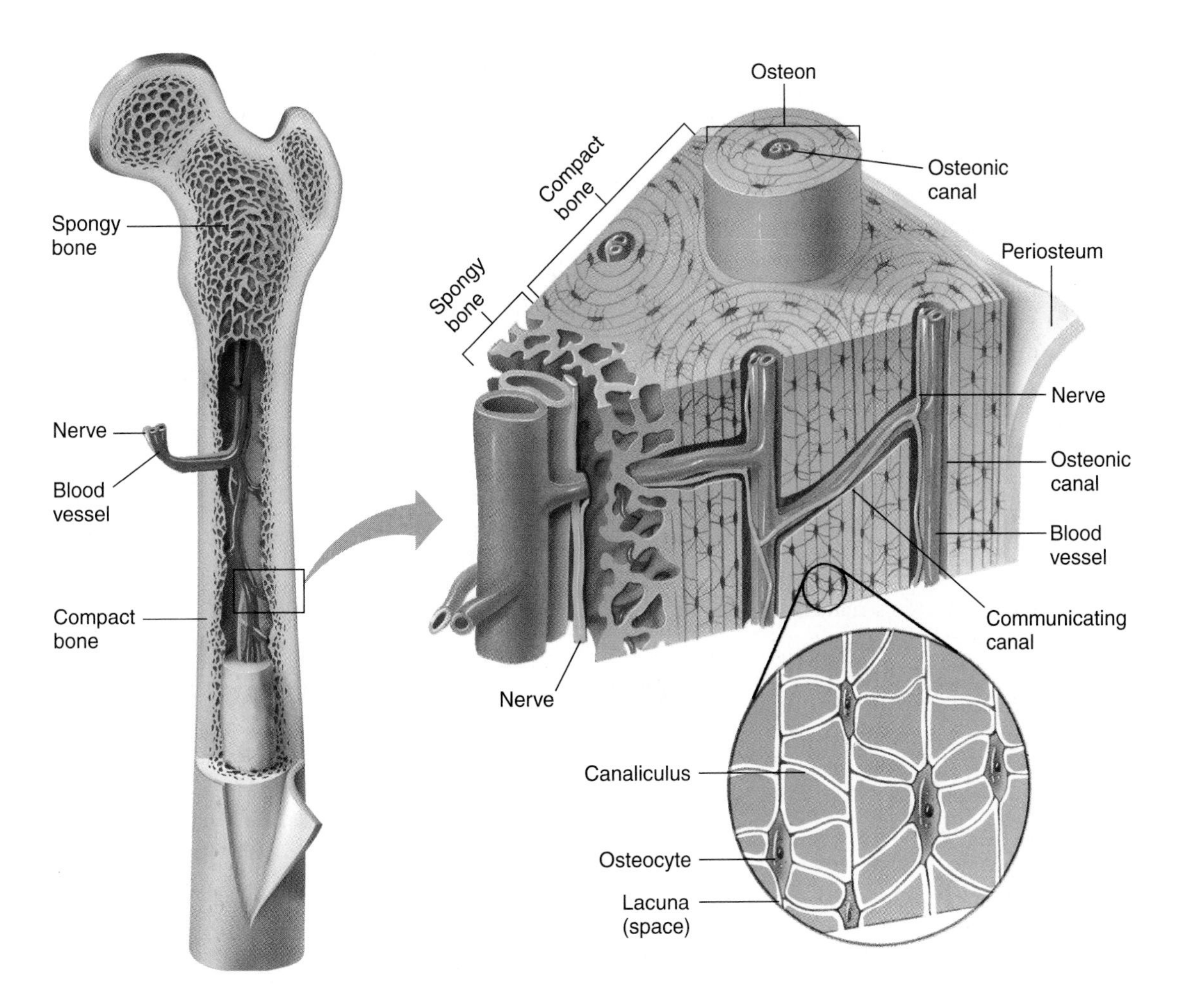

FIGURE 34.8
The Structure of a Long Bone.
The shaft of the bone is hollow and contains yellow marrow. Surrounding this is a layer of spongy bone. The outer coat consists of compact bone. The knobby ends of the bone are spongy bone coated with compact bone.

with a layer of more solid **compact bone.** The weight of long bones, such as limb bones, is further reduced by the **marrow cavity,** a space in the shaft that contains yellow marrow consisting primarily of fat cells. Because a hollow tube is more resistant to certain types of stress than a solid rod of equal mass, this cavity allows for maximum strength with minimum weight.

34.2 MASTERING CONCEPTS

1. How does a braced framework skeleton differ from a hydrostatic skeleton?
2. What are two ways that endoskeletons and exoskeletons differ?
3. What are some functions of the vertebrate endoskeleton?
4. What are the major groups of bones that make up the human skeleton?
5. What are the functions of cartilage and bone in vertebrate skeletons?
6. Describe the organization of bone tissue.

OLC

FIGURE 34.9
Antagonistic Muscles.
Pairs of muscles that act antagonistically surround a joint in a grasshopper's leg, enabling the appendage to be moved.

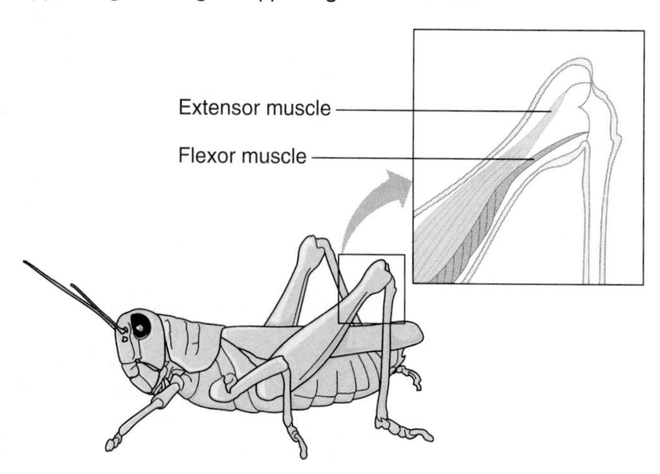

34.3 Muscle Diversity

Muscles contract through organized actions at the molecular level. Muscles in invertebrates are highly coordinated with the exoskeleton to provide rapid adaptive responses to stimuli.

The ability to contract is an ancient characteristic of cells. High-energy molecules such as ATP power the sliding protein filaments that underlie such diverse biological movements as organelle transport within cells, the locomotion of protista such as the amoeba, and such large-scale movements as the graceful gait of a cheetah.

Muscle structure and function among different types of animals are more alike than diverse. At the molecular level, muscles are built primarily of the proteins actin and myosin, with other proteins less abundant but no less crucial. At a cellular and tissue level, muscle can be voluntary (striated muscle) or involuntary (smooth muscle and cardiac muscle). Muscles as organs consist of complex packets of fibers and surrounding tissues.

An individual muscle cell can be in either of two states—contracting or relaxed. A muscle cell cannot expand, or lengthen. However, specific configurations of contracted and relaxed muscle cells provide great versatility in an animal's movements.

Invertebrates use combinations of smooth and striated muscles. A scallop, for example, is a bivalve mollusk that uses striated muscle to snap its shell shut, propelling it forward to escape a hungry predator. But its adductor muscle, which can close the shell for days, is smooth muscle. Figure 34.9 shows the relationship between parts of an insect's exoskeleton and muscle pairs. When the extensor muscle contracts, the limb extends; when the flexor muscle contracts, the limb flexes.

Insect flight illustrates frenetic muscular activity—some flies beat their wings a thousand times a second! Striated muscle produces fast muscle contractions in the insect wing. The elastic tissue surrounding the flight muscles helps sustain a rapid contraction rate. When a group of muscles called depressors lower the wing by contracting, they stretch another group, the elevator muscles. These muscles then contract, raising the wing. When the elevator muscles contract, they in turn stretch the depressor muscles, stimulating them to contract. The result of this oscillation: A single nerve impulse can trigger 20 to 30 contractions of the wing muscles. A fly whizzes by. The opening essay to chapter 32 describes the muscles that enable a blowfly's halteres to balance the animal during flight.

The movements of some invertebrates are so fluid and complex that they are the envy of engineers trying to design machines to explore rugged terrain. A crab, for example, must coordinate the movements of 10 legs to scurry sideways to escape danger. Arthropods are able to move well because of sense organs within their exoskeletons that function as strain gauges, detecting stress on the animals' outsides. These sense organs are linked to sensory neurons beneath the cuticle. When a falling rock slightly dents a crab's leg, for example, the sensory neurons beneath the dent fire action potentials to the central nervous system, which then signals certain muscles to contract in a pattern that enables the crab to quickly leave the danger scene.

34.3 MASTERING CONCEPTS

1. What types of molecules provide movement in animals?
2. In what two states are muscle cells?
3. Describe how insects fly.

OLC

34.4 Vertebrate Muscles

Skeletal muscle cells (fibers) are grouped into bundles encased in connective tissue with associated blood vessels and nerves. Sliding actin and myosin filaments contract the muscle fiber. A nerve cell and the muscle fibers it stimulates form a motor unit. Muscle fibers are of three twitch types.

At the microscopic level, muscle structure and function are similar in different vertebrate species. On a macroscopic level, however, differences become apparent, reflecting the varied pressures of natural selection.

Muscle Cell Types

Recall from chapter 30 that three types of cells form muscle tissues (fig. 34.10). **Smooth muscle** cells are long and tapered at each end. Each cell has a single nucleus. Animals cannot consciously control smooth muscle cells; these cells form involuntary muscles. Smooth muscle is found in many parts of the vertebrate body. In the walls of certain blood vessels, smooth muscle contracts, regulating blood pressure and directing blood flow. In the iris, smooth muscle controls the size of the eye's opening to light. Smooth muscle in the intestinal wall contracts, pushing food through the digestive tract. Sheets of circular and longitudinal smooth muscle line several hollow organs and either push material through or maintain the diameter of a tube. In the bladder this movement makes urination possible.

FIGURE 34.10

Three Types of Muscle Cells and Tissues.

(**A**) Smooth muscle is composed of spindle-shaped cells, each with one nucleus, that often form layers (×250). Smooth muscle is in the walls of the digestive tract; its contraction is involuntary. (**B**) Cardiac muscle is unique to the heart. Its striated cells join at connections called intercalated disks. The cells form a branching network, and their contraction is involuntary (×400). (**C**) Skeletal muscle is under voluntary control, and the cells have many nuclei (×250). The cells are striated due to the orderly organization of contractile proteins.

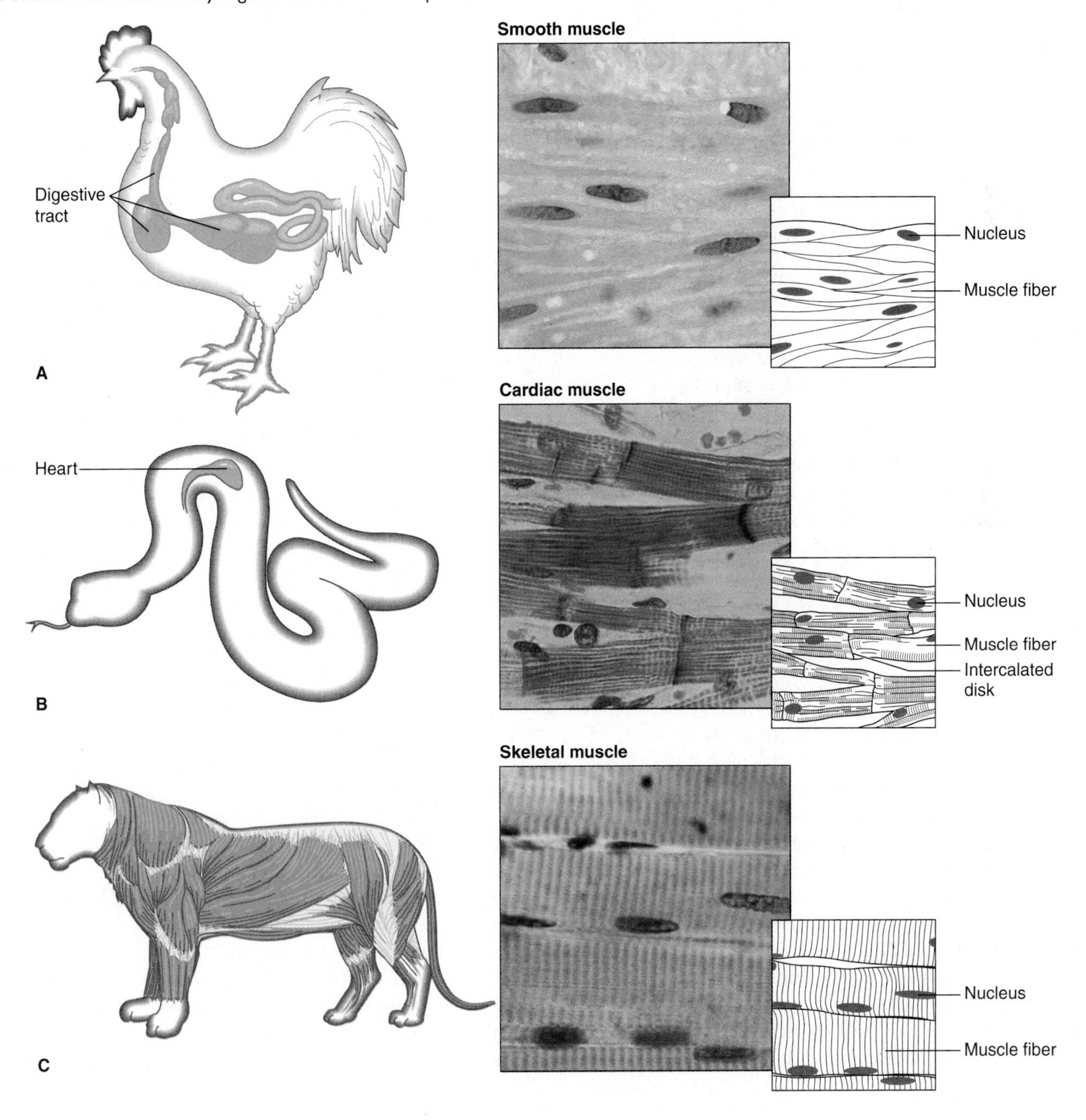

FIGURE 34.11
The Human Muscular System.
The human body has more than 600 skeletal muscles, a few of which are identified here.

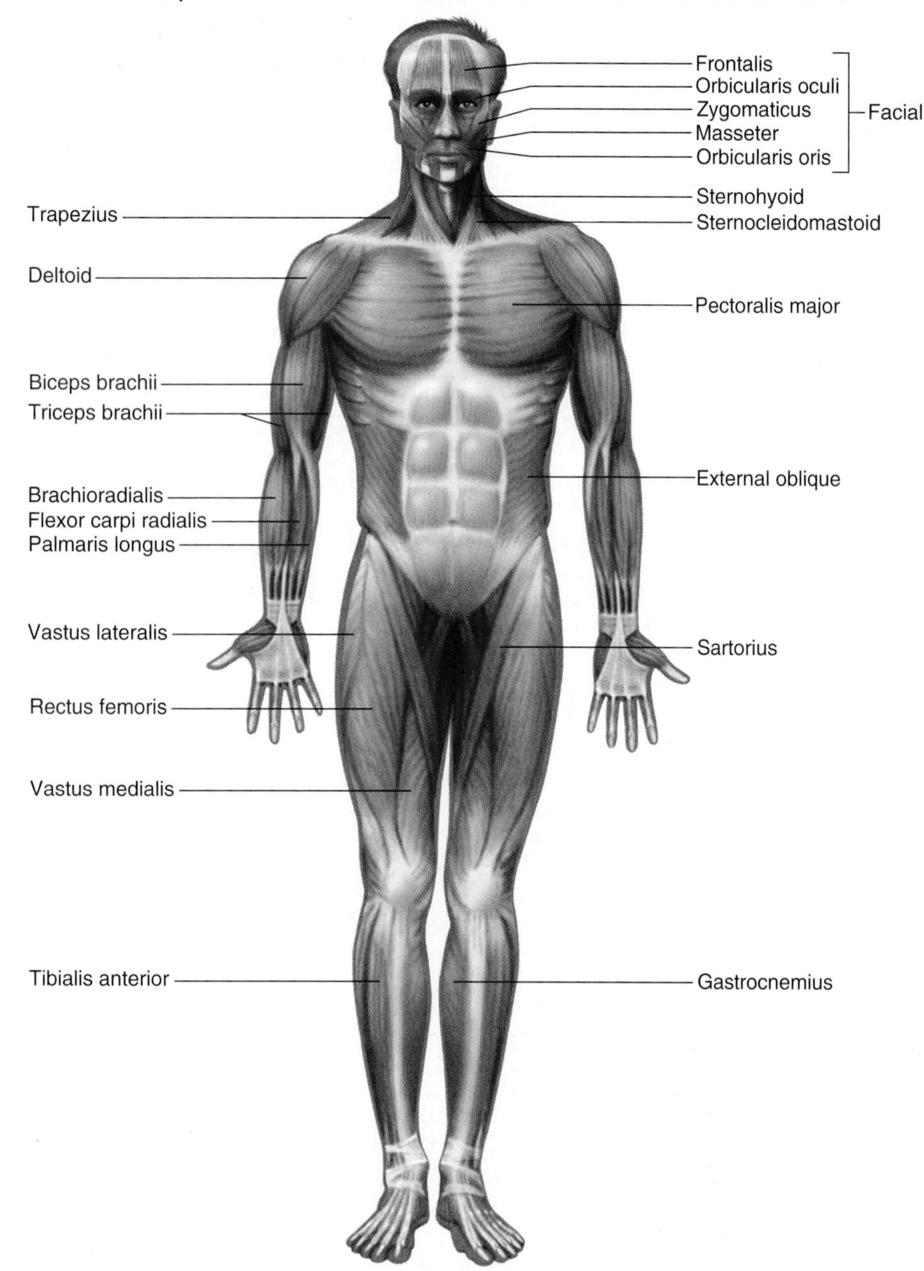

Cardiac muscle cells are unique to the heart, where they contract, propelling blood throughout the body. Cardiac muscle cells appear striated under the microscope because of the pattern of contractile filaments within them. Their control is involuntary, and they have one nucleus per cell. Cardiac muscle cells branch, forming a netlike structure, and strong but permeable membrane foldings called **intercalated disks** join them.

Skeletal muscle cells, so named because they attach to bones, are under voluntary control. Skeletal muscle cells are very long and appear striated. A single skeletal muscle cell has many nuclei, which are pushed to the outer surface. The human muscular system is very complex, and includes more than 600 skeletal muscles. Many small skeletal muscles are responsible for our great variety of facial expressions. These muscles connect the bones of the skull to the skin. Four pairs of skeletal muscles are required to move the jaw in chewing. Eight muscles move a foot. Figure 34.11 identifies a few of the major skeletal muscles in a human.

Skeletal Muscle Consists of Fiber Bundles

The powerful arm muscles that propel a tennis player's racquet can be viewed at several levels (fig. 34.12). The most familiar level is the whole muscle. Each muscle lies along the length of a bone. A heavy band of fibrous connective tissue called a **tendon** attaches each end of the muscle to a different bone. At the microscopic level, the long skeletal muscle cells are called **muscle fibers,** which can be more than 1 foot (0.3 meter) long. These muscle fibers are organized into many larger bundles. A muscle consists of clusters of such bundles. A connective tissue sheath wraps around each fiber, bundle, and whole muscle.

Blood vessels and nerves run between the bundles of muscle fibers. The rich blood supply provides muscle cells with nutrients and oxygen to power contraction and removes cellular wastes. The nerves trigger and control muscle contraction.

Most of the volume of a muscle fiber consists of hundreds of thousands of cylindrical subunits called **myofibrils** (fig. 34.13). The rest of the cell contains the usual organelles. The cell membrane and endoplasmic reticulum (ER) are extensive and have special names—**sarcolemma** for the cell membrane and **sarcoplasmic reticulum** for the ER. Both membrane networks fold against one another at many points. The parts of the outer membrane that jut into the inner membrane are called the **transverse,** or **T, tubules.** These structures carry action potentials to the cell's interior.

Myofibrils are made of even finer "strings," the **myofilaments,** which consist of the contractile proteins. Myofilaments are of two types—thin and thick. Each thin myofilament is composed primarily of two entwined strands of actin and also contains two proteins important in controlling muscle contraction, **troponin** and **tropomyosin.**

The thick myofilaments are composed of myosin. Each myosin molecule is shaped like a golf club. The "shafts" of myosin molecules adhere to one another, forming a bundle that may consist of several hundred molecules. The "heads" protrude from both ends of the

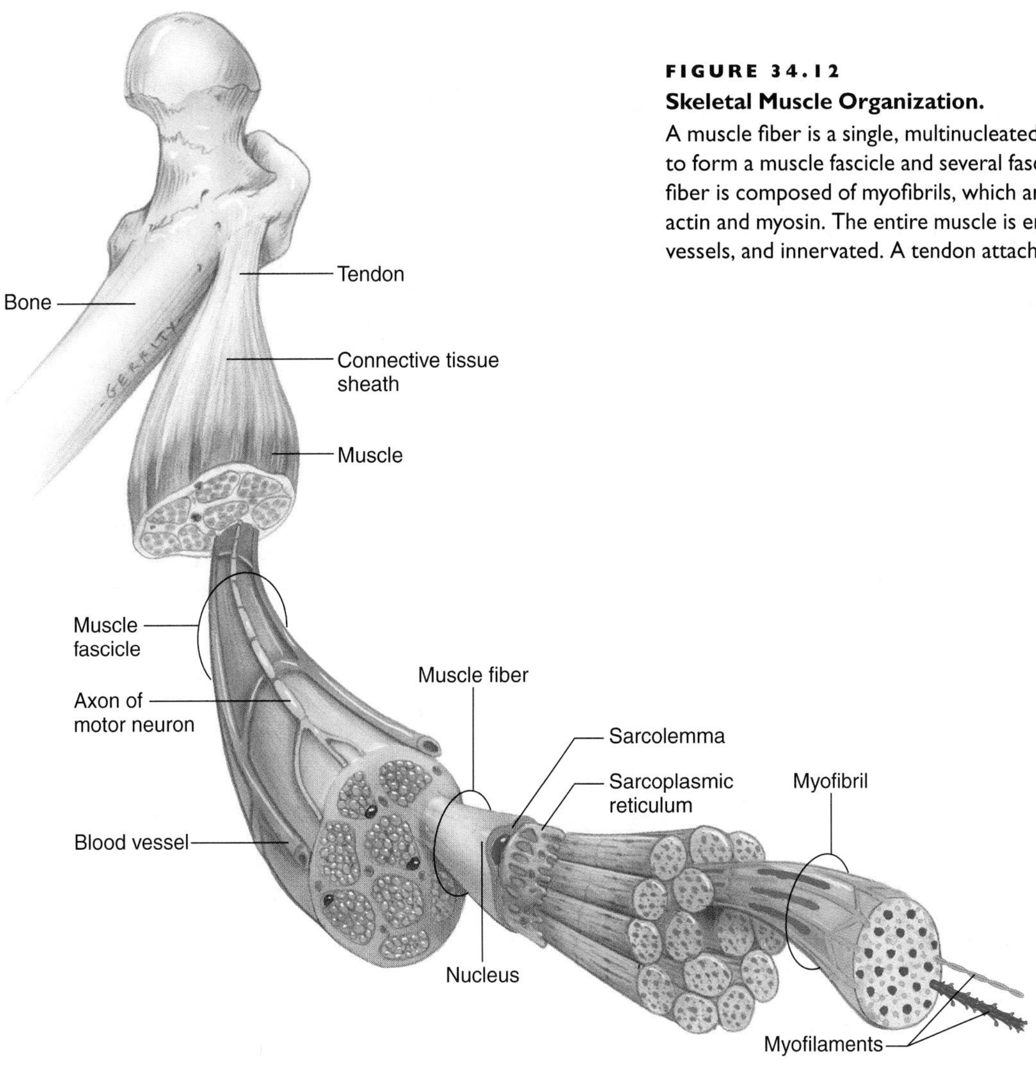

FIGURE 34.12

Skeletal Muscle Organization.

A muscle fiber is a single, multinucleated cell. Several muscle fibers are grouped to form a muscle fascicle and several fascicles make up a muscle. Each muscle fiber is composed of myofibrils, which are in turn composed of myofilaments of actin and myosin. The entire muscle is enclosed in connective tissue, fed by blood vessels, and innervated. A tendon attaches muscle to bone.

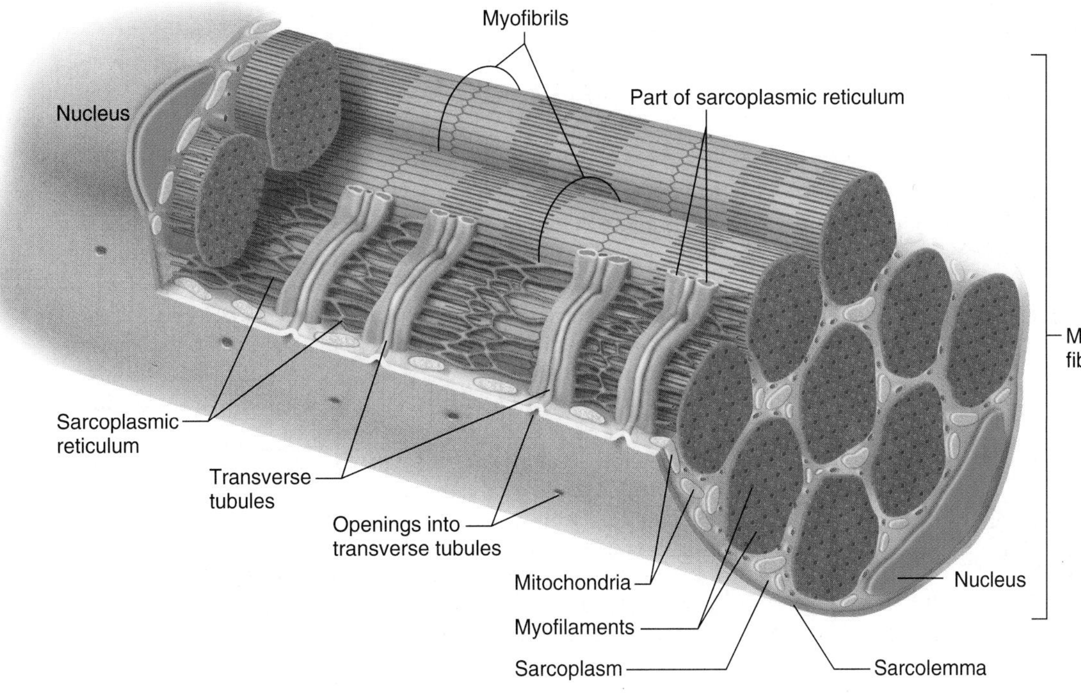

FIGURE 34.13

Anatomy of a Muscle Fiber.

The sarcoplasmic reticulum is a specialized endoplasmic reticulum in a muscle cell. It is highly folded to store the calcium ions that regulate muscle contraction. Transverse tubules are membranes that fold inward between contractile units of the muscle cell. They conduct action potentials to the interior of the cell.

bundle in a spiral pattern. Each thick myosin myofilament is surrounded by six thin actin myofilaments (fig. 34.14). This organization allows connections between myosin and actin to form, break apart, and form again, which is essential for muscle contraction.

Muscle contains other proteins too. A muscle protein need not be very abundant to be vital. For example, Duchenne muscular dystrophy results from lack of or abnormal dystrophin, a protein that makes up only 0.2% of the total muscle protein. Dystrophin normally supports the inside face of muscle cell membranes, enabling them to withstand the force of contraction. When a child lacks dystrophin, muscles gradually fail.

Sliding Filaments Generate Movement

The striations of skeletal muscle derive from the organization of the thick and thin myofilaments. The stripes occur in a pattern of repeated units called **sarcomeres** (muscle units). Figure 34.14 shows a microscopic view of a sarcomere and illustrates how the myofilaments interact to produce the various bands and how the bands change as the muscle contracts.

Different areas within a sarcomere have specific names. Z lines define the boundaries of a sarcomere, and they are membranes to which actin myofilaments attach. A bands are areas occupied by thick (myosin) myofilaments. The central area between actin filaments is called the H zone, and it is occupied by myosin only. Myosin-free I bands extend between adjacent A bands.

Muscle contracts when the thin myofilaments slide between the thick ones. This motion shortens the sarcomere without shortening its protein components, a little like fitting your fingers together to shorten the distance between your hands. The movement of protein myofilaments in a way that shortens skeletal muscle cells is called the **sliding filament model** of muscle contraction.

To understand how actin and myosin are propelled past one another, it is necessary to understand the action of the myosin

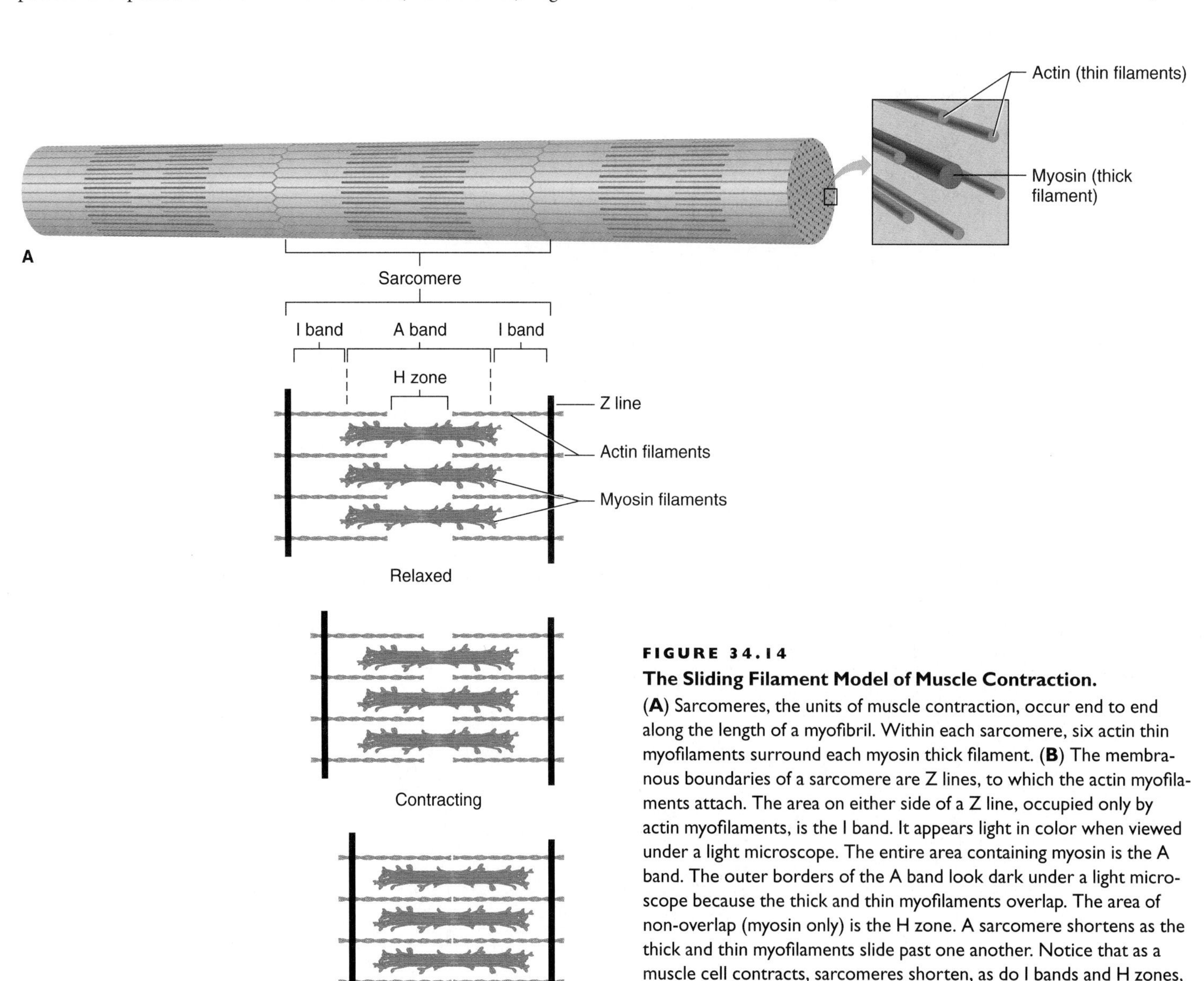

FIGURE 34.14
The Sliding Filament Model of Muscle Contraction.
(**A**) Sarcomeres, the units of muscle contraction, occur end to end along the length of a myofibril. Within each sarcomere, six actin thin myofilaments surround each myosin thick filament. (**B**) The membranous boundaries of a sarcomere are Z lines, to which the actin myofilaments attach. The area on either side of a Z line, occupied only by actin myofilaments, is the I band. It appears light in color when viewed under a light microscope. The entire area containing myosin is the A band. The outer borders of the A band look dark under a light microscope because the thick and thin myofilaments overlap. The area of non-overlap (myosin only) is the H zone. A sarcomere shortens as the thick and thin myofilaments slide past one another. Notice that as a muscle cell contracts, sarcomeres shorten, as do I bands and H zones, but A bands stay the same length. (Note: only two of the six thin myofilaments surround each thick one in this depiction.)

heads. The club-shaped heads attach to the shafts with hingelike connections that allow the head to rock back and forth. This back-and-forth motion is similar to the path of an oar as a stroke propels a boat through the water. The movement pulls on actin and causes it to slide past myosin in the same way the oar's motion moves a boat. Figure 34.15 shows how actin and myosin interact. Myosin heads can bind both actin and ATP. Splitting ATP provides the energy that powers muscle contraction. Muscle contraction also requires calcium ions (Ca^{2+}).

For contraction to occur, actin and myosin must touch. Because a myosin head can swing out to contact an actin molecule it is also called a cross bridge. In a resting muscle, troponin holds tropomyosin in a groove between the actin strands, which blocks myosin cross bridges from binding to actin.

The directive to contract begins when a motor nerve cell that extends from the brain or spinal cord to a muscle cell releases the neurotransmitter acetylcholine at a neuromuscular junction, where a neuron and muscle cell meet (fig. 34.16). When the neurotransmitter binds to the outer membrane of the muscle cell, it produces an electrical wave that races down the T tubules. This wave induces storage sacs within the sarcoplasmic reticulum to release calcium ions, which flood the cytoplasm. The calcium binds to the troponin molecules attached to the actin myofilaments. The troponin then moves, taking the attached tropomyosin molecules to the interior of the thin myofilament, where they no longer block actin from contacting the myosin of the surrounding thick filaments. When tropomyosin moves, myosin cross bridges attach to the exposed actin.

Long before the nerve impulse arrived, ATP slipped into its site on the myosin head, where it was enzymatically split to yield ADP and inorganic phosphate, releasing the energy that activates myosin cross bridges. When the activated cross bridges bind to actin, this energy is released, causing the myosin head to swing back to its original position. This forces the filaments to slide past one another.

After this so-called power stroke, ADP and the inorganic phosphate pop off the myosin head, leaving it free to bind a new ATP. When a new ATP binds, the cross bridge releases from actin even before the ATP splits. Then, as ATP releases its energy, the myosin head flips back to its original position, ready to bind a new actin further along on the chain.

FIGURE 34.15

How a Muscle Moves.

This sequence of drawings shows the molecular interactions of the sliding filament model of muscle contraction.

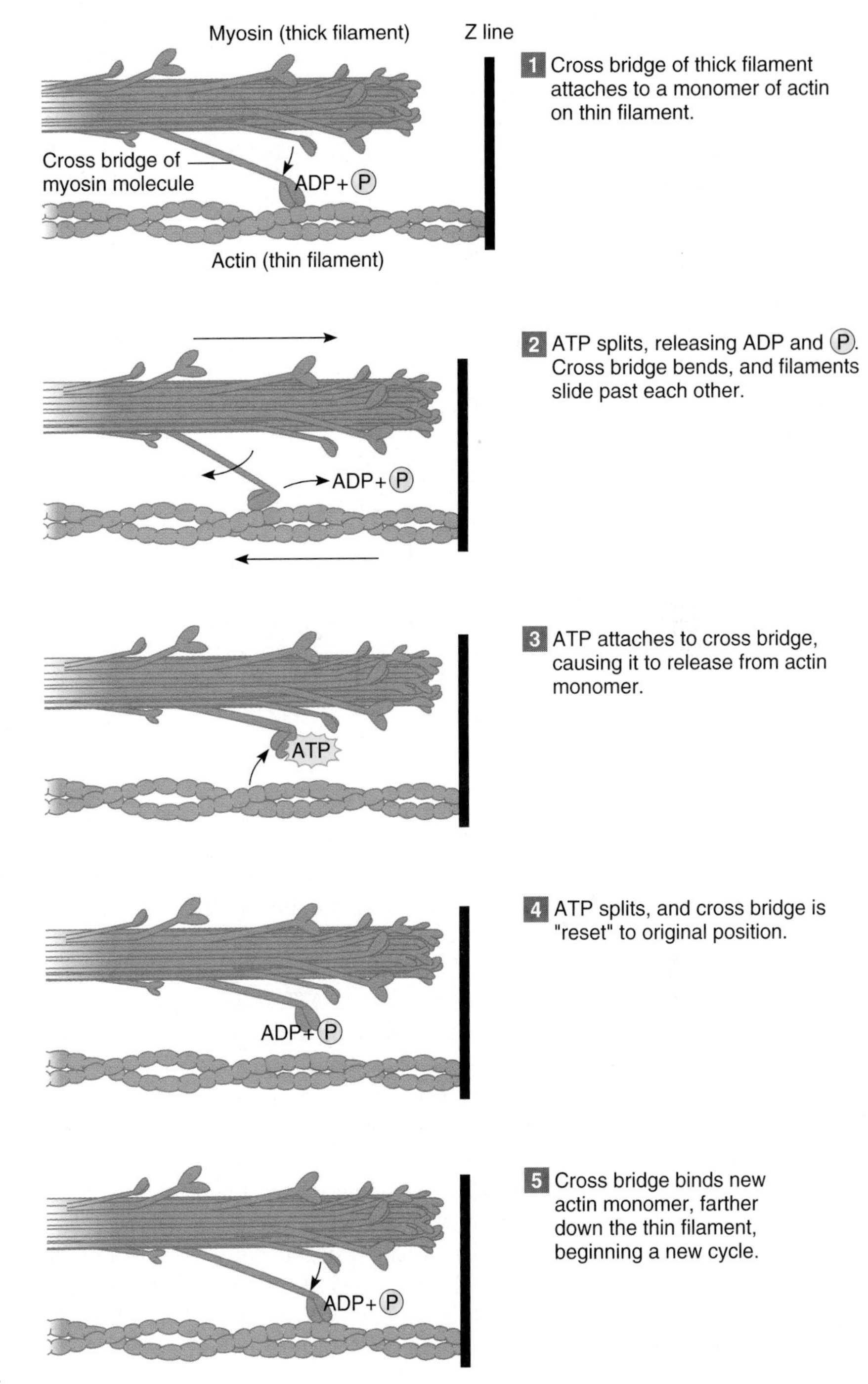

This sliding repeats about a hundred times a second on each of the hundreds of myosin molecules of a thick filament. Although the movement of filaments due to the interactions of single actin and myosin filaments would be no more astounding than the forward progress of a boat with a single person rowing, a skeletal muscle contracts quickly and forcefully due to the efforts of many thousands of "rowers."

Skeletal muscle must also relax on cue. Myofilament sliding stops when calcium ions are no longer available. Shortly after motor nerve stimulation causes the release of calcium ions, they are actively transported back to the sarcoplasmic reticulum. Tropomyosin returns to its original position, blocking actin from interacting with myosin heads, and the sarcomere relaxes.

FIGURE 34.16
A Neuromuscular Junction.
A motor axon branches, forming specialized junctions with muscle fibers. The axon releases a neurotransmitter that triggers electrical changes in the muscle cell membrane, which ultimately triggers release of calcium ions inside the muscle cell, leading to contraction.

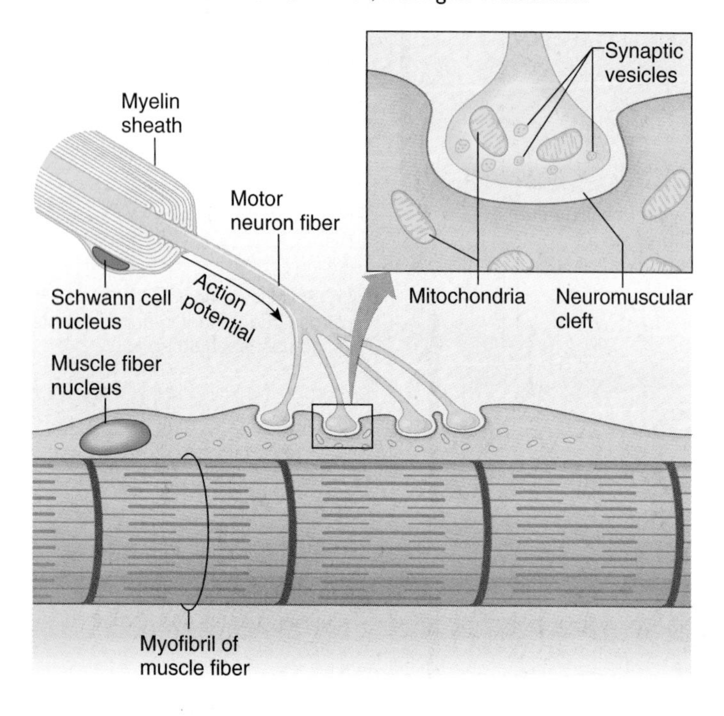

Skeletal muscle contraction requires huge amounts of ATP, both to power the return of Ca^{2+} to the sarcoplasmic reticulum and to break the connection between actin and myosin, allowing a new cross bridge to form. After death, muscles run out of ATP. The cross bridges cannot release from actin, and the muscles remain in a stiff position. This stiffening is called rigor mortis.

An animal's body has several ways to generate ATP. When muscle activity begins, ATP can be replenished as rapidly as it is depleted when a molecule called **creatine phosphate,** which is stored in muscle fibers, donates a high-energy phosphate to ADP to regenerate ATP. Obtaining ATP from creatine phosphate requires only one enzymatic step, which is why this source of ATP is rapid, but it is also depleted first. After the supply of creatine phosphate falls, the body must generate ATP from other sources. As long as enough oxygen can reach the muscle tissue, the mitochondria form ATP by aerobic respiration. The fuel sources are fatty acids and carbohydrates. The muscle stores glycogen, a glucose polymer that can be broken down to provide a ready supply of glucose for metabolism.

When aerobic respiration and creatine phosphate can no longer provide enough ATP, the muscle cell obtains ATP from anaerobic pathways. This is a less efficient metabolic route that rapidly leads to muscle fatigue. Lactic acid buildup from anaerobic pathways causes muscle pain.

A muscle **twitch** is the sequence of contraction and relaxation that follows stimulation. Most skeletal muscles contain fibers of three different twitch types, which are distinguished by how quickly they contract and tire. Slow-twitch (fatigue-resistant) fibers contract slowly because the myosin heads split ATP slowly. They resist fatigue, however, because they are well supplied with oxygen. This oxygen is bound to many molecules of oxygen-carrying myoglobin protein or is delivered by the extensive blood supply serving the muscle fibers. Myoglobin is similar to hemoglobin, the oxygen carrier in the blood supply, and like hemoglobin is reddish. Therefore, these slow-twitch, slow-fatiguing muscles are called red or dark fibers. They are abundant in muscles specialized for endurance. For example, the breast meat (flight muscles) of ducks and geese, which require muscle endurance for long, migratory flights, is reddish "dark meat."

The second type of muscle fiber, fast-twitch (fatigable) fibers, splits ATP quickly but does not have much oxygen. These fibers are white because they lack myoglobin and a rich blood supply. They are specialized for short bouts of rapid contraction. The very white breast muscle of the domesticated chicken, for example, can power barnyard flapping for a short time but cannot support sustained long-distance flight.

Intermediate fibers have the fast ATP-splitting characteristic of fast-twitch fibers but enjoy a better blood supply. The speed and fatigue properties are intermediate between those of the two main muscle twitch types.

The proportion of fast-twitch to slow-twitch fibers within particular muscles affects athletic performance. Most people have about equal numbers of fast- and slow-twitch muscle fibers. Those who have a higher proportion of slow-twitch fibers excel at endurance sports, such as long-distance biking, running, and swimming. Consider a world-class long-distance runner, who holds many first-place finishes in the marathon but does not do nearly as well in the 400-meter dash. This runner's leg muscles might consist of over 90% slow-twitch fibers. By contrast, athletes who have a higher proportion of fast-twitch fibers tend to perform best at short, fast events such as sprinting, weight lifting, and shot putting.

Whole Muscle Responses to Stimuli

The nervous system interacts with the musculoskeletal system to produce smooth, coordinated muscle movements. Two integrated mechanisms are at work. One occurs at the level of the individual muscle cell, and the other controls groups of muscle fibers.

A single muscle cell contracts whenever a motor neuron stimulates it. However, calcium ions are quickly cleared from the cytoplasm of the muscle fiber after each action potential. For a muscle cell to contract to its full range, repeated action potentials are required to sustain calcium availability and maintain cross-bridge formation. Maximal contraction caused by continuous stimulation is called **tetanus** (not to be confused with the disease caused by *Clostridium tetani*). With a high enough rate of action potentials, tetanus ensures smooth, prolonged contractions of individual muscle cells.

The second mechanism that ensures smooth muscle movements involves groups of muscle cells that a single neuron controls. A nerve cell and all the muscle fibers it contacts form a **motor unit** (fig. 34.17). A motor nerve cell that controls only a

FIGURE 34.17
Motor Units.
A motor unit consists of a motor neuron axon and the muscle fibers it innervates. Motor units are of varying sizes. Large motor units can generate large, strong movements because their motor neurons contact and stimulate many muscle cells. Small motor units govern smaller, finer movements where a nerve cell stimulates only a few muscle cells.

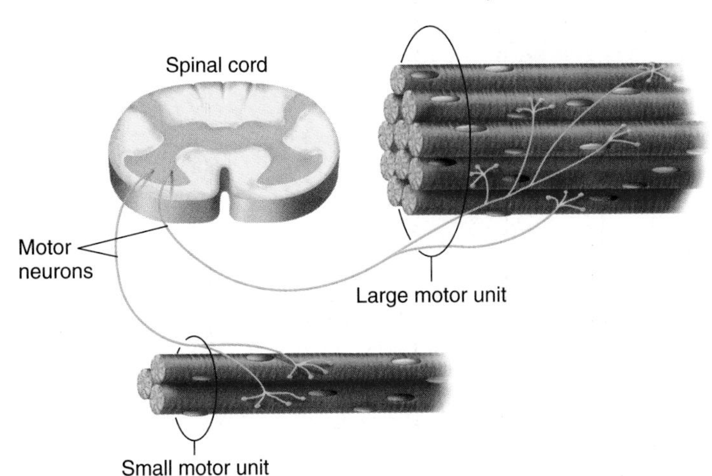

few muscle fibers produces fine, small-scale responses, such as the eye movements required for reading. In contrast, a single motor neuron that innervates hundreds of different muscle cells produces large, coarse movements, such as those required for throwing a ball. Within a given muscle, however, motor units vary in size from tens to hundreds of cells per motor neuron. This is why the same hand can both grip a hammer and pick up a tiny nail.

Scattered among the large muscle fibers are small, modified muscle cells, called **muscle spindles,** that are innervated by stretch-sensitive neurons. Muscle spindles keep muscles always partially contracted, a condition called muscle tone. In the spinal cord, the spindle's stretch-sensitive neurons synapse with all of the muscle's motor neurons. However, the tiny neurons innervating the smallest motor units are most easily depolarized (see chapter 31). When a light object such as a pen is placed into a hand, the spindles are stretched only slightly. The stimulus generates only a few action potentials in the stretch-sensitive neurons, which in turn activate only the smallest motor units. The result is an appropriately minor muscle contraction. In contrast, if someone hands a person a heavy textbook, the muscle spindles are stretched much further, provoking more action potentials in the sensory neurons. These recruit many motor units, both large and small. As a result, the muscle contracts sufficiently to support the book.

Many skeletal muscles occur in antagonistic pairs whose members operate in opposite directions. The muscles in the upper arm, for example, form an antagonistic pair. When a person contracts the biceps (the muscle that bulges to "make a muscle"), the arm bends at the elbow. When a person contracts the triceps, the muscle at the back of the upper arm, the arm straightens. To produce smooth arm movements, these muscles must be precisely coordinated in opposite directions. To accomplish this, sensory neurons from muscle spindles in the biceps not only excite motor neurons that lead back to the biceps, but also signal another neuron in the spinal cord (an interneuron) to inhibit muscle contraction in the triceps. This inhibition is precisely matched to the degree of contraction stimulated in the biceps.

34.4 MASTERING CONCEPTS

1. What are the three muscle cell types?
2. Describe the levels of organization of a vertebrate muscle.
3. Describe how protein interactions move muscles.
4. What is the source of energy that powers muscle contraction?
5. How do nerves control skeletal muscle contraction?

OLC

34.5 Skeletal Muscle and Bone Function Together

A synovial joint is a fluid-filled connective tissue capsule that connects bones and is held in place by ligaments. Tendons attach muscles to bones, forming lever systems that generate movement.

The vertebrate musculoskeletal system is a highly organized framework of bones and muscles and the tissues and structures that connect and support them. All the components work together to produce movement.

Joints—Where Bone Meets Bone

The structures between bones are called joints. Most of the vertebrate skeleton consists of movable bones separated by cartilage, but joined by a capsule of fibrous connective tissue called a synovial capsule (fig. 34.18). Freely movable joints are called **synovial joints.** Their synovial capsules are filled with fluid, which, when combined with the slipperiness of the cartilage on the bone ends, allows bones to move against each other in a nearly friction-free environment. Cartilage on cartilage in a healthy synovial joint generates only 20% of the friction of ice on ice.

Tough bands of fibrous connective tissue called **ligaments** form the joint capsule. Lining the interior of the cavity is the synovial membrane, which secretes the lubricating synovial fluid. The amount of synovial fluid in joints decreases with age, which is one reason why older people have stiffer joints. Within a synovial joint, small membrane-lined packets called **bursae** store the fluid and help to reduce friction between bones and other nearby structures such as skin and muscle. Calcium deposits in the bursae can cause an inflammation known as bursitis. Tennis elbow is a familiar example of bursitis.

Freely movable joints are a prime target for disease and injury. The most common type of joint problem is arthritis, or joint inflammation. The most serious form, rheumatoid arthritis, is an

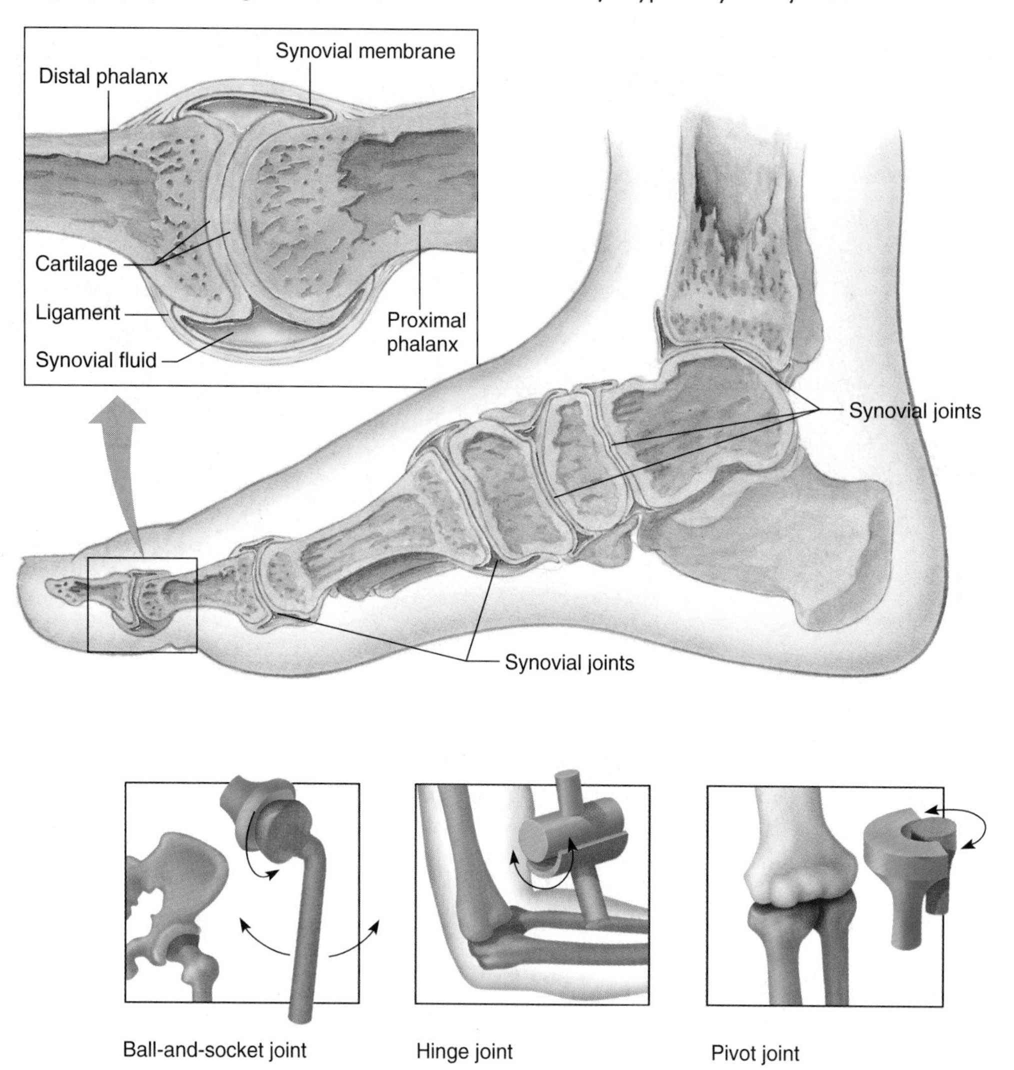

FIGURE 34.18
The Synovial Joint—Where Bone Meets Bone.
A synovial joint is surrounded by a fluid-filled capsule of fibrous connective tissue. This illustration shows the synovial joints in the foot and ankle. The enlargement shows phalanges (bones) of the big toe. The insets show some of the major types of synovial joints.

inflammation of the synovial membrane, usually of the small joints of the hands and feet (fig. 34.19). Rheumatoid arthritis may be caused by a faulty immune system that attacks the synovial membranes. In the more common osteoarthritis, joint cartilage wears away. As the bone is exposed, small bumps of new bone begin to form. Osteoarthritis usually manifests itself as stiffness and soreness in certain joints after age 40.

Muscles, joints, and bones interacting to generate movement form biological lever systems (fig. 34.20). A lever is a structure that can pivot around a point, called the fulcrum, when force is applied. In the body, bones function as levers, joints are fulcrums, and skeletal muscles supply the force.

Skeletal muscles, attached by **tendons,** allow movement between two bones. The muscle's **origin** is the end on the bone that does not move, and the muscle's **insertion** is the end on the movable bone.

Exercise Strengthens Muscles and Bones

Regular exercise strengthens the muscular system and enables it to use energy more efficiently. During the few months after a runner begins training, gradually increasing in distance, leg muscles noticeably enlarge. This exercise-induced increase in muscle mass, or **hypertrophy,** is attributable to an increase in the size of individual skeletal muscle cells rather than to an increase in their number.

Getting into good physical shape causes changes at the microscopic level, too. When an individual runs, muscles consume more than 90% of the total energy generated in the body. The enzymes within a trained runner's muscle fibers are more active and numerous, and mitochondria are more abundant than they are in the skeletal muscle cells of a sedentary person. As a result, the runner's muscles can withstand far more exertion than can the

FIGURE 34.19

Arthritis Inflames Joints.

(**A**) The characteristic gnarled hands of a person who has rheumatoid arthritis make many tasks quite painful, or impossible. (**B**) The inflamed joints are visible in this X ray.

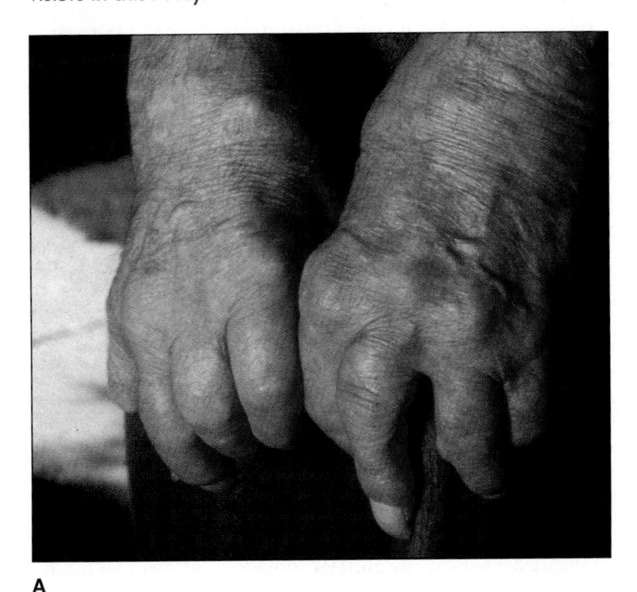

A

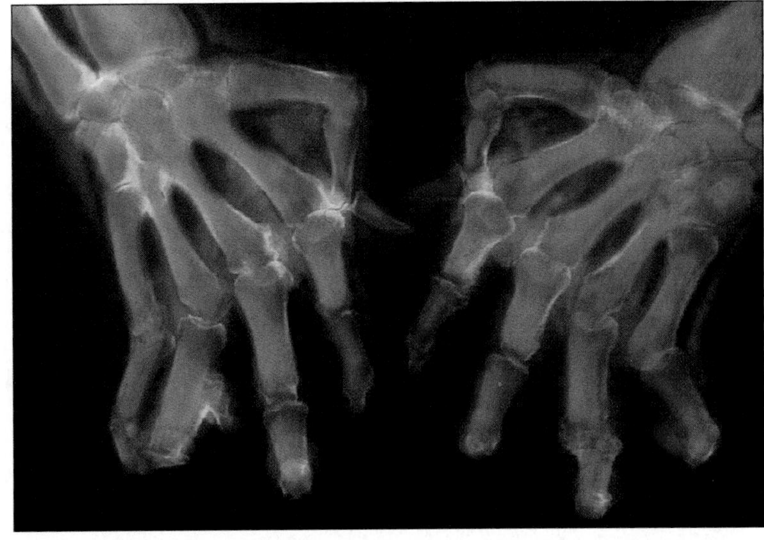

B

FIGURE 34.20

Bones, Joints, and Muscles Work Together as Lever Systems.

When stimulated, muscles contract. To lengthen, a muscle must relax (cross-bridges disengage) so that an opposing force can pull the muscle back to its full length. The arrangement of antagonistic pairs of muscles around a joint (which serves as a type of fulcrum) allows them to act coordinately to stretch each other out as well as provide the force to move a lever (the bone they insert on) in opposite directions.

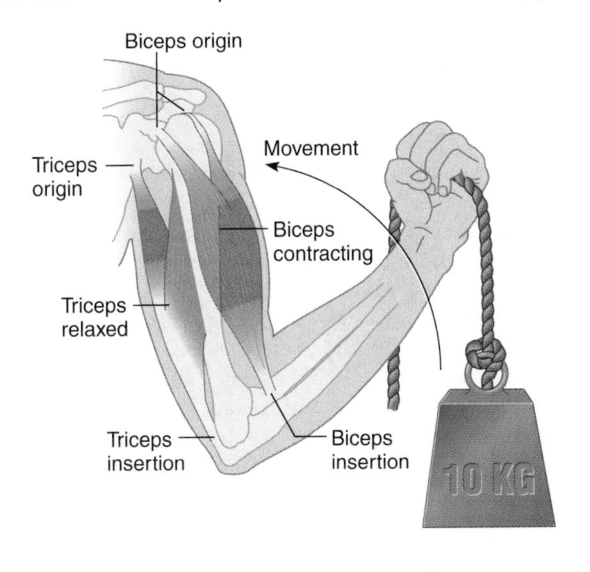

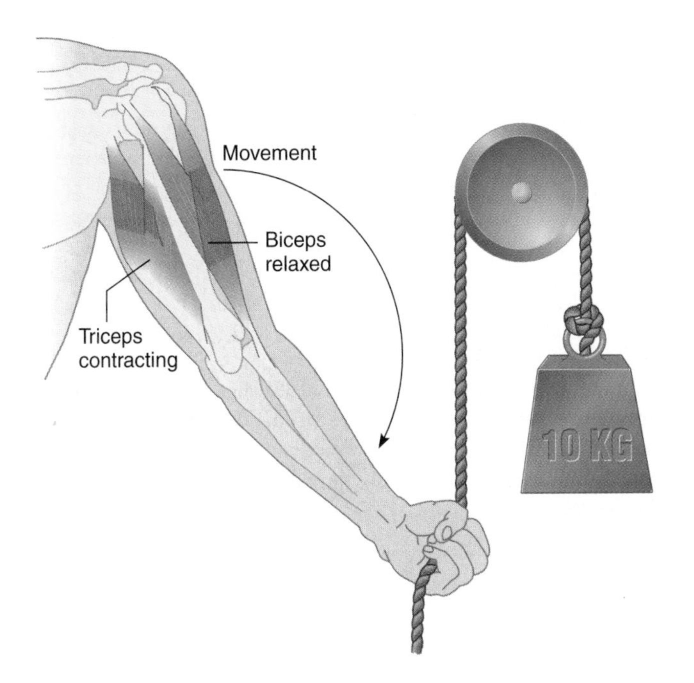

untrained person's muscles before anaerobic metabolism begins. The athlete's muscles also receive more blood and store more glycogen than those of an untrained person.

Exercise-induced hypertrophy is even more pronounced in a weight lifter, because the muscles are greatly stressed by the resistance of the weights. The opposite condition, muscle degeneration resulting from lack of use or immobilization in a cast, is called **atrophy.**

The changes in muscles due to regular exercise disappear quickly if activity stops. After just 2 days of inactivity, mitochondrial enzyme activity drops in skeletal muscle cells. After a week without exercise, aerobic respiration efficiency falls by 50%. The number of small blood vessels surrounding muscle fibers declines, lowering the body's ability to deliver oxygen to the muscle. Lactic acid metabolism becomes less efficient, and glycogen reserves fall. After 2 or 3 months of inactivity, the benefits regular exercise provides have all but disappeared.

In addition to the effects on muscles, strenuous activities such as weight lifting put stress on certain bones. Although the mechanism is not completely understood, such exercise promotes increased bone thickness and strength. Conversely, bone mass is lost with disuse, as the mineral component slowly dissolves.

Along with the muscular and nervous systems, the skeletal system endows us with the capacity to carry out an enormous variety of bodily motions. To ensure that this masterpiece of natural engineering continues to serve its functions throughout the life span, we must use it regularly and provide the nutrients necessary to sustain the continual microscopic ebb and flow of its components.

34.5 MASTERING CONCEPTS

1. What structures form a joint?
2. How do muscles, joints, and bones form levers?
3. How does exercise affect muscle?

Chapter Summary

34.1 A Musculoskeletal System Helps an Animal to Respond to Its Environment

1. The musculoskeletal system helps to maintain homeostasis by enabling an animal to move in response to environmental stimuli.

34.2 Types of Skeletons

2. An animal's skeleton supports its body, protects soft tissues, and enables the animal to move.
3. A **hydrostatic skeleton** consists of tissue containing constrained fluid.
4. A **braced framework,** which has solid components, can be on the organism's exterior as an **exoskeleton** or within the body as an **endoskeleton.**
5. The vertebrate skeleton supports, protects, attaches to muscles, and stores minerals. The **axial skeleton** consists of the skull, **vertebral column,** breastbone, and ribs. The **appendicular skeleton** includes the limbs and the limb girdles (**pectoral** and **pelvic**) that support them.
6. Cartilage and bone make up skeletons. Cartilage entraps a great deal of water, which makes it an excellent shock absorber. Bone has a rigid matrix and derives its strength from collagen and its hardness from minerals. **Spongy bone** has many spaces. In **compact bone,** osteocytes form **osteons.** The central channel for blood supply is an **osteonic canal.** Canaliculi connect lacunae that house bone cells and the osteonic canal. Bone cells extend through canaliculi and pass materials from cell to cell. Bone continually degenerates and builds up.

34.3 Muscle Diversity

7. Many cells and organisms have movement mechanisms based on actin and myosin.
8. Invertebrates have smooth and striated muscles.

34.4 Vertebrate Muscles

9. **Smooth muscle** cells are spindle-shaped and involuntary, and they line organs, push food through the digestive tract, and regulate blood flow and pressure. **Cardiac muscle** cells, in the heart, are striated and involuntary and are joined by **intercalated disks** into a branching pattern. **Skeletal muscle** cells are multinucleate, striated, voluntary, and move bones.
10. Sliding protein filaments in muscles provide movement. **Tendons** attach an intact whole skeletal muscle to bones. Each skeletal **muscle fiber** is a long cylindrical cell composed primarily of thick and thin **myofilaments.** The thick myofilaments are myosin, each with a head (cross bridge) attached to a shaft. The heads project outward from each end of a myosin bundle. The thin myofilaments are composed of actin plus **troponin** and **tropomyosin.**
11. A muscle fiber is a chain of contractile units, called **sarcomeres.** Myofilaments within a sarcomere give the tissue its striated appearance. According to the **sliding filament model,** muscle contraction occurs when the thick and thin myofilaments move past one another so that they overlap more.
12. When a motor neuron stimulates a muscle fiber, acetylcholine is released at a neuromuscular junction. Electrical waves spread along the muscle cell membrane, causing the **sarcoplasmic reticulum** to release calcium ions, which bind to the troponin molecules on myofilaments. As a result, troponin moves and no longer prevents actin from binding to myosin. Once myosin cross bridges touch actin, ATP attached to the myosin head splits and the head moves, causing the actin myofilament to slide past the myosin myofilament. ADP and inorganic phosphate are released. A new ATP binds to the myosin head and the cross bridge to actin breaks, the myosin head returns to its original position, and a new cross bridge forms further along the myofilament.
13. After the nerve impulse, calcium ions are actively pumped back into the sarcoplasmic reticulum and tropomyosin moves to prevent actin and myosin interactions. The muscle relaxes.
14. The energy for muscle contraction comes first from stored ATP, then from **creatine phosphate** stored in muscle cells, and then from aerobic respiration, and finally from anaerobic metabolism. An important source of energy for muscle contraction is glycogen.
15. A nerve cell and the muscle fibers it touches are a **motor unit.**
16. When stimulated, a muscle cell responds in an all-or-none fashion, although not all of the cells in a whole muscle contract. When a muscle cell is stimulated once, it contracts and relaxes. If the stimulation rate increases, the muscle cell does not completely relax and the response strengthens. At a high rate of stimulation, muscle cells reach a sustained state of maximal contraction, **tetanus.**

17. Muscles form **antagonistic pairs,** which enable bones to move in two directions.

34.5 Skeletal Muscle and Bone Function Together

18. Joints attach bones. Some joints, such as those holding the skull bones in place, are immovable. Freely moving **synovial joints** consist of cartilage and connective tissue **ligaments** that contain lubricating synovial fluid.
19. Most voluntary muscles attach to bones, forming lever systems. When a muscle contracts, a bone moves. The muscle end attached to the stationary bone is its **origin.** The end attached to the movable bone is called the **insertion.**
20. A muscle exercised regularly increases in size (**hypertrophies**) because each muscle cell thickens. An unused muscle shrinks (**atrophies**). Regular exercise causes microscopic changes in muscle cells that enable them to use energy more efficiently.

Testing Your Knowledge

1. Distinguish among a hydrostatic skeleton, an exoskeleton, and an endoskeleton. Give an example of an animal with each.
2. What advantages and disadvantages does a jointed exoskeleton have over a shell? What advantage does an endoskeleton have over a jointed exoskeleton?
3. What role does cartilage play in the vertebrate skeletal system?
4. What are the two major components of bone matrix? How do they work together to give bone its characteristics?
5. How is a bone cell nourished?
6. What are the structural differences between spongy bone and compact bone?
7. List four muscle proteins and their functions.
8. How does the muscular system interact with each of the following?
   a. the nervous system
   b. the skeletal system
   c. the circulatory system
9. How do antagonistic muscle pairs move bones? Give an example of such a pair.

Thinking Scientifically

1. How can the musculoskeletal systems of diverse vertebrate species consist of the same molecules, cells, and tissues, and have similar organization, yet also be adapted to a particular species' way of life?
2. What roles does fluid play in hydrostatic skeletons, cartilage, and synovial joints?
3. The aerobics instructor chants, "Just concentrate, and feel your muscles expand and contract." Is her statement an accurate description of muscle action? Why or why not?
4. Why are voluntary muscles in invertebrates termed striated, but in vertebrates called skeletal?
5. In some European nations, the ratio of fast- to slow-twitch fibers is used as a predictor of athletic success in certain events. Athletes give small samples of muscle tissue to test for twitch fiber proportions. How would the ratio probably differ between a champion sprinter and a long-distance runner?
6. A father wants to test his healthy daughter's muscles to determine the percentage of slow-twitch and fast-twitch muscle fibers to decide if she should try out for soccer or cross-country (long-distance) running. Do you think this is a valid reason to test muscle tissue? Why or why not?

References and Resources

Allen, William H. June 1995. Animals and their models do their locomotions. *BioScience,* vol. 45. Researchers study locomotion in crabs, cockroaches, and centipedes running on tiny treadmills.

Dickinson, Michael H., et al. April 7, 2000. How animals move: An integrative view. *Science,* vol. 288. Muscles function as brakes, springs, struts, and motors in animal bodies.

Kale, Sujata, et al. September 2000. Three-dimensional cellular development is essential for *ex vivo* formation of human bone. *Nature Biotechnology,* vol. 18, p. 954. Osteocytes plus transforming growth factor yields bone cell "spheroids" in culture—a model of bone formation.

Petite, Herve, et al. September 2000. Tissue-engineered bone regeneration. *Nature Biotechnology,* vol. 18, p. 959. Hip fractures in sheep are healed with a composite material made of stem cells and coral.

The September 1, 2000 issue of *Science* has several articles on bone remodeling and repair.

Westbroek, Peter, and Frederic Marin. April 30, 1998. A marriage of bone and nacre. *Nature,* vol. 392, p. 861. Why would a human bone accept a molluscan implant? The answer may lie in evolution.

The *LIFE* Online Learning Center provides additional resources and tools for studying this chapter.
www.mhhe.com/life

CHAPTER 35

The Circulatory System

Circulatory System Spare Parts

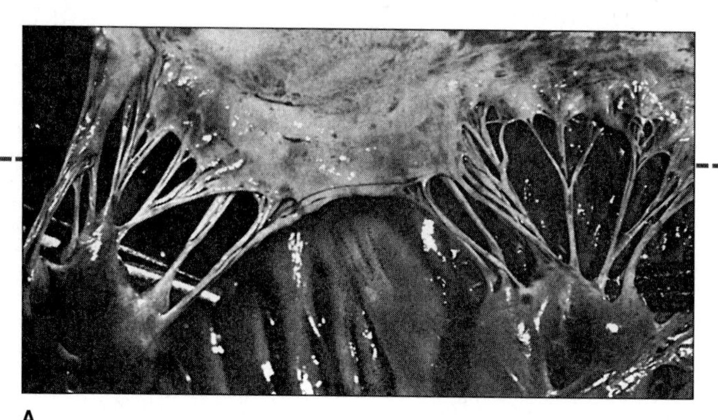

A

A complex circulatory system consists of a pump, vessels, and blood. Malfunction of any component can be deadly, and so medical technology seeks to replace various parts damaged by injury or disease. Described here are a few of them. (The chapter 30 opening essay discusses blood vessel replacements.)

Heart Transplants

When Tina Orbacz was pregnant with her second child, she attributed her fatigue to her pregnant state. A month after her son's birth, she was even more exhausted, and she had lost weight during her pregnancy. She was suffering from heart failure due to a birth defect, called an atrial septal defect, that weakened the tissue between the two upper heart chambers. Because her heart was failing, and a complication of this disorder is lung failure, Tina Orbacz, age 24, would need a heart-lung transplant.

Tina's condition stabilized, and she survived the 3-year wait until a 16-year-old who died in an accident donated a heart and lungs. Because she was young and her cell surface molecules closely matched those of the donated heart and lungs, Tina has done extraordinarily well. Today, she exercises regularly and leads a normal life.

Because of the shortage of donor hearts, and the severity of the illnesses that lead to heart failure, many potential recipients die waiting for a transplant. A mechanical half-heart, called a left ventricular assist device (LVAD), can often maintain cardiac function long enough for a donor heart to become available. In England, a few patients who are too ill to receive a transplant are surviving with permanently implanted LVADs.

Heart Valves

John T. Harrington's heart trouble began when he was 8 years old and had rheumatic fever, which developed into rheumatic heart disease, an inflammation that follows an infection by *Streptococcus* bacteria. Because the cell surfaces of these bacteria resemble those of heart valve cells, the alerted immune system attacks the valve cells, mistaking them for bacteria. (The valves route blood through the heart.) The resulting valve damage can take many years to produce symptoms.

Harrington's aortic valve—the one between the heart and the aorta, the artery that leaves the heart—was indeed damaged. It kept him out of sports in high school, then out of the Vietnam War, and became an object of much cu-

Rheumatic Fever Progresses to Rheumatic Heart Disease When Heart Valves Become Scarred.

The valve in (**A**) is healthy. The thickened and scarred valve leaflets in (**B**) are the result of rheumatic heart disease, a misplaced immune attack triggered by a bacterial infection.

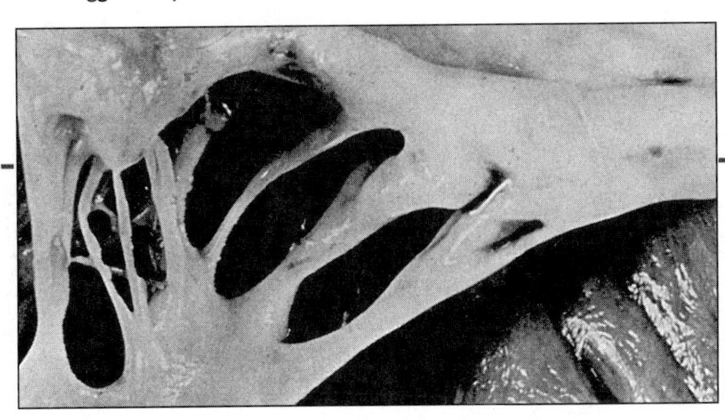

B

riosity to his medical school friends. Although the valve damage produced an interesting sound and abnormalities in various diagnostic tests, he felt no ill effects until 1983.

At that time, Harrington began jogging, just a half mile down a hill from his house, and the half mile back. But late that year, he started feeling pressure in his chest after working out, which would vanish when he rested. In early 1984, at age 47, he saw a cardiologist, who said he needed a "valve job."

Harrington's first replacement valve came from the outer heart muscle layer of a cow. At the time, valves from animal tissues were most practical, since blood clots were likely to form around valves made from synthetic materials.

The cow valve lasted 8 years. Just after Christmas in 1992, while sitting at the breakfast table with his young grandsons, Harrington experienced crushing chest pain. The valve had torn. Waking up in the hospital some time later, Harrington learned that he now had a synthetic valve. Made of new clot-resistant polymers, this one should last about 30 years.

Blood Substitutes

A substitute for human blood must mimic the major functions of this connective tissue. Red blood cells carry oxygen, white blood cells protect against infection, platelets promote clotting, and the fluid portion of blood contains many dissolved substances. Efforts to replace blood in the past have sought to fill in the missing fluid volume, or replicate the oxygen-carrying function. The search for blood substitutes intensified after the two world wars, when injured soldiers desperately needed transfusions, and again when the AIDS pandemic made transfusions dangerous unless blood is properly screened.

A substitute for human blood must mimic the major functions of this connective tissue.

To date, there is no true blood substitute, although several pharmaceutical companies and research labs are working on it. One line of research is developing an artificial red blood cell substitute. A red blood cell substitute must meet several requirements: It must carry oxygen and give it up to tissues, be nontoxic, be storable, function until the body can take over, and not elicit an immune response. Red blood cell substitutes are of two basic types. Perfluorocarbons are synthetic chemicals that carry dissolved oxygen. These were developed in the 1960s, and a famous photo shows a mouse apparently drowning in a beaker of the chemical—even though it is breathing while submerged.

The second type of red blood cell substitute dismantles red blood cells and isolates the oxygen-carrying hemoglobin molecules, which are then linked in various ways. The starting material is usually cow blood, or old stored human blood. A cow hemoglobin preparation saved the life of a young woman whose immune system was attacking her own blood, maintaining her circulation for several days until the illness subsided. In times past, healers used a variety of remedies to replace blood, including wine, ale, milk, plant resins, urine, and opium!

35.1 Circulatory Systems Deliver Nutrients and Remove Wastes

Simple animals exchange nutrients and wastes across readily accessible surfaces. Open circulatory systems have hearts and vessels that lead to spaces; closed circulatory systems contain blood in a system of vessels. Respiratory pigments efficiently deliver oxygen to tissues.

Animals are like towns and cities—they require transportation systems to bring in supplies (nutrients and oxygen for energy) and remove garbage (metabolic wastes) without disturbing their internal environments. Just as small towns can accomplish these tasks with simpler systems than can larger cities, animals with simpler bodies can sustain life with less complex systems. Diffusion across the body surface suffices in some animals; in more complex animals, circulatory systems transport nutrients.

Animals that have flattened bodies can distribute materials to and from cells without the aid of a special circulatory system. In flatworms, for example, a central cavity branches so that all cells lie close to a branch or to the body surface (see fig. 24.14). This simple system enables the animal to use diffusion to meet metabolic demands. Fluid moves through the highly branched digestive tract as muscles within the body wall contract.

A starfish has a fluid-filled body cavity, a coelom, that contacts internal structures. The coelomic fluid washes through the body as cilia lining the coelom beat. Starfishes have an additional system of small channels, called the hemal system, that parallels the water vascular system and appears to assist food distribution.

In an animal thicker than a flatworm or starfish, the increase in size decreases the animal's surface-to-volume ratio. This makes diffusion inefficient for distributing raw materials throughout the organism. Circulatory systems in large, active animals deliver vital materials to cells and remove wastes efficiently enough to permit the animals to take in sufficient energy to move, yet maintain the internal environment.

Circulatory Systems Are Open or Closed

A circulatory system transports fluid in one direction, powered by a pump that forces the fluid throughout the body. Circulatory systems evolved in conjunction with respiratory systems. That is, animals that have organs (lungs or gills) that compartmentalize gas exchange also have systems that transport oxygen throughout the body. Circulatory systems are of two basic types: open and closed.

In an **open circulatory system,** the fluid is not contained in vessels. Most mollusks and arthropods have open circulatory systems. These systems consist of a heart and short vessels that lead to spaces where the fluid, called **hemolymph,** directly bathes cells before returning to the heart (fig. 35.1). Invertebrates with open circulatory systems can be fairly active because their respiratory systems branch in a way that allows the outside environment to come close to internal tissues. Insects, for example, are among the most active animals, and their respiratory systems operate independently of their open circulatory systems.

FIGURE 35.1
Open and Closed Circulatory Systems.
(A) In an open circulatory system, hemolymph bathes cells directly.
(B) In a closed circulatory system, blood flows within vessels.

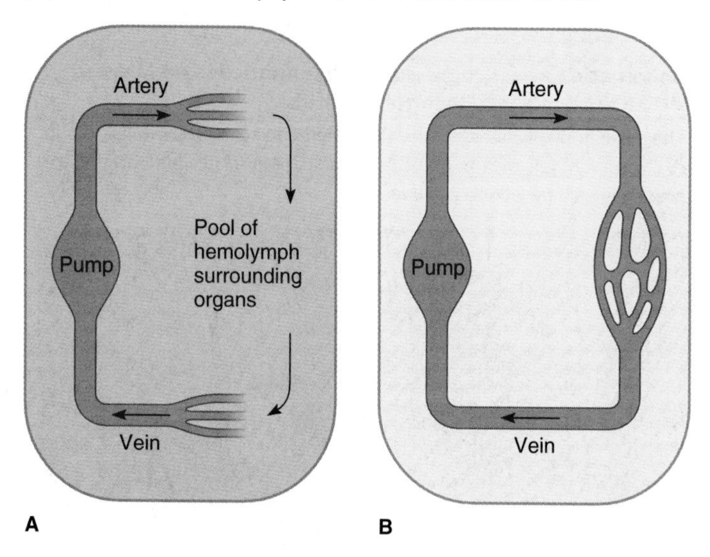

In a **closed circulatory system,** blood remains within vessels. Large vessels called **arteries** conduct blood away from the heart and branch into smaller vessels, called **arterioles,** which then diverge into a network of very tiny, thin vessels called **capillaries.** Materials diffuse between cells and the blood across the walls of capillaries. Blood then collects into slightly larger vessels, called **venules,** which unite to form still larger vessels, the **veins,** which carry blood back to the heart.

Annelids are the simplest animals with a closed circulatory system. Squids, which are mollusks, have an interesting variation on a closed circulatory system. They have two extra hearts in their extensive network of gills, which boost blood circulation through the dense vessels in that region. The extra pumping action and very large gills help the animal to move very quickly.

Respiratory Pigments Carry Additional Oxygen

In many invertebrates, the fluid that transports nutrients, dissolved gases, and wastes is clear and watery. It contains salts, proteins, and phagocytes (protective cells that engulf smaller cells and particles). In larger animals, the circulatory system fluid contains pigment molecules that carry additional oxygen. Such respiratory pigments must be able to bind oxygen where it is plentiful, yet readily release it in oxygen-depleted tissues, where it is required. **Hemoglobin** is the respiratory pigment in mammals, birds, reptiles, amphibians, fishes, annelids, and some mollusks. Hemoglobin enables a milliliter of a mammal's blood to carry 25 times as much oxygen as a milliliter of seawater at a given temperature.

Another respiratory pigment, **hemocyanin,** carries three times as much oxygen as seawater. The horseshoe crab, an arthropod, has hemocyanin. Hemoglobin contains iron, which gives blood its rich red color, and hemocyanin contains copper, which gives the horseshoe crab's hemolymph a metallic blue color. In some animals, such as the annelids, hemoglobin is free in the blood; in vertebrates, cells contain hemoglobin. A single human red blood cell holds about a quarter billion hemoglobin molecules, comprising a third of the cell's mass.

Vertebrate Circulatory Systems Are Closed, with Specialized Blood Cells

Recall from chapter 25 that the vertebrates share phylum Chordata with the more primitive tunicates and lancelets. Comparing their circulatory systems with those of the vertebrates provides a glimpse at the evolution of this organ system.

A tunicate's blood consists of phagocytes carried in fluid plasma. The heart is an expanded section of a blood vessel that is lined on the inside with contractile cells. The heart pushes blood out to the tissues, and then the blood returns over the same pathway. The circulatory system of the lancelets, although it lacks a heart, introduces the unidirectional flow of blood that persists in the vertebrates. The largest blood vessels contract with sufficient force to keep the blood flowing throughout the body.

In vertebrates, the circulatory system has a highly specialized heart, blood containing red and white cells in a closed system, and hemoglobin contained in cells. Circulatory systems become increasingly complex among the fishes, amphibians, reptiles, birds, and mammals, probably because of adaptations to the different physical demands as animals moved onto land. Again, the increasing complexity of the circulatory system parallels the evolution of respiratory structures. Aquatic species exchange gases through gills, which are outfoldings of the integument (skin). Other vertebrates exchange gases through lungs, which are highly folded internal structures.

A fish's blood flows through the heart only once in each circuit around the body (fig. 35.2). The heart pumps the blood through the gills and then to the rest of the body before it returns to the heart. Blood flow slows through the gills, where it is under very low pressure, providing time for gas exchange to occur. A fish's heart has two chambers—one **atrium** (where blood enters) and one **ventricle** (from which blood exits).

In amphibians and reptiles, blood makes two trips through the heart in each circuit around the body. The three-chambered amphibian heart has a left and a right atrium and one ventricle. A reptile's heart also has two atria and one ventricle, but the ventricle is partially divided. In the crocodiles, the division is more pronounced. The ventricular division helps to separate blood that has picked up oxygen (oxygenated blood) from blood depleted of oxygen (deoxygenated blood). An amphibian does not require a divided ventricle, because it obtains much of its oxygen through the mouth surface and skin. A reptile's circulatory system is under greater stress because the animal is fully terrestrial. Its metabolic rate, blood pressure, and oxygen requirements exceed those of amphibians.

Birds and mammals independently evolved four-chambered hearts with two atria and two ventricles. Blood circulates separately through the lungs, in the **pulmonary circulation,** and the rest of the body in the **systemic circulation.** Such double circulation allows the ventricles to divide the work, ensuring that blood circulates through the entire body, including the limbs. Also, separating

FIGURE 35.2
Vertebrate Circulatory Diversity.
(**A**) A fish's two-chambered heart allows blood to flow in a single circuit around the body. (**B**) An amphibian's heart has three chambers—two atria and a large ventricle, which serves as a powerful pump. (**C**) A bird or mammal has a four-chambered heart, supporting two trips through the heart in each circuit. This more powerful pumping system enables birds and mammals to be more active.

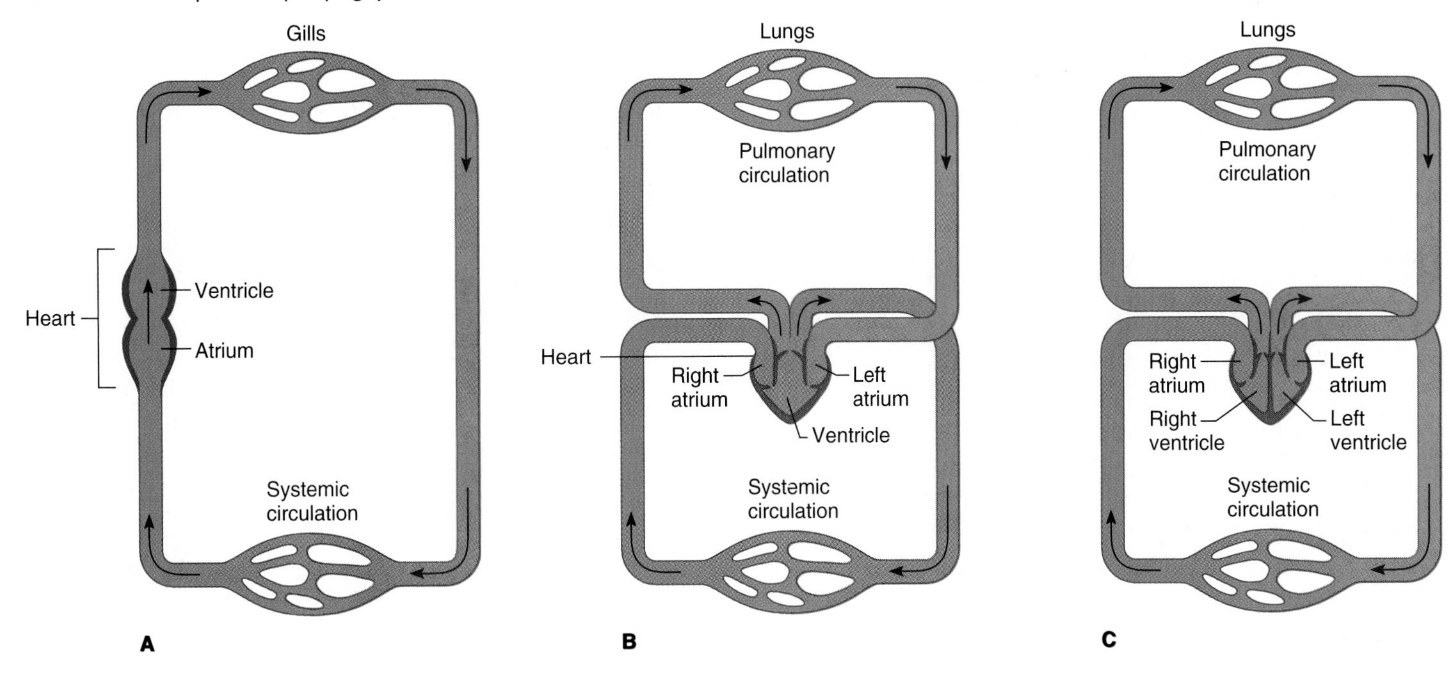

FIGURE 35.3

Phylogenetic Trends in Circulatory Systems.

Increasing complexity of circulatory systems reflects adaptations to different environments.

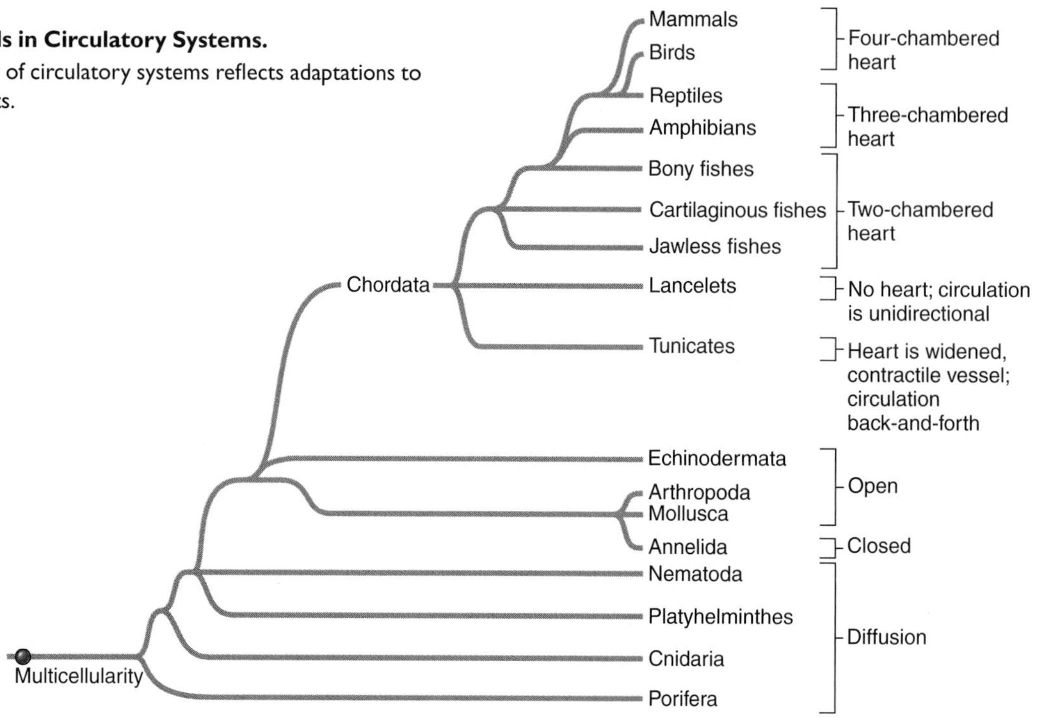

Mammals
Birds
Four-chambered heart
Reptiles
Amphibians
Three-chambered heart
Bony fishes
Cartilaginous fishes
Jawless fishes
Two-chambered heart
Chordata
Lancelets
No heart; circulation is unidirectional
Tunicates
Heart is widened, contractile vessel; circulation back-and-forth
Echinodermata
Arthropoda
Mollusca
Open
Annelida
Closed
Nematoda
Platyhelminthes
Cnidaria
Porifera
Diffusion
Multicellularity

oxygenated from deoxygenated blood maximizes the amount of oxygen reaching tissues. Figure 35.3 depicts the evolution of circulatory systems suggested by modern comparative anatomy.

35.1 MASTERING CONCEPTS

1. How does the organization of cells in a simple multicellular animal enable it to obtain nutrients and dispose of wastes by diffusion?
2. What are the general components of a circulatory system?
3. How do open and closed circulatory systems differ from each other?
4. What is the function of a respiratory pigment?
5. How are the circulatory systems of fishes, amphibians and reptiles, and birds and mammals increasingly complex?

OLC

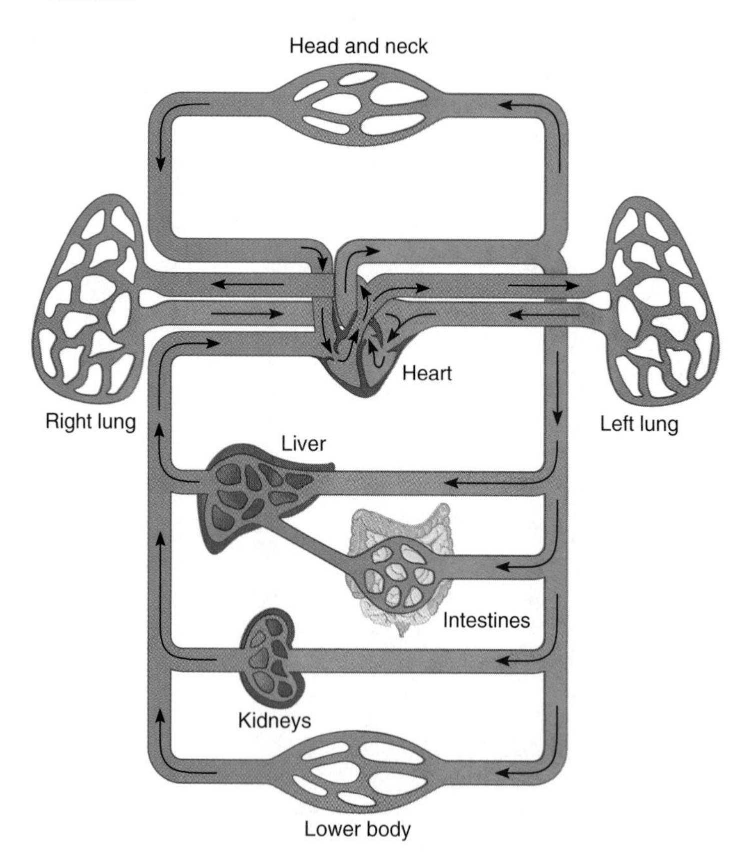

35.2 The Human Circulatory System

The heart's upper chambers receive blood and the two lower chambers propel it. Valves separate the upper and lower chambers and the lower chambers from the arteries. Arteries lead from the heart to arterioles to capillaries, where gases are exchanged, and the capillaries feed into venules and then veins, leading back to the heart. Blood consists of plasma, red blood cells, white blood cells, and platelets.

The human circulatory system consists of a central pump, the **heart,** and a continuous network of tubes, the blood vessels (fig. 35.4). This system is also called the **cardiovascular system**

FIGURE 35.4

The Journey of Blood in the Human Circulatory System.

Red depicts arteries and blue depicts veins, except in the pulmonary circuit, where arteries carry deoxygenated blood and veins carry oxygenated blood. Therefore, the colors for the pulmonary circuit are reversed.

Head and neck
Heart
Right lung
Left lung
Liver
Intestines
Kidneys
Lower body

("cardio" refers to the heart, "vascular" to the vessels). The circulatory system transports blood, a complex fluid that nourishes cells, removes cellular waste, and helps to protect against infection. So extensive is the system that no cell is more than a few cell layers away from one of the smallest vessels. Figure 35.5 shows the major parts of the human circulatory system. Refer back to it when reading this section.

The Human Heart Is a Powerful Pump

Each day, the human heart sends a volume equal to more than 7,000 liters of blood through the body, and it contracts more than 2.5 billion times in a lifetime. This fist-sized muscular pump is specialized to ensure that blood flows in one direction.

Structure of the Pump Contraction of the cardiac muscle that makes up most of the heart walls provides the force that propels blood. Cardiac muscle cells branch, forming an almost netlike pattern. When cardiac muscle contracts, it "wrings" the blood out of the heart. The heart is enclosed in a tough connective tissue sac, the **pericardium** ("around the heart"), that protects the heart but allows it to move even during vigorous beating.

The human heart's four chambers pump blood to and from the lungs to pick up oxygen and then distribute it to the rest of the body. Most of the blood entering the heart moves through the upper chambers (atria) into the relaxed lower chambers (ventricles). The atria are primer pumps that finish filling the much larger ventricles. The thick, muscular walls of the ventricles generate enough force to push blood to even the most remote body

FIGURE 35.5
Human Circulatory System.
The major arteries and veins of the human circulatory system are shown along with the areas each supplies or drains.

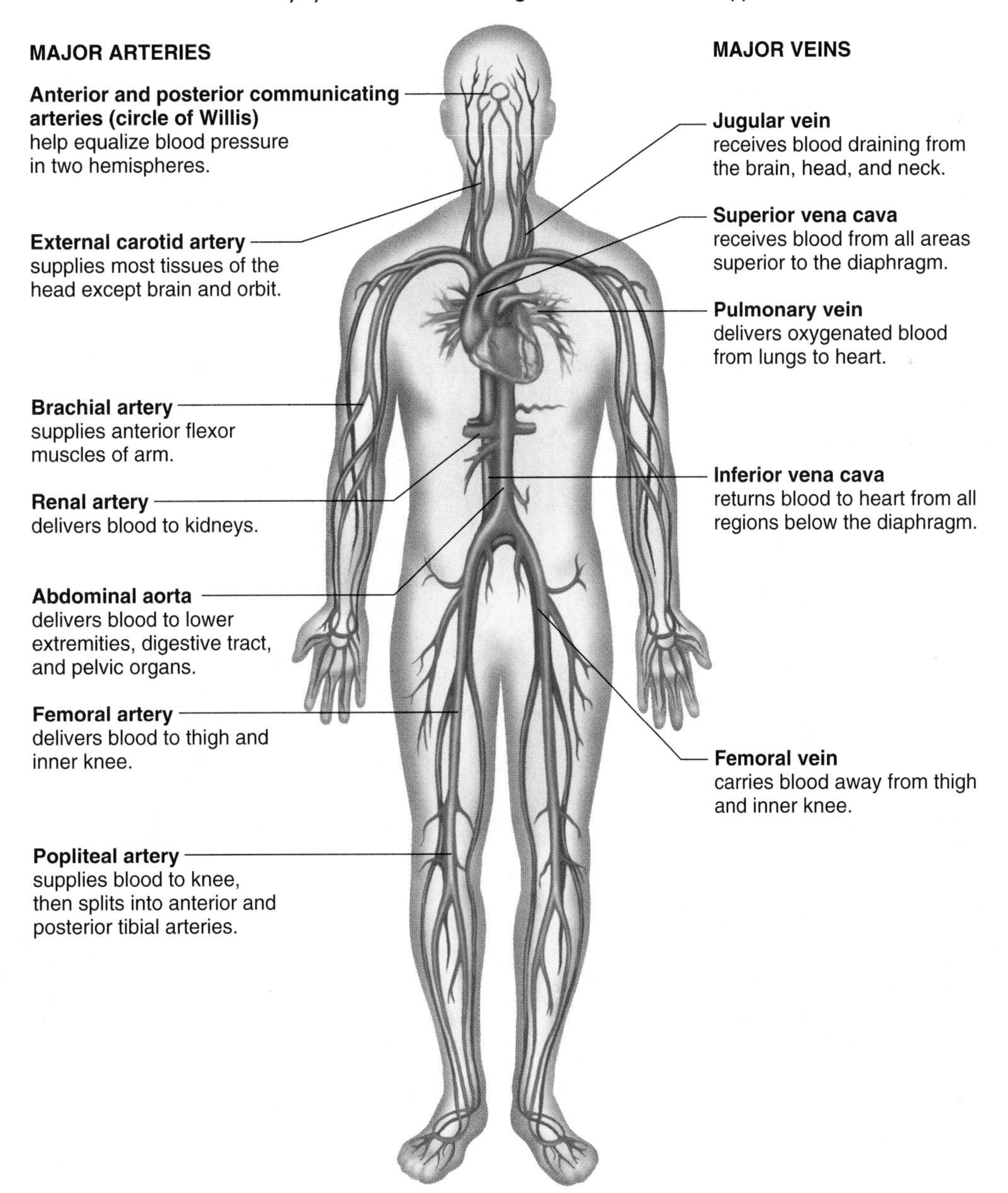

parts. A barrier of connective tissue called the interventricular septum separates the left and right sides of the heart.

Two pairs of heart valves keep blood flowing in a single direction. The **atrioventricular valves** (AV valves) are located between each atrium and ventricle (fig. 35.6). These thin flaps of tissue move in response to pressure changes that accompany the ventricle's contraction. The AV valve on the right side of the heart is the tricuspid valve; the AV valve on the left side is the bicuspid (or mitral) valve. When the ventricle contracts, pressure increases, pushing the flaps up and preventing blood from moving back into the atrium. The valves cannot be pushed into the atria because strings of connective tissue, called chordae tendinae, anchor them to the papillary muscle that extends from the wall of the ventricle.

The other pair of heart valves, the **semilunar valves,** consist of three pocketlike flaps that form a ring in the arteries just outside each ventricle. Blood leaving the ventricle passes these valves easily. Blood moving backward toward the ventricle, however, fills the pockets and causes the valves to bulge, like a parachute filling with air. This blocks the blood from flowing backward.

Two sets of heart valves closing generate the "lub-dup" sound of the heartbeat. The "lub" corresponds to the closing AV valves, and the "dup" to the semilunar valves closing. A heart murmur, which is a variation on the "lub-dup" sound, often reflects abnormally functioning valves. It could mean a dangerous backflow of blood, forcing the heart to work harder, or it could be completely harmless.

Blood Flow in the Heart The heart is sometimes considered to be two pumps that beat in unison because it pumps blood through two closed circuits of blood vessels—the right side sends blood to the lungs and the left side forces blood through the vast systemic circulation in the rest of the body. Both sides of the heart function in the same way, but the blood from each side has a different destination.

Dark red, oxygen-poor blood from the systemic circulation enters the heart at the right atrium and is delivered by the two largest veins in the body, the **superior vena cava** and the **inferior vena cava.** A smaller vein that enters the right atrium, the coronary sinus, returns blood that has been circulating within the walls of the heart.

From the right atrium blood passes through an AV valve into the right ventricle. Next, the blood flows out of the right ventricle, past the pulmonary semilunar valve, and through the **pulmonary arteries** to the lungs, where it picks up oxygen. The oxygen-rich blood, now bright red, returns through four **pulmonary veins** to the heart, where it enters the left atrium. The oxygenated blood then flows from the left atrium through the bicuspid valve and into the left ventricle. The massive force of contraction of the left ventricle, the most powerful heart chamber, sends blood through the aortic semilunar valve and into the **aorta,** the largest artery in the body. From there, the blood circulates throughout the body before returning to the heart. Two branches off of the aorta, the coronary arteries, provide blood to the heart muscle.

FIGURE 35.6
A Human Heart.
(**A**) This illustration depicts the four chambers, the valves, and the major blood vessels of the human heart. (**B**) This cross-sectional view shows the relationship of the four major heart valves, which keep blood flow unidirectional.

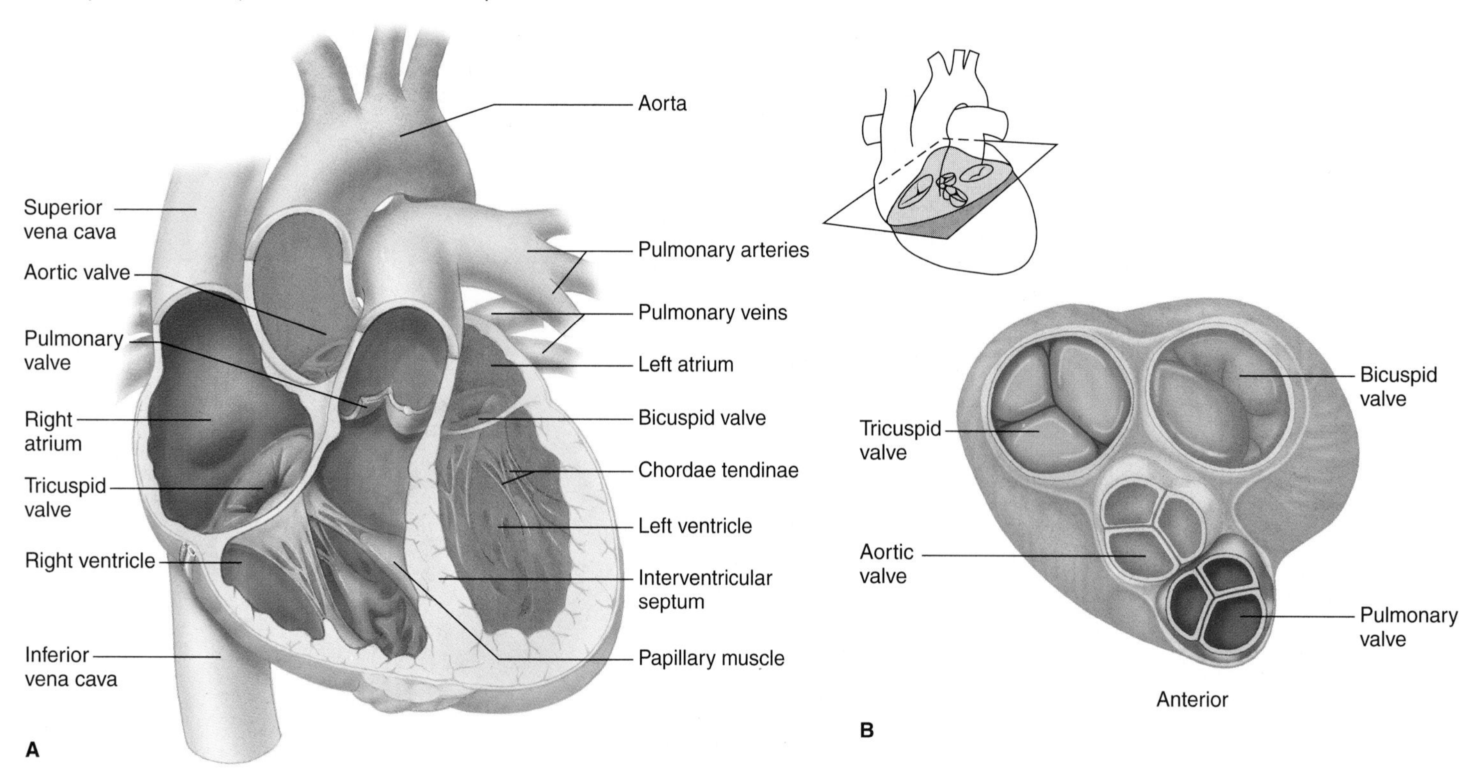

Heartbeat The actions of the two sides of the heart are highly synchronized. Each beat of the heart, called a **cardiac cycle,** consists of a sequence of contraction called **systole,** and relaxation called **diastole.**

The heartbeat originates in the cardiac muscle cells. These cells can contract on their own, without nervous stimulation. The membranes of these cells are leaky to ions that cause the cells to depolarize and contract automatically.

When grown outside the body, isolated heart cells beat at different rates. Unlike skeletal muscle cells, cardiac muscle cells are not electrically isolated. The intercalated disks between cardiac muscle cells help spread electrical impulses, allowing depolarization and repolarization to radiate throughout the heart. Many heart cells contracting in unison generate enough force to powerfully eject blood.

During each heartbeat, muscle cell contractions are coordinated so that excitation reaches each region of the heart and contraction is forceful. The beat begins in a region of specialized cardiac muscle cells called the **sinoatrial (SA) node** in the wall of the right atrium, slightly below the opening of the superior vena cava. Because the SA node sets the tempo of the beat, it is called the **pacemaker.** Once begun, the impulses triggering contraction race across the atrial wall to another region of specialized muscle cells, the **atrioventricular (AV) node.** The AV node branches first into two parts called bundle branches, and these give rise to specialized Purkinje fibers, which are cardiac muscle fibers that conduct electrical stimulation six times faster than other parts of heart muscle. The AV node delays spread of the impulse briefly, which allows the ventricles to fill before they contract. It takes only 0.8 of a second from the time the SA node produces its electrical stimulation until the ventricles contract.

We can observe the electrical changes that accompany the contraction of the heart using a device called an electrocardiograph (fig. 35.7). An electrocardiogram (ECG) can detect an irregular heartbeat, called an arrhythmia. A normal heartbeat is regular and predictable. Highly irregular patterns may result in a dangerous arrhythmic condition called ventricular fibrillation.

Blood Vessels

The structures of blood vessels are well suited to their functions.

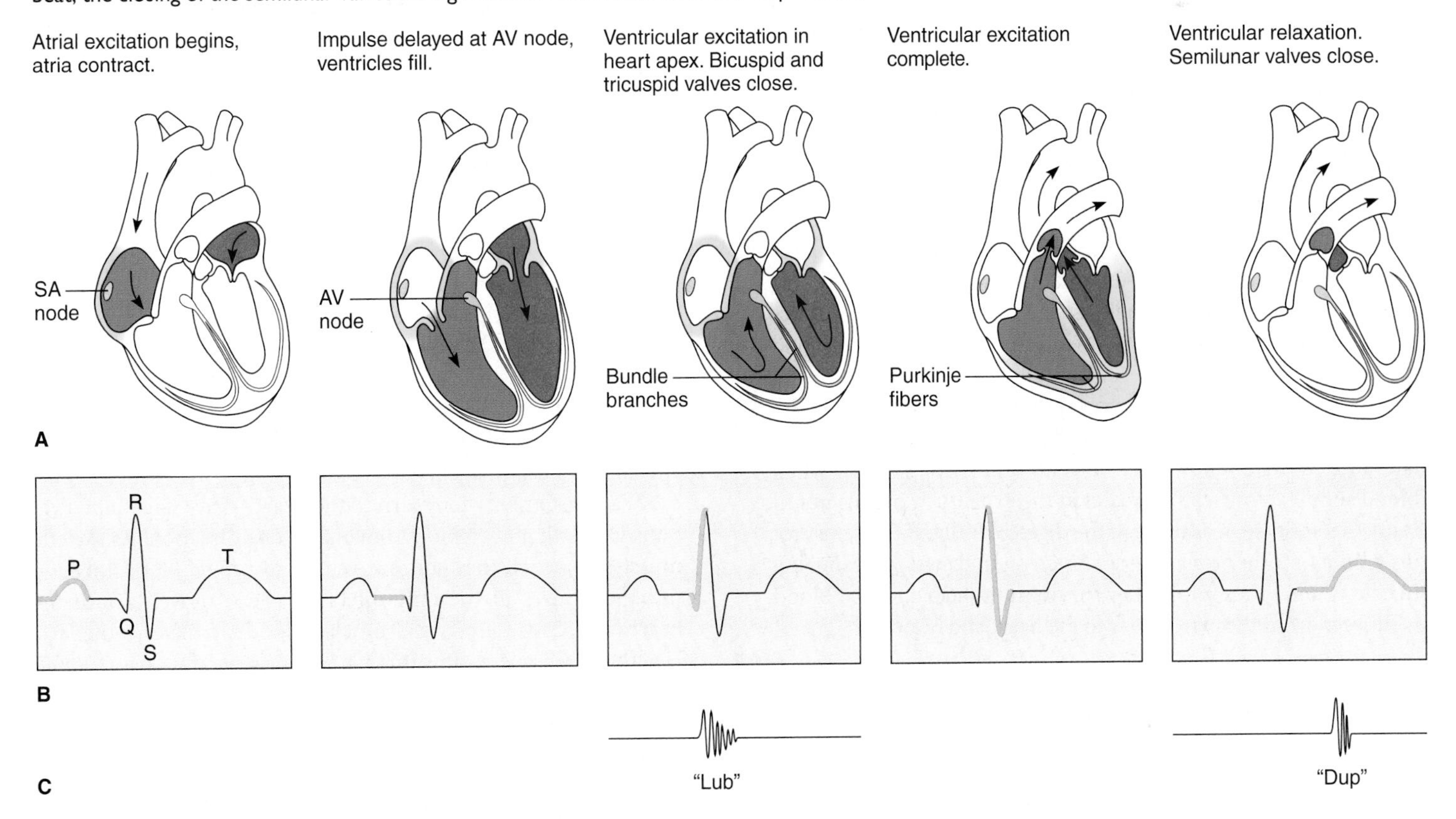

FIGURE 35.7
Electrical Activity in the Heart.
(**A**) Specialized muscle cells form a conduction network through the heart. Electrical changes start in the SA node and travel through the arterial wall to the AV node and then to the ventricle walls through the Purkinje fibers. (**B**) An electrocardiogram (ECG) is a printout of electrical changes on the body's surface in response to electrical activities of the heart. The P wave corresponds to the depolarization of the atria, the QRS complex corresponds to the depolarization of the ventricle, and the T wave corresponds to the repolarization of the ventricles. The wave from the repolarization of the atria is obscured by the massive QRS wave of the ventricles. (**C**) The closing of the AV valves during ventricle contraction produces the "lub" sound of the heartbeat; the closing of the semilunar valves during ventricle relaxation causes the "dup" sound.

FIGURE 35.8
Types of Blood Vessels.
The walls of arteries and veins have three layers: The outermost layer is connective tissue; the middle layer is elastic and muscular; the inner layer is smooth endothelium. Walls of arteries are much thicker than those of veins. The smooth muscle layer is greatly reduced in veins. The capillary wall consists only of endothelium. Many veins have valves, which keep blood flowing in one direction.

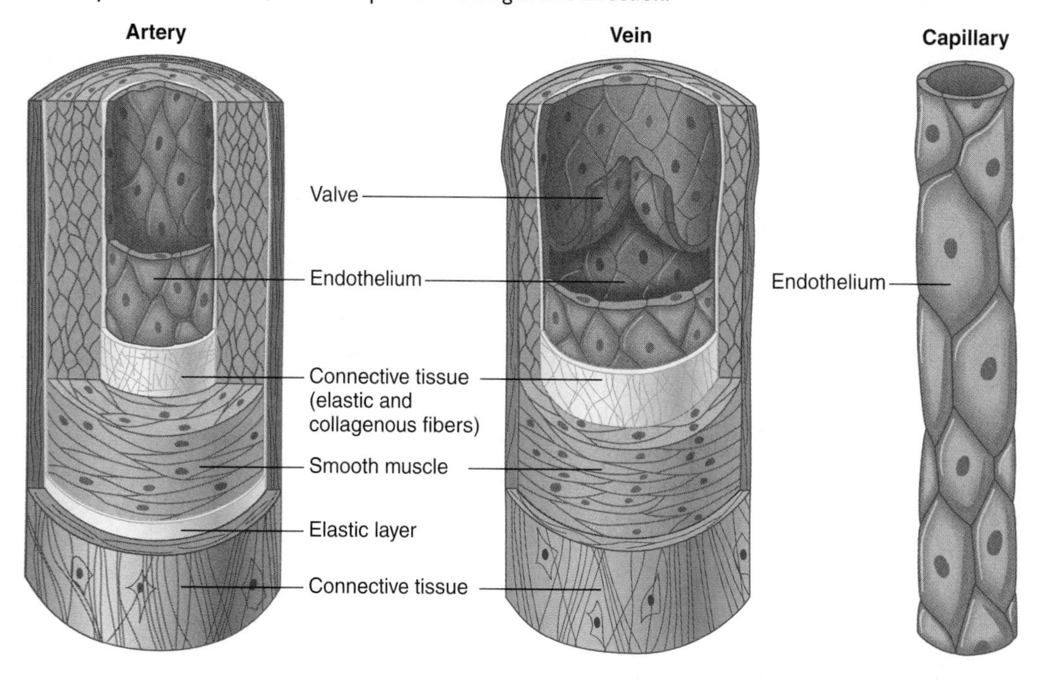

Arteries Lead Away from the Heart The outermost layer of an artery consists of connective tissue, the middle layer is elastic tissue and smooth muscle, and the inner layer is a smooth, one-cell-thick lining of simple squamous epithelium called **endothelium** (fig. 35.8). When heart muscle contracts, a wave of blood pounds against arterial walls, and the arteries dilate, accommodating this surge. Then, while the heart relaxes, the arteries elastically recoil, helping to push the blood along.

The largest arteries have such thick walls that they contain smaller blood vessels to nourish the vessel. As arteries branch farther from the heart, they become thinner and the outermost layer may taper away. In arterioles, the smooth muscle layer is predominant, which enables these small vessels to constrict (**vasoconstriction**) or dilate (**vasodilation**), which helps to regulate blood pressure.

An individual's circulatory requirements vary with activity and environmental conditions. The ability of the arterioles to dilate and constrict allows the body to adjust to exercise, when blood supply to muscles increases at the expense of organs not in immediate use, such as those in the digestive tract. Blood must redistribute for other reasons, too. A person in danger of drowning in cold water, for example, may survive until help comes if blood is shunted from the extremities to the heart and brain.

Capillaries Are the Sites of Gas Exchange Capillaries consist only of endothelium, the single-celled layer that lines the interior of arteries and veins. This simple structure enables capillaries to be the sites of gas exchange. Capillaries are very numerous, providing extensive surface area (fig. 35.9). All the capillaries in a human body laid end to end may extend up to 60,000 miles—about 2.5 times around Earth.

Capillaries form networks called **capillary beds** that connect an arteriole and a venule. Not all capillary beds in the body are open at one time—there simply is not enough blood to fill them all. The smooth muscle in an arteriole can contract to reduce blood flow to the capillary bed it feeds or relax to open the area. Structures called precapillary sphincters can contract like drawstrings, temporarily keeping blood out of capillaries by routing it through areas called thoroughfare channels.

Veins Return Blood to the Heart Veins are thinner and less elastic than arteries, because these characteristics are not critical in a vessel that conducts blood with so little pressure. The middle layer of a vein is thus much reduced or even absent. Veins not filled with blood collapse.

Because the blood pressure in veins is low, some medium and large veins—particularly in the lower legs, where blood must move against gravity to return to the heart—have flaps called venous valves that keep blood flowing in one direction. As skeletal muscles contract, they squeeze veins and propel blood through the open valves in the only direction it can move: toward the heart (fig. 35.10). Gravity also helps blood circulate in the legs. Health 35.1 discusses how this can be a problem for an astronaut.

Blood Vessels Form the Circulation Pathway Blood journeys through the body in a particular pathway through specific blood vessels. Blood leaves the heart through the aorta. Among the first branches of the aorta are vessels that nourish the heart

FIGURE 35.9
Capillaries.
(**A**) A capillary bed is a network of tiny vessels that lies between an arteriole and a venule. Contraction of precapillary sphincters diverts blood away from capillaries and through thoroughfare channels and directly back to venules. (**B**) Capillaries are so small that red blood cells move through them in single file. (**C**) In capillaries, oxygen leaves red blood cells and carbon dioxide enters the circulation. Nutrients are released to the tissues and nitrogenous wastes collected from them.

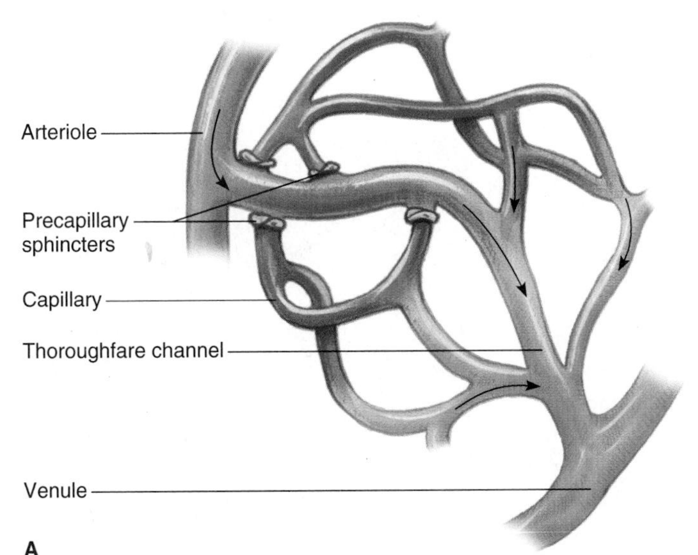

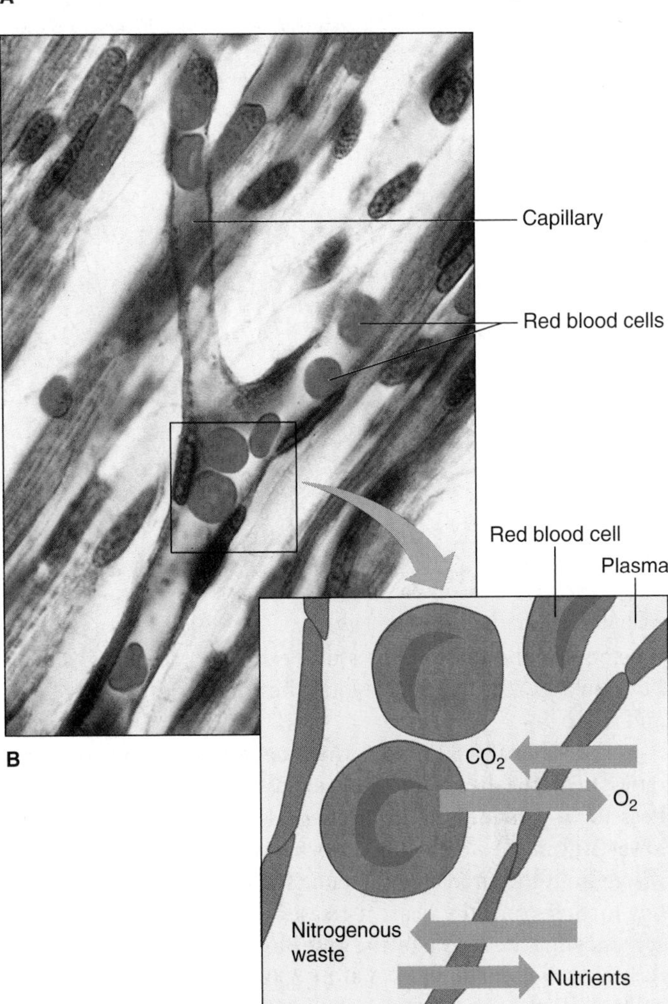

FIGURE 35.10
Valves in Veins Help Return Blood to the Heart.
When skeletal muscles relax, valves prevent blood from flowing backward. When skeletal muscles contract, they thicken and squeeze the veins, propelling blood through the open valves.

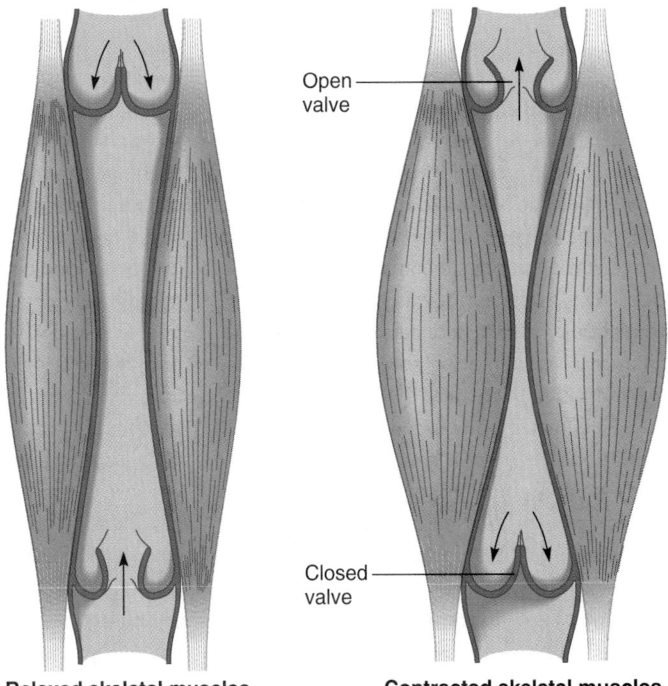

itself, which carry about 4% of the blood leaving the heart. These arteries lead to an extensive network of smaller blood vessels that nourish the active heart muscle cells.

Other branches of the aorta diverge to supply the rest of the body. At rest, about a quarter of the blood travels to the kidneys, and another quarter delivers oxygen and nutrients to muscles. About 15% of the blood goes to the abdominal organs, 10% to the liver, 8% to the brain, and the remainder to other parts of the body.

In the liver, circulating blood detours from the usual route of capillary to venule to vein. The hepatic portal system ("hepatic" means liver) is a special division of the circulatory system that helps quickly harness the chemical energy in digested food. Capillaries from the stomach, small intestine, pancreas, and spleen lead into veins that converge into the hepatic portal vein, which leads to the liver. Here, the hepatic portal vein diverges into venules and then capillaries, and these capillaries reconverge into the hepatic vein. The hepatic vein empties into the inferior vena cava and returns to the heart. The unusual position of the hepatic portal vein between two beds of capillaries allows dissolved nutrients that enter the bloodstream at the digestive organs to reach the liver rapidly, where they are metabolized to supply energy. The liver also removes toxins and microorganisms from blood arriving from the digestive system.

A second, though much smaller, portal system links the hypothalamus of the brain and the anterior pituitary gland. This system delivers releasing hormones from the hypothalamus to their sites of action in the pituitary (see chapter 33).

Space Medicine

When the rescue team approached the space shuttle *Atlantis* just after it landed on September 26, 1996, they brought a stretcher, fully expecting to carry off Mission Specialist Shannon Lucid. The 53-year-old biochemist had just spent 188 days aboard the Russian *Mir* space station, more time in space than any other U.S. astronaut. It was widely known that about 70% of astronauts cannot stand at all upon reencountering gravity—but Lucid walked, albeit unsteadily.

The human body evolved under conditions of constant gravity. So when a body is exposed to microgravity (very low gravity) or weightlessness for extended periods, some very obvious changes occur. The field of space medicine examines anatomic and physiologic responses to conditions in space. Shannon Lucid was expected to require the stretcher because of decreased muscle mass, mineral-depleted bones, and low blood volume. The 400 hours that she logged on the *Mir's* treadmill and stationary bicycle may have helped her stay in terrific physical shape (fig. 35.A).

From the moment she landed, Lucid was poked and prodded, monitored and tested, as medical researchers attempted to learn how 6 months in space affects cardiovascular functioning, respiratory capacity, mood, blood chemistry, circadian rhythms, muscular strength, body fluid composition, and many other aspects of anatomy and physiology. Just minutes after greeting her family, Lucid entered a magnetic resonance imaging tube to record the state of her various parts as they adjusted to return to gravity.

The tendency to feel wobbly and be unable to stand upon returning to Earth is one of the better-studied physiological responses to experiencing low-gravity conditions. It is called orthostatic intolerance. Normally, gravity helps blood circulate in the lower limbs. In microgravity or no gravity, blood tends to pool in blood vessels in the center of the body, registering on receptors there. The body interprets this as excess blood, and in response, signals the kidneys to excrete more fluid. But there really isn't a large blood volume. On return to Earth, the body actually has a pint to a quart less blood than it should, up to a 10 to 20% decrease in total blood volume. If blood vessels cannot constrict enough to counter the plummeting blood pressure, the weakness and wobbliness of orthostatic intolerance results. To minimize the effect, astronauts wear lower-body suction suits, which apply a vacuum force that helps draw blood into the blood vessels of the lower limbs. Maintaining fluid intake helps prevent dehydration.

Shannon Lucid had expected to be in space for 140 days, but problems with the space shuttle and a fierce hurricane extended her stay. Because she spent the longest continuous time in space, her visit is supplying vital information for planning a trip to Mars, which would require that astronauts be in space at least 2 years, round-trip.

FIGURE 35.A
Prolonged Weightlessness.
Shannon Lucid's 188-day stay in space revealed to researchers much about the body's responses to microgravity conditions. While aboard the space station *Mir,* Lucid conducted experiments on quail embryos and growth of protein crystals.

Blood Pressure

The force that drives blood through vessels can be felt as a "pulse" pressing outward on the blood vessel walls. This **blood pressure** results from the heart's pumping action and the degree of vasoconstriction in the peripheral circulation. The peak of the blood pressure pulse occurs when the ventricles contract and is called the systolic pressure. A low point called the diastolic pressure occurs when the ventricles relax. A device called a sphygmomanometer measures how far up a tube the blood pressure can push the heavy liquid mercury, sometimes using a simple pressure gauge (fig. 35.11). A normal blood pressure for a young adult is 120 millimeters of mercury for the systolic pressure and 80 millimeters for the diastolic pressure, expressed as "120 over 80" (written 120/80). "Normal" blood pressure varies with age, sex, and race.

The value 120/80 reflects the arterial blood pressure, in arteries near the heart. Pressure drops with distance from the heart, so it is highest in the arteries, lower in the capillaries, and lower still in the veins. Blood does not flow at the same speed throughout the circulatory system. The branching of blood vessels further from the heart increases their total cross-sectional area, slowing the blood. By the time blood reaches the capillaries, the cross-sectional area is 600 to 800 times that of the aorta. Thus, even though blood pressure is lowest in the veins, blood moves most slowly through the capillaries, which facilitates nu-

FIGURE 35.11

Blood Pressure.

(**A**) A sphygmomanometer, which is an inflatable cuff attached to a pressure gauge, measures blood pressure. The cuff is wrapped around the upper arm and inflated until no pulse is felt in the wrist, which signifies that circulation to the lower arm has been temporarily cut off. A stethoscope placed on the arm just below the cuff detects the sound of returning blood flow when the cuff is slowly deflated. When a thumping is heard, the listener notes the pressure on the gauge. This sound is the blood rushing through the arteries past the deflating cuff as the pressure peaks due to ventricular contraction. (**B**) The value on the gauge when the sound begins is the systolic blood pressure. The sound fades until it disappears, and the pressure reading at this point is the diastolic blood pressure.

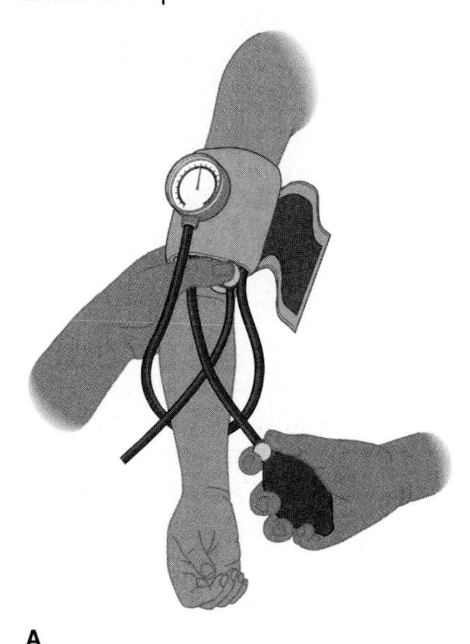

A

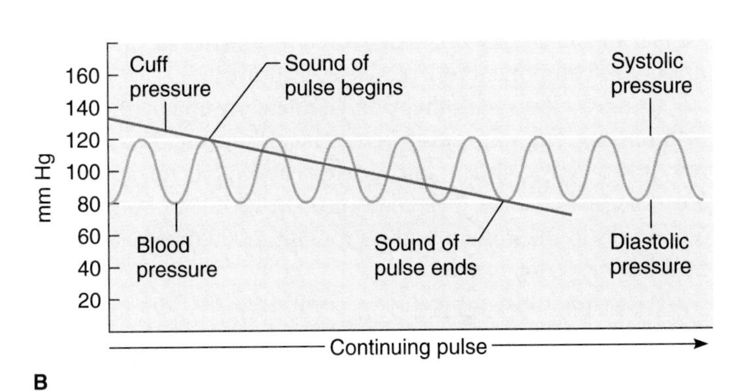

B

FIGURE 35.12

Blood Pressure in Vessels.

(**A**) Blood pressure drops with distance from the powerful left ventricle. The pressure falls so low in veins that muscle contraction, valves, and gravity must assist circulation. (**B**) While blood pressure falls with distance from the heart, blood velocity is very high at the aorta, but falls to a crawl as it moves through the capillaries and picks back up as it moves through the converging veins. (**C**) The total vessel cross-sectional area can explain this behavior at various stages along the course. Even though blood pressure in capillaries is greater than in veins, blood velocity is slower because of the expansion in total cross-sectional area across the capillary beds.

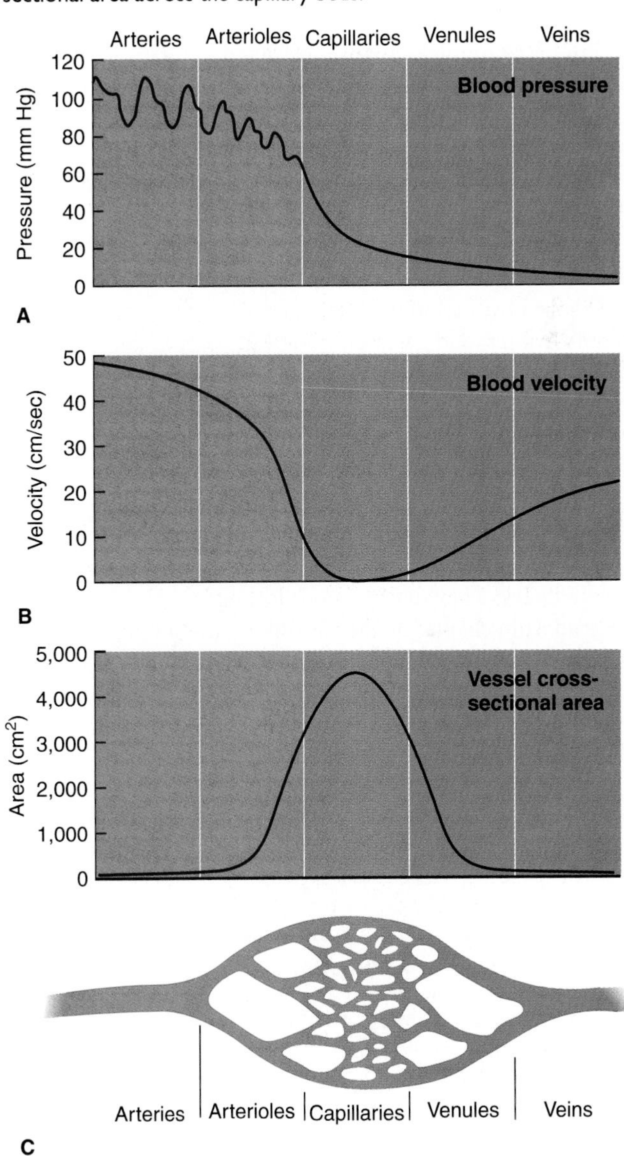

trient and waste exchange. Past the capillaries, veins merge and the total cross-sectional area decreases, helping to speed blood flow back to the heart (fig. 35.12).

Arterial blood pressure rises when blood vessels narrow and lowers when they open. The number of arterioles that are constricted or dilated depends upon several factors, including blood volume, temperature, blood chemistry, and activity level of the tissues surrounding the arterioles. Emotions such as anger and fear can also alter blood pressure. Hypotension, which is blood pressure significantly lower than normal, may cause fainting. Hypertension, which is blood pressure significantly higher than normal, does not produce symptoms directly, but it may severely damage the circulatory system and other organs.

Blood pressure is under constant regulation. The vasomotor center in the medulla of the brain regulates heart rate and the diameters of arterioles, via the autonomic nervous system (see chapter 31). An increased rate of impulses from the sympathetic nervous system constricts arterioles and increases heart rate, raising blood pressure. These actions may occur in a person with low blood pressure, blood loss, or blood that carries excess carbon dioxide. When blood pressure is elevated, brain centers decrease sympathetic nervous system input, causing vasodilation and decreased heart rate, while impulses from the parasympathetic nervous system slow the heart. Pressure receptors within the walls of certain major arteries continually monitor blood pressure, detecting changes and sending information to the medulla. This negative feedback leads to responses that maintain circulatory homeostasis.

Blood Is a Complex Mixture

Blood is a complex mixture that has a number of functions in the body (table 35.1). Blood follows the general connective tissue scheme of cells within a matrix. The matrix of blood is liquid **plasma** that contains many suspended or dissolved biochemicals. Blood cells or cell fragments are called formed elements (fig. 35.13).

Blood plasma, which comprises more than half of the blood's volume, is 90 to 92% water (fig. 35.14). It is about 1% dissolved molecules, including salts, nutrients, vitamins, hormones, metabolic wastes, and gases. Although the concentrations of the dissolved molecules are low, they are critical. For example, the blood is usually about 0.1% glucose; if glucose levels fall to 0.06%, convulsions begin. Plasma also contains about 7 to 8% dissolved proteins of more than 70 different types. Some, such as the albumins, maintain osmotic pressure. The globulins provide immune responses, and some plasma proteins transport lipids or metals, including zinc, iron, and copper. A major plasma protein, fibrinogen, plays a key role in blood clotting.

Recall from chapter 30 that the three varieties of formed elements are red blood cells (erythrocytes), white blood cells (leukocytes), and platelets. A cubic millimeter of blood normally contains about 5 million red blood cells, 7,000 white blood cells, and 250,000 platelets.

Red Blood Cells **Red blood cells** are by far the most numerous of the formed elements. Their precursors form within red bone marrow at the rate of 2 to 3 million per second. As they develop, human red blood cells lose their nuclei, ribosomes, and mitochondria. Mature red blood cells that lack nuclei cannot divide or carry out most cellular metabolism. They do, however, have many glycolytic enzymes that provide ATP for energy to perform their functions.

Mature red blood cells are biconcave discs packed with hemoglobin. During its short life span of about 120 days, a red blood cell pounds against artery walls and squeezes through tiny capillaries. Finally, the cell is destroyed in the liver or spleen, and most of its components are recycled for future use.

The red blood cell's shape and content are adapted to transport oxygen. The thin, easily bent cell can squeeze through narrow passageways, and its biconcave shape increases its surface area for gas exchange. The combined surface area of all the red blood cells in the human body is roughly 2,000 times as great as the body's exterior surface.

A hemoglobin molecule is composed of four polypeptide chains called globins. Each globin is bound to an iron-containing complex called a heme group—hence the name hemoglobin. The iron atom in the center of each heme group can combine with one oxygen molecule. Since there are four hemes per hemoglobin and about a quarter billion hemoglobin molecules per red blood cell, each red blood cell can carry about a billion oxygen molecules.

Hemoglobin exists in two functional forms. Oxyhemoglobin is bright red and compacted, with four globin chains wrapped closely around four iron-containing heme groups that bind oxygen picked up in the lungs. Then the globin chains of oxyhemoglobin unfold, releasing oxygen to the cells. The molecule becomes the deep red deoxyhemoglobin. Conversion of deoxyhemoglobin back to oxyhemoglobin takes only 30 millionths of a second. If all of the oxygen binding sites of the body's hemoglobin molecules are blocked, death follows swiftly. Carbon monox-

FIGURE 35.13
Blood.
Blood consists of formed elements (red blood cells, white blood cells, and platelets) within a watery matrix, plasma.

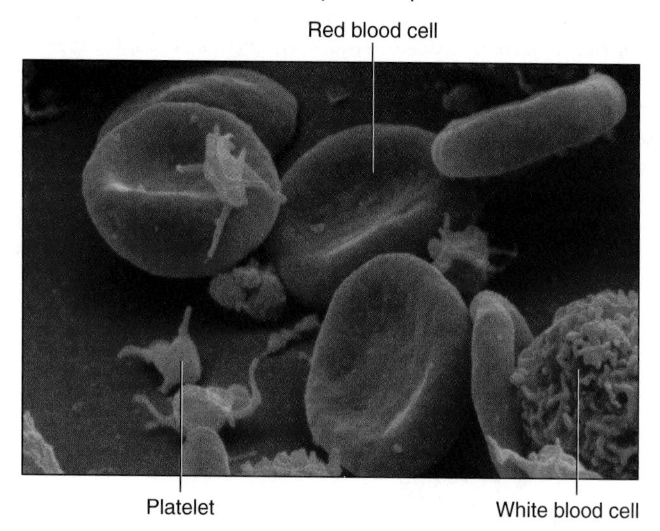

TABLE 35.1 Blood Functions
1. Carries oxygen, nutrients, and hormones to tissues.
2. Carries carbon dioxide (CO_2) as bicarbonate ion (HCO_3^-) to the lungs for exhalation.
3. Transports urea (a waste product of protein metabolism) from the liver, which produces it, to the kidneys, which send it to the bladder to be excreted in urine.
4. Influences the composition of the tissue fluid that surrounds cells. Tissue fluid forms from blood plasma.
5. Maintains homeostasis by helping to regulate pH at 7.4, regulating cells' water content, generating a pressure gradient that keeps plasma in capillaries, and absorbing heat and dissipating it at the body's surface, keeping body temperature constant.
6. Protects from injury by clotting to plug damaged vessels, transporting white blood cells to destroy foreign particles, and sending white blood cells in the inflammatory response, which dilutes toxins at an injury site.

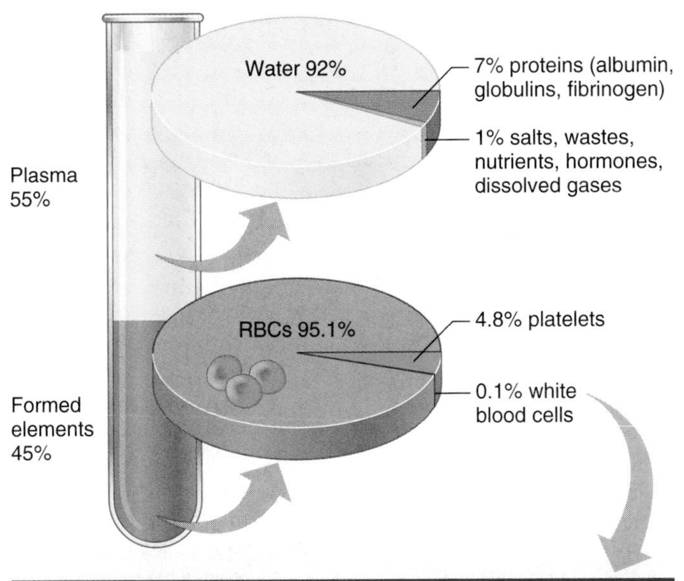

FIGURE 35.14
Blood Composition.
Human blood is a complex mixture of formed elements (platelets, red blood cells, and white blood cells) suspended in a liquid plasma of water, proteins, salts, wastes, nutrients, hormones, and dissolved gases. There are five types of white blood cells: neutrophils surround, engulf, and destroy infectious microorganisms; lymphocytes produce antibodies and cytokines; monocytes attack organisms resistant to neutrophils; eosinophils attack parasites that the neutrophils do not affect; and basophils release chemicals that take part in the inflammatory response.

ide (CO) poisoning occurs because CO binds more strongly to heme groups than oxygen does, although at a different site.

Adequate oxygenation of the body's tissues depends on having a sufficient number of red blood cells carrying enough hemoglobin. The proportion of red blood cells in the blood adjusts to the environment. At high altitudes, where the oxygen content of air is low, red blood cell production increases to maintain an adequate supply of oxygen to the tissues. A similar adjustment occurs as a result of regular aerobic exercise, when the muscles' oxygen use increases. Conversely, a sedentary individual may have fewer than the average number of red blood cells.

The proportion of red blood cells can change because red blood cell maturation rate and hemoglobin synthesis respond to oxygen availability. When certain cells in the kidney do not receive enough oxygen, they produce a substance that combines with a plasma protein to form the hormone **erythropoietin.** This hormone stimulates red blood cell production in red bone marrow.

Anemia is a decrease in the oxygen-carrying capacity of blood, which may reflect a reduction in the number of red blood cells or in the amount of hemoglobin they contain. Because oxygen is necessary to maximize the energy extracted from nutrients to make energy-laden ATP, the symptoms of anemia include fatigue and lack of tolerance to cold. Anemia can result from red blood cells that are too small or too few, that contain too little hemoglobin, or that are manufactured too slowly or die too quickly. The most common cause of anemia is iron deficiency, treated by eating more iron-rich foods or taking iron supplements. Figure 13.17 illustrates sickle cell disease, which is a type of inherited anemia.

White Blood Cells The five types of **white blood cells** in order of abundance in a normal individual are: neutrophils, lymphocytes (B cells and T cells), monocytes, eosinophils, and basophils. These cells protect against toxins, infectious organisms, and to some extent, cancer. White blood cells are larger than red blood cells and retain their nuclei. They originate in the bone marrow and typically live about 1 year, spending only 3 or 4 days in the blood. White blood cells leave the circulation and wander through tissues. The types of white blood cells are distinguished by size, life span, location in the body, nucleus shape, number of granules in the cytoplasm, and staining properties.

In general, white blood cells defend the body against infection or injury. They secrete signals that cause the warmth and swelling of inflammation, surround microbes and destroy them, and produce proteins called antibodies that help attract other white blood cells to destroy foreign microbes or particles. The specific functions and interactions of white blood cells and their products are discussed further in chapter 39.

The proportions of white blood cell types are a window on health, because elevated or diminished numbers of specific types may provide a clue to the type of infection or disease present. For example, the leukemias are cancers of the white blood cells in which abnormal white blood cells greatly predominate because they divide more often than the other blood cells. The abnormal cells form at the expense of red blood cells, so when the patient's "white cell count" rises, the "red cell count" falls. Thus, leukemia also causes anemia.

Having too few white blood cells is also devastating. Acute exposure to radiation or toxic chemicals can severely damage bone marrow, killing many white blood cells. Unless the cells are immediately replaced, death occurs in a day or two from rampant lung and digestive tract infection.

Platelets In mammals, **platelets** are small, colorless cell fragments that live about 1 week and initiate blood clotting. A platelet originates as part of a huge bone marrow cell called a megakaryocyte. This giant cell has rows of vesicles that divide the cytoplasm into distinct regions, like a sheet of stamps. The vesicles enlarge and coalesce, "shedding" fragments that become platelets. Vertebrates other than mammals have small cells with nuclei, called thrombocytes, that carry out clotting.

In a healthy circulatory system, platelets travel freely within the vessels. But when a wound nicks a blood vessel, or a blood vessel's normally smooth inner lining becomes obstructed, platelets catch on the bumps, shattering and releasing biochemicals from secretory granules (fig. 35.15A). These biochemicals combine with plasma proteins to start a complex series of reactions that results in blood clotting.

Clotting begins as soon as an injury occurs, at the same time as the injured vessel constricts, slowing the flow of blood. Within seconds, platelets collect at the site, change shape, and connect to each other. Meanwhile, cells from the blood vessel wall release proteins, called thromboplastins, which activate a series of plasma proteins that eventually convert the blood protein prothrombin into thrombin. Thrombin, in turn, initiates a series of chemical reactions that ultimately converts the plasma protein fibrinogen into **fibrin** (fig. 35.15B). The threadlike fibrin molecules form a meshwork that entraps red blood cells and more platelets. The platelets trapped in the clot release more thromboplastin, perpetuating the clotting cycle in a positive feedback pathway. The clot shrinks as new tissue forms, filling in the wound.

Clotting that occurs too slowly or too rapidly seriously affects health. Inherited bleeding disorders called hemophilias are caused by absent or abnormal clotting factors, the proteins essential for blood clotting. Deficiencies of vitamins C or K can also dangerously slow clotting, because these vitamins are required to synthesize clotting factors.

Blood that clots too readily is extremely dangerous. In atherosclerosis, clots may form as platelets snag on rough spots in blood vessel linings. A clot that blocks a blood vessel is called a thrombus. It can cut off circulation, ultimately causing death. A clot that travels in the bloodstream to another location is an embolus. An embolus that obstructs a blood vessel in a lung, called a pulmonary embolism, can kill.

Blood cells originate in bone marrow from pluripotent stem cells. Recall that "pluripotent" means that these cells have the potential to differentiate into any of a number of specialized cell types. As these stem cells divide, the daughter cells respond to colony-stimulating factors. These are secreted growth factors that turn on some genes and turn off others, sculpting the distinctive characteristics of the different blood cell types.

FIGURE 35.15
Blood Clotting.
(**A**) Blood clotting results from vessel constriction, platelet aggregation, and a biochemical cascade that produces threads of fibrin. (**B**) A scanning electron micrograph of fibrin threads (×2,800).

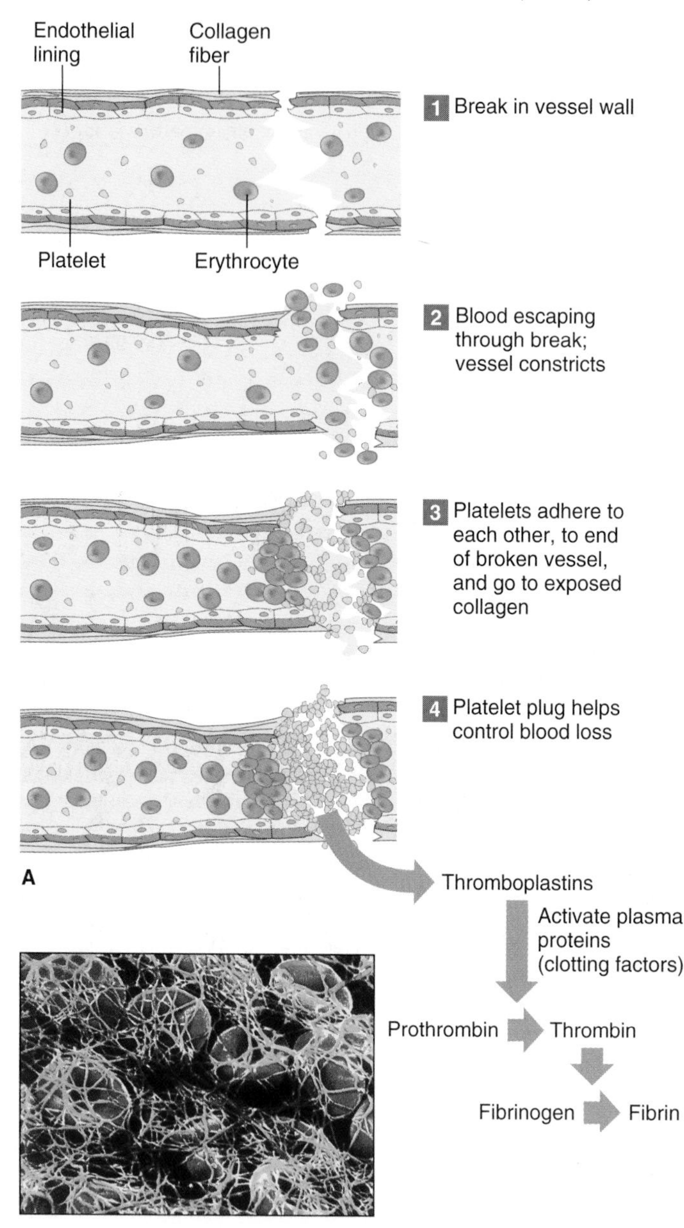

The Effects of Exercise

The normal heart pumps out as much blood as enters it from the veins. If more blood enters the heart, the strength of the next ventricular contraction increases to handle the increased load. Exercise enhances this ability of the heart, so that the amount of blood that the heart pumps each minute, called the **cardiac output,** is maintained at a lower heart rate. Thus, the heart of a person dedicated to exercise performs more efficiently than a sedentary person's heart, pumping the same amount of blood at 50 beats per minute that the sedentary person's heart pumps at 75 beats per minute. (The lowest heart rate ever recorded in an athlete was 25 beats per minute!) The number of red blood cells increases in response to regular exercise, and these cells are packed with more hemoglobin to deliver more oxygen to tissues.

Exercise provides several other cardiovascular benefits. It can lower blood pressure significantly. It elevates the level of high-density lipoproteins (HDLs), the cholesterol carriers that pick up cholesterol and transport it to the liver for elimination. Exercise also spurs the development of collateral circulation, or extra blood vessels (including those within the walls of the heart), which may help prevent a heart attack by providing alternate pathways for blood flow to the heart muscle.

To achieve the benefits of exercise, the heart rate must be elevated to 70 to 85% of its "theoretical maximum" for at least half an hour three times a week. You can calculate your theoretical maximum by subtracting your age from 220. If you are 18 years old, your theoretical maximum is 202 beats per minute. Seventy to 85% of this value is 141 to 172 beats per minute. Tennis, skating, skiing, handball, vigorous dancing, hockey, basketball, biking, and brisk walking can all elevate heart rate to this level.

Health 35.2 considers the unhealthy circulatory system.

The Unhealthy Circulatory System

Coronary Heart Disease

Coronary heart disease affects the coronary arteries that nourish the heart. It often begins with atherosclerosis, when fatty plaques accumulate inside the walls of the coronary arteries and reduce blood flow to the heart muscle. In some people, the fatty plaques build up because their cells cannot remove enough cholesterol from the bloodstream (fig. 35.B). In others, a lifetime of eating fatty food overwhelms the cell surface receptors that transport cholesterol into cells, and the excess backs up into the blood. Blood vessels also become blocked from overgrowth of smooth muscle cells, blood clot formation, and malfunction of inner lining cells.

Although high levels of cholesterol in the blood increase risk of atherosclerosis, total blood cholesterol level does not tell the whole story. Some types of cholesterol are not as harmful as others. Because cholesterol, a lipid, is insoluble in water, a carrier protein must transport it. One carrier, the low-density lipoprotein (LDL), transports cholesterol to artery walls and deposits it. High-density lipoproteins (HDLs) transport cholesterol to the liver for elimination. Exercising and avoiding saturated fats in the diet can raise HDL levels. Guarding against obesity and high blood pressure and not smoking also help maintain heart health.

Atherosclerosis can eventually cause heart attack, angina pectoris, arrhythmia, and congestive heart failure. It can also so weaken the wall of an artery that blood pressure dilates a region of the vessel, forming a pulsating, enlarging sac called an aneurysm (fig. 35.C). An aneurysm that presses on nearby organs can cause symptoms, and if it bursts, a great amount of blood may be lost. Aneurysms may also result from a congenital weakened area of an arterial wall, from trauma, infection, persistently high blood pressure, or an inherited disorder such as Marfan syndrome.

Heart Attack

Blocked blood flow in a coronary artery kills the muscle in the region. This is a heart attack (myocardial infarction), and it is often caused by a blood clot that forms in vessels narrowed by coronary heart disease. The dying portion of the heart develops altered electrical activity. The person may initially feel pressure, fullness, or a squeezing sensation in the chest that spreads to the shoulder, neck, and arms and lasts for at least 2 minutes. He or she may also feel sweaty, short of breath, dizzy, or nauseated.

A heart attack can come on suddenly. The outcome depends to some extent on the availability of help. About two-thirds of heart attack patients die within minutes or hours of the attack. Once help arrives, the person is given a drug to slow the heartbeat and oxygen to keep as much of the heart muscle functioning as possible. In the ambulance, an electrocardiogram locates and assesses the damage while monitoring the heartbeat. If the heart continues to beat erratically, an electric shock to the heart may restore a normal beat.

At the hospital in a coronary care unit, doctors monitor heart rate and blood pressure. Dye is injected into the coronary arteries to see which ones are blocked, blood chemistry is evaluated to see if clots form too

FIGURE 35.B
Atherosclerosis.
Deposits of cholesterol and other fatty materials beneath the inner lining of the coronary arteries cause atherosclerosis.

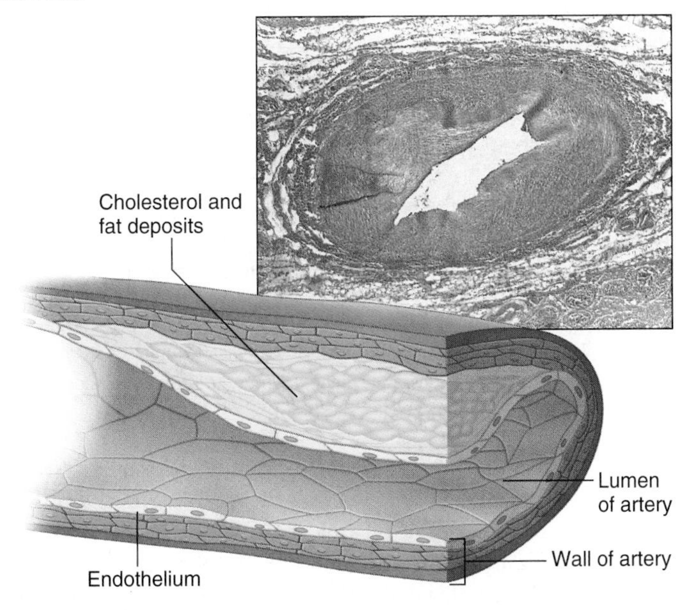

FIGURE 35.C
Aneurysm.
An aneurysm is a ballooning of a weakened part of an arterial wall.

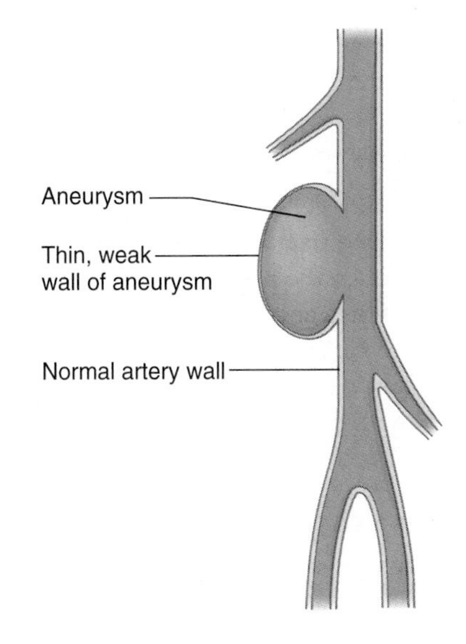

continued

(concluded)

quickly, and enzyme measurements tell the extent of heart muscle damage. Because damage continues over the next 72 hours, the patient begins drug therapy to minimize it and to ward off a second heart attack. A "clot-busting" drug, such as tissue plasminogen activator (tPA) or streptokinase, can save a life if the patient receives it within 4 hours.

If a person survives the few weeks following the initial heart attack, and perhaps takes drugs to prevent a second attack, coronary circulation can be restored as other arteries deliver blood to the damaged part of the heart. If the arteries fail to take over or if atherosclerosis is too severe, a coronary bypass operation may give blood an alternate route to the heart by stitching in a leg vein or another vessel. Otherwise, plaque buildup may continue.

Angina Pectoris

Decreased oxygen flow in the heart due to a partially blocked coronary blood vessel can cause a gripping, viselike pain called angina pectoris. It is often difficult to tell whether such pain is a warning of an impending heart attack, an isolated event, or indigestion. The circumstances of the attack provide clues to its origin. *Angina at rest* occurs during sleep and is due to a calcium imbalance or a blood clot rather than atherosclerosis. *Angina of exertion* follows sudden activity in a sedentary person and also is not usually a result of atherosclerosis. *Unstable angina pectoris* is the dangerous variety, producing chest pain either when the person is at rest but awake or is engaged in minimal exertion.

Arrhythmia

An arrhythmia is an abnormal heartbeat. Some arrhythmias may be transient flutters or racing that lasts only a few seconds. These arrhythmias usually are centered in the atria. They may reflect malfunctioning pacemaker cells, which slow ventricular pumping so that blood flow is too slow to support life. An electronic pacemaker can supplement faulty pacemaker cells. In ventricular fibrillation, pacemaker cells lose control, causing the heart muscle to suddenly twitch wildly. The heart pumps ineffectively, blood pressure plummets and death occurs. Ventricular fibrillation causes 77% of in-hospital deaths following heart attack.

Congestive Heart Failure

In congestive heart failure, the heart is too weak to maintain circulation. Fluid accumulates in tissues, swelling them. People with congestive heart failure also are fatigued, weak, and short of breath. Atherosclerosis or defective heart valves can cause congestive heart failure. Treatment includes drugs—digitalis to strengthen the heart muscle and diuretics to rid the body of excess fluid.

Hypertension

Consistently elevated blood pressure, or hypertension, affects 15 to 20% of adults in industrialized nations. Blood pressure of 140/90 is considered high; 200/100 is dangerously high. Hypertension strains the heart, raising the risk of heart attack or failure. The high pressure may cause a stroke and deposit lipids in arteries, which elevates blood pressure further. Because hypertension often has no symptoms, it is important for medical exams to include blood pressure measurement.

About 10% of the time, a malfunctioning organ or gland, such as the kidneys, pituitary, thyroid, or adrenal cortex, causes hypertension. Narrowed or hardened arteries account for many cases. Rarely, a faulty gene causes too many nerves to surround small blood vessels, constricting them too much, which elevates blood pressure. A fatty diet, obesity, and smoking increase the risk of developing hypertension.

Stroke

In a stroke, a blood clot or bleeding interrupts circulation in the brain, producing numbness, weakness, loss of balance, or paralysis. Ironically, being on a heart-lung machine during bypass surgery can cause strokes by dislodging arterial plaque, which travels in the bloodstream to the brain. Treating a stroke caused by a blood clot within 3 hours with the "clot-busting" drug tPA improves full recovery rate by 50%. Other drugs limit the damaging effects of the response of cells surrounding the site of the stroke.

35.2 MASTERING CONCEPTS

1. How are cells in the walls of the heart organized to produce an efficient blood pumping organ?
2. Describe the path of blood through the heart's chambers and valves, and through the pulmonary and systemic circulations.
3. How does heartbeat originate and spread?
4. How are blood pressure, blood velocity, and vessel diameter across the human circulatory system related?
5. What are the components of blood and their proportions and functions?
6. Where do blood cells originate?
7. How does exercise affect the circulatory system?

OLC

35.3 The Lymphatic System Maintains Circulation and Protects Against Infection

Lymph capillaries collect tissue fluid and pass it to larger lymph vessels that return the fluid to veins. Lymph nodes filter lymph. Lymph nodes, the spleen, thymus, and tonsils produce lymphocytes.

The **lymphatic system** is another transport system (fig. 35.16). **Lymph,** the fluid that fills these capillaries and vessels, is made up of the **interstitial fluid** that bathes cells and forms from blood plasma minus certain proteins too large to leave blood capillaries. Vessels of the lymphatic system begin as open lymph capillaries

FIGURE 35.16

Human Lymphatic System.

The lymphatic system is a network of vessels that collect excess fluid and proteins that leak from the blood capillaries and return them to the blood. At places along the lymph vessels, lymph nodes filter lymph. Lymph vessels have valves, and muscle contraction moves lymph. The inset shows the drainage sites of the lymph vessels into major veins above the heart.

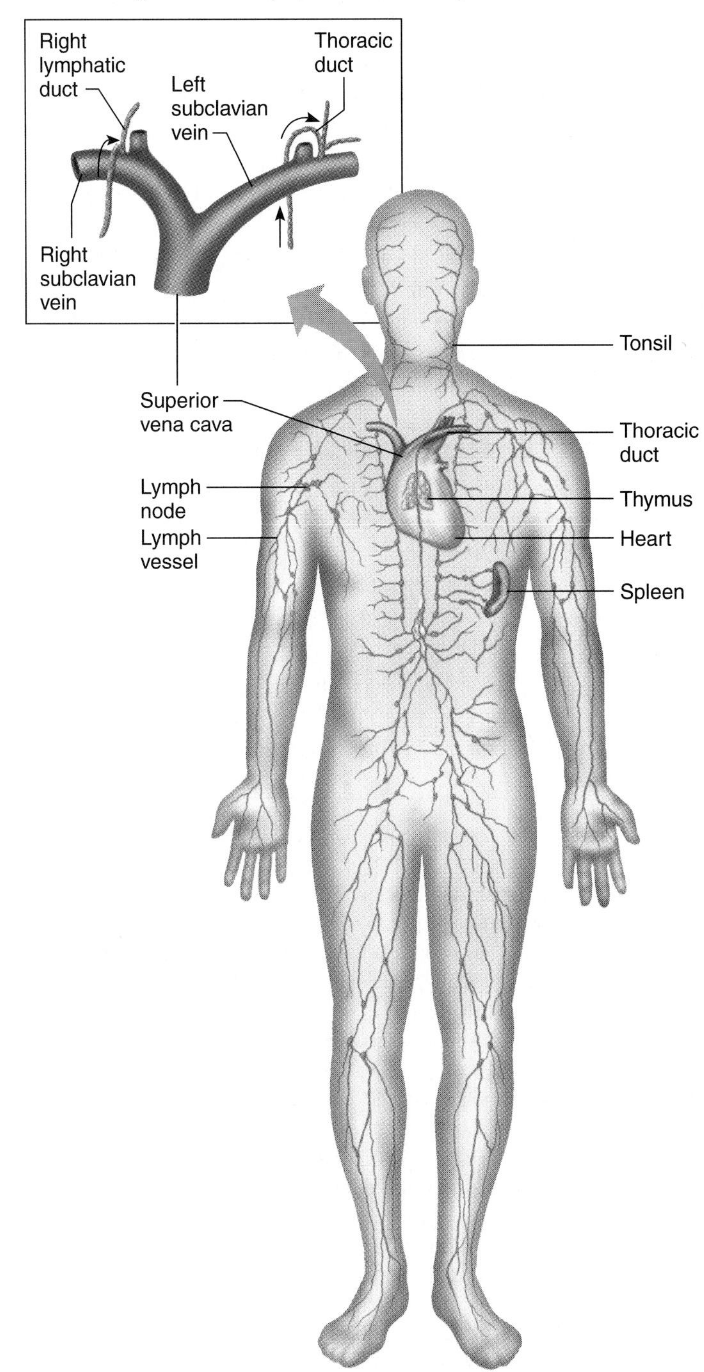

FIGURE 35.17

Lymph Vessels and Lymph Nodes.

(A) Lymph vessels collect interstitial fluid in the tissues of the body and ultimately return it to the circulatory system in large veins. **(B)** Lymph passes through lymph nodes on its way back to veins.

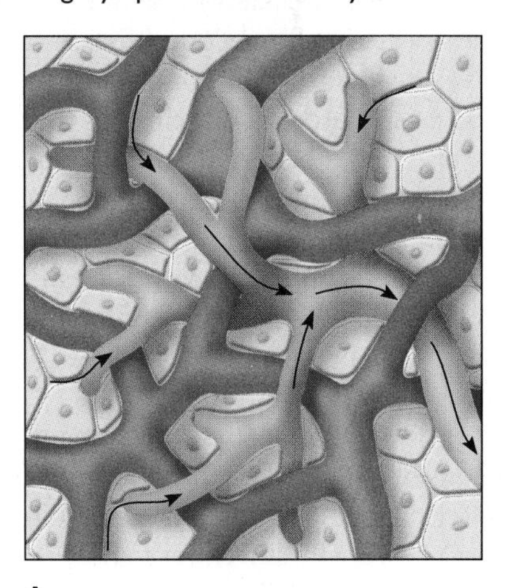

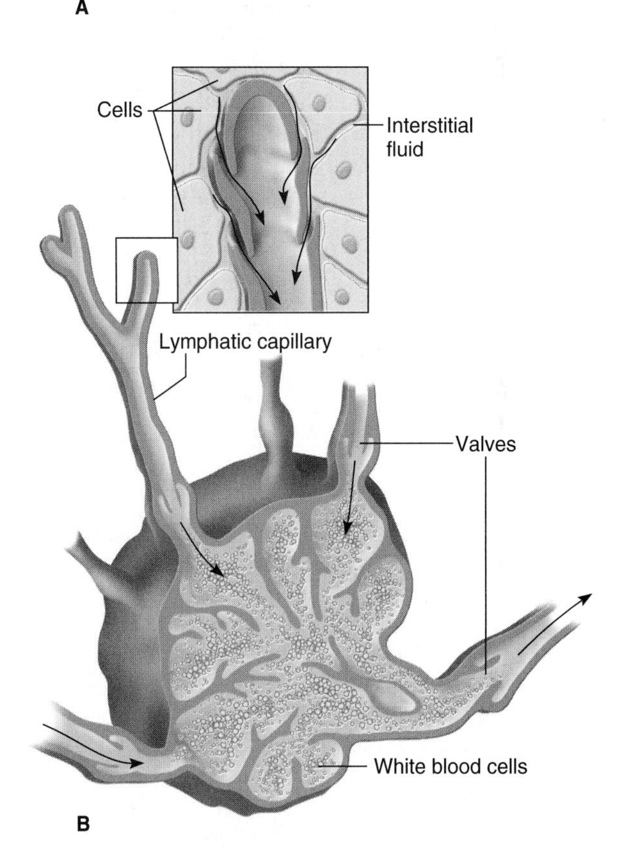

that are highly permeable to interstitial fluid that seeps out of blood capillaries. The lymph capillaries gradually converge to form larger lymph vessels that eventually empty into veins. Therefore, the lymphatic system collects fluid lost from the circulation, and recycles it to the bloodstream (fig. 35.17).

Contractions of surrounding skeletal muscles and valves in lymph vessels help move the sluggish lymph fluid. The lymphatic system normally returns less than 1 ounce (30 milliliters) of fluid per minute to the veins, in contrast to the 4 to 5 quarts (5 or 6 liters) pumped through the circulatory system in the same amount of time. Tissue fluid enters lymphatic vessels and flows toward the chest region, where it returns to the blood. Special lymph capillaries in the small intestine absorb dietary fats.

If the lymphatic system fails, the body swells with the excess tissue fluid, a condition called edema. As the slow leakage of proteins and fluid from the blood circulatory system continues, the swelling may become severe. In an infectious disease called elephantiasis, parasitic worms reside in the lymph vessels, blocking the flow of lymph and causing grotesque swelling of tissues.

The lymphatic system has other important functions. Structures called **lymph nodes** contain many white blood cells that protect against infection. The kidney-shaped lymph nodes are located primarily where the limbs meet the trunk and in the neck. Lymph nodes range in size from microscopic to much larger masses, such as those that can be felt as "swollen glands" along the sides of the neck during respiratory infections.

A lymph node consists of fibrous tissue with many pockets, each filled with millions of lymphocytes that efficiently filter cellular debris and bacteria from the lymph flow. The fluid that enters the nodes is often packed with infectious particles, 99.9% of which are destroyed in the node. Although lymphocytes in the lymph nodes can detect and destroy cancer cells, if a cancer originates in or spreads to a lymph node and is not destroyed, it may spread to the rest of the body in the lymph fluid.

The lymph nodes continually add lymphocytes to the lymph, which transports them to the blood. Lymphocytes are found in other parts of the lymphatic system too—the **spleen,** the **thymus,** and possibly the **tonsils.** B lymphocytes (B cells) secrete antibody proteins. T lymphocytes (T cells) secrete biochemicals called cytokines, discussed further in chapter 39.

The largest component of the lymphatic system is the spleen, which is to the left of and behind the stomach. The thymus is a lymphatic organ in the upper chest near the neck that is prominent in children, but begins to degenerate in early adulthood. The thymus "educates" certain lymphocytes in the fetus to distinguish body cells from foreign cells. Tonsils are collections of lymphatic tissue high in the throat. Tonsils do not filter lymph, but they may help protect against infection in other ways.

The blood circulatory system and the lymphatic system work together. The blood carries a continuous supply of nutrients and oxygen to cells and removes wastes from them, and the lymphatic system cleanses fluid that has leaked from blood capillaries, removing bacteria, debris, and cancer cells and returning the filtered fluid to the blood. The blood circulatory system and the lymphatic system nurture even cells far removed from the body's surface, ensuring that the many trillions of cells of the human body can carry out the functions of life.

35.3 MASTERING CONCEPTS

1. Where does lymph come from?
2. What are the components of the lymphatic system?
3. What do lymph nodes do?
4. How do the blood and lymph circulatory systems work together?

OLC

Chapter Summary

35.1 Circulatory Systems Deliver Nutrients and Remove Wastes

1. A circulatory system consists of fluid, a network of vessels, and a pump. The fluid delivers nutrients and oxygen to tissues and removes metabolic wastes.
2. In simple animals, open body cavities give interior structures direct contact with fluids from the environment so that nutrients and oxygen can diffuse into cells and wastes can diffuse out. In more complex animals, a heart pumps the fluid to the body cells.
3. In an **open circulatory system,** the fluid bathes tissues directly in open spaces before returning to the heart.
4. In a **closed circulatory system,** the heart pumps the fluid through a system of vessels to cells and back.
5. Respiratory pigments increase the oxygen-carrying capacity of blood or **hemolymph.**
6. Tunicates have blood and a rudimentary heart, and a blood flow that goes back and forth. Lancelets lack a heart, but introduce unidirectional circulation.
7. Vertebrates have closed circulatory systems that increase in complexity in fishes, amphibians, reptiles, birds, and mammals, reflecting the challenges of life on land. A double heart and double circulation that separates oxygenated from deoxygenated blood enables land vertebrates to be active.

35.2 The Human Circulatory System

8. The heart is the muscular pump that drives the human circulatory system. The heart has two **atria** that receive blood and two **ventricles** that propel blood in the body. Heart **valves** separate the right atrium and right ventricle, the left atrium and left ventricle, and the site where the **pulmonary circulation** and **systemic circulation** pathways leave the heart. Heart valves ensure one-way blood flow.
9. The **pacemaker** or **sinoatrial (SA) node,** a collection of specialized cardiac muscle cells in the right atrium, sets the heart rate. From there, heartbeat spreads to the **atrioventricular (AV) node** and then along Purkinje fibers through the ventricles.
10. The circulatory system leads to and from the lungs and to and from the rest of the body. Blood leaves the heart through the **aorta** and travels in increasingly narrower **arteries** and **arterioles** to the **capillaries,** where nutrient and waste exchange occur. Blood flows from the capillaries to **venules** and then to **veins,** and it reenters the heart through the **venae cavae.** Arteries have thicker, more elastic walls than veins.
11. The pumping of the heart and constriction of blood vessels produces **blood pressure. Systole** is the pressure exerted on blood vessel walls when the ventricles contract. The low point, **diastole,** occurs when the ventricles relax.
12. Human blood is a mixture of cells and cell fragments (collectively called formed elements), proteins, and molecules that are dissolved

or suspended in **plasma.** The formed elements include **red blood cells, white blood cells,** and **platelets.** Red blood cells carry oxygen. White blood cells protect against infection. Platelets break and collect near a wound, releasing chemicals that trigger blood clotting. Blood cells originate in bone marrow.

13. The circulatory system is controlled to help maintain homeostasis. When the volume of blood entering the heart changes, ventricular contraction changes in response to adjust **cardiac output.** The autonomic nervous system, under the influence of the brain's vasomotor center, speeds or slows heart rate and dilates or constricts blood vessels in a way that adjusts blood pressure.

35.3 The Lymphatic System Maintains Circulation and Protects Against Infection

14. The **lymphatic system** is a network of vessels that collect fluid from the body's tissues, purify it, and return it to the blood. **Lymph nodes** filter **lymph,** and the **spleen, thymus,** and possibly the **tonsils** manufacture lymphocytes. The blood and lymphatic systems continually supply tissues with nutrients and oxygen and remove or destroy wastes.

Testing Your Knowledge

1. Why do large, active animals require circulatory systems? Give three examples of circulatory systems that are adapted to the activity level of the organism.
2. How do open and closed circulatory systems differ?
3. What role do pigments play in the respiratory and circulatory systems of vertebrates?
4. Maintaining the proper proportions of formed elements in the blood is essential for health. What can happen when the blood contains too few or too many red blood cells, white blood cells, or platelets?
5. What process discussed in the chapter illustrates positive feedback? Negative feedback?
6. Name three ways that the circulatory system helps to maintain homeostasis.
7. What causes the heartbeat? Why is the heart sometimes referred to as two hearts that beat in unison?
8. What types of changes in blood vessels would raise blood pressure?
9. Where does lymph originate? What propels it through the lymphatic vessels?

Thinking Scientifically

1. HIV colonizes lymph nodes before it appears in the bloodstream. Why is the presence of HIV in lymph nodes dangerous?
2. Chemotherapy used to treat cancer often severely damages bone marrow. Which biochemical mentioned in the chapter might help patients receiving chemotherapy to recover their bone marrow function?
3. Athletes tend to be slim and strong, have low blood pressure, do not smoke, and alleviate stress through exercise. How might these characteristics complicate a study to assess the effects of exercise on the circulatory system?
4. Explain why women who have had cancerous breasts and the associated lymph nodes removed can develop swollen arms.
5. It usually takes 13 seconds for prothrombin to be converted to thrombin, and thrombin to trigger the formation of fibrin. When a person's diet lacks vitamin K, however, these chemical reactions take 30 seconds or longer. Which circulatory system function does this impair? Which symptoms might a dietary deficiency of vitamin K cause?

References and Resources

Jauhar, Sandeep. September 19, 2000. Saving the heart can sometimes mean losing the memory. *The New York Times,* p. F1. Going on a heart-lung machine during bypass surgery can send dislodged arterial plaque to the brain.

Lewis, Ricki. July 15, 1997. Cloaked RBCs and transgenic tobacco—blood substitutes? *Genetic Engineering News.* Experimental blood substitutes have unusual sources.

Lewis, Ricki. January 24, 2000. Homing in on homocysteine. *The Scientist,* vol. 14, p. 1. A little-known amino acid plays a prominent role in heart health.

Westaby, Stephen, et al. September 9, 2000. First permanent implant of the Jarvik 2000 heart. *The Lancet,* vol. 356, p. 900. Once used only as a "bridge to transplant," the Jarvik Heart LVAD can also stand in for a transplant permanently, according to this preliminary report.

Zimmer, Carl. April 2000. The hidden unity of hearts. *Natural History,* vol. 109, p. 56. The earliest hint of a heart may have been a swelling in a worm's anterior region.

The *LIFE* Online Learning Center provides additional resources and tools for studying this chapter.
www.mhhe.com/life

CHAPTER 36

The Respiratory System

Do Caterpillars Have Lungs?

A New View of Insect Tracheal Systems

Sometimes in science, a closer look can alter a long-standing view. So it is with the way that insects take in oxygen and get rid of carbon dioxide.

Insects exchange gases through a system of narrowing tubes that is continuous with the cuticle. The tubules that lead into the system, called tracheae, open to the outside through holes, called spiracles, that dot the integument, from the thorax (the middle section of the animal) to the end of the abdomen. As they wind through the body, the tracheae split and narrow into smaller vessels called tracheoles, which in turn become yet narrower and wind their way around most of the cells. The smallest tubules supply oxygen to the cells, presumably by diffusion, and return CO_2. The insect's pulsating abdominal muscles are enough to power the tracheal system. Lungs and respiratory pigments aren't necessary—or so it was thought.

Entomologist Michael Locke looked to the tracheae leading from the hindmost spiracles in the caterpillar of the Brazilian skipper butterfly, *Calpodes ethlius.* These tubules are wider than the others, and for a century have been called "aerating tracheae," compared with the other "conducting tracheae." Instead of branching and infiltrating many cells, the aerating tracheae form tufts of very narrow and thin-walled tracheoles. Their role, until Locke's work, wasn't understood. He discovered, in a roundabout way, that circulating hemolymph (the insect's version of blood) picks up oxygen from the waving tuft, which then moves the hemolymph toward the heart.

To show that hemolymph—in particular, the cells it carries, called hemocytes—

Lungs in Caterpillars?
Alice encounters an insect that apparently uses lungs to breathe.

pick up oxygen, Locke decided to see what happened in an oxygen-poor situation. With oxygen plentiful, hemolymph typically contains about 7 million hemocytes per milliliter. But with little oxygen present, the number zooms to 12 million. If hemocytes did not carry oxygen, then why would their number increase when oxygen availability plummeted?

If hemocytes did not carry oxygen, then why would their number increase when oxygen availability plummeted?

Taking an even closer look, Locke found that the hemocytes look different, depending upon the state of oxygen availability. With enough oxygen, the cells are irregularly shaped with extensions and abundant Golgi apparatuses and secretory vesicles. But when oxygen becomes scarce, not only does the number of cells increase, but they become rounder, and lose their extensions and secretory specializations. Some hemocytes are normally in the tracheal system, and Locke further discovered that when oxygen is scarce, their number increases and more of them are in the round form.

In a second set of experiments, Locke traced the aerating tracheae back in another direction from the tufts, to a compartment at the rear of the animal called the tokus where tracheoles project inward. Hemolymph enters this chamber, and the hemocytes dock at the tracheoles, flattening against them, which maximizes the surface area for picking up oxygen. After a while, the cells leave, taking their precious cargo to the heart. The tokus, Locke proposes, functions as a lung—something an insect was not thought to possess, *Alice in Wonderland's* smoking caterpillar to the contrary.

Locke found this close association between the tracheal system and hemolymph in 13 other families of lepidopterans (moths and butterflies), and his work suggests that the insect mode of respiration is not as simple as entomologists have thought. Sums up Locke, "Although as a rule insect tracheae go to tissues . . . hemocytes go to tracheae."

The larger lesson—scientific theories change as we learn more.

36.1 Respiration Exchanges Gases

Respiratory systems enable animals to deliver oxygen to cells and to get rid of carbon dioxide. Oxygen is the final electron acceptor in the pathways that extract energy from nutrients. This gas exchange occurs across respiratory surfaces, which include the body surface, tracheal systems, gills, and lungs.

The first breath of life is the toughest. A baby often emerges from the birth canal with a bluish color that may be frightening to new parents. Soon, as the baby musters up 15 to 20 times the strength needed for subsequent breaths, the millions of tiny sacs in the lungs, each only partially inflated, fill with air for the very first time. As oxygen rapidly diffuses into the bloodstream and reaches the tissues, the infant turns a robust pink and lets out a yowl (fig. 36.1).

The lungs of an adult who has spent years breathing polluted air and smoking cigarettes are quite different from the healthy lungs of a newborn. The passageways to these damaged lungs are dotted with bare patches where dense cilia, which move particles up and out of the respiratory tract, once waved. Deep within the lungs, the pattern of bare patches continues, and sections of air sacs are deflated or altogether gone. While the newborn's lung linings are pure pink, the smoker's are a sooty black; the tissues that capture life-giving oxygen have been ravaged beyond repair. The adult notices the effects with each hacking cough.

What Is Respiration?

Breathing, technically termed **external respiration,** is actually one of three forms of respiration. External respiration is the process by which animals exchange oxygen and carbon dioxide (CO_2) between moist **respiratory surfaces** and the blood. **Internal respiration** exchanges these gases between the blood and cells. Cellular respiration refers to oxygen-utilizing biochemical pathways that store chemical energy as ATP.

FIGURE 36.1
The First Breath of Life Is the Toughest.
A baby must gather great strength to fill the lungs for the first time.

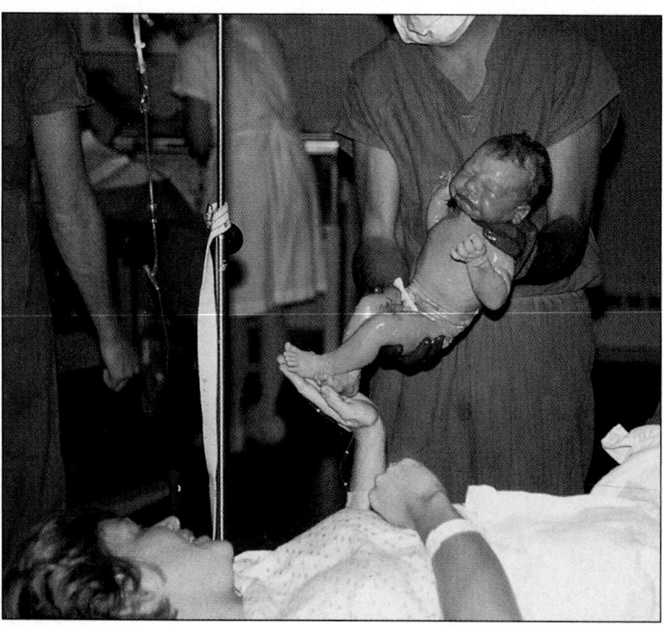

The gas exchange that breathing makes possible enables cells to harness the energy held in the chemical bonds of nutrient molecules. Without gas exchange, cells die.

Recall from chapter 5 that energy is slowly liberated from food molecules when electrons are stripped off and channeled through a series of electron carriers, each at a lower energy level than the previous one. The energy released at each step is used to form ATP. Oxygen is the final acceptor of the low-energy electrons at the end of the chain of acceptors, where it combines with hydrogen to form water. Without oxygen, electrons cannot pass along this series of acceptors; this halts breakdown of organic molecules much earlier and leaves a great deal of energy locked in the nutrient molecule. Oxygen also combines with carbons cleaved from nutrient molecules in the Krebs cycle, producing carbon dioxide.

One-celled organisms can exchange gases using simple diffusion. Larger organisms, however, require respiratory systems with moist membranes that provide enough diffusional surface area to meet metabolic requirements. Because diffusion can only effectively distribute substances over about 0.5 millimeters, more active animals also require a circulatory system to transport gases between the respiratory membrane and cells.

Aquatic animals have respiratory structures that can extract oxygen from a viscous medium holding small quantities of oxygen. Water saturated with oxygen contains only about one-thirtieth of the oxygen present in an equal volume of air, and the warmer or saltier the water, the less oxygen it holds. Furthermore, oxygen takes about 300,000 times longer to diffuse through water than through air, so it takes more time to replenish oxygen used for cellular respiration. This means that to obtain the same quantity of oxygen, an aquatic organism must move a much greater quantity of water over its respiratory surface than the volume of air a land dweller must move across its respiratory surface. In addition, because water is more dense and more viscous than air, an animal in an aquatic environment must use more energy to keep oxygen-rich water flowing over its respiratory system. As a result, an aquatic organism requires a larger respiratory surface than a terrestrial animal of similar size and with similar energy requirements. Land dwellers, however, have their own challenge: keeping the respiratory membrane from drying out. The respiratory surface of terrestrial animals is thus inside the body, where it can be kept moist.

Respiratory surfaces include body surfaces, tracheal systems, gills, and lungs. Each of these specialized structures is an adaptation for obtaining oxygen in a specific environment (fig. 36.2).

Body Surface

In the simplest gas exchange mechanism, all cells are in close enough contact with the environment that gases can diffuse across the cell membrane and reach all cells. A single-celled organism can exchange gases by simple diffusion because its cell membrane provides enough surface area for sufficient oxygen to diffuse into the cell.

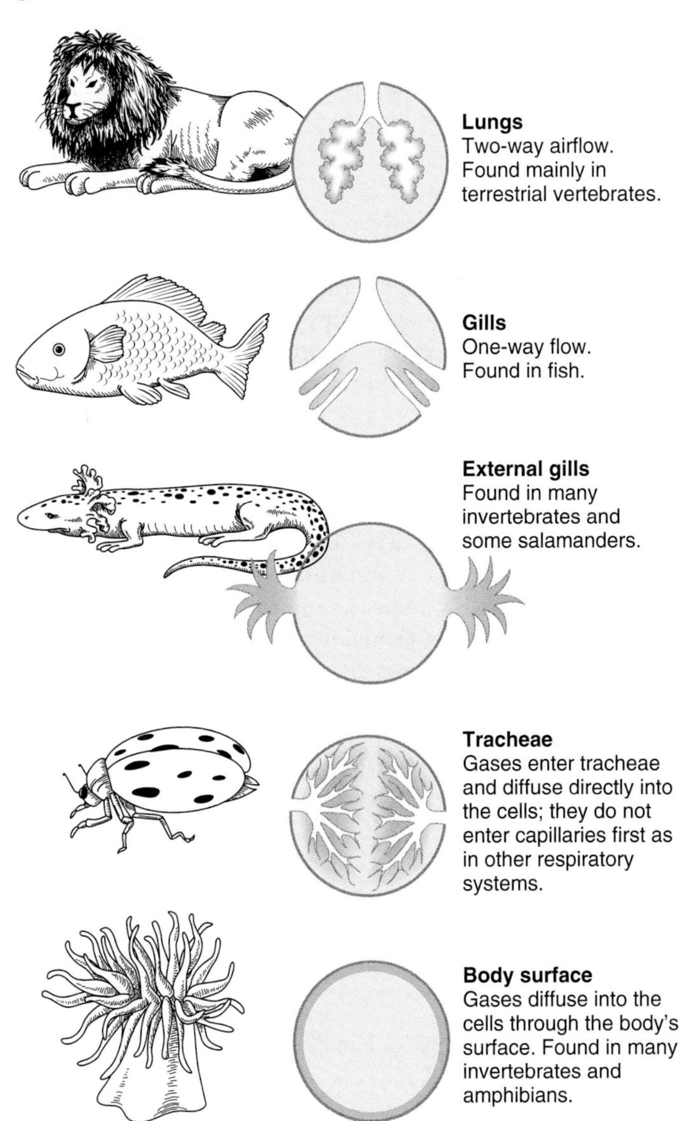

FIGURE 36.2
Diversity in Respiratory Surfaces.
The sea anemone has the simplest type of respiratory surface—diffusion across a thin layer. Many land-dwelling insects exchange gases using an extensive network of tracheae. A mud puppy uses an external gill. A fish uses an internalized gill. Terrestrial vertebrates use lungs.

Multicellular organisms use different strategies for gas exchange. One way to keep cells in contact with the oxygen-rich environment is to have a flat body. For example, cnidaria such as hydra and sea anemones can obtain oxygen by diffusion across their thin, extended body parts. Flatworms (phylum Platyhelminthes) are also thin enough for gas exchange and distribution to occur by diffusion.

In larger animals, diffusion cannot effectively distribute gases. A circulatory system transports gases between body cells and the respiratory membrane. Some organisms, such as earthworms (phylum Annelida) and amphibians, still use the body surface for gas exchange but have a circulatory system to transport gases between cells and the body surface. The respiratory surface and the circulatory system in such animals are intimately and extensively linked; just beneath the skin surface, many tiny blood vessels allow gases to diffuse across the skin into or out of the circulatory system.

Tracheal Systems

In terrestrial arthropods, including insects, centipedes, millipedes, and some spiders, tough, waterproof exoskeletons dip inward to form an extensively branching system of tubules, called **tracheae,** that bring the outside environment close enough to nearly every cell for gases to diffuse (fig. 36.3). The tracheae walls are one cell thick, and covered with cuticle that keeps them from collapsing. The tracheae open to the outside through paired **spiracles** along the animal's segments, as the chapter opening essay describes. In many species, valves guard the spiracles, and can close in dry environments to keep the cells of the tracheae moist. Some arthropods have bristles over the spiracles that can screen out water, dust, and parasites. Tracheal systems can be quite extensive. The silkworm larva, for example, has 1.5 million of the tubules.

Tracheae branch into tiny fluid-filled tubules called **tracheoles,** which branch around individual cells, some hugging the cells so tightly that they indent the cell membrane. This close association may be an adaptation that brings the tracheoles as close to the mitochondria as possible to deliver oxygen.

In some areas of a trachea, the thick, spiral cuticle that normally keeps the tubule open thins. The tubule expands at this site

FIGURE 36.3
Insect Tracheal Systems.
Tracheal systems deliver oxygen directly to an insect's muscle cells by diffusion and also to hemocytes, as highlighted in the chapter opening essay. The enlargement shows how tracheae branch to form tracheoles. In some species the tracheoles also form expanded regions called air sacs.

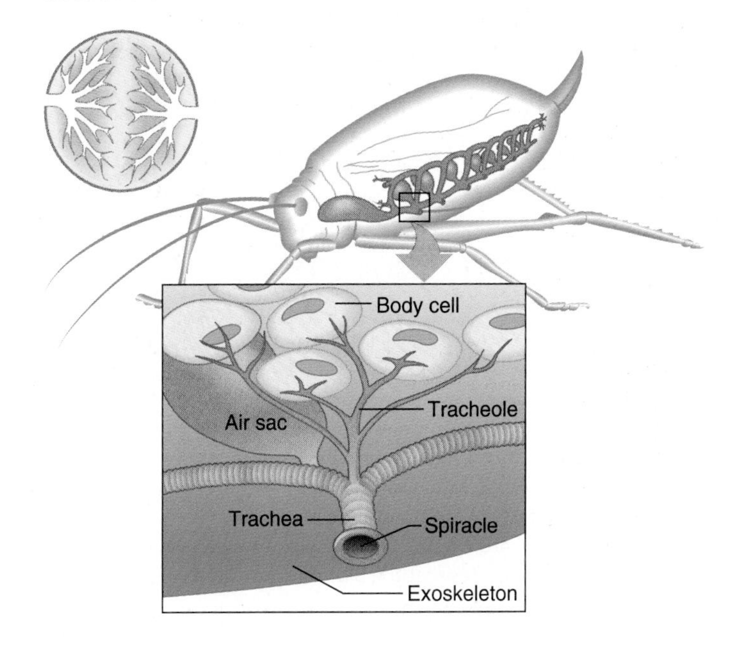

of weakening, forming a localized air sac. These regions are found in the body cavity and in the legs. Air sacs keep the animal lightweight, and also may play a role in moving air through the system, inflating and deflating somewhat like an accordion.

In small arthropods, the tracheae are short enough to effectively deliver oxygen, which follows its concentration gradient into cells. This diffusion pulls on the tracheae, drawing in more air. Large, active arthropods, however, use their abdominal and flight muscles to draw air in. A foraging honeybee, for example, pumps segments of the abdomen back and forth to move air. Insects that are aquatic as larvae can exchange gases through the body wall when they are in that stage, and then through tracheal systems later in development.

Gills

Large, active organisms require a gas exchange surface greater than the skin surface. In aquatic organisms, the vast respiratory surface usually takes the form of **gills.** These structures are extensions of the body wall that are so highly folded and branched that they may superficially resemble feathers. Within each gill is a dense network of capillaries. Oxygen diffuses from the water across delicate gill membranes and thin capillary walls into the blood, which delivers it to cells. Carbon dioxide diffuses in the opposite direction.

Amphibian larvae have external gills that close off as lungs begin working after metamorphosis. Some salamanders retain external gills through adulthood. The fish's complex gills create the extensive surface area necessary to extract maximal oxygen from water and deliver it to cells (fig. 36.4). Gases are exchanged across a very thin respiratory membrane and the single layer of capillary cells, and are transported by the circulatory system. A structure called an operculum protects the gills while pumping water over their surfaces. Maximal gas exchange requires that a continuous supply of fresh water flows from the mouth over the gills. This is more energy efficient than moving water into a gill chamber, only to reverse the direction of flow and push it out again.

Blood flows through the capillaries in the opposite direction that the water flows over the gill membrane. This anatomical arrangement is called **countercurrent flow** and it maximizes the amount of oxygen extracted from the water. Because of the countercurrent flow, blood leaving the gills, which has a relatively high oxygen concentration, moves past water entering the gills, which has the highest available oxygen concentration. Because of this gradient, oxygen diffuses into the blood, bringing oxygen in the water and blood closer to equilibrium as the blood leaves the gills. The oxygen level in the water drops as it continues to flow over the gills. The blood vessels that water encounters have even less oxygen, however, so there is always a gradient for diffusion.

Lungs

As organisms ventured onto the land hundreds of millions of years ago, gills became obsolete—they would simply dry up without a watery environment. Scorpion-like arthropods were among the earliest of terrestrial animals, venturing from the seas during the Silurian period over 400 million years ago. Special respiratory structures called book lungs that contained pagelike surfaces folded into the body allowed arachnids to exchange respiratory gases with the air.

The movement of vertebrates to terrestrial habitats drove the evolution of **lungs** in these organisms. The location of lungs in-

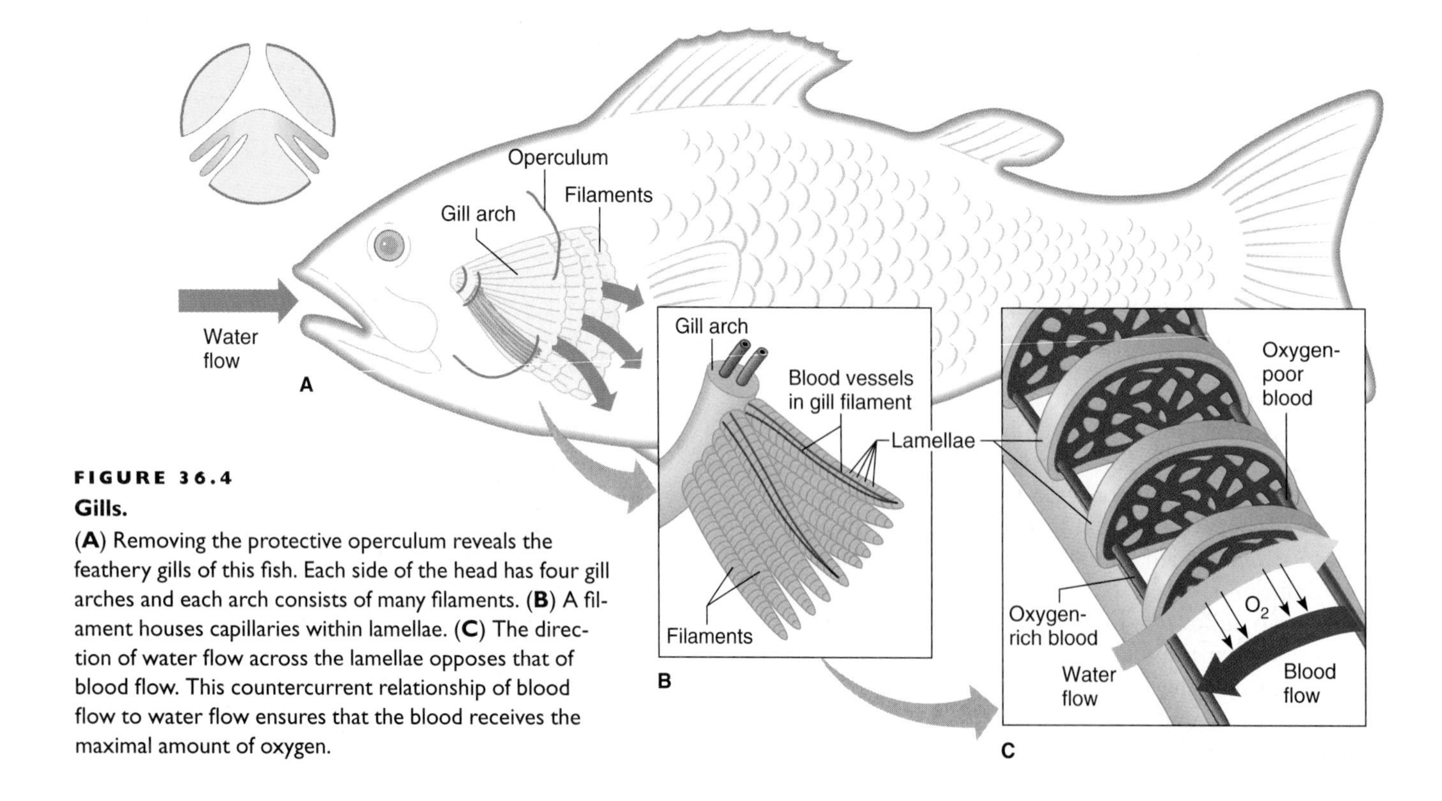

FIGURE 36.4
Gills.
(**A**) Removing the protective operculum reveals the feathery gills of this fish. Each side of the head has four gill arches and each arch consists of many filaments. (**B**) A filament houses capillaries within lamellae. (**C**) The direction of water flow across the lamellae opposes that of blood flow. This countercurrent relationship of blood flow to water flow ensures that the blood receives the maximal amount of oxygen.

FIGURE 36.5

The Evolution of Lungs.

(**A**) The first lungs probably supplemented gills during droughts, and were little more than paired sacs with rich capillary networks in their smooth walls. (**B** and **C**) As vertebrates conquered land, lungs became more elaborate, first developing a few subdivisions in salamanders, then becoming more extensively subdivided in frogs and toads. (**D**) The reptile lung is even more clearly compartmentalized. (**E**) Bird lungs store air in air sacs, which is an adaptation to meet the high metabolic demands of flight. (**F**) The mammalian lung takes maximization of surface area to an extreme, with millions of microscopic air sacs wrapped in capillaries.

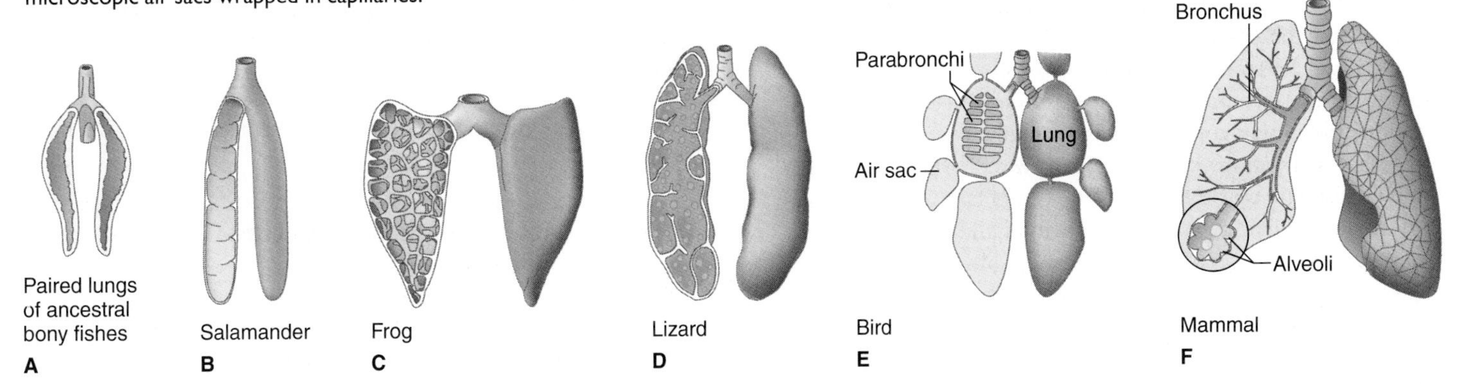

side bodies kept the surfaces for gas exchange moist. A new challenge that lungs introduced, however, was how to move air in and out to renew the oxygen supply, rather than let air simply flow in one direction over the respiratory membrane, as water does over a fish's gills. Lungs and associated structures evolved into **ventilation** machines as well as sites of gas exchange. Ventilation in general refers to any mechanism that increases the flow of air or water across a respiratory surface.

The first lungs appeared in the lobe-finned fishes (see fig. 18.17). Recall that these animals lived about 410 million years ago, during the Devonian "Age of Fishes." They had gills as well as lungs, and would use their lungs when swamps dried up, surfacing briefly to gulp in air. Their direct descendants today are the appropriately named lungfishes.

Looking at lung complexity in modern species gives an idea of how these organs evolved (fig. 36.5). The simplest lungs, of the lungfishes and other primitive fishes, are little more than relatively smooth air-filled pouches that are invested with capillary networks. These lungs supplement gills. The bichir, for example, is a primitive-looking fish that lives in shallow, oxygen-poor waters in Africa, where it occasionally swims to the surface to gulp air. Oxygen diffuses into the lush network of capillaries that pervade its pouchlike lung.

Evolution of lung structure is clearly evident among the amphibians. Salamanders have primitive lungs that resemble those of lungfishes—two paired, rather smooth sacs, with a capillary supply. The lungs of frogs and toads, however, introduce a few segments, an adaptation that increases surface area. Figure 36.6 illustrates how a frog breathes. A reptile's lung has much more extensive subdivisions, and the mammalian lung is truly spectacular in its maximization of surface area—millions of tiny air sacs sit within baskets of capillaries, able to rapidly extract oxygen

FIGURE 36.6

Respiration in Frogs.

(**A**) Unlike most reptiles, birds, and mammals, which suck in air (negative pressure), a frog obtains air with a positive pressure system, forcing it into the lungs. (**B**) The animal draws air into its nostrils with its mouth shut. (**C**) By closing the nostrils and raising the floor of the mouth, the air is forced into the only place it can go—the lungs. (**D**) The body wall contracts and the lungs recoil, expelling air. A frog also obtains oxygen through its moist skin and the lining of its mouth, which are supplementary respiratory surfaces.

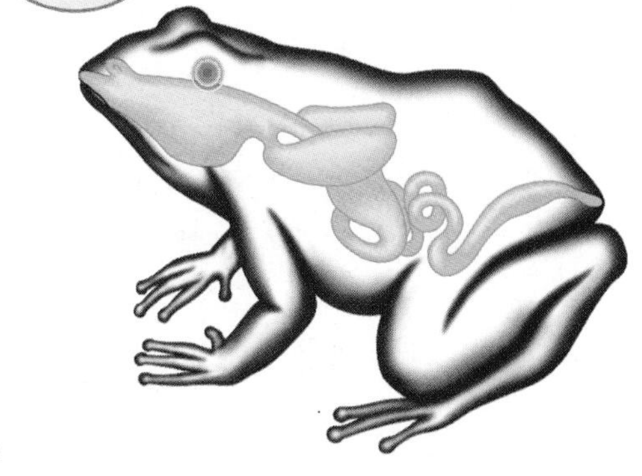

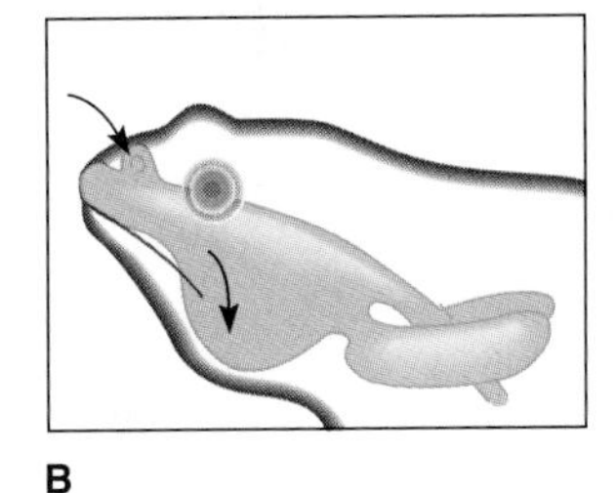

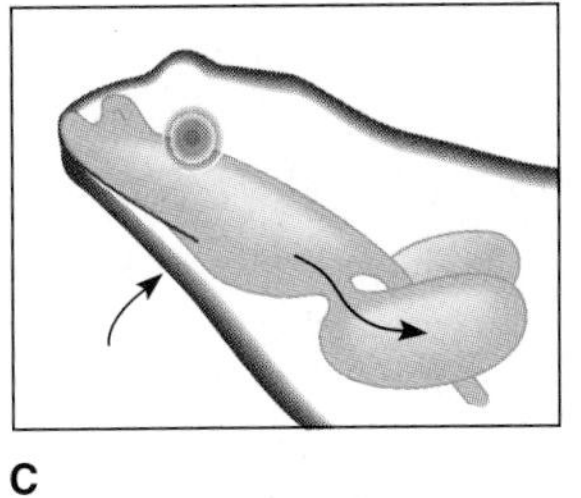

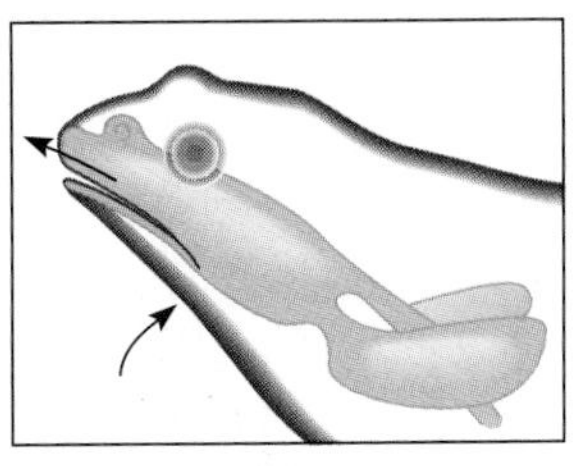

from incoming air and pass it directly to the bloodstream. The complex organization of the mammalian lung supports the high metabolic rate necessary to maintain body temperature and a very active existence.

Birds have an exceptionally extensive respiratory system that provides the continuous oxygen supply required to power flight. A bird's lungs consist of many narrow tubules where gas exchange occurs. These tubules connect to air sacs that are located throughout the bird's body. Because of the added space that the air sacs provide, oxygen remains in the respiratory system even during exhalation, providing a continual oxygen supply (fig. 36.7).

Other adaptations can increase oxygen delivery to cells. An elephant seal and a leatherback turtle can obtain oxygen to support frequent diving because of their abundant, oxygen-rich blood supplies. Many animals use hemoglobin to supply oxygen to tissues by way of the bloodstream, and myoglobin to supply oxygen directly to the muscle.

36.1 MASTERING CONCEPTS

1. What are the three forms of respiration?
2. Why do animals require oxygen?
3. What is the role of a circulatory system in respiration?
4. Why do aquatic animals require a larger respiratory surface than terrestrial animals?
5. Describe how tracheal systems, gills, and lungs participate in gas exchange.
6. How did lungs evolve?

OLC

36.2 Components of the Respiratory System in Humans

Air is cleansed and warmed in the nose, then transported through the pharynx, larynx, and trachea to the lungs. Here, air continues moving through bronchi and bronchioles, ending at the alveoli, where gas exchange occurs.

The human respiratory system is representative of mammalian systems.

Respiratory System Structures Transport Air

The human respiratory system is a continuous network of tubules that delivers the oxygen in air to the circulatory system. Figure 36.8 presents an overview of the system.

The Nose Inhaled air must be cleaned, warmed, and humidified before the lungs can extract oxygen from it. Millions of airborne bacteria and dust particles enter the nose each day. Winter temperatures in many parts of the world would freeze lung tissues, and exposure to dry air would kill lung cells. Clearly, air must be processed. This begins in the nose.

The epithelium in the nose forms a mucous membrane that filters, moistens, and warms incoming air. The nose is subdivided

FIGURE 36.7

Respiration in Birds.

(**A**) A bird's anterior and posterior air sacs enable it to store air and use it even as it exhales. Air remains in the animal for two cycles of inhaling and exhaling and moves through the lungs in a one-way flow. (**B**) With the first inhalation, most of the air goes to posterior air sacs. (**C**) During the first exhalation, the air in the posterior air sacs passes through the lungs, where gas exchange occurs on the surfaces of parallel arrays of tubules. (**D**) The air, depleted of oxygen but having picked up carbon dioxide, goes to the anterior air sacs during the second inhalation. (**E**) Finally, the air from the anterior air sacs is released from the bird during the second exhalation.

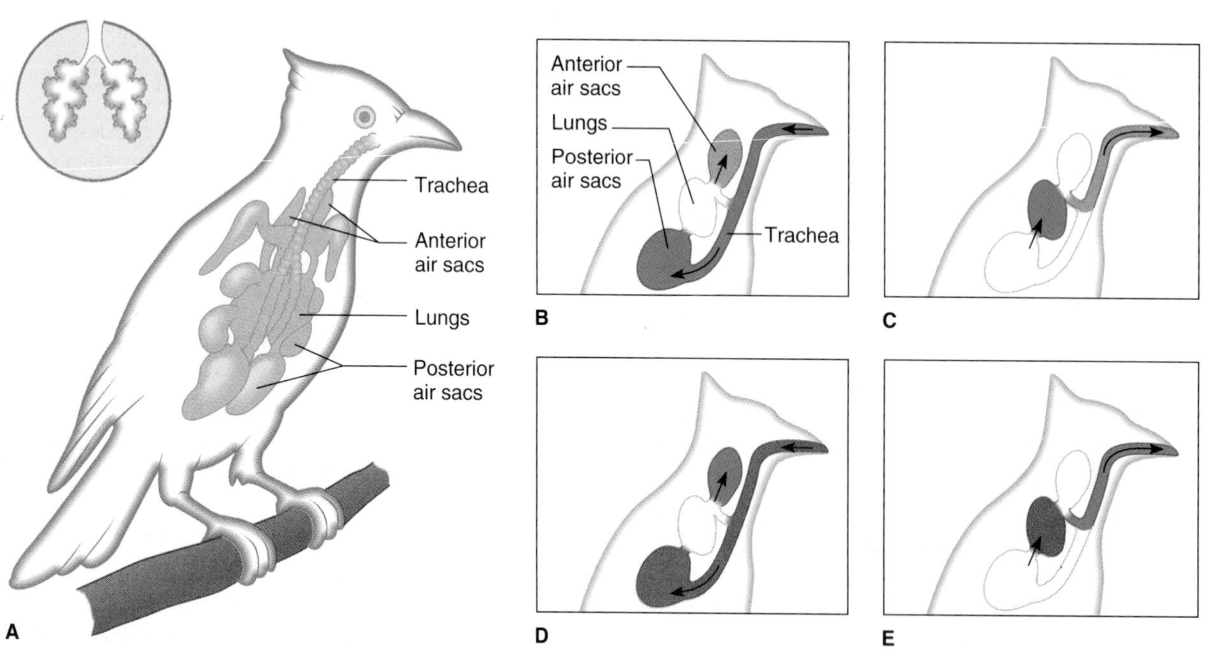

FIGURE 36.8

The Human Respiratory System.

Inhaled air passes through increasingly narrow tubes until it arrives in microscopic air sacs—the alveoli (inset)—where gas exchange occurs. Many other structures assist in the movement of air, including parts of the mouth and surrounding bones and muscles.

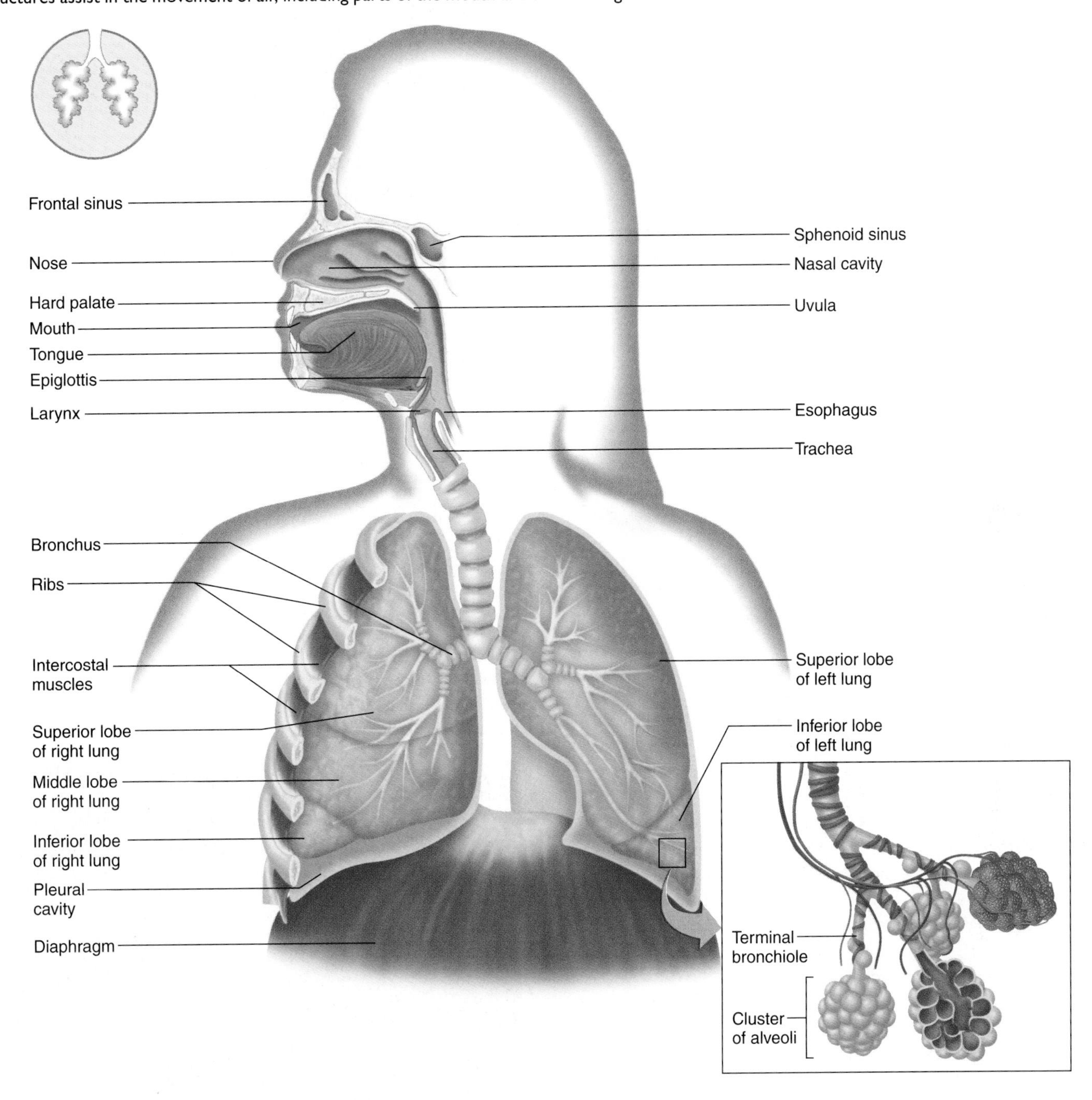

into compartments, forming many surfaces for ciliated cells to cover. Three shelflike nasal bones partition each of the two nasal cavities into channels. The entire surface is also rich with blood vessels. Cells called goblet cells secrete mucus, and the epithelium is fringed with cilia that beat, moving particles along their surfaces (fig. 36.9).

Hairs at the entrance of the nasal cavities first filter incoming air. Most bacteria and particles that pass the hairs catch in sticky mucus or, if inhaled further, are trapped by mucus lower in the respiratory tract and swept back out by waving cilia. A large inhaled particle may trigger a sensory cell in the nose that signals the brain to orchestrate a sneeze, which forcefully ejects the particle. Sneeze-borne particles can travel up to 200 miles (322 kilometers) per hour.

The nose also protects lung tissue by adjusting the temperature and humidity of incoming air. Blood within the many vessels permeating the mucous membrane warms the air, and the mucus moisturizes it. Cigarette smoke disrupts all of these vital

FIGURE 36.9

The Lining of the Upper Respiratory Tract Is Adapted to Expel Particles.

(**A**) Goblet cells secrete mucus, which entraps particles, and (**B**) the cilia on the epithelial cells move matter up and out of the respiratory tract.

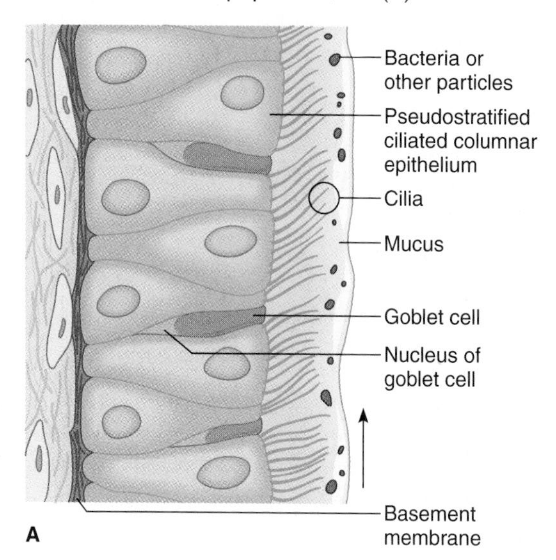

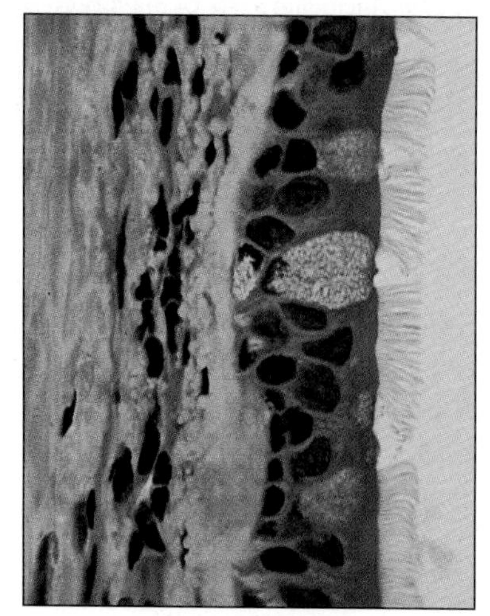

functions of the nose by a variety of mechanisms, including destruction of cilia (see Health 36.1).

The Pharynx and the Larynx The back of the nose leads into the **pharynx,** or throat, which is a 4.5-inch (12-centimeter) tube that conducts food and air. The **larynx,** or Adam's apple, is a box-like structure located below and in front of the pharynx that passes materials and also produces the voice. A reflex action steers food and fluid toward the digestive system during swallowing. Inhaled air passes through the **glottis,** the opening to the larynx. Swallowing moves the larynx upward, flipping down a piece of cartilage called the **epiglottis** that covers the glottis like a trapdoor (see fig. 37.9).

Stretched over the glottis are two elastic bands of tissue, the **vocal cords,** that vibrate when air passes them. The vibrations produce sounds that can be molded into speech. Sound resonates in the pharynx, the nasal cavities, and the mouth and can be molded by moving the tongue and lips. The deepening voice of a male during puberty is caused by expansion of the Adam's apple and lengthening of the vocal cords.

The Trachea, Bronchi, and Bronchioles The larynx sits atop the windpipe, or **trachea,** a 5-inch- (13-centimeter-) long tube about 1 inch (roughly 2.5 centimeters) in diameter. C-shaped rings of cartilage hold the trachea open in spite of the negative pressure that inhalation creates. You can feel these rings in the lower portion of your neck. The inside surface of the trachea is ciliated and secretes mucus. The cilia and mucus filter, warm, and moisten incoming air.

As the trachea approaches the lungs, it branches into two **bronchi.** Each bronchus continues the tracheal organization of C-shaped cartilage rings, an inner ciliated epithelial layer, and an outer layer of smooth muscle. The bronchi branch repeatedly, each branch decreasing in diameter and wall thickness. The finest branches, lacking cilia, are called **bronchioles** (little bronchi). The respiratory passageways within the lung are called the "bronchial tree" because their branching pattern resembles an upside-down tree (fig. 36.10). The 10 million cilia lining the bronchial tree beat hundreds of times a minute to remove inhaled particles.

FIGURE 36.10

Bronchi and Bronchioles Form a Treelike Arrangement in the Lungs.

This plastic cast shows the complex branching pattern formed by the respiratory passages.

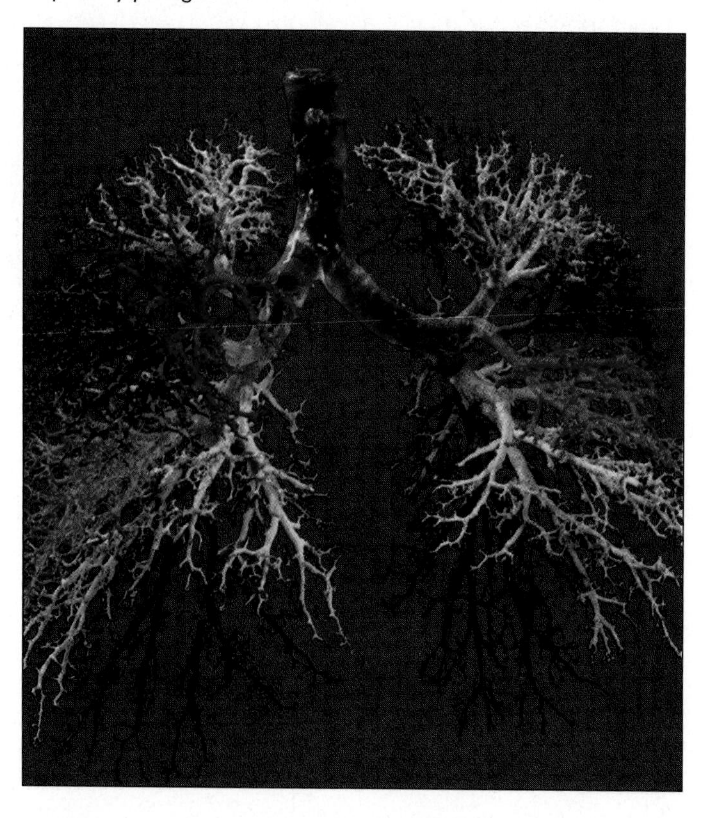

Lung Irritants

The lungs are exquisitely sensitive to the presence of inhaled particles. Such exposures can cause a variety of symptoms that range from a persistent cough to cancer. Causes may be as obscure as the occupational hazards listed below, or as common as cigarette smoking.

A Disorder with Many Names

Repeatedly inhaling dust of organic origin can cause a lung irritation called extrinsic allergic alveolitis. An acute (sudden, of short duration) form of this reaction impairs breathing and causes a fever a few hours after encountering dust. In the chronic (longer-lasting) form, lung changes occur gradually over several years. The condition is associated with several occupations, and has a variety of colorful names:

Bathtub refinisher's lung
Bird breeder disease
Cheese worker's lung
Enzyme detergent sensitivity
Farmer lung
Laboratory technician's lung
Maltworker lung
Maple bark stripper disease
Mushroom picker disease
Snuff taker's lung
Plastic worker's lung
Poultry raiser disease
Wheat weevil disease

Asbestos-Related Disorders

This naturally occurring mineral was once widely used in buildings and on various products because it resists burning and chemical damage. Asbestos easily crumbles into fibers, which, when airborne, can enter human respiratory passages. Asbestos-related problems include:

- asbestosis (shortness of breath resulting from scars in lungs)
- lung cancer
- mesothelioma (a rare cancer of the membranes surrounding the lungs)

Asbestos only causes respiratory illness if it is disturbed, so that fibers break free and become airborne. Sometimes it is safer to encapsulate asbestos in a building and leave it in place, than to move it, which can release the lung-choking fibers. Today, synthetic fiberglass or plastics replace asbestos.

Berylliosis

Beryllium is an element used in fluorescent powders, metal alloys, and in the nuclear power industry. A small percentage of workers exposed to beryllium dust or vapor develop an immune response, which damages the lungs. Symptoms include cough, shortness of breath, fatigue, loss of appetite, and weight loss. Fevers and night sweats indicate the role of the immune system. Radiographs show scars, called granulomas, in the lungs (fig. 36.A). Listening to breath sounds with a stethoscope reveals the impaired breathing characteristic of berylliosis.

Symptoms of berylliosis typically begin about a decade after the first exposure, but this response time can range from several months to as long as forty years. Many people who worked with the element in the Rocky Flats plant in Colorado have developed berylliosis, and they are being monitored at the National Jewish Medical and Research Center in Denver. Affected individuals include those who directly contacted the element frequently, as well as support staff such as secretaries who probably inhaled beryllium.

The tendency to develop berylliosis is inherited, and some workplaces have begun voluntary testing programs to identify susceptible individuals. One test of sensitivity spots a dramatic increase in cell division rates among lymphocytes exposed to beryllium. Genetic tests are in development. Steroid drugs can treat the symptoms of berylliosis, but the condition may ultimately be fatal.

Cigarette Smoking

Damage to the respiratory system from cigarette smoking is slow, progressive, and deadly. Cigarette smoke severely impairs the respiratory system's ability to cleanse itself. With the very first inhalation, the beating of the cilia slows. With time, the cilia become paralyzed and, eventually, vanish. When the cilia can no longer remove mucus, the individual must cough to clear particles from the airways. Smoker's cough is usually worse in the morning because mucus accumulates during sleep.

Smoking produces excess mucus that clogs the air passageways. Disease-causing microorganisms that are normally removed now have easier access to respiratory surfaces, and lung congestion favors their growth. This is why smokers are more susceptible to respiratory infections. In addition, a lethal chain reaction begins. Smoker's cough leads to chronic bronchitis; mucus production increases and the linings of the bronchioles (microscopic tubes that deliver air) thicken, making breathing difficult. As the bronchioles lose elasticity, they can no longer absorb the pressure changes that accompany coughing. As a result, a cough can increase the air pressure within

(continued)

FIGURE 36.A

Exposure to Beryllium Causes Severe Lung Irritation in Genetically Susceptible Individuals.

The workers in (**A**) are sealing a drum that contains beryllium oxide (BeO), a ceramic-grade powder. Respirators, Tyvek coveralls, and gloves prevent both inhalation and skin contact with BeO particles. (**B**) In berylliosis, granuloma scars form in lung tissue.

A

B

(concluded)

the alveoli (microscopic air sacs) enough to rupture their delicate walls, producing a condition called emphysema. Burst alveoli cause a worsening cough, fatigue, wheezing, and impaired breathing. Emphysema is 15 times more common among heavy smokers than among nonsmokers.

As structural changes progress to emphysema, cellular changes may lead to lung cancer. First, cells in the outer border of the bronchial lining begin to divide more rapidly than usual, displacing ciliated cells. If smoking continues, these cells may eventually break through the basement membrane and begin dividing within the lung tissue, forming a tumor that could spread throughout the lung tissue and beyond (fig. 36.B). Eighty-five percent of lung cancer cases occur in smokers.

It pays to quit smoking. Much of the structural damage to the respiratory system can heal. Cilia may reappear, and the thickening of alveolar walls can reverse, although ruptured alveoli are gone forever.

FIGURE 36.B
Long-Term Exposure to the Chemicals in Cigarette Smoke Causes Lung Cancer. Compare the healthy lung (left) to this cancer-ridden lung (right).

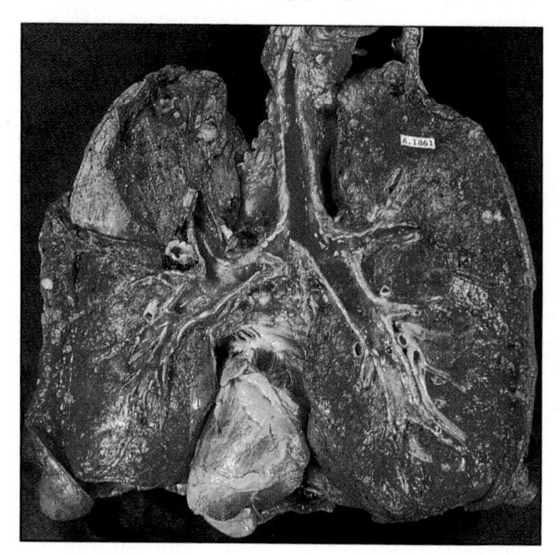

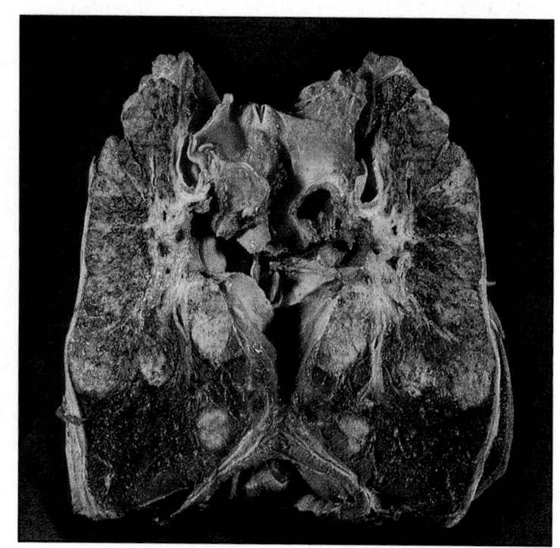

The bronchioles have no cartilage, but their walls contain smooth muscle. The autonomic nervous system controls contraction of these muscles, adjusting airflow to suit metabolic demands. During a stressful or emergency situation, certain muscle cells may require more oxygen to produce the ATP that fuels the emergency response. Under these conditions, the sympathetic nervous system dilates the bronchioles, increasing their diameters.

Contraction of the bronchial wall muscle is usually adjusted to the body's condition. Sometimes, however, as in asthma, the bronchial muscles spasm, making airflow exceedingly difficult. Asthma is chronic inflammation of the bronchial passages and episodic wheezing. An allergy to pollen, dog or cat dander (skin particles), or tiny mites in house dust trigger most asthma attacks. Inhalant drugs that treat asthma usually control chronic bronchial and lung inflammation and relax the bronchial muscles. Some aerosol inhalers spray adrenaline onto the bronchial walls, which mimics the effect of the sympathetic nervous system and dilates the bronchioles.

Gases Are Exchanged in the Lungs at the Alveoli

Each bronchiole narrows into several alveolar ducts, and each duct opens into a grapelike cluster of **alveoli** (fig. 36.11). Each alveolus is a tiny sac with a one-cell-thick wall. Most of the lung is composed of some 300 million alveoli, which makes the lung structure similar to that of foam rubber. The total surface area of the alveoli in one set of human lungs is about 50 times the area of the skin!

A vast network of capillaries surrounds each cluster of alveoli. If all of these capillaries were unwound and laid end to end, they would extend for about 620 miles (1,000 kilometers). Gas diffuses through the thin walls of the alveoli and the neighboring capillaries. Oxygen enters and carbon dioxide leaves the blood within the alveoli.

The alveoli must expand to efficiently use their extensive surface area for gas exchange. A mixture of phospholipid molecules called human lung surfactant keeps the alveoli open by occupying the space between the watery film that lines the inner alveoli surfaces and the air within the alveoli. Without surfactant, the watery lining exerts surface tension, the consequence of cohesive forces between water molecules. This surface tension generates a contractile force that would collapse the alveoli, as the inside of a moist balloon sticks together and is difficult to expand with air. Surfactant counters this force, so the alveoli remain open and available for gas exchange.

Human lung surfactant is especially important at birth, when a baby's lungs must inflate to take the first breath. The infant must overcome surface tension, which makes the first breath 15 to 20 times more difficult than ventilating expanded alveoli. Once the first breath expands the alveoli, however, surfactant, which has been collecting for the preceding few weeks, keeps them open and eases subsequent breaths.

FIGURE 36.11

Alveoli.

Bronchioles narrow to form alveolar ducts, which lead to clusters of alveoli. A lush capillary supply surrounds each cluster. Alveoli greatly increase the surface area available for gas exchange.

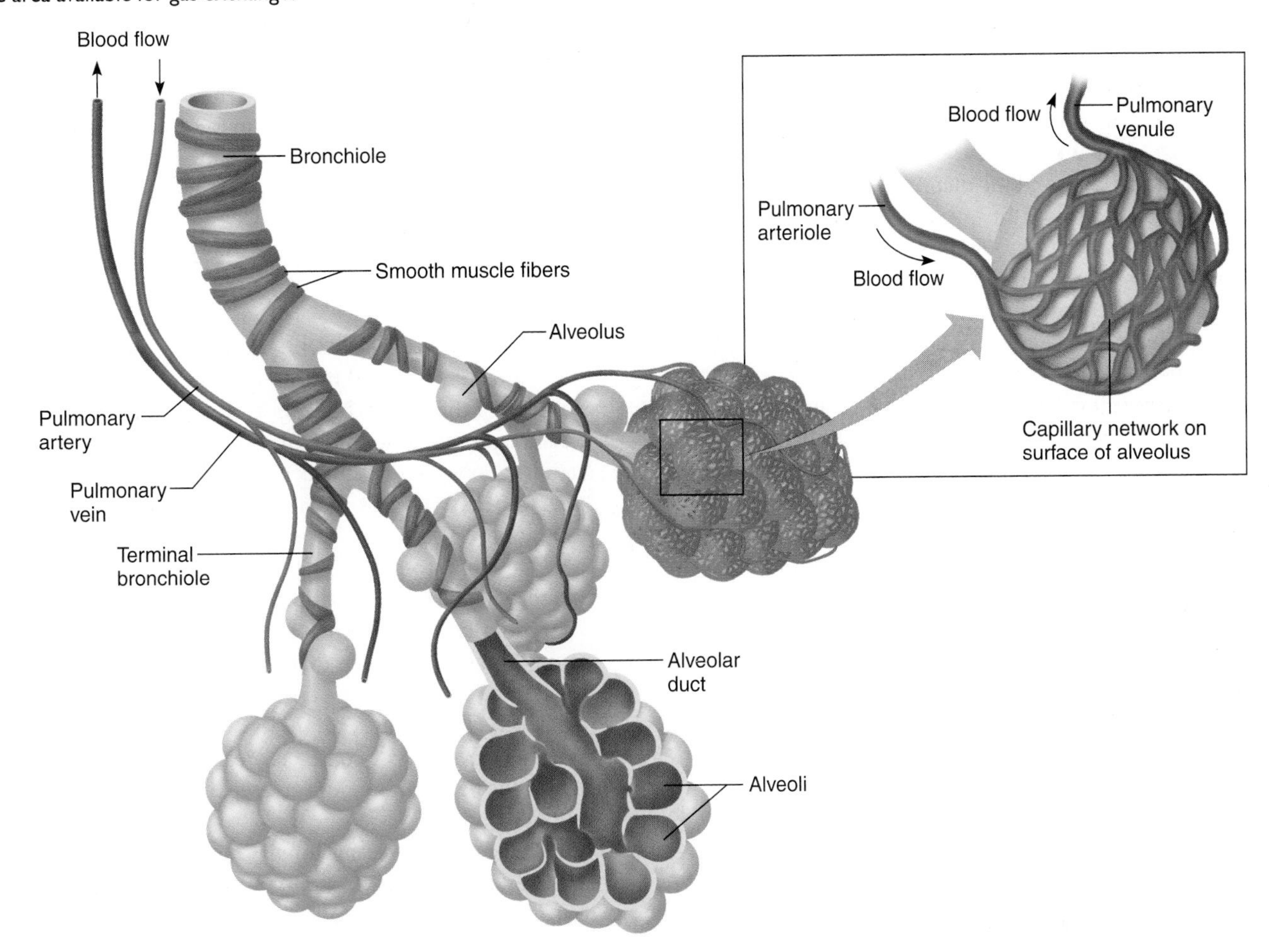

Many premature babies have not yet produced enough surfactant to prevent the alveoli from collapsing after each breath, which causes infantile respiratory distress syndrome. The newborn must fight as hard for every breath as a normal newborn does for the first one. Synthetic surfactant dripped into the respiratory tract can enable these tiny patients to breathe.

The bronchial tree and the alveoli are housed within the paired lungs. Each human lung weighs about 1 pound (454 grams). The left lung is divided into two lobes, and the right is divided into three. The heart lies between the lungs.

36.2 MASTERING CONCEPTS

1. How do structures in the nose clean and warm inhaled air?
2. Which structures prevent food from entering the respiratory system?
3. Describe the structures and functions of the trachea, bronchi, bronchioles, and alveoli.
4. What is the function of lung surfactant?

OLC

36.3 Functions of the Respiratory System in Humans

Breathing occurs because of pressure differences between the lungs and the outside, controlled by muscle action. Hemoglobin in red blood cells delivers oxygen to tissues. Carbon dioxide is removed in plasma, bound to hemoglobin, and carried mostly as bicarbonate ions.

Overall, respiration requires obtaining, filtering, and delivering air, and then the process of gas exchange.

How We Breathe

Breathing moves air between the atmosphere and the lungs in response to pressure differences. Air moves in when the air pressure in the lungs is lower than the pressure outside the body, and air moves out when the pressure in the lungs is greater than the atmospheric pressure.

The anatomy of the thoracic (chest) cavity explains the generation of the pressure changes responsible for inflating and deflating the lungs. The thoracic cavity has no direct connection to the outside of the body. A broad sheet of skeletal muscle, the **diaphragm,** separates it from the abdominal cavity.

Membranes that coat the lungs and thoracic cavity enclose the thin pleural cavity. A collapsed lung occurs when these pleural membranes rupture. Since each lung has its own pleural membranes, the collapse of one lung does not necessarily compromise the other.

Changing the size of the thoracic cavity creates pressure changes that move air in and out of the lungs. When the inspiratory muscles of the rib cage and the diaphragm contract, the thoracic cavity enlarges and draws air into the lungs. This process is called inspiration, or inhalation. When the rib muscles contract, they pull the rib cage upward and forward. Elevation of the rib cage expands the thoracic cavity from front to back, and contraction of the diaphragm elongates it. The increase in size of the thoracic cavity lowers the air pressure within the space between the lungs and the outer wall of the thorax. This causes the lungs to expand, lowering pressure in the alveoli, and air rushes in (fig. 36.12).

When the muscles of the rib cage and the diaphragm relax, the thoracic cavity shrinks, causing expiration, or exhalation. Now the rib cage falls to its former position, the diaphragm rests up in the thoracic cavity again, and the elastic tissues of the lung recoil. As the lungs squeeze back to their resting volume, the pressure within them increases. When the pressure in the lungs exceeds atmospheric pressure, air moves out.

One inspiration and one expiration constitute a respiratory cycle. The volume of air inhaled or exhaled during a single respiratory cycle is called the **tidal volume.** Breathing quietly moves roughly 500 ml of air in and out with each breath. Of this volume, only about 350 ml actually enter the alveoli. The remaining volume of inhaled air is left in the nose, pharynx, trachea, and bronchi. During exercise, tidal volume increases. If you were to take the deepest breath possible and exhale until you could not force any more air from your lungs, you could measure your **vital capacity**—the maximal amount of air you can force in and out of the lungs during one breath. Even after a maximal expiration, a volume of air called residual air remains in the lungs and inflates the alveoli.

FIGURE 36.12
How We Breathe.
When we inhale, the diaphragm flattens and the rib cage rises, expanding the thoracic cavity and lowering the air pressure in the pleural cavity. This expands the lungs, drawing in air. When we exhale, the diaphragm relaxes and the rib cage lowers, reversing the process and pushing air out of the lungs.

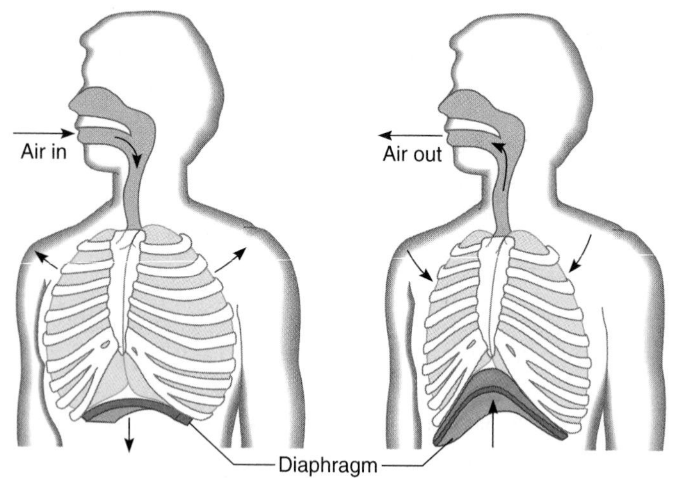

Delivering Oxygen and Removing Carbon Dioxide

Recall that external respiration is the exchange of O_2 and CO_2 between the air and the blood and internal respiration is the exchange of these gases between the blood and the tissues. In external respiration, O_2 enters the bloodstream, where its concentration is lower and CO_2 diffuses from the bloodstream to its lower concentration in the air spaces of the alveoli. In internal respiration, O_2 diffuses from the bloodstream to the lower O_2 concentration around the cells, while CO_2 moves to the bloodstream, where its concentration is lower (fig. 36.13).

From the alveoli, blood carries oxygen throughout the body. About 99% of the oxygen that reaches cells binds to hemoglobin in red blood cells, with the remaining 1% dissolved in plasma. A red blood cell spending only a second or two in the alveolar capillaries becomes almost completely saturated with oxygen. About 98% of the hemoglobin molecules leaving the lungs are saturated with oxygen.

Oxygen delivery is amazingly responsive to the requirements of the cells and helps to maintain homeostasis. The amount of oxygen delivered to the cells can more than triple without increasing blood flow rate. Exercise, for example, causes muscle cells to require more oxygen than usual to produce ATP. Under these metabolic conditions, hemoglobin releases more of its oxygen load to the cells. Hemoglobin does this in response to an exercise-induced rise in acidity in the blood. Strenuous exercise can cause muscle cells to switch to anaerobic pathways and release lactic acid. In addition, a very active muscle cell produces more CO_2 as waste. The CO_2 forms carbonic acid (H_2CO_3) when it dissolves in the water of the tissue fluid and blood, a point we will return to soon.

Active tissues produce heat as a by-product of energy metabolism. Blood flowing through warmer parts of the body releases more of its oxygen. Once again, more oxygen is released to cells that use oxygen at a higher rate.

Besides delivering oxygen, blood must also remove CO_2 produced in cellular respiration. CO_2 transport occurs in three ways. A small amount of CO_2, about 7%, is transported in plasma. Hemoglobin molecules transport slightly less than a quarter (23%) of the CO_2. CO_2 combined with hemoglobin is called carbaminohemoglobin.

About 70% of CO_2 is transported as bicarbonate ion (HCO_3^-) dissolved in plasma. The bicarbonate ion forms when

FIGURE 36.13

Oxygen and Carbon Dioxide Are Exchanged.

(**A**) In external respiration in the lungs, CO_2 diffuses from red blood cells and plasma to the alveoli. Oxygen moves from the alveoli to red blood cells and plasma. (**B**) In internal respiration, oxygen diffuses into tissues, while CO_2 moves into the blood. CO_2 from cells moves in the blood in three ways: dissolved in plasma, carried in hemoglobin, and as bicarbonate ion (HCO_3^-).

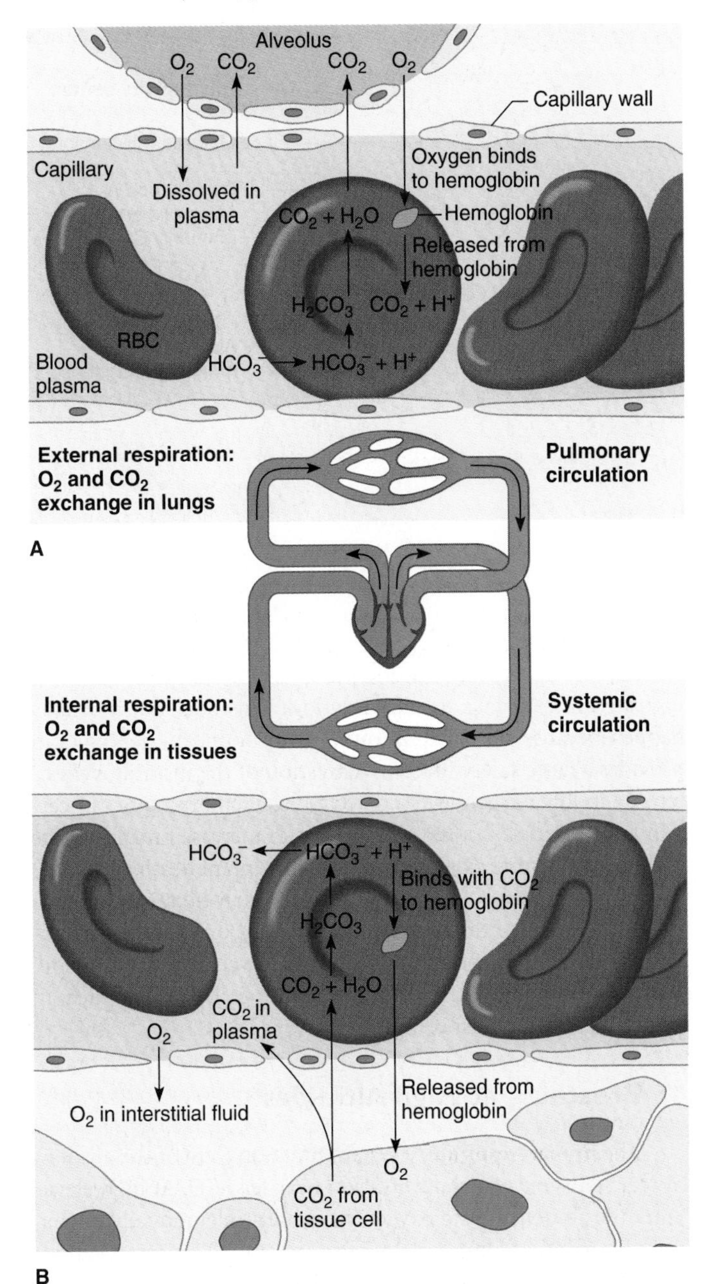

CO_2 that cells produce diffuses into the blood and into the red blood cells. The CO_2 reacts with water to form H_2CO_3, which dissociates to hydrogen ions (H^+) and bicarbonate ions. Within the red blood cell, an enzyme called carbonic anhydrase speeds this reaction. The H^+ then combines with hemoglobin, and the bicarbonate ions diffuse out of the red blood cell into the plasma. When the blood reaches the lungs, the whole process reverses (fig. 36.13). In the presence of carbonic anhydrase within the red blood cell, carbonic acid re-forms CO_2 and H_2O. The CO_2 then diffuses into the alveolar air and is exhaled.

In the tissues, CO_2 enters the bloodstream, where its concentration is lower. In the lungs, CO_2 moves from the bloodstream to the alveoli because the concentration in the bloodstream is higher.

Carbon Monoxide Poisoning

Carbon monoxide (CO) is a colorless, odorless gas that is a by-product of combustion. It is released by campfire smoke, car exhaust, cigarette smoke, kerosene heaters, woodstoves, and home furnaces and can be particularly dangerous in poorly ventilated areas. A few reports cite faulty ice resurfacing machines in public ice skating rinks, construction workers using gas-powered equipment in closed warehouses and garages, and indoor tractor pull events as sources of carbon monoxide poisoning.

Carbon monoxide is poisonous because it binds to hemoglobin more readily than oxygen does. Plus, it is less likely to leave the hemoglobin molecule. As a result, carbon monoxide prevents hemoglobin from binding and carrying oxygen, and the unsuspecting person begins to experience symptoms of oxygen deprivation. In healthy nonsmokers, 2% or fewer of hemoglobin molecules carry CO. When 10 to 20% of an individual's hemoglobin molecules carry carbon monoxide instead of oxygen, the person feels tired, nauseous, and may have a bad headache. A heavy smoker may experience these symptoms at lower carbon monoxide exposures because his or her blood always harbors some CO. When 30% of the hemoglobin molecules carry carbon monoxide, the person loses consciousness and may go into a coma. If the person survives, he or she may be left with permanent confusion, memory loss, or coordination problems.

36.3 MASTERING CONCEPTS

1. What force drives air in and out of the respiratory system?
2. Describe the events of inspiration and expiration.
3. Distinguish among vital capacity, tidal volume, and residual air.
4. What parts of blood deliver oxygen to cells?
5. How does oxygen delivery adjust to energy requirements?
6. What are the three ways that the blood transports CO_2?
7. What is the effect of more than 10% of a person's hemoglobin molecules binding CO?

OLC

36.4 Control of Respiration

The inspiratory center in the medulla causes inhalation, and the expiratory center causes the active exhalations of heavy breathing. Chemoreceptors in the medulla, aorta, and carotid arteries adjust breathing rate to blood pH, which reflects carbon dioxide level.

Awake or asleep, without conscious effort, we breathe an average of 12 times a minute. A rhythmicity center in the brain's medulla controls this steady breathing. During quiet breathing, part of this region, called the inspiratory center, is in control. It periodically triggers the diaphragm to contract and thus initiates inspiration. Expiration is the passive result of the lungs elastic recoil that forces air out of the respiratory passageways.

During strenuous activity, the respiratory cycle changes so that exhalation becomes an active process. The inspiratory center sends impulses to another part of the medulla's rhythmicity center called the expiratory center. This contracts muscles that pull the rib cage down and the sternum inward and also contracts the abdominal muscles, causing the abdominal organs to push the diaphragm farther into the thorax than usual. Air is quickly pushed out of the lungs.

Other brain areas may modify the basic breathing pattern. We can voluntarily alter our breathing pattern through impulses originating in the cerebral cortex. For example, we can control breathing when we speak or play a musical instrument, and we can voluntarily pant like a dog, sigh, or hold our breath.

FIGURE 36.14
Breathing Control.
The medulla has receptors that monitor hydrogen ion concentrations in the cerebrospinal fluid. Also, receptors in the aorta and carotid artery detect carbon dioxide, hydrogen ions, and oxygen in the bloodstream and pass the information to the rhythmicity center in the brain's medulla.

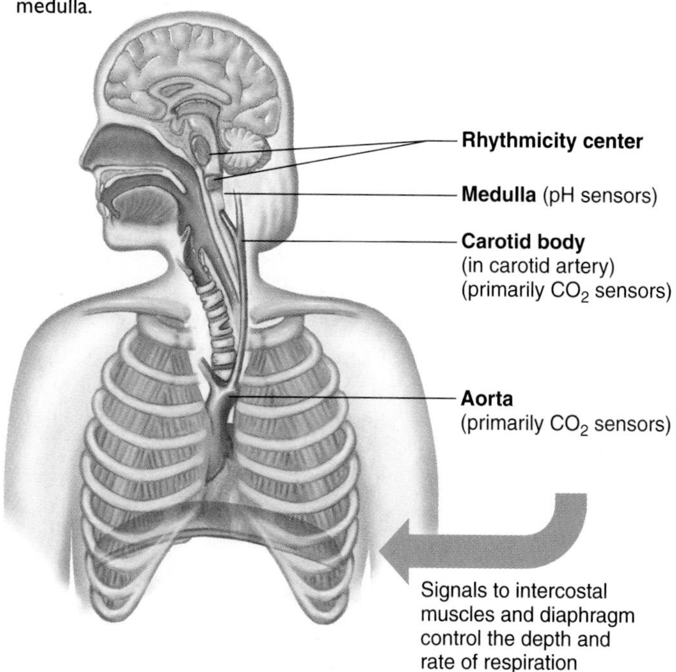

Blood CO_2 Levels Regulate Breathing Rate

The most important factor in regulating breathing rate is the blood CO_2 level, or, more precisely, the number of H^+ formed when CO_2 reacts with H_2O to form carbonic acid. Because under most circumstances CO_2 is a by-product of aerobic respiration, monitoring blood CO_2 levels is a way to determine how quickly cells use oxygen.

Chemoreceptors in the brain's medulla, and to some extent those in the aorta and carotid arteries, monitor blood CO_2 levels (fig. 36.14). These receptors are near the surface of the medulla, where they are bathed in cerebrospinal fluid. Since cerebrospinal fluid lacks the blood's ability to buffer pH changes, dissolved CO_2 changes the pH of cerebrospinal fluid more than that of blood. When the blood CO_2 level increases, it raises the H^+ concentration of the cerebrospinal fluid. Chemoreceptors are then stimulated to send messages to the medulla's inspiratory center. As a result, breathing rate increases. In the opposite situation, rapid depletion of CO_2 in the blood, breathing temporarily ceases, and a person may faint. This is what happens when we hyperventilate—taking rapid deep breaths.

Although the CO_2 concentration in the air we breathe is low, small changes can trigger a tremendous increase in breathing rate. For example, an increase to 5% CO_2 in the air causes us to breath 10 times faster.

Oxygen level is actually less important than CO_2 level in regulating breathing. In fact, oxygen level does not regulate breathing rate unless it falls dangerously low. This does not normally happen because the great amount of O_2 bound to hemoglobin provides a large safety margin. Activation of the inspiratory center due to low oxygen levels, detected by chemoreceptors in certain large arteries, is a last-ditch effort to increase breathing rate and restore normal oxygen levels. Sudden infant death syndrome (SIDS), in which a baby dies while asleep, may be caused by failure of chemoreceptors in major arteries and/or the brain's respiratory centers to respond to low oxygen levels in arterial blood. This cessation of breathing during sleep is called sleep apnea. In adults, it is a cause of snoring.

Breathing at High Altitudes

The human respiratory system functions best at the air density and 21% oxygen concentration near sea level. At high elevations, the air density and oxygen availability fall gradually, so that at an elevation of almost 10,000 feet (about 3,000 meters) above sea level, an individual inhales a third less oxygen with each breath. Each year, 100,000 mountain climbers and high-altitude exercisers experience varying degrees of altitude sickness. More than 150 climbers have died attempting to reach Mt. Everest's 29,028-foot summit, many from the effects of low oxygen.

Altitude sickness occurs when the body's effort to get more oxygen—by increasing breathing and heart rate and stepping up production of red blood cells and hemoglobin—cannot keep pace with declining oxygen concentration. Altitude sickness can

TABLE 36.1 The Effects of High Altitude

CONDITION	ALTITUDE	SYMPTOMS
Acute mountain sickness	5,900 feet (1,800 m)	Headache, weakness, nausea, poor sleep, shortness of breath
High-altitude pulmonary edema	9,000 feet (2,700 m)	Severe shortness of breath, cough, gurgle in chest, stupor, weakness; person can drown in accumulated fluid in lungs
High-altitude cerebral edema	13,000 feet (4,000 m)	Brain swells, causing severe headache, vomiting, altered mental status, loss of coordination, hallucinations, coma, and death

be cured by descending slowly at the first appearance of symptoms, described in table 36.1. A study using brain imaging (magnetic resonance imaging, or MRI) on nine people rescued from mountaintops with the last-stage, high-altitude cerebral edema, found that this condition causes swelling of the white matter in the brain, and that it is reversible, after a few weeks. The study also revealed that the swelling is caused by leakage of fluid through the blood-brain barrier.

36.4 MASTERING CONCEPTS

1. What parts of the brain control breathing?
2. How does strenuous activity alter the respiratory cycle?
3. How is blood CO_2 level detected and responded to?
4. What are the effects of altitude on the respiratory system?

OLC

Chapter Summary

36.1 Respiration Exchanges Gases

1. In **external respiration,** animals exchange oxygen and carbon dioxide between respiratory surfaces and blood. In **internal respiration,** these gases are exchanged between blood and cells. Cellular respiration uses oxygen to store nutrient energy in ATP.
2. CO_2 forms as a by-product of aerobic ATP production and must be eliminated from the body. Gas exchange brings oxygen to and removes CO_2 from cells.
3. In external respiration, oxygen and CO_2 are exchanged by diffusion across a moist membrane. Body size, metabolic requirements, and habitat have affected the evolution of respiratory systems.
4. Simple organisms exchange oxygen and CO_2 directly across the body surface. Larger or more active animals have circulatory systems to transport gases between cells and the environment.
5. Terrestrial arthropods bring the environment into contact with almost every cell through a highly branched system of **tracheae.**
6. Complex aquatic animals exchange gases across **gill** membranes, body surface extensions. In bony fishes, water flows over the gills in the direction opposite blood flow.
7. Vertebrate **lungs** create an extensive, moist internal surface for two-way gas exchange.

36.2 Components of the Respiratory System in Humans

8. In humans, the nose purifies, warms, and moisturizes inhaled air. The air then flows through the **pharynx, larynx,** and **trachea,** which is held open by cartilage rings. The trachea divides to form **bronchi** that deliver air to the lungs. The bronchi branch extensively to form tinier air tubules, **bronchioles,** which end in clusters of thin-walled, saclike **alveoli.**
9. Many capillaries surround the alveoli. Oxygen diffuses into the blood from the alveolar air, while CO_2 diffuses from the blood into the alveolar air. Human lung surfactant, a chemical mixture that reduces surface tension, helps keep alveoli open.

36.3 Functions of the Respiratory System in Humans

10. Breathing brings oxygen-rich air into the lungs and removes air high in CO_2. When the **diaphragm** and rib cage muscles contract, the thoracic cavity expands. This reduces air pressure in the lungs, drawing air in. Expiration results when these muscles relax and the thoracic cavity shrinks, raising air pressure in the lungs and pushing out air.
11. The volume of air moved in and out of the lungs during a respiratory cycle is the **tidal volume.** The amount of air that can be exhaled after a maximal inspiration is the **vital capacity.** Residual air remains in the lungs after expiration.
12. Almost all oxygen transported to cells is bound to hemoglobin in red blood cells. Increase in blood acidity (usually due to a rise in CO_2 level) or elevated temperature due to metabolism increases the amount of oxygen reaching cells.
13. Some CO_2 in the blood is carried bound to hemoglobin or dissolved in plasma. Most CO_2 is transported as bicarbonate ion, generated from carbonic acid that forms when CO_2 reacts with water and ionizes. Carbonic anhydrase speeds this reaction. The process reverses in the lungs, releasing CO_2 for expiration.

36.4 Control of Respiration

14. During quiet breathing, the inspiratory center within the medulla's rhythmicity center spontaneously generates impulses that trigger inhalation. During heavy breathing, the inspiratory center remains active and the expiratory center becomes active, causing active expiration. Other brain regions can alter the basic breathing pattern.

15. Breathing rate adjusts to the body's demands chemically. When CO_2 levels rise, chemoreceptors in the medulla, aorta, and carotid arteries sense the rise in CO_2 concentration and blood acidity and increase breathing rate. Oxygen levels do not change breathing rate unless levels are critically low.

Testing Your Knowledge

1. What are the two major functions of external respiration?
2. What is the connection between external respiration and aerobic, cellular respiration?
3. How does an animal's size, activity level, and environment influence the structure and function of its respiratory system?
4. Trace the path of an oxygen molecule from a person's nose to a red blood cell.
5. To what structures in the human respiratory system do an insect's trachea and trachioles correspond?
6. How is air cleaned before it reaches the lungs?
7. Describe inspiration. Which muscles are active in the process?
8. How does the body transport oxygen to cells?
9. What mechanisms direct more oxygen to metabolically active cells?
10. How does the respiratory system transport most CO_2? In what other ways is CO_2 transported?
11. How does the brain establish breathing rhythm?
12. What chemical change in the blood is most important in altering breathing rate?

Thinking Scientifically

1. It is below 0°F outside, but the dedicated runner bundles up and hits the roads anyway. "You're crazy," shouts a neighbor. "Your lungs will freeze." Why is the well-meaning neighbor wrong?
2. Why does breathing through the mouth instead of the nose dry the throat out?
3. What color might the insides of prehistoric human lungs have been? Why?
4. On December 13, 1799, George Washington spent the day walking on his estate in a freezing rain. The next day, he had trouble breathing and swallowing. Several doctors were called in. One suggested a tracheostomy, or cutting a hole in the president's throat so he could breathe. He was voted down. The other physicians suggested bleeding the patient, plastering his throat with bran and honey, and placing beetles on his legs to produce blisters. Within a few hours, Washington's voice became muffled, his breathing more labored, and he grew restless. For a short time he seemed euphoric; then he died. Washington had epiglottitis, a swelling of the epiglottis to 10 times its normal size. How does this diagnosis explain Washington's symptoms? Which suggested treatment might have worked?
5. Why can't you commit suicide by holding your breath?
6. Why do you think the times of endurance events at the 1968 Olympics, held in 7,218-feet elevation (2,200-meter) Mexico City, were rather slow?
7. Why might it be dangerous for a heavy smoker to use a cough suppressant?
8. The most successful treatment for lung cancer is surgery performed when a tumor is small and confined to one lung. The entire lung or the lobe containing the tumor may be removed. Many physicians, however, will not attempt such surgery on a patient who is a heavy smoker because of a second illness. What might this illness be?

References and Resources

Hackett, Peter H. December 9, 1998. High-altitude cerebral edema evaluated with magnetic resonance imaging. *The Journal of the American Medical Association,* vol. 280, p. 1920. MRI can detect brain changes characteristic of altitude-induced cerebral edema, providing a diagnostic test and revealing the source of swelling.

Marshall, Eliot. July 9, 1999. Beryllium screening raises ethical issues. *Science,* vol. 285, p. 178. Not everyone develops lung disease from exposure to beryllium.

Mill, Peter J. January 8, 1998. Caterpillars have lungs. *Nature,* vol. 391, p. 130. An insect's tracheal system interacts with its circulatory system.

Ridley, Matt. March 2000. Asthma, environment, and the genome. *Natural History,* vol. 109, p. 54. The increasing prevalence of asthma may reflect malfunctioning immunity.

The *LIFE* Online Learning Center provides additional resources and tools for studying this chapter.
www.mhhe.com/life

CHAPTER 37

Digestion and Nutrition

Geophagy

Why Do Animals Eat Soil?

Why do animals as diverse as rabbits, butterflies, reptiles, birds, and pregnant women eat soil? Biologists and anthropologists have noted the practice of geophagy—eating soil—for centuries in many animal species, and in all human societies. Puzzled biologists have hypothesized several explanations for geophagy, but a study of the dirty dietary preferences of blue-headed parrots (*Pionus menstruus*) indicates a reason no one had thought of—minerals in soil deactivate toxins in their other food, which consists of seeds and unripened fruits (see chapter opening figure). The parrots live in the tropical rain forest of Peru, and eat, every morning at the same time, claylike soil from one particular layer of sediment along the banks of the Manu River. They will not consume soil from other layers.

Biologists and anthropologists have noted the practice of geophagy—eating soil—for centuries in many animal species, and in all human societies.

In humans, compulsive eating of nonnutritive substances is called pica, which is Latin for the magpie, a bird that eats almost anything. Pica has variants, defined by the preferred odd food—soil and clay, but also laundry starch, matches, paint, ice, gravel, and hair. In central Georgia, for example, some people regularly consume kaolin, a smooth white clay abundant in the area that is used in papermaking, paint products, fiberglass, ceramics, and over-the-counter stomachache treatments. Grocery stores sell kaolin, and people can dig it up in their backyards. One woman, who consumes three bags of kaolin a week, was introduced to it in childhood, when her perpetually pregnant mother asked her to borrow some from neighbors. The woman tasted it as a child, and has craved it ever since.

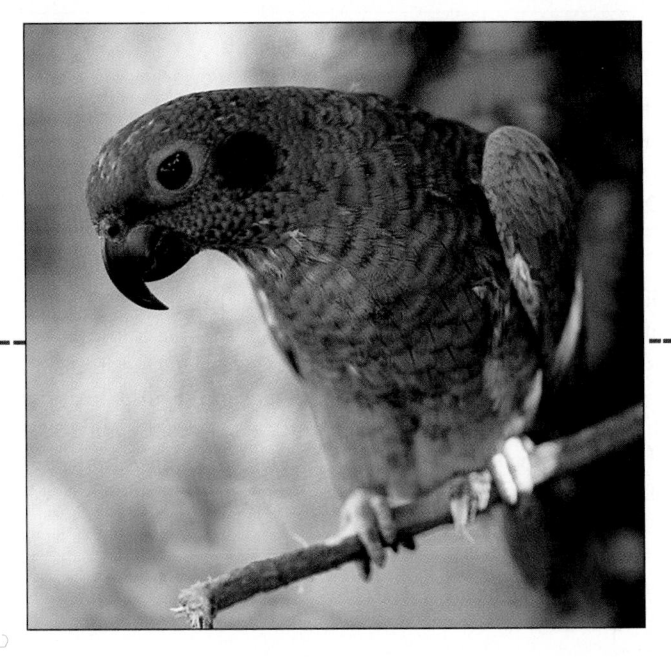

Soil-Eating Birds.
The blue-headed parrot, *Pionus menstruus,* is one of many animal species that eats soil—for a good (adaptive) reason.

Because some people with pica have iron or zinc deficiencies, and because it is most common among pregnant women, who are more likely to have such deficiencies, the hypothesis arose that such behavior in any animal species reflects a diet lacking one or more minerals. But according to the parrots, this may not be the case—the clay that the birds eat isn't especially mineral-laden.

A second hypothesis for the reason to eat soil, based on birds, is that they do so to provide particles for their gizzards, which are muscular parts of the digestive tract that grind food. But the parrots grind their food with their beaks, and their soil food is actually a clay, with particles too fine to be useful in grinding.

Yet a third explanation for the fondness for soil as cuisine, based on the practice among bovines, is that it neutralizes the acidic contents of the rumen, one of a cow's stomachs. Parrots, however, lack rumens. They can't be eating clay for this reason.

Finally, researchers looked more closely at the parrots' diets. They found that the seeds and unripened fruits that the parrots eat contain high levels of alkaloids, chemicals that give them a very bitter taste, not to mention being poisonous. When the researchers fed seed extracts to brine shrimp as a measure of toxicity, most of the shrimp died. But when they mixed the claylike soil in with the extracts, survival rate improved. In a second set of experiments, the researchers fed parrots a known alkaloid, with or without soil. Eating soil along with the alkaloid significantly decreased intestinal absorption of the toxin.

The clay-eating parrots have found a way to undermine the plant's strategy for protecting its immature offspring.

These experiments led the researchers to hypothesize further that the mechanism of soil-mediated detoxification is based on chemical attraction. Clay particles bearing negative charges bind the positively charged alkaloids in the parrots' stomachs. The clay then passes out of the digestive tract, along with its toxic cargo.

The clay-eating parrots have found a way to undermine the plant's strategy for protecting its immature offspring. Seeds are bitter to make animals spit them out or swallow them whole, dispersing them so that they can germinate. Natural selection has similarly made unripened fruits bitter so that animals won't eat them before their seeds are mature. An animal that can eat seeds and unripened fruits that others cannot tolerate, just by nibbling some clay, would have more dietary choices—and therefore a better chance of survival.

37.1 The Art and Evolution of Eating

Digestion mechanically and chemically breaks down nutrient molecules, releasing the energy in the chemical bonds. Different species have different ways to obtain nutrients.

Animals eat to obtain energy. Although the diets of humans and squids, dragonflies, and jellyfishes are certainly different, animals break their varied foods down into the same types of nutrients. Then, the pathways of cellular metabolism extract the energy and use it to produce ATP. Digestive systems accomplish the task of breaking down foods to their constituent nutrients. An American robin's taking a meal illustrates the steps of the process.

The robin (fig. 37.1A) is an animal whose tastes vary with the season. It eats mostly insects, spiders, and worms in the fall, winter, and spring but relies more on fruits and berries in the summer, which makes it a "seasonal frugivore" (table 37.1). The robin's digestive system, like that of many birds, is highly adapted to its eclectic diet. Its digestive biochemicals can break down the protein and fat that make up most of its animal-based meals, yet its stomach is specialized to rapidly digest fruits. In the stomach, large seeds are separated from a fruit meal and regurgitated; the pulp is then digested quickly. The intestine absorbs the nutrients, and the fruit's peel and remaining smaller seeds move on to become a bird dropping.

TABLE 37.1 Types of Feeders

TERM	TYPE OF FOOD
Carnivore	Animals
Frugivore	Fruits and berries
Herbivore	Plants
Insectivore	Insects
Omnivore	Plants and animals
Detritivore	Nonliving organic matter

FIGURE 37.1
Feeding Diversity.
(**A**) A robin's digestive system, including its beak, is adapted to eating fruits and berries, as well as insects, spiders, and worms. It is an omnivore. (**B**) A baleen whale is a filter feeder. It can swim through a school of krill and filter them from 60 tons of water in a few seconds, thanks to a curtain of baleen plates. (**C**) The giant panda has a very specific diet, consuming enormous quantities of bamboo. It is an herbivore.

A

B

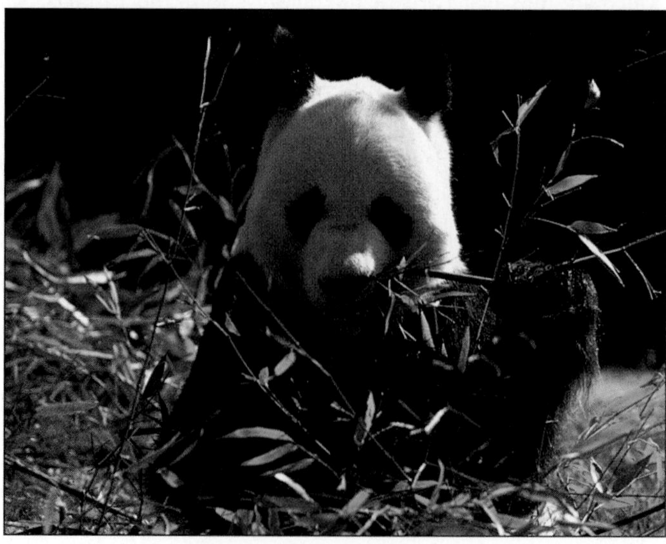

C

For the robin to use the proteins and fats in its worm meal, or the sugars in fruits, its body must dismantle these nutrient molecules into smaller molecules—proteins to amino acids, complex carbohydrates (starches) to simple carbohydrates (sugars), and fats to fatty acids and monoglycerides. Only such small molecules can enter the cells lining the digestive tract and then leave them to enter the circulation. Once the bloodstream delivers nutrients to cells, nutrient molecules are broken down, releasing the energy in their chemical bonds. Nonnutritive parts of food are eliminated from the body as feces.

Animals eat to replace energy they expend in the activities of life and to provide raw material for the structural components required for growth, repair, and maintenance of the body. This process has several major steps. Ingestion is the taking in of food. The next stage, mechanical breakdown, cuts food into smaller pieces, usually by chewing (masticating). Mechanical breakdown exposes sufficient surface area for various digestive enzymes to contact. **Digestion,** the next step, is the chemical breakdown of the food particles. Hydrolytic enzymes add water molecules between the building blocks of large nutrient molecules, splitting them apart. Released nutrients are then absorbed into the bloodstream, typically through the lining of a hollow organ specialized for this function, such as an intestine. Finally, undigested material is eliminated (egested) from the animal's body as feces.

Animals obtain food in diverse ways. Recall from chapter 25 that a hagfish spikes its worm food using tentacles on its mouth, and some lampreys use their oral discs to suck onto fish food. Many animals use teeth or other chewing mouthparts to grab food, and some, such as snakes and lizards, swallow mammal meals whole, as the opening essay to chapter 5 vividly shows. Some animals drink their food—a mosquito alights on human skin to take a "blood meal." Other organisms digest their food alive, as a spider consumes an insect, or a starfish eats a clam.

Several types of aquatic animals are **filter feeders.** They use ciliated tissue surfaces to create waves that usher food particles into their mouths. Filter feeding is particularly adaptive in animals that do not move about much, such as corals, because it brings the food to them, although the menu is limited to organisms that live in the vicinity. Bivalve mollusks such as clams and mussels sweep food towards their mouths using cilia on the feathery surfaces of their gills. Tunicates and lancelets capture particles of food suspended in water in the sticky mucus that lines the pharynx. Cilia move the food to the digestive organs.

Motile filter feeders range from tiny crustaceans to the baleen whale. This behemoth opens its huge mouth and gulps in a few thousand fishes and crustaceans as it moves forward. Smaller dietary fare, such as the plentiful, tiny shrimplike animals called krill, adhere to some 300 horny baleen plates that dangle from the roof of the whale's mouth like a shredded shower curtain (fig. 37.1B). The whale uses its tongue to lick the food off the baleen and then swallows it.

A terrestrial counterpart to the filter feeder is the **deposit feeder.** This type of feeder strains food from soil or other sediments. The earthworm is a familiar deposit feeder.

Digestion breaks food down into nutrients. A nutrient is a food substance used in an organism to promote growth, maintenance, and repair of its tissues. Recall that obtaining nutrients from the environment, as opposed to synthesizing them, is termed heterotrophy.

The earliest heterotrophs probably engulfed organic matter and used it to supply energy and building blocks for proteins and nucleic acids. The evolution of photosynthesis and chemosynthesis, which enabled organisms to manufacture useful biochemicals from solar or geothermal energy, had two important implications for early life-forms. First, some organisms could now synthesize the complex organic molecules they required, freeing them from depending on an uncertain and possibly dwindling food supply. Second, the nutrient molecules obtained from photosynthesis and chemosynthesis, called primary production, eventually built up a reservoir of food that would make heterotrophy a sustainable lifestyle over the long term for many types of organisms.

Heterotrophy may have evolved because many organisms never acquired the ability to synthesize their own biochemicals. In terms of natural selection, heterotrophy is advantageous if it takes less energy to obtain a nutrient from the environment than it does to synthesize it. This may be especially true for a heterotroph that has a nearly constant supply of nutrients, such as an animal that has an abundance of a particular amino acid in its diet. Over time, the species could lose the ability to synthesize this amino acid with no harmful consequence, and divert the energy required for its synthesis to other functions.

Over hundreds of millions of years, considering the great diversity of life, it is not surprising that many animals have become very specialized feeders, relying on one or a few kinds of food to supply nutrients. Consider the giant panda (fig. 37.1C). It has a short digestive tract characteristic of its meat-eating ancestors, with less capacity to absorb nutrients. The panda lacks cellulase to break down bulky plant matter and also lacks an organ to store food. Thus, the panda must eat almost constantly. Even though it consumes 6% of its body weight in bamboo each day (compared with 2% for most plant-eating mammals), it just barely meets its basic nutritional requirements. Animals with more varied diets, such as the robin and human, can live in more widespread habitats.

37.1 MASTERING CONCEPTS

1. Why must digestion break down nutrient molecules?
2. Why do animals eat?
3. How can heterotrophy be adaptive?

37.2 Digestive Diversity

Some protista engulf food and break it down intracellularly. Animals digest nutrients in extracellular compartments. Simpler digestive systems have one opening, whereas more complex systems have two openings. Digestive systems are adapted to the frequency and composition of a species' meals.

Animals' varied ways of obtaining and digesting food reflect specializations for their nutrient requirements, their habitats, the kinds of food available, and each species' evolutionary background. Even the animals in a particular taxonomic group may have diverse feeding habits and digestive adaptations.

Types of Digestion

Because animals and nutrient molecules are composed of the same types of chemicals, digestive enzymes could just as easily attack an animal's body as its food. To prevent this, digestion occurs within specialized compartments. Even in protista such as a paramecium, digestion is separated from other functions. These organisms take in dissolved nutrients by endocytosis and envelop the food in a food vacuole. Some protista have specialized mouthlike openings that food passes through en route to food vacuoles. Once loaded, the food vacuole fuses with another sac containing digestive enzymes that break down large nutrient molecules. Digested nutrients exit the food vacuole and are used in other parts of the cell. The cell extrudes waste. Digestion within a cell's food vacuoles is **intracellular digestion** (fig. 37.2).

Sponges are the only types of animals that rely solely on intracellular digestion. The cells that form the body wall, the choanocytes, capture food particles with their flagella and pass them to amoebocytes, which digest them in food vacuoles, then store the nutrients or pass them to other cells (see fig. 24.8). It isn't surprising that these simplest of animals have protistlike food vacuoles for intracellular digestion, because sponges likely descended from colonial choanoflagellates (see fig. 24.1).

Intracellular digestion can handle only very small food particles. In **extracellular digestion,** hydrolytic enzymes in a cavity or system of tubes connected with the outside world (the vertebrate stomach, for example) dismantle larger food particles outside the cells that will use them. When nutrient molecules are small enough, they enter cells lining the cavity, where chemical breakdown by intracellular digestion may continue. Extracellular digestion eases waste removal and is more efficient and specialized than intracellular digestion. The cavity in which extracellular digestion occurs, along with accessory organs, together constitute a **digestive system.** (Recall that fungi have truly extracellular digestion—it occurs outside their bodies. Fungi secrete enzymes that digest organic molecules in the environment, then take up the liberated nutrients.)

Digestive systems may have one or two openings (fig. 37.3). Some animals, such as hydra and flatworms, have digestive systems with a single opening. Indigestible food exits the digestive system through the same opening it entered. Consequently, food must be digested and the residue expelled before the next meal can begin. This two-way traffic makes any specialized compartments for storing, digesting, or absorbing nutrients impossible. In these organisms, the digestive cavity doubles as a circulatory system, which distributes the products of digestion to the body cells for use. For this reason, the digestive cavity is also called a **gastrovascular cavity.**

The digestive systems of many familiar animals, such as segmented worms, mollusks, insects, starfishes, and vertebrates, have two openings, a separate entrance and exit in a tubelike structure that allows one-way traffic. The roundworms are the simplest animals to have digestive systems with two openings. Hydrolytic enzymes are released into the tube, and the products of digestion are absorbed and delivered to cells via a circulatory system.

FIGURE 37.2
Intracellular and Extracellular Digestion.
(**A**) Intracellular digestion is an ancient form of obtaining nutrients that sponges and single-celled organisms such as the paramecium use as their sole means of digestion. Food is digested when the food vacuole that contains it fuses with an intracellular sac containing digestive enzymes. (**B**) In most multicellular animals such as the hydra, food is broken down in a cavity by extracellular digestion. The cells lining the cavity absorb the nutrients and continue the breakdown process by intracellular digestion.

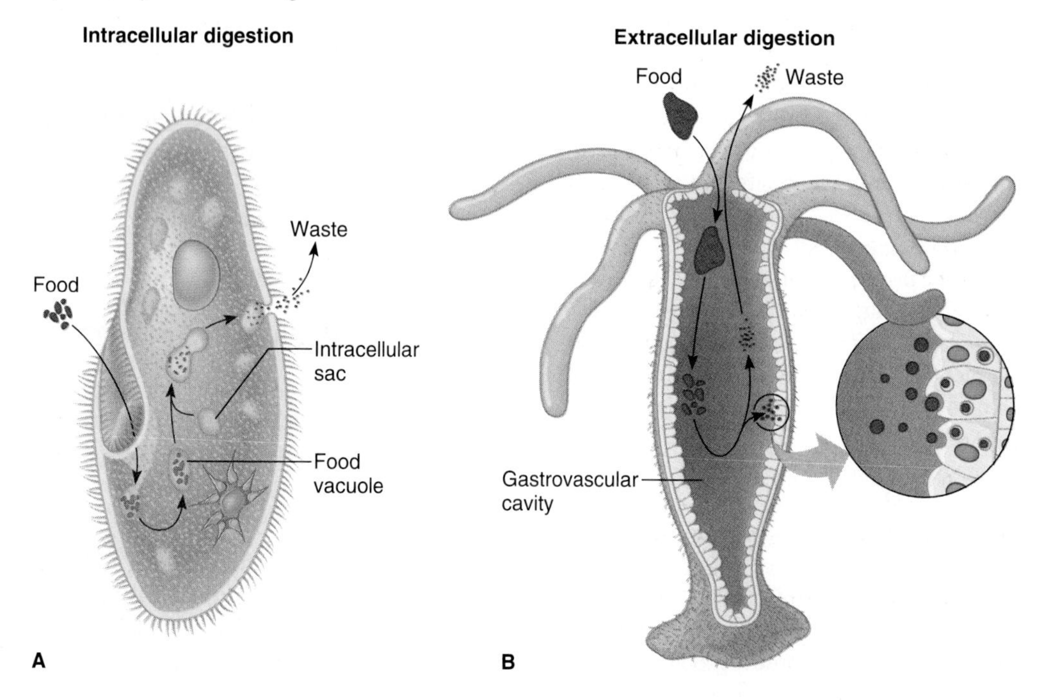

FIGURE 37.3

One-Opening and Two-Opening Digestive Systems.

In a digestive system with a single opening, such as in the gastrovascular cavity of a flatworm (**A**), food and wastes mix. Cells lining the digestive cavity absorb partially digested nutrients. Further digestion occurs in these cells. (**B**) A mosquito's digestive system is open at the mouth and anus. Digestive systems with two openings are often specialized into compartments.

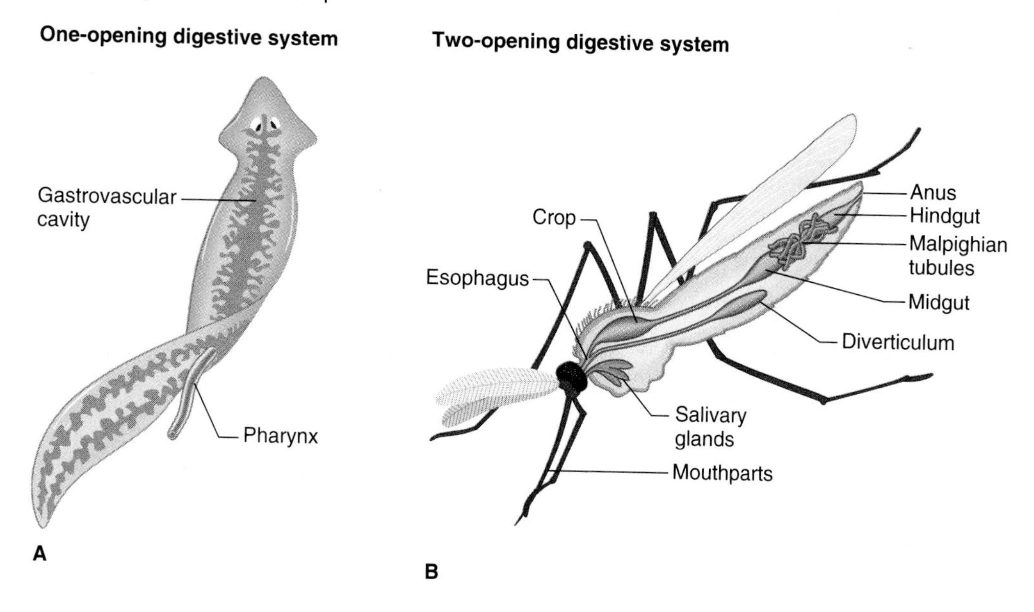

Indigestible food remains in the tube and leaves the body as part of the feces, from an **anus.** A major advantage of a two-opening digestive system is that regions of the tube can become specialized for different functions: breaking food into smaller particles, storage, chemical digestion, and absorption.

Digestive System Adaptations

The vertebrates exhibit great digestive diversity. Birds metabolize nutrients very rapidly, and they must eat continually and digest food quickly to fuel their lightweight bodies. Their mouths have adapted to their diets. The stork's bill easily scoops fish; the strong, short beak of the finch cracks and removes hulls from seeds; the vulture's hooked beak rips carrion into manageable chunks. A bird's esophagus has an enlargement called a **crop,** which temporarily stores food. After the crop, food moves to the proventriculus, a glandular stomach that secretes gastric juice that chemically digests food. Next, food moves to the **gizzard,** a muscular organ lined with ridges. It mechanically digests food with the aid of sand and small pebbles the bird swallows. Nutrients are absorbed in the intestines. The digestive system ends at the **cloaca,** an opening common to the digestive, reproductive, and excretory systems (fig. 37.4).

The intestines of some types of birds show a remarkable adaptation—they enlarge when the animal eats huge amounts of food, such as just before a long migration. In the wren, rufous-sided towhee, rock ptarmigan, and spruce grouse, the intestine lengthens up to 22%. This adaptation enables birds to store more food than usual. Because it takes longer for food to travel through the lengthened intestine, more nutrients are absorbed.

FIGURE 37.4

Digestion in a Bird.

Food is stored temporarily in the crop, then passed through the proventriculus and the gizzard. Absorption occurs at the intestine, and waste exits through the cloaca.

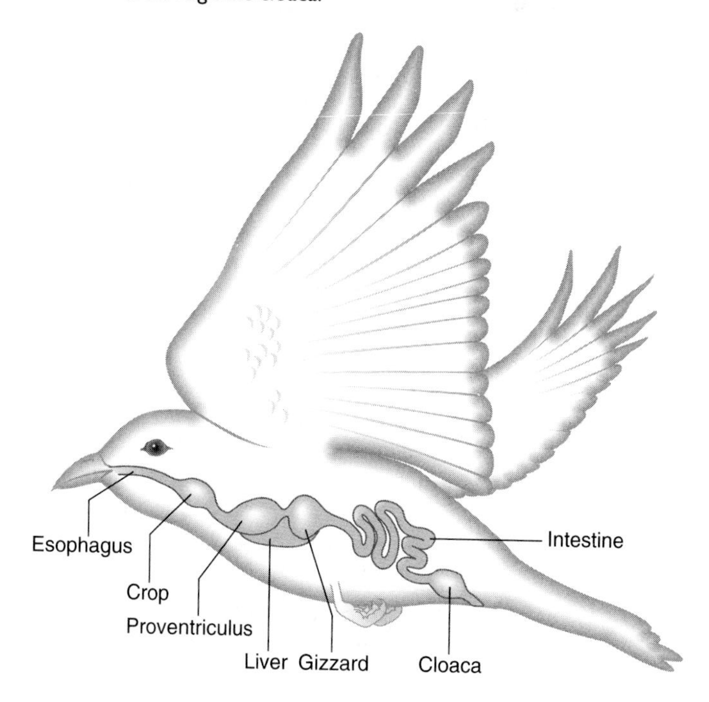

The digestive systems of mammals are adapted to their diets (fig. 37.5). Many mammals have digestive system specializations that enable them to break down the cellulose in the cell walls of plant foods. The cow has a four-sectioned stomach. Grass enters the first section, the **rumen,** where bacteria break it down into balls of cud. The cow regurgitates the cud back up to its mouth, where it chews to mechanically break the food down further. The cow's teeth can mash the cud forward and backward, up and down, and sideways. After chewing the cud, the cow swallows again, but this time the food bypasses the rumen, continuing digestion in the other three sections of the stomach. Sheep, deer, goats, antelope, buffalo, and giraffes also

FIGURE 37.5

Digestive Systems Are Adapted to an Animal's Diet.

(**A**) Small mammals that are insectivores have short, simple digestive systems, because they consume little plant material that would require fermentation. (**B**) The intestines of nonruminant herbivores such as rabbits and rodents harbor anaerobic microorganisms that break down cellulose in plant foods. (**C**) Ruminant herbivores have a characteristic four-part stomach and an extensive gastrointestinal tract. (**D**) The mostly protein diet of carnivores is easier to digest than plants, so the digestive system is much shorter, with a reduced cecum.

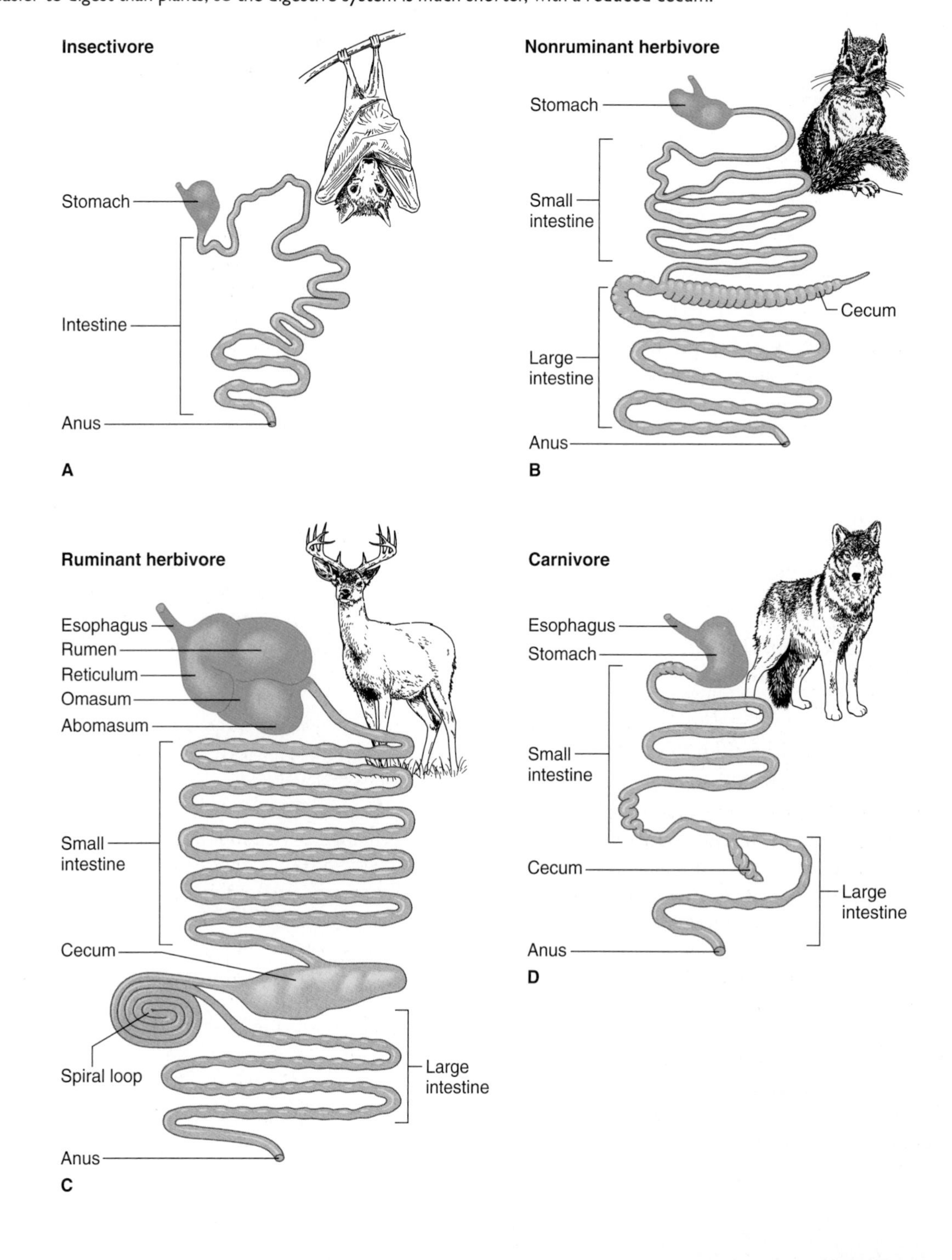

have quadruple stomachs. These animals with large rumens are called **ruminants.**

Elephants have particular difficulty digesting their meals of leaves and twigs. Because they have only tusks and molars, a lot of wood ends up in their massive stomachs. It remains there for over 2 days, churning about amidst digestive juices and cellulose-degrading bacteria. Still, elephant dung contains many undigested twigs. An elephant eats 300 to 400 pounds (135 to 150 kilograms) of food each day.

Like cows and elephants, rabbits have their own way of getting the most out of a meal. The leaves that rabbits eat spend time in their stomachs. Instead of regurgitating the partially digested plant matter, rabbits egest it in moist pellets. Then they eat the moist pellets, which continue digestion in the stomach and intestines. This time, the rabbits egest dry pellets, which they do not eat. The gastrointestinal tracts of rabbits and several other types of mammals have a side compartment, called a **cecum,** where plant matter is fermented and absorbed. Insectivores, like the shrew, and carnivores, like the fox, have mostly protein diets. The digestive tracts of these animals are adapted to this diet by having short intestines and a small or absent cecum (see fig. 37.5).

37.2 MASTERING CONCEPTS

1. Distinguish between intracellular and extracellular digestion.
2. What is the limitation of a one-way digestive system?
3. What are some diverse ways that animals obtain and digest nutrients?

OLC

37.3 The Digestive System in Humans

Salivary amylase begins chemical digestion as the teeth mechanically break down food. The esophagus conducts chewed food to the stomach, propelled by peristalsis, where it liquefies under the action of churning and pepsin. Nutrients are absorbed through small intestinal villi. Digestion is completed in the large intestine, and waste leaves through the rectum and anus. The pancreas secretes digestive enzymes and the liver produces fat-emulsifying bile, which the gallbladder stores and secretes.

The human digestive system consists of a continuous tube called the **gastrointestinal tract** (including the mouth, pharynx, esophagus, stomach, small intestine, and large intestine, or colon) and accessory structures (fig. 37.6). (The gastrointestinal tract is also known as the alimentary canal.) The accessory structures aid food breakdown either mechanically (for example, the teeth and tongue) or chemically by secreting digestive enzymes (the salivary glands and pancreas). Other accessory structures, the liver and gallbladder, produce and store bile, which assists fat digestion. The mesentery, an epithelial (lining) sheet reinforced by connective tissue, supports the digestive organs and glands.

The digestive lining begins at the mouth and ends at the anus. The moist innermost layer secretes mucus that lubricates the tube so that food slips through easily, while also protecting underlying cells from rough materials in food and from digestive enzymes. In some regions of the digestive system, cells in the innermost lining also secrete digestive enzymes. Beneath the lining a layer of connective tissue contains blood vessels and nerves. The blood nourishes the cells of the digestive system and in some regions picks up and transports digested nutrients.

The layers against the innermost layers are muscular. Except for the stomach, which has three layers, two layers produce the movements of the digestive tract (fig. 37.7). The muscles of the inner layer circle the tube, constricting it when they contract, and the muscles in the outer layer run lengthwise, shortening the tube when they contract. These contractions, which occur about three times per minute, churn food, mixing it with enzymes into a liquid. Waves of contraction called **peristalsis** propel food through the digestive tract. When food distends the walls of the tube, the circular muscles immediately behind it contract, squeezing the mass forward.

The Mouth and the Esophagus—Digestion Begins

The thought, smell, or taste of food triggers three pairs of salivary glands near the mouth to secrete saliva. Chemical digestion begins as **salivary amylase,** an enzyme in saliva, starts to break down starch into maltose. Salivary amylase causes a piece of bread in the mouth to taste sweet as the enzyme breaks down starch to sugar.

Mechanical breakdown begins as the teeth chew the food, a process called mastication. Water and mucus in saliva aid the teeth as they tear food into small pieces, thereby increasing the surface area available for chemical digestion. Researchers recently used a computer simulation of chewing to test the hypothesis that each food requires an optimum range of number of chews, so that the food is torn enough to form a smooth lump, called a **bolus,** yet not so softened that it sticks in the throat. Eating raw carrots, for example, requires 20 to 25 chews. Experiments with people confirmed the researchers' predictions.

Tooth structure and shape are adapted to particular functions. The thick **enamel** that covers a tooth is the hardest substance in the human body. Beneath the enamel is the bonelike **dentine,** and beneath that, the soft inner **pulp,** which contains connective tissue, blood vessels, and nerves (fig. 37.8). Two layers on the outside of the tooth, the periodontal ligament and the cementum, anchor the tooth to the gum and jawbone. The visible part of a tooth is the crown, and the part below the surface is the root. Calcium compounds harden teeth, as they do bones.

The tongue rolls chewed food into a bolus and pushes it to the back of the mouth for swallowing. The bolus passes first through the **pharynx** and then through the **esophagus,** a muscular tube leading to the stomach. During swallowing, the epiglottis covers the passageway to the lungs, routing food to the digestive tract (fig. 37.9).

Food does not merely slide down the esophagus due to gravity—contracting esophageal muscles push it along in a wave of peristalsis. This is why it is possible to swallow while standing

FIGURE 37.6

The Human Digestive System.

Food is broken down as it moves through the various chambers of the gastrointestinal tract. Accessory organs deliver digestive enzymes and other chemicals that assist the chemical digestion of food.

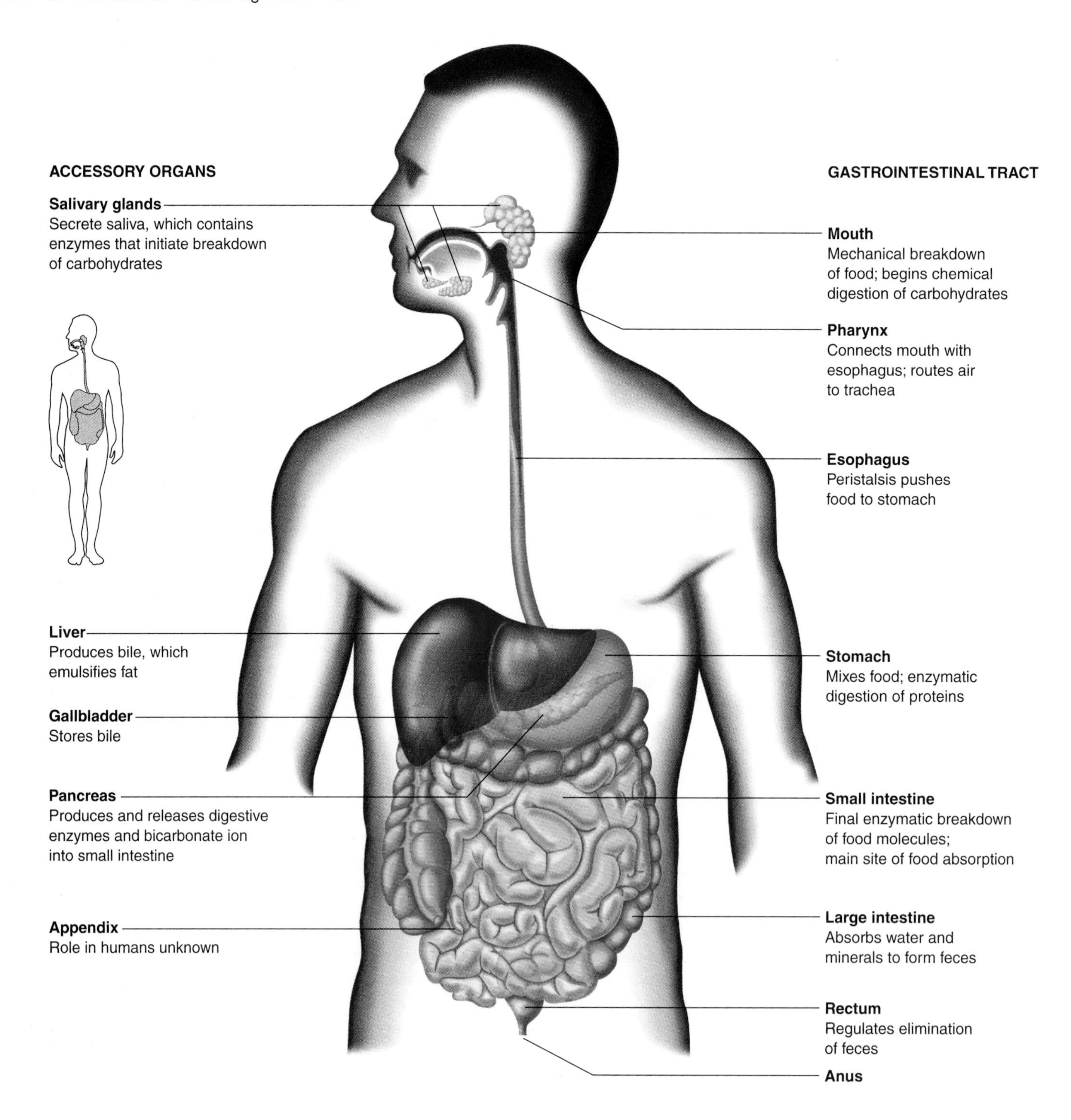

FIGURE 37.7

Muscles Move Food.

The two muscle layers of the digestive tract coordinate their contractions to move food in one direction. Note that the muscle layers are perpendicular to each other. (In the stomach a third layer lies obliquely.)

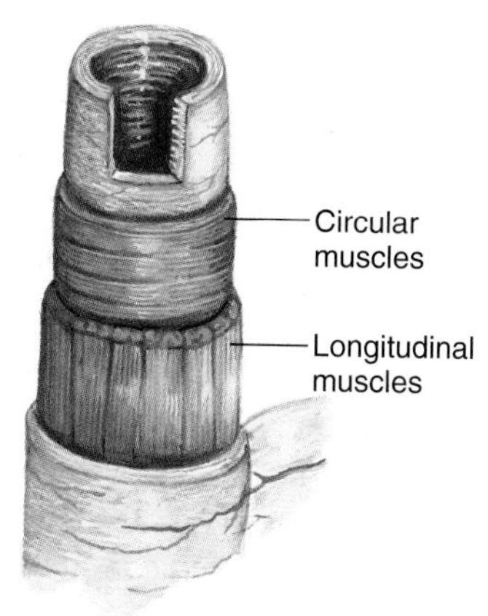

FIGURE 37.8

Anatomy of a Tooth.

For many animals, processing food begins with the teeth, which grasp, cut, tear, shred, crush, and grind food into pieces small enough to be swallowed.

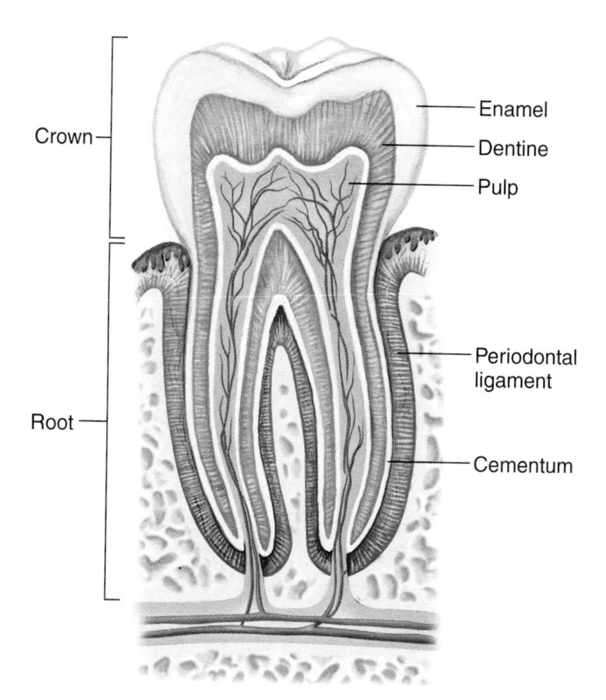

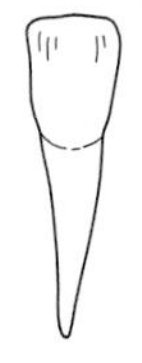

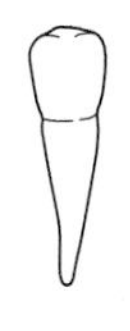

FIGURE 37.9

Separating Air from Food.

During breathing, the tongue and larynx are relaxed and the epiglottis is elevated. When a swallow initiates, the soft palate elevates and the tongue moves food to the rear of the mouth. During the swallowing reflex, the tongue and the larynx elevate and food is diverted down the esophagus by the epiglottis, which now blocks the larynx.

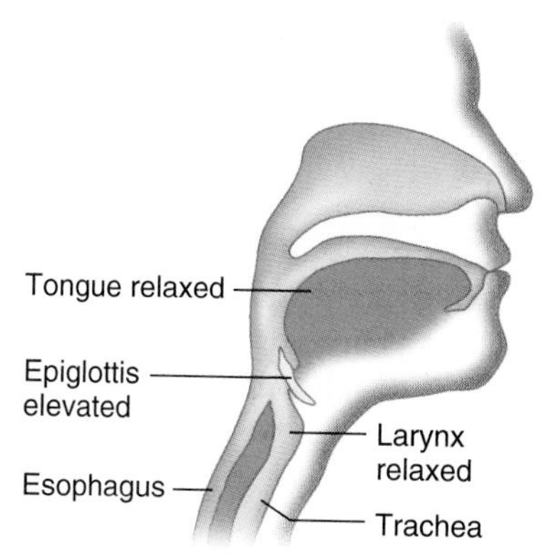

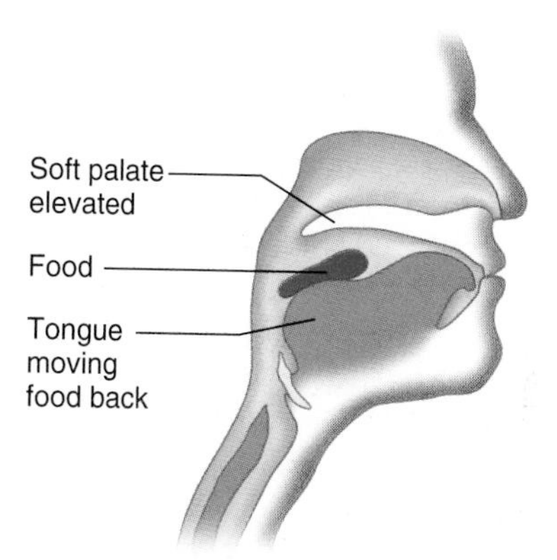

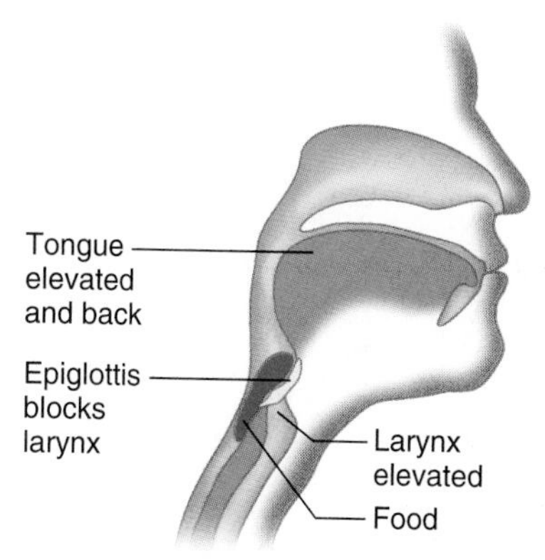

on your head. Muscle contractions continue, propelling the food down the esophagus toward the stomach, where the next stage in digestion takes place.

The Stomach Stores, Digests, and Pushes Food

From the esophagus, food progresses to the **stomach,** a J-shaped bag about 12 inches (30 centimeters) long and 6 inches (15 centimeters) wide. The stomach has three important functions: storage, some digestion, and pushing food into the small intestine.

The stomach is the size of a large sausage when empty, but it can expand to hold as much as 3 or 4 quarts (approximately 3 or 4 liters) of food. Folds in the stomach's mucosa, called **rugae,** can unfold like the pleats of an accordion to accommodate a large meal. Muscular rings called **sphincters** control entry to and exit from the stomach, pinching shut to contain the stomach's contents (fig. 37.10).

Mechanical breakdown and chemical digestion occur in the stomach. Waves of peristalsis push food against the stomach bottom, churning it backwards, breaking it into pieces, and mixing it with gastric juice to produce a semifluid mass called **chyme.** Fats usually remain in the stomach from 3 to 6 hours, proteins for up to 3 hours, and carbohydrates 1 to 2 hours.

Gastric juice is responsible for chemical digestion in the stomach. About 40 million cells lining the stomach's interior secrete 2 to 3 quarts (approximately 2 to 3 liters) of gastric juice per day. Gastric juice consists of water, mucus, salts, hydrochloric acid, and enzymes. The hydrochloric acid creates a highly acidic environment, which activates the enzyme **pepsin** from its precursor, pepsinogen. Pepsin breaks down proteins to yield polypeptides, a first step in protein digestion. Cells called chief cells secrete pepsinogen and parietal cells secrete hydrochloric acid from indentations in the stomach lining called gastric pits.

Nerves and hormones regulate gastric juice secretion. The thought or taste of food initiates nerve impulses that stimulate

FIGURE 37.10

The Stomach.

The stomach is a J-shaped bag that stores, mixes, and digests food until it is fluid enough to move on to the small intestine. (**A**) The stomach has four regions (fundus, cardia, body, and pylorus), two sphincters, and folds called rugae that increase its capacity. (**B**) The lining of the stomach contains gastric pits, where chief cells secrete pepsinogen and parietal cells secrete hydrochloric acid. Other lining cells secrete the abundant mucus that coats the stomach lining, preventing it from digesting itself. The three layers of muscle that make up the stomach wall, running at different angles, enable the organ to move food around, mechanically breaking it into smaller pieces.

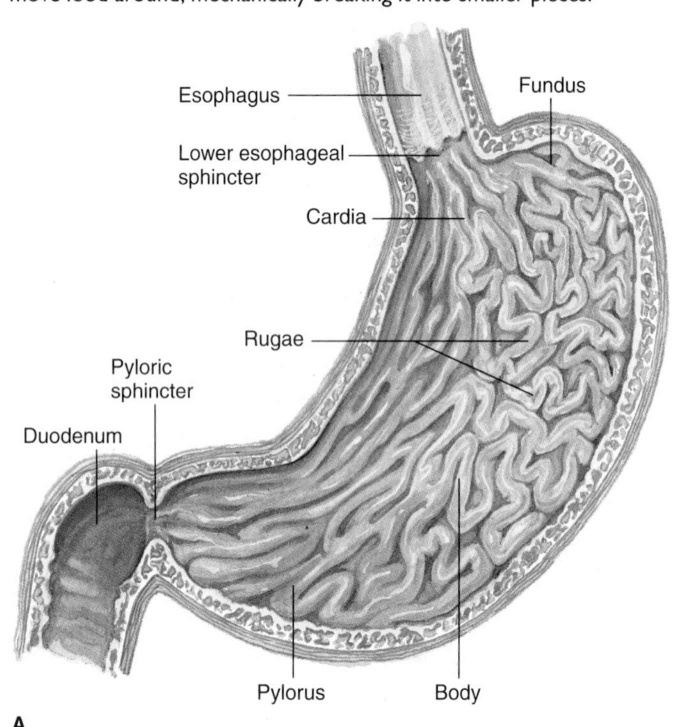

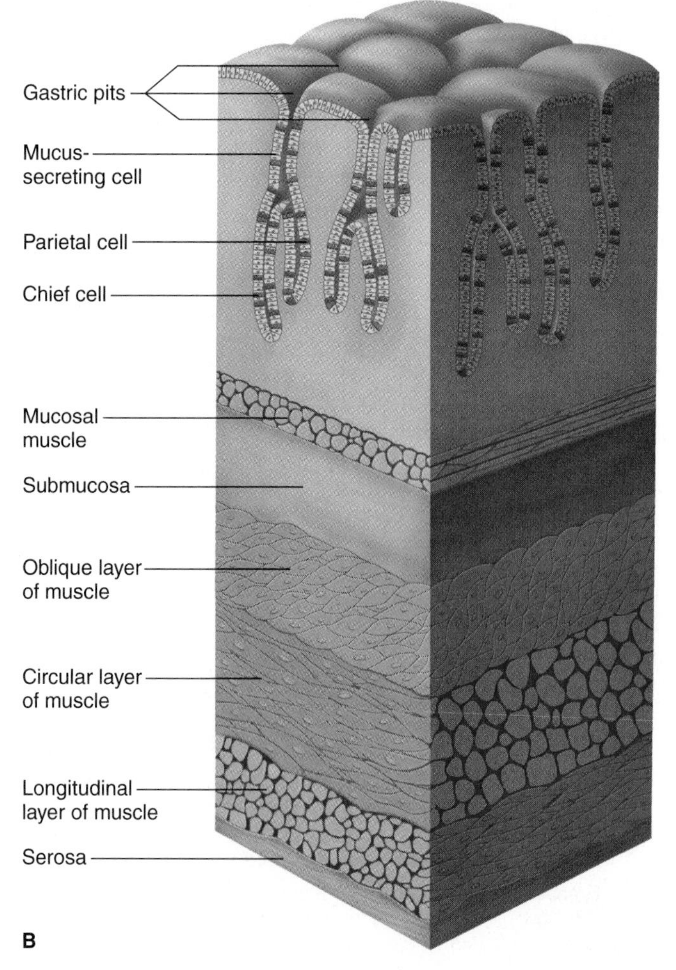

Health 37.1

A New View of Ulcers

Physicians and medical researchers are often hesitant to abandon a long-entrenched idea. So it was with the discovery that gastric ulcers, a common, painful digestive disorder that irritates and erodes the stomach wall, can result from bacterial infection.

Ulcers were thought to be the direct result of excess stomach acid secretion, dating from a German researcher's concise pronouncement in 1910, "No acid, no ulcer." The finding in the 1970s that drugs that block acid production relieve ulcers (albeit temporarily) seemed to confirm the hypothesis that excess acid causes ulcers. Acid buildup was attributed to stress and eating spicy foods. For decades, the standard treatment for ulcers in the stomach or small intestine was a bland diet, stress reduction, acid-blocking drugs, or surgery. Over the many years that ulcers typically persist, cost of this treatment mounts. Considering this background, the idea that an ulcer was actually a bacterial infection that could be easily, quickly, cheaply, and permanently cured seemed preposterous.

Helicobacter pylori—the ulcer bacterium—was identified in the laboratory of J. Robin Warren at Royal Perth Hospital in western Australia. Warren had found bacteria in stomach tissue samples from people suffering from gastritis, an inflammation of the stomach. He didn't know which way to phrase the hypothesis—were the bacteria attracted to inflamed tissue, or did they cause the inflammation?

Warren's assistant, medical resident Barry Marshall, helped choose between the hypotheses. Marshall knew he had a healthy stomach and had never had gastritis or an ulcer. If the bacteria caused the irritation, they would do so in him. So on a hot July day in 1984, Marshall concocted some "swamp water"—a brew of a billion or so of the microbes—and drank it.

Marshall got sick from drinking the bacterial brew, but he was lucky—he suffered for a few days from gastritis, which then cleared up without treatment. A second volunteer, however, was ill for several months. Although a dose of bismuth subcitrate (Pepto Bismol) gave temporary relief, the pain returned. Only when he took two antibiotic drugs for a few weeks did the gastritis vanish completely.

Other researchers, using laboratory animals rather than themselves, soon confirmed the causative link between *H. pylori* and gastritis and then a link to ulcers, too. Still, it took until 1994—a full decade after Marshall's self-experiment—for the National Institutes of Health to advise doctors that a 2-week course of two antibiotic drugs plus an antacid should become the standard treatment for gastritis and ulcers in patients with an infection.

By 1996, with a new subdiscipline in the biomedical sciences devoted to stomach bacteria and medical journals frequently reporting on *H. pylori's* activities in the human digestive tract, the idea that gastritis and ulcers are usually the consequence of bacterial infection had finally been accepted. Although the precise interactions between stress, diet, stomach acid production, and bacterial infection are still not fully understood, for many patients, our knowledge so far may mean that gastritis and ulcers are a pain of the past.

---

the stomach to secrete between 6 and 7 ounces (about 200 milliliters) of gastric juice. This prepares the stomach to receive food. When food arrives in the stomach, endocrine cells in the stomach's lining release a hormone, **gastrin,** that stimulates secretion of another 20 ounces (about 600 milliliters) of gastric juice.

The stomach usually doesn't digest the protein of its own cells along with the protein in food. The organ has built-in triple protection. First, the stomach secretes little gastric juice until food is present for it to work on. Secondly, some stomach cells secrete mucus, which coats and protects the stomach lining from the corrosive gastric juice. Finally, the stomach produces pepsin in an inactive form, pepsinogen, and cannot digest protein until hydrochloric acid is present. Health 37.1 describes the unusual research that led to a better understanding of gastric ulcers, a common condition in which the stomach's protection fails.

The stomach absorbs very few nutrients; most food has not yet been digested sufficiently. However, the stomach can absorb some water and salts (electrolytes), a few drugs (such as aspirin), and, like the rest of the digestive tract, alcohol. This is why aspirin can irritate the stomach lining and why we feel alcohol's intoxicating effects quickly.

After the appropriate amount of time, depending upon the composition of a meal, stretching of the stomach wall or release of gastrin triggers neural messages that relax the **pyloric sphincter** at the stomach's exit. The stomach then squirts chyme in small amounts through the sphincter into the small intestine. Nerve impulses from the first part of the small intestine then control the stomach's actions. When the upper part of the small intestine is full, receptor cells on the outside of the intestine are stimulated and the pyloric sphincter tightens. Once pressure on the small intestine lessens, the sphincter opens, the stomach contracts, and more chyme enters.

The Small Intestine Absorbs Nutrients

The **small intestine,** a 23-foot (approximately 7-meter) tubular organ, completes digestion and absorbs the resulting nutrients. The first 10 inches (25 centimeters) of the organ forms the **duodenum,** the next two-fifths the **jejunum,** and the remainder (almost three-fifths) the **ileum.** Localized muscle contractions carry out mechanical digestion in the small intestine in a process called **segmentation.** During this process, the chyme sloshes back and forth between segments of the small intestine that form when bands of circular muscle temporarily contract. Peristalsis moves the food lengthwise along the intestine.

Digestive secretions from the small intestine, the liver, and the pancreas contribute mucus, water, bile, and enzymes. Chemical digestion in the small intestine acts on all three major types of nutrient molecules. The small intestine continues the protein breakdown the stomach began. Trypsin and chymotrypsin from the pancreas break down polypeptides into peptides, and peptidases secreted by small intestinal cells break peptides into tripeptides, dipeptides, and single amino acids small enough to enter the lining cells. Figure 37.11 summarizes the sites of digestion of major nutrients, and the enzymes that dismantle them.

Fats present an interesting challenge to the digestive system. Lipases, the enzymes that chemically digest fats, are water soluble, but fats are not. Therefore, a lipase can only act at the surface of a fat droplet, where it contacts water. **Bile** from the liver emulsifies fats, breaking them into many droplets, which exposes greater surface area to lipase. Most fats in our diet are triglycerides, which are digested into fatty acids and monoglycerides.

Carbohydrate digestion also continues in the small intestine. The pancreas sends pancreatic amylase to the small intestine, where it acts on starches that salivary amylase missed in the mouth. In addition, the small intestine produces carbohydrases, which chemically break down certain disaccharides into monosaccharides. Sucrase, for example, is a carbohydrase that breaks down the disaccharide sucrose into the monosaccharides glucose and fructose.

Deficiency of a particular carbohydrase can cause digestive distress when a person eats certain foods. Many adults have difficulty digesting milk products because of lactose intolerance, which results from the absence of the enzyme lactase in the small intestine. This enzyme breaks down the disaccharide lactose (milk sugar) into the monosaccharides glucose and galactose. Bacteria ferment undigested lactose in the large intestine, producing abdominal pain, gas, diarrhea, bloating, and cramps. A person with lactose intolerance can avoid these symptoms by consuming fermented dairy products such as yogurt, buttermilk, and cheese instead of fresh dairy products, because lactose has already been broken down. Taking lactase tablets can also prevent symptoms of lactose intolerance.

FIGURE 37.11
Overview of Chemical Digestion.
Digestion gradually breaks down large molecules.

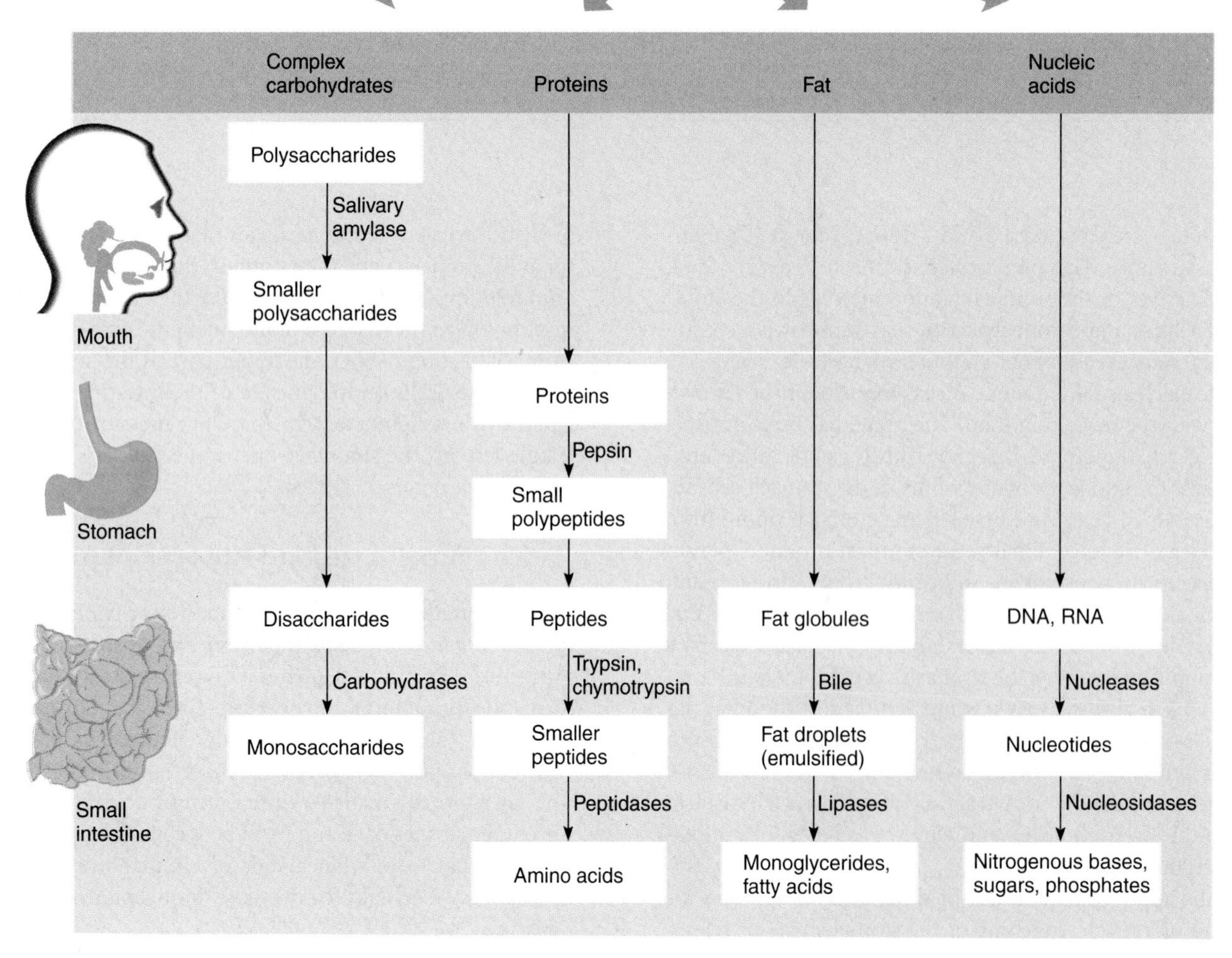

Nucleic acids are not abundant enough to be considered a major nutrient, but they too are digested. Nucleases break RNA and DNA down into nucleotides, and then nucleosidases take apart nucleotides into their constituent bases, sugars, and phosphates.

Hormones coordinate the small intestine's various digestive activities. When the stomach squirts acidic chyme into the small intestine, intestinal cells release the hormone secretin. This triggers the release of bicarbonate from the pancreas, which quickly neutralizes the acidity of the chyme. Secretin also stimulates the liver to secrete bile. Intestinal cells secrete another hormone, cholecystokinin (CCK), in response to chyme that is high in fats. CCK stimulates the liver to continue bile secretion and triggers the gallbladder to contract, sending stored bile to the small intestine. CCK also releases pancreatic enzymes, including lipase, into the small intestine. Secretin and CCK inhibit endocrine cells of the stomach wall from releasing gastrin and temporarily slowing digestion in the stomach.

Hormonal regulation of digestion protects intestinal cells because digestive biochemicals are produced only when food is present. As an additional safeguard, other glands in the small intestine secrete mucus, which protects the intestinal wall from digestive juices and neutralizes stomach acid. Mucus offers limited protection, however. Many intestinal lining cells succumb to the caustic contents. Fortunately, these epithelial cells have a high division rate and replace the lining every 36 hours. Nearly one-quarter of the bulk of feces consists of dead epithelial cells from the small intestine.

At the end of digestion, carbohydrates have been digested to monosaccharides, proteins to amino acids, fats to fatty acids and monoglycerides, and nucleic acids to nitrogenous bases, sugars, and phosphates. Cells lining the small intestine absorb these products, which then enter the circulation.

Nutrient absorption must take place at a surface. The small intestine maximizes surface area. The organ is very long and highly folded, enabling it to fit into the abdomen. In addition, the innermost layer is corrugated with circular ridges almost half an inch high. The surface of every hill and valley of the lining looks velvety due to additional folds—about 6 million tiny projections called **villi** (fig. 37.12). The tall epithelial cells on the surface of

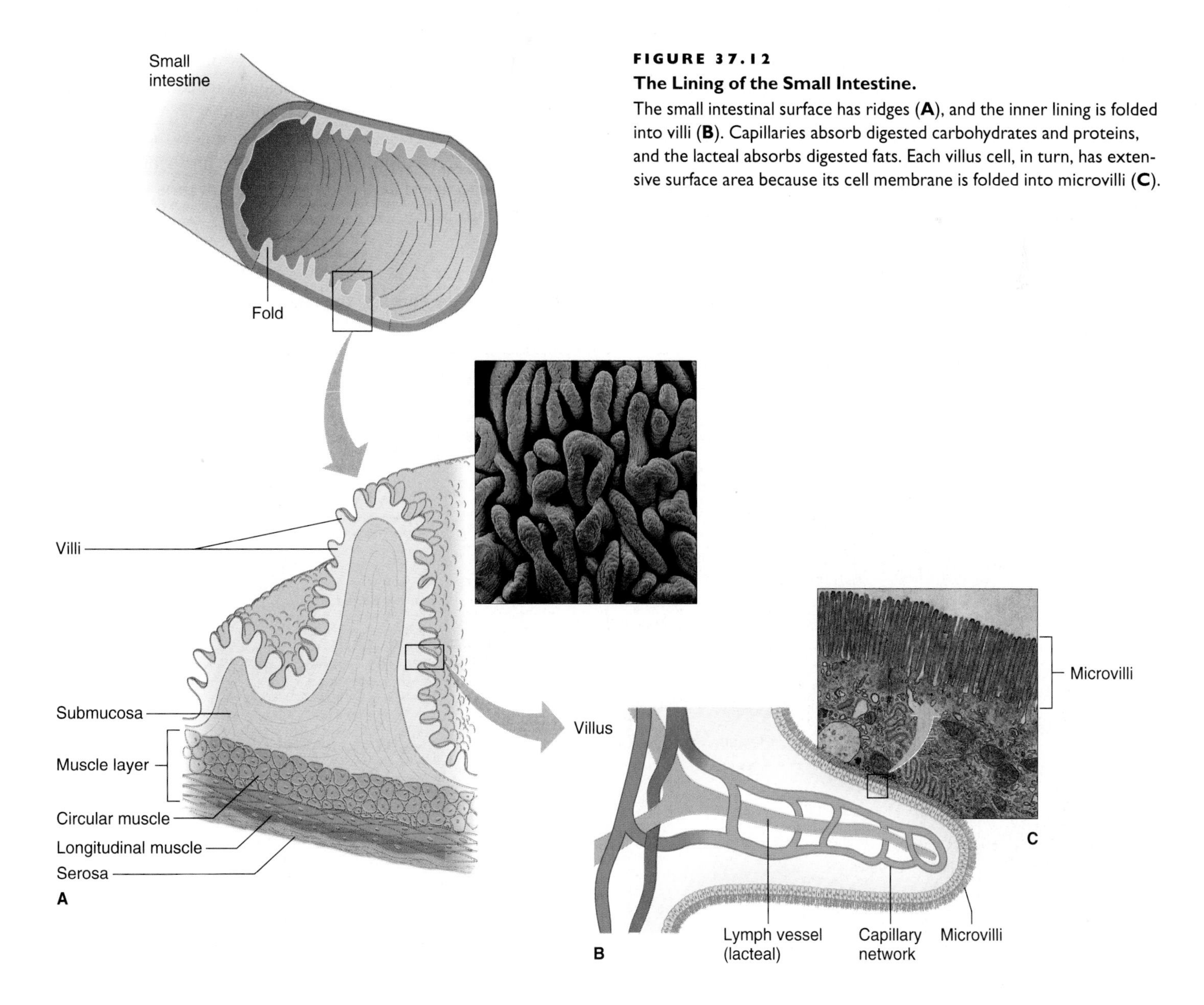

FIGURE 37.12
The Lining of the Small Intestine.
The small intestinal surface has ridges (**A**), and the inner lining is folded into villi (**B**). Capillaries absorb digested carbohydrates and proteins, and the lacteal absorbs digested fats. Each villus cell, in turn, has extensive surface area because its cell membrane is folded into microvilli (**C**).

each villus bristle with projections of their own called **microvilli.** Each villus cell and its 500 microvilli increase the surface area of the small intestine at least 600 times.

Within each villus a capillary network absorbs amino acids, monosaccharides, and water, as well as some vitamins and minerals. These digested nutrients are then distributed to cells. Fatty acids and monoglycerides are reassembled into triglycerides within the small intestinal lining cells, and they are coated with proteins to make them soluble before they enter the lymphatic system. The triglycerides then enter lymph vessels called **lacteals** that are surrounded by capillaries in the intestinal villi. Triglycerides enter the blood where lymphatic vessels join the circulatory system in the upper chest.

The Large Intestine Completes Nutrient Absorption

The material remaining in the small intestine after nutrients are absorbed enters the **large intestine,** or **colon.** The large intestine is shorter than the small intestine, but its 2.5-inch (6.5-centimeter) diameter is greater. The 5-foot (1.5-meter) tube surrounds the convoluted mass of the small intestine, roughly in the shape of a question mark. At the start of the large intestine is a pouch called the cecum (fig. 37.13).

Dangling from the cecum is the **appendix,** a thin wormlike tube. The appendix has an immune function in some vertebrates but its role in humans is not known. In our primate ancestors, the appendix may have helped digest fibrous plant matter. If bacteria or undigested food become trapped in the appendix, the area can become irritated and inflamed and infection can set in, producing severe abdominal pain. If this happens, the appendix must be promptly removed before it bursts and spills its contents into the abdominal cavity and spreads the infection.

The large intestine absorbs most of the water, electrolytes, and minerals from chyme, leaving solid or semisolid feces. Billions of bacteria of about 500 different species are normal inhabitants of the healthy human large intestine. These "intestinal flora" decompose any nutrients that escaped absorption in the small intestine; they produce vitamins B_1, B_2, B_6, B_{12}, K, folic acid, and biotin; and they break down bile and foreign chemicals such as certain drugs.

Intestinal flora produce foul-smelling compounds that cause the characteristic odors of intestinal gas and feces. These bacteria also help prevent infection by other microorganisms. Antibiotic drugs often kill the normal bacteria and allow other microorganisms, especially yeasts, to grow. This alteration in the intestinal flora causes the diarrhea that is sometimes a side effect of taking antibiotic drugs.

The remnants of digestion—cellulose, bacteria, bile, and intestinal cells—collect as feces in the **rectum,** a 6- to 8-inch- (15- to 20-centimeter-) long region. Within the rectum is the 1-inch- (2.5-centimeter-) long anal canal. The opening to the anal canal, the **anus,** is usually closed by two sets of sphincters, an inner smooth-muscle sphincter under involuntary control and an outer skeletal-muscle sphincter, which is voluntary. When the rectum is full, receptor cells trigger a reflex that eliminates the feces.

Figure 37.14 summarizes the lengths of various parts of the human digestive system and the amount of time food spends in each region.

FIGURE 37.13
The Large Intestine.
The large intestine absorbs water, salts, and minerals and temporarily stores feces. Inset: The appendix attaches to the pouchlike cecum, which receives food from the ileum of the small intestine.

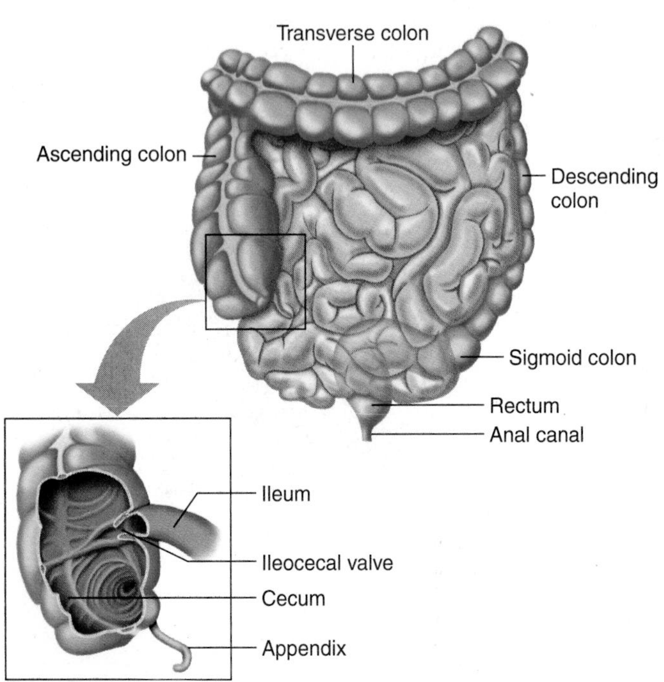

FIGURE 37.14
Food's Journey Through the Human Digestive Tract.
This flowchart indicates the length of each part of the human gastrointestinal tract and the approximate amount of time that food spends in each area.

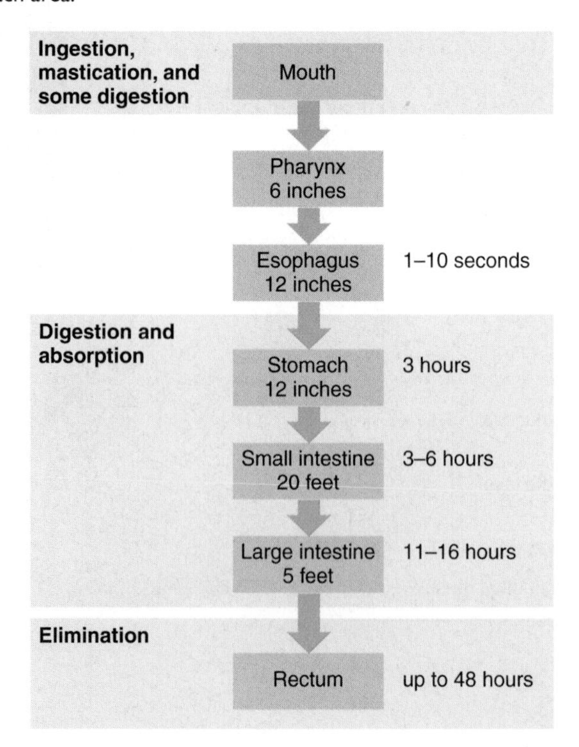

The Pancreas, Liver, and Gallbladder Aid Digestion

Other glands and organs assist digestion. The **pancreas** is a multifunctional structure associated with the digestive tract (fig. 37.15). It sends about a liter of fluid to the duodenum each day, including trypsin and chymotrypsin to digest polypeptides, pancreatic amylase to digest carbohydrates, pancreatic lipase to further break down emulsified fats, and nucleases to degrade DNA and RNA. Pancreatic "juice" also contains sodium bicarbonate to neutralize the acidity that hydrochloric acid produces in the stomach. Recall from chapter 33 that the pancreas also functions as an endocrine gland, regulating blood sugar level.

At a weight of about 3 pounds (1.4 kilograms), the **liver** is the largest solid organ in the body. It has more than 200 functions, including detoxifying harmful substances in the blood, storing glycogen and fat-soluble vitamins, and synthesizing blood proteins. The liver's contribution to digestion is the production of greenish yellow bile. Bile is stored in the **gallbladder** until chyme in the small intestine triggers its release. The cholesterol in bile can crystallize, forming gallstones that partially or completely block the duct to the small intestine. Gallstones are very painful and may require removal of the gallbladder.

Bile consists of pigments derived from the breakdown products of hemoglobin and bile salts from the breakdown of cholesterol. This colorful substance is responsible for the brown color of feces and the pale yellow of blood plasma and urine. It also creates the abnormal yellow complexion of jaundice, a condition that deposits excess bile pigments in the skin.

37.3 MASTERING CONCEPTS

1. Which structures in the mouth and throat participate in digestion?
2. Describe the mechanical and chemical digestion that occurs in the stomach.
3. How is the small intestine specialized to maximize surface area?
4. What are the products of digestion?
5. What occurs in the colon?
6. How do secretions of the pancreas, liver, and gallbladder aid digestion?

OLC

37.4 Human Nutrition

Carbohydrates, proteins, and lipids are required in large amounts and provide energy, which is measured in kilocalories, and raw materials for building bodies. Vitamins and minerals are required in small amounts and have specific functions.

The products of digestion provide energy, which cells use to function and to synthesize compounds. Figure 7.12 depicts the fates of these products—glucose is the immediate energy source, but amino acids and fatty acids enter the energy pathways too. Chapter 2 discusses the functions of the major nutrients.

Types of Nutrients

Carbohydrates, proteins, and lipids are called **macronutrients** because humans require them in large amounts, or energy nutrients because they supply energy. Humans require vitamins and minerals in very small amounts, and hence these nutrients are called **micronutrients.** Water is also a vital part of the diet.

Tables 37.2 and 37.3 provide details on specific vitamins and minerals. Vitamins function as coenzymes and minerals are essential to many biochemical pathways. Vitamins are classified by whether they are soluble in water or fat.

Essential nutrients must come from food because the body cannot synthesize them. Essential nutrients vary among species. Vitamin C is an essential nutrient in humans, guinea pigs, Indian fruit bats, and certain monkeys, but not in most other animals. If a guinea pig eats only rabbit chow, which lacks vitamin C, the guinea pig will develop vitamin C deficiency. Nonessential nutrients come from foods, but are also made in the body. For example, the adult human body synthesizes 11 of the 20 types of amino acids, which are therefore nonessential.

The amount of energy released from a nutrient is measured in **kilocalories (kcal),** often called simply calories. A food's caloric content is determined by burning it in a bomb calorimeter, a chamber immersed in water. When burning food is placed in the chamber, the energy released raises the water temperature, and this energy is measured in kilocalories. One kilocalorie is the energy needed to raise 1 kilogram of water 1°C under controlled conditions. Bomb calorimetry studies have shown that 1 gram of carbohydrate yields 4 kilocalories, 1 gram of protein yields 4 kilocalories,

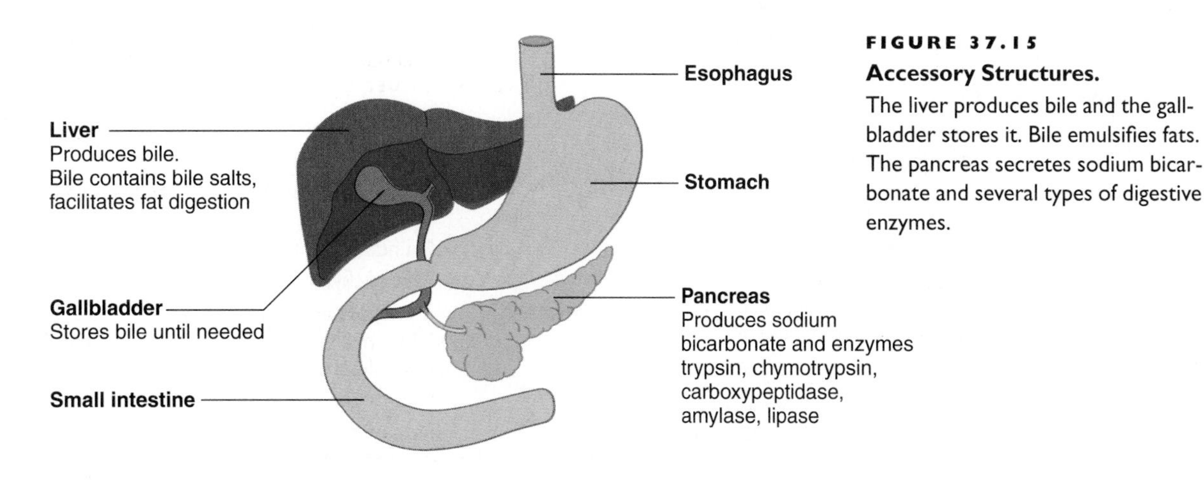

FIGURE 37.15
Accessory Structures. The liver produces bile and the gallbladder stores it. Bile emulsifies fats. The pancreas secretes sodium bicarbonate and several types of digestive enzymes.

TABLE 37.2 Vitamins and Health

VITAMIN	FUNCTION	FOOD SOURCES	DEFICIENCY SYMPTOMS
WATER-SOLUBLE VITAMINS			
Thiamine (vitamin B_1)	Growth, fertility, digestion, nerve cell function, milk production	Pork, beans, peas, nuts, whole grains	Beriberi (neurological disorder), loss of appetite, swelling, poor growth, heart problems
Riboflavin (vitamin B_2)	Energy use	Liver, leafy vegetables, dairy products, whole grains	Hypersensitivity of eyes to light, lip sores, oily dermatitis
Pantothenic acid	Growth, cell maintenance	Liver, eggs, peas, potatoes, peanuts	Headache, fatigue, poor muscle control, nausea, cramps*
Niacin	Growth	Liver, meat, peas, beans, whole grains, fish	Dark rough skin, diarrhea, mouth sores, mental confusion (pellagra)
Pyridoxine (vitamin B_6)	Protein use	Red meat, liver, corn, potatoes, whole grains, green vegetables	Mouth sores, dizziness, nausea, weight loss, neurological disorders*
Folic acid	Manufacture of red blood cells, metabolism	Liver, navy beans, dark green vegetables	Anemia, cancer
Biotin	Metabolism	Meat, milk, eggs	Skin disorders, muscle pain, insomnia, depression*
Cyanocobalamin (vitamin B_{12})	Manufacture of red blood cells, growth, cell maintenance	Meat, organ meats, fish, shellfish, milk	Pernicious anemia
Ascorbic acid (vitamin C)	Growth, tissue repair, bone and cartilage formation	Citrus fruits, tomatoes, peppers, strawberries, cabbage	Weakness, gum bleeding, weight loss (scurvy)
FAT-SOLUBLE VITAMINS			
Retinol (vitamin A)	Night vision, new cell growth	Liver, dairy products, egg yolk, vegetables, fruit	Night blindness, rough dry skin
Cholecalciferol (vitamin D)	Bone formation	Fish-liver oil, milk, egg yolk	Skeletal deformation (rickets)
Tocopherol (vitamin E)	Prevents certain compounds from being oxidized	Vegetable oil, nuts, beans	Anemia in premature infants*
Vitamin K	Blood clotting	Liver, egg yolk, green vegetables	Bleeding, liver problems*

*Deficiencies of these vitamins are rare in humans, but they have been observed in experimental animals.

TABLE 37.3 Minerals in the Human Diet

MINERAL	FOOD SOURCES	FUNCTIONS IN THE HUMAN BODY
BULK MINERALS		
Calcium	Milk products, green leafy vegetables	Bone and tooth structure, blood clotting, hormone release, nerve transmission, muscle contraction
Chloride	Table salt, meat, fish, eggs, poultry, milk	Digestion in stomach
Magnesium	Green leafy vegetables, beans, fruits, peanuts, whole grains	Muscle contraction, nucleic acid synthesis, enzyme activity
Phosphorus	Meat, fish, eggs, poultry, whole grains	Bone and tooth structure
Potassium	Fruits, potatoes, meat, fish, eggs, poultry, milk	Body fluid balance, nerve transmission, muscle contraction, nucleic acid synthesis
Sodium	Table salt, meat, fish, eggs, poultry, milk	Body fluid balance, nerve transmission, muscle contraction
Sulfur	Meat, fish, eggs, poultry	Hair, skin, and nail structure, blood clotting, energy transfer, detoxification
TRACE MINERALS		
Chromium	Yeast, pork kidneys	Regulates glucose use
Cobalt	Meat, eggs, dairy products	Part of vitamin B_{12}
Copper	Organ meats, nuts, shellfish, beans	Part of many enzymes, storage and release of iron in red blood cells
Fluorine	Water (in some areas)	Maintains dental health
Iodine	Seafood, iodized salt	Part of thyroid hormone
Iron	Meat, liver, fish, shellfish, egg yolk, peas, beans, dried fruit, whole grains	Transport and use of oxygen (as part of hemoglobin and myoglobin), part of certain enzymes
Manganese	Bran, coffee, tea, nuts, peas, beans	Part of certain enzymes, bone and tendon structure
Selenium	Meat, milk, grains, onions	Part of certain enzymes, heart function
Zinc	Meat, fish, egg yolk, milk, nuts, some whole grains	Part of certain enzymes, nucleic acid synthesis

and 1 gram of fat yields 9 kilocalories. (A teaspoon of dried food weighs approximately 5 grams.) Although approximate, these values help to explain why a fatty diet may cause weight gain; fats supply more energy than most people can use.

Good nutrition is a matter of balance. When we take in more kilocalories than we expend, weight increases; those who consume fewer kilocalories than expended lose weight—and, if taken to an extreme, may even starve. Balancing vitamins and minerals is important to health, too. Eating a variety of foods helps meet nutritional requirements. The dietary suggestions pictured in a food pyramid (fig. 37.16) can help a person make healthful food choices.

The food pyramid is an improvement over past food group plans. The four food group plan devised in the 1950s suggested that humans need nearly as many daily servings of meat as of grains, dairy products, and fruits and vegetables. In the 1940s, an eight food group plan was followed for a time, including separate groups for butter and margarine and for eggs—foods now associated with the development of heart disease. In the 1920s, an entire food group was devoted to sweets! Clearly, as our knowledge of nutrition grows, our concept of a healthy diet changes.

Nutrient Deficiencies

Obtaining an adequate number of kilocalories each day from a variety of foods is sometimes difficult. A student may eat nothing but pizza during exam week; a busy working person may skip meals or eat on the run; many individuals may not be able to afford enough milk or meat. All these situations may lead to primary nutrient deficiencies, which are caused by diet. Secondary nutrient deficiencies, in contrast, result from inborn metabolic conditions that cause the body to malabsorb, overexcrete, or destroy a particular nutrient.

Micronutrient deficiencies develop more slowly than the obvious weight loss that accompanies macronutrient deficiency. Vitamin or mineral deficiencies cause subtle changes before health is noticeably affected. For example, the blood of a vegetarian who does not eat many iron-containing foods may have an abnormally low number of red blood cells or small red blood cells with little hemoglobin. The person feels fine, temporarily, as the body uses its iron stores. After several weeks, however, the signs of anemia appear—weakness, fatigue, frequent headaches, and a pale complexion.

Health 37.2 looks at a prevalent problem that takes many guises—starvation.

37.4 MASTERING CONCEPTS

1. Why are carbohydrates, proteins, and lipids called macronutrients and vitamins and minerals called micronutrients?
2. What is an essential nutrient?
3. What are the effects of macronutrient and micronutrient deficiencies?

OLC

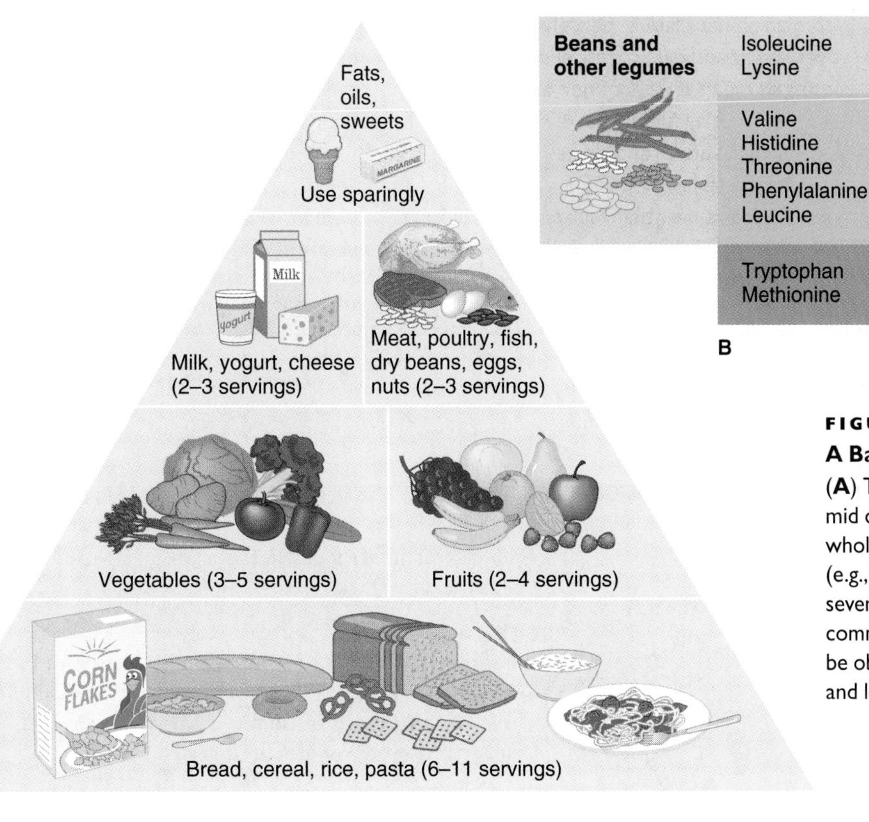

FIGURE 37.16
A Balanced Diet.
(**A**) The U.S. Department of Agriculture's food pyramid of human nutritional requirements emphasizes whole grains, fruits, and vegetables. (**B**) Legumes (e.g., beans) and grains (e.g., rice, corn) are rich in seven amino acids. Because only five amino acids are common to both, all nine essential amino acids can be obtained by eating meals containing both grains and legumes.

Starvation

A healthy human can survive for 50 to 70 days without food. In prehistoric times, this margin allowed survival during seasonal famines. In some areas of the world today, famine is not a seasonal event but a constant condition, and millions of people are starving to death. Hunger strikes, inhumane treatment of prisoners, and eating disorders can also cause starvation.

Whatever the cause, the starving human body begins to digest itself. After only one day without food, the body's reserves of sugar and starch are gone, and it starts to extract energy from fat and then from muscle protein. By the third day, hunger ceases as the body uses energy from fat reserves. Gradually, metabolism slows to conserve energy, blood pressure drops, the pulse slows, and chill sets in. Skin becomes dry and hair falls out as the proteins in these structures are broken down and release amino acids required for the vital functions of the brain, heart, and lungs. When the body dismantles the immune system's antibody proteins for their amino acids, protection against infection declines. Mouth sores and anemia develop, the heart beats irregularly, and bone begins to degenerate. After several weeks without food, coordination deteriorates. Near the end, the starving human is blind, deaf, and emaciated.

Marasmus and Kwashiorkor

Starvation can take different forms. When individuals lack all nutrients, they develop marasmus, a condition that causes a skeletal appearance. Children under the age of two with marasmus often die of measles or other infections; their immune systems are too weak to fight normally mild viral illnesses.

Some starving children do not look skeletal but instead have protruding bellies. These youngsters suffer from a form of protein starvation called kwashiorkor, which in the language of Ghana means "the evil spirit which infects the first child when the second child is born." Kwashiorkor typically appears in a child who has recently been weaned from the breast, usually because of the birth of a sibling. The switch from protein-rich breast milk to the protein-poor gruel that is the staple of many developing nations causes protein deficiency. The children's bellies swell with the fluid that builds up when protein is lacking, and their skin may develop lesions. Infections overwhelm the body as antibodies are broken down.

Anorexia Nervosa

Anorexia nervosa is self-imposed starvation. The condition is reported to affect 1 out of 250 adolescents, 95% of them female. The sufferer, typically a well-behaved adolescent from an affluent family, perceives herself as overweight and eats barely enough to survive, losing as much as 25% of her original body weight. She may further lose weight by vomiting, taking laxatives and diuretics, or engaging in intense exercise.

Anorexia eventually leads to low blood pressure, slowed or irregular heartbeat, constipation, and constant chills. A woman stops menstruating as body fat level plunges. Hair becomes brittle, the skin dries out, and soft, pale, fine body hair called lanugo, normally seen only on a developing fetus, grows to preserve body heat.

A person with anorexia who becomes emaciated may be hospitalized, so that intravenous feedings can prevent sudden death from heart failure due to a mineral imbalance. Psychotherapy and nutritional counseling may help identify and remedy the underlying cause of the abnormal eating behavior. Despite these efforts, 15 to 21% of people with anorexia die from the disease.

Bulimia

A person suffering from bulimia eats, often in huge amounts, and then gets rid of the thousands of extra kilocalories by vomiting, taking laxatives, or exercising frantically. A person with bulimia often binges in private, eating well beyond the point of pain.

Sometimes a dentist is the first to identify bulimia by observing teeth decayed from frequent vomiting. The backs of the hands may bear telltale scratches from efforts to induce vomiting, and the throat may be raw and the stomach lining ulcerated from the stomach acid forced forward by vomiting. The binge-and-purge cycle is very hard to break, even with psychotherapy and nutritional counseling.

Health services at many colleges and universities screen students for eating disorders. One in 10 college students suffers from an eating disorder.

Chapter Summary

37.1 The Art and Evolution of Eating

1. Food is ingested, mechanically broken down, **digested,** absorbed into the bloodstream, and waste eliminated.
2. Animals have varied mechanisms for obtaining food. The advent of heterotrophy allowed organisms to obtain nutrients from the environment.
3. Animals eat diverse foods, but break them down into the same types of nutrients.

37.2 Digestive Diversity

4. Protista and sponges have **intracellular digestion,** in which cells engulf food and digest it in food vacuoles. More complex animals have **extracellular digestion** in a cavity outside cells. The cavity plus accessory organs form the **digestive system.**

5. Digestive cavities can have one opening or two. In a system with two openings, food enters through the mouth and is digested and absorbed; undigested material leaves through the second opening, the **anus.**
6. The length of the digestive tract, number of stomachs, and presence of a **cecum** are adaptations to particular diets. Plant matter is the hardest to digest.

37.3 The Digestive System in Humans

7. In humans, digestion begins in the mouth, where teeth break food into smaller pieces. Salivary glands produce saliva, which moistens food and begins starch digestion. Swallowed food moves through the **esophagus** to the **stomach.** Waves of contraction called **peristalsis** move food along the digestive tract.
8. The stomach stores food, churns it until it liquefies into **chyme,** and mixes it with gastric juice. Hydrochloric acid in the gastric juice activates pepsinogen, forming the protein-splitting enzyme **pepsin** to begin protein digestion.
9. In the **small intestine,** trypsin and chymotrypsin break polypeptides into peptides, and peptidases break these down further. Pancreatic amylase and carbohydrases continue carbohydrate digestion. Nucleases and nucleosidases break down nucleic acids. The small intestine is the main site of nutrient absorption.
10. Material remaining after absorption in the small intestine passes to the **colon,** which absorbs water, minerals, and salts, leaving feces. Many bacteria digest remaining nutrients and produce useful vitamins that are then absorbed. Feces exit the body through the anus.
11. The **pancreas, liver,** and **gallbladder** aid digestion. The pancreas supplies pancreatic amylase, trypsin, chymotrypsin, lipases, and nucleases. The liver supplies **bile,** which emulsifies fat, and the gallbladder stores bile. The nervous and endocrine systems regulate digestive secretion.

37.4 Human Nutrition

12. Nutrients promote growth, maintenance, and repair of body tissues. **Macronutrients,** or energy nutrients, include carbohydrates, proteins, and fats. **Micronutrients** include vitamins and minerals. Water is also vital.
13. **Kilocalories** measure the energy food provides. One gram of carbohydrate or protein yields 4 kilocalories, and 1 gram of fat yields 9.
14. An inadequate diet causes a primary nutrient deficiency. A metabolic abnormality causes a secondary nutrient deficiency.

Testing Your Knowledge

1. What is a nutrient?
2. How do the circulatory, muscular, nervous, and endocrine systems take part in digestion?
3. What are the digestive products of carbohydrates, proteins, and fats?
4. Identify the part of the digestive system that includes the following:
   a. ileum, jejunum, duodenum
   b. cecum, appendix, rectum, anus
   c. villi and microvilli
   d. rugae, body, pylorus, cardia, fundus
5. Name an organism that has each of the following:
   a. extracellular digestion
   b. intracellular digestion
   c. a digestive tract with one opening
   d. a digestive tract with two openings
   e. filter feeding
6. How does mechanical breakdown of food facilitate chemical digestion?
7. How is surface area maximized in the stomach and in the small intestine? Why is it necessary for a digestive system to have extensive surface area?
8. Name three hormones that are associated with digestion and their functions.

Thinking Scientifically

1. How can people consume vastly different diets yet obtain adequate nourishment?
2. Why can't a person gain nutritional benefits by finely grinding meat and injecting it into the bloodstream?
3. Name an organelle in human cells that has a function similar to intracellular digestion in a protistan.
4. The protein in a hamburger is mostly myosin. Why doesn't the human body simply use the cow version of myosin to build muscle tissue? What happens instead?
5. Orlistat is a new weight-loss drug that inhibits the activity of lipases in the small intestine. Why would this be more effective than a drug that blocks absorption of proteins or carbohydrates? What might be a side effect of blocking fat absorption?

References and Resources

Alexander, R. McNeill. January 22, 1998. News of chews: The optimization of mastication. *Nature,* vol. 391, p. 329. Researchers calculated the optimal number of chews for a variety of foods.

Byers, T. April 20, 2000. Diet, colorectal adenomas, and colorectal cancer. *The New England Journal of Medicine,* vol. 342, p. 1206. Two long-term, large-scale clinical trials reveal that a high-fiber diet may not protect against colon cancer.

Diamond, Jared. July 8, 1999. Dirty eating for healthy living. *Nature,* vol. 400, p. 120. Eating soil and clay is adaptive—for parrots and possibly others.

Lewis, Ricki. July 20, 1998. Unraveling leptin pathways identifies new drug targets. *The Scientist,* vol. 13, p. 1. Understanding the controls of body weight suggests new drug strategies to treat obesity.

The *LIFE* Online Learning Center provides additional resources and tools for studying this chapter.
www.mhhe.com/life

CHAPTER 38

Regulation of Temperature and Body Fluids

Beating the Heat

Keeping Cool in the Desert and on the Shore

The Camel's Adaptations to Heat

- Avoids midday sun
- Variable body temperature lowers water lost through evaporation
- Light color reflects radiant heat
- Insulating hair coat
- Concentrated urine
- Dry feces
- Concentrated fat hump
- Heat storage
- Carotid rete
- Eats succulent plants
- Deep, slow breathing reduces water lost through respiration
- Oval-shaped red blood cells

A

Some animals live in environments hostile to cell survival. Here we look at some behavioral and physiological adaptations that have allowed two animals to cope in extreme temperature conditions.

The Camel—Playing It Cool

The camel is a walking definition of adaptation; everything from its hump to its behavior helps the animal conserve water and tolerate heat.

Standing up to 7 feet (2.1 meters) tall and weighing up to 1,000 pounds (454 kilograms), the camel stays cooler longer than a smaller desert dweller. Its characteristic hump, rich in fatty tissue, absorbs heat, preventing some of it from reaching the rest of the body; the long, thin legs radiate heat. The camel's bulk casts a large shadow that provides shade, especially when several camels squat together.

Internal adaptations also help the camel survive in the desert heat. Just be-

B

Adaptations to Heat.
(**A**) The camel has many adaptations that help it tolerate high temperatures. (**B**) Marine iguanas retreat to shade to escape the heat of the noonday sun.

neath the brain, a pool of cool venous blood surrounds arteries that carry warmer blood to the brain in a circulatory specialization called a carotid rete. This mechanism is a lifesaving adaptation because the brain is especially sensitive to heat. A carotid rete is a type of countercurrent exchange system, in which blood in one vessel transfers heat to blood in a nearby vessel.

The camel's blood composition is also adapted to surviving heat. The blood is more watery than the blood of mammals living in cooler climates. The camel's red blood cells are oval, compared with other mammals' rounder cells that shrivel in great heat and block blood flow to the surface that would otherwise dissipate body heat. Oval cells remain oval, allowing blood flow to continue even if the camel loses a third of its body water.

The camel's metabolism is also adapted to the heat, slowing in the summer rather than speeding up as in most mammals. Water loss suppresses activity of the camel's thyroid gland, lowering metabolic rate. Slower metabolism decreases the respiratory rate, conserving water that would otherwise leave through respiration. The urinary system conserves water, too—the concentrated urine is just watery enough to carry out wastes. A camel's droppings are very dry, which also conserves water.

Even the camel's behavior has a cooling effect. The animal sits much of the time, minimizing energy expenditure. Like many other animals, a camel urinates on its legs, enjoying the cooling effects of evaporating liquid.

A Day in the Life of a Marine Iguana

An iguana lizard has a different set of adaptations for surviving intense heat than does a camel. This is because a camel derives body heat mostly from its metabolism, whereas an iguana does so from the environment.

Iguanas spend much of their time regulating their body temperatures, which are about the same as, or slightly higher than, a human's body temperature. Marine iguanas on the Galápagos Islands begin their days basking in the rising sun, draped on boulders and hardened lava, sunning their backs and sides. After about an hour, they turn and raise their bodies and aim their undersides at the sun.

By midmorning, the air temperature is rising rapidly. Iguanas cannot sweat; instead, they must escape the blazing sun. They lift their bodies by extending their short legs, which removes their bellies from the hot rocks and allows breezes to fan them. But by noon, these push-ups are insufficient to stay cool. The iguanas retreat to the shade of rock crevices.

By midday, the animals are hungry. The iguanas dive into the ocean, although it is too cold for them to stay in for more than a few minutes. They eat green algae on the ocean floor or seaweed by the shore, hanging off rocks to reach it. The water is so cold that the lizards' body temperatures rapidly drop. Arteries near their body surfaces constrict to help conserve heat.

After feeding, the iguanas stretch out on the rocks again, warming sufficiently to digest their meal. They continue basking as the day ends, absorbing enough heat to sustain them through the cooler night temperatures until a new day begins.

38.1 How Do Animals Regulate Their Temperature?

An animal's body temperature must remain within a certain range for enzymes to function. Ectotherms obtain heat from the surroundings; endotherms derive heat from metabolism. Body temperature is constant or variable. Adaptations to temperature extremes are anatomical, physiological, metabolic, hormonal, and behavioral.

The diverse ways that animals regulate body temperature, conserve water, and excrete wastes are evolutionary adaptations for life in varied environments. The extremely dry, scorchingly hot home of a camel contrasts sharply with a frigid arctic habitat or with the perpetual humidity of a tropical rain forest. In hot, dry areas, an animal may have difficulty conserving enough water to meet excretory demands and also produce sufficient sweat to tolerate heat. In arctic freshwater habitats, an animal must continually pump excess water out of the body, conserve ions, and survive the cold. Most animals live in environments that are intermediate between these extremes, but each species has specific adaptations that enable it to maintain homeostasis in its environment.

Body Heat Is External or Internal, Body Temperature Constant or Variable

The ability of an animal to balance heat loss and gain within its particular environment is called **thermoregulation.** It is important that an animal's body temperature remain within limits because temperature influences the rate of the chemical reactions that sustain life. An animal's enzymes, for example, work best at its customary body temperature. Should cellular temperatures vary from this optimum, the enzymes function less efficiently, and vital biochemical reactions slow. If enzymes in the same biochemical pathway respond differently to temperature changes, the coordination that enables the pathway to function breaks down.

Extreme temperatures alter biological molecules. A drastic temperature increase alters a protein's three-dimensional shape and disrupts its function. An extreme temperature drop solidifies lipids. Since lipids are an integral component of biological membranes, such a change would impede many functions, especially transport of molecules across membranes.

Animals are classified according to whether their major source of body heat is external or internal (fig. 38.1). **Ectotherms** obtain heat primarily from their external surroundings, such as the iguana described in the chapter opening essay. An ectotherm regulates its body temperature by its behavior, moving to areas where it can gain or lose external heat, depending on circumstances.

An **endotherm** obtains most of its heat internally from its metabolism, such as the camel. The metabolic rate of an endotherm is generally five times that of an ectotherm of similar size and body temperature. Layers of insulation in the form of fat, feathers, or fur help retain body heat in endotherms.

Ectothermic and endothermic ways of life each have advantages and disadvantages. The ectotherm depends upon a continuous ability to obtain or escape environmental heat. An injured iguana that could not squeeze into a crevice to escape the broiling noon sun would cook to death. In contrast, an endotherm's internally generated heat enables it to maintain body heat even in the middle of the night. This internal constancy comes at a cost, however. The endotherm must eat much more food and must use 80% of the energy from that food just to maintain its temperature.

FIGURE 38.1
Ectothermy and Endothermy.
An ectotherm alters its behavior to gain heat from or lose it to the environment, depending upon circumstances. In contrast, most of an endotherm's body heat is generated by its metabolism.

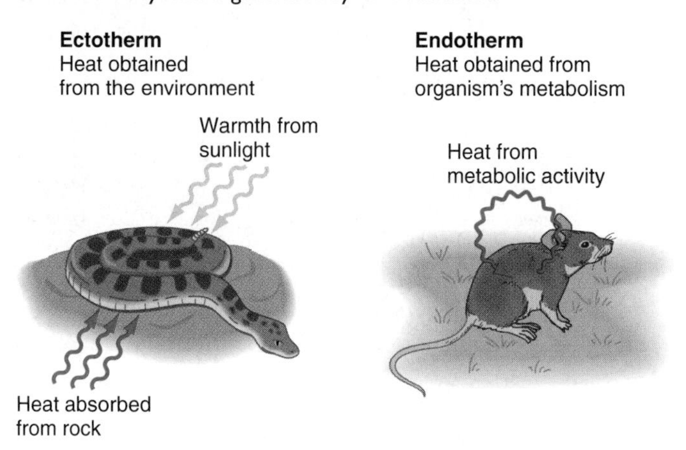

Most endotherms maintain a relatively constant body temperature by balancing heat generated in metabolism with heat lost to the environment, and the body temperature of most ectotherms varies as external conditions change. However, there are exceptions to these trends. Reptiles are ectotherms, but many can maintain their body temperatures within fairly narrow limits by moving around to gain or lose heat. On the other hand, a hummingbird is an endotherm whose body temperature is subject to occasional fluctuations.

Adaptations that enable animals to regulate their temperatures fall into three general categories: anatomical and physiological; metabolic and hormonal; and behavioral. Clusters of neurons in the hypothalamus control many of an animal's thermoregulatory responses by reacting to feedback from sensory cells elsewhere in the body. Thermoregulation depends on the interactions of the musculoskeletal, nervous, endocrine, respiratory, integumentary, and circulatory systems. Animals have adapted to a variety of environmental extremes.

Adaptations to Cold

For many fishes, water temperature determines body temperature. Yet, these animals must maintain body temperatures warm enough to permit survival. To do this, many fishes have biochemicals in their blood that function as antifreeze, lowering the point at which the blood freezes so that it stays liquid even when the water is very cold.

The great white shark and bluefin tuna have another type of adaptation to cold water called a **countercurrent exchange system.** Recall from the chapter opening essay that this is an anatomical arrangement that transfers heat between blood vessels. In these animals, venous blood coming from warm interior muscles warms colder blood in arteries from near the body surface. In another type of circulatory adaptation, marine reptiles such as sea turtles and sea snakes shunt warmer blood to the centers of their bodies, preferentially protecting interior vital organs. Amphibians, which tend to lose what little metabolic heat they generate through their moist skins, conserve heat by restricting their habitats to warm, moist places.

Birds and mammals can function for long periods at low temperatures, thanks to their adaptations to the cold. Feathers retain warmth by creating pockets that trap air near the skin. In many mammals, countercurrent exchange systems transfer the heat in some blood vessels to cooler blood in nearby vessels, retaining heat in the body rather than allowing it to escape to the environment through the extremities. Countercurrent exchange systems allow penguins to spend hours on ice or snow or in frigid water and enable arctic mammals such as wolves to hunt in extreme cold (fig. 38.2).

Blubber helps whales and seals maintain a constant body temperature. In their flippers and tails, which lack blubber, countercurrent heat exchange blood vessels keep the heat in. Mammals also have several behavioral strategies for retaining heat, including migrating, hibernating, and simply huddling together.

Mammals generate much of their metabolic heat in a process called **thermogenesis.** In one type of thermogenesis, fatty acids in brown fat are oxidized to release heat energy. Brown fat is abundant in animals that hibernate, and it also helps many newborn mammals, including humans, to stay warm.

Shivering is another form of thermogenesis. Shivering contracts muscles, releasing heat as ATP splits during actin and myosin filament interaction (see chapter 34). Hormones also provide body heat. Falling body temperatures stimulate the thyroid gland to produce more thyroxine, which increases cell membrane permeability to sodium ions (Na^+). When cell membranes pump Na^+ out of the cells, ATP is split, releasing heat energy. Hormone-directed internal heating along with brown fat utilization is called nonshivering thermogenesis.

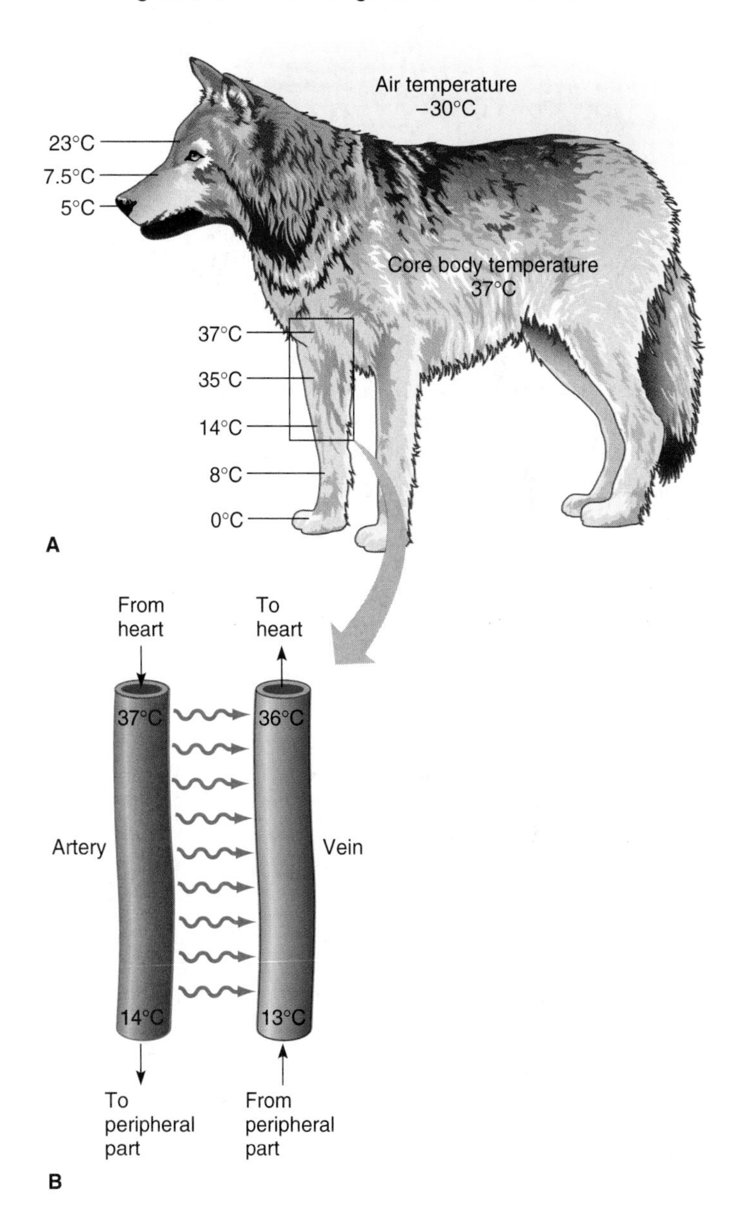

FIGURE 38.2
Countercurrent Heat Exchange.
(A) A mammal's extremities chill more easily than interior body parts. **(B)** In a countercurrent exchange system, arteries carrying warm blood lie in close proximity to veins carrying cooler blood, thereby conserving heat, rather than losing it to the environment.

Adaptations to Heat

Invertebrates and vertebrates share many adaptations for surviving in extreme heat and drought. One common strategy for lowering body temperature is **evaporative cooling,** in which air moves over a moist surface, evaporating the moisture and cooling that part of the body. A coyote pants and an owl flutters loose skin under its throat to move air over moist mouth surfaces. When a honeybee's temperature reaches 113°F (45°C), a heat sensor in its brain stimulates it to regurgitate nectar stored in an organ called a honey-crop. The nectar spills onto a tongue-like structure and the bee licks itself, cooling as its nectar bath evaporates. Australian sawfly larvae smear rectal fluid on their bodies, which evaporates and enables them to survive another hour in 115°F (46°C) heat.

Humans sweat. Some invertebrates sweat in a sense, too. Consider a type of cicada (an insect) that lives in the Sonoran Desert in Arizona, where the temperature can reach 105°F (41°C) in the shade. Water from the cicada's hemolymph moves through tubules to the body's surface, where it exits through ducts and

FIGURE 38.3

Cicadas Living in the Desert "Sweat."

These sweat pores in the cicada's cuticle are magnified 330 times.

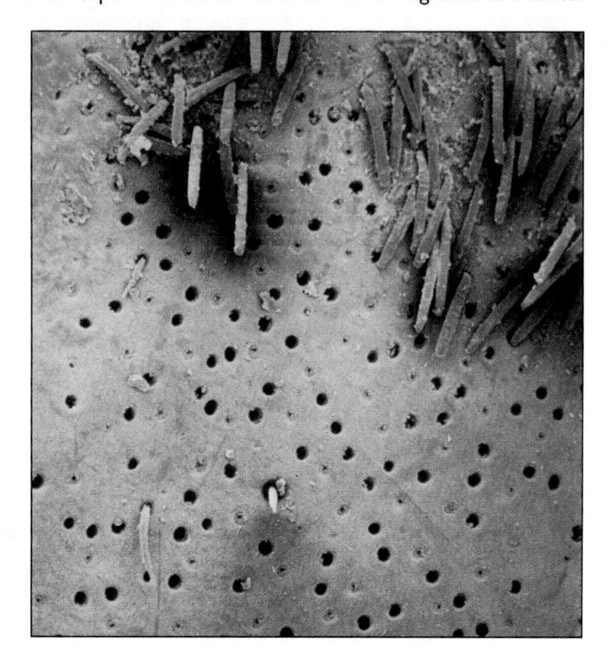

FIGURE 38.4

The Carotid Rete.

The brain of a hoofed mammal or carnivore stays cool thanks to the carotid rete, a network of arteries that cools the blood they deliver to the brain by passing first through a pool of cool venous blood beneath the brain. The venous blood comes from veins in the animal's muzzle.

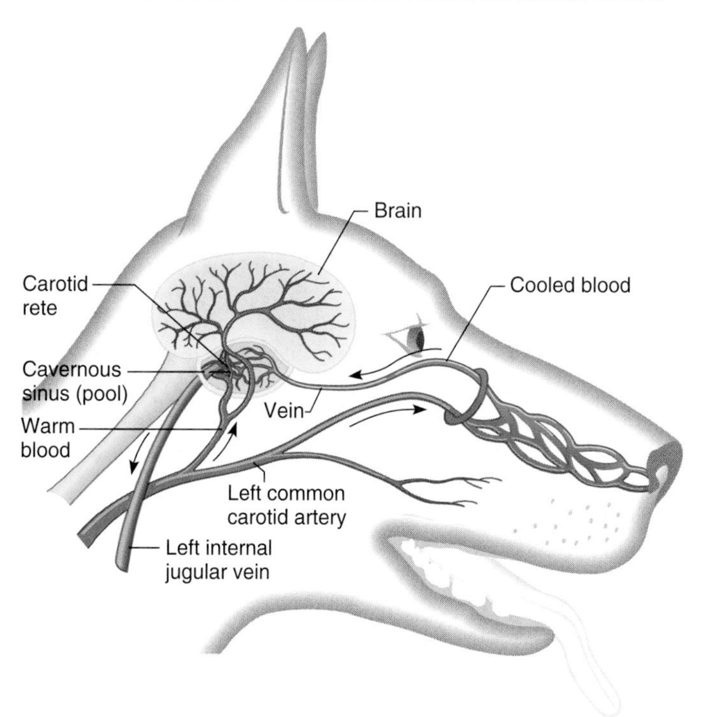

evaporates (fig. 38.3). By "sweating," the cicada can lower its body temperature as much as 18°F to 27°F (10°C to 15°C) below that of surrounding air.

A cicada can lose 20 to 35% of its body weight in fluid when the outside temperature hits 115°F (46°C). Most insects perish at a 20% loss, and humans cannot survive a 10% loss of body fluid. The Sonoran cicada can survive such extreme dehydration because fluid loss is balanced by intake—the insect imbibes tremendous amounts of fluid in its succulent plant food.

In many mammals, evaporative cooling is linked to circulatory system specializations that route cooled blood past warmer blood. An anatomical organization called a carotid rete in many mammals consists of warm-blood-bearing arteries that run through a pool of cool venous blood at the base of the brain (fig. 38.4).

The ancient Greeks, who discovered the carotid rete, named it the *rete mirabile,* which means "wonderful net." Cats, dogs, seals, sea lions, antelopes, sheep, goats, and cows all have them. These animals also have long snouts and tend to pant. The snout interiors have a rich blood supply and many glands that secrete fluids that keep the area moist. Each time the animal pants, water from the secretions evaporates, cooling blood in nearby vessels. Veins take the cooled blood to the pool beneath the brain where the carotid arteries bring warm blood. Cooling just the brain—the area that needs it most—requires less energy than cooling the entire body.

Humans have a different circulatory adaptation for keeping a cool head, which enables us to survive the heat of high fevers. Tiny "emissary" veins in the face and scalp reroute blood cooled near the body's surface through the braincase, which cools brain tissue. This adaptation, called a "cranial radiator," causes the face of a person vigorously exercising to flush.

Behavior and body form provide powerful adaptations to extreme heat and drought, in vertebrates and invertebrates. Consider the Namib ant, which lives in the desert of southwest Africa, where sand temperatures can reach 150°F (66°C). The temperature drops with distance above the sand, even if the distance is just the height of an ant. The Namib ant's long legs place its head high enough that the surrounding temperature is a mere 131°F (55°C). The ant has other behaviors that help it tolerate heat. By clinging to a blade of grass or a leaf for short intervals, or resting atop the dead body of a less lucky insect, an ant can cool off sufficiently to race across another stretch of blisteringly hot sand. This combination of body form and behavior enables the Namib ant to forage for food when individuals of less well-adapted species must seek shade.

Animals combine evaporative, circulatory, body, and behavioral strategies to cool their bodies and retain fluids. Turtles, for example, salivate, urinate on their back legs, and pant. Their heart rates change and their peripheral blood vessels dilate, dissipating heat at the body's surface. A turtle's shell absorbs heat, protecting vital inner organs, and its large bladder stores water. Turtles also burrow into the ground, or seek shade, to avoid heat. Figure 38.5 outlines familiar adaptations of the human body to temperature extremes.

FIGURE 38.5
Thermoregulation in Humans.
Receptors in the skin signal the hypothalamus to keep body temperature within a certain range. Exposure to high environmental temperature causes sweating, dilation of blood vessels to dissipate heat at the body's surface, and shade-seeking behavior. Exposure to cold stimulates shivering, constriction of blood vessels to conserve heat in the body's core, and warmth-seeking behavior.

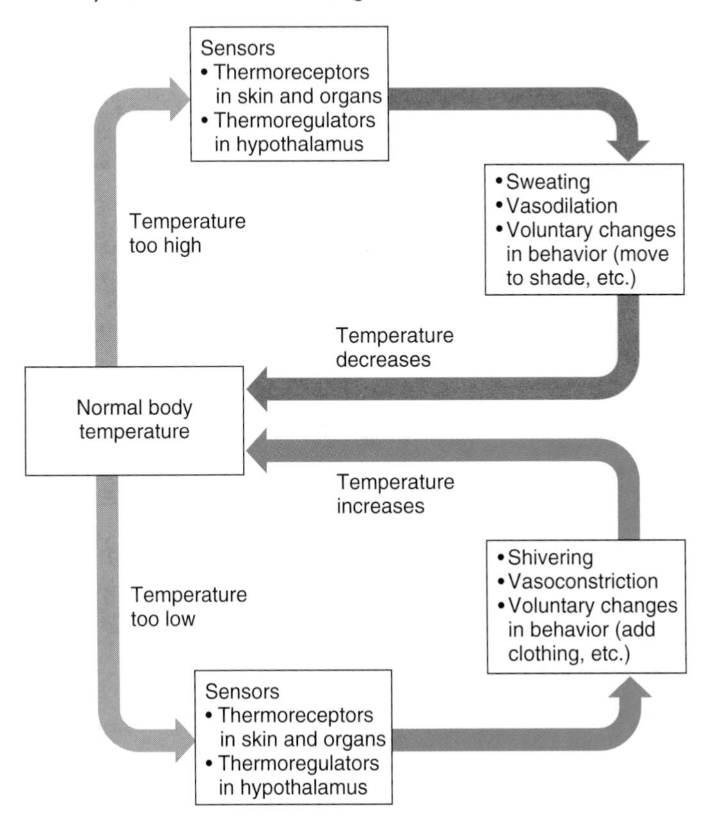

38.1 MASTERING CONCEPTS

1. Why is thermoregulation important?
2. Distinguish between ectotherms and endotherms.
3. What are some adaptations to extreme cold and extreme heat?

OLC

38.2 How Do Animals Regulate Water and Ion Balance?

Animals must maintain specific ranges of concentrations of ions in their body fluids, a process that the environment influences via natural selection. Osmoconformers have body fluid composition similar to that of the environment. Osmoregulators must work to maintain body fluid composition, in response to extreme or changeable environments.

Different environments present different challenges to organisms, not only in regulating body temperature but also in maintaining ion concentrations in body fluids. Osmoregulation is the control of water and ion balance in an organism.

Osmoregulation is highly attuned to the environment. Recall that most cell membranes are permeable to small ions and highly permeable to water molecules. A fish in seawater must handle the influx of salts (ions) into its body and the loss of water to the environment. (Salts diffuse in and water out by following their concentration gradients.) A fish in a lake faces the opposite problem—it must conserve salts, yet rid its body of excess water.

Animals osmoregulate to different extents. Invertebrates that live in the ocean, such as some crustaceans and worms, expend little energy maintaining body fluid composition because they are **osmoconformers**—the ion concentrations in their body fluids are similar to those in the surrounding water. Because their body surfaces are permeable to water and salts, their fluid compositions would change with the composition of their surroundings. But they can survive because the ion concentrations in the ocean are fairly constant.

Invertebrates living in coastal environments, where salinity changes with the tide as fresh water enters from rivers, are **osmoregulators.** They more actively keep their ion concentrations constant. Crabs living in estuaries (where salt water and fresh water meet), for example, have glands in their heads that pump out excess water that enters the body when the surrounding water becomes less salty (fig. 38.6). In the opposite situation, when too many salts exit the crab's body through the gills and in urine, specialized salt-secreting cells in the gills remove needed ions from incoming seawater and transport them to the blood.

FIGURE 38.6
Osmoregulation in a Crab.
This crab is shooting dilute urine from antennal glands near its eyes. The animal uses this technique to maintain fluid volume and ion balance.

FIGURE 38.7

Osmoregulation in Fishes.

(**A**) The marine fish must rid its body of excess salts that diffuse in from the environment and that it drinks in seawater. Secretory cells in the gills actively pump out sodium (Na^+) and chloride ions (Cl^-) and its kidneys excrete magnesium (Mg^{2+}) and sulfate (SO_4^{2-}) ions while reabsorbing most of the water to produce a concentrated urine. (**B**) In contrast the freshwater fish must conserve its body salts, which tend to diffuse out into the more dilute environment. The freshwater fish drinks very little water compared with the saltwater fish. Salt-absorbing cells in the gills actively absorb Na^+ and K^+. Its kidneys shed excess water in dilute urine while returning most of its ions to the bloodstream.

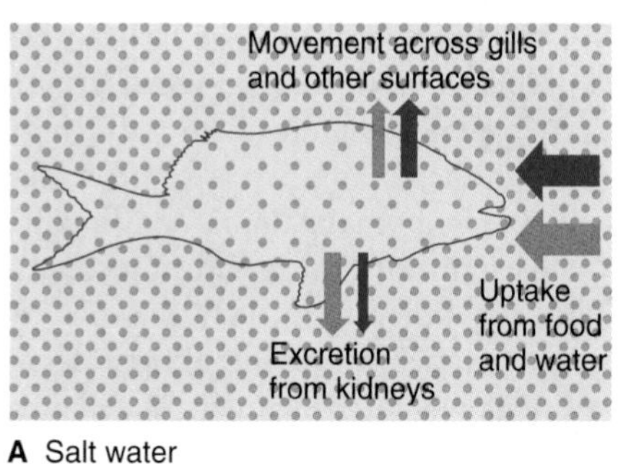

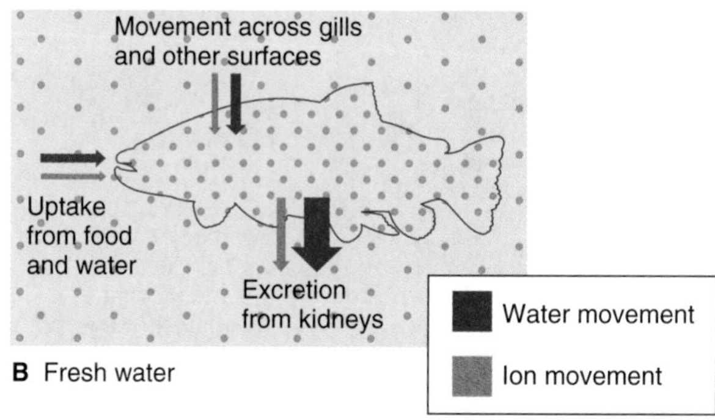

Fishes are osmoregulators, but those that live in the ocean and in freshwater lakes and streams face different challenges in regulating the ion concentration and volume of their body fluids (fig. 38.7). The marine fish must rid its body of excess salts that diffuse in from the environment and that it drinks in seawater. In contrast, the freshwater fish must conserve its body salts, which tend to diffuse out into the more dilute environment. Its kidneys have a more highly developed reabsorbing capacity, returning salts to the bloodstream and shedding excess water in dilute urine.

The marine iguana highlighted in the chapter opening essay also faces special osmoregulatory challenges. Because the iguana's food comes from the ocean, it is very salty. To rid its body of excess dietary salt, the iguana has a multilobed structure called a salt gland in its head (fig. 38.8). Cells lining the interior of the gland secrete a fluid that travels in branching tubules to empty through the nostrils. This fluid carries excess salts with it.

Animals in terrestrial environments must conserve water. They obtain water by eating and drinking and as a by-product of metabolism; they lose water through evaporation from lungs and body surfaces, in feces, and in urine. Different animals use different combinations of strategies to obtain and conserve water. A human takes in most of its water in food and drink. A kangaroo rat, in contrast, derives most of its water from its metabolism (fig. 38.9).

FIGURE 38.8

Salt Gland.

The salt gland of a marine iguana helps the animal rid its body of excess salt.

Salt gland

38.2 MASTERING CONCEPTS

1. What is osmoregulation?
2. Distinguish between osmoconformers and osmoregulators.
3. How does osmoregulation differ between fishes living in oceans and in fresh water?

OLC

FIGURE 38.9

Water Gain and Loss in a Kangaroo Rat and a Human.

A desert resident, the kangaroo rat must conserve water. It gets most of its water from its metabolism and loses little through feces and urine. In contrast, a human gets most of its body water from food and drink, and loses most of it through urine.

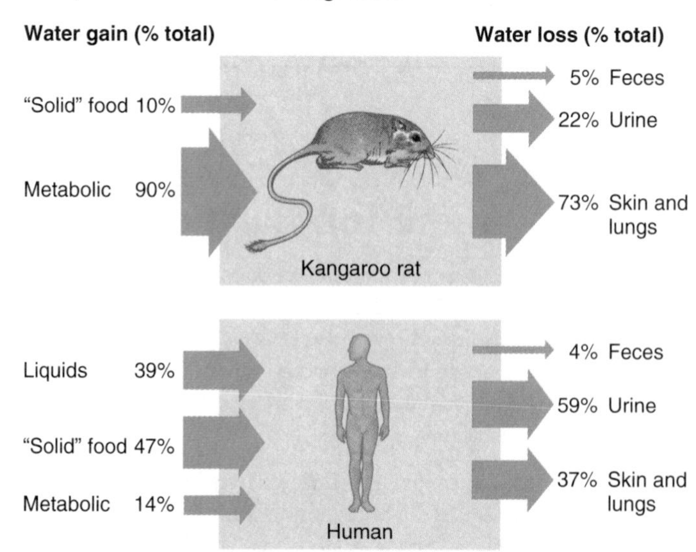

38.3 How Do Animals Rid Themselves of Nitrogenous Wastes?

Breakdown of proteins and nucleic acids releases ammonia, which may be converted to urea or uric acid. Animals must excrete these nitrogenous wastes.

Animals must rid their bodies of nitrogen-containing (nitrogenous) wastes that result when proteins and nucleic acids are broken down. Protein destruction generates three types of nitrogenous wastes: **ammonia, uric acid,** and **urea.** As a protein is degraded, amino groups ($-NH_2$) are released from amino acids. Each amino group then picks up a hydrogen ion and becomes ammonia, NH_3. Excreting ammonia directly is energetically efficient because it requires only one chemical reaction. Because ammonia is very toxic, however, animals must excrete it in a dilute solution.

Some freshwater animals take in water from the environment to dilute ammonia. However, land-dwelling animals, which must conserve water, use energy to convert ammonia to less toxic substances that can be stored and excreted in a relatively concentrated form. Mammals, amphibians, sharks, and some bony fishes convert ammonia to urea as their primary nitrogenous waste. Insects, reptiles, and birds convert most of their nitrogenous wastes to uric acid, which can be excreted in an almost solid form. In birds, uric acid mixes with undigested food in the cloaca to form the familiar "bird dropping."

Humans produce very small amounts of uric acid. Unlike the uric acid of birds and reptiles, which comes from protein breakdown, uric acid in humans comes from breakdown of DNA and RNA purine bases. In the human metabolic disorder called gout, excess uric acid crystallizes in the joints, causing pain.

Many invertebrates have networks of tubules called **nephridia** that form excretory systems. A simple type of nephridium excretory system is the planarian (flatworm) **flame cell system,** which consists of two branched tubules that extend the length of the body (fig. 38.10). Short side branches called flame cells (because their beating cilia resemble leaping flames) project into the body tissues. The cilia on the flame cells generate a negative pressure that draws fluid into the tubules. Cells lining the tubules add some substances, a process called secretion, and remove others, termed reabsorption. The tubules feed into pores that open onto the body surface. Nitrogenous wastes diffuse directly across the animal's body surface.

Some mollusks and segmented worms have a more complex nephridium excretory system. The earthworm has two individual nephridia in most of its segments (fig. 38.11). A single earthworm nephridium consists of a ciliated opening called a nephrostome, which leads to a coiled tubule surrounded by capillaries. The close association between an excretory tubule and a blood supply enables the circulation to reabsorb valuable biochemicals as urine forms. Urine moves to the bladder for storage and then leaves the earthworm's body at an opening called a nephridiopore, located in the adjacent segment.

FIGURE 38.10

The Flame Cell System of a Planarian.

In a planarian flatworm, networks of tubules extend down the body in two columns. Ciliated flame cells wave fluid into the tubule network. Cells lining the tubules add and remove substances, and the remaining waste exits the body through excretory pores.

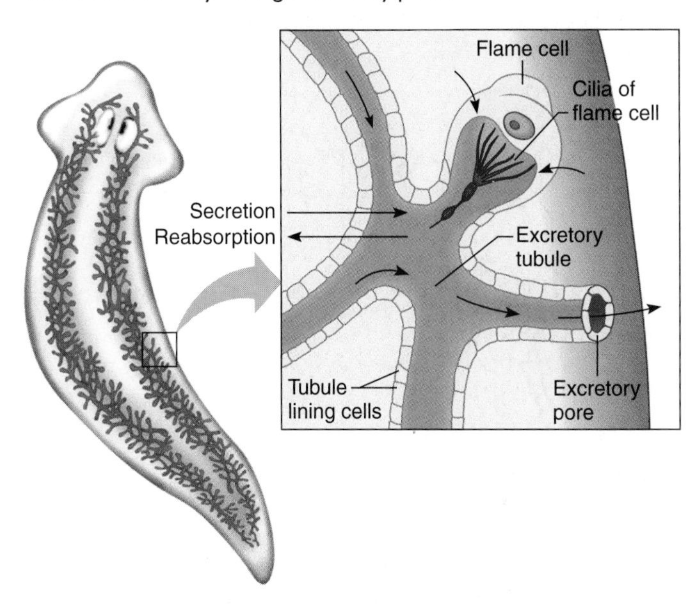

FIGURE 38.11

Earthworm Nephridia.

Earthworms excrete waste through paired nephridia in nearly all body segments. A nephridium begins at an opening, the nephrostome, and then continues as a coiled tubule surrounded by capillaries. A bladder stores waste. The system opens to the outside at a nephridiopore on an adjacent segment.

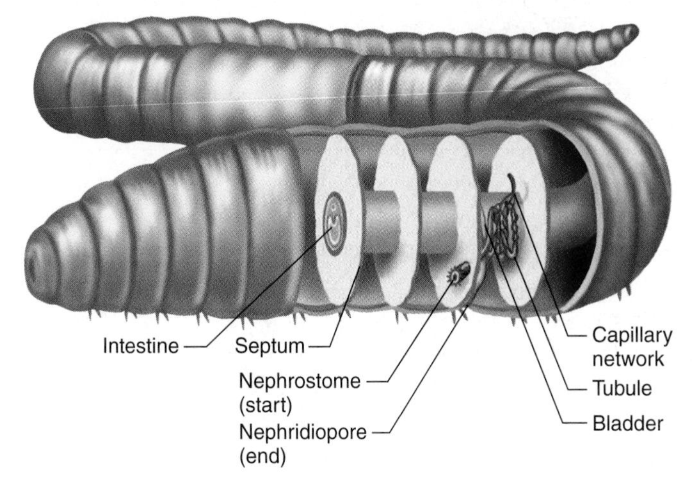

Insects' and spiders' excretory systems consist of **Malpighian tubules** (fig. 38.12). Ions, especially potassium (K^+), are actively secreted into the tubules. When K^+ exits, it draws water as well as nitrogenous wastes with it. The Malpighian tubules join the intestine, where rectal glands reabsorb most of the K^+ and water. The nitrogenous waste leaves the body as uric acid.

FIGURE 38.12

Excretion in Insects.

(**A**) Insect excretory organs, called Malpighian tubules, join the digestive tract. (**B**) Solutes, including sodium (Na^+) and potassium (K^+), are actively transported into the Malpighian tubules, and water, amino acids, sugars, and nitrogenous waste (uric acid) follow passively. Useful substances are reabsorbed in the rectal gland. Uric acid leaves the body through the anus.

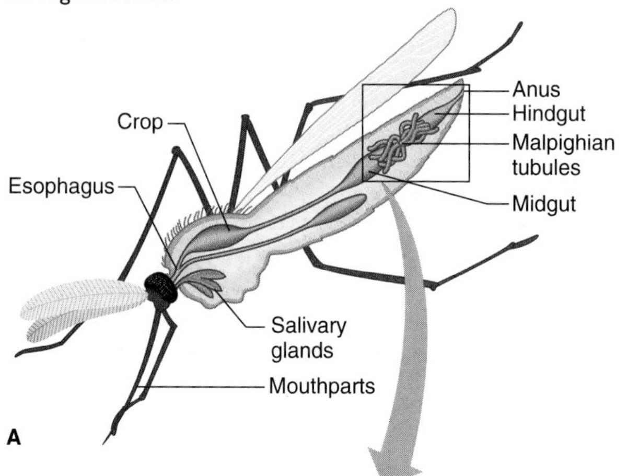

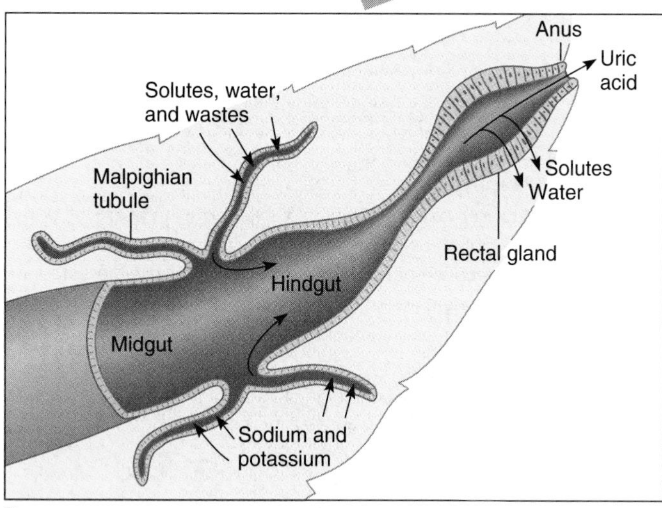

The major excretory structures of vertebrates are **kidneys,** which are paired organs located along the dorsal body wall. Kidneys conserve water and essential nutrients and remove nitrogenous wastes. While the organization of tubules and ducts that remove wastes differs among vertebrate groups, all kidneys share functional units called **nephrons.** We consider the human **urinary system** to illustrate the vertebrate's approach to waste removal.

The human kidneys are the major organs responsible for excretion and osmoregulation (fig. 38.13). Each kidney, located against the wall of the lower back within the abdominal cavity, is about the size of an adult fist and weighs about half a pound. Urine forms within each kidney and drains into a muscular tube about 11 inches (28 centimeters) long called a **ureter.** Waves of muscle contraction squeeze urine along the ureters and squirt it into a saclike, muscular **urinary bladder.** Urine drains from the bladder and exits the body through the **urethra.** In females, this tube is about 1 inch (2 to 3 centimeters) long and opens between the clitoris and vagina. In males, the urethra is about 8 inches (20 centimeters) long and extends the length of the penis.

Two rings of muscle (sphincters) guard the exit from the bladder. Both sphincters must relax before urine can leave the bladder. A spinal reflex involuntarily controls the innermost sphincter. Most people learn to consciously control relaxation of the outer sphincter at about 2 years of age.

The adult bladder can hold about 20 ounces (600 milliliters) of urine; however, as little as 10 ounces (300 milliliters) of accumulating urine stimulates stretch receptors in the bladder. The receptors

FIGURE 38.13

Human Urinary System.

The human urinary system includes the kidneys, ureters, urinary bladder, and urethra.

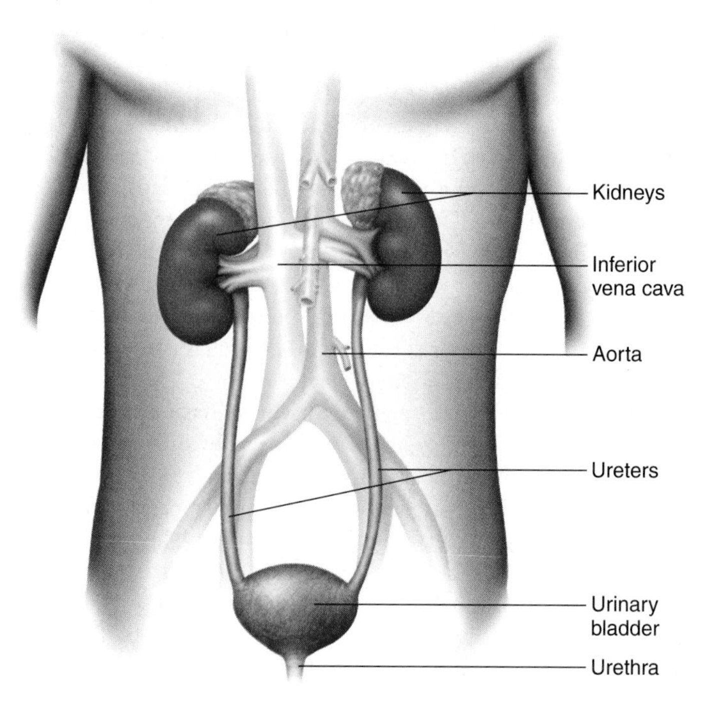

38.3 MASTERING CONCEPTS

1. Where do nitrogenous wastes originate in animal bodies?
2. How do animals detoxify ammonia?
3. Describe three types of structures that animals use to rid their bodies of nitrogenous wastes.

OLC

38.4 The Human Urinary System

Paired kidneys drain into ureters, which enter the bladder, from which urine exits through the urethra. Blood enters the kidney's nephrons, and along their length water and other valuable materials are reabsorbed, and wastes secreted into the forming urine.

send impulses to the spinal cord, which stimulates sensory neurons that contract the bladder muscles, generating a strong urge to urinate. We can suppress the urge to urinate for a short time by contracting the external sphincter. Eventually the cerebral cortex directs the sphincters to relax, and bladder muscle contractions force urine out of the body.

The Kidneys

The kidneys form an incredibly efficient blood-cleansing mechanism by simultaneously adjusting the composition, pH, and volume of the blood. As the kidneys process blood, urine forms.

Each kidney is packed with 1.3 million microscopic tubular nephrons. An individual nephron consists of a tuft of filtering capillaries (the **glomerulus**), a continuous **renal tubule** ("renal" means "of the kidney") plus the **peritubular capillaries** that entwine around it. A stretched-out nephron would be about a half inch (approximately 12 millimeters) long.

The same regions of each nephron lie at similar positions within the kidney (fig. 38.14). The outermost part of the kidney, the **renal cortex,** contains the glomeruli and is grainy in appearance. The middle section, the **renal medulla,** resembles a collection of aligned strings. This region corresponds to the long sections of the renal tubules. The innermost portion of the kidney, the **renal pelvis,** collects the urine each nephron produces.

The kidney illustrates a familiar biological organization: extensive surface area (afforded by the renal tubules) packed into a relatively small volume (the kidneys). The body's entire blood supply courses through the kidney's blood vessels every 5 minutes. At that rate, the equivalent of 425 to 525 gallons (1,600 to 2,000 liters) of blood passes through the kidneys each day. Yet a person excretes only about 0.4 gallons (1.5 liters) of urine daily. This is because most of the blood components that the nephrons process are reabsorbed into the blood rather than excreted in urine.

The kidney selectively retains and recycles important dissolved chemicals and water through three processes (fig. 38.15):

1. It filters wastes, nutrients, and water from the blood into the renal tubule, leaving large structures such as proteins and blood cells in the blood **(filtration).**
2. The kidney reabsorbs salts, water, and nutrients into the peritubular capillaries **(reabsorption).**
3. The tubules transport toxic substances out of the peritubular capillaries for excretion **(secretion).**

Overall, the nephrons extract wastes and recycle valuable nutrients, ions, and water to the bloodstream.

Nephrons

Specific portions of the nephron carry out specific functions.

The Glomerular Capsule Filters Blood Blood approaches a nephron in one of many branches of the renal artery known as an **afferent arteriole.** Each arteriole branches into a ball of capillaries, the glomerulus (Latin for "tiny ball"). These capillaries then join again to form an **efferent arteriole.** The glomerulus is surrounded

FIGURE 38.14
Anatomy of a Kidney.
(**A**) A kidney sliced down the middle reveals an outer area, the renal cortex, and an inner section, the renal medulla. (**B**) The nephrons are aligned; as a result, the same nephron parts are found in the same region of the kidney. (**C**) The nephron is the structural and functional unit of the kidney.

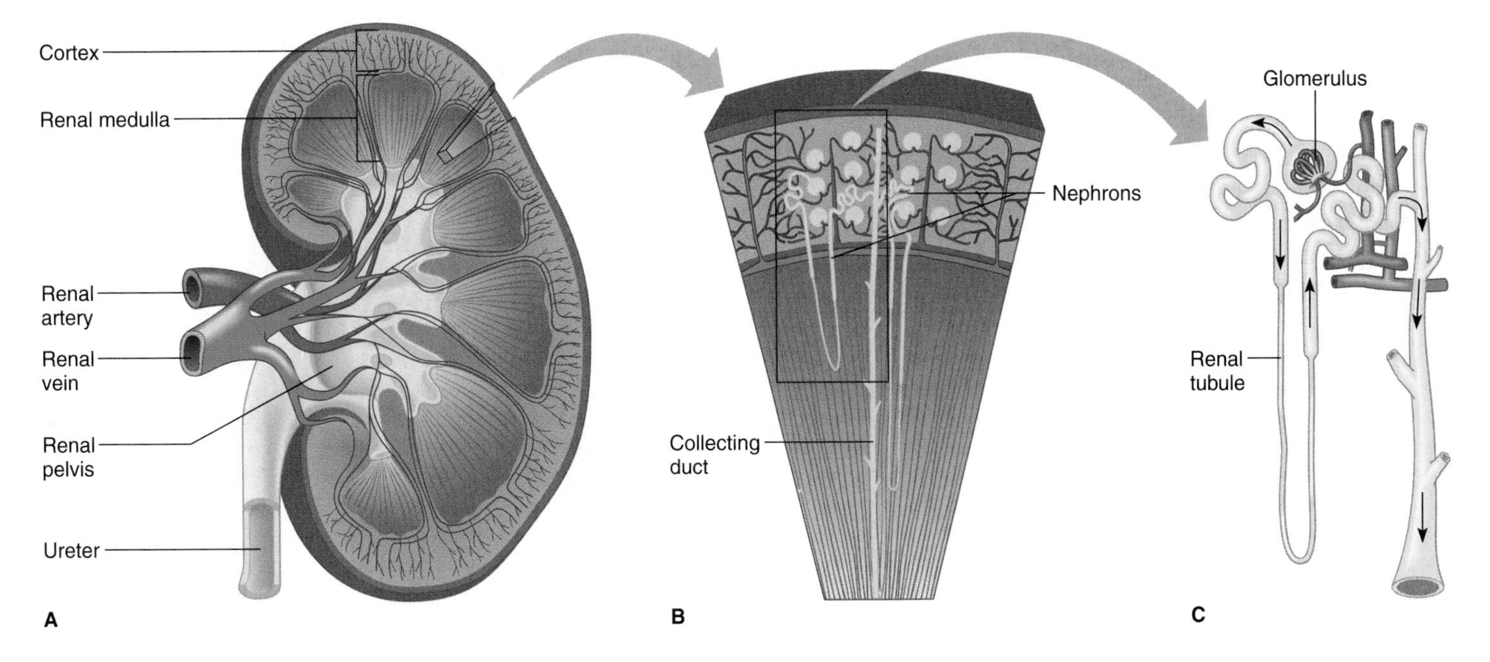

FIGURE 38.15

Overview of Urine Formation in the Nephron.

Urine forms from the processes of filtration, reabsorption, and secretion that selectively move substances between the blood and the nephron tubules. This is a stylized, schematic view that does not reflect nephron anatomy.

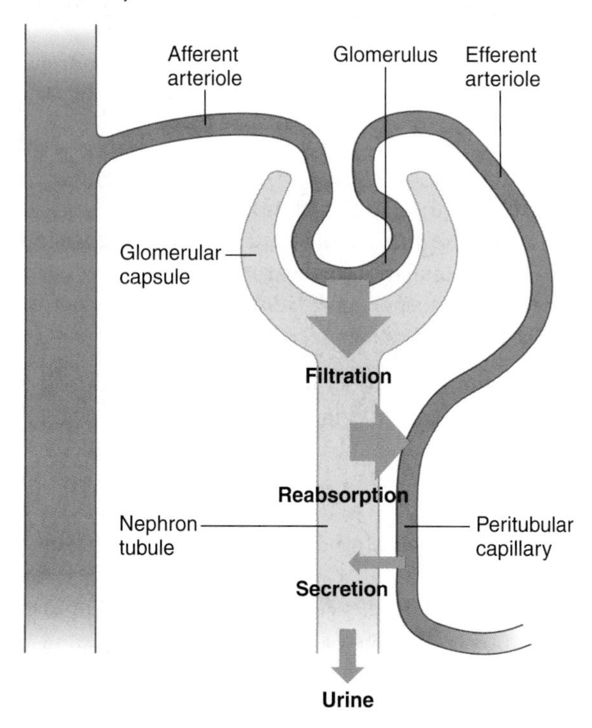

by a cup-shaped structure at the end of the renal tubule known as the **glomerular** (or Bowman's) **capsule** (fig. 38.16). Glomerular capsules, one at the beginning of each tubule, lie in the cortex of the kidney.

The glomerular capillaries filter blood. Anything that fits through the pores in the glomerulus and is not repelled by the charge on the membrane passes into the glomerular capsule. Like a catcher's mitt catching a ball, the glomerular capsule surrounds the glomerulus and captures all of this material.

Blood pressure provides the force that drives substances out of the glomerulus. Plus, the afferent arteriole has a larger diameter than the efferent arteriole. The resulting pressure forces fluid and small dissolved molecules across the capillary walls. Large structures such as plasma proteins, blood cells, and platelets remain in the bloodstream and leave the glomerulus in the efferent arteriole, which ultimately leads into the capillary network surrounding the nephron. These capillaries then empty into a venule, which runs into the renal vein, which finally joins the inferior vena cava.

The material in the glomerular capsule is chemically similar to blood plasma. Because only some substances from the blood enter the nephron here, however, the material passing into the glomerular capsule is called the glomerular filtrate. This is the product of the first step in urine manufacture. The 425 to 525 gallons (1,600 to 2,000 liters) of blood per day that pass through the kidneys produce approximately 45 gallons (180 liters) of glomerular filtrate.

The remainder of the renal tubule consists of a winding passageway with four functional regions: the proximal convoluted tubule, the nephron loop (loop of Henle), the distal convoluted tubule, and the collecting duct (fig. 38.17).

The Proximal Convoluted Tubule Recovers Valuable Substances The glomerular filtrate passes from the glomerular capsule into the **proximal convoluted tubule.** This is an important site for selective reabsorption, a process that returns useful components of the glomerular filtrate to the blood. The proximal convoluted tubule is highly folded, which increases its surface area, and its cells have many mitochondria, which power the active transport of dissolved nutrient molecules back into the blood. All along the tubule, specialized cells actively reabsorb glucose, amino acids, and important ions (electrolytes). Some ions are actively transported, and others tag along passively, attracted by the electrical charge on the transported ion. Water follows the electrolytes by osmosis and is also reabsorbed. Almost two thirds of the water and ions in the filtrate are reabsorbed into the blood in the proximal convoluted tubule.

FIGURE 38.16

The Glomerulus and Glomerular Capsule.

Small molecules dissolved in blood plasma cross the capillary walls, becoming filtrate and leaving some plasma, larger proteins, blood cells, and platelets in the bloodstream to exit through the efferent arteriole. The filtrate passes through filtration slits (enlargement) between extensions of cells called podocytes that wrap tightly around the glomerular capillaries.

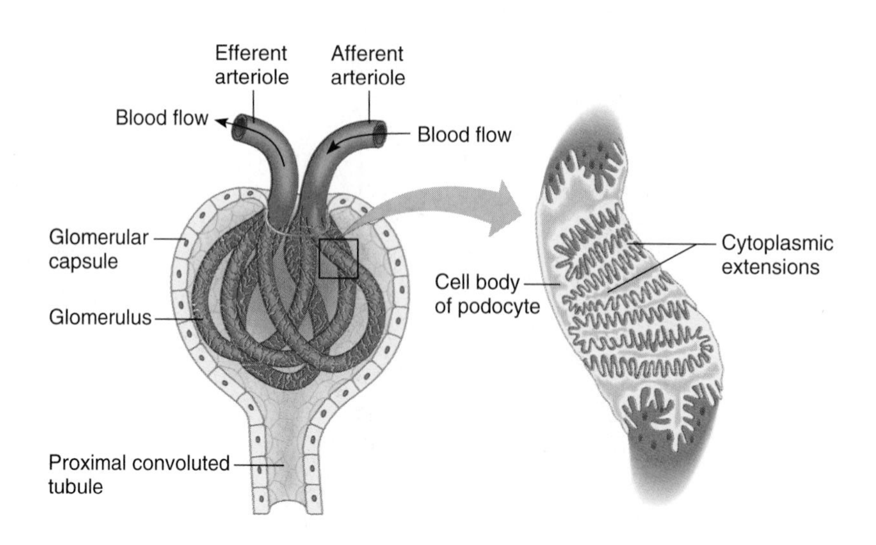

FIGURE 38.17

Activities Along a Nephron.

(**A**) Urine formation begins as the glomerulus filters out the large particles and cells, and passes the filtrate to the proximal convoluted tubule, where glucose, vitamins, ions, and amino acids are reabsorbed into the blood. From there the nephron forms a loop, shown schematically in (**B**). Active transport of Na^+ from the ascending limb creates a briny broth around the loop, which causes water to leave the nephron tubule and be reabsorbed into the bloodstream. Some urea diffuses out of the collecting duct and further adds to the high solute concentration around the nephron loop. The countercurrent flow of fluids in the tubules and adjacent blood vessels improves the exchange of substances. ("Mosm" stands for milliosmole, which is a unit of osmotic pressure.)

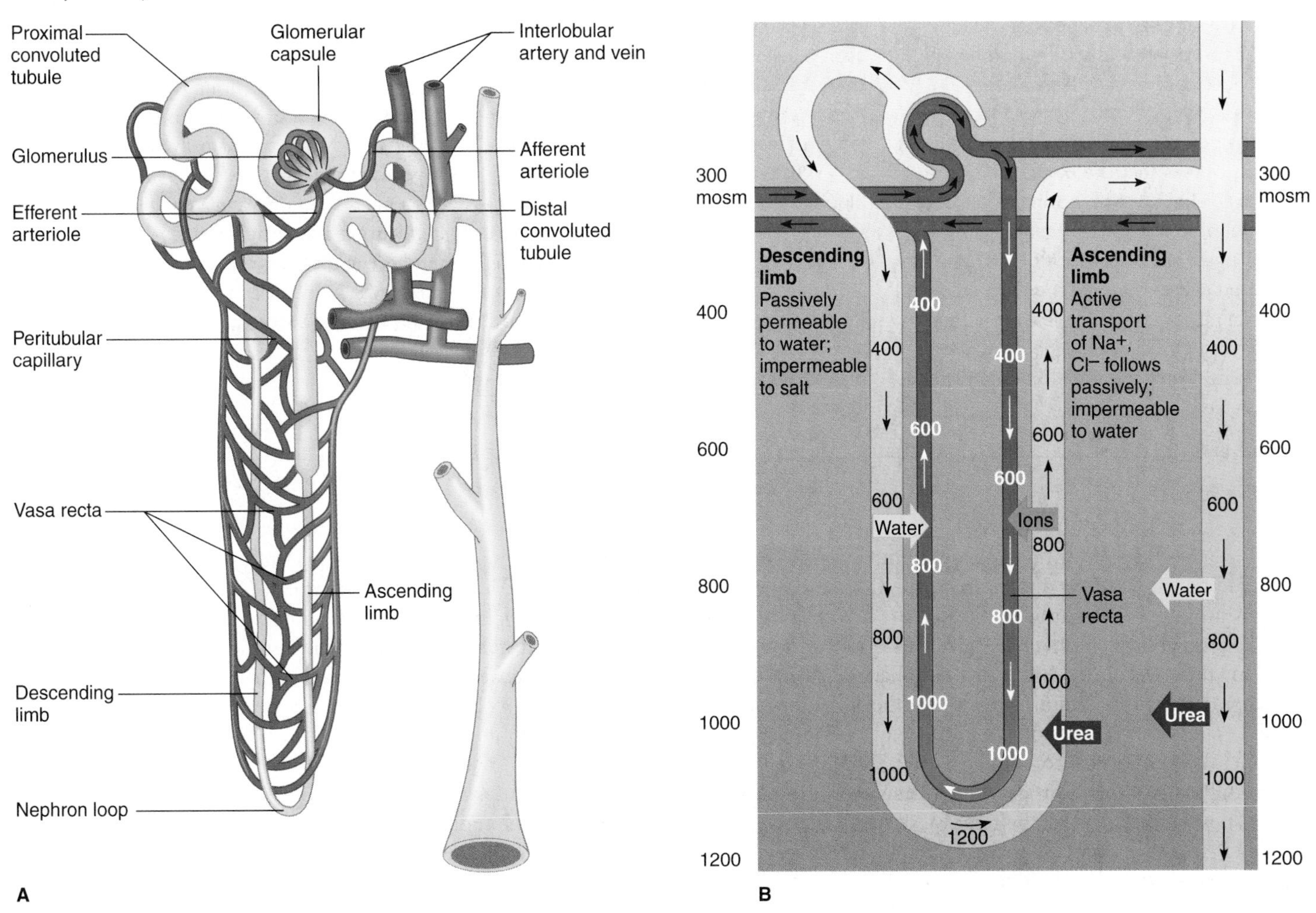

The Nephron Loop Conserves Water and Exchanges Ions with Blood Most of the remaining fluid is reabsorbed into the bloodstream in the **nephron loop,** which conserves water and exchanges ions with blood. The loop consists of a descending limb and then an ascending limb. The descending limb extends from the proximal convoluted tubule in the renal cortex and dips into the medulla. There, it folds back on itself to form the ascending limb, which returns to the cortex, where it becomes the **distal convoluted tubule.** Each nephron loop is closely paralleled by a portion of the peritubular capillaries called the **vasa recta.** Blood in the vasa recta flows in the opposite direction of the filtrate in the nephron loop. These two "antiparallel" fluid streams exchange water and ions in a countercurrent exchange system.

As the descending limb cuts across the kidney, it encounters an increasingly salty environment, which draws water out of the renal tubule and into the vasa recta by osmosis (see chapter 4). The cells of the descending limb are impermeable to ions but permeable to water. As a result, water passively diffuses into the intercellular spaces of the renal medulla and then into the blood of the vasa recta, following its concentration gradient. The vasa recta has a higher solute concentration than the tubule beside it because it is flowing from a saltier environment closer to the kidney's center. As water leaves the descending limb, the Na^+ concentration inside the tubules rises until it reaches its maximum at the bottom of the loop.

The filtrate next moves around the bend and into the ascending limb, which is impermeable to water along its length.

Health 38.1

Urinalysis—Clues to Health

Urine has long been a part of many folk remedies. It has been used as a mouthwash, a toothache treatment, and a cure for sore eyes.

Hippocrates (460–377 B.C.) was the first to observe that the condition of the urine can reflect health, noting that frothy urine denoted kidney disease. During the Middle Ages, health practitioners frequently consulted charts that matched urine colors to certain diseases. In the seventeenth century, British physicians diagnosed diabetes mellitus by having medical students taste sugar in a patient's urine!

Today, urine composition is still used as an indicator of health. The urine of a healthy individual contains water, urea, creatinine, uric acid, ammonia, amino acids, and several salts, in particular proportions. Urine should be pale yellow, with a pH of 4.8 to 8.0. These characteristics may fluctuate slightly due to diet. In some situations, urine is tested for traces of biochemicals from illegal drugs.

What can urine reveal about health? Blood in the urine indicates bleeding in the urinary tract. More than a trace of glucose may be a sign of diabetes mellitus. Carrier molecules actively transport glucose out of the filtrate in the kidney tubules and into the blood. If there are more glucose molecules than carrier molecules can handle, due to deficient insulin, the excess glucose remains in the filtrate and appears in the urine. Excess glucose in the blood may also indicate a high-carbohydrate diet or stress. Stress causes release of excess adrenaline, which stimulates the liver to break down more glycogen into glucose.

Albumin in the urine may be a sign of damaged nephrons, because this plasma protein is too large to normally fit through the pores at the entrance to the kidneys' microscopic tubules. Pus and a complete absence of glucose in the urine indicates infection somewhere in the urinary tract. The pus consists of white blood cells, which fight infection, and the infecting bacteria, which consume all glucose.

Near the bend, the wall of the ascending limb is relatively thin. Here, sodium ions (Na^+) diffuse from the filtrate along their concentration gradient into the more dilute blood of the vasa recta. (Some chloride and potassium ions are transported as well, but they are omitted here for simplicity). At the thick-walled portion of the ascending limb, near the distal convoluted tubule, Na^+ is moved out of the filtrate by active transport, which has two important consequences. First, it helps replenish the medulla's concentration gradient, which would otherwise dissipate over time. Second, the removal of these ions from the tubule makes the filtrate less concentrated than the tissues and blood in the surrounding cortex. This concentration difference between the distal convoluted tubule and the cortex allows for the hormonal regulation of urine composition and volume.

The Distal Convoluted Tubule Regulates Filtrate Composition In the next region of the renal tubule, the distal convoluted tubule in the kidney's cortex, sodium is actively reabsorbed into the peritubular capillaries. The steroid hormone **aldosterone** stimulates Na^+ movement out of the distal tubule. The accumulation of Na^+ outside the distal convoluted tubule stimulates water to diffuse passively out of the tubule into the capillaries. Water reabsorption into the blood from the distal convoluted tubule is also under the control of **antidiuretic hormone,** described in a subsequent section.

Most of the potassium ions (K^+) in the glomerular filtrate are actively reabsorbed into the blood at the proximal convoluted tubule. However, if excess K^+ is present in the blood, it may be secreted into the distal convoluted tubule and collecting duct. K^+ is attracted to negatively charged regions of the tubule that form as Na^+ leaves. Aldosterone also stimulates K^+ secretion, almost as if it were swapping one ion for the other.

The distal convoluted tubule also helps to maintain the pH of blood between 6 and 8. Any variance from this range is deadly. Blood is more likely to be too acidic than too alkaline, largely because most metabolic wastes are acids. The distal convoluted tubules can raise a too-low blood pH by secreting H^+ into the urine. It can lower a too-high blood pH by inhibiting H^+ secretion into the tubules. The distal convoluted tubule also secretes waste molecules such as creatinine (a by-product of muscle contraction) and drugs such as penicillin.

The Collecting Duct Conserves More Water The fourth portion of the renal tubule, the **collecting duct,** descends into the medulla, as does the nephron loop. Several renal tubules drain into a common collecting duct. Some of the urea diffuses out of the collecting duct as it passes through the medulla, where it contributes to the high solute concentration surrounding the nephron loops.

After reabsorption and secretion, the filtrate is urine. Urine contains water, urea, uric acid, creatinine, and several ions (see Health 38.1). From the collecting duct, urine accumulates in the renal pelvis before peristalsis carries it from the ureter to the urinary bladder and it finally moves out of the body through the urethra.

The nephron loop and collecting duct together are adaptations of the mammalian nephron to conserve water. The lengths of these tubes differ among mammalian species. The longer the tubes, the more water is reabsorbed into the bloodstream and the more concentrated the urine. The beaver and the Australian hopping mouse are at two extremes. The beaver lives surrounded by water and therefore does not have to conserve it. This animal has a short nephron loop (and collecting duct) and excretes a watery urine only up to twice as concentrated as its blood plasma. The Australian hopping mouse, in contrast, is a desert-dweller with the longest mammalian nephron loop relative to its body size. Its urine is up to 22 times as concentrated as its blood plasma. In comparison, human urine is up to 4.2 times as concentrated as blood plasma.

38.4 MASTERING CONCEPTS

1. What are the components of the human urinary system?
2. What are the components of a nephron?
3. What are the regions of a kidney?
4. How does urine formation begin?
5. What occurs in the proximal convoluted tubule?
6. What is the function of the countercurrent exchange system in the nephron loop?
7. What happens in the distal convoluted tubule?

38.5 Control of Kidney Function

Antidiuretic hormone controls water reabsorption from the distal convoluted tubule, and aldosterone increases salt retention in the kidneys.

The maintenance of blood pH, volume, and composition within an optimum range is essential to the functioning of each cell. Since the kidney plays a key role in regulating these factors, its activities are continuously monitored and adjusted to meet the body's requirements.

The amount of water reabsorbed from the filtrate influences two important characteristics of blood: the concentrations of plasma solutes and volume. The force generated as water moves by osmosis is called osmotic pressure; the greater the concentration gradient, the greater the osmotic pressure. Osmotic pressure affects many cellular activities, particularly the exchange of materials between cells and blood. Blood volume influences blood pressure and, therefore, affects cardiovascular health.

The solute concentration of the blood remains constant despite variations in the amount of water we consume in food or drink. When we drink too much, our kidneys allow more fluid to pass into the urine. If water is scarce, however, our kidneys conserve it by producing concentrated urine.

Osmoreceptor cells within the hypothalamus determine how much water to retain (fig. 38.18). When blood plasma becomes too concentrated, increasing osmotic pressure, the osmoreceptor cells send impulses to the posterior pituitary gland, which secretes antidiuretic hormone (ADH) in response. ADH increases water permeability in the distal convoluted tubule and the collecting duct. Because of the concentration gradient created by the countercurrent exchange system, increased water permeability increases water reabsorption into the blood.

FIGURE 38.18

Negative Feedback Control of Osmotic Concentration.

A feedback loop connecting the hypothalamus, the posterior pituitary, and the blood controls the amount of water reabsorbed from the kidneys into the bloodstream.

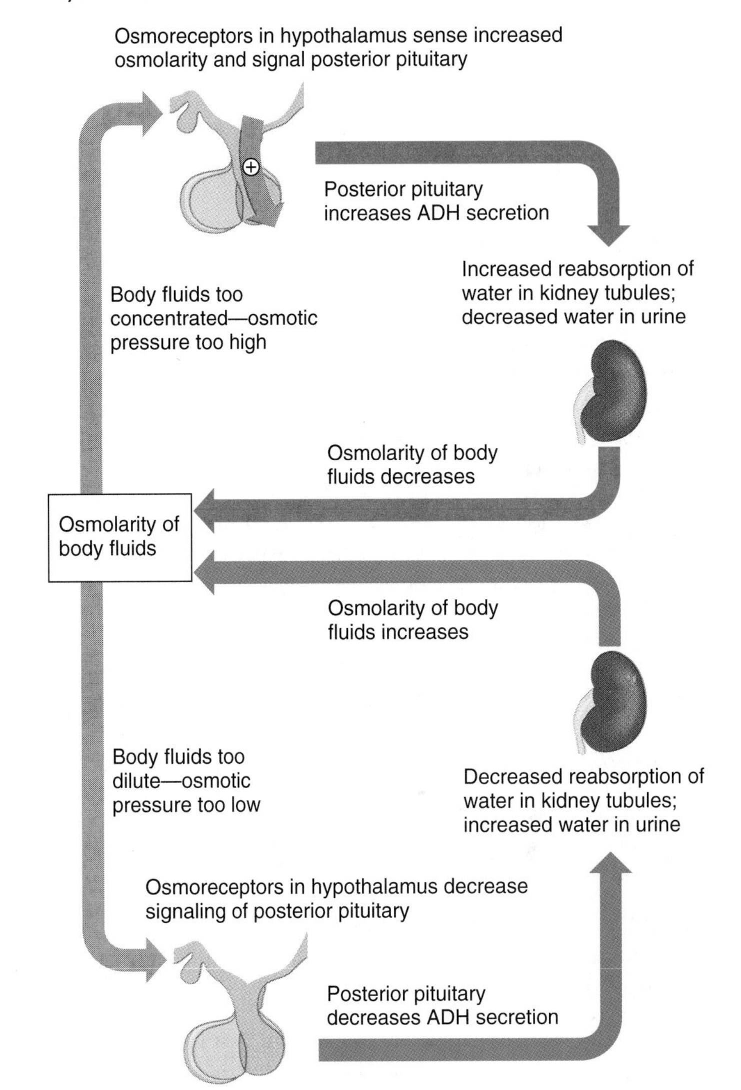

Conversely, if blood plasma is too dilute—if the osmotic pressure is too low—the same cells in the hypothalamus signal the posterior pituitary gland to stop producing ADH. As a result, the distal convoluted tubule and collecting ducts become less permeable to water. More water is excreted in the urine, and the osmotic pressure of the plasma and tissue fluids returns to normal. In a disease called diabetes insipidus, ADH activity is insufficient, and the person urinates from 1 to 2 gallons (4 to 8 liters) per day. Thirst is intense, yet water consumed rapidly leaves through the kidneys.

The ethyl alcohol in alcoholic beverages is a diuretic; this means it increases the volume of urine. Alcohol stimulates urine production by decreasing ADH secretion, thereby decreasing the permeability of the tubules to water. Because it increases water loss to urine, an alcoholic beverage actually intensifies thirst.

Dehydration resulting from drinking too much alcohol causes the discomfort of a hangover. Caffeine is also a diuretic, exerting its effects by decreasing Na^+ (and therefore water) reabsorption from the renal tubule. Similarly, many diuretic drugs increase urine volume by decreasing reabsorption of Na^+ in the proximal tubules. Diuretics, also known as "water pills," lower blood pressure and relieve edema (tissue swelling).

The adrenal hormone aldosterone enhances Na^+ reabsorption in the distal convoluted tubules as well as in the salivary glands, sweat glands, and large intestine. When the sodium level of the blood falls, or when blood pressure or volume declines, aldosterone synthesis increases and more Na^+ enters the bloodstream. Groups of specialized cells in the afferent arterioles also sense decreased blood volume and pressure. They respond by releasing another hormone, **renin,** which initiates a series of chemical reactions that eventually boost aldosterone levels.

38.5 MASTERING CONCEPTS

1. Why must kidney function be constantly monitored?
2. How does the solute concentration of the blood remain constant, despite fluctuations in water intake?
3. How do aldosterone and renin control blood volume and sodium level?

OLC

Chapter Summary

38.1 How Do Animals Regulate Their Temperature?

1. Adaptations permit animals to regulate body temperatures, the amount of water in their bodies, and solute concentrations in body fluids, which enable them to populate diverse habitats.
2. **Ectotherms** regulate body temperature by seeking an environment with the appropriate temperature. **Endotherms** use internal metabolism to generate heat.
3. Animals regulate their body temperatures metabolically, behaviorally, and with anatomical and physiological adaptations.
4. Adaptations to cold include shunting blood to interior organs; **countercurrent exchange systems** of blood vessels; antifreeze-like chemicals in blood; heat-seeking behaviors; blubber, feathers, and oxidation of fatty acids in brown fat; and shivering and non-shivering **thermogenesis.**
5. Heat-resisting adaptations include **evaporative cooling;** circulatory specializations such as the carotid rete; and body forms and behaviors.

38.2 How Do Animals Regulate Water and Ion Balance?

6. **Osmoregulation** is the control of ion concentrations in body fluids. Specific osmoregulatory adaptations depend on habitat.
7. **Osmoconformers** are aquatic organisms that have permeable body surfaces. Their body fluids have solute concentrations similar to those of the surrounding water.
8. **Osmoregulators** include aquatic animals in coastal regions that conserve ions under some conditions and lose excess ions in others. Terrestrial animals have many adaptations to conserve water and regulate ion composition.

38.3 How Do Animals Rid Themselves of Nitrogenous Wastes?

9. Animals must remove wastes. Most nitrogenous wastes come from protein breakdown and include **ammonia, urea,** and **uric acid.**
10. Ammonia formation requires the least energy but releases the most water. Urea requires more energy but is less toxic and helps conserve water. Uric acid requires the most energy to produce but can be excreted in an almost solid form, saving water.
11. Animals have diverse excretory systems to remove nitrogenous wastes, including **nephridia, Malpighian tubules,** and **kidneys.**

38.4 The Human Urinary System

12. The human **urinary system** excretes nitrogenous wastes (mostly urea) and regulates water and electrolyte levels. The kidneys, each of which drains into a **ureter,** produce urine. Ureters drain urine into the **urinary bladder** for storage. Urine leaves the body through the **urethra.**
13. The functional unit of the kidney is the **nephron.** Blood is filtered from the **glomerulus** into the **glomerular capsule.** As the resulting filtrate moves along the renal tubule, its composition is adjusted. Important materials are **reabsorbed** from the filtrate back to the blood and other substances are **secreted** into the urine.
14. Composition adjustment begins in the **proximal convoluted tubule.** The filtrate moves into the **nephron loop** that dips into and then out of the center, or **medulla,** of the kidney. The nephron loop helps to concentrate urine. Water leaves the filtrate because an osmotic gradient forms in the fluid around the nephron loop by a countercurrent exchange system built by Na^+ and urea reabsorption.
15. In the **distal convoluted tubule,** more reabsorption and secretion occur. The reabsorption of Na^+ helps control Na^+ balance. Secretion of H^+ helps regulate blood pH.
16. The filtrate then moves to the **collecting duct** and finally to the central cavity of the kidney, the **renal pelvis,** before draining through the ureter.

38.5 Control of Kidney Function

17. **Antidiuretic hormone** (ADH) regulates the amount of water reabsorbed from the distal convoluted tubule. The posterior pituitary gland secretes ADH in response to signals from the hypothalamus, which senses the osmotic pressure of blood. ADH increases permeability of the distal convoluted tubule and the collecting duct so more water is reabsorbed and urine is concentrated.
18. **Aldosterone** increases the kidneys' salt retention. The adrenal glands release aldosterone in response to either low sodium concentration in the plasma or low blood pressure. Some water reabsorption always follows Na^+ reabsorption.

Testing Your Knowledge

1. How does an endotherm differ from an ectotherm?
2. Cite three adaptations that help keep an animal's feet warm.
3. Explain how a carotid rete cools a mammal's brain.

4. How do humans and iguana lizards differ in body temperature regulation?
5. What are the three types of nitrogenous wastes, and where does each come from?
6. Draw a nephron and label the parts. Indicate which regions of the renal tubule are specialized for each of the three processes that form urine.
7. How is urine concentrated?
8. What are some medical conditions that urinalysis might detect?
9. How do ADH and aldosterone control kidney function?
10. Why can't you live without at least one kidney unless you undergo dialysis?

Thinking Scientifically

1. Cicadas living in the tropics have many more sweat pores than cicadas in cooler climates. Suggest how these differences might have arisen.
2. Imagine you are adrift at sea. Why would you dehydrate more quickly if you drank seawater to quench your thirst instead of fresh water?
3. Which of the substances listed below are excreted, and which are reabsorbed into the bloodstream?

	Concentrations (mg/100mL)		
Substance	**Plasma**	**Glomerular Filtrate**	**Urine**
Glucose	100	100	0
Urea	26	26	1,820
Uric acid	4	4	53
Creatinine	1	1	196

4. Urinary tract infections frequently accompany sexually transmitted diseases, particularly in women. Why?
5. Would an excess or deficiency of renin be likely to cause hypertension (high blood pressure)? Cite a reason for your answer.
6. Why is protein in the urine a sign of kidney damage? What structures in the kidney are probably affected?
7. How could very low blood pressure impair kidney function?

References and Resources

Cloudsley-Thompson, John. August 1993. When the going gets hot, the tortoise gets frothy. *Natural History.* Tortoises use evaporative cooling to maintain body temperature.

Falk, Dean. August 1993. A good brain is hard to cool. *Natural History.* Cranial radiators enable our brains to survive high temperatures.

Marsh, Alan C. August 1993. Ants that are not too hot to trot. *Natural History.* Namib ants scuttle across desert sand by pausing briefly on blades of grass.

Matas, Arthur J. August 10, 2000. Nondirected donation of kidneys from living donors. *The New England Journal of Medicine,* vol. 343, p. 433. Some people donate a kidney to nonrelatives, providing an alternative to dialysis or kidney transplant from a cadaver.

Yagil, Reuven. August 1993. From its blood to its hump, the camel adapts to the desert. *Natural History.* The camel is well adapted to life in dry heat.

The *LIFE* Online Learning Center provides additional resources and tools for studying this chapter.
www.mhhe.com/life

CHAPTER 39

The Immune System

Vaccines Old and New

Defending Against Disease

Vaccines have a long and colorful history. Centuries ago, they consisted of crusts from smallpox sores stuffed up one's nose. Decades ago, youngsters lined up to receive injections of the first polio vaccine, and a few years later, an oral vaccine squirted onto sugar cubes. Today people still receive vaccines as injections, but several new delivery methods will soon be available, including a nasal spray vaccine to protect against influenza, and bananas genetically engineered to arm the immune system against particular infections.

A vaccine is an inactive or partial form of a pathogen that prepares the immune system to respond to any future encounter with that pathogen. A variety of different types of cells and proteins are key to this response. Vaccines range from entire but weakened viruses or bacteria, to their parts.

A vaccine is an inactive or partial form of a pathogen that prepares the immune system to respond to any future encounter with that pathogen.

The First Vaccine

Vaccine technology dates back to the eleventh century in China. Based on the observation that those who recovered from smallpox never got it again, people would collect the scabs of infected individuals and crush them into a powder, which they inhaled or rubbed into pricked skin. In 1796, the wife of a British ambassador to Turkey witnessed the Chinese method of vaccination, and mentioned it to an English country physician, Edward Jenner. Intrigued, Jenner had himself vaccinated the Chinese way, and then thought of a different approach.

Smallpox Vaccine.
Edward Jenner inoculated children with smallpox vaccine in 1798.

It was widely known that people who milked cows contracted a mild illness called cowpox, but did not get smallpox. The cows became ill from infected horses. Since the virus seemed to jump species, Jenner wondered, would exposing a healthy person to cowpox lesions protect against smallpox? Wrote Jenner of the horse ailment that farmers transferred to cows: "It is an inflammation and swelling in the heel, from which issues matter possessing properties of very peculiar kind, which seems capable of generating a disease in the human body . . . which bears so strong a resemblance to the smallpox that I think it highly probable it may be the source of the disease."

A slightly different virus causes cowpox than smallpox, but Jenner's approach would prove successful, leading to development of the first vaccine (from the Latin *vaca* for "cow"). Unable to experiment on himself because he'd already taken the Chinese vaccine, Jenner instead tried his first vaccine on 8-year-old James Phipps. He dipped a needle in pus oozing from a small sore on a milkmaid named Sarah Nelmes, then scratched the boy's arm with it. Young James survived, and the smallpox vaccine was born. Eventually, the vaccine would completely eradicate the disease, although several nations maintain the virus in storage for research purposes.

Plant-Based Vaccines

Recall from Biotechnology 13.3 that transgenic organisms have a "foreign" gene stitched into each of their cells. If edible plants are engineered to encode the antigens that evoke an immune response in the human body, perhaps they can serve as vaccines. The antigens, when released in the digestive system, would activate the immune system in the entire body.

One of the first experimental plant-based vaccines was developed at the Boyce Thompson Institute for Plant Research in Ithaca, New York. A transgenic potato was given an *E. coli* gene that encodes an enterotoxin, which causes diarrhea. A week after mice ate the raw, shredded transgenic potatoes, antibodies against the toxin appeared in the circulation. This vaccine is now being tested on humans, and several other types are being developed.

A plant-based vaccine must be eaten raw. Because raw potato isn't appetizing, future plant-based vaccines may be delivered in bananas. This would certainly be an easy way to vaccinate babies.

TYPES OF VACCINES

Type	Example (some experimental)
Entire weakened (attenuated) pathogen	Polio, influenza
Inactivated toxins	Tetanus
Parts of killed pathogens	Cholera, whooping cough
Recombinant vaccines; subunit vaccine (key pathogen protein only)	Lyme disease, HIV, hepatitis B
"Naked" DNA from pathogen	Influenza, HIV, hepatitis B
Plant-based vaccine	
Potato	Diarrhea
Tomato	Hepatitis B, rabies
Black-eyed peas	Infectious disease in mink
Tobacco	Dental caries

39.1 The Evolution of Immunity

The complexity of the vertebrate immune system has its roots in protective cells and substances that provide immunity in invertebrates. Immunity may have evolved from cell-cell interactions that enabled unicellular forms of life to recognize each other.

Recognition is a theme in biology, and perhaps nowhere is this more apparent than in the functioning of the immune system. The cells of an immune system produce biochemicals that enable an animal cell to recognize the organism of which it is a part, and to destroy any cell or large structure that is not "self," or is an alteration of self, such as a cancer cell. The molecules of immunity distinguish self by recognizing, and binding, nonself antigens. An **antigen** is any molecule or part of a molecule that elicits an immune response. It is usually a carbohydrate or protein.

All animals have some form of immunity, from the simple collection of amoebocytes in a sponge, to the billions of cells of the human immune system. Immunity is a necessity for an animal's survival in a world populated by many other types of organisms. It is likely that this built-in biological protection has its roots in the ancestors of animals, in colonial microorganisms that had to be able to recognize one another and distinguish themselves from others, in order to form the associations that were perhaps the forerunners of multicellular life.

Invertebrate Immunity Is Innate Only

The first line of defense for an organism is often a physical barrier, such as skin. A barrier is part of **innate immunity,** which is defense that is not keyed to any specific target. Mucous membranes, a wide array of biomolecules, and even special cells are part of the innate immune response.

Even sponges, simple animals with only a few cell types, can fight infection. Their amoebocytes function as **phagocytes,** which are scavenger cells that engulf bacteria or viruses that flow in with the seawater, then dispose of the intruders by digesting them. A sponge's sense of self is clearly revealed in experiments that separate two different species of sponge into individual cells, followed by mixing the cells together. Rather than forming a hybrid, the cells rapidly assort into two sponges, one of each type. But a sponge's immune response is not as sophisticated as that of a vertebrate. Like a vertebrate, a sponge rejects grafts of foreign tissue, but it takes just as long to reject a second graft as it will a first. In contrast, a vertebrate has immunological "memory" that enables it to reject a second graft much faster than it did a first graft.

Found in corals, anemones, snails, arthropods, flatworms, and sponges, phagocytes are common components of innate immune systems. In the horseshoe crab *Limulus,* amoebocytes migrate to the site of an injury where bacteria have released a toxin, then initiate pathways that cause the copper-containing blue hemolymph of the crab to clot, entrapping the bacteria and enabling the animal to survive. (*Limulus* hemolymph is the basis of a test used on medical instruments to detect bacterial contamination. If an instrument dipped into a solution containing burst amoebocytes causes clotting, then bacterial contamination is present.) In animals complex enough to have circulatory systems, phagocytes patrol the tissues as they travel in the bloodstream.

The discovery of the protective role of phagocytes, in echinoderms, was a milestone in immunology. In 1882, Russian zoologist Elie Metchnikoff collected transparent starfish larvae on a beach in Sicily, pierced them with rose thorns, and watched what happened (fig. 39.1). A day later, he saw that amoebocytes had

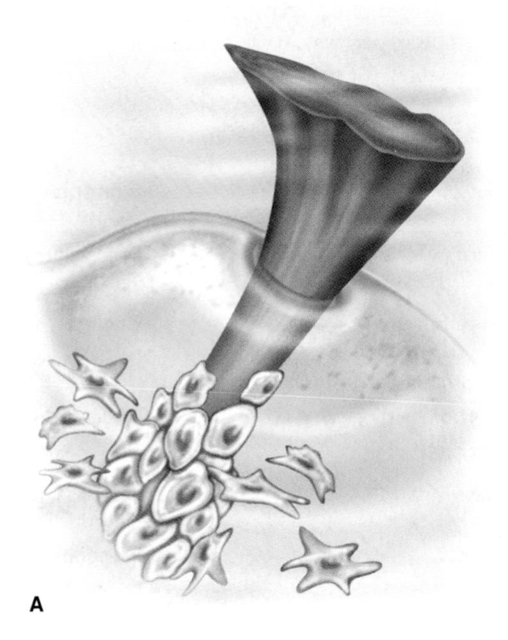

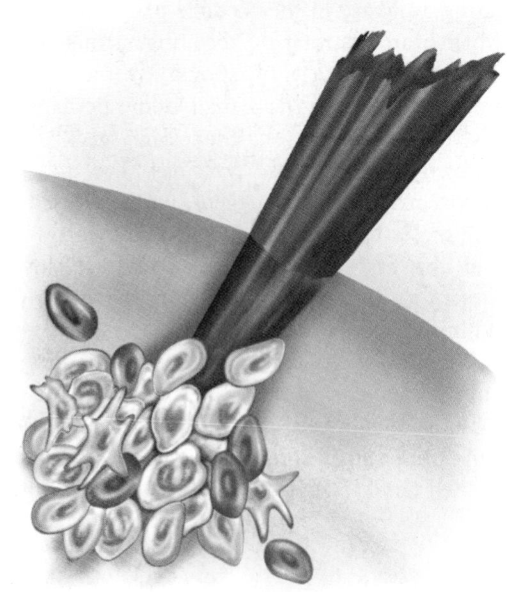

FIGURE 39.1
Responding to Injury.
Most animals respond to the entry of a foreign body, such as a rose thorn, as an initial step in the immune response. Even a simple animal like a starfish larva (**A**) surrounds a thorn with phagocytes just as human skin (**B**) does.

collected at the site of injury. This innate immune response to a thorn in a starfish larva is not unlike that of the human innate immune response to a splinter piercing skin.

Innate immunity is well studied in the arthropods. Insect immunity, for example, takes several forms. Amoebocytes circulate in the hemolymph, detecting and attacking invaders, initiating clotting reactions that immobilize infectious agents, just as in the horseshoe crab (also an arthropod). Insects make many types of antimicrobial peptides that are collectively called **defensins.** Signal transduction pathways that recognize a nonself cell surface or viral particle trigger the expression of genes that encode defensins. A class of defensins called cecropins was discovered in 1979 in the silk moth *Hyalophora cecropia.* Cecropins poke holes in bacterial cell walls and membranes, which are sufficient to cripple the microbes. These peptides are exquisitely sensitive, called into action when only a few bacteria are present. The recent discovery of cecropins in the gastrointestinal tract of pigs is yet another link between the immune systems of invertebrates and vertebrates. Amphibians produce similar proteins called magainins, discovered in 1987 when an astute researcher noticed that frogs recover from surgery performed under nonsterile conditions, even when they swim in bacteria-infested water. The magainins (Hebrew for "shield") protect against bacterial, fungal, and protistan infections, and may provide a new type of antibiotic drug.

The echinoderms are also key players in the evolution of immunity because they introduce molecules very similar to the **cytokines** that vertebrates produce. Cytokines are proteins synthesized in certain activated cells that increase or decrease the activity of other types of cells. The three major types of cytokines in vertebrates are the **interleukins,** the **interferons,** and **tumor necrosis factor.** Amoebocytes of the starfish *Asterias forbesi* produce an interleukin, and tunicates produce both an interleukin and tumor necrosis factor.

Vertebrate Immunity Is Innate and Acquired

The immune system of vertebrates has two parts that are separate but interact in highly coordinated ways—innate and acquired immunity. Innate immunity consists of certain cells and substances that provide a rapid, broad first-line defense against bacteria, viruses, fungi, and other parasites. The term "innate" refers to the fact that these defenses are already present, ready to function if infection occurs.

In contrast, the **acquired immunity** that is a hallmark of just the vertebrates is slower because it must be stimulated into action, taking days compared with the minutes of innate immunity. Acquired immunity is thought to have emerged about 450 million years ago. The first lymphocytes (white blood cells) appeared in the jawless fishes. The earliest jawed fishes—the sharks and rays—had two types of lymphocytes. Somehow, the forerunners of B cells picked up the ability to mix and match parts of genes. Today this ability enables B cells to combine pieces of antibody-encoding genes in a nearly limitless number of ways, to protect against a nearly limitless number of pathogens (disease-causing organisms). In this way the vertebrate immune system evolved the capability to specifically target particular types of invaders.

Acquired immunity is highly specific and more complex than innate immunity, and it leaves a lasting impression—that is, immunological memory. Because of this memory, the immune system of a vertebrate can respond rapidly when it again detects the same foreign antigen any time in the future.

Acquired immunity has two arms, based on the activities of the two types of lymphocytes, B cells and T cells. B cells produce antibodies in the **humoral immune response,** and T cells produce cytokines and can become "killers" in the **cellular immune response.** Innate immunity and then acquired immunity act when pathogens breach physical barriers, such as the skin or mucous membranes (fig. 39.2).

This chapter began with an overview of the vertebrate immune system, because it echoes protective strategies seen among the invertebrates, whose simpler systems are nonetheless incredibly adaptive. The fact that invertebrates greatly outnumber vertebrates attests to the success of their innate immune systems. Many vertebrate immune system components have invertebrate

FIGURE 39.2
Levels of Immune Protection.
Pathogens (disease-causing organisms) such as certain bacteria and viruses first encounter innate responses that prevent their entry into the body of vertebrates. If these barriers are breached, an array of specific cells and molecules attack and eliminate the pathogen from the body. Once activated, these specific cells retain a memory of the pathogen, allowing faster responses to subsequent attacks.

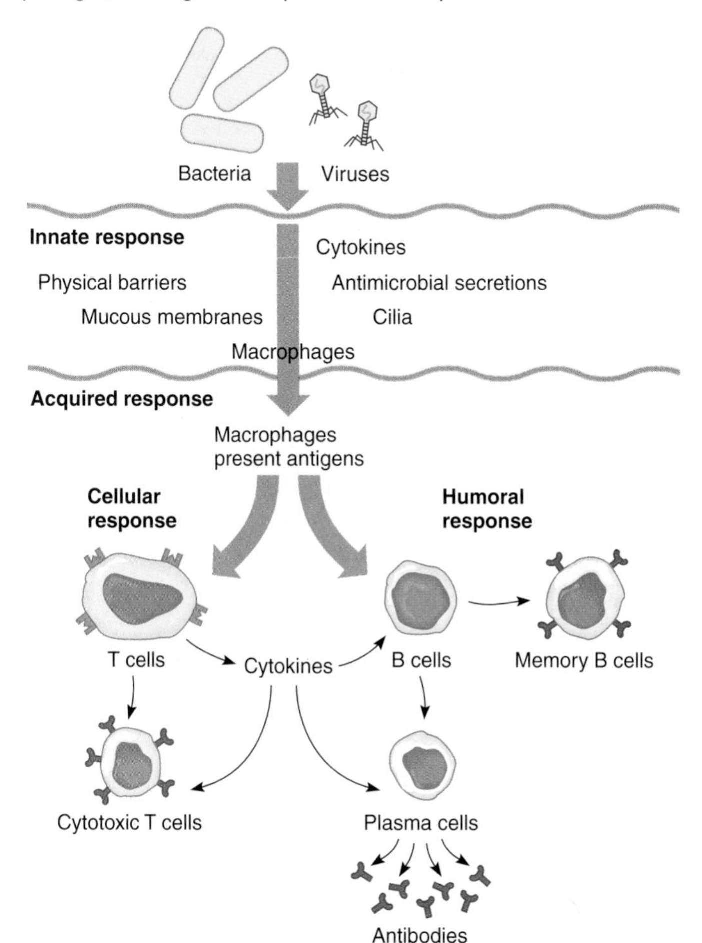

counterparts. For example, insects produce an antibacterial substance called lysozyme—which is also found in human tears. Another aspect of immunity that is traceable up a phylogenetic tree is a group of proteins called **complement.** In the vertebrate innate immune system, complement proteins interact in a series of reactions that perforate bacterial cell membranes, killing the microorganisms. Insects have enzymes very similar to complement proteins that encapsulate microbes. Sea urchins and tunicates have versions of a complement system. Figure 39.3 depicts the probable evolution of immunity, based on our limited current knowledge of when certain capabilities appeared.

The remainder of the chapter considers the human immune system. In many ways, it is quite similar to that of less complex animals.

39.1 MASTERING CONCEPTS

1. How is immunity among vertebrates more complex than immunity among invertebrates?
2. Distinguish between innate and acquired immunity.
3. What is the evidence that vertebrate immunity evolved from invertebrate immunity?

OLC

39.2 Innate Defenses Are Nonspecific and Act Early

Physical barriers and mechanisms to eject microbes are nonspecific, directed generally against any microbial invaders. Inflammation is a complex general response to infection or injury that sends in phagocytes and antimicrobial substances. Collectin proteins recognize nonself cell surfaces.

Infecting a human body is not easy. Barriers nonspecifically prevent microbes from entering the body—that is, they block all microbes, without distinguishing whether they are harmful or not. Unpunctured skin is the most extensive and obvious wall. There are other physical barriers to infection. Mucus in the nose traps inhaled dust particles; tears wash chemical irritants from the eyes and contain lysozyme; wax traps dust particles in the ears. Most microorganisms that pass these barriers and reach the stomach die in a vat of acidic secretions. Bacteria that enter the respiratory system are swept out of the airways by cilia and then swallowed.

If a pathogen breaches the body's outer barriers, the first response is a general process called **inflammation.** This is a nonspecific defense that creates an environment hostile to microorganisms at the site of injury (fig. 39.4). Consider a puncture wound, which triggers the inflammatory response. First, phagocytes infiltrate the injured skin, attacking entering bacteria. Plasma accumulates at the wound site, diluting toxins that bacteria secrete and bringing in antimicrobial substances. Increased blood flow warms the area, turning it swollen and red.

Inflammation is the innate immune response in action. Innate immunity in humans, as in other animals, is based on recognition of patterns—specifically, of the cell surface characteristics unique to pathogens. Phagocytes detect and home in on such surfaces (fig. 39.5). In the human body, some phagocytes are anchored in tissues; others, such as **neutrophils** and **macrophages,**

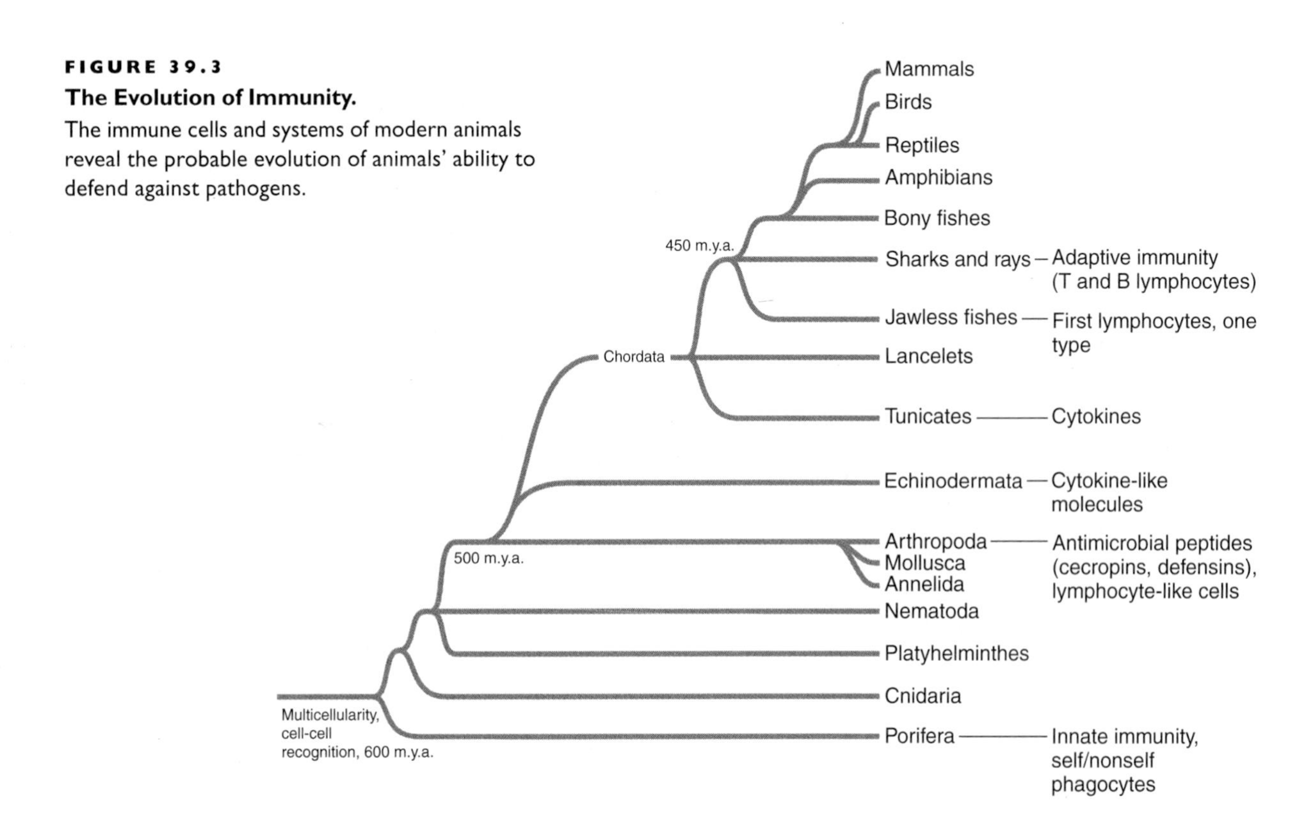

FIGURE 39.3
The Evolution of Immunity.
The immune cells and systems of modern animals reveal the probable evolution of animals' ability to defend against pathogens.

FIGURE 39.4
Inflammation.
An injury sets into motion the several steps of the inflammatory response.

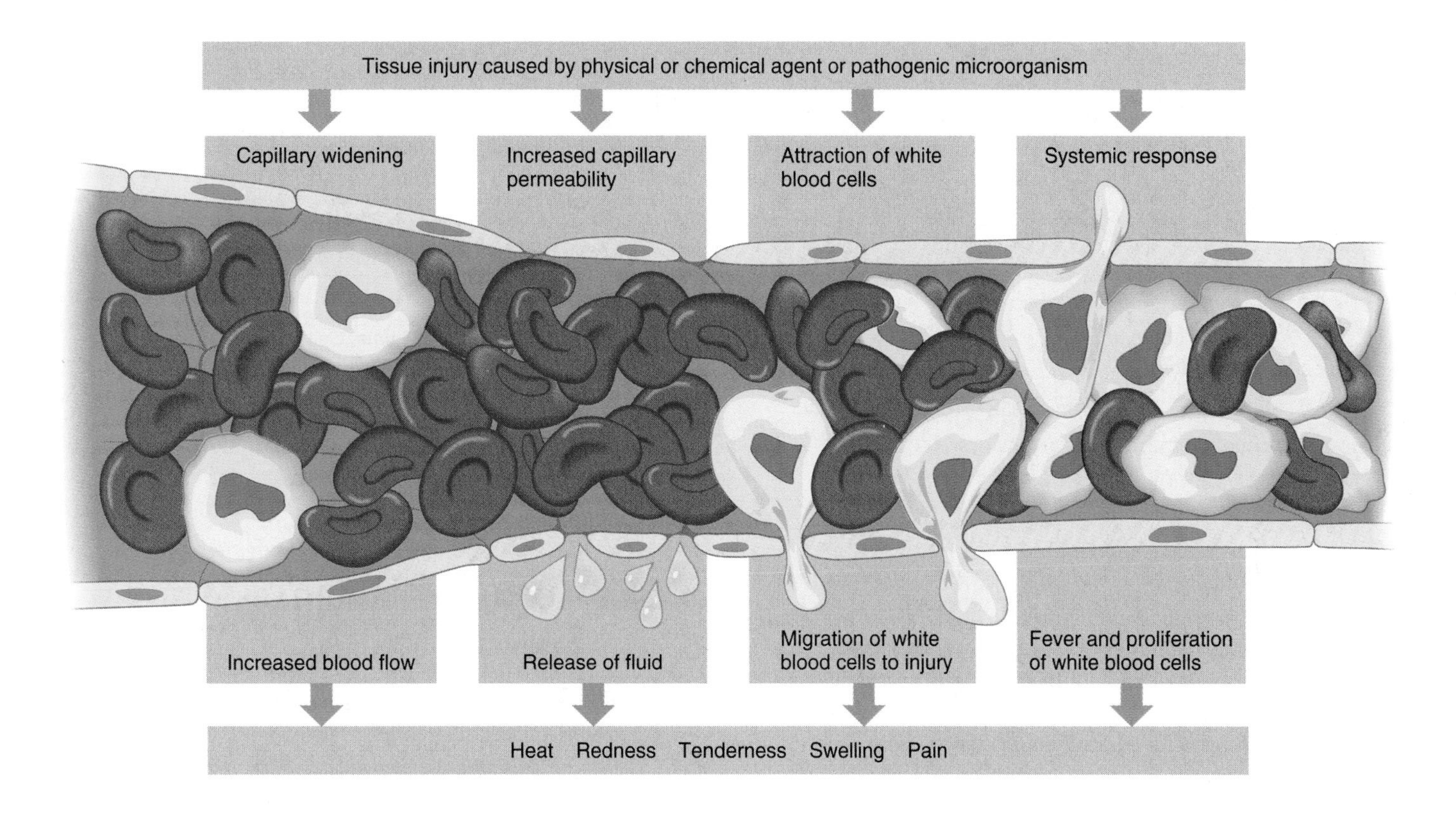

FIGURE 39.5
Nature's Garbage Collectors.
A human phagocyte engulfs a yeast cell.

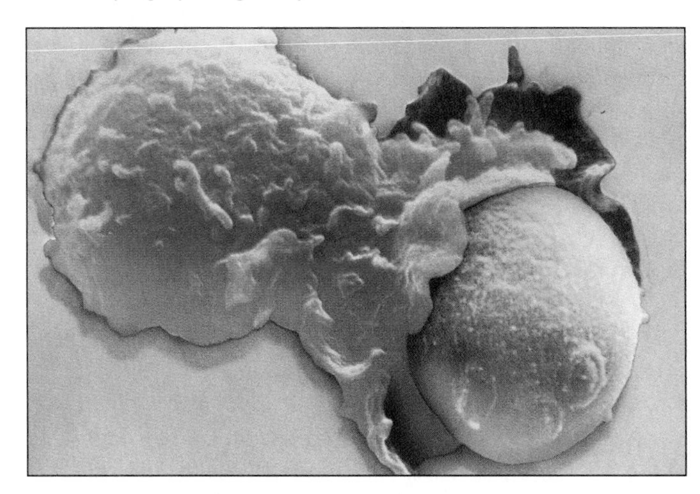

move. A neutrophil—an abundant type of white blood cell—can engulf only about 20 bacteria before it dies, but a macrophage can capture up to 100 bacteria. Dead phagocytes are one component of pus, a sure sign of infection.

Accompanying inflammation is attack by a large and diverse group of antimicrobial biochemicals that act nonspecifically, but are produced and released in response to detection of nonself antigens. One such type of biochemical, the defensins, are proteins produced by neutrophils and other types of white blood cells and in the intestinal epithelium, the urogenital tract, the kidneys, and the skin.

Another type of nonspecific defense, the complement system, is a group of proteins that assist, or complement, several of the body's other defense mechanisms. Some complement proteins trigger a chain reaction that punctures bacterial cell membranes, bursting the cells (fig. 39.6). Other complement proteins assist inflammation by causing **mast cells** to release a biochemical called **histamine.** Histamine dilates (widens) blood vessels, easing entry of white blood cells and fluid to the injured area. Still other complement proteins attract phagocytes to an injury.

Collectins are yet another type of protein that provides broad protection against bacteria, yeasts, and some viruses. These proteins home in on slight differences in the structures

FIGURE 39.6
Complement Kills Bacteria.
Triggered by a bound antibody, complement proteins combine to riddle a bacterium's cell membrane with holes, shattering its physical integrity. The bacterial cell quickly dies.

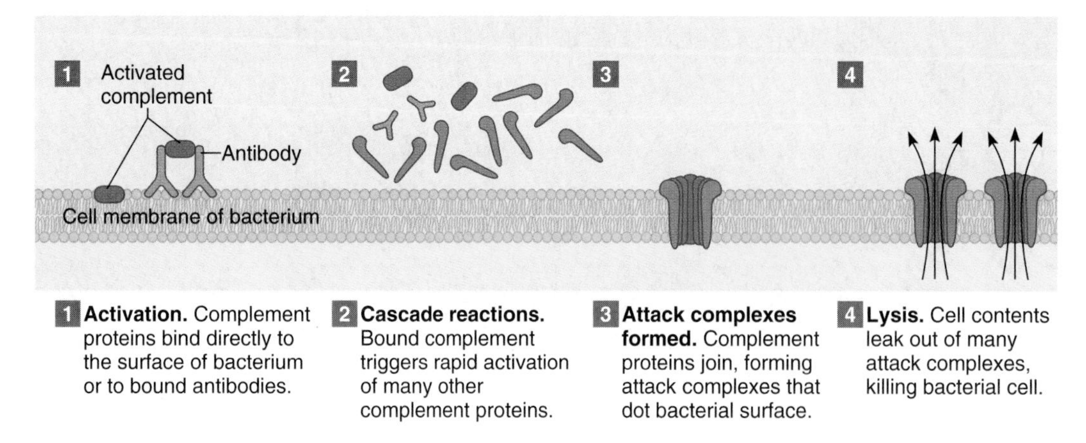

1 Activation. Complement proteins bind directly to the surface of bacterium or to bound antibodies.
2 Cascade reactions. Bound complement triggers rapid activation of many other complement proteins.
3 Attack complexes formed. Complement proteins join, forming attack complexes that dot bacterial surface.
4 Lysis. Cell contents leak out of many attack complexes, killing bacterial cell.

and patterns of sugars that protrude from the surfaces of pathogens. For example, a collectin called mannose-binding protein recognizes subtle differences in the placement of certain OH groups on mannose sugars on the surfaces of certain yeasts, zeroing in on them while being "blind" to similar sugars on self cells. Collectins detect not only the sugar molecules, but the pattern in which they are clustered, grabbing on much like Velcro clings to fabric. The human body's collection of collectins is diverse enough so that groups of them correspond to the surfaces of types of infectious agents. Just as some collectins bind to the mannoses of yeast, others detect lipopolysaccharides (sugar-lipid combinations) of gram-negative bacteria, the glycolipids of mycobacteria, and surface features of a double-stranded RNA virus.

Cytokines, although produced by activated T cells, take part in the innate immune response. Interferons, for example, alert other components of the immune system to the presence of virally infected cells, which are then destroyed to limit the spread of infection. Interleukins are a type of cytokine with many different effects on the body. Some cytokines cause fever, resetting the thermoregulatory center in the hypothalamus to maintain a higher body temperature. The heat kills some infecting bacteria and viruses directly. Fever also counters microbial growth indirectly, because higher body temperature reduces the iron level in the blood. Because bacteria and fungi require more iron as the temperature rises, a fever-ridden body stops their growth. Phagocytes also attack more vigorously when the temperature rises.

39.2 MASTERING CONCEPTS

1. What are some nonspecific barriers to infection in the human body?
2. What happens during an inflammatory response?
3. What are some antimicrobial biochemicals?
4. How is fever protective?

OLC

39.3 Acquired Defenses Are Specific and Act Later

A helper T cell alerted by an antigen-presenting cell activates a B cell to proliferate, differentiate, and secrete antibodies. Cytotoxic T cells kill infected cells, transplanted cells, and cancer cells. Gamma-delta T cells coordinate innate and acquired immune responses.

In the human body, about 2 trillion lymphocytes (B and T cells) and wandering macrophages provide immunity against specific nonself cells and viruses. Stem cells in the bone marrow give rise to T and B cells and monocytes, which mature into macrophages in the bloodstream. B cells migrate from the bone marrow to lymphoid tissues, and then also into the circulation. Upon stimulation by T cells, B cells proliferate explosively and differentiate further into **plasma cells,** which secrete huge numbers of antibodies, and **memory B cells** (fig. 39.7). T cells mature in the thymus gland and in epithelium of the small intestine, the skin, and the lungs. They account for 65 to 85% of all lymphocytes. T cells can act directly on other cells, in addition to secreting cytokines.

Antigen-Presenting Cells Begin the Immune Response

An immune response is a specific sequence of events. One of the first cell types to respond to infection is the macrophage. These scavenger cells alert lymphocytes by displaying antigens from engulfed invaders (fig. 39.8). Each such foreign antigen is attached to the macrophage surface by a self protein that is encoded by a gene that is part of a cluster called the **major histocompatibility complex,** or MHC. All vertebrates have a version of this gene cluster. The MHC encodes two classes of cell surface

FIGURE 39.7

Immune Cells Are Diverse.

T cells, B cells, and macrophages each contribute to an overall immune response. All three types of cells originate in the bone marrow and circulate in the blood.

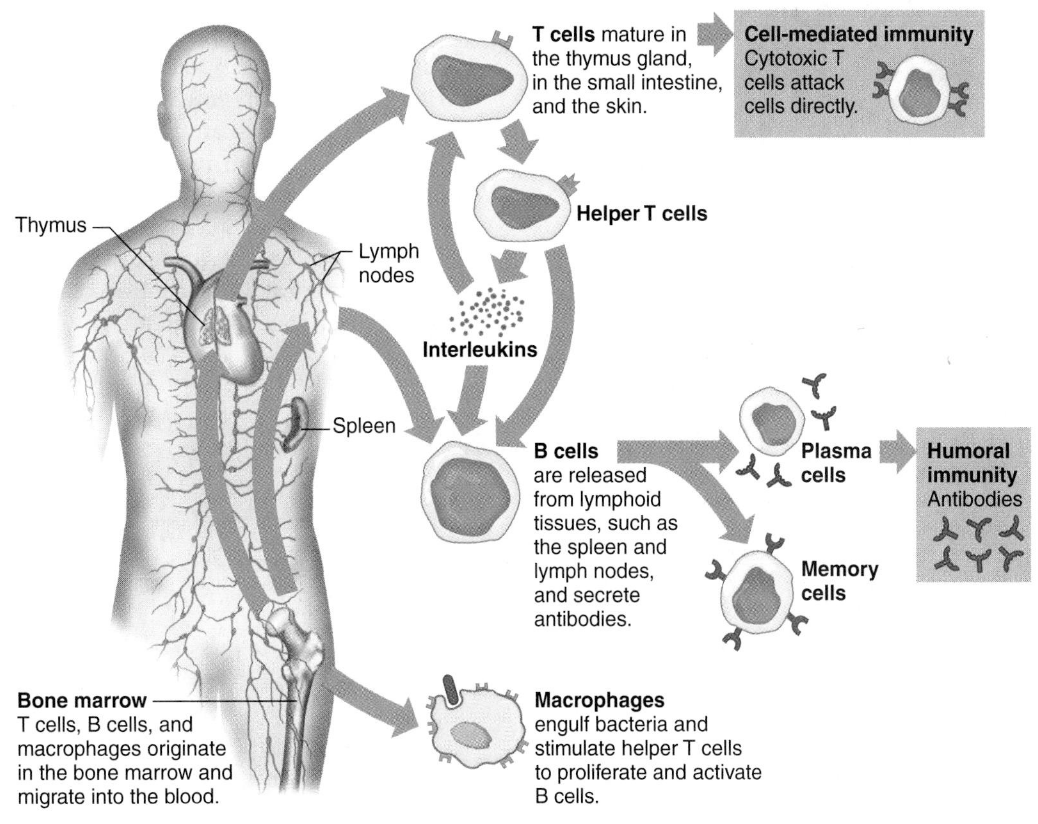

FIGURE 39.8

Macrophages Are Antigen-Presenting Cells.

A macrophage engulfs a bacterium by endocytosis, then displays foreign antigens on its surface, held in place by major histocompatibility complex (MHC) self proteins. This event sets into motion many immune reactions.

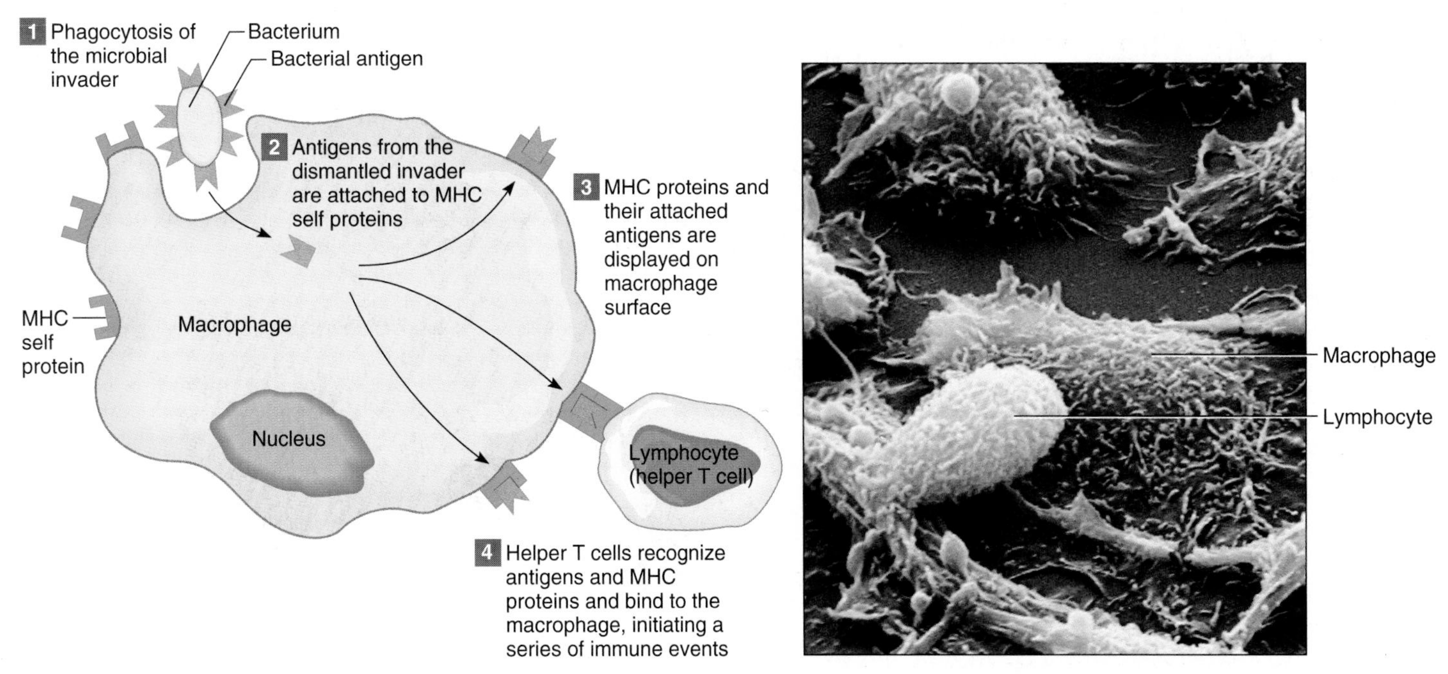

proteins, one class that marks all cells of an individual, and another that marks macrophages and lymphocytes. It is the MHC protein profile that has traditionally been "typed" to determine compatibility in organ transplantation. The MHC marks each cell as "self."

Displaying the foreign antigens like flags, thanks to the MHC proteins, macrophages migrate to the nearest lymph nodes, where they encounter many varieties of T and B cells. When a type of lymphocyte called a **helper T cell** recognizes both the macrophage's antigen flag and the MHC self label, it binds, beginning a chain reaction of events and effects, as the activated T cells secrete cytokines and stimulate B cells. The point where T cell meets antigen-presenting cell is called an "immunological synapse."

B Cells Produce Antibodies

At the heart of humoral immunity, B cells secrete antibodies into the bloodstream in response to activation by a helper T cell reacting to a foreign antigen displayed on a macrophage. Antibodies anchored to the surface of B cells bind to and display the foreign antigen, which is necessary for the helper T cells to recognize them. Interleukins can also indirectly stimulate B cells.

B cells come in an incredible number of varieties. Antibody production begins only when a receptor on the activated T cell binds a surface molecule on a particular B cell (fig. 39.9). If a match is made, the stimulated B cells divide rapidly, producing clones (groups of identical cells) that can identify the foreign antigen that triggered the proliferation. The daughters

FIGURE 39.9
Production of Antibodies.
B cell proliferation and maturation into antibody-secreting plasma cells constitute the humoral immune response. Note that only the B cell that binds the antigen proliferates and develops into memory cells or plasma cells.

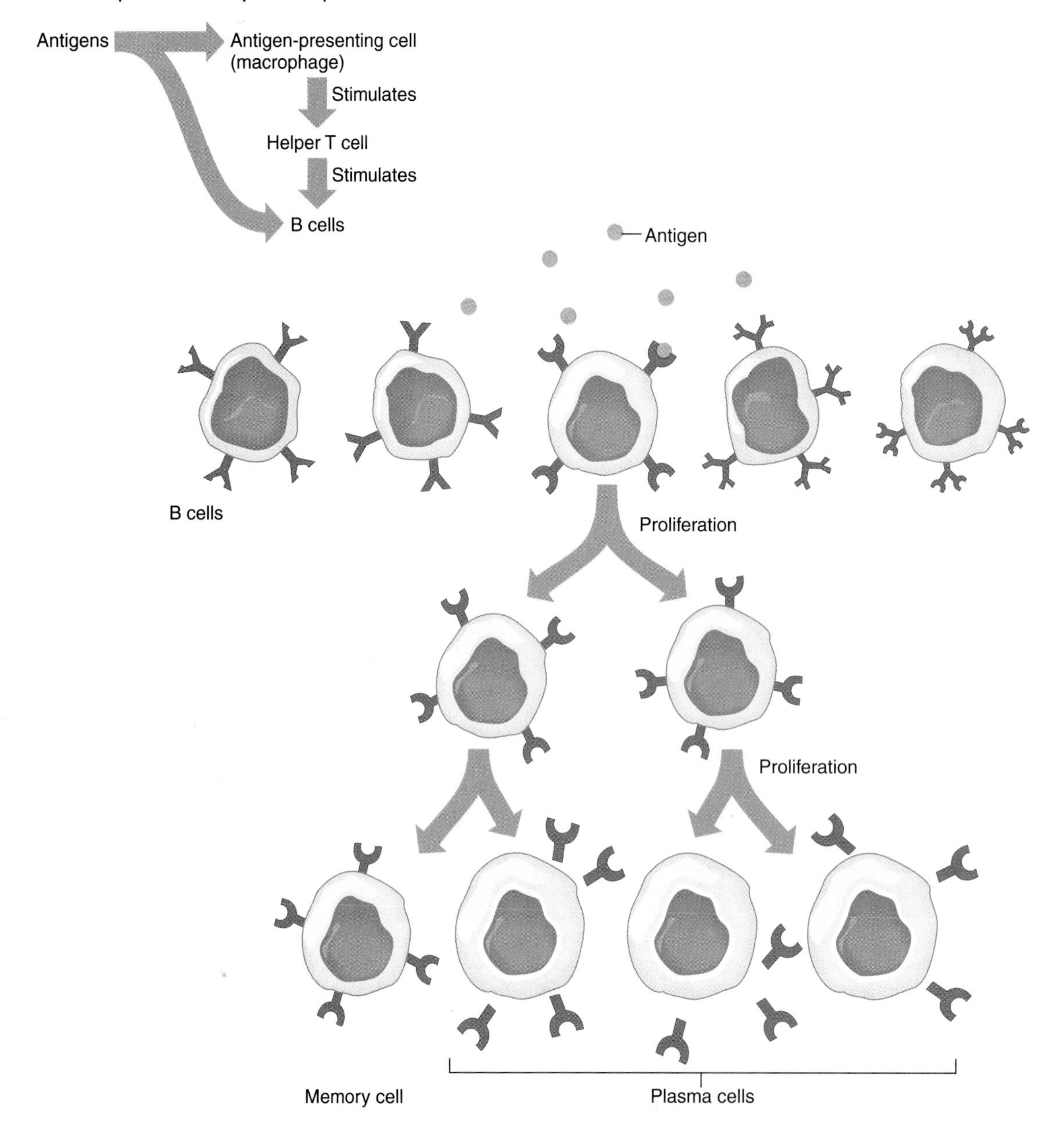

of these stimulated B cells mature into either plasma cells or memory cells.

A plasma cell secretes up to 2,000 identical antibodies per second during its few days of existence. These antibodies surround, bind, and inactivate the foreign antigens that triggered the B cell proliferation. Plasma cells derived from different B cells secrete different antibodies, and each antibody type corresponds to a specific portion of the foreign antigen. The different types of antibodies secreted in response to different portions of an antigen are collectively called a polyclonal antibody response (fig. 39.10).

Memory B cells are usually dormant, but they respond to the antigen quickly and forcefully should it be encountered again in the future. The **primary immune response** is the acquired immune system's reaction to its first meeting with a nonself antigen, and it takes a few days. Subsequent encounters stimulate a **secondary immune response,** which is so much faster that a person may not even be aware of it. Memory B cells function in the secondary immune response. Vaccination produces memory B cells.

The products of B cells, antibodies (also called immunoglobulins), are large, multipolypeptide proteins. Different genes encode the different polypeptides that form an antibody molecule. The simplest antibody molecule consists of four polypeptide chains connected by disulfide (sulfur-sulfur) bonds, forming a shape like the letter Y (fig. 39.11). The two larger polypeptides are called **heavy chains,** and the other two are called **light chains.** The lower portion of each chain is an amino acid sequence that is very similar in all antibody molecules, even in different species. These areas are called **constant regions.** The amino acid sequence of the upper portions of each polypeptide chain, termed **variable regions,** can differ a great deal among antibodies and determines the specificity of each antibody to its target. Thus, an antibody subunit has two

FIGURE 39.10

An Immune Response Recognizes Many Targets.

A humoral immune response is polyclonal, which means that different B cells produce antibody proteins that recognize and bind to different features of a foreign cell's surface.

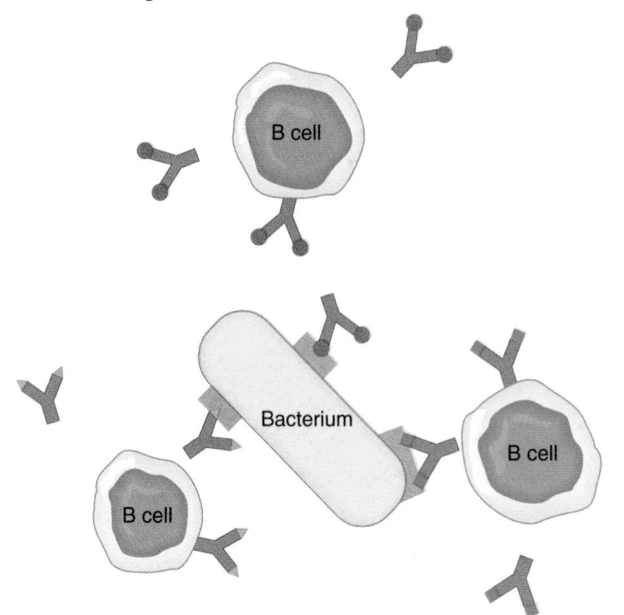

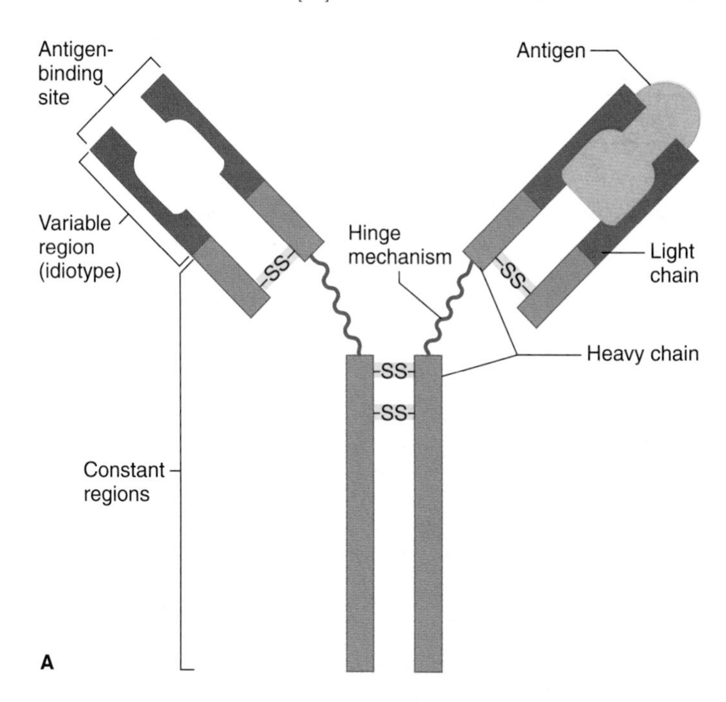

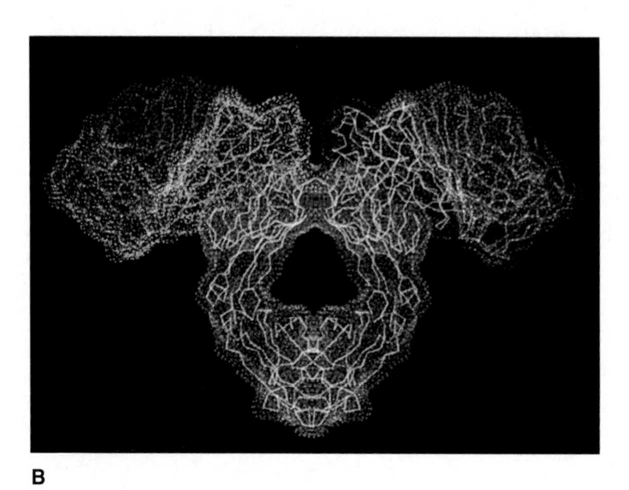

FIGURE 39.11

Antibody Structure.

The simplest antibody molecule **(A)** consists of four polypeptide chains, two heavy and two light, joined by disulfide bonds. Part of each polypeptide chain has a constant sequence of amino acids, and the remainder of the sequence is variable. The tops of the Y-shaped molecules form antigen-binding sites. **(B)** Computer modeling provides a three-dimensional view of an antibody molecule.

TABLE 39.1 Types of Antibodies

TYPE*	LOCATION	FUNCTIONS
IgA	Milk, saliva, and tears; respiratory and digestive secretions	Protects against microorganisms at points of entry into body
IgD	B cells in blood	Stimulates B cells to make other antibodies (little is known about IgD)
IgE	Mast cells in tissues and serum	Provides receptors for antigens that cause mast cells to secrete allergy mediators
IgG	White blood cells and macrophages in blood, serum	Binds bacteria and macrophages at same time, assisting macrophage in engulfing bacteria; activates complement; participates in secondary immune response; passes from pregnant woman to fetus to protect until fetus manufactures antibodies
IgM	B cells in blood, serum	Activates complement in primary immune response

*The letters A, D, E, G, and M refer to the specific conformations of heavy chains characteristic of each class of antibody.

heavy chains and two light chains, and each chain is partly constant and partly variable in its amino acid sequence.

Five major classes of antibodies are distinguished by their location in the body and their functions (table 39.1). Antibody types are also distinguished by their aggregation into complexes. Figure 39.12 shows antibody classes consisting of two and five subunits; the other three classes consist of only one four-polypeptide antibody molecule.

Antibodies can bind to specific antigens because of the three-dimensional shapes of their variable regions. The antibody contorts slightly to form a pocket around the antigen. The specialized ends of the variable regions of the antibody molecule are called **antigen binding sites,** and the particular parts that actually bind the antigen are called **idiotypes** (see fig. 39.11).

It is the binding of antibodies to the antigens on foreign cell surfaces that triggers the various protective actions of the immune response. Antibody-antigen binding inactivates a microbe or neutralizes the toxins it produces. Antibodies can cause pathogens to clump, making them more visible to macrophages, which then destroy the pathogens. Antibodies also activate complement, which destroys microorganisms.

The human body can manufacture an apparently limitless number of different antibodies because different antibody gene segments combine. As B cells develop, sections of their antibody

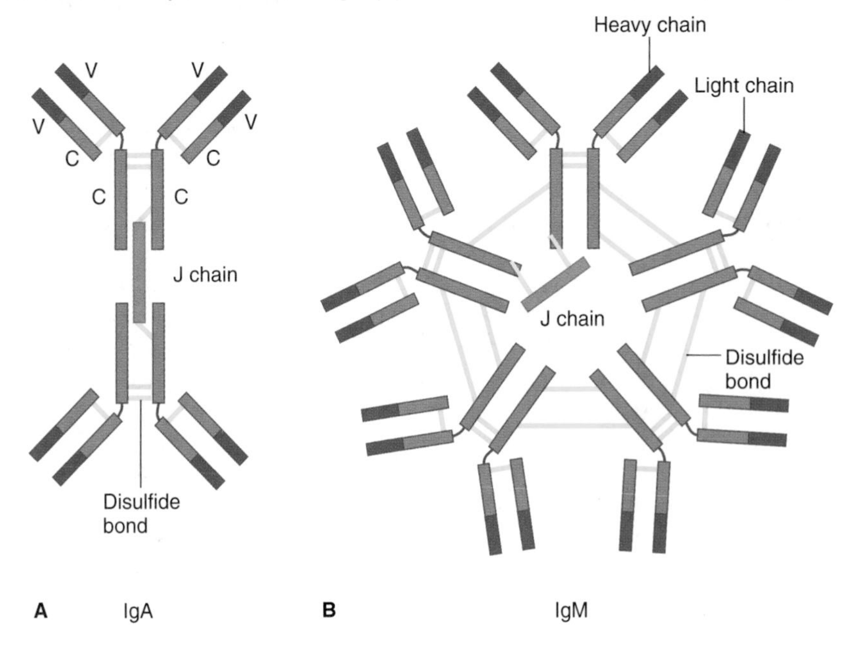

FIGURE 39.12
Classes of Antibodies.
Three of the five classes of antibodies consist of a single four-polypeptide molecule, as in figure 39.11. Each class contains variable (V) and constanct (C) regions. (**A**) IgA, however, consists of two Y-shaped subunits, and IgM (**B**) consists of five subunits. J chain proteins join the units.

genes move to other chromosomal locations, creating new genetic instructions for antibodies. Antibody diversity is analogous to using the limited number of words in a language to compose an infinite variety of stories. Antibody genes combine in many ways to generate an enormous diversity of antibodies. The immune system, as a result, is ready for nearly anything.

T Cells Stimulate, Secrete, Attack, and Coordinate

At the same time that B cells are specializing into plasma cells and secreting antibodies, T cells provide the cellular immune response. It is termed "cellular" because the cells themselves move to where they are required, unlike B cells, which stay where they are and secrete antibodies into the bloodstream. In the thymus, T cells acquire the ability to recognize self cell surfaces and molecules. T cells somehow also "learn" to not attack antigens in food, a characteristic called oral tolerance.

Several types of T cells have distinct functions. Helper T cells are crucial to a multifaceted immune response (fig. 39.13). They stimulate B cells to produce antibodies, secrete cytokines, and activate another type of T cell called a **cytotoxic T cell** (sometimes called a killer T cell). Helper T cells initiate a specific immune response. An immune reaction begins when a helper T cell recognizes a macrophage with a nonself antigen bound to its surface. The stimulated helper T cell then causes B cells to mature and secrete antibodies. The T cell also releases cytokines, including interleukins, interferons, and tumor necrosis factor, which elicit biological responses that destroy the nonself antigen.

FIGURE 39.13
T Cells Are the Heart of an Immune Response.
An activated helper T cell turns on both cellular and humoral immunity.

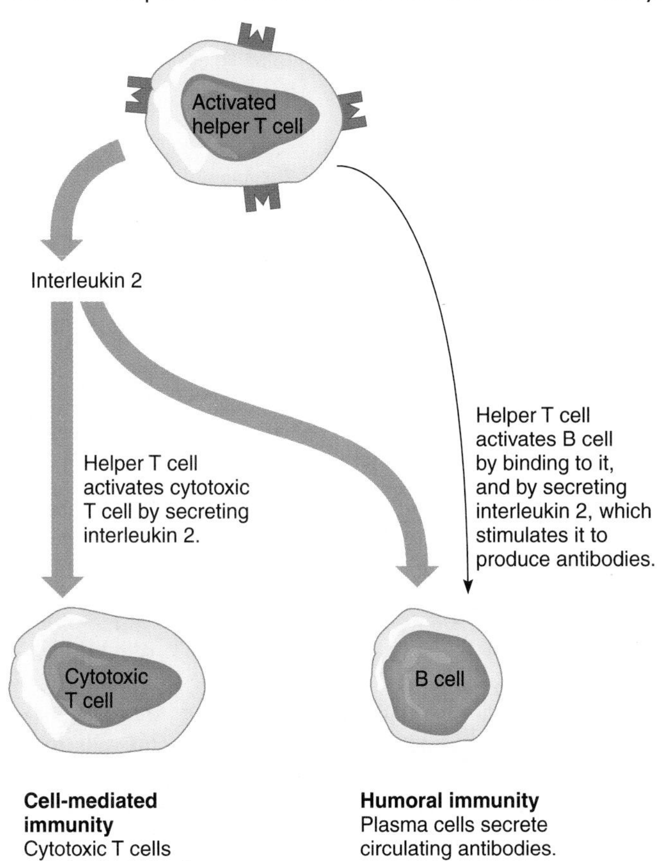

Cytotoxic T cells attack nonself cells by attaching to them and releasing biochemicals. To do so, they use T cell receptors on their surfaces that bind foreign antigens. When a cytotoxic T cell encounters a nonself cell, its T cell receptors bind to the foreign cell. The T cell then releases a biochemical called perforin that cuts holes in the foreign cell's membrane. This disrupts the flow of biochemicals in and out of the foreign cell and kills it (fig. 39.14). Cytotoxic T cells are also attracted by their receptors to body cells infected with viruses. The T cells destroy the infected cells before the viruses can spread. Cancer cells often produce molecules that make them appear "foreign" to the immune system and open them to attack by cytotoxic T cells.

Yet another type of T cell, called a **gamma-delta T cell,** has many functions and oversees the interactions of the innate and acquired immune responses. These cells serve as sentinels, detecting stress and injury, and setting into motion the other parts of the system to counter possible infection. Like other lymphocytes, gamma-delta T cells differentiate from bone marrow stem cells. Their precursor cells wander from the bone marrow to lymphoid patches in the small intestine, where they further specialize. Damaged epithelium activates gamma-delta T cells. Only recently discovered because they make up only a small fraction of the total population of lymphocytes, gamma-delta T cells are known to:

- recognize nonprotein molecules that cells release when they are stressed;
- synthesize epithelial growth factors, which stimulate cell division in epithelium, which is important in wound healing;
- produce biochemicals that contribute to the inflammatory response;
- secrete cytokines that coordinate the functioning of innate and acquired immunity.

Because gamma-delta T cells link the two major branches of the immune system, it will be interesting to see if researchers discover counterparts to them in invertebrates.

Turning off an immune response once an infection has been halted is as important as turning it on, because the powerful immune biochemicals can attack the body's healthy tissues. Certain cytokine combinations shut off the immune response. Special cytotoxic T cells reduce the number of proliferating B and T cells, leaving only memory cells behind.

39.3 MASTERING CONCEPTS

1. How does a macrophage display a foreign antigen?
2. What happens after a B cell is stimulated?
3. Describe the structure and function of an antibody.
4. How are T cells central to the immune response?

OLC

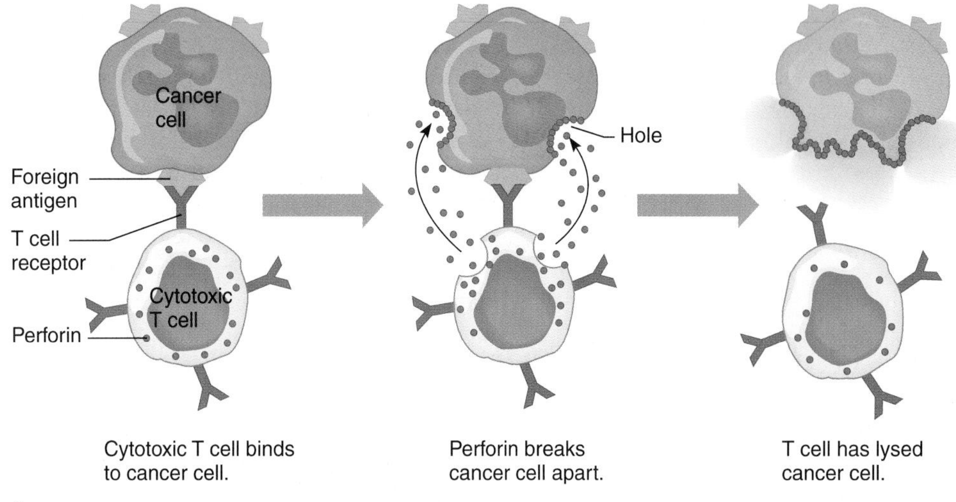

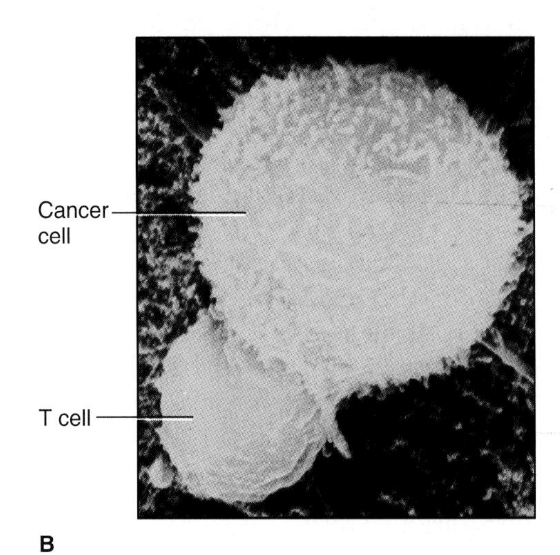

FIGURE 39.14

Death of a Cancer Cell.

(**A**) A cytotoxic T cell binds to a cancer cell and injects perforin, a protein that pokes holes in the cancer cell's membrane. As the holes form, the cancer cell dies, leaving behind debris that macrophages clear away. (**B**) The smaller cell is a cytotoxic T cell, which homes in on the surface of the large cancer cell above it. The cytotoxic T cell will literally break the cancer cell apart, leaving nothing behind but scattered fibers.

39.4 The Human Immune System Throughout Life

A pregnant woman's body does not reject a fetus, but medical complications can result from interactions between the two immune systems. The fetus learns to distinguish self from nonself. A newborn has passive immunity from the mother and soon mounts its own immune response. Immunity begins to decline after adolescence.

The immune system begins to develop in a fetus by cataloging cells destined to differentiate as lymphocytes. These cells will eventually learn to tell self from nonself. Before the lymphocytes have been "educated" to discriminate self from nonself, the fetal body will regard any antigen it encounters as self. A mouse fetus exposed early to another mouse's cells will, as an adult, accept a skin graft from the donor mouse. Similarly, human fraternal twins can have different ABO blood types, yet accept transfusions from each other if each twin's immune system, as it cataloged self antigens before birth, included the blood cell surfaces of the other.

As the fetus is taking inventory of its cells, it is already in a precarious situation. A fetus is totally dependent upon the pregnant woman, whose cell surfaces, although similar to the fetus's, are not identical. The female immune response dampens during pregnancy so that it doesn't reject the embryo and fetus. The woman's body detects the fetus because some of its cells enter her circulation. Pregnant women are more prone to infection while their immune systems rest, but after giving birth, protection returns.

Even though a woman's immune system adjusts to the temporary state of pregnancy, blood type incompatibilities between woman and fetus can cause problems. A blood type is determined by the pattern of antigens on red blood cells. One important antigen is the Rh type—a person can be Rh positive or Rh negative. When an Rh^- woman carries a fetus that is Rh^+, the woman's immune system reacts to the few fetal cells in her bloodstream by manufacturing antibodies against the fetal cells (fig. 39.15). Not enough antibodies form to harm the first fetus, but if she carries a second Rh^+ fetus, the built-up antibody supply attacks the fetal blood. Dying fetal blood cells release biliru-

FIGURE 39.15

Rh Incompatibility.

Fetal cells entering the pregnant woman's bloodstream can stimulate her immune system to make anti-Rh antibodies, if the fetus is Rh^+ and she is Rh^-. A drug called RhoGAM prevents attacks on subsequent fetuses.

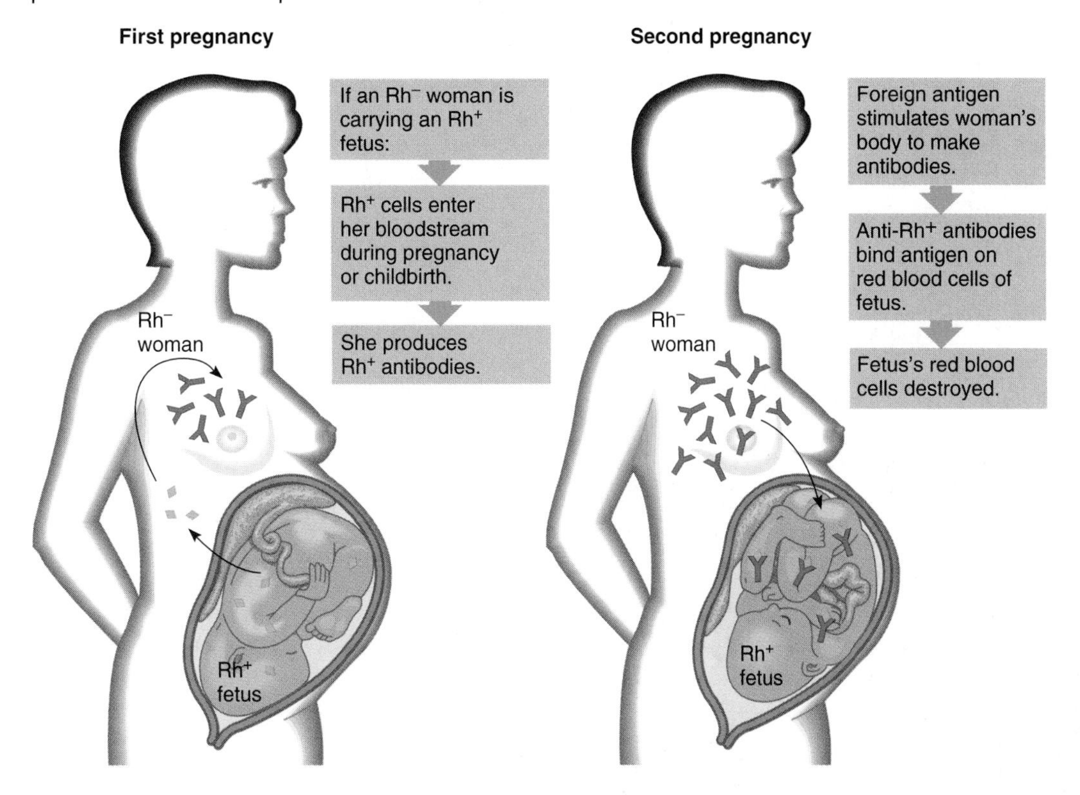

bin, which accumulates and damages the brain. The fetal liver and spleen swell as they struggle to replenish the red blood cell supply. A transfusion of Rh^- blood at birth can save the newborn's life, but this is rarely necessary—couples who face this potential problem are identified with blood tests in advance, and a drug is given that inactivates the woman's anti-Rh^+ antibodies. Health 39.1 describes a medical condition in a woman resulting from fetal cells remaining in her body from a past pregnancy.

A newborn has temporary immunity against certain illnesses from IgG antibodies passed from the woman's body during pregnancy. This is a form of **passive immunity,** because the child doesn't make the antibodies. Human milk provides additional protection. Shortly before and after giving birth, a woman's milk contains a yellow substance called colostrum that is rich in IgA antibodies that protect the baby from certain digestive and respiratory infections. In a few days, mature milk replaces the colostrum, adding antibodies against intestinal parasites.

Just as the maternal antibody supply in milk falls, the infant begins to manufacture its own antibodies. The first antibodies an infant makes are a response to bacteria and viruses transmitted by whoever is in closest physical contact. The mother also makes antibodies to the bacteria and viruses on her baby's skin (faster than the baby can) and passes them to the baby in her milk.

This gradual exposure to foreign antigens seems to be part of the normal development of the immune system. By 6 months of age, a baby can produce a great enough variety of antibodies to overcome most infections. Early exposure to the microbes we share the planet with may be crucial to the continuing development of the immune system.

The immune system begins to decline early in life. The thymus gland reaches its maximal size in early adolescence and then slowly degenerates. By age 70, the thymus is one-tenth the size it was at the age of 10, and the immune system is only 25% as powerful. The declining strength of the immune response is why elderly persons have a higher risk of developing cancer and succumb more easily to infections that they easily fought off at an earlier age. Decline in immune function is mostly due to loss of T cells. B cell activity changes little with age.

39.4 MASTERING CONCEPTS

1. How does a pregnant woman's immune system respond to the presence of a fetus in her body?
2. How does human milk provide immune protection?
3. What stimulates a newborn to manufacture antibodies?
4. At what age does the immune response begin to weaken?

OLC

Retained Fetal Cells Can Cause Scleroderma

"Scleroderma" means "hard skin," and this is an accurate description of this condition that one patient calls "the body turning to stone." Symptoms of fatigue, swollen joints, stiff fingers, and a masklike face typically begin between the ages 45 and 55 (fig. 39.A). In a severe form of the illness, the hardness extends to the endothelium of blood vessels, the lungs, the esophagus, and white blood cells, as well as the dermis.

Scleroderma has long been considered an autoimmune disorder, in which the immune system attacks the body, interpreting "self" as "nonself". But several curiosities about scleroderma alerted immunologists to another cause —fetal cells remaining from past pregnancies lodged in the skin and other affected organs, where they slowly stimulated an immune response. The clues:

- Scleroderma is much more common among women than men.
- Symptoms resemble those of graft-versus-host disease (GVHD), a condition in which transplanted tissue attacks the recipient's body. Antigens on cells in scleroderma lesions are the same as those involved in GVHD.
- Mothers who have scleroderma and their sons have cell surfaces that more closely resemble each other than unaffected mothers and their sons.

Sergio Jimenez and coworkers at Jefferson Medical College in Philadelphia considered retained fetal cells as a cause of scleroderma based on their familiarity with an experimental prenatal diagnostic technique: fetal cell sorting. Described in the opening essay to chapter 11, this test is based on the observation that a woman's blood can contain rare fetal cells for as long as 27 years following the birth. If fetal cells could persist in the circulation, possibly as stem cells, could they find their way to other tissues, such as the skin?

To answer that question, Jimenez sampled cells from scleroderma skin lesions from affected mothers of sons, and found the telltale Y chromosome among some of the cells. (Mothers of daughters can develop scleroderma from their children as well, but the fetal cells cannot be easily distinguished on the basis of chromosomes, as can cells from a male fetus.) The researchers hypothesize that when the fetal and maternal cell surfaces are very similar, then some fetal cells can persist without the immune system quickly destroying them. The cells travel, then differentiate, perhaps into white blood cells, dermis or endothelium. Some years later, possibly due to some triggering event, the woman's immune system becomes aware of the stowaway cells, and attacks. The result is a disorder that appears to be a typically unprovoked autoimmune attack, but is instead a delayed reaction to the retention of cells from a long-ago fetus. Are other autoimmune disorders the consequence of fetal influences?

FIGURE 39.A
Scleroderma Hardens the Skin.
Some cases of the skin-hardening disease scleroderma appear to be caused by a long-delayed reaction of the immune system to cells retained from a fetus—even decades earlier.

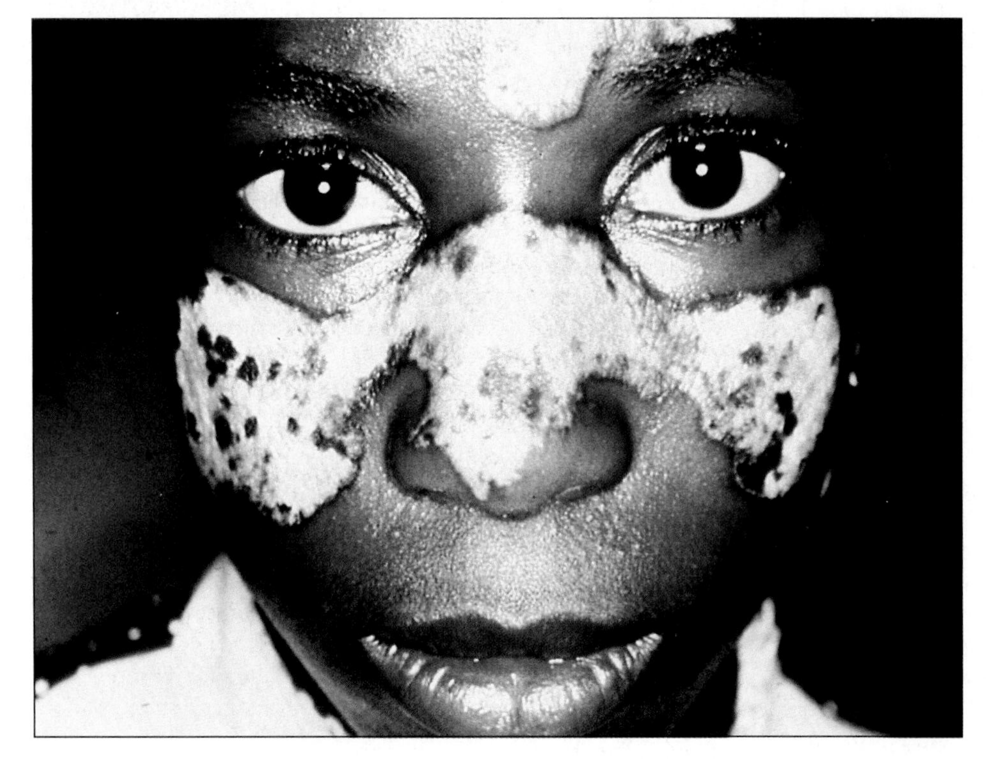

39.5 The Unhealthy Immune System

Immune system malfunction affects health in different ways. Breakdown leads to infection and cancer. An immune response inappropriately directed against self damages tissues. Attack against an innocuous substance produces an allergy. Because the immune system controls the body's response to disease, malfunction can be devastating. The immune response can be deficient, too intense, or misdirected.

Immunity Breakdown

Immune system breakdown devastates health. Such immune deficiency can be acquired or inherited.

HIV Infection Imagine that each day, 3 billion ants appear at your front door. The first day, you kill a billion of them. The next day, 3 billion more ants arrive, but even after you kill another billion, 4 billion remain. Each day you kill about a third of the ants, yet still they replenish their numbers. Soon you tire. You can only kill a fourth of them, then a fifth. Before long, you can no longer wage the battle, and ants overrun your house. As they come through the windows, under the doors, and down the chimney, they open up routes for other unwanted visitors, who damage the house in other ways.

This scenario is not unlike what happens when human immunodeficiency virus (HIV) infects a human body, targeting, at first, helper T cells. The first stage of infection is what one researcher calls "a titanic struggle between the virus and the immune system." For months to a decade or more, the body can produce enough new T cells daily to remove 100 million to a billion viruses. But as the virus destroys helper T cells, immunity falters, and the infections and cancers of AIDS begin. The infections of AIDS are termed "opportunistic" because they do not threaten a person who has a healthy immune system. Kaposi's sarcoma and pneumonia caused by *Pneumocystis carinii* are two common AIDS-associated illnesses in North America, but the infection greatly increases susceptibility to many infectious diseases, some once seen only in nonhuman animals.

Recall from chapter 19 that HIV infects helper T cells by binding to a receptor called CD4 as well as to a coreceptor, with a glycoprotein called gp120 (see fig. 19.4). HIV can evade detection because it is cloaked in a membrane containing molecules that the immune system would recognize as self. In addition, the only exposed viral proteins tend to change with each viral replication, making them nearly impossible for the immune system to recognize. The overall result is that the vulnerable portion of the viral surface—the part that a vaccine might target—is shielded until the precise instant when the virus contacts the helper T cell. An effective treatment for HIV infection must protect this key area on the cell. For now, treatment consists of combining drugs that tackle the virus at different stages of viral replication.

Once HIV infection is established, usually after several years, a viral variant arises that introduces a new weapon—it binds to a receptor on cytotoxic T cells and macrophages, triggering apoptosis (programmed cell death, see chapter 6). This is an indirect attack, because cytotoxic T cells lack the CD4 receptors that provide entrance to helper T cells. Waves of cytotoxic T cells, the ones that normally destroy viruses, die. When this happens, the symptoms of AIDS appear in earnest. The ability of HIV to trigger apoptosis is like an executioner forcing a condemned prisoner to kill himself. Discovery of this effect explained a long-standing mystery surrounding HIV infection—why cytotoxic T cells die if the virus can only enter helper T cells (fig. 39.16).

Not everyone exposed to HIV becomes infected or progresses to AIDS. About 5% of people infected with HIV remain healthy 12 to 15 years after infection. Some prostitutes in Africa who have been exposed to HIV hundreds of times never become infected. Some infants born infected with HIV "lose" the virus. A detailed study of such individuals, called "long-term nonprogressors," revealed a lucky combination of circumstances that may explain their persistent health—a strong immune response plus infection by weakened strains of HIV. They have abundant antibodies and helper T cells. In two of the patients, HIV was missing part of a particular gene, which apparently weakens it. About 1% of people have a mutation that alters certain receptors on CD4 cells that bind HIV. These people cannot become infected because HIV cannot enter their cells.

Although incidence of AIDS is declining in many places, more than 10 million people have died. In some African nations, life expectancy has plummeted due to AIDS (see figure 42.5).

Severe Combined Immune Deficiency AIDS is acquired—it is the consequence of infection. Immune deficiency can also be inherited. Each year, a few children are born defenseless against infection due to severe combined immune deficiency (SCID), in which neither T nor B cells function. David Vetter was one such youngster. Born in Texas in 1971, David had no thymus gland and spent the 12 years of his life in a vinyl bubble, awaiting a treatment that never came. As David reached adolescence, he wanted to leave his bubble. An experimental bone marrow transplant was unsuccessful—soon afterward, David began vomiting and developed diarrhea, signs of infection. David left the bubble, but he died within days of a massive infection.

Today, children with SCID caused by an inherited enzyme deficiency have several treatment options that were unavailable to David. Biotechnology 13.3 tells the story of the first patients to receive gene therapy to treat inherited immune deficiency.

Autoimmunity—Attacking Self

Sometimes the immune system turns against the host, manufacturing **autoantibodies** that attack the body's own cells and cause **autoimmunity.** The specific nature of an autoimmune disorder depends upon the cell types the immune system attacks. Table 39.2 lists some autoimmune disorders.

Why might the immune system turn on itself? Perhaps a virus, while replicating within a human cell, "borrows" proteins from the host cell's surface and incorporates them onto its own surface. When the immune system "learns" the surface of the virus to destroy it, it also learns to attack the human cells that normally bear that particular protein. Another possible explanation of autoimmunity is that T cells never learn to distinguish self

FIGURE 39.16

How HIV Decimates T Cell Populations.

Early in HIV infection, the virus enters helper T cells by first binding to their CD4 receptors. The virus then uses the T cells to reproduce more virus particles, while compromising the host cells' normal functions of coordinating the immune response. Later in infection, cytotoxic T cells, which are distinguished by different receptor molecules (designated CD8 and CCR) and a lack of CD4 receptors, become targets for infection by HIV. A mass suicide of cytotoxic T cells ensues. As populations of both helper and cytotoxic T cells plunge, opportunistic infections and cancers set in. This is AIDS.

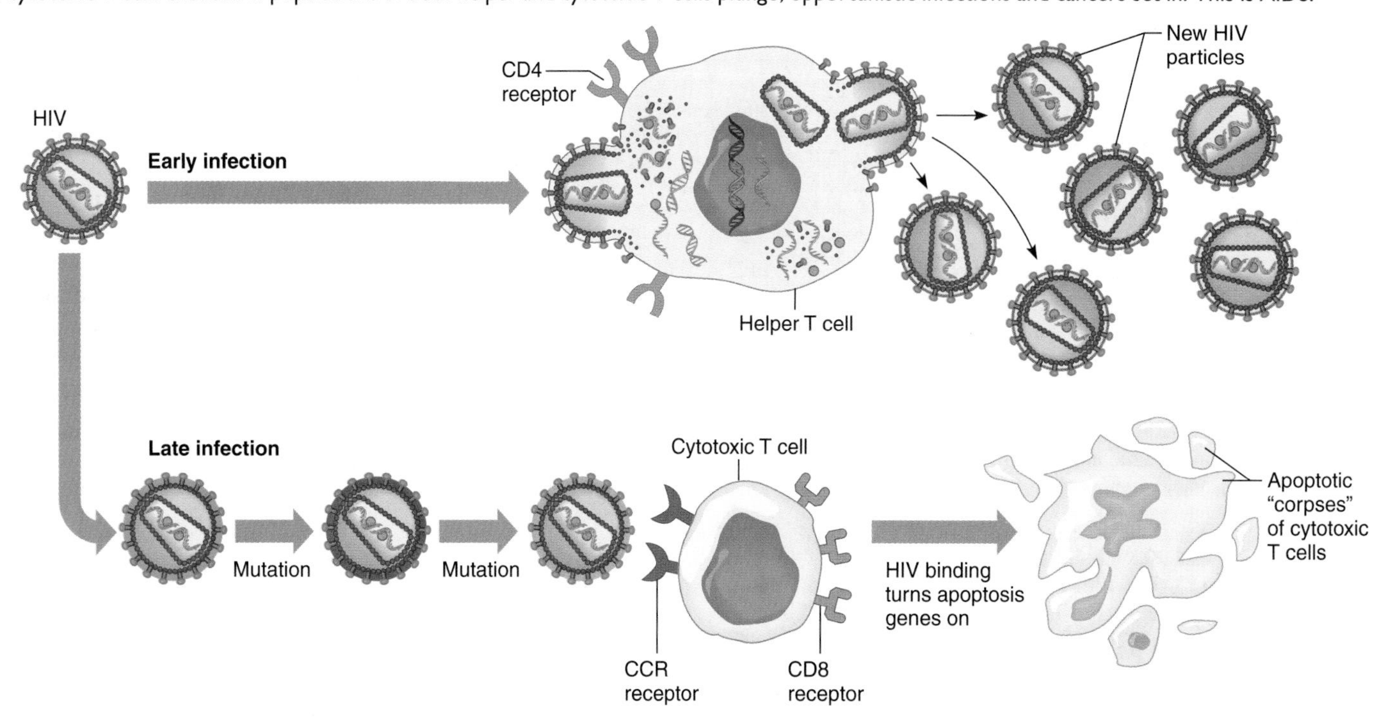

TABLE 39.2 Autoimmune Disorders

DISORDER	SYMPTOMS	TARGETS OF ANTIBODY ATTACK
Glomerulonephritis	Lower back pain	Kidney cell antigens that resemble *Streptococcus* antigens
Grave's disease	Restlessness, weight loss, irritability, increased heart rate and blood pressure	Thyroid gland antigens
Juvenile diabetes	Thirst, hunger, weakness, emaciation	Pancreatic beta cells
Hemolytic anemia	Fatigue and weakness	Red blood cells
Myasthenia gravis	Muscle weakness	Nerve message receptors on skeletal muscle cells
Pernicious anemia	Fatigue and weakness	Binding site for vitamin B on cells lining stomach
Rheumatic heart disease	Weakness, shortness of breath	Heart valve cell antigens that resemble *Streptococcus* antigens
Rheumatoid arthritis	Joint pain and deformity	Cells lining joints
Scleroderma	Thick, hard, pigmented skin patches	Connective tissue cells
Systemic lupus erythematosus	Red rash on face, prolonged fever, weakness, kidney damage	DNA, neurons, blood cells
Ulcerative colitis	Lower abdominal pain	Colon cells

from nonself in the thymus. Normally, T cells that recognize self antigens are weeded out, by apoptosis, in the thymus. If these cells persist, they can attack the body's tissues at a later time.

A third possible route of autoimmunity is when a nonself antigen coincidentally resembles a self antigen. For example, in rheumatic heart disease, which is a complication of rheumatic fever, antibodies attack heart valve cells that have antigens that resemble those on *Streptococcus* bacteria.

Another example of autoimmunity triggered by resemblance of a self antigen to a foreign antigen occurs in Lyme arthritis. In people who have a certain genotype, a protein in their joints resembles an antigen on the spirochete bacterium that causes Lyme disease. When these individuals become infected and their immune systems attack the bacteria, they also attack the joints. Because the source of joint pain is autoimmune, once it begins, antibiotic therapy does not help.

Superantigens Cause an Exaggerated Immune Response

Macrophages trim and display most foreign antigens on their surfaces, held in particular orientations by snugly fitting MHC proteins. A typical displayed antigen might attract and stimulate one in a million or so passing T cells. Certain foreign antigens, however, break these rules. They are called **superantigens.** When displayed, they are unusually large, and hang onto the outsides of the MHC molecules. Because this stance exposes more of the foreign antigen, more T cells respond—many more. Up to 5% of the entire population of T cells may be alerted, some tens of millions of cells. So many T cells secrete such abundant cytokines that harmful symptoms result.

The best-studied superantigens are bacterial toxins. For example, part of a toxin that *Staphylococcus aureus* produces causes some cases of food poisoning because of its role as a superantigen. It is the resulting flood of cytokines that causes vomiting and diarrhea, not the spoiled food or the bacteria. A different *S. aureus* toxin causes toxic shock syndrome. The sudden release of tumor necrosis factor makes blood vessels leaky, dangerously lowering blood pressure and causing a state of shock.

Allergies Misdirect the Immune Response

In an **allergy,** the immune system is overly sensitive and attacks harmless substances. The triggering substances, called **allergens,** activate IgE antibodies, which stimulate mast cells to explosively release **allergy mediators,** which include histamine and heparin (fig. 39.17). Allergy mediators cause symptoms by increasing the permeability of capillaries and venules, contracting smooth muscle, and stimulating mucous glands.

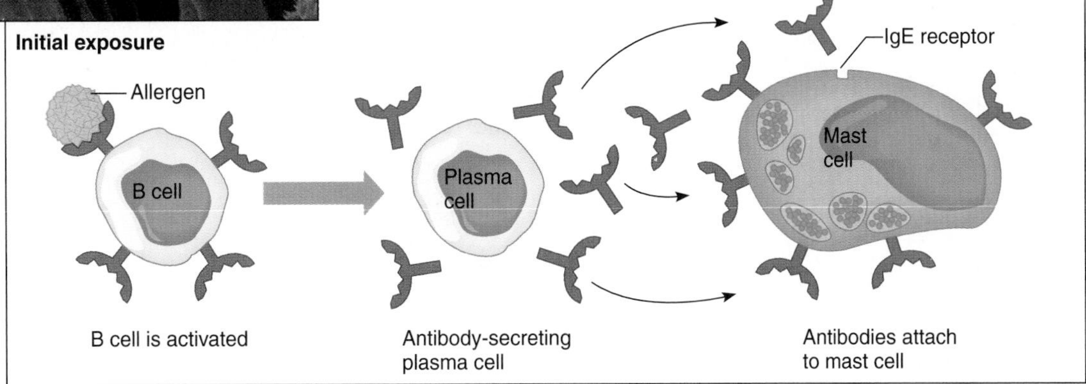

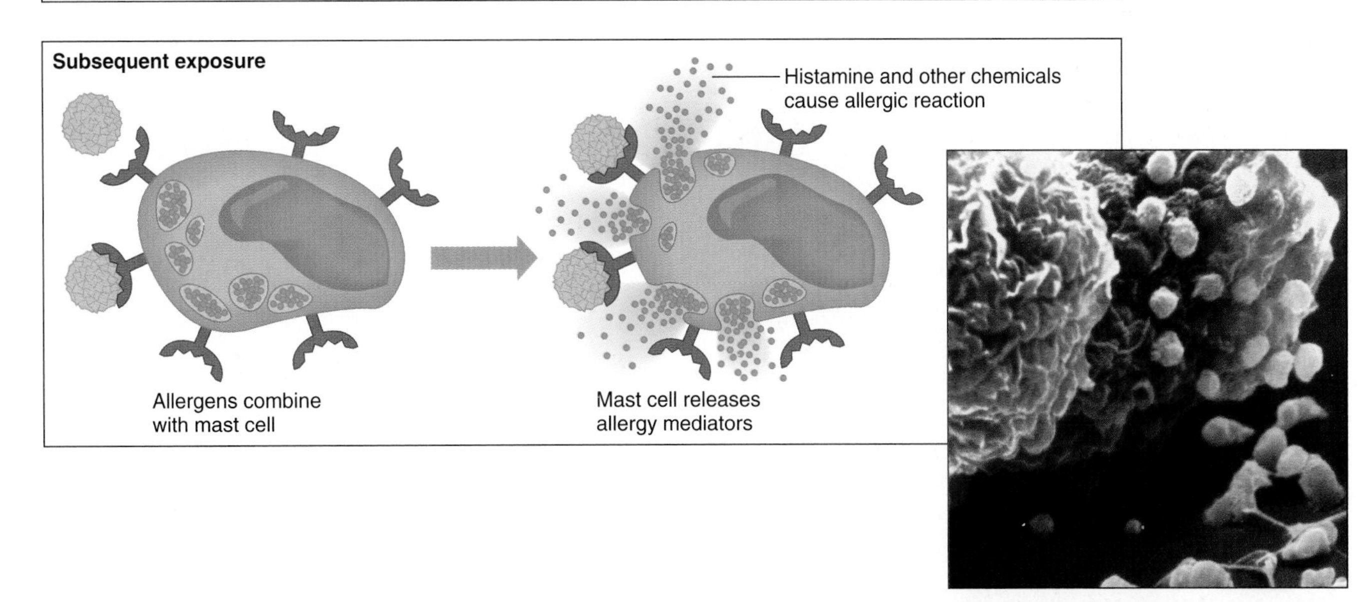

FIGURE 39.17
Allergy.
In an allergic reaction, B cells are activated by an allergen such as pollen (upper inset), and differentiate into antibody-secreting plasma cells. The antibodies attach to mast cells. When allergens are encountered again, they combine with the antibodies on the mast cells. The mast cells burst (lower inset), releasing the chemicals that cause itchy eyes and a runny nose.

Biotechnology 39.1

Immunotherapy

The immune system is remarkably effective at keeping potentially infectious bacteria, viruses, and tumor cells from taking over our bodies. Can we improve on nature? The idea of immunotherapy—amplifying or redirecting the immune response—was born late in the 1890s, when New York surgeon William Coley gave cancer patients killed bacteria. He had noticed that some cancer patients spontaneously recovered following a bacterial infection. Sometimes "Coley's toxins" worked—apparently the immune response against the bacteria also killed the cancer cells.

Today, many immunotherapies are in clinical trials. A few are already part of standard medical practice.

Boosting Humoral Immunity—Monoclonal Antibody Technology

When a single B cell recognizes a single foreign antigen, it manufactures a single, or monoclonal, type of antibody. A large amount of a single antibody type would make a powerful drug because of its great specificity. It could be used to target a particular pathogen or cancer.

In 1975, British researchers Cesar Milstein and Georges Köhler devised monoclonal antibody (MAb) technology, which increases the production of a single, specific antibody. First, they injected a mouse with a preparation of antigens (fig. 39.B). They then isolated a single B cell from the mouse's spleen and fused it with a cancerous white blood cell from a mouse. The fused cell, called a hybridoma, had a valuable pair of talents. Like the B cell, it produced large amounts of a single antibody type. Like the cancer cell, it divided continuously. A hybridoma is a specific antibody-making machine.

Today, MAbs are used in basic research, veterinary and human health care, agriculture, forestry, food technology, and forensics.

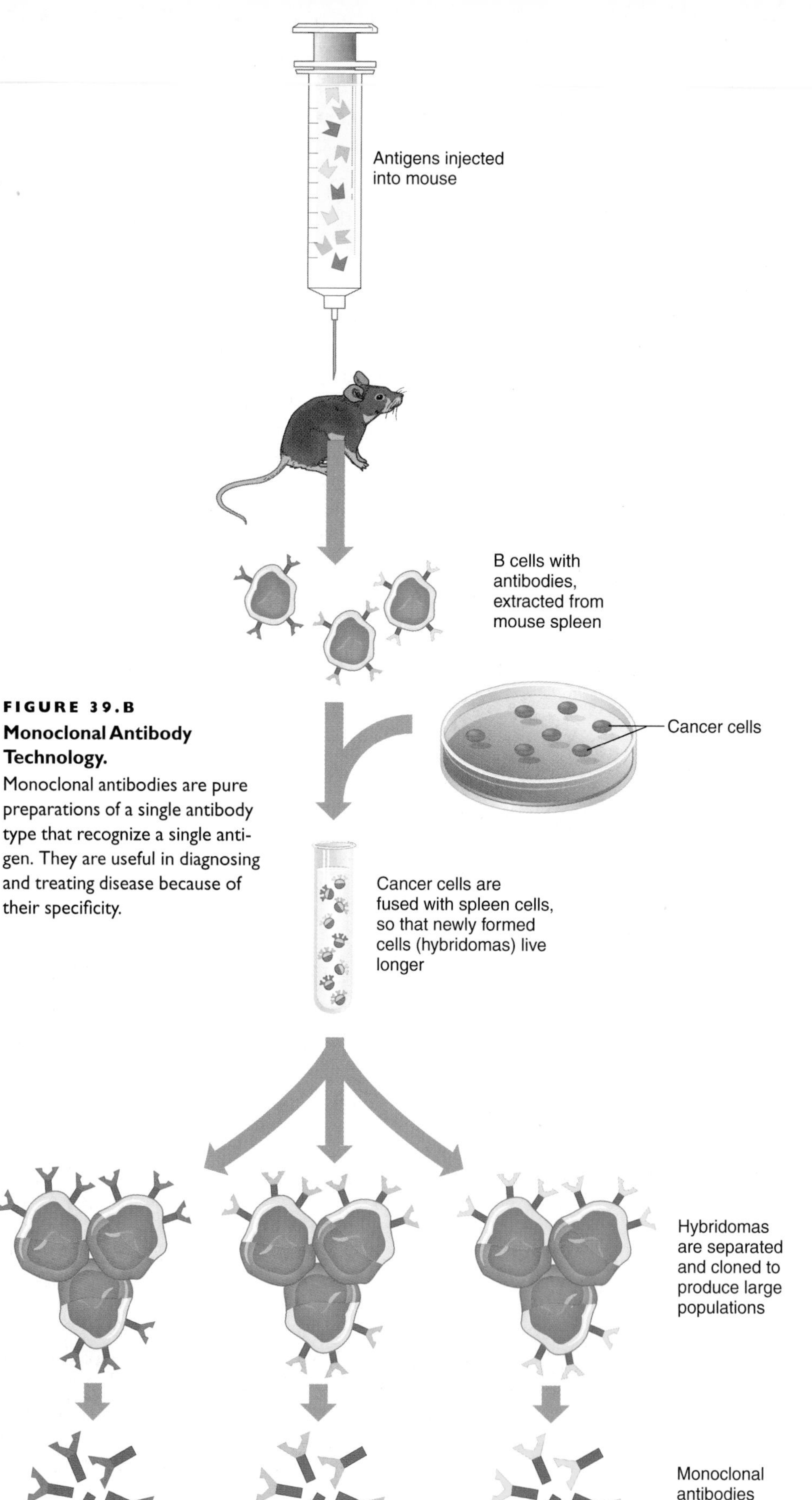

FIGURE 39.B Monoclonal Antibody Technology. Monoclonal antibodies are pure preparations of a single antibody type that recognize a single antigen. They are useful in diagnosing and treating disease because of their specificity.

Researchers have developed techniques to make MAbs more like human antibodies—the original mouse versions cause allergic reactions in many people.

MAb kits that detect tiny amounts of a molecule are used to diagnose everything from strep throat to turf grass disease. One common use is a home pregnancy test. A woman places drops of her urine onto a paper strip containing a MAb that binds to hCG, a hormone present only during pregnancy. A color change ensues if the MAb binds its target—indicating that the woman is pregnant.

MAbs can detect cancer earlier than traditional methods can. The MAb is attached to a fluorescent dye and injected into a patient or applied to a sample of tissue or body fluid. If the MAb binds its target—an antigen found mostly or only on cancer cells—the fluorescence is detected with a scanning technology or fluorescence microscopy. MAbs linked to radioactive isotopes or to drugs can be used in a similar fashion to ferry treatment to cancer cells. Herceptin is a monoclonal antibody-based drug for breast cancer. It blocks receptors for certain growth factors on the cancer cells, preventing them from dividing.

Boosting Cellular Immunity—Harnessing Cytokines

As coordinators of immunity, cytokines are used to treat a variety of conditions, including cancer, multiple sclerosis, genital warts, AIDS, tuberculosis, and parasitic infections. It has been difficult, however, to develop cytokines into drugs for several reasons. They have side effects, remain active only for short periods, and must be delivered precisely where they are needed.

Interferon (IF) was the first cytokine to be tested on a large scale. When researchers discovered it in the 1950s, IF was erroneously hailed as a wonder drug. Although it did not live up to early expectations, IF is used today to treat a few types of cancer, genital warts, and multiple sclerosis.

Interleukin-2 (IL-2) is administered intravenously to treat kidney cancer recurrence. In another approach, IL-2 is applied to T cells removed from a patient's tumor; then the stimulated T cells, along with IL-2, are reinfused into the patient. Much ongoing research is focused on discovering uses for the many other interleukins. Colony stimulating factors, which cause white blood cells to mature and differentiate, are used to boost white blood cell supplies in people with suppressed immune systems. People with AIDS or those undergoing cancer chemotherapy inject themselves with colony stimulating factors to help replenish their white blood cell supplies.

So logical and promising is immunotherapy that cancer treatment is beginning to include combinations of immune system cells and biochemicals. Boosting immunity can enable a patient to withstand higher doses of a conventional drug, or destroy cancer cells that remain after standard treatment.

Because histamine causes many allergy symptoms, drugs with antihistamine action can relieve symptoms.

The particular misery associated with a specific allergic response depends upon the site in the body where mast cells release mediators. Because many mast cells are in the skin, respiratory passages, and digestive tract, allergies often affect these organs, producing hives, runny nose and eyes, asthma, nausea, vomiting, and diarrhea.

Common allergens are foods, dust mites, pollen, and fur. To identify the cause of an allergy, a physician injects extracts of suspected allergens beneath a patient's skin on the upper arm or back. If a red bump develops, the substance injected at that point is an allergen for that person.

A treatment called desensitization can help hay fever and asthma sufferers. A physician periodically injects small amounts of the allergen under the skin. The doses are not enough to stimulate IgE production, but they do stimulate IgG production. When the person next encounters the allergen in a natural setting, IgG binds to the allergen before IgE can. Because only IgE can stimulate mast cells to release allergy mediators, the attack is prevented.

Not all allergic reactions are as benign (although uncomfortable) as hives and watery eyes. Some individuals may react to certain stimuli with a terrifying and potentially life-threatening reaction called anaphylactic shock, in which many mast cells release mediators throughout the body. The person may at first feel an inexplicable apprehension. Then, suddenly, the entire body itches and erupts in hives. He or she may vomit and have diarrhea. The face, tongue, and larynx begin to swell, and breathing becomes difficult. Unless the person receives an injection of adrenaline and sometimes a tracheostomy (an incision into the windpipe to restore breathing), he or she will lose consciousness and die within as little as 5 minutes. Anaphylactic shock most often results from an allergy to penicillin or insect stings.

One theory of allergy origin, and particularly of anaphylactic shock, is that allergies evolved long ago when insect bites and common substances such as penicillin might have threatened human survival. Evidence for this theory is that IgE also protects against roundworm and flatworm infections, which may have once been a greater threat. Perhaps IgE is a holdover from times past, when the environment presented different challenges to the human immune system than it does today.

39.5 MASTERING CONCEPTS

1. How does HIV affect helper T cells and cytotoxic T cells?
2. How does severe combined immune deficiency differ from HIV infection?
3. What events might lead to autoimmunity?
4. What is a superantigen?
5. Which cells and biochemicals participate in an allergic reaction?

OLC

Chapter Summary

39.1 The Evolution of Immunity

1. All animals have immunity, which is based on the ability to distinguish self from nonself **antigens.**
2. All animals have **innate immunity,** but vertebrates alone have **acquired immunity.**
3. Invertebrates' innate immunity involves **phagocytes** and antimicrobial peptides.

39.2 Innate Defenses Are Nonspecific and Act Early

4. Skin, mucous membranes, tears, earwax, and cilia block pathogens.
5. Phagocytes and antimicrobial substances take part in **inflammation,** which is an immediate reaction to injury.
6. **Complement** proteins interact in a cascade that bursts bacterial cells.
7. **Collectins** are proteins that protect against pathogenic bacteria, viruses, and yeasts.

39.3 Acquired Defenses Are Specific and Act Later

8. **Macrophages** are antigen-presenting cells, engulfing, processing, and displaying foreign antigens held in place by **major histocompatibility complex** molecules.
9. A **helper T cell** binding to the antigen-presenting cell triggers the acquired immune response.
10. B cells, when activated by helper T cells, proliferate and differentiate into **plasma cells,** which secrete **antibodies,** and **memory B cells.**
11. An antibody is a Y-shaped protein composed of two **heavy chains** and two **light chains.** Each chain has a **constant** amino acid sequence and a **variable** sequence. The tips form an **antigen binding site.** Antibodies bind antigens and form complexes that attract other immune system components. Antibody molecules are incredibly diverse because DNA segments shuffle during early B cell development. This is the **humoral immune response.**
12. T cells carry out the **cellular immune response.** They are educated in the thymus gland to recognize self. Helper T cells activate other T cells and B cells. **Cytotoxic T cells** release biochemicals that bore into bacteria, kill them, and also destroy cells infected with viruses. T cells secrete **cytokines,** which control communication within the immune system.
13. **Gamma-delta T cells** coordinate responses of innate and acquired immunity.

39.4 The Human Immune System Throughout Life

14. Fetal T cells are "educated" in the thymus to recognize self.
15. A pregnant woman's immune system dampens to accept the fetus, but fetal cells in her bloodstream can cause problems.
16. A baby gets antibodies from the mother, which is **passive immunity,** then begins to make its own. Immune function begins to wane in early adulthood.

39.5 The Unhealthy Immune System

17. HIV enters helper T cells and uses them to reproduce, killing them directly. The virus activates apoptosis in cytotoxic T cells.
18. Immune deficiency can be inherited, and some types are treatable with gene therapy.
19. **Autoimmunity** results when the immune system produces **autoantibodies,** which attack the body's tissues.
20. **Superantigens** are foreign antigens that are displayed abnormally prominently on macrophages, activating many T cells that flood the body with cytokines, which cause symptoms.
21. An allergy is an immune reaction to a harmless substance. An **allergen** activates IgE antibodies, which cause **mast cells** to release **allergy mediators.**

Testing Your Knowledge

1. How do the immune systems of vertebrates differ from those of invertebrates, and how are they similar?
2. What would be a consequence of a lack of:
   a. helper T cells
   b. cytotoxic T cells
   c. B cells
   d. gamma-delta T cells
   e. macrophages
3. State the function of each of the following immune system biochemicals:
   a. defensins
   b. collectins
   c. antibodies
   d. cytokines
   e. complement proteins
4. Describe the immune defenses of three types of organisms.
5. List the cells and biochemicals that provide the humoral and cellular immune responses.
6. What do a plasma cell and a memory cell descended from the same B cell have in common, and how do they differ?
7. What part do antibodies play in allergic reactions and in autoimmune disorders?

Thinking Scientifically

1. Cecropins and gamma-delta T cells have only recently been discovered. Why is it that we know less about the immune system than other organ systems?
2. Why is a vaccine consisting of gp120, the spoke that emanates from HIV, unlikely to be effective? Cite three reasons why developing a vaccine against HIV infection has been so challenging.
3. How might a drug advertised to be a "histamine blocker" relieve allergy symptoms?
4. How can a vaccine cause the illness it is intended to prevent?
5. Why is a polyclonal antibody response valuable in the body, but a monoclonal antibody valuable as a diagnostic tool?

References and Resources

Halim, Nadia. April 17, 2000. New era in vaccine development. *The Scientist,* vol. 14, p. 10. Genome sequences of bacteria are providing information useful in vaccine development.

Hoffmann, Jules A., et al. May 21, 1999. Phylogenetic perspectives in innate immunity. *Science,* vol. 284, p. 1313. Insects and mammals share many mechanisms of innate immunity.

Lewis, Ricki. July 6, 1998. Classic technique reveals HIV in action. *The Scientist,* p. 1. By stripping away HIV's protective sugars and peptides, researchers glimpse the virus touching a helper T cell.

Malissen, Bernard. July 9, 1999. Dancing the immunological two-step. *Science,* vol. 285, p. 207. Where T cell meets antigen-presenting cell is called the immunological synapse.

Walker, Bruce D. and Philip J. R. Goulder. September 21, 2000. Escape from the immune system. *Nature,* vol. 407, p. 313. A changeable HIV evades T cell attack.

The *LIFE* Online Learning Center provides additional resources and tools for studying this chapter
www.mhhe.com/life

CHAPTER 40

Human Reproduction and Development

A Heart in the Wrong Place

In the original *Star Trek* television series, Dr. McCoy often complained when examining Mr. Spock that Vulcan organs weren't where they were supposed to be, based on human anatomy. The good doctor would have had a hard time examining humans with a condition called situs inversus, in which certain normally asymmetrically located organs develop on the wrong side of the body.

In a normal human body, certain organs lie either on the right or the left of the body's midline. The heart, stomach, and spleen are on the left, and the liver is on the right. The right lung has three sections, or lobes; the left lung has two. Other organs twist and turn in either a right or left direction. All these organs originate in the center of an initially symmetrical embryo, and then the embryo turns, and the organs migrate to their final locations.

Which way an organ moves depends upon decisions set in the very early embryo. Experiments show that damaging one end of a fertilized *Xenopus laevis* (African clawed frog) embryo reverses the positions of the adult frog's organs. Similarly, altering the positions of cells in an eight-celled *Caenorhabditis elegans* roundworm embryo—for instance, placing those on top in the middle—causes organs to develop in reverse orientation. Mice show differences in the genes that are expressed on the right and left sides as early as 48 hours after fertilization.

Which way an organ moves depends upon decisions set in the very early embryo.

In humans, misplaced body parts are part of Kartagener syndrome, in which the heart, spleen, or stomach may be on the right, both lungs may have the same number of lobes, the small intestine may twist the wrong way, or the liver may span the center of

Situs Inversus.
The drawing on the left (**A**) shows the normal position of the heart, lungs, liver, spleen, and stomach. In situs inversus, certain organs form on the wrong side of the body (**B**).

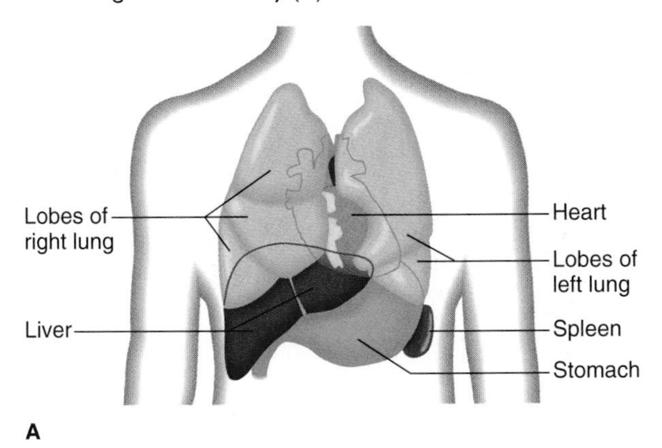

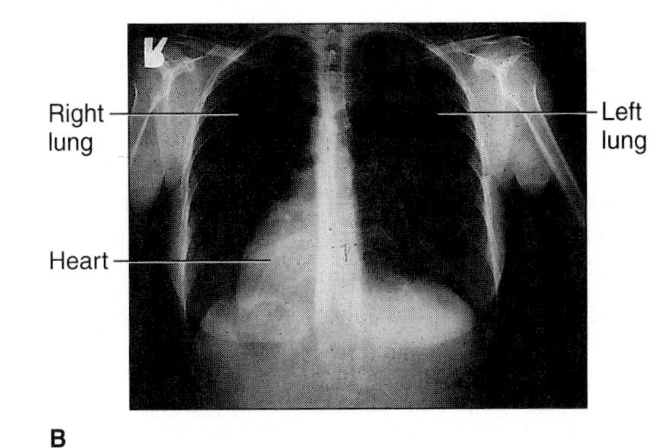

the body. Many people with this syndrome die in childhood from heart abnormalities. Kartagener syndrome was first described in 1936 by a Swiss internist caring for a family with several members who had a strange collection of symptoms—chronic cough, sinus pain, a poor sense of smell, male infertility, and misplaced organs—usually a heart on the right.

Many years later, researchers identified another anomaly in patients with Kartagener syndrome, which would hold a clue to how the heart winds up on the wrong side of the chest. All affected individuals lack dynein. Recall from chapter 4 that dynein is a protein that enables microtubules to slide past one another and generate motion. Without dynein, cilia cannot wave. In the upper respiratory tract, immobile cilia allow debris and mucus to accumulate, which explains the cough, clogged sinuses, and poor sense of smell. Lack of dynein also paralyzes sperm tails, accounting for male infertility. But how could dynein deficiency explain a heart that develops on the right instead of the left?

One hypothesis was based on the fact that dynein helps establish the mitotic spindle, which determines the location of the cleavage furrow in dividing cells of the zygote. The cleavage furrow, in turn, determines where in three-dimensional space a particular daughter cell lies. Therefore, the dynein defect may, early on, set cells on a developmental pathway that diverts migration of the heart from the embryo's midline to the left.

Another explanation for how organs end up in the wrong locations comes from mice genetically altered to lack a gene required for assembling cilia. About 50% of the mice have reversed organs, and all of the animals either lack cilia that are normally present on cells of the early embryo, called node cells, or the cilia cannot move. Node cells are the sites where the first differences between right and left arise, and they set the pattern for placement of organs. Normally, the cilia rotate counterclockwise, which moves fluids to the left. This movement creates a gradient of molecules, called morphogens, that control development. Somehow the differing concentrations of specific morphogens in different parts of the body send signals that guide the development of organs. Without moving cilia, organ placement occurs randomly, on the left or the right. This is why only 50% of the genetically altered mice, and presumably not all people who inherit Kartagener syndrome, have organs in the wrong place. Just by chance, some of them develop normally.

Node cells are the sites where the first differences between right and left arise, and they set the pattern for placement of organs.

40.1 From Embryology to Developmental Biology: A Descriptive Science Becomes Experimental

Before biologists could observe early development, they thought that a fertilized ovum contained a small but complete organism. The idea that cells specialize gradually grew as researchers compared embryos in different species and throughout development. Experiments in the twentieth century revealed that cell specialization is a consequence of differential gene action.

Animal bodies range from simple forms, such as the sponge with its four specialized cell types, to the complex organ systems of a vertebrate that have been detailed throughout this unit. How cells specialize and interact to form the tissues, organs, and organ systems that comprise an animal constitutes the overall process of development. It begins with rapid and profound changes, then stabilizes somewhat, with continuing cell division, death, and growth maintaining the body's form and functions (see fig. 8.1).

Sexual reproduction, on a population or species level, introduces the benefits of genetic diversity in surviving environmental changes, as section 9.1 points out. On an individual organism level, the development of a sexually reproducing animal begins with a single cell, the zygote, which formed when sperm penetrated oocyte, and the nuclei of the parents merged. From this point on, different species develop into their distinctive forms, yet all pass through several similar stages.

Although many of the details of reproduction and development are now understood down to the molecular level, our knowledge has its roots in a lively, colorful, and often controversial process of observation, experimentation, and interpretation.

Does a Fertilized Ovum Contain a Tiny Organism? Preformation Versus Epigenesis

Until the mid-eighteenth century, most biologists supported **preformation,** the idea that a fertilized egg (or even a gamete) contains a tiny, preformed organism, or "germ," that grows, like a homunculus (fig. 40.1). In 1759, German physiologist Kaspar Friedrich Wolff published a revolutionary pamphlet, which stated that a fertilized ovum contains no preformed germ but unspecialized tissue that gradually specializes. For the next 50 years other scientists doubted Wolff's hypothesis, called **epigenesis.**

Through the nineteenth century, biologists continued to observe and sketch embryos. A German scientist, Karl Ernst von Baer, studied dog, chick, frog, and salamander embryos; he was a founder of embryology. Baer watched, fascinated, as a fertilized egg divided into two cells, then four, then eight, gradually forming a solid ball that then hollowed out. The hollow ball would indent, then form two layers, then build up what Baer thought were four layers (later, scientists corrected this to three). Baer's observation—that specialized cell layers arise from nonlayered, unspecialized cells—confirmed Wolff's theory of epigenesis.

FIGURE 40.1
Human Sperm.
(**A**) Scanning electron micrograph of human sperm cells. (**B**) When human sperm were first seen in the microscope, they were thought to be infectious microbes. This 1694 illustration by Dutch histologist Nicolaas Hartsoeker presents another popular hypothesis about the role of sperm—some thought they were carriers of a preformed human called a homunculus.

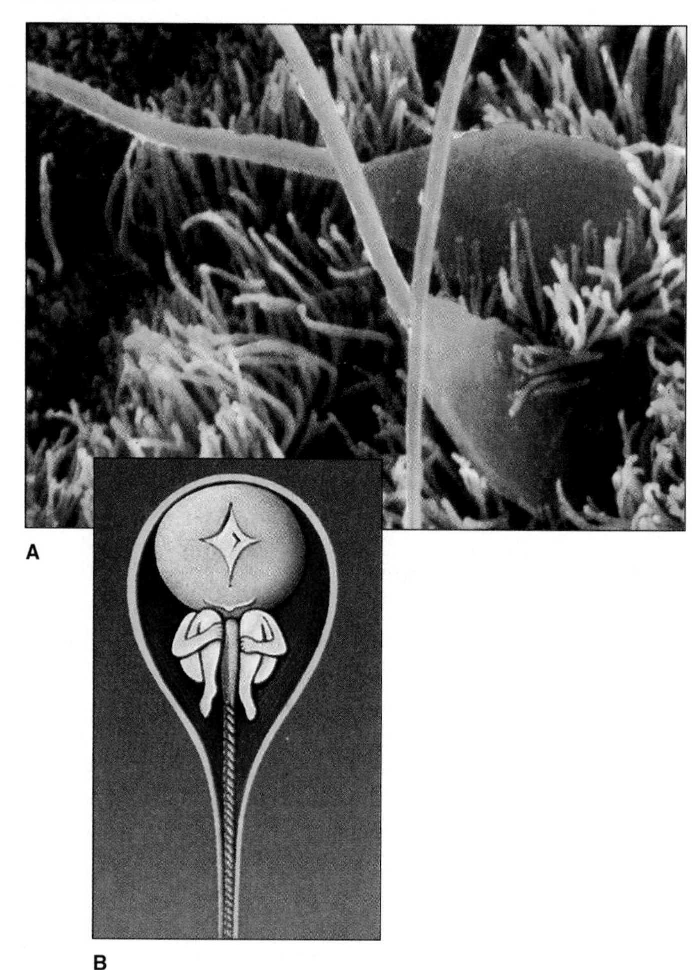

When Baer compared embryos of several vertebrate species, he noted that they were incredibly similar in early stages, but then gradually developed species-distinct characteristics (fig. 40.2). For example, an early appearing bump looked similar on embryos from four species, but one developed as a paw, one an arm, another a wing, and another a flipper. Charles Darwin interpreted Baer's observations as evidence that different species descended from a common ancestor. Baer disagreed vehemently with Darwin on this matter for the rest of his days.

Later, German naturalist Ernst Haeckel extended Darwin's interpretation. Haeckel suggested that vertebrate embryos proceed through a series of stages that represent adult forms of the species that preceded them in evolutionary time. Today, most biologists reject the biogenetic law, as Haeckel's view is called. Structures in early embryos probably resemble each other simply

FIGURE 40.2

Embryonic Resemblances.

In 1828, embryologist Karl Ernst von Baer neglected to label two embryos in an experiment. The embryos were so similar in appearance that without labels, he couldn't tell whether they were embryos of mammals, reptiles, or birds. Notice how similar the early embryos in the top row are—each has gill slits and a tail reminiscent of fish. Later in development, as depicted in the bottom row, the embryos become more distinctive. These sketches were drawn in 1901 and therefore have some inaccuracies.

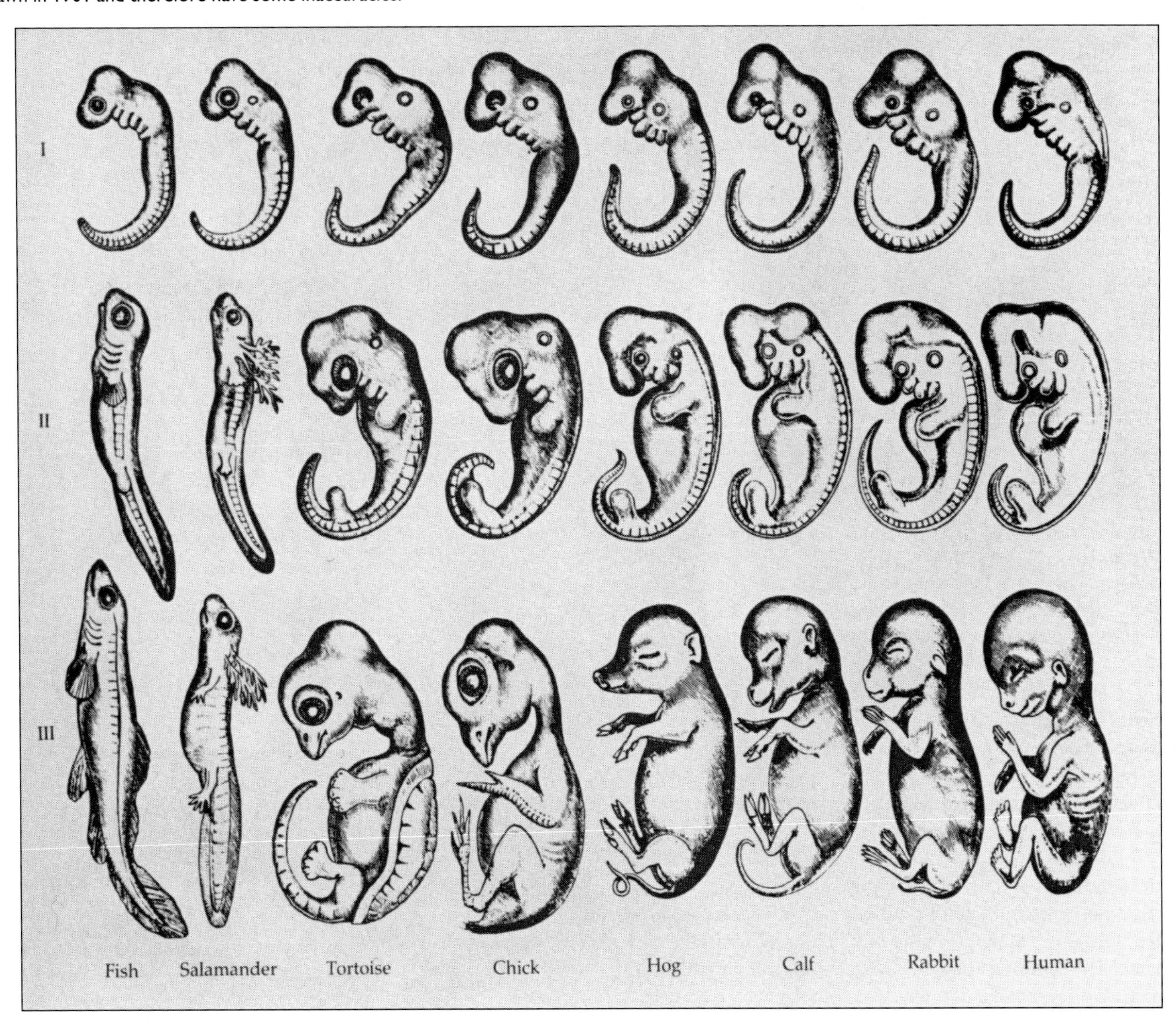

because they have not yet specialized, and early in development, different species go through similar stages as cells organize into tissues and organs. Chapter 17 discussed misrepresentations in Haeckel's famed embryos, and figure 17.13 depicts a modern rendition of the comparisons.

Early Experiments with Embryos

During the late nineteenth and early twentieth centuries, research went in a new direction. Embryologists began intervening in development, disrupting embryonic structures to see what happened. A student of Haeckel's, Wilhelm Roux (1850–1924), was one of the first to alter embryos. He shook them, suspended them upside down, destroyed certain cells with heat or cold, and observed fairly consistent results—disturbed embryos cease developing.

In the 1920s, German zoologist Hans Spemann experimented on newt and frog embryos. He teased apart the cells of a very early embryo and watched them develop, separately, into distinct but identical individuals. Each cell of the original embryo retained the potential to give rise to a complete individual, a capacity called **totipotency.** In the 1980s, scientists repeated

FIGURE 40.3

A Mouse With Six Parents.

The tricolored mouse has six parents—a pair for each of the three original embryos combined to form the one individual. The tricolored mouse was bred to a white male mouse, and she gave birth to the three mice shown above her in the photograph. The fact that the tricolored mouse gave birth to solid-colored mice of the original three colors indicates that some of her gametes descended from each of the original three donor embryos.

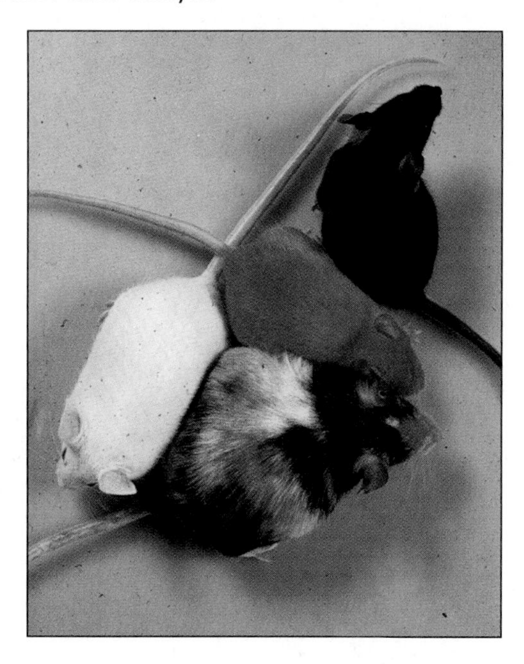

Spemann's experiment with a twist: They separated the cells of three different mouse embryos and reassembled the cells to create one new embryo from cells of all three—which then developed into a mouse with six parents (fig. 40.3)! The mouse experiment only works at early stages of development; by the time a mouse embryo contains 1,000 cells, the cells have shut off the genes that would allow them to generate a complete organism.

Spemann also operated on embryos. In one famous experiment, Spemann's student, Hilde Mangold, removed part of a region called the dorsal lip of the blastopore from a newt embryo. Then she transplanted the dorsal lip to the opposite side of an embryo at the same stage but of a different newt species (fig. 40.4). A new embryo grew at the site where the dorsal lip tissue was grafted onto the second embryo. The dorsal lip, now named Spemann's primary organizer, is therefore a vital trigger in embryonic development.

Transplant experiments on many species helped researchers decipher the fates and functions of embryonic structures. These experimenters sought to determine when a cell commits to follow a specific **developmental pathway,** or series of events that culminates in formation of a particular tissue such as nerve or muscle. If transplanted cells develop in their new location as they would have if not moved, then their developmental pathway was set before the time of transplantation. If, on the other hand, a transplanted cell develops as other cells do in its new location, then the cell was still totipotent when it was transplanted.

FIGURE 40.4

Spemann's Primary Organizer.

When Spemann's student, Hilde Mangold, transplanted the dorsal lip of an early embryo of one newt species to an embryo of a different species, the recipient developed into two embryos. This result indicates that the dorsal lip is very important in early development. The newt species are of different colors, which enabled Mangold to distinguish donor from recipient tissue. The dorsal lip has come to be known as Spemann's primary organizer.

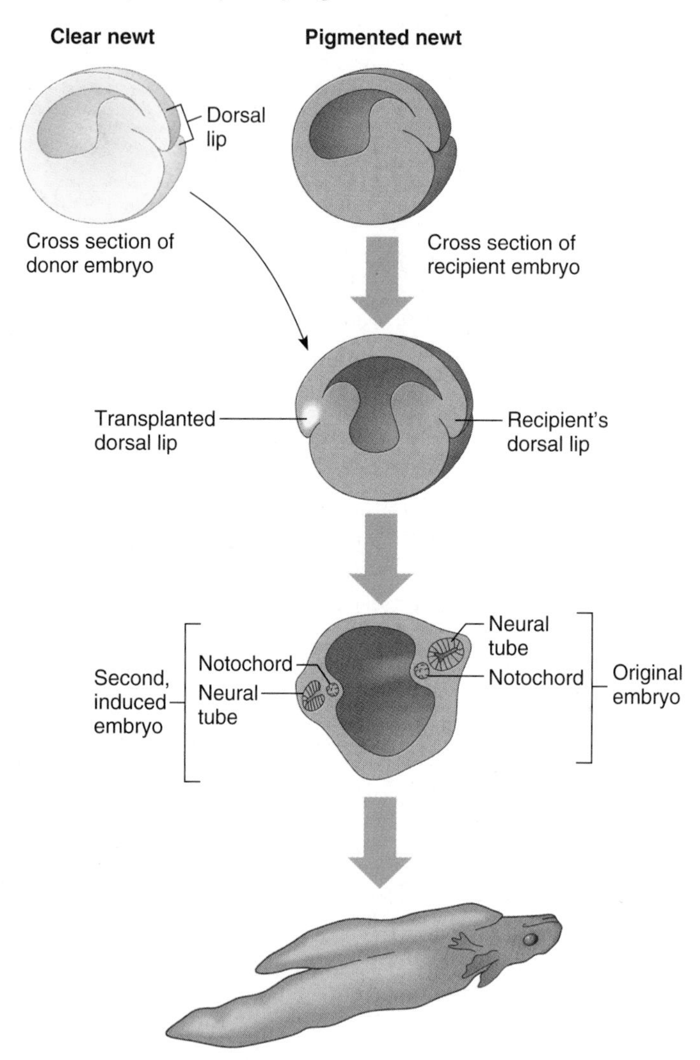

With the discovery of the structure of the genetic material in 1953, embryologists—now known by the broader term developmental biologists—could use the ability to detect changes in gene activity as development proceeds. All cells contain all of the organism's genes, but as cells become specialized, or differentiated, they express only certain genes, whose products endow the cell with its distinctive characteristics. This differential gene expression underlies the transformation of a relatively formless clump of cells into a complex grouping of specialized tissues.

Selective Gene Activation Sculpts Specialized Cell Types

A series of experiments conducted on the African clawed frog, *Xenopus laevis,* in the early 1960s demonstrated that differentiated cells contain a complete genetic package but only use some of the information in it. British developmental biologist J. B. Gurdon designed a way to "turn on" the genes that a differentiated cell normally "turns off."

Gurdon removed the nucleus from a differentiated cell lining the small intestine of a tadpole and injected it into an egg of the same species whose nucleus had been destroyed. Some of the altered eggs developed into tadpoles, and a few developed as far as normal adult frogs (fig. 40.5). The fact that a nucleus from a differentiated cell could support normal development from a single cell proved that such a nucleus contains complete genetic instructions for the species. Genes must therefore be inactivated—not lost, as some investigators once thought—as a multicellular organism develops. In 1997, Scottish researchers conducted a version of Gurdon's experiment with sheep. They successfully cloned a sheep by transferring nuclei from mammary gland cells of a six-year-old ewe to oocytes that had had their nuclei removed (see Biotechnology 40.1). The result was a lamb named Dolly.

Developmental biologists increasingly use molecular techniques to decipher the distributions of proteins that guide totipotent cells to form tissues. This process requires growth (cell division) as well as differentiation. Consider muscle formation in mammals. A "master control gene" turns on dozens of genes, causing the cell to produce muscle-specific proteins. When the master control gene is placed in chick connective tissue cells growing in culture, the cells elongate and then pulsate—they become muscle.

The study of animal development is perhaps most exciting in our own species. But studying human development before birth is difficult for a variety of legal, ethical, and practical reasons. To avoid these problems, and because many developmental processes are very similar in closely related species, researchers often work with model organisms to learn more about human development. Investigating Life 40.1 examines two widely used model organisms. In the pages that follow we focus on human reproduction and development.

40.1 MASTERING CONCEPTS

1. How does epigenesis differ from preformation?
2. What is the significance of the similarity in appearance of early embryos of different species?
3. How did Gurdon determine that genes are deactivated during development as opposed to being lost?

OLC

FIGURE 40.5

Gurdon's Experiment.

Nuclear transplantation shows that the nucleus of a differentiated cell can support complete development. In the first step of the procedure, a frog egg nucleus is inactivated with radiation. A nucleus is removed from a tadpole's differentiated cell. Next, the nucleus from the differentiated cell is transferred to the enucleated egg. The egg, controlled by its transplanted nucleus, develops into a tadpole and then a frog. Today, cloning has been accomplished in several mammalian species, using nuclei from fetal or adult cells.

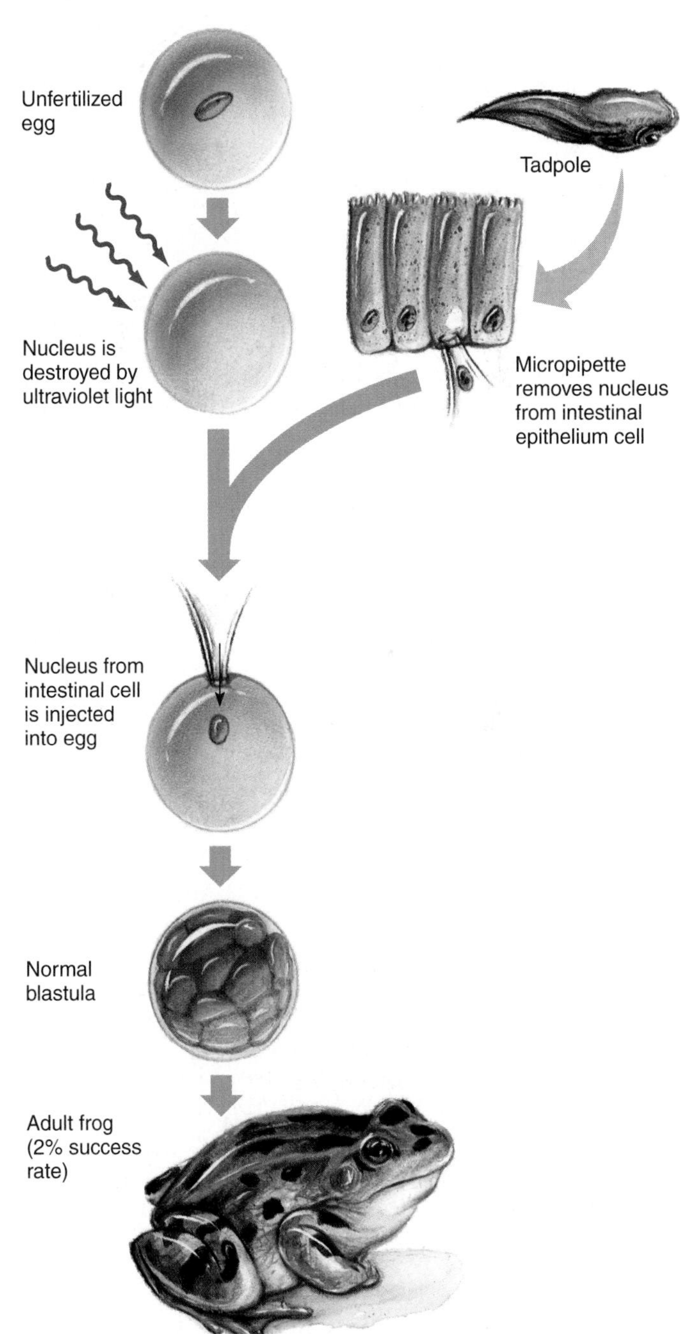

Biotechnology 40.1

Considering Cloning

When the birth of a cloned sheep, Dolly, was reported in 1997, the public was at once fascinated and astonished. Biologists, however, were familiar with the idea of regenerating an animal by transferring the nucleus of a somatic cell into a fertilized ovum whose nucleus had been removed—it had been done in amphibians in the 1960s, but using nuclei from cells of embryos or fetuses. Dolly's donor nucleus, in contrast, came from mammary gland cells of a 6-year-old ewe—actually, of her stored cells. Figure 40.A explains how Dolly came to be.

Although Dolly is a genetic replica of the donor ewe, for subtle genetic reasons, she is not exactly the same. Some of her mitochondria descend from the donor cell, and some from the recipient cell, and mitochondria contain genes. Her chromosomes have a head start in telomere shortening, because the initial cell was already "old"—researchers are carefully watching Dolly for signs of premature aging. So far, she has borne several lambs and seems quite normal, although her chromosome tips are indeed shorter than normal. Another difference is in X chromosome inactivation. In female mammals, early during embryonic development, one X chromosome is turned off in each cell. The pattern of cells with the paternal X inactivated versus the maternal X is probably different in Dolly than it was in her natural mother. She may also be at higher risk for cancer, because the nucleus that she descends from had 6 years to accumulate mutations. Such a somatic mutation might not be noticeable in one of millions of mammary gland cells, but it could be devastating if it programmed development of an entire new individual.

In addition to these genetic reasons why a clone isn't really a clone, another powerful factor is the environment. In humans, environmental influences include experience, diet, stress, and exposure to infectious diseases. Environment can also refer to other changes that occur after the genetic program is set at conception. For example, the pattern of black and white patches on cow clones varies, because cells destined to produce pigment moved about in unique ways as each calf developed.

We probably needn't worry about genetic duplicates among us anytime soon, because cloning isn't easy to do. Dolly was one of 277 attempts; the first cloned mouse, Cumulina, was among 15 liveborn mice from 942 tries. Attempts to commercialize cloning of farm animals from fetal cell nuclei in the 1980s failed, because the newborns were huge and very ill and weak—a problem with cloning that persists. An easier route was to take 8- or 16-celled embryos, tease apart the cells with digestive enzymes, and implant them in surrogates to develop. This method of "embryo splitting" is also used to increase numbers of severely endangered species. For these applications, cloning isn't the most efficient option. Researchers hypothesize that the reason for the difficulty of cloning is that starting development with a somatic cell nucleus is just not the same as starting it with haploid input from both sexes. According to one researcher, "The whole natural order is broken," referring to meiosis. As for humans, stem cell technology is a much more promising route to obtaining replacement tissues than cloning, and if done using adult tissues, would not raise ethical objections (see the opening essay to chapter 30).

If human cloning ever becomes reality, we will probably learn that we are not merely the sum of our genes. This is the essence of objection to cloning: that we are dissecting and redefining our very individuality, reducing it to a biochemistry so supposedly simple that we can duplicate it. Chances are, considering the complexity of biology, that we really can't.

FIGURE 40.A
The First Clone of an Adult Mammal—Dolly.
This diagram depicts the steps involved in cloning a sheep.

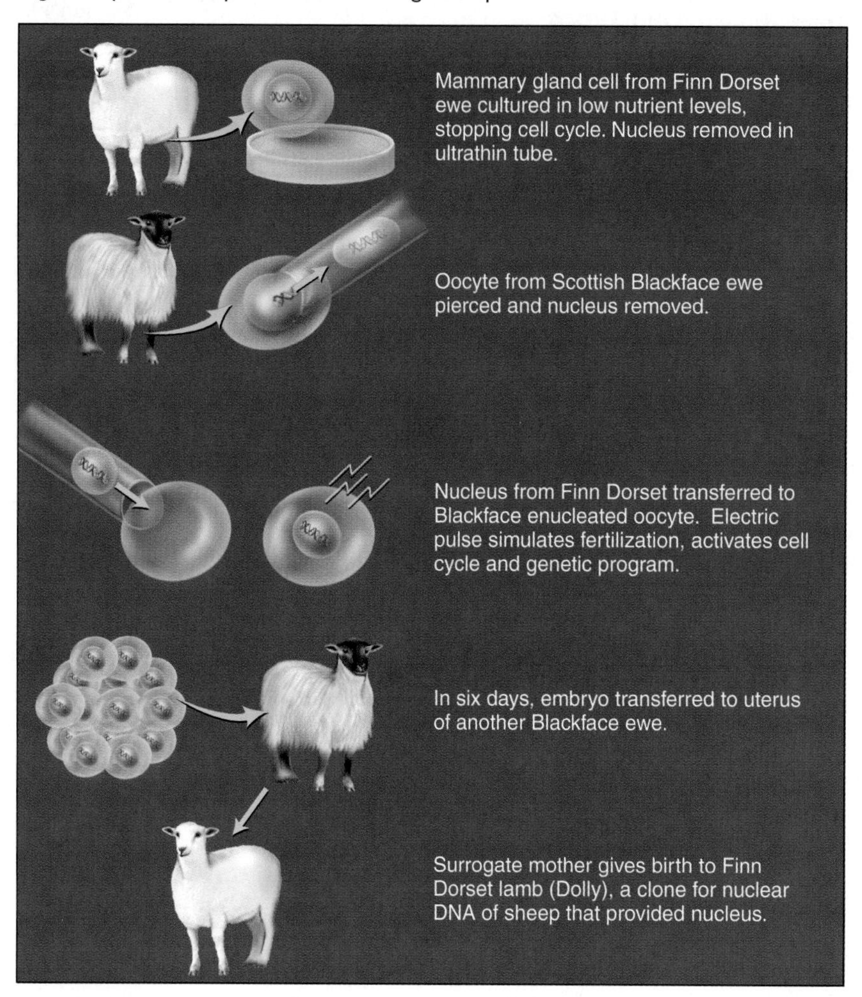

Two Invertebrate Beginnings: Early Development in a Worm and a Fly

A millimeter-long worm, a fruit fly, and a human may not appear to have much in common. However, each animal meets the same challenges in development. It begins as a single cell, it undergoes rapid mitotic cell division, its cells commit to follow particular pathways, and, gradually, distinct structures begin to appear, specialize, interact, and organize.

Watching Development Unfold in a Transparent Worm

The nematode worm *Caenorhabditis elegans,* a 1-millimeter-long soil resident, is in many ways an ideal model organism. An adult worm consists of a mere 959 somatic cells, yet it has differentiated tissues just as complex animals do (fig. 40.B). The worms are easy to maintain in the laboratory and can even be frozen for later use. *C. elegans* grows from fertilized egg to adult in just 3 days and is completely transparent. With these worms, researchers can watch animal development unfold, counting and tracking cells as they divide, migrate, and interact.

That is exactly what Einhard Schierenberg of Germany and John Sulston of England did from 1975 to 1983. They followed the fates of each cell under a light microscope and published a "fate map" showing all cell lineages—which cells give rise to which cells. By slicing the worms into very thin sections and observing them under an electron microscope, the researchers described every aspect of the adult animal's anatomy. For example, they identified 302 nerve cells and 7,600 connections the cells make with each other.

Development begins with fertilization. For the first few cell divisions, one cell in the worm looks much like another. By the fourth mitosis for some cells, and later for others, certain cells commit to follow certain developmental pathways—that is, to become part of the outer covering (cuticle), intestine, muscle, nerve, pharynx, vulva, egg, or gonad. These specialized structures do not actually appear until a certain number of cell divisions later. Sometime during the initial hours of rapid cell division, controls are set to transform the transparent bag of identical-appearing cells into a moving, coordinated multicellular organism.

A half day after the first cell division, the organism is a larva of 558 cells. Further rounds of cell division and the death of precisely 113 cells sculpts the 959-celled adult worm. In addition, 2,000 cells are set aside as germ cells, which will give rise to sex cells.

Thanks to the efforts of Schierenberg, Sulston, and other "worm people," as *C. elegans* researchers call themselves, we now have a complete picture of an animal's development, providing clues to how more complex animals form. Imagine tracking development in the many-trillion-celled human!

FIGURE 40.B
Developmental Biology of *C. elegans*.
Caenorhabditis elegans is an important model organism for developmental biology. The fates of all cells produced from the original fertilized egg have been described and mapped to the final 959 somatic cells that make up the body of this 1-millimeter-long worm. The much simplified fate map shown here indicates the development of major tissues of the worm's body.

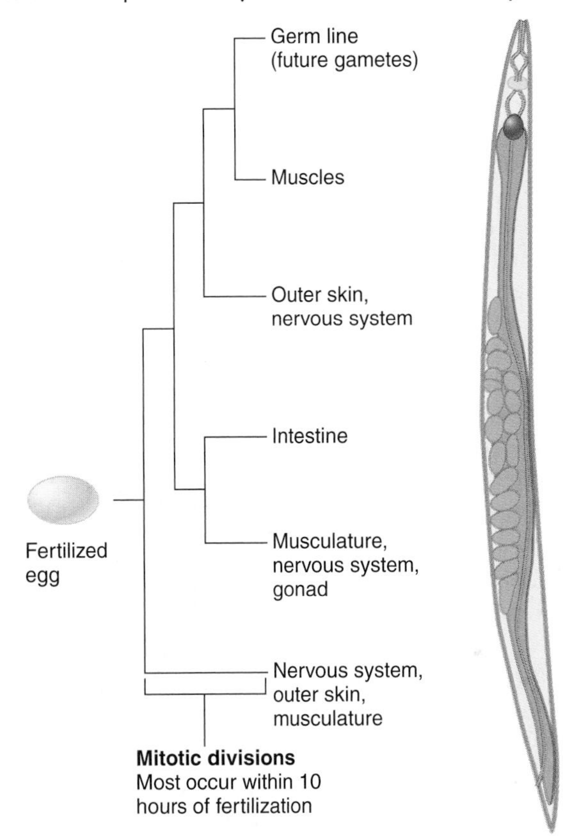

(continued)

Patterns Are Set as a Fly Develops

For an embryo to form, cells must become "aware" of their position—of whether they are located at the anterior, posterior, dorsal, ventral, or left or right side of the forming body. Cells obtain this positional information from protein gradients. Recall that a gradient is a substance or activity that varies in amount or strength continuously over a defined area. In a developing organism, cells at one end contain high levels of a particular protein, cells at the other end contain very little, and those in between have steadily diminishing levels.

The role of genes and the protein gradients they produce is well studied in the fruit fly *Drosophila melanogaster*. For example, a gene in a female fly produces a protein in the egg called bicoid ("two tails"), which instructs the embryo to develop a head end and a tail end. When the gene is mutated, the embryo develops two rear ends! Normally, a greater concentration of bicoid protein in the anterior end of the embryo causes the tissues characteristic of the head to differentiate. At the posterior of the animal, higher concentrations of a protein called nanos signal abdominal tissues to differentiate. Therefore, high levels of bicoid and low levels of nanos indicate "head" formation, and the reverse specifies "tail." Proteins such as bicoid and nanos, present in gradients that influence development, are called morphogens (morphology means "form").

Gradients of several different morphogens distinguish parts of the developing embryo, like using increasing intensities of several colors in a painting (fig. 40.C). The morphogen gradients are signals that stimulate cells to produce yet other proteins, which ultimately regulate the formation of a specific structure. A cell destined to be part of the adult fly's antenna, for example, has different morphogen concentrations than a cell whose future lies in the eye. Sometimes, when morphogen genes mutate, developmental signals go off in the wrong part of the animal—with striking results. The fly in figure 40.D is an example. Called, appropriately, *Antennapedia,* it has legs in place of its antennae. Genes that, when mutated, lead to organisms with structures in the wrong places are called homeotic.

Homeotic genes are widespread. Cows, earthworms, humans, frogs, lampreys, corn, mice, beetles, locusts, bacteria, yeast, chickens, roundworms, and mosquitoes have them. Investigating Life 28.1 depicts homeotic mutations in the mustard relative *Arabidopsis thaliana.* Homeotic mutations cause diseases in humans, including a blood cancer, cleft palate, extra fingers and toes, lack of irises, and thyroid cancer. The fact that homeotic genes are found in diverse species indicates that these controls of early development are both ancient and very important.

FIGURE 40.C

Biochemical Gradients Control Early Development in Fruit Flies.

Very early in the development of a fruit fly, a gradient of bicoid protein distinguishes the embryo's head from its rear. In the first panel of this computer-enhanced image (upper left), different colors represent different concentrations of bicoid. Bicoid protein directs the synthesis of two other proteins, one shown in red and one in green (upper right). A half hour later, another gene directs production of a protein that divides the embryo into seven stripes (lower left). Yet another gene oversees dividing each existing section of the embryo in two (lower right). Overall, genes produce protein gradients that set up distinct biochemical environments in different parts of the embryo, which profoundly influence differentiation.

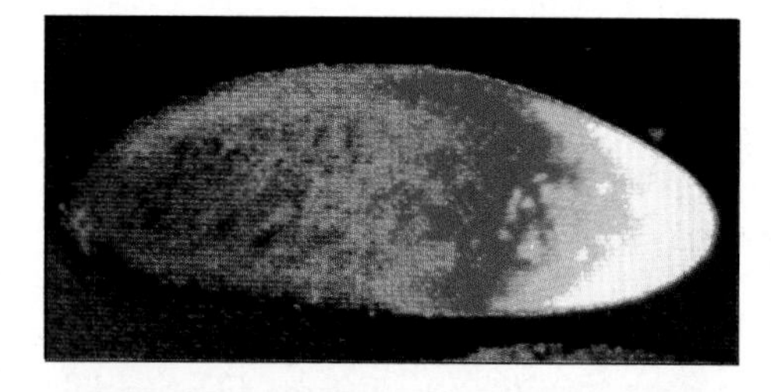
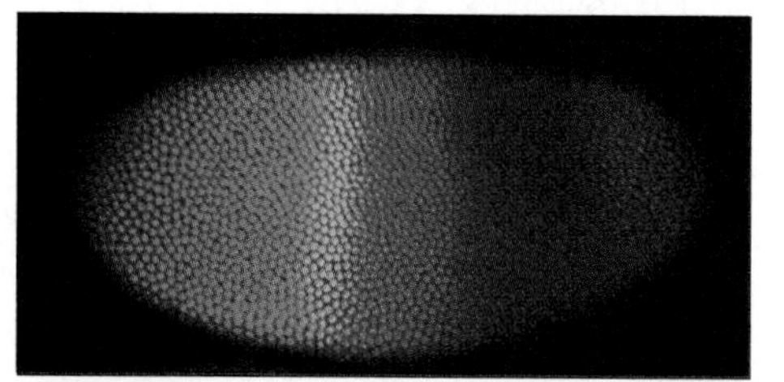
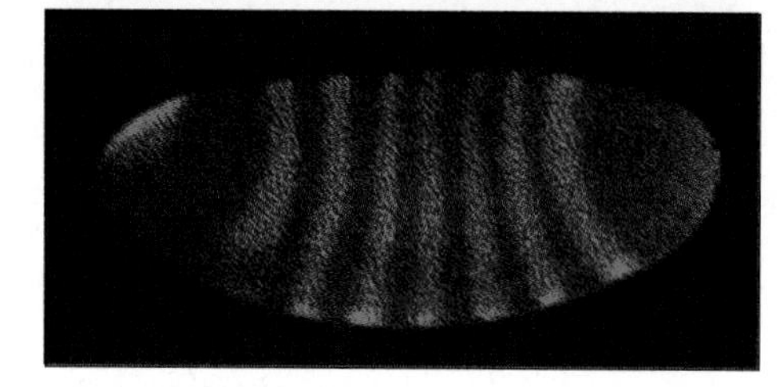
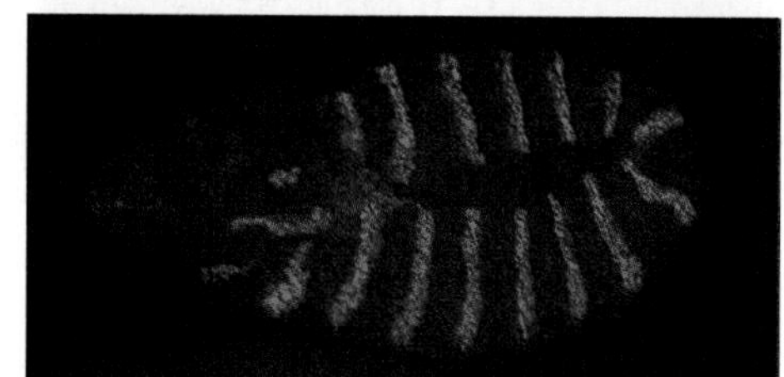

FIGURE 40.D

Homeotic Mutations.

A homeotic mutation sends the wrong morphogen signals to the antenna of a fruit fly, directing the tissue to differentiate as leg. A normal fly (left); a homeotic fly (right).

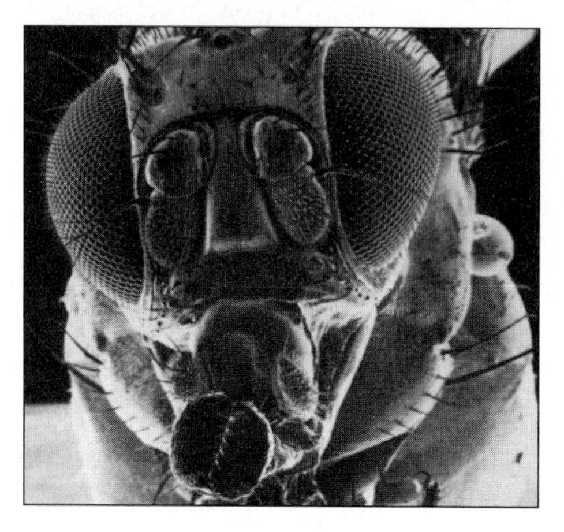
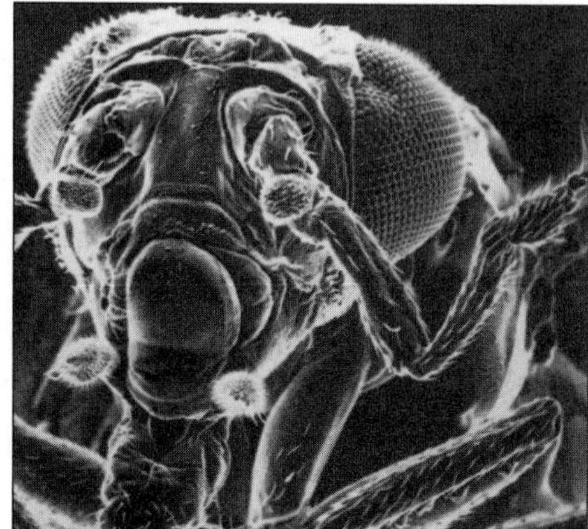

40.2 Setting the Stage for Human Prenatal Development—The Reproductive System

In the reproductive systems of both sexes, gametes are set aside and nurtured, then transported to where they can meet, which occurs following sexual intercourse.

The reproductive systems of the human male and female are similarly organized. Each system has paired **gonads** in which the sperm or oocytes are manufactured; a network of tubes to transport these cells; and hormones and glandular secretions that control the entire process. The details of sperm and oocyte production are covered in chapter 9.

The Human Male Reproductive System

Sperm cells are manufactured within a 410-foot-long (125-meter-long) network of tubes called **seminiferous tubules,** which are packed into paired, oval organs called **testes** (sometimes called testicles) (fig. 40.6). The testes are the male gonads. They lie outside the abdomen within a sac called the scrotum. Their location outside of the abdominal cavity allows the testes to maintain a lower temperature than the rest of the body, which is necessary for sperm to develop properly. Leading from each testis is a tightly coiled tube, the **epididymis,** in which sperm cells mature and are stored. Each epididymis continues into another tubule, the **vas deferens** (plural: vasa deferentia). Each vas deferens bends behind the bladder and joins the **urethra,** the tube that also carries urine out through the **penis.**

Along the sperm's path, three glands contribute secretions. The **prostate gland,** which produces a thin, milky, alkaline fluid that activates the sperm to swim, wraps around the vasa deferentia. Opening into each vas deferens is a duct from the seminal vesicles. These glands secrete the sugar fructose, an energy source, plus hormonelike prostaglandins, which may stimulate contractions in the female reproductive tract that help propel sperm. The bulbourethral glands, each about the size of a pea, attach to the urethra where it passes through the body wall. These glands secrete

FIGURE 40.6
The Human Male Reproductive System.
Sperm cells are manufactured in seminiferous tubules within the paired testes. Sperm mature and are stored in the epididymis and exit through the vas deferens. The paired vasa deferentia join in the urethra, through which seminal fluid exits the body. Secretions are added to the sperm cells from the prostate gland, the seminal vesicles, and the bulbourethral glands. During sexual arousal, three cylinders of erectile tissue fill with blood and cause the penis to become erect.

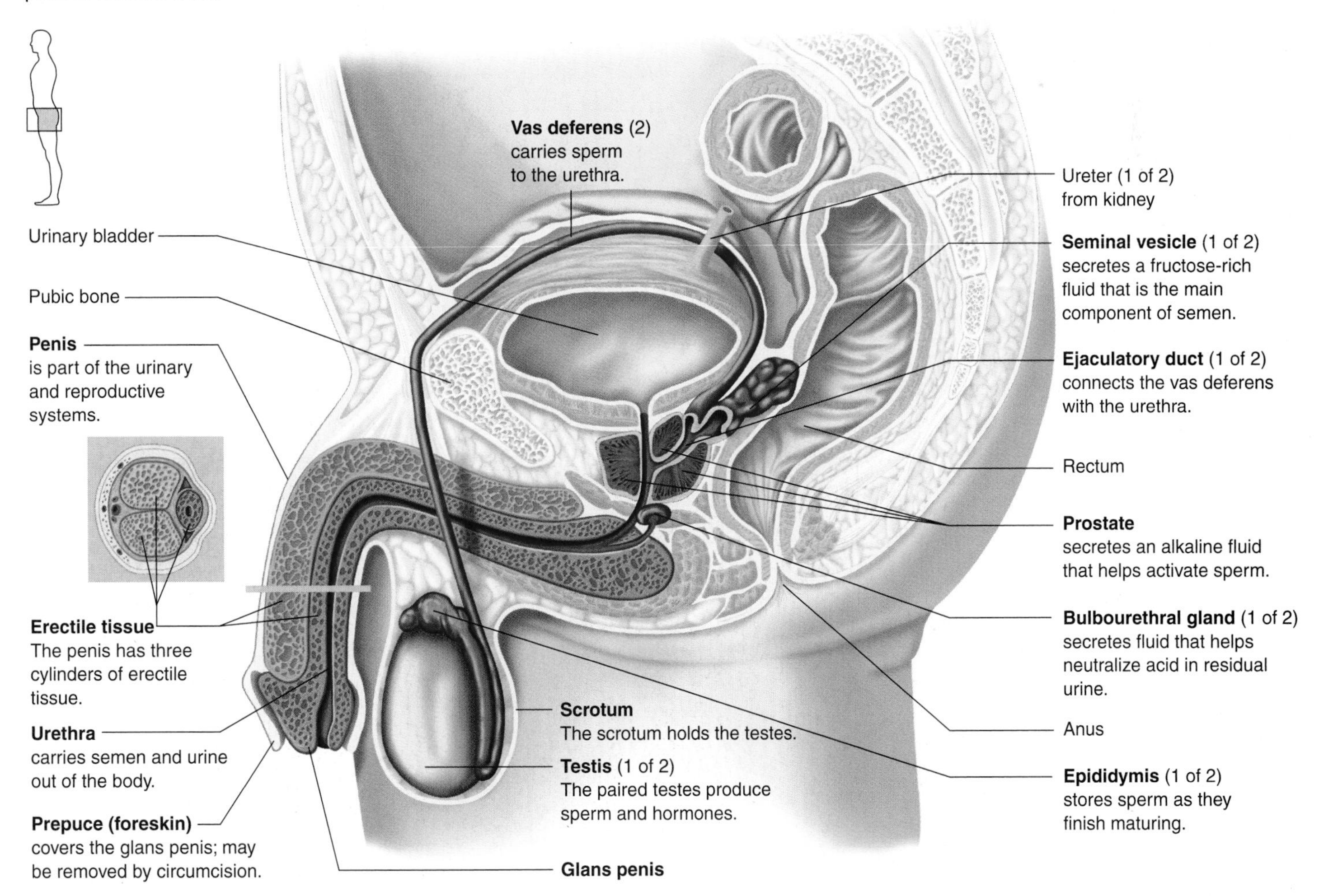

alkaline mucus, which coats the urethra before sperm are released. All of these secretions combine to form the seminal fluid, which carry the sperm cells.

During sexual arousal, the penis becomes erect so that it can penetrate the vagina and deposit sperm in the female reproductive tract. At the peak of sexual stimulation, a pleasurable sensation called **orgasm** occurs, accompanied by rhythmic muscular contractions that eject the sperm from the vasa deferentia through the urethra and out the penis. The discharge of sperm from the penis is called **ejaculation.** One human ejaculation typically delivers about 100 million sperm cells.

The Human Female Reproductive System

The female sex cells develop within paired gonads, called the **ovaries,** in the abdomen (fig. 40.7). Within each ovary of a newborn female are about a million oocytes. Nourishing follicle cells surround each oocyte. Like a testis containing sperm cells in various stages of development, an ovary houses oocytes in different stages of development. Approximately once a month, beginning at puberty, one ovary releases the most mature oocyte. Beating cilia sweep the mature oocyte into the fingerlike projections of one of the two **fallopian tubes.** The tube carries the oocyte into a muscular saclike organ, the **uterus,** or womb.

The lower end of the uterus narrows to form the **cervix,** which opens into the tubelike **vagina,** which opens to the outside of the body. Two pairs of fleshy folds protect the vaginal opening on the outside: the labia majora (major lips) and the thinner, underlying flaps of tissue they protect, called the labia minora (minor lips). At the upper junction of both pairs of labia is a 1-inch-long (2-centimeter-long) structure called the **clitoris,** which is anatomically analogous to the penis. Rubbing the clitoris stimulates females to experience orgasm.

With this anatomy in mind, consider the hormonal fluctuations that regulate the complex timing of oocyte release and the body's preparation for pregnancy. Once released from the ovary, an oocyte can live for about 72 hours, but a sperm can penetrate it

FIGURE 40.7

The Human Female Reproductive System.

Immature egg cells (oocytes) are packed into the paired ovaries. Once a month, one oocyte is released from an ovary and is drawn into the fingerlike projections of a nearby fallopian tube by ciliary movement. If the oocyte is fertilized by a sperm cell in the fallopian tube, it continues into the uterus, where it is nurtured for 9 months as it develops into a new individual. If the ovum is not fertilized, it is expelled along with the built-up uterine lining through the cervix and then through the vagina. The external genitalia consist of the labia minora and majora and the clitoris.

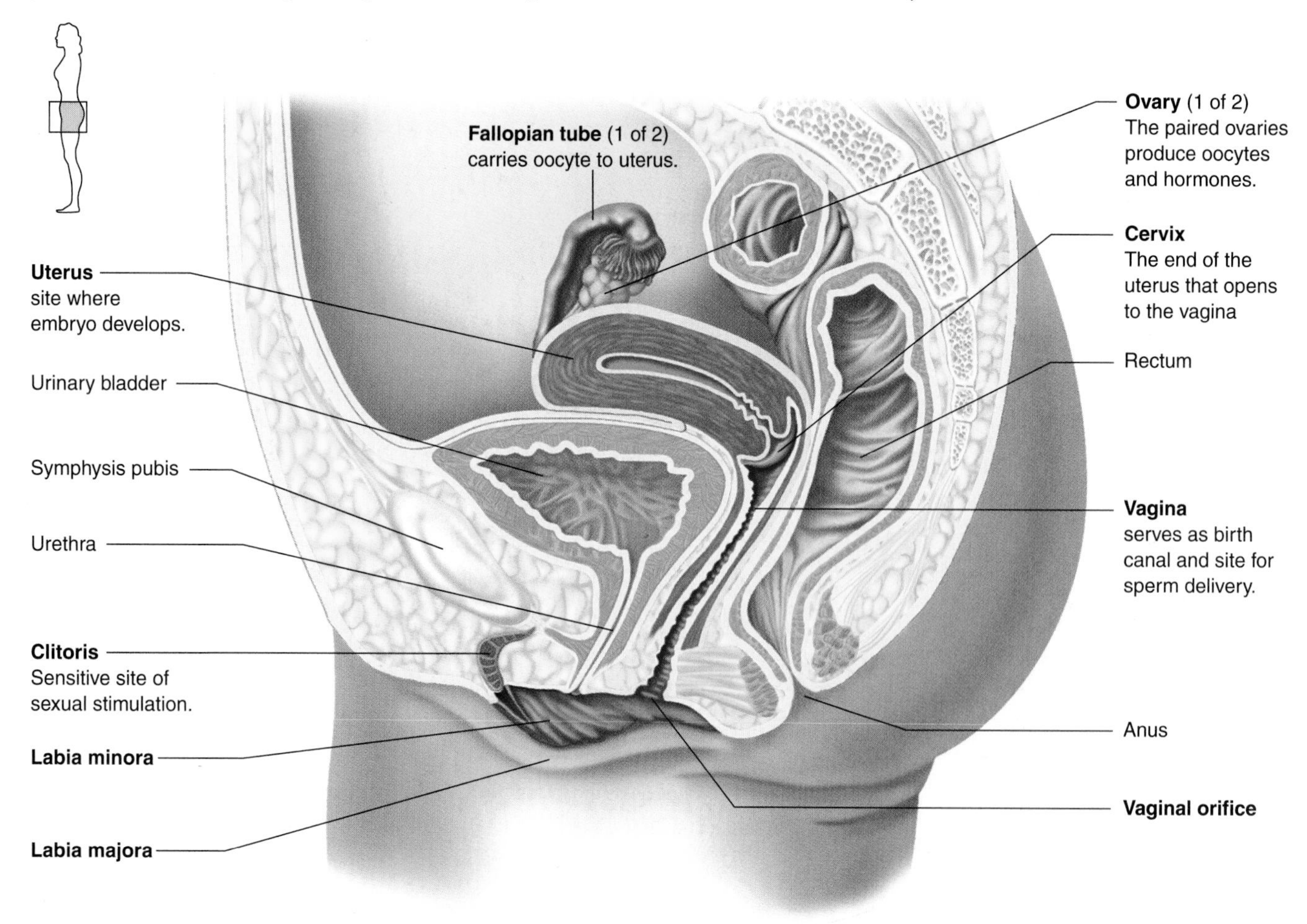

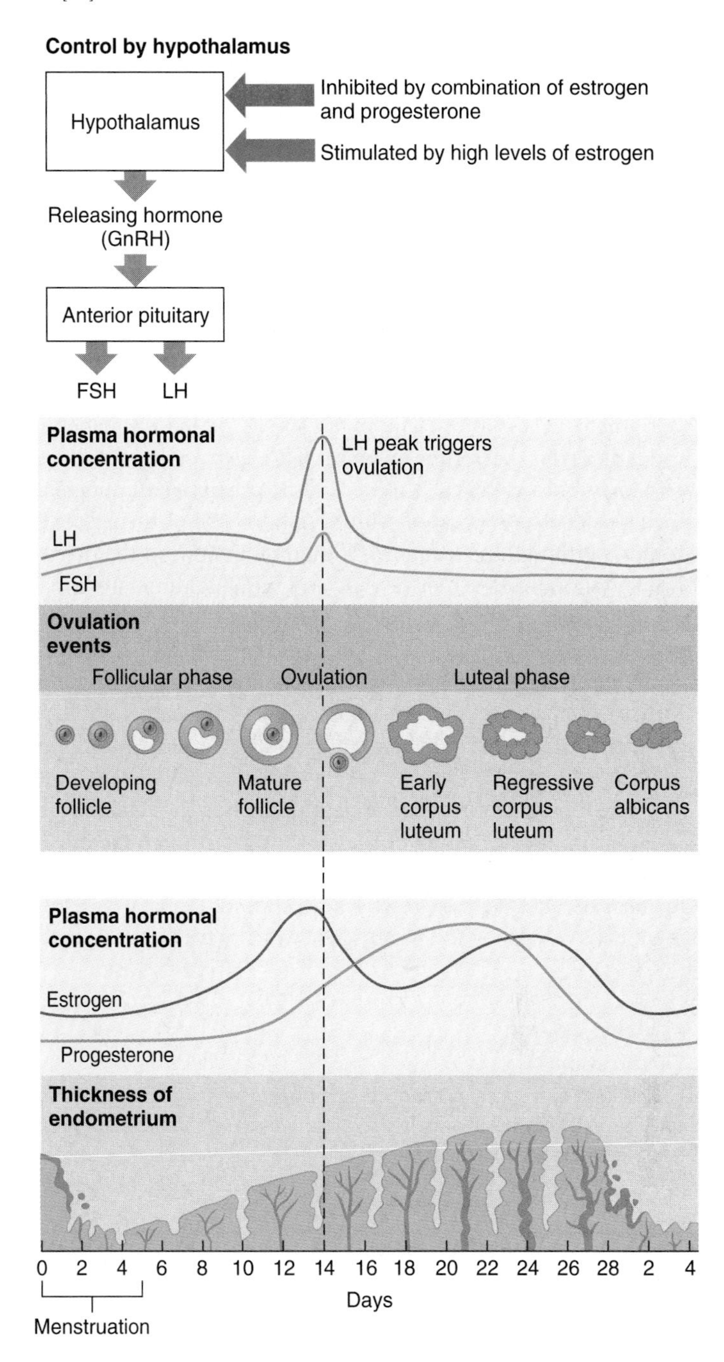

FIGURE 40.8

Three Types of Biological Changes Make Up the Menstrual Cycle.

The middle part of the illustration traces growth of the ovarian follicle. The lower portion depicts the changes in thickness of the uterine lining. The curves in the upper part of each panel indicate changing levels of the hormones estrogen, progesterone, follicle-stimulating hormone (FSH), and luteinizing hormone (LH) throughout the cycle. On the first day of the cycle, the oocyte is within its follicle, the uterine lining is actively being shed, and levels of all four hormones are low. The low hormone levels signal the hypothalamus to release gonadotropin-releasing hormone (GnRH), which increases the anterior pituitary's output of FSH and LH. By the midpoint of the cycle, around day 14, LH level surges, which prompts the oocyte to burst out of its follicle. Meanwhile, the follicle has been producing estrogen to prepare the uterine lining for possible pregnancy. After ovulation, if conception does not occur, the ruptured follicle produces estrogen and progesterone. The uterine lining continues to build as the corpus luteum (the follicle without its oocyte) first enlarges and then degenerates. Once it degenerates, estrogen and progesterone production slow, and by the end of the cycle, the decreasing hormone levels cause the padding of blood and tissue in the uterine lining to deteriorate and exit the body as the menstrual flow.

only during the first 24 hours of this period—possibly even less. If the oocyte encounters a sperm cell in a fallopian tube and the cells combine and their nuclei fuse, the oocyte completes its development and becomes a fertilized ovum, or zygote. It then travels into the uterus and implants in the thick, blood-rich uterine lining that has built up over the preceding few weeks. The uterine lining is called the **endometrium.** Over the next 9 months, the fertilized ovum develops into a new human being. If the oocyte is not fertilized, both endometrium and oocyte are expelled as the menstrual flow.

Hormones control the cycle of oocyte maturation in the ovaries and prepare the uterus to nurture a zygote. On the first day of the cycle, when menstruation begins, low blood levels of estrogen and progesterone signal the hypothalamus to secrete gonadotropin-releasing hormone (GnRH). This prompts the anterior pituitary to release abundant follicle-stimulating hormone (FSH) and small amounts of luteinizing hormone (LH), setting into motion hormonal interactions that thicken the uterine lining in preparation for possible pregnancy. A midcycle surge in the level of LH in the bloodstream triggers the ovary to release an oocyte, an event called **ovulation.** The follicle cells left behind in the ovary undergo changes and are now referred to as the corpus luteum. If pregnancy occurs, the implanted fertilized ovum produces human chorionic gonadotropin (hCG), the hormone that is the basis of pregnancy tests. For a while, hCG keeps the cells of the corpus luteum producing progesterone, which in turn maintains the uterine lining.

If pregnancy does not occur, the corpus luteum degenerates and levels of progesterone and estrogen decline. Reduced levels of these hormones are no longer able to maintain the endometrium, which is then shed through the cervix. Lowered progesterone and estrogen levels also release their inhibition of LH and FSH release in the brain and the cycle begins anew. Figure 40.8 tracks changes in the ovarian follicle, the uterine lining, and the levels of four hormones during the menstrual cycle.

40.2 MASTERING CONCEPTS

1. In which organ are sperm produced?
2. What are the other structures and associated glands of the male reproductive system?
3. In what organ are oocytes produced?
4. What are the other structures of the female reproductive system?

OLC

40.3 The Structures and Stages of Human Prenatal Development

Following fertilization, the zygote divides rapidly, forming a solid ball of cells that hollows out, with a collection of cells set aside that forms three layers. These layers then gradually fold into an embryo. All structures are present by the eighth week, when fetal existence begins. Labor is a series of events that expels a fetus from the uterus.

Developmental biologists recognize three stages of prenatal development, or gestation. The first 2 weeks, which occur before the three layers that define an embryo are distinct, have a variety of names, but we will call this period the **preembryonic stage.** It includes fertilization, rapid mitotic cell division of the zygote as it moves through the woman's fallopian tube toward the uterus; implantation into the uterine wall; and initial folding into layers.

The **embryonic stage** lasts from the end of the second week until the end of the eighth week. Cells of the three layers continue to divide, differentiate, and interact, forming tissues and organs. Structures that support the embryo—the placenta, umbilical cord, and extraembryonic membranes—also develop during this period.

The third stage of prenatal development is the **fetal period,** lasting from the beginning of the ninth week through the full 38 weeks of development. Organs begin to function and coordinate to form organ systems. Growth is very rapid. Prenatal development ends with labor and parturition, which is the birth of a baby.

Fertilization Joins Genetic Packages

The first step in prenatal development is the initial contact between sperm and secondary oocyte. Recall from chapter 9 that the oocyte arrests in metaphase II. A thin, clear layer of proteins and carbohydrates, the **zona pellucida,** encases the oocyte, and a layer of cells called the **corona radiata** surrounds the zona pellucida. The sperm must penetrate these layers to fertilize the ovum (fig. 40.9).

Ejaculation deposits about 100 million sperm cells in the woman's body. (Biotechnology 9.1 describes other routes to fertilization, the assisted reproductive technologies.) A sperm cell can survive here for up to 6 days, but it can only fertilize the oocyte in the 12 to 24 hours after ovulation. A process called **capacitation** activates the sperm inside the woman, altering sperm cell surfaces in a way that enables them to enter an oocyte. In some species, the oocyte secretes a chemical that attracts sperm, although such an attractant has not been found in humans. The female's muscle contractions, the movement of the sperm tails, the surrounding mucus, and the waving cilia on the cells lining the female reproductive tract all propel the sperm toward the oocyte. Despite this help, only 200 or so sperm approach the oocyte, and only one will penetrate it. Even if a sperm touches an oocyte, fertilization may not occur.

When a particular sperm contacts a secondary oocyte, its acrosome bursts, spilling enzymes that digest the zona pellucida and corona radiata. Fertilization (conception) begins when the outer membranes of the sperm and secondary oocyte meet. A wave of electricity spreads physical and chemical changes across

FIGURE 40.9
Fertilization.
Fertilization by a sperm cell induces the oocyte (arrested in metaphase II) to complete meiosis. Before fertilization occurs, the sperm's acrosome bursts, spilling forth enzymes that help the sperm's nucleus enter the oocyte. A series of chemical reactions ensues that help to ensure that only one sperm nucleus enters an oocyte.

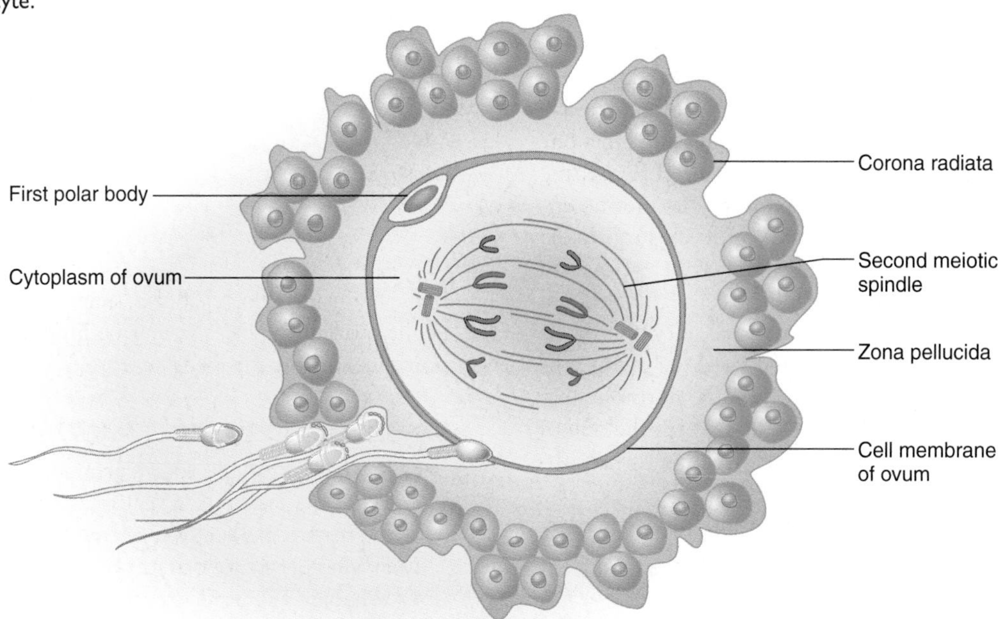

the entire oocyte surface, which keeps other sperm out. If more than one sperm were to enter a single oocyte, the resulting cell would have too much genetic material to develop normally. Very rarely, such fetuses survive until birth, but they have defects in many organs and die within days. When two sperm fertilize two oocytes, fraternal twins result.

As the sperm enters the secondary oocyte, the female cell completes meiosis and becomes a fertilized ovum. Fertilization is not complete, however, until the genetic packages, or pronuclei, of the sperm and ovum meet. Within 12 hours of the sperm's penetration, the nuclear membrane of the ovum disappears, and the two new sets of chromosomes mingle, forming the first cell of the zygote (fig. 40.10). This cell has 23 pairs of chromosomes—one chromosome of each pair from each parent. The cell is still within a fallopian tube.

Contraception is the use of devices or practices that work "against conception." Birth control methods either block the meeting of sperm and oocyte, or they make the female system's environment hostile to sperm or to implantation in the uterine lining. Table 40.1 lists the most common birth control methods available.

Cleavage—Cells Rapidly Divide

About 3 hours after fertilization, the zygote divides for the first time, beginning a period of rapid mitotic cell division called **cleavage.** The daughter cells of cleavage divisions are called **blastomeres.** Sixty hours after fertilization, a second division occurs, and the zygote now consists of four cells. Cleavage continues until a solid ball of 16 or more cells forms. This ball is called a **morula,** Latin for "mulberry," which it resembles (fig. 40.10).

Three days after fertilization, the morula is usually still within the fallopian tube, but it is moving toward the uterus. It is about the same size as the fertilized ovum, because the initial cleavage divisions produce daughter cells that are about half the size of the parent cell. Soon the cellular size levels off, and the zygote begins to enlarge as cells accumulate. During cleavage, organelles and molecules from the secondary oocyte's cytoplasm still control most cellular activities, but some of the zygote's genes become active. The sperm contributes its nucleus, the centrosomes that organize cell division, and proteins that are necessary for cleavage to occur.

The morula reaches the uterus 3 to 6 days after fertilization. It then hollows out, its center filling with fluid that seeps in from the uterus. The preembryo is now also called a **blastocyst,** Greek for "germ bag." The blastomeres are either in an outer layer or on the interior face of the blastocyst. The outer layer of cells is called the trophoblast (Greek for "nourishment of germ"). Certain trophoblast cells will develop into a membrane called the **chorion,** which eventually forms the fetal portion of the placenta, the organ that brings oxygen and nutrients to and removes wastes from the fetus.

The blastomeres inside the blastocyst form the **inner cell mass.** These cells develop into the embryo plus its supportive

FIGURE 40.10
From Ovulation to Implantation.
The zygote forms in the fallopian tube when a sperm nucleus fuses with the nucleus of an oocyte. The first divisions ensue while the zygote moves toward the uterus. By day 7, it begins to implant in the uterine lining.

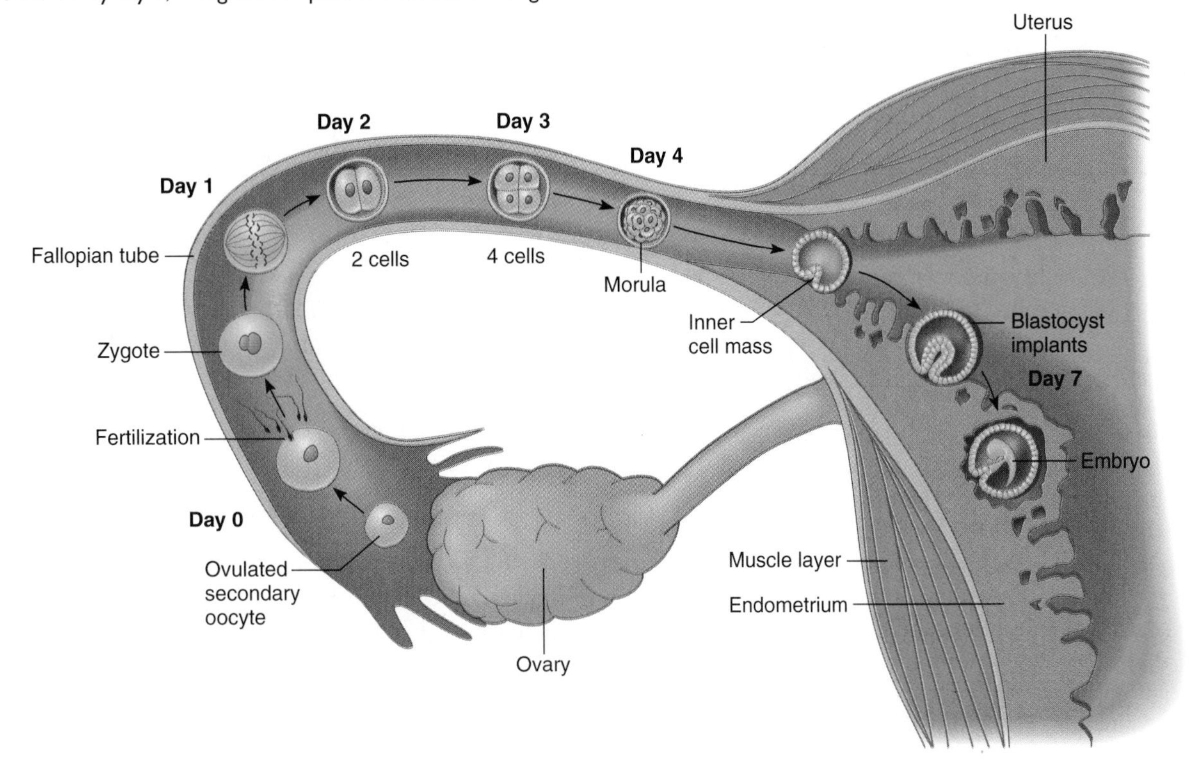

TABLE 40.1 Birth Control Methods

METHOD	MECHANISM	ADVANTAGES	DISADVANTAGES	SUCCESS
BARRIERS AND SPERMICIDES				
Condom and spermicide	Worn over penis or inserted into vagina, keeps sperm out of vagina, and kills sperm that escape	Protects against sexually transmitted diseases	Disrupts spontaneity, reduces sensation	95–98%
Diaphragm and spermicide	Kills sperm and blocks cervix	Inexpensive	Disrupts spontaneity, must be fitted	83–97%
Cervical cap and spermicide	Kills sperm and blocks cervix	Inexpensive, can be kept in for 24 hours	May slip out of place, must be fitted	80–95%
Spermicidal foam or jelly	Kills sperm and blocks cervix	Inexpensive	Messy	78–95%
Spermicidal suppository	Kills sperm and blocks cervix	Easy to use and carry	Irritates 25% of users	85–97%
HORMONAL				
Combination birth control pill	Prevents ovulation and implantation, thickens cervical mucus	Does not interrupt spontaneity, lowers cancer risk, lightens menstrual flow	Raises risk of heart disease in some women, weight gain and breast tenderness	90–100%
Minipill	Blocks implantation, deactivates sperm, thickens cervical mucus	Fewer side effects	Weight gain	91–100%
Depo-Provera	Prevents ovulation, alters uterine lining	Easy to use, lasts 3 months	Menstrual changes, weight gain, injection	99%
Norplant	Prevents ovulation, thickens cervical mucus	Easy to use, lasts 5 years	Menstrual changes, doctor must implant	99.8%
BEHAVIORAL				
Rhythm method	No intercourse during fertile times	No cost	Difficult to do, hard to predict timing	79–87%
Withdrawal	Removal of penis from vagina before ejaculation	No cost	Difficult to do; sperm may leak from penis before it is withdrawn, even without ejaculation	75–91%
SURGICAL				
Vasectomy	Sperm cells never reach penis	Permanent, does not interrupt spontaneity	Requires minor surgery, difficult to reverse	99.85%
Tubal ligation	Oocytes never reach uterus	Permanent, does not interrupt spontaneity	Requires surgery, risk of infection, difficult to reverse	99.6%
OTHER				
Intrauterine device	Prevents implantation	Does not interrupt spontaneity	Severe menstrual cramps, risk of infection	95–99%

structures, called extraembryonic membranes. The fluid-filled center of the ball of cells is called the blastocyst cavity.

Implantation Establishes Pregnancy

Between the fifth and seventh days after fertilization, the blastocyst attaches to the uterine lining, and the inner cell mass within settles against it (fig. 40.10). The trophoblast secretes digestive enzymes that eat through the outer layer of the uterine lining, and ruptured blood vessels surround the blastocyst, bathing it in nutrient-rich blood. This nestling of the blastocyst into the uterine lining is called **implantation,** and it completes by day 14. The trophoblast layer directly beneath the inner cell mass thickens and sends out fingerlike projections into the uterine lining at the site of implantation. These projections develop into the chorion (fig. 40.11).

The trophoblast cells now secrete a hormone, **human chorionic gonadotropin (hCG),** that causes other hormonal changes and prevents menstruation. In this way, the blastocyst helps to ensure its own survival, for if menstruation occurs, it would leave the woman's body along with the tissue in the uterus. The trophoblast cells produce hCG for about 10 weeks.

Gastrulation—Tissues Begin to Form

During the second week of prenatal development, the blastocyst completes implantation, and the inner cell mass changes. A space called the amniotic cavity forms within a sac called the **amnion,** which lies between the inner cell mass and the portion of the trophoblast that has nestled into the uterine lining (fig. 40.11). The inner cell mass then flattens and is called the **embryonic disc.**

FIGURE 40.11
Implantation.
As the inner cell mass settles against the uterine lining, the trophoblast sends out fingerlike extensions that begin to form the chorion, a membrane that develops into the placenta. Meanwhile, the enlarging embryo folds into the three primary germ layers. The yolk sac forms blood cells, immune system stem cells, part of the embryo's digestive system, and becomes incorporated into the umbilical cord.

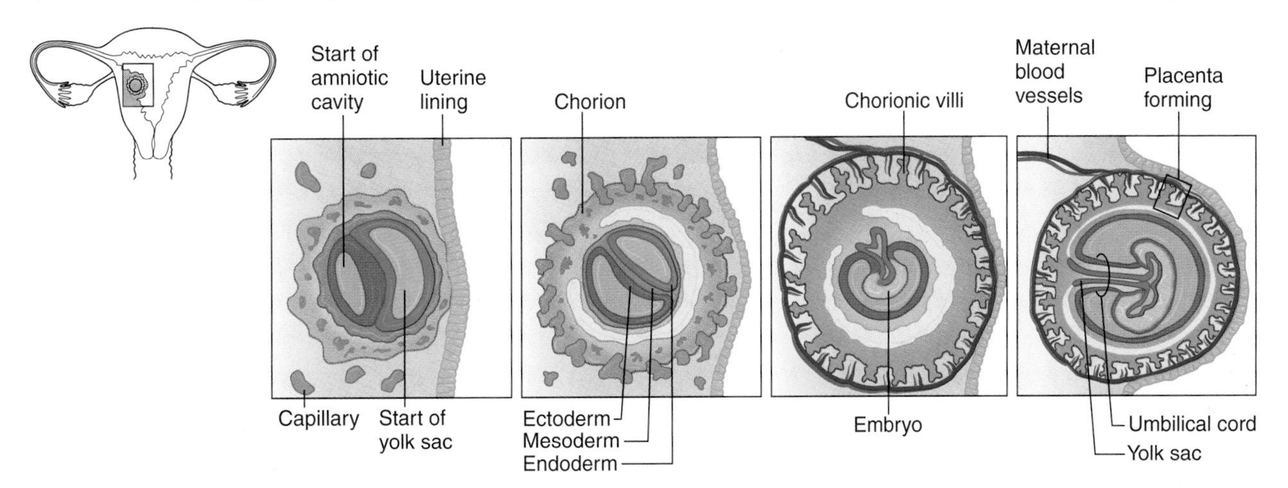

The embryonic disc at first consists of two layers. The outer layer, nearest the amniotic cavity, is ectoderm; the inner layer, closer to the blastocyst cavity, is endoderm. Shortly after, a middle layer, mesoderm, forms between the other two. This three-layered structure is the primordial embryo, or **gastrula.** The process of forming the primordial embryo is called gastrulation, and the layers are called primary germ layers (see chapter 24).

Gastrulation is important in prenatal development because a cell's location in a particular layer determines its fate. Ectoderm cells develop into the nervous system, sense organs, the outer skin layers (epidermis), hair, nails, and skin glands. Mesoderm cells develop into bone, muscle, blood, the inner skin layer (dermis), and reproductive organs. Endoderm cells eventually form the organs and the linings of the digestive and respiratory systems. Gastrulation marks the start of **morphogenesis,** the series of events that forms distinct structures.

The preembryonic stage ends after the second week of prenatal development. Although the woman has not yet missed her menstrual period, she might notice effects of her shifting hormones, such as swollen and tender breasts and fatigue. By now her urine contains enough hCG for an at-home pregnancy test to detect. Highly sensitive blood tests can detect hCG as early as 5 to 7 days after conception.

Organogenesis—The Human Body Takes Shape

During the embryonic stage, organs begin to develop and structures form that will nurture and protect the developing organism. As the days and weeks proceed, different rates of cell division in different parts of the embryo fold tissues into intricate patterns. In a process called embryonic induction, the specialization of one group of cells causes adjacent groups of cells to specialize. Gradually, these changes mold the three primary germ layers into organs and organ systems. **Organogenesis** is the term that describes the transformation of the structurally simple, three-layered embryo into a body with distinct organs. Developing organs are particularly sensitive to damage by environmental factors such as chemicals and viruses.

During the third week of prenatal development, a band called the **primitive streak** appears along the back of the embryonic disc. It gradually elongates to form an axis, which is an anatomical reference point that other structures organize around as they develop. The primitive streak eventually gives rise to connective tissue precursor cells and the **notochord,** a structure that forms the basic framework of the skeleton. The notochord induces overlying ectoderm to differentiate into a hollow **neural tube** (fig. 40.12), which develops into the brain and spinal cord (central nervous system). Formation of the neural tube, or **neurulation,** is a key event in early development because it marks the beginning of organ formation. Soon after neurulation ensues, a reddish bulge containing the heart appears. It begins to beat around day 18. Then the central nervous system begins to elaborate. Figure 40.13 shows some early embryos undergoing these changes.

The fourth week of the embryonic period is a time of rapid growth and differentiation. Blood cells begin to form and to fill primitive blood vessels. Immature lungs and kidneys appear. If the neural tube does not close normally at about day 28, a deformity called a neural tube defect results, which leaves open an area of the spine from which nervous tissue protrudes, causing paralysis. Small buds appear that will develop into arms and legs. The 4-week embryo has a distinct head and jaw and early evidence of eyes, ears, and nose. The rudiments of a digestive system appear as a long, hollow tube that will develop into the intestines. A woman carrying this embryo, which is now only 1/4 inch (0.6 centimeter) long, may suspect that she is pregnant because her menstrual period is about 2 weeks late.

By the fifth week, the embryo's head appears disproportionately large. Limbs extending from the body end in platelike structures. Tiny ridges run down the plates, and by week 6 the ridges deepen as certain cells die, molding fingers and toes. The eyes open, but they do not yet have eyelids or irises. Cells in the brain

FIGURE 40.12
Neurulation.
At a signal from the notochord, ectoderm folds into the neural tube, which will gradually form the brain and spinal cord. The micrograph shows a chick embryo at the neural fold stage.

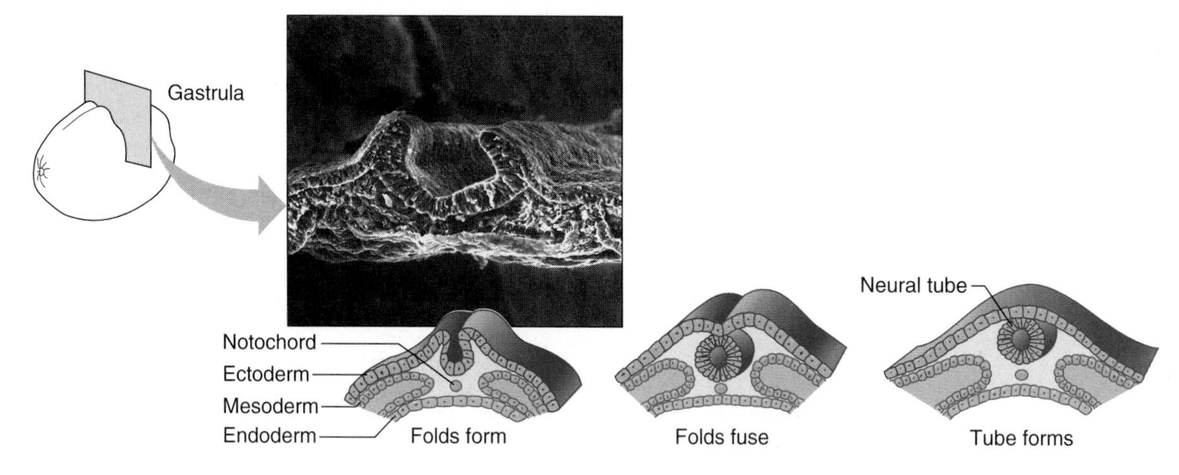

FIGURE 40.13
Early Embryos.
It takes about a month for the embryo to look like a "typical" embryo. At first all that can be distinguished is the primitive streak (**A**), but soon the central nervous system begins to form (**B**). By the 24th day, the heart becomes prominent as a bulge (**C**), and by the 28th day, the organism is beginning to look human (**D**) and (**E**).

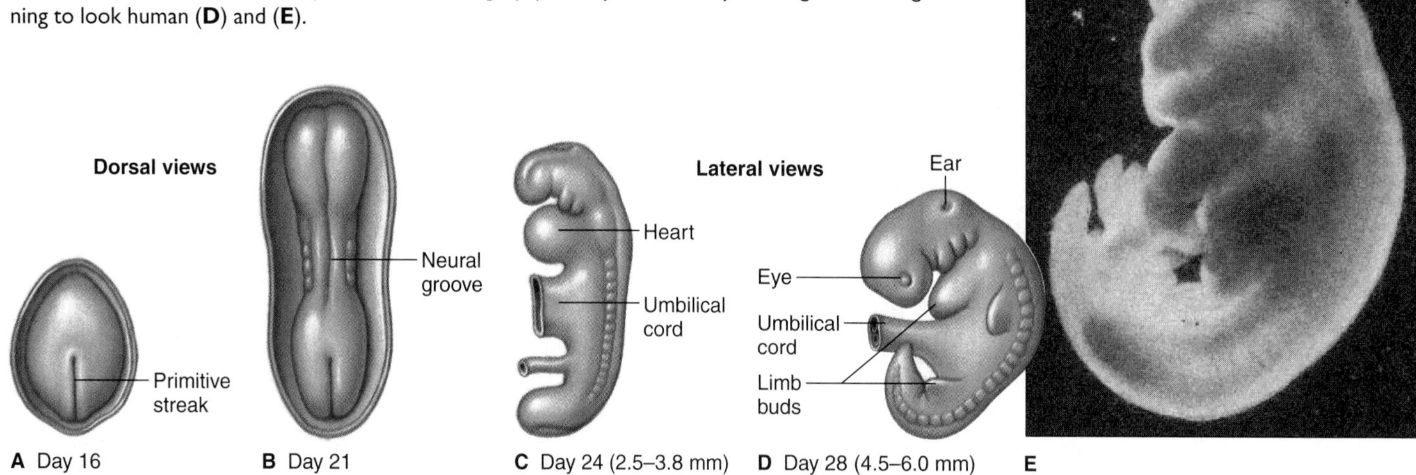

are rapidly differentiating. The embryo is now about 1/2 inch (1.3 centimeters) from head to buttocks.

During the seventh and eighth weeks, a cartilage skeleton appears. The placenta is now almost fully formed and functional, secreting hormones that maintain the blood-rich uterine lining. The embryo is about the size and weight of a paper clip. The eyes now seal shut and will stay that way until the seventh month. The nostrils are closed. A neck appears as the head begins to make up proportionately less of the embryo, and the abdomen flattens somewhat.

Structures That Support the Embryo

By the end of the second week after fertilization, as the embryo folds into three layers, the fingerlike projections from the chorion, called **chorionic villi,** extend further into the uterine lining, establishing the beginnings of the placenta. One side of the placenta—the chorion tissue—comes from the embryo, and the other side consists of blood pools from the pregnant woman's circulation (fig. 40.14). The two blood systems are separate but lie side by side; the chorionic villi lie between the pools of maternal blood. This proximity enables nutrients and oxygen to diffuse from the woman's circulatory system across the chorionic villi cells to the embryo and for wastes to leave the embryo's circulation and enter the woman's, which eventually excretes them.

In addition to providing a lifeline to the embryo and fetus, the placenta secretes hormones that maintain the pregnancy. At 10 weeks, the placenta is completely developed. It is a reddish-brown disc that weighs about 2 pounds (900 grams). Toxins and viruses can pass through the placenta, sometimes damaging the embryo or fetus.

The extraembryonic membranes also form during the early embryonic period. The **yolk sac** begins to appear beneath the embryonic disc at the end of the second week. It manufactures blood cells until about the sixth week, when the liver takes over, and then it starts to shrink. Parts of the yolk sac eventually develop into the intestines and germ cells. Despite its name, the human yolk sac does not actually contain yolk. Similar structures in other animals do contain yolk, which provides nutrients to the developing embryo.

FIGURE 40.14
The Placenta.
In the placenta, chorionic villi that extend from the embryo contact pools of maternal blood. Nutrients and oxygen pass from the maternal blood to the embryo, while wastes diffuse in the opposite direction.

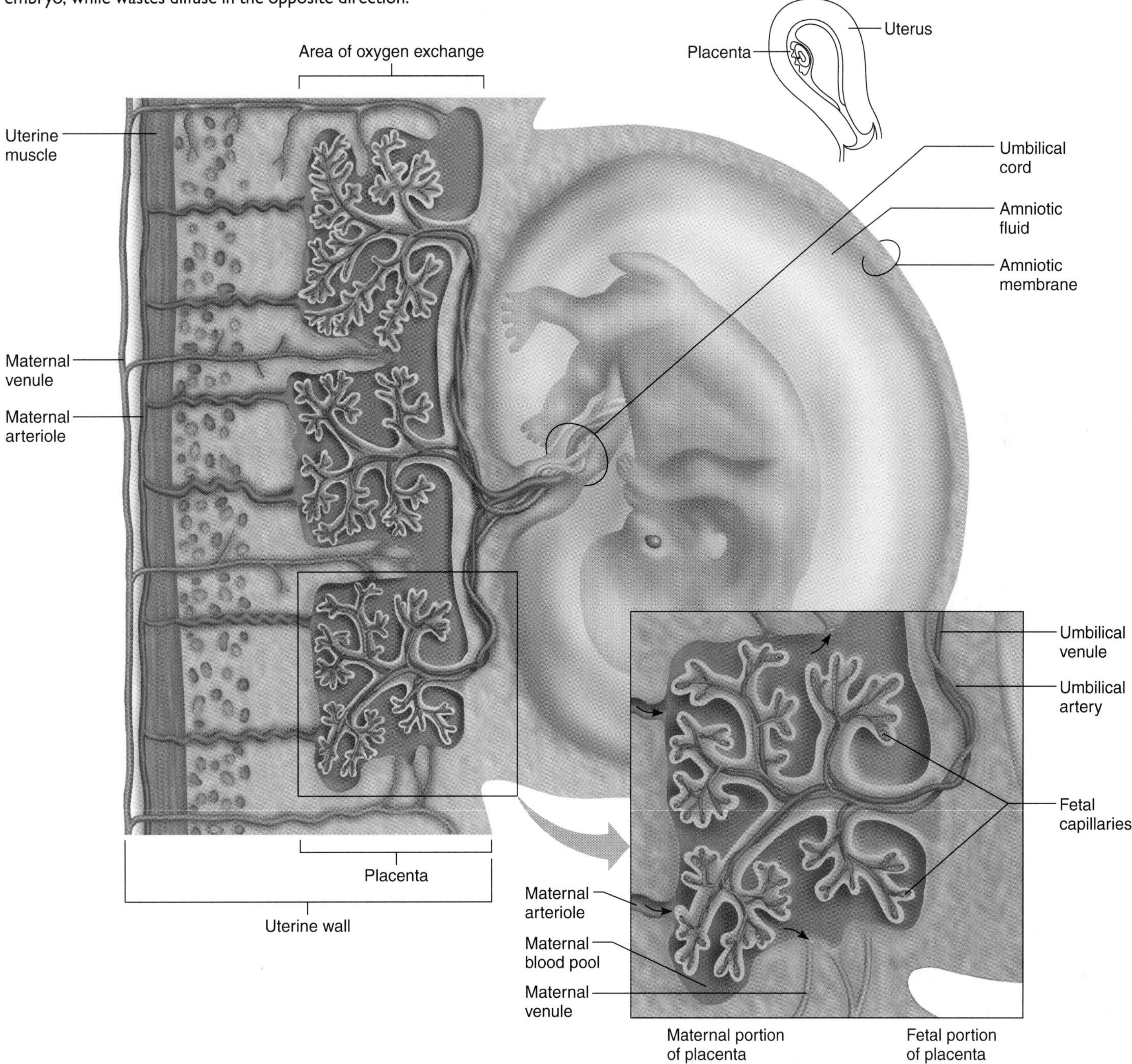

By the third week, an outpouching of the yolk sac forms the **allantois,** another extraembryonic membrane. It, too, manufactures blood cells, and it also gives rise to the fetal umbilical arteries and vein. During the second month, most of the allantois degenerates, but part of it persists in the adult as a ligament supporting the urinary bladder.

As the yolk sac and allantois develop, the amniotic cavity swells with fluid. This "bag of waters" cushions the embryo, maintains a constant temperature and pressure, and protects the embryo if the woman falls. The amniotic fluid comes from the woman's blood. It also contains fetal urine and cells from the amniotic sac, the placenta, and the fetus.

Toward the end of the embryonic period, as the yolk sac shrinks and the amniotic sac swells, the umbilical cord forms. The cord is 2 feet (0.6 meter) long and 1 inch (2.5 centimeters) in diameter. It houses two umbilical arteries, which transport oxygen-depleted blood from the embryo to the placenta, and one umbilical vein, which brings oxygen-rich blood to the embryo. The umbilical cord attaches to the center of the placenta. It twists because the umbilical vein is slightly longer than the umbilical arteries.

Medical technology sometimes uses the structures that support and nourish an embryo. Fetal cells collected from amniotic fluid, or sampled chorionic villi cells, provide information on the genetic health of a fetus and form the basis of many prenatal medical tests, as the opening essay to chapter 11 describes. Umbilical cord blood is a rich source of stem cells used in gene therapy and for bone marrow transplants needed later in life to treat certain blood disorders (see chapter 30's opening essay). Many hair products include extracts from human placentas.

The Fetal Period—Structures Are Elaborated

The body proportions of the fetus appear more like those of a newborn as time goes on. The ears lie low, and the eyes become widely spaced. Bone begins to form and will eventually replace most of the cartilage, which is softer. Soon, as the nerves and muscles coordinate, the fetus will move its arms and legs.

Sex is determined at fertilization, when a sperm bearing an X or Y chromosome meets an oocyte, which always has an X chromosome. A gene on the Y chromosome, called *SRY* (for "sex-determining region of the Y") determines maleness. However, all early embryos have rudiments of both sexes, including unspecialized gonads and sets of tubes. At week 7, if the *SRY* gene is activated, male hormones stimulate differentiation of male reproductive organs and glands from the male precursor structures. If there is no *SRY* gene—that is, if the fetus is female—female reproductive structures develop. Physical differences between the sexes begin to appear after week 7 (fig. 40.15).

By the twelfth week, the fetus is obviously male or female. It now sucks its thumb, kicks, and makes fists and faces, and baby teeth begin to form in the gums. The fetus breathes the amniotic fluid and releases wastes into it. The first trimester (3 months) of pregnancy ends.

FIGURE 40.15
Sexual Development.
Early embryos have rudiments of structures characteristic of both the male and female reproductive systems. If at week 7 the *SRY* gene on the Y chromosome is activated, development continues as a male; if *SRY* does not turn on—because it isn't there—a female develops.

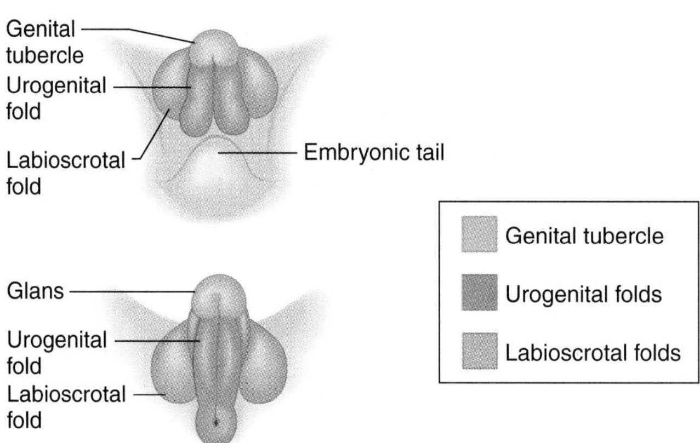

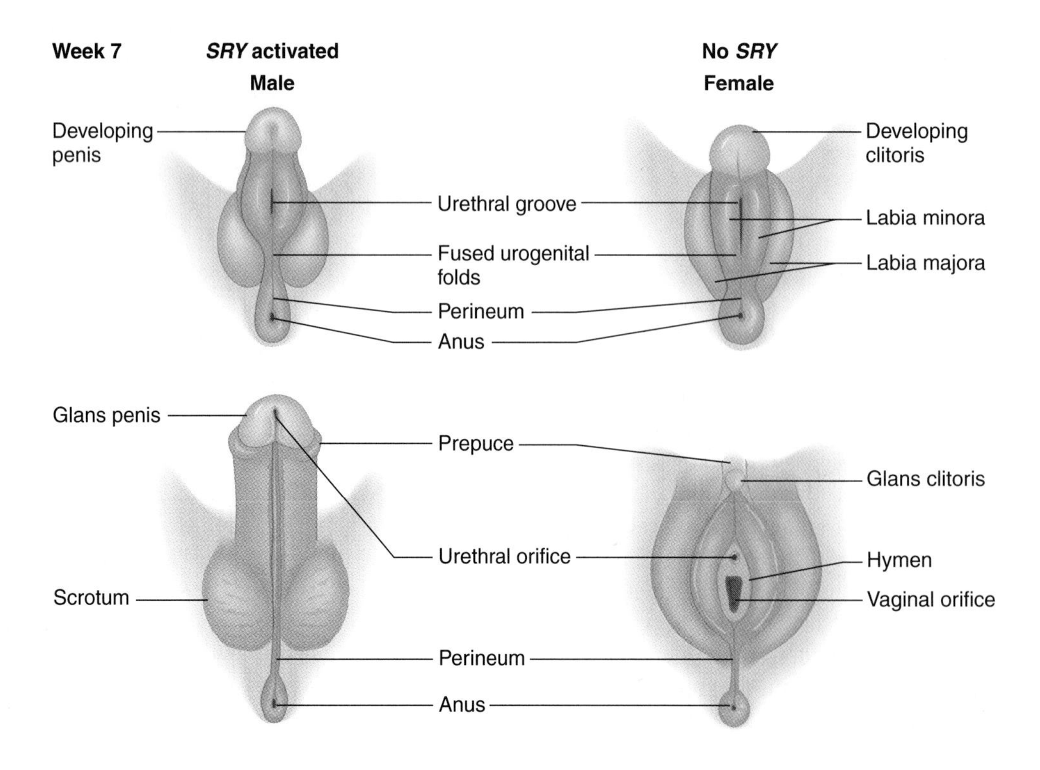

During the second trimester, the body proportions of the fetus become even more like those of a newborn. By the fourth month, it has hair, eyebrows, lashes, nipples, and nails. Some fetuses even scratch themselves before birth. Bone continues to replace the cartilage skeleton. The fetus's muscle movements become stronger, and the woman may begin to feel slight flutterings in her abdomen. By the end of the fourth month, the fetus is about 8 inches (20 centimeters) long and weighs about 6 ounces (170 grams).

During the fifth month, the fetus becomes covered with an oily substance called vernix caseosa that protects the skin. White, downy hair called lanugo holds the vernix in place. By 18 weeks, the vocal cords form, but the fetus makes no sounds because it does not breathe air. By the end of the fifth month, the fetus curls into the classic head-to-knees position. It weighs about 1 pound (454 grams) and is 9 inches (23 centimeters) long.

During the sixth month, the skin appears wrinkled because there isn't much fat beneath it. The skin turns pink as blood-filled capillaries extend into it. By the end of the second trimester, the woman feels distinct kicks and jabs and may even detect a fetal hiccup. The fetus is now about 12 inches (30.5 centimeters) long.

In the final trimester, fetal brain cells rapidly connect into networks, and organs differentiate further and grow. A layer of fat develops beneath the skin. The digestive and respiratory systems mature last, which is why infants born prematurely often have difficulty digesting milk and breathing. About 266 days after a single sperm burrowed into an oocyte, a baby is ready to be born.

Labor and Birth

"Labor" refers, appropriately, to the strenuous work a pregnant woman performs in the hours before giving birth. Labor may begin with an abrupt leaking of amniotic fluid as the fetus presses down and ruptures the sac ("water breaking"). This exposes the fetus to the environment; if birth doesn't occur within 24 hours, the baby may be born with an infection. Labor may also begin with a discharge of blood and mucus from the vagina, or a woman may feel mild contractions in her lower abdomen about every 20 minutes.

As labor proceeds, the hormone-prompted uterine contractions gradually increase in frequency and intensity. During the first stage of labor, the baby presses against the cervix with each contraction. The cervix dilates (opens) a little more each time. At the start of labor, the cervix is a thick, closed band of tissue. By the end of the first stage of labor, the cervix stretches open to about 10 centimeters. The cervix sometimes takes several days to open, with mild labor beginning well before a woman realizes it has started.

The second stage of labor is the delivery of the baby, or parturition (fig. 40.16). It begins when the cervix completely dilates. The woman feels a tremendous urge to push. Within the next 2 hours, the baby descends through the cervix and vagina and is born. In the third and last stage of labor, uterine contractions expel the placenta and extraembryonic membranes from the woman's body.

A baby may be delivered surgically, through a procedure called a cesarean section, for a variety of reasons. The baby may be too large to fit through the woman's pelvis; it may be positioned feet or buttocks down (breech) rather than head down, so

FIGURE 40.16
Labor.
About 2 weeks before birth, the fetus "drops" in the woman's pelvis and the cervix may begin to dilate (**A**). At the onset of labor, the amniotic sac may break (**B**). The baby (**C**) and then the placenta (**D**) and extraembryonic membranes are pushed out of the birth canal.

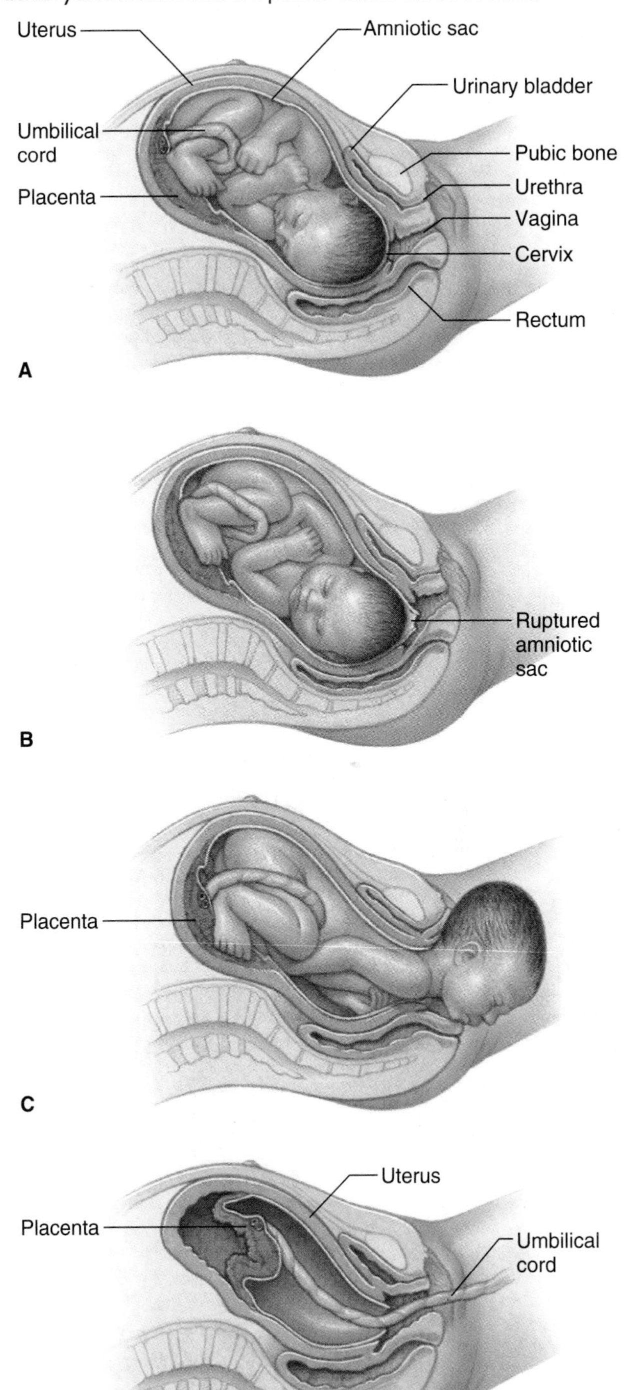

that the head may become caught if the baby is delivered vaginally; or the fetus may be wedged side-to-side. If the umbilical cord wraps around the fetus's neck as it moves into the birth canal, a cesarean section can save the baby's life. About one in five births in the United States is surgical.

The birth of a live, healthy baby seems against the odds, considering the complexity of human development from the beginning. Of every 100 secondary oocytes exposed to sperm, 84 are fertilized. Of these, 69 implant in the uterus, 42 survive 1 week or longer, 37 survive 6 weeks or longer, and only 31 are born alive. Of those that do not survive, about half have chromosomal abnormalities too severe to maintain life. Health 40.1 explores some familiar birth defects and their causes.

40.3 MASTERING CONCEPTS

1. List the events of the preembryonic stage of prenatal human development.
2. How is menstruation prevented when a woman becomes pregnant?
3. What is the relationship among the primitive streak, the notochord, and the neural tube?
4. Which supportive structures and extraembryonic membranes develop during the embryonic period?
5. When do sexual differences appear, and what triggers them?
6. Which events make up the three stages of labor?

40.4 Growth and Development After Birth

The newborn's body must suddenly take over functions fulfilled by the pregnant woman during pregnancy. Childhood is a time of rapid growth and maturation; hormone changes dominate adolescence. Signs of aging become evident in the third decade, but the process is continual, in both active and passive ways, from conception.

Growth and development continue after birth, as the body rapidly enlarges while it retains the specialization of its tissues and organs. Some structures grow and their cells are replenished continually, such as hair, nails, skin, and the lining of the small intestine. Some functions begin at or after birth, peak, and then decline, following specific timetables, yet changing in a coordinated manner so that the body as a whole operates efficiently. A human life can be considered in the stages depicted in figure 40.17.

The Early Years

Birth triggers dramatic changes, as the newborn must suddenly breathe, eat, excrete, and regulate temperature on its own. The blood vessels that linked fetus to pregnant woman close off, and the baby's circulatory system now handles vital gas exchange. The human newborn is helpless, compared with newborns of some other mammals. Llamas, antelopes, and guinea pigs, for example, are born fully furred, and within minutes of birth, they stand and walk about. It is many months before a human infant stands and walks!

Infancy (4 weeks to 2 years) is a time of incredibly rapid growth, as body weight triples. In the first 6 months, adult hemoglobin (the protein that carries oxygen in the blood) gradually replaces the fetal variety, which has a greater affinity for oxygen. The digestive system matures gradually, as evidenced by the increasing variety of foods that a baby can digest. Movements coordinate, immunity begins to build, and the baby learns to communicate.

FIGURE 40.17
Life Stages.
Stages of human postnatal development.

Newborn Birth–4 weeks	Infant 4 weeks–2 years	Child 2–12 years	Adolescent 13–18 years	Adult 18 years–death
• Survival reflexes • Start of respiration, digestion, kidney function, temperature control • Circulatory system changes	• Adult hemoglobin • Senses sharpen • Muscles coordinate • Rapid growth • Teeth erupt • Communication skills	• Skeleton matures • Immune system is completely active • Brain cell division slows and stops • Teeth erupt • Bladder and bowel control	• Sex hormones produced • Secondary sexual characteristics appear • Sex organs mature • Growth spurt • Emotional maturity • Intellectual development	• Organ systems function • Repair of damaged tissue • Muscle strength, senses, hair growth peak • Aging-related changes begin in mid-thirties

Birth Defects

Genetic abnormalities or exposure to toxins can disrupt prenatal development and cause birth defects. However, about 97% of newborns are apparently normal.

The Critical Period

The specific nature of a birth defect depends upon which structures were developing when the exposure occured. The time when genetic abnormalities, toxic substances, or viruses can alter a specific structure is called its critical period (fig. 40.E). Some body parts, such as fingers and toes, are sensitive for short periods of time. In contrast, the brain is vulnerable to damage throughout prenatal development, as well as during the first 2 years of life. Because of the brain's extensive critical period, many birth defect syndromes include mental retardation. The continuing sensitivity of the brain after birth explains why toddlers who accidentally ingest lead-based paint suffer impaired learning.

About two-thirds of birth defects stem from a disruption during the embryonic period. Disruptions during the fetal period cause more subtle defects, such as learning disabilities. For example, damage during the first trimester might cause mental retardation, but in the seventh month of pregnancy, it might interfere with learning to read.

Some birth defects can be attributed to an abnormal gene that acts at a specific point in prenatal development. For example, in a rare inherited condition, phocomelia, an abnormal gene acts during the third to fifth weeks of embryonic development, when the limbs develop. The infant has flipperlike structures where arms and legs should be.

Toxins that pregnant women ingest cause some birth defects. These environmentally caused problems cannot pass to future generations and will not recur unless the exposure does. Chemicals or viruses that cause birth defects are called teratogens. For example, a drug called thalidomide also acts during the third to fifth week of embryonic development and causes severe limb shortening, as the inherited phocomelia does (fig. 40.F).

People may encounter teratogens in the workplace. Women who work with textile dyes, lead, certain photographic chemicals, semiconductor materials, mercury, and cadmium face increased risk of spontaneous abortion and birth defects in their children. Several other teratogens are described below.

Alcohol

Greek philosopher Aristotle noticed problems in children of alcoholic mothers

(continued)

FIGURE 40.E

Sensitive Periods of Development.

The nature of a birth defect resulting from drug exposure depends upon which structures are developing at the time of the exposure. The time when a particular structure is sensitive is called the critical period. Accutane is an acne medication. Diethylstilbestrol (DES) was used in the 1950s to prevent miscarriage. Thalidomide was used to prevent morning sickness.

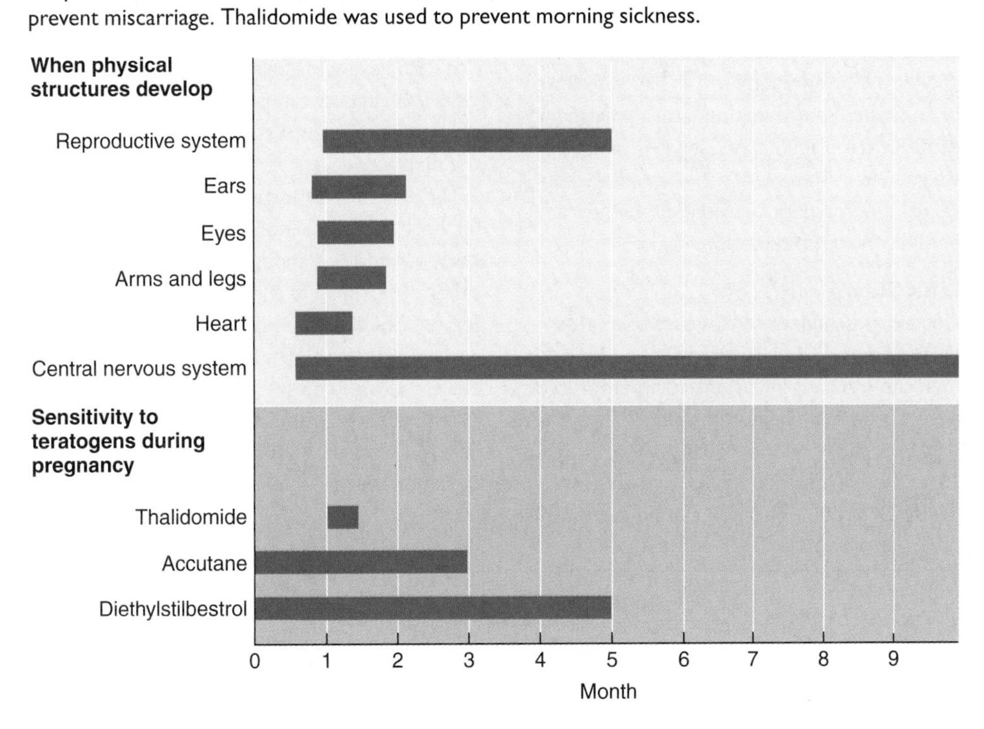

FIGURE 40.F

Teratogenic Effects of Thalidomide.

This child is 1 of about 10,000 children born between 1957 and 1961 in Europe to women who had taken the tranquilizer thalidomide early in pregnancy. The drug is a teratogen that affects developing limbs so that only stumps appear. However, thalidomide is still a very useful drug for certain conditions, such as leprosy and AIDS.

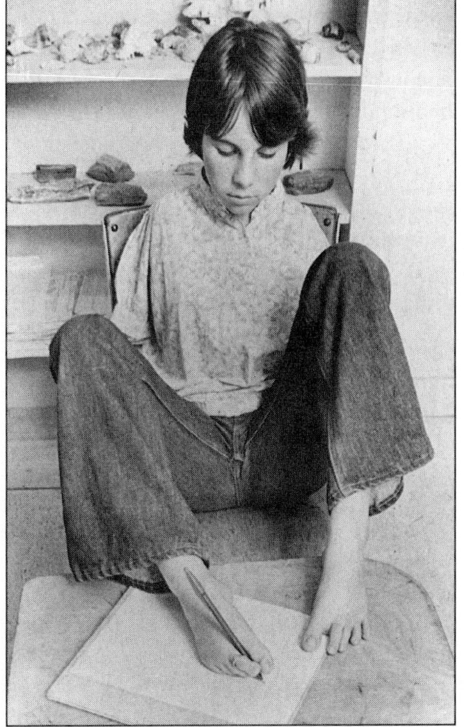

more than 23 centuries ago. In the United States today, fetal alcohol syndrome (FAS) is the third most common cause of mental retardation in newborns, and 1 to 3 in every 1,000 infants has the syndrome—more than 40,000 born each year.

A pregnant woman who has just one or two drinks a day, or perhaps a large amount at a single crucial time in prenatal development, risks FAS. Because each woman metabolizes alcohol slightly differently, physicians advise that pregnant women avoid alcohol when pregnant or when trying to become pregnant.

A child with fetal alcohol syndrome has a small head, misshapen eyes, and a flat face and nose and grows slowly before and after birth. The child has impaired intellect, ranging from minor learning disabilities to mental retardation. Effects of FAS continue beyond childhood. More than 80% of them retain facial characteristics of a young child with FAS. The long-term mental effects of prenatal alcohol exposure are more severe than the physical vestiges. Many adults with FAS function at early grade school level, finding it difficult to learn and to maintain relationships.

Cocaine

Cocaine is very dangerous to the unborn. It can induce a stroke in a fetus and cause spontaneous abortion. Cocaine-exposed infants who survive are distracted and unable to concentrate on their surroundings. Other health and behavioral problems arise as these children grow. A problem in evaluating the effects of cocaine is that affected children are often exposed to other substances and situations that could also account for their symptoms.

Cigarettes

Chemicals in cigarette smoke stress a fetus. Carbon monoxide crosses the placenta and plugs sites on fetal hemoglobin molecules that would normally bind oxygen, robbing rapidly growing fetal tissues of oxygen. Other chemicals in cigarette smoke prevent nutrients from reaching the fetus. The placentas of women who smoke lack important growth factors, thus lowering birth weight. Cigarette smoking during pregnancy increases risk of spontaneous abortion, stillbirth, and prematurity.

Nutrients

Certain nutrients ingested in large amounts, particularly vitamins, act as drugs in the human body. The acne medicine isotretinoin (Accutane), derived from vitamin A, causes spontaneous abortions and defects of the heart, nervous system, and face. The tragic effects of this drug were first noted in the early 1980s, exactly 9 months after dermatologists began prescribing it to young women. Today it is never prescribed to women who could be pregnant.

Another nutrient that can harm a fetus when the pregnant woman takes it in excess is vitamin C. The fetus becomes accustomed to the large doses and, after birth, when the supply drops, the baby develops symptoms of vitamin C deficiency (scurvy). The baby bruises easily and is prone to infection.

Malnutrition during pregnancy threatens a fetus because a pregnant woman requires extra calories. Obstetrical records of pregnant women before, during, and after World War II link inadequate nutrition in early pregnancy to an increase in the incidence of spontaneous abortion. The aborted fetuses had very little brain tissue. Poor nutrition later in pregnancy damages the placenta, causing low birth weight, short stature, tooth decay, delayed sexual development, learning disabilities, and possibly mental retardation.

Infections

Certain viral infections can pass to a fetus during pregnancy or to an infant during birth. Men can transmit infections to an embryo or fetus during sexual intercourse.

HIV, the virus that causes AIDS, infects 15 to 30% of infants born to infected (HIV positive) women. This risk can be cut sharply if an infected woman takes certain drugs used to treat AIDS while pregnant. Fetuses infected with HIV are at high risk for low birth weight, prematurity, and stillbirth.

The virus that causes rubella (German measles) is a powerful teratogen, causing symptoms that are collectively called congenital rubella syndrome. Australian physicians first noted its effects in 1941, and a rubella epidemic in the United States in the early 1960s caused 20,000 birth defects and 30,000 stillbirths. Exposure in the first trimester leads to cataracts, deafness, and heart defects, and later exposure causes learning disabilities, speech and hearing problems, and juvenile-onset diabetes. Widespread vaccination has slashed the incidence of congenital rubella syndrome, and today it occurs only where people are not vaccinated. In 1991, for example, 14 of every 1,000 newborns among the Amish population in rural Pennsylvania had rubella exposure; the incidence in the general U.S. population is 0.006 per 1,000 births.

A herpes simplex viral infection can harm a fetus and newborn, whose immune systems are not yet fully functional. Forty percent of babies exposed to active vaginal lesions become infected, and half of these infants die from it. Of the surviving infants, 25% sustain severe nervous system damage, and 25% have widespread skin sores. A cesarean section can prevent an infected woman from transmitting the virus during the birth process.

Pregnant women are routinely checked for hepatitis B infection, which in adults causes liver inflammation, great fatigue, and other symptoms. Each year in the United States, 22,000 infants are infected with this virus during birth. These babies are healthy, but are at high risk for developing serious liver problems as adults. By identifying infected women, a vaccine can be given to their newborns, which can help prevent complications.

During childhood (2 to 12 years), the bones complete hardening, the immune system becomes fully active, and primary teeth erupt, and some may even be replaced by secondary teeth. Bladder and bowel controls are mastered early in childhood. Adolescence, which may begin as early as age 10 or 11 in a girl and may start as late as 16 in either sex, is another period of rapid and profound change. As sex hormones begin to be produced, the sex organs mature, and secondary sexual characteristics (such as a deepening voice in a male and breast development in a female) appear. Growth is rapid, and a teen may eat voraciously. Social, emotional, and intellectual development occur rapidly too. By early adulthood, all organ systems are fully developed and functioning.

Aging Is Passive and Active

Aging occurs throughout life, as structures break down and functions decline. It is even apparent in a fetus, as cells in the webbing of the hands and feet die as fingers and toes form. Table 40.2 summarizes some of these changes.

The aging process is difficult to analyze because of the intricate interactions of the body's organ systems. One structure's breakdown ultimately affects the way others function. Aging has both passive and active components—existing structures break down, and new biochemicals and structures form.

Aging is passive in the sense that structures break down and functions slow. At the molecular level, passive aging is seen in the degeneration of the connective tissue's elastin and collagen proteins. As these proteins fall apart, skin loses its elasticity and begins to sag, and muscle tissue slackens.

During a long lifetime, DNA sequence errors accumulate. Mistakes in a DNA sequence can occur when DNA replicates in dividing cells. Usually repair enzymes correct this damage immediately. But over many years, exposure to chemicals, viruses, and radiation disrupts DNA repair, so that the burden of fixing errors becomes too great. The cell may die as a result of faulty genetic instructions that go unrepaired.

Highly reactive metabolic by-products called free radicals may stimulate the cellular degradation associated with aging. A free-radical molecule has an unpaired electron in its outermost

TABLE 40.2 Aging in the Human

DECADE (YEARS)	CHANGES ASSOCIATED WITH AGING
0–10	Brain cells begin to die
	Sense of hearing peaks
10–20	Thymus gland begins to shrink
	Peak of male sexuality
20–30	Peak muscle strength
30–40	Peak hair thickness
	Skin begins to lose elasticity, small wrinkles appear
	Decrease in height
	Hearing diminishes
	Peak of female sexuality
	Heart muscle thickens
40–50	Slumping posture as ligaments in back lose elasticity
	Weight gain
	Decrease in height
	Hair grays and thins
	Fewer white blood cells increases susceptibility to infection and cancer
	Farsightedness develops
	Joints may stiffen
50–60	Skin sags and wrinkles appear
	Decline in visual acuity
	Loss of some taste sensation
	Menopause
	Male produces less semen
	Decline in nail growth
	Decline in muscle mass and strength
	Memory wanes
60–70	Decrease in height
	Decrease in lung capacity
70–80	Decrease in height
	Further loss of taste sensation
	Nose, ears, and eyes appear prominent as skin sags but cartilage grows

valence shell, which attracts electrons from other molecules and destabilizes them. This sets into motion a chain reaction of chemical instability that could kill the cell. Exposure to toxins or radiation can also generate free radicals. Enzymes that catalyze reactions that inactivate free radicals before they damage cells diminish in number and activity in the later years. One such enzyme, superoxide dismutase, is promoted as an antiaging remedy at some health food stores. Even though this enzyme is a natural free-radical fighter, no evidence exists that it, or other natural antioxidants such as vitamins C and E and beta carotene, stalls aging on a whole-body level.

Aging can entail the active appearance of new substances or functions, as well as the passive breakdown of structure and function. One active aging substance may be lipofuscin granules that build up in aging muscle and nerve cells. Accumulation of lipofuscin illustrates both passive and active aging. Lipofuscin actively builds up with age, but it does so because of the passive breakdown of lipids. Another example of active aging is autoimmunity, in which the immune system turns against the body and attacks its cells as if they were infectious organisms. Rheumatoid arthritis and some forms of diabetes, both with onset in adulthood, are autoimmune disorders. Apoptosis, a form of programmed cell death, is also an active aging process.

One way to study aging in humans is by examining rare inherited diseases that accelerate the aging timetable. The most severe aging disorders are the progerias (fig. 40.18). In Hutchinson-Gilford syndrome, a child appears normal at birth, but within a few years ages dramatically, developing wrinkles, baldness, and the prominent facial features of advanced age. The body ages on the inside as well, as arteries clog with fatty deposits. The child usually dies of a heart attack or stroke by age 12, although some patients live into their twenties. In an "adult" form of progeria called Werner syndrome, which appears before the twentieth birthday, death from accelerated old age usually occurs when the individual is in his or her forties.

FIGURE 40.18
Progeria.
The Luciano brothers inherited progeria and appear far older than their years.

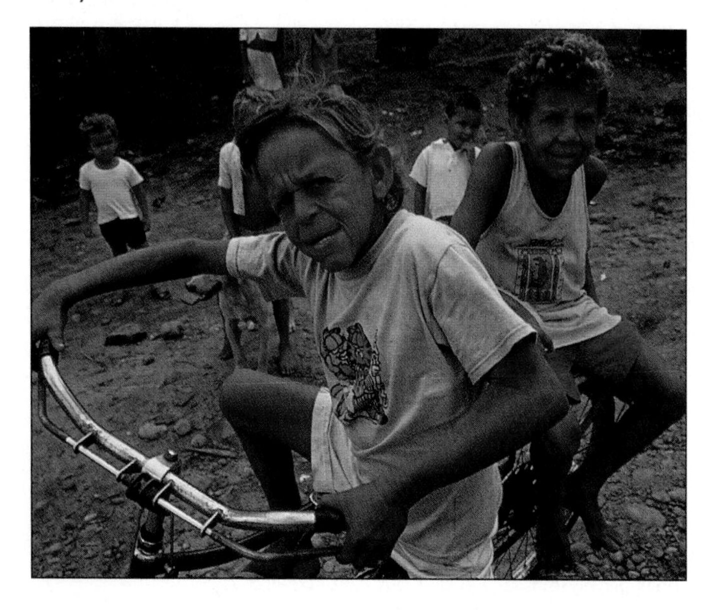

The cells as well as the bodies of progeria patients show aging-related changes. Recall that normal cells in culture divide on average about 50 times before dying. Cells from progeria patients die in culture after only 10 to 30 divisions, as if they were programmed to die prematurely. Certain structures seen in normal cultured cells as they near the 50-division limit (glycogen particles, lipofuscin granules, many lysosomes and vacuoles, and few ribosomes) appear in cells of progeria patients early on. Understanding the mechanisms that cause these diseased cells to race through the aging process may help us better understand the biological aspects of normal aging.

The Human Life Span

Wrote Jonathan Swift, of *Gulliver's Travels* fame, "Every man desires to live long; but no man would be old." In the age-old quest to prolong life while remaining healthy, people have sampled everything from turtle soup to owl meat to human blood. A Russian-French microbiologist, Ilya Mechnikov, believed that a human could attain a life span of 150 years with the help of a steady diet of milk cultured with bacteria. He thought that the bacteria would inhabit the large intestine and somehow increase longevity. He died at 71. More recently, researchers have attempted to use hormone treatments to restore younger functions to older people. A 62-year-old woman has given birth, and men in their sixties have been treated with growth hormone to rebuild sagging tissues.

The human **life span**—theoretically, the longest length of time a human can live—is approximately 120 years. Of course, most people succumb to disease or injury long before they reach such an age. **Life expectancy** is a more realistic projection of how long an individual will live, based on epidemiological information and a person's current age. In the United States, life expectancy from birth is 79 years for males and 83 years for females.

In many nations, women outlive men by a few years. Developmental biologists do not know why this is so, because both males and females suffer from the same life-threatening illnesses. But the longer lives of females may reflect biology, because it is seen in a variety of species, including beetles, flies, mice, fishes, spiders, rats, birds, and butterflies. Differences that enable females to outlive males may begin very early in development. Experiments on human preembryos show that male cleavage embryos divide faster than female cleavage embryos. Perhaps mitotic clocks "tick" faster in males, reaching life's endpoint on average a few years sooner.

Length of life reflects luck, genes, and a lifetime of environmental influences. Families with many very aged members usually have a fortuitous collection of genes, plus shared positive environmental influences, such as sound nutrition, excellent health care, devoted relatives, and a myriad of other nurturing advantages.

Identifying and isolating the inborn and environmental influences on life span and life expectancy is difficult. One approach compares the health of adopted individuals to that of their biological and adoptive parents. In one study, Danish adopted individuals with one biological parent who died of natural causes before age 50 were more than twice as likely to also die before age 50 as were adopted people whose biological parents lived beyond this age. This suggests an inherited component to longevity. Interestingly, adopted people whose biological parents died early due to infection were more than five times as likely to also die early of infection, perhaps because of inherited immune system deficiencies. The adoptive parents' age at death showed no correlation with the longevity of their adopted children, suggesting that heredity may play a more important role than environment in an individual's life expectancy.

Perhaps because many human populations are aging, researchers and physicians are paying more attention to the biological aspects of growing older. The field of gerontology examines the changes of aging at the molecular, cellular, organismal, and population levels. As we learn more about the biology of aging, we can improve not only the length but also the quality of life.

40.4 MASTERING CONCEPTS

1. What changes occur as the newborn adjusts to conditions outside the uterus?
2. What bodily changes characterize adolescence?
3. What are some passive aspects of aging?
4. What are some active aspects of aging?
5. What are some aging-related changes apparent in cells of people suffering with severe aging disorders (progerias)?
6. What is the maximum human life span?
7. What factors influence longevity?

Chapter Summary

40.1 From Embryology to Developmental Biology: A Descriptive Science Becomes Experimental

1. In the mid-eighteenth century, the theory of **preformation** gave way to the theory of **epigenesis** as biologists observed embryos specialize over time. Karl Ernst von Baer noted that early vertebrate embryos look alike but gradually become more distinctive.
2. Cells of the early embryo are **totipotent.** Gradually, biochemically distinct regions of the embryo develop and differentiate down specific **developmental pathways.**
3. A cell commits to a fate at a certain point in development. Before this time, a transplanted cell develops according to its new surroundings; after this point, it retains the specialization other cells exhibit in its original location. Differential gene expression underlies cell specialization.

40.2 Setting the Stage for Human Prenatal Development—The Reproductive System

4. Developing sperm originate in **seminiferous tubules** within the paired **testes.** Sperm mature in the **epididymis** and **vasa deferentia,** and they exit the body through the **urethra** during **ejaculation.** The **prostate gland,** seminal vesicles, and bulbourethral glands add secretions to sperm.
5. Oocytes originate in the female **gonads,** the **ovaries.** Each month after puberty, one ovary releases an oocyte into a **fallopian tube,** which leads to the **uterus.**
6. The ovaries secrete estrogen and progesterone, hormones that stimulate development of female sexual characteristics and together with GnRH, FSH, and LH control the menstrual cycle.

40.3 The Structures and Stages of Human Prenatal Development

7. Human prenatal development begins at fertilization. A single, **capacitated** sperm cell at a secondary oocyte burrows through the **zona pellucida** and **corona radiata.** The two pronuclei join. **Cleavage** ensues, and a 16-celled **morula** forms. Between days 3 and 6, the morula arrives at the uterus and hollows to form a **blastocyst,** made up of individual **blastomeres.** The trophoblast layer and **inner cell mass** form. **Implantation** occurs between days 6 and 14. Trophoblast cells secrete **human chorionic gonadotropin (hCG),** which prevents menstruation.
8. During the second week, the amniotic cavity forms as the inner cell mass flattens, forming the **embryonic disc.** The **primitive streak** appears. Ectoderm and endoderm form, and then mesoderm appears, establishing the germ layers of the **gastrula.** Cells in a particular germ layer develop into parts of specific organ systems.
9. During the third week the **chorion** starts to develop into the **placenta,** and the **yolk sac, allantois,** and umbilical cord form as the **amniotic sac** swells with fluid. **Organogenesis** occurs throughout this embryonic period. Gradually structures appear, including the **notochord, neural tube,** arm and leg buds, heart, facial structures, skin specializations, and skeleton. Structures continue to elaborate during the fetal period.
10. Labor begins as the fetus presses against the cervix. Uterine contractions expel the baby, placenta, and extraembryonic membranes.

40.4 Growth and Development After Birth

11. Postnatal stages include the newborn, infant, child, adolescent, and adult. Drastic changes occur at birth, as the newborn takes on functions that the pregnant woman provided. Infancy and childhood are periods of rapid growth and maturation of organ systems. Hormonal changes dominate adolescence. Organ systems function during adulthood and begin to show signs of aging.
12. In passive aging, structures break down, and DNA repair becomes less efficient. In active aging, lipofuscin accumulates in cells and cells die.
13. The theoretical maximum human **life span** is 120 years. **Life expectancy** is a more realistic measure of how long someone will live, based on age and epidemiological information.

Testing Your Knowledge

1. Why would the scientific method support the hypothesis of epigenesis better than the theory of preformation?
2. How can very different specialized cells have identical genes?
3. How do morphogen gradients cause cells to differentiate?
4. How are the human male and female reproductive tracts similar? How are the structures of the testis and ovary similar?
5. Arrange these prenatal humans in order from youngest to oldest: morula, gastrula, zygote, fetus, blastocyst, and embryo.
6. What events must take place for fertilization to occur?
7. What is the difference between life span and life expectancy?

Thinking Scientifically

1. Is the presence of a complete set of genetic instructions in a fertilized ovum consistent with the hypothesis of preformation or epigenesis?
2. When an isolated sperm is mechanically injected into an oocyte outside the woman's body, and the treated egg is then placed in the woman's body, no embryo develops. Why doesn't development ensue?
3. What technology would be necessary to enable a fetus born in the fourth month to survive in a laboratory setting?
4. What kinds of studies and information would be necessary to determine whether exposure to a potential teratogen during a war can cause birth defects a year later? How would such an analysis differ if it was a man or a woman who was exposed?
5. What factors contribute to longevity?
6. In the year 2050, 1 in every 20 persons in the United States will be over the age of 85. What provisions would you like to see made for older people of the future (especially if you will be one of them)? What can you do now to increase the probability that your final years will be healthy and enjoyable?

References and Resources

Gould, Stephen Jay. March 2000. Haeckel's distortions did not help Darwin. *Natural History,* vol. 109, p. 42. Haeckel misinterpreted the similarities of early embryos.

Hayflick, Leonard. January 27, 2000. New approaches to old age. *Nature,* vol. 403, p. 365. Hayflick of "Hayflick limit" fame discusses aging at the cellular level.

Riddle, Robert D. and Clifford J. Tabin. February 1999. How limbs develop. *Scientific American,* vol. 280, pp. 74–79. Cells communicate in complex ways during prenatal development.

Short, R. V. February 17, 2000. Where do babies come from? *Nature,* vol. 403, p. 705. Centuries ago, William Harvey puzzled over how development proceeds.

Smith, Bradley R. March 1999. Visualizing human embryos. *Scientific American,* vol. 280, pp. 76–83. Spectacular images of human embryos.

Wilmut, Ian, Keith Campbell, and Colin Tudge. 2000. *The Second Creation: Dolly and the Age of Biological Control.* London: Headline. The story of Dolly, as told by her "father."

The *LIFE* Online Learning Center provides additional resources and tools for studying this chapter.
www.mhhe.com/life

UNIT 7 • BEHAVIOR AND ECOLOGY

This baby cowbird is displaying inborn behavior, begging for food at the sight of an adult caregiver. The adult bird pictured, however, is not the cowbird's parent—most adult cowbirds do not build nests or care for their young. Instead, they lay their eggs in the nests of other species, often destroying the host's own eggs in the process. When the cowbird eggs hatch, the surrogate parents feed the babies as though they were their own. Cowbirds thus represent one of many factors that control songbird populations, and both bird species rely on other organisms for food, shelter, and oxygen. Thus, the interacting bird species represent a small part of a larger community of other animals, plants, fungi, and microorganisms that depend on one another for life.

CHAPTER 41

Animal Behavior

A "Green Beard" Gene in Fire Ants Explains Aggressive Behavior

In 1976, evolutionary biologist Richard Dawkins coined the phrase "selfish gene" to denote genes that seemingly cause animals to behave in ways that increase the proportion of a particular allele in a population. According to this theory, genes take on a life of their own, in that they act to increase the probability of their own survival, as if they have a will. One type of selfish gene is what Dawkins dubbed a "green beard" gene. It endows an animal with a phenotype that is very obvious to those bearing the gene, but also compels them to behave in a particular way that increases the chance of the gene's perpetuation. A gene causing a green beard—in any animal—would be quite noticeable.

A gene causing a green beard—in any animal—would be quite noticeable.

But the idea of selfish genes bothered many geneticists, who pointed out that a DNA molecule could hardly dictate behavior in such a willful way. Evidence for selfish genes in general, and green beard genes in particular, was scant. However, observations and experiments with the red imported fire ant *Solenopsis invicta* provide what may be evidence that an allele can indeed select itself. These ants, introduced to the southeastern United States from South America, can deliver painful bites to humans who happen upon their foot-tall mounds.

A gene called *B* seems to determine which fire ants live or die. Fire ants live in two types of colonies. A monogynous ("one female") colony has a single queen of genotype *BB*. In contrast, a polygynous ("many females") colony has several queens. Individuals of genotype *bb*, male or female, die before they reach reproductive age. Workers may be *BB* or *Bb*. But queens that live long enough to reproduce are always genotype *Bb*. What happens to the *BB* queens?

Watching the goings-on in an ant colony reveals that the *BB* queens—and only the *BB* queens—are viciously killed by workers within minutes. The

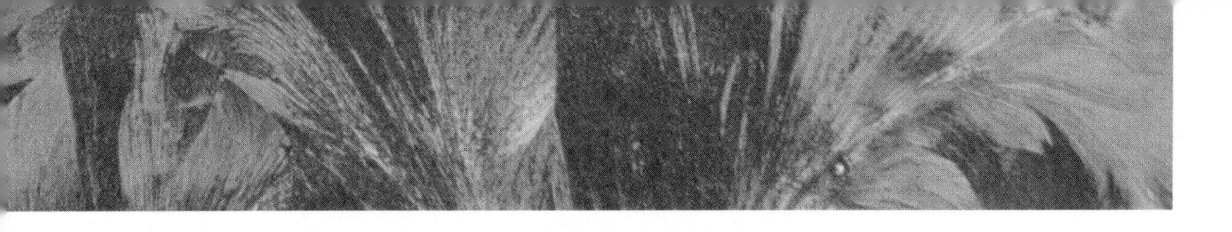

Red Ants with Green Beards?
A red fire ant queen and workers—here as attendants, not executioners.

attack begins when the queens attempt to lay eggs. The murderous workers are mostly of genotype *Bb.* By killing the *BB* queens, the heterozygous males keep the *b* allele in the population, offsetting its removal when *bb* ants die.

To discover how a *Bb* male fire ant "knows" to attack a *BB* queen, Laurent Keller from the University of Lausanne in Switzerland and Kenneth Ross at the University of Georgia set up a series of experiments. After moving entire colonies into Ross's lab, the researchers removed the queens (both *BB* and *Bb*) for 3 days. Upon their return to the 5,000-plus-member colonies, the queens laid eggs, stimulated by the pheromones that the queenless males had emitted. The insect equivalent of a bloodbath (a hemolymph bath) ensued, but the researchers stopped the anticide after 5 minutes. They removed the attackers and victims and determined their genotypes by testing for an enzyme variant whose gene is very closely linked to the *B* gene.

The results were astounding—the attacked queens were all of genotype *BB,* the ignored queens were all *Bb,* and the killers were all *Bb.* When the killer ants were let loose on a different subspecies of ant, they did not kill selectively, as they did their own. Then Keller and Ross noticed that the attackers were being attacked! Did they carry a pheromone from the *BB* queens they had killed? To test this new hypothesis, the researchers rubbed workers against *Bb* queens or young *BB* queens, then put them amid other workers. Sure enough, only those workers rubbed against *BB* queens were attacked, providing evidence that pheromones can direct an animal's aggression against individuals of a particular genotype. This explanation is more scientifically satisfying than endowing genes with human qualities, such as motivation.

41.1 Genes and Experience Shape Behavior

Innate behavior is instinctive and genetically determined, providing responses that are vital for survival. Learned behavior is more a response to environmental stimuli based on experience. Complex behaviors are the result of natural and sexual selection.

The ground squirrel appeared to sense danger as soon as she neared her burrow and saw her neighbor's frenzied tail flagging and darting about. She had been out foraging for food and had left behind three pups. The squirrel's behavior sent an unmistakable message—rattlesnake!

A snake commandeering a burrow to keep itself warm could trap and eat a newborn squirrel. But this squirrel's pups were standing their ground well. Although they were too young to have encountered a rattlesnake before, or to have learned from their mother how to defend themselves, the pups instinctively reacted to the threat. They flagged their tails, pushed out their snouts, jumped back and forth as if anticipating a snake's strike, and vigorously kicked dirt and sand at the reptile. The arrival of the mother sent the snake slithering into the temporarily abandoned burrow. Mother and pups immediately kicked sand and dirt into the burrow. They were safe—for a while (fig. 41.1).

The encounter between the squirrels and the snake vividly illustrates animal behavior. **Ethology,** the study of animal behavior, was officially recognized in 1973, when zoologists Karl von Frisch, Konrad Lorenz, and Niko Tinbergen won a Nobel Prize for their groundbreaking studies over four decades. Ethologists examine behaviors that enable animals to survive and reproduce in their natural habitats. Also in 1973, evolutionary biologist John Maynard Smith compared behaviors that promote survival to strategies in a game in which one individual's behavior depends upon the actions of others. Survival strategies are subject to natural selection.

Long before 1973, biologists painstakingly observed animals in the wild to learn about their behaviors (fig. 41.2). The mostly observational studies that formed the foundation of ethology are today complemented by experiments that examine the components of and driving forces behind particular behaviors. In addition, biologists now use DNA typing to identify and follow particular individuals. In the past, biologists determined how individuals were related from observing animals' interactions. The chapter opening essay describes yet another approach to studying behavior—identifying its genetic underpinnings. All of these approaches are used to reveal how and why animals behave in characteristic ways.

A successful behavioral strategy is the result of inborn actions and learned responses. Although squirrels instinctively can deter a snake, predators still corner and devour many squirrel newborns. Experience teaches a pup to refine its inborn skills to evade snakes and other predators. Heredity (nature) as well as the environment (nurture) determine behaviors, with each influence contributing to particular responses to different degrees.

Behavior is complex. Simple actions may reflect input from several genes. Yet a single gene can control fairly complex behav-

FIGURE 41.1
Defensive Reflex.
A ground squirrel protects itself from a snake with innate reflexes, such as this defensive stance that young do not have to be taught.

FIGURE 41.2
Pioneering Primatologist.
Jane Goodall, a British zoologist, spent many years observing chimpanzees at very close range, discovering much of what we know about their behavior.

ior. In different species of voles (a rodent), for example, different alleles of a gene that encodes the cell surface receptor for the hormone vasopressin determine whether the animal stays with one mate or frequently changes partners. When researchers placed a prairie vole vasopressin gene associated with monogamy into transgenic (genetically engineered) male mice, the normally antisocial mice showed preferences for specific females—apparently, transferring the gene transferred the behavior! Most behaviors, though, are not so simply explained. Heredity and environmental influences can be very difficult to identify and distinguish.

An animal's anatomical systems for detecting stimuli, integrating the input, and responding limit behavior. Genes determine how sophisticated and sensitive these anatomical systems are. On the other hand, learning can be important in mastering complex behaviors, such as reading, speaking a language, or following the rules of a game.

A behavior is usually primarily genetically determined, or **innate,** when it is critical that the animal perform the action correctly the first time. Escape behaviors, such as the squirrel's sand-kicking stance, tend to be under fairly precise genetic control. If an individual fails to respond appropriately in its first encounter with a predator, it is eaten and never gets a second chance to refine the response. Animals that respond correctly to predators live to pass their genes to future generations.

Genetic constraints on learning are many, as animal trainers well know. One group of trainers attempted to teach raccoons, roosters, and pigs to put coins into a piggy bank for a food reward. The animals were not always cooperative. Roosters scratched their feet on the floor, pigs rooted on the bare floor, and raccoons fondled their coins rather than depositing them in the bank. The animals were behaving instinctively to food deprivation—roosters scratch the ground to uncover grain, pigs root around for food, and raccoons wash their food in water, rubbing it with their fingers.

A bird's song reveals the influence genes exert on learning. Young male, white-crowned sparrows learn to sing by hearing the melodies that their fathers sing. If isolated from older males of their species, they sing abnormal songs as adults. Learning doesn't completely control singing, however, because the sparrow can only learn the song of its own species. If a young isolated bird hears recordings of a variety of birdsongs, including a normal white-crowned sparrow song, it learns to sing properly. The bird innately recognizes its species' song and then perfects the performance through experience.

Innate Behavior Is Inborn

Natural selection molds innate behaviors because they are largely genetically determined. Genes that predispose an individual to a behavior that lowers the chances of survival to reproduce will ultimately diminish in the population.

Fixed Action Patterns A **fixed action pattern (FAP)** is an innate, stereotyped behavior, such as a dog digging on the floor as if trying to bury a bone or a kitten pouncing on a rustling leaf as if it were a mouse. Members of a species perform an FAP nearly identically, with little environmental modification.

The egg-rolling response of a female greylag goose is an FAP (fig. 41.3). When a brooding female sees an egg outside her nest, she retrieves it in a characteristic manner: she stretches her bill just beyond the egg and then scoops it toward her, repeating this motion until the egg is back in the nest. She adjusts her movements as the egg wobbles over uneven terrain. This action is adaptive, because an unincubated egg will not hatch.

The goose will retrieve a beer bottle as readily as an egg. Apparently she has only a vague sense of what an egg is and automatically responds to all small, rounded objects outside the nest. If the egg is removed after she has begun rolling it back to the nest, she continues her retrieval motions until the imaginary egg is in the nest. Once an FAP begins, the action continues until completion, even without appropriate feedback.

Releasers The specific factor that triggers an FAP is called a **releaser** (or sign stimulus). An animal's senses are bombarded constantly with many more stimuli than it could possibly respond to. The animal must select and respond to only the few key stimuli that are reliable cues for adaptive behaviors. Even a single object or organism has many aspects to it, such as color, shape, odor, taste, sound, and movement.

To identify the specific releaser for a behavior, ethologists build models that isolate a single stimulus, such as color or scent.

FIGURE 41.3

A Fixed Action Pattern.

Like all fixed action patterns, egg retrieval in the greylag goose is always the same.

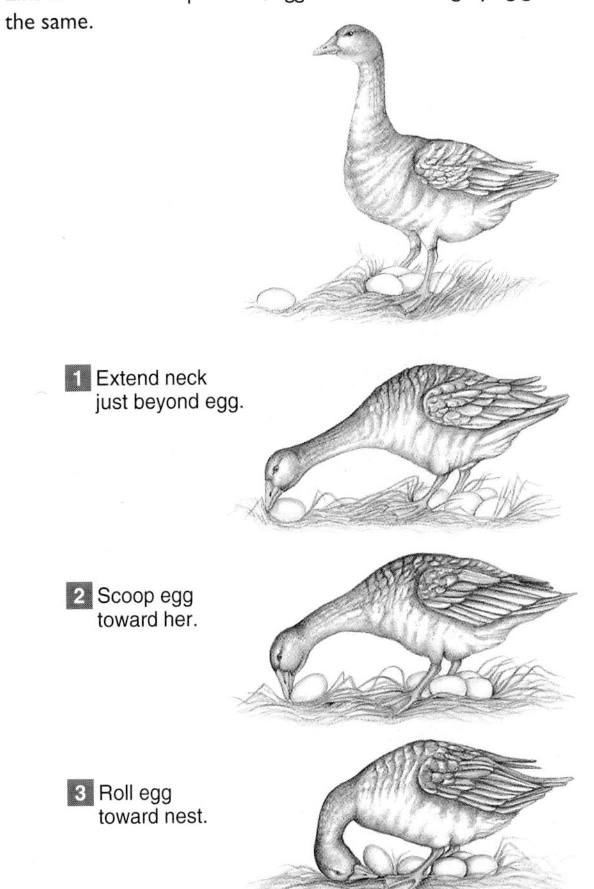

The researcher shows the model to an organism in an appropriate physiological state and observes whether it responds as it would to the natural stimulus. For example, a herring gull chick obtains a regurgitated meal from its mother by pecking a red spot on her bill. Experiments show that a chick will also peck a red-marked, moving stick (fig. 41.4).

Releasers are important in human parent-infant interactions. When an infant gazes and smiles at her exhausted parents, the adults respond with affection. But an infant up to 2 months old will smile at anything with two dots resembling eyes! A model of a face lacking eyes will not elicit a smile.

Many releasers are auditory (sensed by hearing). A mosquito is attracted to his mate by the buzz of her wings. A model of the stimulus, a tuning fork that vibrates at the same frequency as a female's wings, also attracts him. The pulsing chirps of crickets, the clatter of katydids, and the droning of bullfrogs are all auditory signals that release mating behaviors.

Releasers may be tactile (sensed by touch). During the mating ritual of the stickleback fish, the female enters the nest, which stimulates the male to thrust his snout against her rump in a series of quick, rhythmic trembling movements. These thrusts release her spawning behavior. Prodding the female with a glass rod mimics the releaser.

Pheromones are chemical releasers, and they include sex attractants of some insects, crustaceans, fishes, and salamanders. To show that a pheromone is a releaser of mating behavior in the Canadian red-sided garter snake, researchers extracted lipids from the skins of sexually mature female and male snakes. Lipids from the females were absorbed onto paper towels, and the towels placed near males. The snakes became stimulated, some even attempting to mate with the pheromone-soaked paper. When lipids that males normally secrete were added to the towel, the male snakes backed off. Scientists working with pheromones may become releasers themselves, because just a trace of a sex attractant can be powerful. One researcher who worked with moth pheromones was besieged by male moths at a football game, even though he had showered.

A supernormal releaser is a model that exaggerates a releaser and elicits a stronger response than the natural object does. Birds such as the oystercatcher and the herring gull, for example, prefer to sit on large eggs. The female will try to sit on an egg that is more immense than she could have possibly laid, even though her own egg is nearby.

A complex behavior called a **reaction chain** develops when a sequence of releasers connects several behaviors. A reaction chain may be a series of actions an individual takes, such as nest building, or an exchange of releasers between two individuals, such as courtship and mating behaviors. Consider the elaborate mating rituals of the three-spined stickleback fish. The male prepares the nest and attracts a female. He then enters the nest and fertilizes the eggs she has laid. The sequence of behaviors is always performed in the same order, because the completion of each step is the releaser for the next.

FIGURE 41.4

Releasers Elicit Specific Behaviors.

A releaser is a specific stimulus that elicits a specific response from an animal. A red spot on a mother's bill is a releaser for feeding behavior of a herring gull chick. The color red, an oblong shape, and movement comprise the releaser.

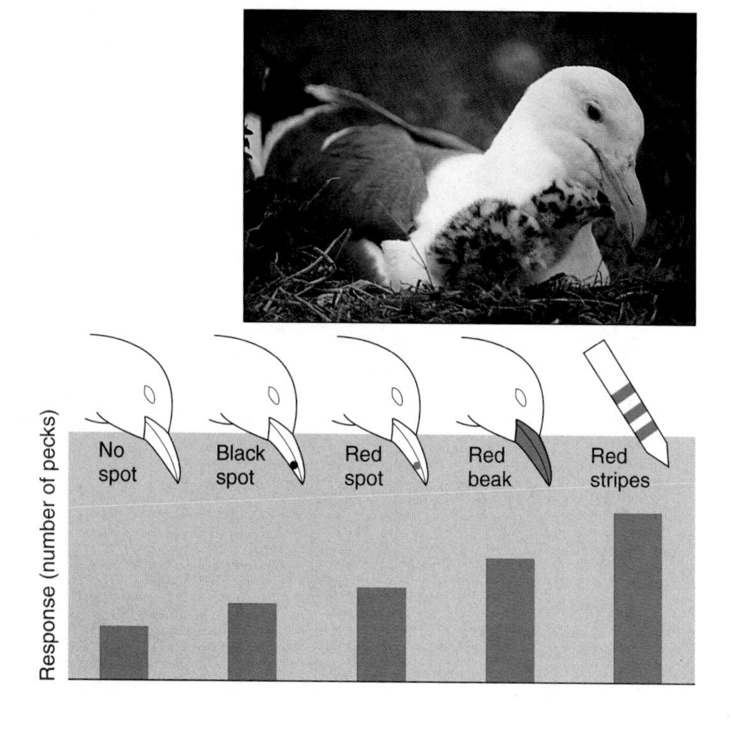

Learned Behavior Results from Experience

Learning is a change in behavior as a result of experience. A young bee innately recognizes a flower pattern of light petals and dark center. An older bee, with experience, recognizes specific flowers. Ethologists recognize several types of learning, and table 41.1 gives familiar examples of learning.

Habituation The simplest form of learning is **habituation,** in which an animal learns to ignore irrelevant stimuli. When a stimulus occurs many times without any consequence, the animal usually decreases its response, perhaps eliminating it completely. For example, young chicks learn not to run from innocuous, common stimuli, such as blowing leaves and nonpredatory birds. Responses to predators are retained because they are rarer and more essential for survival.

Habituation helps ensure that aggression is appropriate and adaptive. If seabirds that often nest within a few feet of one another did not habituate to the presence of neighbors, they would waste much time and energy on aggressive encounters. Instead, they learn not to show aggression toward neighbors as long as they remain outside a certain territorial limit. They remain aggressive against other birds that they encounter less frequently.

Classical Conditioning In **classical conditioning,** an animal learns to respond in a customary way to a new stimulus. A new association forms between a stimulus and a response when the new or **conditioned stimulus** is repeatedly presented immediately before a familiar or **unconditioned stimulus** that normally triggers the response. After the stimuli have been paired in a series of trials, the new stimulus alone elicits the response.

In the most familiar study of classical conditioning, published in 1903, the Russian scientist Ivan Pavlov reported that dogs could be conditioned to associate the sound of a bell with feeding. When fed many times following a bell ring, the dog salivates at the sound of a bell. If, however, the bell rings several times without ensuing food, the dog will no longer salivate in response to the bell. This loss of a conditioned response is called extinction. There are many examples of conditioned responses—guinea pig pets chatter in anticipation when they hear the sound of a refrigerator door opening, ducks approach people expecting to be fed, and park bears associate cars with food (fig. 41.5).

Operant Conditioning **Operant conditioning** is a trial-and-error type of learning in which an animal repeats behavior that is rewarded (positive reinforcement) or avoids a painful stimulus (negative reinforcement). Reinforcement increases the probability that the animal will repeat the behavior. Trial-and-error learning can refine natural behaviors, such as predatory skills. A grizzly bear, for example, might learn that splashing about in a stream does not yield a salmon dinner. Staying still and quiet in one place is much more effective. Figure 41.6 shows another example of operant conditioning.

Ethologist B. F. Skinner designed an apparatus to demonstrate operant conditioning. When a hungry rat is placed in a device called a Skinner box, it explores its surroundings and eventually presses a lever by accident, which releases a food pellet. This food reward increases the probability that the rat will press the lever again.

Animal trainers use operant conditioning. They first reinforce any behavior that vaguely resembles the desired one and then restrict the reward to better approximations. Pigeons learn to play table tennis in this way. Hungry pigeons are taught to use their beaks to hit a ball. When the ball falls in the trough on either side of the table, the trainer delivers a food pellet to the pigeon that won the point.

Imprinting **Imprinting** is a type of learning that occurs quickly, during a limited time (called the critical period) in an animal's life, and usually is performed without obvious reinforcement. Young chicks, goslings, or ducklings are imprinted to follow the first moving object they see, which in the wild is their mother. This "following" response is adaptive, because the mother typically would lead them to a safe place where they are likely to find food.

Ethologist Konrad Lorenz is famous for his studies of imprinting. Lorenz hatched eggs in an incubator and then had the baby birds waddle after him. Thereafter they showed the same responses to him as they would have to their mother (fig. 41.7). A recent film chronicled a flock of geese that imprinted on a 13-year-old girl in Canada and even followed her as she flew in an airplane south for the first migration.

FIGURE 41.5
Classical Conditioning.
This black bear in Yellowstone Park has learned to associate tourist cars with food.

FIGURE 41.6
Operant Conditioning.
Through patient trial-and-error learning, coupled with many tasty rewards, this parrot has been taught to ride a scooter.

Imprinting is important in the development of attachment between mother and newborns in several species, including goats, sheep, and Alaskan fur seals. In the first few critical minutes after a goat kid's birth, the mother learns to identify her offspring by its odor. She will accept and nurse any young that she smelled during the critical period and reject any young she does not recognize.

Insight and Latent Learning **Insight learning,** or reasoning, is the ability to apply prior learning to a new situation without trial-and-error activity. For example, in one experiment, chimpanzees were put into a room with boxes and a banana hanging from the ceiling. The chimpanzees must have reasoned that stacking the boxes would enable them to reach the banana—and that is exactly what they did.

Latent learning occurs without obvious reward or punishment and is not apparent until after the learning experience. Even without reinforcement, a rat that has been allowed to run freely through a maze will master its twists and turns more quickly than a rat with no previous experience in the maze. In the wild, animals may learn the details of their surroundings during daily explorations. This information may not be of immediate use, but knowing where to quickly find hiding places may make the difference between life and death when a predator strikes.

TABLE 41.1 Learning in Humans and Pets

TYPE OF LEARNING	EXAMPLE
Habituation	Being able to concentrate on written work with music playing in the background but jumping when the phone rings
Classical conditioning	Craving popcorn in a movie theater
Operant conditioning	A student learns what to do to earn high grades in school
Imprinting	Puppies and kittens become tame if people handle them in the first few weeks of life
Insight learning	A child places a box on roller skates to build a vehicle
Latent learning	A medical student watches a physician give an injection and then does it herself

41.1 MASTERING CONCEPTS

1. What are some ways that ethologists study animal behavior?
2. What are the roles of heredity and the environment in determining behaviors?
3. What is a fixed action pattern?
4. What is the function of a releaser?
5. Describe the types of learning.

OLC

FIGURE 41.7
Imprinting.
Many baby birds imprint on the first moving object they see. This is usually the mother but may be an ethologist, such as Konrad Lorenz.

41.2 Types of Behavior

Certain types of behavior are seen among diverse animal species. These include migration and homing, aggression, mating behavior, and altruism.

Several familiar types of behaviors have both innate and learned components.

Orientation and Navigation

Each spring, North American skies fill with wings as birds migrate north from the tropics. These animals use the position of the sun or stars, subtle shifts in Earth's magnetic field, sights, sounds, and winds and the aromas they carry as cues to cover vast distances and arrive with great accuracy. Their lives depend on it. Paths of migration take animals past and to areas rich in food, at times when that food is most abundant.

Migration is a regularly repeated journey from one specific geographic region to another. Bird migrations are well studied (fig. 41.8). Shorebirds such as sandpipers, plovers, and oystercatchers may travel 20,000 miles (32,000 kilometers) round-trip, finding their food of worms, snails, clams, and larvae at the water's edge over vast distances. Birds are not the only migrants. Atlantic salmon that begin life in New England rivers migrate through the sea to Greenland to feed and then return to the exact river or stream where they were born to spawn. Migrating animals obviously must be able to **orient** (move in a specific direction) and **navigate** (follow a specific course). However, many nonmigratory animals, even very simple ones, orient to specific environmental cues. Many of those also can navigate to and from specific objects or places.

Orientation and navigation behaviors may be a complex mix of innate and learned responses. Investigating Life 41.1 indi-

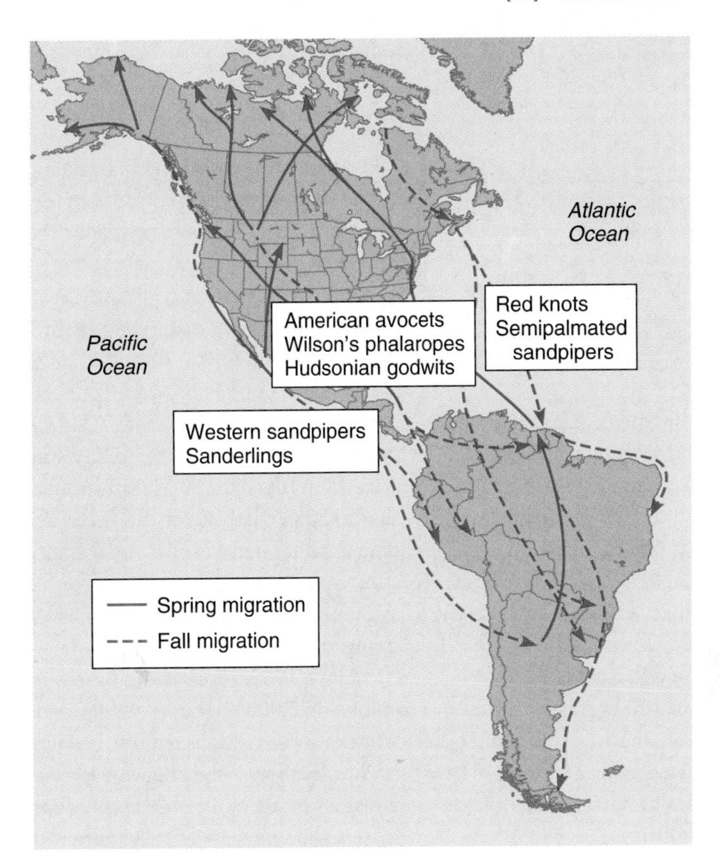

FIGURE 41.8
Bird Migrations.
Some shorebirds fly round-trips of thousands of miles along specific routes to reach their breeding and wintering grounds.

cates that migration among green turtles is learned. However, shorebird chicks can follow the correct routes even when adults are not present, indicating that their migratory behavior is largely innate.

Most migrating species use environmental cues as a compass to orient them in one direction. Caged birds prevented from migrating still orient in the directions that free members of their species do at approximately the right flight times.

The sun can serve as a compass. It rises in the east and moves across the sky at an average rate of 15° an hour, setting in the west. An animal's biological clock tells it the time of day. When the animal combines this information with the position of the sun, it can orient in any direction.

Earth's magnetic field, which runs in a generally north-south direction, can also serve as a compass. A pigeon can find its way home if released up to 1,000 miles (1,609 kilometers) away; however, bar magnets placed on a pigeon's wings impair its ability to find its way home on cloudy days. On sunny days, when the sun can provide cues, the magnets have no effect. Sham magnets (nonmagnetic bars of equal size and weight) never disrupt orientation, indicating that the magnetism, and not the bar itself, disrupts normal navigational cues.

The most complex navigational skill is the ability to home—that is, to return to a given spot after being displaced to an unfamiliar location. Homing requires both a compass sense for direction and a map sense telling the animal where it is relative to home. We know very little about the map sense. It may depend on regular variations in the strength of Earth's magnetic field, which is about twice as strong at the poles as it is at the equator. An animal very sensitive to the strength of a magnetic field would know how far north or south it is. Pigeons become disoriented when sunspots disrupt magnetic fields.

The fact that some mammals can find their way home over great distances suggests that familiar senses may help foster a map sense. A wolf that had spent her whole life in a pen in Barrow, Alaska, was taken 175 miles (282 kilometers) away and released. She found her way home. A deer taken 350 miles (563 kilometers) from its wildlife refuge on the Gulf of Mexico found his way back. Closer scrutiny revealed how the animals may have navigated. The wandering wolf had grown up next to an airport and was accustomed to the sounds of jets taking off and landing. Such loud sounds travel far on the bleak Alaskan landscape, and the animal may have traveled so that the jets became louder. (Another wolf taken on the same trip returned to the wrong airport!) The deer seeking his wildlife refuge may have homed in on scents wafting on gulf breezes.

We simply cannot explain many tales of mammal navigation. Big Mac, a 450-pound (204-kilogram) black bear tracked by a radio collar, wandered far from home, in search of berries to eat.

DNA Testing Sheds Light on the Mystery of Green Turtle Migration

Imagine traveling 1,400 miles (2,253 kilometers) to give birth and then turning around and going back. This is exactly what females in a population of green turtles do every few years (fig. 41.A). Each breeding season, a female turtle migrates from feeding grounds off the coast of Brazil to a tiny elevation in the middle of the Atlantic Ocean called Ascension Island. She digs a trench on the beach, deposits about 100 eggs, swims near shore for 2 weeks, lays another clutch of eggs, and continues this pattern for 3 months. Then she swims 1,400 miles back to Brazil, not eating during the long journey.

Why do these particular turtles do this, when others of their species breed near where they feed? In the 1960s, herpetologist Archie Carr proposed an intriguing hypothesis.

Carr considered the location of Ascension Island. It lies directly above the mid-Atlantic ridge, which is a buckling in the ocean floor where hot rock pours forth from the earth's interior, forming islands that move outwards and are eventually submerged. Ascension Island, then, is constantly being renewed. But it was not always so far from Brazil, nor from Africa. About 80 million years ago, the two continents were connected, and then they began to drift apart. Carr hypothesized that the turtles' ancestors nested on an Ascension Island that was not very far from the Brazilian coast. Over the years, as the continents continued to separate, the turtles instinctively continued to migrate to the island, as they still do today. The turtles haven't changed, but Earth has.

Since 1965, researchers have tagged and followed 28,000 migrating turtles and found that the turtles always return to Ascension Island to lay eggs. If the behavior is innate, implying a genetic basis, then the Ascension turtles should be genetically quite distinct from other populations. In 1994, researchers tested this hypothesis. They collected eggs from the island and analyzed mitochondrial DNA (mtDNA) from the embryos. Recall that mitochondrial DNA traces the maternal lineage. Surprisingly, the studies indicated that the Ascension turtles are not very different from green turtles in other populations. Therefore, the drive to return to the tiny island is probably *not* an instinct handed down from when the ocean was narrower.

In scientific inquiry, if an experiment doesn't support one hypothesis, the next step is to devise another. If the green turtles' homing isn't innate, could it be learned? Many researchers today think so. They hypothesize that baby turtles imprint on some feature of the island, and this is what drives them to return there to nest. This is a more adaptive behavior than being locked into an invariant innate drive, because if some force of nature should destroy the nesting site, the turtles could successfully lay eggs elsewhere.

Mysteries remain concerning these green turtles. Just what do baby turtles recognize on the island that makes such a strong and early impression? What cues do they use to find their way home? The Ascension turtle population has been endangered since seafaring Britons began capturing them to make soup in the last century. Let's hope that the turtles survive and that we can one day understand their migration behavior.

FIGURE 41.A
Green Turtle Migration.
Female green turtles (*Chelonia mydas*) in a certain population migrate 1,400 miles (2,253 kilometers) to a tiny island in the Atlantic to lay their eggs.

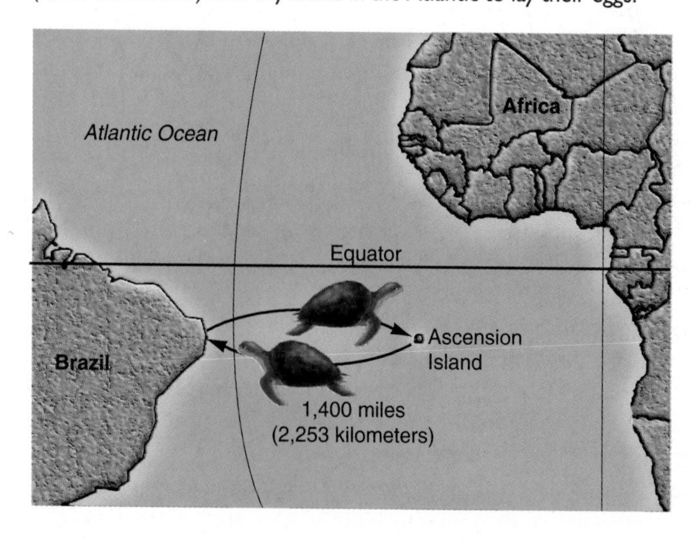

With cold weather imminent, he stopped foraging and tramped directly homeward, through forests, backyards, farms, and roads, without stars or sun for guidance, as snow began to fall. His pace slowed and he started to meander only when he was within 40 miles (64 kilometers) of home, the normal radius a bear travels in. Finally, he arrived—to a tiny crevice in a hillside, 126 miles (203 kilometers) from where his berry hunt had taken him (fig. 41.9).

Aggression

Animals often display aggressive behavior when members of the same species compete for resources, such as mates, food, shelter, and nesting sites. Competition can be fierce because the stakes are high. Winners survive and leave more offspring than losers do.

Territoriality One way that animals acquire resources with minimal aggression is to defend an area of their habitat, or territory, against invasion. Animals claim different types of territories for different functions, including sexual selection, foraging, defense, and protection of young. An animal may defend the territory against other species, against all members of the species, or against only those of the same sex. A territory may be fixed, such as prairie dog burrows, or moving, such as baboon settlements. Individuals hold some territories; groups hold others.

Behavior that defends one's territory, termed **territoriality,** is common. Male birds often occupy territories and repel other males while enticing females to enter. Male insects sometimes take over a food resource that attracts females. A territory may serve groups of animals, who defend it against others of their species. Wolf packs, hamadryas baboons, and pack rats maintain group territories.

Some animals defend territories only during the breeding season, but others hold them year-round. Territory size also may vary tremendously. A flightless cormorant defends only the area it can reach while sitting on its nest. Yet African weaver ants defend territories up to 2,153 square yards (1,800 square meters).

If territorial boundaries are to be respected, an animal must indicate ownership in some way. Figure 41.10 shows a cheetah marking a territorial boundary. Although fighting may accompany territory establishment, boundaries are generally respected without contest once they are set. Animals usually avoid actual combat. An animal becomes less aggressive when it crosses the boundary to leave its territory. The point where an animal is as likely to attack an intruder as to flee from it marks the borderline of the territory.

FIGURE 41.9
Sniffing the Way Home.
Bears provide much information on animal navigation because many are removed from towns as "nuisances," tagged, and transported beyond the 40-mile (64 kilometer) radius of their home territory. Odors may help bears find familiar surroundings. This bear is sniffing the breeze.

FIGURE 41.10
Pheromones Can Be Used to Mark Territories.
This male cheetah is spraying a pheromone onto a tree to mark its territory.

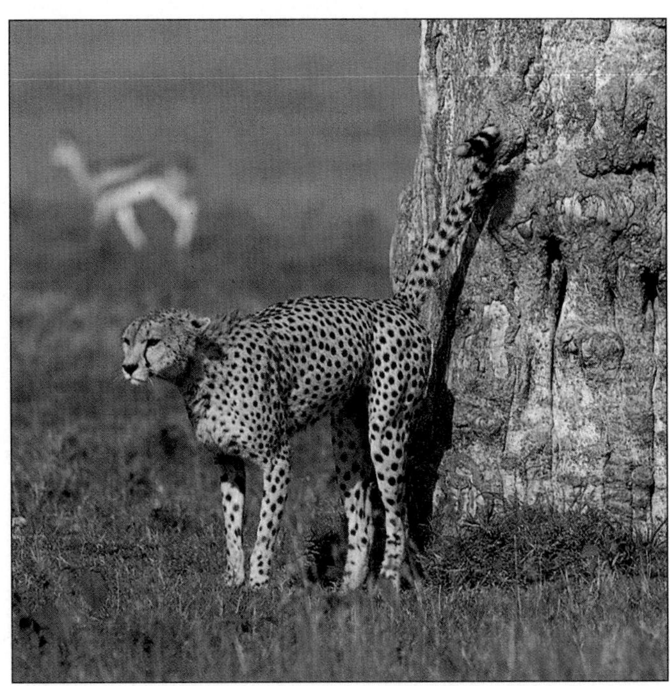

An animal may indicate its territory with a threat posture such as the sight of a male baboon's fangs, or sound, such as a toadfish's rumbling, a bull alligator's roar, or a wolf's howling. Many animals mark their territories with strong scents. Badgers, martins, and mongeese have scent glands at the bases of their tails, and rabbits have chin and anal glands. Urine is a common territorial marker. The giant galago, a primate, urinates on its hands, rubs the urine on its feet, and walks around to mark its territory. A male hippopotamus defines its territory by "dung showering"—rotating its tail to fling solid waste along the perimeter of its area.

When animals ignore territorial markers, threat behavior follows, which is a warning of impending aggression. A horse tosses its head and kicks. A sea anemone withdraws its tentacles and inflates a collarlike structure. If one animal is clearly superior, the other may leave without a challenge.

At any time during a battle, one combatant can end the fight with a display of submission, without provoking retaliation from the victor. Wolves expose their vulnerable throats as a sign of appeasement. In threat postures, animals attempt to appear larger, but appeasing animals try to seem smaller.

Diversion is another response to a threat. The "loser" behaves in a way unrelated to the situation to distract the aggressor. A dog might roll over, paws up, eliciting parental feelings rather than aggression from the other dog.

When competition does lead to combat, it is more like a tournament with a set of rules than a bloody fight to the death. Although roe deer could easily gore an opponent with their antlers, they do not use them in this way against members of their species. Two males face each other, antler to antler, and push in a ritualized fight. In some species, however, battling over a territory can sometimes lead to injury or death.

Dominance Hierarchies A **dominance hierarchy** is another way to distribute resources with minimal aggression. This is a social ranking of each adult group member of the same sex (fig. 41.11). In a linear hierarchy, one animal dominates, a second animal dominates all but the first, a third dominates all but the first two, and so on. Researchers first studied this type of relationship in the pecking order of domestic hens. Each hen knows whom it may peck and who it may be pecked by. Hierarchies in other organisms are often more complex. Female Amboseli baboons, for example, are ranked according to the age of their mothers and the age differences between them and their sisters.

FIGURE 41.11
Top Dog.
Dominance hierarchies can avoid aggression because each individual knows his or her place. The dog on the left is exposing its underside, a sure sign of submission.

A dominance hierarchy forms when members of the group first meet and may involve many threats and an occasional fight. For example, when pairs of Bewick's swans return from Siberia to their home estuary in western England, they flap their wings and make buglelike sounds to quickly establish the dominance hierarchy. Once the status of each individual is set, life within the group is generally peaceful.

Cannibalism More than 1,000 species of invertebrates and vertebrates practice cannibalism—including our own. In Crystal Lake, Vancouver, traveling bands of 300 or so female cannibal stickleback fishes attack lone males guarding nests and devour the eggs. Alligators in shallow rivers in Florida's Everglades become too crowded during droughts as their habitats shrink. When elders eat younger alligators, the remaining population has more food and space.

Cannibalism can be adaptive—for the cannibals, of course, but not for the victims. In some fishes and amphibians, eating one's own kind causes faster growth and development, which enables an individual to have more offspring. In the Costa Rican rain forest after a heavy rain, meadow tree frogs mate frenetically, the females depositing so many clutches of eggs in transient puddles that the habitat very rapidly becomes overcrowded. When the tadpoles hatch, they compete intensely for limited plant food. The crowding and dwindling resources trigger some tadpoles to become cannibals. By the time of metamorphosis, the cannibals are much larger than any remaining vegetarian tadpoles. When the puddles dry up, the larger frogs are more likely to survive because they are more adept at getting food and avoiding becoming food.

To see whether frogs grew faster from eating their own kind, or just from eating meat, researchers set up an experiment. They fed meadow frog tadpoles mashed tadpoles of either their own or a different species. Only the cannibals were bigger than normal at metamorphosis. A possible explanation is that the frogs eat an extra dose of the hormone thyroxine, which regulates metamorphosis, that is tailored to their species. We return to cannibalism later in the chapter (see fig. 41.15).

Mating Behavior

With dominance relationships and territorial boundaries keeping animals apart, how do they come close enough to mate? Courtship rituals are stereotyped, elaborate, and conspicuous behaviors that overcome aggression long enough for mating to occur. For example, a male swan rises up and flaps his wings as a prelude to mating, and king penguins "neck" during their courtship rituals (fig. 41.12). Mating behavior is part of sexual

FIGURE 41.12
Courtship Displays.
(**A**) Bewick's swans pair for life, yet the male performs this courtship display each year. (**B**) Mating in king penguins involves ritualized behaviors such as "necking."

A

B

selection. Recall from chapter 14's opening essay and figure 14.7 that sexual selection is differential reproductive success based on a trait or behavior that increases the likelihood of attracting or competing for mates.

In some species, males behave to reduce aggression in the female. A male orb weaving spider may stand at the edge of a female's web and tug on the suspension cord before venturing in. A male wolf spider slowly approaches a female, then signals his drive to mate by waving specially adorned appendages.

Courtship displays are specific, preventing the costly error of mating with a member of a different species. Male fiddler crabs, for example, wave their large claws to lure females into their burrows, but the exact pattern of movement is unique to different species. Many courtship behaviors hold attractions that we humans simply cannot appreciate. A female jackdaw bird, for example, stuffs regurgitated worms into the mouth of her mate.

Mating Systems Animals have several types of mating systems defined by the number of partners. In **monogamy,** a female and male associate exclusively with one another for a period of time—a few days, a season, or a lifetime. Pairs of golden-rumped elephant shrews are monogamous, but they spend most of their time apart, foraging alone on the forest floor and meeting to mate only every 6 to 8 weeks. In contrast, Bewick's swans also pair for life but are nearly inseparable. Monogamy in a broad biological sense means that the pair spend time together to the exclusion of others, but they may also occasionally mate with others. Only 7% of mammalian species are known to be monogamous. These animals tend to intensely care for and protect their young.

The amount of time and resources a parent spends on producing and raising its offspring is termed **parental investment.** Increased parental investment increases the chances of an individual's offspring surviving. Parental investment may be greater when mating is monogamous because the male somehow knows that he is the only father. Figure 41.13 shows a male caring for young.

Several types of mating systems involve multiple partners. In **polygamy,** a member of one sex associates with several members of the opposite sex. The most common polygamous behavior is polygyny, in which one male mates with several females. Competition among polygynous males is often intense. In polyandry, a female mates with several males. Still other species, including frogs, fruit flies, and antelopes, engage in a communal mating system, called polygynandry, in which both sexes have multiple partners.

FIGURE 41.13
Parental Investment Helps Ensure Reproductive Success.
Males take part in bearing the offspring to different extents in different species. This male yellowhead jawfish carries embryos for their first 4 or 5 days—the female leaves as soon as her eggs are fertilized.

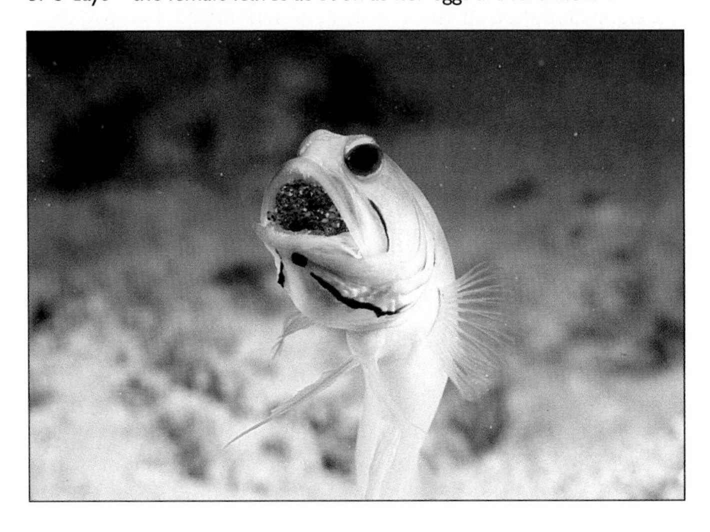

A species' mating system is related to the degree of difference in appearance between the sexes, which is called **sexual dimorphism.** Polygamous species are usually highly sexually dimorphic (very different), whereas the sexes in monogamous species usually look more alike. The predominant mating system in a species is also probably adaptive. For example, in pseudoscorpions, females with two partners have about a third more offspring than do females with single partners.

Diversity of Mating Behavior Mating behavior is diverse among animal species. Consider such behavior in the Sierra dome spider compared to that of the hedge sparrow.

A female Sierra dome spider has several mates, which increases the chances of fertilization. Males seek a female's web for food, protection from predators, and a potential mate, arriving when she secretes a pheromone that indicates that she is sexually receptive. The males battle for entry to her web, and the winner fertilizes 70 to 80% of her eggs. Then she plucks the web to signal his dismissal, and mates with other males.

Female spiders select mates based on their body size and stamina during mating. Spider sex takes 2 to 7 hours and consists of hundreds of couplings of only a few seconds duration. In between couplings, the female eats; the male is too exhausted to compete with her for food. Because males that are large and vigorous mates also tend to be healthy, the female choosing the strongest partners illustrates sexual selection.

Unlike the Sierra dome spider, the hedge sparrow displays all types of mating systems. Females maintain territories in thick hedges, but one territory may house a monogamous pair, another a female with two males, yet another a single male who flits between several females' lairs. Two or three males may mate in groups with two to four females. Parental investment in these birds varies with type of mating behavior. In monogamous pairs of hedge sparrows, the male cares for all of the chicks—and they are all his. In groups of two males and one female, the proportion of time each male spends feeding young is about the same as the proportion of time that he mates with the female.

The diversity of mating styles among hedge sparrows arose, ethologists hypothesize, because the sexes seek different strategies. For a female hedge sparrow, mating is most adaptive when male participation in chick-rearing is maximized, because more chicks will survive. Monogamy or polyandry, where she is the sole female, each accomplishes this. For a male, however, mating with several females (polygyny) maximizes the number of offspring he produces. Polyandry is the least efficient mating style for a male, because he may care for chicks that aren't his.

Male and female hedge sparrows are pitted against each other in a dominance struggle. If the male wins, polygyny proceeds. If the female wins, polyandry rules—and she takes a second mate. If neither male nor female can get a second mate, monogamy is a default option. Finally, polygynandry may represent a stalemate situation, when one female cannot chase away another, and one male can't get rid of another. (Of course, this is just a human view of the birds' mating styles.) Figure 41.14 shows another interesting sexual behavior of hedge sparrows—males can ensure that it is their sperm that fertilizes eggs.

FIGURE 41.14

Fatherhood Assured.

Before a male hedge sparrow mates, he pecks the female on her cloacal opening. In response, she releases a droplet containing sperm stored from previous matings. Only then, with his competition for fatherhood eliminated, does the male complete mating.

Primate species have diverse sexual behaviors too. They form male-female pairs, one-sex bands, mixed-sex troops, harems, homosexual pairs, loners, and social groups with frequently changing members. Devoted monogamous gibbon pairs sing together in the early morning to defend their territory. In bonobo societies, females are dominant and often form intense relationships with each other. Orangutans are polygynous, with a female mating only once every few years.

Primate sexual behavior varies even within species. An aggressive male mountain gorilla mates with and kills an older female who cannot keep up with the troop; yet other mountain gorillas form long-lasting bonds. Some chimpanzee males prefer to mate with an unfamiliar female. Other males take turns mating with a single female. Still other chimpanzees spend their lives in monogamy. Humans have also practiced every mating system observed in nonhuman primates.

Altruism

Dictionaries define altruism as concern for the welfare of others. In the study of animal behavior, altruism has a more specific meaning—increasing another's fitness (ability to pass genes to the next generation) at a cost of one's own fitness. Charles Darwin notes various examples of animals helping their kin to raise young, while not having young of their own. For example, the white-fronted bee-eater lives in extended family groups of five to nine birds. In each group, "cooperative breeders" do not raise their own young, but assist parents, who are usually their siblings or half-siblings, in rearing offspring. Ants, bees, wasps, and naked mole rats also display such altruistic behavior, some individuals even losing their ability to reproduce.

Darwin wondered why natural selection did not act to remove such altruistic individuals from populations, since they leave no progeny. An explanation came with a hypothesis called **kin selection,** proposed by geneticist W. D. Hamilton in 1964. In general, kin selection is the process by which an individual helps a relative, either by assisting it to survive or to reproduce. Hamilton proposed that selection be viewed at the level of genes, instead of individuals. An animal helping a relative to reproduce may not become a parent itself, but nonetheless assists its own genes in staying in the population through nieces, nephews, and cousins, with which it shares genes. Even earlier, evolutionary biologist J. B. S. Haldane said, in 1932, "I would lay down my life for two brothers or eight cousins," referring to the fact that an individual shares half its genes with siblings, and one-eighth with first cousins. Another route to kin selection is cannibalism of nonrelatives, while sparing relatives.

Hamilton developed the concept of **inclusive fitness,** which maintains that two selective mechanisms operate. Direct fitness is the usual route of passing on one's genes through offspring. Indirect fitness, or kin selection, is the transmission of genes by helping relatives to reproduce. A mathematical expression describes the conditions under which kin selection is likely to occur. In words, kin selection occurs if certain genes are more likely to be passed on if the relatives reproduce than if the altruistic individual does. For example, if a naked mole rat does not reproduce, two alleles of a particular gene are lost from the population. If the rat helps a sister to reproduce and she has eight pups, then 16 copies of the gene are passed to the next generation. The gain of 16 alleles exceeds the loss of two alleles.

Inclusive fitness predicts that animals can recognize close relatives. Many observations and experiments have shown that animals recognize kin through pheromones, taste, hormone secretions, appearance, and by somehow knowing that relatives are in a particular place at a particular time. Flour beetles, for example, do not cannibalize other flour beetles if they are in a place where they have just laid eggs, so that they do not kill their relatives.

The theory of kin selection also predicts that the degree of altruism should be directly proportional to the degree of relatedness. The tiger salamander *Ambystoma tigrinum* illustrates the ability of cannibals to recognize kin. Like certain nematodes, rotifers, and wasp larvae, tiger salamander tadpoles can be cannibals or noncannibals. Crowding stimulates certain individuals to develop a wider mouth and prominent teeth and jaws, becoming cannibals (fig. 41.15). (In some other species, the chance ingestion of meat triggers a similar transformation.) The killer tadpoles consume the noncannibals, but preferentially those that are not relatives. Researchers replicated this behavior in tanks in the laboratory, controlling the numbers of relatives and nonrelatives, and tracking who was related to whom. They crowded the tanks, and observed that the more closely related a cannibal is to a particular noncannibal tadpole, the less likely it is to eat it.

FIGURE 41.15
Cannibalism and Kin.
Noncannibal tiger salamanders develop into killers under crowded conditions, but they tend not to eat their relatives.

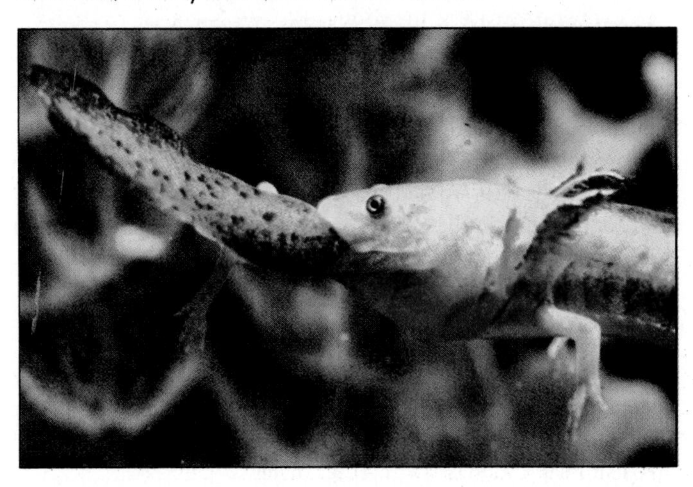

41.2 MASTERING CONCEPTS

1. What types of environmental cues do animals use to navigate?
2. What skills does an animal require to home?
3. How do territoriality and dominance hierarchies limit aggression?
4. What is a courtship ritual?
5. Describe different mating systems in animal species.
6. How might the type of mating system affect parental investment?
7. How is altruism adaptive on a population level but not on an individual level?

OLC

41.3 Animals Living Together

A biological society exhibits cooperative care of young, overlapping generations, and division of labor. Social behavior is adapted to environmental challenges. Communication is vital to the survival of a biological society, and may take diverse forms.

Many animals live in organized groups. Such a group is termed **eusocial,** meaning that it constitutes a biological society, if it exhibits three characteristics:

1. cooperative care of the young;
2. overlapping generations;
3. division of labor.

Communication among members is also a requirement of eusocial groups, but species that do not have such groups also communicate. The best-studied eusocieties are those of ants, termites, and some species of bees and wasps. Fossil evidence of insects in groups suggests that insect societies existed 200 million years ago.

Division of labor may be the key to the evolutionary success of social insect colonies, such as the fire ants described in the chapter opening essay. Coordinated groups of individuals make up the colony, each group with a specific function. One group might locate new food sources, while another assesses the sugar content of prospective foods. Other groups care for young or dispose of dead members. Many individuals each performing only a few tasks minimizes errors. It is similar to the way that a large company operates.

Several factors influence a social insect's role. These include nutrition, the temperature of the nest, pheromones, and age. For many species, younger members stay in the interior of the hive or nest, where they care for the queen and eggs. As the insects age, they move away, first helping to build the nest, then guarding the colony, and finally foraging, the most dangerous activity. The insect society is thus a series of **temporal castes**—groups whose roles change with time.

The proportion of an insect society's population that carries out a specific task changes to meet the colony's needs, even as individual members proceed through the temporal castes. In laboratory studies, if some members of a caste are removed, remaining members compensate by adding tasks to their own jobs. When removed insects are returned, or replaced, everyone resumes their original tasks. It is a little like human workers in a company temporarily taking on more work when a coworker leaves.

Group Living Has Advantages and Disadvantages

Animals may live in small groups that do not meet the criteria of biological societies, or in highly regimented societies whose members number in the thousands. Any degree of grouping presents certain advantages—and some disadvantages.

Groups of animals acquire food, protection, and care for their young more easily than solitary animals. Animal groups have been so successful in securing both individual and species survival that they have evolved independently many times.

More Favorable Surroundings By forming groups, animals may change their environments to their advantage. For example, animals group to conserve moisture or heat. During the winter pigs sleep in heaps. They move and squeal periodically as those on the perimeter seek the warmth of body heat inside the pile, and the central pig becomes too hot and moves to the outside.

FIGURE 41.16
Flocking Helps Flying.
Geese live in groups of a dozen or so individuals. They take turns leading the famous "V" flight pattern, which enables them to conserve strength.

Forming groups is also a way to distribute physical effort, as flying geese illustrate (fig. 41.16).

A group can physically alter its surroundings. Water fleas cannot survive in alkaline water, but when large numbers congregate, carbon dioxide from their respiration decreases the water's alkalinity to acceptable levels. Similarly, fruit flies fare poorly when they lay too few eggs in an area because there are not enough larvae to break up the fruit and make it soft enough to eat.

Better Defense There is safety in numbers. A predator can more easily fell a lone prey than pick one from a group where a few alert individuals warn the others. Vervet monkeys use different calls to warn of the approach of an eagle, a leopard, or a poisonous snake. These different signals are important because the route of escape differs depending upon whether the predator strikes from sky or land.

Group defense can also be passive—that is, it can avoid confrontation. For example, some small fishes swim at specific distances and angles from each other, forming a school. When a predator faces a school, it often suffers a **confusion effect** and is unable to decide which fish to attack. In the presence of a faster-swimming enemy, the school may use a **fountain effect,** splitting in two and regrouping behind the baffled predator.

Ostriches use the **dilution effect** of passive grouping to protect their young. Their societies consist of bonded pairs of a cock and a major hen, as well as minor hens who mate with many males. Minor hens deposit their eggs in the nest of a bonded pair, who sit on them, with their own eggs, for 6 weeks. Why would a bonded pair take in other eggs? Possibly because if not all the eggs are theirs, predators will consume fewer of their eggs. The major hen moves the minor hen's eggs to the edges of the clutch so they are more visible than her own. This is called a dilution effect because the pair's own eggs no longer make up a large percentage of the clutch.

Sometimes the best defense is a good offense, with help from others. Many prey species engage in mobbing behavior, where adults bother a predator that is larger than any individual of the group. Redwing blackbirds make hit-and-run attacks against much larger owls, uttering shrill calls. Baboons and chimpanzees use a similar strategy against leopards. Screaming, they charge and retreat and maybe even throw sticks at the leopard.

Forming a circle with the most formidable parts of the body facing outwards is a common defense strategy. When young catfish are disturbed they mass together, with their large pectoral fins and poisonous spines projecting in all directions. Similarly, adult musk oxen form a ring with their heads pointed outward toward the attacker. The older animals face the threat, while young find shelter within the woolly wall.

Enhanced Reproductive Success Social behavior may increase the probability that an individual will find a mate. An increase in the number of animals in an area may even trigger physiological changes necessary for successful reproduction. For example, a pig's ovaries may not develop normally unless she hears and smells a boar.

The sights, sounds, and scents of other courting individuals enhance and synchronize breeding. Spotted salamanders provide an extreme example of explicitly timed group mating. Each spring in many parts of North America, for two to seven consecutive nights, the salamanders perform courtship dances on pond bottoms, with males depositing bundles of sperm and coaxing females to take them. At other times, the yellow-dotted black amphibians do not seem to interact socially at all.

Reproduction that occurs simultaneously throughout a population can have other benefits. Synchronous breeding also may be adaptive if it coincides with abundance of seasonally available food, for meeting the energy needs of mating animals, or for feeding young.

Improved Foraging Efficiency If food is plentiful in some places and scarce elsewhere, it helps enormously to have others around to locate good feeding sites. When one member of a flock of starlings finds a food source, for example, other birds rapidly change their searching strategies and head for the meal.

Some spider's webs vividly illustrate the advantage of group living in securing a meal. The community spiders of genus *Stegodyphus* that live in South and East Africa build vast silken empires that house hundreds of spiders. At night during the wet early autumn, the spiders repair their netting, torn in the daytime by visitors, rain, and wind. Maintaining the webs is essential, for these structures trap the spiders' food of gnats, termites, ants, stinkbugs, beetles, and grasshoppers. By cooperating to erect this food trap, the spiders enjoy a rich and varied menu that an individual could not subdue on its own. An added benefit is that the web entraps some predators.

Improved Learning Social behavior may enhance learning and the passage of tradition. For example, learning is necessary to acquire the navigational skills for migration. Learning can be adaptive by enabling others to benefit from a behavior that an individual discovers, such as by watching. In Japan, a monkey washed sweet potatoes and wheat in the sea to clean and season the food. Others in her troop imitated her. Evolution of our own species has relied heavily on rapid transmission and assimilation of new behaviors.

Disadvantages of Group Living Group living could be disadvantageous in circumstances where the number of individuals quickly depletes a limited food supply. Colonial living also makes it easier for infections to spread. This is the case for mite infestations in honeybee hives in the United States. The microscopic tracheal mites collect in the breathing tubes of the bees. Mated female mites crawl from the tubes and lie on the tips of the bees' hairs, where they are easily picked up by another bee. Beehives infested with tracheal mites may be morgues by winter's end.

Environmental Conditions Influence Behavior

A species' characteristic social structure is influenced both by the animals' physiology and their surroundings. Consider the three species of zebra, which occupy distinct ranges in Africa (fig. 41.17). The plains zebra (*Equus burchelli*) lives in many areas, including grasslands in East Africa, woodlands in Zambia, and a vast treeless area in South Africa. The mountain zebra (*E. zebra*) lives in the mountains of southern South Africa and Namibia. The Grevy's zebra (*E. grevyi*), the most ancient of the three, occupies rocky, sparse grassland in northern Kenya and Ethiopia.

The social groups of the three types of zebras reflect the availability of resources for females—and the distribution of the males is a response to how the females obtain those resources. Plains and mountain zebras live in groups of one or more adult mares, their youngest calves, and one stallion. The groups need not roam much, because food is abundant in their territory. Powerful social bonds link the stallion to each of the females, and the females to each other. Offspring leave the group as they approach sexual maturity, which may be an adaptation to avoid inbreeding—if a young female stayed, she would end up mating with her father. Often several social groups will join, forming bands hundreds of animals strong. These large groups can effectively ward off other groups of 20 or so "bachelor" males in search of females.

In contrast, social groups are not part of the Grevy's zebra's way of life. The only persistent groups are single mares and their youngest offspring. This organization suits the sparse habitat. The females and their young move from grass patch to grass patch. If the groups were larger, a grassy area could not support them all. Solitary males inhabit these oases, and mate with whichever visiting females happen to be sexually receptive—a fairly rare occurrence in this species whose gestation period is 13 months. The Grevy's male must fight off other males who attempt to take over the territory.

Communication Keeps the Group Together

Communication is obviously necessary for social behaviors. This may include chemical (taste or smell), auditory, tactile, or visual signals. The best way to send a message depends upon the environment. Beneath the sea, for example, light does not travel

FIGURE 41.17

Zebra Social Structure Reflects Habitat.

(**A**) The three typs of zebras have distinct ranges in Africa, reflecting the availability of resources in their various habitats. (**B**) In plains zebras, small social groups coalesce into enormous herds.

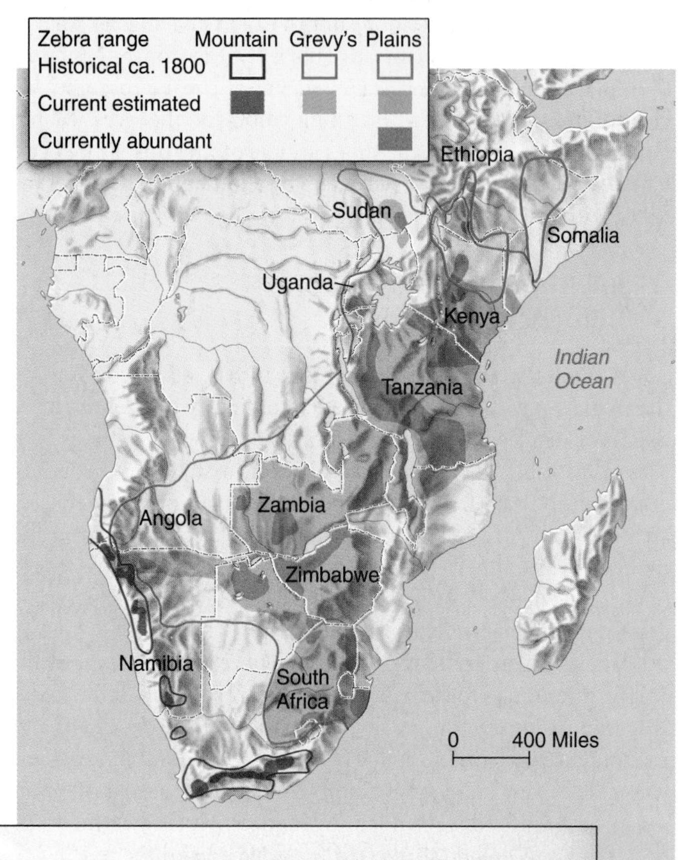

far, so visual communication may not be useful. Sound, however, travels easily under water. The songs of whales can be heard hundreds of miles away.

Chemicals Many animals use chemicals to communicate. Recall that pheromones are chemicals that an organism emits that signal others of the same species (see fig. 32.4). Lobsters, for example, have poorly developed visual systems, but parts of their bodies are covered with millions of tiny hairs that function as chemoreceptors, providing a sense of smell 1,000 to 1 million times more sensitive than ours. Lobster mating depends upon chemical cues (fig. 41.18). When a female enters a crevice and faces its male occupant, the two stand face to face, flicking their antennules to sense whether the other is ready to mate. After several stereotyped motions, the female ejects heavily scented urine. The male senses the stream in his antennules, and, in response, he fans the odor outward. Then, as the female shrinks and sheds her shell, he flips her onto her back, and they mate.

Sound Elephants are masters of auditory communication. They live in matriarchal herds consisting of sisters, cousins, and their offspring. Males roam alone or in groups (see Investigating Life 1.1). Related families greet each other with a cacophony of rumbles, trumpets, and screams. When a female approaches a male to mate, she bellows her intent. Her relatives watch, trumpeting loudly.

Sperm whales live in groups of 10 to 12 relatives, including mothers, daughters, aunts, and young males. They converse in a language of clicks and pauses that resembles Morse code. Researchers describe each pattern of clicks and pauses as a coda. For example, a 6 + 1 + 1 coda corresponds to

click-click-click-click-click-click-pause-click-pause-click

A coda in the sperm whale's language is the equivalent of a letter in our alphabet.

Certain codas often follow others, suggesting that they are not random utterances but meaningful messages. Ethologists observe how sperm whales behave when they use certain codas to try to discern meaning. Sperm whales thousands of miles apart use the same codas. Linguistic differences between sperm whales in the Pacific and in the Caribbean relate to the frequencies of certain codas and not to the "message" of the codas themselves. The clicks and pauses of sperm whale communication are more like human language in their consistency and their association with specific behaviors than are the whistles, moans, grunts, and squeals of other whale species.

Touch Physical contact can cement some social bonds. Touching, for example, can reduce tensions and signify greeting. Members of a wolf pack may surround the dominant male and lick his face and poke his mouth with their muzzles. This ceremony occurs when it is useful to reinforce social ties, such as upon awakening, after separations, and just before hunting.

Touch is common among nonhuman primates, who often sit with their arms about one another. These animals groom each other's fur, using their hands, teeth, and tongues. Grooming removes parasitic insects, prevents infection, and also promotes social acceptance.

Vision Honeybees (*Apis mellifera*) must store large amounts of honey in their hives to provide energy stores to survive the winter, when nectar, the source of the honey, is scarce. Because a hive may house up to 15,000 insects, a division of labor, plus a highly effective communication system, are necessary to collect and store sufficient nectar. The bees' activities must also be highly coordinated

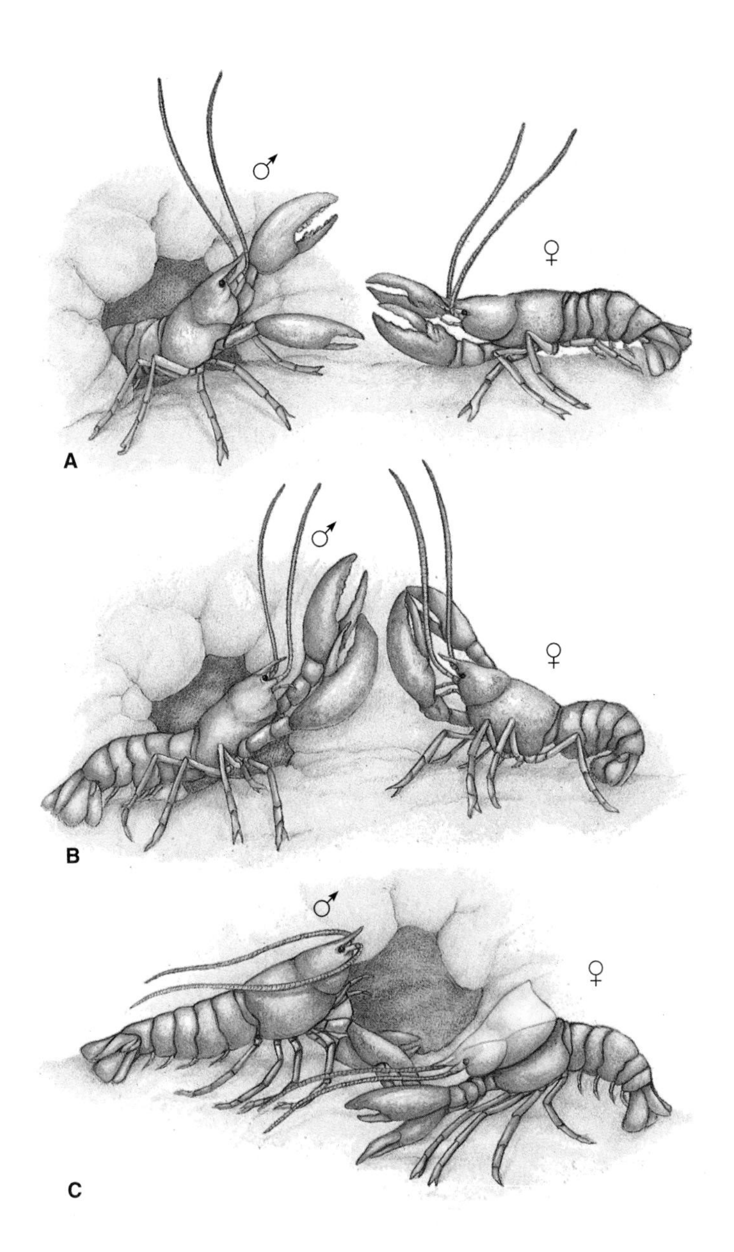

FIGURE 41.18

Chemical Cues are Important in Lobster Mating.

At certain points during courtship behavior, the female emits strong urine, which excites her selected mate—and possibly other males as well. (**A**) Courtship starts when a female approaches a male in his crevice. (**B**) She "knights" him by raising her claws, announcing her intent. (**C**) The female's body shrinks and she sheds her shell. (**D**) The male turns her gently on her back, and they mate. (**E**) For the next week he shelters her as her new shell hardens.

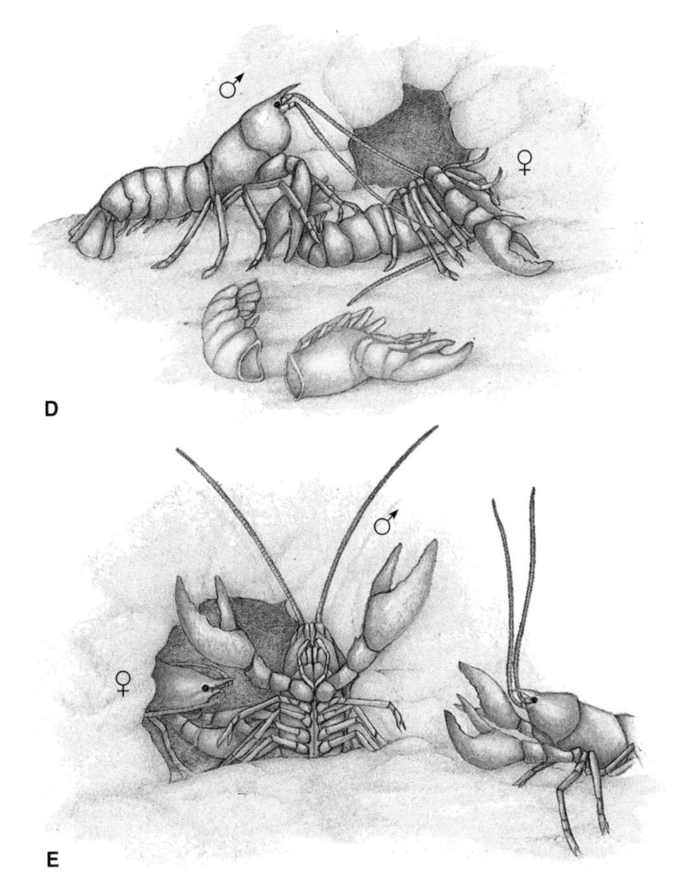

with the life cycles of flowering plants. On a spring day, a bee colony can gather up to 10 pounds of nectar, which ultimately yields half that amount of honey after drying out. About a third of the hive's members are foragers, whose task it is to locate and obtain nectar, and communicate its source. Foragers are infertile daughters of the queen, who spend the last 10 days of their 30-day life spans searching for nectar.

Much of what we know about honeybee foraging behavior comes from the extensive work of Austrian zoologist Karl von Frisch. In 1910, he read a paper stating that bees are color-blind. Why, he wondered, are flowers so brightly colored if not to attract the bees that pollinate them?

To demonstrate color vision in bees, von Frisch placed sugar water on a blue cardboard disc near a hive. The bees drank the water. Next, he placed blue and red discs near the hive; the bees went only to the blue disc. Similar trials with different colors and shades showed that bees not only see color (except red), but they can also detect ultraviolet and polarized light, which humans cannot see.

Von Frisch deciphered the "dances" that bees use to communicate food source locations. As a forager bee dances, recruit bees cluster around her, sensing her movements with their antennae. If the food is close to the hive, the forager dances in tight little circles on the face of the comb. This "round dance" incites the others to search for food close to the hive (fig. 41.19A). The longer the duration of the dance, the sweeter the food. A "waggle dance" signifies that food is farther from the hive (fig. 41.19B). The speed of the dance, the number of waggles on the straightaway, and the duration of the buzzing sound correlate with distance to the food. The orientation of the straight part of the run indicates the direction of the food source relative to the sun. Once other bees have followed the leader to the flower patch, they use odors given off by pollen to home in on particular nectar-rich plants.

When foragers return to the hive with nectar, they regurgitate it. Then another class of bees, called receivers, suck the nectar glistening on the mouthparts of the foragers. These younger bees

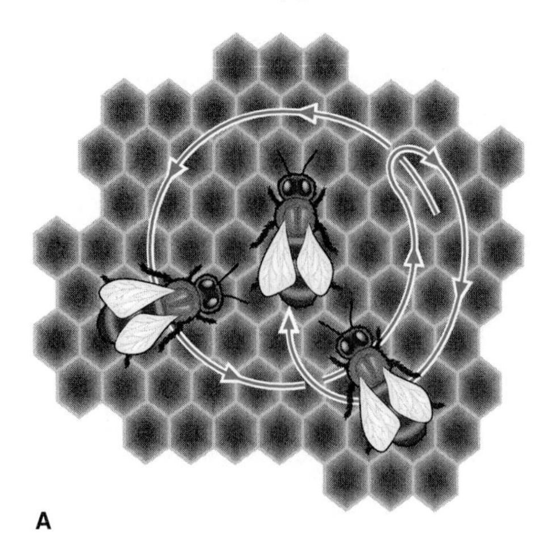

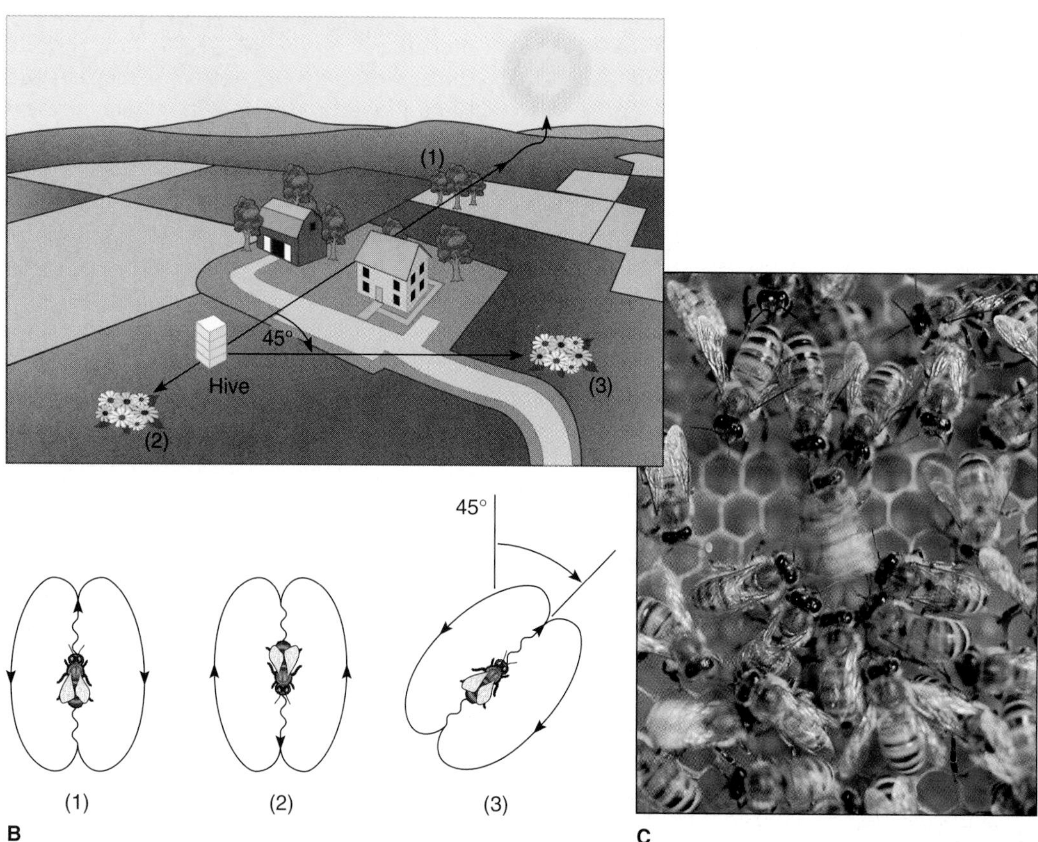

FIGURE 41.19
Bee Dances.
(**A**) A round dance informs recruits that food is close by, but does not point them in the right direction. (**B**) In a hive, honeycombs hang vertically. Bees observing a waggle dance interpret the dancer's movements relative to the vertical—with the angle of the straight run of the dance indicating the angle of the food source relative to the sun. For example, if the food source is straight toward the sun *(1)*, the straight run points directly up. If the food source is in the opposite direction of the sun *(2)*, the straight run of the dance points directly down. If the food source lies at a 45-degree angle relative to the sun *(3)*, then the straight run of the dance is 45 degrees relative to vertical. (**C**) Bees performing a waggle dance move so quickly that they are difficult to see.

deposit the nectar in the waxy cubicles of the honeycomb, where it dehydrates to form honey.

For a beehive to process nectar into honey efficiently, the numbers and rates of work of foragers and receivers must be in sync. Bees use a third type of dance, called the tremble dance, to deal with backups in the production line. Von Frisch described the tremble dance: ". . . they move about on four legs, with the forelegs, themselves trembling and shaking, held aloft approximately in the position in which a begging dog holds its forepaws." However, von Frisch could not discern any meaning in the tremble dance.

More recently, Cornell University biologist Thomas Seeley discovered that bees tremble to signal distress. In experiments conducted in the Adirondack State Park in New York, he removed most of the receiver bees from a hive, and noted that the foragers, when returning with their nectar loads, began to tremble when they couldn't quickly locate receivers. After a few hours, however, receivers were once again present to pick up the nectar, even though Seeley had not returned any of the bees he had removed. He hypothesizes that the bee colony, stressed by having one part of the workforce absent, hastily recruited others to take over their jobs, a little like coworkers at a company temporarily taking over the tasks of those who are absent.

Bees use visual cues to signal distance and direction to a food source. Some species can add a time element to such visual communication. Experiments show that scrub jays (*Aphelocoma coerulescens*) remember which foods they've stored, and where and when they stored them, a skill called episodic memory.

41.3 MASTERING CONCEPTS

1. What are the three requirements for a eusocial group?
2. What are the advantages and disadvantages of group living?
3. How do environmental conditions influence behavior?
4. Which senses are important in communication among animals?

OLC

41.4 Three Animal Societies

Snapping shrimp, naked mole rats, and honeybees are vastly different animals, but all live in highly organized societies.

Animal societies are common among insects and primates. Ethologists are learning more about well-understood social animals, and discovering new ones. Here is a look at a newly described animal society, one known for a few years, and a long-known society.

Snapping Shrimp

A newly recognized social species is the snapping shrimp, *Synalpheus regalis*, named for their powerful fighting claws. These crustaceans occupy cavities within two types of sponges that live among coral reefs in the Caribbean. Shrimp colonies within sponges average 149 individuals but may include up to 300.

Each shrimp colony has only one female that reproduces, which she does very often (fig. 41.20A). Young undergo direct development—there are no larvae. The members of the colony are therefore half or full siblings. The spongy home protects the shrimp and provides food, which floats into the colony from the sponge's internal canal system.

Collecting and dissecting sponges reveals two of the three defining characteristics of an animal society. First, generations overlap, as shown in the groups of young of different size and presumably different age. Second, shrimp colonies have a division of labor, as evidenced in the lone reproductive female. Demonstrating

FIGURE 41.20
Three Animal Societies.
(**A**) Snapping shrimp (*Synalpheus regalis*) live in social groups in sponges in Caribbean coral reefs. Each colony has a single reproductive queen, shown here in the center. Some of the shrimp around her display their characteristic claws, which they use to protect the colony—sponge homes are in high demand. (**B**) Naked mole rats from different colonies attack each other viciously, each trying to drag the enemy into its own territory to sink its teeth into the other animal's flesh. (**C**) Honeybee workers build a chamber called a queen cell in which the potential queen develops.

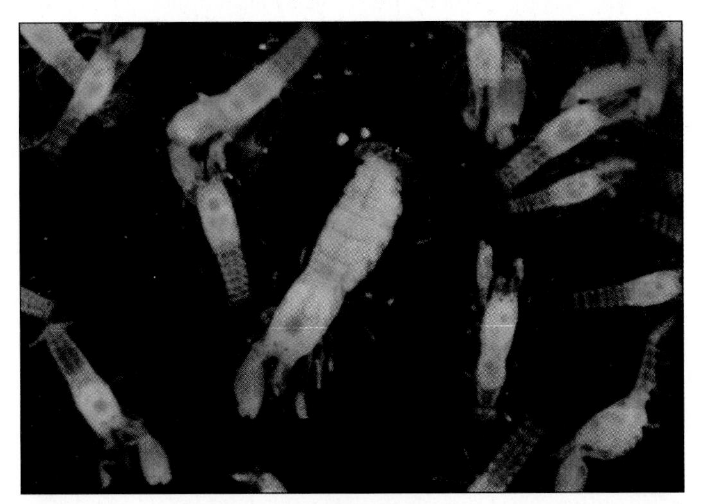

A

B

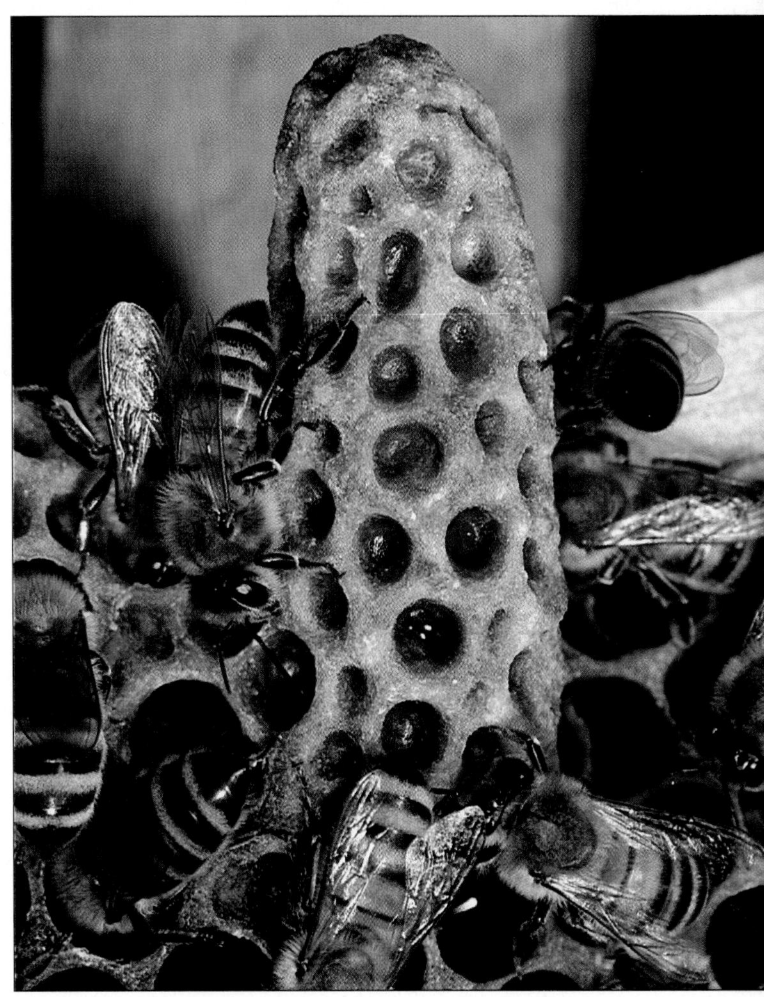

C

the third characteristic of an animal society—cooperative care of young—required more than simply dissecting sponges and observing their crustacean inhabitants. An experiment was necessary. Such behavior might be revealed by identifying a way that adults could help juveniles.

Unlike some other animal societies, young shrimp obtain food without adult assistance—their meals just float in. Nor must adults build a nest—the host sponge provides a secluded home. Biologist J. Emmett Duffy, of the Virginia Institute of Marine Science, hypothesized that adults care for young by protecting their hold on the sponge. He based this hypothesis on the observation that he could not find any sponges that did not house shrimp.

If competition exists for sponge homes, then introducing a threat should provoke protective behavior among resident shrimp. Duffy set up two experimental situations in sponges reared in his laboratory. In the first experiment, a shrimp colony encountered shrimp of another species. In the second experiment, a shrimp colony encountered shrimp of the same species that had been removed from the colony beforehand. Results were striking. When confronted with unrelated shrimp that competed for sponge space, the resident snapping shrimp fought valiantly, chopping up the competition. In contrast, the resident shrimp did not attack their former colony mates. More telling, only the larger individuals defended the colony—which Duffy concluded is evidence of cooperative care of young. The shrimp are the first eusocial invertebrate discovered in the oceans.

Naked Mole Rats

They live in vast, subterranean cities, digging tunnels and chambers in the soils of Kenya, Ethiopia, and Somalia. Roads above them collapse; crops that get in their way, such as yams, they eat. The naked mole rats of Africa live in societies remarkably like those of eusocial paper wasps and sweat bees, with a powerful queen ruling a colony of up to 300 individuals. Unlike most mammalian social groups, in which many females reproduce, the naked mole rat queen is the sole sexually active female.

Other roles in the naked mole rat community are well defined. The "janitors," the smallest and youngest males and females, patrol the colony's tunnels. They keep the walls smooth by rubbing them with their soft, hairless bodies. Janitors eat roots hanging from ceilings and confine excrement to widened dead ends.

The oldest and largest mole rats serve sentinel duty, their senses sharpened to detect intruders (fig. 41.20B). A sentinel will stop and feel the air currents on its nearly sightless eyeballs, a change in current indicating an invader. The sentinel also senses the low-frequency sound of footsteps overhead and detects many odors. If these signs, for example, indicate that a snake has poked its head into a tunnel, the sentinels signal to the others to descend to wider tunnels.

Mole rats dig new tunnels in early morning and late afternoon. "Dirt carriers" form groups; the head animal is a "digger," and a "kicker" works at the back to get rid of the dirt. Food carriers use their long incisors to break off bits of roots and tubers, consuming some of the food before returning to the group.

The queen patrols the tunnels several times an hour. Every 3 months, she gives birth to a litter of up to 27 pups, nursing them for a month while three male workers bring her food. For the next 2 months, these workers care for the young, who then become workers. Apparently the queen's presence is crucial to the integrity of this social colony. In the laboratory, when the queen is removed, chaos ensues. Jobs vanish, and the cooperation and interaction that define the colony break down.

Honeybees

The inside of a honeybee hive is an efficient living machine, with each individual taking a specific role. At the summit of the organizational ladder is the queen, a large, specialized female who lays about 1,000 eggs each day. Eggs fertilized by a male drone develop through larval and pupal stages into female workers. Unfertilized eggs develop into males. Workers secrete a substance called royal jelly onto a very few eggs. These eggs hatch to yield larvae that eat much more royal jelly than do other larvae, which places them on a developmental pathway toward eventual queendom.

The 20,000 to 80,000 workers vastly outnumber the hundred or so drones, and only one queen reigns per hive. If the queen leaves, workers chemically detect her absence and hasten development of the young potential queens (fig. 41.20C). The first new queen to emerge from her pupa case kills the others and then takes a "nuptial flight" to attract drones. She collects and stores their sperm and then later uses it to fertilize eggs. Males die when they inseminate the queen, or if they survive, they are forced out of the hive when food becomes scarce or autumn arrives.

A worker's existence is more complex, including several stages and specializations. Newly hatched workers feed the others. One-week-old females make and maintain the wax cells of the hive, where larvae and pupae develop. Some females are undertakers, ridding the hive of dead bees. Older workers collect food, as described previously.

Orchestrating the overall social order of a beehive is a mixture called "queen mandibular pheromone" that the queen secretes from glands in her jaw. It has different functions in different contexts. The pheromone inhibits workers from developing ovaries and from rearing additional queens. Inside the hive, queen mandibular pheromone inhibits males from mating, yet outside the hive, the pheromone has an opposite effect, stimulating males to mate.

Chapter Summary

41.1 Genes and Experience Shape Behavior

1. **Ethology** considers animal behavior. The field was founded on observational studies, but today it includes experimental approaches.
2. Genes and the environment shape behavior. **Innate** (genetic) influences dominate when it is essential for a behavior to occur correctly the first time. Learning is important in mastering certain complex behaviors. Anatomy and physiology limit possible behaviors for a species.

3. A **fixed action pattern** is an innate, stereotyped response that continues without feedback. A **releaser** is a stimulus that triggers an FAP, and it may be visual, auditory, chemical, or tactile. A supernormal releaser is an exaggerated releaser that is more effective than the natural stimulus in triggering the response. **Reaction chains** are sequences of FAPs.
4. Different types of learning overlap. In **habituation,** an animal ignores a highly repeated stimulus. In **classical conditioning,** an animal gives an old response to a new, **conditioned stimulus** instead of the older, **unconditioned stimulus.** In **operant conditioning,** a behavior changes in frequency because it is positively or negatively reinforced. **Imprinting** occurs during a critical period and doesn't require reinforcement. **Insight learning** applies prior learning to new situations without trial-and-error activity. **Latent learning** uses past observations to perform a new activity.

41.2 Types of Behavior

5. **Orientation** and **navigation** rely on responses to environmental cues. Animals that navigate recognize landmarks, use the sun or stars as a compass, and possibly detect Earth's magnetic field. Homing depends on a compass sense to set direction and a map sense to direct the navigator. The map sense may detect magnetic fluctuations, odors, or other cues.
6. Animals may fight members of their own species when they compete for resources. **Territoriality** and **dominance hierarchies** minimize aggression. Threats and appeasement, diversion, or displacement responses dampen combat. Overcrowding may trigger cannibalism, which improves the habitat for survivors.
7. Courtship rituals calm aggression so that physiologically ready individuals of the same species can mate. Mating systems include devoted pairs (**monogamy**), a male with multiple partners (polygyny), a female with several partners (polyandry), and group mating (polygynandry). **Polygamy** is mating involving multiple partners.
8. **Parental investment** is effort to ensure survival of one's offspring and increases in proportion to how frequently a particular male mates with the mother of his offspring. Polygamous species are highly **sexually dimorphic**—the sexes appear different.
9. Altruism toward a relative, or **kin selection,** increases one's **inclusive fitness,** thereby ensuring that some of an individual's genes persist, if not the individual.

41.3 Animals Living Together

10. A **eusocial** group has cooperative care of young, overlapping generations, and division of labor. Insect societies have **temporal castes,** in which age determines role in the colony.
11. Group living provides the ability to alter the environment, defense, improved reproductive success, more effective foraging, and opportunity to learn. However, group living increases competition for resources and eases spread of infection.
12. A species' social structure reflects environmental constraints. Many behaviors are based on communication, which may include chemical, auditory, tactile, visual or other types of stimuli.

41.2 Three Animal Societies

13. Societies of snapping shrimp have one fertile female and live in sponges. Naked mole rats have large, complex subterranean societies. Honeybee colonies are highly regimented societies.

Testing Your Knowledge

1. Distinguish between an innate and a learned behavior.
2. How can the following behaviors be adaptive (further the likelihood of reproductive success for some individuals)?
   a. territoriality
   b. cannibalism
   c. altruism
   d. sexual selection
   e. dominance hierarchies
3. Define the following terms:
   a. sexual dimorphism
   b. parental investment
   c. inclusive fitness
4. Distinguish among confusion, fountain, and dilution effects in deterring predators.
5. How does altruism differ in its literal and biological senses?
6. What are the advantages of group living?
7. How is communication important in the following situations?
   a. maintaining a eusocial group
   b. minimizing aggression
   c. promoting courtship and mating behavior

Thinking Scientifically

1. An infant who is blind smiles at her mother. Why is this evidence that smiling is an innate behavior?
2. Give examples of humans engaged in threat, appeasement, diversion, and displacement behaviors.
3. The modern feral (wild) horse was domesticated 5,000 years ago but has returned to the wild behavior of its ancestors. Even a modern broken horse will revert to wild behavior if allowed to range freely, with stallions stampeding mares and mock fighting. What type of behavior might explain the primitive actions of the free-ranging modern horse?
4. A dog adopted between 4 and 8 weeks of age often develops closer ties to its owner than one adopted at an older age. What type of behavior accounts for this observation?
5. A taxonomist working in a South American tropical rain forest uncovers a new species of ant. What signs should she look for to determine whether to classify the species as eusocial?
6. In laboratories, naked mole rats live in Plexiglas colonies; bees establish hives in glass-encased honeycombs built into a window; spotted salamanders mate in bathtubs. Do you think that researchers using these setups can accurately assess animal behavior? Can you suggest alternative approaches for observing these organisms?

References and Resources

Hack, Mace A. March 1998. Zebra zones. *Natural History,* vol. 107, pp. 26–31. The social groups of zebras reflect their habitats.

Keller, Laurent, and Kenneth G. Ross. August 6, 1998. Selfish genes: A green beard in the red fire ant. *Nature,* vol. 394, p. 573. Do genotypes underlie selective killing behavior among fire ants?

Parnell, Richard. October 1999. Gorilla exposé. *Natural History,* vol. 108, p. 38. Primates exhibit a diversity of social behaviors.

Rodriguez, Ivan, et al. September 2000. A putative pheromone receptor gene expressed in human olfactory mucosa. *Nature Biotechnology,* vol. 26, p. 18. Humans have counterparts of genes that encode pheromone receptors in rodents.

Seeley, Thomas D. June 1999. Born to dance. *Natural History,* vol. 108, pp. 54–57. New experiments add to von Frisch's classic observations on the dances of honeybees.

The *LIFE* Online Learning Center provides additional resources and tools for studying this chapter.
www.mhhe.com/life

CHAPTER 42

Populations

A Population Out of Control

Snakes Decimate an Island's Biodiversity

A species introduced into a new area can rapidly deplete native species that serve as prey. When this happens on an island, and the new predator is highly adapted to eat many types of organisms, biodiversity can plummet. This fate befell many of the residents of the island of Guam, thanks to the brown tree snake *Boiga irregularis.*

Guam was never exactly thriving in diversity, because of its remote location more than 1,242 miles (2,000 kilometers) from the closest landmasses of New Guinea, Japan, and the Philippines. It is one of 2,100 islands comprising Micronesia. Native animal species are those that can naturally get there, including bats and birds that can fly in, and reptiles, whose eggs float in on vegetation. Humans brought insects such as flies and mosquitoes, and killed off many native bird populations in the search for food. The brown tree snake likely arrived with equipment left over from World War II that was dumped on Guam. Perhaps a lone pregnant female, or a nonpregnant female who had stored sperm, as some snakes do, founded the population. Or perhaps several snakes arrived, slithering from their transport vehicles and mating.

The brown tree snake likely arrived with equipment left over from World War II that was dumped on Guam.

However the brown tree snake population started, it clearly grew explosively. Because this new predator and its prey had not coevolved over time, nature had not selected defensive adaptations that might have enabled more of the island's birds and reptiles to escape. Like children let loose in a well-stocked candy store, the snakes feasted; native species, as well as introduced species that had been there awhile, perished. The forests are silent now, for there are barely any birds left to sing. The snakes also eat lizards and small mammals, even biting and attempting to constrict sleeping human babies.

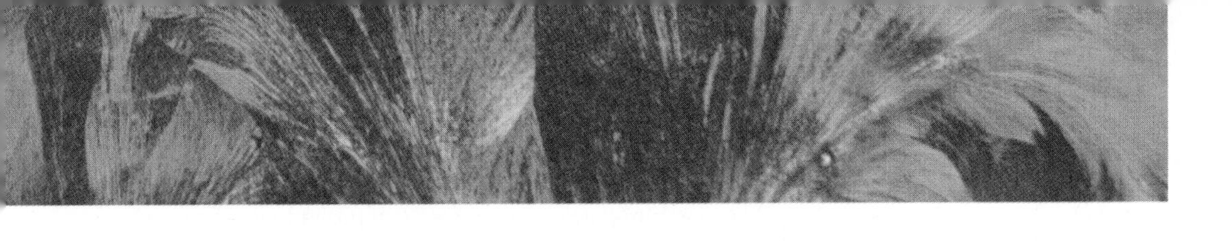

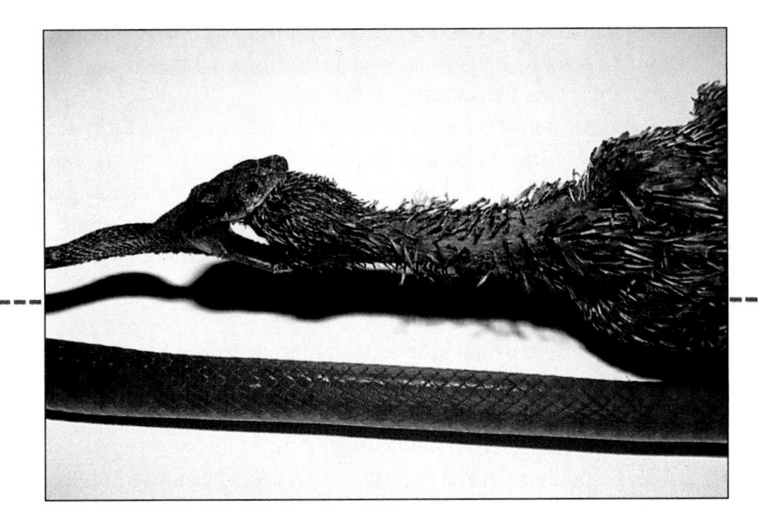

Super-Predators.
Brown tree snakes on the island of Guam will try to eat nearly anything. This snake was electrocuted while trying to swallow a much-too-large pigeon chick on a power pole. The attempt caused an island-wide power outage!

It wasn't until the 1960s that ecologists began to notice that something was amiss on Guam. Birds had vanished, first in the south, but spreading northward. A first hypothesis was that an insecticide was at fault—particularly DDT, known to kill baby birds by thinning eggshells. Alternatively, a disease could have been killing birds, but this would not explain the demise of lizards too. The government assigned an expert in bird diseases, Julie Savidge, to investigate the disappearances on Guam. She implicated the brown tree snake. But many scientists did not at first accept her finding, because it was not consistent with the known behavior of snakes to eat only once in a while. These snakes were apparently eating all the time, devouring nearly anything.

However, evidence that the snake was killing the native animal species of Guam was compelling:

- The snake is extremely common.
- It is well adapted to stalking and killing prey in a variety of ways.
- Areas where snakes were prevalent matched areas where birds disappeared.
- Native birds as well as introduced birds disappeared.
- Birds were not disappearing on nearby islands, which lacked the snakes.
- An insecticide or disease could not be identified.
- Many lizards live on the offshore islands, areas free of snakes.

The snakes also eat lizards and small mammals, even biting and attempting to constrict sleeping human babies.

Once the snake hypothesis was accepted, researchers began surveying the forests of Guam. They found an enormous snake population. Whereas most snakes live in populations of 1 to 10 animals per hectare (2.47 acres), 100 brown tree snakes occupied a hectare! This was a considerably greater population density than the prey species. The birds and lizards of Guam had no chance—they were hopelessly outnumbered by voracious predators.

The snakes also are superbly adapted to killing in several ways. They can pounce on a moving lizard in response to visual cues, or locate an egg meal by smell. The observation that the snakes not only bite human babies but constrict them indicates that the reptiles are actively searching for food, and not just defending themselves. A brown tree snake is far too small to swallow an infant, but unlike other snakes, it seems to have lost the ability to judge prey size. Many experiments reveal that the snakes will follow nearly any cue that suggests a live meal. They also eat dead food, including hamburgers, dog food, corpses, and garbage. Current efforts to kill the snakes entail leaving dead rats laced with acetaminophen, which destroys the snakes' livers. (It is a commonly used painkiller.) Still, Guam is a long way from recovering its biodiversity.

42.1 A Population Is All Individuals of a Single Species in a Particular Place

Ecology considers interactions between organisms and the living and nonliving parts of the environment. Populations, which are composed of members of a single species, exist in communities of interacting organisms. Several parameters are used to describe populations.

A **population** is a group of organisms of the same species in a given geographic location. The term population does not imply a minimum or maximum geographic scale. Instead, the potential for interbreeding defines a population's size. Members of the same population are much more likely to breed with one another than with a member of another population. A population may consist of just a few individuals of a plant species that happen to colonize a newly emerged island, or it may consist of millions of individuals over a large area, as in some insect populations.

This chapter examines the factors that influence populations over time. The next chapter looks at how the interactions between populations translate into patterns of overall biotic composition and nutrient cycling in different types of environments.

Ecology Is the Study of Organisms and Their Environment

A population's habitat is simply where its members live, such as the edge of a pond, a desert, or a rain forest. An organism's habitat might even be another organism, on or in which it lives. Because many different species typically inhabit an area and may compete for resources, populations interact. A **community** includes all the organisms, sometimes hundreds of species, in a given area. The size of a population is often affected by interactions with predators and other members of the same community.

The interactions among individuals in populations and communities are part of the life science of **ecology.** The word ecology comes from the Greek words for "home" (*oikos*) and "study" (*logos*). Ecology is the study of relationships between organisms and their environments—their "homes" in the broadest sense. The environment includes nonliving (abiotic) features and living (biotic) features. These abiotic and biotic features determine the distribution and abundance of species and the kinds of species that coexist in a given place and situation.

The study of population ecology impacts diverse fields, including disease prediction and land management. Typical questions population ecologists ask include: Which weather conditions favor the spread of rodents that transmit human diseases? How large a patch of old-growth forest does a breeding pair of spotted owls require? How many deer should hunters cull to keep the herd healthy? How many humans can Earth support? To answer such questions, ecologists begin by describing the population.

Characteristics of Populations

The term population may refer to all individuals of a species over the species' entire range, or to some of these individuals that live in one part of the range. A population has many descriptive features that reflect its ecological relationships. **Demography** is the statistical study of populations, including vital statistics such as population size, density, and distribution.

Population density is the number of individuals of a species per unit area or unit volume of habitat. Numbers can range from one hare on hundreds of hectares to the millions of bacteria in a single gram of soil. For some species, simply observing the organisms can provide a measure of population density. For others, more sophisticated techniques are required, as Investigating Life 42.1 describes.

The pattern in which individuals are scattered through habitat space is called **population dispersion** (fig. 42.1). Organisms may be dispersed in a random pattern if the environment is relatively uniform (such as the deep shade under a dense forest canopy) and individuals neither strongly attract nor repel one another. A tendency toward uniform spacing may occur if individuals respond negatively to each other's presence. Tree species whose seedlings require abundant sunlight, for example, are more or less uniformly distributed because offspring cannot survive in the shade of their parents. Strongly territorial animals may also spread themselves evenly throughout a habitat. Most often, however, organisms show some degree of aggregation, coexisting where habitat conditions are most favorable, seeking out each other due to social attraction, or clumping in the vicinity of their parents. Recall the three zebra species described in chapter 41. Mountain and plains zebras are fairly uniformly distributed, because their habitats tend to have abundant food. Grevy's zebras, in contrast, dot the landscape in small groups near sparse grassy areas (see fig. 41.17).

A species is rarely distributed continuously and in the same density over broad geographic areas within its range. For example, figure 42.2 shows a range map for the imported fire ant, *Solenopsis* spp., in the southeastern United States. These aggressive, stinging ants nest in open, sunny areas such as pastures and cultivated fields. Their range in the United States corresponds closely to the relatively mild winter climate to which the ants are adapted. Since the ants' preferred habitat type is not distributed continuously throughout the entire region, however, a map of the population viewed at a closer scale would appear much patchier than the range map suggests. These scattered local populations may exchange individuals by migration—immigration is migration into a population and emigration is migration out. Such semi-isolated populations form regional "metapopulations."

42.1 MASTERING CONCEPTS

1. Distinguish between a population and a community.
2. What is ecology?
3. What is population density?
4. Distinguish between the types of dispersion in a habitat.

OLC

Conducting a Wildlife Census

In Washington, D.C.'s Lafayette Park, biologist Vagn Flyger and his helpers place nest boxes in trees and then collect them when gray squirrels wander in. They anesthetize the animals and record their vital statistics—weight, sex, age, and health status. The researchers also tattoo the squirrels so their whereabouts can be monitored and then release them. Once a certain number of squirrels have been tattooed (M), a "recapture" sample of a certain size (R) is taken and the fraction of tattooed animals (f) noted. The number of marked squirrels (M) divided by this fraction (f) estimates the total squirrel population size.

Wildlife biologists also use a variation of this "capture-mark-recapture" approach on deer. Researchers place bright orange collars on a small number of deer. Then, an observer spots a group of deer from a plane and takes an aerial photograph. From the easily seen percentage of collared animals in the photo, biologists calculate the total number of animals in the group.

Counting droppings is another way to take a wildlife census, as is done for deer and moose in the Adirondack Mountains of upstate New York. The numbers of excrement piles are correlated with the number of animals. A wildlife refuge on the Gulf Coast combines the capture-mark-count approach with dropping detection. Scientists capture otters and inject them with a radioactive, harmless chemical, which the animals eliminate in feces. The scientists then collect droppings and extrapolate population size from the proportion of radioactively labeled droppings.

Many wildlife surveys count migrating herds of mammals and flocks of waterfowl from the air (fig. 42.A). Counts are also done from the ground. Every year, the North American Breeding Bird Survey obtains counts along more than 2,000 routes. Perhaps most inventive is a device used in Montana. Rotting meat or fresh sardines are draped in a tree, along with an infrared heat detector attached to a camera. When a bear saunters over to collect the treat, the detector picks up the bear's body heat and snaps its picture (fig. 42.B). Tags are used to trace bear identities.

Describing wildlife populations is useful in evaluating the effects of environmental catastrophes, such as fires, volcanic eruptions, or oil spills, by providing information on population dynamics and habitats before the disaster. Hunting, trapping, and fishing regulations and schedules are based on these data, as well as decisions on where to build—or not to build—houses, dams, bridges, and pipelines. The status of wildlife populations also reflects the overall well-being of an ecosystem.

FIGURE 42.A
Bird Count.
Each spring, U.S. and Canadian waterfowl spotters traverse 1.3 million square miles (3.4 million square kilometers) trying to count the number of animals in flocks, such as these snow geese.

FIGURE 42.B
Bear Count.
A male grizzly bear takes his own picture while stealing food from a tree with a camera triggered by the bear's body heat. From tags or physical features visible in such photos, biologists can obtain information on the overall bear population in a region.

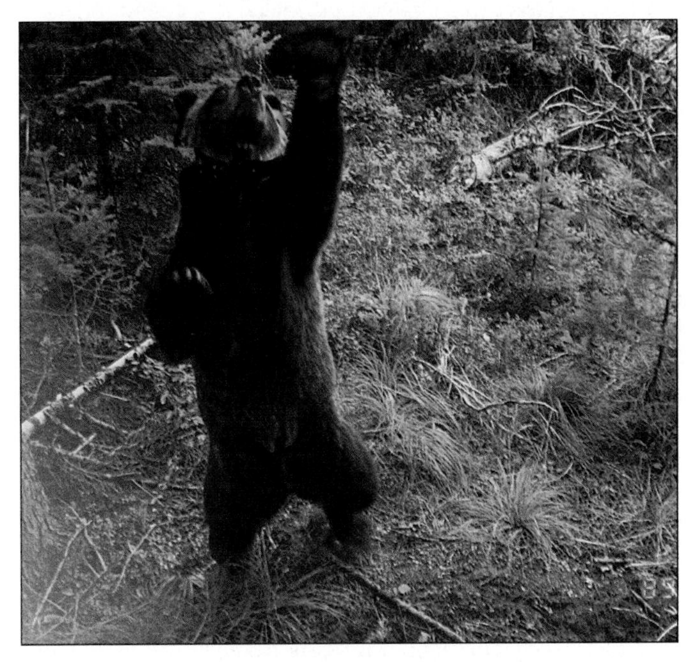

FIGURE 42.1
Population Dispersion Is a Basic Feature of Populations.
Random spacing, as illustrated by these lichens (**A**), is rare because it requires an environment in which resources are very evenly distributed. Social interactions such as territoriality (**B**) can lead to increasing uniformity, whereas an uneven distribution of resources or other social interactions may yield increasing aggregation, as in the human population of the United States (**C**).

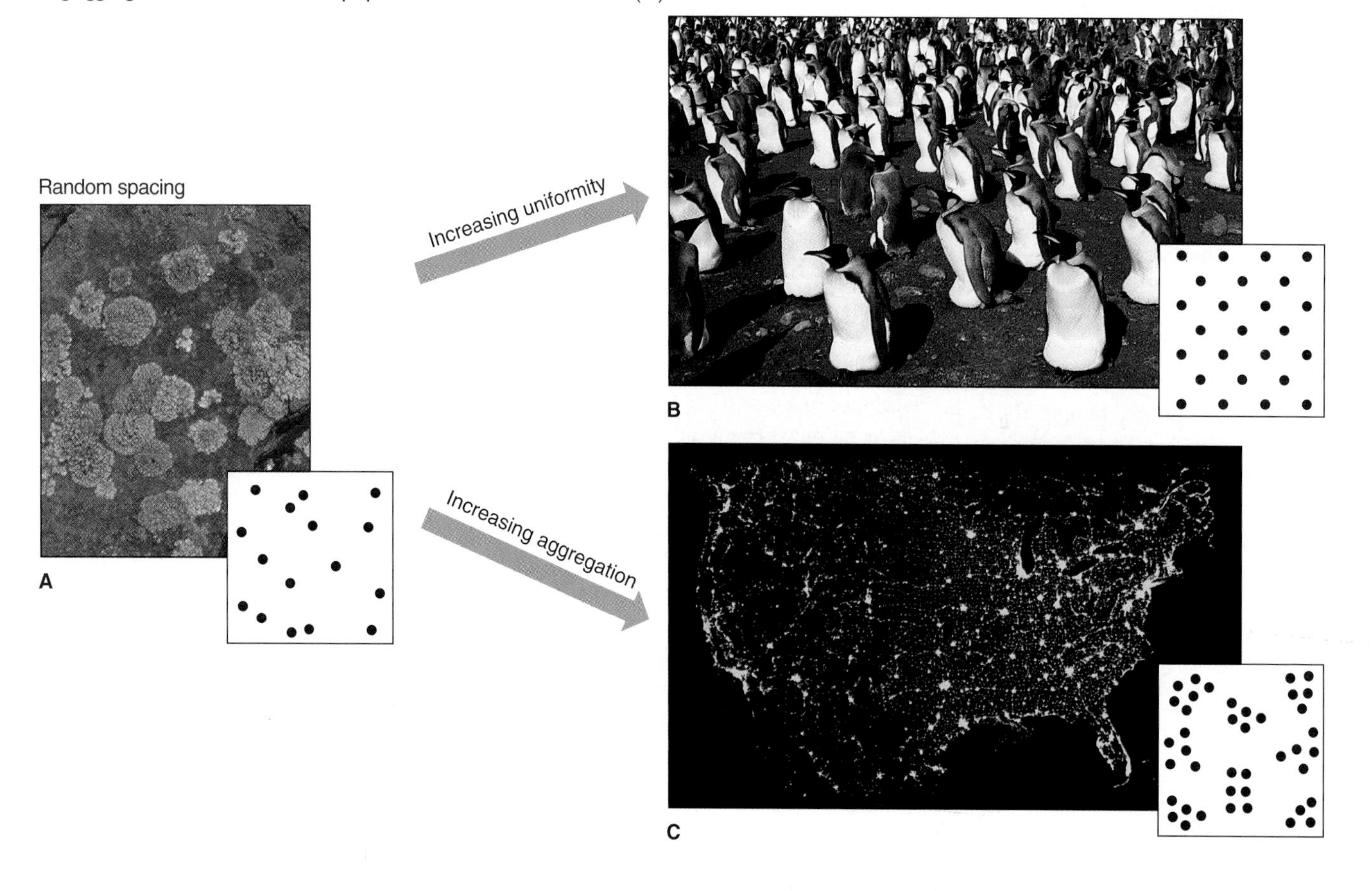

FIGURE 42.2
Fire Ant Range.
Fire ants (*Solenopsis* spp.) occupy open, sunny sites throughout the southeastern United States. Originally from South America, they have steadily expanded their range to 12 southern states over the past 50 years.
Source: Data from National Agriculture Pest Information System, February 28, 1996 and November 8, 1999.

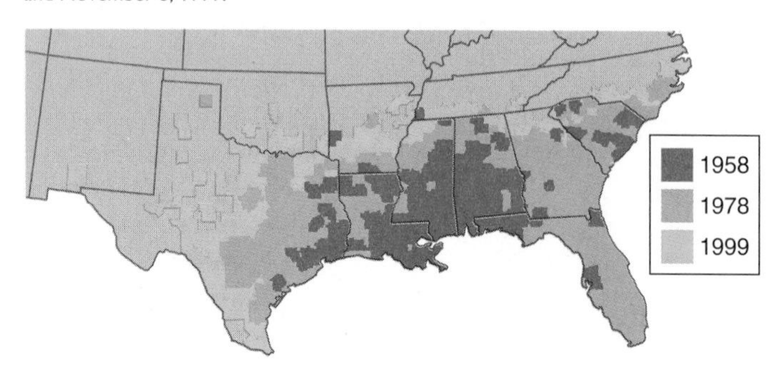

42.2 Several Factors Affect Population Growth

The size of a population reflects birth- and death rates and immigration and emigration. With unlimited resources, population growth is exponential, and can be described with a J-shaped curve. Since an organism's chances of reproducing and dying vary with its age, the growth rate of a population reflects its age structure.

Population density and dispersion measurements provide static "snapshots" of a population at a given time. Two populations of the same initial density, however, may behave very differently if one consists mainly of aging individuals and the other is more youthful. Therefore, to predict whether a population's size will grow, decline, or stay the same, an understanding of the species' reproductive characteristics is essential.

Reproductive Characteristics Influence Population Growth

Simply put, a population will grow if the sum of births plus immigration exceeds the sum of deaths plus emigration. In a local population, organisms moving into or out of an area may

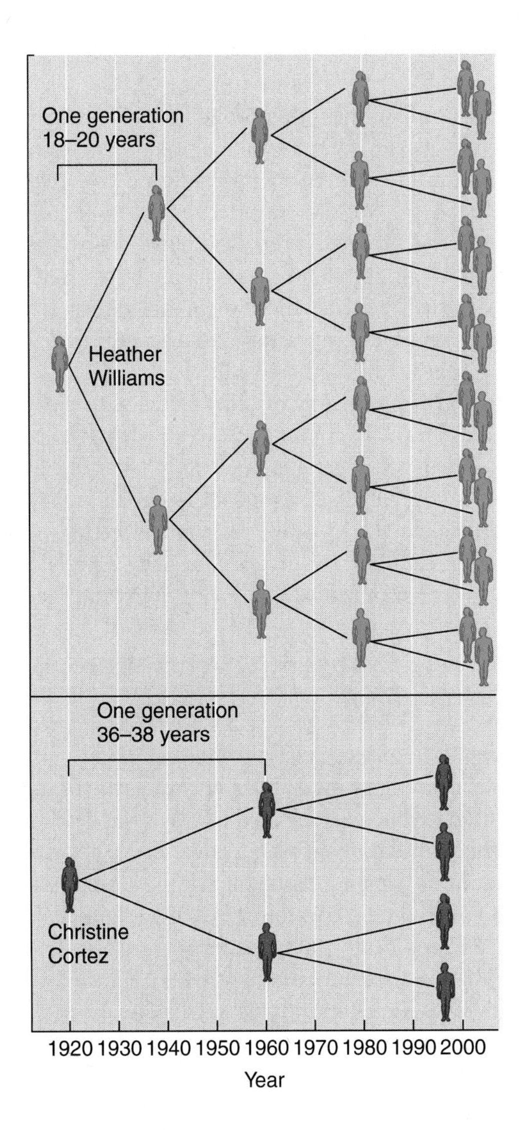

FIGURE 42.3

Early Reproduction Leads to Fast Population Growth.

Women in the hypothetical Williams family, including those marrying into the family, have two children each. The women become mothers when they are between the ages of 18 and 20. The women of the Cortez family do not have children until ages 36 and 38. By the time the matriarch of the Cortez family is a grandmother of 4, the head of the Williams family is already a great-great-grandmother of 16!

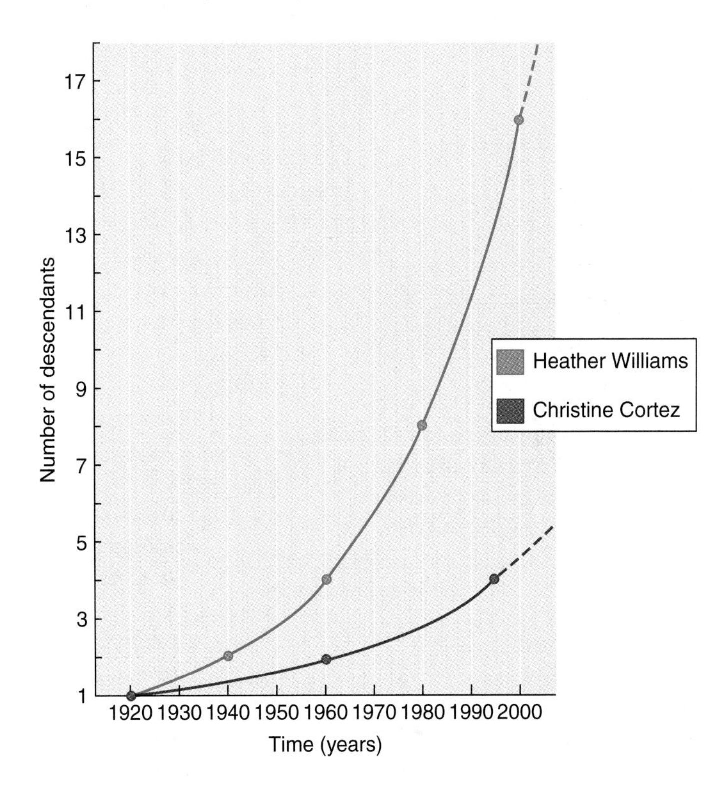

have a substantial effect. For example, immigration has increased the human population in the United States in recent years. Often, however, reproduction and mortality are the most important determinants of population growth.

The likelihood of reproducing and the likelihood of dying both vary with age and species. For any group of individuals that begin life at the same time (a **cohort**), these two probabilities can be tracked over time to generate a **life table.** Although mortality and breeding also clearly depend on other factors, such as temperature and population density, these effects are averaged in calculating the odds of reproducing or dying at a given age.

Fecundity is the number of offspring an individual produces over its lifetime. Fecundity depends on factors such as the number of times an individual reproduces in its lifetime, and the number of offspring resulting from each episode. In addition, all else being equal, the earlier reproduction begins, the faster the population will grow. For example, a multigenerational human family in which women give birth in their teens will be considerably larger than a family in which the women delay motherhood until they are in their mid- to late thirties, even when each woman gives birth to the same number of children (fig. 42.3).

A life table shows the probability of dying at a given age (table 42.1). By graphing survivorship (the proportion of individuals

TABLE 42.1 Age-Specific Life Table for the Desert Sand Scorpion, *Paruroctonus mesaensis.*

AGE (MONTHS)	PROPORTION SURVIVING TO AGE
0	1.000
1	0.849
2	0.688
4	0.612
6	0.589
8	0.567
10	0.542
12	0.500
16	0.451
20	0.419
24	0.326
36	0.044
48	0.007
60	0.001
72	0.000

Source: Data from Polis and Farley, *Ecology* 61(3): 620–629.

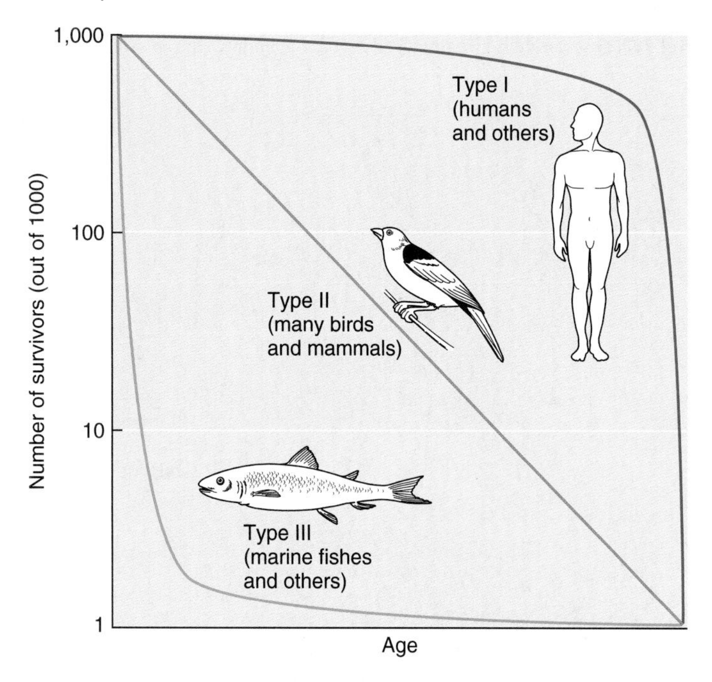

FIGURE 42.4
Survivorship Curves Show Extremes Among Species.
This graph depicts the number of survivors out of a cohort of 1,000 individuals as age increases. The scale of the y axis is logarithmic, which means that straight-line portions of the curves reflect a constant survivorship rate.

that survive) against age for various species, three patterns emerge (fig. 42.4). In Type I species, such as humans and elephants, most offspring survive to old age, when the mortality rate becomes high. Type II species, including many birds and mammals, show a constant survivorship rate throughout life. For Type III species, such as most marine fishes and invertebrates, most insects, and many plants, most offspring die at a very young age. Of course, these generalized examples do not describe all populations. Many species, including the scorpion from table 42.1, have survivorship curves that fall between the theoretical patterns.

Since the probabilities of reproducing and dying change with age, the distribution of age classes of a population, called the **age structure,** determines whether a population is growing, stable, or declining (fig. 42.5). We can consider the members of a population in three age classes: prereproductive, reproductive, and postreproductive. A population with a large fraction of prereproductive individuals, such as those of many countries in Central America and Asia, will grow as individuals enter their reproductive years, assuming they remain healthy and fertile. Conversely, a population consisting mainly of older individuals will decline as members die. This is happening in Spain, where the proportion of females able to bear young is small. Sub-Saharan African populations are sharply declining for a different age-related reason. Many middle-aged HIV-positive males are infecting young girls, who in turn infect their age-mates. Gradually, some populations are coming to consist mostly of young orphans and the elderly.

Growth Is Exponential When Resources Are Unlimited

Any population will grow if the number of individuals added exceeds the number removed. The **per capita rate of increase,** or *r*, is the difference between the birthrate and the death rate. If *r* is negative, the population shrinks. Conversely, a positive *r* means the population is growing. For example, a population in which there are 35 births and 10 deaths per 1,000 individuals is growing at a rate of 25 people per 1,000, or 2.5%.

A population in an environment with unlimited space, shelter, and other resources and that is free of disease and predation will increase by a factor of *r* during each time interval, as the following equation expresses:

$$G = rN$$

G is the number of individuals added per unit time, *r* is the per capita rate of increase, and *N* is the number of individuals at the start of the time interval. Such an increase in population size is called **exponential growth,** and when plotted over time it yields a characteristic J-shaped curve (fig. 42.6). Actual growth

FIGURE 42.5
Population Age Structures Predict the Future.
In age structure diagrams, the width of each bar is proportional to the percent of individuals in that age class. Populations are likely to grow if they have a high proportion of individuals in prereproductive age classes, assuming that the individuals are healthy and fertile. Although the Nigerian profile indicates growth, the numbers of individuals of reproductive age are falling as a result of AIDS. A stable population has roughly equal numbers of people of each age group. A declining population has most of its members in reproductive or postreproductive age classes.
Source: Data from UNAIDS/WHO Working Group on Global HIV/AIDS and STD Surveillance, Switzerland.

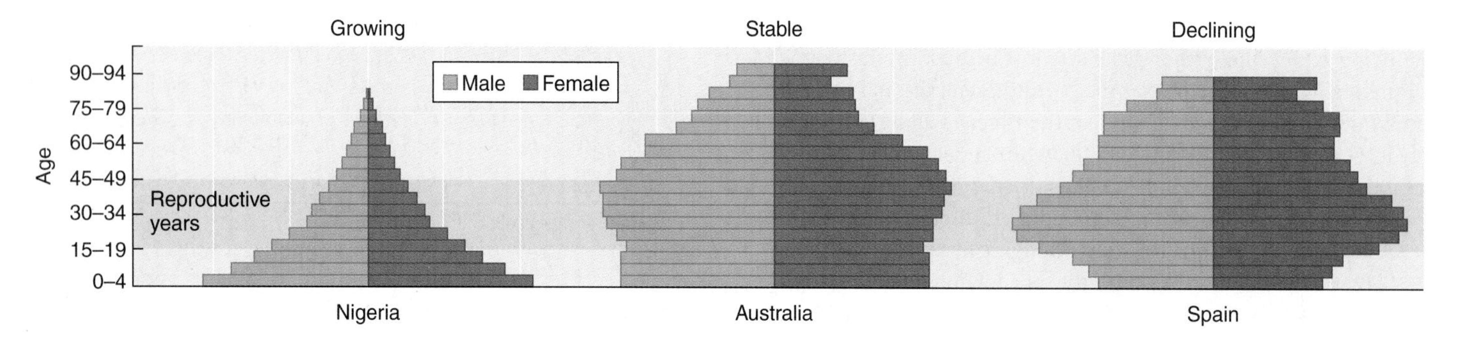

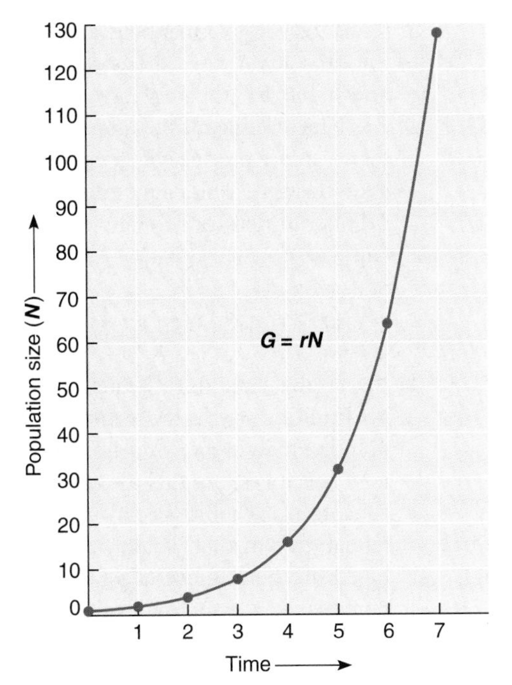

FIGURE 42.6
Exponential Population Growth. During each time interval, the population's growth rate, *G*, is the product of the per capita rate of increase, *r*, and the number of individuals in the population at the beginning of the interval, *N*. With each successive interval, *G* increases.

may be slow at first, but the increase in numbers accelerates continuously. Growth resulting from repeated doubling (1, 2, 4, 8, 16, 32 . . .) such as from binary fission in bacteria, is exponential. Species introduced to an area where they are not native may also proliferate exponentially for a time, since they often have no natural population controls.

The maximum value of *r*, called r_{max} (or the intrinsic rate of increase), occurs in populations that are unrestricted by the environment, so that both reproduction and survival are at their highest possible levels. Under those ideal conditions, a population is growing at the maximum possible rate and fulfilling its **biotic potential.** The growth potential for even the slowest-breeding species is incredibly high, but seldom attained. Although the gestation period for elephants is 22 months, if elephants met their reproductive potential, a single pair could leave 19 million descendants in 750 years! The biotic potential for rapidly breeding organisms is even more astounding. Under ideal conditions, a single *E. coli* could produce more than 500,000,000,000,000,000 offspring in one day! Human biotic potential is also impressive. Imagine a couple who starts a family as soon as physiologically possible and continues having babies as frequently as possible until menopause. If all their children reproduced in the same manner, within five generations (about a century), the couple would have more than half a million descendants.

42.2 MASTERING CONCEPTS

1. How do fecundity, survivorship, and age structure influence population growth?
2. What is the per capita rate of increase?
3. What conditions support exponential population growth?
4. How can a population express its full biotic potential?

OLC

42.3 Factors That Regulate Population Size

Environmental resistance prevents unlimited population growth, and may or may not depend upon population density. The carrying capacity, which may fluctuate over time, is the maximum number of individuals that can be supported indefinitely in a particular environment. Populations may exceed the carrying capacity and crash, or remain near the carrying capacity.

Exponential growth may continue for a short time, but it cannot continue indefinitely because some resource will eventually be depleted. Factors that check population growth by reducing reproduction and immigration or increasing mortality and emigration are termed **environmental resistance.** These factors include shortage of food or other resources, predation, disease, and parasitism. Environmental resistance also includes natural or human-caused environmental disasters, pollution that impairs health or ability to reproduce, and any other factor that reduces population size.

Population Growth Eventually Slows

The fate of reindeer on St. Matthew Island in the Bering Sea, south of Alaska, illustrates how environmental resistance operates (fig. 42.7). In 1944, 29 deer were introduced to the island, which was free of large predators and provided optimal reindeer habitat. By the summer of 1963, the population had exploded to about 6,000 animals, but food was becoming scarce. In the following cold and snowy winter, all but 42 animals starved. The density of reindeer just prior to the population crash was 47 animals per square mile—more than three times the estimated capacity of arctic winter range to support stable populations of reindeer.

FIGURE 42.7
A Population Explosion and Crash.
In 1944, reindeer were introduced to an optimal habitat on St. Matthew Island in the Bering Sea. The population grew exponentially until the winter of 1963–64, when severe weather and a depleted food supply led to mass starvation.
Source: David R. Klein, *Journal of Wildlife Management*, vol. 32, no. 2., April 1968, p. 352.

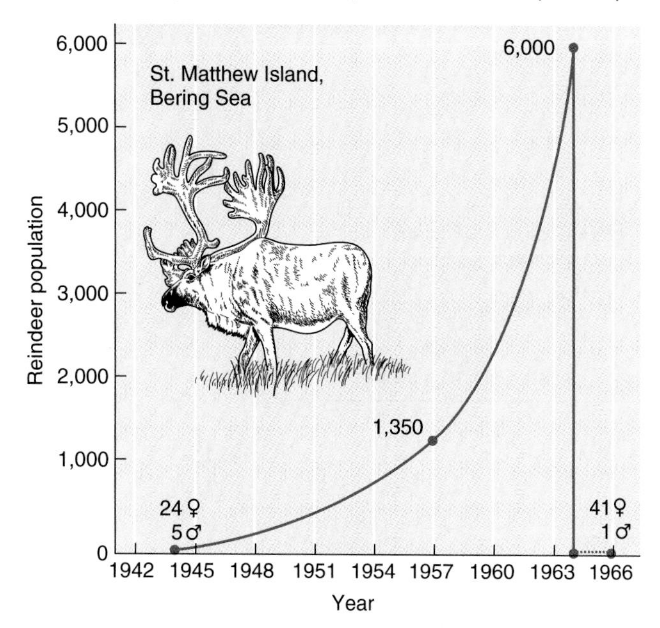

The theoretical maximum number of individuals that an environment can support for an indefinite time period is its **carrying capacity.** A population that is exactly at its carrying capacity has a growth rate of zero, as its birthrate is equal to its death rate. In reality, however, the carrying capacity is not fixed, because the features of the environment that determine population size vary over time, and populations differ in how they respond as they approach the carrying capacity. At one extreme, the population tends to grow exponentially and fluctuate wildly about the carrying capacity. At the other extreme, population growth may slow and stabilize as the carrying capacity nears.

The leveling off of a population in response to environmental resistance produces a characteristic S-shaped curve, or **logistic growth curve.** This curve follows a simple equation that gives the number of new individuals added after a particular time interval (fig. 42.8):

$$G = r_{max} N \left(\frac{K - N}{K} \right)$$

In this equation, G equals the number of individuals added per unit time, r_{max} is the intrinsic rate of increase, N is the number of individuals in the population at a given time, and K is the carrying capacity. The right side of the equation states that growth is equal to $r_{\text{max}}N$ multiplied by $(K - N)/K$. The term $r_{\text{max}}N$ is the potential growth that would occur if resources were not limiting, based on the number, N, of individuals present and the growth potential, r_{max}, associated with each. $(K - N)/K$ applies a "degree of realization" to potential growth. This term simulates the increasing intensity of environmental resistance as the population approaches carrying capacity. When N is very small relative to K, this expression yields a numerical value near 1; most of the potential growth is thus realized. When N is near K, the value of the expression drops to near zero; most of the potential growth is not realized. Thus, the logistic equation can be stated:

$$\text{realized growth} = \begin{matrix}\text{potential}\\ \text{growth}\end{matrix} \times \begin{matrix}\text{degree of}\\ \text{realization.}\end{matrix}$$

Crowding, Competition, Predation, and Pathogens Limit Population Size

Environmental resistance sometimes depends on the size of a population. **Density-dependent factors** are conditions whose growth-limiting effects increase as a local population grows.

FIGURE 42.8
Logistic Growth.
Populations of red deer on the Scottish island of Rum can grow quite large when hunting is not allowed. When food becomes scarce, pregnant does become malnourished, and fetuses die. Population growth slows as it approaches the carrying capacity, K.
Source: Data from Andrew Cockburn, "Deer Destiny Determined by Density," in *Nature,* **399**; 407, June 3, 1999.

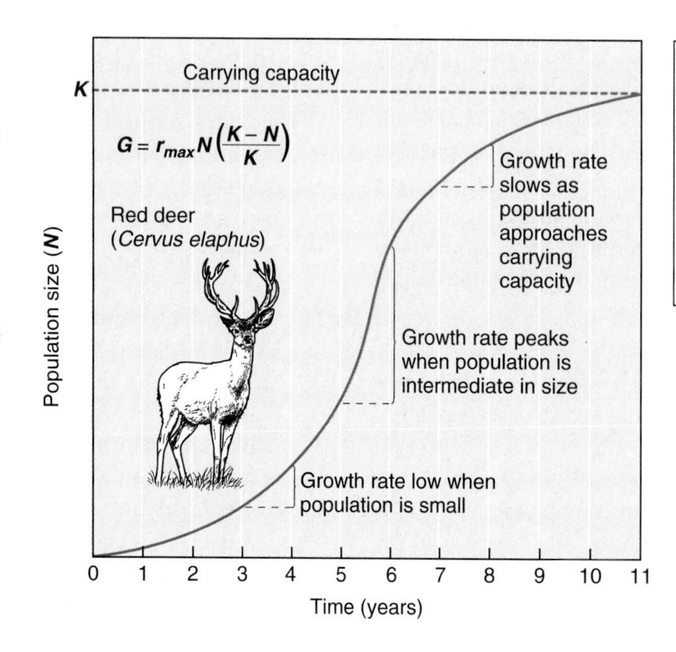

Density-dependent factors include crowding, competition, predation, and infectious disease. In response to crowding, for example, some animals cease mating, neglect their young, and become aggressive. Physiological responses to crowding include increased rate of spontaneous abortion, delayed maturation, and hormonal changes. Such responses ultimately slow population growth. Infectious disease is also considered to be a density-dependent factor, because crowding facilitates the spread of disease.

Competition for space and food is a common density-dependent factor that affects population growth. When many individuals share limited food, none of them may eat enough to be able to reproduce, and population growth slows. For example, if several hundred fruit fly eggs hatch on a very small apple already riddled with larvae, they soon deplete the food supply. Larvae that do not eat enough starve to death or develop into small adults too weak to reproduce. Members of a population may also compete for a resource indirectly. Animals may compete for social dominance or possession of a territory, factors that guarantee the winners an adequate supply of a limited resource. The losers get less of the resource or none at all. Population growth slows because losers of the competition breed less successfully than the winners. Since multiple species often share a habitat, competition between species may also limit population growth, as described in chapter 43.

Predation is another density-dependent factor that limits the population growth of both predator and prey. Effects on the prey population are obvious—their numbers diminish as they are eaten. Predators also influence prey populations by eliminating the weakest individuals, who are more easily captured. Prey populations are often maintained by high reproduction rates, which help compensate for the loss to predation. However, the chapter opening essay describes how populations of the resident species on Guam were decimated after the accidental introduction of the predatory brown tree snake, which has itself flourished in the absence of natural predators.

Removing predators may enable prey populations to swell and expand into new ranges. For example, people in Canada and the northern United States are reporting more frequent encounters with moose (fig. 42.9). In the nineteenth century, many moose died from hunting and conversion of their forest habitat to farms. By the end of World War II, about 350,000 moose lived in Canada and the United States. Moose numbers have recently begun to recover due to a combination of factors—stricter hunting regulations, a return of forest, and fewer wolves and grizzly bears, which are their primary predators. Today the animals top the 30,000 mark in Maine alone, and the metapopulation on the continent exceeds 1 million animals.

From the predator's perspective, inability to find food may mean death, and reproduction among predators is intimately tied to the density of their prey population. Many animals that cannot find sufficient food become temporarily infertile or have small litters. A hungry female bobcat has only one or two kittens, instead of the usual three, and those few may not survive their first 10 months if the mother cannot find enough food for them. Chapter 43 further explores predator-prey interactions in communities.

FIGURE 42.9
Expanding Moose Populations.
A decline in predator populations and changing human activities account for the present resurgence among moose populations. Their range creeps ever southward.

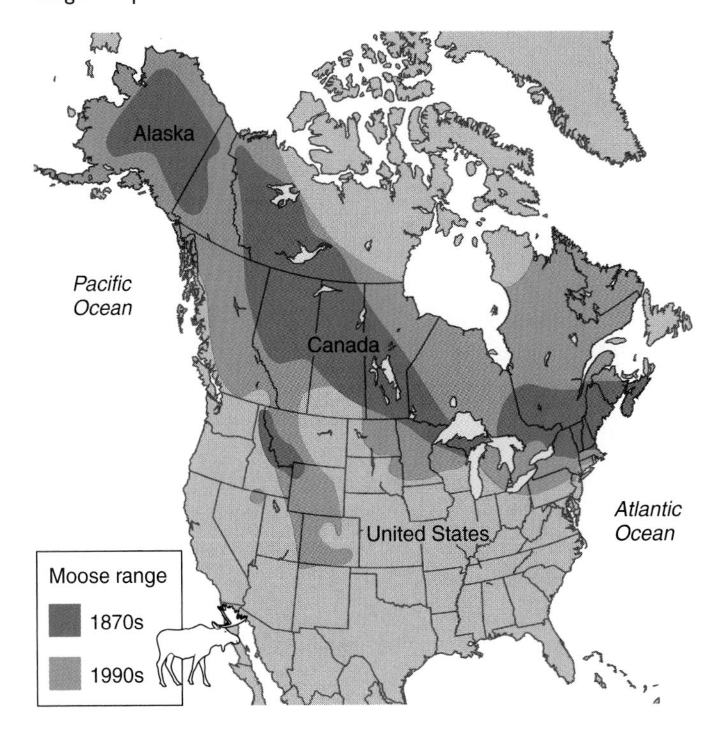

Environmental Disasters Kill Irrespective of Population Size

Density-independent factors exert effects that are unrelated to population density. A severe cold snap, for example, might kill a certain percentage of fishes in two populations, even if the populations differ in size. Natural disasters such as earthquakes and severe weather conditions are typical density-independent factors. Consequences of human-caused environmental disasters are also independent of population density. Oil spills, for example, kill whatever life becomes entrapped.

A spectacular density-independent event occurred on May 18, 1980, when the north face of Mount St. Helens in Washington State began to bulge and rumble. Underground, water superheated by molten rocks was turning to steam. At 8:32 A.M., Mount St. Helens erupted. The 900°F (477°C) blast pulverized rocks and trees, killing nearly everything within a 6-mile (9.7-kilometer) radius and triggering earthquakes and avalanches. A 25-foot- (nearly 8-meter-) high wave of mud and molten rock, joined by melting snow and ice in the higher elevations, buried everything in its path. A forest of five-century-old fir trees disappeared instantly. The animal death toll included some 5,200 elk, 6,000 black-tailed deer, 200 black bears, 11,000 hares, 15 mountain lions, 300 bobcats, 1,400 coyotes, 27,000 grouse, 11 million fishes, and uncountable numbers of insects. The recovery of biotic communities following such disasters occurs by a sequence of events called succession, as chapter 43 explains.

A density-independent event that is disastrous for some species might benefit others. In 1993, a flooding Mississippi River sent deer, raccoons, rodents, skunks, and opossums scurrying to higher ground, but provided rich new shallow-water habitats for microscopic algae and plankton, and the fishes and insects that feed on them. Wading birds ate fishes trapped in the shallower floodwaters and mud. Floodwaters carried wild gourds and zebra mussels to new habitats. Duck populations soared as the ravaged riverbanks provided more nesting sites and places to hide from predators. Other birds weren't as lucky. The least tern, for example, already endangered, became even more so as the waters washed away its nests, eggs, and fledglings.

Sometimes it is difficult to label a population decline as due to density-dependent or density-independent factors. This is the case for the amphibian populations that are being decimated in many places, partly because of a fungal infection called cutaneous chytridiomycosis, as the opening essay to chapter 23 explains. A mutation, perhaps caused by increased exposure to ultraviolet radiation due to thinning of the ozone layer, may have rendered the chytrid, whose relatives attack plants and insects, capable of infecting amphibians. Ozone depletion is a density-independent factor, but infectious disease is considered to be density-dependent, since crowding fosters its spread.

"Boom and Bust" Cycles

In populations with high reproductive rates, exponential growth may overshoot the carrying capacity. Then the population drastically drops or crashes. A population that repeatedly and regularly increases (booms) and decreases (busts) in size over a given period of time exhibits a **boom and bust cycle.** Food supply, predation, and changes in a population's age structure influence boom and bust cycles. These cycles can occur when density-dependent factors are delayed relative to the reproductive rate, which allows population size to reach a level that cannot continue.

Snowshoe hare populations in Canada and Alaska follow a boom and bust cycle, with their numbers peaking every 8 to 11 years. When a snowshoe hare population is at its most dense, up to 10 hares may live on a hectare of land. In the leanest years, one hare may be the only one of its species on 80 hectares. The population densities of animals that eat hares fluctuate along with the changing hare population.

Ecologists have tracked boom and bust cycles among Soay sheep on the island of Hirta, near Scotland (fig. 42.10). In 1932, the Marquis of Bute brought 107 feral (wild) sheep to the island. By 1942, their numbers had grown to 500, and by 1952, to 1,114 animals. A few summers later, when the population exceeded 1,400, the tiny island became littered with sheep corpses. More than a third of the animals had starved to death. Curiously, populations of sheep on other islands, and of other large animals on Hirta, did not decline.

The nature of the Soay sheep's reproductive cycle explained the boom and bust cycle. A combination of factors enables the population to grow much too quickly for its own good. Females can become pregnant before they are a year old, and 20% of them deliver twins. Infant mortality is low. Babies stop nursing in June, when food is abundant, which gives their mothers sufficient time to gain enough weight to conceive in the fall. The Soay sheep also lack predators and cannot leave the island. Although many sheep starve, the population recovers because reproduction rates are high.

FIGURE 42.10
Boom and Bust Cycles.
Soay sheep on the island of Hirta graze in rocky sheltered areas. As their population explodes, due to early and frequent pregnancies, the sheep eat all the vegetation. Every 4 years, many sheep starve, but the population recovers rapidly.

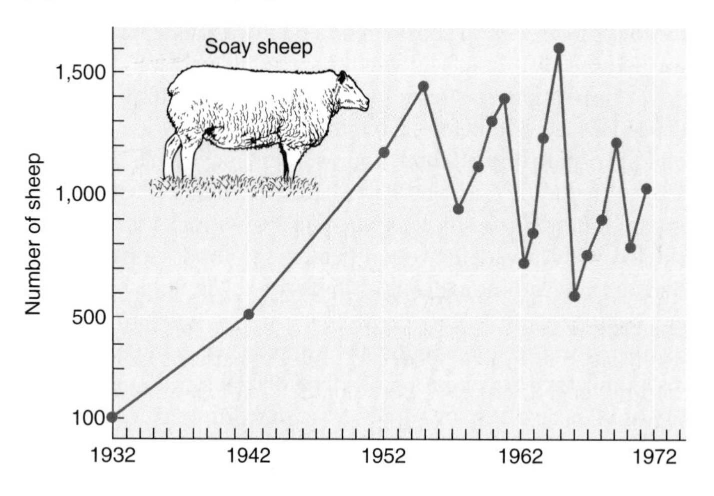

Influences of Natural Selection on Population Growth

Species vary in their life history characteristics. Species with Type III survivorship curves, for example, tend to produce abundant offspring, each of which has a very low probability of surviving to reproduce (see fig. 42.4). Other species, such as those with Type I survivorship curves, produce far fewer young, but parental investment is high so most offspring live and reproduce. Natural selection has adapted the life histories of these species to reflect the trade-off between reproductive output and parental investment. The differences between these types of life histories reflect different population growth rates.

At one extreme are ***r*-selected** species, in which individuals tend to be short-lived, reproduce at an early age, and have many offspring that receive little care. Weeds, crop pests such as insects, and many other invertebrates live in strongly *r*-selected populations. These species are usually excellent colonizers because of their high reproductive rates and effective dispersal strategies.

***K*-selected** species are at the other extreme. Density-dependent factors regulate *K*-selected populations close to the carrying capacity (recall that *K* stands for carrying capacity). Individuals tend to be long-lived, to be late-maturing, and to produce a small number of offspring that receive extended parental care. Many birds and large mammals live in *K*-selected populations. These species tend to be good competitors that exploit relatively stable environments. Whereas many insects that consume crops are *r*-strategists, many parasites and predators are *K*-selected. Perhaps this difference in life history is one reason that the introduction of exotic natural enemies rarely works for control of introduced pests. Table 42.2 compares the characteristics of *K*-selected and *r*-selected species.

TABLE 42.2 Characteristics of *r*-Selected and *K*-Selected Species

r-SELECTION	*K*-SELECTION
Small individuals	Large individuals
Short life span	Long life span
Fast to mature	Slow to mature
Many offspring	Few offspring
Little or no care of offspring	Extensive care of offspring

For decades, ecologists thought that logistic growth and close regulation of populations to a stable carrying capacity was the general rule in nature. Long-term studies of populations of many organisms, however, reveal that such patterns are the exception rather than the rule. Even populations of large animals that show *K*-selected life histories fluctuate greatly in response to changes in their environments that range from gradual and progressive to sudden and catastrophic.

42.3 MASTERING CONCEPTS

1. How does environmental resistance keep populations from growing continuously?
2. What conditions are necessary for logistic (S-shaped) growth to occur?
3. Distinguish between density-dependent and density-independent effects on population size.
4. What is a boom and bust cycle?
5. Distinguish between *K*-selected and *r*-selected species.

OLC

42.4 Human Population Growth

Human population growth surged following acquisition of tool use, the rise of agriculture, and the Industrial Revolution. Today the human population growth rate varies greatly among nations. Human population biology is complex because it reflects not only ecological principles but also diverse cultural, social, and economic influences.

The principles of population ecology described above apply generally to all species, including humans. From the appearance of modern humankind some 40,000 years ago until 1850, our population grew relatively slowly, reaching 1 billion. Yet only 80 years later, by 1930, we doubled our number to 2 billion. In another 30 years, by 1960, 3 billion humans populated Earth. The human population hit 4 billion in 1974, 5 billion in 1987, and 6 billion in October 1999 (fig. 42.11). It is expected to reach nearly 9 billion by 2050. Although the human population continues to grow, the rate of increase may be slowing. In the late 1980s, nearly 90 million people were added each year. By the end of the 1990s, the rate had slowed to 78 million people per year, and by 2050, the United Nations estimates that the human population will grow by 33 million people per year.

Part of the reason for the global slowdown in the overall rate of human population growth is a decline in average fertility. The average number of children has declined from 6 per female in the 1960s to 3 today, partly because of increasing access to contraceptives in developing countries. Still, some populations of Latin America, Asia, and Africa will continue to grow for many years, as each generation more than replaces itself. Today, 85% of human population growth is in developing nations.

Fertility, mortality, and age structure differ worldwide to produce a mosaic of regional human subpopulations—some growing, some stable, and some actually declining (fig. 42.12). The overall growth rate of the human population is 1.3% per year, but it varies from 2.6% per year in the least developed countries to 0.3% per year (or less) in the most developed countries. Unequal growth worldwide has led to sometimes opposite governmental attempts to control reproduction. In Thailand, for example, population growth has fallen in the past 20 years because of increasing availability of contraceptives. China has also controlled runaway population growth, but with somewhat drastic measures, such as rewarding one-child families and revoking the first child's benefits if a second child is born. Yet the governments of France, Canada, and Japan offer financial incentives and extended new parent leaves from work to encourage citizens to have children to bolster aging populations.

Increasing life spans will contribute to the growing population. Average life expectancy is increasing in most parts of the world, thanks in part to medical advances. Life expectancy for the cohort born from 1995 to 2000 ranges from 52 in the least developed countries to 74.5 in most developed countries. In Sweden, for example, the maximum age at death increased from 101 in the 1860s to 108 in the past decade. The current growth of this population, and its increasingly older makeup, is due to several factors—higher birthrates, better survivorship, and lower death rates among those older than 70 years. In Sub-Saharan Africa, however, the AIDS epidemic has significantly affected mortality rates and erased decades of progress in increasing life expectancy. Table 42.3 shows the impact of AIDS on regional human populations.

TABLE 42.3 Estimated Adult and Child Deaths Due to HIV/AIDS from the Beginning of the Epidemic to the End of 1999.

REGION	CUMULATIVE DEATHS
Australia and New Zealand	8,000
Eastern Europe and central Asia	17,000
East Asia and Pacific	40,000
North Africa and Middle East	70,000
Caribbean	160,000
Western Europe	210,000
North America	450,000
Latin America	520,000
South and Southeast Asia	1,100,000
Sub-Saharan Africa	13,700,000

Source: Joint United Nations Programme on HIV/AIDS.

FIGURE 42.11

The Human Population Is Growing Exponentially.

Human population growth is on the rise, as indicated by the J-shaped curve. The most rapid population growth has occurred in the past 200 years. *Source:* Data from U.S. Census Bureau.

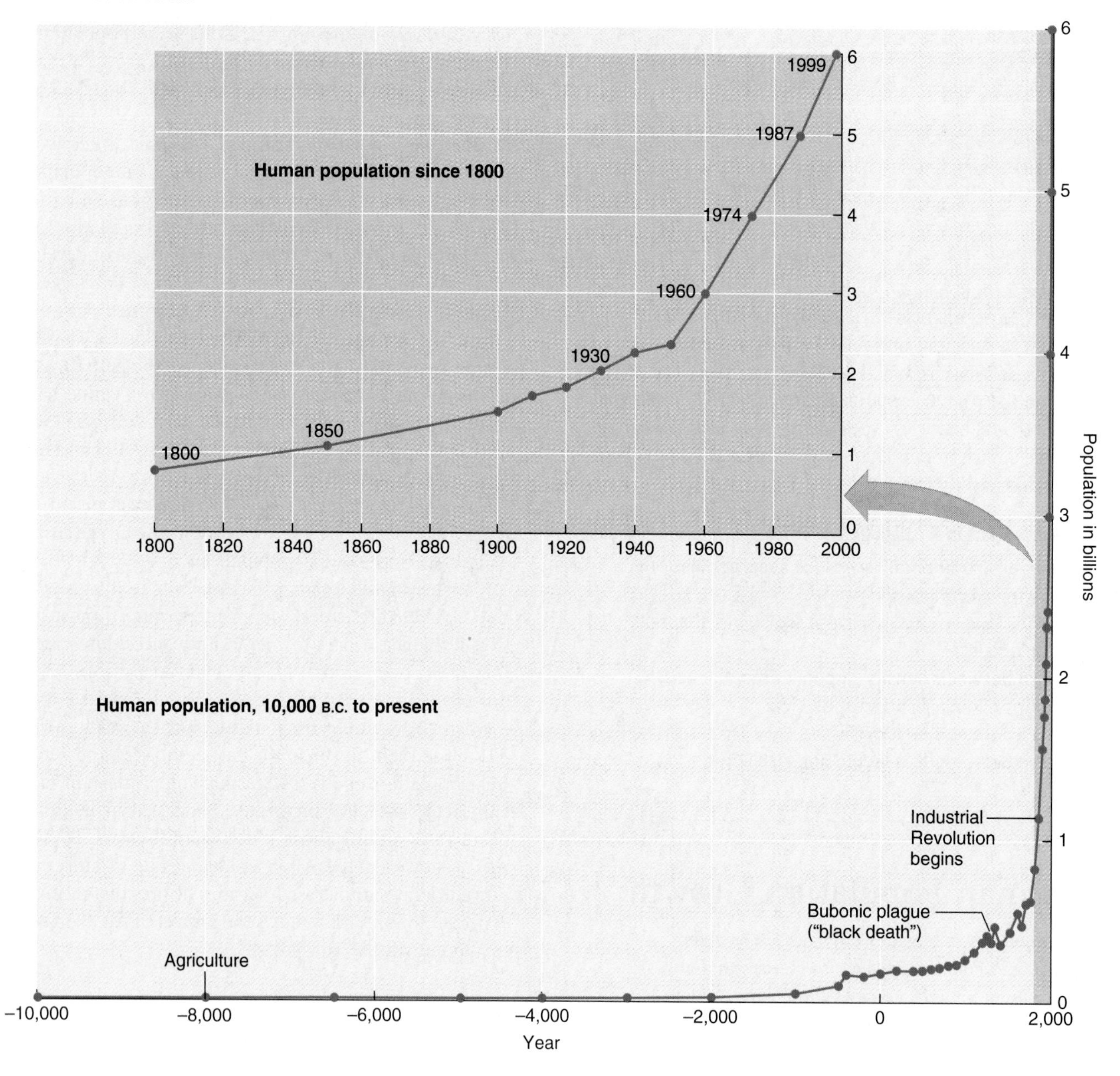

When the study of population dynamics applies to humans, fields other than biology come into play—such as politics, history, economics, anthropology, and sociology. Shifting the focus of studying populations from a global view to a family view helped reveal factors that contribute to rapid human population growth. Following a decade of studies on the structures of families in many societies, attendees at the 1994 United Nations International Conference on Population and Development, held in Cairo, structured a new hypothesis to explain the drive behind burgeoning human populations. In many cultures, they found, families are large because people want many children to help obtain increasingly scarce resources, such as wood for fuel, clean drinking water, and food. Hence, population growth, poverty, and environmental decline (which limits resources) are intricately interrelated.

The key to limiting family size, according to the United Nations conference, is to find ways to give women in developing nations options other than continual reproduction. This suggestion is based on many studies that correlate large families to women who are illiterate and/or work for no pay. Studies also show that as the percent earnings that a mother contributes to the family income rises, the number of children falls. Specific recommendations include educating women for trades, increasing availability of contraceptives, and improving health care.

FIGURE 42.12

Regional Trends in Future Population Growth.

As human population growth continues in the first half of the 21st century, differences in age structure will produce a mix of growing, stable, and declining regional populations. Note that the graphs are plotted on different scales to account for regional differences in population size.

Source: Population Division of the Department of Economic and Social Affairs of the United Nations Secretariat (1999). World Population Prospects: The 1998 Revision. Vol. II: The Sex and Age Distribution World Population. (United Nations publication, Sales No. E.99.XIII.8). The United Nations is the author of the original material.

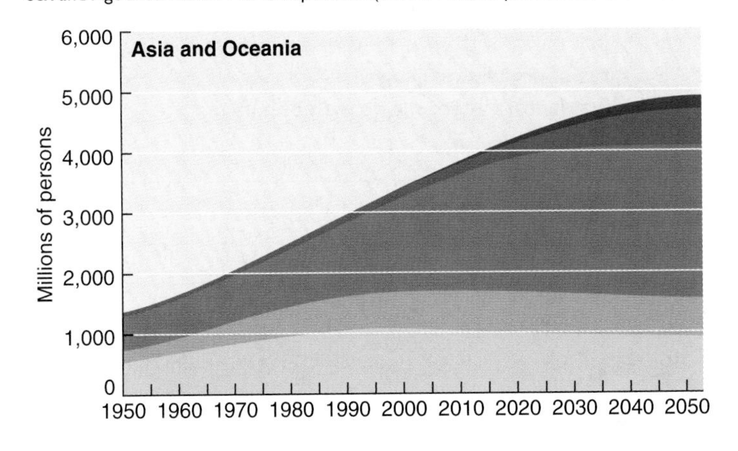

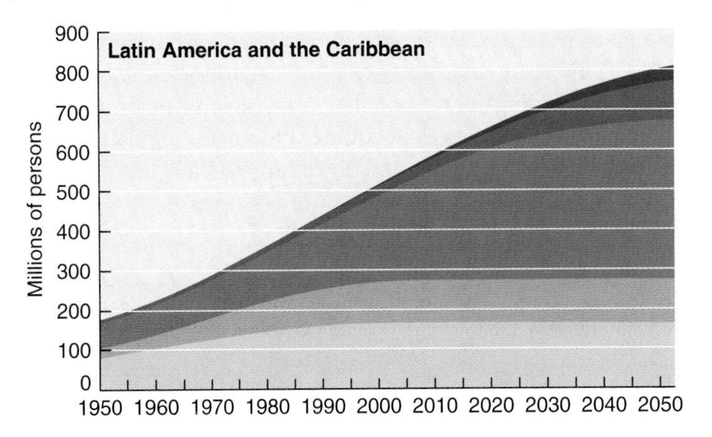

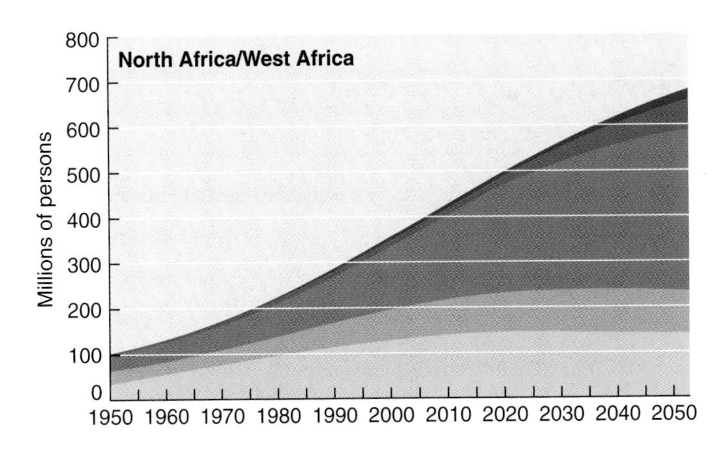

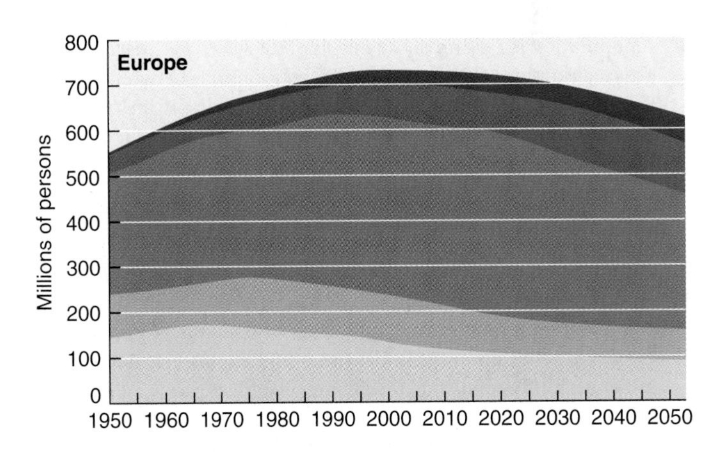

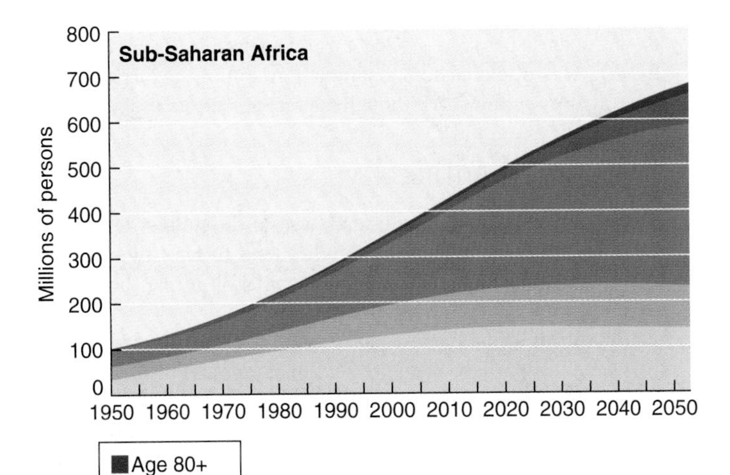

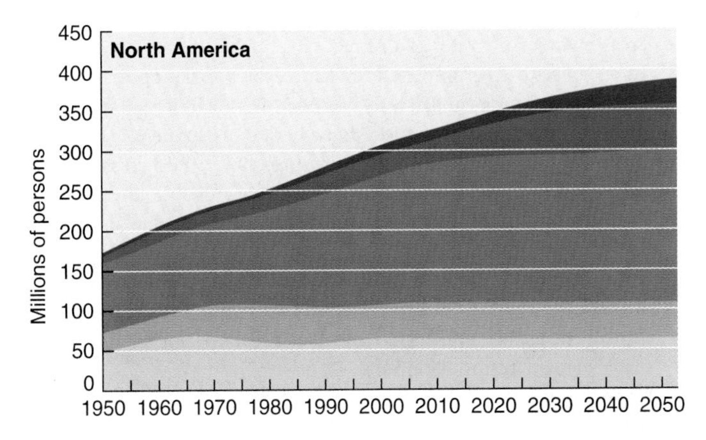

Age 80+
Age 65–79
Age 25–64
Age 15–24
Age 0–14

Birth- and death rates affect the human population, but regionally migration also affects the quality of life. For example, migration from rural to urban areas lacking adequate economic opportunities is producing huge slums in several "megacities" (cities with more than 10 million inhabitants each). Worldwide, increasing numbers of people will mean greater pressure on land, water, and air resources. Deforestation, increased fuel consumption, species extinctions, and other environmental problems resulting from the expanding human population are explored in chapter 45.

Despite the influences of culture on human population dynamics, our population ups and downs are basically like those of other species. Changing numbers reflect birth- and death rates, as well as diverse environmental influences. The next two chapters explore in greater detail the interactions of organisms and the living and nonliving environment—ecology.

42.4 MASTERING CONCEPTS

1. What parts of the world have the highest and lowest rates of population growth?
2. What factors affect the growth of human populations?

OLC

Chapter Summary

42.1 A Population Is All Individuals of a Single Species in a Particular Place

1. A **population** is a group of organisms of the same species living in a geographic region. **Ecology** considers relationships between organisms and their living and nonliving environments. It includes the relations of individuals in populations and their interactions with individuals of other species in **communities.**
2. **Demographic** characteristics such as **population density** and **population dispersion** are static measures of a population.

42.2 Several Factors Affect Population Growth

3. A population grows when more individuals are added through birth or immigration than leave due to death or emigration. Population growth depends upon the initial size of the population, how many individuals are added and at what rate, the age at which individuals begin to reproduce, and the **age structure** of the population.
4. Unrestrained growth is **exponential** and produces a J-shaped curve.

42.3 Factors That Regulate Population Size

5. **Environmental resistance** counters unrestrained growth by increasing mortality rates and/or reducing birthrates.
6. In response to environmental resistance, the population may stabilize at its **carrying capacity,** the number of individuals an environment can indefinitely support. After a period of exponential growth, a population may overshoot the carrying capacity and crash. Alternatively, **density-dependent factors** may slow growth so the population size levels off at the carrying capacity, producing an S-shaped **logistic growth** curve.
7. Environmental resistance includes **density-independent factors,** which kill a fraction of the population regardless of its size. Density-dependent factors, which have a greater effect on large populations, can also regulate population size.
8. Populations that regularly increase and decrease in size have a **boom and bust cycle.**
9. Different life histories reflect a trade-off between the number of offspring and parental investment. ***K*-selected** species invest heavily in rearing relatively few young. In contrast, ***r*-selected** species produce many offspring but do not expend much energy on each.

42.4 Human Population Growth

10. Human population growth has not been steady and occurs unevenly in different parts of the world. Global human population growth will level off towards the end of the 21st century.

Testing Your Knowledge

1. State two descriptive features of population structure.
2. Which resources are necessary for a population to grow?
3. How do the fecundity, survivorship, and age structure of a population affect its growth?
4. Define the following terms:
   a. per capita rate of increase
   b. environmental resistance
   c. carrying capacity
5. What are the differences between the conditions that result in exponential versus logistic population growth?
6. How do density-independent and density-dependent factors affect population growth?
7. Why don't predator populations usually completely decimate prey populations?
8. Distinguish between *K*-selection and *r*-selection.
9. Why will the global human population continue to increase for several more decades even though the rate of increase has peaked?

Thinking Scientifically

1. In some nations, well-educated, well-to-do families tend to have fewer children, and poorer families tend to have more children. How might such a trend affect a population in the short and the long term?
2. Because of China's "one-child" policy, many children growing up in China today will have no brothers or sisters, and their children will have few aunts, uncles, or cousins. The alternative, statistics predict, is mass starvation in the 21st century. Do you think China's "one-child" policy has been an effective way to control population growth? Why or why not?
3. Cite three environmental upheavals that may have had a density-independent impact on wildlife populations.

4. In Pakistan, 43% of the population is under age 15 and 4% is over 65. In the United Kingdom, 19% of the people are under 15 and 15% are over 65. Which population will increase faster in the future? Why?
5. Describe a method to estimate the number of elk in a particular area.

References and Resources

deWaal, Frans B. M., et al. May 2000. Coping with crowding. *Scientific American,* vol. 282, p. 76. Primates can cope with crowding; rodents cannot.

Ezzell, Carol. May 2000. Care for a dying continent. *Scientific American,* vol. 282, p. 96. The human population in Africa will decline in the future, because of AIDS.

Houlahan, Jeff E., et al. April 13, 2000. Quantitative evidence for global amphibian population declines. *Nature,* vol. 404, p. 752. A large-scale study supports anecdotal evidence of declining amphibian populations.

Kaul, Rupert, et al. October 2000. AIDS in Africa: a disaster no longer waiting to happen. *Nature Immunology,* vol. 1, p. 267. In Sub-Saharan Africa, the midsection of the population is vanishing.

Kluger, Jeffrey. Spring 2000. The big crunch. *Time Special Edition.* Every minute, 247 people are born.

Wilmoth, J. R., et al. September 29, 2000. Increase of maximum life span in Sweden, 1861–1999. *Science,* vol. 289, p. 2366. Improved health care can, over several years, change the age structure of a population.

The *LIFE* Online Learning Center provides additional resources and tools for studying this chapter.
www.mhhe.com/life

CHAPTER 43

Communities and Ecosystems

Collecting Black Smokers from Deep-Sea Hydrothermal Vents

In 1977, the submersible *Alvin* took a group of geologists to the sea bottom near the Galápagos Islands, directly above cracks in Earth's crust called deep-sea hydrothermal vents. In these areas of intense pressure and temperature extremes where Earth's crust is born, the researchers were astounded to see abundant life. Tubelike worms waved, anemones clung, crabs crawled and shrimp grazed—most of the animals unknown to marine biologists. Figure 24.20 discusses one such animal, the Pompeii worm, which thrives at the extreme temperature gradients of these underwater ecosystems. Since that time more vent communities were discovered along the ridge of undersea volcanoes that dot the floor of the Pacific Ocean. Deep-sea hydrothermal vents are by no means rare—scientists just didn't think to look on the bottom of the ocean for living communities!

In these areas of intense pressure and temperature extremes where Earth's crust is born, the researchers were astounded to see abundant life.

Too far beneath the surface to use sunlight, it was clear that the producers in these vibrant communities were tapping energy from the chemicals in and of Earth, predominantly hydrogen sulfide. In the early 1980s, researchers discovered chimneylike structures extending from some hydrothermal vents. Black, intensely hot, mineral-laden water shot through the chimneys, earning their descriptive name of "black smokers." The chimneys arise as molten minerals from Earth's mantle crystallize in the cold water.

In 1998, researchers from the University of Washington in Seattle and the American Museum of Natural History in New York added another chapter to the brief history of black smokers—they collected some! Using maps of the sea bottom assembled from sonar and photographic data amassed a year earlier, the team selected chimneys small enough to be removed from the Juan de Fuca Ridge, 180 miles (290 kilometers) off the coast of Washington and British Columbia and

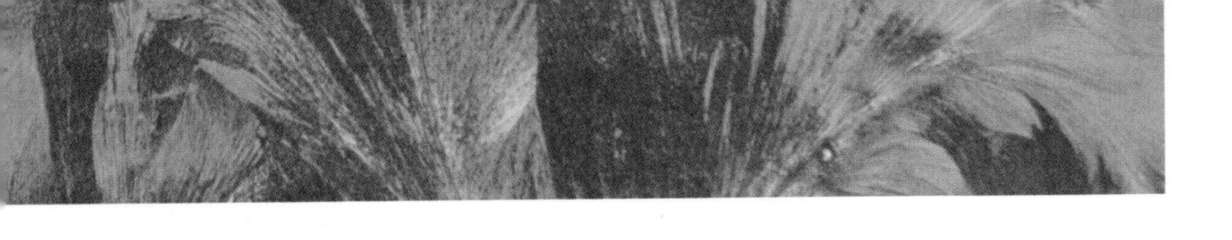

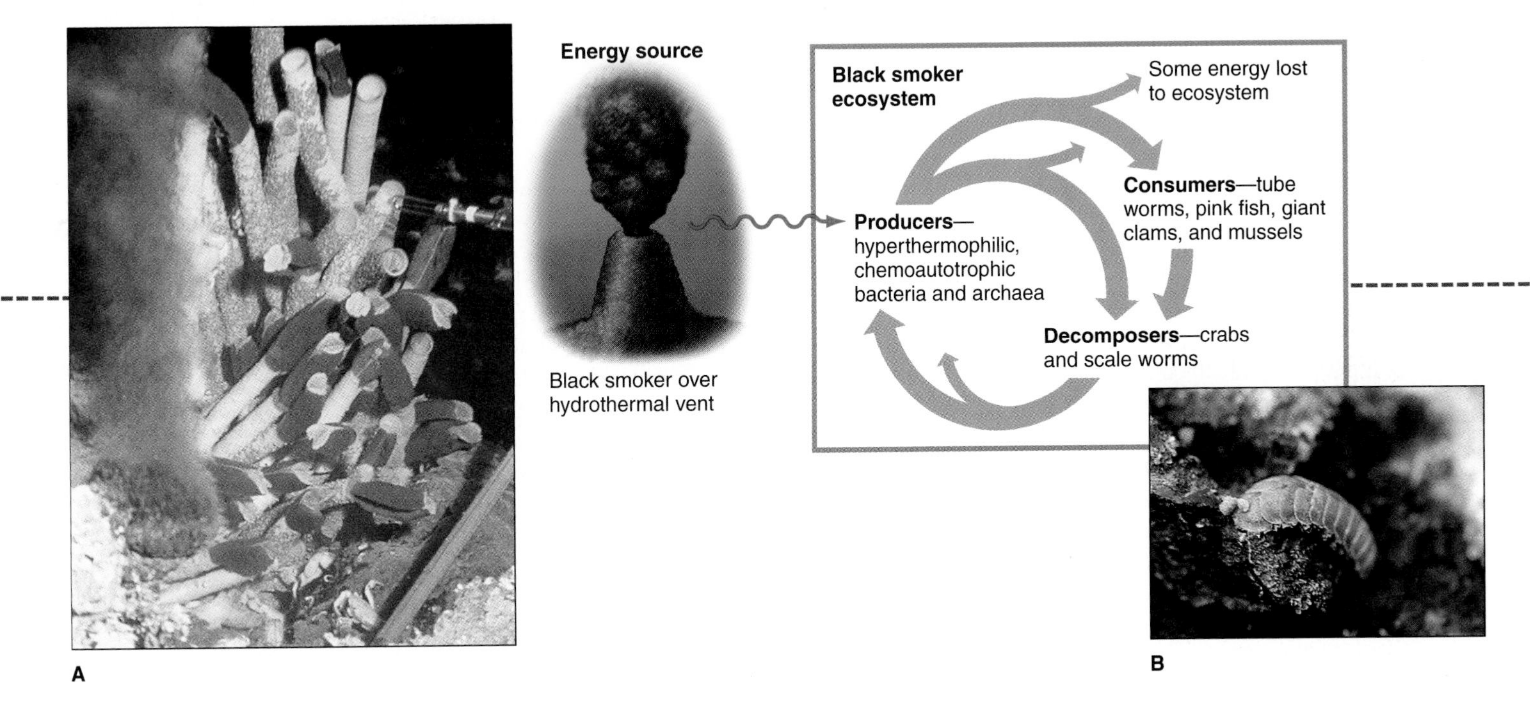

Some Life Thrives Where the Sun Never Shines.
The discovery that life thrives in and around deep-sea hydrothermal vents and the black smokers that rise from them dispelled our human-centric ideas of just what constitutes a habitat. (**A**) Energy in the bonds of inorganic chemicals spewing from Earth's interior supports this ecosystem, in which tube worms are abundant consumers. The "smoke" emerging from the hydrothermal vent is a mixture of minerals. (**B**) Scale worms are among the decomposers in the hydrothermal vent community.

more than 7,000 feet (2,135 meters) beneath the surface. Then they used a remotely operated Canadian vehicle to break off the chimneys. An 8,000-foot (2,440-meter) line anchored to a Canadian coast guard ship enabled the researchers to drag in the dismembered chimneys, layers of life still clinging to them. Each was 5 to 7 feet (1.5 to 2.1 meters) tall, and weighed 1,200 to 1,400 pounds (545 to 636 kilograms). When hauled from the water, each chimney bore a fringe of tube worms, and was still very hot to the touch, despite an hour of travel through ice-cold seawater.

The team collected four chimneys. The first, Phang, was devoid of life—hot water no longer rose through it on the seafloor, and so its living community had died. But three others, named Gwenen, Roane, and Finn (from Irish mythology), now reside in a glass cubicle at the museum. The chimneys, which resemble dripping candles, glisten with black and yellow crystals of zinc, copper, and iron sulfide. Remains of their former fringe of resident tube worms stud their surfaces.

In the next few years, biologists will finally be able to piece together the residents of these undersea ecosystems. Tests of lipid content and DNA sequences have already revealed that the primary producers are chemoautotrophic, hyperthermophilic bacteria and archaea. From 10,000 to 100,000 of these microorganisms inhabit each gram of rock. Consumers include a variety of worms and clams that harbor symbiotic, chemosynthetic prokaryotes. Other organisms prey on the tube worms, and crabs are scavengers and decomposers. So far researchers have identified several hundred of the thousands of types of microorganisms present, but recognized only about 5% of the 300 or so species that are visible to our unaided eyes.

Deep-sea hydrothermal vents and black smokers may hold clues to conditions at the time when life began, as figure 18.4 depicts. More philosophically, the discovery of these ecosystems has helped to dispel the long-held idea that life is restricted to a thin layer of Earth's surface between rock and atmosphere.

Disturbing an ecosystem is a big decision for conservation-minded biologists. Fortunately, the ocean floor has barely noticed the taking of the four black smokers. Within days of the collections, the chimneys had grown anew, up to 6 feet (1.8 meters). Soon, they will again house untold numbers, and varieties, of life.

43.1 A Community Includes All Life in an Area

A species' habitat is where it lives; its niche includes the resources it requires and all other factors that limit its distribution and abundance. Interspecific competition may lead to specialized adaptations that allow a species to exploit a subset of the resources in its habitat. Other interactions among species include symbiosis and predation.

All of the populations in a given area constitute a **biotic community.** One or more communities plus the nonliving, or abiotic, environment constitute an **ecosystem.** A downed Douglas fir, for example, is a bustling biotic community (fig. 43.1). Some resident species prefer dry, shaded branches, others the soggy, rotting roots, and still others, the deepest heartwood. Soon after the tree falls, certain insects invade the inner bark, and then other species enter the sapwood. Still later, different species attack the heartwood. Deep within this decomposing tree, bark beetles whittle a labyrinth-like "egg gallery" where they rear the next generation; the mother beetle guards the entrance. This chamber is also home to more than 100 other types of beetles. Tiny wasps drill in from the outside and lay eggs on the bark beetle larvae. When wasp larvae hatch, they eat the beetle larvae. Scavenging beetles eat any dead bark beetle larvae left over from the wasp's meal. Other insects eat the fungus that grows in indentations in the bark beetles' exoskeletons.

FIGURE 43.1
A Fallen Tree Holds a Thriving Community of Organisms.
Far from lifeless, rotting wood houses a changing community of insects, worms, fungi, and microorganisms.

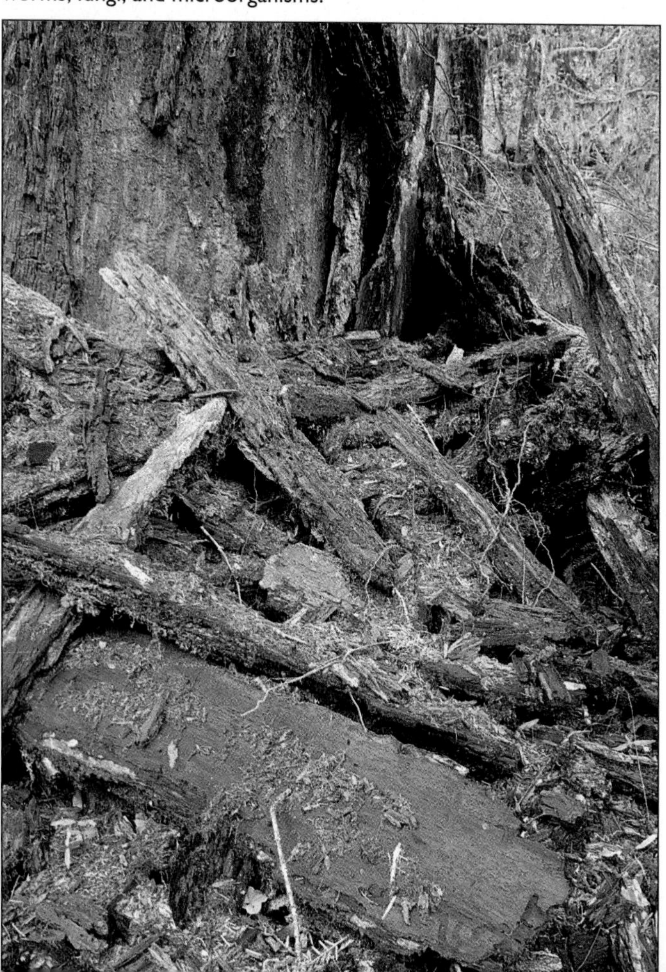

Communities can range greatly in size. Right Whale Bay in the South Georgia region of Antarctica is a community of fur seals, king penguins, macaroni penguins, Dominican gulls, elephant seals, and all the marine organisms that these mammals eat. Yet a dead and decaying squirrel riddled with insects, worms, and microorganisms is also a biotic community. The boundaries that delimit a community or ecosystem are arbitrary, just as they are for populations.

The planet's land and waters house a patchwork of different communities. Biodiversity is highest near the equator, possibly due to the favorable climate, and declines as the climate fluctuates more toward the poles. In general, as described in chapter 44, similar types of biological communities appear at corresponding latitudes because they have similar climates. The steamy heat of South American and African rain forests is likely to be home to species with similar adaptations. Differences among communities at the same latitude in different parts of the world reflect regional climatic characteristics or geographical influences.

Interspecific Competition Prevents Species from Occupying the Same Niche

Each species in a community has a characteristic home and way of life. Recall from chapter 42 that a **habitat** is the physical place where members of a population typically live, such as the bottom of a river or a rain forest canopy. Many different plants, fungi, animals, and microorganisms live together in major habitats, yet species that are very similar often do not live in the same places. To understand the reasons for this, it helps to distinguish between a species' habitat and the way it makes its living, called its ecological **niche.** The niche is the total of all the resources a species exploits for its survival, growth, and reproduction. It also includes the range of abiotic environmental conditions (such as salinity, temperature, and water availability) where the species lives. As part of the resources that comprise a species' niche, the habitat is actually a subset of the niche.

Two terms describe resource use in a niche. A species' **fundamental niche** includes all the resources it could possibly use. If two or more species with similar but not identical fundamental niches live in the same area, the **interspecific competition** between them may restrict each species to only some of the resources available. Or, the species may be forced to use resources in different habitats. Therefore, due to competition, a species' **realized niche,** the one it actually fills, may be smaller than its fundamental niche. Two species of barnacles, for example, live in the intertidal zone along Scotland's shoreline. Both species share the same fundamental niche, the entire intertidal zone. When present alone, either type of barnacle can survive throughout the entire zone. When both are present, however, one species grows faster than the other and crowds out its competitor in the lower, moister region of the intertidal zone. The slower-growing species, however, better tolerates dehydration when exposed to the air while the tide is out and can more efficiently use the resources of the upper region of the intertidal

FIGURE 43.2

Fundamental and Realized Niches.

When the barnacle *Balanus* is absent, *Chthamalus* adults occupy the entire intertidal zone. *Balanus* grows faster than *Chthamalus*, however, and when *Balanus* is present, interspecific competition for space limits *Chthamalus* to the upper intertidal zone.

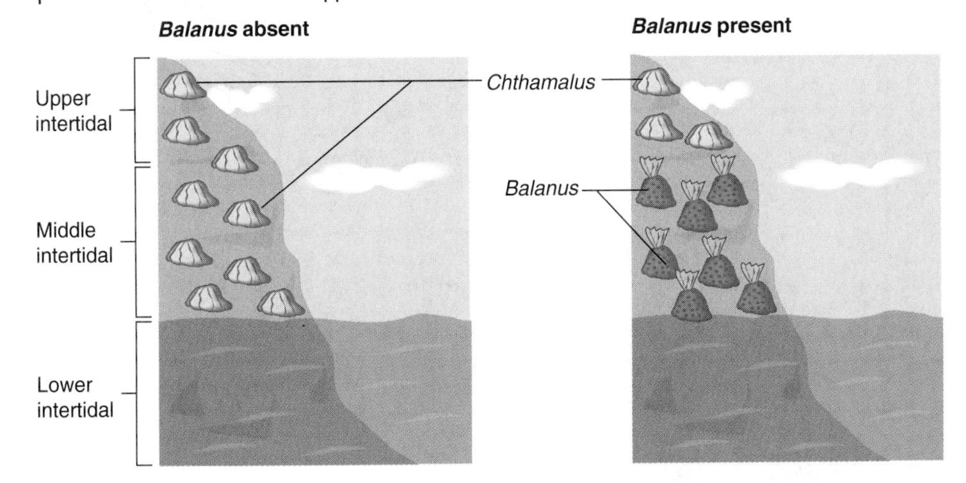

zone. Consequently, the two populations of barnacles have different realized niches (fig. 43.2).

The barnacle example illustrates the **principle of competitive exclusion,** which says that two species cannot coexist indefinitely in the same niche. The two species will continually compete for the limited resources required by both, such as food, nesting sites, or soil nutrients. If one species acquires more of the resources that are in short supply, it will eventually replace the other, less successful species. For example, on Africa's Serengeti Plain, wild dogs and hyenas hunt zebras, wildebeests, and antelopes. Hyenas are larger and more aggressive than wild dogs, and they sometimes displace wild dogs from their kills. The hyena's superior hunting skills have contributed to the decline of the wild dog population (fig. 43.3).

When a new species enters an area, competition may drive a native population, or an entire species, to extinction. This happened when starlings were released in New York City's Central Park in 1891. The starlings robbed the native bluebirds of their nesting sites. The displacement of native species by introduced species occasionally disrupts whole communities. Zebra mussels (*Dreissena polymorpha*), for example, are native to the Caspian Sea in Asia. They were accidentally introduced to the Great Lakes in the 1980s, and since then have spread to many waterways in the United States and Canada. Not only have they crowded out native mussel species along the way, but the tiny filter-feeders have also changed plant communities by greatly increasing the clarity of the waters they inhabit. The new plants provide a different type of habitat than the previous species, so the community of fishes and other organisms has also changed in waterways that zebra mussels inhabit.

The competitive exclusion principle says that species can coexist in a community if they use different resources within the habitat. Alternatively, competing species may use the same resource in a slightly different way or at a different time. This specialization is called **resource partitioning.** For example, in New England forests, five species of warblers, all small insect eaters, coexist in the same trees, each type feeding in different ways and on different parts of the tree (fig. 43.4).

Other Community Interactions Include Symbiosis and Predation

Competition is just one of many ways populations can interact. Some species live in or on another, a phenomenon called **symbiosis** (literally, "living together"). The opening essay to chapter 20 illustrates one example of symbiosis—a squid that glows by the light of its resident bioluminescent bacteria. As table 43.1 shows, the relationship between symbiotic species may be **mutualistic,** or mutually beneficial, as when a cow derives simple sugars from the cellulose-digesting microorganisms living in its rumen. In rare cases, called **commensalism,** one species derives a benefit but the other is unaffected. Most humans, for example, never notice the tiny mites that live, eat, and breed in their hair follicles. More common is **parasitism,** in which one species derives nutrients or other resources at the expense of another. Microorganisms, fungi, plants, and animals may be parasites. Mistletoe, for example, is a parasitic plant that acquires nutrients by tapping into the vascular tissue of a host plant. Whereas parasites rarely kill their hosts outright, another type of symbiotic partner, called a **parasitoid,** does kill. Some wasps are parasitoids, laying their eggs in the larvae of other insects. When the wasp eggs hatch, the tiny larvae eat the host alive from the inside out.

FIGURE 43.3

Interspecific Competition for Food.

This spotted hyena is carrying a scavenged rib cage away from the site of a kill.

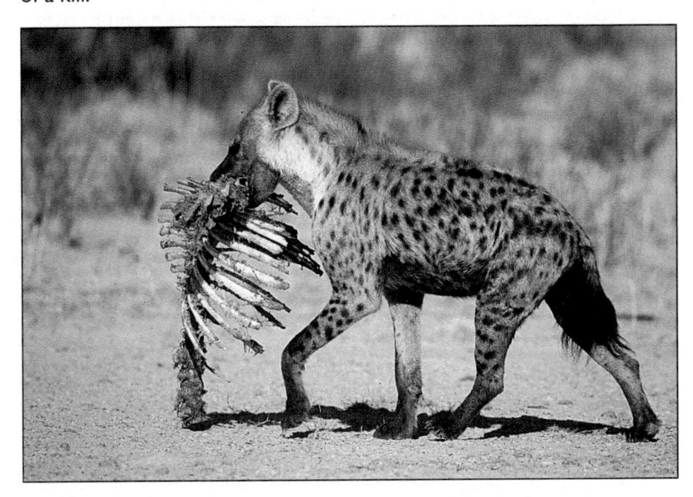

FIGURE 43.4
Populations Coexist by Reducing Competition.
These five species of North American warblers live together in conifer forests because each forages in a different portion of the tree and uses different foraging behavior. Different-colored areas of the tree indicate primary feeding zones for each warbler species.

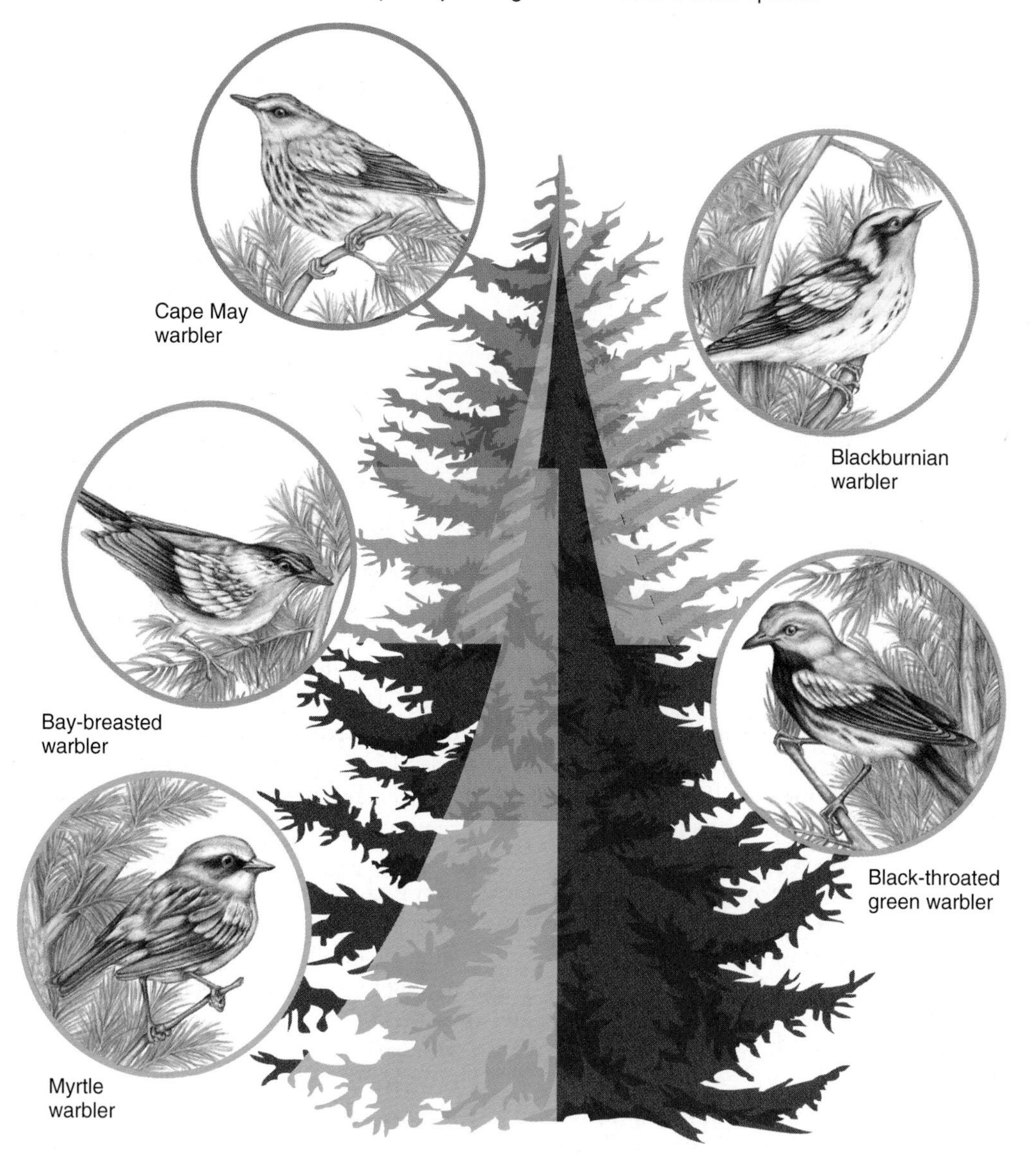

TABLE 43.1 Symbiosis

TYPE	DEFINITION	EXAMPLE
Mutualism	Both partners benefit	Mycorrhizae (see chapter 23)
Commensalism	One partner benefits with no effect on the other	Moss plants on tree bark
Parasitism	One partner benefits to the detriment of the other	Disease-causing organisms (see chapters 20, 21, and 23)
Parasitoidism	One partner benefits and the other dies as a result	Wasps that lay eggs in other insect larvae

Species also interact when one, a **predator,** eats another, the **prey.** Recall from chapter 42 that predator-prey relationships may strongly influence the population size of both the hunter and the hunted. Prey species often have obvious adaptations that help them avoid being eaten (fig. 43.5). **Camouflage,** for example, is an adaptation of shape, color, or behavior that allows a species to "hide in plain sight." Chapter 25's opening essay highlights adaptations among animals in Vietnam that enable them to blend into their surroundings. At the opposite end of the spectrum, some toxic or well-defended prey species,

FIGURE 43.5
Prey Defenses.
Prey species may use camouflage to hide from predators (**A**), or warning coloration to advertise their defenses (**B**). Other species may, in turn, mimic those displays. Although the animals in (**C**) look like ants, each as eight—not six—legs. They are ant-mimicking jumping spiders, with none of the weaponry of the ants they resemble.

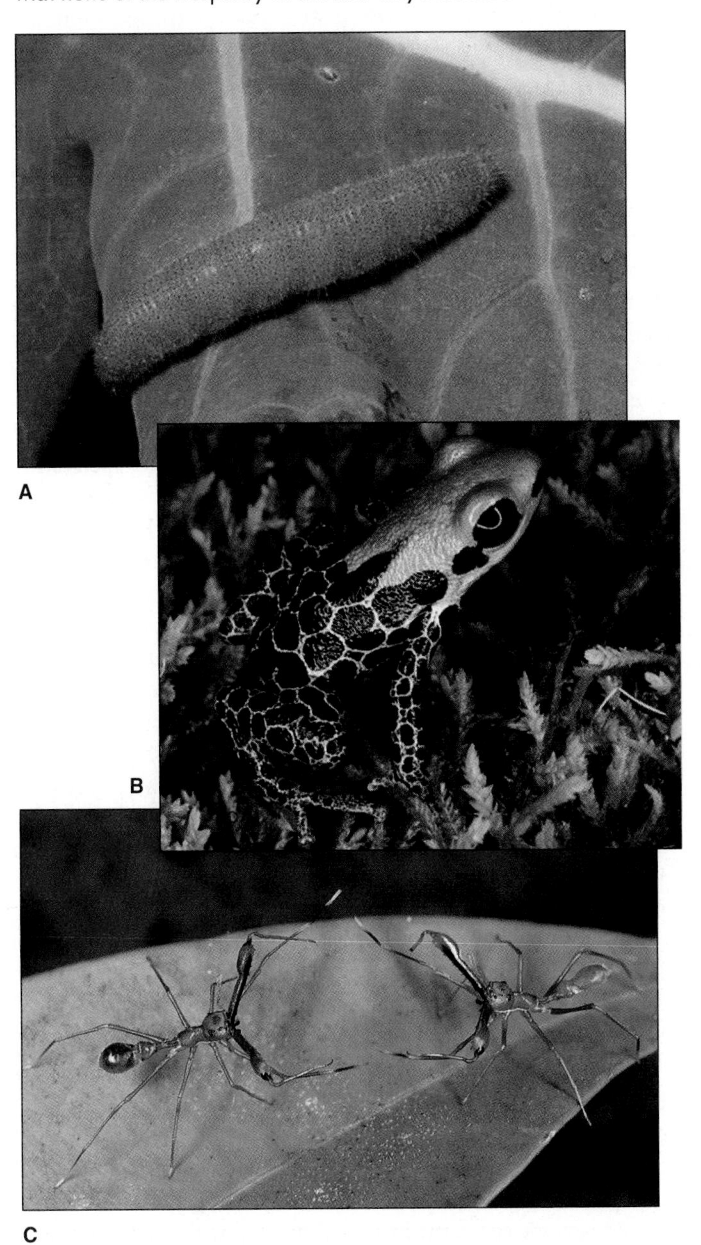

such as stinging bees and poison dart frogs, produce bright or distinctive **warning coloration** to advertise their special defenses. Predators quickly learn to avoid prey with warning coloration. An interesting variation on the theme of warning coloration is **mimicry,** in which different species develop similar appearances. For example, a harmless species may deter predators with coloration similar to that of a noxious or inedible species. In another type of mimicry, several distantly related noxious species may share a similar type of warning coloration, such the yellow and black stripes on many stinging bees and wasps. Such mimicry is mutually beneficial to all species with the coloration.

FIGURE 43.6
Predators Sometimes Cooperate.
The hunt begins when the group of Harris hawks assembles on a lookout. Smaller teams of two or three birds make short flights to nearby perches, scanning the area for prey. After the hunt, the hawks share the meal.

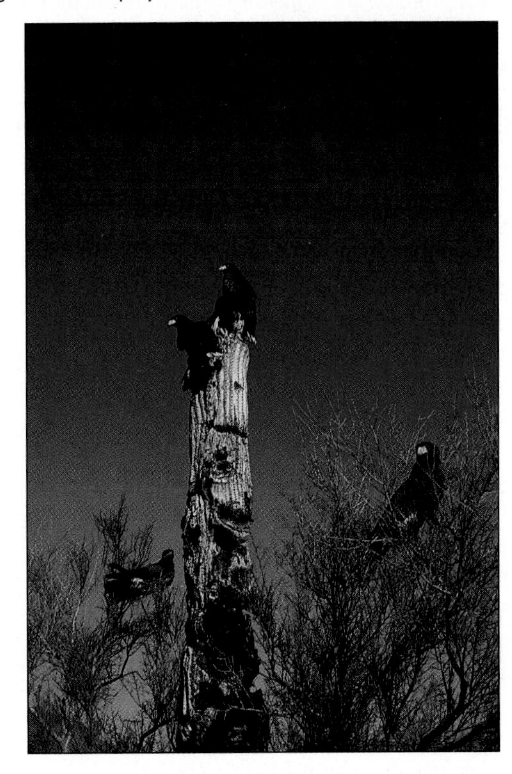

Natural weapons and structural defenses are also commonly used among prey species. Some plant species, for example, produce distasteful or poisonous chemicals that deter most herbivores. Figure 24.10 illustrates the nematocysts that are a defensive trademark of the cnidaria, and figure 24.23 shows the characteristic protective "spiny skin" from which the echinoderms take their name.

Predators are adapted to circumvent the defensive strategies of prey. Monarch butterflies, for example, can tolerate the noxious chemicals in milkweed plants that repel other herbivores. Some predators, such as tigers and other big cats, have markings that hide their shape against their surroundings, which helps them sneak up on their prey. Natural selection may also favor predators with superior senses of smell or vision. Many predators hunt in groups. The Harris hawks perched in a tree in figure 43.6, for example, are scanning the area for prey. When they spot a rabbit, the hawks take turns diving at it. Should the rabbit escape into a burrow, the hawks again take turns in going after it. When the rabbit runs out, the other hawks, waiting in a circle, kill it. Afterwards, the hawk hunting team shares the meal.

Some connections between species are so strong that the species directly influence each other's evolution. In **coevolution,** a genetic change in one species selects for subsequent change in the genome of another species. Of course, all interacting species in a given community have the potential to influence one another, and they are all "evolving together." However, in the strictest sense, such genetic changes are only said to result from coevolution if scientists can demonstrate that the traits in the co-evolving species specifically result from their interactions.

One example of coevolution is the mutualistic relationship between ants (*Pseudomyrmex* spp.) and acacia trees. The ants defend their acacia trees from potential herbivores, seemingly in exchange for food the acacia produces. Each species has special adaptations not found among related species that lack the mutualistic relationship. For example, the acacia produces swollen, hollow thorns in which the ants live. Also, although many plants produce nectar in flowers to attract pollinators, the acacia also produces nectar on its leaves, which the ants consume. These accommodations have, in turn, selected for ant behaviors that defend the acacia tree against potential herbivores.

Sometimes, many species in a community depend on one type of organism, called a **keystone species.** All communities have dominant producer species that are essential to community structure because of their high biomass. **Biomass** is the total dry weight of organisms in an area at a given time. In contrast, keystone species have low biomass but exert a disproportionate influence on community diversity. Many keystone species are versatile predators that maintain competition, such as sea stars that prey on diverse tide-pool invertebrates, or birds that eat herbivorous insects. Mycorrhizal fungi can also act as keystone species by influencing the composition of certain plant communities (see chapter 23).

43.1 MASTERING CONCEPTS

1. Distinguish between habitat and niche.
2. Distinguish between fundamental and realized niche.
3. What is the principle of competitive exclusion?
4. How do the populations that comprise communities interact?
5. Why are keystone species important in communities?

OLC

43.2 Communities Change Over Time

In primary succession, a new community forms; in secondary succession, the types of resident species change after a disturbance to the area. In both types of succession, pioneer species pave the way for changes in the community and the physical environment.

Although communities may appear static, they do change throughout time in response to many influences—climatic change, disturbance, and species invading from other areas. A community also changes as a result of its interaction with its physical environment. This gradual and directional process of change in the community, in which species replace each other and the biotic structure may change considerably, is called **ecological succession.**

Succession often leads toward a **climax community,** one that may remain fairly constant as long as climate does not change and major disturbances do not strike. However, few communities ever reach true climax conditions. Pockets of local disturbance, such as the area affected when a large tree blows over, create a patchy distribution of successional stages across a landscape. Also, such factors as fire, disease, and severe storms can influence successional patterns for centuries. Usually, the rate of change slows late in the successional process, but never ceases. In the Pacific Northwest, for example, old-growth forests are 500 to 1,000 years old, yet they are still changing in their structure and composition.

Primary Succession—A New Community Arises

Ecologists define two major types of succession: primary and secondary. **Primary succession** occurs in an area where no community previously existed. Volcanoes, for example, afford wonderful opportunities to observe primary succession as lava flows obliterate existing life, a little like suddenly replacing an intricate painting with a blank canvas. Road cuts and glaciers that scour the landscape also expose virtually lifeless areas on which new communities eventually arise.

Primary succession can be monitored on a patch of bare rock in New England. The first species to invade, the **pioneer species,** are hardy organisms such as lichens and mosses that can grow on smooth rock. The fungal component of a lichen attaches to the rock and obtains water and minerals, while the algal part photosynthesizes and provides nutrients. As lichens produce organic acids that erode the rock, sand and dust accumulate in the crevices. Decomposing lichens add organic material, eventually forming a thin covering of soil. Microorganisms, worms, and insects colonize the forming soil. Then rooted plants such as herbs and grasses invade. Soil continues to form, and larger plants, such as shrubs, appear. Larger animals move into the area. Next grow aspens and conifers, such as jack pines or black spruces. Finally, hundreds of years after lichens first colonized bare rock, the soil finally becomes rich enough to support other deciduous trees, and a climax community of an oak-hickory forest may develop (fig. 43.7).

Secondary Succession—Replacing a Disturbed Community

Secondary succession occurs where a community is disturbed but not decimated, some soil and life remaining. Because the area isn't completely devastated, secondary succession occurs faster than primary succession. Fires, hurricanes, and agriculture are common triggers of secondary succession.

An abandoned farm in the eastern United States illustrates secondary succession. This "old field" succession begins when the original deciduous forest is cut down for farmland. As long as crops are cultivated, natural succession stops. When the land is no longer farmed, fast-growing pioneer species, such as black mustard, wild carrot, and dandelion, move in, followed by slower-growing, taller goldenrod and perennial grasses. In a few years, trees such as pin cherries and aspens arrive and are eventually replaced by pine and oak. A century or so later, the climax community of beech and maple may again be well developed.

Ecologists have monitored secondary succession in the aftermath of hurricanes. Puerto Rico has seen 15 major hurricanes over the past three centuries. Figure 43.8 depicts the secondary

FIGURE 43.7

Primary Succession Can Yield a Deciduous Forest After Centuries.

First, lichens colonize bare rock and produce acids that begin to break down the rock into a thin layer of soil. Next mosses, fungi, small worms, insects, bacteria, and protista colonize the scanty soil, contributing organic matter as they live and die. Tiny herbs and weeds grow, grasses and shrubs move in, and trees come many years later, after the soil is well developed.

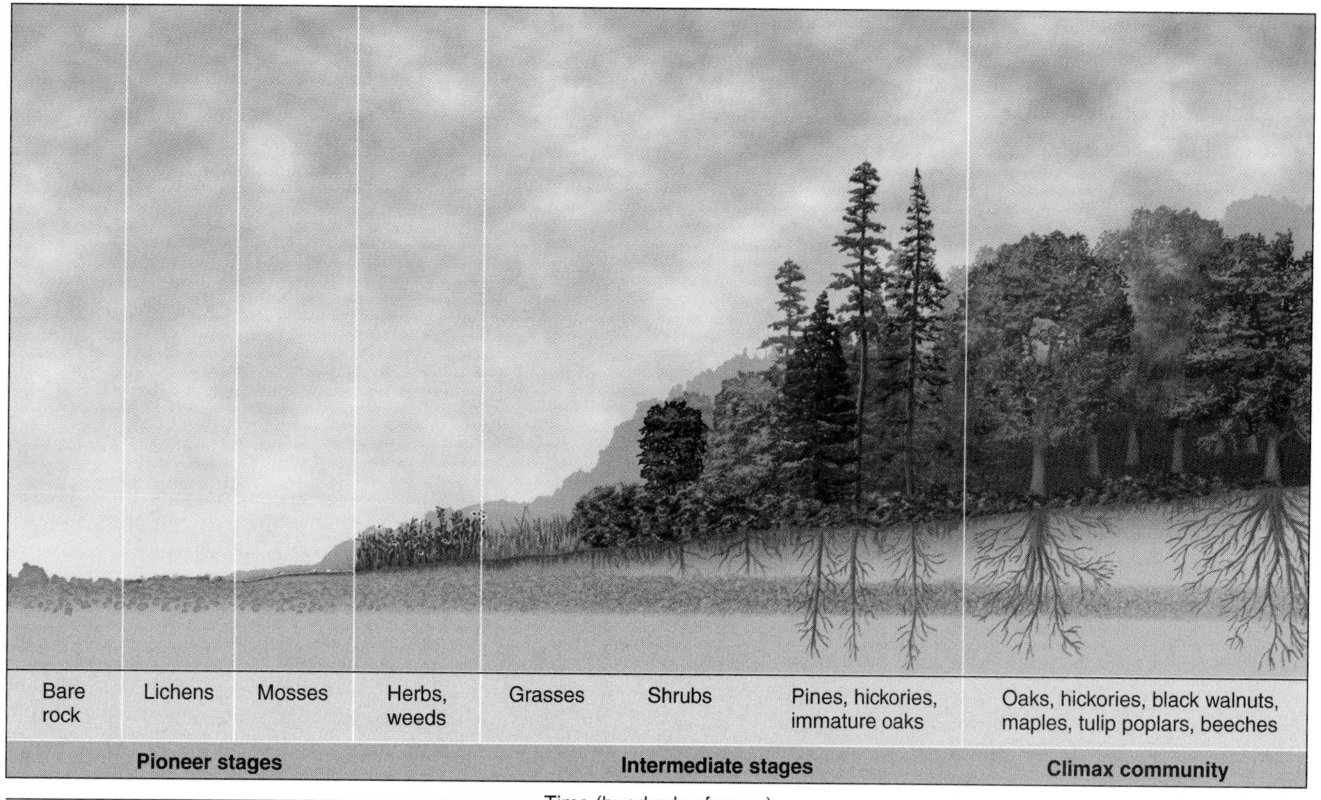

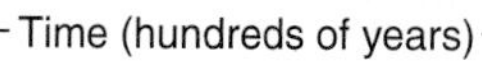

A

B

FIGURE 43.8

Secondary Succession.

Secondary succession occurs in the Luquillo Experimental Forest, Puerto Rico, following Hurricane Hugo. The storm destroyed about half of the forest canopy in 1989 (**A**), but five years later the forest had recovered (**B**).

succession following Hurricane Hugo that struck Puerto Rico on September 18, 1989. The storms there tend to keep the forest clipped down to a low level. If 60 years go by without a major hurricane, a climax community of taller, denser trees grows. Typically, a large hurricane lowers the height of the forest canopy by 50%, the winds stripping leaves and branches, breaking stems, and uprooting trees. Usually fewer than half of the trees die, most of them young and low, their limbs snapped off. Unlike primary succession, which relies on the arrival of new organisms to restart growth, secondary succession occurs in several ways. The Luquillo Experimental Forest in Puerto Rico recovered via resprouting and repopulation of the canopy that had been opened up by tree deaths, as well as from seeds brought in on the wind.

Succession Can Be Complex

Although succession occurs in a variety of different settings, ecologists recognize a common set of processes in both primary and secondary succession. First, pioneer species colonize a bare or disturbed site. Recall from chapter 42 that the hardy pioneer species are usually *r*-selected, with prolific reproduction and efficient dispersal. Interestingly, these early colonists often alter the physical conditions in ways that enable other species to become established. These new arrivals continue to change the environment. Some early colonists do not survive the new challenges, which further alters the community. In contrast to the pioneers, species that appear in the climax community are usually long-lived, late-maturing, *K*-selected species that are strong competitors in a stable environment.

In addition to changes in the physical environment, interactions between species also change the composition of the biotic community. When pine trees invade a site, for example, they shade out lower-growing plants that appeared earlier in succession, and they attract species that grow or feed on pines.

Primary and secondary succession can occur simultaneously in a large area that has faced a major disturbance, such as a volcanic eruption or hurricane. Consider the return of life to Mount St. Helens, a volcano that erupted in Washington State in May 1980. Regions of primary and secondary succession dot the landscape, depending on the extent of devastation. In some areas, scorchingly hot gases and steam killed everything, so primary succession occurred, eventually, when the wind brought in pioneer species and the ground had cooled enough for them to survive and grow. An avalanche that triggered a mudslide also ushered in primary succession, for it buried everything in its path. Secondary succession occurred in the "blowdown zone," where many trees were knocked down but small plants survived. In these areas, insects, birds, and mammals returned relatively rapidly as the forest plants began to regrow.

43.2 MASTERING CONCEPTS

1. What is ecological succession?
2. Distinguish between primary and secondary succession.
3. What processes and events contribute to succession?
4. How common are climax communities?

OLC

43.3 An Ecosystem Is a Community and Its Physical Environment

Ecosystems range greatly in size, and are constantly changing and interacting. Food chains and webs are built on primary producers that pass energy to successive levels of consumers. Pyramid diagrams represent energy flow or numbers or biomass of organisms in an ecosystem.

An ecosystem includes all the living (biotic) organisms and the nonliving (abiotic) environment within a defined area. It may be as large as the whole earth, a small part of it, such as a stretch of creek, or the community and its physical habitat beneath a single large boulder. Even a single leaf is an ecosystem (fig. 43.9). An ecosystem is terrestrial if it is on land and aquatic if it is in water.

Characteristics of Ecosystems

Ecosystems are open systems, which means that they exchange matter and energy with their surroundings (see chapter 5). Some ecosystems, such as lakes, are relatively self-contained—more materials cycle within than enter or leave. In some other ecosystems, such as a section of stream, materials rapidly flow in and out. All ecosystems rely on energy from some outside source. Usually this is the sun (solar energy). Some types of microorganisms, such as the chemosynthetic bacteria at the hydrothermal vents described in the chapter opening essay, derive energy from inorganic chemicals.

Ecosystems can and do change, as the previous examples of succession indicate. Succession leads to change not only in the community, but also in many characteristics of the physical environment.

Ecosystems interact. Insects, mosses, and fungi on the surface of a downed tree interact with organisms in the adjacent soil ecosystem. A terrestrial ecosystem may contain an aquatic one, such as a pond or marsh, or an aquatic ecosystem, such as a bay, may include islands or sandbars. Ecosystems receive wind-dispersed seeds in the air and dissolved nutrients from rivers and streams. Animals continually ferry nutrients and organisms between ecosystems. Even physically distant ecosystems interact through the air and waters. Pollution from smokestacks in the U.S. Midwest, for example, alters the acidity of lakes in the Northeast.

So interconnected are ecosystems that the entire planet can be viewed as one huge, interacting ecosystem, the **biosphere.** This is the part of the planet where life exists, reaching about 6.2 miles (10 kilometers) into the atmosphere and down about the same distance into the deepest ocean trenches. The biosphere can be divided into **biomes,** which are major ecosystem types, such as desert, grassland, or deciduous forest. Each terrestrial biome has a characteristic group of predominant plant species, corresponding to the climate of the region. Aquatic ecosystems have characteristic physical and chemical conditions. The next chapter discusses biomes.

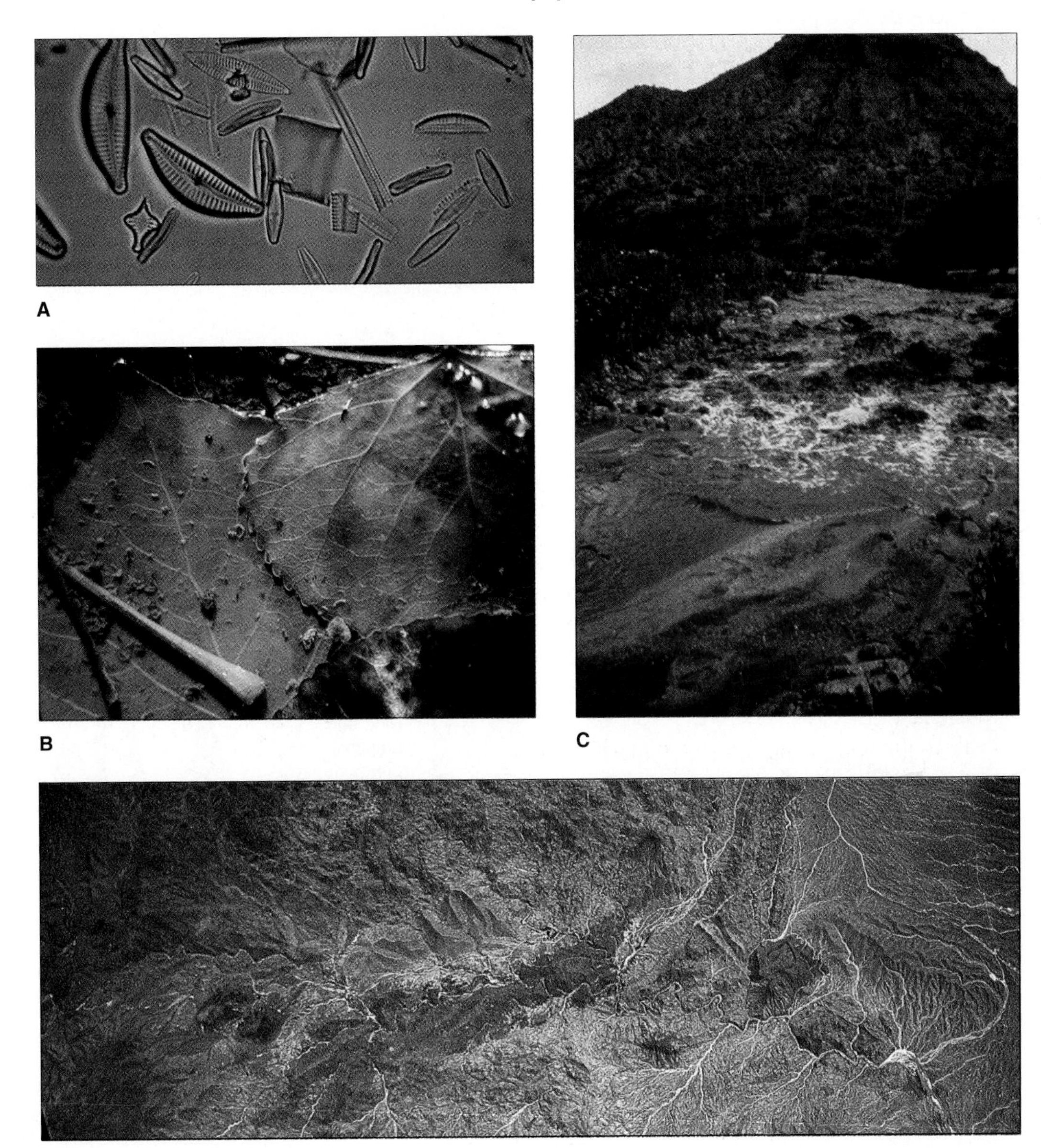

FIGURE 43.9
An Ecosystem Can Be Studied at Several Scales.
Fourteen streams feed into the Sycamore Creek drainage basin near Phoenix, Arizona. (**A**) The water abounds with microscopic diatoms. (**B**) Fallen leaves provide food for invertebrates. (**C**) A flash flood is not necessarily a catastrophe—invertebrates survive under rocks and in crevices, fishes swim into eddies and protected pools, and decimated insect populations recolonize the stream quickly because of their high reproductive rates. An aerial photograph (**D**) reveals the pattern of stream channels and the relation of the stream to vegetation of its floodplain and watershed. In this photo, dense vegetation appears bright red.

Energy Flows Through an Ecosystem

Energy flows through an ecosystem in one direction, beginning usually with solar energy that photosynthetic organisms transduce to chemical energy (see chapter 6). This stored energy then passes through a **food chain,** which is a series of organisms that successively eat one another. Eventually, all the stored energy is dissipated as heat, so all ecosystems require continual energy input to function indefinitely.

An organism's **trophic level** is its position in the food chain, relative to the ecosystem's energy source. **Primary producers,** or **autotrophs** ("self-feeders"), form the first link in the food chain and the base of the trophic structure. Primary producers ultimately provide energy for all other organisms; they use inorganic materials and energy to produce all the organic material they require. Familiar primary producers include plants and algae, but many prokaryotes are primary producers too, deriving energy from the sun or inorganic chemicals.

The other levels in a food chain are **consumers,** or **heterotrophs** ("other eaters"), which obtain energy from the producers. Herbivores (plant-eaters) are primary consumers and form the second trophic level. Carnivores (meat-eaters) are secondary consumers, animals that eat herbivores. Carnivores that eat other carnivores form a fourth trophic level; they are tertiary consumers. These animals expend a great deal of energy capturing their prey. Scavengers, such as vultures, eat remains of another's meal, and **decomposers,** such as certain fungi, bacteria, insects, and worms, break down **detritus** (dead organisms and feces).

Food chains interconnect, forming complex **food webs,** such as the Antarctic web in figure 43.10. Webs form when one species eats or is eaten by several other species and when one species functions at more than one trophic level. As described in Investigating Life 43.1, food webs can unravel as a result of the loss of a keystone species.

FIGURE 43.10

The Antarctic Web of Life.

As in most food webs, the multiple interactions among Antarctic residents form a very complex network. Note, however, that producers, consumers, and scavengers are all present. Decomposers return inorganic nutrients that producers require.

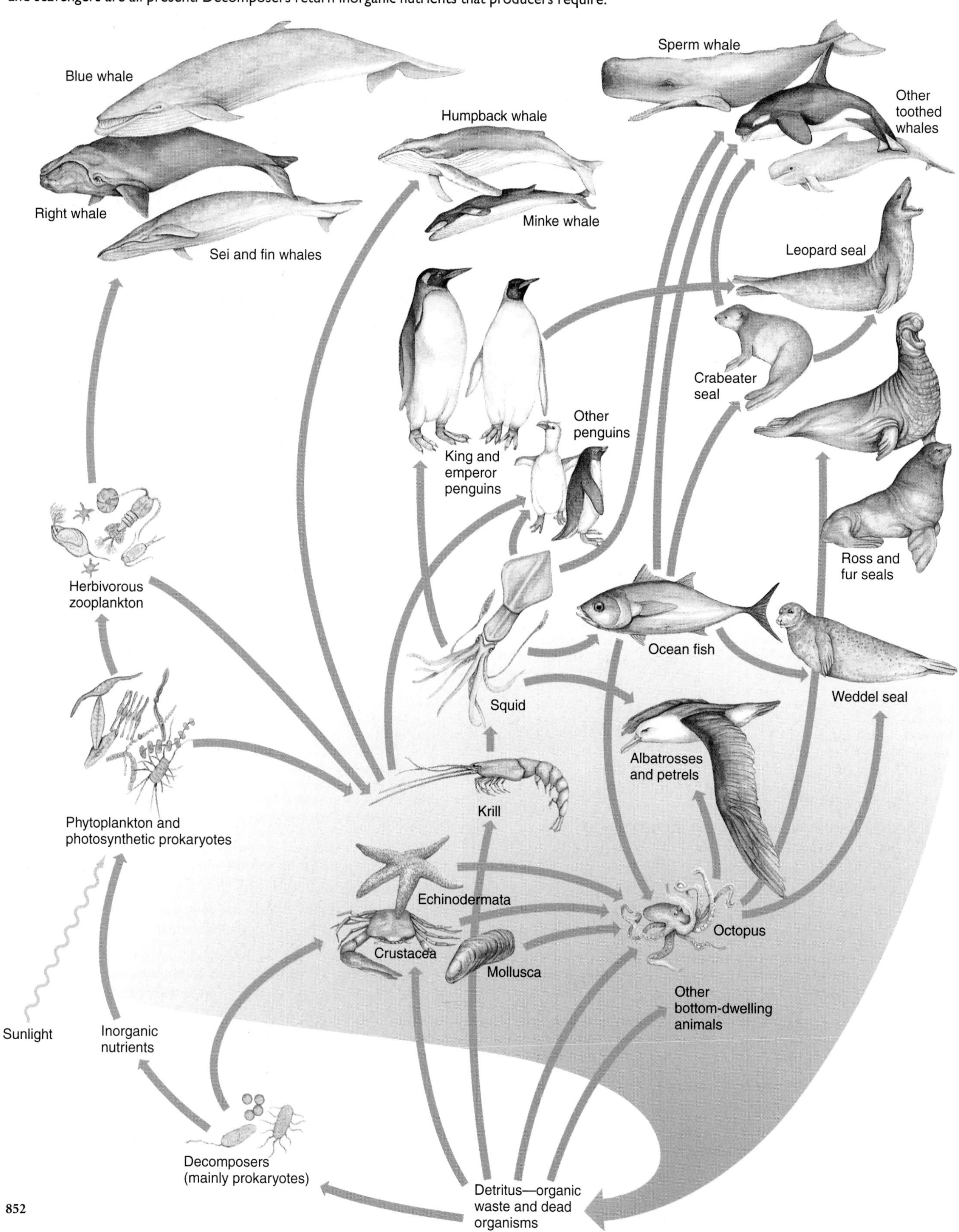

Investigating Life 43.1

Otter Deaths Topple a Food Web in the Pacific Northwest

The removal of a keystone species can slowly dismantle a food web, as the links of who-eats-whom change or fall apart. This is happening in ocean and coastline ecosystems near the Aleutian Islands in the Pacific Northwest. The keystone species in the vast underwater kelp forests is the sea otter, which eats sea urchins that would otherwise devour the kelp.

James Estes, a marine ecologist with the U.S. Geological Survey, and his coworkers first noticed a decline in the sea otter population in 1990. By tracking otters with radio collars, they ruled out disease, pollution, or reproductive problems as the cause of the growing scarcity of otters. A clue was the lack of dead otters—a new type of predator could have been responsible.

The next clue came from an observant researcher in the group, who saw an orca (killer whale) kill an otter, an oddity because the two species had lived together for decades. Then others noticed similar attacks. A natural experiment provided evidence for the hypothesis that whales were killing the otters. Adak Island had two lagoons with otters—one where killer whales could enter, and one where they could not. Where orcas had access, the otter population plunged by two-thirds, but where they didn't, only 12% of the otter population perished.

By considering these clues and taking into account other members of the ocean and coastline ecosystems, the researchers pieced together a plausible hypothesis to explain the shifting food web. They began with the sea otters and worked backwards, then forwards to predict repercussions of their declining population (fig. 43.A).

The series of events leading to the otters' demise began with a decrease in the sizes of populations of oily, nutrient-rich fishes such as herring and ocean perch. It isn't known what caused this to happen, but the researchers have three suggestions: overfishing; a warming trend; or whaling, which would have released many plankton from predation, which in turn would have overfed pollock, a less-nutrient-packed fish. When Steller sea lions and harbor seals ate pollock instead of their usual oilier fish fare, they began to die from malnutrition. The orcas that normally eat sea lions and seals needed another food source, and they turned to the coastal ecosystem and its abundant otters. As a result, over a 500-mile (805-meter) expanse of coastline, a population of otters 53,000 strong in the 1970s now numbers 6,000 animals.

Without otters, the sea urchin population mushroomed, and they ate away at the kelp. Ecologists are predicting the effects of the vanishing kelp forest. Mussels and fishes that live among the undersea fronds will lose their food source and be more visible to predatory birds. When the fishes starve, so too will some birds. At the same time, numbers of starfishes will rise, because their predators, otters, have vanished.

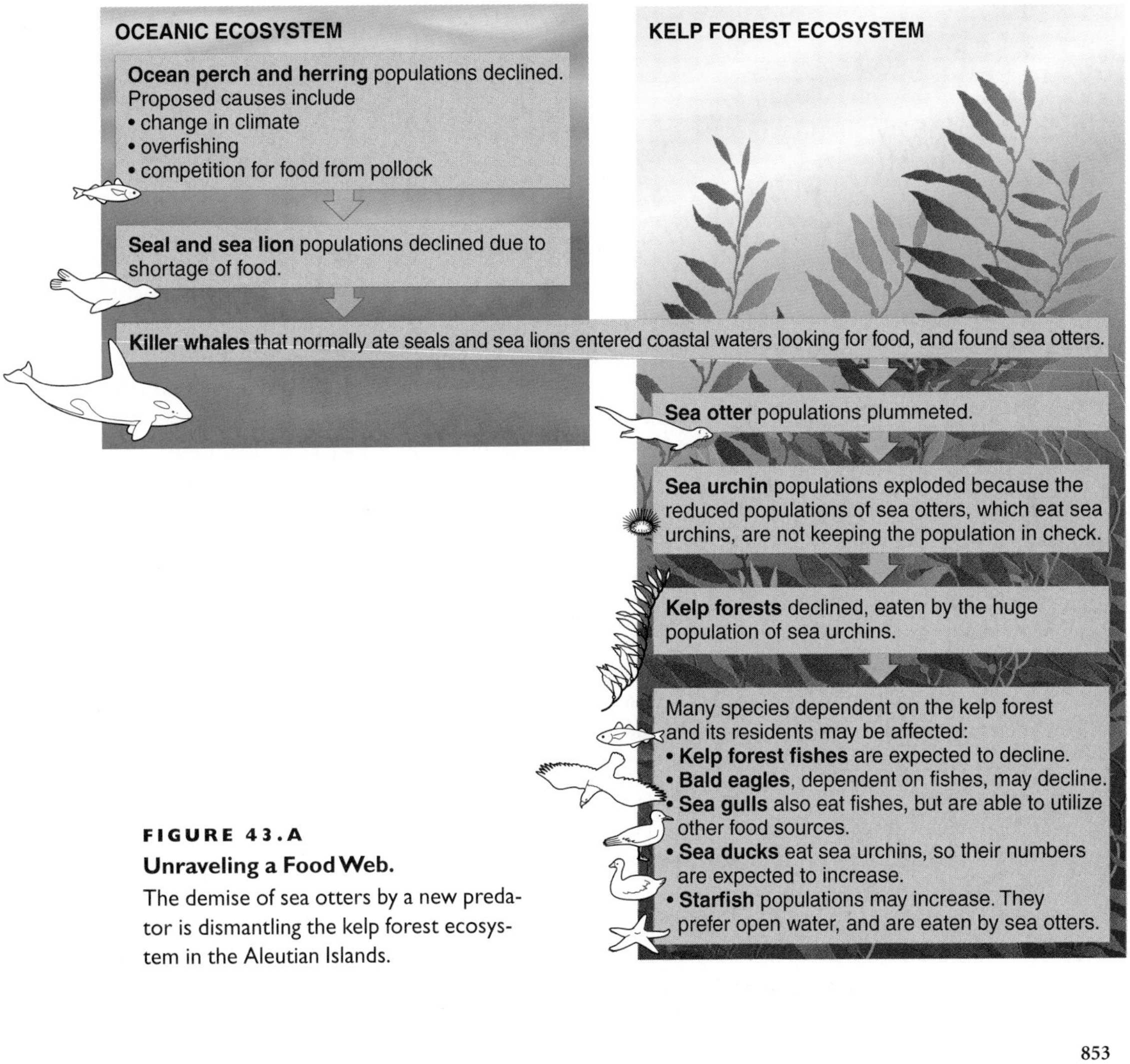

OCEANIC ECOSYSTEM

Ocean perch and herring populations declined. Proposed causes include
- change in climate
- overfishing
- competition for food from pollock

Seal and sea lion populations declined due to shortage of food.

Killer whales that normally ate seals and sea lions entered coastal waters looking for food, and found sea otters.

KELP FOREST ECOSYSTEM

Sea otter populations plummeted.

Sea urchin populations exploded because the reduced populations of sea otters, which eat sea urchins, are not keeping the population in check.

Kelp forests declined, eaten by the huge population of sea urchins.

Many species dependent on the kelp forest and its residents may be affected:
- **Kelp forest fishes** are expected to decline.
- **Bald eagles**, dependent on fishes, may decline.
- **Sea gulls** also eat fishes, but are able to utilize other food sources.
- **Sea ducks** eat sea urchins, so their numbers are expected to increase.
- **Starfish** populations may increase. They prefer open water, and are eaten by sea otters.

FIGURE 43.A
Unraveling a Food Web.
The demise of sea otters by a new predator is dismantling the kelp forest ecosystem in the Aleutian Islands.

FIGURE 43.11
Energy Flow Through an Ecosystem.
(**A**) The cat eats the bird that eats the beetle that eats the tomato plant. The pyramid of energy shows that only a small percentage of the energy stored at one trophic level per unit time is transferred and stored in new growth and reproduction at the next level. (**B**) Humans who derive energy by eating meat are getting only a small fraction of the energy originally present in grain.

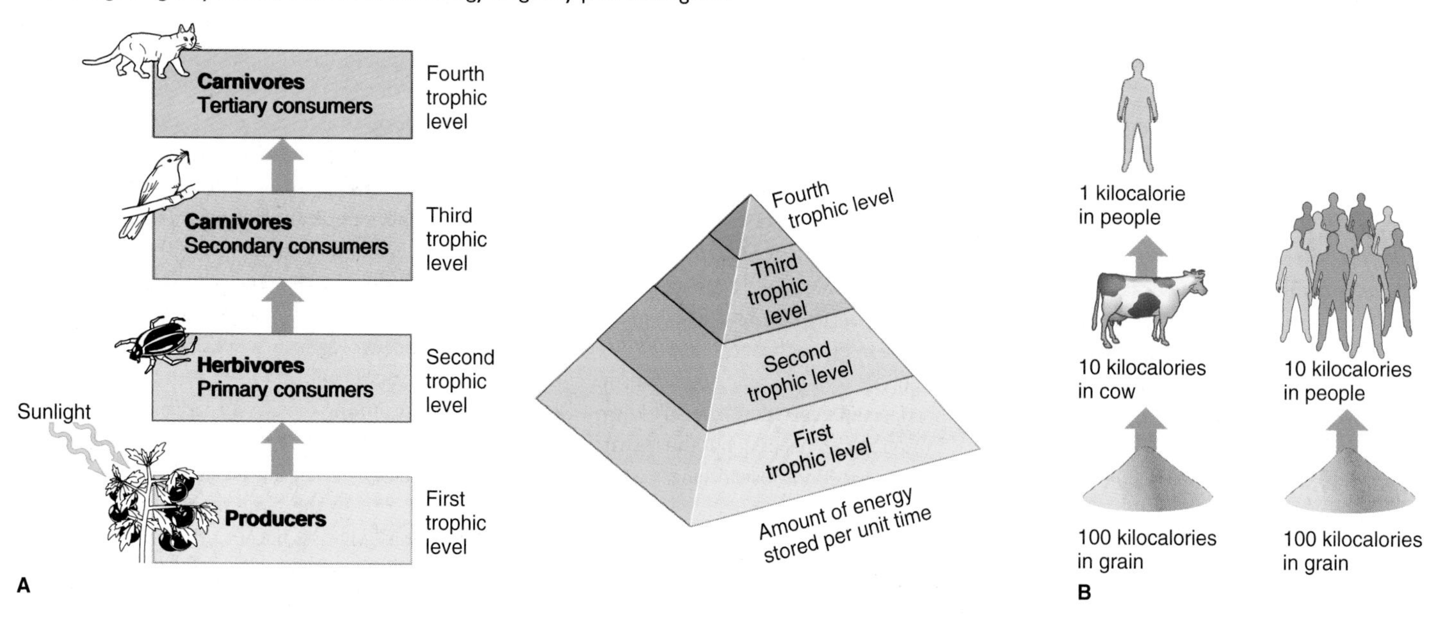

Ecological Pyramids Describe Ecosystem Characteristics

The total amount of energy that is trapped, or "fixed," by all autotrophs in an ecosystem is called gross primary production. However, the autotrophs use some of this energy to generate ATP for their own needs. **Net primary productivity** is the amount of energy left over after respiration by the autotrophs; it is the amount of energy that is available to consumers.

Consumers cannot completely digest all the food they eat; they also lose energy to heat. Because of these inefficiencies, only about 2 to 30% of the potential energy stored in the bonds of organic molecules at one trophic level fuels growth and reproduction of organisms at the next trophic level (fig. 43.11). If a food chain consists of four organisms—such as a cat eating a bird that has eaten a beetle, which ate a tomato leaf—the cat can only use a tiny fraction of the energy in the leaf for its growth and development. This inefficiency of energy transfer is why food chains rarely extend beyond four trophic levels. One way to maximize energy obtained from producers is to lower the number of trophic levels. A person can do this by getting protein from beans instead of meat.

If each trophic level is represented by a volume directly proportional to the energy stored in new tissues per unit time, these volumes can be stacked to form a steep-sided **pyramid of energy** (see fig. 43.11). Consider a simplified example. In Cayuga Lake in New York, algae store 1,500 kilocalories of energy in a period of time. The algae feed small aquatic animals, which store about 150 kilocalories as new tissues. When a smelt fish eats these small aquatic animals, it derives perhaps 15 kilocalories for growth and development. A human eating the smelt would convert only about 1.5 kilocalories of the original 1,500 kilocalories to new growth.

Other types of pyramids describe other aspects of ecosystems. A **pyramid of numbers** shows the number of organisms at each trophic level. The shape of this type of pyramid depends on the size and number of the producers and consumers. The pyramid of numbers in a grassland community has a broad base because the producers—grasses—are small and numerous (fig. 43.12A). In a forest, in contrast, the pyramid of numbers stands on a very narrow base, because a single tree can feed many herbivores. Such an inverted pyramid can also represent a single large animal that supports a large number of parasites, such as the field mouse and its ticks depicted in figure 43.12B.

A **pyramid of biomass** takes into account the weight of organisms at different trophic levels. Many pyramids of biomass are wide at the bottom and narrow at the top, because energy is lost as heat at each trophic level. In some aquatic ecosystems, however, the pyramid is inverted, because the biomass of the primary producers (phytoplankton) is smaller than that of the primary consumers (zooplankton). This is because biomass is measured at one time. Phytoplankton reproduce quickly, but zooplankton eat them almost immediately so few are present at any given time. In contrast, zooplankton live longer, so more of them are present at one time. If we consider the biomass of all the phytoplankton that zooplankton consume during their life span, the pyramid would be upright.

FIGURE 43.12
Pyramids of Numbers.
(**A**) A pyramid of numbers that represents a grassland has a broad base. (**B**) Parasitic organisms illustrate an inverted pyramid of numbers. A field mouse may carry several deer ticks in its fur, which in turn each harbor many spirochete bacteria (*Borrelia burgdorferi*). The bacteria cause Lyme disease in humans.

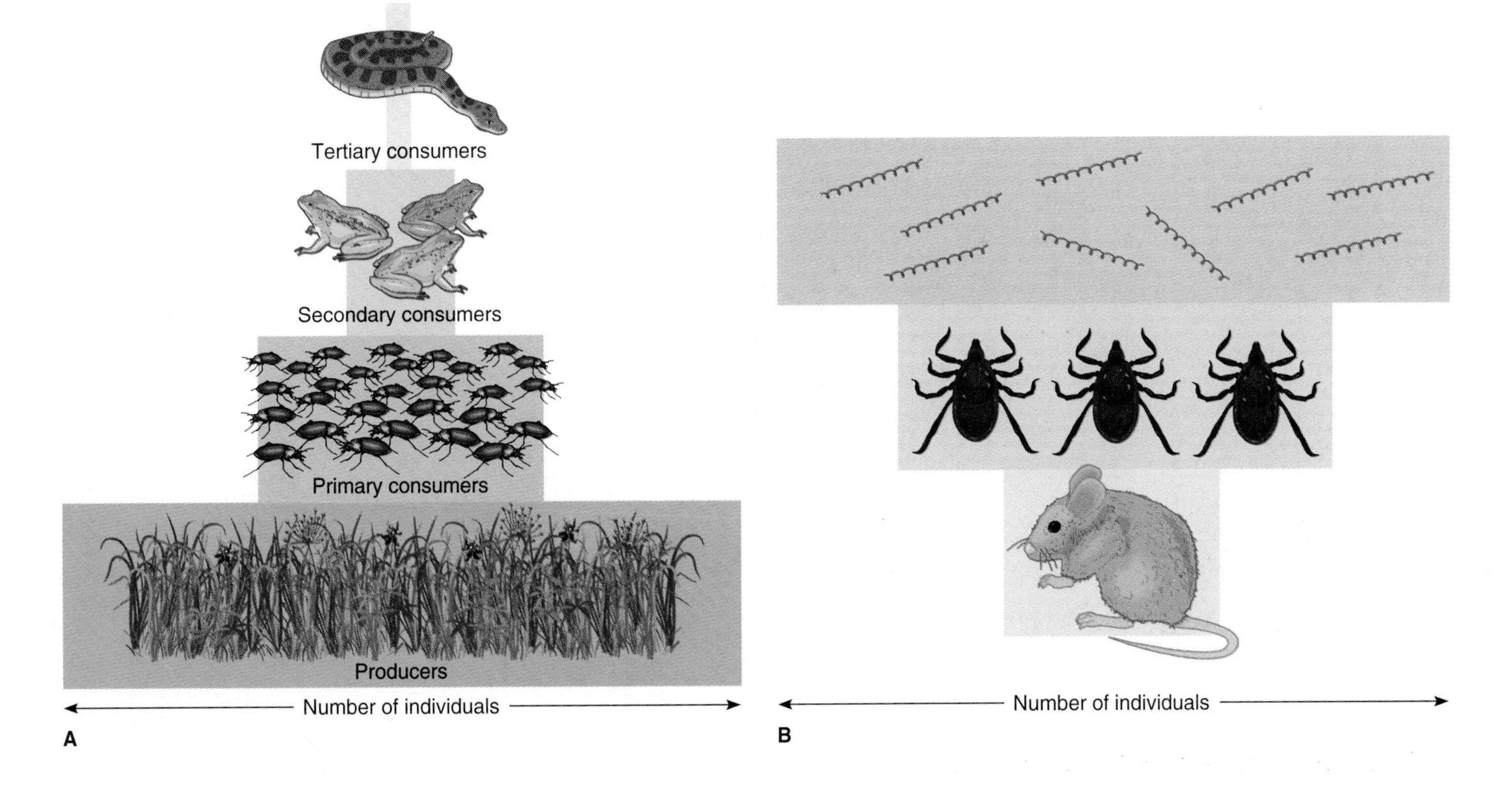

43.3 MASTERING CONCEPTS

1. How do ecosystems range in size?
2. Identify the types of organisms that form trophic levels.
3. Distinguish between gross and net primary production.
4. How efficient is energy transfer between trophic levels in food webs?
5. List three types of pyramids that are used to describe ecosystems.

OLC

43.4 How Chemicals Cycle Within Ecosystems—Biogeochemical Cycles

Biogeochemical cycles describe how elements move between organisms and the physical environment. Elements from the environment ascend food webs, and return to the physical environment when an organism decomposes. Some chemicals are concentrated as they ascend food webs, with toxic effects.

All life, through all time, must use the elements present when Earth formed. These elements continuously recycle through the interactions of organisms and their environments. If not for this constant recycling, these elements would have been depleted as they became bound in the bodies of organisms that lived eons ago. Because recycling chemicals essential to life involves both geological and biological processes, these pathways are called **biogeochemical cycles.**

Each chemical element in an ecosystem has a characteristic biogeochemical cycle, but all cycles have steps in common. Generally, an inorganic form of the element is taken from the environment and incorporated into the organic molecules of autotrophic organisms, such as plants. If an animal eats the plant, the element may become part of animal tissue. If another animal eats this animal, the element may be incorporated into the second animal's body. All organisms die, and decomposers break their tissues down, which releases the elements back into the environment in inorganic form.

The Water Cycle

Water covers much of Earth's surface, primarily as oceans but also as lakes, rivers, streams, ponds, swamps, snow, and ice. Water is also below the land surface as groundwater. The ability of water to exist in all three states of matter—solid (ice), liquid, and gas

FIGURE 43.13

The Water Cycle.

Water falls to Earth as precipitation. Organisms use some water, and the remainder evaporates, runs off into streams, or enters the ground. Animals return water to the environment by respiring and excreting, and plants do so by transpiring.

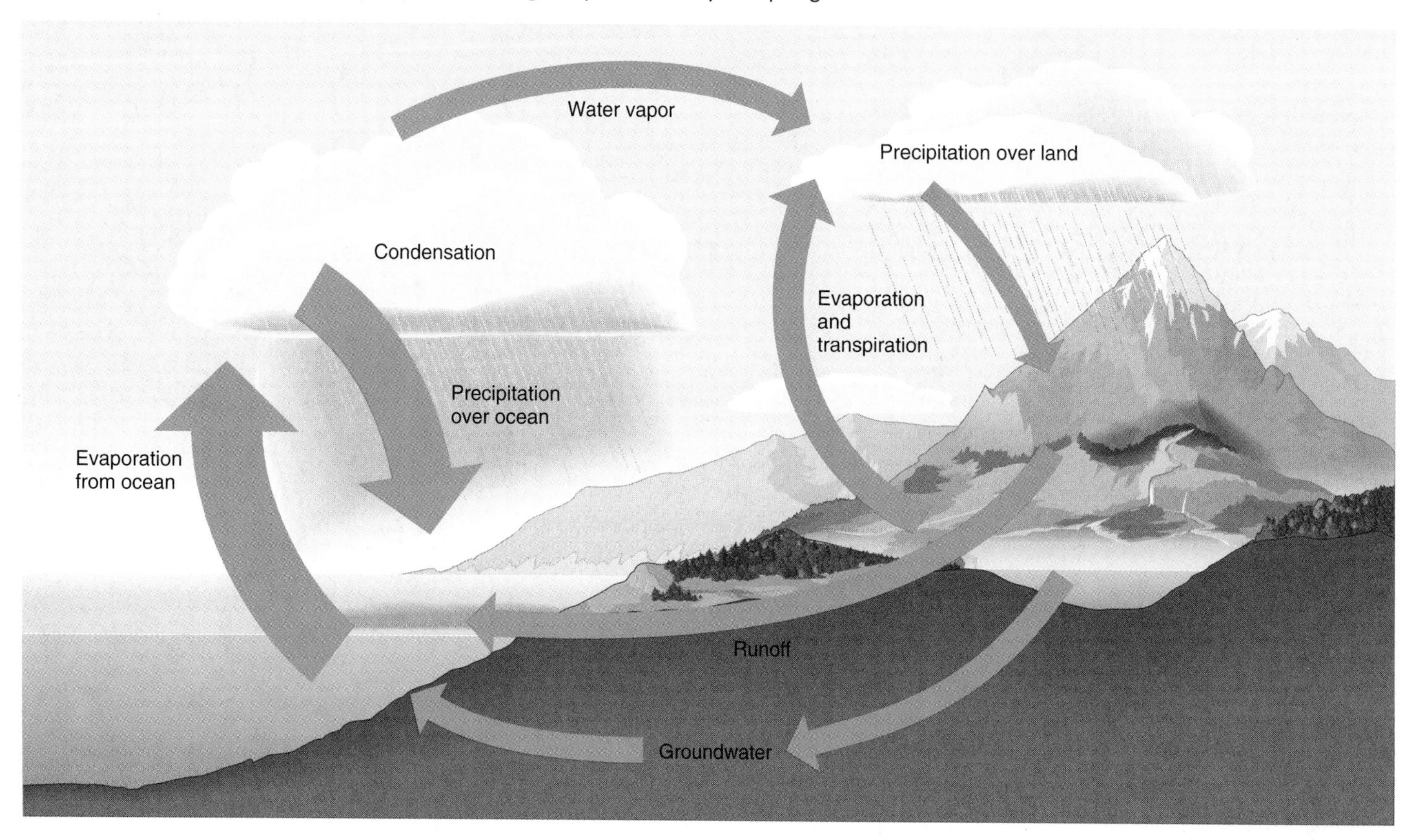

(vapor)—in the temperature range of the biosphere allows it to cycle efficiently from Earth's surface to the atmosphere and back again. (fig. 43.13)

The sun's heat causes evaporation of water. Plants absorb water from soil, use it, and release some of it from their leaves in transpiration (see chapter 27). Animals return water to the environment through evaporation and excretion. Water vapor may rise on warm air currents, cooling and forming clouds. If air currents carry this moisture higher or over cold water, more cooling occurs, and the vapor condenses into water droplets. Depending on the temperature and atmospheric pressure, the droplets fall as rain, snow, hail, fog, sleet, or freezing rain.

Most precipitation enters oceans or other bodies of water, but some also falls on land, where it either soaks into the ground and porous rock to restore soil moisture and groundwater or runs along the surface following the natural contours of the land. Rivulets join and form streams that unite into rivers. Most rivers eventually lead back to the ocean, where the sun's energy again heats the surface, evaporating the water and continuing the cycle. Water trapped in porous rock between the ground surface and impervious rock far below creates an aquifer. This underground water supplies wells and feeds springs that many species use. Spring water may evaporate or flow into streams, linking the groundwater system to the overall water cycle.

The Carbon Cycle

In the carbon cycle, autotrophs such as plants capture the sun's energy and use it with atmospheric carbon dioxide (CO_2) to synthesize organic compounds that are incorporated into their tissues (fig. 43.14). Cellular respiration releases carbon to the atmosphere as CO_2. Dead organisms and excrement also return carbon to soil or water. Invertebrates feed on and fragment the dead, and bacterial and fungal decomposers complete the breakdown of these organic compounds to release simple carbon compounds into the soil, air, and water.

Certain geological deposits contain carbon from past life. Limestone, for example, consists mostly of exoskeletons and shells of ancient sea inhabitants. Fossil fuels, such as coal and oil, form from the remains of long-dead organisms. When these fuels burn, carbon returns to the atmosphere as CO_2.

FIGURE 43.14

The Carbon Cycle.

Carbon dioxide in the air and water enters ecosystems through photosynthesis and then passes along food chains. Respiration and combustion return carbon to the abiotic environment. Carbon can be retained in geological formations and fossil fuels for long periods of time.

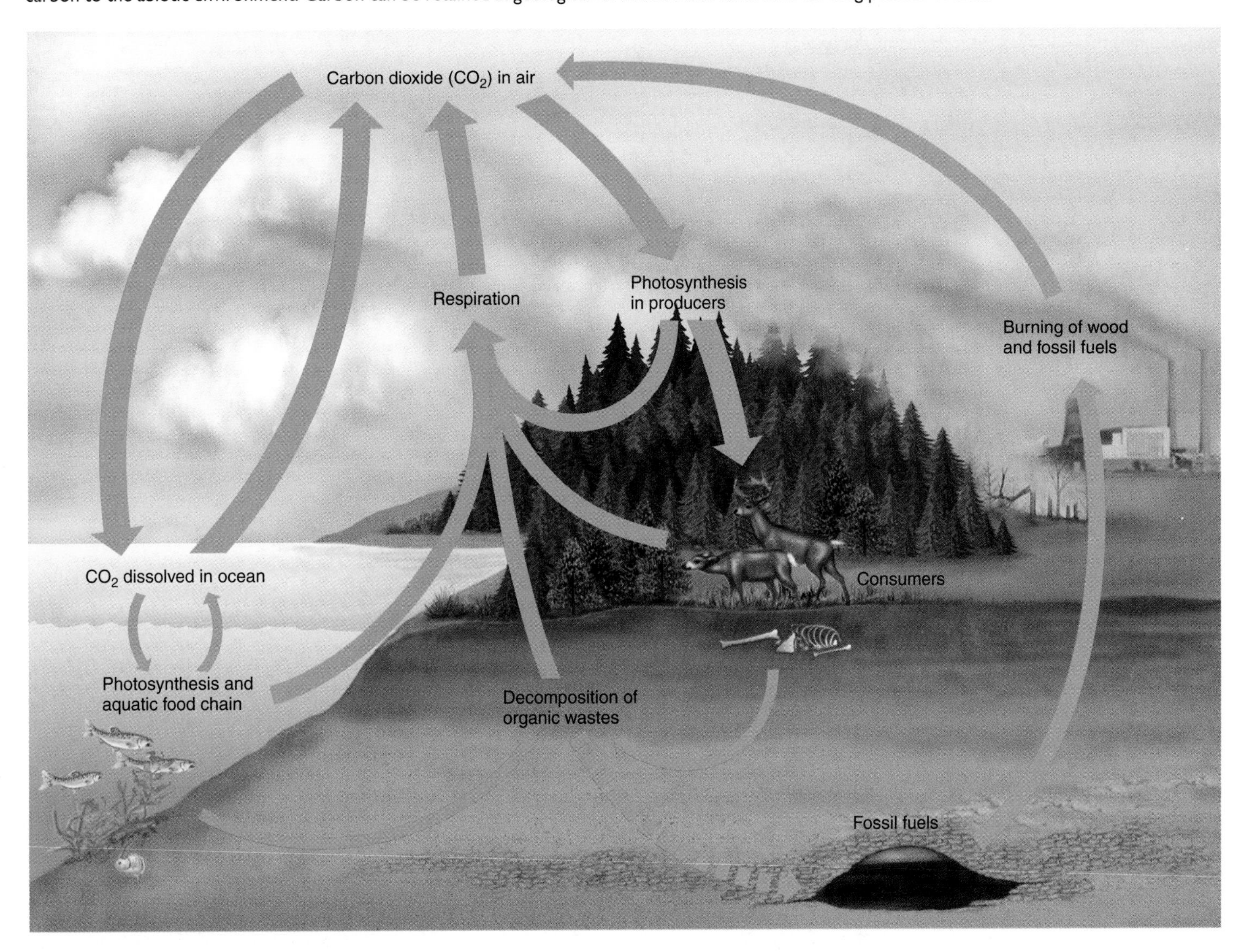

The Nitrogen Cycle

Nitrogen is an essential component of proteins and nucleic acids, as well as other parts of living cells. Although the atmosphere is 79% nitrogen gas (N_2), most organisms cannot use this nitrogen to manufacture biochemicals. They depend on free-living or symbiotic **nitrogen-fixing bacteria** that convert N_2 into ammonia, NH_3 (present as ammonium ion, NH_4^+), which can be incorporated into plant tissue. Decomposers also release some ammonia. **Nitrifying bacteria** convert ammonia to nitrites (NO_2^-) and eventually to nitrates (NO_3^-), which plants can incorporate and then pass to animals. (Some nitrate is also produced when lightning fixes atmospheric nitrogen.) Finally, N_2 returns to the atmosphere when **denitrifying bacteria** convert nitrites and nitrates to nitrogen gas (fig. 43.15).

The enzyme nitrogenase enables nitrogen-fixing bacteria to convert N_2 to other nitrogen-containing compounds. Because oxygen inactivates this enzyme, nitrogen-fixing microbes are typically anaerobic or shield nitrogenase from oxygen. For example, *Rhizobium* bacteria live in nodules on the roots of legumes such as beans, peas, and clover. Farmers rotate legumes with nonleguminous crops, such as corn, so the soil is continually enriched with biologically fixed nitrogen.

FIGURE 43.15

The Nitrogen Cycle.

Nitrogen-fixing bacteria convert atmospheric nitrogen gas (N_2) to organic forms and ammonium ion, which plants can use. Plants use nitrogen to synthesize amino acids, nucleic acids, and chlorophyll. They pass these biochemicals along food chains. Nitrogen returns to the abiotic environment in urine and feces and by decomposition of dead organic matter. Specific groups of bacteria convert ammonium ion to nitrate (another form plants can use), and nitrate to nitrogen gas, completing the cycle.

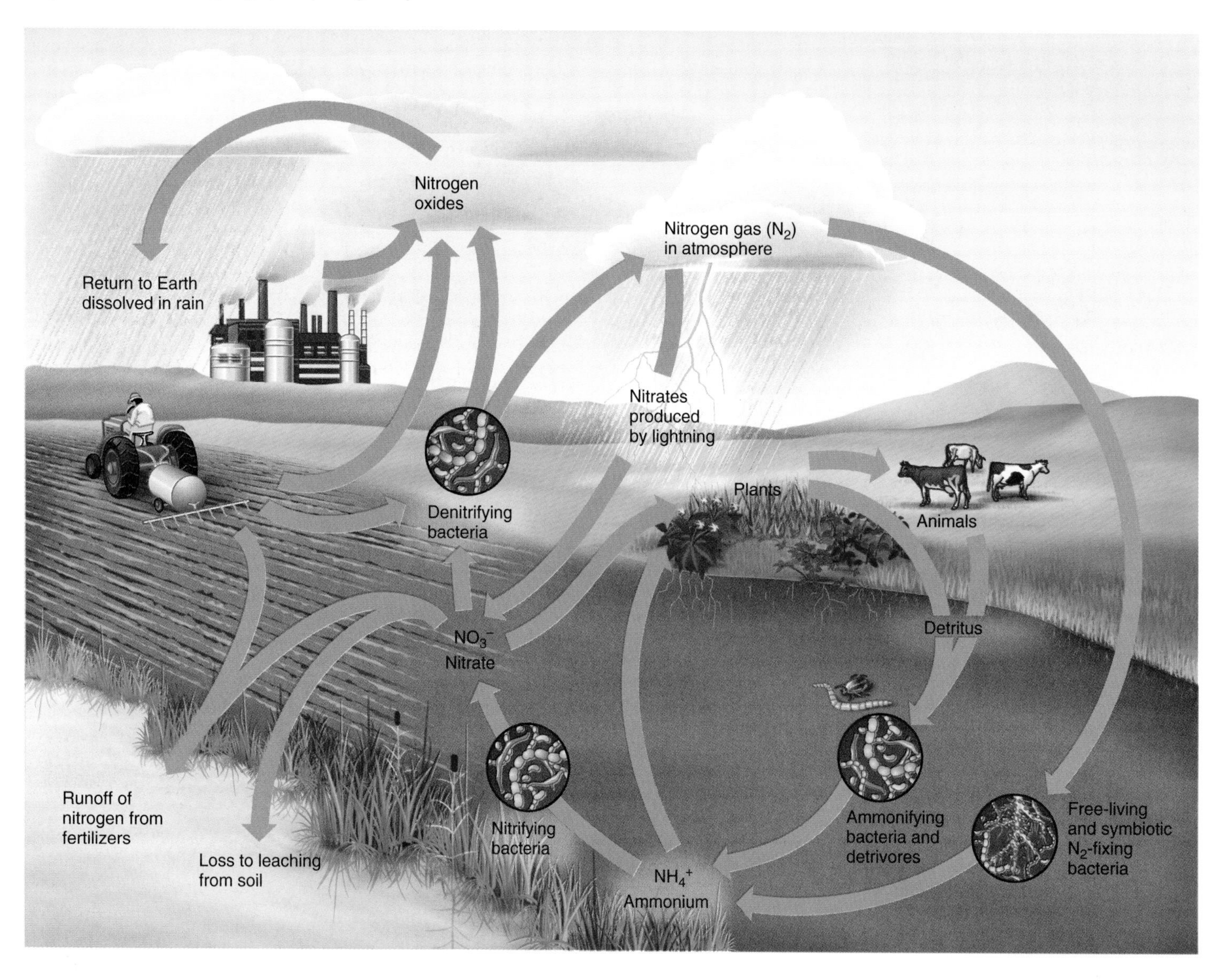

The Phosphorus Cycle

Phosphorus is a vital component of nucleic acids, ATP, and membrane phospholipids and is a structural component of many organisms. Unlike the other biogeochemical cycles, that for phosphorus is based mostly in sediments and rocks, rather than the atmosphere (fig. 43.16). These geological sources gradually release phosphorus in the form of usable phosphate (PO_4^{-3}) to ecosystems as they erode. Organisms assimilate some of this phosphorus, but much of it ultimately returns to the oceans and other bodies of water, where it becomes part of sediments. Geological uplift eventually returns sedimentary rock to the land. Decomposers return phosphorus assimilated by living organisms to soil and water.

Some Chemicals Become Concentrated as They Ascend Food Webs

Because cells admit some chemicals but not others, certain chemicals become more concentrated within cells than in the surrounding environment. This process, termed **bioaccumulation,** can concentrate particular elements or compounds in

FIGURE 43.16
The Phosphorus Cycle.
Phosphorus comes from rock, which slowly erodes. Soluble phosphates are taken up by plants (often with the help of mycorrhizal fungi) and passed up food chains. Decomposers return phosphorus to the abiotic environment. Phosphorus mining and its distribution in fertilizer (not shown) has increased phosphorus availability to both terrestrial and aquatic organisms.

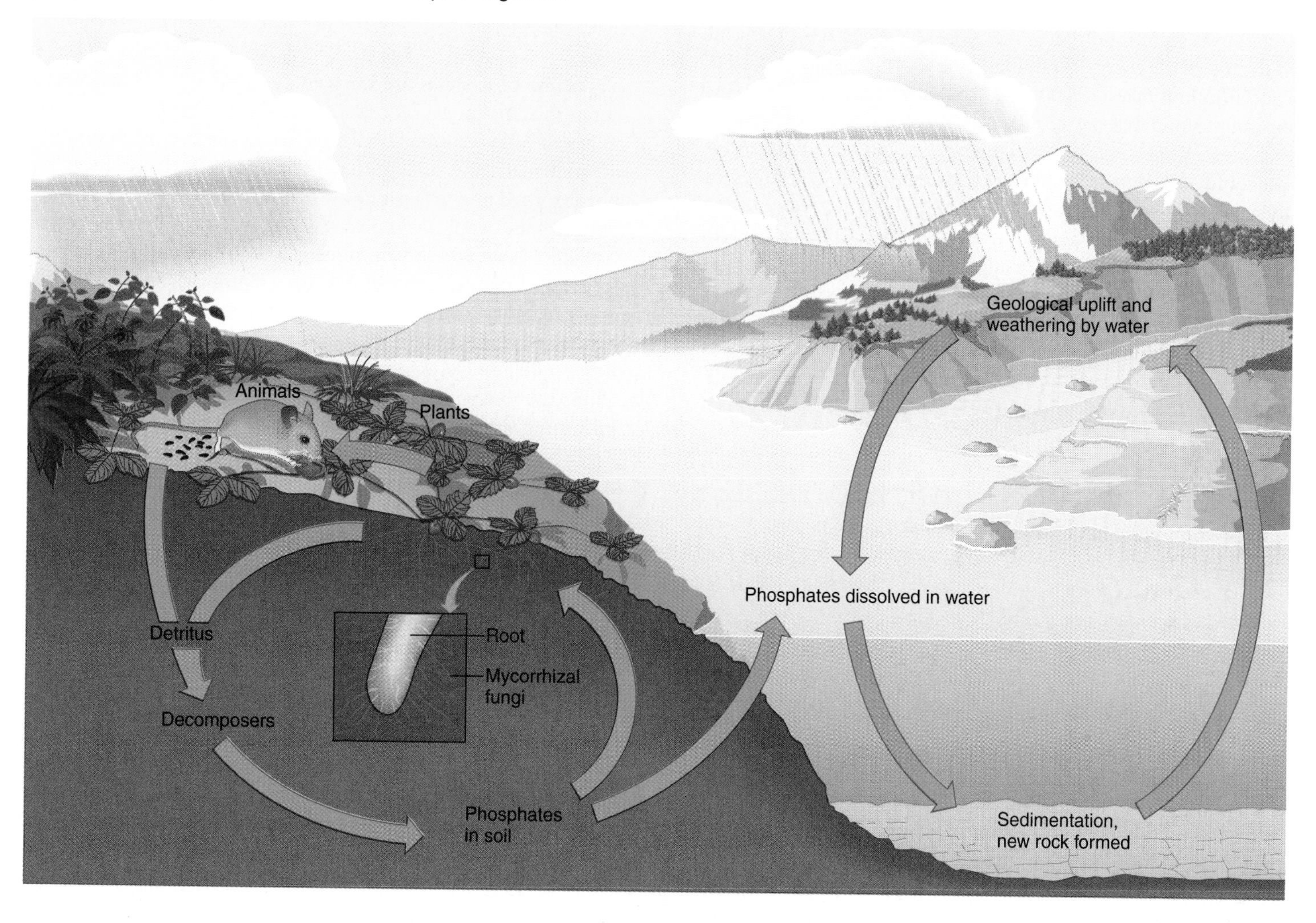

cells to thousands or millions of times their concentrations in the environment. Some substances, including certain synthetic compounds not normally found in ecosystems, become increasingly concentrated in organisms at successively higher trophic levels. This process, called **biomagnification,** occurs because the chemical passes to the next consumer rather than being metabolized or excreted. Two classic examples of biomagnification concern the pesticide DDT, and mercury contamination of fish.

DDT was once widely used to kill insect pests because it damages their nervous systems. But DDT soon proved to harm many other species too. The United States banned use of DDT in 1971, after much evidence showed it to cause cancer, disrupt wildlife, bioaccumulate, and become biomagnified (fig. 43.17). DDT is still

FIGURE 43.17
DDT Accumulates in Top Predators.
The corpse of this about-to-hatch peregrine falcon, discovered amidst broken, empty eggs in Scotland in 1971, is testament to the concentration of DDT at the top of the food web, a position the animal occupies as a predatory bird. The pesticide caused birds' livers to abnormally metabolize the hormones required to secrete a firm, calcium-rich eggshell.

used in some countries and it has touched nearly all life on Earth. DDT and its breakdown products have even been found in the fat of Antarctic penguins, who live where DDT was never used. Most of the DDT entering an animal remains in its fat and then passes through successive trophic levels. The pesticide becomes concentrated with each trophic level and most severely affects organisms at the top of the energy pyramid. By the fourth trophic level, DDT concentrations may be 2,000 times greater than in organisms at the base of the food web (fig. 43.18).

Mercury is a naturally occurring element that bioaccumulates and is biomagnified. Plants growing near volcanoes or thermal springs, such as in parts of California, Hawaii, and Mexico, assimilate airborne mercury particles through leaves and mercury in the soil through roots. Phytoplankton bioaccumulate mercury from fresh water or ocean water. Mercury can also be a pollutant. Biomagnification of mercury along food chains may create levels in fishes that are hazardous to humans and top carnivores such as birds of prey and marine mammals. Health 43.1 discusses some effects of the bioaccumulation of mercury.

43.4 MASTERING CONCEPTS

1. What are the basic steps of a biogeochemical cycle?
2. Describe the steps of the water, carbon, nitrogen, and phosphorus cycles.
3. What is bioaccumulation?
4. What is biomagnification?

OLC

43.5 A Sampling of Ecosystems

Earth has many different ecosystems. Here is a look at a few of them.

The deep-sea hydrothermal vents described in the chapter opening essay are but one of the diverse types of ecosystems. Here are a few others (fig. 43.19).

The Sea Surface Microlayer

The sea surface, perhaps the richest ecosystem on Earth, covers approximately 71% of the planet. Life is diverse and abundant in a microlayer only half an inch (1.27 centimeters) thick. The microlayer is an invisible organic film, rich in lipids and fatty acids. Bacteria and fungi are concentrated in the microlayer, as are microscopic algae and other protista. The microlayer is also an important nursery for diverse species of fishes and crustaceans, whose eggs and larvae concentrate there.

The sea surface microlayer is incredibly thin compared to the vast depth of the ocean, but it is of great ecological importance. It is a transitional zone between ocean and atmosphere and is the base of a huge food web, with birds from above and larger animals from below feeding on its abundant life. The sea surface microlayer is also a reservoir for pollutants from oil spills, pesticide residue, sewage, by-products of combustion from planes and cars, and wastes from incinerators and power plants.

Vernal Pools

Some ecosystems are transient, existing only under certain conditions that prevail at certain times of the year. In many places in North America, small bodies of water called vernal pools form as snow melts and flows into depressions in the land. If water is not absorbed into the ground and cannot exit other than by evaporation, the pool may persist until summer (vernal means "spring"). That is plenty of time for a vibrant community of bacteria, algae, plants, invertebrates, and amphibians to arrive. Without predatory fishes, which cannot live in such temporary habitats, the mix of species that forms the vernal pool community flourishes. Decaying plant matter in the still water provides abundant nutrients to fuel food chains, and soon the pool is brimming with assorted larvae.

The yellow-spotted salamander is one species that instinctively seeks the vernal pool of its birth, where it will mate and lay eggs. If a pool vanishes, they will usually not mate. To protect the animals from being crushed by traffic on roads that they must traverse in their migration, some human communities post "salamander crossing" signs or even dig tunnels to assist them.

When the heat of summer dries up a vernal pool, its residents may either move on, or display other adaptations to drying. Most insects lay eggs in the drying mud of summer and leave, their offspring hatching the next spring. Midges and mayflies wait until spring to lay their eggs. Snails, small invertebrates called fairy shrimp, and tiny clams can enter a dormant stage and survive until the pool forms again, even if that is years later. Amphibians simply walk away, much as they arrived.

Niagara Escarpment

A sheer 98-foot (30-meter) cliff face doesn't seem a likely habitat, but several hardy species occupy the Niagara Escarpment in Ontario, Canada. The climate there is much more like the dry cold of the Arctic tundra than the surrounding wet and temperate Canadian forest.

Eastern white cedars of species *Thuja occidentalis* emerge from cracks in the cliff face. These trees are superbly adapted to their unusual environment. All of the trees are stunted, so that they cannot easily blow over. They grow very slowly, which may be a response to scarce nutrients and little space for root growth. A tree may widen only 0.002 inches (0.05 millimeter) in a year.

Life exists not only between the rocks of the escarpment but within them. Colored layers just beneath the rock surface consist of algae, fungi, and lichens. This sparse, cryptoendolithic ("hidden inside rocks") ecosystem survives because the translucent rock lets sufficient sunlight through to allow photosynthesis to occur.

FIGURE 43.18

Biomagnification in a Sampling of Wetland Organisms.

The concentration of DDT in organisms' bodies increases up the food web. Values represent concentrations of DDT per unit of tissue, measured in parts per million (ppm).

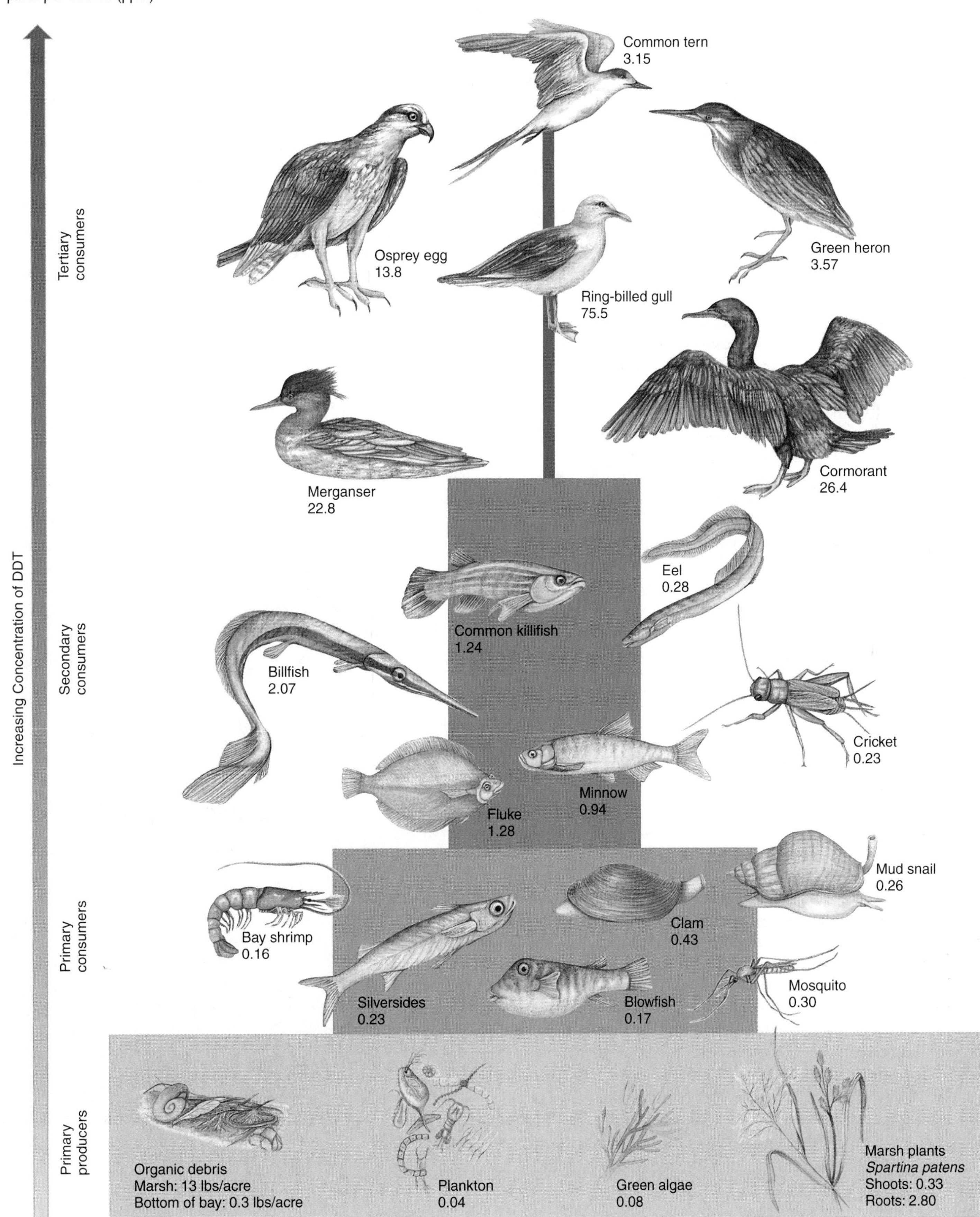

Biomagnification: "Cat's-Dancing Disease" Revealed Mercury Disaster

The world learned about the bioaccumulation of mercury from the small fishing village of Minamata, Japan. The pollution that would eventually cause severe nervous system damage in people who ate tainted fish began in the 1930s, when a chemical production facility moved to the area and began releasing mercury into the bay. There, the element was converted into toxic methyl mercury, which microorganisms passed up the food chain through fishes and crustaceans, to mammals such as cats and humans. The Minamata disaster is unusual, and perhaps as devastating as it was, because the bay ecosystem was relatively closed, delivering very high levels of mercury to higher trophic levels.

In the years following World War II, the mercury pollution in the water increased, and by the mid 1950s, dead fishes floated in the bay. Even stranger was an epidemic of "cat suicides." The animals would "dance" about, some falling into the bay and drowning. People were beginning to show symptoms too, subtle at first—poor coordination, shaking, and impaired hearing. Symptoms progressed to complete lack of control over movements, partial paralysis, inability to speak, and convulsions, as their brains were being slowly destroyed. As the opening essay to chapter 31 describes, mercury compounds concentrate in neural tissue.

The mysterious condition became known as "cat's-dancing disease." What the cats and the people had in common was eating fish—mercury-contaminated fish. Fishermen's families were the earliest victims; the first reported cases were five children who had collected seafood.

By the end of the 1950s, the source of the disease was traced to wastewater from the chemical company, although this fact was not publicized. Twenty adults had died, and at least four times that many were ill. The company made compensatory payments to affected families and tried to control the pollution, but the worst damage had already been done—children soon began to be born with the illness, which for a time was called Minamata disease. When pregnant women ate the heavily contaminated fish, mercury concentrated in the fetal brain, causing children to be born with symptoms of cerebral palsy, mental retardation, and deafness. Some children were severely deformed, a disaster that was vividly documented in *Life* magazine in the 1970s (fig. 43.B).

The Public Health Service officially identified the cause of the disease in 1963, and from then until the present time, great effort and expense have been applied to cleaning up the area and removing the contaminated sediments. A memorial statue now stands beside the bay, which reopened for use in 1997.

The link between industrial pollution and biomagnification was so stark in the Minamata case that Rachel Carson highlighted it in her famous book, *Silent Spring*. Yet other outbreaks of mercury poisoning stemming from pollution have been more difficult to assess. Complicating factors include multiple chemical contaminants in fishes, variation of the concentration of mercury from fish to fish, the amount of fishes that a population normally eats, how closed the ecosystem is, and the genetic background of the population. Reports have been conflicting. A 1997 study identified neurological deficits among children in the Faroe Islands, whose mothers had eaten fish diets high in mercury during pregnancy, as indicated by hair analysis for the element. The Faroe Islands problems were not as severe as in Minamata, consisting mostly of difficulties with memory, language skills, and attention span. Yet a 1998 investigation of similarly exposed children in the Republic of Seychelles (in the Indian Ocean) did not reveal neurological problems.

While the Minamata tragedy of "cat's-dancing disease" and its human counterpart illustrated biomagnification of a toxin, it is clear that assessing effects of pollution can be very complex.

FIGURE 43.B
Mercury Poisoning.
This young man was exposed to methyl mercury as a fetus, because his mother ate fish contaminated with waste released from a chemical manufacturing facility. The mercury was biomagnified up the food chain.

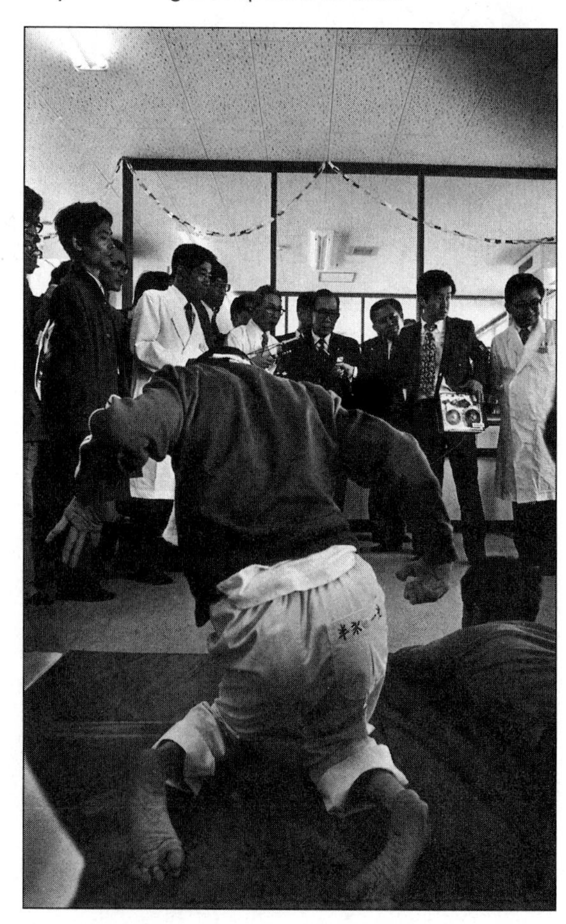

FIGURE 43.19
A Sampling of Ecosystems.
(**A**) The sea surface microlayer brims with tiny adult animals, eggs, larvae, and nutrients. (**B**) This yellow-spotted salamander is migrating to its ancestral pond as the snow melts. (**C**) Eastern white cedar trees are adapted to life on the Niagara escarpment. (**D**) Life in a fjord tolerates the unusual condition of salt water meeting fresh water. This is New Zealand's Milford Sound.

A

B

C

D

Fjords—Where Fresh Water Meets Seawater

In the southwest corner of New Zealand's national park, narrow glacier-cut valleys called fjords (or fiords) extend from the Tasman Sea 10 miles (16.1 kilometers) distant, ending between the Southern Alps. The fjords flood, but in an unusual way—seawater pours in, but near-daily rains send rivulets of fresh water down the steep mountain slopes. The result is a body of water that is fresh water for 10 to 12 feet (3.1 to 3.7 meters) on top but seawater below. Fresh water rises because it is less dense and because the fjords are so narrow that waves, which would mix the layers, cannot form.

The spectrum of life in the fjords ranges from a freshwater to a seawater ecosystem. In the surface layer, freshwater mussels and barnacles abound. The water here is stained yellow-brown from leaf litter. The color blocks much sunlight, so the sealike portion of the fjord lacks the light-dependent seaweeds and other algae common in the ocean. But dark-adapted sea life is plentiful and includes sea slugs, corals, and sponges. Brachiopods, crinoids ("sea lilies"), and other shelled animals cling to the cliff walls.

Predatory starfishes hover at the interface between fresh and salty water and shift position as the interface moves. When a day without rain raises the boundary, freshwater animals are trapped in salt water—giving the waiting starfishes an easily captured meal. As soon as the rain returns, the starfishes move down, for they cannot tolerate fresh water.

The next chapter continues our look at the environment, from a broader perspective.

43.5 MASTERING CONCEPTS

1. What kinds of organisms live in the sea surface microlayer?
2. Why do invertebrates and amphibians flourish in vernal pools?
3. What types of species are adapted to life on cliffs?
4. What is unique about fjords?

Chapter Summary

43.1 A Community Includes All Life in an Area

1. **Biotic communities** are made up of coexisting species that are specialized in where, when, and how they live. Communities are most diverse near the equator. An **ecosystem** consists of biotic communities and the abiotic environment.
2. Each species has characteristic conditions where it lives **(habitat)** and resources necessary for its life activities **(niche).** Slight differences in niche allow different species to share surroundings. Because of **interspecific competition,** a species' **realized niche** is often smaller than its **fundamental niche.**
3. **Symbiotic** relationships and **predation** influence community structure. In **coevolution,** the interaction between species is so strong that genetic changes in one population select for genetic changes in the other. **Keystone species** are members of communities that disproportionately affect community composition.

43.2 Communities Change Over Time

4. As species interact with each other and their physical habitats, they change the community, a process called **ecological succession. Primary succession** occurs in a previously unoccupied area; **secondary succession** occurs after a disturbance. Succession leads toward a stable **climax community,** but complete stability is rare—most communities continue to change.

43.3 An Ecosystem Is a Community and Its Physical Environment

5. An ecosystem can range from a very small area to the entire **biosphere.** Ecosystems interact and change. Major terrestrial ecosystem types are called **biomes.**
6. A **food chain** begins when **primary producers** harness energy from the sun or inorganic chemicals, forming the first **trophic level.** The total amount of energy converted to chemical energy is gross primary production. The energy remaining after producers' metabolism is **net primary production.**
7. **Consumers** comprise the next trophic levels. Primary consumers (herbivores) eat the primary producers. A secondary consumer may eat the primary consumer, and a tertiary consumer may eat the secondary consumer. **Decomposers** break down nonliving organic material **(detritus)** into inorganic nutrients.
8. Food chains rarely extend beyond four trophic levels because only a small percentage of the energy in one trophic level transfers to the next level. Ecological pyramids measure energy, numbers of organisms, or **biomass** in a food chain. Food chains interact, forming **food webs.**

43.4 How Chemicals Cycle Within Ecosystems—Biogeochemical Cycles

9. **Biogeochemical cycles** are geological and chemical processes that recycle chemicals essential to life.
10. Water cycles from the atmosphere as precipitation over land or water, then into organisms that release water in transpiration, evaporation, or excretion.
11. Autotrophs use atmospheric carbon in CO_2 to manufacture carbohydrates. Cellular respiration and burning fossil fuels release CO_2. Decomposers release carbon from once-living material.
12. **Nitrogen-fixing bacteria** convert atmospheric nitrogen to ammonia, which plants can incorporate into their tissues. Decomposers convert the nitrogen in dead organisms back to ammonia. **Nitrifying bacteria** convert the ammonia to nitrites and nitrates. **Denitrifying bacteria** convert nitrates and nitrites to nitrogen gas.
13. As rain falls over land, rocks release phosphorus as useable phosphates. Decomposers return phosphorus to the soil.
14. **Bioaccumulation** concentrates chemicals in cells relative to their surroundings. **Biomagnification** concentrates chemicals to a greater degree at successive trophic levels as the chemical passes to the next consumer rather than being metabolized or excreted.

43.5 A Sampling of Ecosystems

15. The top half-inch of the ocean contains abundant life.
16. Vernal pools exist for only a few months, but in that time house thriving communities.
17. Hardy trees and microorganisms live on cliff faces.
18. Fjords include layers of fresh water and seawater.

Testing Your Knowledge

1. How does a community differ from an ecosystem?
2. How does a habitat differ from a niche? How does a fundamental niche differ from a realized niche? Describe your habitat and niche.
3. How can two species share the same habitat?
4. Describe and give examples of four types of symbiotic relationships.
5. What adaptations do prey use to avoid predators?
6. What is a keystone species?
7. Distinguish between primary and secondary succession.
8. What are the components of ecosystems?
9. How do organisms return water, carbon, nitrogen, and phosphorus to the abiotic environment?

Thinking Scientifically

1. Why are true climax communities rarely reached?
2. How is the inside of the human mouth an ecosystem?
3. Some people have suggested that we "farm" the krill in Antarctica and use it to feed people who are starving elsewhere. What effects might krill farming have on the Antarctic ecosystem?
4. After fires destroyed much of Yellowstone National Park in 1988, forest managers suggested humans could help the areas recover by feeding deer, bringing in plants, and planting trees. What are some of the advantages and disadvantages of intervening in recovery from a natural disaster?
5. How is natural selection apparent in ecological succession?
6. Identify a consumer in
   a. a black smoker
   b. the coastline ecosystem of the Aleutian Islands
   c. the blowdown zone around Mount St. Helens
   d. an ecosystem not mentioned in the text
7. In one type of mimicry, a harmless species such as a jumping spider physically resembles a noxious species such as an aggressive type of ant. Explain why the spiders must be less abundant than the ants for this system to benefit the spiders.
8. How can an organism be both a producer and a consumer?

References and Resources

Allchin, Douglas. June 1999. The tragedy and triumph of Minamata. *The American Biology Teacher,* vol. 61, p. 413. Biomagnification of mercury caused an environmental disaster in Japan in the 1950s.

Bange, Hermann W. November 16, 2000. It's not a gas. *Nature,* vol. 408, p. 301. Nitrous oxide is a greehouse gas that depletes ozone.

Brown, Kathryn. October 6, 2000. Ghost towns tell of tales of ecological boom and bust. *Science,* vol. 290, p. 35. Ecologists study the soils of desert towns devastated by mining efforst in the last century.

Falkowski, P., et al. October 13, 2000. The global carbon cycle: a test of our knowledge of Earth activity as a system. *Science,* vol. 290, p. 291. Increasing atmosphere carbon dioxide, as a result of human activity, will continue to affect the climate.

Stevens, William K. January 5, 1999. Search for missing otters turns up a few surprises. *The New York Times,* p. F1. When sea otters began dying, the ecosystem dominated by great kelp forests began to unravel.

Tyson, Peter. June 1999. Neptune's furnace. *Natural History,* vol. 108, p. 42. A participant describes what it was like to collect a black smoker from the ocean floor.

Vitousek, Peter. June 1997. After the volcano. *Natural History,* vol. 106, p. 48. A volcano to wipe the biological slate clean and a sparse ecosystem create perfect conditions under which to study primary succession.

The *LIFE* Online Learning Center provides additional resources and tools for studying this chapter.
www.mhhe.com/life

CHAPTER 44

Biomes and Aquatic Ecosystems

Ecosystems in Flux

At the Epicenter of El Niño

Sea Surface Temperatures Change During El Niño Years.
An El Niño brings warm waters to the Peruvian and Ecuadoran coasts. In these maps, red areas represent warm water in non-El Niño (**A**) and El Niño (**B**) conditions.

For the people who fish along the coasts of Peru and Ecuador, El Niños are part of life. An El Niño is a periodic global climate shift in which a slack in the trade winds sends unusually warm water from the west Pacific eastward, setting into motion an unstoppable ecological chain reaction. Widespread changes in wind patterns, rainfall, and temperature drastically alter food webs. Some populations swell, while others starve. Along the Peruvian coast, scallop and shrimp populations explode, yet many types of fishes become so scarce that for a time, the fishing industry collapses. With the fishes gone, birds, seals, and sea lions starve.

"El Niño" is Spanish for "the child," and refers to the Yuletide beginnings of the phenomenon. The event usually occurs every 2 to 7 years, and typically lasts a year or two. Ten have occurred since about 1960. An El Niño is often followed by a reversal of conditions called a La Niña.

The 1982–83 event was supposed to be the "El Niño of the century." In some places, the wind shifted direction, causing powerful storms and raising sea surface temperature along a 5,000-mile (8,050-kilometer) belt in the Pacific around the equator. Fishes that were adapted to tropical and subtropical waters followed the warmth toward the poles. Many shore populations starved as changing sea levels and temperatures drastically disrupted ecosystems.

Severe weather patterns also affected the distribution of life. Monsoons in the central Pacific, typhoons in Hawaii, and forest fires raging through Indonesia and Australia decimated many local populations. In northern Peru and Ecuador, torrential rains, caused by the unusually warm Pacific waters, turned deserts into muddy grasslands. Here, insects flourished, providing food for burgeoning toad and bird populations. Fish populations suffered, however, as the animals migrated from the swollen ocean to pockets of salinity in the transient lakes that appeared in what was usually desert. When conditions began to reverse, fishes became trapped and died as the pools dried up.

In the United States, the 1982–83 El Niño caused flooding in southern California, blizzards in the Rockies, a warm winter in the Northeast, and rising chicken prices, as the fishes used in chicken feed migrated south from their

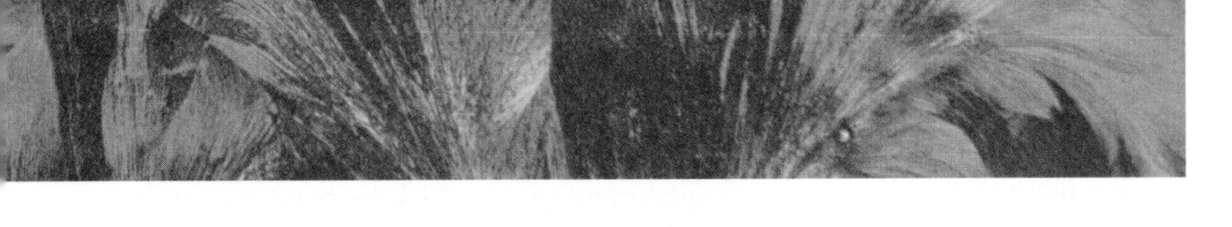

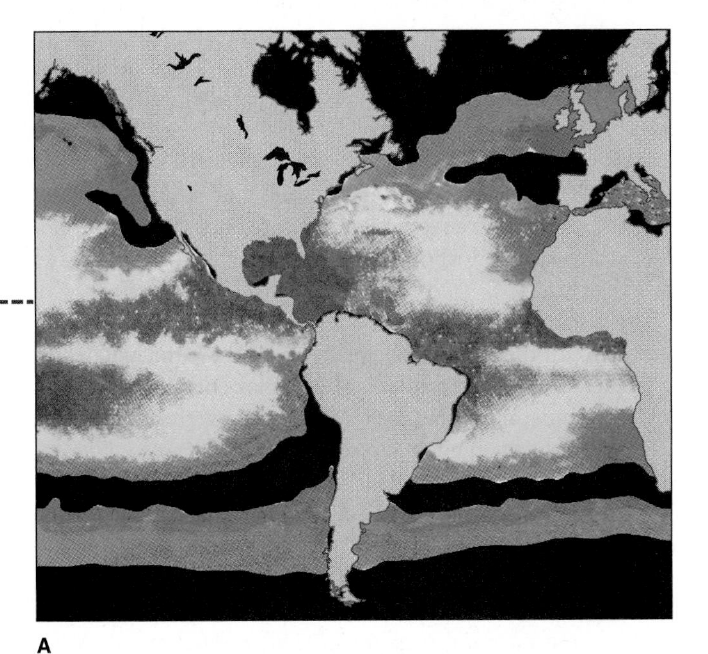

A

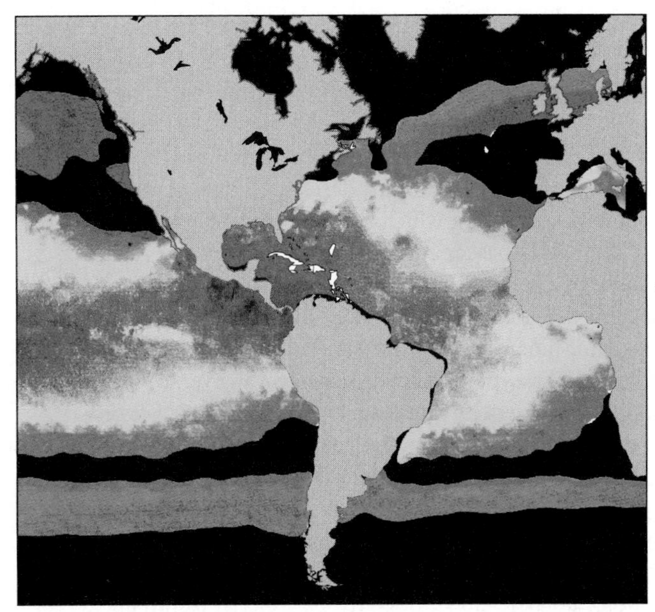

B

usual habitats in Peru and Ecuador and escaped fishing nets. Yet at the same time, areas of Mexico, India, the Philippines, Australia, and southern Africa lost countless plants and animals to severe droughts.

The El Niño that began in December 1996 eclipsed the 1982 event. After a year-long period of starvation, the beaches of Peru became littered with sea lion pups, born prematurely because their mothers were starving. Adult carcasses washed ashore weeks later. Dead Peruvian boobies, a guano-producing bird, appeared too, as seabird nests went empty. Humboldt penguins were especially hard hit, an entire generation of nestlings wiped out by a 52-hour rainstorm in an area that hardly ever sees rain. Ecologists canvassed the area to assess the damage to populations.

Ecologists deduced the chain reaction of climatic and biological events comprising this most recent El Niño. An atmospheric event called the Madden-Julian oscillation (MJO) boosted the Pacific warming that was already beginning in December 1996, and again in March 1997. An MJO is a 1- to 2-month-long wind burst near the equator that forms a patch of clouds that moves quickly over the Pacific from west to east. The MJO also increased evaporation in the western Pacific, which cooled conditions there. When the western Pacific cooled, the temperature difference between it and the eastern Pacific lessened, which diminished the winds that normally keep warmer water in the west. The overall effect: warming of the eastern Pacific. By January 1998, the average temperature of the eastern Pacific was 72°F (22°C). That was 13°F greater than the previous 30-year average and three times the temperature increase predicted by models that did not take MJO into account.

The rising sea surface temperature pushed down the thermocline, which is a zone of rapid temperature change where the cold nutrient-rich water from below meets the surface water heated by the sun. When the thermocline fell under the influx of warm water, the huge numbers of anchovies and sardines that live in the colder waters moved down. When this happened, the birds and seals that normally scoop these smaller fishes from the water could no longer reach them. The predators starved.

This most recent El Niño began to subside by May 1998. A year later, many of the areas in the Americas overwhelmed by El Niño–induced floods began to experience droughts caused by La Niña. This event begins in the eastern tropical Pacific Ocean, as low barometric pressure combines with a pool of westward-moving warm surface water. Despite these changes, the fishing families off the coast of Peru know that it is just a matter of time until the oceans warm again.

EL NIÑO-INDUCED CHANGES IN SELECTED ANIMAL POPULATIONS

	Number of Living Animals	
Population	**January 1997**	**May 1998**
Humboldt penguins	5,000	50
Fur seals	40,000	2,800
Sea lions	150,000	28,000

44.1 Biomes

We can view the planet as a patchwork of large geographic areas characterized by specific temperature and precipitation ranges, soils, and plant communities, which determine communities of other organisms. Biomes include several types of forests, savannas, grasslands, deserts, and tundra.

Ecology can be studied at several levels, from populations, to communities, to ecosystems. Scientists recognize several distinctive major types of terrestrial ecosystems, called **biomes,** that occupy large geographic areas. Forests, deserts, and grasslands are examples of biomes, each of which supports characteristic communities. The water-based equivalents of biomes are called aquatic ecosystems. They occur in fresh water, in the oceans, and in the regions where salt and fresh water meet.

Plants define biomes because they are primary producers, and therefore form the bases of food webs. They form an overall pattern of vegetation that influences which microorganisms, fungi, and animals can exist in the biome. The types of plants reflect long-term interactions with regional climate and soils, although the taxa in a given biome may differ on different continents. Animals also contribute to the structure of a biome. Humans have drastically altered many natural biomes, replacing forests, for example, with farmland, suburban housing, and cities. Finally, regional disturbances such as periodic fires influence the communities in some biomes.

Climate Influences Biomes

The planet can be divided into major climatic regions based on temperature, within which biomes can be distinguished by the availability of moisture, which determines the spectrum of species present. The major climatic regions, in order of decreasing average temperature, are the tropics, the temperate zone, the subarctic, and the arctic. For example, tropical biomes range from the wettest, the tropical rain forest, to the tropical dry forest, to the still drier savanna, to semidesert and the hot, dry desert. Major changes in temperature also occur with elevation within biomes. Figure 44.1 depicts these temperature-defined regions in terms of latitude and altitude. Conditions generally become drier and colder at higher latitudes and increasing altitudes, and biodiversity tends to decrease (table 44.1). Resident populations must be adapted to specific ranges of temperature, moisture, and wind conditions, and also to how these aspects of climate fluctuate during the year.

The great variety of physical environments on Earth is a consequence of the fact that the planet is a sphere that rotates at an angle as it orbits the sun (fig. 44.2). Solar energy is most intense at the equator, where the sun is directly overhead. In other areas the sun's rays hit the surface at oblique angles. Due to Earth's curvature, the same amount of solar energy is distributed over a larger area, and so the average temperature falls with distance from the equator. Different latitudes receive the strongest solar rays at different times of the year because of the tilt to the axis of rotation. The northern and southern hemispheres take turns receiving maximal sunlight, receiving equal amounts on about March 21 and about September 23. Therefore from the middle to high latitudes, temperature and day length change throughout the year.

Unequal heating of Earth's surface causes the air movements that distribute moisture. When intense sunlight heats the air over

FIGURE 44.1
Latitude and Altitude Influence Plant Communities.
With increasing elevation on a mountain, the dominant types of plants parallel many of the types that appear as distance from the equator (latitude) increases.

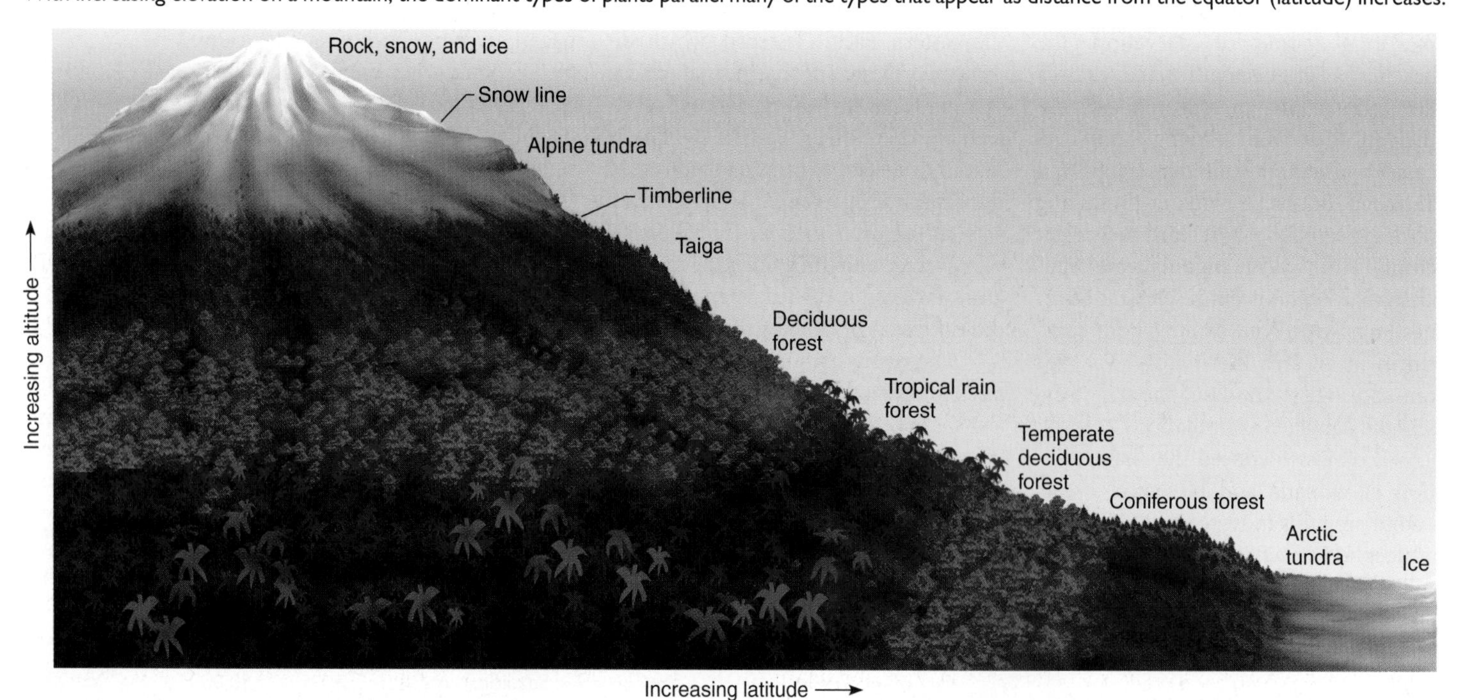

TABLE 44.1 Butterfly Species Diversity

LOCATION	LATITUDE	AREA (HECTARES)	TOTAL SPECIES
Garza Cocha, Ecuador	0° 29′	500	676
La Selva, Costa Rica	10° 26′	1,000	442
Estacion Los Tuxlas, Mexico	18° 35′	700	212
H. J. Andrews Forest, Oregon	44° 14′	6,400	62

Source: P. J. DeVries, "Diversity of Butterflies" in S. Levin (ed.), *Encyclopaedia of Biodiversity,* 2000, Academic Press.

FIGURE 44.2
Earth's Seasons.
The tilt of Earth's axis produces distinct seasons in the northern and southern hemispheres as Earth travels around the sun.

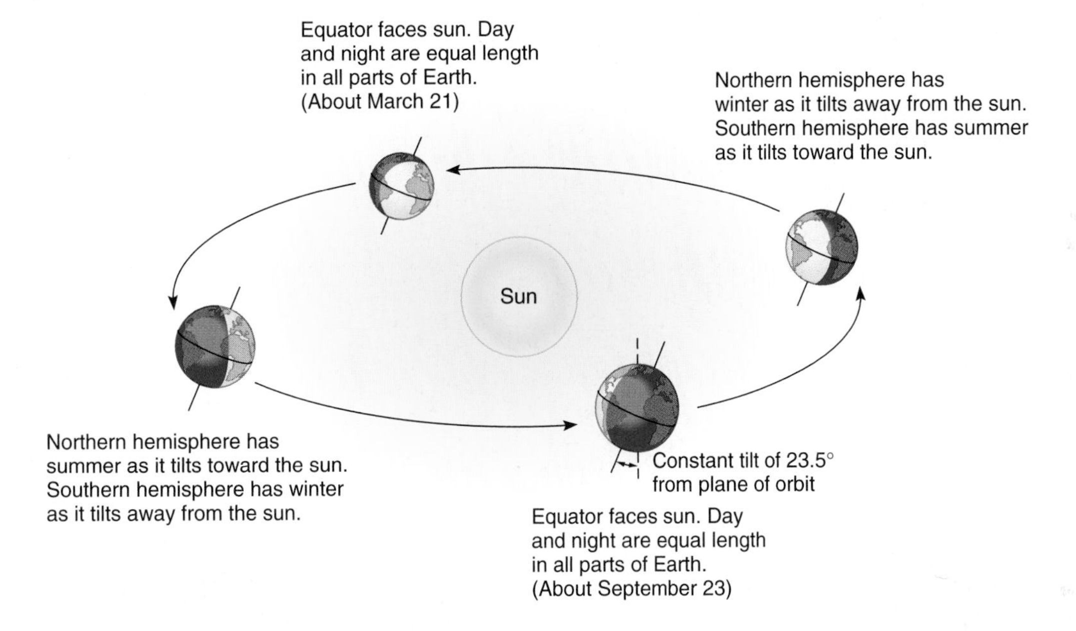

the equator, the air rises and expands. As the air rises, it cools. Because cool air cannot hold as much moisture as warm air, the excess water vapor condenses, forming the thick cloud cover that pours near-constant rain over the tropics. Equatorial air also travels north and south, rising and drying. As it cools at higher latitudes and elevations, the air's density increases, and it sinks back down, at about 30° north and south latitude. Here it absorbs moisture from the land, creating the vast deserts of Asia, Africa, the Americas, and Australia. Some of the air continues poleward, rising and cooling at about 60° north and south latitude, causing rain in these areas, which supports temperate forests. This air rises and some continues toward the poles, where precipitation is quite low, and some returns to the equator. Near the equator the air is heated again, and the cycle begins anew. A cycle of heating and cooling, rising and falling air is called a cell. Rising, moist air loses water as it cools; falling dry air heats up and retains moisture. The planet has three such cells above the equator and three below (fig. 44.3).

Overall, Earth's tilt, resulting in uneven solar energy distribution, causes major winds. The slight winds over the equator are called the doldrums. Trade winds lie above and below the doldrums, blowing from the east-northeast in the northern hemisphere and from the east-southeast in the southern hemisphere. Westerlies blow from west to east from 30° to 60° south and north latitude. The polar areas have weak winds called easterlies. The trades, westerlies, and easterlies blow from the east or west because of Earth's rotation.

Soils form the framework of biomes, for they directly support the defining plant life. Recall from figure 27.4 that soil is a complex mixture of nonliving and living matter, rich with fragmented and pulverized rocks and microbial life.

Climate influences soil development in many ways. Heavy precipitation may leach soluble materials from surface layers and deposit them in deeper layers, or it may remove them entirely from the soil system. Temperature and moisture conditions may also determine the fate of organic matter from living organisms. Under a warm, moist climate, rapid decomposition may leave little humus (organic material) in the soil. In cold, damp climates, undecomposed peat may accumulate in the soil.

The following discussion considers some of the world's major biomes, as shown in figure 44.4. There are others, and they tend to be more continuous than the lines on the map imply.

FIGURE 44.3

Patterns of Air Circulation and Moisture.

(**A**) Air heated over the equator rises, cools, and drops its moisture. The air falls at about 30° north and south latitude, absorbs moisture from the land, then some spreads farther north and south. Six patterns of air circulation (cells) form. The winds are deflected to the right in the northern hemisphere and to the left in the southern hemisphere because of Earth's rotation. (**B**) Clouds form over the tropics as equatorial air cools and releases moisture. Light winds here are called doldrums.

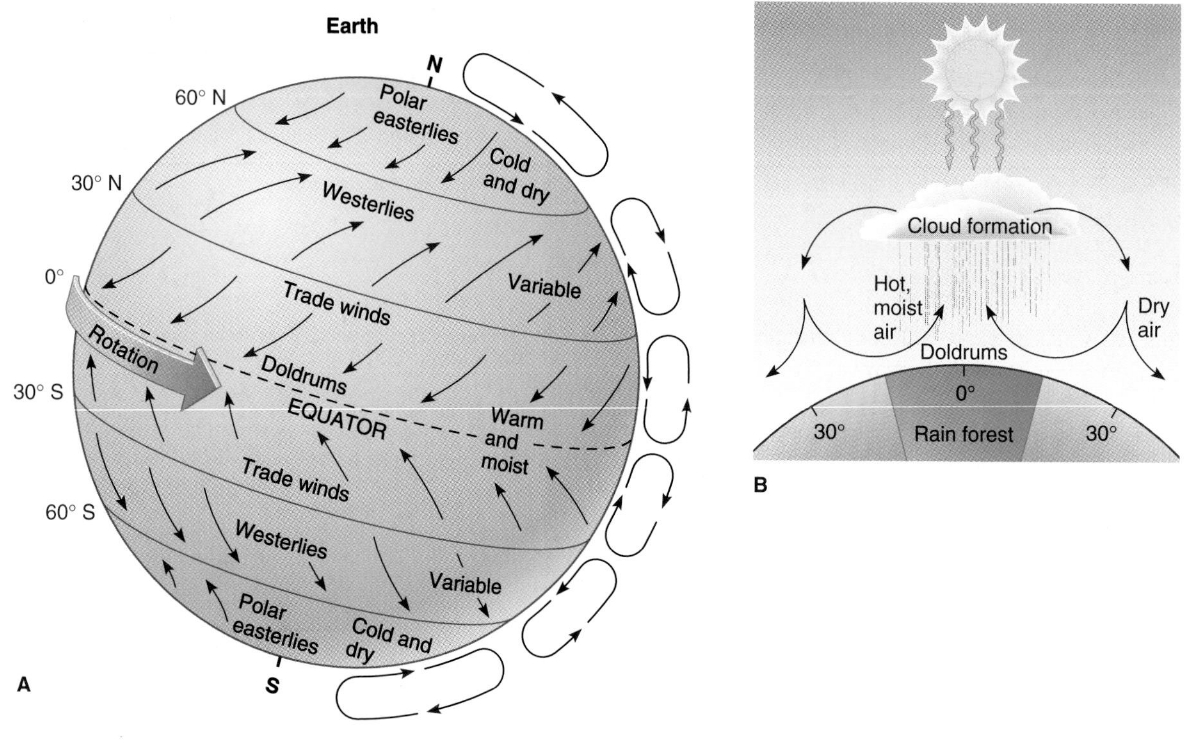

FIGURE 44.4

Earth's Major Biomes.

Biomes are large terrestrial areas that have characteristic plant communities.

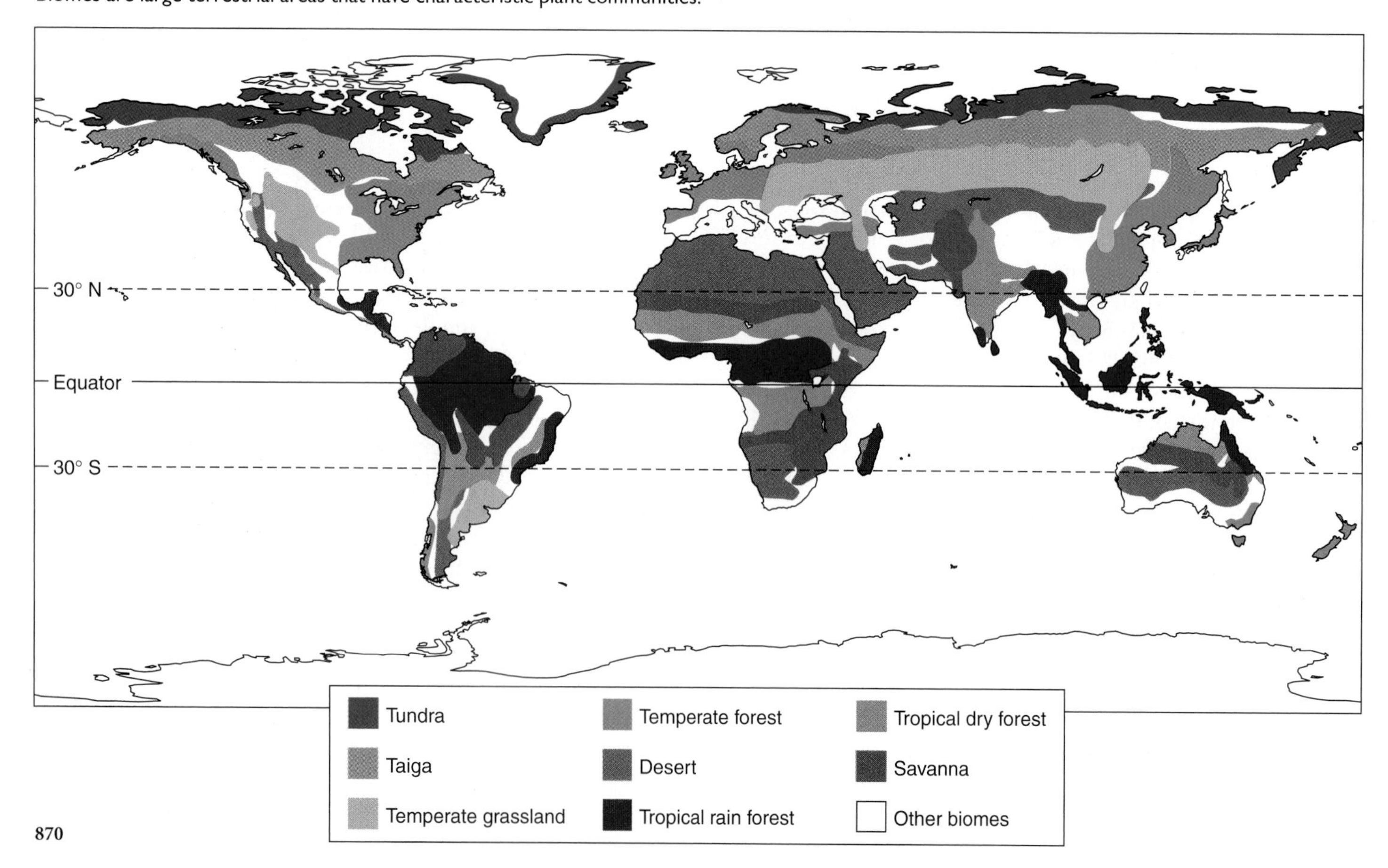

Tropical Rain Forest

Tropical rain forests occur where the climate is almost constantly warm and moist, within 10° north and south of the equator in West Africa, Southeast Asia, and Central and South America. Rainfall is typically between 79 and 157 inches (200 and 400 centimeters) per year, and the soils are often poor in nutrients because water leaches them away. The tropical rain forest of the Amazon Basin in South America is incredibly vast; its meandering rivers and dense foliage cover an area 90% the size of the continental United States. Within the lush maze of intertwined branches and moss-covered vines live a staggering diversity of species. An area of tropical rain forest 4 miles square (10.4 kilometers square) is likely to house, among its 750 tree species, 60 species of amphibians, 100 of reptiles, 125 of mammals, and 400 of birds. One tree alone may support thousands of insect species.

From the air, a tropical rain forest appears as a solid, endless canopy of green, consisting of treetops 50 to 200 feet (15 to 66 meters) above the forest floor. Plants beneath the canopy compete for sunlight and form layers of different types of organisms. This layering of different plant species from the forest floor to the canopy is called **vertical stratification** (fig. 44.5).

Plants in the tropical rain forest are adapted in different ways to capture sunlight. Very tall trees poke through the canopy. Other trees have broad, flattened crowns that maximize their sun exposure in this equatorial region where the sun's rays are almost perpendicular to Earth's surface. Vines and epiphytes (small plants that grow on the branches, bark, or leaves of another plant) grow on tall trees. Epiphytes may have aerial roots that penetrate masses of rich soil on branches. The ground of the tropical rain forest looks rather bare beneath the overwhelming canopy, but it is a habitat for countless shade-adapted species. Many tree saplings grow slowly in the deep shade, but growth surges when a gap opens in the canopy.

The lush vegetation feeds a variety of herbivores. Insects, sloths, and tapirs devour leaves. Small deer and peccaries, monkeys, rodents, bats, and birds primarily eat fruit. Many carnivores, such as large cats, birds of prey, and snakes, consume the herbivores. Most animals are small or medium in size, but a few are enormous. The pirarucu fish of the Amazon River is more than 7 feet (2.1 meters) long, and the wingspans of many insects are several inches.

Nutrients cycle rapidly in the tropical rain forest. Frequent torrential rains wash away nutrients that bacteria and fungi release as they decay dead organisms. The heat and humidity speed decomposition. Termites recycle nutrients with the help of protista in their guts, which break down cellulose and release its atoms back to the environment. Roots of giant trees absorb nutrients so efficiently that most decaying plant material is saved. Essential in a tree's recycling program are fine roots that permeate the upper 3 inches (7.6 centimeters) of ground and attach to dead leaves. These roots absorb nutrients from falling dust and rain. Fungi on rootlets quickly absorb nutrients released from dead leaves. Animals recycle nutrients too. Certain leaf-cutting ants, for example, farm gardens of fungi that hasten decomposition (see chapter 23).

FIGURE 44.5
The Tropical Rain Forest.
Plants in the tropical rain forest form distinct layers, from the tallest trees emerging from the canopy to the tiniest residents of the shady forest floor.

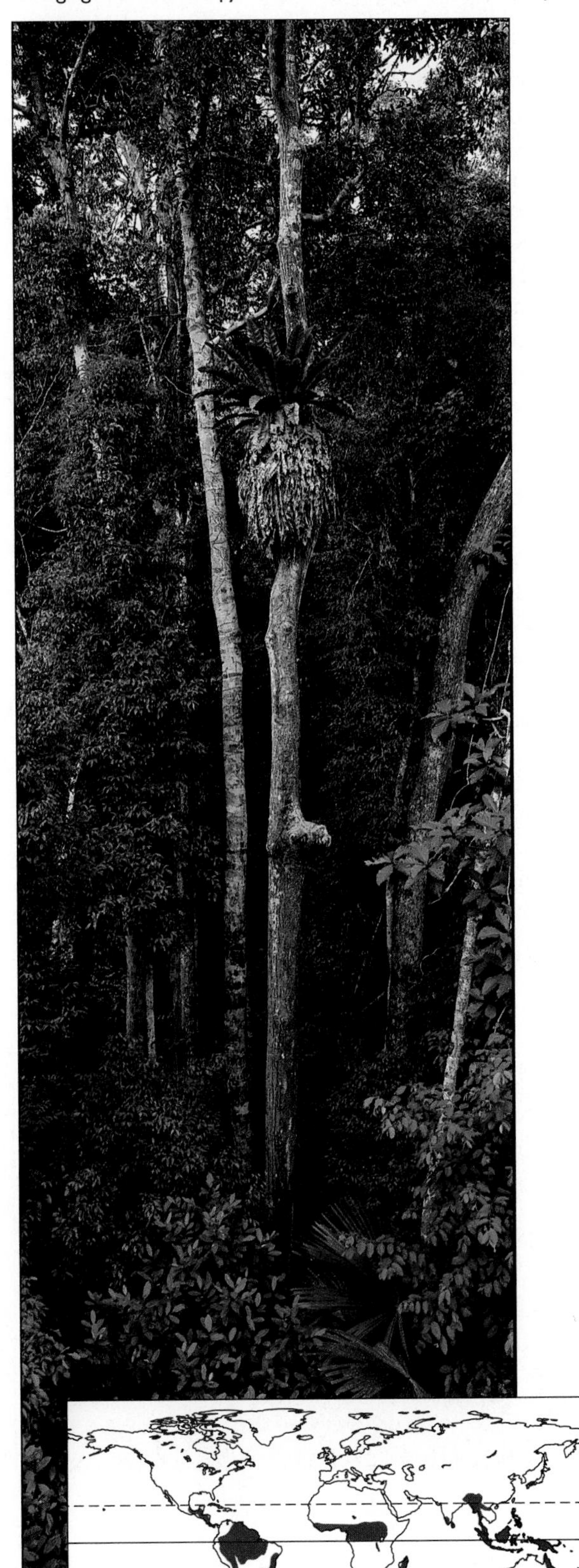

Tropical Dry Forest

From 10° to 25° latitude lie **tropical dry forests,** which correspond to the diminishing moisture level with distance from the equator. These biomes hug the rain forests of central Africa and the Amazon. They cross Australia, and run along Mexico's west coast. In these forests, life is adapted to yearly dry and rainy seasons and fluctuating temperatures. The soil is richer in nutrients than rain forest soils, although the torrential rains of the wet season may quickly erode them away.

The residents of a tropical dry forest are adapted to the seasons. Trees enter states of dormancy during the 6-month dry season, as figure 44.6 vividly illustrates by comparing the same forest in different seasons. Animals, able to move, simply depart when conditions become too dry, migrating to wetter habitats such as the nearby rain forests or river ecosystems. In the wet season, the plants produce abundant flowers and fruits that attract pollinators and herbivores. Unfortunately, the rich soil makes these areas prime agricultural land, and many of these forests have been lost to cultivation by humans. In Central America and Mexico, most of the tropical dry forests have become cotton fields, cattle ranches, and farms where grain is grown. Because of the dry season, these forests are easier to burn and clear than are tropical rain forests.

Costa Rica's Guanacaste National Park is a tropical dry forest that is being preserved. It is named for the guanacaste tree, which produces abundant disc-shaped fruits. Today domesticated animals such as cattle and horses eat the fruits and disperse the undigested seeds in their feces, but these animals are nonnatives. In times past, large populations of wild horses, camels, ground sloths, and other herbivores did the job, and the trees flourished. These animals became extinct about 10,000 years ago. Introducing more horses into the park is helping to restore this biome.

Tropical Savanna

At 10° to 20° latitude, neighboring tropical dry forests are the drier tropical **savannas.** Rainfall is about 12 to 20 inches (30 to 50 centimeters) annually, but certain tropical savannas may exceed this. A distinct dry season lasts at least 5 months and alternates with a wet season. Savannas are grasslands with scattered trees or shrubs and bands of woody vegetation along stream courses (fig. 44.7). Tropical savannas are in Africa south of the Sahara; in South America in south-central Brazil, Venezuela, and Colombia; in northern Australia; and in northwest India and eastern Pakistan.

The animal communities of the tropical savanna differ in different parts of the world. They include many migrating populations, an adaptation to the seasonality of this biome. Australian savanna, for example, is home to many birds and kangaroos, whereas the grassland in Africa is covered with herds of zebra, giraffes, wildebeests, gazelles, and elephants, with lions, cheetahs, wild dogs, and hyenas preying on them and vultures and other scavengers eating the leftovers. In many tropical savannas, termites are major detritivores, and their huge nests dot the landscape. It was from the African savanna that hominids and hominoids—our ancestors—evolved, emerged, and spread (see chapter 18).

A complex interaction of soils, rainfall, grazing, and fire determines the particular vegetation pattern of a tropical savanna. Fires tend to start at the very end of the dry season, when lightning easily ignites dried plants. Grasses, however, can resprout following fire, but often trees cannot. The frequent fires help maintain the grasses.

Desert

Deserts receive less than 8 inches (20 centimeters) of rainfall per year. The days can be searingly hot because few clouds block or filter the sun's strongest rays. Nights are cool, sometimes 86°F

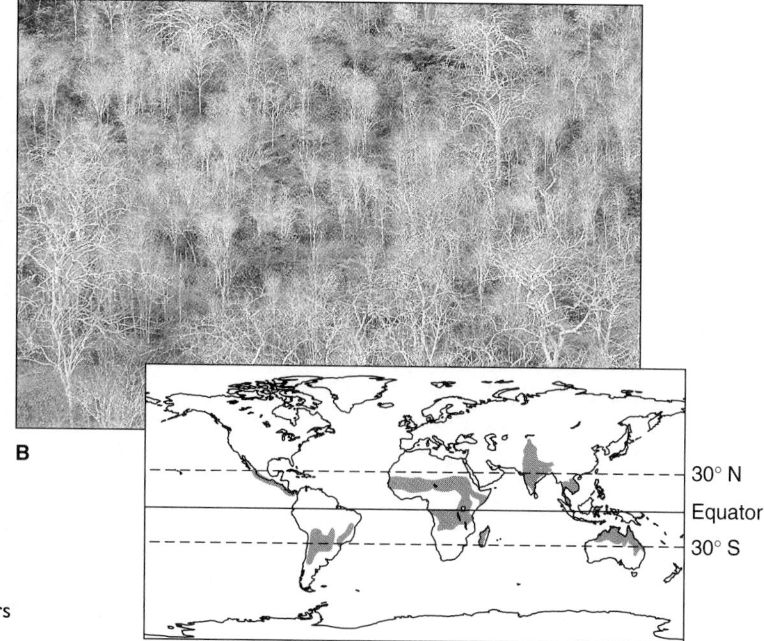

FIGURE 44.6

Tropical Dry Forest.

It's easy to distinguish the wet from the dry season in a tropical dry forest. With the rains, trees spring back to activity, turning the landscape a lush green (**A**). This forest in the Galápagos Islands appears dead during the dry season (**B**).

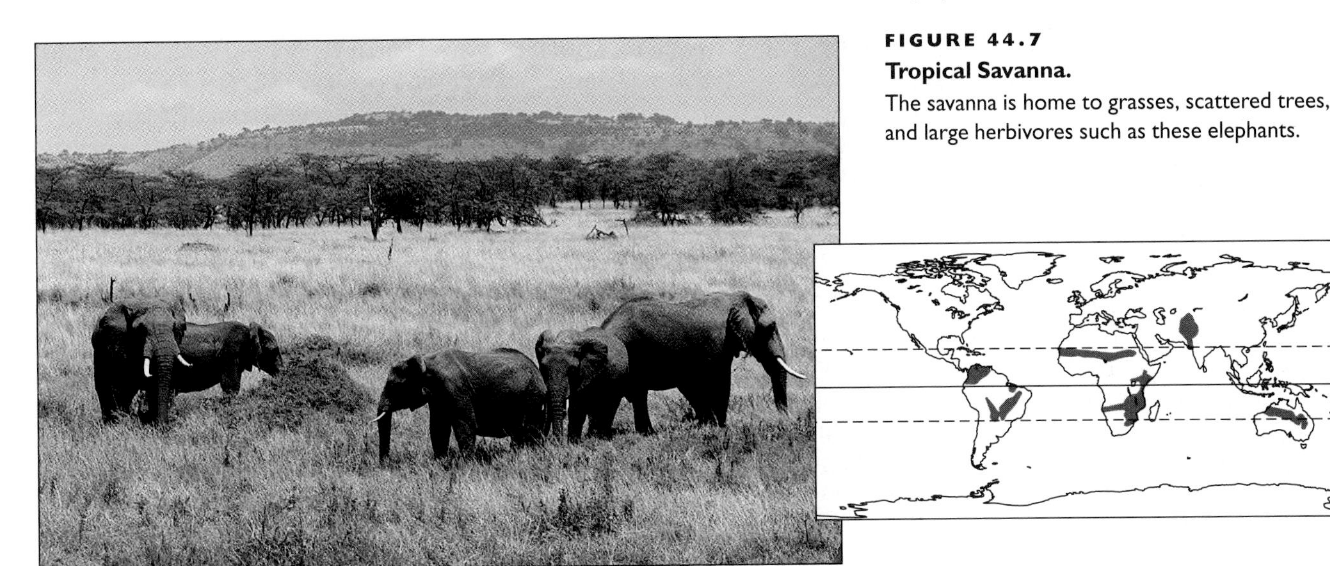

FIGURE 44.7
Tropical Savanna.
The savanna is home to grasses, scattered trees, and large herbivores such as these elephants.

(30°C) below the daytime temperature, because heat radiates rapidly to the clear night sky. Deserts ring the globe at 30° north and south latitude, corresponding to where the circulating atmosphere soaks up moisture.

Deserts vary in biodiversity. Although many desert habitats support only a few species, others, such as the upland desert of southern Arizona, are rich in species (fig. 44.8). In the driest and largest desert in the world, Africa's great Sahara, rainfall is less than 1 inch (2 centimeters) per year. Many areas are almost devoid of life.

Desert plants and animals show many striking adaptations to heat and drought. Some shrubs, such as mesquite, have extremely long taproots that exploit groundwater far below the surface. Annual plants grow quickly, squeezing their entire life cycles into wet periods between droughts. Most of their seeds germinate only after a soaking rain, which rinses growth inhibitors from their coats. The thick-skinned, fleshy leaves of succulent perennials help hold precious water. In rainy times, annual and perennial flowers bloom magnificently.

The reduced leaves of cacti minimize water loss. After a rainfall, the stems of barrel and saguaro cacti expand as they take up and store water. Spines guard their succulent tissues, although some desert animals still eat them. The root system of a cactus is shallow but widespread—that of a large cactus extends up to 55 to 65 yards (50 to 60 meters). Some desert plants produce chemicals that inhibit the growth of other plants nearby, decreasing competition for water.

Desert animals also cope with water scarcity. Body coverings, such as a scorpion's exoskeleton or a reptile's leathery skin, minimize water loss. Some small mammals, such as the kangaroo rat, have very concentrated urine, which saves water. Few animals face the midday sun. Most burrow or seek shelter during the day and become active when the sun goes down and the risk of water loss lessens.

Particularly harsh areas in the deserts of North and South America, Africa, and Australia are salt flats, where runoff collects in basins and evaporates, leaving behind extremely salty soil. Most plants cannot tap into the water table beneath the flats, even though

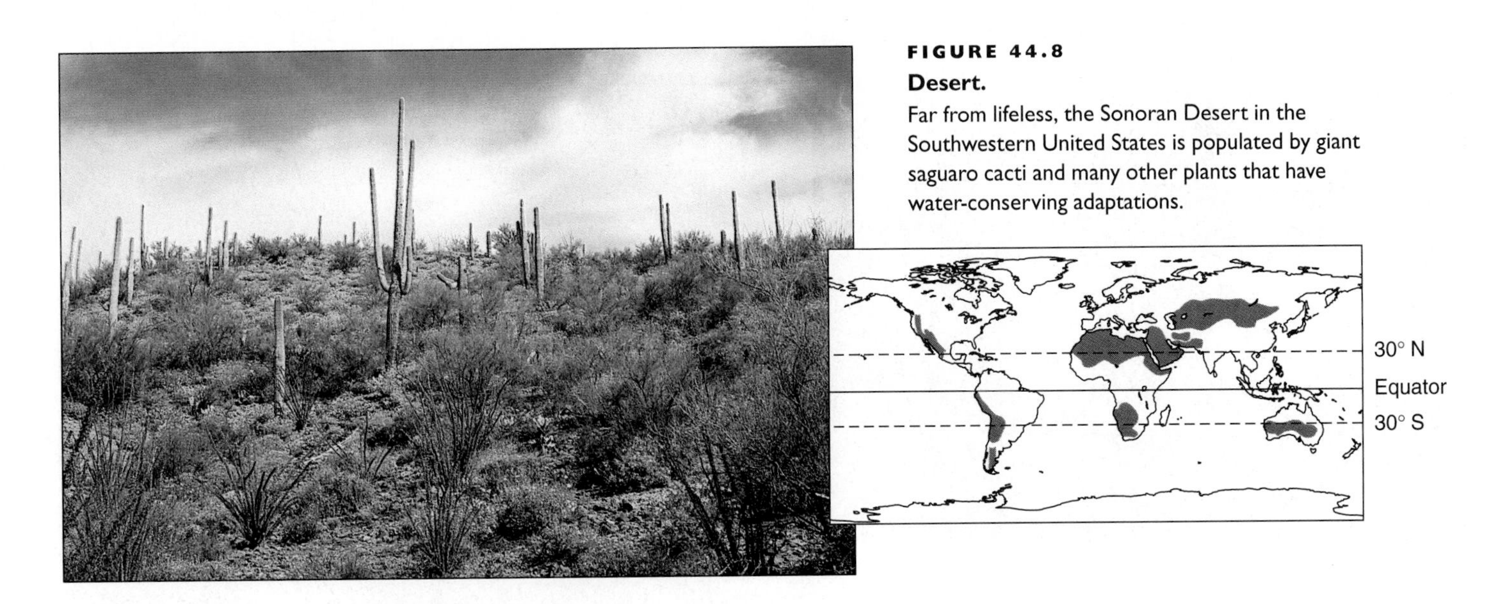

FIGURE 44.8
Desert.
Far from lifeless, the Sonoran Desert in the Southwestern United States is populated by giant saguaro cacti and many other plants that have water-conserving adaptations.

it is shallow, because the high salinity would actually draw water by osmosis from their cells. However, the members of one family of plants, the Chenopodiaceae, are adapted to these salty surroundings (fig. 44. 9). They can sequester salt within their cells at higher concentrations than are found in the salt flats, enabling them to extract water from these areas. These plants are called "halophytes," which means salt-loving. They thrive where others cannot grow.

The saltbush plant *Atriplex* is one such salt-loving desert resident. It concentrates environmental salt in the outermost cells of its leaves. Some of the cells burst, showering the leaf surface with spiky salt crystals that deter herbivores as well as shield the plant from sunlight. Because the saltbush plant can readily obtain water from beneath the salt flats, its leaves remain green year-round. The high salt content of the saltbush plant deters herbivores too—because any fluid animals obtained from the plant would be saltier than their cells. The animals would actually lose water by eating them, much as a person dying of thirst on the ocean cannot survive by drinking seawater.

Natural selection tends to find ways around obstacles, and in a striking example of convergent evolution, three distantly related types of rodents can indeed feed on *Atriplex*. The chisel-toothed kangaroo rat hails from the Great Basin Desert in North America; the red vizcacha rat from the Monte Desert in Argentina; and the fat sand rat from the Sahara Desert. All three rodent species have unusually shaped teeth that enable them to peel off the outer leaf cells, enabling them to eat only the less salty tissues within.

Temperate Grasslands

Temperate grasslands have 10 to 40 inches (25 to 100 centimeters) of rainfall annually, which is often not sufficient to support trees. These biomes usually have one or two severe dry seasons, when the vegetation becomes dry and flammable, so that fire is an important ecological factor.

The North American prairie is a temperate grassland (fig. 44.10). The height of the grasses reflects local moisture. Grasses reach 4 to 8 feet (1.2 to 2.4 meters) around the Mississippi Valley, where moisture from the Great Lakes and Gulf of Mexico contributes to an annual rainfall of about 39 inches (100 centimeters). In this tallgrass prairie, grazing and fire suppress the growth of trees. Westward, toward the Rocky Mountains, annual rainfall decreases, and shorter grass species dominate the landscape. Unlike many trees, grasses can do quite well with little water. Varied blade texture, surface, and shape are adaptations that enable grasses to conserve water. Root systems are extensive, with some species sending roots 6 feet (1.8 meters) underground to reach water. The mat of roots holds soil together and prevents it from blowing away during drought. Growth response to rain is rapid.

Unlike trees, perennial grasses easily survive damage caused by fire and herbivores. The perennial buds of grasses lie below the soil surface and are protected from flames. Because these plants grow from below, removal of the blade does not hinder growth. Even chunks of grass kicked up by a grazing animal can reroot. In contrast, the growing regions of trees—the tips of branches—are the first structures that fire or herbivores destroy.

Grasslands are rich in plant and animal species. In North America, bison, elk, and pronghorn antelope were originally the important grazing herbivores. Their predators were the gray wolf and coyote. Antelopes used speed to escape these predators. The size and power of elk and bison were sufficient protection against natural predators, but they were no match for humans with guns. Other herbivores include rodents, prairie chickens, and insects. Many rodents, such as prairie dogs, retreat into burrows to escape predators. Mouse and grasshopper populations can grow particularly large in a temperate grassland.

FIGURE 44.9
Adaptations in the Desert.
(**A**) The saltbush plant *Atriplex* stores salt in the outer cells of its leaves, enabling it to extract water from beneath salt flats. When some of the cells burst, they coat the leaves with salt crystals. Three species of rodents, from three continents, have teeth and bristles that enable them to strip the salty layer from *Atriplex,* enabling them to eat the less-salty interior. The red vizcacha rat (**B**) has the characteristic leaf-shredding incisors, but also has bristle bundles in the mouth (arrow) that aid in stripping the plants of their salty coverings.

A

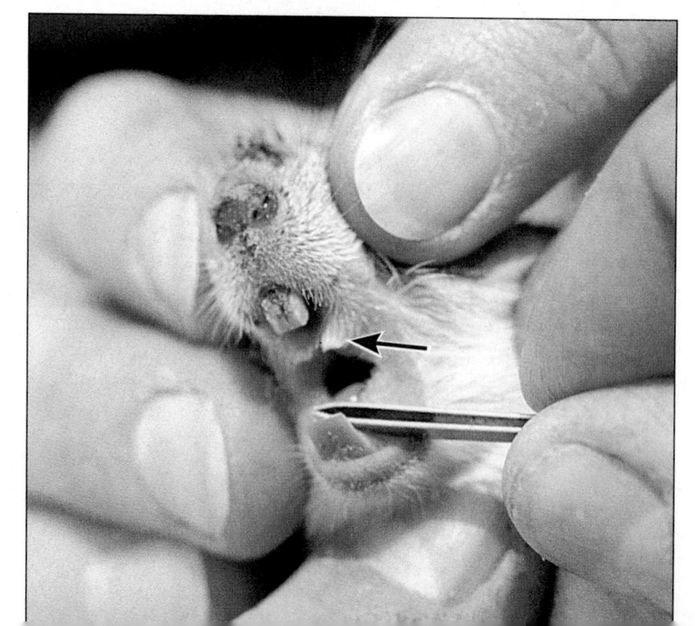

B

FIGURE 44.10
Temperate Grassland.
Bison once dominated the North American prairie, a temperate grassland.

Today the once expansive North American prairie has been shattered into thousands of pieces called, appropriately, "remnants." Farmland has replaced the prairie, wheat and corn taking the place of diverse grasses. The plant survivors in the remnants peek out at the edges of housing developments and graveyards, along railroad tracks and roadsides, and at the borders of cultivated fields. Some remnants are so small that they partition off small groups from populations, leading to genetic drift and population bottlenecks that decrease genetic diversity and locally endanger the species. Prairie remnants are usually too small to sustain the herds of large herbivores that once roamed the plains.

In Prairie City, Iowa, the U.S. Fish and Wildlife Service is attempting to resurrect a larger, more self-sustaining prairie. A handful of researchers and volunteers combed the state, collecting seeds from different places and planting them on a 5,000-acre site that was slated to become a nuclear power plant. The project began in the early 1990s, and the area now boasts native grasses exceeding 12 feet in height and a few dozen other plant species—not close to the 150 to 200 plant species native to the prairie, but a start. Buffalo, elk, birds, bats, and of course many insects have been reintroduced or have returned on their own.

Temperate Forests

Covering much of Earth between 30° and 55° latitude, temperate forests are either deciduous (composed of trees that shed their leaves seasonally) or coniferous, although a walk in the woods reveals that many such forests contain both types of trees. Deciduous trees tend to be predominant in temperate forests that have mild winters and a moist growing season lasting at least 4 months, rainfall ranging from 26 to 118 inches (65 to 300 centimeters), and soil rich in nutrients. In contrast, dry summers and long, harsh winters favor the conifers. Vertical stratification occurs in temperate forests, although species diversity is lower than in tropical forests.

Usually one or two species of trees predominate in the **temperate deciduous forest,** as in oak-hickory or beech-maple forests (fig. 44.11A). The dominant trees, victors in the competition for light, are also best suited for the average rainfall. Trees in the deciduous forest often have ball-shaped tops that maximize light absorption for the angle of the sun's rays at these latitudes. Shrubs grow beneath the towering trees. Below them, herbaceous flowering plants grow mostly in early spring, when light penetrates the leafless tree canopy. Mosses and liverworts coat damp spots on the forest floor.

Decomposers in temperate forests—nematodes, earthworms, bacteria, and fungi—break down leaf litter and create a rich soil. In the North American deciduous forest, herbivores include whitetail deer, ruffed grouse, and gray squirrels. Red fox and raccoon are common carnivores. All residents of the deciduous forest must cope with seasonal changes. In winter, some insects enter an inactive state called diapause, and some mammals hibernate. Some birds migrate to warmer climates. Many mammals survive winter by fattening up, growing thicker coats, and recovering hidden stores of food.

The remains of the once-great temperate deciduous forest that stretched from the Great Lakes to the Atlantic and from Canada to the southern states is currently changing, becoming strikingly redder. The red maple is taking over the niche of oaks and hickories, thriving because it can live under a variety of conditions. In the past, fires—natural events in this biome—preferentially killed the thin-barked red maples, while at the same time clearing the land in a way that enabled oak seedlings to take root. With the coming of humans, fires became less frequent. The red maples flourished on land that people had cleared for farming and then abandoned as they moved westward.

Red maples are well adapted to take over, because they are generalists. They grow under a wide range of shade and moisture conditions and can tolerate the acid deposition that kills other trees. The red maples drop their fruit in the spring, giving their offspring a head start on those of oaks and hickories, which disperse their seeds in the fall. Plus, red maples produce alkaloid compounds that render their leaves unpalatable to gypsy moths and deer, which instead devour the leaves of other tree species. As the red maple thrives, many other types of trees are diseased, including sycamores, sugar maples, dogwoods, elm, chestnuts,

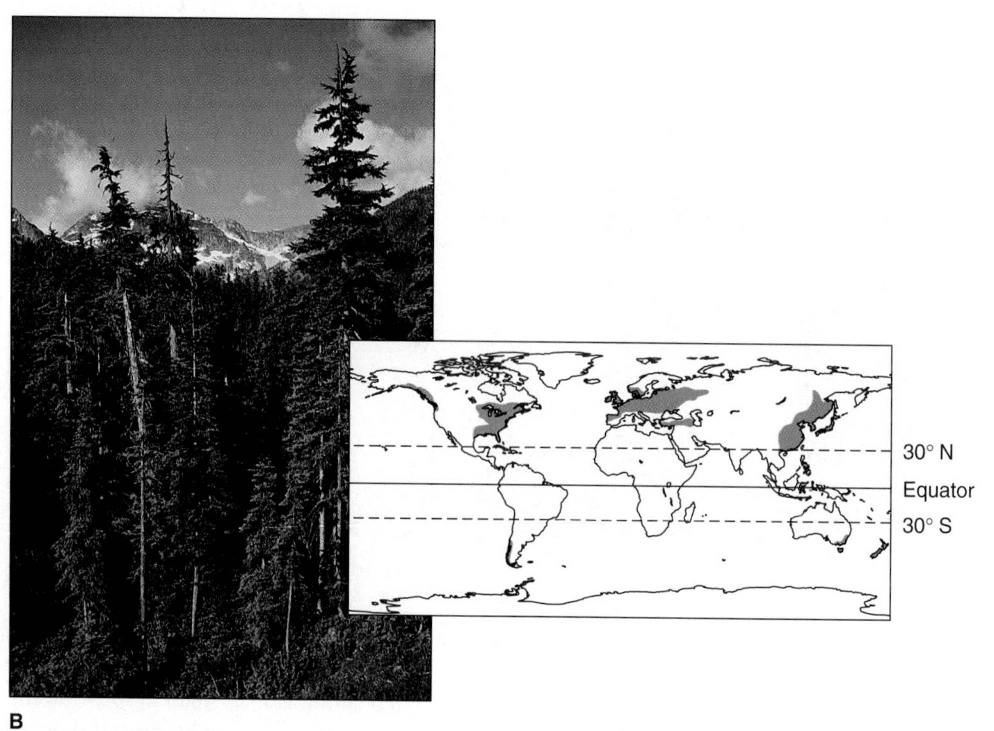

A

B

FIGURE 44.11
Temperate Forests.
Temperate deciduous forests (**A**) are dominated by trees that lose their leaves each fall. Temperate coniferous forests (**B**) have evergreen trees such as these western hemlocks in Washington State.

walnuts, hemlocks, white ash, and balsam fir. In parallel to the shifting plant species will be changes in the communities of animal species that are adapted to living with mostly oaks. Affected populations include beetles and other insects that live in and on bark, and the many types of birds and small mammals that eat acorns.

In contrast to the deciduous forest, most trees in the **temperate coniferous forest** are evergreen (fig. 44.11B). They lose leaves a few at a time, so most conifers are green all year. Many coniferous forest soils are thin, acidic, and poor in nutrients. Spruces, pines, firs, and hemlock are the dominant trees. The shortage of nutrients favors evergreen plants over deciduous species, which require these resources to replace all their leaves in a short time.

The coniferous forest also has a dense understory of shrubs such as alder and hazelnut. In North America, herbivores include the whitetail deer, red squirrel, spruce grouse, and moose. The bobcat and black bear are typical carnivores. Legions of tiny invertebrates, fungi, and bacteria recycle nutrients.

Conifers are adapted to more severe winters and drier summers than are deciduous trees. They are also adapted to recurring fires. Fire recycles nutrients and selects against some species, opening new niches. The thick bark on certain pine trees resists flames, and some pine cones liberate their seeds only after exposure to the extreme temperature of a forest fire.

Parts of New Jersey's Pine Barrens near Atlantic City burn about every 30 years. The sandy soil, the pine needles and branches that litter the ground, and the "pitch" that seals pine cones shut all create conditions just right for fire. In the last big fire in the Pine Barrens, in 1994, squirrels, deer, reptiles, and rabbits fled the affected area. The fire cleared the forest floor of three decades of accumulated plant debris, leaving space for future growth. The seeds that the fire liberated soon yielded abundant seedlings against the backdrop of blackened trees. Soon, the forest will look as if a fire had never happened.

Taiga

North of the temperate zone in the northern hemisphere lies the cold, snowy **taiga,** also called the boreal forest for the Greek word for "north" (fig. 44.12). From above, the taiga appears as alternately heavily treed areas, regions of sparser trees where shrubs and bushes such as juniper and blueberry grow, and lakes and rivers. The taiga extends across central Canada, central Alaska, Siberia and Russia, to Scandinavia, between 50° and 65° north latitude, and also through parts of the Rocky Mountains and on mountains in Asia and south-central Europe. Because of the harsh conditions, biodiversity is lower than it is in the temperate zone.

The species that inhabit the taiga are adapted to the biome's long winters and large annual temperature fluctuations. In central Siberia, for example, summer temperatures may reach 86°F (30°C), but can plunge to –94°F (–70°C) in winter. Annual precipitation is 8 to 24 inches (20 to 60 centimeters), most of it snow and ice.

Soils in the taiga are cold, damp, acidic, and nutrient-poor. The topsoil is thin, and the subsoil may be frozen. The low temperature and pH slows decomposition of dead organisms, and nutrients tend to stay in the leaf litter above the soil, rather than entering the topsoil. Mycorrhizal fungi on the shallow and abundant roots of trees help the plants maximize nutrient uptake from the leaf litter (see fig. 23.14).

Spruce, fir, pine, larch, and tamarack are the dominant trees in the taiga, but aspen and birch grow after fires, and willows ring lakeshores. Conifers are well adapted to an environment so cold that water is usually frozen throughout the winter. With a scant amount of sap, a conifer has little liquid to freeze. The needle

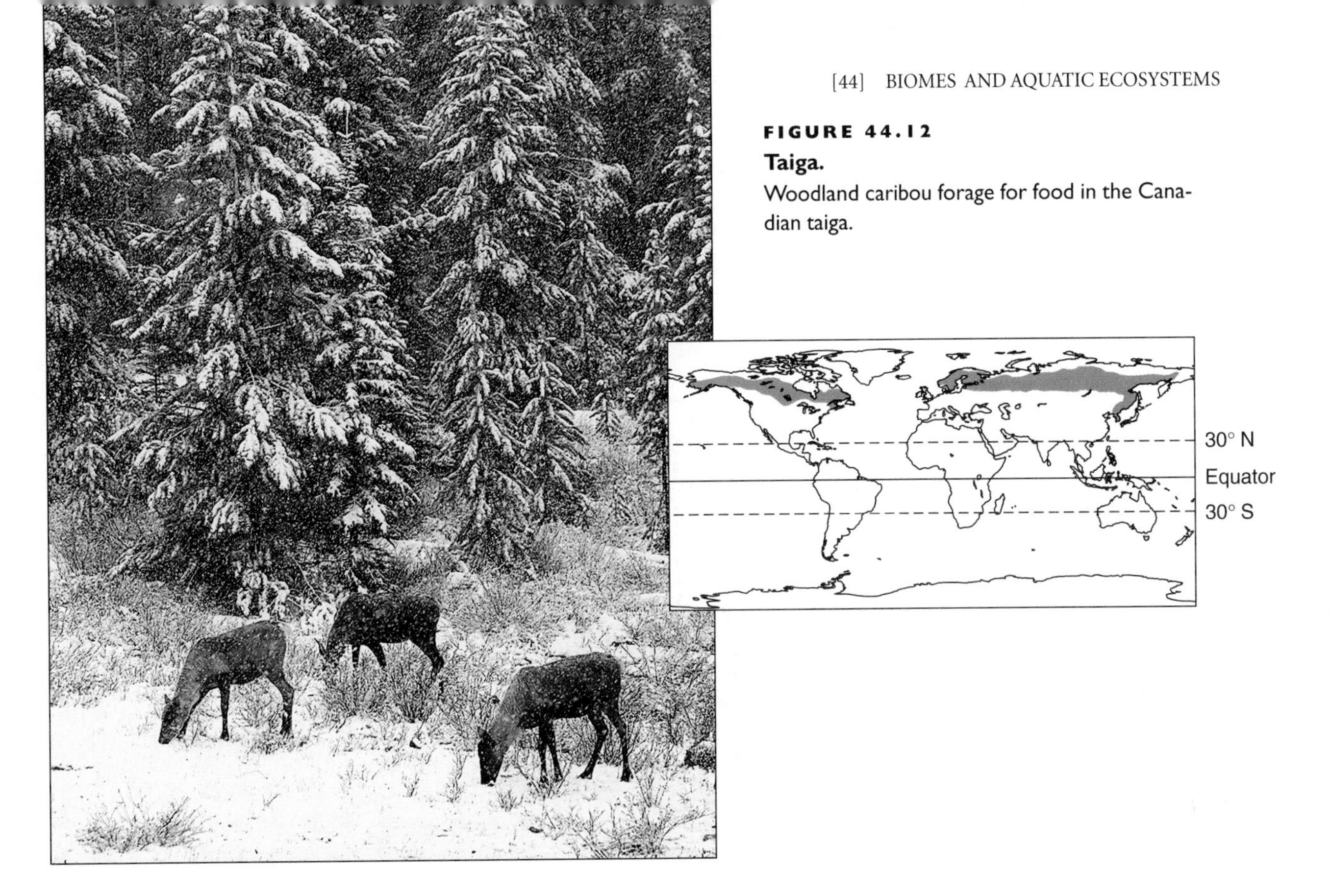

FIGURE 44.12
Taiga.
Woodland caribou forage for food in the Canadian taiga.

shape, waxy coat, and minimal stomata set in deep pits are adaptations of conifer leaves to conserve water. The conical tree shape helps capture the oblique rays of light in northern latitudes and prevents damaging snow and ice buildup.

The taiga forest floor is often boggy, with scattered shrubs and mats of mosses and lichens. Fungi are common decomposers in the taiga. Typical herbivores include the woodland caribou, porcupines, red squirrels, snowshoe hares, moose, and spruce grouse. Carnivores include lynx, gray wolves, and wolverines. As is true nearly everywhere else on the planet, insect populations can at times grow very dense.

Tundra

A band of **tundra** runs across the northern parts of Asia, Europe, and North America. Winter is bitterly cold, with typical temperatures about –26°F (–3°C). Precipitation is 8 to 24 inches (20 to 60 centimeters) per year. The ground remains frozen, in a zone called **permafrost,** even during summer, when temperatures range from 40°F to 70°F (4.5°C to 21°C). Permafrost begins at 18 inches (46 centimeters) below the surface and extends 300 feet (91.5 meters) down. Because the permafrost blocks water infiltration, spring runoff from ice and snow drains rapidly into rivers or accumulates, forming bogs and small, stagnant ponds. Permafrost also limits rooting depth, which prevents large plants from becoming established.

The shallow tundra soil supports dwarf shrubs, low-growing perennial plants such as sedges and broad-leafed herbs, and reindeer lichens (fig. 44.13). It is difficult for annual plants to germinate, grow, and flower in the short growing season, which may be

FIGURE 44.13
Tundra.
In the tundra, permafrost limits plant life to shallow-rooted shrubs and abundant lichens.

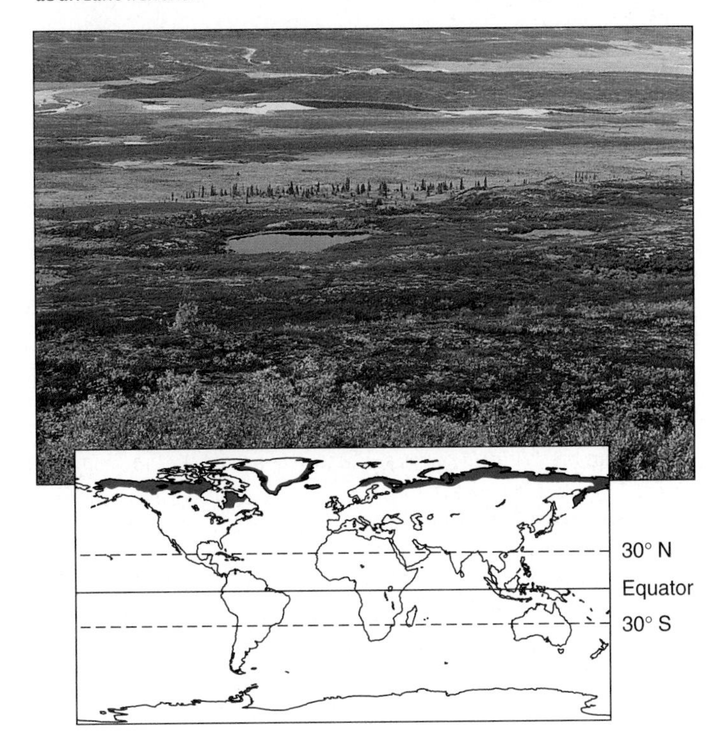

less than 60 days, so most plants are perennials. Tundra plants tend to be low and flat, a shape that lets the wind blow over them and protects them under snow. The plants often clump together, which helps to conserve moisture, and they are often buried amid rocks, which block the wind. Some plants have protective hairs that insulate them and help break the wind. The dark green color of many plants allows them to absorb more light for photosynthesis.

Animal inhabitants of the tundra include caribou, musk oxen, reindeer, lemmings, snowy owls, foxes, and wolverines. Polar bears sometimes visit coastal areas of the tundra to den. In the summer, migratory birds stop in tundra areas to raise their young and feed on the insects that flourish in the tundra and its ponds.

Like plants, the animals of the tundra have adapted to the harsh climate. Both the hunters and the hunted benefit from camouflage. White winter colors make the arctic fox, ptarmigan, ermine, and arctic hare as inconspicuous against snow as their brown summer colors make them against the snow-free landscape. These animals often have short extremities, a form that helps to conserve heat. The snowshoe hare's big feet are natural snowshoes. The shallow soil, short growing season, and slow decomposition of the tundra make it a very fragile environment.

44.1 MASTERING CONCEPTS

1. What are the major climatic regions of the world?
2. How do climate and soil composition determine characteristics of biomes?
3. Describe the rainfall and temperature patterns, nutrient cycling, and inhabitants of each of the major terrestrial biomes.

OLC

44.2 Freshwater Ecosystems

Life in lakes, rivers, and streams must be adapted to water velocities, changing nutrient and oxygen levels, and drought and flooding conditions.

Earth's waters house diverse species adapted to the temperature, light, current, and nutrient availability of their surroundings. Aquatic ecosystems are distinguished by physical and chemical factors such as current pattern and degree of salinity. Two types of freshwater ecosystems are standing water, such as lakes, swamps, and ponds, and running water, such as rivers and streams.

Lakes and Ponds

Light penetrates the regions of a lake to differing degrees. These differences determine the types of organisms that live in particular areas (fig. 44.14).

The **littoral zone** is the shallow region along the shore where light reaches the bottom sufficiently for photosynthesis to occur. Some photosynthetic organisms in this region are free-floating; others are rooted to the bottom and have submerged, floating, or emergent leaves. The littoral zone is the richest area of a lake or pond. Producers here include free-floating and attached cyanobacteria, green algae, and diatoms. Floating plants such as water lilies and emergent plants such as cattails, reeds, and rushes are also part of the flora. Animal life is diverse and includes damselfly and dragonfly nymphs, crayfish, rotifers, flatworms, hydra, snails, snakes, turtles, frogs, and the young of some species of deep-water fishes.

The **limnetic zone** is the layer of open water where light penetrates. Plankton (mostly protista, such as diatoms, ciliates, and algae) and fishes inhabit this area. The **profundal zone** is the deep region beneath the limnetic zone where light does not penetrate. Organisms here, which rely on falling organic material from above, include mostly scavengers and decomposers such as insect larvae and bacteria. The sediment at the lake bottom comprises the **benthic zone.**

Oxygen and mineral nutrients in a lake are distributed unevenly. The concentration of oxygen is usually greater in the upper layers, where it comes from the atmosphere and from photosynthesis. As dead organic matter sinks to the bottom, decomposers consume oxygen and release phosphates and nitrates into the lower layers of the lake. In a shallow lake, wind blowing across the surface mixes the water, redistributes nutrients, and restores oxygen to bottom waters.

Deeper lakes in temperate regions often develop layers with very different water temperatures and densities. This thermal stratification prevents the free circulation of nutrients and oxygen in the lake. The degree of thermal stratification varies with the season.

In the summer, the sun heats the surface layer of the lake, but the deepest layer remains cold. Between these two layers is a third region, the thermocline, where water temperature drops quickly. In the fall, the temperature in the surface layer drops as the air cools. Gradually, water temperature becomes the same throughout the lake. Wind then mixes the upper and lower layers, creating a **fall turnover** that redistributes nutrients and oxygen throughout the lake. During winter, surface water cools. When water cools to 39°F (4°C), the temperature at which it is most dense, it sinks. Water colder than this floats above the 39°F layer and may freeze, giving the lake an ice cover. In the spring, when the surface layer warms to 39°F, a **spring turnover** occurs, again redistributing nutrients and oxygen. After the spring turnover, algae thrive in the warming, nutrient-rich surface water.

Lakes age. Younger lakes are often deep, steep-sided, and low in nutrient content. The deep zone of bottom water stores a large quantity of oxygen, which is rarely depleted. These lakes are termed **oligotrophic,** which means they are low in fertility and productivity. They are clear and sparkling blue, because phytoplankton aren't abundant enough to cloud the water. Lake trout and other organisms that thrive in cold, oxygen-rich deep water are numerous.

As a lake ages, organic material from decaying organisms and sediment begins to fill it in, and nutrients accumulate. These lakes are termed **eutrophic,** which means they are nutrient-rich and high in productivity. The rich algal growth turns the water green and murky. Decomposing organisms in the deeper waters deplete oxygen during the summer. Fish and plankton communities

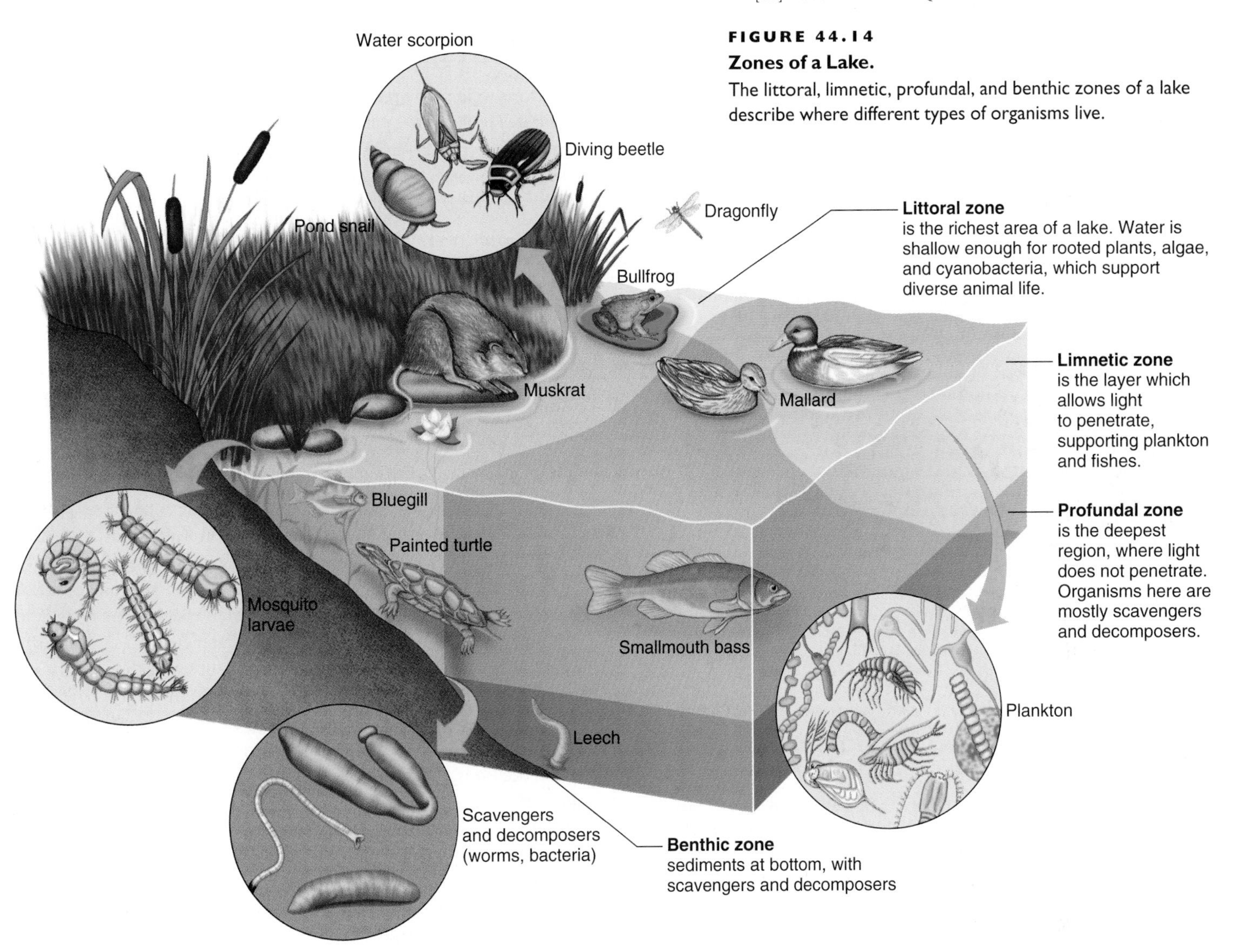

FIGURE 44.14
Zones of a Lake.
The littoral, limnetic, profundal, and benthic zones of a lake describe where different types of organisms live.

change, and fishes that can tolerate low oxygen conditions replace species such as lake trout. In time, the lake becomes a bog or marsh and, eventually, dry land. Discharge of nutrient-rich urban wastewater and runoff carrying phosphate-rich fertilizers from cultivated lands can speed conversion of oligotrophic lakes to eutrophic lakes. This transformation is termed eutrophication. In extreme cases, the nutrients promote excessive algal growth. When the algae die, they sink to the lake bottom, where decomposers deplete the water of oxygen. Fish kills and unpleasant odors often result.

Rivers and Streams

The rivers and streams that flow across the terrestrial landscape carry rainwater, groundwater, snowmelt, and sediment from all portions of the land toward the ocean or an interior basin (such as the Great Salt Lake). The flow is not constant, however. Where the landscape flattens, the water may slow to a virtual standstill, forming pools. Elsewhere, the water flows in straight runs or in bends called riffles. Rapids are the fastest-moving parts. Along the way, rivers provide moisture and habitat to a variety of aquatic and streamside organisms, which are adapted to both flooding and drying.

Rivers change, physically and biologically, as they move toward the ocean (fig. 44.15). At the headwaters, the water is relatively clear, and the channel is narrow. Where the current is swift, turbulence mixes air with water, so the water is rich in oxygen. In fast-moving streams, some organisms cling to any available stationary surface, such as rocks or logs. Algae, diatoms, mosses, and snails that graze on them, live here. Larval and adult insects burrow into sediments or adhere to the undersides of rocks with hooks or suckers. Many of these invertebrates eat decaying plant material that drops in from streamside vegetation, which often provides the bulk of the energy that fuels the headwater stream food chain.

As the river flows toward the ocean, it continues to pick up sediment and nutrients from the channel. Tributaries contribute to water flow, so the river widens, and as the land flattens the

FIGURE 44.15
Rivers Change Along Their Course.
A narrow, swift stream in the mountains becomes a slow-moving river as it accumulates water and sediments and approaches the ocean.

current slows. The river is murky, restricting photosynthesis to the banks and water surface. As a result, the oxygen content is low relative to the river upstream. Such slower-moving rivers and streams support more diverse life, including crayfish, snails, bass, and catfish. Worms burrow in the muddy bottom, and plants line the banks.

Rivers and streams depend heavily on the land for water and nutrients. Dead leaves and other organic material that fall into a river add to the nutrients that resident organisms recycle. Rivers also return nutrients to the land. Many rivers flood each year, swelling with meltwater and spring runoff and spreading nutrient-rich silt onto their floodplains. When a river approaches the ocean, its current diminishes, which deposits fine, rich soil that forms new delta lands. The opener to the next chapter describes the incredible disruption to ecosystems that occurred when people altered a river's course in southern Florida.

44.2 MASTERING CONCEPTS

1. Describe the types of organisms that live in each zone of a lake or pond.
2. How are oxygen and nutrients distributed (and redistributed) in lakes?
3. What is eutrophication?
4. What adaptations enable organisms to survive in moving water?
5. Describe the ways a river changes from its headwaters to its mouth.

OLC

44.3 Marine Ecosystems

In areas where salt water meets fresh water, organisms are adapted to fluctuating salinity. In the intertidal zone, the ebb and flow of the tide challenges organisms. Life in the oceans is abundant and diverse, but we know little about it because oceans are vast and mostly inaccessible.

The ocean, covering 70% of Earth's surface and running 7 miles (11.2 kilometers) deep in places, is the largest and most stable aquatic ecosystem. Specific regions are based on proximity to land.

Coasts

Several types of aquatic ecosystems border shorelines. Figure 44.16 illustrates these coastal areas.

Estuaries At the margin of the land, where the fresh water of a river meets the salty ocean, is an **estuary.** Life in an estuary must be adapted to a range of chemical and physical conditions. The water is brackish, which means that it is a mixture of fresh water and salt water; however, the salinity fluctuates. When the tide is out, the water may not be much saltier than water in the river. The returning tide, however, may make the water nearly as salty as the sea. As the tide ebbs and flows, nearshore areas of the estuary are alternately exposed to drying air and then flood.

Organisms able to withstand these environmental extremes enjoy daily deliveries of nutrients, from the slowing river as well as from the tides. Photosynthesis occurs in shallow water. An estuary houses a very productive ecosystem, its rocks slippery with algae, its shores lush with salt marsh vegetation, and its water teeming with plankton. Almost half of an estuary's photosynthetic products go out with the tide and nourish coastal communities.

Estuaries are nurseries for many sea animals. More than half the commercially important fish and shellfish species spend some part of their life cycle in an estuary. Migratory waterfowl feed and nest here as well. These important ecosystems, however, are currently threatened by human activities, as described in chapter 45.

Mangrove Swamps Another type of aquatic ecosystem where salinity varies is a **mangrove swamp,** which is distinguished by characteristic salt-tolerant plants. The general term "mangrove" refers to plants that are adapted to survive in shallow, salty water, typically with aerial roots. About 40 species of trees are considered to be mangrove. Mangrove swamps mark the transitional zone between forest and ocean and are located in many areas of the tropics. Within them, salinity varies from the salty ocean, to the brackish estuary region, to the fresh water of the forest.

A mangrove swamp is home to a diverse assemblage of species because it provides a variety of microenvironments, from its treetops to deeply submerged roots in its own version of vertical stratification. Life is least abundant in the treetops, where sun exposure is greatest and water availability the lowest. Snakes, lizards, birds, and many insects live here. A hollow elevated mangrove branch may house a thriving community of scorpions, termites, spiders, mites, roaches, beetles, moths, and ants.

FIGURE 44.16

Coastal Ecosystems.

Coastal ecosystems include estuaries (**A**), where salt water and fresh water meet. Some coastal areas have mangrove swamps (**B**), rocky intertidal zones (**C**), or coral reefs (**D**).

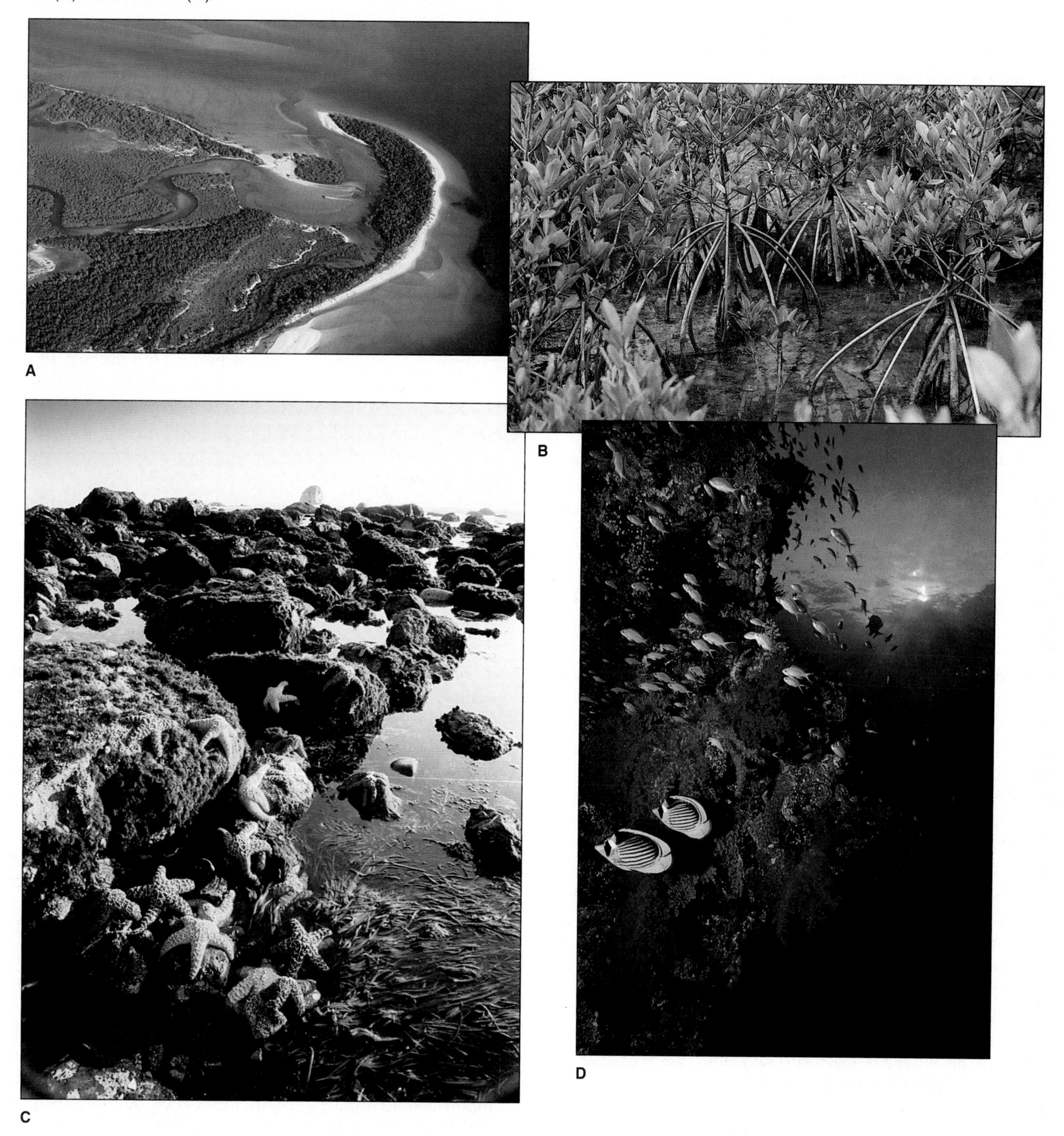

Aerial roots of mangroves provide the middle region of the swamp's vertical stratification. Here, roots are alternately exposed and submerged as the tide goes in and out. Barnacles, oysters, crabs, and red algae cling to the roots. Lower down lies the root region of the mangrove swamp, populated by sea anemones, sponges, crabs, oysters, algae, and bacteria. The algal slime that coats roots discourages hungry animals.

Submerged roots form the lowest region of the mangrove swamp. Here live sea grasses, polychaete worms, crustaceans, jellyfishes, the ever-present algae, and an occasional manatee. Ecologists estimate that up to 30% of the resident species here are unknown.

Unfortunately, many mangrove swamps are in prime vacation spots for *Homo sapiens*—which means habitat destruction. When people cut down mangrove trees, small shrubs that can tolerate salt grow in the area, and trees cannot grow back. The diverse mangrove ecosystem vanishes.

The Intertidal Zone Along coastlines, in the littoral zone, lie the rocky or sandy areas of the **intertidal zone.** This region is alternately exposed and covered with water as the tide ebbs and flows.

The organisms in a rocky intertidal zone often attach to rocks, which prevents wave action from carrying them away. Holdfasts attach large marine algae (seaweeds) to rocks. Threads or suction fastens mussels to rocks. Sea anemones, sea urchins, snails, and sea stars live in pools of water that form between rocks as the tide ebbs. The organisms of the sandy beach, such as mole crabs, burrow to escape the pounding waves that would wash them away. Sandy beaches have very little primary production.

Coral Reefs Colorful and highly productive **coral reef** ecosystems border some tropical coastlines. Coral reefs are vast, underwater structures of calcium carbonate whose nooks and crannies collectively provide habitats for a million species of plants and animals, and an unknown variety of microorganisms. The Great Barrier Reef of Australia, for example, is composed of some 400 species of coral, and supports more than 1,500 species of fishes, 400 of sponges, and 4,000 of mollusks. Other residents include algae, snails, sea stars, sea urchins, and octopuses. Food is abundant because the sun penetrates the shallow water, allowing photosynthesis to occur, and constant wave action brings in additional nutrients.

Coral animals have colorful popular names based on their varied forms—brain, staghorn, lace, vase, bead, button, and organpipe corals are just a few. Recall from chapter 24 that individual animals, called polyps, build the reefs and house symbiotic algae that are essential for the coral's, and the ecosystem's, survival (see fig. 24.12). The living coral is but a thin layer atop the remains of ancestors. A coral reef, then, is at the same time an immense graveyard, and a thriving ecosystem. It is rich in biodiversity, yet fragile. Chapter 45 considers threats to coral reefs.

The Oceans

The oceans cover 70% of Earth's surface, but we know less about biodiversity there than we do about biodiversity in a single tree in a tropical rain forest. The reason for our scant knowledge of life in the oceans may be simply the vastness of this aquatic ecosystem. Populations are sometimes small, usually very dispersed, and nearly always difficult for us to observe. Biologists have explored only 5% of the ocean floor, and 1% of the huge volume of water above. Yet by the year 2010, 65% of the world's human population will reside within 10 miles (16.1 kilometers) of an ocean.

We can, however, describe the ocean's physical characteristics, which determine the nature of its biological communities. The planet's five oceans and many seas, plus the bridging waters that interconnect them, hold about 95 trillion gallons (360 cubic kilometers) of water. The temperature ranges from 35°F (–1.5°C) in the Antarctic Ocean to 81°F (27°C) near the equator. Sunlight quickly dissipates with depth. Within the first 10 meters, the water absorbs 80% of incoming sunlight, reflecting the blue wavelengths. The blue deepens with depth, and from about 600 meters and lower, everything is black, except for the occasional glow from bioluminescent organisms (see the opening essay to chapter 20).

The sun heats the surface water, causing its molecules to move faster than molecules below. This warm upper layer is separated from denser, colder water below by a thin thermocline layer where the temperature changes rapidly. Tropical oceans and seas have a thermocline year round, but it only appears in the summer in temperate waters.

Very productive ocean environments arise where cooler, nutrient-rich bottom layers move upward in a process called **upwelling.** The resulting sudden influx of nutrients causes phytoplankton to "bloom," and with this widening of the food web base, many ocean populations grow. Upwelling generally occurs on the western side of continents, where wind pushes surface waters offshore, such as along the coasts of southern California, South America, parts of Africa, and the Antarctic.

Like other aquatic ecosystems, the ocean is considered in zones (fig. 44.17). These designations are horizontal and vertical (depth). The horizontal zones describe the relationship between the ocean and the land. The intertidal zone, discussed in the previous section, is the shoreline. The **neritic zone** is the area from the coast to the edge of the continental shelf, and its waters reach a depth of 200 meters. The remainder of the ocean is the **oceanic zone,** which is in turn subdivided according to depth.

Two general divisions of the oceanic zone that refer to habitats are the benthic or bottom zone, and the waters above, collectively called the **pelagic zone.** The pelagic zone is in turn subdivided according to depth. From about 200 meters to the surface lies the epipelagic zone, the only area where photosynthesis can occur. Beneath that is the mesopelagic zone (200 to

FIGURE 44.17
Zones of the Ocean.
Zones in the ocean describe the relationship between regions of water and the land.

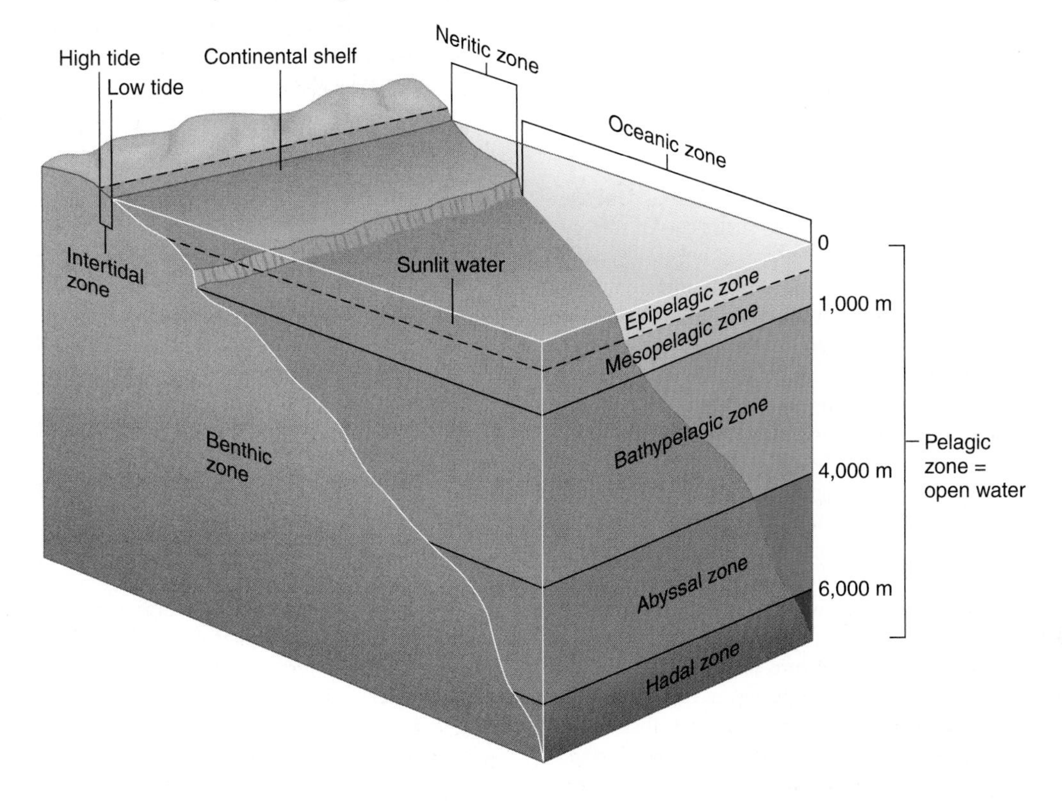

1,000 meters), and then the bathypelagic zone (1,000 to 4,000 meters). Beneath these zones lies the abyssal zone, from 4,000 to 6,000 meters. The hadal zone describes the areas of the ocean bottom that dip down even more, as far as 10 kilometers below sea level.

Chapter 43 considered life at the very top of the ocean—the surface microlayer—as well as life in deep-sea hydrothermal vents. The very top of the epipelagic zone is rich in microscopic species that photosynthesize and form the bases of the great oceanic food webs. Other microorganisms and small animals feed on them. These top-dwelling organisms die and provide a continual rain of nutrients to the species below. The diverse living communities of the hydrothermal vents are fueled by chemosynthetic bacteria that harness the chemical energy inside Earth. In between lies a universe of diverse life-forms, including great numbers and varieties of fishes, mollusks, echinoderms, crustaceans, and species yet to be discovered and described. One estimate of biodiversity in the benthic zone is 10 million species; it may be even higher in the enormous pelagic zone.

We have much to learn about the oceans and seas, the environmental descendants of the aquatic setting where life probably arose. But we may never do so unless the current state of the oceans changes, and soon. Already the oceans include 50 "dead zones," areas devoid of life. In some cases the cause of the dead zone is obviously human intervention, or population shifts such as algal blooms or the effects of dinoflagellate toxins. But in others, the trigger for the unraveling of food webs and ecosystems is a mystery. The final chapter explores some of the ways that the oceans, and other biomes and aquatic ecosystems, are struggling.

44.3 MASTERING CONCEPTS

1. Describe and distinguish among types of coastal aquatic ecosystems.
2. What are some adaptations of organisms to life where water meets land, or salt water meets fresh water?
3. What are the major zones of the ocean?

OLC

Chapter Summary

44.1 Biomes

1. **Biomes** are major types of terrestrial ecosystems. The equivalent in water is an aquatic ecosystem. Each biome or aquatic ecosystem has a characteristic group of species. Biomes occupy large geographic areas.
2. Temperature and rainfall define the major climatic regions. Uneven heating due to the angle of solar rays hitting Earth's curved surface generates wind and moisture patterns.
3. The **tropical rain forest** is hot and wet, with diverse life. Competition for light leads to **vertical stratification.** Nutrients cycle rapidly.
4. **Tropical dry forest** borders tropical rain forest, with rich soil and distinct dry and wet seasons.
5. Tropical **savannas** have alternating dry and wet seasons, and are dominated by grasses, with sparse shrubs and woody vegetation and migrating herds of herbivores.
6. **Deserts** have less than 8 inches (20 centimeters) of rainfall a year. Desert plants are well-adapted for obtaining and storing water, with rapid life cycles, deep roots, or succulent tissues. Animals minimize water loss with tough integuments, and they are active at night. Some desert organisms are adapted to living in high salt conditions.
7. **Temperate grasslands** receive less water than deciduous forests and more water than deserts. The more moisture, the taller the grasses.
8. **Temperate deciduous forests** require a growing season of at least 4 months, are vertically stratified, and have less diverse life than tropical rain forests. Tree shapes maximize sun exposure. Decomposers form soil from leaf litter. **Temperate coniferous forests** have poor soil and a cold climate. Periodic fires occur in these areas.
9. The **taiga** is a very cold northern coniferous forest. Adaptations of conifers include needle shapes, year-round leaf retention, and conical tree shape.
10. The **tundra** has very cold and long winters. A layer of frozen soil called **permafrost** lies beneath the surface. During the spring and summer, meltwater forms rivers and pools. Lichens are common in the treeless tundra, and animals include caribou, reindeer, lemmings, and snowy owls.

44.2 Freshwater Ecosystems

11. Freshwater ecosystems include standing water (lakes and ponds) and running water (rivers and streams).
12. The **littoral zone** of a lake is the shallow area where light reaches the bottom; the **limnetic zone** is the lit upper layer of open water; the **profundal zone** is the dark deeper layer. The lake bottom is the **benthic zone.** In the littoral zone, most producers are rooted plants. In the limnetic zone, phytoplankton predominate. Nutrients fall from the upper layers and support life in the profundal and benthic zones.
13. Deep lakes in the temperate zone rely on **fall turnover** and **spring turnover** to mix oxygen and nutrients. Young, deep, **oligotrophic** lakes are clear blue, with few nutrients to support algae. Nutrients gradually accumulate, and algae tint the water green. The lake becomes a productive, or **eutrophic,** lake.
14. In rivers, organisms are adapted to local current conditions. Near the headwaters the channel is narrow and the current is swift. As the river accumulates water and sediments, the current slows and the channel widens.

44.3 Marine Ecosystems

15. In **estuaries** rivers empty into oceans. Life here is adapted to fluctuating salinity.
16. **Mangrove swamps** have changing salinity, and are defined by characteristic salt-tolerant plant species.
17. Residents of the **intertidal zone** are adapted to stay in place as the tide ebbs and flows.
18. **Coral reefs** support many thousands of species in and around 400 or so types of coral.
19. The region of ocean near the shore is the **neritic zone.** Open water is the **oceanic zone** and includes the benthic zone (the bottom), and the **pelagic zone** (open water above the ocean floor). The most productive areas are in the neritic zones where **upwelling** occurs.

Testing Your Knowledge

1. How does the fact that Earth is a sphere tilted on its axis influence the distribution of life?
2. Describe the zones of a(n)
   - **a.** lake
   - **b.** ocean
   - **c.** mangrove swamp
3. What are the sources of energy in the various zones of the ocean?
4. List adaptations that enable organisms to survive conditions in the following biomes:
   - **a.** tropical rain forest
   - **b.** tropical dry forest
   - **c.** savanna
   - **d.** temperate grassland
   - **e.** tundra
   - **f.** desert
   - **g.** taiga
5. How can the tropical rain forest support diverse and abundant life with such poor soil?
6. What is permafrost?
7. How does photosynthetic activity differ in the zones of a lake?

Thinking Scientifically

1. Excess nutrients in a lake or in the ocean can greatly disrupt biological communities. Organisms die. How can this happen, if life depends upon a supply of nutrients?
2. Researchers and citizens in Prairie City, Iowa, are reconstructing the prairie by collecting seeds from remnants and reintroducing animals. Which other biomes and aquatic ecosystems discussed in the chapter might be possible to reconstruct, and which not?
3. Cite two biomes or aquatic ecosystems where population bottlenecks might occur, and describe how this might happen.
4. How can a devastating fire be a natural part of a biome's dynamics? Give an example of a biome in which fires occur regularly.
5. Some scientists are currently attempting to catalog all of the world's biodiversity. What are some of the technical problems they may encounter?

References and Resources

Kerr, Richard A. July 16, 1999. Does a globe-girdling disturbance jigger El Niño? *Science*, vol. 285, p. 322. The Madden-Julian Oscillation may worsen an El Niño.

Samuels, Sam Hooper. September 21, 1999. In Iowa, restitching the torn fabric of the prairie. *The New York Times*, p. F1. Will collecting seeds and reintroducing animals be enough to rebuild part of the once-great prairie?

Stevens, William K. April 27, 1999. Eastern forests change color as red maples proliferate. *The New York Times*, p. F1. Superb adaptations and a generalist lifestyle give the red maple a competitive advantage in the temperate deciduous forest of North America.

Wuethrich, Bernice. February 4, 2000. How climate change alters rhythms of the wild. *Science*, vol. 287, p. 793. Shifting climate affects animal behavior and alters population dynamics.

The *LIFE* Online Learning Center provides additional resources and tools for studying this chapter.
www.mhhe.com/life

CHAPTER 45

Environmental Challenges

The Endangered Everglades

If ecology and environmentalism have taught us anything, it is that nature is unpredictable. Nature also operates over long time periods. Making decisions and taking actions that drastically change ecosystems based on information collected over a relatively short period of time can lead to disaster. Perhaps nowhere is this better illustrated than in the Everglades, the "sea of grasses" in southern Florida.

Prior to the end of the nineteenth century, much of the state was a continuous waterway. When the Kissimmee River up north swelled, it fed into Lake Okeechobee, which would flood its banks, sending water into the Everglades, which would in turn pass the water south to the Floridian Bay System and through a series of rivers to the east coast. This vast area included several types of estuaries, saw grass plains, mangrove swamps and other wetlands, and tropical hardwood forests. The Everglades were home to marsh grasses and cypress trees; shorebirds such as egrets, eagles, and herons; and panthers and alligators. The region was naturally unstable, prone to droughts, floods, fires, and hurricanes. Today, only half of the area bears any resemblance to what it once was, and only one-fifth of it is protected. What is left of the once-lush Florida ecosystem is currently called a "degraded remnant".

The destruction of the Everglades began at the turn of the twentieth century. From 1903 until 1917, the state's "Everglades reclamation project" drained the area to prevent flooding, via four large canals. In 1930, following disastrous hurricanes in 1926 and 1928, levees were built around Lake Okeechobee. Farmers, reassured of their safety from the floodwaters, doubled their sugarcane crop.

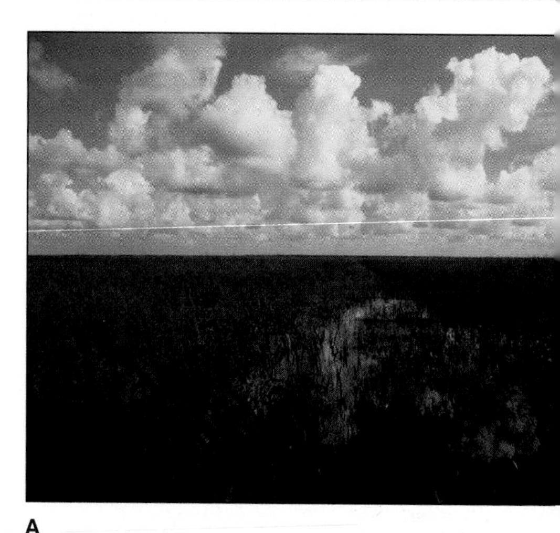

A

Then an organization of biologists called the Soil and Crop Science Society reported that flood control was changing the ecosystem too much. As a result, a "rewatering" plan began in 1939, in an attempt to revert some of the drained land back into wetlands.

But the years 1947 and 1948 doomed the Everglades. Hurricanes dumped 108 inches (274 centimeters) of rain, nearly twice the normal rainfall, submerging many farms and homes for months. Congress decided to intervene to control flooding by installing a water supply system. The plan, called the "Central and Southern Florida Project for Flood Control and other Purposes," divided the state into three major land use areas: agricultural (27%), water conservation (33%), and Everglades National Park (21%). Of the remainder, 12% went to urban development and 7% remained undeveloped but not protected. Then they built a system of canals, levees, and pumps that covered nearly 1,000 miles (1,600 kilometers), to direct water to agricultural and urban areas. As a result, 6.4 billion liters of fresh water began to be pumped into the Atlantic Ocean and Gulf of Mexico every day. As the life was being pumped

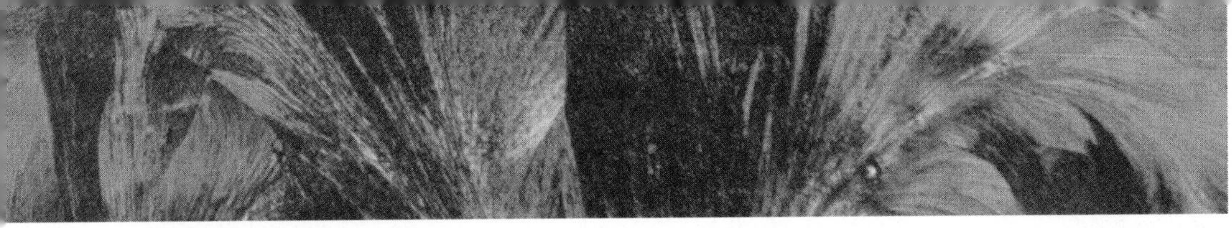

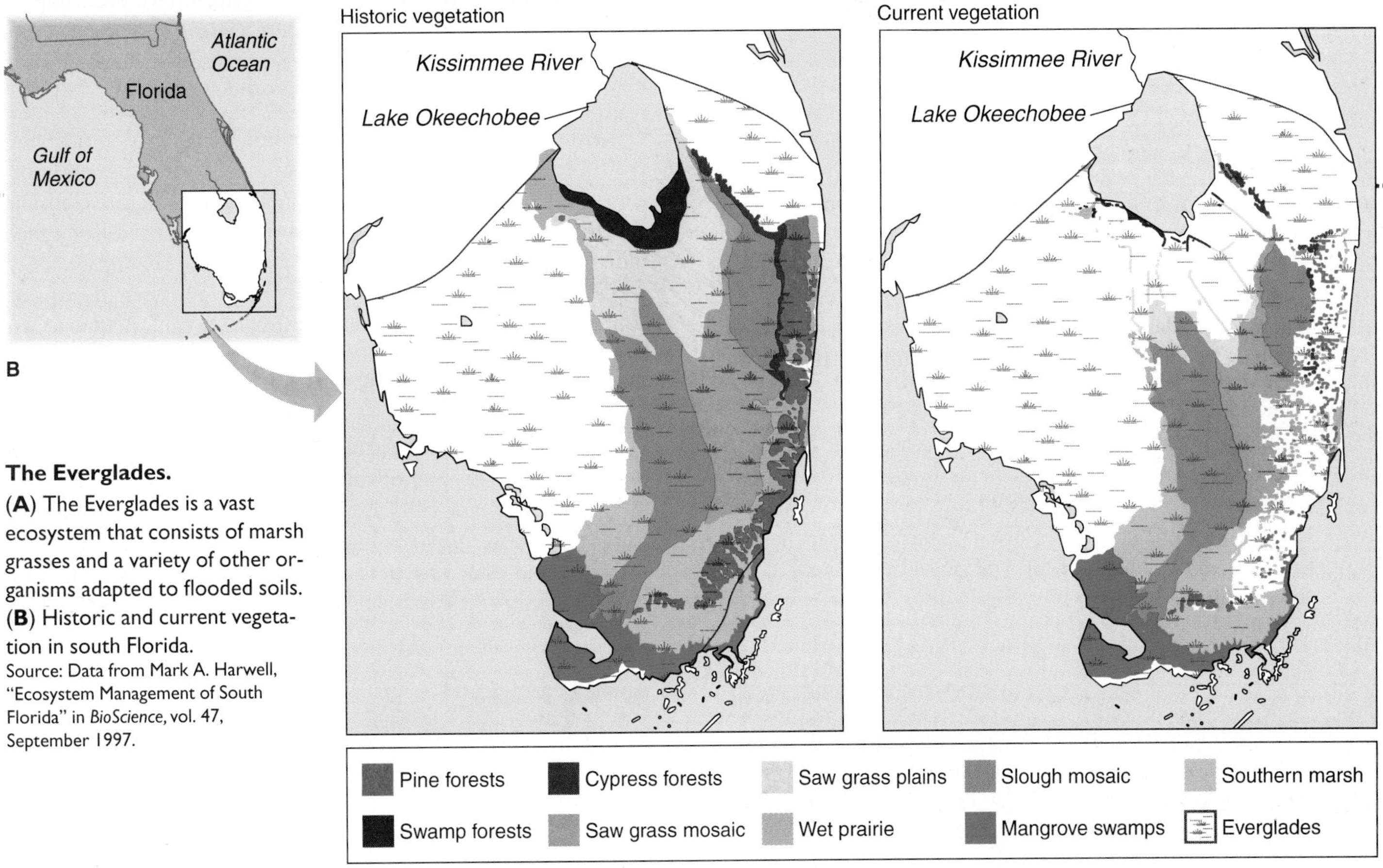

The Everglades.

(**A**) The Everglades is a vast ecosystem that consists of marsh grasses and a variety of other organisms adapted to flooded soils. (**B**) Historic and current vegetation in south Florida.

Source: Data from Mark A. Harwell, "Ecosystem Management of South Florida" in *BioScience*, vol. 47, September 1997.

from the already burdened Florida lands and waters, the human population was swelling, from three-quarters of a million to more than 4 million.

In 1950, an engineer working on the project became concerned, and reported to officials that the efforts would rob the southern Everglades of half of the natural water supply, and that this would drastically affect the distribution of life. He was ignored. Then came disaster—nature changed. In the early 1960s a drought started that would last until 1975. Nearly 741,000 acres (300,000 hectares) of land burned. The tables had turned, and now all of the human effort to prevent flooding in years past created a terrible water shortage.

The situation worsened in the 1980s. A drought occurred in 1981, followed by the torrential rains of the 1982–83 El Niño (see the opening essay to chapter 44). In 1986, a fifth of Lake Okeechobee fell under an algal bloom, the result of nitrogen and phosphorus from fertilizer and dairy operations washing into the water. Eutrophication had set in.

Meanwhile, the living communities were responding to these changes to the environment. The native saw grass gave way to species that could tolerate the nutrient pollution. The influx of fresh water killed mangroves by robbing them of salt water, which led to declines in populations of sea grasses and various animals. Wading bird populations plummeted to 10% of their levels a century earlier. Dozens of species tottered on the brink of extinction.

Since the mid 1980s, efforts have been under way to undo the damage to the vast Florida ecosystem. In 1984, flooding was restored to a small area of the historic Kissimmee River floodplain. The results were encouraging. Waterfowl, aquatic invertebrates, and fishes quickly moved into the area as native vegetation returned. In response to the success of the demonstration project, state and federal agencies have agreed on a more ambitious plan to restore historic channels and floodplain wetlands to the central one-third of the Kissimmee River. Canals and levees will be removed, and water levels will be manipulated to simulate historic floods. The project will cost hundreds of millions of dollars in labor and land acquisition, and take about 15 years to complete.

Ecologists do not yet know whether this larger project will bring back the original character of the Kissimmee River, or whether the effort is too little, too late. Ecologists do know, however, that it is never easy to restore environmental quality after decades or centuries of decline. This final chapter outlines some of the other challenges facing the natural world.

45.1 Sustainability

As we alter habitats, we may be causing irreversible changes to ecosystems whose consequences we cannot foresee. The concept of sustainability holds that we should use natural resources in a way that does not deplete them for future generations.

The science of biology, and much of this book, considers *Homo sapiens* as but one of millions of species. Like other forms of life, we acquire energy and use it to power our cells; we sense and react to stimuli; we use an informational system based on DNA to construct the proteins that carry out the activities of life. Also like other species, ours adapts. How we deviate from others, however, is in the impact that humans have on the environment (fig. 45.1). It is profound. Consider some facts:

- Humans have altered nearly 50% of the land. We have replaced grasslands with wheat fields, wetlands with suburbs, forests with grazing areas.
- By burning fossil fuels and using fertilizer on farmland, humans have doubled the amount of available nitrogen in the environment.
- Overfishing, introducing nonnative species, and destroying habitats has caused or hastened many extinctions.
- Half of the world's mangrove swamps have been cleared for building or for aquaculture (farming seafood, especially shrimp). Ocean fishes are used to feed farmed shrimp, depleting marine food webs.

Humans are part of the environment. Like other heterotrophs, we eat other types of organisms. We also use other organisms to build and power our homes, to manufacture clothing, and to develop drugs. When our use of organisms kills enough individuals to deplete populations and reduce genetic diversity, species can become endangered—at first locally, then perhaps on a growing and global scale. Alternatively, we can attempt to use living resources in a way that does not irreversibly destroy them for the future. This approach of preserving populations of organisms while using them is called **sustainability.** It is a guiding principle of modern agriculture. The Everglades ecosystems described in the chapter opening essay, and many areas discussed in this chapter, are vivid examples of how misguided human activity is countering sustainability.

A famous essay written in 1968 has proven timeless in describing the concept of sustainability. In "The Tragedy of the Com-

A

C

B

FIGURE 45.1
Sustainability Means Ensuring That Natural Resources Do Not Become Depleted.
(A) A housing development in California replaces a natural ecosystem. **(B)** Until the mid 20th century, demand for whale oil to manufacture soaps, candles, margarine, and a variety of other products led to depletion of many populations, which have yet to recover. These casks are filled with sperm whale oil. **(C)** This cattle pasture in western Brazil was not so long ago a lush tropical rain forest.

mons," Garrett Hardin used a metaphor of a common (shared) area, such as a water supply. If each person takes from the resource without concern for the negative effects on others, eventually the resource will disappear. Hardin's scenario envisioned a town. Today the idea of sustainability has taken on a more global meaning. This final chapter explores specific challenges to sustainability that affect some of the major biomes and aquatic ecosystems discussed in the previous chapter.

45.1 MASTERING CONCEPTS

1. What is sustainability?
2. How have human activities countered sustainability?

OLC

45.2 The Air

Atmospheric pollution contributes to acid deposition. Release of certain chemicals has thinned the ozone layer, and carbon dioxide trapped near Earth's surface has caused a greenhouse effect, which may contribute to global warming.

Air pollution, acid deposition, depletion of the ozone layer, and global warming are some of the problems affecting the atmosphere.

Air Pollution

Pollution is any chemical, physical, or biological change in the environment that adversely affects living organisms. Pollution impacts air and water worldwide.

People have been polluting the air since ancient times. Analysis of lake sediments in Sweden and peat deposits in England reveal local air pollution during the Roman Empire, from 500 B.C. to A.D. 300. At that time, lead mining and metal smelting were as common as during the Industrial Revolution, a period previously credited with introducing air pollution (fig. 45.2). Analysis of Greenland ice reveals that the smelters of the Roman Empire polluted the air with lead on a global scale.

December 5, 1952, stands out as a more recent time when we became particularly aware of air pollution. On this day, a temperature inversion occurred over London. A high-level layer of warm air blanketed the city, trapping cold air at ground level and preventing the particulates and gases from burning coal and other fossil fuels from dispersing upward. Sulfur dioxide (SO_2), which reacts in fog to form sulfuric acid, was the main pollutant. This "London fog" sent hundreds of people to hospitals with respiratory symptoms, and the death rate soared. London-type air pollution, also called **industrial smog,** occurs in urban and industrial regions where power plants, industries, and households burn sulfur-rich coal and oil.

Another form of air pollution is **photochemical smog,** which results from vehicle emissions and the chemical reactions they undergo in the presence of light. Photochemical smog forms in warm, sunny areas where automobile use is heavy. The ultraviolet radiation in sunlight causes nitrogen oxides and incompletely burned hydrocarbons to react to form ozone (O_3) and other oxidants. These compounds poison plants and cause severe respiratory problems in some animals, including humans.

FIGURE 45.2
Air Pollution.
(**A**) During the Industrial Revolution in England, soot-spewing smokestacks were a status symbol. (**B**) Air pollution continues to plague many cities. This is Mexico City.

A

B

Acid Deposition

Patterns of air circulation and geography can interact to harm organisms far from the source of air pollution. In 1852, British scientist Angus Smith coined the phrase "acid rain" to describe one effect of the Industrial Revolution on the clean British countryside. In the mid 1950s, studies on clouds in the northeastern United States, farmlands in Scandinavia, and lakes in England rediscovered the effects of **acid deposition.**

Today, acid deposition in the United States originates mostly from coal-burning power plants in the Midwest that release sulfur and nitrogen oxides (SO_2 and NO_2) into the atmosphere (fig. 45.3). Gasoline and diesel fuel burned in internal combustion engines and fumes from heavy metal smelters add to the problem. In the atmosphere, these oxides join water and fall over large areas as sulfuric and nitric acids (H_2SO_4 and HNO_3), forming acid rain, snow, fog, and dew. These acids can also return to the earth as dry particles.

Because the atmosphere contains carbon dioxide (CO_2) and water, all rainfall includes some carbonic acid and is therefore slightly acidic (pH about 5.6). Burned fossil fuels, however, have made the average rainfall in the eastern United States 25 to 60 times more acidic than normal. Acid deposition also affects the Pacific Northwest, the Rockies, Canada, Europe, East Asia, and the former Soviet Union.

Acid deposition impacts lake life dramatically, because it may lower the pH range by 3.0 to 5.0 units. The acid can leach toxic metals such as aluminum or mercury from soils and sediments. In this abnormal environment, fish eggs die or hatch to yield deformed offspring. Amphibian eggs do not hatch at all. Crustacean exoskeletons fail to harden. In very acidified lakes, most aquatic invertebrates and protista die, and lake-clogging mosses, fungi, and algae replace aquatic flowering plants. Clinging algae outcompete photosynthetic microorganisms and coat shallow nearshore areas, preventing trout from spawning. Organisms that feed on the doomed species must seek alternate food sources or starve, which disrupts or topples food webs. Over time, these conditions reduce lake life to a few species that can tolerate increasingly acidic conditions.

The effects of acid rain on lakes are often difficult to study because of ecosystem complexity and lack of information on a lake's condition prior to acidification. In an experimental approach, since 1968 researchers in Winnipeg, Manitoba, Canada, have added sulfuric acid to a lake 200 miles (322 kilometers) southeast of the city. They have cataloged changes in the biotic, abiotic, and energy components of this experimental ecosystem. In that time, the lake's pH has dropped from 6.8 to 5.0.

Canada's experimentally acidified lake shows that life continues, but biodiversity plummets. Before acidification, the lake housed large populations of 80 species of midges, a type of insect. Today, 95% of the midge community is of one previously unknown species. A formerly rare midge variant able to thrive in low pH conditions survived as natural selection removed competing species.

The Canadian experimental lake shows that acid deposition alters food webs and biogeochemical cycles, but it does not seem to affect primary production, nutrient levels, and decomposition rates. Natural events also counter acidification. Calcium compounds in the earth have a buffering effect, just as similar compounds do in stomach antacids. Bacteria on the needles of jack pine trees neutralize acid rain, and sphagnum peat mosses ringing lakes absorb and buffer acid rain. Bacteria deep within lakes break down nitrogen and sulfur compounds. Researchers at the Canadian lake hypothesize that if pollution would stop, lakes would recover in just 5 to 10 years.

FIGURE 45.3
Acid Deposition.
(**A**) In the United States the average pH of precipitation drops in an eastward direction. Sulfur and nitrogen oxides from industry in the U.S. midwest cause the acid deposition in the East. (**B**) Brown conifer needles are an early sign of acidified soil. (**C**) Years of acid deposition has defaced this statue in Herten, Germany.
Source: Data from Thomas Graedel and Paul Crutzen, "Atmosphere, Climate and Change" in *Scientific American Library*, 1995.

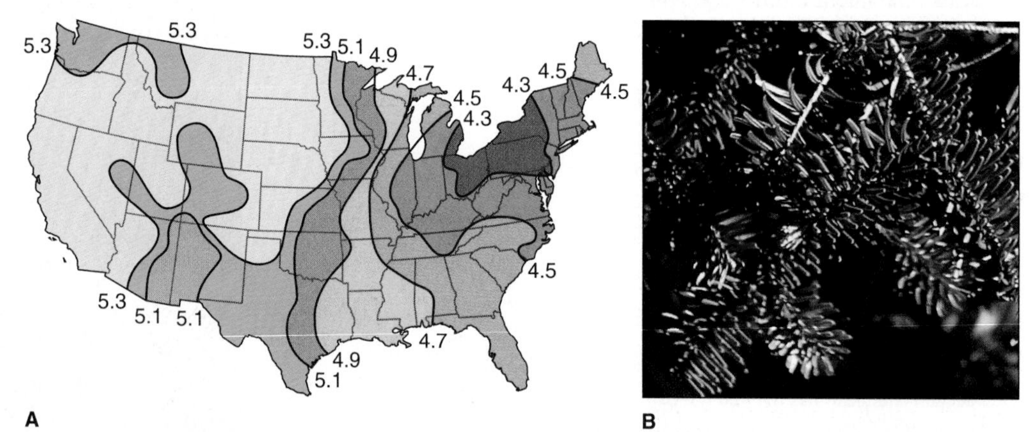

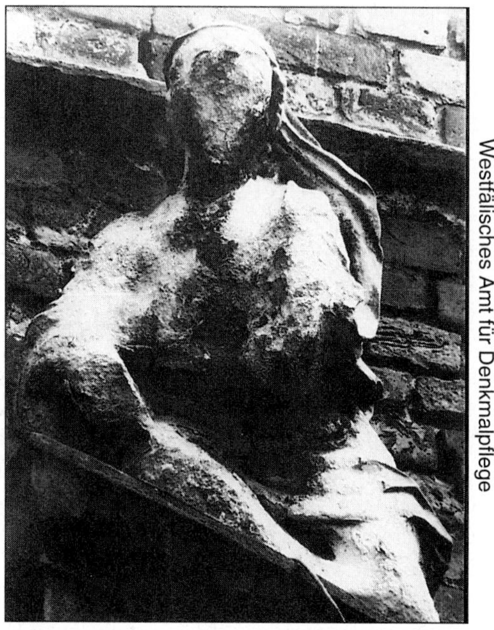

Westfälisches Amt für Denkmalpflege

Acid deposition alters forests too. In a coniferous forest high in the mountains of Fichtelgebirge, Germany, acid rain and snow thins trees and yellows needles. Here, nitric and sulfuric acids trigger a cascade of effects. As soil pH drops, aluminum ions percolate toward tree roots, where they displace nutrients such as calcium ions (required for growth of twigs, leaves, stems, and trunks) and magnesium ions (part of chlorophyll). Loss of essential nutrients stunts tree growth. Yet the constant input of nitrogen, in nitric acid, stimulates trees to grow. These mixed environmental signals stress trees and make them less able to resist infection or to survive harsh weather. As a result, acid deposition is thinning high-elevation forests throughout Europe and on the U.S. coast from New England to South Carolina.

Although smog and acid deposition still exist, air quality has improved in the United States in recent decades. The Clean Air Act, originally passed in 1970 and amended several times since, sets minimum air quality standards for many types of air pollutants. Among other provisions, the law mandates reductions in sulfur and nitrogen oxides from automobiles and industrial sources such as power plants and smelters. Since 1970, emissions of nitrogen and sulfur oxides, lead, carbon monoxide, particulates, and other pollutants have declined, leading to significant improvements in regional air quality.

Thinning of the Ozone Layer

In photochemical smog at Earth's surface, ozone is a harmful pollutant. In the upper atmosphere (stratosphere), however, this gas forms a protective **ozone layer** about 12 to 30 miles (20 to 50 kilometers) above Earth's surface. This ozone forms when ultraviolet radiation reacts with oxygen gas (O_2), and it absorbs much of the harmful ultraviolet (UV) radiation that would otherwise strike Earth and harm organisms.

The **ultraviolet** portion of the electromagnetic spectrum includes wavelengths shorter than 400 nm. UV-B radiation (wavelength 290 to 320 nm), along with radiation of even shorter wavelengths, can damage biological molecules such as DNA. In humans, the result of exposure to UV-B can be skin cancer or cataracts. UV-B can also harm phytoplankton, the base of the food web in many aquatic ecosystems.

Chlorine, fluorine, and bromine—some of which enter the atmosphere as **chlorofluorocarbon (CFC) compounds**—can accelerate the breakdown of stratospheric ozone. These compounds were once used in refrigerants such as Freon, as propellants in aerosol cans, and to produce foamed plastics. They can persist for decades in the upper atmosphere, catalyzing chemical reactions that break down ozone. The ozone layer has thinned so much since 1985 that a "hole" has appeared in satellite images over Antarctica (fig. 45.4). The ozone layer has also thinned over parts of Asia, Europe, North America, Australia, and New Zealand. The Montreal Protocol, signed in 1987 and subsequently amended, is an international treaty that calls for the phaseout of most ozone-depleting chemicals early in this century.

Natural events, such as the 1991 eruption of Mount Pinatubo, thin the ozone layer too. Chemical reactions that deplete ozone occur readily on the dust particles that the volcano sent into the atmosphere. Still, by 1995, when researchers had expected Pinatubo's influence to have waned, the ozone hole over Antarctica was still present. The persistence may be due to lingering effects of the volcano, increased use of chlorine and bromine-containing chemicals, or a cold wave in the stratosphere above Antarctica—or to a combination of these influences. Researchers

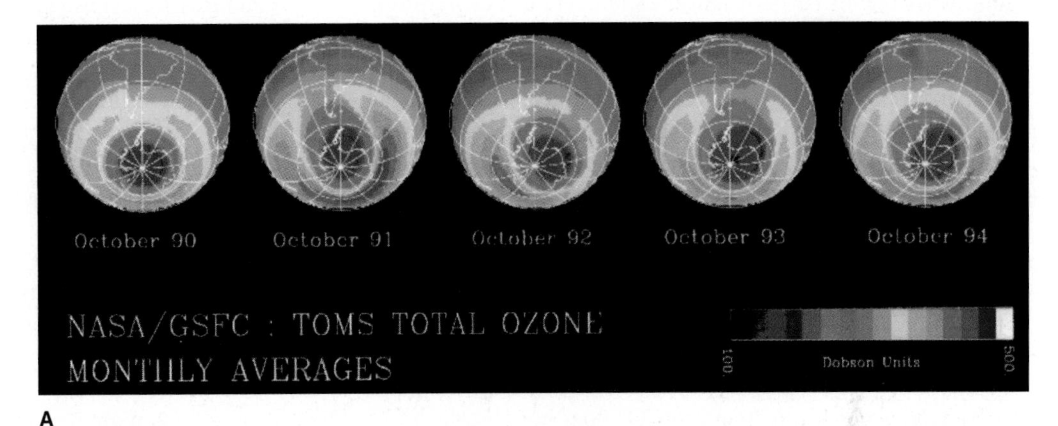

A

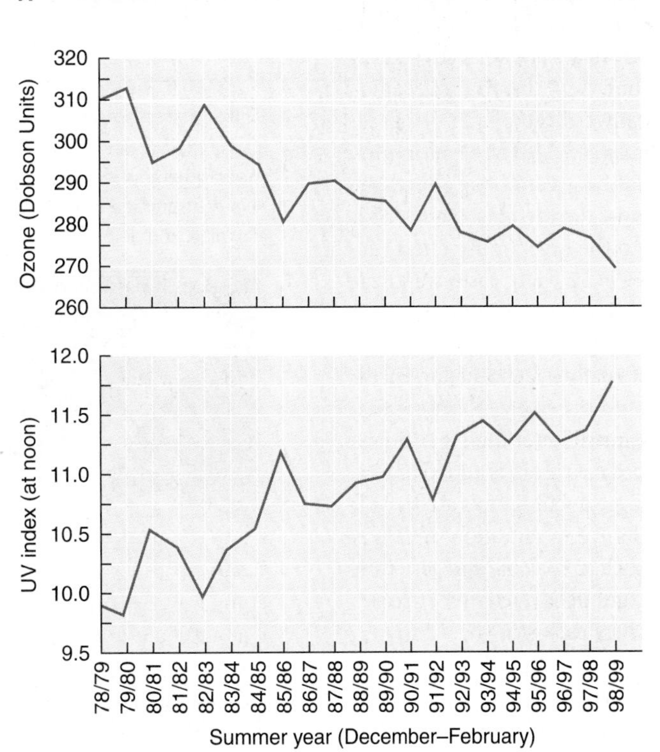

B

FIGURE 45.4
The Ozone Hole and Ultraviolet Radiation.
(**A**) Satellite images taken over time reveal the ozone hole over Antarctica. In these images, purple and blue represent thinned areas of the ozone layer.
Source: NASA
(**B**) These two charts clearly show that as ozone level decreases, UV intensity increases.
Source: Data from Richard McKenzie, et al., "Increased Summertime UV Radiation in New Zealand in Response to Ozone Loss" in *Science*, vol. 285, September 10, 1999.

do not know whether the ozone hole is temporary and localized, or the beginning of global ozone depletion.

Correlating ozone depletion to rising UV-B levels has been difficult because many factors can reflect or filter out UV-B, including snow, pollution, clouds, and volcanic ash. A decade-long study, however, recently strengthened the connection. Researchers monitored midday UV-B levels at a rural part of South Island in New Zealand that is unusually free of these confounding factors. Results were astonishing—UV-B levels are 12% higher today than they were in 1989! That figure is close to the 15% predicted based on the rate of depletion of ozone in the area since 1978. As a control, the level of UV-A remained constant over the past decade, which is consistent with the insensitivity of these longer wavelengths to ozone. To put this information in a human health perspective, a decrease in ozone of 1% is associated with a 3% increase in the incidence of skin cancers caused by UV exposure.

Global Warming

Recall from chapter 44 that climate greatly influences the distribution of life on Earth. Changes in temperature or moisture can greatly alter populations, killing some organisms outright, stressing others, or propelling those that can move to migrate in search of the conditions to which they are adapted. Average global temperatures appear to be increasing, with calculations projected from existing data predicting a rise in temperature of 2.5°F (1.4°C) by 2050.

The primary cause of global warming is elevation in the level of carbon dioxide (CO_2) in the stratosphere. Carbon dioxide is a colorless, odorless gas present at a concentration of 350 parts per million. Although a minor atmospheric constituent, CO_2 has major effects on life by influencing temperature.

CO_2 that blankets Earth prevents radiation from exiting the atmosphere as quickly as it enters. CO_2 does not absorb short-wave solar radiation, including light waves. Thus, light reaches Earth's surface, where it is converted to heat. The surface radiates long-wave heat radiation outward, but CO_2 absorbs this radiation and reradiates some of it back toward the planet's surface. This traps heat near Earth's surface (fig. 45.5). The resulting increase in surface temperature is called the **greenhouse effect,** because CO_2 blocks heat escape much as do the glass panes of a greenhouse. Other gases contribute to the greenhouse effect, including methane, nitrous oxide, and CFCs. These gases trap heat much more efficiently than CO_2 does, but because they are scarcer, they contribute only half as much to global warming.

In a sense, the greenhouse effect actually benefits life on Earth, because without its blanket of CO_2 and other greenhouse gases, Earth's average temperature would be much lower than it is. But as CO_2 accumulates in the atmosphere because more is added than is removed, scientists predict that average global temperatures will climb. Today, the increase in atmospheric concentration of CO_2 is largely caused by burning fossil fuels such as coal, oil, and gas. This, plus tropical deforestation and other combustion activities, releases 5 million tons of CO_2 into the atmosphere each year (fig. 45.6).

The first prediction that increasing stratospheric CO_2 could raise temperature came from Svante Arrhenius, of panspermia fame (see chapter 18's opening essay). In 1896, he calculated that a 50% increase in CO_2 level would cause a 7.4°F (4.1°C) increase in temperature from 30° to 40° north latitude, and in the ocean of 5.9°F (3.3°C)—results similar to today's computer-derived predictions. The industrialization of the Western world prompted Arrhenius's interest in global warming.

Physical effects of global warming were obvious before biological effects. Precipitation has increased at mid-latitudes and decreased at low latitudes over the past 30 to 40 years. Average global temperatures have increased since 1900, and especially since 1980. Ice sheets in Greenland and glaciers in Alaska are shrinking. Growing seasons in temperate areas are lengthening. Computer models predict that these trends will continue, and will shift the distributions of species.

Evidence that species' ranges are shifting is beginning to mount. For example, the northern boundaries of the ranges of at least 34 species of butterflies are moving northward. At the southern ends of the ranges, where temperatures are climbing too high, some species have become locally extinct. Also, the northern limits of many bird populations in the United Kingdom have moved northward. Among birds in the United Kingdom and the United States, comparing census records from 1971 through 1995 shows that birds are laying eggs about 9 days earlier in the spring.

Predictions of the future effects of global warming sound like fictional doomsday scenarios, but they may prove to be accurate. A United Nations Intergovernmental Panel on Climate Change warns countries to prepare for possible consequences. If the warming continues, the U.S. south may become too dry to sustain many traditional crops, with expected drops in the wheat harvest of 20%, corn by 30%, and soybeans by 40%. Water shortages worldwide may affect more than a billion people, coastal flooding 23 million, and hunger 22 million. Infectious disease patterns may also shift. Outbreaks of paralytic shellfish poisoning and algal blooms are already spreading and appearing where they haven't before. Tropical diseases, such as malaria, African sleeping sickness, and river blindness may come to more northern areas. The United Nations committee

FIGURE 45.5

The Greenhouse Effect.

Solar radiation heats Earth's surface. Some energy returns to the atmosphere as infrared radiation (heat), and some of this heat is trapped near the surface by CO_2 and other gases. Respiration, industrial processes, farming, and deforestation produce these greenhouse gases.

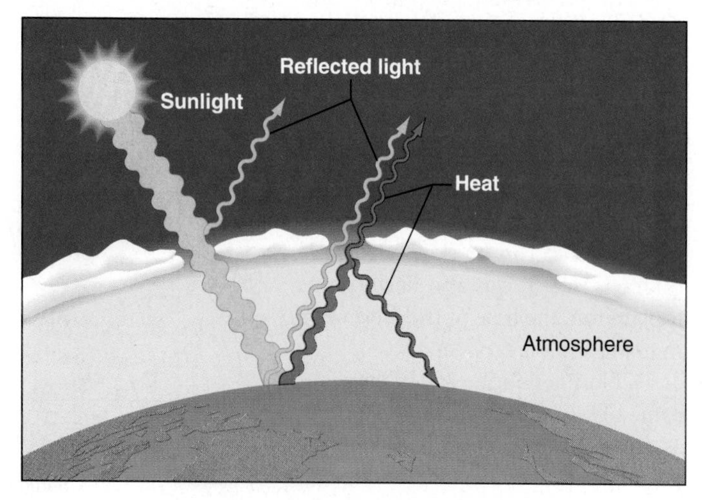

FIGURE 45.6

CO_2 Level and Global Average Temperature Are Correlated.

Atmospheric CO_2 levels continue to rise annually, as measured in an observatory at Mauna Loa, Hawaii. As CO_2 accumulates, the average global temperature may also be increasing.

Source: Data from U.S. Environmental Protection Agency (atmospheric CO_2 concentration) and NASA Goddard Institute for Space Studies (temperature data).

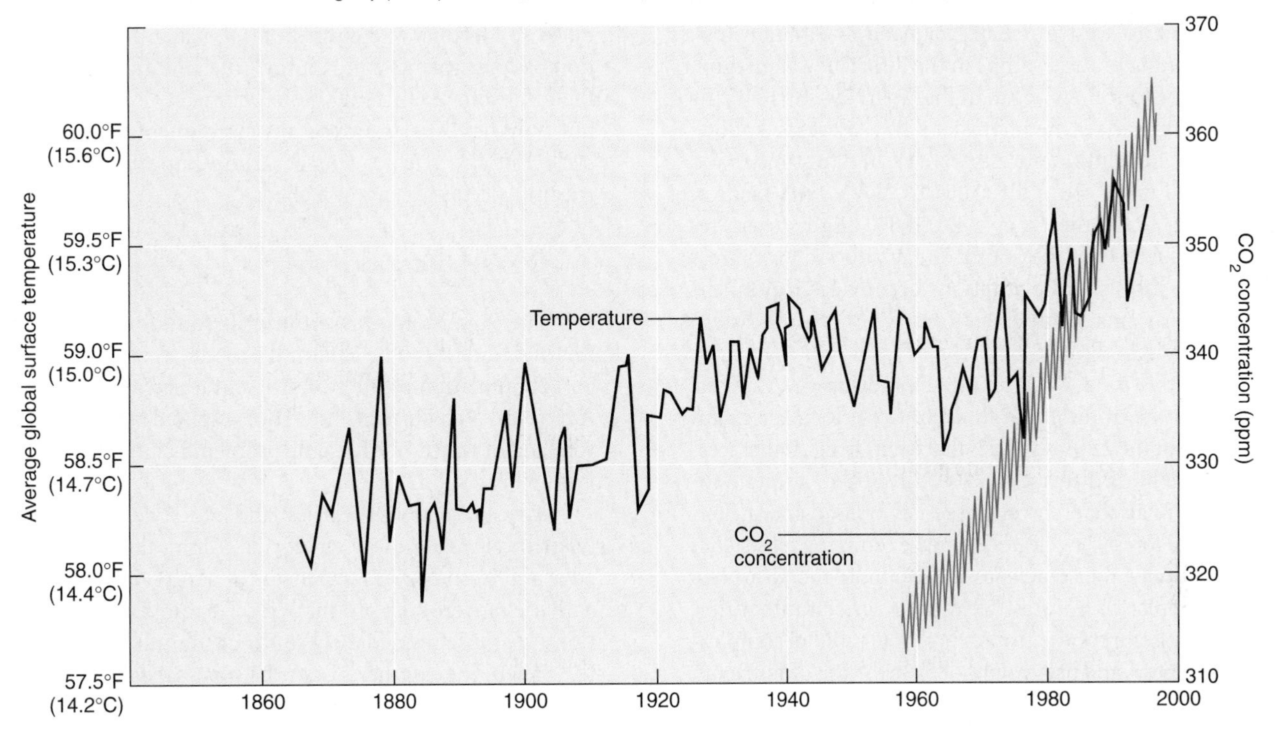

suggests that to limit the effects of the warming trend, nations learn to desalinate water; train people for jobs other than agriculture; encourage people to move inland; and stop destroying habitats that could protect against flooding, such as mangrove swamps.

45.2 MASTERING CONCEPTS

1. What are major sources of air pollution?
2. How does acid deposition form?
3. What effects do air pollution, acid deposition, thinning of the ozone layer, and global warming have on life?

OLC

45.3 The Land

Short-sighted agricultural practices combined with natural weather events and climatic conditions have destroyed forests and extended deserts.

On the land, habitat destruction is threatening the world's forests and grassland ecosystems.

Deforestation

The removal of all tree cover from a forested area is called **deforestation.** Both tropical and temperate forests are disappearing, at an average of about 1% annually. Satellite data indicate that tropical deforestation affects more than 17.4 million acres (7 million hectares) per year (fig. 45.7). Tropical rain forests are

FIGURE 45.7

Tropical Deforestation.

Dark green areas in this satellite image of a forest in Brazil represent natural forest, whereas the light yellowish and pink areas indicate areas of deforestation.

shrinking in South America, Africa, and Southeast Asia. Nearly half of the world's moist tropical forests have already been cleared.

The disappearance of the native North American temperate forest has clearly paralleled settlement by Europeans. In the first half of the eighteenth century, European settlers began clearing the land from east to west to create farmland and obtain fuel. Logging and draining of the swamps in the lush Tidewater region of Virginia and the Carolinas began in the late 1760s. By 1840, the path of destruction had reached Louisiana and Arkansas. Following the Industrial Revolution, forests were cleared for timber and turpentine and to make room for railroads, towns, and fields of soybeans, cotton, and other crops. By 1880, virgin forest remained in less than 35% of the South. Today, although vast areas of managed pine forests and plantations occupy the region, less than 1% of the original temperate forest of the southeastern United States survives (fig. 45.8).

When trees die, food webs topple. Destroying the *Casearia corymbosa* tree in the tropical rain forest of Costa Rica, for example, starves many of the 22 bird species that eat its fruits. Other trees whose seeds these birds disperse are also threatened, as are monkeys and other animals that rely on these trees for food and shelter.

Most tropical forests are cleared to make room for subsistence agriculture. Ironically, the same soils that support the lush plant growth of the tropical rain forest yield poor agriculture returns. The warm temperatures of the tropics promote rapid decomposition of organic matter, and heavy rains deplete the soil of nutrients. Therefore, most of the nutrients in the intact forest are bound up in the plant growth. Once these are cleared to make room to plant crops or graze animals, the nutrient-poor soils rapidly harden into a cementlike crust that cannot support plants.

FIGURE 45.8

Temperate Deforestation in the United States.

Since 1620, nearly all the native forests in the continental United States (dark green areas on the map) have been cleared.

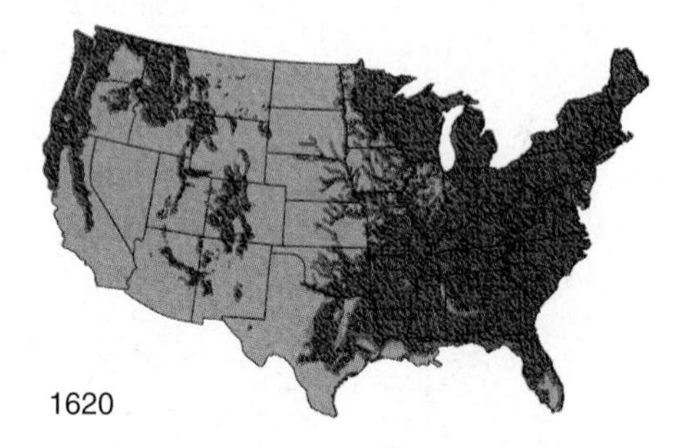

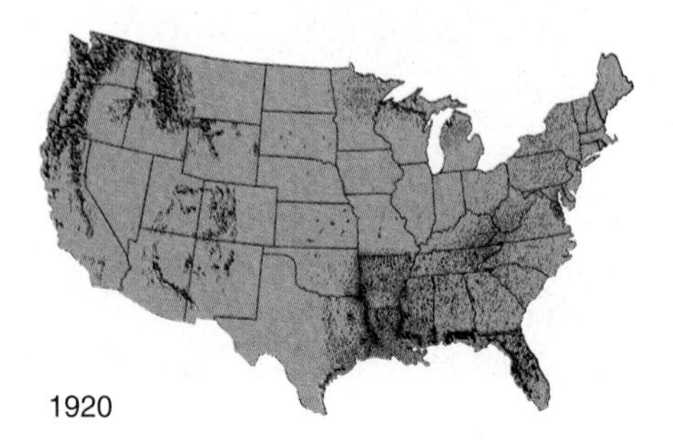

Deforestation contributes to diverse social and environmental problems. Tropical forest destruction threatens many native people. Forests also harbor a tremendous diversity of plant, animal, and microbial life, which may become extinct as the forests are destroyed. Deforestation promotes soil erosion, which reduces soil fertility and contributes to water pollution. Transpiration by forest plants contributes large amounts of water vapor to the atmosphere, affecting global climates. Burning these forests not only removes an important component of the global water cycle, but also releases stored carbon into the atmosphere, contributing to the greenhouse effect.

Desertification

The expansion of desert into surrounding areas is called **desertification.** It can be a natural process, or a human-driven one, or a combination of the two. Current desertification problems in Africa and Asia illustrate how short-sighted agricultural practices rob land of water. In Africa the problem is cattle grazing; in Asia it is growing cotton.

Africa's Sahara is the driest and largest desert in the world. Its dust travels as far away as Florida. Along the Sahara's southern edge and in countries farther south, natural drought and human activities are eroding productivity. The most severely affected nations are those immediately south of the Sahara in an area called the Sahel. The soil here is so dry that seeds cannot germinate. Wells are dry. Birds, reptiles, and desert shrubs are rare, and some parts of the Sahel appear to be lifeless.

As in many ecological events, desertification of the Sahara is the culmination of several steps. After 10 years of plentiful rain, a great drought began in 1968. As grasses shriveled and trees died in the Sahel, farmers let their cattle browse farther to find food. The animals ate everything growing near oases, where grasses have deep roots that tap into the low water table. When their leaves were eaten, the plants redirected their energy into replacing the leaves rather than extending their roots. The grasses died of lack of water. In place of the doomed grasses grew plants that cattle would not eat, presumably because of their taste. The short roots of the new plants, plus the movements of the cattle, broke apart and eroded the soil. At first, patches of desert appeared where the plants died and grazing cattle compacted the soil. Then the patches enlarged and joined, and soon, the desert had spread. Where would the cattle find food? The farmers cut down more forest to create more grazing land, which is destined to become desert too, if these practices continue.

Drought is not unusual in this part of the world, and desertification is not limited to Africa. In many nations, overgrazing, improper irrigation, unregulated vehicle activity, and other human impacts are degrading arid land. In the former Soviet Union, the catalyst for desertification was irrigating land to grow cotton. As a result, the city of Nukus, in the Republic of Karakalpakstan, unexpectedly turned into a dust bowl. Residents hurry indoors when the wind picks up, to avoid the toxic dust storms that carry residues of years of uncontrolled fertilizer and pesticide use. Rates of respiratory illness and anemia are sky high.

The death of central Asia's Aral Sea—actually once the world's fourth-largest lake—began in the 1920s, when its water was first used to irrigate cotton. In the 1950s, the program accelerated, with the Soviet government building a system of canals to tap and divert the water from the two rivers that fed the sea. But the water was being removed much faster than it was replenished by rainfall in this naturally dry area, and by the early 1960s, the Aral Sea was clearly shrinking. Today it is but three small remnants of what it once was, with 9 million acres (3.6 million hectares) of seabed exposed (fig. 45.9). Only a fifth of the original volume of water remains, and it has grown so salty from the fertilizer and pesticides used on nearby crops that all 24 native fish species have become locally extinct. Salt deposits are visible. Boats have become stranded.

The entire Aral Sea ecosystem and its surrounding lands are on their way toward lifelessness, or selection for salt- and drought-tolerant species. Efforts are under way to correct the damage, or at least prevent it from worsening. Total restoration of the Aral Sea to a productive aquatic ecosystem would take 50 years of no irrigation. Because some farming must continue to feed human residents, smaller amounts of water are being diverted from the tributaries. The hope is that the water table will rise and the Aral Sea might begin to refill, albeit slowly, with rainwater, which will also eventually dilute the saltiness.

45.3 MASTERING CONCEPTS

1. What events have led to tropical and temperate deforestation worldwide?
2. What factors promote desertification?

OLC

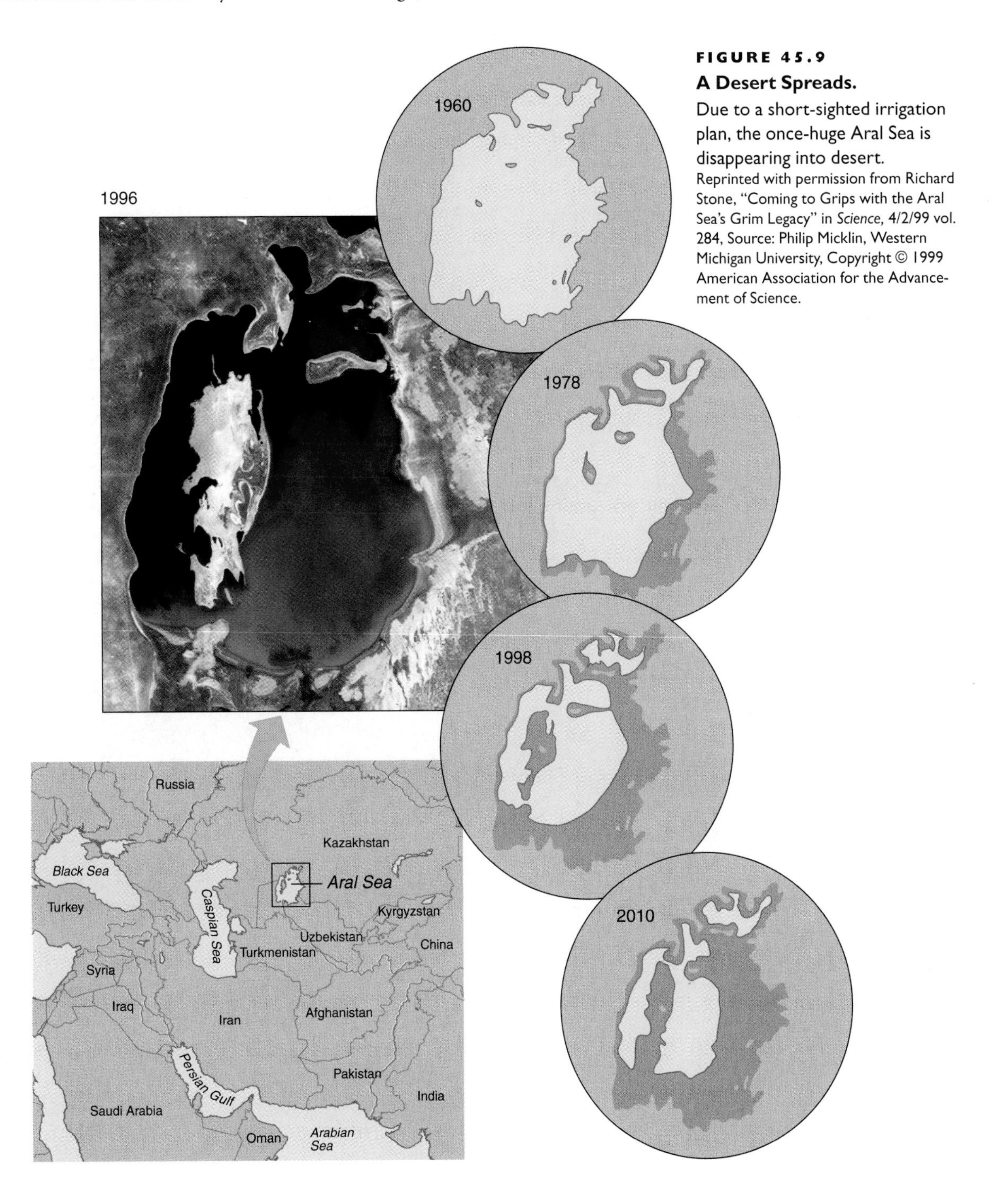

FIGURE 45.9
A Desert Spreads.
Due to a short-sighted irrigation plan, the once-huge Aral Sea is disappearing into desert.
Reprinted with permission from Richard Stone, "Coming to Grips with the Aral Sea's Grim Legacy" in *Science*, 4/2/99 vol. 284, Source: Philip Micklin, Western Michigan University, Copyright © 1999 American Association for the Advancement of Science.

45.4 The Waters

Human activities are fouling the waters and destroying and displacing populations before we have a chance to fully understand the scope of biodiversity in aquatic ecosystems.

Like the air, the waters face human-induced challenges. Pollution with toxic chemicals, excessive nutrient input, and overharvest of aquatic resources threaten ocean and freshwater ecosystems.

Chemical Pollution

Water can become polluted with chemicals that adversely affect organisms. Factories, municipal wastewater treatment plants, the shipping industry, and agriculture are important sources of surface water pollution. Groundwater contamination often comes from landfills, leaking storage tanks, and agricultural operations.

The chemical pollutants affecting rivers, lakes, and groundwater are very diverse. Poisons such as cyanide and organic mercury compounds from mining operations can alter ecosystems by killing key members of food webs. Over a longer term, some organic chemicals such as PCBs (polychlorinated biphenyls) and PAHs (polycyclic aromatic hydrocarbons) may cause cancer and disturb reproduction in some species. An essential nutrient such as phosphorus may become a pollutant if too much of it is available; the effects of lake eutrophication were described in chapter 44. On land, nitrates from cattle feedlots seep into the soil and move readily into groundwater; high nitrate concentrations in drinking water can kill babies. Table 45.1 lists some examples of chemical water pollutants.

Pollutants may also be as seemingly innocuous as sediments, which may reduce photosynthesis by blocking light penetration into water. Even heat can be a pollutant. Hot water discharged from power plants reduces the ability of a river to carry dissolved oxygen, affecting fishes and other aquatic organisms.

TABLE 45.1 Examples of Chemical Water Pollutants

ORGANIC	INORGANIC
Sewage	Chloride ions
Detergents	Heavy metals (mercury, lead, chromium, zinc, nickel, copper, cadmium)
Pesticides	Nitrogen from fertilizer
Wood-bleaching agents	Phosphorus from fertilizer and sewage
Hydrocarbons from petroleum	Cyanide
Humic acids	Selenium
Polychlorinated biphenyls (PCBs)	
Polycyclic aromatic hydrocarbons (PAHs)	

FIGURE 45.10
The Polluted Rhine River.
Efforts to fight the 1986 fire in a chemical warehouse in Switzerland caused tons of poisons to flow into the already polluted Rhine, leading to massive fish kills. This man is scooping up dead eels.

Organisms living in polluted areas are often exposed to several toxins. The Rhine River, which flows through western Europe, has until recently been a chemical soup of industrial and agricultural pollutants that have flowed in since Europe recovered from World War II. The Rhine was contaminated with detergents, chlorine-containing pesticides and bleaching products,

FIGURE 45.11
Oil Spills Can Cause Massive Destruction.
The crude oil fouling this beach in the Kenai Fjords area of Alaska was spilled hundreds of miles away when the Exxon *Valdez* ran aground in Prince William Sound.

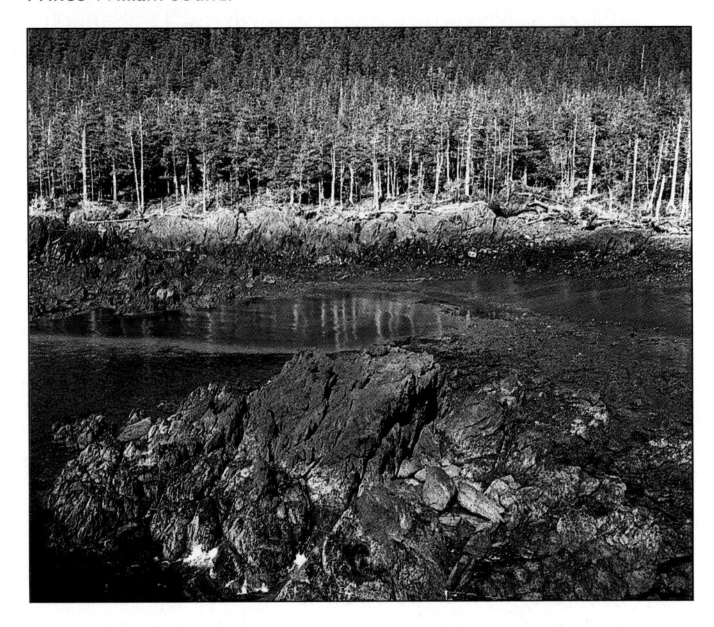

petroleum products, sewage, heavy metals, nitrogen, phosphorus, and other chemicals. In 1986, following a fire in a Swiss chemical warehouse, tons of poisonous liquids flowed into the Rhine and killed hundreds of thousands of fishes (fig. 45.10). In response, the Rhine Action Program identified 30 types of pollutants and then developed regulations and sewage treatments that had, by 1995, lowered the levels of discharge of all the pollutants by at least 50%. Salmon are back in the Rhine for the first time since the 1950s, and it appears that the river may be on its way to recovery.

Perhaps the most devastating recent example of chemical pollution was the oil spill which occurred in Alaska's Prince William Sound on March 24, 1989. The sound once harbored millions of birds, land mammals, marine mammals, invertebrates, and fishes. An oil tanker, the Exxon *Valdez,* ran aground on Bligh Reef, spilling 11 million gallons (more than 50 million liters) of oil (fig. 45.11). Unfortunately, some well-meant attempts to clean the mess backfired. For example, of 14,000 sea otters, 2,800 perished directly from the oil. Of the 1,200 others that were captured, tranquilized, and scrubbed, only 197 were alive 2 months later to be released into a still-polluted inlet. Some of the released otters carried a deadly herpesvirus to healthy wild animals. Scrubbing oil-covered rocks with very hot water under high pressure also had unexpected repercussions. Although the treatment cleansed rocky habitats, it harmed many organisms that had survived the oil pollution. This was the fate of *Fucus,* a light brown alga also called rockweed, that had dominated parts of the sound. *Fucus* has grown back sporadically in some places, but it has yet to cover many areas in the shaggy carpet that it did before the oil spill.

Just as with air quality, the United States has seen improved water quality in recent decades. Among other provisions, the Clean Water Act of 1972 required virtually every city to build and maintain a sewage treatment plant, dramatically curtailing the discharge of raw sewage into rivers and lakes. The 1987 Water Quality Act followed up on the Clean Water Act, providing for regulation of water pollution from industry, agricultural runoff, sewage overflows during storms, and runoff from city streets. Although water quality problems still exist in the United States, many of the nation's surface waters have largely recovered from past unregulated discharge of phosphorus, other nutrients, and toxic chemicals.

Physical Alteration of Waterways

Because a river provides diverse habitats and can cover a large area, human intervention can have profound effects on life. For example, along the banks of the Mississippi River, levees built to prevent damage from flooding alter the pattern of sediment deposition. Nutrients and sediments that once spread over the floodplain during periodic floods are now confined to the river channel, which carries them to the Gulf of Mexico. There, the nutrients feed protista, causing their populations to bloom. Red and brown tides, the result of such blooms, may kill fishes, manatees, and other sea life.

Damming alters a river ecosystem too. Changes in water temperatures and nutrient levels alter populations of protista, aquatic insects, and fishes, which creates great gaps in food webs. Channelization, or straightening the course of a meandering river, increases the water's flow rate, which erodes channel sediments and carries them downstream, where they can choke out stream communities.

Endangered Estuaries

Estuaries, as links between fresh water and salt water, are pivotal in environmental quality. Many fishes and shellfishes spend part of their lives in estuaries, and a diverse array of algae and flowering plants supports the food web. Yet humans have drained and filled estuaries to build houses and dumped garbage and other pollutants in their waters. Laws now protect estuaries in some parts of the nation.

Pollutants pour into the estuaries from rivers and streams that drain the surrounding watersheds. Animal droppings, human sewage and medical waste, motor oil, fertilizer from large farms and millions of lawns, plus industrial waste, oil spills, and garbage, all may end up in estuaries. In salt water, chemical reactions encapsulate pollutants into particles, making them heavy enough to sink. Here they remain, unless disturbed by human action or extreme weather.

Once contaminated sediments are disturbed, a deadly chain reaction ensues. Released nitrogen and phosphorus trigger algal and dinoflagellate blooms (see fig. 21.10), which sometimes produce toxins that poison aquatic animals. Decomposers feeding on dead algae and dinoflagellates deplete the water's dissolved oxygen.

The 64,000-square-mile (166,000-square-kilometer) Chesapeake Bay basin, on the eastern coast of the United States, touches six states and houses more than 200 species of fishes and 75 species of birds. The ecological changes of the past few decades here are ominous. The bay's food web base is in upheaval, in response to influx of sewage, agricultural fertilizers, heavy metals, pesticides, and oil. Each day millions of gallons of waste flow into the bay.

In response to the high nutrient loads, phytoplankton populations have bloomed and the composition of their communities has changed. Dinoflagellate and green algae populations are growing, while the numbers of diatoms are falling. Reduced light penetration has killed eelgrass. Eutrophication has sharply reduced the oxygen content of the deep water. As a result of these changes in food sources, the fish and crustacean populations are declining.

The Chesapeake Bay oyster is one victim of pollution and overharvest. In colonial times, oysters lived up to 10 years and grew to a foot or more in length. Today, they barely live past 3 or 4 years and are about the size of a golf ball. In the 1890s, 15 to 18 million bushels of oysters were harvested a year; today the number is under a million. Blue crabs were once so plentiful that at certain times of the year a wader could not avoid stepping on them. The crabs are fast losing their feeding and molting grounds.

Cleanup efforts may help restore some of the Chesapeake's lost diversity. New sewage treatment facilities are being built and old ones repaired. Floating plants introduced in some areas use the excess nitrogen. So far, however, efforts to reduce the phosphorus and nitrogen flowing into the estuary have been successful only in some areas.

Coral Reefs

Besides being home to a staggering number of species, coral reefs play other roles in the environment. They serve as barriers between powerful ocean currents and beaches, halting erosion. Some coral reefs filter outgoing material from mangrove swamps, thereby controlling nutrient flow to the oceans. And the residents of coral reefs feed millions of people living in tropical areas. On another front, chemical compounds derived from reef organisms serve as the basis for many new drugs.

Despite their recognized great value, coral reefs are in dire trouble. Unless recently begun conservation and restoration programs help, the world may lose up to two-thirds of its reefs within the next four decades.

Threats to reef ecosystems are both natural and human-induced. Several violent hurricanes in recent years caused much destruction, as has an increase in infectious diseases of reef residents. In the Caribbean, the local extinction of the sea urchin *Diadema antillarum* in the 1980s led to overgrowth of the algae that the sea urchins normally eat. The algae clogged and killed many coral reefs. Also in the Caribbean, elkhorn and staghorn corals have died in large numbers from at least 12 types of viral infections and other diseases.

Three specific types of human activities have harmed coral reefs:

- In "cyanide fishing," divers squirt cyanide into the crevices within reefs, momentarily stunning the fishes within. When the fishes emerge, they are easily caught (fig. 45.12). But the reef—and some divers too—may die in the process, for cyanide is a potent poison. The practice has become a popular way to provide tropical fishes for collectors and for diners in Asia who like to select their meals from tanks displayed in restaurants.
- "Dynamite fishing" uses underwater explosives to blast fishes out of the water. Some biologists have had to abandon experiments for fear of being bombed out along with the fishes.
- Overbuilding of resort facilities along shorelines pumps municipal waste into coastal waters, joining nutrient-packed sediment and runoff from agriculture. As populations of algae explode, the coral polyps in the living layer of the reef become smothered in algae and die. In Phuket Island, Thailand, such nutrient pollution invited a new predator to the local coral reef, the crown-of-thorns starfish. The algal bloom may have provided abundant food for the starfish's larvae. Once it has developed into an adult, this animal everts its stomach through its mouth, releases digestive enzymes that dissolve the polyps, then absorbs the coral animals' remains.

Since 1975, conservation at the Great Barrier Reef has served as a model to other coral reef areas. The Barrier Reef is divided into four sectors, and each of these is subdivided into three areas: open only to scientists, open to divers and snorkelers who can observe but not touch the coral, and general use areas where some other activities are permitted. But in other parts of the world, coral reefs and their surroundings are overfished to feed people starving today, even though this threatens sustainability for tomorrow. To help solve this dilemma, the International Coral Reef Initiative be-

FIGURE 45.12
Cyanide Fishing Is Killing Corals.
In the Philippines, divers use cyanide to stun fishes that live in coral reefs, so that the fishes can be caught. The corals may die.

gan in 1995 to try to coordinate efforts, from local to global levels, to preserve these ecosystems.

Polluted Oceans

Many species in the vast oceans of the world are harmed by the destruction of estuaries and other coastal habitats by urbanization, housing, tourism, dredging, mining, and agriculture. Many marine animals live at least parts of their lives in coastal areas, and the loss of these habitats can threaten populations of commercially important species. These populations are further depleted as fishing crews harvest the adults faster than the fishes can reproduce. In addition, many marine mammals, seabirds, sea turtles, and nontarget fish species are killed accidentally as they are caught up in the nets set for the target species (fig. 45.13).

Toxic chemicals and nutrients from shipping, urban and agricultural runoff, oil drilling, and industrial and municipal wastewater discharges also pollute oceans. In the top layer of the ocean, which supports vast webs of marine life, petroleum-based pollutants choke out life from above. Fishing nets and plastic also entrap and kill millions of seabirds and marine mammals, including 50,000 Alaskan fur seals. Some birds build their nests with plastic and feed their nestlings plastic bits. Sea turtles mistake floating blobs of plastic for their natural food, jellyfishes, and swallow them. The plastic lodges in their intestines and kills them.

Human activities, El Niño, and global warming have contributed to an apparent increase in infectious disease among ocean occupants. Victims include kelp, sea mammals, fishes, corals, sea grass, clams, and sea urchins, and the infectious agents include viruses, bacteria, protista, fungi, and nematodes. Some infections appear to be new, while others have spread because exposure to pollutants impairs host immune systems. Infections have also been transferred to new marine hosts. For example, a canine distemper virus, transmitted from sled dogs, has killed seals in Antarctica and Siberia. Global warming has shifted the migration routes of birds whose excrement passes influenza viruses, infecting seals and whales (see fig. 19.13).

Like other biomes, oceans have natural defenses. Many toxins remain sequestered if undisturbed. The sheer volume of the oceans dilutes some pollutants, and microorganisms naturally degrade certain compounds. People are attempting to aid this natural healing. Cardboard devices are replacing plastic beverage carriers, ships are no longer supposed to dump waste at sea, cattle are being kept from rivers, and many people are more aware of what they discard and pour down the drain.

45.4 MASTERING CONCEPTS

1. What are some pollutants of aquatic ecosystems?
2. What types of changes adversely alter river ecosystems?
3. Why is pollution in estuaries particularly devastating?
4. What climatic events and human activities have damaged coral reefs?
5. How does pollution affect oceans?

OLC

FIGURE 45.13
Ocean Problems.
This green turtle is trapped in a fishing net.

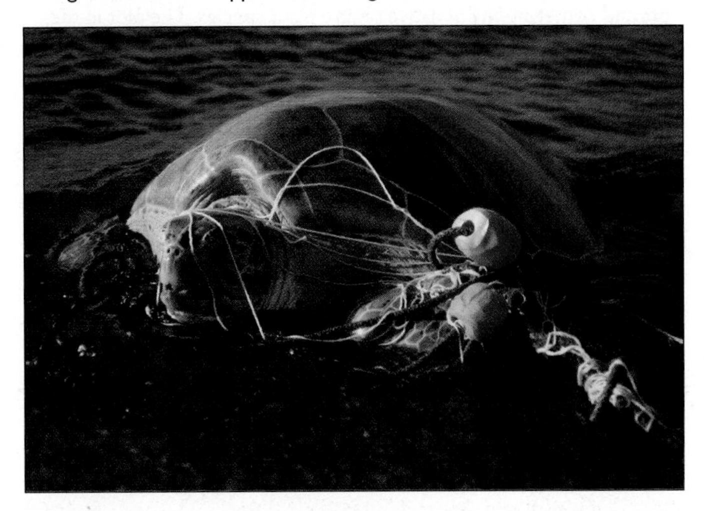

45.5 Loss of Biodiversity

Human-directed habitat destruction and invasion of nonnative species are hastening the shrinking of the biosphere into a disjointed collection of species-poor remnants of a once-diverse living landscape. As we teeter on the verge of a major extinction, it may be too late to stop plummeting biodiversity.

Genetic diversity in a population provides insurance against disaster—should the environment change, individuals with certain genotypes are likely to survive. On a broader scale, species diversity in ecosystems provides some assurance that at least some species will survive environmental change.

Biodiversity and Ecosystem Stability

The **diversity-stability hypothesis** holds that the more diverse and the greater the number of species in an ecosystem, the more it will be able to withstand a threat. Long-term experiments conducted in a Minnesota grassland support the hypothesis, but not all ecologists agree that the experiment accurately mirrors a natural ecosystem.

Grassland threats include fire, frost, hail, drought, herbivores, and extreme temperatures. To assess an ecosystem's response to disaster, researchers measure the biomass of plants and catalog the number of species. Starting in 1982, researchers monitored 207 plots of land in the Minnesota grassland and determined plant biomass to measure ecosystem stability (fig. 45.14). They controlled the number of species by supplying certain amounts of nitrogen in fertilizer—the more nitrogen, the more plant species a particular plot supports.

After a severe drought in 1987–88, the researchers measured the plant biomass in the plots. Plots with the most species lost half their biomass, but plots with the fewest species lost seven-eighths of their biomass! The conclusion based on this experimental ecosystem: When ecosystems are stressed, biodiversity

FIGURE 45.14
The Diversity-Stability Hypothesis.
Each plot in this biodiversity experiment is planted with different numbers and combinations of native prairie plant species. The plot in the foreground has four plant species; the plot just to the right has 16. Mowed areas are walkways between the plots.

provides stability. Under favorable environmental conditions, however, a species-poor area will also produce a large biomass.

Even though other experiments using controlled ecosystems have confirmed the finding that species diversity promotes ecosystem stability, and the idea of a "diversified portfolio" or "not putting all your eggs in one basket" seems logical, the hypothesis remains controversial. One criticism of the Minnesota grassland work is that applying nitrogen to control the number of species selects for those that grow the fastest. This might occur at the expense of slower-growing plants that have drought-resistance genes and that might have a better chance in a natural ecosystem.

Another criticism is that the conclusion that species number alone ensures ecosystem survival is an oversimplification. An ecosystem remains stable not necessarily because of a large number of species but because of their diverse roles. A forest of plants of very different heights, for example, might withstand fire better than a forest of trees of similar height, even if both areas include the same number of species. Another oversimplification of equating biodiversity with ecosystem stability is that species do not equally influence composition of a community. Recall from chapter 43 that keystone species influence community composition more than other members of the same community.

Will the Twenty-First Century End in Mass Extinction?

The last mass extinction occurred 65 million years ago, and it wiped out two-thirds of all terrestrial species, with unknown numbers vanishing in the waters. Recall from chapter 16 that past mass extinctions were caused by changing physical conditions, such as a meteor impact, drastic temperature shifts, or volcanic eruptions.

The next mass extinction, which we may be on the brink of, will almost certainly be related to human activities. Some researchers estimate that extinctions are occurring at 100 to 1,000 times the background rate. Given that biologists are unsure how many species exist on Earth, accurate extinction estimates are hard to obtain. However, extrapolations based on known rates of habitat destruction predict that one-third to two-thirds of all species will be lost during the second half of the twenty-first century, mostly in the tropics.

The cause of this next mass extinction will be less cataclysmic than meteor impacts, volcanic eruptions, or glacial movement. It will come from the slow and silent fragmenting of the biosphere into "remnants"—many of which have already been discussed. Grasslands are turning from seas of grasses to patches; the inland Aral Sea is giving way to desert; on a local level, streams dry up into pools that become biological ghettos for the survivors; forests become grazing pastures which then dry out; and the oceans have at least 50 documented pockets of dead zones. Instead of the biosphere being a blanket of diverse ecosystems that interconnect and overlap to cover Earth in life, it may become a series of isolated remnants, each housing gene pools too restricted for the species to survive much longer. One researcher calls these organisms "the living dead," for their populations may be doomed.

Extinctions directly caused by humans are known. Passenger pigeons, for example, were hunted to extinction in the late 1800s. In the Hawaiian Islands, Europeans brought cats and avian diseases that killed off 70 of the 140 native bird species. If anyone doubts that human encroachment condemns biodiversity, just compare a landscape of the New York City area from the year 1500 to a cityscape today. Our urban and agricultural ecosystems change the environment so radically that many native species must move on or perish. Extinctions are forever. They are the antithesis of sustainability. Once a type of organism is gone, there is no way to get it back.

Invasion Biology and Introduced Species

Habitat destruction is the primary cause of diminishing biodiversity. The second cause is biological invasion—the introduction of nonnative species, which then displace the native species. The opening essay to chapter 42 describes the effect of introducing brown tree snakes to the island of Guam, and chapter 43 describes the invasion of zebra mussels in the Great Lakes. Similarly, 75% of Florida's current plant life is not part of the lush ecosystem described in the chapter opening essay, but invaders introduced from agriculture and urbanization.

Humans are the conduits of species invasions because we tend to take plants and animals that we like with us when we settle new areas—and also some that aren't as popular, such as microorganisms on and in our crops, pets, and livestock, and stowaways such as rodents and insects on ships. A successful invasion, though, isn't highly likely. Of every 100 species introduced, about 10 survive in the short term, competing with native species, and only 1 persists to take over a niche. Invasions are more likely to occur in areas that have been disturbed—which suggests a dangerous link to habitat destruction. They are also more common on islands, where diversity is naturally low.

A successful invasion can have severe repercussions on an ecosystem. One famous case of species invasion is that of Australian rabbits. In 1859, wealthy Thomas Austin imported a few rabbits to his estate so that he might enjoy hunting them. He apparently did not shoot them all, and the survivors did what rabbits do—they made more rabbits. Soon they had taken over and spread, eating much of the vegetation and displacing native animals. The government took action in the 1950s by unleashing the myxomatosis virus, which kills only rabbits. When that virus became ineffective by 1995, they substituted another. Today the arid lands of Australia show signs of recovery from the rabbit attack.

A more recent example of species invasion is occurring now in Europe. In the early 1990s, a plane from the United States unintentionally brought the western corn rootworm, actually a beetle larva, to war-torn Yugoslavia. By 1995, the rootworms had spread to Croatia and Hungary, and today they are in Bulgaria and Italy too. By the time the people noticed that their corn crops were disappearing, it was too late to eradicate the pest.

Plants may also be invasive. Scientists estimate that 5,000 nonindigenous plant species are established in U. S. ecosystems. One example is hydrilla (*Hydrilla verticillata*), which chokes waterways in Florida, alters nutrient cycles, affects aquatic animals, and reduces recreational use of lakes and rivers (fig. 45.15).

In Hawaii, the state government is attempting to fight the species invasion problem by creating refuges for native species. At the Kulani correctional facility, for example, conservationists have gathered the Mauna Loa silversword plant and tree ferns, protecting them from voracious leaf-eating nonnative pigs. The refuge also houses several species of songbirds, whose numbers are plummeting from a triple attack—diminishing habitat, avian malaria, and invading species that eat their eggs.

The dangers of introduced species are 2-fold. On an ecosystem-by-ecosystem basis, natural selection will favor those species that grow and reproduce the fastest, and are generalists. This is the case for the red maple that is taking over the temperate deciduous forest in the United States because it can tolerate a wide range of conditions (see chapter 44). More philosophically, introducing species is homogenizing the biosphere, gradually transforming it from a collection of distinctive interacting areas whose species arrived by their own means—such as birds flying to islands—to a human-directed sameness. In natural ecosystems, the combination of sameness and shrinking habitats may prove devastating to the diversity of life.

FIGURE 45.15

Biological Invaders.

Hydrilla verticillata clogs this waterway in central Florida. This species came to the United States from southeast Asia as an aquarium plant, and quickly became a nuisance when residents put the imported plant into local waters.

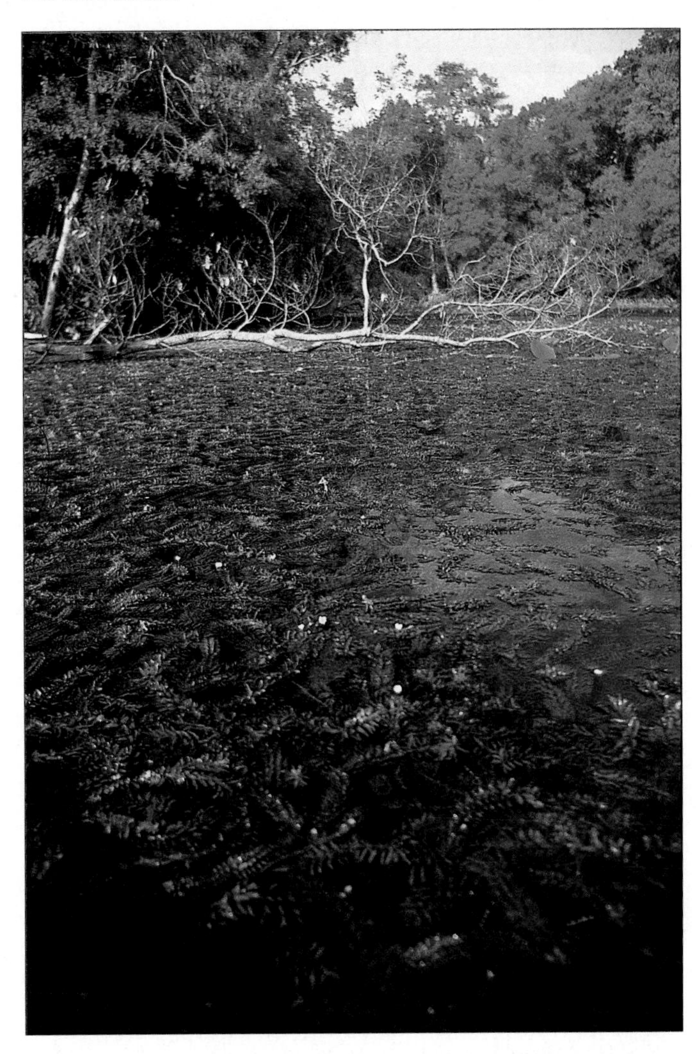

The Resiliency of Life

Life on Earth has had many millions of years to adapt, diversify, and occupy nearly every part of the planet's surface. It would be very difficult to halt life on Earth completely, short of a global catastrophe such as a meteor collision or a nuclear holocaust. The biosphere has survived mass extinctions in the past, and it probably will in the future.

Life has prevailed through many localized challenges—from natural events such as hurricanes, volcanic eruptions, and forest fires to human-caused garbage heaps, oil spills, and pollution. Organisms may perish—many of them—but the plasticity of genetic material ensures that, in most cases, some individuals will inherit what in one environment is a quirk but in another is the key to survival. Even when biodiversity is severely challenged, some species usually persist. Life does not completely end; it changes.

Can we continue to rely on adaptation and diversity to maintain life on Earth? Life may be resilient in an overall sense, yet it is fragile in its interrelationships. Intricate food webs easily collapse if a single member species declines or vanishes, or if a new one is introduced.

Maybe we have just been lucky, so far, that no single event has decimated so many key players in the game of life that other species could not replace them. But could multiple stresses combine to make Earth too inhospitable for life as we know it to continue? Will lifeless patches of oceans grow and coalesce, as ozone-depleted areas over the poles and stretches of dead desert do? A single eutrophied or dried-up lake, an acid deposition–ravaged forest, a grassland turned cornfield and shopping mall, an oil-soaked shore—alone they are terrible; together they could begin to unravel the tangled threads that tie all life together.

On the other hand, it is important to recognize that humans may also have the power to undo some of our past mistakes. For example, because of the Endangered Species Act of 1973, some species that were in danger of extinction, such as the bald eagle, are slowly recovering. The last several decades have seen other major pieces of environmental legislation (such as the Clean Air and Clean Water Acts) that have yielded dramatic improvements in environmental quality in the United States. Worldwide, governments and private organizations have set aside natural areas to protect them from agriculture and urbanization. Major restoration projects such as the Everglades plan described in the chapter opening essay show that recovery, while costly and difficult, may yet be able to reverse some of the sobering trends described in this chapter.

As the human population continues to grow, pressure on Earth's natural resources will certainly increase. Ironically, in developing countries, families may respond by having more children to help obtain increasingly scarce basic necessities such as fuel, food, and clean water. In all parts of the world, local depletion of natural resources, pollution, and occasional natural disasters will continue to damage the environment. For these reasons, one key to reversing environmental decline will be to slow the growth of the human population.

Scientists and politicians, as well as ordinary citizens, carry the heavy burden of protecting the remaining resources for the future, while maintaining a reasonable standard of living for all people. Part of the solution lies within you and how you choose to live. This book has shown you the wonder of life, from its constituent chemicals to its cells, tissues, and organs, all the way up to the biosphere. Do whatever you can to preserve the diversity of life, for in diversity lies resiliency and the future of life on Earth.

45.5 MASTERING CONCEPTS

1. What is the evidence that biodiversity fosters ecosystem stability?
2. What are the most important human-related causes of species extinctions worldwide?
3. How do introduced species disrupt ecosystems?
4. Why will the next mass extinction probably not mean the end of life on Earth?

OLC

Chapter Summary

45.1 Sustainability

1. **Sustainability** is the careful use of natural resources in a way that does not deplete them for future use.

45.2 The Air

2. **Pollution** is any change in the environment that harms living organisms.
3. Air pollutants include heavy metals, particulates, and emissions from fossil fuel combustion in automobiles and industries. Some of these pollutants react in light to form **photochemical smog.**
4. **Acid deposition** forms when nitogen and sulfur oxides react with water in the upper atmosphere to form nitric and sulfuric acids. These acids return to earth as dry particles or in precipitation. Acidification of lakes changes aquatic communities. In terrestrial ecosystems, acid deposition kills leaves and releases toxic aluminum from soils.
5. The Clean Air Act improved air quality in the United States.
6. Use of **chlorofluorocarbon compounds** has thinned the stratospheric **ozone layer,** which protects life from **ultraviolet radiation.**
7. The **greenhouse effect** results from CO_2 and other gases trapping heat near Earth's surface. Agriculture, burning fossil fuels, and destruction of tropical rain forests generate greenhouse gases. The greenhouse effect may contribute to global warming. Shifting vegetation patterns, species ranges, and egg-laying schedules are responses to global warming.

45.3 The Land

8. **Deforestation** is the destruction of tree cover from forested areas. Subsistence-level agriculture contributes to the removal of tropical rain forests.
9. Draining lakes to irrigate crops and cattle grazing cause and hasten **desertification,** the expansion of desert into surrounding areas.

45.4 The Waters

10. Organic and inorganic toxins, sediments, heat, heavy metals, and excessive nutrient levels pollute aquatic ecosystems. Water quality in the United States has improved in recent decades.
11. Dams and levees alter river ecosystems.
12. Preserving estuaries is important because they are breeding grounds for many species.
13. Fishing with cyanide and dynamite, excessive nutrients, and hurricanes threaten coral reefs.
14. Loss of coastal habitat and overharvest contribute to declines of populations of valuable ocean species.
15. Plastic and oil contaminate surface waters.
16. Human activities and climatic fluctuations contribute to the spread of infectious disease among many ocean inhabitants.

45.5 Loss of Biodiversity

17. The **diversity-stability hypothesis** proposes that an ecosystem with many diverse species is better able to survive stresses than a species-poor ecosystem.
18. Reduction of the biosphere to similar ecosystem remnants may put Earth on the brink of a mass extinction.
19. After habitat destruction, biological invasion is the top cause of decreasing biodiversity as nonnative species displace native organisms.

Testing Your Knowledge

1. What is sustainability?
2. Cite biological evidence of global warming.

3. In what ways is the greenhouse effect both beneficial and detrimental?
4. What is being done to restore or prevent further destruction of:
   a. the Everglades
   b. the Aral Sea
   c. grasslands in the midwestern United States (see chapter 44)
   d. the Great Barrier Reef
   e. oceans
   f. the Rhine River
   g. Prince William Sound
5. What natural protections do ecosystems have against the destructive effects of each of the following?
   a. acid deposition
   b. global warming from the greenhouse effect
   c. ocean pollution
6. How can too much of a nutrient alter an ecosystem?
7. Give an example of an environmental problem that had an immediate effect and one that had a delayed effect.

Thinking Scientifically

1. Several polar bears were recently discovered that are hermaphrodites—they have organs of both sexes. These bears also have high levels of PCBs in their blood, but researchers do not yet know whether exposure to PCBs is related to the disturbed sexual development. Given that many heavily polluted ecosystems are tainted with several types of pollutants, how could you link a particular chemical to a particular biological effect?
2. An approach to combat species invasion is to kill the invaders. In Hawaii, officials shoot feral cats, goats, and pigs. In Australia, the government added chlorine and copper to a bay to kill zebra mussels, killing everything living in the water. Do you think that this is an effective approach? Suggest an alternative.
3. Chapter 23 explained that if the number of species of mycorrhizal fungi in an ecosystem increases, then the number of plant species that they support increases. Does this finding support or challenge the diversity-stability hypothesis?
4. In the Aral Sea and in the Everglades, humans diverted natural water flow. Compare and contrast the effects of these actions in these two very different ecosystems.
5. How does desertification of the Sahara support the hypothesis that species invasion favors organisms that reproduce rapidly?
6. Select an ecosystem or biome from chapters 43 or 44, and list three ways that an earlier spring and later fall resulting from global climate change might affect its spectrum of species.

References and Resources

Alewell, C., et al. October 19, 2000. Is acidification still an ecological threat? *Nature,* vol. 408, p. 856. Some ecosystems are recovering from acid deposition.

Covington, William Wallace. November 9, 2000. Helping western forests heal. *Nature,* vol. 408, p. 135. U.S. forest ecosystems are in trouble.

Daily, Gretchen C., and Brian H. Walker. January 20, 2000. Seeking the great transition. *Nature,* vol. 403, p. 243. Scientists and business leaders must cooperate to achieve sustainable economies.

Hardin, Garrett. 1968. The tragedy of the commons. *Science,* vol. 162, pp. 1243–1248. A classic metaphor for destruction of freely available resources.

Kaiser, Jocelyn. August 25, 2000. Rift over biodiversity divides ecologists. *Science,* vol. 289, p. 1282. Do data on the relationship between ecosystem diversity and stability from experimental plots reflect nature?

McNeill, J. R. 2000. *Something new underneath the sun: an environmental history of the twentieth-century world.* Allen Lane/W. W. Norton, London. A look at humanity's impact on the atmosphere, waters, and lands during the past century.

Normile, Dennis. November 17, 2000. Reelf migrations, bleaching effects stir the air in Bali. *Science,* vol. 290, p. 1282. Biodiversity is plummeting among fishes in coral reef fish ecosystems, as the bleaching problem continues.

Ostrom, Elinor, et al. April 9, 1999. Revisiting the commons: Local lessons, global challenges. *Science,* vol. 284, p. 278. A reassessment of Garrett Hardin's famous essay, "The Tragedy of the Commons."

Ruiz, G. M., et al. November 2, 2000. Invasion biology: global spread of microorganisms by ships. *Nature,* vol. 408, p. 49. Ballast water has introduced many pathogens around the world.

Wilson, E. O. Spring 2000. Vanishing before our eyes. *Time,* Special Edition. Mass extinctions are ongoing.

The *LIFE* Online Learning Center provides additional resources and tools for studying this chapter.
www.mhhe.com/life

UNITS OF MEASUREMENT
METRIC/ENGLISH CONVERSIONS

APPENDIX A

Length

1 meter = 39.4 inches = 3.28 feet = 1.09 yards

1 foot = 0.305 meter = 12 inches = 0.33 yard

1 inch = 2.54 centimeters

1 centimeter = 10 millimeters = 0.394 inch

1 millimeter = 0.001 meter = 0.01 centimeter = 0.039 inch

1 fathom = 6 feet = 1.83 meters

1 rod = 16.5 feet = 5 meters

1 chain = 4 rods = 66 feet = 20 meters

1 furlong = 10 chains = 40 rods = 660 feet = 200 meters

1 kilometer = 1,000 meters = 0.621 mile = 0.54 nautical mile

1 mile = 5,280 feet = 8 furlongs = 1.61 kilometers

1 nautical mile = 1.15 miles = 1,853.25 meters

Area

1 square centimeter = 0.155 square inch

1 square foot = 144 square inches = 929 square centimeters

1 square yard = 9 square feet = 0.836 square meter

1 square meter = 10.76 square feet = 1.196 square yards = 1 million square millimeters

1 hectare = 10,000 square meters = 0.01 square kilometer = 2.47 acres

1 acre = 43,560 square feet = 0.405 hectare

1 square kilometer = 100 hectares = 1 million square meters = 0.386 square mile = 247 acres

1 square mile = 640 acres = 2.59 square kilometers

Volume

1 cubic centimeter = 1 milliliter = 0.001 liter

1 cubic meter = 1 million cubic centimeters = 1,000 liters

1 cubic meter = 35.3 cubic feet = 1.307 cubic yards = 264 U.S. gallons

1 cubic yard = 27 cubic feet = 0.765 cubic meter = 202 U.S. gallons

1 cubic kilometer = 1 million cubic meters = 0.24 cubic mile = 264 billion gallons

1 cubic mile = 4.166 cubic kilometers

1 liter = 1,000 milliliters = 1.06 quarts = 0.265 U.S. gallon = 0.035 cubic foot

1 U.S. gallon = 4 quarts = 3.79 liters = 231 cubic inches = 0.83 imperial (British) gallon

1 quart = 2 pints = 4 cups = 0.94 liter

1 acre foot = 325,851 U.S. gallons = 1,234,975 liters = 1,234 cubic meters

1 barrel (of oil) = 42 U.S. gallons = 159 liters

Mass

1 microgram = 0.001 milligram = 0.000001 gram

1 gram = 1,000 milligrams = 0.035 ounce

1 kilogram = 1,000 grams = 2.205 pounds

1 pound = 16 ounces = 454 grams

1 short ton = 2,000 pounds = 909 kilograms

1 metric ton = 1,000 kilograms = 2,200 pounds

Temperature

Celsius to Fahrenheit
$°F = (°C \times 1.8) + 32$

Fahrenheit to Celsius
$°C = (°F - 32) \div 1.8$

Energy and Power

1 erg = 1 dyne per square centimeter

1 joule = 10 million ergs

1 calorie = 4.184 joules

1 kilojoule = 1,000 joules = 0.949 British Thermal Units (BTU)

1 kilocalorie = 1,000 calories = 3.97 BTU = 0.00116 kilowatt-hour

1 BTU = 0.293 watt-hour

1 kilowatt-hour = 1,000 watt-hour = 860 kilocalories = 3,400 BTU

1 horsepower = 640 kilocalories

1 quad = 1 quadrillion kilojoules = 2.93 trillion kilowatt-hours

METRIC CONVERSION

APPENDIX B

	METRIC QUANTITIES	METRIC TO ENGLISH CONVERSION	ENGLISH TO METRIC CONVERSION
LENGTH	1 kilometer (km) = 1,000 (10^3) meters 1 meter (m) = 100 centimeters 1 centimeter (cm) = 0.01 (10^{-2}) meter 1 millimeter (mm) = 0.001 (10^{-3}) meter 1 micrometer* (μm) = 0.000001 (10^{-6}) meter 1 nanometer (nm) = 0.000000001 (10^{-9}) meter *formerly called micron	1 km = 0.62 mile 1 m = 1.09 yards = 39.7 inches 1 cm = 0.394 inch 1 mm = 0.039 inch	1 mile = 1.609 km 1 yard = 0.914 m 1 foot = 0.305 m = 30.5 cm 1 inch = 2.54 cm
AREA	1 square kilometer (km^2) = 100 hectares 1 hectare (ha) = 10,000 square meters 1 square meter (m^2) = 10,000 square centimeters 1 square centimeter (cm^2) = 100 square millimeters	1 km^2 = 0.3861 square mile 1 ha = 2.471 acres 1 m^2 = 1.1960 square yards = 10.764 square feet 1 cm^2 = 0.155 square inch	1 square mile = 2.590 km^2 1 acre = 0.4047 ha 1 square yard = 0.8361 m^2 1 square foot = 0.0929 m^2 1 square inch = 6.4516 cm^2
MASS	1 metric ton (t) = 1,000 kilograms 1 metric ton (t) = 1,000,000 grams 1 kilogram (kg) = 1,000 grams 1 gram (g) = 1,000 milligrams 1 milligram (mg) = 0.001 gram 1 microgram (μg) = 0.000001 gram	1 t = 1.1025 ton (U.S.) 1 kg = 2.205 pounds 1 g = 0.0353 ounce	1 ton (U.S.) = 0.907 t 1 pound = 0.4536 kg 1 ounce = 28.35 g
VOLUME (SOLIDS)	1 cubic meter (m^3) = 1,000,000 cubic centimeters 1 cubic centimeter (cm^3) = 1,000 cubic millimeters	1 m^3 = 1.3080 cubic yards = 35.315 cubic feet 1 cm^3 = 0.0610 cubic inch	1 cubic yard = 0.7646 m^3 1 cubic foot = 0.0283 m^3 1 cubic inch = 16.387 cm^3
VOLUME (LIQUIDS)	1 liter (l) = 1,000 milliliters 1 milliliter (ml) = 0.001 liter 1 microliter (μl) = 0.000001 liter	1 l = 1.06 quarts (U.S.) 1 ml = 0.034 fluid ounce	1 quart (U.S.) = 0.94 l 1 pint (U.S.) = 0.47 l 1 fluid ounce = 29.57 ml
TIME	1 second (sec) = 1,000 milliseconds 1 millisecond (msec) = 0.001 second 1 microsecond (μsec) = 0.000001 second		

PERIODIC TABLE OF ELEMENTS

APPENDIX C

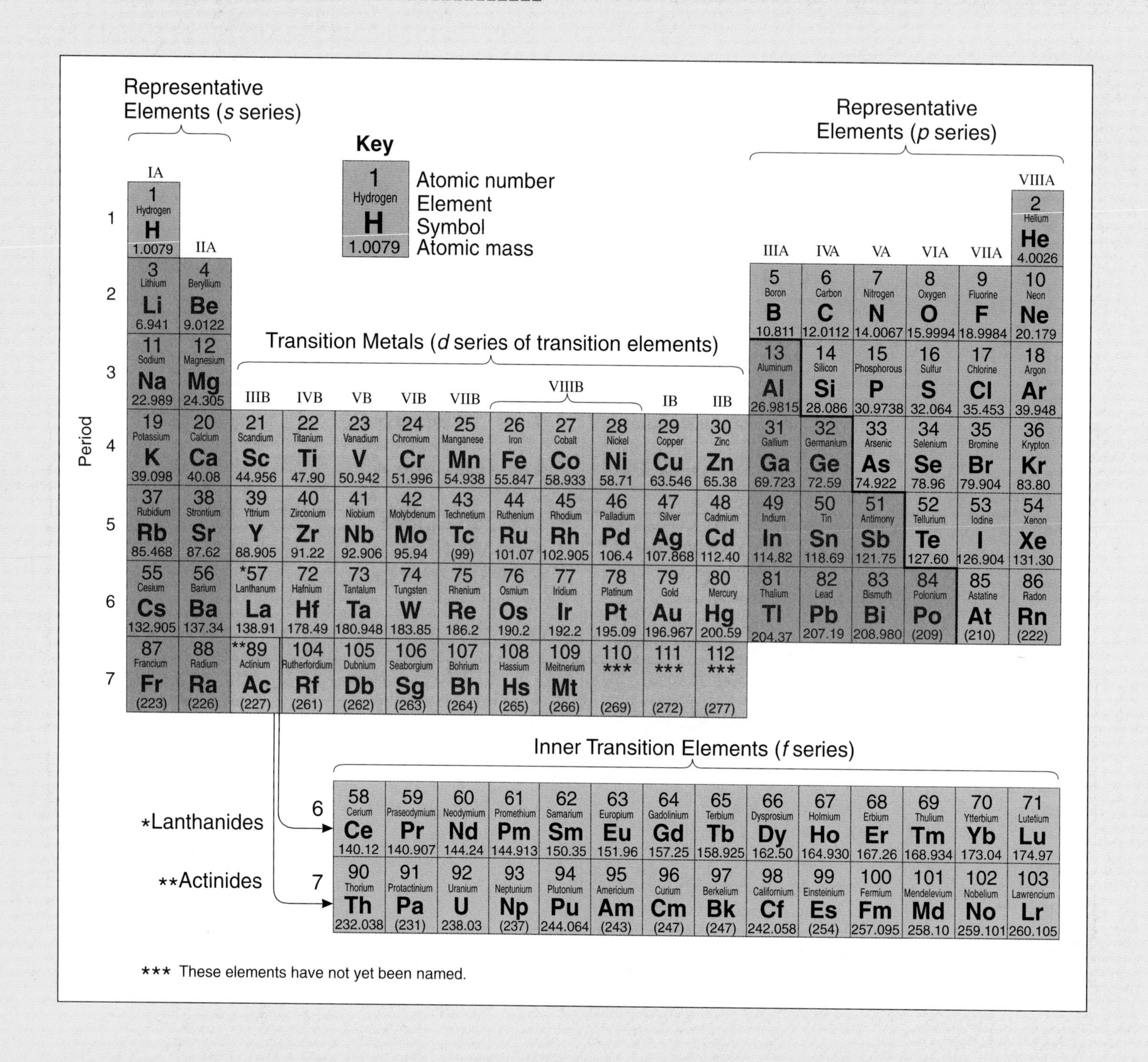

TREE OF LIFE

APPENDIX D

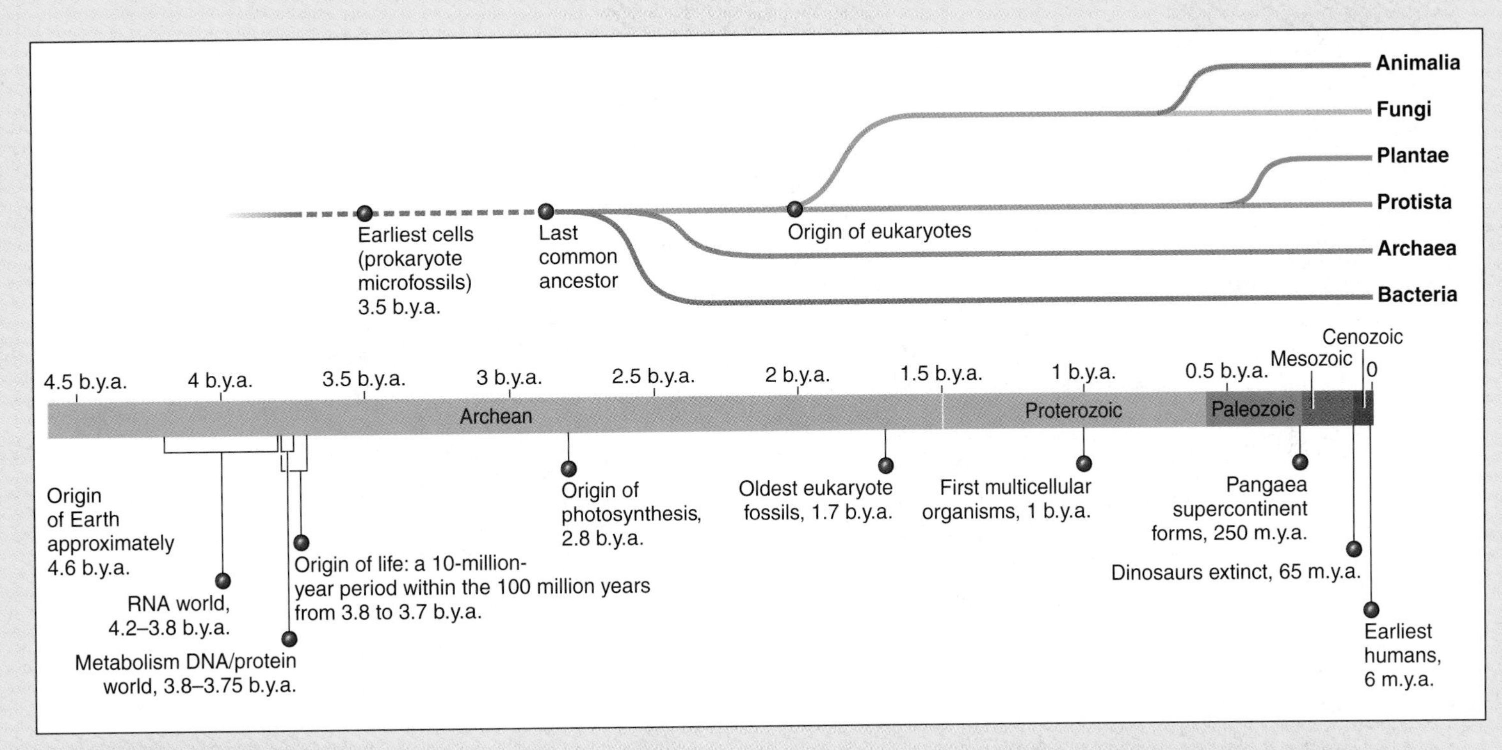

A tree of life showing possible evolutionary relationships among the major groups of organisms. Life scientists recognize three domains (Bacteria, Archaea, and Eukarya) and either five or six, (or sometimes more) kingdoms. The six-kingdom scheme counts Archaea as a kingdom. The timescale is based on fossil evidence, and on evidence from protein amino acid and DNA sequence differences among modern species. The precise evolutionary relationships among Bacteria, Archaea, and Eukarya are still unclear. Based on molecular evidence, the Archaea and Eukarya are more closely related to each other than either is to the Bacteria.

TAXONOMY

APPENDIX E

Kingdom Animalia

Kingdom Fungi

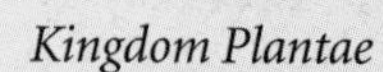
Kingdom Plantae

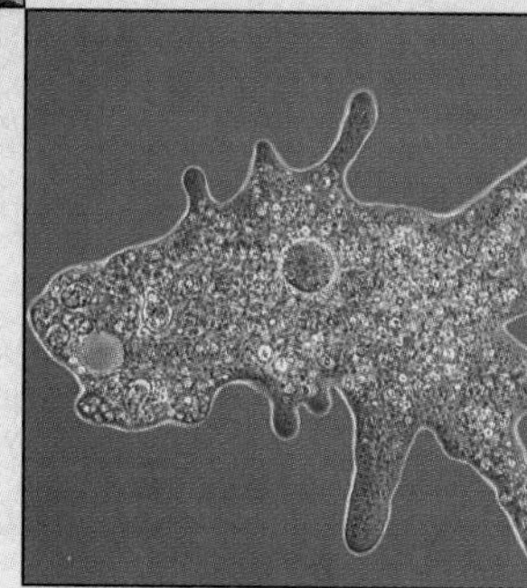
Kingdom Protista

We can group the millions of living and extinct species that have dwelled on Earth according to many schemes. Taxonomists traditionally grouped organisms to reflect morphological and anatomical similarities. More recently, the ongoing sequencing of gene sequences (and entire genomes) has simplified the task of grouping organisms by descent from a common ancestor. This new information occasionally conflicts with traditional groupings. As a result, the classification of organisms remains in flux. Two-, three-, four-, five-, and six-kingdom classifications have been proposed, as well as a system of three domains that supercede kingdoms.

The three-domain system outlined here contains many (but by no means all) of the known phyla and divisions of life; those in bold type are discussed in Unit 4. In recognition of the technological advances that enable constant revisions of these groups, we urge the reader to visit other web sites specializing in the classification of life.

Tree of Life: www.phylogeny.arizona.edu/tree/phylogeny.html
University of California, Berkeley, Museum of Paleontology Phylogeny Exhibit: www.ucmp.berkeley.edu/exhibit/phylogeny.html
National Center for Biotechnology Information (NCBI) GenBank: www.ncbi.nlm.nih.gov/

Domain Bacteria

Domain Bacteria is one of two domains containing prokaryotic organisms, and includes many organisms that cause disease in humans. The importance of bacteria extends far beyond human pathogens, however, for they play an essential role in global nutrient cycling. Bacteria exhibit a diversity of metabolic lifestyles—they may be heterotrophs or autotrophs, aerobes or anaerobes, photosynthesizers or chemosynthesizers. One scheme for grouping bacteria, adapted from *Bergey's Manual of Systematic Bacteriology*, 1984, appears below. Note that this scheme is based on observable characteristics of cells, and does not necessarily reflect shared ancestry among the bacteria.

Spirochetes
Motile, helical or vibrioid gram-negative bacteria
Nonmotile, gram-negative curved bacteria
Gram-negative aerobic rods and cocci
Facultatively anaerobic gram-negative rods
Anaerobic gram-negative rods

Sulfate- or sulfur-reducing bacteria
Anaerobic gram-negative cocci
Rickettsias and chlamydias
Mycoplasmas
Endosymbionts of protista, fungi, arthropods, and other invertebrates
Gram-positive cocci
Endospore-forming gram-positive rods and cocci
Nonsporing gram-positive rods
Mycobacteria
Nocardiforms
Gliding, non-fruiting bacteria
Anoxygenic phototrophic bacteria
Budding and/or appendaged bacteria
Sheathed bacteria
Gliding, fruiting bacteria
Chemolithotrophic bacteria
Cyanobacteria
Actinomycetes
Streptomycetes

Domain Archaea

Like domain Bacteria, domain Archaea contains organisms whose cells lack nuclei and membrane-bounded organelles. They are, however, distinct from bacteria in important aspects of cell structure and metabolism. In fact, archaea are more similar to eukaryotes than they are to bacteria in many ways. Researchers first discovered archaea in habitats with extreme salt, acid, or heat, but have since also found them in more moderate conditions. Like bacteria, archaea may be aerobic or anaerobic, autotrophs or heterotrophs.

Crenarchaeota (many are extreme acidophiles and thermophiles)
Euarchaeota (many are extreme halophiles or generate methane gas)
Korarchaeota (uncultured organisms from terrestrial hot springs)

Domain Eukarya

Domain Eukarya contains all organisms with eukaryotic cells (those with membrane-bounded organelles such as nuclei and mitochondria). The classification of the Eukarya is in flux; here we use a traditional four-kingdom system.

Kingdom Protista This kingdom contains an extremely diverse assemblage of organisms that do not fit neatly into the other three eukaryotic kingdoms. They can be unicellular or multicellular, autotrophs or heterotrophs, free-living or obligate parasites of other organisms. The list below is by no means complete, as there are dozens of separate groups of protista whose relationships to each other remain unknown. Some scientists classify the green algae in kingdom Plantae, but we have used the traditional classification of the green algae with the protista. Also, some scientists classify the stramenopiles as a separate kingdom (Chromista) within domain Eukarya.

Protista lacking mitochondria
- **Diplomonadida**
- **Parabasalia**
- **Entamoebae**

Mitochondrial Protista
- **Jakobids**
- **Euglenida and Kinetoplastida**
- **Acrasiomycota (cellular slime molds)**
- **Myxomycota (plasmodial, or acellular, slime molds)**
- Crown eukaryotes
  - **Choanoflagellates**
  - **Alveolates (dinoflagellates, apicomplexa, ciliates)**
  - **Stramenopiles (water molds, diatoms, brown and golden algae)**
  - **Rhodophyta (red algae)**
  - **Chlorophyta (green algae)**

Kingdom Plantae The plants are multicellular, autotrophic (by photosynthesis), and their life cycles include an alternation of multicellular haploid and diploid generations. Their cells have cellulose cell walls. The major groups of plants are distinguished by the presence or absence of vascular tissue, seeds, and flowers and fruits.

Nonvascular plants
- **Division Anthocerotophyta (hornworts)**
- **Division Hepaticophyta (liverworts)**
- **Division Bryophyta (mosses)**

Vascular plants without seeds
- **Division Pterophyta (ferns)**
- **Division Psilophyta (whisk ferns)**
- **Division Lycophyta (club mosses)**
- **Division Sphenophyta (horsetails)**

Nonflowering vascular plants with seeds
- **Division Cycadophyta (cycads)**
- **Division Ginkgophyta (ginkgos)**
- **Division Coniferophyta (conifers)**
- **Division Gnetophyta (gnetophytes)**

Division Anthophyta, or Magnoliophyta (flowering vascular plants)

Kingdom Fungi This kingdom includes mostly multicellular eukaryotic organisms that are heterotrophic, consuming organic matter by secreting digestive enzymes and absorbing the remaining nutrients. Their cells have cell walls of chitin. The major groups are distinguished by the presence or absence of flagella and the type of sexual spores produced.

Division Chytridiomycota (flagellated fungi; chytrids)
Division Zygomycota (bread molds)
Division Ascomycota (yeasts, morels, truffles, many molds)
Division Basidiomycota (mushrooms, puffballs, rusts, smuts)

Kingdom Animalia Animals include multicellular, eukaryotic organisms that are heterotrophic, consuming organic matter by ingestion. Their cells lack cell walls. Animals have specialized tissues and organs, and most have nervous and locomotive systems. Phyla are distinguished largely on the basis of body form and symmetry, characteristics generally established in the early embryo.

Parazoa
- **Phylum Porifera (sponges)**

Eumetazoa
- Radiata
  - **Phylum Cnidaria (jellyfishes, anemones, coral)**
  - Phylum Ctenophora (comb jellies)
- Bilateria
  - Acoelomates
    - **Phylum Platyhelminthes (flatworms)**
    - Phylum Nemertea (ribbon worms)
  - Pseudocoelomates
    - Phylum Rotifera (rotifers)
    - Phylum Gastrotricha (gastrotrichs)
    - Phylum Kinorhyncha (kinorhynchs)
    - **Phylum Nematoda (roundworms)**

- Phylum Nematomorpha (horse-hair worms)
- Phylum Acanthocephala (spiny-headed parasitic worms)
- Phylum Entoprocta (entoprocts)
- Phylum Priapula (priapulid worms)
- Phylum Loricifera (loriciferans)

Eucoelomates
- Protostomes
  - **Phylum Mollusca (snails, clams, squids)**
  - **Phylum Annelida (segmented worms)**
  - **Phylum Arthropoda (insects, spiders, crabs)**
  - Phylum Echiuria (marine worms)
  - Phylum Sipunculida (bottom-dwelling marine worms)
  - Phylum Tardigrada (water bears)
  - Phylum Pentastomida (tongue worms)
  - Phylum Onychophora (velvet worms)
  - Phylum Pogonophora (beard worms)
- Deuterostomes
  - Lophophorates
  - **Phylum Echinodermata (starfishes, urchins, sea cucumbers)**
  - **Phylum Chaetognatha (arrow worms)**
  - **Phylum Hemichordata (acorn worms)**
  - **Phylum Chordata (vertebrates and relatives)**

GLOSSARY

A

aboral end *AB-or-al end* The end of an animal's body opposite its mouth. 453

abscisic acid *ab-SIS-ik AS-id* One of five classes of plant hormone. 561

absolute dating Determining the age of a fossil by direct measurement, usually involving radioisotope decay. 324

absorption The process of taking in and incorporating nutrients or energy. 729

accessory pigment Photosynthetic pigment other than chlorophyll that extends the range of light wavelengths useful in photosynthesis. 102

accommodation Moving or changing the shape of the lens to bring an object into focus. 630

acellular slime mold A protist that can form a multinucleated mass called a plasmodium. 398

acetyl CoA formation *AS-eh-til Co-A FOR-MAY-shun* The first step in aerobic respiration. In the mitochondrion, pyruvic acid loses a carbon dioxide and bonds to coenzyme A to form acetyl CoA. 123

acetyl coenzyme A The molecule that enters the Krebs cycle. 123

acid A molecule that releases hydrogen ions into water. 27

acid deposition Low pH precipitation or particles that form when air pollutants react with water in the upper atmosphere, forming nitric and sulfuric acids. 889

acidophile Organism that lives in extremely acidic environments. 385

acoelomate An animal that lacks a coelom (body cavity). 454

acquired immunity Slower-acting defense system in vertebrates that requires a previous encounter with a foreign antigen. 757

acrocentric *ak-ro-SEN-trik* A chromosome whose centromere divides the chromosome into a long arm and a short arm. 202

acrosome *AK-ro-som* A protrusion on the anterior end of a sperm cell containing digestive enzymes that enable the sperm to penetrate layers around the oocyte. 168

actin A type of protein in the thin myofilaments of skeletal muscle cells. Also part of cytoskeletons. 581

action potential An all-or-none electrochemical change across the cell membrane of a neuron; the basis of a nerve impulse. 598

active immunity Immunity generated by an organism's production of antibodies and cytokines. 767

active site The part of an enzyme that provides catalysis. 36

active transport Movement of a molecule through a membrane against its concentration gradient, using a carrier protein and energy from ATP. 66

adaptation An inherited trait that enables an organism to survive a particular environmental challenge. 7

adaptive Referring to traits that confer some advantage in reproduction or survival. 307

adaptive radiation The divergence of several new types of organisms from a single ancestral type. 307

adenine One of two purine nitrogenous bases in DNA and RNA. 37

adenosine triphosphate *ah-DEN-o-seen tri-FOS-fate* (ATP) A molecule whose three high-energy phosphate bonds power many biological processes. 90

adhesion The tendency of water to hydrogen bond to other compounds. 26

adipocyte *a-DIP-o-site* Fat cell. 579

adipose tissue A type of connective tissue consisting of cells laden with lipid. 579

adrenal cortex The outer portion of an adrenal gland that secretes mineralocorticoids, glucocorticoids, and sex hormones. 652

adrenal glands Paired glands atop the kidneys that produce catecholamines, mineralocorticoids, glucocorticoids, and sex hormones. 651

adrenal medulla The inner portion of an adrenal gland that secretes epinephrine and norepinephrine. 652

adrenocorticotropic hormone (ACTH) A hormone of the anterior pituitary that stimulates the adrenal cortex to secrete hormones. 651

adventitious root *ad-ven-TISH-shus ROOT* Root that forms on stems or leaves. 517

aerial root *AIR-ee-al ROOT* Adventitious root that grows in the air. 519

aerobe Organism that requires oxygen. 119

aerobic respiration Extraction of energy from glucose in the presence of oxygen. 119

afferent arteriole Branch of the renal artery that delivers blood to glomerular capillaries. 747

age structure The distribution of ages of individuals of a population. 832

alcoholic fermentation *AL-ko-HALL-ik FER-men-TAY-shun* An anaerobic step that yeast use after glycolysis that breaks down pyruvic acid to ethanol and carbon dioxide. 129

aldosterone Mineralocorticoid that increases sodium ion reabsorption in the kidneys. 750

alga (pl. algae) *AL-ga* Unicellular or multicellular photosynthetic eukaryote that generally lacks roots, stems, leaves, vascular tisssue, and complex sex organs. 396

allantois An extraembryonic membrane that forms as an outpouching of the yolk sac. 793

allele *ah-LEEL* An alternate form of a gene. 163

allergen Harmless substance that triggers an immune attack. 771

allergy mediator Biochemical that mast cells release when contacting an allergen, causing allergy symptoms. 771

allopatric speciation The formation of new species due to the physical separation of two groups. 300

allopolyploid *AL-lo-POL-ee-ploid* An organism with multiple chromosome sets resulting from mating of individuals of different species. 304

alternation of generations The life cycle of plants and many green algae, which alternates between a diploid sporophyte stage and a haploid gametophyte stage. 159, 409, 414

altricial Describes young that hatch helpless and totally dependent on their parents. 492

alveolar duct *al-vee-O-ler DUCT* In lungs, the narrowed ending of bronchioles that opens into a cluster of alveoli. 710

alveolate Group of protista whose cells have flattened sacs called alveoli beneath the cell membrane. 401

alveolus (pl. alveoli) Microscopic air sac in the lungs where gas exchange occurs between the air and the blood. In some protista, a flattened sac beneath the cell membrane. 401, 710

amino acid *a-MEEN-o AS-id* An organic molecule consisting of a central carbon atom bonded to a hydrogen atom, an amino group, a carboxyl group, and an R group. 33

amino group *a-MEEN-o GROOP* A nitrogen atom single-bonded to two hydrogen atoms. 34

ammonia A nitrogenous waste generated by deamination of amino acids. 745

amnion An extraembryonic membrane in which the amniotic cavity forms. 790

amniote A vertebrate that has extraembryonic membranes (amnion, chorion, and allantois); includes reptiles, birds, and mammals. 482

amniote egg An egg containing an internal source of fluid and nutrients surrounded by several membranes. 352, 491

amoeba Protist that moves when its cytoplasm flows into projections called pseudopods. 398

amoebocyte *a-MEE-bo-site* Mobile cell in a sponge. 456

amphibian A tetrapod vertebrate that can live on land, but requires water to reproduce. 488

ampulla (pl. ampullae) *am-PULL-ee* The enlarged base of a semicircular canal in the inner ear, lined with hair cells that detect fluid movement and transduce it into action potentials. 634

anabolism *an-AB-o-liz-um* Metabolic reactions that use energy to synthesize compounds. 87

anaerobe *AN-air-robe* Organism that can live in an environment lacking oxygen. 119

anaerobic respiration *an-air-RO-bic res-per-A-shun* Cellular respiration in the absence of oxygen. 128

analogous structure *ah-NAL-eh-ges STRUC-cher* Body part in different species that is similar in function but not in structure that evolved in response to a similar environmental challenge. 328

anamniote A vertebrate that lacks extraembryonic membranes (amnion, chorion, allantois); includes the fishes and the amphibians. 482

anaphase *AN-ah-faze* The stage of mitosis when centromeres split and two sets of chromosomes part. 142

anaphase I *AN-ah-faze I* Anaphase of meiosis I; homologs separate. 162

anaphase II *AN-ah-faze II* Anaphase of meiosis II, when chromatids separate. 163

anapsid Reptile lacking holes on the side of its skull. 491

androecium *an-dro-EE-shee-um* The innermost whorl of a flower's corolla, consisting of male reproductive structures. 543

anemia A reduction in the number of red blood cells or the amount of hemoglobin. 693

aneuploid *AN-you-ploid* A cell with one or more extra or missing chromosomes. 211

angiosperm *AN-gee-o-sperm* A group of plants that produce flowers and whose seeds are borne within a fruit. 415

Animalia The kingdom that includes the animals. 9

annelid A segmented worm. 466

annulus (pl. annuli) *AN-yew-li* Demarcation between the segments of leeches. 467

antagonistic muscles *an-tag-o-NIS-tik MUS-uls* Two muscles or muscle groups that flank a bone and move it in opposite directions. 675

antenna complex An interconnected series of membrane proteins in the thylakoid membranes that harvests light energy. 104

anterior pituitary The front part of the pituitary gland; secretes FSH, LH, ACTH, GH, TSH, and prolactin. 648

anther *AN-ther* Pollen-producing body at the tip of a stamen. 543

antheridium *an-ther-ID-ee-um* A sperm-producing structure in algae, bryophytes, and seedless vascular plants. 417

anthocyanin *an-tho-SI-ah-nin* One of several types of photosynthetic pigment. 102

antibody Protein that B cells secrete that recognizes and binds to foreign antigens, disabling them or signaling other cells to do so. 763

anticodon *AN-ti-ko-don* A three-base sequence on one loop of a transfer RNA molecule that is complementary to an mRNA codon and joins an amino acid and its mRNA instructions. 245

antidiuretic hormone (ADH) A hypothalamic hormone released from the posterior pituitary that acts on kidneys and smooth muscle cells of blood vessels to maintain the composition of body fluids. 651, 750

antigen The specific part of a molecule that elicits an immune response. 756

antigen binding sites Specialized ends of antibodies that bind specific antigens. 764

antiparallelism *AN-ti-PAR-a-lel-izm* The head-to-tail relationship of the two rails of the DNA double helix. 225

anus Exit of digestive tract, where feces is released. 723, 732

aorta The largest artery; it leaves the heart. 686

apical complex Microtubular structure that helps apicomplexa attach to or penetrate host cells. 402

apical dominance In plants, the suppression of growth of lateral buds by an intact terminal bud. 559

apical meristem The meristem (dividing cells) at the tip of a root or shoot. 505

apicomplexa A group of protista whose cells have apical complexes of specialized organelles. 402

apolipoprotein *APE-o-LIP-o-PRO-teen* The protein parts of cholesterol-carrying molecules. 34

apoptosis *ape-o-TOE-sis* Programmed cell death. 138

appendicular skeleton *AP-en-DEK-u-lar SKEL-eh-ten* In vertebrates, the limb bones and the bones that support them. 663

appendix A thin sac extending from the cecum in the human digestive system. 732

aqueous humor A fluid between the cornea and the lens that helps focus incoming light rays and maintains the shape of the eyeball. 630

aqueous solution *AWK-kwee-us so-LEW-shun* A solution in which water is the solvent. 27

arachnid A terrestrial, eight-legged, chelicerate arthropod, such as a spider. 469

Archaea *ar-KEE-a* One of the three domains of life. 380

archegonium *arch-eh-GO-nee-um* In bryophytes and some vascular plants, a multicellular egg-producing organ. 417

arteriole Small artery in which the smooth muscle layer predominates; important in regulation of blood pressure. 682

artery Vessel that carries blood away from the heart. 682

arthropod Segmented protostome with a jointed exoskeleton. 468

artificial selection Selective breeding. 269

ascending limb *ay-SEN-ding LIM* The distal portion of the loop of the nephron, which ascends from the kidney's medulla. 749

ascomycete *ass-ko-mi-SEET* Fungus that produces spores in sacs called asci. 434

ascospore A spore produced after meiosis in an ascus. 421

ascus (pl. asci) Saclike structure in which ascospores are produced. 441

asexual reproduction Any form of reproduction which does not require the fusion of gametes. 6, 540

aster Collection of microtubules that anchor the centriole and spindle apparatus to the cell membrane. 140

atom *AT-um* A chemical unit, composed of protons, neutrons, and electrons, that cannot be further broken down by chemical means. 21

atomic number *a-TOM-ic NUM-ber* The number in an element's box in the periodic table that indicates number of protons. 20

atomic weight Also atomic mass, the mass of an atom. 20

ATP synthase *ATP SIN-thaze* An enzyme complex that allows protons to move through the mitochondrial membrane and trigger phosphorylation of ADP to ATP. 106

atriopore *AYT-ree-o-pore* The hole in a lancelet through which water exits the body. 481

atrioventricular (AV) node Specialized autorhythmic cells involved in the control of heartbeat. 687

atrioventricular valve Flap of heart tissue that prevents the backflow of blood from a ventricle to an atrium. 686

atrium An upper heart chamber that receives blood from veins. 683

atrophy *AT-tre-fee* Muscle degeneration resulting from lack of use or immobilization. 677

auditory canal The ear canal; funnels sounds from the pinna to the tympanic membrane. 633

auditory nerve Nerve fibers from the cochlea in the inner ear to the cerebral cortex. 634

autoantibody Antibody that attacks the body's tissues, causing autoimmune disease. 769

autoimmunity An organism's immune system attacking its own body. 769

autonomic nervous system Motor pathways that lead to smooth muscle, cardiac muscle, and glands. 607

autopolyploid *aw-toe-POL-ee-ploid* An organism with multiple chromosome sets. 302

autosome *AW-toe-soam* A nonsex chromosome. 161

autotroph Also known as a producer, an organism that produces its own nutrients by acquiring carbon from inorganic sources. 84, 384

auxin *AWK-zin* One of five classes of plant hormone. 558

axial skeleton In a vertebrate skeleton, the skull, vertebral column, ribs, and sternum. 663

axil *AX-el* The region between a leaf stalk and stem. 510

axon An extension of a neuron that transmits messages away from the cell body and toward another cell. 581, 597

B

bacillus (pl. bacilli) *ba-SILL-us* Rod-shaped bacterium. 384

background extinction rate The steady, gradual loss of a small percentage of existing species through natural competition or loss of diversity. 308

Bacteria *bac-TEAR-e-a* One of the three domains of life. 380

bacteriophage A virus that infects bacteria. 366

balanced polymorphism *BAL-anced POL-ee-MORF-iz-um* Stabilizing selection that maintains a genetic disease in a population because heterozygotes resist an infectious disease. 286

bark In woody plants, all tissues outside the vascular cambium. 520

Barr body *BAR BOD-ee* The dark-staining body in the nucleus of a female mammal's cell, corresponding to the inactivated X chromosome. 211

basal eukaryote Diverse protista that evolved earlier than crown eukaryotes. 396

base A molecule that releases hydroxide ions into water. 27

basement membrane A thin, noncellular layer that anchors epithelial tissues to other tissues. 578

basidiomycete *bass-ID-eo-mi-SEET* Fungus that has basidia as reproductive structures. 434

basidiospore A spore produced, after meiosis, on a basidium.442

basidium *bass-ID-ee-um* Club-shaped structure on which basidiospores are produced externally. 441

basilar membrane *BA-sill-ar MEM-brane* The membrane beneath hair cells in the cochlea of the inner ear that vibrates in response to sound. 634

B cell Lymphocyte that produces antibodies. 762

behavioral isolation A form of reproductive isolation based upon differences in behavior, such as mating rituals. 299

benign tumor *bee-NINE TOO-mer* A noncancerous tumor. 136

benthic zone *BEN-thick ZONE* The bottom of an ocean or lake. 878

bilateral symmetry A body form in which only one plane divides the animal into mirror image halves. 453

bile A digestive biochemical that emulsifies fats. 730

binary fission *BI-nair-ee FISH-en* A type of asexual reproduction in which a prokaryotic cell divides into two identical cells. 158, 387

bioaccumulation Higher concentration of a substance in cells compared to the surrounding environment. 858

biochemical Molecule that is important in biological systems. 29

biodiversity *bi-o-di-VER-city* The spectrum of different life-forms. 8

bioenergetics *bi-o-en-er-JET-ix* The study of energy in life. 82

biogeochemical cycle *bi-o-gee-o-KEM-i-kal SI-kull* Geological and biological processes that recycle chemicals vital to life. 855

biogeography *bi-o-gee-OG-grah-fee* The physical distribution of organisms. 268

biological evolution The process by which the genetic structure of populations changes over time. 297

bioluminescence *bi-o-loom-in-ES-ents* A chemical reaction that causes an organism to glow. 378

biomagnification *bi-o-mag-nif-i-KAY-shun* Increasing concentrations of a chemical with higher trophic levels. 859

biomass Total dry weight of organisms in a given area at a given time. 854

biome *BI-ohm* One of several major types of terrestrial ecosystems. 850

bioremediation *bi-o-ree-meed-e-AY-shun* Use of organisms that metabolize toxins to clean the environment. 392

biosphere *BI-o-sfere* The ecosystem of the entire planet. 6

biotic community The interacting populations in a given area. 844

biotic potential The potential growth of a population under ideal conditions and with unlimited resources.833

bipedalism *by-PEED-a-liz-m* The ability to move on two limbs. 354

biramous *bi-RAYM-us* A double-branched appendage in arthropods. 468

bird A tetrapod vertebrate that has feathers, wings, lungs, a four-chambered heart, and endothermy. 492

bivalve A mollusk that has a two-part shell. 465

blade The flattened region of a leaf. 511

blastocyst Stage of human prenatal development in which the morula hollows out and fills will fluid. 789

blastomere A cell in a preembryonic animal resulting from cleavage divisions. 789

blastopore An indentation in an animal embryo that develops into the mouth in protostomes and the anus in deuterostomes. 455

blastula The stage of early animal embryonic development that consists of a hollow ball of cells. 453

blind spot Point where the optic nerve exits the retina; it is devoid of photoreceptors. 630

blood A complex mixture of cells suspended in a liquid matrix that delivers nutrients to cells and removes wastes. 579, 692

blood-brain barrier Close-knit cells that form capillaries in the brain, limiting the substances that can enter. 617

blood pressure The force that blood exerts against blood vessel walls. 690

Bohr model A classic representation of an atom by Neils Bohr wherein electrons occupy circular orbits around a nucleus. 22

bolus *BO-lus* Food rolled into a lump by the tongue. 725

bond energy The energy required to form a particular chemical bond. 88

bone A connective tissue consisting of osteoblasts, osteocytes, and osteoclasts, embedded in a mineralized matrix. 579

bony fish A fish with a skeleton reinforced with mineral deposits to form bone. 485

boom and bust cycle The repeated and regular increases and decreases in size of some populations. 836

braced framework A skeleton built of solid structural components strong enough to resist collapsing. 661

bract *BRAK* Floral leaf that protects a developing flower. 514

brain A distinct concentration of nervous tissue at the anterior end of an animal. 607

brain stem Part of the vertebrate brain closest to the spinal cord; controls vital functions. 610

bronchiole Microscopic branch of the bronchi within the lungs. 708

bronchus (pl. bronchi) *BRON-kus* One of a pair of tubules that branch from the trachea as it reaches the lungs. 708

brown alga Multicellular photosynthetic protista (stramenopiles) with unique pigments. 407

bryophyte *BRI-o-fite* Collective term for plants that lack vascular tissue, including liverworts, hornworts, and mosses. 415

budding Formation of a small daughter cell from a yeast cell. 436

buffer system Pairs of weak acids and bases that maintain body fluid pH. 28

bulbourethral gland *BUL-bo-u-REE-thral GLAN* Small gland near the male urethra that secretes mucus. 785
bulk element An element that an animal requires in large amounts. 20
bundle-sheath cell *BUN-dull SHEETH SEL* Thick-walled plant cell surrounding veins that functions in C_4 photosynthesis. 111
bursa (pl. bursae) *BURR-sa* Small packet in joint that secretes lubricating fluid. 675

C

C_3 plant Plant that uses only the Calvin cycle to fix carbon dioxide. 108
C_4 photosynthesis In plants, a biochemical pathway that helps prevent photorespiration. 110
caecilian A type of limbless amphibian. 488
calcitonin A thyroid hormone that decreases blood calcium levels. 652
calorie The energy required to raise the temperature of 1 gram of water 1°C under standard conditions. 82
camouflage Adaptation that helps an organism blend in with its surroundings. 846
CAM photosynthesis Reactions that reduce the effect of photorespiration by storing carbon during nighttime for use during the day in hot, arid climates. 112
canaliculus (pl. canaliculi) *can-al-IK-u-lus* Passageway in bone that connect lacunae. 579
capacitation Activation of sperm cells in the female reproductive tract. 788
capillary Tiny vessel that connects an arteriole with a venule. 682
capillary bed Network of capillaries. 688
capsid The protective protein container of the genetic material of a virus. 363
capsomer The individual proteins that comprise a viral capsid. 363
capsule Firm, sticky layer surrounding some prokaryotic cells. 384
carbohydrase *KAR-bo-HI-drase* Enzyme that breaks down certain disaccharides into monosaccharides. 730
carbohydrate *KAR-bo-HI-drate* Compound containing carbon, hydrogen, and oxygen, with twice as many hydrogens as oxygens; sugar or starch. 29
carbon reactions Also known as the Calvin-Benson cycle, the reactions of photosynthesis that synthesize glucose from carbon dioxide. 104
carboxyl group *kar-BOX-ill GROOP* A carbon atom double-bonded to an oxygen and single-bonded to a hydroxyl group (OH). 34
cardiac cycle The sequence of contraction and relaxation that makes up the heartbeat. 687
cardiac muscle Type of muscle composed of striated, involuntary, single-nucleated contractile cells. 581, 670
cardiac muscle cell Striated, involuntary, single-nucleated contractile cell in the mammalian heart. 670
cardiac output The volume of blood the heart ejects in 1 minute. 694
cardiovascular system The system of vessels and muscular pump that transports blood throughout the body. 584, 684
carnivore An animal that feeds on animals. 720
carotenoid (carotenoids) *kare-OT-in-oyd* Yellow or orange plant pigment. 102
carotid rete *care-OT-id REE-tee* A configuration of blood vessels that cools the brain. 742
carpel *KAR-pel* Structure in a flower that encloses one or more ovules. 544
carrying capacity The theoretical maximum number of individuals an environment can support indefinitely. 834
cartilage A supportive connective tissue consisting of chondrocytes embedded in collagen and proteoglycans. 579, 675
cartilaginous fishes Group of jawed fishes that have a skeleton made of cartilage rather than bone; includes sharks, skates, and rays. 484
Casparian strip *kas-PAHR-ee-an STRIP* A waxy region of suberized cell walls in the endodermis of roots. 517, 534
caspase Enzyme that triggers apoptosis. 150
catabolism *cah-TAB-o-liz-um* Metabolic degradation reactions, which release energy. 88
catalysis *kat-AL-i-sis* Speeding a chemical reaction. 36
catastrophism A theory of geological change championed by Cuvier, that stated that new life comes to an area damaged or destroyed by a catastrophe. 266
catecholamine *kat-ah-KOLE-ah-meen* Hormone of the adrenal medulla. 652
cavitation A break in the water column in a xylem tube. 533
cecum Compartment in digestive tract where plant matter is absorbed. 725
cell The structural and functional unit of life. 42
cell body The enlarged portion of a neuron that contains most of the organelles. 581, 591
cell cycle The life of a cell, in terms of whether it is dividing or in interphase. 138
cell membrane Proteins embedded in a lipid bilayer, which forms the boundary of cells. 62
cell plate The dividing region between two daughter plant cells undergoing the last stages of mitosis. 144
cell population A group of cells with characteristic proportions in particular stages of the cell cycle. 149
cell theory The ideas that all living matter consists of cells, cells are the structural and functional units of life, and all cells come from preexisting cells. 43
cellular adhesion molecule (CAM) A protein that enables cells to interact with each other. 76
cellular immune response The actions of T and B cells in the immune system. 757
cellular respiration Biochemical reactions that extract energy in mitochondria. 82
cellular slime mold A protist that is unicellular when food is available, but forms a mobile, multinucleated mass when food is scarce. 398
cell wall A rigid boundary surrounding many prokaryotic, plant, and fungal cells. 48, 382
central nervous system (CNS) The brain and the spinal cord. 598
centriole *SEN-tre-ole* One of a pair of oblong structures consisting of microtubules in animal cells that organize the mitotic spindle. 140
centromere *SEN-tro-mere* A characteristically located constriction in a chromosome. 140
centrosome *SEN-tro-soam* A region near the cell nucleus that contains the centrioles. 140
cephalization Development of an animal body with a head end. 453
cephalopod A type of marine mollusk with a reduced or absent shell and well-developed brain and eyes; includes octopuses and squids. 465
cerebellum An area of the brain that coordinates muscular responses. 610
cerebral cortex The outer layer of the cerebrum. 613
cerebral hemisphere One of the two halves of the cerebrum. 613
cerebrospinal fluid Fluid similar to blood plasma that bathes and cushions the CNS. 617
cerebrum The region of the brain that controls intelligence, learning, perception, and emotion. 610
cervical vertebra Vertebra of the neck. 663
cervix The opening of the uterus into the vagina. 786
chaperone protein *shap-ER-one PRO-teen* A protein that stabilizes a growing amino acid chain. 251
charophyte Group of green algae thought to be most closely related to terrestrial plants. 415
checkpoints Points in the cell cycle where cell division is halted in response to control signals. 138
chemical bond Attachments atoms form by sharing or exchanging electrons. 22
chemical equilibrium *KEM-e-cal e-kwil-IB-ree-um* When a chemical reaction proceeds in both directions at the same rate. 88
chemical reaction Interactions in which atoms exchange or share electrons, forming new chemicals. 22
chemiosmosis *KEM-ee-oss-MOE-sis* Phosphorylation of ADP to ATP occurring when protons that are following a concentration gradient contact ATP synthase. 126
chemiosmotic phosphorylation Also known as oxidation phosphorylation, the reactions that produce ATP using the energy of a proton gradient. 106

chemoautotrophs Organisms that obtain carbon from inorganic sources and energy from chemical bonds. 84

chemoreception *KEEM-o-ree-SEP-shun* Smell and taste. 625

chemoreceptor Receptor cell responsive to chemicals. 624

chemotaxis Directed movement toward or away from a chemical. 382

chemotroph *KEEM-o-trofe* An organism that obtains energy by oxidizing chemicals. 385

chlorofluorocarbon (CFC) compounds Persistent chemicals that accelerate the breakdown of ozone in the stratosphere. 891

chlorophyll *a* *KLOR-eh-fill A* A green pigment plants use to harness the energy in sunlight. 100

chloroplast *KLOR-o-plast* An organelle housing the reactions of photosynthesis in eukaryotes. 55

choanocyte *cho-AN-o-site* Collar cell in sponges that produces a current that draws food and water into the central cavity. 457

chondrocyte *KON-dro-site* A cartilage cell. 579

chordate Animal that at some time during its development has a notochord, hollow nerve cord, gill slits, and postanal tail. 478

chorion A membrane that develops into the placenta. 789

chorionic villus Fingerlike projection extending from the chorion to the uterine lining. 792

choroid The middle layer of the eyeball that is rich in blood vessels. 629

chromatid *KRO-mah-tid* A continuous strand of DNA comprising an unreplicated chromosome or one-half of a replicated chromosome. 140

chromatophore *kro-MAT-o-for* Structure in an amphibian's skin that provides pigmentation. 488

chromosome *KRO-mo-soam* A dark-staining, rod-shaped structure in the nucleus of a eukaryotic cell consisting of a continuous molecule of DNA wrapped in protein. 202

chyme Semisolid food in the stomach. 728

chymotrypsin *KI-mo-TRIP-sin* A pancreatic enzyme that digests protein in the small intestine. 730

chytridiomycete *KI-trid* A microscopic fungus that produces motile zoospores. 434

ciliary body *SIL-ee-air-ee BOD-ee* A highly folded, specialized structure in the center of the choroid coat of the human eye that houses the ciliary muscle, which alters the shape of the lens. 629

ciliate Group of protista whose cells are covered with cilia. 404

cilium *SIL-ee-um* Protein projection from cells that beats, moving cells and substances. 71

circadian rhythm *sir-KA-dee-en RITH-um* Regular, daily rhythm of a biological function. 570

clade A group of similar organisms sharing a set of traits derived from a common ancestor. 307

cladogram A diagram representing the evolutionary relationships of a group of organisms. 317

classical conditioning A form of learning in which an animal responds to a formerly irrelevant stimulus. 808

cleavage A period of rapid cell division following fertilization but before embryogenesis. 789

cleavage furrow The initial indentation between two daughter cells in mitosis. 144

climax community A community that persists unless it is disturbed by environmental change. 848

clitellum *cli-TELL-um* Part of the outside of earthworms that secretes mucus during copulation and a cocoon early in development. 467

clitoris A small, highly innervated bit of tissue that is the female anatomical equivalent of the penis. 786

cloaca An opening common to the digestive, reproductive, and excretory systems in some animals. 723

closed circulatory system A circulatory system in which blood is confined to vessels. 467, 682

club moss A type of seedless vascular plant. 421

cluster-of-differentiation (CD4) antigen A protein on a helper T cell that recognizes a nonself antigen on a macrophage, initiating an immune response. 770

Cnidaria *NID-air-ee-a* An animal phylum whose members have a hollow two-layered body, with a jellylike interior, cnidocytes, and radial symmetry. 458

cnidocyte Cell in cnidaria that contains nematocysts. 459

coccus (pl. cocci) *KOK-sus* Spherical prokaryote. 384

cochlea The spiral-shaped part of the inner ear, where vibrations are translated into nerve impulses. 634

codominant *ko-DOM-eh-nent* Alleles that are both expressed in the heterozygote. 192

codon *KO-don* A triplet of mRNA bases that specifies a particular amino acid. 243

coefficient of relationship *co-eff-FISH-ent uv re-LAY-shun-ship* A measurement of how closely related two individuals are, based upon the proportion of genes they share. 195

coelom A fluid-filled animal body cavity that is completely lined by tissue derived from mesoderm. 454

coelomate An animal with a coelom. 455

coenzyme Organic molecule required as part of a functional enzyme complex. 92

coevolution *ko-ev-eh-LU-shun* The mutual, direct influence of two species on each others' evolution. 847

cofactor Inorganic molecule required for activity of an enzyme. 92

cohesion The attraction of water molecules to each other. 26

cohesion-tension theory Theory that explains the movement of water in xylem from roots through stems to leaves. 532

cohort A group of individuals that begin life at the same time. 831

collagen *COLL-a-jen* A connective tissue protein. 578

collectin Protein that provides broad protection against bacteria, yeasts, and some viruses. 759

collecting duct A structure in the kidney into which nephrons drain urine. 750

collenchyma *kol-LEN-kah-mah* Plant tissue composed of elongated, living cells with thickened walls that support growing regions of leaves and shoots. 506

colon The large intestine. 732

colony-stimulating factor Secreted growth factor that stimulates blood cells to specialize. 773

columnar Tall cells, as in epithelium. 579

commensalism *co-MEN-sal-izm* A symbiotic relationship in which one member benefits without affecting the other member. 845

community All the interacting organisms in a given area. 6

compact bone A layer of solid, hard bone that covers spongy bone. 668

companion cell In phloem, cell that transfers carbohydrates to and from sieve tube members. 509

complement A group of proteins that assist other immune defenses. 758

complementary base pairs Bonding of adenine to thymine and guanine to cytosine in the DNA double helix. 225

complex trait A trait caused by genes and the environment. 193

compound A molecule including different atoms. 21

compound eye An eye formed from several ommatidia. 468, 628

compression fossil A fossil made by pressing an organism in layers of sediment. 323

concentration gradient Difference in solute concentrations between two adjacent compartments. 64

concordance *kon-KOR-dance* A measure of the inherited component of a trait. The number of pairs of either monozygotic or dizygotic twins in which both members express a trait, divided by the number of pairs in which at least one twin expresses the trait. 196

conditioned stimulus A new stimulus coupled to a familiar or unconditioned stimulus, so that an animal associates the two. 809

cone A pollen- or ovule-bearing structure in many gymnosperms. 442

cone cell Specialized receptor cell in the center of the retina that detects colors. 630

conformation *KON-for-MAY-shun* The three-dimensional shape of a protein. 35

confusion effect The indecision caused in a predator when viewing a tight grouping of prey. 818

conifer One of four divisions of gymnosperms. 423

conjugation *con-ju-GAY-shun* A form of gene transfer in prokaryotes. 160, 389

connective tissue Tissue type consisting of widely spaced cells in a matrix; includes loose and fibrous connective tissues, cartilage, bone, and blood. 578

conservative Referring to DNA replication that retains the complete, original parental DNA while making an entirely new copy of that molecule. 228

constant regions Sequences of amino acids in the heavy and light chains that are the same for all antibodies. 763

consumer An organism that must obtain nutrients by consuming other organisms. 6

contact inhibition A property of most noncancerous eukaryotic cells that inhibits cell division when they come in contact with one another. 146

contractile vacuole *KON-tract-till VAK-u-ol* An organelle that pumps water out of the cell. 404

convergent evolution *kon-VER-gent ev-o-LU-shun* Organisms that have similar adaptations to a similar environmental challenge but that are not related by descent. 268

coral reef Underwater deposits of calcium carbonate formed by colonies of animals. 882

cork cambium A type of lateral meristem that produces the periderm of woody plants. 518

cork cell Waxy cell on mature stems and roots. 520

cornea A modified portion of the human eye's sclera that forms a transparent curved window that admits light. 629

corona radiata Layer of cells around the oocyte. 788

corpus callosum Thick band of nerve fibers that interconnect the cerebral hemispheres. 614

corpus luteum *KOR-pis LU-te-um* A gland formed from an ovarian follicle that has recently released an oocyte; produces estrogen and progesterone. 787

cortex The ground tissue between the epidermis and vascular tissue in stems and roots. 510

cotransport Two substances crossing the cell membrane together through a single channel complex. 67

cotyledon *KOT-ah-LEE-don* Embryonic seed leaf in flowering plants. 551

countercurrent exchange system A system of parallel vessels in which fluid flows in opposite directions and maximizes the exchange of substances or heat. 741

countercurrent flow Fluid flow in different parts of a continuous tubule in opposite directions, which maximizes the amount of a particular substance that diffuses out of the tubule. 704

coupled reactions Two chemical reactions that occur simultaneously and have a common intermediate. 86

covalent bond Atoms sharing electrons. 23

cranial nerve Peripheral nerve that exits the vertebrate CNS from the brain. 607

creatine phosphate *KRE-ah-tin FOS-fate* A molecule stored in muscle fibers that can donate its high-energy phosphate to ADP to regenerate ATP. 674

crista (pl. cristae) *KRIS-tay* Fold of the inner mitochondrial membrane along which many of the reactions of cellular respiration occur. 54

critical period The time during prenatal development when a structure is vulnerable to damage. 810

crop A part of the digestive tract in birds and some invertebrates that stores or digests food. 723

crossing over Exchange of genetic material between homologous chromosomes during prophase I of meiosis. 163

crown eukaryote Eukaryotic organisms, including plants, animals, fungi, and many protista, thought to have diverged after basal eukaryotes. 396

crustacean A segmented, aquatic arthropod with gills, two-part appendages, mandibles, and antennae. 470

cuticle *KEW-tah-kal* A waxy layer covering the aerial parts of a plant. 507

cutin *KEW-tin* A fatty material that a plant's epidermal cells produce that forms a cuticle. 507

cyanobacterium *si-an-o-bak-TEAR-ee-um* A type of photosynthetic prokaryote. 384

cycad *SI-kad* One of four divisions of gymnosperms. 423

cyclic adenosine monophosphate (cAMP) A second messenger formed from ATP by the activation of the enzyme adenyl cyclase. 77, 642

cyclin *SI-klin* A type of protein that controls the cell cycle. 148

cyst *SIST* A dormant form of a protist that has a cell wall and lowered metabolism. 397

cytochrome *SI-to-krome* An iron-containing molecule that transfers electrons in metabolic pathways. 92

cytogenetics *si-to-jen-ET-ix* Correlation of an inherited trait to a chromosomal anomaly. 203

cytokine Protein synthesized in certain immune cells that influences the activity of other immune cells. 757

cytokinesis *SI-toe-kin-E-sis* Distribution of cytoplasm, organelles, and macromolecules into two progeny cells in cell division. 138

cytokinin *SI-toe-KI-nin* One of five classes of plant hormone. 559

cytoplasm *SI-toe-PLAZ-um* The jellylike fluid in which cell structures are suspended. 42

cytosine *SI-toe-seen* One of the two pyrimidine nitrogenous bases in DNA and RNA. 225

cytoskeleton *SI-toe-SKEL-eh-ten* A framework of arrays of protein rods and tubules in animal cells. 62

cytotoxic T cell Immune system cell that kills nonself cells by binding them and releasing chemicals. 765

D

day-neutral plant Plant that does not rely on photoperiod to flower. 566

dead space The pharynx, the trachea, and the upper third of the lungs, which contain air not used in gas exchange. 708

deciduous tree *dah-SID-u-us TREE* Tree that sheds leaves at the end of a growing season. 514

decomposer An organism that consumes feces and dead organisms. 6

defensin Antimicrobial peptide in arthropods. 757

deforestation Removal of tree cover from a previously forested area. 893

degenerate codons *de-JEN-er-at KO-donz* Different codons that specify the same amino acid. 255

dehydration synthesis *de-hi-DRA-shun SYN-theh-sis* Formation of a covalent bond between two molecules by loss of water. 29

deletion A type of mutation in which genetic material is removed from a chromosome. 214

demography The statistical study of populations. 828

dendrite Thin neuron branch that receives neural messages and transmits information to the cell body. 581, 597

denitrifying bacterium *de-NI-tri-fy-ing bak-TEAR-e-um* A bacterium that converts nitrites and nitrates to nitrogen gas. 857

dense connective tissue Connective tissue with dense collagen tracts. 579

density-dependent factor Condition that limits population growth when populations are large. 834

density-independent factor Population-limiting condition that acts irrespective of population size. 835

dentine Bonelike substance beneath a tooth's enamel. 725

deoxyhemoglobin *de-OX-ee-HEEM-o-GLO-bin* Hemoglobin that is deep red after releasing its oxygen to tissues. 692

deoxyribonucleic acid *de-OX-ee-RI-bo-nu-KLAY-ic AS-id* (DNA) A double-stranded nucleic acid composed of nucleotides containing a phosphate group, a nitrogenous base (A, T, G, or C), and deoxyribose. 37

deoxyribose *de-OX-ee-RI-bose* A five-carbon sugar that is part of DNA. 223

deposit feeder Animal that eats soil and strains out nutrients. 721

derived character The common distinguishing feature of a clade. 317

dermal tissue *DER-mal TISH-ew* Tissue that covers a plant. 507

dermis The layer of skin, derived from mesoderm, that lies beneath the epidermis in vertebrates. 588

descending limb The proximal portion of the loop of the nephron, which descends into the kidney's medulla. 749

desert One of several types of terrestrial biomes; very low precipitation. 872

desertification The expression of desert into surrounding areas. 894

desmosome A junction that anchors intermediate filaments of two adjoining cells in a single spot on the cell membrane. 75

determinate cleavage Condition in an embryo in which the fate of a cell is predetermined. 455

detritus Collective term for feces and dead organic matter. 851

deuteromycete *doo-ter-o-mi-SEET* Fungus that lacks a sexual phase. 414

deuterostome A coelomate lineage with radial cleavage, an anus that forms from the blastopore, indeterminate cleavage, and a coelom that forms from outpocketings of the archenteron. 455

developmental pathway Series of events that forms a particular anatomical structure. 780

diabetes mellitus Disease resulting from an inability of the body to produce or use insulin. 653

diaphragm A sheet of muscle separating the thoracic and abdominal cavities. 712

diapsid Animal with two openings behind each eye orbit in its skull. 491

diastole Relaxation of the heart. 687

diatom *DIE-a-tom* A photosynthetic (protist) with distinctive silica walls. 408

dicotyledon(dicot) An angiosperm that has two cotyledons, or seed leaves. 426

differentiate *diff-er-EN-shee-ate* Specialize. 149

diffusion *de-FUZE-jhun* Movement of a substance from a region where it is highly concentrated to an area where it is less concentrated without energy input. 64

digestion Chemical breakdown of food. 721

digestive system System of tubes where food is broken down into nutrient molecules that are absorbed into capillaries. 584, 722

dihybrid cross *DI-HI-brid KROS* Mating between individuals heterozygous for two particular genes. 187

dikaryon *DI-kari-on* A cell having two genetically distinct haploid nuclei. 435

dilution effect A behavior that decreases the chance of an event occurring. 818

dinoflagellate *di-no-FLADJ-el-et* Group of aquatic protista with flagellated cells. 401

dioecious In plants, having separate male and female individuals of the same species. 545

diploblastic Describing an animal whose adult tissues arise from two germ layers in the embryo. 453

diploid cell *DIP-loid SEL* A cell with two copies of each chromosome. Also known as 2*n*. 158

Diplomonadida Group of basal eukaryotes whose cells lack mitochondria. 397

direct development The gradual development of a juvenile animal into an adult, without an intervening larval stage. 454

directional selection Changes in the prevalence of a characteristic that reflects differential survival of individuals better adapted to a particular environment. 285

disaccharide *di-SAK-eh-ride* A sugar that consists of two bonded monosaccharides. 29

dispersive Referring to DNA replication that results in two daughter molecules that each contain fragments of the original parent molecule. 228

disruptive selection A population in which two extreme expressions of a trait are equally adaptive. 286

distal convoluted tubule The region of the kidney distal to the nephron loop and proximal to a collecting duct. 749

disulfide bond Attraction between two sulfur atoms within a protein molecule. 35

diversity-stability hypothesis Proposes that the more species in a community, the better the community will resist environmental change. 899

dizygotic twins *di-zi-GOT-ik TWINZ* Fraternal (non-identical) twins. 196

DNA polymerase *DNA po-LIM-er-ase* An enzyme that inserts new bases and corrects mismatched base pairs in DNA replication. 231

domain A taxonomic designation that supercedes kingdom. 8

dominance hierarchy A social ranking of members of a group of the same sex, which distributes resources with minimal aggression. 814

dominant Describing an allele that masks the expression of another allele. 178

dormancy *DOR-man-see* A temporary state of lowered metabolism and arrested growth. 570

dorsal *DOOR-sal* The back side. 453

dorsal hollow nerve cord One of the four characteristics of chordates; derived from a plate of embryonic ectoderm. 479

double-blind An experimental protocol where neither participants nor researchers know which subjects received a placebo and which received the treatment being evaluated. 14

double fertilization In angiosperms and gnetophytes, the fusion of one sperm cell with an egg, and another sperm cell with the polar nuclei to yield a triploid endosperm. 549

duodenum The first section of the small intestine. 729

duplication A repeated portion of DNA in a chromosome. 214

dynamic equilibrium A state where equal concentrations of a substance exist across a semipermeable membrane. 64

E

echinoderm A deuterostome with a five-part body plan, radial symmetry, and spiny outer covering. 470

echolocation Ability to locate objects by bouncing sound waves off them. 632

ecological succession Change in the species composition of a community over time. 848

ecology The study of relationships between organisms and their environments. 828

ecosystem All organisms and their nonliving environment in a defined area. 6

ectoderm In an animal embryo, the outermost germ layer, whose cells become part of the nervous system, sense organs, and the outer skin layer. 453

ectotherm An animal that obtains heat primarily from its external surroundings. 482, 740

edema *eh-DEEM-ah* Swelling of a body part from fluid buildup. 752

efferent arteriole Arteriole that receives blood from the glomerular capillaries of a nephron. 747

ejaculation Discharge of sperm through the penis. 786

elastin *e-LAS-tin* A type of connective tissue protein. 579

electrolyte *e-LEK-tro-lite* Solution containing ions.26

electromagnetic spectrum *e-LEK-tro-mag-NET-ik SPECK-trum* A spectrum of naturally occurring radiation. 99

electron *e-LEK-tron* A negatively charged subatomic particle with negligible mass that orbits the atomic nucleus. 20

electronegativity *e-LEK-tro-neg-a-TIV-it-ee* The tendency of an atom to attract electrons. 24

electron orbital *e-LEK-tron OR-bit-al* The volume of space where a particular electron is 90% of the time. 20

electron transport chain Membrane-bounded molecular complex that transfers energy from electrons to form gradients and energy-carrying molecules. 89

element A pure substance consisting of atoms containing a characteristic number of protons. 20

El Niño *l NEEN-yo* A periodic slack in the trade winds that prevents upwelling along Peru's coast and changes global weather patterns. 866

embryo A prenatal stage of development when all structures form. 791

embryonic disc Flattened inner cell mass of the embryo. 790

embryonic induction *EM-bree-ON-ik in-DUK-shun* In an embryo, the ability of a group of specialized cells to stimulate neighboring cells to specialize. 792

embryonic stage In humans, the stage of prenatal development from the second through eighth weeks, when tissues and organs begin to differentiate. 788

embryo sac The mature female gametophyte in angiosperms. 547

emergent property A quality that appears as biological complexity increases. 4

emigration Migration out of a population. 828

empiric risk *em-PEER-ik RISK* Risk calculation based on prevalence. 194

enamel Hard substance covering a tooth. 725

endergonic reaction *en-der-GONE-ik re-AK-shun* An energy-requiring chemical reaction. 88

endocrine gland A concentration of hormone-producing cells in animals. 642

endocrine system Glands and cells that secrete hormones in animals. 584, 642

endocytosis *EN-doe-si-TOE-sis* The cell membrane's engulfing extracellular material. 68

endoderm In an animal embryo the germ layer, whose cells become the organs and linings of the digestive and respiratory systems. 453

endodermis *en-do-DER-mis* The innermost cell layer in a root's cortex. 515

endometrium The inner uterine lining. 787

endoplasmic reticulum *EN-doe-PLAZ-mik reh-TIK-u-lum* Interconnected membranous tubules and sacs that wind from the nuclear envelope to the cell membrane, along which proteins are synthesized (in rough ER) and lipids synthesized (in smooth ER). 52

endorphin Pain-killing protein produced in the nervous system. 648

endoskeleton An internal scaffolding type of skeleton in vertebrates and some invertebrates. 470, 662

endosome A membrane-bounded compartment containing the products of endocytosis. 68

endosperm *EN-do-sperm* In angiosperms, a triploid tissue that nourishes the embryo in a seed. 550

endospore *EN-doe-spor* A walled structure that enables some prokaryotic cells to survive harsh environmental conditions. 382

endosymbiont theory *EN-doe-SYM-bee-ont THER-ee* The idea that eukaryotic cells evolved from large prokaryotic cells that engulfed once free-living bacteria. 58

endothelium Layer of single cells that lines blood vessels. 688

endotherm An animal that uses metabolic heat to regulate its temperature. 492, 740

energy The ability to do work. 82

energy nutrients Fats, proteins, and carbohydrates. 88

energy of activation Energy required for a chemical reaction to begin. 93

energy shell Levels of energy in an atom formed by electron orbitals. 22

entrainment *en-TRANE-ment* The resynchronization of a biological clock by the environment. 570

entropy *EN-tro-pee* Randomness or disorder. 86

envelope The lipid layer around some viruses. 363

environmental resistance Factors that limit population growth. 833

enzyme *EN-zime* A protein that catalyzes a specific type of chemical reaction. 36

enzyme pathway A series of enzymes that carries out a complex series of chemical reactions. 94

enzyme-substrate complex *EN-zime SUB-strate COM-plex* A transient structure that forms when a substrate binds an enzyme's active site. 36

epidemiology *EP-eh-dee-mee-OL-o-gee* The analysis of data derived from real-life, nonexperimental situations. 10

epidermis *ep-ee-DERM-is* The outer integumentary layer in several types of animals; also the outermost cell layer of young roots, stems, and leaves. 507, 588

epididymis In the human male, a tightly coiled tube leading from each testis, where sperm mature and are stored. 168, 785

epigenesis The idea that specialized tissue arises from unspecialized tissue in a fertilized ovum. 778

epiglottis Cartilage that covers the glottis, routing food to the digestive tract and air to the respiratory tract. 708

epinephrine (adrenaline) A hormone produced in the adrenal medulla that raises blood pressure and slows digestion. 652

epipelagic zone *epi-pe-LAG-ic ZONE* Part of the ocean where photosynthesis occurs. 882

epistasis *eh-pis-tah-sis* A gene masking another gene's expression. 190

epithelial tissue Tightly packed cells that form linings and coverings. 578

epoch *EH-poc* Time within a period. 322

equational division *ee-QUAY-shun-el deh-VISZ-un* The second meiotic division, when four haploid cells form from two haploid cells that are the products of meiosis I. 161

era Very long period of time of biological or geological activity. 322

erythrocyte Red blood cell. 692

erythropoietin A hormone produced in the kidneys that stimulates red blood cell production when oxygen is lacking. 693

esophagus A muscular tube that leads from the pharynx to the stomach. 725

essential nutrient Nutrient that must come from food because the body cannot synthesize it. 733

estrogen A steroid hormone produced in ovaries of female vertebrates that helps regulate reproductive cycles. 655

estuary *ES-tu-air-ee* An area where fresh water in a river meets salty water of an ocean. 880

ethology Study of how natural selection shapes adaptive behavior. 806

ethylene *ETH-eh-leen* One of five classes of plant hormone. 561

etiolated seedling *E-ti-o-LAY-tid SEED-ling* Seedling that has abnormally elongated stems, small roots and leaves, and a pale color, because it was grown in the dark. 569

euchromatin *u-KROME-a-tin* Light-staining genetic material. 202

Euglenida Group of motile basal eukaryotes (protista), some of which are photosynthetic. 399

Eukarya *yoo-KAR-ee-a* One of the three domains of life, including organisms that have eukaryotic cells. 9

eukaryotic cell *u-CARE-ee-OT-ik SEL* A complex cell containing membrane-bound organelles. 50

euploid *U-ployd* A normal chromosome number. 211

eusocial Describing a population of animals that communicate, cooperate in caring for young, have overlapping generations, and divide labor. 818

eustachian tube (auditory tube) Tube that connects the middle ear with the air passageways; allows for the adjustment of pressure on the inside of the tympanic membrane. 634

eutrophic *yu-TRO-fik* A lake containing many nutrients and decaying organisms, often tinted green with algae. 878

evaporative cooling Loss of body heat by evaporation of fluid from the body's surface. 741

evolution Changing gene frequencies in a population over time. 264

excision repair *ex-SIZ-jhun ree-PARE* Cutting pyrimidine dimers out of DNA. 233

excurrent siphon *EX-cur-ent SI-fon* A structure that shoots water from a tunicate when it is disturbed. 479

exergonic reaction *ex-er-GONE-ik re-AK-shun* An energy-releasing chemical reaction. 88

exocrine gland Structure that secretes substances through ducts. 642

exocytosis *EX-o-si-TOE-sis* Fusing of secretion-containing organelles with the cell membrane. 68

exon *EX-on* The bases of a gene that code for amino acids. 245

exoskeleton A braced framework skeleton on the outside of an organism. 468, 661

experiment A test to disprove a hypothesis. 10

experimental control An extra test that can rule out causes other than the one being investigated. 14

exponential growth Population growth in which numbers double at regular intervals. 832

external respiration Exchange of gases between respiratory surfaces and the blood. 702

extinction *ex-TINK-shun* Disappearance of a type of organism. 308

extracellular digestion Dismantling of food by hydrolytic enzymes in a cavity within an organism's body. 722

extracellular matrix A nonliving complex of substances that surrounds cells of connective tissue. 75

extraembryonic membrane *EX-tra-EM-bree-on-ik MEM-brane* Structure that supports and nourishes the mammalian embryo and fetus, including the yolk sac, allantois, and amnion. 792

F

facilitated diffusion *fah-SIL-eh-tay-tid dif-FU-shun* Movement of a substance down its concentration gradient with the aid of a carrier protein. 66

facultative anaerobe Organism that can live in the presence or absence of oxygen. 385

fallopian tube In the human female, one of a pair of tubes leading from near the ovaries to the uterus. 786

fall turnover The seasonal mixing of the upper and lower layers of a lake. 878

fatty acid A hydrocarbon chain that is a part of a triglyceride. 32

feather follicle An extension of the epidermis from which a feather extends. 590

fecundity *fee-KUN-dit-ee* The number of offspring an individual produces in its lifetime. 831

fermentation *fur-men-TAY-shun* An energy pathway in the cytoplasm that extracts limited energy from the bonds of glucose. 129

fertilization The union of two gametes of eukaryotic origin. 159

fetal period The final stage of prenatal development, when structures grow and elaborate. 788

fibrin A threadlike protein that forms blood clots. 694

fibrinogen *fi-BRIN-o-jen* A protein that is the precursor of fibrin. 694

fibroblast A connective tissue cell that secretes collagen and elastin. 578

fibrous root system A root system composed of many similar-sized roots, as in grasses and many other monocots. 515

fiddlehead *FID-ul-hed* A leaf formation in a true fern. 441

filter feeder An aquatic animal that uses ciliated tissue surfaces to feed on small food particles. 721

filtrate Material that passes from the glomerulus into the glomerular capsule. 748

filtration Filtering of substances across a filtration membrane, as occurs in the glomerulus. 747

first filial generation *FURST FILL-e-al gen-er-A-shun* (F_1) The second generation in a genetics problem. 180

fixed action pattern (FAP) An innate, stereotyped behavior. 807

flagellum (pl. flagella) A long whiplike appendage a cell uses for motility and composed of microtubules. 71, 382

flame cell system A simple excretory system in flatworms. 745

flatworm An unsegmented worm lacking a body cavity; phylum Platyhelminthes. 461

flavin adenine dinucleotide *FLAY-vin AD-e-neen di-NUKE-lee-o-tide* (FAD) An electron carrier molecule that functions in certain metabolic pathways. 92

flower The reproductive structure in angiosperms. 543

fluorescence Absorbing light energy at one wavelength and emitting it at another wavelength. 104

follicle cell *FOL-ik-kel SEL* Nourishing cell surrounding an oocyte. 787

follicle-stimulating hormone (FSH) A pituitary hormone that controls oocyte maturation, development of ovarian follicles, and their release of estrogen. 648

food chain The linear series of organisms that successively eat each other. 851

food vacuole *FOOD VAK-you-ole* A membrane-bounded sac that simple animals use for digestion. 722

food web A network of interconnecting food chains. 851

forebrain The front part of the vertebrate brain. 610

formed elements Blood cells and platelets. 692

founder effect Genetic drift when small groups found new settlements, partitioning a subset of the original population's genes. 284

fountain effect Splitting and regrouping of a school of fish, which confuses a predator. 818

fovea centralis An indentation in the retina opposite the lens that has only cones and provides visual acuity. 630

frameshift mutation *FRAME-shift mew-TAY-shun* A mutation that adds or deletes one or two DNA bases, altering the reading frame. 255

free energy The usable energy in the bonds of a molecule. 88

free radical Highly reactive by-product of metabolism that can damage tissue. 799

frond *FROND* Leaf of a true fern. 422

frugivore An animal that feeds on fruits. 720

fruit Seed-containing structure in flowering plants. 543

fruiting body Spore-bearing organs in fungi. 436

frustules *FRUS-tools* Two-part silica walls of diatoms. 407

fucoxanthin *foo-ko-ZAN-thin* A photosynthetic pigment that gives golden and brown algae their distinctive color. 407

fundamental niche *fun-da-MEN-tal NEESH* All the resources that a species could possibly use in its environment. 844

G

G_0 phase Resting phase of the cell cycle where cells continue to function, but do not divide. 139

G_1 phase The gap stage of interphase when proteins, lipids, and carbohydrates are synthesized. 138

G_2 phase The gap stage of interphase when membrane components are synthesized and stored. 140

gallbladder An organ beneath the liver that stores bile. 733

gamete *GAM-eet* A sex cell. The sperm and ovum. 158

gametogenesis *ga-MEET-o-gen-i-sis* Meiosis and maturation; making gametes. 160

gametophyte *gam-EET-o-fite* The haploid, gamete-producing stage of the plant life cycle. 414, 543

gamma-delta T cell A type of T cell that oversees interactions of the innate and acquired immune responses. 765

ganglion Cluster of neuron cell bodies. 606

ganglion cell The first cell type in the visual pathway to generate action potentials. 630

gap junction A connection between two cells that allows cytosol to flow between both cells. 75

gastric juice *GAS-trik JUICE* The fluid that stomach cells secrete that carries out chemical digestion. 725

gastrin A hormone that stomach cells secrete that stimulates secretion of more gastric juice. 729

gastrodermis *gas-tro-DERM-is* The inner tissue layer of a cnidarian. 458

gastrointestinal tract A continuous tube along which food is physically and chemically digested. 725

gastropod A mollusk with a broad flat foot for crawling (snails and slugs). 465

gastrovascular cavity A digestive chamber with a single opening found in cnidarians and flatworms. 458, 722

gastrula A three-layered embryo immediately following gastrulation. 453, 791

gemma (pl. gemmae) Asexual reproductive structure in liverworts. 418

gene A sequence of DNA that specifies the sequence of amino acids of a particular polypeptide. 37

gene flow A change in a gene pool due to the random loss or addition of alleles through migration. 280

gene pool All the genes in a population. 280

genetic code Correspondence between specific DNA base triplets and amino acids. 246

genetic drift Changes in gene frequencies caused by separation of a small group from a larger population. 284

genetic heterogeneity Containing nonidentical components of genetic information. 193

genetic load Collection of deleterious alleles in a population. 285

genetic marker A detectable piece of DNA closely linked to a gene of interest whose precise location is unknown. 205

genotype *JEAN-o-type* Genetic constitution of an individual. 181

genotypic ratio *jean-o-TIP-ik RAY-shee-o* Proportions of genotypes among offspring of a genetic cross. 182

geological timescale A division of time into major eras of biological and geological activity, then periods within eras, and epochs within some periods. 321

germ cell Gamete or sex cell. 160

germinal mutation *GER-min-al mew-TAY-shun* A mutation in a sperm or oocyte. 254

germination The beginning of growth in a seed, spore, or other structure. 552

gibberellin *GIB-ah-REL-in* One of five classes of plant hormone. 559

gill A highly folded respiratory surface for gas exchange in aquatic animals. 479, 704

ginkgo *GEENG-ko* One of four divisions of gymnosperms. 423

gizzard A muscular part of the digestive tract that grinds food in some animals. 723

glomerular or Bowman's capsule The cup-shaped proximal end of the renal tubule that surrounds the glomerulus. 748

glomerulus A ball of capillaries between the afferent arterioles and efferent arterioles in the proximal part of a nephron. 747

glottis Opening from the pharynx to the larynx. 708

glucagon A pancreatic hormone that breaks down glycogen into glucose, raising blood sugar levels. 653

glucocorticoid Hormone that the adrenal cortex secretes that enables the body to survive prolonged stress. 651, 653

glycerol *GLI-sir-all* A three-carbon alcohol that forms the backbone of triglyceride fats. 32

glycocalyx *gli-ko-CAY-lix* A sticky layer outside a prokaryotic cell wall that consists of proteins and/or polysaccharides. 384

glycolysis *gli-KOL-eh-sis* A catabolic pathway occurring in the cytoplasm of all cells. One molecule of glucose splits and rearranges into two molecules of pyruvic acid. 118

gnetophyte *NEE-to-fit-e* One of four divisions of gymnosperms. 424

goiter Swelling of the thyroid gland caused by lack of iodine in the diet. 652

golden alga Photosynthetic protist with unique pigments. 408

Golgi body *GOL-gee bod-EE* A system of flat, stacked, membrane-bounded sacs where sugars are polymerized to starches or bonded to proteins or lipids. 52

gonad Organ that manufactures gametes in animals. 654, 785

gonadotropic hormone *go-NAD-o-TRO-pik HOR-moan* A hormone made in the anterior pituitary that affects the ovaries or testes. 655

gradualism Slow evolutionary change. 305

Gram stain Technique for classifying major groups of bacteria, based on cell wall structure. 382

granum A stack of flattened thylakoid discs that forms the inner membrane of a chloroplast. 55

gravitropism *grav-eh-TROP-izm* A plant's growth response toward or away from the pull of gravity. 562

gray matter Nervous tissue in the CNS consisting of neuron cell bodies, unmyelinated fibers, interneurons, and neuroglial cells. 610

green alga Photosynthetic protist that has pigments, starch, and cell walls most like plants. 409

greenhouse effect Elevation in surface temperature caused by carbon dioxide and other atmospheric gases. 892

ground tissue The tissue that makes up most of the primary body of a plant; consists of parenchyma, collenchyma, and sclerenchyma cells. 506

growth hormone (GH) A pituitary hormone that promotes growth and development of tissues by increasing protein synthesis and cell division rates. 648

guanine *GWAN-een* One of the two purine nitrogenous bases in DNA and RNA. 225

guard cell One of a pair of epidermal cells that open and close stomata in plants by gaining and losing turgor pressure. 507

gustation The sense of taste. 627

gymnosperm *JIM-no-sperm* A seed-producing plant whose seeds are not enclosed in an ovary. 415

gynoecium *guy-no-EE-see-um* The second innermost whorl of a flower, consisting of female reproductive structures. 543

H

habitat The physical place where an organism lives. 844

habituation The simplest form of learning, in which an animal learns not to respond to irrelevant stimuli. 809

hagfish A jawless fish with a cranium and lacking supportive cartilage around the nerve cord. 482

hair cell Mechanoreceptor that initiates sound transduction in the cochlea. 634

hair follicle An epidermal structure anchored in the dermis, from which a hair grows. 589

half-life The time it takes for half the isotopes in a sample of an element to decay into a second isotope. 324

halophile *HAL-o-file* Organism that tolerates extremely salty surroundings. 385

haploid cell A cell with one copy of each chromosome. Also called 1*n*. 158

hardwood Wood in dicots, such as oak, maple, and ash. 520

Hardy-Weinberg equilibrium Maintenance of the proportion of genotypes in a population from one generation to the next. 280

heart Muscular organ that pumps blood or hemolymph. 684

heart attack A blockage of blood flow in a coronary artery. 695

heart valve Flap in the heart that keeps blood flow in one direction. 686

heartwood Nonfunctioning wood in the center of a tree. 520

heat capacity The amount of heat necessary to raise the temperature of a substance. 28

heat of vaporization *HEET of VA-por-i-ZA-shun* The amount of heat required to convert a liquid to vapor (gas). 28

heavy chain Large polypeptide of an antibody subunit. 763

helper T cell Lymphocyte that produces cytokines and stimulates activities of other immune system cells. 762

heme group *HEEM GROOP* An iron-containing complex that forms the oxygen-binding part of hemoglobin. 253

hemizygous *HEM-ee-ZY-gus* A gene on the Y chromosome in humans. 207

hemocoel *HEEM-o-seal* The body cavity of an arthropod. 468

hemocyanin A respiratory pigment in mollusks that carries oxygen. 682

hemoglobin The iron-binding protein that carries oxygen in mammals. 682

hemolymph The "blood" in animals with open circulatory systems. 468, 682

hepatic portal system *heh-PAH-tik POR-tel SIS-tum* A division of the circulatory system that enables the liver to rapidly harness chemical energy in digested food. 689

herbivore An animal that feeds on plants. 720

heritability *herr-it-a-BILL-it-ee* The proportion of a trait attributable to heredity. 194

hermaphrodite Organism that produces sperm and eggs in the same individual. 453

heterochromatin *het-er-o-KROME-a-tin* Dark-staining genetic material. 202

heterogametic sex *HET-er-o-gah-MEE-tik SEX* The sex with two different sex chromosomes. 205

heterotherm *HET-er-o-therm* An animal with variable body temperature. 740

heterotroph *HET-er-o-TROFE* An organism that obtains carbon by eating another organism. 384

heterozygous *HET-er-o-ZI-gus* Possessing two different alleles for a particular gene. 181

highly conserved A protein or nucleic acid sequence that is very similar in different species. 332

hindbrain The lower portion of the vertebrate brain, which includes the brain stem and controls vital functions. 610

hippocampus Area of the cerebral cortex important in long-term memory. 615

Hirudinea *hear-EW-din-ay* A class of annelids that includes the leeches. 467

histamine An allergy mediator that dilates blood vessels and causes allergy symptoms. 759

homeostasis The ability of an organism to maintain constant body temperature, fluid balance, and chemistry. 6, 585

homeotherm *HOME-e-o-therm* An animal with constant body temperature. 740

homeotic *home-ee-OT-ik* A mutant in which body parts form in the wrong places. 781

homing The returning to a given spot. 811

hominid *HOM-eh-nid* Animal ancestral to humans only. 354

hominoid *HOM-eh-noid* Animal ancestral to apes and humans. 354

homogametic sex *HO-mo-gah-MEE-tik SEX* The sex with two identical sex chromosomes. 205

homologous pairs *ho-MOL-eh-gus PAIRZ* Chromosome pairs that have the same sequence of genes. 161

homologous structures *ho-MOL-eh-gus STRUK-churs* Similar structures in different species that have the same general function, indicating descent from a common ancestor. 328

homozygous *HO-mo-ZI-gus* Possessing two identical alleles for a particular gene. 181

horizontal gene transfer Transfer of genetic information between possibly unrelated organisms. 386

hormone A chemical synthesized in small quantities in one part of an organism and transported to another, where it affects target cells. 558, 642

hornwort *HORN-wart* A type of bryophyte. 418

horsetail A type of seedless vascular plant. 421

host range The organisms that can be infected by a virus. 370

human chorionic gonadotropin (hCG) A hormone secreted by the preembryo and embryo that prevents menstruation. 790

humoral immune response Secretion of antibodies by B cells in response to a foreign antigen. 757

humus Partially decomposed organic matter in soil. 529

hybridoma *hi-bri-DOME-ah* An artificial cell created by fusing a B cell with a cancer cell that secretes a particular antibody. It mass-produces the antibody. 772

hydrocarbon *HI-dro-kar-bon* A molecule containing carbon and hydrogen. 24

hydrogen bond *HI-dro-gen BOND* A weak chemical bond between negatively charged portions of molecules and hydrogen ions. 25

hydrolysis *hi-DROL-eh-sis* Splitting a molecule by adding water. 30

hydrophilic *HI-dro-FILL-ik* Attracted to water. 25

hydrophobic *HI-dro-FOBE-ik* Repelled by water. 25

hydrostatic skeleton The simplest type of skeleton, consisting of flexible tissue surrounding a constrained liquid. 661

hydroxide ion *hi-DROX-ide I-on* (OH–) A molecule consisting of one oxygen and one hydrogen. 27

hypertension Condition in which blood pressure is significantly higher than normal. 691

hypertonic *hi-per-TON-ik* The solution on one side of a membrane where the solute concentration is greater than on the other side. 65

hypertrophy *hi-PER-tro-fee* Increase in muscle mass, usually due to exercise. 676

hypha *HI-fa* A fungal thread, the basic structural unit of a multicellular fungus. 436

hypodermis *hi-po-DERM-is* An outer layer of a roundworm that secretes a protective cuticle. 464

hypothalamus A small structure beneath the thalamus that controls homeostasis and links the nervous and endocrine systems. 613, 647

hypothesis *hy-POTH-eh-sis* An educated guess based on prior knowledge. 10

hypotonic *hi-po-TON-ic* The solution on one side of a membrane where the solute concentration is less than on the other side. 65

I

ileum The last section of the small intestine. 729

imbibition *IM-bah-BISH-un* The absorption of water by a seed. 26

immigration Migration into a population. 828

immune system System of specialized cells that defends the body against infections. 584

impact theory Proposed theory for mass extinctions caused by impacts of extraterrestrial origin. 309

implantation Nestling of the blastocyst into the uterine lining. 790

impression fossil The preserved evidence of an organism, such as a footprint. 323

imprinting A type of learning that usually occurs early in the life and is performed without obvious reinforcement. 810

inclusive fitness Fitness defined by reproductive success of an individual and its relatives. 817

incomplete dominance A heterozygote whose phenotype is intermediate between the phenotypes of the two homozygotes. 192
incomplete penetrance A genotype that does not always produce a phenotype. 189
incus *INK-us* A small bone in the middle ear. 633
independent assortment The random organization of homologs during metaphase of meiosis I. 163
indeterminate cleavage Condition in an embryo in which the fate of a cell is not predetermined. 455
industrial melanism *in-DUS-tree-al MEL-an-iz-um* Coloration that is adaptive in a polluted area. 286
industrial smog Air pollution resulting directly from industrial and urban emissions. 889
inferior vena cava The lower branch of the largest vein that leads to the heart. 686
inflammation Increased blood flow and accumulation of fluid and phagocytes at the site of an injury, rendering it inhospitable to bacteria. 758
inhibiting hormone A hormone produced by the hypothalamus that inhibits another gland. 648
initiation complex *in-ish-e-A-shun COM-plex* A small ribosomal subunit, mRNA, and an initiator tRNA, joined. 243
initiation site *in-ish-e-A-shun SITE* The site on a chromosome where DNA replication begins. 231
innate Instinctive; developing independently of experience. 807
innate immunity Cells and substances that provide preexisting defenses against infection without prior exposure to an antigen. 756
inner cell mass The cells in the blastocyst that develop into the embryo. 789
inner ear Fluid-filled chamber that houses both balance and hearing structures. 634
insect An arthropod in the class Insecta; adults have a head with one pair of antennae, a thorax with wings and six legs, and an abdomen. 470
insectivore An animal that feeds on insects. 720
insertion *in-SER-shun* The end of a muscle on a movable bone. 676
insight learning Ability to apply prior learning to a new situation without trial-and-error activity. 810
insulin A pancreatic hormone that lowers blood sugar level by stimulating body cells to take up glucose from the blood. 653
integrin A key protein that anchors cells to connective tissue. 76
integument (integumentary system) Outer covering of an animal's body. 584
intercalary meristem *in-TER-kah-LARE-ee MER-eh-stem* Dividing tissue between mature regions of grass stem. 505
intercalated disks *in-TER-kah-LAY-tid DISKS* Tight foldings in cardiac muscle cell membranes that join adjacent cells. 670
intercellular junctions Structures that join cells, including tight junctions, gap junctions, desmosomes, and plasmodesmata. 74
interferon A polypeptide produced by a T cell infected with a virus that stimulates surrounding cells to manufacture biochemicals that halt viral replication. 757
interleukin A class of immune system biochemicals. 757
intermediate filament Cytoskeletal element intermediate in size between a microtubule and a microfilament. 70
intermembrane compartment *in-ter-MEM-brain cum-PART-ment* The space between a mitochondrion's two membranes. 120
internal respiration Exchange of gases between the blood and the body cells. 702
interneuron A neuron that connects one neuron to another to integrate information from many sources and to coordinate responses. 598
internode *IN-ter-node* Part of stem between nodes. 510
interphase *IN-ter-faze* The period when the cell synthesizes proteins, lipids, carbohydrates, and nucleic acids. 138
interspecific competition The struggle between members of a different species for vital resources. 844
interstitial fluid Liquid that bathes cells in a vertebrate's body. 696
intertidal zone *IN-ter-TI-dal ZONE* The region along a coastline where the tide recedes and returns. 882
intestinal cecum *in-TES-tin-al SEE-kum* The site of digestion and absorption in flatworms. 461
intracellular digestion Digestion within food vacuoles in cells. 722
intrinsic rate of natural increase Population growth rate, the difference between birth and death rates. 833
intron *IN-tron* Bases of a gene that are transcribed but are excised from the mRNA before translation into protein. 245
inversion *in-VER-shun* An introverted gene sequence. 214
invertebrate Animal lacking a vertebral column. 452
ion *I-on* An atom that has lost or gained electrons, giving it an electrical charge. 25
ionic bond *i-ON-ik bond* Attraction between oppositely charged ions. 25
ionizing radiation *I-o-nize-ing rade-e-A-shun* Radiation that ejects electrons from atoms. 254
iris Colored part of the eye that regulates the size of the pupil. 629
irritability *IR-eh-tah-BIL-eh-tee* An immediate response to a stimulus. 7
islet of Langerhans Cluster of cells in the pancreas that secretes hormones that control nutrient use. 653
isotonic *ice-o-TON-ik* When solute concentration is the same on both sides of a membrane. 65
isotope *I-so-tope* A differently weighted form of an element. 20

J

jejunum The middle section of the small intestine. 729
joint Where a bone contacts another bone. 663
juvenile hormone An insect hormone produced in larvae that controls metamorphosis. 645

K

karyokinesis *KAR-ee-o-kah-NEE-sus* Division of the genetic material. 138
karyotype *KAR-ee-o-type* A size-order chart of chromosomes. 202
keratin *KERR-a-tin* A hard protein that accumulates in the integument of many animals and forms specialized structures. 35
keystone species Species that exert an effect on community structure that is disproportionate to their biomass. 848
kidney Organ consisting of millions of tubules that excrete nitrogenous waste and regulate ion and water levels. 746
kilocalorie (kcal) The energy required to raise 1 kilogram of water 1°C. One food calorie. 733
kinase *KI-nase* A type of enzyme that activates other proteins by adding a phosphate. 148
kinetic energy *kin-ET-ik EN-er-gee* The energy of motion. 84
kinetochore Microtubule fibers anchored in the centromere that connect chromosomes to the spindle apparatus. 140
kinetoplast Unique, DNA-containing structure that forms part of the mitochondrion in Kinetoplastida. 400
Kinetoplastida Group of protista whose cells contain kinetoplasts. 400
kin selection Process by which an individual helps a relative survive or reproduce. 817
Koch's postulates Rules used to verify that an organism causes a particular disease. 390
Krebs cycle *KREBS SI-kle* The stage in cellular respiration that completely oxidizes the products of glycolysis. 119
K-selection Selection for individuals that are long-lived, late-maturing, and have few offspring that each receive heavy parental investment. 836

L

lacteal A lymph vessel that absorbs fat in the small intestine. 732

lactic acid fermentation *LAK-tik AS-id fer-men-TAY-*shun Conversion of pyruvic acid into lactic acid, occurring in some anaerobic bacteria and tired mammalian muscle cells. 131

lacuna (pl. lacunae) *la-KEW-na* Space in cartilage and bone tissue. 579

laminin *LAM-i-nin* A protein that binds skin layers. 589

lamprey The first jawless fish with cartilage around the nerve cord; the simplest true vertebrate. 484

lancelet One of three types of primitive invertebrate chordate. 480

large intestine Part of the digestive tract that extends from the small intestine to the rectum. 732

larva An immature stage of an animal that usually does not resemble the adult of the species. 454

larynx The "voice box" and a conduit for air. 708

latency A stage in some viral reproduction in which the viral genome is hidden within the host chromosome. 368

latent learning Learning without reward or punishment; not apparent until after the learning experience. 810

lateral line system A network of canals that extends along the sides of fishes and houses receptor organs that detect vibrations. 485

lateral meristem *LAT-er-al MER-ee-stem* A meristem that gives rise to secondary plant tissue. 505

law of segregation Allele separation during meiosis. 179

laws of thermodynamics Physical laws that describe energy and energy use. 85

leader sequence A short sequence at the start of each mRNA that enables the mRNA to hydrogen bond with part of the rRNA in a ribosome. 250

leaf The primary photosynthetic organ of most plants. 510

leaf abscission *LEAF ab-SCISZ-on* Shedding of a tree's leaves as part of its life cycle. 514

lens The structure in the eye through which light passes and is focused. 629

leukocyte White blood cell. 692

lichen *LI-ken* An association of a fungus and an alga or cyanobacterium. 442

life expectancy Prediction of how long an individual will live, based on current age and epidemiology. 800

life span The longest a member of a species can live. 800

life table Data summarizing the probability of reproducing and dying at a given age. 831

ligament Tough band of fibrous connective tissue that connects bone to bone across a joint. 675

ligand *LIG-and* A messenger molecule that binds to a cell surface protein. 68

ligase *LIG-aze* An enzyme that catalyzes formation of covalent bonds in the DNA sugar-phosphate backbone. 232

light chain Small polypeptide chain in an antibody subunit. 763

light reactions Also known as the Hill reactions, the reactions in photosynthesis that harvest light energy and store that energy in molecules of ATP or NADPH. 104

lignin *LIG-nin* A complex polymer in plant cell walls that adds support. 415, 433

limnetic zone *lim-NET-ik ZONE* The layer of open water in a lake or pond where light penetrates. 878

linkage map Diagram of gene order on a chromosome based on crossover frequencies. 205

linked genes Genes on the same chromosome. 203

lipase *LI-payse* Enzyme that chemically digests fats. 730

lipid *LIP-id* Compound that contains carbon, hydrogen, and oxygen, but with less oxygen than carbohydrates have. 31

littoral zone *LIT-or-al ZONE* The shallow region along the shore of a lake or pond where sufficient light reaches to the bottom for photosynthesis. 878

liver The organ that detoxifies blood, stores glycogen and fat-soluble vitamins, synthesizes blood proteins, and monitors blood glucose level. 733

liverwort *LIV-er-wart* A type of bryophyte. 418

lobe-finned fish A type of fish with limblike fins. 486

logistic growth Population growth that levels off as the carrying capacity is approached. 834

long-day plant Plant that requires dark periods shorter than some critical length to flower. 566

long-term memory A memory that can persist for years as a result of enhanced synaptic transmission and/or altered synaptic connection patterns. 614

loose connective tissue Connective tissue with widely spaced fibroblasts and a few fat cells. 579

low-density lipoprotein *LO DEN-sit-ee LIP-o-PRO-teen* A molecule that carries cholesterol. 34

lumbar vertebra Vertebra of the small of the back. 663

lungfish A type of fish with air bladders adapted as lungs. 486

lungs Paired structures that house the bronchial tree and the alveoli; the sites of gas exchange in some vertebrates. 704

luteinizing hormone (**LH**) A hormone made in the anterior pituitary that promotes ovulation. 648

lymph Blood plasma minus some large proteins, which flows through lymph capillaries and lymph vessels. 696

lymphatic system A circulatory system that consists of lymph capillaries and lymph vessels that transport lymph. 696

lymph capillary *LIMF CAP-eh-lair-ee* Dead-end, microscopic vessel that transports lymph. 697

lymph node Structure in the lymphatic system that contains white blood cells and fights infection. 698

lymph vessel *LIMF VES-sel* Vessel that transports lymph and eventually empties into a vein. 697

lysis In viral replication, the release of virus particles by bursting the host cell. 367

lysogenic A bacteriophage infection in which the phage DNA is incorporated in the host chromosome. 367

lysosome *LI-so-soam* A sac in a eukaryotic cell in which molecules and worn-out organelles are enzymatically dismantled. 52

M

macroevolution *MAK-ro-ev-eh-LU-shun* Large-scale evolutionary changes, such as speciation and extinction. 264

macromolecule *MAK-ro-MOL-e-kuel* A very large molecule. 22

macronucleus *MAK-ro-NEW-klee-us* A large diploid nucleus in a protistan cell. 404

macronutrients *MAK-ro-NU-tree-entz* Carbohydrates, fats, and proteins obtained from food in large amounts. 733

macrophage A phagocyte that destroys bacteria and cell debris and presents antigen to T cells. 758

major histocompatibility complex (**MHC**) A cluster of genes that code for cell surface proteins. 760

malignant tumor A cancerous tumor. 151

malleus *MAL-e-us* A small bone in the middle ear. 633

Malpighian tubule An insect excretory structure. 468, 745

mammal A tetrapod vertebrate with hair, mother's milk, endothermy, four-chambered heart, and a muscular diaphragm. 495

mammary gland Milk-producing sweat gland derivative in mammals. 495

mangrove swamp *MAN-growv SWAMP* Tropical wetland dominated by salt-tolerant trees. 880

mantle A dorsal fold of tissue that secretes a shell in mollusks. 465

marrow cavity Space in a bone shaft that contains yellow marrow. 668
marsupial A pouched mammal. 495
marsupium A pouch in which the immature young of marsupial mammals nurse and develop. 495
mass extinctions The abrupt loss of many species over a wide area. 308
mast cell Immune system cell that releases allergy mediators when stimulated. 759
matrix *MAY-trix* The inner compartment of a mitochondrion. The nonliving part of connective tissue. 120
maturation The process of development toward an adult, or final, stage. 799
mechanoreceptor Receptor cell sensitive to mechanical energy. 624
medulla oblongata The part of the brain stem nearest the spinal cord; regulates breathing, heartbeat, blood pressure, and reflexes. 612
medusa The free-swimming form of a cnidarian. 459
megagametophyte *MEG-ah-gah-MEE-toe-fight* The female gametophyte in a plant. 546
megakaryocyte *MEG-ah-KAR-ee-o-site* A huge bone marrow cell that breaks apart to yield platelets. 693
megaspore *MEG-ah-spor* Structure in plants that develops into the female gametophyte. 546
meiosis *mi-O-sis* Cell division that halves the genetic material. 159
melanocyte Cell that produces melanin pigment. 588
melanocyte-stimulating hormone (MSH) A pituitary hormone that controls skin pigmentation in some vertebrates. 651
melatonin A hormone produced in the pineal gland that may control other hormones by sensing light and dark cycles. 654
memory B cell Mature B cell, specific to an antigen already met, that responds quickly by secreting antibodies when that antigen is encountered again. 760
meninge One of three membranes that cover and protect the CNS. 617
meristem Undifferentiated plant tissue which gives rise to new cells. 505
mesoderm The middle embryonic germ layer, whose cells become bone, muscle, blood, dermis, and reproductive organs. 453
mesoglia *mez-o-GLEE-a* A jellylike substance in the middle of a cnidarian body. 458
mesohyl *MEZ-o-hile* A part of a sponge that contains amoebocytes. 456
mesophyll cell *MEZ-o-fill SEL* Thin-walled plant cell that takes part in photosynthesis. 110, 514
messenger RNA *MESS-en-ger RNA* (mRNA) A molecule of ribonucleic acid that is complementary in sequence to the sense strand of a gene. 243
metabolic pathway *met-a-BOL-ik PATH-way* A series of connected, enzymatically catalyzed reactions in a cell. 86
metabolism *meh-TAB-o-liz-um* The biochemical reactions that acquire and use energy. 4
metacentric *met-a-SEN-trik* A chromosome whose centromere divides it into two similarly sized arms. 202
metamerism *MET-a-mer-izm* A body consisting of repeating segments. 453
metamorphosis A developmental process in which an animal changes drastically in body form between the juvenile and the adult. 454
metaphase *MET-ah-faze* The second stage of cell division, when chromosomes align down the center of a cell. 142
metaphase I *MET-ah-faze I* In meiosis I, when homologs align down center of cell. 162
metaphase II *MET-a-faze II* When replicated chromosomes align down the center of the cell in meiosis II. 163
metapopulation *met-a-pop-eu-LAY-shun* Local populations that frequently exchange members. 828
metastasis *meh-TAH-stah-sis* Spreading of cancer. 136
methanogen *meth-AN-o-gen* Organism that produces methane as a by-product of its metabolism. 391
microevolution *MIKE-ro-ev-eh-LU-shun* Subtle, incremental single-trait changes that underlie speciation. 264
microfilament *MI-kro-FILL-ah-ment* Tiny actin rod in cells, especially contractile cells. 70
microgametophyte *MIK-ro-gah-MEE-toe-fight* The male gametophyte in a plant. 546
micronucleus *mi-kro-NUKE-lee-us* A small, haploid nucleus in a protistan cell. 404
micronutrients *MIKE-ro-NU-tree-entz* Vitamins and minerals, which are required in small amounts. 733
microspore *MIKE-ro-spor* Structure in plants that develops into the male gametophyte. 546
microtubule *MIKE-ro-TU-bule* Long, hollow tubule of tubulin protein that moves cells. 68
microvillus (pl. microvilli) Tiny projection on the surface of epithelial cell that comprises intestinal villi. 732
midbrain Part of the brain between the forebrain and hindbrain; important in vision and hearing. 610
middle ear Chamber that lies between the tympanic membrane and the inner ear; contains the maleus, incus, and stapes. 633
migration A regularly repeated journey from one specific geographic region to another. 810
mimicry Similar appearances of different species. 847
mineralocorticoid Adrenal hormone that helps maintain blood volume and electrolyte balance. 653
mismatch repair Any DNA repair system that recognizes and corrects a defect comprising the wrong nucleotide as part of a base pair. 235
missense mutation *MISS-sentz mu-TAY-shun* A mutation that changes a codon specifying a certain amino acid into a codon specifying a different amino acid. 254
mitochondrion *MI-toe-KON-dree-on* Organelle within which the reactions of cellular metabolism occur. 54
mitosis *mi-TOE-sis* A form of cell division in which two genetically identical cells form from one. 138
mitotic spindle *mi-TOT-ik SPIN-del* A structure of microtubules that aligns and separates chromosomes in mitosis. 140
mode of inheritance Whether a trait is autosomal or sex-linked, and dominant or recessive. 183
molecular clock Using differences in molecules as markers for the amount of time that has passed since two species were the same species. 333
molecular evolution Estimating degrees of evolutionary relatedness by comparing informational polymer sequences among species. 332
molecular tree diagram A depiction of hypothesized relationships between species based on molecular sequence data. 317
molecular weight The sum of the weights of the atoms that make up a molecule. 22
molecule *MOL-eh-kuel* Combined atoms. 21
mollusk An unsegmented protostome with a mantle, muscular foot, and visceral mass. 465
molting hormone An insect hormone produced in the late larva that triggers molting. 645
Monera *mone-AIR-ah* The kingdom that includes prokaryotic organisms. 9
monoclonal antibodies *MON-o-KLON-al AN-tee-bod-eez* Identical antibodies provided by a single B cell. 772
monocotyledon (monocot) *MON-o-kot* An angiosperm that has one cotelydon, or seed leaf. 426
monoecious In plants, having male and female reproductive parts on the same individual. 544
monogamy Formation of a permanent male-female pair. 815
monohybrid cross *MON-o-HI-brid CROSS* Mating of individuals for a particular gene. 182
monomer *MON-o-mer* A single link in a polymeric molecule. 29
monosaccharide *MON-o-SAK-eh-ride* A sugar that is one five- or six-carbon unit. 30
monosomy *MON-o-SOAM-ee* Absence of one chromosome. 212
monotreme An egg-laying mammal. 495

monounsaturated (fat) *MON-o-un-SAT-yer-a-tid* A fatty acid with one double bond. 32

monozygotic twins *mon-o-zi-GOT-ik TWINZ* Identical twins resulting from the splitting of a fertilized ovum. 196

morphogen *MORF-o-gen* Protein that forms gradients that influence development. 784

morphogenesis The series of events during embryonic development that leads to the formation of distinct structures. 791

morula The preembryonic stage consisting of a solid ball of cells. 789

moss A type of bryophyte. 419

motif *moe-TEEF* A common part of a transcription factor. 241

motor neuron A neuron that transmits a message from the CNS toward a muscle or gland. 598

motor unit A neuron and all the muscle fibers it contacts. 674

mouth The beginning of gastrointestinal tract. 725

M phase The portion of the cell cycle where nuclear material is divided; mitosis or meiosis. 140

mucigel *MYUS-eh-gel* A slimy substance that root cap cells produce. 515

multicellular *mull-tee-SEL-u-lar* Consisting of many cells. 6

multifactorial *mull-tee-fac-TORE-e-al* Traits molded by one or more genes and the environment. 194

muscle fascicle *MUS-sel fah-SIK-ul* Bundle of muscle fibers. 671

muscle fiber Skeletal muscle cell. 670

muscle spindle Receptor in skeletal muscle fiber that monitors tension. 675

muscle tissue Tissue consisting of contractile cells that provide motion. 578

muscular foot A ventral organ in mollusks that is used in locomotion. 465

muscular system Organ system of skeletal muscles whose contractions form the basis of movement in vertebrates. 584

mutagen *MUTE-a-jen* An agent that causes a mutation. 253

mutant *MU-tent* A phenotype or allele that is not the most common for a certain gene in a population or that has been altered from the "normal" condition. 181

mutation *mu-TAY-shun* A change in a gene or chromosome. 252

mutualism *MU-chu-a-lism* Symbiosis that benefits both partners. 845

mycelium *mi-SELL-ee-um* An assemblage of hyphae that forms an individual fungus. 436

mycorrhiza An association of a fungus and the roots of a plant. 442, 529

myelin sheath A fatty material that insulates some nerve fibers in vertebrates, allowing rapid nerve impulse transmission. 601

myoepithelial cell *mi-o-ep-ee-THEEL-ee-al SEL* Cell that can contract and be part of a layer. 588

myofibril A cylindrical subunit of a muscle fiber. 670

myofilament Actin or myosin "string" that is part of a myofibril. 670

myosin The protein that forms thick filaments in muscle tissue. 581, 670

myriapod *mi-REE-a-pod* Centipedes and millipedes. 470

N

nastic movement Plant growth or movement that is not oriented toward the provoking stimulus. 563

natural selection The differential survival and reproduction of organisms whose genetic traits better adapt them to a particular environment. 8

navigation The following of a specific course. 810

necrosis *neck-RO-sis* A form of cell death. 150

negative feedback An action that counters an existing condition; important in homeostatic responses. 94, 585, 644

nematocyst A stinging structure contained within cnidocytes of cnidarians. 459

neonatology *NE-o-nah-TOL-eh-gee* Study of the newborn. 797

nephridium Network of tubules in some invertebrates that has an excretory function. 745

nephron A microscopic tubular subunit of a kidney, consisting of a renal tubule and peritubular capillaries. 746

nephron loop (loop of Henle) Part of a nephron that lies between the proximal and distal convoluted tubules, where water is conserved and urine concentrates by a countercurrent exchange system. 749

Neptunism *NEP-tune-iz-um* The idea that a single great flood organized the features of Earth's surface present today. 266

neritic zone *ner-IT-ik ZONE* The region of an ocean from the coast to the edge of the continental shelf. 882

nerve Bundle of axons. 597

nerve net Diffuse network of neurons, as in cnidarians. 606

nervous system Interconnected network of neurons and supportive neuroglia that transmits information rapidly throughout the body. 583

nervous tissue A tissue whose cells (neurons and neuroglia) form a communication network. 578

net primary production Energy available to consumers in a food chain, after cellular respiration by producers. 854

neural tube Embryonic precursor of the CNS. 791

neuroglia Cells associated with neurons in the nervous system. 581, 601

neuromodulator Peptide that alters a neuron's response to a neurotransmitter or blocks the release of a neurotransmitter. 602

neuromuscular junction A chemical synapse of a neuron onto a muscle cell. 673

neuron A nerve cell, consisting of a cell body, a long "sending" projection called an axon, and numerous "receiving" projections called dendrites. 581, 597

neurosecretory cell Cell that functions as a neuron at one end but as an endocrine cell at the other by receiving neural messages and secreting hormones. 645

neurotransmitter A chemical passed from a neuron to receptors on another neuron or on a muscle or gland cell. 602

neurulation Interaction between the notochord and nearby ectoderm that triggers formation of the nervous system. 791

neutron *NEW-tron* A particle in an atom's nucleus that is electrically neutral and has one mass unit. 20

niche All resources a species uses for survival, growth, and reproduction. 844

nitrifying bacterium *NI-tri-fy-ing bac-TEAR-e-um* Bacterium that converts ammonia into nitrites and nitrates. 857

nitrogen fixation A microbial process that reduces atmospheric nitrogen gas to ammonia. 380, 531, 857

nitrogenous base A nitrogen-containing compound that forms part of a nucleotide, giving it individuality. 37

node Area of leaf attachment on a stem. 510

node of Ranvier A short region of exposed axon between Schwann cells on neurons of the vertebrate peripheral nervous system. 601

nodule Swelling, inhabited by symbiotic nitrogen-fixing bacteria, on the roots of certain types of plants. 531

nondisjunction *NON-dis-JUNK-shun* Unequal partition of chromosomes into gametes during meiosis. 212

nonpolar covalent bond *non-POE-lar co-VAY-lent BOND* A covalent bond in which atoms share electrons equally. 24

nonsense mutation *NON-sents mu-TAY-shun* A point mutation that alters a codon that encodes an amino acid to one that encodes a stop codon. 254

nonseptate hypha A multicellular hypha lacking crosswalls, or septa. 436

nonshivering thermogenesis *non-SHIV-er-ing ther-mo-GEN-i-sis* Hormone-directed internal heat. 741

norepinephrine An adrenal hormone that raises blood pressure, constricts blood vessels, and slows digestion. 652

notochord A semirigid rod running down the length of a chordate's body. 478, 791

nuclear envelope *NEW-klee-ar EN-vel-ope* A two-layered structure bounding a cell's nucleus. 53

nuclear pore *NEW-klee-ar POOR* A hole in the nuclear envelope. 53

nucleic acid *new-CLAY-ic AS-id* A biochemical that encodes an amino acid sequence. 37

nucleoid *NEW-klee-oid* The part of a prokaryotic cell where the DNA is located. 48, 382

nucleolus *new-KLEE-o-lis* A structure within the nucleus where RNA nucleotides are stored. 53

nucleosome *NEW-klee-o-some* DNA wrapped around eight histone proteins as part of chromosome structure. 225

nucleotide *NEW-klee-o-tide* The building block of a nucleic acid, consisting of a phosphate group, a nitrogenous base, and a five-carbon sugar. 37

nucleus *NEW-klee-is* The central region of an atom, consisting of protons and neutrons. A membrane-bounded sac in a eukaryotic cell that contains the genetic material. 20

nutrient Any element that is required for metabolism, growth, and reproduction. 526, 733

O

obligate aerobe An organism that requires oxygen to live. 385

obligate anaerobe An organism that must live in the absence of oxygen. 385

obligate photoperiodism *OB-lah-get fo-toe-PER-ee-o-diz-um* A plant's requirement of a particular photoperiod in order to flower. 566

oceanic zone *O-shee-AN-ik ZONE* The open sea beyond the continental shelf. 882

ocellus *o-SELL-us* Eyespot; visual organ in flatworms. 628

octet rule *OC-tet ROOL* The tendency of an atom to fill its outermost shell. 23

olfaction The sense of smell. 626

olfactory epithelium *ol-FAK-tore-ee ep-e-THEEL-e-um* Neurons specialized to detect odors, in a small patch of tissue high in the nostrils. 626

oligochaete *OL-eh-go-keet* A class of annelids that includes the earthworms. 467

oligodendrocyte Cell that produces myelinated neurons in the brain. 601

oligotrophic *OL-ah-go-TRO-fik* A lake with few nutrients, usually very blue. 878

ommatidium The visual unit of a compound eye. 628

omnivore An animal that feeds on plants and animals. 720

oncogene *ON-ko-jean* A gene that normally controls cell division but when overexpressed leads to cancer. 152

oocyte *OO-site* The female sex cell before it is fertilized. 169

oogenesis *oo-GEN-eh-sis* The differentiation of an egg cell from a diploid oogonium, to a primary oocyte, to two haploid secondary oocytes, to ootids, and finally, after fertilization, to a mature ovum. 169

oogonium *oo-GO-nee-um* The diploid cell where egg formation begins. 169

oomycete Filamentous, heterotrophic protista, also called water molds. 407

open circulatory system A circulatory system in which hemolymph circulates freely through the body cavity. 468, 682

operant conditioning Trial-and-error learning, in which an animal voluntarily repeats any behavior that brings success. 809

operculum *o-PER-cu-lum* A flap of tissue that protects gills. 704

operon *OP-er-on* A series of genes with related functions and their controls. 241

optic nerve Nerve fibers that connect the retina to the brain; formed of ganglion cell axons. 630

oral end The mouth. 453

organ A structure of two or more tissues that functions as an integrated unit. 4, 578

organelle *OR-gan-NELL* Specialized structure in eukaryotic cells that carries out specific functions. 42

organogenesis *or-GAN-o-gen-eh-sis* Development of organs in an embryo. 791

organ system System of physically or functionally linked organs. 4, 578

orgasm A pleasurable sensation associated with sexual activity. 786

orientation Movement in a specific direction. 810

origin The end of a muscle on an immobile bone. 676

osculum *OS-kew-lum* A large hole at the top of a sponge that expels water and wastes. 457

osmoconformer An organism whose ion concentrations match those of its surroundings. 743

osmoregulator *OZ-mo-reg-u-LAY-tor* An organism that actively controls its ion concentrations in a changing environment. 743

osmosis *oz-MO-sis* Passive diffusion of water through a semipermeable membrane. 64

ossicle *OSS-i-kull* Calcium-containing structure in echinoderms' coverings. 470

osteoblast A bone cell that secretes matrix. 579

osteoclast A bone cell that degrades matrix. 579

osteocyte A mature bone cell in a lacuna. 579

osteon Concentric circles of osteocytes in bone. 665

osteonic canal *oss-tee-ON-ik ca-NAL* Portal that houses blood vessels in bone. 665

osteoprogenitor cell *OSS-tee-o-pro-GEN-i-tor SEL* A cell lining a bone passageway that can differentiate into an osteoblast in the event of injury or growth. 579

ostium *OSS-tee-um* Surface pore on a sponge. 457

ostracoderm *os-TRAK-o-derm* A type of extinct, jawless, bottom-dwelling filter-feeding fish. 484

otolith *O-toe-lith* Calcium carbonate granules in the vestibule of the inner ear whose movements provide information on changes in velocity. 636

oval window A membrane between the middle ear and the inner ear. 634

ovary One of the paired female gonads that house developing oocytes. Also, in a flowering plant, the enlarged basal portion of a carpel. 170, 544, 654, 786

oviparous Egg-laying (in animals). 482

ovoviviparous Retaining eggs inside the birth canal and giving birth to live offspring (in animals). 482

ovulation The release of an oocyte from an ovarian follicle. 787

ovule *OV-yul* In seed plants, a structure that contains a megagametophyte and egg cell. 544

oxidation *OX-e-DAY-shun* The loss of one or more electrons by a participant in a chemical reaction. 82

oxidative phosphorylation *ox-i-DAY-tiv fos-for-e-LAY-shun* Phosphorylation of ADP to ATP coupled to protons moving down their concentration gradient across a membrane. 126

oxyhemoglobin *ox-ee-HEEM-o-GLO-bin* Bright red hemoglobin that has just picked up oxygen in the lungs. 692

oxytocin A hypothalamic hormone released from the posterior pituitary that stimulates muscle contraction in the mammary glands and the uterus. 646, 651

ozone layer Atmospheric zone rich in ozone gas, which absorbs the sun's ultraviolet radiation. 891

P

pacemaker Specialized cells in the wall of the right atrium that set the pace of the heartbeat. 687

Pacinian corpuscle *pah-SIN-ee-en KOR-pus-el* A receptor in the skin that senses touch. 636

paedomorphosis When adults of a species have features present in larval stages of their ancestors. 489

pain receptor Specialized receptor cell that serves a protective function in detecting mechanical damage, temperature extremes, etc. 636

paleontology *PAY-lee-on-TOL-ah-gee* Study of evidence of past life. 316

pancreas An organ with an endocrine part that produces somatostatin, insulin, and glucagon and a digestive part that produces pancreatic juice. 733

Parabasalia Group of protista whose cells lack mitochondria. 396

parapatric evolution The formation of a new species at the boundary zone between two species. 301

parapodia *par-a-PODE-ee-a* Pairs of fleshy appendages that polychaete worms use to locomote. 467

parasitism *par-a-si-tizm* A symbiotic relationship in which one member derives nutrients or resources at the expense of the other. 845

parasitoid A symbiotic organism that lives within its host and kills it. 845

parasympathetic nervous system Part of the autonomic nervous system that controls vital functions such as respiration and heart rate. 607

parathyroid gland One of four small groups of cells behind the thyroid gland that secretes parathyroid hormone. 650

parathyroid hormone (PTH) A hormone involved in maintaining calcium levels in the blood. 652

parenchyma *pah-REN-kah-mah* Plant tissue composed of living, thin-walled cells of variable function. 506

parental generation (P) The first generation in a genetic cross. 204

parental investment The amount of time and resources a parent spends on producing and raising its offspring. 815

parsimony analysis *PAR-si-mone-ee eh-NAL-eh-sis* A statistical method to identify the most realistic evolutionary tree based on molecular data. 319

parthenogenesis Female reproduction without fertilization. 158

passive immunity Immunity generated when an organism receives antibodies from another organism. 767

passive transport Movement of substances across a membrane down a concentration gradient. 66

pectoral girdle Collarbones plus shoulder blades in the vertebrate skeleton. 663

pedicellariae *ped-ee-SELL-ar-ay* Echinoderm pincers. 470

pedigree *PED-eh-gree* A chart showing relationships of relatives and which ones have a particular trait. 183

pelagic zone *pah-LA-gik ZONE* Water above the ocean floor. 882

pelvic girdle Bones that support a vertebrate's hind limbs. 663

penis Male organ of copulation and in mammals, urinary excretion. 785

pentaradial symmetry *pent-a-RAY-de-al SIM-it-ree* A body form with five arms, characteristic of echinoderms. 470

pepsin A stomach enzyme that chemically digests protein. 728

peptide bond *PEP-tide BOND* A chemical bond between two amino acids resulting from dehydration synthesis. 35

peptide hormone A water-soluble, amino acid–based hormone that cannot freely diffuse through a cell membrane. 643

peptidoglycan A complex, cross-linked polysaccharide in a bacterial cell wall. 382

per capita rate of increase The difference between the birthrate and death rate in a population 832

perception An animal's interpretation of a sensation. 622

pericardium Connective tissue sac that houses the heart. 685

pericycle *PEAR-ee-si-kel* A ring of cells in a root's cortex that produces branch roots. 517

periderm *PEAR-ee-derm* Outer covering of woody stems and roots. Includes cork, cork cambium, and phelloderm. 520

period Time period within an era. 322

periodic table *peer-ee-OD-ic TA-ble* Chart that lists naturally occurring elements according to their properties. 20

peripheral nervous system Neurons that transmit information to and from the CNS. 602

peristalsis Waves of muscle contraction that propel food along the digestive tract. 725

peritubular capillaries Capillaries that surround renal tubules in kidney nephrons. 747

permafrost *PER-mah-frost* Permanently frozen part of the ground in the tundra. 877

peroxisome *per-OX-eh-soam* A membrane-bounded sac that buds from the smooth ER and that houses enzymes important in oxygen use. 56

petal A flower part often colored to lure pollinators. 543

petiole *PET-ee-ol* The stalklike part of a leaf. 511

petrification The preservation of tissue structure by replacing biomolecules with minerals. 324

phagocytes White blood cells that engulf and digest foreign material and cell debris. 756

phagocytic vacuole *FAG-o-sit-ik VAK-u-ole* A space that forms within a protistan cell when it engulfs food particles. 404

pharyngeal gill slit *farr-in-GEE-al GILL SLIT* Chordate structure that functions in feeding or respiration. 479

pharyngeal pouch Where endoderm and ectoderm grow toward each other in the throat region of a chordate embryo. 479

pharynx A muscular tube that connects the mouth and esophagus. 479, 708, 725

phenocopy *FEEN-o-kop-ee* An environmentally caused trait that resembles an inherited trait. 192

phenotype *FEEN-o-type* Observable expression of a genotype. 181

phenotypic ratio *feen-o-TIP-ik RAY-shee-o* Proportions of different phenotypes among offspring of a genetic cross. 182

pheromone Biochemical an organism secretes that elicits a response in another member of the species. 623, 642

phloem *FLOW-m* Plant tissue that transports photosynthetic products. 415, 508

phloem sap Solution of water, simple sugars, and other dissolved substances in phloem. 535

phospholipid *FOS-fo-LIP-id* A molecule consisting of a lipid and a phosphate that is hydrophobic at one end and hydrophilic at the other end. 62

phosphorylation The addition of a phosphate to a molecule. 91

photochemical smog *fo-to-KEM-i-kal SMOG* Air pollution resulting from chemical reactions among pollutants in sunlight. 889

photon *FOE-ton* A packet of light energy. 82

photonasty *FO-toe-nas-tee* A nastic response in plants to light and dark. 565

photoperiodism *fo-toe-PER-ee-o-diz-um* A plant's ability to measure seasonal changes by the length of day and night. 566

photophosphorylation *FO-toe-FOS-for-eh-LAY-shun* A photosynthetic reaction in which energy released by the electron transport chain linking the two photosystems is stored in the high-energy phosphate bonds of ATP. 106

photoreactivation *fo-to-re-ac-ti-VAY-shun* A type of DNA repair in which an enzyme uses light energy to break pyrimidine dimers. 233

photoreceptor Receptor cell sensitive to light energy. 628

photorespiration *fo-to-res-per-A-shun* A process that reverses photosynthesis. 110

photosynthate The product of photosynthesis. 98

photosynthesis *FO-toe-SIN-the-sis* The series of biochemical reactions that enable plants to harness sunlight energy to manufacture organic molecules. 82

photosystem *FO-toe-SIS-tum* A cluster of pigment molecules that enables photosynthetic organisms to harness solar energy. 103

phototroph *FO-toe-trofe* An organism that derives energy from the sun. 385

phototropism *fo-toe-TROP-iz-um* A plant's growth towards unidirectional light. 562

pH scale A measurement of how acidic or basic a solution is. 27

phylogeny *fi-LODJ-ah-nee* Depiction of evolutionary relationships among species. 298

phylum *FI-lum* A major taxonomic group, just beneath kingdoms. 8

phytochrome *FI-toe-krome* A pale blue plant pigment involved in timing flowering, seed germination, and other processes. 568

phytoplankton *FITE-o-PLANK-tun* Microscopic photosynthetic organisms that drift in water. 854

pilus (pl. pili) Short projection on bacterial cells used to attach to objects or other cells. 382

pinacocyte *pine-AK-o-site* Cell in the body wall of a sponge. 456

pineal gland A small structure in the brain that produces melatonin in response to light and dark periods. 654

pinocytosis *pine-o-si-TOE-sis* A protistan cell pinches inward and brings fluid into the cell. 68

pioneer species The first species to colonize an area devoid of life. 848

pistil *PIS-til* The female reproductive structure in a flower. 543

pith *PITH* Ground tissue in the center of a stem or root, within the vascular cylinder. 510

pituitary gland A pea-sized gland attached to the vertebrate brain that releases several types of hormones. 647

placebo *pla-SEE-bo* An inert substance used as an experimental control. 14

placenta A structure that connects the developing fetus to the maternal circulation in many mammals. 495

placental mammal A mammal in which the developing fetus is nourished by a placenta. 495

placoderm An extinct line of giant fishes with jaws, paired fins, and notochord with some bone. 484

plankton *PLANK-tun* Microscopic organisms that drift in water. 878

Plantae *PLAN-tay* The plant kingdom. 9

plaque With respect to viral infections, a zone of killed cells in a layer of host cells. 362

plasma A watery, protein-rich fluid that forms the matrix of blood. 579, 692

plasma cells Mature B cells that secrete large quantities of a single antibody type. 760

plasmid Small, circular DNA apart from an organism's chromosome. 10, 382

plasmodesma Connection between plant cells that allows cytoplasm to flow between them. 76

plasmodium *plaz-MO-dee-um* A multinucleated mass of an acellular slime mold. 398

plastid *PLAS-tid* A plant organelle that encapsulates photosynthetic membranes. 102

platelet A cell fragment that is part of the blood and orchestrates clotting. 579, 693

plate tectonics The movement of landmasses resting on plates floating on molten rock. 301

Platyhelminthes *plat-ee-hel-MEN-theez* Flatworms. 461

pleiotropic *PLY-o-TRO-pik* A genotype with multiple expressions. 190

pluripotency *plur-e-POE-ten-see* A cell that retains the potential to specialize in any way. 694

pneumatophore *new-MAT-o-for* The floating portion of a cnidarian. 458

point mutation *POYNT mu-TAY-shun* A change in a single DNA base. 254

polar body A small cell generated during female meiosis, enabling cytoplasm to be partitioned into just one of the four meiotic products, the ovum. 169

polar covalent bond *PO-lar co-VAY-lent BOND* A covalent bond in which electrons are attracted more toward one atom's nucleus than to the other. 24

polar nuclei *PO-lar NU-klee-i* The two nuclei in a cell of a plant's megagametophyte that fuse with sperm nucleus to form the triploid endosperm nucleus. 547

pollen grain Immature male gametophyte in seed plants. 441, 546

pollen sac One of four cavities in an anther, in which pollen grains are produced. 546

pollen tube A tube, formed upon germination of a pollen grain, that carries sperm to the ovule. 426

pollination Transfer of pollen from an anther to a receptive stigma. 547

pollution Any change in air, land, or water that adversely affects organisms. 889

polyandry *pol-ee-AN-dree* A mating system in which one female mates with several males. 815

polychaete *POL-ee-kete* Marine segmented worm. 467

polygamy A mating system in which a member of one sex associates with several members of the opposite sex. 815

polygenic *pol-ee-JEAN-ik* A trait caused by more than one gene. 194

polygynandry *POL-ee-gine-AN-dree* Communal mating displays, involving several males and several females. 815

polygyny *pol-IJ-ah-nee* A mating system in which one male mates with several females. 815

polymer *POL-eh-mer* A long molecule composed of similar subunits. 29

polyp The sessile form of a cnidarian. 459

polypeptide chain *pol-ee-PEP-tide CHAYN* A long polymer of amino acids. 33

polyploidy *POL-ee-PLOID-ee* A cell with extra chromosome sets. 211

polyunsaturated (fat) *POL-ee-un-SAT-yer-a-tid FAT* A fatty acid with more than one double bond. 32

pons An oval mass in the brain stem where white matter connects the medulla to higher brain structures and gray matter helps control respiration. 612

population A group of interbreeding organisms living in the same area. 6, 828

population bottleneck A type of genetic drift. An event kills many members of a population, and a small number of individuals restore its numbers, restricting the gene pool. 284

population density The number of individuals of a species per unit area or volume of habitat. 828

population dispersion The pattern in which individuals are scattered throughout their habitat. 828

Porifera *pore-IF-er-a* Sponges. 456

positive feedback Mechanism by which the products of a process further activate the process. 94, 585, 645

postanal tail One of the four characteristics of chordates in which the notochord plus muscles extend posteriorly beyond the anus. 479

posterior pituitary The back part of the pituitary gland; secretes ADH and oxytocin. 651

postsynaptic cell *post-sin-AP-tik SEL* One of two adjacent neurons that receives a message. 603

postzygotic reproduction isolation The separation of species due to nonviability of an embryo or offspring. 299

potential energy The energy stored in the position of matter. 84

prebiotic simulation Experiment that attempts to recreate the conditions of an early Earth that gave rise to a living cell. 339

precocial Describes babies that are born capable of independent behavior. 492

predator An organism that kills another for food. 846

preembryonic stage Prenatal development before the organism folds into layers. 788

preformation The idea that a gamete or fertilized ovum contains an entire preformed organism. 778

pressure flow theory Theory that explains how phloem sap moves from photosynthetic tissues to nonphotosynthetic plant parts. 535

presynaptic neuron Neuron that releases neurotransmitters into a synaptic cleft. 603

prey An organism killed by another for food. 846

prezygotic reproductive isolation The separation of species due to factors that prevent the formation of a zygote. 299

primary growth Lengthening of a plant due to cell division in the apical meristems. 505

primary immune response The immune system's response to its first encounter with a foreign antigen. 763

primary nutrient deficiency Too little of a particular nutrient due to inadequate diet. 735

primary oocyte *PRI-mare-ee OO-site* An intermediate in ovum formation. 169

primary producer Species forming the base of a food web or the first link in a food chain. 851

primary spermatocyte *PRI-mare-ee spur-MAT-o-site* An intermediate in sperm formation. 167

primary (1°) structure The amino acid sequence of a protein. 35

primary succession Appearance of life in an area previously devoid of life. 848
primitive streak The pigmented band along the back of an embryo that develops into the notochord. 791
principle of competitive exclusion The idea that two or more species cannot indefinitely occupy the same niche. 845
principle of superposition The idea that lower rock layers are older than those above them. 266
prion *PRI-on* Infectious protein particle. 373
producer Also known as an autotroph, organism that produces its own nutrients from inorganic carbon sources. 6
product The result of a chemical reaction. 87
product rule The chance of two events occurring equals the product of the chances of either event occurring. 188
profundal zone *pro-FUN-dal ZONE* The deep region of a lake or pond where light does not penetrate. 878
progenote *pro-JEAN-note* Collections of nucleic acid and protein that was a forerunner to cells. 343
progesterone A steroid hormone produced by the ovaries that controls secretion patterns of other reproductive hormones. 655
proglottid *pro-GLOT-id* Structure in tapeworms that contains reproductive structures. 463
prokaryotic cell *pro-CARE-ee-OT-ik SEL* A cell that lacks organelles. Bacteria and Archaea. 45
prolactin A pituitary hormone that stimulates milk production. 648
promoter *pro-MOW-ter* A control sequence near the start of a gene that attracts RNA polymerase and transcription factors. 241
pronucleus *pro-NU-kle-us* The genetic material of a gamete when it has become part of a fertilized ovum. 789
prophase *PRO-faze* The first stage of cell division, when chromosomes condense and become visible. 140
prophase I *PRO-faze I* Prophase of meiosis I, when synapsis and crossing over occur. 162
prophase II *PRO-faze II* Prophase of the second meiotic division. 163
prostaglandin Lipid released locally and transiently at the site of a cellular disturbance. 656
prostate gland A small gland that produces a milky, alkaline fluid that activates sperm. 785
protein *PRO-teen* A polymer of amino acids. 33
Protista *pro-TEES-ta* The kingdom that includes mostly unicellular, eukaryotic organisms. 396
proton *PRO-ton* A particle in an atom's nucleus carrying a positive charge and having one mass unit. 20
protonephridium (pl. protonephridia) *pro-toe-nef-RID-ee-um* Structure in flatworm that helps maintain internal water balance. 463
proton gradient A difference in concentration of hydrogen ions across a membrane.106
protoplasm *PRO-tow-plaz-m* Living matter. 570
protostome A coelomate lineage with spiral cleavage, a blastopore that develops into a mouth, determinate cleavage, and a coelom that forms from splits in blocks of mesoderm. 455
proximal convoluted tubule Region of the nephron, adjacent to the glomerular capsule, where selective reabsorption of useful components of the glomerular filtrate occurs. 748
pseudocoelom A fluid-filled animal body cavity lined by endoderm and mesoderm. 464
pseudocoelomate An animal with a pseudocoelom. 455
pseudoplasmodium *su-doe-plaz-MO-dee-um* The multicellular, mobile stage of the life cycle of a cellular slime mold. 398
pseudopod *SU-doe-pod* The portion of an amoeba that extends outward, moving the organism. 398
pseudostratified epithelium *su-doe-STRAT-if-eyed ep-eh-THEEL-e-um* A single layer of epithelium whose nuclei are at different levels, giving the appearance of more than one layer. 579
pulmonary artery The artery that leads from the right ventricle to the lungs. 686
pulmonary circulation Blood circulation through the heart and the lungs. 683
pulmonary semilunar valve *PULL-mo-nair-ee SEM-ee-LOON-er VALVE* The valve that leads from the right ventricle to the pulmonary artery. 686
pulmonary vein Vein that leads from the lungs to the left atrium. 686
pulp The soft inner part of a tooth, consisting of connective tissue, blood vessels, and nerves. 725
punctuated equilibrium *PUNK-chew-ate-ed ee-kwa-LIB-ree-um* The view that life's history has had periods of little change interrupted by bursts of rapid change. 305
Punnett square A device used to diagram the various possible genetic results of combining gametes. 182
pupil The opening in the iris that admits light into the eye. 628
purine *PURE-een* A type of organic molecule with a double ring structure, including the nitrogenous bases adenine and guanine. 224
Purkinje fiber *per-KIN-gee FI-ber* Muscle fiber that branches from the atrioventricular node and transmits electrical stimulation rapidly in the heart. 687
pyloric sphincter Circular muscle at the stomach's exit. 729
pyramid of biomass A diagram depicting dry weight of organisms at each trophic level. 854
pyramid of energy A diagram depicting energy stored at each trophic level at a given time. 854
pyramid of numbers A diagram depicting number of organisms at each trophic level. 854
pyrenoid *PIE-re-noyd* A starch-containing chloroplast in a green alga. 409
pyrimidine *pie-RIM-eh-deen* A type of organic molecule with a single ring structure, including the nitrogenous bases cytosine, thymine, and uracil. 224
pyruvic acid *pi-ROO-vic AS-id* The products of glycolysis. 119

Q

quaternary (4°) structure *QUAT-eh-nair-ee STRUK-sure* The organization of polypeptide chains of a protein. 36

R

radial cleavage The pattern of directly aligned blastomeres in the early deuterostome embryo. 455
radial symmetry An animal body form in which any plane passing from one end to the other divides the body into mirror images. 453
radicle *RAD-eh-kil* The first root to emerge from a seed. 552
radiole *RAY-de-ole* Ciliated structure on the anterior end of polychaete worms that filters small organisms and organic particles from seawater. 467
radiometric dating *RAD-ee-o-MET-rik DAY-ting* Using measurements of natural radioactive decay as a clock to date fossils. 324
radula *RAD-yew-la* A chitinous, tonguelike structure that mollusks use to eat. 465
ray-finned fish Group of bony fishes with fins supported by parallel bony rays connected by webs of thin tissue. 486
reabsorption *re-ab-SORP-shun* The kidney's return of useful substances to the blood. 745
reactant *re-AK-tant* A starting material in a chemical reaction. 22
reaction center Clusters of chlorophyll and proteins that receive photon energy in photosynthesis. 104
reaction chain A sequence of releasers that joins several behaviors. 808
reading frame The DNA nucleotide corresponding to the first codon position in mRNA. 255

realized niche The resources in a species' environment that it can actually use, considering competition and other limitations. 844

receptacle The area where a flower attaches to a floral stalk. 551

receptor-mediated endocytosis *re-CEP-ter ME-dee-a-ted en-do-ci-TOE-sis* Binding of a ligand by a cell surface protein that draws the ligand into the cell in a vesicle. 68

receptor potential A change in membrane potential in a neuron specialized as a sensory receptor. 623

recessive *re-SESS-ive* An allele whose expression is masked by the activity of another allele. 178

reciprocal translocation *re-SIP-ro-kal tranz-lo-CAY-shun* Two nonhomologous chromosomes exchanging parts. 214

recombinant Containing genetic material from two of more sources. 204

rectum A storage region leading from the large intestine to the anus. 732

red alga Photosynthetic protist with unique pigments. 408

red blood cell A disc-shaped cell, lacking a nucleus, that contains hemoglobin. 579

red marrow Immature blood cells and platelets in cavities in spongy bone. 665

reduction The gain of one or more electrons by a participant in a chemical reaction. 89

reduction division Meiosis I, when the diploid chromosome number is halved. 161

reflex arc A neural pathway that links a sensory receptor and an effector. 610

regulatory enzyme An enzyme that controls the activity of a biochemical pathway because it can be activated or deactivated by binding a compound other than its substrate under certain conditions. 94

relative dating Determining the age of a fossil by comparisons to known ages of adjacent fossils in rock strata. 324

releaser The specific factor that triggers a fixed action pattern. 807

releasing hormone A hormone produced by the hypothalamus that stimulates another gland. 648

renal cortex The outer portion of a kidney. 747

renal medulla The middle part of a kidney. 747

renal pelvis The inner part of a kidney, where urine collects. 747

renal tubule The tubule part of a nephron that contains filtrate. 747

replication fork A locally unwound portion of DNA where replication occurs. 231

repressor A regulatory protein that inhibits transcription. 241

reproductive system A system of organs that produces and transports gametes and may nurture developing offspring. 584

reptile A tetrapod vertebrate with an amniote egg, ectothermy, and a tough scaly body covering; the first vertebrate to reproduce independent of water. 490

reservoir With respect to viral infections, an organism that can harbor a virus that infects a different species. 370

resource partitioning Specialization of resource use that allows species with similar niches to coexist. 845

respiratory chain A series of electron-accepting enzymes embedded in the inner mitochondrial membrane. 124

respiratory surface Site of external respiration. 702

respiratory system Organ system that acquires oxygen gas and releases carbon dioxide. 584

resting potential The electrical potential inside of a neuron not conducting a nerve impulse. 598

reticular activating system *rah-TIK-u-lar AK-tah-vay-ting SIS-tem* (RAS) A diffuse network of cell bodies and nerve tracts that extends through the brain stem and into the thalamus; screens sensory input to the cerebrum. 613

reticular formation A diffuse network of nerve tracts that extends through the brain stem and into the thalamus; screens sensory input to the cerebrum. 613

retina *RET-eh-na* A sheet of photoreceptors at the back of the human eye. 629

retinal Photo-sensitive portion of rhodopsin molecule. 631

reverse transcriptase *re-VERS tran-SCRIPT-aze* An enzyme that constructs a DNA molecule from an RNA molecule. 342

R group An amino acid side chain. 34

rhizoid *RI-zoyd* Rootlike extension on gametophytes of some nonvascular plants that anchors the plant and absorbs water and minerals. 421

rhizome *RI-zome* Fleshy, horizontal underground stem. 417

rhodopsin A light-sensitive pigment molecule stored in rod cells of the retina. 628

ribonucleic acid *RI-bo-nu-KLAY-ik AS-id* (RNA) A single-stranded nucleic acid consisting of nucleotides containing a phosphate, ribose, and nitrogenous bases adenine, guanine, cytosine, and uracil. 37

ribose *RI-bose* The five-carbon sugar that is a structural component of RNA. 223

ribosomal RNA *RI-bo-SOAM-el RNA*(rRNA) RNA that, along with proteins, forms a ribosome. 243

ribosome *RI-bo-soam* A structure built of RNA and protein upon which mRNA anchors during protein synthesis. 48

ribozymes *RI-bo-zimes* Small RNAs that function as enzymes. 342

ribulose bisphosphate The 5-carbon intermediate of the carbon reactions of photosynthesis. 109

RNA polymerase *RNA poe-LIM-er-ase* An enzyme that takes part in DNA replication and RNA transcription. 231

RNA primer *RNA PRI-mer* A small piece of RNA, inserted at the start of a piece of DNA to be replicated, which attracts RNA polymerase and is later removed. 231

RNA world The theory that the first genetic material was composed of RNA. 341

rod cell Specialized receptor cell in the retina that provides black-and-white vision. 630

root The underground part of a plant. 515

root cap A thimble-shaped mass of cells that protects a growing root tip. 515

root hair Outgrowth of root epidermal cell that increases the surface area for absorbing water and minerals. 508

rough endoplasmic reticulum The organelle where secreted proteins are synthesized. 52

roundworm An unsegmented worm with a pseudocoelom; phylum Nematoda. 464

r-selection Selection for individuals that are short-lived, reproduce early and have many offspring that each get little parental investment. 836

rubisco The enzyme that adds carbon to ribulose bisphosphate. 109

ruga (pl. rugae) Fold in the mucosa of the stomach. 728

rumen A bacteria-rich part of the stomach of large grazing animals. 724

ruminant An animal that has a rumen. 725

S

saccule *SAK-yul* A pouch in the vestibule of the inner ear, containing a jellylike fluid and lined with hair cells that contain calcium carbonate granules that move in response to changes in velocity, firing action potentials. 636

sacrum Fused pelvic vertebrae. 663

salivary amylase An enzyme produced in the mouth that begins chemical digestion of starch. 725

salt A molecule composed of cations and anions. 25

saltatory conduction Jumping of an action potential between nodes of Ranvier in myelinated axons. 601

sample size The number of individuals in an experiment. 14

saprotroph An organism that absorbs nutrients from dead organic matter. 435

sapwood *SAP-wood* Outermost wood that actively transports water and dissolved nutrients within a plant. 520

sarcolemma Cell membrane of a skeletal muscle cell. 670

sarcomere A pattern of repeated bands in skeletal muscle. 670

sarcoplasmic reticulum The endoplasmic reticulum of a skeletal muscle cell. 670

satellite The end of a chromosome attached to the rest of the chromosome by a bridge. 202

saturated (fat) A triglyceride with single bonds between the carbons of its fatty acid tails. 32

savanna One of several terrestrial biomes; a grassland with scattered trees. 872

scavenger Animal that eats the remains of other animals' meals. 851

scent gland Sweat gland derivative in different body regions in different species and produces fluids that evaporate to yield a distinctive odor. 592, 814

Schwann cell A type of neuroglia that forms a sheath around certain neurons. 601

scientific method A systematic approach to interpreting observations, involving reasoning, predicting, testing, and drawing conclusions, and then putting them into perspective with existing knowledge. 10

sclera The outermost layer of the eye; the white of the eye. 629

sclerenchyma *sklah-REN-kah-mah* Supportive plant tissue composed of elongated cells with thick, nonstretchable secondary cell walls. 506

scolex *SKO-lex* The holdfast organ of a tapeworm. 461

scrotum *SKRO-tum* The sac of skin containing the testes. 785

sebaceous gland *se-BAY-shis GLAND* Gland in human skin that secretes a mixture of oils that soften hair and skin. 592

secondary growth Thickening of a plant due to cell division in lateral meristems. 505

secondary immune response The immune system's response to subsequent encounters with a foreign antigen. 763

secondary nutrient deficiency Too little of a particular nutrient due to an inborn metabolic condition. 252

secondary oocyte *SEC-un-derry OO-site* A haploid cell that is an intermediate in ovum formation. 169

secondary spermatocyte *SEC-un-derry sper-MAT-o-site* A haploid cell that is an intermediate in sperm formation. 167

secondary (2°) structure The shape a protein assumes when amino acids close together in the primary structure chemically attract. 35

secondary succession Change in a community's species composition following a disturbance. 848

second filial generation *SEK-und FIL-e-al jen-er-A-shun* (F_2) The third generation in a genetics problem. 180

second messenger A biochemical activated by an extracellular signal that transmits a message inside a cell. 78

secretion A cell's release of a biochemical. Also, the addition of substances to the material in a kidney tubule. 745

seed In seed plants, a dormant sporophyte within a protective coat. 551

seed coat A tough outer layer of a seed that protects a dormant plant embryo and its food supply. 551

segmentation In digestion, localized muscle contractions in the small intestine that provide mechanical digestion. Also, division of an animal body into repeated subunits. 729

segregation Mendel's principle that one of each gene pair is placed in a separate gamete. 179

selectively permeable *sah-LEK-tive-lee PERM-ee-ah-bul* A biological membrane that admits only some substances. 64

semicircular canal Fluid-filled structure in the inner ear that provides information on the position of the head. 634

semiconservative replication Mode of DNA replication. Each double helix has one parental and one new strand. 228

semilunar valve Tissue flaps in the artery just outside each ventricle that maintain one-way blood flow. 686

seminal fluid *SEM-in-el FLEW-id* Secretions that carry sperm. 785

seminal vesicle *SEM-in-el VES-eh-kel* In the human male, one of a pair of structures that adds fructose and prostaglandins to sperm. 785

seminiferous tubule Tubule within the testis where sperm form and mature. 168, 785

senescence The cessation of growth and subsequent aging of tissues. 570

sensory adaptation Lessening of sensation with prolonged exposure to the stimulus. 623

sensory neuron A neuron that transmits information from a stimulated body part to the CNS. 598

sensory receptor Sensory neuron or specialized epithelial cell that detects and passes stimulus information to a sensory neuron. 622

sepal *SEE-pel* A leaflike floral structure that encloses and protects inner floral parts. 543

septate hypha A hypha with septa that partition off individual cells. 436

seta (pl. setae) *SEE-ta* Bristle on the side of an earthworm; also, hairlike, vibration-sensitive structure at base of insect antenna. 467, 632

sex chromosome A chromosome that carries genes that determine sex. 161

sex hormone Hormone that controls development of secondary sexual characteristics and prepares an animal for sexual reproduction. 655

sex-linked A gene on the X chromosome or a trait that results from activity of such a gene. 207

sex pilus *SEX PILL-us* A prokaryotic cell outgrowth that transfers DNA from one cell to another. 389

sex ratio The proportion of males to females in a population. 206

sexual dimorphism The difference in appearance between males and females of the same species. 816

sexual reproduction The combination of genetic material from two individuals to create a third individual. 6

sexual selection Natural selection of traits that increase an individual's reproductive success. 271

shoot The aboveground part of a plant. 504

short-day plant Plant that requires dark periods longer than some critical length to flower. 567

short-term memory A memory that persists only as long as the stimulation continues. 614

sieve cell Relatively unspecialized phloem cell, found in gymnosperms and seedless vascular plants. 508

sieve plate End wall of a sieve tube member in phloem. 508

sieve tube member Specialized conducting cell in phloem of flowering plants. 508

signal transduction The biochemical transmission of a message from outside the cell to inside. 77

signature sequence DNA sequence unique to the members of specific taxonomic groups. 385

simple epithelium *SIM-pel ep-eh-THEEL-ee-um* A single layer of epithelium. 579

sink In plants, areas where products of photosynthesis are unloaded from phloem. 535

sinoatrial (SA) node Specialized cells in the wall of the right atrium that set the pace of the heartbeat; the pacemaker. 687

skeletal muscle Voluntary muscle type composed of unbranched, multinucleated cells that connect to skeletal elements. 581, 670

skeletal muscle cell Single, multinucleated cell that contracts when actin and myosin filaments slide. Makes up voluntary, striated muscle. 670

skeletal system System of bones and ligaments that support body structures and attach to muscles. 584

sliding filament model Sliding of protein myofilaments past each other to shorten skeletal muscle cells, leading to contraction. 672

slime layer A loose sticky layer surrounding some prokaryotic cells. 384

slime molds Protistan that forms mobile masses that enable cells to collectively migrate. 398

small intestine Connects the stomach with the colon; important site of chemical digestion and absorption. 729

smooth endoplasmic reticulum The organelle of lipid synthesis and detoxification. 52

smooth muscle Type of muscle consisting of involuntary, nonstriated, spindle-shaped cells. 581, 669

smooth muscle cell Cell that makes up involuntary, nonstriated contractile tissue that lines the digestive tract and other organs. 669

sodium-potassium pump A mechanism that uses energy released from splitting ATP to transport Na^+ out of cells and K^+ into cells. 67

softwood Wood of gymnosperms. 520

solute *SOL-yoot* A chemical that dissolves in another, forming a solution. 26

solution A homogenous mixture of a substance (the solute) dissolved in water (the solvent). 26

solvent *SOL-vent* A chemical in which others dissolve, forming a solution. 26

somatic mutation *so-MAT-ik mew-TAY-shun* A mutation in a body cell. 254

somatic nervous system Motor pathways that lead to skeletal muscles. 607

somatostatin *so-MAT-owe-STAT-in* A pancreatic hormone that controls the rate of nutrient absorption into the bloodstream. 653

sorus *SOR-us* Dark spot on the underside of fern fronds that is a collection of spore-producing structures. 440

source In plants, areas where products of photosynthesis are loaded into phloem. 535

speciation *SPE-she-AY-shun* Appearance of a new type of organism. 297

species A group of similar individuals that breed in nature only among themselves. 8

sperm The male sex cell. 168

spermatid *sper-ma-TID* An intermediate stage in sperm development. 167

spermatogenesis *sper-MAT-o-JEN-eh-sis* The differentiation of a sperm cell from a diploid spermatogonium, to primary spermatocyte, to two haploid secondary spermatocytes, to spermatids, and finally to mature spermatozoa. 167

spermatogonium *sper-mat-o-GOWN-e-um* A diploid cell that divides, yielding progeny cells that become sperm cells. 167

spermatozoa *sper-mat-o-ZO-a* Mature sperm. 168

S phase The synthesis phase of interphase, when DNA replicates and microtubules assemble from tubulin. 140

sphincter Muscular ring that controls passage of a substance from one part of the body to another. 728

spicule Glassy or limy material that makes up sponge skeletons. 458

spinal cord Tube of nervous tissue that extends through the vertebral column. 607

spinal nerve Peripheral nerve that exits the vertebrate CNS from the spinal cord. 607

spiracle Opening in the body wall of arthropods; used for breathing. 703

spiral cleavage Pattern of early cleavage cells in protostomes, resembling a spiral. 455

spirillum (pl. spirilla) A spiral-shaped prokaryotic cell. 384

spleen An abdominal organ that produces and stores lymphocytes and red blood cells. 698

sponge A simple, asymmetrical or radially symmetrical animal lacking true tissues and gastrulation; phylum Porifera. 456

spongin *SPUNJ-in* Organic material in sponge skeletons. 458

spongocoel *SPUNJ-o-seel* The central cavity of a sponge. 458

spongy bone Flat bones and tips of long bones with large spaces between a web of bony struts. 665

spontaneous generation The idea, proven untrue, that life can arise from nonliving matter. 336

spontaneous mutation A change in DNA sequence not caused by a mutagen, usually resulting from a DNA replication error. 253

sporangium A structure in which spores are produced. 418

spore Reproductive structure in prokaryotes and fungi that may be specialized for survival or dissemination. In plants, the spore develops into a gametophyte. 414, 436

sporophyte *SPOR-o-fite* Diploid, spore-producing stage of the plant life cycle. 414, 543

squamous *SKWAY-mus* Flat, as in epithelium. 579

spring turnover The seasonal mixing of upper and lower layers of a lake. 878

stabilizing selection When extreme phenotypes are less adaptive than an intermediate phenotype. 286

stamen *STA-men* Male reproductive structure in flowers; consist of a stalklike filament with a pollen-producing anther at its tip. 543

stapes *STAY-peez* A small bone in the middle ear. 634

statocyst *STAT-o-sist* A fluid-filled cavity that contains minerals and sensory hairs that control balance and equilibrium. 633

statolith *STAT-o-lith* Structure in plant cells that detects gravity. 563

stem Central axis of a plant's shoot system. 510

stem cell An undifferentiated cell that divides to give rise to specialized cells. 578

steroid hormone A lipid hormone that can cross a target cell's membrane and bind to intracellular receptors. 643

sterols Lipid molecules based on a complex molecule of four interconnected carbon rings. 32

stigma *STIG-mah* The tip of a flower's style that receives pollen grains and on which they germinate. 544

stipe *STYP* Stemlike region of a brown alga. 408

stolon *STOL-on* Stem that grows along the soil surface. 510

stoma (pl. stomata) *STO-mah* Pore in a plant's epidermis through which gases exchange between the plant and the atmosphere. 108, 508

stomach J-shaped compartment that receives food from the esophagus. 728

storage leaf Fleshy leaf that stores nutrients. 514

stramenopile Diverse group of protista including brown algae, diatoms, and water molds. 406

stratified epithelium *STRAT-eh-fyed ep-eh-THEEL-ee-um* Layered epithelium. 579

stroke Interrupted blood circulation in the brain. 696

stroma *STRO-ma* The nonmembranous inner region of the chloroplast. 55

style A stalk that arises from the top of an ovary in a flower. 544

subcutaneous layer *sub-kew-TAYN-ee-us LAY-r* In human skin, the layer of fat beneath the dermis. 589

submetacentric Referring to the location of the centromere as midway between bisecting the chromatid and the telomere. 202

substrate A reactant an enzyme acts upon. 36

substrate-level phosphorylation ATP formation from transferring a phosphate group to ADP. 120

succulent stem Fleshy plant tissue that stores water. 510

superantigen Larger than usual antigen on the outside of an MHC molecule that promotes an exaggerated immune response. 771

superior vena cava The upper branch of the largest vein that leads to the heart. 686

supernormal releaser A model that exaggerates a releaser and elicits a stronger response than does the natural object. 808

superposition The principle that older fossils lay in lower layers of rock strata. 266

survivorship The fraction of a group of individuals that survive to a particular age. 831

suspension feeder Animal that captures food particles suspended in the water. 467, 470

sustainability Use of resources while preserving them for the future. 888
sweat gland Epidermal invagination that produces sweat. 592
swim bladder An organ that allows bony fishes to adjust their buoyancy. 485
symbiosis *sim-bi-O-sis* One type of organism living in or on another. 845
sympathetic nervous system Part of the autonomic nervous system that mobilizes the body to respond to environmental stimuli. 607
sympatric speciation The formation of a new species within the boundaries of a parent species. 302
synapse Area where two neurons meet and transmission of electrochemical information occurs. 602
synapsids Vertebrate with a single opening behind each eye orbit; mammals and their immediate ancestors. 491
synapsis *sin-AP-sis* The gene-by-gene alignment of homologous chromosomes during prophase of meiosis I. 163
synaptic cleft The space between two neurons at a chemical synapse. 603
synaptic integration A cell's overall response to many incoming neural messages. 605
synaptic knob Enlarged tip of an axon that contains synaptic vesicles. 602
synovial joint *sin-OV-ee-el JOYNT* A capsule of fluid-filled fibrous connective tissue between freely movable bones. 675
synovial membrane *sin-OV-ee-el MEM-brane* Lining of the interior of a joint capsule, which secretes lubricating synovial fluid. 675
synteny *SIN-ten-ee* Comparison of gene order on chromosomes between species. 331
systematics *sis-te-MAT-ix* Study of the evolutionary relationships among species. 317
systemic circulation Blood circulation through the heart and the body. 683
systole The contraction of the ventricles of the heart. 687

T

taiga *TI-e-gah* One of several terrestrial biomes; the northern coniferous forest. 876
taproot system A root system in which lateral root branches arise from a tapering main root. 515
target cell A cell that a hormone binds and directly affects. 642
taste receptors Specialized neurons that detect taste. 627
taxonomy *tax-ON-o-mee* Classification of organisms on the basis of evolutionary relationships. Taxonomic levels include, in order: domain, kingdom, phylum (or division), class, order, family, genus, and species. 8
T cell T lymphocyte; a component of the immune system. 757
T cell receptor The antigen-binding molecules on the surface of T lymphocytes. 765
tectorial membrane *tek-TOR-ee-al MEM-brane* The membrane above hair cells in the cochlea of the inner ear that is pressed by hair cells responding to the basilar membrane's vibration in the presence of sound waves. 634
tectum Thickened area of gray matter at the roof of the midbrain. 612
tegument *TEG-yew-ment* The protective body covering of certain flatworms. 461
telocentric *tell-o-SEN-trik* A chromosome with the centromere at the tip. 202
telomerase *tell-OM-er-ase* An enzyme that extends chromosome tips using RNA as a template. 145
telomere *TELL-o-meer* A chromosome tip. 145
telophase *TELL-o-faze* The final stage of cell division, when two cells form from one and the spindle is disassembled. 142
telophase I *TELL-o-faze I* Arrival of homologs at opposite poles in meiosis I. 162
telophase II *TELL-o-faze II* Nuclear envelope formation around meiotic products. 163
temperate coniferous forest One of several terrestrial biomes; coniferous trees dominate. 876
temperate deciduous forest *TEMP-er-et de-SID-u-us FOR-est* One of several terrestrial biomes; dominated by deciduous trees. 875
temperate grassland One of several terrestrial biomes; grazing, fire, and drought restrict tree growth. 874
temporal caste Group in a society whose role changes with time. 818
temporal isolation A form of reproductive isolation based on differences in the timing of interactions. 299
tendon A band of fibrous connective tissue attaching a muscle to a bone. 670
tendril *TEN-dril* Shoot or modified leaf that supports plants by coiling around objects. 514
teratogen *teh-RAT-eh-jen* Something that causes a birth defect. 797
territoriality Behavior that defends one's area. 813
tertiary (3°) structure *TER-she-air-ee STRUK-sure* The shape a protein assumes when amino acids far apart in the primary structure chemically attract one another. 36
test cross Breeding an individual of unknown genotype to a homozygous recessive individual to reveal the unknown genotype. 182
testis Male gonad containing seminiferous tubules, where sperm are manufactured. 168, 654, 785
testosterone A male steroid hormone involved in regulation of sperm production and development of male characteristics. 655
tetanus Maximal muscle contraction caused by continual stimulation. 674
tetrapod A four-limbed vertebrate. 478
thalamus A tight grouping of nerve cell bodies beneath the cerebrum that relays sensory input to the cerebrum. 613
thallus *THALL-us* Relatively undifferentiated body form lacking roots, stems, and leaves. 436
theory An explanation for observations and evidence of natural phenomena. 10
thermodynamics *THERM-o-di-NAM-ix* Study of energy transformations in nature. 85
thermogenesis Generating heat metabolically. 741
thermophile *THERM-o-file* Organism that lives under conditions of high heat. 562
thermoreceptor A receptor cell that responds to temperature changes. 636
thermoregulation Ability of an animal to balance heat loss and gain with the environment. 740
thigmonasty *THIG-mo-NAS-tee* A nastic response to touch. 565
thigmotropism *THIG-mo-TRO-piz-um* A plant's growth response toward touch. 563
thoracic vertebra Vertebra of the upper back. 663
thorn Stem modified for protection. 510
threshold potential Membrane potential at which an action potential is initiated. 599
thrombin *THROM-bin* A blood-clotting protein. 694
thromboplastin *THROM-bo-PLAS-tin* A protein released from blood vessel walls following injury that converts prothrombin to thrombin. 694
thrombus *THROM-bus* A blood clot that blocks a blood vessel. 694
thylakoid *THI-lah-koyd* Disclike structure that makes up the inner membrane of a chloroplast. 55
thylakoid membrane The membrane of the thylakoid, the location of the light reactions of photosynthesis. 102
thylakoid space The inner compartment of the thylakoid. 103
thymine *THI-meen* One of the two pyrimidine bases in DNA. 37
thymus A lymphatic organ in the upper chest where T cells learn to distinguish foreign from self antigens. 698
thyroid gland In humans, a gland in the neck that manufactures thyroxine, a hormone that increases energy expenditure. 650
thyroid-stimulating hormone (TSH) A pituitary hormone that stimulates the thyroid gland to release two hormones. 651
thyroxine A thyroid hormone that increases the rate of cellular metabolism. 225, 651

tidal volume Volume of air inhaled or exhaled during a normal breath. 712

tight junction A connection between two cells that prevents fluid from flowing past the cells. 74

tissue Groups of cells with related functions. 578

tonsil Collection of lymphatic tissue in the throat. 698

topsoil The uppermost layer of soil, often rich in organic matter. 529

totipotent Describing cells that retain the potential to specialize in any way. 779

trace element An element an animal requires in small amounts. 20

trachea The respiratory tube just beneath the larynx; the "windpipe." 708

tracheae Branching system of tubules that brings the external environment in close contact with an animal's cells so that gas exchange can occur. 703

tracheid *TRAY-kee-id* Relatively unspecialized conducting cell in xylem. 508

tracheole Tiny fluid-filled branches of arthropod tracheae that contact individual cells. 703

transcription *tranz-SKRIP-shun* Manufacturing RNA from DNA. 240

transcription factor *tranz-SCRIPT-shun fac-TOR* A protein that turns on and off different genes in a particular cell. 241

transcytosis Movement of substances from one side of a cell through to the matrix on the opposite side. 69

transduction Conversion of energy from one form to another. Also a method of horizontal gene transfer in which a virus transfers DNA from one host cell to another. 389

transfer RNA (tRNA) A small RNA molecule that binds an amino acid at one site and an mRNA codon at another site. 245

transformation Method of horizontal gene transfer in which an organism takes up naked DNA from the environment. 388

translation *tranz-LAY-shun* Assembly of an amino acid chain according to the sequence of base triplets in a molecule of mRNA. 240

translocation *TRANZ-lo-KAY-shun* Exchange of genetic material between nonhomologous chromosomes. 214

transmissible spongiform encephalopathy A disease caused by the ingestion of prions. 374

transpiration The movement of water vapor from plant parts to the atmosphere through open stomata. 531

transplantation *tranz-plan-TAY-shun* Replacing a diseased or damaged organ with one from a donor. 586

transposable element *tranz-POSE-a-bull EL-e-ment* Jumping gene. 255

transverse (T) tubule *TRANZ-verse TU-bule* Part of the sarcolemma that juts from the sarcoplasmic reticulum of a skeletal muscle cell. 670

triglyceride *tri-GLI-sir-ide* A type of fat that consists of one glycerol and three fatty acids. 32

triiodothyronine A thyroid hormone that increases the rate of cellular metabolism. 651

trilobite *TRI-low-bite* Extinct arthropod that was once very abundant. 469

triploblastic Describing an animal whose adult tissues arise from three germ layers in the embryo. 453

trisomy *TRI-som-mee* A cell with one extra chromosome. 212

trochophore *TRO-ko-for* A mollusk larva. 465

trophic level *TRO-pik LEV-l* An organism's position along a food chain. 851

trophoblast *TRO-fo-blast* A layer of cells in the preembryo that develops into the chorion and then the placenta. 789

tropical dry forest One of several terrestrial biomes; yearly dry and rainy seasons in tropics. 872

tropical rain forest One of several terrestrial biomes; year-round high temperatures and precipitation. 871

tropic hormone A hormone that affects another hormone's secretion. 648

tropism *TRO-piz-um* Plant growth toward or away from an environmental stimulus. 562

tropomyosin *TRO-po-MI-o-sin* A type of protein in thin myofilaments of skeletal muscle cells. 670

troponin *tro-PO-nin* A type of protein in thin myofilaments of skeletal muscle cells. 670

true fern A type of seedless vascular plant.422

trypanosome Protist (Kinetoplastida) that causes human diseases transmitted by biting flies. 400

tube foot Cuplike structure on echinoderms connected to the water vascular system and involved in locomotion. 470

tuber *TU-ber* Swollen region of stem that stores nutrients. 510

tubulin *TOOB-u-lin* Protein that makes up microtubules. 70

tumor necrosis factor An immune system biochemical with varied functions in cancer, infection, and inflammation. 757

tumor suppressor gene *TOO-mer sup-PRESS-er JEAN* A gene which, when inactivated or suppressed, causes cancer. 153

tundra *TUN-drah* One of several terrestrial biomes; low temperature and short growing season. 877

tunicate *TOON-i-cat* A type of primitive invertebrate chordate. 479

turgor pressure *TER-ger PRESH-er* Rigidity of a cell caused by water pressing against the cell wall. 66

twitch Rapid contraction and relaxation of a muscle cell following a single stimulation. 674

tympanal organ *tim-PAN-al OR-gan* A thin part of an insect's cuticle that detects vibrations and therefore sound. 633

tympanic membrane The eardrum, a structure upon which sound waves impinge. 633

U

ultratrace element An element vital to an organism in very small amounts. 20

ultraviolet (UV) radiation Portion of the electromagnetic spectrum with wavelengths shorter than 400 nm. 891

umbilical cord *um-BIL-ik-kel KORD* A ropelike structure that contains one vein and two arteries that connects a pregnant female placental mammal to unborn offspring. 793

unconditioned stimulus A stimulus that normally triggers a particular response. 809

unicellular Organisms consisting of single cells. 6

uniformitarianism A theory of geological change, championed by Lyell, that stated that changes were the result of steady, gradual processes whose overall rates remain constant. 266

uniramous *uni-RAY-muss* A single-lobed arthropod appendage. 468

unsaturated *un-SAT-yur-RAY-tid* (fat) A triglyceride with double bonds between some of its carbons. 32

upwelling Upward movement of cold, nutrient-rich lower layers of a body of water. 882

uracil *YUR-eh-sil* One of two pyrimidine bases in RNA. 37

urea A nitrogenous waste derived from ammonia. 745

ureter A muscular tube that transports urine from the kidney to the bladder. 746

urethra Tube that transports urine from the bladder out of the body. 746, 785

uric acid A nitrogenous waste derived from ammonia. 745

urinary bladder A muscular sac where urine collects. 746

urinary system Organ system that filters blood and helps maintain concentrations of body fluids. 584, 746

uterus The muscular, saclike organ in a female placental mammal where the embryo and fetus develop. 786

utricle *U-trah-kel* A pouch in the vestibule of the inner ear filled with a jellylike fluid and lined with hair cells that contain calcium carbonate granules that move in response to changes in velocity, firing action potentials. 636

V

vaccine A killed or weakened form of, or part of, an infectious agent that initiates an immune response so that when the real agent is encountered, antibodies are already available to deactivate it. 755

vacuole A storage sac in a cell. 51

vagina Passage from the uterus to the outside of the body in female mammals. 784

valence electron *VAY-lense e-LEC-tron* Electron in the outermost shell. 23

van der Waals attractions *VAN dur WALLS a-TRAC-shuns* Dynamic attractions within or between molecules when oppositely charged regions attract. 25

variable A portion of an experiment that is changed to test a hypothesis. 14

variable region Amino acid sequence that forms the upper part of heavy and light chains, which varies for different antibodies. 763

variably expressive A phenotype that varies in intensity in different individuals. 189

vasa recta Portion of peritubular capillary in which blood flows in opposite direction of the filtrate in the nephron loop. 749

vascular bundle *VAS-ku-ler BUN-del* A strand of vascular tissue containing primary xylem and primary phloem and associated cells. 510

vascular cambium *VAS-ku-ler KAM-bee-um* A type of lateral meristem in plants; produces secondary xylem and phloem. 518

vascular tissue *VAS-ku-ler TISH-ew* Specialized conducting tissue in plants; xylem and phloem. 508

vas deferens In the human male, a tube from the epididymis that joins the urethra in the penis. 168, 785

vasoconstriction Decrease in the diameter of a vessel. 688

vasodilation Increase in the diameter of a vessel. 688

vein In plants, a strand of vascular tissue in leaves. In animals, a vessel that returns blood to the heart. 682

venous valve *VEEN-is VALVE* Flaplike structure in a vein that keeps blood flowing in one direction. 688

ventilation Any mechanism that increases the flow of air or water across a respiratory surface. 705

ventral *VEN-tral* The belly or underside. 453

ventricle Space in the brain into which cerebrospinal fluid is secreted. Also, a muscular heart chamber beneath the atria. 683

ventricular fibrillation When pacemaker cells lose control of heart contraction; heart muscle twitches wildly. 687

venule Vessel that drains blood from capillaries. 682

vertebra One unit of the vertebral column. 482

vertebral column *ver-TE-bral COL-um* Bone or cartilage that supports and protects the spinal cord. 663

vertebrate An animal that has a vertebral column. 452, 482

vertical gene transfer Inheritance of genetic information from one generation to the next. 386

vertical stratification *VER-ti-kal strat-if-i-KAY-shun* Layering of different plant species in the forest canopy. 871

vesicle *VES-i-cal* A membrane-bounded sac in a cell. 52

vessel element *VES-el EL-eh-ment* Specialized conducting cell in xylem. 508

vestibule *VES-teh-bule* A structure in the inner ear that provides information on the position of the head with respect to gravity and changes in velocity. 635

vestigial organ *ves-TEEG-e-el OR-gan* A structure that seems not to have a function in an organism but that resembles a functional organ in another species. 328

villus (pl. villi) Tiny projection on the inner lining of the small intestine that greatly increases surface area for nutrient absorption. 731

virion A complete, infectious virus particle. 363

viroid *VIE-royd* Infectious genetic material. 372

virus An infectious particle consisting of a nucleic acid (DNA or RNA) wrapped in protein. 362

visceral mass Part of the molluscan body that contains the digestive and reproductive systems. 465

vital capacity The maximal amount of air that can be forced in and out of the lungs during one breath. 712

vitreous humor *VIT-ree-us HU-mer* A jellylike substance behind the lens that makes up most of the volume of the eye. 630

viviparous Bearing live young without retaining eggs in the birth canal (in animals). 482

vocal cord Elastic tissue band that covers the glottis and vibrates as air passes, producing sound. 708

W

warning coloration Bright or distinctive display that advertises an organism's defenses. 847

water mold Filamentous, heterotrophic protista, also called oomycetes. 407

water vascular system System of canals in echinoderms; involved in locomotion and osmotic balance. 470

wavelength The distance a photon moves during a complete vibration. 99

whisk fern A type of seedless vascular plant. 422

white blood cell A cell that helps fight infection. 579, 693

white matter Nervous tissue in the CNS consisting of myelinated fibers. 610

wild-type The most common phenotype or allele for a certain gene in a population. 181

X

X inactivation *X IN-ak-tah-VA-shun* Turning off one X chromosome in each cell of a female mammal at a certain point in prenatal development. 210

X-inactivation center *X IN-ak-tah-VA-shun SEN-ter* The part of the X chromosome that inactivates the X. 210

X-linked Referring to traits, usually recessive, whose alleles reside on the X chromosome. 207

xylem *ZI-lem* Plant tissue that transports dissolved ions and water. 415, 508

xylem sap Dilute solution of water and dissolved minerals in xylem. 532

Y

yeast A unicellular fungus. 436

yolk sac An extraembryonic membrane that forms beneath the embryonic disc and manufactures blood cells. 792

Z

zona pellucida A thin, clear layer of proteins and sugars that surrounds a secondary oocyte. 788

zoonosis An infection in humans that originates in another animal. 370

zooplankton *ZO-O-PLANK-tun* Abundant heterotrophic microorganisms in bodies of water. 854

zoospore A flagellated spore produced by chytrids and water molds. 419, 437

zygomycete *zi-go-mi-SEET* A fungus that produces zygospores. 434

zygospore *zi-go-SPOOR* Diploid resting spore produced by fusion of haploid gametangia in zygomycetes. 437

zygote *ZI-goat* The fused egg and sperm that develops into a diploid individual. 160

CREDITS

LINE ART

Chapter 3

Fig. 3A(5): Schematic Courtesy Michael Schmid, Institut F. Allgemeine Physik, TU Wien.

Chapter 12

Fig. 12.13: Reproduced with permission of Quadrant Holdings, Cambridge, LTD.

Chapter 15

Fig. 15.5: From Dr. Stephen J. O'Brien and Dr. Michael Dean/NYT Pictures.

Chapter 16

Fig. 16.12: From Jonathan B. Losos and Kevin de Queiroz, *Darwin's Lizards.*

Chapter 17

Fig. 17.10: From Kevin Padian and Luis M. Chiappe, "The Origin of Birds and Their Flight" in *Scientific American,* vol. 278, No. 2, February 1998. Reprinted by permission of Ed Heck, Illustrator.
Fig. 17.11: From *Patterns in Evolution* by Roger Lewin © 1997 by *Scientific American Library.* Used with the permission of W.H Freeman and Company.

Chapter 18

Fig. 18.13B: From R. Raff, *The Shape of Life.* Copyright © The University of Chicago Press. Reprinted by permission.

Chapter 24

Opener B: Reprinted by permission from *Nature* from Peter Funch and Reinhardt M. Kristensen, "Cycliophora is a New Phylum with Affinities to Entoprocta and Ectoprocta" in *Nature,* vol. 378, pp. 711–714, 1995. Copyright © 1995 MacMillan Magazine, Ltd., www.nature.com

Chapter 25

Fig. 25.5: Reprinted by permission from *Nature* from S. Kumar and S.B. Hedge, "A Molecular Time Scale for Vertebrate Evolution" in *Nature,* vol. 392, p. 919, 4/30/98, fig. 3. Copyright © 1998 MacMillan Magazine, Ltd., www.nature.com **Fig. 25.21:** From Kenneth V. Kardong, Vertebrates: Comparative, Anatomy, Function, Evolution, 2nd edition. Copyright ©1998 The McGraw-Hill Companies. All Rights Reserved. Reprinted by permission.

Chapter 37

Fig. 37.5: After E. Rogers, *Looking at Vertebrates,* Longman Group Essex, England, 1986 from Hickman, et al., *Integrated Principles of Zoology,* 10th edition. Copyright © 1997 The McGraw-Hill Companies, Inc. All Rights Reserved. Reprinted by permission.

Chapter 40

Fig. 40.A: From Ricki Lewis, *Human Genetics,* 3rd edition. Copyright © 2000 The McGraw-Hill Companies, Inc. All Rights Reserved. Reprinted by permission.

Chapter 41

Fig. 41.17: From Mace A. Mack and Daniel Rubenstein, "Zebra Zones" in *Natural History,* 3/98 p. 79, American Museum of Natural History. Reprinted by permission of Joe LeMonnier, Illustrator. **Fig. 41.19:** Reproduced with permission from *Animal Communication* by Judith Goodenough; Carolina Biological Reader Series No. 143. Copyright © 1984, Carolina Biological Supply Co., Burlington, North Carolina, U.S.A.

Chapter 42

Fig. 42.9: Based on study by Joel Berger, University of Nevada-Reno/NYT Pictures.

Chapter 43

Fig. 43.B: Data from James A. Estes, U.S. Geological Survey.

PHOTOGRAPHS

Preface

Ricki Lewis photo courtesy of Rich Lannon. Doug Gaffin and Mariëlle Hoefnagels photos courtesy of Coral McCallister. Bruce Parker photo © Glen Ricks Photography, Inc. Photo of Precision Graphics Illustration Team provided by Precision Graphics.

Graphic for chapters in unit one: Digital Stock/Natural World.
Graphic for chapters in unit two: Corbis/Microscape Backgrounds.
Graphic for chapters in unit three: PhotoDisc/vol 6/ Nature, Wildlife, Environment.
Graphic for chapters in unit four: PhotoDisc/vol. 29/ Modern Technologies.
Graphic for chapters in unit five: Digital Stock/Natural World.
Graphic for chapters in unit six: Digital Stock/Natural World.
Graphic for chapters in unit seven: PhotoDisc/vol. 29/Modern Technologies.
Front and end matter graphics: Photodisc/vol. 36/ Nature Scenes.

Unit Openers

1: © Jane McAlonan / Visuals Unlimited; **2:** © Tony Stone Images; **3:** © Ed Reschke / Peter Arnold, Inc.; **4:** NSSDC / Dr. Richard J. Allenby, Jr. Id# 69-059A-01A; **5:** Professor James Miller and The Geo-Images Project, Department of Geography, University of California, Berkeley; **6:** Philip Brownell, Oregon State University; **7:** Maslowski Photo

Chapter 1

Opener: Corbis #WST0049 Western Scenics; **1.1(Biosphere):** Corbis Eye on Earth CD #EOE0013; **1.1(Ecosystem):** © Lynn Stone / Animals Animals; **1.1(Community):** © Carl R. Sams II / Peter Arnold, Inc.; **1.1(Population):** © Rod Planck / Photo Researchers, Inc.; **1.1(Multicellular organism):** © Rod Planck / Photo Researchers, Inc.; **1.1G:** © Dwight Kuhn; **1.1H:** Ed Reschke; **1.4A:** © R. Kessel - G. Shih / Visuals Unlimited; **1.4B:** © Runk / Schoenberger / Grant Heilman Photography; **1.4C:** Corbis Animals in Action CD #AIA0064; **1.5A,B:** © Michael and Patricia Fogden; **1.6(all):** © Steven Vogel; **1.7(1):** © Kwangshin Kim / Photo Researchers, Inc.; **1.7(2):** James King-Holmes / Science Photo Library / Photo Researchers, Inc.; **1.7(3):** © Ken W. David / Tom Stack & Assoc.; **1.7(4):** © Dwight Kuhn; **1.7(5):** © David Dennis / Animals Animals; **1.7(6):** © Michael Fogden/ Animals Animals; **1.9A:** © Inga Spence / Tom Stack & Associates; **1.9B:** Dr. James G. Sikarskie, DVM / Michigan State University; **1.9C:** Planet Earth Pictures; **1.9D:** © Scimat / Photo Researchers, Inc.; **1.10 A-1:** © Philip Gould / Corbis; **1.10 A-2:** © Galen Rowell / Corbis; **1.A:** © Katy Payne; **1.11:** Courtesy National Library of Medicine

Chapter 2

Opener A: © SIU / Visuals Unlimited;
Opener B: © Hans Reinhard / Okapia / Ovis Ammon Aries / Photo Researchers, Inc.; **2.5B:** © Dan McCoy / Rainbow; **2.7A:** Corbis #PN007203; **2.7B:** Corbis #CB003879; **2.7C:** Corbis #IL001283; **2.12:** Courtesy of Diane R. Nelson; **2.14A:** © BioPhoto Associates / Photo Researchers, Inc.; **2.14B:** © Darylyne A. Murawski / Peter Arnold, Inc.; **2.17:** © Stephen J. Krasemann / DRK Photo; **2.19A:** © Richard R. Hansen / Photo Researchers, Inc.; **2.19B:** © Gerald Lacz / Peter Arnold, Inc.; **2.19 C:** © John Cancelosi / Peter Arnold, Inc.; **2.22B:** Div. of Computer Research & Technology, National Institute of Health / Science Photo Library/ Photo Researchers, Inc.

Chapter 3

Opener (left): ©Boehringer Ingelheim International GmbH Photo: Lennart Nilsson; **Opener (right):** ©Nancy Kedersha / Immunogen / SPL / Photo Researchers, Inc.; **3.1A:** © Manfred Kage / Peter Arnold, Inc.; **3.1B:** © Edwin A. Reschke / Peter Arnold, Inc.; **3.2A:** © Kathy Talaro / Visuals Unlimited; **3.2B:** Science VU / Visuals Unlimited; **3.A(top, left):** © Doug Martin / Photo Researchers, Inc.; **3.A (top, right):** © BioPhoto Associates / Photo Researchers, Inc.; **3.A(middle, left):** © SIU / Visuals Unlimited; **3.A(middle, right):** © Don Fawcett /

Visuals Unlimited; **3.A (bottom):** © Science VU-IBMRL / Visuals Unlimited; **3.CA:** © Fred Hossler / Visuals Unlimited; **3.CB:** ©Boehringer Ingelheim International GmbH Photo: Lennart Nilsson "The Body Victorious" Dell Publishing; **3.CC:** © Aaron H. Swihart and Robert C. MacDonald; **3.6B:** CNRI / Science Photo Library / Photo Researchers, Inc.; **3.7A:** © David M. Phillips / Visuals Unlimited; **3.7B:** © Thomas Tottleben / Tottleben Scientific Company; **3.7C:** © T.E. Adams / Visuals Unlimited; **3.8:** From J.T. Staley, M.P. Bryant, N. Pfennig, and J.G. Holt, (Eds.), *Bergey's Manual of Systematic Bacteriology,* vol. 3. p. 224, fig. 25.37 © 1989 Williams and Wilkins Co., Baltimore Reprinted with permission of Bergey's Manual Trust; **3.9(nucleus):** © David M. Phillips / The Population Council / Science Source/ Photo Researchers, Inc.; **3.9(golgi):** © Biophoto Assoc. / Science Source / Photo Researchers, Inc.; **3.9(mitochondrion)** © K.R. Porter / Photo Researchers, Inc.; **3.10:** © G. Chapman - P. Devadoss / Visuals Unlimited; **3.12C:** © David M. Phillips / The Population Council / Science Source/ Photo Researchers, Inc.; **3.13:** © R. Bolender - D. Fawcett / Visuals Unlimited; **3.14:** Prof. P. Motta & T. Naguro / Science Photo Library / Photo Researchers, Inc.; **3.15:** © Bill Longcore / Photo Researchers, Inc.; **3.16:** © Biophoto Associates / Photo Researchers, Inc.; **3.17:** Prof. P. Motta & T. Naguro / Science Photo Library / Photo Researchers, Inc.; **3.18A:** © D. Friend - D. Fawcett / Visuals Unlimited; **3.18B:** From S.E. Frederick and E.H. Newcomb, *Journal of Cell Biology,* 4:343–53, 1969

Chapter 4

Opener A: Corbis #BEO58072; **4.1:** © NIBSC / SPL / Photo Researchers, Inc.; **Opener B:** Gordon Leedale / BioPhoto Associates; **4.6A:** © Stanley Flegler / Visuals Unlimited; **4.6B:** © Veronika Burmeister / Visuals Unlimited; **4.6C:** © Science VU / Visuals Unlimited; **4.7B,C:** © Cabisco / Visuals Unlimited; **4.13A:** © Biology Media / Photo Researchers, Inc.; **4.13B:** © M. Schliwa / Visuals Unlimited; **4.13C(all):** The Company of Biologists, Ltd; **4.14:** K.G. Murti / Visuals Unlimited; **4.15A:** © D.W. Fawcett / Photo Researchers, Inc.; **4.15B:** © CNRI / SPL / Photo Researchers, Inc.; **4.16B:** © I. Gibbons, Don Fawcett / Visuals Unlimited; **4.16C:** © David M. Phillips; **4.17:** Mary Reedy; **4.18B:** AP / Wide World Photos; **4.19(top):** © David M. Phillips / Visuals Unlimited; **4.19(middle):** Courtesy Camillo Peracchia; **4.19(bottom):** © D. Allbertini - D. Fawcett / Visuals Unlimited; **4.20B:** © C. Gerald Van Dyke / Visuals Unlimited.

Chapter 5

Opener: © Gunter Ziesler / Bruce Coleman Collection; **5.1:** © Stephen Dalton / Photo Researchers, Inc.; **5.2(top):** © Jeremy Walker / Tony Stone Images; **5.2(bottom):** Des & Jen Bartlett / NGS Image Collection; **5.3:** © D. Foster / Visuals Unlimited; **5.4A:** © Bryan Yablosky / Duomo; **5.4B:** © Stephen Dalton / NHPA; **5.6:** © Manoj Shah / Tony Stone Images; **5.7:** Dr. E.R. Degginger; **5.9A:** © Blair Seitz / Photo Researchers, Inc.; **5.9B:** © Gamma Liaison; **5.A:** © Ivan Polunin / Bruce Coleman Inc.; **5.A(inset):** © E.R. Degginger / Photo Researchers, Inc.

Chapter 6

Opener: Dr. Rainer H. Koehler; **6.1:** Steve Raymer / NGS Image Collection; **6.7:** Courtesy of Dr. Eldon Newcomb, University of Wisconsin, Madison; **6.13A:** © BioPhot; **6.14A:** © Geoff Bryan 1996 / Photo Researchers, Inc.; **6.14B,C:** © Colin Weston / Planet Earth Pictures

Chapter 7

Opener: Willard R. Chappell; **7.1A:** © Lori Adamski Peek / Tony Stone Images; **7.1B:** Ed Reschke; **7.4:** © SPL / Custom Medical Stock Photos; **7.8:** PhotoDisc vol. 06 #6066 NA004439; **7.10A:** Courtesy of Peter Hinkle; **7.12:** PhotoDisc Food & Dining vol 12 #12282; **7.A:** © Auril Ramage / Oxford Scientific Films / Animals Animals / Earth Scenes; **7.13A:** © Vance Henry / Nelson Henry

Chapter 8

8.1: © Lennart Nilsson; **8.3B:** AP / Wide World Photos; **8.4B:** From Dr. A.T. Sumner, *"Mammalian Chromosomes from Prophase to Telophase,"* Chromosoma 100:410-418, 1991. © Springer Verlag; **8.5:** BPS; **8.6:** Ed Reschke; **8.7(1-5):** Ed Reschke; **8.7(6):** Peter Arnold Inc. / © Ed Reschke; **8.7(7):** Ed Reschke; **8.8B:** Dr. Mark W. Kirschner Dept. of Cell Biology Harvard Med. School; **8.9:** © R. Calentine / Visuals Unlimited; **8A:** From L. Chong. *"A Human Telomeric Protein,"* SCIENCE, 270:1663-1667, © American Association for the Advancement of Science. Photo Courtesy, Dr. Titia DeLange; **8.15(top):** © Lennart Nilsson; **8.15A:** Robert Maier / Animals Animals; **8.15B:** © Gordon & Cathy Illg / Animals Animals; **8.16:** Micrograph provided by Dr. John Heuser of Washington University School of Medicine, St. Louis

Chapter 9

Opener A,B: © Dwight Kuhn; **9.1A(all):** © Carolina Biological Supply Company / PhotoTake; **9.1B:** © Francis Gohier / Photo Researchers, Inc.; **9.3:** Dr. L. Caro / Science Photo Library / Photo Researchers, Inc.; **9.4B:** © R. Kessel - G. Shih; **9.5:** CNRI / Science Photo Library / Photo Researchers, Inc.; **9.8B:** © John / Cabisco / Visuals Unlimited; **9.11:** © Francis LeRoy / BioCosmos / SPL / Photo Researchers, Inc.; **9.12B:** © David M. Phillips / Visuals Unlimited; **9.12C:** The Granger Collection, New York; **9.14:** Larry Johnson, Dept. of Veterinary Anatomy and Public Health, Texas A&M University. Originally appeared in Biology of Reproduction, 47:1091-1098, 1992; **9.16:** © David M. Phillips / Visuals Unlimited; **9.17:** Courtesy of Dr. Y. Verlinsky; **9.D:** © CNRI / Phototake

Chapter 10

Opener A(both): © Archive Photos; **Opener B:** Courtesy of W. Dorsey Stuart, University of Hawaii; **10.7A:** Courtesy of Dr Myron Neuffer, University of Missouri; **10.7B:** © Joe McDonald / Visuals Unlimited; **10.A:** George McCord; **10.B(top):** Courtesy James R. Poush; **10.B(bottom):** © Courtesy James. R. Poush; **10.14A:** © Porterfield - Chickering / Photo Researchers; **Table 10.5(gray):** Norvia Behling / Animals Animals / Earth Scenes; **Table 10.5(chinchilla):** © Grant Heilman / Grant Heilman Photography; **Table 10.5(light gray):** © Robert Maier / Animals Animals /Earth Scenes; **Table 10.5(himalayan):** © Hans Burton / Bruce Coleman; **Table 10.5(albino):** © Jane Burton / Bruce Coleman; **10.15:** © Lester Bergman; **10.16A:** North Wind Picture Archives; **10.20(all):** Courtesy Thomas Kaufman. Photo by Phil Randazzo and Rudi Turner; **10.22:** Library of Congress

Chapter 11

11.1: © Science VU / Visuals Unlimited; **11.8:** © BioPhoto Associates / Photo Researchers, Inc.; **11.11B:** © Cabisco / Visuals Unlimited; **11.12A:** © Horst Schaefer / Peter Arnold, Inc.; **11.12B:** © William E. Ferguson; **11.14A:** Courtesy Colleen Weisz; **11.14B:** Courtesy of Integrated Genetics; **11.16B:** Courtesy of Lawrence Livermore National Laboratory; **11.17A:** David M. Phillips / Visuals Unlimited; **11.17B(all):** From R. Simensen, R. Curtis Rogers, *"Fragile X Syndrome,"* American Family Physician, 39:186 May 1989, © American Academy of Family Physicians; p. 305; Courtesy of Integrated Genetics

Chapter 12

Opener: Stock Montage, Inc.; **12.1:** © Dr. Gopal Murti / SPL / Photo Researchers, Inc.; **12.4B:** © Oliver Meckes / MPI - Tubingen / Photo Researchers, Inc.; **12.6:** © A.C. Barrington Brown / Photo Researchers, Inc.; **12.7C:** Div. of Computer Research & Technology, National Institute of Health / Science Photo Library / Photo Researchers, Inc.; **12.10C:** © 1948 M-C Escher Foundation-Baarn- Holland, All Rights Reserved; **12.11(top):** © 1979 Olins and Olins / BPS; **12.11(bottom):** Visuals Unlimited; **12.A(B):** Courtesy of Cellmark Diagnotics, Germantown, Maryland

Chapter 13

Opener(E-coli): © R. Kessel - G. Shih / Visuals Unlimited; **Opener(face):** © Harvey Lloyd / Peter Arnold, Inc.; **Opener(round worm):** © A.M. Siegelman / Visuals Unlimited; **13.9C:** © Tripos Associates / Peter Arnold, Inc.; **13.C:** © K.G. Murti/ Visuals Unlimited; **13.16B:** © Kiseleva - Fawcett / Visuals Unlimited; **13.17(both):** © Bill Longcore / Photo Researchers, Inc.; **13.F(sheep):** Corbis #CB011809; **13.F(salmon):** © Norbert Wu / Peter Arnold, Inc.; **13.F(human):** Corbis #CB038761; **13.F(rabbit):** PhotoDisc vol. Series 44 Image #44104 WL000339

Chapter 14

Opener A: © J.C. Carton/ Bruce Coleman, Inc.; **14.2B:** © Jeff Greenberg / Peter Arnold, Inc.; **14.3:** © Galen Rowell / Peter Arnold, Inc.; **14.4(both):** © Tom McHugh / Photo Researchers, Inc.; **14.4B(top):** © Jeff Rotman / Peter Arnold, Inc.; **14.4B(bottom):** © Tom McHugh / Photo Researchers, Inc.; **14.4C(left):** © A.J. Copley / Visuals Unlimited; **14.4C(middle):** Peter Scoones / © Planet Earth Pictures; **14.4C(right):** © D. Holden Bailey / Tom Stack & Associates; **14.5(both):** © Peter Grant / Princeton University; **14.6:** BioPhot; **14.7A,B:** © Tom McHugh / Photo Researchers, Inc.; **14.7C:** © Robert and Linda Mitchell; **14.7D:** © C. Allan Morgan / Peter Arnold, Inc.; **14.B(left):** © Science VU / Visuals Unlimited; **14.B(right):** © Hans Gelderblom / Visuals Unlimited; **14.8A:** © David M. Phillips / Visuals Unlimited; **14.8B:** © A.B. Dowsett / SPL / Photo Researchers, Inc.; **14.8C(top):** © Stanley F. Hayes, National Institutes of Health / SPL / Photo Researchers, Inc.; **14.8C(bottom):** © Ken Greer / Visuals Unlimited; **14.8(inset):** © Cabisco / Phototake; **14.9:** The Bettmann Archives

Chapter 15

Opener(both): Reprinted by permission from Nature (vol. 394 2 July 1998 p. 69) Macmillan

Magazines Ltd.; **15.A(bulldog):** © Gerard Lacz / Peter Arnold, Inc.; **15.A(basset):** © D. Cavagnaro / Visuals Unlimited; **15.A(paws):** © Scott Camazine / Photo Researchers, Inc.; **15.A(cat):** © Barb Zurawski; **15.2B:** Courtesy of Dr. Victor A. McKusick / Johns Hopkins Hospital

Chapter 16

Opener A(left): © Milton W. Tiemey, Jr. / Visuals Unlimited; **Opener A(right):** PhotoDisc #WL003910; **16.1A(left):** S. Lowry / Univ. Ulster / © Tony Stone Images; **16.1A(middle):** Corbis id #CB030324; **16.1A(right):** © Robert Fried / Stock Boston; **16.1B(left):** Corbis AWI029 Animals & Wildlife; **16.1B(middle):** Corbis #CB009677; **16.1B(right):** PhotoDisc # WL003983; **16.1C:** David Liittschwager and Susan Middleton / © 1973 Reprinted with permission of Discover Magazine; **16.3A:** © Tom McHugh / Photo Researchers, Inc.; **16.4(frog / left):** © John R. MacGregor; **16.4(frog / right):** © C. Allan Morgan / Peter Arnold, Inc.; **16.4(moose):** © Sandy Macys / Gamma Liaison; **16.4(abalone):** © Tom McHugh / Photo Researchers, Inc.; **16.4(cat / dog):** © Renee Lynn / Photo Researchers, Inc.; **16.4(liger):** © Porterfield / Chickering / Photo Researchers, Inc.; **16.6A,B:** Courtesy of Dr. Tom Smith, San Francisco State University; **16.A(top / both):** © Porterfield / Chickering / Photo Researchers, Inc.; **16.A(bottom):** Photo by L. Rieseberg and G. Seiler, Biology Department, Indiana University; **16.7:** W.L. Perry, DM Lodge, and JL Feder; Dept. of Biology, University of Notre Dame; **16.8(both):** © Melvin Grey; **16.10A(top):** © Kjell B. Sandved / Visuals Unlimited; **16.10A(bottom):** E.R. Degginger / Photo Researchers, Inc.; **16.10B(both):** © J. Wm. Schopf / UCLA; **16.11:** © Discover Magazine; **16.12(1-5):** © Kevin de Queiroz; **16.12(6):** Jonathan B. Losos; **16.14A:** Photo by Mario Maier, Erlangen / Germany

Chapter 17

Opener A: photo courtesy of Terry D. Jones and John A. Ruben; **Opener B:** O. Louis Mazzatenta / NGS Image Collection; **17.1A:** Reprinted with permission from SCIENCE vol. 282 No. 5394 pages 1601–1772 November 27 © 1998 / David Dilcher and Ge Sun; **17.1B:** Reprinted with permission from SCIENCE vol. 283 No. 5399 pages 133–288 January 8 © 1999 / Grifford H. Milla; **17.1C:** Photo by Dr. Hendrik Poinar / Max-Planck Institute; **17.3B(left):** photo courtesy of Terry D. Jones and John A. Ruben; **17.3B(middle):** O. Louis Mazzatenta / NGS Image Collection; **17.3(right):** Corbis / Animals in Action #AIA0079; **17.4A-C:** Dr. Luis M. Chiappe; **17.7A,B:** © William E. Ferguson; **17.7C:** © Bob Gossington / Bruce Coleman; **17.7D:** © John Reader / SPL / Photo Researchers, Inc.; **17.7E:** Dr. Shuhai Xiao / Botanical Museum Harvard University; **17.C:** Courtesy Department Library Services - American Museum of Natural History Neg. No. 320496; **17.12A:** © E.R. Degginger / Animals Animals; **17.12B:** Visuals Unlimited / © Science VU; **17.13(top - all):** Courtesy Michael Richardson; **17.13(bottom - all):** © Bodleian Library; **17.14A,B:** Courtesy Dr. H. Hameister, Universitatulm Ulm, Abteilung Medizinische Genetik; **17.17(left):** From James H. Asher and Thomas B. Friedman, *Journal of Medical Genetics* 27:618–626, 1990; **17.17(middle):** © Vickie Jackson; **17.17(right):** From James H. Asher and Thomas B. Friedman, Journal of Medical Genetics 27:618–626, 1990

Chapter 18

Opener: NASA; **18.A:** CORBIS / Bettmann; **18.10B:** © M. Abbey / Visuals Unlimited; **18.10C:** © W. Burgdorfer / Visuals Unlimited; **18.10D:** Charles O'Kelly, Bigelow Laboratory for Ocean Sciences; **18.11:** © Dr. Malcolm Walter; **18.12, 18.13:** J.G. Gehling, all rights reserved; **18.14(all):** Bruce Parker ; **18.15:** Photo by W.A. Taylor / permission NATURE 9/4/97 vol. 389 p. 35 fig. 2d; **18.17A:** Field Museum of Natural History Chicago (Geo 75400C); **18.17B:** Milwaukee Public Museum Silurian reef diorama; **18.23A:** © G. Hinter Leitner / Gamma Liason; **18.23B:** © Burt Silverman / Silverman Studios, Inc.

Chapter 19

Opener: K.P. Hennes and C.A. Suttle; **19.1A:** © Runk / Schoenberger / Grant Heilman Photography, Inc.; **19.1B:** © Dennis Kunkel / Phototake; **19.2:** © E.C.S Chan / Visuals Unlimited; **19.3(bacterium):** Science VU / Visuals Unlimited; **19.3(mosaic):** © Dr. O. Bradfute / Peter Arnold, Inc.; **19.3(adenovirus):** © E.O.S. / Gelderblom/ Photo Researchers, Inc.; **19.3(herpesvirus):** Peter Arnold, Inc.; **19.3(poxvirus):** © Hans Gelderblom / Visuals Unlimited; **19.5:** © Oliver Meckes / MPI - Tubingen / Photo Researchers, Inc.; **19.9:** © Science Source / Photo Researchers, Inc.; **19.10:** © R.F. Ashley / Visuals Unlimited; **19.A:** Culver Pictures, Inc.; **19.11:** Theodore Diener / USDA / Plant Virology Laboratory; **19.13:** © Ralph Eagle Jr. / Photo Researchers, Inc.

Chapter 20

Opener B: Prof. Margaret J. McFall-Ngai; **Opener C:** Photograph Hans Reichenbach, GBF, Braunschweig, Germany; **20.1A:** © Sylvan Wittwer / Visuals Unlimited; **20.1B:** © C.P. Vance / Visuals Unlimited; **20.3A:** © G. Murti / SPL / Photo Researchers, Inc.; **20.3B:** American Society for Microbiology; **20.4B:** © Raymond B. Otero / Visuals Unlimited; **20.5A:** Courtesy of Charles C. Brinton, Jr. and Judith Carnaham; **20.5B:** © Fred Hossler / Visuals Unlimited; **20.5C:** © George Musil / Visuals Unlimited; **20.6A:** © David M. Phillips / Visuals Unlimited; **20.6B:** © George J. Wilder / Visuals Unlimited; **20.6C:** © Thomas Tottleben / Tottleben Scientific Company; **20.7:** Esther R. Angert & Norman R. Pace; **20.D:** Dr. Walther Stoeckenius; **20.8B:** Dr. Kari Lounatmaa / Science Photo Library / Photo Researchers, Inc.; **20.11:** Courtesy of E.L. Wollman; **20.12A:** © Arthur M. Siegelman / Visuals Unlimited; **20.12B:** © Dr. Dennis Kunkel / Phototake; **20.13:** © Joe Munroe / Photo Researchers, Inc.

Chapter 21

Opener A: Whirling Disease Foundation; **Opener B:** © Ken Lucas / Visuals Unlimited; **Opener C:** © Larry S. Roberts / Visuals Unlimited; **21.1A:** Serguei Karpov; **21.2:** © A.M. Siegelman / Visuals Unlimited; **21.3:** © M. Abbey / Visuals Unlimited; **21.4A,B:** From M. Schaechter, G. Medoff, and D. Schlessinger (Eds.) *Mechanism of Microbial Disease,* 1989 Williams and Wilkins; **21.5:** © Tony Stone Images / Robert Brons / BPS; **21.6:** © B. Beatty / Visuals Unlimited; **21.7(both):** © Carolina Biological Supply / Phototake; **21.9B:** © Arthur M. Siegelman / Visuals Unlimited; **21.9C:** Armed Forces Institute of Pathology; **21.10A:** © David M. Phillips; **21.10C:** © Bill Bachman / Photo Researchers, Inc.; **21.11A:** Courtesy of Dr. JoAnn Burkholder; **21.11B:** NCSU Aquatic Botany Lab; **21.11C:** Courtesy of Dr. JoAnn Burkholder; **21.12:** DJP Ferguson, Oxford University with permmision from The Biologist Ref; **21.A:** NCSU Aquatic Botany Lab; **21.C:** © Lennart Nilsson, *The Body Victorious,* Dell Publishing Company; Unlimited; **21.14:** © M. Abbey / Visuals; **21.15:** © Biophoto Associates / Science Source / Photo Researchers, Inc.; **21.16A:** M.S. Fuller; **21.16B:** W.E. Fry, Plant Pathology, Cornell University; **21.17:** © John D. Cunningham / Visuals Unlimited; **21.18:** © Gregory Ochocki / Photo Researchers, Inc.; **21.19:** © Daniel W. Gotshall / Visuals Unlimited; **21.20:** © Andrew J. Martinez / Photo Researchers, Inc.

Chapter 22

Opener(left): © Roland Seitre / Peter Arnold, Inc. **Opener(right):** © Ray Pfortner / Peter Arnold, Inc.; **22.3(left):** Brigitte Meyer-Berthaud permission NATURE 398 p. 700 4/22/99; **22.3(middle):** Reprinted with permission from SCIENCE vol. 282 No. 5394 pgs. 1601–1772 November 27 © 1998 / David Dilcher and Ge Sun; **22.3(right):** Maria A. Gandolfo; **22.5:** © Jane Burton / Bruce Coleman, Inc.; **22.6A,B:** Courtesy of Howard Crum, University of Michigan Herbarium; **22.6C:** © John D. Cunningham / Visuals Unlimited; **22.7:** © William E. Ferguson; **22.8A:** © William E. Ferguson; **22.8B:** © Dave Schiefelbein / Tony Stone Images; **22.8C:** © Kingsley Stern; **22.10(left):** © Kjell B. Sandved / Bufferfly Alphabet; **22.10(right):** © David M. Dennis / Tom Stack & Associates; **22.10(bottom):** © Stan Elems / Visuals Unlimited; **23.11A:** © Bud Lehnhausen / Photo Researchers, Inc.; **23.11B:** © Joe McDonald / Visuals Unlimited; **23.11C:** © W. Ormerod / Visuals Unlimited; **23.11D:** © Rod Planck / Tom Stack & Associates; **22.12A,B:** © Robert & Linda Mitchell; **22.14A:** Dr. Dennis Stevenson; **22.14B:** © Biophoto Associates / Photo Researchers, Inc.; **22.14C:** © Jeff Gnass; **22.14D:** © Doug Sokell / Visuals Unlimited; **22.17A:** © Richard Weiss / Peter Arnold, Inc.; **22.17B:** © Dwight Kuhn; **22.17C:** © James L. Castner; **22.17D:** © Hans Reinhard / OKAPIA / Photo Researchers, Inc.; **22.A:** © Fred Bavendam; **22.B:** © Michael Ederegger / Peter Arnold, Inc.; **22.C:** © Kingsley Stern; **22.D-G:** © W. Barthlott

Chapter 23

Opener A: Lee Berger; **23.1A:** © R.S. Hussey / Visuals Unlimited; **23.1B:** Thomas J. Volk / University of Wisconsin-La Crosse (www.wisc.edu/botany/fungi/volkmyco.html); **22.3:** Courtesy of N. Allin and G.L. Barron, University of Gue; **22.5:** © Dwight Kuhn; **23.5B(left):** Runk / Schoenberger From Grant Heilman; **23.5B(right):** Courtesy of Dr. Garry T. Cole / The Univ. of Texas at Austin; **23.6A:** © Carolina Biological / Phototake; **23.7A:** © J. Robert Waaland /

Biological Photo Service; **23.7B:** M.S. Fuller; **23.8(left):** Runk / Schoenberger From Grant Heilman Photography, Inc.; **23.8(right):** Ed Reschke; **23.9:** © Dwight Kunn; **23.10A:** © John D. Cunningham / Visuals Unlimited; **23.10B:**© Doug Sherman / Geofile; **23.11B:** © Bruce Iverson; **23.12A:** © Bill Keogh / Visuals Unlimited; **23.12B:** © Hans Reinhard / Bruce Coleman, Inc.; **23.12C,D:** Courtesy of G.L. Barron, The University of Guelph; **23.13:** © Stanley Flegler / Visuals Unlimited; **23.14A:** John Dennis / Canadian Forest Service; **23.14B:** Biological Photo Service; **23.15B:** © V. Ahmadjian / Visuals Unlimited; **23.16:** © Steve & Sylvia Sharnoff / Visuals Unlimited; **23.A:** © E. Chan / Visuals Unlimited; **23.B:** © Holt Studios, Ltd. / Animals Animals / Earth Scenes

Chapter 24

Opener A: © Biofoto / P. Funch / R.M. Kristensen; **24.7B:** © James C. Amos / Photo Researchers, Inc.; **24.9B:** H. Armstrong Roberts; **24.9C:** © Science VU / Visuals Unlimited; **24.12A:** © L Newman & A. Flowers / Photo Researchers, Inc.; **24.12B:** © David Mechlin / Phototake NYC; **24.13B:** © CNRI / SPL / Photo Researchers, Inc.; **24.14:** © Carolina Biological / Phototake; **24.15B:** © A.M. Siegelman / Visuals Unlimited; **24.16B:** © Franklin J. Viola; **24.16C:** © Chesher / Photo Researchers, Inc.; **24.16D:** © Fred Bavendam / Peter Arnold, Inc.; **24.18B:** © Kjell B. Sandved / Visuals Unlimited; **24.18C:** E.R. Degginger / Bruce Coleman, Inc.; **24.18D:** © Bill Beatty / Visuals Unlimited; **24.20:** Craig Cary/Univ. of Delaware; **24.21:** © William Ferguson; **24.22B:** © Stephen Krasemann / Photo Researchers, Inc.; **24.22C:** © Luis C. Marigo / Peter Arnold, Inc.; **24.22D:** © Zigmund Leszczynski / Animals Animals; **24.22E:** © G.C Merker / Visuals Unlimited; **24.22F:** © Dr. James L. Castner; **24.22G**(beetle): Davies & Starr / © Tony Stone Images; **24.22G**(grub): © George K. Bryce / Animals Animals; **24.23B:** © Andrew Martinez / Photo Researchers, Inc.; **24.23C:** © Norbert Wu / Peter Arnold, Inc.; **24.23D:** © Rick M. Harbo; **24.24A:** © David B. Fleetham / Visuals Unlimited; **24.24B:** © Nancy Sefton / Photo Researchers, Inc.; **24.25A:** © Peter Parks / Animals Animals; **24.25B:** © E.R. Degginger / Animals Animals

Chapter 25

Opener(both): © Mark Moffett / Minden Pictures; **25.3B:** © Nancy Sefton / Photo Researchers, Inc.; **25.4B:** © Runk / Schoenberger / Grant Heilman Photography; **25.6B:** Illustration by: Rob Wood - Wood Ronsaville Harlin, Inc.; **25.6C:** © Russ Kinne / Photo Researchers, Inc.; **25.8B:** © Hal Beral / Visuals Unlimited; **25.8C:** © W. Gregory Brown / Animals Animals; **25.10B:** © Tom McHugh / Photo Researchers, Inc.; **25.10C:** Peter Scoones / Planet Earth Pictures; **25.10D:** © Dr. E.R. Degginger; **25.12B:** © Mark Moffett / Minden Pictures; **25.12C:** © Suzanne L. Collins & Joseph T. Collins / Photo Researchers, Inc.; **25.12D:** E.D. Brodie, Jr.; **25.14B:** © Ed Reschke / Peter Arnold, Inc.; **25.14C,D:** © Joe McDonald / Animals Animals; **25.15:** © Joe McDonald / Bruce Coleman, Inc.; **25.18B:** © Meckes / Ottawa / Photo Researchers, Inc.; **25.20B:** © Fritz Prenzel / Animals Animals; **25.20C:** © Dalton, S. / Animals Animals; **25.20D:** © Tim Davis / Photo Researchers, Inc.; **25.20E:** © Ansell Horn / Phototake; **25.22B:** © Tom Ulrich / Visuals Unlimited; **25.22C:** © Roland Seitre / Peter Arnold, Inc.; **25.22D:** © John Glustina / GIUST / Bruce Coleman, Inc.

Chapter 26

Opener(left): Image provided by DHP Papermill & Press; **Opener(right):** From the collection of the Robert C. Williams American Museum of Papermaking; **26.2:** © Jack M. Bostrack / Visuals Unlimited; **26.3A:** © Randy Moore / BioPhot; **26.3B:** © George Wilder / Visuals Unlimited; **26.3C:** © BioPhot; **26.3D:** © Bruno P. Zehnder / Peter Arnold, Inc.; **26.4A:** © BioPhot; **26.4B:** © Dwight Kuhn; **26.4C:** From Yolanda Heslop-Harrison, *"SEM of Fresh Leaves of Pinguicula"*, SCIENCE 667:173, © 1970 by the AAAS; **26.4D:** © Oliver Meckes / Eye of Science / Photo Researchers, Inc.; **26.4E:** Animals Animals / Earth Scenes © Patti Murray; **26.5A,B:** © Ray Simon / Photo Researchers, Inc.; **26.6A:** © John D. Cunningham / Visuals Unlimited; **26.6B:** © George J. Wilder / Visuals Unlimited; **26.7:** © George Wilder / Visuals Unlimited; **26.8A(left):** © Cabisco / Visuals Unlimited; **26.8A(right):** Dwight Kuhn; **26.8B(right):** © Science VU / Visuals Unlimited; **26.9A:** © Charles Gurche; **26.9B:** © Kenneth W. Fink / Photo Researchers, Inc.; **26.9C:** © G.C. Kelley / Photo Researchers, Inc.; **26.9D:** © Franz Krenn / Photo Researchers, Inc.; **26.9E,F:** © Dwight R. Kuhn; **26.10A,B:** © Kjell Sandved / Butterfly Alphabet; **26.10C:** © Robert Maier / Earth Scenes / Animals Animals; **26.11A,B:** © Dwight R. Kuhn; **26.12:** © M.I. Walker / Photo Researchers, Inc.; **26.13:** © Ed Reschke; **26.15A:** © John D. Cunningham / Visuals Unlimited; **26.15B,C:** © Stan Elems / Visuals Unlimited; **26.15E:** © John D. Cunningham / Visuals Unlimited; **26.16:** © Runk / Schoenberger / Grant Heilman Photography; **26.18A,B:** © W. H. Hodge / Peter Arnold, Inc.; **26.18C:** © William Ferguson; **26.20B:** © Manfred Kage / Peter Arnold, Inc.; **26.20C:** © A.J. Copley / Visuals Unlimited

Chapter 27

Opener B: © Dennis Flaherty / Photo Researchers, Inc.; **27.3A,B:** © 1991 Regents University of California Staewide IPM Project; **27.4:** © Deborah A. Kopp / Visuals Unlimited; **27.9A:** Dr. E.R. Degginger

Chapter 28

Opener(left): © M.T. Frazier / PSU / Photo Researchers, Inc.; **Opener(middle):** Dr. E. R. Degginger; **Opener(right):** © Stephen Dalton / Photo Researchers, Inc.; **28.1A:** © Nardin / Jacana / Photo Researchers, Inc.; **28.1B:** © David Sieren / Visuals Unlimited; **28.1C:** Dr. Jeffry B. Mitton / University of Colorado; **28.C&D(all):** © David Littschwager, Susan Middleton / Discover Magazine; **28.6A:** © William E. Ferguson; **28.6B(both):** © Robert A. Tyrrell; **28.6C:** © Joe McDonald / McDonald Wildlife Photography; **28.6D:** © Merlin D. Tuttle / Bat Conservation International / Photo Researchers, Inc.; **28.7A,B:** © Leonard Lessin / Photo Researchers, Inc.; **28.8:** © Angelina Lax / Photo Researchers, Inc.; **28.10B:** © John D. Cunningham / Visuals Unlimited; **28.10C:** Biodisc.com; **Table 28.1(1):** © Inga Spence / Visuals Unlimited; **Table 28.1(2):** © Alan & Linda Detrick / Photo Researchers, Inc.; **Table 28.1(3):** © Zig Leszczynski / Earth Scenes / Animals Animals; **Table 28.1(4):** © Dwight Kuhn; **Table 28.1(5):** © Phil Degginger; **Table 28.1(6):** © Dwight Kuhn; **Table 28.1(7):** © Dr. E.R. Degginger; **Table 28.1(8):** © R.J. Erwin / Photo Researchers, Inc.; **28.12A:** © Adam Hart-Davis / SPL / Photo Researchers, Inc.; **28.12B:** © O.S.F. / Earth Scenes / Animals Animals; **28.12C:** © Rod Planck / Photo Researchers, Inc.; **28.12D:** © W.H. Hodge / Peter Arnold, Inc.; **28.12E:** © Richard Shiell / Earth Scenes / Animals Animals; **28.13A:** © Dwight Kuhn; **28.13B:** © Ed Reschke

Chapter 29

Opener B: Courtesy of University of California, Statewide Integrated Pest Management Project, Photo by Jack Kelly Clark; **29.3A:** © Robert E. Lyons / Visuals Unlimited; **29.3B:** © Sylvan H. Wittwer / Visuals Unlimited; **29.5:** © David G. Clark; **29.6:** © Leonard Lessin / Peter Arnold, Inc.; **29.7A:** © C. Calentine / Visuals Unlimited; **29.7B:** © BioPhoto; **29.A-1&2:** © Dr. Randy Moore / BioPhoto; **29.9:** © William E. Ferguson; **29.10A:** © John Kaprielian / Photo Researchers, Inc.; **29.10B:** © Richard H. Gross; **29.10C:** © John Kaprielian / Photo Researchers, Inc.; **29.11(both):** © Tom McHugh / Photo Researchers, Inc.; **29.16:** © John D. Cunningham / Visuals Unlimited; **29.17:** © L. West / Photo Researchers, Inc.; **29.18:** © Tim Thompson / Tony Stone Images; **29.B:** ©Joanne Chory

Chapter 30

30.B(left): Vaughn Youtz / Sipa Press; **30.B(right):** Jewish Hospital, Kleinert, Kutz and Associates Hand Core Center, and University of Louisville; **30.7(zebra):** © Roland Seitre / Peter Arnold, Inc.; **30.7(toucan):** © Michael Sewell / Peter Arnold, Inc.; **30.7(frog):** © Rob & Ann Simpson **30.7(reptile):** © Werner H. Muller / Peter Arnold, Inc. **30.9B:** © CNRI / SPL / Photo Researchers, Inc.

Chapter 31

Opener: Dr. David W. Nierenberg; **31.1:** © Leonard Lee Rue III; **31.2B:** ©Secchi-League / Roussel - UCLAF / CNRI / SPL / Photo Researchers, Inc.; **31.6B:** © Fawcett / Coggeshall / Photo Researchers, Inc.; **31.8B:** © Don Fawcett / Photo Researchers, Inc.; **31.10:** © E.R. Lewis / University of California, Berkeley; **31.14D:** © Manfred Kage / Peter Arnold, Inc.

Chapter 32

Opener A: W. G. Evans. University of Alberta; **Opener C:** © Dave Watts / Tom Stack & Associates; **Opener D:** © Larry Miller / Photo Researchers, Inc.; **32.1A,B:** © Thomas Eisner; **32.4A:** © Hans Pfletschinger / Peter Arnold, Inc.; **32.4B:** micrograph R.A. Steinbrecht; **32.7B:** © Thomas Eisner; **32.7C:** © Fred Bavendam / Peter Arnold, Inc.; **32.8C:** © Frank S. Werblin

Chapter 33

Opener A,B: Courtesy of Eli Lilly & Company; **33.6A:** © Dan Kline / Visuals Unlimited; **33.9A-C:** Clinical Pathological Conference on Acromegaly, Diabetes, Hypermetabolism, Protein

Use and Heart Failure, American Journal of Medicine 20:133, (1956)

Chapter 34

Opener: Photography by Runk / Schoenberger / Grant Heilman Photography; **34.1A:** Photo © Lois Greenfield; **34.1B:** © Dwight Kuhn; **34.3A:** © David Scharf / Peter Arnold, Inc.; **34.3B:** © Zigmund Leszezynski / Animals Animals; **34.4A:** © Nancy Stefton / Photo Researchers, Inc.; **34.4B:** © Gregory Ochocki / Photo Researchers, Inc.; **34.4C:** © J&B Photo / Animals Animals / Earth Scenes; **34.5:** From Kenneth R. Gordon, Ph.D., *"Adaptive Nature of Skeletal Design,"* BioScience 39(11): 784–790, Dec. 1989; **34.7A:** © Chuck Brown / Photo Researchers, Inc.; **34.7B:** © Ed Reschke; **34.B:** © Michael Klein / Peter Arnold, Inc.; **34.10A:** © Ed Reschke; **34.10B:** © Manfred Kage / Peter Arnold, Inc.; **34.10C:** © Ed Reschke; **34.19A:** © Custom Medical Stock Photos; **34.19B:** © B.S.I.P. / Custom Medical Stock Photos

Chapter 35

Opener(both): From Leonard V. Crowley, *Introduction to Human Disease,* 3rd. ed. © 1992 Jones and Bartlett Publishers, Boston, Reprinted by Permission.; **35.9B:** © Lennart Nilsson, *Behold Man,* Little Brown & Co.; **35.A:** © NASA; **35.13:** © ©Boehringer Ingelheim International GmbH Photo: Lennart Nilsson, The Body Victorious, Dell Publishing; **35.15B:** SPL / Photo Researchers; **35B:** J&L Weber / Peter Arnold

Chapter 36

Opener: Mary Evans Picture Library; **36.1:** © Douglas M. Munnecke / Biological Photo Service; **36.9B:** © D.W. Fawcett / Photo Researchers; **36.10:** © John Watney Photo Library; **36.A:** Brush Wellman photograph; **36.B:** © D.J. Staats MD, Custom Medical Stock Photo; **36.B(both):** © Martin Rotker / Martin Rotker Photography

Chapter 37

Opener: © Barbara Magnuson / Visuals Unlimited; **37.1A:** © Dwight Kuhn; **37.1B:** © James D. Watt / Planet Earth Pictures; **37.1C:** © Tom McHugh / Photo Researchers, Inc.; **37.12A:** © David Scharf / Peter Arnold, Inc.; **37.12C:** Courtesy of David H. Alpers, M.D.

Chapter 38

Opener A: © Richard Nowitz; Opener **38B:** © Fred Bavendam / Peter Arnold; **38.3:** Courtesy of Eric Toolson; **38.6:** © Roger Eriksson

Chapter 39

Opener: Science VU / Visuals Unlimited; **39.5:** © Biology Media / Photo Researchers, Inc.; **39.8:** © Manfred Kage / Peter Arnold, Inc.; **39.11:** Courtesy D. Arthur J. Olson, Scripps Institute; **39.14B:** © Dr. A Liepins / SPL / Photo Researchers, Inc.; **39.A:** © Ken Greer / Visuals Unlimited; **39.17(top):** © David Scharf / Peter Arnold, Inc.; **39.17(bottom):** © Phil Harrington / Peter Arnold, Inc.

Chapter 40

Opener B: Dr. Brett Casey; **40.1A:** © Professor P. Motta / Dept. of Anatomy / University of La Sapienza, Rome / SPL / Photo Researchers, Inc.; **40.3:** Furnished through the Courtesy of C.L. Markert; **40.C(1):** Courtesy of Christine Nusslein-Volhard; **40.B(2-4):** Courtesy of Jim Langeland, Steve Paddock, Sean Carroll / Howard Hughes Medical Institute, University of Wisconsin; **40D(both):** Courtesy of F.R. Turner, Indiana University, Bloomington, IN; **40.12:** Courtesy Kathryn Tosney; **40.13E:** © Petit Format / Nestle / Science Source / Photo Researchers, Inc.; **40.17B:** © Erika Stone / Peter Arnold, Inc.; **40.F:** © Topham / The Image Works; **40.18:** © J. L. Bulcao / Gamma Liaison

Chapter 41

Opener: Photo by Walter R. Tschinkel; **41.1:** © J.A.L. Cooke; **41.2:** © K&K Ammann / Bruce Coleman; **41.4:** © Frans Lanting / Minden Pictures; **41.5:** © Willard Luce / Animals Animals; **41.6:** Bonnie Bird; **41.7:** Nina Leen / Life Magazine; **41.A:** © Marty Snyderman; **41.9:** © Lynn Rogers; **41.10:** © Diana and P. Sullivan / Bruce Coleman, Inc.; **41.11:** © Tom & Pat Leeson / Photo Researchers, Inc.; **41.12A:** © Lady Philippa / Scott / NHPA; **41.12B:** Tom Brakefield / Bruce Coleman, Inc.; **41.13:** © Franklin J. Viola; **41.15:** David W. Pfennig; **41.16:** E.R. Degginger / Color-Pic. Inc.; **41.17B:** © Gregory G. Dimijian / Photo Researchers, Inc.; **41.19C:** © Scott Camazine / Photo Researchers, Inc.; **41.20A:** © Emmett Duffy, College of William & Mary, Virginia Institute of Marine Science; **41.20B:** © Raymond A. Mendez / Animals Animals; **41.20C:** © Donald Specker / Animals Animals

Chapter 42

Opener: Gordon Rodda / USGS; **42.A:** © C. Gable Ray ; **42.B:** Courtesy of Montana Dept. of Fish, Wildlife, and Parks; **42.1A:** Wayne P. Armstrong; **42.1B:** © John A. Novak / Animals Animals; **42.1C:** NOAA / NGDC DMSP Digital Archive / Science Source / Photo Researchers Inc.

Chapter 43

Opener A: © Ken Smith / Scripps / OSF / DRK Photo; **Opener B:** Peter Tyson / NOVA; **43.1:** Janis Burger / Bruce Coleman, Inc.; **43.3:** © J. Sneesby / B. Wilkins / Tony Stone Images; **43.5A:** © Edward S. Ross; **43.5B:** © James L. Castner; **43.5C:** © Simon D. Pollard / Photo Researchers, Inc.; **43.6:** © Norm Smith and Jim Dawson; **43.8A,B:** USDA Forest Service - IITF Library; **43.9all:** Courtesy, Dr. Stuart Fisher, Department of Zoology, Arizona State University; **43.17:** © R.T. Smith / Ardea; **43.B:** © Aileen M. Smith, Courtesy Black Star, Inc. New York; **43.19A:** © Peter David / Photo Researchers, Inc.; **43.19B:** © Gustav W. Verderber / Visuals Unlimited; **43.19C:** © Uta Matthes-Sears / Univ. of Guelph; **43.19D:** G.R. Roberts

Chapter 44

44.5: © Michael. J. Doolittle / Peter Arnold, Inc.; **44.6 A,B:** Minden Pictures; **44.7:** Dr. E.R. Degginger; **44.8:** Ed Reschke; **44.9A,B:** Michael A. Mares; **44.10:** PhotoDisc #WL002106; **44.11A:** Larry Lefever / Grant Heilman; **44.11B:** E.R. Degginger; **44.12:** © Thomas Kitchin / Photo Researchers, Inc.; **44.13:** © Frank Krahmer / Bruce Coleman, Inc.; **44.16A:** © Joe / Arrington / Visuals Unlimited; **44.16B:** © James Castner; **44.16C:** © D. Brown / Animals, Animals; **44.16D:** © Mike Bacon / Tom Stack & Assoc.

Chapter 45

Opener A: Photodisc #AA002992; **45.1A:** © Stouffer Prod. / Animals Animals; **45.1B:** Old Dartmouth Historical Society - New Bedford Whaling Museum; **45.1C:** © Michael Nichols / Magnum; **45.2A:** Erich Lessing / Art Resource, NY; **45.2B:** © Ulrike Welsch / Photo Researchers, Inc.; **45.3B:** © Ray Pfortner / Peter Arnold, Inc.; **45.3C(both):** "Ornament figure at the Herten Castle. Westfälisches Amt für Denkmalpflege. Salzstrasse 38, 4400 Münster, BRD"; **45.4A:** Courtesy Arlin J. Krueger / Goddard Space Flight Center / NASA; **45.7:** NRSC LTD / Science Photo Library / Photo Researchers, Inc.; **45.9:** Dr. Rainer Ressl German Aerospace Center (DLR) German Remote Sensing Data Center (DFD); **45.10:** Sygma; **45.11:** Ken Graham / Bruce Coleman, Inc.; **45.12:** © Lyn Funkhouser / Peter Arnold, Inc.; **45.13:** PhotoDisc vol. Series 31 Environmental Concerns image #31186 WL001177; **45.14:** David Tilman; **45.15:** © M.T. O'Keefe / Bruce Coleman Inc.

INDEX

Page numbers followed by an f indicate figures; numbers followed by a t indicate tables; numbers in italics indicate illustrations and photos.

A

B

D

E

F

G

I

J

K

L

P

Q

R

S

T

U

X

Y

Z